MATHEMATICAL REVIEW

Area of a circle of radius R	$A = \pi R^2$
Circumference of a circle	$C = 2\pi R$
Surface area of a sphere	$A = 4\pi R^2$
Volume of a sphere	$V = \frac{4}{3}\pi R^3$
Area of a triangle	$A = \frac{1}{2}bh$
Volume of a circular cylinder of length l	$V = \pi R^2 l$
Pythagorean Theorem	$C^2 = A^2 + B^2$

$\sin \theta = A/C$
$\cos \theta = B/C$ $\qquad \tan \theta = \dfrac{\sin \theta}{\cos \theta}$
$\tan \theta = A/B$

Quadratic Equation: Where $ax^2 + bx + c = 0$

$$x = \frac{-b \pm \sqrt{b^2 - 4ac}}{2a}$$

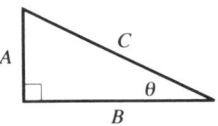

PHYSICAL CONSTANTS

Quantity	Symbol	Value
Gravitation constant	G	$6.672\,59 \times 10^{-11}$ N·m^2/kg^2
Speed of light in vacuum	c	$2.997\,924\,58 \times 10^8$ m/s
Electron charge	e	$1.602\,18 \times 10^{-19}$ C
Planck's Constant	h	$6.626\,076 \times 10^{-34}$ J·s
		$4.135\,669 \times 10^{-15}$ eV·s
Universal gas constant	R	$8.314\,510$ J/mol·K
Avogadro's number	N_A	$6.022\,137 \times 10^{23}$ mol^{-1}
Boltzmann Constant	k_B	$1.380\,66 \times 10^{-23}$ J/K
		$8.617\,39 \times 10^{-5}$ eV/K
Coulomb force constant	k_0	$8.987\,55 \times 10^9$ N·m^2/C^2
Permittivity of free space ($1/\mu_0 c^2$)	ε_0	$8.854\,19 \times 10^{-12}$ C^2/N·m^2
Permeability of free space	μ_0	$1.256\,64 \times 10^{-6}$ T·m/A
Permeability constant	$\mu_0/4\pi$	10^{-7} T·m/A
Electron mass	m_e	$9.109\,39 \times 10^{-31}$ kg
Electron rest energy	$m_e c^2$	$0.510\,999$ MeV
Electron magnetic moment	μ_e	$9.284\,77 \times 10^{-24}$ J/T
Electron charge/mass ratio	e/m_e	$1.758\,82 \times 10^{11}$ C/kg
Electron Compton wavelength	λ_c	$2.426\,31 \times 10^{-12}$ m
Proton mass	m_p	$1.672\,623 \times 10^{-27}$ kg
		$1.007\,276$ u
Proton rest energy	$m_p c^2$	938.272 MeV
Proton magnetic moment	μ_p	$1.410\,608 \times 10^{-26}$ J/T
Neutron mass	m_n	$1.674\,929 \times 10^{-27}$ kg
		$1.008\,66$ u
Neutron rest energy	$m_n c^2$	939.566 MeV
Neutron magnetic moment	μ_n	$9.662\,37 \times 10^{-27}$ J/T
Bohr magneton	μ_B	$9.274\,015 \times 10^{-24}$ J/T
Stefan-Boltzmann Constant	σ	$5.670\,51 \times 10^{-8}$ W/m^2·K^4
Rydberg constant	R	$1.097\,373 \times 10^7$ m^{-1}
Bohr radius	r_1	$5.291\,77 \times 10^{-11}$ m
Faraday constant	F	$9.648\,53 \times 10^4$ C/mol

PHYSICS
ALGEBRA/TRIG

PHYSICS
ALGEBRA/TRIG
THIRD EDITION

EUGENE HECHT
Adelphi University

THOMSON

BROOKS/COLE

Australia • Canada • Mexico • Singapore • Spain
United Kingdom • United States

THOMSON

BROOKS/COLE

™

Editor: *Angus McDonald*
Editorial Assistant: *Emily Levitan*
Technology Project Manager: *Samuel Subity*
Marketing Manager: *Kelley McAllister*
Marketing Assistant: *Sandra Perin*
Advertising Project Manager: *Stacey Purviance*
Project Manager, Editorial Production: *Tom Novack*
Print/Media Buyer: *Kristine Waller*
Permissions Editor: *Sue Ewing*
Production Service: HRS *Interactive*

Text Designer: HRS *Interactive*
Photo Researcher: *Jennifer Burke*
Photo Acquisitions: *Carolyn Hecht*
Copy Editor: *Betty Pesagno*
Illustrator: HRS *Interactive*
Cover Designer: *Denise Davidson*
Cover Image: *Courtesy of NASA*
Cover Printer: *Lehigh Press*
Compositor: HRS *Interactive*
Printer: *Quebecor/World–Versailles*

Printed in the United States of America

2 3 4 5 6 7 06 05 04 03

For more information about our products, contact us at:
Thomson Learning Academic Resource Center
1-800-423-0563

For permission to use material from this text, contact us by:
Phone: 1-800-730-2214   **Fax:** 1-800-730-2215
Web: http://www.thomsonrights.com

Library of Congress Control Number: 2002107407

ISBN 0-534-37729-7

Brooks/Cole–Thomson Learning
511 Forest Lodge Road
Pacific Grove, CA 93950
USA

Asia
Thomson Learning
5 Shenton Way #01-01
UIC Building
Singapore 068808

Australia
Nelson Thomson Learning
102 Dodds Street
South Melbourne, Victoria 3205
Australia

Canada
Nelson Thomson Learning
1120 Birchmount Road
Toronto, Ontario M1K 5G4
Canada

Europe/Middle East/Africa
Thomson Learning
High Holborn House
50/51 Bedford Row
London WC1R 4LR
United Kingdom

Latin America
Thomson Learning
Seneca, 53
Colonia Polanco
11560 Mexico D.F.
Mexico

Spain
Paraninfo Thomson Learning
Calle/Magallanes, 25
28015 Madrid, Spain

Preface

Physics is the study of the material Universe—all there *is*. And that's a bold and wonderful agenda. The Universe is incredibly awesome and mysterious, and we, after all, are just beginning to understand it. Almost 3000 years in the making, physics—incomplete as it is—stands as one of the great creations of the human intellect. It has been a privilege and an unending joy to have spent much of my life studying physics, and it is out of gratitude and admiration that this book takes its form. If this work transmits a sense of the grandeur, unity, and vitality of the subject, it will have met its most important objective.

To the Student

Most people will find this course to be a rather different and demanding experience. It's a lot like learning a foregn language: there's a whole new vocabulary to be memorized and a subtle grammar to be mastered. What you learn today forms the groundwork for what you'll learn tomorrow, so the effort must be sustained. And it's not enough just to take notes in class and skim the book at home—you must develop a deeper understanding by doing problems, lots of problems.

Once class is over for the day, you're essentially on your own and that's when much of the learning has to take place. To help that process, this third edition has been integrated with an accompanying **CD-ROM**, created by the author to be **your personal tutor**. This CD is a major extension of previous versions that have been developed over the last ten years. It's the latest in an ongoing effort to create a powerful electronic tutor that will be there for you, whenever you need it.

For some, the hardest part of this course will be mastering problem solving. The Text-CD package was designed to provide everything you need to know to get the job done. In each chapter, "play" with the simulations called INTERACTIVE EXPLORATIONS, making sure you understand the operative principles; go over the interactive multiple choice questions,

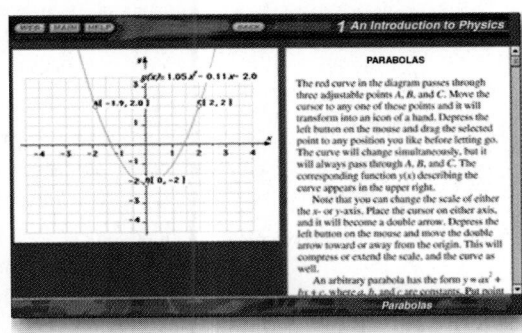

or WARM-UPS, until you feel comfortable with the basics; and then study the detailed step-by-step problem analyses, the WALK-THROUGHS. Working at your own pace, you will learn how to approach any problem: where to begin, how to organize your thoughts, how to proceed, what the pitfalls are, and even what you should be thinking along the way. After doing all of that for a given topic, revisit the appropriate Examples in the textbook, making sure that you completely understand them. If you're not totally confident about your math and calculator skills, redo every calculation in each Example as you go through it. Once you've worked the CD and studied the Examples, go on to the Problems in the text. First focus on the I-level ones; begin with the Coordinated Problems (printed in magenta), then do the Progressive Problems (in blue), and then tackle the regular problems. That done, move on to the II-level problems—there are answers to *all* odd-numbered problems, and complete solutions to many of them, in the back of the book.

The Text-CD package is a proven, effective learning system—make use of it!

To the Instructor

The Basis of This Revision

During the last several years, the second edition has been extensively reviewed by faculty and students to help determine both its strong points and its shortcomings. In addition, hundreds of users have e-mailed suggestions and comments. This third edition was created in response to all of that feedback. On the positive side, we learned that: The illustrations (all conceived by the author) supported the physics in a uniquely effective and appealing way. The writing style, humor, and candor were enjoyed. The clarity and rigor of the exposition, as well as the error-free presentation, were important. The Problem Sets got high grades, particularly because of the inclusion of "real-life" data and situations. The **Five-Step Problem-Solving Approach** used in every Example, and the **Quick Check** at the end of each of them, were touted by instructors, many of whom have long made use of similar schemes in their classrooms. The **Selected Solutions** in the back of the book were more appreciated by students. They also liked having arrows over the symbols for vectors. The **Multiple Choice Questions** proved to be a big help to those taking exams in that format (e.g., the MCAT). The **Exploring Physics on Your Own** inserts, which consist of easy-to-do home experiments that can be performed without any special equipment (e.g., adiabatic cooling using a rubber band), were very well received. The vast majority of respondents agreed that the history of ideas made the physics far more approachable and the book more engaging. The integration of modern physics throughout the text was seen as a necessary and welcomed advance.

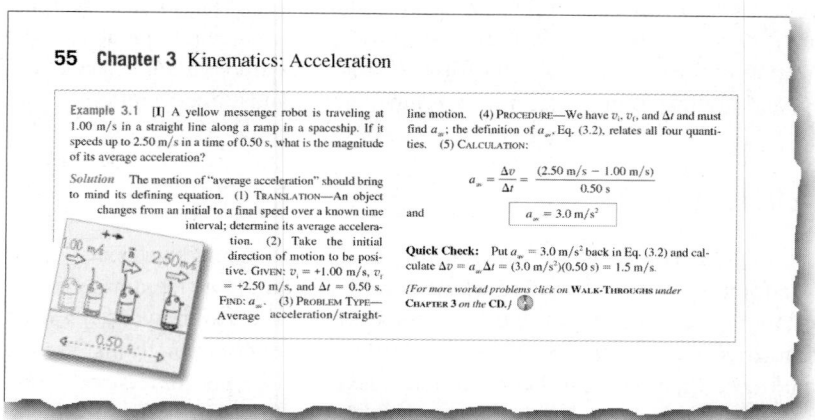

On the other hand, students wanted more illustrations for the Examples and more worked-out problems. They felt that the solutions to some Examples needed to be made clearer. Many were confounded by the concept of the free-body diagrams and needed to see them in a wider range of applications. They also requested help in determining "what was important" for them to know. It was suggested by faculty that the Five-Step Problem-Solving Approach would benefit from the addition of an introductory sentence or two. And finally the author, having used the second edition in class continuously (except for a blessed sabbatical), shared in the universal frustration associated with trying to teach *everyone* in the room how to solve physics problems and actually get correct numerical answers.

With all of these and numerous other concerns in mind, the basic agenda for the revision took form: Every Example in the book was fine-tuned, and a substantial number of new illustrations were added. To help students determine "what's important" and see the "big picture," the text now contains a number of **Study Guide** inserts that integrate, explain, remind, direct, and caution. Charts of *free-body diagrams* were created for statics and dynamics (pp. 124 and 125) and for rotational motion (p. 266). A fair number of students often remain uncomfortable with their calculators and are still making rudimentary errors well into the semester. To address that situation, a new device called **Calculator Tips** was introduced.

For most instructors, teaching Problem Solving is a preeminent concern and, more often than not, a long-standing source of disappointment. There's simply not enough class time to do the job as well as one would like. As we'll see presently, this new edition, along with its accompanying CD-ROM, addresses that issue more robustly than has ever been done before.

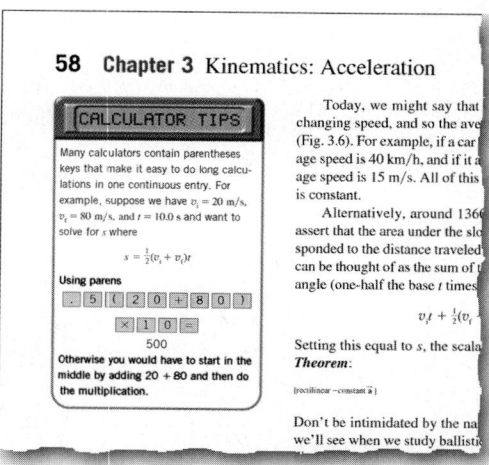

Pedagogy

The present work has assimilated many of the important educational research findings of the last several decades. Its numerous innovations (e.g., in notation, graphics, topic order, problem handling, and photography) have been extensively tested in classrooms throughout the world.

This text is rich in small but important teaching advances that are difficult to notice without actually using the book. For example, most textbooks write the scalar components of a two-dimensional vector $\vec{A} = \vec{A}_x + \vec{A}_y$ as

$$A_y = A \sin \theta \quad \text{and} \quad A_x = A \cos \theta$$

where A is the always positive *magnitude* of $\vec{A}$. That's fine—we usually make sure to tell students that lightface symbols are the magnitudes of the corresponding vectors. Here, however, A_x and A_y can be positive or negative (depending on θ) and are **not** the magnitudes of the component vectors $\vec{A}_x$ and $\vec{A}_y$, all of which is rather clumsy and confusing. In fact, it's so confusing that one introductory physics text defines A_x and A_y as is done above, and yet on the facing page actually calls them "magnitudes." Just think of what must be going on in the mind of a student reading this for the first time.

The problem only gets worse in kinematics. After all, **speed** (v) is the magnitude of the **velocity** ($\vec{v}$), which is a vector quantity. One cannot have a negative speed, and so what shall we call -10 m/s? And what symbol shall we give it? It isn't v, and by definition, without an explicit direction, it certainly isn't $\vec{v}$. Surely, -10 s is a scalar, just as -10 m/s is a scalar. The majority of books deliberately ignore the issue, but inevitably write the equations of constant acceleration and present formulas such as $v_f = v_i + at$. Here v, which can be positive or negative, isn't the *speed*, despite the fact that it's a lightfaced letter (and therefore supposedly the magnitude of $\vec{v}$). At this point, without addressing (or even acknowledging) the problem, texts simply call v the "velocity." The student, who is new to all of this, may well conclude that since vectors are boldfaced, "velocity" is now a scalar.

The muddle continues with angular velocity. Another text rightly tells the reader that ω "can be either positive or negative," but five pages later it calls v, in the seminal equation $v = r\omega$, "the linear speed (the magnitude of the linear velocity)." Obviously, v is not linear *speed*! Ah, but by the time we get to wave motion, where we traditionally write $y = f(x \pm vt)$, all is well because now v is linear *speed*! Remember that the original rationale for calling v "velocity" was that the motion was one-dimensional. Well, it's one-dimensional here where v is speed. What the student will make of this logical conundrum is anyone's guess; most will quietly assume the confusion is theirs. Perhaps this is partly why physics has a reputation for being so "hard."

Clearly, this whole matter is a pedagogical morass that merits much more careful management than it's gotten thus far. Although in the classroom we can fly past these awkward little moments, a textbook must patiently and rigorously sort it all out—that much is long overdue!

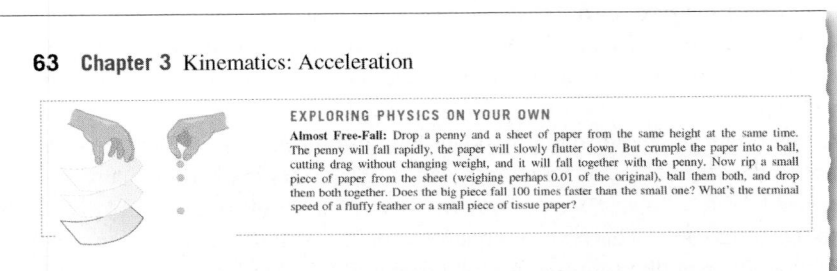

63 **Chapter 3** Kinematics: Acceleration

EXPLORING PHYSICS ON YOUR OWN

Almost Free-Fall: Drop a penny and a sheet of paper from the same height at the same time. The penny will fall rapidly, the paper will slowly flutter down. But crumple the paper into a ball, cutting drag without changing weight, and it will fall together with the penny. Now rip a small piece of paper from the sheet (weighing perhaps 0.01 of the original), ball them both, and drop them both together. Does the big piece fall 100 times faster than the small one? What's the terminal speed of a fluffy feather or a small piece of tissue paper?

284 **Chapter 9** Solids, Liquids, & Gases

STARDUST & YOU

An adult human being is a collection of over 10^{27} atoms bound together as $\approx 10^{14}$ cells. We are a constantly changing aggregate of atoms, themselves created thousands of millions of years ago. The living cells are recent enough, though almost without exception the atoms that form us are at least as old as the Solar System and often much older than that. They have circulated around the Earth for the last four-and-a-half thousand-million years, through air and water, through fish and fowl and trees and Whopper burgers and dung, and back to the soil and then, for the moment, to you. Formed in the thermonuclear fires of stars long gone, the atoms you ate for breakfast extend your lineage back in time to the dawn of creation. We *are* stardust, borrowers in the ancient ritual of "ashes to ashes."

As another example of how this book incorporates small but important teaching advances, consider a related kinematical issue. Obviously, it would be wise to develop kinematics in a way that takes advantage of the student's familiarity with motion, a familiarity predicated on automobiles with odometers and speedometers (neither of which ever displays a negative value). Thus Chapter 2 first treats **motion along a curved path**—just the kind of motion we all know intimately from walking, or traveling on bikes, cars, trains, boats, and planes. This small shift in sequencing has proven to be quite significant pedagogically.

By contrast, out of purely mathematical concerns, the standard texts start kinematics, *as they have for well over half a century*, with the special case of one-dimensional motion (something that would be unrealistic even on a flat Earth). Though it was meant as a simplification, it turns out to be a poor idea. Locating an object as it moves in a straight line using its *displacement* from the origin is a far more subtle matter than treating the overall *distance* it traveled. That approach leads to negative displacements and negative velocities, which, to the modern traveler, are counterexperiential—students do very little traveling backwards and never use displacement or velocity meters. As we've seen, because they're treating one-dimensional motion, these texts go on to completely blur the difference between speed and velocity. In the effort to immediately develop a sophisticated vector description, much of the physics often gets lost. It is far better and more intuitive to begin with two-dimensional motion specified by the scalar path-length along a curve (using distance and speed), and once that's mastered, to move on to the vector notions of displacement and velocity.

Another important pedagogical improvement is in **notation**. Consider the needless confusion engendered by the choice of symbols in most introductory texts. How many different symbols are used for force—F, f, w, W, N, T, R—in any one book? How could that not be confusing to the uninitiated? This text uses one symbol for force, F. There are tensile forces F_T, normal forces F_N, reaction forces F_R, weight forces F_W, friction forces F_f, elastic forces F_e, electrical forces F_E, and so on. When a student sees F representing a physical quantity anywhere in this book, it is force. In the same way, W is work, nothing else.

This small list only touches on some of the important pedagogical concerns dealt with in this edition; there are dozens more. For example, what's the surprising problem with the Work-Energy Theorem as it's usually applied in contemporary textbooks? When you jump, does the floor do work on you to change your KE? If work is force times displacement, is zero work done against friction when you drag a load back to where you started and $s = 0$? If planes fly via Bernoulli, how do they fly upside down? How does the E-field manage to propel an electric current through bends in a wire? Why is most of the sky actually blue? If photons can only exist at c, what is the index of refraction? Is mass really a function of speed? And so on.

Pacing & Support

The Introductory Physics course makes a wide range of demands on the student, and the experienced instructor knows that the first few weeks of the semester can be crucial. Accordingly, Chapter 1 contains a preparatory section (p. 14) on graphs. And that's backed up on the companion CD-ROM with a great interactive simulation called **PARABOLAS & STRAIGHT LINES**, and a set of **WARM-UPS**. The material of kinematics has been rearranged to introduce the physics more gradually and allow time for the ideas to be assimilated. Consequently, Chapter 2 deals only with the physical concepts of speed, displacement, and velocity. The explanations are elaborate, and there are many examples, graphs, and illustrations. Acceleration comes later, in Chapter 3.

The concept of vectors appears for the first time as it relates to displacement (p. 29). It is justified, made logically appealing, and illustrated with

42   Chapter 2 Kinematics: Speed and Velocity

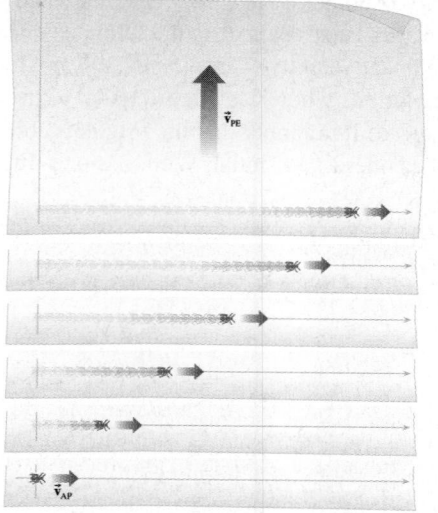

Figure 2.33 An ant walking across a sheet of paper that is itself being moved at a velocity $\vec{v}_{PE}$. The bug is carried along with the paper so that it moves northeast with respect to the Earth at a velocity $\vec{v}_{AE}$.

explanatory diagrams you'll see nowhere else (e.g., p. 31 and p. 43). To help establish a basic working knowledge, the CD-ROM contains a set of **Warm-Ups**, entitled **Vectors**. Moreover, there are two especially helpful interactive simulations—**Vector Components** and **Vector Addition**—on the **CD**.

Only after the student has presumably worked out dozens of problems and has begun to learn how to learn physics do we turn to Chapter 3 and acceleration. By the time the equations of uniform acceleration are reached, the typical reader is much better able to deal with what's involved. With a bit of trigonometry and vector algebra in place, and with a more realistic understanding of the demands of the experience, the serious student is then ready to move ahead more rapidly.

The Utility of History

Historical materials are incorporated in the book for a variety of reasons. For instance, when details of the lives of any of the great scientists are given, it is with an eye toward making these larger-than-life figures less intimidating (e.g., p. 63 on Galileo, or p. 85 on Newton) and their work a little more approachable. In a similar vein, *the book is responsive to the significant contribution made by women in physics*. It highlights the accomplishments of such outstanding scientists as Amalie Noether, Maria Goeppert Mayer, and Lise Meitner, among many others.

Most important, the text follows a historical approach whenever doing so allows the physics to unfold more clearly. The **history of ideas** is examined in order to make those ideas more immediately accessible. Once the student learns what Buridan (1330) was thinking about when he conceived the notion of momentum, the concept instantly becomes understandable: mv makes sense. The brilliant idea of *Conservation of Momentum* came out of Descartes' metaphysical musings in a way that's perfectly reasonable, though no one would call it physics. Having read why Huygens was unhappy with Descartes' momentum, and why he squared the v as an alternative (thereby discovering a new conserved quantity), the concept of kinetic energy comes alive.

A Modern Approach

The central glory of twentieth-century physics is the discovery of an overarching unity in Nature. All matter is composed of myriad identical clones of a small number of fundamental interacting particles. Everything physical is presumably understandable within that context. Thus, to treat the subject as if it were an encyclopedic collection of unrelated ideas is to miss the whole point. *From the beginning of this text to its end we study the unity of natural phenomena, the various manifestations of matter interacting with matter via the fundamental Four Forces.* Even as we explore concepts conceived centuries ago, we bring to bear the perspective of contemporary physics.

Practical Emphasis

A driving force shaping this text is the belief that students are motivated to learn about what directly affects their lives and concerns. The perspective of the book is therefore largely practical: How do we walk? Why do bones break? How does a speedometer work? How can we make a battery? Throughout the text there are photographs of commonplace phenomena that usually go unnoticed (e.g., standing-wave patterns in a wash bucket, p. 401). The work is rich in life-science applications (see p. xxvii), although it strives to be broadly interdisciplinary. If the textbook does its job in this regard, the instructor should never hear that forlorn and exasperating question, "Why do I have to take this course?"

Problem Solving

Every physicist who has ever taught Introductory Physics has heard the universal student lament, "I understand everything; I just can't do the problems." Nonetheless, most instructors believe that "doing" problems is the culmination of the entire experience. This edition (along with its CD-ROM) goes considerably further than any other text has ever gone in systematically addressing this challenge. The end-of-chapter material is now much more than just a source of homework problems.

Five-Step Approach to Problem Solving   Every worked **EXAMPLE** in the book and every **WALK-THROUGH** on the CD follow the same Five-Step Approach, which is based on contemporary pedagogical research and discussions with hundreds of teachers. (It's described in detail on page 17.) This scheme carries students from one necessary phase of the analysis to the next, providing the organizational framework that so many desperately need. Classroom testing during the last five years has proven the efficacy of the method and shown that it is tremendously helpful for the book to approach every solution in the same systematic way. The Five-Step Approach certainly isn't foolproof, but it does help even the poorest student, especially if it's made a requirement on exams involving problem solving.

Quick Checks   All of the **EXAMPLES** in the text and the **WALK-THROUGHS** on the CD end with a **QUICK CHECK**. This teaches a wide variety of verification methods and establishes the habit of checking one's work.

Suggestions on Problem Solving   Every chapter includes this section that explores techniques applicable to those particular problems. It may also contain approximation methods as well as discussions of the pitfalls and common errors specific to the material at hand. For example, a common error in kinematics is to compute the average speed of a uniformly accelerating object using $v_{av} = \frac{1}{2}(v_f - v_i)$ rather than $v_{av} = \frac{1}{2}(v_f + v_i)$; students are appropriately cautioned.

Problems   An extensive selection of problems, most built on real data and referring to actual situations, is provided at the end of each chapter. An instructor will find a wealth of choices from which to assign homework. The Problem Sets are arranged in three levels of increasing difficulty (indicated by the symbols [I], [II], and [III], respectively) and always include a large selection of single-concept problems that explore one idea at a time (from several perspectives) to help students establish a strong foundation of competence and familiarity with the basics. The student is carefully guided through some of the more demanding problems by a variety of **Hints**. This is an important and unique feature of this book. It allows someone who is less well prepared to nonetheless tackle some of the harder problems.

778   Chapter 21 AC and Electronics

29. [I] A 0.15-H coil with a negligible resistance has a reactance of 10 Ω when wired into a sinusoidal-ac circuit. Determine the frequency of the source. [*Hint: Inductive reactance is frequency dependent.*]

30. [I] What is the inductive reactance of a 0.200-H coil in a 100-Hz ac circuit?

31. [I] We are to design a coil having an inductive reactance of 0.50 kΩ to be used in a 60.0-Hz ac circuit. What should be its inductance?

32. [I] Determine the inductive reactance of a 10.0-mH coil connected to a 100-Hz ac generator.

780   Chapter 21 AC and Electronics

76. [I] THIS PROBLEM EXPLORES INDUCTIVE AND CAPACITIVE REACTANCE. Consider a 60-mH inductor and a 25-μF capacitor. We want to determine at what frequency they will have the same reactance. (a) Write an expression for the inductive reactance. (b) Write an expression for the capacitive reactance. (c) Show that these reactances are equal at the angular frequency $\omega = 1/\sqrt{LC}$. (d) Determine the numerical value of the frequency f for which the reactances are equal.

Coordinated Problems   Every end-of-chapter Problem Set contains several special magenta-colored groupings, each consisting of three very similar problems. To the uninitiated these three **Coordinated Problems** read quite differently, but despite that, as far as the physics is concerned, they are really only slight variations on the same theme, and are all solved in much the same way. The first problem in each such group has an elaborate hint and, if that's not enough, there's a complete solution in the back of the book. Once that lead problem is mastered, the next two in the group should be manageable by even the weakest student. This is an important confidence-building mechanism.

Progressive Problems   Every Problem Set contains a selection of so-called Progressive Problems that are introduced in blue. These unfold step-by-step carrying along the analysis in a much more suggestive way than is customary nowadays. In effect, the problem is divided into a number of questions that take the student by the hand through the solution as a whole. Not surprisingly, students generally respond very positively to these problems, which help to develop analytic skills while sustaining self-assurance.

Solutions ***Approximately 15 percent of the end-of-chapter problems are worked out*** ***succinctly at the back of the book in order to encourage and strengthen independent*** ***study.*** (These problems are indicated by a boldface numeral.) Another 10 percent of the solutions are provided in a student solutions manual, which is available at the instructor's request to the bookstore. (These problems are indicated by an italic numeral.) In response to student requests for more solutions, the end-of-chapter Problem Sets now include a small selection of new representative problems immediately followed by their complete **Solutions**.

The CD-ROM: Your Personal Tutor

Every text has limited resources; there are just so many pages. Discussions must be cut short and topics have to be excluded. Although it would be wonderful if the book could take all students by the hand and walk them through the analysis of two or three hundred additional problems—exploring the logic and pointing out the pitfalls in each—that's simply not possible in the textbook format. Ergo, the CD-ROM.

This third edition comes with a free CD that complements and extends the book. If you'd like to see an in-depth discussion of the gyroscope, or read a complete derivation of the Thin-Lens Equation, or learn how a refrigerator works (and a great deal more), it's there, fully illustrated; just click on FURTHER DISCUSSIONS on the CD.

Even more important, the CD continues the problem-solving job with hundreds of WARM-UPS, interactive multiple choice questions (with complete solutions) that establish the basic foundation of knowledge. These are followed by hundreds of completely worked out substantial problems, called WALK-THROUGHS. Students are taught how to "read" a problem, how to draw an appropriate diagram, and how to analyze that problem.

The CD-ROM also contains over forty marvelous simulations of key phenomena. These INTERACTIVE EXPLORATIONS are like having an elementary lab at your disposal. Students are encouraged to "play" with the simulations until they understand the principles involved. All references to the CD in the text are marked with a icon.

Additional Improvements and New Features

Sticker Art To encourage students to draw diagrams and, indeed, to teach them how to do so, the text and the CD-ROM contain hundreds of pieces of "sticker art," many of them new to this edition. These illustrate how to transform a verbal statement into a simple visual one.

Discussion Questions All chapters end with a selection of discussion questions designed to develop and extend the conceptual understanding of the material. Most chapters in this edition contain one or more new questions.

Multiple Choice Questions *A group of roughly 20 to 25 multiple choice questions similar to those found on national medical (MCAT) and optometry (OATP) school entrance exams is included in each chapter.* Among other types, these comprise single-concept calculational questions as well as probing conceptual questions. In this edition groups of five or six related MCAT-like questions pertaining to a single diagram or graph have been added.

Core Material & Study Guide Every chapter ends with a section called CORE MATERIAL & STUDY GUIDE. Each is an in-depth summary of what's important if one is to be able to do the problems: it outlines what should be studied and reread, which EXAMPLES should be given special attention, and so forth. All of these have been revised with the intent of making them more useful.

Key Terms To help the reader determine "what's important," the end-of-chapter Core Material & Study Guide sections now contain a list of **Key Terms**. These are the new words, introduced in the chapter, that must be learned.

Example Problems Every Example Problem in the book has been revised.

Problems Many Problems have been clarified, and hundreds of new ones have been added.

Exploring Physics on Your Own These are small experiments done with everyday materials that students are encouraged to perform on their own. Several new ones were added in this edition (e.g., p. 383).

Calculator Tips Because some students have trouble making full and accurate use of their calculators, especially in the early weeks of the course, this edition now contains a margin insert called Calculator Tips. It anticipates and addresses the usual screwups.

Margin Call-Outs To help the student recognize the key points and definitions within the discussion, there are margin call-outs (with arrows leading back to the appropriate text) where the idea is restated in a succinct summary fashion. This edition contains dozens of new call-outs.

Internet Sites There are some great physics-related Internet sites, and the web addresses of several of the best are provided in this edition.

Study Guide Notes To assist the reader in figuring out "what's important" and to help in seeing the "forest for the trees," this edition now contains a number of **Study Guide** margin inserts.

Textual Material This edition contains some new textual material [on the collapse of the Twin Towers (p. 342), the Doppler Effect (p. 405), circuit boards (p. 639), medical applications of isotopes (p. 1078), and Radiation Damage: Dosimetry (p. 986), for example].

Biomedical & Bioengineering Material By popular request, the range of applications of physics to other disciplines has been enriched, especially when it comes to the life sciences and bioengineering.

Drawings & Photographs The third edition contains dozens of new drawings and photographs. There are charts of free-body diagrams (p. 124 and p. 266), and a number of photos of actual mechanical systems to be analyzed in the Problem Sets (e.g., pp. 132, 135, 268, and 275).

Integrated CD-ROM The text and CD-ROM are now completely integrated. The student is advised to consult the CD at appropriate places throughout the book. Over forty wonderful simulations allow the reader to interactively explore many of the key physical principles. In addition, the CD-ROM contains many new WARM-UPS and WALK-THROUGHS, all with complete solutions. No problem in the text is duplicated on the CD. Moreover, the CD now contains an extensive math review. In addition to all of that new content, the CD has been completely restructured so that it runs easier and faster.

A Complete Ancillary Package

The following comprehensive teaching and learning package accompanies this book.

For the Student

Your Personal Tutor CD-ROM, Version 2003 Free with every student copy of the text! Each copy of the text is available with a FREE tutorial CD-ROM, Your Personal

---

374 Chapter 11 Waves & Sound

The **period** is the time it takes for a wave to go through one complete oscillation.

To Find the Ve

A periodic wave (wh as the endless repeti → **period** (T) of such

---

6 Chapter 1 An Introduction to Physics

Internet

For more on symmetry and conservation go to the website dedicated to teaching symmetry in the introductory physics curriculum:

www.emmynoether.com

---

756 Chapter 21 AC and Electronics

STUDY GUIDE

As was the case with dc analysis, when solving ac circuit problems the goal is often to find the equivalent impedance and then use Ohm's Law.

The quantities X_C and X_L, which depend on frequency, usually need to be determined. (1) **Compute those individual reactances.** Next, X_C and X_L must be combined, not algebraically, but by adding them as time-independent phasors (see Fig. 21.19). (2) Thus, **compute the combined reactance** $X = (X_L - X_C)$. This has the form it does because the phasors are always colinear and oppositely directed (180° out-of-phase). (3) Then **compute the impedance of the entire circuit** using the Pythagorean Theorem: $Z = (R^2 + X^2)^{1/2}$. Notice that the sign of X is of no significance here. (4) Knowing either the voltage or the current, **use the ac version of Ohm's Law** ($V = IZ$) **to compute** I or V. As we'll see, this overall scheme works for both series and parallel circuits.

Tutor, written by the author specifically to correlate with the textbook. Hecht has written hundreds of problems—many new to this edition—designed to tutor a student progressing through the course. These problems include **WARM-UPS**—interactive multiple choice questions that establish a basic level of understanding; **WALK-THROUGH EXAMPLES**—in which a typical homework problem is broken down into pieces and analyzed step-by-step in an interactive manner; 50 **INTERACTIVE EXPLORATIONS**—state-of-the-art physics simulations that allow students to do everything from firing a ball out of a virtual cannon to learning about projectile motion, to varying the parameters in a Carnot Cycle; and **FURTHER DISCUSSIONS**—additional textual material available only on the CD-ROM.

Student Solutions Manual Written by Jerry Shi and Eugene Hecht, this manual includes answers to selected odd-numbered discussion questions, answers to odd-numbered multiple choice questions, and solutions to selected odd-numbered problems (approximately 10%) not already solved in the book.

For the Instructor

Complete Solutions Manual Written by Jerry Shi and Eugene Hecht, this manual contains answers to all discussion questions, answers to all multiple choice questions, and solutions to all end-of-chapter problems in the text.

Test Items This feature includes more than 1000 multiple choice and short-answer questions. The notation of the test items carefully follows that of the main text.

Instructor's Manual This manual contains suggested demonstrations, homework, chapter/lecture outlines, teaching hints, and a correlation chart of problems mapped to those in the second edition.

ExamView® Computerized Testing (ISBN: 0-534-39694-1) Create, deliver, and customize tests and study guides (both print and online) in minutes with this easy-to-use assessment and tutorial system. *ExamView* offers both a Quick Test Wizard and an Online Test Wizard that guide you step-by-step through the process of creating tests, while the unique "WYSIWYG" capability allows you to see the test you are creating on the screen exactly as it will print or display online. You can build tests of up to 250 questions using up to 12 question types. Using *ExamView's* complete word processing capabilities, you can enter an unlimited number of new questions or edit existing questions.

Transparencies Acetate transparencies in full color include over 100 illustrations from the text, enlarged for use in the classroom and lecture halls.

Multimedia Manager: A Microsoft® PowerPoint® Link Tool (ISBN: 0-534-39696-8) This easy-to-use multimedia lecture tool allows you to quickly assemble art and database files with notes to create fluid lectures. The CD-ROM includes a database of animations and images from Brooks/Cole physics titles. The simple interface makes it easy for you to incorporate graphics, digital video, animations, and audio clips into your lectures.

MyCourse 2.0 Ask us about our new FREE online course builder! Whether you want only the easy-to-use tools to build it or the content to furnish it, Brooks/Cole offers you a simple solution for a custom course Web site that allows you to assign, track, and report on student progress; load your syllabus; and more. Contact your Thomson–Brooks/Cole representative for details or visit http://mycourse.thomsonlearning.com for a demonstration.

The Brooks/Cole Physics Resource Center, http://physics.brookscole.com When you adopt a Thomson–Brooks/Cole physics text, you and your students can have access to a rich array of teaching and learning resources that you won't find anywhere else. This outstanding site features everything from online quizzing to tutorials, and more! It's the ideal way to make teaching and learning an interactive and intriguing experience. The site includes Suggested Course and Lecture Outlines; a Syllabus Builder; Downloadable PowerPoint presentation slides; topic-by-topic Internet links for every chapter in many Brooks/Cole physics texts; and so much more!

Core Concepts in College Physics, Version 2.0 and Workbook (Microsoft® Windows®/Macintosh® CD-ROM; ISBN: 0-03-033701-1) The *Core Concepts in College Physics CD-ROM* set applies the power of multimedia to the introductory physics course, offering full-motion animation and video, engaging interactive graphics, clear and concise text, and guiding narration. Hundreds of screens provide in-depth coverage of the abstract and often difficult principles of physics, making clear connections between physical concepts and mathematics. *Core Concepts* contains more than 350 video clips, both animated and live, including laboratory demonstrations, real-world examples, and more. Many of these videos are highly interactive, allowing students to manipulate key objects within the frame. The accompanying *Workbook* contains practical physics problems coordinated with the CD-ROM, along with worked solutions. *Core Concepts* can be purchased as a stand-alone product or packaged with the text for significant savings. To package *Core Concepts in College Physics, Version 2.0 and Workbook* with Hecht's *PHYSICS: Algebra/Trig*, Third Edition, please use ISBN: 0-534-45627-8. Contact your local Thomson–Brooks/Cole representative for details or call (800) 423-0563.

WebAssign This is a Web-based homework delivery, collection, grading, and recording service developed at North Carolina State University. Instructors who sign up for WebAssign can assign homework to their students, using questions and problems taken directly from *PHYSICS: Algebra/Trig*, Third Edition. Details about and a demonstration of WebAssign are available at http://webassign.net/info. For more information about ordering this service, contact WebAssign at webassign@ncsu.edu.

Acknowledgments

Over the years, I have received comments, suggestions, articles, and photos from hundreds of colleagues, and I most sincerely thank them all, especially

William Achor, *Western Maryland College*, David Adamson, *Plainview High School*, Elise Adamson, *Wayland Baptist University*, W. Ariyasinghe, *Baylor University*, Andrew Baden, *University of Maryland*, John Barrere, *Apex High School*, Henry J. Bartol, *Newark Academy*, Fred Becchetti, *University of Michigan*, James R. Benbrook, *University of Houston*, Stan Bergkamp, *Maize High School*, Susan Berrend, *Intermountain High School*, Randy Booker, *University of North Carolina at Asheville*, Robert Boughton, *Bowling Green State University*, Michael Browne, *University of Idaho*, Gary Buckwalter, *Daytona Beach Community College*, Anthony Buffa, *California Polytechnic State University, San Luis Obispo*, Phillips B. Burnside, *Ohio Wesleyan University*, Myron Campbell, *University of Michigan*, The Physics Class of 1994, *Wilcox High School*, James A. Castiglione, *Kean University*, Lawrence B. Coleman, *University of California, Davis*, Joe Conti, *Piscataway High School*, Dennis B. Crossley, *University of Wisconsin—Sheboygan*, Jose D'Arruda, *University of North Carolina at Pembroke*, J. R. Dennison, *Utah State University*, Eric Dietz, *California State University, Chico*, Paul Dooher, *Freeport High School*, Jess Dowdy, *Northeast Texas Community College*, Miles Dresser, *Washington State University*, Larry Dukerich, *Dobson High School*, Mark Edwards, *Hofstra University*, Terry L. Ellis, *Jacksonville University*, John Erdei, *University of Dayton*, David Ernst, *Vanderbilt University*, Mark Fischer, *College of Mount St. Joseph*, Roger Freedman, *University of California, Santa Barbara*, Raymond E. Frey, *University of Oregon*, Sherman Frye, *Northern Virginia Community College*, Carl L. Gibson, *North Central High School*, John

Eric Goff, *Oberlin College,* Martin Goodson, *Delta College,* Mark Gross, *California State University, Long Beach,* Robert Hallock, *University of Massachusetts,* Teodoro Halpern, *Ramapo College,* Grant W. Hart, *Brigham Young University,* Lory E. Heron, *Hillsboro High School,* Michael Hughes, *Bear River High School,* Fred Inman, *Mankato State University,* Bill Jameson, *DeForest Area High School,* Fred Jarka, *Stark State College of Technology,* John Gordon Jones, *College of Charleston,* Michael R. Kelley, *William Henry Harrison High School,* Sanford Kern, *Colorado State University,* James Kettler, *Ohio University, Eastern Campus,* Terence M. Kite, *Pepperdine University,* Paul Kramer, *State University of New York at Farmingdale,* Victor Kriss, *University of San Diego,* Brian Lane, *Hillsborough Community College,* Paul Lee, *Louisiana State University,* Stan Linman, *Exeter Union High School,* Richard Mancuso, *College of Brockport, SUNY,* Mark Manley, *Kent State University,* Robert March, *University of Wisconsin,* David Markowitz, *University of Connecticut,* Donald H. Martins, *University of Alaska Anchorage,* Robert Marx, *Edward R. Murrow High School,* Troy J. Massey, *Poland Seminary High School,* George Matous, *Indiana University of Pennsylvania,* Lorin Matthews, *Baylor University,* Susan Matts, *Mary Washington College,* David K. McDaniels, *University of Oregon,* Rich McNamara, *Washington-Lee High School,* Marvin Morris, *San Jose State University,* Anthony J. Nicastro, *West Chester University of Pennsylvania,* Andrew Njaa, *Falmoth High School,* Kari Oakes, *Charles E. Smith Jewish Day School,* Melvin Oakes, *University of Texas, Austin,* Dale Wilson Olson, *University of Northern Iowa,* Mary Ethel Parrott, *Notre Dame Academy,* Robert Payton, *Enterprise High School,* Terry Jay Phillips, *The University of Texas at Brownsville and Texas Southmost College,* Channon Price, *University of Alaska,* Jon Pumplin, *Michigan State University,* G. N. Rao, *Adelphi University,* Kurt Reibel, *Ohio State University,* Marc Reif, *Fayetteville High School,* Diane Riendeau, *Barrington High School,* Dawn Rowden-Brett, *Cochise College,* Sema'an I. Salem, *California State University, Long Beach,* Kandula Sastry, *University of Massachusetts, Amherst,* Pauline Seales, *Pioneer High School,* William Sebastian, *Northgate High School,* Dwight Sederholm, *North Idaho College,* Tom Speicher, *Martin County High School,* Chuck Stone, *Forsyth Technical Community College,* Joseph Vanderway, *Granada Hills High School,* Ted Walker, *Diablo Valley College,* Eric Weeks, *Emory University,* Richard Witt, *Ladue Horton Watkins High School,* Nancy A. Woods, *DesMoines Area Community College,* and Jiang Yu, *Fitchburg State College.*

I am particularly grateful to Prof. Zvonimir Hlousek of California State University, Long Beach; Prof. Jerry Shi of Pasadena City College; and Prof. Mark Headlee of Armand Hammer United World College for their many contributions. And I'd also like to thank my students, Andrew Gennawey, Kevin Hansen, Maria Rivera, and Gregory Schoellig, for all their help.

I am indebted to all those people at Brooks/Cole and beyond who made the project possible: Sean Wakely, Michael Johnson, Angus McDonald, Marcus Boggs, Lisa Moller, and Beth Wilbur; and to Assistant Editor Emily Levitan, Marketing Manager Kelley McAllister, Art Director Vernon Boes, Production Supervisor Tom Novack, and Media Editor Sam Subity, each of whom contributed their special talents and I thank them. I also thank Kelly Shoemaker, whose commitment to excellence will always be remembered. Jerry Shi of Pasadena City College did a brilliant job of upgrading the solutions manuals. I remain most appreciative of the contribution of Connie Jirovsky who championed this project from its outset.

The book was produced by HRS *Interactive,* which did a splendid job of putting it all together despite having to deal with an uncompromising, relentless, picky author. Lorraine Burke watched over every aspect of the process with exceptional skill, extraordinary attention to detail, and sustained equanimity in the face of every sling and arrow; Ed Burke's design was strong, innovative, and beautiful, his high standards an inspiration; Alan Wiener took care of a range of production issues with inexhaustible patience and good humor; Hilda Espreo performed wonders leading the team of compositors; Pat Hannagan produced incomparable art; Chris Burke created the marvelous chapter-opener collages; and Jennifer Burke skillfully managed the photographs. It was a joy to work with such gifted, creative people, and they all have my deepest appreciation.

Finally, I bow appreciatively to my Personal Editor, friend and wife, Carolyn Eisen Hecht (*eshet chayil*). She tracked down hundreds of photographs, researched innumerable obscure topics, efficiently wrote hundreds of e-mail messages, spelled thousands of words, more or less patiently tolerated my papers lying scattered, stacked, and stashed everywhere throughout the house, and in general watched the store. Knowing full well the tremendous personal sacrifices demanded by this project, I gratefully thank her for coping with one more edition of one more book.

If you have any comments about this edition, suggestions for the next edition, or favorite problems you'd like to share, send them to E. Hecht, Adelphi University, Physics Department, Garden City, N.Y. 11530 (or genehecht@aol.com).

Eugene Hecht
Freeport, N.Y.

To Ca, b. w. l.

Brief Contents

1   An Introduction to Physics   1

2   Kinematics: Speed & Velocity   20

3   Kinematics: Acceleration   53

4   Newton's Three Laws   85

5   Centripetal Force & Gravity   141

6   Energy   171

7   Momentum & Collisions   209

8   Rotational Motion   234

9   Solids, Liquids, & Gases   283

10   Elasticity & Oscillations   329

11   Waves & Sound   371

12   Thermal Properties of Matter   423

13   Heat & Thermal Energy   456

14   Thermodynamics   490

15   Electrostatics: Forces   529

16   Electrostatics: Energy   569

17   Direct Current   607

18   Circuits   635

19   Magnetism   667

20   Electromagnetic Induction   709

21   AC & Electronics   743

22   Radiant Energy: Light   783

23   The Propagation of Light: Scattering   812

24   Geometrical Optics & Instruments   845

25   Physical Optics   888

26   Special Relativity   933

27   The Origins of Modern Physics   968

28   The Evolution of Quantum Theory   996

29   Quantum Mechanics   1027

30   Nuclear Physics   1056

31   High-Energy Physics   1094

Contents

1 An Introduction to Physics 1

1.1 Law and Theory 1
1.2 The Modern Perspective 2

Measurement 6
1.3 Length 7
1.4 Mass and Weight 9
1.5 Time 10
1.6 Significant Figures 11

The Language of Physics 13
1.7 Equations 13
1.8 Graphs 14
1.9 Approximations and Checks 15
Core Material & Study Guide 16
Discussion Questions 16
Multiple Choice Questions 16
Suggestions on Problem Solving 17
Problems 17

2 Kinematics: Speed & Velocity 20

Speed 20
2.1 Average Speed 20
2.2 Constant Speed 24
2.3 Delta Notation: The Change in a Quantity 26
2.4 Instantaneous Speed 26

Velocity 28
2.5 The Displacement Vector 29
2.6 Some Vector Algebra 30
2.7 Instantaneous Velocity 32
2.8 Components and Vector Addition 34

Relative Motion 38
2.9 Velocity with Respect to... 39
Core Material & Study Guide 44
Discussion Questions 44

Multiple Choice Questions 45
Suggestions on Problem Solving 46
Problems 47

3 Kinematics: Acceleration 53

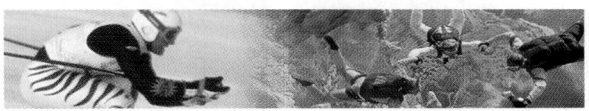

The Concept of Acceleration 53
3.1 Average Acceleration 53
3.2 Instantaneous Acceleration 55

Uniformly Accelerated Motion 56
3.3 Constant Acceleration 56
3.4 The Mean Speed 58
3.5 The Equations of Constant Acceleration 60

Free-Fall 62
3.6 Air Drag 62
3.7 Acceleration Due to Gravity 63
3.8 Straight Up & Down 65
3.9 Two-Dimensional Motion: Projectiles 68
Core Material & Study Guide 76
Discussion Questions 76
Multiple Choice Questions 77
Suggestions on Problem Solving 78
Problems 79

4 Newton's Three Laws 85

The Three Laws 85
4.1 The Law of Inertia 85
4.2 Force 88
4.3 The Second Law 92
4.4 Interaction: The Third Law 96

Dynamics & Statics 97
4.5 The Effects of Force: Newton's Laws 98
4.6 Weight: Gravitational Force 101

4.7 Coupled Motions 106
4.8 Friction 107
4.9 Equilibrium: Statics 115
Core Material & Study Guide 123
Discussion Questions 125
Multiple Choice Questions 126
Suggestions on Problem Solving 128
Problems 128

5 Centripetal Force & Gravity 141

Centripetal Force 141
5.1 Centripetal Acceleration 142
5.2 Center-Seeking Forces 144

Gravity 147
5.3 The Law of Universal Gravitation 147
5.4 Terrestrial Gravity 152

The Cosmic Force 155
5.5 The Laws of Planetary Motion 155
5.6 Satellite Orbits 157
5.7 Effectively Weightless 160
5.8 The Gravitational Field 162
Core Material & Study Guide 163
Discussion Questions 163
Multiple Choice Questions 165
Suggestions on Problem Solving 166
Problems 166

6 Energy 171

The Transfer of Energy 171
6.1 Work 172

Mechanical Energy 180
6.2 Kinetic Energy 180
6.3 Potential Energy 183

Conservation of Mechanical Energy 187
6.4 Mechanical Energy 188
6.5 Applying Conservation of Energy 192
6.6 Power 195
6.7 Energy Conservation & Symmetry 198
Core Material & Study Guide 199
Discussion Questions 200

Multiple Choice Questions 201
Suggestions on Problem Solving 202
Problems 202

7 Momentum & Collisions 209

Linear Momentum 209
7.1 Impulse & Momentum Change 210
7.2 Varying Force 212
7.3 Jets and Rockets 215
7.4 Conservation of Linear Momentum 216
7.5 Collisions 219
7.6 Linear Momentum & Symmetry 226
Core Material & Study Guide 227
Discussion Questions 228
Multiple Choice Questions 228
Suggestions on Problem Solving 229
Problems 230

8 Rotational Motion 234

The Kinematics of Rotation 234
8.1 Angular Displacement 234
8.2 Angular Speed 236
8.3 Angular Acceleration 239
8.4 Equations of Constant Angular
Acceleration 241

Rotational Equilibrium 243
8.5 Torque 244
8.6 Second Condition of Equilibrium 246
8.7 Extended Bodies & the c.g. 249

The Dynamics of Rotation 253
8.8 Torque & Rotational Inertia 253
8.9 Rotational Kinetic Energy 258
8.10 Angular Momentum 259
8.11 Conservation of Angular Momentum 260
Core Material & Study Guide 265
Discussion Questions 266
Multiple Choice Questions 268
Suggestions on Problem Solving 270
Problems 270

9 Solids, Liquids, & Gases 283

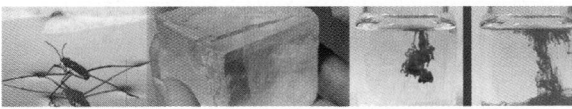

Atoms & Matter 283
9.1 Atomism 283
9.2 Density 285
9.3 The States of Matter 287
Fluid Statics 290
9.4 Hydrostatic Pressure 290
9.5 Pascal's Principle 296
9.6 Buoyant Force 298
Fluid Dynamics 302
9.7 Fluid Flow 303
9.8 The Continuity Equation 305
9.9 Bernoulli's Equation 306
Core Material & Study Guide 316
Discussion Questions 317
Multiple Choice Questions 319
Suggestions on Problem Solving 320
Problems 321

10 Elasticity & Oscillations 329

Elasticity 329
10.1 Hooke's Law 329
10.2 Stress & Strain 334
10.3 Strength 336
10.4 Elastic Moduli 339
Harmonic Motion 344
10.5 Simple Harmonic Motion 344
10.6 Elastic Restoring Force 349
10.7 The Pendulum 352
10.8 Damping, Forcing, & Resonance 354
Core Material & Study Guide 359
Discussion Questions 360
Multiple Choice Questions 362
Suggestions on Problem Solving 363
Problems 363

11 Waves & Sound 371

Mechanical Waves 371
11.1 Wave Characteristics 371

11.2 Transverse Waves: Strings 376
11.3 Compression Waves 379
Sound 382
11.4 Acoustics: Sound Waves 382
11.5 Wavefronts & Intensity 387
11.6 The Speed of Sound in Air 389
11.7 Hearing Sound 390
11.8 Intensity-Level 393
11.9 Sound Waves: Beats 395
11.10 Standing Waves 397
11.11 The Doppler Effect 404
Core Material & Study Guide 411
Discussion Questions 412
Multiple Choice Questions 413
Suggestions on Problem Solving 415
Problems 415

12 Thermal Properties of Matter 423

Temperature 423
12.1 Thermodynamic Temperature & Absolute Zero 425
Thermal Expansion 429
12.2 Linear Expansion 429
12.3 Volumetric Expansion: Solids & Liquids 432
The Gas Laws 435
12.4 The Laws of Boyle, Charles, & Gay-Lussac 435
12.5 The Ideal Gas Law 438
12.6 Phase Diagrams 440
12.7 Kinetic Theory 441
Core Material & Study Guide 446
Discussion Questions 447
Multiple Choice Questions 448
Suggestions on Problem Solving 449
Problems 450

13 Heat & Thermal Energy 456

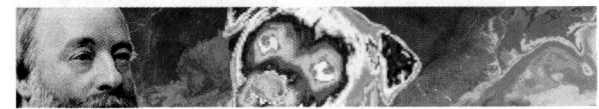

Thermal Energy 456
13.1 Heat & Temperature 457
13.2 Quantity of Heat 459
13.3 The Mechanical Equivalent of Heat 461
13.4 Specific Heat: Calorimetry 463
Change of State 467
13.5 Melting & Freezing 467

13.6   Vaporization 469
13.7   Boiling 471

The Transfer of Thermal Energy 473
13.8   Radiation 473
13.9   Convection 475
13.10 Conduction 476
       Core Material & Study Guide 480
       Discussion Questions 481
       Multiple Choice Questions 482
       Suggestions on Problem Solving 484
       Problems 484

14   Thermodynamics 490

The First Law of Thermodynamics 490
14.1   Conservation of Energy 491
14.2   Thermal Processes & Work 493
14.3   Isothermal Change of an Ideal Gas 497
14.4   Adiabatic Change of an Ideal Gas 500

Cyclic Processes: Engines & Refrigerators 503
14.5   The Carnot Cycle 504
14.6   Refrigeration Machines & Heat Pumps 507

The Second Law of Thermodynamics 510
14.7   Entropy 513
       Core Material & Study Guide 519
       Discussion Questions 520
       Multiple Choice Questions 521
       Suggestions on Problem Solving 522
       Problems 523

15   Electrostatics: Forces 529

Electromagnetic Charge 529
15.1   Positive and Negative Charge 530
15.2   Insulators and Conductors 532

The Electric Force 537
15.3   Coulomb's Law 538

The Electric Field 544
15.4   Definition of the E-Field 546

15.5   Gauss's Law 557
       Core Material & Study Guide 561
       Discussion Questions 562
       Multiple Choice Questions 563
       Suggestions on Problem Solving 564
       Problems 565

16   Electrostatics: Energy 569

Electric Potential 569
16.1   Electrical-PE & Potential 569
16.2   Potential of a Point-Charge 575
16.3   The Potential of Several Charges 579
16.4   Conservation of Charge 582

Capacitance 584
16.5   The Capacitor 584
16.6   Capacitors in Combination 588
16.7   Energy in Capacitors 592
       Core Material & Study Guide 595
       Discussion Questions 596
       Multiple Choice Questions 597
       Suggestions on Problem Solving 598
       Problems 599

17   Direct Current 607

Flowing Electricity 607
17.1   Electric Current 607

Resistance 614
17.2   Ohm's Law 615
17.3   Resistivity 618
17.4   Voltage Drops & Rises 622
17.5   Energy and Power 623
       Core Material & Study Guide 625
       Discussion Questions 626
       Multiple Choice Questions 628
       Suggestions on Problem Solving 629
       Problems 629

18 Circuits 635

Circuit Principles 635
18.1 Sources and Internal Resistance 635
18.2 Resistors in Series & Parallel 639
18.3 Ammeters and Voltmeters (Optional) 645
18.4 *R-C* Circuits 647

Network Analysis (Optional) **649**
18.5 Kirchhoff's Rules 650
     Core Material & Study Guide 654
     Discussion Questions 654
     Multiple Choice Questions 656
     Suggestions on Problem Solving 658
     Problems 659

19 Magnetism 667

Magnets & the Magnetic Field 667
19.1 Permanent Magnets 667
19.2 The Magnetic Field 670

Electrodynamics 677
19.3 Currents and Fields 677
19.4 Ampère's Law 684

Magnetic Force 687
19.5 The Force of a Moving Charge 687
19.6 Forces on Wires 691
     Core Material & Study Guide 696
     Discussion Questions 697
     Multiple Choice Questions 699
     Suggestions on Problem Solving 702
     Problems 702

20 Electromagnetic Induction 709

Electromagnetically Induced emf 709
20.1 Faraday's Induction Law 710

20.2 Motional emf 716

Generators 720
20.3 The AC Generator 721
20.4 The DC Generator 722

Self-Induction 725
20.5 Inductance 725
20.6 The *R-L* Circuit: Transients 728
20.7 Energy in the Magnetic Field 730
     Core Material & Study Guide 731
     Discussion Questions 732
     Multiple Choice Questions 734
     Suggestions on Problem Solving 736
     Problems 736

21 AC & Electronics 743

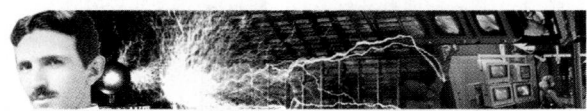

Alternating Current 743
21.1 AC and Resistance 745
21.2 AC and Inductance 748
21.3 AC and Capacitance 750

***R-L-C* AC Networks** (Optional) **753**
21.4 *R-L-C* Circuits 753
21.5 The Transformer 761
21.6 Domestic Circuits and Hazards 762

Electronics (Optional) **765**
21.7 Semiconductors 765
     Core Material & Study Guide 772
     Discussion Questions 773
     Multiple Choice Questions 775
     Suggestions on Problem Solving 776
     Problems 776

22 Radiant Energy: Light 783

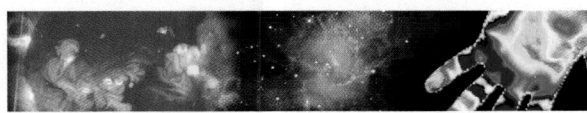

The Nature of Light 783
22.1 Waves and Particles 783
22.2 Electromagnetic Waves 784
22.3 Waveforms and Wavefronts 786
22.4 Energy and Irradiance 790
22.5 The Origins of EM Radiation 791

22.6   Energy Quanta 793
22.7   Atoms & Light 795
        The Electromagnetic-Photon Spectrum 796
22.8   Radiowaves 797
22.9   Microwaves 799
22.10   Infrared 800
22.11   Light 800
22.12   Ultraviolet 802
22.13   X-Rays 804
22.14   Gamma Rays 805
        Core Material & Study Guide 805
        Discussion Questions 806
        Multiple Choice Questions 807
        Suggestions on Problem Solving 808
        Problems 808

23 The Propagation of Light: Scattering 812

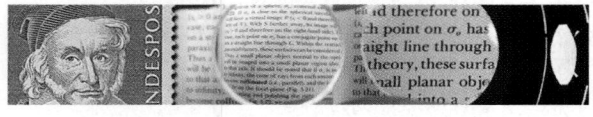

Scattering 812
23.1   Rayleigh Scattering: Blue Skies 812
        Reflection 816
23.2   Internal and External Reflection 816
23.3   The Law of Reflection 817
        Refraction 822
23.4   The Index of Refraction 822
23.5   Snell's Law 824
23.6   Total Internal Reflection 829
        The World of Color 832
23.7   White, Black, and Gray 832
23.8   Colors 833
        Core Material & Study Guide 835
        Discussion Questions 836
        Multiple Choice Questions 837
        Suggestions on Problem Solving 839
        Problems 839

24 Geometrical Optics & Instruments 845

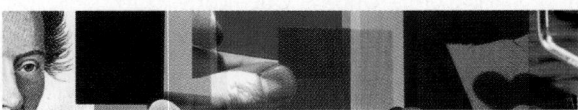

Lenses 845
24.1   Aspherical Surfaces 846
24.2   Spherical Thin Lenses 848
24.3   Focal Points and Planes 851
24.4   Extended Imagery: Lenses 852
24.5   A Single Lens 857
24.6   Thin-Lens Combinations (Optional) 863
        Mirrors 871
24.7   Curved Mirrors 871
        Core Material & Study Guide 878
        Discussion Questions 879
        Multiple Choice Questions 880
        Suggestions on Problem Solving 881
        Problems 882

25 Physical Optics 888

Polarization 888
25.1   Natural Light 889
25.2   Polarizers 892
25.3   Polarizing Processes 895
        Interference 899
25.4   Young's Experiment 901
25.5   Thin-Film Interference 905
25.6   The Michelson Interferometer 908
        Diffraction 910
25.7   Single-Slit Diffraction 913
25.8   The Diffraction Grating 915
25.9   Circular Holes and Obstacles 917
25.10   Holography 920
        Core Material & Study Guide 922
        Discussion Questions 923
        Multiple Choice Questions 924
        Suggestions on Problem Solving 925
        Problems 925

26 Special Relativity 933

Before the Special Theory 933
26.1   The Michelson—Morley Experiment 933
        The Special Theory of Relativity 936
26.2   The Two Postulates 936
26.3   Simultaneity and Time 939
26.4   The Hatter's Watch: Time Dilation 942

26.5 Shrinking Alice: Length Contraction 945
26.6 Tweedledee & Tweedledum: The Twin Effect 947
26.7 Addition of Velocities 949
      Relativistic Dynamics 951
26.8 Relativistic Momentum 951
26.9 Relativistic Energy 954
      Core Material & Study Guide 959
      Discussion Questions 960
      Multiple Choice Questions 961
      Suggestions on Problem Solving 963
      Problems 963

27 The Origins of Modern Physics 968

      Subatomic Particles 968
27.1 The Quantum of Charge 968
27.2 Cathode Rays: Particles of Charge 970
27.3 X-Rays 973
27.4 The Discovery of Radioactivity 976
      The Nuclear Atom 979
27.5 Rutherford Scattering 980
27.6 Atomic Spectra 982
27.7 The Proton 984
27.8 The Neutron 985
27.9 Radiation Damage: Dosimetry 986
      Core Material & Study Guide 989
      Discussion Questions 990
      Multiple Choice Questions 991
      Suggestions on Problem Solving 992
      Problems 993

28 The Evolution of Quantum Theory 996

      The Old Quantum Theory 996
28.1 Blackbody Radiation 996
28.2 Quantization of Energy:
      The Photoelectric Effect 1002
28.3 Bremsstrahlung 1006
28.4 The Compton Effect 1007
      Atomic Theory 1009
28.5 The Bohr Atom 1010

28.6 Stimulated Emission: The Laser 1014
28.7 Atomic Number 1017
      Core Material & Study Guide 1019
      Discussion Questions 1020
      Multiple Choice Questions 1021
      Suggestions on Problem Solving 1023
      Problems 1023

29 Quantum Mechanics 1027

      **The Conceptual Basis of
      Quantum Mechanics 1027**
29.1 De Broglie Waves 1027
29.2 The Principle of Complementarity 1029
29.3 The Schrödinger Wave Equation 1031
      Quantum Physics 1035
29.4 Quantum Numbers 1035
29.5 The Zeeman Effect 1036
29.6 Spin 1038
29.7 The Pauli Exclusion Principle 1038
29.8 Electron Shells 1039
29.9 The Uncertainty Principle 1044
29.10 QED and Antimatter 1047
      Core Material & Study Guide 1050
      Discussion Questions 1051
      Multiple Choice Questions 1052
      Suggestions on Problem Solving 1053
      Problems 1053

30 Nuclear Physics 1056

      Nuclear Structure 1056
30.1 Isotopes: Birds of a Feather 1056
30.2 Nuclear Size, Shape, & Spin 1059
30.3 The Nuclear Force 1062
30.4 Nuclear Stability 1064
      Nuclear Transformation 1068
30.5 Radioactive Decay 1068
30.6 The Weak Force 1072
30.7 Gamma Decay 1073
30.8 Half-Life 1074
30.9 Induced Radioactivity 1078
30.10 Fission & Fusion 1079

Core Material & Study Guide 1086
Discussion Questions 1087
Multiple Choice Questions 1088
Suggestions on Problem Solving 1089
Problems 1090

31   High-Energy Physics 1094

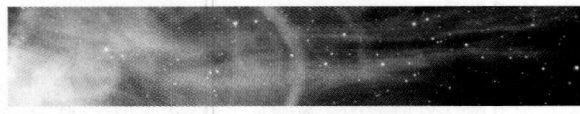

Elementary Particles 1094
31.1   Leptons 1095
31.2   Hadrons 1097

Quantum Field Theory 1100
31.3   Gauge Theory 1102
31.4   Quarks 1104
31.5   Quantum Chromodynamics 1108
31.6   The Electroweak Force 1109
31.7   GUTs & Beyond:
       The Creation of the Universe 1112
       Core Material & Study Guide 1115
       Discussion Questions 1115
       Multiple Choice Questions 1117
       Suggestions on Problem Solving 1118
       Problems 1118

Appendixes:
A Brief Mathematical Review A-1

Appendix A: Algebra A-1
A-1   Exponents A-1
A-2   Powers of Ten: Scientific Notation A-2
A-3   Logarithms A-3
A-4   Proportionalities & Equations A-3
A-5   Approximations A-4
A-6   The Change in a Quantity A-5

Appendix B: Geometry A-5

Appendix C: Trigonometry A-6

Appendix D: Vectors A-6

Appendix E: Dimensions A-6

Complete Math Review CD-ROM

Answer Section AN-1

Credits C-1

Index I-1

Applications to the Life Sciences and Bioengineering

1. The relationship between the amount of creatinine excreted and human body mass   *p. 16*

2. The physiological effects of acceleration on the human body   *p. 104*

3. Friction and the knee joint; how synovial fluid lubricates the joint   *p. 115*

4. The foot as a lever, showing the relationship of the ankle joint and the Achilles tendon   *p. 116*

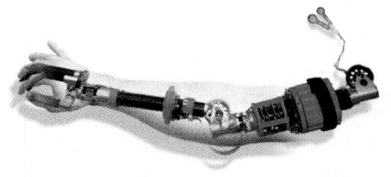

5. A modern prosthetic arm that incorporates a shoulder that rotates overhead and an elbow that bends   *p. 116*

6. The human body as a structure and how the weights of its various parts are supported by the musculoskeletal system   *p. 116*

7. The Thomas leg splint   *p. 118*

8. The application of statics to the treatment of congenital hip dislocation   *p. 121*

9. Analogy between vertebrates and a trusswork structure, discussing the roles of the vertebrae and rib compression members, and the intercostal muscles as tension members   *p. 123*

10. The application of orthodontal wire braces as a statics problem   *p. 136*

11. The action of the quadriceps muscle and the patellar tendon at the knee   *p. 139*

12. An analysis of the forces on the hip in a standing male showing the action of the femur on the pelvis   *p. 139*

13. A discussion of how muscles do work; the contraction of muscle fibers and the rate at which work is done   *p. 177*

14. Human power and its relationship to oxygen consumption: the metabolic rate   *p. 197*

15. The basal metabolic rate, BMR   *p. 207*

16. Human mobility and the torques generated by muscles acting on bones   *p. 246*

17. The concurrent force system acting on the head of a person leaning forward   *p. 247*

18. The center-of-gravity of the human arm   *p. 249*

19. The center-of-gravity of the extended leg as a composite system   *p. 250*

20. The center-of-gravity of the human body   *p. 250*

21. The human arm as an end-pivoted lever system, showing the action of the humerus and biceps   *p. 251*

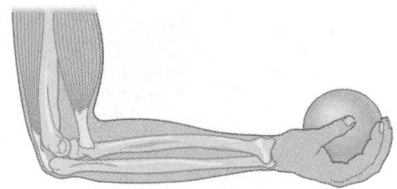

22. A discussion of bipedal stability and balance as related to the location of the center-of-gravity   *p. 252*

23. The moment of inertia of the human body   *p. 255*

24. The detection of rotation of the head via the action of the semicircular canals in the inner ear   *p. 257*

25. The long jump   *p. 262*

26. The operation of the arm via torques created by the triceps and biceps   *p. 266*

27. The action of the knee in relation to torques   *p.274*

28. Measurement of the location of the c.g. of a person's body   *p. 280*

29. The body as a collection of atoms and the human being as star dust   *p. 284*

30. The explosion of a staphylococcus bacterium exposed to antibiotic   *p. 295*

31. Fluid pressures at various locations within the human body   *p. 295*

32. Intravenous liquid infusion and pressure   *p. 295*

33. Muscle and fat ratios of the body as they relate to floating   *p. 300*

34. Vascular flutter, snoring, and arteriosclerosis and their relationship to fluid flow and the Venturi Effect   *pp. 311, 312*

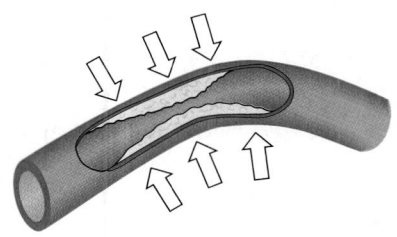

35. The operation of the human heart and blood pressure   *p. 314*

36. Measuring blood pressure   *p. 315*

37. The cardiovascular system   *p. 315*

38. The operation of the aspirator for removing unwanted body fluids   *p. 322*

39. The behavior of biological materials such as bone, collagen, and elastin under stress   *pp. 333, 335*

40. The rupture of the human femur   *pp. 338, 341*

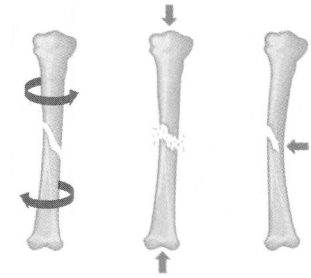

41. Fatigue of medical sutures and bone braces   *pp. 338, 339*

42. The strengths of muscles and tendons   *p. 338*

43. Torques and the breaking angles of human bones. Common bone breaks   *p. 341*

44. The resonant frequencies of internal

organs and the effects of vibrations with high input energy levels *p. 381*

45. *Ultrasound and its application to producing sonograms* **p. 381**

46. *The operation of the human ear as a detector of sound* **p. 390**

47. *Pitch, timbre, and loudness as characteristic responses of the human auditory system* **pp. 390, 391**

48. *The frequency range of the human voice* **p. 394**

49. *The perception of loudness and its relation to intensity and sound level* **p. 392**

50. *Percussion of the human body* **p. 404**

51. *The vocal tract—larynx, pharynx, and nasal and oral cavities* **p. 405**

52. *The use of Doppler-shifted ultrasound waves to measure blood flow in the diagnosis of deep-vein thrombosis* **p. 409**

53. *Lowering the temperature of the body during an operation* **p. 428**

54. *Pressure in the chest cavity and the operation of the lungs* **p. 436**

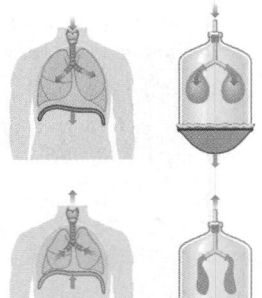

55. *The energy content of various foods* **p. 460**

56. *Energy "consumed" per hour by the human body in various activities* **p. 461**

57. *Skin color & radiation* **p. 474**

58. *Thermal conduction; skin, fat, and countercurrent exchange* **p. 477**

59. *The electrophoretic separation of biological molecules such as DNA* **p. 547**

60. *Electrically responsive cells in the shark* **p. 548**

61. *The analysis of blood samples via an applied electric field* **p. 576**

62. *A map of equipotentials in the brain of an epileptic* **p. 578**

63. *Equipotentials across the chest due to heart activity* **p. 580**

64. *Electrocardiography* **p. 580**

65. *Electrical stimulation of the heart* **p. 593**

66. *Propagation of an electrical signal, the action potential pulse along an axon* **pp. 603, 605**

67. *Electrocardiography* **p. 622**

68. *Strain gauges measuring axial loading of the jaw* **p. 636**

69. *The cardiac pacemaker and R-C circuits* **p. 649**

70. *Magnetotactic bacteria* **p. 670**

71. *Levitation of a live frog in an inhomogeneous magnetic field* **p. 674**

72. *Magnetic fields generated by currents in the brain* **p. 678**

73. *The stimulation of paralyzed muscles resulting from spinal cord injury* **p. 715**

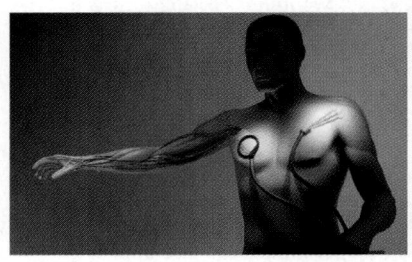

74. *Effects of various currents on the human body* **p. 764**

75. *Abnormal blood flow visualized via IR thermography* **p. 800**

76. *A broad band view of the human arm* **p. 801**

77. *Sensitivity of the human eye* **p. 906**

78. *Using fiberoptics to explore within the human body* **p. 831**

79. *The human eye* **p. 860**

80. *Farsightedness (hyperopia), nearsightedness (myopia)* **pp. 866, 867**

81. *Polarized light reflected from a pygmy octopus* **p. 890**

82. *Seeing your eye from the inside* **p. 913**

83. *Fraunhofer diffraction from cancer cells* **p. 917**

84. *Radiation damage: dosimetry and the radiosensitivity of organisms* **pp. 986–989**

85. *Laser surgery* **p. 1016**

86. *Electron micrograph of a neuron* **p. 1028**

87. *Positron Emission Tomography (PET)* **p. 1049**

88. *Magnetic resonance imaging (MRI)* **p. 1062**

89. *Natural radioactivity of foods* **p. 1069**

90. *You are radioactive. The absorption of radioactive materials by the human body* **p. 1071**

91. *Gamma-ray images of the bones of a cancer patient* **p. 1074**

92. *The amount of radioactive potassium in the human body* **p. 1077**

93. *Technetium-99m used to detect cancer of the bone* **p. 1078**

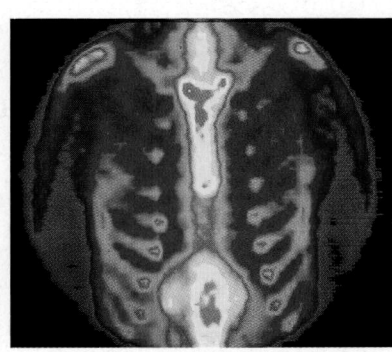

94. *Radioactive isotopes used in nuclear medicine* **p. 1078**

Chapter 1
An Introduction to Physics

*P*hysics is the study of the material Universe. And though we all know what the words material Universe refer to, it would take volumes to attempt to explain what they mean. The problem is that the basic underlying concepts—such as matter, space, and time—are difficult if not impossible to define conceptually, a shortcoming we share with the philosophers.

The primary property of matter is that it is observable; it interacts in ways that result in changes we can detect. **Physics is the study of matter, interaction, and change.**

Practically speaking, physics is a guide to action in the complex world of natural phenomenon: What do you do if you find yourself in an open field during a thunderstorm? What's the maximum height you can jump from without breaking your legs?

1.1 Law and Theory

The aim of physics is to understand the natural events that we are party to, to understand the Universe—what it is, how it works, what it's doing, and maybe even how it came to exist, and how it might end. If that bold agenda is possible at all, it's only because natural phenomena occur in reproducible ways; there are rules and rhythms in the seeming chaos. Accordingly, we'll study motion, heat, sound, electricity, light, and so forth, all the categories of our understanding.

The doing of physics usually involves a record of phenomena: observation and the collection of **data**—information objectively perceived. An event is observed, either deliberately or not, and things are recorded, *measured* (How much? How long? How many? How big?). *Physics quantifies; it associates numbers with its concepts.* Within the body of observations, patterns are sought that reveal relationships among the data. A **law** is a description of a relationship in Nature that manifests itself in recurring patterns of events. It can be a prescription for how things change, or it might be a statement of how things remain immune to change.

How does our basic underlying knowledge, our world view, of space and time, of matter interacting with matter, account for the patterns of events? How can we understand what is observed? **Theory** is the explanation of phenomena in terms of more basic natural processes and relationships. To explain phenomena, we draw on intuition and imagination and guess at what is happening. We propose **hypotheses** and leap beyond what we know to what might be. *A construct of definitions, hypotheses, and laws that explain some*

The question "What is matter?" is of the kind that is asked by metaphysicians and answered in vast books of incredible obscurity.

BERTRAND RUSSELL (1959)
BRITISH PHILOSOPHER

... in Science it is when we take some interest in the great discoverers and their lives that it becomes endurable, and only when we begin to trace the development of ideas that it becomes fascinating.

JAMES CLERK MAXWELL (1831–1879)
BRITISH PHYSICIST

Read Sections 1.1 and 1.2 and just try to enjoy the ideas. This is an introduction to the subject, and you are certainly not expected to remember it all.

Learning physics is similar, in some ways, to learning a foreign language; every chapter will have a bunch of new words with very specific meanings, that you must know (p. 16). And, almost without exception, every chapter will build on the previous one.

As you go through this chapter, keep your eye on the big concepts (e.g., atoms, the Four Forces, length, mass, weight, time, and significant figures). There's lots of information that's here just to help you grasp those central ideas, but it's information you should not memorize, like the mass of the Empire State Building.

Pay close attention to the discussion of units (p. 7) in Sections 1.3, 1.4, and 1.5 and to the distinction between weight and mass (p. 9). Make sure you understand Section 1.6 on significant figures.

Insofar as the propositions of mathematics refer to reality they are not certain, and insofar as they are certain they do not refer to reality.

ALBERT EINSTEIN

---

observed order in Nature is the essence of theory. A powerful theory allows us to deduce already known laws as well as to predict new occurrences and relationships that, once tested and confirmed, may become new laws.

All theory is tentative. The formalisms of physics cannot be proven absolutely true or even absolutely false. Our current world view may be wrong and all the proofs based on it equally wrong, our hypotheses may be in error, our laws may only be approximate. Newtonian Mechanics worked amazingly well for 200 years before it began to deviate even slightly from the most refined observations. Then Albert Einstein proved that the world view on which it was based is wrong; Newtonian Mechanics is a brilliant approximation of a more complete truth.

Physics must be continuously tested against Nature. Our ideas must correspond in every detail to the observed workings of the Universe, but even that does not prove them true. The testing never ends.

At this moment, there is no single significant theory in physics that can be said to be both finished and completely satisfactory. We have a magnificent understanding, but it's only the beginning. The game is wonderfully wide open and the real surprise, as Einstein suggested, is that we can do so much and know so little.

1.2 The Modern Perspective

Physics has evolved to its present state as a result of about 2500 years of effort. Simple early theories were displaced by increasingly more effective later ones, and these, in turn, were subsumed by still more powerful, far-reaching, and complex contemporary treatments.

Classical Physics

The discipline, as it developed up until the 1920s, is known as Classical Physics. It rests on three theoretical pillars (Fig. 1.1): Newtonian Mechanics, Thermodynamics, and Electromagnetic Theory. The completion of Classical Physics was accomplished by Einstein. His Special Theory of Relativity (1905) reformulates our conception of space, time, and motion. His General Theory of Relativity (1915) considers gravity in terms of curved space-time and allows us to begin to comprehend such cosmic phenomena as black holes, pulsars, and the Big Bang creation of the Universe. **Classical Physics represents the basic conceptual material that must be understood if we are to interact effectively with the physical environment on an everyday level.**

Physicists were supremely confident throughout the last decades of the 1800s. But, as time went by, the atomic domain seemed to become more and more inaccessible to classical analysis.

Figure 1.1 Classical Physics encompasses Newtonian Mechanics (the study of motion and gravity), Electromagnetic Theory (the synthesis of electricity and magnetism), Thermodynamics (the study of thermal energy), and Relativity. It corresponds to physics prior to the 1920s.

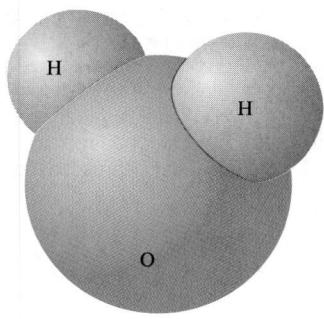

Figure 1.2 A water molecule consists of two hydrogen (H) atoms each sharing its single electron with an oxygen (O) atom. The angle made by these "Mickey Mouse" ears is 104°, and the distance between O and H nuclei is 0.096 nm. The radius of a single hydrogen atom is about 0.05 nm.

Atomic Theory

One of the great revelations of physics is that ***all the ordinary things in our world (mountains, frogs, TV sets, etc.) are composed of combinations of tiny specks of matter called atoms***. There are roughly 110 different basic kinds of atoms [oxygen (O), hydrogen (H), uranium (U), etc.]. This is a tremendously important idea; with a few simple insights (which will be explored later) we can understand much of how the material Universe works.

(a) ***Atoms experience a short-ranged attraction that operates effectively over distances of only one or two atomic diameters.*** When they are close together, they are powerfully drawn to one another. In fact, atoms usually bond strongly in small groupings called ***molecules***; there are oxygen molecules (O_2), water molecules (H_2O), sugar molecules, and so forth (Fig. 1.2). This same short-ranged attraction bonds molecules to one another (although less strongly because of the larger distances) to form all the familiar bulk substances from plastic to peanut butter. In the end, it's this interatomic attraction that keeps your nose stuck to your face and holds most things together. {For an introduction to what holds atoms and molecules together click on CHEMICAL BONDS under FURTHER DISCUSSIONS on the CD.} 💿

(b) ***When atoms and/or molecules are far apart (a few atomic diameters), they barely interact.*** That's why you can walk through air and, to some extent, why liquids pour (Fig.1.3).

(c) ***When atoms and/or molecules come very close to one another, they powerfully repel.*** Consequently, when atoms attract to form bulk matter, there will be a limit to how close together they come. This is in part why ordinary matter isn't very dense; in fact, like all solids, your own body is mostly empty space (p. 287). If the atoms in the Empire State Building (331×10^6 kg) were crushed, removing all the space, it would end up smaller than a one-inch sewing needle. Because atoms repel, you stand *on* the floor rather than passing through it.

(d) ***Atoms and molecules are in never-ending motion.*** Open a bottle of perfume and in seconds you can smell it across the room.

We've learned in the last hundred years or so that atoms are complex systems made up of still smaller grains, all in perpetual motion. An atom consists of a minuscule core or ***nucleus*** encircled by a distant whirling cloud of ***electrons***. The nucleus is formed of two varieties of composite particles: ***neutrons*** and ***protons***.

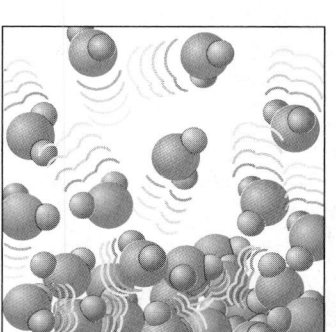

(a) Water and steam

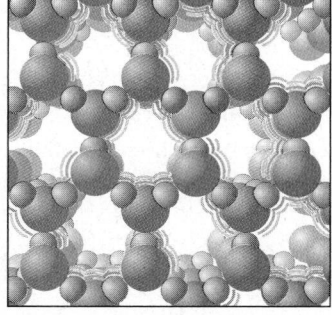

(c) Ice

Figure 1.3 Water molecules enlarged about 50 million times. (a) steam; (b) liquid water; and (c) ice.

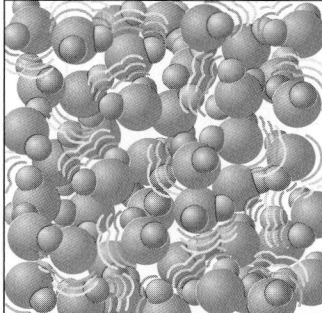

(b) Water

Quantum Physics

In 1905 Einstein showed that light was emitted and absorbed in minute blasts he called *quanta*. These "particles" of light, the most tenuous form of matter, are now known as ***photons***. Today, the material Universe is considered to be composed of myriads of identical clones of a handful of elementary particles—the ***quarks***, ***leptons***, and ***photons***. Neutrons and protons are made of quarks (Fig. 1.4), whereas the electron is a fundamental particle, a lepton. All of us, all of the trees, the planets, and the stars are basically exquisitely organized swarms of quarks surrounded by leptons, absorbing and emitting photons.

Contemporary physics has extended its domain from the subnuclear to the entire Universe. Though it seems presumptuous, Genesis has become a branch of physics. Particle accelerators can replicate conditions that existed in the first moments of creation. All of this, and more, is the fruit of a new theoretical approach that began to develop in the 1920s called ***Quantum Mechanics***. We know now that physical processes, although they appear

Light is, in short, the most refined form of matter.
LOUIS DE BROGLIE (1892–1987)
FRENCH PHYSICIST

These are the tracks left by an upward stream of high-speed subatomic particles. It is from millions of pictures like this that we have learned much of what we know about the subatomic domain. Every aspect of the tumult in the photo will be explained by the time we get to the end of this book. For more details now see Goronway T. Jones, "A Simple Estimate of the Mass of the Positron," *The Phys. Teacher* **31**, 95 (1993).

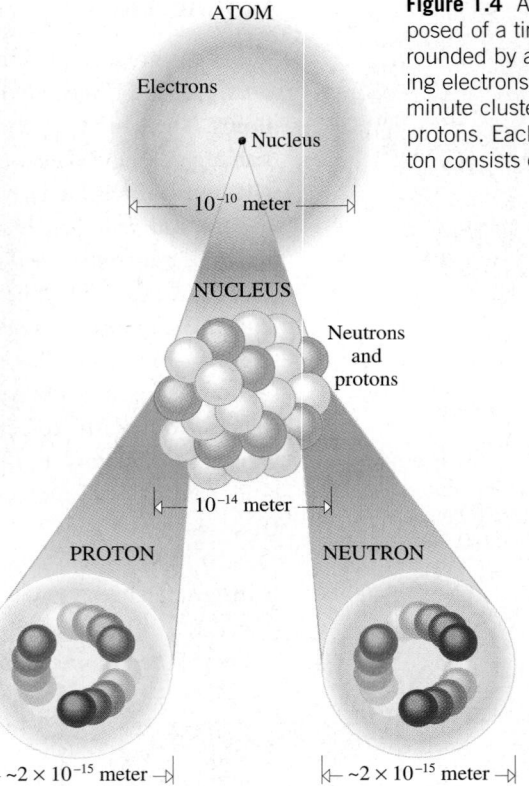

Figure 1.4 An atom is composed of a tiny nucleus surrounded by a cloud of fast-moving electrons. The nucleus is a minute cluster of neutrons and protons. Each neutron and proton consists of three quarks.

to be continuous on a macroscopic level, are fundamentally discrete. In the subatomic microworld *change occurs in jumps*; the landscape is quantized.

The Four Forces

Matter, whatever it is, interacts. Contemporary theoretical physics maintains that in the Universe as it exists today, there are four distinct basic interactions, or forces: gravitational, electromagnetic, strong, and weak. These determine the entire range of observable occurrences (Table 1.1), and we will study each in turn.

When a change in the state of something occurs, we assume a cause and logically presume that the change is carried out by a force. When an apple falls, or a grasshopper leaps, or a supernova explodes, or a neutron decays, these very different events unfold because of the involvement of the Four Forces. In this modern context, we initially define **force** broadly as **the agent of change**.

The **gravitational force** keeps you, the atmosphere, and the seas fixed to the surface of the planet. Though gravity is the weakest of all the interactions, it is also the least selective—it acts between all particles. Because its range is unlimited and because it is only attractive, gravity rules the cosmos on a grand scale. It holds the Earth in orbit around the Sun, keeps the Sun locked within our galaxy of a hundred thousand million stars, and reaches all across the thousands of millions of galaxies that constitute the Universe. If it were much stronger than it is, it would quickly halt the present expansion of the Universe and send all the galaxies collapsing back down into oblivion.

The **electromagnetic force** gives rise to the short-range interatomic attraction. As such, it binds together the smaller

QUANTUM MECHANICS & THE LOSS OF CERTAINTY

Perhaps the most profound philosophical distinction between Classical and Modern Physics is the **loss of certainty**. The Universe may grind along like a great machine, but its tiniest parts are inherently elusive. We cannot possibly know everything we would like to know. There is a soft edge to certainty, and the future has yet to be written. Despite its baffling text and strange version of reality, Quantum Mechanics has never once failed an experimental test. It is extremely reliable, though not transparently comprehensible. It is likely true that "no one understands Quantum Mechanics" (Sect. 29.9), although it is equally true that in some wonderful way Quantum Mechanics understands the Universe.

To be sure, we say that an area of space is free of matter; we call it empty, if there is nothing present except a gravitational field. However, this is not found in reality, because even far out in the universe there is starlight, and that is matter.

ERWIN SCHRÖDINGER (1887–1961)
AUSTRIAN PHYSICIST

Table 1.1

The Four Forces

Force	Couples with	Strength*	Range
Strong	Quarks and particles composed of them	10^4	$\approx 10^{-15}$ meter
Electromagnetic	Electrically charged particles	10^2	Unlimited
Weak	Most particles	10^{-2}	$\approx 10^{-17}$ meter
Gravitational	All particles	10^{-34}	Unlimited

*Strengths listed are the forces (in newtons) between two protons separated by a distance equal to their diameter ($\approx$2 femtometers).

things—atoms, molecules, trees, buildings, and you. As with the gravitational force, we can perceive its consequences directly (just hold two magnets near one another). The electromagnetic force acting on a microscopic level is responsible for a variety of what once were thought to be different forces. It produces contact forces, as for example, between a fist and a punching bag, or a hamburger and teeth. It generates friction and drag, produces adhesion and cohesion, and is responsible for the elastic force. The range of the electromagnetic force is unlimited, but it can be either attractive or repulsive, which generally restricts its influence to only short distances. Electromagnetism governs chemistry and biology, rules life and death, and keeps the Earth and everything on it from being crushed by gravity.

The **strong force** binds quarks together to form neutrons and protons and binds these together to form the nuclei of atoms (Fig. 1.4). It's an extremely powerful, exceedingly short-range force whose influence extends only a few times the diameter of a proton. That tiny range accounts for its not obtruding directly into our normal experience, and, as a result, not being discovered until the twentieth century. Without it, familiar matter, from planets to puppies, would all disintegrate into a fine quark dust.

The **weak force** is a million times fainter than the strong force and a hundred times shorter in range. Among other things, it transforms one type of quark into another. It can change protons into neutrons and is thereby responsible for the slow decay of certain radioactive atoms. A star like the Sun derives its energy from the thermonuclear furnace at its core. There, it "burns" hydrogen into helium, a process that relies on the gradual transformation of protons via the weak force; no weak force, no sunshine, no life on Earth.

The great agenda of physics is to describe the Universe in terms of a minimum number of basic concepts. Contemporary physics operates under the assumption that at its most fundamental level, Nature is superbly concise; that what appears to be a great diversity of existence and activity is the outward manifestation of an inner simplicity. Guided by the commitment to *unification*, physicists have established that at exceedingly high energies, now available only in the laboratory, the electromagnetic and weak forces coalesce into a single **electroweak force**. Carrying the logic to its natural end, it is widely believed that the Four Forces sprang from a single superforce that once pervaded the primordial fireball of creation.

Amalie (Emmy) Noether (1882–1935) was an outstanding mathematician who did most of her work in abstract algebra. After a long struggle she won the right as a woman to lecture, without pay, at Göttingen University in Germany. It was in 1918 that she presented the results of an analysis dealing with symmetry that became a guiding principle for contemporary physics. Noether taught at Göttingen until 1933 when she came to the United States after the Nazis learned that she was Jewish and expelled her from Germany.

Conservation & Symmetry

Physicists in the last quarter of the twentieth century have come to a remarkable revelation: on a fundamental level, Nature possesses a range of symmetries that are associated with the most basic laws of physics. By understanding these symmetries, we can produce theories that presumably have the correct internal logic.

There are many kinds of symmetry, the simplest of which is geometrical. Imagine a perfectly uniform disc mounted on a central axle. Suppose you close your eyes and someone rotates the disc into a new position. When you open your eyes, everything is as it was before—you can't tell that anything has occurred. *When a change in a physical system leaves some aspect of the system unchanged, the system possesses a corresponding sym-*

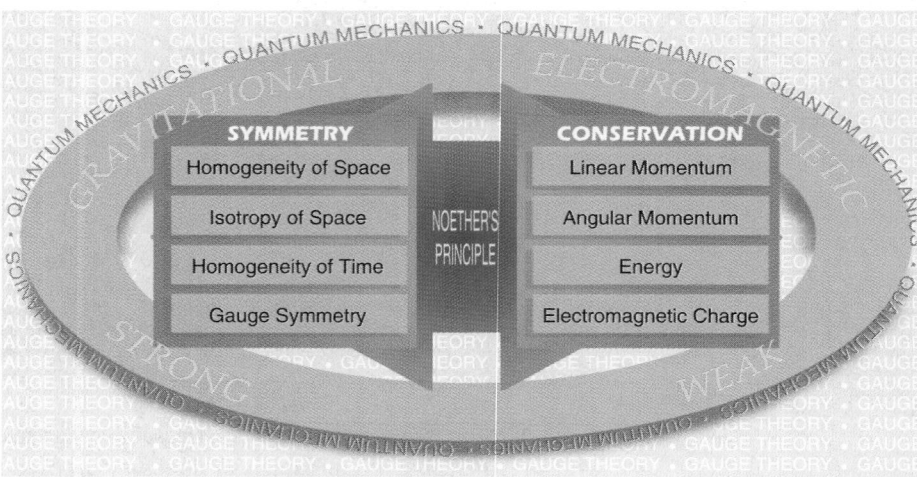

Figure 1.5 Some of the theoretical ideas we'll study in future chapters. Don't worry about the details now—it will all become clear in time.

metry. The existence of a symmetry means that a feature of the system is changeless, or *invariant*.

In Nature, there are important invariances associated with more subtle symmetries. The first person to grasp the implications of this was Amalie Noether. Among the central conceptual pillars of physics are three conservation laws, powerful insights that took hundreds of years to develop: Conservation of Linear Momentum (p. 216), Conservation of Angular Momentum (p. 260), and Conservation of Energy (p. 187). Each law maintains that a specific quantity is constant at every moment as an isolated system undergoes change. As a result of Noether's work (1918), we now know that *for every conservation law there is a corresponding symmetry, and vice versa.* When we study these conservation laws (Fig. 1.5), we'll learn how each is a direct consequence of a specific symmetry of space and time. {For more on this topic click on SYMMETRY under FURTHER DISCUSSIONS on the CD.} Many physicists believe that the wildly energetic, incredibly dense infant Universe was a place of unparalleled symmetry and undifferentiated simplicity. The superforce that drove creation ultimately broke into the Four Forces as the Universe cooled, calmed, and shattered into complexity (Fig. 31.23).

Internet

For more on symmetry and conservation go to the website dedicated to teaching symmetry in the introductory physics curriculum:

www.emmynoether.com

Measurement

Physics is founded on experimentation: we ground it directly to Nature through observations that entail the measurement of physical quantities. Today, the scientific community follows the **Système International** (SI), a program of weights and measures based, for the most part, on the following units: for length, the *meter* (m); for mass, the *kilogram* (kg); for time, the *second* (s); for electric current, the *ampere* (A); and for temperature, the *kelvin* (K).

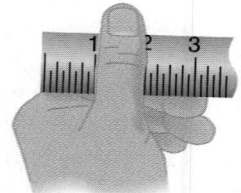

EXPLORING PHYSICS ON YOUR OWN

Body Measurements: The human body was the most convenient measuring device in early times. Then the *inch* was the width of a man's thumb at its widest. The *foot* began as the length of a Roman sandal. Legend has it that in the twelfth century Henry I of England fixed the *yard*, or double-cubit, to be the greatest length from his nose to his extended outermost fingertip. Compare each of these with your own body measures. A yard is about 3.5 inches less than a meter, and that's about the width of a *hand*, which is still the way we measure horses (1 hand = 10.16 cm).

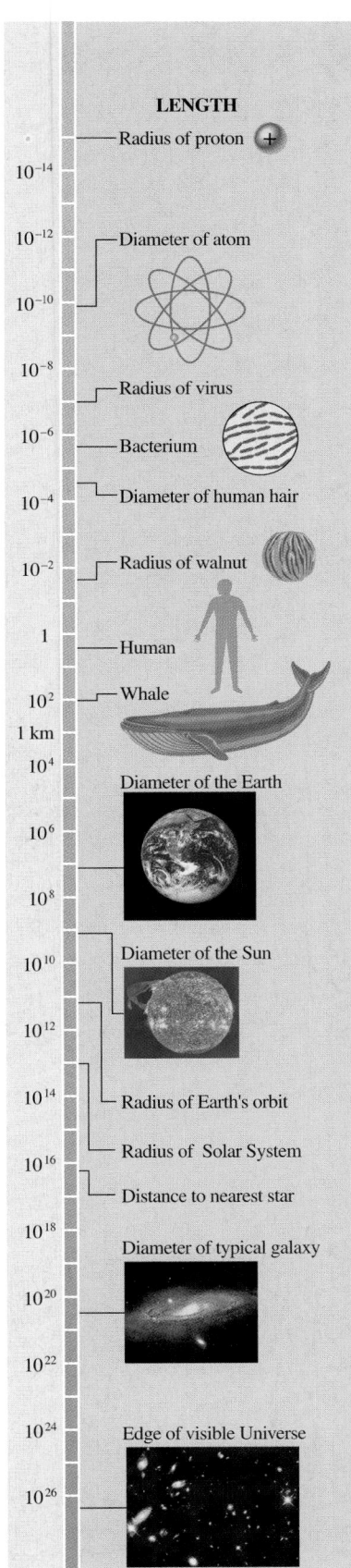

Figure 1.6 The sizes (in meters) of objects from protons to the Universe itself.

1.3 Length

Length is a distance or extent in space (whatever distance is, or for that matter, whatever space is). Stick out your arm and announce that the distance from elbow to farthest out-stretched fingertip is hence to be known as a *cubit*. That's the way it was done over 4000 years ago in Egypt and Mesopotamia. *An agreed-upon measure by which we reckon any physical quantity is a* **unit**; a cubit is a unit of length. The Great Pyramid was built to cubit specifications, as was Noah's Ark. Of course, you'd have a hard time building a large precise structure if everyone's forearms are different. An advanced society must evolve *an unchanging embodiment of each unit to serve as a primary reference, or* **standard**. The black granite master cubit was just such a standard against which cubit sticks in Egypt were regularly checked.

For centuries, primitive body measures were used throughout Europe. But the rise of science and the growth of commerce both suffered from the lack of a standardized system. Though decimal schemes for weights and measures were proposed elsewhere (e.g., by Thomas Jefferson) the French were the first (1799) to adopt the approach in their **Metric System**, dividing all quantities into 10, 100, 1000, and so on, equal parts. Their new unit of length was the **meter**, abbreviated m. It was to be one ten-millionth of the distance from the North Pole to the Equator along a meridian line passing through Paris. However patriotic, errors in the difficult measurement made the whole exercise quite arbitrary. They could have just as well stayed home and used the length of Napoleon's sword instead. Be that as it may, in 1889 the distance between two lines inscribed in a bar of platinum-iridium alloy became the world's standard meter (Fig. 1.6).

The Metric System, often called the *mks system*, is based on the *m*eter, *k*ilogram, and *s*econd; it's the foundation of the SI. Although shunned by the British for over a century, the new weights and measures were soon adopted throughout Europe. The United States, in happy distant oblivion, continued to use the dreadful British units and is still trying to make the transition. The system used today in the United States is called **U.S. Customary Units**. It's based on length, weight, and time expressed in feet, pounds, and seconds, respectively (Table 1.2).

Because a billion (1 000 000 000) in the United States is a thousand times smaller than a European billion (1 000 000 000 000), the common names for numbers beyond a million are rarely used in scientific work. Instead, **prefixes** (Table 1.3) are added to the units. Greek prefixes like kilo, mega, and giga are multiples; the Latin ones like centi, milli, and micro are subdivisions.

Table 1.2

Length and Volume Equivalents	
1 light-year (5.88 × 10^{12} mi)	9.461×10^{15} m
1 mile (5280 ft)	1.609 km
1 yard (yd; 3 ft)	0.914 m
1 foot	0.304 8 m
1 inch	2.54 cm
1 meter	3.281 ft
1 meter	39.37 in.
1 meter	1.094 yd
1 gallon U.S. (4 quarts)	3.785×10^{-3} m^3
1 quart (2 pints)	9.464×10^{-4} m^3
1 ounce	2.957×10^{-5} m^3
1 tablespoon (1/2 oz)	1.479×10^{-5} m^3
1 cubic foot	28 320 cm^3
1 cubic foot	0.028 m^3
1 cubic meter	1.308 yd^3
1 cubic meter	61 023 in.^{3}

CALCULATOR TIPS

Your calculator, not having much imagination, carries out instructions in the sequence they're given. For example, compute

$$x = \frac{4+6}{2}$$

If you enter [4] [+] [6] [÷] [2] [=]
the answer will be $4 + 6/2 = 7$, which is wrong. The 4 and 6 must be combined before dividing by 2. One way to handle that is to enter [4] [+] [6] [=] get 10 and then divide that by 2. Alternatively, if your calculator has parens, you can enter

[(] [4] [+] [6] [)] [÷] [2] [=]

Table 1.3

Common Unit Prefixes

Power of ten	Prefix	Pronunciation	Symbol	Example
10^{15}	peta-	pe′ta	P	petasecond, Ps
10^{12}	tera-	ter′a	T	terahertz, THz
10^{9}	giga-	jig′a[a]	G	gigavolts, GV
10^{6}	mega-	meg′a	M	megawatt, MW
10^{3}	kilo-	kil′oe	k	kilogram, kg
10^{2}	hecto-	hek′toe[b]	h	
10	deka-	dek′a[b]	da	
10^{-1}	deci-	des′i	d	decibel, dB
10^{-2}	centi-	sen′ti	c	centimeter, cm
10^{-3}	milli-	mil′i	m	millimeter, mm
10^{-6}	micro-	my′kroe	μ	microgram, μg
10^{-9}	nano-	nan′oe	n	nanometer, nm
10^{-12}	pico-	pee′koe	p	picofarad, pF
10^{-15}	femto-	fem′toe	f	femtometer, fm

[a]Although the given pronunciation is supposed to be standard, many physicists use gig′a instead. [b]The prefixes hecto- and deka- are avoided in physics.

Example 1.1 [I]* How many centimeters are there in 25.00 meters?

Solution This is a unit-conversion problem going from meters to centimeters. (1) TRANSLATION—Convert a known number of meters to centimeters. (2) GIVEN: A length of 25.00 m. FIND: Equivalent in cm. (3) PROBLEM TYPE—Unit conversion. (4) PROCEDURE—The most systematic way to do all unit conversions is to form a ratio equal to 1, with the unit you want on top and the one you want to replace on the bottom. (5) CALCULATION—Since 1 m is exactly 100 cm,

$$1 = \frac{100.0 \text{ cm}}{1.000 \text{ m}}$$

Then multiply the original number by the unit ratio

$$25.00 \text{ m} \times \frac{100.0 \text{ cm}}{1.000 \text{ m}} = \boxed{2500 \text{ cm}}$$

Units can be treated as algebraic quantities and so meters (m) cancel, leaving centimeters (cm).

Quick Check: There are 100 cm per 1 m, hence 25 m × 100 cm/m = 2500 cm.

* Examples are labeled [I], [II], or [III], according to increased difficulty.

{*For more worked problems click on* **WALK-THROUGH EXAMPLES** *in* **CHAPTER 1** *on the CD.*}

Example 1.2 [I] How many cubic centimeters are in a cubic meter?

Solution This is a unit-conversion problem going from cubic meters to cubic centimeters. (1) TRANSLATION—Convert 1 cubic meter into cubic centimeters. (2) GIVEN: 1 m³. FIND: corresponding number of cm³. (3) PROBLEM TYPE—Unit conversion. (4) PROCEDURE—We know that 1 m = 100 cm exactly, hence a cube 1 m by 1 m by 1 m is 100 cm by 100 cm by 100 cm. (5) CALCULATION—From Fig. 1.7 {or see **MATH REVIEW: PART B** on the CD }, the volume (V) is given by the length times width times height:

$$V = (100 \text{ cm})(100 \text{ cm})(100 \text{ cm}) = \boxed{1\,000\,000 \text{ cm}^3}$$

There are exactly one million cubic centimeters in a cubic meter.

Quick Check: Imagine a big cube made up of 1 cm³ blocks. There will be 100 layers, each 1-cm tall and 100-cm long by 100-cm wide. Each layer contains 100×100 or 10 000 little cubes, and there are 100 such layers, yielding $100 \times 10\,000$, which equals 10^6 cm³.

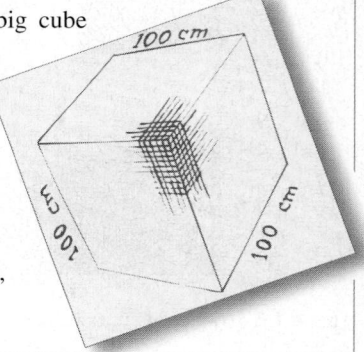

People who are new to physics often have trouble converting the units for areas and volumes. Remember that a 1.00-m² area, A, corresponds to

$$A = (100 \text{ cm})(100 \text{ cm}) = 10\,000 \text{ cm}^2$$

and a 1.00-m³ volume, V, corresponds to

$$V = (100 \text{ cm})(100 \text{ cm})(100 \text{ cm})$$
$$V = 1\,000\,000 \text{ cm}^3$$

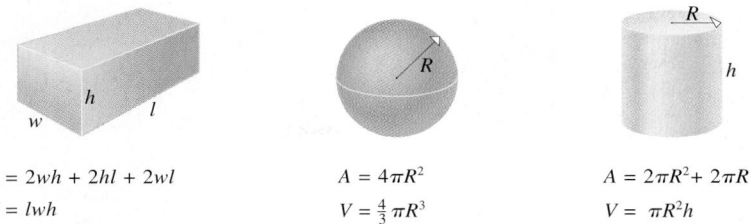

$$A = 2wh + 2hl + 2wl$$
$$V = lwh$$

$$A = 4\pi R^2$$
$$V = \tfrac{4}{3}\pi R^3$$

$$A = 2\pi R^2 + 2\pi Rh$$
$$V = \pi R^2 h$$

Figure 1.7 The areas and volumes of several geometric figures.

A related approach that's used, especially where things are on a small scale, as is often the case in the life sciences, is the **cgs system** (*c*entimeter, *g*ram, *s*econd). Here, a *centimeter* (cm) is a one-hundredth part of a meter (1 cm = 0.01 m) and 1 inch (in.) is defined to be *exactly* 2.54 cm. The centimeter, gram, and second are SI units, but all the others in the cgs system (the units for force, pressure, density, etc.) are not.

With the advent of stable lasers, the speed of light was measured with extraordinary precision, and so in 1983 it became the basis of the last and final definition of the meter. The speed of light in vacuum (c) is now *defined* to be exactly 299 792 458 meters per second. Consequently, ***the meter is the distance light travels in vacuum during a time of exactly 1/299 792 458 second***. This standard will never need to be revised. Improvements in measuring time will simply improve the measurement of the meter.

1.4 Mass and Weight

The concepts of weight and volume were conceived out of the necessities of commerce in earliest times. So it was remarkable when the thirteenth-century theologian Aegidius Romanus proposed yet another measure of matter. His insight came from a quandary: The Catholic church maintained that the sacramental bread and wine were literally transformed into the body and blood of Christ. How could that be, given the obvious discrepancies in the weights and volumes of each before and after?

Aegidius suggested that there was another measure of substance in addition to weight and volume. That measure, the **quantity-of-matter**, indicates *how much matter* there is in a material object. This was the origin of one of the most basic ideas in physics, the concept of **mass** (*m*). Almost 400 years later, Newton used the terms *mass* and *quantity-of-matter* interchangeably. Indeed, he was among the first to use the word mass in this modern way.

Historically, **weight is the downward force exerted on (or by) an object at the surface of the Earth**, and it was assumed to be constant. The fact that the weight of an object varies with location was discovered inadvertently in 1671 by the astronomer J. Richer, who found that his pendulum clock, when brought from Europe to French Guiana, ran slow, losing $2\frac{1}{2}$ minutes each day. Newton promptly explained this surprising effect by distinguishing between *weight* and *mass* (see Discussion Question 4).

According to Newton, **mass is a fixed property of a sample of matter.** It is constant everywhere on Earth. In fact, we believe that the mass of an object will be found constant no matter where in the Universe it is. A thing (a frog, a shoe, a can of peas) has mass because the subatomic particles (quarks and leptons) that compose it have mass. **Masses interact gravitationally, and that results in the force called weight.** The weight of an object on Earth is essentially due to the gravitational interaction between it and the planet. At a far distance from all material bodies (planets, moons, stars, etc.), the weight of an object effectively vanishes. Mass never vanishes.

The unit of mass is the **kilogram** (kg). Today the standard kilogram is a platinum-iridium cylinder watched over by the International Bureau of Weights and Measures. It's a little thing about the size of a whisky shot glass, 39 mm (1.5 in.) in both height and diameter.

Kilogram Number 20 is the standard mass of the United States.

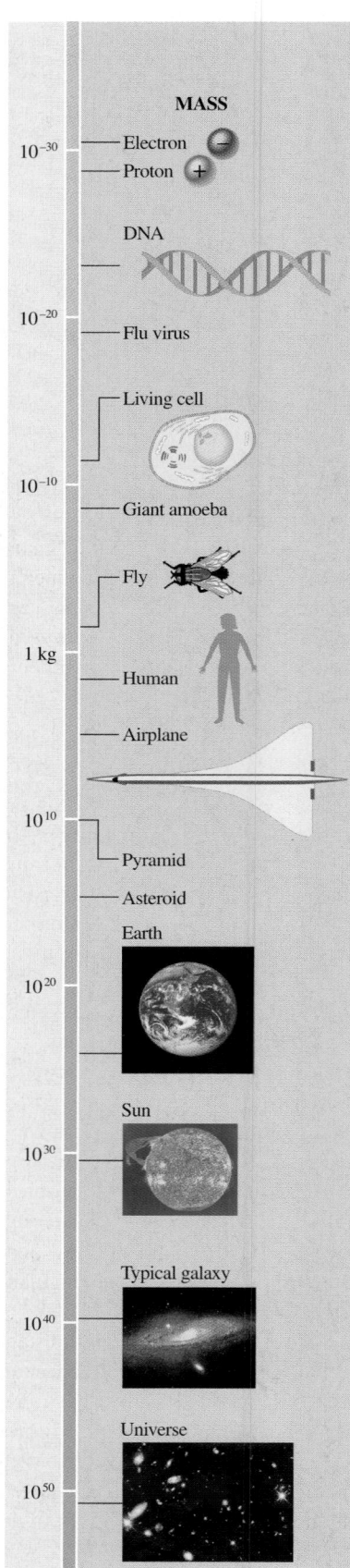

Figure 1.8 Masses (in kilograms) of objects from the electron to the Universe.

On Earth an object with a mass of 15.9 kg has a weight of 35 lb.

Table 1.4	
Weight-Mass Correspondence on Earth*	
1 gram	0.035 273 40 ounce
1 kilogram	2.204 622 pounds
1 ounce	28.349 52 grams
1 pound	0.453 592 4 kilogram

*Mass and weight are very different features of matter. The values given here are the weights that would be measured for the corresponding masses at the surface of the Earth.

A **gram** is a one-thousandth part of a kilogram. Typically, a gram of any solid corresponds to about 10^{21} atoms. Once the International Kilogram was agreed upon and copies made, a pan balance could then be used anywhere on the planet to compare the mass of an object with a set of standard masses (Fig. 1.8). Since gravity is the same at each pan, if the balance is level, then the masses are equal, independent of what they might weigh at that location.

Forty nearly identical kilogram masses were made in London in 1884. The standard for the United States is Kilogram Number 20, which sits in a basement vault at the National Institute of Standards and Technology (NIST). Whenever you buy anything in the U.S.A. by the gram, kilogram, ounce, grain, carat, pound, or ton, that measure, of either mass or weight, should be traceable directly back through a long series of calibrations to old Number 20 (Table 1.4).

1.5 Time

The rhythms of life—the passing of successive days, the pounding of a heart—suggest a progression, a sequencing of occurrences. Time is a measure of the unfolding of events, and it can be reckoned against uniform change. That's often done using some cyclic phenomenon like the rising of the Sun. More than 3000 years ago the Egyptians divided the day and night into 12 equal hours. Babylonian arithmetic had 60 as its number base, and there subsequently developed a tradition of subdividing things into 60 parts. It was probably in the fourteenth century, after the arrival of the mechanical clock, that the hour (h) got divided into 60 minutes (min). The word minute was abbreviated from the Latin for *first minūte part* (which is why the words minute and minūte are spelled the same). Later, when the *second minūte part* could be measured, it was natural to take 60 seconds (s) to a minute. {For more on the subject click on **SPACE & TIME** under **FURTHER DISCUSSIONS** on the **CD**.}

The SI unit of time, the **second**, is defined as the interval corresponding to 9 192 631 770 oscillations of the radiation from cesium-133 atoms. At present, the primary time standard in the U.S.A. is NIST F-1, an atomic fountain clock that drifts less than 1 second in 20 000 000 years.

Operationally, time is that which is measured by a clock. Conceptually, time is a measure of the rate at which change occurs. Facetiously, time is that which keeps everything from happening all at once (Fig. 1.9).

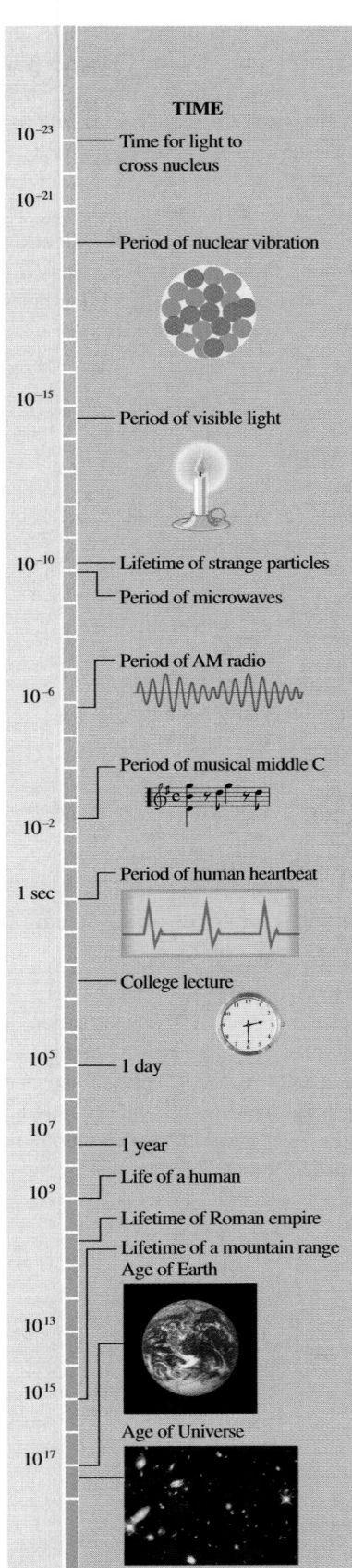

Figure 1.9 Intervals (in seconds) from the time it takes light to traverse a nucleus to the age of the Universe.

If we ever communicate with extraterrestrials, they surely won't use meters, kilograms, and seconds; but they will certainly measure length, mass, and time. We should have no trouble making the necessary conversions. Our physics, to the extent that it's correct, must be their physics.

1.6 Significant Figures

Measurement is very different from counting, even though both associate numbers with notions. We can count the number of beans in a jar and know it exactly. But we cannot measure the height of the jar exactly. ***There is no such thing as an exact measurement.***

Figure 1.10 shows a rod whose length is being measured with a ruler. Assuming one end of the ruler and rod are lined up precisely, the object is between 4.1 and 4.2 cm long. Because there are no finer divisions than 1 mm on the scale, fractions of a millimeter will have to be estimated. Doing that, we judge that the rod is about 4.15 cm long. It's inappropriate to say that the length is exactly 4.15 cm. The length seems closer to 4.15 cm than to either 4.14 cm or 4.16 cm, which is what we mean when we write 4.15, although it can be stated explicitly as 4.15 ± 0.01. The 4 is certain, as is the 1, but the 5 might be in error by as much as ± 1, and trying to arrive at any more figures with our ruler would be meaningless. A digit in a number is said to be a **significant figure** when it's known with some reliability, like the 4, the 1, and even the 5. When a result is presented, such as 4.15 cm, it can be assumed that all the figures are significant and that the 5 is the least significant.

To three significant figures, the length of the rod is 4.15 cm, or 41.5 mm, or 0.041 5 m, or 0.000 041 5 km. Each of these is equivalent, and each has three significant figures. Moving the decimal point and changing the unit prefix have no effect on the accuracy of the number. Hence, because there are *only zeros* and no other digits to the left of the 4, those zeros just locate the decimal point and are not significant figures. Zeros to the right of the 4, however, are significant. If you use a micrometer and measure the rod to have a length of 4.150 cm, there are four significant figures.

The number 1001 has four significant figures, as does 1000; and the number 1000.0 has five significant figures. On the other hand, if a quantity is said to have a value of a thousand but is only known to three significant figures, it must be written as 1.00×10^3 or 10.0×10^2. Consequently, suppose the distance to the Sun is given as 146 million kilometers (i.e., three significant figures). Writing it as 146 000 000 km, which implies nine significant figures, is inappropriate. Only significant figures should be provided, and the decimal should be located using scientific notation {see **MATH REVIEW: PART A-2** on the **CD** 💿 }. The distance to the Sun is then 146×10^9 m, or 146×10^6 km, or 14.6×10^7 km, and so forth.

In summary, the quantities 1 m, 0.7 g, 0.04 km, 0.02×10^{25} s, and 9×10^8 kg all have one significant figure, whereas 5.00 m, 600 nm, 0.300 cm, 0.020 0 kg, and 4.00×10^{12} s all have three significant figures.

Processing Significant Figures

Computations frequently involve measured quantities, and we need to establish how to deal with the multiplication, division, addition, and subtraction of significant figures. When processing physical quantities with units, convert everything to SI and use the same prefixes throughout a calculation; for example, *don't mix meters and centimeters*.

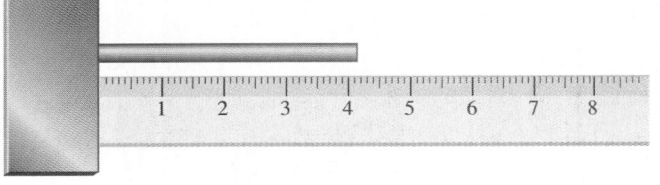

Figure 1.10 This rod measures somewhere between 4.1 cm and 4.2 cm, closest to 4.15 cm, but that last digit is not certain.

$$
\begin{array}{r}
10.75 \\
\times\ 3.54 \\
\hline
430\ 0 \\
5\ 375 \\
32\ 25 \\
\hline
38.055\ 0
\end{array}
$$

Multiplication and Division: Consider the area of a rectangle with sides of 10.75 cm and 3.54 cm (to help keep track of things, the doubtful digits are boldfaced). The area {see **MATH REVIEW: PART B** on the CD 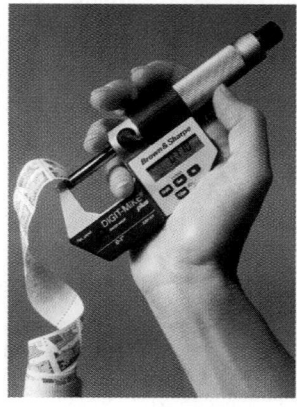 } is the product of length and width, (10.75 cm)(3.54 cm) = 38.**055 0** cm². As a rule, we *keep only one doubtful digit and round off the answer accordingly. If the first of the insignificant figures being dropped is 5 or more, raise the last significant figure being retained by 1; if it's less than 5, leave it unchanged.* Here 38.**055 0** cm² is rounded up to 38.1 cm². Rather than computing the number of significant figures in each case, use this rule: **The result of multiplication and/or division should be rounded off so that it has as many significant figures as the least precise quantity used in the calculation.** The rule is rough; it can inadvertently reduce the precision, but regardless of those rare exceptions, we will stick with it.

Internet

For more on SI units, go to the website of the National Institute of Science and Technology:

physics.nist.gov/cuu/index.html

Example 1.3 **[I]** Someone is 2.00-yd tall. Using the fact that one inch is exactly 2.54 cm, how tall is the person in centimeters?

Solution This is a unit-conversion problem going from yards to centimeters, but from now on we have to watch out for significant figures. (1) TRANSLATION—Convert height in yards to centimeters. (2) GIVEN: $h = 2.00$ yd, and 1.00 in. = 2.54 cm. Note that the height has three significant figures. FIND: h in cm. (3) PROBLEM TYPE—Unit conversion/significant figures. (4) PROCEDURE—We have a relationship between inches and centimeters, but h is in yards. If we had h in inches, it would be easy to find it in centimeters. (5) CALCULATION—Compute height in inches. Since there are 3.00 feet per yard,

$$1.00 = \frac{3.00\ \text{ft}}{1.00\ \text{yd}}$$

The height in feet is

$$h = 2.00\ \text{yd}\ \frac{3.00\ \text{ft}}{1.00\ \text{yd}} = 6.00\ \text{ft}$$

Note how the yd unit cancels leaving the result in feet. Similarly, since there are 12.0 inches per foot,

$$1.00 = \frac{12.0\ \text{in.}}{1.00\ \text{ft}}$$

and

$$h = 6.00\ \text{ft}\ \frac{12.0\ \text{in.}}{1.00\ \text{ft}} = 72.0\ \text{in.}$$

Inasmuch as there are 2.54 centimeters per inch,

$$1.00 = \frac{2.54\ \text{cm}}{1.00\ \text{in.}}$$

hence

$$h = 72.0\ \text{in.}\ \frac{2.54\ \text{cm}}{1.00\ \text{in.}} = 182.88\ \text{cm}$$

and to three significant figures $\boxed{h = 183\ \text{cm}}$.

Quick Check: 1 yd ≈ 0.9 m; 2 yd ≈ 1.8 m ≈ 180 cm.

$$
\begin{array}{r}
0.140 \\
+\ 2.0 \\
\hline
2.140
\end{array}
$$

Addition and Subtraction: The significance of digits in addition and subtraction is determined in relation to the location of the decimal point. Suppose a rod of length 140 mm is added to one 2.0-m long, and we want the total length. It might be assumed that (0.140 m) + (2.0 m) = 2.140 m, but we know nothing about the digits in the second and third decimal places of the rod that is 2.0∗∗ m long and cannot know the sum accurate to three decimal places. It makes sense to round off the sum to the least number of decimal places of any number appearing in the summation: the length of the combined rods is

---

CALCULATOR TIPS

The EXP key on your calculator allows you to work with scientific notation. Using a single-line (old fashioned) calculator the number 2.5×10^{-2} is entered as

[2] [.] [5] [EXP] [+/-] [2]

Whereas on a two-line calculator you'll enter

[2] [.] [5] [EXP] [–] [2]

A fine micrometer can measure distances in divisions of 0.001 mm.

God taking the measure of the Universe, painted by the poet and artist William Blake (1757–1827).

The sight of day and night, of months and the revolving years, of equinox and solstice has caused the invention of number and bestowed on us the notion of time.

PLATO

Table 1.5

Mathematical Symbols

Symbol	Meaning
$=$	equal to
$\neq$	not equal to
$\equiv$	defined as
$\pm$	plus or minus
$\parallel$	parallel
$\perp$	perpendicular
$\div$	divide by
$\rightarrow$	approaches
$\approx$	approximately equal to
$<$	less than
$>$	greater than
$\ll$	much less than
$\gg$	much greater than
$\leq$	less than or equal to
$\geq$	greater than or equal to
$|z|$	absolute value of z
$\propto$	proportional to
∞	infinity
$\sum$	sum of

2.1 m, and both the 2 and the 1 are significant figures. As a rule: **The result of addition and/or subtraction should be rounded off so that it has the same number of decimal places (to the right of the decimal point) as the quantity in the calculation having the least number of decimal places.** Here are a few examples of how this procedure works: (275 s − 270 s) = 5 s and (120 kg − 40.0 kg) = 80 kg. Keep in mind that *you can only add or subtract physical quantities with the same units. Watch out for those unit prefixes; always convert all numbers to the same SI units (including prefixes) before adding them*: 1.000 kg + 1.0 g = 1.000 kg + 0.001 0 kg = 1.001 kg.

Sines and Cosines   In time, we will work with sines, cosines, and tangents. These are usually evaluated using a scientific calculator; for example, sin 35.1° = 0.575 005 252. The number whose sine is being taken, the *argument* of the function, is 35.1°. The calculator, trying to please, provides us with a full register of figures, most of which are meaningless. What it actually gives us is sin 35.100 000 00°. The following (admittedly crude) rule is a lot more realistic: **The values of trigonometric functions have the same number of significant figures as their arguments.** Thus, sin 35.**1**° = 0.57**5**. The 35 in the argument is certain, but there is some doubt about the .1; consequently, on the right we are certain about the 0.57 and have doubt about the .005. In agreement with the rule, sin 35.0° = 0.574 and sin 35.2° = 0.576. If the values of trigonometric functions are computed in the range from −90° to +90°, the rule will work well enough for all our needs without any further refinement.

Because rounding off intermediate numbers within a calculation can produce cumulative errors, **one or two insignificant figures should be carried through intermediate calculations and only the final answers should be properly rounded off.** It's appropriate to run through an entire analysis on a calculator and round off only the final answer.

The Language of Physics

Once we learn the scientific vocabulary, we can transpose the experiences of everyday life into the carefully defined terminology of physics. For example, How much space is occupied by a big round red balloon 10 meters across? This question translates into, What is the volume of a sphere 10 m in diameter? The scientific words *volume*, *sphere*, and *diameter* have exact meanings, and there is a specific relationship between them—we know how to calculate the volume of a sphere in terms of its diameter (Fig. 1.7). No such relationship exists for the ordinary words *space*, *round*, and *across*.

1.7 Equations

Newton was the last great physicist to do his work without routinely representing physical quantities symbolically and writing equations describing their interrelationships {see **MATH REVIEW: PART A-4** on the **CD** 💿 }. Today, that approach would be unthinkable. Each physical quantity is assigned a symbol; t for time, a for acceleration, V for volume, R for radius, and so forth. We try to pick symbols that will be easy to remember, though that's not always possible. Along with these, there are standardized mathematical signs that act as a conceptual shorthand (Table 1.5).

The reason for transforming verbal scientific statements into symbolic ones is twofold. First, they are concise. A sentence such as "The volume of a sphere is equal to four-thirds of the product of pi times the radius cubed" becomes $V = \frac{4}{3}\pi R^3$. Certainly you have to know the definitions of volume, radius, and pi for this equation to make any sense. Once the technique is mastered, a whole page of symbols can be read like a story. Second, there exists a mathematical symbolic logic called algebra that's immediately applicable to physics. If you know the volume of a sphere and want to find its radius, the rules of algebra tell us that $3V = 4\pi R^3$, $R^3 = 3V/4\pi$, and $R = \sqrt[3]{3V/4\pi}$. Try thinking that through in sentences!

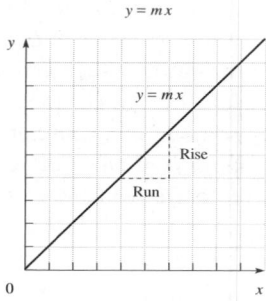

Figure 1.11 A plot is linear when it graphs as a straight line, which means that the slope, the rise over the run, is constant and equal to m. This plot passes through the origin when $x = 0$, $y = 0$.

1.8 Graphs

Complex relationships can be represented algebraically, but it's often hard to get a feeling for the interplay of the variables in an equation. That's a great virtue of graphs; we can plot an equation and literally see what happens to one quantity as another varies, and take it all in at a glance.

The simplest situation is when one variable (y) depends *directly* on another (x): $y = mx$, where m is a constant. The larger m is, the faster y changes with changes in x. For the moment, let x and y be general variables that can stand for anything. For example, if you get a 10% discount ($m = 0.10$) on the price of any shirt (x) in the store, you save ($y = mx$) dollars. Figure 1.11 is a plot of y versus x, and m is the slope of the line. Because the graph is a straight line, $y = mx$ is called a *linear equation*. There are many physical relationships, such as Ohm's Law and Hooke's Law, that are linear. Here, the line passes through zero so that at $x = 0$, $y = 0$, but that needn't be the case. Figure 1.12 shows a graph of the function $y = mx + b$, where m is again the slope and b is the point where the line now crosses the y-axis when $x = 0$.

Frequently, a quantity varies as the square of some other parameter. Mathematically, that situation appears as $y = kx^2$, where k is a constant. Since the variable is raised to the second power, this is a second-degree equation. For example, the area (A) of a square whose sides are of length l is given by $A = l^2$ (where $k = 1$). Double the side length and the area of the square increases by a factor of four. An equation of this kind is said to be *quadratic*, and the graph of y versus x is a *parabola* (Fig. 1.13). {To see how all of this works in a really lovely way, click on **PARABOLAS & STRAIGHT LINES** under **INTERACTIVE EXPLORATIONS** on the **CD.**}

Another common situation occurs when one quantity (y) varies *inversely* with some other quantity (x): $y = k/x$ or $yx = k$, where k is constant. The bigger x gets, the smaller y becomes, but yx is constant. For example, suppose you are designing a rectangular floor that must have a fixed area (k). The longer (x) the floor is made, the narrower (y) it must be, since $yx = k$. The graph of y versus x (Fig. 1.14) is a hyperbola that gradually approaches zero as x increases. At a constant temperature, a plot of pressure against volume for a variety of gases is also hyperbolic.

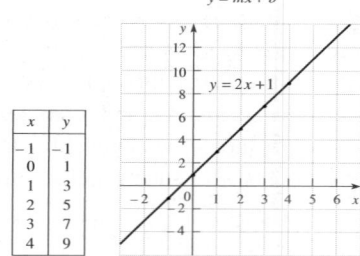

Figure 1.12 Here $y = 2x + 1$, the slope m is $+2$, and the y-intercept (b) is 1. When $x = 0$, $y = 1$. Between $x = 0$ and $x = 4$, the rise is 8, the run is 4, and the slope, 8/4, is 2.

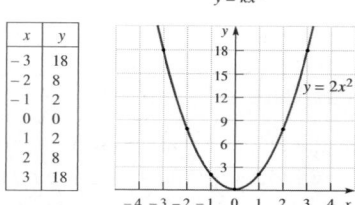

Figure 1.13 A plot of $y = kx^2$ (where k is a constant) is a curve that is symmetric about the y-axis called a *parabola*. At $x = 0$, $y = 0$, and the curve passes through the origin. There are many physical relationships that are parabolic.

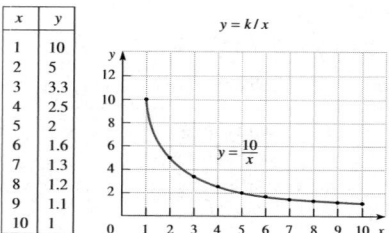

Figure 1.14 A plot of $y = k/x$ (where k is constant) is a hyperbola that approaches $y = 0$ as $x \rightarrow \infty$. For the gas in an air pump, the pressure times the volume is constant at a fixed temperature. Double the pressure, and you halve the volume.

The volume of gasoline pumped into the tank is directly proportional to the time (t) during which it is pumped. If V is the volume of gas in the tank at any moment, b is the amount in the tank at the start, and m is the constant flow rate, then $V = mt + b$. A plot of V versus t is a straight line.

1.9 Approximations and Checks

There are about 10^{11} galaxies in the Universe. No one has ever counted them, but we have photographed the sky, counted the number of galaxies in a small region, and then estimated the total. Making approximations like that is part of the everyday doing of physics.

How accurate an approximation is depends on the accuracy of the data that goes into it and the level of detail we devote to the calculation. For example, if all the people in the United States were placed head-to-foot, how far would they extend? There are about 250×10^6 people, each roughly 2 m tall; that's $\approx 5 \times 10^8$ m, which is greater than the distance to the Moon (3.8×10^8 m). That's not a very good approximation because 2 m is a poor guess for the average height—it's too tall. So what we calculated is an extreme. We know that the exact answer (there probably isn't one, with people being born and dying) is less than 5×10^8 m. Things could be improved by assuming there are $\approx 186 \times 10^6$ adults, ≈ 1.6-m tall, and $\approx 64 \times 10^6$ children, averaging ≈ 1-m tall. That yields a distance of $\approx 4 \times 10^8$ m, which can't be too far off—after all, if everyone in the country averaged only 1-m tall, the distance would still be $\approx 2.5 \times 10^8$ m. Notice how looking at the extremes of a situation puts things in perspective.

It's also good practice to check the solutions of problems by making quick approximations {see **MATH REVIEW: PART A-5** on the **CD** }. For example, suppose we had a water storage tank 3.107 9-m deep, 4.921-m wide, and 18.689-m long. What's its volume (V)? Since the volume is the length times width times depth (Fig. 1.7):

$$V = (18.689 \text{ m})(4.921 \text{ m})(3.107\ 9 \text{ m}) = 285.8 \text{ m}^3$$

To check this result, round each number off to one significant figure and redo the calculation, as follows:

$$V = (2 \times 10^1 \text{ m})(5 \text{ m})(3 \text{ m}) = 3 \times 10^2 \text{ m}^3$$

Because we rounded up more than down, this approximation can be expected to be on the large side. Still, it provides a quick check to show us that 285.8 m^3 is roughly the right size.

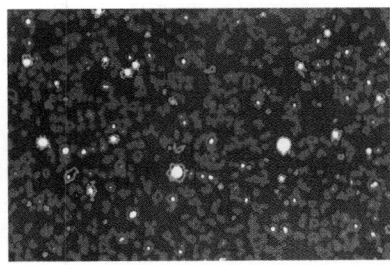

A long-exposure image of a small region of the southern sky. Here galaxies appear red and stars green. Clearly, there are a lot of galaxies in the Universe.

Core Material & Study Guide

MEASUREMENT

The SI system of units is, in part, based on the meter, kilogram, and second. U.S. Customary Units use the foot, pound, and second and are in general to be avoided. Reread Sections 1.3 (Length), 1.4 (Mass and Weight), and 1.5 (Time) and make sure you understand Examples 1.1 and 1.2. **To make sure you've understood the basics, work through all the Warm-Ups on the CD. Once that's done go on to the Walk-Through Examples.**

The result of multiplication and/or division should be rounded off so that it has as many significant figures as the least precise quantity used in the calculation. Reread Section 1.6 (Significant Figures) and make sure you understand Example 1.3.

The result of addition and/or subtraction should be rounded off so that it has the same number of decimal places as the quantity in the calculation having the least number of decimal places.

Always convert all numbers to the same SI units (getting rid of prefixes) before working with them.

The values of trigonometric functions have the same number of significant figures as their arguments.

Reread Section 1.8 (Graphs) and familiarize yourself with the curves.

Key Terms

atom	Metric System
nucleus	SI
electron	meter
quark	weight
lepton	mass
force	kilogram
interaction	second
conservation law	prefixes
symmetry	scientific notation
units	significant figures

Discussion Questions

1. Which of the fundamental Four Forces is responsible for a stamp adhering to an envelope? Explain.

2. If gravity is pulling you downward, which of the fundamental Four Forces is keeping you from passing into the chair you are sitting on?

3. When scientists take a large number of healthy normal people of all ages and sizes and measure the amount of creatinine (an alkaline constituent of blood and urine) passed out through the kidneys each day, they get a broad range of values. And yet when the data is plotted, the result is rather simple (Fig. Q3). Determine the slope of the curve and discuss its units. What can you conclude in general? Why might one suggest, "the more muscle the more creatinine"? Explain.

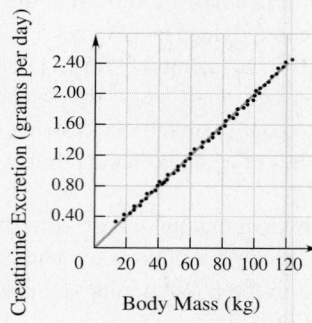

Figure Q3

4. Explain the meaning of each of the Key Terms at the top of this page.

5. Gregor Mendel (1865) pointed out that when a recessive gene occurs in some percentage, x, of all observations and the corresponding trait it determines occurs in some percentage, y, of all observations, y varies directly as the square of x. What would it look like if y was plotted against x?

6. Given that a law says that some relationship among physical quantities will be found to *always* exist, how does that transcend experience? Implicit in the statement of law is the understanding that it applies *everywhere*. Comment on the practicalities of this notion.

7. After a series of tests using pendulums, Newton wrote: "By experiments made with the greatest accuracy, I have always found the quantity-of-matter in bodies to be proportional to their weight." Explain what he was talking about.

8. What did Einstein mean when he wrote:

> The aim of science is, on the one hand, a comprehension, as *complete* as possible, of the connection between the sense experiences in their totality and, on the other hand, the accomplishment of this aim *by the use of a minimum of primary concepts and relations*.

What's the point of specifying "a minimum of primary concepts"?

9. The realization that sound and heat are both manifestations of the mechanical behavior of matter and are governed by the same basic laws is an example of one of the central themes of contemporary physics. Identify it and discuss it.

10. Give a brief description of an atom and list a few of its main behavioral characteristics.

11. Ohm's Law is $V = IR$ where V is voltage, I is current, and R is a constant resistance. What will be the shape of the curve of V versus I? Does the curve pass through the origin?

Multiple Choice Questions

1. Of the following sets of units, which are all SI? (a) cm, s, kg, lb, μm (b) mm, μm, g, s, in. (c) fm, ns, kg, mm, μs (d) km, s, kg, μm, ft (e) none of these.

2. Which length is the largest? (a) 10^1 cm (b) 10^{-10} m (c) 1×10^2 mm (d) 1 m (e) none of these.

3. The diameter of your eyeball is about (a) 2.0×10^2 cm (b) 3.5×10^{-10} m (c) 1.5×10^2 mm (d) 2.5 cm (e) none of these.

4. The length of an unsharpened wooden pencil is about (a) 2×10^2 cm (b) 2×10^{-2} m (c) 2×10^3 mm (d) 2×10^3 nm (e) none of these.

5. Which mass is the smallest? (a) 10^5 μg (b) 10^2 g (c) 1 kg (d) 10^3 mg (e) none of these.

6. Given are four masses: (1) 10 mg (2) 1000 μg (3) 10^2 kg (4) 10^{-4} kg. These are ordered in ascending size as (a) 1, 2, 3, 4

(b) 2, 1, 4, 3   (c) 4, 3, 2, 1   (d) 2, 1, 3, 4   (e) none of these.

7. Which of the following is longest?   (a) 1×10^4 cm   (b) 100×10^2 mm   (c) 10^6 μm   (d) 10^9 nm   (e) none of these.

8. A day has roughly   (a) 86×10^2 s   (b) 8640 s   (c) 9×10^4 s   (d) 1.44×10^3 s   (e) none of these.

9. A year has roughly   (a) 8.77×10^2 h   (b) 5×10^5 min   (c) 3.7×10^3 days   (d) 32×10^5 s   (e) none of these.

10. A cube 1000 cm on a side has a volume of   (a) 10^2 cm^2 (b) 10^2 cm^3 (c) 10^6 cm^3 (d) 10^9 cm^3 (e) none of these.

11. A rectangular floor is 6.6 m by 12 m. Its area is   (a) 79 m^2 (b) 18.6 m^2 (c) 7.92 m^2 (d) 79.2 m   (e) none of these.

12. A femtosecond is   (a) 10^{-12} s   (b) -15 s   (c) 10^{15} s   (d) 10^{-15} s   (e) none of these.

13. A 20.0-in. bar is   (a) 20.0-cm long   (b) 508-mm long   (c) 51-m long   (d) (2.54/20)-cm long   (e) none of these.

14. One pound has an equivalent mass of exactly 453.592 37 g. To four significant figures, that's   (a) 453.5 g   (b) 453.592 3 g   (c) 400.0 g   (d) 453.6 g   (e) none of these.

15. The product of 12.4 m and 2 m should be written as   (a) 24.8 m   (b) 24.8   (c) 25 m^2 (d) 0.2×10^2 m   (e) none of these.

16. The product of 15.0 cm and 5 cm should be written as   (a) 75 cm^2 (b) 7.5×10^1 cm^2 (c) 0.75×10^2 cm^2 (d) 0.8×10^2 cm^2 (e) none of these.

17. The weight of 1 kg on Earth is about   (a) 1 lb   (b) 1000 g   (c) $2\frac{1}{4}$ lb   (d) 0   (e) none of these.

18. If a bag of screws costs 10¢ per pound, a kilogram of them will cost about   (a) 100¢   (b) 22¢   (c) 4.5¢   (d) $22   (e) none of these.

19. If coffee is $12 a kilogram, roughly how much will it be by the pound?   (a) $26   (b) $12   (c) $5.5   (d) 55¢   (e) none of these.

20. A liter is 1000 cm^3, which means that a cube 100 cm on a side has a volume of   (a) 1000 liters   (b) 0.001 m^3 (c) 100 liters   (d) 1000 liters^3 (e) none of these.

21. A kilometer is   (a) just under half a mile   (b) just over half a mile   (c) about 1000 ft   (d) roughly 5280 ft   (e) none of these.

For more Multiple Choice Questions with answers click on WARM-UPS in CHAPTER 1 on the CD.

Suggestions on Problem Solving

1. TRANSLATION—Problems are often posed in nonscientific prose, for example, "A young man runs 20 meters from his front door to the street in 10 seconds. On average, how fast was he going?" The physics to handle this problem will be developed in Chapter 2; here, we are only concerned with general principles of problem solving. Clearly, the question is full of extraneous information—it doesn't matter that it's a young man, and it doesn't matter that he goes from door to street. The following is a good way to proceed. (a) *Decide on what's happening as far as the physics is concerned and lift that information from the inconsequential background*. An object travels 20 m in 10 s—how fast did it move? (b) *Translate the statement into the language of physics*, as you continue to learn that language. An object travels a known *distance* (20 m) in a known *time* (10 s); determine its *average speed*.

2. GIVEN/FIND—*Symbolically represent the variables*: An object travels a known *distance* ($d = 20$ m) in a known time ($t = 10$ s); find its *average speed* (v_{av}). Once the problem has been so restated, *write down what is* given ($d = 20$ m and $t = 10$ s) *and what you must find* (v_{av}).

3. PROBLEM TYPE—Determine the type of problem you are dealing with. That is, what's the central idea (e.g., average speed, constant acceleration, temperature)? This is usually easy to do for textbook problems because you know what chapter you're in, but it may not be so easy on an exam.

4. PROCEDURE—*Determine a relationship between the given quantities and the one(s) to be found*. It is the business of physics to produce such relationships in the form of laws and definitions. As a rule, *it's always a good idea to* draw a diagram. The process helps to organize your thoughts.

5. CALCULATION—In a numerical problem the last step is to substitute in numbers and carry out the calculation. As you become more confident, you'll work more with the symbols and only plug in numbers at the end. That allows symbols to cancel and avoids multiplying by some numerical quantity only to find that you divide by it three steps later. *Show all of your work throughout the analysis*. Doing so will let you, and anyone else attempting to understand your efforts at some later date, follow your logic.

When solving a problem involving quantities with units having prefixes such as ms, or km, or μm, replace the prefixes by scientific notation before carrying out the calculation. Notice that $(1.0 \text{ km})^3 = (1.0 \times 10^3 \text{ m})^3 = 1.0 \times 10^9$ m^3, not 1.0×10^9 m.

Problems ✦ Coordinated Problems ✦ Progressive Problems ✦ Solutions

STUDY GUIDE **1. Coordinated Problems:** The three problems within each magenta-colored grouping are solvable in similar ways. Note that the first of these always has a hint; moreover, its solution is provided in the back of the book. *Work out each of these sets; they'll strengthen technique and build confidence.* **2. Progressive Problems:** The problems introduced in blue unfold step-by-step carrying along the analysis in a more suggestive way than is customary. *Work out all of these; they'll guide you through the analytic process and help develop problem-solving skills.* **3. Worked-Out Solutions:** Studying worked-out solutions is an important part of learning how to solve problems. Accordingly, additional *solutions* to a number of model problems are given below. *Make sure you understand each of them before you go on to the next problem.* **4.** Also provided in the back of the book are the *Answers* to all odd-numbered problems, as well as worked-out *solutions* to those with boldface numbers. Problem numbers in italic indicate that a solution appears in the Student Solutions Manual.

SECTION 1.3: LENGTH

SECTION 1.4: MASS AND WEIGHT

SECTION 1.5: TIME

1. [I] The human brain has about 10 (U.S.) billion nerve cells, more tightly packed than in any other tissue. Write that number out and then put it in scientific notation.

2. [I] It's been estimated that the human brain can store 100 (U.S.) trillion bits of information. Write that number out, put it in scientific notation, and then express it by adding the proper prefix to the word *bits*.

3. [I] Express each of the following quantities in seconds in decimal form: (a) 10.0 ms; (b) 1000 μs; (c) 10.000 ks; (d) 100 Ms; (e) 1000 ns. (Don't worry about significant figures in this problem.)

4. [I] Express each of the following quantities in micrograms: (a) 10.0 mg; (b) 10^4 g; (c) 10.000 kg; (d) 100×10^3 g; (e) 1000 ng. (Don't worry about significant figures.)

5. [I] How many millimeters are in 10.0 km?

> SOLUTION: There are 1000 mm in 1 m, so first convert 10.0 km to meters and then to millimeters. $(10.0 \text{ km})(1000 \text{ m/km}) = 10.0 \times 10^3$ m and $(10.0 \times 10^3 \text{ m})(1000 \text{ mm/m}) = 1.00 \times 10^7$ mm.

6. [I] A U.S. nickel coin has a diameter of 2.1 cm; how much is that in inches?

7. [I] The *Titanic* (1912-1912) was 882 feet long; how long is that in meters?

8. [I] A red blood cell lives for about four months and travels roughly 1.0×10^3 mi through the body. How far is that in kilometers?

9. [I] How many millimeters are in 3.00 in.? [*Hint: Change inches to centimeters using 1 in. = 2.54 cm. Then change centimeters to millimeters.*]

10. [I] What is the equivalent in millimeters of 1.00 ft?

11. [I] How many inches are the equal of 200.0 mm?

12. [I] The unit known as an angstrom (Å), exactly equal to 0.1 nm, is still widely used. It's named after the nineteenth-century physicist Anders J. Ångstrom (Å is the letter that comes after Z in the Swedish alphabet). Given that an atom is about 1 Å in diameter, how much is that in centimeters?

13. [I] The length of a lightwave is about 5×10^3 Å (see previous problem). How much is that in nanometers?

> SOLUTION: 1.0 Å equals 0.1 nm. Therefore the number of nanometers is 5×10^3 Å $= (5 \times 10^3$ Å$)(0.1$ nm/Å$)$ and to one significant figure 5×10^3 Å $= 0.5 \times 10^3$ nm.

14. [I] About how many centimeters tall is a stack of 25 CDs without their cases?

15. [I] The distance light travels in a year is called a *light-year* (ly). Given that 1.00 ly $= 5.88 \times 10^{12}$ mi, how far is that in meters? [*Hint: We first need to find the number of meters to 1 mile. The most rudimentary way to do this is to use 1 mi = 5280 ft. Then there are 5280×12 inches per mile and $5280 \times 12 \times 2.54$ centimeters per mile.*]

16. [I] The Moon is, on average, 2.39×10^5 miles away. How much is that in meters?

17. [I] The nearest star beyond the Sun is Alpha Centauri. It's 4.2 light-years away (1 ly $= 5.88 \times 10^{12}$ mi). How far is that in kilometers?

18. [I] By what number would you multiply a given number of centimeters to convert it to millimeters?

19. [I] There are an average of 32 million bacteria on each square inch of the human body. Given a total skin area of 1.7 m^2, how many bacteria are you carrying around (exclude the bacteria that are internal)?

20. [I] How many square centimeters are in a square inch?

21. [I] A nickel (U.S. 5-cent coin) has a mass of about 5 g. How many nickels correspond to a kilogram? [*Hint: We need to determine how many 5-g objects are equivalent to 1 kilogram, which equals 1000 g.*]

22. [I] A stack of 4.00-g buttons is placed on one pan of a balance and a standard 1.00-kg mass is placed on the other pan. How many buttons will it take to balance the scale?

23. [I] Suppose that you have a bag of identical marbles each of which has a mass of 20.0 g. How many marbles will it take to match the mass of an exactly 1/2-kg banana?

24. [I] In a human nose, the total area of the region that detects odors is about 3/4 in.^{2}. Compare that to the sensory organ of a hunting dog, whose nose has an active area of about 65 cm^2.

25. [I] It takes your brain about one five-hundredth of a second to recognize a familiar object once the light from that object enters your eye. Express that time interval in milliseconds, microseconds, and nanoseconds, each to one significant figure.

26. [I] How many seconds are there in an exactly 24-h day?

27. [I] Each second the human brain undergoes 1×10^5 different chemical reactions. At that rate, how many will it experience in 10 h?

28. [II] A GM diesel locomotive pulling 40 to 50 loaded freight cars at 70 mi/h uses 1.0 gallon of fuel every 632 yd. How many meters can it travel on 10 gallons of fuel?

29. [II] By what number would you multiply a given number of centimeters to convert it into inches? Give the answer to four significant figures and show your work.

> SOLUTION: There are exactly 2.54 cm per inch and we want the number of inches per centimeter, which is one over that:
> $$\frac{1}{2.54 \text{ cm/in.}} = 0.393\,7 \text{ in./cm}$$

30. [II] THIS PROBLEM DEALS WITH UNIT CONVERSIONS. A Boeing 747 jumbo jet carrying 385 people while cruising at 39 000 ft travels 280 yd on a gallon of aviation fuel. We want to find out how many gallons it will need to travel 2000 km. a) How many meters does 280 yd equal? b) How many meters does 2000 km equal? c) How many meters does the plane travel per gallon of fuel? d) How many gallons will the plane need to travel 2000 km?

31. [II] Most people lose about 45 hairs per day out of a typical headful of 125 000. Suppose each hair averages 10-cm long. If you placed a year's lost hairs end-to-end, how far would they extend?

32. [II] What's the SI equivalent of 1.000 in.^{2}?

33. [II] Express the equivalent of 1.000 ft^2 in SI units.

34. [II] In an average lifetime, a human inhales roughly 5.0×10^5 yd^3 of air. How many cubic meters is that?

35. [II] A one-cup measure is equivalent to 237 milliliters. How much is that in the preferred SI unit of cubic meters? [*Hint: There are 1000 cm^3 to a liter, so each 1 cm^3 is a milliliter. Moreover, (p. 8) there are 10^6 cm^3 per 1 m^3.*]

36. [II] How many liters of water would fill a cube-shaped tank whose inner dimensions are 1.00 m on each side?

37. [II] Considering the tank in the previous problem, if it contained 20.0 liters of water, how deep would the liquid be?

38. [II] Roughly what size cube of water would have a mass of 1.0 kg? [*Hint: 1 cm^3 of water has a mass of 1 g.*]

39. [II] Each hour a large man sheds about 6×10^5 particles of skin. In a year, that amounts to about 1.5 lb (on Earth). Roughly how much is the mass of each such particle? What mass of skin will be shed in 50 years of adulthood?

40. [II] We are advised to keep our total blood cholesterol level at less than 200. What that actually means is 200 mg of cholesterol per deciliter of blood. Physicists usually shun the *deci* prefix and often avoid the liter as well, so convert this quantity to SI units.

SECTION 1.6: SIGNIFICANT FIGURES
SECTION 1.9: APPROXIMATIONS AND CHECKS

41. [I] It's been estimated that the mass of junk mail delivered each and every second in the United States is 4.4×10^1 kg. How much arrives in an hour?

42. [I] The potato spindle virus with a diameter of 2×10^{-8} m is among the smallest viruses. If we place three million of them end-to-end, how far will the column reach?

43. [I] Can the quantity 100.0 kg be expressed in grams in decimal form with the correct number of significant figures? Explain.

44. [I] Express each of the following in grams in decimal form with the correct number of significant figures: (a) 1.00 μg; (b) 0.001 ng; (c) 100.0 mg; (d) 10 000 μg; and (e) 10.000 kg.

> **SOLUTION:** Keep in mind that we can write each answer in many different ways. (a) 1.00 μg has three significant figures, and 1 μg = 10^{-6} g; therefore 1.00 μg = 1.00×10^{-6} g. (b) 0.001ng has one significant figure, and 1 ng = 10^{-9} g; therefore 0.001 ng = 0.001×10^{-9} g or 1×10^{-12} g. (c) 100.0 mg has four significant figure, and 1 mg = 10^{-3} g; therefore 100.0 mg = 100.0×10^{-3} g or 0.100 0 g. (d) 10 000 μg has five significant figures, and 1 μg = 10^{-6} g; therefore 10 000 μg = $10 000 \times 10^{-6}$ g or $1.000 0 \times 10^{-2}$ g. (e) 10.000 kg has five significant figures, and 1 kg = 10^3 g; therefore 10.000 kg = 10.000×10^3 g or $1.000 0 \times 10^4$ g.

45. [I] Determine the number of significant figures for each of the following: (a) 0.002 0; (b) 0.99; (c) 1.75 ± 0.02; (d) 1.001; (e) 4.44×10^4; and (f) 0.01×10^{34}.

46. [I] Write the speed of light in vacuum (299 792 458 m/s) to one, three, four, and eight significant figures.

47. [I] State the number $\pi = 3.141 592 65\ldots$ to one, three, four, and five significant figures.

48. [I] To four significant figures, how many square meters correspond to a square foot?

> **SOLUTION:** A foot is 12 inches and each inch is 2.54 cm, so each foot is 30.48 cm. One square foot corresponds to (30.48 cm)(30.48 cm) = $9.290 3 \times 10^2$ cm^2. Moreover, 1.000 m^2 = (100 cm)(100 cm) = 1.000×10^4 cm^2. Thus
>
> $$1.000 \text{ ft}^2 = 9.290 3 \times 10^2 \text{ cm}^2 = \frac{9.290 3 \times 10^2 \text{ cm}^2}{1.000 \times 10^4 \text{ cm}^2/\text{m}^2}$$
>
> $$1.000 \text{ ft}^2 = 9.290 \times 10^{-2} \text{ m}^2$$

49. [I] Liters are sometimes used for liquid and gas measurement, although the preferred unit is the cubic meter. How much is 1.000 liter in cubic meters?

50. [I] The radius of a hydrogen atom is $5.291 77 \times 10^{-11}$ m. How many of them, lined up one "touching" the next, would stretch a length of 1.0 m?

51. [I] The radius of the Moon ($R_☾$) is 1.738×10^3 km. Assume the Moon is a sphere and determine its volume to two significant figures. [*Hint: The volume of a sphere is $V = \frac{4}{3}\pi R^3$. Compute with four figures and round down to two.*]

52. [I] A spherical tank contains 4.19 m^3 of water. What's the diameter of the tank?

53. [I] A solid sphere having a volume of 1.00 m^3 fits tightly in a cubical box. What's the volume contained within the box?

54. [II] Compute the circumferences of circles having diameters of (a) 5.42 μm, (b) 0.542 0 nm, (c) 5.420 000 mm, (d) 542.0 m, and (e) 0.542 km.

55. [II] Compute the sum of the following quantities: 0.066 m, 1.132 m, 200.1 m, 5.3 m, and 1600.22 m. [*Hint: Here all the numbers have the same units (including prefixes), so this is a straight addition. Watch out for significant figures.*]

56. [II] Add the following quantities: 0.10 ms, 20.2 s, 6.33 s, 18 μs, and 200.55 ms.

57. [II] Determine the sum of the following: 1.00 g, 1.00 mg, 1.00 kg, and 1.00 μg.

58. [II] What is the product of the following quantities: 0.002 1 g, 655.1 kg, and 4.41 μg?

59. [II] THIS PROBLEM DEALS WITH MAKING APROXIMATIONS. We want to find out roughly how many grains of sand would form a sphere the size of the Earth? a) Approximately how big is a grain of sand, assuming first that it's a cube; and then that it's a sphere? b) What are the two volumes one might compute for such a grain of sand? c) The planet's diameter is about 1.3×10^7 m; what's the volume of the Earth? d) Roughly how many grains of sand would fit inside a sphere the size of the Earth? Don't worry about empty spaces between grains.

60. [II] We wish to uniformly paint a house having an area A. Write an expression approximating the thickness, τ, of the resulting layer of paint when one gallon is used. There are 3785 cubic centimeters per gallon. Determine τ when the area is a reasonable 8×10^2 m^2.

61. [II] Write expressions that crudely estimate the area and volume of a human body by representing it as a cylinder of radius r and height h. (Take the person to be a rather large 2-m tall.) Approximate a cell to be a sphere 30 μm in diameter; roughly how many cells are in a human body?

62. [III] A *pipe* is 2 *hogsheads,* and a hogshead is 63 gallons (or 2 barrels). Given that 1 U.S. gallon is 231 in.^{3}, what's the SI equivalent of a pipe to three significant figures?

63. [III] It has been estimated that a drop (0.1 cm^3) of light oil spreads out, on the calm surface of a pond, into a film with an area of about 40 m^2. Assuming this to be a single molecule thick, approximate the size of the molecule.

Chapter 2

Kinematics: Speed and Velocity

E very atom in every solid object, including the human body, is jiggling wildly. All the galaxies are rushing outward; the entire Universe is expanding. Everything is moving. The branch of physics that describes motion *with no concern for what is causing the motion* is known as **kinematics**. There are two methods of description (scalar and vector) that have evolved as our mathematical techniques developed. One is based on the concepts of distance and speed, the other on displacement and velocity. The first is simpler and more intuitive but rather limited; the second is mathematically demanding but much more powerful. We'll begin with the familiar ideas of distance and speed—the concepts measured every day by odometers and speedometers.

Speed

The faster a thing moves, the farther it goes in any interval of time. We have meter sticks to measure *distance* and clocks for *time*; we only need to relate these two ideas to quantify *speed*, to determine "how fast." The concept goes back to the ancient Greeks, who specified speed as *the distance traveled in a given amount of time*. Over 1000 years later, if you asked a traveler how fast he was moving you would still get an answer like "I've gone 100 miles in 2 days."

2.1 Average Speed

Today, we carry the logic one step further and define the **average speed** as *the distance traveled (along any path) divided by the time it took to do the traveling*:

$$\text{Average speed} = \frac{\text{distance traveled}}{\text{time elapsed}}$$

Dividing by the time in specific units—seconds, minutes, or hours—yields the *distance traversed **per unit time***: per second, per minute, and so on. The division creates a new

Average speed is distance traveled over time elapsed.

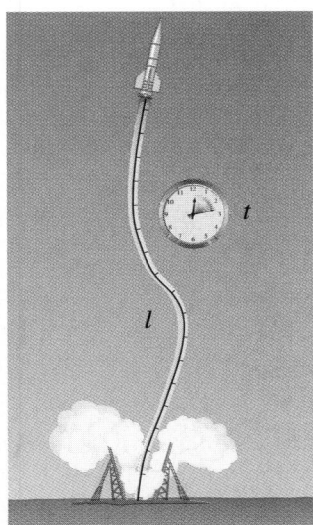

Figure 2.1 The rocket travels a distance measured along its actual path through space. After a time t, having gone a distance l, its average speed is $v_{av} = l/t$.

Table 2.1

Typical Speeds

Speed, m/s	Motion	Speed, mi/h
300 000 000	Light, radiowaves, X-rays, microwaves (in vacuum)	669 600 000
210 000	Earth-Sun travel around the galaxy	481 000
29 600	Earth around the Sun	66 600
1000	Moon around the Earth	2300
980	SR-71 reconnaissance jet	2200
333	Sound (in air)	750
267	Commercial jet airliner	600
62	Commercial automobile (max.)	140
37	Falcon in a dive	82
29	Running cheetah	65
10	100-yd dash (max.)	22
9	Porpoise swimming	20
5	Flying bee	12
4	Human running	10
2	Human swimming	4.5
0.01	Walking ant	0.03
0.000 045	Swimming sperm	0.000 1

concept: a single quantity—speed—with a single numerical value and its own units (Table 2.1).

The above definition can be expressed symbolically as

[along any path]

$$v_{av} = \frac{l}{t}$$ (2.1)

where v_{av} is the average speed,* l is the path-length traveled, and t is the time it took to travel it (Fig. 2.1). The units of speed are always those of *length over time* (kilometers per hour, meters per second, etc.). An automobile traversing a total along-the-road distance of 30 km in 2.0 h has an average speed of 15 km/h. It could have stopped a few times on the way and then sped up and still covered the 30 km in the same 2.0 h—that's why we call this concept *average* speed.

Example 2.1   **[I]** A student drives from school to home at an average speed of 25.3 km/h. If it takes 4.72 h to get to her destination, how far did she travel?

Solution   The words "average speed" in the statement of the problem should immediately bring to mind the definition of v_{av} in terms of distance traveled and time elapsed. (1) TRANSLATION— An object moves at a known average speed for a specified time; determine the distance traveled. (2) GIVEN: $v_{av} = 25.3$ km/h and $t = 4.72$ h. FIND: l. (3) PROBLEM TYPE—Average speed. (4) PROCEDURE—We are provided v_{av} and t and asked to find l; Eq. (2.1) is the definition of v_{av} and it relates all three quantities. We can convert to m/s or more easily stay with km/h. (5) CALCULATION—Solving Eq. (2.1) for distance we get

$$l = v_{av}t = (25.3 \text{ km/h})(4.72 \text{ h})$$

Since the least number of significant figures is three,

$$l = 119 \text{ km}$$

Quick Check: Recalculate one of the given terms (e.g., v_{av}) using the computed value of l: $v_{av} = l/t$ = (119.4 km)/(4.72 h) = 25.297 km/h = 25.3 km/h, so all is well.

{For more worked problems click on **WALK-THROUGHS** *under* **CHAPTER 2** *on the CD.}*

---

*We do not use s for speed because it has often been associated with the displacement in "space" (see p. 29). The symbol v derives from the word velocity (see p. 33).

Example 2.2 **[II]** The Moon moves in a nearly circular orbit around the Earth with an average radius R of 3.84×10^8 m. If it takes 27.3 days to complete one revolution, determine its average orbital speed in meters per second.

Solution The words "average orbital speed" in the statement of the problem should call to mind the definition of v_{av}. The reference to "circular orbit" and "radius" suggests, that we are probably going to have to compute the distance traveled. (1) TRANSLATION—An object moves a known circular distance in a specified time; determine its average speed. (2) GIVEN: $R = 3.84 \times 10^8$ m and a once-around time of $T = 27.3$ days (this is a particular time interval, so give it its own sym-

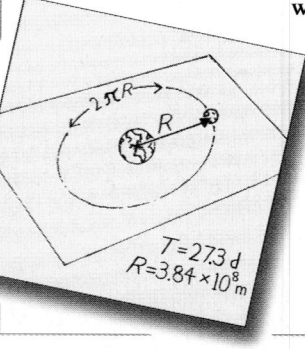

bol). FIND: v_{av} in m/s. (3) PROBLEM TYPE—Average speed. (4) PROCEDURE—Compute the motion over one circular orbit. The Moon's orbital circumference is $2\pi R$ {see **MATH REVIEW: PART B** on the **CD** for a geometry review ⊕ }. Thus $v_{av} = l/t$ where $l = 2\pi R$ and $t = T$; consequently, $v_{av} = 2\pi R/T$. (5) CALCULATION—Get the time in seconds because we want the speed in m/s: $T = 27.3$ d $\times$ 24 h/d $\times$ 60 min/h $\times$ 60 s/min $= 2.359 \times 10^6$ s. The average speed is then

$$v_{av} = \frac{2\pi R}{T} = \frac{2\pi(3.84 \times 10^8 \text{ m})}{2.359 \times 10^6 \text{ s}} = \boxed{1.02 \times 10^3 \text{ m/s}}$$

Quick Check: *Always make sure your answer is the correct order-of-magnitude* {see **MATH REVIEW: PART A-5** on the **CD** ⊕ }. The quantity $2\pi \approx 6$, and the circumference is about $6 \times 4 \times 10^8$ m $\approx 24 \times 10^8$ m; that value divided by the time (about 24×10^5 s) yields 10^3 m/s.

An average speed in one set of units can be expressed in any other set of units (Table 2.2). To carry out such a conversion (p. 8), use multiplicative factors equal to 1 so that the size of the physical quantity never changes. For example, to transform 15 km/h into m/s, first convert kilometers to meters by multiplying by (1000 m/1.0 km), which equals 1 exactly:

$$\left(\frac{15 \text{ km}}{1.0 \text{ h}}\right)\left(\frac{1000 \text{ m}}{1.0 \text{ km}}\right) = \frac{15\,000 \text{ m}}{1.0 \text{ h}}$$

Convert this to m/s by multiplying by whatever it takes to transform the hours into seconds, namely (1.0 h/3600 s) = 1.0:

$$\left(\frac{15\,000 \text{ m}}{1.0 \text{ h}}\right)\left(\frac{1.0 \text{ h}}{3600 \text{ s}}\right) = 4.2 \text{ m/s}$$

Table 2.2

A Rough Comparison of Speeds in Different Units			
mi/h	ft/s	km/h	m/s
0	0	0	0
10	15	16	4
20	29	32	9
30	44	48	13
40	59	64	18
50	73	80	22
60	88	97	27
70	103	113	31
80	117	129	36
90	132	145	40
100	147	161	45
200	293	322	89
500	734	805	224
1000	1467	1609	447

Most of the world uses km/h for the units of speed. This sign is in Mexico.

Example 2.3 **[II]** The 1980 speed record for human-powered vehicles was set on a measured 200-m run by a sleek machine called Vector. Pedaling back-to-back, its two drivers averaged 62.92 mi/h. This awkward mix of units is the way the data appeared in a newspaper article reporting the event. (a) Determine the average speed in m/s and (b) then compute how long the record run lasted.

Solution The phrase "average speed" in the statement of the problem should remind us of the definition of v_{av} in terms of l and t. (1) TRANSLATION—An object moves a known distance at a known average speed; determine the trip time. (2) GIVEN: $l = 200$ m and $v_{av} = 62.92$ mi/h. FIND: v_{av} in m/s and t. (3) PROBLEM TYPE—Unit conversion/average speed. (4) PROCEDURE—The quantities that are given (l and v_{av}), and to be found (t), are related by Eq. (2.1), $v_{av} = l/t$. (5) CALCULATION—(a) First convert v_{av} into m/s. Changing miles

to feet, $v_{av} = (62.92 \text{ mi/h})(5280 \text{ ft/mi}) = 3.322 \times 10^5$ ft/h. Since 1.00 m = 3.281 ft, $v_{av} = (3.322 \times 10^5 \text{ ft/h})/(3.281 \text{ ft/m}) = 10.12 \times 10^4$ m/h. To convert hours to seconds, divide by (3600 s/h); so $(10.12 \times 10^4 \text{ m/h})/(3600 \text{ s/h}) = \boxed{28.1 \text{ m/s}}$. (b) The time can be gotten by rewriting Eq. (2.1) for t:

$$t = \frac{l}{v_{av}} = \frac{200 \text{ m}}{28.1 \text{ m/s}} = \boxed{7.12 \text{ s}}$$

Quick Check: Recalculate t using v_{av} in mi/h; $l = 200$ m = (200 m)(6.214×10^{-4} mi/m) = 0.124 mi. $t = l/v_{av} = (0.124$ mi)/(62.92 mi/h) $= 1.98 \times 10^{-3}$ h = 7.1 s. Alternatively, it's useful to remember that 60 mi/h = 88 ft/s and so $v_{av} \approx 60$ mi/h ≈ 88 ft/s ≈ 29 m/s and therefore $l/v_{av} = (200 \text{ m})/(29 \text{ m/s}) \approx$ 7 s. Of course, you could always use a conversion factor for the speed; $(62.92 \times 0.447\ 0$ m/s) = 28.1 m/s.

Example 2.4 **[III]** A 600-km cross-country automobile race is won by a team of two drivers, each of whom had the wheel for half the distance of the trip. If one averaged 60 km/h and the other 20 km/h, what was their overall average speed?

Solution The words "overall average speed" ought to call to mind the definition of v_{av} in terms of l (the overall distance) and t (the overall time). (1) TRANSLATION—Two objects move known distances at known average speeds; determine their net average speed. (2) GIVEN: Introduce a subscript notation to keep track of the two sets of data. Total length $l = 600$ km; distances traveled by drivers 1 and 2, $l_1 = l_2 = 300$ km (half the total distance); average speeds $(v_{av})_1 = 60$ km/h and $(v_{av})_2 = 20$ km/h. FIND: v_{av} net for the whole trip. (3) PROBLEM TYPE—Average speed. (4) PROCEDURE—The overall average speed, Eq. (2.1), is the *total length l* divided by the *total time t*. We have l but need $t = t_1 + t_2$. Problems will often be encountered that deal with portions of a total distance or time— *the sum of the parts equals the whole* is one more equation to be used in the analysis. The times t_1 and t_2 can be computed from the individual journeys. (5) CALCULATION—Determine t_1 and t_2 using Eq. (2.1), $v_{av} = l/t$,

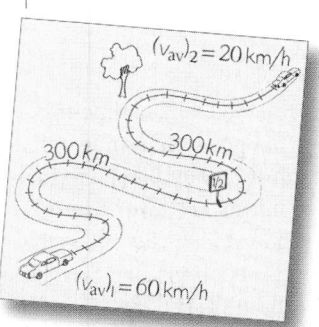

$(v_{av})_2 = 20$ km/h

300 km   300 km

$(v_{av})_1 = 60$ km/h

$$t_1 = \frac{l_1}{(v_{av})_1} = \frac{300 \text{ km}}{60 \text{ km/h}} = 5.0 \text{ h}$$

and

$$t_2 = \frac{l_2}{(v_{av})_2} = \frac{300 \text{ km}}{20 \text{ km/h}} = 15 \text{ h}$$

The total time elapsed is 20 h, and therefore the average speed for the total trip is

$$v_{av} = \frac{l}{t} = \frac{600 \text{ km}}{20 \text{ h}} = \boxed{30 \text{ km/h}}$$

Notice that this is *not* equal to the average of the two average speeds, which is

$$\frac{(v_{av})_1 + (v_{av})_2}{2} = \frac{60 \text{ km/h} + 20 \text{ km/h}}{2} = 40 \text{ km/h}$$

The two are not the same because one person drove three times longer than the other and has a correspondingly greater influence on v_{av}. If that bothers you, imagine two rooms, one containing 100 basketball players all 6-ft tall and the other containing 2 kids each 3-ft tall. Find the average height of the people in each room and then find the average height of all the people. If you got 4.5 ft, you missed the point.

Quick Check: Because the second driver traveled three times longer, that person's speed must be three times more weighty in the average. One person drives for 5 h out of 20 h, and the other for 15 h out of 20 h, hence

$$v_{av} = \frac{5(60 \text{ km/h}) + 15(20 \text{ km/h})}{20} = \frac{120 \text{ km/h}}{4}$$

and $v_{av} = 30$ km/h.

Table 2.3

Speed Conversion Table*

m/s	km/h	mi/h	ft/s	
1.000	0.277 8	0.447 0	0.304 8	m/s
3.600	1.000	1.609	1.097	km/h
2.237	0.621 4	1.000	0.681 8	mi/h
3.281	0.911 3	1.467	1.000	ft/s

*The units being converted head up the columns. The units into which they are transformed via multiplication are on the right of each row. Example: To go from km/h (second column) to mi/h (third row), multiply by 0.621 4. Thus, 1.00 km/h = (1.00)(0.621 4) mi/h.

To go from U.S. Customary Units to SI, or vice versa, requires knowledge of a numerical constant that relates the two systems. For instance, for distance 1.000 in. = 2.540 cm, or 1.000 0 mi = 1.609 3 km. The conversion of 60 mi/h to SI is

$$60 \text{ mi/h} = \left(\frac{60 \text{ mi}}{1.0 \text{ h}}\right)\left(\frac{1.609 \text{ km}}{1.0 \text{ mi}}\right)\left(\frac{1000 \text{ m}}{1.0 \text{ km}}\right)\left(\frac{1.0 \text{ h}}{3600 \text{ s}}\right) = 27 \text{ m/s}$$

The three terms multiplying 60 mi/h can be compressed into a single conversion factor; that is, (0.447 0 m/s) = (1.000 mi/h); see Table 2.3. *It is not recommended that such specific conversion factors be memorized*; 1.000 mi ≈ 1.609 km should be all you need.

2.2 Constant Speed

All of us have traveled in automobiles and have some intuitive sense of what **constant** or **uniform speed** is—just keep the speedometer locked at, say, 55 km/h and you have it (Fig. 2.2). If v_{con} is the constant speed with which some object moves, then that's also its average speed, and Eq. (2.1) becomes $v_{av} = v_{con} = l/t$. The path-length traveled in a given time at a fixed speed is

[constant speed] $$l = v_{con}t$$ (2.2)

   If we use time as an additional dimension and plot distance-traveled versus time-traveled, we gain a new perspective. The resulting graph, which fans out position in time, tells us about the motion. Figure 2.3 is just such a **distance-time graph** for a bumblebee flying at a constant 10 m/s. The bee traverses 10 m during each second; its *speed* or *rate-of-change of distance* is constant, and the curve is a straight line. The *slope* of the line is the ratio of its *rise* to its *run* (the vertical distance between any two points on the line divided by their horizontal separation) {see **MATH REVIEW: PART A-4** on the **CD** }. For the

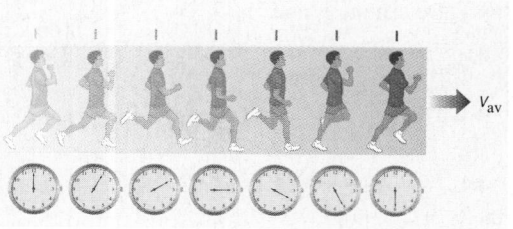

Figure 2.2 An object traveling at a constant speed covers equal distances in equal intervals of time.

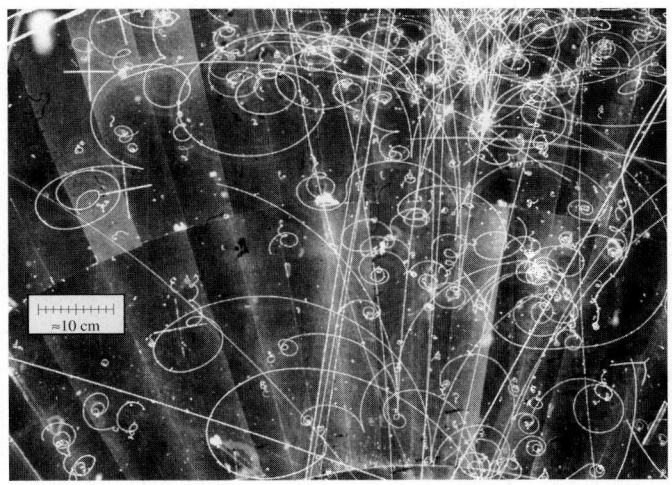

Tracks left by a stream of subatomic particles pouring in at the bottom of the photo. The trails were created over a time interval of only a few nanoseconds and are actually quite long. The particles that made them were initially traveling at a substantial fraction of the speed of light ($c = 2.998 \times 10^8$ m/s).

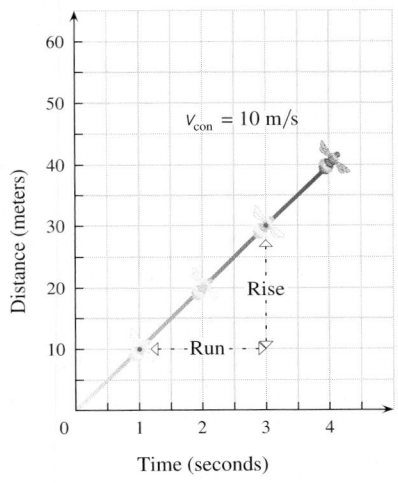

Figure 2.3 Distance-time graph of a bumblebee traveling at a constant speed (v_{con}). Note that the angle (although not the slope) of the line depends on the scale used to plot the curve. Here the slope is 10 m/s and $l = 10t$.

two red points shown in Fig. 2.3, the rise is $(30 \text{ m} - 10 \text{ m}) = 20$ m, the run is $(3.0 \text{ s} - 1.0 \text{ s}) = 2.0$ s, and the slope is 20 m/2.0 s = 10 m/s. ***The slope of the straight line equals the constant speed***: a faster object has a greater slope (Fig. 2.4).

In this case the **speed-time graph** is a horizontal straight line (Fig. 2.5); the speed of the bee is a constant 10 m/s. The rectangular area under the line within, say, the first 3.0 s of flight equals the height (i.e., the speed, 10 m/s) multiplied by the length (i.e., the time, 3.0 s), which yields 30 m, the distance. The bee traveled 30 m in those 3.0 s. ***The area under the speed-time curve bounded by any two moments in time equals the distance traveled during that interval.*** This statement may not seem like much of a revelation in this instance, but it applies to all speed-time graphs.

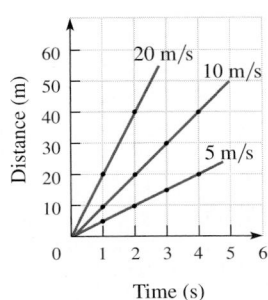

Figure 2.4 For a uniformly moving object, the distance-time graph is a straight line and its slope is the speed. The greater the slope, the greater speed.

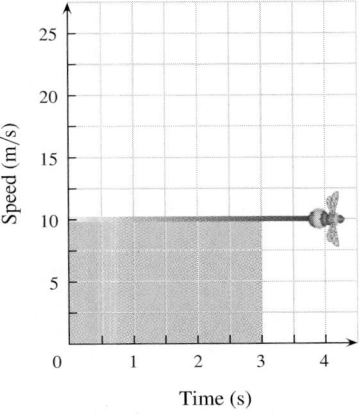

Figure 2.5 A speed-time graph for a bee traveling at a constant 10 m/s. The area under the curve between any two times is the height (i.e., the speed) multiplied by the length (i.e., the time); the result equals the distance traveled.

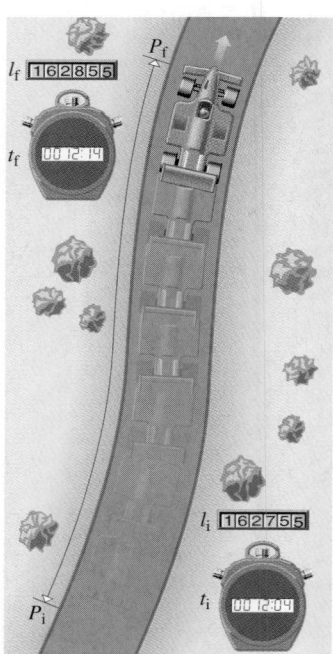

Figure 2.6 Measuring the average speed of a car over a run from point P_i to point P_f. The distance traveled is $l_f - l_i = 10.0$ km, and the time elapsed is $t_f - t_i = 10$ min. The average speed is $v_{av} = (10.0$ km$)/(10$ min$) = 60$ km/h.

2.3 Delta Notation: The Change in a Quantity

Imagine that you decide to measure the average speed of your new Ferrari (Fig. 2.6). As you pass some arbitrary initial point, P_i, on the road, your watch reads an initial time, t_i, of 12:04. At that moment, the odometer displays an initial distance, l_i, of 16 275.5 km. A far-off tree marks the final point of the course, P_f. As the car passes it, the odometer reads a final distance, l_f, of 16 285.5 km and the clock a final time, t_f, of 12:14. Here, $l_f - l_i = 10$ km and $t_f - t_i = 10$ min.

The Greek capital letter delta, Δ, placed before the symbol for some quantity means *the change in that quantity*. Thus, Δl (read delta "el") is the *change in the distance* or the length traversed: $\Delta l = l_f - l_i$. Similarly, Δt (read delta "tee") is the *change in time*, or the duration elapsed: $\Delta t = t_f - t_i$, and

$$v_{av} = \frac{\Delta l}{\Delta t} = \frac{l_f - l_i}{t_f - t_i} \tag{2.3}$$

In this case, where $\Delta l = 10$ km and $\Delta t = 10$ min or $\frac{1}{6}$ h, $v_{av} = 60$ km/h. This form of Eq. (2.1) is useful when the data are presented in terms of initial and final values, as opposed to overall lengths and times. Had you used a stopwatch and a resettable odometer or tripmeter, you could have started the run at $t_i = 0$ and $l_i = 0$. In that instance, $\Delta l = l_f = l = 10$ km, the whole distance; $\Delta t = t_f = t = 10$ min, the whole time; and Eq. (2.3) becomes identical to Eq. (2.1).

2.4 Instantaneous Speed

While in a car or on a bicycle, if you ask "How fast am I going NOW?" you want the *instantaneous speed*—the speed at that very moment. Today, that's usually read from a speedometer. Still, what are we really talking about?

The bumblebee flying in Figs. 2.3 and 2.5 has a constant speed, and it doesn't matter when you yell NOW; its speed at any moment is 10 m/s. More realistically, Fig. 2.7a depicts a nonlinear distance-time curve. The bee leaps into the air at $t = 0$, lands on a flower after flying 2.0 m, rests, and takes off again at $t = 3.0$ s. It flies past your nose at $t = 5.0$ s and alights on a branch at $t = 7.0$ s. *The average speed over any part of a trip is the slope of the line from the beginning to the end of that portion of the distance-time curve.* In Fig. 2.7b, for the interval from say $t = 2.0$ s to $t = 7.0$ s, $v_{av} = \Delta l/\Delta t = 8.0$ m/5.0 s $= 1.6$ m/s.

How do we specify the speed at some instant—NOW—like the one when the bee flew past the point P, on the tip of your nose, at $t = 5.0$ s? To do that, evaluate v_{av} using as an

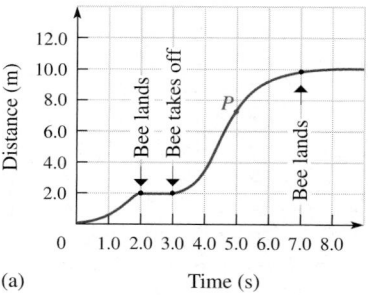

(a)

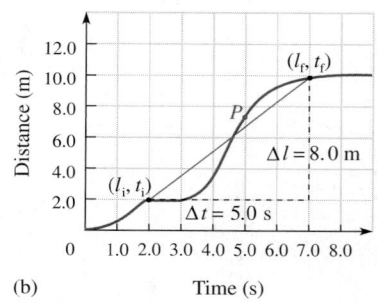

(b)

Figure 2.7 (a) The distance-time curve for a bee. The bee flew to a flower at $t = 2.0$ s, rested until $t = 3.0$ s, and then flew past your nose at point P, alighting on a branch at $t = 7.0$ s. (b) The average speed of the bee during the interval from $t = 2.0$ s to $t = 7.0$ s is $v_{av} = \Delta l/\Delta t = (8.0$ m$)/(5.0$ s$) = 1.6$ m/s.

interval Δl, a small test course that extends along the flight path from a little before to a little after point P (Fig. 2.8a). Because it takes a time Δt for the bee to traverse Δl, the average speed is again $\Delta l/\Delta t$. Our scheme will become increasingly better as Δl straddling P is narrowed, and that narrowing will occur if we reduce the time interval of the measurement (Fig. 2.8b). Therefore, suppose the bee flew from 0.000 005 m before P to 0.000 005 m beyond P in 0.000 002 s. Its v_{av} over that tiny distance is (0.000 010 m)/(0.000 002 s) = 5 m/s. This is the average speed measured during a small but finite time N-O-W as opposed to an instantaneous NOW. It's not necessarily the speed exactly at P, but it's very close to it. By further shrinking Δt, and thereby Δl, we can get as close as we like to the speed NOW.

There is a geometrical way to envision the instantaneous speed of the bee at P. As Δt is made ever smaller, Δl also shrinks, and $\Delta l/\Delta t$ (the average speed) approaches the slope of the distance-time curve at P. ***The slope of any curve at any point is the slope of the tangent to the curve at that point*** (Fig. 2.9a). The slope precisely at P is the instantaneous speed at P, just as the slope anywhere along the curve of Fig. 2.3 is the constant instantaneous speed 10 m/s. From Fig. 2.9b, we see that the bee flew past your nose at 2.9 m/s. Alternatively, suppose you drive a car—up and down hills, and around curves—and every minute or so you record the odometer reading and time. *The slope of the resulting distance-time curve at any point in time equals the instantaneous speed at that moment.*

The central notion here is that Δl will become infinitesimally small as Δt becomes infinitesimally small, and yet their ratio, $\Delta l/\Delta t$, will approach a finite limiting value: a tiny number divided by another tiny number need not itself be tiny. This limiting value approached by $\Delta l/\Delta t$ as $\Delta t \rightarrow 0$ (i.e., as Δt gets infinitesimally small) is called the **instantaneous speed**, v. Mathematically, that's written as

[instantaneous speed]
$$v = \lim_{\Delta t \to 0} \left[\frac{\Delta l}{\Delta t} \right]$$
(2.4)

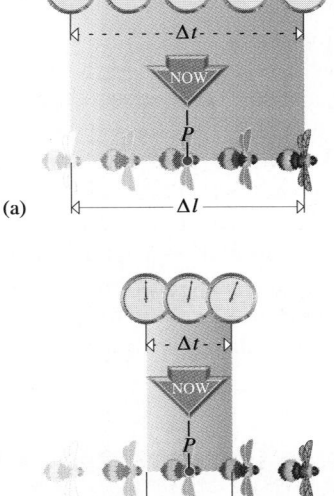

Figure 2.8 To find the instantaneous speed of the bee at point P, determine the average speed over a tiny interval straddling P. Then shrink Δl until v_{av} remains constant at the instantaneous speed v.

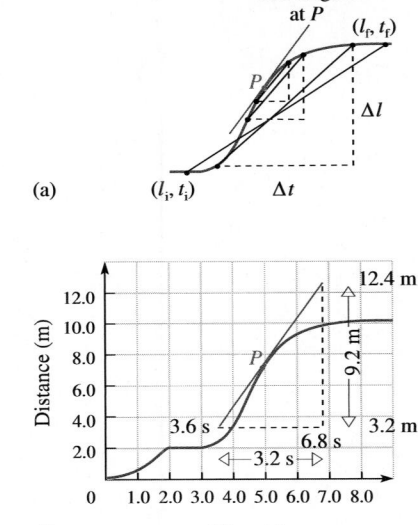

Figure 2.9 (a) As Δt shrinks, Δl shrinks and the line from t_i to t_f approaches the tangent to the curve at P. In this case, both t_i and t_f approach $t = 5.0$ s. (b) The slope of the tangent at P is $v = (9.2 \text{ m})/(3.2 \text{ s}) = 2.9$ m/s, and this value is the instantaneous speed at $t = 5.0$ s.

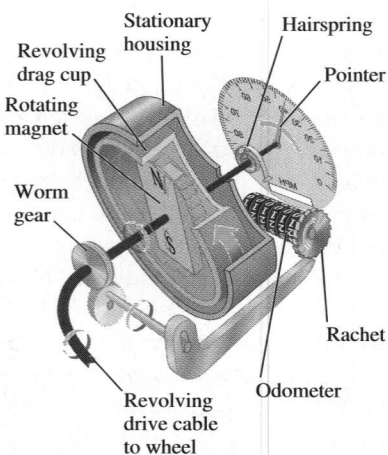

Figure 2.10 A speedometer and odometer. As the wheels of the car turn, the flexible cable turns and the magnet at its end rapidly revolves. That produces a tug on the drag cup that tries to follow the magnet but is held back by the spring. The pointer is attached to the cup and turns with it.

A modern speedometer in the United States shows speed in both mph (miles per hour) and km/h (kilometers per hour).

Instantaneous speed is the limiting value of the average speed $(\Delta l/\Delta t)$ *determined as the interval over which the averaging takes place* (Δt) *approaches zero.* Conceptually, this is an important definition, even though we will not use it directly to make numerical calculations. {For more on calculus and speed, click on **Derivative** under **Further Discussions** on the **CD**.}

Henceforth, when we talk about "speed," instantaneous speed is what we have in mind. Notice that $\Delta l/\Delta t$ is to be determined as Δt *approaches*, rather than reaches, zero. It would make no sense to ask for the average speed over a zero time interval; nothing moves a real distance in no time, and the definition of average speed falls apart when $\Delta t = 0$. This subtlety was a point of confusion for centuries. For us, it will suffice to think of v as v_{av} taken over a tiny interval Δt centered on NOW, where Δt is small enough so that v is effectively constant within it.

Velocity

A wooden horse on a carousel can swing around on its pedestal through 40 m in 20 s, and though it's forever coming back to the same spot in space, it still travels with an average speed of (40 m)/(20 s) = 2.0 m/s. *Speed is independent of the direction of motion*, and like all such quantities that have nothing to do with spatial orientation (length, time, etc.), it has only a size or *scale* and no direction. For that reason it's called a **scalar quantity**.

To describe the motion of an object completely, we must specify both the rate of travel (i.e., the speed) and its direction. The concept that embraces both speed and direction is **velocity**, and velocity can't be fully understood without the notion of **displacement**. We now develop the mathematics needed to treat all such directional quantities, the mathematics of *vectors*.

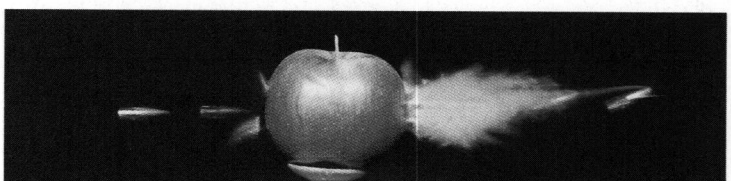

Here a single bullet traveling at 800 m/s is captured four times in a quadruple exposure. During each 500-ns exposure, the bullet certainly moved, but not enough to appreciably blur its image. The time intervals between the first and second, second and third, and third and fourth are 50 μs, 200 μs (to give the apple time to explode), and 50 μs, respectively. For all practical purposes the average speed over the 0.50 μs exposure time is the bullet's instantaneous speed.

2.5 The Displacement Vector

Suppose we put a little turtle on a piece of paper at some point P_o and leave for a while. On our return, the turtle is discovered at a new spot, P_f. Nothing is known about the route it took, but we do know the turtle has moved. ***Its displacement is the straight-line shift in position from P_o to P_f, specifying both length and direction.*** In Fig. 2.11, the meandering reptile ended up 10 cm northeast of where it started. Any concept such as this that can be stated completely only if we provide both its magnitude (size) and direction is a **vector quantity**. To distinguish vectors from scalars in handwritten material, a small arrow is usually placed just above the symbol. With that in mind, we will designate a vector by a bold-faced letter with an arrow above it (e.g., $\vec{\mathbf{A}}$).

The displacement of an object from its starting point P_o is the vector extending from that point to the position where the object is. Displacement is often important; think of the displacement of a football from the person who threw it. The *displacement vector* is designated as $\vec{\mathbf{s}}$, a symbol that comes from the notion of the "space" between two points.

In Fig. 2.11, the vector $\vec{\mathbf{s}}$ is represented pictorially by an arrow 10-cm long pointing northeast. And if 10 cm is inconveniently long to draw, simply scale it down and just label it 10 cm. The *length* of the arrow corresponds to the **magnitude** (the ***positive numerical value***) of that vector. The magnitude of the displacement vector $\vec{\mathbf{s}}$ is written as $|\vec{\mathbf{s}}|$; it's specified by a single number, in this case $|\vec{\mathbf{s}}| = 10$ cm. Although it's a terrible idea, for reasons we'll soon see (p. 38), the magnitude of a vector such as $\vec{\mathbf{s}}$ is often written as just s. We'll avoid this usage wherever possible.

Imagine that the turtle takes off on a new trip, this time traveling 3 cm east from point P_o, undergoing its first displacement $\vec{\mathbf{s}}_1$, as shown in Fig. 2.12. It then turns 90 degrees left, going north for 4 cm to point P_f, undergoing a displacement $\vec{\mathbf{s}}_2$. The **resultant** displacement (with respect to P_o) is the vector $\vec{\mathbf{s}}$, which is independent of the actual path taken. Its magnitude can be computed from the right triangle using the Pythagorean Theorem {see **MATH REVIEW: PART B** on the **CD** 💿}: $|\vec{\mathbf{s}}| = \sqrt{3^2 + 4^2} = \sqrt{25}$. Fortunately, the turtle walked two sides of a 3-4-5 right triangle, so $\vec{\mathbf{s}}$, the hypotenuse, is 5-cm long—P_f is 5 cm from P_o in a direction somewhat east of north.

There can be a big difference between *distance* and *displacement*. The odometer on a car measures distance traveled (l) not displacement ($\vec{\mathbf{s}}$). If you make a 100-km round trip, the odometer will record a 100-km journey even though you return to the starting place and your displacement is zero. *Only when the path traveled is a straight line, and there is no doubling back, will distance equal the magnitude of the displacement.*

We represent a physical **vector quantity** by an arrow whose length corresponds to the size of the quantity, pointing in the direction in which it occurs.

The **displacement** of an object with respect to some reference point is the vector drawn from that point to the object.

A **vector quantity** is charaterized by both a magnitude, or size, and a direction.

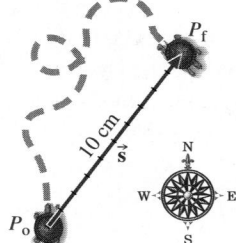

Figure 2.11 The displacement vector $\vec{\mathbf{s}}$ of a turtle that meandered from point P_o to point P_f. The vector quantity $\vec{\mathbf{s}}$ tells us where the turtle ended up in relation to where it started.

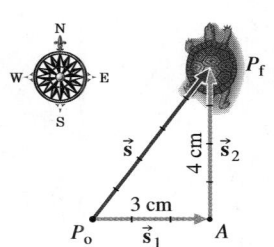

Figure 2.12 The total displacement $\vec{\mathbf{s}}$ of a turtle that traveled from P_o to A to P_f via displacements $\vec{\mathbf{s}}_1$ and $\vec{\mathbf{s}}_2$. The total displacement equals the vector sum of the individual displacements: $\vec{\mathbf{s}} = \vec{\mathbf{s}}_1 + \vec{\mathbf{s}}_2$.

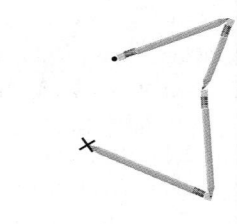

2.6 Some Vector Algebra

A vector is not completely defined until the rules for its behavior (e.g., **vector addition**) are established. We don't have a satisfactory idea of what numbers, or scalars, are until we know how to manipulate them, and the same is true with vectors.

Envision somehow adding the intermediate displacements $(\vec{s}_1 + \vec{s}_2)$ in Fig. 2.12 to arrive at the total displacement $\vec{s}$; it's evident that we cannot simply add in the usual algebraic way. The total path-length traveled is $|\vec{s}_1| + |\vec{s}_2| = 7$ cm, whereas the magnitude of the total displacement is $|\vec{s}| = 5$ cm; $|\vec{s}| \neq |\vec{s}_1| + |\vec{s}_2|$. The illustration itself suggests a rule for adding vectors known as the **Tip-to-Tail Method**. *Two vectors can be added together by arranging them such that the tail of the second is at the tip of the first; the sum of these, also called the **resultant vector**, extends from the tail of the first to the tip of the second.* In the process of adding two vectors, the sum will not change if either one (or both) is moved parallel to its original direction.

Had the turtle walked north first and then east (inverting the order of the displacements in Fig. 2.12), it would still have arrived at P_f, which means that $\vec{s} = \vec{s}_1 + \vec{s}_2 = \vec{s}_2 + \vec{s}_1$. As with ordinary scalars, *the order of addition is irrelevant*, and you can put either tip to either tail when adding two vectors.

Figure 2.13 makes the same point, this time with vectors that form an oblique triangle. It suggests another technique for graphically adding two vectors. Beginning with the Tip-to-Tail scheme and realizing that the order of addition is irrelevant, the resultant vector can be constructed in two ways which, taken together, form a parallelogram. Draw the two vectors we wish to add, $\vec{s}_1$ and $\vec{s}_2$, emerging from a *common origin*. Using these as adjacent sides, construct a parallelogram. *The diagonal emerging from the origin is then the resultant of the two.* This procedure is called the **Parallelogram Method**. It was devised by Isaac Newton in order to deal with the addition of forces (which is also a vector quantity) long before the mathematical concept of "vector" was introduced.

The line along which a vector lies is called its **line-of-action**. *When the lines-of-action of several vectors intersect at the same point, they are said to be **concurrent**.* The two vectors in Fig. 2.13 intersect and are therefore concurrent. That circumstance will become especially important when we consider a number of forces acting on a body.

Once we have a rule for adding two vectors, we have a rule for adding any number of vectors. The four vectors, $\vec{A} + \vec{B} + \vec{D} + \vec{F}$, in Fig. 2.14, can be added two at a time following the Tip-to-Tail Method. In that way, $\vec{A}$ plus $\vec{B}$ is $\vec{C}$, to which we add $\vec{D}$, such that $\vec{C} + \vec{D} = \vec{E}$, to which we add $\vec{F}$, such that $\vec{E} + \vec{F} = \vec{G}$, so that $\vec{A} + \vec{B} + \vec{D} + \vec{F} = \vec{G}$. It's not necessary to bother with the intermediate sums of $\vec{C}$ and $\vec{E}$. The figure shows that we could have

> When **adding several vectors** you can slide them around as long as you don't change the direction in which any of them points.

Figure 2.13 (a) Two vectors that are to be added in two ways. (b) The tip-to-tail addition of two vectors $(\vec{s}_1 + \vec{s}_2 = \vec{s}_2 + \vec{s}_1 = \vec{s})$. (c) The parallelogram formed with $\vec{s}_1$ and $\vec{s}_2$ as its sides.

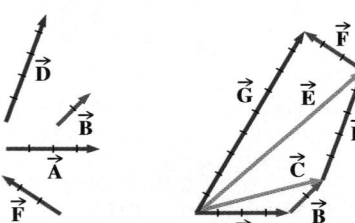

Figure 2.14 The addition of several vectors $(\vec{A} + \vec{B} + \vec{D} + \vec{F})$ via the tip-to-tail method. Here $\vec{C} = \vec{A} + \vec{B}$, $\vec{E} = \vec{C} + \vec{D}$, and $\vec{G} = \vec{E} + \vec{F} = \vec{A} + \vec{B} + \vec{D} + \vec{F}$. Knowing how to add two vectors allows us to add any number of vectors.

Example 2.5 **[I]** A treasure map was torn into six pieces by its maker, a clever one-eyed gentleman with a parrot. The first piece shows a large recognizable tree and simply reads "start here." Although in no particular order, each of the other five fragments reads: "3 m–EAST," "5 m–NORTH," "5 m–SOUTHEAST," "2.5 m–SOUTH," and "6.5 m–WEST." Find the treasure.

Solution Each term, such as "5 m–NORTH," specifies a displacement vector. (1) TRANSLATION—Add five known displacement vectors. (2) GIVEN: $\vec{s}_1 = 3$ m–EAST, $\vec{s}_2 = 5$ m–NORTH, $\vec{s}_3 = 5$ m–SOUTHEAST, $\vec{s}_4 = 2.5$ m–SOUTH, and $\vec{s}_5 = 6.5$ m–WEST. FIND: $\vec{s}$. (3) PROBLEM TYPE—Vector addition: displacement. (4) PROCEDURE—Use the tip-to-tail method knowing that vector addition is independent of order. (5) CALCULATION—From Fig. 2.15, the treasure is buried 1 m due south of the old tree, or $\vec{s} = 1$ m–SOUTH. Note that **any numerical representation of a vector should always have a direction explicitly specified.**

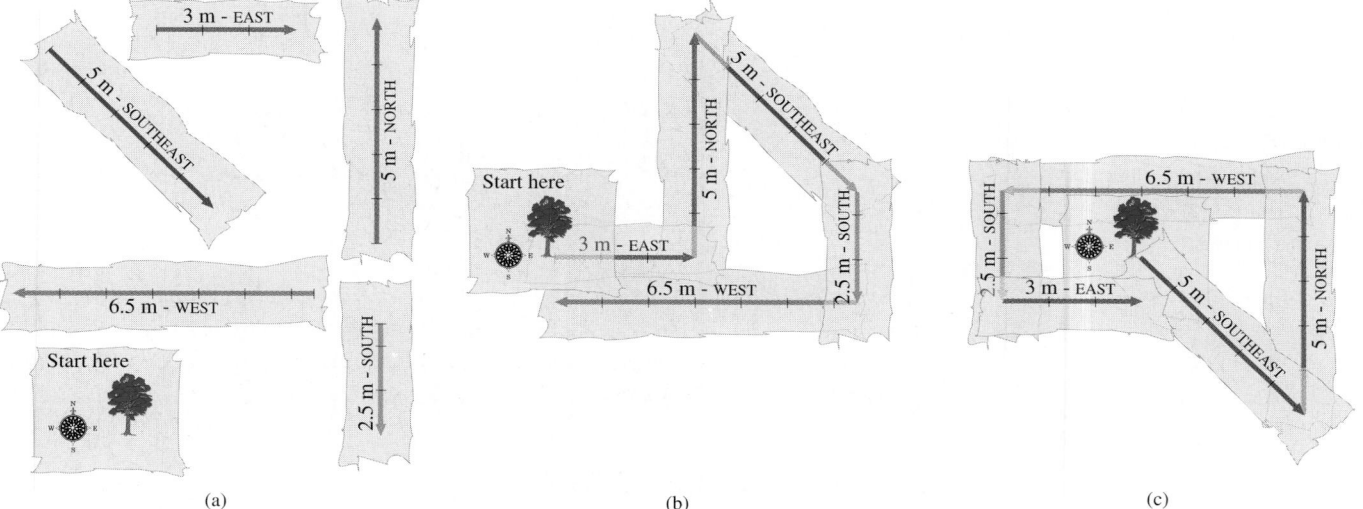

Figure 2.15 The five vectors in (a) can be added in any order to find the treasure. In (b), we start at the tree and first move east 3 m. In (c), that 3-m displacement is last, but we still arrive at the same spot where the treasure is buried.

just as well tip-to-tailed $\vec{A}$, $\vec{B}$, $\vec{D}$, and $\vec{F}$ directly and, for that matter, we could have done it in any order at all.

Vectors that are either parallel or antiparallel (i.e., oppositely directed) are added in the usual Tip-to-Tail way, and the results are simple and highly useful (Fig. 2.16). In the case of two parallel vectors, $\vec{s}_1$ and $\vec{s}_2$, the magnitude of the resultant $|\vec{s}|$ equals the sum of the magnitudes of the scalar distances.

Suppose now that we add to a vector $\vec{s}$ an identical vector $\vec{s}$. The resultant of $(\vec{s} + \vec{s})$ lies in the same direction as either constituent $\vec{s}$ and is twice as long as either one. It would seem reasonable from the usual procedures of algebra to write the resultant as $2\vec{s} = \vec{s} + \vec{s}$, which defines the process of **multiplication of a vector by a scalar**. For example, if $\vec{s}$ is a displacement of 10 cm–EAST, $6\vec{s}$ is a displacement of 60 cm–EAST. *Multiplying a vector by a positive scalar multiplies the vector's magnitude in the usual way, leaving its direction unchanged.*

If two vectors are antiparallel (Fig. 2.16b), when $|\vec{s}_1| > |\vec{s}_2|$, the magnitude of the resultant equals the difference of the individual magnitudes, $|\vec{s}_1| - |\vec{s}_2|$. Vectors that are oppositely directed subtract algebraically. A displacement of 10 m–EAST, followed by a displacement of 10 m–WEST, yields a total displacement of zero. If we interpret the symbol $(-\vec{s})$ to be a vector equal in magnitude but antiparallel to $\vec{s}$, then the sum $\vec{s} + (-\vec{s}) = \vec{s} - \vec{s} = 0$, so the two cancel in the usual algebraic way. This suggests that **vector subtraction** can be defined as follows: *any vector $\vec{B}$ can be subtracted from any vector $\vec{A}$ by reversing $\vec{B}$ to form $-\vec{B}$ and then adding the two in the usual Tip-to-Tail way* (Fig. 2.17).

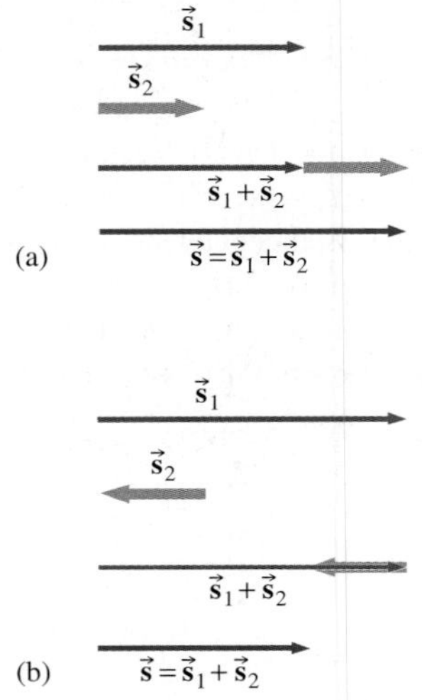

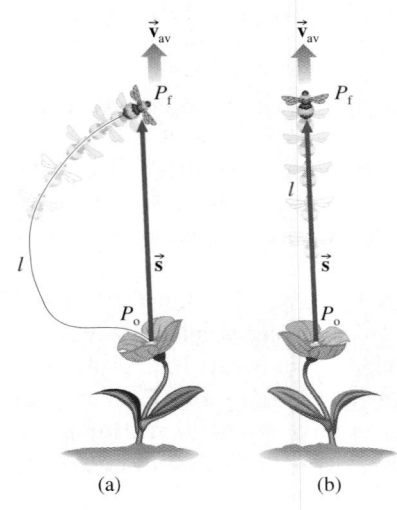

Figure 2.16 (a) The sum of two parallel vectors. Vectors that point in the same direction can be added like scalars in that $\vec{s} = \vec{s}_1 + \vec{s}_2$. (b) The sum of two antiparallel vectors.

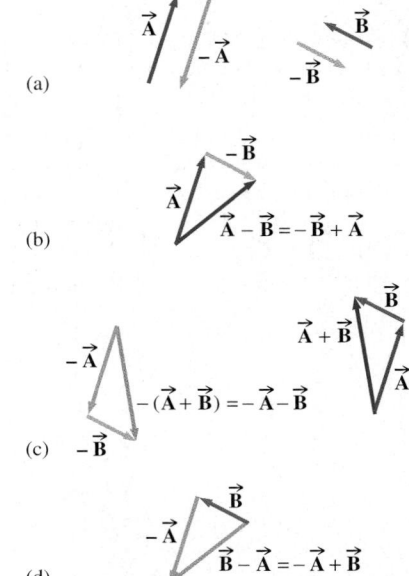

Figure 2.17 The sum and differences of two vectors. In (a) we see vectors $\vec{A}$ and $\vec{B}$ and their negatives $-\vec{A}$ and $-\vec{B}$; (b) shows $\vec{A} - \vec{B}$; (c) shows $-\vec{A} - \vec{B}$ and $\vec{A} + \vec{B}$; and (d) adds $-\vec{A}$ and $\vec{B}$ to get $\vec{B} - \vec{A}$. Notice that $-\vec{B} + \vec{A}$ is the negative of $-\vec{A} + \vec{B}$.

Multiplying a vector by a negative scalar multiplies the vector's magnitude but also reverses its direction. If $\vec{s}$ is a displacement of 10 cm–EAST, $-6\vec{s}$ is a displacement of 60 cm–WEST. {Make sure to work with the wonderful simulation called **VECTOR ADDITION & SUBTRACTION** under **INTERACTIVE EXPLORATIONS** on the **CD**. }

2.7 Instantaneous Velocity

Back to our bumblebee, which in a time t now flies from P_o to P_f in an arc shown in Fig. 2.18a. The bee's average speed is $v_{av} = l/t$; with this expression as a guide, define the **average velocity** to be the ***vector formed by the displacement, measured from the starting point of the motion, divided by the time elapsed***, or

$$\vec{v}_{av} = \frac{\vec{s}}{t} \qquad (2.5)$$

Since t is a positive scalar, multiplying $\vec{s}$ by $1/t$ has no effect on its direction; $\vec{v}_{av}$ ***is parallel to*** $\vec{s}$. Average velocity is an intermediate idea on the way to the far more important concept of instantaneous velocity. By itself $\vec{v}_{av}$ is not very useful for anything other than straight-line motion—a racecar roaring around a track will have zero displacement and zero average velocity whenever it returns to its starting point, no matter how fast it's going.

Consider the bee in Fig. 2.19. While it flies from P_i to P_f in a time Δt, the vector $\vec{s}$, measured from P_o, changes as it tracks the bee instant-by-instant. Starting at P_i the displacement with respect to P_o is $\vec{s}_i$. This plus the change in the displacement, $\Delta\vec{s}$, equals the final displacement with respect to P_o, namely, $\vec{s}_f$ where $\vec{s}_i + \Delta\vec{s} = \vec{s}_f$. Therefore, $\Delta\vec{s} = \vec{s}_f - \vec{s}_i$, which is reminiscent of Eq. (2.3). The displacement of the bee on going from P_i to P_f is $\Delta\vec{s}$.

Using the definition [Eq. (2.5)], the average velocity over the route from P_i to P_f becomes

$$\vec{v}_{av} = \frac{\Delta\vec{s}}{\Delta t} = \frac{\vec{s}_f - \vec{s}_i}{t_f - t_i} \qquad (2.6)$$

Earlier, we transformed the notion of average speed into instantaneous speed by causing the time interval over which the averaging was happening to shrink toward zero. Using the same reasoning, we define the **instantaneous velocity**, $\vec{v}$, ***at any point*** P_i ***to be the lim-***

Figure 2.18 The distance traveled (l) and the displacement (s) in the two different cases of (a) motion along a curve and (b) motion along a straight line. For the special case of straight-line motion in a fixed direction, the magnitude of the average velocity s/t equals the average speed l/t. That is generally not the case.

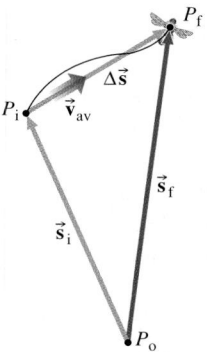

Figure 2.19 The change in the displacement of the bumblebee P_i to P_f occurring in a time Δt is $\Delta \vec{\mathbf{s}}$. Notice that $\vec{\mathbf{v}}_{av}$ is in the direction of $\Delta \vec{\mathbf{s}}$.

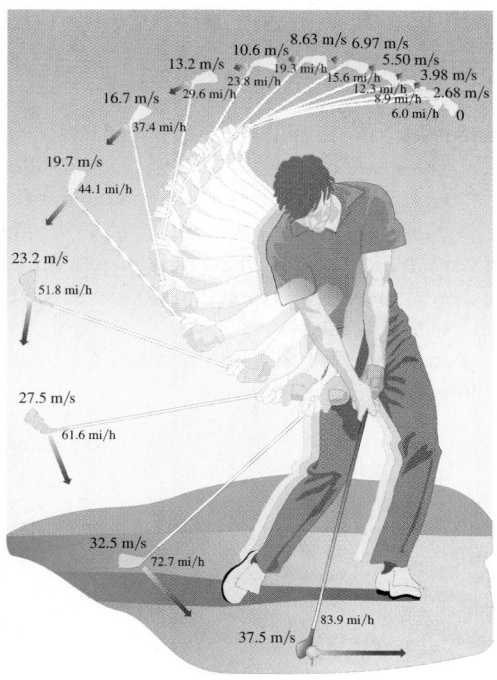

Figure 2.20 A typical swing of a golf club. The velocity vector for the head is shown every 10 milliseconds.

it as $\Delta t \to 0$ of the displacement of the object (from that starting point) divided by the time interval

[instantaneous velocity]

$$\vec{\mathbf{v}} = \lim_{\Delta t \to 0} \left[\frac{\Delta \vec{\mathbf{s}}}{\Delta t} \right] \tag{2.7}$$

We'll simply call this the **velocity**. [Equation (2.7) is an important definition, even though it will not be used directly in any numerical calculations.]

The magnitude of the velocity equals the instantaneous speed: $|\vec{\mathbf{v}}| = v$. That's why we used the symbol v for speed in the first place. *The velocity vector at any location is tangent to the path and always points in the direction of motion* (Fig. 2.20). *When the velocity of an object is constant, it travels in a straight line at a constant speed.* {For a proof of the last several points, click on **VELOCITY** under **FURTHER DISCUSSIONS** on the **CD**.}

Example 2.6 **[II]** A large clock mounted on a wall in a railroad station (Fig. 2.21) has a second hand 1.0-m long. Assuming that the hand sweeps smoothly around, find the velocity of a point on its very tip at exactly 15 s after noon.

Figure 2.21 A clockface at 15 s after 12:00. At the instant shown, the point on the very end of the second hand is moving downward with a velocity $\vec{\mathbf{v}}$.

Solution "Velocity" is a vector quantity with magnitude and direction. (1) TRANSLATION—An object moves in a circle at a constant speed; determine its velocity at a specific moment. (2) Notice that some of the needed information is not stated explicitly. GIVEN: radius $R = 1.0$ m, and once-around time $T = 60$ s. FIND: $\vec{\mathbf{v}}$

(both its speed and direction) at time $t = 15$ s. (3) PROBLEM TYPE—Velocity/constant speed. (4) PROCEDURE—The tip's constant speed equals its instantaneous speed, and that suggests Eq. (2.2), $v_{con} = v = l/t$, where we first compute l. (5) CALCULATION—l is the total distance swept out by the very tip of the second hand in one revolution (see Example 2.2); $l = 2\pi R = 2(3.14)(1.0$ m$) = 6.28$ m and

$$v = \frac{l}{T} = \frac{6.28 \text{ m}}{60 \text{ s}} = 0.10 \text{ m/s}$$

Since $|\vec{\mathbf{v}}| = v = 0.10$ m/s, at 15 s after noon

$$\boxed{\vec{\mathbf{v}} = 0.10 \text{ m/s—STRAIGHT DOWNWARD}}$$

Quick Check: The velocity is always tangent to the circle (i.e., perpendicular to the radius), pointing in a direction determined by the clockwise motion. $2\pi R \approx 6$ m, $v \approx 6$ m/60 s ≈ 0.1 m/s.

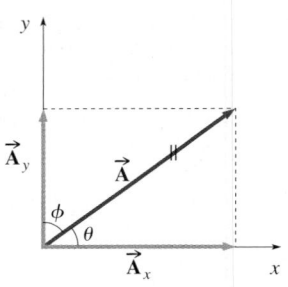

Figure 2.22 Any vector $\vec{A}$ can be thought of as equivalent to its two perpendicular components $\vec{A}_x$ and $\vec{A}_y$. Notice that the x-component of the vector can be written in terms of either $\cos\theta$ or $\sin\theta$ and similarly the y-component of the vector can be written in terms of either $\sin\theta$ or $\cos\theta$. Be careful with those angles.

2.8 Components and Vector Addition

We now develop a powerful analytic technique for adding vectors, making use of trigonometry and the concept of *vector components*. Figure 2.22 shows two perpendicular vectors (acting along the x- and y-directions) $\vec{A}_x$ and $\vec{A}_y$; these might be displacements, or velocities, or any other vector quantity. Using the parallelogram method (p. 30), it's evident that $\vec{A} = \vec{A}_x + \vec{A}_y$. Conversely, the same diagram can be interpreted as showing that *any vector $\vec{A}$ in the xy-plane may be resolved into two perpendicular component vectors $\vec{A}_x$ and $\vec{A}_y$ which together are equivalent in every regard to the original vector*. Given a specific value of $\vec{A}$ (i.e., magnitude, A, and direction, θ), the most accurate way to determine these components is to use trigonometry.

Recall {see **Math Review: Part C** on the **CD** 💿 } that, for a right triangle, the sine of either acute angle is the ratio

$$\sin\theta = \frac{\text{side opposite }\theta}{\text{hypotenuse}}$$

Similarly, the cosine of either acute angle of a right triangle is the ratio

$$\cos\theta = \frac{\text{side adjacent }\theta}{\text{hypotenuse}}$$

The triangle in Fig. 2.22 has sides equal to the lengths (i.e., magnitudes) of the vectors: $|\vec{A}_x|$, $|\vec{A}_y|$, and, $|\vec{A}|$. The ratios of these are

$$\sin\theta = \frac{|\vec{A}_y|}{|\vec{A}|} \qquad \text{and} \qquad \cos\theta = \frac{|\vec{A}_x|}{|\vec{A}|}$$

Multiplying both sides of each of these equations by $|\vec{A}|$ yields

$$|\vec{A}_y| = |\vec{A}|\sin\theta \qquad \text{and} \qquad |\vec{A}_x| = |\vec{A}|\cos\theta \qquad (2.8)$$

Example 2.7 **[I]** An aircraft in a 45.0° dive is traveling at a constant 800 km/h. Determine its speed, in m/s, in the vertical direction—that is, the rate at which the plane is losing altitude.

Solution The problem provides the velocity vector and asks for its vertical component. (1) Translation—An object moves with a known constant velocity at an angle θ; determine the vertical component. (2) The plane's *constant* velocity vector $\vec{v}$ points downward at an angle θ. Given: $v = 800$ km/h at $\theta = 45.0°$. Find: v_y. (3) Problem Type—Speed/components. (4) Procedure—Equations (2.8) pro-

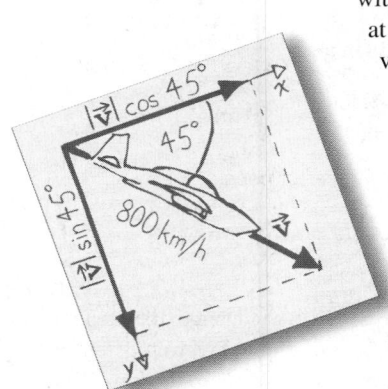

vide the needed relationships between any vector and its two perpendicular components. (5) Calculation—First get v in m/s; $v = (800 \text{ km/h})(1000 \text{ m/km})/(3600 \text{ s/h}) = 222.2$ m/s. Then

$$v_y = v\sin\theta = (222.2 \text{ m/s})\sin 45.0° = \boxed{157 \text{ m/s}}$$

Because both acute angles of this velocity triangle are equal (both are 45°), it is isosceles and the two sides, v_x and v_y, are equal. The plane's horizontal and vertical speeds are therefore both 157 m/s. As the aircraft advances horizontally, it simultaneously descends at the same speed vertically.

Quick Check: You should know that $\sin 45° = 0.707 = \cos 45°$, and so either component is $\approx(0.7)(800 \text{ km/h}) \approx 5.6 \times 10^2$ km/h ≈ 156 m/s. *Always make a rough check of the numerical computation to be sure you have pushed the correct calculator keys.* Check that your calculator is set for degrees.

Example 2.8 **[I]** A firefighter dashes up a 26-m ladder (making an angle of 67.4° with the ground). What is the vertical displacement of her feet at the top?

Solution We're essentially given the displacement vector and asks to find its vertical component. (1) TRANSLATION—A body is displaced at an angle θ through a known distance; determine the vertical component. (2) GIVEN: $|\vec{s}| = 26$ m and $\theta = 67.4°$. FIND: $\vec{s}_y$. (3) PROBLEM TYPE—Vector components. (4) PROCEDURE—The ladder is the hypotenuse of a right triangle having a base angle of 67.4°. The displacement vectors form the same triangle. (5) CALCULATION—From Eq. (2.8)

$$|\vec{s}_y| = |\vec{s}| \sin \theta = (26 \text{ m})(\sin 67.4°) = (26 \text{ m})(0.923)$$

and $|\vec{s}_y| = 24$ m; therefore $\boxed{\vec{s}_y = 24 \text{ m—STRAIGHT UP}}$

Quick Check: The displacement triangle {see MATH REVIEW: PART B on the CD for a geometry review} is a 5-12-13 right triangle. Its sides are 10-24-26.

In Fig. 2.22 $\vec{A}$ is in the first quadrant and everything is positive, but for $\vec{A}$ in any other quadrant, $\sin \theta$ and $\cos \theta$ may each be negative. That means these equations must be generalized since, as is, the left side of each is always positive and the right side isn't. Accordingly, define the **scalar components of a vector** $\vec{A}$ as

$$A_y = |\vec{A}| \sin \theta \qquad \text{and} \qquad A_x = |\vec{A}| \cos \theta \qquad (2.9)$$

where* $A_x = \pm|\vec{A}_x|$ and $A_y = \pm|\vec{A}_y|$; *the scalar component can be positive or negative, the magnitude is always positive*. When $\vec{A}_x$ points in the positive x-direction A_x is positive and when it points in the negative x-direction A_x is negative, and the same is true for A_y. The quantities A_x and A_y are the *scalar x- and y-components* of $\vec{A}$. {For an extraordinary animated look at how a vector projects its components click on VECTOR COMPONENTS under INTERACTIVE EXPLORATIONS on the CD.}

To get a feel for the vector nature of velocity, consider the police cruiser parked on the side of the road in Fig. 2.23. The officer inside is using radar to continuously measure the speed of approaching traffic. The speedometer in any oncoming car displays the speed along the road, v, which is the magnitude of the car's velocity at any instant, $|\vec{v}|$. By contrast, what's being determined by the radar gun is the speed at which a car is approaching the cop, and that's not v. The speed of approach is the magnitude of the velocity component along the line connecting the two vehicles; at any instant that's $|\vec{v}| \cos \theta$. In other words, the car, traveling at v, is moving toward the radar gun at a speed of $|\vec{v}| \cos \theta$. Thus the gun measures less then the speedometer, but when the car is far away, θ is very small, and

$\theta = 20°$
speedometer: 65 mph
radar gun: 61 mph

Figure 2.23 The speedometer reading as seen by the driver inside any approaching car is always greater than the radar-gun reading by a factor of $\cos \theta$.

*Frequently you'll see Eq. (2.9) written as $A_y = A \sin \theta$ and $A_x = A \cos \theta$ where we then have to remember that although A is the magnitude of $\vec{A}$, A_x and A_y are not the magnitudes of $\vec{A}_x$ and $\vec{A}_y$; all of which is rather clumsy.

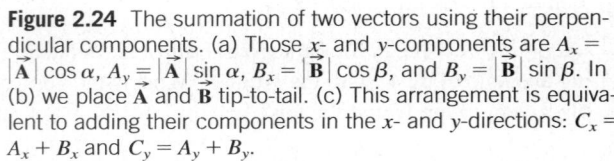

Figure 2.24 The summation of two vectors using their perpendicular components. (a) Those x- and y-components are $A_x = |\vec{A}| \cos \alpha$, $A_y = |\vec{A}| \sin \alpha$, $B_x = |\vec{B}| \cos \beta$, and $B_y = |\vec{B}| \sin \beta$. In (b) we place $\vec{A}$ and $\vec{B}$ tip-to-tail. (c) This arrangement is equivalent to adding their components in the x- and y-directions: $C_x = A_x + B_x$ and $C_y = A_y + B_y$.

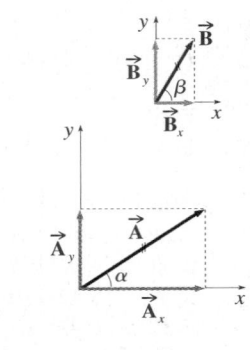

(a)

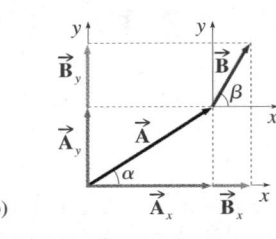

(b)

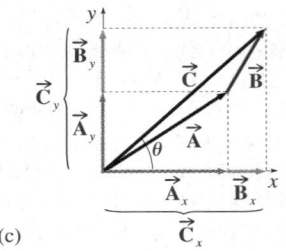

(c)

$\cos \theta \approx 1$. This example underscores the power of the vector approach—how else would you explain the cosine dependence?

Figure 2.24*a* shows two arbitrary vectors $\vec{A}$ and $\vec{B}$ that are added graphically in Fig. 2.24*b* using the Tip-to-Tail technique. The resultant vector $\vec{C}$ is constructed in Fig. 2.24*c*. Notice that each of the components of the resultant is equal to the sums of the components of the constituents. In other words,

$$C_x = A_x + B_x \qquad \text{and} \qquad C_y = A_y + B_y \qquad (2.10)$$

Find the scalar x-components of the vectors to be summed, **give them proper signs** (Fig. 2.25), and then add them algebraically. That total is the scalar x-component of the resultant, and the same procedure applies to the y-component. To reconstruct $\vec{C}$ from C_x and C_y, use the Pythagorean Theorem to compute C, the magnitude of the resultant,

$$C = \sqrt{C_x^2 + C_y^2} \qquad (2.11)$$

Now determine the direction of $\vec{C}$ by finding θ. The most convenient way to do that is via the **tangent** function for right triangles

$$\tan \theta = \frac{\text{side opposite } \theta}{\text{side adjacent } \theta}$$

The angle θ is the *inverse tangent* ($\tan^{-1}$) of the ratio of the two sides

$$\theta = \tan^{-1} \frac{|C_y|}{|C_x|} \qquad (2.12)$$

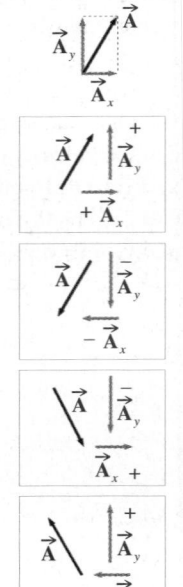

Figure 2.25 The vector $\vec{A}$ has components in the x- and y-directions of $\vec{A}_x$ and $\vec{A}_y$. Here we see the signs that must be associated with the scalar quantities A_x and A_y if they are to be correctly added algebraically to the scalar components of other vectors.

where $|C_y|$ and $|C_x|$ are the positive values of the components as shown Fig. 2.26. For example, suppose $C_x = -2.0$ and $C_y = +4.0$. The vector $\vec{C}$ is then in the second quadrant, pointing a little more upward than to the left. From Eq. (2.12), $\theta = \tan^{-1}(4.0/2.0) = 63°$, that is, $\vec{C}$ lies 63° up from the negative x-axis or $90° + 63° = 153°$ up from the positive x-axis {see **MATH REVIEW: PART D** on the CD for a geometry review 🔘 }.

To compute **inverse trig. functions** first make sure your calculator is in DEG mode. Then enter the number which we know is the sine, cosine, or tangent of some unknown angle which you must find. That done, the inverse functions are usually accessed by pressing the 2nd key (which is often color coded). For example, let's determine the angle whose sine is 0.50, that is

$$\theta = \sin^{-1} 0.50:$$

enter

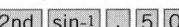

and 30, which is 30°, appears as the answer.

If you have one of the newer two-line calculators that does things in normal math sequence, enter

2nd sin-1 . 5 0

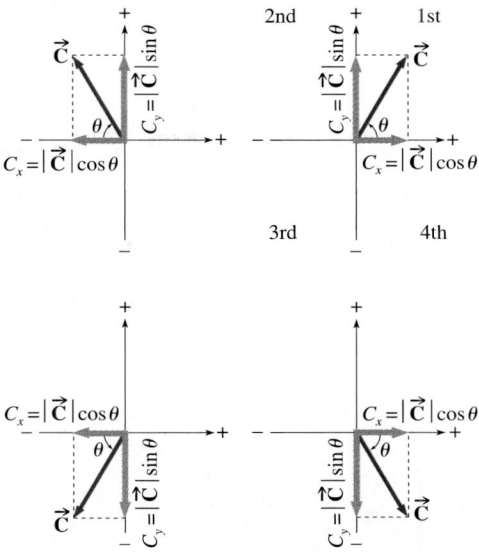

Figure 2.26 In the process of adding any two vectors $\vec{A}$ and $\vec{B}$ to get $\vec{C}$ use Eq. (2.9) to determine C_x and C_y. Make sure to associate the proper signs with A_x and B_x, and A_y and B_y before adding them (see Fig. 2.25). The signs of C_x and C_y determine in which quadrant $\vec{C}$ will be and also determine θ. Thus, if C_x is negative and C_y is positive, $\vec{C}$ is in the second quadrant (see the upper left diagram).

Example 2.9 **[II]** Two vectors lie in the *x-y* plane, as shown in Fig. 2.24. Vector $\vec{A}$ is 10 units long and points upward 30° above the positive *x*-axis. Vector $\vec{B}$ is 15 units long and points 45° above the positive *x*-axis. What is the resultant of the two vectors?

Solution We're being called on to analytically add two vectors, which means find their scalar components and add them. (1) TRANSLATION—Add known vectors $\vec{A}$ and $\vec{B}$.   (2) GIVEN: $\vec{A}$ = 10 units–UP 30° ABOVE *X*-AXIS and $\vec{B}$ = 15 units–UP 45° ABOVE *X*-AXIS. FIND: $\vec{C} = \vec{A} + \vec{B}$.   (3) PROBLEM TYPE—Analytic vector addition.   (4) PROCEDURE—From Eqs. (2.9) compute A_x, A_y, B_x, and B_y and then use Eqs. (2.10) to (2.12).   (5) CALCULATION:

$$A_x = |\vec{A}| \cos \alpha = 10 \cos 30° = 8.66$$

$$A_y = |\vec{A}| \sin \alpha = 10 \sin 30° = 5.00$$

$$B_x = |\vec{B}| \cos \beta = 15 \cos 45° = 10.61$$

$$B_y = |\vec{B}| \sin \beta = 15 \sin 45° = 10.61$$

Therefore,

$$C_x = 8.66 + 10.61 = 19.27 \quad \text{and} \quad C_y = 5.00 + 10.61 = 15.61$$

And so,

$$C = \sqrt{C_x^2 + C_y^2} = \sqrt{(19.27)^2 + (15.61)^2} = 24.8$$

To two significant figures, $C = 25$ units. To find the angle θ that $\vec{C}$ makes with the *x*-axis, use

$$\tan \theta = \frac{|C_y|}{|C_x|} = \frac{15.61}{19.27} = 0.810\,1$$

and $\tan^{-1} 0.810\,1 = \theta = 39°$. Both $\vec{A}$ and $\vec{B}$ point above the *x*-axis in the first quadrant and so

$$\boxed{\vec{C} = 25 \text{ UNITS–UP } 39° \text{ ABOVE } X\text{-AXIS}}$$

Quick Check: A_x and B_x are positive as are A_y and B_y, so C_x and C_y must be positive; that is, $90° \geq \theta \geq 0°$. From the parallelogram method, $\vec{C}$ should lie between $\vec{A}$ and $\vec{B}$ and be greater than either, and it is. The smaller the angle between $\vec{A}$ and $\vec{B}$, $(45° - 30°)$, the more C approaches $A + B = 25$ as a maximum and $C = 24.8$ is reasonable.

Graphing Straight-Line Motion

At any moment an object moving in a plane, like the firefighter in Example 2.8, can be located with respect to some origin by the vector $\vec{s}$, which has scalar components s_x and s_y along the *x*- and *y*-axes. In the simpler case of motion in a straight line along, say, the *x*-axis, $\vec{s}$ again locates the object and s_x is its scalar value. To be precise, s_x is equal to the numeri-

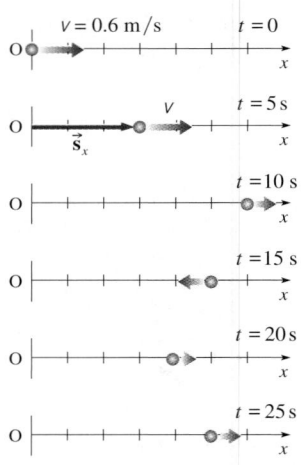

Figure 2.27 The motion of an object in one dimension at successive 5-s intervals. Notice that sometime between $t = 10$ s and $t = 15$ s, the object stops and reverses direction. It can't move along a line and change direction without stopping, if only for an instant.

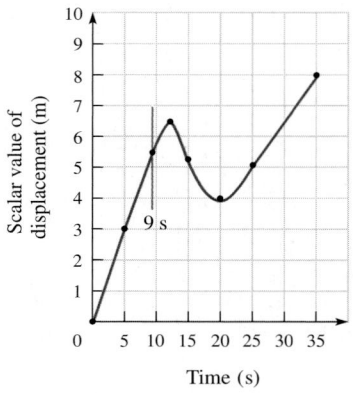

Figure 2.28 A plot of the scalar value of the displacement versus time for the object in Fig. 2.27 as it moves along its straight path.

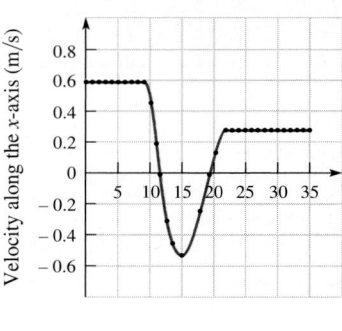

Figure 2.29 The velocity (magnitude and direction) along the x-axis determined each second from the slope of the curve of Fig. 2.22. At $t = 11.5$ s and $t = 19.6$ s, the slope of the displacement versus time curve is zero and the velocity is zero.

cal value of the displacement measured from the origin along the x-axis. We associate positive numbers ($s_x > 0$) with displacements in the positive x-direction and negative numbers ($s_x < 0$) with displacements in the negative x-direction. Therefore, whereas l, the total distance traveled (e.g., an odometer reading), never decreases and is never negative, s_x can increase, decrease, or even be negative. This is a bit inconsistent with the previous use of a light-faced symbol as the *magnitude* of a vector—magnitudes are always positive. Nonetheless, we will be able to avoid any confusion by referring to $s_x = \pm|\vec{s}_x|$ as the *scalar value of the displacement*, or just the *scalar displacement*.

In the scalar formulation of the first part of this chapter, we found that *for motion in general, the slope of the distance-time curve at any point in time equals the speed at that moment* (p. 27). Similarly, in the vector formulation, **for straight-line motion, the slope of the displacement-time curve at any point in time provides us with both the magnitude and direction (sign) of the velocity.** To illustrate, suppose an object moves along the x-axis (Fig. 2.27) and we are given a plot of s_x against t (Fig. 2.28) to interpret. The object is already moving at 0.6 m/s in the positive x-direction at $t = 0$, and it continues at that rate (shown by the fact that s_x versus t has a constant slope beginning at O) until $t = 9$ s, when the slope of the curve, and therefore the speed, starts to decrease (Fig. 2.28). At $t = 11.5$ s, the tangent has zero slope and so the speed is zero. Thereafter s_x decreases, the curve has a negative slope, and the object moves back toward O. At $t = 19.6$ s, the object momentarily stops and then again moves in the positive x-direction (the curve again has a positive slope).

The velocity vector is always in the direction of motion. When the slope in Fig. 2.28 is positive, $\vec{v}$ points in the $+x$-direction; and when the slope is negative, $\vec{v}$ points in the $-x$-direction. Figure 2.29 is a graph of the instantaneous velocity (magnitude and direction, via the sign) versus time, determined from the tangents to the curve in Fig. 2.28. Note that *a change in the direction of one-dimensional motion must always be marked by a point where* $\vec{v} = 0$. The object initially moves in the positive x-direction, stops at $t = 11.5$ s, and turns around, moving back toward the origin until $t = 19.6$ s. **The net area under the curve (area above, minus area below) as of time t equals the net displacement from the origin at t.**

Relative Motion (Optional)

Most people believe that they can know with certainty whether a thing is moving or not; they think that motion is absolute. But the Earth is orbiting the Sun (at about 66 000 mi/h), and the Sun is whirling along with the galaxy (at about 480 000 mi/h). Nothing is at rest;

A midair refueling. Even though both the tanker and the fighter are traveling at hundreds of miles per hour, their relative speed (v_{FT}) is zero.

Figure 2.30 A bee flying a straight path in a room. A vector from the origin, *O*, to the bee tracks the insect's motion. The velocity of the bee with respect to the room is $\vec{\mathbf{v}}_{BR}$. If the room happens to be in a moving mobile home, the bee's motion with respect to the ground can get rather complicated.

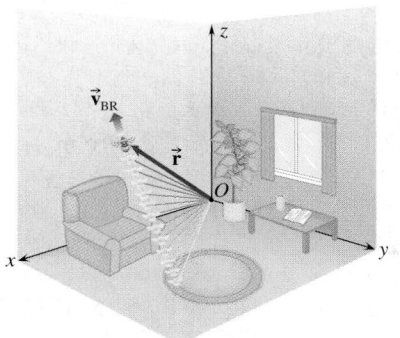

besides, there is no way to measure the difference between rest and uniform motion. There is no physical difference between rest and uniform motion. That's why it feels as though you are at rest at this very moment, sitting reading this book.

We learn from the Special Theory of Relativity that motion is relative, not absolute. Motion is always measured *with respect to* (w.r.t.) something; a car moves at 80 km/h with respect to the road or a parked police cruiser. We usually hold the Earth to be at "rest" and think about objects moving relative to it (Fig. 2.30). Even so, two things moving with respect to the Earth may be moving relative to each other, and that latter motion might even be the more important one. A front-runner in a race should be more concerned about her velocity relative to the person in second place than with her velocity relative to the finish line.

The central idea in relative motion is *velocity*. To combine the effects of several simultaneous motions, we need only add the corresponding velocity vectors.

2.9 Velocity with Respect to...

Consider the journey of a little turtle meandering eastward along the moving ruler of Fig. 2.31. At the moment shown, the ruler has been displaced with respect to the Earth by $\Delta\vec{s}_{RE}$. The turtle has undergone a displacement with respect to the end of the ruler of $\Delta\vec{s}_{TR}$. The total displacement of the turtle with respect to the Earth (i.e., with respect to the starting point P_o fixed on the ground) is given by the vector sum

$$\Delta\vec{s}_{TE} = \Delta\vec{s}_{TR} + \Delta\vec{s}_{RE} \qquad (2.13)$$

A stationary observer sees the turtle travel from P_o to P_f in some time interval Δt. If we divide both sides of the equation by Δt and take the limit as $\Delta t \rightarrow 0$, we get, via Eq. (2.7), the corresponding velocity

$$\vec{\mathbf{v}}_{TE} = \vec{\mathbf{v}}_{TR} + \vec{\mathbf{v}}_{RE} \qquad (2.14)$$

The rockets are traveling at relatively low speeds with respect to the plane. By comparison, the speed of the rockets with respect to the grounds is much higher.

If, for example, the velocity of the turtle with respect to the ruler is 0.01 m/s–EAST and the velocity of the ruler with respect to the Earth is 0.50 m/s–EAST, then $\vec{\mathbf{v}}_{TE} = 0.51$ m/s–EAST. The velocity vectors add to give the resultant, just as the displacement vectors do. The equation is quite general even though the subscripts may change; it applies as well to turtles, rulers, and Earths as to trains, rubberbands, and elephants.

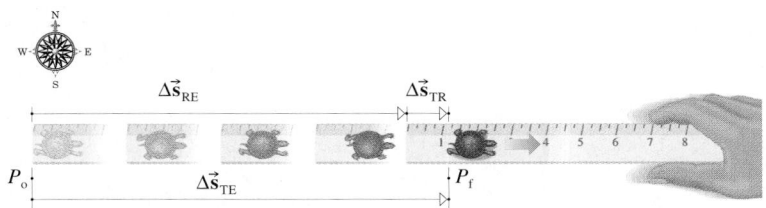

Figure 2.31 The displacement of a turtle walking on a moving ruler. The turtle walks to the right with respect to the ruler, while the ruler moves right with respect to the Earth. As a result, the little traveler ends up a rather large distance Δs_{TE}, from the point on Earth, P_o, where it started.

Example 2.10 **[II]** A big, black dog saunters down the aisle of an Amtrak Pullman at a velocity of 5.00 km/h–EAST with respect to the car. The train, all the while, is traveling at 10.00 km/h–EAST. A resident flea heading rumpward along the hound's back moves at 0.01 km/h–WEST with respect to the dog. Find the velocity of the flea with respect to the Earth.

Solution   We have to sum three separate motions in such a way as to get the velocity of the flea w.r.t. the Earth, $\vec{v}_{FE}$. (1) TRANSLATION—The *F*lea moves w.r.t. the *d*og, the *d*og moves w.r.t. the *t*rain, the *t*rain moves w.r.t. the *E*arth; determine the velocity of the *f*lea w.r.t. the Earth. (2) GIVEN: $\vec{v}_{FD}$ = 0.01 km/h–WEST, $\vec{v}_{DT}$ = 5.00 km/h–EAST; $\vec{v}_{TE}$ = 10.00 km/h–EAST. FIND: $\vec{v}_{FE}$. (3) PROBLEM TYPE—Relative velocity. (4) PROCEDURE— Extend the concept of Eq. (2.13) to a three-vector sum and arrange the first and last subscripts on the right to be F and E.   (5) CALCULATION:

$$\vec{v}_{FE} = \vec{v}_{FD} + \vec{v}_{DT} + \vec{v}_{TE} \qquad (2.15)$$

The D's and T's effectively cancel, yielding the desired resultant. In particular, $\vec{v}_{FE}$ = (0.01 km/h–WEST) + (5.00 km/h–EAST) + (10.00 km/h–EAST). Because parallel and antiparallel vectors combine algebraically, we simply add the speeds of similarly directed velocities and subtract those of oppositely directed ones. The easterly speed is greater, and so we subtract 0.01 km/h–WEST from 15.00 km/h–EAST to get

$$\boxed{\vec{v}_{FE} = 14.99 \text{ km/h–EAST}}$$

Quick Check: Even though the flea is running due west, it is going east with respect to the ground—still, it is getting right to the end it wants. Both the dog and the train are going east, so $\vec{v}_{FE}$ ≈ 15 km/s–EAST is reasonable. Some keen-eyed observer outside at rest watching this circus will see the train, the dog, and the flea all moving east at different speeds.

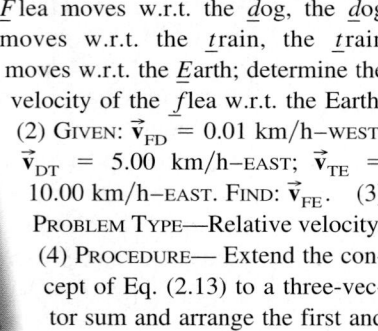

The order of the subscripts in Eq. (2.14) is crucial to the logic, even though the order of the vector addition is not. Arrange the subscripts so that the first and last on one side are the same as the first and last on the other side. Imagine that identical subscript letters cancel when they come one after the other in the sequence on the right. For instance, the vector addition effectively removes the R subscripts in Eq. (2.14): $\vec{v}_{TR} + \vec{v}_{RE} = \vec{v}_{TE}$. Keeping that in mind we'll know how to set up Eq. (2.14).

Now, for a slight variation, imagine the midnight mail *t*rain carrying a safe full of gold heading due north at v_{TE} = +20 km/h relative to the *E*arth. A *r*obber, riding on a horse at $\vec{v}_{RE}$ = 25 km/h–NORTH next to the caboose, is about to leap onto the train. What is his speed relative to the train (v_{RT})? This speed is his main concern because he wants to land on the train, not the ground. Unless he misses, he couldn't care less about either $\vec{v}_{RE}$ or $\vec{v}_{TE}$.

We want $\vec{v}_{RT}$ and can construct an appropriate vector equation for it, namely,

$$\vec{v}_{RT} = \vec{v}_{RE} + \vec{v}_{ET} \qquad (2.16)$$

which is fine except we don't have $\vec{v}_{ET}$. ***Reversing the order of the subscripts reverses the direction of the vector.*** For exam-

TURTLES, TREADMILLS, & TRAIN STATIONS

Suppose that the turtle (at P_f) in Fig. 2.31 turns around and heads *west* at a speed, with respect to the ruler, of 0.01 m/s (Fig. 2.32). Equation (2.14) can again be used; in fact, it applies regardless of the directions of the vectors. Now, $\vec{v}_{TR}$ and $\vec{v}_{RE}$ are antiparallel. The resulting speed is the *difference* between their speeds, and subtracting the smaller from the larger yields $\vec{v}_{TE}$ = 0.49 m/s–EAST. The turtle approaches the left edge of the ruler, but because $v_{RE} > v_{TR}$, it is nonetheless carried east of the fixed point P_f.

If instead the turtle again began moving west from P_f, but this time *with a constant speed equal to that of the ruler* (0.50 m/s), it would end up at the left end of the ruler and yet be trudging along "in place" at P_f. Its speed relative to the Earth would be zero. This is what happens when someone aboard a train pulling out of a station runs to the rear of the car trying to linger face-to-face with a friend standing on the platform. It is the classic treadmill scheme well known to rodents and cardiologists.

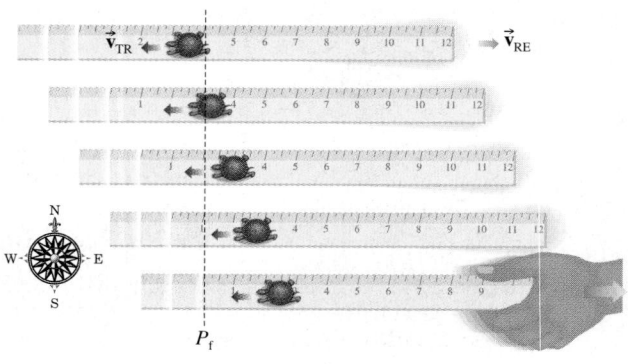

Figure 2.32 With respect to the Earth, the turtle travels east even though it's heading toward the west end of the ruler. The situation is much like that of a person walking to the back of a jet airliner while flying at 600 mi/h. If the walk takes 1 min, the plane travels 10 mi in the process.

ple, $\vec{\mathbf{v}}_{ET} = -\vec{\mathbf{v}}_{TE}$. The motion of the *train* with respect to (the sheriff standing still on) the *E*arth is 20 km/h–NORTH (that's $\vec{\mathbf{v}}_{TE}$), but to the robber, once he makes his jump onto the caboose, the sheriff will be receding *southward* at 20 km/h from him (that's $\vec{\mathbf{v}}_{ET}$). Therefore Eq. (2.16) can be rewritten as

$$\vec{\mathbf{v}}_{RT} = \vec{\mathbf{v}}_{RE} + (-\vec{\mathbf{v}}_{TE})$$

$$\vec{\mathbf{v}}_{RT} = (25 \text{ km/h–NORTH}) + (-20 \text{ km/h–NORTH}) \qquad (2.17)$$

Because parallel and antiparallel vectors combine algebraically, we simply add the speeds of similarly directed velocities and subtract those of oppositely directed ones. This gives the robber a velocity, just before he leaps, of 5.0 km/h–NORTH with respect to the train. This analysis applies as well to spaceships docking, airplanes midair-refueling, relay runners baton-passing, and police cars hotly pursuing.

Example 2.11 **[II]** Two knights with lances are about to joust. Sir John the Slow gallops south at 5.0 km/h, while Sir Peter the Fast races north at 25 km/h. What is the velocity of John with respect to (i.e., as seen by) Peter? At what speed do they close the gap between one another?

Solution We have to sum two separate motions in such a way as to get the velocity of John w.r.t. Peter, $\vec{\mathbf{v}}_{JP}$. (1) TRANSLATION— John moves w.r.t. *E*arth, *P*eter moves w.r.t. *E*arth. Determine the velocity of *J*ohn w.r.t. *P*eter. (2) GIVEN: $\vec{\mathbf{v}}_{JE} = (5.0 \text{ km/h–SOUTH})$ and $\vec{\mathbf{v}}_{PE} = (25 \text{ km/h–NORTH})$. FIND: $\vec{\mathbf{v}}_{JP}$.

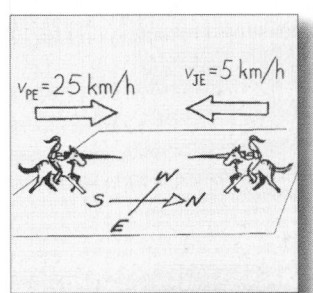

(3) PROBLEM TYPE Relative velocity. (4) PROCEDURE Use the idea embodied in Eq.(2.13). (5) CALCULATION:

$$\vec{\mathbf{v}}_{JP} = \vec{\mathbf{v}}_{JE} + \vec{\mathbf{v}}_{EP}$$

We don't have $\vec{\mathbf{v}}_{EP}$, but it equals $(-\vec{\mathbf{v}}_{PE})$, and we have that. Therefore,

$$\vec{\mathbf{v}}_{JP} = \vec{\mathbf{v}}_{JE} - \vec{\mathbf{v}}_{PE}$$

$$\vec{\mathbf{v}}_{JP} = (5.0 \text{ km/h–SOUTH}) - (25 \text{ km/h–NORTH})$$

The minus sign means that, if we do the vector addition graphically, we must draw the second vector reversed, that is, pointing south. The resultant is then

$$\boxed{\vec{\mathbf{v}}_{JP} = 30 \text{ km/h–SOUTH}}$$

and the riders close at a speed of $\boxed{30 \text{ km/h}}$.

Quick Check: Peter sees John coming south at 30 km/h and John sees Peter ($\vec{\mathbf{v}}_{PJ}$) galloping north toward him at 30 km/h, where $\vec{\mathbf{v}}_{JP} = -\vec{\mathbf{v}}_{PJ}$. The distance between them is being covered from both ends at a net speed of 30 km/h.

Figure 2.33 shows an *a*nt scampering with a velocity of $\vec{\mathbf{v}}_{AP}$ east across a piece of *p*aper that is simultaneously being pulled north at a velocity of $\vec{\mathbf{v}}_{PE}$. The ant, as seen from above by an observer at rest on the *E*arth, has a diagonal velocity of $\vec{\mathbf{v}}_{AE}$. In exactly the same way, we can think of this situation in terms of a *s*tream rushing north ($\vec{\mathbf{v}}_{SE}$) being crossed by a *b*oat bearing due east ($\vec{\mathbf{v}}_{BS}$) with respect to the *s*tream. The resultant motion of the *b*oat with respect to the *E*arth (i.e., the shore) is diagonal ($\vec{\mathbf{v}}_{BE}$)—the stream carries the boat somewhat northward even though its bow is heading east.

Alternatively, imagine a *j*et plane heading east through the *a*ir ($\vec{\mathbf{v}}_{JA}$), which, in turn, is flowing north as a strong wind ($\vec{\mathbf{v}}_{AE}$) with respect to the *E*arth's surface (Fig. 2.34*a*). If the air speed of the jet is a low 120 km/h and the wind speed a constant 50.0 km/h, the Pythagorean Theorem gives a hypotenuse of $v_{JE} = 130$ km/h. The plane has a ground speed in excess of its air speed. The pilot must get the plane to its destination, in spite of winds, even if it means pointing the aircraft more toward the south in order not to be carried off course by a sustained wind blowing north. The southerly component of the plane's velocity (Fig. 2.34*b*) must cancel the northerly velocity of the wind, leaving only an easterly motion of the craft.

The angle of flight, θ, can be determined via the definition of the tangent: $\tan \theta = (50.0 \text{ km/h})/(120 \text{ km/h}) = 0.4167$. We want the angle ($\theta$) whose tangent is known, and that angle is obtained using the *inverse* or *arc tangent*; $\tan^{-1}(0.417) = \theta = 22.6°$, which here is north of east.

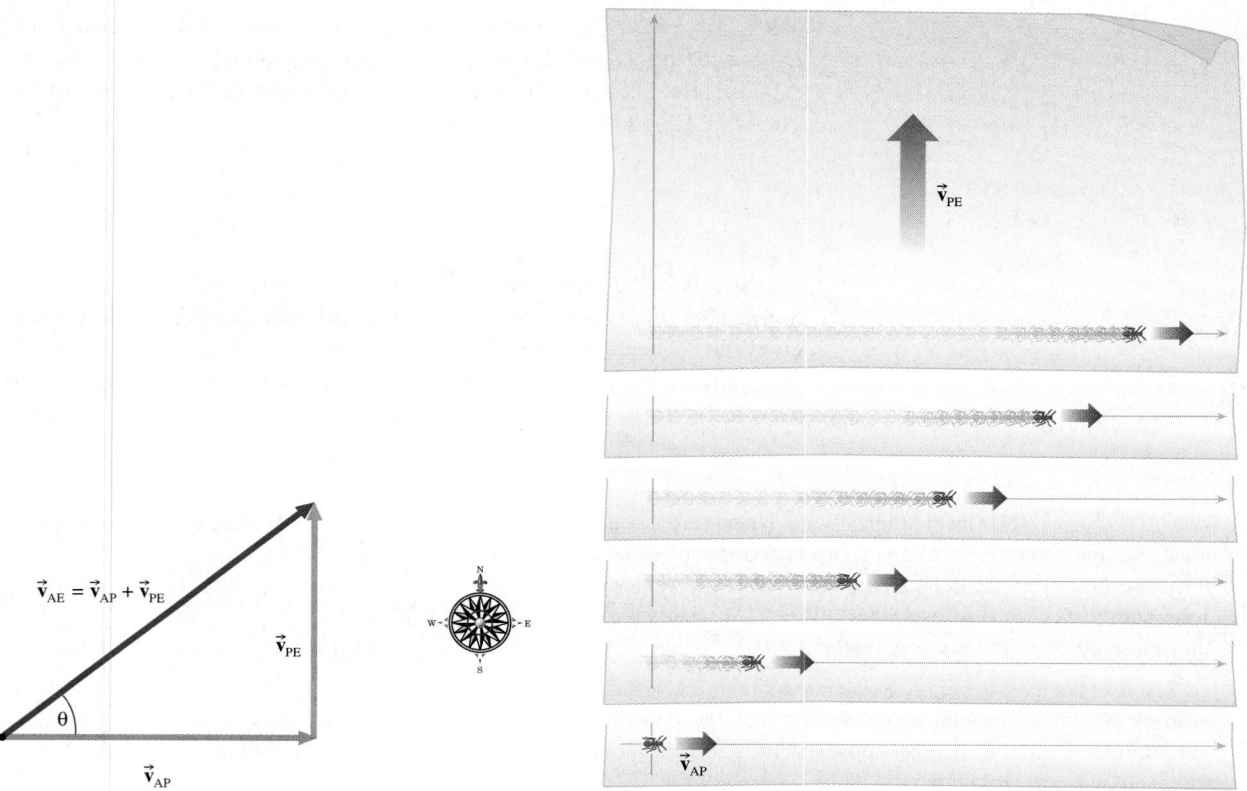

$\vec{\mathbf{v}}_{AE} = \vec{\mathbf{v}}_{AP} + \vec{\mathbf{v}}_{PE}$

$\vec{\mathbf{v}}_{PE}$

θ

$\vec{\mathbf{v}}_{AP}$

Figure 2.33 An ant walking across a sheet of paper that is itself being moved at a velocity $\vec{\mathbf{v}}_{PE}$. The bug is carried along with the paper so that it moves northeast with respect to the Earth at a velocity $\vec{\mathbf{v}}_{AE}$.

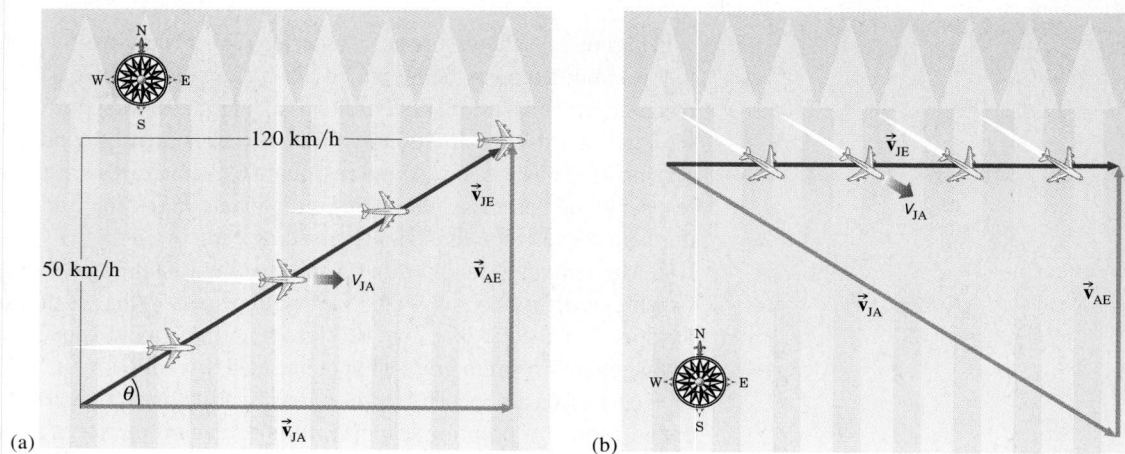

(a)
120 km/h
50 km/h
$\vec{\mathbf{v}}_{JE}$
V_{JA}
$\vec{\mathbf{v}}_{AE}$
θ
$\vec{\mathbf{v}}_{JA}$

(b)
$\vec{\mathbf{v}}_{JE}$
V_{JA}
$\vec{\mathbf{v}}_{JA}$
$\vec{\mathbf{v}}_{AE}$

Figure 2.34 (a) A plane heading east across a wind blowing north. The plane is carried N of E at an angle θ. (b) To cross the air stream due east, the plane must point south-of-east such that its southerly velocity component effectively cancels the northerly velocity of the wind. Someone hitting a golf ball in a crosswind has the same problem as the pilot. When you are moving across a medium that's moving, you can't head straight toward your destination.

Example 2.12 **[III]** The blimp in Fig. 2.35*a* is cruising with a constant air speed of 180 km/h due north against a steady 100 km/h wind blowing out of the northwest (i.e., along the 45.0°-diagonal between north and west). Compute its ground speed, its actual direction of flight, and the distance traveled after 3.00 h of cruising.

Solution We have to sum two separate motions in such a way as to get the velocity of the blimp w.r.t. the ground, $\vec{v}_{BG}$. (1) TRANSLATION—A *b*limp moves w.r.t. the *a*ir with a known velocity, and the *a*ir moves w.r.t. the *g*round with a known velocity; determine the velocity of the *b*limp w.r.t. the *g*round. (2) GIVEN: $\vec{v}_{BA} = 180$ km/h–NORTH; $\vec{v}_{AG} = 100$ km/h–SOUTH-EAST; and $t = 3.00$ h. FIND: $\vec{v}_{BG}$ and *s*. (3) PROBLEM TYPE—Relative velocity. (4) PROCEDURE—Determine $\vec{v}_{BG}$ using the concept of relative velocity, and analytically add vectors; then $s = v_{BG}t$. (5) CALCULATION—Working from Eq. (2.14),

$$\vec{v}_{BG} = \vec{v}_{BA} + \vec{v}_{AG}$$

Figure 2.35*b* represents this equation graphically. To carry out the addition, resolve $\vec{v}_{BA}$ and $\vec{v}_{AG}$ into their *x*- and *y*-components. This is easily done for $\vec{v}_{BA}$; it has no *x*-component. **Associate a sign with each component** using Fig. 2.26. With the signs taken care of, every numerical value is entered as a positive number. The scalar *x*-component of $\vec{v}_{AG}$ is $v_{AG} \cos 45.0°$, and the *y*-component is $-v_{AG} \sin 45.0°$.

The scalar *y*-component of $\vec{v}_{BG}$ is therefore

$$v_{BA} - v_{AG} \sin 45.0° = 180 \text{ km/h} - (100 \text{ km/h})(0.707) = 109 \text{ km/h}$$

while the scalar *x*-component of $\vec{v}_{BG}$ is

$$v_{AG} \cos 45.0° = (100 \text{ km/h})(0.707) = 70.7 \text{ km/h}$$

The magnitude of $\vec{v}_{BG}$ is gotten from the Pythagorean Theorem:

$$v_{BG} = \sqrt{(109 \text{ km/h})^2 + (70.7 \text{ km/h})^2} = \boxed{130 \text{ km/h}}$$

The angle $\vec{v}_{BG}$ makes with the *x*-axis (or east-west line) is θ where

$$\tan \theta = \frac{109 \text{ km/h}}{70.7 \text{ km/h}} = 1.54$$

and so $\tan^{-1}(1.54) = \boxed{\theta = 57.0° \text{ N of E}}$. In 3.00 h, the blimp travels a distance

$$s = v_{BG}t = (130 \text{ km/h})(3.00 \text{ h}) = \boxed{390 \text{ km}}$$

Quick Check: There are many problems that can be checked using the Law of Sines {see **MATH REVIEW: PART C** on the **CD** for a geometry review 💿 }. This law maintains that for any triangle the ratio of any one side to the sine of the angle opposite that side equals the ratio of any other side to the sine of the angle opposite it. For the triangle in Fig. 2.35*a*, $v_{BG}/\sin 45.0° = (130$ km/h)/0.7071 = 184 m/s and this quantity should equal $v_{AG}/\sin(90° - 57°) = 184$ m/s, and all is well.

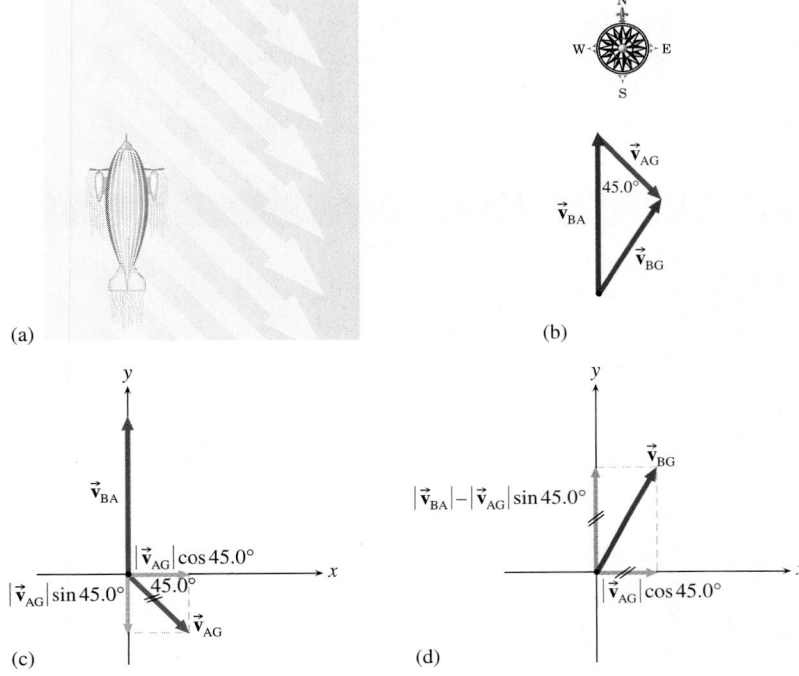

(a)

(b)

(c)

(d)

Figure 2.35 (a) A blimp heads north through the air with a velocity $\vec{v}_{BA}$ while the air is moving with a velocity $\vec{v}_{AG}$ with respect to the ground. Thus, the blimp is blown off course and travels northeast with a velocity $\vec{v}_{BG}$ with respect to the ground. (b) To find its velocity $\vec{v}_{BG}$, vectors $\vec{v}_{BA}$ and $\vec{v}_{AG}$ are resolved into their components along the *x*- and *y*-axes. (c) All the components along each axis are added, yielding two perpendicular vectors that are themselves the *x*- and *y*-components of the resultant $\vec{v}_{BG}$.

Core Material & Study Guide

SPEED

Average speed is *the distance traversed in one unit of time*, or

$$v_{av} = \frac{l}{t} \qquad [2.1]$$

When the speed is constant

$$l = v_{con}t \qquad [2.2]$$

Reread Sections 2.1 (Average Speed) and 2.2 (Constant Speed), and make sure you understand Examples 2.1 through 2.4. **Look at the CD WARM-UPS, then study the WALK-THROUGH EXAMPLES.** 💿 .

Study Section 2.4 (Instantaneous Speed). The slope of the distance-time curve at any point is the speed at that point. The area under the speed-time curve (bounded by any two moments in time) equals the distance traveled during that interval (p. 25). The **instantaneous speed** is

$$v = \lim_{\Delta t \to 0} \left[\frac{\Delta l}{\Delta t} \right] \qquad [2.4]$$

VELOCITY

The important mathematical concept of vectors is introduced in Section 2.5 (The Displacement Vector) and elaborated in Section 2.6 (Some Vector Algebra). Any physical quantity that can be specified completely only if both its magnitude and direction are provided is known as a **vector quantity**. Two vectors can be added by arranging them such that the tail of the second is at the tip of the first; the sum of these, also called the **resultant vector**, extends from the tail of the first to the tip of the second (p. 29). These ideas are applied to velocity in Section 2.7 (Instantaneous Velocity). The **average velocity** is

$$\vec{v}_{av} = \frac{\vec{s}}{t} \qquad [2.5]$$

and the **instantaneous velocity** is

$$\vec{v} = \lim_{\Delta t \to 0} \left[\frac{\Delta \vec{s}}{\Delta t} \right] \qquad [2.7]$$

Reread Section 2.8 (Components and Vector Addition) and make sure you understand Examples 2.7 through 2.9. The scalar x- and y-components of any vector $\vec{A}$ are

$$A_y = |\vec{A}| \sin \theta \qquad \text{and} \qquad A_x = |\vec{A}| \cos \theta \qquad [2.9]$$

If $\vec{C}$ is the sum of the two vectors $\vec{A}$ and $\vec{B}$, then

$$C_x = A_x + B_x \qquad \text{and} \qquad C_y = A_y + B_y \qquad [2.10]$$

and

$$C = \sqrt{C_x^2 + C_y^2} \qquad [2.11]$$

where

$$\theta = \tan^{-1} \frac{|C_y|}{|C_x|} \qquad [2.12]$$

RELATIVE MOTION

Figure 2.32 illustrates the concept of relative motion. In this case, the velocity of the turtle with respect to the Earth is

$$\vec{v}_{TE} = \vec{v}_{TR} + \vec{v}_{RE} \qquad [2.14]$$

Make sure that you can do all the I-level problems. Do the magenta-colored problems next. Each grouping contains three very similar problems where the first one has both a hint and a solution in the back of the book.

Key Terms

kinematics	displacement
average speed	magnitude
constant speed	resultant
distance-time graph	Tip-to-Tail Method
speed-time graph	Parallelogram Method
delta notation	concurrent
instantaneous speed	average velocity
slope	instantaneous velocity
scalar quantity	scalar value
vector	relative motion

Discussion Questions

1. Suppose that the average speed over some time interval is zero. Is it possible for the average speed over a still smaller segment of that interval to be nonzero? Suppose that the average velocity over some time interval is zero. Is it possible for the average velocity over a still smaller segment of that interval to be nonzero?

2. If the tangent at a given point on the distance-time graph of some object is horizontal, what is the object's instantaneous speed at that moment? What does it mean if the tangent to the graph is vertical? Can that actually occur?

3. Is it possible to travel from one place to another with some average speed without ever having had an instantaneous speed equal to it somewhere along the journey? Explain.

4. Figure Q4 is a plot of distance versus time for the World Championship race run by Carl Lewis in Rome 1987. Discuss his speed over the entire race. Does his speed vary in a way you might have expected?

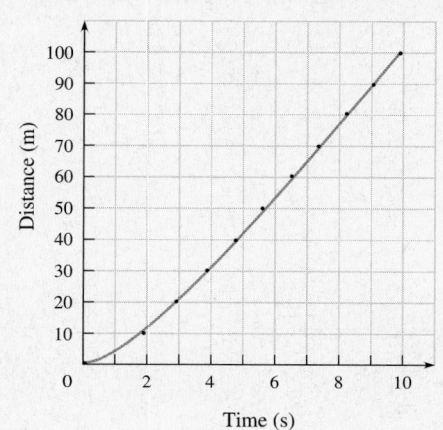

Figure Q4

5. Imagine that you leave the base of a mountain at 9:00 A.M. and ascend along a path reaching the top at 11:00 A.M. You sleep there and descend along the same path starting at 9:00 A.M. the next day. Will there be some point along the trail that you reach at exactly the same time you were there the day before, regardless of the variations in your speeds during the two journeys?

6. Can the change in displacement of an object in some interval of time exceed the distance traveled in that time?

7. Figure Q7 shows four different distance-time curves. For each one describe what, if anything, happens to the corresponding speed as time progresses.

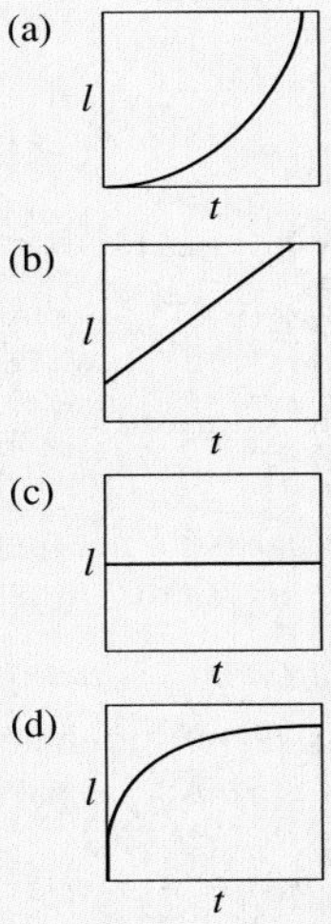

8. When we talk about displacement vectors for travel on our planet, we assume a flat Earth, or at least trips short enough so it's approximately flat. Just for fun, suppose you were standing at the North Pole and walked on the surface (no burrowing) 10 km–SOUTH, 20 km–EAST, and 10 km–NORTH, where would you end up?

9. Figure Q9 is a plot of the total distance traveled by a dog running in a field. Describe the motion. What can be said about the dog's displacement at any time? What was the total distance traveled? When was he moving the fastest?

Figure Q7

10. Is it possible during a given interval for a graph of distance (l) versus time for the same object to be different from a graph of the magnitude of the displacement (s) versus time? Explain.

11. Suppose we have a vector representing some physical quantity. Will the numerical value of the magnitude change if we express it in different units? Will its orientation change?

12. Figure Q12 is a plot of the displacement of a mouse running through a straight, narrow tunnel, the entrance to which is taken as the origin. Describe the mouse's motion in general terms throughout

its journey. Where and when did it start moving? Where did it finish? Did it ever rest?

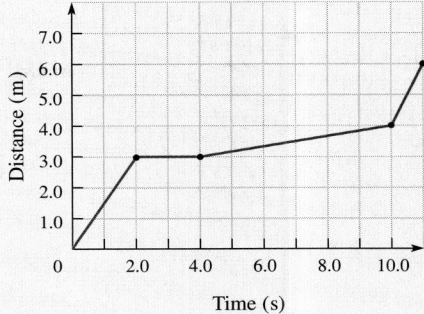

Figure Q9

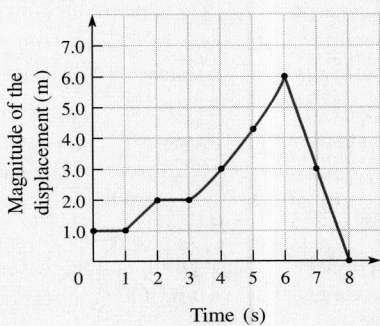

Figure Q12

13. Suppose we multiply a vector $\overrightarrow{\mathbf{M}}$ representing some physical quantity by a scalar n representing yet another physical quantity. Will $n\overrightarrow{\mathbf{M}}$ have the same units as $\overrightarrow{\mathbf{M}}$?

14. If the magnitude of the displacement of a body is given by $s = (At^2 + Bt)/D(C + t)$ where A, B, C, and D are constants, (a) determine the displacement at $t = 0$; (b) find the approximate value of s when t is very much larger than C—that is, when $t \gg C$. What is the approximate displacement when $t \ll C$?

15. Explain the meaning of each of the Key Terms on page 44.

Multiple Choice Questions

1. If $[L]$ represents the dimensions of length and $[T]$ that of time, then the dimensions of speed are   (a) $[L + T]$ (b) $[T/L]$ (c) $[L/T^2]$ (d) $[L/T]$ (e) none of these.

2. When converting a given number of kilometers per hour to kilometers per second, you can expect the resulting number to be   (a) always smaller (b) the same   (c) sometimes smaller (d) never smaller (e) none of these.

3. How far does light travel in one year moving at 3×10^8 m/s? (a) 3×10^8 m   (b) 9.5×10^{15} m   (c) 3×10^8 km   (d) 9.5×10^8 km   (e) none of these.

4. A speed of 300 m/s is equivalent to   (a) 0.300 km/h   (b) 984 ft/s   (c) 134 mi/h   (d) 83.3 km/h   (e) none of these.

5. The three-person crew of *Apollo 11* made the 3.8×10^5 km journey to the Moon in exactly three days with an average speed of   (a) 5.3×10^3 km/d   (b) 53×10^3 km/h   (c) 5.3×10^3 km/s   (d) 5.3×10^3 km/h   (e) none of these.

6. If a body moves at a constant speed v along a closed path, its average speed in comparison to v is   (a) greater   (b) less   (c) the same   (d) sometimes greater, sometimes less   (e) none of these.

7. On a distance-time graph, the slope at any point is   (a) the distance traveled   (b) the time elapsed   (c) the instantaneous speed (d) the average speed   (e) none of these.

8. Figure MC8 on p. 46 is a portion of a distance-time curve for a jogger. At $t = 14$ s the instantaneous speed is about   (a) 0.50 m/s (b) 0.27 m/s   (c) 6.5 m/s   (d) 0.46 m/s   (e) none of these.

9. During a journey over which the speed varies, the average speed in comparison to the maximum attained speed is always   (a) one-half of it   (b) greater than it   (c) less than it   (d) equal to it   (e) none of these.

10. The *Mariner 2* spacecraft lifted off from Cape Canaveral on August 27, 1962. After a successful flight, it passed out of tracking range at about 87 million kilometers (54×10^6 mi) on January 4,

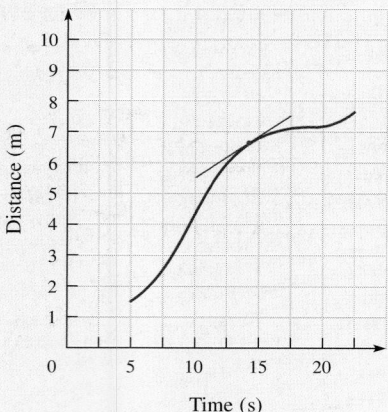

Figure MC8

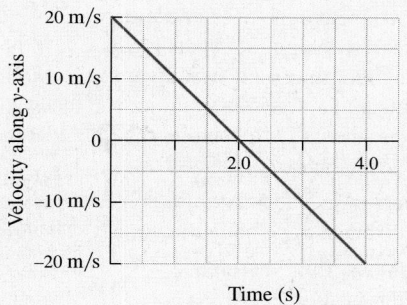

Figure MC16

1963. The magnitude of its average velocity up until that point was roughly   (a) not enough information to calculate   (b) 1.7×10^4 km/h   (c) equal to the average speed   (d) 28×10^3 km/h   (e) none of these.

11. When is the resultant that arises when $\vec{A}$ is subtracted from $\vec{B}$ equal to that when $\vec{B}$ is subtracted from $\vec{A}$?   (a) always   (b) never   (c) not enough information   (d) only when $\vec{A} = \vec{B}$ (e) none of these.

12. It is possible that two vectors of magnitude 8.0 and 3.0 can be added so as to produce a third vector of magnitude   (a) 12   (b) 15   (c) 8.0   (d) 3.0   (e) none of these.

13. A vector 10 units long pointing northeast is added to a vector 24 units long pointing northwest. The magnitude of the resultant is   (a) 26 units   (b) 14 units   (c) 34 units   (d) 0 units   (e) not enough information.

14. The inhabitants of the mythical planet Mongo measure distance in units of "glongs," one of which is the length of their leader's breathing tube. A Mongoian displacement vector 90.0 glongs long points south from the truffle tree to the methane fountain, while a vector 120 glongs long points west from the fountain to the main reflector. The displacement from tree to reflector is   (a) 150 glongs–49° S of W   (b) 150 m–37° S of W   (c) 150 glongs–37° S of W   (d) 80.0 glongs–41° S of E   (e) not enough information.

15. In general, if a vector $\vec{A}$ is to be added to a vector, $\vec{B}$ the magnitude of the resultant when $A \geq B$ must be between   (a) $2A$ and B (b) A and $A + B$ (c) $A + B$ and $A - B$ (d) $B - A$ and B (e) none of these.

The following questions pertain to Fig. MC16, which is a velocity-time curve for a projectile fired straight up into the air.

16. At what speed was the projectile launched?   (a) 5.0 m/s   (b) 10 m/s   (c) 20 m/s   (d) 0.0 m/s   (e) none of these.

17. What was its minimum speed and when did it occur?   (a) 5.0 m/s at $t = 0.5$ s   (b) 10 m/s at $t = 1.0$ s   (c) 20 m/s at $t = 4.0$ s   (d) 0.0 m/s at $t = 4.0$ s   (e) none of these.

18. What was its maximum speed and when did it occur?   (a) 20 m/s at $t = 1.0$ s   (b) 10 m/s at $t = 2.0$ s   (c) 20 m/s at $t = 0.0$ s and at $t = 4.0$ s   (d) 20 m/s at $t = 2.0$ s   (e) none of these.

19. How far did it travel during its first 2.0 s of flight?   (a) 40 m   (b) 20 m   (c) 0.0 m   (d) 10 m   (e) none of these.

20. What is its net displacement as of 4.0 s?   (a) 0.0 m   (b) 40 m   (c) 20 m   (d) 10 m   (e) none of these.

The following questions pertain to Fig. MC21.

21. As seen from the car, the man on foot approaches with a velocity of   (a) $\vec{v}_{ME}$ (b) $\vec{v}_{EC}$ (c) $\vec{v}_{MC}$ (d) $\vec{v}_{CM}$ (e) none of these.

Figure MC21

22. As seen by the man on foot, the car approaches with a velocity of   (a) $-\vec{v}_{ME}$ (b) $\vec{v}_{EM}$ (c) $\vec{v}_{EC}$ (d) $-\vec{v}_{MC}$ (e) none of these.

23. As seen from the car, the man on foot moves with a velocity of   (a) $\vec{v}_{MC} + \vec{v}_{CE}$ (b) $\vec{v}_{ME} + \vec{v}_{EC}$ (c) $\vec{v}_{EM} + \vec{v}_{MC}$ (d) $\vec{v}_{CM} + \vec{v}_{ME}$ (e) none of these.

24. If the car, as seen by the man on foot, approaches with a velocity of $\vec{v}_{CM}$, then the velocity of the man with respect to the road is   (a) $\vec{v}_{ME} + \vec{v}_{CE}$ (b) $\vec{v}_{ME} + \vec{v}_{EC}$ (c) $\vec{v}_{EM} + \vec{v}_{MC}$ (d) $-\vec{v}_{CM} + \vec{v}_{CE}$ (e) none of these.

For more Multiple Choice Questions with answers click on WARM-UPS in CHAPTER 2 on the CD.

Suggestions on Problem Solving

1. Reread the Suggestions on p. 17. If one idea, like distance, has several values, call them distance-1, distance-2, and so on. When several distances or times are involved, determine their relationship to the total distance or time. List what is *given* and what is to be

found. This list is the best summary of the exercise. It will suggest which equations will be useful in the solution.

2. Read the problem several times and determine the broad subject to which it pertains—the subject may be obvious from the location

in a chapter, but it might not be so obvious in the uncategorized real world or even on a final exam or MCAT. Determine the subgrouping of ideas to which the problem most clearly belongs, such as average speed, instantaneous speed, displacement, and so on. Look for key ideas and phrases that distinguish one group of notions from another, such as references to direction. These indications tell you that you are dealing with displacement or velocity. Watch for expressions like *straight road*, *steady motion*, *constant speed*, and so forth—they provide vital information.

3. Draw sketches that illustrate the problem being analyzed. Put into these all the known information and, where appropriate, draw in the vectors to represent directional quantities.

4. For most simple single-concept problems there will be one unknown that can be determined from one formula. Write down the equation that contains the desired unknown (and only that one unknown) and solve for it algebraically in terms of the remaining knowns. Problems involving two unknowns require solving two equations simultaneously {see **MATH REVIEW: PART A-4** on the **CD** for a geometry review 💿 }. Remember you will need the same number of equations as there are unknowns to be found.

5. **Unless otherwise required, work all problems in SI units**. Convert the given numerical information into appropriate units

before substituting into equations. It's best to convert as soon as possible to avoid accidentally using the wrong values (10 cm is *not* 10 m). Carry the units for each term through the whole calculation as a double check that you haven't messed up the algebra: an answer of 10 km/h for how *far* the chicken traveled suggests something is amiss; an answer of just 10 tells you a good deal less and is incomplete. Always include units with your numerical answers.

6. Always try to check your answers by recomputing the problem in a different way.

7. Watch out for the number of significant figures. If you travel 10.2 m (three significant figures) in 2.1 s (two significant figures), your average speed is 4.9 m/s (two significant figures) and *not* 4.857 142 9 m/s.

8. To foster some independence from the calculator, a number of problems are given with "nice" numbers so they are easier to do by hand than by machine. Avoid using the calculator to divide 100 by 2, wondering all the while why you bothered. Some important national examinations do not allow calculators to be used. You should know what the curves of the sine, cosine, and tangent functions look like. Also know their values for 0°, 30°, 45°, 60°, and 90°. Become familiar with the 3-4-5 and 5-12-13 right triangles.

Problems ✛ Coordinated Problems ✛ Progressive Problems ✛ Solutions

STUDY GUIDE **1. Coordinated Problems:** The three problems within each magenta-colored grouping are solvable in similar ways. Note that the first of these always has a hint; moreover, its solution is provided in the back of the book. *Work out each of these sets; they'll strengthen technique and build confidence.* **2. Progressive Problems:** The problems introduced in blue unfold step-by-step carrying along the analysis in a more suggestive way than is customary. *Work out all of these; they'll guide you through the analytic process and help develop problem-solving skills.* **3. Worked-Out Solutions:** Studying worked-out solutions is an important part of learning how to solve problems. Accordingly, additional *solutions* to a number of model problems are given below. *Make sure you understand each of them before you go on to the next problem.* **4.** Also provided in the back of the book are the *Answers* to all odd-numbered problems, as well as worked-out *solutions* to those with boldface numbers. Problem numbers in italic indicate that a solution appears in the Student Solutions Manual.

SECTION 2.1: AVERAGE SPEED

1. [I] THIS PROBLEM WILL HELP US BETTER UNDERSTAND AVERAGE SPEED. During a trip to the supermarket a car covers 5.0 km in 0.25 h.   (a) What's the total distance traveled in meters?   (b) What's the total time in seconds that it took to travel that distance? (c) What's the car's average speed? Give your answer in m/s.

2. [I] THIS PROBLEM WILL HELP US BETTER UNDERSTAND AVERAGE SPEED. A kid on a bike traveled 500 m around a circular track in 20 min.   (a) What was the total distance traveled in meters?   (b) What was the total time in seconds that it took to travel that distance?   (c) How fast, on average, was she moving? Give your answer in m/s.

3. [I] In the next 30.0 s, the Earth will travel roughly 885 km (i.e., 550 mi) along its orbit around the Sun. Compute its average orbital speed in km/s. Draw a diagram. [*Hint: Average speed is defined by Eq. (2.1).*]

4. [I] A baseball is thrown and in 2.0 s its shadow on the ground travels 180 m in a straight line. What's the average speed of the shadow? Is that more, less, or the same as the average speed of the ball? Draw a diagram.

5. [I] Standing on the roof of a building, a kid drops a plastic bag filled with water at a height of 100 m, and 4.5 s later it strikes the ground. Determine the bag's average speed.

6. [I] Hair grows at an average rate of 3×10^{-9} m/s. How long will it take to grow a 10-cm strand?

> **SOLUTION:** The speed at which a hair grows is 3×10^{-9} m/s. The time it takes for the end to travel 10 cm = 0.10 m is
>
> $$t = \frac{l}{v_{av}} = \frac{0.10 \text{ m}}{3 \times 10^{-9} \text{ m/s}} = 3 \times 10^7 \text{ s}$$

7. [I] Given that a glacier creeps along at an average speed of 1×10^{-6} m/s, how long will it take to advance 1.0 km?

8.[I] How long does it take the sound of thunder to go 1.000 mile (1.609 km) traveling at an average speed of $3.314\ 5 \times 10^2$ m/s?

9. [I] Referring to Table 2.3 (p. 24), derive each conversion factor in the first column.

10. [I] Referring to Table 2.3 (p. 24), derive each conversion factor in the second column.

11. [I] If the odometer in a car at the beginning of a trip read 12 723.10 km and 2.00 h later it read 12 973.10 km, what was the average speed during the journey?

> **SOLUTION:** $\Delta l = 12\ 973.10$ km $- 12\ 723.10$ km $= 250$ km, hence $v_{av} = \Delta l / \Delta t = (250$ km$)/(2.00$ h$) = 125$ km/h.

12. [I] Having traveled 900 km in 9.00 h, how fast should a driver go if he is to finish the entire 1000-km journey at an average speed of 100 km/h?

13. [I] A 30-km trip is negotiated in two equal-length segments. The first, on highways taking 15 min; the second, through heavy traffic and taking 45 min. Compute   (a) the average speed along each leg and   (b) the average speed over the whole trip. Draw a diagram.

14. [I] The SR-71 strategic reconnaissance aircraft, the *Blackbird*, set a world speed record by flying from London to Los Angeles, a distance of 8790 km (5463 mi), in 3 h 47 min 36 s.   (a) Compute its average speed in m/s.   (b) It recaptured the 1000-km closed-circuit-course record (previously held by the Russian MIG-25 *Foxbat*)

at an average speed of 2092 mi/h (i.e., 935.1 m/s). How long in time was that flight? Draw a diagram.

15. [II] A driver covered 42.0 km in 6.0 h (during that time she stopped for 30.0 min for lunch). She then speeds up, traversing another 56.0 km in 4.0 h. What was her average speed for the entire journey? What was her average speed while actually drving?

16. [II] The evil Dr. X secretly leaves Space Port L4 in a warship capable of traveling at an average speed of $(v_{av})_X$. Two hours later her escape is noticed and our hero blasts off after her at a speed that will effect rendezvous in 6.4 hours. Write an expression for his speed, $(v_{av})_H$, in terms of hers, $(v_{av})_X$. Draw a diagram.

17. [II] The bad guys come roaring onto a highway at 100 km/h headed for the Mexican border 300 km away. The cops, in hot pursuit, arrive at the highway entrance one-half hour later. What must be their minimum average speed if they are to intercept the crooks this side of the border? Draw a diagram.

18. [II] Having traveled halfway to the end of a journey at an average speed of 15 km/h, how fast must you traverse the rest of the trip in order to average 20 km/h? Draw a diagram.

19. [III] A swimmer crosses a channel at an average speed of 10 km/h and returns at half that rate. What was his overall average speed for the round trip? Draw a diagram. [*Hint: Let L be the one-way distance. Write expressions for the time-out and time-back. The overall average speed is 2L divided by the total time.*]

20. [III] To test a small rocket motor it's fired up a long vertical tube. It rises to a height of 100 m in 5.0 s and then falls back to the ground at an average speed of 10.0 m/s. How long did the whole trip take, and what was the net average speed? Draw a diagram.

21. [III] A three-lap relay race is run at average speeds for each lap of 10 km/h, 12 km/h, and 14 km/h. What is the average speed for the entire race? Guess at the answer and then calculate it. Draw a diagram. Redo the problem changing 10 km/h to 1.0 km/h.

SECTION 2.2: CONSTANT SPEED
SECTION 2.4: INSTANTANEOUS SPEED

22. [I] You know how you close your eyes when you sneeze? Suppose you are driving along at a constant 96.5 km/h (i.e., 60 mi/h) and you experience a 1.00-s-long, eyes-closed, giant sneeze. How many meters does the car travel while you are out of control?

23. [I] THIS PROBLEM DEALS WITH THE GRAPHIC REPRESENTATION OF SPEED. Figure P23 depicts a speed versus time curve for a toy airplane. (a) Approximately, over which time intervals did the speed increase? (b) Approximately, over which time intervals did the speed decrease? (c) What was its average speed over the interval from 10 s to 15 s?

24. [I] THIS PROBLEM DEALS WITH THE GRAPHIC REPRESENTATION OF SPEED. Figure P23 depicts a speed versus time curve for a toy airplane (a) When did the speed reach its maximum value? (b) What's the plane's instantaneous speed at 20 s? (c) What's the plane's instantaneous speed at 0 s? (d) What's the plane's instantaneous speed at 12 s?

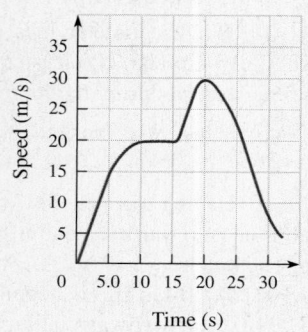

Figure P23

25. [I] Light travels in vacuum at a fixed speed of roughly 2.998×10^8 m/s, and its speed in air is only negligibly (0.03%) slower than that. (a) How long does it take light to traverse 1.0 ft? (b) That means that when you look at something 1000 m away, you are seeing it as it was _____ second (s) back in time.

26. [I] If an alien power straight out of a comic book were to cause the Sun to vanish *now*, we would still be bathed in sunshine for the next 8.3 min. We would see our star blazing in the sky as usual for all that time. Taking the speed of light to be roughly 3.0×10^8 m/s, compute the average Earth–Sun distance in meters.

> **SOLUTION:** $v_{av} = \Delta l / \Delta t$. We need the time in seconds: (8.3 min)(60 s/min) = 498 s. $\Delta l = v_{av} \Delta t = (3.0 \times 10^8 \text{ m/s})(498 \text{ s}) = 1.49 \times 10^{11}$ m = 1.5×10^{11} m.

27. [I] Figure P27 is a plot of the speed of a cat versus time. How far did the cat travel during the third second of its journey? What were its maximum and minimum speeds? When, if ever, was its speed nonzero and constant?

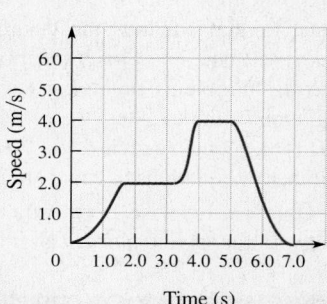

Figure P27

28. [I] The Earth rotates once around its spin axis in 23 h 56 min, and its equatorial diameter is 1.276×10^7 m (i.e., 7927 mi). At what speed would you be traveling with respect to the stationary stars while sitting still in Mbandaka, Zaire, on the Equator? Draw a diagram.

29. [I] Use Fig. 2.5 (p. 25) to calculate the distance traveled by the bee (whose speed-time graph is plotted) in the time interval from 1.33 s to 2.83 s.

30. [I] The driver of a car sets a tripmeter to zero and starts a stopwatch while en route along a highway. The following are a series of pairs of readings taken during the run: 6.0 min, 2.75 km; 30 min, 13.8 km; 45 min, 20.6 km; 1.0 h, 27.5 km; 1.00 h 15 min, 34.4 km; 1.00 h 30 min, 41.3 km. Draw a distance-time graph and describe the car's speed.

31. [I] Figure P31 shows the distance traveled versus time curve for a toy car. What was the toy car's average speed during the time interval from $t = 2.0$ s to 8.0 s?

32. [II] Suppose you fire a rifle bullet (1600 m/s) in a shooting gallery and hear the gong on the target ring 0.731 s later. Taking the speed of sound to be 330 m/s and assuming the bullet travels straight downrange at a constant speed, how far away is the target? Draw a diagram.

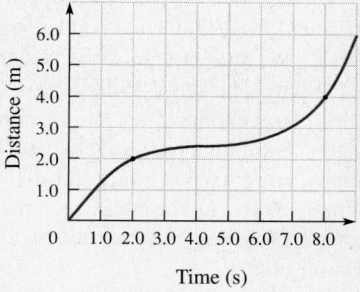

Figure P31

33. [II] The speed of sound in air (at 0 °C) is a constant 3.3×10^2 m/s, while the speed of light (i.e., all electromagnetic radiation) is about 3.0×10^8 m/s. If you see a flash of lightning and then 5.0 s later hear its roll of thunder, how far away did the bolt strike? [*Hint:*

The speed of light is a million times faster than sound. Just the fact that you can hear the thunder means the lightning occurred fairly close by.]

34. [II] A test explosion occurs at ($t = 0$) while you are in a distant bunker. The instruments in the bunker record the flash 0.02 ms later. Roughly when will you experience the shock wave, which travels at a constant 3×10^2 m/s? Draw a diagram.

35. [II] While the Apollo astronauts were on the Moon, 3.844×10^8 m away, they communicated using radio-waves, which travel at the speed of light. The Earth's atmosphere, which is about 160-km thick, affects the speed of the signal by less than 0.1 percent. About how long will the ground controller have to wait from the end of a question to the start of an answer, given that an astronaut takes 1 s to begin to respond?

36. [II] Figure P27 is a plot of the speed of a cat versus time. Approximately, what was its instantaneous speed at each of the following times, 0, 1.0 s, 2.0 s, 4.5 s, 6.0 s, and 7.0 s? During what time intervals was its speed increasing? When was its speed decreasing?

37. [II] While driving in a car along a winding mountain road a passenger makes a plot of the tripmeter's readings against time. Placing a dot on the distance versus time curve every 10.0 s for 5 min he gets a straight line passing through the $l = 0$, $t = 0$ origin and having a slope of 16.0 m/s. (a) What is the instantaneous speed of the car 45 s into the exercise? (b) How far does the car travel in the time between $t = 86$ s and $t = 186$ s?

> **SOLUTION:** (a) The slope of the curve is the speed and it's constant at 16.0 m/s $= v_{av} = \Delta l / \Delta t$. (b) We need the distance, $\Delta l = v_{av} \Delta t$ and therefore the time interval, $\Delta t = 186$ s $- 86$ s $= 100$ s. $\Delta l = v_{av} \Delta t = (16.0 \text{ m/s})(100 \text{ s}) = 16.0 \times 10^2$ m.

38. [III] If the equation describing the motion of a rocket in SI units is $l = 10 + 5t^2$, find its average speed during the first 5.0 s of flight. [*Hint: At a time $(t + \Delta t)$, the object is at $(l + \Delta l)$.*] Write an expression for the instantaneous speed of the rocket [see Eq. (2.4)].

39. [III] Given the equation describing the motion of a falling object $l = Ct^2$ (where C is a constant), show that its average speed during the interval from t to $(t + \Delta t)$ is

$$v_{av} = 2Ct + C\Delta t$$

[*Hint: At a time $(t + \Delta t)$, the object is at $(l + \Delta l)$.*] Now determine an equation for the instantaneous speed from Eq. (2.4).

SECTION 2.5: THE DISPLACEMENT VECTOR
SECTION 2.6: SOME VECTOR ALGEBRA

40. [I] A mouse runs straight north 1.414 m, stops, turns right through 90°, and runs another 1.414 m due east. Through what distance is it displaced? Draw a diagram.

41. [I] A green frog with a body temperature of 18 °C at rest at the edge of a 1.0-m-high table jumps off perpendicularly and lands on the ground 1.0 m out from the edge. Determine the length of its displacement vector. Draw a diagram.

42. [I] A youngster on the roof of a 19.0-m-tall building stands at the edge and throws a paper plane from a height of 1.00 m above the roof. It sails around before landing directly in front of him 15.0 m from the building. What is the magnitude of the displacement of the plane from its launch point? Draw a diagram.

43. [I] A robot on the planet Mongo leaves its storage closet and heads due east for 9.0 km. It then turns south for 12 km and stops. Neglecting the curvature of the planet, what is the magnitude of the robot's displacement? Draw a diagram. [*Hint: We need to determine the sum of two displacement vectors. The resulting triangle has sides of 9 km and 12 km; that suggests a 3-4-5 right triangle.*]

44. [I] After lifting off its launchpad, a rocket is found to be 480 m directly above an observer who is 360 m due east of the pad. What is the displacement of the rocket from the pad at that moment? Draw a diagram.

45. [I] While meandering on a blanket on the grass at a picnic, an ant takes 603 steps north from a cookie crumb. It then walks 804 steps due west and stops at a raisin. What is the displacement of the ant from the crumb? Give your answer in ant-steps and provide the direction. Draw a diagram.

46. [I] A jogger in the city runs 4 blocks north, 2 blocks east, 1 block south, 4 blocks west, 1 block north, 1 block east, and collapses. Determine the magnitude of the jogger's displacement. Draw a diagram.

47. [I] A stationary wooden horse on a merry-go-round is carried along a 31.4-m circumference each time the machine turns once around. Determine the magnitude of the horse's displacement for half a turn. Draw a diagram.

48. [I] Practicing golf at the Grand Canyon, someone taps a ball that rolls in a long arc coming to the edge of a cliff. The point the ball reaches at the edge is a straight 20.0 m away at the same elevation as the tee from which it was hit. The ball drops over the edge, falling into a bird's nest on a ledge 32.0 m down. Determine the magnitude of the displacement of the ball from the tee. Draw a diagram.

> **SOLUTION:** The displacement from the tee is the hypotenuse of the right triangle formed by the two displacements of 20.0 m and 32.0 m. Thus $s = \sqrt{(20.0 \text{m})^2 + (32.0 \text{m})^2} = 37.7$ m.

49. [I] THIS PROBLEM WILL HELP US BETTER UNDERSTAND THE NOTION OF DISPLACEMENT. G3, a maintenance android, leaves his closet and heads due east for 15.5 m. He stops for 310 s to make a repair, and then travels 22.1 m east to get a tool. He then wheels around 180° and walks 7.6 m west, at which point he determines the magnitude of his displacement from his closet. (a) Draw a diagram showing each displacement vector sequentially placed tail-to-tip. (b) What is the significance of the vector drawn from the first tail to the last tip? (c) What is the magnitude of his displacement?

50. [I] THIS PROBLEM WILL HELP US BETTER UNDERSTAND THE DIFFERENCE BETWEEN DISTANCE AND DISPLACEMENT. Someone holding a ball out of a window 20.0 m above ground drops it. The ball hits the sidewalk and on its first bounce climbs 10.0 m straight up. (a) What was the ball's displacement from the person's hand at the moment it struck the ground? (b) What was its displacement from the hand after it rose 10.0 m? (c) How far had it traveled up until that moment?

51. [I] A boy scout troop marches 10 km east, 5.0 km south, 4.0 km west, 3.0 km south, 6.0 km west, and 8.0 km north. What is their net displacement from their starting point? How far have they marched? Draw a diagram.

52. [I] A marble, on the first shot of a tournament game in 1942 in Brooklyn, was displaced by $\vec{s}$ equal to 3.0 m in a direction 60° N of E. On the next shot, it was displaced from its new location by an amount $-4.0 \vec{s}$. Where did it end up? Draw a diagram.

53. [I] Someone runs up four zigzagging flights of stairs, each one 10-m long and rising 6.0-m high. At the end of the climb, the person stands directly above the starting point. Determine the runner's total displacement.

54. [I] Draw the vector $\vec{A}$, 10 units long pointing due east, and graphically add to it the vector $\vec{B}$, 15 units long pointing northeast at 45°. Measure both the magnitude of the resultant and its direction.

55. [I] Graphically subtract $\vec{A}$ (5.0 units–EAST) from $\vec{B}$ (7.1 units–45° N OF E) to get $\vec{C}$.

56. [I] Graphically determine $\vec{C}$, which is the resultant of $\vec{A}$ = 20 units–30° S OF E and $\vec{B}$ = 10 units–30° N OF E.

57. [I] Graphically add $\vec{A}$ = 5.0 units–EAST to $\vec{B}$ = 7.1 units–45° N OF W to get $\vec{C}$.

58. [II] An athlete dashes past the starting point in a straight-line run at a fairly constant speed of 5 m/s for the first 2 s and then at 10 m/s for the next 3 s. At that moment she abruptly stops, turns around, waits 1 s, and runs back to the start with a constant speed in 3 s. Draw a graph of the magnitude of her displacement versus time.

59. [II] A cannonball fired from a gun at ground level located 20 m away from a castle rises high into the air in a smooth arc and sails down, crashing into the wall 60 m up from the ground. Determine the projectile's displacement.

60. [II] While playing catch a young woman throws a ball. It leaves her hand 1.00 m above the ground, sails through the air, and strikes a building that is 5.00 m away from her. The ball hits the wall at a height of 13.0 m. What was the magnitude of its displacement from its launch point at the moment it struck the wall? Draw a diagram.

61. [II] A kid at a window 20.0 m up from the street throws a ball high into the air at an angle somewhere around 50°. It lands at a spot directly in front of and beneath him, 30.0 m from the building. What is the magnitude of the displacement of the point of impact with respect to him? Draw a diagram.

62. II] Two soccer players are 20.0 m apart when one kicks a ball into the air. In a moment the ball is midway between them and 10.0 m high. What is the magnitude of the displacement of the ball, at that instant, from each player's foot? Draw a diagram.

63. [II] At t = 0 a car traveling south at 20.0 m/s on a straight track passes inches from a stationary observer. What is the magnitude of the displacement of the car with respect to the observer 10.0 s later if she (who happens to be a world-class runner), in the meantime, has raced 100 m north along the road?

64. [II] Standing still sniffing the air, a dog observes a kid on a bike pass him at 10.0 m/s heading due north. Fifteen seconds later the bike reaches an intersection and turns due east without slowing down. What is the magnitude of the displacement of the bike with respect to the dog after an additional 10.0 s? Draw a diagram.

SECTION 2.7: INSTANTANEOUS VELOCITY
SECTION 2.8: COMPONENTS AND VECTOR ADDITION

65. [I] A bumblebee flew 43 m along a twisting path only to land on a flower 3.0 m due south of the point on its hive from which it started. If the entire journey took 10 s, what was its average speed and average velocity? [*Hint: Average speed is total distance traveled (a scalar) over time; average velocity is net displacement (a vector) over time.*]

66. [I] While on vacation a tourist left the center of town in a rent-ed car having an odometer reading of 26 725.10 km. He traveled south of west for 6.00 h and then swung northeast for 14.0 h, ending up 420 km due east of the center of town. At that point the odometer reading was 27 725.10 km. Compute his average velocity and average speed.

67. [I] The city of Kisirkaya on the Black Sea is 27.0 km due north of Istanbul. A party of photographers leaves Istanbul heading west at 11:00 in the morning. They stop for lunch on the road and arrive at Kisirkaya at 8:30 in the evening. What was their average velocity over the journey? Can you compute their average speed?

68. [I] A driver in a cross-country race heads due east, attaining a speed of 30 km/h in 10 s. Maintaining that speed for another 30.0 min, he then makes a hard 90° left turn and heads north. After 50.0 min at that speed and heading, he veers right and slows, moving northeast at 20 km/h for the next 40.0 min, at which point he crosses the finish line. What is the car's velocity at each of the following times: (a) 40 s, (b) 30.0 min, (c) 50 min, and (d) 100 min?

69. [I] Determine the two acute angles of an exactly 5-12-13 right triangle accurate to two decimal places (four significant figures).

70. [I] What is the ratio of the sides of a right triangle whose acute angles are 30° and 60°? *This ratio is a good one to remember.* Which side is opposite the 30° angle?

71. [I] A telescope pointed directly at the top of a distant flagpole makes an angle of 25° with the ground. If the scope is low to the ground and 25 m from the base of the pole, how tall is the pole? Draw a diagram.

72. [I] A Roman catapult fires a boulder with a launch speed of 20 m/s at an angle of 60° with the ground. What are the projectile's initial horizontal and vertical speeds? Draw a diagram. (See the previous problem.)

73. [I] A skier is moving at 85 km/h straight down a tall mountain having a 60.0° slope. At what rate is his altitude decreasing? Draw a diagram.

74. [I] A pintail duck is flying northwest at 10 m/s. (a) At what rate is it progressing north? (b) What is the magnitude of its westward velocity component? Draw a diagram.

75. [I] A bullet is fired from a 1910 9-mm Luger pointed up at 32.0° with respect to the horizontal. If the muzzle speed—that is, the speed at which the projectile leaves the gun—is 300 m/s, find the magnitudes of: (a) the horizontal and (b) the vertical components of the bullet's initial velocity. (c) Check your results using the Pythagorean Theorem.

76. [I] A rope tied to the top of a flagpole makes an angle of 45° with the ground. If a circus performer walks up the $10\sqrt{2}$-m rope, what is her altitude at the top?

77. [I] A toy electric train runs along a straight length of track. Its displacement versus time curve is shown in Fig. P77. Is the train's velocity ever constant, and if so, when? What is its instantaneous velocity at t = 2.0 s and at t = 6.5 s? Did it change direction, and if so when?

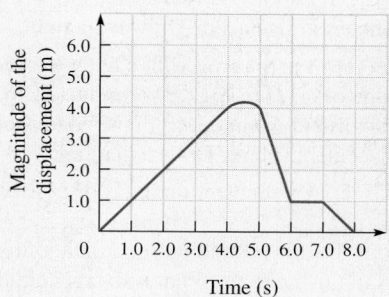

Figure P77

78. [II] Figure P78 shows the displacement-time graph of a gerbil running inside a straight length of clear plastic tubing. (a) How far did it travel during the interval from 15 s to 20 s? (b) What distance did it traverse in the first 35 s? (c) What is its average speed during the first 35 s? (d) What is its instantaneous speed at $t = 20$ s?

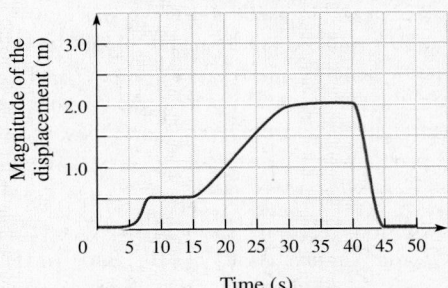

Figure P78

79. [II] Refer to Fig. P78 and assume the gerbil starts out heading north. (a) What is his velocity at $t = 12$ s; at $t = 38$ s? (b) What is his velocity at 42.5 s?

80. [II] A toy train travels around a closed circular loop of track having a 12.57-m circumference, at a constant speed of 0.50 m/s. The kid playing with the train starts a stopwatch when the engine passes its most northerly point (heading east). Is the velocity of any point on the train ever constant? What's the maximum displacement of the engine from the starting point? What's the velocity at the point of maximum displacement?

> **SOLUTION:** No, the velocity is not constant anywhere on the train. The magnitude of the maximum displacement is one diameter. The circumference equals $2\pi R = 12.57$ m and therefore $R = 2.00$ m, and so the maximum displacement is 4.001 m–SOUTH. The velocity is 0.50 m/s–WEST.

81. [II] THIS PROBLEM WILL HELP US BETTER UNDERSTAND THE NOTION OF VELOCITY. Just as it comes out of a cannon a projectile has a vertical speed of 200 m/s and a horizontal speed of 100 m/s. (a) What is the relationship between the magnitudes of any two perpendicular vector components and the magnitude of their resultant? (b) At what speed did the projectile leave the gun?

82. [II] THIS PROBLEM WILL HELP US BETTER UNDERSTAND THE NOTION OF VELOCITY. Fired, at a speed v, from a pipe making an angle θ of 60.0° up from the ground, a tennis ball rises into the air along a path lying in a vertical east-west plane. (a) Write an expression in terms of θ and v for the horizontal scalar component of the velocity. (b) If the ball is launched with a horizontal speed of 20.0 m/s, at what net speed, v, did it leave the pipe?

83. [II] Ideally, a bullet fired straight up at 300 m/s will constantly slow down as it rises to a height of 4588 m in 30.6 s, at which point it will come to a midair stop. It then plummets back down again (overlooking friction losses) and will speed up, reaching the gun barrel at the original 300 m/s after falling 30.6 s. Determine its average speed and average velocity on both the upward leg and on the round trip. Draw a diagram.

84. [II] Referring to the previous problem: (a) What is the velocity of the bullet just as it leaves the gun? (b) What is the bullet's velocity at $t = 30.6$ s; at $t = 61.2$ s?

85. [II] A chicken is resting at a location 3.0 m north of a stationary farmer who weighs 185 lb. The chicken then meanders to a new location 3.0 m east of the farmer in a time of 2.0 s. Compute the bird's average velocity during its little journey.

86. [II] A trolley travels along a straight run of track and Fig. P86 is a plot of its velocity versus time. Approximately how far did it travel in the first 3.0 s of its journey? How far from its starting point is it at $t = 6.0$ s?

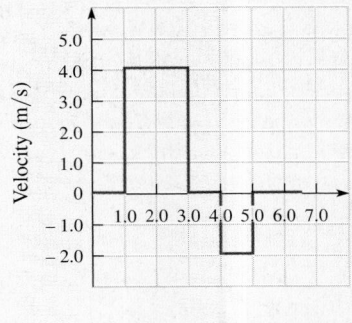

Figure P86

87. [III] A cannon is fired due north with an elevation of 45°. The projectile arcs into the air and then descends, crashing to the ground 600 m downrange 3 s later. With the Sun directly overhead, the projectile's shadow races along the flat stretch of Earth at a fairly constant speed. (a) Compute the instantaneous velocity of the shadow at the moment of impact. (b) What was the average velocity of the shell for the entire flight?

88. [III] An observer on a golf course, at 2:00 in the afternoon, stands 60 m west of a player who drives a ball due north down the fairway. If the ball lands 2.0 s later, 156 m from the observer, what was its average velocity? Draw a diagram.

SECTION 2.9: VELOCITY WITH RESPECT TO …

89. [I] Each of two runners at either end of a 1000-m straight track jogs toward the other at a constant 5.00 m/s. How long will it take before they meet?

90. [I] The jet stream is a narrow current of air that flows from west to east in the stratosphere above the temperate zone. Suppose a passenger plane capable of cruising at an air speed of 965 km/h (i.e., 600 mi/h) rode the jet stream on a day when that current remained at a constant 483 km/h (i.e., 300 mi/h) with respect to the Earth. Find the plane's ground speed.

91. [I] Suppose that a fly were to go back and forth, essentially without a pause, from one runner to the other in Problem 89 at an average speed of 10 m/s. How far will it have traveled in total by the time the athletes meet? [*Hint: How long is the fly in the air?*]

92. [I] A deckhand on a ship steaming north walks toward the rear of the vessel at 5.00 km/h carrying a horizontal wooden plank. A ladybug on the plank scampers away from the human at 0.01 km/h–SOUTH. If the ship cuts through the calm sea at 15.00 km/h: (a) What is the speed of the ladybug with respect to the ship? (b) What is the speed of the plank with respect to the Earth? (c) What is the velocity of the bug with respect to the shore?

93. [I] While on a bike heading east at 5.0 m/s a kid throws a ball due east at 10.0 m/s toward a stationary friend. At what speed does the ball approach her friend? Now suppose the friend is running west at 3.0 m/s. At what speed is the ball approaching her? If at the instant the ball is launched the two kids are 9.0 m apart, how long would the ball be in flight before it's caught?

94. [I] A pack of hounds running at 15.0 m/s is 50.0 m behind a mechanical rabbit, itself sailing along at 10.0 m/s. (a) How long will it take before the rabbit is caught? (b) How far will it travel before being overtaken?

95. [I] Pennsylvania Avenue and Prince Street intersect at right angles. If Mary is running at 8.0 m/s northwest toward the intersection along the avenue, and John is heading northeast at 6.0 m/s toward the intersection along the street, at what speed are the two runners approaching one another? [*Hint: The two velocity vectors, which are perpendicular, add such that* $\vec{v}_{ME} + \vec{v}_{EJ} = \vec{v}_{MJ}$. *The Pythagorean Theorem yields the relative speed.*]

96. [I] At 12:00 noon a speedboat is heading due west at 12.0 m/s while a cruise ship 300 m to the west and 400 m north is traveling due south at 9.0 m/s. Given that the water is still, how fast are they closing the gap between one another?

97. [I] While heading north at 100 m/s toward the star Polaris a spaceship sends out a probe due east. If the probe is launched at 200 m/s with respect to the ship, what is its speed with respect to the star?

98. [II] A jet fighter heading directly north toward a flying target drone fires an air-to-air missile at it from 1000 m away. The drone has a constant speed (due north) with respect to the ground of 700 km/h. The fighter maintains a ground speed of 800 km/h, and the missile travels at 900 km/h with respect to the fighter. (a) What is the speed of the plane with respect to the drone? (b) What is the ground speed of the missile? (c) How long will it take the missile to overtake the target?

99. [II] At the moment this sentence was written, I was in New York at a latitude of just about 41° N. Using Fig. P99, compute the circumference of the circle (the *parallel* of latitude) swept out by a point at that location as the planet rotates. Take the Earth to be a sphere of 6400-km radius turning once in 24 h. At what speed was I moving about the spin axis while sitting here at "rest"? Where are you and how fast are you moving?

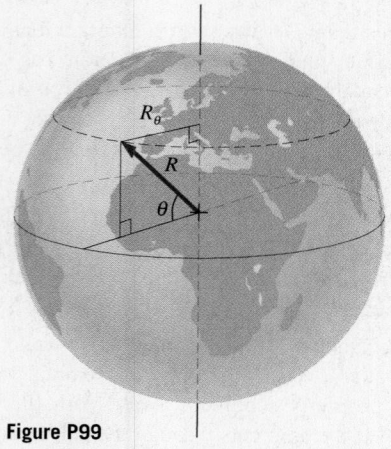

Figure P99

100. [II] A ferryboat has a still-water speed of 10 km/h. Its helmsman steers a course due north straight across a river that runs with a current of 5.0 km/h—EAST. If the trip lasts 2.0 h, how wide is the river and where does the ferry dock?

101. [II] A hawk 50 m above ground sees a mouse directly below running due north at 2.0 m/s. If it reacts immediately, at what angle and speed must the hawk dive in a straight line, keeping a constant velocity, to intercept its prey in 5.0 s? Incidentally, the mouse escaped by jumping in a hole.

102. [II] A ship, in an old black-and-white war movie, is steaming 45° S of W at 30 km/h. At that moment 50 km due south of it, an enemy submarine heading 30° W of N spots it. At what speed must a torpedo from the sub travel if it's to strike the ship? Assume that the sub fires straight ahead from its forward tubes, and overlook the fact that this problem is a bit unrealistic. [*Hint: What can be said about both westerly displacements?*]

103. [III] At times of flood, the Colorado River reaches a speed of 48 km/h (i.e., 30 mi/h) near the Lava Falls. Suppose that you wanted to cross it there perpendicularly, in a motorboat capable of traveling at 48 km/h (i.e., 30 mi/h) in still water. (a) Is that possible? Explain. (b) If on a quieter day the water flows at 32 km/h (i.e., 20 mi/h), at what angle would you head the boat to cut directly across the river? (c) At what speed (in SI units) with respect to the shore would you be traveling?

104. [III] A train is traveling east along a straight run of track at 60 km/h (i.e., 37 mi/h). Inside, two children 2.0 m apart are playing catch directly across the aisle. The kid wearing a Grateful Dead T-shirt throws the ball horizontally north. The ball crosses the train and is caught in 1.0 s by her little brother. Ignoring any effects of gravity or friction, (a) find the ball's velocity with respect to the little brother. (b) What is the velocity of the ball as seen by someone outside standing still?

105. [III] A cruise ship heads 30° W of N at 20 km/h in a still sea. Someone in a sweatsuit dashes across the deck traveling 60° E of N at 10 km/h. What is the jogger's velocity with respect to the Earth?

106. [III] A ship bearing 60° N of E is making 25 km/h while negotiating a 10-km/h current heading 30° N of E. (a) What is the velocity of the ship with respect to the Earth? (b) How fast is it traveling northward?

107. [III] The Kennedy Space Center on the east coast of Florida is located at a latitude of about 28.5°. Using Problem 99, we can show that the Space Center is moving eastward at 1470 km/h as the Earth rotates. (a) Suppose a rocket is fired so it arcs over and travels horizontally westward at 1470 km/h. What's its speed with respect to both the center and the surface of the Earth? (Incidentally, it's the former speed that determines the orbit of a satellite.) (b) What would be the rocket's speed with respect to the center of the Earth if it were launched at 1470 km/h eastward? (c) Which way do you think we usually fire rockets from Kennedy? (d) Why, overlooking the weather, was Florida rather than Maine originally chosen for the Kennedy Space Center?

Chapter 3
Kinematics: Acceleration

*V*ariations in motion are commonplace; change is the rule rather than the exception. We travel from one place to the next, speed up, slow down, turn, pause, and move again. How can we quantify this kind of varying motion? This chapter does precisely that using the concept of *acceleration*, the time rate-of-change of velocity.

The Concept of Acceleration

The idea of acceleration comes from the notion of going faster and faster, but the modern definition is more inclusive. The change in velocity must be part of a careful formulation of acceleration, but it's only a part. The *rate* at which change occurs is crucial. Waiting in a car, at rest on the entrance ramp to a highway, we know what's needed. The car must move from an initial speed $v_i = 0$ to a final speed $v_f = 88.5$ km/h (i.e., 55 mi/h) to merge into a gap in the flowing ribbon of traffic. The change in velocity ($\Delta \vec{\mathbf{v}}$) must occur in a specific amount of time (Δt)—it cannot take 10 h, or 10 min, but more like 10 s. **Acceleration is the time rate-of-change of velocity** (Fig. 3.1).

STUDY GUIDE

Thus far we've learned about distance, speed, displacement, and velocity. Moreover, at this point we know something about slopes and how to think about the instantaneous values of physical quantities. With these ideas to build on, we now explore the vector quantity known as *acceleration* (p. 53).

Because we're limited to a noncalculus treatment, we'll focus on situations where the acceleration is constant (p. 56).

3.1 Average Acceleration

The **average acceleration** ($\vec{\mathbf{a}}_{av}$) of a body is defined as *the ratio of the change in its velocity over the time elapsed in the process*:

$$\text{Average acceleration} = \frac{\text{change in velocity}}{\text{time elapsed}}$$

$$\vec{\mathbf{a}}_{av} = \frac{\Delta \vec{\mathbf{v}}}{\Delta t} = \frac{\vec{\mathbf{v}}_f - \vec{\mathbf{v}}_i}{t_f - t_i} \tag{3.1}$$

A vector divided by a scalar is a vector; both *velocity and acceleration are vector quantities*. Something accelerates in a specific direction—it changes its velocity in that direction.

As a vector, acceleration is a two-part concept that arises either because of a change in the *direction* of the velocity or because of a change in the *magnitude* of the velocity (i.e., its

Figure 3.1 Uniform acceleration. Notice how the distances traveled in equal time intervals increase. In fact, those distances go as the odd integers 1, 3, 5, 7,... something we'll be able to prove presently—see Eq. (3.9).

At present it is the purpose of our Author merely to investigate and to demonstrate some of the properties of accelerated motion (whatever the cause of this acceleration may be).

GALILEO
DIALOGUES CONCERNING TWO NEW SCIENCES (1638)

speed), or both. A racehorse running faster and faster on a straightaway is accelerating because its speed is changing. A merry-go-round stallion revolving at a constant speed is also accelerating because the direction of its motion (the direction its nose is pointing) is forever changing. A colt chasing butterflies in a field is likely to be accelerating because of changes in both speed and direction.

Whenever an object speeds up or slows down, its velocity changes size and it has an acceleration along, or tangent to, the path traveled. Whenever an object moves in a curve, its velocity changes direction and it has an acceleration perpendicular to the path traveled. (We'll come back to this *centripetal acceleration* in Chapter 5.) When you read about a car's ability to accelerate or decelerate, it is this *tangential acceleration* they're talking about. And when a car's accelerometers detect a value of the tangential acceleration that's too large, they lock up the seat belts and fire the front air bags. To keep things simple, much of this chapter will deal with rectilinear (i.e., straight-line) motion where the acceleration occurs along (i.e., tangent to the path of) the motion.

AUTOMOBILE ACCELEROMETERS

It would be great if cars had an accelerometer gauge that showed the component of $\vec{a}$ in the forward direction, that is, tangent to the path—call it a_T. A device of this sort would alternately display positive and negative values of a_T, depending on whether you were stepping on the gas, coasting, or hitting the brake. In this context, a_T is the *scalar value of the tangential acceleration vector*. On a straight run $a_T = a$; we'll deal with curved motion later.

There are a great many practical occurrences where the motion is rectilinear (a falling rock is just one of them). Then $\vec{v}_i$, $\vec{v}_f$, and $\vec{a}$ are *colinear*; they all act along the same line. As we have seen, colinear vectors can be treated algebraically by working with their scalar values [Eq. (2.9) p. 35], that is, by associating appropriate signs with their magnitudes (which are positive by definition). In all of physics there's probably only one case where we have given a special name to the magnitude of a vector and that's the **speed**, so we have to be especially careful with it when it comes to signs. Let v be the scalar value of the velocity, as distinct from the speed ($v = |v|$), which is provided by a speedometer and is always positive. Thus v, *which depends on the direction of motion*, can be less than, greater than, or equal to zero. For example, if a car going east (which we take as the positive direction) has a speed of $v = 20$ m/s, then $v = +20$ m/s $= v$. On the other hand, if it's traveling west at a speed of $v = 20$ m/s, then $v = -20$ m/s.

Usually we'll take the direction in which the object first moves to be positive: $v_i > 0$. With such a sign convention, Eq. (3.1) can be written using scalar values:

Displacement ($\vec{s}$), velocity ($\vec{v}$), and acceleration ($\vec{a}$) are vector quantities. Nonetheless, we can deal with them using our simple scalar equations, provided the motion is along a straight line. Then all we have to do is pick a direction to be positive and enter the scalar values of the various quantities in the equations with the proper signs: + if the corresponding vector is in the positive direction, − if it's not.

[straight-line motion]
$$a_{av} = \frac{\Delta v}{\Delta t} = \frac{v_f - v_i}{t_f - t_i} \tag{3.2}$$

When $v_f < v_i$ the moving object has slowed, the average tangential acceleration is negative, and the object has **decelerated**. The dimensions of acceleration are those of *speed over time* or *distance over time, over time*. The preferred SI units of acceleration are m/s², read "meters per second squared," although (m/s)/s is more informative—an object changes speed by so many meters per second, every second. Incidentally, the range of acceleration for a typical car is between −6.6 m/s² and +6.4 m/s².

Example 3.1 **[I]** A yellow messenger robot is traveling at 1.00 m/s in a straight line along a ramp in a spaceship. If it speeds up to 2.50 m/s in a time of 0.50 s, what is the magnitude of its average acceleration?

Solution   The mention of "average acceleration" should bring to mind its defining equation.   (1) TRANSLATION—An object changes from an initial to a final speed over a known time interval; determine its average acceleration.   (2) Take the initial direction of motion to be positive. GIVEN: $v_i = +1.00$ m/s, $v_f = +2.50$ m/s, and $\Delta t = 0.50$ s. FIND: a_{av}.   (3) PROBLEM TYPE— Average acceleration/straight-

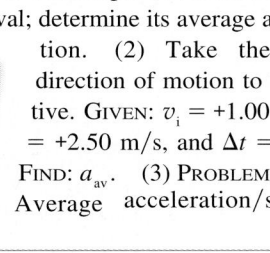

line motion.   (4) PROCEDURE—We have v_i, v_f, and Δt and must find a_{av}; the definition of a_{av}, Eq. (3.2), relates all four quantities.   (5) CALCULATION:

$$a_{av} = \frac{\Delta v}{\Delta t} = \frac{(2.50 \text{ m/s} - 1.00 \text{ m/s})}{0.50 \text{ s}}$$

and

$$\boxed{a_{av} = 3.0 \text{ m/s}^2}$$

Quick Check: Put $a_{av} = 3.0$ m/s² back in Eq. (3.2) and calculate $\Delta v = a_{av}\Delta t = (3.0 \text{ m/s}^2)(0.50 \text{ s}) = 1.5$ m/s.

{For more worked problems click on **WALK-THROUGHS** *under* **CHAPTER 3** *on the CD.}*

3.2 Instantaneous Acceleration

If an object at rest increases its speed, for example, by 5 m/s during each 1 s of travel, the magnitude of its acceleration will be constant at 5 m/s². Because the motion is along a straight line, we graph *velocity versus time* (Fig. 3.2a), with the initial direction of motion positive. (In this simple case it's possible to effectively graph a vector quantity because we can plot both the magnitude and sign of the velocity, knowing the axis along which the object moves.) The curve is a diagonal *straight line*, where $v = 0$ at $t = 0$. At the end of successive 1-s intervals, each point is plotted 5 m/s higher than the previous point, and so the graph is a straight line. The **slope**, *the rise* (Δv) *over the run* (Δt), is constant at 5 m/s over 1 s. The greater the constant acceleration, the greater the slope of the line. If instead the speed was *decreasing* ($\Delta v < 0$), the rise over the run would be *negative* (Fig. 3.2b). A negative slope, one progressing downward, corresponds to an acceleration vector pointing opposite to the initial velocity—the object is decelerating. By contrast, a constant speed graphs as a horizontal line, a slope of zero, and no acceleration (Fig. 2.5, p. 25).

Figure 3.3 on the next page is a plot of the speed-time curve for a classic 12-cylinder Jaguar XJ12L sedan. Although it's fairly straight for the first 3 s or 4 s, the curve leans over, leveling off as time goes on. Its slope decreases, slowly at first, but more so as time pro-

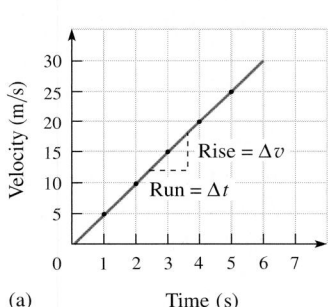

(a)   Time (s)

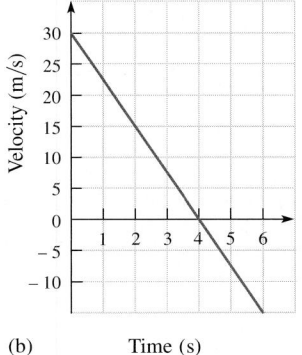

(b)   Time (s)

Figure 3.2 Graphs of the motions of two objects moving along straight lines. The slope of the velocity-time plot is the average acceleration over the interval. In (a), the slope is positive and constant, as is the acceleration. In (b), the slope is constant and negative. The body decelerates to $v = 0$ and continues to accelerate in the negative direction such that, after $t = 4$ s, it has a negative velocity.

Speed skiing can take a person from 0 to 60 mi/h in 3 s and up to 100 mi/h in 6 s. The men's world record is 139 mi/h.

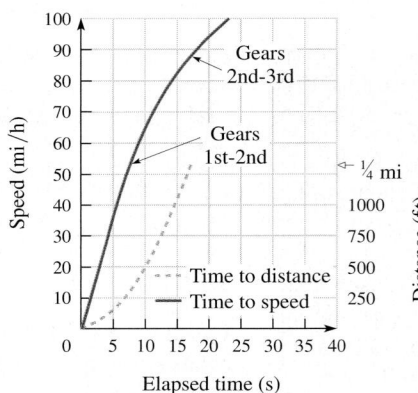

Figure 3.3 The speed-time graph (solid line) for a classic Jaguar XJ12L sedan traveling along a straight path, as it appeared in *Road & Track* magazine. The speed versus time curve gradually levels off, ultimately becoming horizontal.

gresses—the car's speed is increasing, ever more slowly (the acceleration is decreasing, which does not mean that the car is decelerating). It cannot accelerate forever because of air drag (which increases in proportion to v^2). The Jaguar will never hit 1000 km/h, even with the gas pedal floored for a week.

Any driver is concerned with the acceleration at each moment and usually doesn't care about a_{av}. What's important is the acceleration NOW: the acceleration at *any instant* determined over a *tiny time interval*. The limiting value approached by the average acceleration—Eq. (3.1)—as the time interval (over which the averaging occurs) shrinks toward zero is the **instantaneous acceleration**,

> **Acceleration** is the time rate-of-change of velocity.

$$\vec{\mathbf{a}} = \lim_{\Delta t \to 0} \left[\frac{\Delta \vec{\mathbf{v}}}{\Delta t} \right] \qquad (3.3)$$

The slope of the tangent to the velocity-time curve at any moment is the instantaneous acceleration at that moment. Even though Eq. (3.3) is an important basic definition, it's not very useful for the kinds of calculations we'll be doing (see Problem 21).

Uniformly Accelerated Motion

Although in the real world, acceleration is rarely if ever precisely constant, there are circumstances where it can be treated as if it were, at least over some limited time interval. The first few seconds of an automobile's acceleration is one such case (Fig. 3.3).

3.3 Constant Acceleration

Without calculus, it's appropriate to stay with situations in which $\vec{\mathbf{a}}$ is constant and therefore equal to $\vec{\mathbf{a}}_{av}$. Then, no matter how large Δt is,

> As a rule we'll take the initial direction of motion to be positive.

$$[\text{constant } \vec{\mathbf{a}}] \qquad \vec{\mathbf{a}} = \frac{\Delta \vec{\mathbf{v}}}{\Delta t} = \frac{\vec{\mathbf{v}}_f - \vec{\mathbf{v}}_i}{t_f - t_i} \qquad (3.4)$$

In this chapter we'll only deal with objects that speed up or slow down uniformly while traveling in a straight line; in other words, objects that have a constant **tangential acceleration** (a_T). Then, by definition, $a_T = \Delta v / \Delta t$, where we'll use the subscript T to remind us that this is an acceleration tangent to (i.e., along) the path of the motion. If the path is curved, as it usually is, there'll also be a radially directed centripetal acceleration (a_c), but we'll treat that later.

The notation can be simplified a little if we start our clocks so that $t_i = 0$. The final time t_f is then the entire trip-time, represented now as t. Thus

$$a_T = \frac{\Delta v}{\Delta t} = \frac{v_f - v_i}{t_f - t_i} = \frac{v_f - v_i}{t} \qquad (3.5)$$

Figure 3.4 For a uniformly accelerating body, the final velocity (v) equals the sum of the initial velocity (v_i) and the subsequent change in the velocity (at). In (a) $v_i > 0$ and $a > 0$; in (b) $v_i < 0$ and $a > 0$; in (c) $v_i > 0$ and $a < 0$; in (d) $v_i < 0$ and $a < 0$.

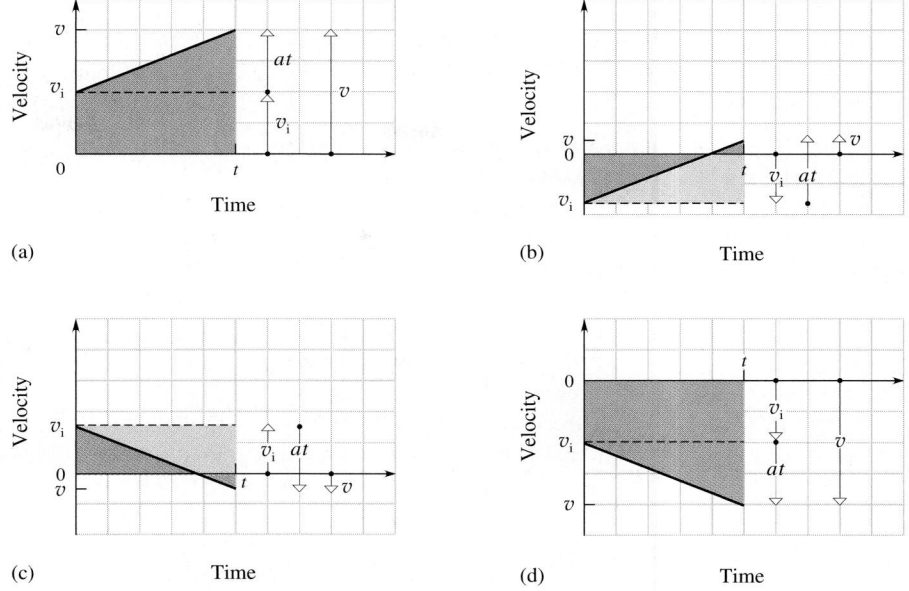

(a)   (b)   (c)   (d)

Multiplying both sides by t and bringing v_i over to the other side, we arrive at one of the basic equations of *uniformly accelerated motion*,

[constant $\vec{a}$]

$$v_f = v_i + a_T t \qquad (3.6)$$

Incidentally, taking the forward direction as positive, this equation can be applied to a car driving up and down hills and around curves; it's not restricted to rectilinear motion. Four quantities are interrelated by Eq. (3.6); given any three, the remaining one can be calculated, and it doesn't have to be v_f. ***Numerical values entered into Eq. (3.6) for v_f, v_i, and a_T can be positive or negative; t is always positive*** (Fig. 3.4).

Depending on the context, the one word—*acceleration*—customarily either means the vector acceleration or the scalar value of the acceleration. {For an animated look at how velocity changes with uniform acceleration click on **CONSTANT ACCELERATION** under **INTERACTIVE EXPLORATIONS** on the **CD**.}

Example 3.2 **[I]** A red Jaguar XJ12L can screech to a stop from 96.54 km/h (i.e., 60.00 mi/h) in about 3.7 s. Compute the scalar value of its tangential acceleration, assuming it to be constant.

Solution The mention of "acceleration" should remind us of its defining equation. (1) TRANSLATION—A moving object changes from an initial to a final speed (of zero) during a known time interval; determine its constant scalar tangential acceleration. (2) Select a direction of motion. We'll usually take the initial direction of motion to be positive. GIVEN: v_i = +96.54 km/h, $v_f = 0$, and $t =$ 3.7 s. FIND: a_T. (3) PROBLEM

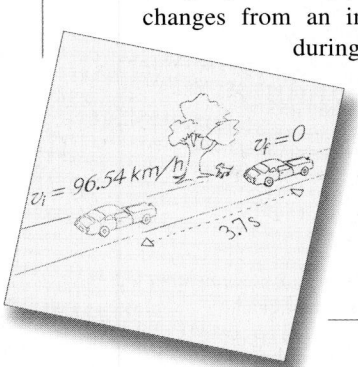

TYPE—Constant acceleration. (4) PROCEDURE—We have v_i, v_f, and t and must find a_T; either Eq. (3.5) or (3.6) relates all four quantities. Notice that the Jag is decelerating and so a_T must be negative. (5) CALCULATION—Determine the initial speed in m/s: v_i = (96.54 km/h)(1000 m/1 km)(1 h/3600 s) = +26.82 m/s. Since

$$a_T = \frac{v_f - v_i}{t}$$

$$a_T = \frac{0 - 26.82 \text{ m/s}}{3.7 \text{ s}} = \boxed{-7.2 \text{ m/s}^2}$$

Quick Check: Recompute using different units; 60 mi/h = 88 ft/s and so a_T = −23.78 ft/s², multiplying by 0.304 8 m/ft, yields −7.2 m/s².

Example 3.3 **[I]** A bicyclist pedaling along a straight road at 25.0 km/h uniformly accelerates at +3.00 m/s² for 3.00 s. Find her final speed.

Solution The phrase "uniformly accelerates" should remind us of the definition of a_T. (1) TRANSLATION—An object moving in a straight line accelerates at a given rate from an initial speed during a known time interval; determine its final speed. (2) Select the initial direction of motion as positive. GIVEN: $v_i =$ +25.0 km/h, $a_T = +3.00$ m/s², and $t = 3.00$ s. FIND: v_f. (3) PROBLEM TYPE—Constant acceleration. (4) PROCEDURE—Because a_T is positive, the bicyclist is going faster and faster.

We have v_i, a_T, and t and must find v_f; either Eq. (3.5) or (3.6) relates all four quantities. (5) CALCULATION—First put everything in the same SI units:

$$v_i = \frac{(25.0 \text{ km/h})(1000 \text{ m/km})}{3600 \text{ s/h}} = 6.94 \text{ m/s}$$

Then
$$v_f = v_i + a_T t$$

$$v_f = (6.94 \text{ m/s}) + (+3.00 \text{ m/s}^2)(3.00 \text{ s})$$

and to three significant figures, $\boxed{v_f = 15.9 \text{ m/s}}$.

Quick Check: Use the answer to compute a_T; $\Delta v/\Delta t = (15.9$ m/s $- 6.94$ m/s$)/(3$ s$) = 3$ m/s².

3.4 The Mean Speed

The most important contribution to the theory of motion that came from the Middle Ages (ca. 1335) was the Mean-Speed Theorem. Having successfully defined constant acceleration, scholars next wanted to relate a_T to the distance traveled (Fig. 3.5). They knew that $s = v_{av}t$; could it then be possible to determine a v_{av} that would produce the same displacement in the same time as would the constant acceleration in question? In other words, what is v_{av} for an object that is uniformly accelerating along a straight line from an initial scalar velocity v_i to a final scalar velocity v_f? The answer is the **mean speed*** given by

[constant $\vec{a}$]
$$v_{av} = \tfrac{1}{2}(v_i + v_f) \tag{3.7}$$

Today, we might say that a constant tangential acceleration produces a uniformly changing speed, and so the average value is midway between the initial and final speeds (Fig. 3.6). For example, if a car accelerates uniformly from 30 km/h up to 50 km/h its average speed is 40 km/h, and if it accelerates uniformly from 20 m/s down to 10 m/s its average speed is 15 m/s. All of this can happen on either a straight or curved road provided a_T is constant.

CALCULATOR TIPS

Many calculators contain parentheses keys that make it easy to do long calculations in one continuous entry. For example, suppose we have $v_i = 20$ m/s, $v_f = 80$ m/s, and $t = 10.0$ s and want to solve for s where

$$s = \tfrac{1}{2}(v_i + v_f)t$$

Using parens

| . | 5 | (| 2 | 0 | + | 8 | 0 |) |

| × | 1 | 0 | = |

500

Otherwise you would have to start in the middle by adding 20 + 80 and then do the multiplication.

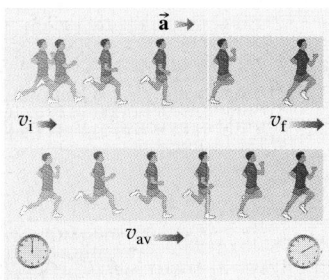

Figure 3.5 A body that accelerates uniformly from v_i to v_f over a certain straight-line distance will cover exactly the same distance in the same time traveling at a fixed speed of v_{av}, known as the *mean speed*.

---

*It should really be the *mean scalar velocity* because it can be positive or negative. Remember that the word "scalar" here is necessary because by itself the term "velocity" is a vector quantity and must have an explicitly stated direction.

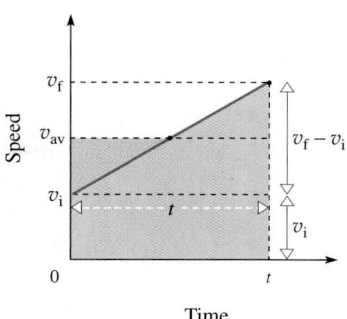

Figure 3.6 For a body that is uniformly accelerating, the mean speed, v_{av}, is the height of the midpoint of the straight-line speed-time graph.

Figure 3.7 The area under the speed-time curve is the distance traveled. The area under the curve (that is, the tilted line) equals the area under the v_{av} line. In other words, $v_{av}t = v_i t + \frac{1}{2}(v_f - v_i)t$.

Alternatively, around 1360, Nicolas Oresme used a primitive speed-time graph to assert that the area under the sloping straight line representing uniform acceleration corresponded to the distance traveled (Fig. 3.7). The area in question has two distinct pieces and can be thought of as the sum of the areas of the rectangle (base t times height v_i) and the triangle (one-half the base t times the altitude $v_f - v_i$). The total area is then

$$v_i t + \tfrac{1}{2}(v_f - v_i)t = (v_i - \tfrac{1}{2}v_i + \tfrac{1}{2}v_f)t = \tfrac{1}{2}(v_i + v_f)t$$

Setting this equal to s, the scalar value of the displacement, we arrive at the **Mean Speed Theorem**:

[rectilinear − constant $\vec{a}$]
$$s = \tfrac{1}{2}(v_i + v_f)t \tag{3.8}$$

Don't be intimidated by the name; this is simply $s = v_{av}t$ with v_{av} given by Eq. (3.7). As we'll see when we study ballistics (p. 68), there are good reasons to want an expression for displacement rather than path-length. Once we go the route of using s we have to stay with straight-line motion.

Example 3.4 [I] A bullet fired from a 38-caliber handgun with a 15.2-cm (6.00-in.) barrel attains a muzzle speed of 330 m/s. Assuming a constant acceleration, how much time does it take the bullet to travel down the barrel?

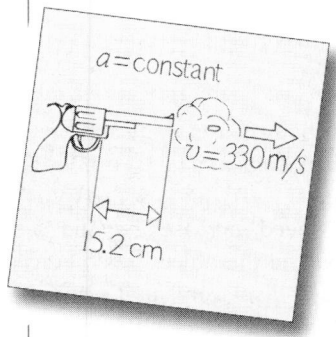

Solution At this point, the phrase "constant acceleration" should suggest both the definition of a_T and the Mean Speed Theorem. (1) TRANSLATION— An object moving in a straight line accelerates from zero to a known final speed, over a known distance; determine the time. (2) Select the initial direction of motion as positive. GIVEN: $v_i = 0$, $v_f = +330$ m/s, and $s = 15.2$ cm. FIND: t. (3) PROBLEM TYPE—Constant acceleration. (4) PROCEDURE—All these quantities are related by $s = \frac{1}{2}(v_i + v_f)t$. Incidentally, we are not given a_T, and that should immediately call to mind either Eq. (3.7) or (3.8). (5) CALCULATION—Change the distance from centimeters to meters and solve for t:

$$s = \tfrac{1}{2}(v_i + v_f)t$$
$$s = 0.152 \text{ m} = \tfrac{1}{2}(0 + 330 \text{ m/s})t$$

and $t = 2(0.152 \text{ m})/(330 \text{ m/s}) = \boxed{9.21 \times 10^{-4} \text{ s}}$.

Quick Check: In U.S. Customary Units, $\frac{1}{2}$ ft $= \frac{1}{2}(1083$ ft/s$)t$; $t = 0.92$ ms.

Example 3.5 [I] That classic 12-cylinder Jaguar can go from rest to 48.3 km/h (i.e., 30 mi/h) in 3.8 s, accelerating uniformly (in a straight line) at a rate of 3.54 m/s². Find its average speed and determine how much of a runway it will use in the process.

Solution The phrases "accelerating uniformly" and "average speed" should call to mind the definition of a_T and the Mean Speed Theorem. (1) TRANSLATION—An object moving in a straight line accelerates at a known rate, from zero to a known final speed, in a specified time; determine its average speed and distance traveled. (2) Select the direction of motion as positive. GIVEN: $v_i = 0$, $v_f = +48.3$ km/h, $t = 3.8$ s, and $a_T = +3.54$ m/s². FIND: v_{av} and s. (3) PROBLEM TYPE—Constant accelera-

continued

tion. (4) PROCEDURE—Having the initial and final speeds, we can compute the average speed, which in this case equals $v_{av} = \frac{1}{2}(v_i + v_f)$. Knowing v_{av} the distance is just $s = v_{av}t = \frac{1}{2}(v_i + v_f)t$. (5) CALCULATION—First convert v_f to m/s:

$$v_f = \frac{(48.3 \text{ km/h})(1000 \text{ m/km})}{3600 \text{ s/h}} = 13.4 \text{ m/s}$$

Then (to three significant figures)

$$v_{av} = \frac{1}{2}(v_i + v_f) = \frac{1}{2}(0 + 13.4 \text{ m/s}) = \boxed{6.70 \text{ m/s}}$$

which equals 15 mi/h. The distance traveled (to two significant figures) is

$$s = v_{av}t$$

$$s = (6.70 \text{ m/s})(3.8 \text{ s}) = \boxed{25 \text{ m}}$$

or about 84 ft.

Quick Check: From Table 2.3, $v_f = (48.3)(0.277\ 8)$ m/s = 13.4 m/s, so that piece is correct. Since 60 mi/h = 88 ft/s, 30 mi/h = 44 ft/s; hence, $v_{av} = 22$ ft/s = 6.7 m/s.

Whenever you have a constant-acceleration problem in which the final speed is neither given nor requested, it's very likely that the solution can be arrived at via Eq. (3.9). It relates displacement, initial scalar velocity, and acceleration. Given any two of these quantities, you can find the third.

To solve Eq.(3.9) for t you'll have to be familiar with the Quadratic Equation {see **MATH REVIEW: PART A-4** on the CD 💿 }. When $v_i = 0$ solving Eq. (3.9) for t becomes much easier.

3.5 The Equations of Constant Acceleration

There are five motion variables (v_i, v_f, a_T, s, t) and five equations that are traditionally used to interrelate them. These *constant-$\vec{a}$* equations derive from the basic definitions of speed and acceleration, and so there are actually only two independent equations and two variables that can be solved for. The five equations are developed primarily as a convenience in problem solving. We already have the definition of acceleration,

$$v_f = v_i + at \qquad [3.6]$$

the average speed

$$v_{av} = \frac{1}{2}(v_i + v_f) \qquad [3.7]$$

and the Mean Speed Theorem $\qquad \boxed{s = v_{av}t = \frac{1}{2}(v_i + v_f)t} \qquad [3.8]$

Any uniform-acceleration problem that we can solve, can be solved with just Eqs. (3.6) and (3.8). Still, we can make life a little easier by anticipating the kinds of situations we're likely to encounter over and over again, and work through some of the details now, once and for all. Accordingly, it would be nice to have an expression for s that's not explicitly dependent on v_f, which can be difficult to measure directly. To derive such an equation, replace v_f in Eq. (3.8) by $v_f = v_i + a_T t$:

$$s = \frac{1}{2}[v_i + (v_i + a_T t)]t = \frac{1}{2}(2v_i t + a_T t^2)$$

[rectilinear−*constant-$\vec{a}$*]

$$\boxed{s = v_i t + \frac{1}{2}a_T t^2} \qquad (3.9)$$

The first term ($v_i t$) is the scalar value of the displacement as the body coasts along at its initial speed, even if it's not accelerating. If $v_i > 0$ the plot of displacement against time is a straight line with a positive slope (Fig. 3.8), since the slope equals v_i. To this is added (provided $a_T > 0$) another displacement ($\frac{1}{2}a_T t^2$) traversed as the speed increases beyond v_i. Alternatively, if a_T is negative, the term ($\frac{1}{2}a_T t^2$) is negative and the object is slowing down—it will not travel as far as ($v_i t$). In fact, since s is the scalar displacement, it's possible for the negative acceleration to carry the object back to where it started ($s = 0$), or even

---

CALCULATOR TIPS

Calculators usually carry out the command to multiply or divide before they execute the next command to add. That allows you to do a sequence of steps just as they appear in the equation (without entering parens). For example, suppose we have an object for which v_i = 250 m/s, $t = 8.00$ s, and $a_T = 9.81$ m/s^2 and we want to solve for s using

$$s = v_i t + \frac{1}{2}a_T t^2$$

2313.92

As soon as you hit $+$ the calculator computes the first product (2000) and is ready to add to it the next product, no matter how many terms it has. Only if that product contains a sum will you need parens to tell the calculator to do the addition first.

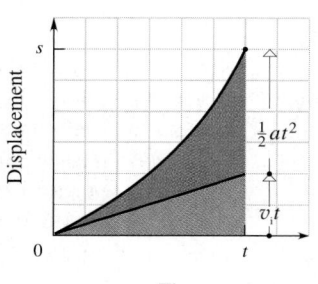

Figure 3.8 The displacement of a uniformly accelerating body is the sum of a linear term ($v_i t$) and a parabolic term ($\frac{1}{2}at^2$).

to where $s < 0$. {You can study these various possibilities via a fun simulation; click on **CONSTANT ACCELERATION** under **INTERACTIVE EXPLORATIONS** on the **CD**. See what happens when $a_T < 0$ with $v_i > 0$ or alternatively when $a_T > 0$ and $v_i < 0$. Observe the difference between l and s in the graph.}

Example 3.6 **[II]** According to *Road & Track* magazine, a Lotus Esprit super-coupe can travel a straight 100.0 ft (i.e., 30.48 m) from rest in 3.30 s. Assume the acceleration is constant, as it nearly is, and determine it in m/s^2.

Solution The phrase "acceleration is constant" should call up the definition of a_T and the Mean Speed Theorem. That's all we really need, but to save time the equations of constant acceleration should also come to mind. (1) TRANSLATION—An object moving in a straight line accelerates from rest covering a known distance in a known time; determine its constant acceleration. (2) Take the initial direction of motion as positive. GIVEN: $v_i = 0$, $s = +30.48$ m, and $t = 3.30$ s. FIND: a_T. (3) PROBLEM TYPE—Constant acceleration. (4) PROCEDURE—Recall that the car is speeding up and therefore a should turn out to be positive. The list of variables (v_i, s, t, a_T) suggests Eq. (3.9), $s = v_i t + \frac{1}{2}a_T t^2$, which contains all, and only, these quantities.

(5) CALCULATION:

$$s = v_i t + \tfrac{1}{2}a_T t^2$$

$$(30.48 \text{ m}) = 0 + \tfrac{1}{2}a_T(3.30 \text{ s})^2$$

Multiplying both sides by 2 and dividing by $(3.30 \text{ s})^2$ yields

$$a_T = \frac{60.96 \text{ m}}{(3.30 \text{ s})^2} = +5.598 \text{ m/s}^2$$

and to three figures, $\boxed{a_T = 5.60 \text{ m/s}^2}$. Note that s and a_T are both positive quantities.

Quick Check: Find v_{av} using the computed value of a_T and then calculate s: $v_f = v_i + a_T t = 0 + (5.6 \text{ m/s}^2)(3.3 \text{ s}) = 18.5$ m/s; therefore $v_{av} = 9.25$ m/s, which in 3.30 s means a distance of 30.5 m.

The last of the five expressions we want relates s, v_i, v_f, and a_T *independent of time*. First, solve Eq. (3.6) for the time, $t = (v_f - v_i)/a_T$. Then substitute this into $s = \frac{1}{2}(v_i + v_f)t$ so that t no longer appears explicitly:

$$s = \left(\frac{v_i + v_f}{2}\right)\left(\frac{v_f - v_i}{a_T}\right) = \frac{v_f^2 - v_i^2}{2a_T}$$

STUDY GUIDE

If you encounter a problem involving s, $v_i \neq 0$, and a_T, and need to find t, the natural choice is Eq. (3.9). Still, if you wish to avoid dealing with the Quadratic Equation you can always use Eqs. (3.8) and (3.10) instead.

Multiplying both sides by $2a_T$ and adding v_i^2 to both sides yields

[rectilinear – *constant*-$\vec{a}$]

$$v_f^2 = v_i^2 + 2a_T s \qquad (3.10)$$

Whenever we have a problem concerning uniform acceleration in which time does not appear, Eq. (3.10) will likely be the key formula.

Equations (3.6), (3.7), (3.8), (3.9), and (3.10) are the five *constant*-$\vec{a}$ expressions we set out to derive. Each represents a distinct relationship among a specific group of variables, and depending on the problem at hand, one or more of these equations will provide a route to its solution. Nonetheless, they are nothing more than a convenient restatement of two basic ideas: the definitions of velocity and acceleration. *Any problem that can be solved with the five constant-$\vec{a}$ equations can be solved (perhaps with a bit more effort) using only the two definitions.*

{To see how all of this applies to braking a car, considering the reaction time of the driver, click on **STOPPING DISTANCE** under **FURTHER DISCUSSIONS** on the CD.}

Example 3.7 **[II]** The fastest animal sprinter is the cheetah, reaching speeds in excess of 113 km/h (i.e., 70 mi/h). These animals have been observed on a straight run to bound from a standing start to 72 km/h in 2.0 s. (a) Use this data to compute the magnitude of the cheetah's acceleration, assuming it to be constant. (b) What minimum distance is required for the cheetah to go from rest to 17.9 m/s (i.e., 40 mi/h)?

Solution Reading that the acceleration is constant should bring to mind the definition of a_T and the Mean Speed Theorem. To save time, the equations of constant acceleration should also be recalled. (1) TRANSLATION—(a) An object moving in a straight line accelerates from rest to a known speed, in a known time; determine its constant acceleration and (b) the distance needed to reach a certain speed. (2) GIVEN: (a) $v_i = 0$, $v_f = 72$

continued

km/h, and $t = 2.0$ s. FIND: a_T. GIVEN: (b) $v_i = 0$, $v_f = 17.9$ m/s, and a. FIND: s. (3) PROBLEM TYPE—Constant acceleration. (4) PROCEDURE—(a) The list of variables (v_i, v_f, t, a_T) suggests that we can compute a_T using Eq. (3.6), which contains all, and only, these quantities. (b) Having determined a_T the variables (v_i, v_f, a_T, s) suggest that Eq. (3.10) will provide s. (5) CALCULATION—(a) Determine a_T from $v_f = v_i + a_T t$ where $v_f = 72$ km/h = 20 m/s,

$$a_T = \frac{v_f - v_i}{t} = \frac{20 \text{ m/s}}{2.0 \text{ s}} = \boxed{10 \text{ m/s}^2}$$

Part (b) calls for s without specifying t and that suggests $v_f^2 = v_i^2 + 2a_T s$. Rearranging terms and using $v_f = 17.9$ m/s and $v_i = 0$, we get

$$s = \frac{v_f^2}{2a_T} = \frac{(17.9 \text{ m/s})^2}{2(10 \text{ m/s}^2)} = \boxed{16 \text{ m}}$$

The cheetah can hit 17.9 m/s from a dead stop in a mere 16 m.

Quick Check: Use the average speed to confirm the last answer. $v_{av} = \frac{1}{2}(72 \text{ km/h} + 0) = 10$ m/s; and it takes a distance $s = (10 \text{ m/s})(2.0 \text{ s}) = 20$ m to reach 20 m/s. Therefore $s = 16$ m for $v_f = 17.9$ m/s is reasonable.

a = constant
$v_i = 0$
$v_f = 72$ km/h
2.0 s

Insofar as air drag can be ignored, all objects fall with the same **constant acceleration**.

Free-Fall

Probably the most important situation where acceleration is effectively constant is free-fall. Drop a bowling ball at the Earth's surface and Eqs. (3.6) through (3.10) will describe the motion quite accurately.

A falling object ordinarily moves through air, and the essential nature of the motion is often obscured by fluid friction. Nonetheless, we now know that **on Earth all objects falling through a vacuum accelerate downward at the same fairly constant rate, regardless of their weight**.

3.6 Air Drag

Ever since Aristotle (384−322 B.C.E.), most people have erroneously believed that a falling body descends at a rate that's proportional to its weight. Though that's fundamentally wrong, it is a fairly accurate picture of the special case of extremely light objects, such as feathers and snowflakes, dropping in air. It applies, as well, for heavy bodies descending

If Galileo Galilei were alive he probably would be tugging at his beard with pleasure about the elementary physics experiment **Apollo 15** astronaut David R. Scott performed today on the moon.

Scott dropped a hammer and a feather from waist high to illustrate that both objects accelerated equally by the moon's gravity and that both would hit the surface at the same time despite their differences in mass or weight…

"In my left hand I have a feather. In the right hand a hammer," Scott said, standing in front of the camera mounted on the lunar Rover…

Then he dropped both objects and, sure enough, they struck the lunar surface simultaneously.

ASSOCIATED PRESS NEWS RELEASE,
JULY, 1971

Galileo Galilei at the age of about 60, drawn by his contemporary Ottavio Leoni. Galileo was born in 1564, just three days before Michelangelo died, and within months of Shakespeare's birth.

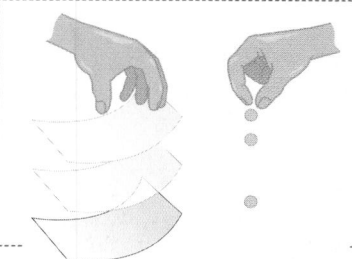

through thick fluids, like oil or honey, but it misses the central point. A stone 10 times heavier than another does *not* fall through air 10 times faster!

The person who contributed most to our modern understanding is Galileo Galilei, the hero of free-fall. Galileo was born in Italy in 1564. He was vigorous and robust, built short and solid with wavy red hair, a man who fathered three more children than perhaps was proper for a lifelong bachelor. He was an arrogant, argumentative, self-assured spokesman of modernity.

Galileo really was not saying anything new when he asserted that, ***in vacuum, where there is no air resistance, all bodies fall at the same rate***. Others had guessed as much long before him. By contrast, when an object falls through the air, moving ever more rapidly, the resistance to its motion, the *drag*, also increases. (Riding on a motorcycle gives a sense of how formidable the effect can be—at 100 mi/h the force exerted on a 1.0 ft^2 perpendicular area is about 25 lb, and it increases to 100 lb at 200 mi/h.) The greater the rate of descent, the greater the resistance, as more air must be pushed aside per second. Finally, a balance is reached where no further increase in falling speed can occur. This point is called the **terminal speed**, and it depends on the shape, surface, and weight of the object (see Table 3.1).

A sky diver, after dropping about 620 m (in the thick atmosphere below an altitude of ≈3000 m) with arms and legs extended spread-eagle, will reach a top speed of roughly 200 km/h (i.e., 120 mi/h).* In a head-down dive, the same person will have a terminal speed up around 300 km/h (i.e., 185 mi/h). With an open parachute, drag increases tremendously, and the terminal speed drops to around 30 km/h (i.e., 20 mi/h). On the other hand, a compact heavy object, such as a smooth stone or an aerial bomb, might descend 150 m to 200 m before even beginning to deviate from free-fall.

3.7 Acceleration Due to Gravity

The philosopher-physicist Strato (ca. 300 B.C.E.) knew that ***objects in free-fall accelerate***. Rainwater running off a roof begins its descent in a continuous stream that ultimately breaks up into a flutter of separate splashes. "This could never happen," Strato maintained,

Table 3.1

Approximate Terminal Speeds

Object	Speed (m/s)
Fluffy feather	0.4
Sheet of paper (flat)	0.5
Snowflake	1
Parachutist	7
Penny	9
No. 6 shot	9
Mouse	13
Sky diver spread-eagle (reached in 10 s to 12 s)	58
Bullet (high caliber)	100
Large rock	200
Cannonball	250

The terminal speed of each person (with arms and legs outstretched) is about 200 km/h. Notice how this flat, belly-down orientation is stable. Throw some playing cards into the air. Do they fall edge-first?

*Five seconds is the record for a 310-ft free-fall onto an air bag. The stuntman, who reached 80 mi/h and burst the bag on impact, walked away smiling.

THE RECORD FREE-FALL

Lieutenant I. M. Chisov of the former Soviet Union was flying his *Ilyushin 4* on a bitter cold day in January 1942, when it was attacked by 12 German *Messerschmitts*. Convinced that he had no chance of surviving if he stayed with his badly battered plane, Chisov bailed out at 21 980 ft. With the fighters still buzzing around, Chisov cleverly decided to fall freely out of the arena. It was his plan not to open his chute until he was down to only about 1000 ft above the ground. Unfortunately, he lost consciousness en route. As luck would have it, he crashed at the edge of a steep ravine covered with 3 ft of snow. Hitting at about 120 mi/h, he plowed along its slope until he came to rest at the bottom. Chisov awoke 20 min later, bruised and sore, but miraculously he had suffered only a concussion of the spine and a fractured pelvis. Three and one-half months later he was back at work as a flight instructor. The reason he survived is embodied in Eq. (7.3).

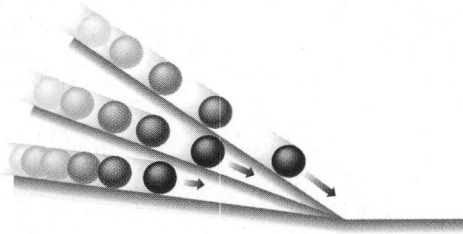

Figure 3.9 To see if the acceleration of a falling body was constant, Galileo rolled balls down inclined planes of increasing tilt. As the angle increased, the acceleration increased, but for each position it was constant. He argued that at a tilt of 90° the object would free-fall, also with a constant acceleration.

"unless the water was falling more swiftly through each successive distance than it had through the earlier ones." The droplets accelerate, and those falling longer move faster and pull away from the slower ones behind them. On some level, you already knew that falling bodies accelerate—although you might volunteer to catch a baseball dropped from a few meters, you would surely decline if it were to come down from the top of the Sears Tower in Chicago (443 m). The ball would then descend at a tremendous rate, which could only mean that it accelerated.

It was Galileo who first designed and carried out an experiment to confirm that the acceleration is constant. Two practical obstacles made it impossible for him to simply drop an object, measure v_f at several values of t, and then compute a_T for each. The first was that free-fall happens rapidly and clocks in his era were extremely crude. He surmounted this problem by slowing things down—instead of dropping balls, he rolled them along inclined planes (Fig. 3.9). He argued that if a_T was constant for each incline, it would continue to be constant, even when the track was vertical and the ball fell freely.

The second difficulty Galileo encountered was that he had no way of measuring the instantaneous final speed at the end of each run. To handle this obstacle, he reformulated the problem, replacing v_f by quantities that were directly observable. In so doing he arrived at a statement of $s = v_i t + \frac{1}{2}a_T t^2$. *If the acceleration was constant, starting from rest ($v_i = 0$), the distance traveled by the ball would depend on the time squared ($s \propto t^2$).* Doubling the time a ball was allowed to move would quadruple the distance it traveled. "In such experiments," he wrote, "repeated a full hundred times, we always found that the spaces traversed were to each other as the squares of the times, and this was true for all inclinations of the plane."

On free-fall. For all things that fall... the empty void cannot on any side, at any time, support anything, but rather, as its own nature desires, it continues to give place; wherefore all things must needs be borne on through the calm void, moving at equal rate with unequal weights.

LUCRETIUS (CA. 94–55 B.C.)
ON THE NATURE OF THINGS

Regarding terminal speeds. You can drop a mouse down a thousand-yard mine shaft and, on arriving at the bottom, it gets a slight shock and walks away. A rat is killed, a man is broken, a horse splashes.

J. B. S. HALDANE
BRITISH GENETICIST, 1892–1964

NOT QUITE FREE-FALL

All else being constant, terminal speed increases as the object's weight increases and decreases as its cross-sectional area increases. The terminal speed of a BB (≈ 9 m/s) is lower than that of a cannonball (≈ 250 m/s). The terminal speed of a child is less than that of an adult. This fact accounts, in part, for small animals and insects surviving falls that are immense in proportion to their size. Air friction spares us from being hurt by raindrops and hailstones, which ordinarily strike the Earth at only about 25 km/h to 30 km/h but might otherwise fall at several hundreds of kilometers per hour (Problem 89, p. 83).

Galileo knew that for all falling bodies, air resistance would "render the motion uniform," provided the descent was long enough. His ideas were confirmed some thirty years later, when Robert Boyle pumped the air out of a long cylinder and, for the first time, dropped various objects in a vacuum. No amount of theorizing is as convincing as seeing a feather and a lead ball come slamming down together.

A cannon ball weighing one or two hundred pounds, or even more, will not reach the ground by as much as a span ahead of a musket ball weighing only half a pound, provided both are dropped from a height of 200 cubits.

GALILEO GALILEI

Table 3.2

Acceleration Due to Gravity*

Location	m/s^2	ft/s^2
Equator	9.780	32.09
Panama	9.782	32.09
Honolulu	9.789	32.12
Key West	9.790	32.12
Austin	9.793	32.13
Tokyo	9.798	32.15
San Francisco	9.800	32.15
Princeton	9.802	32.16
New York	9.803	32.16
Chicago	9.803	32.16
Munich	9.807	32.18
Winnipeg	9.810	32.18
Leningrad	9.819	32.22
Arctic Red River	9.824	32.23
North Pole	9.832	32.26

*The measured values of g are actually due to both gravity and the rotation of the Earth.

The free-fall acceleration, due as it is to gravity, is represented by its own symbol, g. There is some small variation in g over the planet (Table 3.2), but we can generally take g to be a constant equal to its average value of 9.806 65 m/s^2 or 9.81 m/s^2 (i.e., 32.2 ft/s^2). {For the dependence of g on both altitude and the Earth's spin click on **GRAVITATIONAL ACCELERATION** under **FURTHER DISCUSSIONS** on the **CD**.}

3.8 Straight Up & Down

If an object is moving freely near the surface of the Earth, its gravitational acceleration is along a vertical line and is assumed to be constant. All the equations of uniformly accelerated motion—Eqs. (3.6), (3.7), (3.8), (3.9), and (3.10)—then apply, as is, with $a_T = g$. In that case g is treated as just another acceleration,* and like a_T, its numerical value can be positive or negative (depending on the details of the analysis); g is the scalar value of $\vec{g}$. On the way up a projectile has a negative acceleration. On the way down it has a positive acceleration. In any event, **in free-fall the acceleration is always $\vec{g}$ and it's always straight downward regardless of the motion**.

Example 3.8 **[I]** A salmon is dropped by a hovering eagle. How far will the fish fall in 2.5 s? Ignore air drag.

Solution Anything that free-falls has a constant acceleration g, and that should remind us of the *constant-$\vec{a}$* equations. (1) TRANSLATION—An object falls freely from rest for a known time; determine the distance it descends. (2) The initial motion is down, so take *down* as positive. Because $\vec{g}$ always points down, selecting down to be positive makes $a_T = g$ positive. Similarly, the displacement vector (drawn from the initial to the final position) points down and s too will be positive. GIVEN: $a_T = g = +9.81$ m/s^2, $v_i = 0$, and $t = 2.5$ s. FIND: s. (3) PROBLEM TYPE—Constant acceleration/free-fall. (4) PROCEDURE—We have a_T, v_i, and t and need s, which suggests Eq. (3.9), $s = v_i t + \frac{1}{2} a_T t^2$. (5) CALCULATION:

$$s = 0 + \tfrac{1}{2}(+9.81\,\text{m/s}^2)(2.5\,\text{s})^2$$

The fish falls a distance $s = +30.6$ m or to two figures,

$$\boxed{s = 31\ \text{m}}$$

Quick Check: Its final speed is $v_f = v_i + gt = 24.5$ m/s, and since $v_f^2 = v_i^2 + 2a_T s$, $s = (24.5\,\text{m/s})^2/2g = 30.6$ m.

---

*Alternatively, some people prefer to keep g a positive quantity. Unfortunately, that requires changing the sign in front of every term containing g in the *constant-$\vec{a}$* equations from + to ±, an approach that was popular around the 1920s; we won't be doing that.

Example 3.9 [II]   A ball is thrown straight down from the roof of a dormitory at 10.0 m/s. If the building is 100-m tall, at what speed will the ball hit the ground? How long will the trip take?

Solution   Anything that free-falls has a constant acceleration ($a_T = g$), and that should bring to mind the equations of constant acceleration. (1) TRANSLATION—An object having an initial *nonzero speed* falls freely for a known time; determine the distance it descends.   (2) Take *down* as positive. Then $a_T = g$, v_i, s, and v_f will all be positive because the corresponding vectors all point downward. GIVEN: $a_T = g = +9.81$ m/s^2, $v_i = +10.0$ m/s, and $s = +100$ m. FIND: v_f and t. (3) PROBLEM TYPE—Constant acceleration/free-fall. (4) PROCEDURE—We are given a_T, v_i, and s, and need v_f, but don't have t. That suggests

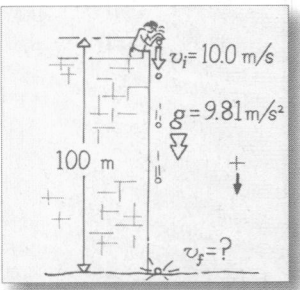

using $v_f^2 = v_i^2 + 2a_T s$, which is independent of t, to get v_f. Then, having v_f, we can find t using $v_f = v_i + a_T t$. (5) CALCULATION:

$$v_f^2 = v_i^2 + 2gs = (+10.0 \text{ m/s})^2 + 2(+9.81\text{m/s}^2)(+100 \text{ m})$$

and $\boxed{v_f = 45.4 \text{ m/s}}$. The time can most easily be gotten from $v_f = v_i + gt$ where

$$t = \frac{v_f - v_i}{g} = \frac{45.4 \text{ m/s} - 10.0 \text{ m/s}}{+9.81 \text{ m/s}^2} = \boxed{3.61 \text{ s}}$$

Quick Check: Using $s = v_i t + \frac{1}{2}a_T t^2$; $\frac{1}{2}(9.81 \text{ m/s}^2)t^2 + (10.0$ m/s$)t - 100$ m $= 0$. From the quadratic equation {see **MATH REVIEW: PART A-4** on the CD 🌑 }

$$t = \frac{-10.0 \text{ m/s} \pm 45.4 \text{ m/s}}{+9.81 \text{ m/s}^2}$$

ignoring the negative solution $t = 3.6$ s.

If a ball is thrown upward, its speed continuously diminishes at a rate of –9.81 m/s per second as it climbs. Always downward, $\vec{\mathbf{g}}$ has the effect of increasing the speed of descending objects and decreasing the speed of ascending objects. The rising ball, moving slower and slower, will momentarily stop (the acceleration is still $\vec{\mathbf{g}}$) and then plunge downward faster and faster. The height of that stopping point, the maximum or **peak altitude** (s_p), is the vertical distance it takes to go from v_i to $v_f = 0$.

Ideally, the up-and-down journey is symmetrical in space and time around the peak altitude (Fig. 3.10). The speeds will be equal for any amount of time before and after the peak altitude is reached. The ascent unfolds as if it were a movie of the descent run backward. And *the descent is exactly the same as if the object were dropped from the peak altitude*. At high projectile speeds, air friction limits the motion, causing the trajectory to be lower and asymmetrical (see p. 75). For example, typical antiaircraft fire rises to less than about 1500 m (see Example 3.10).

Consider an object thrown vertically upward with a scalar velocity v_i. At any time t thereafter, we measure its displacement from the point of launch. That displacement is always given by $s = v_i t + \frac{1}{2}gt^2$ regardless of whether the projectile happens to be going up or down at t. Just put in the right signs at the start, and this equation will take care of all possibilities, including the one where the projectile has gone past s_p and is on the way down. Indeed, that's the primary reason displacement was used in this chapter rather than path-length. To illustrate how these two ideas differ, consider a projectile thrown straight up that reaches peak altitude, and then comes halfway back down. The displacement at that moment is $\frac{1}{2}s_p$ whereas the total distance traveled is $l = \frac{3}{2}s_p$.

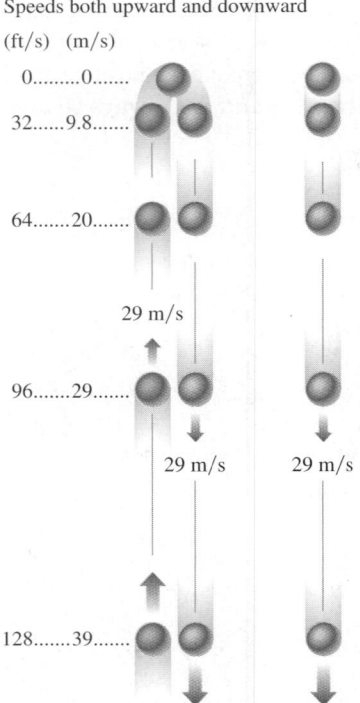

Speeds both upward and downward

(ft/s)   (m/s)

0.........0.......

32......9.8.......

64.......20.......

29 m/s

96.......29.......

29 m/s   29 m/s

128.......39.......

Figure 3.10   A ball thrown straight up at some speed—in this case 39 m/s (i.e., 128 ft/s)—will (neglecting air friction) return to its launch height at that same speed. In fact, the speeds moving up and down are equal at equal heights. In this instance, the ball takes 4.0 s to reach maximum altitude and 4.0 s to get back.

Example 3.10 [II] A .32-caliber bullet fired from a revolver with a 3-in.-long barrel will have a relatively low muzzle speed of about 200 m/s. If it's shot straight up, neglecting air resistance, (a) what is the peak height the bullet will reach? (b) How fast will it be moving when it returns to the height of the gun? (c) How long will the whole trip take?

Solution Anything that free-falls has a constant acceleration ($a_T = g$), and that should bring to mind the *constant-$\vec{a}$* equations. Even physicists, who generally like to derive what they need, have these memorized. (1) TRANSLATION—An object ascends with an initial speed; determine (a) its peak altitude, (b) return speed, and (c) total flight time. (2) Select *up* as the positive direction. GIVEN: $a_T = g = -9.81$ m/s^2 and $v_i = +200$ m/s. FIND: (a) s_p; (b) final speed; and (c) total time of flight t_T. (3) PROBLEM TYPE—Constant acceleration/free-fall. (4) PROCEDURE—For the upward leg of the trip we have a, v_i, and $v_f = 0$ and need s_p, which suggests $v_f^2 = v_i^2 + 2a_T s$. (5) CALCULATION—With $a_T = g$ and $v_f = 0$,

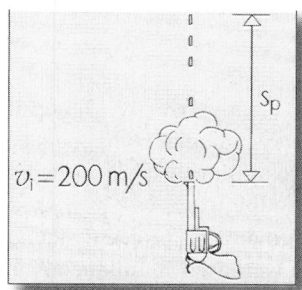

$$0 = v_i^2 + 2g s_p$$

and $$s_p = -\frac{v_i^2}{2g} \quad (3.11)$$

The bullet is decelerating, $\vec{g}$ *is downward, and so we enter* g *as a negative number.* The answer to part (a) is

$$s_p = -\frac{(200 \text{ m/s})^2}{2(-9.81 \text{ m/s}^2)} = \boxed{+2.04 \times 10^3 \text{ m}}$$

(b) Because the flight is symmetrical, the bullet ideally returns to the height of the gun at a speed of $\boxed{200 \text{ m/s}}$. Table 3.1 tells us that the actual speed would be closer to 100 m/s, which is still very dangerous. (c) The time for the trip *up*, the ***time to peak altitude*** (t_p), can be computed using Eqs. (3.6), (3.8), or (3.9)—Eq. (3.6) is the easiest:

$$v_f = v_i + gt$$

$$0 = 200 \text{ m/s} + (-9.81 \text{ m/s}^2)t_p$$

and $t_p = -(200 \text{ m/s})/(-9.81 \text{ m/s}^2) = 20.4$ s, which is half the total flight time and thus $\boxed{t_T = 40.8 \text{ s}}$.

Quick Check: Compute the down-trip time starting with $s = v_i t + \frac{1}{2}gt^2$ where $v_i = 0$, and the displacement is -2040 m measured from the top of the trajectory downward to the initial height of firing. Then

$$t_p = \left(\frac{2s}{g}\right)^{1/2} = \left[\frac{2(-2040 \text{ m})}{-9.81 \text{ m/s}^2}\right]^{1/2} = 20.4 \text{ s}$$

Alternatively, factor Eq. (3.9) and set $s = 0$; $t(v_i + \frac{1}{2}gt) = 0$, hence $t = 0$ and $(v_i + \frac{1}{2}gt) = 0$, and so $t = t_T = -2v_i/g = 40.8$ s.

Example 3.11 [III] A ball is hurled straight up at a speed of 15.0 m/s, leaving the hand of the thrower 2.00 m above ground. Compute the times and the ball's speeds when it passes an observer sitting at a window in line with the throw 10.0 m above the point of release.

Solution Anything that free-falls has a constant acceleration of g, and that should call to mind the *constant-$\vec{a}$* equations. (1) TRANSLATION—An object ascends freely with an initial speed from an initial height; determine (a) its speeds and (b) times when at a later height. (2) Because nothing about ground level is to be computed, the height of the hand at launch is taken as the origin. Make *up* positive. GIVEN: $a = g = -9.81$ m/s^2, $v_i = +15.0$ m/s, and $s = +10.0$ m. FIND: the *two* times and the *two* final speeds. (3) PROBLEM TYPE—Constant acceleration/free-fall. (4) PROCEDURE—We could use $s = v_i t + \frac{1}{2}a_T t^2$ to find t directly (Problem 101). Instead we will find v_f first, via $v_f^2 = v_i^2 + 2a_T s$. (5) CALCULATION:

$$v_f^2 = (15.0 \text{ m/s})^2 + 2(-9.81 \text{ m/s}^2)(+10.0 \text{ m}) = 28.9 \text{ m}^2/\text{s}^2$$

and $$\boxed{v_f = \pm 5.37 \text{ m/s}}$$

The ball passes the window traveling at +5.37 m/s on the way

up and -5.37 m/s on the way down. The corresponding times (t_u and t_d) can be gotten from $v_f = v_i + a_T t$. On the way up

$$v_f = v_i + gt$$

$$+5.37 \text{ m/s} = +15.0 \text{ m/s} + (-9.81 \text{ m/s}^2)t_u$$

$$\boxed{t_u = 0.982 \text{ s}}$$

On the way down, $v_f = -5.37$ m/s, and so

$$-5.37 \text{ m/s} = +15.0 \text{ m/s} + (-9.81 \text{ m/s}^2)t_d$$

$$\boxed{t_d = 2.08 \text{ s}}$$

The peak altitude was reached when $v_f = 0 = v_i + (-9.81$ m/s$^2)t_p$, that is, when $t_p = 1.53$ s. The ball passed the window on the way up 0.55 s earlier than t_p and on the way down 0.55 s later. This symmetry around the peak can be useful in problem solving.

Quick Check: Use t_u to find the distance; $s = v_i t + \frac{1}{2}a_T t^2 = (15.0 \text{ m/s})(0.982 \text{ s}) + \frac{1}{2}(-9.81 \text{ m/s}^2)(0.982 \text{ s})^2 = 10.0 \text{ m}$.

At the surface of the Earth, in situations where air friction is negligible, objects fall with the same acceleration regardless of their weights. Notice that the larger ball is experiencing a little bit more drag.

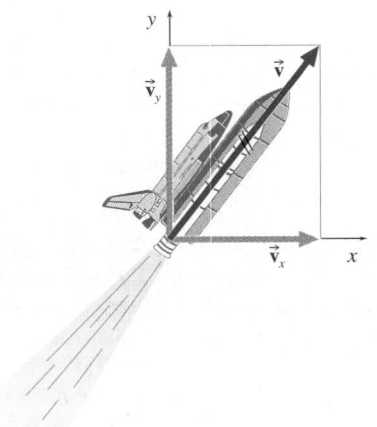

Figure 3.11 The Shuttle goes from rest to 17×10^3 mi/h in 8 min. At any instance its velocity, $\vec{\mathbf{v}}$, is in the direction of motion. As depicted here the craft is traveling in a vertical plane and has velocity components of $\vec{\mathbf{v}}_x$ and $\vec{\mathbf{v}}_y$.

3.9 Two-Dimensional Motion: Projectiles

Consider an object (e.g., a car, or the Space Shuttle in Fig. 3.11) moving in two dimensions. At any moment its velocity vector can be resolved into two perpendicular components. The motion unfolds as if it were composed of these two separate motions superimposed: $\vec{\mathbf{v}} = \vec{\mathbf{v}}_x + \vec{\mathbf{v}}_y$. And that's quite generally the case for all types of motion.

Galileo was probably the first to recognize that the movement of a projectile launched into the air at some arbitrary angle can be imagined as if it were two simultaneous *independent* motions: a horizontal flight (ideally at constant speed) and a vertical gravity fall (ideally at constant acceleration). The gravitational interaction with the planet causes a vertical acceleration, but no such influence exists horizontally, and the horizontal acceleration (overlooking air drag) is zero.

Now imagine a ball fired horizontally. There is no initial component of the velocity in the vertical direction ($v_{iy} = 0$); the ball sails straight off horizontally at a *constant speed* equal to its initial speed, v_{ix}. But on Earth, heavy objects fall freely downward with a constant acceleration. The projectile will continuously descend, faster and faster, as it progresses laterally, sweeping out a smooth arc that curves increasingly downward (Fig. 3.12). Figure 3.13 shows much the same motion for a heavy ball (negligible drag) dropped from

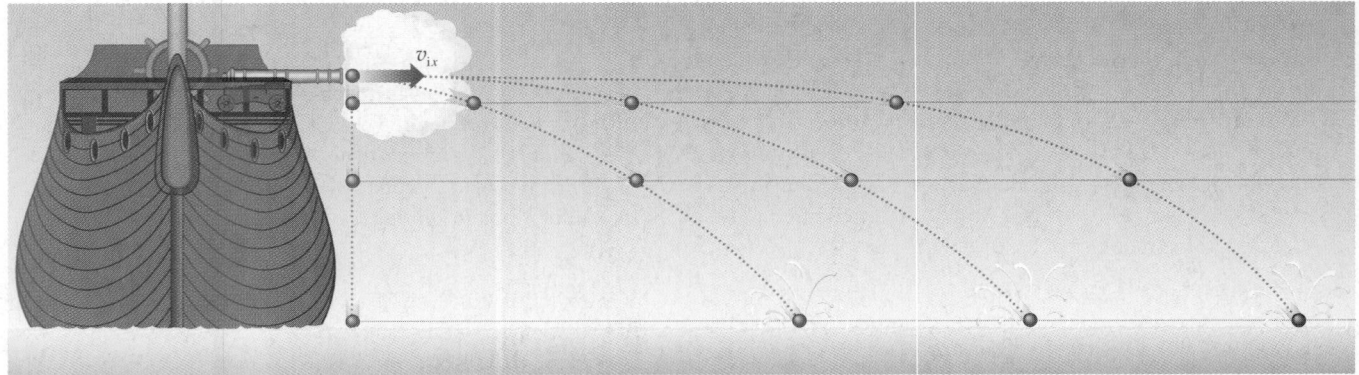

Figure 3.12 The faster each cannonball is fired, the farther it will go, but all fall at the same rate and hit the water after the same flight times. The horizontal speed, v_{ix}, is constant, provided air friction is negligible. The vertical speed increases at the rate g.

Figure 3.13 A plane traveling at a constant velocity. When friction is negligible, the horizontal speed of the ball equals the speed of the plane and it is constant. As a result, the ball is always under the plane.

a plane traveling on a straight horizontal path at a fixed speed. The ball's horizontal speed is ideally constant and equal to the speed of the plane, which is why it hits the water directly below the aircraft. *To the extent that both the curvature of the Earth and friction can be neglected, the path of a free, or ballistic, projectile is parabolic and it lies in a vertical plane.* {For a proof click on **PARABOLIC TRAJECTORIES** under **FURTHER DISCUSSIONS** on the **CD**.}

Example 3.12 [**III**]   A youngster hurls a ball horizontally at a speed of 10 m/s from a bridge 50 m above a river. Ignoring air resistance: (a) How long will it take for the ball to hit the water? (b) What is the velocity of the ball just before it lands (Fig. 3.14)? (c) How far from the bridge will it strike?

Solution   The vertical part of the motion is free-fall, and that should remind us of the equations of constant acceleration. The independent horizontal motion is at a constant speed.   (1) TRANSLATION—A projectile has a known horizontal initial speed and initial height; determine (a) its total time of flight, (b) final or impact velocity, and (c) horizontal distance traveled. (2) Take *down* as positive. GIVEN: $v_{ix} = 10$ m/s, $v_{iy} = 0$, and $s_y = +50$ m. FIND: (a) t_T, (b) $\vec{v}_f$, and (c) s_x.   (3) PROBLEM TYPE— Projectile motion/horizontal launch.   (4) PROCEDURE—(a) The time it takes to hit the water is just the free-fall time (i.e., the time it takes to fall 50 m) given by Eq. (3.9), $s = v_i t + \frac{1}{2} a_T t^2$, applied to the vertical motion as if it alone were occurring.   (5) CALCULATION—With $a_T = g$ and $v_{iy} = 0$,

$$s_y = \tfrac{1}{2} g t_T^2$$

With $g = +9.81$ m/s²,

$$(50 \text{ m}) = \tfrac{1}{2}(+9.81 \text{ m/s}^2) t_T^2$$

and

$$t_T = \sqrt{\frac{2(50 \text{ m})}{9.81 \text{ m/s}^2}} = 3.19 \text{ s} = \boxed{3.2 \text{ s}}$$

(b) The velocity at which the ball hits the water is the vector resultant of its final horizontal, $\vec{v}_{fx}$, and vertical, $\vec{v}_{fy}$, motions: $\vec{v}_f = \vec{v}_{fx} + \vec{v}_{fy}$. The horizontal speed is constant at $v_{ix} = 10$ m/s; now find $\vec{v}_{fy}$. With $v_{iy} = 0$, Eq. (3.6), $v_f = v_i + a_T t$, leads to

Figure 3.14 A projectile thrown horizontally from a bridge hits the water with a velocity $\vec{v}_f$ that has both horizontal, $\vec{v}_{fx}$, and vertical, $\vec{v}_{fy}$, components. With negligible air-friction losses, the initial horizontal scalar velocity v_{ix} equals v_{fx}.

continued

$$v_{fy} = gt_T = (+9.81 \text{ m/s}^2)(3.19 \text{ s}) = 31.3 \text{ m/s}$$

The resultant speed (p. 36) is then

$$v_f = \sqrt{v_{fx}^2 + v_{fy}^2} = \sqrt{1079} \text{ m/s} = 33 \text{ m/s}$$

Notice that this quantity is larger than the largest of the two components, as it should be. The angle, θ, made by $\vec{v}_f$ and the horizontal (Fig. 3.11) is such that

$$\tan \theta = \frac{v_{fy}}{v_{fx}} = \frac{31.3 \text{ m/s}}{10 \text{ m/s}} = 3.13$$

therefore $\theta = 72°$ below the horizontal. We could use either sin θ or cos θ instead of tan θ but that would involve v_f, and if an error was made in computing v_f, it would make θ wrong, too.

Finally then,

$$\boxed{\vec{v}_f = 33 \text{ m/s–DOWN } 72° \text{ BELOW HORIZONTAL}}$$

(c) Because the ball's horizontal speed is constant and because it is in the air for a time t_T,

$$s_x = v_{ix} t_T = (10 \text{ m/s})(3.2 \text{ s}) = \boxed{32 \text{ m}}$$

Quick Check: The average vertical speed is $\frac{1}{2}v_{fy} = 15.65$ m/s; the time of flight is then $t_T = (50 \text{ m})/(15.65 \text{ m/s}) = 3.19$ s. Observe that v_{fy} is about 3 times v_{fx}, so a fairly large value of θ is reasonable.

Now for the more general case. Imagine a cork popping out of a champagne bottle at some arbitrary angle θ, with an initial speed v_i, as in Fig. 3.15a. [This could just as well be a leaping frog, an athlete executing a broad jump (p. 262), or a kicked football—any unpowered, i.e., ***ballistic flight***.] The motion occurs as if it were two independent motions—one vertical, with an initial scalar velocity of

$$v_{iy} = v_i \sin \theta$$

the other horizontal, with an initial scalar velocity of

$$v_{ix} = v_i \cos \theta$$

The displacement vector $\vec{s}$ drawn from the point of launch to the projectile at any moment has a scalar horizontal component of

$$s_x = v_{ix} t$$

because $a_x = 0$, and a scalar vertical component of

$$s_y = v_{iy} t + \frac{1}{2} g t^2$$

because $a_y = g$.

Sparks fly into the air, arcing along parabolic paths.

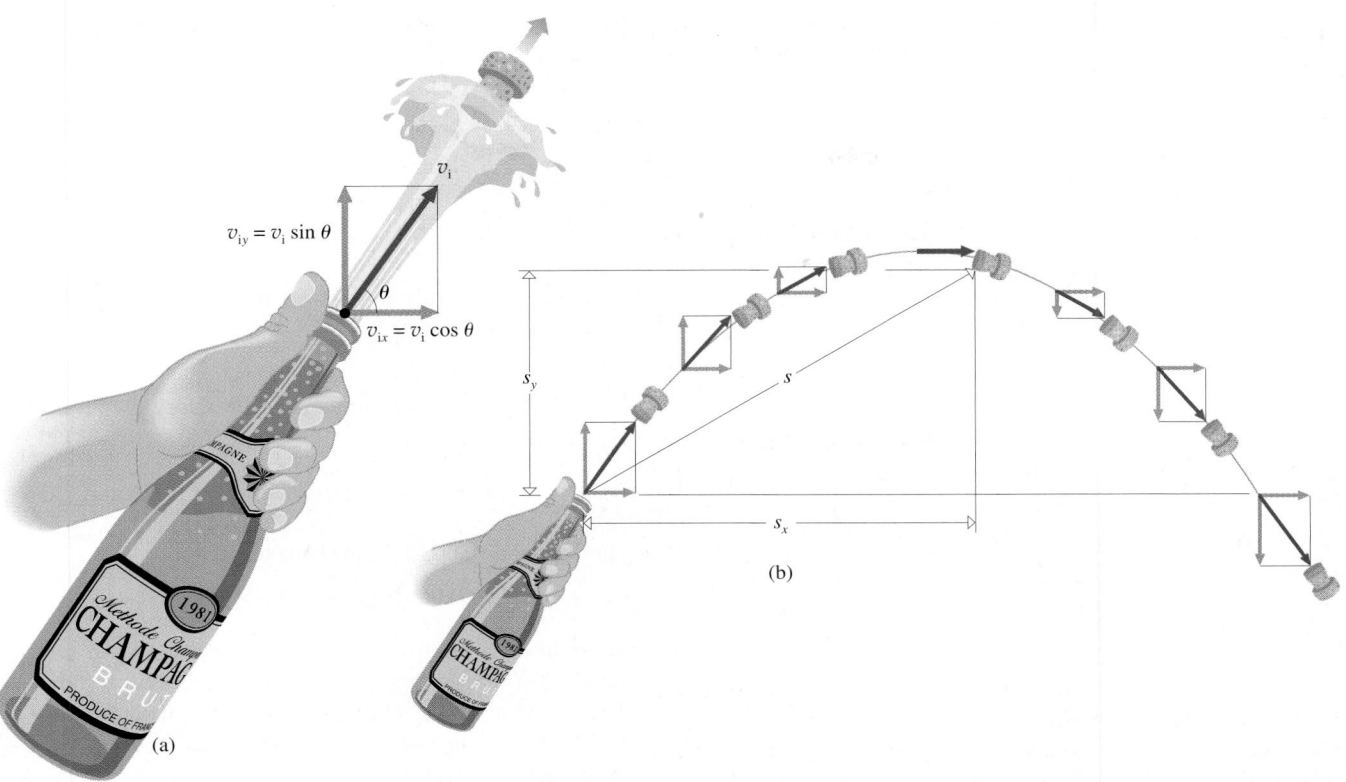

Figure 3.15 (a) Launching a projectile into the air at an angle θ. (b) Notice how the vertical component of the velocity varies along the trajectory. A longstanding record was set in Reno, Nevada on July 4, 1981, when a cork from a champagne bottle was "fired" 105 ft 9 in. Because of air friction, corks are launched at 40° to cover more distance than height.

The cork rises (take up to be positive), slowing under the influence of a downward acceleration, $\vec{\mathbf{g}}$, until it reaches its peak altitude. At that point, its vertical speed is zero, although it continues uninterruptedly to move horizontally (Fig. 3.15*b*). Thereafter, the cork descends to the launch elevation at very nearly the speed with which it was first fired.*

The equation for s_y describes both the trip up and down; s_y is the scalar vertical *displacement* from the launch point, *not* the total vertical distance traveled. On the way up, s_y increases; on the way down, it decreases. When the projectile drops back to the launch height, $s_y = 0$. If it should fall into a hole or over a cliff so that it lands beneath the level of launch, s_y would become negative.

The Peak Height

Provided friction losses are negligible, the path is a parabola, symmetrical around the peak altitude. But if you ran along at a constant speed staying directly under the cork, it would have no horizontal velocity with respect to you; it would be seen simply to rise up, stop at peak altitude, and come straight down. The time that it takes (t_p) to reach peak altitude (s_p) is again obtained from $v_{fy} = v_{iy} + gt_p$, where $v_{fy} = 0$. Consequently, $t_p = -v_{iy}/g$; when *up* is positive so that v_{iy} is positive, g will be negative and vice versa. In any event, ***the time is always positive***.

*When the Palestine Liberation Organization left Beirut, Lebanon, in 1982, they fired so many shots into the air to celebrate that at least half a dozen people were seriously wounded by falling bullets.

In the simplest case, the journey ends when the projectile returns to the height at which it was launched. Then, because of the symmetry, twice the time-to-peak equals the **total flight time** t_T:

[symmetrical trajectory]

$$t_T = \frac{-2v_{iy}}{g} = -\frac{2v_i \sin \theta}{-9.81 \text{ m/s}^2} = \frac{2v_i \sin \theta}{9.81 \text{ m/s}^2} \qquad (3.12)$$

The **peak height** attained is

$$s_p = \frac{-v_{iy}^2}{2g} = -\frac{(v_i \sin \theta)^2}{2(-9.81 \text{ m/s}^2)} = \frac{(v_i \sin \theta)^2}{2(9.81 \text{ m/s}^2)} \qquad (3.13)$$

which applies as well to basketballs or water fountains. These are very useful notions, nonetheless they should be rederived as needed rather than memorized.

Example 3.13 **[III]** The person in the gondola shown in Fig. 3.16 throws a ball at 40.0° to the horizon at 10.0 m/s. If the ball is launched at a height of 100 m, where will it land? Ignore air friction.

Solution The vertical part of the motion is free-fall, and that should remind us of the equations of constant acceleration. The independent horizontal motion is at a constant speed. (1) TRANSLATION—An object is launched at a known angle with a known initial speed and initial height; determine the horizontal distance traveled. (2) Take *up* as positive. GIVEN: $v_i = +10.0$ m/s, $\theta = 40.0°$, and the launch height is 100 m. FIND: (a) s_x. (3) PROBLEM TYPE—Projectile motion. (4) PROCEDURE—To use $s_x = v_{ix}t_T$ we need t_T, but this path from hand to ground is not symmetrical around the peak and $t_T \neq 2t_p$, which means that Eq. (3.12) does not apply. Alternatively, get t_T by finding how long the total vertical motion lasts using $s = v_i t + \frac{1}{2}a_T t^2$. With the launch point as the origin, $s_{fy} = -100$ m. (5) CALCULATION:

$$s_y = v_{iy}t_T + \frac{1}{2}gt_T^2$$

where

$$v_{iy} = (10.0 \text{ m/s}) \sin 40.0° = 6.428 \text{ m/s}$$

and so $-100 \text{ m} = (6.428 \text{ m/s})t_T + \frac{1}{2}(-9.81 \text{ m/s}^2)t_T^2$

To solve for t_T, rearrange this equation so that it has the standard form of the quadratic equation {see **MATH REVIEW: PART A-4** on the **CD** 💿 }, namely,

$$(4.905 \text{ m/s}^2)t_T^2 + (-6.428 \text{ m/s})t_T + (-100 \text{ m}) = 0$$

whereupon t_T equals

$$\frac{6.428 \text{ m/s} \pm \sqrt{(-6.428 \text{ m/s})^2 - 4(4.905 \text{ m/s}^2)(-100 \text{ m})}}{2(4.905 \text{ m/s})}$$

The one positive solution is $t_T = 5.22$ s. The ball hits the ground

at a horizontal distance from the launch point of

$$s_x = v_{ix}t_T = (10.0 \text{ m/s})(\cos 40.0°)(5.22 \text{ s}) = \boxed{40.0 \text{ m}}$$

Quick Check: The time it takes to get back down to the launch height is $2t_p = 1.31$ s. At that point, the ball is moving down at 6.428 m/s and, if the answer is right, it will take (5.22 s − 1.31 s) = 3.91 s to descend the 100 m. Hence, $s_y = v_{iy}t + \frac{1}{2}gt^2$ = (6.428 m/s) × (3.91 s) + $\frac{1}{2}$(9.81 m/s²)(3.91 s)² = 100 m and all's well.

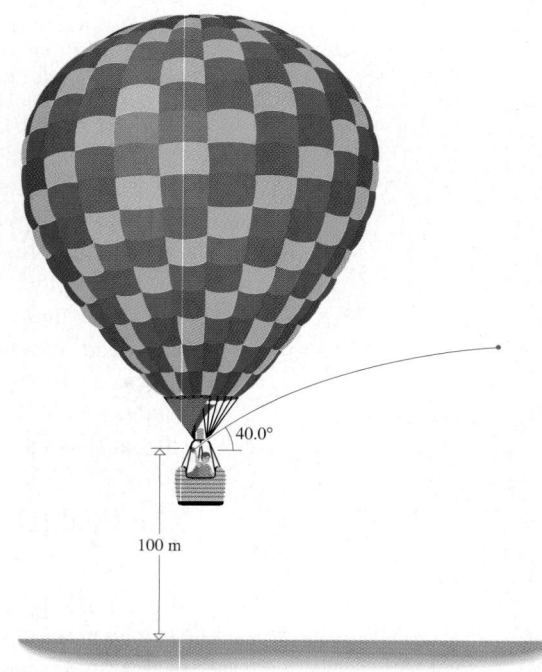

Figure 3.16 A ball is thrown at 40.0° from a gondola 100 m above the ground. Where will it land?

The Range of a Projectile

Range s_R is the total horizontal distance that a projectile travels returning to the same height from which it was launched. The only way this notion makes sense is if the object is fired in some direction upward. Since v_{ix} is constant,

$$s_R = v_{ix}t_T = (v_i \cos \theta)t_T$$

Using $t_T = -2v_{iy}/g$ for the total-flight time, the range becomes

$$s_R = -\frac{2v_i^2}{g} \cos \theta \sin \theta = \frac{2v_i^2}{9.81 \text{ m/s}^2} \cos \theta \sin \theta \qquad (3.14)$$

Neglecting friction and lift, a projectile has **maximum range** when launched at 45°.

This horizontal displacement is nonzero and positive (to the right of the launch point) when $\theta < 90°$, whereupon both $\sin \theta$ and $\cos \theta$ are positive. When $\theta = 90°$, the projectile goes straight up and down, $\cos 90° = 0$, and $s_R = 0$; moreover, when $\theta = 0°$, $s_R = 0$. Neglecting aerodynamic effects, maximum range occurs when ($\cos \theta \times \sin \theta$) has its maximum value. A simple plot of that product reveals that it reaches a maximum when $\theta = 45°$. Altering the launch angle by the same amounts, either increasing or decreasing it from 45°, produces the same range (Fig. 3.17a). *The projected horizontal distance is the same for each member of any pair of firing angles that add up to 90°.* Interestingly, if a projectile is launched in a fixed vertical plane with the same initial speed at varying angles, the envelope of all the possible trajectories is itself a parabola (Fig. 3.17b).

The latest thing in cannons, the Crusader fires a 155-mm diameter projectile weighing between 95 and 107 lbs at about 958 m/s. The barrel is 7.5 m long, and the gun has a maximum range of more than 40 km.

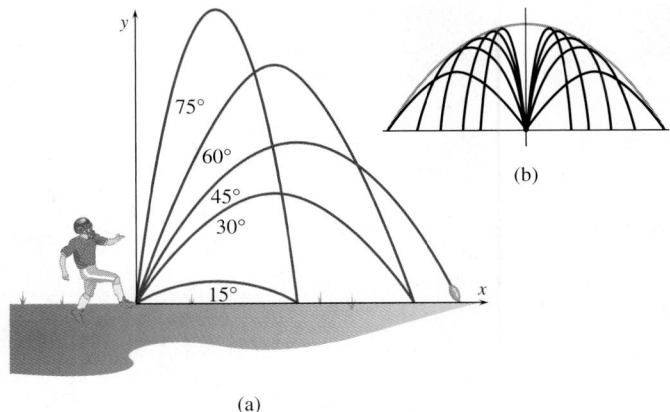

(a)

(b)

Figure 3.17 (a) Ideally, the maximum range of a projectile corresponds to a launch angle of 45°. Notice that if we change the angle by going up or down some amount from 45°, the range for each angle is the same. For example, firing at 45° + 15° or 45° − 15° yields the same range. (b) The envelope of all of the parabolic trajectories is itself parabolic.

Example 3.14 **[I]** A baseball recoiling from a bat soars into the air at an angle of 40.0° above the ground traveling at 45.7 m/s (i.e., 150 ft/s). Assuming it's caught at the same height at which it's hit, calculate the ball's theoretical range ignoring aerodynamic effects.

Solution The mention of "range" should call to mind the equation for s_R, which you can either derive as needed, or memorize. (1) TRANSLATION—An object is launched at a known angle with a known initial speed; determine its range. (2) Take *up* as positive. GIVEN: $\theta = 40.0°$ and $v_i = +45.7$ m/s. FIND: (a) s_R. (3) PROBLEM TYPE— Projectile motion. (4) PROCEDURE—Use Eq.

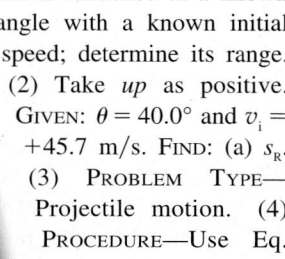

(3.14) for s_R, which can be derived from $s_R = v_{ix}(2t_p)$. (5) CALCULATION:

$$s_R = -\frac{2v_i^2}{g}\cos\theta\sin\theta$$

$$s_R = \frac{2(45.7 \text{ m/s})^2}{9.81 \text{ m/s}^2}\cos 40.0°\sin 40.0°$$

or $\boxed{s_R = 210 \text{ m}}$ (i.e., 688 ft). In reality, a baseball experiences both drag and lift. At the speed given, it's not likely to go much farther than about 500 ft. Moreover, maximum range is actually attained when the ball is hit at 40° rather than 45°.

Quick Check: Use $s_R = v_{ix}t_T$ directly: $t_T = 2t_p = -2v_{iy}/g = -2(45.7 \text{ m/s})\sin 40°/(-9.8 \text{ m/s}^2) = 6.0$ s and $s_R = v_{ix}t_T = [(45.7 \text{ m/s})\cos 40°](6.0 \text{ s}) = 210$ m.

On September 25, 1998 Sammy Sosa executed a powerful upward arcing swing that sent the ball flying at around 35° to 40° for his 66th home run.

Keep in mind that this treatment of ballistics is only an approximation; air friction can have a profound effect on projectiles (see Table 3.3). This is especially true for small-mass high-speed objects (Fig. 3.18). In an environment where there's appreciable drag, maximum range actually occurs at angles somewhat less than 45° (Fig. 3.19). For example, a batted baseball will attain maximum range when hit at around 35°. Conclusion—ignoring friction works better for basketballs than bullets.

{All of the above ideas can be studied via a really nice simulation; click on **PROJECTILE MOTION** under **INTERACTIVE EXPLORATIONS** on the **CD**.}

Table 3.3

Actual Projectile Data*

	Projectile Characteristics			Maximum Range	
Weapon	**Mass** (kg)	**Diameter** (mm)	**Muzzle Speed** (m/s)	**Computed** (km)	**Actual** (km)
Grenade launcher	0.23	40.0	76	0.59	0.40
Mortar	4.2	81.0	240	5.90	3.74
Pistol 0.45 caliber	16.2×10^{-3}	11.4	262	7.00	1.50
Mortar	11.8	106.7	299	9.10	5.61
Cannon/howitzer M198	43.0	155.0	376	14.4	9.87
Cannon/howitzer 8 in. M1	90.7	203.0	594	36.0	16.6
Rifle M-14	10.1×10^{-3}	7.62	853	74.0	3.7
Rifle M-16	3.6×10^{-3}	5.56	991	100.0	2.6

*There's a fair amount of variation in muzzle speed and range depending on such factors as environmental conditions and powder charge.

EXPLORING PHYSICS ON YOUR OWN

Projectiles and Air Friction: When there's an appreciable amount of drag on a lightweight projectile, it will dramatically affect the trajectory. Instead of sailing off along a symmetric parabolic arc, the flight path terminates abruptly and the projectile descends much more steeply. Take a 2- or 3-in. square of tissue paper and lightly ball it up so that it forms a fluffy sphere about 0.5 in. in diameter. Throw it into the air at around 45° at as high a launch speed as possible (drag is proportional to the speed squared). The wad will reach a peak height and then fall almost straight down. Tighten the ball and decrease the speed. What happens?

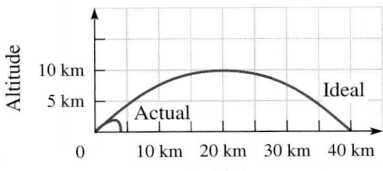

Figure 3.18 Actual and ideal paths of a relatively slow cannonball in free flight through the air, and a high-speed projectile like a rifle bullet with a muzzle speed of 0.6 km/s.

BALLISTICS WITH REAL-WORLD DRAG

The above analysis is fine for cannoneers on the Moon where there is no atmosphere, but on Earth it fails noticeably when we go beyond a gentle game of catch. Air friction has an appreciable effect on projectiles, especially on those that are *light and fast*; drag is then proportional to the *speed squared.* A well-struck baseball, in the air for a rather long time, can lose as much as half its initial speed and travel only a bit more than half as far as it would have had there been no friction. A basketball is usually thrown at a low speed (around 7.6 m/s or 25 ft/s), which keeps drag down, but the ball's light weight and large surface area nonetheless make air resistance a factor; there's a deceleration of ≈ 1 m/s² tangent to the flight path. A player must compensate for that by throwing the ball with a speed about 5% greater than would be necessary if the court were in vacuum.

In contrast, a rifle bullet (with a mass of only about 150 g) fired at a substantial 0.6 km/s will experience a good deal of drag. If unimpeded, it would have a tremendous maximum range of about 40 km according to Eq. (3.14). As a result of air drag the bullet is not likely to travel much beyond 4 km (Fig. 3.18). By comparison, because it fires a massive projectile, the 210-mm (diameter) howitzer (used by Iraq in the 1991 Gulf War) has the greatest range of all field artillery, 57 km (Table 3.3). When a 12-bore shotgun fires No. 6 shot straight up, the pellets reach a height of only about 110 m. Shot at 405 m/s they have a theoretical zero-friction peak height of 8.4 km.

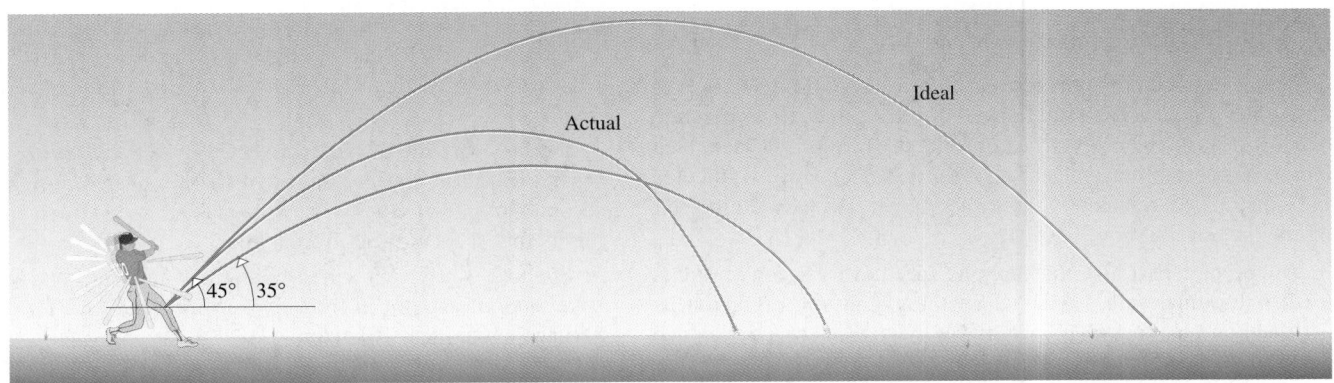

Figure 3.19 The maximum range of a batted ball depends on air drag, and it generally corresponds to a launch angle of around 35°. An 85-mph "fastball" hit by a bat traveling at 80 mph will have a range of about 430 ft.

THE CONCEPT OF ACCELERATION

Average acceleration is defined as

$$\vec{a}_{av} = \frac{\Delta\vec{v}}{\Delta t} = \frac{\vec{v}_f - \vec{v}_i}{t_f - t_i} \qquad [3.1]$$

The scalar form of this provides us with an expression for the tangential acceleration:

[straight-line motion] $$a_{av} = \frac{\Delta v}{\Delta t} = \frac{v_f - v_i}{t_f - t_i} \qquad [3.2]$$

Reread Section 3.1 (Average Acceleration) and study Example 3.1. As $\Delta t \to 0$, this ratio approaches the instantaneous acceleration

$$\vec{a} = \lim_{\Delta t \to 0}\left[\frac{\Delta\vec{v}}{\Delta t}\right] \qquad [3.3]$$

Work to understand the basic idea of Section 3.2 (Instantaneous Acceleration) without getting overly concerned with the details of Eq. (3.3). **Look at the CD WARM-UPS, then study the WALK-THROUGH EXAMPLES.**

UNIFORMLY ACCELERATED MOTION

Sections 3.3 (Constant Acceleration) and 3.4 (The Mean Speed) are preliminary to Section 3.5 (The Equations of Constant Acceleration) where it all comes together. Study Examples 3.1– 3.5. Restricting the treatment to *straight-line motion at constant acceleration*, there are five useful equations:

$$v_f = v_i + a_T t \qquad [3.6]$$

$$v_{av} = \tfrac{1}{2}(v_i + v_f) \qquad [3.7]$$

$$s = \tfrac{1}{2}(v_i + v_f)t \qquad [3.8]$$

$$s = v_i t + \tfrac{1}{2}a_T t^2 \qquad [3.9]$$

and

$$v_f^2 = v_i^2 + 2a_T s \qquad [3.10]$$

These equations can all be derived as needed, but it's more practical to just memorize them. Study Examples 3.6 and 3.7.

FREE FALL

Objects in free-fall at the surface of the Earth (for which air friction is inconsequential) descend with a nearly *uniform acceleration* (g) equal to 9.81 m/s^2 or 32.2 ft/s^2. For vertical motion study Section 3.8 (Straight Up & Down) and Examples 3.8–3.11.

The above set of equations can be applied to projectiles (p. 68) where the constant-speed horizontal motion and the constant-acceleration vertical motion occur independently. Such quantities as the peak altitude (s_p), the time to peak altitude (t_p), the total time of flight (t_T), and the range (s_R) are important ideas that can easily be calculated when friction is negligible. Reread Section 3.9 (Two-Dimensional Motion: Projectiles) and study Examples 3.12–3.14.

Key Terms

average acceleration	terminal speed
tangential acceleration	gravitational acceleration
instantaneous acceleration	peak altitude
mean speed	projectile
Mean Speed Theorem	parabolic trajectory
constant-$\vec{a}$ equations	ballistic flight
free-fall	total time of flight
air drag	range

1. Does a car's speedometer measure a vector or a scalar? How about the odometer? What does it measure? Is it possible for a car to accelerate and yet have its speedometer read a constant value? Can two airplanes have the same acceleration and different speeds? the same speeds and different accelerations? Can either plane at a given instant have a large acceleration and a small velocity? A small acceleration and a large velocity? Explain.

2. Galileo used a "thought experiment" from the Middle Ages to justify the conclusion that all things fall at the same rate in vacuum. See if you can reconstruct it from the following: consider three identical objects (three balls of clay, for instance). How would they fall separately in vacuum? Stick two together. How would the pair of clay masses fall now? Explain.

3. An airplane in a dive increases its speed every second by one-tenth its original value. What, if anything, can you say about its acceleration? Is it possible for a rocket to have an acceleration due east and a velocity due west? Explain.

4. Can the average acceleration of a body over a finite time ever equal the instantaneous acceleration for more than an instant during the journey? Can it ever not equal the instantaneous acceleration for at least an instant? Explain.

5. Two identical cars are in a drag race. The first starts and continues to uniformly accelerate up to the finish line some time later. The second car follows the first out of the starting box 1 s afterward traveling exactly as the first did. Will the cars be the same distance apart throughout the race? If not, when will they be closest to each other? How much time will there be between their successive crossings of the finish line?

6. A rocket can accelerate at an increasing rate as its fuel is consumed and it gets lighter and lighter. If the rocket starts from rest, what can we say about the *average speed* as compared with the *final speed*?

7. Explain why the distance traversed by a body during its tenth second of fall is greater than the distance covered during its first second—if, in fact, it is. What's the average speed of an apple during its first second of fall from a tree?

8. Suppose you are in a free-falling elevator and you hold your keys motionless right in front of your face and then let go. What will happen to them? Explain.

9. Consider a free-falling sky diver descending straight down. Make a graph of v versus t showing (a) the straight-line terminal speed v_t; (b) the zero-drag free-fall with $v_i = 0$; and (c) your approximation of the actual speed-time descent curve.

10. To get a feeling for how extremes of a variable affect an equation, suppose that the acceleration of a body is given by $a = [(At^2 - Bt)/(t + C)D] + Et^2/(t - C)D$ where A, B, C, D, and E are con-

stants. (a) Determine the value of a when $t = 0$. (b) Find the value of a when $t \gg C$. (c) Find the value of a when $t \ll C$.

11. Imagine an object thrown straight upward into the air. Ignoring friction, draw a rough plot of the vertical velocity versus time. Does the velocity ever become negative, and if so, when? What can you say about the object's acceleration?

12. A portion of the flight of a toy rocket that travels straight up from the surface of the Earth is depicted in Fig. Q12. Describe the motion as completely as you can.

13. Suppose a hunter fires a blast of No. 6 shot straight up into the air. Given that the pellets have a terminal speed of about 9 mys, describe the flight of a typical shot and comment on the danger to the hunter. The terminal speed of a .22-caliber bullet (nose first) is about 67 mys. What does that suggest about shooting into the air anywhere near people? (See the footnote on p. 71 and comment on the physics.)

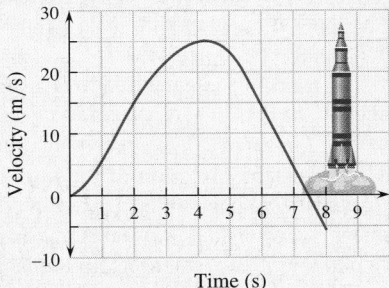

Figure Q12

14. An object is thrown straight up at a speed v_i. At that instant an identical object, directly above the first, is dropped. Discuss their relative motion. What is their relative speed?

15. In 1917-1918 the U.S. War Department conducted a series of experiments on falling bombs. The data in Table 3.4 corresponds to a bomb that dropped 5539 ft from an airplane. (a) Make a plot of the distance fallen versus the horizontal distance traveled (b) Make a table of the horizontal distances traveled in successive 1.0 s intervals. For how long was the horizontal speed of the bomb more or less constant? (c) Make a table of the vertical distance fallen in each successive second. Given that $g = 32$ ft/s^2 the average vertical speed during each successive second should have increased by 32 ft/s, did it? (d) For how many seconds was the downward acceleration constant?

16. Explain the meaning of each of the Key Terms on page 76.

Table 3.4 The Fall of a Bomb Dropped From an Airplane

Time (s)	Horizontal travel (ft)	Distance fallen (ft)
0.000	0	0
1.000	98	16
2.000	196	64
3.000	294	144
4.000	392	256
5.000	490	400
6.000	587	576
7.000	684	782
8.000	780	1019
9.000	875	1287
10.000	970	1585
11.000	1065	1912
12.000	1158	2268
13.000	1251	2652
14.000	1343	3064
15.000	1433	3502
16.000	1523	3965
17.000	1611	4453
18.000	1698	4965
19.000	1783	5499
19.075	1790	5539

Multiple Choice Questions

The first five questions pertain to Fig. MC1 and Table 3.5. The data is for a Space Shuttle launch and was taken from a video of the liftoff. The origin was at the corner of the picture frame, and a painted spot on the side of the vehicle served as a reference point.

1. During the first 6.4 s the vehicle (a) rose precisely vertically (b) slowly drifted horizontally as it rose (c) only accelerated appreciably horizontally (d) ascended with a constant vertical velocity (e) none of these.

2. At $t = 0$ the reference point locating the vehicle was at a distance from the origin of (a) 102.4 m (b) 0 m (c) 73.2 m (d) 71.536 m (e) none of these.

Table 3.5*

t (s)	x (m)	y (m)	v_x (m/s)	v_y (m/s)
0.000	71.536	15.649		
0.400	71.536	16.109	0.549	3.452
0.800	71.975	18.410	0.549	5.753
1.200	71.975	20.711	0.000	5.753
1.600	71.975	23.013	0.000	6.904
2.000	71.975	26.234	1.097	8.630
2.400	72.853	29.916	1.646	10.356
2.800	73.292	34.519	1.097	11.506
3.200	73.730	39.121	1.097	12.082
3.600	74.169	44.184	1.646	13.808
4.000	75.047	50.167	1.646	16.109
4.400	75.486	57.071	2.194	16.684
4.800	76.803	63.515	1.646	17.259
5.200	76.803	70.879	1.646	19.561
5.600	78.119	79.163	3.292	21.862
6.000	79.436	88.638	3.292	21.862
6.400	80.752	96.653	2.743	23.588
6.800	81.630	107.238		

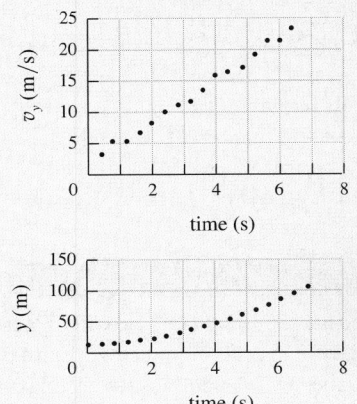

Figure MC1

*The craft had already begun to ascend when the clock was started at $t = 0$.

3. From the plot of y-versus-t we can conclude that the vertical component of the Shuttle's velocity   (a) is clearly constant throughout the first 6.4 s of liftoff   (b) is increasing at a rate that seems to be parabolic   (c) is decreasing at a rate that seems to be parabolic   (d) is increasing at a rate that might even be constant   (e) none of these.

4. Using the plot of v_y-versus-t and approximating a_y to be constant, the vertical acceleration of the vehicle turns out to be about   (a) -9.81 m/s^2 (b) zero   (c) 3.8 m/s^2 (d) 13.6 m/s^2 (e) none of these.

5. From the table of data it follows that the instantaneous velocity at $t = 6.40$ s is   (a) 23.7 m/s–UP 83.4° IN XY-PLANE   (b) 23.588 m/s–UP 90° IN XY-PLANE   (c) 2.743 m/s–UP 6.6° IN XY-PLANE   (d) 26.33 m/s–UP 90° IN XY-PLANE   (e) none of these.

6. The expression $s = vt$ is applicable when the   (a) speed is constant   (b) acceleration is constant   (c) distance is constant   (d) acceleration is linear   (e) none of these.

7. A body moving with an acceleration having a constant magnitude must experience a change in   (a) velocity   (b) speed   (c) acceleration   (d) weight   (e) none of these.

8. Which of the following pairs of concepts cannot both simultaneously be constant and nonzero for a given body?   (a) the speed and velocity   (b) the distance and displacement   (c) the magnitude of the acceleration and the acceleration   (d) the velocity and acceleration   (e) none of these.

9. A mouse runs along a straight narrow tunnel. If its velocity-time curve is a straight line parallel to the time axis, then the acceleration is   (a) a nonzero constant   (b) zero   (c) varying linearly   (d) quadratic   (e) none of these.

10. Figure MC10 shows the velocity versus time curve for a vehicle traveling along a straight track. What is the vehicle's maximum distance from its starting point, and when is it there?   (a) 20 m, at $t = 0$ (b) 30 m, at $t = 9.0$ s   (c) 30 m, at $t = 5.0$ s   (d) 60 m, at $t = 4.0$ s   (e) none of these.

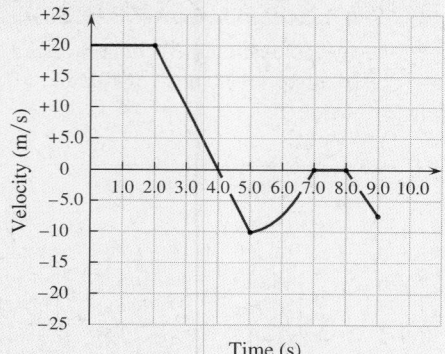

Figure MC10

11. Figure MC10 shows the velocity versus time curve for a vehicle traveling along a straight track. When did it have a positive acceleration? Between   (a) 0 s and 2.0 s   (b) 2.0 s and 4.0 s   (c) 2.0 s and 5.0 s   (d) 5.0 s and 7.0 s   (e) none of these.

12. Figure MC10 shows the velocity versus time curve for a vehicle traveling along a straight track. When was it at rest? Only between   (a) 0 s and 2.0 s   (b) 8.0 s and 9.0 s   (c) 7.0 s and 8.0 s   (d) 5.0 s and 7.0 s   (e) none of these.

13. Figure MC10 shows the velocity versus time curve for a vehicle traveling along a straight track. When was the acceleration zero? Only   (a) between 0 s and 2.0 s   (b) between 8.0 s and 9.0 s   (c) between 7.0 s and 8.0 s   (d) at 4.0 s   (e) none of these.

14. The speed of a body traveling in a straight line with a constant positive acceleration increases linearly with   (a) distance   (b) time   (c) displacement   (d) distance squared   (e) none of these.

15. If [L] represents the dimension of length and [T] that of time, then the dimensions of acceleration are   (a) $[L + T^2]$ (b) $[L/T]$ (c) $[L^2/T]$ (d) $[L/T^2]$ (e) none of these.

16. Taking the direction of the initial motion to be positive, the slope of the velocity-time curve of a body moving in a straight line is negative when the   (a) acceleration is in the opposite direction to the initial motion   (b) acceleration increases the speed   (c) acceleration decreases in time   (d) acceleration is constant and in the same direction as the initial motion   (e) none of these.

17. If the displacement of a body is a quadratic function of time, the body is moving with   (a) a uniform acceleration   (b) a nonconstant acceleration   (c) a uniform speed   (d) a uniform velocity   (e) none of these.

18. A driver traveling at 20 km/h abruptly stops the car in 3 m. Later, while moving at 40 km/h, the driver again stops the car with the same deceleration, bringing it to a halt after   (a) 6 m   (b) 9 m   (c) 60 m   (d) 12 m   (e) none of these.

19. Two statues of Kermit the Frog, one made of aluminum and one of brass, are the same size, although the former is 3.2 times lighter than the latter. Both are dropped at the same moment from the same height of 2 m. They hit the ground   (a) at nearly the same time with very different speeds   (b) at very different times with nearly the same speeds   (c) at nearly the same times with nearly the same speeds   (d) at very different times with very different speeds   (e) none of these.

20. A ballast bag of sand dropped from a hot-air balloon hits the ground at a certain speed and the craft slowly rises and soon comes to a stop. If a second identical bag is then dropped and it hits the ground twice as fast as the first, how high was the balloon when it was dropped in comparison to when the first bag fell?   (a) 1/2 as high   (b) 2 times as high   (c) 4 times as high   (d) 8 times as high   (e) none of these.

21. The average speed of a coconut during a 2-s fall from a tree, starting at rest, is   (a) 19.6 m/s   (b) 9.8 m/s^2 (c) 39.2 m/s   (d) 9.8 m/s   (e) none of these.

22. An alien space traveler exploring Earth records that his phasor pistol, dropped from a high cliff, fell a distance of 1 glong in a time of 1 tock. How far will it fall in two tocks?   (a) 1.5 glongs   (b) 2 glongs   (c) 3 glongs   (d) 4 glongs   (e) none of these.

For more Multiple Choice Questions with answers click on WARM-UPS in CHAPTER 3 on the CD.

Suggestions on Problem Solving

1. Watch units. When making conversions, anticipate whether the answer will be larger or smaller than the original number *before* doing the calculation. Never mix different sets of units (e.g., cm, m,

km) in the same problem, let alone the same equation. Be careful not to take 32 as the numerical value of g in a problem using SI units.
2. Be consistent with the signs of the vector quantities ($\vec{s}$, $\vec{v}$, and $\vec{a}$),

particularly in free-fall and projectile problems. Take the initial direction of motion as positive and stick with it throughout the problem. This approach is not always necessary, but in the beginning it is certainly advisable. ***Draw diagrams***.

3. Consider Eq. (3.9), which varies with t and t^2 both. When t is given, it's easy to find s, v_i, or a_T. When t is the unknown to be determined, the most straightforward approach is to use the quadratic equation. Rearranging Eq. (3.9) yields $\frac{1}{2}a_T t^2 + v_i t - s = 0$, which has the general form $At^2 + Bt + C = 0$, where $A = \frac{1}{2}a$, $B = v_i$, and $C = -s$. The solution is then $t = (-B \pm \sqrt{B^2 - 4AC})/2A$ {see **MATH REVIEW: PART A-4** on the CD 💿 }. The quadratic equation can be avoided by using combinations of other equations: Eq. (3.10) to find v_f, and then Eq. (3.6) or (3.8) to get t. The wise choice of equations will save time and effort.

4. Think about your answers. Often an error will produce a result that is unrealistically large or small. If you think the problem through so that you can anticipate the size, direction, sign, and so on, most computational errors will be spotted. We all make computational errors; not all of us find them. A reindeer falling off the roof of a one-story building is not going to land on the ground at 5000 km/h. Given the real-world physical situation of a problem, ask yourself if your answer is reasonable.

5. ***Always check your answers***. The best way to do that (although it may not always be practical) is to recalculate the answer using a different approach. At least go over your calculations several times.

6. When a problem requests the distance traveled during a certain time interval, such as the fifth second of motion, this is not the same as asking how far the body moved in 5 s. The fifth second is the one that begins at $t = 4$ s and ends at $t = 5$ s.

7. It is not necessary to remember the whole set of equations for projectile motion: t_p is easily derived from the definition of a, where $v_y = 0$; t_T is just $2t_p$ and s_R is simply $v_x t_T$. The fact that $v_y = 0$ (and $v_x \neq 0$) at the peak altitude is important and should always be kept in mind.

8. The Mean Speed Theorem applies to straight-line motion in one direction and so is appropriate for dealing with the uniformly accelerated vertical component of a projectile's movement. *Do not apply it to the overall motion* (as for example, in an attempt to determine the total final speed v_f knowing s and t)—the projectile travels along an arc and $l \neq s$.

Problems ✦ Coordinated Problems ✦ Progressive Problems ✦ Solutions

STUDY GUIDE   **1. Coordinated Problems:** The three problems within each magenta-colored grouping are solvable in similar ways. Note that the first of these always has a hint; moreover, its solution is provided in the back of the book. *Work out each of these sets; they'll strengthen technique and build confidence.* **2. Progressive Problems:** The problems introduced in blue unfold step-by-step carrying along the analysis in a more suggestive way than is customary. *Work out all of these; they'll guide you through the analytic process and help develop problem-solving skills.* **3. Worked-Out Solutions:** Studying worked-out solutions is an important part of learning how to solve problems. Accordingly, additional *solutions* to a number of model problems are given below. *Make sure you understand each of them before you go on to the next problem.* **4.** Also provided in the back of the book are the *Answers* to all odd-numbered problems, as well as worked-out *solutions* to those with boldface numbers. Problem numbers in italic indicate that a solution appears in the Student Solutions Manual.

SECTION 3.1: AVERAGE ACCELERATION

1. [I] THIS PROBLEM WILL HELP US BETTER UNDERSTAND THE RELATIONSHIP BETWEEN SPEED AND AVERAGE ACCELERATION. Figure P1 depicts a speed versus time curve for a toy airplane. (a) What is the significance of the slope of the curve between any two points? (b) When, if ever, was the plane at rest? (c) What's the plane's average acceleration over the interval from 12 s to 14 s?

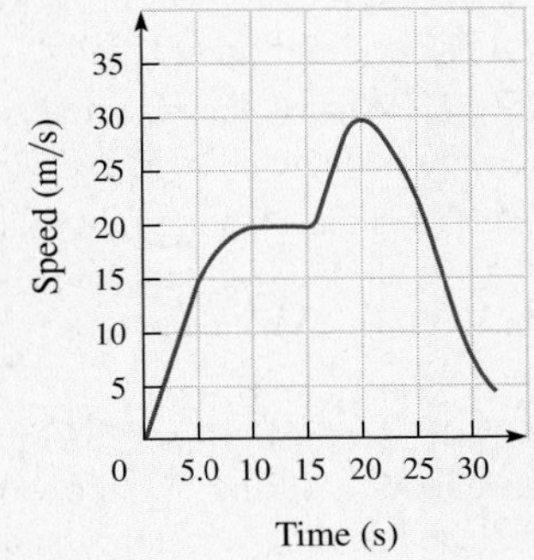

Figure P1

2. [I] THIS PROBLEM WILL ASSIST US IN UNDERSTANDING THE RELATIONSHIP BETWEEN SPEED AND AVERAGE ACCELERATION. Figure P1 depicts a speed versus time curve for a toy airplane. (a) At what time did the plane have its maximum speed? (b) Over what interval, roughly, did it have its maximum positive average acceleration? (c) When did it start to decelerate?

3. [I] A rocket lifts off its launchpad and travels straight up attaining a speed of 100 m/s in 10 s. Calculate its average acceleration.

4. [I] A canvasback duck heading south at 50 km/h at 2:02 A.M. is spotted at 2:06 A.M. still traveling south but at 40 km/h. Calculate its average acceleration over that interval—magnitude and direction.

5. [I] An android on guard duty in front of the Institute of Robotics is heading due south at 1:07 P.M. at a speed of 10 m/s when it receives a command to alter course. At 1:09 P.M. it is recorded to be moving at 10 m/s due north. Compute its average acceleration over that interval—magnitude and direction.

6. [I] The 1997 Corvette Sport Coupe (16-valve V8, 5.7-liter engine) goes from 0 to 60 mph (i.e., 26.8 m/s) in 4.8 s. What's its average acceleration in SI units?

7. [I] A finalist in the Soap Box Derby starts with a push down a long straight run at an initial speed of 1.0 m/s. At the bottom, 1.0 min 2.0 s later, it reaches a speed of 15.0 m/s. Find its average acceleration.

8. [I] Pushing backward, a sprinter leaves the blocks at 3.0 m/s. If 1.0 s later he is moving at 5.2 m/s, what was his average acceleration during that 1.0-s interval?

9. [I] A piston-engine dragster set a world record by starting from rest and hitting a top speed of 244 mi/h in 6.2 s over a straight measured track of 440 yd. Compute the scalar value of its average acceleration in m/s². [*Hint: 1 mi/h = 0.447 0 m/s; the average scalar acceleration is the change in the speed divided by the time, regardless of the distance traveled.*]

10. [I] During a typical launch a Space Shuttle goes from a vertical speed of 5.75 m/s at $t = 1.20$ s to a vertical speed of 6.90 m/s at $t = 1.60$ s, while rising 2.30 m. Determine the average acceleration—magnitude and direction—over that interval.

11. [I] A bag of sand drops from a hot-air balloon; 12.0 s later having fallen 700 m it's traveling straight down at 116 m/s. Determine the average acceleration vector for the bag during that descent.

12. [II] THIS PROBLEM WILL HELP US BETTER UNDERSTAND ACCELERATION. Operating on an automatic program, the belt on a treadmill moving at 2.5 m/s increases its speed to 3.7 m/s in 2.4 min. (a) What was the change in its speed. (b) Over what time interval did

that change occur? (c) Determine the belt's average acceleration.

13. [II] THIS PROBLEM WILL HELP US BETTER UNDERSTAND ACCELER-ATION. A kid coasting along at 12.0 m/s, holds down the brake on a bicycle. With a resulting average tangential acceleration of −0.60 m/s², she soon comes to rest. (a) During the braking, what was her initial speed? (b) During the braking, what was her final speed? (c) How long did it take her to come to a stop? (d) Does it matter that she's traveling a curved path?

14. [II] A VW Rabbit can go from rest to 80.5 km/h (50.0 mi/h) in a modest 8.20 s. How long will it take to speed up from 48.3 km/h to 64.4 km/h, along a straight run, if the average acceleration is the same as before?

15. [II] During a baseball game a runner traveling at 4.0 m/s slides into second base. Given that it takes her 0.50 s to come to rest, what was her average acceleration? Approximately how fast was she moving 0.30 s into the slide?

16. [II] Having been kicked, a soccer ball rolls in a straight line past a kid holding a stopwatch. At the moment the ball passes the kid it has an instantaneous speed of 4.0 m/s and the watch reads 10.0 s. If the watch reads 23.3 s when the ball comes to rest, what's its average acceleration?

> **SOLUTION:** The initial speed is 4.0 m/s at $t = 10.0$ s. At $t = 23.3$ s, $v = 0$. By definition $a_{av} = \Delta v/\Delta t = (0 - 4.0$ m/s$)/(23.3$ s $- 10.0$ s$) = (-4.0$ m/s$)/(13.3$ s$) = -0.30$ m/s².

17. [II] Videos taken of a typical male sprinter show that he goes from 0 up to 3.0 m/s in the first step, reaches 4.2 m/s in the next step, and 5.0 m/s in the third step. Given that each step takes essentially the same amount of time, what can be said about his acceleration?

18. [II] A motorboat starting from a dead stop accelerates at an average rate of 2.0 m/s² for 3.0 s, then very rapidly roars up to 4.0 m/s² and holds it constant for 4.0 s. What is its approximate average acceleration over the first 5.0 s of motion?

19. [II] Superman slams head-on into a locomotive speeding along at 60 km/h, bringing it smoothly to rest in an amazing 1/1000 s and saving Lois Lane, who was tied to the tracks. Calculate the average deceleration of the train in m/s².

20. [III] Two motorcycle stuntpersons are driving directly toward one another, each having started at rest and each accelerating at an average rate of 5.5 m/s². At what speed will they be approaching each other 2.0 s into this lunacy?

SECTION 3.2: INSTANTANEOUS ACCELERATION

21. [I] THIS PROBLEM DEALS WITH THE RELATIONSHIP BETWEEN SPEED AND INSTANTANEOUS ACCELERATION. Figure P1 depicts a speed versus time curve for a toy airplane. (a) What is the significance of the slope of the curve at any point? (b) Did the plane's instantaneous acceleration change between 0 and 3 s? (c) What's its instantaneous acceleration at 12 s? (d) What's its instantaneous acceleration at 20 s?

22. [I] THIS PROBLEM EXAMINES THE RELATIONSHIP BETWEEN SPEED AND INSTANTANEOUS ACCELERATION. Figure P1 depicts a speed versus time curve for a toy airplane. (a) What's its average acceleration over the interval from 0 s to 5.0 s? (b) What's its instantaneous acceleration at 2.5 s? (c) Compare these two answers and explain why they are as they are.

23. [I] What is the instantaneous acceleration of the object whose motion is depicted in Fig. 3.2a (p. 55) at a time of 1.58 s?

24. [I] Figure P24 is a velocity-time graph for a test car on a straight track. The test car initially moved backward in the negative x-direction at 20 m/s. It slowed, came to a stop, and then moved off in the positive x-direction at $t = 2.0$ s. What was its average acceleration during each of the time

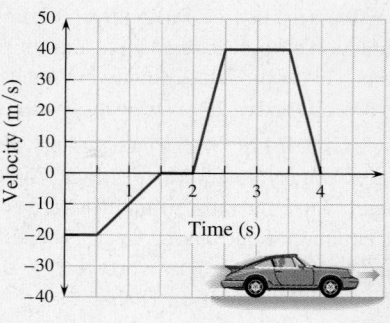

Figure P24

intervals 0 to 0.5 s, 1.5 to 2.0 s, and 2.0 to 2.5 s? What was its instantaneous acceleration at $t = 2.25$ s?

25. [I] In Fig. P24, what was the car's instantaneous acceleration at $t = 3.0$ s? Is the instantaneous acceleration positive or negative at $t = 3.7$ s? How about at $t = 1.1$ s? What is the instantaneous acceleration of the car at $t = 0.25$ s?

26. [I] Figure P26 shows the speed-time curves of three cyclists traveling a straight course. What are their respective instantaneous speeds at $t = 2$ s? Which if any starts out at $t = 2$ s with the greatest instantaneous acceleration?

27. [I] Figure P26 shows the speed-time curves of three cyclists traveling a straight course. Which one has the greatest instantaneous acceleration at each of the following times, $t = 2.1$ s, 3.3 s, and 6.5 s?

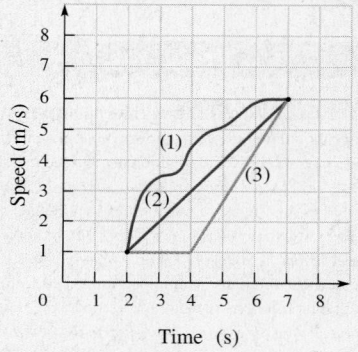

Figure P26

28. [I] Figure P26 still shows the speed-time curves of three cyclists traveling a straight course. What are their respective instantaneous speeds at $t = 7$ s? Which, if any, has the greatest instantaneous acceleration at $t = 7$ s?

29. [I] What is the instantaneous acceleration of the object whose motion is depicted in Fig. 3.2b (p. 55) at $t = 4$ s?

30. [I] A car on a straight road goes from rest to 10 km/h in 10 s; at the end of 20 s it's moving at 20 km/h; at the end of 30 s it reaches 30 km/h. What can you say about its acceleration? Make a graph of a versus t.

31. [II] Referring back to Fig. P24, what was the car's average acceleration during the interval from $t = 0$ to $t = 3.0$ s? What was its instantaneous acceleration at $t = 0$ and $t = 3.0$ s?

32. [II] Referring back to Fig. P26, which shows the speed-time curves for three cyclists traveling a straight course, describe their motions and compute their average accelerations over the entire interval shown. Which of the cyclists has the greater average acceleration over the interval from 2.0 s to 4.6 s? Which one has the greatest instantaneous acceleration at $t = 4.6$ s?

33. [II] Using Fig. 3.3 (p. 56), graphically determine the car's approximate acceleration at 3.8 s into the run.

34. [II] At time $t = 0$, a body located at $s = 0$ has a scalar velocity of $v = 0$ and an instantaneous acceleration of 1 m/s². It travels in a straight line maintaining that acceleration for 4 s and then

immediately ceases accelerating. That condition lasts for 1 s at which point it decelerates at 2 m/s^2 for 2 s and then again ceases accelerating. Draw graphs of a-versus-t and v-versus-t.

35. [II] Suppose that in Problem 18, after holding at 4.0 m/s^2 for 4.0 s, the boat next decelerates to a stop 20 s later at a rate of 1.1 m/s^2. (a) What is its average acceleration during the first 27 s of the run? (b) What is its instantaneous acceleration 10 s into the trip?

36. [II] A rocket accelerates straight up at 10 m/s^2 toward a helicopter that is descending uniformly at 5.0 m/s^2. What is the relative acceleration of the missile with respect to the aircraft as seen by the pilot who is about to bail out?

37. [III] The speed of a flying saucer in ascent mode-4 as reported by an alleged eyewitness (who got it straight from the talkative alien navigator) is given by the expression $v = At + Bt^2$. Amazingly, A and B are constants in SI units, but the little green navigator would not reveal their values. Determine the instantaneous acceleration of the craft. [*Hint: At $(t + \Delta t)$ the speed is $(v + \Delta v)$.*]

SECTION 3.3: CONSTANT ACCELERATION

SECTION 3.4: THE MEAN SPEED

38. [I] An R75 maintenance robot on a spaceship is standing in front of the bathroom when it begins to move down the straight passageway. It accelerates at a constant 2.0 m/s^2. Find its speed at the end of 5.0 s.

39. [I] With the previous problem in mind, how far did the robot travel in the 5.0 s?

40. [I] During a swing the head of a golf club is in contact with the ball for about 0.5 ms and the ball goes from zero to 70 m/s. Assuming the acceleration is constant, determine its value.

41. [I] If a truck traveling 40.0 km/h uniformly accelerates up to 60.0 km/h, what's its average speed in the process?

42. [I] Slamming on the brakes, a driver decelerates her car from 25.0 m/s to 15.0 m/s in 3.5 s. Find her average speed assuming the acceleration was uniform.

43. [I] If during a race, a horse accelerates fairly constantly up to 16 m/s, what's its average speed?

44. [I] How far does the car in Problem 42 travel while dropping in speed from 25.0 m/s to 15.0 m/s in 3.5 s?

45. [II] According to the *New York Times*, a 1997 Corvette Sport Coupe, starting from rest, can travel 1/4 mile (i.e., 402 m) in 13.3 s. What's its average speed in SI units? Assuming the acceleration is constant (which it isn't), what would be its maximum speed?

46. [II] THIS PROBLEM WILL HELP US BETTER UNDERSTAND THE RELATIONSHIP BETWEEN SPEED AND ACCELERATION. An elevator accelerates upward from rest at 0.98 m/s^2. (a) What was its initial speed? (b) How fast is it moving after 3.0 s? (c) How high has it risen in those 3.0 s?

47. [II] THIS PROBLEM DEALS WITH THE RELATIONSHIP BETWEEN SPEED AND ACCELERATION. A robot accelerates in a straight line at a constant rate from 1.20 m/s to 6.20 m/s while traveling 30.0 m. (a) What was its average speed? (b) How long did that portion of its journey take? (c) Determine its acceleration.

48. [II] The Corvette in Problem 45 can come to a stop from 60 mph (i.e., 26.8 m/s) in 116 ft (i.e., 35.4 m). Determine the deceleration (in SI units) assuming it to be constant.

49. [II] A good male sprinter can run 100 m in 10 s. What's his average speed? He will typically reach a peak speed of 11 m/s at about 5 s and slow down toward the finish. Assuming his acceleration is fairly constant for the first 5 s, how fast will he be going 3 s into the race?

50. [II] If a van moving at 50.0 km/h uniformly accelerates up to 70.0 km/h in 20.0 s, how far along the straight road will it travel in the process?

> **SOLUTION:** The acceleration is constant, and we know the initial and final speeds so we can compute the average speed and with that the distance. Convert to m/s using 1 km/h = 0.2778 m/s; 50.0 km/h = 13.89 m/s and 70.0 km/h = 19.45 m/s. $v_{av} = \frac{1}{2}(19.45 \text{ m/s} + 13.89 \text{ m/s}) = 16.67$ m/s therefore $l = v_{av}t = 333$ m.

51. [II] Supposing that the acceleration of the 1997 Corvette Sport Coupe in Problem 6 is constant (which it really isn't) how much road will it travel in going from 0 to 60 mph (i.e., 26.8 m/s) in 4.8 s?

SECTION 3.5: THE EQUATIONS OF CONSTANT ACCELERATION

52. [I] THIS PROBLEM WILL HELP US BETTER UNDERSTAND THE RELATIONSHIP BETWEEN DISTANCE TRAVELED AND ACCELERATION. During takeoff a small plane has an average tangential acceleration of 5.0 m/s^2 and travels for 20 s before becoming airborne. (a) What is the initial speed of the plane? (b) How long must the runway be?

53. [I] THIS PROBLEM EXPLORES THE RELATIONSHIP BETWEEN DISTANCE TRAVELED AND ACCELERATION. A driver stops his test car at a rate of 6.0 m/s^2 in a distance of 410 m. (a) What was the car's final speed? (b) What was the car's acceleration? (c) How fast was the car going when he applied the brakes?

54. [I] A locust, extending its hind legs over a distance of 4 cm, leaves the ground at a speed of 340 cm/s. Determine its acceleration, presuming it to be constant.

55. [I] A wayward robot, R2D3, is moving along at 1.5 m/s when it suddenly shifts gears and roars off, accelerating uniformly at 1.0 m/s^2 straight toward a wall 10 m away. At what speed will it crash into the wall? [*Hint: We have the initial speed, the constant acceleration, and the distance traveled. To find the final speed use a constant-$\bar{a}$ equation that relates these quantities.*]

56. [I] The driver of a car traveling at 10.0 m/s along a straight road depresses the accelerator and uniformly increases her speed at a rate of 2.50 m/s^2. How fast will the car be moving as it passes a parked police cruiser 100 m away?

57. [I] The length of a straight tunnel through a mountain is 25.0 m. A cyclist heads directly toward it, accelerating at a constant rate of 0.20 m/s^2. If at the instant he enters the tunnel he is traveling at a speed of 5.00 m/s, how fast will he be moving as he emerges?

58. [I] Regarding Problem 19, Lois was a mere 10 cm down the tracks from the point where Superman struck the train. How far away from her did the "Man of Steel" finally stop the engine? (Thank goodness he arrived in time!)

59. [I] A Jaguar in an auto accident in England in 1960 left the longest recorded skid marks on a public road: an incredible 290-m long. As we will see later, the friction force between the tires and the pavement varies with speed, producing a deceleration that increases as the speed decreases. Assuming an average acceleration of −3.9 m/s^2 (that is, −0.4 g), calculate the Jag's speed when the brakes locked.

60. [I] A modern supertanker is gigantic: 1200- to 1300-ft long with a 200-ft beam. Fully loaded, it chugs along at about 16 knots (i.e., 30 km/h or 18 mi/h). It can take 20 min to bring such a monster to a full stop. Calculate the corresponding deceleration in m/s² and determine the stopping distance.

61. [I] A bullet traveling at 300 m/s slams into a block of moist clay, coming to rest with a fairly uniform acceleration after penetrating 5 cm. Calculate its acceleration.

62. [I] Consider an automobile collision in which a vehicle traveling at 30.0 m/s is brought to rest in a distance of 50.0 cm. What is the deceleration, assuming it to be constant?

63. [II] A driver traveling at 60 km/h sees a chicken dash out onto the road and slams on the brakes. Accelerating at −7 m/s², the car stops just in time 23.3 m down the road. What was the driver's reaction time (i.e., the time that elapsed before he engaged the brake)?

64. [II] The longest passenger liner ever built was the *France*, at 66 348 tons and 315.5-m long. Suppose its bow passes the edge of a pier at a speed of 2.50 m/s while the ship is accelerating uniformly at 0.01 m/s². At what speed will the stern of the vessel pass the pier?

65. [II] A swimmer stroking along at a fast 2.2 m/s ceases all body movement and uniformly coasts to a dead stop in 10 m. Determine how far she moved during her third second of unpowered drift.

66. [II] A little electric car having a maximum speed of 40.0 km/h can speed up uniformly at any rate from 1.00 m/s² to 4.00 m/s² and slow down uniformly at any rate from 0.00 to −6.00 m/s². What is the shortest time in which such a vehicle can traverse a distance of 1.00 km starting and ending at rest?

67. [II] Two trains heading straight for each other on the same track are 250 m apart when their engineers see each other and hit the brakes. The Express, heading west at 96 km/h, slows down, decelerating at an average of 4 m/s² while the eastbound Flyer, traveling at 110 km/h, slows down, decelerating at an average of 3 m/s². Will they collide?

68. [II] The drivers of two cars in a demolition derby are at rest 100 m apart. A clock on a billboard reads 12:17:00 at the moment they begin heading straight toward each other. If both are accelerating at a constant 2.5 m/s², at what time will they collide?

> **SOLUTION:** The accelerations are constant, and we know their initial separation. How long will it take for either car to cover 50.0 m? $l = \frac{1}{2}at^2 = 50.0$ m $= \frac{1}{2}(2.5$ m/s²$)t^2$ and so $t = 6.3$ s.

69. [II] A rocket-launching device contains several solid-propellant missiles that are successively fired horizontally at 1-s intervals. (a) What is the horizontal separation of the first and second missiles just at the moment the second one is fired, if each has an initial speed of 60 m/s and a constant acceleration of 20 m/s² lasting for 10 s? (b) What is the horizontal separation of the first and second just as the third is launched?

70. [II] Referring to the previous problem, what is the relative acceleration of the first missile with respect to the second missile once both are in the air and accelerating? What will be their horizontal separation 6 s into the flight?

71. [II] Imagine that you are driving toward an intersection at a speed v_i just as the light changes from green to yellow. Assuming a response time of 0.6 s and an acceleration of −6.9 m/s², write an expression for the smallest distance (s_S) from the corner in which you could stop in time. How much is that if you are traveling 35 km/h?

72. [II] Considering the previous problem, it should be clear that the yellow light might reasonably be set for a time t_Y, which is long enough for a car to traverse the distance equal to both s_S and the width of the intersection s_I. Assuming a constant speed v_i equal to the legal limit, write an equation for t_Y, which is independent of s_S.

73. [II] The driver of a pink Cadillac traveling at a constant 26.8 m/s (i.e., 60 mi/h) in a 55-mi/h zone is being chased by the law. The police car is 20 m behind the perpetrator when it too reaches 26.8 m/s (i.e., 60 mi/h), and at that moment the officer floors the gas pedal. If her car roars up to the rear of the Cadillac 2.0 s later, what was the scalar value of her acceleration, assuming it to be constant? [*Hint: This problem really begins when the cop gets to a point 20 m behind the Cadillac. At that moment the two cars are at rest with respect to one another. The cop then takes 2.0 s to cover the 20-m distance.*]

74. [II] While making a movie, a cowboy on a horse rides up to a moving train traveling at 5.0 km/h along a long straight length of track. After running next to the last car for a while, he charges ahead toward the engine 100 m away and gets there in 1.10 min. Assuming it was constant, determine the scalar value of his acceleration.

75. [II] Ed, the leading runner in a race taking place along a straight track, is traveling at his top speed, a constant 10.0 m/s. He has been 4.0 m in front of Harry, his chief rival, for the last 10.0 s. But aware that Ed will hit the finish line in 20.0 s, Harry puts on a burst of speed. At what minimum constant rate must he accelerate if he's not to lose the race?

76. [II] With Problems 71 and 72 in mind, how long should the yellow light stay lit if we assume a driver-response time of 0.6 s, an acceleration of −6.9 m/s², a speed of 35 km/h, and an intersection 25-m wide? Which of the several contributing aspects requires the greatest time?

77. [II] In Problem 63, how far does the police car travel in the process of closing the Cadillac's 20-m lead?

78. [III] Having taken a nap under a tree only 20 m from the finish line, Rabbit wakes up to find Turtle 19.5 m beyond him, grinding along at 1/4 m/s. If the bewildered hare can accelerate at 9 m/s² up to his top speed of 18 m/s (40 mi/h) and sustain that speed, will he win?

79. [III] Superman is jogging alongside the railroad tracks on the outskirts of Metropolis at 100 km/h. He overtakes the caboose of a 500-m-long freight train traveling at 50 km/h. At that moment he begins to accelerate at +10 m/s². How far will the train have traveled before Superman passes the locomotive?

80. [III] A motorcycle cop, parked at the side of a highway reading a magazine, is passed by a woman in a red Ferrari 308 GTS doing 90.0 km/h. After a few attempts to get his cycle started, the officer roars off 2.00 s later. At what average rate must he accelerate if 110 km/h is his top speed and he is to catch her just at the state line 2.00 km away?

SECTION 3.8: STRAIGHT UP & DOWN

81. [I] THIS PROBLEM EXAMINES FREE-FALL. A boulder on the mythical planet Mongo drops off a cliff and falls from rest 1000 m in 10.0 s. (a) What's the initial speed of the boulder? (b) Determine the acceleration due to gravity on Mongo. Ignore friction.

82. [I] THIS PROBLEM IS ABOUT FREE-FALL. It takes 3.75 s for a bowling ball to strike the sidewalk when dropped from the window

of a building. (a) What's the initial speed of the ball? (b) How high up is the point of release? Ignore friction.

83. [I] EXPLORING PHYSICS ON YOUR OWN: Ask a friend to hold his or her thumb and forefinger parallel to each other in a horizontal plane. The fingers should be about an inch apart. Now you hold a 1-ft ruler vertically in the gap just above and between these fingers so that it can be dropped between them. Have your friend look at the ruler and catch it when you, without warning, let it fall. Calculate the corresponding response time. Now position a dollar bill vertically so that Washington's face is between your friend's fingers—is it likely to be caught when dropped?

84. [I] A kangaroo can jump straight up about 2.5 m—what is its takeoff speed?

85. [I] At what speed would you hit the floor if you stepped off a chair 0.50-m high? Ignore friction. Express your answer in m/s, ft/s, and mi/h.

86. [I] If a stone dropped (not thrown) from a bridge takes 3.7 s to hit the water, how high is the rock-dropper? Ignore friction.

87. [I] Ignoring air friction, how fast will an object be moving and how far will it have fallen after dropping from rest for 1.0 s, 2.0 s, 5.0 s, and 10 s?

88. [I] A cannonball is fired straight up at a rather modest speed of 9.81 m/s. Compute its maximum altitude and the time it takes to reach that height (ignoring air friction).

> SOLUTION: We have the initial vertical speed and need the peak altitude and corresponding time. On the way up g is negative:
> $$l_p = -v_i^2/2g = -(9.81 \text{ m/s})^2/2(-9.81 \text{ m/s}^2) = 4.91 \text{ m}$$
> $$t_p = -v_i/g = -(9.81 \text{ m/s})/(-9.81 \text{ m/s}^2) = 1.00 \text{ s}$$

89. [I] Calculate the speed at which a hailstone, falling from a height of $0.914\ 4 \times 10^4$ m (i.e., 3.00×10^4 ft) out of a cumulonimbus cloud, would strike the ground, presuming air friction is negligible (which it certainly is not). Give your answer in m/s.

90. [I] A circus performer juggling while standing on a platform 15.0-m high tosses a ball directly upward into the air at a speed of 5.0 m/s. If it leaves his hand 1.0 m above the platform, what is the ball's maximum altitude? If the juggler misses the ball, at what speed will it hit the floor? Ignore air friction.

91. [I] Draw a curve of s in meters versus t in seconds for a free-falling body dropped from rest. Restrict the analysis to the first second of fall, and make up a table of values of t, t^2, and s at 1/10-s intervals. Draw a curve of s versus t^2. Explain your findings.

92. [I] The acceleration due to gravity on the surface of the Moon is about $g/6$. If you can throw a ball straight up to a height of 25 m on Earth, how high would it reach on the Moon when launched at the same speed? Ignore the minor effects of air friction.

93. [II] THIS PROBLEM DEALS WITH FREE-FALL. While falling toward the ground at 100 km/h a skydiver releases a bag of rocks and then opens his parachute. (a) What was the initial downward speed (in m/s) of the bag at the moment it was released, at a height of 800 m? (b) At what speed will the bag hit the ground? Ignore air friction.

94. [II] THIS PROBLEM EXAMINES FALLING AT THE EARTH'S SURFACE. (a) Make a sketch of how far an object falls from rest in a total of 1 s, 2 s, 3 s, etc. (b) Show that the distances traversed by a body in free-fall during consecutive equal intervals of time (e.g., the first second, the second second, the third second, etc.) are in the ratios of 1:3:5:7:9 etc.

95. [II] A young kid with a huge baseball cap is playing catch with himself by throwing a ball straight up. How fast does he throw it if the ball comes back to his hands a second later? At low speeds air friction is negligible.

96. [II] An arrow is launched vertically upward from a crossbow at 98.1 m/s. Ignoring air friction, what is its instantaneous speed at the end of 10.0 s of flight? What is its average speed up to that moment? How high has it risen? What is its instantaneous acceleration 4.20 s into the flight?

> SOLUTION: To find out if it is still going up determine $t_p = -v_i/g = -(98.1 \text{ m/s})/(-9.81 \text{ m/s}^2) = 10.0$ s. In fact, it is at peak altitude and its speed is zero. $l_p = -v_i^2/2g = -(98.1 \text{ m/s})^2/2(-9.81 \text{ m/s}^2) = 491$ m. Its average speed is $v_{av} = \frac{1}{2}(98.1 \text{ m/s} + 0 \text{ m/s}) = 49.1$ m/s. Its acceleration is always g.

97. [II] A lit firecracker is shot straight up into the air at a speed of 50.00 m/s. How high is it above ground level 5.000 s later when it explodes? How fast is it moving when it blows up? How far has it fallen, if at all, from its maximum height? Ignore drag and take g equal to 9.800 m/s^2.

98. [II] The observation deck of a skyscraper is 1377 ft above ground. Ignoring air friction, how long will it take for a massive object to free-fall that far? Use $g = 9.81$ m/s^2. Considering the discussion in the text, is it reasonable to ignore air drag in this problem? Explain.

99. [II] A human being performing a vertical jump generally squats and then springs upward, accelerating with feet touching the Earth through a distance s_a. Once fully extended, the jumper leaves the ground and glides upward, decelerating until the feet are a maximum height s_{max} off the floor. Assuming the jumping acceleration a to be constant, derive an expression for it in terms of s_a and s_{max}. Ignore friction.

100. [II] A small rocket is launched vertically, attaining a maximum speed at burnout of 1.0×10^2 m/s and thereafter coasting straight up to a maximum altitude of 1510 m. Assuming the rocket accelerated uniformly while the engine was on, how long did it fire and how high was it at engine cutoff? Ignore air friction.

101. [II] The illustrative Example 3.11 did not use the equation $s = v_i t + \frac{1}{2}at^2$ because it required the quadratic formula. Carry out that calculation, solving it for the two values of t. Notice the symmetry around the time of peak altitude.

102. [III] Imagine that someone dropped a firecracker off the roof of a building and heard it explode exactly 10 s later. Ignoring air friction, taking $g = 9.81$ m/s^2 and using 330 m/s as the speed of sound, calculate how far the cherry bomb had fallen at the very moment it blew up.

103. [III] A bag of sand dropped by a would-be assassin from the roof of a building just misses Tough Tony, a gangster 2-m tall. The missile traverses the height of Tough Tony in 0.20 s, landing with a thud at his feet. How high was the building? Ignore friction.

SECTION 3.9: TWO-DIMENSIONAL MOTION: PROJECTILES

104. [I] A shoe is flung into the air such that at the end of 2.0 s it is at its maximum altitude, moving at 6.0 m/s. How far away from the thrower will it be when it returns to the height from which it was tossed? Ignore air friction.

105. [I] THIS PROBLEM DEALS WITH PROJECTILE MOTION. A ball is thrown horizontally at a height of 4.91 m above the ground. It has

an initial speed of 16.0 m/s. (a) What's its initial vertical speed? (b) How long will it take to drop to the ground? (c) How far, measured horizontally from the launch point, will the ball strike the ground?

106. [I] THIS PROBLEM EXPLORES HORIZONTAL PROJECTILE MOTION. Fired at 60.0 m/s from a horizontal compressed-gas gun, a tennis ball strikes the ground 80.0 m away. (a) What's its initial vertical speed? (b) How long will the ball be in the air? (c) How high above the ground is the gun?

107. [I] Show that the range of a projectile can be expressed as

$$s_R = \frac{-2v_{ix}v_{iy}}{g}$$

Ignore air friction.

108. [I] Suppose you point a rifle horizontally directly at the center of a paper target 100 m away from you. If the muzzle speed of the bullet is 1000 m/s, where will it strike the target? Assume aerodynamic effects are negligible.

109. [I] A raw egg is thrown horizontally straight out of the open window of a fraternity house. If its initial speed is 20 m/s and it hits ground 2.0 s later, at what height was it launched? At such low speeds air friction is negligible. [*Hint: Even though it's been thrown horizontally, the egg falls vertically for 2.0 s exactly as if it were simply dropped.*]

110. [I] While rolling marbles on a horizontal window sill a youngster accidentally shoots one at 3.0 m/s out the open window. He sees it land in a flower pot on a neighbor's fire escape 3.0 s later. How far beneath the sill is the pot?

111. [I] Several clowns in a circus are performing high up in the riggings of the tent. One throws a plastic bag full of water (at a height of 20.0 m) directly at a companion who is 10 m away and also 20.0 m above the ground. The bag just misses and 1.5 s later lands on the head of a third clown. Ignoring air friction, how high is his wet head above the ground?

112. [I] A golfer wishes to chip a shot into a hole 50 m away on flat level ground. If the ball sails off at 45°, what speed must it have initially? Ignore aerodynamic effects.

113. [I] Check the dimensions of both sides of the equation $v_i s_y^2 = tv_i^2 \cos\theta + \frac{1}{2}gv_i t^2$ to see if they are the same; if not, the equation is wrong. This technique was introduced by the French mathematician and physicist, J.B.J. Fourier, around 1822. Is it possible that the above equation is correct?

114. [II] THIS PROBLEM DEALS WITH HORIZONTAL PROJECTILE MOTION. Hurled horizontally through an open window at a height of 20.0 m above the ground, a rock leaves the thrower's hand at 60.0 m/s. (a) What are its initial vertical and horizontal speeds? (b) How long will it take to fall to the ground? (b) At what net speed will it land?

115. [II] THIS PROBLEM IS ABOUT PROJECTILE MOTION. Making an angle of 50.0° with the horizontal, a cannon fires a ball with a muzzle speed of 100 m/s. (a) What are its initial vertical and horizontal speeds? (b) What's its peak altitude? (c) On the way up, how long will it take to reach an altitude of 100 m?

116. [II] A flea jumps into the air and lands about 8.0 in. away,

having risen to an altitude of about 130 times its own height (that's comparable to you jumping 650 ft up). Assuming a 45° launch, compute the flea's take-off speed. Make use of the mathematical fact that $2\sin\theta\cos\theta = \sin 2\theta$ and ignore air friction.

SOLUTION: Its range is (8.0 in.)(2.54 cm/in.) = 20.32 cm where,

$$s_R = \frac{2v_i^2}{9.81 \text{ m/s}^2}\cos\theta\sin\theta = \frac{v_i^2}{9.81 \text{ m/s}^2}\sin 2\theta$$

$$v_i^2 = \frac{(20.32 \times 10^{-2}\text{ m})(9.81 \text{ m/s}^2)}{\sin 90°} \text{ and } v_i = 1.4 \text{ m/s}$$

117. [II] Two diving platforms 10-m high terminate just at the edge of each end of a swimming pool 30-m long. How fast must two clowns run straight off their respective boards if they are to collide at the surface of the water midpool? Ignore friction.

118. [II] A golf ball hit with a 7-iron soars into the air at 40.0° with a speed of 54.86 m/s (i.e., 180 ft/s). Overlooking the effect of the atmosphere on the ball, determine (a) its range and (b) when it will strike the ground.

119. [II] A silver dollar is thrown downward at an angle of 60.0° below the horizontal from a bridge 50.0 m above a river. If its initial speed is 40.0 m/s, where and at what speed will it strike the water? Ignore the effects of the air. [*Hint: Find the final vertical speed knowing the height; get the time of flight from the Mean Speed Theorem. Use that to compute the horizontal distance. And finally from the Pythagorean Theorem get the total speed.*]

120. [II] Someone at a third-floor window (12.0 m above the ground) hurls a ball downward at 45.0° at a speed of 25.0 m/s. How fast will it be traveling when it strikes the sidewalk?

121. [II] A small rocket is fired in a test range. It rises high into the air and soon runs out of fuel. On the way down it passes near an observer (sitting in a 20.0-m-high tower) who sees the rocket traveling at a speed of 30.0 m/s and moving in a vertical plane at an angle of 80.0° below horizontal. How fast will it be traveling when it hits the ground?

122. [II] Using Fig. 3.17 (p. 73) and extending it where necessary, show that an angle less than 45° with respect to the ground will result in a greater horizontal displacement when the landing point is lower than the launch point. Neglect friction.

123. [II] Show that at any time t, where friction is negligible, the velocity of a projectile (launched at an angle θ with a speed v_i) makes an angle θ_t with the horizontal given by the expression

$$\theta_t = \tan^{-1}\left(\tan\theta + \frac{gt}{v_i\cos\theta}\right)$$

where $g < 0$.

124. [III] A baseball is hit as it comes in, 1.30 m over the plate. The blast sends it off at an angle of 30° above the horizontal with a speed of 45.0 m/s. The outfield fence is 100 m away and 11.3-m high. Ignoring aerodynamic effects, will the ball clear the fence?

125. [III] A burning firecracker is tossed into the air at an angle of 60° up from the horizon. If it leaves the hand of the hurler at a speed of 30 m/s, how long should the fuse be set to burn if the explosion is to occur 20 m away? Ignoring friction, just set up the equation for t.

Chapter 4
Newton's Three Laws

*T*his chapter is about Newton's Three Laws—(1) the Law of Inertia; (2) the relationship between force and the accompanying change in motion; (3) and what we might call the Law of Interaction. These insights initiated the study of how motion changes under the influence of force, a discipline called **dynamics** (p. 97). When the forces exerted on a system cancel one another so that there is no change in velocity, no acceleration, the system is in *equilibrium*—there is a balance, the study of which is called **statics** (p. 115). The idea of force is central to the Newtonian formulation, and in this century we have come to appreciate the fundamental role played by the concept of momentum in understanding force.

STUDY GUIDE

The last two chapters provided the vocabulary of motion: distance, speed, displacement, velocity, acceleration, and so forth. And we were introduced to the underlying vector nature of motion. Putting it all together, we learned how to describe uniformly accelerated motion. We'll be relying on that basic knowledge here, so if you need to, go back and review Chapters 2 and 3.

This chapter is about forces—tension (p. 90), weight (p. 101), friction (p. 107), etc. The general principles are Newton's Three Laws. The study becomes tremendously practical because it treats the response of everyday objects to various forces.

This is the same fundamental theoretical formulation that allows us to analyze everything from the structure of an artificial limb, or bridge, or building, to a car crash, or Space Shuttle flight.

The Three Laws

Isaac Newton was born on Christmas Day, 1642, within a year of Galileo's death. Newton was a short man, nearsighted, already silver-gray in his thirties, inattentive to personal appearance, incredibly forgetful, and probably forever virginal. This nervous, hypochondriacal, sensitive, pious, vulnerable man was one of the greatest geniuses to ever live. Among Newton's many achievements was a theory of motion that withstood every test for over 200 years.

4.1 The Law of Inertia

The first of Newton's Three Laws was proposed, in an incomplete form, by Galileo: *a body once set in motion and thereafter undisturbed will continue at a constant speed forever, all by itself.* This was totally opposite to the prevailing Aristotelian view that without a force pushing it, a moving body will immediately come to rest. Aristotle's theory appears to agree with experience, but it's wrong. We live on a planet where gravity and friction obscure what's really going on.

The experiment that led Galileo to his Law of Inertia used two inclined planes set end-to-end (Fig. 4.1). A ball rolled from rest, swept past the bottom and up the second incline. Regardless of the tilt of that second track, the ball always reached a vertical height that was just a little lower than it started from. Rather than overlook this slight difference, Galileo

The ability of matter to resist any change in its motion is called **inertia**.

Isaac Newton, 1642–1727.

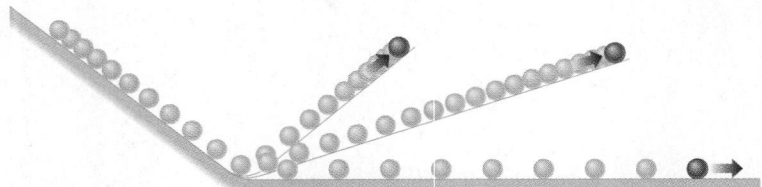

Figure 4.1 A ball rolled down an inclined track always climbs back to nearly the same height at which it was released, before it stops. Here a succession of tracks is shown with decreasing tilt. Galileo argued that if the second track were horizontal and frictionless, the ball would roll forever.

The First Law: A body at rest stays at rest, and a body in motion stays in motion with a constant velocity, all by itself.

searched for its cause. He polished the ball and smoothed the tracks, and finally concluded that the loss was due to friction.

Next, Galileo gradually reduced the angle of the second track so that the ball rolled farther before coming to rest. Then he asked the right question: What would happen if the second incline were made *perfectly frictionless* and *perfectly horizontal*? The ball would roll forever with "a motion which was uniform and perpetual." We ordinarily must keep pushing objects to maintain them in motion because of friction (p. 107), not because of the fundamental nature of the process. Had Aristotle lived aboard a spaceship, the Law of Inertia would have been obvious.

In 1687, Newton set out his "laws of motion" in the book we now call the *Principia* (pronounced prin-<u>sip</u>-pee-uh). His First Law was the **Law of Inertia**:

> **Every body continues in a state of rest or of uniform motion in a straight line except insofar as it is compelled to change that state by forces impressed upon it.**

An equivalence is drawn here between rest and uniform motion. Altering either requires a force, but once either state is established, it persists forever in the absence of force.

The *Apollo* astronauts, in 1969, shut down their engines and coasted in unpowered flight for much of the way to the Moon. At this very instant, the space probe *Pioneer 10* (launched March 2, 1972) is hurtling toward the star Aldebaran in Taurus at tens of thousands of kilometers per hour, its rockets burned out decades ago. The famous trick of whisking a tablecloth out from under a setting of dishes is a grandiose First-Law gesture.

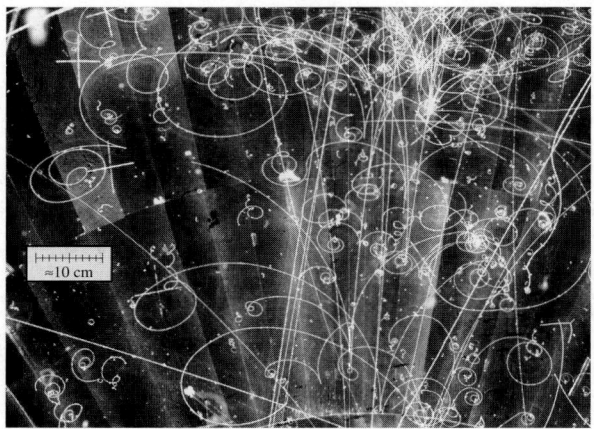

A stream of high-speed subatomic particles enters at the bottom of the photo (see p. 4). The tracks that are visible are curved paths, thereby indicating that the particles that made them were acted upon by an external force. In this case, there was an applied magnetic field and a resulting magnetic force (Sect. 19.5). Every track was made by a charged particle, and every track was curved by a magnetic force. (The two parallel, straight, slightly tilted lines running top to bottom near the middle of the picture are reference wires stretched across the chamber.)

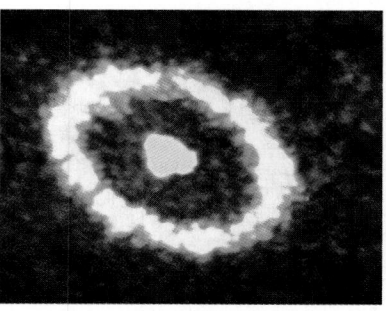

The pink central region is the remnant of a star that was seen to explode in 1987 (supernova 1987A). This 1990 photo shows a yellow ring of glowing gas about 12×10^{12} mi across. It was blown off by the star some 20 000 years ago and is still hurtling out into space in accord with Newton's First Law. The central debris of the supernova is moving outward much faster and will overtake and tear apart the ring within the next 100 years.

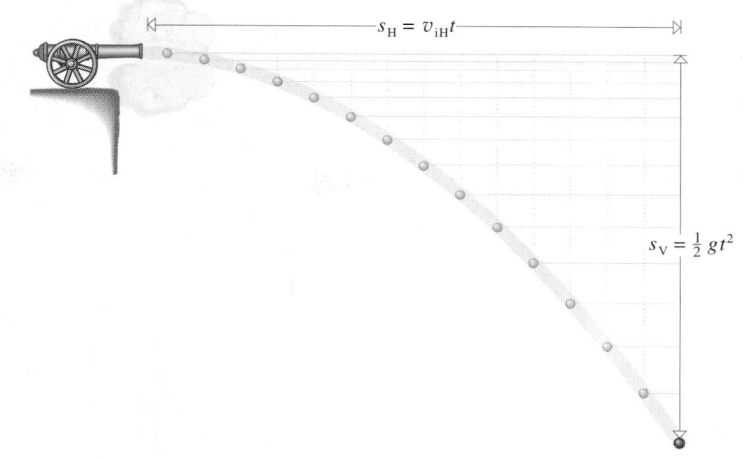

Figure 4.2 Without friction, a projectile fired horizontally falls as it progresses, sustaining its constant horizontal inertial motion (v_{iH} = constant) while experiencing a vertical acceleration of g.

And the purpose of seat belts becomes clear when the body in motion tending to stay in motion after slamming on the brakes is yours.

Dynamics and Inertia

To see how all of this works theoretically, let's reconsider the flight of a projectile (p. 68). The Law of Inertia maintains that left on its own, a cannonball should sail straight off in the direction fired (Fig. 4.2) at a ***constant velocity*** equal to its initial velocity. This is what would happen in intergalactic space far from anything else where gravity is negligible and there is no air friction. But on Earth the ball's trajectory curves downward, and there must be a downward force acting on it. The gravitational attraction between the ball and the planet results in a downward force on the ball known as its weight (p. 101). And so a projectile's motion is a combination of a horizontal (force-free) constant-velocity inertial motion, added to a constant-$\vec{\mathbf{g}}$ free-fall.

As one more application, consider an object in free-fall inside a uniformly moving environment, like the train in Fig. 4.3. We know from experience that a dropped bundle of keys doesn't slam into the rear wall of the car. With respect to the dropper, the keys initially have zero horizontal velocity ($\vec{\mathbf{v}}_{iH} = 0$), and the Law of Inertia maintains that that will

---

An Unappreciated Anticipation. No body begins to move or comes to rest of itself.

ABU ALI IBN SINA, KNOWN AS AVICENNA
(980–1037)

On the Motion of Bodies Before Galileo. Once moving they are never at rest unless impeded. Once in motion only an outside force can restrain them.

G. B. BENEDETTI (1585)

CAUSE & EFFECT AND THE LAW OF INERTIA

It's impossible to confirm the Law of Inertia directly; there is no place in this Universe where an object can be completely free of external influences, and the idea of undisturbed unending straight-line motion is unrealistic in a cosmos cluttered with galaxies. Still, the Law of Inertia allows us to understand a tremendous range of observations—and that's its real power. Moreover, the law introduces ***cause and effect*** into the discourse. *If a moving body does not maintain a constant speed in a straight line it must be experiencing a force, and we can look for the source of that force somewhere external to the body.*

The First Law: A body in motion tends to stay in motion. This train was supposed to have stopped at the platform. Instead it continued on after the brakes were applied and crashed through the terminal wall.

Were it not for the accidental impediment of the air, I verily believe that, if at the time of the ball's going out of the piece [that is, a cannon] another were let fall from the same height directly downwards, they would both come to the ground at the same instant.

GALILEO

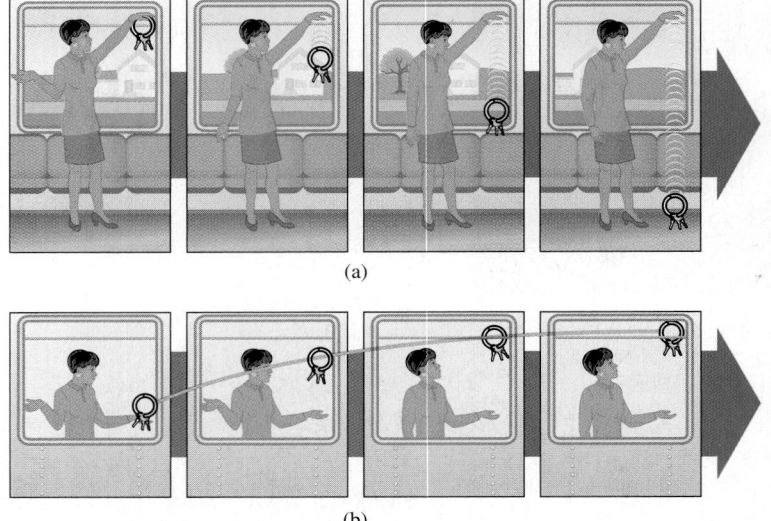

(a)

(b)

Figure 4.3 An experiment performed inside a uniformly moving train. (a) As seen from inside, the dropped keys fall straight down. Someone outside, however, would see them fall along a descending parabola. (b) The same is true when the keys are tossed straight up. As seen from outside, the keys appear to move along an ascending parabola.

continue throughout their descent. The keys fall straight down and land directly under the hand that released them. To someone outside at "rest" looking in, the keys (even before they're released) have a constant horizontal velocity equal to that of the train ($\vec{v}_{iH} = \vec{v}_T$). Because there's no horizontal force acting on them, the Law of Inertia tells us that the keys will continue to travel horizontally at $\vec{v}_T$ with respect to the outside observer. As the person in the train is moved forward, the keys (in midair) move forward at the same speed, and that continues throughout the descent until the keys land directly below the hand that dropped them. The outside observer sees the person in the train move forward in a straight line as the keys travel along a forward parabolic arc (Fig. 4.3a). The First Law makes it all understandable!

THE ORIGINS OF INERTIA

Newton felt that **inertia**—that is, *the resistance to a change in motion*—is an intrinsic property of matter itself, independent of the environment. Inertia is a manifestation of mass; the more mass a body has, the more inertia it has. Yet that opinion was soon challenged by the philosopher George Berkeley. Many today believe, like Berkeley, that inertia arises from the interaction of all matter in the Universe, an idea that's called Mach's Principle. Still, there are other possibilities. Physicists have certainly not yet resolved the fundamental question of the origin of inertia.

4.2 Force

The First Law states that a body at rest will remain at rest, and a body in uniform motion will remain in uniform motion *except insofar as it is compelled to change that state by a force impressed upon it*. But what is a force? Force is often naively defined in terms of pushes and pulls (reasonably enough, we

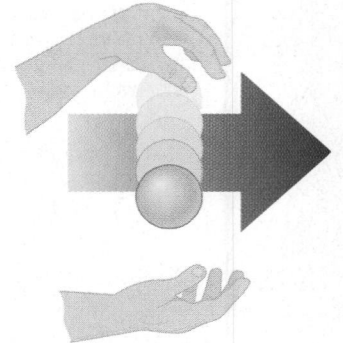

EXPLORING PHYSICS ON YOUR OWN

Free-Fall and the Law of Inertia: Hold a ball high in your right hand and position your left hand a few feet directly below it. Now walk with a constant velocity and, somewhere along the way, drop the ball without breaking stride. Where will it land? Where will the ball land if, just as you drop it, you begin to decelerate? Try throwing the ball straight up while walking at a constant velocity—it should come down directly into your hand. Place a dime at the very edge of a table and slide a nickel directly at it as fast as you can so that it knocks the dime off and follows it over the edge. The nickel will fall almost straight down, while the dime will travel a meter or so horizontally, but *they will both strike the floor simultaneously*—why?

tend to go back to our physiological experiences as a starting point), but that just replaces one ill-defined word by two undefined words. Moreover, the sensation of force is as subjective as the sensations of taste or smell. If we were ghostlike and viewed the world remotely without being able to feel, what would we say about force? A force is not something that's directly observable; what is perceived is the effect, that is, the change produced by the force (e.g., something is compressed, or twisted, or moved).

Newton defined force as "an action exerted upon a body, in order to change its state, either of rest or of moving uniformly…" **Force is the agent of change**. In the particular case of dynamics, *it is that which alters motion*. In the Universe as it exists today, there are four fundamental forces: gravitational, electromagnetic, strong, and weak. In one way or another, all of physics is concerned with these Four Forces. *

As with most scientific concepts, we try to define force in a way that allows us to measure it. The easiest procedure is to take a standard mass (p. 9), such as a kilogram, hang it on a spring scale, and call whatever the arrow points to "so many units of force." After all, ordinary weight is the downward gravitational force. With a beam balance, we can divide a kilogram mass into fractional pieces that could then be used to calibrate the spring scale. A 0.453 592 37-kg piece, at a place where $g = 9.806\,65$ m/s^2, has a weight equal to precisely 1 lb. Once appropriately labeled, the spring scale provides a convenient, reproducible way to measure forces. Unfortunately, this scheme uses gravity and has little to do with Newton's definition of force as the changer of motion. Many operational definitions have been put forth over the years, but they all fall short for one practical reason or another. In the end, *force is a fundamental concept, and it defies being completely satisfactorily defined*.

Later, when we define the unit of force in the SI system, the *newton* (N), it will be by way of the change in motion of an object, independent of gravity. For now, 1 N is the weight on Earth of a mass of about 0.10 kg. In terms of the outmoded pound, 1 N = 0.224 809 lb or 4.448 22 N = 1 lb. We will treat force in newtons, keeping in mind that 1 N $\approx \frac{1}{4}$ lb—the weight of a medium-sized apple.

Force as a Vector

Two oppositely directed forces acting on the same object work against each other, in effect, partially or totally canceling one another, just as two vectors would. Moreover, several forces applied to a point on a body act as if they were a single force. The man on the left in

An apple of medium size weighs about 1 N.

*The concept of force has taken on a broadened meaning in contemporary physics. Force not only changes motion, it transforms matter (e.g., altering quarks).

The application of a **force** to a body (one that is free to move) changes the body's state of motion; that is, it changes its velocity.

If you insist upon a precise definition of force, you will never get it!

RICHARD FEYNMAN
NOBEL LAUREATE (1963)

Table 4.1

A Comparison of Masses

Approximate Masses	kg
Our galaxy	2×10^{41}
Sun	2×10^{30}
Earth	6×10^{24}
Moon	7×10^{22}
Lake Erie	10^{15}
Great Pyramid at Giza	10^{10}
Super tanker	10^8
Boeing 747	3.5×10^5
Blue whale	1.4×10^5
Elephant, African	6×10^3
Car	10^3
Human—adult	0.7×10^2
Dog	10^1
Chicken	2
Baseball	1.5×10^{-1}
Paperclip	0.5×10^{-3}
Ant	10^{-5}
Pencil mark	10^{-10}
DNA molecule	10^{-20}
Oxygen atom	3×10^{-26}
Proton	2×10^{-27}
Electron	9×10^{-31}

Figure 4.4 The four forces to the right (which all pass through a single point in the ring) act as a single net force equal to their vector sum.

Fig. 4.4 experiences one large force pulling him backward even though he has four individual adversaries.

If you're asked to "push on someone with a force of 10 N," it could be done in a lot of different ways: up, down, right, left. Force is specified completely only when both its magnitude and direction are given; **force is a vector quantity**. *It's the net force, the resultant of all the forces acting, that changes the motion of a body*.

The forces in Fig. 4.4 are applied to ropes and not directly to any object. A rope, chain, or tendon transmits the applied force via electromagnetic interactions between its intervening atoms. The rope stretches slightly and the displaced atoms pull on the atoms behind them, and so on until the force is communicated to the object. The force vector acts along the taut rope, and this direction is called the **line-of-action** of the force. Because the force tends to stretch the rope, it's known as a **tensile force** ($\vec{\mathbf{F}}_T$). Imagine that we cut the rope and splice in a spring scale of negligible mass. The scale will read the magnitude of the applied tensile force. That always-positive quantity (F_T) is called the **tension** (study Fig. 4.5).

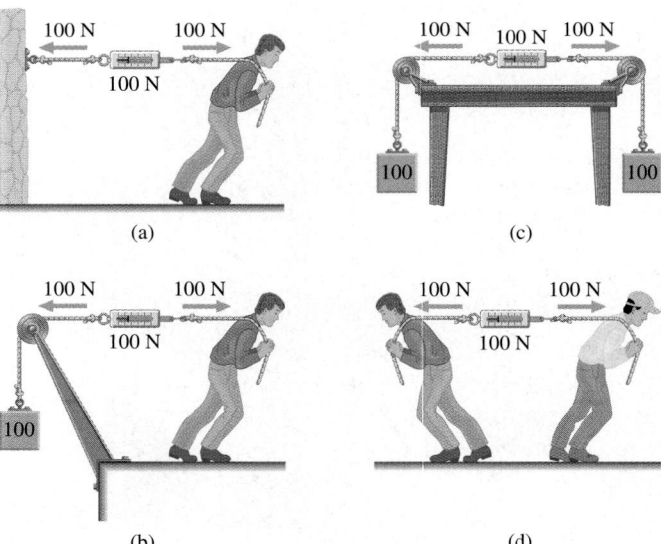

Figure 4.5 A spring scale being used to measure force. Shown are several situations that look different but are equivalent as far as the scale is concerned.

Example 4.1 [I] Determine the resultant, or net force, exerted on the stationary elephant by the two clowns in Fig. 4.6. What is the tension in the rope attached to the elephant?

Solution Force is a vector quantity, and we're reminded of that by the word "resultant." (1) TRANSLATION—Add two known force vectors. (2) GIVEN: $\vec{F}_1 = 300$ N–NORTH and $\vec{F}_2 = 400$ N–EAST. FIND: The resultant $\vec{F} = \vec{F}_1 + \vec{F}_2$. (3) PROBLEM TYPE—Vector addition. (4) PROCEDURE—Because the two vectors are perpendicular, $\vec{F}$ can be determined using the Pythagorean Theorem to find F and the tangent to get θ. (5) CALCULATION—Since the forces (300 N and 400 N) form a 3-4-5 right triangle, we needn't calculate F; it equals 500 N. To find

the angle $\vec{F}$ makes with the west-to-east axis, use

$$\theta = \tan^{-1}\frac{300 \text{ N}}{400 \text{ N}} = \boxed{36.9° \text{ north of east}}$$

and so $\boxed{\vec{F} = 500 \text{ N–AT } 36.9° \text{ NORTH OF EAST}}$. The net force the clowns exert is transmitted to the rope attached to the elephant, and so the tension in that rope is $\boxed{500 \text{ N}}$.

Quick Check: For the clown pulling east (500 N) cos 36.9° = 400 N.

[For more worked problems click on **WALK-THROUGHS** *under* **CHAPTER 4** *on the CD.]*

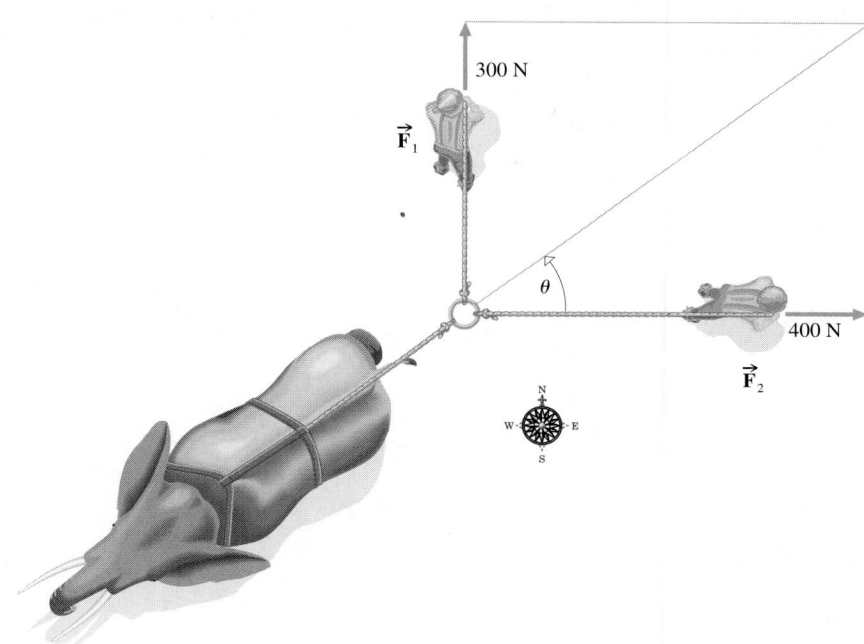

300 N

$\vec{F}_1$

θ

400 N

$\vec{F}_2$

Figure 4.6 Each clown pulls on the ring with a force as shown. The stationary elephant feels a net force pulling him north of east at an angle θ.

Example 4.2 [I] Determine the net force exerted on the ring by the three people in Fig. 4.7a on page 92.

Solution The phrase "net force" means add vectors, and here we'll add them analytically. (1) TRANSLATION—Add three known force vectors. (2) GIVEN: $|\vec{F}_1| = 707$ N, $|\vec{F}_2| = 500$ N, $|\vec{F}_3| = 966$ N, and their angles. FIND: The resultant $\vec{F} = \vec{F}_1 + \vec{F}_2 + \vec{F}_3$. (3) PROBLEM TYPE—Vector addition. (4) PROCEDURE—Resolve the three vectors into their components along the x- and y-axes, as in Fig. 4.7b. Add all the scalar x-components (taking to the *right* as positive) to get F_x, and all the

scalar y-components (taking *up* as positive) to get F_y. Next $\vec{F}$ is determined using the Pythagorean Theorem to find F and the tangent to get θ. (5) CALCULATION—(*Because this is our first example of the kind, we're going to do this one putting in every single step.*) The vector y-component of the resultant is

$$\vec{F}_y = \vec{F}_{1y} + \vec{F}_{2y} + \vec{F}_{3y}$$

Thus the scalar y-component of the resultant is

$$F_y = F_{1y} + F_{2y} + F_{3y}$$

continued

where $F_{1y} = F_1 \cos 30.0°$, $F_{2y} = F_2 \cos 45.0°$, and $F_{3y} = F_3$

Let's now put in the numbers, being especially careful about the signs. Here F_{1y} and F_{2y} correspond to upward forces and are positive, whereas F_3 corresponds to a downward force and is negative. Consequently,

$$F_{1y} = (707 \text{ N}) \cos 30.0°, \quad F_{2y} = (500 \text{ N}) \cos 45.0°,$$
$$F_3 = -966 \text{ N}$$

$$F_y = (707 \text{ N})(0.866) + (500 \text{ N})(0.707) + (-966 \text{ N})$$

and therefore $F_y = 0$

Similarly, in the x-direction

$$\vec{F}_x = \vec{F}_{1x} + \vec{F}_{2x} + \vec{F}_{3x}$$

and $$F_x = F_{1x} + F_{2x} + 0$$

where $F_{1x} = F_1 \sin 30.0°$ and $F_{2x} = F_2 \sin 45.0°$

Here F_{1x} corresponds to a force to the left and is negative, where-

as F_{2x} corresponds to a force to the right and is positive. Consequently,

$$F_{1x} = -(707 \text{ N})(0.500) \quad \text{and} \quad F_{2x} = (500 \text{ N})(0.707)$$
$$F_x = -(707 \text{ N})(0.500) + (500 \text{ N})(0.707)$$

and $$F_x = 0$$

Since both components are zero, $\boxed{\vec{F} = 0}$. The people heave and strain, but if the situation began with the ring at rest, it remains at rest because the net force acting on it is zero!

Quick Check: If the three vectors actually add up to zero, they must form a closed triangle, as in Fig. 4.7c. The angles inside this *force triangle* do add up to 180°, which is a good sign. To be sure things are right, we check that the Law of Sines applies:

$$\frac{707}{\sin 45°} = \frac{500}{\sin 30°} = \frac{966}{\sin 105°} = 10^3$$

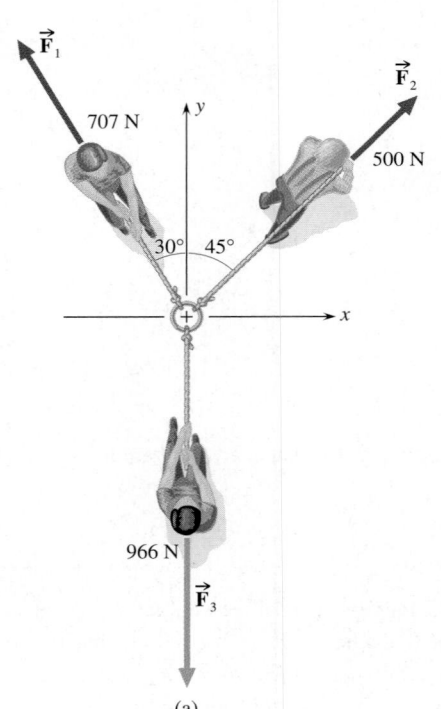

(a)

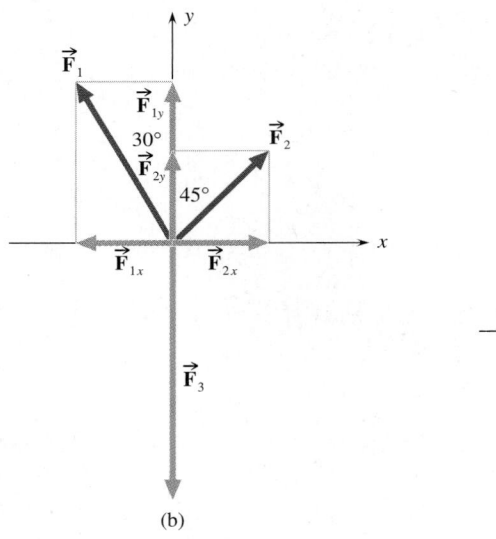

(b)

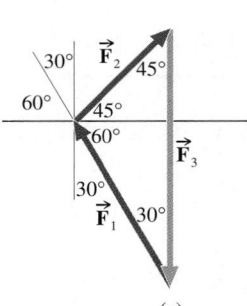

(c)

Figure 4.7 (a) Three forces passing through a point are applied in different directions. (b) The forces are resolved in the two perpendicular directions x and y. (c) Because the net force is zero, the three force vectors form a closed triangle called the *force triangle*.

4.3 The Second Law

The state of motion of an object changes with the application of a net force—the First Law tells us as much. But by how much does it change? The answer is in Newton's Second Law, which is a quantified restatement and extension of the First Law. At its heart, the Second Law is about a new concept known as *momentum*; the law tells us how force changes momentum.

Momentum

The realization that speed alone fails to give us some essential aspect of the motion of an object goes back to Jean Buridan, working around 1330. He was trying to understand why an iron ball of the same size as a wooden one would travel much farther when both were launched at the same speed. Or, equivalently, why you would rather be hit by a firefly traveling at 60 km/h than by a fire engine at the same speed. Buridan reasoned that the crucial concept was the product of the mass (m) and the speed (v). The "drive" that a moving body has—that ability to plow along—cannot be attributed to its speed alone (the firefly has the same speed as the fire engine). The "true measure of motion" was a new basic quantity, the product of *mass* and *speed*. The more mass a body has the greater its inertia, and the more difficult it is to alter the way it's moving; it's a lot harder to change the motion of a fire engine than a firefly.

> **Momentum ($\vec{p}$)** is mass times velocity.

The idea, although somewhat muddled, was picked up by Galileo who called it *momento* and later by Descartes who called it the *quantity-of-motion*. When Newton set out the central concept of *his* theory—the measure of motion—he defined the **momentum ($\vec{p}$)** as *the product of the object's mass and velocity*: $\vec{p} = m\vec{v}$. This was long before the concept of "vector" was formalized and so he never wrote it in this concise way, but he was well aware of the directional nature of momentum.

The Time Rate-of-Change of Momentum

> **The change of motion is proportional to the force impressed; and is made in the direction of the straight line in which that force is impressed.**

This is the Second Law translated from Newton's original Latin. Even though the wording is a little obscure, it's well worth the effort to try to appreciate his thinking. We know from the First Law that force is the changer of motion. Moreover, "the measure of the motion" is the *momentum*. The Second Law says that an applied force (F), which for simplicity we'll take to be constant, produces a proportional change in the momentum of the body (Δp), that is, $F \propto \Delta p$.

We know from experience that time is involved here, too. The longer a rocket fires the greater its final speed, and the greater will be Δp. There is no explicit reference to time in the original statement of the Second Law, but it's clear that Newton was speaking about the change in motion *per unit time*. *The force applied equals the resulting change in momentum per unit time.* A modern version of **the Second Law** is

> **The Second Law:** The net force applied to an object equals the resulting time rate-of-change of the object's momentum.

> **The force exerted on a body equals the resulting change in its momentum divided by the time elapsed in the process.**

Newton understood the directional nature of the phenomenon—the change in motion is in the direction of the force impressed. In other words, $\Delta\vec{p}$ is parallel to $\vec{F}$, and putting it all together in vector notation,

> **Force** is associated with the flow of momentum. On a macroscopic level, it is that which changes momentum.

[constant force]

$$\vec{F} = \frac{\Delta\vec{p}}{\Delta t}$$

(4.1)

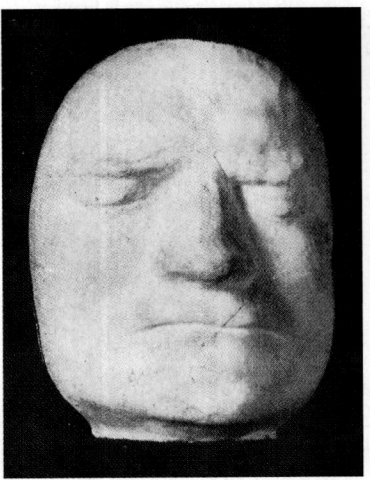

Isaac Newton's death mask (a plaster casting of his face), made soon after he died at about 1:00 A.M. on March 20, 1727.

This is Newton's Second Law in all its glory, and it's one of the most important insights in physics. *An event (a change) occurs when two or more entities interact; they interact by exchanging momentum.* The way we know that something happens (e.g., the phone rings) is by observing change. Change is the direct result of the transfer of momentum (e.g., sound waves are created that carry momentum to your ear). Every event is connected to its observers by the action of forces, or equivalently, by the flow of momentum. Contemporary theory maintains that **the Four Forces operate through the exchange of force-mediating momentum-carrying *virtual particles***. Any two entities (e.g., you and the Earth) interact by invisibly absorbing and emitting a torrent of virtual particles (gravitons).

Not long after Newton's work, Jacob Hermann (1716) formalized an alternative statement of the Second Law in terms of acceleration. Although it's far less profound, that simplistic version is easier to deal with—so much so that nowadays it's customary to study it before exploring the direct consequences of Newton's Second Law as he gave it. And so we'll do just that now, coming back to Newton's superior insight later (Chapter 7).

Force Equals Mass Times Acceleration

Today apparatus exists (e.g., air tracks and motion detectors) that can be used to directly measure the speed of a moving object (see Discussion Question 1). After making a few reasonable assumptions, it becomes an easy matter to apply a constant force (F) to a known mass (m) and determine the resulting acceleration (a). What we find is that $F = ma$: **A net force applied to an object causes it to accelerate at a rate that is inversely proportional to the mass of the object:** $a = F/m$.

This alternative form of the Second Law was first derived from Eq. (4.1): $\vec{\mathbf{F}} = \Delta(m\vec{\mathbf{v}})/\Delta t$. The change in momentum $\Delta(m\vec{\mathbf{v}})$ equals $(m_f\vec{\mathbf{v}}_f - m_i\vec{\mathbf{v}}_i)$, but if the mass of the object is constant, which it most often is, $m_i = m_f = m$, and $\Delta(m\vec{\mathbf{v}}) = m(\vec{\mathbf{v}}_f - \vec{\mathbf{v}}_i) = m\Delta\vec{\mathbf{v}}$. It follows that $\vec{\mathbf{F}} = m\Delta\vec{\mathbf{v}}/\Delta t$ and, using $\vec{\mathbf{a}} = \Delta\vec{\mathbf{v}}/\Delta t$

[constant force]
$$\vec{\mathbf{F}} = m\vec{\mathbf{a}} \qquad (4.2)$$

> The **net force** applied to an object equals its mass times the resulting acceleration.

As long as it's remembered that the force and acceleration are in the same direction, this equation* can be written in scalar form as $F = ma$. Though Newton was aware of this rela-

---

*When we study Relativity we'll find that at immense speeds (upwards of 10 000 times faster than anyone has ever traveled) traditional theory increasingly departs from reality. The fault lies in the fact that momentum must be redefined ($\vec{\mathbf{p}} = \gamma m\vec{\mathbf{v}}$). In the high-speed relativistic regime, even though Eq. (4.1) is true, Eq. (4.2) is not; F is not rigorously equal to ma, and in general $\vec{\mathbf{F}}$ is not even parallel to $\vec{\mathbf{a}}$. Nonetheless, at ordinary speeds, and a good deal beyond, Eq. (4.2) works perfectly well.

THE ELUSIVE OPERATIONAL DEFINITION

It's easy to imagine a perfect experiment in which various constant forces are successively applied to a mass and the resulting accelerations are measured. Ideally, a plot of F versus a would then be a straight line having a slope of m. In real life, however, it's all but impossible to isolate the mass completely from extraneous forces, such as gravity and friction. And so even though experiments of this sort can work extremely well, they cannot be used as the basis of a rigorous operational definition of either force or mass.

tionship, it's a credit to his genius that he appreciated the fundamental nature of momentum.

A constant force of 1.0 N will cause a mass of 1.0 kg to accelerate at a constant rate of 1.0 m/s², from which it follows that

$$1\ \text{N} = 1\ \text{kg}\cdot\text{m/s}^2$$

Table 4.2 lists the units of force in the several systems that are likely to be encountered. If, while in space (Fig. 4.8), we push on a floating weightless $\frac{1}{2}$-kg mass with a constant 1-N force, it will immediately accelerate in the direction of the force, at a rate of 2 m/s² *all the while the force acts.*

Table 4.2

Units of Force, Mass, and Acceleration in Various Systems

	SI	cgs*	U.S. Customary Units
Force	$1\ \text{N} \equiv 1\ \text{kg}\cdot\text{m/s}^2$	$1\ \text{dyne} \equiv 1\ \text{g}\cdot\text{cm/s}^2$	1 lb
Mass	1 kg	1 g	$1\ \text{slug}^\dagger \equiv 1\ \text{lb}\cdot\text{s}^2/\text{ft}$
Acceleration	$1\ \text{m/s}^2$	$1\ \text{cm/s}^2$	$1\ \text{ft/s}^2$

*The centimeter-gram-second system, which we will avoid, uses the dyne as the unit of force. $1\ \text{N} = 10^5\ \text{dynes} = 0.225\ \text{lb}$.

†Although the slug is still used in engineering as the unit of mass, we will make no use of it.

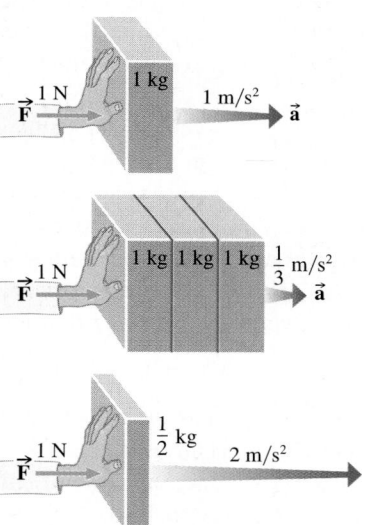

Figure 4.8 Several examples of how $\vec{\mathbf{F}} = m\vec{\mathbf{a}}$ works. In each case the force—at the moment shown—produces a proportional instantaneous acceleration. For as long as the force is applied, the body accelerates in the direction of $\vec{\mathbf{F}}$.

Example 4.3 **[II]** A ball (0.142 kg) left a player's hand at a speed of 20.0 m/s. If the straight throw lasted 0.020 s, determine the magnitude of the force exerted on the ball, assuming it to be constant.

Solution Whenever you see the word "force" in a problem, think of Newton's Second Law. (1) TRANSLATION—Determine the constant force needed to accelerate a body of known mass from rest to a specified final speed in a known time interval.

(2) Before you go any further, pick a direction to be positive. Let's take the direction of motion to be positive. Then enter the numbers into the GIVEN & FIND with the proper signs. GIVEN: $m = 0.142$ kg, $v_\text{i} = 0$, $v_\text{f} = +20.0$ m/s, and $\Delta t = 0.020$ s. FIND: F. (3) PROBLEM TYPE—Newton's Laws/constant force/acceleration. (4) PROCEDURE—We will use the scalar form of Eq. (4.2), $F = ma$, but we'll first need to determine a via its definition. (5) CALCULATION—Since F is constant, a is constant. Remembering that $\Delta v = v_\text{f} - v_\text{i}$,

continued

$$a = \frac{\Delta v}{\Delta t} = \frac{+20.0 \text{ m/s}}{0.020 \text{ s}} = \boxed{1000 \text{ m/s}^2}$$

From Eq. (4.2),

$$F = ma = (0.142 \text{ kg})(1000 \text{ m/s}^2) = \boxed{1.4 \times 10^2 \text{ N}}$$

Quick Check: From Newton's Second Law as he gave it [Eq. (4.1)], $F\Delta t = (1.4 \times 10^2 \text{ N})(0.020 \text{ s}) = 2.8 \text{ N·s}$ and this should equal $\Delta p = (0.142 \text{ kg})(20.0 \text{ m/s}) = 2.8 \text{ kg·m/s}$, and so everything looks okay (Fig. 4.9).

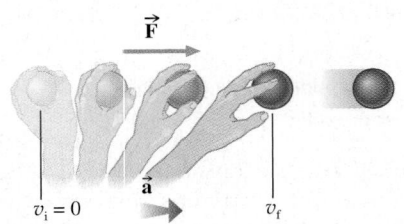

Figure 4.9 A ball being launched along a straight path. The constant force accelerates the ball from $v_i = 0$ to v_f at a constant rate $\vec{\mathbf{a}}$.

4.4 Interaction: The Third Law

Newton's Third Law completes the logical picture of the concept of force. A change of motion arises out of a net impressed force. Thus far, when we talked about a force applied to a body, it was an **external force**, *one whose source is external to the body*. An isolated body obeys the Law of Inertia. It cannot alter its own uniform straight-line motion because that requires some outside intervention via one or more external entities.

We can assume that when two bodies interact, *both* will be affected by the encounter; the motions of both will be altered, both will experience an external force. Two identical bodies moving toward each other at the same speed must experience the same alteration of motion on impact and, therefore, must exert the same force on each other.

Experiments on gases, fashionable when Newton was a young man, demonstrated that when a gas is compressed within a container by a piston pushing down on it, there is a backward reaction—the piston tends to pop back up. The "spring of the air" was a common phrase coming out of those studies. Indeed, a spring was a perfect image to use—push on it and it pushes back on you; pull it in one direction and it pulls you in the opposite direction.

In Newton's own words the Third Law is:

To every action there is always opposed an equal reaction.

The interaction of two bodies always occurs by way of a force and an equal-magnitude oppositely directed counterforce. There is no such thing as a single force exerted by some active entity on a passive one; **there is only interaction**. A hand cannot simply push on a chair—the chair must always push back on the hand. Entities interact, one upon the other: *force is a thing of pairs.*

RIGID BODIES & ATOMIC INTERACTIONS

A chisel struck by a hammer transmits a force to a piece of wood, just as the balloon transmitted a force from the hand to the door. One layer of atoms is displaced, pushing (electromagnetically) on the next layer, and so on down to the end of the chisel. During such a process some distortion occurs, even in a hard steel chisel. By contrast, the distortion of the balloon was obvious. In the real world, a perfectly rigid body cannot exist. ***Even a flea deforms the rock it lands on*** (granted, it's only by a minuscule amount, but to some extent the rock is squashed). The shifting of atoms within the rock, and the subsequent change in their electrical interaction, supports the flea.

The Third Law: Any force is always part of an interaction pair.

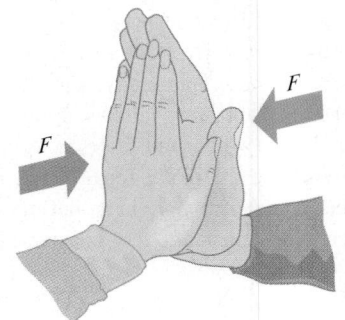

EXPLORING PHYSICS ON YOUR OWN

Newton's Third Law: To experience the Third Law, lean on a wall allowing gravity to press you against it. You'll feel the wall pushing back, supporting your weight. Now have someone position her hand, palm forward, fingers up. Place your hand against hers, palm to palm just touching. Ask her to push on your hand, but don't push back; that is, keeping in contact, withdraw your hand just as hers advances. It will feel a bit weird, but she can't push on you unless you push on her. Stretch a rubber band—you pull on it, it pulls on you; you pull to the right on it, it pulls to the left on you. The forces are equal in magnitude and opposite in direction.

An object offers as much resistance to the air as the air does to the object. You may see that the beating of an eagle's wings against the air supports its heavy body in the highest and rarest atmosphere ...

LEONARDO DA VINCI (1452–1519)
CODEX ATLANTICUS

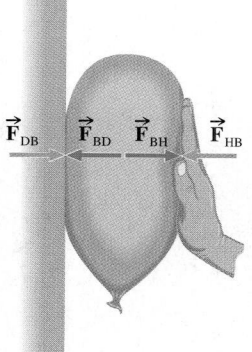

Figure 4.10 A balloon pressed against a door. $\vec{\mathbf{F}}_{HB}$ and $\vec{\mathbf{F}}_{BH}$ are an interaction pair, as are $\vec{\mathbf{F}}_{DB}$ and $\vec{\mathbf{F}}_{BD}$. All forces are the result of interactions, and all come in pairs comprising equal and opposite influences.

Subatomic particles interact with each other, and all the familiar forces that are manifest in our everyday macroscopic world (weight, tension, friction, etc.) arise from those interactions. Your feet hurt after you've been standing for a while because all the quarks and electrons in your body are gravitationally attracting all the quarks and electrons in the Earth, and vice versa. The planet is pulled up, and you're pulled down, and that flattens your feet. ***When we refer to a specific force, we are focusing on half of an interaction***; we are then only looking at what's happening to one of the two interacting entities. Whenever we see some object deviate from the Law of Inertia, we must be able to find another object interacting with it.

Consider, for example, a balloon being pressed against a closed door. You could argue that the balloon is squashed by the pusher alone, without any help from the door. Yet the balloon is flattened on both sides, just as it would be if the door had opened out to where the balloon was being held fixed in hand and proceeded to squash it. The hand pushes on the balloon (Fig. 4.10) with a force $\vec{\mathbf{F}}_{HB}$, and the balloon pushes back with an equal and opposite force $\vec{\mathbf{F}}_{BH}$. The balloon, acting as if it were an extension of the hand, transmits the force that is applied to it to the door ($\vec{\mathbf{F}}_{BD} = \vec{\mathbf{F}}_{HB}$). The door, in turn, reacts with an equal-magnitude, oppositely directed counterforce $\vec{\mathbf{F}}_{DB}$. The balloon is squashed in on both sides by the forces *applied to it*, $\vec{\mathbf{F}}_{DB}$ and $\vec{\mathbf{F}}_{HB}$.

WHY YOU CAN'T LIFT YOURSELF UP AND FLY

A system may have several interacting parts that push and pull on each other with *internal* forces that don't affect the overall motion. The reason you can't grab your belt, pull up, and fly is that you and the belt are the system and the forces exerted are all internal. Your hand pulls up on your belt and your belt pulls down on your hand—that's an interaction pair. Meanwhile, your belt pulls up on your waist and your waist pushes down on your belt—that's an interaction pair. And all the forces are numerically equal. The net force on the belt is zero, and the net force on you is zero—you go nowhere. For the same reason you have to get out and push if you want to move a stalled car: inside, you push forward on the dashboard and it pushes you backward against the seat with an equal and opposite force. Your back pushes to the rear of the car with the same force your hands push forward with—while inside, the net force you exert on the car is zero, and the net force it exerts on you is zero.

Interaction between entities *always* means ***two equal-magnitude, oppositely directed forces (one on each participant) forming the pair***. Here, each interaction pair is composed of two forces having reverse-ordered subscripts. **The forces of such a pair are always in opposite directions and never act on the same body.** Thus, $\vec{\mathbf{F}}_{HB}$ and $\vec{\mathbf{F}}_{BH}$ are an interaction pair; $\vec{\mathbf{F}}_{HB}$ is exerted on the balloon, while $\vec{\mathbf{F}}_{BH}$ is exerted on the hand. Two external horizontal forces are acting on the balloon, $\vec{\mathbf{F}}_{DB}$ and $\vec{\mathbf{F}}_{HB}$. These are not an interaction pair, even though they happen to be equal in magnitude and opposite in direction. As far as the motion of the balloon as a whole is concerned, the two external horizontal forces cancel, yielding a net applied force of zero; $a = 0$, and the balloon stays put at rest, squashed up against the door.

Dynamics & Statics

A body experiencing a net external force will accelerate in the direction of that force. It is the relationship between various kinds of forces and the resulting accelerations that will be examined next.

Both teams pull with all their might in opposite directions. But when the teams are equally matched $\sum \vec{F} = 0$. The net force on the rope is zero, the accelaration of the rope is zero, and it remains at rest. Of course, the tension (p. 90) in the rope is far from zero.

4.5 The Effects of Force: Newton's Laws

The force in the expression $\vec{F} = m\vec{a}$ is the **net applied force**, *the vector sum of all of the forces acting on the object*. For example, the elephant in Fig. 4.6 experiences a net force of $\vec{F} = 500$ N–AT 36.9° NORTH OF EAST. If its mass is 6.3×10^3 kg and if it happens to have greased feet so that it's frictionless, it will accelerate at a rate of $a = F/m = (500 \text{ N})/(6.3 \times 10^3 \text{ kg}) = 0.079 \text{ m/s}^2$. The two clowns pull, and the elephant accelerates in the direction of that net force such that $\vec{a} = 0.079 \text{ m/s}^2$–AT 36.9° NORTH OF EAST.

Because $\vec{F}$ represents the *net* applied force, it's helpful to write the equation as

$$\sum \vec{F} = m\vec{a} \tag{4.3}$$

where $\sum$ (Greek capital sigma) stands for "the sum of" and reminds us that we are dealing with the vector sum of all the forces acting on the body. The vector $\vec{F}$ can be resolved into components along any three perpendicular axes, and Eq. (4.3) then applies independently to each of the three component motions. We'll restrict ourselves to the simpler case of two-dimensional motion (Fig. 4.11). Consequently, for a body moving in a plane, for example, the xy-plane, the scalar component equations are

$$\sum F_x = ma_x \qquad \text{and} \qquad \sum F_y = ma_y \tag{4.4}$$

These two expressions are applicable independently and simultaneously. Always keep in mind that $\sum F$ means the sum of the scalar force components, or the net scalar force, in a particular direction; it may be positive or negative.

The Free-Body Diagram

There is a wonderful conceptual device called a **free-body diagram** that's quite helpful in analyzing all sorts of mechanics problems. To create such a diagram, isolate the object of concern (which might be a car, a knot in a rope, or a piece of a bridge) from the rest of the system; that is, (1) remove anything in contact with it and (2) replace each source of inter-action, be it a pushing hand or a pulling string, by a force vector. That will create a picture of the object all by itself, with a bunch of force vectors acting on it. It's then possible to apply $\sum \vec{F} = m\vec{a}$ to the isolated piece of the system. For example, Fig. 4.12a depicts a man pushing a refrigerator that is on frictionless wheels. The free-body diagram of the "fridge" (Fig. 4.12b) shows it by itself, with all the forces acting on it: the applied 100-N force, the weight, and the supporting forces exerted by the floor. Because the fridge is at rest in the vertical direction $a_y = 0$ and $\sum F_y = ma_y = 0$, the force of the floor up cancels the weight of the object down, and the net vertical force is zero. By contrast, the 100-N horizontal force results in a horizontal acceleration ($\sum F_x = ma_x$) of 1.00 m/s^2. When the sum of the external horizontal forces acting on an object is zero (i.e., $\sum F_x = 0 = ma_x$), as in Fig. 4.13, there will be no horizontal acceleration ($a_x = 0$).

It is the net **external force** that causes a body to accelerate; that is, the vector resultant of all forces applied to the body by agents outside of itself.

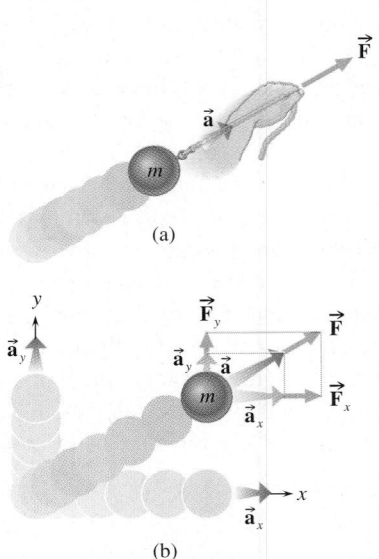

Figure 4.11 (a) An object accelerating in a straight line under the influence of a force $\vec{F}$ (b) The motion can be thought of as occurring along two perpendicular axes. Newton's Second Law applies independently along both axes.

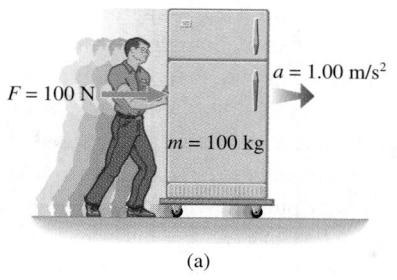

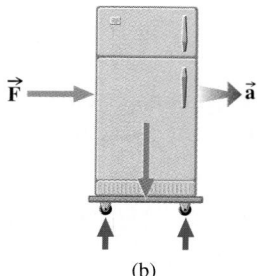

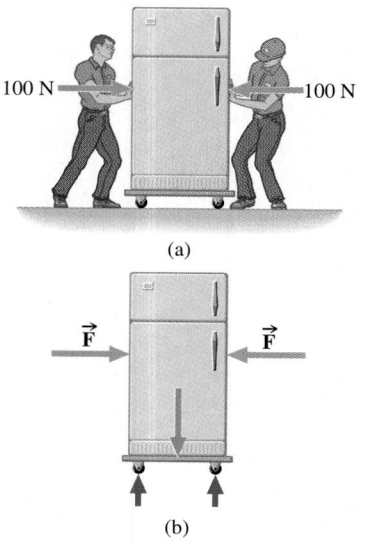

Figure 4.12 (a) If the only horizontal force acting on a body of mass m is F, then it will accelerate horizontally such that $F = ma$. This motion is independent of the vertical forces that we'll deal with shortly. (b) The free-body diagram shows the net horizontal force.

Figure 4.13 (a) Two people apply equal, oppositely directed forces to an object. Because force is a vector quantity, these two forces cancel each other and $\sum F_x = 0$, which means there cannot be any horizontal acceleration. (b) The free-body diagram makes the same point more directly.

Example 4.4 **[II]** The kid in Fig. 4.14a pulls a loaded wagon having a total mass of 100 kg. He applies a constant force of 100 N along the handle at 30.0°. Ignoring friction, compute the horizontal force on the wagon and the resulting acceleration. By the way, the wagon is much too heavy to lift off the ground, so don't worry about that.

Solution When you see the words "force" and "acceleration" in a problem, think of $F = ma$. (1) TRANSLATION—A known force is applied at a specified angle to a body of known mass; determine the horizontal force component and the horizontal acceleration. (2) Take the direction of motion to be positive. GIVEN: $m = 100$ kg, $F = 100$ N, and $\theta = 30.0°$. FIND: F_x and a_x. (3) PROBLEM TYPE—Newton's Laws/constant force and acceleration. (4) PROCEDURE—Draw a free-body diagram as is done in Fig. 4.14b. Compute the component of $\vec{F}$ in the direction of interest (horizontal), and then use $\sum F = ma$ to determine a_x. (5) CALCULATION:

$$\overset{+}{\rightarrow}\sum F_x = ma_x$$

where, because the motion is to the right, we make that the positive direction (*the arrow and plus sign are included as aids for the student—if you don't need the reminder, leave them off*). The only force acting horizontally is the component of $\vec{F}$;

$$F_x = F\cos\theta = F\cos 30.0° = \boxed{+86.6 \text{ N}}$$

Therefore

$$\overset{+}{\rightarrow}\sum F_x = +86.6 \text{ N} = ma_x$$

and

$$\boxed{a_x = +0.866 \text{ m/s}^2}$$

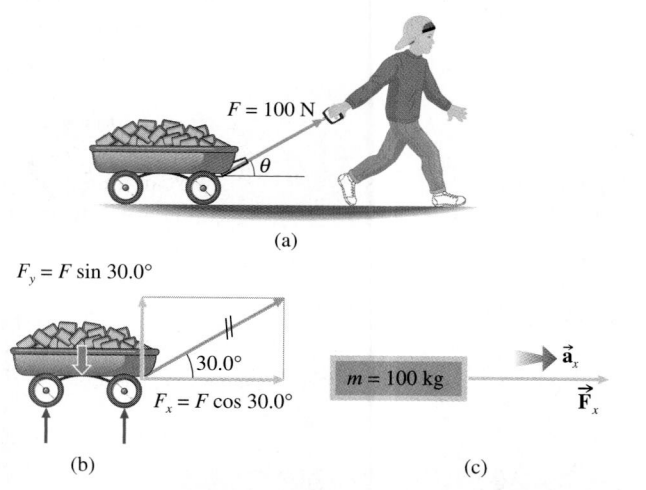

Figure 4.14 (a) The wagon experiences a force directed upward along the handle at an angle θ. (b) Its horizontal component $\vec{F}_x$ is the only horizontal force, and it will determine the horizontal acceleration. (c) Ignoring the vertical forces for the moment, the free-body diagram is simple.

Quick Check: If the total 100-N force acted horizontally, it would have produced an acceleration of 1.00 m/s², so the size of this answer is reasonable.

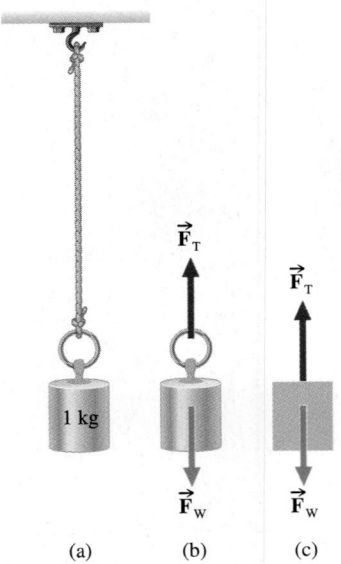

(a)   (b)   (c)

Figure 4.15 (a) A hook supporting a rope and a 1-kg brass cylinder. Suppose we wish to study the forces acting on the cylinder. In that case, draw a free-body diagram of the object of interest. (b) The weight ($\vec{F}_w$) acts down from the middle of the mass. (c) A simple free-body diagram of the brass cylinder.

In Fig. 4.15*a* a cylindrical 1-kg mass hangs on a rope. To study the forces on the cylinder we isolate it in a free-body diagram (Fig. 4.15*b*). Gravity tugs on every one of its atoms, but it will suffice to draw the overall weight vector $\vec{F}_w$ acting straight down along the center line (we'll come back to this when we consider the center-of-gravity or *c.g.*, p. 249). The magnitude of $\vec{F}_w$ is the weight of the cylinder, F_w. *The weight of the rope can be ignored because we will only deal with situations in which that weight is negligible in comparison to the load.* Gravity pulls down on the mass, and the mass pulls down on the rope with the same force, F_w. The rope, in turn, pulls up on the mass with a force whose magnitude is the tension F_T. Applying the Second Law in the vertical direction to the cylinder

$$\sum \vec{F}_y = \vec{F}_T + \vec{F}_w = m\vec{a}_y = 0$$

Taking *up* as positive, we can sum the scalar values of the forces, but since both weight and tension are inherently positive quantities we'll have to enter them into the equation with the proper signs:

$$+\uparrow \sum F_y = (+F_T) + (-F_w) = 0$$

and we conclude that $F_T = F_w$. Notice that there is a + sign and an arrow in front of the summation to serve as a reminder that *up* is positive — if you wish, you can leave it out.

The tension (F_T) *at any two points in a rope will be the same* (provided no tangential forces act on the rope between those two points). We can assume that **the tensile force exerted *by* any rope has a magnitude of F_T and is directed along the rope's length toward its center**.

Average Force

More often than not, the net force applied to an object (during a substantial time interval Δt) is not constant, nor is the resulting acceleration. Yet if the initial and final velocities of the object are known, $a_{av} = \Delta v / \Delta t$ and

[straight-line motion] $$F_{av} = ma_{av} \qquad (4.5)$$

We can think of F_{av} as the value of a constant force that produces the same change in motion in the same time interval as does the actual varying force (Fig. 4.16).

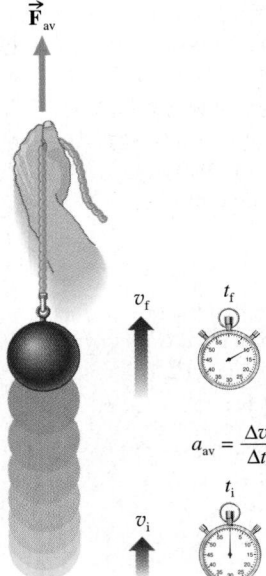

Figure 4.16 The application of a net force F_{av} for a time Δt results in a change in momentum and an average acceleration, $a_{av} = (v_f - v_i)/\Delta t$.

A spider web is a total tension structure. Every fiber is in tension.

Example 4.5 [II]   An old Checker cab having a mass of 1741.7 kg cruises along a straight level road at 35.8 m/s (Fig. 4.17). The driver lets the car coast in neutral, while air drag and friction decelerate it at a nonconstant rate down to 22.4 m/s in 24 s.   (a) Calculate the size of the average deceleration during the period.   (b) Determine the average scalar retarding force acting on the cab.

Solution   When you read the words "force" and "acceleration" in a problem think of $F = ma$. (1) TRANSLATION—(a) A body of known mass goes from one known speed to another in a specified time; determine the average deceleration. (b) Compute the average applied force.   (2) Take the direction of motion to be positive. GIVEN: $m = 1741.7$ kg, $v_i = 35.8$ m/s, $v_f = 22.4$ m/s, and $\Delta t = 24$ s.   FIND: (a) a_{av} and (b) F_{av}.   (3) PROBLEM TYPE—Newton's Laws/average force and acceleration. (4) PROCEDURE—Draw a free-body diagram; Figs. 4.17b and c are successively simpler versions. The retarding force opposes the

motion and slows the vehicle. Since F_{av} is called for, we'll use $F_{av} = ma_{av}$. But first, knowing Δt, v_i, and v_f, determine a_{av} via its definition.   (5) CALCULATION—Remembering that $\Delta v = v_f - v_i$

$$a_{av} = \frac{\Delta v}{\Delta t} = \frac{(22.4 \text{ m/s} - 35.8 \text{ m/s})}{24 \text{ s}} = -0.558 \text{ m/s}^2$$

or $\boxed{a_{av} = -0.56 \text{ m/s}^2}$. (b) Consequently

$$F_{av} = ma_{av} = (1741.7 \text{ kg})(-0.558 \text{ m/s}^2) = -971.9 \text{ N}$$

and so $\boxed{F_{av} = -0.97 \text{ kN}}$. The minus sign indicates deceleration.

Quick Check: Using momentum, from Eq. (4.1), $F_{av}\Delta t = \Delta p$, $(0.97 \text{ kN})(24 \text{ s}) = 23.3$ kN·s, whereas $\Delta p = (1742 \text{ kg})(13.4$ m/s$) = 23.3 \times 10^3$ kg·m/s, and all is well.

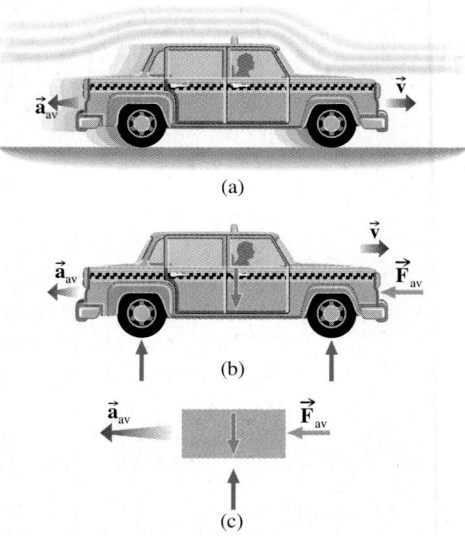

(a)

(b)

(c)

Figure 4.17   (a) An old cab coasting down from a speed v_i to a speed v_f. Air drag produces an average force $\vec{F}_{av}$, and the cab decelerates. Overlooking vertical forces for the moment, (b) is a free-body diagram of the cab. Notice how $\vec{a}_{av}$ is in the same direction as $\vec{F}_{av}$ but opposite to the motion. (c) When we begin to do a problem, the free-body diagram can be simple.

4.6 Weight: Gravitational Force

At this very moment the Sun, the Moon, all the planets, and a fair number of stars are gravitationally interacting with you. (Gravity propagates at the speed of light and if you are 100 years old, only those stars within a distance of 100 light-years of Earth can interact with you.) And so for the sake of completeness, we might say that **your weight is the net gravitational force exerted on you by the Universe**. Practically, however, for earthlings, the influence of our planet is so overwhelmingly dominant that everything else can be ignored. Moreover, the customary definition of weight evolved while we were stuck on the Earth without a thought of ever standing on some other celestial body. **The weight of an object on Earth is the downward force experienced by that object (*usually at the surface of***

The **weight** of an object is the downward force exerted on it by the Earth (or any other celestial object large enough and near enough to produce a substantial gravitational interaction).

By pushing back on a fixed block the runner causes the block to push forward on her. It is this force that initially accelerates the athlete in the forward direction.

the planet) **as a result of the Earth-object gravitational interaction**. Of course, if you're an alien living on the mythical planet Mongo, you just substitute the word Mongo for Earth.*

Because the Earth is revolving, a scale in your bathroom is actually moving in a circle and therefore accelerating, and its readings will be affected accordingly (Sect. 5.4). Let's call the "weight" it reads, the **effective weight**. In general, effective weight is slightly lower than the true weight as defined above, but the differences are usually negligible.

The Second Law provides a relationship between the concepts of weight ($\vec{F}_W$) and mass (m). Because the object's Earth-weight acts downward, causing the object to accelerate at g, $\vec{F} = m\vec{a}$ becomes

A body behaves gravitationally as if all of its mass were located at a single point called the **center-of-gravity** or *c.g.* (p. 249). Weight always acts down along a line passing through the *c.g.*

$$\vec{F}_W = m\vec{g} \qquad (4.6)$$

the more mass, the more weight. ***Whenever you are given a body's weight you can find its mass, and vice versa;*** $F_W = mg$. The weight of an object numerically equals the force that must be exerted upon it (e.g., by the floor or a scale) in order to keep the object from accelerating downward because of gravity. The weight on Earth of a 2.0-kg chicken is (2.0 kg)(9.81 m/s^2) = 20 N.

ON PUSHING ELEPHANTS, WEIGHTLESS AND OTHERWISE

Suppose you meet the 61.78-kN bull elephant (from Fig. 4.6) now wearing ice skates and stranded in the middle of a frozen pond and you decide to push him to shore. Neglecting friction, how hard must you constantly shove to uniformly accelerate the pachyderm from rest up to 6.7 m/s (i.e., 15 mi/h) in 10 s? The required acceleration [$a = \Delta v/\Delta t = (6.7$ m/s)/(10 s)] is 0.67 m/s^2, which is not very much. Then

$$F = ma = \frac{F_W}{g} a = \frac{(61.78 \times 10^3 \text{ N})}{9.81 \text{ m/s}^2}(0.67 \text{ m/s}^2) = 4.2 \text{ kN}$$

or 0.95×10^3 lb. Well, forget that idea. If you were even more ambitious and wanted the elephant to accelerate over the ice at $a = 9.81$ m/s$^2 = g$, then a would cancel g in the equation and $F = F_W$. *You would have to push with a force equal to the elephant's weight.* That's true on a frictionless ice pond or out in space far from any celestial body where gravity is negligible and the object is essentially weightless—the needed force depends on the mass and the acceleration. The moral of this tale is that you *can* get squashed out in space between two colliding weightless elephants; inertia has to do with mass, not weight.

The Physics of Standing Still

The force exerted by one object back on another is called a **reaction force**. It may have both a tangential component, often due to **friction** ($\vec{F}_f$), and a perpendicular component (i.e., normal to the surface) known as the **normal force** ($\vec{F}_N$). Let's first deal with the normal force. Because we live on a planet where gravity is appreciable, objects most often rest on horizontal surfaces and $\vec{F}_N$ usually points straight up (Fig. 4.18). That need not be the case: $\vec{F}_N$ must be perpendicular to the surface, but the latter could be tilted in any direction (a ping-pong ball flattened against a vertical paddle experiences a horizontal normal force). The magnitude of the normal force is the positive quantity F_N.

*It might surprise you to learn that the physics community has not as yet formalized a complete definition of "weight," but we are slowly hashing it out in the literature.

The floor pushes up and to the left with a reaction force $\vec{\mathbf{F}}_R$. It has components $\vec{\mathbf{F}}_f$ (a friction force keeping his foot from sliding) and $\vec{\mathbf{F}}_N$ (the normal force equal to his weight).

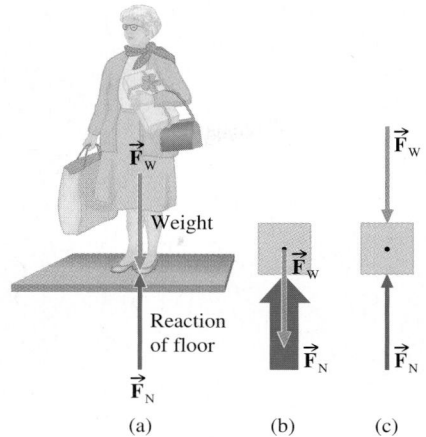

Figure 4.18 (a) The weight of an object is supported by an equal normal force (actually distributed over both feet) exerted upward by the floor. (b) The free-body diagram shows the two forces acting on the standing person. Because $\vec{\mathbf{F}}_W$ and $\vec{\mathbf{F}}_N$ are colinear and tend to overlap, we can draw them as shown in (b). Alternatively, just slide either vector along its line-of-action as in (c).

Weight is an *external force* acting *on* an object. The Earth pulls down on the object, and the object pulls up on the Earth—that's the interaction pair. As a rule, we don't go plummeting down to the center of the planet, so there must be some other force stopping us. The weight of the person in Fig. 4.18 causes her feet to press on the ground and the ground pushes back on her—that's also an interaction pair that is basically electromagnetic; it's the same phenomenon that keeps you from walking through walls. **When two colinear forces acting on a body point toward each other, the body is under compression**; the woman in Fig. 4.18 is under compression via $\vec{\mathbf{F}}_W$ and $\vec{\mathbf{F}}_N$. **An object in tension pulls, an object in compression pushes.**

Call the force exerted perpendicularly on a supporting surface the **load**. The weight of a brick resting on a floor numerically equals the downward load *on the floor*, and the equal-and-opposite force *on the brick* exerted by the floor is the upward normal force. If you lift up slightly on the brick, decreasing the load, then the normal force will be reduced accordingly. If instead you step onto the brick, the normal force will numerically equal the combined weight, the total load.

When you stand still, the net vertical force *on* you (weight down, normal force up) equals zero. The Second Law then demands that you continue to remain at rest in the vertical direction. If, for structural reasons, the floor cannot exert an upward force equal to your weight, the nonzero net force acting on you would be down, and down you would accelerate through the floor, just as if you were trying to stand on a paper box or walk on water.

The Physics of Jumping

Have you ever seen $\approx 1\frac{1}{2}$-year-old kids (who have not yet learned to jump) trying desperately to leap into the air? What they do not know is that a net *external* vertical force must be applied if a mass is to accelerate upward. The Third Law is the key: ***push down on the floor and it will push up on you***. Push with your leg muscles downward ($\vec{\mathbf{F}}_M$) on the floor (Fig. 4.19). The total load is then ($\vec{\mathbf{F}}_M + \vec{\mathbf{F}}_W$) down. The upward reaction, the normal force ($\vec{\mathbf{F}}_N$), is equal to $-(\vec{\mathbf{F}}_M + \vec{\mathbf{F}}_W)$. But only two forces act *on you*, $\vec{\mathbf{F}}_W$ down and $\vec{\mathbf{F}}_N$ up. $\vec{\mathbf{F}}_M$ is a force

SAGGING FLOORS AND ATOMIC INTERACTIONS

Floors do not usually collapse underfoot. They exactly match your weight with a reaction force, and one might wonder how they manage that. Materials, no matter how rigid, give under the influence of a body pushing on them, regardless of how lightly. When two objects exert a contact force on each other, atoms in both must be somewhat displaced by the interaction. The greater the load, the more a floor stretches, thereby building up a counterforce via springlike atomic interactions (that are electromagnetic) within the floor. The material ceases to distort any further when the upward reaction of the floor exactly cancels the load on it. Wooden floors often betray the sagging by squeaking as one board rubs past another, but all floors sag underfoot.

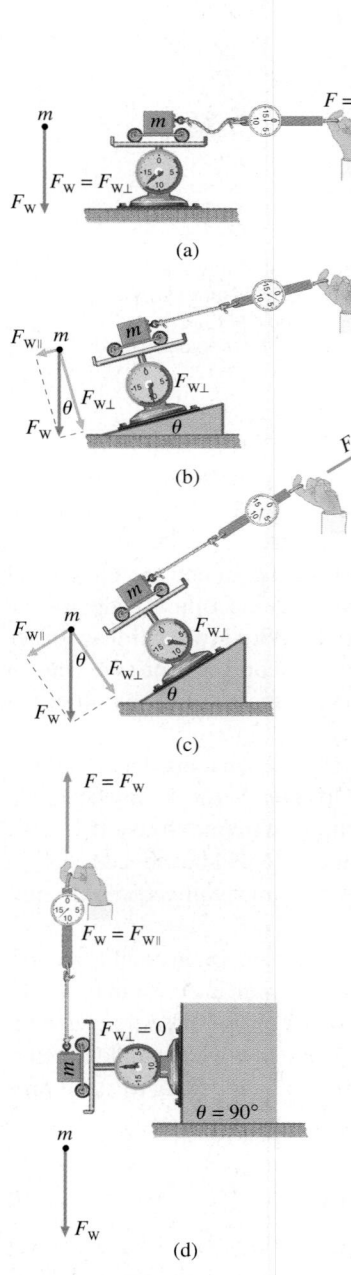

Figure 4.20 (a) A cart of mass m resting on a scale exerts a downward force equal to its weight F_W. (b) When the plane is inclined, the motionless (frictionless) cart (held at rest by the hand's force F) experiences a constant vertical weight F_W. (c) As θ increases, the component of the weight $F_{W\perp}$ decreases as $F_{W\parallel}$ increases. (d) When $\theta = 90°$, $F_{W\perp} = 0$ and $F_{W\parallel} = F_W$.

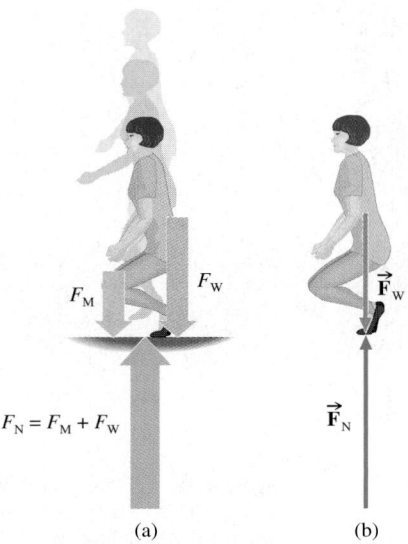

Figure 4.19 (a) When an individual jumps, two forces act *on* the person: F_W down and the normal force F_N, acting up. (b) The floor experiences a net downward force of $(F_M + F_W)$, and it exerts an upward equal and opposite reaction F_N. The net force on the jumper is F_M upward.

exerted *by you down on the floor* (Fig. 4.19*b*). The sum of the vertical forces on you equals your mass times your acceleration:

$$\sum \vec{\mathbf{F}} = m\vec{\mathbf{a}} = \vec{\mathbf{F}}_W + \vec{\mathbf{F}}_N = \vec{\mathbf{F}}_W + [-(\vec{\mathbf{F}}_M + \vec{\mathbf{F}}_W)] = -\vec{\mathbf{F}}_M$$

which is a net upward force, and up you go ($a \neq 0$). The harder you push down, the greater the resulting upward acceleration. Skeptics are encouraged to leap up from a bathroom scale to see it read ($F_W + F_M$) during the accelerated ascent. For much of that time the downward force averages about $2.3F_W$, but this decreases greatly before the feet lift off. An ordinary athlete will be able to leap about 45 to 60 cm (1.5 to 2 ft) straight up, whereas the very best male jumpers can just about double that height (Table 4.3).

The Inclined Plane

Although the force of gravity acts straight down, when an object is supported by an inclined plane (Fig. 4.20), components of its weight exist both parallel ($\vec{\mathbf{F}}_{W\parallel}$) and perpendicular ($\vec{\mathbf{F}}_{W\perp}$) to the surface. As you can see in Fig. 4.21, the only possible motion is along the incline, and it's $F_{W\parallel}$ that drives that motion. As we'll learn presently, $F_{W\perp}$ influences friction

Table 4.3

The Physiological Effects of Acceleration		
Acceleration	**Body orientation**	**Effect**
$2g$	Upright parallel to a	Walking becomes strenuous
$3g$	Upright parallel to a	Walking is impossible
$4g-6g$	Upright parallel to a	Progressive dimming of vision due to decrease of blood to retina, ultimate blackout
$9g-12g$	Reclining perpendicular to a	Chest pain, fatigue, some loss of peripheral vision, but one is still conscious and can move hands and fingers

The roof of this building is constructed from strongly tilted inclined planes to allow rain and snow to slide off rather than pile up and load the structure.

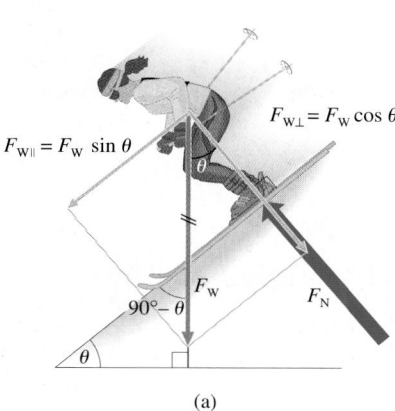

$F_{W\|} = F_W \sin \theta$

$F_{W\perp} = F_W \cos \theta$

$90° - \theta$

F_W

F_N

θ

(a)

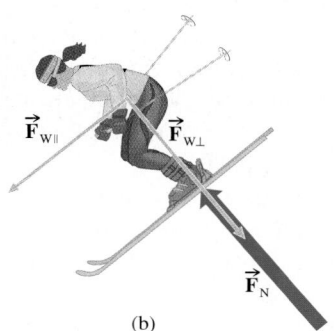

$\vec{F}_{W\|}$

$\vec{F}_{W\perp}$

$\vec{F}_N$

(b)

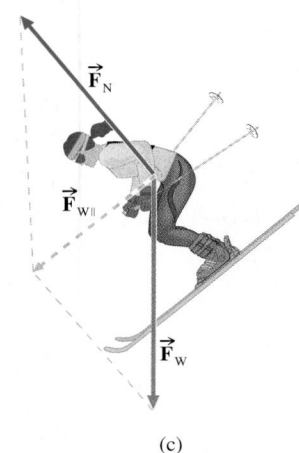

$\vec{F}_N$

$\vec{F}_{W\|}$

$\vec{F}_W$

(c)

Figure 4.21 A body on an inclined plane. (a) The component of weight acting down the inclined plane drives the body downhill. The component of weight acting perpendicular to the plane is matched by the normal force. (b) Applying the Second Law down the incline yields $\sum F_\| = F_{W\|} = ma_\|$. (c) Notice how $\vec{F}_W + \vec{F}_N = \vec{F}_{W\|}$.

and will be of interest as well. The angle between the line-of-action of the weight vector ($\vec{F}_W$) and the normal to the incline equals θ, the incline angle. Consequently

$$F_{W\|} = F_W \sin \theta \qquad \text{and} \qquad F_{W\perp} = F_W \cos \theta \qquad (4.7)$$

$F_W \sin \theta$ pushes the object down the incline, and $F_W \cos \theta$ pushes it into the surface. The Second Law applies independently in each perpendicular direction. As shown in the free-body diagram (Fig. 4.21b), the force $F_{W\|}$ will accelerate the object down the slope; $\sum F_\| = ma_\|$, from which we can compute $a_\|$. There is no acceleration perpendicular to the slope; $\sum F_\perp = ma_\perp = 0$, which just means that $F_{W\perp} = F_N$.

Example 4.6 **[II]** The 50.0-kg skier in Fig. 4.21 coasts along the surface of a snow-covered slope (assumed to be frictionless) tilted at an angle of 30.0°. Compute (a) the magnitude of the normal force acting on her; (b) the magnitude of the force tending to drive her down the inclined plane; and (c) the resulting acceleration ignoring air drag.

Solution When you see the words "force" and "acceleration" in a problem, think of $F = ma$. (1) TRANSLATION—A known mass slides down an inclined plane; determine (a) the normal force, (b) the component of the weight down the slope, and (c) the acceleration. (2) GIVEN: $m = 50.0$ kg and $\theta = 30.0°$. FIND: (a) F_N, (b) $F_{W\|}$, and (c) $a_\|$. (3) PROBLEM TYPE—Newton's Laws/inclined plane. (4) PROCEDURE—Draw a free-body diagram (Fig. 4.21b); apply $\sum \vec{F} = m\vec{a}$ in each direction. (5) CALCULATION: The component of the skier's weight pressing

her into the surface is

$$F_{W\perp} = mg \cos \theta = (50.0 \text{ kg})(9.81 \text{ m/s}^2) \cos 30.0° = 425 \text{ N}$$

The skier doesn't leave the surface of the incline, and so $a_\perp = 0$. Taking *up* (normal to the surface) as positive, the sum of the forces perpendicular to the incline is zero;

$$+\nwarrow \sum F_\perp = 0 = F_N + (-F_{W\perp})$$

$F_{W\perp}$ points into the surface in the negative direction—we enter it with a minus sign. Therefore, $\boxed{F_N = 425 \text{ N}}$. (b) The magnitude of the driving force down the incline is

$$F_{W\|} = mg \sin \theta = (50.0 \text{ kg})(9.81 \text{ m/s}^2) \sin 30.0° = \boxed{245 \text{ N}}$$

(c) To compute the down-plane acceleration (taking the direction of motion *down the slope* as positive),

continued

$$+\swarrow\sum F_\parallel = ma_\parallel = F_{W\parallel}$$

and $\quad a_\parallel = \dfrac{F_{W\parallel}}{m} = \dfrac{(mg\sin\theta)}{m} = g\sin 30.0°$ $\quad$ (4.8)

or $\boxed{a_\parallel = \frac{1}{2}g}$. This result is independent of the mass and applies to any body sliding down a frictionless incline at $\theta = 30°$.

Quick Check: We can think of the inclined plane as a kind of "gravity reducer." With $\sin 30° = \frac{1}{2}$ in Eq. (4.7), the body behaves as if it were in free-fall (down the slope) at $\frac{1}{2}g$. Since the skier's weight is 490 N, in this tilted world it's driven down by $\frac{1}{2}$490 N. A normal force of 425 N at a small incline is also reasonable.

4.7 Coupled Motions

The two masses in Fig. 4.22 are attached together by an unstretchable rope. The pulleys are weightless and frictionless, and the tensions are therefore constant throughout each rope. For the moment, we'll require that the surfaces be frictionless as well. Suppose the motion takes place in the direction shown by the arrow in each case. The leading mass m_1 pulls the connecting rope along, and the rope pulls the trailing mass m_2. The trailing mass can never overtake and slacken the rope, nor can it lag behind, accelerating slower than the rope; each mass must accelerate at the same rate. With two masses, we can write two coupled Second-Law equations and solve for two unknowns, usually F_T and a, although any two parameters could be unknown.

If there is any ambiguity about how the system moves, *guess at the direction of the overall motion and make that positive*, even if it means down is positive for one part and up for another. If you guess wrong, the values of the unknowns will just come out negative.

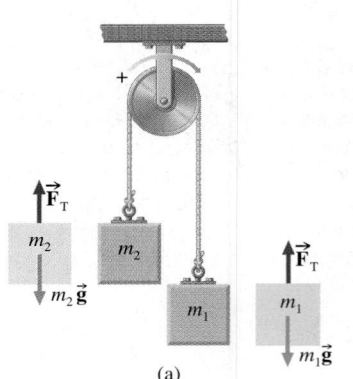

(a)

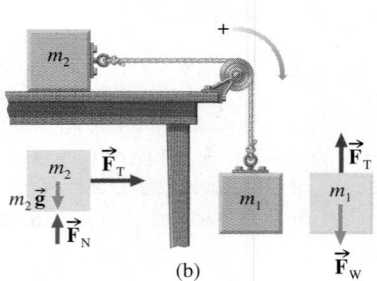

(b)

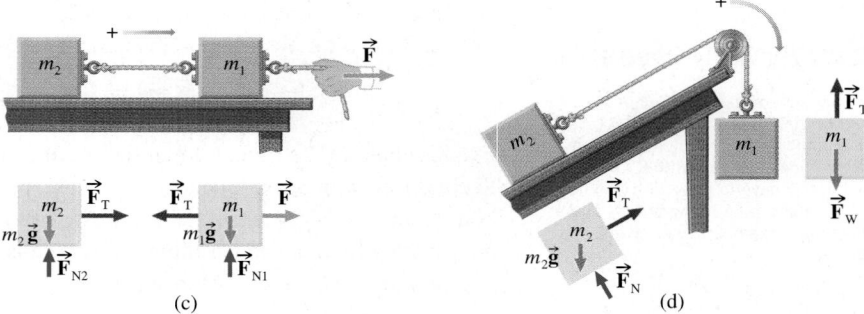

(c)

(d)

Figure 4.22 Four examples of coupled arrangements where two masses are attached together by a rope. The free-body diagram for every mass is drawn. In each of the four examples, the same tension acts on both masses and both accelerate at the same rate.

Example 4.7 **[II]** Mary ($m_M = 50$ kg) and her boyfriend, Don ($m_D = 70$ kg), are tied together (we won't ask why) by a rope of negligible mass. She is standing on a frictionless horizontal sheet of wet ice when Don, who is not too smart, accidentally steps off a cliff (Fig. 4.23*a*). Assume the unlikely possibility that the tree limb is frictionless and that her length of the rope is horizontal. Determine (a) the tension in the rope and (b) the accelerations of the ill-fated lovers. What would happen if she cut the rope?

Solution Whenever you see the words "tension" and "acceleration" in a problem, think of $F = ma$. Here there are two moving objects, and so we'll have to apply $F = ma$ to them separately. (1) TRANSLATION—Two known masses are tied together such that the falling one pulls the other; determine (a) the tension and (b) the accelerations. (2) GIVEN: $m_M = 50$ kg, and $m_D = 70$ kg. FIND: (a) F_T, (b) a_M, and a_D. (3) PROBLEM TYPE—Newton' Laws/coupled motion. (4) PROCEDURE—Draw free-body diagrams (Figs. 4.23*b* and *c*). Apply $\sum F = ma$

continued

to each body, in the direction of motion. As long as the rope does not stretch, $a_D = a_M = a$. Take clockwise to be positive as in Fig. 4.22b. (5) CALCULATION—There are two unknowns, a and F_T, and we must construct two equations. Mary and the Second Law (for the horizontal motion) yield

$$\xrightarrow{+}\sum F_{xM} = F_T = m_M a \qquad (4.9)$$

while Don's vertical contribution provides

$$+\downarrow\sum F_{yD} = F_{WD} + (-F_T) = m_D a \qquad (4.10)$$

Solve these two equations simultaneously {see **MATH REVIEW: PART A-4** on the **CD** 🌐} for either unknown; let's do it for a first. Substituting Eq. (4.9), $F_T = m_M a$, into Eq. (4.10) leads to

$$F_{WD} - m_M a = m_D a$$

Since we know m_D, replace Don's weight F_{WD} by $m_D g$, via Eq. (4.6),

$$m_D g = m_D a + m_M a$$

and

$$a = \frac{m_D}{(m_D + m_M)g}$$

Putting this expression into Eq. (4.9) produces the desired equation for the tension:

$$F_T = \frac{m_M m_D}{(m_D + m_M)} g$$

The numerical values are $\boxed{a = 0.58g}$ and $\boxed{F_T = 0.29 \text{ kN}}$. The tension is the magnitude of the force accelerating Mary. If the rope were to be cut, the tension would drop to zero and she would subsequently coast at a fixed speed. Don would then descend in free-fall.

Quick Check: When $m_D \gg m_M$, $(m_D + m_M) \approx m_D$, $a \approx g$, which means Don is approximately in free-fall, and that's reasonable. Moreover, $F_T \approx m_M g$, and that too is to be expected. When $m_M \gg m_D$, $a \approx 0$ and $F_T \approx F_{WD}$, as it would be if Don were tied to a massive boulder. There is then no acceleration and he just hangs over the edge. Both equations produce sensible results at the extremes, suggesting that they may be correct. Furthermore, both have correct units.

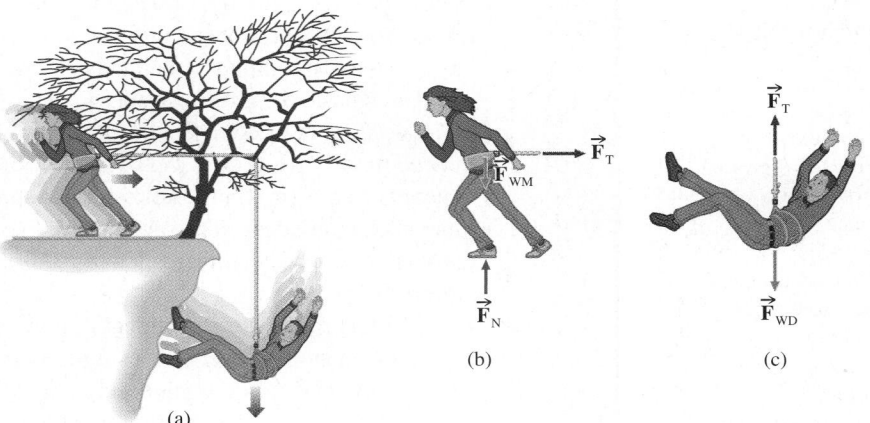

(a)

(b)

(c)

Figure 4.23 (a) Mary sliding along the ice, after Don falls off the cliff. Free-body diagrams of (b) Mary and (c) Don.

4.8 Friction

Our common experience is that objects in motion do not usually stay in motion, despite the unrefuted wisdom of the First Law. In practice we move, or try to move, one thing against another, and there are interactions that resist the motion. A force of this sort that *opposes an impending or actual motion* is said to be a ***frictional force*** ($\vec{\mathbf{F}}_f$). When friction impedes a motion that is in progress, it's *kinetic friction*; when it prevents motion from occurring altogether, it's *static friction*.

Our present concern is with dry friction and its influence on motion. We limit the discussion to solids sitting, sliding, and rolling on other solids. Even so, our knowledge is rather rudimentary, the analysis is incomplete, and the experimental data (because of contaminants) are often imprecise and difficult to reproduce.

A gecko can walk up a vertical glass wall using the millions of tiny hairs or setae on the bottoms of its feet. These "hairs" get so close to the atoms of the wall that each seta electromagnetically attracts the wall with a force of about 200×10^{-6} N. This adhesive force (which arises from the minute interactions between nearby atoms and molecules) is known as the Van der Waals force. The setae release when tipped at 30°, which accounts for the gecko peeling its toes off the surface as it walks (see p. 115)

Static Friction

Figure 4.24 shows a block on a table that supports it with a normal force F_N. A small horizontal force F is applied to the right—the spring scale registers that force, but the block doesn't accelerate—it doesn't seem to move at all. The Three Laws explain what's happening: Because $a = 0$, the net force is zero, and there must be a force to the left equal to F. *This force acting parallel to the surface, resisting the motion, is the force of* **static friction**, F_f. Increase F and, if the object remains at rest, F_f must also have increased. If F is made larger and larger, *the block will ultimately break loose and move when F exceeds the maximum possible value of the static friction, $F_f(\text{max})$, by even the tiniest amount*. When there is no overall motion of the block, the static friction force is equal and opposite to the driving force F, and that equality continues right up to $F_f(\text{max})$, just as a rope can react with any size tension right up to the moment when it breaks.

Three basic insights regarding $F_f(\text{max})$ come from simple observations (Leonardo da Vinci was aware of them in the mid-fifteenth century). (1) $F_f(\text{max})$ *is proportional to the normal force*: $F_f(\text{max}) \propto F_N$. (2) The same block in Fig. 4.24 sitting on another surface will experience an $F_f(\text{max})$ proportional to F_N, but it will likely be different. Some surfaces are more slick than others. The statement $F_f(\text{max}) \propto F_N$ can be turned into an equality using a constant of proportionality, namely μ_s. This **coefficient of static friction** depends on the two materials in contact.

We get around our theoretical limitations by compressing the details of what is happening on an atomic level into a single coefficient that, because it cannot yet be calculated from theory, must be determined experimentally (Table 4.4). Accordingly

$$F_f(\text{max}) = \mu_s F_N \tag{4.11}$$

Table 4.4

Approximate Friction Coefficients*

Material	μ_s	μ_k
Steel on ice	0.1	0.05
Steel on steel—dry	0.6	0.4
Steel on steel—greased	0.1	0.05
Rope on wood	0.5	0.3
Teflon on steel	0.04	0.04
Shoes on ice	0.1	0.05
Climbing boots on rock	1.0	0.8
Leather-soled shoes on carpet	0.6	0.5
Leather-soled shoes on wood	0.3	0.2
Rubber-soled shoes on wood	0.9	0.7
Auto tires on dry concrete	1.0	0.7−0.8
Auto tires on wet concrete	0.7	0.5
Auto tires on icy concrete	0.3	0.02
Rubber on asphalt	0.60	0.40
Teflon on Teflon	0.04	0.04
Wood on wood	0.5	0.3
Ice on ice	0.05−0.15	0.02
Glass on glass	0.9	0.4

*The first column lists values of various coefficients of static friction. The second gives the corresponding values of the kinetic coefficients of friction, a concept that will be discussed presently.

Figure 4.24 (a) A horizontal force F acting through a spring scale is applied to a motionless block of weight F_W. Static friction opposes the impending motion with a force F_f that can be as much as $F_f(\text{max}) = \mu_s F_N$. (b) The free-body diagram indicates that no horizontal acceleration will occur until $F > F_f$.

FRICTION, THE TWO-FACED FORCE

Friction arises via the electromagnetic interaction between atoms that can be in any bulk state—solid, liquid, or gas. *Liquid-gas friction* saves us from being pelted by high-speed raindrops; it allows the wind to make the sea choppy and retards the rising bubbles in a glass of beer. *Solid-liquid friction* slows the flow of petroleum in pipelines and blood in vessels; it helps to make bullets fairly useless under water and oil slicks fairly deadly on highways. *Solid-gas friction* hinders the flow of air in pumps, pipes, and lungs; it slows skaters, cyclists, baseballs, and fire engines (at 70 mi/h a typical car expends 70% of its fuel just pushing air out of its way); it burns up meteors, threatens returning spaceships, and is the unflagging joy of parachutists. *Solid-solid friction* stops your car whenever you use your brakes, and it wastes about 20% of the engine's power; without it, you could neither walk nor hold up your socks; it keeps cloth, carpets, ropes, baskets, and wicker furniture from unraveling; it allows us to write with pencils and turn pages. Friction is the great two-faced force of nature, at once the bane and the backbone of modern terrestrial technology.

At this county fair, teams of horses attempt to drag heavy loads of concrete across the ground. Here a team strains to overcome static friction.

where F_f is the magnitude of $\vec{\mathbf{F}}_f$. A 40-N dictionary on a surface where $\mu_s = 0.3$ will require a minimum horizontal force of 12 N to overcome static friction and make it slide. For modern rubber-soled running shoes on artificial grass $\mu_s \approx 1.5$.

(3) The third insight is the surprising observation that F_f (max) *is independent of the apparent size of the contact area between the two solid surfaces*. Place this book on a table resting on either its narrow spine or its broad face, and the maximum-static-friction force will be the same.

Example 4.8 **[II]** A climber (Fig. 4.25) stands on the rock face of a mountain. The soles and heels of her boots have a static-friction coefficient equal to 1.0. (a) What is the steepest slope she can stand on without slipping? (b) Assuming her pants have a static-friction coefficient for rock of 0.3, what happens if she sits down to rest?

Solution This problem is about static friction, so we should be thinking about μ_s and F_N. (1) TRANSLATION—An object is on an inclined plane; knowing the coefficient of static friction, determine (a) the greatest angle before slipping occurs and (b) what happens when the coefficient decreases. (2) GIVEN: (a) $\mu_s = 1.0$ and (b) $\mu_s = 0.3$. FIND: The corresponding maximum slope θ_{max} for each. (3) PROBLEM TYPE—Newton's Laws/static friction. (4) PROCEDURE—Draw a free-body diagram (refer back to Fig. 4.20). The climber's weight F_W has two components: $F_{W\perp} = F_W \cos \theta$, which determines F_N, and $F_{W\parallel} = F_W \sin \theta$, which acts down the slope. Static friction opposes the impending motion due to $F_{W\parallel}$ and acts *up* the incline, parallel to it. (5) CALCULATION—Because she is standing still, $a_\perp = 0$

$$+\nearrow \sum F_\perp = 0 = F_N - F_W \cos \theta$$

or

$$F_N = F_W \cos \theta$$

Inasmuch as $a_\parallel = 0$

$$+\nwarrow \sum F_\parallel = 0 = F_f - F_W \sin \theta$$

and

$$F_f = F_W \sin \theta$$

To combine the two functions of θ into one tangent expression,

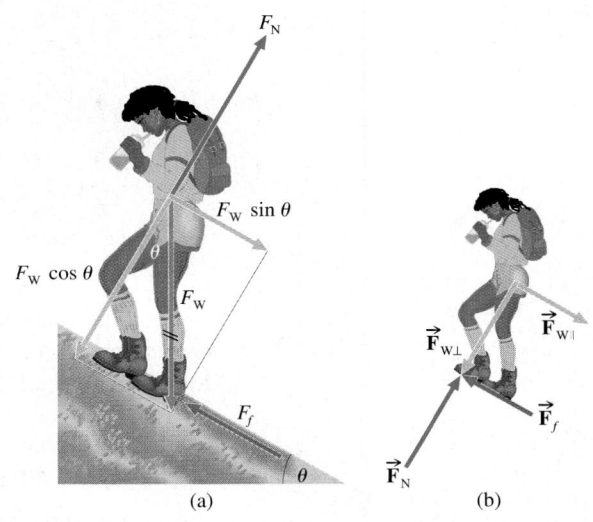

Figure 4.25 (a) A climber at rest on an inclined surface, where friction opposes her impending downward motion. (b) The free-body diagram is similar to Fig. 4.24b except that it's tilted.

continued

divide F_f by F_N, yielding

$$\frac{F_f}{F_N} = \tan \theta$$

This ratio has its maximum value when $F_f = F_f(\text{max}) = \mu_s F_N$, whereupon $\theta = \theta_{\text{max}}$, thus

$$\frac{F_f(\text{max})}{F_N} = \tan \theta_{\text{max}} = \mu_s$$

and $\theta_{\text{max}} = \tan^{-1} \mu_s$ (4.12)

independent of the mass. Measuring the tilt angle at which a body starts to slide down an incline is a convenient way to determine μ_s. With $\mu_s = 1$, $\boxed{\theta = 45°}$. If she were to sit, $\tan^{-1}(0.3) = \boxed{17°}$, and she would slide down any hill tilted at more than 17°.

Quick Check: Both sides of Eq. (4.12) are unitless, which is appropriate. At the extremes, when $\mu_s = 0$, $\theta_{\text{max}} = 0$; when $\mu_s = \infty$, $\theta_{\text{max}} = 90°$, and that makes sense.

Both the proportionality between F_f and F_N and the area independence of F_f are working principles rather than fundamental relationships. They generally hold true; and even when they are off, it's usually by no more than about 10%. As we will see, however, there are a few exceptions. Similarly, the idea that μ_s is always constant is not exactly true either—it can change as the contact time increases. Have you ever tried to unscrew a nut that has been in place for a few years or move a vase that has been sitting on a painted surface for several months?

Moving via Static Friction

We walk by cleverly arranging for an external force to propel us. Newton's Third Law is at the center of the process—push backward on the floor and it will react with a forward force that propels you. But only if there is friction can you push tangentially backward on the

The bicycle moves forward when the rear tire pushes back on the ground via friction. The ground pushes the tire, and hence the bike, forward.

floor. And that push cannot exceed $F_f(\text{max})$ or your foot will slide out from under you. Friction, opposing the backward motion of your foot, propels you forward—*friction is the driving force*.

A car traveling at a constant speed leaves unsmeared tire marks in damp sand because there is no horizontal motion of the tread with respect to the ground. Provided there is no slipping, *it's the static-friction force between tires and road that propels a car*, even when it is in motion. It's the same static-friction force that accelerates it. Skidding with the brakes locked and the tires not turning or attempting to accelerate so rapidly that the wheels spin (burning rubber) results in something different called *kinetic friction*.

EXPLORING PHYSICS ON YOUR OWN

STATIC FRICTION: Push on an empty chair, first gently, then harder and harder until it slides. Have someone sit in it and try again. With the added load, $F_f(\text{max})$ is increased tremendously. Now place this book flat on a table and put something (e.g., a calculator or a wallet) on the cover. Slowly raise the cover until the object slides. Notice the angle (θ) that the cover makes with the horizontal; $\tan \theta = \mu_s$ as per Example 4.8.

Example 4.9 [II] A washing machine in a wooden crate has a total mass of 100 kg. It is to be dragged across an oak floor by tugging on a rope (Fig. 4.26), making an angle of 30° with the horizontal. What minimum force will be needed to get the thing moving? Will it be more or less when $\theta = 0$?

Solution This is a maximum-static-friction problem, so we should be thinking about μ_s and F_N. (1) TRANSLATION—An object is made to overcome static friction; knowing the angle at which the force acts, determine its minimum value. (2) GIVEN: $m = 100$ kg, $\theta = 30°$, and from Table 4.4 $\mu_s = 0.5$. FIND: F when $\theta = 30°$ and when $\theta = 0°$. (3) PROBLEM TYPE— Newton's Laws/static friction. (4) PROCEDURE—Draw a free-body diagram. The central equation is $F_f(\max) = \mu_s F_N$, but here $F_N \neq F_W$. Use the Second Law to determine F_N. $\vec{F}$ has two scalar components, one parallel ($F \cos \theta$) and one perpendicular ($F \sin \theta$) to the direction of impending motion. (5) CALCULATION— Since the crate is at rest, taking vertically *up* as positive yields

$$+\uparrow \sum F_y = 0 = F_N + F \sin \theta - F_W$$

and $F_N = F_W - F \sin \theta$. Taking the direction of impending motion horizontally as positive, we get

$$\xrightarrow{+} \sum F_x = 0 = F \cos \theta - F_f$$

and $F_f = F \cos \theta$. The crate begins to move when the driving force just exceeds F where $F \cos \theta = F_f(\max)$. Furthermore

$$F_f(\max) = \mu_s F_N = \mu_s(F_W - F \sin \theta)$$

Figure 4.26 The washing machine crate experiences an applied force $\vec{F}$ that both lifts upward, lightening the load, and pulls forward to overcome static friction. The normal force equals the load, which now is $(F_W - F_y)$. The crate will move when $F_x > F_f(\max)$, where $F_x = F \cos \theta$.

and setting these last two equations for $F_f(\max)$ equal to each other yields

$$F \cos \theta = \mu_s F_W - \mu_s F \sin \theta$$

and $$F = \frac{\mu_s m g}{\cos \theta + \mu_s \sin \theta} = \frac{490.3 \text{ N}}{(0.866 + 0.250)} = \boxed{0.4 \text{ kN}}$$

The force must exceed 0.4 kN . When $\theta = 0$, $F = \mu_s F_W = 490$ N, or $\boxed{0.5 \text{ kN}}$.

Quick Check: If F is horizontal, $\theta = 0$, $\cos 0 = 1$, $\sin 0 = 0$, and the expression for F becomes $F = \mu_s m g = \mu_s F_N$, which suggests that it might be right.

Kinetic Friction

Look at Fig. 4.24, where we pulled on a body at rest to observe its static-friction force. As the driving force increased, F_f increased until it reached a maximum value of $F_f(\max)$. Applying a still greater force will cause the body to break loose and move in the direction of the applied force. The **kinetic (or sliding) friction force**, *the retarding force exerted on a sliding body in contact with a surface, is equal and opposite to the driving force that's needed to maintain the body in uniform motion.* Once again, experiments show that the three principles of friction apply, this time with a *coefficient of kinetic friction* μ_k, where

$$F_f = \mu_k F_N \tag{4.13}$$

On Friction. Generally it is easier to further the motion of a moving body than to move a body at rest.

THEMISTIUS (390–320 B.C.E.)
PHYSICA

EXPLORING PHYSICS ON YOUR OWN

WALKING AND FRICTION: To begin to walk you must accelerate and, therefore, must exert a proportionately large force backward on the ground; the greater is a, the greater must be F. Try to accelerate as fast as you can and the physics will become obvious. By contrast, it takes just a few ounces of force to continually overcome air drag as someone walks along at a constant ($a = 0$) pace. To experience that, have a person "walk in place." As she stands there marking time, push forward on her back to get her moving. The greater the a you wish to impart, the greater the force you must apply. Once started, however, you need only push with a tiny force (enough to overcome air friction) to keep her moving at a *constant* speed. That's why people barely notice that they're pushing back as they walk forward. Why do you walk very slowly on wet ice? Why do runners on dirt tracks wear shoes with cleats?

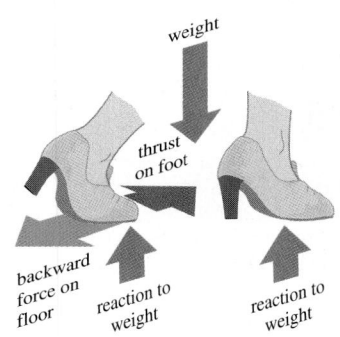

FOOTPRINTS, TREAD MARKS, & STATIC FRICTION

When you walk or run at a uniform rate, you gently push backward, but your feet don't move horizontally while in contact with the ground. That's why you make neat footprints trotting on snow. While traveling at a constant speed, a runner's foot actually moves through space with a motion that closely matches that of a rolling wheel (Fig. 4.27), and static friction propels both! Imagine a dot of red paint on the rim of a uniformly rolling wheel. At the instant the dot hits the ground, it will be moving backward with respect to the wheel's center at the same speed the entire wheel is moving forward. With respect to the ground, just as the dot comes down and touches, it will be motionless.

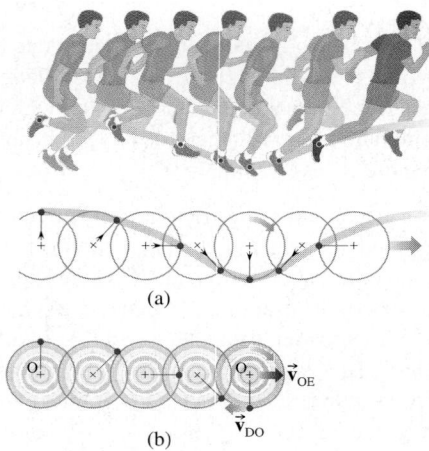

Figure 4.27 (a) The motion of a running foot resembles that of a point on a uniformly rolling wheel. (b) At the instant shown, the red dot is moving backward at a velocity with respect to the center of the wheel of $\vec{\mathbf{v}}_{DO}$. This is equal in magnitude and opposite in direction to the wheel's velocity with respect to the Earth, $\vec{\mathbf{v}}_{OE}$. Hence $\vec{\mathbf{v}}_{DE} = \vec{\mathbf{v}}_{DO} + \vec{\mathbf{v}}_{OE} = 0$ and the dot is at rest, at that instant, with respect to the Earth.

Though it has to be applied with care, this remarkably simple relationship holds true for a wide range of sliding situations both lubricated and not. Table 4.4 lists experimentally determined values of μ_k; it makes it clear that static friction is usually greater than kinetic friction (a fact apparent to anyone dragging a heavy mass). The observation that kinetic friction is, more or less, independent of speed is attributed to Charles Augustin de Coulomb (1785) who was quite aware of the limitations of that notion. Indeed, for automobile tires μ_k varies with temperature, load, and sliding speed.

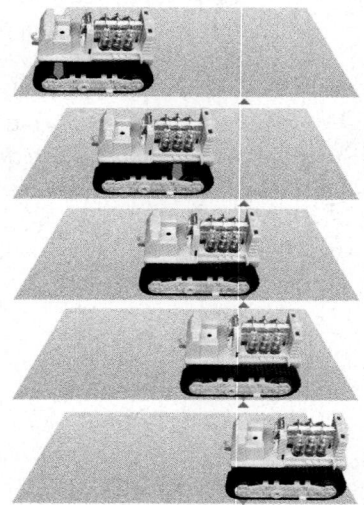

Just as with the rolling wheel in Fig. 4.27, any point on a tractor's tread that touches the ground (e.g., the green mark fixed to the tread) is motionless with respect to the Earth. In effect, the tractor lays the tread down on the ground where it remains at rest as the vehicle drives over it. Thus it's static friction that propels both cars and tractors as they travel without slipping.

Example 4.10 **[II]** Someone wants to push a (100-kg) box full of books along the floor by exerting a constant horizontal force of 600 N. Given that the coefficient of static friction is 0.6 and the coefficient of kinetic friction is 0.1, determine the resulting motion of the box.

Solution The first thing to figure out is if the box moves. (1) TRANSLATION—An object of known mass experiences a known

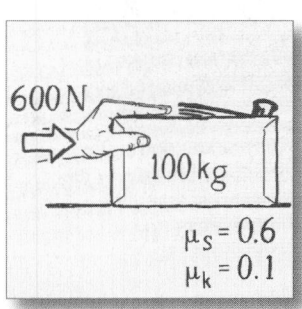

applied horizontal force in the presence of friction; determine the resulting horizontal acceleration, *if there is any*. (2) GIVEN: $m = 100$ kg, $F = 600$ N, $\mu_s = 0.6$, and $\mu_k = 0.1$. FIND: a. (3) PROBLEM TYPE—Newton's Laws/static and kinetic friction. (4) PRO-CEDURE—Draw a free-body

diagram. First use $F_f(\text{max}) = \mu_s F_N$ to see if the box breaks free and moves; if it doesn't $a = 0$, if it does use $\overset{+}{\rightarrow}\sum F_x = ma$ to determine a. (5) CALCULATION—To see if $F > F_f(\text{max})$ we need $F_N = F_W = mg = 981$ N. Then

$$F_f(\text{max}) = \mu_s F_N = (0.6)(981 \text{ N}) = 588.6 \text{ N}$$

which is less than F; therefore, the block moves. Take the direction of motion as positive and realize that kinetic friction ($F_f = \mu_k mg$) opposes the motion; it's in the negative direction and must be entered as a negative quantity,

$$\overset{+}{\rightarrow}\sum F_x = ma = F - \mu_k mg = 600 \text{ N} - (0.1)(981 \text{ N}) = 501.9 \text{ N}$$

and $\quad a = (F - \mu_k mg)/m = (501.9 \text{ N})/(100 \text{ kg}) = \boxed{5 \text{ m/s}^2}$

Quick Check: If $a = \sum F/m \approx 5$ m/s², then $\sum F \approx 500$ N. Since $\mu_k = 0.1$, a force of $1/10$ of the weight is needed to overcome kinetic friction and the remainder accelerates the body. Thus the applied force must be ≈ 98 N larger than $\sum F$, or ≈ 598 N.

Rolling Friction

Wheels were used in transportation 5000 years ago, and it has been obvious for all that time that it's easier to roll a load than to slide it. Ideally, overlooking air drag, if a rigid wheel is rolled along a flat inflexible surface, it should go on forever. That doesn't happen, and we call the retarding force **rolling friction**. The effect can be described in the same form as were static and kinetic friction, namely,

$$F_f = \mu_r F_N \qquad (4.14)$$

where μ_r is the *coefficient of rolling friction*. The friction force is equal in magnitude to the force required to keep the object rolling at a uniform speed.

The effect arises from the deformation of both the wheel and the surface, which creates an obstacle to free rolling. When the surfaces are hard, there is little deformation, as when steel wheels roll on steel tracks, whereupon μ_r is only about 0.001. By contrast, a flexible rubber tire on concrete will have a coefficient around 0.01 to 0.02 (with radials lower than cross-ply). Ideally, a car tire should have high sliding friction, but low rolling losses. The rolling coefficient for tires actually varies with speed, increasing as much as 15% between 0 and 120 km/h. However small μ_r is, the average auto at 80 km/h uses about 30% of its power to overcome rolling friction. At low speeds it exceeds air drag. That situation changes at around 55 km/h (i.e., 34 mi/h), depending on the shape of the car, after which air drag predominates. The admonition to maintain proper tire pressure translates into lower rolling friction and better fuel economy.

The Causes of Friction

Even under heavy loads rigid steel wheels and tracks experience little deformation, and therefore have a low coefficient of rolling friction.

Today tribology, the study of friction, is an active area of research. Even though friction has been a practical concern for thousands of years, we still don't have a complete understanding of what causes it. Coulomb suggested that friction had two components: one dependent

Burned rubber due to rapid deceleration. A tire's surface heats up and the rubber burns, leaving marks on the road as the vehicle accelerates. These tracks were left by the two tires on the right side of a truck. Variations show that the driver pumped the brakes. Keep in mind that you don't get much friction sliding along on melted rubber. It's much better not to lock up the wheels.

The flattening of the tire contributes to rolling friction. Every time the wheel turns, energy is lost reshaping the rubber, and the tire heats up.

on the normal force (the load), and another arising from adhesion. This view seems to match the latest experimental results where friction, on an atomic scale, has been observed even with no load.

The attraction between the atoms and molecules that holds solid matter together is electromagnetic in nature. Without it, there would be no such thing as textbooks, toenails, or teacups, nor would there be ordinary friction. But only when the atoms are close to one another will they experience an appreciable attractive force (i.e., adhesion); the effect is short-ranged. Such forces become negligible at distances equivalent to only about 4 or 5 atomic diameters (each of which is $\approx 1 \times 10^{-10}$ m to 2×10^{-10} m). Friction is in part due to the shearing of many tiny regions where the two materials in contact have bonded; these are known as adhesive junctions. An additional mechanism that can contribute to friction is the mechanical plowing of ridges on one surface across the other surface.

We've talked about static friction for a block on a surface (Fig. 4.24), and said that the two objects were at rest up to the point where the applied force exceeds $F_f(\text{max})$. Well, that's not quite true; the atoms in the face of the block actually do move by microscopic amounts under the influence of F. *The application of a force to an object always distorts the object.* The electromagnetic forces on the displaced atoms are springlike and tend to return the atoms to their unshifted positions. It is this process that accounts for the fact that $F_f = F$ right up until the "springs" are overwhelmed and the block breaks loose and moves as a whole.

Common objects may seem smooth to the eye and touch, but on a microscopic level they are jagged and rough. Even a polished surface is a craggy affair with features about 3×10^{-8} m across, each hill corresponding to perhaps 100 atomic diameters. When two ordinary surfaces are in contact, they only touch at a few high spots. Any material will

STOPPING A MOVING CAR WITH STATIC FRICTION

In practice, kinetic friction decreases as the sliding speed increases. Drivers know from experience that they must ease off the brakes if they are to bring an automobile to a smooth stop—the kinetic friction on the brake shoes increases as the wheels slow, and after a few jolting stops, the new driver learns to lighten the pressure appropriately. Thus, while μ_k for a sliding tire is roughly 0.8 at 8 km/h, it is likely to be less than about 0.5 at 130 km/h, and that can be crucial if your car is skidding. The best way to stop a car is to engage the brakes so that the wheels are just on the threshold of locking up but are still turning. This technique will ensure a maximum retardation via *static* friction, which is greater than the kinetic friction of skidding rubber. The process can be done manually—you can feel the wheels locking—but it takes practice, especially if it is to be accomplished under stress. That's why computer-operated antilocking brakes are a good idea, and why we're always told to pump the brakes for a fast stop.

Gauge blocks are precise, smooth steel blocks that are used to calibrate measuring tools. When they are squeezed together they interact so strongly that they stick to each other. The adhesive force at work here is the same force that gives rise to friction. It's also the same force that allows a gecko to walk up a vertical glass wall (see p. 108).

deform if the *force per unit area*, or *pressure*, on it is great enough. With the load supported on a few tiny prominences, each experiences a crushing pressure and subsequently flattens out. The contact area thereby increases, decreasing the pressure until the material squashes no further. Still, only a small fraction of the total surface area is in contact, while the remainder is separated by gaps of 10 to 50 atomic diameters.

Suppose a book is oriented to give the greatest apparent contact area with a surface (Fig. 4.28). Resting on a flat face may mean more contact points, but each will experience less pressure and will deform less. The overall effect is that the actual total contact area is the same, independent of the apparent area. That's equivalent to the third principle of friction. Moreover, the interatomic forces, and thereby the adhesion and the friction, also depend on the kinds of atoms involved, which is equivalent to the second principle of friction.

Static friction arises out of the need to rip apart the areas of bonded contact. Two clean pieces of metal, such as gold, pressed together, will literally fuse at the contact points, a process known as **cold welding**. To move the two surfaces, the welds must be torn. Keeping the surfaces separated usually reduces adhesion and friction—that's what oil and grease do, and that's why baby powder protects the skin against chafing (Fig. 4.29). Adhesion is lessened if the two surfaces are already moving past one another, and usually $\mu_s > \mu_k$. If no relative motion occurs, the surfaces enmesh, coming into close contact. When there is sliding, the surfaces ride across each other mostly on the high prominences. Still, in everyday situations the fact that static friction is greater than kinetic friction is probably due to impurities (i.e., "dirt") on the surfaces as much as anything else. By contrast, in rolling friction the welds are peeled apart rather than sheared or ripped, and that takes less effort. Moreover, during sliding, surface layers of dirt, oxides, and grease will be cut through and some good contacts can be made. This action doesn't happen with rolling objects and $\mu_k > \mu_r$.

4.9 Equilibrium: Statics

Consider the special case where, even though two or more external forces act on a body, their effects cancel and there is no change in motion. Such a body is said to be in **equilibrium**—it may or may not be moving, but its motion is unchanging ($\vec{a} = 0$). Here we'll deal only with *translational equilibrium* where there is no rotation.

Structurally, the human body shares remarkable similarities with high-rise buildings and trestle bridges. About 500 years ago, Leonardo da Vinci discovered that the bones and

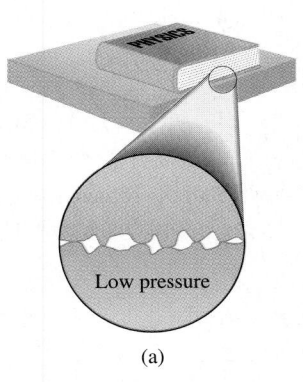

Low pressure

(a)

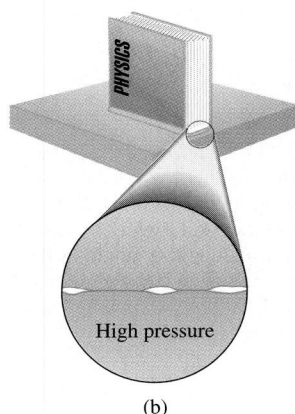

High pressure

(b)

Figure 4.28 (a) When the load is spread out over a large area, the pressure is low, and there are many small contact areas. (b) When the load is concentrated, the pressure is greater and each place of contact is larger, but there are fewer of them. Thus, the total contact area is the same in (a) as in (b).

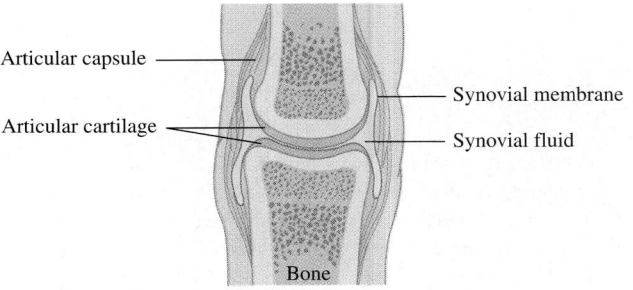

Cross-sectional view

Figure 4.29 A human bone joint lubricated with synovial fluid. As the bones move, the contact area shifts, squeezing fluid out of the porous cartilage. The result is a remarkably small coefficient of friction of about 0.000 3.

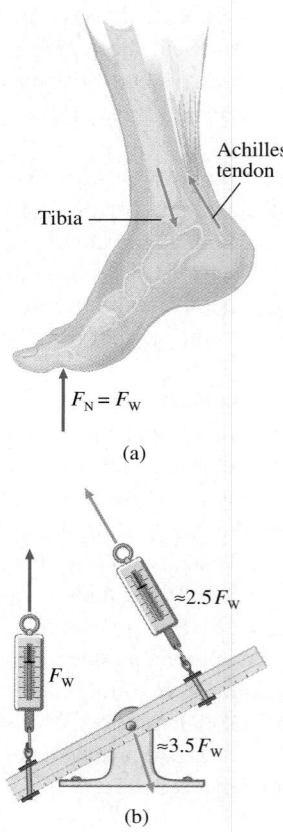

$F_N = F_W$

(a)

$\approx 2.5 F_W$

F_W

$\approx 3.5 F_W$

(b)

Figure 4.30 (a) The foot on tiptoe. Ligaments hold the bones together, and tendons attach the muscles to the bone. Think of the foot pivoted at the ankle joint, with the Achilles tendon pulling upward. (b) The foot can be modeled, and, as we'll see presently, the forces in the tibia and tendon can be computed.

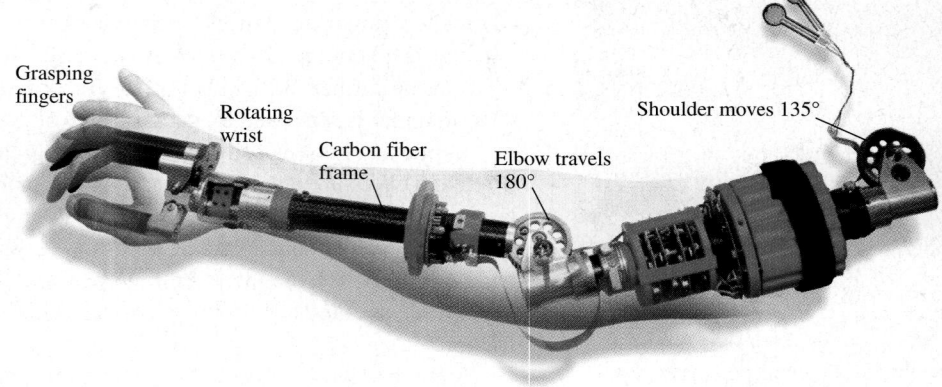

Figure 4.31 All sorts of prosthetic devices are created using the basic principles of mechanics. This artificial arm, designed in Edinburgh, Scotland, is the first to incorporate a shoulder that rotates overhead, an elbow that bends, a wrist that twists, and fingers that grasp.

muscles of vertebrates form a system of levers (Fig. 4.30). We move most of our body parts (fingers, arms, legs, etc.) using muscles to pivot one bone against another (Fig. 4.31).

For simplicity we'll only study systems of forces that all act in a single plane. It follows from $\vec{\mathbf{F}} = m\vec{\mathbf{a}}$ that equilibrium ($\vec{\mathbf{a}} = 0$) occurs when

$$\sum \vec{\mathbf{F}} = 0 \qquad (4.15)$$

and this is called the **First Condition of Equilibrium**. If we resolve all the forces acting on a body into their components along *any* two perpendicular axes, the equivalent scalar statement is

[forces in a plane] $\qquad \sum F_x = 0 \qquad$ and $\qquad \sum F_y = 0 \qquad (4.16)$

Keep in mind that $\sum F$ means the sum of the scalar force components, or the net scalar force, in a particular direction.

THE HUMAN SKELETON AS A STRUCTURE

Most buildings are constructed on a frame of wood, reinforced concrete, or steel, and that structure carries the loads of weight and wind down to the ground. The human body is constructed on a similar frame of bones, ligaments, and muscles. Loads in the upper body are carried by the spine to the pelvis and then down through the legs to the ground (Fig. 4.32). The bones of the skeleton are primarily compressive members that carry both body weight and the forces exerted by the muscles. The skeletal muscles are attached at both ends via tendons to different bones. These muscles can only contract, pulling on the bones and putting them in compression.

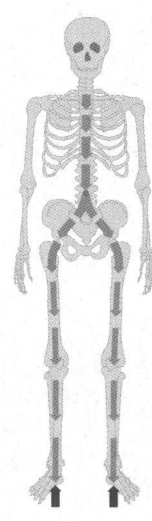

Figure 4.32 The weights of the various parts of the human body are supported by the musculoskeletal system. The hanging arms are held in place by tendons and muscles in the shoulders. That load is carried back to the spine, which transmits the weight of the entire upper body to the pelvis. The bone arch of the pelvis splits the downward force, directing it through the legs to the ground.

Archimedes, who was kinsman and a friend of King Hieron of Syracuse, wrote to him that with any given force it was possible to move any given weight, and emboldened, as we were told, by the strength of his demonstration, he declared that, if there were another world, and he could go to it, he could move this. Hieron was astonished, and begged him to put his proposition into execution, and show him some great weight moved by a slight force. Archimedes therefore fixed upon a three-master merchantman of the royal fleet, which had been dragged ashore by the great labors of many men, and after putting on board many passengers and the customary freight, he seated himself at a distance from her, and without any great effort, but quietly setting in motion with his hand a system of compound pulleys, drew her towards him smoothly and evenly, as though she were gliding through the water.

PLUTARCH (CA. 46–120 A.D.)
LIFE OF MARCELLUS

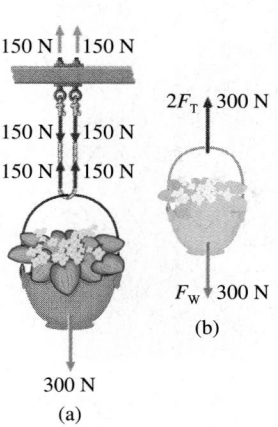

Figure 4.33 (a) A load supported by two lengths of rope. Since the system is in equilibrium, $\sum F_y = 0$. (b) The net upward force ($2F_T$) equals the net downward force (F_w).Here you count the number of ropes, just as you would count the number of legs supporting a table.

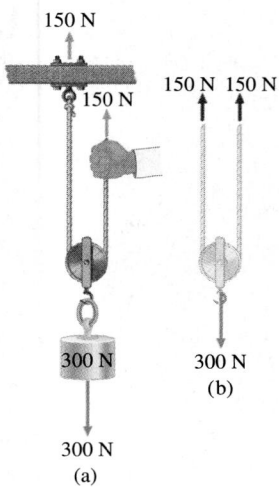

Figure 4.34 (a) This pulley arrangement acts as a force multiplier. Count the number of ropes supporting the load. In this case, there are two, and the system applies twice the force applied to it. The hand lifts with 150 N and the system lifts with 300 N. (b) The tension in each supporting length of rope is 150 N. The total upward force therefore is 300 N.

Parallel Force Systems

Return to Fig. 4.15, which depicts a mass attached to an essentially weightless rope hanging from a ceiling hook. The mass, the rope, and the hook are separately in equilibrium and, for each, $\sum F_y = 0$. A single length of rope supporting a load is called a **hanger**. *When there is only one hanger, the tension in it equals the load*: $F_T = F_w$.

If two lengths of light rope support a load, as with the 300-N pot in Fig. 4.33, the tension in each length is 150 N and the force on each hook is 150 N. If three ropes equally share the 300-N load, the tension in each is 100 N and the force on each ceiling hook to which a rope is attached is 100 N. Figure 4.34 incorporates an essentially weightless, frictionless pulley that distributes the load equally on the two rope segments; other than that, the situation is identical to Fig. 4.32. We apply a force of 150 N, and the system delivers a force of 300 N. This arrangement is a *force-multiplier*; it's one of the most important simple machines ever devised.

These people have reached terminal speed. The weight of each diver acting down exactly equals the drag force acting up, or $\sum F_y = 0$. Hence, $a = 0$ and each person is in equilibrium even though he or she is falling at about 160 km/h.

A contemporary pulley system used to redirect force.

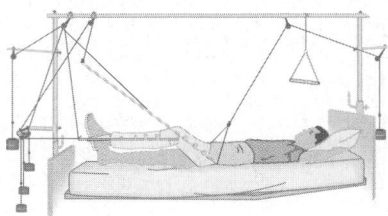

Figure 4.36 It's been common practice since the fourteenth century to use ropes and pulleys to put broken limbs in traction. Shown here is a Thomas leg splint with a Pearson attachment.

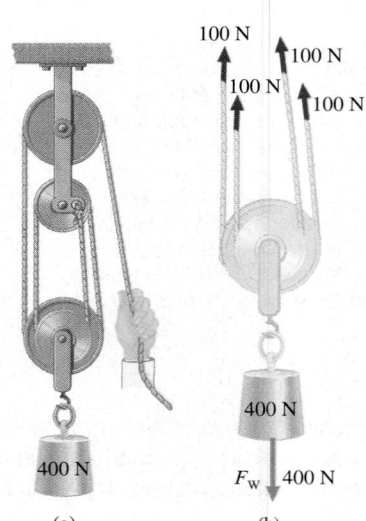

Figure 4.35 Four ropes support the load, and the hand need only pull down with 100 N.

When all the forces on a **pinned bar** are applied at its ends the bar will either be in pure tension or compression—the two resultant forces will act along the length of the bar.

Any number of pulleys can be used, mounted next to or just below one another. In Fig. 4.35 four ropes pull upward on the load (the rope in the hand pulls down and doesn't support the load) and so this arrangement multiplies the applied force by 4. Pulleys are used extensively in applications where long distances are involved such as elevators, scaffolds, and cranes.

Each of the pulleys in Fig. 4.36 serves to change the direction of the applied force rather than act as a multiplier. Used in this way it provides a convenient means of applying a constant force (equal to the hanging weight) to a system.

Equilibrium of Two-Force Systems. When you hold a steering wheel on each side and push with one hand and pull equally with the other, the sum of these two oppositely directed forces is zero, but the wheel turns nonetheless; it's not in complete equilibrium, certainly not in rotational equilibrium. Accordingly, it's possible to have a rigid structure acted on by forces where $\sum \vec{\mathbf{F}} = 0$, and yet the thing can rotate (and maybe even tumble over) even though it can't translate. In the case of the steering wheel, the lines-of-action of the two parallel forces are separated, and that causes them to twist the body. Clearly then, **if a rigid body is in equilibrium under the influence of two forces, those forces must be equal in magnitude, oppositely directed** (so that $\sum \vec{\mathbf{F}} = 0$), **and colinear** (so there's no twisting). The hanging mass in Fig. 4.15 and the woman standing in Fig. 4.18 (p. 103) are examples of the rule.

Many modern structures are made using a framework of lightweight, short, rigid members or **bars**. These are often attached together by pins (bolts or rivets) at their ends. Because there can't be any twisting (as there could be if they were welded rigidly together), every bar in the structure experiences a single net force at each end. The only way such a two-force member can be in equilibrium is if the forces at its ends are colinear; **when loaded at their ends, the bars are in pure tension or compression** (see Problem 132).

If two colinear forces acting on a body point toward each other, the body is under compression; the woman in Fig. 4.18 is under compression via F_W and F_N. She's drawn downward gravitationally (F_W) and pushed upward electromagnetically (F_N). *An object in tension pulls; an object in compression pushes.*

Concurrent Force Systems

In the remainder of this chapter we focus on **concurrent force systems**, that is, situations where *the lines-of-action of all the applied forces acting on an object pass through the same point* (as they did with the rings in Figs. 4.4, 4.6, and 4.7). This situation is easy to analyze because the forces, being concurrent, act with a common origin and can be combined into a single resultant $(\sum \vec{F})$. For complete equilibrium the only requirement is that the resultant be zero, $\sum \vec{F} = 0$; there can't then be any twisting. It may at first seem as though this is a very special situation, but it really isn't. In fact, **whenever an object is maintained in equilibrium by three forces in a plane acting along nonparallel lines, those forces must be concurrent** (see e.g., Figs. 4.30 and 4.37). {For a formal proof of this, read Section 8.6 and click on CONCURRENT FORCES under FURTHER DISCUSSIONS on the CD.} 💿

When three coplanar, nonparallel forces maintain an object in equilibrium, those forces must all meet at a single point; in other words, they must be **concurrent**.

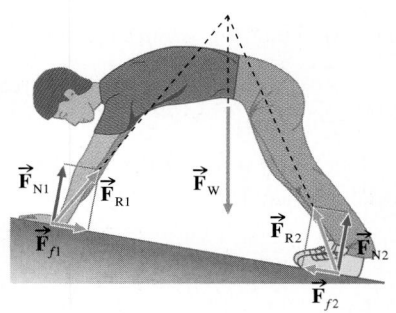

Figure 4.37 The floor pushes up on the hands and feet with reaction forces $\vec{F}_{R1}$ and $\vec{F}_{R2}$, respectively. These each comprise both friction and a normal force: $\vec{F}_{R1} = \vec{F}_{f1} + \vec{F}_{N1}$ and $\vec{F}_{R2} = \vec{F}_{f2} + \vec{F}_{N2}$. Since the person is in equilibrium, $\vec{F}_{R1}$, $\vec{F}_{R2}$, and $\vec{F}_{W}$ are concurrent. As we'll see, the weight acts through the center-of-gravity which is more or less behind the navel.

Example 4.11 [**II**] The engine hanging motionlessly in Fig. 4.38a weighs 800 N. If θ is 20.0°, compute (a) the tension in each length of rope and (b) the horizontal force tending to pull out the support pins. Use Table 4.5 to pick an appropriate rope for the job. To avoid failure from high momentary forces (e.g., sudden jerks), it's customary to use ropes at no more than about one-sixth their breaking strength.

Solution This is an equilibrium problem, and that means the sum-of-the-forces-equals-zero. (1) TRANSLATION—A load is supported in equilibrium by two ropes making known angles; determine (a) the tensions and (b) the horizontal forces on the supports. (2) GIVEN: $F_{W} = 800$ N and $\theta = 20.0°$. FIND: tensions F_{T1}, F_{T2}, F_{T3}, and the pull-out force. (3) PROBLEM TYPE—Equilibrium/concurrent forces. (4) PROCEDURE—**We**

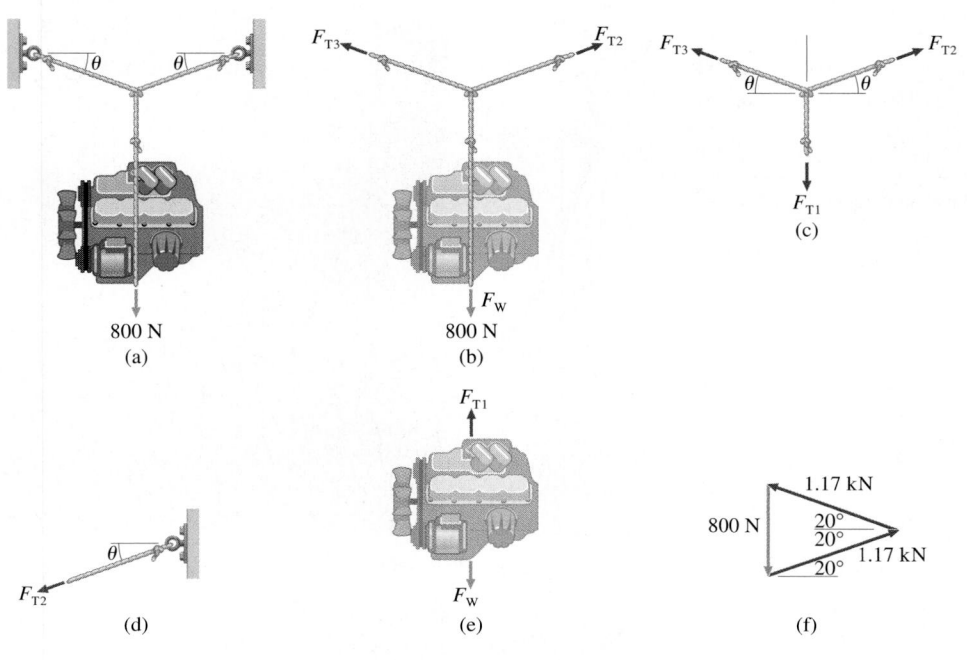

Figure 4.38 (a) The engine is supported by a system of ropes, and all the forces acting on that arrangement are concurrent. (b), (c), (d), and (e) are free-body diagrams. (f) Because $\sum \vec{F} = 0$, the three forces must form a closed triangle such that the resultant is zero.

continued

are to find the three tensions and so must isolate one or more parts of the system where the desired forces act. The equilibrium equations for those parts, analyzed via free-body diagrams, will contain the sought-after unknowns — three unknowns, three equations. The nice thing about ropes, cables, and muscles is that they always act in tension, and the forces applied to and by them act along them. Figure 4.38b shows $\vec{F}_{T2}$, $\vec{F}_{T3}$, and $\vec{F}_W$ acting on the rope-engine system. They must be concurrent because the knot at the center is in equilibrium; indeed, they all pass through the knot. Draw free-body diagrams (Figs. 4.38c, d, and e), one for each part of the system that contains a force you wish to study. (5) CALCULATION—A good place to start is with $\vec{F}_{T1}$; the engine is supported by a single hanger, so the tension in that rope equals the load

[Equation-1] $$\boxed{F_{T1} = F_W = 800 \text{ N}}$$

Because of the symmetry, the tensions on either segment of the V-shaped support rope must be equal ($F_{T2} = F_{T3}$). Whatever happens, happens identically to both segments. Let's not take advantage of that fact now but just press on with the analysis step by step. We need F_{T2} and F_{T3} and could use either Fig. 4.38b for the whole system or Fig. 4.38c for just the knot. Each must be in equilibrium, and each contains the two forces we want.

Take *up* and to the *right* as positive. The sum-of-the-horizontal-forces equals zero on the knot, and

[Equation-2] $$\overset{+}{\rightarrow}\sum F_x = F_{T2} \cos \theta - F_{T3} \cos \theta = 0$$

confirming that $F_{T2} = F_{T3}$. The sum-of-the-vertical-forces equals zero on the knot, and

[Equation-3] $$\overset{+\uparrow}{\sum} F_y = F_{T2} \sin \theta + F_{T3} \sin \theta - F_W = 0$$

Hence $$2F_{T2} \sin 20.0° = 800 \text{ N}$$

and $\boxed{F_{T2} = 1.17 \text{ kN}}$. At each wall pin (Fig. 4.38d), the horizontal force is

$$F_{T2} \cos 20.0° = \boxed{1.10 \text{ kN}}$$

Since $F_{T1} = 800$ N and $F_{T2} = 1.17$ kN, none of the 3/16-in. ropes will suffice. The braided nylon and Dacron™ 1/4-in. lines will do nicely.

Quick Check: Make a sketch of the three force vectors acting at the knot. They must add up to zero and form a closed triangle, and they do (Fig. 4.38f). ***Use the Law of Cosines on the vector triangle***; $\sqrt{2(1170 \text{ N})^2 - 2(1170 \text{ N})^2 \cos 40.0°} = F_{T1} = 800$ N.

Table 4.5
Various Line Strengths

Diameter (inches)	Breaking strength*					
	Manila hemp (kN)	Nylon filament (kN)	Dacron™ filament (kN)	Nylon braid (kN)	Dacron™ braid (kN)	Cable (kN)
3/16	1.8	4.0	4.0	4.6	4.6	19
1/4	2.4	6.6	6.7	7.1	8.2	31
5/16	4.0	9.6	10.3	12.0	12.8	44
3/8	5.4	14.9	15.4	16.9	18.4	64
7/16	7.0	20.0	20.0	23.1	25.0	78
1/2	10.6	27.1	28.3	32.0	32.7	101

*As a rule, working loads should not exceed one-sixth of the breaking strength for ropes and one-fifth for cable.

This person is acted upon by five nonconcurrent forces. The gravitational interaction with the Earth produces a downward force $\vec{F}_W$. The wall ($\vec{F}_1$), table ($\vec{F}_2$), and chair ($\vec{F}_3$ and $\vec{F}_4$) each exert normal forces. Since the person is in equilibrium, the sum-of-the-forces equals zero.

SAGGING ROPES AND TENSIONS

Notice in Fig. 4.38 that, if F_W has any non-zero value, θ cannot equal zero because sin 0 equals zero and **an undeflected rope exerts no upward force**. The rope stretches and must sag as it interacts with the object being supported. In fact, the rope will sag under its own weight. *No amount of horizontal force can pull a rope so that it straightens and hangs totally horizontal*. Provided $\theta < 30°$, the tension ($F_{T2} = F_{T3}$) in either supporting rope will be greater than the load and the smaller is θ, the greater the tension.

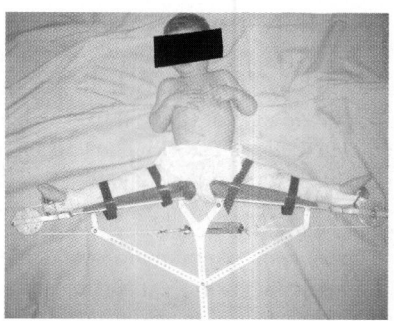

An interesting application of statics for the treatment of congenital hip dislocation. The spring scale reads the tension applied to each leg. The little turnbuckle (on the left of the scale) allows the tension to be adjusted as needed. The double-Y-shaped frame keeps the legs apart and allows $\sum \vec{F} = 0$.

Example 4.12 **[II]** Determine the angle θ in the arrangement shown in Fig. 4.39. Assume the pulley is weightless and frictionless.

Solution This is an equilibrium problem, and that means the sum-of-the-forces-equals-zero. (1) TRANSLATION—A load is supported in equilibrium by two ropes making different angles, one of which is known; determine the other. (2) GIVEN: The angle is 20°, the leg weight 150 N, and the counterforce 200 N. FIND: θ. (3) PROBLEM TYPE—Equilibrium/concurrent forces. (4) PROCEDURE— Draw the free-body diagram of the ring (Fig. 4.39b), which is where θ is formed. With no friction on the weightless pulley, the tension in the rope on the right is uniform throughout. It should be evident then that $F_{T2} = 200$ N, $F_{T1} = 150$ N, and, because θ may not equal 20°, F_{T3} may not equal F_{T2}. The three forces on the ring must be concurrent because the ring is in equilibrium; therefore $\sum F_x = 0$ and $\sum F_y = 0$ is the way to go. (5) CALCULATION—Starting at the ring,

$$\xrightarrow{+} \sum F_x = (200\ \text{N}) \cos 20° - F_{T3} \cos \theta = 0$$

and

$$+\uparrow \sum F_y = (200\ \text{N}) \sin 20° + F_{T3} \sin \theta - 150\ \text{N} = 0$$

Then, $F_{T3} \cos \theta = 187.9$ N and $F_{T3} \sin \theta = 81.60$ N. These are the scalar components of $\vec{F}_{T3}$, and if we divide the second by the first, the tension cancels:

$$\frac{\sin \theta}{\cos \theta} = \tan \theta = 0.434$$

and to two significant figures

$$\boxed{\theta = 23°}$$

Quick Check: Knowing θ we can compute the tension: $F_{T3} = (187.9\ \text{N})/\cos \theta = 204.9$ N, and again, $F_{T3} = (81.60\ \text{N})/\sin \theta = 204.9$ N. Figure 4.39c is the force triangle, and it correctly has a zero resultant. Alternatively, *use the Law of Cosines on the vector triangle*;

$$\sqrt{(205\ \text{N})^2 + (200\ \text{N})^2 - 2(205\ \text{N})(200\ \text{N}) \cos 43.5°} =$$
$$F_{T1} = 150\ \text{N}$$

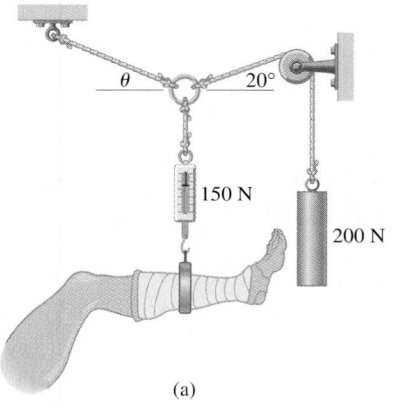

(a)

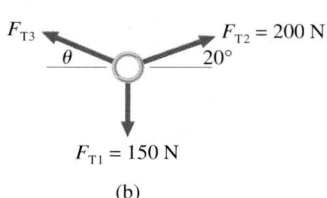

(b)

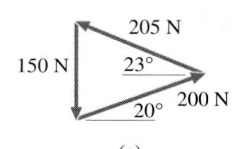

(c)

Figure 4.39 (a) A concurrent force system. (b) The free-body diagram of the ring showing the three forces that meet at its center. (c) The force triangle corresponding to equilibrium.

Example 4.13 **[II]** A 400-N sign is suspended from the end of a 2.0-m-long, negligibly light horizontal bar, which is attached to a wall via a pivot (Fig. 4.40). The bar is held up by a rope, making an angle of 50°, which is experiencing 0.70 kN of tension. The bar is 2.00 m long, and the rope is attached 0.50 m from its end. Find the reaction force on the pivot at the wall.

Solution This is an equilibrium problem, and that means the sum-of-the-forces-equals-zero. (1) TRANSLATION—A bar is in equilibrium under the action of three coplanar forces: a known tensile force, an unknown reaction, and a known load. Determine the reaction force. (2) GIVEN: $F_W = 400$ N, and $F_T = 0.70$ kN at 50°. FIND: $\vec{F}_R$ on the pivot. The three forces on the bar must be concurrent because the bar is in equilibrium. (3) PROBLEM TYPE—Equilibrium/concurrent forces. (4) PROCEDURE—Draw (Fig. 4.40) a free-body diagram of the bar since it experiences the desired force, $\vec{F}_R$. *You can draw $\vec{F}_R$ in any direction you like, acting at the end of the bar.* But the vectors $\vec{F}_T$ and $\vec{F}_W$ intersect at a point and $\vec{F}_R$ must pass through that point (see the Quick Check). Moreover, if the bar is to be in equilibrium, the horizontal component of $\vec{F}_R$ must cancel the horizontal component of $\vec{F}_T$ (which is to the left); therefore, $\vec{F}_R$

acts to the right and down. *Had you drawn it incorrectly, it would simply turn out negative when you solve for it.* (5) CALCULATION— $\sum F_x = 0$ and $\sum F_y = 0$; two equations two unknowns, F_R and θ:

$$+\uparrow \sum F_y = F_T \sin 50° - F_W - F_{Ry} = 0$$

$$\overset{+}{\rightarrow} \sum F_x = F_{Rx} - F_T \cos 50° = 0$$

Knowing that $F_T = 0.70$ kN it follows that $F_{Ry} = 0.136$ kN and $F_{Rx} = 0.450$ kN. Therefore

$$F_R = \sqrt{(0.136 \text{ kN})^2 + (0.450 \text{ kN})^2} = \boxed{0.47 \text{ kN}}$$

The angle that $\vec{F}_R$ makes with the horizontal is then

$$\theta = \tan^{-1} \frac{F_{Ry}}{F_{Rx}} = \boxed{17°}$$

Quick Check: $\vec{F}_T$ crosses $\vec{F}_W$ at a distance d_T below point E where $d_T = (\tan 50°) \times (0.50 \text{ m}) = 0.6$ m. Similarly, $\vec{F}_R$ crosses $\vec{F}_W$ at a distance d_R below point E where $d_R = (\tan 17°)(2.00 \text{ m}) = 0.6$ m. *The vectors are concurrent.*

The pulley system on the left is the force multiplier in this crane. Notice the lightweight truss boom.

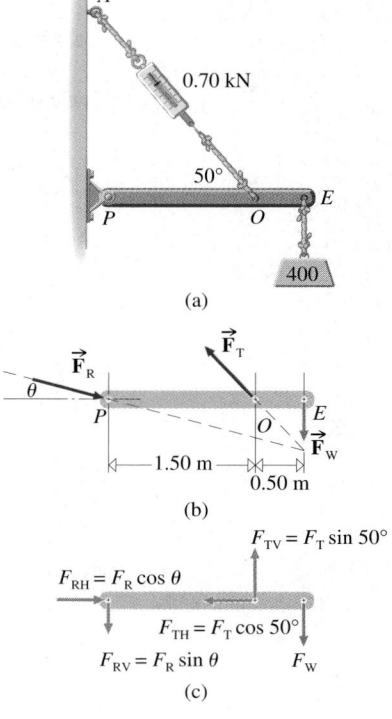

(a)

(b)

(c)

Figure 4.40 (a) The problem is to calculate the reaction at P. (b) Since the bar must be in equilibrium, the three forces (which are not parallel) must be concurrent. (c) The forces resolved into horizontal and vertical components. Notice that since all the forces do not act only at the ends of the bar the reaction force at P will not act along the length of the bar. If the rope's tie point O were moved to the end of the bar (at E), the reaction force at P would then have to be horizontal.

TRUSSES, BRIDGES, AND BONES

A *truss* is a structural framework of members joined together at their ends in triangular patterns. Trusses are widely used in aircraft, TV towers, bridges, buildings, and so on. Made of lightweight components of moderate length, the truss provides strength and great space-spanning capability.

In the early days of the railroads, a variety of truss bridges were developed. In the United States, where iron was expensive, timber beams were used in compression, with thin iron rods as the tension members. Figure 4.41 shows one of these truss bridges, and it immediately brings to mind the musculoskeletal system of quadrupedal vertebrates, with the spine spanning the space between the shoulder girdle and the pelvis.

In humans and other animals, the compression members—vertebrae and ribs—are made of bone. The tension members are the intercostal muscles that crisscross between the ribs, pulling them together. For the bridge to sag under a load, the vertical beams would have to spread apart at their lower ends, an action prevented by the tension crossbars. Likewise, the spine-rib-muscle structure is the bridge supporting the weight of the trunk of the body. The legs are the compression columns on which the entire load rests.

Figure 4.41 (a) An old railroad trusswork bridge made of timber and braced with iron tension rods. (b) Vertebrates, such as this lizard, are supported by a similar system of bone in compression, crisscrossed by muscles in tension.

Core Material & Study Guide

THE THREE LAWS

The **Law of Inertia** states that: Every body continues in a state of rest or in uniform motion in a straight line except insofar as it is compelled to change that state by forces impressed upon it. The *weight* of an object on Earth is the downward force arising from the gravitational interaction between the object and the planet. The SI unit of force, the **newton** (N), is approximately equal on Earth to the weight of a mass of 0.10 kg, or $1 \text{ N} \approx \frac{1}{4}$ lb. Reexamine Sections 4.1 (The Law of Inertia) and 4.2 (Force) and review Examples 4.1 and 4.2.

The measure of motion in Newton's theory (p. 93) is **momentum ($\vec{p} = m\vec{v}$)**. **Newton's Second Law** says that

[constant force] $$\vec{F} = \frac{\Delta \vec{p}}{\Delta t}$$ [4.1]

and $1 \text{ N} = 1 \text{ kg·m/s}^2$. Equation (4.1) will be revisited later, don't worry about it now. The all-important notion now is

[constant force] $$\sum \vec{F} = m\vec{a}$$ [4.3]

When the applied force (p. 100) varies in time

[straight-line motion] $$F_{av} = ma_{av}$$ [4.5]

Review Section 4.3 (The Second Law) and study Example 4.3. The **Third Law** maintains that two objects interact such that for every force exerted on one, there is an equal and oppositely directed force exerted on the other.

DYNAMICS AND STATICS

For applications of the three laws see Section 4.5 (The Effects of Force: Newton's Laws) and Section 4.7 (Coupled Motion); study Examples 4.4–4.6.

The weight (p. 101) of an object is

$$\vec{F}_W = m\vec{g}$$ [4.6]

The force exerted by a surface perpendicular to itself and equal to the load supported is the **normal force** (F_N). The relationship $F_W = mg$ comes up all the time—know it!

The maximum static friction force (p. 107) is

$$F_f(\max) = \mu_s F_N$$ [4.11]

where μ_s is the *coefficient of static friction*. For surfaces in relative motion, kinetic friction is given by

$$F_f = \mu_k F_N$$ [4.13]

where μ_k is the *coefficient of kinetic friction*. For the friction encountered when one object rolls across another

$$F_f = \mu_r F_N$$ [4.14]

where μ_r is the *coefficient of rolling friction*. Focus on the subsections *Static Friction* and *Kinetic Friction*; then study Examples 4.8–4.10.

The notion of the free-body diagram (p. 98) is important, and so we illustrate a number of dynamical situations here in Table 4.6 on p. 124. Make sure you understand each of them.

For *coplanar concurrent forces*, equilibrium occurs when (p. 115)

$$\sum \vec{F} = 0 \qquad [4.15]$$

which is the **First Condition of Equilibrium**. A system acted upon by forces that all pass through a single point is in equilibrium when

$$\sum F_x = 0 \qquad \text{and} \qquad \sum F_y = 0 \qquad [4.16]$$

Review Section 4.9 (Equilibrium: Statics) and study Examples 4.11–4.13. Table 4.7 provides free-body diagrams for a number of equilibrium situations; study each of them.

Read all the Suggestions on Problem Solving and go over the Examples in the text. **Look at the CD WARM-UPS, then study the WALK-THROUGH EXAMPLES.** Do the odd-number Multiple Choice Questions in the textbook and check your answers. You should then be ready to try the problems. Do all the odd I-level Problems first. The complete solutions for many of them (those with boldfaced numbers) are provided. Once you feel confident go on to the II-level Problems.

Key Terms

inertia
Law of Inertia
force
line-of-action
tensile force
tension
momentum
Second Law
interaction
Third Law
sum-of-the-forces
free-body diagram
weight
reaction force

normal force
static friction
kinetic friction
rolling friction
cold welding
equilibrium
First Condition of Equilibrium
Second Condition of Equilibrium
force-multiplier
tension
compression
concurrent forces
truss

Table 4.6 Free-body diagrams for objects in motion.

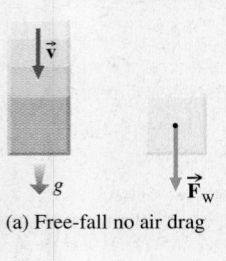

(a) Free-fall no air drag

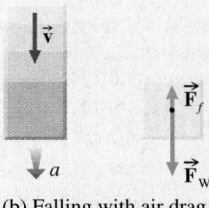

(b) Falling with air drag $a \neq g$

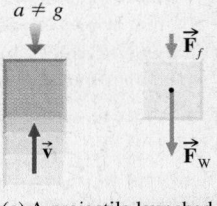

(c) A projectile launched upward with air drag

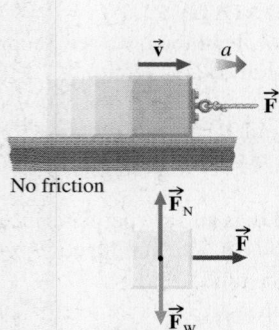

(d) An object speeding up as it's pulled to the right: $F_N = F_W$

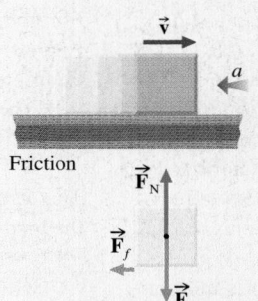

(e) An object moving to the right slowing down due to friction: $F_N = F_W$

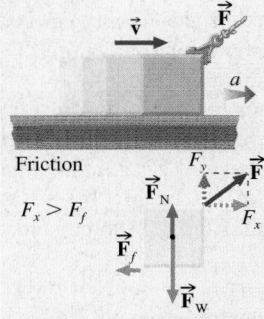

(f) An object accelerating to the right: $F_N + F_y = F_W$

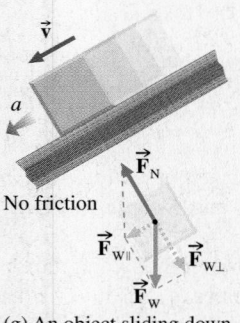

(g) An object sliding down an incline: $\vec{F}_{W\parallel} = \vec{F}_N + \vec{F}_W$

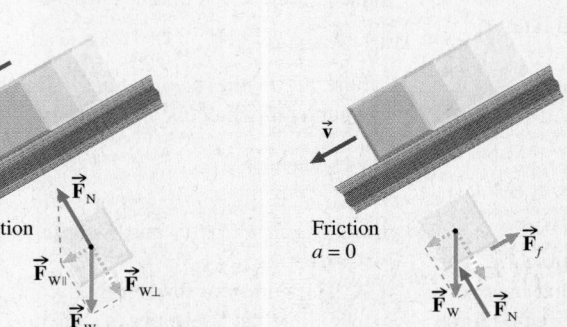

(h) Translational equilibrium. An object sliding down at a constant speed: $\sum \vec{F} = 0$

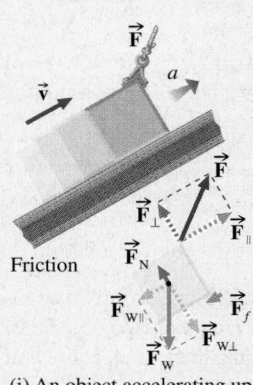

(i) An object accelerating up an incline: $F_\parallel > F_{W\parallel} + F_f$

Table 4.7 Free-body diagrams for objects at rest.

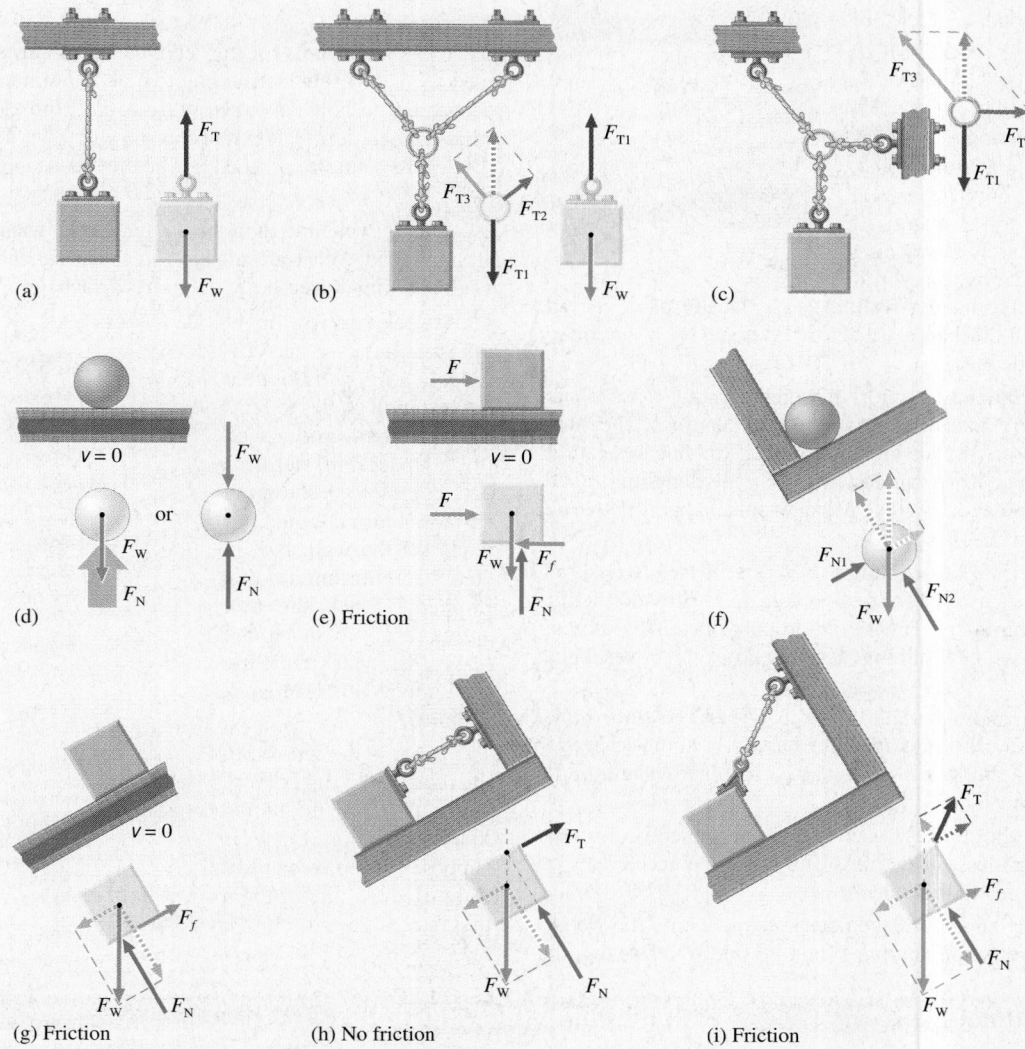

(a)

(b)

(c)

(d)

(e) Friction

(f)

(g) Friction

(h) No friction

(i) Friction

Discussion Questions

1. Ubaldi, in his book *Mechanicorum Liber* (1577), was the first to point out that the screw, wedge, and inclined plane are related devices. Explain this conclusion. Use diagrams.

2. Show that when three or more concurrent forces act on a body in equilibrium, any one force must be equal in magnitude and opposite in direction to the sum of all the other forces.

3. Examine the air track depicted in Fig. Q3. This is a straight hollow track through which air is pumped so that it escapes in jets from hundreds of tiny holes. A glider then floats almost frictionlessly above the track on a cushion of air. Explain how this setup can be used to confirm that $F = ma$. Make a list of all the assumptions being made. The glider has a small flag of width L; and as it passes through the photogate, the latter measures the time at which it arrives, t_1, and the interval it takes to pass by, Δt_1. It does the same at the second photogate (measuring t_2 and Δt_2). Come up with two ways to determine a. [For some actual data see Satinder Sidhu, "Pristinely Pure Law in the Lab," *The Phys. Teacher* **32**, 282 (1994).]

4. In a science-fiction TV show several years ago, the hero spent the entire 40 min fretting about running out of fuel before he could

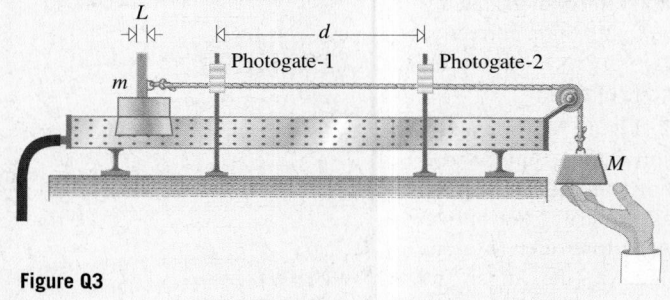

Figure Q3

pilot his spaceship back to base. Time and again the camera cut from his worried face to the blasting rockets, to the trembling fuel gauge, and back to his pained expression as the engines roared and the fuel diminished. What should he have done, and why would you say the writers were Aristotelians?

5. A car stopped at a red light is hit in the rear by another car. Describe what will happen to the two drivers as a result of the impact. What kind of injuries might each driver sustain?

6. The bombardier in the World War II B-17 shown in Fig. Q6 has evidently just dropped a "stick" of bombs. Explain why they form a nice neat column in the picture.

Figure Q6

7. EXPLORING PHYSICS ON YOUR OWN: Ride a bike at a constant leisurely rate in a straight line. Is it possible to bounce a ball (thrown "straight downward") off the ground and catch it coming "straight up," on the move? Try it. If you don't have a bike handy, walk at a fast steady pace instead. Draw a diagram of the ball's trajectory as seen by a stationary bystander as you ride past.

8. There is a classic demonstration in which a heavy mass is suspended from a fairly fine thread and another length of the same thread is attached to its underside and allowed to hang beneath it. Someone then gives a sharp tug downward on the dangling thread. Which string will break and why? What would happen if the tug were gradual and strong? Explain.

9. Describe the process of swimming in terms of Newton's laws. Swimmers sometimes use rubber fins on their feet. How do the fins work? While floating in space, can you propel yourself by moving your hands and feet as if swimming? Explain how one moves a boat by rowing.

10. A weight lifter and a barbell are both on a large floor scale. While the weight lifter lifts and holds the barbell, will the scale ever momentarily exceed and/or be less than the combined weight of the two bodies? Explain.

11. Two equal-weight monkeys each hang from the ends of a rope passing over a weightless, frictionless pulley. If one accelerates up the rope, what happens to the other?

12. A truck carrying a load of caged canaries pulls onto the weighing scales at a highway toll station. While no one was looking, the driver banged on the side of the cab so that the birds would fly around inside. What happens to the weight of the load?

13. EXPLORING PHYSICS ON YOUR OWN: What physical mechanism accounts for the "stickiness" of adhesive tape? Attach a piece of tape a few inches long to a solid horizontal surface leaving a little tab to hold on to. Now try to pull the tape off horizontally (that is, shear it off). Next, peel it off with a vertical force. Compare the results and discuss the significance as it applies to rolling friction.

14. Draw a picture of a horse pulling a wagon over a smooth straight road. What are all the forces acting on the horse? On the wagon? Draw a free-body diagram of each.

15. A rock thrown straight up stops at its peak altitude. Is it in equilibrium at the instant it comes to rest? A ball thrown into a wall on a spaceship deforms and comes to rest for an instant before bouncing back, off the wall. Is it in static equilibrium at that moment? Once the ball springs back to shape and leaves the wall (assuming zero gravity), is it in equilibrium?

16. What is the tension in the rope on the right in Fig. Q16a? Each of the pulley systems in Fig. Q16b is supposed to be in equilibrium. Redraw any of them that may be in error.

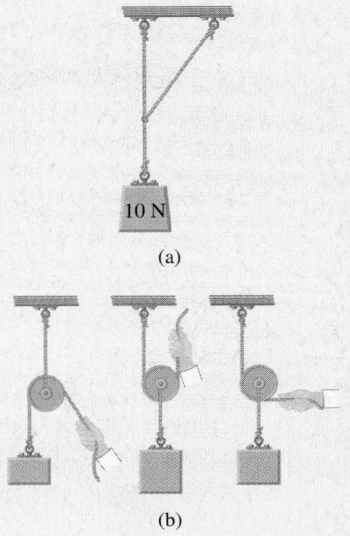

Figure Q16

Multiple Choice Questions

1. Figure MC1 shows a ball rolling down a curved incline. Which diagram depicts the resulting motion correctly? (a), (b), (c), (d), or (e) none of these.

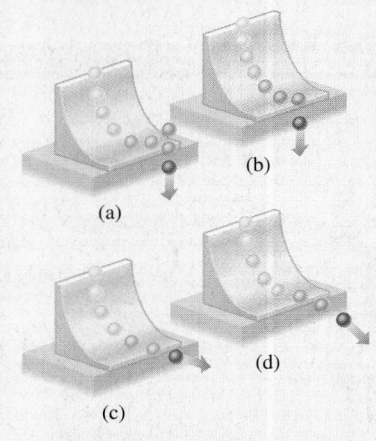

Figure MC1

2. Figure MC2 depicts a spiral tube lying at rest horizontally on a table, as viewed from above. A ball is rolled into one end of the tube and emerges from the other end. Ignoring friction, which diagram correctly represents the motion? (a), (b), (c), (d), or (e) none of these.

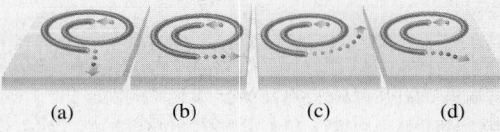

Figure MC2

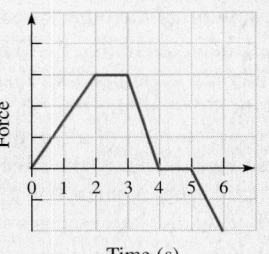

Figure MC3

3. Refer to the force-time curve for an object of constant mass m in Fig. MC3. When was the object at rest? (a) 0 s to 1 s (b) 4 s to 5 s (c) 2 s to 3 s (d) 3 s to 4 s (e) not enough information given.

4. When did the object in Fig. MC3 have a constant velocity? (a) 3 s to 4 s (b) 2 s to 3 s (c) 4 s to 5 s (d) 5 s to 6 s (e) not enough information given.

5. Which interval in Fig. MC3 corresponds to the greatest change in the speed of the body? (a) 0 s to 1 s (b) 1 s to 2 s (c) 2 s to 3 s (d) 3 s to 4 s (e) 5 s to 6 s.

6. During which time interval in Fig. MC3 did the body decelerate? (a) 0 s to 1 s (b) 2 s to 3 s (c) 3 s to 4 s (d) 5 s to 6 s (e) none of these.

7. If L stands for length, T for time, and M for mass, the dimensions of force are (a) $[ML^2]$ (b) $[ML/T]$ (c) $[ML/T^2]$ (d) $[LT/M]$ (e) none of these.

The next seven questions refer to Fig. MC8 where the man pushes on the box with a force at 53.1°. He pushes harder and harder until at 200 N the box, which weighs 400 N, is on the verge of sliding.

8. What is the vertical component of the force he exerts on the box? (a) 200 N, in the $-y$-direction (b) 160 N, in the $-y$-direction (c) 120 N, in the $-y$-direction (d) 400 N, down (e) none of these.

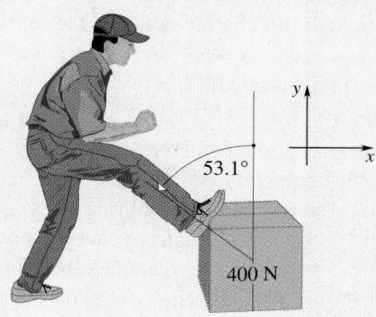

Figure MC8

9. What is the horizontal component of the force he exerts on the box? (a) 200 N, in the $+x$-direction (b) 160 N, in the $+x$-direction (c) 120 N, in the $+x$-direction (d) 400 N, in the $+x$-direction (e) none of these.

10. What is the horizontal component of the force exerted by the box on him? (a) 200 N, in the $-x$-direction (b) 160 N, in the $-x$-direction (c) 120 N, in the $-x$-direction (d) 400 N, in the $-x$-direction (e) none of these.

11. What is the normal force on the box? (a) 200 N, in the $+y$-direction (b) 560 N, in the $+y$-direction (c) 520 N, in the $+y$-direction (d) 400 N, in the $+y$-direction (e) none of these.

12. What is the static friction force exerted on the box? (a) 200 N, in the $-x$-direction (b) 160 N, in the $-x$-direction (c) 120 N, in the $-x$-direction (d) 400 N, in the $-x$-direction (e) none of these.

13. What is the coefficient of static friction between the box and the surface? (a) 0.03 (b) 3.3 (c) 0.40 (d) 0.31 (e) none of these.

14. Given that his back foot doesn't move, what is the friction force exerted on that foot? (a) 200 N, in the $-x$-direction (b) 160 N, in the $-x$-direction (c) 120 N, in the $-x$-direction (d) 400 N, in the $-x$-direction (e) none of these.

15. Is it possible to devise a technique to push on a table without it pushing back on you? (a) Yes, out in space. (b) Yes, if someone else also pushes on it. (c) A table never pushes in the first place. (d) No. (e) None of these.

16. If a nonzero constant net horizontal force is acting on a body sitting at rest on a frictionless table, the body will (a) sometimes accelerate (b) always move off at a constant speed (c) always accelerate at a constant rate (d) accelerate whenever the force exceeds its weight (e) none of these.

17. If (with no friction) a force F results in an acceleration a when acting on a mass m, then tripling the mass and increasing the force sixfold will result in an acceleration of (a) a (b) $a/2$ (c) $2a$ (d) $a/6$ (e) none of these.

18. A bubble level can be used as an accelerometer. If the level is accelerated due east while in normal operating position aligned east-west, the bubble will (a) move west (b) move east (c) move north (d) remain at rest (e) none of these. Try it.

19. A 250-lb man holding a 30-lb bag of potatoes is standing on a scale in an amusement park. He heaves the bag straight up into the air, and before it leaves his hands, a card pops out of a slot with his weight and fortune. It reads (a) 250 lb (b) 280 lb (c) less than 250 lb (d) more than 280 lb (e) none of these.

20. Why does it take more force to start walking than to continue walking? (a) It doesn't. (b) Because kinetic friction is less than static friction. (c) Because air drag at walking speeds is much less than the needed inertial force, ma. (d) People get tired walking and prefer to stand still. (e) None of these.

21. Imagine that you are standing on a cardboard box that just supports you. What would happen to it if you jumped into the air? It would (a) collapse (b) be unaffected (c) spring up as well (d) move sidewise (e) none of these.

22. Imagine a flat, lightweight wheeled cart that is low to the ground and has well-oiled bearings. What will happen to it if, while standing at rest on it, you begin to walk along its length? It will (a) remain stationary (b) advance along with you (c) not enough information to say (d) move in the opposite direction (e) none of these.

23. With the previous question in mind, what would happen if you approached the cart, stepped onto it, and walked its length at a constant speed? It would (a) remain nearly stationary (b) advance along with you (c) move rapidly in the opposite direction (d) move forward then backward (e) none of these.

24. An 800-N acrobat holds a 40-N chicken in one hand while his 480-N assistant Jane sits on his shoulders holding a 4.0-N box of cigars. The acrobat stands motionlessly on one foot on a 36-N bathroom scale. The scale reads (a) 1360 N (b) 1324 N (c) 662 N (d) 680 N (e) none of these.

25. Two forces, each of 100 N acting at a point such that they are 120° apart, are equivalent to a single force of (a) 100 N (b) zero (c) 200 N (d) 86.6 N (e) none of these.

26. A light rope is looped over a weightless pulley and tied to a 50-N rock. The other end is held by a 50-N monkey who climbs up the rope a bit and then stops. The tension in the rope at that moment is (a) 100 N (b) zero (c) 50 N (d) 25 N (e) none of these.

27. A 100-N load hangs from the ceiling on a rope. Someone pushes on the load with a continuing horizontal force, and it moves off to the side a little, coming to rest at a bit of an angle with the vertical. As a result, the tension in the rope (a) is unchanged (b) is zero (c) is decreased (d) is increased (e) none of these.

28. A 10-m length of rope, weighing 10 N/m, is hanging straight down from a ceiling hook. The tensions in the rope at the free end, at 5.0 m up from the free end, and at the hook are, respectively, (a) 0, 100 N, and 100 N (b) 0, 50 N, and 100 N (c) 100 N, 100 N, and 100 N (d) 100 N, 50 N, and 0 (e) none of these.

29. In order to hold the weight in Fig. MC29 in equilibrium, the hand and the ceiling exert forces, respectively, of (a) 100 N and 300 N (b) 300 N and 100 N (c) 100 N and 400 N (d) 300 N and 600 N (e) none of these.

Figure MC29

For more Multiple Choice Questions with answers click on WARM-UPS in CHAPTER 4 on the CD.

Suggestions on Problem Solving

1. Begin problem solving with a sketch. Watch out for the signs. Take the direction of initial motion as positive and stick with it for the whole analysis. Put the signs in your diagram at the outset. Sometimes you will have to guess at the direction of the overall motion of the system (actual or impending); call that positive. If a complicated system is moving clockwise, make that the positive direction and draw a big arrow in that direction from one end of the system to the other. Part of the system may be moving up, part right, and part down. Provided each portion follows the overall clockwise motion, up, right, and down are positive for the corresponding portions of the system.

2. When adding force vectors, keep in mind that the x- and y-axes may not be the simplest choice of directions along which to resolve your vectors. Watch for symmetries that tell you that force components are equal or perhaps cancel. Two equal and oppositely directed vectors cancel each other no matter what the angle of their common line-of-action is.

3. The more experience you have, the further along you will go in the analysis before substituting in numbers. It makes no sense to multiply by some number only to find that you have to divide by it later on. Also, do not forget to square the quantity when you have a problem involving t^2 or v^2.

4. Isolate each object of interest via its own free-body diagram. Then apply $\sum F = ma$ to each body, remembering that if the body is at rest or moving at a constant velocity, $a = 0$. Apply the Second Law in the direction of actual or impending motion and (if you need normal forces) perpendicular to it. Each such application will provide an independent equation, and the number of equations must equal the number of unknowns.

5. If forces act that are neither perpendicular nor parallel to the actual or impending motion, they must be resolved into components in those directions. The motions in these two directions are independent. For an inclined plane, we apply $\sum F = ma$ both down the incline and perpendicular to it—the former tells us about the acceleration, the latter about the normal force.

6. The tensions acting at each end of a rope are equal and oppositely directed (provided no other tangential forces act). If two masses are connected by a rope, the same tension will act (in opposite directions) on the two bodies.

7. Don't forget that $F_w = mg$, which means that if you know either the weight or the mass, you can calculate the other.

8. For a body at rest, make sure that the net force can overcome the static friction before you go on to calculate the effects of kinetic friction.

9. Draw a free-body diagram of each segment of the setup to be analyzed. Guess at the directions of the internal forces, and if you are wrong, they will turn out negative. In time, you should be able to guess correctly by recognizing what the directions of the forces must be to sustain equilibrium. If you are to solve for the force in a given rope or bar, part of that member must be included in one of these diagrams. You cannot solve for a quantity unless it appears in one of the equations.

10. Consider the geometrical symmetries of a given setup. Such symmetries usually mean that the internal forces will be symmetrical if the load is symmetrical, as in Fig. 4.38. If the arrangement is asymmetrical, the internal forces are also likely to be different.

11. When dealing with pulleys, determine how many lengths or segments of a single continuous rope support the load. A set of pulleys may be fixed to a load and move with it; if so, determine the number of rope segments supporting it (as in Fig. 4.34), because they support the load. If the rope is continuous and there are no other tangential forces on it, the tension is the same at every point in it. If there are N supporting rope segments, the tension in each is $1/N$th of the load. When several different ropes support the load, each may have a different tension in it.

Problems ✦ Coordinated Problems ✦ Progressive Problems ✦ Solutions

SECTION 4.1: THE LAW OF INERTIA

SECTION 4.2: FORCE

1. [I] A youngster sitting in a bus gently throws a ball straight up at a speed of 1.00 m/s. The bus is traveling south at a constant rate of 10 m/s, and it passes someone sitting on a fence. With what velocity does that person outside see the ball moving when it reaches maximum altitude?

2. [I] The hero in a Western movie is standing on top of a train that is moving due north at 30 km/h, while a pack of bad guys chases in hot pursuit. Cleverly, the hero tosses a lit bomb straight up at a speed of 9.8 m/s. When and where will it land (neglecting air friction)?

3. [I] Marc Antony, stretched out on a couch, is waiting, mouth upward and open. Cleopatra, carrying a bunch of grapes, is dashing across the palace floor straight toward him at a speed of 2.213 6 m/s. She holds one grape 1.000 0 m above his face. How far from him should she release her projectile if it is to land in the middle of his mouth? Neglect air friction and take the acceleration of gravity at the palace to be 9.800 0 m/s².

4. [I] A girl running at a constant speed of 2.0 m/s in a straight line throws a ball directly upward at 4.9 m/s. How far will she travel before the ball drops back into her hands? Ignore air friction.

SOLUTION: The ball is in the air for a time $2t_p$. She runs at a constant horizontal speed v_H and travels a distance $l = v_H 2t_p$, where $t_p = -v_i/g$. Hence $l = v_H 2t_p = -v_H 2v_i/g = -(2.0 \text{ m/s})2(9.4 \text{ m/s})/(-9.81 \text{ m/s}^2)$
$l = 3.8$ m.

5. [I] This problem will help us understand the action of forces. (a) How much force does the scale on the right exert on the scale on the left? (b) Observing that both scales read 100 N, what is the force on the wall in Fig. P5? (c) What force does the hand pull with?

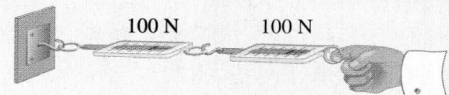

Figure P5

6. [I] This problem will help us understand the action of forces. A force of 2.00 kN in the xy-plane acts at 35.0° up from the positive x-axis. (a) Draw a diagram. (b) What is the angle θ between the force and the x-axis? (c) Resolve the force into components in the x- and y-directions.

7. [I] What is the force exerted on the wall by the person in Fig. 4.5a (p. 90)?

8. [I] What is the net force exerted on the scale in Fig. 4.5a?

9. [I] What is the net force exerted on the person on the left by the person on the right in Fig. 4.5d?

10. [I] There are five forces acting on the truck in Fig. P10. What is the net external force on it?

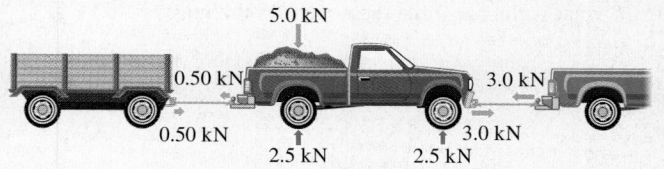

Figure P10

11. [I] The two ropes attached to the hook in Fig. P11 are pulled on with forces of 100 N and 200 N. What size single force acting in what direction would produce the same effect?

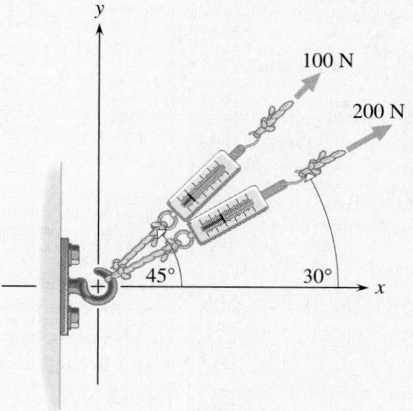

Figure P11

12. [II] The scale in Fig. P12 is being pulled on via three ropes. What net force does the scale read?

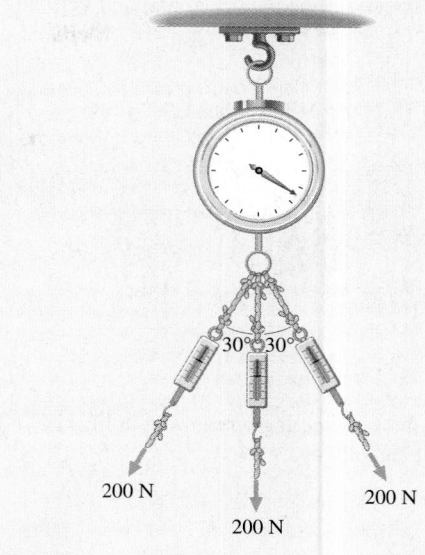

Figure P12

13. [II] What is the net force acting on the ring in Fig. P13?

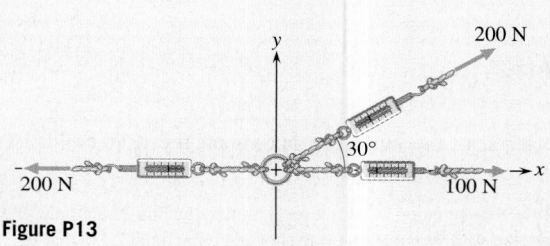

Figure P13

14. [II] What is the net force acting on the ring in Fig. P14?

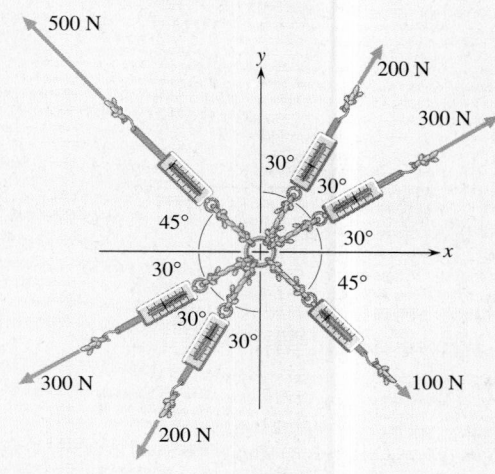

Figure P14

15. [II] Here's a classic: Suppose a conservationist points a dart gun directly at a monkey in a tree. Show that if the monkey drops from its branch just when it sees the gun go off, it will still get hit.

16. [II] Write expressions for the net horizontal and vertical forces due to the hands acting on the block in Fig. P16. The hand on the left pushes downward diagonally.

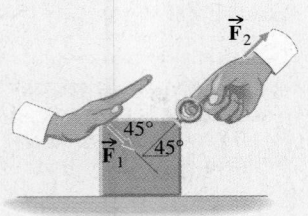

Figure P16

17. [II] What is the net force acting on the ring in Fig. P17? Try to solve this one in an elegant way.

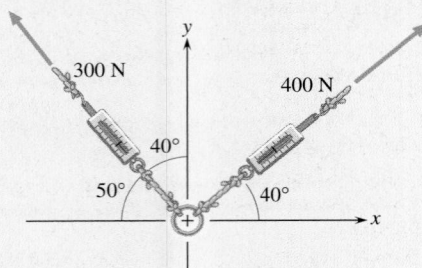

Figure P17

18. [II] Determine the net force acting on the ring in Fig. P18. Again, try to do this one elegantly.

SOLUTION: The situation is symmetrical around the line-of-action of the 200-N force so we resolve all the forces in and perpendicular to that line. Taking up and to the right as positive $\nearrow\sum F = 200$ N $- 100$ N cos 30° $- 100$ N cos 30° $= 27$ N. In the perpendicular direction, the two components arising from the hands pulling cancel $\nwarrow\sum F_\parallel = 0$ and the total force is 27 N at 45° upward.

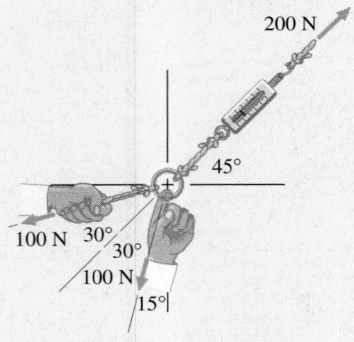

Figure P18

19. [II] THIS PROBLEM ENHANCES THE WAYS WE CAN DEAL WITH FORCES. Two forces, one of which is 60.0 N pointing somewhat north of east and the other is 50.0 N due east, are separated by

50.0°. (a) Draw a diagram. (b) Slide the force vectors parallel to themselves, thereby forming a parallelogram. What is the value of the large angle between the two vectors? (c) Determine the magnitude of the resultant force using the Law of Cosines.

20. [II] THIS PROBLEM PROVIDES PRACTICE IN DEALING WITH FORCES. At the moment of launch, the force exerted by each of the two rubber bands in Fig. P20 is 120 N acting along the band. (a) What are the two force components for each band in the forward direction and perpendicular to it? (b) What is the net force due to the slingshot on the pellet?

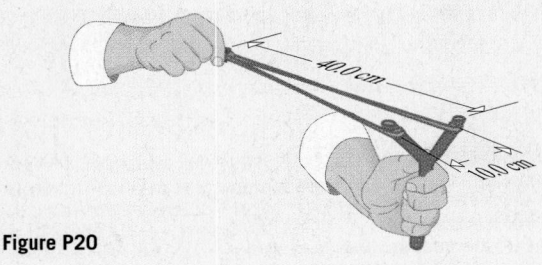

Figure P20

21. [II] A ring is fixed via a harness to a block of stone, and three ropes are attached to the ring. Each rope is to be pulled in the same horizontal plane with a force of 2 kN. How should they be arranged so as to produce a net force due east of 4 kN on the block?

22. [III] Figure P22 shows three people pulling down via ropes, each with 200 N, on the top of a mast 20-m tall. If they stand at equal distances of 20 m from the base of the mast along lines 120° apart, what is the net force they exert on the mast?

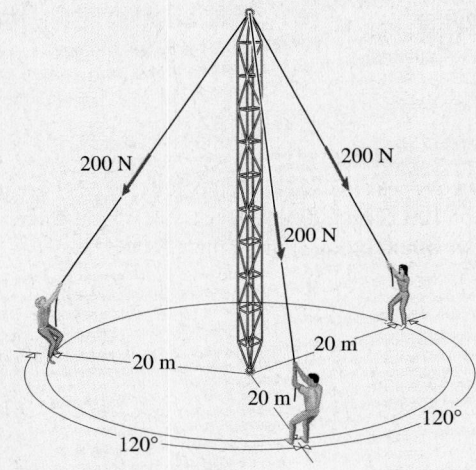

Figure P22

23. [III] Four firefighters hold a square net, one at each corner. Each person exerts a force of 200 N whose line-of-action passes through a point just below the center of the net and makes an angle with the vertical of 60°. What is the net force the firefighters exert on the net?

SECTION 4.3: THE SECOND LAW

SECTION 4.5: THE EFFECTS OF FORCE: NEWTON'S LAWS

24. [I] Draw a free-body diagram of a car rolling frictionlessly down an inclined plane.

25. [I] Draw a free-body diagram of a ball dropped by an astronaut just above the surface of the Moon.

26. [I] By pumping air up through thousands of tiny holes we can make a horizontal air table that will support several pucks so that they can move around on cushions of air with very little friction. Suppose that one such puck, having a mass of 0.25 kg, is pushed along by a 10.0-N force for 10.0 s. Determine its acceleration.

> **SOLUTION:** We have the force and the mass and need the acceleration; the time is irrelevant. $\sum F = ma$, 10.0 N = (0.25 kg)a; a = (10.0 N)/(0.25 kg) = 40 m/s^2.

27. [I] A 60-kg ice skater holds up a large sheet of cardboard that can catch the wind and drive her (frictionlessly) across the ice. While she's moving at 0.5 m/s, a wind that is constant and horizontal for 5.0 s exerts a forward force on the cardboard of 2.0 N. What is the skater's initial acceleration once the wind begins to blow?

28. [I] Two bar magnets of mass 1.0 kg and 2.0 kg have their like poles pressed together so that there is a repulsive interaction between them. In a frictionless environment the two fly apart when released such that the more massive one has an initial acceleration of 10.0 m/s^2 due north. What is the initial acceleration of the other magnet?

29. [I] A person pushes a shopping cart with a force of 100 N acting down at 45°. If the cart travels at a constant speed of 1.0 m/s, what is the value of the combined retarding force acting horizontally on the cart due to friction and air drag? [*Hint: It is the horizontal component of the force that drives the cart forward. The constant speed tells you that* $\sum F_x = ma = 0$.]

30. [I] A rope is tied around a large box of books (of mass m) and someone attaches a spring scale to the rope and pulls up and to the right on it (with a force F). The scale then makes an angle θ with the horizontal. Write expressions for the horizontal and vertical components of the applied force. Write an expression for the normal force exerted on the box by the floor. Once set in motion the box is pulled at a constant speed. If F is then the constant applied force as indicated by the scale, write an expression (in terms of F and θ) for the net frictional retarding force acting horizontally on the box.

31. [I] A kid on an old dirt bike is being pushed along at a constant speed by her father who exerts a constant downward force on the child of 10.0 N at an angle of 30.0° below the horizontal in the forward direction. If the combined mass of the kid and the bike is 30.0 kg, what is the total retarding force (e.g., air friction, friction on the tires, etc.) acting on her and the bike?

32. [I] Studies show that a male lion (170 kg) accelerates toward prey at about 10 m/s^2, which is about the same rate a human sprinter can achieve (compare that to 3.8 m/s^2 for a Porsche). How much force must the lion exert horizontally on the ground during such a charge?

33. [I] A 100-kg gentleman standing on slippery grass is pulled by his two rather unruly children. One tugs him with 50 N north,

toward the ice cream stand, while the other hauls with 120 N toward the bathroom, due east. Overlooking friction, compute Pop's resulting acceleration.

34. [II] The ballistocardiograph is a device used to assess the pumping action of the human heart. A patient lies on a horizontal 2.0-kg platform "frictionlessly" suspended on air-bearings. The rush of blood pumped in one direction will be accompanied by a counterforce on the body and table in the opposite direction. The resulting acceleration of the platform is recorded by an extremely sensitive accelerometer able to measure values as small as 10^{-5} m/s^2. For a young, healthy adult the acceleration over an interval of 0.10 s during the pumping cycle may be as great as 0.06 m/s^2. (a) Compute the force exerted by the heart if the patient has a mass of 70 kg. (b) What is the corresponding change in momentum?

35. [II] A bullet is fired from a handgun with a 24.0-cm-long barrel. Its muzzle speed is 350 m/s and its mass is 6.00 g. Compute the average force exerted on the bullet by the expanding gas in the barrel.

36. [II] Suppose a car stopped on the road is hit from behind by a bus so that it accelerates, in a straight line, up to 4.47 m/s (i.e., 10 mi/h) in 0.10 s. If the driver of the car has a mass of 50 kg and her front-seat passenger has a mass of 80 kg, what average net force must the seat exert on them? (Car seats have been known to collapse under this sort of treatment, though they obviously shouldn't.)

> **SOLUTION:** We know the initial and final speeds, and the time interval over which that change occurred, so we can determine the average acceleration: $a_{av} = \Delta v/\Delta t = $ (4.47 m/s)/(0.10 s) = 44.7 m/s^2. Therefore $\sum F_{av} = ma_{av} = $ (50 kg + 80 kg)(44.7 m/s^2) = 5.8 kN.

37. [II] When a golf club strikes a 0.046-kg ball the latter may attain a speed of 70 m/s during the 0.50-ms collision. Find the average force exerted by the club on the ball.

38. [II] During a particular rocket-powered sled run, Colonel J. P. Stapp of the U.S. Air Force decelerated from a speed of 286.5 m/s (i.e., 940 ft/s) to a dead stop in 1.40 s. Assuming he weighed 175 lb (i.e., has a mass of 79.38 kg), compute the average force, in newtons, exerted on him. Incidentally, the primary dangers in this sort of test are things like blood vessels tearing away from the lungs or the heart, and even retinal detachment, via the First Law.

39. [II] A bullet fired into wet clay will decelerate fairly uniformly. If a 10-g bullet hits a block of clay at 200 m/s and comes to rest in 20 cm, what average force does it exert on the block? [*Hint: From the fact that the bullet goes from v_i to 0 in a distance s, you can find a_{av} and with that F_{av}.*]

40. [II] Imagine a car involved in a head-on crash. The driver, whose mass is m, is to be brought uniformly to rest within the passenger compartment by compressing an inflated air bag through a distance s_c. Write an expression for the average force exerted on the bag in terms of m, v_i, and s_c. Compute that average force for a 26.8-m/s (i.e., 60-mi/h) collision, where the driver's mass is 60 kg and the allowed stopping distance in the bag is 30 cm. Assume the car deforms only negligibly.

41. [II] A 4265-lb Jaguar XJ12L requires 164 ft of straight runway as a minimum stopping distance from a speed of 60.0 mi/h. Assuming its deceleration is uniform, compute the average stopping force exerted on the car. Sorry about the units, but that's the

way the manufacturer gave the data and we should know how to handle it.

42. [II] Suppose a 6.00-g bullet traveling at 100 m/s strikes a bulletproof vest and comes to rest in about 600 μs. What average force will it impart to the happy wearer?

43. [III] During a parachute exercise over Alaska in 1955, a United States trooper jumped from a C-119 at 1200 ft, but his chute failed to open. He was found flat on his back at the bottom of a $3\frac{1}{2}$-ft-deep crater in the snow, alive and with only an incomplete fracture of the clavicle. Compute the average force that acted on him as he plowed into the snow. Assume the deceleration was constant; take his mass to be 90 kg and his terminal speed to be 120 mi/h.

SECTION 4.6: WEIGHT: GRAVITATIONAL FORCE

44. [I] What is the smallest force needed to lift a 0.50-kg bullfrog up from the ground, and under what circumstances would that answer be applicable?

45. [I] Using a force platform, it's been found that when people jump straight up as high as they can they exert a net downward force for much of the time equal to about 2.3 times their weight. Approximately what acceleration do they achieve?

46. [I] THIS PROBLEM EXAMINES THE RELATIONSHIP BETWEEN FORCE AND ACCELERATION. An astronaut floating in space next to a 10.0-kg "weightless" beacon wants to push on it so that the beacon accelerates away at 9.81 m/s². (a) What force must she apply? (b) Compare that to the beacon's weight on Earth. (c) If she just pushes with her arms, what will happen to her?

47. [I] THIS PROBLEM EXAMINES THE RELATIONSHIP BETWEEN FORCE AND ACCELERATION. The roller in Fig. P47 weighs 5.00 N and the scale reads 3.0 N. (a) What is the tension in the string? (b) What is the mass of the roller? (c) At what rate will the roller accelerate at the moment it is released, assuming it slides frictionlessly? Compare your result with that of Problem 73.

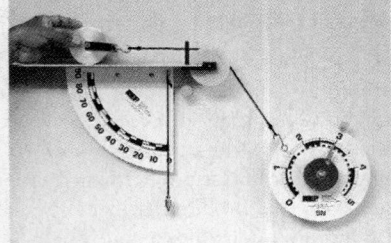

Figure P47

48. [I] The gravitational acceleration on the surface of Mercury is 0.38 times its value on Earth. What is the weight of a 1.0-kg mass on that planet?

> **SOLUTION:** GIVEN $g_M = 0.38g$, the weight will also be 0.38 times smaller: $F_w = mg_M = m0.38g = (1.0\ \text{kg})(0.38)(9.81\ \text{m/s}^2) = 3.7\ \text{N}$.

49. [I] A youngster who weighs 392.4 N on Earth is standing weightless in a space station. She jumps into the air with an average vertical acceleration of 5.00 m/s². Determine the average force exerted by the floor on the jumper during the leap. [*Hint: We have the average acceleration and need the propelling force. That can be found from the Second Law. For that you'll need the mass, but you have the weight.*]

50. [I] A maintenance robot 1.70-m tall, weighing 752 N when constructed on Earth, is found floating 400 m from an orbiting power station. The human crew decides to tie a light rope to it and haul it aboard. How much force must be exerted if the robot is to be uniformly accelerated for 1.00 s at 9.81 m/s²?

51. [I] A rocket that weighed 98.1 N when put together on Earth is attached to a 1.00×10^3-kg signal relay in intergalactic space. The rocket develops a constant thrust (i.e., it pushes with a force) of 10.0 kN. Compute the resulting acceleration.

52. [I] An *F-14* twin-engine supersonic fighter weighs 69 000 lb. Its two Pratt and Whitney TF30-P-412 engines can each deliver a peak thrust (i.e., a forward force), using afterburners, of 20 200 lb. Compute its maximum acceleration (in terms of *g*) during horizontal flight (neglect air friction). Forgive the units, but this is still the way the data is provided by the military. Clearly it's important to practice unit conversions.

53. [II] Someone of mass 100 kg is standing on top of a steep cliff with only an old rope that he knows will support no more than 500 N. His plan is to slide down the rope using friction to keep from falling freely. At what minimum rate can he accelerate down the rope in order not to break it?

54. [II] A parachutist of mass *m* lands with legs bent, coming to rest with an upwardly directed average acceleration of 4.0g. What is the average force exerted on him by the ground? Write your answer in terms of his weight.

> **SOLUTION:** The force needed to bring her to rest is $F_{av} = ma_{av} = m4.0g = 4.0mg = 4.0F_w$, but the ground also supports her weight so the net force is $F_w + 4.0F_w = 5.0F_w$.

55. [II] The all-time champion jumper is the flea. This tiny pest can attain a range of up to 12 in. or about 200 times its own body length. (That's the same as you jumping five city blocks.) Assuming a 45° liftoff angle, a push-off time of 1.0 ms, and a mass of 4.5×10^{-7} kg, compute (a) the initial acceleration of the flea, assuming it to be constant, and (b) the force it exerts on the floor in addition to its weight.

56. [II] On its first flight in 1981, the *Columbia* spaceship was part of a 4.5×10^6-lb (i.e., 20 MN) 18-story-high launch assembly that developed a total thrust of about 6.4×10^6 lb (i.e., 28.5 MN). (a) What was its initial acceleration at full power? (b) According to newspaper reports, it was traveling at 33.5 m/s (i.e., 75 mi/h) 6.0 s after blastoff when it cleared the 347-ft (i.e., 105.8-m) support tower. Are these observations consistent? To what average acceleration does that correspond? (c) Compare these values of acceleration and explain any differences.

57. [II] A 2000-kg car in neutral at the top of a 20° inclined 20-m-long driveway slips its parking brake and rolls downward. At what speed will it hit the garage door at the bottom of the incline? Neglect all retarding forces.

58. [II] A rescue helicopter lifts two people from the sea on an essentially weightless rope. Jamey (100 kg) hangs 15 m below Amy (50.0 kg), who is 5.0 m below the aircraft. What is the tension (a) on the topmost end of the rope and (b) at its middle while the helicopter hovers? (c) Compute those answers again, this time with the aircraft accelerating upward at 9.8 m/s².

59. [II] A pink pom-pom of mass *m* hangs on a chain from the rearview mirror of a Corvette L82. That machine can go from rest to 26.8 m/s (60 mi/h) in 6.8 s. The pom-pom serves as a pendulum accelerometer. Assuming a constant maximum acceleration, compute the angle the pom-pom makes with the vertical.

60. [II] It's known from laboratory studies (Problem 45) that when people jump straight up as high as they can, they exert a net downward force for much of the time equal to about 2.3 times their weight. Starting with the knees bent, the vertical distance over

which the force is applied is about 0.4 m. At approximately what speed will they leave the ground and how high will they rise?

61. [II] Suppose that a person of mass m steps off a ladder at a height s_h and lands on the ground without bouncing. If the total compression of the body and the soil during impact is s_c, and if the deceleration is assumed constant, show that the force exerted by the ground is given by

$$F_{av} = mg\,(s_h/s_c)$$

Notice that bending the knees extends s_c considerably, decreasing F_{av} accordingly.

62. [II] When a person jumps and lands stiff-legged on the heels of the feet, a considerable force can be exerted on the long leg bones. The greatest stress occurs in the tibia, or shin bone, a bit above the ankle where the bone has its smallest cross section. If a force in excess of about 50 000 N is applied upward on the heel, the tibia will probably fracture. Keeping in mind the results of Problem 61, what is the minimum height of a fall above which a 60-kg person is likely to suffer a tibial fracture? Assume the body decelerates uniformly through a distance of 1.0 cm and the landing occurs squarely on both feet.

63. [II] In Problem 36, how much force must the driver's neck exert to keep her head in line with her body during the collision if her head weighs 44.5 N (10 lb)? Think about "whiplash."

64. [III] Navy jets are hurled off the deck of a modern carrier by a combination of catapult and engine thrust. The catapult (e.g., the C-7) has a 250-ft stroke (that is, runway) and will yank a 70 000-lb F-14A jet, with its engines developing a net average thrust of 16 000 lb, from rest to a speed of 200 ft/s in just 2.4 s.   (a) Compute the average force exerted by the catapult.   (b) Is the acceleration of the plane uniform? [*Hint: Try several of the equations for constant a.*]

SECTION 4.7: COUPLED MOTIONS

65. [I] Referring to Fig 4.22b, suppose m_2 equals 2.00 kg and it's on a frictionless surface. If $m_1 = 10.0$ kg, someone holds m_2 at rest, what will be the tension in the string?

66. [I] The pulley in Fig. 4.22a is essentially weightless and frictionless. If $m_1 = 10.0$ kg, the m_2 weighs 300 N, and someone holds on to m_2 so that the system is motionless, what is the tension in the rope and the acceleration of m_1?

67. [I] The pulley in Fig. 4.22a is essentially weightless and frictionless. If m_1 weighs 100 N and the m_2 weighs 300 N, and someone holds on to m_1 so that the system is motionless, what is the tension in the rope and the acceleration of m_1? How much force must the person exert and in what direction?

68. [I] Referring to Fig 4.22b suppose m_2, which weighs 300 N, rests on a frictionless surface. If the tension in the string is measured to be 100 N, what is the acceleration of m_1?

69. [I] The pulley in Fig. 4.22a is essentially weightless and frictionless. If $m_1 = 10.0$ kg and the second mass m_2 weighs 98.1 N, what is the tension in the rope and the acceleration of m_1?

70. [I] The two masses in Fig. 4.22c are $m_1 = 10.0$ kg and $m_2 = 20.0$ kg. What must be the tension in the rope on the right if the two blocks accelerate at 0.50 m/s^2 in a straight line over a frictionless surface?

71. [I] The pulley in Fig. 4.22a is essentially weightless and fric-

tionless. Suppose that someone holds onto $m_2 = 10.0$ kg and accelerates it upward at 4.905 m/s^2. What will then be the tension in the rope given that $m_1 = 10.0$ kg?

72. [I] Suppose that the blocks in Fig 4.22b are $m_1 = 10.0$ kg and $m_2 = 20.0$ kg and the surfaces are frictionless.   (a) If someone holds on to m_1 and pulls it down with an acceleration of 9.81 m/s^2, what is the tension in the rope?   (b) What is the value of the force (F_A) applied to m_1 by the person?

> **SOLUTION:** (a) The mass m_2 experiences a net horizontal force such that $\sum F_H = F_T = m_2 a = (20.0\text{ kg})(9.81\text{ m/s}^2) = 196$ N; that's the tension. (b) The mass m_1 experiences a net force such that $\sum F = m_1 a = (10.0\text{ kg})(9.81\text{ m/s}^2)$; moreover, taking down as positive $\sum F = F_W + F_A - F_T = m_1 a = m_1 g = 98.1$ N; hence $F_A = 98.1\text{ N} - F_W + F_T = 98.1\text{ N} - 98.1\text{ N} + F_T$ and $F_A = F_T = 196$ N.

73. [I] THIS PROBLEM WILL HELP US LEARN ABOUT FORCES AND ACCELERATION. In Fig. P73 the cylindrical roller weighs 5.0 N and each of the hanging masses weighs 1.0 N. Assume the pulley is weightless and frictionless, and the roller slides freely.   (a) What is the tension in the rope when everything is at rest?   (b) After the roller is released, describe the motion of the system.   (c) Is the tension now the same as it was before?   (d) Draw a free-body diagram of both the roller and the hanging mass.   (e) When released, at what rate will the cylinder accelerate? (f) Compare your result with that of Problem 47 and make sure you understand the difference.

Figure P73

74. [II] THIS PROBLEM WILL HELP US LEARN ABOUT FORCES AND ACCELERATION. Figure P74 depicts four identical 20.0-kg cartons being pushed forward by a 80.0-N force on a frictionless horizontal plane.   (a) What is the total mass being accelerated?   (b) Determine the acceleration of each block.   (c) What is the net force exerted on each block?   (d) What force does the block second from the left exert on the block third from the left?

Figure P74

75. [II] A tug pulls two small barges (of mass $m_1 = 4.00 \times 10^3$ kg and $m_2 = 3.50 \times 10^3$ kg) tied together, one behind the other. The tug exerts 1.00 kN on the line to the first barge (m_1) and accelerates at 0.100 m/s^2. Compute the tension in the two ropes and, knowing that friction opposes the motion, determine the friction on each barge, assuming it's the same for both. [*Hint: There are two unknowns, the tension in the connecting rope and the friction. Take the sum of the forces on the first barge and the sum of the forces on the second barge, and solve the two equations simultaneously.*]

76. [II] Examine the air track depicted in Fig. Q3 and suppose that when let loose the hanging mass creates a tension of 0.260 N in the string that's attached to a 0.280-kg glider. Now imagine that a second glider of mass 0.200 kg is placed on the track in front of (i.e., to the right of) the first one and the two are attached by a very light spring. The hanging mass is released; determine the acceleration of the gliders and the force compressing the spring. Ignore friction.

77. [II] Two crates rest on the floor one next to the other. A man pushes with a horizontal force of 400 N on the larger (50.0 kg), which, in turn, pushes the other (25.0 kg) so that both slide as he walks along. Each crate experiences a friction force opposing the motion equal to 40% of its weight. Determine the acceleration of each crate. What is the force exerted by one crate on the other?

78. [II] Refer to Fig. P78. Compute the acceleration of the 1.3×10^3-N weight. Neglect friction and the mass of the rope and pulleys. Now remove the 1.8×10^3-N weight, replace it by a constant downward tug of 1.8×10^3 N acting on the end of the rope, and recalculate the acceleration of the first weight.

Figure P78

79. [II] An 80.0-kg man inside a 40.0-kg dumb-waiter (Fig. P.79) pulls down on the rope. At that moment the scale on which he is standing reads 200 N. Determine the elevator's acceleration.

80. [II] In 1784, George Atwood published the description of a device for "diluting" the effect of gravity, thereby facilitating the accurate measurement of g. Figure P80 shows the apparatus: two masses tied together by a length of essentially massless rope slung over an essentially massless free-turning pulley. With $m_2 > m_1$, prove that both masses accelerate at a rate of

$$a = \frac{(m_2 - m_1)}{(m_2 + m_1)} g$$

Show that the tension in the rope is

$$F_T = \frac{2m_1m_2}{(m_2 + m_1)} g$$

If $m_2 = 2m_1$, what is a? When is a equal to zero? When $m_2 \gg m_1$, find a.

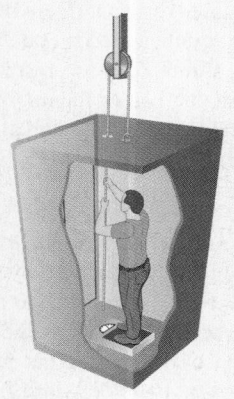

Figure P79

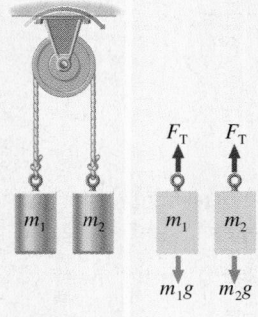

Figure P80

81. [III] Figure P81 shows three masses attached via massless ropes over weightless, frictionless pulleys on a frictionless surface. Compute (a) the tension in the ropes and (b) the acceleration of the system. Draw all appropriate free-body diagrams. (c) Discuss your results in terms of internal and external forces.

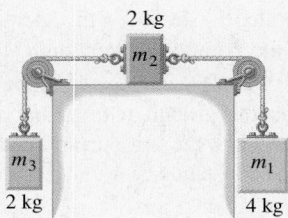

Figure P81

82. [III] A physicist on planet Mongo is using a device equivalent to Atwood's machine (Fig. P80) to measure the Mongoian gravitational acceleration g_M. He fixes one of the two 0.25-kg masses at each end of the rope. While both are at rest, he places a 0.025-kg gronch (a toadlike creature) on one of the masses. That body and its wart-covered passenger descend 0.50 m before the gronch hops off. The body continues traveling downward another 1.2 m in the next 3.0 s. Compute g_M.

SECTION 4.8: FRICTION

83. [I] A dog weighing 300 N harnessed to a sled can exert a maximum horizontal force of 160 N without slipping. What is the coefficient of static friction between the dog's foot pads and the road?

84. [I] Assume that your mass is 70.0 kg and that you are wearing leather-soled shoes on a wooden floor. Now walk over to a wall and push horizontally on it. How much force can you exert before your feet start sliding away?

85. [I] What is the maximum acceleration attainable by a four-wheel-drive vehicle with a tires-on-the-road static coefficient of μ_s?

86. [I] Mass m_1 sits on top of mass m_2, which is pulled along at a constant speed by a horizontal force F. If $m_1 = 10.0$ kg, $m_2 = 5.0$ kg, and μ_k for all surfaces is 0.30, find F.

87. [I] Someone wearing leather shoes is standing in the middle of a wooden plank. One end of the board is gradually raised until it makes an angle of 17° with the floor, at which point the person begins to slide down the incline. Compute the coefficient of static friction.

88. [I] A 30.0-kg youngster is dragged around the living room floor at a constant speed (giggling all the while) via a 60-N horizontal force. What was the appropriate pants-carpet friction coefficient?

89. [I] A wooden crate containing old $1000 bank notes has a total mass of 50.0 kg. It is transported on a flatbed truck to a facility to be burned. If the coefficient of static friction between the bed and the crate is 0.3 and the truck begins to climb a 20° incline at a constant speed, will the crate begin to slide?

> **SOLUTION:** The maximum friction force that the crate can experience is $F_f(\text{max}) = \mu_s F_N = 0.3 \, F_w \cos 20° = 0.3 \, (491 \text{ N}) \cos 20° = 138.4 \text{ N}$. The driving force down the incline is $F_w \sin 20° = (491 \text{ N}) \sin 20° = 167.9 \text{ N}$ so it slides.

90. [I] A garbage can partly filled with sand weighs 100 N, and it takes a force of 40 N to drag it down to the street at a uniform speed. How much force will it take to drag a full can weighing 150 N?

91. [II] A crate is being transported on a flatbed truck. The coefficient of static friction between the crate and the horizontal bed is

0.50. What is the minimum stopping distance if the truck, traveling at 50.0 km/h, is to decelerate uniformly and the crate is not to slide forward on the bed?

92. [II] What is the steepest incline that can be climbed at a constant speed by a four-wheel-drive vehicle having a tires-on-the-road coefficient of static friction of 0.90?

93. [II] Assume an even weight distribution on all four tires of a car with four-wheel antilock brakes. What is the minimum stopping time from 27 m/s if the coefficients of static and kinetic friction are 0.9 and 0.8, respectively?

94. [II] Place a book flat on a table and press down on it with your hand. Now suppose the hand-book and table-book values of μ_k are 0.50 and 0.40, respectively; the book's mass is 1.0 kg and your downward push on it is 10 N. How much horizontal force is needed to keep the book moving at a constant speed if your hand is stationary with respect to the table?

> SOLUTION: The normal force between your hand and the book is 10 N; the friction force between your hand and the book is $F_{f_{HB}} = \mu_k F_N = 0.50(10 \text{ N}) = 5.0$ N. The normal force between the book and the table is $(10 \text{ N} + g \times 1.0 \text{ kg}) = (10 \text{ N} + 9.81 \text{ N}) = 19.81$ N; the friction force between the book and the table is $F_{f_{BT}} = \mu_k F_N = 0.40(19.81 \text{ N}) = 7.92$ N. The net friction force opposing the motion is 12.9 N or, to two significant figures, 13 N. This is the required force.

95. [II] THIS PROBLEM WILL HELP US LEARN ABOUT FORCES AND EQUILIBRIUM. The soda can with the wad of clay on it in Fig. P95 has a net mass of 404.2 g and rests on a surface where there is friction. The incline angle is 20.0°, and the two hanging weights are each 1.00 N. (a) What is the component of the weight acting down the incline? (b) What is the normal force acting up on the can? (c) Draw a free-body diagram of the can and clay together. (d) If the can is just on the verge of sliding up the incline, what is the value of the coefficient of static friction between it and the incline?

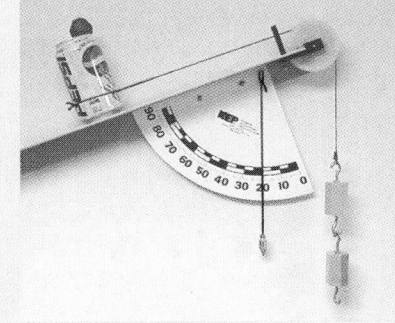

Figure P95

96. [II] THIS PROBLEM WILL HELP US LEARN MORE ABOUT FORCE AND ACCELERATION. The two blocks, each of mass *m*, shown in Fig. P96 are being accelerated at a rate *a* across a surface that has a coefficient of friction of μ_k by a horizontal force *F*. (a) Write an expression (in terms of *m, g, μ_k*, and *a*) for the tension in the rope that ties the blocks together. Ignore the weight of both the rope and the chain. (b) Write an expression for the force *F* (in terms of *m, g, μ_k*, and *a*).

Figure P96

97. [II] THIS PROBLEM DEALS WITH FORCE AND ACCELERATION. Using Fig. P96 and with the previous problem in mind, suppose each block has a mass of 10.0 kg. If the connecting rope breaks at a tension of 20.0 N, (a) write an expression for the acceleration that will break the rope. (b) The force *F* applied via the chain to the first

block on the right must overcome both friction and inertia. What value of *F* will cause that rope to break given that friction exists? (c) What is the effect of friction on the breaking of the rope?

98. [II] THIS PROBLEM WILL HELP US LEARN MORE ABOUT FORCE AND ACCELERATION. Figure P98 shows three objects on a frictionless surface. (a) Write an expression for F_{T3} given that the system accelerates as the mass *m* descends. (b) Write expressions for F_{T2} and F_{T1}. (c) Write an expression for the acceleration, *a*, of the system in terms of *g* and the masses. Does your answer agree with the solutions for Problem 73.

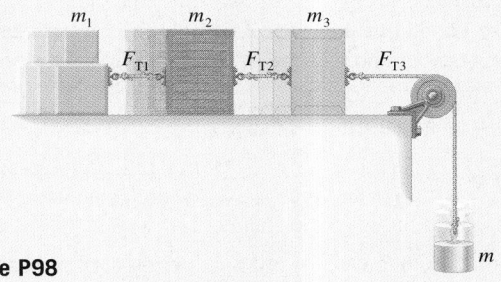

Figure P98

99. [II] A skier on a 4.0° inclined, snow-covered run skied downhill at a constant speed. Compute the coefficient of friction between the waxed skis and snow on that day when the temperature was around 0°C. Why is it that if the temperature were to drop to −10°C, μ_k would rise to around 0.22? Neglect air friction.

100. [II] A youngster shoots a bottle cap up a 20° inclined board at 2.0 m/s. The cap slides in a straight line, slowing to 1.0 m/s after traveling some distance. If $\mu_k = 0.4$, find that distance.

101. [II] A 100-kg bale of dried hay falls off a truck traveling on a level road at 88.0 km/h. It lands flat on the blacktop and skids 100 m before coming to rest. Assuming a uniform deceleration, compute the coefficient of kinetic friction between the bale and the road.

102. [II] A 100-kg trunk loaded with old books is to be slid across a floor by a young woman who exerts a force of 300 N down and forward at 30° with the horizontal. If $\mu_k = 0.4$ and $\mu_s = 0.5$, compute the resulting acceleration.

> SOLUTION: The maximum static friction force that the trunk can experience is $F_f(\text{max}) = \mu_s F_N$ where $F_N = F_W + 300 \text{ N} \sin 30° = (100 \text{ kg})(9.81 \text{ m/s}^2) + 300 \text{ N} \sin 30° = 1.131$ kN. Thus $F_f(\text{max}) = 0.5(1.131 \text{ kN}) = 566$ N. The horizontal force she exerts is $(300 \text{ N}) \cos 30° = 259.8$ N and the trunk does not move: $a = 0$.

103. [II] Suppose the woman in Problem 102 puts aside some of the books so that the mass of the load is 50 kg. (a) Pushing with the same force, what will the acceleration be now? (b) A bit annoyed, she squirts some oil under the trunk so that $\mu_s = 0.4$ and $\mu_k = 0.3$. What now?

104. [II] One of two identical 200-kg crates is dragged along the floor at a constant speed by a horizontal force of 200 N. How much force would it take to pull them both at some constant speed if (a) they are tied together one behind the other? (b) the second is stacked on top of the first?

105. [III] What is the steepest incline that can be climbed at a constant speed by a rear-wheel-drive auto having a tires-on-the-road μ_s of 0.9 and 57% of its weight on the front axle? Ignore all other forms of friction and remember that the wheels are turning.

106. [III] Determine the force (F) needed to keep the blocks depicted in Fig. P106 moving at a constant speed—m_2 to the right and m_1 to the left. Assume both the frictionless pulley and the rope to be massless: m_1 has a weight of 5.0 N; m_2 has a weight of 10 N; and $\mu_k =$ 0.40 for all surfaces.

Figure P106

107. [III] A greasy flatbed truck is carrying a crate weighing 2.0 kN. The truck accelerates uniformly from rest to a speed of 30 km/h in a distance of 30.0 m. In that time, the crate slides 1.0 m back toward the end of the truck. Compute the coefficient of friction between bed and box.

SECTION 4.9: EQUILIBRIUM: STATICS

108. [I] A 200-N chandelier is hung from a ceiling hook by a chain weighing 50 N.   (a) What force is exerted on the hook?   (b) What force acts on the bottom of the chain?

109. [I] What are the tension forces acting on each of the weightless ropes in Fig. P109?

Figure P109

110. [I] THIS PROBLEM WILL HELP US LEARN MORE ABOUT FORCES AND EQUILIBRIUM. There are three strings tied to the small motionless ring in Fig. P110. The one on the right makes an angle θ with the horizontal, while the one on the left makes an angle ϕ. Given that each of the six hanging masses weighs 1.00 N,   (a) Draw a free-body diagram of the ring. (b) Write expressions for the horizontal and vertical components of each force exerted by a string on the ring. (c) Use the sum-of-the-forces-equals-zero to determine θ and ϕ.

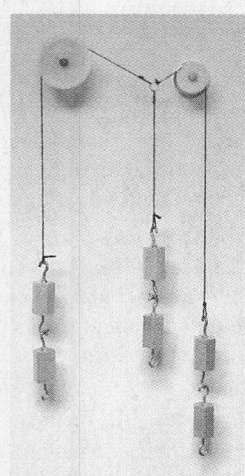

Figure P110

111. [I] The 10-kg block in Fig. P111 is held at rest by the four ropes shown. If the tension in the ropes on the right and top are each 98 N, what is the tension in the remaining two ropes?

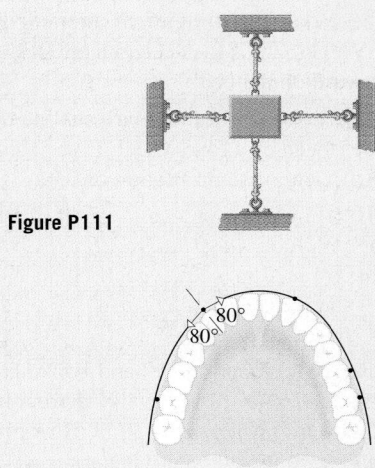

Figure P111

112. [I] The orthodontal wire brace in Fig. P112 makes an angle of 80.0° with the perpendicular to the protruding tooth. If the tension in the wire is 10.0 N, what force is exerted on the tooth by the brace?

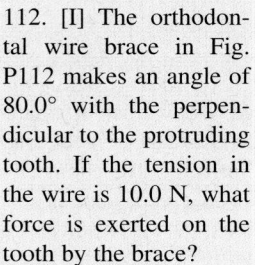

Figure P112

113. [I] Referring to Fig. P113, determine the scale reading in the right arm of the suspension and the angle θ. The scale was set to read zero with no suspended load.

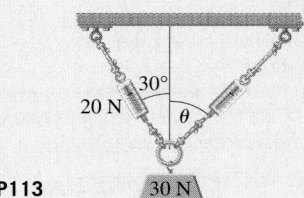

Figure P113

114. [I] The steel beam that hangs horizontally in Fig. P114 weighs 8.00 kN, and the lower cables hook on to it at angles of 70.0°. What are the tensions in the lower cables?

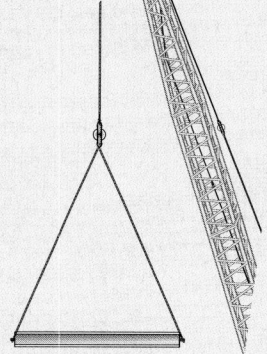

Figure P114

115. [I] In the static arrangement shown in Fig. P115, the pulleys and ropes are essentially weightless. If weight-1 is 15.0 N, weight-2 is 31.0 N, and the two angles are measured to be $\theta = 45.0°$ and $\phi = 20.0°$, determine the value of weight-3. Check that the system is in equilibrium.

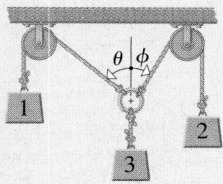

Figure P115

116. [I] The hand in Fig. P116 exerts a downward force that holds the length of pipe at rest. The pulleys are essentially weightless. (a) Compute that force. (b) What is the tension in the rope that is strung over the pulley? (c) What force supports the upper hook?

SOLUTION: (a) The load is 300 N. The bottom pulley is attached to the load and both are supported by two vertical lengths of rope. The tension in each length of rope is $(300\,\text{N})/2 = 150\,\text{N}$. (b) There is one continuous rope and the tension everywhere within it is 150 N. (c) The hook supports the load (300 N) and the downward pull of the hand (150 N), and so the ceiling exerts an upward force of 450 N.

Figure P116

300 N

117. [I] Look at Fig. P117. If the hand pulls down so that the 100-N weight remains at rest, what is the tension in the long rope? What force is exerted down on the rightmost ceiling bracket?

100 N

Figure P117

118. [I] What is the tension in each length of rope in Fig. P118, given that the load is at rest?

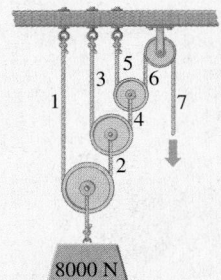

Figure P118 8000 N

119. [I] The cylinder (and the several masses attached to it) photographed in Fig. P119 is at rest, although it can roll frictionlessly along the 30°-inclined plane. The scale reads 2.7 N. Compute the net weight of the cylinder and the attached masses.

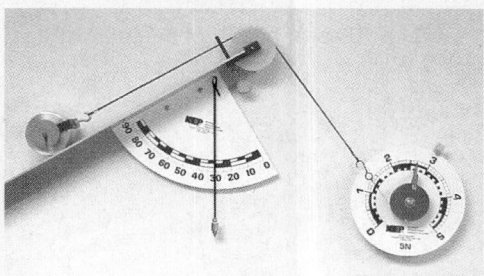

Figure P119

120. [I] What downward force on the rope in Fig. P120 will hold the 300-N block at rest assuming the pulleys are essentially weightless? Notice that there are two different ropes holding up the load.

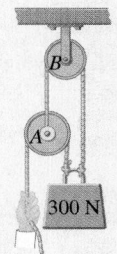

Figure P120

300 N

121. [II] Determine the weight of the motionless mass *m* in Fig. P121. Assume the pulleys and ropes are all essentially weightless.

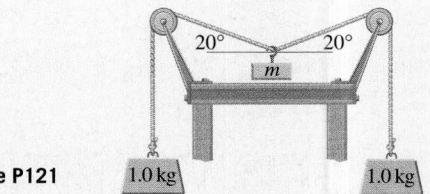

20°   *m*   20°

Figure P121   1.0 kg   1.0 kg

122. [II] An essentially weightless rope is strung nearly horizontally over a light pulley, as shown in Fig. P122. The spring balance is adjusted to read zero. A 100-N weight is then hung at the midpoint of the span and the rope sags, descending 10 cm from the horizontal. Determine the scale reading.

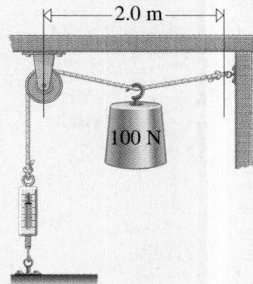

2.0 m

100 N

Figure P122

123. [II] A 533.8-N tightrope walker dances out to the middle of a 20-m-long wire stretched parallel to the ground between two buildings. She is wearing a pink tutu, of negligible weight; the wire sags, making a 5.0° angle on both sides of her feet with the horizontal. Find the tension.

124. [II] If a force of 100 N in Fig. P124 holds the load motionless, what is its mass? What is the tension in rope 1?

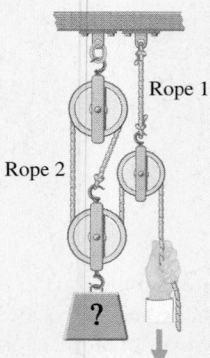

Figure P124

125. [II] Determine both the angle at which the pulley hangs and the tension in the hook supporting it in Fig. P125.

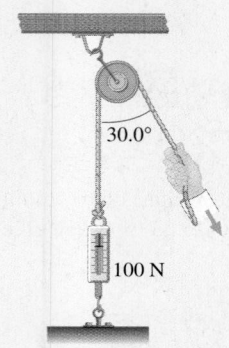

Figure P125

126. [II] THIS PROBLEM WILL HELP US LEARN MORE ABOUT FORCES AND EQUILIBRIUM. The spring in Fig. P126 exerts a horizontal force of 200 N (45.0 lb) on the ring, which is at rest. (a) Draw a free-body diagram of the ring. (b) What is the horizontal force exerted by cable-AB on the ring? (c) What is the tension in cable-AB? (d) What is the vertical force exerted by cable-AB on the ring? (e) What is the value of the load?

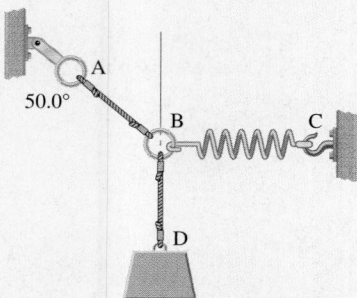

Figure P126

127. [II] THIS PROBLEM WILL HELP US LEARN ABOUT FORCES AND EQUILIBRIUM. The system shown in Fig. P127 is in equilibrium. (a) Draw a free-body diagram of the ring. (b) What is the vertical force exerted by rope-2 on the ring? (c) Determine the angle θ. (d) What is the horizontal force exerted by rope-2 on the ring? (e) What is the tension in each rope? (f) How much does the mass on the left weigh?

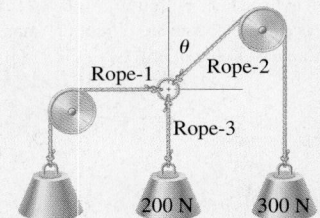

Figure P127

128. [II] THIS PROBLEM IS ABOUT FORCES AND EQUILIBRIUM. In Fig. P128 a 500-g mass sits on top of a 397-g spool of solder. The string makes an angle of 29° with the horizontal, and the two hanging masses each weigh 1.00 N. (a) What are the horizontal and vertical components of the tensile force exerted by the string on the spool? (b) Determine the normal force on the bottom of spool. (c) What is the coefficient of static friction between the spool and platform, given that the system is just on the verge of moving?

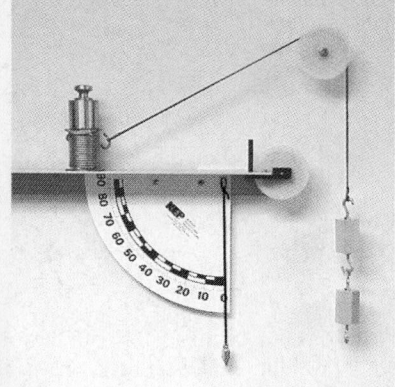

Figure P128

129. [II] THIS PROBLEM WILL HELP US LEARN MORE ABOUT FORCES AND EQUILIBRIUM. The crane in Fig P129 supports a 20.0-kN (4.5 × 10³ lb) load. To simplify things take the cables to be represented by two separate bundles: cable-AB and cable-BC. (a) Draw a free-body diagram of point-B putting in the forces exerted by the boom, cable-AB, and cable-BC. (b) Determine the horizontal and vertical components of each force. (c) Using $\sum F_x = 0$ and $\sum F_y = 0$ determine the compressive force on the boom.

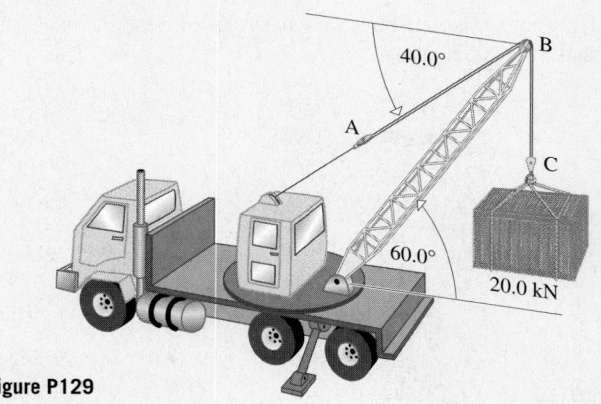

Figure P129

130. [II] Determine the net compressive force acting on the patella as a result of the action of the quadriceps muscle (F_m upward) and the patellar tendon (F_t downward) as shown in Fig. P130. Take the angle between the muscle group and tendon to be 160°, and suppose the leg is bent symmetrically so that $F_m = F_t =$ 100 N. Check your answer using the vector triangle and the Law of Cosines. What happens to the net force as the knee is further bent?

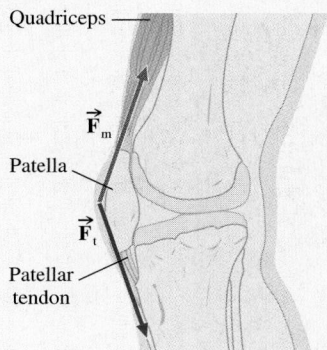

Quadriceps

$\vec{F}_m$

Patella

$\vec{F}_t$

Patellar tendon

Figure P130

131. [II] Figure P131 depicts a rod of negligible mass pinned to a wall at one end and supported by a rope at the other end. The load is 2.00 kN and the scale reads 1.00 kN. (a) Compute the horizontal and vertical reaction forces on the rod at the wall. (b) Make a sketch showing that the forces on the bar are concurrent. *Notice that the forces do not all act at the ends of the bar and so the bar will not be in pure compression or tension—the reaction force will not be parallel to the bar.* [*Hint: The bar is in equilibrium under the three forces that act on it;* $\sum F_x = 0$ *and* $\sum F_y = 0$. *The reaction force must act to the right and upward (since the forces are concurrent); the left end of the rod would drop downward unless supported.*]

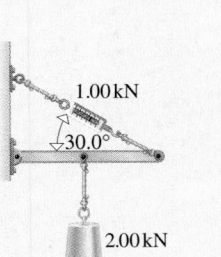

1.00 kN
30.0°
2.00 kN

Figure P131

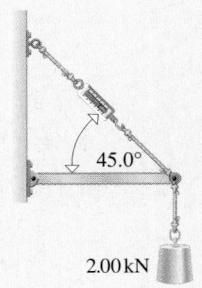

45.0°
2.00 kN

Figure P132

132. [II] A very lightweight beam is attached to a wall via a pinned bracket (Fig. P132). If the load is 2.00 kN, determine the reading of the spring scale and the force exerted by the wall bracket on the beam.

133. [II] A streetlight weighing 2.40 kN is hung from a pipe of negligible weight as shown in Fig. P133. (a) Determine the value of the reaction force exerted by the wall on the pipe and the angle that force makes with the horizontal. (b) Make a sketch showing

that the forces on the pipe are concurrent. *Notice that the forces do not all act at the ends of the bar and so the bar will not be in pure compression or tension—the reaction force will not be parallel to the bar.*

2460 N

2.40 kN

Figure P133

134. [II] The three bodies in Fig. P134 are at rest. If a fly lands on the 80.0-kg mass and the system then begins to move, what is the static coefficient of friction for the box and surface? The pulleys are weightless and frictionless.

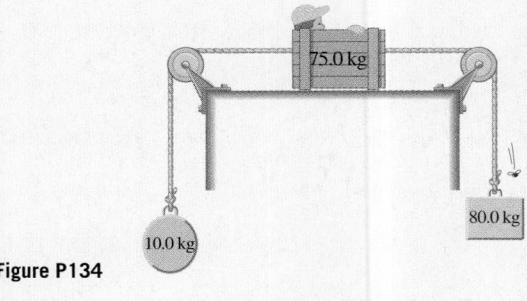

75.0 kg

10.0 kg

80.0 kg

Figure P134

135. [III] While standing on two legs, the hips support about 66.64% of the weight of an adult male. Given that the hip abductor muscles (on each leg) exert a force (F_M) at an angle with the horizontal of 70° (Fig. P135) and that the upward reaction force exerted by the head of the femur on the pelvis (F_R) makes an angle of 75°, show that $F_M = 0.99F_W$ whereas $F_R = 1.3F_W$.

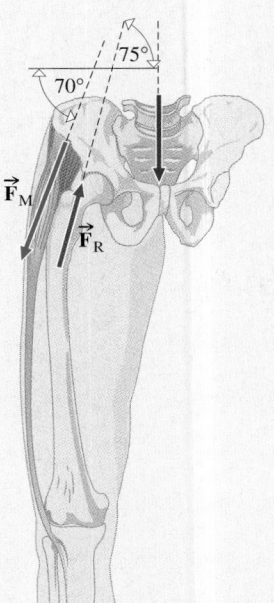

75°
70°

$\vec{F}_M$

$\vec{F}_R$

Figure P135

136. [III] Consider the arrangement depicted in Fig. P136. Now, compute the internal forces in each of the ropes and in the boom.

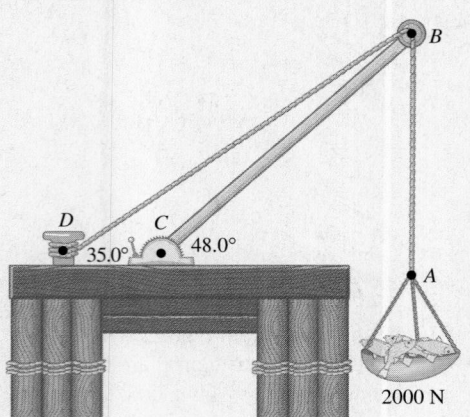

Figure P136

137. [III] Determine the reaction forces at the four contact points *A*, *B*, *C*, and *D*, in Fig. P137. The hollow balls, which are homogeneous and made of the same material, have masses of 1.02 kg and 2.04 kg, and the chamber they sit in is 55.98-cm wide. Everything is smooth, so that friction is negligible.

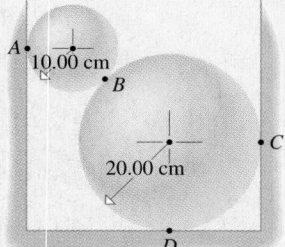

Figure P137

Chapter 5
Centripetal Force & Gravity

*T*he central problem in Newton's work, indeed the central question of the age, was "How do the planets move?" The motion of the heavens had long been one of the great issues of Western science, philosophy, and theology. It was Newton who finally welded dynamics and astronomy into a mathematical understanding that swept away Aristotelian scholarship as if that genius had been little more than an amusing schoolboy. At the heart of the Newtonian synthesis is the concept of a center-seeking gravitational force holding the planets in their orbits.

Centripetal Force

The common wisdom 350 years ago (though it's wrong!) was that revolving objects are thrown radially outward by a "centrifugal" force. Even so, G. B. Benedetti (1585), while considering the motion of a rock in a sling, correctly pointed out that an object revolving in a circle and suddenly unleashed sails off in a straight line *tangent* to the curve at the point of release. (David knew as much intuitively. Perhaps Goliath did, too.) The person who first solved the problem, and in so doing gave impetus to the mathematization of physics, was Christiaan Huygens (1658). Newton probably independently came to the correct understanding of motion along a curve much later in the early 1680s, though his analysis was superior. It was Newton who coined the phrase "centripetal force."

Whirl a ball tied to a string in a circle (Fig. 5.1). Its velocity vector continuously changes as it's whipped around, even at a constant speed, because the direction of $\vec{v}$ keeps changing. That means there must be an acceleration and, by the Second Law, a force causing that acceleration. The hand pulls the string inward toward the center of the motion, and the string continuously pulls the ball off its straight-line inertial course. *A center-seeking or **centripetal** force must be exerted on an object if it is to move in a curved path.* That centripetal force might be due to the pull of a string, or gravity, or magnetism, or friction; but whatever its cause, there must be an external force acting toward the center of the circular motion. Remove that inward centripetal force and the motion instantly becomes straight-line inertial (and that's tangential, *not* outwardly radial).

STUDY GUIDE

This chapter contains two independent primary ideas: centripetal force (p. 144),

$$F_C = \frac{mv^2}{r}$$

which relates to motion along a curved path, and the Law of Universal Gravitation (p. 147)

$$F_G = G\frac{mM}{r^2}$$

Everything else is an application of these principles. They're together in this chapter only because we need both ideas to understand planetary motion.

There's lots of historical material in this chapter. Don't worry about all the details, just concentrate on the physical ideas and how they developed.

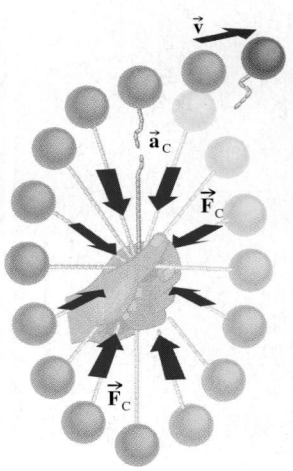

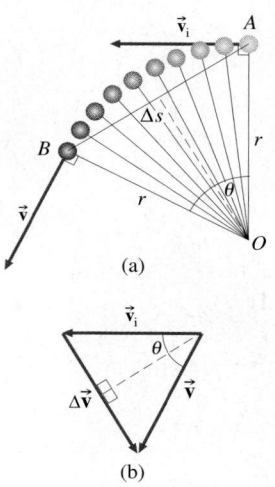

Figure 5.1 An object tied to a string revolves in a circle under the influence of a constant inward tug. The object is continuously accelerated toward the center of the circle. If the string breaks as it did here at the top of the swing, the object flies off tangentially.

Figure 5.2 The geometry of circular motion. In this simple derivation the speed is constant, but that need not be the case. At any instant, $a_C = v^2/r$.

5.1 Centripetal Acceleration

Instead of flying off tangentially, the ball is tugged inwardly and forced to turn inwardly; it accelerates toward the center of the arc along which it is traveling. We can derive an expression for this **centripetal acceleration**, $\vec{a}_C$, for an object that moves uniformly in a circle (Fig. 5.2). The body travels from A to B in a time interval Δt, and as it does, the radius r sweeps through an angle θ. The velocity $\vec{v}_i$ at A changes direction, becoming $\vec{v}$ at B. Since both these vectors are perpendicular to r, $\vec{v}_i$ must also sweep through an angle θ as it moves into $\vec{v}$ (see Fig. 5.2b). The initial velocity plus the change $\Delta\vec{v}$ equals the final velocity ($\vec{v} = \vec{v}_i + \Delta\vec{v}$). Inasmuch as $v = v_i$, since the speed is constant, the velocity triangle is isosceles, just as is triangle ABO. Indeed, these two are similar triangles, both having the same vertex angle θ. It follows that the ratios of any two of their corresponding sides are equal:

$$\frac{\Delta v}{v} = \frac{\overline{AB}}{r}$$

The side $\overline{AB}$ is the magnitude of the displacement Δs as the body moves from A to B, and so

$$\Delta v = \frac{v}{r}\Delta s$$

Dividing by Δt yields

$$\frac{\Delta v}{\Delta t} = \frac{v}{r}\frac{\Delta s}{\Delta t}$$

In the limit as Δt approaches zero, $\Delta s/\Delta t$ becomes v, by way of Eq. (2.7), and $\Delta v/\Delta t$ becomes a, by way of Eq. (3.3). This particular instantaneous acceleration is known as the **centripetal acceleration**. For a body in circular motion, a_C is given by

$$a_C = \frac{v^2}{r} \tag{5.1}$$

where the units are m/s².

The acceleration vector $\vec{a}_C$ points radially inward toward the center of the motion. The direction of $\vec{a}_C$ is the direction of $(\Delta\vec{v}/\Delta t)$ in the limit as $\Delta t \to 0$. There are several

The hammer has a mass of 7.26 kg and is whirled around several times before being released. Ultimately, the range (87 m is the record) is determined by the launch speed, and that is limited by the centripetal force the athlete can exert.

ways to see that $\Delta\vec{v}$ points inward perpendicular to the cord of the circle, Δs; for instance, in the two similar triangles, $\overline{OA}$ is perpendicular to $\vec{v}_i$, $\overline{OB}$ is perpendicular to $\vec{v}$, and therefore Δs must be perpendicular to $\Delta\vec{v}$. {For another proof that $\vec{a}_C$ points radially inward click on **CENTRIPETAL ACCELERATION** under **FURTHER DISCUSSIONS** on the **CD**.}

Example 5.1   [II] A beetle standing on the edge of an antique 12-in. vinyl record of "Sergeant Pepper's Lonely Hearts Club Band" is whirling around at $33\frac{1}{3}$ rotations per minute. Compute the magnitude of the creature's centripetal acceleration.

Solution   Centripetal acceleration depends on speed and radius. (1) TRANSLATION—An object is moving in a circle of known radius at a known rate; determine its centripetal acceleration.   (2) GIVEN: $r = 6.0$ in. $= 0.152$ m and a rotation rate of $33\frac{1}{3}$ rpm. FIND: a_C. (3) PROBLEM TYPE—Circular motion/centripetal acceleration. (4) PROCEDURE—There is one defining expression for a_C, namely, $a_C = v^2/r$. We must

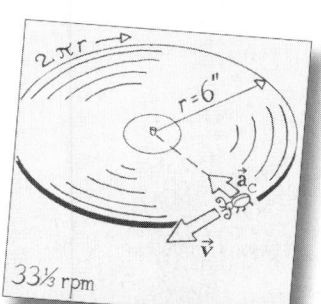

first determine v from the rotation rate ($33\frac{1}{3}$ rotations per minute) and the circumference ($2\pi r$).   (5) CALCULATION—The total distance traveled in 1 min is $33\frac{1}{3} \times 2\pi r = 31.9$ m; therefore, $v = (31.9 \text{ m})/(60 \text{ s}) = 0.532$ m/s and

$$a_C = \frac{v^2}{r} = \frac{(0.532 \text{ m/s})^2}{(0.152 \text{ m})} = \boxed{1.9 \text{ m/s}^2}$$

Quick Check: At a speed of 1 m/s and a radius of 1 m, a_C would be 1 m/s^2. Here, v is more than one-half that and r about one-sixth, so the answer should be a bit more than $(6/4)$ $(1 \text{ m/s}^2) \approx 3/2$ m/s^2.

{For more worked problems click on **WALK-THROUGHS** *in* **CHAPTER 5** *on the* **CD**.*}*

Equation (5.1) applies to paths that are not completely circular (Fig. 5.3) provided that, at any instant, we use for r the radius of the circular segment being traveled, the *radius of curvature*.

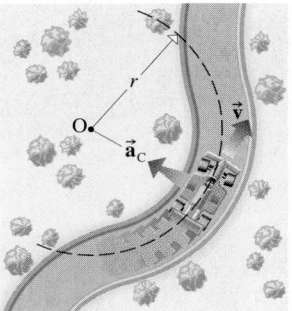

Figure 5.3 The centripetal acceleration of a car on a curved road at the moment when the radius of curvature is r.

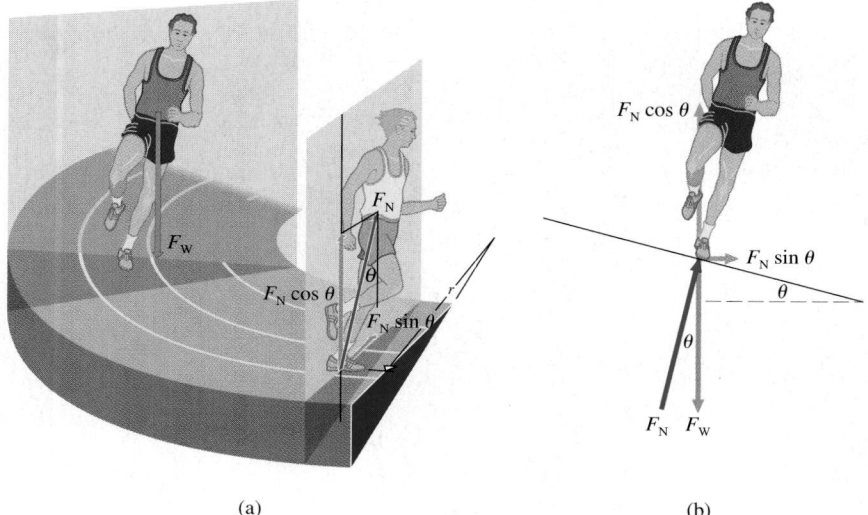

The reaction force F_R acting up along the line of the skate blade has two components: F_N (which is vertical and equal to the weight, and supports the skater) and F_C (which is the needed centripetal force). Here F_C is due to friction— if this were greased glass (where the blades could not dig in), that skater could never make the turn.

5.2 Center-Seeking Forces

If a body of mass m is accelerating, it must be experiencing a net force given by $F = ma$. Any object moving in a circle must be restrained by an inwardly directed **centripetal force** $\vec{F}_c$, having a magnitude of

$$F_C = ma_C = \frac{mv^2}{r} \qquad (5.2)$$

This is the force tugging the body off its otherwise straight-line inertial course. And the tighter the curve (the smaller is r), the more the path deviates from straight-line motion and the greater the needed force (Fig. 5.4).

Banked Turns

It's possible to bank a roadway so that the normal force exerted by the road provides the centripetal force rather than relying on friction. Figure 5.5 shows how the normal force $\vec{F}_N$ (perpendicular to the track's surface) acting on a runner has components both vertical (supporting the weight) and horizontal (providing the centripetal force). With no vertical acceleration

$$^{+\uparrow}\sum F_V = F_N \cos \theta - F_W = 0$$

> When a force such as gravity or friction continues to act perpendicular to the direction of motion of an object (which therefore moves in a circular arc), we call that a **centripetal force.**

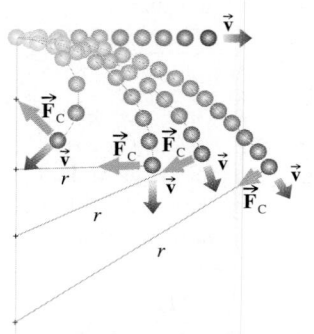

Figure 5.4 It takes no force for an object to move along a straight line at a uniform speed. Such a path can be thought of as a circle having an infinite radius. If the object is made to follow an arc of finite radius r, a centripetal force must be applied. The smaller the radius, the tighter the curve, and the greater is the force needed to pull the object away from its inertial path.

(a)   (b)

Figure 5.5 On a banked road the normal force F_N, which is always perpendicular to the road, is not aligned with the weight F_W, which is always straight down. The normal force has a radial component, $F_N \sin \theta$, that can be made equal to F_C.

CENTRIPETAL FORCE, AMTRAK, & THE BULGING EARTH

Centripetal force is not a new physical interaction; it's simply the name given to any force directed toward the center of the motion. For the Moon revolving about the Earth, $\vec{F}_C$ is gravitational. For a car rounding a turn on a flat highway, $\vec{F}_C$ is frictional. There are trains derailed all over the world simply because they're driven too fast around curves; Amtrak's Silver Meteor was derailed in December 1991 by an engineer oblivious to Eq. (5.2).

It was Newton who first showed that the bulge of the Earth (about 20 km at the equator) was due to the planet's rotation. Any piece of spinning Earth tends to fly off tangentially and the planet essentially stretches out. Given a high enough v, the dirtball would rip apart. It's the equatorial region that stretches most because v is greatest there.

Notice that $F_N > F_W$. There is a horizontal or radial acceleration, and so

$$\overset{\text{\tiny +}}{\rightarrow}\sum F_H = ma_H$$

with $a_H = a_C$

$$F_N \sin\theta = F_C = \frac{mv^2}{r}$$

The first equation contains $\cos\theta$ and the second, $\sin\theta$. One way to solve for θ is to combine the two into a single expression in terms of $\tan\theta$. Accordingly, write the first equation as $F_N \cos\theta = F_W = mg$ and divide the second one by it, thus

$$\frac{F_N \sin\theta}{F_N \cos\theta} = \frac{mv^2/r}{mg}$$

and

[banked turns]
$$\tan\theta = \frac{v^2}{gr} \qquad (5.3)$$

This expression gives the proper banking angle for any one speed, and it applies equally well to cars on roads, trains on tracks, and even to birds and planes banking during turns in level flight. Any car—Eq. (5.3) is independent of mass—traveling at the posted speed can safely make the tilted turn even if the road is slippery, which is the reason highway ramps are banked. It follows from Eq. (5.3) that the smaller the radius and the greater the speed, the larger is $\tan\theta$ and, hence, the greater the banking angle. Motorcyclists and tobogganers are forever careening at high speeds around incredibly tilted tracks.

Example 5.2 **[II]** A circular track with a 20-m radius is to be banked at an angle θ appropriate for a "4.0-min mile," which is equivalent to 1.609×10^3 m in 240 s. Compute θ.

Solution Whenever you see the term "banked" in relation to a track, $\tan\theta$ should come to mind. (1) TRANSLATION—An object is moving in a circle of known radius at a known rate; determine the proper banking angle. (2) GIVEN: $r = 20$ m and v corresponding to a speed of 1 mi per 4 min. FIND: θ. (3) PROBLEM TYPE—Circular motion/centripetal acceleration/banked track. (4) PROCEDURE—The banking angle is found using $\tan\theta = v^2/gr$. (5) CALCULATION—

First find v in SI units:

$$v = \frac{1.609 \times 10^3\,\text{m}}{240\,\text{s}} = 6.71\,\text{m/s}$$

Then

$$\tan\theta = \frac{v^2}{gr} = \frac{(6.71\,\text{m/s})^2}{(9.81\,\text{m/s}^2)(20\,\text{m})} = 0.229$$

and

$$\boxed{\theta = 13°}$$

The runner will lean into the curve balancing automatically and quite comfortably at a 13° tilt.

Quick Check: Calculate v using the above determined value of θ: $v = \sqrt{gr\tan\theta} = \sqrt{(9.8\,\text{m/s}^2)(20\,\text{m})(\tan 13°)} = 6.7\,\text{m/s}$.

Internet

For a great video of the ultimate banked track, showing a car speeding around a cylindrical vertical wall, go to

http://www.mut.ac.th/~physics

Click on *Physics Magic*, and then on *The Wall of Death*.

Circular Motion

Imagine a frog sitting in the bottom of a bucket that's being whirled in a horizontal plane (Fig. 5.6). The frog, left to its own, would move in a straight tangential line, but the bottom of the bucket, constrained to a circular path, advances and continuously intercepts that line. The bucket is forever nudging its passenger out of the way, thereby exerting an inward centripetal force on it. This is a normal force exerted *on the frog* by the bucket. The frog, in turn, exerts a radially outward reaction force *on the bucket*. {For a discussion of circular motion in a vertical plane click on VERTICAL ARCS under FURTHER DISCUSSIONS on the CD.}

Figure 5.6 A frog being whirled around in a horizontal circle. The bottom of the bucket exerts a normal force that is the centripetal force.

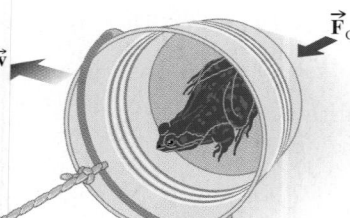

ARTIFICIAL GRAVITY, THE CENTRIFUGE, & SHARP TURNS

Suppose the frog and bucket of Fig. 5.6 were in deep space, where there is no appreciable gravity. If $a_c = v^2/r = 9.81$ m/s², the "frogonaut" would experience a normal force equal to its Earth-weight, as though it were sitting on the ground back at the old mudhole. It has been suggested that great doughnut-shaped, spinning space stations (where "up" is radially inward toward the central axis) could provide a more natural long-term environment for humans.

The **centrifuge** is a practical device using the same frog-in-the-bucket principle. The idea is simple—for a sample at rest, the sedimentation rate of particles suspended in a liquid (like blood corpuscles in serum) is proportional to g. The centrifuge revolves the sample, creating a simulated gravity of as much as $500\,000g$ at perhaps $60\,000$ rpm. Particles denser than the surrounding liquid end up in the bottom of a test tube, much as the frog ended up pressing on the bottom of the bucket.

A driver colliding with the door in a turning car (Fig. 5.7) experiences the same thing as a cell in a centrifuge or a frog in a pail. The wall of the car intercepts the driver's linear motion, forcing her radially inward toward the center of the turn. For ordinary turns, there is usually enough friction via the seats to supply F_c and keep people from crashing into doors.

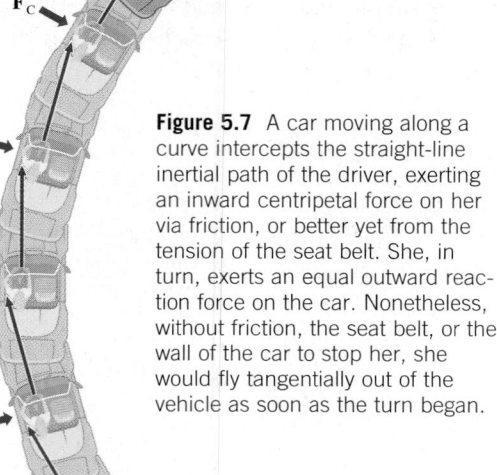

Figure 5.7 A car moving along a curve intercepts the straight-line inertial path of the driver, exerting an inward centripetal force on her via friction, or better yet from the tension of the seat belt. She, in turn, exerts an equal outward reaction force on the car. Nonetheless, without friction, the seat belt, or the wall of the car to stop her, she would fly tangentially out of the vehicle as soon as the turn began.

ON ROUNDING CURVES & SCREECHING TIRES

Ordinarily when a car, runner, bicycle, or skater makes a turn, it is friction with the ground that provides the centripetal force (see photo on p. 144). The runner usually leans into the turn, pushing horizontally—radially outward—on the ground. The ground pushes back on him with a friction force that acts radially inward toward the center of the curved path such that $F_f = F_c$. When a car makes a tight turn (i.e., r is small) at a high speed (i.e., v is large) it can happen that the needed centripetal force exceeds that which can be provided by road-tire friction. That's especially true if you happen to move onto an oil slick or a patch of ice. In that unfortunate instance the car will tend to leave the road tangentially; there are battered concrete highway dividers at every sharp turn that prove the point. When the tire tread pulls free, overcoming friction on a sharp turn, it vibrates, producing a characteristic screech (see p. 359 for a discussion of stick-slip friction) that warns of impending danger.

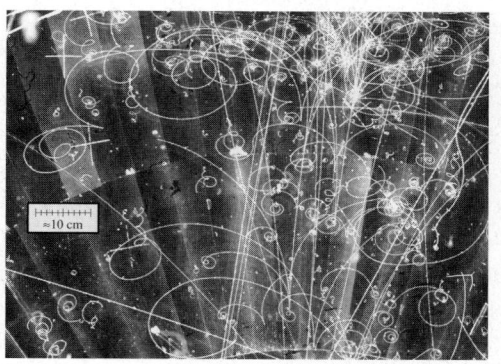

When a charged object traverses a magnetic field, it experiences a centripetal force. Low-mass particles, like electrons and positrons, have little momentum and follow tightly curved paths. These particles, traveling through a liquid mixture of neon and hydrogen, quickly lose energy; they rapidly spiral inward as they slow down.

Gravity

An amusement-park ride that makes use of gravity to supply centripetal force. At the top, the seat exerts a normal force (F_N) downward on a passenger such that $F_C = F_N + F_W$. Therefore $F_N = F_C - F_W$ and when the ride is designed so that $F_C = F_W$, $F_N = 0$ and the passenger is effectively weightless, floating upside down in the seat.

After 2000 years of thinking about gravity, we have a powerful, albeit still incomplete, understanding of the phenomenon. That understanding comes from several different theoretical perspectives. First, there is Newtonian Theory, which is reliable, practical, and simple. It was the formulation that guided us to a precise landing on the Moon (1969); yet, on a grand scale, it simply fails. Then there is Einstein's General Theory of Relativity, which relates gravity to the fabric of space and time. It subsumes Newton's Theory and goes far beyond it, allowing us to analyze events on a vast cosmic scale. But Einstein's formulation is highly mathematical and quite unnecessary for most earthly considerations. Today we believe that all interactions arise from the continuous exchange of force-carrying particles (Chapter 31). Unhappily, General Relativity seems to be irreconcilable with this principle, and so we must conclude that it too is incomplete. In this chapter we focus on Newtonian theory, exploring how it applies on Earth and beyond.

On gravity. For my part I think that gravity is nothing but a certain natural striving with which parts have been endowed...so that by assembling in the form of a sphere they may join together in their unity and wholeness.

NICOLAUS COPERNICUS

5.3 The Law of Universal Gravitation

Newton's **Law of Universal Gravitation** states that the gravitational force F_G between any two bodies of mass m and M, separated by a distance r, is described by

$$F_G \propto \frac{mM}{r^2} \tag{5.4}$$

It is a center-to-center attraction between all forms of matter. Yet Newton's achievement was not so much in formulating the law as it was in applying it with supreme mathematical skill. As we will see, the pieces to the gravity puzzle were already there before Newton's great work of 1685.

Evolution of the Law

The force of gravity between any two objects varies directly in proportion to the product of their masses and inversely with their separation squared.

The Universal Law: The "universal" aspect of Newton's law alludes to the fact that *all* matter interacts via gravity. That insight, which still surprises some people today, developed gradually. Nicolaus Copernicus suggested that gravity was the tendency of *similar* matter to come together. He explained that the Sun, Moon, and Earth were all spherical because each kind of substance tends to pull itself inward into a ball. That idea is for the most part correct, but there is no suggestion here that Earth-stuff might attract Moon-stuff. A century later, Galileo was still committed to the same parochial gravity, in which only like substances attract.

Nicolaus Copernicus (1473–1543), pictured here on a stamp as a young man in Torun, Poland.

In 1600, the physician William Gilbert wrongly suggested that gravity was due to magnetism. (Hold two magnets close together and you immediately know the wonder of an invisible force reaching across space.) Johannes Kepler made the distinction between gravity and magnetism and argued, correctly, that both the Sun and Moon attract the "waters of the sea," causing the tides. The Earth, the Moon, the Sun, the waters—all interact gravitationally. Extending the concept, Roberval (1644) hinted at an attraction between all "worldly matter." In 1670, Hooke stated that gravity was an interaction between "All Celestial Bodies." And, finally, Newton proposed that the gravitational interaction exists between all material objects: "We must... universally allow that all bodies whatsoever are endowed with a principle of mutual gravitation."

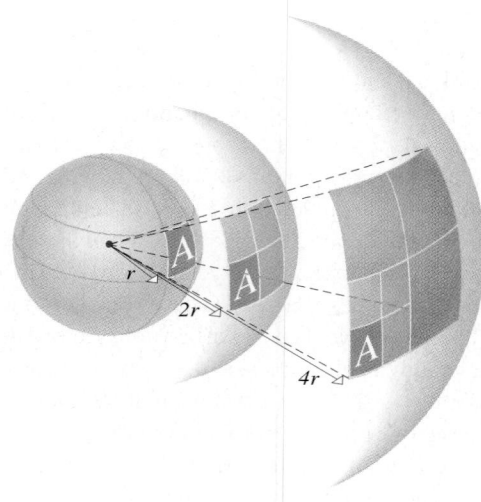

Figure 5.8 When the radius of a sphere r increases, the area of the sphere increases as r^2. Thus, if we double the radius, the little area A projects into an area of $4A$. Similarly, if the radius goes from r to $4r$, the area A projects out to an area of $16A$. The total area of the sphere, $4\pi r^2$, expands to $4\pi(4r)^2$ $= 16(4\pi r^2)$, or 16 times the original area.

The Moon causes the movement of the waters and the tides of the ocean.

The Sun (chief inciter of action in nature)… causes the planets to advance in their course.

The Moon alone of all planets directs its movements as a whole toward the Earth's centre, and is near of kin to Earth, and as it were held by ties to Earth.

The force which emanates from the Moon reaches to the Earth, and, in like manner, the magnetic virtue of the Earth pervades the region of the Moon.

WILLIAM GILBERT (1540–1603)

On gravity. If the attractive force of the Moon reaches down to the Earth, it follows that the attractive force of the Earth, all the more, extends to the Moon and even farther.

J. KEPLER

The Earth pulls every subatomic particle in your body downward, producing a net gravitational force called your *weight*, just as it pulls downward on apples, rocks, and frogs giving them weight as well. You, in turn, gravitationally attract every particle of the Earth with a net equal and opposite force; that's the interaction pair. Matter somehow draws on matter, and you attract (weakly, to be sure) all the apples, rocks, and even the frogs—everything attracts everything else.

As the Product of the Two Masses: The realization that the force of gravity, Eq. (5.4), is proportional to the interacting masses comes from a suggestion by Gilbert, even though it was little more than a lucky guess. He observed that the force between two magnets depended on their sizes and weights—which just happened to be the case for the natural magnets available at the time. Carrying the idea over to gravity, Kepler (1609) maintained that "two stones… placed anywhere in space" would gravitationally attract and "come together… each approaching the other in proportion to the other's mass."

Inversely with the Square of the Distance: The force of gravity decreases as the separation increases, Eq. (5.4), and we might also wonder how that insight came to be. In fact, it was Kepler who initiated it, albeit in a confused context. Kepler (without the Law of Inertia) wrongly thought that the planets were pushed along in orbit by a force spreading out from the Sun. Since the farther a planet is from the Sun, the slower it moves, he proposed that this force decreased with distance. After all, the magnetic force between two bodies decreases as their separation increases.

When anything diffuses outward in all directions from a source (Fig. 5.8), it spreads out, becoming less concentrated as it advances. Imagine all of the flow passing successively through surrounding concentric spheres centered on the source, any one of which has a radius r and an area $4\pi r^2$. The same quantity of emanation (whether it's hairspray or gravity) that initially passes densely through a tiny spherical surface will later pass thinly through a much larger one. The concentration (the amount per unit area) decreases inversely with the area of the sphere—it *decreases inversely with the distance squared* $(1/r^2)$. Everyone knew you can only read by a candle flame if you stand near it. And Kepler once wrote that the solar force "weakened through spreading from the Sun in the same manner as light." If gravity spreads out from a body uniformly in all directions, an inverse-square dependence is reasonable to suppose.

A Center-to-Center Force: In 1645, the astronomer Ismaël Bullialdus asserted that the Sun's attractive action on each planet *was along the center-to-center line joining the two and dropped off inversely with the distance squared*. The idea that gravity acts along a line connecting the "centers" of the interacting bodies was not startling either. The Medieval scholar Albert of Saxony had defined weight as the tendency of the center-of-gravity (*c.g.*) of a body to approach the center-of-gravity (p. 249) of the Earth. Ever since Archimedes (Section 9.6), it had been common to think of the weight of a body acting at some single central point. The vision of a center-to-center gravitational interaction was quite natural.

Newton's Confirmation of $1/r^2$

Early in the 1660s, Hooke, already convinced of the variation of gravity with distance, attempted to arrive at its form experimentally. He carried out a series of measurements to determine the minute differences in the weight of a body at various altitudes, within deep

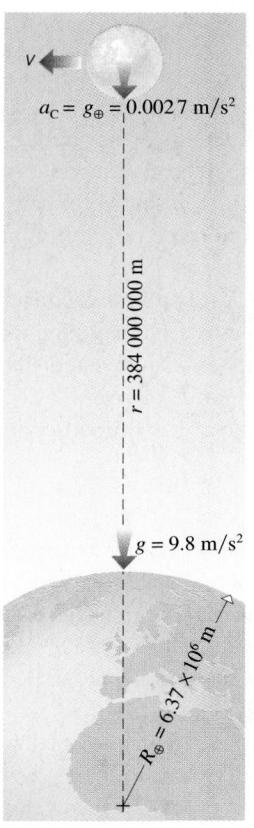

Figure 5.9 The acceleration due to gravity drops off inversely with distance squared. Since g at the Earth's surface is 9.8 m/s², it should be equal to 0.002 7 m/s² at the distance of the Moon. And that is precisely equal to the centripetal acceleration we know the Moon must have to maintain its orbit.

wells and high "on Paul's steeple and Westminster Abbey, but none that were fully satisfactory." Years later, Newton claimed to have made his own first test of the inverse-square dependence just around that same time, himself a young man of twenty-four. As he put it, I "compared the force requisite to keep the Moon in her orb with the force of gravity at the surface of the Earth; and found them answer pretty nearly."

Although its orbit is slightly elliptical, we can envision the Moon circling the Earth at a center-to-center distance of 38.4×10^7 m under the influence of a centripetal force provided by gravity (Fig. 5.9). Its centripetal acceleration, $a_c = v^2/r$, can be easily determined numerically. The lunar orbital speed is the distance traveled in one revolution ($2\pi r$) divided by the time it takes—the *period*. The orbit's circumference equals $2\pi(38.4 \times 10^7$ m), while the period is 27.32 days (or 2.360×10^6 s), which yields a speed of 1.02 km/s and an acceleration of

$$a_c = \frac{v^2}{r} = 0.002\,7 \text{ m/s}^2$$

Any object at that distance from Earth, whether orbiting or not, should experience this same downward acceleration via gravity.

Newton next compared this value with the acceleration at the planet's surface (6.371×10^6 m from its center). An object *on* the Earth experiences an acceleration due to gravity of 9.81 m/s². Assuming that the downward gravitational force trails off with distance from the planet as $1/r^2$, the resulting acceleration will do the same ($F = ma$). *Let $g_\oplus$ be this distance-dependent gravitational acceleration associated with the Earth*. The Moon is 60.33 Earth-radii away, so it (and anything else at that distance) should experience a gravitational acceleration toward Earth that is about 60^2 times smaller than does an object at the planet's surface:

$$g_\oplus = \frac{9.81 \text{ m/s}^2}{60^2} = \frac{9.81 \text{ m/s}^2}{3600} = 0.002\,7 \text{ m/s}^2$$

Amazing! The Moon's centripetal acceleration equals the gravitational acceleration due to the Earth at that altitude: $a_c = g_\oplus$. The whole calculation, simple and profound, is wonderful.

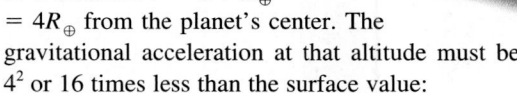

Example 5.3   [II] An astronaut out in space descends straight down toward the Earth in free-fall. Compute her acceleration at the moment she reaches an altitude of 1.911×10^4 km from the planet's surface. Neglect the presence of any other celestial object.

Solution This problem mentions a large "altitude" and the corresponding acceleration, and that brings to mind the dependence of g on r. (1) TRANSLATION—An object is at a known distance from the Earth; determine the gravitational acceleration at that location. (2) GIVEN: $h = 1.911 \times 10^4$ km, measured from the *surface*. FIND: $g_\oplus$. (3) PROBLEM TYPE—Gravitational acceleration/inverse square. (4) PROCEDURE—For the moment (p. 153), we don't have an expression for $g_\oplus$ in terms of r. We'll have to arrive at a value using the acceleration at the surface of the planet and the fact that $g_\oplus$ drops off as $1/r^2$. Remember that r is the center-to-center distance.   (5)

CALCULATION—Letting $R_\oplus$ be the Earth's radius (6.371×10^3 km), $h = 3R_\oplus$, and the astronaut is at a distance $r = h + R_\oplus = 4R_\oplus$ from the planet's center. The gravitational acceleration at that altitude must be 4^2 or 16 times less than the surface value:

$$g_\oplus = \frac{9.81 \text{ m/s}^2}{16} = \boxed{0.613 \text{ m/s}}$$

Quick Check: $r = 25\,481$ km and $g_\oplus/(9.8 \text{ m/s}^2) = [1/(4R_\oplus)]^2/(1/R_\oplus)^2 = R_\oplus^2/(4R_\oplus)^2 = (6.4 \times 10^3 \text{ km})^2/(25 \times 10^3$ km$)^2$; hence, $g_\oplus = 0.6$ m/s².

The asteroid Ida and its tiny moon Dactyl. The picture was taken in 1994 by the *Galileo* spacecraft from a distance of 10 500 km. Ida has a maximum diameter of about 50 km, which is much too small to crush itself gravitationally into a sphere. It is orbited by Dactyl, which is just 1.5 km across. By studying the motion of this moonlet, astronomers will be able to determine the mass of Ida (see Section 5.9).

By all their influences, you may as well Forbid the sea for to obey the moon....

W. SHAKESPEARE (1564–1616)
The Winter's Tale

The Dependence on Mass

That all terrestrial objects at the surface of the planet accelerate downward at the same rate of $g = 9.81$ m/s^2 is both remarkable and suggestive. From Newton's Second Law ($F = ma$), we know that a body of mass m propelled by a force equal to its own weight F_w accelerates at a rate g such that $F_w = mg$. Consequently, $F_w/m = g$, but this equivalence is true for *all* masses, which can only be the case if F_w is proportional to m so that the ratio can be independent of m. Inasmuch as the weight of an object is the force of gravity pulling it down toward the planet ($F_w = F_G$), it follows that $F_G \propto m$.

Arguing from Newton's Third Law, we know that, if the Earth pulls down on an object, the object must tug *up* on the Earth with an equal force. Imagine two interacting masses m and M. Since the force of gravity acting on any body depends on its mass, and since there is an equal and opposite force of magnitude F_G on each of them, $F_G \propto m$ and $F_G \propto M$. This implies that the gravitational force on either one of the two objects is proportional to both m and M:

$$F_G \propto \frac{mM}{r^2}$$

The force being proportional to the product mM agrees with our everyday experiences. We know that if one apple of mass m in a bag on a scale at the market has a weight of 1 N, then two identical apples will have a weight of 2 N. Doubling the mass in the bag (m) doubles the weight ($F_w = mg$) despite the immense mass of the Earth ($M_\oplus = 5.975 \times 10^{24}$ kg). This would not occur, for instance, if F_G were in proportion to the sum of the masses ($m + M_\oplus$) rather than their product ($mM_\oplus$).

To be rigorous, Newton's Law—Eq. (5.4)—ought to be applied only to idealized point masses. We would then, by mathematical means, determine the gravitational force due to a body of any shape by adding up the forces from all its constituent particles. Developing the ability to perform that calculation and then proving that the law as it stands applies between spherical masses (such as planets) was one of the main problems Newton had to overcome. {For Newton's solution click on **The Gravity of a Sphere** under **Further Discussions** on the **CD**.} 💿

The Gravitational Constant

The relationship between F_G, m, M, and r^2 can be recast as an equality (something Newton never had the need to do) by introducing a constant of proportionality G {see **Math Review: Part A-2** on the **CD** 💿 }. This constant also serves to make the resulting equation dimensionally correct—the units on both sides of the proportionality are not the same. **Newton's Law of Universal Gravitation** takes its final form as

$$F_G = G \frac{mM}{r^2} \tag{5.5}$$

Here, G has the units of N·m^2/kg^2, and the force is properly in newtons. The numerical value of this **Universal Gravitational Constant** has to be determined experimentally.

That measurement was first done in 1798 by Lord Henry Cavendish, one of the richest men of his age and one of the most peculiar. Incredibly withdrawn, he rarely spoke to men and never, if he could help it, interacted with women.

Eccentricities aside, determining G was a formidable technical accomplishment. To do it, Cavendish attached a small lead sphere about 2 in. in diameter to each end of a 6-ft-long, light, rigid rod that was suspended horizontally from its middle on a thin vertical wire (Fig. 5.10). He then brought two lead balls, roughly 8 in. in diameter, into stationary positions, one in front of and one behind the small spheres. The resulting gravitational attraction caused the suspended rod-assembly to turn. The smaller balls swung very slowly toward the larger ones until the twisting wire brought them to rest. Having previously determined the force that would twist the wire through any angle, Cavendish arrived at a value for F_G.

Guidance systems for missiles and aircraft require a detailed knowledge of the Earth's gravity. Here U.S. Air Force technicians very precisely measure variations in F_G as the altitude increases.

A modern version of the Cavendish experiment. The two large lead balls are easily visible, one in front, one in back. One of the small balls can be seen just to the right of the red laser beam.

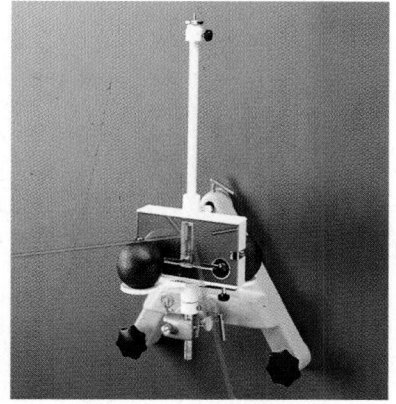

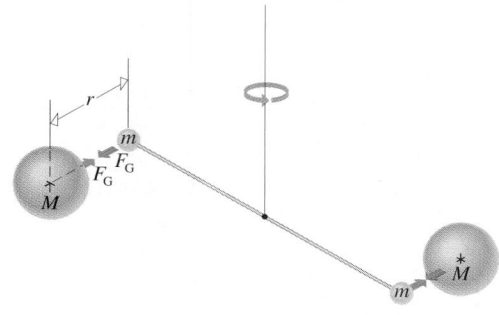

Figure 5.10 The Cavendish torsion pendulum for determining G. Drawn by their gravitational attraction, the two small balls move toward the large stationary balls, thereby twisting the wire.

Universal Gravitation on a Celestial Scale.
All Celestial Bodies whatsoever have an attraction or gravitating power towards their own Centres, whereby they attract not only their own parts…but that they do also attract all the other Celestial Bodies that are within the sphere of their activity.

ROBERT HOOKE (1670)

Measuring m, M, r, and F_G, the constant G could then be computed; the presently accepted value is

$$G = 6.672\,59 \times 10^{-11} \ \text{N} \cdot \text{m}^2/\text{kg}^2$$

(Cavendish got 6.75×10^{-11} N·m²/kg².) It's interesting that of all the fundamental constants we will encounter, G is the least accurately known.

Example 5.4 **[I]** Compute the gravitational attraction between two 100-kg uniform spheres separated center-to-center by 1.00 m. (That's the rough equivalent of two 220-lb football players at arm's length.)

Solution The phrase "compute the gravitational attraction between" suggests a direct application of Newton's Law. (1) TRANSLATION—Two objects of known mass are separated by a known distance; determine their gravitational interaction. (2) GIVEN: $m = 100$ kg, $M = 100$ kg, and $r = 1.00$ m. FIND: F_G. (3) PROBLEM TYPE—Gravitational force. (4) PROCEDURE—Use the Law of Universal Gravitation. (5) CALCULATION:

$$F_G = G \frac{mM}{r^2}$$

$$F_G = (6.67 \times 10^{-11} \ \text{N} \cdot \text{m}^2/\text{kg}^2) \frac{(100 \ \text{kg})(100 \ \text{kg})}{(1.00 \ \text{m})^2}$$

and

$$\boxed{F_G = 6.67 \times 10^{-7} \ \text{N}}$$

which equals 0.15×10^{-6} lb, or just about the weight of a baby flea. No wonder we're not noticeably drawn to each other via gravity.

Quick Check: The force between two 1-kg masses separated by 1 m is 6.7×10^{-11} N, and the force in our problem is 100×100 times greater.

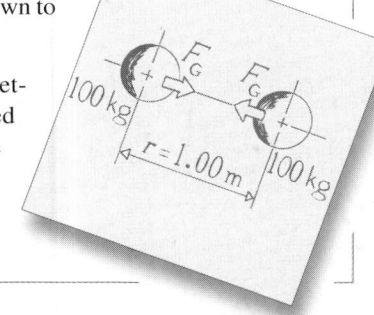

Example 5.5 **[I]** The mass of the Moon is 7.35×10^{22} kg, and its distance from Earth is 384×10^3 km. Taking the planet's mass to be 5.98×10^{24} kg, compute what Newton called "the force requisite to keep the Moon in her orb."

Solution When you have to compute the attraction between celestial objects, always think of the Law of Universal Gravitation. (1) TRANSLATION—Two objects of known mass are separated by a known distance; determine their gravitational interaction. (2) GIVEN: $m_{\text{☾}} = 7.35 \times 10^{22}$ kg, $M_{\oplus} = 5.98 \times 10^{24}$ kg, and $r_{\oplus\text{☾}} = 384 \times 10^3$ km. FIND: F_G. (3) PROBLEM TYPE—Gravitational force. (4) PROCEDURE—Use Newton's Universal Law of Gravitation. (5) CALCULATION:

$$F_G = G \frac{m_{\text{☾}} M_{\oplus}}{r_{\oplus\text{☾}}^2}$$

continued

$$F_G = (6.67 \times 10^{-11} \text{ N·m}^2/\text{kg}^2)$$

$$\times \frac{(7.35 \times 10^{22} \text{ kg})(5.98 \times 10^{24} \text{ kg})}{(384 \times 10^6 \text{ m})^2}$$

and $\boxed{F_G = 1.99 \times 10^{20} \text{ N}}$

Incidentally, this value is approximately 4.5×10^{19} lb, or the equivalent of the thrust of 6 million *Saturn 5* Moon rockets.

Quick Check: From p. 149, $g_{\oplus} = 0.0027$ m/s²; hence, $F_G = m_{\mathbb{C}}a = (7.4 \times 10^{22}$ kg$)(0.0027$ m/s²$) = 2 \times 10^{20}$ N.

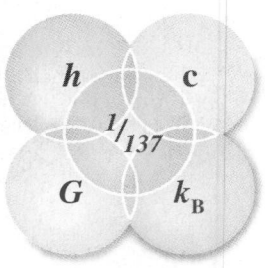

Figure 5.11 The gravitational constant G is the first of several fundamental constants that will be encountered as we continue the study of physics. The number 1/137 is the dimensionless fine structure constant.

The quantity G is a measure of the strength of the gravitational force. Since it's an extremely small number, F_G will be appreciable only when at least one of the interacting masses is very large. Still, because it has unlimited range (dropping to zero when $r = \infty$), and because it is purely attractive (not being weakened by an opposing repulsive aspect), gravity rules the Universe on a grand scale (Fig. 5.11).

5.4 Terrestrial Gravity

Cavendish had an ulterior purpose when he performed his famous experiment—he wanted to determine the density of the Earth. An apple of mass m, hanging on a scale at the Earth's surface a distance $R_{\oplus} = 6.37 \times 10^6$ m from the planet's center, weighs $F_w = mg$. The scale and the apple are motionless on the Earth, and therefore both are revolving with the planet. If, for a moment, we remove the small effect of that rotation, we must replace g by g_0, the nonspinning, **absolute acceleration due to gravity**, whereupon $F_w = mg_0$. This equation represents the "true" weight of an object due only to its gravitational interaction with the Earth and is therefore identical to $F_w = GmM_{\oplus}/R_{\oplus}^2$. Setting the two weight expressions equal and canceling m results in *the absolute gravitational acceleration at the Earth's surface,*

[at surface of spherical Earth] $$g_0 = \frac{GM_{\oplus}}{R_{\oplus}^2}$$ (5.6)

There is no mention of m in this equation; *in the vicinity of the planet, the acceleration due to gravity is the same for all bodies independent of their masses*. Moreover, by appropriately substituting for M and R, the acceleration due to gravity can be determined for any celestial body (Table 5.1).

Table 5.1

Field Events in the Solar System: Long Jump, High Jump, and Shot Put on the Moon and Nine Planets*

Planet	Gravitational acceleration (m/s²)	Long jump[1] distance (m)	High jump[2] height (m)	Shot put[3] distance (m)
Mercury	3.7	24	4.7	56
Venus	8.9	9.9	2.5	25
Earth	9.8	8.9	2.4	22
Moon	1.6	54	9.4	126
Mars	3.7	23	4.6	56
Jupiter	26	3.3	1.5	9.4
Saturn	12	7.5	2.2	19
Uranus	11	7.6	2.2	19
Neptune	12	7.3	2.1	19
Pluto	2	45	7.9	105

* Adapted from *Scientific American*, August 1992. Here we assume that the athlete can run, jump, and throw in the various gravitational environments at the following speeds:
1. The athlete runs at 10 m/s and then jumps at 71° at 4.08 m/s.
2. The athlete jumps at 5.20 m/s.
3. The athlete throws the shot at 14.2 m/s at slightly less than 45°.

EXPLORING PHYSICS ON YOUR OWN

Banked Tracks, Wine Glasses & Orbits: Put a marble in a paper or Styrofoam cup, the kind that's slightly conical. Now revolve the cup, thereby setting the marble moving around the inner wall. Study what happens to the marble as you increase the speed. Now try it again, gently, with an inexpensive wine glass. What makes it rise up the wall? How is this related to the banked track? The walls of most wine glasses change their slope halfway up and start to bend inward. Consequently, you can find a spot where the marble will orbit horizontally quite stably. For a wonderful strobe effect, look through the glass at a white computer screen as you whirl the marble. Check out your conclusions with an empty cylindrical jar. What happens now? You can get the marble to orbit if the jar has an indented lip and it's upside down—how is that possible?

Knowing both the value of g_0 (which differs from g by less than 0.4%) and $R_\oplus$, and having just determined G experimentally, Eq. (5.6) can be solved for the mass of the Earth:

$$M_\oplus = \frac{R_\oplus^2 g_0}{G} = 6.0 \times 10^{24} \text{ kg}$$

and that's a splendid thing to be able to compute. Taking the planet to be a sphere of volume $V_\oplus = \frac{4}{3}\pi R_\oplus^3 = 1.1 \times 10^{21} \text{ m}^3$, Cavendish found that the **average density**, or ***mass per unit volume*** (ρ), was

$$\rho_\oplus = \frac{M_\oplus}{V_\oplus} = 5.5 \times 10^3 \text{ kg/m}^3$$

which is 5.5 times greater than that of water (whose density is $1.0 \times 10^3 \text{ kg/m}^3$).

Newton never had access to a measured value of G, but he could have guessed at the Earth's density and then just run the above calculation backward to find G, had he so wished. We know that he made just such an extraordinary guess in the *Principia*, where he wrote: "Since...the common matter of our Earth on the surface thereof is about twice as heavy as water, and a little lower, in mines, is found about three or four, or even five times more heavy, it is probable that the quantity of the whole matter of the Earth may be five or six times greater than if it consisted all of water..."

Example 5.6 **[II]** By what percentage would your weight change if you ascended a height equivalent to that of the Sears Tower in Chicago (1454 ft or 4.432×10^2 m)? The Earth's diameter is 1.274246×10^7 m.

Solution This example explores the distance dependence of the gravitational interaction. (1) TRANSLATION—By what percentage does the weight of an object change as its height above the Earth changes by a specified amount? (2) GIVEN: $h = 4.432 \times 10^2$ m and $R_\oplus = 6.37123 \times 10^6$ m. FIND: F_{Wh}/F_W. (3) PROBLEM TYPE—Gravitational force/weight. (4) PROCEDURE—This is a percentage problem, so we use ratios and therefore

don't need several of the constants. Use the Universal Law of Gravitation. (5) CALCULATION—The ratio of the weights is

$$\frac{F_{Wh}}{F_W} = \frac{1/(R_\oplus + h)^2}{1/R_\oplus^2} = \frac{(6.37123 \times 10^6 \text{ m})^2}{(6.37167 \times 10^6 \text{ m})^2}$$

which to four significant figures is $\boxed{99.99\%}$. The reduction in the weight, and therefore the reduction in $g_\oplus$, is a mere 0.01%.

Quick Check: The ratio equals $[R_\oplus/(R_\oplus + h)]^2 = [1/(1 + h/R_\oplus)]^2 = 0.9999$.

The Apple Myth. One day in the year 1666 Newton had gone to the country, and seeing the fall of an apple, as his niece told me, let himself be led into a deep meditation on the cause which thus draws every object along a line whose extension would pass almost through the center of the Earth.

VOLTAIRE (1738)

Newton showed that *as far as a point outside was concerned, a sphere behaves gravitationally as if all its mass were at its center*. {For Newton's proof click on **THE GRAVITY OF A SPHERE** under **FURTHER DISCUSSIONS** on the CD.} Consequently, using the same logic that led to Eq. (5.6), at a distance $r > R_\oplus$ from the center of the Earth the acceleration due to gravity is

[beyond the Earth]
$$g_\oplus = \frac{GM_\oplus}{r^2} \tag{5.7}$$

Note that for an approximately spherical Earth g_0 is a constant, whereas $g_\oplus$ is a function of r.

When Newton saw an apple fall, he found
In that slight startle from his contemplation—
'Tis said (for I'll not answer above ground
For any sage's creed or calculation)—
A mode of proving that the earth turn'd round
In a most natural whirl, called "gravitation";
And this is the sole mortal who could grapple,
Since Adam, with a fall, or with an apple.
　　　　　　LORD BYRON (1788–1824)

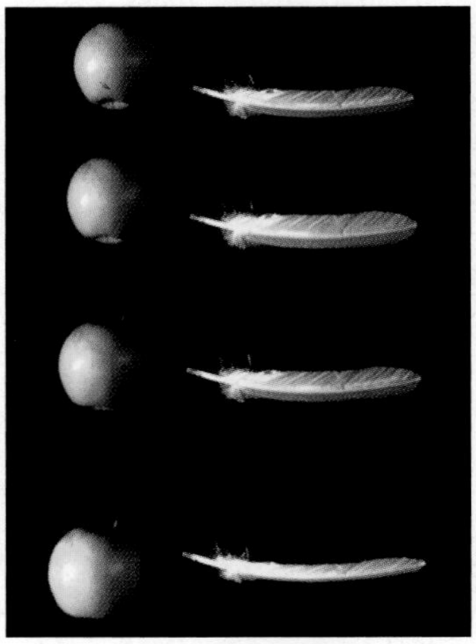

A feather and an apple fall at the same rate. The apple is more massive and must experience a greater force ($F_W = mg$) if it is to accelerate at a rate equal to that of the feather. But it has more mass and does weigh more ($F_W \propto mM_\oplus/R_\oplus^2$). Accordingly, the acceleration due to gravity is independent of the mass of the falling object. Here air drag is slowing the feather more than the apple.

An Imperfect Spinning Earth

The Earth is not a sphere, and it is not uniform. Excluding its hills and valleys, the planet is slightly flattened at the poles, with a little bulge at the north giving the whole thing a pear shape ($R_\oplus$ at the equator is about 21.5 km greater than at the poles). At close range, the deviation in the gravitational force from an ideal $1/r^2$ dependence is small (about 0.1%), and yet it markedly affects the orbits of artificial Earth satellites. Farther out, at the distance of the Moon, the interaction with the Earth deviates from the inverse square by only about 0.000 03%. In general, *any two mass configurations will interact in an approximately $1/r^2$ fashion when r is large in comparison to the sizes of the interacting objects.*

The planet's oblate shape, combined with the fact that it's rotating, produces a variation in the measured surface value of g that's about five parts in a thousand (0.5%) as one moves from pole to equator (Table 3.2). To determine the effect of the Earth's rotation on g, consider an object of mass m supported by a scale situated at the *equator*. The two external forces acting on the object are its downward interaction with the planet (F_G) and the upward scale-force equal in magnitude to the object's **measured weight** (F_w). But since the object rotates as the planet turns, it must experience a centripetal acceleration. Hence ($+\downarrow \sum F = ma$),

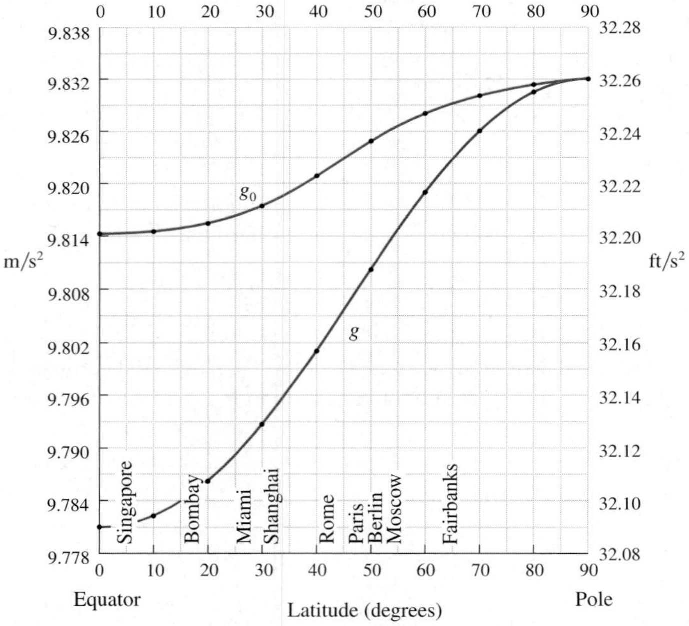

$$F_G - F_w = ma_c$$

$$\frac{GmM_\oplus}{R_\oplus^2} - mg = ma_c$$

and so

$$g = \frac{GM_\oplus}{R_\oplus^2} - a_c = g_0 - a_c$$

Figure 5.12 A plot of both the measured (g) and the absolute (g_0) values of the acceleration due to gravity on Earth. The former includes the rotational effects of the spinning planet; the latter does not.

At the poles, $a_c = 0$, $g = g_0$, and the measured weight mg equals the **true weight** mg_0. The planet's spin produces almost two-thirds of the observed change in g with latitude (Fig. 5.12). {For more on gravity click on **GRAVIMETERS** under **FURTHER DISCUSSIONS** on the **CD**.}

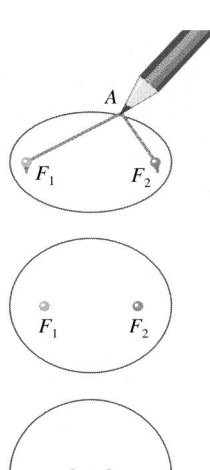

Figure 5.13 The ellipse. Point A moves along an ellipse when the sum of the distances $\overline{F_1 A}$ and $\overline{F_2 A}$ is constant. Imagine a slack string tacked down at its ends, F_1 and F_2. A pencil holding the string taut will sweep out an ellipse, since $(\overline{F_1 A} + \overline{F_2 A})$ always equals the constant length of the string. As the two foci, F_1 and F_2, are set closer together, the figure increasingly resembles a circle.

The Cosmic Force

Johannes Kepler, at twenty-nine, was sensitive, provincial, and an ardent believer in the notion that the Earth and all the planets orbited the Sun. The young mystic, who came from a family of misfits and witches, was a little-known teacher of mathematics, a pauper, a caster of uncanny horoscopes, and a genius. On the death of the astronomer Tycho Brahe (1601), Kepler, who had been working for him, absconded with Tycho's vast hoard of observational data on the motions of the planets. Struggling over the next two decades, Kepler ultimately formulated his famous Three Laws of Planetary Motion.

5.5 The Laws of Planetary Motion

Unlike almost everyone before him,* who insisted that the planetary orbits must have the mystic perfection of the circle, **Kepler's First Law** maintains that *the planets move in elliptical orbits with the Sun at one focus* (Fig. 5.13). Granted, the paths are nearly circular—but the difference is crucial.

Kepler also noticed that each planet progressed at varying speeds over the course of its journey around the Sun. Relying on the mathematics of his era—geometry—he succeeded in describing the motion in detail. **Kepler's Second Law** states that *as a planet orbits the Sun it moves in such a way that a line drawn from the Sun to the planet sweeps out equal areas in equal time intervals* (Fig. 5.14). The Earth-Sun line will sweep out the same area in a week no matter where it is in orbit, whether it's a week in June or a week in January. Near the Sun the speed is great, the arc is long, and the slice of orbital pie is short and broad. Far from the Sun, the speed is low, the slice is narrow but tall, and still the area swept out is the same.

Armed with the physics of gravity, we can appreciate what's happening on another level. The planet, attracted to the Sun, swoops "down," accelerating as it rushes in. Reaching a maximum speed at its point of closest approach, it sails around and past. As it "climbs" against the Sun's gravity-pull, it decelerates. The force gradually weakens, and the planet continues to slow down until it reaches its minimum speed at the point farthest away, only to sweep around and begin the plunge once again. {For a discussion of Newton's proofs of Kepler's first and second laws click on **GRAVITY & KEPLER'S LAWS** under **FURTHER DISCUSSIONS** on the **CD**.}

The larger a planetary orbit is, the slower the planet moves—that much was already known. In 1618, Kepler found the relationship between a planet's orbital period (T), the time it takes to travel once around the Sun, and its average distance from the Sun ($r_\odot$). **Kepler's Third Law** states that *the ratio of the average distance from the Sun cubed to the period squared is the same constant value for all planets*:

$$\frac{r_\odot^3}{T^2} = C_\odot \qquad (5.8)$$

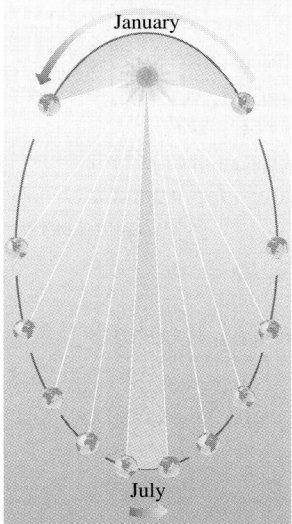

Figure 5.14 A line drawn from the Earth to the Sun sweeps out equal areas in equal times. Here each segment corresponds to a one-month interval and all have the same area. The Earth reaches its perihelion in January where it travels at maximum speed.

*Perhaps the one exception is the Muslim al-Zarqali (1081), "the blue-eyed one," who wrote about the 'oval' orbit of Mercury.

where $C_\odot$ is a to-be-determined constant that depends on the specific units and is associated with the Sun as the central body. This equation with a different constant would describe the motion of all satellites about the Earth, and the same is true for any central body. The fact that all the planets orbiting the Sun share the same constant suggests that they are all ruled by the same underlying mechanism—gravity.

Example 5.7 **[II]** Given that the mean solar distance for the Earth is 149.6×10^6 km, compute the corresponding value for Venus. The periods (in Earth-days) of the Earth ($\oplus$) and Venus (ς) are 365.256 days and 224.701 days, respectively.

Solution We've got astronomical distances and periods, and that means Kepler's Third Law. Still, you can always begin an orbit problem by setting the gravitational force equal to the centripetal force and deriving what you need. (1) TRANSLATION— Knowing the radius of a planet's orbit and its period, find the radius of another planet's orbit knowing its period. (2) GIVEN: $r_{\odot\oplus} = 149.6 \times 10^6$ km, $T_\oplus = 365.256$ d, and $T_\varsigma = 224.701$ d. FIND: $r_{\odot\varsigma}$. (3) PROBLEM TYPE—Gravitational force/Kepler's Third Law. (4) PROCEDURE—Using Kepler's Third Law and the Earth's data, determine the constant $C_\odot$ for the Solar System. With that, apply the law again to get $r_{\odot\varsigma}$. (5) CALCULATION— From Eq. (5.8)

$$\frac{r_{\odot\oplus}^3}{T_\oplus^2} = \frac{r_{\odot\varsigma}^3}{T_\varsigma^2} = C_\odot$$

Solving for the Sun–Venus distance,

$$r_{\odot\varsigma} = \left(T_\varsigma^2 \frac{r_{\odot\oplus}^3}{T_\oplus^2} \right)^{\frac{1}{3}} = r_{\odot\oplus} \left(\frac{T_\varsigma^2}{T_\oplus^2} \right)^{\frac{1}{3}} = r_{\odot\oplus} \left(\frac{T_\varsigma}{T_\oplus} \right)^{\frac{2}{3}}$$

$$r_{\odot\varsigma} = (149.6 \times 10^6 \text{ km}) \left(\frac{224.701 \text{ d}}{365.256 \text{ d}} \right)^{\frac{2}{3}} = \boxed{108.2 \times 10^6 \text{ km}}$$

Notice that the division canceled the units of time, so no conversions were needed.

Quick Check: We use the data for the Earth to determine $C_\odot = 2.509\ 6 \times 10^{19}$ km³/d² and from that value and T_ς find $r_{\odot\varsigma} = \sqrt[3]{C_\odot T_\varsigma^2} = 108.2 \times 10^6$ km, which at least shows that we did the math correctly.

Johannes Kepler (1571–1630).

Kepler's solar constant $C_\odot$ has its simplest value (viz., 1.00) when distance is measured in *astronomical units* (1 AU $= r_{\odot\oplus} \approx 1.5 \times 10^8$ km $\approx 93 \times 10^6$ mi) and time in *Earth-years*. Since $r_\odot^3/T^2 = C_\odot$, we can reevaluate $C_\odot$ using data for the Earth, and that same value will apply to all the planets orbiting the Sun. Thus, $C_\odot = (1 \text{ AU})^3/(1 \text{ Earth-year})^2 = 1$. Table 5.2 gives results for the planets known to Kepler. {For a wonderful animated look at orbital motion, click on **KEPLER'S LAWS** under **INTERACTIVE EXPLORATIONS** on the **CD.**} 💿

Newton's Proof of Kepler's Third Law

The last of Kepler's relationships is easy to understand in terms of Newtonian theory. A planet of mass m moving in a circle at a distance $r_\odot$ from the Sun must be under the influ-

Table 5.2

Kepler's Third Law*

Planet	Average distance to Sun (astonomical units) $r_\odot$	Period (Earth-years) T	Period squared T^2	Distance cubed $r_\odot^3$
Mercury	0.39	0.24	0.058	0.059
Venus	0.72	0.62	0.38	0.37
Earth	1.00	1.00	1.00	1.00
Mars	1.53	1.88	3.53	3.58
Jupiter	5.21	11.9	142	141
Saturn	9.55	29.5	870	871

*Compare the last two columns.

ence of a centripetal force given by

$$F_c = ma_c = \frac{mv^2}{r_\odot}$$ [5.2]

And if gravity is to provide such a force ($F_G = F_c$), then

$$F_G = G\frac{mM_\odot}{r_\odot^2} = \frac{mv^2}{r_\odot} = F_c$$

where $M_\odot$ is the mass of the central body. Dividing the circumference ($2\pi r_\odot$) by the period ($2\pi r_\odot/T$) yields the orbital speed. Canceling m and substituting $v = 2\pi r_\odot/T$

$$\frac{GM_\odot}{r_\odot^2} = \frac{(2\pi r_\odot/T)^2}{r_\odot}$$

and

$$\frac{r_\odot^3}{T^2} = \frac{GM_\odot}{4\pi^2}$$ (5.9)

which is Kepler's Third Law, where the right side equals $C_\odot$. *Only the mass of the central body remains in the expression*; the mass of the orbiting object cancels out (see Problem 78 for a slightly more accurate version). Exchange $M_\oplus$ for $M_\odot$, and replace $r_\odot$ with $r_\oplus$ (the distance to our planet's center), and the same equation applies to the circling Moon or to any artificial satellite in Earth-orbit.

5.6 Satellite Orbits

Long before *Sputnik,* the first artificial Earth satellite, Newton had observed that any projectile launched horizontally is, in a sense, a *satellite* (a word coined by Kepler). A rock thrown from a tall building sails in a modest orbit that soon intersects the Earth not far from its point of launch (Fig. 5.15). Galileo maintained that this trajectory should be parabolic, but that assumed a flat world (p. 69). On a spherical Earth, a horizontally hurled baseball (neglecting friction) theoretically arcs along an elliptical orbit, with its far focus at the planet's center. All the planet's mass acts as if it were at its center, and the ball attempts to orbit that point. If the Earth were squashed down small enough, the baseball would become a moonlet.

If the ball were fired more swiftly to start with, it would travel farther; the ellipse would flatten out, becoming less elongated (Fig. 5.16). Further increasing the speed would result in ever larger, rounder elliptical paths and more distant impact points. Finally, at one particular launch speed, the ball would glide out just above the planet's surface all the way around to the other side without ever striking the ground. Like a little leather moon, it would wheel around the globe in a circular orbit until something (air friction, collision with birds or buildings) brought it down.

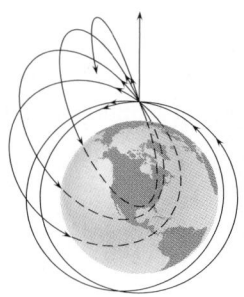

Figure 5.15 Various orbits for projectiles with the same speeds launched in different directions. In each case, the major or long axis is unchanged.

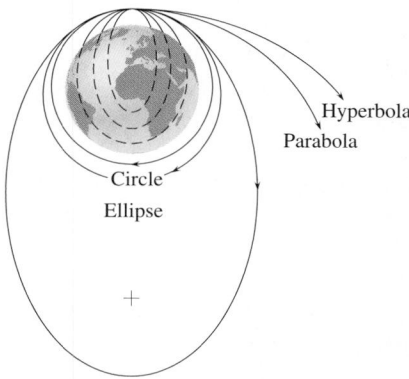

Figure 5.16 The orbits of projectiles fired horizontally close to the Earth's surface at various speeds (based on a drawing by Newton). The circular orbit occurs at a launch speed of about 8 km/s. From 8 to 11.2 km/s, the orbit is elliptical. At 11.2 km/s, it's parabolic, and beyond that, it's hyperbolic.

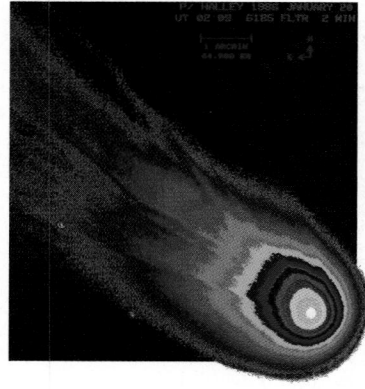

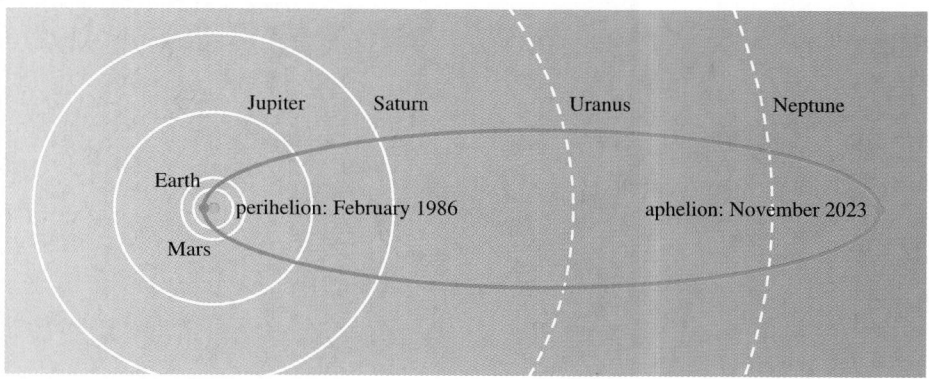

Halley's comet as it appeared in its most recent approach in 1985. The comet is held by gravity in an elongated elliptical orbit that causes it to swing near the Sun roughly every 75 years.

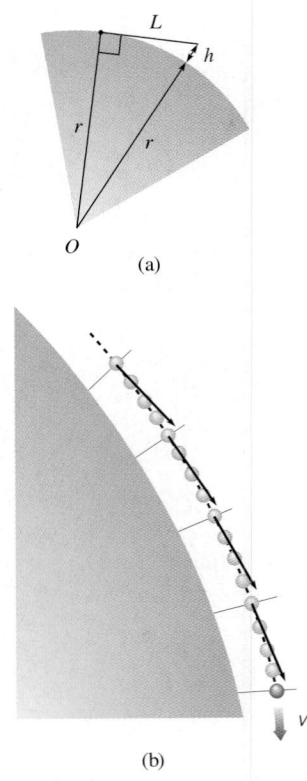

Figure 5.17 (a) The Earth drops away by an amount h below the tangent line of length L. (b) If a projectile is fired with a speed such that it travels a distance L in the time it falls a height h, it will remain in orbit.

A low, circular flight path means that the satellite follows the planet's curvature. In effect, *it continually drops toward the center-of-force, but since the ground "falls away" at the same rate, it never gets any closer to the surface*. Figure 5.17 shows how the curvature of the Earth's surface drops a distance h beneath a tangent of length L such that the Pythagorean Theorem yields

$$(r + h)^2 = r^2 + L^2$$

$$r^2 + 2rh + h^2 = r^2 + L^2$$

and

$$2rh + h^2 = L^2$$

Here we are looking at a very small segment of a large curve, and so h is tiny and h^2 by comparison with $2rh$ (since r is big) is quite negligible. Then $2rh \approx L^2$ and

$$L \approx \sqrt{2rh}$$

Now, suppose we are dealing with an orbit just above the planet's surface and the small segment corresponds in time to 1 s of motion. Then, L is the straight-line distance the object would travel if there were no gravity and h is the height it would fall due to gravity alone (see Fig. 4.2). From Eq. (3.9), $h = \frac{1}{2}gt^2 = \frac{1}{2}g(1\text{ s})^2 = 4.9$ m and, with $r \approx R_\oplus$,

$$L \approx \sqrt{2R_\oplus h} \approx 7.9 \text{ km}$$

This means that if an object were initially fired horizontally with a speed such that it traveled 7.9 km in 1 s (that is, 18×10^3 mi/h), it would subsequently orbit the Earth at the *lowest possible altitude*, effectively dropping 4.9 m each second.

An alternative way of seeing this, which is less graphic but more general, comes from equating F_G and F_C. An object of mass m at a distance r from the center of a large body of mass M will travel a circular path with an **orbital speed** v_o when the gravitational force exactly equals the centripetal force, whereupon

$$G\frac{mM}{r^2} = \frac{mv_o^2}{r}$$

and

[circular orbital speed] $$v_o = \sqrt{\frac{GM}{r}} \qquad (5.10)$$

Any satellite (Fig. 5.18) *regardless of its mass* (provided $M \gg m$) will move in a circular orbit of radius r, as long as its speed obeys Eq. (5.10). A launch vehicle initially rises

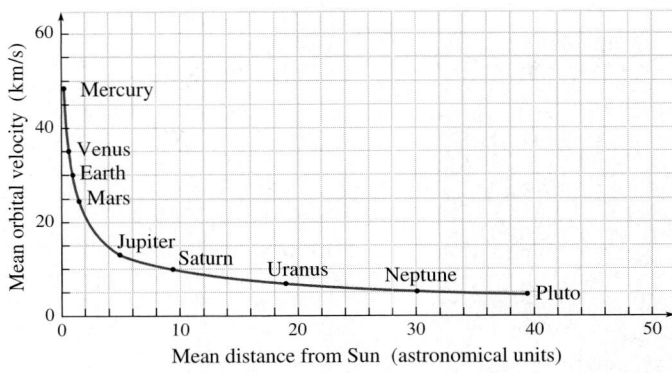

Figure 5.18 Because the Sun represents 99.9% of the mass of the Solar System, each planet essentially "sees" only the Sun as the central gravitating body. Consequently, for all planets $v_o \propto 1/\sqrt{r}$; the larger the orbit, the slower the orbital speed.

Saturn and its ring system. The rings are composed of billions of individual particles that typically range in size from pebbles to boulders. Gravity holds them in orbit and also, by its tidal action, keeps them from coalescing. Any moon orbiting closer than a certain distance will be ripped into pieces by its gravitation interaction with Saturn.

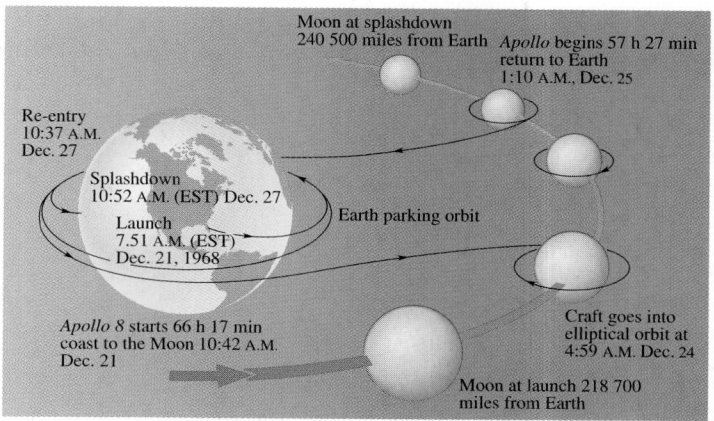

Figure 5.19 This drawing summarizes the *Apollo 8* journey to the Moon in 1968. The spacecraft first went into a parking orbit around the Earth. After the crew briefly fired its rockets and sped it up, the ship coasted off to the Moon.

> *Y*ᵉ suns conversion doth turn the planet out of this line framing its motion into a circular, but the former desire of *y*ᵉ planet to move in a streight line hinders the full conquest of *y*ᵉ Sun, and forces it into an Ellipticke figure.
> JEREMIAH HORROCKS (1618–1641)

vertically, gradually rolling over. If on separation the payload is moving tangentially at a speed v_o, it will be injected into a circular orbit. If the speed is less than v_o, the craft will descend in an elliptical orbit (Fig. 5.16). If the speed is greater than v_o but less than $\sqrt{2}v_o$, it will ascend into a large elliptical orbit. At $\sqrt{2}v_o$, it will escape into an open parabolic path that carries it out and away forever (p. 193). At speeds in excess of $\sqrt{2}v_o$, the escape trajectory becomes hyperbolic and increasingly flattened (Fig. 5.19). {For more on satellites click on **ORBITS** under **FURTHER DISCUSSIONS** on the CD.}

Example 5.8 **[I]** At what speed must a spacecraft be injected into orbit if it is to circle the Earth at treetop-height?

Solution The mention of "speed," "orbit," and "circle" should have you thinking about the equation for the orbital speed of a satellite. Still, you can always begin an orbit problem by setting the gravitational force equal to the centripetal force and deriving what you need. (1) TRANSLATION—An object is in the lowest possible circular Earth orbit; determine its speed. (2) GIVEN: $r \approx R_\oplus = 6.4 \times 10^6$ m and $M = M_\oplus = 6.0 \times 10^{24}$ kg. FIND: v_o. (3) PROBLEM TYPE—Gravitational force /circular orbital motion. (4) PROCEDURE—Use Eq. (5.10) with $r \approx R_\oplus$. (5) CALCULATION:

$$v_o = \left(\frac{GM_\oplus}{R_\oplus} \right)^{\frac{1}{2}}$$

and

$$v_o = \left[\frac{(6.67 \times 10^{-11} \text{ N·m}^2/\text{kg}^2)(6.0 \times 10^{24} \text{ kg})}{6.4 \times 10^6 \text{ m}} \right]^{\frac{1}{2}}$$

$$\boxed{v_o = 7.9 \times 10^3 \text{ m/s}}$$

Quick Check: This speed is the same one we obtained from the analysis where we used Fig. 5.17 and the Earth's curvature. Alternatively, at the surface of the Earth, from Eq. (5.6) for g_0,

$v_o = \sqrt{g_0 R_\oplus} = [(9.8 \text{ m/s}^2)(6.4 \times 10^6 \text{ m})]^{1/2} = 7.9 \times 10^3 \text{ m/s}$.

{To confirm this calculation you can go to the CD under CHAPTER 5, choose INTERACTIVE EXPLORATIONS, and open KEPLER'S LAWS. Select KEPLER'S 1ST LAW and enter the lowest possible orbit.}

Example 5.9 **[II]** Compute the altitude above the planet's surface and the necessary injection speed for a geostationary circular orbit, that is, one for which the satellite as seen from the ground appears to be stationary in the sky.

Solution You can always begin an orbit problem by setting the gravitational force equal to the centripetal force and deriv-

ing what you need. (1) TRANSLATION—An object is in a geostationary orbit; determine its altitude and speed. (2) GIVEN: $M_\oplus = 5.98 \times 10^{24}$ kg and the most important fact (though not stated explicitly) $T = 24$ h. FIND: the altitude h and v_o. (3) PROBLEM TYPE—Gravitational force /circular orbital motion. (4) PROCEDURE—Remember that $h = r - R_\oplus$, the difference between the distance from the center of the Earth to the satellite,

continued

and the Earth's radius. Determine r by setting $F_G = F_C$, using $v = 2\pi r/T$.   (5) CALCULATION—Canceling m,

$$\frac{GM_\oplus}{r^2} = \frac{(2\pi r/T)^2}{r}$$

$$r = \left(\frac{M_\oplus T^2 G}{4\pi^2}\right)^{\frac{1}{3}}$$

and $r = (7.544 \times 10^{22}\,\text{m}^3)^{\frac{1}{3}} = 42.25 \times 10^6\,\text{m}$

We want $h = r - R_\oplus = \boxed{3.59 \times 10^7\,\text{m}} \approx 5.6 R_\oplus \approx 22 \times 10^3$ mi. Do you suppose this is why the nation of Colombia, which straddles the equator, claims air space above it up to 23 000 miles? As for the speed—in one transit of the orbit, the satellite travels a distance of $2\pi r$ in a time T. Hence

$$v_o = \frac{2\pi r}{T} = \frac{2\pi (42.25 \times 10^6\,\text{m})}{(24\,\text{h})(60\,\text{min/h})(60\,\text{s/min})}$$

and $\boxed{v_o = 3.1 \times 10^3\,\text{m/s}}$

or 6.9×10^3 mi/h. By the way, *Syncom II* was the first of the species successfully launched in July 1963.

Quick Check: Using Eq. (5.10), $v_o = [(6.67 \times 10^{-11}\,\text{N·m}^2/\text{kg}^2)(5.975 \times 10^{24}\,\text{kg})/(42.25 \times 10^6\,\text{m})]^{1/2} = 3.1 \times 10^3$ m/s.

DISHES & SATELLITES

One of the most desirable orbital arrangements, especially for communications purposes, is the **geostationary** or **geosynchronous orbit**. While parked in one of these, a satellite, having been lofted directly above the equator, will circle once around in a day, revolving in synchronization with the rotating Earth. As seen from anywhere on the globe, the craft will remain in a fixed location in the sky and therefore in continuous line-of-sight communication with a ground station or a satellite dish on the roof of a house. Well over 100 military and civilian spacecraft are crowded into the geosynchronous band with more being added every year.

5.7 Effectively Weightless

Nothing in this Universe of $\approx 10^{22}$ stars is gravityless: there's always something around to interact with. Yet a body falling freely behaves as if it were without weight, and so we say it's **effectively weightless**.

Imagine that you're in an elevator falling freely, accelerating downward at g. Down you would go, and a bundle of keys or a book resting in your open hand would fall just as you fall. The falling book would hover "weightlessly" in your falling hand. In fact, if you were standing on a scale, it would drop out from beneath you; not supporting your weight, it would read zero. It is only when a scale pushes up with a normal force equal to your weight, thereby keeping you from accelerating downward, that it will read your weight.

The floor, dropping out from beneath your feet, would exert no upward response unless you pushed down on it with some muscle-generated force (Fig. 5.20). To be precise, the gravitational force still acts; in fact, *gravity is the only force that does act*. According to our definition of weight, an object in free-fall is not weightless; $F_W = mg$ is not zero. However, if we define the **effective weight** of an object as *the force it exerts on a scale*, then bodies in free-fall have an effective weight of zero, which is what the popular literature calls "zero-g" or "zero gravity." Because of the rotation of the planet, your bathroom scale is also accelerating downward, so it reads effective weight. {For more on this subject click on **VERTICAL ACCELERATIONS** under **FURTHER DISCUSSIONS** on the **CD.**} 🔘

The **effective weight** of an unconstrained object is the net downward force on the object measured by an ordinary scale.

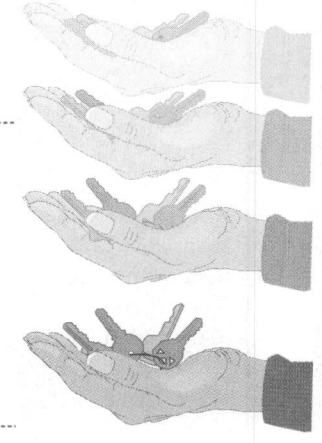

$a = g$

EXPLORING PHYSICS ON YOUR OWN

Weightlessness & Free-Fall: Hold your palm facing up, and place a bunch of keys in it, keeping your hand open. Now drop your arm, accelerating it at $9.8\,\text{m/s}^2$. The keys descend at this rate all by themselves—let them fall freely and simply have your hand precede them down. At the point where your hand reaches $a = g$, you will no longer feel the keys pressing on your palm; your hand will fall away as the keys fall, and the keys will become *effectively weightless*. Put the keys back in your open "stationary" hand and now step off a chair. As you descend, the keys will again be weightless and you'll feel them thump down into your hand when you hit the floor.

WEIGHTLESS IN ORBIT & THE VOMIT COMET

Return to Eq. (5.10), $v_o = \sqrt{GM/r}$, which describes the speed of a satellite in a circular orbit. The independence of v_o on m makes the point that an astronaut inside an orbiting spaceship is also orbiting at that same speed. And so, too, is all the loose paraphernalia in the vehicle—all of which is in orbit within the confines of the craft independent of it. The pilot will float around inside the ship just as she would float around outside of it, moving along at speed v_o in orbit. And since the only force acting is gravity, the crew will be *effectively weightless*. Indeed, orbiting astronauts are in *free-fall* toward the Earth at a rate $g_\oplus = a_C$. The concept of "falling" is generally thought of from a flat-Earth perspective, where the thing that falls gets closer to the surface in the process, which need not be the case on a spherical planet, the surface of which also "falls away."

The National Aeronautics and Space Administration (NASA) routinely flies a padded research plane called the "Vomit Comet" in orbital arcs so that trainee astronauts can float effectively weightless for about 40 s at a time. Gravity acts uniformly on all parts of the body, and so in free-fall one feels nothing—no perception of endless descent—just effortless floating. By the way, you too can be "weightless," just leap into the air; any object in unpowered flight (experiencing only the force of gravity, once you leave the floor) is effectively weightless.

Figure 5.20 Floating around within a free-falling elevator after pushing off from its floor. The elevator moves downward more rapidly than the person who slowed his descent by pushing on the floor.

Suppose that our elevator (at the surface of the Earth) is now hurled upward, at first accelerating at a rate a and then let loose to continue up in free unpowered flight like a spent rocket. If you stepped off a chair inside the elevator during the powered ascent, you would accelerate down at g while the floor accelerated up to meet you at a. The combined result would be a relative acceleration of you with respect to the floor of $(g + a)$. Once you alight on the floor, you would have an effective weight of $m(g + a)$. During the subsequent unpowered portion of the ascent, our elevator-turned-rocketship will continue to climb, slowing down as it rises because of the ever-present tug of gravity. Step off the chair now and you will accelerate earthward at g as usual, but so will the elevator, even as

Astronauts orbiting Earth are effectively weightless in free-fall.

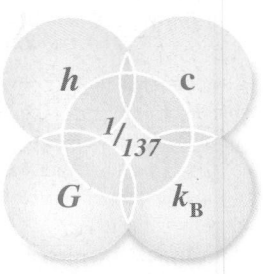

Figure 5.21 The speed of light (c) is the upper limit on the speed of all interactions. It's one of the fundamental constants of Nature.

When an object experiences forces of a particular kind over a continuous range of locations in space, we say that a **force field** exists in that region.

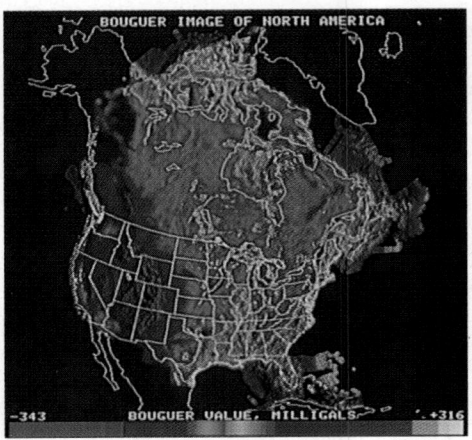

Variations in the strength of the gravitational field across North America. The units are milligals, where 1 gal = 1 $cm/s^2 = 10^{-2}$ m/s^2. The field is strongest (red) in the Atlantic basins off the continental shelf and weakest in the high reaches of the Rocky Mountains (purple).

All things by immortal power,
Near or far,
Hiddenly
To each other linked are,
That thou canst not stir a flower
Without troubling of a star.
FRANCIS THOMPSON (1859–1907)

it continues to climb. Both passenger and craft are in free-fall, although heading upward, and you again hover effectively weightless in midair right where you left the chair. The Apollo astronauts floated around in their spaceships on the way to and from the Moon. They were not gravityless, or weightless, just effectively weightless in free-fall.

5.8 The Gravitational Field

An apple loosed from a branch drops earthward under the influence of gravity. But how does the apple get the message? How does it know which way is down and how hard it's supposedly being pulled?

We are content with the idea of *direct contact*—two things pressed against one another, force transmitted by touching. But matter is composed of separate, spaced atoms and the idea of *direct* contact becomes meaningless. Atoms certainly interact, yet they never touch in the familiar way we think of macroscopic objects touching each other. Even here, in the very essence of contact, there is *action-at-a-distance*, the idea of objects exerting influences across an intervening void. When your finger "touches" this book, it approaches close enough for the electromagnetic repulsion between the electron clouds of finger atoms and book atoms to become palpable.

Each particle of matter has associated with it a surrounding field of influence, and it is that field that carries the interaction between separated bits of matter. A **field of force** *exists in a region of space when an appropriate object placed at any point therein experiences a force.* We think of every object that has mass permeating the surrounding space with a gravitational field that extends out indefinitely, its strength dropping off with distance.

It is from the study of the microworld that we have learned to expect that *all* forces are mediated by streams of special particles. Contemporary Quantum Field Theory maintains that the forces between objects arise from the exchange of these field particles. The carrier of the gravitational interaction is the still hypothetical massless **graviton**. These particles, traveling at the speed of light (c) from Earth to apple and vice versa, carry the interaction from one to the other (Fig. 5.21).

In comparison to the other kinds of forces (Fig. 5.22) to be encountered in later chapters (viz., electromagnetic, weak, and strong), gravity is by far the faintest. Not surprisingly, then, the graviton remains elusive, its existence neither confirmed nor denied. For the time being, the graviton is a theoretical entity in a marvelous scheme that has replaced the invisible tentacles of classical gravity with the equally invisible particles of quantum gravity. {To read about gravity in the early universe click on **INFLATION & THE BIG BANG** under **FURTHER DISCUSSIONS** on the **CD**.}

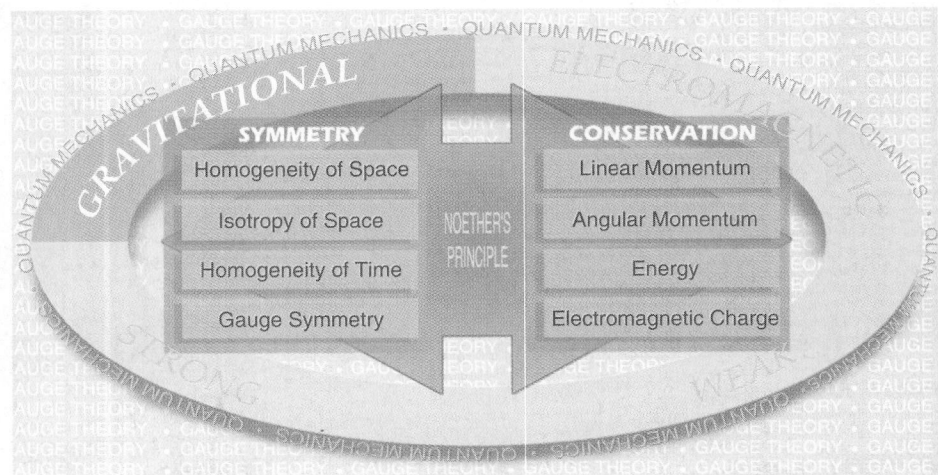

Figure 5.22 This summary illustration will help us keep track of where we've been and where we've yet to go in the theoretical landscape of physics. The diagram now shows that we have studied the gravitational force field.

Core Material & Study Guide

CENTRIPETAL FORCE

An inwardly directed force (gravitational, frictional, electrical, whatever) called a **centripetal force** must be exerted on an object if it is to travel in a curved path. The *centripetal acceleration*, a_C, which occurs when any object uniformly moves in a circle, is

$$a_C = \frac{v^2}{r} \qquad [5.1]$$

Such an object must experience an inwardly directed centripetal force given by

$$F_C = ma_C = \frac{mv^2}{r} \qquad [5.2]$$

Reread Sections 5.1 (Centripetal Acceleration) and 5.2 (Center-Seeking Forces), and make sure you understand Example 5.1.

When a track is banked at an angle θ such that

$$\tan \theta = \frac{v^2}{gr} \qquad [5.3]$$

the normal force will supply the centripetal force rather than friction. Study Example 5.2. Read all the Suggestions on Problem Solving. Work through the **WARM-UPS** in **CHAPTER 5** on the CD to establish that you understand the basics; then do as many I-level problems as you can. After that study the **WALK-THROUGHS** in **CHAPTER 5** on the **CD**.

GRAVITY

Newton's **Law of Universal Gravitation** is

$$F_G = G\frac{mM}{r^2} \qquad [5.5]$$

where $G = 6.672\,59 \times 10^{-11}$ N·m²/kg². This expression holds exactly for point masses or uniform spheres. *As far as a point outside is concerned, a sphere behaves gravitationally as if all its mass were at its center*. In general, any two mass configurations will interact in an approximately $1/r^2$ fashion when r is large in comparison to the sizes of the objects. Reread Section 5.3 (The Law of Universal Gravitation), and make sure you understand Examples 5.4 and 5.5.

On Earth's surface, the *absolute acceleration due to gravity* is

[spherical Earth] $\qquad g_0 = \frac{GM_\oplus}{R_\oplus^2} \qquad [5.6]$

At a distance $r > R_\oplus$ from the center of the planet the acceleration due to gravity is

[beyond the Earth] $\qquad g_\oplus = \frac{GM_\oplus}{r^2} \qquad [5.7]$

Reread Section 5.4 (Terrestrial Gravity).

THE COSMIC FORCE

Kepler's *Three Laws of Planetary Motion* are:

1. The planets move in elliptical orbits with the Sun at one focus (and nothing at the other).

2. Any planet moves in such a way that a line drawn from the Sun to its center sweeps out equal areas in equal time intervals.

3. The ratio of the average distance from the Sun cubed to the period squared is the same constant value for all planets. That is,

$$\frac{r_\odot^3}{T^2} = C_\odot \qquad [5.8]$$

A satellite in a circular orbit about a large mass M, at a distance r from its center, moves with a speed

$$v_o = \sqrt{\frac{GM}{r}} \qquad [5.10]$$

Reread Sections 5.5 (The Laws of Planetary Motion) and 5.6 (Satellite Orbits), and review Examples 5.7, 5.8, and 5.9.

It's useful to imagine that any mass is accompanied by a *gravitational field* that extends out into the surrounding space.

Key Terms

centripetal force	Kepler's Laws
centripetal acceleration	astronomical unit
radius of curvature	altitude
Law of Universal Gravitation	circular orbital speed
inverse square	effectively weightless
period	gravitational field
Gravitational Constant	field of force
absolute gravitational acceleration	graviton
Principia	geostationary orbit
measured weight	geosynchronous orbit
true weight	

Discussion Questions

1. There is an amusement-park ride where people stand in a bowl-shaped cage that revolves about a vertical axis. At a certain speed the floor drops away and everyone is pinned to the outer wall. How does that happen? What role does friction play?

2. Discuss the tension in any one of the ropes in the accompanying photo taken not far from the Sun Pyramid in Teotihuacan, Mexico. Is the tension greater or less than the weight of the man it supports? Explain.

3. Can an object ever move in a direction other than that of the net applied force? Can it accelerate in a direction different from that of the force?

4. A 160-lb gymnast doing a giant swing on a high bar experiences a force directed inward toward the bar of about 120 lb at the top and roughly 790 lb at the bottom. Explain what is happening.

Figure Q2

5. What are all the forces acting on the cyclist in Fig. Q5? What, if anything, must F_N equal? What, if anything, must F_C equal?

Figure Q5

6. What will happen when the string holding the wrench handle in Fig. Q6 burns through?

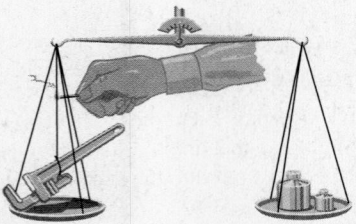

Figure Q6

7. Suppose that an astronaut in space is in a high circular Earth orbit moving eastward (that is, counterclockwise looking down on the North Pole). Imagine that she releases an apple into space. What will be the flight path of the apple? What if the apple is hurled directly forward (eastward)? What if it's thrown in the backward direction? What do you think would happen to it if it were thrown radially toward the planet? (This last question goes beyond the material we have studied so far and is only meant as a test of your intuition.)

8. The following quote is from *Bioastronautics Data Book*, 2nd ed., NASA SP-3006 (1973), p. 149. Why is it unmitigated nonsense?

> In its third form, acceleration occurs as a component of the attraction between masses. The resulting force is directly proportional to the product of the masses and inversely proportional to the square of the distance between them. The proportionality constant is the gravitational constant *g*, which represents an acceleration of 24 feet per second (fps) within the terrestrial field of reference. This is the accepted unit of measurement of acceleration.

9. Why is it easier to launch Earth satellites in an easterly direction than in a westerly direction? The John F. Kennedy Space Center is on which coast of Florida? Why? Why is it in Florida?

10. Why do you lose weight when you enter a tunnel passing through a mountain or walk inside a skyscraper?

11. Figure Q11 shows a spaceship changing from a low to a high circular orbit. Explain how it's being done. Compare the initial and final orbital speeds. Assume an ordinary chemical rocket that can be fired in limited duration bursts.

12. The Sun seems to move across the star field more rapidly in the winter than it does in the summer. What does this phenomenon imply about the Earth-Sun distances at these two times?

Changing orbits

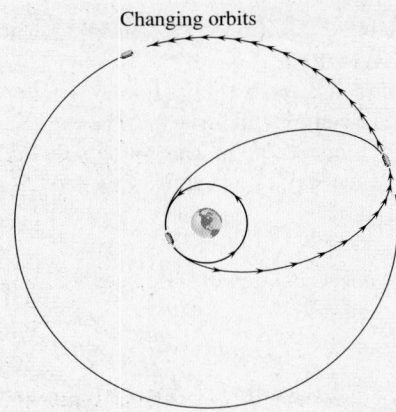

Figure Q11

13. Figure Q13 shows a maneuver that a rocket can use to escape from a circular orbit about a planet into a hyperbolic unpowered flight path. What must happen at point *A*? If the process is run backward and the craft free-falls toward the planet, what must happen at *A* to effect capture?

14. Figure Q14 depicts a double-thrust hyperbolic departure of a rocket from a circular orbit. Explain how it happens. Incidentally, for high speeds at great distances, this maneuver is more fuel-efficient than that considered in the previous question.

15. It has become commonplace to see large (≈9-ft-diameter) parabolic TV antennas on the roofs of buildings, especially bars and hotels. In what direction do they point, and at what are they pointed?

16. The Sun gravitationally attracts the Moon. Does the Moon orbit the Sun? Explain.

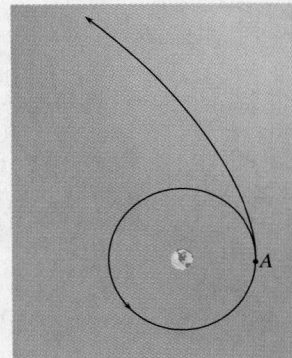

Figure Q13

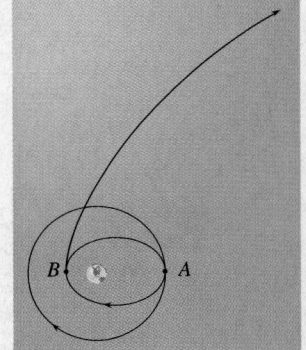

Figure Q14

17. In 1978, the *International Sun-Earth Explorer 3* (*ISEE-3*) satellite was launched into a remarkable orbit. What's so extraordinary is that it orbits a *libration point* in space where no central mass exists. This point lies on the Earth-Sun center-to-center line and is the place where the gravitational forces of these two bodies exactly counterbalance each other. In other words, the gravitational field there is essentially zero. The plane of the orbit is perpendicular to the center line and contains the libration point. Explain, qualitatively, how *ISEE-3* could be held in orbit this way.

18. Imagine that all the mass of the Earth were somehow compressed into a sphere < 200 m in diameter. It can be shown that a *solid steel* satellite 1.0 m in diameter descending toward the planet would be pulled apart as soon as it approached to within 100 m of the center of this miniworld. Similarly, if the Moon were to orbit

Earth at a center-to-center distance of less than $2.9R_\oplus$, it, too, would be torn apart. Explain.

19. In April 2001 drivers practicing for the first CART race at the Texas Motor Speedway began complaining about lightheadedness, blurred vision, and dizziness. They had been tooling around the relatively short (1.7 mi) highly banked (24°) track at speeds up to 230 mph. Consultants from NASA determined that the drivers were experiencing accelerations of as much as $5.7g$, substantially higher than the sustained safe limit for fighter pilots. The race was subsequently canceled. Considering blood flow to the brain, explain what was happening. What was the significance of the fact that the track was short? What was the significance of the fact that the track was highly tilted? What was the significance of the fact that the cars were reaching 230 mph?

Multiple Choice Questions

The first three of the following questions pertain to Fig. MC1, which shows a vehicle of mass m, traveling a level road at a constant speed v.

1. Where along the path is the vehicle's net acceleration the largest? (a) A to B (b) B to C (c) C to D (d) D to E (e) none of these.

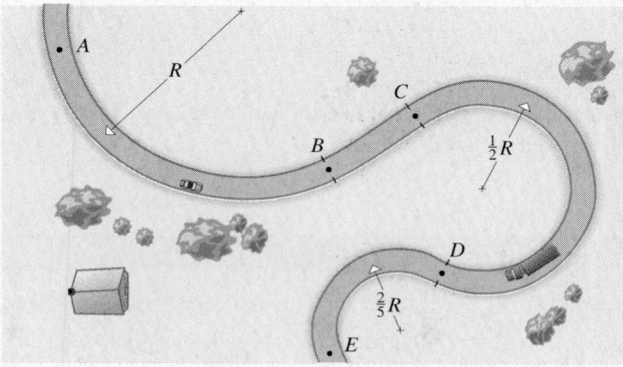

Figure MC1

2. Where along the path is the vehicle's net acceleration the smallest? (a) A to B (b) B to C (c) C to D (d) D to E (e) none of these.

3. What is the magnitude of the vehicle's net acceleration between C and D? (a) $mv^2/2R$ (b) $2mv^2/R$ (c) $4mv^2/R$ (d) $mv^2/4R$ (e) none of these.

4. For an object moving along a curved path at an increasing speed (a) its acceleration is perpendicular to its instantaneous velocity (b) its velocity is constant (c) its acceleration vector is constant (d) its acceleration is zero (e) none of these.

5. For an object moving along a curved path at a constant speed (a) its acceleration is perpendicular to its instantaneous velocity (b) its velocity is constant (c) its acceleration vector is constant (d) its acceleration is zero (e) none of these.

6. Suppose you are whirling a ball tied to the end of a string in a horizontal circle and you triple its speed. (a) Its centripetal acceleration triples. (b) Its velocity goes up by a multiplicative factor of nine. (c) Its acceleration vector is constant. (d) Its centripetal acceleration goes up by a multiplicative factor of nine. (e) none of these.

7. A toy airplane is traveling in a circle at the end of a guide wire. It is made to go faster as more wire is played out until both the speed and the radius of the circle are constant at double their original values. (a) The plane's centripetal acceleration is unchanged. (b) The magnitude of its centripetal acceleration is doubled. (c) Its acceleration decreases by a factor of two. (d) The plane's centripetal acceleration is zero (e) none of these.

8. Imagine a man on a motorcycle (together with a net weight F_W) driving around on the inside wall of a large vertical cylinder. (a) $F_C = F_W$ and $F_N = F_W$ (b) $F_C = F_f$ and $F_N = F_W$ (c) $F_C = F_W$ and $F_N = F_f$ (d) $F_C = F_N$ and $F_f = F_W$ (e) none of these.

9. If the dimensions of length, time, and mass are L, T, and M, respectively, then the dimensions of G are (a) $[L^2/MT^2]$ (b) $[L^3/MT]$ (c) $[L^3/MT^2]$ (d) $[L^3/M^2T]$ (e) $[L^2/M^3T]$.

10. A 1.00-kg chicken weighs 9.8 N on the surface of the Earth. At a distance of one Earth-radius above the planet's surface (a) its weight is 4.9 N (b) its mass is 0.50 kg (c) its weight is 19.6 N (d) its mass is 2.00 kg (e) none of these.

11. Call the gravitational attraction between you and the planet, your *true* weight. Because of the rotation of Earth, the reading that you get when you stand on a scale (a) will be more than your *true* weight (b) will everywhere be less than your *true* weight (c) will always be your *true* weight (d) will only be your *true* weight at the two poles (e) will only be your *true* weight at the Equator.

12. Suppose you were transported to the mythical planet Mongo, which is four times as massive as Earth and has twice the diameter. Your Mongoian weight as compared to your present weight would be (a) 4 times larger (b) the same (c) 2 times smaller (d) 4 times smaller (e) none of these.

13. Mars has a mass of $0.107\,4M_\oplus$ and is at a mean distance from the Sun that is 1.52 times larger than that of Earth. By comparison to the gravitational force exerted on Mars by our world, the force exerted on Earth by Mars is (a) 0.107 4 times smaller (b) 0.107 4 times larger (c) the same (d) 1.52 times less (e) none of these.

14. The acceleration of a meteor at a height above the Earth of $1R_\oplus$ is (a) about 2.5 m/s² (b) 9.8 m/s² (c) not enough information to say (d) can be anything from 9.8 m/s² to 0 (e) none of these.

15. The asteroid Geographos (one of the Apollo group, each of which crosses the Earth's orbit on the way around the Sun) has a radius of $2.4 \times 10^{-4}R_\oplus$ and a mass of $8.4 \times 10^{-12}M_\oplus$. How does the gravitational acceleration on its surface compare to the corresponding value g_0 on the Earth? It equals (a) $2.4 \times 10^{-4}g_0$ (b) $8.4 \times 10^{-12}g_0$ (c) $1.5 \times 10^{-4}g_0$ (d) $3.5 \times 10^{-8}g_0$ (e) none of these.

16. An astronaut on the Moon has a mass that by comparison to his mass on Earth is (a) unchanged (b) six times greater (c) six times less (d) not enough information to say (e) none of these.

17. The acceleration due to gravity, as measured by a spring-balance determination of the weight of an object ($F_W = mg$), varies

from place to place on Earth because (a) the mass changes (b) g is affected by the rotation of the planet only (c) g depends on the shape of the planet only (d) g depends on both the rotation and shape of the planet (e) none of these.

18. If *Martian Orbiter 1* is sailing about that planet in a circle with an orbital radius nine times that of *Orbiter 2*, whose speed is v_2, what is the speed of *Orbiter 1*? (a) $\frac{1}{3}v_2$ (b) $3v_2$ (c) v_2 (d) $81v_2$ (e) none of these.

19. Figure MC19 shows a spaceship in orbit about a star. If its speeds at the four points shown are v_A, v_B, v_C, and v_D, respectively, then (a) $v_A < v_B < v_C < v_D$ (b) $v_A > v_B > v_C > v_D$ (c) $v_A > v_B = v_D > v_C$ (d) $v_A < v_B = v_D < v_C$ (e) none of these.

20. A spacecraft is in a circular orbit about a planet located at point O in Fig. MC20. When it reaches point A on its orbit, the First Mate throws a canister out the front port straight ahead; the canister goes into a new orbit shown in (a) part a (b) part b (c) part c (d) part d (e) none of these.

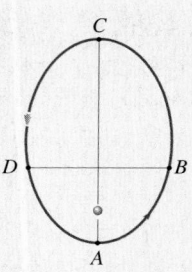

Figure MC19

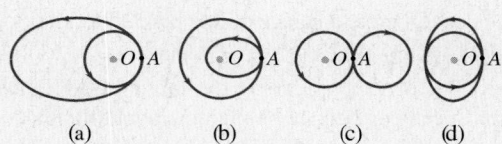

Figure MC20

21. It is desired that a spacecraft initially in a circular orbit drop straight down to the planet below. Figure MC21 shows the planned double-thrust maneuver, which supposedly will save fuel. The craft (a) slows down at B and brakes to zero speed at D (b) speeds up at A and brakes to zero speed at E (c) brakes slightly at A and then again at E (d) the maneuver is impossible (e) none of these.

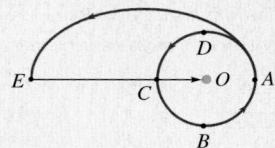

Figure MC21

For more **Multiple Choice Questions** with answers click on WARM-UPS in CHAPTER 5 on the CD.

Suggestions on Problem Solving

1. Make sure that your answers are reasonable. Compare your results with known values to see if they fit within recognized extremes. A value of $g_\oplus$ computed anywhere inside or outside the Earth should be between 0 and roughly 9.81 m/s². A value of 98 m/s² must be wrong!

2. Don't confuse G (the universal constant), g (the measured free-fall acceleration at the surface of Earth), g_0 (the acceleration due only to gravity at the surface of Earth), and $g_\oplus$ (the acceleration anywhere due only to Earth's gravity).

3. Consider units—don't enter a number in *kilometers* when

you want *meters*, or use 32.2 m/s² when you should use 9.81 m/s².

4. Remember that F_G varies inversely with the *square* of the distance. An object moved from the surface of the Earth to a point 100 Earth-radii from the planet's *center* drops in weight to 1/10 000th its original value, *not* 1/100th.

5. Many of the equations in the chapter call for the radial distance. When given the numerical value of a diameter, remember to divide by 2 before substituting. Keep in mind that the altitude above the surface of a planet is not the radial distance to its center.

Problems + Coordinated Problems + Progressive Problems + Solutions

STUDY GUIDE **1. Coordinated Problems:** The three problems within each magenta-colored grouping are solvable in similar ways. Note that the first of these always has a hint; moreover, its solution is provided in the back of the book. *Work out each of these sets; they'll strengthen technique and build confidence.* **2. Progressive Problems:** The problems introduced in blue unfold step-by-step carrying along the analysis in a more suggestive way than is customary. *Work out all of these; they'll guide you through the analytic process and help develop problem-solving skills.* **3. Worked-Out Solutions:** Studying worked-out solutions is an important part of learning how to solve problems. Accordingly, additional *solutions* to a number of model problems are given below. *Make sure you understand each of them before you go on to the next problem.* **4.** Also provided in the back of the book are the **Answers** to all odd-numbered problems, as well as worked-out *solutions* to those with boldface numbers. Problem numbers in italic indicate that a solution appears in the Student Solutions Manual.

SECTION 5.1: CENTRIPETAL ACCELERATION

SECTION 5.2: CENTER-SEEKING FORCES

1. [I] Determine the acceleration of a kid on a bike traveling at a constant 10.0 m/s around a flat circular track of radius 200 m.

2. [I] A supersonic jet diving at 290 m/s pulls out into a circular loop of radius R. If the craft is designed to withstand forces accompanying centripetal accelerations of up to $9.0g$, compute the minimum value of R.

3. [I] THIS PROBLEM EXAMINES CENTRIPETAL ACCELERATION. A small ball rolling at a constant speed in a horizontal circular groove having a diameter of 2.00 m experiences a centripetal acceleration of 0.500 m/s². (a) What is the radius of the ball's orbit? (b) How does the centripetal acceleration depend on the speed? (c) Determine that speed.

4. [I] THIS PROBLEM DEALS WITH CENTRIPETAL FORCE. While moving in a horizontal arc having a radius of 2.00 m, a 28-g mouse travels at a constant speed of 1.60 m/s. (a) What force allows the mouse to travel in a circular arc? (b) What is the mouse's mass in kilograms? (c) What is the total friction force acting on its little feet at that instant? Neglect air friction.

5. [I] A 25-g pebble is stuck in the tread of a 28-in.-diameter tire. If the tire can exert an inward radial friction force of up to 20 N on the pebble, how fast will the pebble be traveling with respect to the center of the wheel when it flies out tangentially? [*Hint: We have the mass, radius, and centripetal force, and need the speed. The defining equation for F_C should do the trick.*]

6. [I] Determine the maximum speed that can be reached by a 19.61-N (as measured on Earth) steel ball tied to the end of a thin thread. The ball, which is essentially weightless, is being whirled in a circle of 2.00-m radius by an astronaut out in space. The thread has a breaking strength of 16.0 N.

7. [I] A 10.0-kg mass is tied to a 3/16-in. Manila line, which has a breaking strength of 1.80 kN. What is the maximum speed the mass can have if it is whirled around in a horizontal circle with a 1.0-m radius and the rope is not to break?

8. [I] A youngster on a carousel horse 5.0 m from the center revolves at a constant rate, once around in 15.0 s. What is her acceleration?

9. [I] A baseball player rounds second base in an arc with a radius of curvature of 4.88 m at a speed of 6.1 m/s. If he weighs 845 N, what is the centripetal force that must be acting on him? Notice how this limits the tightness of the turn. What provides the centripetal force?

10. [I] Compute the Earth's centripetal acceleration toward the Sun. Take the time it takes to go once around as 365 d and the radius on average as 1.50×10^8 km.

> **SOLUTION:** The centripetal acceleration is given by $a_C = v^2/r$, but we don't know v. Still, $v = l/t$ and we can compute l, which is the circumference. $l = 2\pi r = 2\pi(1.50 \times 10^{11}$ m$) = 9.425 \times 10^{11}$ m. The time to go once around, the period, is $(365$ d$)(24$ h/d$)(60$ min/h$)(60$ s/min$) = 3.154 \times 10^7$ s. Thus $v = l/t = (9.425 \times 10^{11}$ m$)/(3.154 \times 10^7$ s$) = 2.988 \times 10^4$ m/s. **And finally,** $a_C = v^2/r = (2.988 \times 10^4$ m/s$)^2/(1.50 \times 10^{11}$ m$) = 5.95 \times 10^{-3}$ m/s².

11. [I] A hammer thrower at a track-and-field meet whirls around at a rate of 2.0 revolutions per second, revolving a 16-lb ball at the end of a cable that gives it a 6.0-ft effective radius. Compute the inward force that must be exerted on the ball.

12. [I] A test tube in a centrifuge is pivoted so that it swings out horizontally as the machine builds up speed. If the bottom of the tube is 150 mm from the central spin axis, and if the machine hits 50 000 revolutions per minute, what would be the centripetal force exerted on a giant amoeba of mass 1.0×10^{-8} kg at the bottom of the tube?

13. [I] A 100-mm-long test tube is held rigidly at 30° with respect to the vertical in a centrifuge with its top lip 5.0 cm from the central spin axis of the machine and its bottom somewhat farther. If it rotates at 40 000 revolutions per minute, what is the centripetal acceleration of a cell at the bottom of the tube?

14. [II] A circular automobile racetrack is banked at an angle θ such that no friction between road and tires is required when a car travels at 30.0 m/s. If the radius of the track is 400 m, determine θ.

15. [II] A front-loading clothes washer has a horizontal drum that is thoroughly perforated with small holes. Assuming it to spin dry at 1 rotation per second, have a radius of 40 cm, and contain a 4.5-kg wet Teddy bear, what maximum force is exerted by the wall on the bear? What happens to the water?

16. [II] EXPLORING PHYSICS ON YOUR OWN: Drop a string of length L through the hole in a vertically held spool of thread (Fig. P16). Tie 10 paper clips to each end and twirl the string so that one bunch of clips moves in a horizontal circle at a speed v while the other hangs vertically. Neglecting friction, write an expression (a) for the distance d as a function of v and L where $L = (r + d)$ and (b) for g in terms of v and r, the radius of the circle.

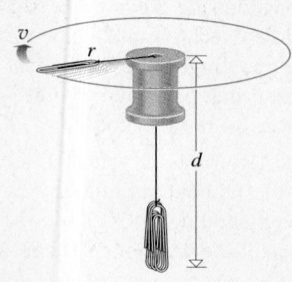

Figure P16

17. [II] A skier, of mass m, comes down a slope that has the shape of a vertical segment of a circle (of radius r, with the center above the path), ending in a tangential, flat, horizontal run. Write an expression in terms of v, m, r, and g for the normal force exerted by the snow on the skis at the bottom just before the skier leaves the circular portion. [*Hint: At the bottom of the arc the ground must supply the centripetal force and support the weight.*]

18. [II] While whirling around in a vertical circle with a radius of 1.50 m, a 2.00-kg mass is held on a rope attached to a very light spring scale. What value does the scale read when the mass is moving at 4.00 m/s at the lowest point of its orbit?

19. [II] A scale is fitted into the seat of a roller coaster car, and a person weighing 800 N sits down on it. The car then descends along a path that has the shape of a 100.0-m-radius vertical circle with its lowest point at the bottom where the car reaches its greatest speed of 40.0 m/s. What is the maximum reading of the scale?

20. [II] THIS PROBLEM DEALS WITH CENTRIPETAL ACCELERATION. An object moves in a circular orbit at a constant speed. It takes a time T, called the *period*, to go once around. We want to show that

$$a_C = \frac{4\pi^2 r}{T^2}$$

But first, (a) write an expression for the circumference of the circular orbit. (b) Knowing how long it takes to go once around, write an expression for the orbital speed in terms of r and T. (c) Now derive the above expression for a_C.

21. [II] After a few seconds a toy car on a 2.00-m diameter circular track reaches and thereafter maintains a constant speed. If it then makes 1200 revolutions in 1.00 h (a) How fast is it moving? (b) What is its centripetal acceleration?

22. [II] A 1000-kg car traveling on a road that runs straight up a hill reaches the rounded crest at 10.0 m/s. If the hill at that point has a radius of curvature (in a vertical plane) of 50 m, what is the net downward force acting on the car at the instant it is horizontal at the very peak?

> **SOLUTION:** The net downward force equals the net upward force which is the normal force. On a flat road $F_W = F_N$, but here at the top of a vertical curved road there must be a net downward force equal to F_C; otherwise the car would fly off tangentially. Thus, taking down as positive, $F_W - F_N = F_C$. The centripetal force is given by $F_C = ma_C = mv^2/r$ and so $F_N = F_W - F_C = mg - mv^2/r = (1000$ kg$)(9.81$ m/s²$) - (1000$ kg$)(10.0$ m/s$)^2/(50$ m$) = 7.8$ kN.

23. [II] At a given instant, someone strapped into a roller coaster car hangs upside down at the very top of the circle (of radius 25.0 m) while executing a so-called loop-the-loop. At what speed must he be traveling if at that moment the force exerted by his body on the seat is half his actual weight? Assume that at the start of the ride the straps were fairly loose.

24. [II] Suppose you wish to whirl a pail full of water in a vertical circle without spilling any of its contents. If your arm is 0.90 m-long (shoulder to fist) and the distance from the handle to the surface of the water is 20.0 cm, what minimum speed is required?

25. [II] A cylindrically shaped space station 1500 m in diameter is to revolve about its central symmetry axis to provide a simulated 1.0-g environment at the periphery. (a) Compute the necessary spin rate. (b) How would "g" vary with altitude up from the floor (which is the inside curved wall of the cylinder)?

26. [II] A stunt pilot flying an old biplane climbs in a vertical circular loop. While upside down, the force acting upward normally on the seat is one third of her usual weight. If the plane is traveling at that moment at 300 km/h, what is the radius of the loop? (A World War I pilot did this little trick without fastening his seat belt and—you guessed it—fell out. What can you say about his speed at the top of the loop?)

27. [III] Take the Earth to be a perfect sphere of diameter 1.274 × 10^7 m. If an object has a weight of 100 N while on a scale at the South Pole, how much will it weigh at the Equator? Take the equatorial spin speed to be $v = 465$ m/s.

28. [III] Design a carnival ride on which standing passengers are pressed against the inside curved wall of a rotating vertical cylinder. It is to turn at most at $\frac{1}{2}$ revolution per second. Assuming a minimum coefficient of friction of 0.20 between clothing and wall, what diameter should the ride have if we can safely make the floor drop away when it reaches running speed?

Miscellaneous data: $M_\oplus = 5.975 \times 10^{24}$ kg, $M_\odot = 1.987 \times 10^{30}$ kg, $M_\mathbb{C} = 7.35 \times 10^{22}$ kg, $R_\odot = 6.97 \times 10^8$ m, $R_\oplus = 6371.23$ km, $R_\mathbb{C} = 1.74 \times 10^6$ m, $r_{\odot\oplus} = 1.495 \times 10^{11}$ m, $r_{\oplus\mathbb{C}} = 3.844 \times 10^8$ m.

SECTION 5.3: THE LAW OF UNIVERSAL GRAVITATION
SECTION 5.4: TERRESTRIAL GRAVITY

29. [I] What would happen to the weight of an object if its mass was doubled while its distance from the center of the Earth was also doubled?

30. [I] The gravitational attraction between a 20-kg cannonball and a marble separated center-to-center by 30 cm is 1.48 × 10^{-10} N. Compute the mass of the marble.

31. [I] Suppose that two identical spheres, separated center-to-center by 1.00 m, experience a mutual gravitational force of 1.00 N. Compute the mass of each sphere.

32. [I] At what center-to-center distance from the Earth would a 1.0-kg mass weigh 1.0 N?

33. [I] Suppose the Earth were compressed to half its diameter. What would happen to the acceleration due to the gravity at its surface?

34. [I] Compare the gravitational force of the Earth on the Moon to that of the Sun on the Moon.

35. [I] If the average distance between Uranus (♅) and Neptune (Ψ) is 4.9 × 10^9 km, and $M_♅ = 14.6M_\oplus$ while $M_\Psi = 17.3M_\oplus$, compute their average gravitational interaction.

36. [I] Imagine two uniform spheres of radius R and density ρ in contact with each other. Write an expression for their mutual gravitational interaction as a function of R, ρ, and G.

37. [I] If you can jump 1.00 m high on Earth, how high can you jump on Venus, where $g_♀ = 0.88g_\oplus$? Assume the same takeoff speed.

38. [I] What fraction of what you weigh on Earth would you weigh in a rocket ship firing its rockets so that it was stationary with respect to the center of the planet $4R_\oplus$ from its surface?

> SOLUTION: The ship is five Earth-radii from the center of the Earth. Given that your weight on Earth is F_w, since the force drops off as the distance square, your weight in the ship is $F_w/25$.

39. [I] Consider two subatomic particles, an electron and a proton, which have masses of 9.1 × 10^{-31} kg and 1.7 × 10^{-27} kg, respectively. When separated by a distance of 5.3 × 10^{-11} m, as they are in a hydrogen atom, the electrical attraction (F_E) between them is 8.2 × 10^{-8} N. Compare this with the corresponding gravitational interaction. How many times larger is F_E than F_G?

40. [I] Considering the Earth as a sphere, when does $g_0 = g_\oplus$? Which of these is a constant and which is a function of r? What would have to happen for $g_0 = g$?

41. [I] The acceleration due to gravity on the surface of Mars is 3.7 m/s^2. If the planet's diameter is 6.8 × 10^6 m, determine the mass of the planet and compare it to Earth.

42. [I] THIS PROBLEM IS ABOUT THE ACCELERATION DUE TO GRAVITY. A spacecraft of mass m is at a distance of $4R_\oplus$ from the center of the planet. (a) What is the formula for the gravitational force on the vehicle at that distance? (b) What is the weight of the vehicle at a distance of $4R_\oplus$? (c) Alternatively, express that weight in terms of the acceleration due to gravity at $4R_\oplus$. (d) Write an expression for the gravitational acceleration at $4R_\oplus$, in terms of $R_\oplus$, G, and $M_\oplus$.

43. [II] What is the value of the acceleration due to the Moon's gravity 100 km above its surface?

44. [II] Imagine an astronaut having a mass of 70 kg floating in space 10.0 m away from the center-of-gravity (the point where all the mass may be imagined to act gravitationally, p. 249) of an *Apollo* Command Module whose mass is 6.00 × 10^3 kg. Determine the gravitational force acting on, and the resulting accelerations (at that instant) of both the ship and the person.

45. [II] Venus has a diameter of 12.1 × 10^3 km and a mean density of 5.2 × 10^3 kg/m^3. How far would an apple fall in one second at its surface? [*Hint: We need the acceleration due to gravity at Venus's surface, and to get that we must first determine the planet's mass.*]

46. [II] Imagine a great sphere of water (of density 1.00 × 10^3 kg/m^3) floating in space. If it has a radius of 10.0 × 10^3 km (about 1.57 times the size of Earth), what would be the acceleration due to gravity at its surface? Check your answer using the Earth's mean radius (6.35 × 10^6 m) and density (5.5 × 10^3 kg/m^3).

47. [II] Gold has density of 19.3 × 10^3 kg/m^3. How big would a solid gold sphere have to be if the acceleration due to gravity at its

surface is to be 9.81 m/s^2? Check your answer against the radius of the Earth, which has a mean density of 5.5 × 10^3 kg/m^3.

48. [II] Given that $M_{\mathbb{C}}/M_{\oplus} = 0.01230$ and $R_{\mathbb{C}}/R_{\oplus} = 0.2731$, compute the ratio of an astronaut's Moon-weight ($F_{w\mathbb{C}}$) to Earth-weight ($F_{w\oplus}$).

SOLUTION: $F_{w\mathbb{C}}/F_{w\oplus} = (GM_{\mathbb{C}}m/R_{\mathbb{C}}^2)/(GM_{\oplus}m/R_{\oplus}^2) = (M_{\mathbb{C}}/M_{\oplus})/(R_{\mathbb{C}}/R_{\oplus})^2 = 0.01230/(0.2731)^2 = 0.1649.$

49. [II] Taking the surface value of $g_{\oplus}$ (see p. 153) to be g_0, show that

$$g_{\oplus} = g_0(R_{\oplus}/r)^2 \quad \text{for } r \geq R_{\oplus}$$

50. [II] Locate the position of a spaceship on the Earth-Moon center line such that, at that point, the tug of each celestial body exerted on it would cancel and the craft would literally be weightless.

51. [II] Mars has a mass of $M_{\male} = 0.108 M_{\oplus}$ and a mean radius $R_{\male} = 0.534 R_{\oplus}$. Find the acceleration of gravity at its surface in terms of $g_0 = 9.8$ m/s^2.

52. [II] Three very small spheres of mass 2.50 kg, 5.00 kg, and 6.00 kg are located on a straight line in space away from everything else. The first one is at a point between the other two, 10.0 cm to the right of the second and 20.0 cm to the left of the third. Compute the net gravitational force it experiences.

53. [II] It is believed that during the gravitational collapse of certain stars, such great densities and pressures will be reached that the atoms themselves will be crushed, leaving only a residual core of neutrons. Such a *neutron star* is, in some respects, very much like a giant atomic nucleus with a tremendous density of roughly about 3 × 10^{17} kg/m^3. Compute the surface acceleration due to gravity for a one-solar-mass neutron star.

54. [III] Figure P54 shows two concentric, thin, uniform spherical shells of mass m_1 and m_2, at the center of which is a small ball of lead of mass m_3. Write an expression for the gravitational force exerted on a particle of mass m at each point A, B, and C located at distances r_A, r_B, and r_C from the very center. [*Hint: There is no force inside a spherical mass shell due to that mass.*]

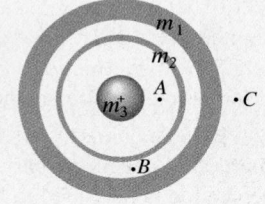

Figure P54

55. [III] Write an expression for the gravitational force on a small mass m imbedded in a uniform spherical cloud of mass M and radius R. Take the particle to be at $r < R$. [*Hint: There is no force inside a spherical mass shell due to that mass.*]

56. [III] Two 2.0-kg crystal balls are 1.0 m apart. Compute the magnitude and direction of the gravitational force they exert on a 10-g marble located 0.25 m from the center-to-center line as shown in Fig. P56.

57. [III] A neutron star (see Problem 53) can be envisioned as an immense nucleus held together by its self-gravitation.

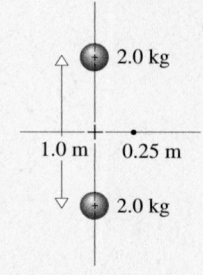

2.0 kg

1.0 m 0.25 m

2.0 kg

Figure P56

What is the shortest period with which such a star could rotate if it's not to lose mass flying off at the equator? Take $\rho = 1 \times 10^{17}$ kg/m^3. It is widely believed that *pulsars*, strange celestial emitters of precisely pulsating radiation, are rapidly rotating neutron stars (see p. 261).

58. [III] Draw a graph of the weight of an object of mass m due to the Earth versus the height above the surface, out to roughly 700 km. What can you say about the curve (as long as $R \gg h$)?

59. [III] Let M be the mass of a uniform spherical planet of radius R. If h is the height above its surface, show that the absolute gravitational acceleration g_p varies with h as

$$g_p = \frac{GM}{R^2}\left(1 + \frac{h}{R}\right)^{-2}$$

This expression can be approximated using the binomial expansion. {See **MATH REVIEW: PART A-5** on the CD.} 🔘

$$(a + x)^n = a^n + na^{n-1}x + \tfrac{1}{2}n(n-1)a^{n-2}x^2 + \cdots$$

where $x^2 < a^2$. Here, $a = 1$, $x = h/R$, $n = -2$, and we limit the calculation to the case where $h \ll R$. Show that

$$g_p \approx \frac{GM}{R^2}\left(1 - \frac{2h}{R}\right)$$

Notice that GM/R^2 is the surface value of g_p occurring when $h = 0$, as in Eq. (5.6).

60. [III] Calculate the acceleration due to gravity 10000 m above the Earth's surface in the following two ways: (1) using $g_{\oplus} = GM_{\oplus}/r^2$; and (2) using the approximation of Problem 59. Compare your results.

61. [III] In light of Problem 59, determine the acceleration due to gravity experienced by the Lunar Module when it was 100 m above the Moon's surface. Is it appreciably different from the surface value?

SECTION 5.5: THE LAWS OF PLANETARY MOTION

62. [I] Using the data for the Earth's orbit, compute the mass of the Sun.

63. [I] Determine the approximate speed of a Lunar Orbiter revolving in a circular orbit at a height of 62 km. Take the Moon's radius as 1738 km.

64. [I] Each of the *Apollo* Lunar Modules was in a very low orbit around the Moon. Given a typical mass of 14.7 × 10^3 kg, assume an altitude of 60.0 km and determine the orbital period.

65. [I] For any Earth satellite in a circular orbit, show that its period (in seconds) and its distance from the center of the planet (in meters) are related by way of

$$T = 3.15 \times 10^{-7}(r_{\oplus})^{\frac{3}{2}}$$

66. [I] *Sputnik 1*, the first artificial satellite to circle the planet (October 1957) had a mean orbital radius of 6950 km. Compute its period.

67. [I] A satellite is to be raised from one circular orbit to another twice as large. What will happen to its period?

68. [I] Referring to Problem 61, compare the two orbital speeds v_1 and v_2.

69. [I] What is the acceleration due to the gravity of the Moon at the center of the Earth?

70. [I] Find the gravitational acceleration 1.00 m from a small sphere whose mass is 1.00 kg.

71. [I] What is the gravitational acceleration due to the Sun at the Earth?

72. [II] By definition, the Earth is a distance of 1.0000 AU from the Sun. Using the fact that Jupiter is, on average, 5.2028 AU from the Sun, compute its period in Earth-years.

73. [II] Determine the period in Earth-years of a satellite placed in a circular solar orbit with a radius of 371.6 million miles.

74. [II] Imagine a central body of mass M_B (be it a star, planet, or moon) about which another object is orbiting such that its Keplerian constant is C_B. Show that C_B/M_B is a universal constant, the same for all bodies.

> **SOLUTION:** From Eq. (5.9) $r_B^3/T^2 = GM_B/4\pi^2 = C_B$ and so $C_B/M_B = G/4\pi^2$ which is constant.

75. [II] THIS PROBLEM WILL HELP US UNDERSTAND KEPLER'S THIRD LAW. The average distance to the Sun for the planet Mercury is 57.9×10^6 km, and its period in Earth-years is 0.241. (a) What is the value of Kepler's solar constant in units of $km^3/Earth-year^2$? (b) If the average distance to Venus is 108.2×10^6 km, determine the approximate period of Venus.

76. [II] THIS PROBLEM WILL HELP US UNDERSTAND KEPLER'S THIRD LAW. The average distance to the Sun for the planet Jupiter is 778.3×10^6 km, and its period in Earth-years is 11.86. (a) What is the value of Kepler's solar constant in units of $km^3/Earth-year^2$? (b) If the period of Saturn is 29.5 Earth-years, determine the average distance of Saturn from the Sun.

77. [II] Imagine the two comparable masses m_1 and m_2 of Fig. P77 orbiting their so-called barycenter O at distances r_1 and r_2, respectively, with a common period T. Since their mutual gravitational interaction provides their individual centripetal forces F_{C1} and F_{C2}, these must be equal. Show that

$$\frac{r_1}{r_2} = \frac{m_2}{m_1}$$

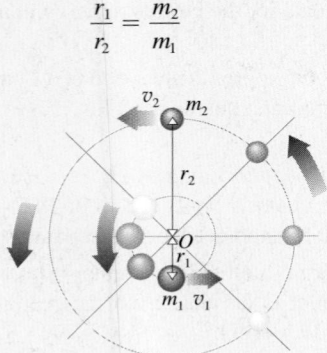

Figure P77

78. [II] Referring to Problem 77 and Fig. P77, write explicit expressions for $F_G = F_{C1}$ and $F_G = F_{C2}$, cancel out whatever you can, and then combine them to get Kepler's Third Law in the more accurate Newtonian version, namely,

$$\frac{(r_1 + r_2)^3}{T^2} = \frac{G(m_1 + m_2)}{4\pi^2}$$

Unlike Eq. (5.8), which applies when the central body (here the Sun) has essentially infinite mass, this expression is more realistic. Notice that the equations are identical when $m_1 \gg m_2$, so that m_2 is negligible; of course, $r = r_1 + r_2$.

79. [II] For an elliptical orbit, the *mean* or *average distance* from the central body to the orbiter is just half the main symmetry axis. This *semi-major axis* is also equal to the average of the perihelion and aphelion distances (i.e., the nearest and farthest approaches to the Sun). The tiny asteroid Icarus (which has a mass of 5.0×10^{12} kg and a radius of just 0.7 km) orbits the Sun in an elongated elliptical path. It crosses the Earth's orbit as it sweeps in to a perihelion of only 0.186 AU and out to an aphelion at 1.97 AU. Compute its mean distance from the Sun and its orbital period in Earth-years.

80. [II] The Andromeda galaxy, known as M31, is a great spiral star-island 2.2×10^6 light-years away. Measurements show that a star out at its extremities—5×10^9 AU from the center—orbits the nucleus at a speed of 200 km/s. Approximate the mass of M31 by assuming that all the mass within the confines of the star's orbit can be taken to act at the center of the galaxy.

81. [III] What is the value of the gravitational acceleration halfway down inside of a solid uniform sphere of mass M and radius R? (See Problem 55.)

82. [III] Determine the gravitational acceleration at the center of a uniform ring of mass M and radius R.

83. [III] Galileo discovered the four major moons of the planet Jupiter in 1610. The nearest one, Io, has a period of 1.7699 d and is 5.578 Jovian radii ($R_{2\!\!\!/}$) from the center of the planet. Use this information to calculate the mean density of Jupiter.

84. [III] The binary stars Sirius A and Sirius B each orbit a common point called the barycenter with a period of 50 years. Their separation is measured to be 20.0 AU (or 2.99×10^{12} m), with the much fainter star, Sirius B, being twice as far from the barycenter as is Sirius A. Compute both their masses. [*Hint: Go back and look at Problems 77 and 78.*]

85. [III] We wish to put an artificial Earth satellite in a circular orbit halfway out to the Moon. Compute its period and the necessary orbital speed.

Chapter 6
Energy

*T*he Newtonian formulation of Mechanics was unrivaled for over a hundred years, but by the beginning of the 1800s a powerful alternative, predicated on the concept of energy, was already taking shape. One of the great achievements of that era was the Law of Conservation of Energy (p. 187).

The emphasis here is on Mechanics, although the idea of energy is all-embracing. In one way or another, it influences our thinking about every branch of physics. Yet there is no completely satisfactory definition of energy. Even so, we will quantify its various manifestations as we struggle to define it.

The Transfer of Energy

Over the centuries, the word energy has had different meanings. The nontechnical usage derives from the Greek *en* (which means in) and *ergon* (which means *work*). Energy is the capacity to do work, an inherent vigor. The word has been used in this way at least since the late 1500s. Galileo (1638) employed the term *l'energia*, though he never defined it. Only in the last 200 years has the idea taken on a scientific meaning.

In very general terms, energy describes the state of a system in relation to the action of the Four Forces. It is a property of all matter and is observed indirectly through changes in speed, mass, position, and so forth. There is no universal energy meter that measures energy directly. The change in the energy of a system, which is all we can ever determine experimentally, is a measure of the physical change in that system. *Force is the agent of change; energy is a measure of change.* Because a system can change through the action of different forces in different ways, there are several distinct manifestations of energy. Energy is a measure of the change that has already occurred, and the change that is yet to occur. In that way it is intimately related to time (see p. 198).

Energy is a scalar quantity associated in various amounts with all the "things" that exist, from minute massless particles to immense whirling galaxies. By observing the changing behavior of matter, we infer the presence of one form or another of energy. *Energy is not an entity in and of itself—there is no such thing as pure energy* (just as there is no such thing as pure momentum). Matter is the vehicle of energy; just as there cannot be wind without air, there cannot be energy without matter.

STUDY GUIDE

This chapter is about energy, which is a new way to quantify change. Newtonian physics deals with interactions and how momentum changes as a result of them. Now we'll examine a complementary perspective: interactions and how energy changes as a result of them.

In succession the chapter describes several kinds of energy: work (p. 172) $W = Fl \cos \theta$, kinetic energy (p. 180) $KE = \frac{1}{2} mv^2$, and gravitational potential energy (p. 183) $PE_G = mgh$.

As we'll see later (p. 216), Newton's formulation is really based on the principle of Conservation of Momentum. Similarly, here everything comes together in the section on Conservation of Mechanical Energy (p. 187).

The rate of transfer of energy is power (p. 195): $P_{av} = \Delta W/\Delta t$ and $P = Fv$.

ON THE VARIOUS MANIFESTATIONS OF ENERGY

We buy *electrical energy* "packaged" from the hardware store and "on tap" from the power company. *Chemical energy* can be gotten from a slice of pizza or a tank of gasoline. Rubber bands, ligaments, and girdles all store *elastic energy*. Snowflakes fall because they have *gravitational energy*. We bake apple pies with *thermal energy* and defend them with *nuclear energy*. You don't play on railroad tracks because trains have lots of *kinetic energy*. *Radiant energy* floods in from the Sun to warm us and from TV stations to entertain us.

It is important to realize that in physics today, we have no knowledge of what energy is.

R. P. FEYNMAN
NOBEL LAUREATE

The concept of energy provides a means of quantitatively accounting for physical change. When a material system (one comprising quarks, leptons, and/or photons) manifests an observable change due to some interaction, we associate an amount of energy with the extent of that change. Interaction is crucial; if matter did not interact, the concept of energy would be superfluous. Even though we speak about electrical energy, chemical energy, thermal energy, and so forth, they're all basically the same; they all arise from the action of the Four Fundamental Forces.

Perhaps the most important characteristic of energy is that it is transferred from one entity to another such that **the total amount of energy always remains unchanged**. Thermal energy can be converted into electrical energy, and some of that turned into light, and back again into thermal energy, but the net amount of energy is always the same—**energy is conserved**. This aspect of the Universe, which arises out of the equal-and-opposite symmetry of action and reaction, allows us to keep track of the effects of all of the interactions going on in a material system, and that's an extraordinary accomplishment.

To see this from a somewhat different perspective, imagine an isolated system consisting of a single point-mass in a vast void. There can be no interactions and so nothing much will happen. Now suppose instead that there are two or more interacting particles present. Under the action of forces, the particles will change both their positions and speeds. As we'll soon see, the loss or gain in the *potential energy* (p. 183) of any particle is associated with the change in its relative position, while the loss or gain in its *kinetic energy* (p. 180) is associated with the change in its relative speed. When two systems of particles (e.g., a bat and a ball) interact, they change each other; energy is transferred from one to the other, and we say that the provider of energy does *work* on the recipient of energy.

6.1 Work

The ancient Greeks seem to have had a vague concept of work; it appears just beneath the surface in their explanations of how a large weight could be lifted by exerting a small force on a lever. In the early 1600s, Galileo was only beginning to grope toward the essential idea. He considered the behavior of a pile driver and recognized that the combination of the weight of the hammer and the distance through which it fell determined its effectiveness. Distance and force are linked in some crucial way.

We learned earlier that the product of *force* and the *time* over which it acts equals the *change in momentum* (p. 93). Now, we will see that *the product of force and the distance over which it acts is a measure of the change in energy*. When that idea was formalized by Gaspard Coriolis (1829), he called the product of force and distance **work**. That word is still used, although *energy mechanically transferred* would be more to the point. **Work is the change in the energy of a system resulting from the application of a force acting over a distance.** At the heart of the concept of work is the notion of *movement against resistance*, be it the resistance produced by gravity, or friction, or inertia, or whatever.

Work is the energy transferred into or out of a system through the action of a force.

Work Along a Straight Line

Let's begin with the simplest case of a point-mass, or equivalently, a finite object that is rigid so that every part of it moves in the same way. Figure 6.1 shows a ***constant external force*** (F) exerted on a "rigid" object, moving it (him) through a horizontal distance (l). The applied force is parallel to the straight-line motion, and in that case, a preliminary definition of work (W) is

[straight-line motion: $\vec{\mathbf{F}} \parallel \vec{\mathbf{s}}$] $$W = \pm Fl \qquad (6.1)$$

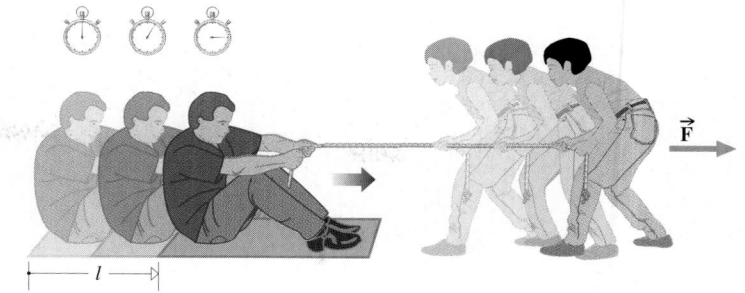

Figure 6.1 A man sitting on a sheet of plastic is pulled a distance *l* along a smooth floor by a woman exerting a constant force $\vec{\mathbf{F}}$. She does an amount of work $W = +Fl$ in the process of overcoming kinetic friction.

Energy cannot be imparted to either a point-mass or a hypothetical rigid object if it remains stationary—there's no way to change such objects—they must move if work is to be done on them.

As we cannot give a general definition of energy, the principle of the conservation of energy simply signifies that there is something which remains constant. Well, whatever new notions of the world future experiments may give us, we know beforehand that there will be something which remains constant and which we shall be able to call energy.

HENRI POINCARE (1854–1912)

As long as the object moves along the line-of-action of the force while the force acts on it, work is being done. Here $\vec{\mathbf{F}}$ is in the direction of motion; it sustains the motion and does positive work on the object. If $\vec{\mathbf{F}}$ was opposite to the direction of motion, it would oppose the motion and do negative work on the object (Fig. 6.2). Of course, force means interaction. The woman in Fig. 6.1 pulls on the rope in the direction of the motion and does positive work. The man pulls back on the rope, in the opposite direction to the motion, and does an equal amount of negative work. As we'll see presently, this effectively means that energy goes from the woman ($W > 0$) to the man ($W < 0$).

The SI unit of work is the *newton-meter*. To be more concise and to honor J. P. Joule (Sect. 13.3), the work done by a 1-N force moving a body through 1 m is defined as 1 *joule* (J), or

$$1 \text{ J} = 1 \text{ N} \cdot \text{m}$$

If the force in Fig. 6.1 is 20 N and the displacement 2.0 m, the work done on the man is 40 N·m, or 40 J. In U.S. Customary Units, still used in engineering, work is expressed in *foot-pounds*, where

$$1 \text{ J} = 0.737 \text{ 6 ft} \cdot \text{lb} \qquad \text{and} \qquad 1 \text{ ft} \cdot \text{lb} = 1.356 \text{ J}$$

Positive work is done on an object when the point of application of the force moves in the direction of the force. If the object experiencing a force does not move, there is no work done. Raise a 100-kg barbell into the air against its downward weight and you do work. Hold it at rest above your head, and you do none.

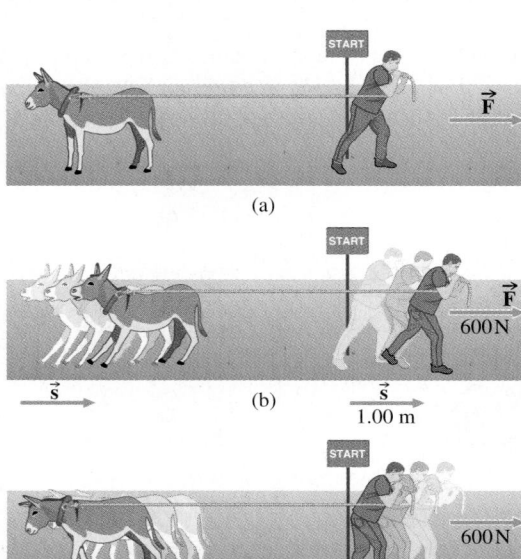

(a)

(b)
1.00 m

(c)

Figure 6.2 (a) The man and the donkey interact via a rope. (b) The man does a positive amount of work (+600 J) in dragging the donkey to the right 1.00 m. (c) Provided he keeps pulling with 600N, he does a negative amount of work (−600 J) as the donkey drags him back to where he started.

Example 6.1 **[I]** A locomotive exerts a constant forwardly directed force of 400 kN on a train that it pulls for 500 m along a straight run. (a) How much work does the engine do on the train (W_{et}) during that period? Coming toward a station, the locomotive slows the train, applying a constant force of 100 kN in the opposite direction. (b) How much work does it then do on the train over a distance of 1000 m?

Solution This problem is about work, where the applied force is parallel to the distance moved. (1) TRANSLATION—An object moves a known distance under the influence of a known force, initially in the direction of motion and later opposite to it; determine the work done. (2) GIVEN: (a) $F = +400$ kN, $l = 500$ m and (b) $F = 100$ kN, $l = 1000$ m. FIND: W_{et} in both cases. (3) PROBLEM TYPE—Energy/work. (4) PROCEDURE—So far there is one defining expression for work, $W = \pm Fl$. (5) CALCULATION—

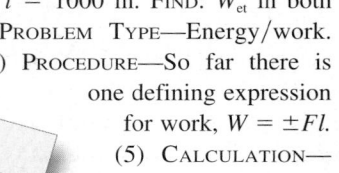

$F = 400$ kN

$l = 500$ m

Part (a)

(a) $\vec{F}$ is in the direction of motion and

$$W_{et} = +Fl = +(400 \text{ kN})(500 \text{ m}) = \boxed{+200 \times 10^6 \text{ J}}$$

The locomotive does 200 MJ of positive work *on* the train. (b) Now $\vec{F}$ is opposite to the direction of motion and

$$W_{et} = -Fl = -(100 \text{ kN})(1000 \text{ m}) = \boxed{-100 \times 10^6 \text{ J}}$$

The engine opposes the motion and does negative work on the train. Equivalently, the train does work ($W_{te} = -W_{et}$) on the engine.

Quick Check: (a) $W_{et} = +4 \times 5 \times 10^7$ J $= 200$ MJ. (b) $W_{et} = -1 \times 1 \times 10^8$ J $= -100$ MJ.

{Make sure to study the WARM-UPS on WORK in CHAPTER 6 on the CD. You should also go over the WALK-THROUGHS before attempting to do the Problems.}

Work in General

Imagine a "rigid" object being pulled along by a person tugging on a rope that makes an angle θ with the direction of motion, as in Fig. 6.3a. Part of the force $\vec{F}$ acts upward (tending to lift the object and lessen the load), and part acts in the direction of the motion and does work. The force is a vector; it has two perpendicular components, one vertical ($F \sin \theta$) and the other horizontal ($F \cos \theta$). The situation is identical to two people pulling on the object (i.e., the seated man) at the same time via two ropes, one up and one forward. Assuming he doesn't lift off the floor, the man only moves horizontally and only the horizontal puller does work.

Equation (6.1) can be generalized by requiring that *the work done on a body by a constant applied force is the product of the component of the force in the direction of the motion multiplied by the distance over which it acts*:

Work can only be done by a force if it is applied to an object that subsequently begins to move or is already moving.

[straight-line motion: $\vec{F} \parallel \vec{s}$]

$$W = (F \cos \theta)l = Fl \cos \theta \tag{6.2}$$

When $\theta = 0$, $\cos 0 = 1$ and this expression reduces to Eq. (6.1).

A working force must to some extent be in, or opposite to, the direction of the motion. While walking around on a flat floor at a constant speed with a barbell above your head

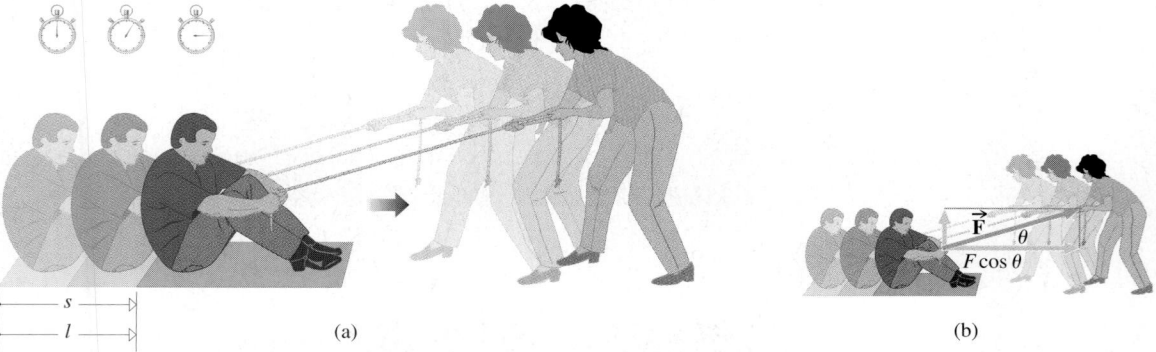

(a)

(b)

Figure 6.3 A man sitting on a sheet of plastic is pulled a distance l along a smooth floor by a woman exerting a constant force $\vec{F}$ at an angle θ with respect to the direction of motion. She does an amount of work $W = +Fl \cos \theta$.

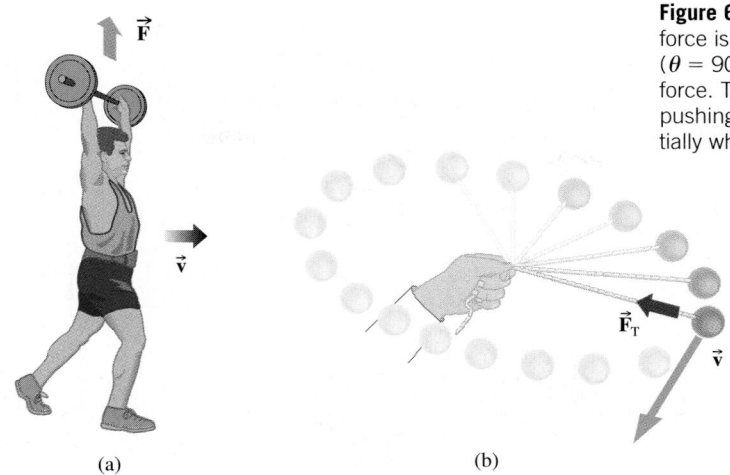

Figure 6.4 Two situations in which a force is perpendicular to the motion ($\theta = 90°$) and no work is done by that force. The man walks horizontally while pushing upward. The ball moves tangentially while being pulled inward.

(a)   (b)

(Fig. 6.4*a*), you do no work on it because the supporting force ($\vec{\mathbf{F}} = -\vec{\mathbf{F}}_w$) is vertical and the motion ($\vec{\mathbf{v}}$) at any moment is always horizontal ($\theta = 90°$). Similarly, Fig. 6.4*b* shows a ball moving in a circle at a constant speed. The tensile force ($\vec{\mathbf{F}}_T$) is perpendicular to the motion ($\vec{\mathbf{v}}$) and $\theta = 90°$; the rope does no work on the ball.

Example 6.2   [I] The truck in Fig. 6.5 is dragging a stalled car up a 20° incline. The tensile force on the towline is constant, and the two vehicles accelerate at a constant rate. If the cable makes an angle of 30° with the road and the tension is 1600 N, how much work was done by the truck on the car in pulling it 0.50 km up the incline?

Solution   This one's about work, where the applied force is not parallel to the distance moved. (1) TRANSLATION—An object moves a known distance under the influence of a specified force at a known angle to the direction of motion; determine the work done. (2) GIVEN: $F_T = 1600$ N, $\theta = 30°$, and $s = 500$ m; the angle of the incline is irrelevant here. FIND: W_{tc}.

(3) PROBLEM TYPE—Energy/work. (4) PROCEDURE—From the definition $W = Fl \cos \theta$. The motion is along the incline, and the component of the force acting in the direction of the displacement is $F_T \cos 30°$, no matter what the tilt of the hill. (5) CALCULATION:

$$W_{tc} = (F_T \cos 30°)l = (1600 \text{ N})(0.866)(500 \text{ m})$$

and

$$\boxed{W_{tc} = 6.9 \times 10^5 \text{ J}}$$

Quick Check: The working force is 1386 N acting over 500 m; hence, $W_{tc} \approx (1.4 \times 10^3 \text{ N})(5 \times 10^2 \text{ m}) \approx 7 \times 10^5 \text{ N} \cdot \text{m}$.

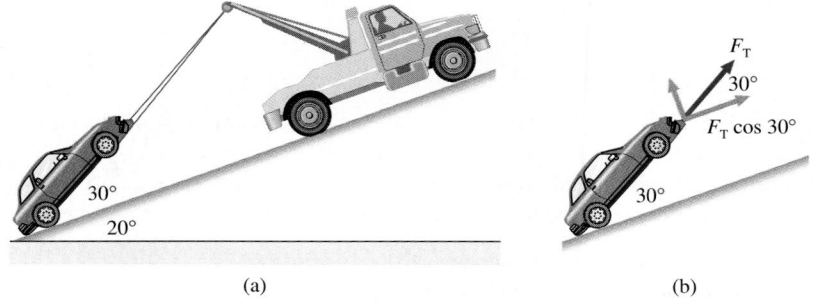

(a)   (b)

Figure 6.5 (a) A tow truck pulling a car up an inclined plane. Only the component of the force parallel to the displacement does work. (b) That working component is $F_T \cos 30°$.

Work Done Overcoming Gravity

In general, work is done to overcome some force, and one of the most common situations involves simply raising a mass against gravity. Figure 6.6 depicts a block being lifted at a

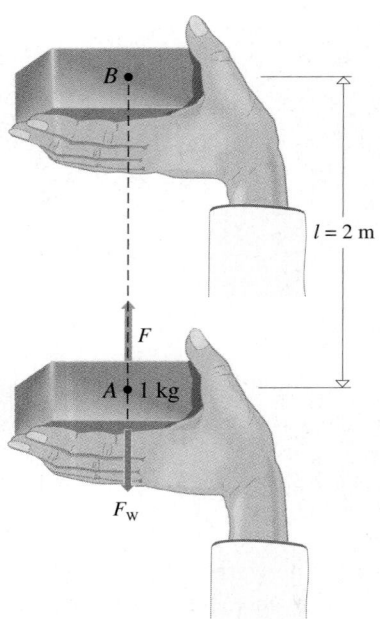

Figure 6.6 The hand slowly raises the block by exerting an upward force F negligibly greater than its weight F_w. Work is done to overcome gravity—that is, to move against the block-Earth gravitational interaction.

$l = 2$ m

constant speed through a vertical distance at the Earth's surface. To raise the block so that $a = 0$, the sum-of-the-forces acting vertically must be zero ($\sum F = ma = 0$), meaning that the applied force $\vec{F}$ is equal in magnitude and opposite in direction to the weight of the block $\vec{F}_w$. (To begin the upward motion, $|\vec{F}|$ must momentarily exceed $|\vec{F}_w|$, but thereafter $|\vec{F}| = |\vec{F}_w|$.) The force is straight up, as is the displacement; $\theta = 0$, and $W = Fl \cos \theta = Fl$. Call the height through which the block is raised h. Then $l = h$ and the work done by the hand on the block, ***the work done against gravity***, is

[overcoming gravity] $$W_g = Fh = F_w h = mgh \qquad (6.3)$$

assuming g to be constant.

Although it's traditional to say that this is the work done on the block, it's best to keep in mind that it really is the work done on the block-Earth system; it arises out of the block-Earth gravitational interaction. In effect, the hand "stretches" the system of interacting entities.

Example 6.3 **[I]** A rigid 1.0-kg block is raised at a constant speed (Fig. 6.6) through a vertical distance of 2.0 m at the Earth's surface. Determine the work done on the block (i.e., on the block-Earth system) by the hand in overcoming gravity.

Solution This problem deals with the work done against gravity. (1) TRANSLATION—An object moves a known distance under the influence of a specified force in the direction of motion; determine the work done. (2) GIVEN: $m = 1.0$ kg, $h =$ 2.0 m, and $g = 9.81$ m/s^2. FIND: W_g. (3) PROBLEM TYPE— Energy/work/gravity. (4) PROCEDURE—By definition $W = Fl \cos \theta$, and since $\theta = 0$, $W = Fl = F_w h$. (5) CALCULATION:

$$W_g = mgh = (1.0 \text{ kg})(9.81 \text{ m/s}^2)(2.0 \text{ m}) \qquad [6.3]$$

and $\boxed{W_{hb} = 20 \text{ J}}$. This is the work done in overcoming gravity.

Quick Check: $W_g / F_w = h = 20$ N·m/9.8 N ≈ 2 m.

In Example 6.3, the hand pushes up on the block and the block pushes down on the hand. The latter force comes from the block-Earth gravitational interaction. The work done by the hand on the block ($W_{hb} = +20$ J $= W_g$) is positive because the force is in the direc-

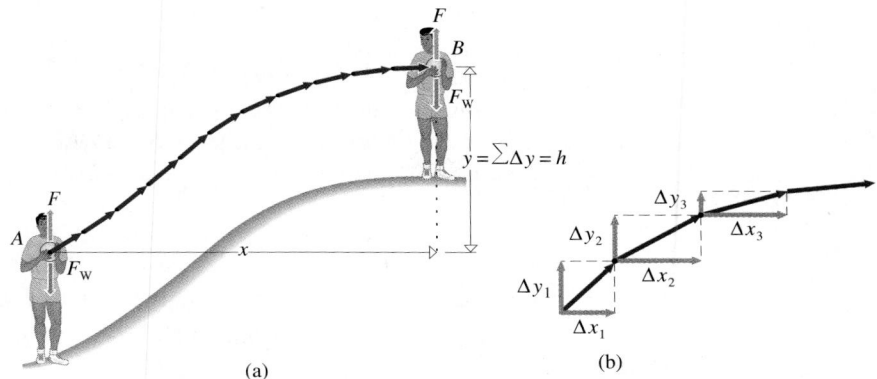

Figure 6.7 The curved path from A to B is divided into tiny straight segments. The small amount of work done during each minute step has the form $\Delta W_g = mg\,\Delta y$. Thus, $W_g(A{\rightarrow}B) = mg\sum\Delta y = mgh$.

tion of the motion; that's the *work done against gravity*. While the block is moved upward, the force it exerts on the hand acts downward; $\theta = 180°$, $\cos\theta = -1$, and the work done by the block on the hand is $W_{bh} = -20\ \text{J} = -W_g$; that's the *work done by gravity*. It's negative because the weight of the block resists the upward motion.

Meandering in the Gravitational Field: Imagine a person holding a bowling ball with a constant upward force $F = F_w$. Suppose that the ball is to be brought up a hill along a smooth path (Fig. 6.7) so that it's at rest at both the initial (A) and final locations (B). Each step carries the ball both upward a small distance Δy (doing a tiny amount of work $\Delta W_g = mg\,\Delta y$) and forward horizontally (doing no work). The total work done in overcoming gravity (going from A to B) is the sum of all these little contributions $W_g(A{\rightarrow}B) = mg\sum\Delta y$. And when the net height raised ($\sum\Delta y$) is h, then $W_g(A{\rightarrow}B) = mgh$. The work depends on the net vertical distance traversed, regardless of what happens horizontally. **In going from A to B in a gravitational field, the work done is independent of the path taken—it is determined solely by the weight of the body and its vertical displacement**. The ball could be carried from A up, down, and around the hill in any zigzag path to B; negative work will subtract from positive work and, in the end, the net work will always be the same: mgh. {For a more rigorous proof click on **WORK & GRAVITY** under **FURTHER DISCUSSIONS** on the **CD**.}

WORK & MUSCLES

Despite the beads of sweat on your brow after hours of trying to lift a barbell, if it hasn't been raised, you have done no work *on it*. Still, you will certainly have generated a good deal of "heat"; your metabolic rate and oxygen consumption will both have gone up. This seeming contradiction can be understood by examining how muscles function. When a muscle contracts, exerting a force over a distance, it does work, but a muscle maintaining a constant tension is doing something as well. The skeletal muscles are composed of bundles of elongated cells known as *fibers*. In response to nerve signals, these cells individually contract (generating a brief pulse of tension) and then relax, all in a matter of a few milliseconds. Each time a fiber contracts it does work on the muscle. The net result of all of these tiny, short-lived contributions is the apparently constant muscle force that may or may not be doing work on some other body.

As a rule, the faster a muscle is called on to respond, the less work it can perform—more work can be delivered to the ball while putting a massive 7.26-kg shot than pitching a baseball.

The Work Done Overcoming Friction

Think about an object (Fig. 6.8) being pushed at a constant speed in a straight line along the floor by an applied force F. Because $a = 0$, the sum of the forces must be zero and F must be equal in magnitude and opposite in direction to the force of kinetic friction, F_f. The latter is the macroscopic average of countless microscopic interactions between the block and the floor—neither is rigid. Regions in contact abrade each other, tiny welds

Figure 6.8 A force F is applied to a block that also experiences a friction force F_f. An amount of work $W_f = F_f l_{cm}$ goes into overcoming friction in the block-floor system, and that energy ultimately appears as thermal energy.

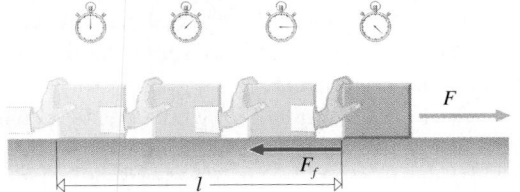

are formed and pulled apart, and work is done. In the process, atoms are jostled and energy is transferred internally to *both* the block and the floor. *There is no way to calculate the work done on just the block by the friction force.* However, **the net work done on the block-floor system is** Fl. This is the total amount of mechanical energy delivered to the ***block-floor*** system, and since $F = F_f$, it follows that $W_f = F_f l$. Insofar as F_f is constant, the ***work done in overcoming kinetic friction*** is, quite generally,

[overcoming friction] $$W_f = F_f l = \mu_k F_N l$$

As we will soon see, the energy transferred into the system by the action of F_f appears primarily as thermal energy in *both* the block and the floor; both get warmer.

Example 6.4 **[I]** A youngster weighing 250 N is sitting on the grass holding on to a large dog via a leash stretched horizontally. The dog pulls on the leash with a force of 100 N and drags the kid at a constant speed 20 m straight across the yard and then stops. How much work did the dog do on the child and the ground in overcoming kinetic friction?

Solution This problems deals with work done against friction. (1) TRANSLATION—An object moves a known distance under the influence of a specified force in the direction of motion; determine the work done.   (2) GIVEN: $F = F_f = 100$ N and $l = 20$ m. FIND: W_f.   (3) PROBLEM TYPE—Energy/work/friction.   (4) PROCEDURE—By definition $W = Fl \cos \theta$, and so because $\theta = 0$, $W = Fl$.   (5) CALCULATION:

$$W_f = F_f l = (100 \text{ N})(20 \text{ m}) = \boxed{2.0 \times 10^3 \text{ N} \cdot \text{m}}$$

Quick Check: $W_f = F_f l = 10^2 \text{ N} \times 2 \times 10^1 \text{ m} = 2 \times 10^3 \text{ N} \cdot \text{m}$.

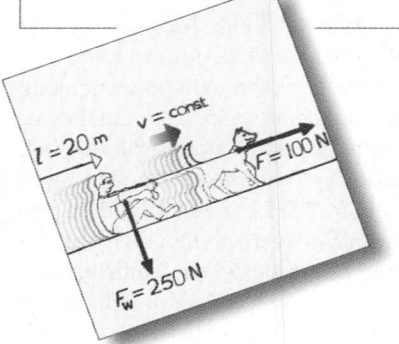

When positive work is done on one system by another, an amount of energy equal to W is transferred to the first from the second. In Example 6.4, the dog was the source of 2.0 kJ that ended up warming the child and the ground.

Since friction usually opposes the motion, $\vec{\mathbf{F}}_f$ ***changes direction*** whenever the motion changes direction, and *the work done against a friction force that opposes the motion is always positive.* If it takes +2.0 kJ to drag the youngster 20 m to the right, it will take another +2.0 kJ to slide him back. *The net work done equals the force applied to overcome friction, times the total path-length (l) traveled.* Work is sometimes defined in terms of displacement, but that can be misleading at this level: If a body experiencing kinetic friction returns to its starting point so that its displacement is zero, the work done in overcoming friction is not zero!

Forces That Change

When we defined work ($W = Fl \cos \theta$), it was assumed that the driving force F was constant. Although there are some important situations where that *is* true, there are many more where it is not. For instance, the force of gravity actually varies as $1/r^2$; the force exerted by a hammer on a nail, or a fist on a nose, will vary with contact distance.

Picture an object being displaced along the x-axis by a varying force $\vec{\mathbf{F}}$ that has a scalar component F_x that does the work. Figure 6.9 is a plot of F_x versus x. We cannot simply plug into Eq. (6.2) to find the work done because the force varies from point to point, but we can use Fig. 6.9, which is a plot of exactly how F_x changes. The area under the curve (F_x times x) is the desired work. Therefore, divide that curve into a number of very narrow intervals: $\Delta x_1, \Delta x_2, \Delta x_3$, and so forth. Assume that the force is constant at some average value over each of these segments.

We have a definition for work done by a constant force; hence, for any segment $\Delta W = F_x \Delta x$. The net work done in going from $x = A$ to $x = B$ is approximately the sum of all these contributions. And that approximation improves as the Δx intervals get narrower and

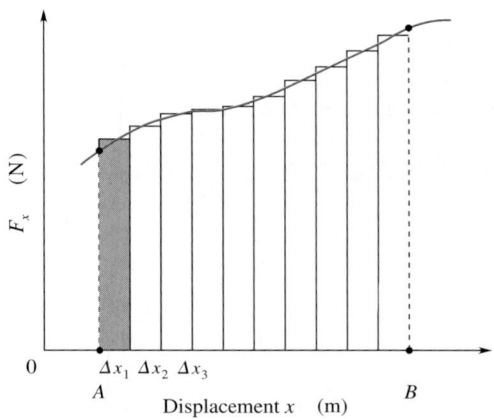

Figure 6.9 A force that varies from point to point in space plotted against distance. The area under the curve is the work done in moving from A to B.

narrower because the average force in each case then more closely approaches the instantaneous force—that is, the corresponding point on the F_x-curve. As these intervals approach zero in width and the number of them increases toward infinity, *the sum approaches the area under the curve between the two points, which is the total work done by the variable force in moving the object from A to B*. If the applied force is positive, the corresponding area is positive; if the force is negative, the area under that portion of the curve is negative. **The work done in going from one point to another equals the total area under the force-distance curve.**

If we can plot the curve, the *work diagram*, we can determine the area and therefore the work. The area can be found directly, by counting boxes on graph paper; mechanically, by tracing the curve with a gadget called a planimeter; or electronically, using a computer. It's common practice when evaluating all kinds of devices, from mechanical hearts to steam engines, to automatically produce work diagrams.

Study the **WARM-UPS** and **WALK-THROUGHS** on the **CD**.

Example 6.5   [[I]] Figure 6.10a shows a mass attached to a spring in a zero-gravity environment. The mass is initially at a height of 0.3 m. It is subsequently pulled down to 0.0 m and released. The force-distance curve for the spring is shown in Fig. 6.10b, where *up* is positive.   (a) Describe the varying force and   (b) compute the work done by the spring on the mass in moving from 0.0 to 0.3 m.   (c) How much work is done by the spring as the mass goes from 0.0 to 0.6 m?

Solution   This example deals with the work done against a spring force.   (1) TRANSLATION—An object moves under the influence of a nonconstant force; determine the work done from the force-distance curve.   (2) GIVEN: The curve. FIND: *W* in the ranges 0.0 to 0.3 m and 0.0 to 0.6 m.   (3) PROBLEM TYPE— Energy/work/force-distance curve   (4) PROCEDURE—Determine the area under the curve between the specified points. (a) The spring-force acts along a vertical line such that it is parallel to the displacement ($\theta = 0$). It follows from the curve that the force varies linearly with the distance from the unstretched

equilibrium position (0.3 m). The force is up and maximum (3.0 N) at the very bottom (0.0 m); zero at equilibrium (0.3 m); and down and maximum (-3.0 N) at the top (0.6 m). From 0.0 to 0.3 m, the spring-force drives the motion, and from 0.3 to 0.6 m, it opposes it.   (5) CALCULATION—(b) The triangular area under the force-distance curve from 0.0 to 0.3 m equals $\frac{1}{2}$(base)(altitude); the work done by the spring on the mass is

$$W_{\text{sm}} = \tfrac{1}{2}(0.3 \text{ m})(3.0 \text{ N}) = 0.45 \text{ N·m}$$

or

$$\boxed{W_{\text{sm}} = +0.5 \text{ J}}$$

(c) Once the mass passes the equilibrium point on the way up, the spring-force opposes the motion. The work done by the spring is negative; the mass does work on the spring. In this case, the area between 0.3 and 0.6 m is negative and $W_{\text{sm}} = -0.5$ J. The total amount of work done by the spring in the complete upward journey is zero.

Quick Check: The total area under the curve is zero.

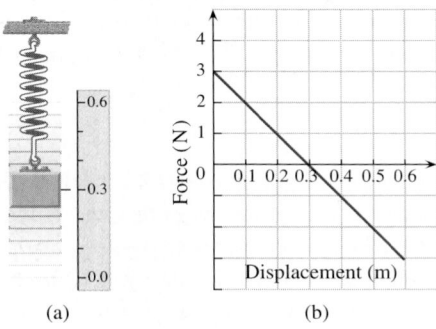

(a)   (b)

Figure 6.10 (a) A mass in a zero-gravity environment attached to a vertical spring. It is pulled down to point 0.0 and released. (b) The plot is a graph of the variable force exerted *by the spring* versus displacement. We'll study the spring-force in detail later; for the moment, notice that it's linear.

Christian Huygens (1629–1695), Dutch mathematician, physicist, and astronomer.

Mechanical Energy

There are four basic forces in nature as it exists today, and when work is done by or against any of them—and that's the only way work can be done—change occurs and energy is transferred.*

All matter interacts gravitationally, resulting in a web of force that binds every entity in the Universe together. It's been suggested that inertia arises from that invisible web (Mach's Principle, 1872). If that's true (and the idea is quite speculative), we can anticipate that a change in the distribution of matter effected by a moving body will correspond to an energy of motion. As a body moves it readjusts its relationship to everything it interacts with.

6.2 Kinetic Energy

Huygens never cared for the idea that the basic measure of motion was momentum. Nonetheless, that view was widely held in the seventeenth century by the followers of Descartes. To be meaningful, momentum ($m\vec{v}$) had to be directional. Two identical cannonballs flying toward each other at the same tremendous speed have momenta that are equal in magnitude and opposite in direction; the net momentum of the two is zero. What kind of essential quality of motion equals zero when the bodies being described are hurtling through the air? That thought did not sit well with Huygens, who searched to find a different measure independent of direction, one that would vanish only when motion ceased. His studies of rigid colliding balls led him to conclude that there was something special about the product of mass and velocity-squared. Remarkably, adding the values of mv^2 for each ball prior to a collision yielded a total that was essentially the same after the collision, even though all the velocities had changed. Squaring the velocity removed any dependence on the sign of v (i.e., on the direction of $\vec{v}$); mv^2 is always positive and only vanishes when v vanishes.

Leibniz picked up on the idea and showed that, for a falling body, mv^2 was proportional to Galileo's product of weight and height (see Discussion Question 19). It was not until 1807 that Thomas Young, an English physicist and physician, spoke of mv^2 for the first time as *energy*. He concluded that "labour expended in producing any motion, is proportional to the energy which is obtained." In other words, work that causes motion equals the resulting change in energy.

Work and the Change in KE

Today, we call the energy associated with motion **kinetic energy** (KE), a term introduced (1849) by Lord Kelvin. Imagine a *rigid* body moving under the influence of a constant force acting parallel to the motion (Table 6.1). Given that it travels a distance l, the work done on the body is $W = Fl$ and, since $F = ma$, $W = mal$. The acceleration is constant and so from Eq. (3.10), $v_f^2 = v_i^2 + 2al$. Therefore, $al = \frac{1}{2}v_f^2 - \frac{1}{2}v_i^2$, and multiplying both sides by m yields

[overcoming inertia]
$$W = \tfrac{1}{2}mv_f^2 - \tfrac{1}{2}mv_i^2 \qquad (6.4)$$

When Coriolis first carried out this analysis, he knew he had come upon something important because, except for the $\frac{1}{2}$, this was Huygens's mv^2. Under the influence of a net force ($\theta = 0$), a body accelerates as positive work is done on it to overcome inertia. It increases its speed and gains kinetic energy (KE) in the process. *We define the kinetic energy of any object of mass m traveling at speed v as*

$$\mathrm{KE} = \tfrac{1}{2}mv^2 \qquad (6.5)$$

Wind blows against the sails, exerting a force F and driving the ship forward. Air and water resistance produce a net drag F_f on the craft. When $F > F_f$, $\sum F = F - F_f = ma$, the ship accelerates, and its KE increases.

Kinetic energy is the energy of motion of any moving entity.

---

* The common dictionary definition, energy is the ability to do work, is at best misleading (see Discussion Question 11). We will learn that the First Law of Thermodynamics maintains that energy is conserved, while the Second Law tells us that the availability of energy to do work tends to decrease as processes occur.

Table 6.1
Rough Values of the Energies of Various Occurrences

Occurrence	Energy (J)
Creation of the Universe	10^{68}
Emission from a radio galaxy	10^{55}
Supernova explosion	10^{44}
Yearly solar emission	10^{34}
Earth moving in orbit	10^{33}
Earth spinning	10^{29}
Earth's annual sunshine	10^{25}
Yearly U.S. sunshine	10^{22}
Annual tidal friction	10^{20}
Exploding volcano (Krakatoa)	10^{19}
Severe earthquake (Richter 8)	10^{18}
100-megaton H-bomb	10^{17}
Burning a million tons of coal	10^{16}
Hurricane	10^{15}
Thunderstorm	10^{15}
Atomic bomb (Hiroshima)	10^{14}
Saturn V rocket	10^{11}
Lightning bolt	10^{10}
Burning a cord of wood	10^{10}
One gallon of gasoline	10^{8}
One day of heavy manual labor	10^{7}
Explosion of 1 kg of TNT	10^{6}
Woman running for 1 h	10^{6}
Burning match	10^{3}
Hard-hit baseball	10^{3}
Apple falling 1 m	1
Human heartbeat	0.5
Depressing typewriter key	10^{-2}
Cricket chirrup	10^{-3}
Hopping flea	10^{-7}
Proton in supercollider	10^{-7}
Fission of 1 uranium nucleus	10^{-11}
Electron mass-energy	10^{-13}
Electron in atom	10^{-18}
Photon of light	10^{-19}
KE of room-temperature air molecule	10^{-21}

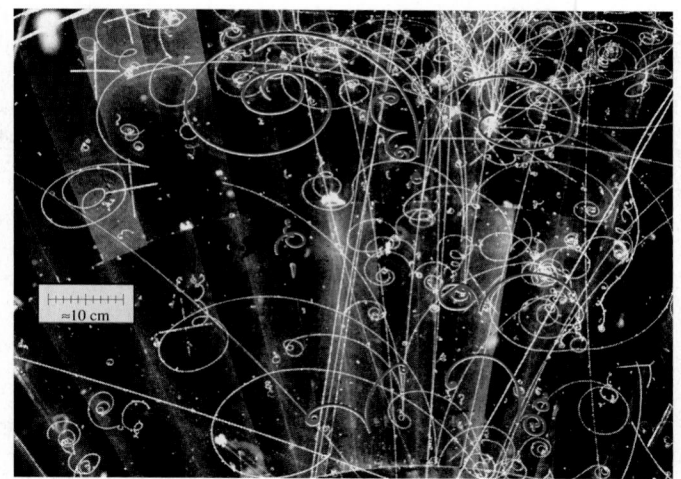

The liquid filling this chamber is a mix of neon and hydrogen which forms tiny bubbles along the paths of charged particles (p. 535). A powerful magnetic field causes these particles to move in circles. Photons (most streaming in from the bottom) leave no tracks, but occasionally, they create an electron-positron pair (several of which are highlighted in red and blue). These low-mass particles immediately collide with the liquid neon atoms and lose energy rapidly. As their **KE** diminishes, they spiral inward (electrons winding clockwise, positrons counterclockwise). Sometimes a photon strikes an atom and knocks out a single electron (several of which are highlighted in red). They also lose energy, slow down and spiral inward. Keep in mind that you are looking at tracks in three dimensions.

In that way Eq. (6.4) is the difference between the final and the initial KE:

$$W = \text{KE}_f - \text{KE}_i = \Delta\text{KE} \tag{6.6}$$

From the derivation, the units of KE are the units of W, namely, *joules*. That can be confirmed by the following: $F = ma$, $1\text{ N} = 1\text{ kg·m/s}^2$, and $W = Fl$; hence, $1\text{ J} = 1\text{ N·m} = 1\,(\text{kg·m/s}^2)\text{m} = 1\text{ kg(m/s)}^2$. *All forms of energy will be measured in joules* (Table 6.1).

Example 6.6 suggests a fundamental point: KE *is a relative quantity*; we effectively choose the zero of KE by selecting the coordinate system with respect to which v is measured. Here the Earth was the motionless reference frame, and ground speed was used in the calculation. If you were in that jet sitting at "rest" next to your 20-kg suitcase, it would have no KE with respect to you and yet, to someone at "rest" on the ground watching, your bag would flash past with almost 3×10^3 J. *As a rule, we are not interested in the total energy*

Example 6.6 **[I]** A Boeing 747 airliner, weighing 2.2×10^6 N (i.e., 5.0×10^5 lb) at takeoff, is cruising at a ground speed of 268 m/s (i.e., 600 mi/h). Compute its kinetic energy. If 1 kg of TNT yields 4.6×10^6 J, how much TNT is the plane's KE equivalent to?

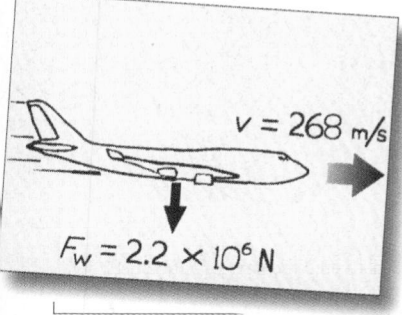

$v = 268$ m/s

$F_w = 2.2 \times 10^6$ N

Solution The mention of "kinetic energy" should bring to mind its definition in terms of mass and speed. (1) TRANSLATION—An object of known mass moves at a known speed; determine its kinetic energy. (2) GIVEN: $F_w = 2.2 \times 10^6$ N and $v =$ 268 m/s. FIND: KE. (3) PROBLEM TYPE—Energy/kinetic energy. (4) PROCEDURE—By definition, $\text{KE} = \frac{1}{2}mv^2$. (5) CALCULATION—First compute the mass; $m = F_w/g = (2.2 \times 10^6\text{ N})/(9.81\text{ m/s}^2) = 2.24 \times 10^5$ kg. Then

$$\text{KE} = \tfrac{1}{2}mv^2 = \tfrac{1}{2}(2.24 \times 10^5\text{ kg})(268\text{ m/s})^2$$

and

$$\boxed{\text{KE} = 8.0 \times 10^9\text{ J}}$$

This is equivalent to the amount of energy given off when $\boxed{1.7 \times 10^3\text{ kg}}$ of TNT explode.

Quick Check: Run the calculation backwards; $\text{KE} = 8.0 \times 10^9\text{ J} = \frac{1}{2}mv^2$; $v^2 = 2(8.04 \times 10^9\text{ J})/(2.24 \times 10^5\text{ kg})$; $v = 268$ m/s.

NO WORK DONE, NO CHANGE IN KE

There can be forces acting and yet no work done—we saw this with a ball whirling in a circle at the end of a string (Fig. 6.4). The speed is constant, and the string is perpendicular to the motion ($\vec{F}_T \perp \vec{v}$); $\theta = 90°$, $W = 0$, and $\Delta KE = 0$. To increase the ball's speed there must be a component of $\vec{F}_T$ parallel to $\vec{v}$; $\theta \neq 90°$, $W \neq 0$, $\Delta KE \neq 0$, and the string pulls the ball forward ($W = \Delta KE$).

of a system but in the energy added to or taken from it. That's appropriate because the total energy of a system is a relative quantity; it depends on the observer (Table 6.2).

The Work-Energy Theorem

Consider a rigid body that does work or has work done on it to overcome only inertia (i.e., to accelerate it); it doesn't experience friction, nor does it rise or fall in a gravitational field.

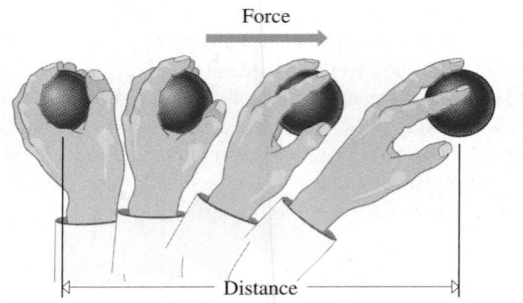

Under these conditions the net work done equals the body's change in kinetic energy (Eq. 6.6). This rather restricted and fairly unremarkable conclusion is, for historical reasons, widely referred to as the **Work-Energy Theorem** (Fig. 6.11).

Your arm can do work on a bowling ball and thereby increase the ball's kinetic energy ($W = \Delta KE$). The ball, going from rest to some launch speed, travels down the lane, crashes into the pins, does work on them, and loses a corresponding amount of KE. A moving object can do a maximum amount of work equal to $\frac{1}{2}mv^2$. Work → KE → Work, the transformation goes either way. Every system that moves transports energy in the form of KE from one place to another.

Figure 6.11 A force acting over a distance results in a change in kinetic energy.

The Work-Energy Theorem is really an intermediate step in the grand scheme of Conservation of Energy (p. 187) and so, by itself, can sometimes be misleading—we'll just have to be careful using it. *The problem with the Work-Energy Theorem is that it strictly pertains only to rigid bodies and so cannot indiscriminantly be applied to ordinary objects that are more or less deformable.* For example, throw a wad of soft clay against a stone wall. The clay comes to a stop, ΔKE can be quite large, but W_{wc} is, for all practical purposes, zero. The wall exerts forces on the clay across the essentially stationary clay-wall interface. The wall is not perfectly rigid, and there is a tiny motion of the points of application of force as the wall distorts, but $W_{wc} \approx 0$. All of this aside, the Work-Energy Theorem works well for fairly rigid things and that's how we'll apply it.

We know that the range of a projectile depends on its initial speed. So this athlete will try to give the shot the highest launch speed she can; that means the largest initial **KE** possible. To produce the largest Δ**KE** she will have to maximize the work she does on the shot; she will exert as much force as she can deliver, over as great a distance as possible.

Table 6.2

Approximate Kinetic Energy and Impact Likelihood of Earth-Striking Meteors		
Expected frequency (once every)	**Meteor radius** (meters)	**Impact energy** (kilotons of TNT)*
Month	1	0.74
	2	5.9
Year	3	20
	4	47
	5	93
Decade	10	7.4×10^2
Century	20	5.9×10^3
	30	20×10^3
10^3 years	50	93×10^3
10^4 years	100	74×10^4
10^6 years	1000	7.4×10^8

*The atom bomb dropped on Hiroshima in 1945 delivered the energy of about 12.5 kilotons of TNT (where 1 kiloton $\approx 4 \times 10^{12}$ J).

The dogs pull on the sled and do work on it, W_{ds}, over some displacement. This energy goes into overcoming drag due to both the air and snow, W_f. If the sled accelerates over the run, $W_{ds} = W_f + \Delta KE$; if it continues at a constant speed, $W_{ds} = W_f$.

Example 6.7 **[II]** According to the record books, Aleksandr Zass (known to his admirers as "Samson") would, when he wasn't bending iron bars, catch a 104-lb (i.e., 463-N) woman fired from a cannon at around 45 mi/h. Assuming that Samson brought her to rest uniformly in a distance of 1.00 m, compute the average force he exerted on our heroine. As a guess, take her "landing speed" to be 8.94 m/s (i.e., 20 mi/h). Only a negligible amount of energy went into overcoming friction.

Solution This example talks about force, distance, and speed, and so it relates work and KE. (1) TRANSLATION—An object of known mass moving at a known speed is brought to rest over a specified distance by an average force, which is to be determined. (2) GIVEN: $F_w = 463$ N, $v = 8.94$ m/s, and $l = 1.00$ m. FIND: F_{av}. (3) PROBLEM TYPE—Energy/kinetic/work. (4) PROCEDURE—Use the Work-Energy Theorem. To find the force,

we require the work done in stopping her, which equals her KE at impact. (5) CALCULATION—To find her KE, we need her mass: $m = F_w/g = (463\ \text{N})/(9.81\ \text{m/s}^2) = 47.2$ kg. Then

$$\text{KE} = \tfrac{1}{2}mv^2 = \tfrac{1}{2}(47.2\ \text{kg})(8.94\ \text{m/s})^2 = 1.89\ \text{kJ}$$

Samson had to do 1.89 kJ of work on the woman to bring her to rest; hence, $W = F_{av}l = \Delta \text{KE}$ and

$$F_{av} = \frac{\Delta \text{KE}}{l} = \frac{1.89 \times 10^3\ \text{J}}{1.00\ \text{m}} = \boxed{1.89\ \text{kN}}$$

Quick Check: Let's rework the problem in different units. $\text{KE} = \tfrac{1}{2}[3.3\ \text{lb}/(\text{ft}/\text{s}^2)](29\ \text{ft/s})^2 = 1.4 \times 10^3\ \text{ft·lb}$. Converting that to joules, $(1.4 \times 10^3\ \text{ft·lb})(1.356\ \text{J/ft·lb}) = 1.9$ kJ. Alternatively, using Eq. (3.10), $a_{av} = -v_0^2/2l = -39.96\ \text{m/s}^2$ and $F_{av} = ma_{av}$.

Note that doubling the speed of an object (the young lady fired from the cannon or perhaps the family car on the highway) quadruples the kinetic energy and also quadruples the work needed either to get it up to speed or to stop it. Insofar as the braking force on a car is fairly constant once the brakes are engaged, it will take about four times the distance to stop an auto traveling at 40 km/h as it takes for one moving at 20 km/h.

6.3 Potential Energy

Imagine a force that is *continuously* exerted on a body, for example, the gravitational interaction producing the body's weight. To move upward against that downward pull requires the application of a counterforce (provided by leg muscles or elevators) and the doing of work. The crucial point is that the force will continue to act after the displacement. A body raised in a gravitational field still experiences a downward force while held up there at rest. When let loose, it will fall; it will be driven back down toward where it came from, and kinetic energy will be imparted to it in the process.

All of that is fine; work done on the system is ultimately converted into KE. But what's happening while the body is held motionless, high in the gravitational field? KE is not "liberated" until the body is released. Apparently, it is possible to do work on a system and not have it immediately appear as KE. Yet the potential for generating that energy is there because of the ever-present force. *Energy is stored in the system of interacting objects,*

What is matter? The paper on which I am writing is matter, the air is still matter, but even light has become matter now, due to *Einstein's* discoveries. It has mass and weight; it is not different from ordinary matter, it too having both energy and momentum. The only difference is that light is never at rest, but always moving with the same characteristic velocity.

WOLFGANG PAULI
NOBEL LAUREATE 1945

waiting to be let loose. This *retrievable stored energy, energy by virtue of position or configuration in relation to a force*, is known as **potential energy** (PE), a name suggested by W. Rankine in the mid-1800s. The concept of PE gives continuity to the idea of energy; W done on a body against a force such as gravity goes into changing its PE.

The change in the potential energy of a body incurred in moving from one point to another equals the work done in overcoming the interaction (gravitational, electromagnetic, strong, or weak) that stores the energy. PE exists only in relation to objects that interact in a sustained way—no interaction, no PE.

Potential energy is energy stored through the interaction of two or more material objects.

Gravitational-PE

When a painter climbs a flagpole, she does work to overcome the downward pull of gravity. Work corresponds to a change in energy, and we associate this change with the *gravitational*-PE; inasmuch as she is motionless, high on her perch, $W = \Delta PE_G$. This is the work she does in overcoming gravity and it equals the product of the force she exerts (i.e., a force insignificantly greater than her weight F_W) and the vertical height through which she ascends, Δh:

$$\Delta PE_G = F_W \Delta h \tag{6.7}$$

Given that the painter has a mass m, we can write Eq. (6.7) in terms of her initial (h_i) and final (h_f) heights independent of the path taken (p. 177);

$$\Delta PE_G = mg\Delta h = mg(h_f - h_i) \tag{6.8}$$

Raising his mass, this climber increases the *gravitational*-PE stored via the interaction between him and the Earth.

Potential energy, like KE, is a relative quantity; even the idea of "height" is relative. How high is your own nose? Above what—your lip, the ground, or perhaps sea level? Any of these would do; **the zero-reference level of PE_G is arbitrary** and can be taken wherever it's convenient. We simply say that the height at some level is zero and take the PE_G with respect to that level. Usually, we are concerned only with changes in PE_G due to changes in altitude, which then removes the need to deal with the zero-reference level. In fact, **the only thing we are able to measure is the change in** PE_G, **the change in energy.** That tells us something very profound about the nature of the Universe (p. 198).

It's useful to have an expression for the PE_G of a body at any height h above the chosen zero-reference, especially if that reference is a commonly used one, such as the surface of the planet. Equation (6.8) can be rewritten as

$$\Delta PE_G = PE_{Gf} - PE_{Gi} = mg(h_f - h_i)$$

which suggests that $PE_{Gf} = mgh_f$ and $PE_{Gi} = mgh_i$. Make h_i the reference level (i.e., $h_i = 0$ and $PE_{Gi} = 0$). Simplify the notation by setting $PE_G = PE_{Gf}$ (at any elevation $h = h_f$ measured from $h = 0$). The desired expression is then

The ***gravitational*-PE** of a body at the surface of the Earth arising from its interaction with the planet.

[g assumed constant]

$$PE_G = mgh \tag{6.9}$$

The work done raising a mass equals the increase in its potential energy [Eq. (6.3)].

PE, RELATIVITY, & "REALITY"

Many physicists and philosophers struggled with the concept of potential energy during the past century. The question is, "Is PE 'real'?" or is it just an accounting device? As J. W. N. Sullivan (1933) put it, "Potential energy, it must be admitted, is a somewhat mysterious notion. Other forms of energy, such as energy of motion and heat energy are obviously 'energetic.' But potential energy is undetectable until it is transformed." If it is indeed "undetectable," is PE no more than a mental construct? The solution to this quandary seems to lie in the Special Theory of Relativity, which maintains that the mass of a system depends on its potential energy. If an object is raised into the air by some external agency, the mass of the Earth–object system increases; that increase is (at the present) undetectably small, but it is presumably real enough. On the other hand, if you raise the object, energy goes from chemical to gravitational, but there is no net change in the energy of the Earth-you-object system and no change in its overall mass. Still, your mass will decrease as the mass of the object increases. Perhaps whenever we talk about PE we should be thinking Δm.

Don't let any of this bother you; on the contrary, it's part of the delight in studying Physics that we're still sorting out the basics.

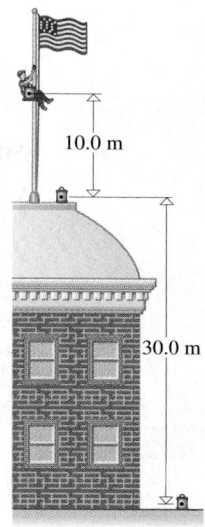

Figure 6.12 A flagpole painter has *gravitational*-PE given by $PE_G = mgh$. Here, h is measured with respect to any level taken as zero. Only the change in PE_G is measurable, so the location of the zero level is arbitrary.

10.0 m

30.0 m

Example 6.8 **[II]** The flagpole painter in Fig. 6.12 is carrying a 2.00-kg can of white paint. She has climbed (raising the can 10.0 m) up the pole from its base, where she just finished having lunch on the roof of a 30.0-m-tall tower. (a) What is the increase in the potential energy of the can just as a result of her climbing the pole? (b) What is the total increase in the can's potential energy above the value it had while it sat on the ground?

Solution The mention of "potential energy" should bring to mind its definition in terms of mass and height. (1) TRANSLATION—An object of known mass is raised a specified distance; determine the increase in *gravitational*-PE. (2) GIVEN:

Heights of 30.0 m and 10.0 m, and $m = 2.00$ kg. FIND: Appropriate PEs. (3) PROBLEM TYPE—Energy/gravitational potential energy. (4) PROCEDURE—Use Eq. (6.8) to compute the change in the PE_G with height. (5) CALCULATION—(a)

$$\Delta PE_G = mg\Delta h = (2.00 \text{ kg})(9.81 \text{ m/s}^2)(10.0 \text{ m}) = \boxed{196 \text{ J}}$$

(b) For a rise of 40.0 m

$$\Delta PE_G = mg\Delta h = (2.00 \text{ kg})(9.81 \text{ m/s}^2)(40.0 \text{ m}) = \boxed{785 \text{ J}}$$

Quick Check: Raising the can from the ground ($PE_G = 0$) to the roof increases its PE_G by $(2.00 \text{ kg})(9.81 \text{ m/s}^2)(30.0 \text{ m}) = 588.6$ J. That plus another 196.1 J equals 785 J.

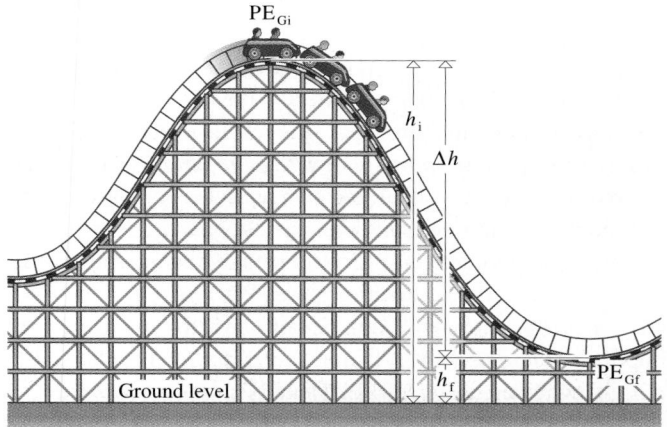

PE_{Gi}

h_i

Δh

h_f

PE_{Gf}

Ground level

Figure 6.13 A roller coaster showing a car at its highest point, where the potential energy is PE_{Gi}. The zero-PE_G level can be located anywhere. When the car drops to the lowest point, it has a potential energy of PE_{Gf}. If PE_{Gi} is set equal to zero, the PE_G anywhere on the slope is negative; thus, $PE_{Gf} = -mg\Delta h$.

Depending on where the zero-reference is chosen to be, PE_G can be positive or negative. Figure 6.13 shows a car on a roller coaster. It starts at the top and drops to the first low point. If we take the ground level as the zero of *gravitational*-PE, $\Delta PE_G = (mgh_f - mgh_i) = -mg\Delta h$, where Δh is positive. PE_G decreases, ΔPE_G is negative, but each value of PE_G is positive. If, however, we take the very top to be the zero-PE_G level, then $PE_{Gi} = 0$ and $\Delta PE_G = PE_{Gf} = -mg\Delta h$. The PE_G is negative everywhere on the ride, but the change in potential energy is the same as before, when the ground was the zero level.

But Gravity Varies with Distance: Equation (6.9) is true provided that g is constant; yet as seen in Eq. (5.7) it isn't! So $PE_G = mgh$ is an approximation, a fine one when h is small compared to the radius of the Earth, but an approximation all the same. This circumstance arises because an object's weight is a function of distance ($F_G = GmM/r^2$). The work done in moving against the nonconstant gravity force can be determined from the area under an appropriate portion of the force-distance curve. Such an analysis yields

$$PE_G = -\frac{GmM}{r} \tag{6.10}$$

A descending roller-coaster car converts *gravitational*-PE into KE. An ascending car does just the reverse.

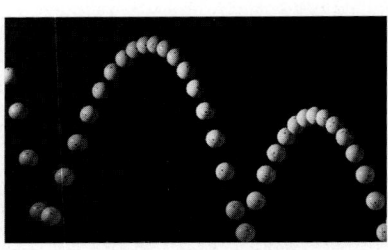

The loss of energy generally runs a system down—in this case, a bouncing golf ball. With each impact, a certain amount of energy goes into random thermal motion, warming the ball and floor; less is returned as organized KE to the ball, so it rises more slowly and doesn't bounce as high. The closer the images of the ball, the slower it is moving.

which is the *gravitational*-PE of a mass m at a distance r from the center of a mass M, which might be a planet, or star, or anything else. It's customary when dealing with an attractive force to choose the zero of PE to be at $r = \infty$, and that was done here. The farther m is from M, the larger is r and the larger is the PE_G, inasmuch as it gets to be a less negative number. Since PE_G is negative, zero is its greatest value. {For a rigorous derivation and a proof that $\Delta PE_G \approx mgh$ click on ***GRAVITATIONAL*-PE** under **FURTHER DISCUSSIONS** on the **CD**.}

Internal Energy & Conservative Forces

It is convenient to distinguish two basic forms of energy: KE (energy of motion) and PE (energy of position). For objects that have structure (that is, interacting parts), it's helpful to further recognize two manifestations of each of these forms called *internal* and *external*. A bowling ball at rest has all its atoms randomly vibrating about their equilibrium positions at fairly high speeds (typically as much as 0.4 km/s). ***That disorganized internal kinetic energy is known as* thermal energy.** A bowling ball flying through the air has an organized motion superimposed on all its jiggling atoms, and we have called this external energy, *kinetic energy*. Similarly, a bowling ball in the air has "external" potential energy by virtue of its interaction with the Earth. When the ball is distorted (for example, during a collision) it has *elastic energy*, one of several kinds of internal PE.

Thermal energy is the random KE of the individual atoms and molecules of a system.

Imagine two structureless particles interacting gravitationally. As we have seen, the work done in bringing a mass from any point A to any point B is independent of the path taken. It follows that moving a particle from A to B and back to A results in zero work done. That feature, namely, that ***zero work is done in moving a particle around a closed path***, defines a **conservative force**. To be conservative, the force on a particle at any given location in space must be constant, no matter how or *when* the particle gets there. Whatever energy goes into such an interaction on moving from A to B, comes out on moving from B to A; the process is reversible. The conservative interaction stores PE.

A force is **conservative** if it does no net work in the process of moving a particle once around a closed path, returning it to the point at which it started.

As we continue our study, we will encounter a number of ways in which the Four Forces act to store PE both internally and externally. A stretched garter retains *elastic*-PE; a bright red stick of dynamite is gift-wrapped *chemical*-PE; a battery supplies *electrical*-PE; two magnets pulled apart wait to reunite with *magnetic*-PE; and all those neatly painted warheads hiding in their silos threaten with *nuclear*-PE.

Internal Energy Sources & Work

The work done on passive objects (crates, blocks, balls, and sailboats) by externally applied forces has been our primary concern thus far. Now, consider self-propelled objects (cars, people, helicopters, frogs, etc.) that have their own internal energy sources. Each can be accelerated by a net external force ($F = ma$), a reaction force, arising from the object's interaction with the environment. *As a rule, such a force does no net positive work on the active nonrigid body, and $W \neq \Delta$KE.* We'll see presently that no energy is transferred to the body from the environment via the reaction force, even though that force accelerates the body.

A swimmer's hand (Fig. 6.14*a*) pushes back on the water, and the water pushes forward on the hand, accelerating the person. But the hand is moving backward all the while. The hand does positive work on the water; the force it exerts is in the same direction as the displacement. By contrast, the water pushes in the forward direction on the hand. It does negative work on the swimmer: $W_{ws} < 0$ even though ΔKE > 0. The water gets energy from the swimmer—it gets warmer. The swimmer is self-powered and uses the water to generate a reaction force so that she can accelerate. Just imagine a motorboat suspended above the water with its engine running at full speed, going nowhere. All the energy needed is provided by the fuel, but without the water to push on, the boat cannot accelerate.

The energy to walk, climb (Fig. 6.14*b*), skate, or jump comes from internal energy stored in the person (see Discussion Questions 4 and 5). When you jump (Fig. 4.19), the upward reaction force that accelerates you acts at the stationary foot-floor interface. If the floor is rigid, *there is no motion of the point of application of the force, and no work is done by the floor on you*: $W_{fu} = 0$ even though ΔKE > 0. Actually, the floor will sag slightly while it exerts a normal force on you, and W_{fu} is positive though very small, since *l* is very small.

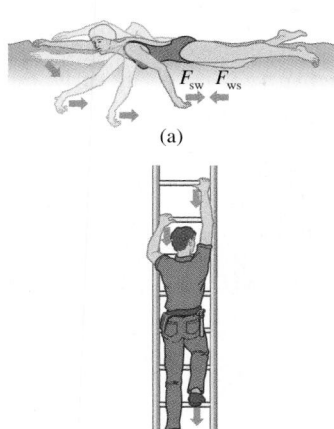

Figure 6.14 (a) A swimmer pulls on the water with a force F_{sw}, drawing it back a distance *x* and doing positive work on it. The forward reaction of the water on the swimmer's hand (F_{ws}) is opposite to the displacement of the hand, and the work done on the swimmer is negative. Energy, supplied by her food, is transferred to the water. (b) A person climbing a ladder exerts forces down on the rungs, but the point of application does not move and no work is done on the ladder or on the climber. The increase in PE_G comes from internal energy.

Conservation of Mechanical Energy

It was Descartes (1644) who boldly proclaimed that the total amount of momentum in the Universe was constant and could only be changed by God (p. 216). And so when Huygens sought to replace *mv* as the essential measure of motion, it was natural for him to look for a quantity (mv^2) that was likewise conserved (i.e., constant in time). The realization that there were various interchangeable forms of an otherwise unchanging total energy had already been proposed by several people before Rankine (1853) provided us with the ringing phrase "Conservation of Energy." That era gave rise to the study of Thermodynamics with its premier insight that heat was another form of energy. Yet it was not until 1905 that Einstein added the final piece to the construct. He showed that mass is equivalent to energy; that by virtue of its very existence, a particle of mass possesses a store of energy. Because an immense amount of energy is equivalent to a minuscule amount of mass, that equivalence was only first observed in the twentieth century. Accordingly, although we really ought to be talking about the conservation of *mass-energy,* we needn't worry about that until later when we get into atomic physics.

The grandest generalization in all of physics is the fully developed statement of the Law of Conservation of Energy:

Conservation of Energy maintains that energy can be transformed, but it can neither be created nor destroyed.

> **The total energy of any system that is isolated from the rest of the Universe remains constant, even though energy may go from one form to another within the system.**

Assuming that *the energy of the Universe is constant*, the energy of any portion of the Universe isolated from the rest must also be constant. If there is no flow of energy into or out of a system (and that's what is meant by *isolated*), then there can be no change in energy either there or in the remainder of the Universe.

6.4 Mechanical Energy

Define the **mechanical energy** (E) of a system as *the sum of the* KE *and gravitational*-PE *of all its parts*. Now suppose the system does an amount of work (W_{out}) in overcoming some applied force (other than gravity), such as friction. The mechanical energy is diminished by an amount $W_{out} > 0$, and therefore $E_i > E_f$.

$$W_{out} = \Delta E = \Delta KE + \Delta PE_G \tag{6.11}$$

$$W_{out} = \Delta E = E_i - E_f$$

Alternatively, if no additional forces are applied, W_{out} is zero, no energy is transferred into or out of the system, $E_f = E_i$, and *mechanical energy is conserved*. This is an early, limited formulation of what would later develop into the concept of conservation of all forms of energy.

Assuming that g is constant, Eq. (6.11) becomes

[a nongravitational force] $$W_{out} = (\tfrac{1}{2}mv_f^2 - \tfrac{1}{2}mv_i^2) + (mgh_f - mgh_i)$$

When the only force acting is gravity (which is accounted for via ΔPE_G), $W_{out} = 0$ and

[only gravity] $$\tfrac{1}{2}mv_f^2 + mgh_f = \tfrac{1}{2}mv_i^2 + mgh_i \tag{6.12}$$

or $$KE_f + PE_{Gf} = KE_i + PE_{Gi}$$

and $$KE + PE = E = \text{constant} \tag{6.13}$$

When the only external force acting on a system is gravity, the **mechanical energy** of the system is constant.

➤ *At every instant in the motion, the total mechanical energy is constant*; if the body's KE increases, its PE_G must decrease, and vice versa.

Fire a ball straight up into the air by doing work on it, giving it a kinetic energy of 100 J (Fig. 6.15a). Let's focus on the launched projectile. Once fired (ignore air friction), no

Modern race cars are designed to break up in a high-speed crash. The flying pieces carry away much of the original **KE**. With less energy available to mangle it, the driver's compartment will be more likely to come to rest intact.

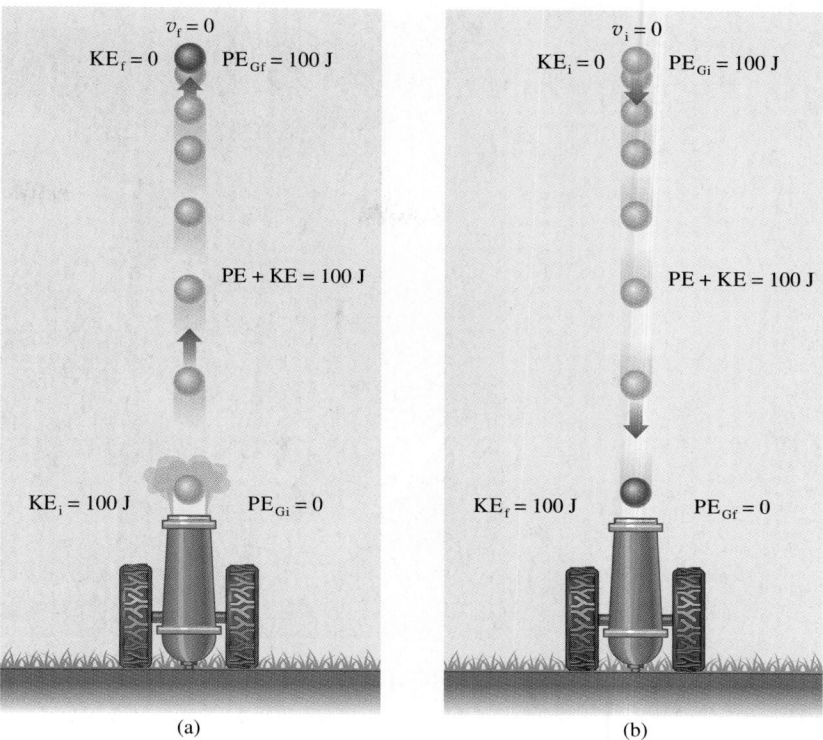

Figure 6.15 (a) A cannonball fired upward loses KE as it rises and gains PE$_G$. Fired with a KE$_i$ = 100 J, it will always have a total energy of 100 J. Of course, in real life, some energy will be lost to air friction. (b) When the ball drops from rest, it loses PE$_G$ and gains KE.

forces other than gravity act and no work is expended overcoming them, $W_{out} = 0$. At $h_i = 0$, PE$_G$ = 0, while the KE, equal to 100 J, is a maximum (equal to E). As the ball rises, increasing in PE$_G$, it slows down, decreasing in KE such that at any moment KE + PE$_G$ = E = 100 J. At peak altitude, where the ball momentarily comes to rest, KE = 0, and PE$_G$ = 100 J is a maximum (equal to E). As it falls back (Fig. 6.15b), PE$_G$ decreases to zero, its speed increases, and the KE attains a maximum value of 100 J (equal to E) as it returns to the point from which it was launched. The similar situation of a swinging pendulum is depicted in Fig. 6.16. {To examine this point in an animated environment click on **THE PENDULUM** under **INTERACTIVE EXPLORATIONS** on the **CD**.}

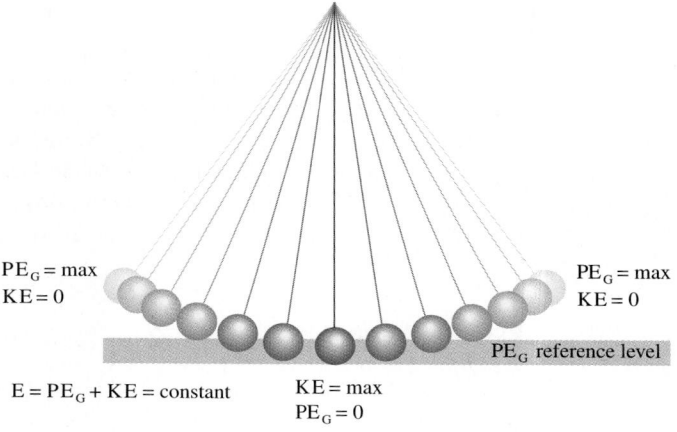

Figure 6.16 The transfer of energy from one form to another in the swinging pendulum.

A vaulter transfers his run-up KE into *elastic*-PE as the pole bends. When the pole straightens out, it imparts energy to the vaulter. At maximum altitude, he's motionless and subsequently drops straight down.

Example 6.9 [II] A vaulter carrying a graphite-fiberglass pole (a thin-walled tube weighing around 4 lb) is about to make a jump. At what speed must he run in order to clear the 20.0-ft (i.e., 6.096-m) mark? Neglect all possible energy losses, take his "*center*" to be 1.00 m above the floor while standing, and presume that it just clears the bar.

Solution The mention of no "energy losses" suggests Conservation of Energy. (1) TRANSLATION—A moving object converts all its kinetic energy into rising a specified height against gravity; determine its initial speed. (2) GIVEN: $h_i = 1.00$ m, $h_f = \Delta h + h_i = 6.10$ m. FIND: v_i when $v_f = 0$. (3) PROBLEM TYPE—Energy/Conservation of Mechanical Energy: $E_f = E_i$. (4) PROCEDURE—Since $KE_f + PE_{Gf} = KE_i + PE_{Gi}$, the final PE_G equals the initial KE. *Because we are not given the mass, it probably cancels out of the equations.* (5) CALCULATION—*Select the zero-PE level* at the initial height of

the jumper's center (i.e., his *c.g.*). If that point rises an additional $\Delta h = 6.096$ m $- 1.00$ m, then

$$0 + mg\Delta h = \tfrac{1}{2}mv_i^2$$

The mass cancels, yielding

$$v_i^2 = 2g\Delta h = 2(9.807 \text{ m/s}^2)(5.096 \text{ m}) = 99.95 \text{ m}^2/\text{s}^2$$

and $\boxed{v_i = 10.0 \text{ m/s}}$. This is a considerable speed for someone running with a pole. Actually, at the last moment, high in the air, as he approaches the bar, the athlete will straighten his arms, thereby exerting a force on the pole, lifting himself above his hand position and even above the end of the pole.

Quick Check: When launched at v_i, the peak height (p.71) reached by a projectile is $v_i^2/2g \approx 10^2/(<20) \approx >5$ m above the 1-m liftoff height.

During a running jump, a person can only get as high as his vertical velocity will carry him. However, using a pole to vault into the air provides a means of transforming nearly all of a runner's KE into PE_G. At the start, the person runs as fast as possible to gain as much KE as possible. Once the athlete jumps off the ground and is pole-borne, his run-up KE is gradually converted into both *gravitational*-PE and *elastic*-PE (Chapter 10) as the pole continues to bend while he rises. It soon straightens out, gives up its *elastic*-PE, and hurls him even higher.

Relativity—Energy & Mass

The Special Theory of Relativity (Chapter 26) expresses the total energy (E) of a system as the sum of the energy when it's at rest, the so-called **rest energy** (E_0), plus the kinetic energy: $E = E_0 + KE$. One of the most profound conclusions of the theory (one that we will derive later) is that $E_0 = mc^2$. Thus, when the system is motionless KE = 0, E = E_0 and

EXPLORING PHYSICS ON YOUR OWN

KE, PEG, Work, & Marbles: Find something long and straight with a groove in it, like a ruler. Hold it at the edge of a table so that a marble can roll down the groove and onto the table without bouncing. Get a sense of the speed (and therefore the KE) that the marble has as it hits the table. Increase the height of release or the tilt of the track (i.e., the PE_G) and observe the effect on the KE at the bottom. Have the marble slam into something light and movable (e.g., a stack of Post-it®s or a sugar cube). Study how far that object is pushed (which tells you about W_f) as the tilt of the track is increased. Now try it all again with a more massive marble or a steel ball, if you have one.

[rest energy]
$$E = mc^2$$

Except for a multiplicative constant, the mass of a system at rest is equal to the energy of that system. In part, this suggests that mass can be converted into energy and vice versa.* Whatever else might be implied, we can conclude that **potential energy and mass are equivalent**.

When a process takes place (whether chemical, mechanical, nuclear, whatever), if energy is liberated, that ΔE ultimately comes from a reduction in mass. Similarly, if energy is gained, mass increases; stretching a spring causes it to gain mass, just as raising it in the Earth's gravitational field causes it to gain mass. The constant c^2 is an immense number, 8.988×10^{16} m²/s², and therefore even a substantial change in energy corresponds to only a minute change in mass.

Turn your head, explode an atomic bomb, or drive around the block, and you convert mass into energy. Ignite a firecracker, and there's a blast of high-speed atoms and a blaze of photons; both carry off KE. Certainly we can say that *chemical*-PE (stored in the interatomic bonds of the explosive) was transformed into KE—that's what a chemical explosion is all about—but if we could somehow collect everything, all the atoms and all the photons, there would be less mass after the explosion than before it. For example, burn 1 kg of hydrogen with 8 kg of oxygen ($H_2 + O_2 \rightarrow H_2O + O$) and the gigantic explosion that results liberates about 1.2×10^8 J (the equivalent of 26 kg of TNT). The corresponding change in mass is $m = E/c^2 = 1.3 \times 10^{-9}$ kg; that's a mere one thousandth of a millionth of a kilogram. And it's a change in mass a thousandfold smaller than can be measured by ordinary chemical techniques.

The gravitational interaction is the weakest of all, and so we shouldn't expect to be able to measure changes in the mass of an object as its PE_G changes. The chemical transformation that bursts the firecracker is all electromagnetic, and the electromagnetic interaction has a relatively modest strength (at least compared to the strong nuclear force, p. 5). That's why you get over a million times more energy fissioning uranium than burning gasoline and why it's the strong force that we must look to if we are to observe Δm experimentally.

A system of particles that comes together under the influence of an attractive force loses PE and therefore loses mass. For example, an alpha particle consists of two neutrons and two protons, so powerfully bound by the strong force that it has 0.75% less mass together than apart—that's a substantial difference. By comparison, in the course of coming together under the influence of gravity, the Earth decreases in mass by only about 0.5×10^{-7}%.

The first direct experimental confirmation of $E = mc^2$ was in 1932, and it came, not surprisingly, via the strong force. At the time J. D. Cockcroft and E. T. S. Wilson fired a beam of protons, each with a KE of about 8.0×10^{-14} J, at a thin film of the light metal lithium (composed of three protons and four neutrons). When a nuclear reaction occurred a lithium atom absorbed a proton forming an unstable nucleus (four protons, four neutrons) that immediately disintegrated into two alpha particles; these flew off with a net KE, on average, of 2.9×10^{-12} J. The masses of all the particles were known, and remarkably, the total mass before and after the reaction was not the same; moreover, thirty-six times more

Internet

For a fun simulation where you design your own roller coaster using energy principles, go to

http://themeparks.about.com/tr
avel/themeparks/gi/dynamic/off
site.htm?site=http%3A%2F%2F
www.Funderstanding.com%2Fk
12%2Fcoaster%2F

---

* There is still some debate about the subtleties of interpretation here (see Chapter 26). Does mass transform into energy or is every change in E simply accompanied by a change in *m*, as some people maintain?

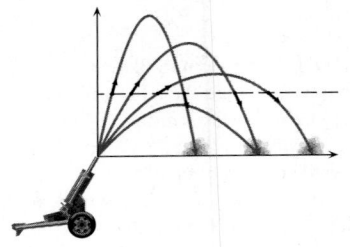

Figure 6.17 Cannonballs fired with the same muzzle speed will all be traveling at the same speed at the same altitude.

KE came out of the reaction than went into it. In effect, the loosely bound unstable nucleus broke into two smaller pieces that were very much more tightly bound and therefore less massive by an amount Δm. Had Relativity not been available to show that the energy difference, $\Delta E = 2.9 \times 10^{-12}$ J $- 8.0 \times 10^{-14}$ J, exactly equaled $\Delta m\, c^2$ there would have been a major crisis in Physics. We would have either dispensed with Conservation of Energy or somehow discovered that $E = mc^2$.

6.5 Applying Conservation of Energy

Unlike Newton's Laws, which require us to know the details of the motion in order to carry out the analysis, the energy formulation provides an overview. With it we can determine the final state of a system from its initial configuration, without looking at the intermediate stages through which it passes. For instance, knowing that the identical cannonballs in Fig. 6.17 were launched with the same initial speed, and therefore with the same KE, we can be sure that at any one height (where their potential energies are the same) they will all be moving with equal speed regardless of the path taken. Similarly, knowing the height of a snowy slope, we can compute a skier's speed at the bottom of a run without any knowledge of the route she took, or even the shape of the mountain, so long as friction is negligible.

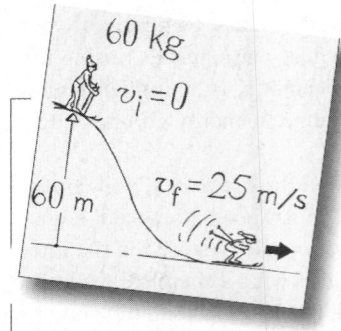

Example 6.10 **[II]** A 60-kg skier starts from rest at the top of a 60-m-high slope. She descends without using her poles. (a) What is her initial *gravitational*-PE with respect to the level ground at the bottom? (b) Assume friction is negligible and compute at what speed she will ideally arrive at the bottom. (c) Now if she reaches the bottom of the run, traveling at 25 m/s, what is the net energy transferred via friction (i.e., "lost" to warming her, the skis, the slope, and the air)?

Solution The mention of "*gravitational*-PE" followed by "speed" should call to mind KE and Conservation of Energy. (1) TRANSLATION—An object at rest at a given height descends along some path: determine its ideal final speed. It arrives at a different (lesser) speed; determine the energy "lost" to friction. (2) GIVEN: $m = 60$ kg, $h_i = 60$ m, $v_i = 0$, and $v_f = 25$ m/s. FIND: (a) PE_{Gi}; (b) v_f, no friction; and (c) W_f. (3) PROBLEM TYPE—Energy/Conservation of Mechanical Energy: $E_f = E_i$ / friction. (4) PROCEDURE—Use $KE_f + PE_{Gf} = KE_i + PE_{Gi}$, where $KE_f = PE_{Gi}$. (5) CALCULATION—*Select the zero-PE level at the bottom.* (a) At the top

$$PE_{Gi} = mgh_i = (60 \text{ kg})(9.81 \text{ m/s}^2)(60 \text{ m})$$

and

$$\boxed{PE_{Gi} = 3.5 \times 10^4 \text{ J}}$$

(b) With no work done against friction, $E_f = E_i$ and from Eq. (6.12), with $v_i = 0$ and $h_f = 0$,

$$\tfrac{1}{2}mv_f^2 + 0 = 0 + mgh_i = 3.5 \times 10^4 \text{ J}$$

And so

$$v_f = \sqrt{\frac{2(3.5 \times 10^4 \text{ J})}{60 \text{kg}}} = \boxed{34 \text{ m/s}} = 76 \text{ mi/h}$$

(c) Earlier (p. 188) we let W_f be the work done overcoming friction (i.e., the work done *by* the system); therefore $W_f = E_i - E_f$ and

$$W_f = mgh_i - \tfrac{1}{2}mv_f^2$$

$$W_f = (60 \text{ kg})(9.81 \text{ m/s}^2)(60 \text{ m}) - \tfrac{1}{2}(60 \text{ kg})(25 \text{ m/s})^2$$

and

$$W_f = 35\,280 \text{ J} - 18\,750 \text{ J} = \boxed{1.7 \times 10^4 \text{ J}\,.}$$

Quick Check: Working backward, $\tfrac{1}{2}mv_f^2 = \tfrac{1}{2}(60 \text{ kg})(34 \text{ m/s})^2 = 35 \text{ kJ} = PE_{Gi}$.

Reconsider Eq. (6.10), $PE_G = -GmM/r$. At $r = \infty$, PE_G is zero, while everywhere else it's negative. A negative PE is characteristic of attractive forces. Had the interaction been repulsive, as between two magnets with like poles, the PE would have been positive at finite distances. When the force is repulsive, work must be done on one body to bring it from infinity "down" to the vicinity of the other body, and so the change in PE (from zero) must be positive. By contrast, if the force is attractive, the bodies will come together on their own. The force field will do the work, and the change in PE (from zero) will be negative. *If a system can return to its zero-PE configuration by doing positive work, then the PE stored in the system is positive.* This is the case with two magnets with like poles repelling and separating out to infinity (they lose mass apart). Squeeze them together and

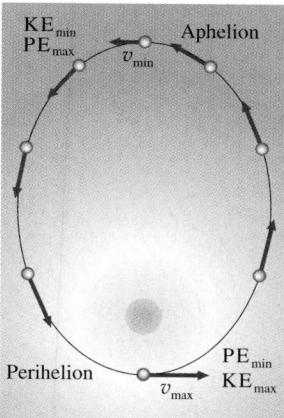

Figure 6.18 A planet in an elliptical orbit around the Sun. As it gets closer to the Sun, it speeds up. The speeds are the same for equal distances, either moving toward or away from the Sun, since there the PEs must be the same and therefore the KEs must also be the same.

In 1992 the comet Shoemaker-Levy 9 passed within 16×10^3 km of Jupiter and was ripped into about 23 pieces by the huge gravitational (tidal) force of the planet. In July of 1994 the string of fragments, traveling at 60 km/s (134 000 mph), returned to crash into Jupiter. Fragment G, which at 3 to 4 km across was among the largest, blasted into the atmosphere with a kinetic energy of about 2×10^{22} J or the equivalent of the explosion of 6 million megatons of TNT (that's about 100 million Hiroshima-sized atom bombs). On impact it created a fireball about 8000 km across.

WHY A PLANET IN ORBIT SPEEDS UP AS IT APPROACHES THE SUN

We know how to find the answer from Newton's force picture, but now we have an alternative. A planet's speed and KE both increase because its *gravitational*-PE decreases as it approaches the Sun. Any energy losses are negligible, and the total mechanical energy is everywhere constant. At the point of closest approach and smallest PE_G, the planet must have its maximum KE and be traveling with maximum speed (Fig. 6.18).

---

they gain energy and mass. *If positive work must be done on the system to bring the system back to the zero-PE state, then the PE stored is negative.* This is the case with two attracting masses being separated (they gain mass apart). The pieces of the Earth attract, and as they come together, they lose PE_G and lose mass.

The Escape Velocity

Suppose we wish to fire a craft into deep space with a single short-lived burst of energy. The minimum blast-off velocity, *the lowest velocity with which we could fire a projectile straight up and never have it fall back to the planet from which it was launched, is the* **escape velocity**, $\vec{v}_{esc}$. When given that speed, the object will have enough KE to overcome the planet's gravity. It will sail on forever, gradually slowing down but nonetheless arriving at infinity (whatever that means) as the final speed becomes zero. It follows from Eq. (6.10) that at launch, at the planet's surface (a distance R from its center),

$$E_i = KE_i + PE_{Gi} = \tfrac{1}{2}mv_{esc}^2 + \left(-\frac{GmM}{R}\right)$$

and no matter what this equals, it must be forever constant, provided nothing external acts on the craft. Ideally, the object will reach infinity, where $PE_{Gf} = 0$, traveling at zero speed such that $KE_f = 0$. Consequently, $E_f = 0$ at $r = \infty$, and since E is everywhere constant, it must be that $E_i = 0$. Therefore

$$\tfrac{1}{2}mv_{esc}^2 - \frac{GmM}{R} = 0$$

and

$$v_{esc} = \sqrt{\frac{2GM}{R}} \qquad (6.14)$$

At the surface of the Earth, where, from Eq. (5.6), $mg_0 = GmM_\oplus/R_\oplus^2$

[from the Earth's surface]
$$v_{esc} = \sqrt{2g_0 R_\oplus} \qquad (6.15)$$

Although the escape speed is independent of the mass of the projectile, the needed KE certainly is not. That's why it takes big rockets to raise big payloads. Notice that this is also the speed with which an object would collide with the planet, having fallen from a very

---

HOW THE UNIVERSE COULD POP INTO BEING

It's been suggested that since *gravitational*-PE is negative, if the total of all other forms of energy is equal in magnitude to the total PE_G then the net energy of the entire Universe is zero. It follows that such a system could have spontaneously appeared out of "nothingness" without violating energy conservation.

Table 6.3
Escape Speeds from the Surfaces of
Celestial Bodies

Body	Escape speed (km/h)
Eros asteroid	55
Moon	8×10^3
Mercury	15×10^3
Mars	18×10^3
Venus	37×10^3
Earth	40×10^3
Neptune	83×10^3
Saturn	130×10^3
Jupiter	216×10^3
Neutron star*	$\approx 0.8 \times 10^9$

*That's about 80% c.

THE EXPANDING UNIVERSE

The Universe is an explosion that began $\approx 13 \times 10^9$ years ago. Its galactic star-islands are still rushing outward to this day; the Universe—all there is—is expanding. If it has enough KE to override its own gravitational field, if the galaxies, and whatever else is out there, have the needed escape speed, the Universe will go on expanding forever. If not, it will ultimately stop and fall back upon itself thousands of millions of years from now.

ESCAPING THE SOLAR SYSTEM

Because gravity drops off inversely with r^2, most of the energy required to travel away from a large body like the Earth is needed to get beyond the first few hundred thousand miles. While the engine-cutoff speed to escape to deep space from Earth is 1.1×10^4 m/s or 25×10^3 mi/h, it takes a launch at about 99% of that speed just to get as far as the Moon. The Earth orbits the Sun, and attempting to leave our star requires a still greater speed. By taking advantage of the planet's orbital motion and sailing off in a judicious direction, this speed can be reduced to 16.4 km/s or 37×10^3 mi/h. The *Pioneer 10* space probe passed beyond the orbit of Pluto in 1983 (coasting along at 30613 mi/h) to become the first human-made object to leave the Solar System.

great distance. Recall Eq. (5.10) for the speed of a body in a circular orbit; namely, $v_o = \sqrt{GM/r}$. The escape speed from *any altitude* [where R becomes r in Eq. (6.14)] is equal to $\sqrt{2}$ times the orbital speed at that altitude. The higher the orbit, the slower the speed, and the easier the escape (Table 6.3).

Example 6.11 **[I]** In 1865, Jules Verne published a story called *From the Earth to the Moon* in which a gigantic cannon fires a capsule carrying three men to the Moon. Incidentally, the cannon was located at "27°7′N. lat. 5°7′W. long. of the meridian of Washington." (That turns out to be in Florida about 100 mi south of the Kennedy Spaceflight Center.) Compute the muzzle speed needed to escape the Earth. Use $R_\oplus = 6.4 \times 10^6$ m. Why is this technique impractical?

Solution The phrase "escape the Earth" suggests the escape speed. (1) TRANSLATION—An object is to be fired from the surface of the Earth; determine its escape speed. (2) GIVEN: $R_\oplus = 6.4 \times 10^6$ m and $g_0 = 9.81$ m/s². FIND: v_{esc}. (3) PROBLEM TYPE—Energy/escape speed. (4) PROCEDURE—Use the defining equation, and the fact that the projectile leaves the

Earth's surface. (5) CALCULATION:

$$v_{esc} = \sqrt{2g_0 R_\oplus} \qquad [6.15]$$

$$v_{esc} = [2(9.81 \text{ m/s}^2)(6.4 \times 10^6 \text{ m})]^{1/2} = \boxed{1.1 \times 10^4 \text{ m/s}}$$

This is equal to 25×10^3 mi/h, which is formidable. In fact, enough friction would be encountered in the dense lower atmosphere at that speed to burn up the ship. This is one reason why we use rockets that build up speed gradually, only reaching v_{esc} in the thin upper atmosphere. Besides this, the acceleration within the cannon would be lethal.

Quick Check: From Table 6.3, $v_{esc} = 40 \times 10^3$ km/h = $(40 \times 10^6/3600)$ m/s = 1.1×10^4 m/s.

Example 6.12 **[II]** In his *Exposition of the System of the World* (1798), the great French mathematician Laplace pointed out that an object of the same density as the Earth with a diameter 250 times that of the Sun would have so large a gravitational attraction that its escape speed would exceed the speed of light. He concluded that such a body, which could prevent light from leaving its surface, would be invisible to us. (a) Show that for a body of constant density ρ, the escape speed is proportional to its radius R. (b) Verify Laplace's assertion.

Solution This one's clearly about escape speed. (1) Translation—An object is 250 times the size of the Sun and has a known density; determine its escape speed. (2) Given: $R = 250R_\odot$. Find: v_{esc}. (3) Problem Type—Energy/escape velocity. (4) Procedure—Use the defining equation, Eq. (6.14), $v_{esc} = \sqrt{2GM/R}$. (5) Calculation—(a) The mass equals the volume times the density, $M = V\rho = \frac{4}{3}\pi R^3\rho$. Therefore

$$v_{esc} = \left(\frac{2G\frac{4}{3}\pi R^3\rho}{R}\right)^{1/2} = R\left(\frac{8}{3}G\pi\rho\right)^{1/2}$$

(b) Because the object has the same density as the Earth, the ratio of their escape speeds equals the ratio of their radii. The ratio of the diameter of the Sun to the Earth is $(1.393 \times 10^9 \text{ m})/(1.274 \times 10^7 \text{ m}) = 109$, which means that an object 250 times the size of the Sun is $(109)(250)$, or 2.7×10^4 times the size of the Earth. Hence, its escape speed is 2.7×10^4 times 11.2 km/s, or 3.1×10^8 m/s , which slightly exceeds c. Collapsed stars that are so dense and so powerful gravitationally that light cannot escape from them actually exist. Nowadays, they are known as **black holes**. Although Laplace had the wrong theory for both gravity and light, he ended up predicting a real, albeit fantastic, entity.

Quick Check: With any derivation it's always good to check the units; $v_{esc} = R(\frac{8}{3}G\pi\rho)^{1/2}$ has units of $m[(N\cdot m^2/kg^2) \times (kg/m^3)]^{1/2} = m[(kg\cdot m/s^2)(1/kg\cdot m)]^{1/2} = m/s$.

6.6 Power

When you want a day's labor and are concerned about getting every minute's worth, you buy a certain amount of work per hour. That's the practical language of industry, and it was in the beginning of the Industrial Revolution that the idea of *power* became quantified. **Power** was defined as ***the rate of doing work***:

Power is the time rate-of-change of energy.

$$\text{Power} = \frac{\text{work done}}{\text{time interval}}$$

More generally, ***power is the rate at which energy is transferred into or out of a system***. An engine that will do the same amount of work as another in half the time is twice as powerful. If the time interval Δt is finite, we are talking about the *average power*, or

$$P_{av} = \frac{\Delta W}{\Delta t} \tag{6.16}$$

To express the moment-by-moment power, the **instantaneous power**, take the limit of this

In 1979, the human-powered Gossamer Albatross flew across the English Channel. The pedaling pilot averaged 190 **W**, or about 0.25 horsepower.

Table 6.4

Power Equivalencies

Quantity	Equivalent
1 Btu per hour	0.293 W
1 joule per second	1 W
1 foot-pound per second	1.356 W
1 kilocalorie per hour	1.16 W
1 kilowatt-hour per day	41.7 W
1 kilocalorie per minute	69.77 W
1 horsepower	745.7 W
1 kilowatt	1000 W
1 Btu per second	1054 W
1 gallon of gasoline per hour	39 kW
1 gallon of oil per minute	2.5 MW
1 million barrels of oil per day	73 GW

ratio as the time interval goes toward zero, whereupon

$$P = \lim_{\Delta t \to 0}\left(\frac{\Delta W}{\Delta t}\right) \tag{6.17}$$

In many cases, the power will be constant and Eq. (6.16) will describe P.

The idea of standardizing the measure of power using the output of a horse had been around for a while before James Watt (1783) formalized it. Watt found that a large draft horse could exert a pull of about 150 lb while walking at about $2\frac{1}{2}$ mi/h for a considerable amount of time. That rate of doing work is equivalent to 33 000 foot-pounds per minute or 550 ft·lb/s, which he called 1 **horsepower** (hp). Even though *horsepower* is now an outdated measure, its use persists.

In the SI system, the unit of power is the **watt** (W), equal to 1 joule of work done per 1 second: 1 W = 1 J/s. Almost all electrical devices are rated in watts to indicate how much power they require. Since a watt is a fairly small amount of power, the more convenient kilowatt (1 kW = 1000 W) is often used:

$$1 \text{ hp} = 550 \text{ ft·lb/s} = 746 \text{ W} = 0.746 \text{ kW}$$

Presumably, someday all forms of power will be specified in SI units so that automobile engines will be rated in kilowatts along with toasters (Table 6.4).

Example 6.13 **[II]** The express elevator in the Sears Tower in Chicago averages a speed of 9.144 m/s (i.e., 1800 ft/min) in its climb to the 103rd floor, 408.4 m (i.e., 1340 ft) above ground. Assuming a load of 1.0×10^3 kg, what average power must the lifting motor supply?

Solution Work is done and power is expended. (1) TRANSLATION—An object of known mass is to rise a specified distance at a known rate; determine the power needed. (2) GIVEN: $v_{av} = 9.144$ m/s, $\Delta l = 408.4$ m, and $m = 1.0 \times 10^3$ kg. FIND: P_{av}. (3) PROBLEM TYPE—Energy/work/mechanical power. (4) PROCEDURE—Use the defining equation: $P_{av} = \Delta W/\Delta t$. (5) CALCULATION—We need ΔW, which, in turn, equals the weight (F_w) times the vertical displacement (Δl):

$$\Delta W = F_w \Delta l = mg\, \Delta l$$

$$\Delta W = (1.0 \times 10^3 \text{ kg})(9.81 \text{ m/s}^2)(408.4 \text{ m}) = 4.0 \times 10^6 \text{ J}$$

To find Δt, remember that $v_{av} = \Delta l/\Delta t$ and so

$$\Delta t = \frac{408.4 \text{ m}}{9.144 \text{ m/s}} = 44.66 \text{ s}$$

Finally

$$P_{av} = \frac{\Delta W}{\Delta t} = \frac{4.0 \times 10^6 \text{ J}}{44.66 \text{ s}} = 89.57 \text{ kW} = \boxed{90 \text{ kW}}$$

Quick Check: 90 kW will lift the 9.8×10^6-N load a distance of 408 m in a time of about $(9.81 \times 10^3 \text{ N})(408 \text{ m})/(90 \text{ kW}) = 44.5$ s. Whereas $\Delta l/\Delta t = (408 \text{ m})/(9.144 \text{ m/s}) = 44.6$ s.

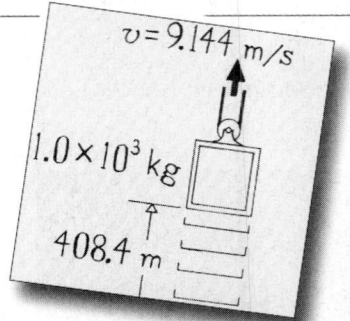

Power and Motion

When a constant force F acts on a body that, in the process, moves through a distance Δl, an amount of work is done, $\Delta W = (F \cos \theta)\, \Delta l$. If this occurs in a time interval Δt

$$P_{av} = \frac{\Delta W}{\Delta t} = (F \cos \theta)\frac{\Delta l}{\Delta t} = (F \cos \theta)v_{av}$$

or, in the limit, as $\Delta t \to 0$

$$P = Fv \cos \theta \tag{6.18}$$

The power delivered to a moving object equals the product of the component of the force in the direction of motion and the speed. At any instant, it depends on the instantaneous speed at which the point of application of the force is moving. In many cases, $\vec{F}$ and $\vec{v}$ are parallel, $\theta = 0$, and the expression becomes

$[\vec{F} \parallel \vec{v}]$

$$P = Fv \tag{6.19}$$

HUMAN POWER

A person in good physical condition can deliver work at a fairly continuous rate of roughly $\frac{1}{10}$ hp, or about 75 W. Working at a steady pace, the human body develops power in proportion to the amount of oxygen it consumes. A rate of consumption of about 1 liter (10^3 cm^3) per minute is what's required for an output of 75 W. An athlete can develop continuous power at around 300 W. A sustained power level of about twice this amount is the limit that a human can maintain for a period in excess of a minute or so. Over short time intervals (e.g., the time it takes to leap into the air or throw a ball), much higher power levels can be attained; sprinters can produce 1200 W in 6-s bursts. When muscles are rested, they have an additional short-term supply of oxygen that may also be called upon. This limited reserve can supply bursts of power of, say, 450 W for a minute or so, or as much as several kilowatts for a fraction of a second. Once exhausted, that reserve supply takes some time to replenish, and any athlete is aware of its one-shot nature.

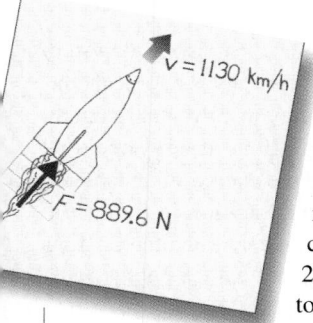

$v = 1130$ km/h

$F = 889.6$ N

Example 6.14 **[I]** In 1935, R. H. Goddard, the American rocket pioneer, launched several A-Series sounding rockets. Given that the engine had a constant thrust of 889.6 N (i.e., 200 lb), how much power did it transfer to the rocket while traveling at its maximum speed of 1130 km/h?

Solution One of the things to keep in mind when you see the word "power" is that the power developed by a moving system depends on the force exerted and the speed. (1) TRANSLATION—An object experiencing a known force is propelled at a known speed; determine the power delivered. (2) GIVEN: $F = 889.6$ N and $v = 1130$ km/h. FIND: P. (3) PROBLEM TYPE—Energy/mechanical power/speed. (4) PROCEDURE—We have force and velocity ($\theta = 0$) and need power; that means $P = Fv$. (5) CALCULATION:

$$P = Fv = (889.6 \text{ N})(313.9 \text{ m/s}) = \boxed{279 \text{ kW}}$$

Quick Check: Let's recalculate using U.S. Customary Units: (1130 km/h)(0.9113 ft·h/km·s) = 1029 ft/s; hence, P = (200 lb)(1029 ft/s) = 20.6 × 10^4 ft·lb/s. Using Table 6.3, multiply this value by 1.356 W/(ft·lb/s), yielding 279 kW.

This makes it clear that exerting the same force on an object that's moving increasingly rapidly requires the application of a proportionately greater amount of power—it's easy to accelerate when jogging at 1 mph, but it's a lot harder to attain the same acceleration when you're already running at 10 mph (Table 6.5).

Table 6.5

Human* Power and Oxygen Consumption

Activity	Power (watts)	Oxygen Consumption (liters O$_2$/min)
Sleeping	83	0.24
Sitting at rest	120	0.34
Sitting in class	210	0.60
Walking slowly (4.8 km/h)	265	0.76
Cycling (13–18 km/h)	400	1.14
Shivering	425	1.21
Playing tennis	440	1.26
Swimming breaststroke	475	1.36
Climbing stairs (116/min)	685	1.96
Cycling (21 km/h)	700	2.00
Running cross-country	740	2.12
Playing basketball	800	2.28
Cycling, professional racer	1855	5.30
Sprinting	2415	6.90

*Normal 76-kg male.

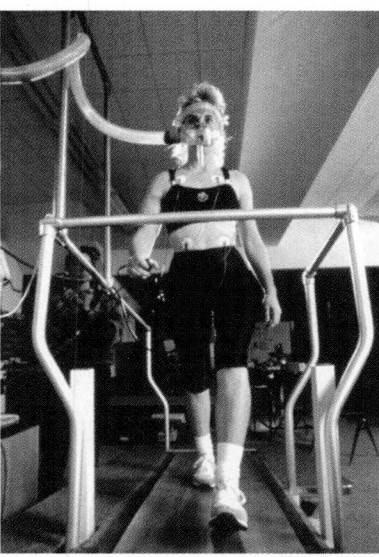

The rate of oxygen consumption is proportional to the power developed. That power level, for any animal, is called the metabolic rate, and it's an important indicator of overall health and physical conditioning.

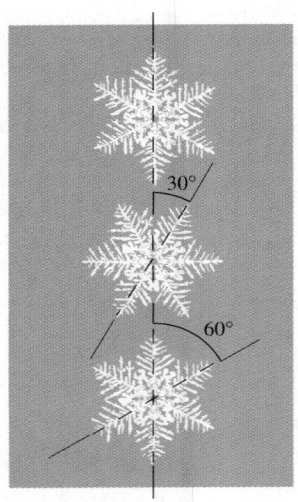

60° snowflake symmetry.

6.7 Energy Conservation and Symmetry

In 1918 Emmy Noether (pronounced like the writer J. W. van Goethe but with an N) made a discovery about the fundamental nature of physical law that has since become one of the most important guiding insights in theoretical physics (p. 5). Her conclusions have been generalized into what is now known as **Noether's Principle**:

For every symmetry of Nature there is a corresponding conservation law, and vice versa.

There are obvious geometrical symmetries like those of snowflakes and plus signs; rotate a snowflake through 60° and it appears unchanged. But the idea of symmetry can be extended to conceptual symmetries: *When a change in a physical system leaves some aspect of the system unchanged, the system possesses a corresponding symmetry.* The existence of a symmetry means that a feature of the system is changeless, or *invariant*.

A conservation law maintains that a specific quantity is constant as an isolated system undergoes change. What Noether showed was that Conservation of Energy is a logical consequence of the fact that the behavior of the Universe (as reflected by its physical laws and constants) is unchanging in time. (As we'll see again later—Section 29.9—energy and time are concepts that are linked in a fundamental way.) If we measure the energy of an isolated system in some laboratory, the results will be independent of the location in time (Fig. 6.19). We can move from one point in time to another, and the results are unaltered. If, as we believe is the case, physical law is timeless throughout the Universe, then energy must be conserved.

Though the proof of this assertion is beyond the level of our treatment, we can make the conclusion plausible. Imagine an isolated lab in which you separate two gravitationally interacting masses, thereby storing a certain amount of PE_G, say 1000 J, in that system. Assume, for the sake of argument, that the laws of physics change in time. If you hold the objects motionless and simply wait for gravity to decrease (or increase), a time will come when the PE_G is perhaps 900 J. At that moment, you let the objects fall back to where they were to start with. The lab is returned to its original configuration, but there is now a 100-J-energy imbalance—Conservation of Energy is violated.

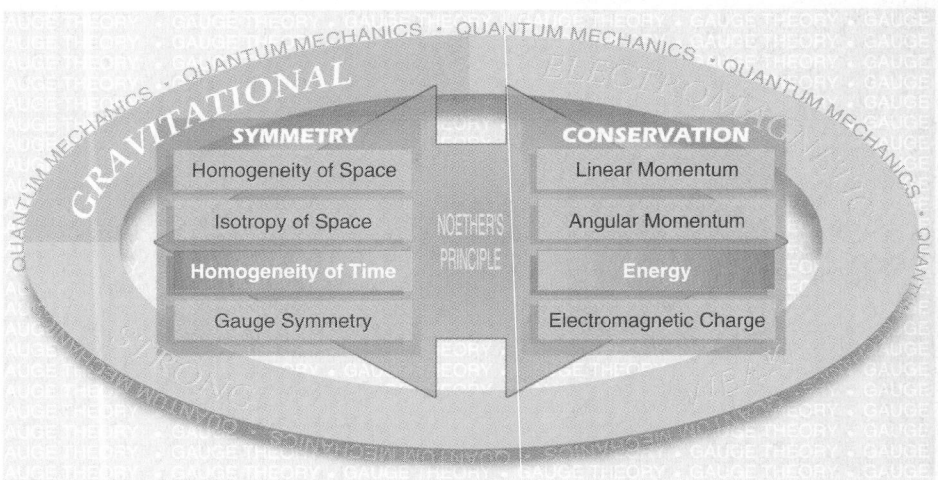

Figure 6.19 This is our summary diagram—it reminds us of where we've gone in the world of theoretical physics. Now we know that energy is conserved and it's conserved because the Universe exhibits temporal displacement symmetry.

Internet

For more on symmetry and conservation go to the website dedicated to teaching symmetry in the introductory physics curriculum:

www.emmynoether.com

The nature of the fundamental interactions (Newton's Third Law) ensures that the transfer of energy everywhere, while the laws of physics are constant, will be such as to maintain a fixed total energy. When any two systems interact, both will change; one will lose the energy the other gains. Insofar as our Universe exhibits temporal displacement symmetry, energy is and will be conserved.

Core Material & Study Guide

THE TRANSFER OF ENERGY

The measure of the amount of energy transferred into or out of a system via an applied force acting over a distance is known as **work**, where

$$W = Fl \cos \theta \qquad [6.2]$$

The unit of work is the *newton-meter* or *joule*. Section 6.1 (Work) is important, especially the part called *Work in General*. Make sure you understand Examples 6.1–6.4. Study both the **WARM-UPS** and **WALK-THROUGHS EXAMPLES** in **CHAPTER** 6 on the **CD**.

MECHANICAL ENERGY

When work is done on a system to overcome inertia, it goes into **kinetic energy** (KE), where

$$\text{KE} = \tfrac{1}{2}mv^2 \qquad [6.5]$$

Reexamine Section 6.2 (Kinetic Energy) and study Example 6.6; the subsection *The Work-Energy Theorem* is useful for solving certain problems (e.g., Example 6.7).

The energy *stored in a system of interacting objects* is known as **potential energy** (PE). Assuming *g* is constant

$$\text{PE}_\text{G} = mgh \qquad [6.9]$$

The potential energy of a body of mass *m* at a distance *R* from the center of a mass *M* (e.g., a planet) is

$$\text{PE}_\text{G} = -\frac{GmM}{r} \qquad [6.10]$$

A force is **conservative** if *the work done in moving an object from one point to another against the force depends only on the locations of the points and not on the path taken* (p. 186). Study Section 6.3 (Potential Energy) and make sure you understand Example 6.8.

CONSERVATION OF MECHANICAL ENERGY

The Law of *Conservation of Energy* (p. 187) maintains that The total energy of any system that is isolated from the rest of the Universe remains constant, even though energy may be transformed from one kind to another within that system.

In the absence of friction and any other force but gravity, mechanical energy is conserved;

$$\tfrac{1}{2}mv_\text{f}^2 + mgh_\text{f} = \tfrac{1}{2}mv_\text{i}^2 + mgh_\text{i} \qquad [6.12]$$

Review Section 6.4 (Mechanical Energy) and study Example 6.9.

Section 6.5 (Applying Conservation of Energy) brings together the ideas of work, KE and PE$_\text{G}$. The lowest speed at which we could fire a projectile so that it never falls back to the place from which it was launched is the **escape speed**

$$v_\text{esc} = \sqrt{\frac{2GM}{R}} \qquad [6.14]$$

Makes sure you can follow Examples 6.10–6.12.

Power is *the rate of doing work*, the rate at which energy is transferred. *Average power* is

$$P_\text{av} = \frac{\Delta W}{\Delta t} \qquad [6.16]$$

The SI unit of power is the **watt**: 1 W = 1 J/s. When an object is in motion at a speed *v* under the influence of a force

$$P = Fv \cos \theta \qquad [6.18]$$

Reread Section 6.6 (Power) and go over Examples 6.13 and 6.14. Work as many I-level problems as possible, then do the groups of three problems that are printed in color.

Key Terms

force	Conservation of Energy
energy	rest energy
work	escape velocity
joule	power
work diagram	average power
kinetic energy	instantaneous power
Work-Energy Theorem	horsepower
potential energy	watt
gravitational-PE	Noether's Principle
thermal energy	temporal displacement
conservative force	symmetry

1. On page 71 of *Unit Outlines In Physics* (1935), published by the College Entrance Book Company Inc., one finds the statement "Work is done when a force moves an object." Is that completely true? What about the work done by Superman stopping a speeding train? If you came up behind a walking elephant and amble along while pushing on its butt, would you be doing work on it even though it was already moving without you? Explain.

2. Figure Q2 is a plot of speed squared versus braking distance for a typical automobile (without computer-assisted brakes) stopping on a good road surface. Assuming the brakes are uniformly applied, explain the physics underlying the diagram. Different brakes and/or road surfaces will result in a set of straight lines all with different slopes and all passing through the origin. Explain the reasons for that. Determine the coefficient of friction.

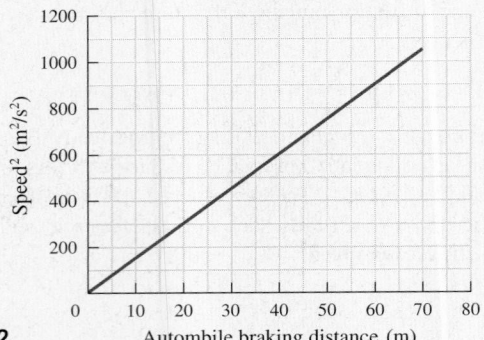

Figure Q2

3. A rocket is held in place as its engines are fired up. Is any work done on the rocket? Once released, the rocket rises into the air. What force accelerates it? Considering both the vehicle and the exhaust, talk about the KE of each and the nature of the forces doing work. Where does the energy that ends up as KE come from? How is this different from swimming? Even though the thrust is constant, the longer the engine fires, the faster the rocket goes. What does that say about the power transferred to the craft as a function of time?

4. In the process of rowing, what external force propels the boat? Describe the work done on the boat and on the water. Where does the change in KE of the boat come from? How is this similar to the operation of a propeller on an airplane?

5. A person walks at a constant speed by pushing backward on the ground with a force F. If he travels a distance l in a straight line on level ground, (a) what force is he overcoming? (b) Suppose his feet slide a little as he pushes back—does the reaction force of the ground do work on him? (c) What if his feet don't slip as he walks; is work done on him by the floor? (d) Where does the energy expended come from?

6. A lump of soft clay of mass m falls from a height h above the floor, where it crashes and comes to rest. What force brings the clay to a stop? How much work is done by the floor on the clay? What is the net amount of work done on the clay during the fall? Where does that energy end up?

7. A youngster on a bicycle starts at rest and pedals a straight path on level ground at a constant speed, coming to a stop some distance away. (a) Is it reasonable to assume that he is the energy source for the journey? (b) What external force propels the bicycle and rider? (c) Does the ground do positive work on the bike?

8. Work is done against resistance. In each of the following cases, describe the forces overcome: (a) hammering a nail into a piece of wood; (b) stretching a piece of rubber; (c) cutting a loaf of bread; (d) putting a book on a high shelf; (e) opening the door on a modern refrigerator; (f) pulling a piece of masking tape off a roll.

9. Two identical balls travel along frictionless grooves in a flat surface. One groove is straight and runs horizontally along the surface, while the other dips down below the surface and up again smoothly in a vertical plane. The two runways begin and end parallel and adjacent to each other. Assume that neither ball leaves its track and that both are given the same initial speed. Describe their motions in a race where the starting and finish lines are across horizontal portions of the grooves.

10. A rock at the bottom of a well has a negative amount of *gravitational*-PE with respect to the ground level. What does that mean? What will happen if it's launched upward with an equal positive amount of KE? Can a body have a negative KE? Discuss the significance of the situation in which a body has a positive KE and a negative PE, where $|PE| >$ KE. What happens when $|PE| <$ KE?

11. The late physicist and Nobel laureate Richard P. Feynman once wrote, "It is important to realize that in physics today, we have no knowledge of what energy *is*." What do you think he meant by that? The usual definition of energy is *the ability to do work*. Discuss the circular nature of that definition. Consider the alternative definition: energy is *the ability to impart vis viva* (mv^2); is this definition any better? Here's a rather esoteric definition: *energy is that measure of the physical change of a system that is conserved as a result of temporal displacement symmetry*. Explain it.

12. Potential energy belongs to the entire system of interacting objects and not to any one of them. Explain this statement and discuss how it applies to the *gravitational*-PE of a frog sitting on a rock on the planet Earth.

13. A ball on a string whirls around in a circle at a constant speed. Is there a net force on the ball? Does it do work? Is there a corresponding change in either PE or KE? How can the ball be made to accelerate tangentially? Will there then be work done on it and a change in either PE or KE?

14. The cars of a roller coaster are usually tugged up to the top of the first hill by a chain drive and then released to coast the rest of the way around. Consider that friction is acting all along the journey. (a) Where is the *gravitational*-PE maximum? (b) How should the ride be configured to produce the greatest value of the maximum speed? (c) Having once attained its maximum *gravitational*-PE, can it reach that value again? (d) Having once attained its maximum KE (which may well not be the maximum possible KE for a structure of that height), can it reach that value again? (e) Must each successive hill be lower than the previous one? (f) Must the second hill be lower than the first?

15. A crate of mass m rests on a flatbed truck. The truck accelerates from rest to a speed v, during which time the crate does not slide on the bed. How much work is done on the crate by friction? Is that work positive or negative? What is the work done against? In what final form is the energy that was provided to the crate? Now suppose the truck comes to a gradual stop. Given this situation, answer the same questions again.

16. During the first Moon-landing flight (*Apollo 11*, July 1969), the spacecraft was put into a circular parking orbit at an altitude of 119

mi above the Earth. The engines were later fired, sending the ship off to the Moon at 24 245 mi/h. Compare this speed to the escape speed and explain the difference. When we calculated the formula for the escape speed for a planet, we neglected the fact that our particular dirtball is spinning. Discuss the effect, if any, of the spin if launch is from some parking orbit about the Earth. What if liftoff is from the surface?

17. A water skier is pulled along at a constant speed (see photo).

Figure Q17

Discuss what's happening in the picture from an energy perspective. What forces act on him? What if anything does he do work on? Considering Conservation of Energy, what's the significance of the water shooting into the air? Where does the energy transferred to the skier by the boat end up? What's the source of all the KE of the water, skier, and boat?

18. Explain the meaning of each of the "Key Terms" on page 199.

19. The Dutch scientist Willem 'sGravesande (1688–1742) performed a series of experiments in which he dropped various objects onto soft clay. He found that when the impact speed was doubled, the penetration depth went up by a multiplicative factor of four, and when the speed tripled, it went up by a factor of nine. Explain his results, first using Newton's Laws ($F = ma$), and then using the ideas of work and energy. (Incidentally, the 's in his name means "of the.")

Multiple Choice Questions

1. A large locomotive exerts an average force of 5×10^5 N while pulling a train from rest a distance of 1 km. How much work does it do on the train during this acceleration? (a) 5×10^5 J (b) 5×10^8 J (c) 1×10^3 J (d) 5×10^8 J (e) none of these.

2. The gravitational force of the Sun on the Earth is about 3.5×10^{22} N and the Earth is in a near circular orbit at a mean distance of 1.5×10^{11} m. How much work does the Sun do on the Earth during one orbit? (a) 9.4×10^{11} J (b) 5.3×10^{33} J (c) 1.5×10^{11} J (d) zero. (e) none of these.

3. A 10-kg mass is held 1.0 m above a table for 25 s. How much work is done during that period? (a) none (b) 10 J (c) 250 J (d) 9.8 J (e) none of these.

4. The net work done on an object moving along a closed path in a force field is zero when it returns to the origin. The force is (a) conservative (b) nonconservative (c) impossible (d) liberal (e) none of these.

5. A typical adult male's heart pumps about 160 milliliters of blood per beat. It beats around 70 times per minute and does roughly 1 J of work per beat. How much work does it do in a day? (a) 10^5 J (b) 10^6 J (c) 70 J (d) 70×10^5 J (e) none of these.

6. Which of the following is not a measure of the same quantity as the others? (a) newton-meter per second (b) kilogram-meter^2 per second^3 (c) joule per second (d) watt (e) none of these.

7. If a 20-kW engine can raise a load 50 m in 10 s, how long will it take for it to raise that same load 100 m? (a) 20 s (b) 40 s (c) 5.0 s (d) not enough information (e) none of these.

8. If a 25-hp motor can raise an elevator 10 floors in 20 s, how long will it take a 50-hp motor to do the same? (a) 40 s (b) 10 s (c) 20 s (d) 5.0 s (e) not enough information.

9. Which of the following is not a measure of the same quantity as the others? (a) foot-pound (b) newton-meter (c) watt (d) joule (e) none of these.

10. If this book is placed on an ordinary table and slid along a path that brings it back to where it started, (a) no net power will have been required (b) work will certainly have been done (c) assuming a conservative gravitational field, no net work will be done (d) not enough information is given to say anything about the work done (e) none of these.

11. A fairly small asteroid (1000 kg) out in deep space is to be accelerated from rest up to 10 m/s. Inasmuch as it is weightless, will work have to be done on it during the acceleration and, if so, how much? (a) no (b) yes, 10 000 J (c) yes, 50×10^3 J (d) yes, 10 000 N (e) yes, 50×10^3 N.

12. Work is done on an object far out in space where it has negligible *gravitational*-PE. If in the process there is no net change in its KE, we can conclude (a) that friction *may* have been operative (b) that this situation is impossible (c) that the energy of the object has decreased (d) that the object's speed decreased (e) none of these.

13. A rocket coasting along in space at some speed v fires its engines thereupon doubling its speed, but at the same time it jettisons some cargo, reducing its mass to half its previous value. In the process, its KE is (a) doubled (b) tripled (c) quadrupled (d) unchanged (e) none of these.

14. A kid in a wagon rolls from rest down a hill reaching the bottom at 12 m/s. On the next run, she gets a push and starts down at 5.0 m/s. At what speed does she now arrive at the bottom? (a) 12 m/s (b) 17 m/s (c) 7 m/s (d) 13 m/s (e) none of these.

15. An arrow is fired, via a bow, straight up. It rises for a while and then drops back to the ground. The process, taking the arrow from loading to just prior to touch down, can best be described by a series of energy transformations corresponding to (a) work, *elastic*-PE, KE, *gravitational*-PE, KE (b) work, KE, *elastic*-PE, KE (c) KE, *gravitational*-PE, work (d) *elastic*-PE, *gravitational*-PE, KE (e) none of these.

16. How fast must a 1.0-kg mass be moving if its kinetic energy is 1.0 J? (a) 1.0 m/s (b) 9.8 m/s (c) 1.414 m/s (d) 10 m/s (e) 0.102 m/s.

17. How high above the Earth's surface must a 1.0-kg mass be for it to have a *gravitational*-PE of 1.0 J with respect to that surface? (a) 9.8 m (b) 1.0 m (c) 0.10 m (d) 0.01 m (e) 32 m.

18. While a ball rolls down the circularly curved track shown in Fig. MC18, its speed, acceleration, and kinetic energy, respectively (a) increase, increase, increase (b) decrease, decrease, decrease (c) increase, decrease, decrease (d) decrease, increase, increase (e) none of these.

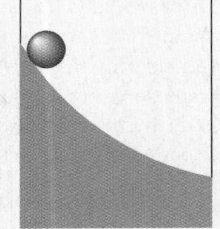

Figure MC18

19. A 458 Winchester magnum cartridge has a 500-grain (1 gr = 0.0648 g) bullet that attains a muzzle speed of 644.9 m/s (2120 ft/s). After traveling 456 m (500 yd), it would still be moving at 365.6 m/s (1202 ft/s). What fraction of its original KE does it have at that distance? (a) 23% (b) 50% (c) 32% (d) 68% (e) 77%.

20. Taking infinity as the zero of *gravitational*-PE, through what height at the Earth's surface must a 1.0-N body be raised if its PE_G is to be increased by 1.0 J? (a) ∞ (b) 10 m (c) 1.0 m (d) −1.0 m (e) none of these.

21. If Superman really is "more powerful than a locomotive," then with regard to freight trains (a) he can pull more cars at the same speed (b) he can pull the same number of cars faster (c) he can pull the same number of cars up steeper hills (d) all of these (e) none of these.

For more Multiple Choice Questions with answers click on WARM-UPS in CHAPTER 6 on the CD.

Suggestions on Problem Solving

1. A common error in *work* problems involving forces that are not parallel to the displacement is to forget to use the component of F, namely ($F \cos \theta$), in computing W. If the force is not parallel to the displacement, check whether or not the perpendicular component ($F \sin \theta$) is affecting the system. For example, ($F \sin \theta$) can increase or decrease the normal force and thereby affect the friction and W.

2. When a body moves under the influence of a constant force, the power developed is given by $P = Fv$. If you know the power provided by a rocket motor traveling at a given speed, you can determine its thrust. Or if you know the maximum power of a locomotive and the force exerted on it, you can calculate its maximum speed.

3. Remember to *square* the speed when doing a KE calculation. When considering the change in KE, bear in mind that $v_f^2 - v_i^2$ is *not* equal to $(v_f - v_i)^2$.

Problems ✦ Coordinated Problems ✦ Progressive Problems ✦ Solutions

STUDY GUIDE 1. **Coordinated Problems:** The three problems within each magenta-colored grouping are solvable in similar ways. Note that the first of these always has a hint; moreover, its solution is provided in the back of the book. *Work out each of these sets; they'll strengthen technique and build confidence.* 2. **Progressive Problems:** The problems introduced in blue unfold step-by-step carrying along the analysis in a more suggestive way than is customary. *Work out all of these; they'll guide you through the analytic process and help develop problem-solving skills.* 3. **Worked-Out Solutions:** Studying worked-out solutions is an important part of learning how to solve problems. Accordingly, additional *solutions* to a number of model problems are given below. *Make sure you understand each of them before you go on to the next problem.* 4. Also provided in the back of the book are the *Answers* to all odd-numbered problems, as well as worked-out *solutions* to those with boldface numbers. Problem numbers in italic indicate that a solution appears in the Student Solutions Manual.

SECTION 6.1: WORK

1. [I] While floating out in space, a constant force of 500 N is applied to a 542.3-kg robot by a small rocket motor. The robot moves along a straight line in the direction of the thrust of the motor. How much work is done on it by the rocket for every 10.0 meters traveled?

2. [I] A 0.50-kg glider on an air track (Fig. Q1, p. 125) is pulled along frictionlessly by a constant tensile force of 15.0-N weight. Over what distance will it travel in order that 10.0 J of work be done on it?

SOLUTION: The force is constant at 15.0 N. The work done, $W = Fl$ is 10.0 J = (15.0 N)l, l = (10.0 J)/(15.0 N) = 0.67 m.

3. [I] Workers load a 500-kg safe onto an elevator that lifts it 90 m to the twentieth floor of an office building. How much work did the elevator do on the safe?

4. [I] What is the work done by gravity when a 2.0-kg ball falls to the floor from a height of 1.50 m? Is it positive or negative? Explain.

5. [I] A book is slowly slid in a straight line, at a constant speed, 1.5 m across a level table by a 15-N horizontal force. How much work is done on the book-table system and what is it done against? [*Hint: Use the definition of work where the force is in the direction of the motion.*]

6. [I] During a lab experiment, a 2.00-kg block of wood is pulled along the floor at a constant speed of 1.00 m/s over a distance of 4.00 m by a 2.00-N horizontal force. How much work was done by the source of the applied force in overcoming friction?

7. [I] A force gauge is used to push horizontally on the back of a student who is sitting on a table. When the gauge reads 400 N the student is sliding along at a constant speed in a straight line. How much work will be expended in moving the student a distance of 2.00 m under those conditions?

8. [I] A 1.0-kg mass is raised 10 m into the air by a 10-N force exerted vertically. Determine the amount of work done by the force. Now suppose a 10-N force pulling horizontally moves a 1.0-kg mass through 10 m over a frictionless floor. How much work is done here? What is the work done against in both instances?

SOLUTION: (a) $W = Fh$, $W = (10 \text{ N})(10 \text{ m}) = 1.0 \times 10^2$ J. (b) $W = Fl$, $W = (10 \text{ N})(10 \text{ m}) = 1.0 \times 10^2$ J. (c) In the first part the work is done to overcome gravity and also to give the body a small acceleration. In the second part the work is done to give the body a larger acceleration.

9. [I] A delicatessen owner holds a knife horizontally above a 10-cm-thick cheese. With one smooth motion, exerting a downward force of 20 N, the cheese is cut in two. How much work was done by the owner?

10. [I] A 2224-N piano is rolled on little wheels a distance of 3.1 m across a horizontal floor by an 890-N piano mover. If, while lifting upward on one end with 111 N, he pushes horizontally with 445 N, how much work will he do?

11. [I] A nurse pushes someone in a wheelchair 100 m doing 400 J of work in the process. What average horizontal force did the nurse exert in the direction of motion?

12. [I] THIS PROBLEM APPLIES THE PHYSICAL CONCEPT OF WORK. A constant horizontal push of 4.0 N is applied to a 5.0-kg bag of marbles which travels 8.2 m at a constant speed across a horizontal table in the direction of the force. (a) What is the value of the horizontal acceleration of the bag? (b) How much work is done in overcoming friction?

13. [I] THIS PROBLEM APPLIES THE PHYSICAL CONCEPT OF WORK. A 20.0-kg dog stands on a frictionless inclined plane. A constant upward 60.0-N force parallel to the incline is applied to the butt of the dog. As a result the dog slides upward 12.0 m along the surface. (a) What is the component of the applied force in the direction of the motion? (b) How much work is done on the dog by that force? (c) What was this work done against?

14. [I] A newspaper delivery boy pulls horizontally on a rubber-wheeled cart that has a coefficient of rolling friction of 0.02 and a mass of 25 kg. It travels 10 km on level streets. How much work does he do in overcoming friction between the cart and the ground?

15. [I] When a solid-fuel rocket burns, the flame front advances into the material in a direction perpendicular to the ignited surface. By configuring the fuel core in various ways and lighting it along the entire length, one can obtain all sorts of performance characteristics. Figure P15 shows a starred-cavity motor. How much work was done by the device in the first 90 s of flight if the rocket rose vertically to a height of 2.0 km in that time?

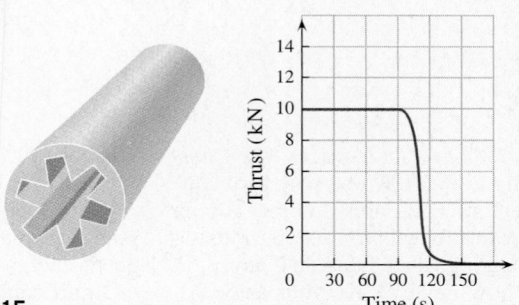

Figure P15

16. [II] A rigid steel crowbar is rested on an upright brick standing near the back of a car. The end of the bar touches the bottom of the bumper 30 cm from its pivot point on the brick. Someone pushes straight down on the other end of the bar, 270 cm from the pivot point, and raises the car's chassis 5.0 cm. Given ideally that the work-in equals the work-out, if it takes 3200 N to raise the car, how much force did the person exert on the bar?

> **SOLUTION:** The two distances to the pivot are 30 cm and 270 cm which are in the ratio of 1 to 9. Accordingly, the geometry is such that if one end goes up 5.0 cm, the other goes down 9 times as much, or 45.0 cm. The work done on the car is $W = Fh = (3200 \text{ N})(5.0 \times 10^{-2} \text{ m}) = 1.6 \times 10^2$ J and that must be the work done by the person. Thus 1.6×10^2 J $= Fl$, $F = (1.6 \times 10^2 \text{ J})/(45.0 \times 10^{-2} \text{ m}) = 3.6 \times 10^2$ N.

17. [II] Assuming no friction and weightless pulleys, how much rope will have to be drawn off if the weight in Fig. P17 is to be raised 1.0 m? Remember: ideally, work-in equals work-out.

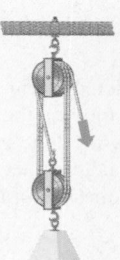

Figure P17

18. [II] A pickup truck is hauling a barge along a canal at a constant speed. The truck, driving parallel to the waterway, is attached to the barge by a cable tied to the bow making a 30° angle with the forward direction. If the truck exerts a force of 1000 N on the cable, how much work is done in overcoming friction as the barge is moved 10 km?

19. [II] A person having a mass of 59.1 kg stands before a flight of 30 stairs each of which is 25.0 cm high. He runs up 20 stairs, turns around, walks down 10, changes his mind, and goes up the remaining 20. How much work did he do in overcoming gravity?

20. [II] While testing a model of a cannon, a 1.0-kg ball is fired straight up into the air. It rises 22.5 m and falls back to the height at which it was launched. What is the net amount of work done on the ball by gravity?

> **SOLUTION:** Gravity does the same amount of negative work on the way up as positive work on the way down, and so the net work it does is zero.

21. [II] A 100-N box, which is on the ground, is slid along a 13-m-long ramp up to a platform 5.0 m above the ground. How much work is done if friction is negligible? How much work would have been done if the box were lifted straight up to the platform? [*Hint: Use the definition of work where it is the component of the weight down the incline ($F_w \sin \theta$) that must be overcome. Notice that $\sin \theta = 5/13$.*]

22. [II] Several crates, having a total weight of 400 N, are loaded into a 10.0-kg wagon that is then pulled up a wooden ramp 10.0-m long making an angle of 30.0°. Knowing that friction was negligible, how much work was done?

23. [II] Starting from rest, a 25.0-kg kid runs up a 5.0-m long slide. At the end he is standing still 3.0 m higher than at the start; how much work did he do?

24. [II] A 50-kg keg of beer slides upright down a 3.0-m-long plank leading from the back of a truck 1.5 m high to the ground. Determine the amount of work done on the keg by gravity.

> **SOLUTION:** The keg descends 1.5 m in the gravitational field. $W_g = F_w h = mgh = (50 \text{ kg})(9.81 \text{ m/s}^2)(1.5 \text{ m}) = 7.4 \times 10^2$ J.

25. [II] The newspaper boy in Problem 14 pushes his cart 25 m along a road inclined at 10°. How much work does he do to overcome road friction?

26. [II] Five fat dictionaries, each 10-cm thick and each 2.5 kg, are resting side by side flat on a table 1.0 m high. How much work would it take to stack them one atop the other?

27. [II] A constant force of 100 N is applied to an object over a straight distance of 2.0 m parallel to the displacement. Draw a work diagram (F vs. l) with a scale of 1.0 cm = 10 N and 1.0 cm = 0.25 m. (a) What was the total work done? (b) How much work does an area of 1.0 cm^2 equal? (c) What is the area under the curve?

28. [II] Efficiency is defined as work-out/work-in. Muscles operate with an efficiency of about 20% in converting energy internally into work externally. Accordingly, how much energy will be expended

by a 60-kg person in the process of ascending several flights of stairs to a height of 25 m?

29. [II] A variable force, depicted in Fig. P29, acts on a 10-kg body. How much work does it do in moving the body from $x = 2$ to $x = 6$? What is the total amount of work done in going from $x = 0$ to $x = 10$? Over which 1-m interval was the smallest amount of work done? How much work is done in going from $x = 2$ to $x = 12$?

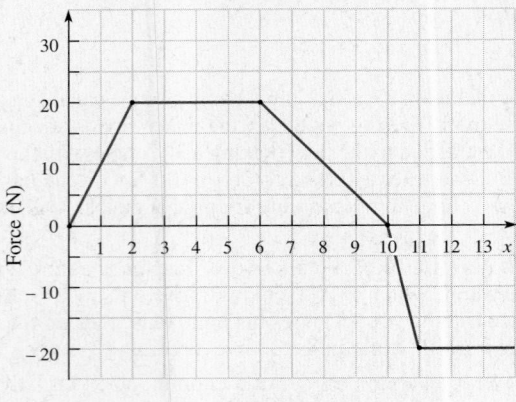

Figure P29

30. [II] THIS PROBLEM EXAMINES WORK DONE OVERCOMING FRICTION. A locked car weighing 6.0 kN has its parking brake on. The coefficient of kinetic friction for rubber on concrete is 1. (a) What will be the value of the friction force on the car once it is set in motion? (b) What force must be exerted horizontally to slide the car at a slow constant speed? (c) How much work will be expended in pushing the car, very slowly, 10.0 m horizontally?

31. [II] THIS PROBLEM EXAMINES WORK DONE OVERCOMING FRICTION. A string making an angle of 30.0° up from the horizontal is attached to a bust of Newton which is being dragged at a constant speed across the floor. (a) What is the significance of the fact that the speed is constant? (b) If the tension in the string is 10.0 N, what is the magnitude of the component of the tensile force parallel to the floor? (c) How much work is done in overcoming friction while the bust moves 3.0 m?

SECTION 6.2: KINETIC ENERGY

32. [I] Which graph in Fig. P32 best represents the kinetic energy of a mass as a function of speed?

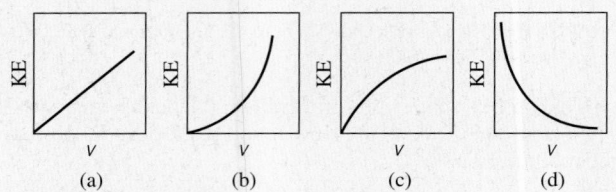

Figure P32

33. [I] Rockets fire, and a 3.20×10^4 kg spaceship moves straight away from a docking station reaching a top speed of 200 m/s. With respect to an observer on the station, what's the ship's maximum kinetic energy?

34. [I] The record average speed for the men's 10-km walk is 4.4 m/s. How much kinetic energy would a 70-kg athlete have at that speed?

35. [I] The men's world swimming record for the 50-m freestyle corresponds to an average speed of about 2.29 m/s. If the swimmer has a mass of 75.0 kg, what's his average kinetic energy during the race?

36. [I] A major concern in the design of spacecraft is the presence in space of tiny, high-speed meteoroids. Micrometeoroids, as they're called, have been detected traveling as fast as 70 km/s. Compute the kinetic energy of a 1.0-g mass moving at that rate.

37. [I] THIS PROBLEM DEALS WITH FORCE AND KINETIC ENERGY. Figure P37 is the force versus distance curve for a changing force acting on an object that was at rest at $t = 0$. The direction of motion, across a frictionless horizontal surface, is always along the line-of-action of the force. (a) When is the object's acceleration constant? (b) When is the object's acceleration not constant? (c) When is the object's acceleration zero? (d) Is the object moving and/or accelerating at each of the following points: point-A, point-B, point-C, point-E, point-F, and point-H? (e) During which interval, if any, was the kinetic energy nonzero and constant?

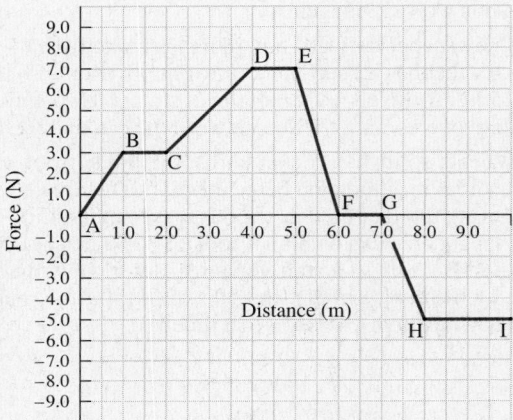

Figure P37

38. [I] THIS PROBLEM DEALS WITH FORCE AND KINETIC ENERGY. Referring to Fig. P37 and with Problem 37 in mind, (a) compare the object's KE at point-A to that at point-B, which is larger? (b) Compare the object's KE at point-B to that at point-C, which is larger? (c) During which interval, if any, did the kinetic energy decrease and by how much? Give your answer to two significant figures.

39. [I] It has been suggested that a controlled fusion reaction (a mini-H-bomb) could be achieved by causing a small (on the order of 1 mm), extremely high-speed projectile to hit a stationary target, both made of the appropriate materials. Determine the kinetic energy of a 0.5-g mass traveling at 200 km/s.

40. [I] One ton of uranium-235 can provide about 7.4×10^{16} J of nuclear energy. If that much energy went into accelerating a 3.5×10^6-kg spaceship (that's the size of a fully loaded *Saturn V* Moon rocket) from rest, what would its final speed be?

> **SOLUTION:** There are 7.4×10^{16} J available that will go into KE and so KE $= \frac{1}{2}mv^2 = \frac{1}{2}(3.5 \times 10^6 \text{ kg})v^2 = 7.4 \times 10^{16}$ J: solving for v: $v^2 = 2(7.4 \times 10^{16} \text{ J})/(3.5 \times 10^6 \text{ kg}) = 4.229 \times 10^{10} \text{ m}^2/\text{s}^2$: $v = 2.1 \times 10^5$ m/s.

41. [I] A 6.5-g bullet is fired from a 2.0-kg rifle with a speed of 300 m/s. What is the kinetic energy of the bullet?

42. [II] Suppose that a 0.149-kg baseball is traveling at 40.0 m/s. How much work must be done on the ball to stop it? If it's brought to rest in 2.0 cm, what average force must act on the ball?

43. [II] The shot used by male shot-putters has a mass of 7.26 kg. A

good throw (which might go roughly 23 m) would correspond to a launch at about 14 m/s. Determine the shot's kinetic energy and compare that to the kinetic energy of a 149-g baseball thrown at a record pitching speed of 45 m/s. Can you explain the difference, that is, where does that energy difference go when a baseball is thrown? Surely the pitcher is trying just as hard as the shot-putter.

44. [II] During a throw (Problem 43) a shot is initially swung around (see Fig. P44) in a circle reaching a speed of about 3.5 m/s. It is then accelerated, more or less, in a straight line over a distance of about 1.7 m, leaving the hand at roughly 14 m/s. (a) How much kinetic ener-gy does the shot initially get in the turning phase? (b) How much kinetic energy does it pick up in the straight-line portion of the launch? (c) What's the average force exerted on the shot during this latter part of the launch? Compare this to the measured peak force of 600 N. Is that reasonable?

Figure P44

45. [II] While swinging a golf club, about 30% of the work done by the player goes into the KE of his arms and body, 20% becomes the KE of the club's shaft, and the remaining 50% ends up as the KE of the head. Typically, a 0.20-kg club head attains a top speed of 50 m/s. Determine the total amount of work done by the swinger.

46. [II] A video of a 70.0-kg male sprinter shows him going from zero to 3.0 m/s on the first step, reaching 4.2 m/s on the second step, and 5.1 m/s on the third step. Compare the speed he gained in each step and the energy gained in each. What can be said about the work he was doing? Remember too that the faster a muscle must act, the less work it can perform.

47. [II] Laboratory studies have shown that a runner at a speed of 6 m/s in still air uses about 7.5 % of his total energy output in over-coming air drag. And that increases to 13% at about 10 m/s. Ignoring all other losses, how much metabolic energy, or work, must a 70-kg sprinter do to reach a speed of 10 m/s?

48. [III] A runner accelerates to a top speed of 9.9 m/s, taking a number of strides (N). If the average horizontal force exerted on the ground is 1.5 times body-weight, and if it acts during each stride over a length of $\frac{1}{3}$ m, determine the value of N.

49. [III] A 0.046-kg golf ball is driven from rest to 70 m/s in about 1.0 ms. Determine the kinetic energy of the ball, and approx-imate the distance over which the club acted on the ball.

SECTION 6.3: POTENTIAL ENERGY

50. [I] What is the kinetic and potential energy of a Boeing 747 air-liner weighing 2.22×10^6 N, flying at 268 m/s (i.e., 600 mi/h) at an altitude of 6.1 km (i.e., 20×10^3 ft)?

    SOLUTION: $KE = \frac{1}{2}mv^2 = \frac{1}{2}[(2.22 \times 10^6 \text{ N})/g](268 \text{ m/s})^2 = 8.13 \times 10^9$ J: $PE_G = mgh = (2.22 \times 10^6 \text{ N})/g)(g)(6.1 \times 10^3 \text{ m}) = 1.35 \times 10^{10}$ J.

51. [I] A single barrel of oil contains the equivalent *chemical*-PE of about 6×10^9 J. How high in the air could that much energy raise a million kilogram load, assuming it is all converted to *gravitation-al*-PE? [*Hint: Use the definition of* PE_G *assuming g is constant.*]

52. [I] The energy content of beer is about 1.8×10^6 J/kg. If that energy could be turned completely into *gravitational*-PE, how much beer would we need to raise a 1.0-kg mass 1.0 km into the air?

53. [I] The daily food intake for an adult male is equivalent to about 1.3×10^7 J. Assuming 100% efficiency in its utilization,

roughly how high a mountain could an 80-kg man climb on that much energy?

54. [I] A uniform rod 6.0-m long, weighing 60 N, is pivoted about a horizontal axis 1.0 m up from its center so that it hangs vertical-ly. Neglecting friction, (a) what's the net work done on the rod by gravity in the process of raising its end, making it horizontal? (See if you can come up with two ways to do this.) (b) What change in potential energy, if any, has the rod experienced?

55. [I] THIS PROBLEM EXAMINES THE RELATIVE NATURE OF POTENTIAL ENERGY. Figure P55 shows a car of mass m at the start of a roller-coaster ride. (a) Write an expression for the weight of the car. (b) What are the heights of point-A, point-B, and point-D above the ground? (b) Write expressions for the car's gravitational potential energy at point-A, point-B, and point-D with respect to the ground as the zero of PE_G.

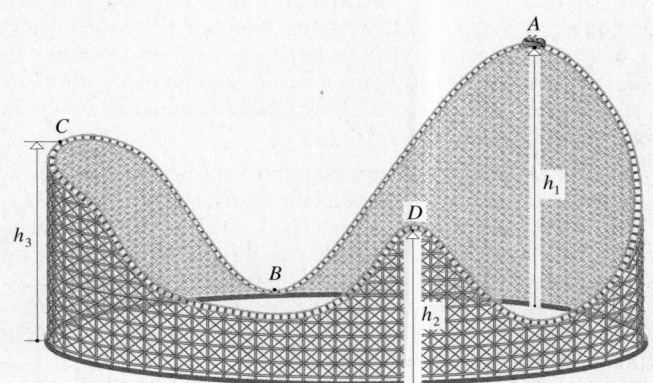

Figure P55

56. [I] THIS PROBLEM EXAMINES THE RELATIVE NATURE OF POTEN-TIAL ENERGY. Figure P55 shows a car of mass m at the start of a rollercoaster ride. (a) What change in height does it experience in going from point-A to point-D? (b) Write an expression for the change in the car's gravitational potential energy when it goes from point-A to point-D. Take the ground as the zero of PE_G: $h_1 > h_2 > h_3$.

57. [I] THIS PROBLEM EXAMINES THE RELATIVE NATURE OF POTEN-TIAL ENERGY. Figure P55 shows a car of mass m at the start of a roller-coaster ride. (a) What is the height of point-A, measured with respect to point-C? (b) Write an expression for the car's gravita-tional potential energy at point-A with respect to point-C as the zero of PE_G.

58. [I] THIS PROBLEM EXAMINES THE RELATIVE NATURE OF POTEN-TIAL ENERGY. Figure P55 shows a car of mass m at the start of a roller-coaster ride. (a) What are the heights of point-A and point-D measured with respect to point-C? (b) Write an expression for the change in the car's gravitational potential energy when it goes from point-A to point-D. Take point-C as the zero of PE_G: $h_1 > h_2 > h_3$.

59. [I] A car with a mass of 1000 kg is at rest at the base of a hill. It accelerates up the incline reaching a speed of 20 m/s at a height of 100 m. What is the total increase in mechanical energy (KE plus PE_G) at that point?

60. [II] A 60-kg person is slowly doing chin-ups, during each one of which her mass (i.e. her c.g., p. 249) can be considered to rise just about $\frac{1}{2}$ m. If the biceps contract roughly 4.0 cm in the process of each lift, how much tension is there, on average, in the muscles of each arm?

61. [II] A 1000-kg car at rest at the top of a hill accelerates down the road, reaching a speed of 20 m/s after descending a height of

100 m. What was its total change in mechanical energy (KE plus PE_G) as of that moment?

62. [II] A 10.0-kg package is raised from rest by an elevator at a constant acceleration of 2.00 m/s^2 for 20.0 s. (a) What is its KE at $t = 10.0$ s? (b) What is its increase in gravitational potential energy after 10.0 s? (c) Assuming no losses, how much work was done on it by the elevator in 10.0 s?

> **SOLUTION:** (a) First find the speed at $t = 10.0$ s: $v = at = (2.00$ m/s^2)(10.0 s) = 20.0 m/s and so KE $= \frac{1}{2}mv^2 = \frac{1}{2}(10.0$ kg)(20.0 m/s)^2 $= 2.00$ kJ. (b) We need the height: $l = \frac{1}{2}at^2 = \frac{1}{2}(2.00$ m/s^2)(10.0 s)$^2 =$ 100 m. Hence, $PE_G = mgh = (10.0$ kg)$g(100$ m$) = 9.81 \times 10^3$ J. $W =$ $\Delta KE + \Delta PE_G = 2.00$ kJ $+ 9.81$ kJ $= 11.81$ kJ.

63. [II] A 100-N weight sitting on the floor is tied to a light rope that passes over a very light frictionless pulley 20 m above it. The other end of the rope hangs down to the floor where it is being held by a 10-N monkey named George. Suppose George now climbs 10 m above the floor. (a) How much work does he do? (b) How much rope ends up on the floor? (c) What is the total change in *gravitational*-PE of the system?

64. [II] During a vertical jump, a person crouches down to lower his or her center and then leaps straight up. The leg muscles essentially do all the work, accelerating the body over a push-off distance of about $\frac{1}{3}$ m. Typically, people can support an additional load using the legs equal to their own weight, but only with considerable effort. Let's suppose then that our person can push off with a muscle-force equal to 1.5 times body-weight. Neglecting losses due to friction, how high can our friend jump?

65. [II] Imagine that you have a large number (N) of thin metal plates each of mass m that are to be stacked one on top of the other to make a vertical column. (a) Approximately how much work must be done to raise the last plate to the top if the height is h? (b) Roughly how much *gravitational*-PE is stored in the column?

66. [II] Consider a vertical cylindrical storage tank, sealed at the bottom and rising to a height h. With the last problem in mind, derive an equation for the exact increase in PE_G that results when M kilograms of water are pumped up from a stream at the level of the base, filling the tank.

67. [II] There are three identical flat stone blocks each 0.50 m high lying on the ground. What is the increase in gravitational potential energy when the blocks are stacked one on top of the other to a height of 1.50 m? Each block weighs 1000 N.

68. [III] A vertical cylinder 2.0-m tall with an inner diameter of 0.305 m is fitted with a piston having a mass of 68.0 kg that can move vertically frictionlessly. A valve at the bottom of the cylinder admits water under pressure and the piston rises 1.22 m up from the bottom. What is the resulting increase in the potential energy of the system given that water has a density of 1.00×10^3 kg/m^3? [*Hint: Remember Problem 66.*]

69. [III] (a) Determine the value of the gravitational acceleration at the surface of the Moon. (b) What is the change in the *gravitational*-PE of a 1000-kg spacecraft when it has risen 100 m above the surface? Take the radius of the Moon to be one-fourth that of the Earth and the mass to be about 100 times less.

70. [III] A locomotive exerts a maximum force of 22.7×10^4 N while pulling a freight train up a $\frac{1}{2}$% grade (i.e., the rise in the roadbed is $\frac{1}{2}$ m per 100 m). The train is 45 cars long, and each one loaded weighs 4.0×10^5 N. The total friction varies with the load and speed, but in this case, 35 N per 10^4 N of weight is a reasonable number. Using energy considerations, how far will the train travel while accelerating from a speed of 6.7 m/s to twice that?

SECTION 6.4: MECHANICAL ENERGY
SECTION 6.5: APPLYING CONSERVATION OF ENERGY

71. [I] What is the kinetic energy of a 10.0-kg piece of concrete after it has fallen for 2.00 s from the facade of an old building?

72. [I] A 60-kg stuntperson runs off a cliff at 5.0 m/s and lands safely in the river 10.0 m below. What was the splashdown speed?

> **SOLUTION::** Using Conservation of Energy $E_i = \frac{1}{2}mv_i^2 + mgh =$ $E_f = \frac{1}{2}mv_f^2$: $v_f^2 = v_i^2 + 2gh$. Solving this for the final speed; $v_f =$ $\sqrt{v_i^2 + 2gh} = \sqrt{(5.0 \text{ m/s})^2 + 2(9.81 \text{ m/s}^2)(10.0 \text{ m})} = 15$ m/s.

73. [I] Two automobiles of weight 7.12 kN and 14.24 kN are traveling along horizontally at 96 km/h when they both run out of gas. Luckily, there is a town in a valley not far off, but it's just beyond a 33.5-m-high hill. Assuming that friction can be neglected, which of the cars will make it to town? [*Hint:* $KE_f + PE_{Gf} = KE_i + PE_{Gi}$.]

74. [I] While traveling along at 96 km/h, a 14.2-kN auto runs out of gas 16 km from a service station. Neglecting friction, if the station is on a level 15.2 m above the elevation where the car stalled, how fast will the car be going when it rolls into the station, if in fact it gets there?

75. [I] A kid in a wagon is traveling at 10 m/s just as she reaches the bottom of a hill and begins to climb a second hill. How high up it will she get before the wagon stops, assuming negligible friction losses?

76. [I] A ball having a mass of 0.50 kg is thrown straight up at a speed of 25.0 m/s. (a) How high will it go if there is no friction? (b) If it rises 22 m, what was the average force due to air friction?

77. [I] An athlete whose mass is 55.0 kg steps off a 10.0-m-high platform and drops onto a trampoline, which, while stretching, brings her to a stop 1.00 m above the ground. Assuming no losses, how much energy must have momentarily been stored in the trampoline as she came to rest? How high will she rise?

78. [I] What is the potential energy of a 1.0-kg mass sitting on the surface of the Earth if we take the zero of PE_G at infinity?

79. [I] Figure P55 shows a car of mass m at the start of a roller coaster ride where its speed is just about zero. Write an expression for its speed at point-D.

80. [II] A constant force is applied to a 2.5-kg mass which travels from rest up a frictionless 30.0°-inclined plane. The force is 20.0 N parallel to the incline. If the mass moves 10.0 m along the surface of the incline, how much work was done on the object? By how much did its potential energy increase? What's its final speed? Check your answer using $F = ma$.

81. [II] Referring to Problem 78, compute the *gravitational*-PE of a 1.0-kg object at distances of 1, 2, 3, 4, 5, and 10 Earth-radii. Draw a plot of PE_G against r measured from the center of the planet in units of Earth-radii and $r > 1$. Now suppose the object is fired upward with a KE of 5.2×10^7 J. What is its initial total mechanical energy? Draw a horizontal line on your diagram representing E. Where does it cross your curve, and what is the significance of that point?

82. [II] A 1000-kg car racing up a mountain road runs out of gas at a height of 35 m while traveling at 22 m/s. Cleverly, the driver shifts into neutral and coasts onward. Neglecting all friction losses, will he clear the 65-m peak? Would it help to throw out any extra weight or even jump out and run alongside the car? Not having any brakes, at what speed will he reach the bottom of the mountain?

SOLUTION: The car has to coast a height $\Delta h = 65$ m $- 35$ m $= 30$ m. At the highest point the car reaches $\Delta PE_G = KE_i$ and so $mg\Delta h = \frac{1}{2}mv_i^2$: using $\Delta h = 30$ m the car needs an initial speed such that $(9.81$ m/s$^2)$ $(30$ m$) = \frac{1}{2}v_i^2$ and so $v_i = 24.26$ m/s. Thus, traveling at 22 m/s the car will not get to the top. It will go up to some height, stop, and roll back to a height of 35 m again traveling at 22 m/s. Descending 35 m, it will reach a speed v_f such that $mg\Delta h + \frac{1}{2}mv_i^2 = \frac{1}{2}mv_f^2$: $2g\Delta h + v_i^2 = v_f^2$: $v_f = 34$ m/s.

83. [II] If the mass of the Moon is 7.4×10^{22} kg and its radius is 1.74×10^6 m, compute the speed with which an object would have to be fired in order to sail away from it, completely overcoming the Moon's gravity pull.

84. [II] A pendulum consists of a small spherical mass attached to a rope so that its center hangs down a distance L beneath the suspension point. The bob is displaced so that the taut string makes an angle of θ_i with the vertical, whereupon it is let loose and swings downward. Show that its maximum speed is given by

$$v_{max}^2 = 2gL(1 - \cos \theta_i)$$

85. [II] It's been suggested that we mine either the Moon or some of the asteroids in order to get raw material from which to build space stations. The idea is that removing material from the Moon "would consume only five percent of the energy needed to lift the same payload off Earth." Show that this conclusion is roughly true.

86. [II] At what speed should a space probe be fired from the Earth if it is required to still be traveling at a speed of 5.00 km/s, even after coasting to an exceedingly great distance from the planet (a distance that is essentially infinite)?

87. [II] A satellite is in a circular orbit about the Earth moving at a speed of 1500 m/s. It is desired that by firing its rocket, the craft attain a speed that will allow it to escape the planet. What must that speed be?

88. [III] An inclined plane at 30.0° is 6.40-m long. A book, which has a kinetic coefficient of friction with the incline of 0.20, is placed at the top and immediately begins to slide. Using energy considerations, how long will it take for the book to reach the bottom of the incline?

89. [III] Meteorites typically strike Earth's atmosphere at speeds ranging from 1.4×10^4 m/s, or 9 mi/s, to about 2.5×10^4 m/s, or 16 mi/s. On passing through, they lose both mass and speed so that a chunk weighing several newtons (a few pounds) is likely to be traveling between 120 and 240 m/s when it hits ground. Determine a good theoretical number for the minimum speed with which a meteorite will enter the atmosphere.

90. [III] The coefficient of restitution of two colliding bodies is defined as the ratio of their "relative speeds" (after the impact to before the impact). Imagine that we drop a sphere made of some material of interest from an initial height onto a test anvil and measure the final height to which the ball bounces. Derive an expression for the coefficient of restitution in terms of these two heights. If the coefficient for glass on steel is 0.96, to what height will a glass marble bounce off a steel plate when dropped from 1.0 m?

91. [III] A light rope is passed over a weightless, frictionless pulley, and masses m_1 and m_2 are attached to its ends. The arrangement is called Atwood's Machine. The starting configuration corresponds to both masses held at rest at the same height. The two are then released. During some arbitrary interval of time, the heavier one falls a distance y while the lighter one rises a distance y. Derive an expression for the speed of either body in terms of g, y, and the masses.

SECTION 6.6: POWER

92. [I] THIS PROBLEM EXAMINES ENERGY AND POWER. The maximum power that can be developed by an athlete is 2×10^2 W. (a) How much energy can that athlete develop in 1 s? (b) How much energy does such a person expend in 20 s of exertion?

93. [I] THIS PROBLEM EXAMINES ENERGY AND POWER. The energy expended in doing one push-up is about 3×10^2 J. (a) How much power is needed to do 1 push-up per second? (b) How much power must be developed if someone is to do 10 push-ups in 5.0 s?

94. [I] How much power does it take to raise an object weighing 100 N a distance of 20.0 m in 50.0 s?

95. [I] A runner traverses a 50-m-long stretch on a horizontal track in 10 s at a fairly constant speed. All the while she experiences a retarding force of 1.0 N due to air friction. What power was developed in overcoming that friction?

96. [I] A small hoist can raise 100 kg of bricks to the top of a construction project 30 m above the street in half a minute. Determine the power provided.

SOLUTION: The work done by the hoist is $W = \Delta PE_G = mgh = (100$ kg$)(9.81$ m/s$^2)(30$ m$) = 2.94 \times 10^4$ J and $P = \Delta W/\Delta t = (2.94 \times 10^4$ J$)/(30$ s$) = 9.8 \times 10^2$ W.

97. [I] The per capita power consumption in the United States is around 10 kJ each and every second. At what speed would you have to push a car exerting a force of 1.0 kN on it all year, day in and day out, to be equivalent to your share?

98. [I] Prove that 1 hp $= 746$ W.

99. [I] In Problem 39 we talked about inducing nuclear fusion via a high-speed collision. If the projectile has a mass of 0.5 g and is traveling at 200 km/s, how much power would be provided to the target if the collision lasted 10 ns?

100. [I] A 17.8-kN car is to be accelerated from 0 to 96.5 km/h (60 mi/h) in 10 s. Neglecting all frictional losses, how much power must be supplied in the process (independent of the exact nature of the acceleration)? Incidentally, air friction increases rapidly with speed, becoming a major concern above 64 km/h (40 mi/h).

101. [II] A well has water 20.0 m down from ground level. How much power must a motor supply to a pump if it is to raise 180 liters $(180 \times 10^{-3}$ m$^3)$ of water per minute to the surface? The density of water (mass per unit volume) is $(1.00 \times 10^3$ kg/m$^3)$.

102. [II] The Zambesi River in Africa rushes over Victoria Falls at a rate of 25×10^6 gallons per minute. The falls are 108 m high, 1 gallon equals 3.785×10^{-3} m^3 (or 0.133 7 cubic foot), and water has a density of 1.00×10^3 kg/m^3 (each cubic foot weighs 62.4 pounds). Determine the power developed by the water in SI units.

103. [II] A 2.5-hp motor drives a hoist that can raise a load of 50 kg to a height of 20 m. At full power, how long will the hoist take to do it?

104. [II] The belt connecting an auto engine to an air conditioner is moving at 40 m/min and has an effective tension of 20 N. How

much work does it do in an hour? How much power does it transmit?

> SOLUTION: In 1 hour the distance moved is $l = (40 \text{ m/min})(60 \text{ min/h})$
> $= 2400$ m and in an hour $W = Fl = (20 \text{ N})(2400 \text{ m}) = 48 \times 10^3$ J. $P = \Delta W/\Delta t = (48 \times 10^3 \text{ J})/(3600 \text{ s}) = 13$ W.

105. [II] An animal's power level is known as its **metabolic rate**. If a man typically uses 10 MJ in the course of a day, what is his metabolic rate? While resting but awake, an average person has a *basal* power level of about 90 W. This figure, however, depends on the size of the person, being larger for larger people, who have greater surface areas and thus lose more energy. So to standardize the notion, making it independent of body size, it is customary to divide by the skin surface area (typically 1.5 m²)—the number thus arrived at is the *basal metabolic rate* or BMR. What is the BMR for our average person?

106. [II] The oxygen taken in by the body reacts with fats, carbohydrates, and protein, liberating energy *internally* at a rate of about 2.0×10^4 joules per liter. If a 70-kg man requires 77 W of power even while sleeping, what is his rate of oxygen consumption?

107. [II] Refer to Problems 105 and 106. What is the BMR of a person who (according to the standard charts) has a surface area of 1.8 m² and is consuming oxygen at a rate of 0.40 liters per minute?

108. [II] (a) What is the KE of a 17.8-kN car traveling at 64 km/h? (b) Assuming the car gets 20 miles to the gallon at that speed, and each gallon of gasoline provides 1.3×10^8 J of chemical energy, what's its power consumption? Incidentally, of this total, only about 20% gets into the mechanical system (pumps, transmission, etc.). About 10^4 J/s goes into propelling the car, of which half is dissipated in overcoming tire and road friction and half in air friction. Older cars typically lose about 10^3 J/s in unburned fuel that evaporates from the carburetor and never even gets into the engine.

109. [II] A 2×10^3-kg spacecraft has an ion engine that delivers 150 kW of propulsive power via its six thrusters. Assuming it to be far from any massive object, how much time would it take to accelerate from "rest" to 268 m/s (i.e., 600 mi/h)?

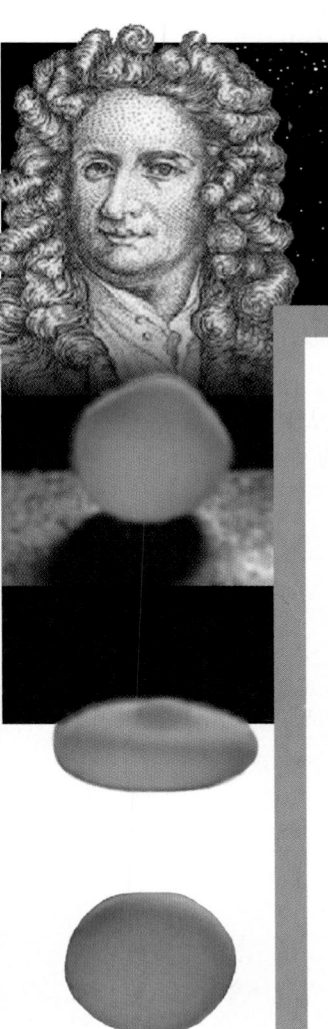

Chapter 7
Momentum & Collisions

*W*e return now to the concept of momentum. In the twentieth century, we learned that, along with energy, momentum is one of the premier notions of physics. When there is no net external force acting on a system, its momentum remains unchanged and we have the all-important Law of Conservation of Linear Momentum. Indeed, it can be argued that Newton's Laws are the direct result of Conservation of Momentum. In all of physics there is only a handful of these wonderful conservation laws, and each is related to some fundamental symmetry of the Universe (p. 226).

Linear Momentum

Newton formulated his version of dynamics (p. 85) based on the concept of **momentum** ($\vec{\mathbf{p}}$), which he defined as ***the product of mass and velocity***:

Momentum ($\vec{\mathbf{p}}$) is mass times velocity.

$$\vec{\mathbf{p}} = m\vec{\mathbf{v}} \qquad\qquad (7.1)$$

and because velocity is a vector quantity, so too is momentum. The direction of $\vec{\mathbf{p}}$ is the direction of $\vec{\mathbf{v}}$; the scalar value of $\vec{\mathbf{p}}$ is mv with SI units of kg·m/s. When $\vec{\mathbf{p}}$ is constant, there is straight-line motion, and so it's more precisely called *linear momentum*.

A critical feature of momentum is that oppositely directed momenta cancel, which is a key quality of vectors. If a train travels with a velocity of 20 km/h–EAST, and you jog along on its flatcars with a velocity of 20 km/h–WEST, you can run right off the end, motionless with respect to the ground. You will drop *straight down* as if you had stepped off a chair in your living room. Your horizontal velocity with respect to the Earth up there in midair will be zero, and your momentum will be zero as well. Like velocity, linear momentum is relative. The momentum of an object moving at $\vec{\mathbf{v}}$ with respect to some observer is $m\vec{\mathbf{v}}$, with respect to that observer. The momentum of a pilot flying at 800 km/h is zero with respect to the plane. {There's a set of interactive multiple choice **WARM-UPS** that deal with **MOMENTUM** in **CHAPTER 7** on the **CD**.}

Example 7.1 **[I]** Rich Gossage set a fastball record by hurling a 0.14-kg baseball at a speed of 46.3 m/s (i.e., 153 ft/s). What was the magnitude of the ball's momentum as it left his hand? Here everything is measured with respect to the Earth.

Solution The mention of "momentum" should bring to mind its definition. (1) TRANSLATION—An object has a known mass and speed; determine its linear momentum. (2) GIVEN: $m_B = 0.14$ kg and $v_B = 46.3$ m/s. FIND: p_B. (3) PROBLEM TYPE—Linear momentum. (4) PROCEDURE—Use the definition of momentum. (5) CALCULATION—The magnitude of the momentum is provided by the scalar form of Eq. (7.1):

$$p_B = m_B v_B = (0.14 \text{ kg})(46.3 \text{ m/s}) = \boxed{6.5 \text{ kg} \cdot \text{m/s}}$$

Quick Check: A baseball thrown at a casual 7.2 m/s has a momentum of $(0.14 \text{ kg})(7.2 \text{ m/s}) = 1$ kg·m/s, so the above momentum at a speed ≈ 6.4 times faster is correct.

[For more worked problems click on **WALK-THROUGHS** *in* **CHAPTER 7** *on the CD.]*

Example 7.2 **[II]** A 0.149-kg baseball traveling at 28 m/s due south approaches a waiting batter. The ball is hit and momentarily crushed; it springs back, sailing away at 46 m/s due north. Determine the magnitudes of its initial and final momenta and the change in its momentum.

Solution Keep in mind that momentum is a vector quantity; the ball has to be stopped and turned around. (1) TRANSLATION—An object with a known mass and velocity has its velocity changed; determine the change in its linear momentum. (2) GIVEN: $m_B = 0.149$ kg, $\vec{v}_i = 28$ m/s–SOUTH, and $\vec{v}_f = 46$ m/s–NORTH. FIND: p_i, p_f, and $\Delta\vec{p}$. (3) PROBLEM TYPE—Linear momentum. (4) PROCEDURE—Use the definition of momentum, keeping in mind that it's a vector quantity. (5) CALCULATION—***Taking north as positive***

$$p_i = m_B v_i = (0.149 \text{ kg})(-28 \text{ m/s}) = \boxed{-4.2 \text{ kg} \cdot \text{m/s}}$$

$$p_f = m_B v_f = (0.149 \text{ kg})(+46 \text{ m/s}) = \boxed{+6.9 \text{ kg} \cdot \text{m/s}}$$

with $$\Delta\vec{p} = \vec{p}_f - \vec{p}_i$$

we can either visualize the subtraction vectorially, reversing $\vec{p}_i$ and adding it tip-to-tail to $\vec{p}_f$, or subtract the scalar values, whereupon

$$\Delta p = (+6.9 \text{ kg} \cdot \text{m/s}) - (-4.2 \text{ kg} \cdot \text{m/s}) = +11 \text{ kg} \cdot \text{m/s}$$

Either way, $\boxed{\Delta\vec{p} = 11 \text{ kg} \cdot \text{m/s–NORTH}}$.

Quick Check: $p_i/p_f = v_i/v_f = 28/46 = 4.2/6.9 = 0.61$, so the numbers look right. The ball has had its direction reversed, as if a momentum of 11 kg·m/s–NORTH was added to its original 4.2 kg·m/s–SOUTH by the impact. ***This kind of large change occurs when the motion of an object is turned around in a collision*** (Fig. 7.1).

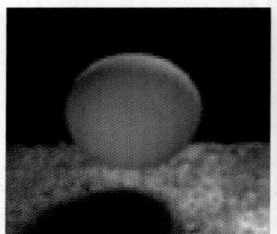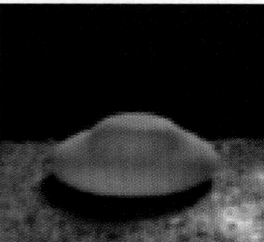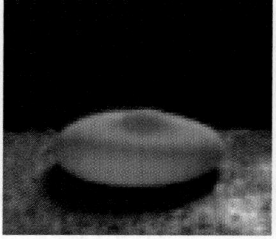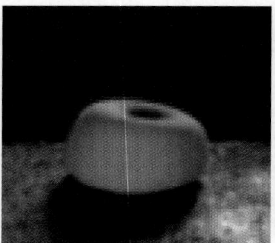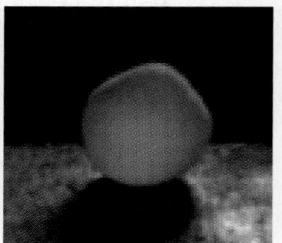

Figure 7.1 A racquetball strikes a surface, compresses, springs back, and moves off in the opposite direction such that $\Delta\vec{p} = \vec{p}_f - \vec{p}_i$ where $\vec{p}_f \approx -\vec{p}_i$.

7.1 Impulse and Momentum Change

When a force is applied to a body, there will be a resulting proportional change in its motion. Earlier we studied that behavior via $\vec{F} = m\vec{a}$; now we'll concentrate on momentum and how it changes. Newton's Second Law states that an applied force (F) produces a proportional change in the momentum of the body (Δp): $F \propto \Delta p$. Since momentum is relative, the significant quantity is not p, which depends on the motion of the observer, but Δp, which does not.

STUDY GUIDE

Momentum is a vector quantity, so you will again have to be careful about signs. Before writing any scalar equations, pick a direction to be positive. The numerical value of the momentum opposite to that direction must be negative. Suppose we have a chicken running at 1.0 m/s and you take its direction to be negative. Then $v_c = -1.0$ m/s, and the chicken's momentum $p_c = m_c v_c$ will also be negative.

Newton's Second Law states that *the net force applied to an object equals the resulting change in its momentum per unit time* [Eq. (4.1)]. Over a finite time interval (Δt), during which the force may change, Newton's Second Law is

[variable $\vec{F}$]
$$\vec{F}_{av} = \frac{\Delta \vec{p}}{\Delta t}$$

The **average net force equals the resulting time rate-of-change of momentum**. For motion along a straight line the scalar form will suffice:

[variable F, straight-line motion]
$$F_{av} = \frac{\Delta p}{\Delta t} \qquad (7.2)$$

Example 7.3 **[I]** A rocket fires its engine, which exerts an average force of 1000 N for 40 s in a fixed direction. What is the magnitude of the rocket's momentum change?

Solution The word "force" should bring to mind Newton's Second Law. (1) TRANSLATION—An object experiences a known force for a specified time interval; determine the resulting change in its momentum. (2) GIVEN: $F_{av} = 1000$ N and $\Delta t = 40$ s. FIND: Δp. (3) PROBLEM TYPE—Linear momentum /

Newton's Second Law. (4) PROCEDURE—The momentum change occurs in the direction of the force, and so we can use the scalar formulation, namely, Eq. (7.2). (5) CALCULATION—$F_{av} = \Delta p / \Delta t$ and

$$\Delta p = F_{av} \Delta t = (1000 \text{ N})(40 \text{ s}) = \boxed{40 \times 10^3 \text{ N} \cdot \text{s}}$$

Quick Check: 1 kN $\times$ 40 s = 40 kN·s.

Newton's early studies had been done around the same period as important work on collisions by Hooke, Wren, Wallis, and Huygens. Billiards was new and very popular at the time, and there was a lot of interest in momentary impacts. In such cases, neither $\vec{F}_{av}$ nor Δt is known very well, but the change in momentum is easily determined. Consequently, it's useful to combine the force and time into a single notion that equals the known momentum change. That single concept ($\vec{F}_{av}\Delta t$) is called the **impulse** of the force, and an alternative statement of the Second Law becomes

Impulse = change in momentum

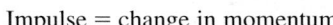

$$\vec{F}_{av} \Delta t = \Delta \vec{p} \qquad (7.3)$$

For motion along a straight line the scalar form will meet our needs:

[straight-line motion]
$$F_{av} \Delta t = \Delta p \qquad (7.4)$$

A given impulse will produce a specific change in momentum no matter what the mass or speed of the recipient body. An object originally at rest will take off in the direction of the net applied force, acquiring a momentum $\Delta p = F_{av} \Delta t$, which is what happens when you throw a dart, discharge a syringe, or strike a golf ball (Fig. 7.2). As long as the club is in contact with the ball, applying force, there is an ongoing gain in momentum in the direction of the force (Table 7.1). Once the ball leaves the club, it flies off under the First Law, forceless and changeless. The longer the barrel of a gun, or the greater the pitcher's throwing motion—the larger will be the time (Δt) during which the propelling force acts and the more the change in momentum (Δp) of the projectile (Fig. 7.3).

In order to turn a pitched baseball arriving at 90 mph (i.e., 40 m/s) into a homerun slam leaving at 110 mph (i.e., 49 m/s), a bat must apply a force of up to about 8000 lb (i.e., 36 kN). During the impact, which lasts only ≈1.25 ms, the ball is crushed to half its diameter (see photo on p. 210).

Force applied to a body already in motion may either increase or decrease its momentum depending on whether $\vec{F}_{av}$ acts parallel or antiparallel to the initial velocity. To slow the *Lunar Excursion Module* as it plunged toward the Moon's surface in 1969, a downward-

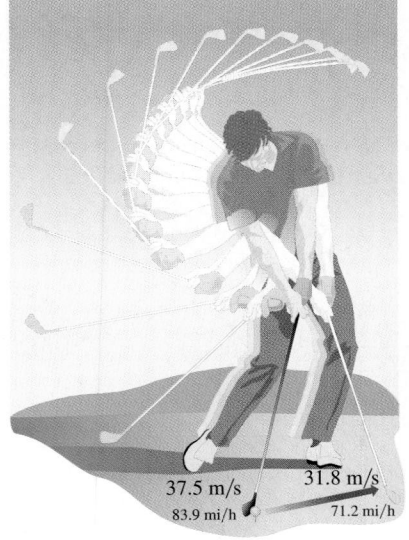

37.5 m/s
83.9 mi/h

31.8 m/s
71.2 mi/h

Figure 7.2 On colliding with the ball, the club lost about 16% of its momentum. The face of the head is tilted, and so the ball sails up and away. Incidentally, because of the First Law, while high in the air at the start of the swing, the shaft bends as the head tends to remain at rest.

Figure 7.3 The pitcher exerts a force on the ball over as long a distance and for as long a time as possible. The greater is $F\Delta t$, the larger is Δp, and the higher the release speed. Here the ball is shown every 1/100th of a second. The greater the distance between images of the ball, the greater the speed.

Table 7.1

TYPICAL PARAMETERS FOR BALLS HIT FROM REST BY THE APPROPRIATE OBJECT

Ball	Mass (kg)	Speed imparted (m/s)	Impact time (ms)
Baseball	0.149	39	1.25
Football (punt)	0.415	28	8
Golf ball (drive)	0.047	69	1
Handball (serve)	0.061	23	12.5
Soccer ball (kick)	0.425	26	8
Tennis ball (serve)	0.058	51	4

An X-ray photo of a 45-caliber bullet fired from a pistol. In effect, the gun pushes on the bullet, and the bullet pushes back on the gun. Some of that reaction goes into operating the slide, exposing the barrel and ejecting the spent shell.

pointing retro-rocket was fired that exerted an upward force on the craft, decreasing its downward momentum and speed. {There's a set of **WARM-UPS** that deal with **IMPULSE** in **CHAPTER 7** on the **CD**.}

Example 7.4 **[II]** On September 12, 1966, a *Gemini* space-craft piloted by astronauts Pete Conrad and Richard Gordon met and docked with an orbiting *Agena* launch vehicle. With plenty of fuel left in the spacecraft, NASA decided to determine the mass of the *Agena*. While coupled, *Gemini*'s motor was fired, exerting a constant thrust of 890 N in a fixed direction for 7.0 s. As a result of that little nudge, the *Gemini-Agena* sped up by 0.93 m/s. Assuming *Gemini*'s mass as a constant 34×10^2 kg, compute the mass of the *Agena*.

Solution The word "thrust" means force, and that should bring to mind Newton's Second Law. (1) TRANSLATION—An object experiences a known force for a known time interval and increases speed by a specified amount; determine its mass. (2) GIVEN: $F_{av} = 890$ N, $\Delta t = 7.0$ s, $\Delta v = +0.93$ m/s, and $m_G = 34 \times 10^2$ kg. FIND: m_A. (3) PROBLEM TYPE—Linear momentum/Newton's Second Law. (4) PROCEDURE—Compute Δp, and from that, knowing Δv, find the mass. The calculation will be inexact to the degree that both the thrust and m_G are assumed constant. (5) CALCULATION—The basic

relationship is $F_{av} \Delta t = \Delta p$, and because $m = m_A + m_G$ is constant,

$$\Delta p = \Delta(mv) = m\Delta v = (m_A + m_G)\Delta v$$

and so

$$F_{av}\Delta t = (m_A + m_G)\Delta v$$

The rocket's impulse produces a well-defined change in speed, regardless of the initial speed. Substituting in the numbers yields

$$(890 \text{ N})(7.0 \text{ s}) = (m_A + 34 \times 10^2 \text{ kg})(0.93 \text{ m/s})$$

Solving for m_A,

$$m_A = \frac{(890 \text{ N})(7.0 \text{ s})}{0.93 \text{ m/s}} - (34 \times 10^2 \text{ kg})$$

and $m_A = 3299$ kg, or to two figures, $\boxed{33 \times 10^2 \text{ kg}}$

Quick Check: Assume the answer is correct and calculate the force; $F_{av} = m\Delta v/\Delta t = (67 \times 10^2 \text{ kg})(0.93 \text{ m/s})/(7.0 \text{ s}) = 890$ N, as given.

7.2 Varying Force

Force may change from moment to moment (the force on your feet as you walk around certainly does). To obtain a more realistic treatment, envision a force that varies in time. At each and every moment, there is a value of the **instantaneous force**; that is, *the limit*

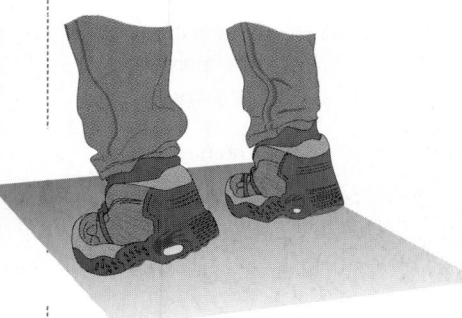

EXPLORING PHYSICS ON YOUR OWN

Impulse & Momentum Change: You can leap into the air as high as you can and still have a soft landing even on a hard floor. That's provided you hit with the balls of your feet (the region just behind the toes) and your knees bent. It then takes lots of time for the feet to collapse before your heels make contact with the floor. Moreover, the trunk of your body continues to descend as your knees bend further: Δt, the impact time, is big, and for the same Δp, F_{av} is small—try it. By contrast, if your knees are locked and you land on your heels, only the flesh pad on the bottom of each heel is there to collapse: Δt is small, and for the same Δp, F_{av} is dangerously large. Even a tiny jump of 1 or 2 cm on concrete will hurt. Don't even try to go any higher! Jump with shoes on and compare that to sneaker landings. Why can you dive into a pool from 10 m up and not get hurt (provided it's filled)?

CATCHING BASEBALLS & BUMPING CARS

It follows from Eq. (7.3) that for a given change in momentum (either an increase or a decrease), the longer the force acts (Δt), the smaller it will have to be. Your hands provide the stopping force for an incoming ball when you catch it; that is, when you change $\vec{p}_i$ to $\vec{p}_f = 0$. If the ball is soft and deforms as it comes to rest, the stopping time will be large and the stopping force will be comparatively small. Even with the same incoming momentum, so that $\Delta \vec{p}$ is the same when the balls are brought to rest, a soft ball will take longer to stop and so require less force to catch than a hard ball. Furthermore, if you wear a padded mitt (which will extend Δt) and draw the glove back toward you just as the ball strikes it, Δt will be still longer and the stopping force still smaller (Fig. 7.4). Automobile bumpers that compress and airbags both extend the stopping time (Δt) and thereby decrease the force of impact.

approached by the average force as the time interval (*over which the averaging is taking place*) *shrinks to zero*. Push on something; the force exerted initially builds, the flesh on your hand compresses and distorts, the force peaks, and then decreases to zero. That's what happens when you shove a book or hurl a shot put. Similarly, a quick blow with a hammer or the blast of an explosive charge will usually produce a force that even more rapidly rises, peaks, and falls off to zero.

Figure 7.5 is a force-time graph for an impact of some sort, a plot of the instantaneous force versus time. The width of the curve and the details of its shape will vary from one situation to another, but the area under the curve is the net impulse imparted to the body. To quantify the impulse, divide the region under the curve into a number of subintervals, each with a little rectangular area whose base is an interval of time: Δt_1, Δt_2, Δt_3, and so forth (Fig. 7.6). Next, fix the height of each rectangle at a value equal to the average force exerted over that particular time interval. For example, the area of the second rectangle is $(F_{av})_2 \Delta t_2$, which corresponds to the *impulse* associated with that subinterval. The sum of all the little subareas is exactly equal to the total impulse exerted by the force.

If the number of subintervals is made increasingly large while the width of each approaches zero ($\Delta t \rightarrow 0$), the tops of the rectangles will blend ever more smoothly into the curve. In other words, the value of the average force for each subinterval will approach the value of the instantaneous force, and these, in turn, are the corresponding points on the curve. Just as the area under any portion of the speed-time curve is the distance traveled during a given period, *the area under the force-time curve is the impulse exerted during that interval and it therefore equals the resulting change in momentum*.

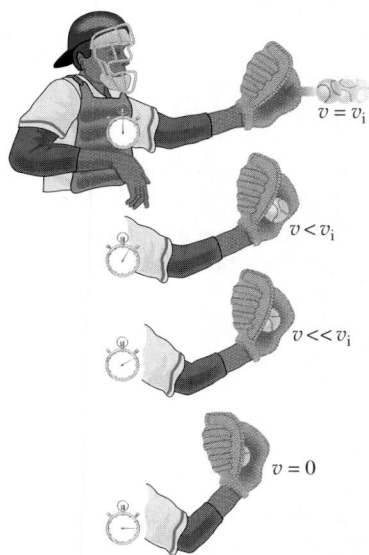

Figure 7.4 By drawing back the glove as the ball arrives, the catcher extends Δt and reduces F_{av} for a given Δp.

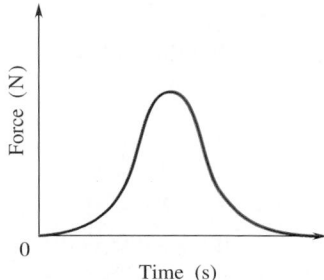

Figure 7.5 The force-time graph for a body undergoing an impact. The force rises, peaks, and falls to zero.

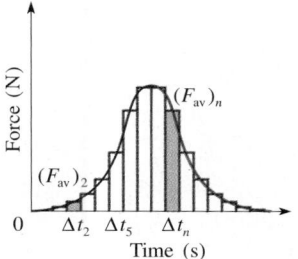

Figure 7.6 The area under the force-time graph is the net impulse exerted, and that equals the net change in momentum imparted to the body in question.

Example 7.5 **[II]** A golfer's club hits a 47.0-g golf ball from rest to a speed of 60.0 m/s in a collision lasting 1.00 ms. The force on the ball rises to a peak value of F_{max} and then drops to zero as it leaves the club. Compute a rough value for this maximum force by approximating the force-time curve, with a triangle of altitude F_{max}.

Solution We have a time-varying force, and that should suggest the area under the force-time curve. (1) TRANSLATION—An object of known mass experiences a force that rises and falls over a known time interval. As a result, the object increases speed by a specified amount; approximate the maximum force. (2) GIVEN: $m = 47.0 \times 10^{-3}$ kg, $v_i = 0$, $v_f = 60.0$ m/s, and $\Delta t = 1.00 \times 10^{-3}$ s. FIND: F_{max}. (3) PROBLEM TYPE— Impulse & Momen-tum. (4) PROCEDURE—Find the change in

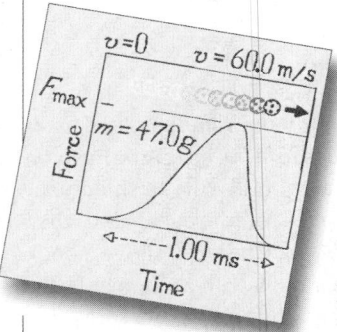

momentum that equals the net impulse. The net impulse equals the area under the force-time curve, which can be approximated by a triangle with an altitude of F_{max}. (5) CALCULATION:

$$F_{av}\Delta t = \Delta(mv) = p_f - p_i = (47.0 \times 10^{-3}\,\text{kg})(60.0\,\text{m/s}) - 0$$

and the impulse is 2.82 kg·m/s, or 2.82 N·s. Representing the force-time curve by a triangle of altitude F_{max} and base Δt, the area encompassed is $\frac{1}{2}\Delta t(F_{max})$, which equals the impulse:

$$2.82\,\text{N·s} = \tfrac{1}{2}(1.00 \times 10^{-3}\,\text{s})F_{max}$$

Therefore, $\boxed{F_{max} = 5.64\,\text{kN}}$ or 1.27×10^3 lb.

Quick Check: Whenever the force-time curve is approximated by a triangle, $F_{av}\Delta t = \frac{1}{2}(\Delta t)F_{max}$ and the peak force is twice the average force. As a consequence, F_{max} may break bones, even though F_{av} is not large enough to do damage. Since force is impulse over time, $F_{av} = (2.82\,\text{N·s})/(1.00\,\text{ms})$ and $2F_{av} = 5.6 \times 10^3\,\text{N} = F_{max}$.

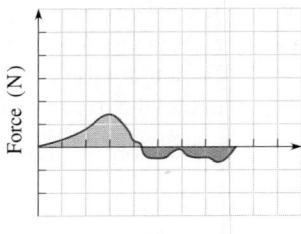

Figure 7.7 Instantaneous force versus time. The total impulse is the net area under the curve; take the area above the axis as positive and that below as negative. Oppositely directed forces produce impulses with opposite signs, and adding the areas above and below the axis takes this opposition into account.

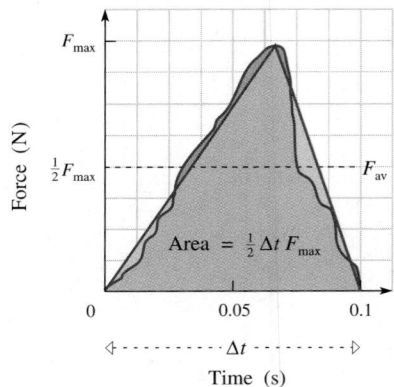

Figure 7.8 A triangular approximation to the area under the force-time curve.

A real force-time curve can have a complex shape (Fig. 7.7). A spaceship with a variable-thrust engine can generate a force-time curve with many bumps and wiggles, but the area encompassed in a given time interval will always be equal to the resulting change in momentum. If the ship turns 180° around, the same engine can fire in the opposite direction, against the motion. That would generate a negative impulse, a negative change in momentum, and a decrease in speed. The total impulse is, then, the area of the force-time curve above the axis minus the area below the axis. If those two portions happen to be equal, the net change in momentum will be zero. When you throw a punch, your fist starts with $p_i = 0$ and ends with $p_f = 0$. It accelerates up to a maximum speed and then decelerates to zero when your arm is extended. Given this fact, a karate blow is always aimed at a point inside the target so that it makes contact when p is maximum.

Car Crashes

When a car crashes into a brick wall, its front end deforms as it slows to a crushing stop. On average automobiles compress roughly 1 inch for every mile per hour of speed just prior to impact; smaller cars typically collapse a little less than this, larger cars a little more. The process is similar to a struck baseball, except the car does not spring back to its original shape.

If we assume that the collision-deceleration is fairly uniform, the crush-distance s_c divided by $v_{av} = \frac{1}{2}(v_i + v_f)$, with $v_f = 0$, is the impact time. Thus

$$\Delta t = \frac{s_c}{v_{av}} = \frac{2s_c}{v_i}$$

Since tests show that s_c is proportional to v_i, the impact time for cars of comparable stiffness should be independent of the speed of impact. A typical head-on, brick-wall collision lasts around 100 milliseconds (airbags inflate in ≈55 ms). By the way, a driver would not be much better off hitting an oncoming car. In fact, crashing head-on into an identical car traveling toward you at the same speed is effectively the same as hitting a stationary brick wall. To appreciate that, just imagine a large movable sheet of steel mounted on wheels. Now, if both cars plow into this steel sheet on opposite sides, the autos will be quite destroyed, but

Most of the moving car is transformed into thermal energy during the crash. The yellow plastic bumper extends the collision time and therefore decreases the force exerted on the car.

the sheet will remain in the middle of the melee unmoved, as if it were unmovable, just like the brick wall.

We can expect the force-time curve for a car colliding with a stationary barrier to have a base width of about 100 ms. An 1800-kg car impacting at 26.8 m/s (i.e., 60 mi/h) will be crushed about 1.5 m (i.e., 60 in.). The engine will end up in the passenger compartment as the remainder of the car moves forward, past it. The momentum change of the car is $\Delta p = 48.2 \times 10^3$ kg·m/s, which equals the area under the force-time curve in Fig. 7.8. If we again approximate that curve by a triangle, the peak force exerted on the car is approximately 0.96×10^6 N, or over 100 tons.

The impulse exerted by the punch changes the momentum of the recipient's head. The brain, which is surrounded by a liquid, slams into the skull. The gloves are heavily padded to increase Δt, thereby decreasing F_{av}, but it often doesn't help.

Example 7.6 **[II]** A 70-kg passenger riding in a typical automobile is involved in a 17.9-m/s (i.e., 40 mi/h) head-on collision with a concrete barrier. Taking the stopping time as 100 ms, compute the average force exerted by the seat belt and shoulder strap on the person.

Solution We have force and the time over which it acts, and that calls to mind the impulse. (1) TRANSLATION—An object of known mass having a known speed comes to rest in a specified time; determine the average force F_{av} experienced. (2) GIVEN: $m_p = 70$ kg; taking the initial direction of motion as positive, $\Delta v = -17.9$ m/s, and $\Delta t = 0.100$ s. FIND: F_{av}. (3) PROBLEM TYPE—Impulse

& Momentum. (4) PROCEDURE—This problem involves mass, speed change, time interval, and average force and suggests impulse-momentum. (5) CALCULATION:

$$F_{av}\,\Delta t = m_p\,\Delta v$$

$$F_{av} = \frac{(70\text{ kg})(-17.9\text{ m/s})}{0.100\text{ s}}$$

and $\boxed{F_{av} = -1.3 \times 10^4\text{ N}}$ or -2.8×10^3 lb. The minus sign means the force is in the opposite direction to the initial motion. Obviously, you can't simply brace yourself with your arms—you need seat belts.

Quick Check: Assume the force is correct and calculate the time; $\Delta t = m_p\,\Delta v/F_{av} = (70\text{ kg})(-17.9\text{ m/s})/(-1.3 \times 10^4\text{ N}) = 0.096$ s = 0.10 s.

70 kg F_{av}

$v_i = 17.9$ m/s $\quad v_f = 0$

100 ms

0

7.3 Jets and Rockets

Imagine yourself on roller skates holding a bag of oranges. Now throw one of them due north, and away you go due south. You are thrust backward just as a gun recoils backward when it fires a projectile. You push on the orange in the forward direction during the throw, it pushes back with the same impulse on you, and back you go. Hurl several oranges and

ROCKETS IN SPACE

Rockets are not propelled by pushing against either the ground or the atmosphere, though that erroneous opinion is widely held even today. In fact, a *New York Times* editorial of 1920 advised Robert H. Goddard, the American who launched the first liquid-fuel rocket, to give up any thoughts of space travel. After all, even a schoolboy knows that rockets obviously cannot fly in space because a vacuum is devoid of anything to push on.

On flying rockets in space. That Professor Goddard … does not know the relation of action to reaction, and of the need to have something better than a vacuum against which to react—to say that would be absurd. Of course, he only seems to lack the knowledge ladled out daily in high schools.

(THE NEW YORK TIMES, JAN. 13, 1920)

Further investigation and experimentation have confirmed the findings of Isaac Newton in the 17th century, and it is now definitely established that a rocket can function in a vacuum as well as in an atmosphere. The *Times* regrets the error.

(THE NEW YORK TIMES, JAN. 17, 1969)

A jet plane sucks in air and blasts out a high-speed exhaust that drives the craft forward. The engines push the exhaust backward, and the exhaust pushes the engines forward.

you have an orange-rocket, because that's exactly the way a rocket works. Rather than throwing a few large objects slowly, a rocket engine hurls out a tremendous number of tiny high-speed (3- or 4-km/s) objects—molecules. For example, during launch, the solid-fuel boosters for the Space Shuttle expel 8.5 tons of fiery exhaust each second. The engines blast exhaust downward, and the escaping gas, in turn, pushes up on the rocket (interaction). That's why rockets can be used in space.

Example 7.7 **[II]** A rocket engine testing a low-power fuel expels 5.0 kg of exhaust gas per second. If these molecules are ejected at an average speed of 1.2 km/s, what is the thrust of the engine?

Solution We have a flow of momentum, and that calls to mind Newton's Second Law. (1) TRANSLATION—Each second a known amount of mass is given a known speed; determine the average force required. (2) GIVEN: 5.0 kg of exhaust gas are ejected per 1.00 s, $v_i = 0$, and $v_f = 1.2$ km/s. FIND: The thrust, F_{av}, exerted by the motor. (3) PROBLEM TYPE—Impulse & Momentum. (4) PROCEDURE—This problem involves F_{av}, Δt, m, and Δv and suggests $F_{av} \Delta t = m\Delta v$. (5) CALCULATION— We know that each second, 5.0 kg of exhaust gas experiences

an increase in speed of $\Delta v = 1.2$ km/s. There's an average force exerted on the gas:

$$F_{av} = \frac{m\Delta v}{\Delta t} = \frac{(5.0 \text{ kg})(1.2 \times 10^3 \text{ m/s})}{1.00 \text{ s}}$$

$$F_{av} = 6.00 \times 10^3 \text{ kg·m/s}^2$$

The engine pushes the gas backward with a force of $\boxed{F_{av} = 6.0 \text{ kN}}$, and the gas exerts an equal reaction force forward on the rocket.

Quick Check: Assuming the force is correct, calculate the mass: $m = (F_{av})\Delta t / \Delta v = 5$ kg.

7.4 Conservation of Linear Momentum

Conservation of Momentum means that the net momentum of an isolated system cannot change.

One of the great guiding ideas of physics is the law of Conservation of Linear Momentum: **When the resultant of all the external forces acting on a system is zero, the linear momentum of the system remains constant** ($p_i = p_f$). Descartes, long before Newton had refined dynamics, wrote of his Creator of the Universe:

> He set in motion in many different ways the parts of matter when He created them, and since He maintained them with the same behavior and with the same laws as He laid upon them in their creation, He conserves continually in this matter an equal quantity-of-motion.

In other words, the total momentum (i.e., quantity-of-motion) of the Universe persists unchanged and will continue to be preserved forever.

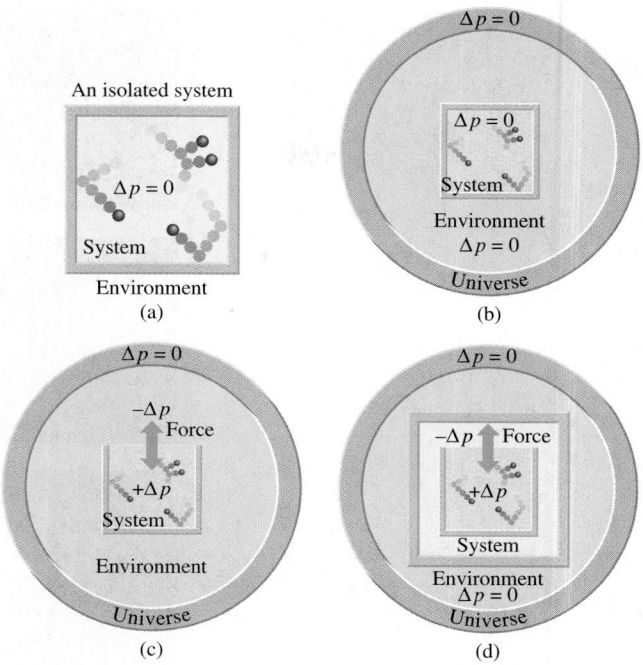

Figure 7.9 (a) A system isolated from the environment conserves momentum; that is, $\Delta p = 0$. (b) Here, we see an isolated system within the Universe. In all cases, Δp for the Universe is zero. (c) When the system communicates with its environment, it may experience a net momentum change, $+\Delta p$, but the environment will then experience an equal and opposite change, $-\Delta p$. (d) If the system is enlarged and yet isolated, its parts can experience momentum changes, but its net change will be zero.

THE EARTH & YOU AS A SYSTEM

If while hovering motionlessly ($p_i = 0$) far off in space you throw a ball, your final momentum will be equal in magnitude and opposite in direction to the projectile's momentum. On Earth, things can be quite different. The ball gets thrown, but you usually just stand there. In the process, you brace yourself by pushing on the ground with a force (F_{YG}) and the ground pushes back on you with an equal and opposite force (F_{GY}). If you and the ball are the system, F_{GY} is an *external* force acting on that system and momentum is properly *not* conserved. If, on the other hand, we take the ball, you, and Earth as the system, F_{GY} and F_{YG} are internal forces and momentum is conserved—the Earth must move with a momentum equal and opposite to the ball's momentum (an effect that's much too small to measure directly).

In the process of throwing a bouquet or catching a cigar, a force is applied by the hand to the projectile. Simultaneously, there is an equal and opposite force back on the hand. The *impulse* exerted on the cigar, bringing it to rest, is matched by an equal and oppositely directed impulse on the hand that catches it. The interaction times are the same, and the forces are equal in magnitude and opposite in direction; the resulting momentum changes must also be equal and opposite.

A particle or a group of particles constitutes a *system*. Forces whose sources are within the system are *internal forces* (Fig. 7.9*a*). **The total momentum of a system of interacting masses must remain unaltered, provided that no net external force is applied.** Just try pushing a car off the road while sitting inside it (p. 97).

The momenta of individual interacting members of a system may certainly change, but each change is accompanied by an equal and opposite change in momentum of the interaction partner. If the whole of our Universe is taken as the system, there are no *external forces* whose sources lie outside the system, and the total momentum must be conserved forever (Fig. 7.9*b*). If a system is not isolated and external forces do act on it (Fig. 7.9*c*), its momentum will change, say, by an amount $+\Delta p$. But the momentum of the communicating environment will then change by $-\Delta p$. If that system is part of a still larger isolated system (Fig. 7.9*d*), then the change in momentum of everything going on inside the large system will be zero, as if it were a little universe unto itself. Figure 7.10 is an example of how this all works in practice.

Figure 7.10 Take the two astronauts playing catch with a little moon as the system. There will then be no external forces acting, and momentum will be conserved. (a) Everything starts out at rest with a net momentum of zero: $p = 0$. (b) After the moonlet is thrown, the net momentum of the system is still zero. (c) And it's zero after the moonlet is caught and (d) thrown again. Alternatively, take the moon and/or either astronaut as the system. External forces will now be applied, and the momentum of this less inclusive system will change. Note that we do not know how much momentum the moonlet has in (c).

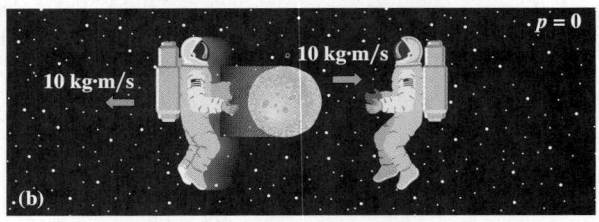
(a)

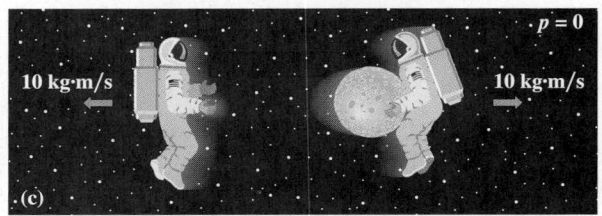
(b)

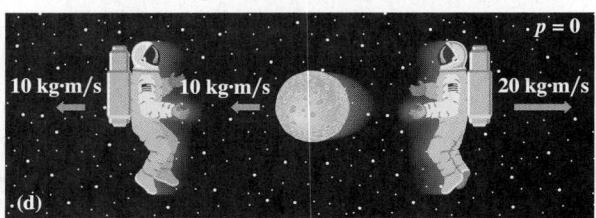
(c)

(d)

A problem involving Conservation of Momentum ($\vec{p}_i = \vec{p}_f$) is really a vector problem, so you will again have to be quite careful about signs. Before you write the scalar equation corresponding to Conservation of Momentum, first pick a direction to be positive. Any scalar velocity opposite to that direction must be entered numerically with a minus sign. Study Example 7.8 before pressing on.

Example 7.8 **[I]** According to published figures, a bullet fired from a standard 9-mm Luger pistol has a mass of 8.0 g and a muzzle speed of 352 m/s (i.e., 1155 ft/s). If the mass of the gun is 0.90 kg, what is its recoil speed when fired horizontally? Studies of handgun recoils for ordinary low-speed bullets confirm that the escaping gases can be ignored.

Solution The word "recoil" should suggest Conservation of Momentum. (1) TRANSLATION—A known small mass is given a known speed by an object whose large mass is known; determine the recoil speed of the large mass. (2) Take the direction of motion of the bullet as positive. GIVEN: $m_B = 8.0$ g, $m_G = 0.90$ kg, $v_{Bi} = v_{Gi} = 0$, and $v_{Bf} = +352$ m/s. FIND: v_{Gf}. (3) PROBLEM TYPE—Linear Momentum/Conservation of Momentum. (4) PROCEDURE— Taking the gun and bullet as the system, there are no external

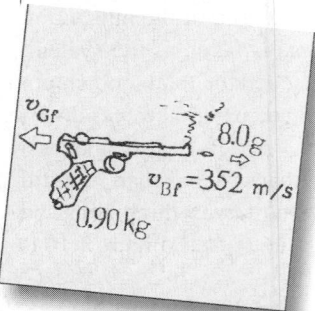

forces and therefore p_i (before firing) equals p_f (after firing). (5) CALCULATION— $p_i = 0 = p_f$

$$0 = m_B v_{Bf} + m_G v_{Gf}$$

and so

$$v_{Gf} = -\frac{m_B}{m_G} v_{Bf}$$

$$v_{Gf} = -\frac{0.008\ 0\ \text{kg}}{0.90\ \text{kg}} (352\ \text{m/s})$$

and

$$\boxed{v_{Gf} = -3.1\ \text{m/s}}$$

The minus sign means that the gun moves in the opposite direction of the bullet. This is the scalar velocity; the speed is just 3.1 m/s.

Quick Check: $|\vec{p}_{Gf}| = |\vec{p}_{Bf}|$; the bullet is roughly 100 times less massive than the gun, and its speed is roughly 100 times greater.

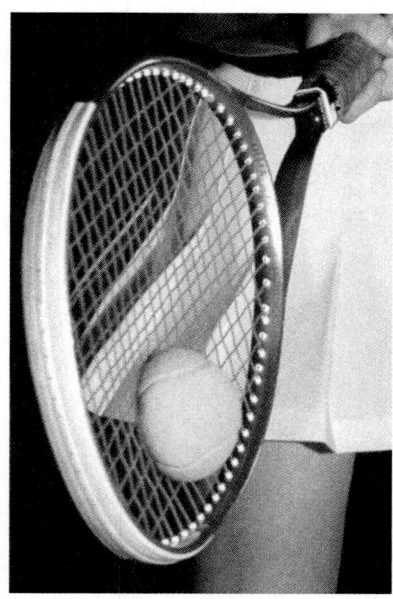

The force of the impact crushes a tennis ball. The racket and ball are in contact for a duration known as the *collision time*.

In the absence of externally applied forces, the total momentum of the objects undergoing any kind of collision is always constant—**momentum is always conserved**.

A collision is **inelastic** when $KE_i \neq KE_f$.

MORE FUNDAMENTAL THAN THE THIRD LAW

No experiment can prove that the law of Conservation of Linear Momentum is true, although observations confirm it all the time: There are no places in this Universe free of outside influences. Still, subatomic particles appear to scrupulously conserve momentum. Indeed, Conservation of Momentum is more fundamental than the Third Law. That's true because interactions between separated objects take time to propagate, and there can be an appreciable delay between the observed action and the observable reaction. Modern Quantum Field Theory (Ch. 31) maintains that all the basic interactions (gravitational, electromagnetic, strong, and weak) arise from the exchange of momentum-carrying particles that travel at finite speeds. Although momentum is conserved at all times, we may have to wait a while after the action for the reaction.

7.5 Collisions

Much of what we know about the atomic and subatomic domains has been learned from the observation of collisions. Beams of particles are hurled against targets, and the motion of the scattered fragments is analyzed. Similar relationships govern both these exotic events and the more mundane ones, such as clubs hitting golfballs or cars crashing. *A collision is marked by the transfer of momentum between objects in relative motion resulting from their interaction via at least one of the Four Forces.* **In all cases where there are no external forces acting, the total momentum of the colliding objects is conserved**. In the macroscopic world of balls and cars and hammers, some of the organized KE of the colliding objects will always end up as internal energy (as disorganized atomic KE, or if the object's shape is changed, as internal PE). Disorganized atomic KE corresponds to random motion that has no net momentum, and so the momentum of the objects must still be conserved.

Inelastic Collisions in One Dimension

An **inelastic collision** *is one where the final KE of the system is different from the initial KE.* A tennis ball dropped against a concrete floor distorts, momentarily comes to rest, and then springs back, popping into the air. But the squashing produces some internal heating as the molecules shift position (and even the crack of sound carries off a little energy), and so the ball only returns about two-thirds of the way back up. One-third of the initial KE has been converted into other energy forms. If, after an impact between two macroscopic objects, either one increases in temperature or remains distorted, the collision is inelastic. *All collisions between macroscopic objects are more or less inelastic.*

The **completely inelastic** collision is at one extreme—where the impacting objects stick together—and the maximum amount of KE is transformed (Fig.7.11). A wad of oatmeal landing on a table, a bug hitting the windshield of a moving bus, or a neutron absorbed by a uranium nucleus are all representative of inelastic collisions. Less drastic is the case where the colliding objects bounce apart and yet lose KE in the process. A lead ball rebounding from a steel anvil will not even retain 1% of its original KE, whereas a soft cork sphere might retain around 35%. **The more rigid the colliders, the less KE is transformed** (i.e., lost), which is why billiard balls are not soft. A glass marble will bounce off the anvil with about 95% of its initial KE (Fig. 7.12). Our purposes will be served by examining only the two extremes of elastic and completely inelastic collisions.

Early in its existence (1663), the British Royal Society sponsored a scientific study of collisions. Among the leading scholars who participated were

An example of an inelastic collision. It took a good deal of energy to reshape this parked car and that energy came from the KE of the vehicle that hit it.

Figure 7.11 Inelastic collisions. (a) Here the skaters flying toward each other have equal masses and speeds. The total momentum is initially zero and remains zero after the collision, as they simply come to rest and stay that way. (b) Here number 9 initially has all of the system's momentum. After impact, since the total momentum is constant and they stay together, the moving mass is twice what it was, and therefore the speed must be halved. The two sail off at half the speed with which number 9 came in. (c) Taking revenge, number 2 comes flying at number 9, at twice 9's speed. The total momentum is $m(2v) + mv$. After the collision, the two skaters sail off together ($2m$) with a scalar velocity of ($3v/2$) that conserves the initial momentum ($3mv$).

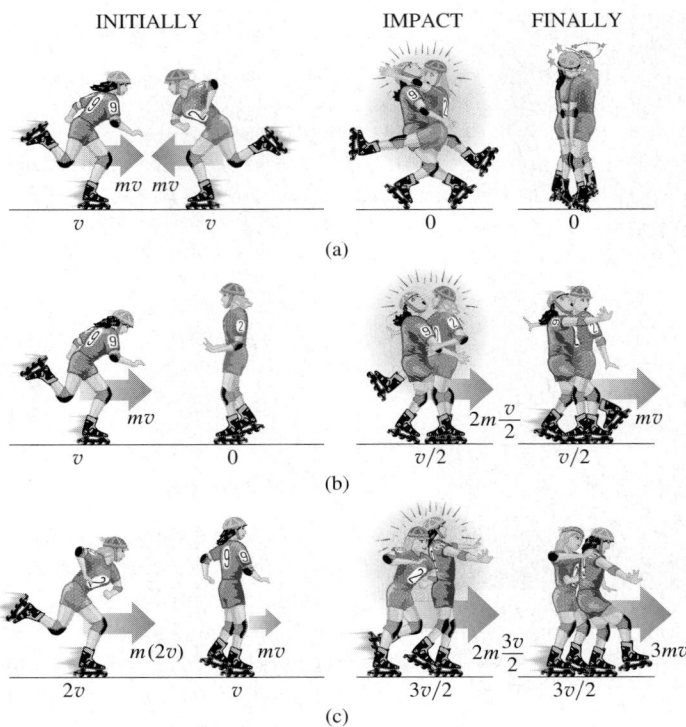

Huygens and the English mathematician John Wallis. Wallis, in a paper in 1668, was probably the first to apply the Law of Conservation of Momentum correctly. When two objects of mass m_1 and m_2, initially moving at scalar velocities v_{1i} and v_{2i}, collide head-on and fly off at v_{1f} and v_{2f}, initial momentum equals final momentum:

[head-on, inelastic] $$m_1 v_{1i} + m_2 v_{2i} = m_1 v_{1f} + m_2 v_{2f} \tag{7.5}$$

and that's all we need in order to solve any problem relating to inelastic collisions. The head-on restriction requires that the motion of both bodies at the moment of impact be along a common straight line connecting the two centers. As a result, the final motion is also along that line.

Now let's see how the kinetic energy is distributed by the collision. For completely inelastic collisions between two bodies, only one of which (m_1) is initially moving, the ratio of the total final KE to the total initial KE is

$$\frac{\text{KE}_f}{\text{KE}_i} = \frac{\frac{1}{2}(m_1 + m_2)v_f^2}{\frac{1}{2}m_1 v_{1i}^2}$$

Although it's not obvious, we can get rid of the scalar velocities by expressing v_f in terms of v_{1i}, and that can be done using momentum conservation; $m_1 v_{1i} = (m_1 + m_2)v_f$. Then, $v_f = m_1 v_{1i}/(m_1 + m_2)$, and squaring and substituting yield

[head-on, inelastic] $$\text{KE}_f = \left(\frac{m_1}{m_1 + m_2}\right)\text{KE}_i$$

This suggests that $\text{KE}_f \approx \text{KE}_i$ and little kinetic energy will be lost when m_1, the moving mass, is much greater than m_2. Then $(m_1 + m_2) \approx m_1$

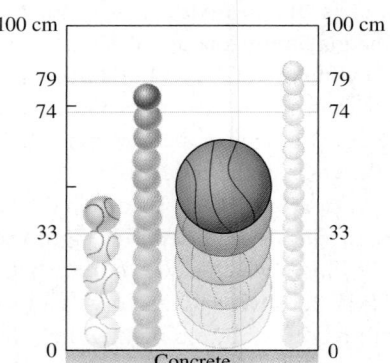

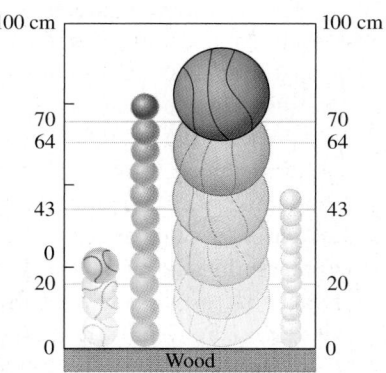

Figure 7.12 Rebound heights for a baseball, racquetball, basketball, and golf ball, each dropped from 1.00 m. The basketball is in contact with the floor for a relatively long time, allowing the wood to spring back.

Example 7.9 **[II]** During a rainy day football game, a 854-N (i.e., 192-lb) quarterback is standing holding the ball looking for a receiver when he's unkindly hit by a 1281-N (i.e., 288-lb) tackle charging in at 6.1 m/s (i.e., 20 ft/s).   (a) At what speed do the two men, tangled together, initially sail off on the wet field? Assume friction is negligible and that the impact is head-on.   (b) How much mechanical energy is lost to friction?

Solution   The men are "tangled together" so we've got an inelastic collision.   (1) TRANSLATION—A known large mass moving at a specified speed collides with a known smaller mass at rest; given that they stick together, determine the final speed. (2) Take the initial direction of motion of the tackle to be positive. GIVEN: $F_{wq} = 854$ N, $F_{wt} = 1281$ N, $v_{qi} = 0$, and $v_{ti} = 6.1$ m/s. FIND: the final speed of the two, v_f, and the lost mechanical energy. (3) PROBLEM TYPE—Linear Momentum / Conservation of Momentum /inelastic collision. (4) PROCEDURE—This collision is completely inelastic because the two bodies slide away together. Considering the two men to be the system, since there are no external forces acting (e.g., no friction), momentum is conserved. On the other hand, because the collision is inelastic, the net KE will not remain constant.   (5) CALCULATION—(a) Initial momentum equals final momentum,

$$p_i = p_f$$

$$m_q v_{qi} + m_t v_{ti} = (m_q + m_t)v_f$$

Using $m = F_w/g$,

$$0 + \frac{1281\ \text{N}}{9.81\ \text{m/s}^2}6.1\ \text{m/s} = \left(\frac{854\ \text{N}}{9.81\ \text{m/s}^2} + \frac{1281\ \text{N}}{9.81\ \text{m/s}^2}\right)v_f$$

$$796.8\ \text{N/s} = (217.7\ \text{N} \cdot \text{s}^2/\text{m})v_f$$

and

$$\boxed{v_f = 3.7\ \text{m/s}}$$

This problem has one unknown and can be solved using only Conservation of Momentum.

The two players move off in the positive direction. Without ever coming to rest, the tackle captures the quarterback and carries him off. This, of course, is why tackles are designed with so much mass. (b) No change in PE_G occurs provided no one falls, but there is a change in KE since

$$KE_i = \tfrac{1}{2}m_t v_{ti}^2 = 2.4\ \text{kJ}$$

whereas

$$KE_f = \tfrac{1}{2}(m_q + m_t)v_f^2 = 1.5\ \text{kJ}$$

Apparently, $\boxed{0.9\ \text{kJ}}$ was transformed, mostly to thermal energy.

Quick Check: Using U.S. Customary Units, $p_i = p_f$,

$$0 + \frac{288\ \text{lb}}{32\ \text{ft/s}^2}20\ \text{ft/s} = \left(\frac{192\ \text{lb}}{32\ \text{ft/s}^2} + \frac{288\ \text{lb}}{32\ \text{ft/s}^2}\right)v_f$$

and $v_f = 12$ ft/s. Similarly, $KE_i = 1.8 \times 10^3$ ft·lb and $KE_f = 1.1 \times 10^3$ ft·lb; $\Delta KE = 0.7 \times 10^3$ ft·lb = 0.9 kJ.

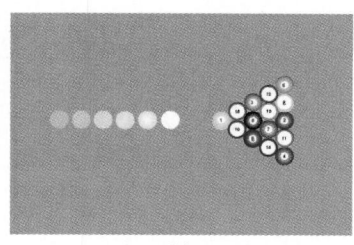

(a)

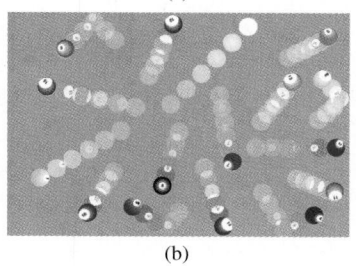

(b)

Figure 7.13 Billiard balls are quite rigid and undergo collisions that are nearly elastic. (a) Here KE_i equals the kinetic energy of the cue ball before the break. (b) That amount of KE is distributed among all the balls ($KE_i \approx KE_f$) immediately after the break, although it's ultimately lost to friction.

and the term in parentheses becomes 1. If a sports car locks bumpers with the back of a stopped garbage truck, the two will roll slowly ahead—KE_i will be greater than KE_f, and the difference will go into mangling the participants. On the other hand, if the truck did the crashing with the same KE_i, the two would roll away at nearly the same speed as that at which the truck came in, and the collision would be less destructive. A small, high-speed projectile in an inelastic collision with a movable target will do more damage than a massive, slow one with the same KE. {You can study a wide variety of collision possibilities via a great simulation; click on **ELASTIC & INELASTIC COLLISIONS** under **INTERACTIVE EXPLORATIONS** on the **CD**.} 💿

Elastic Collisions in One Dimension

A collision is **elastic** *when* **KE is constant** (Fig. 7.13). Particles on the submicroscopic level collide elastically, presumably because they don't suffer any permanent rearrangement of their component parts (if there are any). Suppose a head-on collision takes place between mass-1 and mass-2 on a horizontal plane with m_2 at rest, where this time the impact is completely elastic (Fig. 7.14). Let's determine a general expression for each of the final, post-collision, scalar velocities v_{1f} and v_{2f}. Conservation of Momentum provides

$$p_i = p_f$$

$$m_1 v_{1i} + 0 = m_1 v_{1f} + m_2 v_{2f}$$

but now the participants move off independently. Since $\Delta PE_G = 0$, Conservation of Energy

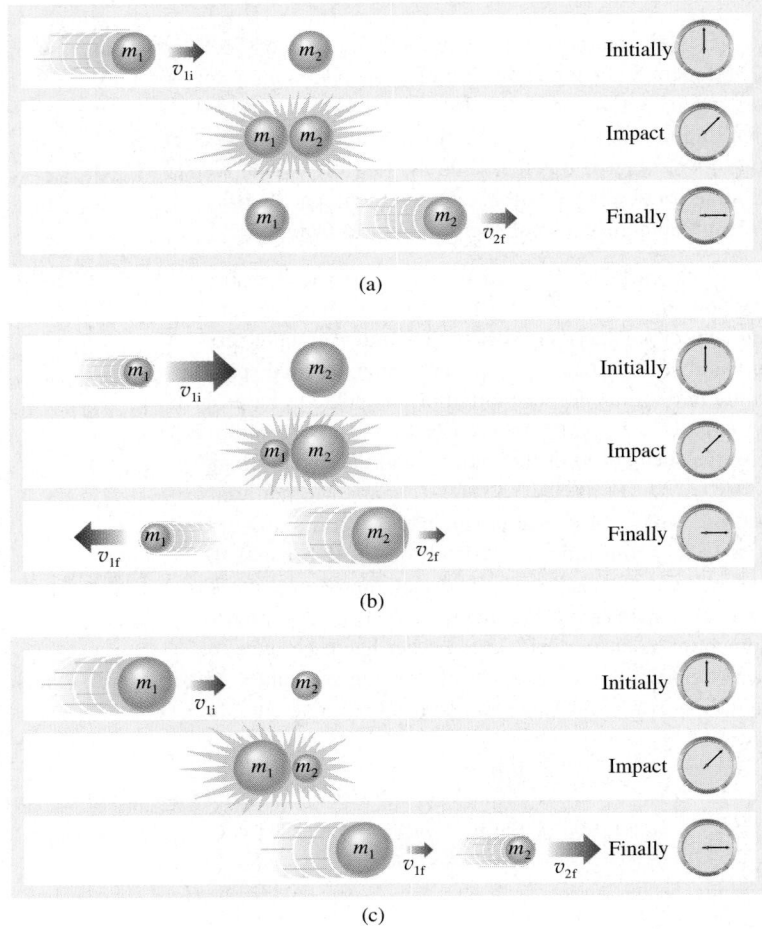

Figure 7.14 Three elastic collisions in which a moving mass m_1 crashes into a stationary mass m_2. In (a) $m_1 = m_2$ and they essentially exchange velocities. In (b) $m_1 < m_2$, and the balls move off in opposite directions. In (c) $m_1 > m_2$, and both balls move off in the direction in which m_1 was originally traveling.

A collision is **elastic** when $\text{KE}_i = \text{KE}_f$.

leads to

$$\text{KE}_{1i} + \text{KE}_{2i} = \text{KE}_{1f} + \text{KE}_{2f}$$

$$\tfrac{1}{2}m_1 v_{1i}^2 + 0 = \tfrac{1}{2}m_1 v_{1f}^2 + \tfrac{1}{2}m_2 v_{2f}^2$$

There can be two unknowns, usually the two final speeds, because there are two independent equations. Accordingly, we can solve for v_{1f} and v_{2f} in terms of v_{1i}. First rearrange the above two equations as

$$m_1(v_{1i} - v_{1f}) = m_2 v_{2f} \tag{7.6}$$

and

$$m_1(v_{1i}^2 - v_{1f}^2) = m_2 v_{2f}^2$$

Factoring this last expression yields

$$m_1(v_{1i} + v_{1f})(v_{1i} - v_{1f}) = m_2 v_{2f}^2 \tag{7.7}$$

Dividing the right side of Eq. (7.7) by the right side of Eq. (7.6) and the left by the left gets rid of the masses:

$$v_{1i} + v_{1f} = v_{2f} \tag{7.8}$$

$$v_{1i} = v_{2f} - v_{1f}$$

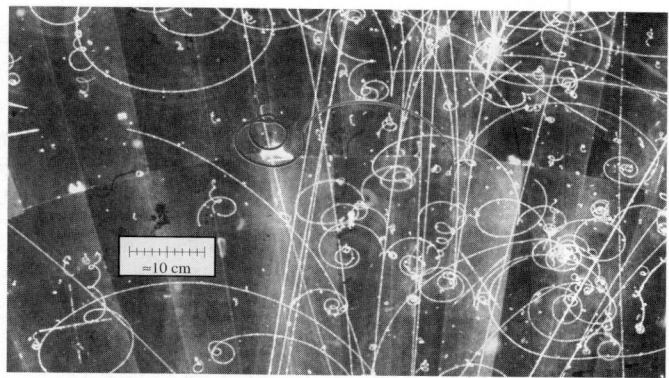

All sorts of collisions take place in this bubble chamber photo, but one is especially rare and interesting. Positrons and electrons have opposite charge but the same mass. Here a positron (shown in blue) arcs to the left and collides head-on with a more or less stationary electron (shown in red). The electron flies off (spiraling clockwise) with essentially all the momentum delivered by the positron, which presumably comes to rest. This is just what happens with two elastically colliding equal-mass balls; the moving one transfers all its momentum to the stationary one (see Fig. 7.14a). (For discussions of a variety of aspects of this same photo see pages 5, 25, 86, 144, and 181.)

Mark McGwire swung his 35 oz (0.99 kg) bat at about 85 mph in order to create a 1.0-ms collision with an incoming 90 mph fastball hurled by Montreal pitcher Carl Pavano. Experiencing around 8000 lb of force, the ball sailed off at about 110 mph, on September 27, 1998, for McGwire's 70th home run. This kind of collision is never completely elastic.

The relative speed with which the objects approached each other before impact was v_{1i}, and the speed with which they separate after impact is still v_{1i}. **The "relative speeds" of the two bodies before and after an elastic impact are equal.** The relative velocity is reversed by the collision, but its magnitude is unchanged. *This is true for all elastic collisions* (see Problem 72). Substituting v_{2f} from Eq. (7.8) into Eq. (7.6) leads to the sought-after expression for the final speed of mass-1:

[head-on, elastic, $v_{2i} = 0$]
$$v_{1f} = \frac{m_1 - m_2}{m_1 + m_2} v_{1i} \tag{7.9}$$

Substituting this into Eq. (7.8), we get $v_{2f} = [v_{1i}(m_1 - m_2)/(m_1 + m_2)] + v_{1i}$, and the expression for the final speed of mass-2 becomes

[head-on, elastic, $v_{2i} = 0$]
$$v_{2f} = \frac{2m_1}{m_1 + m_2} v_{1i} \tag{7.10}$$

If we take the direction of $\vec{v}_{1i}$ as positive, the last two equations describe the resulting motion for the three possible cases.

(1) When $m_1 = m_2$, as with two billiard balls, $v_{1f} = 0$ and $v_{2f} = v_{1i}$. The moving number-1 ball slams to a stop, and the struck number-2 ball slides (no rolling, please) away with the initial speed. The balls, in effect, exchange velocities (Fig. 7.14a). (2) When $m_1 < m_2$, v_{1f} is negative, and the incoming number-1 ball rebounds in the opposite direction while the initially stationary number-2 ball moves off in the positive direction, as in Fig. 7.14b.

Exploring Conservation of Momentum. A cannon-ball is fired at an equal-mass ball at rest on the platform. The balls exchange momentum and the ball which was at rest flies away. (From Johann Marcus Marc, *De proportione motus*, 1648.)

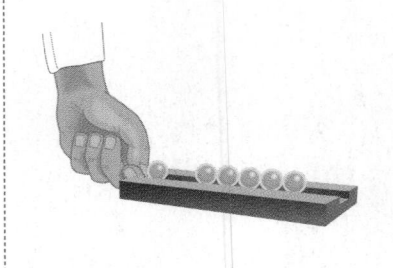

EXPLORING PHYSICS ON YOUR OWN

Elastic Collisions: Newton's cradle is the familiar setup in Fig. 7.15. When one ball hits on the left, one pops up on the right; if two swing down on the left, two pop up on the right. All of this is understandable via Conservation of Momentum and Energy taken together. With one ball hitting, only one can rebound. If instead two pop up, the speed would have to be reduced by a factor of 1/2 to conserve momentum and simultaneously by $1/\sqrt{2}$ to conserve energy and that's impossible. Use a grooved horizontal ruler and five or six marbles to make an equivalent of the cradle. Keep increasing the percentage of marbles launched at the remaining others—how many pop out? What happens when more than half the total number of balls are shot in?

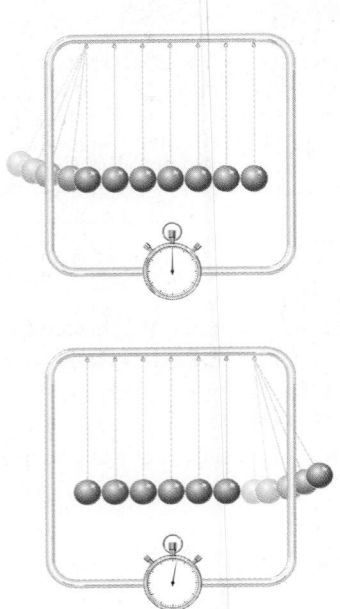

Figure 7.15 Newton's cradle. A ball slams down on the left, and one flies out on the right. The collision is nearly elastic and both energy and momentum are conserved.

Indeed, when $m_1 \ll m_2$, $v_{1f} \approx -v_{1i}$ and $v_{2f} \approx 0$, which is like bouncing a ball (m_1) off the back of a stationary bus (m_2). (3) Finally, when $m_1 > m_2$, v_{1f} is positive as is v_{2f}; both move off in the initial direction of motion (Fig. 7.14c). When a tennis racket slams into an essentially motionless ball during a serve, it continues to move in the forward direction but at a diminished speed. In the extreme, when $m_1 \gg m_2$, $v_{1f} \approx v_{1i}$, and since m_2 is negligible in the denominator of the last equation for v_{2f}, it follows that $v_{2f} \approx 2v_{1i}$. Hit a Ping Pong ball with a massive paddle, and it will fly off with twice the paddle's speed (*and still the relative speeds will be unchanged*). {To study elastic collisions in one dimension using a fun simulation, please click on **ELASTIC & INELASTIC COLLISIONS** under **INTERACTIVE EXPLORATIONS** on the CD.}

Collisions in Two Dimensions (Optional)

When two objects with masses m_1 and m_2 elastically slam into one another off-center (Fig. 7.16), in what's known as a *glancing collision*, they subsequently move away at angles θ and ϕ to the original direction, respectively. Any pool player knows how to strike a ball on the side so that it sails off at an angle. Provided there are no external forces acting, momentum is always conserved ($\vec{p}_i = \vec{p}_f$). Because of the two-dimensionality of the collisions, we'll make use of the fact that the scalar momentum components in the x- and y-directions are conserved independently:

$$p_{ix} = p_{fx} \qquad \text{and} \qquad p_{iy} = p_{fy} \qquad (7.11)$$

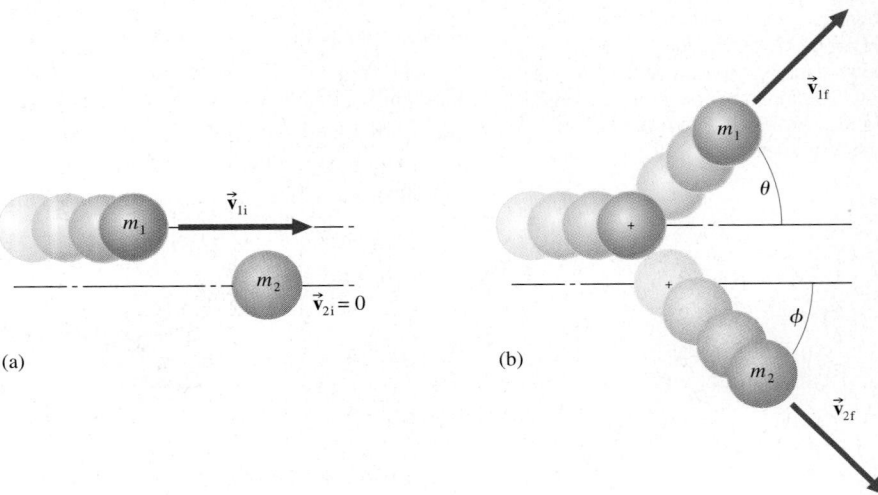

Figure 7.16 A glancing elastic collision. When $m_1 = m_2$ and $v_{2i} = 0$, then $\theta + \phi = 90°$. See the two photos on the next page.

A beam of protons enters from the left. One strikes a proton in the medium, and two scatter away, making an angle of 90°.

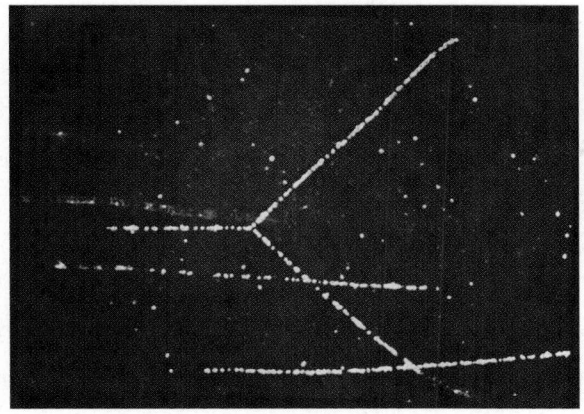

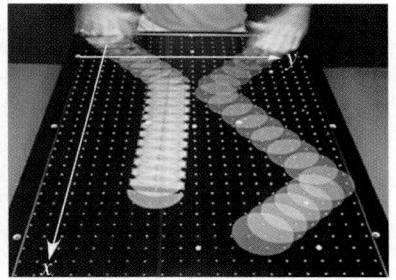

Two identical pucks on an air table. The blue puck has a positive y-component of momentum mv_{By}. The yellow puck has a negative y-component of momentum $-mv_{Yy}$, where $mv_{By} > mv_{Yy}$. After colliding, the blue puck has no y-component and the yellow puck moves off with a positive y-component equal to the initial net y-component $(mv_{By} - mv_{Yy})$. Both the x- and y-components of momentum are conserved independently.

A collision between two billiard balls each of 173 g. The flashes were every 1/30 s. The ball with the dot entered from the left; the other was at rest initially. And again they scatter at 90°.

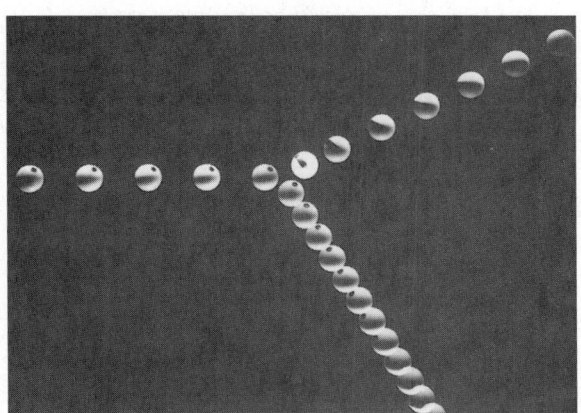

We ambitiously might set out to find expressions for $\vec{\mathbf{v}}_{1f}$ and $\vec{\mathbf{v}}_{2f}$ in terms of the masses and $\vec{\mathbf{v}}_{1i}$ and $\vec{\mathbf{v}}_{2i}$, but that's not possible. To find these two vectors, we would have to determine the x- and y-components for each, and that's four unknowns and we have only three equations to play with: $p_{ix} = p_{fx}$, $p_{iy} = p_{fy}$, and $\mathrm{KE}_i = \mathrm{KE}_f$. Interestingly, it's possible to prove that **when two equal-mass objects undergo a glancing elastic collision, provided one of them is at rest to begin with, the two will always move off at right angles to each other** (Fig. 7.16b).

The difficulty of having too many unknowns (encountered above) is no longer a problem when the colliding bodies stick together. Then there's only one final velocity with two unknown components, and we only need two equations [viz., Eq. (7.11)]. {To study these various possibilities via a simulation, click on **TWO-DIMENSIONAL COLLISIONS** under **INTERACTIVE EXPLORATIONS** on the **CD**. Play with this piece for a while, and you'll surely get a sense of how the various quantities contribute to the process.}

Example 7.10 **[III]** Two cars enter an icy intersection and skid into each other (Fig. 7.17). The 2.50×10^3-kg sedan was originally heading south at 20.0 m/s, whereas the 1.45×10^3-kg coupe was driving east at 30.0 m/s. On impact, the two vehicles become entangled and move off as one at an angle θ in a southeasterly direction. Determine θ and the speed at which they initially skid away after crashing.

Solution The cars are "entangled," so we've got an inelastic collision. (1) TRANSLATION—Two known masses, traveling with known velocities, collide and stick together; determine their final speed and direction of motion. (2) Take directions south and east to be positive and call them y and x. GIVEN: $m_s = 2.50 \times 10^3$ kg, $m_c = 1.45 \times 10^3$ kg, $v_{si} = 20.0$ m/s, and $v_{ci} = 30$ m/s. FIND: the final speed of the two, v_f, and θ. (3) PROBLEM

continued

TYPE—Linear momentum /Conservation of Momentum/two-dimensional inelastic collision. (4) PROCEDURE—This collision is completely inelastic because the two bodies slide away together. Considering the two cars to be the system, since there are no external forces acting (e.g., no friction), momentum is conserved. On the other hand, because the collision is inelastic, the net KE will not remain constant. This situation is two-dimensional, and so we will use both $p_{ix} = p_{fx}$ and $p_{iy} = p_{fy}$. (5) CALCULATION—In the x-direction (where east is positive)

$$p_{ix} = p_{fx}$$

$$m_c v_{ci} + 0 = (m_c + m_s) v_{fx}$$

Using $v_{fx} = v_f \cos \theta$,

$$v_{fx} = \frac{m_c v_{ci}}{m_c + m_s}$$

$$v_{fx} = \frac{(1.45 \times 10^3 \text{ kg})(30.0 \text{ m/s})}{(1.45 \times 10^3 \text{ kg} + 2.50 \times 10^3 \text{ kg})} = 11.01 \text{ m/s}$$

and so $\qquad v_f \cos \theta = 11.01 \text{ m/s}$ (i)

Similarly, in the y-direction (where south is positive)

$$p_{iy} = p_{fy}$$

$$m_s v_{si} + 0 = (m_c + m_s) v_{fy}$$

Using $v_{fy} = v_f \sin \theta$,

$$v_{fy} = \frac{m_s v_{si}}{m_c + m_s}$$

$$v_{fy} = \frac{(2.50 \times 10^3 \text{ kg})(20.0 \text{ m/s})}{(1.45 \times 10^3 \text{ kg} + 2.50 \times 10^3 \text{ kg})} = 12.66 \text{ m/s}$$

and so $\qquad v_f \sin \theta = 12.66 \text{ m/s}$ (ii)

To get rid of the v_f terms divide Eq. (ii) by Eq. (i):

$$\tan \theta = \frac{12.66 \text{ m/s}}{11.01 \text{ m/s}} = 1.1499$$

and $\theta = 48.988°$, or to three significant figures, $\boxed{\theta = 49.0°}$. Putting this into $v_f \cos \theta = 11.01 \text{ m/s}$ yields

$$v_f = \frac{11.01 \text{ m/s}}{\cos 48.988°} = \boxed{16.8 \text{ m/s}}$$

Quick Check: Using $\cos^2 \theta + \sin^2 \theta = 1$, square and add Eqs. (i) and (ii) to get $v_f^2 = (11.01 \text{ m/s})^2 + (12.66 \text{ m/s})^2$ and $v_f = 16.8 \text{ m/s}$.

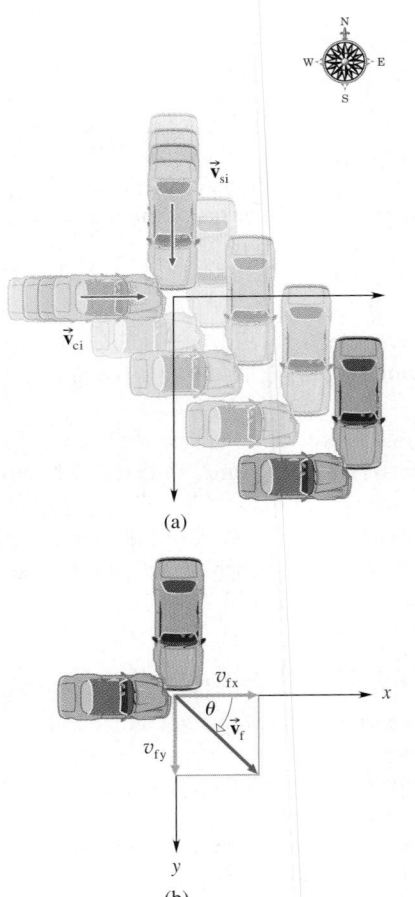

(a)

(b)

Figure 7.17 An inelastic collision.

7.6 Linear Momentum and Symmetry

When a change in a physical system leaves it essentially unchanged, that system possesses a corresponding symmetry (p. 5). We believe that the Universe is homogeneous with regard to its governing physical principles. Empty space is the same everywhere. The same laws of physics apply everywhere, and the fundamental constants (e.g., the speed of light, the gravitational constant) are independent of position.

Noether's Principle maintains that for every conservation law there is a corresponding symmetry, and vice versa. In particular, Conservation of Linear Momentum is a consequence of the invariance of Nature with respect to place. Insofar as no conceivable experiment can distinguish between one point in empty space and another (on the basis of variations in physical behavior), space is homogeneous and linear momentum must be conserved (Fig. 7.18). Noether's proof of this is beyond our means, but we can show the general idea.

Assume that the physical Universe is *not* homogeneous, that the laws of physics are *not* the same everywhere. Now hold two identical objects at separate distant points (A and B) in space. Suppose that they attract each other either electrically, magnetically, or gravitationally. Given the "inhomogeneity," assume the force law is different at the two locations. Then the force on object A due to object B will be different from the force on object B due to object A. These forces are not an equal-and-opposite interaction pair. Suppose the objects are now released with a total initial momentum of zero. They will immediately accelerate toward each other, increasing in momentum but at different rates, thereby instantly violating Conservation of Linear Momentum. The converse is as follows: space *is* homogeneous, the force laws are the same everywhere, the change in momentum of the two objects is equal in magnitude and opposite in direction, and linear momentum is conserved throughout the Universe.

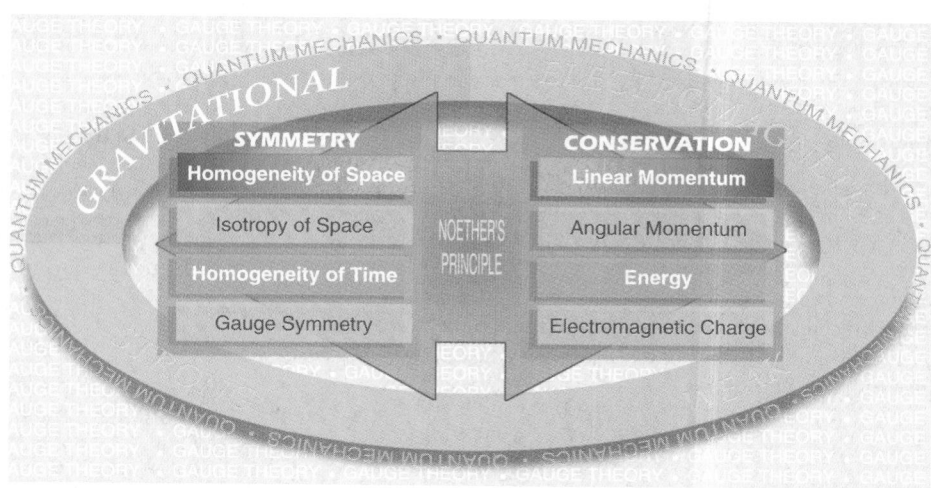

Figure 7.18 Thus far, of the Four Fundamental Forces, we've studied the gravitational force. Moreover we've found that Conservation of Energy arises from the fact that the laws and constants of Nature do not change in time—there is temporal displacement symmetry. Now we learn that the homogeneity of space gives rise to Conservation of Linear Momentum.

Core Material & Study Guide

LINEAR MOMEMTUM

Recall (p. 93) that the measure of motion in Newton's theory is **momentum ($\vec{p}$)**, defined as

$$\vec{p} = m\vec{v} \qquad [7.1]$$

The scalar form $p = mv$ is often all we'll need, especially for straight-line motion. Study Example 7.1 and be sure you understand the concept. {Work out the **WARM-UPS** in **CHAPTER 7** on the **CD**. That will ensure that you've got the basics. Then study the **WALK-THROUGHS**.} Newton's Second Law says that

$$F_{av} = \frac{\Delta p}{\Delta t} \qquad [7.2]$$

Alternatively, *the change in the momentum of a body equals the net applied impulse*:

$$\vec{F}_{av}\,\Delta t = \Delta\vec{p} \qquad [7.3]$$

For motion in a straight line the scalar formulation will do:

[straight-line motion] $F_{av}\,\Delta t = \Delta p \qquad [7.4]$

Reread Sections 7.1 (Impulse and Momentum Change) and 7.2 (Varying Force) and study Examples 7.2–7.6.

Momentum is of interest primarily because it's conserved. The law of **Conservation of Linear Momentum** states that *when the resultant of all the external forces acting on a system is zero, the momentum of the system remains constant* (p. 216). Examine Section 7.4 (Conservation of Linear Momentum) and make sure you understand Example 7.8. The ideas of Conservation of Momentum and Energy are then applied to the analysis of collisions in Section 7.5 (review Examples 7.9 and 7.10).

Section 7.5 is all about collisions. *An inelastic collision is one where the final KE of the system is different from the initial KE* (p. 219). All collisions between macroscopic objects are more or less inelastic. The **completely inelastic** collision is at one extreme—where the impacting objects stick together—and the maximum amount of KE is transformed (i.e., lost) into internal energy. Then only Conservation of Momentum holds:

[head-on, inelastic] $m_1 v_{1i} + m_2 v_{2i} = m_1 v_{1f} + m_2 v_{2f} \qquad [7.5]$

A collision is **elastic** *when* $KE_i = KE_f$ (p. 221):

$$KE_{1i} + KE_{2i} = KE_{1f} + KE_{2f}$$

The **"relative speeds"** of the two bodies before and after an elastic impact are equal (p. 223). In two dimensions the scalar momentum components in the x- and y-directions are conserved independently:

$$p_{ix} = p_{fx} \qquad \text{and} \qquad p_{iy} = p_{fy} \qquad [7.11]$$

Conservation of Linear Momentum arises from the homogeneity of space (p.226).

Key Terms

linear momentum	inelastic collision
impulse	completely inelastic
rockets	elastic collision
Conservation of Linear Momentum	relative speed
internal forces	glancing collision
external forces	homogeneity of space

Discussion Questions

1. During the Korean War in the 1950s, American pilots flying the F86 reported that they lost 30 to 40 miles per hour in air speed whenever they fired their guns. Explain.

2. A 500-g cart is sent along a straight horizontal track toward a stationary bumper. Magnets mounted on the cart and bumper produce a repulsive force that extends the collision time. An ultrasonic detector measures the scalar value of the velocity of the cart, and a force transducer on the cart reads out the force on impact. A computer then determines that the area under the force-versus-time curve is 0.37 N·s. Figure Q2 shows plots of the data taken—explain every aspect of it that you can.

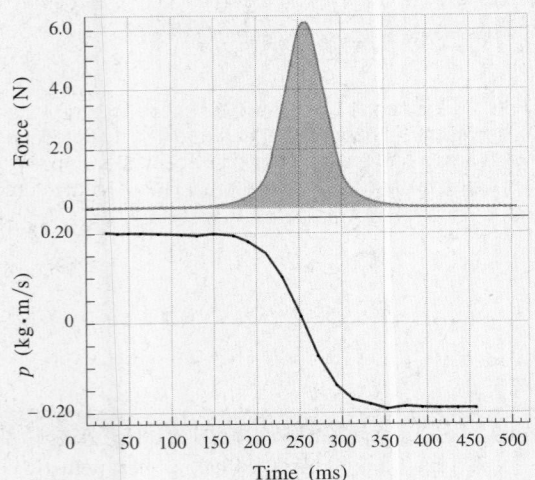

Figure Q2

3. If you dive into the water and land flat on your chest in what is referred to in some circles as a "belly flop," you are likely to remember it for a while. Explain the physics behind what's happening.

4. Why should a batter pay more attention to follow-through when hitting a softball as opposed to a hardball? Every now and then a player will hit a ball so hard that the bat itself shatters. Explain how this happens in terms of Newton's Laws. Drop a ball and watch it hit the floor and rebound back upward. According to the First Law, there must have been an externally applied upward force on the ball. What was it?

5. It is fairly easy to catch a 20-N steel weight dropped from a height of $\frac{1}{4}$ m or so. But if you attempt to catch the weight while resting your hand flat on a table, you are likely to be injured. Explain.

6. A human being can survive a feet-first impact at speeds up to roughly 12 m/s (i.e., 27 mi/h) on concrete; 15 m/s (i.e., 34 mi/h) on soil; and 34 m/s (i.e., 75 mi/h) on water. Explain the spread in these values.

7. Which would be more difficult to throw at a given speed while hovering in space: a "weightless" tennis ball or a "weightless" bowling ball? Explain. Can astronauts floating in orbit tell which objects around them would be heavy or light on Earth even though everything is weightless? How?

8. Which stings the hand more (if either)—catching a ball in flight while running toward it or away from it? Explain.

9. What would happen to two astronauts if, while floating stationary with respect to one another, they played catch with a baseball?

10. Suppose that while working out in space an astronaut accidentally hits her thumb with a "weightless" hammer. Will it hurt? Explain.

11. Two wads of clay with identical momenta but different masses are thrown at a wall. Which will produce more thermal energy when they slam to a stop?

12. The number One billiard ball moving with a speed v_{1i} strikes the Two ball, which is at rest, hitting it off-center and sailing away in an elastic collision. With respect to the initial line of motion, the two balls move off at angles θ and ϕ, respectively. Draw a diagram showing that $\vec{\mathbf{p}}_{1i}$, $\vec{\mathbf{p}}_{1f}$, and $\vec{\mathbf{p}}_{2f}$ form a closed triangle. Explain your reasoning.

13. Given two moving bodies with identical kinetic energies but different masses, will their momenta be equal? If not, which will have the greater momentum?

14. It is possible to hold a large rock or a massive lump of soft clay and slam it with a hammer and yet not hurt the hand holding the object? Explain what is happening in terms of energy and momentum.

15. A glass marble is dropped from some height onto wet sand and penetrates to a depth d. The experiment is repeated, but this time from twice the original height. How deep will the sphere penetrate and why? What assumptions are made?

16. You wish to hit a nail into a board with a hammer. It is your intention to slam the nail so that the hammer remains in contact with the nail without rebounding. Would you do better delivering the same energy with a light or a heavy hammer? Explain.

17. A mass is hoisted into the air and dropped onto a stake, driving it into the ground. Which will have more momentum on impact, a heavy object falling through a small distance or a light object fall through a great distance, provided each is lifted to its drop point with the same expenditure of work? And if there is no rebounding, which is the more efficient driver? Explain.

18. Explain the meaning of each of the Key Terms on page 227.

Multiple Choice Questions

1. If we represent the dimensions of mass, length, and time by M, L, and T, respectively, then the dimensions of impulse are (a) $[ML/T]$ (b) $[ML^2/T^2]$ (c) $[ML/T^2]$ (d) $[LT/M]$ (e) none of these.

2. Suppose a projectile's speed and mass are both doubled. Its momentum will then be (a) doubled (b) unchanged (c) quadrupled (d) halved (e) none of these.

3. A 20-N force acts on a body and its momentum changes by 80.0 kg·m/s. How long does the force act? (a) 8.0 s (b) 4.0 s (c) 2.0 s (d) 1.0 s (e) none of these.

4. As a rule (which is true up to a point) the longer the barrel of a gun the greater the muzzle velocity. This is the case because (a) on average there's less friction (b) the mass of the projectile decreases as it flies down the barrel (c) the potential energy of

the projectile is greater (d) the time over which the force acts is greater (e) none of these.

5. An object in motion must have (a) an acceleration (b) momentum (c) potential energy (d) charge (e) none of these.

6. The units of momentum can be (a) $kg \cdot s^2$ (b) $kg \cdot m/s^2$ (c) $N \cdot s^2$ (d) $N \cdot s$ (e) none of these.

7. A firecracker explodes in midair. Considering all the fragments upon explosion (a) the total KE remains constant (b) the total momentum decreases (c) the total KE decreases (d) the total momentum remains constant (e) none of these.

8. While floating motionlessly in a spacestation a 20-kg girl pushes on a 40-kg boy. He sails away at 1.0 m/s and she (a) remains motionless (b) moves in the opposite direction at 2.0 m/s (c) moves in the same direction at 1.0 m/s (d) moves in the opposite direction at 1.0 m/s (e) none of these.

9. Light carries momentum, so when a beam impinges on a surface, it will exert a force on that surface. If the light is reflected rather than absorbed, the force will be (a) less (b) greater (c) equal either way (d) not enough information given (e) none of these.

10. A 1.0-kg body at rest experiences a force of 5.0 N exerted in the positive x-direction for 2.0 s followed by a force of 10 N exerted in the negative x-direction for 1.0 s. Its resulting speed will be (a) +10 m/s (b) −10 m/s (c) +20 m/s (d) −20 m/s (e) none of these.

11. A 75.0-kg astronaut wearing a 25.0-kg spacesuit while floating in space fires a little gas jet that shoots out a blast of 10.0 g of gas at 100 m/s. He in turn (a) moves in the opposite direction at 100 m/s (b) moves in the opposite direction at 10.0 m/s (c) moves in the opposite direction at 1.00 m/s (d) moves in the opposite direction at 0.100 m/s (e) none of these.

12. A car traveling at a certain speed undergoes a head-on collision with an identical vehicle traveling at the same speed, and both come to rest in 10 ms. Now suppose, instead, that the car crashes into a brick wall and again comes to a rest in 10 ms. From the perspective of the original driver, which would be less dangerous? (a) the two-car crash (b) the car-wall crash (c) both are identical (d) not enough information (e) none of these.

13. A garbage truck crashes head-on into a Volkswagen and the two come to rest in a cloud of flies. Which experiences the greater impact force? (a) the truck (b) the Volkswagen (c) both experience the same force (d) not enough information given (e) none of these.

14. Two equal-mass bullets traveling with the same speed strike a target. One of the bullets is rubber and bounces off; the other is metal and penetrates, coming to rest in the target. Which exerts the greater impulse on the target? (a) the rubber bullet (b) the metal bullet (c) both exert the same (d) not enough information (e) none of these.

15. An open railroad car filled with coal is coasting frictionlessly. A girl on board starts throwing the coal horizontally backward straight off the car, one chunk at a time. The car (a) speeds up (b) slows down (c) first speeds up and then slows down (d) travels at constant speed (e) none of the above.

16. A tank car coasting frictionlessly horizontally along the rails has a leak in its bottom and dribbles several thousand gallons of water onto the roadbed. In the process it (a) speeds up (b) slows down (c) gains momentum (d) loses momentum (e) none of the above.

17. What happens to the momentum of a body of constant mass if while it's traveling its kinetic energy is doubled? (a) it doubles (b) it remains the same (c) it increases by a multiplicative factor of $\sqrt{2}$ (d) it decreases by a multiplicative factor of $\sqrt{2}$ (e) none of the above.

18. A bomb hanging from a string explodes into pieces of different sizes and shapes. After the explosion (a) the vector momentum of each piece is identical (b) the total momentum is increased (c) the momentum of all the pieces, exhaust, and smoke adds up to zero (d) not enough information to comment (e) none of the above.

19. A can of whipped cream floating in space develops a hole in the bottom from which it squirts backward a mess of gas and cream at a constant speed with respect to the can. The can thereupon (a) accelerates forward throughout the squirting (b) moves forward at a constant speed (c) remains at rest (d) first speeds up, and then slows down when the gas runs out (e) none of the above.

20. A 9.5-g wad of chewing gum traveling at 1.0 m/s horizontally crashes into a 9.5-g puck floating frictionlessly on an air table. They stick together and sail away at (a) 1.0 m/s (b) 2.0 m/s (c) 0.0 m/s (d) 0.50 m/s (e) none of these.

21. Two balls collide head-on elastically. Ball-1 (of mass m_1) has an initial speed and ball-2 (of mass m_2) is at rest. The transfer of energy from ball-1 to ball-2 will be maximum when (a) $m_1 > m_2$ (b) $m_2 > m_1$ (c) $m_1/m_2 \approx 0$ (d) $m_1/m_2 = 1$ (e) none of these.

22. Two cars, each of 1570 kg, collide. One is heading north at 22.0 m/s while the other is traveling east at 22.0 m/s. If on impact they mash together and skid off as one, in what direction will the wreckage travel? (a) 45° north of west (b) 45° north of east (c) east (d) 22.5° north of east (e) none of these.

For more Multiple Choice Questions with answers click on WARM-UPS in CHAPTER 7 on the CD.

Suggestions on Problem Solving

1. Begin problem solving with a sketch. Watch out for the signs. Take the direction of initial motion as positive and stick with it for the whole analysis. Put the signs in your diagram at the outset.

2. Think in terms of Conservation of Momentum whenever you have and want masses and speeds before and after some interaction. Newton's Laws will provide an alternative means of solving such problems, and either approach can be used to check the other.

3. Remember to *square* the speed when doing a KE calculation. When considering the change in KE, bear in mind that $v_f^2 - v_i^2$ is *not* equal to $(v_f - v_i)^2$.

4. Momentum is a vector quantity. If the velocity vectors of two bodies lie along a line connecting their centers, a collision problem can be treated via a single scalar momentum expression. This is the case provided that you first assign a sign to the appropriate direc-

tions. Then colinear vectors will properly add (or subtract) as scalars. If the initial velocity vectors are not colinear, their separate component momenta in, say, the x- and y-directions will be conserved individually. Be very careful with the signs in momentum problems.

5. We are restricted to the two extreme classes of collision problems, and that's helpful to remember. If the colliding bodies stick together, the collision involves only Conservation of Momentum and there will be one equation and only one unknown. When the two bodies move off together, there will be one final velocity for both. By contrast, in one dimension, if the collision is elastic, we can also use Conservation of Energy, and there will be two equations and there can be two unknowns. The final velocities of the bodies will be different.

6. Proceed cautiously when confronted with a collision problem where it's not explicitly stated that mechanical energy is conserved. A bullet that squashes against a steel plate or imbeds in a phone book or passes right through a wooden board will "lose" an appreciable amount of energy. Nonetheless, provided no external forces are acting on the bullet-board system, momentum will be conserved. Watch out for situations where any of the objects taking part suffer deformation or generate "heat" via friction.

Problems ✦ Coordinated Problems ✦ Progressive Problems ✦ Solutions

STUDY GUIDE **1. Coordinated Problems:** The three problems within each magenta-colored grouping are solvable in similar ways. Note that the first of these always has a hint; moreover, its solution is provided in the back of the book. *Work out each of these sets; they'll strengthen technique and build confidence.* **2. Progressive Problems:** The problems introduced in blue unfold step-by-step carrying along the analysis in a more suggestive way than is customary. *Work out all of these; they'll guide you through the analytic process and help develop problem-solving skills.* **3. Worked-Out Solutions:** Studying worked-out solutions is an important part of learning how to solve problems. Accordingly, additional *solutions* to a number of model problems are given below. *Make sure you understand each of them before you go on to the next problem.* **4.** Also provided in the back of the book are the **Answers** to all odd-numbered problems, as well as worked-out *solutions* to those with boldface numbers. Problem numbers in italic indicate that a solution appears in the Student Solutions Manual.

SECTION 7.1: IMPULSE AND MOMENTUM CHANGE
SECTION 7.2: VARYING FORCE

1. [I] A 400-kg L75-domestic robot is flying due north down the emergency shaft of a space freighter at 12.0 m/s. Determine its momentum.

2. [I] If the magnitude of the momentum of a 47-g golf ball is 2.8 kg·m/s, what's its speed?

3. [I] From how high (in meters) must a car fall if it's to have the same momentum it would have driving at 60 mi/h (i.e., 60×0.4470 m/s)?

4. [I] What is the linear momentum of the Earth ($m_\oplus = 5.975 \times 10^{27}$ g) as it moves through space if its orbital speed is about 6.66×10^4 mi/h (i.e., $6.66 \times 10^4 \times 0.447\,0$ m/s)?

5. [I] A 1.0-kg wad of clay is slammed straight into a wall at a speed of 10 m/s. If the clay sticks in place, what was the impulse that acted on it via the wall? [*Hint: Take the direction of motion as positive, and compute the change in momentum of the clay.*]

6. [I] What is the impulse provided by the racket on a tennis ball (0.058 kg) served at 50 m/s?

7. [I] During a game a soccer ball having a mass of 0.425 kg is kicked from rest to 26 m/s in a collision with a player's foot lasting 8 ms. What is the impulse imparted to the ball?

8. [I] A dried pea fired from a plastic drinking straw has a mass of 0.50 g. If the force exerted on the pea is an average 0.070 lb (i.e., 0.070×4.448 N) over the 0.10-s flight through the straw, at what speed will it emerge?

9. [I] A 1.0-kg body initially traveling in the positive x-direction at 10 m/s is acted upon for 2.0 s by a force in that same direction of 20 N. It then experiences a force acting in the negative direction for 20 s equal to 2.0 N. Draw a force-time curve and determine the final momentum.

10. [I] A 10-kg asteroid is traveling through space at 2 m/s toward a spaceship. To avoid a collision, an astronaut, with the help of a backpack rocket, exerts a force of 20 N on the asteroid in the opposite direction for 5 s. (a) Find the asteroid's final speed. (b) How long should she have pushed in order to just stop the asteroid?

11. [I] If a 0.061-kg handball goes from rest to 20 m/s during a serve and the player feels an average force of 100 N, how long did the impact last?

> **SOLUTION:** $F_{av}\Delta t = \Delta p$ and $\Delta p = (0.061 \text{ kg})(20 \text{ m/s}) = 1.22 \text{ kg·m/s}$ hence $\Delta t = \Delta p/F_{av} = (1.22 \text{ kg·m/s})/(100 \text{ N}) = 1.2 \times 10^{-2}$ s.

12. [I] Figure P12 shows the force-time curves for two idealized blows delivered during a martial arts contest. Which of the two is likely to correspond to a boxer's punch as opposed to a karate chop? Compute an approximate value for the corresponding impulse in each case. Which is more likely to break bones, and why?

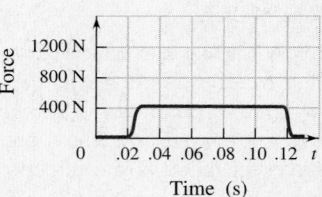

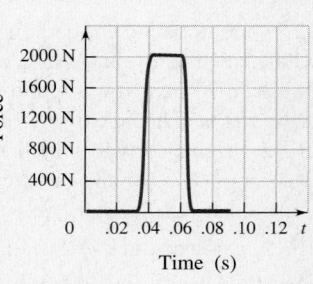

Figure P12

13. [I] Figure P13 is a force-time curve for a small frictionless cart. When, if ever, was its momentum constant?

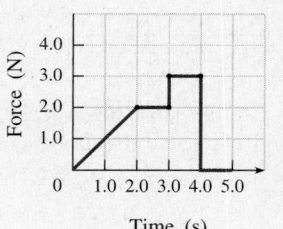

Figure P13

14. [I] Figure P13 is a force-time curve for a small frictionless cart. By how much did its momentum change during the interval from 0 to 2.0 s?

15. [II] THIS PROBLEM EXAMINES THE SPEED OF AN OBJECT WHOSE FORCE-TIME CURVE IS GIVEN. Figure P13 depicts the force applied to a frictionless cart, which was at rest at $t = 0$, and which has a mass of 0.50 kg. (a) What is the physical significance of the area under the curve? (b) Determine the area under the curve between $t = 0$ and $t = 4.0$ s. (c) How fast is the cart moving at $t = 4.0$ s?

16. [II] THIS PROBLEM EXAMINES THE SPEED OF AN OBJECT WHOSE

FORCE-TIME CURVE IS GIVEN. Figure P16 depicts the force applied to an ion engine (i.e., the thrust) mounted in a test vehicle whose total mass is 200.0 kg. (a) What is the physical significance of the area under the curve? (b) What is the significance of the fact that some of the area is below the axis? (c) Determine the area under the curve between $t = 0$ and $t = 9.54$ s. (d) If the craft is traveling through space at 10.0 m/s at $t = 0$, how fast will it be moving at $t = 9.54$ s?

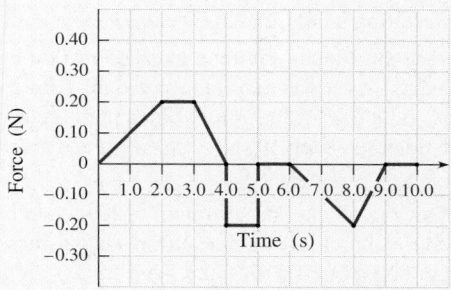

Figure P16

17. [II] A youngster having a mass of 50.0 kg steps off a 1.00-m high platform. If she keeps her legs fairly rigid and comes to rest in 10.0 ms, what is her momentum just as she hits the floor? What average force acts on her during the subsequent deceleration? [*Hint: Use the fact that she falls with a constant acceleration to compute her impact speed.*]

18. [II] A 20.0-kg ball of wet clay drops off a workman's scaffold and falls for 4.0 s before hitting the ground. If it comes to rest in 5.0 ms, what average force did the floor exert on the clay? Use momentum considerations to solve the problem.

19. [II] A 47-g golf ball is hit into the air at 60 m/s. It lands in sand at the same elevation and comes to rest in 10 ms. Ignoring air friction, what was the average force the sand exerted on the ball?

20. [II] A golf ball with a mass of 47.0 g can be blasted from rest to a speed of 70.0 m/s during the impact with a clubhead. Taking that impact to last only about 1.00 ms, (a) calculate the scalar value of the change in momentum of the ball. (b) What is the average force on the ball during the collision?

> **SOLUTION:** Take the direction of motion of the club to be positive. (a) $\Delta p = (0.047 \text{ kg})(70.0 \text{ m/s}) = 3.29 \text{ kg·m/s}$. (b) $F_{av}\Delta t = \Delta p$; hence $F_{av} = \Delta p / \Delta t = (3.29 \text{ kg·m/s})/(1.00 \times 10^{-3} \text{ s}) = 3.29 \times 10^3$ N.

21. [II] A 1-kg hammer slams down on a nail at 5 m/s and bounces off at 1 m/s. If the impact lasts 1 ms, what average force is exerted on the nail?

22. [II] A person can just survive a full-body collision (either to the front, back, or side) at roughly 9 m/s (20 mi/h) with an impact time of approximately 10 ms. At greater speeds or shorter times, fatal brain damage will likely occur. Could someone survive a fall from 4.0 m landing flat on his back on soft soil so that he decelerates to rest through a distance of 10 cm (that's the total compression of body and soil)? If his mass is 70 kg, what's the impulse exerted on his body by the ground? Assume the deceleration is constant.

23. [II] The human heart pumps about 2 ounces of blood into the aorta on each stroke, which lasts roughly 0.1 s. During that time, the pulse of blood is accelerated from rest to about 50 cm/s. Compute the corresponding average propulsion force exerted on the blood by the heart.

24. [II] During a 40 mi/h (i.e., 40×0.447 0 m/s) car crash, a 50-

kg passenger comes to rest in 100 ms. What is the area under the force-time curve for that person? Assuming a seat belt was worn, which would smooth out the curve so there were no devastating peaks, what average force would be experienced?

25. [II] A modern nuclear aircraft carrier weighs in at around 90 000 tons and is capable of traveling in excess of 30 knots (15.4 m/s). Suppose one of these giants plowed into a pier at 30 knots, coming to rest in 0.50 min. What average force would it exert in the process? (One ton, as used in the United States and Canada, equals 2 000 pounds.)

26. [II] A beam consisting of around 1.0×10^{15} electrons per second, each of mass 9.1×10^{-31} kg, paints pictures on the face of a TV tube. Assuming that the electrons are absorbed by the screen and exert an average net force on it of 1.0×10^{-7} N, what is their average speed at impact?

27. [II] An F-14A jet weighing 70 000 lb (i.e., 31.75×10^3 kg) has a typical liftoff distance of 3000 ft (i.e., 3000×0.304 8 m) and a corresponding liftoff speed of 135 knots (69.5 m/s). Assuming its acceleration is constant, calculate the net average thrust of its engines in newtons.

28. [III] A typical tubular aluminum arrow 70-cm long has a mass of 25.0 g. The force exerted by a bow varies in a fairly complicated fashion from the initial draw force of, say, 175 N (39 lb) to zero, as the arrow leaves the string. Assume a full draw (70 cm) and a launch time of 16.0 ms. If the arrow when fired straight up reaches a maximum height of 143.3 m above the point of launch, then (a) What is the average force exerted on it by the bow? (b) What's the area under the force-time curve? (c) What impulse is applied to the shooter?

29. [III] The batter in a baseball game is ready to swing at a ball (of mass 0.149 kg) traveling toward him at 35.0 m/s horizontally. The bat strikes the ball, which flies upward at 45.0° at a speed of 39.0 m/s. Determine the impulse applied to the ball.

SECTION 7.3: ROCKETS

SECTION 7.4: CONSERVATION OF LINEAR MOMENTUM

30. [I] During the launch of a space rocket, fuel is consumed at a constant rate of 7.5 kg/s. If the thrust is to be 18 kN, determine the necessary exhaust speed.

31. [I] A rocket engine blasts out 1000 kg of exhaust gas at an average speed of 2 km/s every second. Calculate the resulting average thrust developed by the engine.

32. [I] Water squirting from a hose emerges with a speed of 10 m/s at a rate of 100 kg/s. Compute the average reaction force exerted on the hose. Have you ever seen firefighters working to control a large hose?

33. [I] THIS PROBLEM DEALS WITH CONSERVATION OF LINEAR MOMENTUM. Two gliders, one of mass 6.0 kg and the other 2.0 kg, are on a frictionless air track. They are locked together, thereby compressing a spring which is positioned between them. (a) Discuss the forces exerted by the spring on gliders. (b) When released they fly apart. What can be said about the impulse imparted to each glider? (c) What can be said about the momentum imparted to each glider? (d) The 6.0-kg glider travels at 3.0 m/s due east. What is the magnitude of its momentum? (e) What is the final velocity of the 2.0-kg glider?

34. [I] THIS PROBLEM CONCERNS CONSERVATION OF LINEAR MOMENTUM. A toy cannon having a mass of 4.0 kg sits on a frictionless level surface. It fires a 250-g projectile horizontally and immediately recoils westward at 1.5 m/s.   (a) Find the change in momentum of the cannon.   (b) Determine the change in momentum of the projectile.   (c) What was the muzzle velocity of the projectile?

35. [I] A 50-kg person at the southernmost end of a 150-kg rowboat at rest in the water begins to walk to the northern end. If, at a given instant, her speed with respect to the water is 10 m/s, what is the velocity of the boat with respect to the water at that moment? [*Hint: The initial momentum of the system (boat and person) is zero and momentum is conserved.*]

36. [I] While constructing a space platform a 100-kg robot finds himself standing on a 30-m long 200-kg steel beam that is motionless with respect to the platform. Using his magnetic feet he walks along the beam traveling south at 2.00 m/s. What is the velocity of the beam with respect to the platform as the robot walks?

37. [I] The glider on an air track (Fig.Q1 in Chapter 4) is 30.0-cm long and has a mass of 0.500 kg. It floats frictionlessly on compressed air and is free to move along an east-west axis. A 100-g wind-up toy car is placed on the glider and it immediately rolls due east at 0.50 m/s with respect to the lab. Does the glider move and if so, how fast and in what direction?

38. [I] While floating in space a 100-kg robot throws a 0.800-kg wrench at 12.0 m/s toward his partner working on the spaceship. How fast will the robot move away from the ship?

39. [I] A 60-kg cosmonaut jumps from her 5000-kg spaceship in order to meet her copilot who is floating at rest a few hundred meters away. If she sails toward him at a speed of 10 m/s, what is the resulting motion of the ship?

40. [I] A kid holding a baseball bat stands at rest in a small boat. The total mass of the kid, bat, and boat is 250 kg. Someone on shore throws a 0.149-kg baseball at the boat. It flies straight down the length of the craft, reaching the batter at 20.0 m/s. He hits it with the bat and the ball sails back the way it came at 20.0 m/s. What's the speed of the boat immediately after the ball is hit?

41. [I] Two kids in a small boat, which is at rest on the water, play catch with a 1.00-kg lead ball. The total mass of the boat, ball, and kids is 250 kg. What is the speed of the boat just as the ball leaves the hand of the kid in the front at a speed of 5.00 m/s? In what direction does the boat move?

> **SOLUTION:** Call the momentum of the ball $\vec{p}_B$ and of the rest of the system $\vec{p}_R$. Since the initial momentum before the throw is zero, the net momentum after the throw is also zero. The ball travels toward the rear (take that to be the positive direction) and the boat moves forward. So $\vec{p}_B = -\vec{p}_R$ after the throw; $(1.00 \text{ kg})(5.00 \text{ m/s}) = -(249 \text{ kg})v_R$ and so $v_R = -2.01 \times 10^{-2}$ m/s.

42. [I] With Problem 41 in mind what is the speed of the boat just after the second kid catches the ball?

43. [II] Water with a density of 1.00×10^3 kg/m³ streams from a hose in a horizontal jet 2.4 cm in diameter at 22 m/s. It strikes a nearby vertical sponge-like absorbing wall, thereupon transferring all its kinetic energy to the wall so there is no splashing. What average force does the water exert on the wall?

44. [II] An ion engine is a rocket designed to produce small thrusts for very long periods of time. The SERT 2 engine develops 30 mN of thrust by expelling mercury ions at speeds of around 22 000 m/s (about 50 000 mi/h). How many kilograms of exhaust emerge per second?

45. [II] An ice skater (with a mass of 55.0 kg) throws a snowball while standing at rest. The ball has a mass of 200 g and moves straight out with a horizontal speed of 20.0 km/h. Neglecting friction, and assuming the skate blades are parallel to the direction of the throw, describe the skater's resulting motion in detail.

46. [II] A 90-kg astronaut floating out in space is carrying a 1.0-kg TV camera and a 10-kg battery pack. He's drifting toward his ship but, in order to get back faster, he hurls the camera out into space at 15 m/s and then throws the battery at 10 m/s in the same direction. What's the resulting increase in his speed after each throw?

47. [II] A railroad flatcar having a mass of 10 000 kg is coasting along at 20 m/s. As it passes under a bridge, 10 men (having an average mass of 90 kg) drop straight down onto the car. What is its speed as it emerges with its new passengers from beneath the bridge?

48. [II] A fireworks rocket rises straight up to its maximum altitude of 50.0 m. At that point it explodes into two equal-mass pieces; one heads straight down at 20.0 m/s, the other travels straight upward. What's the maximum altitude attained by the second upwardly-moving fragment?

49. [II] Two girls on roller skates, face-to-face, push on each other as hard as they can. Girl-1, whose mass is 20.0-kg, moves off west at 2.00 m/s. Write a general expression for the speed of the second body in such a situation and then calculate the specific speed of girl-2, whose mass is 40.0 kg.

50. [III] During a test a rocket is launched into a ballistic trajectory that ordinarily would have a range of 40.0 km. At the top of its parabolic arc across the sky it blows up, blasting into two equal-mass pieces. One drops straight down, while the other is hurled forward horizontally—where does it land?

51. [III] Two astronauts floating at rest with respect to their ship in space decide to play catch with a 0.500-kg asteroid. Neil (whose mass is 100 kg) heaves the asteroid at 20.0 m/s toward Sally (whose mass is 50.0 kg). She catches it and heaves it back at 20.0 m/s (with respect to the ship). Before Neil catches it a second time, how fast is each person moving, and in what direction?

SECTION 7.5: COLLISIONS

52. [I] Two identical cars of mass 2000 kg each drive toward one another, both traveling at 20.0 m/s. They collide head-on and smash together; what is the final speed of the wreckage?

53. [I] THIS PROBLEM DEALS WITH CONSERVATION OF LINEAR MOMENTUM. Two blocks on a frictionless air table slide directly toward each other. The first, traveling east has a momentum of 10.0 kg·m/s; the second traveling west has a momentum of 15.0 kg·m/s. They collide and more or less bounce off one another.   (a) Taking east to be positive, what is the initial momentum of each block? (b) What is the total momentum of the two blocks before the collision?   (c) What is the total momentum of the two blocks after the collision?

54. [I] THIS PROBLEM IS ABOUT CONSERVATION OF LINEAR MOMENTUM. Two carts, one of mass 9.0 kg and the other 3.0 kg, move frictionlessly on a horizontal surface. The 9.0-kg cart traveling north at 2.0 m/s crashes head-on into the 3.0-kg cart heading south and both slam to a stop.   (a) Taking north to be positive, what is the initial momentum of the 9.0-kg cart?   (b) What was the total momentum of the two carts after the collision?   (c) What was the total momentum of the two carts before the collision? (d) What was the momentum of the 3.0- kg cart prior to the collision? (e) Determine the speed of the 3.0- kg cart prior to the collision. (f) Was this an elastic collision?

55. [I] The glider on an air track has a mass of 1.00 kg and floats on compressed air, so it can move frictionlessly. Someone shoots a 20.0-g lump of clay at it. The clay strikes the glider, sticks to it, and both move away with a speed of 20.0 cm/s. What was the speed of the clay as it hit the glider?

> SOLUTION: Momentum is conserved for the clay (C) and glider (G). The collision is totally inelastic. Take the direction of motion of the clay as positive. Initially, $p_i = m_C v_{Ci}$; finally, $p_f = (m_C + m_G)v_f$ and since $p_i = p_f$, $m_C v_{Ci} = (m_C + m_G)v_f$. Putting the numbers in, $(20.0 \times 10^{-3} \text{ kg})v_{Ci} = (1.02 \text{ kg})(20.0 \times 10^{-2} \text{ m/s})$ and $v_{Ci} = 10.2 \text{ m/s}$.

56. [I] Two wads of putty are propelled horizontally directly toward each other. Wad-1 has a mass of 5.00 kg and is traveling at 21.0 m/s south. Wad-2, which is 6.00 kg, is moving at 12.0 m/s north. They collide and stick together. What is the velocity of the joint mass of putty?

57. [I] A 90-kg signal relay floating in space is struck by a 1000-g meteoroid. The latter imbeds itself in the craft and the two sail away at 5.0 m/s. What was the initial speed of the meteoroid?

58. [I] A glass sphere of mass 100 g falls toward a thick horizontal hard-steel plate, striking it at 10.0 m/s. Given that the collision is completely elastic, at what speed will the sphere rebound?

59. [I] A steel ball (call it ball-1) is fired at 20.0 m/s at an identical steel ball (call it ball-2). What is the speed of each ball immediately after the collision if no kinetic energy is lost?

60. [I] A glass ball of mass 200 g is traveling directly at a steel ball of 400 g. An observer at rest in the lab sees the glass ball moving east at 10.0 m/s and the steel ball moving west at 20.0 m/s. What will be the speed at which the two balls approach or recede from each other after undergoing an elastic collision?

61. [II] A 1263-kg (i.e., 2785-lb) Triumph TR-8 sports car traveling south at 40 km/h (i.e., $40 \times 0.277 \, 8$ m/s) crashes head-on into a 1742-kg (i.e., 3840-lb) Checker cab moving north at 90 km/h (i.e., $90 \times 0.277 \, 8$ m/s). If the two cars remain tangled together but free to coast, describe their motion immediately after the collision. What is their final speed? How much kinetic energy is lost?

> SOLUTION: Momentum is conserved for the Triumph (T) and cab (C). Take south to be positive, in which case $v_{Ti} = (40 \times 0.277 \, 8$ m/s) and $v_{Ci} = -(90 \times 0.277 \, 8$ m/s). The collision is totally inelastic. Initially, $p_{Ci} = m_C v_{Ci}$ and $p_{Ti} = m_T v_{Ti}$. Finally, $p_f = (m_C + m_T)v_f$ and since $p_i = p_f$, $m_C v_{Ci} + m_T v_{Ti} = (m_C + m_G)v_f$. Putting the numbers in (1263 kg)(11.1 m/s) + (1742 kg)(−25.0 m/s) = (3005 kg)v_f and $v_f = -9.8$ m/s. $KE_i - KE_f = \frac{1}{2}m_T v_{Ti}^2 + \frac{1}{2}m_C v_{Ci}^2 - \frac{1}{2}(m_T + m_C)v_f^2 = 4.8 \times 10^5$ J.

62. [II] In an arrangement for measuring the muzzle velocity of a rifle or pistol, the bullet is fired up at a wooden mass, into which it imbeds. The wood is blasted straight up into the air to a measured height h. Assuming negligible losses to friction, write an expression for the velocity in terms of the known masses and height. If a 100-grain (6.48-g) 25-06 Remington rifle bullet is fired into a 5.00-kg block that then rises 4.0 cm into the air, what was the muzzle speed of that bullet?

63. [II] An 8.0-kg puck floating on an air table is traveling east at

15 cm/s. Coming the other way at 25 cm/s is a 2.0-kg puck on which is affixed a wad of bubble gum. The two slam head-on into each other and stick together. Find their velocity after the impact. How much kinetic energy is lost?

64. [II] Two identical blocks, each of mass 10.0 kg, are to be used in an experiment on a frictionless surface. The first is held at rest on a 20.0° inclined plane 10.0 m from the second, which is at rest at the foot of the plane. The one descends the incline, slams into and sticks to the second, and they sail off together horizontally. Calculate their speed immediately after impact.

65. [II] A 2000-kg car is traveling east at 20.0 m/s when it's rammed in the rear by a 1000-kg car that was traveling at 30.0 m/s just before impact. The two cars tangle together and move off together at a speed that we now wish to determine—please do so.

66. [II] As seen from the window of a space station, a 100-kg satellite sailing along at 10.0 m/s collides head-on with a small 300-kg asteroid, which was initially at rest. Taking the collision to be elastic, and neglecting their mutual gravitational interaction, what are the final velocities of the two bodies? What were their "relative speeds" before and after the collision?

67. [II] Two billiard balls, one heading north at 15.0 m/s and one heading south at 10 m/s, collide head-on. Take the collision to be perfectly elastic. What is the post-impact speed of each ball? [*Hint: It will help if you remember that for completely elastic collisions, the "relative speeds" before and after are equal.*]

68. [II] A skater with a mass of 75 kg is traveling east at 5.0 m/s when he collides with another skater of mass 45 kg heading 60° south of west at 15 m/s. If they stay tangled together, what is their final velocity?

69. [III] A soft clay block is suspended so as to form a so-called ballistic pendulum, as shown in Fig. P69. A bullet is fired point-blank into the block, imbedding itself therein and raising the latter to a height h. Write an expression for the muzzle speed of the bullet in terms of g, h, and the masses. Remember that although friction on the pendulum is negligible, clay-bullet friction losses are not, and *mechanical energy is not conserved* in the collision.

Figure P69

70. [III] Referring to Problem 69, derive an expression for the percentage of the kinetic energy converted into internal energy during the bullet-clay impact.

71. [III] Two identical hard spheres, one at rest and the other moving, slam into one another in a *non-head-on* elastic collision, as shown in Fig. P71. Prove that they will always fly apart at 90° to each other.

72. [III] Earlier (p. 223) we saw that $v_{1i} = v_{2f} - v_{1f}$ for elastic collisions where body-2 was initially at rest. To show that this is true in general, first establish that $m_1(v_{1i} + v_{1f})(v_{1i} - v_{1f}) = m_2(v_{2f} + v_{2i})(v_{2f} - v_{2i})$; then obtain $m_1(v_{1i} - v_{1f}) = m_2(v_{2f} - v_{2i})$ and finally show that $v_{1i} - v_{2i} = -(v_{1f} - v_{2f})$.

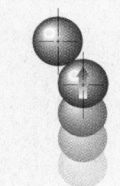

Figure P71

Chapter 8
Rotational Motion

Motion occurs in two basic forms: **translational** and **rotational** (Fig. 8.1). If a rigid body is moving such that a line drawn between any two of its internal points remains parallel to itself, the body is translating. If, on the other hand, that line does not remain parallel to itself, the body is rotating. The Earth, the Sun, and even our galaxy of 10^{11} stars are all spinning through space. At the other extreme, the atoms that form matter are composed of far smaller parts that are in perpetual rotation.

The Kinematics of Rotation

Imagine an object revolving about some point; it might be a wheel, a line of hand-holding skaters, or a string of beads as in Fig. 8.2. Whirled in a circle, each bead sails along a different arc, but *each sweeps through the same angle θ*. Until now, we've measured angles in *degrees,* 360° to a circle. Yet, there is nothing special about "degrees"—they are a remnant of the ancient Babylonian number system, whose base was 60 rather than 10.

8.1 Angular Displacement

STUDY GUIDE

Rotation Kinematics is conceptually very similar to Translational Kinematics. The motion of an object moving around an axis, circular motion, is described in terms of the angular displacement (θ), which is in units of radians. Make sure that you understand what θ is and what a radian is before pressing on (p. 234). We'll then define angular speed (ω) in radians per second (p. 236), and angular acceleration (α) in radians per second squared (p. 239). The familiar equations of constant translational acceleration can easily be transformed into the equations of constant angular acceleration via the substitutions $s \rightarrow \theta, v \rightarrow \omega, a_\mathrm{T} \rightarrow \alpha$, and $t \rightarrow t$. We are then in a position to analyze all sorts of problems concerning rotation, and that's the point of all of this.

The angle θ can be specified in a less arbitrary way by relating it back to an essential aspect of the circle, namely, its circumference. Each bead (Fig. 8.2b) moves an arc-length (ℓ) at some radial distance (r) from the center. The larger r is, the larger ℓ is. In each case, θ is the same, suggesting that we might formulate θ in terms of the ratio of ℓ to r. Therefore, let

$$\theta = \frac{\ell}{r} \quad \text{or} \quad \ell = r\theta \tag{8.1}$$

The angle θ is now specified as a distance divided by a distance and is *unitless*—it's a pure number. The degree is not a physical unit in the strict sense either—it's not referenced back to meters, kilograms, or seconds. When $\ell = r$, $\theta = 1$ and, as a reminder that we are measuring angles via Eq. (8.1), we call this 1 *radian* or 1 rad. If we ever meet up with aliens from some other world, they will probably use the equivalent of radians, too.

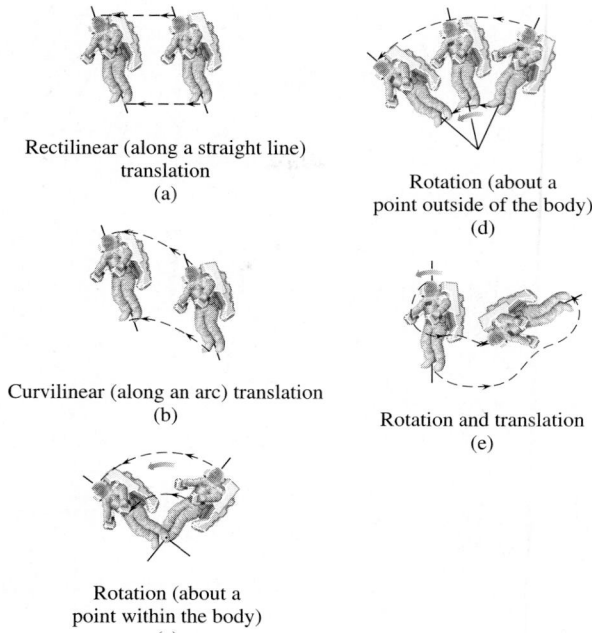

Figure 8.1 The movement of all rigid bodies in space can be described in terms of rotational and translational motion.

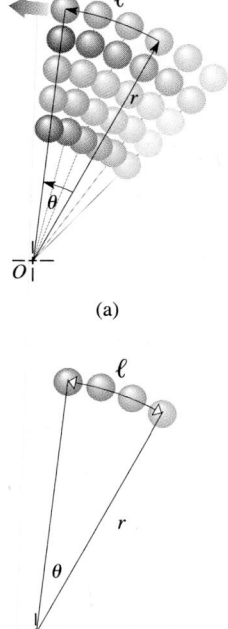

Figure 8.2 A revolving string of beads all moving around their respective circles in step with each other.

Draw a circle of radius r (Fig. 8.3a). Now lay off arc-lengths equal to r around the circle. You will end up with six such segments plus an additional fraction of one. That's because the circumference equals $2\pi r$, which is $(6.283\,185\,...)r$. Figure 8.3b shows the relationship between an arc-length ℓ and the angle θ subtended by it at point O. Once around the circle ($360°$) corresponds to $\ell = 2\pi r$ or $\ell/r = 2\pi$ rad $= \theta = 360°$. Any angle θ in radians can be transformed into degrees by simply taking proportions

$$\frac{\theta\,(\text{radians})}{2\pi(\text{radians})} = \frac{\theta\,(\text{degrees})}{360(\text{degrees})} \tag{8.2}$$

which means that π rad $= 180°$, $\pi/2$ rad $= 90°$, and so forth (see Table 8.1).

The Moon has a diameter $D_{\mathbb{C}}$ equal to about 3.5×10^6 m, and it is a distance $r = 3.8 \times 10^8$ m from the Earth's surface. If we approximate its straight-line diameter as an arc-

Table 8.1

Angles in Degrees and Radians

Degrees	Radians in terms of π	Radians in decimals
0	0	0
15	$\pi/12$	0.262
30	$\pi/6$	0.524
45	$\pi/4$	0.785
57.3	1.00	1.00
60	$\pi/3$	1.047
90	$\pi/2$	1.571
120	$2\pi/3$	2.094
180	π	3.142
270	$3\pi/2$	4.712
360	2π	6.283

A solar eclipse—the Moon passes directly in front of the Sun.

This photo, made at a location well below the equator, was taken by holding the camera's shutter open for several hours. As the Earth revolved, the stars left circular streaks about the spin axis. A similar photo, taken in the Northern Hemisphere, would show Polaris, the North Star, as a bright tiny arc very near the center of the star traces.

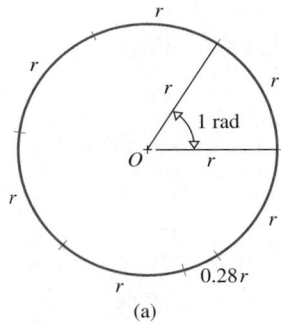
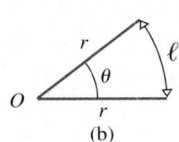
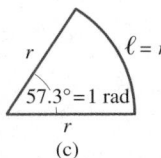
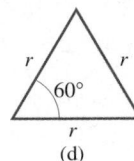

Figure 8.3 The relationship between arc-length, radius, and angle. (a) The circumference of a circle equals $2\pi r$. (b) As r moves through the angle θ, its end point sweeps out a length ℓ. (c) When $\ell = r$, $\theta = 1$ rad. (d) This shape is close to that of an equilateral triangle where $\theta = 60°$.

length, then the angle θ subtended at the Earth (Fig. 8.4) by the Moon is

$$\theta = \frac{\ell}{r} = \frac{D_{\leftmoon}}{r} = \frac{3.5 \times 10^6 \text{m}}{3.8 \times 10^8 \text{m}} = 0.009 \text{ rad}$$

or 0.5°. By comparison, the diameter of the Sun (1.4×10^9 m) is immensely larger, but its distance to the Earth (1.5×10^{11} m) is also much greater. Coincidentally, the Sun also subtends an angle of 0.009 rad. That's why the Sun and Moon seem to be the same size.

8.2 Angular Velocity

Imagine the last bead in Fig. 8.2 moving in a circle of radius r. As it travels along the circular segment ℓ, the radius sweeps through an angle θ, the **angular displacement**. If the bead revolves through six complete counterclockwise turns and comes back to its starting point, we will take $\theta = 6 \times (2\pi)$ rad rather than $\theta = 0$. **Define θ swept out counterclockwise as positive and clockwise as negative**, and the same for ℓ. Consequently, four turns in one direction followed by four turns in the other brings the bead back to where it started, $\theta = 0$ (and $\ell = 0$); in that sense, we are dealing with angular displacements.

If the bead travels the distance $\ell = r\theta$ in a total time t, divide both sides by t to get

$$\frac{\ell}{t} = r\frac{\theta}{t} \tag{8.3}$$

The ratio on the left is reminiscent of the average speed [Eq. (2.1), p. 21] measured along the path, but here it can be positive (counterclockwise motion) or negative (clockwise motion) and cannot properly be referred to as speed. With that in mind, let $v_{av} = \ell/t$ and call it the scalar value of the tangential velocity. The ratio on the right is the scalar value of the **average angular velocity**, denoted by the lowercase Greek letter *omega* (ω) such that

$$\omega_{av} = \frac{\theta}{t} \tag{8.4}$$

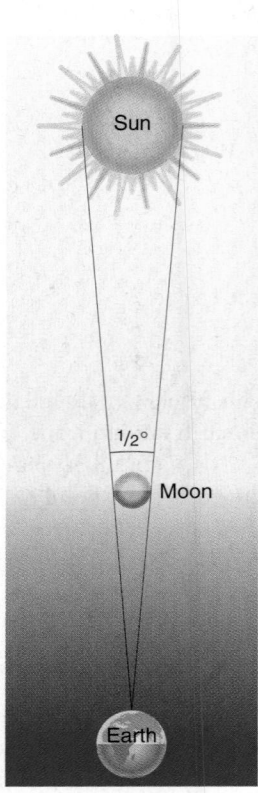

Figure 8.4 The Moon is closer to us than the Sun. Therefore, even though it is much smaller than the Sun, it happens to subtend very nearly the same angle when viewed from Earth.

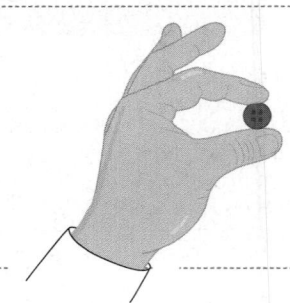

EXPLORING PHYSICS ON YOUR OWN

The Sun, the Moon, Buttons, & Aspirins: The reason we see nice neat solar eclipses is that the Sun and Moon subtend about the same angle: 0.009 rad (i.e., 0.5°). You remember how big the Moon looks, so it should be an easy matter to find something that you think will be the same size when held at arm's length. Amazingly, a man's shirt button (1.0 cm) at arm's length (≈ 70 cm) is "bigger" than the Moon (i.e., ≈ 0.014 rad)—try it. An aspirin (1.1 cm) at arm's length subtends a bit less than 1°, and the width of your index finger subtends a bit more than 1°; that's twice the size of the Moon.

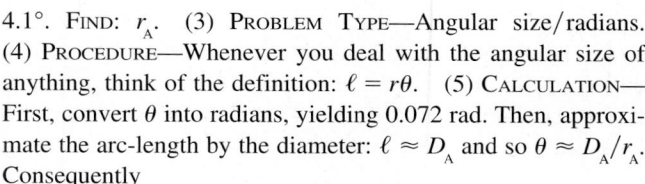

Example 8.1 **[I]** The Andromeda galaxy is a giant spiral cluster of stars whose mass is that of 300 thousand million Suns. You can see it with the naked eye as a faint elongated cloud in the night sky. Inasmuch as it subtends an angle of 4.1° and is known to be larger than our own galaxy [163×10^3 light-years (ly) in diameter for Andromeda as compared to 100×10^3 ly for our galaxy], how far away is it in light-years?

Solution When you see the phrase "subtends an angle," it should call to mind θ and radians. (1) TRANSLATION—An object of known size subtends a known angle; how far away is it? (2) GIVEN: The diameter of Andromeda $D_A = 163 \times 10^3$ ly and its subtended angle $\theta =$

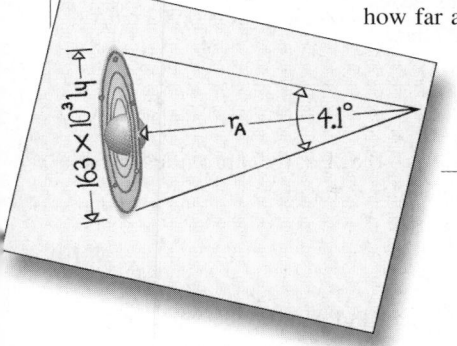

4.1°. FIND: r_A. (3) PROBLEM TYPE—Angular size/radians. (4) PROCEDURE—Whenever you deal with the angular size of anything, think of the definition: $\ell = r\theta$. (5) CALCULATION—First, convert θ into radians, yielding 0.072 rad. Then, approximate the arc-length by the diameter: $\ell \approx D_A$ and so $\theta \approx D_A/r_A$. Consequently

$$r_A = \frac{D_A}{\theta} = \frac{163 \times 10^3 \text{ ly}}{0.072} = \boxed{2.3 \times 10^6 \text{ ly}}$$

Quick Check: Use this result to calculate D_A. When the arm r_A-long sweeps through θ, its end point traces an arc of length $(4.1°\pi/180°)(2.3 \times 10^6 \text{ ly}) = 1.6 \times 10^5 \text{ ly}$.

[For more worked problems click on **WALK-THROUGHS** *in* **CHAPTER 8** *on the* **CD**.*]*

The units of angular velocity are rad/s, although *revolutions per second* (rev/s) and *rotations per minute* (rpm) are often used. Notice that each revolution swings through 2π rad, so that 1 rev/s = 2π rad/s.

It's common practice to avoid the ponderous but precise scalar-value terminology and just refer to ω as the "angular velocity" or the "angular speed," even though it can be negative and speed is never negative. And so, when there is no chance of confusion we may be less than fastidious and speak of ω as speed.

Only when the angular velocity is specified in rad/s will Eq. (8.3) become

$$v_{av} = r\omega_{av} \tag{8.5}$$

Figure 8.5 depicts the slightly different circumstance in which both θ and ℓ are measured from some reference line, in this case, the x-axis. Here, the intervals $\Delta\ell = \ell_f - \ell_i$ and $\Delta\theta = \theta_f - \theta_i$, as before, are related by

$$\Delta\ell = r\Delta\theta \tag{8.6}$$

If this is traversed in a time Δt,

$$\frac{\Delta\ell}{\Delta t} = r\frac{\Delta\theta}{\Delta t} \tag{8.7}$$

which provides the alternative statements

$$v_{av} = \frac{\Delta\ell}{\Delta t} \quad \text{and} \quad \omega_{av} = \frac{\Delta\theta}{\Delta t} \tag{8.8}$$

The advantage of this notation is that we can define the scalar value of the **instantaneous angular velocity** (ω) as the limit of the average value as the averaging time interval becomes smaller and smaller:

$$\omega = \lim_{\Delta t \to 0}\left[\frac{\Delta\theta}{\Delta t}\right] \tag{8.9}$$

The tachometer in a car indicates the rate at which the crankshaft is rotating.

SUBTENDED ANGLES & PERSPECTIVE

The Muslim scientist Alhazen (A.D. 1000) correctly suggested that visual perspective arises because the angle an object subtends gets smaller as it recedes from us—a nearby tree appears large, whereas a distant one, the same size, seems minute. As we'll learn later, if the angle subtended is smaller, the image on the retina is also smaller. Why do you have to get your eye very close to use a peep hole?

Example 8.2 [I] On August 24, 1968, a very fast horse named Dr. Fager finished a 1.00-mi race in 1.00 min 32.2 s with an amazing average speed of 62.8 km/h or 17.4 m/s. Assuming he ran once around a circular track, what was his average angular speed?

Solution The words "angular speed" should call to mind $v = r\omega$. (1) TRANSLATION—An object travels with a known average linear speed in a circle of specified size; determine its average angular speed. (2) GIVEN: $v_{av} = 17.4$ m/s, $\ell = 1.00$ mi on a circular path, and time elapsed $t = 1.00$ min 32.2 s. FIND: ω_{av}. (3) PROBLEM TYPE—Angular motion/average angular speed.

(4) PROCEDURE—We have v_{av} and must determine ω_{av}, which suggests $v_{av} = r\omega_{av}$, where we first find r. (5) CALCULATION—To get the radius, use the fact that $2\pi r = 1.00$ mi; hence, $r = 0.159$ mi $= 0.256$ km. From Eq. (8.5)

$$\omega_{av} = \frac{v_{av}}{r} = \frac{0.017\,4 \text{ km/s}}{0.256 \text{ km}} = \boxed{0.067\,9 \text{ rad/s}}$$

Quick Check: You could use Eq. (8.4) instead. Since $\theta = 2\pi$ rad,

$$\omega_{av} = \frac{\theta}{t} = \frac{2\pi \text{ rad}}{1.537 \text{ min}} = \frac{6.28 \text{ rad}}{1.537 \text{ min}} = 4.09 \text{ rad/min} = 0.068 \text{ rad/s}.$$

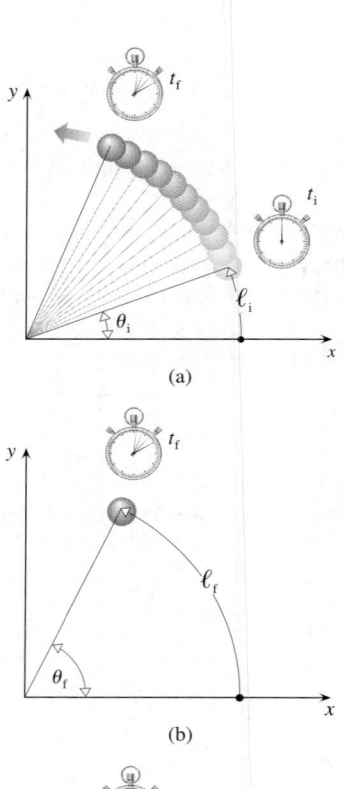

(a)

(b)

(c)

Figure 8.5 As the ball moves from angular position θ_i at ℓ_i to angular position θ_f at ℓ_f on the arc, the angle changes by $\Delta\theta$ and the arc-length measured up from the x-axis changes by $\Delta\ell$.

ω *is the time rate-of-change of* θ. Taking the limit as $\Delta t \to 0$ of Eq. (8.7) provides the relationship between the instantaneous linear and angular scalar velocities:

$$v = r\omega \qquad (8.10)$$

where ω is in rad/s. At any instant, if we know ω, we can find v, and vice versa. Keep in mind that the velocity vector is always tangent to the curved path and v is the magnitude of that vector along with the appropriate sign ($v = \pm v = \pm |\vec{v}|$).

It's evident from Fig. 8.6 that the more distant a bead is from the center of rotation, the more rapidly it must be moving. Once around for bead-1 is a small circle, while bead-2 travels an appreciably larger path in the same time. If they stay abreast, $v_2 > v_1$ even though ω is the same for each of them. That's why the outermost skater in a long revolving line has to be the fastest and why the inner rail position in a race is most advantageous.

When a rigid body rotates about an axis at some angular speed ω, *that rate characterizes the motion and is the same for the entire object*. Nonetheless, each point on the body moves with a linear speed determined by its perpendicular distance from the spin axis.

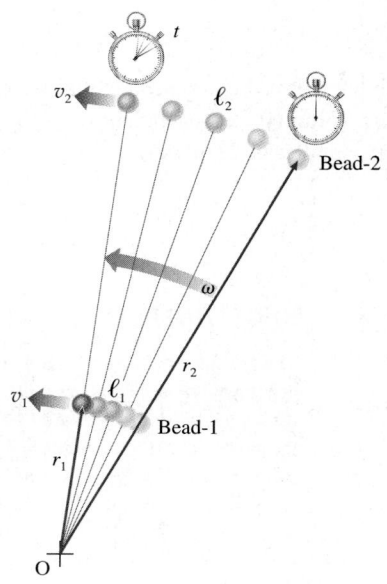

Figure 8.6 Here, beads -1 and -2 move in step, swinging around their respective circles together. Bead-2 travels a larger path and must therefore be moving faster ($v_2 > v_1$), even though both beads have the same w.

Example 8.3 **[I]** The music on a compact disc (CD) is encoded on tiny raised ridges lying on a spiraling path (up to 5.4 km long) that winds outward from the center of the disc (Fig. 8.7). A laserbeam tracks along the path at a constant linear speed of 1.2 m/s, and the fluctuating light reflected off the ridges carries the information lifted from the CD. Typically, the track begins 2.3 cm from the center and ends about 5.9 cm from the center. To keep the linear speed of the laser readout unit constant, the angular speed at which the disc turns is continuously varied. Determine the angular speed of the disc at the beginning and end of the data path. Give your answers in rad/s and rev/s.

Solution The mere mention of "angular speed" should bring to mind $v = r\omega$. (1) TRANSLATION—A point revolves with a known linear speed in a circle of known radius; determine its angular speed. (2) GIVEN: $v = 1.2$ m/s = constant, $r_i = 2.3$ cm, and $r_f = 5.9$ cm. FIND: ω_i and ω_f. (3) PROBLEM TYPE—Angular motion/changing angular speed. (4) PROCEDURE—

We have the constant v and must determine ω, which suggests $v = r\omega = r_i\omega_i = r_f\omega_f$. (5) CALCULATION:

$$\omega_i = \frac{v}{r_i} = \frac{1.2 \text{ m/s}}{2.3 \times 10^{-2} \text{ m}} = 52.2 \text{ rad/s}$$

and

$$\omega_f = \frac{v}{r_f} = \frac{1.2 \text{ m/s}}{5.9 \times 10^{-2} \text{ m}} = 20.3 \text{ rad/s}$$

Each revolution corresponds to 2π rad, and so dividing each of these by 2π and keeping two significant figures, we get

$$\boxed{\omega_i = 52 \text{ rad/s} = 8.3 \text{ rev/s}}$$

and

$$\boxed{\omega_f = 20 \text{ rad/s} = 3.2 \text{ rev/s}}$$

Quick Check: Because v is constant, the ratio of the angular speeds must equal the inverse ratio of the radii: $\omega_i/\omega_f = r_f/r_i$ and so (8.3 rev/s)/(3.2 rev/s) = 2.6 should equal (5.9 cm)/(2.3 cm) = 2.6.

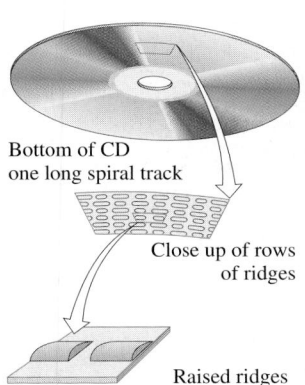

Bottom of CD
one long spiral track

Close up of rows
of ridges

Raised ridges

Figure 8.7 The data stored on the bottom surface of a CD is in the form of one long spiraling trail of tiny raised ridges. It begins at the center and winds out to the rim.

8.3 Angular Acceleration

Variations in ω are as commonplace as are variations in v. Just think about a CD (Example 8.3) spun so that its ω gradually decreases as it plays. If a rotating body changes its angular speed by an amount $\Delta\omega$ in a time interval Δt, its **average angular acceleration** (denoted by the lowercase Greek letter *alpha*) is

$$\alpha_{av} = \frac{\Delta\omega}{\Delta t} = \frac{\omega_f - \omega_i}{t_f - t_i} \tag{8.11}$$

The unit of angular acceleration is radians per second-per second or rad/s². The **instantaneous angular acceleration** (α) is the time rate-of-change of ω at any instant. For simplicity, we restrict our study to situations where α is constant, in which case $\alpha = \alpha_{av} =$ constant.

Since $v = r\omega$, it follows that a change in angular speed (with r constant) will be accompanied by a change in linear speed, $\Delta v = r\Delta\omega$ and

$$\alpha_{av} = \frac{\Delta\omega}{\Delta t} = \frac{\Delta v}{r\Delta t} = \frac{1}{r}a_{av} \tag{8.12}$$

where $a_{av} = \Delta v/\Delta t$. Here Δv, whether an increase or decrease, is along (i.e., tangent to) the path of the motion. Since $\alpha_{av} = \alpha =$ constant, it follows that $a_{av} = a =$ constant. To distinguish this **tangential acceleration** from the centripetal acceleration (p. 142) we'll put a subscript T on it:

$$\alpha = \frac{a_T}{r} \tag{8.13}$$

(see Table 8.2). Clearly, a_T doesn't exist if v is constant; that is, if $\Delta v = 0$. It's the scalar value of the acceleration vector drawn from the moving object tangent to the path of the motion at any instant (Fig. 8.8):

[$\alpha =$ constant]

$$a_T = r\alpha \tag{8.14}$$

Centripetal acceleration a_C arises from a change in the direction of the motion (not a change in speed): v may or may not be constant. As long as the direction of $\vec{v}$ changes, there will be an a_C, and that always occurs for circular motion. When both the direction and magnitude of $\vec{v}$ vary in time, $\vec{a}_T$ will exist acting perpendicular to $\vec{a}_C$. This is true for a runner who speeds up around a turn or a monkey who slows while swinging on a vine (Fig. 8.9).

Table 8.2
Linear and Angular Kinematic Parameters

Equation	Units	
$\ell = r\theta$	ℓ (m)	θ (rad)
$v = r\omega$	v (m/s)	ω (rad/s)
$a_T = r\alpha$	a_T (m/s²)	α (rad/s²)

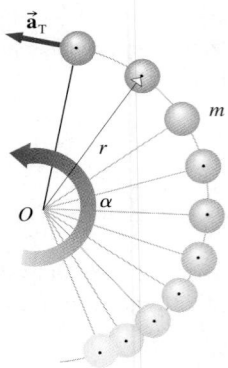

Figure 8.8 When an object moving in a circle goes faster (or slower), it accelerates tangentially. The angle that marks its position (θ) changes at a rate (ω) which is not constant. The object has an angular acceleration (α) such that $a_T = r\alpha$.

Figure 8.9 As the speed of the club head increases, $a_C = v^2/R$ increases even more rapidly (as does $F_C = ma_C$). Because the material of the club is not perfectly rigid, its head actually bends away from the shaft during the bottom of the swing.

Example 8.4 **[III]** At a particular moment in a race, a car roaring around a turn with a radius of 50 m had an angular speed of 0.60 rad/s and an angular acceleration of 0.20 rad/s². Compute its (a) linear speed, (b) its centripetal acceleration, (c) its tangential acceleration, and (d) its *total* linear acceleration at that moment (Fig. 8.10).

Solution Again, the mention of "angular speed" should bring to mind $v = r\omega$. In addition, this problem talks about a_C, a_T, and the resultant of the two, a. (1) TRANSLATION—An object moves in a circle of known radius with a known angular speed and angular acceleration; determine its linear speed, centripetal acceleration, tangential acceleration, and *total* linear acceleration. (2) GIVEN: $r = 50$ m, $\omega = 0.60$ rad/s, and $\alpha = 0.20$ rad/s². FIND: (a) v, (b) a_C, (c) a_T, and (d) the total acceleration $\vec{a}$. (3) PROBLEM TYPE—Angular motion/constant angular acceleration. (4) PROCEDURE—(a) First, we'll find v from ω using $v = r\omega$. (5) CALCULATION—The linear speed is

$$v = r\omega = (50 \text{ m})(0.60 \text{ rad/s}) = \boxed{30 \text{ m/s}}$$

(b) From which it follows that

$$a_C = \frac{v^2}{r} = \frac{(30 \text{ m/s})^2}{50 \text{ m}} = \boxed{18 \text{ m/s}^2}$$

(c) The tangential acceleration is

$$a_T = r\alpha = (50 \text{ m})(0.20 \text{ rad/s}^2) = \boxed{10 \text{ m/s}^2}$$

To find the total linear acceleration, refer to the figure, which shows the relationship between $\vec{a}_C$, $\vec{a}_T$, and $\vec{a} = \vec{a}_C + \vec{a}_T$. Accordingly,

$$a = \sqrt{a_C^2 + a_T^2} = \sqrt{(18 \text{ m/s}^2)^2 + (10 \text{ m/s}^2)^2} = \boxed{21 \text{ m/s}^2}$$

at an angle of

$$\phi = \tan^{-1}\frac{a_C}{a_T} = \boxed{61°}$$

Quick Check: It's always best to avoid making a calculation that depends on a previous one. Since we computed v, we should have used $a_C = r\omega^2 = (50 \text{ m})(0.60 \text{ rad/s})^2 = 18 \text{ m/s}^2$ to get a_C. Moreover, $a_C = a \sin \phi = (21 \text{ m/s}^2)(\sin 61°) = 18 \text{ m/s}^2$.

{For a selection of interactive multiple choice questions go to **WARM-UPS** *in* **CHAPTER 8** *on the CD.}*

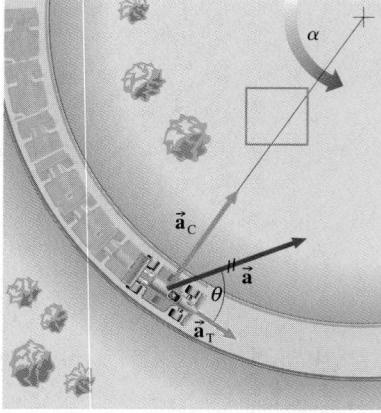

Figure 8.10 A car moving faster and faster around a curve has both a tangential ($\vec{a}_T$) and a centripetal ($\vec{a}_C$) acceleration. Its total acceleration is the vector $\vec{a} = \vec{a}_T + \vec{a}_C$. The two little slashes on $\vec{a}$ are to remind you that it's been resolved into its components $\vec{a}_C$ and $\vec{a}_T$.

8.4 Equations of Constant Angular Acceleration

In Chapter 3, we considered linear motion at constant acceleration and arrived at five important equations. With the proper interpretation, these same expressions also apply to curvilinear motion. If we take ℓ to be the arc-length traveled,

[fixed axis—constant a_T] $$v_f = v_i + a_T t \qquad (8.15)$$

[fixed axis—constant a_T] $$v_{av} = \tfrac{1}{2}(v_i + v_f) \qquad (8.16)$$

[fixed axis—constant a_T] $$\ell = v_{av}t = \tfrac{1}{2}(v_i + v_f)t \qquad (8.17)$$

[fixed axis—constant a_T] $$\ell = v_i t + \tfrac{1}{2}a_T t^2 \qquad (8.18)$$

[fixed axis—constant a_T] $$v_f^2 = v_i^2 + 2a_T \ell \qquad (8.19)$$

where the direction in which ℓ is positive is the direction in which v and a_T are both positive. Here ℓ is the distance traveled along the circumference. It's also called the *curvilinear displacement*, and (like s) it represents the distance through which the body is ultimately displaced rather than the total distance traversed: if a horse runs 10 times around a 1.0-km circular track going clockwise and then 10 times around going counterclockwise, it's back where it started and $\ell = 0$ (whereas the total path-length traveled is 20 km). Of course, if the rotation takes place in only one direction, without any doubling back, the arc-length (ℓ) equals the path-length (l).

These equations of curvilinear motion can most easily be transformed into an equivalent angular description when the path is circular. We know that

$$\ell = r\theta \qquad [8.1]$$

$$v = r\omega \qquad [8.10]$$

and $$a_T = r\alpha \qquad [8.14]$$

It's then a simple matter of substitution into Eqs. (8.15) through (8.19) to arrive at

[constant α] $$\omega_f = \omega_i + \alpha t \qquad (8.20)$$

[constant α] $$\omega_{av} = \tfrac{1}{2}(\omega_i + \omega_f) \qquad (8.21)$$

[constant α] $$\theta = \omega_{av}t = \tfrac{1}{2}(\omega_i + \omega_f)t \qquad (8.22)$$

[constant α] $$\theta = \omega_i t + \tfrac{1}{2}\alpha t^2 \qquad (8.23)$$

[constant α] $$\omega_f^2 = \omega_i^2 + 2\alpha\theta \qquad (8.24)$$

where the direction in which θ is positive is the direction in which ω and α are both positive.

Example 8.5   [II] Mounted in a bus is a 2.0-m-diameter flywheel, a massive disk often used to store rotational energy. If it is accelerated from rest at a constant rate of 2.0 rpm per second, what will be the angular speed of a point on the rim of the flywheel after 5.0 s? Through what angle will that point have rotated?

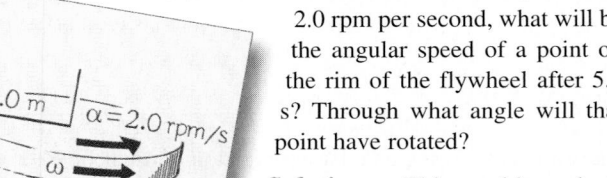

Solution   This problem deals with uniformly accelerated angular motion.   (1) TRANSLATION—A rotating object of known radius angularly accelerates from rest at a known rate for a known time; determine its final angular speed and the angle through which it moves.   (2) GIVEN: $r = 1.0$ m, $\alpha = +2.0$ rpm/s, and $\omega_0 = 0$. FIND: ω and θ at $t = 5.0$ s.   (3) PROBLEM TYPE—Angular motion/constant angular acceleration.   (4) PROCEDURE—Since we have ω_0, α, and t, use $\omega = \omega_0 + \alpha t$ to determine ω.   (5) CALCULATION—Let's first get α in rad/s²:

$$\alpha = \frac{(2.0 \text{ rpm/s})(2\pi \text{ rad/rot})}{(60 \text{ s/min})} = 0.209 \text{ rad/s}^2$$

continued

Then

$$\omega = \omega_0 + \alpha t = 0 + (0.209 \text{ rad/s}^2)(5.0 \text{ s}) = 1.045 \text{ rad/s}$$

or, to two figures, $\boxed{\omega = 1.0 \text{ rad/s}}$. Equation (8.24) provides θ:

$$\theta = \tfrac{1}{2}\alpha t^2 = \tfrac{1}{2}(0.209 \text{ rad/s}^2)(5.0 \text{ s})^2 = \boxed{2.6 \text{ rad}}$$

Quick Check: Notice that at $\alpha = 2.0$ rpm/s, the flywheel's angular speed should pick up 10 rpm in 5 s; that is, $(2\pi 10 \text{ rad/min})/(60 \text{ s/min})$, or $\omega = 1.05$ rad/s, which gives the flywheel an ω_{av} of $(1.05/2)$ rad/s, so it will revolve through $\theta = \omega_{av}t = (1.05/2)(5)$ rad = 2.6 rad.

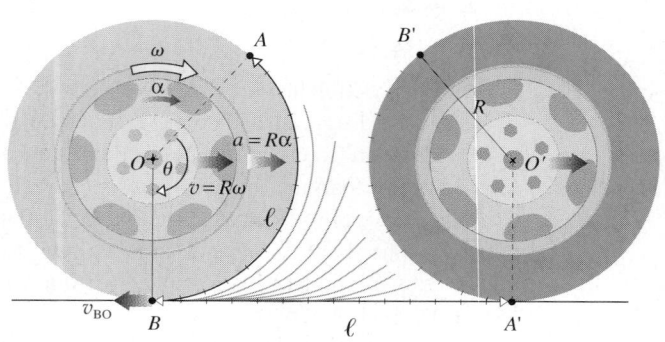

Figure 8.11 As a body rolls, revolving about its center at a rate ω, with an angular acceleration α, point O moves at a speed $v = R\omega$, with an acceleration at that instant of $a = R\alpha$.

A Free Rolling Wheel

In most cases, as a body moves, rotating and translating, the linear displacement of a point within it cannot be easily correlated with its angular displacement. The important exception is a freely rolling disk, cylinder, or sphere (Fig. 8.11). *The term* **freely rolling** *means there is no slipping at the point of contact with the ground*—no skidding and no spinning in place. As we'll see, for that to happen there must be sufficient friction to hold the lowest point on the wheel, B, so that it is instantaneously at rest just as it descends and touches the surface. Rolling, in the usual sense of the word, requires an interaction between the object and the surface; a ball cannot "roll" on a frictionless surface. If it were possible to put a spinning ball down onto a perfectly frictionless surface, it would just sit there spinning. On the other hand, when a billiard ball is hit off-center and too fast, it initially slides and spins at the same time; given enough distance, friction (which opposes the forward slide) will slow the ball until it stops skidding and rolls freely.

The wheel in the figure rolls to the right, and point O on the axis moves to O' as A moves to A' and B moves to B'. The arc-length from B to A equals ℓ just as length $\overline{BA'} = \overline{OO'} = \ell$. The straight-line distance traveled by the centerpoint is $\ell = R\theta$, and so $v_0 = R\omega$ and $a_0 = R\alpha$. **The center of a freely rolling wheel has a linear speed and acceleration equal in magnitude to that of any point on its rim** (w.r.t. O). Notice that the velocity of point B, while in contact with the ground, measured with respect to O, namely $\vec{\mathbf{v}}_{BO}$, is to the left. The velocity of O with respect to the ground, $\vec{\mathbf{v}}_{OG}$, is to the right. Furthermore, $|\vec{\mathbf{v}}_{BO}| = |\vec{\mathbf{v}}_{OG}| = R\omega$; and

$$\vec{\mathbf{v}}_{BG} = \vec{\mathbf{v}}_{BO} + \vec{\mathbf{v}}_{OG} = 0$$

The velocity of the lowest point on a freely rolling ball or wheel, with respect to the ground, at the moment of touching is zero—the contact point doesn't slip across the surface (see Fig. 4.27, p. 112).

Example 8.6 [III] A cyclist traveling at 5.0 m/s uniformly accelerates up to 10.0 m/s in 2.0 s. Each tire of the bike has a 35-cm radius, and a small pebble is caught in the tread of one of them. (a) What is the angular acceleration of the pebble during those two seconds? (b) Through what angle does the pebble revolve? (c) How far around the wheel does the pebble travel during that accelerating interval?

continued

Solution On the first reading, it should be clear that we'll need the equations of constant angular acceleration. (1) TRANSLATION—A rolling object uniformly accelerates from a known initial to a known final linear speed, in a specified time. Knowing its radius, determine (a) its angular acceleration, (b) the angle through which it revolves, and (c) how far a point on the circumference travels. (2) GIVEN: $v_i = 5.0$ m/s, $v_f = 10.0$ m/s, $t = 2.0$ s, and $R = 0.35$ m. FIND: (a) α, (b) θ, and (c) ℓ. (3) PROBLEM TYPE—Angular motion/ constant angular acceleration. (4) PROCEDURE—Since we have the linear speeds and time, compute a and from that α. (5) CALCULATION—(a) The bike moves along with the center of each wheel, accelerating at a linear rate of

$$a = \frac{(v_f - v_i)}{t} = \frac{(10.0 \text{ m/s}) - (5.0 \text{ m/s})}{2.0 \text{ s}} = 2.5 \text{ m/s}^2$$

This, in turn, equals $R\alpha$ and therefore

$$\alpha = \frac{a}{R} = \frac{2.5 \text{ m/s}^2}{0.35 \text{ m}} = \boxed{7.1 \text{ rad/s}^2}$$

which is the angular acceleration of each wheel and, hence, of the pebble. (b) Inasmuch as the linear speed of the pebble equals the linear speed of the bike, $v_i = r\omega_i = 5.0$ m/s, $\omega_i = 14.29$ rad/s, and so

$$\theta = \omega_i t + \tfrac{1}{2}\alpha t^2 = (14.29 \text{ rad/s})(2.0 \text{ s}) + \tfrac{1}{2}(7.1 \text{ rad/s}^2)(2.0 \text{ s})^2$$

and $\boxed{\theta = 43 \text{ rad}}$. (c) Both bike and pebble travel distances of $\ell = R\theta = \boxed{15 \text{ m}}$.

Quick Check: At an average speed of 7.5 m/s, moving for 2.0 s, the cyclist traverses 15 m, and everything checks.

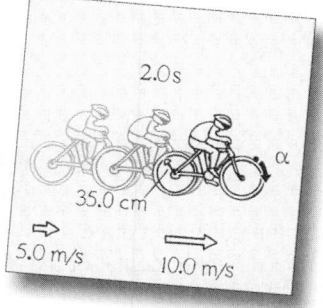

Rotational Equilibrium

Over 2000 years ago Aristotle attempted to analyze the practical problem of balances, levers, and seesaws. When a pivoted beam is balanced horizontally so that it doesn't revolve, we say that it's in *rotational equilibrium*. Archimedes (287 B.C.E.–212 B.C.E.) correctly extended Aristotle's efforts, and their rudimentary *Law of the Lever* states that *unequal forces, acting perpendicularly on a pivoted bar, balance each other, provided $F_1 r_1 = F_2 r_2$* (Fig. 8.12). One must consider the sizes of the forces *as well as the distances at which they act from the pivot*—that much can easily be confirmed experimentally.

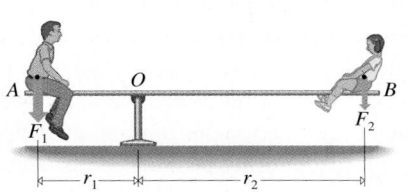

Figure 8.12 Unequal parallel forces acting at unequal distances from the pivot. The system is in rotational equilibrium.

Example 8.7 **[I]** A boy (mass 30 kg) wishes to play on a centrally pivoted seesaw with his dog Irving (mass 10 kg). When the dog sits 3.0 m from the pivot, where must the boy sit if the 6.5-m-long board is to be balanced horizontally?

Solution This problem is about the balance of an object that's free to rotate. (1) TRANSLATION—Two known masses are balanced on a horizontal center-pivoted beam; knowing the position of one, locate the other. (2) GIVEN: $m_B = 30$ kg, $m_D = 10$ kg, the dog-pivot distance $r_D = 3.0$ m, and the board length is

6.5 m. FIND: The boy-pivot distance r_B. (3) PROBLEM TYPE—Angular motion/rotational equilibrium. (4) PROCEDURE—Use the so-called Law of the Lever. (5) CALCULATION—The forces are the weights $F_{WB} = m_B g$ and $F_{WD} = m_D g$; therefore,

$$(m_B g)r_B = (m_D g)r_D$$

and

$$r_B = \frac{(10 \text{ kg})(3.0 \text{ m})}{(30 \text{ kg})} = \boxed{1.0 \text{ m}}$$

The boy must sit 1.0 m from the pivot.

Quick Check: $m_B r_B = m_D r_D = 30$ kg·m.

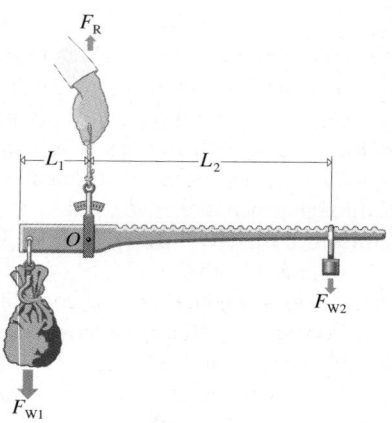

Figure 8.13 An unequal-arm balance, or steelyard. The small weight, the *poise*, is moved down the arm until a balance is reached with the heavy object being weighed. The balanced steelyard is a system of separated forces whose lines-of-action are parallel and therefore nonconcurrent.

8.5 Torque

Figure 8.14 shows situations in which a force $\vec{F}$ acts on a wrench along different directions and at different points. The axis about which rotation will occur runs down the center of the bolt through point *O* perpendicular to the plane of the diagram. The ***position vector $\vec{r}$ is drawn from O to the point of application of the force***. In each case, the ability of $\vec{F}$ to produce a rotation of the wrench will be different. It is this turning effect, or *moment*, that da Vinci appreciated as crucial. When the line-of-action of the force (i.e., the line lying along and extending beyond $\vec{F}$) passes through the pivot, no rotation will occur at all (Fig. 8.14*a*). You cannot close a door by pushing on its edge *toward* the hinge. By contrast, the force applied at the very end of the wrench, where the line-of-action is far from *O*, as in Fig. 8.14*c*, produces the greatest twist. In fact, it's precisely for that reason that the wrench was invented.

The **moment-arm** *of the force with respect to the axis passing through O is defined as the perpendicular distance* $(r_\perp)$ *drawn from O to the line-of-action of* $\vec{F}$. In each part of Fig. 8.14, notice that, although the magnitude of the force is unchanged, the moment-arm increases from $r_\perp = 0$ in (a), to $r_\perp = r \sin \theta$ in (b), to $r_\perp = r$ in (c). The **moment-of-the-force** *about O is defined as the product of F and the moment-arm*. We symbolize the moment-of-the-force by the Greek letter tau (τ) and particularize it with respect to point *O* as τ_0. In this way

$$\tau_0 = r_\perp F$$

This quantity is called the **torque** (from the Latin *torquere*, to twist). The term "moment-of-the-force" is also used. ***The torque is a measure of the twist produced by a force around a particular axis that may be located anywhere.*** Torque has the dimensions of force multiplied by distance (N·m).

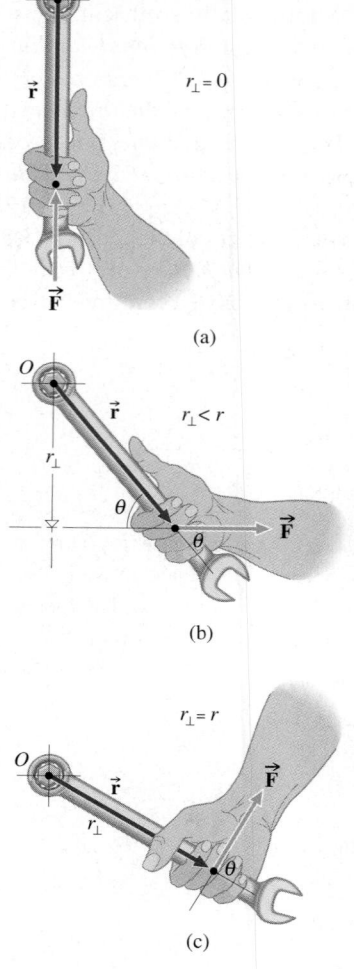

(a)

(b)

(c)

Figure 8.14 The amount of twist, or torque, one can get from a wrench depends on how big the force is, in what direction it acts, and where it is applied. (a) No torque is exerted when the force goes through the pivot point *O*. (b) Here, torque about *O* exists. It can be increased (c) by increasing θ, which increases $r_\perp$.

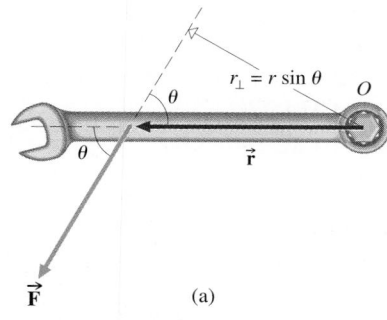

$r_\perp = r \sin\theta$ O

$\vec{\mathbf{r}}$

$\vec{\mathbf{F}}$ (a)

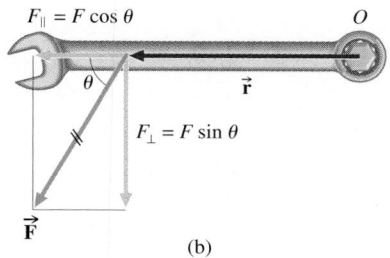

$F_\parallel = F \cos\theta$ O

$\vec{\mathbf{r}}$

$F_\perp = F \sin\theta$

$\vec{\mathbf{F}}$ (b)

Figure 8.15   Torque can be thought of as $Fr_\perp$ or $F_\perp r$, since $Fr_\perp = F_\perp r = Fr \sin\theta$.

In Fig. 8.15*a*, the line-of-action of the force vector $\vec{\mathbf{F}}$ makes an angle of θ with the line-of-action of the position vector $\vec{\mathbf{r}}$. Here, $r_\perp = r \sin\theta$ and

$$\tau_0 = Fr_\perp = Fr \sin\theta$$

Alternatively, resolve $\vec{\mathbf{F}}$ into its components ($\vec{\mathbf{F}}_\parallel$ parallel to $\vec{\mathbf{r}}$, and $\vec{\mathbf{F}}_\perp$ perpendicular to $\vec{\mathbf{r}}$ as in Fig. 8.15*b*). **When the line-of-action of a force passes through any point *O*, its moment-arm with respect to *O* is zero and its torque about that point is also zero.** This is the case with $\vec{\mathbf{F}}_\parallel$ and so with respect to *O*, the torque produced by $\vec{\mathbf{F}}_\perp$ must equal the torque produced by $\vec{\mathbf{F}}$ as a whole. Since $F_\perp$ (which equals $F \sin\theta$) has a moment-arm of r, its torque about *O* is

$$\tau_0 = rF_\perp = Fr_\perp = rF \sin\theta \tag{8.25}$$

and we use whichever form is the more convenient. **The direction or, better yet, the sense of the torque is either clockwise or counterclockwise depending on which way the torque tends to rotate the object around the reference axis.**

When you look down on the diagram in Fig. 8.15, you can see that the force rotates the wrench counterclockwise about *O*. Some other force could act on the wrench in a different direction to produce an opposing clockwise twist (as when two people arm wrestle, for instance). The resulting twist produced by opposing torques would then be the sum of the two: $\sum\tau_0$, taking one positive and the other negative.

Example 8.8   [I]   The exercise illustrated in Fig. 8.16 consists of swinging the lower leg out until it's horizontal and then lowering it, all the while supporting a mass *m*. Write an expression for the torque about the knee due to the strapped-on mass in terms of θ, m, g, and the knee-heel length r. For the time being, ignore the mass of the leg. What is the value of the torque when $m = 1.0$ kg, $r = 50$ cm, and $\theta = 30.0°$?

Solution   This one's about "torque," and that's force times moment-arm.   (1) TRANSLATION—A known force acts with a known moment arm w.r.t. a pivot point; determine the torque around that point.   (2) GIVEN: $m = 1.0$ kg, $r = 50$ cm, and $\theta = 30.0°$. FIND: τ_0 in general and then specifically, where *O* is at the knee joint.   (3) PROBLEM TYPE—Angular motion/torque.   (4) PROCEDURE—Use the definition of torque.   (5) CALCULATION—The downward force producing a tendency for the leg to rotate clockwise (i.e., a clockwise torque) is the weight *mg*:

$$\tau_0 = r_\perp F = r_\perp mg$$

The moment-arm is the perpendicular distance from *O* to the line-of-action of the force, that is,

$$r_\perp = r \sin\theta$$

where θ is the angle between $\vec{\mathbf{r}}$ and $\vec{\mathbf{F}}$. Finally,

$$\boxed{\tau_0 = mgr \sin\theta}$$

and

$$\tau_0 = (1.0 \text{ kg})(9.81 \text{ m/s}^2)(0.50 \text{ m})(\sin 30.0°) = \boxed{2.5 \text{ N·m}}$$

Quick Check: The torque varies with $\sin\theta$; reasonably, it's zero when $\theta = 0$ and maximum, $\tau_0(\text{max}) = mgr$, when $\theta = 90°$.

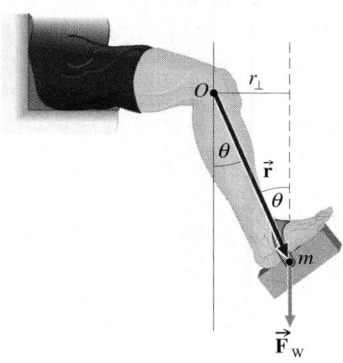

O $r_\perp$

θ $\vec{\mathbf{r}}$

θ

m

$\vec{\mathbf{F}}_w$

Figure 8.16   The weight exerts a torque about *O* that is countered by an equal and opposite muscle-generated torque.

Human mobility arises from muscles attached to bones, pulling on other bones that are pivoted so they rotate under the action of torques. That's how you walk, bend, chew, and even hold your head up. The torque that can be generated by a muscle, and therefore its ability to rotate the associated body segment, depends on the moment-arm between the muscle's line-of-action and the joint center. As the limb in Fig. 8.17 passes through its range of action, the moment-arm changes; that directly affects the torque that can be generated. {For an introduction to the vector nature of torque and the cross product, click on **TORQUE VECTORS** under **FURTHER DISCUSSIONS** on the CD.}

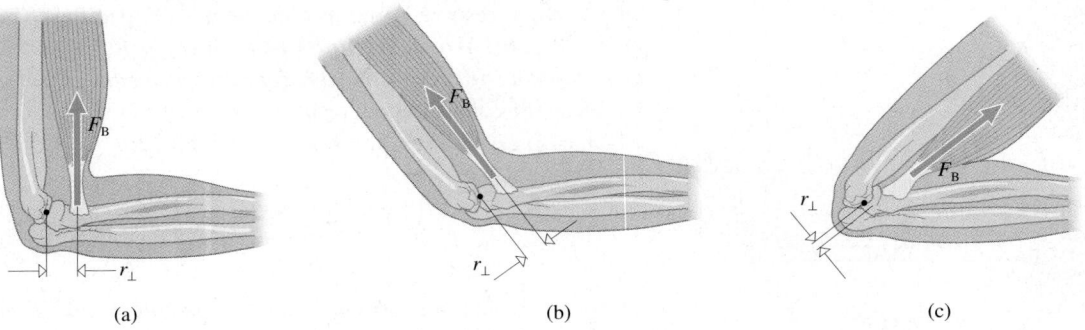

(a) (b) (c)

Figure 8.17 As the forearm moves, the biceps force F_B changes its moment-arm $r_\perp$ about the elbow pivot; that changes the torque even when F_B is constant.

Example 8.9 **[II]** At the moment depicted in Fig. 8.18, the hand exerts a 200-N force at the end of a pivoted 1.0-m-long lever. The return spring fixed to the lever's midpoint pulls back on it with a horizontal force of 80 N. What is the net torque acting on the lever about the pivot?

Solution This one's about "torque," and that's force times moment-arm, but it also asks for the sum-of-the-torques. (1) TRANSLATION—Two known forces act with known moment arms w.r.t. a pivot point; determine the net torque around that point. (2) GIVEN: Referring to Fig. 8.18b, $F_h = 200$ N, $r_h = 1.0$ m, $F_s = 80$ N, and $r_s = 0.50$ m. FIND: τ_0 where O is at the pivot. (3) PROBLEM TYPE—Angular motion/torque. (4) PROCEDURE—Use the definition of torque. (5) CALCULATION—The moment-arm of the hand-force (the perpendicular distance from O to the line-of-action of F_h) is just r_h, while the moment-arm of the spring-force is given by $r_s \cos 60°$. Arbitrarily taking clockwise as positive, we have

$$\circlearrowleft + \sum \tau_0 = r_h F_h + (-r_s F_s \cos 60°)$$

$$\circlearrowleft + \sum \tau_0 = (1.0 \text{ m})(200 \text{ N}) - (0.50 \text{ m})(80 \text{ N})(0.50)$$

and
$$\boxed{\circlearrowleft + \sum \tau_0 = +0.18 \text{ kN} \cdot \text{m}}$$

Quick Check: The angle between $\vec{r}_s$ and $\vec{F}_s$ is 150°; hence, the torque that is contributed by $\vec{F}_s$ is $-r_s F_s \sin 150°$, which equals $-r_s F_s \cos 60°$.

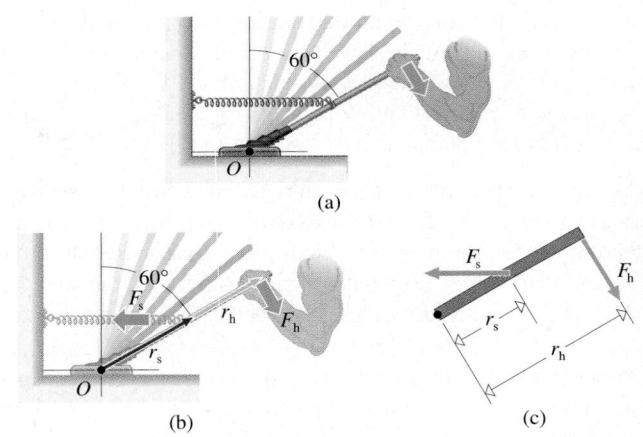

Figure 8.18 (a) A pivoted bar with two forces acting on it. (b) The forces and their moment-arms about point O. (c) A simplified version of the apparatus.

8.6 Second Condition of Equilibrium

Real bodies deform to some extent under the influence of applied forces, but many change so little that we can take them to be rigid. Assume such a body is made up of a large number of separate, stationary, interacting particles (atoms or molecules). If the body is in equilibrium under the action of several externally applied forces, all of its constituent particles are also in equilibrium. The net force ($\vec{F}$) acting on any one particle must be zero. Now, choose some point O—anywhere you like—and determine the resultant torque about O due to the net force acting on that arbitrary particle. But here, $\vec{F} = 0$, so the torque about O is

Rotational equilibrium in Afghanistan.

zero, too. That's true for each and every particle in the body, so the net torque on the body due to all the internal and external forces is zero. Because of the colinear equal-and-opposite nature of the internal interaction pairs holding the body together, the torque about any point due to each pair must cancel. Consequently, the sum of all the torques due to the internal forces must be zero. Therefore, **the sum-of-the-torques about any point due to all the externally applied forces acting on a rigid body in equilibrium must be zero**. Here we have only the suggestion of a derivation, but it makes the point that this powerful insight is based squarely on Newton's Laws and is not some new physical principle.

The fact that

[translational equilibrium]
$$\sum \vec{F} = 0$$
[4.15]

is the First Condition of Equilibrium. For a rigid body

[rotational equilibrium]
$$\sum \tau_0 = 0$$
(8.26)

> By using both the **First and Second Conditions** together, all sorts of force systems in equilibrium can be analyzed, including those that are neither concurrent nor parallel.

> **Varignon's Theorem:** The torque produced by a force is equivalent to the sum of the torques produced by its components—which goes without saying, or proving, here.

is the **Second Condition of Equilibrium**. Note that the point O about which the torques are taken is arbitrary.

When the forces acting on a body pass through a common point (i.e., when they are concurrent) the First Condition suffices for equilibrium. (The Second Condition is then automatically true because the torque produced by a system of *concurrent forces* equals the torque produced by its resultant, which is zero.) **When dealing with concurrent forces, there will be only two independent equations (either sum-of-the-forces or sum-of-the-torques), and no more than two unknowns can be determined.** Figures 4.37, 4.38, and 4.39 depict concurrent force systems; and although we solved those problems using the sum-of-the-forces, we could have used the sum-of-the-torques instead.

You know you've got a concurrent force system when an object is maintained in equilibrium by three coplanar nonparallel forces. Still, that can at times be less than obvious; witness Fig. 8.19*a*. Here a person is leaning over, the head is supported at the pivot P (the occipital condyles), and held in place by a muscle force F_M which provides a counterclockwise torque about P. As we'll soon see (p. 249), the weight of the head acts vertically through a point called the center-of-gravity which lies somewhat forward of the pivot point. That's not surprising because if you relax your neck muscles, your head naturally slumps forward; the weight of the head produces a clockwise torque around P. The two other forces acting on the head are a shear force F_S (across the spine, keeping the head from sliding off) and a normal force F_N (along the spine, keeping the head from further compressing the spine). If we combine these last two into a single spinal reaction force F_R, then there are three nonparallel forces keeping the head in equilibrium and they must meet at a single point (Fig. 8.19*b*).

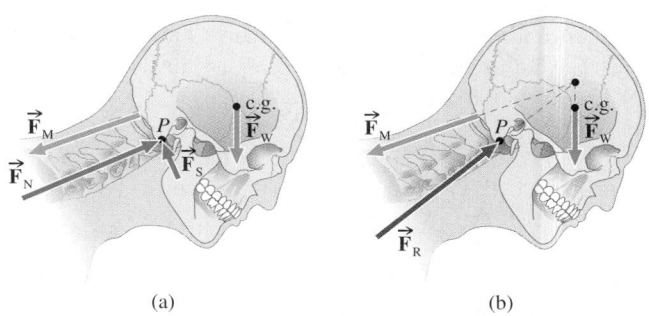

(a)   (b)

Figure 8.19 A concurrent force system. The head of a person leaning over is supported by a shear force $\vec{F}_S$ and a normal force $\vec{F}_N$, both supplied by the spine, and a muscle force $\vec{F}_M$. Since there are three forces, $\vec{F}_M$, $\vec{F}_W$, and $\vec{F}_R = \vec{F}_N + \vec{F}_S$, acting on the head in equilibrium, those forces must meet at a point.

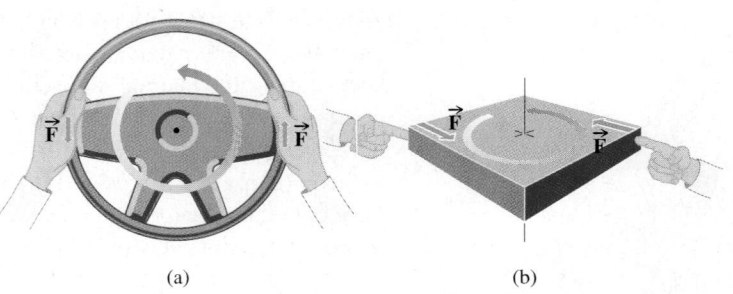

(a)   (b)

Figure 8.20 (a) Two equal, oppositely directed forces acting at different points on a body do not produce equilibrium even though $\sum \vec{F} = 0$. There is a torque, and the system cannot be in rotational equilibrium. (b) The block will rotate without translating.

Nonconcurrent Forces

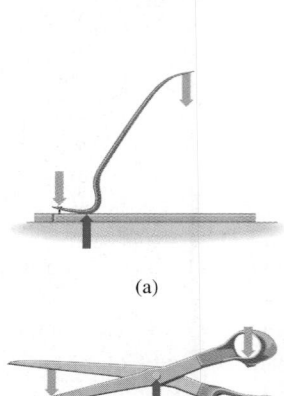

(a)

(b)

Figure 8.21 Two tools that are the same kind of lever as the seesaw. Both have fulcrums or pivot points between the applied force and the force exerted by the device. Shown are the forces acting on the tools. In (b) we see the forces acting on the lower blade as the sharp edge rotates upward in the act of cutting something.

When two equal but antiparallel forces act at different points on a body (Fig. 8.20), equilibrium will not be established despite the fact that $\sum \vec{F} = 0$. There will be a net torque, and the body will revolve increasingly rapidly—it will accelerate. Although the body is in translational equilibrium, it will only be in rotational equilibrium when $\sum \tau = 0$. *Any number of parallel separated forces acting on a body in equilibrium can give rise to only two independent equations (either one force and one torque, or two torque), and only two unknowns can be determined.*

The steelyard and the seesaw, the crowbar and the scissor (Fig. 8.21), and the human foot (Fig. 4.30) are all the same kind of simple machine—a lever of the class of levers that have the fulcrum, or pivot, located *between* the two forces acting on it.

The wind blowing to the right on the sail produces a clockwise torque that must be balanced by a net counterclockwise torque produced by the sailors.

Example 8.10 **[I]** The steelyard in Fig. 8.13 consists of a nonuniform horizontal bar that is balanced without any added weights when supported at point O. If the weights F_{W1} and F_{W2} are attached and the system is balanced as shown, write an expression for L_2 in terms of L_1, F_{W1}, and F_{W2}. What is the magnitude of the reaction force F_R supporting the system? Ignore the weight of the steelyard itself. Compute L_2 when $F_{W1} = 5000$ N, $F_{W2} = 250$ N, and $L_1 = 10.0$ cm.

Solution The system is "balanced," and that means the sum-of-the-torques-equals-zero. (1) TRANSLATION—Two known forces act on a balanced beam; if the moment-arm w.r.t. the pivot is known for one, determine the other moment-arm and the reaction force. (2) GIVEN: $F_{W1} = 5000$ N, $F_{W2} = 250$ N, and $L_1 = 10.0$ cm. FIND: The moment-arm L_2 and the reaction F_R. (3) PROBLEM TYPE—Angular motion/rotational equilibrium. (4) PROCEDURE—The forces are nonconcurrent, and we will have to apply both conditions of equilibrium. (5) CALCULATION—The sum-of-the-torques about *any* point must be zero, but not all such points are equally convenient. It would be advantageous

not to have to deal with the reaction force in the first part of the problem. Since F_R passes through O, it has no torque around O and consequently does not enter the equation, which therefore contains only one unknown. Weight-2 tends to swing the system clockwise around O, whereas weight-1 tends to rotate it counterclockwise. With clockwise as positive,

$$\curvearrowleft +\sum \tau_0 = 0 = F_{W2}L_2 - F_{W1}L_1$$

and

$$L_2 = \frac{F_{W1}L_1}{F_{W2}} = \frac{(5000 \text{ N})(0.100 \text{ m})}{(250 \text{ N})} = \boxed{2.00 \text{ m}}$$

As for the reaction force

$$+\uparrow \sum F_y = F_R - F_{W1} - F_{W2} = 0$$

not surprisingly $\boxed{F_R = F_{W1} + F_{W2} = 5250 \text{ N}}$

Quick Check: The problem can be redone using the Law of the Lever, which is a special case of Eq. (8.26). Note that the units check.

8.7 Extended Bodies & the Center-of-Gravity

Any ordinary object can be thought of as composed of a large number of point-masses. Each atom within you experiences a downward gravitational force, all of which are essentially parallel and combine to produce a single resultant force, the *weight* ($\vec{F}_W$) of the body;

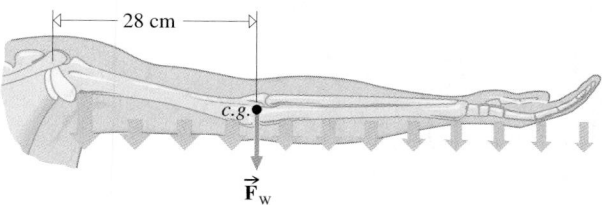

but where on the body does $\vec{F}_W$ act? We define the **center-of-gravity (c.g.) of an object to be that point where the total weight, $\vec{F}_W$, can be imagined to act**. Having a single force $\vec{F}_W$ act at the *c.g.* produces exactly the same mechanical result as having gravity act on all the point-masses that constitute the body. Gravity acts on every atom of the outstretched arm (Fig. 8.22), and yet the effect is exactly equivalent to having the total weight act at the *c.g.* located near the elbow, about 28 cm from the shoulder joint. The torque produced at the shoulder is the weight of the arm multiplied by the distance from the joint to the *c.g.*

Figure 8.22 The weight of the arm is distributed along its entire length. Nonetheless, it can be represented by a single force $\vec{F}_W$ acting at a single point, the *c.g.*

The weight of the balanced athlete in the photograph is straight down, and if we think of it as acting through a single point, the *c.g. must be somewhere on the line-of-action of the normal force*. Since the line-of-action of the weight passes through the *c.g.*, **the force of gravity generates no torque about the center-of-gravity**. A body suspended from its *c.g.* is in equilibrium in any orientation whatsoever and will stay where it is put. **The c.g. is the balance point.** If a body of any shape is freely suspended from *any* point, it will so orient itself in reaching equilibrium that the line-of-action of $\vec{F}_W$ will pass through the support point (Fig. 8.23), producing a no-torque situation. In that case, $\vec{F}_R = -\vec{F}_W$ and the line-of-action of $\vec{F}_R$, which is vertical, passes through the *c.g.* To find the center-of-gravity experimentally, one need only suspend the object from two different points. The lines-of-action of the two successive reaction forces will intersect at the *c.g.*

For simplicity, suppose we have a flat body made up of a large number of tiny point-masses: $m_{\bullet1}, m_{\bullet2}, m_{\bullet3}, \ldots$, as shown in Fig. 8.24. We can compute the location of its center-of-gravity in the following way. Each particle experiences a torque (due to its weight) about the *arbitrary* point O. To find the moment-arm for each of these tiny torques, extend the line-of-action of the weight of each particle and drop a perpendicular to it from O—that x-distance is the moment-arm. The torques have the form $F_{W1}x_1$, $F_{W2}x_2$, and so forth. The total weight-generated torque about O produced by all the individual particles, taking clockwise as positive, is

> As far as its gravitational interaction is concerned, a body behaves as if all of its mass is at a point called the **center-of-gravity**.

Because the human body is flexible, the location of the *c.g.* is not fixed. In equilibrium, however, it must be located on the line-of-action of the normal force.

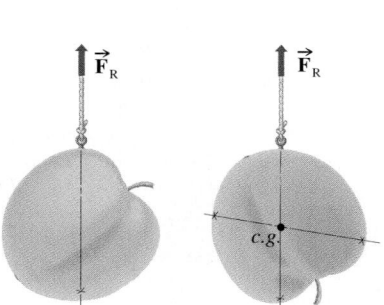

Figure 8.23 The *c.g.* of an object can be located by suspending it from several different points. The *c.g.* is always on the line-of-action of the force supporting the object.

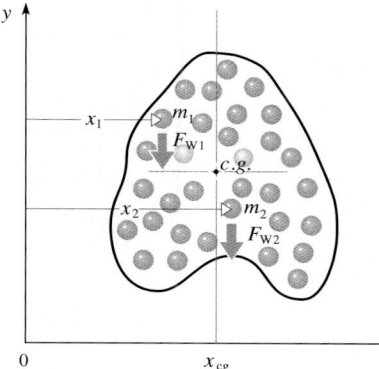

Figure 8.24 The sum of all the torques about any point O caused by the weights of the point-masses constituting a body is equal to the torque produced by the total weight of the body acting at the *c.g.* Measured along the x-axis, the *c.g.* is at x_{cg}. Measured along the y-axis, it's at y_{cg}.

This mechanical arm is driven by two hydraulic pistons that can both push and pull, so each does the work of a pair of muscles like those in Fig. 8.26.

$$F_{W1}x_1 + F_{W2}x_2 + \cdots = \sum F_W x$$

It follows from the definition of the *c.g.* that the total weight, written as

$$F_{W1} + F_{W2} + \cdots = \sum F_W$$

must produce that same torque when it acts as a single force at the *c.g.*; that is,

$$\left(\sum F_W\right)x_{cg} = \sum F_W x$$

and

$$x_{cg} = \frac{\sum F_W x}{\sum F_W} \tag{8.27}$$

Take each point-mass, multiply its weight by its distance from the $x = 0$ axis, add them, divide by the total weight, and you obtain x_{cg}, measured from the axis no matter where the origin was chosen to be. If the object and the coordinate system are both rotated 90° so that the y-axis is horizontal, we get an expression of exactly the same form for y_{cg}.

The c.g. of a regularly shaped body of uniform composition lies at its geometric center. This applies to spheres, cubes, cylinders, rings, and so forth. Bear in mind that the *c.g.* of an object need not reside within the space occupied by material. For example, the *c.g.* of a doughnut is at the center of the hole. *When a body is composed of several parts, the overall c.g. can be located by treating each part as a point-mass located at its own c.g., acted upon by its own weight.*

Example 8.11 **[I]** For a man of mass *m* and height *h*, the *c.g.* of the lower leg and foot, and the *c.g.* of the upper leg are shown in Fig. 8.25 along with the corresponding weights. Find the *c.g.* of the whole leg in the extended position measured from the sole of the foot. This kind of information is important in physical therapy.

Solution Keep in mind that each segment of a compound object behaves as if all of its mass was a point-mass located at its center-of-mass. (1) TRANSLATION—Several objects with known *c.g.*s combine to form a single object; determine its *c.g.* as a whole. (2) GIVEN: $F_{W1} = 0.059mg$, $x_1 = 0.19h$, $F_{W2} = 0.097mg$, and $x_2 = 0.42h$. FIND: x_{cg}. (3) PROBLEM TYPE—Center-of-gravity. (4) PROCEDURE—We can treat the two leg parts as point-masses located at their respective *c.g.*s. (5) CALCULATION—Apply the definition, Eq. (8.27):

$$x_{cg} = \frac{F_{W1}x_1 + F_{W2}x_2}{F_{W1} + F_{W2}}$$

$$x_{cg} = \frac{(0.059mg)(0.19h) + (0.097mg)(0.42h)}{0.097mg + 0.059mg}$$

$$x_{cg} = \frac{0.052mgh}{0.156mg} = \boxed{0.33h}$$

Quick Check: The torque about the heel due to the whole leg is $(0.33h)(0.156mg) = (0.05mgh)$; the torque due to the separate parts $\approx (0.06mg)(0.2h) + (0.1mg)(0.4h) \approx 0.05mgh$, and they *are* equal.

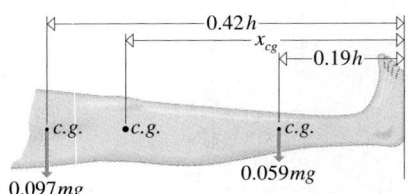

Figure 8.25 The locations of the various *c.g.*s for several parts of a man's leg.

Example 8.12 **[I]** The biceps muscle (Fig. 8.26) is connected from the shoulder (scapula) to the radius bone at a point 5.0 cm from the elbow. Its contractions flex the arm. Taking the mass of the hand and forearm together to be 5.5% of the total body mass (with a *c.g.* as shown), compute the force exerted by the biceps of a 70-kg man holding a 2.0-kg sphere. Assume the force of the biceps acts vertically.

Solution Because the forces aren't concurrent, you can expect to deal with torques. (1) TRANSLATION—A specified system is in equilibrium acted on by four nonconcurrent forces, two of which are unknown; determine one of them. (2) GIVEN: The mass of the arm $m_a = 5.5\%(70\text{ kg})$, $m_s = 2.0$ kg, and all the appropriate distances as drawn. FIND: F_B, the force exerted by the biceps. (3) PROBLEM TYPE—Angular motion/rotational

continued

equilibrium.   (4) Procedure—Use $\sum \tau_0 = 0$.   (5) Calculation—Take the torque about the elbow joint to avoid dealing
with the force F_H exerted by the humerus (see the free-body diagram in Fig. 8.26b). With clockwise as positive,

$$\circlearrowright + \sum \tau_0 = 0 = m_s g (0.34 \text{ m}) + m_a g (0.16 \text{ m}) - F_B (0.050 \text{ m})$$

$$F_B = g \frac{(2.0 \text{ kg})(0.34 \text{ m}) + (3.85 \text{ kg})(0.16 \text{ m})}{(0.050 \text{ m})}$$

and $\boxed{F_B = 2.5 \times 10^2 \text{ N}}$ or about 57 lb.

Quick Check: Rounding off the numbers, the clockwise torque
is roughly $g(2 \text{ kg})(0.3 \text{ m}) + g(4 \text{ kg})(0.2 \text{ m}) = 1.4g \approx 14 \text{ N·m}$,
and dividing by 0.05 m yields $F_B \approx 0.28 \text{ kN}$.

THE C.G. OF THE HUMAN BODY

The *c.g.* of a person varies with posture
and limb position. For someone standing
at attention, it lies on a vertical centerline
about 3 cm forward of the ankle joint and
up around the height of the second sacral
vertebra, or roughly a few inches behind
the navel. The *c.g.* also varies from person to person—in men with large shoulders, it's usually a little higher than in
women with large pelvises. The location
of the body's *c.g.* is of interest to everyone who stands or walks, but it's of particular concern to dancers and athletes.
For example, Dick Fosbury devised a
backward high-jump technique in which
his body, face upward, arcs over the bar,
while his *c.g.* passes beneath it. How high
the *c.g.* is raised determines the amount
of effort that will have to be expended in
the jump.

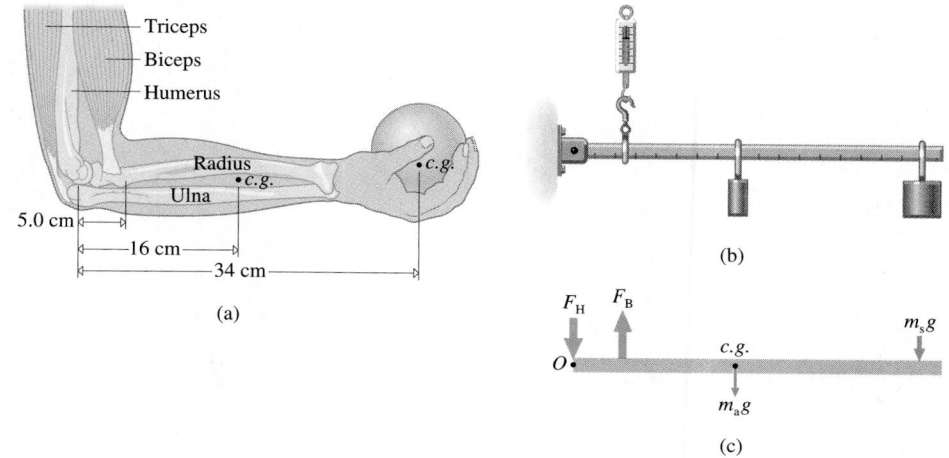

Figure 8.26   (a) Skeletal muscles can only contract. Thus, to raise (flex) the arm, the biceps is
contracted. The pivot is at one end, the load at the other, and the applied force is between the
two. This kind of end-pivoted lever is common in the limbs of mammals and is basically the
same as encountered in the use of the shovel and rake, or a pair of tweezers. (b) A model of
the arm. (c) The four forces acting are shown in the free-body diagram.

Stability and Balance

*A body is said to be in stable equilibrium if, after receiving a small momentary displace-
ment, it returns to its original condition.* Each object in Fig. 8.27 is in stable equilibrium
because, once slightly tilted, the resulting torque produced by its weight acting through the

The Fosbury flop. The jumper extends the head and
legs downward so the *c.g.* actually passes below the bar.
As we've seen, raising the *c.g.* takes energy, and having
it slip under the bar means the bar can be higher.

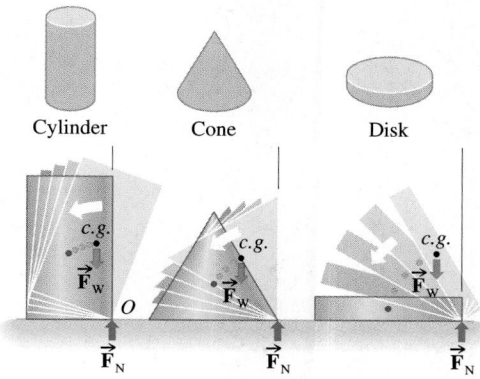

Figure 8.27   Bodies such as these are in stable
equilibrium because they fall back to their original
orientations even after a large displacement.

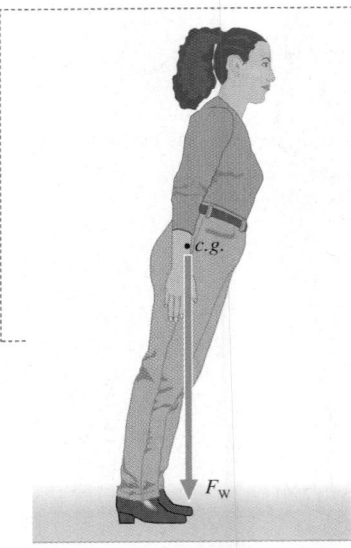

F_w

EXPLORING PHYSICS ON YOUR OWN

Your Center-of-Gravity: (a) Sit back upright in a straight chair with your legs touching the floor just in front of the chair. Where is your *c.g.*? Without shifting your position, can you stand up? Explain. (b) Now stand, feet flat on the ground, with your back and heels against a wall. Try to touch your toes. Explain why you cannot do it. (c) Face the edge of an open door. Place one foot on each side of the door so that both feet extend somewhat beyond the edge against which you are resting your nose. Try to stand up on your toes. Explain what's happening. (d) To crudely locate your *c.g.*, stand rigid, hands at sides, and slowly lean forward until the upward normal force can just pass through your *c.g.*—any farther and you fall over. Now do it again leaning right or left. Your *c.g.* is where the lines-of-action of the normal forces intersect.

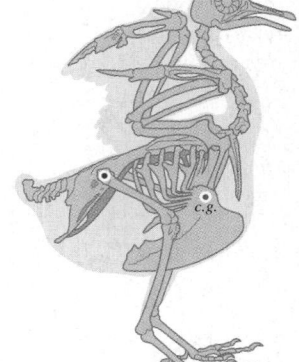

Figure 8.28 Birds are stable on two legs because those legs pivot at the pelvis above the *c.g.*, which is low because of the massive breast bone that anchors the wing muscles. In flight, the *c.g.* is supported by, and located under, the wings.

STUDY GUIDE

Always take the sum-of-the-torques about a point through which pass forces that you do not want to determine. That will keep those unknown forces out of your equation and thereby lessen the complexity of the math. For example, if a balanced beam has several forces acting on it and you're not interested in the reaction force at the pivot, take the sum-of-the-torques about the pivot.

Remember that you will need one independent equation for each unknown.

c.g. tends to rotate the body back to its original orientation. In each such case the initial displacement causes a raising of the *c.g.*, which, when released, tends naturally to fall back downward. If, instead, the cone was balanced on its point, it would be in *unstable* equilibrium—any slight shift would cause it to move farther from its original position. As a rule, *an object supported at a point above the c.g. will be stable, while one supported at a point below the c.g. will be unstable*. A cane hanging from its crook on a finger is stable, but one balanced upright on a fingertip is unstable.

KILLER COKE MACHINES

According to Walter Reed Army Medical Center, between 1975 and 1977, 3 young men were killed and 12 severely injured by falling soda vending machines at U.S. army bases. The victims had apparently deliberately tipped the machines, which, when fully loaded, weigh between 800 and 1000 pounds. Since that study was published, another 48 victims (of whom 11 died) were reported. All were male, ranging in age from 10 to 33; 7 were civilians.

Shenyang Acrobatic Troupe. The combined *c.g.* of the two horizontal men and the pole lies on a vertical line passing through the bottom of the pole. That line then passes between the feet of the man on whose shoulder the pole rests.

This athlete's weight acts straight down through his *c.g.* The two nonparallel reaction forces ($\vec{F}_{R1}$ and $\vec{F}_{R2}$) acting upward on his hands combine to produce a net force that's equal and opposite to his weight. The three forces ($\vec{F}_W$, $\vec{F}_{R1}$, and $\vec{F}_{R2}$) acting on him must be kept coplanar and concurrent. If they are not, there will be a torque about the line passing through his hands and he will rotate around it.

A typical SUV has a high *c.g.*, and a relatively narrow wheel-to-wheel stance. With the *c.g.* high above the axles, the vehicle is not very stable. Particularly poorly designed SUVs tend to roll over with little provocation. It is predicted that in the United States during the next year 70,000 SUVs will roll over and 2000 people will die in the process.

When we bipeds walk, our *c.g.* must remain somewhere over the one supporting foot in contact with the floor, and that requires us to do a little side-to-side wiggling. It's not by accident that we are creatures with the big feet. In fact, the *c.g.* of the upper portion of the human body resides well above the support structure of the hips, and that's not a particularly good design. Birds, which have been bipedal a lot longer than we, are far better suited to that purpose, having their *c.g.* hanging below the hip line (Fig. 8.28). {For more on the stability of football players, robots, buildings, and automobiles click on **STABILITY** under **FURTHER DISCUSSIONS** on the **CD**.}

The Dynamics of Rotation

Newton's First Law has its rotational equivalent in the concept that *a body at rest tends to stay at rest, while a body in uniform rotational motion tends to stay in such motion, except insofar as it is acted upon by a torque*. Remember (p. 247) the two determining conditions of nonaccelerating systems: $\sum \vec{F} = 0$ and $\sum \tau = 0$. Force is the changer of all motion. Specifically, the *moment-of-the-force* is the *torque*, and **torque is the changer of rotational motion**.

The resistance to the change in motion of an object is called *inertia*, and that resistance is physically embodied in the *inertial mass* (or just *mass*, for short). This same property of matter can appear as a resistance to the change in rotational motion, where it is called *rotational inertia*. This rotational resistance is associated with both the amount of mass and its distribution with respect to the axis of rotation. Euler called that composite concept the *moment-of-inertia*.

8.8 Torque & Rotational Inertia

In swinging closed a heavy door or turning a massive wheel, it's evident that if the thing is to accelerate rotationally, a force must be applied, and it must be applied with some moment-arm. There must be a net torque.

Envision a point-mass m. constrained to move on a circle of radius r about an axis passing through O under the influence of an applied tangential force F (Fig. 8.29). The particle will experience a tangential acceleration via the Second Law

$$F = m.a_T$$

or, using $a_T = r\alpha$,

$$F = m.r\alpha$$

The torque about O arising from F is rF, and multiplying both sides of the above equation by r yields

$$\tau_0 = rF = m.r^2\alpha \tag{8.28}$$

By comparison with $F = m.a$, this equation suggests that the rotational equivalent of mass

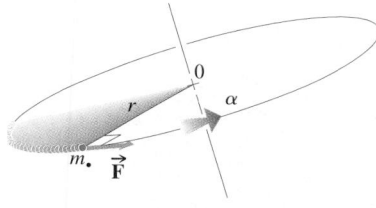

Figure 8.29 A point-mass m. revolving in a circle of radius r about point O. It is accelerated (α) by a tangential force $\vec{F}$.

The **moment-of-inertia of a point-mass** is given by $I_{\bullet} = m_{\bullet}r^2$.

(i.e., of *inertia*) is the quantity $m_{\bullet}r^2$. This is the **moment-of-inertia of a point-mass** about a given axis; it's designated as $I_{\bullet} = m_{\bullet}r^2$, and so $\tau_0 = I_{\bullet}\alpha$ just as $F = m_{\bullet}a$.

A rigid body consists of a great many interacting particles. When such a body is set into rotation, Eq. (8.28) describes any one of those component particles. The sum of all the torques acting on all the constituent point-masses results in an overall angular acceleration of the body such that

$$\sum \tau_0 = \left(\sum m_{\bullet}r^2\right)\alpha \qquad (8.29)$$

Moment-of-inertia is a measure of a body's resistance to changes in its angular motion.

The summation on the right is the **moment-of-inertia of the body** about the axis of rotation passing through O, hence

$$I = \sum m_{\bullet}r^2 \qquad (8.30)$$

Each particle (1, 2, 3, …) with its own mass and perpendicular distance from O has its own moment-of-inertia. All these can be added so that $I = m_{\bullet 1}r_1^2 + m_{\bullet 2}r_2^2 + m_{\bullet 3}r_3^2 + \cdots$.

Calculus is usually used to compute moments-of-inertia, but we'll rely on Table 8.3, which lists values of I for some uniform symmetrical bodies about various axes. ***The more mass and the farther it is from the axis, the greater will be I, and the greater will be the resistance to the change in the rotational motion*** (Fig. 8.30). That's why the thin-walled ring in Table 8.3 has twice the moment-of-inertia of the solid disk with the same mass and why the moment-of-inertia of the hollow sphere is 5/3 that of the solid sphere. The expression

Table 8.3

Moments-of-Inertia

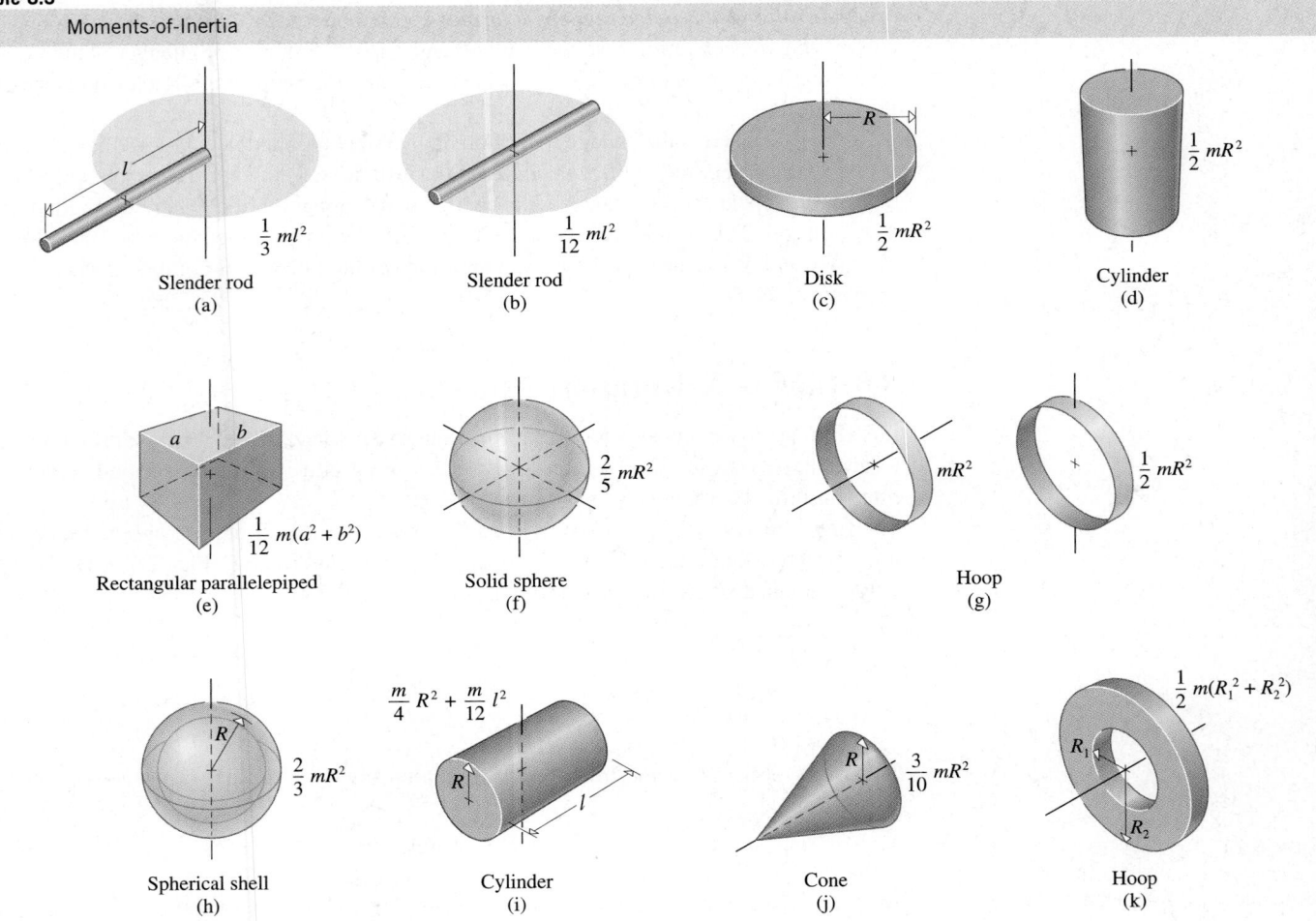

Slender rod
(a) — $\frac{1}{3}ml^2$

Slender rod
(b) — $\frac{1}{12}ml^2$

Disk
(c) — $\frac{1}{2}mR^2$

Cylinder
(d) — $\frac{1}{2}mR^2$

Rectangular parallelepiped
(e) — $\frac{1}{12}m(a^2 + b^2)$

Solid sphere
(f) — $\frac{2}{5}mR^2$

Hoop
(g) — mR^2 ; $\frac{1}{2}mR^2$

Spherical shell
(h) — $\frac{2}{3}mR^2$

Cylinder
(i) — $\frac{m}{4}R^2 + \frac{m}{12}l^2$

Cone
(j) — $\frac{3}{10}mR^2$

Hoop
(k) — $\frac{1}{2}m(R_1^2 + R_2^2)$

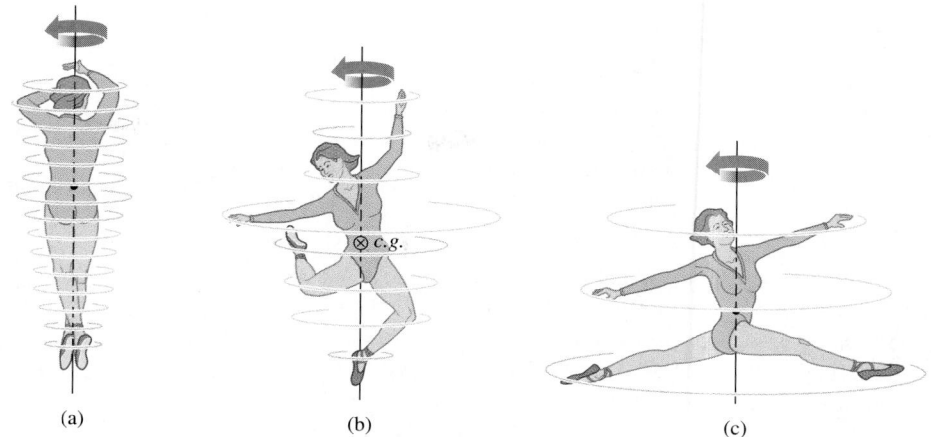

Figure 8.30 With the limbs drawn in near the spin axis as in (a), the moment-of-inertia is comparatively small. With bent knees I increases, as in (b), and reaches a maximum with arms and legs extended perpendicular to the axis (c).

[fixed-axis rotation]
$$\sum \tau_0 = I\alpha$$
(8.31)

deliberately resembles Newton's Second Law and is the primary dynamical equation of rotating systems. The farther a bit of mass is from the axis, the larger will be I, and the more torque must be applied to attain a given α. "Choking up" on a baseball bat (that is, gripping it in from the end) decreases I about the pivot and allows the same torque to produce a larger α. If a youngster can't swing the bat fast enough to meet the incoming pitch, a larger α will help.

WALKING (TIGHTROPE & OTHERWISE)

Equation (8.31) governs all sorts of angular motion including, for example, that of a pole-carrying tightrope walker. The long balance pole has a large moment-of-inertia about its center. Since the acrobat cannot topple over sideways off the wire without rotating the pole, and since such angular acceleration is resisted by the large rotational inertia of the pole, it can be quite handy. Any kid walking on a narrow curb knows to stick both arms out for "balance."

Legs swinging about hip joints also have rotational inertia. Long, heavy legs require large torques to accelerate them. That's why animals with four light-weight legs can usually run faster than erect creatures with two necessarily big feet and relatively massive legs. One thing to do to reduce the rotational inertia about the hip joint when high accelerations are needed is to bend the legs. Most people must know that, since you hardly ever see anyone running stiff-legged.

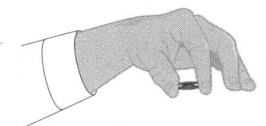

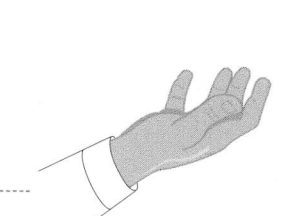

EXPLORING PHYSICS ON YOUR OWN

Omega & the Nickel-Quarter Flip: Place a nickel between two quarters and hold the stack horizontally with your thumb and index finger. Place your other hand about a foot below and release the lower quarter and the nickel; to your utter amazement, they'll land with the quarter on top! Now drop them from around two feet and the nickel will land on top. What's happening? [Answer: You inadvertently release one edge of the quarter first. While the other edge is still supported by your finger, the weight vector down causes a net torque and the nickel-quarter combination immediately begins to rotate. Continuing to spin about its *c.g.*, it revolves half a rotation in the 1/4 s it takes to fall 1 ft. That corresponds to $\omega_{av} = 120$ rpm.]

Example 8.13 **[III]** The evil Dr. Doom, involved in an extortion threat to despin the Earth, plans to mount a series of surplus rockets tangentially all along the Equator. Taking the planet to be a uniform sphere of radius 6.37×10^6 m and mass 5.98×10^{24} kg, how much continuous total thrust would the rockets need to apply to accomplish the heinous deed in the course of 12 hours?

Solution   The problem talks about despinning a rotating body, and that means torques and angular acceleration. (1) TRANSLATION—A sphere of known mass and angular speed (revolving about a central axis) is to have its speed reduced to zero in a specified time; determine the required force applied tangentially at the surface. (2) GIVEN: $R_\oplus = 6.37 \times 10^6$ m, $M_\oplus = 5.98 \times 10^{24}$ kg, and $\omega_f = 0$ when $\Delta t = 12$ h. FIND: F. (3) PROBLEM TYPE—Angular motion/rotational dynamics. (4) PROCEDURE—The necessary torque is given by $\sum \tau_0 = I\alpha$, so we must first determine I and α. (5) CALCULATION—According to Table 8.3, the moment-of-inertia of a solid sphere having the mass of the Earth (about a central axis) is

$$I = \tfrac{2}{5}mR_\oplus^2 = \tfrac{2}{5}(5.98 \times 10^{24} \text{ kg})(6.37 \times 10^6 \text{ m})^2$$

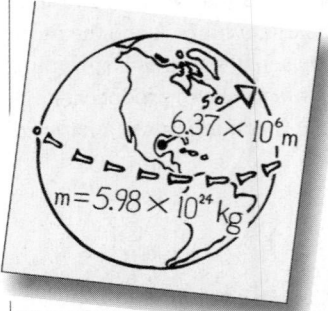

or $I = 9.71 \times 10^{37}$ kg·m². It must decelerate from its present angular rate to zero in 12 h. The former (ω_i) can be computed accurately enough assuming the planet now spins through 2π rad in 24 h; that is, $\omega_i = 0.727 \times 10^{-4}$ rad/s, and

$$\alpha = \frac{\omega_f - \omega_i}{\Delta t} = \frac{0 - 0.727 \times 10^{-4} \text{ rad/s}}{12 \text{ h} \times 60 \text{ min/h} \times 60 \text{ s/min}}$$

$$\alpha = -1.68 \times 10^{-9} \text{ rad/s}^2$$

The total required torque is then

$$\sum \tau_0 = I\alpha = (9.71 \times 10^{37} \text{ kg·m}^2)(-1.68 \times 10^{-9} \text{ rad/s}^2)$$

$$\sum \tau_0 = -1.63 \times 10^{29} \text{ N·m}$$

Since each rocket acts about the spin axis with a moment-arm of $R_\oplus$, it follows that

$$\sum \tau_0 = R_\oplus \sum F$$

and the total thrust must be

$$\sum F = \frac{\sum \tau_0}{R_\oplus} = \frac{-1.63 \times 10^{29} \text{ N·m}}{6.37 \times 10^6 \text{ m}} = \boxed{-2.6 \times 10^{22} \text{ N}}$$

The minus sign shows that the thrust opposes the positive ω_i. The value 2.6×10^{22} N is a realistically large force. Compare it to the 3.4×10^7 N thrust of a *Saturn* Moon rocket. Dr. Doom would have to fire over 700 million million such rockets continuously over the whole 12 hours! So, not to worry.

Quick Check: Recompute an approximate value of the moment-of-inertia; $I \approx 0.4(6 \times 10^{24} \text{ kg})(36 \times 10^{12} \text{ m}^2) \approx 9 \times 10^{37}$ kg·m². We can check ω_i by confirming that $\omega_i(24 \text{ h} \times 60 \text{ min/h} \times 60 \text{ s/min}) = 6.28$ rad $= 2\pi$, it does and so $\omega_i = 0.727 \times 10^{-4}$ rad/s is okay. Using $\theta = \omega_{av}t = (3.6 \times 10^{-5}$ rad/s$)(4.3 \times 10^4$ s$) = 1.6$ rad, and from Eq. (8.24) $\alpha = -\omega_i^2/2\theta = -1.7 \times 10^{-9}$ rad/s², which also checks.

Example 8.14 **[III]** A 10.0-kg mass hangs on a rope wrapped around a freely rotating 2.00-kg cylinder of radius 10.0 cm as shown in Fig. 8.31. Determine the tension in the rope and the accelerations of both the cylinder and the mass. Use $g = 9.81$ m/s².

Solution   The cylinder is rotating and accelerating, and that should bring to mind the sum-of-the-torques-equals $I\alpha$. (1) TRANSLATION—A known mass, attached by a rope around a free-turning cylinder, falls; determine the tension and accelerations. (2) GIVEN: $m = 10.0$ kg, $M_c = 2.00$ kg, and $R = 10.0$ cm $= 0.100$ m. FIND: F_T, α, and a. (3) PROBLEM TYPE —Angular motion/rotational dynamics. (4) PROCEDURE—The central equation is $\sum \tau_0 = I\alpha$. There are three unknowns, and we'll need three equations (the sum-of-the-torques, the sum-of-the-forces, and the relation between a and α). (5) CALCULATION—The tension of the rope produces a torque on the cylinder such that

$$\circlearrowleft + \sum \tau_0 = F_T R = I\alpha$$

and so $F_T = I\alpha/R$. Furthermore, for the descending mass traveling in the y-direction

$$\downarrow + \sum F_y = mg - F_T = ma$$

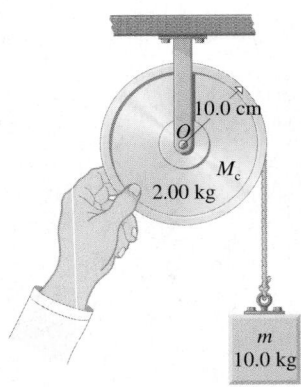

Figure 8.31 Up to this point, we have neglected the motion of the pulley in this sort of problem. Now, however, we can take into account that it really isn't massless. The pulley has a moment-of-inertia and will reduce the acceleration of m.

continued

The rope accelerates at a rate equal to that of a point on the cylinder, $a = R\alpha$. Substituting this last relationship into the above two equations yields

$$mg - F_T = ma = mR\alpha$$

$$mg - \frac{I\alpha}{R} = mR\alpha$$

and so

$$\alpha = \frac{mg}{mR + I/R}$$

Since $\quad I = \frac{1}{2}M_cR^2 = \frac{1}{2}(2.00\ \text{kg})(0.100\ \text{m})^2 = 0.010\ \text{kg}\cdot\text{m}^2$

$$\alpha = \frac{98.1\ \text{kg}\cdot\text{m/s}^2}{(1.00\ \text{kg}\cdot\text{m} + 0.100\ \text{kg}\cdot\text{m})} = \boxed{89.2\ \text{rad/s}^2}$$

This result corresponds to $a = R\alpha = \boxed{8.92\ \text{m/s}^2}$ and a tension of $F_T = m(g - a) = \boxed{8.9\ \text{N}}$

Quick Check: Let's examine the expression for α at its extremes; namely, when $m >> M_c$ and $m << M_c$. In the first case $(m >> M_c)$, I/R would be negligible compared to mR. Thereupon $\alpha \approx g/R$; that is, $a \approx g$; $F_T = 0$, and we would have free-fall. When $m << M_c$, $I/R >> mR$, and $\alpha \approx mgR/I$, which is vanishingly small—hardly any motion occurs. Both conclusions are reasonable.

The Center-of-Mass

As far as its translational motion is concerned, a body under the influence of external forces behaves as if all of its mass were at a point called the **center-of-mass**.

➤ It is possible to find a point, the **center-of-mass** (*c.m.*), *where all the mass* (*m*) *of an object can be imagined concentrated*, as the object translates in compliance with Newton's Second Law: $\sum \vec{\mathbf{F}} = m\vec{\mathbf{a}}_{cm}$. Suppose the body is acted upon by several external forces. If the sum of those forces is nonzero, the body must accelerate in the direction of the resultant $\sum \vec{\mathbf{F}}$. We could replace all the separate forces by the resultant, but where on the body should it act? The answer is, through the *c.m.* The body would then behave as if its entire mass were at the *c.m.*

It turns out that *when the gravitational field is uniform over the body, and that's usually the case, the c.m. and the c.g. are the same point*. Nonetheless, there are important situations where the differences can be exploited. Several artificial satellites maintain their orientations in space using the $1/r^2$ drop in the Earth's gravity. A long tube extending down from a satellite experiences a nonuniform field—the *c.g.* is displaced downward from the *c.m.*, allowing torques to develop that maintain a stable alignment. {For a more mathematical treatment of the *c.m.* click on CENTER-OF-MASS under FURTHER DISCUSSIONS on the CD.}

Motion of & Around the c.m.: If $\sum \vec{\mathbf{F}} = 0$ and $\sum \tau \neq 0$—the body will rotate, accelerating about the *c.m.*—the motion will be purely rotational. Adding a force will superimpose a translational acceleration *of* the *c.m.* onto the rotation *about* the *c.m.* A wrench or a base-

HOW WE, EYES CLOSED, DETECT ROTATION

Humans sense rotation using three perpendicular loops or semicircular canals in the inner ear (Fig. 8.32). These are each filled with a watery fluid. When the head turns, one or more of the canals revolves with it, but the fluid therein has rotational inertia and tends to stay at rest. A little hinged flap (called a cupola) is attached to the inner wall of each canal and extends into the liquid. When there is a relative motion of the fluid with respect to the canal, the flap swings and a sensory nerve at its base signals rotation.

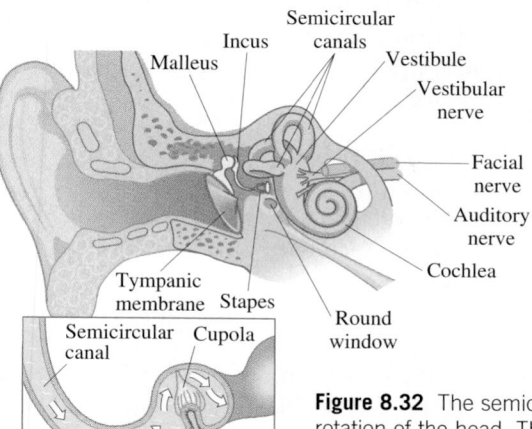

Figure 8.32 The semicircular canals detect any rotation of the head. They lie in three perpendicular planes and are filled with a fluid that tends to stay at rest as the head rotates.

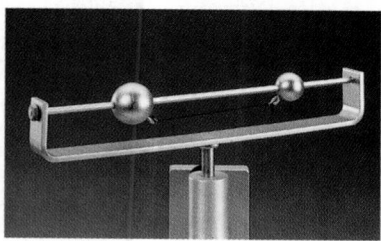

Two unequal-mass balls fastened together with a string are free to slide along the central horizontal rod. The system can be spun about the vertical shaft. Only when that vertical axis passes through the *c.m.* of the two balls will they remain where they are as the system rotates. A slight displacement to either side will unbalance the centripetal forces and cause the balls to slide to that side.

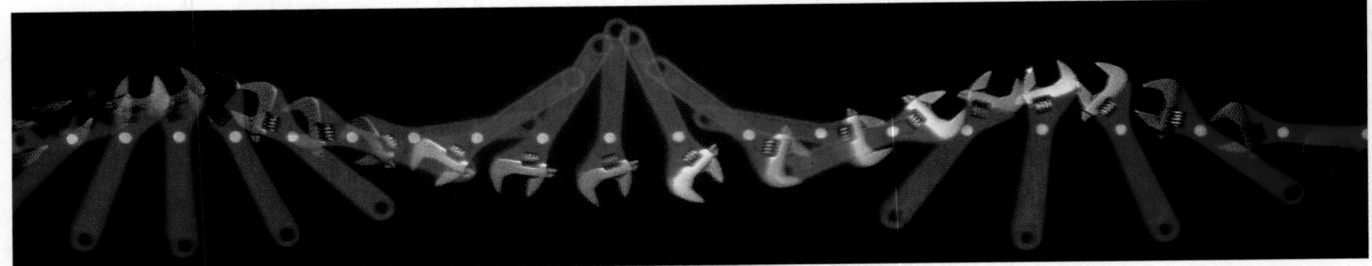

A tumbling wrench on a horizontal surface shown every 1/30 of a second. The center-of-mass (indicated by a spot of light paint) moves along a straight path, while the wrench revolves around it.

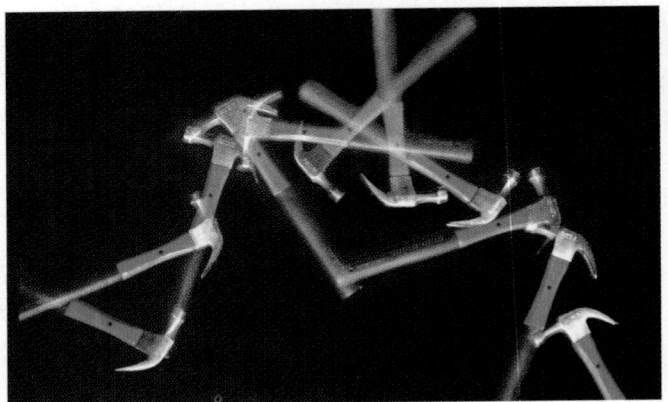

The center-of-mass (black dot) of this hurled hammer flies along a ballistic parabola as if all of the mass were at that point.

Figure 8.33 Since the motion of the *c.m.* can only change with the application of an external force, and no new force is applied, the *c.m.* moves in a ballistic arc regardless of the diver's twisting.

ball bat hurled through the air may tumble around (if $\sum \vec{\mathbf{F}}$ does not pass through its *c.m.* at launch), but the center-of-mass will sail along a parabola as if it marked the motion of a point-mass. Only if you hit a cue ball along a line through its *c.m.* will it move away spinless.

A diver in the air (Fig. 8.33) can tumble and twist, changing the location of her *c.m.* with respect to her body from one instant to the next via *internal forces*, and still her *c.m.* will free-fall along a parabola. Her *c.m.* will move in a ballistic arc as she rotates around it.

8.9 Rotational Kinetic Energy

Imagine an object rotating around some axis passing through it but not otherwise moving. Almost every atom of the object is revolving around that axis at some distance from it. Because each of these atoms is literally moving through space, each has translational KE as a result of the rotation. It's therefore reasonable to conceptualize a KE for the body as a whole and we call it the **rotational kinetic energy** (KE_R). This is equal to the sum of all the rotational kinetic energies of all the point-masses,

$$KE_R = \sum \left(\tfrac{1}{2} m_i v^2\right) = \sum \left(\tfrac{1}{2} m_i r^2 \omega^2\right) = \tfrac{1}{2}\left(\sum m_i r^2\right)\omega^2$$

we can factor out the ω^2 because it's the same for each point-mass. The term in parentheses is the moment-of-inertia of the body: $I = \sum m_i r^2$. Therefore

[rotational KE]
$$KE_R = \tfrac{1}{2} I \omega^2 \qquad (8.32)$$

When a rigid body translates without rotating, its total $KE = \tfrac{1}{2} m v^2$. When it rotates without translating, its total $KE = \tfrac{1}{2} I \omega^2$. When the body simultaneously rotates and translates, $KE = \tfrac{1}{2} m v^2 + \tfrac{1}{2} I \omega^2$. There's really nothing new happening here; all of this energy is the translational KE of the atoms expressed in two component forms, and, as ever, energy is conserved.

Example 8.15 **[III]** A uniform solid ball of radius R and mass m is at rest at a height h atop an inclined plane making an angle θ, as shown in Fig. 8.34. Write an expression for the linear speed of the sphere at the bottom of the incline assuming it rolls without slipping. Compare that to the speed a hollow sphere of the same mass and size would attain.

Solution Knowing how far the ball descends vertically, we know its potential energy change, and that suggests using Conservation of Energy to solve the problem. (1) TRANSLATION—A sphere of known size and mass at a given height rolls down an inclined plane; determine its speed at the bottom. (2) GIVEN: R, m, h, and θ. FIND: v at the bottom. (3) PROBLEM TYPE—Angular motion/Conservation of Energy/rotational KE. (4) PROCEDURE—Since we are asked to find the final speed and don't care about the forces, this suggests an energy approach, namely, $E_i = E_f$. Notice that v and ω are not independent here

Figure 8.34

because the ball is rolling rather than flying through the air. (5) CALCULATION—Initially, transitional KE = 0, $KE_R = 0$, and all the energy is *gravitational*-PE; this ultimately transforms into translational and rotational KE such that

$$PE_{Gi} = KE_f + KE_{Rf}$$

$$mgh = \tfrac{1}{2}mv^2 + \tfrac{1}{2}I_{cm}\omega^2$$

The right side of the equation is the KE of the c.m. plus the KE about the c.m. From Table 8.3, $I_{cm} = \tfrac{2}{5}mR^2$ and since $v = R\omega$

$$mgh = \tfrac{1}{2}mv^2 + \tfrac{1}{2}(\tfrac{2}{5}mR^2)(v/R)^2 = \tfrac{7}{10}mv^2$$

and $$\boxed{v = \sqrt{10gh/7}}$$

Had the sphere been hollow, its moment-of-inertia ($I_{cm} = \tfrac{2}{3}mR^2$) would have been larger, and therefore both ω and $v = \sqrt{6gh/5}$ would be smaller. In a no-slip race, a rolling solid sphere will always beat a hollow one.

Quick Check: Let's at least check the units; $\sqrt{gh}$ has units of $\sqrt{(m/s^2)(m)}$ or m/s, which is OK.

8.10 Angular Momentum

The rotational equivalent of force is torque, the *moment of the force*. Likewise, the rotational equivalent of linear momentum (p) is **angular momentum** (L), the *moment of momentum*.

Figure 8.35 depicts a particle of mass $m_\bullet$ and speed v moving along a straight line past a point O. We define the magnitude of the angular momentum with respect to O as

$$L_0 = r_\perp p = r_\perp m_\bullet v \tag{8.33}$$

with units of $kg \cdot m^2/s$. It might seem surprising that a particle moving in a straight line has *angular* momentum. Yet experience tells us that a straight-blowing wind can impart angular motion to a windmill and that a door can be swung closed (imparting angular momentum to it) by bouncing a ball straight off it.

Now, examine the flat extended object made up of point-masses in Fig. 8.36. It's revolving about a perpendicular axis through point O. Each mass point has a distinctive angular momentum even though they all have the same angular speed ω. Since, for every particle, $\vec{r}$ is perpendicular to $\vec{v}$, $r = r_\perp$ and the magnitude of the body's total angular momentum L_0 about O is

$$L_0 = \sum r m_\bullet v = \left(\sum m_\bullet r^2\right)\omega$$

since $v = r\omega$, and ω is the same for all point-masses. The term in parentheses is the body's moment of inertia about O [Eq. (8.30)], and so

$$L_0 = I_0\omega \tag{8.34}$$

Here point O, which locates the axis of rotation, happened to lie on the body, but it need not have. The O subscript tells us that L is referenced to that point; it's really not crucial and is often omit-

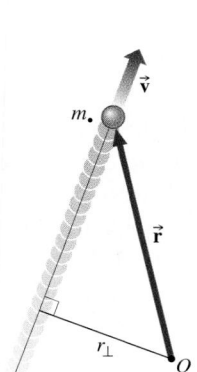

Figure 8.35 A particle of mass $m_\bullet$ has an angular momentum L_0 with respect to point O equal to $r_\perp m_\bullet v$.

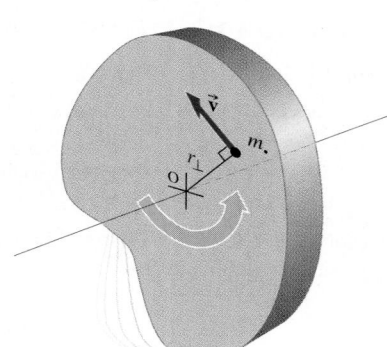

Figure 8.36 An object rotating about an axis passing through point O.

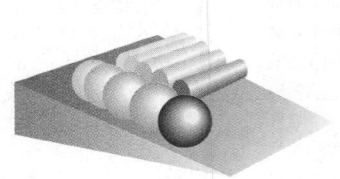

EXPLORING PHYSICS ON YOUR OWN

Free Rolling: Notice that the answer in Example 8.15 is independent of the mass and size of the objects rolling down the incline. Isn't that surprising? It would seem that a marble and a cannonball should roll together, whereas a BB pellet should beat a can of soup or a hollow rubber ball every time. Make an inclined plane (e.g., tilt a table) and run some experiments. What happens with a roll of tape and a piece of chalk? Now try a Ping-Pong ball and an empty jar, and if you can anticipate that, how about a Ping-Pong ball and a piece of chalk? Make sure your inclined plane is smooth and not tipped to either side. The closer the mass distribution is to the *c.m.*, the faster it rolls.

Example 8.16 **[I]** Taking the Earth to be a uniform sphere of radius 6.37×10^6 m and mass 5.98×10^{24} kg, compute its angular momentum about its spin axis.

Solution Here the Earth is just a sphere rotating about its central axis. (1) TRANSLATION—A sphere of known radius and mass is spinning at a known rate; determine its angular momentum. (2) GIVEN: $R_\oplus = 6.37 \times 10^6$ m and $M_\oplus = 5.98 \times 10^{24}$ kg. FIND: L. (3) PROBLEM TYPE—Angular motion/angular momentum. (4) PROCEDURE—By definition $L = I\omega$, so we'll need I and ω. (5) CALCULATION—As we saw earlier, $I = \frac{2}{5}mR_\oplus^2 = 9.71 \times 10^{37}$ kg·m^2, whereas $\omega = 0.727 \times 10^{-4}$ rad/s;

therefore,

$$L = I\omega = (9.71 \times 10^{37} \text{ kg·m}^2)(7.27 \times 10^{-5} \text{ rad/s})$$

$$\boxed{L = 7.06 \times 10^{33} \text{ kg·m}^2/\text{s}}$$

Incidentally, looking down onto the North Pole, the Earth rotates counterclockwise.

Quick Check: Check the units: kg·m^2/s = m(kg·m/s), which *are* the units of *rmv*. Furthermore, $L \approx (10^{38}$ kg·m$^2)(7 \times 10^{-5}$ rad/s) $\approx 7 \times 10^{33}$ kg·m^2/s.

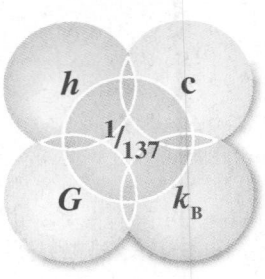

Figure 8.37 Fundamental constants of Nature. Planck's Constant, which is of tremendous importance in the atomic microworld, goes unnoticed in the macroscopic domain of everyday life.

In the remainder of this chapter, we'll learn that Rotation Dynamics is very similar to Translational Dynamics, in that except for a change in symbols, the two can be formulated in much the same way. To that end, we define monent-of-inertia (I), torque (τ), rotational kinetic energy (KE$_R$), and angular momentum (L). To save us the need to memorize a whole new set of equations, add to the kinematical substitutions, $s \to \theta$, $v \to \omega$, $a_T \to \alpha$, and $t \to t$ the dynamical ones, $m \to I$, $F \to \tau$, KE $\to$ KE$_R$, and $p \to L$. The equations of translational motion can then be carried over into rotational motion.

ted. The last equation shows angular momentum to be analogous to linear momentum: $p = mv$. In transforming to the rotational regime, m becomes I, v becomes ω, and p becomes L.

On an Atomic Level

A variety of experiments performed during the twentieth century revealed a surprising fact: *angular momentum is an intrinsic property of matter on the atomic and subatomic levels.* Each of the elementary particles that constitutes ordinary matter (the leptons and quarks) has a characteristic angular momentum, just as it has a characteristic mass and electric charge. The building blocks of the atom—electrons, protons, and neutrons—each have an intrinsic angular momentum. The discrete measure of angular momentum associated with these particles turns out to be the smallest observed amount—namely, $h/4\pi = 5.27 \times 10^{-35}$ kg·m^2/s. Because it comes in specific doses, we say that angular momentum is *quantized*— it exists only in whole-number multiples of $h/4\pi$. That's even true for light where each photon has an angular momentum of $h/2\pi$. The quantity h is the fundamental constant of Quantum Mechanics known as **Planck's Constant** (Fig. 8.37). The angular momentum of macroscopic objects is so large compared to $h/4\pi$ that we never have to worry about its inherently quantized nature.

8.11 Conservation of Angular Momentum

Suppose that a constant net torque τ acts on a body for a time interval Δt. There will then be a constant angular acceleration such that

$$\tau = I\alpha = I\frac{\Delta\omega}{\Delta t}$$

Forasmuch as I too is constant, any change in the angular momentum is a result of a change in ω; that is, $\Delta L = I\Delta\omega$, and therefore

$$\tau = \frac{\Delta L}{\Delta t} \tag{8.35}$$

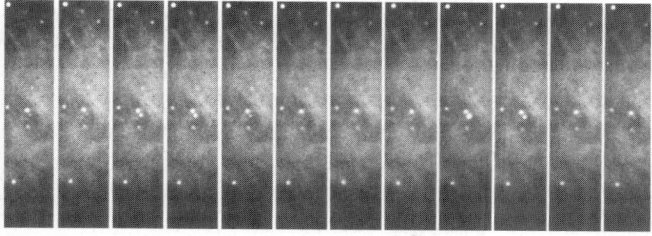

The Crab Nebula is the remnant of a star that exploded (a supernova) in A.D. 1054. At its center are the crushed remains now in the form of a spinning neutron star. It's called a pulsar because it emits radiant energy like a lighthouse beam that sweeps past, flashing on and off. You can see it appear and disappear in the center of this sequence of close-up photos.

PIROUETTES & PULSARS

The same mechanism that governs a pirouetting skater governs a whirling hurricane or a spinning star. A fuel-spent star will collapse as it dies, gravitationally crushing itself to immense densities. The tiny remnant of this cosmic compression is a neutron star. If the thing is spinning at the outset, as it collapses, I decreases tremendously and ω increases in proportion. The end result is a rapidly rotating source of radiant energy—a *pulsar*. One such object, perhaps only 25 km (15 mi) in diameter and having about the same mass as the Sun, is at the center of the Crab Nebula. Whirling around 30 times each second it emits regular pulses of light and X-rays that sweep the sky every 33 milliseconds.

The rider leans into a turn so that $\vec{\mathbf{F}}_f = \vec{\mathbf{F}}_C$. The total weight of the bike and rider ($\vec{\mathbf{F}}_W$) acts down through the combined *c.g.* He doesn't fall over, thereby rotating about the *c.g.* and gaining angular momentum, because the sum of the torques ($F_N r_{\perp N} - F_f r_{\perp f}$) w.r.t. the *c.g.* equals zero, but only if he does it right.

The torque equals the time rate-of-change of the angular momentum. For simplicity we are limiting the discussion to symmetrical objects revolving around symmetry axes. Equation (8.35) is the rotational equivalent of Newton's Second Law, and, as before, it points to a grand underlying insight. **In the absence of a net external torque acting on a system, its angular momentum remains constant.** This is the law of **Conservation of Angular Momentum**.

The **angular momentum** of an isolated system (i.e., one with no external torques acting on it) is conserved.

An upwelling of air, here above warm water, causes a drop in pressure that draws in air from all sides. If there is any circulation of the surrounding air, it spirals inward, moving faster and faster as its distance from the center decreases. Thus, a great vortex forms, conserving angular momentum all the while.

Example 8.17 **[I]** By pushing on the ground, a skater with arms extended (as in Fig. 8.38*a*) manages to whirl around at a maximum angular speed of 1.0 rev/s. In that configuration, her moment-of-inertia about the spin axis is 3.5 kg·m². The point-masses making up her arms and outstretched leg are, for the most part, fairly far from the vertical spin axis, and therefore her moment-of-inertia is relatively large. What will happen to her spin rate as she draws her arms in and stands upright in a pirouette (Fig. 8.38*b*), whereupon her moment-of-inertia is only 1.0 kg·m²? Assume the friction forces acting on her are negligible.

Solution There are no external torques acting on the spinning skater so she conserves angular momentum. (1) TRANSLATION—An object with a known moment-of-inertia and angular speed takes on a new moment-of-inertia; determine its new angular speed. (2) GIVEN: $\omega_i = 1.0$ rev/s, $I_i = 3.5$ kg·m², and $I_f = 1.0$ kg·m². FIND: ω_f. (3) PROBLEM TYPE—Angular motion/Conservation of Angular Momentum. (4) PROCEDURE—With no externally applied torques, the angular momentum of the system is constant. By definition $L = I\omega$. (5) CALCULATION:

$$L_i = L_f$$

$$I_i \omega_i = I_f \omega_f$$

and so

$$\omega_f = \frac{I_i}{I_f} \omega_i = \frac{3.5 \text{ kg·m}^2}{1.0 \text{ kg·m}^2} (1.0 \text{ rev/s}) = \boxed{3.5 \text{ rev/s}}$$

Quick Check: The skater speeds up as her moment-of-inertia decreases, all the while conserving her angular momentum of $(3.5)(1.0)(2\pi)$ kg·m²/s = 22 kg·m²/s.

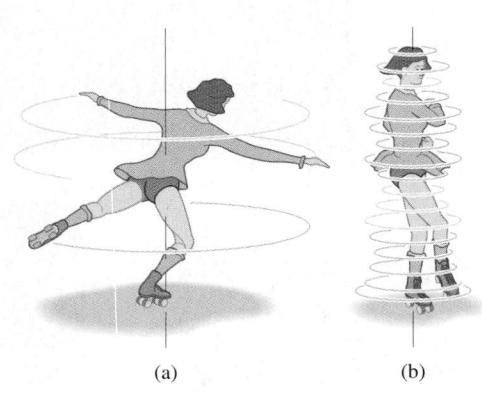

(a)   (b)

Figure 8.38 (a) The skater revolves as rapidly as possible, maximizing ω_i while maintaining a large I about the spin axis, which produces a substantial $L_i = I_i\omega_i$. (b) By pulling in her arms and legs, her moment-of-inertia decreases ($I_f < I_i$). Since there are no appreciable external torques, $L_i = L_f$ and $\omega_f > \omega_i$; her spin rate increases.

In this old-fashioned long jump, the athlete rotates his legs forward while pulling his arms backward. Once airborne, his net angular momentum cannot change because there are no external torques on him. The angular momentum of his arms rotating one way must cancel the angular momentum of his legs rotating the other way.

The tendency for matter to sustain rotational motion is utilized in a number of practical devices. The modern automobile engine is fitted with a flywheel that is set in high-speed rotation by the engine. Mounted on the end of the crankshaft, it stores angular momentum and keeps the shaft turning smoothly between successive strokes of the pistons.*

A high diver leaping into a pool can exert a torque about his *c.m.* while still in contact with the board and so begin to rotate. Once in the air and "torqueless," L is constant no

*The flywheel is geared via a disengaging mechanism to the starting motor, and if you've ever heard a horrible grinding sound when turning the ignition key for the second or third time, it's because the flywheel was still spinning.

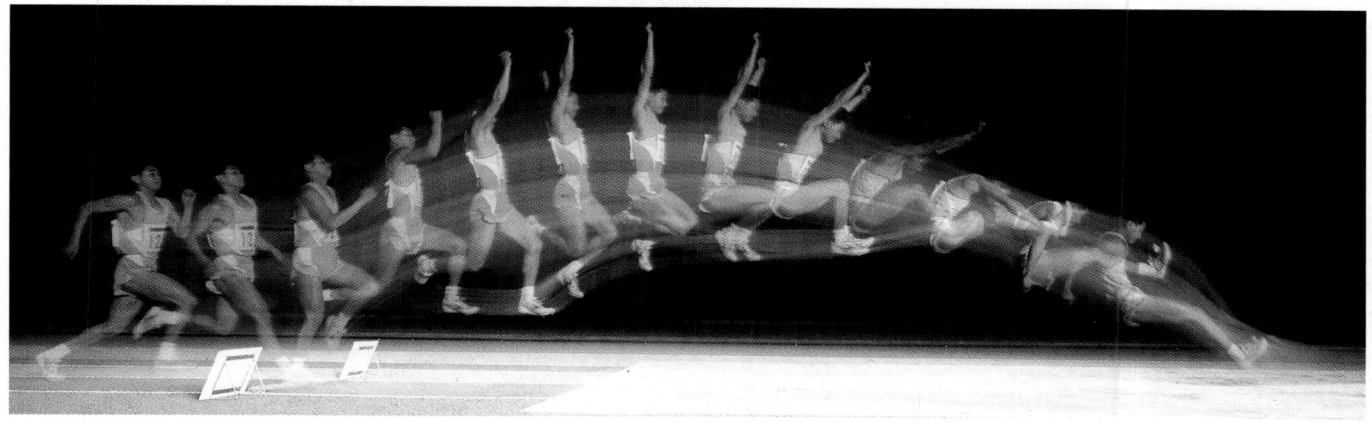

Once the jumper is airborne, no external torques are applied and angular momentum is conserved: the angular momentum due to a rotation of his arms one way is canceled by the angular momentum due to a rotation of his legs the other way. As with torque, we select a direction of rotation to be positive. Since he wants to land legs extended, he jumps with arms up. Then, while in midair, he pulls his arms down (producing angular momentum we'll call positive). At the same time, he draws his legs up (producing an equal amount of negative angular momentum). The net change in L is zero, and the jumper lands feet first.

Example 8.18 [II] A very small sphere of mass 0.20 kg is being whirled in a horizontal circle at the end of a 2.0-m-long string at a constant speed of 1.0 m/s. (a) Determine its orbital angular momentum about the axis of rotation. Suppose the string is quickly reeled in, leaving a radius of only 1.0 m. (b) How fast will the sphere be moving then? (c) Talk about the tension in the string.

Solution Are there any external torques acting on the sphere about O? (1) TRANSLATION—An object of known mass revolves in a circle of known radius at a constant speed; (a) determine its angular momentum and (b) its speed when there is a new reduced radius. (2) GIVEN: $m = 0.20$ kg, $v_i = 1.0$ m/s, $r_i = 2.0$ m, and $r_f = 1.0$ m. FIND: (a) L w.r.t. the axis of rotation, and (b) v_f. (3) PROBLEM TYPE—Angular Motion/ Conservation of Angular Momentum. (4) PROCEDURE—With no externally applied torques, the angular momentum of the system is constant. By definition $L = I\omega$. (5) CALCULATION:

$$L_i = L_f$$

(a) The angular momentum can be obtained by approximating the sphere as a point located at its *c.m.* whereupon

$$L = rm_{\cdot}v = (2.0 \text{ m})(0.20 \text{ kg})(1.0 \text{ m/s}) = \boxed{0.40 \text{ kg·m}^2/\text{s}}$$

(b) The string exerts a centripetal force on the ball that passes through O, and no torque about the axis is produced. No net torque means no change in the ball's angular momentum—the initial angular momentum ($m_{\cdot}r_i v_i$) before the string is reeled in must equal the final value ($m_{\cdot}r_f v_f$) afterward:

$$m_{\cdot}r_i v_i = m_{\cdot}r_f v_f$$

or

$$I_i \omega_i = I_f \omega_f$$

and

$$v_f = \frac{v_i r_i}{r_f} = \frac{(1.0 \text{ m/s})(2.0 \text{ m})}{(1.0 \text{ m})} = \boxed{2.0 \text{ m/s}}$$

(c) The tensile force equals $F_C = ma_C = mv^2/r$. Since v doubles when r is halved, v^2 quadruples. That means the tension goes up by a factor of eight. One might wonder which force causes the sphere to accelerate from v_i to v_f. Figure 8.39 shows that as the body is pulled out of its circular orbit by an increased tensile force $\vec{\mathbf{F}}_T$, the latter can be resolved into two components, one parallel ($\vec{\mathbf{F}}_{T\parallel}$) and one perpendicular ($\vec{\mathbf{F}}_{T\perp}$) to the motion. While $\vec{\mathbf{F}}_{T\perp}$ changes the direction, it is $\vec{\mathbf{F}}_{T\parallel}$ that tangentially accelerates the body. The *net* torque about O is always zero, and L is constant throughout.

Quick Check: The moment-of-inertia decreases, and as a result the angular speed increases, as does the linear speed. The radius is halved; the speed must double.

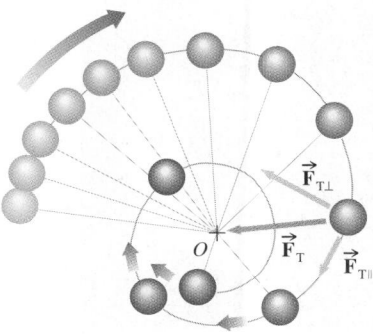

Figure 8.39 A body spiraling inward due to the tension in an attached string.

matter how he tumbles, and tumble he must. Just as a translating body will continue to move with a constant $\vec{v}$ unless acted upon by a force, a rotating body will continue to rotate forever except insofar as it is acted upon by an *external* torque. {For a derivation of Kepler's Second Law via angular momentum, click on **PLANETARY MOTION** under **FURTHER DISCUSSIONS** on the **CD**. For a discussion of gyroscopic stability, click on **GYROS** under **FURTHER DISCUSSIONS** on the **CD**.}

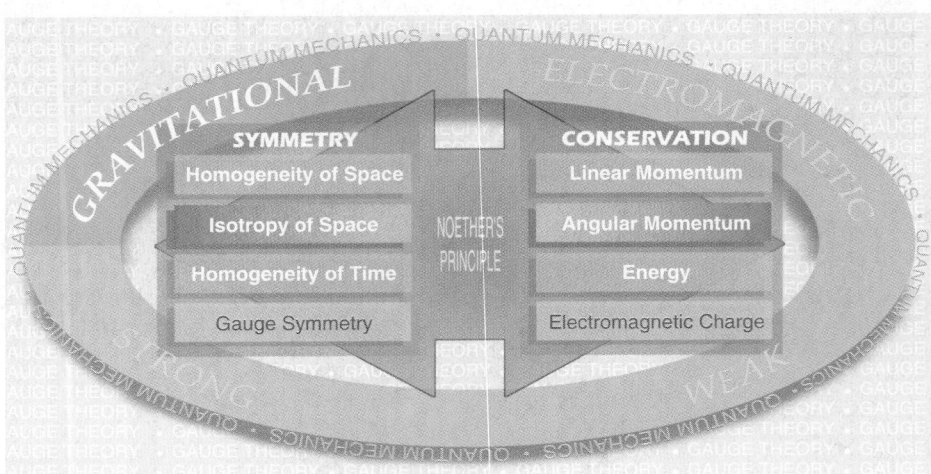

Figure 8.40 Conservation of Angular Momentum is one of the fundamental principles of physics. It arises from the fact that the laws and constants of physics are independent of orientation in space.

Angular Momentum and Symmetry

Noether showed (p. 5) that every conservation law is linked to a symmetry of some kind (Fig. 8.40); we change an aspect of the physical system and some characteristic remains unchanged or conserved—that's what a symmetry is.

The Earth and Moon form a single system bound together by gravity. This photograph was taken by the *Galileo* spacecraft from a distance of 6.2×10^6 km.

THE TIDES & TIME

The Moon gravitationally draws the Earth's crust, waters, and atmosphere, generating tidal motions in each. These create frictional forces that gradually slow the planet's spin rate. The effect is quite small, an increase in period of roughly 25×10^{-9} s each day. Still, one thousand million years from now, a day may be approximately 3 h longer as a result (see Problem 50). It's been suggested that 1.5×10^9 years ago a day may have been only 9 or 10 hours long, with roughly 900 such days per year.

The source of the torque causing the despin is the gravitational interaction of the Moon, which is *external* to the Earth; hence, the planet's angular momentum must change. But if this same torque is envisioned as *internal* to the Earth-Moon system, the angular momentum of that larger system must be conserved. As the planet loses angular momentum, its moon, to which it is coupled gravitationally, must gain angular momentum. The Moon is receding from the Earth and thereby increasing its distance, r, at a rate of 1 foot in about 30 years and, consequently, increasing its orbital angular momentum ($F_C = F_G$, therefore $mv^2/r \propto m/r^2$ and $v^2 \propto 1/r$, and since $mvr = L$ it follows that $L \propto m\sqrt{r}$).

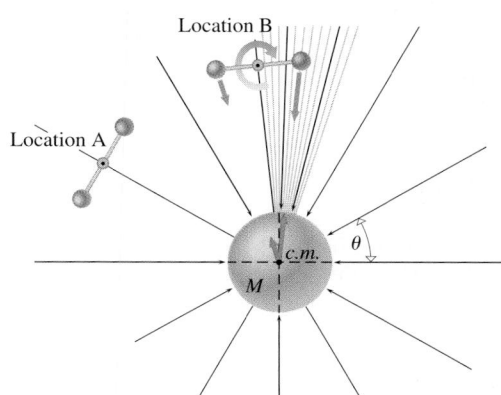

Location B

Location A

θ

c.m.

M

Figure 8.41 A dumbbell at two locations in the anisotropic gravity field of a uniform sphere in an imaginary anisotropic space. At location B, the dumbbell will have an angular acceleration and a nonzero *L*. The torque on the dumbbell is now not equal to the torque on the sphere, and, as a result, Conservation of Angular Momentum will be violated.

Conservation of Angular Momentum is a consequence of the isotropy of empty space—space is the same in all directions; the constants and the laws of physics are independent of orientation. Imagine a laboratory in a spaceship far from any other objects. If the ship is rotated through some arbitrary angle, none of the experiments performed in the lab will change. That this invariance demands Conservation of Angular Momentum is not easy to prove, but we can at least make it reasonable.

Suppose space were anisotropic and therefore the laws of physics were dependent on spatial orientation. Imagine a uniform spherical mass M having a hypothetical gravitational field that varies with angle (i.e., θ in Fig. 8.41). A dumbbell with two identical balls, one red and one blue, is pivoted at its center so that it can turn freely. Holding it at rest at location A in a region of uniform field, the gravitational attraction on both small masses is equal, and the dumbbell remains motionless. But holding it at location B is like putting a paddlewheel in a waterfall. Although initially $L = 0$, as soon as the dumbbell is allowed to, it will start to revolve clockwise. The red ball in the strong-field region will descend, swing past the vertical, and enter the weak-field region, whereupon the blue ball will fall through the strong field. The dumbbell will accelerate around, and Conservation of Angular Momentum is violated.

We assume that in our Universe the converse is true. **Empty space is isotropic (physics is indifferent to orientation), and therefore angular momentum is conserved.**

<hr>

Core Material & Study Guide

THE KINEMATICS OF ROTATION

The basic measure of **angular displacement** θ is defined by

$$\theta = \frac{\ell}{r} \quad \text{or} \quad \ell = r\theta \quad\quad [8.1]$$

where ℓ is the *arc-length*. The units of θ are *radians* such that π rad = 180°. The *average angular speed* ω_{av} equals $\Delta\theta/\Delta t$ and in the limit as $\Delta t \to 0$ that ratio approaches ω, the **instantaneous angular speed**, where

$$v = r\omega \quad\quad [8.10]$$

and ω has units of rad/s. Reread Sections 8.1 (Angular Displacement) and 8.2 (Angular Velocity) and study Examples 8.1–8.3. **Try the WARM-UPS on the CD and then study the WALK-THROUGHS.** The instantaneous angular acceleration, α, equals the limit of $\Delta\omega/\Delta t$ as $\Delta t \to 0$, and

$$a_T = r\alpha \quad\quad [8.14]$$

The units of a_T are m/s^2, whereas those of α are rad/s^2 (p. 245). Study Section 8.3 (Angular Acceleration). Example 8.4 is complicated, but it's something of a summary in itself. Make sure you understand it.

When α *is constant,*

$$\omega_f = \omega_i + \alpha t \quad\quad [8.20]$$

$$\omega_{av} = \tfrac{1}{2}(\omega_i + \omega_f) \quad\quad [8.21]$$

$$\theta = \omega_{av}t = \tfrac{1}{2}(\omega_i + \omega_f)t \quad\quad [8.22]$$

$$\theta = \omega_i t + \tfrac{1}{2}\alpha t^2 \quad\quad [8.23]$$

$$\omega_f^2 = \omega_i^2 + 2\alpha\theta \quad\quad [8.24]$$

These equations are identical to those of translational motion in Chapter 3, where $s \to \theta$, $v \to \omega$, $a_T \to \alpha$, and $t \to t$. Review Section 8.4 (Equations of Constant Angular Acceleration); the piece called "A Free Rolling Wheel" will come up in problems. Examples 8.5 and 6.6 are typical. Remember that there are lots more examples worked out on the CD.

ROTATIONAL EQUILIBRIUM

The **torque** about point O is defined as

$$\tau_0 = Fr_\perp = F_\perp r = Fr \sin \theta \qu\quad [8.25]$$

Examples 8.8 and 8.9 in Section 8.5 (Torque) are designed to help establish what torques are and how to add them. For a rigid body in equilibrium, the sum of the torques about any point O must be zero,

$$\sum \tau_0 = 0 \qu\quad [8.26]$$

which is the **Second Condition of Equilibrium**. Reexamine Section 8.6 (Second Condition of Equilibrium)—Eq. (8.26) is the primary result here. Example 8.10 begins the process of learning how to analyze a simple system in **rotational equilibrium**.

The **center-of-gravity** (*c.g.*) *of an object is that point where the total weight of an object can be imagined to act* (p. 249):

$$x_{cg} = \frac{\sum F_W x}{\sum F_W} \qu\quad [8.27]$$

Section 8.7 (Extended Bodies & the Center-of-Gravity) introduces an idea that will allow us to realistically treat more complicated objects, as in Example 8.12.

Table 8.4 provides free-body diagrams for a number of situations of rotational equilibrium. In every case the beam weighs F_{Wb}. Study each diagram and make sure that you understand it completely.

Table 8.4

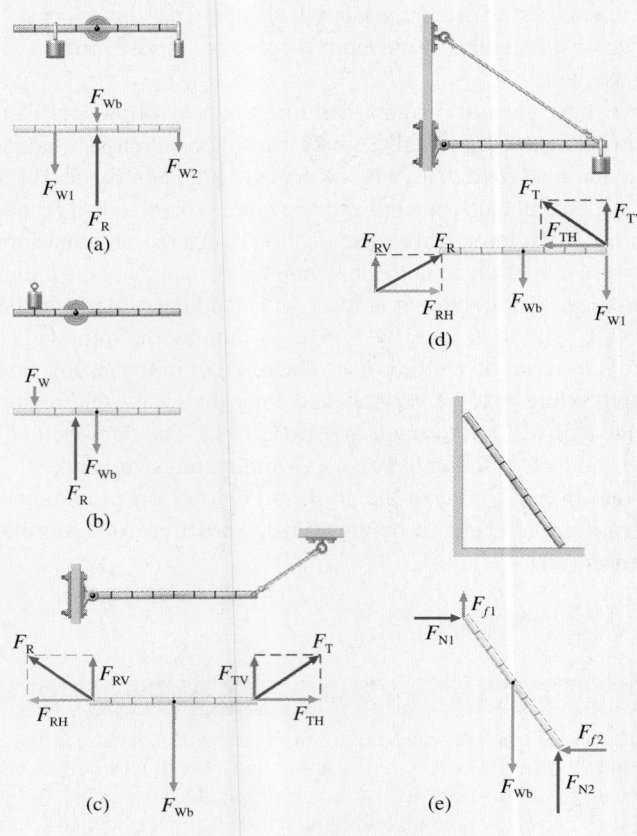

THE DYNAMICS OF ROTATION

Section 8.8 (Torque & Rotational Inertia) begins the study of **rotational dynamics**. For a body composed of a number of particles of mass m_*, the moment-of-inertia about an axis passing through some point O is

$$I = \sum m_* r^2 \qquad [8.30]$$

Table 8.3 lists moments-of-inertia for homogeneous bodies. The rotational equivalent of $F = ma$ is

$$\sum \tau_0 = I\alpha \qquad [8.31]$$

Examples 8.13 and 8.14 explore how torque effects rotation. It is possible to find a point, the **center-of-mass**, where, as far as its motion is concerned, all of the mass of a body can be thought of as concentrated.

Because of the rotation of a body, it possesses rotational kinetic energy

$$KE_R = \tfrac{1}{2}I\omega^2 \qquad [8.32]$$

Example 8.15, in Section 8.9 (Rotational Kinetic Energy), is typical of how this idea is applied.

Section 8.10 introduces the important idea of **angular momentum**, the moment of momentum, defined as

$$L_0 = r_\perp p = r_\perp m. v \qquad [8.33]$$

For symmetrical bodies spinning around symmetry axes

$$L_0 = I_0 \omega \qquad [8.34]$$

The rotational equivalent of the Second Law is

$$\tau = \frac{\Delta L}{\Delta t} \qquad [8.35]$$

All of these equations are identical to those of translational motion in Chapter 4 provided that $m \to I$, $F \to \tau$, and $p \to L$.

In the absence of a net torque, the angular momentum of the system remains constant; this is the law of **Conservation of Angular Momentum** (p. 260). Study Section 8.11 (Conservation of Angular Momentum) and go over Examples 8.17 and 8.18.

Key Terms

arc-length	moment-arm
angular displacement	moment-of-the-force
radian	Second Condition of Equilibrium
angular velocity	sum-of-the-torques
rpm	nonconcurrent forces
angular acceleration	center-of-gravity
tangential acceleration	stable equilibrium
centripetal acceleration	rotational inertia
free rolling	moment-of-inertia
rotational equilibrium	center-of-mass
Law of the Lever	rotational kinetic energy
torque	angular momentum

Discussion Questions

1. Figure Q1 shows a forearm and hand exerting a downward force on a book. Explain the physics of what's happening. How would it be different if the arm were supporting the book? Try it yourself, checking your own muscles.

2. EXPLORING PHYSICS ON YOUR OWN: Why is it easier to hold a long stick horizontally at its midpoint rather than at its end? Position a stick horizontally so that it rests on your two outstretched index fingers held about 2 ft apart. Now slowly move your fingers toward one another. Observe the movement of the stick over your fingers: Why does it alternate, sliding first over one finger and then the other? What is the significance of the point where your fingers finally meet?

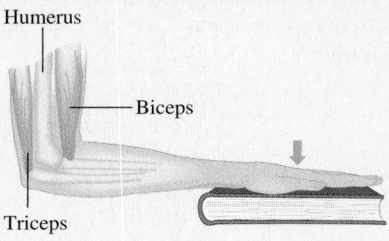

Figure Q1

Figure Q3

3. About 100 years ago, there was a popular toy called a tumbler (Fig. Q3). When toppled, the toy immediately righted itself. The modern equivalents are great inflatable plastic things that kids punch. How do such toys work?

4. EXPLORING PHYSICS ON YOUR OWN: Try to balance an ordinary kitchen bowl or paper cup on the eraser end of a pencil, first at its middle on the inside and then on the bottom outside of the bowl. Explain the difference.

5. When a car turns, do the wheels on its right side have the same angular velocity as those on its left side? What does the answer suggest about mounting the wheels rigidly on a common axle?

6. Is a truck loaded with several tons of feathers more likely to tip over in a high wind than one carrying the same weight of bricks? Explain. Occasionally, one sees a narrow van filled with cargo carrying a heavy load on its roof. Why is that dangerous, and how can the risk be minimized? Explain the significance of the narrowness of the van.

7. People have been predicting that a rise in the temperature of the planet would result in the melting of the polar ice caps. Would such massive melting, in turn, have any effect on the length of the day? Explain. Apparently the storage of large amounts of water in reservoirs in the Northern Hemisphere has already had an effect.

8. Figure Q8 shows a gymnast in three postures, spinning about the frontal axis passing through her *c.m.* Which has the greatest moment of inertia about that axis and which has the least? Explain. Draw a rectangular parallelepiped in the shape of a deck of playing cards. Fix the *c.m.* at its geometric center. Draw, passing through the *c.m.*, the three axes perpendicular to the three pairs of faces, numbering them 1, 2, and 3 as the area of each increases. About which axis is the moment of inertia a maximum? A minimum?

9. A diver shown in Fig. Q9 leaves the springboard an instant after being projected by it into the air. Resolve the normal force on his feet into two components, one along the line through his *c.m.* and one perpendicular to it. Discuss the motion that results from each component.

Figure Q8

(a)

(b)

(c)

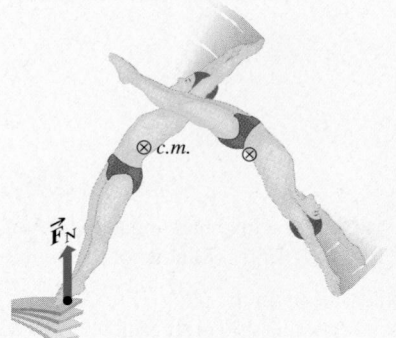

Figure Q9

10. EXPLORING PHYSICS ON YOUR OWN: Imagine that you spin two eggs, one raw and one hard-boiled, on a smooth flat surface at the same rate. Which would come to a stop first and why? Again, spin the raw egg but now stop it just for an instant with a fingertip and then let it go. Why does it start spinning again?

11. Consider the extraordinary situation of a polished sphere rolling without slipping on a flat horizontal surface with a constant angular speed. What can you say about the force of friction on the sphere?

12. Suppose you were playing pool with balls floating in space. How would you hit a motionless cue ball to propel it forward in a straight line, making it rotate as if it were in contact with a table? (By the way, on Earth where the game is only played *on* a table, this is known as a "follow shot" because the cue ball follows after the ball it hits.)

13. Explain why most helicopters have a small additional propeller at the tail end that revolves in a vertical plane.

14. Imagine Superman in a black leather jacket sitting on a motorcycle floating out in space. Describe what would happen as he gunned the engine (presuming it was modified to run out there).

15. If a person throws a ball, its angular momentum is increased, but the person doesn't rotate. Is angular momentum conserved in the ball-person system? What, if any, external torque acts on the person? Is the angular momentum of the ball-person-Earth system conserved? What, if anything, happens to the motion of the planet? What would happen to someone floating in space who threw a ball in an overhand pitch?

16. An astronaut floating in space has zero angular momentum initially. He then proceeds to rotate both his extended arms around an axis passing across his chest through his two shoulders. While looking at his right side, you see his arms revolving counterclockwise. What motion, if any, will be executed by the rest of his body? What will happen when he stops?

17. One of the best ways of transporting the magnetic tape in a recorder is by a direct-drive system where a motor turns the capstan that moves the tape (see Problem 34). Such motors are often fitted with heavy flywheel housings that revolve with the shaft. Why do you think this might be done? (See Fig. Q17.)

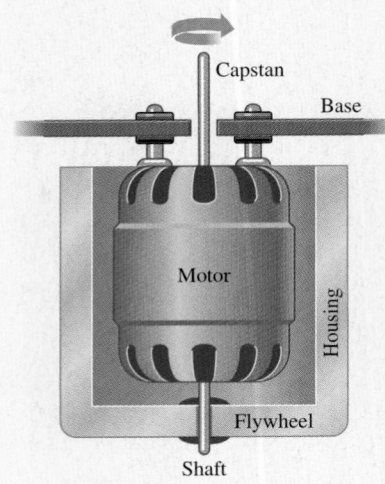

Figure Q17

18. When a rear-wheel drive car or motorcycle rapidly accelerates, it tends to rotate about its *c.m.* so that the front end noses upward. What forces are acting *on* the vehicle? Make a sketch and explain what is happening. What would happen if the front brakes engaged first and locked during a quick stop?

19. A motorboat generally noses up out of the water, and the more it does so, the faster it travels. Make a sketch locating the *c.m.* of the motorboat and the level of its propeller. Draw in all the forces acting on the boat while it is traveling at a constant speed—don't forget the upward force of the water on the boat and the friction on the hull bottom. Explain what's happening.

20. Imagine a gradual, inwardly spiraling groove cut just into the surface of a flat tabletop. Now, suppose we propel a small sphere into the channel with a speed v so that it rolls toward the center point O (with no appreciable losses due to friction). What, if anything, will happen to its speed and why?

21. A physics student, with arms stretched out to either side, holds a weight in each hand. While sitting on a bar stool, he is swung around by a helpful friend. What will happen to his angular speed, if anything, when he lets the weights fall to the floor? Think this one through carefully.

22. In the past, many satellites have had to be put into orbit spinning more rapidly than was operationally desirable. A cute despin system consists of two masses each (known as a yo-yo) tied to a long cable. These are wound around the craft, and the weights are held in place on opposite sides. On command, the weights are released, and they unwind as the satellite spins until the ends of the cables unhook and each weight, trailing its cable, flies away into space. Explain how this system works to reduce the satellite's spin rate—usually by a factor of about 10.

23. Explain the meaning of each of the Key Terms on page 266.

24. The motionless beam in Fig. Q24 is supported at its center on a frictionless pin. Is it in equilibrium? Does the fact that the beam is tilted have any effect on the state of equilibrium? How does it affect the sum-of-the-torques equation? What would happen if you lowered the right side of the beam through, say, 60° or 70° bringing the system to rest at a new location? Would it stay there or swing back to where it was originally?

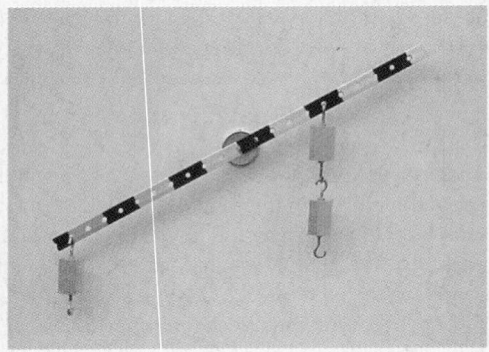

Figure Q24

Multiple Choice Questions

1. What is the equivalent of 25 degrees in radians and 25 radians in degrees? (a) 1.4×10^3 radians and 0.44 degrees (b) 0.44 radians and 1.4×10^3 degrees (c) 25 radians and 1.4×10^3 degrees (d) 0.44 degrees and 0.44 radians (e) none of these.

The next three questions pertain to Fig. MC2 which shows a uniform 100-cm long bar of weight F_w balanced in equilibrium on a frictionless pivot.

2. Where does the weight of the bar act? (a) 50 cm to the right of the pivot (b) 10 cm to the left of the pivot (c) 30 cm to the right of the pivot (d) at the pivot (e) none of these.

3. What is the value of the upward reaction force on the bar at the pivot? (a) F_w (b) $2F_w$ (c) $3F_w$ (d) $4F_w$ (e) none of these.

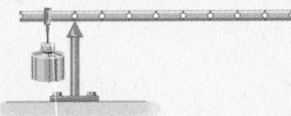

Figure MC2

4. What is the value of the load hanging at the left of pivot? (a) F_w (b) $2F_w$ (c) $3F_w$ (d) $4F_w$ (e) none of these.

5. The motionless beam in the photo (Fig. MC5) is 55 cm long and is supported at a frictionless pivot. If each of the two hanging objects

on the left weighs 1.0 N, how much does the wad of clay weigh? (a) 1.0 N   (b) 2.0 N   (c) 3.0 N   (d) 4.0 N   (e) none of these.

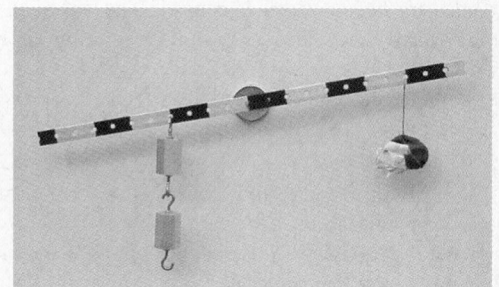

Figure MC5

6. The beam in the photo (Fig. MC5) weighs 0.8 N. Using the results of the previous question, what is the value of the upward normal force exerted by the pivot on the beam?   (a) 1.8 N   (b) 2.8 N   (c) 3.8 N   (d) 4.8 N   (e) none of these.

7. The motionless beam in the photo (Fig. MC7) is supported at a frictionless pivot. If each of the two hanging objects on the left weighs 1.0 N, how much does the wad of clay weigh?   (a) $\frac{2}{8} \times 5$ N   (b) $\frac{8}{2} \times 5$ N   (c) $\frac{5}{2} \times 8$ N   (d) $\frac{2}{5} \times 8$ N   (e) none of these.

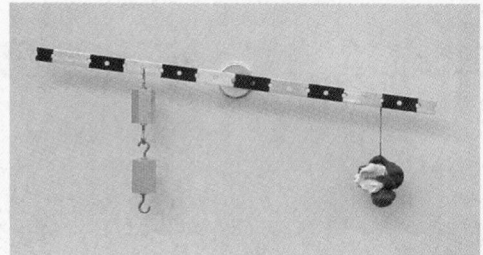

Figure MC7

8. The lightweight pivoted bar in Fig. MC8 will be in rotational equilibrium when a 200-N force acts   (a) down at D (b) up at B (c) down at E or up at C (d) up at C only   (e) none of these.

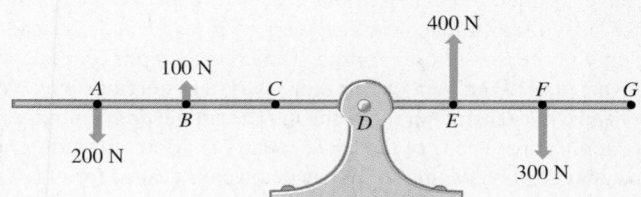

Figure MC8

9. Given a body acted upon by several forces, if the sum of those forces is exactly zero, the body   (a) may be in both translational and rotational equilibrium   (b) must be in both translational and rotational equilibrium   (c) cannot be in both translational and rotational equilibrium   (d) not enough information to come to any conclusion   (e) none of these.

10. In the photo of Chinese acrobats on p. 252, the combined $c.g.$ of the pole and the two suspended people   (a) is on a vertical line passing through the third man's left shoulder   (b) is to the left of the third man   (c) is to the right of the third man   (d) is on a horizontal line passing through the third man's shoulder   (e) none of these.

11. If an object is divided into two parts by slicing it vertically straight through its $c.g.$, the two pieces   (a) must have the same weight   (b) must have different weights   (c) can have different weights   (d) must have the same mass   (e) none of these.

12. Two astronauts push on either end of a satellite floating in space. If the craft is then in translational and rotational equilibrium, it is required that   (a) both forces must pass through the satellite's $c.g.$ (b) the two forces be equal, coplanar, and pass through the $c.g.$ (c) there be two forces, they need not be equal, but they must be antiparallel   (d) the forces must be equal, colinear, antiparallel, and pass through the $c.g.$ (e) none of these.

13. Someone places a 6.0-m-long steel rod on a rock so that one end is under a baby moose weighing 2.0 kN. The person pushes down on the other end of the rod with a force of 400 N, and the moose is held in the air at rest. The rock was   (a) 1.0 m from the moose   (b) 5.0 m from the moose   (c) 1.0 m from the person   (d) 6.0 m from the person   (e) none of these.

14. In analyzing the condition of rotational equilibrium, one takes the sum of the torques about an axis   (a) which must pass through the body's $c.g.$ (b) through which the line-of-action of all the forces must pass   (c) passing through the center of the body   (d) located anywhere   (e) none of these.

15. Every point-mass on a rigid body that is rotating has   (a) the same angular speed and angular acceleration   (b) a different angular speed but the same angular acceleration   (c) the same linear and angular speed   (d) the same angular speed and the same linear acceleration   (e) none of these.

16. At what latitude on Earth would you have the greatest angular velocity?   (a) 0°   (b) 45° N   (c) 45° S   (d) 60° N   (e) none of these.

17. A body free of support and experiencing a zero net applied force   (a) must be accelerating   (b) must only be translating at a constant speed   (c) must only be rotating about its $c.m.$ (d) may be at rest or moving at a constant speed and may be rotating about its $c.m.$ as well   (e) none of these.

18. A flying saucer in a TV show banks and just blocks out the disk of the Moon (which subtends an angle of 0.009 rad) when a nearby radar device determines its range to be 10 km away. How big was the craft?   (a) 9 m   (b) 0.09 km   (c) 0.9 km   (d) 0.9 m   (e) none of these.

19. Two slender uniform rods have the same cross section and the same mass. If one is light plastic and the other lead, what can you say regarding their respective moments of inertia about an axis through the very center perpendicular to each rod's long axis?   (a) Nothing.   (b) The moments of inertia of the rods are equal.   (c) The lead rod must have a greater moment of inertia.   (d) The plastic rod must have a greater moment of inertia.   (e) None of these.

20. The moment-of-inertia of a spinning body about its spin axis depends on its   (a) angular speed, shape, and mass   (b) angular acceleration, mass, and spin-axis position   (c) mass, shape, spin-axis position, and size   (d) mass, size, shape, and speed   (e) none of these.

21. When a single force is applied to a free body floating in space, it is possible that there will be a resulting change in   (a) only its

translational motion   (b) both its translational and rotational motions   (c) both (a) and (b) are possible   (d) only its rotational motion   (e) none of these.

22. For the net force acting on a body to result in purely linear motion   (a) it must be zero   (b) it must pass through the *c.m.*   (c) it must be less than the weight of the body   (d) it must not pass through the *c.m.*   (e) none of these.

23. Two straight inclined grooved tracks are lined up next to one another. Number 1 track is coated with a slippery film and number 2 is not; otherwise, they are identical. Two steel balls are placed one on each track and released so that ball-1 slides down and ball-2 rolls all the way.   (a) Both reach the bottom at the same time.   (b) What happens depends on the exact tilt of the tracks and so we cannot say who wins.   (c) Ball-1 wins always.   (d) Ball-2 wins always.   (e) Ball-1 wins as often as ball-2.

24. A solid cylinder, a solid sphere, and a hoop all of the same mass but different radii, roll without sliding down an inclined plane. The body that gets to the bottom first will invariably be   (a) the hoop   (b) the sphere   (c) the cylinder   (d) they all arrive together   (e) not enough information.

25. The dimensions of angular momentum are   (a) $[ML^2T]$ (b) $[M^{-1}L^2T]$ (c) $[ML^{-2}T]$ (d) $[ML^2T^{-1}]$ (e) $[ML^{-2}T^{-1}]$.

26. Low-altitude artificial Earth satellites traversing wisps of atmosphere are initially slowed down by air-drag so that they cannot remain in their original orbits and inevitably fall back to the planet. Spiraling down ever closer to the Earth, they usually break up and burn up before hitting ground. The speed of such a device   (a) increases as it descends and its angular momentum is not conserved   (b) decreases as it descends and its angular momentum is conserved   (c) increases and then decreases, conserving its angular momentum   (d) must remain constant, conserving its angular momentum   (e) none of these.

27. At a given moment, the net angular momentum of a system of two particles about a certain point is zero. Accordingly, we know that the particles   (a) must be at rest   (b) must be moving on the same circle in opposite directions   (c) must be moving on straight perpendicular paths   (d) must have the same linear momentum   (e) none of these.

28. A particle of mass *m* and speed *v* has zero angular momentum with respect to a point *A* in space. We know that the particle   (a) has already passed through *A*   (b) will never pass through *A*   (c) is moving along a line that is perpendicular to the plane containing it and *A*   (d) is moving along a line that passes through *A*   (e) will go somewhere, but we don't have enough information to comment.

For more Multiple Choice Questions with answers click on WARM-UPS in CHAPTER 8 on the CD.

Suggestions on Problem Solving

1. When the forces acting on a body in equilibrium are nonconcurrent, at least one application of the sum-of-the-torques equation will be required. Use $\sum \tau_0 = 0$ sparingly to avoid nonindependent torque equations.

2. Always draw a picture of the entire system. Then draw a free-body diagram of each segment of the setup to be analyzed. Guess at the directions of the internal forces, and if you are wrong, they will turn out negative. In time, you should be able to guess correctly by recognizing what the directions of the forces and torques must be to sustain equilibrium.

3. When finding torques, consider the point about which they will be computed. *Avoid the unknown forces you are not interested in by taking torques about points through which those forces pass.*

4. To determine the signs in a sum-of-the-torques equation, assign a direction (clockwise or counterclockwise) as positive. Then imagine an axis perpendicular to your drawing running through the point about which you are taking the torques. For each force, determine how it, acting alone, would rotate the system about that axis.

5. Watch out for those unit conversions! Do not confuse rpm with rev/s or rad/s. Remember the factor of 2π. And be especially careful about signs. A decreasing ω means a negative α.

6. Remember that F, τ, p, and L are dynamical variables; seeing any of these in a problem should call to mind certain specific equations from the second half of the chapter. For example, it's not very likely that you will ever come across a numerical problem that talks about angular momentum and does not require either Eq. (8.34) or Eq. (8.35). Similarly, the mere mention of the word *torque* suggests angular acceleration and vice versa, although a time-varying angular momentum should come to mind as well. It's certainly possible to have a torque problem without ever seeing the word explicitly. Such "disguised" torque problems work by giving $\vec{F}$ and *r* instead—when you see these, think torque. The crucial step is to relate the "Given" and "Find" via one or more familiar equations once you recognize the general type of problem. The simple problems require one application of one of the basic equations. More difficult problems may require solving for things that weren't asked for.

Problems   ✦   Coordinated Problems   ✦   Progressive Problems   ✦   Solutions

STUDY GUIDE   **1. Coordinated Problems:** The three problems within each magenta-colored grouping are solvable in similar ways. Note that the first of these always has a hint; moreover, its solution is provided in the back of the book. *Work out each of these sets; they'll strengthen technique and build confidence.* **2. Progressive Problems:** The problems introduced in blue unfold step-by-step carrying along the analysis in a more suggestive way than is customary. *Work out all of these; they'll guide you through the analytic process and help develop problem-solving skills.* **3. Worked-Out Solutions:** Studying worked-out solutions is an important part of learning how to solve problems. Accordingly, additional *solutions* to a number of model problems are given below. *Make sure you understand each of them before you go on to the next problem.* **4.** Also provided in the back of the book are the *Answers* to all odd-numbered problems, as well as worked-out *solutions* to

those with boldface numbers. Problem numbers in italic indicate that a solution appears in the Student Solutions Manual.

SECTION 8.1: ANGULAR DISPLACEMENT

1. [I] What does 1.000 0° equal in radians?

2. [I] How many radians are swept out by the minute hand of a clock as it goes from 12:00 noon to each of the following times: 1:45 P.M., 2:30 P.M., and 3:17 P.M.?

3. [I] What does 3.141 6 rad equal in degrees?

4. [I] What does 30.0° equal in radians?

> **SOLUTION:** $30.0°/180° = x/(\pi \text{ rad})$; $x = \frac{\pi}{6}$ rad $= 0.524$ rad, which agrees with the fact that $90° = \frac{\pi}{2}$ rad.

5. [I] (a) What is the equivalent of 45.0° in radians? (b) Express your answer as a fraction in terms of π.

6. [I] (a) What is the equivalent of 240° in radians? (b) Express your answer as a fraction in terms of π.

7. [I] What is the equivalent of $3\pi/2$ rad in degrees?

8. [I] What is the equivalent of $13\pi/6$ rad in degrees?

9. [I] If a point on the circumference of a rotating wheel passes through 200 rad, how many times has it revolved? [*Hint: Once around is 2π rad.*]

10. [I] A race car on a circular track goes 2.88 times around in 15 minutes. Through how many radians did it pass in that time, as measured by an observer at the center of the track?

11. [I] A model plane at the end of a control line circles at a constant speed 10.6 times around in 50.0 s. Through how many radians does it fly in 25.0 s?

12. [I] A 25-cent coin is roughly 2.5 cm in diameter. How far away should it be held if it is to subtend the same angle as does the Moon?

13. [I] A well-adjusted laserbeam is emitted as an exceedingly narrow diverging cone of light with a typical spread angle of about 8×10^{-4} rad. How large a circle will be illuminated on the surface of the Moon by such a device ($r_{\oplus ☽} = 3.8 \times 10^8$ m)?

14. [I] The 10-point center-circle bull's eye on the official slow fire pistol target is $\frac{7}{8}$ in. in diameter. What angle does it subtend as seen by a shooter 50 ft away? How does this compare with the apparent size of the Moon?

15. [II] THIS PROBLEM EXAMINES THE RELATIONSHIP BETWEEN ANGULAR DISPLACEMENT AND ARC-LENGTH. A carousel has a diameter of 10.0 m and is revolving clockwise. In a certain time interval a youngster sitting at the very edge is carried through an angular displacement of −160°. (a) What is the radius of the circular arc along which she moves though space? (b) Show that during that interval the carousel whirls her along an arc-length of −14.0 m.

16. [II] THIS PROBLEM EXAMINES THE RELATIONSHIP BETWEEN ANGULAR DISPLACEMENT AND ARC-LENGTH. A passenger in a fixed car on a Ferris Wheel is 12.0 m from the center of rotation. At that moment his girlfriend on the ground, looking at the Wheel as if it were the face of a clock, sees him to be at the 2 o'clock position. And he's moving counterclockwise on the last turn of the ride. (a) How many degrees does each "hour" correspond to? (b) Verify that that's equivalent to $\frac{\pi}{6}$ rad. (c) Through how many radians must he travel to get to the ground? (d) What is the distance (arc-length) he must traverse before the ride ends?

17. [II] Assume the human eye can resolve details separated by about 1.0 minute of arc (where there are 60 minutes per degree). How far away must you sit from a TV set (whose picture is 30-cm high) in order to no longer distinguish the horizontal scan lines? (Typically, 525 horizontal lines make up a modest TV picture.)

18. [II] An ant positioned on the very edge of a Beatles record that is 26 cm in diameter revolves through an angle of 100° as the disk turns. How far does he travel?

19. [II] If a disc 30 cm in diameter rolls 65 m along a straight line without slipping, (a) How many revolutions would it make in the

process? (b) Through what angular displacement would a speck of gum on its rim be carried?

> **SOLUTION:** (a) $N = L/2\pi r = (65 \text{ m})/2\pi(0.15 \text{ m}) = 69$ turns. (b) For each complete turn, the wheel revolves through 2π rad; therefore $\theta = 4.3 \times 10^2$ rad.

20. [II] The bob at the end of a pendulum 100-cm long swings out an arc 15.0 cm in length. Find the angle in radians and degrees through which it moves. Check your answer by determining the fraction of the complete circle to which this corresponds and then taking that fraction of 2π.

21. [II] Determine, via geometry, the angle in radians subtended by a 2.00-m-wide car seen head-on from a distance of 1.00 m away. What would that angle be as seen standing 100 m? Compute this last part exactly (using geometry and the straight-line width) and approximately (taking the width as an arc-length). Now, go back to the first part and do it with the same approximation. Compare your results.

SECTION 8.2: ANGULAR VELOCITY
SECTION 8.3: ANGULAR ACCELERATION

22. [I] What is the equivalent of 1.00 rev/min in rad/s?

23. [I] If a point on a wheel turns through 45.0° in 0.050 s, what is its angular speed? [*Hint: How many radians does it turn through per second?*]

24. [I] A radial line is drawn on a disc that is subsequently set revolving about its center. If the line sweeps through 480° in 0.25 s, how fast is the disc spinning in rad/s?

25. [I] The angular speed of a wheel is to be determined by affixing a tiny mirror to the circumference and recording the returning light bounced off it as it spins past a laserbeam. If 100 return pulses are detected in 0.020 0 s, what is the angular speed of the wheel?

26. [I] What is the average angular speed of the Earth in its orbit? Take a year to be 365.24 days. Give your answer to three significant figures.

27. [I] Take the Earth to be moving in a circular orbit ($r_{\oplus \odot} = 1.495 \times 10^{11}$ m) about the Sun. Compute its average linear speed.

28. [I] When hit with a driver (having a head sloped at 10°), a golf ball sails off rotating at about 50 rev/s. Since the ball has a diameter of 4 cm, how fast is a point on its surface moving just as a result of the spin?

29. [I] A well-trained runner can swing a leg forward in the vertical plane around a horizontal axis through the knee at a rate of about 700° per second. How much is that in rad/s?

30. [I] What is the angular speed of the 10-cm-long second hand of a clock? What is ω for the same-length minute hand?

31. [I] THIS PROBLEM DEALS WITH AVERAGE ANGULAR ACCELERATION. A large horizontal disc, which can revolve about a vertical axis, begins at rest and attains an angular speed of 25.0 rad/s. It does this in a time of 2.00 s. (a) What is the disc's initial angular speed? (b) What is its average angular acceleration?

32. [I] THIS PROBLEM DEALS WITH AVERAGE ANGULAR ACCELERATION. The propeller on an old airplane starts from rest and accelerates at an average rate of 4.0 rad/s² for 1 minute and 20 seconds. (a) What is the propeller's initial speed? (b) How long, in seconds, was it

accelerating?   (c) How fast will it be turning after 1 minute and 20 seconds?

33. [I] Thirty seconds after the start button is pressed on a big electric motor, its shaft is whirling around at 500 rev/s. Determine its average acceleration.

34. [I] A golf ball hit by the tilted face of a number 7 iron flies off spinning at 130 rev/s. If the impact lasts 1.0 ms, what is the average angular acceleration?

SOLUTION: $\alpha_{av} = \Delta\omega/\Delta t$

$$\alpha_{av} = \frac{(130\ \text{rev/s})(2\pi\ \text{rad/rev})}{1.0 \times 10^{-3}\ \text{s}} = 8.2 \times 10^5\ \text{rad/s}^2$$

35. [I] A steam engine is running at 200 rpm when the engineer shuts it off. The friction of its various parts produces torques that combine to decelerate the machine at 5.0 rad/s/s. How long will it take to come to rest?

36. [I] What average angular acceleration is required to stop a turbine blade spinning at 20 000 rpm in 50.0 s?

37. [II] When vinyl records were in their prime, most good players were fitted with a band of equal-spaced radial lines on the edge of the rotating platter on which the record sat. As the platter turned, a small lamp flickered at an exact rate. When the lines appeared to stand still, the platter was turning at the right speed. Suppose the light flashed on 60 times per second and the radial lines occurred every 3.33°. What's the proper angular speed of the platter in rpm?

38. [II] The motor in Fig. P38 is mounted with a 20-cm-diameter pulley and revolves at 100 rpm. We would like to drive the attic fan so that no point on each of its four 1.0-m-long blades exceeds a speed of 7.0 m/s. What size pulley should it have?

Figure P38

39. [II] A train rounds a turn 304.8 m (i.e., 1000 ft) in radius while traveling at a speed of 8.9 m/s (i.e., 20 mi/h). Determine its angular speed and centripetal acceleration. Give the answer in SI units.

40. [II] Imagine two ordinary gears of different diameters meshed together, with the larger being the driver. If the larger gear has 100 teeth around its circumference and rotates at 5.0 rad/s, the smaller gear, which has only 25 teeth, will rotate at what speed?

41. [II] The pulley on the left of Fig. P41 is 0.6 m in diameter and rotates at 1.0 rpm. It is attached by a twisted belt to the 0.2-m-diameter hub of a compound pulley whose outer diameter is 0.8 m. If the driver turns clockwise, find the velocity of the suspended body.

42. [II] Figure P42 shows the structure of a magnetic tape cassette.

The tape is transported across the heads at a constant speed of $1\frac{7}{8}$ in/s. Contrary to popular belief, this is not done by driving the feed and take-up hubs. Why not? Actually, the tape is moved by being squeezed between a precisely rotating shaft—the capstan—and a soft pinch roller. If the capstan has a diameter of 3/32 in., how fast must it turn? Give the answer in SI units.

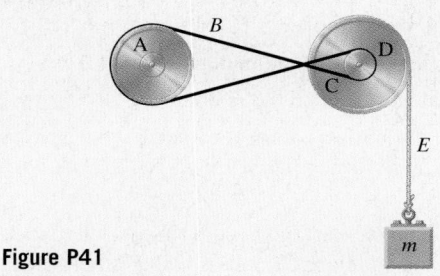

Figure P41

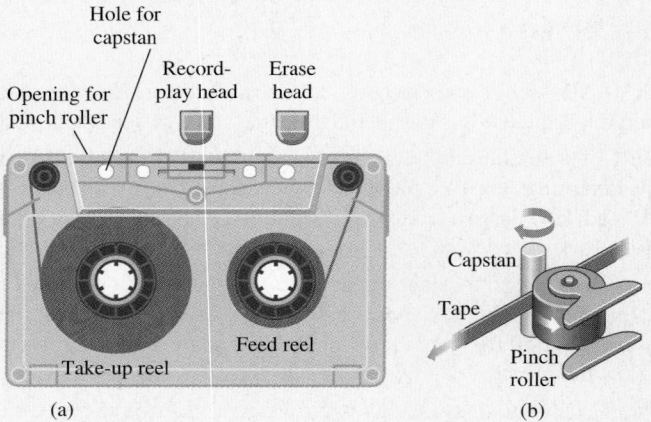

Figure P42

43. [II] New York City is traveling around at a tangential speed of about 353 m/s (i.e., 790 mi/h) as the Earth spins. Assuming the planet is a sphere of radius 6371 km, how long is the perpendicular from the city to the spin axis? Compute the city's latitude (i.e., the angle measured above or below the equator).

44. [II] Two spacecraft are freely coasting in circular orbits, one 2 km beneath the other. Referring to the conclusions of Chapter 5, compare their angular speeds and show that the lower ship will pull ahead of the upper one.

45. [II] A gear train consists of five meshed gears arranged so that the first (20 teeth) drives the second (80 teeth), which drives the third (40 teeth), and so on. If the first is mounted on the shaft of a 1500-rpm motor rotating clockwise, at what speed and in what direction will the fourth (25 teeth) and fifth (75 teeth) gears rotate? Figure out a general rule for the speed and direction of the last gear in such a train.

46. [II] The pegged wheel in Fig. P46 was a precursor of the modern cam. Here it's shown operating a trip hammer the same way it was used to crush ore and forge metal 1000 years ago in Europe. At what speed should the central drive shaft be turned, presumably by a waterwheel, for the hammer to pound away at one smash every 3 s?

Trip hammer

Drive shaft

Pegged wheel

Figure P46

47. [III] A cylindrical wheel is rotating with a constant angular acceleration of 4.0 rad/s². At the instant it reaches an angular speed of 2.0 rad/s, a point on its rim experiences a total acceleration of 8.0 m/s². What is the radius of the wheel?

48. [III] A cord is wrapped around a 0.24-m-diameter pulley and the free end is tied to a weight. That weight is allowed to descend at a constant acceleration. A stopwatch is started at the instant when the weight reaches a speed of 0.02 m/s. At a time of $t = 2.0$ s, the weight has dropped an additional 0.10 m. Take $y = 0$ to occur at $t = 0$ and show that

$$y = (0.02t + 15 \times 10^{-3}t^2) \text{ m}$$

Prove that for any point on the circumference

$$a_C = (33 \times 10^{-4} + 10 \times 10^{-3}t + 75 \times 10^{-4}t^2) \text{ m/s}^2$$

and $$a_T = 0.03 \text{ m/s}^2.$$

49. [III] If the hand in Fig. P49 pulls down on the rope moving it at a speed of 2.0 m/s, what will be the resulting velocity of the hanging mass?

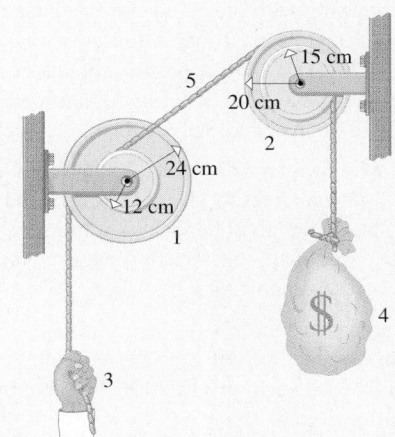

15 cm

5

20 cm

24 cm

12 cm

2

1

$

4

3

Figure P49

50. [III] Because of tidal friction, the Earth is despinning. Its period is increasing at a rate of roughly 25×10^{-9} s per day. Approximate the corresponding angular acceleration in rad/s². Compute the decrease in the Earth's angular speed that would exist one thousand million years from now, assuming α is constant. What will its new period be?

SECTION 8.4: EQUATIONS OF CONSTANT ANGULAR ACCELERATION

51. [I] A videodisc revolves at 1800 rpm beneath a laser read-out head. If the beam is 12 cm from the center of the disc, how many meters of data pass beneath it in 0.10 s?

52. [I] A variable speed electric drill motor turning at 100 rev/s is uniformly accelerated at 50.0 rev/s² up to 200 rev/s. How many turns does it make in the process?

53. [I] Placed on a long incline, a wheel is released from rest and rolls for 30.0 s until it reaches a speed of 10.0 rad/s. Assume its acceleration is constant, what angle did it turn through?

54. [I] A large motor-driven grindstone is spinning at 4 rad/s when the power is turned off. If it rotates through 100 rad as it uniformly comes to rest, what is its angular acceleration?

55. [II] A ring-shaped space satellite is revolving at 100 rev/min when its despin rockets are fired and it decelerates at a constant 2.0 rad/s². How long must the rockets fire in order to bring the craft's rotation to a stop? Through what angle will it turn in the process? [*Hint: The change in the speed over the time is the acceleration. And that will give you the time. There are several ways to get the angular displacement. Watch out for signs; t is positive.*]

56. [II] An antique spring-driven Victrola phonograph plays recordings at 78 rpm. At the end of each record the arm hits a lever that activates a brake that brings the platter to rest in about 1 s. Through how many radians does it turn in the process of stopping?

57. [II] A wheel is revolving at 20.0 rad/s when a brake is engaged and the wheel is brought to a uniform stop in 15.92 revolutions. How long did it take to stop the wheel and what was the deceleration?

58. [II] A bicycle with 24-in.-diameter wheels is traveling at 10 mi/h. At what angular speed do the wheels turn? How long do they take to turn once around?

59. [II] THIS PROBLEM DEALS WITH ANGULAR ACCELERATION. A large horizontal disc, which can revolve about a vertical axis, begins at rest and attains a speed of 25.0 rad/s. It does this while turning around exactly 10 times. (a) What was its average angular speed? (b) What was its angular displacement? (c) How long did it take to go the 10 turns? (d) Show that the disc had an average angular acceleration of 4.98 rad/s².

60. [II] THIS PROBLEM EXAMINES ANGULAR DISPLACEMENT. An electric motor attached to a set of fan blades starts from rest and accelerates at an average rate of 4.01 rad/s². After exactly 1 minute and 2 seconds the fuse blows and the motor coasts to a stop 120 s later. (a) What was the maximum angular speed it reached? (b) What was its average speed during the positive acceleration phase? (c) What was its average speed during the negative acceleration phase? (d) What was its angular displacement during the positive acceleration phase? (e) What was its angular displacement during the negative acceleration phase? (f) How many revolutions did the blades make in total?

61. [II] A 1.0-m-diameter nontranslating disk is made to accelerate from rest up to 20 rpm at a rate of 5 rad/s². Through how many turns will it revolve in the process? How far will a point on its rim travel while all this is happening?

62. [II] A bicycle has its gears set so that the front sprocket, which is driven by the pedals, has 52 teeth whereas the rear one, to which

it is attached via the chain, has 16 teeth. If it's pedaled at a most efficient 50 rpm and has 24-in. wheels, how fast is the bicycle moving? Give the answer in SI units.

63. [II] A chimp sitting on a yellow unicycle with a wheel diameter of 20 in. is pedaling away at 100 rpm. How fast does it travel? Give the answer in SI units.

64. [II] A wheel of radius R is freely rolling along a flat surface at an angular rate of ω. Determine, analytically, the linear speed (with respect to the ground at any instant) of   (a) the center of the wheel   (b) the topmost point, and   (c) the point in contact with the ground.

65. [II] An electric circular saw reaches an operating speed of 1500 rpm in the process of revolving through 200 turns. Assuming the angular acceleration is constant, determine its value. How long does it take to get up to speed? Now redo the problem using the average angular speed.

66. [III] If the hand in Problem 49 uniformly accelerates the rope downward from rest at $1\frac{1}{3}$ m/s^2, how fast will the hanging mass be moving 1.0 s later?

67. [III] A 30-cm-diameter turntable platter of a record player is attached, by a belt (to reduce vibrations), to a motor-driven pulley 2.0 cm in diameter (Fig. P67). Determine the motor's angular acceleration if the platter is to reach $33\frac{1}{3}$ rpm in 6.0 s. How many rotations will it make before reaching operating speed?

Figure P67

68. [III] An antique rim-drive record player has a 6.0-cm-diameter rubber wheel rotating in contact with the inner surface of the 30-cm-diameter platter (Fig. P68). If the turntable is to go from $33\frac{1}{3}$ rpm to 78 rpm in 3.0 s, what must be the angular acceleration of the drive wheel?

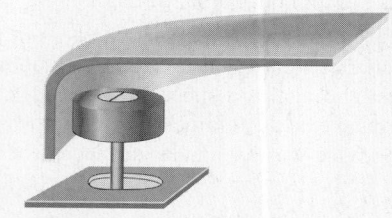

Figure P68

SECTION 8.5: TORQUE
SECTION 8.6: SECOND CONDITION OF EQUILIBRIUM

69. [I] A force of 100 N is applied perpendicularly to the middle of a meter stick; the stick lies north-south, the force points east. What is the size and sense of the torque around each end of the stick?

70. [I] Determine the torques about the elbow and shoulder produced by the 20-N weight in the outstretched hand in Fig. P70.

71. [I] Compute the torque about the pivot O, generated by the 100-N force on the gearshift lever in Fig. P71.

72. [I] Figure P72 shows a tendon exerting an 80-N force on bones in the lower leg. Assume the tendon acts horizontally 0.06 m from the knee pivot. Draw a simplified version of the system, and compute the tendon's torque about the pivot.

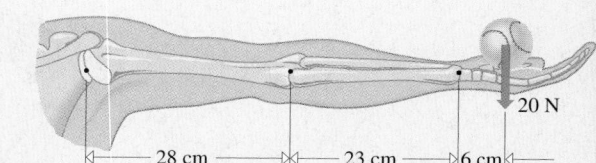

Figure P70

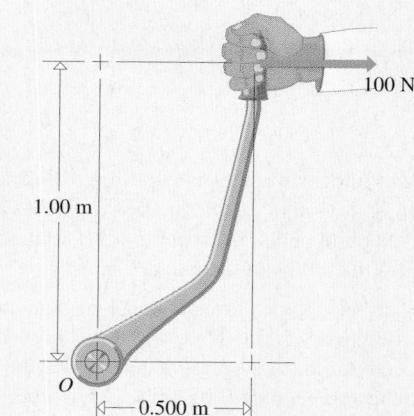

Figure P71

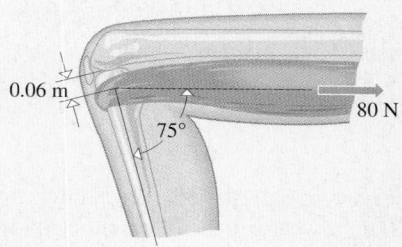

Figure P72

73. [I] While working with precision components, especially out in space, it is often necessary to use a torque wrench, a device that allows the user to exert only a preset amount of torque. Having dialed in a value of 35.0 N·m, what maximum perpendicular force should be exerted on the handle of a wrench 25.0 cm from the bolt?

74. [I] Harry, who weighs 320 N, and 200-N Gretchen are about to play on a 5.00-m-long seesaw. He sits at one end and she at the other. Where should the pivot be located if they are to be balanced? Neglecting the weight of the seesaw beam, what is the reaction force exerted by the support on it?

75. [I] THIS PROBLEM DEALS WITH ROTATIONAL EQUILIBRIUM. The uniform beam (whose weight acts at its center), and the mass on the left in Fig. P75, each weighs 20 N. The central pivot is essentially frictionless, and the system is in equilibrium.   (a) Draw a free-body diagram of the beam.   (b) How much does the mass on the right weigh?   (c) Determine the reaction force exerted on the beam by the pivot.

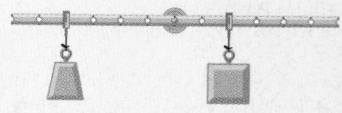

Figure P75

76. [I] THIS PROBLEM DEALS WITH ROTATIONAL EQUILIBRIUM. The uniform beam in Fig. P76 (whose weight acts at its center) is completely at rest. The scale reads 90.0 N and the hanging mass on the right weighs 45.0 N. (a) Draw a free-body diagram of the beam. (b) How much does the mass on the left weigh? (c) Determine the weight of the beam.

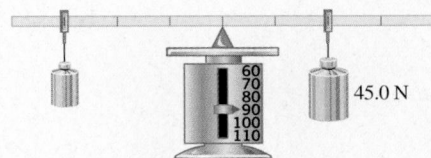

Figure P76

77. [I] An 8000-N automobile is stalled one-quarter of the way across a bridge (see Fig. P77). Compute the additional reaction forces at supports A and B due to the presence of the car. Take the length of the bridge to be $\overline{AB}$.

SOLUTION: Let L be the length of the bridge;

$$\text{Eq.(i)} \quad +\uparrow \sum F_y = 0 = F_{RA} + F_{RB} - 8000 \text{ N}$$

$$\text{Eq.(ii)} \quad (+\sum \tau_A = 0 = (8000 \text{ N})(\tfrac{3}{4}L) - F_{RB}L$$

Using Eq.(ii), $F_{RB} = (8000 \text{ N})\tfrac{3}{4} = 6000 \text{ N}$ and using Eq.(i)

$$F_{RA} = 2000 \text{ N}$$

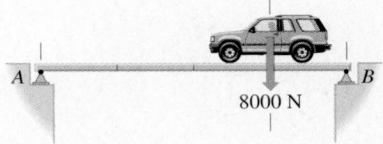

Figure P77

78.[I] The little bridge in Fig. P78 has a weight of 20.0 kN, which acts at its center. Calculate the reaction forces at points A and B.

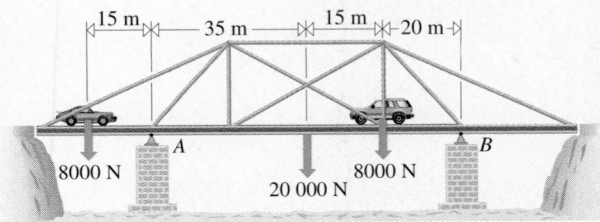

Figure P78

79. [I] The scale in Fig. P79 reads 2.1 N. Neglecting the weight of the beam and assuming the pivot, which is in the first hole on the left (see Fig. Q24), is frictionless, determine the hanging load.

80. [I] With the previous problem in mind, completely specify the reaction force exerted on the beam by the pivot.

81. [I] THIS PROBLEM DEALS WITH AN OBJECT IN EQUILIBRIUM. Each of the hanging objects on the right in Fig. P81 weighs 1.0 N. Assume the pivot (which is in the notch at the far left end) is frictionless. If the weight of the uniform beam is neglected, and the scale reads 5.0 N, (a) write an expression for the sum of the torques about the pivot and set it equal to zero. (b) What is the value of the angle θ? (c) What is the value of the horizontal component of the reaction force on the pivot? (d) What is the value of the vertical component of the reaction force on the pivot?

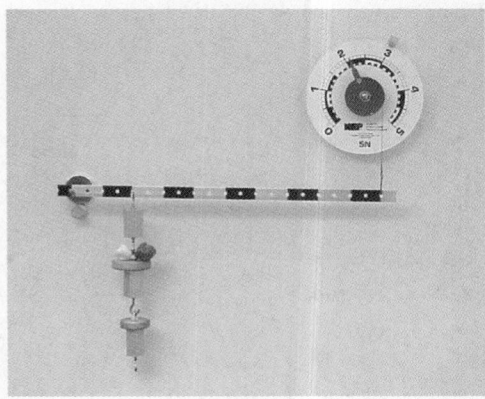

Figure P79

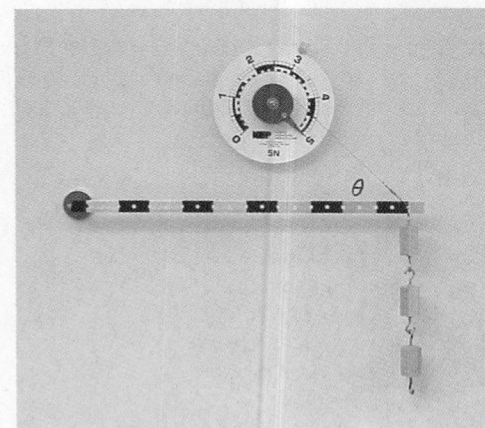

Figure P81

82. [II] Determine the magnitude and direction of the reaction force of the pivot on the beam in the previous problem. Refer to the previous problem. Notice that the pivot rod does not go through a whole, but instead sits in a half-circle groove at the end of the beam. Explain how the beam stays in place.

83. [II] Two campers carry their gear (90.72 kg) on a light, rigid horizontal pole whose ends they support on their shoulders 1.829 m apart. If Selma experiences a compressive force of 533.8 N, where is the load hung on the pole, and what will Rocko feel?

84. [II] The weightless ruler in Fig. P84 is in equilibrium. Determine both the unknown mass M and the left-hand scale reading.

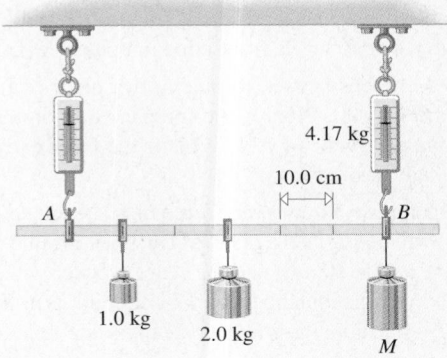

Figure P84

85.[II] If the beam in Fig. P85 is of negligible mass, what value does the scale read (in kg)?

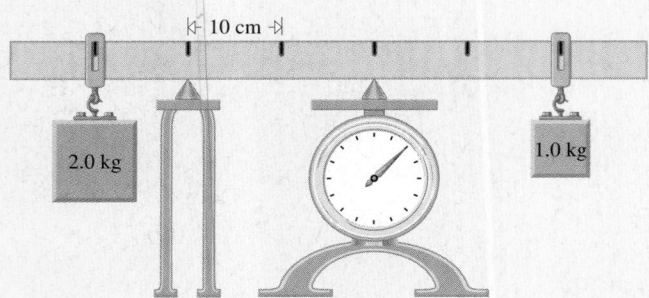

10 cm

2.0 kg   1.0 kg

Figure P85

86. [II] Figure P86 shows a laboratory model of a roof truss constructed of two very light rods pinned at points B and C. Because of this construction, the forces act only along the lengths of the rods. A light cable tightened by a turnbuckle runs from A to C to complete the truss. What do the two spring scales read in newtons? Determine the compression in rod AB. [*Hint: Find out the reading of scale-1 and then draw a free-body diagram of A.*]

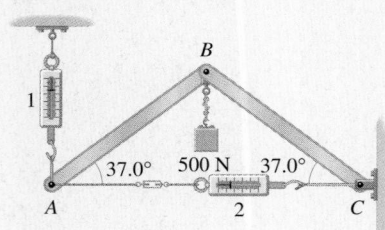

B
1
37.0° 500 N 37.0°
A   2   C

Figure P86

87. [II] The little temporary bridge in Fig. P87 consists of two very light planks. The 480-N person stands 3.00 m from the end of the upper board, which itself rests on rollers. What are the reaction forces at points A, C, and D?

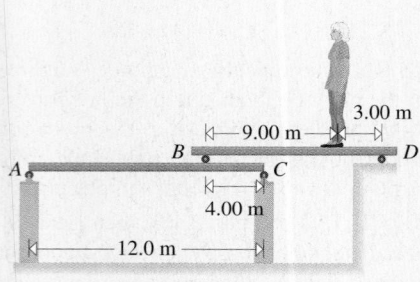

3.00 m
9.00 m
B   D
A   C
4.00 m
12.0 m

Figure P87

88. [II] Find the force F_R in Fig. P88 by taking torques about both point A and the point where the lines-of-action of the two tensile forces on the ropes intersect.

89. [II] The hand in Fig. P89 exerts a force of 300 N on the hammer handle. Compute the force acting on the nail, just as the head begins to pivot about its front edge in contact with the board.

90. [II] A forceps is used to pinch off a piece of rubber tubing, as shown in Fig. P90. What is the force exerted on the rubber if each finger squeezes with 10.0 N? What is the force exerted on the pivot by each half of the forceps?

91. [III] Suppose a downward force of 80 N acts on the pedal of the bike shown in Fig. P91. The chain goes around the front chain wheel (of radius 10 cm) and the rear sprocket wheel (of radius 3.0 cm). The pedal is attached to a 17-cm crank arm that turns around point A. What impelling force will the scale read (the bike, chain,

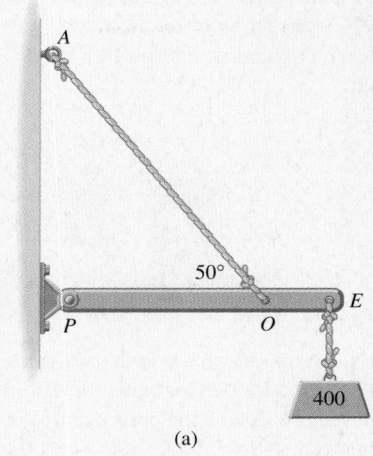

A
50°
P   O   E
400

(a)

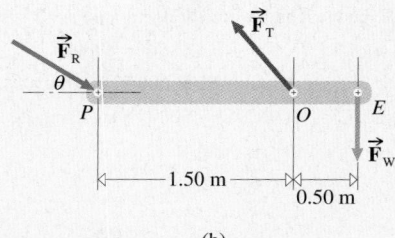

$\vec{F}_R$ $\vec{F}_T$
θ
P   O   E
$\vec{F}_W$
1.50 m
0.50 m

Figure P88   (b)

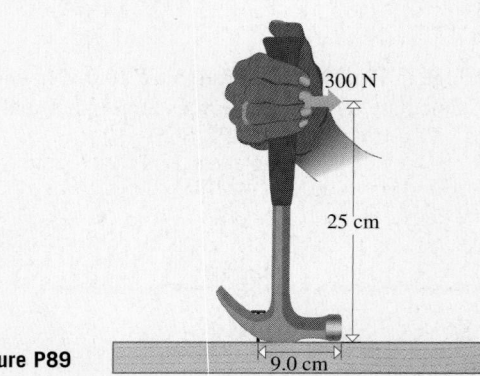

300 N
25 cm
Figure P89   9.0 cm

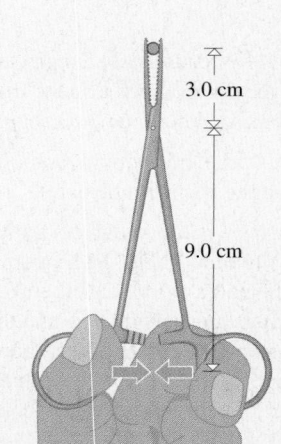

3.0 cm
9.0 cm

Figure P90

and wheel are motionless at the moment depicted when the crank arm is horizontal)? What provides that force?

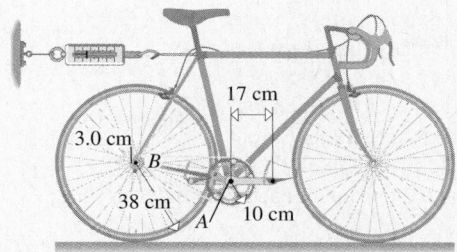

Figure P91

92. [III] A sphere of mass 10.0 kg rests in a groove, as shown in Fig. P92. Assuming no friction and taking the weight of the sphere to act at its center, compute the reaction forces exerted by the two surfaces.

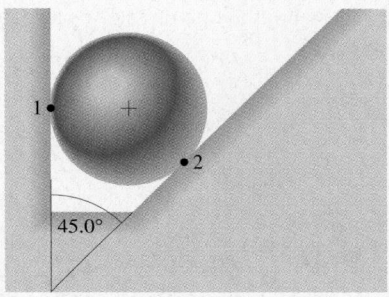

Figure P92

93. [III] Determine the internal forces acting at point A in Fig. P93 when a 1000-N weight is hung from the hook. All of the structural members can be taken as weightless. [*Hint: Because of the pinned construction, the forces act only along the lengths of the rods.*]

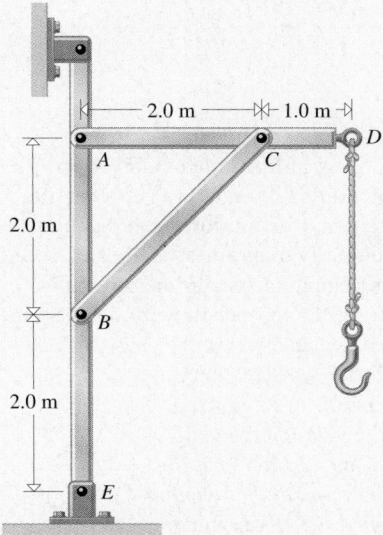

Figure P93

94. [III] Figure P94 shows a smooth rod resting horizontally inside a bowl. Compute the force F that will maintain the rod in position, given negligible friction.

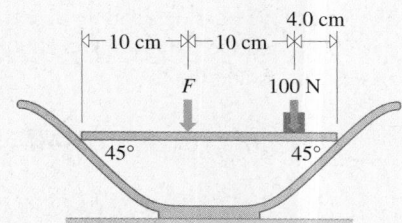

Figure P94

SECTION 8.7: EXTENDED BODIES & THE CENTER-OF-GRAVITY.

95. [I] A uniform cylindrical broom handle is 2.0 m long and weighs 8.0 N. Locate its center-of-gravity.

96. [I] Two identical 60-cm diameter hollow copper spheres are attached to each other with superglue. Locate the center-of-gravity of the system.

97. [I] Two identical uniform 50-cm long rods are attached to each other at their centers so that they lie in the same plane making an angle of 90° like a + sign. One of four identical 10-N lead spheres is then fixed at each end of the rods. Locate the center-of-gravity of the system.

98. [I] A 40-cm diameter 20-N metal sphere is glued to the surface of a 120-cm diameter 20-N plastic sphere. Locate the center-of-gravity of the system.

99. [I] Two lead spheres weighing 20.0 N and 10.0 N are separated 30.0 cm, center-to-center, by a horizontal weightless rigid rod. Locate the center-of-gravity of the system.

> **SOLUTION:** Place the spheres on the positive x-axis with the origin at the center of the 20-N sphere. Using Eq. (8.27), with the total weight being 30.0 N, $x_{cg}(30.0 \text{ N}) = (10.0 \text{ N})(0.300 \text{ m})$ and $x_{cg} = 0.100$ m. The c.g. is 10.0 cm to the right of the center of the larger sphere.

100. [I] THIS PROBLEM DEALS WITH THE CENTER-OF-GRAVITY OF A COMPOUND SYSTEM. To each end of a lightweight (horizontal) straight rod is affixed a horizontal metal ring. Each end of the rod is attached to the outer rim of a ring and the rod lies along the line connecting the centers of the two rings. One ring weighs 60.0 N, the other 15.0 N. (a) Locate the c.g. of each ring. (b) If their center-to-center separation is 100 cm, locate the center-of-gravity of the system.

101. [I] Two uniform blocks of wood 4 cm × 4 cm × 12 cm, each weighing 1.0 N, are glued together as shown in Fig. P101. Find the c.g. of the system.

102. [I] A uniform piece of soft, thin copper wire 60-cm long and weighing 1.2 N is bent into three equal-length segments to form two right angles, so that it has the shape of an angular letter "C." Locate the folded wire's c.g.

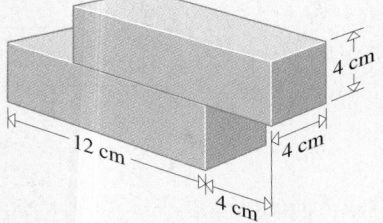

Figure P101

103. [I] A 0.50-m-long thin steel rod with a mass per unit length of 0.50 kg/m is bent in half (making a right angle) into the shape of a letter "L." Locate its c.g.

104. [I] The glider in Fig. P104 is descending at a constant speed. If the drag (F_D) is 600 N and the plane weighs 6000 N, determine both the angle of descent, θ, and the lift, F_L.

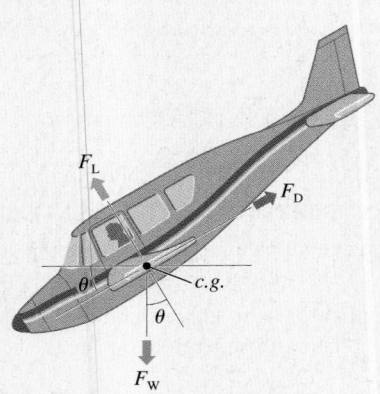

Figure P104

105. [I] Scale-1 in Fig. P105 reads 100 N. What does scale-2 read? How much does the suspended body weigh? Discuss the location of the vertical line passing through the *c.g.*

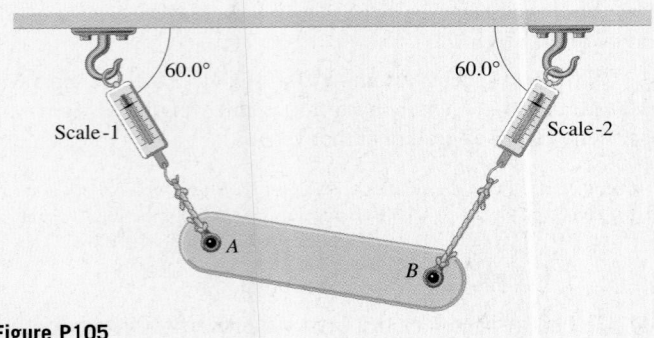

Figure P105

106. [I] Refer to the mobile in Fig. P106. If the star weighs 10.0 N, how much does the sphere weigh?

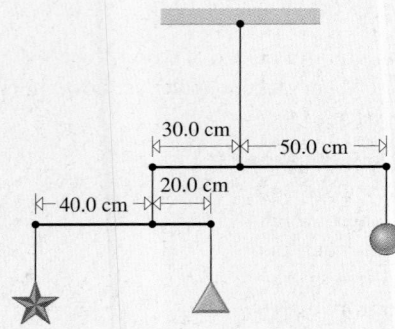

Figure P106

107. [I] Now suppose we displace the sphere in Problem 106 downward, whereupon the top rod tilts down at 30.0° with the horizontal. Will the mobile be in equilibrium?

108. [I] This problem explores rotational equilibrium. The heavy motionless uniform steel beam hanging in Fig. P108 weighs an unspecified amount F_W. (a) Where does the weight of the beam act? (b) Draw the free-body diagram of the beam. (c) What is the value of the net upward force due to the supporting ropes? (d) Which two forces produce torques about the left end of the beam? (e) Determine the tension in each rope in terms of F_W.

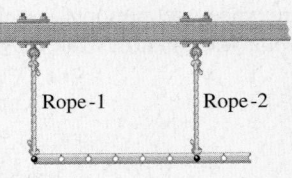

Figure P108

109. [I] This problem explores rotational equilibrium. The uniform beam in Fig. P109 weighs 0.8 N, and the scale reads 2.5 N. Assume the pivot in the first hole at the right (see p. 269) is frictionless. (a) Where does the weight of the beam act? (b) Draw the free-body diagram of the beam. (c) List all the forces acting on the beam and specify the directions (clockwise or counterclockwise) of the torques they produce about the pivot. (d) What is the tension in the rope at the far left? (e) Write an expression for the sum-of-the-torques about the pivot. (f) Determine the net weight of the hanging load.

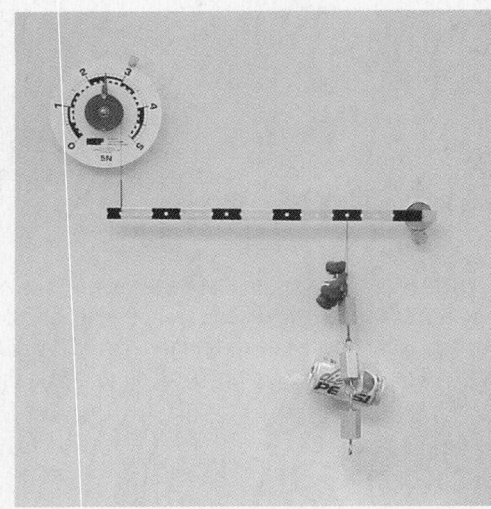

Figure P109

110. [I] This problem explores rotational equilibrium. The heavy motionless uniform beam shown in Fig. P110 is frictionlessly pivoted at its left end. (a) In what directions are the horizontal and vertical reaction forces on the pivot? Explain why. (b) Draw the free-body diagram of the beam. (c) List all the forces acting on the beam and specify the directions (clockwise or counterclockwise) of the torques they produce about the pivot. (d) If the beam has the same weight (F_W) as the load it supports, what is the value of the vertical reaction force at the pivot, in terms of F_W? (e) What is the value of the vertical component of the tension? [*Hint: To find the direction of the reaction force at the pivot (without making any calculations), imagine that the pivot pin was removed; the beam would then move left while simultaneously rotating clockwise around the end of the rope.*]

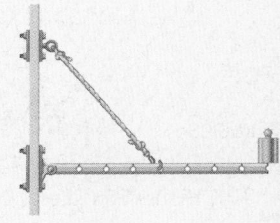

Figure P110

111. [I] The tilted motionless uniform heavy rod shown in Fig. P111 is frictionlessly pivoted at its right end. Draw the free-body diagram of the rod. [*Hint: To find the direction of the reaction force at the pivot (without making any calculations), imagine that the pivot pin was removed; the upper portion of the rod would rotate counterclockwise around the end of the rope.*]

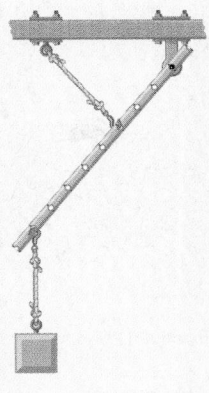

Figure P111

112. [II] A pole vaulter carries a 6.0-m uniform pole horizontally. He holds one end with his right hand pushing downward and 1.0 m away pushes upward with his left hand. If the pole weighs 40 N, compute both forces he exerts. What is the net force he applies? What would these forces be if the pole were vertical? What does this answer suggest about the carrying angle?

113. [II] We wish to stack five uniform wooden blocks so that they extend as far right as possible and still remain stable. How should each be positioned? Can the top block have its entire length beyond the edge of the bottom block (Fig. P113)?

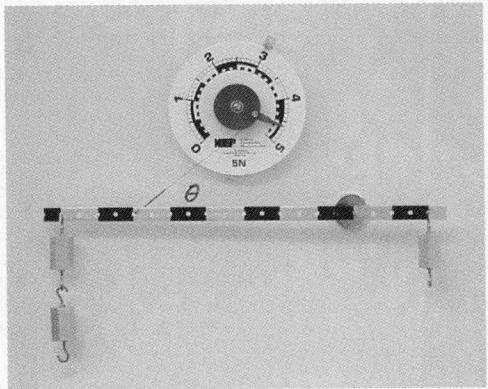

Figure P113

114. [II] THIS PROBLEM DEALS WITH ROTATIONAL EQUILIBRIUM. We want to compute θ, the smaller angle between the string and the beam, in the photo shown in Fig. P114. Each of the hanging objects weighs 1.0 N, and the scale reads 4.6 N. Assume the pivot, in the fifth hole from the right, is frictionless and the weight of the uniform beam is 0.80 N. (a) Where does the weight of the beam act? (b) What are the horizontal and vertical forces exerted on the beam by the scale? (c) What is the value of the horizontal component of the reaction force at the pivot. (d) List all the forces acting on the beam and specify the directions (clockwise or counterclockwise) of the torques they produce about the pivot. (e) Find θ.

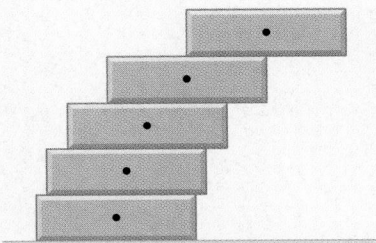

Figure 114

115. [II] THIS PROBLEM DEALS WITH ROTATIONAL EQUILIBRIUM. We want to find the reaction force at the pivot in the photo shown in

Fig. P81. Each of the hanging objects on the right weighs 1.0 N and the scale reads 5.0 N. Assume the pivot is frictionless and the weight of the uniform rod is 0.8 N. (a) Where does the weight of the rod act? (b) In what directions are the horizontal and vertical reaction forces on the pivot? Explain why. (c) Draw the free-body diagram of the rod. (d) List all the forces acting on the rod and specify the directions (clockwise or counterclockwise) of the torques they produce about the pivot. (e) Compute θ, this time including the weight of the rod. (f) What is the value of the vertical component of the reaction force at the pivot? (g) What is the value of the horizontal component of the reaction force at the pivot? Refer to Problem 81.

116. [II] An essentially weightless 10.0-m-long beam is supported at both ends, as in Fig. P116. A 300-N child stands 2.00 m from the left end, and a 6.00-m-long stack of newspapers weighing 100 N per linear meter is uniformly distributed at the other end. Determine the two reaction forces supporting the beam.

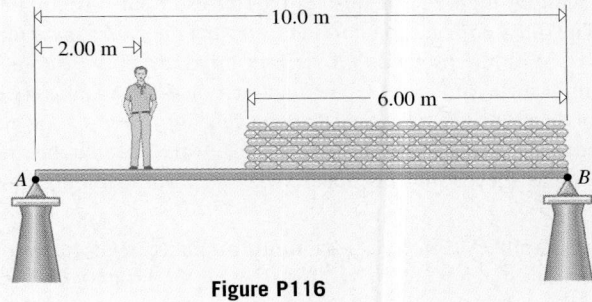

Figure P116

117. [II] Figure P117 shows two ropes supporting a body. (a) Prove that the resultant of the two tensile forces, whose magnitudes are indicated by the scale readings, has the same magnitude as the weight. (b) Show that the plumb line passes through the body's *c.g.*

SOLUTION: (a) The sum of the horizontal forces must be zero, which means that the horizontal components of the tensile forces must cancel each other, leaving only their vertical components. These must combine to form a resultant that is equal in magnitude and opposite in direction to the weight, since the sum of the vertical forces must also be zero. (b) The entire system (ropes and body) experiences two forces: a force up (along the plumb line) exerted by the ring, and the weight of the system down. If these were not colinear, the body would experience a nonzero torque and rotate.

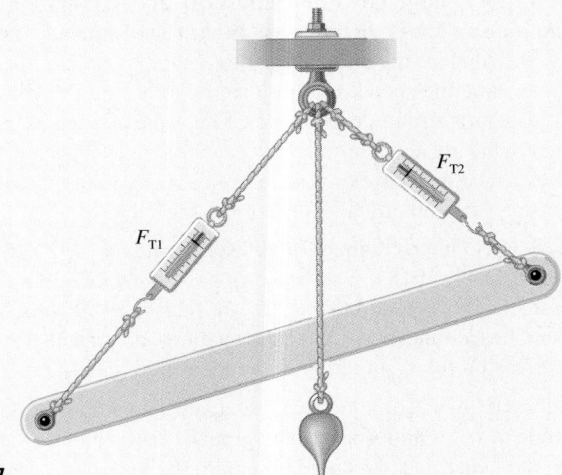

Figure P117

118. [II] A 55.0-kg woman is standing "rigid"—straight upright, hands at sides, feet together. At their widest, her two feet have a breadth of 20.0 cm and her *c.g.* is 95.0 cm above the ground. What minimum horizontal force in the plane of her body (i.e., perpendicular to the forward direction) will tilt her over if it acts along a shoulder-to-shoulder line 145-cm high?

119. [II] THIS PROBLEM EXAMINES THE FORCES ACTING ON AN OBJECT IN ROTATIONAL EQUILIBRIUM. A heavy uniform ladder leans motionlessly against a smooth, essentially frictionless wall. It rests on a rough floor that keeps it from sliding out. (a) Identify the two components of the reaction force of the floor on the ladder. (b) Identify the single reaction force of the wall on the ladder. (c) What is the relationship between the friction force at the ground and the normal force exerted by the wall? (d) Draw the free-body diagram of the ladder.

120. [II] THIS PROBLEM EXAMINES THE FORCES ACTING ON AN OBJECT IN ROTATIONAL EQUILIBRIUM. A 200-N uniform board leans motionlessly with its high end against a smooth, essentially frictionless, wall. The other end rests on a rough floor that keeps it from sliding out. (a) With the previous problem in mind, draw the free-body diagram of the board. (b) Given that the board-floor static coefficient of friction is 0.50, show that the reaction force of the wall must be ≤100 N if the board is not to slip. (c) Show that the minimum angle the board can make with the floor without sliding is 45°.

121. [II] An 8.0-m-long, 30-kg uniform plank leans against a smooth wall, making an angle of 60.0° with the ground. Compute the reaction force, $\vec{F}_{RG}$, of the ground on the plank.

122. [II] In Problem 74, Harry and Gretchen were sitting on a weightless seesaw. Suppose now that the beam is uniform and weighs 200 N—find the new pivot point.

123. [II] A 65-kg woman is horizontal in a pushup position in Fig. P123. What are the forces acting on her hands and feet?

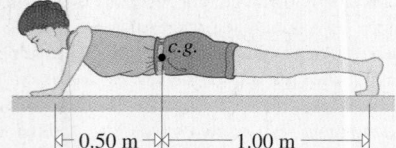

Figure P123　|← 0.50 m →|←　　1.00 m　　→|

124. [II] A large box of breakfast cereal is 6.0-cm × 35-cm × 46-cm high and has a net mass of 1.0 kg. The contents are a uniformly distributed material. The box is next to an open window. Determine the speed of a uniform wind hitting the large surface if it just starts tilting the box over. The force exerted by a perpendicular wind in newtons on each square meter is given roughly by $0.6v^2$, wherein v is expressed in m/s. Assume the distributed force acts at the geometrical center of the face.

125. [III] One technique for measuring the *c.g.* of a person is illustrated in Fig. P125. The board is positioned according to the person's height (*h*), and the scales are then reset to zero. The participant lies down, and the scale readings are taken. Determine an expression for x_{cg} in terms of the measured quantities.

126. [III] A uniform ladder leans against a smooth wall, making an angle of θ with respect to the ground. The ground's reaction force on the ladder is at an angle of ϕ with the horizontal. Determine the relationship between θ and ϕ. Which is larger?

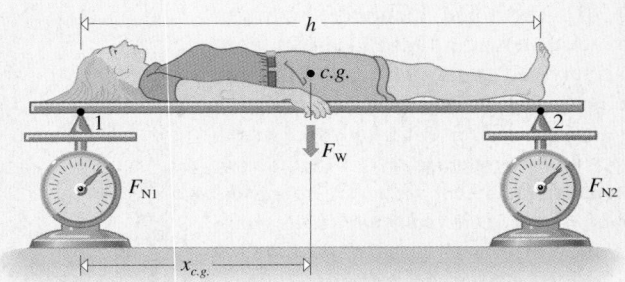

Figure P125

127. [III] The uniform plank in Fig. P127 is 5.00-m long and weighs 100 N. The cord that attaches to the plank 1.00 m from the bottom end holds it from sliding, there being no friction. Find the tension in the cord.

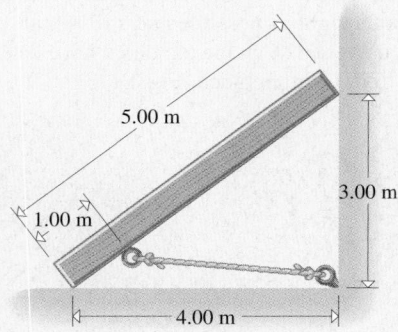

Figure P127

SECTION 8.8: TORQUE & ROTATIONAL INERTIA

128. [I] Determine the moment-of-inertia of a solid disk 0.50 m in diameter, having a mass of 4.8 km.

129. [I] Consider a thin ring or hoop of mass *M* and radius *R* of negligible thickness. Compute the moment-of-inertia about its *c.m.* by imagining it to be made up of a large number of small segments.

130. [I] Two solid 3.0-kg cones, each with a base radius of 0.30 m, are glued with their flat faces together. Determine the moment-of-inertia of this device about the symmetry axis passing through both vertices.

131. [I] The **Parallel-Axis Theorem** states that if I_{cm} is the moment-of-inertia of a body about an axis through the *c.m.*, then I, the moment-of-inertia about any axis parallel to that first one, is given by

$$I = I_{cm} + md^2$$

Here, *m* is the object's mass, and *d* is the perpendicular distance through which the axis is displaced. Accordingly, use the theorem for a thin rod of length *l*; $I_{cm} = \frac{1}{12}ml^2$ to compute *I* about one end.

132. [I] Show that a small object (such as a sphere) that orbits a distant axis can be approximated by a point-mass when you want to compute its moment-of-inertia. Begin with the theorem of the previous problem.

133. [I] Envision a solid cylindrical wheel of radius *r* and mass *m* resting upright on its narrow edge on a flat plane. With Problem 101 in mind, compute the wheel's moment-of-inertia about the line of contact with the surface.

134. [I] Two small rockets are mounted tangentially on diametrically opposite sides of a cylindrically shaped artificial satellite. The spacecraft has a 1.0-m diameter and a moment-of-inertia about its central symmetry axis of 25 kg·m². The rockets each develop a thrust of 5.0 N, are oppositely directed, and are aligned to produce a maximum spin-up of the craft. What's the resulting angular acceleration when they are both fired?

135. [II] A 26.0-in.-diameter bicycle wheel supported vertically at its center is spun about a horizontal axis by a torque of 50.0 ft·lb. If the rim and tire together weigh 3.22 lb, determine the approximate angular acceleration of the wheel. Ignore the contribution of the spokes. Of course, these ugly units should cancel out. [*Hint: The net applied torque equals the moment-of-inertia times the angular acceleration. This wheel is essentially a hoop.*]

136. [II] Determine the angular acceleration that would result when a torque of 0.60 N·m is applied about the central spin axis of a hoop of mass 2.0 kg and radius 0.50 m.

137. [II] How much torque must be applied to a hoop having a moment-of-inertia about its central symmetry axis of 1.50 kg·m² if it is to uniformly accelerate from rest to 10.0 rad/s in 10.0 s?

138. [II] Derive an expression for the acceleration of a hollow sphere of mass m and radius R rolling, without slipping, down an incline of angle θ. Discuss the implications of the absence of m and R in this formula.

139. [II] A block of mass m is tied to a light cord that is wrapped around a vertical pulley of radius R and moment-of-inertia I. Assuming the bearings are essentially frictionless, write an expression for the rate at which the block will accelerate when released.

140. [II] A 10-kg solid steel cylinder with a 10-cm radius is mounted on bearings so that it rotates freely about a horizontal axis. Around the cylinder is wound a number of turns of a fine gold thread. A 1.0-kg monkey named Fred holds on to the loose end and descends on the unwinding thread as the cylinder turns. Compute Fred's acceleration and the tension in the thread.

141. [II] How would Fred manage in Problem 140 if the bearings exerted a frictional torque of 0.060 N·m?

142. [II] A water turbine (having a moment-of-inertia of 1000 kg·m²) is essentially a wheel with a lot of blades attached to it much like its predecessor, the waterwheel. When a high-speed blast of water hits the blades, it drives the turbine into rotation. Suppose the input valve is closed and the turbine winds down from its operating speed of 200 rpm, coming to rest in 30 minutes. Determine the frictional torque acting.

143. [II] A hoop of mass m and radius R rolls without slipping down an incline at an angle of θ. Write an expression for the acceleration of its *c.m.* and find its numerical value when $m = 1.0$ kg, $R = 1.0$ m, and $\theta = 30°$.

144. [III] Figure P144 shows two masses strung over a 0.50-m-diameter pulley whose moment-of-inertia is 0.035 kg·m². If there is a constant frictional torque of 0.25 N·m at the bearing, compute the acceleration of either mass. (The tension in the rope is not the same on either side of the pulley.)

145. [III] Consider a long slender rod of mass M and length l pivoted at its far end at

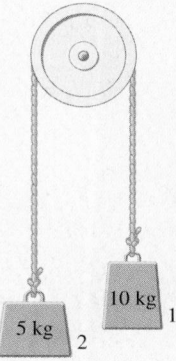

Figure P144

point O. Now compute I_0 directly by dividing the rod into 100 equal-mass parts and summing up the moments of inertia for all of these. You will need to know that

$$1^2 + 3^2 + 5^2 + \cdots + (2n - 1)^2 = \frac{n}{3}(2n + 1)(2n - 1)$$

146. [III] A solid cylindrical pulley of mass m and radius R is held upright, and a number of circular turns of light cord are wrapped around it within the groove. The loose end of the cord is held vertically, and the pulley is released so it falls straight downward, unwinding like a yo-yo. Please compute its linear acceleration, in terms of g, and the tension in the cord, in terms of m and g, before it reaches the end of its rope.

147. [III] A solid cylinder is placed at rest on an inclined plane at a vertical height h and allowed to roll down without slipping. Using forces and torques show that its speed at the bottom is given by $v = 2\sqrt{gh/3}$. Look at Problem 156.

148. [III] A cylinder of radius R and mass m mounted with an axle is placed on a horizontal surface. It's pulled forward via a mechanical arrangement so that the applied horizontal force F passes through the *c.m.* Assuming the cylinder rolls without slipping, determine both its acceleration and the friction force at the surface in terms of F.

SECTION 8.9: ROTATIONAL KINETIC ENERGY

149. [I] Determine the kinetic energy of a nontranslating disk that is spinning around its central symmetry axis at 300 rpm. The disk has a moment-of-inertia of 1.00 kg·m².

150. [I] A spherical space satellite having a moment-of-inertia of 250 kg·m² is to be spun up from rest to a speed of 12 rpm. How much energy must be imparted to the satellite?

151. [I] Calculate the angular speed of a uniform solid disk about its central symmetry axis if its rotational kinetic energy is 100 J and its moment-of-inertia is 10.0 kg·m².

152. [I] A small lead sphere of mass 0.250 kg is whirling in a horizontal circle at the end of a 2.00-m long string at a speed of 10.0 rad/s. Approximately what is the value of its rotational KE? If the string breaks, what will be its linear KE immediately thereafter?

153. [I] If the dimensions of mass, length, and time are M, L, and T, what are the dimensions of angular kinetic energy?

154. [I] A solid rubber ball of radius R and mass m is thrown with a speed v. If it leaves the pitcher's hand spinning at a rate ω, write an expression for its total KE.

155. [II] A hollow cylinder, or hoop, of mass m rolls down an inclined plane from a height h. If it begins at rest, show that its final speed is given by

$$v = \sqrt{gh}$$

156. [II] A solid cylinder of mass 2.0 kg rolls without slipping down a long curved track from a height of 10.0 m. Calculate the linear speed with which it exits the track at the bottom. Look at Problem 147.

157. [II] What is the total kinetic energy of a solid ball of mass m rolling along a horizontal surface at a constant speed V?

158. [II] With Problem 157 in mind, what fraction of the total energy of a solid sphere rolling at a constant speed exists as translational KE?

159. [II] A long pencil is balanced straight up on its point on a horizontal surface. Without slipping, the pencil topples over. Show that

the speed at which the eraser end strikes the surface is

$$v = \sqrt{3gL}$$

[*Hint: Take the pencil to be a uniform rod and apply Conservation of Energy to the c.g.*]

SECTION 8.10: ANGULAR MOMENTUM
SECTION 8.11: CONSERVATION OF ANGULAR MOMENTUM

160. [I] A uniform disk of mass 800 kg and radius 0.50 m is rotating around its central symmetry axis at a rate of 60 rpm. Determine its angular momentum.

161. [I] A tiny sphere of mass 50.0 g is a vertical distance of 1.00 m from a point P. If at that instant the particle is moving horizontally at a speed of 2.50 m/s, what if any is its angular momentum with respect to P?

162. [I] Determine the angular momentum of a hoop having a mass of 2.00 kg and a diameter of 2.00 m rolling down an incline at 2.50 m/s.

163. [I] A gymnast in the process of performing a forward somersault in midair increases his angular velocity by 450% while going from an arms-overhead layout position to a tight tuck. What can you say about the change, if any, in the moment-of-inertia about the frontal axis (parallel to the shoulder-to-shoulder line) passing through his *c.m.* (in one side and out the other)?

164. [I] An astronaut of mass 70 kg rotates his arms backward about a shoulder-to-shoulder axis at a rate of 1 rev/s. *Both of his arms together* have a mass of 12.5% of his body mass and a total moment-of-inertia about the shoulders of 1 kg·m^2. Compute the angular momentum imparted to the rest of his body in the process.

165. [I] A tiny hummingbird with a mass of only 2×10^{-3} kg is circling around a flower in a 1.0-m-diameter orbit. If it travels once around in 1.0 s, approximate its angular momentum with respect to the blossom (look at Problem 132).

166. [II] Compute the orbital angular momentum of Jupiter ($M_{2\!\!4} = 1.9 \times 10^{27}$ kg, $r_{2\!\!4} = 7.8 \times 10^{11}$ m, and $v_{av} = 13.1 \times 10^3$ m/s) and then compare it to the spin angular momentum of the Sun ($M_\odot = 1.99 \times 10^{30}$ kg, $R_\odot = 6.96 \times 10^8$ m). Assume the Sun, whose equator rotates once in about 26 days, is a rigid sphere of uniform density. It would seem that most of the angular momentum of the Solar System is out there with the giant planets that also rotate quite rapidly.

167. [II] It is possible for a large star (one greater than 1.4 solar masses) to gravitationally collapse, crushing itself into a tiny neutron star perhaps 40 km in diameter. Suppose such a thing could happen to the Sun, which is 1.39×10^9 m in diameter and spinning at a rate of about once around every 27 days. What would be the spin rate for the resulting neutron star? See the photos on p. 261.

168. [II] A small mass m is tied to a string and swung in a horizontal plane. The string winds around a vertical rod as the mass revolves, like a length of jewelry chain wrapping around an outstretched finger. Given that the initial speed and length are v_i and r_i, compute v_f when $r_f = r_i/10$.

169. [II] A very thin 1.0-kg disk with a diameter of 80 cm is mounted horizontally to rotate freely about a central vertical axis. On the edge of the disk, sticking out a little, is a small, essentially massless, tab or "catcher." A 1.0-g wad of clay is fired at a speed of 10.0 m/s directly at the tab perpendicular to it and tangent to the disk. The clay sticks to the tab, which is initially at rest, at a distance of 40 cm from the axis. What is the moment-of-inertia   (a) of the clay about the axis?   (b) of the disk about the axis?   (c) of both clay and disk about the axis?   (d) What is the linear momentum of the clay before impact?   (e) What is the angular momentum of the clay with respect to the axis just before impact?   (f) What is the angular speed of the disk after impact?

170. [II] Write an expression for the orbital angular momentum of a small artificial satellite of mass m in a circular flight-path of radius r about the Earth. Check the units. What happens to angular momentum as the orbit increases?

171. [III] An astronaut is working out at the far reach of a 100-m tether, the opposite end of which is attached to a pivoting ring on the nose of the space station. She does not notice that an air hose on her backpack has developed a leak. The little blast of gas produces a tangential thrust and a corresponding acceleration of $a_T = 1.0 \times 10^{-3}g$. After two minutes, she realizes what has happened and shuts off the leak. At that point, she and her life support system (total mass of 150 kg) are sailing along at a tangential speed of ———with an angular momentum of ———. Annoyed, she decides to return to the craft by pulling hand-over-hand on the tether. If she manages to get 5.0 m from the ring on the ship, her tangential speed would be ———. Considering that the centripetal force she would have to hold against is then ———, it's unlikely she would ever get that close.

Chapter 9
Solids, Liquids, & Gases

One of our continuing concerns in studying the physical Universe is *matter*: What is it? How does it behave, and why? This chapter is the start of an exploration into the mechanical, thermal, electrical, magnetic, and optical properties of bulk matter. It begins with a sketch of the atomic landscape, explores the various states of matter, and then goes on to focus on fluids.

Atoms and Matter

It might seem easy to hold a thing in hand and know what **matter** is—it's stuff, substance, the material aspect of the Universe. But language fails us if we try to go much beyond that. A century ago, one could have responded that matter possesses mass and fills the space it resides in. Yet, we now know that light is material; it transports energy and momentum. And the particle of light, the *photon*, which is the most tenuous form of matter, is totally massless. Moreover, there is a tiny chargeless speck called the *neutrino* whose mass is so small it's never been detected. (If indeed it has any mass at all, it's less than 1/180 000 that of an electron, the lightest particle whose mass has been measured.) The old maxim that "matter is that which is impenetrable and occupies space" is also not very useful. A cannonball may appear solid enough, but it's mostly empty. Bulk matter is composed of separated atoms and does not fill space in any ordinary sense. Indeed, the particles that make up the atoms are fluttery, elusive things that recoil from localization quite unlike the imagery of tiny billiard balls. The idea of atoms as impenetrable specks of solid matter is an ancient mirage.

9.1 Atomism

The concept of the *atom* dates back to the Greek Leucippus (ca. 450 B.C.E.). If a substance, a piece of gold, say, is divided into ever smaller portions, could the process go on forever, or would some last indivisible fragment—the *atomos (that which cannot be further cut)*— be reached? Robert Boyle embraced the concept as an explanation of the "springiness of air" (in a balloon, for example). In 1665, Hooke proposed that crystalline structures resulted from the packing of tiny spherical particles—atoms stacked like cannonballs.

After a brief introduction to atoms and the various states of matter (p. 283), this chapter focuses on fluids. It treats both fluid statics and fluid dynamics. If you're a sailor or a plumber, the relevance of this material will be apparent. So too if you're a human being—dependent for life on pumped blood and breathed air—you might find this chapter of more than passing interest.

Fluid statics is mostly about pressure. First, we'll deal with incompressible fluids open to the atmosphere (p 290–302), like lakes and bathtubs; then—because of its tremendous practical importance—we'll treat fluids confined to sealed systems (Pascal's Principle). Statics ends with the notions of buoyancy and floating (p. 298).

Fluid dynamics rests on the Continuity Equation (which arises from the incompressibility of liquids) and Bernoulli's Equation (which is an expression of Conservation of Energy).

A stack of plums. Hooke was right; many elements (e.g., zinc and nickel) arrange their atoms like this.

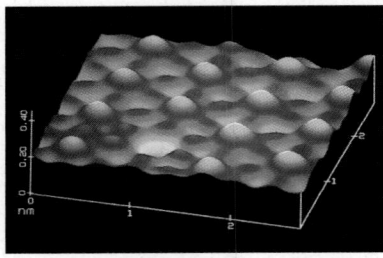

This regular pattern of bumps is formed by iodine atoms absorbed on a platinum surface.

Salt dissolves into water, which suggests that the liquid, instead of being continuous, is rife with empty spaces. A cup of alcohol and a cup of water will combine to form less than two cups of "good cheer"—the atoms seem to be sliding in between one another, filling holes. The smell of perfume can fill a room in minutes, suggesting a mingling of colliding particles.

The breakthrough in atomic research came in the early 1800s at the hands of an unknown country schoolmaster, a gruff, awkward man named John Dalton. It had already been shown that *elements combine to form compounds in certain definite proportions.* From this he deduced the relative masses of the atoms involved. Suspecting that hydrogen was the lightest of the atoms, he set its mass equal to 1 and so on down the list of elements.

The table of relative atomic masses that Dalton drew up had many errors, but they were soon corrected. Nowadays, we take carbon (^{12}C) as the standard instead of hydrogen, set its mass equal to exactly 12.000 00 unified atomic mass units (u), and measure the other atoms relative to it (Table 30.1). In this way a hydrogen atom is said to have a mass of 1.007 825 u. {For more details click on **ATOMISM** under **FURTHER DISCUSSIONS** on the **CD**.}

Avogadro's Number

In 1811, Amedeo Avogadro, Count of Quaregna, made two brilliant proposals (coining the word *molecule* in the process): (1) *The gaseous elements can exist in molecular form.* (2) *Equal volumes of molecular gases* (*under the same conditions of temperature and pressure*) *contain the same number of molecules.*

Ordinary hydrogen exists as a molecule (H_2) with a relative mass of 2. Suppose we admit 2 g of it into an expandable chamber that can be maintained at **standard temperature (0°C) and pressure (1 atm)**—STP. The gas is found to occupy a volume of 22.4 liters. According to Avogadro, 22.4 liters of oxygen, or of any gas, would contain the same number of molecules. Since each O_2 molecule has a mass of 2×16 u = 32 u, 22.4 liters of oxygen should have a mass of 32 g. Indeed, ***22.4 liters of any gas at STP has a mass in grams numerically equal to its molecular mass***. This measure is called a *gram-molecular mass*, or **mole**. For instance, one molecule of carbon dioxide (CO_2) has a relative mass of 12 u + 2×16 u = 44 u, and 1 mole of it has a mass of 44 g.

Similarly, *a **kilomole** is the mass of any substance in kilograms that is numerically equal to the molecular mass of that substance.* A kilomole (abbreviated kmol) of any gas at STP occupies 22.4 m^3. One kmol of CO_2 has a mass of 44 kg.

We could start out with 22.4 liters of a gas and cool it down to a liquid or freeze it into a solid and still have a mole. It follows that a mole of uranium, a shiny 238-g block, has the same number of atoms as there are molecules in 44 g of carbon dioxide. To honor the good Count, the number of molecules in a mole is known as **Avogadro's number** (N_A), though he himself had no idea of its numerical value:

$$N_A = 6.022\,14 \times 10^{23} \text{ molecules/mole}$$

> The number of grams in a **mole** of a substance is numerically equal to the molecular mass of that substance in units of u.

> The number of kilograms in a **kilomole** of a substance is numerically equal to the molecular mass of that substance in units of u. 1 kilomole = 1000 moles

EXPLORING PHYSICS ON YOUR OWN

Atomism: Get a large, clear glass bottle or beaker and fill it with cold water. Let a few drops of ink or food coloring fall onto its surface. Watch how the dye slowly diffuses throughout the liquid. Time how long it takes for the entire quantity to become uniformly colored. Now try it again with much warmer water; the difference will be quite dramatic. What's your interpretation of what's happening?

Example 9.1 **[I]** What is the mass of a carbon atom in kilograms? Provide your answer to four significant figures.

Solution The mention of "mass," "carbon atom," and "kilogram" suggests the conversion of atomic mass units. (1) TRANSLATION—Knowing that the mass of a carbon atom is 12.00 u, determine that atom's mass in kg. (2) GIVEN: A mole of carbon corresponds to 12.00 g and contains N_A atoms. FIND: m_C in kg. (3) PROBLEM TYPE—Mass conversion/Avogadro's Number. (4) PROCEDURE—The mass of a carbon atom equals the mass of a mole of carbon divided by the number of atoms in a mole. (5) CALCULATION:

$$m_C = \frac{12.00 \text{ g/mole}}{6.022 \times 10^{23} \text{ atoms/mole}}$$

$$m_C = 1.993 \times 10^{-23} \text{ g} = \boxed{1.993 \times 10^{-26} \text{ kg}}$$

Quick Check: In SI units, $N_A = 6.022 \times 10^{26}$ molecules/kilomole and $m_C = (12.00 \text{ kg})/(6.022 \times 10^{26}) = 1.993 \times 10^{-26}$ kg.

[For more worked problems click on **WALK-THROUGHS** *in* **CHAPTER 9** *on the* **CD***. There's also a whole section of* **WARM-UPS** *on* **MOLES & AVOGADRO'S NUMBER** *in* **CHAPTER 9***.]*

9.2 Density

One of the more evident features of bulk matter is that different substances are not packed with equal density—a kilogram of lead has a much smaller volume than a kilogram of feathers. When dealing with the behavior of materials rather than individual objects, it's often convenient to use quantities that are *independent of volume*. Consequently, define the **density** of any material as the ***mass per unit volume***:

$$\text{Density} = \frac{\text{mass}}{\text{volume}}$$

The symbol for density is the lowercase Greek letter ρ (rho), and so

$$\rho = \frac{m}{V} \tag{9.1}$$

Density is a characteristic of a substance; mass is a characteristic of an object (Table 9.1).

At one time, the centimeter-gram-second (cgs) unit of g/cm³ was widely used; it still persists when dealing with small quantities of substances, as in medicine. Since

$$1 \text{ g/cm}^3 = \frac{1 \times 10^{-3} \text{ kg}}{(1 \times 10^{-2} \text{ m})(1 \times 10^{-2} \text{ m})(1 \times 10^{-2} \text{ m})} = 10^3 \text{ kg/m}^3$$

the numerical values differ by a factor of 10^3. Because the gram was defined as the mass of a 1-cm³ volume of water, the density of water is 1 g/cm³, or 10^3 kg/m³. The pungent-smelling metal osmium, whose atoms are massive at 190 u (though not the most massive), is so closely packed that it is the densest (22.5×10^3 kg/m³) substance on Earth.

The Size of Atoms

By the 1880s, a number of experiments had established a rough size for atoms and molecules. One of the easiest to appreciate takes a known volume of oil and allows it to spread

The pale blue, almost transparent solid resting on a mound of shaving cream is silica aerogel. Its density is only about 3 times that of air, and yet it can support a load ≈1600 times its own weight.

A gram-mole of mercury, water, glass, salt, and iron. The glass marble can be used to establish the scale. A mole of almost anything solid or liquid will occupy a volume of about (2 or 3 cm)³.

Table 9.1

Densities of Some Materials

Substance	Density (kg/m³)	Substance	Density (kg/m³)
Interstellar space	10^{-18} to 10^{-21}	Chloroform	1.53×10^3
Hydrogen*	0.090	Sugar	1.6×10^3
Oxygen	1.43	Magnesium	1.7×10^3
Helium	0.178	Bone	$(1.5-2.0) \times 10^3$
Air, dry (30°C)	1.16	Clay	$(1.8-2.6) \times 10^3$
Air, dry (0°C)	1.29	Ivory	$(1.8-1.9) \times 10^3$
Styrofoam	0.03×10^3	Glass	$(2.4-2.8) \times 10^3$
Balsa wood	0.12×10^3	Cement	$(2.7-3.0) \times 10^3$
Cork	$(0.2-0.3) \times 10^3$	Aluminum	2.7×10^3
Pine wood	$(0.4-0.6) \times 10^3$	Marble	2.7×10^3
Oak wood	$(0.6-0.9) \times 10^3$	Diamond	$(3.0-3.5) \times 10^3$
Ether	0.736×10^3	Moon	3.34×10^3
Ethyl alcohol	0.791×10^3	Planet Earth,	
Acetone	0.792×10^3	average	5.25×10^3
Turpentine	0.87×10^3	Iron	7.9×10^3
Benzene	0.899×10^3	Nickel	8.8×10^3
Butter	0.9×10^3	Copper	8.9×10^3
Olive oil	0.92×10^3	Silver	10.5×10^3
Ice	0.92×10^3	Lead	11.3×10^3
Water (0°C)	$0.999\,87 \times 10^3$	Mercury	13.6×10^3
Water (3.98°C)	$1.000\,00 \times 10^3$	Uranium	18.7×10^3
Water (20°C)	$1.001\,80 \times 10^3$	Gold	19.3×10^3
Tar	1.02×10^3	Tungsten	19.3×10^3
Seawater	1.025×10^3	Platinum	21.5×10^3
Blood plasma	1.03×10^3	Osmium	22.5×10^3
Blood, whole	1.05×10^3	Pulsar	$10^8 - 10^{11}$
Ebony wood	$(1.1-1.3) \times 10^3$	Nuclear matter	$\approx 10^{17}$
Rubber, hard	1.2×10^3	Neutron star, core	$\approx 10^{18}$
Brick	$(1.4-2.2) \times 10^3$	Black hole	
Sun, average	1.41×10^3	(1 solar mass)	$\approx 10^{19}$

*Gases are at 0°C and 1 atm unless otherwise indicated.

out on the surface of water. There are good reasons to assume that the resulting film would thin out until it was only one molecule thick. Measuring the area of the slick gives us the thickness of the film and, therefore, the diameter of the oil molecule. Nowadays, we know that *atoms of every kind are nearly the same size, varying in diameter from roughly 0.1 nm to about 0.3 nm*.

Example 9.2 **[II]** Determine the mass of a gold sphere 0.10 m in diameter. How many atoms are in it? The atomic mass of gold is 197 u.

Solution On first reading, the mention of "mass" and "diameter" suggests volume and density, and talk about the number of atoms calls to mind Avogadro's Number. (1) TRANSLATION—A sphere of known radius is composed of a material of known density and atomic mass; determine (a) its mass and (b) the number of atoms present. (2) *Use Table 9.1 when you need a density.* GIVEN: $D = 0.10$ m, $\rho = 19.3 \times 10^3$ kg/m³, and the atomic mass is 197 u. FIND: The mass m and the number of atoms N. (3) PROBLEM TYPE—Density/Avogadro's Number. (4) PROCEDURE—Whenever you're asked about the number of atoms, N_A is usually involved. Use the definition $\rho = m/V$ to find m. (5) CALCULATION—(a) First compute V,

$$V = \tfrac{4}{3}\pi R^3 = \tfrac{4}{3}(3.141\,6)(0.050 \text{ m})^3 = 0.524 \times 10^{-3} \text{ m}^3$$

then

$$m = \rho V = (19.3 \times 10^3 \text{ kg/m}^3)(0.524 \times 10^{-3} \text{ m}^3)$$

and

$$\boxed{m = 10 \text{ kg}}$$

(b) A kilomole of gold is 197 kg, so this sphere is the equivalent of 0.051 27 kmol and

$$N = N_A\, 0.051\,27 = (6.022 \times 10^{26} \text{ atoms/kmol})(0.051\,27 \text{ kmol})$$

and so

$$\boxed{N = 3.1 \times 10^{25} \text{ atoms}}$$

Quick Check: $(3.1 \times 10^{25} \text{ atoms})(197 \text{ kg·kmol}^{-1})/(6.022 \times 10^{26} \text{ atoms/kmol}) = 10 \text{ kg}$.

We can get an estimate of the dimensions of atoms in a solid or liquid by assuming that they "fill" the volume of the material. They don't, but there can't be a great deal of interatomic space since neither solids nor liquids can be compressed very much. If we take each atom as a little cube of volume L^3, the volume V of a mole would then be $N_A L^3$. If the mass of that mole is M_m,

$$\rho = \frac{M_m}{V} = \frac{M_m}{N_A L^3}$$

and
$$L = \left(\frac{M_m}{N_A \rho}\right)^{\frac{1}{3}} \tag{9.2}$$

This means that a water molecule should have a size of very roughly

$$L = \left[\frac{18 \text{ g}}{(6.022 \times 10^{23} \text{ atoms/mole})(1.0 \text{ g/cm}^3)}\right]^{\frac{1}{3}}$$

which is 3×10^{-10} m, or 0.3 nm. Interestingly, if we do this same calculation for most simple solids and liquids, they generally turn out to have atomic dimensions within a factor of 2 or so of water. For example, $L = 0.5$ nm for ethyl alcohol.

The atom is a complex structure consisting of a massive, minute core—the *nucleus*—surrounded by a tenuous mist of electrons (Fig. 1.4). The nucleus, whose diameter is only about one ten-thousandth the diameter of the atom, is a ball of rapidly moving, positively charged protons and neutral neutrons. The electrons, which tumble around at tremendous speeds in the outer reaches of the atom, are retained by the electromagnetic force (Fig. 9.1). We no longer envision well-defined electron orbits. Instead, the atomic landscape is calculated in terms of the patterns in which the electrons are likely to reside, patterns known as electron clouds.

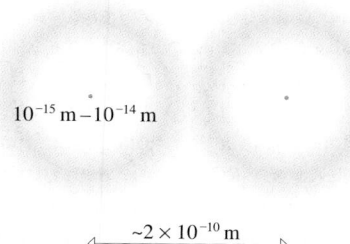

10^{-15} m $- 10^{-14}$ m

$\sim 2 \times 10^{-10}$ m

Figure 9.1 The atoms in a solid are roughly 0.2 nm apart, center-to-center, and have almost all their mass concentrated in the tiny 10^{-15} m to 10^{-14} m nuclei.

9.3 The States of Matter

Under ordinary planet-Earth conditions, matter exists in bulk as solid, liquid, gas, or plasma. If the interatomic forces are strong enough, a collection of particles independently *maintains its shape and volume,* and that ability is the distinguishing feature of a **solid**. A **liquid** is characterized by weaker binding, so that *it flows, assuming the shape of its container, although it maintains a constant volume regardless of that shape.* If the forces are still weaker, the material exists as a **gas**; the atoms or molecules tend to disperse, and *the substance flows, assuming both the shape and volume of its container.* If some of the atoms of a gas are *ionized* (i.e., electrons are removed or added), *the mix of atoms, ions, and electrons is* called a **plasma**.

This polycrystalline mass of calcium carbonate ($CaCO_3$) grew in a professor's kidney until it rather painfully found its way down and out. Fortunately this photo is about six times life-size.

CHERRY-PIT NUCLEI & ALICE'S POTION

Atoms in solids and liquids can be imagined as tightly packed (i.e., cloud to cloud), whereas in gases they are quite a bit farther apart—gases under ordinary conditions are about a million times more compressible than solids. The regions between the nuclei of adjacent atoms in all forms of bulk matter are comparatively immense. If you could drink Alice in Wonderland's growth potion and swell up until the core of an atom in your body was the size of a cherry pit, you would stand about a thousand million kilometers tall! And most of your huge volume would be void, with hundreds of meters of emptiness separating the pit-sized concentrations of matter.

Some materials can exist as solid, liquid, or gas and readily shift from one state to another and back, depending on the prevailing conditions of temperature and pressure. Most solids become less dense by a few percent on melting. The space between molecules increases but not by very much. On the other hand, a tremendous decrease in density results when a liquid is transformed into a gas. Table 9.2 shows the effect in terms of the number of molecules per cubic meter, which provides a sense of the intermolecular spacing.

Table 9.2

Number of Molecules per Cubic Meter

Substance	State	(Number of molecules)/m³
Gold	solid	5.9×10^{28}
Ice	solid	3.1×10^{28}
Water	liquid	3.3×10^{28}
Mercury	liquid	4.2×10^{28}
Oxygen	gas	2.7×10^{25}
Air	gas	2.7×10^{25}

Solids

Many more materials in our moderate-temperature environment are in the solid state than in either the liquid or gaseous states. One way to classify these solids is as either *crystalline, quasi-crystalline, amorphous,* or *composite.* The atoms of a solid are in perpetual motion, but it's a small jiggling, vibratory motion about fixed, closely spaced positions. When the atoms arrange themselves in an *orderly, three-dimensional pattern that repeats itself,* the solid is **crystalline**. Most minerals and all metals and salts exist in at least one crystalline configuration. Many materials are made up of tiny crystal grains, each no more than a fraction of a millimeter in size. Indeed, ordinary metal objects (knives, forks, and hubcaps) are all **polycrystalline** (many-crystalled). {For more on this topic click on **CRYSTALS** under **FURTHER DISCUSSIONS** on the **CD**.}

A distinguishing property of most crystalline solids is their sharp transition to the liquid state at a constant temperature. Crystals melt suddenly. Because the distances between atoms are well defined, the interatomic bonds that hold the crystal together are all nearly equal in strength, and all rupture almost simultaneously.

If a substance is composed of atoms that are *not arranged in any orderly and repetitive array,* the solid is said to be **amorphous**. Rubber, pitch, resin, plastics, and various glassy materials are examples. The atoms of glasses are strung out in an irregular mesh that lacks any *long-range* order (Fig. 9.2). What order there is, is patchy, localized, and nothing like the neat repetitive array of row upon row of atoms in a crystal. In contrast, the long-chain molecules of rubber and many plastics resemble the entwined, but still not total, chaos of a bowl of spaghetti. Noncrystalline materials (e.g., glass or tar) melt gradually over a broad range of increasing temperature.

Between these two long-recognized traditional configurations—crystalline and amorphous—there is a new form of solid matter that was only discovered in 1983. When the atoms arrange themselves in an *orderly three-dimensional pattern that nonetheless does not repeat itself with a regular periodicity,* the solid is **quasi-crystalline**.

And lastly, **composite** solids such as wood, concrete, fiberglass, bone, and blood vessel walls are composed of several different materials bonded together.

The Transition from One State to Another. When homogeneous solids (such as ice or steel) are heated, their atoms and/or molecules pick up KE in the form of random vibrational motion: they oscillate more vigorously about their equilibrium locations. Once enough energy is obtained to overcome the intermolecular forces, a solid melts. Small groupings of associated molecules still persist in the *liquid,* but as the patterns shift, they link and unlink. There is order, but it's local and changing. A web of cohesive, fairly long-ranged, intermolecular force remains, though the molecules are now energetic enough to easily move against it. They stay relatively near to each other, still interacting appreciably, but the locked, powerful bonding of the solid is gone.

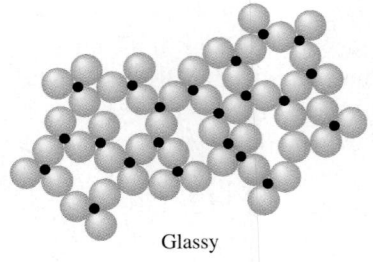

Glassy

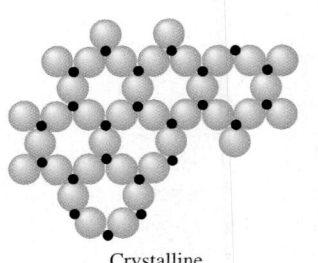

Crystalline

Figure 9.2 Glassy and crystalline solids—short- and long-range order.

Raising the temperature further brings the liquid to the boiling point, where bonds between molecules are overcome. For some molecules, their random thermal KE exceeds the cohesive potential energy, and they escape the liquid altogether. The short-range groupings disperse and the liquid vaporizes—it turns into a *gas*. {For an introduction to what holds atoms and molecules together click on CHEMICAL BONDS under FURTHER DISCUSSIONS on the CD.}

Liquids

Very few liquids occur naturally in any great quantity. Apart from water and petroleum, most others, such as gasoline, are manufactured. Not surprisingly, living organisms are typically composed of between 65% and 95% water. In fact, many of the familiar liquids—from blood and beer to berry juice—are primarily water.

The balance between bonding energy and thermal energy determines the state of an atomic system. The liquid is a transition state, existing between the random violence of the gas and the relative calm of the solid. At ordinary temperature and pressure, the liquid state is most effectively formed of either highly polarized molecules with electrically positive and negative ends (like water) or large heavy molecules (like petroleum). As with Goldilocks's porridge, the temperature of the liquid has to be "just right" with respect to the intermolecular forces—too hot and it's a gas, too cold and it's a solid.

The ability to flow, the hallmark of the fluid, varies with the cohesive force from one substance to the next—from acetone to water to motor oil to molasses to tar. This notion was first treated analytically by Newton (ca. 1687). **Viscosity** is the internal resistance, or friction, offered to an object moving through a fluid. Usually small molecules, such as water and benzene, move around easily and manifest little viscosity compared to large complex molecules, such as tar. *Much of our discussion will be simplified by treating* **ideal liquids that are incompressible and nonviscous**—characteristics well approximated by water and many other real liquids under ordinary conditions.

Gases

The alchemist van Helmont (ca. 1620) transmuted the Greek word for *chaos* into Flemish and came up with the term *gas*—and a raging chaos it is. At this very instant, each square inch of your face, and every other exposed surface at sea level, is being bombarded by 2×10^{24} air molecules every second! They pelt, ricochet, hit other molecules, and rebound, over and over in a tireless fury (powered, for the most part, by energy from the Sun). At its greatest, the density of the atmosphere at our planet's surface is only 1/800 that of water. Still, at STP there are roughly 3×10^{19} molecules careening around in every cubic centimeter. Although the distance between them is constantly changing, a typical separation is about 10 times the size of the molecules themselves or around 3 nm, which is relatively large.

An average air molecule (Table 9.3) travels at a remarkable 450 m/s, or about 1000 mi/h! This aimless bombardment goes on for each molecule in a gas at a rate of roughly 6×10^9 collisions per second. Such a jostling mass of molecules will rapidly spread out in a container until it encounters and rebounds from the walls. Swelling of its own internal violence, a gas more or less uniformly fills whatever vessel it's in.

Zinc crystals on a galvanized parking lot guardrail. Crystals like these can be seen on streetlamps and garbage cans. Old brass doorknobs etched by sweat and years of handling also show a fine-grain polycrystalline structure.

Nitrogen, the main constituent of air, becomes a liquid at a very low temperature (−198 °C). It looks and pours like water but evaporates much more quickly.

Table 9.3

The Composition of the Earth's Atmosphere	
Gas	**Percent by volume**
Dry air	
Nitrogen	78.08
Oxygen	20.95
Argon	0.9
Carbon dioxide	0.03
Neon	0.002
Helium	0.000 5
Methane	0.000 1
Krypton	0.000 1
Nitrous oxide	0.000 05
Hydrogen	0.000 05
Ozone	0.000 007
Xenon	0.000 009
Typical air	
Nitrogen	76.9
Oxygen	20.7
Water vapor	1.4
All others	1.0

No physicist would have predicted the existence
of a liquid state from our present knowledge of
atomic properties.
VICTOR WEISSKOPF (1972)
PHYSICIST

Gravity can function on a large scale to restrain a gas much as a container does. It is the
internal gravity that causes an interstellar gas cloud to contract into a star, and it is gravity
that holds down the Earth's atmosphere. A balance of molecular speed against strength of
attraction keeps that thin layer of air (99.999% lies below 90 km) from escaping and leaving us all breathless.

Fluid Statics

Liquids and gases flow, and we call both **fluids**. Their atoms and/or molecules can move
around fairly freely, and that contributes to a range of shared properties. The mobility inherent in fluids makes them crucial to all known life forms. The human body itself is a fluid-
dynamical system; we breathe, drink, bleed, and excrete fluids.

9.4 Hydrostatic Pressure

Rather than forces that act at a point on a body, we now treat ***distributed forces*** that act over
an extended area. Accordingly, **pressure** (P) is defined as *the scalar value of the force acting perpendicular to, and distributed over, a surface, divided by the area of that surface*:

For an object experiencing compression, **pressure** is the magnitude of the
perpendicular force, per unit area,
over which it acts.

$$P = \frac{F_\perp}{A} \tag{9.3}$$

Pressure is a scalar quantity; at any point, it has magnitude but no direction (Table 9.4).
The units of P are N/m^2 or pascals (Pa): $1\ N/m^2 = 1\ Pa$.

Imagine a container filled with a liquid—a cup of tea, say, or a barrel of wine. Gravity
acts on the substance and very slightly compresses the fluid, pushing molecule against molecule and ultimately transmitting the force to the bottom of the container, which pushes
upward on the liquid. Like a mound of dry sand, the liquid would run off sideways were it
not for the walls of the tank. A fluid exerts an outward push on the walls of the container,

This is the carton box for an air conditioner. What's wrong with it?

Table 9.4

A Sampling of Pressures

Source			Pressure (Pa)
Sun, at the center			2×10^{16}
Earth, at the center			4×10^{11}
Laboratory, highest sustained			1.5×10^{10}
Ocean, deepest point			1.1×10^{8}
Spiked heels			$\approx 10^{7}$
Atmospheric, Venus			90×10^{5}
		Earth	1.0×10^{5}
		Mars	≈ 700
Air, Earth	1-km altitude		90×10^{3}
	10-km altitude		26×10^{3}
	100-km altitude		≈ 0.1
Air, top of Sears Tower, Chicago			95×10^{3}
Air, top of Mt. Everest			30×10^{3}
Air, commercial jet altitude (11 km)			23×10^{3}
Solar radiation, at Earth			5×10^{-6}
Laboratory vacuum (very good)			10^{-12}

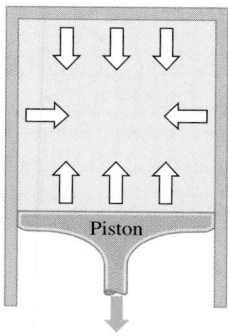

Figure 9.3 A liquid in a cylinder being stretched as the tight-fitting piston is pulled down. This is resisted by the liquid's intermolecular (electromagnetic) cohesive forces pulling inward. Pure water can exert an inward force per unit area of up to 30 MN/m².

which restrains it with a counterforce—a counterpressure. This outward pushing of the liquid is observed when water spurts from a hole in the side of a container.

The force exerted by a fluid at rest acting on any rigid surface is always perpendicular to that surface. It cannot be otherwise since the fluid has no rigidity. A wooden stick certainly could be used to push on a wall in a direction that is not perpendicular, and the same thing could also be done with a stream of water from a hose, but a fluid *at rest* only pushes perpendicularly. The forces exerted by a fluid at rest give rise to what's called **hydrostatic pressure**.

A fluid at rest within a container is typically in static equilibrium under the perpendicular compressive forces exerted by the walls. It is important to realize, however, that a liquid, by virtue of its internal (electromagnetic) cohesive force, can also sustain a tensile force (Fig. 9.3). Although liquids usually push outward, they can also pull inward. *The pressure exerted by a fluid will be taken as positive when the fluid is under compression*, as is most often the case. (Discussion Question 12 deals with negative pressures.) *We will only be concerned with positive pressures.*

The mobility of the molecules is the mechanism that transmits pressure *independent of direction*. The value of P measured at any point in a liquid is independent of the orientation of the measuring gauge. Under water or in a jet plane you can turn your head any way you like and your ears will still feel stuffy.

Gravity and Hydrostatic Pressure

Gravity is the principal cause of hydrostatic pressure. To see this, suppose there is a tank of liquid with a postage stamp in it having an area A parallel to and below the surface at a depth h (Fig. 9.4). The downward normal force on the top face of the stamp due to the liquid must be equal to the weight of the column of fluid above the stamp. Assuming the density ρ is constant (because the fluid is essentially incompressible), the mass of the column is its volume $V = Ah$ times its density; that is, $m = \rho Ah$. The weight ($F_w = mg$) of the column is ρAhg, and the average pressure on the stamp due to the liquid *only* is

$$P_l = \frac{F_w}{A} = \frac{\rho Ahg}{A}$$

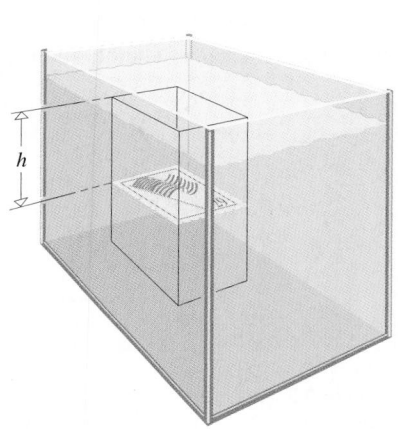

Figure 9.4 The weight of the column of liquid above the stamp, of height h, determines the pressure at that level.

or

$$P_l = \rho gh \qquad (9.4)$$

Example 9.3 [I] What pressure (due to only the water) will a swimmer 20 m below the surface of the ocean experience?

Solution The words "pressure," "water," and "below" bring to mind hydrostatic pressure. (1) TRANSLATION—Determine the pressure at a known depth in seawater due only to the liquid. (2) GIVEN: $h = 20$ m and (from Table 9.1) for seawater $\rho = 1.025 \times 10^3$ kg/m³. FIND: P_l. (3) PROBLEM TYPE—Hydrostatics/pressure. (4) PROCEDURE—So far there's only one relationship dealing with hydrostatic pressure and the depth in a liquid; $P_l = \rho g h$. (5) CALCULATION:

$$P_l = \rho g h = (1.025 \times 10^3 \text{ kg/m}^3)(9.81 \text{ m/s}^2)(20 \text{ m})$$

$$\boxed{P_l = 2.0 \times 10^5 \text{ N}}$$

or, since 1 Pa = 1.45×10^{-4} lb/in.², this is 29 lb/in.², or 29 psi, which is about the pressure in an automobile tire.

Quick Check: The pressure increase per 1 m of depth is roughly $(1 \times 10^3 \text{ kg/m}^3)(10 \text{ m/s}^2)(1 \text{ m}) = 10^4$ Pa; at a depth of 20 m, the pressure must be $\approx 20 \times 10^4$ Pa.

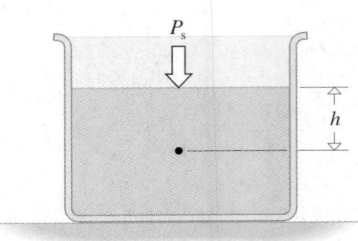

Figure 9.5 In a single liquid, the total pressure at a point a distance h beneath the surface is the sum of the pressure on the surface P_s and $\rho g h$.

By supporting a load at a number of "points," the normal force at each is small. The pressure at each "point" is then also relatively small.

When a uniform pressure P_s exists on the surface of the liquid (Fig. 9.5), **the total pressure at any point within the liquid** is the sum of the contributing pressures:

$$P = P_s + P_l = P_s + \rho g h \qquad (9.5)$$

The legendary Dutch boy who held back the North Sea by sticking his finger in a hole in the dike could really have accomplished the feat. The pressure that he had to overcome depended on h, the depth in the liquid. The force on the end of his finger acted perpendicularly, tending to pop it out of the hole like a cork in a champagne bottle. But it didn't matter whether he was holding back a leak from an ocean or a bathtub; there is nothing in Eq. (9.5) about the total amount of water present.

Pressure in a Container

It follows from successive applications of Eq. (9.3) that **the pressure at every point at a given horizontal level in a single body of fluid at rest is the same** (Fig. 9.6). Consequently, the pressure at any point on the flat horizontal bottom of a container of fluid is the same, regardless of how weirdly shaped the container may be, since P only depends on the depth beneath the open surface (Fig. 9.7). The validity of this conclusion was proven experimentally by Simon Stevin in 1586 (Fig. 9.8).

What's happening in Figs. 9.7 and 9.8 is that the walls of the containers are interacting with the liquid and participating in the process. The pressure in Fig. 9.9 at point A differs from that at point B by an amount determined by the weight of the column of liquid between A and B, namely, $\rho g \Delta h$. The pressure at B equals that at D, and so the pressure at C, also a height Δh above D, must equal that at A, even though C is under the glass shoulder. Anywhere at the level of point A, the pressure in the fluid is the same whether there is

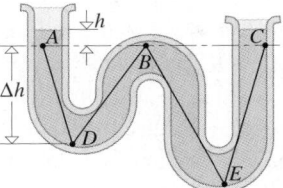

Figure 9.6 The pressure at point A is $P_s + \rho g h$. It increases in going down to D by $\rho g \Delta h$ but decreases by the same amount rising back up to B. Again, it increases even more in descending to E but drops back that same amount in rising the same vertical distance to point C. Hence, the pressure at A, B, and C is the same.

Figure 9.7 The pressure at a given depth is independent of the shape of the vessel.

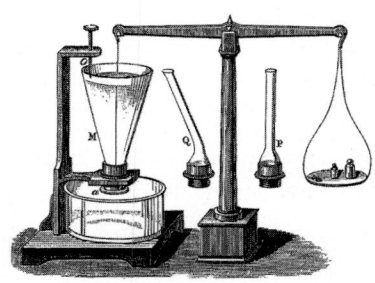

Figure 9.8 Stevin's experiment to determine the pressure on the bottoms of differently shaped containers. The balance holds up the bottom of each test vessel with an identical force equal to the load on the right-hand pan.

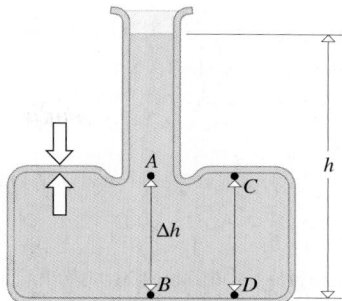

Figure 9.9 The walls of the container contribute to the process that leads to the pressure being dependent only on the depth in the liquid.

liquid or glass directly above the point. The pressure generates an upward normal force on each unit area of the horizontal shoulder, and the glass, in turn, pushes down on the liquid with the same force per unit area. The pressure is precisely equal to that produced by the weight of the column of fluid above *A* in the neck of the vessel. A guppy swimming around could not tell from the pressure whether it was under the shoulder or the neck.

Atmospheric Pressure and the Barometer

A problem plaguing the European mining community in the seventeenth century was the seepage of groundwater into their deep mine shafts. The real frustration arose from the mysterious inability of even the best suction pumps to raise water any more than 34 ft above the flood level. In Italy, Galileo was consulted about the strange phenomenon. The maestro knew that air had weight; he had determined that much experimentally by weighing a glass bulb sealed under room conditions and then reweighing it after forcing in air under pressure. Yet it was only after Galileo died that his assistant, Evangelista Torricelli, was able to recognize that the two apparently unrelated ideas were cause and effect (1643).

Being 13.6 times denser than water, mercury would presumably rise only about 30 in. instead of 34 ft and so would be far more convenient to work with in the laboratory. Torricelli sealed one end of a glass tube about 2-m long, filled it with mercury, and covering the opening with a finger, upended the tube into a bowl of mercury (Fig. 9.10). On removing his finger from the submerged end, some of the liquid poured out of the tube into the bowl, and the shiny column dropped to a height of about 76 cm (i.e., 30 in.) above the open mercury level. Furthermore, since no air had entered the tube, the space above the mercury column was apparently empty and this was essentially the first man-made vacuum. {For more details click on **VACUUM** under **FURTHER DISCUSSIONS** on the **CD**.}

"On the surface of the liquid which is in the bowl," wrote Torricelli, "there rests the weight of a height of fifty miles of air." Air pressure produces a normal force that pushes down on the open liquid surface, driving the mercury up the tube. The column settles with a height such that the downward force it exerts, at the level of the air-mercury interface, equals the upward force beneath and supporting the column. In other words, *the pressure is equal everywhere at the level of the open surface, both inside the liquid in the tube and out* (Fig. 9.11). The column of mercury effectively takes the place of a much taller, though

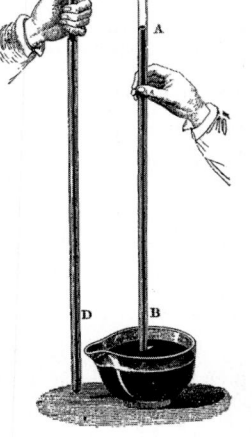

The pressure at the surface of each liquid column must be the same. Therefore, the level of each surface must be independent of the shape of the column, and all must stand at the same height.

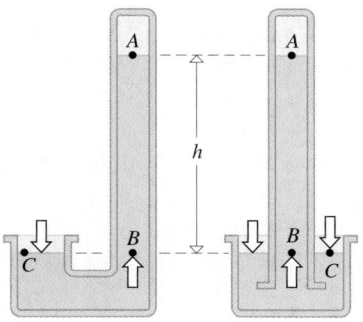

Figure 9.10 The atmosphere pressing down on the open mercury surface can support a 30-in. column.

Figure 9.11 Two equivalent versions of the barometer. The pressure at *A* is essentially zero. The pressure at *B* equals that at *C*, which is atmospheric. Hence $\rho g h$ must equal atmosheric pressure.

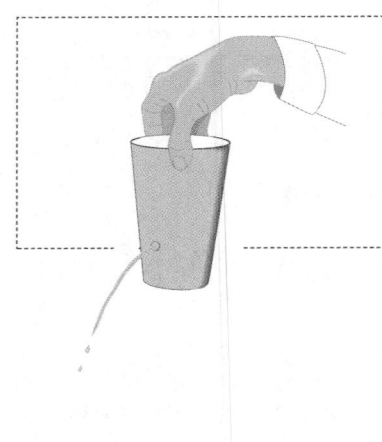

EXPLORING PHYSICS ON YOUR OWN

"Weightless" & Pressureless: For a weightless fluid, one far out in space, no internal pressure exists. (That's true provided the amount of fluid is modest, and its self-gravity is negligible. Such is not the case for something huge like a star.) Take a paper cup filled with water, punch a small hole in the side near the bottom, and watch the water pour out in a thin steam. Now drop the cup. What happens to the stream while the cup falls? What would happen if it accelerated upward?

equally weighty, column of air, producing the same pressure inside as the air produces outside. Clearly, the height of the mercury is a direct measure of the **atmospheric pressure** acting on the device. Torricelli had constructed the first mercury *barometer*.

"We live immersed at the bottom of a sea of... air," he observed—a fluid sea wherein pressures are generated just as they are in a liquid. Approximately 5×10^{18} kg of air press down on the planet, giving rise to an average sea-level atmospheric pressure (P_A) of 1.013×10^5 N/m² or, in proper SI units, 1.013×10^5 Pa. There are a number of older units for pressure that are still in use (Table 9.5). An *atmosphere* (abbreviated atm) is defined as 1 atm = $1.013\,25 \times 10^5$ Pa, which equals 14.7 lb/in². The *torr* (for Torricelli) was formerly the "millimeter of mercury" (or mm Hg), a pressure corresponding to a barometric mercury height of 1 mm and equal to 133.3 Pa. The mm Hg still survives, especially in medical literature.

One standard *atmosphere* corresponds to a barometric reading of 29.9 in. of mercury

Table 9.5

Pressure Conversions

Quantity	Equivalent in pascals
1 atm	1.013×10^5
1 lb/in². (psi)	6.895×10^3
1 mm Hg (torr)	1.333×10^2
1 in. Hg	3.386×10^3
1 in. H₂O	2.491×10^2
1 bar	1.000×10^5

Equivalent of 1 atmosphere
1.013×10^5 Pa
14.70 lb/in².
7.60×10^2 mm Hg (0 °C)
29.92 in. Hg (0 °C)
1.013 bar
2.117×10^3 lb/ft²
4.068×10^2 in. H₂O

EYE DROPPERS, STRAWS, & BUBBLES

The medicine dropper and its larger culinary cousin, the meat baster, are devices on the family tree of the barometer. The low-pressure region created by squeezing the rubber bulb and allowing it to expand back to size corresponds to the near vacuum in the top of the barometer. Air acting on the open surface of the liquid forces it up the medicine dropper tube. A good bit of residual air left in the bulb pushes down on the top of the liquid column, effectively adding to its weight. The result is that a much smaller amount of liquid is forced up the tube, which is fine, since one rarely wants a 34-ft-long eye dropper. The pipette and the drinking straw are the same machine, with the rubber bulb replaced by a mouth. A drop in mouth pressure of a few percent will operate the thing nicely, while an excess of mouth pressure will force the liquid in the straw below the open-surface level, allowing the operator to "make bubbles."

Example 9.4 **[II]** One *atmosphere* is defined as the pressure equivalent to that produced at 0°C by exactly 76 cm of mercury of density $13.595\,0 \times 10^3$ kg/m³ under standard gravitational conditions, where $g = 9.806\,65$ m/s². Show that a mercury barometer column 76.00-cm (i.e., 29.921-in.) high corresponds to an air pressure of 1.013×10^5 Pa.

Solution This is a barometer problem concerning hydrostatic pressure produced by mercury. (1) TRANSLATION— Determine the pressure equivalent to a column of mercury of known height. (2) GIVEN: Hg column 76.00-cm high. FIND: P, pressure beneath it. (3) PROBLEM TYPE—Hydrostatics/pressure. (4) PROCEDURE —The pressure is due to the weight of the column of mercury: $P_l = \rho g h$. (5) CALCULATION: Consider the barometer in Fig. 9.11. The height of the column above the

open-surface level is $h = 0.760\,0$ m. The pressure at point B is

$$P = P_s + P_l = 0 + \rho g h$$

Since there is vacuum above the mercury in the tube ($P_s = 0$), the pressure at B, at the open-surface level, is

$$P = \rho g h$$

$$P = (13.595\,0 \times 10^3 \text{ kg/m}^3)(9.806\,65 \text{ m/s}^2)(0.760\,0 \text{ m})$$

$$\boxed{P = 1.013 \times 10^5 \text{ Pa}}$$

Quick Check: This is also the pressure outside at the open surface of the mercury—namely, at point C where $P_l = 0$. Since $h = 0$, $P = P_s = P_A = 1.013 \times 10^5$ Pa. It's helpful to remember that 1 MPa $\approx$ 10 atm.

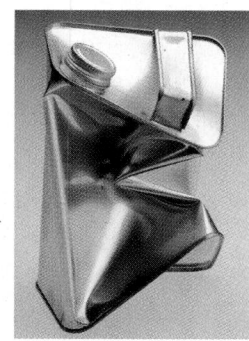

(76 cm), although air pressure ordinarily varies from 29 to 30 in. of mercury. In the central eye of a hurricane, it might be as low as 27 in. of mercury. Incidentally, water vapor is less dense than air. (A liter of moist air has a mass of 1.18 g, whereas a liter of dry air has a mass of 1.29 g.) As the air becomes drier, it becomes heavier, and a rising barometer usually portends fair weather.

Gauge Pressure

In the practical world of pumps, automobile tires, and compressed gas tanks, pressure is measured with a gauge that's set to read zero in the open air and neglects the effect of the atmosphere. This pressure above or below atmospheric is the **gauge pressure** (we will distinguish it with a subscript G). The pressure referenced to the perfect vacuum as zero is the **absolute pressure** (P). Until now we have only considered absolute pressure. It's given as the sum of the gauge and atmospheric pressures:

$$P = P_A + P_G \qquad (9.6)$$

There is nothing profound about this relationship. If we agreed to limit ourselves to absolute pressure, P_G would be superfluous. But people measure the pressure in veins, chests, and scuba tanks using gauge pressure, which makes Eq. (9.6) a practical necessity.

For the fluid in Fig. 9.12 to flow from the collapsible pouch into the vein, the gauge pressure at the needle ($P_G = \rho g h$) must exceed the gauge blood pressure in the arm. The gauge pressure in a vein (Table 9.6) is less than ≈ 2 kPa (i.e., 15 mm Hg), and so the pouch must be raised at least 20 cm above the arm.

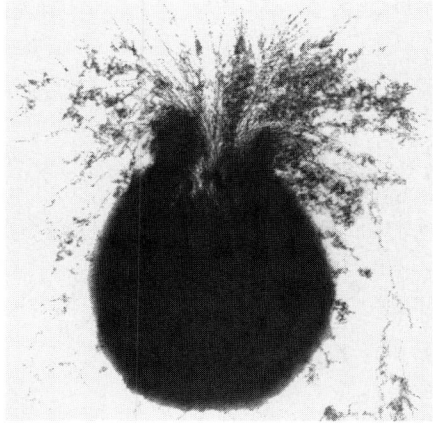

Magnified about 1.5×10^5 times, this staphylococcus bacterium explodes after being exposed to a low-level dose of antibiotic. With an internal pressure of between 25 and 30 times atmospheric pressure, the cell wall ruptured at its weakest point.

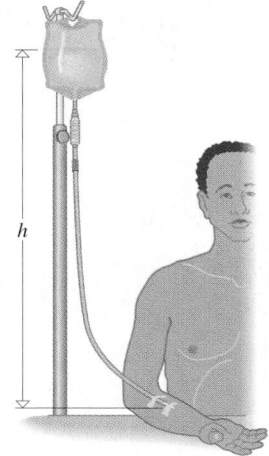

Figure 9.12 The liquid's gauge pressure at the needle is $P_G = \rho g h$, where ρ is the density of the fluid being injected.

Table 9.6

Some Fluid Pressures in the Human Body

Location	Gauge pressure	
	mm Hg	kPa
Brain, surrounding fluid	5−12	0.7−1.6
Eye, aqueous humor	12−24	1.6−3.2
Gastrointestinal	10−20	1.3−2.7
Lungs		
inhalation	−2	−0.3
exhalation	3	0.4
Chest (intrathoracic)		
inhalation	−6	−0.8
exhalation	−2.5	−0.3
Venous blood		
venules	8−15	1−2
veins	4−8	0.5−1
major veins	4	0.5
Arterial blood at heart level		
maximum (systolic)	100−140	13−19
minimum (diastolic)	60−90	8−12
Bladder		
average	0−25	0−3
during urination	15−30	2−4

This house exploded during a hurricane. The pressure outside dropped below the inside pressure, which raised the roof. Had the occupants left several windows open, the explosion would not have happened.

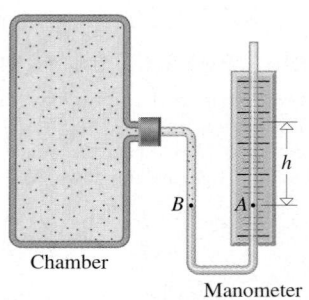

Figure 9.13 A manometer. It is usually filled with mercury, but water is also used in medical applications, such as measuring lung pressure, because mercury vapor is poisonous. Here, the gas pressure in the chamber is being measured. It is assumed that the weight of the column of gas above B is negligible. As shown, the chamber pressure exceeds atmospheric by ρgh.

Chamber

Manometer

For several centuries the only accurate practical pressure gauge was the mercury manometer (Fig. 9.13), a U-shaped open tube filled with mercury. The gauge pressure at point A equals the pressure at point B, equals the chamber pressure. When h is above or below point B, the chamber pressure is above or below atmospheric: $P_G = \pm\rho gh$. Because of the manometer, blood pressure is measured in mm Hg to this day.

Example 9.5 **[II]** A straw is to be used to raise water from a glass to a mouth 15.0 cm above the liquid's surface (at sea level). Determine the absolute pressure that must be maintained in the mouth. What is the corresponding gauge pressure?

Solution The circumstances of this problem are very much like that of a barometer. (1) TRANSLATION—Determine the pressure above a column of water needed to sustain its known height at sea level. (2) GIVEN: $h = 15.0$ cm and $\rho = 1.00 \times 10^3$ kg/m³. FIND: P_s in the mouth at the surface of the water at the top of the straw. (3) PROBLEM TYPE—Hydrostatics/pressure. (4) PROCEDURE—The column is to rise 0.150 m above the open-surface level where the pressure is P_A. In the liquid within the straw at that open-surface level $P = P_s + P_l = P_A$. (5) CALCULATION—Knowing that $P_l = \rho gh$, where h is the vertical height of liquid in the straw,

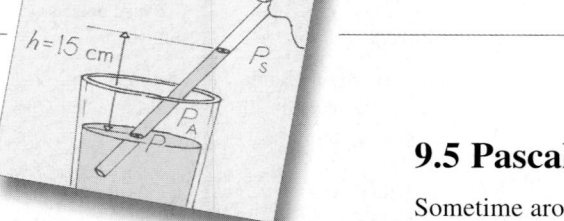

$$P_s = P_A - \rho gh$$

$$P_s = 1.013 \times 10^5 \text{ Pa} - (1.00 \times 10^3 \text{ kg/m}^3) \times$$

$$(9.807 \text{ m/s}^2)(0.150 \text{ m})$$

and the pressure at the surface of the liquid, high in the straw (i.e., in the mouth), is $\boxed{P_s = 0.998 \times 10^5 \text{ Pa}}$. That's 98.5% of atmospheric pressure. To find the gauge pressure just subtract atmospheric pressure:

$$P_G = P_s - P_A = \boxed{-1.47 \times 10^3 \text{ Pa}}$$

Quick Check: The weight of the water column of cross-sectional area A is $\rho ghA \approx (10^3 \text{ kg/m}^3)(10 \text{ m/s}^2)(0.15 \text{ m})A$, and the pressure is ≈ 1.5 kPa: air pressure must exceed mouth pressure by this amount.

9.5 Pascal's Principle

Sometime around 1651, Pascal wrote a treatise entitled *On the Equilibrium of Liquids*. It contained the first precise statement* of what has come to be known as **Pascal's Principle:**

> **An external pressure applied to a fluid confined within a closed container is transmitted undiminished throughout the entire fluid.**

When pressure is put on some region of a confined liquid (as, for example, when a piston pushes down on the liquid in a cylinder), the fluid compresses slightly and distributes the pressure uniformly everywhere therein. This process is quite different from the internal pressure generated by gravity and would exist even in a weightless liquid. Pascal's syringe (Fig. 9.14) illustrates the point nicely, as does the aerosol spray can (Fig. 9.15).

On work—long before the notion was formalized. And it is truly admirable that there is encountered in this new [hydraulic] machine the constant rule which appears in all the older machines, such as the lever, the wheel and axle, the endless screw, etc., which is, that the path is increased in the same proportion as the force.

BLAISE PASCAL
Traité de l'Équilibre des Liqueurs

---

*Archimedes had come close to it 2000 years before.

Figure 9.14 Pascal's syringe is a bottle punctured with holes and fitted with a tight piston. Pushing down on the piston increases the pressure on the fluid, which blows out the holes uniformly in all directions. This phenomenon suggests that the applied pressure is distributed equally throughout the liquid.

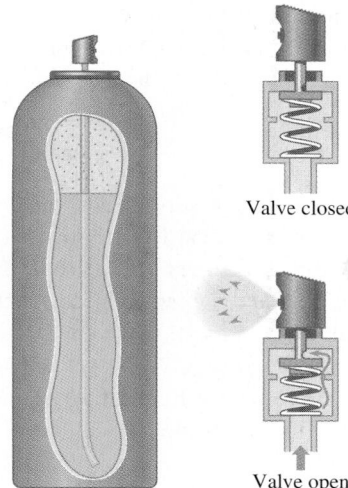

Valve closed

Valve open

Figure 9.15 An aerosol spray can contains a gas under pressure called the propellant. It pushes down on the surface of the liquid that is to be sprayed. When the valve is opened, the top end of the long tube is at atmospheric pressure, and the bottom end is at a pressure well above that. The difference propels the liquid up and out.

ON FULL BLADDERS & EARTHWORMS

One reason you don't let people sit on your belly when your bladder is full is Pascal's Principle. In the same way, pressure on the abdomen of a pregnant woman is transmitted to the fetus via the amniotic fluid. This mechanism is also responsible for the motion of soft-bodied animals, like the earthworm, that have hydrostatic skeletons. Using a mesh of perpendicular muscles, a worm squeezes itself into shape, becoming long and thin or short and fat as needed.

Hydraulic Machines

The same amount of pressure can be produced within a liquid by pistons of different sizes acting with proportionately different forces (Fig. 9.16)—the larger the cross-sectional area of the piston, the larger the force needed to create a given pressure. It was at this juncture that Pascal recognized the tremendous practical significance of his principle. For the first time since antiquity, a new class of force multipliers known as *hydraulic* machines (from the Greek for *water* and *pipe*) was possible (although a practical device was not built until Bramah devised a functioning pressure-seal in 1796).

If two chambers fitted with different-sized pistons are connected so that they share a common working fluid, the pressure generated by one will be transmitted undiminished to the other (Fig. 9.17). Nowadays, automobiles in gas stations are unceremoniously lofted into the air on hydraulic lifts. Most are activated by compressed air pressing on oil, but the simplest arrangement is a U-tube—narrow on one side, wide on the other—with sealed movable pistons at both ends. A downward input-force F_i acting over the small input-area

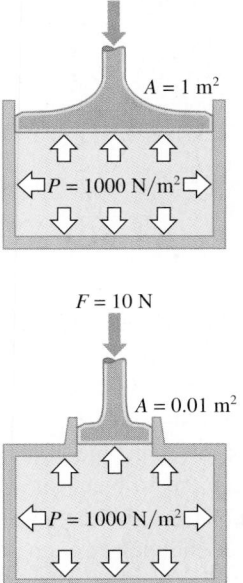

$F = 1000$ N

$A = 1$ m^2

$P = 1000$ N/m^2

$F = 10$ N

$A = 0.01$ m^2

$P = 1000$ N/m^2

Figure 9.16 The force divided by the area of the piston determines the pressure. Different piston areas and forces produce the same pressure.

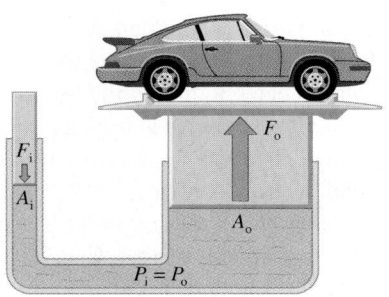

F_i

A_i

F_o

A_o

$P_i = P_o$

Figure 9.17 A hydraulic lift. The input force F_i creates a pressure F_i/A_i that is transmitted to the output cylinder, where that same pressure equals F_o/A_o.

Today, hydraulic devices that push and pull are commonplace.

A_i of the narrow piston generates an input-pressure $P_i = F_i/A_i$. But this pressure is distributed uniformly and so equals the output-pressure ($P_i = P_o$), which is given by $P_o = F_o/A_o$. Therefore, $F_i/A_i = F_o/A_o$, and

$$\frac{F_o}{F_i} = \frac{A_o}{A_i} \qquad (9.7)$$

Although the pressures in and out are equal, the forces are certainly not. When the output piston under the car has 100 times the area of the input piston, it will experience an upward force 100 times greater than the input-force. We exert a force of 200 N on a hydraulic jack, and it exerts a force of 20 000 N on the car. Like all machines, this is not a *work* multiplier: at best, when energy losses (e.g., those due to friction) are negligible, work-in equals work-out. In this example, the small piston will have to descend a distance $y_i = 100$ cm for every 1 cm the large one rises (y_o). In other words, the volume of liquid displaced at the input side is $A_i y_i$, which is equal to the volume displaced at the output $A_o y_o$. Accordingly,

$$\frac{A_o}{A_i} = \frac{y_i}{y_o} \qquad (9.8)$$

One person doing a lot of pumping over a lot of distance (y_i) can jack a car up (y_o) using a relatively small force, but *work-in ≥ work-out*.

Keep an eye out for hydraulic machines. They can be found on garbage compacters, forklifts, robots, cherry pickers, plows, tractors, airplane landing gear, elevators, and even in the brake system of the family car.

Example 9.6 **[I]** A barber's chair rests on a hydraulic piston 10 cm in diameter. The input side has a piston with a cross-sectional area of 10 cm^2, which is pumped on using a foot pedal. If the chair and the client together have a mass of 160 kg, what force must be applied to the input piston?

Solution The mention of a "hydraulic piston" suggests that this will be a Pascal's Principle problem. (1) TRANSLATION—In a hydraulic system the input and output cylinder areas are known, as is the mass of the output load; determine the input force. (2) GIVEN: $A_i = 10$ cm$^2 = 0.001\ 0$ m^2, $r_o = 5.0$ cm $= 0.050$ m, and $m_o = 160$ kg. FIND: F_i. (3) PROBLEM TYPE—

Hydrostatics/hydraulic pressure. (4) PROCEDURE—Pascal's Principle is the primary notion and $F_i/A_i = F_o/A_o$. (5) CALCULATION:

$$F_i = \frac{F_o A_i}{A_o} = \frac{m_o g (0.001\ 0 \text{ m}^2)}{\pi (0.050 \text{ m})^2}$$

$$F_i = (1569 \text{ N})(0.127\ 3) = \boxed{2.0 \times 10^2 \text{ N}}$$

Quick Check: $A_i/A_o = \pi(5 \text{ cm})^2/(10 \text{ cm}^2) = 7.85 = F_o/F_i = 1.6$ kN$/F_i$ and $F_i = 0.2$ kN.

9.6 Buoyant Force

It can be argued that the study of hydrostatics was begun by Archimedes in the third century B.C.E. The greatest physicist of ancient times, Archimedes was apparently a kinsman of Hieron II, Tyrant of Syracuse. Legend has it that the king ordered a solid gold crown to be made. But when the piece was delivered, Hieron suspected that his jeweler had substituted silver for gold in the hidden interior. Archimedes was given the challenge of determining the truth without damaging the royal treasure. After pondering the problem for some time, its solution came to him while he was musing in a warm tub at the public baths. The distracted philosopher leaped from the water and ran home naked, shouting through the streets, *"Heureka! Heureka!"* I have found it! I have found it! What he found was far more valuable than Hieron's crown.

A completely submerged body displaces a volume of liquid equal to its own volume. Experience also tells us that when an object is submerged, it appears lighter in weight; the water buoys it up, pushes upward, partially supporting it somehow. That much would be

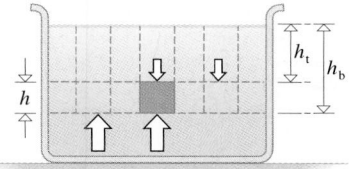

Figure 9.18 The buoyant force on a cube is the difference between the downward force on its top face and the larger upward force on its bottom face.

obvious to anyone who ever tried to submerge an inflated tire tube or a beach ball. Archimedes quantified the phenomenon. His **Buoyancy Principle** asserts that

> **an object immersed in a fluid will be lighter (i.e., it will be buoyed up)**
> **by an amount equal to the weight of the fluid it displaces.**

The upward force exerted by the fluid is known as the **buoyant force**. A 10-N body that displaces an amount of water weighing 2 N will itself "weigh" only 8 N while submerged.

Buoyant force is caused by gravity acting on the fluid. It has its origin in the pressure difference occurring between the top and bottom of the immersed object. Imagine a solid cube, each face of which has an area A, somewhere below the surface of a fluid of density ρ_f, as depicted in Fig. 9.18. The gauge pressure on the bottom, $P_b = \rho_f g h_b$, is greater than the gauge pressure on the top, $P_t = \rho_f g h_t$, and that difference $\Delta P = \rho_f g(h_b - h_t) = \rho_f g h$ gives rise to the *buoyant force* F_B. The force pushing up on the bottom is greater than the force pushing down on the top by an amount

$$F_B = A\Delta P = \rho_f g A h$$

but $Ah = V$, the volume of the body or equivalently the volume of the fluid displaced. It follows, since the mass of fluid displaced is $m_f = \rho_f V$, that

$$F_B = g\rho_f V = m_f g \tag{9.9}$$

The buoyant force equals the weight of the fluid displaced. {To see how Archimedes might have solved Hieron's problem click on **BUOYANT FORCE** under **FURTHER DISCUSSIONS** on the **CD**.}

Example 9.7 [II] An empty spherical weather balloon with a mass of 5.00 kg has a radius of 2.879 m when fully inflated with helium. It is supposed to carry a small load of instruments having a mass of 10.0 kg. Taking air and helium to have densities of 1.16 kg/m³ and 0.160 kg/m³, respectively, will the balloon get off the ground?

Solution   The balloon is to float in the air, so this is a buoyancy problem. (1) TRANSLATION—A sphere of known size is filled with a gas of known density; compare its total weight with its buoyant force.   (2) GIVEN: The bag plus the load has a mass of 15.0 kg, $\rho_{He} = 0.160$ kg/m³, $\rho_f = 1.16$ kg/m³, and $R = 2.879$ m. FIND: The buoyant force compared to the weight.   (3) PROBLEM TYPE—Hydrostatics/buoyancy.   (4) PROCEDURE— Determine the net weight and the total buoyant force.   (5) CALCULATION—To find the weight of fluid (air) displaced, we need the volume of the spherical body: $V = \frac{4}{3}\pi R^3 = 100$ m³. The buoyant force is therefore the weight of 100 m³ of air:

$$F_B = \rho_f Vg = (1.16 \text{ kg/m}^3)(100 \text{ m}^3)(9.81 \text{ m/s}^2)$$

$$\boxed{F_B = 1.14 \times 10^3 \text{ N}}$$

By comparison, the weight of the helium is

$$\rho_{He}Vg = (0.160 \text{ kg/m}^3)(100 \text{ m}^3)(9.81 \text{ m/s}^2) = 156.9 \text{ N}$$

The total weight of the helium, balloon, and load is

$$F_W = (15.0 \text{ kg})(9.81 \text{ m/s}^2) + 156.9 \text{ N} = 304 \text{ N}$$

The buoyant force exceeds the weight, and the balloon will accelerate upward rapidly.

Quick Check: $F_B \approx (100 \text{ m}^3)(1 \text{ kg/m}^3)(10 \text{ m/s}^2) \approx 10^3$ N.

Specific Gravity

The notion of density can be applied more conveniently by forming the ratio of the density of any material to that of water. Suppose an object of mass m is immersed in water such that the ***mass of fluid displaced*** is m_w. The buoyant force is $F_B = gm_w$ and if the body is weighed while submerged in water it would "weigh" that much less. Similarly, using a balance (Fig.

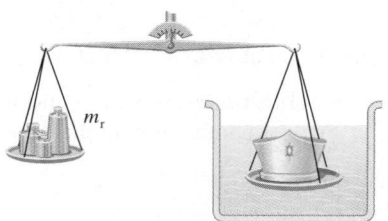

Figure 9.19 A crown of mass m, immersed in water, has a reduced mass of $m_r = m - m_w$, where m_w is the mass of water displaced. We can measure m and m_r directly and then calculate $m/(m - m_r)$, which equals the specific gravity.

9.19) to determine mass, the water would effectively reduce the measured mass by m_w. We would then directly measure the object's apparent or **reduced mass** (m_r) defined as $m_r = m - m_w$. This provides a measure of the relative density ρ/ρ_w:

> The **specific gravity** of a substance is the unitless ratio of its density to the density of water.

$$\frac{\rho}{\rho_w} = \frac{m/V}{m_w/V} = \frac{m}{m_w} = \frac{m}{m - m_r} = \text{sp. gr.} \qquad (9.10)$$

The ratio of the mass of a body to the mass of an identical volume of water is equal to the relative density, or, as it's been called for centuries, the **specific gravity**. More practically, the mass of an object in air divided by the difference between its mass in air and its measured mass in water is its specific gravity. The idea was introduced by the Persian al-Biruni sometime around A.D. 1025 in order to use unitless quantities and so avoid the muddle of different units existing throughout the Muslim Empire. The relative density tells us how many times more or less dense a material is than water. Equation (9.10) provides a convenient method for measuring the average density of a human body by submerging a person, sitting on a scale, in what is fondly called a "fat tank."

Floating

When an object weighs more than the total volume of fluid it can displace, it sinks. If the fluid is water, this is equivalent to saying that an object whose relative density exceeds 1.0 will sink. A solid cubic foot of gold, copper, concrete, or glass weighs more than a cubic foot of water and will sink in spite of the (278-N, or 62.4-lb) buoyant force.

When an object weighs less than the total volume of fluid it can displace, it will settle downward until the buoyant force equals the weight and it floats partially submerged. In water, an object whose relative density is less than 1.0 will float partially submerged. A steel ship can encompass a great deal of empty space and so have a large volume and a relatively small average density. Provided it weighs less than the maximum amount of water it can displace (i.e., provided it weighs less than the maximum buoyant force), a

A swimming pool used as a "Fat Tank" provides a measure of a person's average density.

ON FLOATING THE HUMAN BODY

People are slightly less dense than water, especially when the lungs are filled with air, and so float partially submerged. The average specific gravity varies from person to person and breath to breath. Body fat, which is ≈18% for a male and ≈28% for a female, has a specific gravity of ≈0.8. Muscle has a specific gravity of ≈1.0 and bone about 1.5–2.0. A lean, muscular body will tend to sink. In general, young people and women have lower average specific gravities, but 0.98 is a typical lungs-filled value. Thus, a person can float with at most 2% of the body out of the water. In seawater, the relative density is $(0.98 \times 10^3$ kg/m$^3)/(1.025 \times 10^3$ kg/m$^3) = 0.956$, and so about 4% of the body will be above the surface. The problem in swimming for humans is keeping the dense head above the water in such a way as to be able to breathe.

The density of the salt-rich water of Great Salt Lake is appreciably greater than that of pure water. Consequently, the swimmer can float with much of her body out of the water.

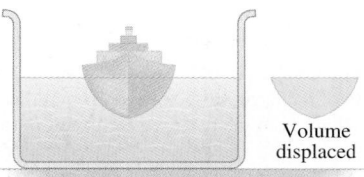

Figure 9.20 A floating object displaces its own weight of liquid.

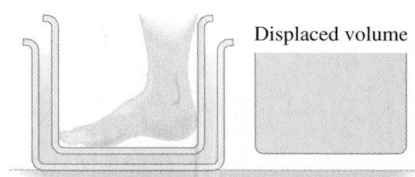

Displaced volume

Figure 9.21 A way to displace more liquid than is actually present..

A **floating object** displaces its own weight of liquid.

Most ships have markings on their hulls showing how deep they are resting in the water and therefore how heavy their cargos are.

ship—cargo and all—will float. *The weight of the water displaced by the submerged portion of a floating object equals the weight of the object* (Fig. 9.20).

The greater the density of the fluid, the more the buoyant force. That's why it's easier to swim in the ocean than in a salt-free pool and why a penny will float on mercury. A fresh egg, one that has developed little or no gas, will be relatively dense and will sink in a glass of tap water. Dissolving a few tablespoons of salt into the water will float the egg up to the surface.

Surprisingly, *it is possible to displace more fluid than might be present*, and it is the *displaced* fluid that counts with regard to floating. However paradoxical that sounds, it really isn't, and Fig. 9.21 should make the business clear. Again, the walls of the container play a crucial role as the fluid transmits the load to them. It doesn't matter how much water is surrounding the inner beaker.

When an object's weight equals the weight of the total amount of fluid it can displace—that is, equals the maximum buoyant force—it neither sinks nor rises but hovers beneath the surface in static equilibrium. Some fish accomplish this balance by storing gas in a bladder, thereby being able to remain at a fixed depth without the need to move about. To rise, or dive, or hover even without operating its engines, a submarine takes on and discharges seawater into ballast tanks.

Example 9.8 **[I]** An object of volume V floats in water with a volume V_u up above the surface. (a) Write an expression for its average density in terms of the density of water. (b) Apply this solution to find the average density of a loaded barge with 20% of its volume above the waterline.

Solution The object floats, so this is likely to be a buoyancy problem. (1) TRANSLATION—An object of known volume floats in water with a known portion above the liquid; determine its density. (2) GIVEN: The total volume of the object is V, and the volume above the water is V_u. FIND: ρ. (3) PROBLEM TYPE—Hydrostatics / buoyancy / floating. (4) PROCEDURE—Determine the buoyant force, in terms of V_u, and set that equal to the weight, in terms of V. (5) CALCULATION—(a) The volume of water displaced is $(V - V_u)$, and the buoyant force is

$$F_B = (V - V_u)\rho_w g$$

But this equals the weight of the object, which is also ρgV; therefore,

$$\rho gV = (V - V_u)\rho_w g$$

as a result

$$\rho = \frac{V - V_u}{V}\rho_w$$

(b) With $V_u = 0.20V$

$$\rho = (1 - 0.20)\rho_w = \boxed{0.80\rho_w}$$

Quick Check: It is always wise to check the extremes. To hover below the surface, $V_u = 0$ and $\rho = \rho_w$, which makes sense. For the entire object to be above water, $V_u \to V$ and $\rho \to 0$, which also makes sense.

Surface Tension

In the absence of gravity, or in an effectively weightless environment, a glob of water would draw itself into a sphere. Free-falling raindrops are roughly spherical—though only roughly because of the air friction. If we shake some mercury in a jar, thousands of tiny,

EXPLORING PHYSICS ON YOUR OWN

"Floating" Paper Clips: An old-fashioned double-edged razor or a plastic-coated paper clip can be supported on top of water by surface tension even though either one is much denser and should otherwise sink. The secret is to gently lay the clip flat onto the surface without breaking or penetrating it. Fill the cup to the very top and gently slide the paper clip onto the surface. Is the clip actually floating in the water? With the clip "floating," dip the end of a toothpick in liquid soap and touch it to the surface far from the clip. What happens almost immediately?

Astronauts in Earth orbit in the *Space Shuttle Columbia* examine a ball of water. In the effectively weightless environment, surface tension pulls the liquid into a minimum energy sphere.

A water strider resting on the surface of a pond. Its "feet," or tarsi, work something like snowshoes. They are covered with fine hairs that increase the area in contact with the liquid.

almost spherical beads will be formed on the glass, beads quite unwilling to conform to the shape of the container's walls. These familiar liquid "tricks" are the outward manifestations of internal interactions that give rise to what is called **surface tension**.

Any molecule within the body of a liquid is surrounded by other molecules. These pull on it more or less uniformly in all directions via the cohesive interaction that binds the substance together. By contrast, a molecule on a liquid's *free surface* (i.e., one bounded by gas or vacuum) will have none of its comrades above it. It will be drawn more strongly to its neighbors on the side and below. Consequently, the surface sheet of molecules reacts as if it were an elastic membrane confining the liquid, being drawn in, assuming the least area it can. About 200 years ago, Dr. Thomas Young suggested this stretchy membranous imagery—it's helpful *but should not be taken literally*.

Generally, a system under the influence of forces moves toward an equilibrium configuration that corresponds to a minimum PE. The sphere contains the most volume for the least surface area and therefore minimizes surface PE. That's why there are no cubic raindrops. Lead shot used to be manufactured by dropping molten metal within tall temperature-controlled towers. One NASA project may lead to the commercial production of perfect ballbearings in facilities orbiting the planet.

Surface tension is responsible for a range of familiar occurrences from soap bubbles to teardrops. The bristles of a fine paintbrush tend to stand apart in a bushy tuft until they are wet with paint and the surface tension pulls them together. That's why your hair, wet after a shower, is a matted clump. {For a quantitative treatment click on **SURFACE TENSION** under **FURTHER DISCUSSIONS** on the **CD**.}

Fluid Dynamics

A moment doesn't go by without each of us somehow interacting with fluids in motion. We walk, drive, and fly through the air, all the while breathing at least 6 quarts of it per minute. With blood pumping in our veins, we ourselves are *hydrodynamic* systems. On a larger

The sloped fairing above the driver's compartment channels air up and over the body of the truck, cutting drag by about 20%.

Streamlines mark the stable paths along which a fluid flows.

scale, every thriving city on Earth has a lifeline of water pouring through its veins; New York City alone consumes 1.5×10^9 gallons per day. How do liquids flow through pipes, pumps, and arteries? What happens to the pressure as air blows over the wing of a plane or up the face of a skyscraper? How can we quantify fluid flow?

9.7 Fluid Flow

Experiments by O. Reynolds (1883) on the motion of fluids in pipes showed that there are two distinct flow regimes: *laminar* and *turbulent*. Gently blow air through your lips, and the well-defined stream resembles idealized smooth flow. Cough, and the burst of air is a complex swirling turbulence that represents the other extreme.

Laminar Flow

When a fluid moves such that the velocity at any point within it is fixed, we have *steady-state motion*. In the real world where fluids have internal friction, steady-state motion usually means slow flow. The velocity may be different at different points, but consecutive specks of fluid arriving at a given spot will have a fixed velocity at that location. We can inject a trickle of dye at several points within a liquid (or smoke in a gas) and trace the paths taken by the particles of fluid as they move in orderly procession, each following the route of the particle that arrived just before it. These unchanging lines of flow (Fig. 9.22) are called **streamlines**. At every point on any one of them, the tangent to the curve is in the direction of the velocity of the fluid at that point. **The streamlines crowd together as the speed gets higher.** There is no flow perpendicular to the streamlines. No two streamlines ever cross, since that would mean that at the point of intersection a particle of fluid would have two different velocities. This kind of streamline flow is also called **laminar flow** because successive layers of fluid molecules move smoothly past one another. The boundary of a bundle of streamlines defines a *tube of flow*, wherein the fluid travels as if it were contained within an invisible pipe (Fig. 9.23).

For a normal person, about half the volume of blood is made up of cells, and so blood flow, which is smooth in most vessels, cannot be perfectly laminar, though it usually comes close.

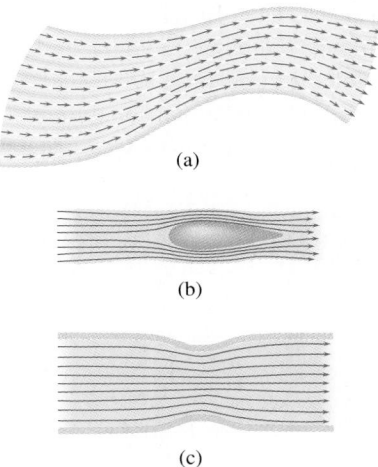

(a)

(b)

(c)

Figure 9.22 (a) The laminar flow pattern revealed by drawing the velocity vector at each point in the stream. (b) Streamlines in the laminar flow around an obstacle and (c) in a pipe. Note that the closer the lines, the higher the speed.

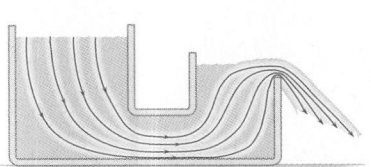

Figure 9.23 The fluid in a tube of flow travels as if in an invisible pipe, neither leaving the confines nor mixing with the contents of adjacent tubes.

This is a clear example of turbulent flow.

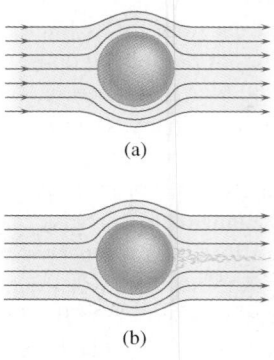

(a)

(b)

Figure 9.24 (a) An ideal fluid flows smoothly around an obstruction. (b) A real fluid cannot always follow the solid surface and forms a jumble of whirlpools that corresponds to turbulence.

Turbulent Flow

Turbulent flow corresponds to nonsteady, chaotic, changing motion (Fig. 9.24). With increasing speed, the molecules of a real fluid meeting discontinuities or obstructions swirl into little shifting whirlpools, energy-carrying vortices that are spun off as curls in the flow lines (Fig. 9.25). At high enough speeds the fluid has sufficient momentum to sail past obstructions, so the flow no longer simply conforms to the shape of those obstructions.

A real fluid streaming through pipes and around objects is influenced by its viscosity. Moreover, the molecules of a fluid easily come into intimate contact with those of a solid, and the short-range (electromagnetic) adhesive force produces appreciable effects. *The layer of fluid in contact with a solid will adhere to that surface and remain at rest with respect to it.* The speed of flow increases from zero up to the free-stream value as the distance from the surface increases (Fig. 9.26). That takes place across a relatively thin region, depending on the medium, called the **boundary layer**.

When the flow over an object is very rapid, the fluid may not be able to make a smooth transition from zero speed at the surface layer to some high value nearby. The result is instability and turbulence. This state is characterized by a diminished flow rate (there is a lot of backward swirling), substantial mixing (indeed, turbulence is required if efficient mixing is to occur), and noise (it is the turbulence at the heart valves that makes that familiar thumping sound). A healthy person breathes quietly, but obstructions in the air passages cause turbulence and noise that can be heard with a stethoscope.

For an object moving through a liquid or gas, the resistance offered by the medium increases with the turbulence. While the viscous drag force varies directly with speed, the turbulent drag force is proportional to the speed-squared. Since the power expended in overcoming drag is force times speed, that power varies as the speed-cubed! That's why

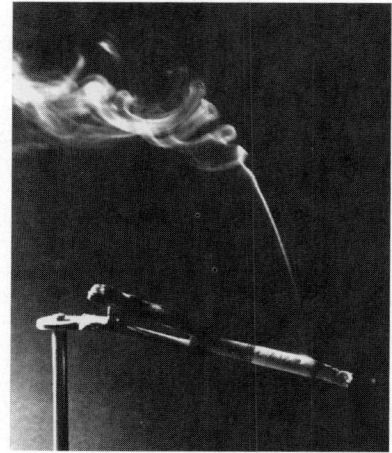

The smoke from a cigarette first rises in a laminar column and then breaks into turbulent flow.

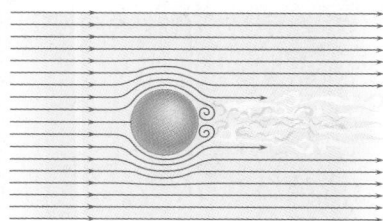

Figure 9.25 As the flow speed of a real fluid increases, its ability to follow the contours of a solid obstacle decreases. It tears away from the surface and forms a wake of turbulence that carries away energy.

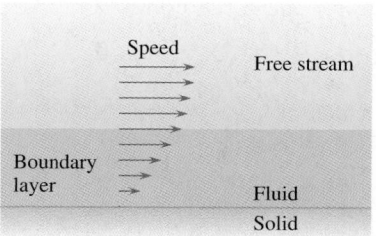

Figure 9.26 The fluid in contact with the solid surface is at rest. Its speed increases, depending on the viscosity, until it reaches the unencumbered free-stream speed, after which it's constant. The transition region is the boundary layer.

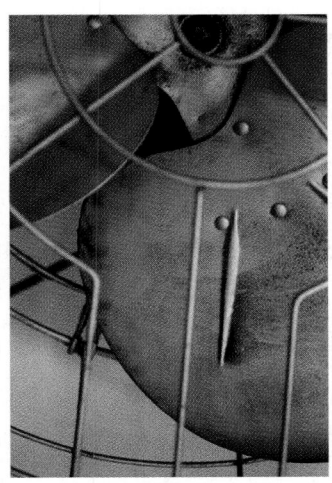

Dust caught in the boundary layer sticks to the fan blade despite its rapid motion.

auto manufacturers are finally streamlining their cars and why big square-front trucks are being retrofitted with wind deflectors, or fairings. The price of fuel demands that less energy be wasted making vortices. A fairing can reduce drag by 20% and save thousands of dollars a year. One wonders why it took the industry so long to figure out what kids, crouching on bicycles, have known all along.

9.8 The Continuity Equation

The constancy of the density of a flowing liquid is the basis of a fundamental relationship that allows us to understand how liquids progress in pipes and veins. Envision a tube of flow in a liquid (Fig. 9.27). There are no *sources* within the volume of the tube (no nozzles squirting in fluid), and there are no *sinks* (no drains tapping off fluid). The fluid enters at boundary-1, where area A_1 is perpendicular to the flow lines, and emerges at boundary-2, where A_2 is perpendicular to the flow. Let v_1 and v_2 be the average speeds of the fluid over A_1 and A_2. In a tiny time Δt, during which the particles of fluid entering the tube travel a distance $v_1\Delta t$, the particles leaving it travel a distance $v_2\Delta t$. Since the volume entering equals the volume leaving, we have

$$A_1 v_1 \Delta t = A_2 v_2 \Delta t$$

and
$$A_1 v_1 = A_2 v_2 \qquad (9.11)$$

DUST, FAN BLADES, & DIRTY CARS

Although we might expect fan blades running at high speeds to stay clean, they certainly don't. The dust particles low to the surface come into the static-air layer and are dragged along with the blades. Instead of the wind blowing the whirling blades clean, the dust remains at rest in the film of calm air adhering to the solid surface. Similarly, one might imagine that driving a car at 80 km/h would keep it clean, but it doesn't.

This is the **Continuity Equation**, and it reveals that, *as the cross-sectional area increases, the speed decreases and vice versa* (Fig. 9.28). A river flowing leisurely along a broad bed speeds up as the bed narrows and all the water rushes through a smaller cross-sectional area. If instead, the river branches into a number of identical tributaries so that the *total* cross-sectional area is increased, the speed in each narrow branch will be slower than in the river. In the human body, as the blood-carrying tubing branches from the single aorta (net area $\approx 2.5 \times 10^2$ mm^2) into the many arteries (net area $\approx 2 \times 10^3$ mm^2) into the far more numerous arterioles (net area $\approx 4 \times 10^3$ mm^2) into the billions of capillaries (net area $\approx 2.5 \times 10^5$ mm^2), the speed of flow decreases proportionately even though the individual blood vessels are getting narrower.

The quantity Av, which is constant within any tube, is denoted by the symbol J and called the *discharge rate*, *flow rate*, or **volume flux**—the latter name applies because at

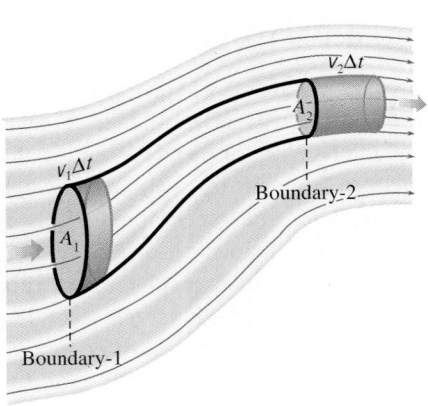

Figure 9.27 A fluid in laminar flow passes from left to right. Assuming incompressibility, the volume entering the flow tube per second at boundary-1 must equal the volume per second leaving the boundary-2, or anywhere else along its length. The flow tube, which is part of a broader stream, could also be envisioned as a pipe with a changing cross section.

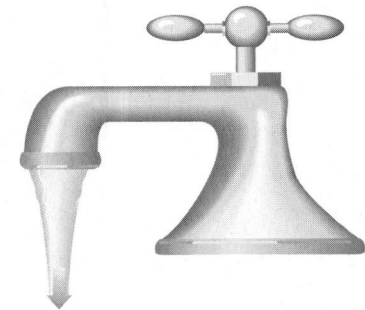

Figure 9.28 As the water falls, its speed increases, and so its cross-sectional area decreases as mandated by the Continuity Equation.

both boundaries $v = \Delta l/\Delta t$ and $\Delta V = A\,\Delta l$; therefore,

$$J = Av = \frac{\Delta V}{\Delta t}$$ (9.12)

This quantity is the volume of fluid flowing past a point in the tube per interval of time Δt and has SI units of m^3/s. Multiplying J by the density of the fluid provides the *mass flux*, or rate of mass flow in kg/s. Envision a system pumping water into a hose at a fixed rate. If a nozzle at its end reduces the area of the opening by a factor of 2, the exit speed must go up by a factor of 2—if the hose carries 0.000 5 m^3/s, the nozzle must discharge 0.000 5 m^3/s. Similarly, in the human circulatory system, the amount of blood pumped out of the heart's ventricles equals the amount returning to the atria.

Example 9.9 **[II]** Blood is pumped out of the heart via the thick-walled (2-mm) tube known as the aorta (inside diameter about 18 mm) at an average speed, for an adult at rest, of 0.33 m/s. (a) Compute the discharge rate. The aorta branches into about 32 major arteries that are each roughly the same size— namely, 4-mm inside diameter. (b) Determine the speed of the blood through these arteries. The smallest branches of the system are the capillaries, about 8×10^{-6} m in inner diameter. (c) Given that the net cross-sectional area of capillaries is 2.5×10^5 mm^2, what is the speed of flow in a capillary?

Solution Because blood is an incompressible fluid, any reference to speeds and cross-sectional dimensions should suggest the Continuity Equation. (1) TRANSLATION—A fluid moves at a known speed through a channel of known cross-sectional area; determine its flow rate. The channel branches in a known way; determine its new speed. (2) GIVEN: $v_A = 0.33$ m/s, $D_A = 18$ mm, 32 arteries with $D_a = 4$ mm, and $A_c = 2.5 \times 10^5$ mm^2. FIND: (a) J_A, (b) v_a, (c) v_c. (3) PROBLEM TYPE—Fluid dynamics/volume flux. (4) PROCEDURE—Use the Continuity Equation. (5) CALCULATION—(a) For the aorta, which we subscript with an A,

$$J_A = Av_A = \pi\left(\frac{D_A}{2}\right)^2 v_A = \pi(9.0 \times 10^{-3}\ \text{m})^2(0.33\ \text{m/s})$$

and

$$\boxed{J_A = 8.4 \times 10^{-5}\ \text{m}^3/\text{s}}$$

(b) The aorta branches into 32 arteries. To find the speed of the blood, we need the *net cross-sectional area of all of these arteries*: $A_a = 32\pi(D_a/2)^2 = 4.02 \times 10^{-4}\ m^2$. Since the flow rate from the aorta equals that in all the arteries taken together,

$$J_A = J_a = A_a v_a$$

$$8.4 \times 10^{-5}\ \text{m}^3/\text{s} = (4.02 \times 10^{-4}\ \text{m}^2)v_a$$

and

$$\boxed{v_a = 0.21\ \text{m/s}}$$

(c) Similarly

$$J_A = J_c = A_c v_c$$

$$8.4 \times 10^{-5}\ \text{m}^3/\text{s} = (2.5 \times 10^{-1}\ \text{m}^2)v_c$$

and

$$\boxed{v_c = 3.4 \times 10^{-4}\ \text{m/s}}$$

Quick Check: $A_a/A_c = (4 \times 10^{-4}\ m^2)/(0.25\ m^2) = 16 \times 10^{-4}$, which should equal $v_c/v_a = 16 \times 10^{-4}$.

9.9 Bernoulli's Equation

The first modern treatment of hydrodynamics was a work completed in 1734 by the Swiss mathematician, physicist, and physician Daniel Bernoulli. He was a disciple of Leibniz and an intellectual adversary of the Newtonian school. Shunning Newton's Laws, Bernoulli used the ideas of Huygens (p. 180) to derive the central formula of fluid dynamics. That derivation, though it came long before the concept of *energy* was formalized, turned out to be equivalent to the principle of Conservation of Energy.

The first point to consider is that a pressurized fluid must contain energy by virtue of the work done on it to establish that pressure. Pop the cap off a bottle of soda that has been well shaken, and fluid will come blasting out with plenty of KE. Clearly, potential energy is stored in the pressurized system and *a fluid that undergoes a pressure change undergoes an energy change*.

Daniel Bernoulli, 1700–1782

Now, imagine an ideal incompressible fluid for which there are no losses due to the conversion of mechanical energy into rotational, thermal, or any other form of energy. The pressure acting on a moving sample of such a fluid does work on it that appears as a net change in the kinetic and/or potential energy of the system; that is,

$$\Delta W = \Delta KE + \Delta PE_G$$

Let's first determine ΔW for a sample of fluid.

Figure 9.29 depicts a narrow tube of flow where the specified values of pressure, speed, displacement, and height are to be thought of as averages. Now to find the net work done on the disk-shaped sample of fluid by the surrounding medium. The pressure-force $F_1 = P_1 A_1$ acting on the sample, pushing it in the direction of motion, does an amount of work *on* it of $F_1 \Delta l_1$. In the process, molecules throughout the tube are shifted to the right with the effect that an equal-volume sample of length Δl_2 is displaced into region-2. This time, the fluid external to the tube pushes to the left with a force $F_2 = P_2 A_2$, which is opposite in direction to the displacement. Here, the liquid in the tube is doing work pushing on the surrounding fluid, and so the work done *on* it is $-F_2 \Delta l_2$. We want the net work done *on* the fluid sample in terms of the pressure difference. But it should also be independent of the dimensions of the arbitrary sample:

$$\Delta W = F_1 \Delta l_1 - F_2 \Delta l_2 = P_1 A_1 \Delta l_1 - P_2 A_2 \Delta l_2$$

Using the fact that $\Delta l_1 = v_1 \Delta t$ and $\Delta l_2 = v_2 \Delta t$, along with the Continuity Equation, $A_1 v_1 = A_2 v_2 = A v$, we get

$$\Delta W = P_1 A_1 v_1 \Delta t - P_2 A_2 v_2 \Delta t = A v \Delta t (P_1 - P_2)$$

Since the mass of the sample is $\Delta m = \rho \Delta V = \rho (A v \Delta t)$,

$$\Delta W = \frac{\Delta m}{\rho} (P_1 - P_2) \tag{9.13}$$

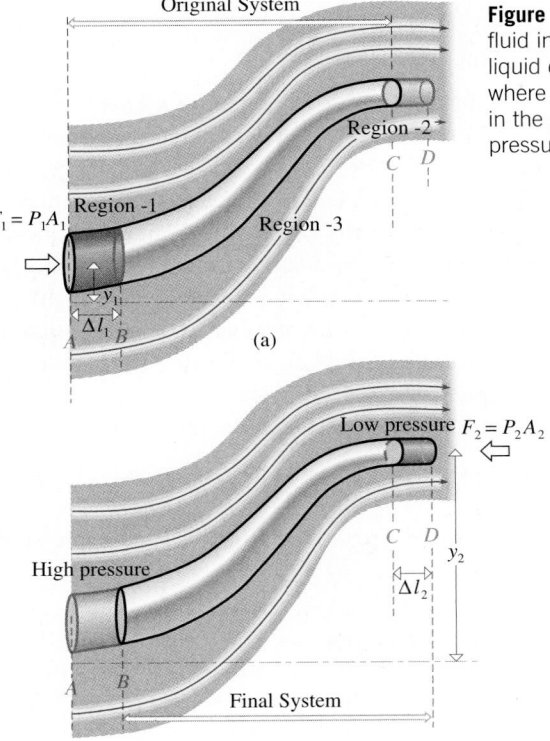

Original System

Region -2

$F_1 = P_1 A_1$ Region -1

Region -3

(a)

Low pressure $F_2 = P_2 A_2$

High pressure

Final System

(b)

Figure 9.29 The energy analysis for an incompressible fluid in laminar flow. In a time interval Δt, the tube of liquid essentially shifts from where it was in (a) to where it is in (b). A mass of fluid Δm has been raised in the gravitational field; there's been a change in pressure and a change in speed.

The sample, in moving from a region of higher pressure to a region of lower pressure, has positive work done on it by the surrounding fluid.

Now for the accompanying change in KE. The system of interest originally is the tube of liquid (Fig. 9.29a) extending from position A to C. The sample of fluid Δm is moving at v_1 in region-1 (A to B). After a time, Δt (Fig. 9.29b), that sample has moved into region-3, and an equivalent mass of fluid that was in region-3 has moved into region-2. The system now extends from B to D. In effect, Δm has gone from region-1 at speed v_1 to region-2 at v_2. The change in kinetic energy of the fluid in the tube is

$$\Delta \text{KE} = \tfrac{1}{2}\Delta m(v_2^2 - v_1^2) \tag{9.14}$$

The change in the *gravitational*-PE arises because the shifting of all the molecules in the tube has effectively brought a mass Δm from region-1 to region-2. Measuring the y values up from some arbitrary reference level,

$$\Delta \text{PE}_\text{G} = \Delta mg(y_2 - y_1) \tag{9.15}$$

Using the last three equations,

$$\Delta W = \Delta \text{KE} + \Delta \text{PE}_\text{G}$$

$$\frac{\Delta m}{\rho}(P_1 - P_2) = \tfrac{1}{2}\Delta m(v_2^2 - v_1^2) + \Delta mg(y_2 - y_1)$$

and

$$(P_1 - P_2) = \tfrac{1}{2}\rho(v_2^2 - v_1^2) + \rho g(y_2 - y_1) \tag{9.16}$$

Rearranging terms, we get **Bernoulli's Equation**

$$P_1 + \tfrac{1}{2}\rho v_1^2 + \rho g y_1 = P_2 + \tfrac{1}{2}\rho v_2^2 + \rho g y_2 \tag{9.17}$$

Because this is true at any two points on a streamline, it follows that both sides of this equation must be constant. Therefore, *in the steady flow of an ideal liquid, at all points along a streamline*

$$P + \tfrac{1}{2}\rho v^2 + \rho g y = \text{constant} \tag{9.18}$$

which is often referred to as **Bernoulli's Theorem**.

Each term has the dimensions of energy per unit volume, or energy density. There is a *kinetic-energy density* ($\tfrac{1}{2}\rho v^2$) associated with the bulk motion of the fluid, a *potential-energy density* ($\rho g y$) associated with changes in location within the gravitational field, and a *pressure-energy density* (P) arising from internal forces on the moving fluid. This latter energy is similar to that stored in a compressed spring in the sense that it's a manifestation of the interatomic electromagnetic interaction. In effect, Eq. (9.18) says that, as long as no energy enters or leaves, *the net energy density contained in the fluid is constant all along a given flow tube*. Though it was derived for an incompressible fluid, Bernoulli's Equation can be applied to gases, provided the pressure changes are only a few percent.

Notice that when the fluid is at rest, $v_1 = v_2 = 0$, and Eq. (9.17) becomes

$$P_1 - P_2 = \rho g(y_2 - y_1)$$

which is equivalent to the result (p. 292) for the pressure difference between any two points in a liquid.

Torricelli's Result

Consider the closed tank of liquid shown in Fig. 9.31 that has an orifice of area A_2, out of which is pouring the fluid. The rate at which the fluid leaves, the *speed of efflux* (v_2), is the quantity we want to determine. The vessel is taken to be very large so that neither the liquid level nor the pressure above it (P) changes significantly. Since the flow tube goes from

STUDY GUIDE

When analyzing a problem concerning the flow of an incompressible fluid, the presence of information about *heights* and *pressures* usually suggests the application of **Bernoulli's Equation**.

Draw the streamlines showing the flow from region-1 (where you have data) to region-2 (where you want to determine P, v, or h).

Apply the Continuity Equation if you need to compute the speed at region-1 from that at region-2, or vice versa.

The pressure just inside a liquid at a free surface (i.e., one bounded by air) is atmospheric. That's true whether the liquid is at rest in an open container or squirting into the air.

Set Bernoulli's Theorem [Eq. (9.18)] for region-1 equal to that for region-2.

Example 9.10 **[II]** Figure 9.30 depicts an uncovered vat of brewing beer and the narrow pipe used to take samples from it. The cross-sectional area of the vat is 1.50 m². At a given instant, the liquid level is falling at 1.0 cm/s, while the beer is traveling at 50 cm/s past the gauge. Determine the absolute pressure reading at that point within the pipe, at that moment. Take the density of beer to be 1.0×10^3 kg/m³.

Solution The problem talks about a flowing liquid, and so we think about the Continuity Equation; in addition, any mention of pressure calls to mind Bernoulli's Equation. (1) TRANSLATION —A fluid (of known density) at one point on a streamline at a known height and pressure moves at a known speed. It arrives at a second point on the streamline at a known height and speed; determine the pressure at that point. (2) GIVEN: Gauge 2.0 m below surface, area of tank 1.50 m², speed of surface 1.0 cm/s, and speed in pipe 0.50 m/s. FIND: Pressure in pipe. (3) PROBLEM TYPE—Fluid dynamics. (4) PROCEDURE—Use Bernoulli's Equation. (5) CALCULATION—Take the influx end of a tube of flow as region-1 located at the open surface of the liquid in the vessel. Take region-2 in the pipe near the pressure gauge. Since the reference level is arbitrary when dealing with potential energy, let $y_2 = 0$. The height of the liquid surface (above the gauge) is $y_1 = h = 2.0$ m. Then

$$P_1 + \tfrac{1}{2}\rho v_1^2 + \rho g y_1 = P_2 + \tfrac{1}{2}\rho v_2^2 + \rho g y_2$$

becomes $\quad P_A + \tfrac{1}{2}\rho v_1^2 + \rho g h = P_2 + \tfrac{1}{2}\rho v_2^2 + 0$

and $\quad P_2 = P_A + \tfrac{1}{2}\rho(v_1^2 - v_2^2) + \rho g h$

$$P_2 = 1.013 \times 10^5 \text{ Pa} + \tfrac{1}{2}(1.0 \times 10^3 \text{ kg/m}^3)[(1.0 \times 10^{-2} \text{ m/s})^2 - (0.50 \text{ m/s})^2] + (1.0 \times 10^3 \text{ kg/m}^3)(9.81 \text{ m/s}^2)(2.0 \text{ m})$$

$$P_2 = 1.013 \times 10^5 \text{ Pa} - 1.25 \times 10^2 \text{ Pa} + 1.96 \times 10^4 \text{ Pa}$$

and $\qquad \boxed{P_2 = 1.2 \times 10^5 \text{ Pa}}$

Quick Check: The pressure at a point that is 2 m down must be $P_A \approx 100$ kPa, plus the contribution due to the weight of the column of fluid $\rho g h \approx (10^3 \text{ kg/m}^3)(10 \text{ m/s}^2)(2 \text{ m}) \approx 20$ kPa, plus a negligible drop in pressure (≈ 0.1 kPa) due to an increase in speed, for a grand total of ≈ 120 kPa.

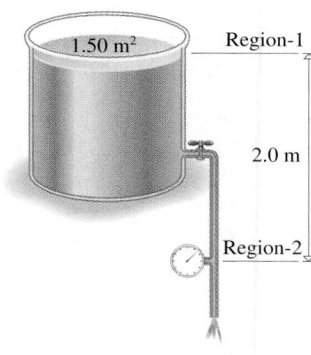

Figure 9.30 A vat of beer is sampled via a narrow tube with a pressure gauge mounted 2.0 m below the surface.

region-1 out to region-2, which is open in the air, let $P_1 = P$. The emerging stream in region-2 is surrounded by air at atmospheric pressure, and nothing is there to sustain a pressure difference. *A free fluid jet must be at the same pressure as the enveloping air*, $P_2 = P_A$. Since $y_1 - y_2 = h$, Bernoulli's Equation is then

$$P + \tfrac{1}{2}\rho v_1^2 + \rho g h = P_A + \tfrac{1}{2}\rho v_2^2$$

and $\qquad v_2^2 = v_1^2 + \dfrac{2(P - P_A)}{\rho} + 2gh \qquad\qquad (9.19)$

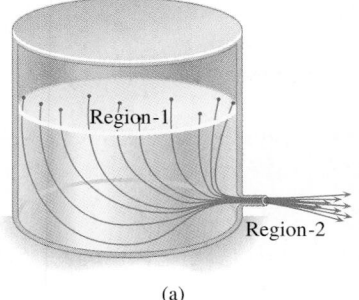

(a)

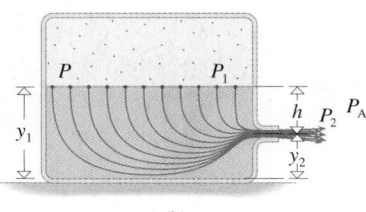

(b)

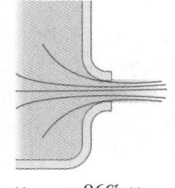

$v_{\text{actual}} \approx 96\% \; v_{\text{ideal}}$

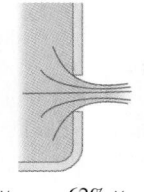

$v_{\text{actual}} \approx 62\% \; v_{\text{ideal}}$

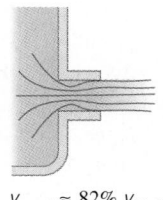

$v_{\text{actual}} \approx 82\% \; v_{\text{ideal}}$

(c)

Figure 9.31 (a) The streamlines for a fluid flowing out of a closed vessel, where the pressure on the liquid's surface is *P*. (b) To apply Bernoulli's Equation, we take the top of the liquid as region-1, the outside of the nozzle as region-2, and follow a streamline from 1 to 2. (c) In actuality, the shape of the orifice affects the shape of the jet and its speed.

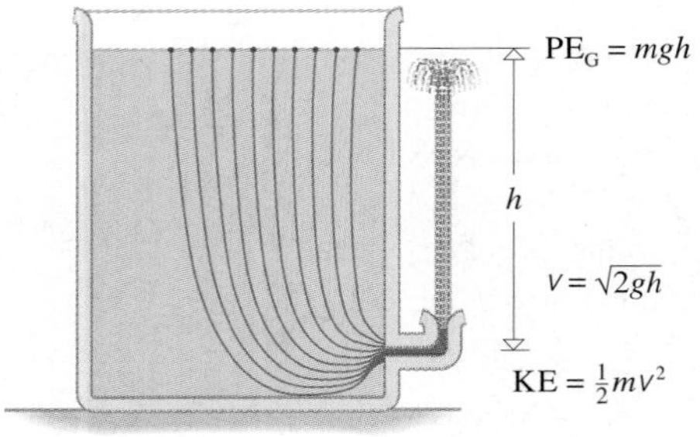

Figure 9.32 The jet reaches almost the height of the surface level. The difference is due to friction.

From the Continuity Equation ($v_1 A_1 = v_2 A_2$), it follows that, when $A_1 \gg A_2$ and $v_2 \gg v_1$, then v_1^2 is negligible compared to v_2^2. In the special, though common, case where the tank is open to the air ($P = P_A$), the pressure-energy density vanishes—the fluid runs out via gravity without also being pushed out by a pressure difference, and

$$v_2 = \sqrt{2gh} \tag{9.20}$$

This relationship is **Torricelli's Result**.

    Being frictionless, in theory at least, the liquid spurts from the hole with a speed equal to that which it would have gained in free-falling through the height h. Torricelli observed that the emerging jet, if directed straight up, would almost reach the original level (Fig. 9.32). The reason the jet doesn't quite attain the initial height is that some energy is invariably converted to thermal energy via friction.

Example 9.11 [III] A boat on a lake crashes into a submerged rock, ripping a hole 40 cm² in its hull 1.00 m below the waterline. At what speed does the water pour in? If the craft can take on 10.00 m³ of water before getting its cargo drenched, roughly how much time does the crew have to do something clever? Assume the water flows straight in the hole and neglect any frictional effects.

Solution   The problem mentions a flowing liquid, and we therefore think about the Continuity Equation; any suggestion of a pressure difference also calls to mind Bernoulli's Equation. (1) TRANSLATION—A fluid (of known density) at one point on a streamline at a known height and pressure moves at a known speed. It arrives at a second point on the streamline at a known height and pressure; determine the speed at that point and flow rate through a hole. (2) GIVEN: Height below waterline $h = 1.00$ m and area of hole 40×10^{-4} m². FIND: The speed of the water pouring in (v_2) and time to take on 10.00 m³. (3) PROBLEM TYPE—Fluid dynamics. (4) PROCEDURE—Use Bernoulli's Equation. (5) CALCULATION—Let region-1, the input end of a flow tube, be at the surface of the lake and region-2, the output end, be at the hole—water flows down and into the boat. Assuming there is atmospheric pressure in the boat at the hole,

$$P_1 + \tfrac{1}{2}\rho v_1^2 + \rho g y_1 = P_2 + \tfrac{1}{2}\rho v_2^2 + \rho g y_2$$

and $P_1 + \tfrac{1}{2}\rho v_1^2 + \rho g(y_1 - y_2) = P_2 + \tfrac{1}{2}\rho v_2^2$

Since the level of the lake is not going to fall very rapidly,

$$P_A + 0 + \rho g h = P_A + \tfrac{1}{2}\rho v_2^2$$

which gives us Torricelli's Result

$$v_2 = \sqrt{2gh} \tag{9.20}$$

We should have anticipated this to begin with—the container with the leak is the whole lake pouring water into the boat, so

$$v_2 = \sqrt{2(9.81 \text{ m/s}^2)(1.00 \text{ m})} = \boxed{4.43 \text{ m/s}}$$

Now $J = v_2 A_2 = (4.43 \text{ m/s})(40 \times 10^{-4} \text{ m}^2)$

and so $J = 1.77 \times 10^{-2} \text{ m}^3/\text{s}$

It will take on 10.00 m³ of water in a time of

$$\frac{10.00 \text{ m}^3}{1.77 \times 10^{-2} \text{ m}^3/\text{s}} = 565 \text{ s} = \boxed{9.4 \text{ min}}$$

This solution assumes the boat doesn't sink appreciably lower in the water in the process, thereby changing h and so J.

Quick Check: The water effectively falls 1 m; hence, $y = \tfrac{1}{2}gt^2$ and $t = 0.45$ s; thus, $v = gt = 4.43$ m/s. $J \approx 160 \times 10^{-4} \text{ m}^3/\text{s}$; 10 m³/$J \approx 10$ min.

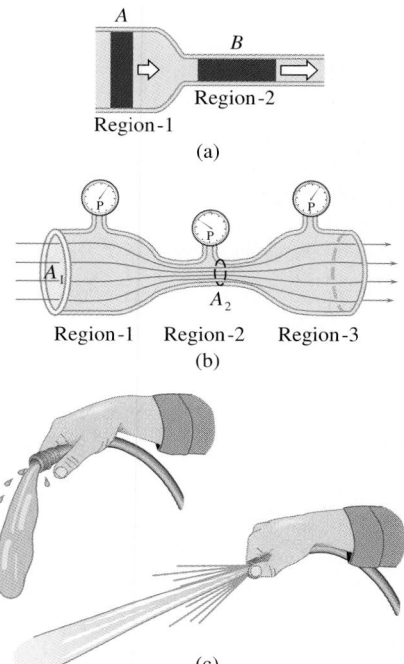

Figure 9.33 (a) In region-1, the cross-sectional area is large, the speed is low, and the pressure is high. In region-2, the area is small, the speed is high, and the pressure is low. (b) In region-3, the area is again large, the speed is small, and the pressure is high. (c) If you've ever played with a garden hose you've experienced this effect first hand.

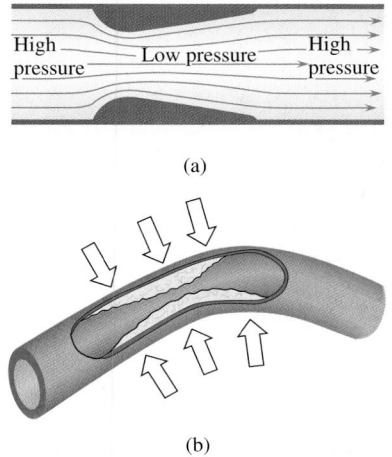

Figure 9.35 (a) The flow of a fluid through a constricted channel. (b) The same thing happens when plaque builds up in an artery.

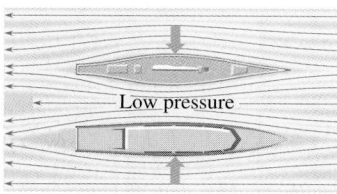

Figure 9.34 Two ships traveling side by side or moored in moving water. The deflection of the flow between the two vessels causes a Venturi drop in pressure, and the ships experience a force pushing them together.

The Venturi Effect

Some of the most practical applications of hydrodynamics (a word coined by Bernoulli) arise from the interdependence of pressure and speed. For example, consider Fig. 9.33a where a slug of water travels from the wide region into the narrow one. The liquid is incompressible so the volume of the slug is constant even though it elongates. And because it elongates, the water must be traveling faster in the narrow tube than in the wide one. Clearly, the slug of water must have accelerated, and that could only happen if it experienced a net force. Hence, there must be a higher pressure in the wide region than in the narrow one.

To see this more analytically examine Fig. 9.33b. Since changes in *gravitational*-PE along any streamline are ignorably small, Bernoulli's Equation relates differences in pressure to differences in KE and, therefore, speed. With transverse areas $A_1 > A_2$, and a horizontal pipe, $y_1 = y_2$, Bernoulli's Equation becomes

$$P_1 + \tfrac{1}{2}\rho v_1^2 = P_2 + \tfrac{1}{2}\rho v_2^2$$

We know from the Continuity Equation that the fluid speeds up at the constriction, $v_2 > v_1$, and therefore $P_2 < P_1$:

Rapidly moving fluids sustain less pressure than slowly moving fluids.

This important result is known as the *Venturi Effect*, named for the Italian researcher who first studied it (1791). As a large truck or bus passes alongside a car, the pressure between the two will often drop from the rushing air, giving the driver of the auto the disturbing feeling of being pulled toward the larger vehicle (Fig. 9.34).

It is possible to mechanically increase the speed of a fluid, as for example, with a pump. The KE thereby added comes from the work done on the fluid. Thus, the pressure in a high-speed free jet produced by a pump is still atmospheric. That's not to be confused with the Venturi Effect, where no additional energy enters the system and an increase in KE must come from a decrease in pressure energy, and vice versa. {For a more detailed analysis click on **VENTURI EFFECT** under **FURTHER DISCUSSIONS** on the **CD**.}

Airfoils

An **airfoil** is an object whose shape is such that it generates a desired reaction force from the fluid through which it moves. For instance, the wing-shaped object shown in Fig. 9.36 is an asymmetrical airfoil, and so, unlike Fig. 9.35a, the streamline pattern is asymmetrical. Air moves with respect to the long, curved upper surface, at a more rapid speed than the airstream beneath the wing. With an appropriately shaped wing, the corresponding drop in pressure above, as compared to below, creates a net upward force known as **lift**. Lift makes possible the level flight of powered airplanes, holds gliders in the air, and supports soaring birds.

Still, it's not clear at this point why the air moves at a higher speed above the wing than below. Most popularized versions suggest that the two streams must take the same time to arrive at the rear edge and so travel at different speeds, but that explanation doesn't account for how the air finds out that it has a scheduled arrival time to meet, nor, in fact, is it even true—the traversal times need not be the same. What actually happens is affected by the viscosity of the air and its adhesion to the wing. The airfoil induces a circulation of

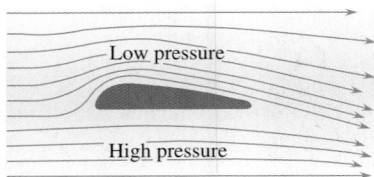

Figure 9.36 Flow past a horizontal asymmetrical airfoil. Note the high-speed, low-pressure regions where the flowlines are closer together. Compare this flow pattern with that of Fig. 9.35*a*.

VASCULAR FLUTTER & SNORING

Arteriosclerosis arises when plaque builds up on the inner walls of the arteries, restricting the flow of blood. Any such obstruction results in a Venturi pressure drop (Fig. 9.35). In an advanced stage of the disease, the pressure difference, inside and out, can be great enough to cause the artery to collapse momentarily. Backed-up blood pressure will soon force it open, only to collapse again and so on, producing a vascular flutter that can be heard with a stethoscope. The Venturi Effect in airstreams is responsible for other flutterings, such as the sound of snoring and the more melodious tones of reeded woodwinds.

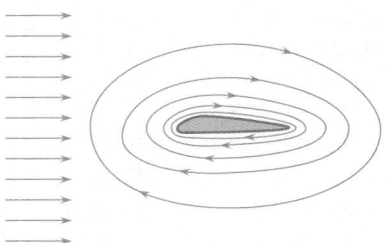

Figure 9.37 Circulation around an airfoil, in cross section. The sum of the circulation and the free-air flow produces the pattern in Fig. 9.36, where the speed is higher above the airfoil.

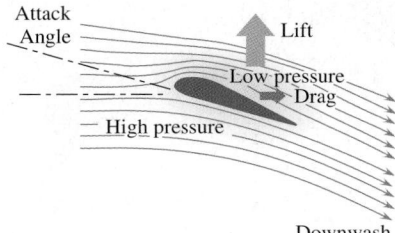

Figure 9.38 The downwash tells us that there is an upward reaction force exerted on the airfoil by the deflected air stream. The time rate-of-change of the momentum of the air must equal the magnitude of the upward force.

air (Fig. 9.37) around the wing. This circulation combines with the free-air flow to produce a higher speed stream above than below the wing.

The amount of circulation generated depends on the shape of the airfoil, its speed, and its orientation with regard to the air flow. For example, a plane can fly horizontally, nose up, with its wings tilted at the so-called *attack angle*. The pressure above drops and the pressure below rises even more, thereby increasing the lift (Fig. 9.38). There is a practical limit to the process (typically around 15°): too much tilt and the boundary-layer air tears away from the top of the wing, becoming turbulent—the wing *stalls*, lift vanishes, and the plane falls.

There's a popular engineering adage that if a plane's engines have enough power, they can drag almost anything into the air. That being the case, the curvature of a plane's wings becomes less important. Wings that are symmetrical airfoils get their lift primarily by being pushed through the air at a nonzero attack angle—in effect, the air moving at the underside is deflected downward and transfers an upward momentum to the wing. That's how planes manage to fly upside down—if the only thing occurring was a reduced pressure due to wing curvature, the craft would be driven into the ground by the inverted "lift," but that doesn't happen.

A simple argument can be made, via Bernoulli's Equation (Problem 123), showing that *lift is proportional to the product of air speed-squared and wing area*. This relationship turns out to be true, and it explains why big planes and big flying birds have relatively big wings. It's also why big birds have a higher minimum air speed that they must reach to take off and must maintain to stay airborne.

The skier in midair becomes an airfoil.

Figure 9.39 City winds and vortices. Poorly designed and positioned buildings can create dangerous windstorms at street level. The average person can be blown off balance by a 65-km/h (i.e., 40-mi/h) wind, and turbulent winds of 16 km/h (i.e., 10 mi/h) impede walking. A gust of wind in Boston's Copley Square once blew over a half-ton mail truck. High winds can also be created by building-lined streets via the Venturi Effect..

A satellite photo of vortices in the cloud cover formed in the wake behind an island in the ocean.

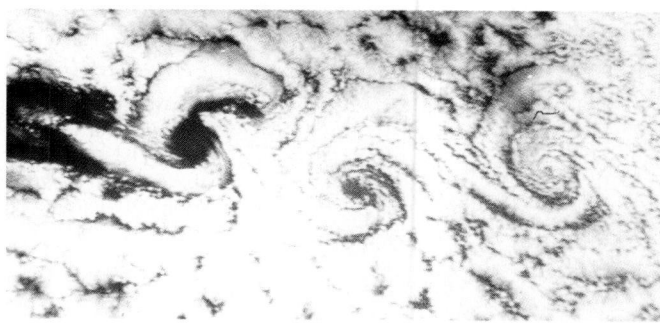

A vortex street.

This photo of the Great Red Spot on the surface of Jupiter was taken by *Voyager 2*. The spot is a giant stable vortex that has been sustained in the deep atmosphere of that rotating planet for at least 300 years.

Vortices

A **vortex** is a whirling mass of fluid bounded by a region that is not rotating. Tornadoes and hurricanes are large-scale powerful vortices; the whirling spiral of water draining from a bathtub and the great mushroom cloud of an atomic bomb are also vortex structures. Whenever an object moves with respect to a fluid (a plane through the air or an oar through the water), it tends to spawn vortices. Small invisible whirlwinds are everywhere—behind moving ships and cars and planes and baseballs. Stirring a cup of coffee creates vortices, which is why it's done—to churn up turbulence that will disperse the ingredients.

Airfoils create vortices at their two ends, due to the difference in pressure at top and bottom. Air flows outward away from the fuselage, up and around the wing tips from the high-pressure region beneath to the low-pressure region above the wings. Figure 9.39 shows how different city situations can also spawn troublesome vortices. These vary from the street-level variety that steals hats to the high-altitude kind that blows out windows.

Imagine a cylindrical body immersed in a moving fluid (Fig. 9.40). Generally, there will be a more or less irregular flow pattern behind the cylinder. But if conditions are right (the speed of the fluid is high enough, its viscosity low enough), two parallel rows of equally spaced alternating eddies will be created—known as a *von Kármán vortex street*. The periodic shedding of vortices by an obstruction results in an oscillating pressure on the body that often tends to set it vibrating in a direction *perpendicular* to the flow. During World War II, the vibration of machine-gun barrels jutting from airplane turrets was a frequent problem; skyscrapers are driven into oscillation by the same vortex mechanism (p. 358).

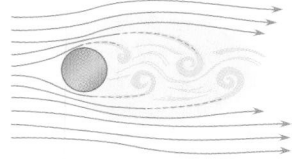

Figure 9.40 A fluid moving past a cylindrical obstruction. If conditions are right, an alternating series of vortices will be formed in the wake.

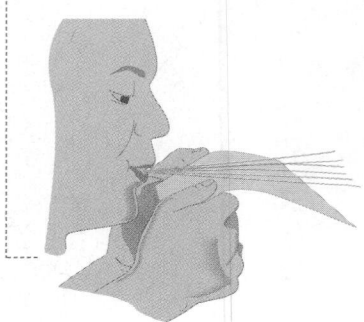

EXPLORING PHYSICS ON YOUR OWN

Making Vortices: Rows of vortices are easy to generate by moving a cylindrical object such as a pen or a finger at various speeds across the surface of a deep pan of water. Blow continuously and gently across the top of a bent downwardly hanging piece of paper. There'll be a pressure drop in the high-speed stream above the paper, and it will smoothly rise up. Now blow in hard bursts and the paper will violently flutter. What's happening?

Tip vortices made visible during crop spraying. The presence of powerful vortices behind large aircraft, such as the Boeing *747*, determines the maximum rate at which planes can follow one another down an airport runway.

Blood Pressure

The human heart is actually two twin-chambered pumps (Fig. 9.41) mounted in a single unit: a low-pressure device (the right atrium and ventricle) and a high-pressure device (the left atrium and ventricle). The right half delivers oxygen-depleted blood to the nearby lungs. With each *systole* or rhythmic contraction of both ventricles, the right half discharges 70 to 80 cm³ of blood at a mean gauge pressure of 2 kPa (i.e., 15 mm Hg) into the pul-

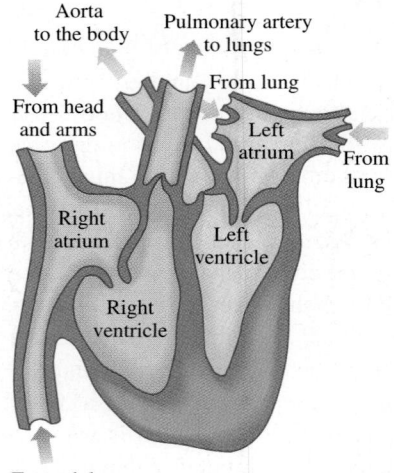

Figure 9.41 Both atria contract at the same time, forcing blood into their corresponding ventricles. Thereafter, the ventricles contract, forcing blood into the arteries and out to the body and lungs. The bluish regions carry oxygen-depleted blood; the reddish regions carry oxygen-rich blood.

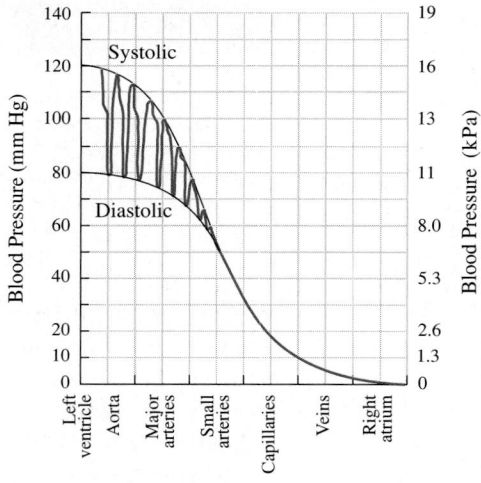

Figure 9.42 Blood pressure in the adult human male body. As the heart goes through its pumping cycle, the pressure in the main vessels fluctuates. The systolic and diastolic pressures are the maximum and minimum values developed in the arteries. Typical female pressures are slightly lower (110 mm Hg/70 mm Hg).

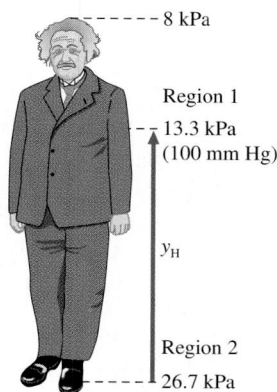

-------- 8 kPa

Region 1
-- 13.3 kPa
(100 mm Hg)

y_H

Region 2
------ 26.7 kPa

Figure 9.43 Blood pressure varies over the body, in part due to gravity. A person who jumps up from a prone position can feel momentarily dizzy because of the sudden drop in blood pressure. Similarly, standing still can cause blood to pool in the legs, and dizziness can again occur. Fainting usually lowers the head.

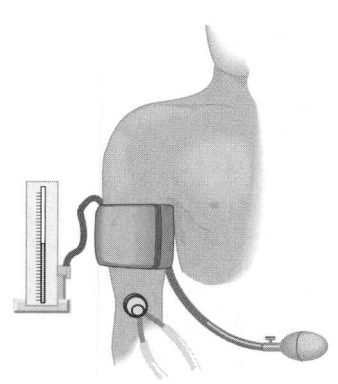

Figure 9.44 The mercury manometer reads the gauge pressure in the cuff wrapped around the arm. The small rubber bulb pumps up the cuff until it cuts off the blood flow in the brachial artery in the upper arm. The pressure is released gradually, and with the use of a stethoscope, the blood can be heard to start flowing.

monary artery on the way to the lungs. An equal amount of oxygen-rich blood is returned from the lungs to the left half of the heart (at ≈8 mm Hg) for delivery to the rest of the body. The left ventricle pumps the blood out, maintaining a pulsating pressure that rises and falls 70 to 80 times per minute. The peak, or *systolic pressure*, developed during pumping (Fig. 9.42) is around 16 kPa (i.e., 120 mm Hg), and the minimum or *diastolic pressure*, which occurs when the heart is essentially relaxed and refilling, is around 11 kPa (i.e., 80 mm Hg).

Blood is a viscous fluid, and there will be some loss in energy and a corresponding drop in pressurization along the flow route—a real liquid drops in pressure as it negotiates even a smooth channel. The effects of viscous drag become less troublesome as the pipe gets larger and more of the fluid is moving farther from the walls. Arteries have relatively large cross sections, and blood progresses through them with little frictional loss and little diminution of pressure energy. That's why the pulses of blood pressure are most easily detected with arteries, especially ones close to the surface at the neck and wrist.

Insofar as viscous effects in the main arteries are not significant, Bernoulli's Equation applies. Since the flow rates and cross-sectional areas in the main arteries are fairly uniform, the speeds of flow are approximately equal as well. Let's compare the pressure at the heart (P_H) and feet (P_F). Take these to be regions-1 and -2, respectively (Fig. 9.43). For someone standing, $y_1 = y_H$ and $y_2 = 0$ and $P_H + \rho g y_H = P_F$.

In effect, the weight of the blood causes an increase in pressure at the feet, just as it would in the static case. This is why it's so civilized to have a nap after lunch—the heart doesn't have to work quite as hard pumping against gravity.

The most common way to measure arterial blood pressure, though it's not very accurate, is with an inflatable bag or cuff attached to a manometer (Fig. 9.44). The cuff is tightly wrapped around the upper arm *at the level of the heart* (so there is no $\rho g y$ contribution to P) and then inflated with a rubber bulb pump. Squeezing the arm, it cuts off blood flow in the brachial artery. The overpressure in the cuff is then allowed to gradually drop while the observer listens with a stethoscope placed downstream on the artery. When the pressure in the cuff (as indicated by the manometer) equals the peak or systolic gauge pressure, blood begins to spurt past the constriction, and the turbulent, noisy flow is easily heard in the stethoscope. Further reduction of the cuff pressure is accompanied by a changing sound until at the diastolic pressure the flow is again laminar and the characteristic noise of turbulence vanishes (Table 9.7).

Table 9.7

The Human Cardiovascular System	
Blood density (37 °C)	$1.059\,5 \times 10^3$ kg/m^3
Number of red cells per mm^3 of blood	5×10^6
Number of white cells per mm^3 of blood	8×10^3
Flow rate at heart	
resting	8×10^{-5} m^3/s (5 liters/min)
intense activity	40×10^{-5} m^3/s (25 liters/min)
Time for tracer to circulate	18−24 s
Time for complete circulation	54 s
Blood volume (70-kg male)	5.2×10^{-3} m^3 (5.2 liters)
Gauge pressure, right ventricle, peak	3 kPa (25 mm Hg)
Gauge pressure, left ventricle, peak	16 kPa (120 mm Hg)
Mean gauge pressure, large veins	1.07 kPa
Mean gauge pressure, large arteries	12.8 kPa
Mean gauge pressure, at the heart	13.3 kPa
Pressure drop per cm of aorta	4.3 Pa (3.2×10^{-2} mm Hg)
Power required by heart at rest	≈10 W

Core Material & Study Guide

ATOMS AND MATTER

The carbon atom (^{12}C) is the mass standard, equal to 12.000 00 *unified atomic mass units* (u). A volume of 22.4 liters of any gas at STP has a mass in grams numerically equal to its molecular mass, which constitutes a **mole** (p. 284). The number of molecules per mole is Avogadro's number: $N_A \approx 6.022 \times 10^{23}$ molecules/mole.

Density is defined as

$$\rho = \frac{m}{V} \qquad [9.1]$$

Solids can be classified as either *crystalline, quasi-crystalline, amorphous,* or *composite* (p. 288). The above ideas are considered in Sections 9.1 (Atomism), 9.2 (Density), and 9.3 (The States of Matter). Calculations of the sorts done in Examples 9.1 and 9.2 are typical of what you should know. **Look at the CD WALK-THROUGH EXAMPLES.**

FLUID STATICS

Pressure (P) is *the scalar value of the total force acting perpendicular to a surface divided by the area of that surface:*

$$P = \frac{F_\perp}{A} \qquad [9.3]$$

The units of pressure are N/m^2 or *pascals*. The pressure a distance h beneath the surface of a homogeneous fluid is

$$P_l = \rho g h \qquad [9.4]$$

and if there is an additional pressure on its surface P_s, then

$$P = P_s + P_l = P_s + \rho g h \qquad [9.5]$$

The pressure at every point at a given horizontal level in a single fluid at rest is the same.

An **atmosphere** is defined as 1 atm = 1.013×10^5 Pa, which equals 14.7 lb/in^2. Absolute pressure (P) is the sum of the gauge and atmospheric pressures:

$$P = P_A + P_G \qquad [9.6]$$

All of these ideas [Eqs. (9.3) – (9.6)] are found in Section 9.4 (Hydrostatic Pressure), and they constitute the basic computational material you should know. Reexamine Examples 9.3 – 9.5 and work out as many I-level problems as you can.

Pascal's Principle states that an external pressure applied to a fluid that is confined within a closed container is transmitted undiminished throughout the entire fluid. Hence, for two connected pistons

$$\frac{F_o}{F_i} = \frac{A_o}{A_i} \qquad [9.7]$$

Study Section 9.5 (Pascal's Principle), make sure you understand Example 9.6, which applies Eq. (9.7), and then work out as many I-level problems as you can.

The **Buoyancy Principle**—*an object immersed in a fluid will be lighter (i.e., it will be buoyed up) by an amount equal to the weight of the fluid that it displaces*—is an important basic insight. Reread Section 9.6 (Buoyant Force) and make sure you understand how things float; check out Examples 9.7 and 9.8. The *relative density*, ρ/ρ_w, is called the **specific gravity**.

FLUID DYNAMICS

The analysis of fluid motion begins with Section 9.7 (Fluid Flow), and there are only two basic ideas at work: (1) the **Continuity Equation**

$$A_1 v_1 = A_2 v_2 \qquad [9.11]$$

and (2) **Bernoulli's Equation**

$$P_1 + \tfrac{1}{2}\rho v_1^2 + \rho g y_1 = P_2 + \tfrac{1}{2}\rho v_2^2 + \rho g y_2 \qquad [9.17]$$

or

$$P + \tfrac{1}{2}\rho v^2 + \rho g y = \text{constant} \qquad [9.18]$$

Study Section 9.8 (The Continuity Equation) and Section 9.9 (Bernoulli's Equation) focusing on how to apply these two formulas. An example of what can be determined is the speed of efflux from a hole in an open container a distance h below the surface

$$v_2 = \sqrt{2gh} \qquad [9.20]$$

which is **Torricelli's Result** (p. 308).

Key Terms

atom	barometer
STP	gauge pressure
mole	absolute pressure
kilomole	Pascal's Principle
Avogadro's number	buoyant force
density	specific gravity
solid	surface tension
liquid	laminar flow
gas	turbulent flow
plasma	streamlines
crystalline	boundary layer
polycrystalline	Continuity Equation
amorphous	volume flux
quasi-crystalline	flow tube
viscosity	Bernoulli's Equation
fluid	Bernoulli's Theorem
pressure	Torricelli's Result
hydrostatic pressure	Venturi Effect
total pressure	vortex
atmospheric pressure	

Discussion Questions

1. Explain the meaning of each of the Key Terms on page 316.

2. Where in your home might you find each of the following?

aluminum	tungsten	carbon
oxygen	mercury	zinc
chlorine	chromium	copper
sodium	iron	nickel
lead	magnesium	cobalt

3. EXPLORING PHYSICS ON YOUR OWN: Figure Q3 shows a spherical (the distortion is optical, not actual) glob of olive oil (0.9 g/cm^3) suspended in a solution of roughly 50% alcohol (0.8 g/cm^3) and water. Explain what you see. Incidentally, by using a mix of bromobenzene (1.497 g/cm^3) and xylene (0.867 g/cm^3), we can determine the specific gravity of blood in this same way.

Figure Q3

4. Imagine that you are in an elevator with a tank of water and three goldfish. A device in the gravel measures gauge pressure at the bottom of the tank. (a) Suppose it reads 4000 N/m^2 with the elevator car at rest. Will it increase, decrease, or stay the same when the car accelerates upward? Explain. (b) What will it read if the car drops at 9.8 m/s^2?

5. Consider the mercury barometer (a word coined by Boyle). What, if anything, would happen to the reading (i.e., the difference in height between the two surface levels) if someone floated a coin on the open mercury? Explain.

6. Imagine a closed bottle filled with water into which is sealed a drinking straw. Could you "suck" the liquid up the straw? Why is the idea of sucking up fluids misleading? What would happen if we now punched a hole in the lid? How does the rubber suction cup on a dart stick to a wall? Why is it a common practice to wet the suction cups first? NASA is developing suction-cup shoes for use in space stations. Will they be of any use outside, for walking on the hull?

7. In Fig. Q7, the tube is filled, pinched off, and then its ends are positioned as shown and released. The liquid is *siphoned* until the vessel on the left is emptied or the two levels become equal. What makes it function? [*Hint: A siphon can work in vacuum.*]

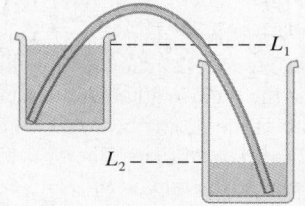

Figure Q7

8. Picture yourself on a raft floating in a large tank of water. On board with you is a tin cup, a pine bust of Napoleon, a banana, and a chicken. What happens to the level of the water in the tank if: (a) you reach out, fill your cup with water, and drink it? (b) you toss the bust overboard? (c) you eat the banana? (d) the chicken flies away?

9. Imagine a metal vessel completely closed except for a small threaded hole on top. It is partially filled with water and placed on a scale that then reads 400 N. Holding on to a solid rod, you lower it down into the water. (a) Will the scale reading change? (b) If the rod were weightless, would that alter your answer? (c) What will happen to the scale reading if the rod is fixed in place by a very light collar that screws into the hole and holds the rod so that you can let go?

10. The photo in Fig. Q10 shows a glass tube open on top and bottom standing in a quantity of trichloroethylene (cleaning fluid) of density 1.44×10^3 kg/m^3. Water (with a little black ink in it for contrast) was subsequently poured into the tube. Explain what the photo shows.

Figure Q10

11. The underwater living chamber depicted in Fig. Q11 is wide open on the bottom for easy entrance. What keeps the water from rushing in? What's the pressure inside? What would happen if it were punctured at the top? Describe what happens to the volume of gas inside as the thing is hauled up to the surface.

Figure Q11

12. Figure Q12 shows an arrangement for producing negative pressures (P_T) at the top of the liquid column. Write an expression for P_T in terms of the pressure at the level of the liquid surface, the density, and the height of the column. Explain what is happening. Pressures of up to -300 atm have been achieved using water. ***Why is it incorrect to define pressure as the magnitude of the applied force per unit area***, as so many texts do?

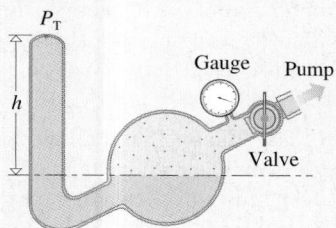

Figure Q12

13. Pascal performed a rather startling demonstration: breaking a barrel using water pressure. He fit the cask with a very tall narrow tube sealed tightly into the lid (Fig. Q13). The barrel and then the tube were filled with water. Explain what happened.

14. EXPLORING PHYSICS ON YOUR OWN: Sprinkle some pepper on the surface of a cup of water and then touch the center of the distribution with a freshly washed and dried finger (no skin oil should be left). Now rub a little soap onto your fingertip and do it again. What happens and why? Put a little oil on your finger and do it again.

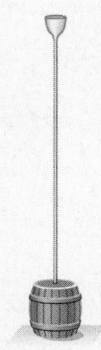

Figure Q13

15. EXPLORING PHYSICS ON YOUR OWN: Rubbing alcohol and acetone both reduce the surface tension of water. With this fact in mind, float two wooden matchsticks in a bowl of water parallel to one another about 2 cm apart. What do you think will happen if you very gently introduce a drop of water between the sticks? Now do that again, this time with a drop of acetone or vodka—what happens?

16. A Ping-Pong ball can be supported in the airstream from a hair dryer even when it is tilted at an appreciable angle (Fig. Q16). Explain.

Figure Q16

17. EXPLORING PHYSICS ON YOUR OWN: Concoct a bubble-making solution of 1 part water to 1 part dishwashing liquid detergent. If you have some glycerin (available at drugstores), add it in (3 parts detergent, 2 parts water, 1 part glycerin). Construct a wire ring about 2 or 3 in. in diameter and tie a small loop of thread to it. Now dip the ring in the solution so that a film stretches across the whole ring and the loop "floats" around within it. Puncture the film within the sagging thread-loop (use a hot pin if necessary). What happens? See Fig. Q17.

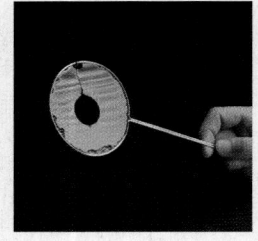

Figure Q17

18. The spinning baseball in Fig. Q18a carries along a boundary layer of air that revolves with it. If the spin axis is horizontal and the ball is thrown as shown in the figure, what will its flight path look like? Figure Q18b shows the flow pattern around a rotating cylinder. An experimental ship built in the 1920s crossed the Atlantic using two large rotating vertical columns and the wind as its source of power. Explain.

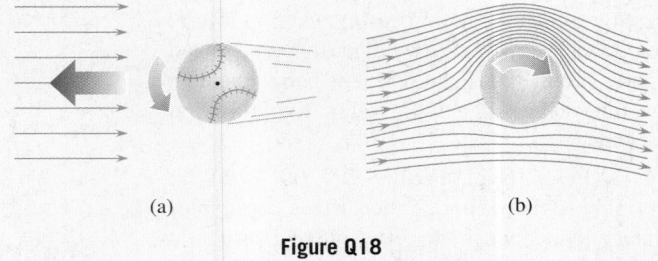

(a)   (b)

Figure Q18

19. Figure Q19 shows two sets of spheres in streams of ideal fluid. Describe the forces, if any, that appear to be acting between them.

An ideal fluid flows parallel to an essentially infinite smooth wall that serves as a boundary. Immersed in the fluid is a solid sphere. Describe the force, if any, that appears to exist between the sphere and the wall.

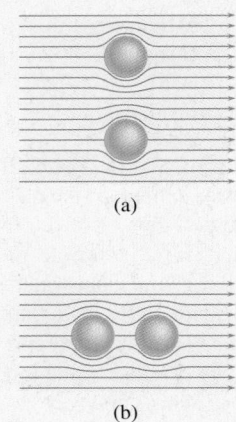

(a)

Figure Q19   (b)

20. Figure Q20 shows a wing that has recently begun to move and around which has just developed a clockwise circulation of air giving rise to lift. Behind the wing we see a so-called *starting vortex*, which always accompanies the initiation of air circulation around a wing. Once formed, the starting vortex washes away downstream. Discuss why you might expect such an eddy to exist solely from momentum considerations. What do you think would happen if the wing was accelerated from rest and then rapidly stopped?

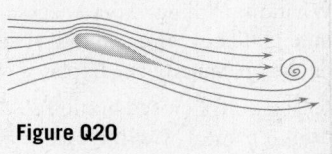

Figure Q20

21. If you cut your finger badly, why might it be wise to hold it high above your head until the bleeding stops?

22. When a helium balloon soars into the sky, it gains KE and *gravitational*-PE. Where does that energy come from?

23. Why does a wide slow river become a fast torrent when it enters a narrow gorge?

24. Figure Q24 shows a gasoline-pump nozzle. We want to examine the built-in automatic system for shutting off the flow when the end of the nozzle becomes immersed in liquid. In Fig. Q24a valve-D is open and gasoline passes through the nozzle. In Fig. Q24b, because the tank is full, valve-D has slammed closed interrupting the flow, and we see the last few ounces of gasoline leaving the nozzle. (Diaphragm-F shuts off the nozzle when the liquid pressure drops because you've gotten all the gas you paid for at a self-service pump; don't worry about it.) Diaphragm-K instantly shuts off the nozzle when it buckles and pulls the pin-E with it. What's the purpose of the constriction at B? Air is sucked in at the little hole at A (you can see the bubbles at B)—why does that happen? What can you say about the pressure at A, C and J in Fig. Q24a? What happens to the pressure in chamber-C when A is immersed (Fig. Q24b) in gasoline? Explain. So, all-in-all, how does the nozzle shut-off system work? For a complete discussion see N. R. Greene and M. R. Dworsak, "Bernoulli at the Gas Pump," *The Phys. Teacher* **39**, 346 (2001).]

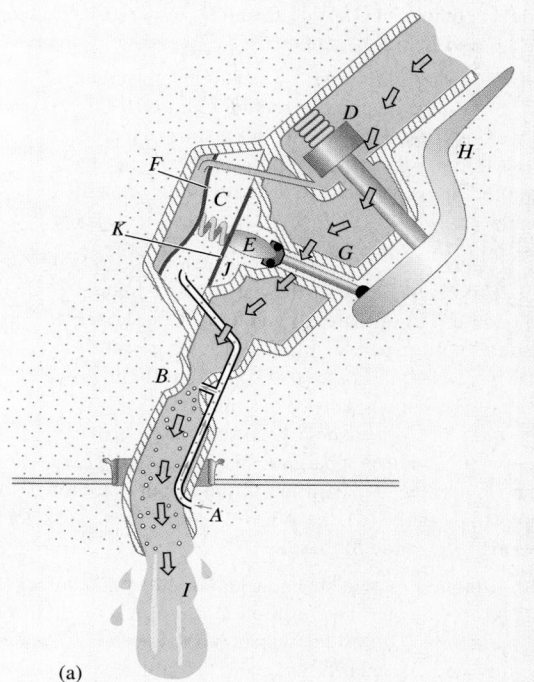

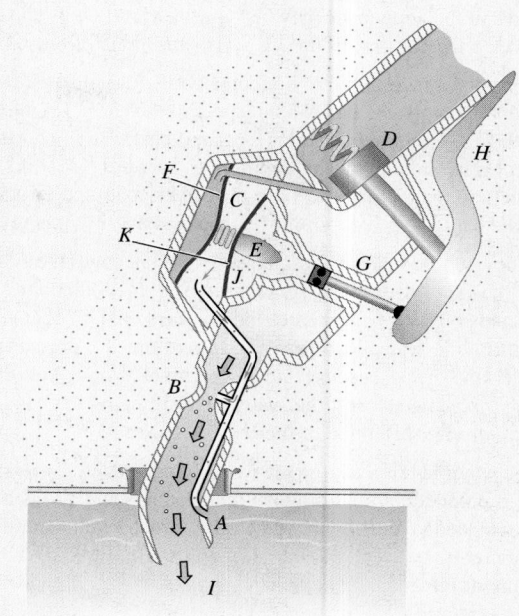

Figure Q24   (a)   (b)

Multiple Choice Questions

1. The dimensions of density are   (a) $[L/M]$ (b) $[L^3/M]$ (c) $[M^3/L]$ (d) $[M/L^3]$ (e) $[M/L]$.

2. If we totally decompose 11 g of carbon dioxide, how many grams of oxygen would be liberated?   (a) 11   (b) 8   (c) 3   (d) 44   (e) none of these.

3. Figure MC3 shows a nail that was left in a concentrated solution of sugar for several days. What has grown on the nail can best be described as (a) a single crystal   (b) an amorphous mass   (c) a glassy solid   (d) a polycrystalline mass   (e) none of these.

4. A simple molecule such as water has a size of about   (a) 0.3 nm   (b) 3.0 nm   (c) 3×10^{-6} m   (d) 0.3 mm   (e) none of these.

Figure MC3

5. The absolute pressure measured at a point within a liquid stored in a large open tank is independent of   (a) atmospheric pressure   (b) the depth below the surface   (c) the density of the liquid   (d) the shape of the tank   (e) none of these.

6. In the case of a liquid barometer, the height of the column is independent of   (a) the density of the liquid   (b) atmospheric pressure   (c) the diameter of the column   (d) the altitude of the barometer   (e) none of these.

7. Suppose we fit a mercury barometer with a snug washer-shaped weightless disk with a hole in the middle that sits on the open surface like a piston. What would happen to the pressure reading if a bird landed on the disk?   (a) nothing   (b) the pressure would exceed atmospheric and the column would rise with respect to the open surface   (c) the pressure would fall slightly, and so the central column of mercury would drop a little   (d) the weight of the bird would press on the mercury, increasing the pressure, and decreasing the height of the column   (e) none of these.

8. Is there a buoyant force acting on you at this very moment?   (a) no   (b) not enough information to say   (c) only if you are floating around in water   (d) yes, the atmosphere exerts a buoyant force   (e) none of these.

9. EXPLORING PHYSICS ON YOUR OWN: Float an ice cube in a glass of water filled to the very brim. Will it overflow as the ice, jutting above the rim, melts?   (a) no, it simply fills the displaced volume and the level is unchanged   (b) yes, but only when it melts rapidly   (c) no, but that is only provided not much ice is in the glass   (d) yes, no matter how fast it melts   (e) none of these.

10. Figure MC10 shows glass tubing, a rubber bulb, and two bottles. What has taken place in order to produce the situation depicted?   (a) nothing, the arrangement is impossible   (b) the tubing was filled with a single liquid, turned over, and inserted into the bottles   (c) the bulb was squeezed while the tubing was in place as shown, and the two different-density liquids rose   (d) the bulb was squeezed more tightly on one side than the other, so the single liquid rose as shown   (e) none of these.

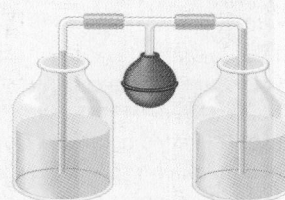

Figure MC10

11. EXPLORING PHYSICS ON YOUR OWN: Suppose you stick a drinking straw into a deep cup of liquid, tightly cover the upper end with a finger, and raise the straw up and out.   (a) some liquid stays in the straw, but its level initially drops, the gas pressure above the liquid in the straw drops, and the straw works like a poor barometer   (b) the

liquid simply runs out of the straw, there being nothing to keep it up (c) the liquid stays in the straw by surface tension alone (d) the liquid stays in the straw primarily because of friction with the long walls, but surface tension contributes as well (e) none of these.

12. Consider a helium-filled balloon floating around in a car. The driver suddenly stops the car. With respect to the car, (a) the driver, the air in the car, and the balloon are pushed backward (b) the driver is pushed backward, but the balloon moves forward (c) only the driver moves—and that is forward (d) the driver is pushed forward as is the air in the car, so the balloon moves backward (e) none of these.

13. A device that reads gauge pressure is suspended in a fixed position above the bottom in a large swimming pool. A helicopter with pontoons lands on the surface. Does the pressure reading change, and if so, how? (a) the water level rises and therefore the pressure rises (b) the pressure drops because the water level drops (c) the pressure remains unchanged (d) the water level remains unchanged, but the pressure drops (e) none of these.

14. The water level in a large tank supplying a small town is 37 m high. Water is pumped up for storage in the evenings and allowed to free-fall down as needed during the day. The street-level water pressure is (a) undetermined (b) 37 Pa (c) 36×10^4 Pa (d) 53×10^4 Pa (e) none of these.

15. Figure MC15 shows a hollow, narrow, open cylinder immersed in a beaker of water. Before being pushed into the fluid, a light plastic disc with a string glued to it was positioned on the bottom end of the cylinder and held in place by pulling up on the thread. Once immersed, the string could be let loose and the disc held in place by the water. That done, a liquid can now be poured into the cylinder from above until the bottom drops out. What is the main point of this experiment? (a) that it's fun to play with water (b) that liquids are more dense than air (c) that the pressure in a fluid is greater than the pressure outside (d) that the pressure in a liquid is due to the weight of the column of fluid above (e) none of these.

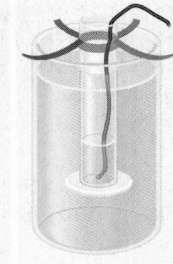

Figure MC15

16. Consider Fig. MC16, which shows a filled fish tank in the atmosphere. At which of the labeled points must the pressure always be equal? (a) A and E (b) B and C (c) A and B (d) C and D (e) none of these.

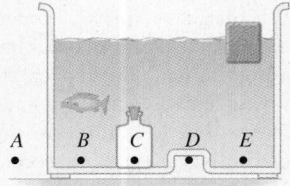

17. In the fish tank of Fig. MC16, it is possible for the pressure to be

Figure MC16

equal at points (a) A and B, and B and E (b) B and C and E (c) A and C and E (d) A and C and B (e) none of these.

18. The long glass tube of a barometer weighs 1.0 N, and the quantity of mercury almost filling it weighs 10 N (Fig. MC18). If the tube (which is very thin, so the mercury-buoyant force can be ignored) is supported by a spring scale, it will read (a) 11 N (b) 9 N (c) zero (d) 1.0 N (e) none of these.

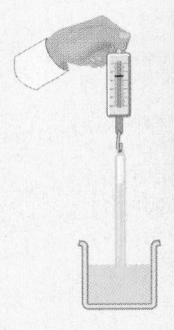

Figure MC18

19. The elderly lady in the park was running a once-in-a-lifetime sale, so you bought the 1.5-kg "solid" gold paperweight. When lowered into a fish tank back at home, it displaced 100 cm³ of water. The known density of gold is 1.93×10^4 kg/m³. (a) you made a great purchase (b) it may not have been a bargain, but at least it's got the right mass (c) bad luck—the density is wrong, so it's definitely not solid gold (d) the density is too high, so something must be wrong (e) none of these.

20. A stone tied to a string is supported from a spring scale that reads 10 N. A pail of water rests on a platform scale that reads 80 N. The stone is lowered into the water, and the spring scale then reads 5 N. It follows that the platform scale (a) still reads 80 N (b) now reads 70 N (c) now reads 90 N (d) now reads 75 N (e) none of these.

21. A block of wood floats in a pot of water on the floor of an elevator. Ignoring surface tension, how is the depth at which it floats changed when the elevator accelerates downward at a rate less than g? (a) the block rises because the water sinks (b) the block lowers because the water rises (c) nothing happens to it (d) the depth decreases because the water gets effectively lighter (e) none of these.

22. Water traveling at a speed of 0.1 m/s enters a pipe with a 1.0-cm radius. It emerges from a section of pipe with a radius of $\frac{1}{2}$ cm at a speed of (a) 0.4 m/s (b) 0.2 m/s (c) 1.0 m/s (d) 0.04 m/s (e) none of these.

23. A filled milk container has two holes punched in its side, one at point 1, 7.5 cm down from the liquid level, and the other at point 2, 15 cm down from that level. The two speeds of efflux are related by (a) $v_1 = v_2$ (b) $v_2 = 2v_1$ (c) $v_1 = 2v_2$ (d) $v_2 = \sqrt{2}\, v_1$ (e) none of these.

24. Two tall glass containers are connected by a narrow tube closed off with a stopcock. One container is filled with water, and when the stopcock is opened, the water passes into the other container until both levels are equal. At the end of this process, in which the "fluid seeks its own level," (a) the *gravitational*-PE has increased to a maximum (b) the KE has increased (c) the *gravitational*-PE has become a minimum (d) the *gravitational*-PE is unchanged (e) none of these.

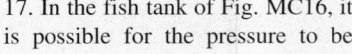

Suggestions on Problem Solving

1. We will work with the abbreviated notation of mega- and giga-, as for example MPa and GPa, which is often the way data is given in the literature. Get in the habit of entering the numbers in your calculations in scientific notation right away. It's easier to substitute the numbers in MPa, but it's also easier to lose the factor of 10^6 that way.

2. When doing buoyancy problems with balloons and the like, don't forget to add in, along with the load, the weight of the gas doing the lifting—it is part of the weight of the craft.

3. Some people are accustomed to working in the cgs system when it comes to density and since $\rho_w = 1.00$ g/cm³. Moreover, since the specific gravity of water is 1.00, there's a tendency to lose the factor of 10^3 when substituting $\rho_w = 1.00 \times 10^3$ kg/m³. Remember, 1.00 g/cm³ = 1.00×10^3 kg/m³. Also remember that 1 cm² = (1 cm)(1 cm) = $(1 \times 10^{-2}$ m$)(1 \times 10^{-2}$ m$) = 10^{-4}$ m². Many problems deal with pressure beneath the water in an open container—***don't forget atmospheric pressure acts as well***. It's also helpful to keep in mind that 1 MPa ≈ 10 atm.

4. There are a number of situations in this chapter where we have the same quantity on both sides of an equation, or equivalently, where there are ratios of the same quantity (e.g., sp. gr.). In such cases we can take some short cuts with the units. For example, if both pressures in the expression $P_i V_i = P_f V_f$ are given, say, in mm Hg, they need not be converted to Pa before substitution because the units, whatever they are, cancel. This can save a lot of effort, but it should go without saying that you must be careful. *For the time being, use such short cuts only as checks.*

5. There are only two basic equations in this treatment of fluid dynamics, the Continuity Equation and Bernoulli's Equation. To these are added the derived expressions for Torricelli's Result and the equation for the Venturi Effect. Read the problem carefully and determine which of these concepts is involved. If the problem deals with fluid flowing through a hole or a constricted channel, one of the last two equations might be called for rather than going back to Bernoulli's Equation to rederive those results each time.

6. If you are considering a pipeline of constant diameter, the Continuity Equation requires that the speed of an ideal fluid be constant regardless of twists and turns, rises and falls.

7. When using Bernoulli's Equation, **draw a diagram**. If possible, locate region-1, where the fluid pressure, elevation, and speed are known. At a place where the section area is much larger than elsewhere in the flow, the fluid speed there may be negligible. The pressure in a fluid jet is the same as that of its surroundings. Locate region-2, where the unknown quantity is to be determined. Pick a zero y-level. The values of y_1 and y_2 may turn out positive or negative. If there are two unknowns, use the Continuity Equation as well. You can use either gauge or absolute pressure, but don't mix them. Bernoulli's Equation is applicable to the flow from one region to another. It applies *along a tube of flow*; in general, you cannot use the equation for *any* two points in the fluid and get exact results.

Problems ✦ Coordinated Problems ✦ Progressive Problems ✦ Solutions

STUDY GUIDE **1. Coordinated Problems:** The three problems within each magenta-colored grouping are solvable in similar ways. Note that the first of these always has a hint; moreover, its solution is provided in the back of the book. *Work out each of these sets; they'll strengthen technique and build confidence.* **2. Progressive Problems:** The problems introduced in blue unfold step-by-step carrying along the analysis in a more suggestive way than is customary. *Work out all of these; they'll guide you through the analytic process and help develop problem-solving skills.* **3. Worked-Out Solutions:** Studying worked-out solutions is an important part of learning how to solve problems. Accordingly, additional *solutions* to a number of model problems are given below. *Make sure you understand each of them before you go on to the next problem.* **4.** Also provided in the back of the book are the *Answers* to all odd-numbered problems, as well as worked-out *solutions* to those with boldface numbers. Problem numbers in italic indicate that a solution appears in the Student Solutions Manual.

SECTION 9.1: ATOMISM

SECTION 9.2: DENSITY

1. [I] How many molecules of trichloromonofluoro-methane are there in 22.4×10^3 cm^3 of the gas at STP?

2. [I] What quantity of water will be produced when 5.0 g of hydrogen are burned in the presence of 40 g of oxygen? What if 6.0 g of hydrogen are burned in the presence of the same amount of oxygen?

3. [I] White dwarf stars are very small and very massive. A 1-in.^{3} chunk of such star-stuff brought to Earth would weigh about 1 ton (2000 lb). Determine its density in SI units.

4. [I] THIS PROBLEM EXAMINES THE UNITS OF DENSITY. Human muscle has a density 1.04 g/cm^3. (a) How many grams are there in exactly 1 kg? (b) How many cm^3 are there in exactly 1 m^3? (c) What is the density of muscle in kg/m^3?

5. [I] THIS PROBLEM DEALS WITH DENSITY AND ITS SI UNITS. It is found that 74.0 cm^3 of a liquid has a mass of 94.0 g (a) What is its density in g/cm^3? (b) What is its density in kg/m^3?

6. [I] The mass of the Earth is 5.975×10^{24} kg, and its mean diameter is $1.274\ 246 \times 10^7$ m. Determine its average density. Given that the density of nuclear matter is about 2×10^{17} kg/m^3, if the Earth were squashed into a small sphere, how big would it have to be to have a comparable density?

7. [I] A typical human being contains roughly 13 gallons of water, where 1.00 gallon = 3.785 liters. What is the corresponding mass of water?

8. [I] How many grams of salt are in a gram-mole of NaCl?

9. [I] Imagine that a kilomole of a substance undergoes some chemical change that liberates 3.2×10^{-19} J (i.e., 2.0 eV) per reaction, per molecule. How much energy will be released by the entire kilomole?

SOLUTION: Number of molecules in a kilomole is $N = 10^3 N_A = 6.02 \times 10^{26}$. Each liberates 3.2×10^{-19} J for a total of 0.19 GJ.

10. [I] A cubic centimeter of water has a mass of 1.00 g. How many *molecules* does it contain? How many atoms are in a 1.00-g ice cube?

11. [I] How many molecules of pure sugar ($C_{12}H_{22}O_{11}$) are in 684 g of the stuff?

12. [I] A ring of pure gold has a mass of 19.7 g. How many atoms are in it?

13. [I] Benzoylmethylecgonine (cocaine) is $C_8H_{13}N(OOCC_6H_5)(COOCH_3)$. What is its atomic weight?

14. [I] Assuming that you weigh 150 lb (i.e., 667 N) and that your body is roughly 70% water, how many H_2O molecules are you?

15. [I] Make a rough estimate of the number of atoms in all the stars in the known Universe. [*Hint: Assume an average atomic mass of, say, 10 u; a star-mass of 10^{33} g; 10^{11} stars in a galaxy; and 10^{11} galaxies in the known Universe.*]

16. [II] Derive a general expression for the number of molecules there are in a cubic meter of any substance at STP.

17. [II] Using the results of Problem 16, determine the number-density for water—that is, the number of molecules there are per cubic meter. (Check your answer with Table 9.2.)

18. [II] The mass of a proton is 1.673×10^{-27} kg, and it can be considered to be a sphere of roughly 1.35×10^{-15}-m radius. Determine its density and compare it with that of most solids. How much is that in tons per cubic inch (1 ton = 2000 lb)?

19. [II] A cubic foot of water weighs 62.4 lb. How many kilograms of hydrogen are in that much water?

20. [II] What is the mass of a single water molecule, and how many of them are in 1.000 g of pure water? The atomic masses of hydrogen and oxygen are 1.008 u and 16.00 u, respectively.

21. [II] Ethyl alcohol, C_2H_5OH, has a density of 0.789×10^3 kg/m³. Estimate the linear size of the molecule.

22. [II] Assume that a water molecule is a sphere of diameter 0.3 nm. Approximately how many of these spherical molecules will be in 18 cm³ of water? How does this result compare with the correct answer? Why is it in error?

23. [III] What is the approximate separation between gas molecules at STP? How does this compare to the separation between molecules in a solid?

SECTION 9.4: HYDROSTATIC PRESSURE

24. [I] How high should a vertical pipe filled with water be if a gauge at its bottom is to read a pressure of 400 kPa due only to the liquid?

25. [I] A swimming pool 5-m wide by 10-m long is filled to a depth of 3-m. What is the pressure on the bottom due only to the water?

26. [I] A rectangular tank 2.0 m by 2.0 m by 3.5 m high contains gasoline, with a density of 0.68×10^3 kg/m³, to a depth of 2.5 m. What is the gauge pressure anywhere 2.0 m below the surface of the gasoline?

27. [I] An oxygen tank sitting in the corner of a laboratory has an internal gauge pressure 5.00 times atmospheric pressure. What outward force is exerted per square centimeter on the inner wall of the tank?

28. [I] At about 10 miles up in the atmosphere, air pressure drops to roughly 2 lb/in.² (down from 14.7 lb/in.² at sea level). Using the fact that 1.000 lb = 4.448 N and 1.000 in. = 2.540 cm, what is the air pressure at that altitude in *atmospheres* and *pascals*?

29. [I] Prove that 1.000 lb/in.² = 6.895×10^3 N/m², using the fact that 1.000 lb = 4.448 23 N.

30. [I] A typical automobile tire has a gauge pressure of around 30 lb/in.² How much is that in pascals? If a car weighs 8897 N (i.e., 2000 lb), how much area (in SI units) on each tire is in contact with the road? See the previous problem.

31. [I] The U.S. Navy bathyscaphe *Trieste* in 1960 descended to a record depth of 10 920 m in the Pacific Ocean near the island of Guam. To two significant figures, determine the maximum gauge pressure (in both Pa and psi) it encountered. Assume the density of the sea is constant. [*Hint: The gauge pressure depends on the depth and the density of the fluid.*]

Figure P31

32. [I] In a large apartment house water is stored in a tank on the roof 30.5 m above a faucet in a first-floor kitchen. What is the gauge pressure at the faucet?

33. [I] How deep must you dive in fresh water before the gauge pressure equals 1.00 atmosphere (1.00 atm = 1.013×10^5 N/m²)?

34. [I] A swimming pool 5.0-m wide by 10-m long is filled to a depth of 3.0 m. What is the absolute pressure on the bottom?

35. [I] When Pascal's brother-in-law climbed to the top of the 3200-ft-high Mont Puy-de-Dôme, the barometer he carried showed a drop in mercury of about 3.0 in. What was the corresponding pressure drop?

36. [I] A swimming pool 5.0-m wide by 10-m long is filled to a depth of 3.0 m. What is the total force exerted on the bottom due to the water?

37. [I] THIS PROBLEM WILL HELP US LEARN ABOUT THE HEIGHT DEPENDENCE OF HYDROSTATIC PRESSURE. A water tank is 310 m above the ground on top of a skyscraper, and a pipe runs straight from it all the way down to street level. (a) What is the meaning of the term *gauge pressure*? (b) Compute the gauge pressure at street level in the pipe. (c) If the pipe has a cross-sectional area of 0.50 m², what is the weight of the water in the pipe?

38. [I] THIS PROBLEM EXAMINES HYDROSTATIC PRESSURE AND FORCE. The *Titanic* sunk on its maiden voyage from Southampton, England to New York on April 14, 1912. It rests in the North Atlantic in water 3.0 km deep. (a) What is the density of seawater? (b) What is the meaning of absolute pressure? (c) Neglecting variations in water density, what was the maximum external pressure (both gauge and absolute) that the craft experienced? (d) What was the value of the force exerted by the sea on every square centimeter of the hull?

39. [II] The medical literature uses yet another pressure unit: the cm of water (water columns are convenient for measuring modest gauge pressures). Thus, one reads that a newborn baby is capable of developing a "momentary intrathoracic negative pressure of the order of 40 cm of water." How much is that in pascals?

40. [II] To remove unwanted fluids from the body, an aspirator such as that shown in Fig. P40 is used. A small electric pump provides the suction (typically at gauge pressure levels of −90 to −120 mm Hg), while the bottle containing water provides the control. Explain how the depth beneath the water surface of the control tube establishes the pressure in the drain tube. What happens to the suction pressure as the drain tube in the bottle becomes submerged in liquid aspirated from the patient? When will the system shut itself off? Suppose the procedure starts with the control tube 10 cm into the water— what is the suction pressure to the patient?

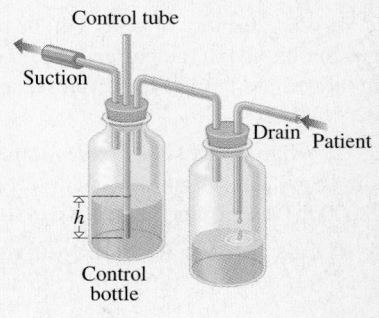

Figure P40

41. [II] Given that most people cannot "suck" water up a straw any higher than about 1.1 m, what's the lowest gauge pressure they can create in the lungs?

42. [II] The gauge pressure in a basement water pipe supplying a tall building is 3.00×10^5 Pa. In an apartment several floors up, the pressure is half this value. How high up is the apartment?

43. [II] A rubber pipe is attached to one end of an open U-tube water-filled manometer (such as the one in Fig. 9.13). A patient exhales strongly into the rubber pipe, and a difference in height

between the two columns of 61 cm results. What absolute pressure was developed by the lungs?

44. [II] The piston in equilibrium, shown in Fig. P44, has a face area of 0.10 m² and experiences a downward force of 1.00 kN. The mercury in the vessel is 10.0-cm deep. What is the gauge pressure at the very bottom of the chamber?

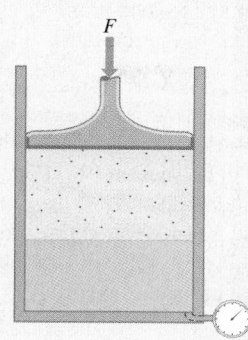

Figure P44

45. [II] THIS PROBLEM EXPLORES HYDROSTATIC PRESSURE DUE TO A COMBINATION OF LIQUIDS. Suppose you take a tall vertical tube, sealed at the bottom, and pour in water to a height of 24 cm, and then gently add another 12 cm of olive oil (with a density of 0.92 g/cm³).   (a) What is the gauge pressure on the bottom before the oil is added to the water?   (b) Does the oil float on the water? How do you know?   (c) What is the gauge pressure at the water-oil interface?   (d) What is the gauge pressure on the bottom of the tube after the oil is added?

46. [II] THIS PROBLEM TREATS HYDROSTATIC PRESSURE DUE TO A COMBINATION OF FACTORS. A tank 10 m by 5.0 m is filled with lubricating oil (having a density of 0.91 g/cm³) to a height of 4.0 m. A 10 000 lb mahogany boat is then floated on the oil.   (a) What is the weight of the oil in the tank?   (b) What is the area of the bottom of the tank?   (c) What is the gauge pressure on the bottom before the boat arrives?   (d) How much does the boat weigh in newtons? (e) After the boat arrives, what's the total weight, of oil and boat, acting on the bottom of the tank?   (f) What is then the gauge pressure at the bottom?

47. [II] A submarine rests in 20.0 m of water. How much force must a diver exert against the pressure of the sea in order to pull open a hatch 1.0 m × 0.50 m, assuming the internal pressure in the boat is 90% of atmospheric?

> **SOLUTION:** Outside the sub $P_{out} = P_A + \rho_w gh$; inside the sub $P_{in} = 90\% P_A$; the difference in pressure is $P_{out} - P_{in} = (0.10P_A + \rho_w gh)$. Thus $F = A(0.10P_A + \rho_w gh) = (0.50 \text{ m}^2)(0.10P_A + \rho_w gh) = 1.0 \times 10^5$ N.

48. [II] The open U-tube in Fig. P48 contained some water before a less dense liquid was poured in on the right side. If the density of the unknown liquid is ρ_x, show that

$$\rho_x = \frac{\rho_w h_w}{h_x}$$

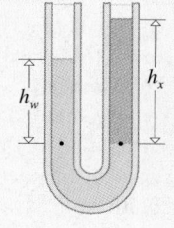

Figure P48

49. [II] Return to the manometer in Fig. 9.13, but this time suppose it's filled with mercury and the level on the right is 4.0 cm lower than the level on the left. What is the pressure in the chamber, taking atmospheric pressure to be 0.101 MPa?

50. [III] A glass plate 0.12 m × 0.20 m is mounted on the top of a submerged camera box to make an angle of 30° with the horizontal. The uppermost 0.12-m edge is 0.20 m below the surface of the water. The box is sealed, and there is air inside at atmospheric pressure. Please compute the net force acting on the plate due to the water.

51. [III] Figure P51 deals with the problem of showing that the pressure within a fluid is independent of orientation. Shown is the end of a wedge-shaped volume of fluid where the average pressure on each of its faces is P_a, P_b, and P_c. The lengths of the sides of the top face are both L, and it makes an angle of ϕ with the bottom. Find the net force on each face, and show that the fluid wedge is in equilibrium when

$$P_a L^2 \sin \phi = P_c L^2 \sin \phi$$

and

$$P_b L^2 \cos \phi = P_c L^2 \cos \phi + \tfrac{1}{2}\rho g L^3 \sin \phi \cos \phi$$

Now, let the wedge shrink and show that $P_a = P_b = P_c$ in the limit as $L \to 0$.

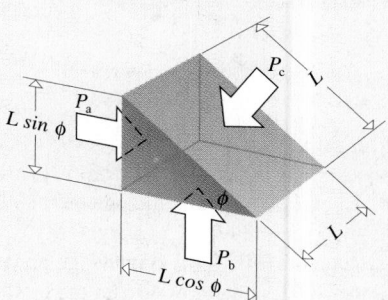

Figure P51

52. [III] The diving bell shown in Fig. P52 allows swimmers to comfortably remain under water for long intervals. Show that

$$d = \frac{P_A h}{(H - h)\rho g} + h$$

where P_A is the original pressure in the bell (usually atmospheric). [*Hint: Don't forget atmospheric pressure on the water.*]

53. [III] A swimming pool 5-m wide by 10-m long is filled to a depth of 3 m. What is the total force on either narrow wall due only to the water?

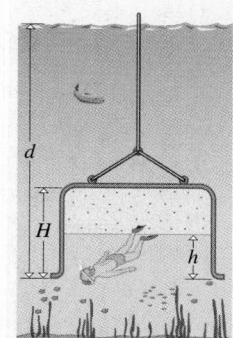

Figure P52

SECTION 9.5: PASCAL'S PRINCIPLE

54. [I] Consider a narrow horizontal cylindrical chamber permanently sealed at one end and closed off at the other end with a tightly fitted movable piston having an area of 0.050 m². The chamber is filled with oil, and a compressive force of 1000 N is applied to the liquid via the piston. Determine the pressure read by a sensor   (a) at the flat far end and   (b) halfway in along the wall.

55. [I] The digging bucket on a large mechanical excavator is moved by a piston sliding within a single hydraulic cylinder. According to the manufacturer, the hydraulic fluid in the cylinder is pressurized to 1000 psi (6.895 × 10⁶ Pa or 68 atm) by a pump. The cylinder has a diameter of 0.165 m and a stroke of 1.36 m. Determine the maximum force that can be exerted by the cylinder.

> **SOLUTION:** $F = PA = (6.895 \times 10^6 \text{ Pa})(\tfrac{1}{4}\pi D^2) = 147$ kN.

56. [I] The pressure that operates a hydraulic elevator is provided by an electric pump. If the elevator and its cargo have a combined mass of 4200 kg and the piston diameter is 22.0 cm, what pressure must be maintained on the fluid?

57. [I] THIS PROBLEM HELPS US BETTER UNDERSTAND THE TRANSMIS-SION OF FORCE THROUGH A CONFINED INCOMPRESSIBLE LIQUID. The face of the piston on the left in Fig. P57 has an area of 21.2 cm^2, and the one on the right has an area of 63.6 cm^2 (a) What is the ratio of the areas of the two piston faces?   (b) If the mass on the left is 26.2 kg, how much mass should be placed on the right to balance the system?

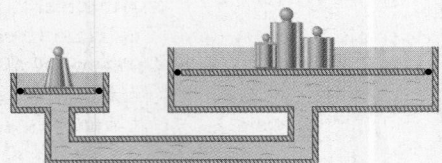

Figure P57

58. [I] THIS PROBLEM EXAMINES THE TRANSMISSION OF FORCE IN A CONFINED INCOMPRESSIBLE FLUID. The face of the piston on the left in Fig. P57 has an area of 21.2 cm^2, and the one on the right has an area of 63.6 cm^2. The mass on the left is 24.0 kg.   (a) What is the force acting on the piston on the left?   (b) What is the pressure on the piston on the left?   (c) What is the pressure on the piston on the right?

59. [I] A hydraulic press consists of two connected cylinders—one 8.00 cm in diameter, the other 20.00 cm in diameter. The cylinders are sealed with movable pistons, and the whole system is filled with oil. If a force of 600 N is then applied to the smaller piston, what force will be exerted on the larger piston? Ignore friction. [*Hint: The pressure everywhere in the sealed system is equal.*]

60. [I] The area of the face of the small piston of a hydraulic press is 10.0 cm^2. An input force of 100 N is applied to that piston, and we wish to have the large piston exert a corresponding output force of 9600 N. What must be the area in cm^2 of the face of the large piston?

61. [I] Imagine a hydraulically operated dentist's chair having a mass of 200.0 kg, and in it is sitting a 54.8-kg patient. The large piston beneath the chair has a diameter of 5.00 cm, whereas the small piston, moved by a pedal on which the dentist steps, has a diameter of 1.00 cm. What is the pressure in the interconnecting fluid when the dentist operates the chair? How much force must he exert on the small piston?

62. [I] Figure P62 is a simplified drawing of a brake arrangement on a car; the master cylinder actually connects to all four brakes via two separate systems. Suppose a force of 100 N is applied to the piston in the master cylinder as a result of someone stepping on the brake.

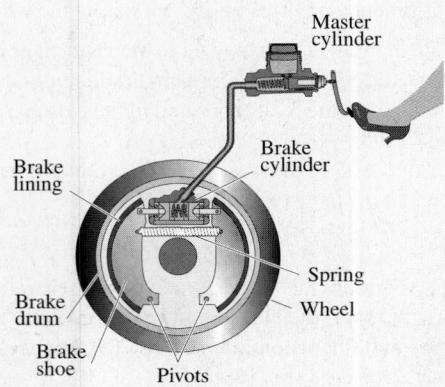

Master cylinder

Brake lining

Brake cylinder

Spring

Wheel

Brake drum

Brake shoe

Pivots

Figure P62

That cylinder has a face area of 6.45 cm^2 as compared to the 19.35-cm^2 face area of the brake cylinder. How much force will be transmitted to each brake shoe pushing the two stationary linings outward against the rotating drum?

63. [II] A small hand-operated hydraulic jack has a rated maximum lifting capacity of 4500 lb (where 1.000 lb = 4.448 N). As can be seen in Fig. P63, the piston has a diameter of 2.00 in. (where 1.000 in. = 2.540 cm). Determine the fluid pressure in the jack when operating under a maximum load.

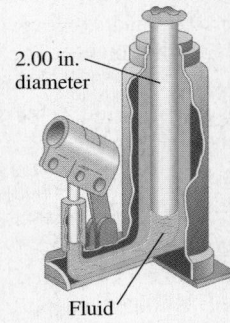

2.00 in. diameter

Fluid

Figure P63

SOLUTION: $P = F/A = (4500 \times 4.448 \text{ N})/(\frac{1}{4}\pi D^2) = (4500 \times 4.448 \text{ N})/\frac{1}{4}\pi(0.050\ 8 \text{ m})^2 = 9.88 \text{ MPa}.$

64. [II] THIS PROBLEM TAKES INTO ACCOUNT HYDROSTATIC PRESSURE EVEN THOUGH THE SYSTEM IS SEALED. Figure P64 shows two pistons (small and large, with masses m_S and m_L, and areas A_S and A_L) in equilibrium. The connecting liquid is oil having a density ρ_o. The mass of the small piston is unknown, so we want an equation for m_S in terms of the other parameters, including h, the column height. (a) Write an expression for the absolute pressure just below the large piston.   (b) Write an expression for the absolute pressure at that same level, but below the small piston. Do not neglect the weight of the oil.   (c) What can be said about these two pressures? (d) Write an expression for m_S in terms of M, m_L, A_S, A_L, ρ_o and h.

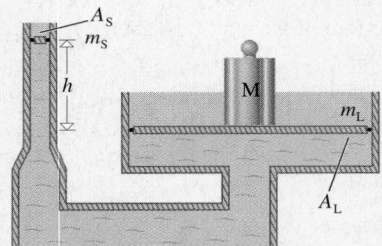

A_S

m_S

h

M

m_L

A_L

Figure P64

65. [II] THIS PROBLEM TREATS THE TRANS-MISSION OF FORCE IN A CONFINED INCOM-PRESSIBLE LIQUID. We want to study the forces (input and output) indicated in the cross-sectional view shown in Fig. P65, keeping in mind that the large piston has a substantial mass m_L. The system, which supports an external load equal to F_o, is in equilibrium.   (a) Write an expression for the gauge pressure just below the large piston.   (b) Neglecting the weight of the fluid, write an expression for the gauge pressure just above the small piston.   (c) What can be said about these two pressures?   (d) Write an expression for F_o in terms of F_i, m_L, A_S, and A_L.

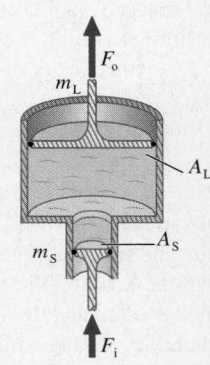

F_o

m_L

A_L

m_S

A_S

F_i

Figure P65

66. [II] A hydraulic lift consists of two interconnected pistons filled with a common working liquid. If the areas of the piston faces are 64.0 cm^2 and 3200 cm^2 and if a 900-kg car rests on the latter, how

much force must be exerted to raise the vehicle very slowly? If the car is to be raised 2.00 m, how far must the input piston be depressed?

67. [II] A hydraulic cylinder on a large machine operates at a pressure of 1000 psi (i.e., 6.895×10^6 Pa), and has a diameter of 0.20 m and a stroke length of 1.36 m. What's the maximum amount of power that it can deliver if a complete stroke takes 6.0 s?

68. [II] Emergency rescue teams often carry small hydraulic devices (like the one in Fig. P68) to cut or bend debris in order to remove victims from wreckage. Compute the output force exerted by the device when a force of 360 N is applied to the handle, which is pivoted at the far right. The small piston has a diameter of 2.00 cm; the large one has a diameter of 6.00 cm.

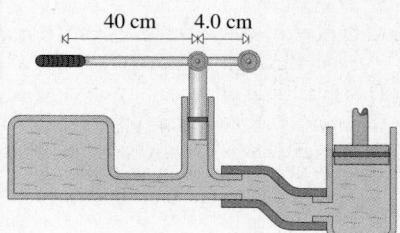

40 cm   4.0 cm

Figure P68

SECTION 9.6: BUOYANT FORCE

69. [I] What is the buoyant force exerted on a sunken treasure chest that has come to rest on a few small rocks at the bottom of a fresh-water lake? The chest is 1.00-m long, 0.50-m wide, and 0.60-m high and contains 10 kg of pure gold.

70. [I] The density of uranium is 18.7×10^3 kg/m^3. What is its specific gravity? How much more dense than water is uranium?

71. [I] The fluid in a lead storage battery for a car has a specific gravity of 1.30 when it's fully charged and 1.15 when it's discharged. By how much does the density of the fluid change when the battery is completely run down?

72. [I] Imagine a rectangular solid having a height of 1.50 cm and a density of 10.5×10^3 kg/m^3 floating in an unknown liquid. Given that the block is submerged to a depth of 1.16 cm, find the density of the liquid.

73. [I] An ancient coin, which X-rays show is solid and homogeneous, has a mass of 0.010 0 kg. When submerged, it displaces 0.952 g of water. What is its specific gravity, and what is it probably made of? [*Hint: Use the definition of specific gravity, ρ/ρ_w.*]

74. [I] Determine the specific gravity of a statue if its mass is 25.0 kg in air and appears to be only 15.0 kg when measured while submerged in water.

75. [I] A metal object is hung from a scale and found to weigh 10.0 N. It is then lowered into a tank of water and its "weight," measured while it's submerged, is found to be 8.00 N. What is its specific gravity?

76. [I] Envision a thin-walled jar having a mass of 10.0 g containing 0.100 g of hydrogen at atmospheric pressure. Estimate the difference we can expect to observe if we weigh the filled jar in dry air at 0°C as opposed to weighing it in vacuum. The density of hydrogen is 90×10^{-6} g/cm^3.

77. [II] THIS PROBLEM EXAMINES THE DENSITY OF NUCLEAR MATTER.

The nucleus of a hydrogen atom has a diameter of about 2.4 fm = 2.4×10^{-15} m and a mass of 1.673×10^{-27} kg.   (a) Determine the volume of the nucleus.   (b) What is the density of the nucleus? (c) What is the specific gravity of nuclear matter?

78. [II] THIS PROBLEM DEALS WITH THE SPECIFIC GRAVITY OF AN UNKNOWN LIQUID. The measured mass of a chunk of material in air is 60.0 g, in water it's 34.0 g, and in the unknown liquid it's 14.0 g. (a) What is the reduced mass of the chunk in water?   (b) What is the mass of the water displaced?   (c) What is the volume of water displaced?   (d) What is the volume of the chunk of material?   (e) What is the mass of the unknown liquid displaced?   (f) What is the density of the unknown liquid?   (g) What is the specific gravity of the unknown liquid?

79. [II] The buoyant force may be thought of as acting upward at a point sometimes called the center-of-volume (Fig. P79). For a person, because the lungs are light and yet voluminous, the center-of-volume is usually above the center-of-gravity. On allowing the body to descend fully into the water, a net torque will initially exist and the floating person will rotate until the line-of-action of the

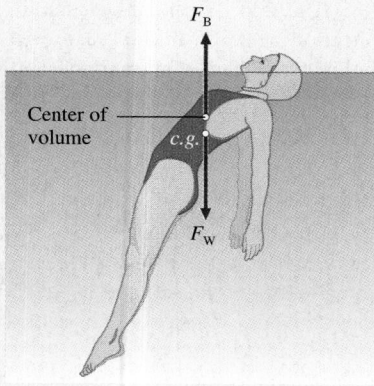

F_B

Center of volume

c.g.

F_W

Figure P79

buoyant force passes through the *c.g.* Studies show that the average density of a young woman with her lungs inflated is about 980 kg/m^3. What percentage of her body will remain in the air when she's floating in fresh water ($\rho_w = 1.000 \times 10^3$ kg/m^3)?

80. [II] A 227-kg block of cement, of density 2.8×10^3 kg/m^3, rests on a pedestal in front of the Al Capone Memorial Library. How much did it weigh submerged while being hauled out of the river (freshwater)?

SOLUTION: $F'_w = F_w - F_B = m_c g - g \rho_w V_c = m_c g - g \rho_w m_c / \rho_c = m_c g (1 - \rho_w / \rho_c) = (227 \text{ kg})(9.81 \text{ m/s}^2)(1 - 1/2.8) = 1.4 \text{ kN}.$

81. [II] A ring weighs 6.327×10^{-3} N when measured in air and 6.033×10^{-3} N when submerged in water. What is its volume (to three significant figures)? What is it likely made of? What is its relative density (to three significant figures)?

82. [II] A woman weighing 500 N jumps into a swimming pool 10 m × 10 m by 5 m deep and floats around. By how much does the water level change as a result of her arrival?

83. [II] Icebergs float in the ocean with much of their huge volumes hidden below the surface. What fraction is visible above the water? What portion of an ice cube floats above the surface of a glass of tap water?

84. [II] A person floats on the Great Salt Lake, which has a density of 1.15×10^3 kg/m^3. Approximately how much of the swimmer's body is above the water?

85. [II] THIS PROBLEM EXPLORES FLOATING IN SEAWATER. A steel can has a mass of 390 g and a capacity of 1000 cm^3.   (a) What's the average density of the can as a whole?   (b) Will it float when empty?   (c) How much does it weigh?   (d) What's the value of the

buoyant force acting on it? (e) How much does it weigh when filled with acetone? (f) Will it float when filled with acetone?

86. [II] THIS PROBLEM TREATS THE BUOYANCY OF A GAS-FILLED BALLOON IN AIR. A hovering balloon filled with hydrogen is tied to a spring scale, which is fixed to the ground. The balloon (when empty) has a mass of 30.1 kg, and now it contains 100 m^3 of gas. Take the density of the air to be 1.29 kg/m^3. (a) Draw a free-body diagram. (b) What is the weight of the balloon? (c) What is the weight of the hydrogen gas? (d) What is the buoyant force on the balloon? (e) What is the value of the force that the scale reads?

87. [II] Determine the mass of helium needed to provide enough buoyancy (in dry air at 0°C) to lift a balloon and its load having a net mass of 454 kg. The load has a negligible volume.

88. [II] A 226.7-kg polar bear standing 6-ft tall walks onto a floating sheet of ice 0.305-m thick. How big is the ice sheet if it sinks just below the surface while supporting the bear?

89. [II] Given a pine raft, of density 0.50 × 10^3 kg/m^3 and dimensions 3.05 m by 6.10 m by 0.305 m, how much of a load can this raft take on for each 2.54 cm it settles into fresh water? How deep does it sink into the water when unloaded?

90. [III] An object has a weight measured in air of F_{Wa}; and when immersed in water, its effective weight is F_{Ww}. It is then submerged in an unknown liquid, and its effective weight is found to be F_{Wl}. Determine the density of the unknown liquid in terms of these measured quantities and the density of water.

SECTION 9.8: THE CONTINUITY EQUATION
SECTION 9.9: BERNOULLI'S EQUATION

91. [I] Petroleum flows along a 10-cm diameter pipeline at a speed of 2.0 m/s. Determine the flow rate, assuming the liquid is ideal.

92. [I] A pipe 5.0 cm in diameter carries gasoline with a density of 0.68 × 10^3 kg/m^3 at a speed of 2.5 m/s. Calculate the mass-flow rate, assuming the liquid is ideal.

93. [I] Gasoline (with a density of 0.68 kg/m^3) is flowing in a pipeline having a 0.50-m diameter. Taking the fluid to be ideal, what pressure change results when the pipe descends 4.0 m down an embankment?

94. [I] Determine the blood pressure at the brain, taking its height above the heart to be 40 cm.

95. [I] Someone cuts a finger and, to control the bleeding, raises the wound 85 cm above the level where the hand was being bandaged. What is the change in the pressure of the blood? Compare that to the pressure at the heart.

96. [I] A horizontal water main with a cross-sectional area of 200 cm^2 necks down to a pipe of area 50 cm^2. Meters mounted in the flow on each side of the transition coupling show a change in gauge pressure of 80 kPa. Determine the flow rate through the system, taking the fluid to be ideal.

97. [I] Derive an expression for the speed of efflux of an ideal liquid from a hole in a large tank filled with fluid in terms of the gauge pressure at the level of the orifice and the density of the liquid. [*Hint: Use Torricelli's Result.*]

98. [I] The drain on a bath tub is 0.50 m below the level of the water when the plug is removed. At approximately what speed will the water emerge from the tub? Don't worry about the shape of the orifice.

99. [I] Water is flowing into a below-deck compartment of a ship through a hole in its side at a speed of 10.0 m/s. How far beneath the water's surface is the hole?

100. [I] A length of tubing is used as a siphon to empty a large upright cask of wine. At a given moment, the lower opening of the tube from which the wine emerges is 20 cm below the liquid level in the cask. Compute the speed at which the wine then emerges.

101. [I] A large vat of clear chicken soup is about to spring a leak from a small hole a distance *h* below the surface of the broth. Knowing that the hole is a height (*y*) above the floor, at what distance (*x*) away from the vat should you place a bowl on the floor to catch the flow, assuming it to be ideal?

102. [I] THIS PROBLEM EXAMINES THE FLOW OF AN INCOMPRESSIBLE LIQUID. Water moves through a horizontal pipe as shown in Fig. P102. (a) Does the fluid undergo any change in *gravitational*-PE in traveling from one section to the other? (b) If the speed of the water in the large section is 3.00 m/s, how fast is it moving in the narrow section?

150 kPa

65.0 kPA

Figure P102

103. [II] THIS PROBLEM EXAMINES THE RATE OF FLOW OF AN INCOMPRESSIBLE LIQUID. A rectangular open tank is 152 cm wide, 305 cm deep, and 610 cm long. We wish to fill the tank using a 2.54-cm diameter hose that delivers water at a speed of 305 cm/s. (a) Determine the volume of the tank in cm^3. (b) What is the cross-sectional area of the hose in cm^2? (c) Compute the volume of water delivered per second by the hose. (d) How long will it take to fill the tank?

104. [II] Refer to Fig. 9.43 on p. 315. Suppose a person standing erect or sitting upright is accelerated upward at a rate *a*. Write an expression for the resulting blood pressure in the brain in terms of the pressure at the heart and discuss your results.

105. [II] A Texas pipeline carrying natural gas (ρ = 0.90 kg/m^3) with a mass-rate of flow of 1.0 kg/s is 35 cm in diameter. Determine the average speed at which the gas is moving along.

106. [II] THIS PROBLEM EXAMINES LAMINAR FLOW OF AN INCOMPRESSIBLE LIQUID. Water moves through a horizontal pipe as shown Fig. P102, but this time the gauge on the left reads 130 kPa and the one on the right is broken. The cross-sectional areas of the pipe are 4.49 cm^2 and 1.00 cm^2. If the speed of the water in the large section is 2.00 m/s, (a) how fast is it moving in the narrow section? (b) What is the pressure in the narrow section?

107. [II] THIS PROBLEM EXAMINES LAMINAR FLOW OF AN INCOMPRESSIBLE LIQUID. A pipe carries glycerine (having a density of 1.26

g/cm³) under pressure from the floor of a processing plant up 20 m to a bottling machine. The liquid enters the pipe, which has a cross-sectional area of 6.25×10^{-2} m², at 14.0×10^5 Pa. Up at the bottling machine, the end of the pipe necks down and attaches to a horizontal delivery tube that has a cross-sectional area of 100 cm². Assume there is little or no friction impeding the flow.   (a) Draw a diagram showing streamlines from the lower pipe up into the delivery tube. (b) If the liquid initially enters the pipe at 5.00 m/s, at what speed is it traveling in the delivery tube?   (c) What is the pressure inside the delivery tube?

108. [II] The pressure of blood in a main artery at the level of the heart (Table 9.6) reaches a maximum of about 120 mm Hg. Suppose the artery is cut, how high will blood spurt vertically? [*Hint: Take the speed of the blood in the artery to be essentially zero.*]

109. [II] THIS PROBLEM EXAMINES LAMINAR FLOW OF AN INCOMPRESSIBLE LIQUID. A large open vertical tank containing a liquid, springs a small leak 2.0 m beneath the surface of the fluid.   (a) Draw a diagram showing streamlines from the open surface, down though the hole into the exiting stream.   (b) What is the pressure at points at the start and end of each of these lines?   (c) Use Bernoulli's Equation to determine the speed at which the fluid leaves the tank.

110. [II] A cylindrical water tank 2.0 m in diameter has a spigot with a hole whose area is 5.0 cm² located 4.0 m below the surface. Show that for a point at the surface the $\frac{1}{2}\rho v^2$ term in Bernoulli's Equation is properly ignorable compared to the corresponding term outside of the spigot in the escaping water jet. Assume no contraction of the stream as it leaves the tank.

> **SOLUTION:** Point-1 is at the surface of the water, point-2 is at the hole. At the surface $A_1 = \frac{1}{4}\pi D^2 = 3.142$ m². At the hole $A_2 = 5.0 \times 10^{-4}$ m². Since $A_1 v_1 = A_2 v_2$; $v_1 = (A_2/A_1)v_2 = 1.6 \times 10^{-4} v_2$. And $v_1^2 = 2.5 \times 10^{-8} v_2^2$, hence $\frac{1}{2}\rho v_1^2$ is ignorable compared to $\frac{1}{2}\rho v_2^2$.

111. [II] A Pitot tube, shown in Fig. P111, is a fluid-speed measuring device that has applications in many fields, including physiology. Often used on airplanes to measure the airspeed of the craft, it can be seen mounted on the wings or fuselage. At the front opening, the streamlines part and there is a **stagnation point** there. The fluid rushes away from that point in all directions, and *its speed there is zero.* Accordingly, the pressure (P) there and all along the inner tube is relatively high. On the other hand, the speed of the fluid over any of the small holes in the outer cylinder is the undisturbed freestream speed (v). Derive an expression for v in terms of the pressure difference indicated by the attached manometer, keeping in mind that the air within the device is at rest. By the way, the device was first used by Henri Pitot to measure water speeds in the Seine River (1730).

112. [II] A small plane, having a mass of 3000 kg, is flying in air of density 1.0 kg/m³. Air moves over the top and bottom surfaces of the wings at 160 m/s and 130 m/s, respectively. What is the net minimum wing area needed? Use the results of Problem 103.

113. [II] The instrument depicted in Fig. P113 is inserted in a pipeline to determine flow rates and speed. Known as a Venturi meter, it consists of a throat and two open manometer tubes. Derive an expression for the speed of flow in the pipe in terms of Δy, the difference in the column heights, and the known cross-sectional areas of pipe and throat. In actuality, the manometer tubes are quite narrow and short, and the fluid within them is at rest.

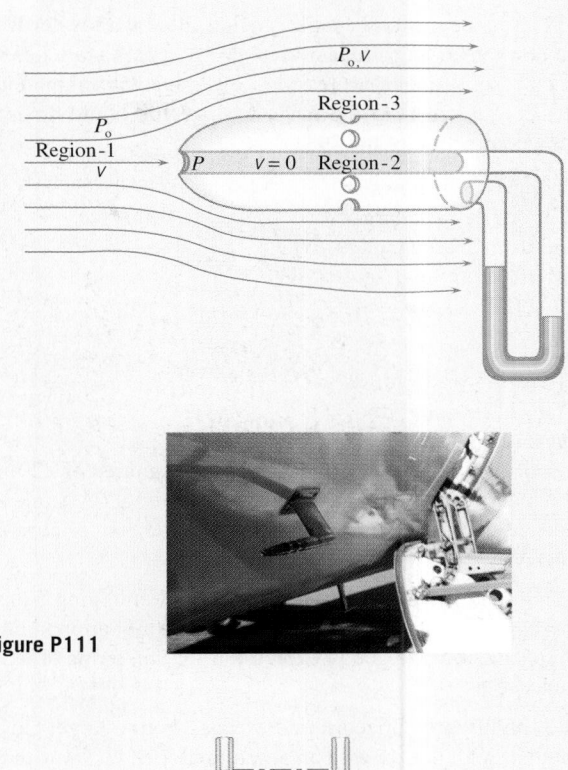

Figure P111

Figure P113

114. [II] The flowmeter in Fig. P113 shows a difference in height of 5.0 cm on an oil pipeline 200 cm² in cross section. If the throat of the device has a diameter of 10 cm, what is the pipeline flow rate? (Assume the liquid to be ideal.)

115. [II] A vaccination gun forces vaccine through a small aperture a few thousandths of an inch in diameter, at high pressures (550 psi), and so does away with the need for hypodermic needles. Compute the speed at which the fluid leaves the gun. Take the flow speed of the vaccine ($\rho = 1.1 \times 10^3$ kg/m³) in the reservoir within the body of the gun to be negligible.

116. [II] A cardboard milk container has two holes, one above the other, punched in its side. At a given moment, the two escaping streams of milk strike the table the container is standing on, at the same point. If the heights of the upper and lower holes are, respectively, y_u and y_l, write an expression for y, the level of milk at that instant, in terms of these heights.

117. [II] A large-diameter open cylindrical storage tank stands on a high platform. It has a small horizontal spigot at its very bottom, a height Y above the ground and a depth h below the surface of the water. Use Bernoulli's Equation to follow a tube of flow from point-1 at the surface, to point-2 just outside the spigot, to point-3 at the point of impact with the ground. Write expressions for v_2 and for v_3 (the speeds at the spigot and at the point of impact with the ground) in terms of g, h, and Y. Did you expect these results?

118. [II] A large, open storage tank is being filled with water from a pipeline at a rate J. Unfortunately, it springs a leak via a hole of area A, at its base. Write an expression for y, the equilibrium height to which the water rises in the tank. Assume the liquid emerges without any contraction of the jet.

119. [II] Figure P119 shows a submerged orifice discharging a liquid from a large tank. Determine an expression for the ideal efflux speed at point 2.

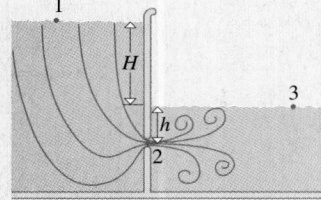

Figure P119

120. [III] During a storm, a 50-m/s (i.e., 112-mi/h) wind blows horizontally across the flat level roof of a supermarket. If the roof has an area of 220 m², what is the most force that's likely to be exerted on it? Assume the place is locked up tight for the night. (Take the density of air to be 1.1 kg/m³.)

121. [III] Figure P121 shows a stream of liquid emerging from a tube in the base of an open tank. Use Bernoulli's Equation applied between points 2 and 3 to get an expression for y in terms of θ and h. Can y exceed h?

122. [III] A highly pressurized liquid escapes from a large, closed tank through a small nozzle with an aperture of area A. Show that a

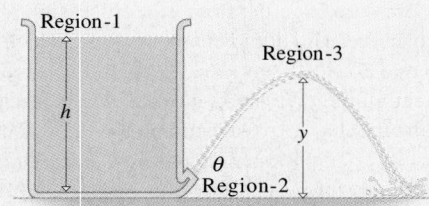

Figure P121

thrust results, given by

$$F = \rho A v^2$$

which is essentially only dependent on the orifice area and the gauge pressure in the tank. There are toy rockets that you fill with water and then pressurize with compressed air via a handpump—they blast out the water and lift off quite effectively.

123. [III] Show that the lift on the wing of an airplane of area A is given by

$$F_L = \tfrac{1}{2}\rho(v_\alpha^2 - v_\beta^2)A$$

where v_α is the speed of the air above, v_β the speed below, and ρ is the density of air. Although it's not clear how to predict these two speeds, it is reasonable to assume that each is proportional to the air-speed of the craft, and hence $(v_\alpha^2 - v_\beta^2) \propto v^2$.

M E C H A N I C S

Chapter 10
Elasticity & Oscillations

*A*lthough collapsing buildings and falling bridges are a rarity, these things happen, especially with some provocation like an earthquake (p. 373). Boats still snap in half on the high seas, and airplanes occasionally come apart in awkward ways. This chapter begins with an introduction to the mechanical properties of materials: how they stretch and compress, fatigue, break, and shear. But its real focus is on one of the most interesting properties, **elasticity**—*the ability of a system once distorted to spontaneously return to its original configuration*, and in so doing, more often than not, to **oscillate** (p. 344). Systems that oscillate run the gamut from violins and jiggling flesh to skyscrapers and stars. The atom itself is the most fundamental of all such systems. Atoms vibrate much as springs vibrate, and so as we study the common spring, we are laying the groundwork for far grander things to come.

Elasticity

Galileo began his book, *Two New Sciences* (1638), with a discussion "treating of the resistance which solid bodies offer to fracture." By the middle of the 1600s, there were some other tentative studies of the strengths of wires and rods, but it remained for that remarkably homely, and rather improper genius, Robert Hooke, to really get things going in what they called "*ye* science of *elasticity*"—that is, the analysis of the behavior of materials and structures under the influence of applied forces.

10.1 Hooke's Law

In 1676, wishing to establish priority for some new discoveries, Hooke published two cryptic anagrams. Two years later he decoded them, announcing the invention of the spring balance and the discovery of what has come to be known as Hooke's Law. When an object is acted on by a force, it may be compressed, stretched, or bent. If, after the force is removed, the object rapidly returns to its original configuration, we say that it is *elastic*. If the force is not too large, materials such as steel, bone, rock, tendon, glass, flesh, and rubber are elastic. By contrast, solids or quasi-solids, such as chewing gum, wax, lead, moist clay, and putty, do not return to their original configurations once distorted. These materials are *plastic*.

Hooke's Law maintains that for a linearly elastic object, the distortion resulting from an applied force is directly proportional to that force.

What Hooke found was that, beyond being elastic, **many materials deform in proportion to the load they support**. That's **Hooke's Law**, and it applies to *linearly elastic* or *Hookean* objects—to steel bars, rods, and wires; to springs, and diving boards; and to some extent to rubber bands, among other things. Thus, if an ordinary coiled (helical) spring in equilibrium ($\sum F = 0$) is either stretched or compressed by an applied force F, the spring will be deformed (extended or contracted) by a distance s that is proportional to the force: $F \propto s$ (Fig. 10.1).

The relationship $F \propto s$ can be transformed into an equality by introducing a constant-of-proportionality. This constant, k, takes care of the differences between units on both sides of the equation. *It introduces the physical characteristics of a particular spring into the formula.* Presumably, if we knew more physics, we could express k in terms of the atomic parameters, but that's still beyond our means. **Hooke's Law** is then

[linearly elastic object]
$$F = ks \tag{10.1}$$

This is the force that must be exerted *on* the spring to produce either an extension or compression of length s: when F is positive, s is positive. In other words, the force applied to the spring causes the displacement and, naturally enough, is always in the direction of the displacement. From Newton's Third Law, just the opposite is true for the force exerted *by* the spring (Fig. 6.10, p. 179); it's always opposite to the displacement. Because this force always tends to return the system to its equilibrium configuration, it's called a ***restoring force***.

The **spring constant** or **elastic constant** k, as defined by Eq. (10.1), **is a measure of the stiffness of the object being deformed**. It has the dimensions of force over distance and the units of N/m. *For a given force, the larger the value of k, the smaller the extension of the spring—that is, the more rigid the spring* (Fig. 10.2). The greater the slope (k) of the force-displacement curve, the stiffer the spring, rod, or diving board. The elastic constant varies with the size and shape of the object as well as with the material: the element, compound or alloy, its impurities, heat treatment, crystal structure, and so on.

You may have seen a Slinky mercilessly stretched until portions of it became distorted, unable to return to its original unstretched condition. Any rod, girder, spring, or wire can be bent or stretched beyond its **elastic limit** so that it becomes permanently deformed, losing its springiness. In such a sorry state, it will no longer obey Eq. (10.1); it will no longer be *linearly* elastic. Accordingly, those nice straight-line curves in Fig. 10.2 don't go on forever. {Before you try any problems, go over the selection of multiple choice **Warm-Ups** called **Hooke's Law** in **Chapter 10** on the CD.}

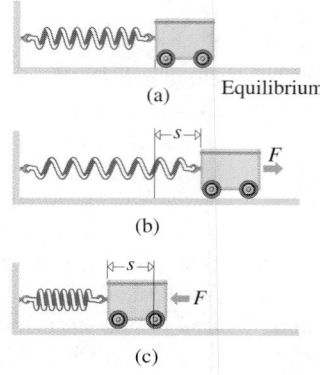

Figure 10.1 (a) A spring in equilibrium. A force F applied to the spring can either (b) stretch or (c) compress it by a distance s. Provided s is not large, $F = ks$.

Heavy trucks use leaf springs and shock absorbers to smooth out the ride. The leaf spring is a stack of thick steel bars of decreasing length strapped together. They are slightly bent so that an impulsive vertical force tends to straighten them out.

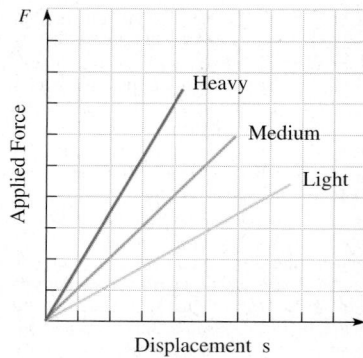

Figure 10.2 The force F applied to a spring stretches it a distance s. This figure is a plot of F versus s for three different springs. The slope of each curve is the spring's elastic constant, k. The less rigid the spring (heavy, medium, and light), the smaller the value of k.

Example 10.1 **[II]** A spring balance (Fig. 10.3) has a 10-cm graduated scale that can read mass from zero to 500 g. (a) How does the device work? Would this balance be useful on the Moon? (b) Determine the spring constant. (c) What displacement would the pointer experience if you pulled down on the support hook with 10 N?

Solution   The mere mention of a spring suggests Hooke's Law. (1) TRANSLATION—An object stretches a known amount under a specified load; determine its elastic constant.   (2) GIVEN: (*i*) the stretch 0–10 cm and mass load range 0–500 g, and (*ii*) a load of 10 N. FIND: (b) *k* and (c) *s* for 10 N.   (3) PROBLEM TYPE—Elasticity/Hooke's Law. (4) PROCEDURE—Use the load, the extension, and Hooke's Law to find *k*. (5) CALCULATION—(a) The spring scale indicates *s*, which is proportional to *F*. On Earth, that force is more or less constant and proportional to the constant mass of the load. If calibrated on

Figure 10.3 A spring scale much like the one invented by Robert Hooke. A load either stretches or compresses a spring, changing its length proportionately.

Earth, it can be used to measure mass. If the scale were set to read in newtons, it would function perfectly on the Moon. When set to read grams, it is assumed the device is on Earth, where $F_W/(9.81 \text{ m/s}^2) = m$.

(b) The weight of a 500-g mass is

$$F_W = mg = (0.500 \text{ kg})(9.81 \text{ m/s}^2) = 4.91 \text{ N}$$

producing a displacement of 10 cm, or 0.10 m; since the applied force is F_W

$$k = \frac{F_W}{s} = \frac{4.91 \text{ N}}{0.10 \text{ m}} = \boxed{49 \text{ N/m}}$$

(c) If you pulled down on the balance with a force of 10 N, a displacement of

$$s = \frac{F}{k} = \frac{10 \text{ N}}{49 \text{ N/m}} = 0.20 \text{ m}$$

would have to occur, which is impossible in this case because the maximum *s* is 10 cm.

Quick Check: Each 1 cm shift corresponds to 500 g/10 = 0.050 kg; hence, $k = F/s = mg/s = (0.050 \text{ kg})(9.81 \text{ m/s}^2)/(0.010 \text{ m}) = 49 \text{ N/m}$.

[For more worked problems click on **WALK-THROUGHS** *in* **CHAPTER 10** *on the* **CD.]**

These massive springs beneath a railroad car absorb energy whenever the car moves vertically.

Elastic Potential Energy

One of the practical functions served by springs is to store energy (Fig. 10.4). The mainspring in a clock (a device invented by Hooke) replaces the *gravitational*-PE of a falling weight or pendulum with the *elastic*-PE of a spring (PE$_e$). As we saw in Chapter 6, ***the work done on a system against a nonconstant force, such as the force of a spring, is the area under the force-distance curve, which equals the change in the potential energy of the system.*** For any of the plots in Fig. 10.2, the *elastic*-PE stored in stretching the spring a distance *s* is the area of the triangle under the line—namely, $\frac{1}{2}$ the base (*s*), times the altitude (*F* = *ks*):

$$\Delta PE_e = \tfrac{1}{2}s\,(ks)$$

and so

$$\Delta PE_e = \tfrac{1}{2}ks^2 \tag{10.2}$$

This is the same as saying that the work done by the displacement-dependent spring-force is its average value, $F_{av} = \frac{1}{2}(F_f + F_i) = \frac{1}{2}(F_f + 0) = \frac{1}{2}ks$ times the displacement *s*: $W = F_{av}s = \frac{1}{2}ks^2 = \Delta PE_e$.

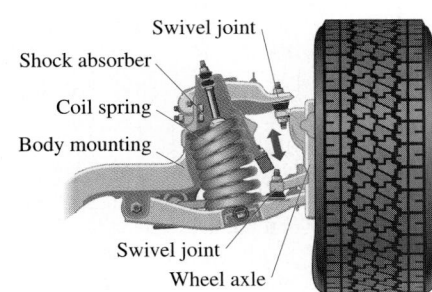

Swivel joint
Shock absorber
Coil spring
Body mounting
Swivel joint
Wheel axle

Figure 10.4 The suspension system of an automobile. When the wheel of a car is jolted upward by a bump in the road, a coil spring is compressed, slowing and smoothing out the vertical motion. The shock absorbers convert the stored mechanical energy to thermal energy, keeping the car from bouncing on the springs.

Example 10.2 **[II]** Four people, each having a mass of 70 kg, enter a car of mass 1100 kg—two in front and two in back. The vehicle settles down in a little while on its four identical springs, compressing each of them a distance of 2.5 cm. (a) Determine the elastic constant of each spring and the amount of energy stored in each. (b) What is the equivalent elastic constant of the suspension taken as a single unit? (c) Where does the stored energy come from?

Solution This problem is about springs, and that suggests Hooke's Law. (1) TRANSLATION—Four springs compress a known amount under a known load; determine (a) the elastic constant for each, (b) the energy stored, and (c) the equivalent elastic constant. (2) GIVEN: Mass of each person, 70 kg; mass of car, 1100 kg; and $s = 2.5$ cm. FIND: (a) each k, (b) each ΔPE_e, and (c) the net spring constant k'. (3) PROBLEM TYPE—Elasticity/Hooke's Law. (4) PROCEDURE—The spring system can be taken to be in equilibrium when the car is empty, so the car's mass is irrelevant. Use the known load, compressed distance, and Hooke's Law to find the spring constant. (5) CALCULATION—(a) The total mass of people supported by the four springs is 4×70 kg, and this weighs

$$F_w = (280 \text{ kg})(9.81 \text{ m/s}^2) = 2747 \text{ N}$$

Each spring compresses the same amount (0.025 m) and supports the same load; hence, the compressive force exerted by each is $\frac{1}{4}$ the total load

$$F = \tfrac{1}{4}(2747 \text{ N}) = ks = k(0.025 \text{ m})$$

and $\boxed{k = 2.7 \times 10^4 \text{ N/m}}$. (b) Taking the springs together as one, exerting 2747 N upward, the equivalent force constant is

$$k' = \frac{2747 \text{ N}}{0.025 \text{ m}} = \boxed{1.1 \times 10^5 \text{ N/m}}$$

which is four times the individual spring constant. (c) The *elastic*-PE stored in each spring is

$$\Delta PE_e = \tfrac{1}{2}ks^2 = \tfrac{1}{2}(2.7 \times 10^4 \text{ N/m})(0.025 \text{ m})^2 = \boxed{8.6 \text{ J}}$$

while the whole system stores four times that. The energy stored elastically is a fraction of the decrease in the *gravitational*-PE—the remainder is lost as thermal energy.

Quick Check: The total stored PE_e is $\frac{1}{2}k's^2 = 34.4$ J, which equals 4(8.6 J).

Elastic Materials

Suppose we take a sample of some material, apply an increasing force F to it, and plot the resulting change in length ΔL, as in Fig. 10.5b for animal tissue and 10.5c for synthetic stretchy polymer. For moderate extensions, the data goes up and back along the two curves, and both represent elastic behavior. For some elastic materials there's a linear relationship between F and ΔL that corresponds to Hooke's Law

$$F = k\Delta L \tag{10.3}$$

However, a linear response only occurs over a limited range of applied force. Removing a small load from a distorted sample allows the atoms, which were only slightly shifted, to return to their equilibrium configuration and the material comes back to its original length.

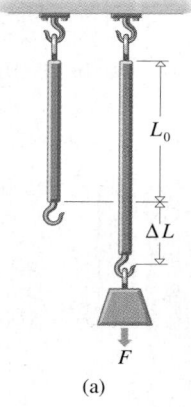

(a)

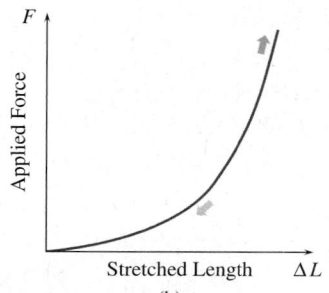

(b)

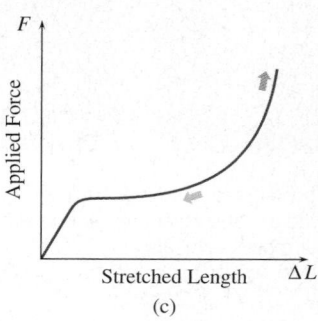

(c)

Figure 10.5 (a) A length (L_0) of material stretches an amount ΔL under a load F. (b) Soft animal tissue stretches elastically, forming a J-shaped curve and contracting back along the same arc as the force is reduced. (c) Rubbery synthetic solids stretch elastically along an S-shaped curve. They also return to $\Delta L = 0$ along the curve as F goes to zero.

The athlete bends the pole, storing energy in it in the form of *elastic*-PE. The pole springs back, returning its energy to him and increasing his *gravitational*-PE.

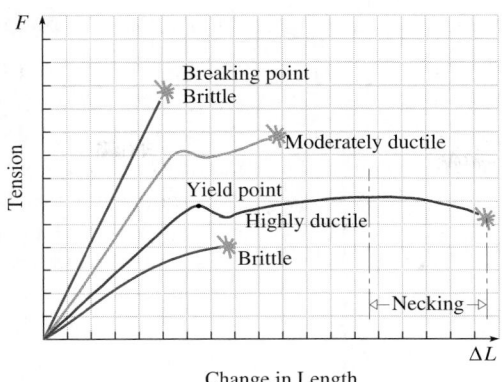

Figure 10.6 Load or tension versus elongation for several different engineering materials. Brittle substances rupture suddenly without much elongating. Ductile substances elongate via plastic flow. A ductile metal might be Hookean up to 0.25%–0.5% elongation and fracture at around 50%. Some nonmetallic materials such as wood and fiberglass fracture after elongations of 1%–3%.

INTERATOMIC BONDS & RESTORING FORCES

The elastic energy of a spring arises from pulling against the interatomic bonds holding the material together—it's really *electromagnetic*-PE. In fact, whenever a contact force acts on any kind of body, the interaction is accompanied by a distortion of the body and the resulting electrical restoring forces among the many millions of atoms involved. There is no such thing as an absolutely rigid body.

Consider what happens to a sample as the load on it gets very large. Depicted in Fig. 10.6 is the behavior of several different solid rods under increasing tension. The longest curve with the greatest value of ΔL corresponds to a material like pure iron or aluminum. Such a sample will elongate a little linearly until its elastic limit, or yield point, is reached and the bonds between atoms begin to be overcome. Rows of atoms then slide past one another in a process known as *slip*, and the material plastically deforms in an unrecoverable way. Materials (e.g., chewing gum, copper, gold, and silver) that can be stretched thin are *ductile*. Once the plastic mode is initiated, the elongation occurs more dramatically until ultimately the specimen ruptures at the breaking point. You have probably never stretched a steel rod to the breaking point, but the same thing happens with a piece of taffy or a Tootsie Roll. {For more discussion, and to find out why a bridge in Canada broke into pieces when a train went over it, click on **BRITTLE MATERIALS** under **FURTHER DISCUSSIONS** on the **CD**.}

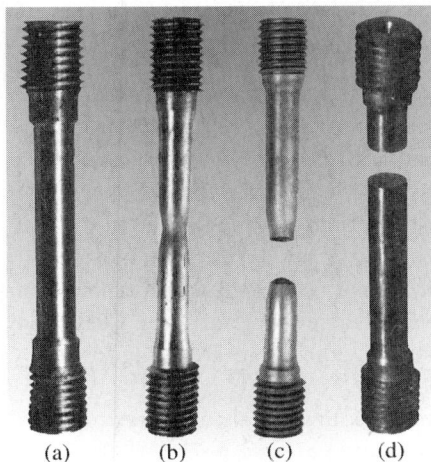

(a) When a sample is put under tension, it elongates linearly by less than ≈0.5%. If it's ductile, it continues to elongate several hundred times as much under slightly increased loads. (b) After a maximum load is reached, the sample begins to neck down until (c) it fails. (d) If the material is brittle, it fractures without deforming plastically.

STICKS AND STONES...

Bone is a composite structure made of crystals of hydroxyapatite (a calcium compound good in compression) and collagen fibers (good in tension). Bone, which is varyingly brittle, remains elastic almost to the point of rupture and cannot be either permanently compressed or elongated unless the loads are applied for a very long time. Infants' bones are high in collagen and very flexible. By young adulthood, hydroxyapatite has built up to the point (≈67%) where bones are both flexible and strong. As adults age, collagen becomes less flexible and calcium tends to be reabsorbed from the hydroxyapatite, making bones brittle and more breakable.

STRESS & STRAIN

Remarkably, about a century and a half elapsed before Hooke's ideas were further developed by the French mathematician Augustin Louis Cauchy. It's likely that this long period of idleness was in no small part due to Newton, who detested Hooke and, outliving him by 25 years, had ample opportunity to denigrate the latter's practical science. Sir Isaac was a loner—chaste, sensitive, and vulnerable—while Hooke was argumentative, vain, and a womanizer, a testy, earthy sort who took pleasure in controversy as much as Newton shrank from it. It can be argued that Hooke contributed to Newton's nervous breakdown, and it can also be argued that Sir Isaac had his revenge, at least on the theory of elasticity. In any event, Cauchy (who himself was widely held to be selfish and narrow-minded) introduced the central notions of *stress* and *strain* (1822).

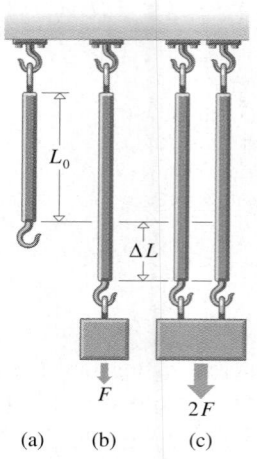

Figure 10.7 (a) A sample of an elastic material. (b) Under a load F, it elongates by an amount ΔL. (c) Two identical samples loaded with $2F$ also elongate by the same ΔL.

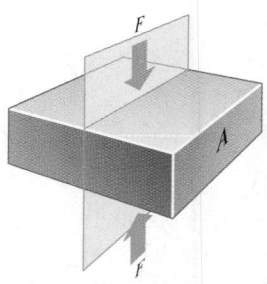

Figure 10.8 A pair of equal-magnitude, parallel forces shearing a sample. Think of how you tear a piece of cardboard or snip a twig or wire.

10.2 Stress & Strain

The application of a force to an elastic object will cause it to deform. The concentration of force is called **stress**. The measure of the resulting deformation is called **strain**. As we'll soon see, the fact that for a specific object stress and strain are proportional, is reflected in the **elastic constant** and **Hooke's Law** (p. 330). Even more remarkably, the fact that for a specific kind of material stress and strain are proportional, is reflected in the **Young's Modulus** (p. 339).

What severely limits the scope of Hooke's Law is that it deals with the specimen or spring in its entirety. The elastic constant depends on the details of the object under test (whether it's a wire or a tree trunk) and not just on the material of which it's made. What we need is a reformulation of Hooke's Law that focuses on the internal processes at work that are independent of the external dimensions.

The notion of *stress* was introduced to describe the distribution of force within a solid acted upon by a load. Ordinary interactions between bodies are always distributed over some contact area; the idea of a force acting at a point is an idealization. ***The stress experienced within a solid is defined as the magnitude of the force acting, divided by the area over which it acts***:

$$\text{Stress} = \frac{\text{force}}{\text{area}}$$

Using the Greek letter sigma (σ) for stress, this equation becomes

[stress] $$\sigma = \frac{F}{A} \qquad (10.4)$$

The idea is very much like the concept of pressure, and like pressure it has the SI units of N/m^2, or Pa. Since a pascal is actually very small, stress is more often given in megapascals (MPa = 10^6 Pa) or even gigapascals (GPa = 10^9 Pa). For example, when all your weight is on one leg as you walk, the stress on your knee can be more than 1 MPa (i.e., 10 atm).

As we are about to discover, there are several kinds of stress. The hanging bar in Fig. 10.7b experiences both a downward force and the upward supporting force provided by the ceiling. Imagine a plane cutting perpendicularly across the bar. The stress in that cross section is given by Eq. (10.4), and it is known as the **normal stress**—*the forces are normal or perpendicular to the area*. Notice in Fig. 10.7 that while a load F is supported by a single rod (which stretches a distance ΔL in the process), two such rods will support twice that load ($2F$), each elongating by that same ΔL. Now if each rod has a cross-sectional area A, the stress (F/A) in the single rod exactly equals the stress in the double-rod suspension ($2F/2A$), even though the elastic constants in Hooke's Law are not the same. That's the great virtue of the idea of stress—it reflects how strongly the atoms are being forced to interact.

In addition to normal stress, there is also something known as **shear stress** (σ_s) that exists when ***the force-pair acts parallel to the cross-sectional area A*** (Fig. 10.8). Shear arises when the applied forces cause two regions of an object to move laterally in opposite directions. A pair of scissors cuts paper by applying a pair of equal-magnitude, oppositely directed forces parallel to the cross section to be severed. It produces a shearing stress that is great enough to sever the material. The shear stress, again given by Eq. (10.4), is a measure of the effort needed to cause the atoms in one part of a solid to slide past an adjacent part—like a pushed deck of cards that edges over at a slant or one that's fanned out. The latter reflects the fact that shear stress also occurs when the sample (e.g., a wrench, or wrist) is twisted.

Strain

Until now, we've talked about rods and the way they elongate or shorten under a tensile or compressive load. But that change in length depends on the original dimensions of the rod. Imagine that a tensile force pulling on the atoms stretches their bonds and increases the interatomic separation by, say, 0.01%. Then, with all its atoms 0.01% farther apart axially, the rod must be 0.01% longer, no matter how long it was originally. Rods of different length will deform by different amounts, but the percentage change will be the same, and it's the change in the interatomic spacing that is crucial here. It's that change that generates the counterforce to the load. Consequently, we define the **normal strain** *under an axial load as the change in the length over the original length*:

$$\text{Strain} = \frac{\text{change in length}}{\text{original length}}$$

which can be compressive or tensile. Commonly, the Greek letter epsilon (ε) represents the strain, which becomes

[normal strain]

$$\varepsilon = \frac{\Delta L}{L_0} \qquad (10.5)$$

and it is assumed that the sample has a constant cross-sectional area. Strain is a measure of how much the atoms are shifted and, therefore, how much the bonds are being stretched or compressed. It is a *dimensionless quantity*. In engineering applications, strain is generally kept within $\pm 1.0\%$ and rarely even exceeds $\pm 0.1\%$. The columns holding up a high-rise building 300-m tall, whether made of concrete or steel, will be designed to compress by little more than about 3 cm. A 100-m elevator cable that stretches 10 cm under a load has a strain of $(10 \times 10^{-2}\,\text{m})/(100\,\text{m}) = 1.0 \times 10^{-3} = 0.1\%$.

There must be a strain as a solid—any solid—yields under the influence of a stress, however small. That's true for normal stress, as well as shear stress, and there must be a **shear strain** (ε_s), a corresponding change in shape. The solid block shown in Fig. 10.9 is being twisted. The larger the side l_0, the more the edge is displaced, Δl, but the shifting on an atomic level will be the same no matter what l_0 is, provided the angle γ is constant. *We define the shear strain as the angle of distortion, γ (Greek lowercase gamma), measured in radians, which makes it*

Human skin is under tension like a rubber glove. As a person ages, the skin wrinkles because it loses its elasticity.

ELASTICITY & BIOLOGICAL TISSUE

Figure 10.10c depicts the behavior of a typical biological tissue sample—pull on your lip or ear lobe. Soft connective animal tissue often contains a mix of two different protein materials: elastin and collagen. The elastin fibers, abundant in the vessels of the human circulatory system, are easily stretched, requiring little stress to produce considerable strain. These are frequently reinforced with a network of quasi-Hookean collagen fibers, as in a human artery. Collagen is the strongest of the connective tissues. It's the collagen that gives flesh its toughness so that it doesn't tear easily. Collagen makes meat tough, although at high temperatures ($\approx 100\,°\text{C}$) it breaks down to soft gelatin, leaving the other more tender constituents like elastin behind—that's one reason why we cook meat.

Some biological tissue can stretch elastically well beyond the 800% value that's typical of soft materials such as rubber.

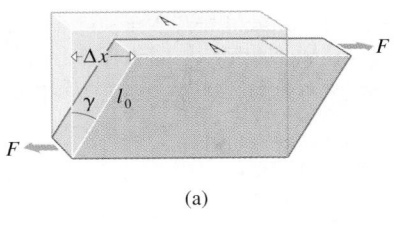

(a)

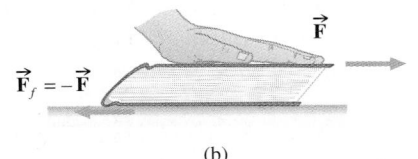

(b)

Figure 10.9 Deformation due to shear. (a) Two equal-magnitude, oppositely directed forces twist the block through an angle γ that corresponds to the shear strain. The shear stress is F/A. (b) The pages of a book slide past one another much as layers of atoms move over one another in a solid that is being sheared.

Figure 10.10 (a) The stress-strain curves for several Hookean materials. The ductile samples, such as soft steel, stop behaving linearly at their yield points (σ_Y). (b) When stressed, rubbery polymers first elongate by straightening out their molecules and thereafter by tugging on the chemical bonds. (c) Most biological materials are under tension even when not strained. Your skin is like a rubber glove stretched over your body. (d) Elastin is usually reinforced with collagen in biological systems such as arteries. Tendon is made principally of collagen.

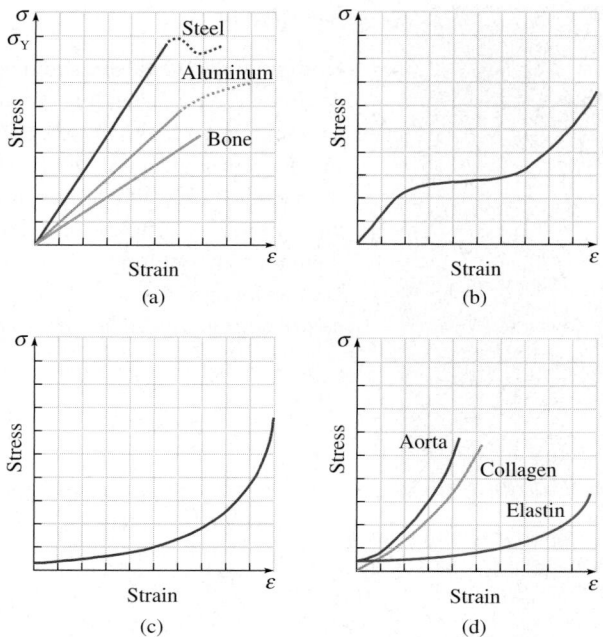

Floor tiles sheared by spiked heels. The downward force increases as a person shifts back onto the heels just before sitting.

THE HIGH-STRESS MENACE

Several decades ago, women's high-heeled shoes with a bare steel shank were fashionable. Ordinarily, a shoe might have roughly 20 in.² of contact area with the ground and a 140-lb woman, her weight shifted to one foot, would produce a meager stress of about 48 kN/m² (7 psi). But when that weight was concentrated on the spiked heel of one shoe (with an area of about 0.12 in.²), the stress would soar to 8 MN/m² (i.e., 1.1×10^3 psi). The resulting dents, holes, and chipped tiles prompted some shopkeepers to ban the menace.

dimensionless. Since the angle of distortion is generally small,

[shear strain]
$$\varepsilon_s = \gamma \approx \frac{\Delta x}{l_0} \tag{10.6}$$

An elastic material is one that returns to its original shape after an initial deformation, and that applies to shear strain as well. Metals, bone, stone, and concrete display *elastic* shear strain of less than about 1°, which is typical of many hard solids. Still, stone bends more than one might imagine; indeed, every building bends in the wind. The towers of the World Trade Center (p. 343) used to heave over as much as 3 ft in a strong wind (you could even hear them creak) and they would bend away from the vertical by 6 or 7 ft in a hurricane. By comparison, stretchy things such as soft biological tissue can recover from shear strains of upwards of 40°. Take hold of the skin on your belly and pull; the way it returns depends on such things as dehydration and, of course, age. Figure 10.10 is a *stress-strain diagram* for several materials (not surprisingly, part *a* resembles Fig. 10.2).

10.3 Strength

In broad terms, *strength* means the ability to withstand a load without failure. More specifically, the **strength** *of a material is the stress, in newtons per meter-squared, that will cause failure in a sample of that substance*. Strength is inherently linked to the atomic behavior of the material and tells us how any sample should behave, all else being equal (Table 10.1).

Figure 10.11 is the stress-strain curve for a sample of an aluminum alloy. *The stress at which slip is first observed, at which the curve stops being linear, is the* **yield strength.** Any mechanical part that will have to support a load without plastically deforming will

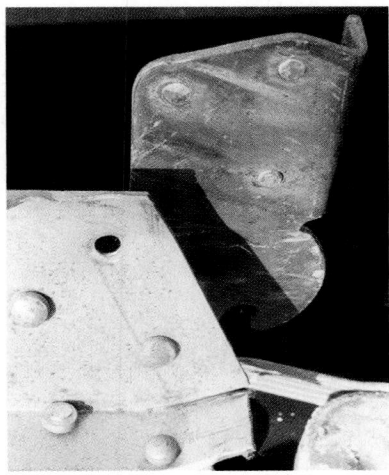

The top of a failed extension ladder. Each side had a steel holding plate fastened to the body with three aluminum rivets. The steel plates are much harder than the rivets, and when there was enough torque, they sheared the rivets in half. The plate twisted upward, still holding its half of the three severed rivets. One disembodied head has dropped out of its hole in the body and rests on the flange. The ladder collapsed, and the man on it luckily broke only his wrist.

Table 10.1

Approximate Ultimate Strengths of Various Materials

Material	Tensile (MPa)	Compressive (MPa)	Shear (MPa)
Muscle	0.1		
Bladder wall	0.2		
Cartilage	3		
Concrete	4	30−40	
Brick	5.5	10−21	
Skin	10		
Marble	10	110	22
Lead	12		12
Granite	20	240	35
Glass	40−175	50	
Polystyrene	48	90	55
Fir, Douglas	50	50	8
Tendon	65−82		
Pine	100	27−50	9
Bone, compact	110	150	
Brass	120−400		
Femur, horse	121	145	
Femur, human	124	170	
Iron, cast-gray	170	650	240
Hair	190		
Spider's silk	240		
Catgut	350		
Cotton	350		
Aluminum	300−570		200−330
Silk	350		
Steel, structural	400		
Nylon thread	1100		
Steel piano wire	3100		

have to have an appropriately high-yield strength—we can't make the girders supporting a building out of soft steel.

The **tensile strength,** *or* **ultimate strength,** *is the stress corresponding to the greatest applied force*—it's the high point on the stress-strain curve. Usually a ductile sample will deform nonuniformly over its cross section, tending to narrow or **neck** down. The start of that process also marks the ultimate strength (see photo on p. 333). By the time a piece of material has reached its point of ultimate strength, it has already been irreversibly elongated. We don't often want a ladder that will permanently sag when the first person climbs it.

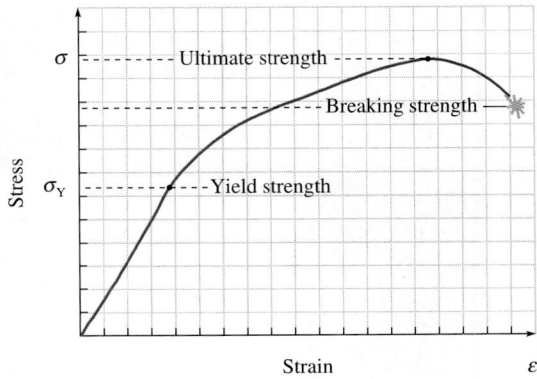

Figure 10.11 The stress-strain curve for an aluminum alloy. There are several different strengths that are of interest to someone using the material: (1) The yield strength tells us when it will cease being elastic and begin to permanently deform. (2) The ultimate strength is the maximum stress it can experience. (3) The breaking strength is the stress at which it fails.

The stress corresponding to the rupture of the sample is the **breaking strength**, and that applies whether the sample is being stretched, compressed, or sheared. Brittle materials show little or no necking, and their ultimate strengths equal their breaking strengths (see photo on p. 333). {For more, especially about both structural steel & Kevlar, click on **STRENGTH OF MATERIALS** under **FURTHER DISCUSSIONS** on the CD.}

Example 10.3 **[II]** The human thigh bone, the femur, at its narrowest point resembles a hollow cylinder with an outer radius of roughly 1.1 cm and an inner radius of just about half of that. Taking the compressive strength of the bone to be 170 MPa, how much force will be required to rupture it?

Solution This problem is about the breaking strength of bone. (1) TRANSLATION—Consider an object of known cross section and compressive strength; determine the maximum load it can support. (2) GIVEN: $R_o = 1.1$ cm, $R_i = 0.50 R_o$, and $\sigma_c(\text{max}) = 170$ MPa. FIND: max F. (3) PROBLEM TYPE—Elasticity/compressive strength. (4) PROCEDURE—The stress corresponding to the rupture of the sample is the breaking strength, which is 170 MPa $= \sigma_c(\text{max}) = F/A$. (5) CALCULATION—To find the stress we need the area. The cross-sectional

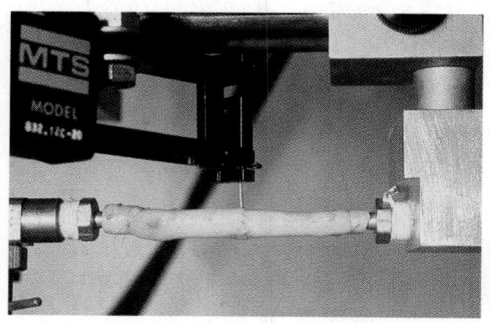

area of bone matter is

$$A = \pi(R_o^2 - R_i^2) = \pi R_o^2(1.00 - 0.50^2)$$
$$A = \tfrac{3}{4}\pi(1.1 \times 10^{-2}\,\text{m})^2 = 2.85 \times 10^{-4}\,\text{m}^2$$

Since $\sigma_c(\text{max}) = F/A$

$$F = \sigma_c(\text{max})A = (170 \times 10^6\,\text{Pa})(2.85 \times 10^{-4}\,\text{m}^2)$$

$$\boxed{F = 4.8 \times 10^4\,\text{N}}$$

or about 11×10^3 lb. An adult who steps off a kitchen table and lands on the heel of one bare foot with the knee locked will experience a force large enough to shatter the femur.

Quick Check: $A = A_o - A_i = 3.8 \times 10^{-4}\,\text{m}^2 - 0.79 \times 10^{-4}\,\text{m}^2 = 3 \times 10^{-4}\,\text{m}^2$; $F/A = \sigma_c \approx (50\,\text{kN})/(3 \times 10^{-4}\,\text{m}^2) \approx 0.17$ GPa.

Fatigue

Many ordinary situations exist in which a metal component can experience a cyclical application of stress. Even though the applied stress is well below the yield strength of the metal, it is still possible for the component to rupture after repeated applications of stress

THE STRENGTH OF MUSCLE

Our forearms and calves house massive concentrations of muscle. Muscle tissue has a very low strength (0.1 MPa), and there has to be a good deal of it to maintain the needed cross-sectional area. Hands and feet, which must move around quickly, need to be light, agile, and still strong. The solution is to bunch our muscles fairly far away and link them to our fingers and toes by strong (82 MPa), and therefore affordably thin, light tendons. Tendon, though its strength is less than that of mild steel ($\approx$450 MPa), is about seven times less dense and so is actually stronger pound for pound.

The fuselage of this Aloha Airlines jet ripped off during a flight in 1988. Experts attribute the failure to metal fatigue. Clearly, this was not a case of elastic behavior.

Fatigue testing of medical sutures under conditions resembling the pulsative pressure loading found in the human body.

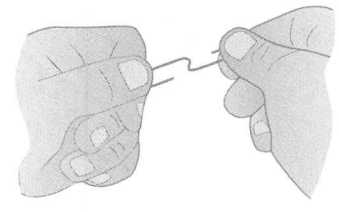

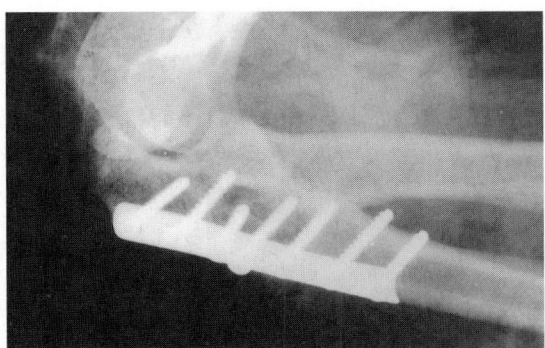

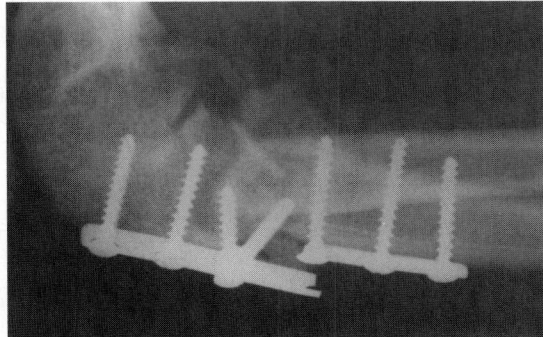

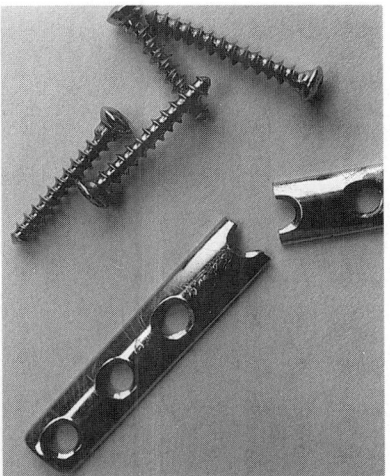

An X-ray of a steel plate used to bridge a broken bone. The bone had not yet knit, and as a result of excessive cyclic loading during physical therapy the plate fatigued and broke. The break occurred at a hole, right where you would expect.

Thomas Young (1773–1829), physicist and physician, has a long list of accomplishments to his credit, including his introduction of the Modulus of Elasticity. He is best known for his work in optics.

(see the photo on p. 338). This failure mode, induced by a relatively small, cyclic stress, is known as **fatigue**.

The accompanying photo is an X-ray of a broken ulna in the forearm of a man showing the steel brace used to hold the two segments together. Unfortunately, the bone did not knit, and so whenever the patient used his arm, especially during physical therapy, the brace carried the load and experienced cyclic stresses. It should have been expected that the fairly thin brace would fail sooner or later due to fatigue, but no one knew the bone had not fused. Finally, the brace ruptured at the place where the steel was narrowest and the stress the greatest, right across a screw hole. {For more, click on **METAL FATIGUE** under **FURTHER DISCUSSIONS** on the **CD**.}

10.4 Elastic Moduli

In 1807, Dr. Thomas Young introduced a notion he called "the modulus of the elasticity" (*modulus* is Latin for "a small measure"). But it was the French engineer C. Navier who (1826) provided the modern formulation of what has come to be known as **Young's Modulus** (Y) or the Elastic Modulus. Young had found a basic measure of the stiffness of materials.

Table 10.2
Approximate Values of Young's Modulus
for Various Solids

Material	Young's Modulus (GPa)
Biological tissue, soft	0.000 2
Rubber	0.007
Cartilage, human	0.024
Collagen	0.6
Tendon, human	0.6
Nylon	2
Spider thread	3
Catgut	3
Nylon fiber	5.5
Plywood	7
Hair	10
Fir, Douglas	13
Oak	14
Brick	14
Lead	16
Compact bone, compression	10
tension	22
Concrete	25–30
Marble	50
Aluminum	56–77
Glass	65
Iron, cast, gray	70–145
Granite	70
Gold	79
Copper	120
Bronze	120
Iron, wrought	180–210
Steel, stainless	190
Steel, structural	200
Tungsten	360
Diamond	1200

Young's Modulus

The normal stress-strain graphs for many materials begin with a linear region where the substance, at least at low strain, is Hookean (Fig. 10.10a); the slope of the curve is a constant. This **elastic stress per unit strain** defines Young's Modulus:

$$\text{Young's Modulus} = \frac{\text{stress}}{\text{strain}}$$

or symbolically

$$Y = \frac{\sigma}{\varepsilon} \tag{10.7}$$

which corresponds to either tension or compression (Table 10.2). The units of Y are the same as those for stress—namely, N/m^2 or Pa. The steeper the slope, the less distortion the sample experiences for a given stress, the stiffer it is, and the greater is Y. A stiff material more nearly maintains its size and shape under a load in the elastic regime.

Referring to the loaded rod in Fig. 10.7b, since $\sigma = F/A$ and $\varepsilon = \Delta L/L_0$, Eq. (10.7) rearranged, $\sigma = Y\varepsilon$, leads to $F/A = Y\varepsilon$, and

$$F = \frac{YA}{L_0}\Delta L \tag{10.8}$$

which is the equivalent of Hooke's Law, $F = k\Delta L$, where $k = YA/L_0$.

The stronger the interatomic bonding, the greater the force that must be applied to separate the atoms and, in turn, to stretch the material. We should expect that for materials with larger and larger bonding forces, as indicated by increasing melting points, Young's Modulus would increase as well (Table 10.3). {For a wonderful approximation of Y using atomic considerations, click on **YOUNG'S MODULUS** under **FURTHER DISCUSSIONS** on the **CD**.} 💿

Table 10.3
Young's Modulus and the Melting Point
for Various Solids

Material	Young's Modulus (GPa)	Melting point (°C)
Lead	16	327
Magnesium	42	650
Aluminum	56–77	660
Copper	120	1083
Tungsten	360	3410

Example 10.4 **[II]** Return to Example 10.3, which deals with a broken femur, and calculate the fractional change in length of the bone at the moment of collapse, assuming Y is constant right up to failure. Use Table 10.2.

Solution This problem is about stress and strain. (1) TRANSLATION—An object of known cross section and compressive strength is compressed until it fails; determine the fractional change in length. (2) GIVEN: $\sigma_c(\text{max}) = 170$ MPa. FIND: $\varepsilon_c = \Delta L/L_0$. (3) PROBLEM TYPE—Compressive strength/Young's Modulus. (4) PROCEDURE—The fractional change in length is the strain and $\varepsilon = \sigma/Y$, and we can look up Y. (5) CALCULATION—From Eq. (10.7) and Table 10.2

$$\varepsilon_c = \frac{\sigma_c}{Y} = \frac{170\text{ MPa}}{10\text{ GPa}} = 17 \times 10^{-3} = \boxed{1.7\%}$$

Quick Check: Using Example 10.3, the force is $F = YA\varepsilon_c = (10\text{ GPa})(2.85 \times 10^{-4}\text{ m}^2)(17 \times 10^{-3}) = 4.8 \times 10^4$ N.

Table 10.4

Shear and Bulk Moduli

Material	Shear Modulus (GPa)	Bulk Modulus (GPa)
Rubber	0.000 8–0.001 6	
Ethyl alcohol		0.9
Water		2
Seawater		2.1
Pine	0.6	3
Glycerin		4.5
Lead	5	8
Mercury		25
Quartz	8–30	35
Compact bone	3.5	12
Marble		70
Aluminum	30	70
Glass	8–30	40
Iron, cast	50	95
Granite		50
Brass	40	70
Copper	50	140
Iron, wrought	85	60
Steel, stainless	85	70
Steel, structural	80	160
Tungsten	150	200
Diamond		620

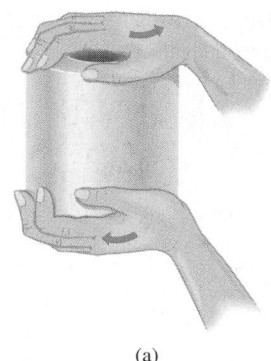

(a)

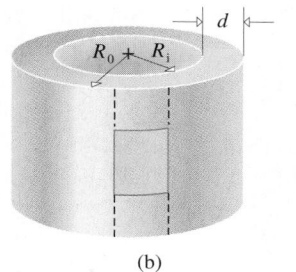

(b)

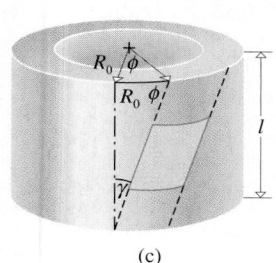

(c)

Figure 10.12 As an example of the theoretical Shear Modulus, consider a hollow thin-walled tube: $S = (F/A)/\gamma$. (a) A torque is applied and the rod is twisted through an angle γ. A small region in the untwisted state (b) experiences shear (c) in the twisted state.

Shear and Bulk Moduli

The stress-strain curves for shear are also often linear initially and therefore an elastic **Shear Modulus** (S) can be introduced:

$$\text{Shear Modulus} = \frac{\text{shear stress}}{\text{shear strain}}$$

that is

$$S = \frac{\sigma_s}{\varepsilon_s} = \frac{F/A}{\gamma} \qquad (10.9)$$

where (Fig. 10.9) the force acts parallel to the area, and we have made use of Eq. (10.4). The Shear Modulus is typically two or three times less than the Young's Modulus for the same material (Table 10.4).

When an ordinary helical spring is "stretched," the coiled wire itself does not elongate. Rather than being in tension or compression, the wire is actually twisted by a load that tends to pull it out into a straight length. The spring constant is actually dependent on the Shear Modulus.

Figure 10.12 shows the geometry for a tube, resembling a hollow human bone, being twisted by a torque τ. The resulting angle $\gamma \approx (R_0\phi)/l$ is proportional to τ/S. If some structure or component is twisted (a bone, for example) through a large enough angle ϕ, it may fracture, which is what often happens to the tibia when a skier's leg is twisted in a fall (Fig. 10.13). Table 10.5 lists a few values of both the breaking torque and the corresponding value of ϕ for some human bones.

Table 10.5

Torques and Breaking Angles for Human Bones

Bone	τ (N·m)	ϕ (degrees)
Ulna or radius (forearm)	20	15
Humerus (upper arm)	60	6
Femur (thigh)	140	1.5
Tibia (lower leg)	100	3

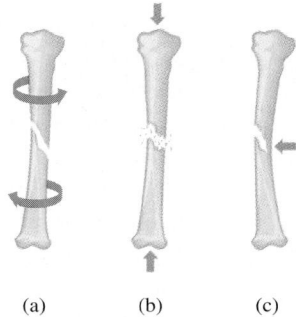

(a)   (b)   (c)

Figure 10.13 Common bone breaks. (a) A spiral fracture results from a torque that twists the bone. This type of break is often associated with skiing accidents. (b) Under compression, the bone may fail and the fractured ends break into pieces. (c) An applied transverse force bends the bone. The side away from the force is stretched, and since bone is weak in tension, it shatters. The near side is in compression, and if the bone is flexible, it might not break there at all.

When you have a narrow object (e.g., a rod, beam, or pipe) under *axial* compression or tension, the object is either shortened or elongated without changing its general shape. The forces act normal to the cross-sectional area. In such cases (where volume changes are ignored), the Elastic (i.e., **Young's) Modulus** applies.

When an object is acted upon by forces tangent to the area, the object distorts laterally. One layer of atoms slides over another. In such cases (where volume changes are ignored), the **Shear Modulus** applies.

When an object is acted upon by forces in all directions, tending to increase or decrease its volume, the **Bulk Modulus** applies.

Similarly, we can define a **Bulk Modulus** (B) as the ratio of the volumetric stress (the pressure P) to the volumetric strain:

$$\text{Bulk Modulus} = -\frac{\text{volumetric stress}}{\text{volumetric strain}}$$

that is

$$B = -\frac{\sigma_v}{\varepsilon_v} = -\frac{F/A}{\Delta V/V_0} \tag{10.10}$$

The negative sign makes the modulus positive since an increase in stress causes a decrease in volume and therefore a negative strain (Table 10.4). Notice that the larger the value of B, the smaller the change in the volume ΔV, and the less compressible is the material.

Clearly, solids do not compress very much. By comparison, water is about 25 times more compressible than granite, but that still isn't a great deal. Nonetheless, estimates indicate that if water were truly incompressible, sea level would be roughly 30 m higher than it is.

Example 10.5 **[I]** Determine the change in the volume of a 1.00-m³ block of granite when it's submerged about 3 km in the ocean, where the pressure on all its surfaces is about 300 times ordinary atmospheric pressure (1.013×10^5 Pa).

Solution This problem is about the volume change of a solid and probably involves the Bulk Modulus. (1) TRANSLATION— An object of known initial volume experiences a specified pressure; determine the resulting change in volume. (2) GIVEN: $P = F/A = 300(1.013 \times 10^5 \text{ Pa})$ and $V_0 = 1.00$ m³. FIND: The change in volume ΔV. (3) PROBLEM TYPE—Elasticity/Bulk

Modulus. (4) PROCEDURE—From Table 10.4, $B = 50$ GPa. Knowing B we can calculate ΔV from its definition. (5) CALCULATION—From Eq. (10.10)

$$\Delta V = -\frac{V_0 F/A}{B} = -\frac{(1.00 \text{ m}^3)(3.039 \times 10^7 \text{ Pa})}{50 \times 10^9 \text{ Pa}}$$

$$\boxed{\Delta V = -6.1 \times 10^{-4} \text{ m}^3}$$

Quick Check: $B = -(F/A)/(\Delta V/V_0) \approx (-3 \times 10^7)/(-6 \times 10^{-4})$ Pa $\approx 0.5 \times 10^{11}$ Pa.

The Collapse of the World Trade Center

The Twin Towers of the World Trade Center in New York were designed to enclose a vast amount of open floor space (10×10^6 ft²). Previously, a typical skyscraper was built around a cluttered framework of horizontal steel beams and vertical columns that obstructed the internal spaces. By contrast, the steel columns supporting each of the Twin Towers were located only around the outer surface of the building and within its central core (Fig.10.14). Between the exterior (208-ft × 208-ft) walls and the inner (79-ft × 139-ft) core was a 60-ft-deep column-free interior. The external walls were composed of a total of two hundred and thirty six ≈14-in. square hollow steel box columns, spaced 39 in. apart center-to-center, which ran the entire height of each building. These were welded to steel spandrels that ringed the buildings at every floor, forming the world's highest load-bearing walls. The perimeter columns also provided wind and seismic bracing (wind forces on a single face could be as much as ≈11 × 10⁶ lb). Any tendency for the building to topple was resisted by this exterior steel lattice, which exerted restoring forces that had large moment arms (about the bottom edges of the tower) and could generate substantial counter torques (p. 244). Each building's concrete and steel core contained 103 elevators and 4 stairwells. It was formed of 47 massive steel columns that supported about 60% of the building's weight (≈2.76 × 10⁵ ton above the plaza). Extending from the core to the exterior columns were light 60-ft-long horizontal trusses (p. 123), made largely of ≈1-in. diameter steel rods. These supported the ≈4-inch-thick reinforced concrete floors. These trusses also stiffened the outside walls against buckling forces due to wind. During construction each prefabricated floor system, weighing in total about 3.2 × 10⁶ lb, was lowered into place, bolted and welded to the exterior lattice and interior core columns.

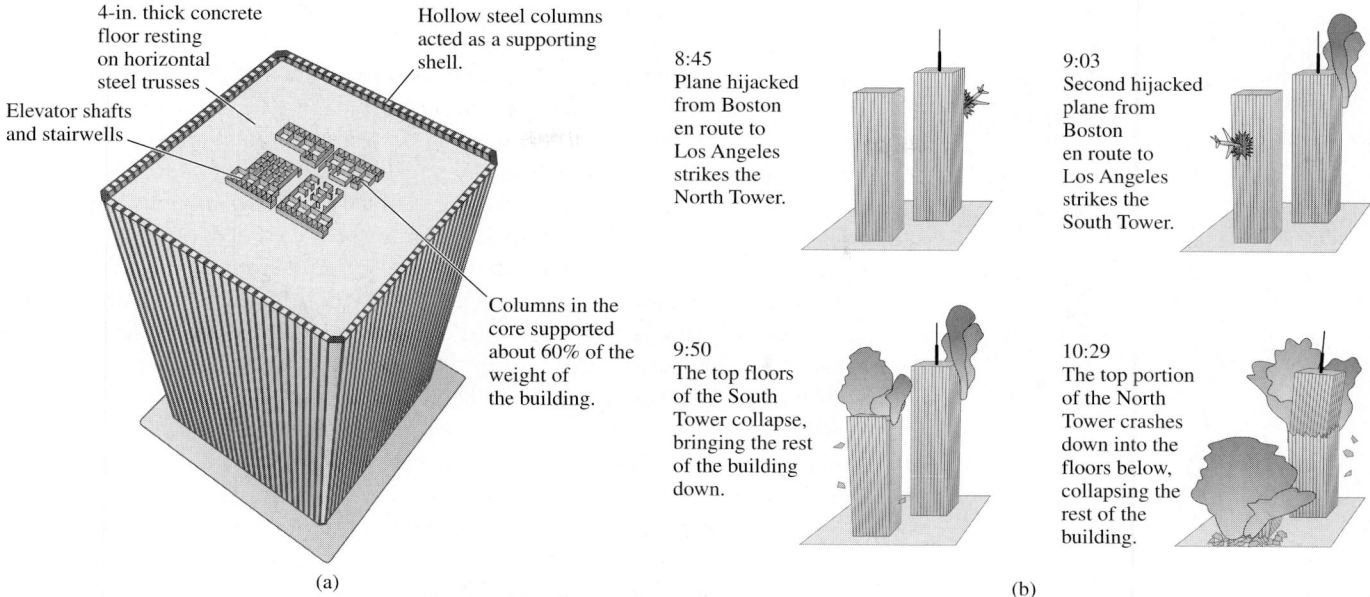

Figure 10.14 (a) A cutaway view of one of the Twin Towers showing the steel exoskeleton and central core. Each tower was essentially a strong core, surrounded by a thin steel lattice shell. And between these were hung 110 light-weight concrete floors. Once the collapse began there was little to keep these floors from tearing away and descending, pulling the shell down with them. In the basement, deep below ground, the last half dozen floors came to rest, one tight upon the next, like a stack of broken dishes.

On the morning of September 11, 2001, each 110-story building was struck by a 200-ton (1.8×10^5-kg) Boeing 767 airliner traveling at least 300 mph (134 m/s) and carrying about 60 000 lbs of jet fuel. Each plane, having a wing span of 156 ft and impacting with a tremendous amount of kinetic energy ($\approx 2 \times 10^9$ J), sheared (p. 341) about three dozen external support columns as it blasted into the building. Even though each plane momentarily exerted a sideways force of perhaps 25×10^6 lb, neither building tipped over. Roughly 3000 gallons of fuel (the equivalent of 90 tons of TNT) detonated on impact. The plane's thin aluminum wings and fuselage were immediately shredded into a hail of small chunks. But the engines, landing gear, and other massive parts, carrying tremendous amounts of momentum, tore into the interior of the building where they probably had enough energy to severely damage almost half of the core columns. Despite that extensive damage, North Tower 1, which was hit between the 94th and 99th floors at 8:45 A.M. remained standing until 10:29 A.M. South Tower 2, which at 9:03 A.M. was struck lower down between the 78th and 84th floors, collapsed at 9:50 A.M. Understanding why those buildings disintegrated provides a poignant review of much of what we've learned thus far.

The great Swiss mathematician Leonard Euler was the first person (1744) to analyze how a column buckles under a load—he had recently invented the calculus of variations and needed a problem to try it out on. His work led to the modern understanding that the load (F) under which a column will buckle is given by $F = K\pi^2 Y \mathscr{I}/L^2$, where K is a constant (e.g., 1, 4, 1/4) determined by how the ends are held in place; Y is Young's Modulus (p. 340); $\mathscr{I}$ is a quantity that depends on the cross-sectional area; and L is the unsupported length of the column. It was to increase $\mathscr{I}$, and so resist buckling, that the columns of the Twin Towers were made hollow like bamboo and leg bone. Welding the columns to spandrels and floor trusses at every story decreased L and further reduced the tendency to buckle. The stiffness of a column depends on Y, and it's essential to have a high Young's Modulus. That's why steel ($Y = 200$ GPa) is the material of choice, but that modulus markedly decreases with temperature, and structural steel begins to lose its strength at ≈ 800 °F. Unfortunately, jet fuel burning in a confined space can produce temperatures well over 1800 °F, far in excess of what would cause the columns to soften and buckle (steel melts at between $\approx$2500 °F and $\approx$2700 °F and some of the wreckage showed unmistakable signs of melting).

Amazingly, the undamaged columns in the region of the 80th floor of South Tower 2 were still able to support the 30-story structure above them. The crash no doubt knocked loose large portions of the fireproofing that had been sprayed on to protect all the steel in the building. With perhaps 10 000 gallons of jet fuel burning unchecked for almost 50 minutes, the temperature rose dramatically, and many of the remaining columns lost their stiffness and gave way. Whether the lightweight floor trusses failed first, or not, may never be known. But inevitably, the upper $\approx$$10^5$ tons of steel, concrete, furniture, and file cabinets came crashing down on the already weakened floors in the affected area. Under that gigantic dynamic load, which was at least an order of magnitude larger than the designed capacity, the floors disintegrated one after the other. That initial blast of energy sent an explosive pulse (p. 379) down into Tower 2. And just behind the billowing cloud of concrete dust, the 30-story shaft plunged into the still-standing portion of the building. As they bent, the floor trusses may have pulled the outer columns inward before they tore away, confining the debris. In any event, each floor was successively hit by an ever greater weight of falling material. The resulting progressive collapse roared toward the ground at an accelerating rate very nearly equal to g, grinding up the 30-story structure as it descended into the remains of the Tower, striking the ground at roughly 120 mph.

It had taken seven years and an immense amount of work (p. 175), roughly 10^{12} J, to haul that vast amount of material high into the air, and in about 12 seconds it all came crashing down. Approximately 30% of that stored energy went into pulverizing much of the concrete, and twisting and tearing steel. The rest of the PE_G was transformed into the KE of the falling debris. When that $\approx$3 $\times$ 10^5 tons of material (minus a fair amount of dust and paper) came crashing down to the street below, part of the associated KE was transformed into thermal energy (p. 456); part went into further rupturing and distorting the debris, and part went into penetrating and deforming the building's substructure and the ground. Roughly 0.1 % of the KE subsequently appeared as seismic energy, creating a modest-sized magnitude 2 earthquake (p. 373) measured 30 km away. Though the destruction was initiated by terrorists, the tremendous *gravitational*-PE (the equivalent of $\approx$120 tons of TNT) stored via the Earth-Tower interaction ultimately powered its collapse.

North Tower 1, with only about ten stories above its crash site, managed to survive a little longer before it too vanished from the skyline of Manhattan. In all, more than three thousand souls perished that day.

Harmonic Motion

There are all sorts of vibrating systems: the balance wheel in a watch oscillates back and forth; puckered lips blowing a trumpet, or a kiss, vibrate; a walker's swinging arms oscillate; so does a singing vocal cord. The first person to really study oscillations was Pythagoras of Samos. And because he was only interested in its relationship to music, vibratory periodic motion came to be known as *harmonic motion*.

10.5 Simple Harmonic Motion

A single sequence of moves that constitutes the repeated unit in a periodic motion is called a *cycle* (Fig. 10.15). During one cycle, a system progresses in some way, returning on completion to its initial configuration and motion. The child on the swing leaves her father's raised hands, swoops down, rises high into the air, stops, descends backward, and comes up again to meet his hands and finish the cycle. ***The time it takes for a system to complete a cycle is a* period** (T). The period of the Earth's rotation about its spin-axis is 24 h (it's closer to 23 h 56 min). The period is *the number of units of time per cycle*; the reciprocal of that—*the number of cycles per unit of time*—is known as the **frequency** (f), after Galileo who called it the *frequenza*:

The **period** is the time it takes a system to progress through one complete cycle returning to its original state.

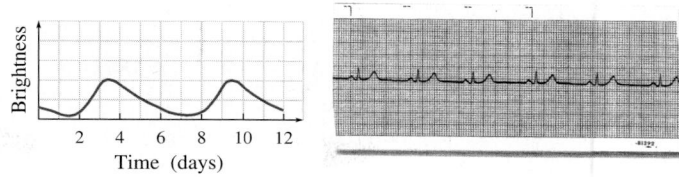

Figure 10.15 (a) The giant star Delta Cephei swells and then shrinks every 5.4 days. The plot of its brightness is therefore a regularly rising and falling curve. (b) This graph is periodic, too, but it's just an electro-cardiogram of a physicist's heart.

> The **frequency** of a periodic motion is the number of cycles per second.

$$f = \frac{1}{T} \tag{10.11}$$

The SI unit of frequency is the *hertz* (Hz), so named to honor Heinrich Hertz, where 1 Hz = 1 cycle/s = 1 s^{-1}. The appropriate units of *cycles per second* got messed up in common usage during the first half of this century. Radio people, especially, talked about kilocycles and megacycles, forgetting the "per second," so today, to avoid that difficulty, we have the hertz. For example, once set swaying by a wind, the Empire State Building oscillates through one cycle in about 8 s, that is, $T = 8$ s and $f = 1/T = 0.1$ s^{-1} = 0.1 Hz (Fig. 10.16).

Imagine an object moving in a circular orbit (Fig. 10.17) at a constant rate: its angular speed ω is constant. Each time it makes one complete orbit, one cycle, the object sweeps through an angle of 2π rad. Since the number of cycles it makes per second is f, *the number of radians it moves through per second* is $2\pi f$, and that's exactly what angular speed ω is:

$$\omega = 2\pi f = \frac{2\pi}{T} \tag{10.12}$$

It's common practice to refer to ω as the **angular frequency**.

Most real oscillatory phenomena wiggle about in a complicated fashion with lots of different frequencies occurring at different strengths, all at once. Even so, many important systems vibrate with a dominant frequency that far outweighs all the others. *Ideally, when the motion is sinusoidal (i.e., harmonic) with a single frequency*, it is known as **simple harmonic motion**, or SHM for short.

Uniform rotational motion, simple harmonic motion, and periodic wave motion are all intimately related and can be analyzed with the same mathematics. We can get a sense of that interrelationship from Fig. 10.18, which depicts an arrangement called a Scotch yoke. As the disk turns uniformly, the rod oscillates linearly in SHM. The frequencies and periods of the two motions are the same.

We next derive the general expressions that describe SHM. These results will later be applied to specific systems, such as vibrating springs and pendulums.

Figure 10.16 The 1250-tall Empire State Building set oscillating by a strong wind. Even though the displacement as drawn is exaggerated, tall buildings do sway up to several meters. In fact, on a windy day, the Twin Towers of the World Trade Center in Manhattan would shift 6 or 7 ft, and the elevators had to be slowed down.

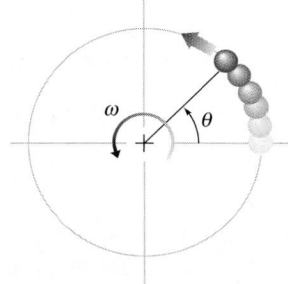

Figure 10.17 An object moving in a circle with a constant angular speed ω. Its angular position $\theta = \omega t$ changes at a constant rate.

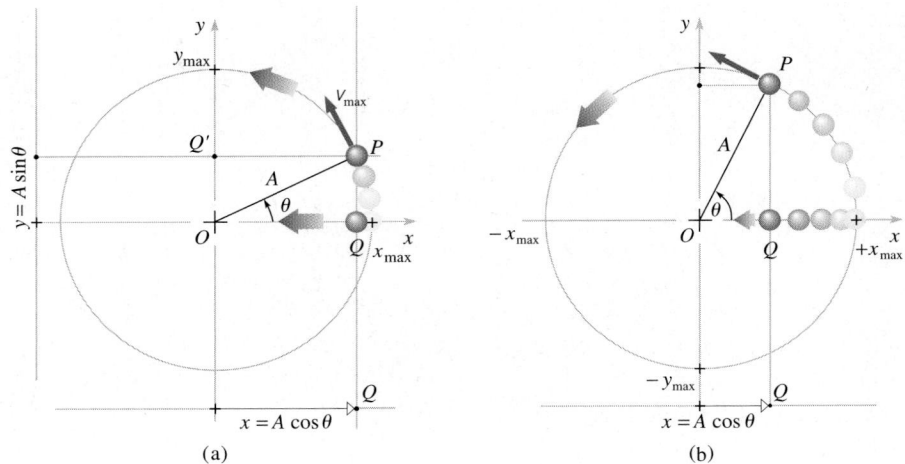

Figure 10.19 A particle P revolving around a circle at a speed v_{max}. Its projection on the x-axis is point Q at a distance $x = A \cos \theta$ from the nondisplaced origin O. As P revolves, Q oscillates between $+x_{max}$ and $-x_{max}$ in SHM. (a) and (b) show two locations as the motion progresses.

Displacement in SHM

An object oscillating in SHM can be located at any instant with respect to a fixed point O called the *origin*. To do that, imagine a particle P moving at a uniform speed v_{max} counterclockwise along a circle of radius A (Fig. 10.19) centered on O. At any moment, the position of the particle is given by θ. Since the linear speed is constant, the corresponding angular speed ω is also constant, since $v = r\omega$. After a time t, it follows from Eq. (8.4) that $\theta = \omega t$. The projection of P down onto the x-axis locates the point Q; and as P circles around with a constant speed, Q oscillates from $+x_{max}$ to $-x_{max}$ and back at a constant frequency in SHM. The projection onto the y-axis locates Q', and it will also display SHM. The displacement of Q from O is given by

$$x = A \cos \theta = x_{max} \cos \omega t \tag{10.13}$$

where $\omega = 2\pi f$. The location of Q provided by Eq. (10.13) can be thought of as describing the top of the Empire State Building wavering in the wind, or anything else oscillating in SHM about its undeflected ($x = 0$) position. {For more of the mathematical details, click on **INITIAL PHASE** on the **CD**.}

The maximum displacement of Q from O is the length $A = x_{max}$, the radius of the reference circle, which is called the **amplitude** of the oscillation. At any instant, the value of x is the size of the displacement from O. The oscillator starts out ($t = 0$) at its maximum displacement ($\cos 0 = 1$); that is, $x = A$. The relationship between uniform circular motion, SHM, and the sinusoidal function is explored in Figs. 10.20, 10.21, and 10.22. {If you'd like to check your understanding of SHM thus far, click on **CHAPTER 10** on the **CD**. Go to **WARM-UPS** and study the first few questions under **OSCILLATIONS**.}

Velocity and Acceleration in SHM

The displacement of a system in SHM varies sinusoidally with time. As that's occurring, what happens to the system's velocity and acceleration? These, too, must change from moment to moment.

Velocity. Point P moves around its circular path ($r = A$) with a constant linear speed $v_{max} = A\omega = 2\pi f A$, while its projection, point Q, always remains just below it on the

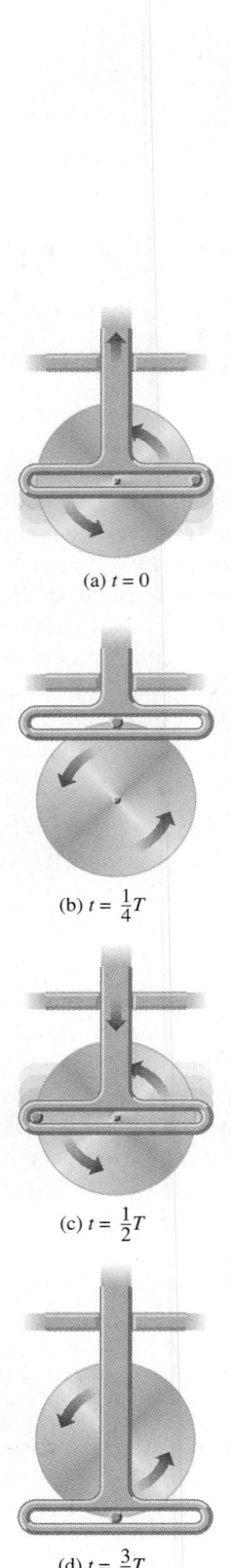

(a) $t = 0$

(b) $t = \frac{1}{4}T$

(c) $t = \frac{1}{2}T$

(d) $t = \frac{3}{4}T$

Figure 10.18 This Scotch yoke produces a vertical oscillation that is SHM as the wheel revolves uniformly.

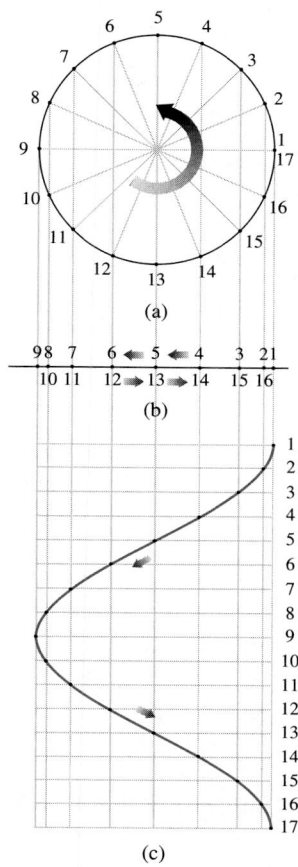

Figure 10.20 (a) Uniform circular motion projected onto a line (b), resulting in linear SHM projected on a "moving" axis (c), which results in a sinusoidal oscillation.

The **maximum speed** of a simple harmonic oscillator equals the amplitude times the angular frequency: $v_{max} = A\omega$.

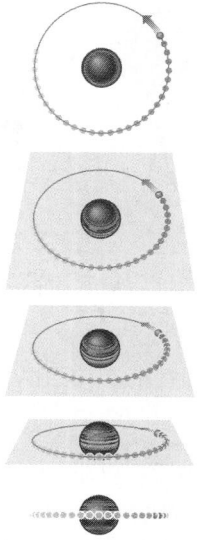

Figure 10.21 A schematic representation of a moon circling around Jupiter. As the viewing angle gets narrower, the orbit flattens, and the moon is seen to oscillate in SHM along a line in the plane of the motion.

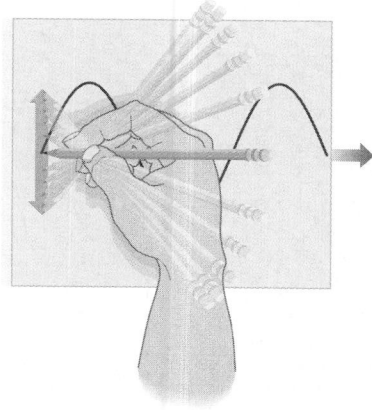

Figure 10.22 The pencil oscillates in SHM along a vertical line. Pulling the paper to the right at a constant speed results in a sinusoidal curve.

x-axis. This means that the x-component of the velocity vector ($\vec{v}_{max}$) for P must equal the velocity of Q, namely, $\vec{v}_x$. Figure 10.23 shows the geometry. To write the scalar equation for the speed of Q, we must be careful about the signs. As shown, $\vec{v}_x$ points in the negative x-direction ($0 < \theta < \pi$), and so v_x is negative: $v_x = -v_{max}\sin\theta = -A\omega\sin\theta$.

Therefore
$$v_x = -A\omega\sin\omega t \qquad (10.14)$$

The scalar value of the maximum velocity, $v_x(max) = \mp A\omega$, must depend on the frequency—the point cannot oscillate at a high frequency without having a high speed.

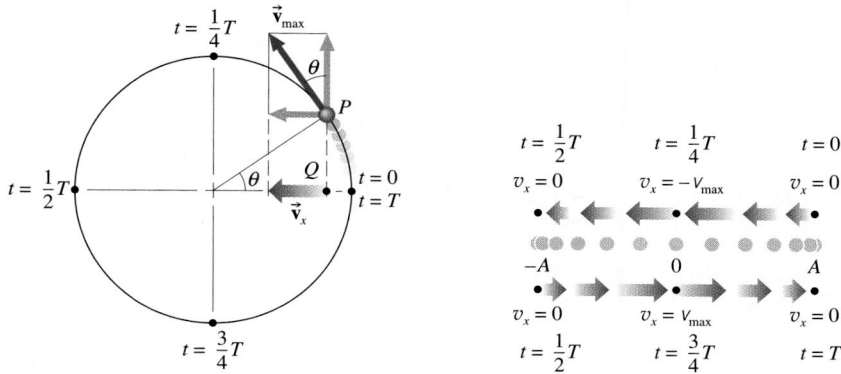

Figure 10.23 The point Q oscillates horizontally from $x = +A$ to $x = -A$, moving from $v_x = 0$, to a maximum of $v_x = -v_{max}$ at $x = 0$, to $v_x = 0$ back at $x = -A$. (a) This reference circle shows the vector $\vec{v}_{max}$ and its component v_x. (b) Here we see the time variation of the speed of a harmonic oscillator.

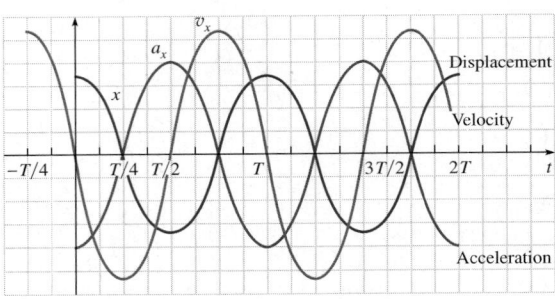

Figure 10.24 Plots of the velocity, displacement, and acceleration for a simple harmonic oscillator. Note that the velocity leads the displacement by $\pi/2$, while the acceleration leads (or equivalently lags) the displacement by π.

The oscillator is momentarily at rest at its two maximum displacements ($x = \pm x_{max}$), where $t = 0$ or $t = \frac{1}{2}T$ and, correspondingly, $\theta = 0$ or π, whereupon $\sin\theta = 0$. As it begins a cycle ($t = 0$), the particle in Fig. 10.23b heads left, picking up speed that rises to a maximum at $x = 0$ when $t = \frac{1}{4}T$ and $\theta = \frac{1}{2}\pi$, whereupon $v_x = -A\omega = -V_{max}$. Sailing past $x = 0$, the speed of Q diminishes until it is again zero, when $t = \frac{1}{2}T$ at $x = -x_{max}$. Like a ball thrown straight up in Earth's gravity field, it stops for an instant at its maximum displacement.

We can also derive an expression for the scalar velocity of Q in terms of x. From Fig. 10.19 and the Pythagorean Theorem, $\overline{PQ} = y = \sqrt{A^2 - x^2}$ and so $\sin\theta = \sqrt{A^2 - x^2}/A$. Using $V_{max} = A\omega$, we obtain

$$v_x = -A\omega\sin\omega t = \mp V_{max}\sqrt{1 - (x/A)^2} \tag{10.15}$$

which is the scalar velocity of a harmonic oscillator at any position x in terms of its maximum speed, V_{max}.

The trigonometric identity $\cos(\alpha \pm \beta) = \cos\alpha\cos\beta \mp \sin\alpha\sin\beta$ tells us that $\cos(\omega t + \pi/2) = -\sin\omega t$, which, in turn, means that Eq. (10.15) could be written as $v_x = -A\omega\cos(\omega t + \pi/2)$. Therefore, as can be seen in Fig. 10.24, the speed leads the displacement by 90°. The speed, which is a cosine function shifted one-quarter period ($t = \frac{1}{4}T$ or $\omega t = \frac{1}{2}\pi$) to the left, results in a curve that is an inverted sine function—that is, minus sine.

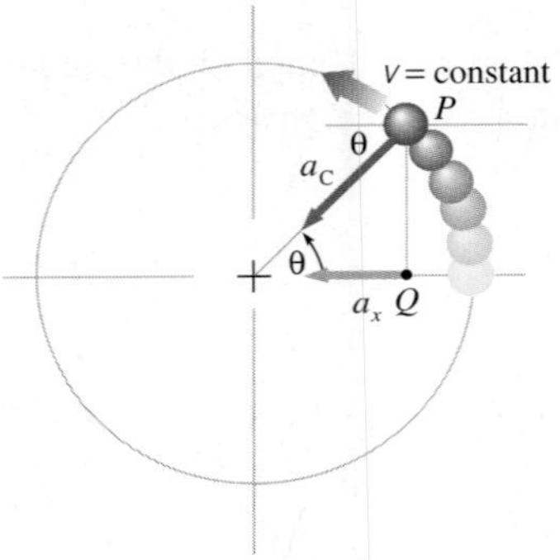

Acceleration. Particle P experiences a constant centripetal acceleration, $a_C = v^2/r = r\omega^2 = A\omega^2$, directed toward the origin (Fig. 10.25). Because Q always follows along, the horizontal component of the centripetal acceleration has to equal Q's acceleration. As before, to write $\vec{a}_x$ as a scalar we must include a negative sign because it points in the negative x-direction:

$$a_x = -a_C\cos\theta = -A\omega^2\cos\theta$$

and

$$a_x = -A\omega^2\cos\omega t \tag{10.16}$$

The oscillator has its maximum acceleration at the limits of its motion ($x = \pm A$) where the speed is zero: $a_x(max) = -A\omega^2$. Prior to this chapter, we studied systems where the acceleration was either zero or constant, but that's not the case here. Moreover, because the acceleration of P is centripetal, the acceleration of Q is also center-seeking, pointing toward the equilibrium position, $x = 0$.

By comparing Eqs. (10.13) and (10.16), it is evident that

$$a_x = -\omega^2 x \tag{10.17}$$

independent of time. **The acceleration of a simple harmonic oscillator is proportional to its displacement.** That's the hallmark of SHM.

Figure 10.25 As an object moves in a circle, there is a centripetal acceleration a_C that has a horizontal component a_x. The latter is the acceleration of the oscillator.

In **SHM** the acceleration at any moment is proportional to the displacement.

Example 10.6 **[II]** A spot of light on the screen of a computer is oscillating to and fro along a horizontal straight line in SHM with a frequency of 1.5 Hz. The total length of the line traversed is 20 cm, and the spot begins the process at the far right. Determine (a) its angular frequency, (b) its period, (c) the magnitude of its maximum velocity, and (d) the magnitude of its maximum acceleration. (e) Write an expression for x and find the location of the spot at $t = 0.40$ s.

Solution   This problem is about SHM. (1) TRANSLATION— An object oscillates in SHM with a known frequency, amplitude, and starting position; determine (a) its angular frequency, (b) period, (c) maximum speed, and (d) maximum acceleration. (e) Write an expression for x.   (2) GIVEN: $f = 1.5$ Hz and $A = 10$ cm.   FIND: (a) ω, (b) T, (c) $|v_x(\text{max})|$, (d) $|a_x(\text{max})|$, (e) x in general, and x at $t = 0.40$s.   (3) PROBLEM TYPE—SHM.   (4) PROCEDURE—Determine ω and T from f. Then use $v_x(\text{max}) = -A\omega$ and $a_x(\text{max}) = -A\omega^2$.   (5) CALCULATION—From Eq. (10.12)

(a) $\omega = 2\pi f = 2\pi(1.5 \text{ Hz}) = 9.4 \text{ rad/s} = \boxed{3.0\pi \text{ rad/s}}$

(b) $T = 1/f = 1/1.5 \text{ Hz} = \boxed{0.67 \text{ s}}$

(c) $|v_x(\text{max})| = V_{\text{max}} = A\omega = 2\pi f A$

$V_{\text{max}} = 2\pi(1.5 \text{ Hz})(0.10 \text{ m}) = \boxed{0.94 \text{ m/s}}$

(d) From Eq. (10.17),

$|a_x(\text{max})| = A\omega^2 = A(2\pi f)^2 = (0.10 \text{ m})(2\pi1.5 \text{ Hz})^2$

$|a_x(\text{max})| = \boxed{8.9 \text{ m/s}^2}$

(e) $\boxed{x = A \cos \omega t = (0.10 \text{ m}) \cos (9.4 \text{ rad/s})t}$

At $t = 0.40$ s

$x = (0.10 \text{ m}) \cos (3.77 \text{ rad}) = (0.10 \text{ m})(-0.81) = \boxed{-8.1 \text{ cm}}$

Quick Check: From (e) at $t = 0$, $x = +A$, as it should be. The value of x at $t = 0.40$ s; namely, -8.1 cm is reasonable since the period is 0.67 s, and so after $\frac{1}{2}T = 0.33$ s, the spot is on the negative side at $-A = -10$ cm heading right. At 0.07 s later, it's not surprising that it be at $x = -8.1$ cm.

10.6 Elastic Restoring Force

When a system oscillates naturally (without being driven by some external source of energy), it does so by moving against a *restoring force* that tends to return it to its undisturbed equilibrium condition. The system in equilibrium is first distorted. An additional quantity of potential energy is thereby stored in it, and then it's let loose to return toward equilibrium. Invariably, it overshoots (there being no tendency to stop at its undisplaced configuration since the system carries an excess of energy and still possesses momentum when it reaches that equilibrium point). It sails past its undistorted configuration only to be displaced in the opposite direction, to again become distorted and begin the process anew. As the system vibrates, PE goes into KE and back to PE, and so on (with no losses) indefinitely. That "lossless" single-frequency ideal vibrator is known as a **simple harmonic oscillator**.

An Oscillating Spring

If a spring with a mass attached to it is slightly stretched or compressed and then let loose, it will oscillate in a way very closely resembling SHM. Many elastic systems (buildings, flagpoles, airplane wings, etc.) behave in a similar fashion, and so the spring merits a detailed examination.

Reconsider Hooke's Law, $F = ks$: the force that must be exerted on an object—a spring or a wire or a tree—to make it distort *slightly* is proportional to the displacement from equilibrium that the object undergoes. Provided the distortion (stretching, compressing, twisting, bending, whatever) is not too great, $F \propto s$, and the object behaves elastically. From Newton's Third Law, the force exerted *by* an elastically stretched spring or a bent pole (the *elastic restoring force* pulling back against that which is doing the stretching) is $F_e = -ks$.

If a body of mass m, attached to a helical spring of negligible mass (as in Fig. 10.26), is displaced a small distance $s = x$ from equilibrium and then let go, the body will thereafter experience a force $F_e = -kx$, exerted on it by the spring. Yet since $F = ma$, the resulting acceleration of the body is $a_x = -(k/m)x$. *The acceleration of the body is proportional to the negative of its displacement*—the mass oscillates about $x = 0$ in SHM. This description

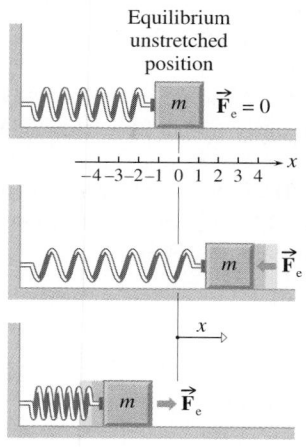

Figure 10.26 A mass on a spring vibrating horizontally in simple harmonic motion. Here, $\vec{F}_e$ is the force exerted *by* the spring, and there is no friction.

assumes no energy is lost to friction (either in the spring, on the table, or through the air); otherwise the motion is *damped harmonic*, not SHM, and it will die out. The restriction that the spring, diving board, or suspension bridge (whatever it is that's being stretched) must behave in a Hookean fashion is equivalent to saying that F is linear in x, which is equivalent to saying a is linear in x, which is the hallmark of SHM. Still, precise linear behavior is more an idealization than a reality. To the degree that a system behaves in a way that is approximately linear, it will oscillate in a way that is approximately SHM. To that end, the spring must only be stretched slightly and the string plucked gently—nature may love simplicity, but twist it too far and it gets very complicated.

Example 10.7 **[III]** Use energy considerations to derive an expression for the maximum speed of the mass oscillating in Fig. 10.26 in terms of k, m, and A. That done, confirm Eq. (10.15) for v_x as a function of the displacement of a simple harmonic oscillator.

Solution This problem deals with SHM and maximum speed therein. (1) TRANSLATION—A mass attached to a spring is oscillating in SHM; using energy considerations, write an expression for the maximum speed it reaches. Confirm Eq. (10.15). (2) GIVEN: k, m, and A. FIND: an expression for v_{max}. (3) PROBLEM TYPE—SHM/energy (4) PROCEDURE—We know that the total mechanical energy (E) is conserved and, moreover, that *elastic*-PE is given by $\Delta PE_e = \frac{1}{2}kx^2$. Find an expression for E and equate it to PE_e when KE = 0. (5) CALCULATION—Let's look at the extremes: At $x = 0$, the KE is maximum and the PE is zero; hence, the total mechanical energy is

$$E = \frac{1}{2}mv_{max}^2$$

At $x = \pm A$, the KE is zero ($v_x = 0$), the PE_e is maximum, and

$E = \frac{1}{2}kA^2$. Since E is conserved,

$$\frac{1}{2}mv_{max}^2 = \frac{1}{2}kA^2$$

and the maximum speed is

$$v_{max} = A\sqrt{\frac{k}{m}}$$

In general

$$E = \frac{1}{2}mv_x^2 + \frac{1}{2}kx^2 = \frac{1}{2}kA^2$$

Solving for v_x^2, we obtain

$$v_x^2 = \frac{k}{m}(A^2 - x^2) = \frac{k}{m}A^2\left(1 - \frac{x^2}{A^2}\right) = v_{max}^2\left(1 - \frac{x^2}{A^2}\right)$$

which is equivalent to Eq. (10.15).

Quick Check: We have already seen that $v_{max} = A\omega$, and we will show presently that $\omega = \sqrt{k/m}$. Now let's just check that the units are right. Using $F = ks$, it follows that k has units of N/m. Thus, k/m has units of $(N/m)/(N \cdot s^2/m) = 1/s^2$, and $\sqrt{k/m}$ correctly has the units of $1/s$.

Frequency and period. When the mass on the spring is initially displaced some distance $\pm x_{max}$ (i.e., via compression or elongation) and released from rest, that displacement will remain the maximum value attainable, and the mass will oscillate between $+x_{max}$ and $-x_{max}$. Of course, this is the amplitude of the oscillation: $x_{max} = A$. Now, compare the body's acceleration $a_x = -(k/m)x$ with Eq. (10.17) for the acceleration of an object in SHM; namely, $a_x = -\omega^2 x$. The angular frequency equals the square root of k/m, but here we will use the symbol ω_0 instead of ω to remind us that it is the **natural angular frequency**, *the specific frequency at which a physical system oscillates all by itself once set in motion* (the spring can be made to vibrate at other frequencies by driving it). Thus

> The **displacement** of an ideal oscillator at the moment of release equals the amplitude of the oscillation.

[natural angular frequency]
$$\omega_0 = \sqrt{\frac{k}{m}} \qquad (10.18)$$

and since $\omega_0 = 2\pi f_0$,

[natural linear frequency]
$$f_0 = \frac{1}{2\pi}\sqrt{\frac{k}{m}} \qquad (10.19)$$

Since $T = 1/f_0$,

The balance wheel of a mechanical clock oscillates about its central axis.

Equilibrium
$x = -x_{max}$ $x = 0$ $x = +x_{max}$

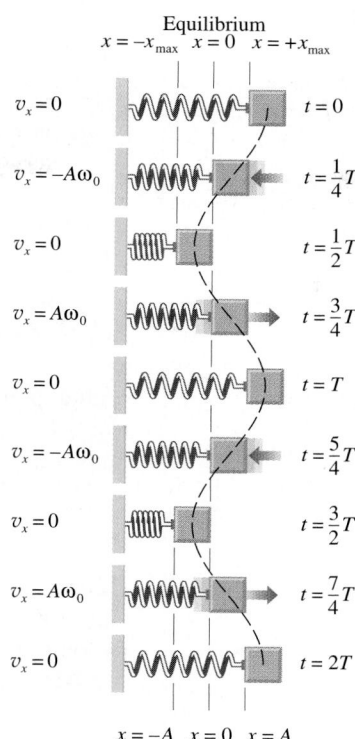

$v_x = 0$ $t = 0$

$v_x = -A\omega_0$ $t = \frac{1}{4}T$

$v_x = 0$ $t = \frac{1}{2}T$

$v_x = A\omega_0$ $t = \frac{3}{4}T$

$v_x = 0$ $t = T$

$v_x = -A\omega_0$ $t = \frac{5}{4}T$

$v_x = 0$ $t = \frac{3}{2}T$

$v_x = A\omega_0$ $t = \frac{7}{4}T$

$v_x = 0$ $t = 2T$

$x = -A$ $x = 0$ $x = A$

Figure 10.27 A simple harmonic oscillator shown every quarter cycle for two cycles. Indicated are the corresponding speeds and displacements.

ON OSCILLATING TOWERS

A skyscraper can be made as stiff as the builder is willing to pay for, up to a point at least. It is the financial as much as the structural concerns that dictate how rigid a tall building will be. As a rule, such structures are designed so that the top never displaces more than about 1/400 of the height when acted upon by a sustained pressure of 1.4 kPa (i.e., 30 lb/ft²), which is equivalent to a 150-km/h (i.e., 95-mi/h) wind. The period with which the building will rock is then determined by its mass and stiffness (k). The World Trade Center towers swayed with a period of 10 s, whereas the smaller Citicorp building, also in Manhattan, has a period of only 7 s.

[period]

$$T = 2\pi\sqrt{\frac{m}{k}}$$ (10.20)

Figure 10.27 shows the relationship among x, v_x, t, and T.

The stiffer the coil spring or wire, the larger the elastic force constant k and, therefore (with a given mass), the higher the vibrational frequency and the shorter the period. The springs on a car must have a large k to keep the up-and-down excursions small, but because the mass of the car is large, the frequency will nonetheless be low. For a given spring (i.e., a given k), if the oscillating mass is increased, the inertia of the system will be increased, causing it to accelerate more slowly and increasing the period. {For a delightful vibrational experience click on **CHAPTER 10** on the **CD**. Go to **INTERACTIVE EXPLORATIONS** and select **THE OSCILLATOR**.}

Example 10.8 **[III]** The cart in Fig. 10.28 has a mass of 1.00 kg, and someone reaches down and displaces it 5.00 cm to the right with an axial horizontal force of 10.0 N. (a) Assuming no friction, what is the period of the resulting simple harmonic oscillation when the cart is released? (b) Where will it be 0.200 s after release? (c) What is the elastic force constant for the system if one of the two identical springs is removed? (d) Determine the new frequency.

Solution   This problem is about SHM. (1) TRANSLATION— A mass attached between two springs is displaced a known amount by a known force and released. It oscillates in SHM;

Figure 10.28 To displace the cart horizontally, energy must be provided to the system. That energy is conserved and ideally transforms back and forth from PE$_e$ to KE as the cart oscillates in SHM. In reality, the oscillation will continue to diminish until all the energy ends up as thermal energy via friction and internal spring losses.

continued

determine (a) its period, (b) its location at some known time after release, (c) the elastic constant with one spring and (d) the new frequency. (2) GIVEN: $m = 1.00$ kg, $A = 0.0500$ m, and $F = 10.0$ N. FIND: (a) T, (b) x at $t = 0.200$ s, (c) k with one spring, and (d) new f. (3) PROBLEM TYPE—SHM/elastic. (4) PROCEDURE—We have an oscillating mass moving as part of an elastic system. (a) The period can be gotten from Eq. (10.20) once we have k for the system. (5) CALCULATION— (a) Inasmuch as the applied force produces a displacement x such that $F = kx$,

$$k = \frac{F}{x} = \frac{10.0 \text{ N}}{0.0500 \text{ m}} = 200 \text{ N/m}$$

This is the elastic force constant of the system—both springs acting at once, and both undergoing the displacement x as if they were attached side by side to the same end of the cart. Consequently

$$T = 2\pi\sqrt{\frac{m}{k}} = 2\pi\sqrt{\frac{1.00 \text{ kg}}{200 \text{ N/m}}} = \boxed{0.444 \text{ s}}$$

(b) To find where the cart will be at $t = 0.200$ s, we go back to the equation of x as a function of time,

$$x = A \cos\theta = A \cos\omega_0 t \qquad [10.13]$$

where the motion began at $x = A$, at $t = 0$. We need either ω_0 or f_0, and since we have T,

$$f_0 = \frac{1}{T} = 2.25 \text{ Hz}$$

Therefore,

$$x_{0.2} = A \cos 2\pi f_0 t$$

$$x_{0.2} = (0.0500 \text{ m}) \cos 2\pi(2.25 \text{ Hz})(0.200 \text{ s}) = \boxed{-0.0476 \text{ m}}$$

(c) If a spring is removed, displacing the body will require only half the force it did before. Now $\boxed{k = 100 \text{ N/m}}$, half its previous value. (d) The resulting frequency, Eq. (10.19), will be $1/\sqrt{2}$ of what it was—namely, $\boxed{1.59 \text{ Hz}}$.

Quick Check: The answer to (b) is reasonable since 0.200 s is very nearly $\frac{1}{2}T$, at which time $x = -A = -0.050$ m. (c) The new period is $T = (0.444 \text{ s})(1.414) = 0.63$ s; $f = 1/T = 1/0.63$ s $= 1.6$ Hz.

Example 10.9 **[III]** A 2.0-kg bag of candy is hung on a vertical, helical, steel spring that elongates 50.0 cm under the load, suspending the bag 1.00 m above the head of an expectant youngster. The candy is pulled down an additional 25.0 cm and released. How long will it take for the bag to return to a height of 1.00 m above the child?

Solution This problem deals with elasticity and SHM. (1) TRANSLATION—A known mass load stretches a spring by a known amount. The mass is then displaced a specified distance and released. It oscillates in SHM; determine the time to return to its initial equilibrium position. (2) GIVEN: $m = 2.0$ kg, $\Delta L = 50.0$ cm, and $A = 25.0$ cm. FIND: t when $y = 0$. (3) PROBLEM TYPE—SHM/elastic. (4) PROCEDURE—k is gotten from the initial elongation ΔL due to the weight of the bag. The loaded system is in equilibrium when the mass is 1.00 m above the small person. It returns there from its lowest point, the starting

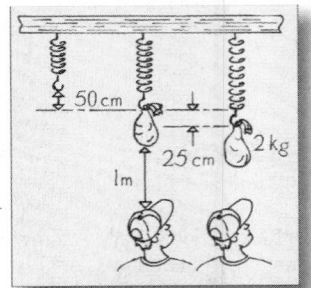

position which is $x = A$, after $\frac{1}{4}$ cycle—that is, at $t = \frac{1}{4}T$. (5) CALCULATION—To find T, we must first find k. Initially the load $(F = F_w)$ stretches the spring [Eq. (10.3)] such that

$$F = k(\Delta L) = mg$$

and so $k = \dfrac{mg}{\Delta L} = \dfrac{(2.0 \text{ kg})(9.81 \text{ m/s}^2)}{0.50 \text{ m}} = 39.2 \text{ N/m}$

We know from Eq. (10.21) that

$$T = 2\pi\sqrt{\frac{m}{k}} = 2\pi\sqrt{\frac{2.0 \text{ kg}}{39.2 \text{ N/m}}} = 1.4 \text{ s}$$

and so

$$\tfrac{1}{4}T = \boxed{0.35 \text{ s}}$$

Quick Check: $F = k\,\Delta y = mg$, $m/k = \Delta y/g$, $T = 2\pi\sqrt{\Delta y/g}$ $= 2\pi\sqrt{0.5 \text{ m}/g} = 1.4$ s.

10.7 The Pendulum

Legend has it that Galileo, while an undergraduate at Pisa, was in the cathedral (1581) when an attendant pulled a candelabrum hanging from the ceiling off to one side to light it. After it was let loose, it swayed to and fro, and Galileo sensed a rhythm in the smooth, repetitive motion. At first the lamp rushed quickly along large arcs, and then as the swing diminished,

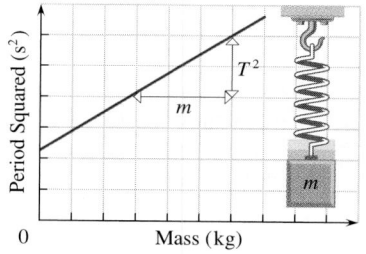

Figure 10.29 Since $T = 2\pi\sqrt{m/k}$, it follows that $T^2 = 4\pi^2 m/k$, and a plot of T^2 versus m must be a straight line with a slope of $T^2/m = 4\pi^2/k$. The reason it doesn't pass through the origin is that the spring itself has mass, which adds inertia to the system and increases the period.

BUT THE SPRING HAS MASS

Continuing from Eq. (10.20) for the period, namely, $T = 2\pi\sqrt{m/k}$, it follows that $T^2 \propto m$. A plot of T^2 versus m, for a number of different masses hung on a vertical coil spring, should be a straight line with a slope of $4\pi^2/k$ passing through the origin. In practice, for small values of A, the graph is straight—the motion is SHM—but it *does not* pass through the origin (Fig. 10.29). That's because the spring itself is not massless. The complete analysis is complicated by the fact that the spring's mass oscillates along with the load, but each segment of it has a different amplitude. It turns out that the effective mass of the system is actually the mass of the load plus one-third the mass of the spring.

the speed slowed so that the rhythm remained strangely constant. Being a student of medicine, Galileo naturally used the beat of his own pulse to time the swing (there were no watches then). To his astonishment, the duration of each cycle (what he called the *periodo*) was constant.

Galileo subsequently performed a series of experiments using balls of different weight hung on various lengths of string. What he soon discovered was remarkable: ***the period of a pendulum is independent of the mass and is determined by the square root of its length***. The implication was clear: the string deflected the ball along a curved path, but it was otherwise free to fall, and since all objects fall at the same rate, two pendulums of the same length should, indeed, swing in step.

The pendulum in Fig. 10.30 is depicted in flight at some angle θ. The bob is displaced a positive distance l measured to the right along the arc of the path from the vertical zero-θ axis. Taking θ in radians, $l = L\theta$. At that moment, the weight of the bob (mg) is, as always, straight down, but it has a component tangent to the arc. It is this component, this unbalanced force

$$F = -mg\sin\theta$$

that drives the pendulum back to equilibrium ($\theta = 0$). The minus sign is needed because $\vec{F}$ points in the negative l-direction, toward decreasing θ. Since $F = ma_\text{T}$, the tangential acceleration is

$$a_\text{T} = -g\sin\theta \qquad (10.21)$$

which is proportional to the *sine* of θ and not to l, and so it is *not* the condition for SHM. Still, for small angles, the value of θ in radians is very nearly the value of $\sin\theta$; they differ

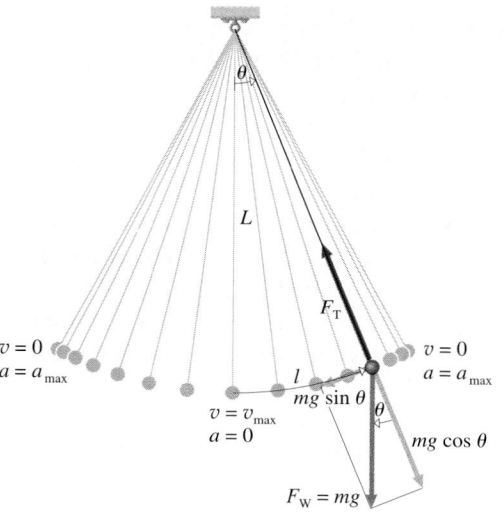

Figure 10.30 A pendulum is driven into oscillation by the nearly horizontal component of its weight $mg\sin\theta$. Provided θ is small, the mass will move in SMH.

by less than 2% out to about 20°. Hence, $\sin \theta \approx \theta = l/L$ and

$$a_{\mathrm{T}} \approx - \left(\frac{g}{L} \right) l \tag{10.22}$$

which is SHM, provided θ is small. It follows from $a_x = -\omega_0^2 x$ and Eq. (10.22) that $\omega_0 = \sqrt{g/L}$ and so

$$f_0 \approx \frac{1}{2\pi} \sqrt{\frac{g}{L}} \tag{10.23}$$

and

$$T \approx 2\pi \sqrt{\frac{L}{g}} \tag{10.24}$$

At any location on Earth, the period of a pendulum is dependent only on the square root of its length. For small initial angular displacements, less than 23°, the actual periods vary by less than 1% from the predictions of Eq. (10.24), and the motion is approximately SHM. The agreement improves tremendously for still smaller starting angles.

Example 10.10 **[I]** How long should a pendulum be if it is to have a period of 1.00 s at a place on Earth where the acceleration due to gravity is 9.81 m/s²?

Solution This problem deals with the period of a pendulum. (1) TRANSLATION—Determine the length of a 1-s pendulum. (2) GIVEN: $T = 1.00$ s and $g = 9.81$ m/s². FIND: L. (3) PROBLEM TYPE—SHM/pendulum. (4) PROCEDURE—The period of a pendulum depends on the square root of its length.

(5) CALCULATION—From Eq. (10.24), $T^2 = 4\pi^2(L/g)$; hence

$$L = \frac{T^2 g}{4\pi^2} = \frac{(1.00 \text{ s})^2 (9.81 \text{ m/s}^2)}{4\pi^2} = \boxed{0.248 \text{ m}}$$

Quick Check: A pendulum 1 m long is sometimes called a "seconds pendulum" because one swing (half a period) takes 1 s: $T = 2\pi\sqrt{L/g} = 2.0$ s. The pendulum in this problem has half that period and therefore should have $\frac{1}{4}$ the length.

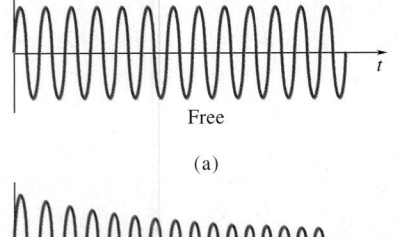

Free

(a)

Weakly damped

(b)

More strongly damped

(c)

Figure 10.31 (a) An ideal harmonic oscillator, free of all forms of energy loss, will vibrate harmonically forever with no decrease in amplitude. (b) When friction is present, the oscillator will be damped, and its amplitude will decrease in time. (c) The more the damping, the more repressed is the oscillation.

10.8 Damping, Forcing, and Resonance

There are usually external forces acting on an oscillator in addition to the restoring force. These forces may either impede (i.e., damp) the motion or drive it so that its amplitude is sustained, or even grows. Sometimes this latter effect can cause catastrophic failures, as when large buildings are toppled during earthquakes.

Damping

In practice, oscillating mechanical systems lose energy in a variety of ways via friction, and the sinusoidal behavior of SHM is invariably replaced by a damped motion that gradually dies away (Fig. 10.31). For example, a common friction mechanism is **viscous damping** (we are, after all, surrounded by air). At low speeds, viscous damping is proportional to the speed of the body in the damping fluid (Fig. 10.32). Any undriven system oscillating in the air must inevitably come to rest.

With weak damping, a system can continue to oscillate a relatively long time before stopping at its zero displacement position. This is what happens to an ordinary pendulum losing energy via friction. Its motion slowly decays, just as the vibrations of a tuning fork or a gong slowly decay. With still more friction, *the amplitude decreases rapidly, and the mass undergoes fewer oscillations before coming to rest*. Such a system is **underdamped** (Fig. 10.33). It's still periodic, although the frequency is lower than it would be without damping.

If friction is further increased, a displaced system can return to equilibrium without overshooting, in which case there will be no oscillation at all. The shock absorbers on a car should damp out any oscillations in less than a cycle. You don't want your car bobbing up

INERTIAL & GRAVITATIONAL MASS

To get Eq. (10.21), we set ma_T equal to $-mg \sin \theta$ and then canceled the mass. Well, that really was more significant than it might seem. The mass in $F = ma$ is the **inertial mass**, the mass associated with the object's tendency to resist changes in its motion. It appears to have nothing to do with gravity and perhaps might even be labeled m_i to underscore the difference in its conceptual origins. On the other hand, the weight of a body is determined by a physical property it possesses called **gravitational mass**, m_g. That property is proportional to the gravitational interaction between objects and seemingly has nothing to do with inertia. Thus, assuming these masses to be different, $T \approx 2\pi\sqrt{m_i L/m_g g}$. But experiments going back to Newton himself all confirm (nowadays to within 1 part in 10^{12}) that the period is independent of the mass of the bob, which implies that $m_i = m_g$. The assertion of the equality of inertial and gravitational mass is known as the *Equivalence Principle*, and it is central to the General Theory of Relativity. Perhaps gravity—the Universe acting gravitationally on each mass—*is* the cause of inertia.

and down long after it hits a bump in the road. When this nonvibratory motion occurs in the shortest amount of time, such that *once let loose the system sweeps right back to its undisturbed configuration*, it is said to be **critically damped**. It would be nice to have the swinging door at the kitchen of a restaurant or the entrance of a supermarket close swiftly without overshooting and slamming back to hit you from behind.

An earthquake caused the collapse of the double-decked Nimitz Freeway in Oakland, California, in 1989. This portion of the highway rests on fine-grained sediments and mud that transmitted frequencies of around 2 Hz. This value is close to the natural frequency of the highway, which experienced violent resonant vibrations. Fifty-three cars were crushed when the supports for the top deck gave way.

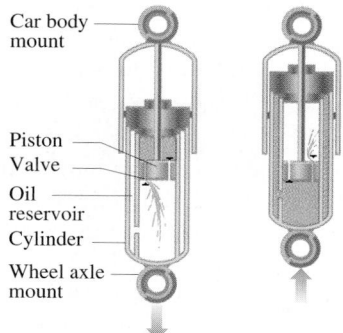

Figure 10.32 The shock absorber (or dashpot) uses viscous damping, which is proportional to the speed of the piston. A rapid jerk of the piston, as might be produced when a car hits a pothole, is met with a large damping force. On the other hand, a gradual depression of the piston is hardly resisted.

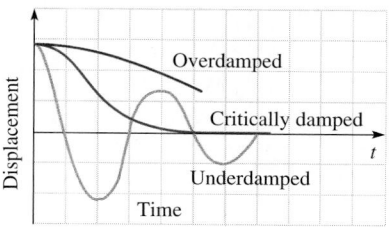

Figure 10.33 When a damped oscillating system experiences still more friction, it will stop vibrating and the motion will gradually decay. The system is then overdamped. When the motion decays as quickly as possible, the system is critically damped.

The tuned-mass damper at the top of the Citicorp tower. (See Example 10.11.)

DAMPERS

A variety of damping devices are used to suppress vibrations of all kinds in physical structures and machines. Olympic gold medals have been won using skis equipped with vibration dampers—sandwiches of lead plates and spongy absorbent foam. There are space satellites that contain little carts that ride back and forth along tracks, set into motion by any inadvertent wobbling of the vehicle. Eventually, the wobble is damped out and the carts come to rest a bit warmer via friction. The World Trade Center in New York contained viscoelastic dampers: large plates coated with a sticky polymer that dragged across one another whenever the towers swayed in the wind. They converted troublesome shear energy into harmless thermal energy.

With even more damping, ***the system again does not oscillate, but it takes much longer to return to equilibrium***. Heavy doors in public buildings almost all have hydraulic devices at their tops to keep them from slamming, and if you have ever waited interminably for one to close, you have experienced the results of **overdamping**.

Forced Oscillation and Resonance

One way to sustain the oscillations of a system suffering damping is to periodically pump in energy via a force that does positive work on the system. Pushing a playground swing can keep it going indefinitely, despite friction. When mom pushes, it's best done after junior has reached the peak of the swing and is moving away from her so that she does positive work on the system. Energy is efficiently transferred to the system when it's pumped in in step with the natural oscillation.

The application of an external alternating driving force to a system capable of vibrating produces **forced oscillation.** Specific physical causes are many. Pulsations of pressure in lines carrying fluids will cause vibrations—some houses have "singing" plumbing, and refrigerators and air conditioners often buzz.

Forced oscillation is easily seen with either a Slinky or a mass on a long elastic band (Fig. 10.34). Displace either and let it loose to oscillate at its natural frequency (f_0), which for the Slinky will be around 1 Hz. That done, stop the oscillation and now move your supporting hand up and down with a small amplitude (≈ 2 cm) at a very low frequency ($f \approx 0.3$ Hz, $T \approx 3.0$ s), thereby driving the elastic system into motion. The Slinky will follow your hand (the driver), moving upward when your hand moves up, and downward when it moves down. Moreover, it will oscillate at the same frequency (f) as the driver but with a relatively small amplitude. Little energy is transferred from the driver to the Slinky as the process continues.

Similar behavior occurs when the driving frequency is much greater than the natural frequency ($f \gg f_0$), except now when your hand goes down, the Slinky will move up, and vice versa. One thing will become obvious when you further vary the driving frequency—*as f approaches the natural frequency of the system, the resulting oscillation will dramatically increase in amplitude*. The vibrational amplitude will reach a maximum when $f = f_0$, a condition known as **resonance**. At that special driving frequency, which is also known as the **resonant frequency**, energy is most efficiently transferred to the system, whether it's a hand-pumped Slinky, a windblown skyscraper, or a human internal organ set into vibration by the overamplified tumult of a rock band.

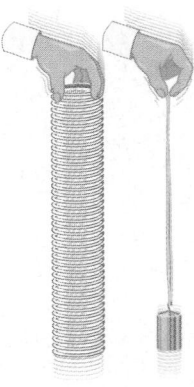

Figure 10.34 By moving up and down, the hand pumps energy into the system, forcing it to oscillate at the driving frequency.

WINE GLASSES & BRIDGES

Tap a good quality wine glass and it will ring, undriven, at its natural frequency, somewhere around 900 Hz. If it's now forced at that frequency, for example, by a blast of vibrating air (i.e., sound) up near the rim, the glass will efficiently absorb energy and begin to vibrate. When the note is sustained and carries enough energy to overcompensate for internal damping, the rim of the glass will oscillate with increasing amplitude and ultimately will shatter because glass is brittle. The same resonance mechanism causes the ashtray in your car to clatter annoyingly to the beat of the radio, and it's the reason why troops marching over bridges are usually ordered to break step. At least two large bridges collapsed in the nineteenth century, one near Manchester, England (1831), the other over the Maine in Anjou, France (1849), both having been set into resonance by the pounding, low-frequency cadence of marching soldiers.

A man playing a tune on a street in New Orleans by resonating (via stick-slip friction) different sized and pitched glasses. (See the photo on p. 401.)

At resonance, most of the small amount of energy available during each cycle is stored in both the elastic medium and the moving mass; none is returned to the driver, though a bit is invariably lost to friction. The less damping there is, the greater the resulting amplitude of the vibration. The phenomenon of resonance is at the heart of a tremendous diversity of occurrences from the atomic absorption and emission of light to the tuning of a TV set, from the swaying of bridges to the rattling of china cabinets when a low-flying plane passes overhead.

Example 10.11 **[I]** A room at the top of the Citicorp building houses a 400-ton (i.e., 3.629×10^5 kg) block of concrete that slides "floating" on a layer of oil pumped under pressure. One end of the block is attached to pneumatic springs fixed to one side of the building while the other end is attached to a hydraulic piston that acts like a large shock absorber, mounted to the other end of the building. The device is called a "tuned-mass damper" because it's tuned to oscillate at the natural frequency of the building. As the tower begins to sway, the block goes into resonant oscillation out-of-phase with the building. The wind's energy is transferred to the tower and then to the block, and then, by moving against the shock absorber, it's finally dissipated as thermal energy. Take the period of oscillation of the building to be 7.0 s and calculate the elastic force constant of the pneumatic spring that will produce resonance with the damping removed. (See the photo on page 356)

Solution This problem deals with the elastic constant and SHM. (1) TRANSLATION—An object of known mass (attached to pneumatic springs) is oscillating with a known period; determine the elastic force constant. (2) GIVEN: $T = 7.0$ s and $m = 3.629 \times 10^5$ kg. FIND: k. (3) PROBLEM TYPE—SHM/elastic. (4) PROCEDURE—The natural period of the spring-mass system must match that of the building—namely, 7.0 s. And the period is that of a simple harmonic oscillator. (5) CALCULATION—From $T = 2\pi \sqrt{m/k}$, it follows that

$$k = \frac{4\pi^2 m}{T^2} = 4\pi^2 \frac{3.63 \times 10^5 \text{ kg}}{49 \text{ s}^2} = \boxed{2.9 \times 10^5 \text{ N/m}}$$

Quick Check: Compute T using the value of k we just determined; $\omega_0 = \sqrt{k/m} = \sqrt{(2.924 \times 10^5 \text{ N/m})/(3.63 \times 10^5 \text{ kg})} = 0.9$ rad/s; $f_0 = \omega_0/2\pi = 0.14$ Hz; $T = 1/f_0 = 7$ s.

Self-Excited Vibrations

Remarkably, it is possible to initiate and sustain the vibration of a particular kind of system even though the energy source is nonoscillatory. In such situations, the response of the vibrating system itself produces alternations in applied force. These **self-excited vibrations** are commonplace and often quite dangerous, though blowing a kiss (sucking in a stream of air so that the lips vibrate) can be harmless enough.

There are many *self-excited vibrational* phenomena that are driven by constant streams of fluid, singing being one of them. Blowing across the mouth hole of a flute causes vortices to peel off periodically, creating a fluctuating pressure and setting the instrument vibrating. Similarly, a constant wind streaming over a banner stretched across a street can set it vibrating.

The first bridge across the Tacoma Narrows at Puget Sound, Washington, had a main span 2800-ft long and 39-ft wide with 8-ft-tall steel stiffening girders. Opened for traffic on July 1, 1940, it soon became infamous for oscillating wildly whenever the wind was

The Tacoma Narrows Bridge oscillating and ultimately collapsing. The stalled car belonged to a reporter who crawled off the bridge, leaving the car behind. The only fatality was a dog initially trapped in the car. Professor Farquharson, who was studying the bridge, risked his own life to free the dog, who then chose not to leave the scene (the dog is the small spot left of center).

blowing. Thrill-seeking motorists were lured from miles around for the ride of a lifetime. On the morning of November 7, 1940, a wind of 40 to 45 mph set the span rippling as usual, at a frequency of 36 vibrations per minute. When the amplitude became too large, the bridge was cleared. At around 10:00 A.M., the north cable became loose in its collar, allowing the main deck to abruptly begin to vibrate in a twisting resonant mode ($f_0 = 0.2$ Hz) around the yellow centerline of the roadway.

Although there are still questions to be answered, the basic mechanism that caused the catastrophe seems to have been vortex-induced self-excited vibrations. Air rushing toward the tall windward stiffening girder broke into two streams, shedding vortices alternately above and below the deck. Once the structure began to oscillate, that motion led to the formation of other motion-induced vortices (Fig. 10.35). These were formed as the deck rose and fell, and so matched the natural vibrational frequency. The rate at which energy was absorbed from the wind soon exceeded the frictional loss, and a state existed in which the oscillations quickly grew in amplitude. Shortly after 11:00 A.M., the center span tore into shreds like so much cotton ribbon.

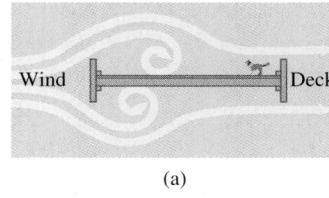

(a)

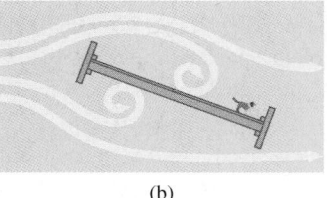

(b)

Figure 10.35 (a) Cross-sectional view of the roadway deck of the Tacoma Narrows Bridge, initially set into oscillation by alternating vortices. (b) As the deck oscillated, it caused other vortices to be induced that drove the bridge at its resonant frequency. These so-called motion-induced vortices were created in step with the twisting deck and furthered the motion.

Internet

For a short video showing the oscillation and collapse of the Tacoma Narrows Bridge go to:

http://www.enm.bris.ac.uk/research/nonlinear/tacoma/tacnarr.mpg

http://instruction.ferris.edu/loub/media/BRIDGE/tacoma.mpg

http://instruction.ferris.edu/loub/media/BRIDGE/Bridge.htm

STICK-SLIP FRICTION

Take a bow and a violin and attempt the business of making music—or at least sound. We wish to set the strings vibrating; they will set the air vibrating, and that's what a concerto is, more or less. Moving in one direction, say, to the right, the bow is made to brush over a string. Initially at rest with respect to the bow, the string is drawn to the right, and bent. The displacement continues until the growing restoring force of the string overcomes the static friction force, and it breaks loose from the bow. The string vibrates at a natural frequency, oscillating almost freely for a little while (losing energy in the form of sound). It encounters a markedly increased friction force when its relative speed with respect to the bow is low. Thus, the moving bow will recapture the string, again displacing it to the right, pumping energy in, and beginning the process over once more. The energy source is the constantly moving bow, but the motion of the system—the string—is vibratory.

This so-called **stick-slip friction** is the process responsible for the screech of a fingernail drawn across a blackboard. The high-pitched creak of an unoiled door is due to a torsional vibration of the short, stiff hinge pin. As Galileo pointed out, a wine glass can be made to ring at its resonant frequency in much the same way. Just run a moist, clean finger around the upper edge of the lip, gently pressing downward—stick-slip friction will do the rest.

Core Material & Study Guide

ELASTICITY

Hooke's Law for elastic objects (such as springs, rods, wires, etc.) is

$$F = ks \qquad [10.1]$$

where s is either the distance the object is elongated or compressed and k is the elastic constant with units of N/m (p. 329). The elastic potential energy stored in such an object is

$$\Delta PE_e = \tfrac{1}{2}ks^2 \qquad [10.2]$$

Many materials manifest a linear relationship between the load F they support and the resulting change in length of the sample (ΔL). Accordingly, Eq. (10.1) can be restated as

$$F = k\Delta L \qquad [10.3]$$

Section 10.1 (Hooke's Law) is important material; make sure you understand it as manifested by the above three equations and study Examples 10.1 and 10.2. Work out as many I-level problems as you can. Look at the **CD WALK-THROUGH EXAMPLES** throughout the chapter.

Stress (p. 334) is defined as the force acting divided by the area over which it acts:

$$\sigma = \frac{F}{A} \qquad [10.4]$$

Strain is the fractional change in some geometrical measure of a sample under stress. For a specimen experiencing an axial load, the *normal strain*, compressive or tensile, is

$$\varepsilon = \frac{\Delta L}{L_0} \qquad [10.5]$$

Similarly, the *shear strain* equals the angle of distortion γ:

$$\varepsilon_s = \gamma \approx \tan \gamma \approx \frac{\Delta x}{l_0} \qquad [10.6]$$

Section 10.2 (Stress & Strain) provides the above basic definitions which will be used in the analysis of problems in later sections. The **strength** (p. 336) of a material is the stress, in N/m^2, that will cause failure of some sort in a sample.

The *elastic stress per unit strain* defines **Young's Modulus**:

$$Y = \frac{\sigma}{\varepsilon} \qquad [10.7]$$

in either tension or compression (p. 339). The units of Y are the same as those for stress—namely, Pa. Alternatively

$$F = \frac{YA}{L_0}\Delta L \qquad [10.8]$$

is the equivalent of Hooke's Law. The elastic **Shear Modulus** (p. 341) is the shear stress over the shear strain:

$$S = \frac{\sigma_s}{\varepsilon_s} = \frac{F/A}{\gamma} \qquad [10.9]$$

Similarly, the **Bulk Modulus** (p. 341) is the ratio of the volumetric stress (the pressure P) to the volumetric strain:

$$B = -\frac{\sigma_v}{\varepsilon_v} = -\frac{F/A}{\Delta V/V_0} \qquad [10.10]$$

Section 10.4 (Elastic Moduli) deals with these issues. Example 10.4 brings together several ideas and is a good one to re-read. Work out as many I-level problems as you can.

HARMONIC MOTION

The time it takes for a system to complete a cycle is its **period** (T). The reciprocal of that is the **frequency** (f)

$$f = \frac{1}{T} \qquad [10.11]$$

The SI unit of frequency is the *hertz* (Hz). The **angular frequency** is

$$\omega = 2\pi f = \frac{2\pi}{T} \qquad [10.12]$$

Simple harmonic motion (SHM) corresponds to the situation where the periodicity has a single frequency and the motion is simply sinusoidal. In that case, the displacement is

$$x = A \cos \theta = x_{max} \cos \omega t \qquad [10.13]$$

The speed of a harmonic oscillator is

$$v_x = -A\omega \sin \omega t \qquad [10.14]$$

where $v_x(\max) = \mp A\omega$. Furthermore

$$v_x = \mp \omega \sqrt{A^2 - x^2} = \mp v_{\max} \sqrt{1 - (x/A)^2} \qquad [10.15]$$

which is the speed of the oscillator at any position x in terms of its maximum speed $v_{\max}$. Its acceleration is

$$a_x = -A\omega^2 \cos \omega t \qquad [10.16]$$

$$a_x = -\omega^2 x \qquad [10.17]$$

Reexamine Section 10.5 (Simple Harmonic Motion) and be sure you understand Example 10.6.

The *natural frequency* of a spring carrying a load (m) is

$$f_0 = \frac{1}{2\pi} \sqrt{k/m} \qquad [10.19]$$

and the period is

$$T = 2\pi \sqrt{m/k} \qquad [10.20]$$

while the natural frequency of a *simple pendulum* is

$$f_0 \approx \frac{1}{2\pi} \sqrt{g/L} \qquad [10.23]$$

Study Sections 10.6 (Elastic Restoring Force) and 10.7 (The Pendulum) and review Examples 10.7–10.11.

Key Terms

elasticity	harmonic motion
Hooke's Law	period
spring constant	frequency
elastic constant	hertz
elastic limit	angular frequency
elastic-PE	SHM
stress	amplitude
normal stress	simple harmonic oscillator
shear stress	natural frequency
strain	pendulum
strength	underdamped
yield strength	critically damped
tensile or ultimate strength	overdamped
breaking strength	forced oscillation
fatigue	resonance
Elastic or Young's Modulus	resonant frequency
Modulus	self-excitation
Bulk Modulus	

Discussion Questions

1. Explain the meaning of each of the Key Terms above.

2. Figure Q2 shows the stress-strain curves for several kinds of steel. (a) Which is the most ductile? (b) Which is the most brittle? (c) What can be said about their Young's Moduli? (d) Which has the greatest ultimate strength? (e) Which has the greatest rupture strength? (f) Which has the least yield strength? (g) Which has the greatest stiffness in the elastic region? (h) Which has the greatest strain at rupture?

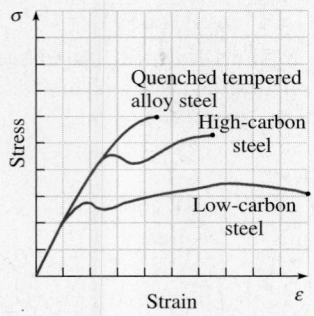

Figure Q2

3. Human bones in the embryo stage are mostly collagen, and only later are they gradually reinforced with calcium and phosphorus. Compare the corresponding values of Young's Modulus and discuss the implications in regard to the mechanical properties of children.

4. The area under the force-displacement curve is called the **strain energy**. (a) Discuss the physical meaning of this quantity. Similarly,

the area under a stress-strain curve is called the **strain-energy density**. (b) Discuss its physical meaning. What are the units of strain-energy density? (c) Figure Q4 shows a σ-ε curve for a material strained to a value of ε_F beyond the yield point. It is unloaded and returns to zero stress, but not zero strain. There is a permanent strain ε_p. What is the significance of the colored area?

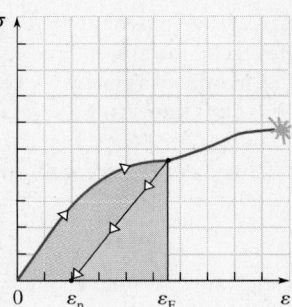

Figure Q4

5. Imagine two vertical hanging ropes that are identical except that one is considerably longer than the other. A gradually increasing load is applied to each of their bottom ends until one or both break. (a) What can you say about the two breaking loads? (b) Compare the amounts of strain energy stored—that is, the amount of energy pumped in—in order to break the ropes. (See Question 4.)

6. Mountains are typically shaped like a cone. Why is that configuration more able to reach great heights than a straight vertical columnlike structure? Explain why a column of marble about 4.2

km high would collapse under its own weight. Given that granite has about the same density as marble, how tall a column could be made of it before it crushed itself?

7. There is a general principle (attributed to Galileo) that deals with scaling structures—that is, increasing or decreasing their sizes proportionately. For example, King Kong was a scaled-up gorilla. The *square-cube rule* points out that while the volume of a thing being scaled-up increases as the cube of the dimensions, the cross-sectional area of the load-supporting members only increases as the square of the dimensions. How does the weight of the structure increase? What happens to the stress in the load-bearing members? Assuming King Kong was made out of normal gorilla-stuff, why would he have to be very careful when standing up?

8. The strengths and rigidities of animal bones do not vary all that much from one type of mammal to the next. Human bone (femur), for instance, has a relatively high compressive strength of 170 MPa; at the other extreme, pig bone has a low compressive strength of 100 MPa—not that much different. With Question 7 in mind, what does this suggest about the relative safety of hopping rabbits as compared to jumping elephants? What advice would you give to a horse about prancing around?

9. Rubber at a working elastic strain of 300% has a stress of 7 MPa, and the area under its nonlinear stress-strain curve is 10 MJ/m^3. What is the significance of this area (which, by the way, is about 10 times the corresponding value for spring steel)? Why do we make the heels of shoes out of rubber? (See Question 3.)

10. For a Hookean material, the area under the linear portion of the stress-strain curve is called the **resilience**. It's a measure of how much energy the material or structure can store without permanently distorting. The area under the curve all the way out to rupture is known as the **toughness**. It's a measure of the ability of a material to withstand a blow without rupturing. In other words, it's a measure of the energy it takes to break the object. What can you say about the resilience of a human tendon or a bridge or a steel spring? Is a glass bowl or a china plate tough? Explain. Suppose you were designing a bow for shooting arrows or a pole for vaulting. Do you want these to be tough or resilient?

11. Figure Q11 shows a spring with a mass attached to it resting on a belt. The belt is then set into constant motion counterclockwise. Assuming ordinary dry friction, describe the circumstances under

which the mass will go into unstable oscillation, cycling back and forth in a way resembling a sinusoid *growing* in amplitude. Remember that friction increases as the relative speed decreases.

12. Draw a simple swinging pendulum and label the locations of the maxima and minima of (a) speed, (b) acceleration, (c) kinetic energy, and (d) potential energy.

13. Why is it perfectly reasonable for the period of a pendulum to decrease as *g* increases? Why is it also reasonable for the period of a pendulum to increase as *L* increases?

14. When a person in a parachute descends, vortices are shed alternately up over one side and then the other. This phenomenon periodically changes the pressure on the chute. How might that become dangerous? [*Hint: What resonance effect is of concern? Incidentally, some chutes have central holes that are there precisely to break up the vortices.*]

15. EXPLORING PHYSICS ON YOUR OWN: Figure Q15 shows a horizontal wire (or string) supporting several simple pendulums. When the bob on the far left is displaced and set oscillating, it twists the wire a little with each swing. What do you think will happen? Try it.

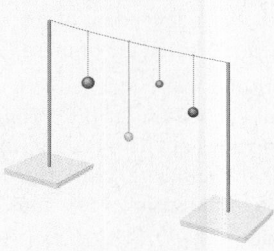

Figure Q15

16. What is the general relationship between the mass of someone jumping up and down on a trampoline (or a mattress) and the frequency of the oscillation?

17. Imagine an antiroll device for a ship: a mass on horizontal springs or a tank of water tuned to resonate at the natural rolling frequency of the ship. If waves begin to rock the ship near its resonant frequency, a very dangerous situation will develop. How does the device avert catastrophe?

18. The mythical planet Mongo is a homogeneous, nonrotating sphere whose inhabitants have drilled a hole running through the center from one side to the other. Show that a ball dropped into the hole would execute SHM.

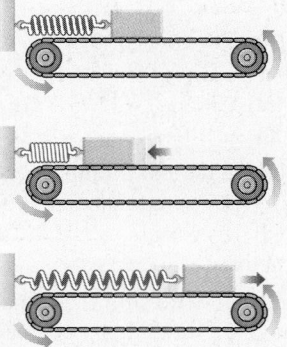

Figure Q11

1. The elastic constant in Hooke's Law is a measure of (a) the stiffness of the object being stressed. (b) the breaking strength of the object being stressed (c) the hardness of the object being stressed (d) the density of the object being stressed (e) none of these.

2. Hooke's Law applies to (a) all objects that are plastic regardless of what they are made of (b) any elastic system (c) only to springs (d) all stretchable objects (e) none of these.

3. The period of an object vibrating in SHM depends only on (a) the material it is made of (b) the elastic constant of the system (c) the length of the oscillator and the mass of the object (d) the mass of the object and the elastic constant of the oscillator (e) none of these.

4. Two wires are identical except wire-1 is twice as long as wire-2. Both are identically loaded. We can expect (a) both wires to stretch the same amount (b) wire-1 to stretch half as much as wire-2 (c) wire-1 to stretch twice as much as wire-2 (d) wire-1 to stretch four times as much as wire-2 (e) none of these.

5. An oscillator consists of a mass attached to a spring. The mass oscillates on a horizontal frictionless surface. The force exerted on the mass by the spring (a) is constant (b) is always directed away from the equilibrium position (c) is always directed toward the equilibrium position (d) is independent of the amount of stretch or compression of the spring (e) none of these.

6. An oscillator consists of a mass attached to a spring. The mass oscillates on a horizontal frictionless surface. The frequency (a) is independent of the spring constant (b) is independent of the mass (c) depends on the length of the spring (d) depends on the amount of stretch or compression of the spring (e) none of these.

7. The dimensions of stress are (a) $[L/M]$ (b) $[L^3T/M]$ (c) $[M^3T/L]$ (d) $[M/T^2L]$ (e) $[M/T^3L]$.

8. A metal part fails via fatigue (a) when it gets tired (b) as a result of the repeated application of a cyclic stress (c) when the stress exceeds the breaking strength (d) when the metal is very old and the load very large (e) none of these.

9. An ordinary piece of blackboard chalk is (a) ductile (b) quite tough (c) brittle (d) highly resilient (e) none of these.

10. A rod of tungsten is quite stiff, and so one can say that tungsten (a) is soft (b) is tough (c) has a large conductivity (d) has a large Young's Modulus (e) none of these.

11. A spring with a large elastic force constant can be said to be fairly (a) soft (b) rigid (c) long (d) short (e) none of these.

12. A spring balance with a linear scale can be used to measure mass because (a) springs are strong (b) metal springs are nearly weightless (c) a spring-force is linear because it has a large Young's Modulus (d) the spring-force is linear in displacement via Hooke's Law (e) none of these.

13. A piece of wire is stretched by a certain amount and then allowed to return to its original configuration. It is then stretched twice that initial amount, without exceeding its elastic limit. Compared to the first stretching, the second elongation stored (a) twice as much energy (b) four times as much energy (c) half as much energy (d) no additional energy (e) none of these.

14. A concrete block supported horizontally at its two ends can be broken with a downward karate chop to the middle because (a) the bottom of the block will be in tension and concrete is weak in tension, so it will crack on the bottom face (b) the top will be in compression and concrete is weak in compression, so it will crack at the top (c) the bottom will be in compression, so it will crack on the bottom (d) the top will be in tension, so it will crack on top (e) none of these.

15. It is possible, while tightening the nuts on the wheel of your car, to use too much torque and break off one of the bolts. This happens when the (a) shear stress exceeds the breaking strength (b) compressive stress becomes too large (c) the screw has too low an elastic constant (d) volumetric stress on the bolt is too great (e) none of these.

16. The units of strain are (a) m/s (b) N/m (c) m^2 (d) N/m^2 (e) none of these.

17. The breaking strength of a ductile material is generally (a) lower than the yield strength (b) very low (c) higher than the yield strength (d) higher than the ultimate strength (e) none of these.

18. A steel wire stretches by 10 cm when a load is hung from it. If a new wire made of the same material with twice the cross-sectional area supports the same load, it will stretch (a) the same amount (b) not at all (c) half as much (d) twice as much (e) not enough information given.

19. A dress made of cloth will hang nicely, conforming to the shape of the body beneath, because cloth has both (a) a large Young's Modulus and a low Bulk Modulus (b) a low Shear Modulus and a low Modulus of Elasticity (c) a high Young's Modulus and a low Modulus of Toughness (d) a low Bulk Modulus and a high Shear Modulus (e) none of these.

20. Of the several moduli, the only one applicable to liquids and even to gases is (a) Young's Modulus (b) the Modulus of Toughness (c) the Shear Modulus (c) the Elastic Modulus (d) the Bulk Modulus (e) none of these.

21. The end of a whip antenna is oscillating in SHM. Its maximum acceleration is given by the expression (a) $2\pi f_0^2 A$ (b) $4\pi^2 f_0^2 A$ (c) $4\pi^2 f_0^2 A^2$ (d) $4\pi f_0 A^2$ (e) none of these.

22. Two coiled springs are made of the same wire such that they are identical in all respects, except the first is half the length of the second. Compared to the natural frequency of the first (f), the natural frequency of the second is (a) $2f$ (b) $\frac{1}{2}f$ (c) f (d) $f/\sqrt{2}$ (e) none of these.

23. The free end of a clamped saw blade vibrates 12.8 times in 19 s. Its frequency is (a) 0.67 Hz (b) 12.8 Hz (c) 19 Hz (d) 1.48 Hz (e) none of these.

24. The tension in the wire of a simple pendulum is at maximum when the bob is at (a) its highest point (b) its lowest point (c) halfway down (d) any location since the tension is actually constant (e) none of these.

25. An adult and a child are sitting on adjacent identical swings. Once they get moving, the adult, by comparison to the child, will necessarily swing with (a) a much greater period (b) a much greater frequency (c) the same period (d) the same amplitude (e) none of these.

26. A pendulum clock is set to run accurately at sea level. It is then brought to the top of a high mountain, where it is found to (a) function unchanged (b) run slow (c) run fast (d) stop running (e) none of these.

27. A tuning fork is struck and so set vibrating. Its handle is then touched to the handle of an identical fork held near the ear, and this second one immediately begins to vibrate at the same frequency. This transfer of energy is an example of (a) contact potential (b) critical damping (c) resonance (d) radiation (e) none of these.

28. Sometimes a really heavy truck will go by and set the windows of the house vibrating for a moment with a low-frequency buzz. They oscillate (a) in-phase with the truck due to critical damping (b) at one of their natural frequencies (c) at a random changing frequency (d) in an undamped mode (e) none of these.

29. An oscillator at a frequency of 1.25 Hz will make 100 vibrations in (a) 125 s (b) 12.5 s (c) 8.0 s (d) 80 s (e) none of these.

30. A pendulum clock in a free-falling elevator (a) runs normally (b) runs a little fast (c) runs a little slow (d) runs very fast (e) none of these.

31. A pendulum clock in an elevator traveling up at a constant speed (a) runs normally (b) runs a little fast (c) runs a little slow (d) runs very fast (e) none of these.

32. The period of a pendulum swinging in an elevator that is accelerating upward at $g/100$, as compared to one at rest, is (a) unchanged (b) a great deal larger (c) a great deal smaller (d) slightly larger (e) none of these.

33. Astronauts in space took a light, coiled spring of known elastic spring constant, attached a mass to it, and set it oscillating. Measuring the period, they could determine (a) the time of day (b) the acceleration due to gravity (c) the mass of the bob (d) the weight of the bob (e) none of these.

For more Multiple Choice Questions with answers click on WARM-UPS in CHAPTER 10 on the CD.

Suggestions on Problem Solving

1. Area conversions (for example, cm^2 to m^2) are common in this chapter, so be careful with them. Don't mix up diameters and radii when calculating areas.

2. We deal extensively with the abbreviated notation of mega- and giga-, as for example MPa and GPa, which is often the way data is given in the literature. Get in the habit of entering the numbers in your calculations in scientific notation right away. It's easier to substitute the numbers in MPa, but it's also easier to lose the factor of 10^6 that way.

3. Young's Modulus provides an insight into the meaning of the spring constant by way of Eq. (10.8). When confronted with a problem dealing with the physical parameters that describe the object being loaded (length, cross section, etc.) think in terms of Y and that equation.

4. A system undergoing oscillation is not moving with a constant acceleration, and the kinematic equations in Chapter 3 *do not apply*!

5. When dealing with SHM, the basic notions to keep in mind are the definition of the spring constant ($F = ks$), the reference circle, and the fact that $\omega_0 = \sqrt{k/m}$ for a spring and $\omega_0 = \sqrt{g/L}$ for a pendulum; most everything else can be quickly derived from these ideas, as required. If you remember an equation for ω, there's no need to also store one for $f = \omega/2\pi$ or $T = 1/f$. In general, we know that $v = R\omega$, so the maximum speed, v_{max}, corresponds to the maximum radius $R = A$; hence, $v_{max} = A\omega$. Physicists don't attempt to commit to memory every equation they encounter—it's much easier to *understand and retain the analysis* than the individual formulas derived from it.

6. Don't get upset by the minus signs in Eqs. (10.14), (10.16), or (10.17)—they only indicate the direction of the vectors. Remember that A is always a positive quantity.

7. Some spring problems (of middle-level difficulty) ask for f or T, thereby requiring that you have k, even though it's not explicitly given. The idea is to first determine k from the initial conditions ($F = ks$).

Problems ✛ Coordinated Problems ✛ Progressive Problems ✛ Solutions

STUDY GUIDE **1. Coordinated Problems:** The three problems within each magenta-colored grouping are solvable in similar ways. Note that the first of these always has a hint; moreover, its solution is provided in the back of the book. *Work out each of these sets; they'll strengthen technique and build confidence.* **2. Progressive Problems:** The problems introduced in blue unfold step-by-step carrying along the analysis in a more suggestive way than is customary. *Work out all of these; they'll guide you through the analytic process and help develop problem-solving skills.* **3. Worked-Out Solutions:** Studying worked-out solutions is an important part of learning how to solve problems. Accordingly, additional *solutions* to a number of model problems are given below. *Make sure you understand each of them before you go on to the next problem.* **4.** Also provided in the back of the book are the *Answers* to all odd-numbered problems, as well as worked-out *solutions* to those with boldface numbers. Problem numbers in italic indicate that a solution appears in the Student Solutions Manual.

SECTION 10.1: HOOKE'S LAW

1. [I] THIS PROBLEM EXPLORES THE BEHAVIOR OF A SPRING. The spring-loaded handle of a pinball machine is pulled out 15 cm and held there. The spring has an elastic constant of 140 N/m. (a) Is the force that has to be applied to the handle while pulling it out, constant? (b) By how much, in meters, is the spring's length changed? (c) Determine the maximum force applied to the handle by the player.

2. [I] THIS PROBLEM WILL HELP US BETTER UNDERSTAND HOOKE'S LAW. A jumper on a pogo stick compresses the spring, whose elastic constant is 3.0 kN/m, by 15 cm on each jump. (a) What is the

compression distance in meters? (b) How much vertical force does the jumper apply to the pogo stick? (c) How much vertical force does the pogo stick exert on the jumper?

3. [I] A helical spring 20-cm long extends to a length of 25 cm when it supports a load of 50 N. Determine the spring constant. [*Hint: Hooke's Law governs the extension of springs.*]

4. [I] When an object weighing 200 N is hung from a vertical spring, the spring stretches 10.0 cm. Calculate the spring's elastic constant.

5. [I] A steel spring is suspended vertically from its upper end, and a monkey weighing 10.0 N grabs hold of its bottom end and hangs motionlessly from it. If the elastic constant of the spring is 500 N/m, by how much will the monkey stretch it?

6. [I] Given a spring with an elastic constant of 500 N/m, how much will it contract when pushed on by an axial force of 10 N?

7. [I] An ordinary helical spring having an elastic constant of 2.00 N/cm is stretched 10.0 cm and held in that configuration. How much work was thereby done on the spring?

> **SOLUTION:** The work done equals the energy stored, and that's $\Delta PE_e = \frac{1}{2}ks^2 = \frac{1}{2}(2.00 \times 10^2 \text{ N/m})(0.100 \text{ m})^2 = 1.00$ J.

8. [I] A long metal wire hangs from the roof truss in a factory building. A 1000-kg machine is attached to it so that the load is suspended above the floor. If the wire stretches 0.500 cm, what is its elastic constant?

9. [I] When a bowstring is pulled back in preparation for shooting an arrow, the system behaves in a Hookean fashion. Suppose the string is drawn 0.700 m and held with a force of 450 N, what is the elastic constant of the bow?

10. [I] How much energy is stored in a helical spring with an elastic constant of 50 N/m when compressed 0.05 m?

11. [II] THIS PROBLEM EXPLORES THE STATIC RESPONSE OF A LOADED SPRING. A helical spring with an elastic constant of 1000 N/m is suspended vertically from its top. A mass of 7.99 kg is attached to the spring's bottom and the system comes to equilibrium. (a) What is the value of the force applied to the bottom of the spring? (b) Does Hooke's Law apply? (c) By how much is the spring stretched by the load? (d) How much work was done on the spring in the process of stretching it?

12. [II] THIS PROBLEM DEALS WITH THE BEHAVIOR OF AN IDEAL SPRING. A jack-in-the-box is built around a spring. It takes 4.4 N to hold the lid down on his little head after he's compressed 15 cm from his sprung equilibrium length. (a) What is the value of the force applied by the jack on the lid? (b) By how much in meters is the spring compressed while in the box? (c) What's the value of Jack's spring constant? (d) How much work was done on Jack in the process of getting him into the box?

13. [II] A helical spring is 55-cm long when a load of 100 N is hung from it and 57-cm long when the load is 110 N. Find its spring constant.

14. [II] Two identical helical springs are attached to one another, end-to-end, making one long spring. If $k = 500$ N/m is the elastic constant of each separate (essentially weightless) spring, what is the constant for the combination?

15. [II] A Hookean spring is suspended vertically, and a mass of 2.00 kg is hung from its end. The spring then stretches 10.0 cm. How much more will it elongate if an additional 0.50 kg is attached to the first mass?

16. [II] Two identical wires each of length L_0 are attached to one another producing a single long wire. If $k = 500$ N/m is the elastic constant of each separate (essentially weightless) wire, what is the constant for the combination?

> **SOLUTION:** Suppose the load on the combined wire is F. Each wire is then under a tension of F, and each stretches an amount s as if it were supporting the load all by itself. Each length of wire elongates by $s = F/k$, giving a total elongation of 2s. Hence, for the composite wire $k' = F/2s = k/2 = 250$ N/m.

17. [II] Imagine a horizontal spring with an elastic constant of 25.0 N/cm, held in place just above an air table. If a 2.00-kg mass traveling at 1.50 m/s slams axially into the end of the spring, by how much will it be compressed in bringing the mass to rest?

18. [III] During the filming of a movie a 100.0-kg stunt man steps off the roof of a building and free-falls. He is attached to a safety line 50.0-m long that has an elastic constant of 1000 N/cm. What will be the maximum stretch of the line at the instant he comes to rest, assuming it remains Hookean?

SECTION 10.2: STRESS & STRAIN
SECTION 10.3: STRENGTH

19. [I] A vertical iron rod, having a known length and diameter, is to support a given load. If the rod is replaced by a new rod twice as long, and all else is kept constant, compare the stresses on both rods.

20. [I] Hanging vertically, a steel wire of a length L and radius R is to support a given load. If its cross-sectional area is halved and all else is the same, what will happen to the stress on the wire?

21. [I] An aluminum bar of known length and diameter is supported vertically, and a given load is attached to its lower end. If this bar is replaced by one having half the previous diameter, and everything else is unchanged, compare the stresses on the two bars.

22. [I] Standing vertically a copper cylinder of a length L and diameter D supports a compressive load placed on its top. An identical cylinder having half the diameter supports half the original load. If all else is the same, compare the stresses on the two cylinders?

23. [I] When a mass of 4.00 kg is suspended from the end of a narrow metal rod having a radius of 0.707 mm, the rod stretches by 0.060 cm. Determine the normal stress in the rod.

> **SOLUTION:** $\sigma = F/A = mg/\pi R^2 = (4.00 \text{ kg})(9.81 \text{ m/s}^2)/(3.142)(0.707 \times 10^{-3} \text{ m})^2 = 2.50 \times 10^7$ Pa.

24. [I] When a mass of 4.00 kg is suspended from the end of a narrow metal rod 60.0-cm long having a diameter of 0.707 mm, the rod stretches by 0.060 cm. Determine the normal strain on the rod.

25. [I] An axial force of 200 N is applied to a rod with a cross-sectional area of 10^{-4} m^2. What is the normal stress within the rod?

26. [I] A man leans down on a cane with a vertical axial force of 100 N. The narrowest part of the cane has a cross-sectional area of 1.0 cm^2. What is the maximum compressive stress in the cane?

27. [I] A 10-m-long wire strung between two trees has a "Curb Your Dog" sign weighing 10 N hung from it. If the wire stretches 1.5 cm, what strain is it under?

28. [I] A rubber band is stretched to twice its original length by a kid who is going to shoot it across a room. What strain is it experiencing?

29. [I] If the normal strain experienced by a rod in compression is 5.000×10^{-4} and its original length is 40.00 cm, what's its final length?

30. [I] Given that collagen has an ultimate tensile strength of 60 MPa, what maximum load could be supported by a fiber with a cross-sectional area of 10^{-6} m²?

31. [I] A length of tendon L_0-mm-long stretches an amount ΔL under a load F. A second piece twice as long, with twice the cross-sectional area, experiences the same load. By how much will it elongate in comparison?

32. [I] Pure natural rubber has a tensile strength of 21 MPa. When carbon particles are added to reinforce it, the tensile strength rises to 31 MPa (which is one reason why you see so much black rubber around). If a rod of natural rubber can support a maximum load of 1000 N, how much will an identical rod of carbon-reinforced rubber support?

33. [I] A rod made of structural steel, 4-m long and 2 mm in diameter, supports a 1.0-lb microphone hanging 3 m above the floor of a gym. A 181.4-kg (400-lb) Sumo wrestler reaches up, grabs the mike, and hangs from it. How much will the rod stretch, assuming it doesn't break?

34. [II] THIS PROBLEM IS A PRACTICAL APPLICATION OF THE IDEA OF STRESS. Figure P34 shows the square base of the Washington Monument, and we want to determine if it's overstressed at all. The structure, which is made of marble and granite, is 555 ft tall and weighs 181.7×10^6 lb and rests on the base.   (a) Determine the surface area (in SI units) of the top of the base on which the monument rests.   (b) What is the weight of the monument in newtons?   (c) Determine the stress in both psi and GPa.   (d) Is this stress large for a stone base?

(a)   (b)

Figure P34

35. [II] THIS PROBLEM EXPLORES THE IDEAS OF STRESS AND STRAIN. Attached to a ceiling hook is a 0.019 7-in. diameter 30.10-in. long wire. When a load of 2.11 kg is attached to the bottom end of the wire, it stretches to a new length of 30.16 in.   (a) What is the cross-sectional area of the wire in SI units?   (b) Determine the normal stress on the wire?   (c) By how much did the wire change its length?   (d) Compute the strain on the wire.

36. [II] A thin rod of structural steel 5.0-m long is to carry a tensile load such that it's just about to begin to deform plastically. Given that the yield strength in tension of steel is 250 MPa, how much will the rod stretch?

37. [II] A rubber motor belt with a circular cross section of radius 1.25 cm breaks at a tensile load of 7363 N. If a new belt was made of the same material with twice the cross-sectional radius, at what load would it break?

38. [II] What is the largest working strain you should design for when dealing with a structural steel member that is not to be permanently deformed by the tensile load? The yield strength of the steel in tension is 250 MPa as compared to its tensile strength of 400 MPa.

39. [II] Marble crushes under a compressive stress of 110 MPa. What maximum strain will it experience?

40. [II] A certain kind of brick has a density of 2×10^3 kg/m³ and a compressive strength of 40 MPa. What's the highest vertical parallel-walled tower that can be built that will not crush the bottom bricks?

41. [II] Two 1.00-cm-thick steel plates on a bridge are riveted to one another with five 4.00-cm-diameter steel rivets. Given that the yield strength in shear of structural steel is 145 MPa, what maximum force parallel to the plates can they be expected to handle without permanent deformation?

42. [II] A certain kind of brick has a density of 2×10^3 kg/m³ and a compressive strength of 40 MPa. What's the highest vertical parallel-walled tower that can be built that will not crush the bottom bricks?

43. [II] A 1.5-m-long brass wire is to support a 200-N "Eat At Joe's" sign without permanently elongating. Given that the yield strength of this variety of yellow cold-rolled brass is 435 MPa, what minimum diameter wire is called for?

44. [II] Show that the *elastic*-PE stored in a normally strained material of uniform cross-sectional area A and original length L_0 is given by

$$PE_e = \frac{F^2 L_0}{2AY}$$

where Y is the stress over the strain.

> **SOLUTION:** $PE_e = \frac{1}{2}k(\Delta L)^2$; but $\sigma = F/A = Y\varepsilon = Y\,\Delta L/L_0$; from Hooke's Law $F = k\Delta L$, and $k = F/\Delta L = AY/L_0$; rewriting this, $\Delta L = FL_0/AY$ and so $PE_e = \frac{1}{2}k(\Delta L)^2 = \frac{1}{2}(AY/L_0)(FL_0/AY)^2 = \frac{1}{2}(F^2L_0/AY)$

45. [III] The bar in Fig. P45 is made from a single piece of material. It consists of two segments of equal length $\frac{1}{2}L_0$ and cross-sectional areas of A and $2A$. Show that the *elastic*-PE stored under the action of an axial force F is

$$PE_e = \frac{3F^2 L_0}{8AY}$$

[*Hint: Look at Problem 44.*]

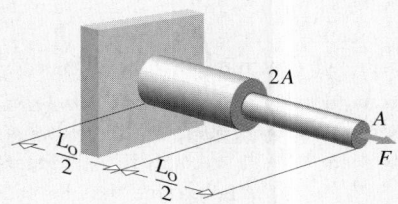

Figure P45

SECTION 10.4: ELASTIC MODULI

46. [I] A piece of fresh human cartilage, with a 1.0-cm² cross-sectional area, is being studied in a laboratory. It is found that when loaded with 100 N, its length increases by 4.2%. Determine its Elastic Modulus.

47. [I] THIS PROBLEM EXPLORES THE RELATIONSHIP BETWEEN STRESS AND STRAIN. We want to find the strain on a 1.00-m long rod of structural steel that experiences a compressive loading of 0.050 MN. The rod has a cross-sectional area of 0.10 m². (a) What is the compressional stress on the rod? (b) What's the relationship between stress and strain for an elastic material? (c) Find the strain on the rod.

48. [I] THIS PROBLEM EXPLORES THE RELATIONSHIP BETWEEN STRESS AND STRAIN. In the process of tuning a piano, a 1.0-m long steel string is stretched by the application of an additional force of 1000 N. (a) If the wire has a cross-sectional area of 1.0 mm², how much is that in m²? (b) Determine the stress on the wire. (c) What's the relationship between stress and strain for an elastic material? (d) Knowing that the wire is made of ordinary steel, by how much did the wire elongate?

49. [I] When a vertical structural steel column is loaded so that it experiences a compressive stress of 0.50 MPa, by what percentage will it compress? [*Hint:Use the tabulated value of Young's Modulus.*]

50. [I] A brick column having a horizontal flat 1.00-m² area at its top is to support a distributed load of 6.00 MN. By what percent can we expect the column to compress?

51. [I] An upright rectangular glass rod 10.0 cm by 10.0 cm by 10.0 m is to support an axial compressive load. If the glass is to squash no more than 0.010 mm, what is the maximum load it can handle?

52. [I] A cube of marble 1.00 m on each edge is submerged in the sea so that there is a distributed force F acting perpendicularly inward on each of its faces. What must that force be if the block is to decrease in volume by 1.00%?

53. [I] Imagine a 20.0-cm cubic block of cast iron being sheared like the one in Fig. 10.9*a*. What horizontal displacement will result when the two antiparallel forces are each 500 N?

54. [I] An aluminum cylinder with a volume of 1.00 m³ is lowered into the ocean to a depth (around 1 km) where the gauge pressure is 10.0 MPa. Calculate the reduction in its volume that results.

55. [II] When a mass of 4.00 kg is suspended from the end of a metal wire 60.0-cm long having a radius of 0.707 mm, the wire stretches by 0.060 cm. Determine Young's Modulus for the wire.

56. [II] THIS PROBLEM EXPLORES SHEAR. The rubber sole of a sneaker is 1.8 cm thick, has a surface area of 0.020 m², and a Shear Modulus of 0.001 5 GPa. (a) If while running, a horizontal friction force of 520 N is applied, what is the shear stress on the rubber? (b) What is the relationship between the shear strain and the angle of distortion? (c) Determine the shear strain. (d) By how much is the sneaker's inner sole (touching the foot) displaced horizontally?

57. [II] THIS PROBLEM EXPLORES THE ELONGATION UNDER STRESS OF DIFFERENT OBJECTS. Two wires made of the same material are subject to the same load. Wire-1 has half the diameter and half the length of wire-2. We want to determine the elongation of wire-1 compared to wire-2. (a) Write expressions for the strain on each wire in terms of Y. (b) Write expressions for the quantity F/Y for each wire. (c) Now compare the changes in the two lengths.

58. [II] THIS PROBLEM EXPLORES THE BEHAVIOR OF AN OBJECT UNDER SHEAR STRESS. A 4.0-cm on-each-side block of gelatin sitting on a plate experiences a horizontal force of 1.2 N acting at and parallel to its top surface. (a) If the block distorts without slipping, as shown in Fig. P58, what is the value of the angle γ? (b) What is the area of the top face? (c) Determine the Shear Modulus of the gelatin.

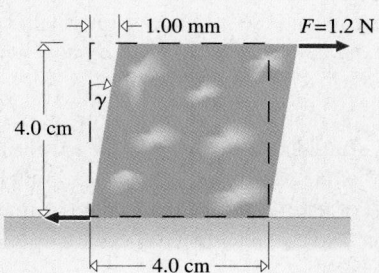

Figure P58

59. [II] THIS PROBLEM HELPS US BETTER UNDERSTAND THE BEHAVIOR OF AN OBJECT UNDER COMPRESSIVE LOADING IN ALL DIRECTIONS. A cast iron anchor falls to the bottom of the ocean, dropping 10 km. (a) What is the pressure at that depth? (b) What does the change in volume of the anchor depend on? (c) What is the Bulk Modulus of cast iron? (d) By what percent does the anchor's volume decrease? In other words, what is the ratio of the change in volume to the original volume?

60. [II] The compressive strength and maximum strain of the crown of a human tooth are 150 MPa and 2.3%, respectively. Assuming the crown to behave elastically until it ruptures, determine its Young's Modulus and (remembering Discussion Question 10) its resilience.

61. [II] The cable on a crane is to be made of high-strength, low-alloy steel with a yield strength in tension of 345 MPa. If the maximum load is to be 25 000 lb, how big in diameter should the cable be? What maximum strain can it experience given that its Modulus of Elasticity is 200 GPa?

62. [III] A solid lead sphere is placed in a pressure chamber so that the force per unit area acting on the sphere is hundreds of times atmospheric pressure (1 atm = 1.013×10^5 Pa). What pressure will produce a 1% increase in the density of the lead?

63. [III] If a uniform rod of cross-sectional area A and length L_0 can sustain a maximum stress of σ_R without rupture, show that it stores an amount of elastic energy given by

$$\text{PE}_e = \frac{\frac{1}{2} A L_0 \sigma_R^2}{Y}$$

in getting to that stress level.

64. [III] A robot's main body is supported on two legs, each containing two vertical rods one above the other jointed at the knee in human fashion. Each rod has a cross-sectional area of 5 cm² and a length of 50 cm and is made of a bonelike material that can sustain

a maximum stress of 150 MPa without rupturing and has a Young's Modulus (in compression) of 10 GPa. How much elastic energy can be stored in these two leg supports (as a result of jumping in a gravity field) without having to go back to the shop for repair? If the robot has a mass of 70 kg, how high a drop can it sustain on Earth?

SECTION 10.5: SIMPLE HARMONIC MOTION

65. [I] What is the period of an old fashioned phonograph record that turns through $33\frac{1}{3}$ rotations per minute?

66. [I] The respiratory system of a medium-sized dog resonates at roughly 5 Hz so that it can pant (in order to cool off) very efficiently at that frequency. How many breaths will the dog be taking in a minute? For comparison, the dog would ordinarily breathe at around 30 breaths per minute.

67. [I] An antique phonograph record is turning uniformly at 78 rpm while an ant sitting at rest on its rim is being viewed by a child whose eyes are in the plane of the record. Describe the ant's motion as seen by the child. What is the frequency of the ant? What is its angular frequency?

68. [I] A hovering fair-sized insect rises a little during the downstroke of its wings and essentially free-falls slightly during the upstroke. The end result is that the creature oscillates in midair. If it typically falls about 0.20 mm per cycle, what is the wingbeat period and frequency? Compare that to the ≈ 10 Hz oscillation of a butterfly, which cannot hover.

69. [I] A point at the end of a spoon whose handle is clenched between someone's teeth vibrates in SHM at 50 Hz with an amplitude of 0.50 cm. Determine its acceleration at the extremes of each swing.

70. [I] A light hacksaw blade is clamped horizontally, and a 5.0-g wad of clay is stuck to its free end, 20 cm from the clamp. The clay vibrates with an amplitude of 2.0 cm at 10 Hz. (a) Find the speed of the clay as it passes through the equilibrium position. (b) Determine its acceleration at maximum displacement and compare that with g.

71. [I] A small mass is vibrating in SHM along a straight line. Its acceleration is 0.40 m/s^2 when at a point 20 cm from the zero of equilibrium. Determine its period of oscillation.

72. [I] A point on the very end of one-half of a tuning fork vibrates approximately as if it were in SHM with an amplitude of 0.50 mm. If the point returns to its equilibrium position with a speed of 1.57 m/s, find the frequency of vibration.

73. [I] A body oscillates in SHM according to the equation

$$x = 5.0\cos(0.40t + 0.10)$$

where each term is in SI units. What is (a) the amplitude, (b) the frequency, and (c) the initial phase at $t = 0$? (d) What is the displacement at $t = 2.0$ s?

74. [I] A body oscillates in SHM according to the equation

$$x = 8.0\cos(1.2t + 0.4)$$

where each term is in SI units. What is the period?

75. [I] This problem uses Figure 10.23 to prove that in SHM

$$T = 2\pi A/v_{\max}$$

(a) What is the speed at which the little object travels around the circle? (b) What is the distance once around the circle? (c) Write an expression for the time it takes for the object to go once around the circle.

76. [II] This problem explores SHM for an ideal mass-spring system. A 0.802-kg bob hangs in equilibrium from a lightweight spring. It is pulled downward by a positive amount and released. It subsequently oscillates in SHM according to the equation

$$y = 0.760\cos(5.50t)$$

where each term is in SI units. (a) Write a general expression for the displacement of an oscillator whose bob is at $+A$ at $t = 0$. What is (b) the amplitude of the oscillation, (c) the angular frequency, (d) the linear frequency, and (e) the period? (f) What is the bob's displacement at $t = (\frac{1}{2}\pi/5.50)$ s? (g) How fast is it moving at that time? (h) How fast is it moving at $t = (\pi/5.50)$ s? (i) What's the value of its acceleration at that moment?

77. [II] Suppose the general equation $x = A\cos(\omega t + \epsilon)$ is to be applied to the mass in Fig. 10.26, where the cycle is to begin at $t = 0$ with m released with zero speed at -3 m. What must ϵ equal?

> **SOLUTION:** At $t = 0$, $x = A\cos(0 + \epsilon) = A\cos\epsilon = -3$ m. Whereas $v_x = -A\omega\sin(\omega t + \epsilon) = -A\omega\sin(0 + \epsilon) = -A\omega\sin\epsilon = 0$ and so $\sin\epsilon = 0$ and $\epsilon = 0$ or π. To see which is the case, recall that $A\cos\epsilon = -3$ m and that's < 0. Since A is always > 0 this means $\cos\epsilon < 0$. But $\cos\epsilon < 1$ so $\epsilon = \pi$.

78. [II] Suppose the general equation $x = A\cos(\omega t + \epsilon)$ is to be applied to the mass in Fig. 10.25, where the timing clock is started arbitrarily and it is found that at $t = \frac{1}{4}T$ the mass is at $x = +\frac{1}{2}A$. Find ϵ.

79. [II] A uniform block with sides of length a, b, and c floats partially submerged in water. It is pushed down a little and let loose to oscillate. Given that the vertical edge has length b and the density of the block is ρ, show that the motion is SHM and determine its period. Check the units of your result.

80. [II] A 5.0-kg block of wood is floating in water. It is found that a downward thrust of 10.0 N submerges the block an additional 10 cm, at which point it is let loose to oscillate up and down. Determine the frequency of the oscillation. Assume SHM.

81. [II] Show that a value of $\epsilon = +\frac{3}{2}\pi$ is equivalent to a value of $\epsilon = -\frac{1}{2}\pi$ in the general equation $x = A\cos(\omega t + \epsilon)$. Draw a sketch of the functions.

82. [II] A spot of light on a computer screen oscillates horizontally in SHM along a line 20-cm long at 50 Hz. The spot reaches the center of the line at $t = T/8$. Show that $v_x = -10\pi\sin(100\pi t + \frac{1}{4}\pi)$ is the expression for the speed of the spot as a function of time.

83. [II] A particle is oscillating with SHM along the z-axis with an amplitude of 0.50 m and a frequency of 0.20 Hz. If it is at $z = +0.50$ m at $t = 0$, where will it be at $t = 5.00$ s, $t = 2.50$ s, and $t = 1.25$ s?

84. [III] A vibration platform oscillates up and down with an amplitude of 10 cm at a controlled variable frequency. Suppose a small rock of mass m is placed on the platform. At what frequency will the rock just begin to leave the surface so that it starts to clatter?

85. [III] Use Fig. P85 to prove that the reciprocating motion of the piston is oscillatory but not SHM.

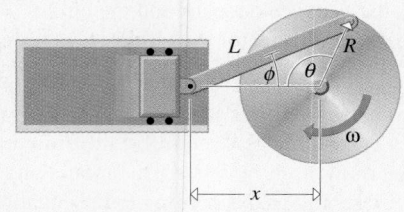

Figure P85

SECTION 10.6: ELASTIC RESTORING FORCE

86. [I] A 1.0-N bird descends onto a branch that bends and goes into SHM with a period of 0.50 s. Determine the effective elastic force constant for the branch.

87. [I] A bug having a mass of 0.20 g falls into a spider's web, setting it into vibration with a dominant frequency of 18 Hz. Find the corresponding elastic spring constant.

88. [I] A 1.0-N weight is attached to a vertical spring and it's then set oscillating up and down. Later three additional 1.0-N weights are added on to the bottom of the spring and it's again set oscillating vertically. Compare the periods of the two oscillations.

89. [I] THIS PROBLEM IS ABOUT A SPRING VIBRATING IN SHM. A hanging scale in a vegetable market oscillates at 3.0 Hz after a 1.20-kg bag of potatoes is dropped on it. If the pan of the scale has a mass of 0.50 kg, (a) what is the total mass supported by the spring? (b) What is the natural frequency of the system? (c) Determine the elastic constant of the spring.

90. [I] THIS PROBLEM IS ABOUT AN ELASTIC SYSTEM VIBRATING IN SHM. A 85.0-kg man tied to a 25.0-m bungee cord steps of a bridge. He just misses the water, at which point the cord is stretched to 45.0 m in length. We want to find out how much time will elapse from when he leaves the bridge to when he nearly returns on the first bounce. (a) What's the maximum stretch of the cord? (b) How much *gravitational*-PE did he have before he left the bridge? (c) What was the maximum amount of elastic energy stored in the cord? (d) Determine the cord's elastic constant. (e) Compute the period of the oscillation.

91. [I] THIS PROBLEM APPLIES ENERGY CONSIDERATIONS TO SHM. A mass m on a horizontal frictionless table is attached to the end of a spring with an elastic constant of k. The mass has a scalar velocity of v_1 when it's at x_1. (a) Write an expression for the total energy of the oscillator. (b) Write an expression for the maximum potential energy of the oscillator in terms of A, the amplitude of the oscillation. (c) Write an expression for A in terms of m, k, v_1, and x_1.

92. [I] A 5.0-kg mass resting on a frictionless airtable is attached to a spring with an elastic constant of 50 N/m. If this mass is displaced 10 cm (compressing the spring) and is then released, find its maximum speed.

93. [I] Two kilograms of potatoes are put on a scale that is displaced 2.50 cm as a result. What is the elastic spring constant? If the scale is pushed down a little and allowed to oscillate, what will be the frequency of the motion?

94. [I] A 250-g mass is attached to a light helical spring with an elastic force constant of 1000 N/m. It is then set into SHM with an amplitude of 20 cm. Determine the total energy of the system.

95. [I] Consider the frictionless cart in Fig. 10.27. If the elastic spring constants are k_1 and k_2, respectively, determine the frequency of vibration in terms of these quantities.

96. [I] What is the speed of the mass in Problem 92 as it reaches a point 5.0 cm from its undisplaced position?

97. [II] A light 3.0-m-long helical spring hangs vertically from a tall stand. A mass of 1.00 kg is then suspended from the bottom of the spring, which lengthens an additional 50 cm before coming to a new equilibrium configuration well within its elastic limit. The bob is now pushed up 10 cm and released. Write an equation describing the displacement y as a function of time.

98. [II] A 200-g mass hung on the bottom of a light helical spring stretches it 10 cm. This mass is then replaced with a new 500-g bob that is set into SHM. Compute its period.

99. [II] An object is hung from a light vertical helical spring that subsequently stretches 2.0 cm. The body is then displaced and set into SHM. Compute the frequency at which it oscillates.

100. [II] Derive an expression for the period of oscillation of the frictionless system shown in Fig. P100. Remember that the displacement of the mass m is the sum of the displacements of the two springs.

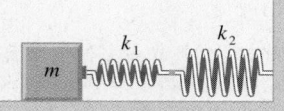

Figure P100

101. [II] An essentially weightless helical spring hangs vertically. When a mass m is suspended from it, it elongates an amount ΔL. When let loose, it oscillates with a measured period of T. Show how you might use this arrangement to determine g.

102. [II] THIS PROBLEM APPLIES ENERGY CONSIDERATIONS TO SHM. A 0.70-kg mass at rest on a horizontal frictionless table is attached to the end of an ideal spring having an elastic constant of 90 N/m. A second 0.30-kg mass is launched toward the first at a speed of 18.0 m/s. The two masses collide and stick together. We wish to write an expression describing the resulting motion. Assuming momentum is conserved in the collision, (a) at what speed does the combined mass move immediately after impact? (b) How much KE is imparted to the composite oscillator via the collision? (c) What is the amplitude of the resulting oscillation? (d) Determine the angular frequency of the system. (e) Keeping in mind that $x = 0$ at $t = 0$, write an expression for the oscillatory displacement.

103. [II] THIS PROBLEM DEALS WITH THE ENERGY OF A MASS-SPRING SYSTEM IN SHM. Considering the system in Problem 76 (a) What is the value of the spring constant? (b) What is the system's maximum potential energy? (c) Assuming it to be lossless, what is its total energy at any moment? (d) Determine the bob's displacement 3.0 s after release. (e) What is its potential energy at that time? (f) What is the KE of the bob at that time? (g) What is the speed of the bob at that moment?

104. [II] A 5.0-g bullet is fired horizontally into a 0.50-kg block of wood resting on a frictionless table. The block, which is attached to a horizontal spring, retains the bullet and moves forward, compressing the spring. The block-spring system goes into SHM with a frequency of 9.0 Hz and an amplitude of 15 cm. Determine the initial speed of the bullet.

105. [II] Using energy considerations, derive an expression for the period of oscillation of a mass m. It is fixed to a horizontal spring having an elastic constant k and is free to move on a frictionless table.

SOLUTION: For SHM the maximum KE equals the maximum PE: $PE_{max} = \frac{1}{2}kA^2 = KE_{max} = \frac{1}{2}mv_{max}^2$ and so $A/v_{max} = \sqrt{m/k}$. Using the results of Problem 75 namely, $T = 2\pi A/v_{max} = 2\pi\sqrt{m/k}$

SECTION 10.7: THE PENDULUM

106. [I] What happens to the period of a simple pendulum if we increase its length so that it's four times longer?

107. [I] What happens to the frequency of a simple pendulum if we increase its length so that it's four times longer?

108. [I] What happens to the period of a simple pendulum if we increase its mass so that it's four times greater?

109. [I] Find the frequency of a simple pendulum of length 10.0 m.

110. [I] A small lead ball is attached to a light string so that the length from the center of the ball to the point of suspension is 1.00 m. Determine the natural period of the ball swinging through a small displacement in a vertical plane.

111. [I] How long must a simple pendulum be if it is to have a period of 10.0 s? [*Hint: The period of a simple pendulum is proportional to the square root of the length.*]

112. [I] While standing on a distant planet, it is found that a 1.00 m long pendulum has a period of 10.00 s. What is the acceleration due to gravity on that planet?

113. [I] How long must a simple pendulum be if it is to have a period of 10.00 s on Mars where the acceleration due to gravity is 39% of its value on Earth?

114. [I] One might guess that the period of a pendulum is proportional to the mass m, the length L, and the acceleration due to gravity g. Consequently, assume that

$$T = Cm^aL^bg^c$$

where C is a unitless constant. Now solve for a, b, and c using the fact that the dimensions on both sides of the equation must be the same. Of course, if this calculation produces nonsense, C wasn't unitless. Although you cannot get it from this analysis, what is the correct value of C?

115. [I] What happens to the frequency of a simple pendulum if we move it to the Moon's surface where the acceleration due to gravity is 1/6 that at the Earth's surface?

116. [II] THIS PROBLEM WILL HELP US TO FURTHER UNDERSTAND THE DEPENDENCE OF A PENDULUM'S MOTION ON GRAVITY. We need an equation for the period of a pendulum at any distance r from the center of the Earth. (a) What does the period of a pendulum

depend on? (b) How does g vary with r? (c) Write an equation for T as a function of r.

117. [II] THIS PROBLEM EXAMINES THE DEPENDENCE OF A PENDULUM'S MOTION ON GRAVITY. A pendulum at the surface of the Earth has a period T_s. It is taken to an altitude h where its period is then T_h. We want to prove that

$$h = r_\oplus\left(\frac{T_h}{T_s} - 1\right)$$

(a) With the previous problem in mind, write an expression for T_h in terms of r. (b) Express r in terms of h and the Earth's radius $r_\oplus$. (c) Write an expression for T_s in terms of $r_\oplus$. (d) Write an equation for T_h/T_s. (e) Now derive the above formula for h.

118. [II] THIS PROBLEM TREATS THE MOTION OF A PENDULUM UNDER VARIOUS CIRCUMSTANCES. A pendulum of length L_1 has a period T_1, at a location where the acceleration due to gravity is $\frac{1}{2}g$. We want to find out how long a second pendulum should be if its period is to be $4T_1$ at a location where the acceleration due to gravity is g? (a) Write an expression for T_1 in terms of L_1 and g. (b) Write an expression for T_2 in terms of L_2 and g. (c) Write an expression for T_2 in terms of T_1. (d) Write an expression for L_2 in terms of L_1.

119. [II] THIS PROBLEM TREATS THE MOTION OF A PENDULUM FROM THE ENERGY PERSPECTIVE. The mass of the bob in a simple pendulum has a maximum speed of 3.0 m/s and we wish to compute the height (h) to which it rises at the end of each swing. (a) Write an expression for the maximum KE of the bob. (b) Write an expression for the maximum PE_G of the bob. (c) What can you say about these last two expressions? (d) Now create an expression for the maximum height. (e) Compute the height at each end of its swing.

120. [II] Consider a simple pendulum of length L. Show that it gains an amount of potential energy

$$\Delta PE_G \approx \frac{1}{2}mgL\theta^2$$

when the bob is raised through an angle θ. [*Hint: You may need to know that $1 - \cos\theta = 2\sin^2\frac{1}{2}\theta$.*]

121. [II] If the period of a simple pendulum is T, what will its new period be if its length is increased by 50%?

122. [II] Extending the ideas of Problem 120, show that the pendulum's maximum angular speed is given by $\omega_{max} = \theta_{max}\sqrt{g/L}$.

123. [II] A pendulum consists of a 2.00-cm-diameter lead sphere attached to a frictionless pivot via a long very light cable. It is to swing from its maximum displacement on one side to its maximum displacement on the other side in 1.00 s. If it's to do this at a place where the acceleration due to gravity is 9.80 m/s², how long must the cable be?

124. [II] THIS PROBLEM EXPLORES THE DEPENDENCE OF A PENDULUM'S PERIOD ON ITS ACCELERATION. We want to derive an expression for the period of a pendulum swinging vertically in an elevator that is itself accelerating vertically at a rate a. (a) Imagine the elevator to be accelerating uniformly upward. Would you seem to weigh more or less in such an environment? (b) If we imagine a to be produced by a change in the gravitational field acting on a stationary elevator should we increase or decrease g? (c) Now suppose the elevator to be accelerating uniformly downward. Would you seem to weigh more or less? (d) If we imagine a to be pro-

duced by a change in the gravitational field acting on a stationary elevator, should we now increase or decrease g? (e) Prove that

$$T = 2\pi\sqrt{\frac{L}{g-a}}$$

125. [II] THIS PROBLEM WILL HELP US TO MORE DEEPLY UNDERSTAND THE BEHAVIOR OF THE PENDULUM. We need an equation for the period of a pendulum at any distance r from the center of the Earth. (a) What does the period of a pendulum depend on? (b) How does g vary with r? (c) Write an equation for T as a function of r.

126. [III] A small mass is attached to the end of a long vertical string. It is displaced through an angle θ and then swung so that the mass moves in a horizontal circle. Show that if you looked at the resulting conical pendulum in the plane of motion, you would see the bob moving back and forth with a period given by

$$T = 2\pi\sqrt{\frac{L\cos\theta}{g}}$$

Chapter 11
Waves & Sound

*T*here are only two fundamental mechanisms for transporting energy and momentum: a streaming of particles and a flowing of waves. And even these two seemingly opposite conceptions are subtly intertwined—there are no waves without particles and no particles without waves, but we'll come back to that later (Chapter 29).

After a general introduction to wave motion, the second part of this chapter deals with a particular kind of wave—sound. The word "sound" has two meanings of interest here. There is the psychological and physiological concept, namely, that which humans hear. And there is the physical concept of sound as any compression wave with its frequency in the range from 20 Hz to 20 kHz.

Mechanical Waves

In general, a wave is a moving self-sustained disturbance of a medium, and that medium can be either a field (e.g., the gravitational field) or a substance (a solid or fluid). Here, the focus is on waves in material media, and these are known as **mechanical waves**.

11.1 Wave Characteristics

Consider an object—a bell, a rope, or even the Earth. Each is a vast collection of atoms forming an essentially continuous elastic medium. Within limits, if the atoms in a material are pushed together, they repel and, if separated, they attract (p. 3). Electrically interacting, the atoms behave as if they are connected to one another via springs. As a result of this restoring force, once displaced, the medium returns to its equilibrium configuration, that is, its normal stable state. Imagine that a driving force acting on the medium displaces some of its atoms; for example, the bell is struck with a hammer. Energy is imparted to these impacted atoms, which are forced off their equilibrium positions, thereby interacting more strongly with neighboring atoms, doing work on them and displacing them. These newly shifted atoms interact with still others, and so the displacement process continues, with energy transferred from atom to atom (Fig. 11.1). *The state of being displaced moves through the medium as a wave.* The disturbance of a medium under the influence of a restoring force is common to all mechanical waves.

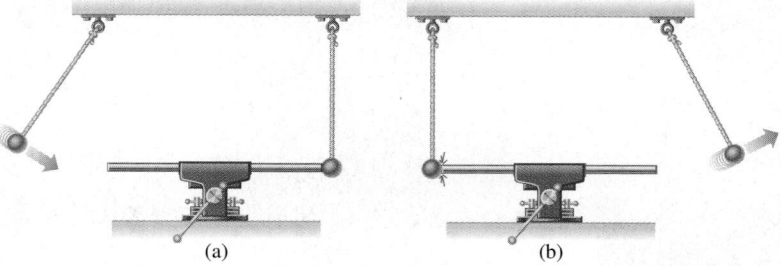

Figure 11.1 Energy is pumped into the horizontal rod by the impact of the ball on the left. A compression wave travels down the rod and slams into the ball on the right. The wave of displaced atoms transports energy and momentum from one ball to the other.

> Once launched, a **wave** is a self-sustaining disturbance of a supporting medium.

A **progressive** *or* **traveling wave** *is a self-sustaining disturbance of a medium that propagates from one region to another, carrying energy and momentum.* There is a great variety of progressive mechanical waves, among which are waves on strings, surface waves on liquids, sound waves in the air, and compression waves in both solids and fluids. In all cases, although the energy-carrying disturbance advances through the medium, the individual atoms remain in the vicinity of their equilibrium positions: *the disturbance advances, not the material medium*. That's the crucial aspect of a wave that distinguishes it from a stream of particles. The wind blowing across a field sets up "waves of grain" that sweep by, even though each stalk only sways in place. Da Vinci seems to have been the first person to recognize that a wave does not transport the medium through which it travels, and it is precisely for this reason that waves are capable of traveling very rapidly.

Longitudinal and Transverse Waves

There are two basic forms of waves: **longitudinal** and **transverse**. In addition to these, there are also torsion waves, which are a variation of the latter, and water waves, which are a combination of the two.

As the medium for a wave, envision a long horizontal spring, like a Slinky (Fig. 11.2). Compress several of its coils and release them all at once. The region of compression, the disturbance from equilibrium, will rapidly advance along the length of the spring. *When the sustaining medium is displaced parallel to the direction of propagation, the wave is longitudinal* (meaning "lengthwise"). The compression wave in a bell, sound waves, and certain seismic waves are all longitudinal.

If the end of the spring is displaced up and down, a hump will be produced (Fig. 11.3) and this disturbance will move, effectively as a pulse of energy, along the length of the

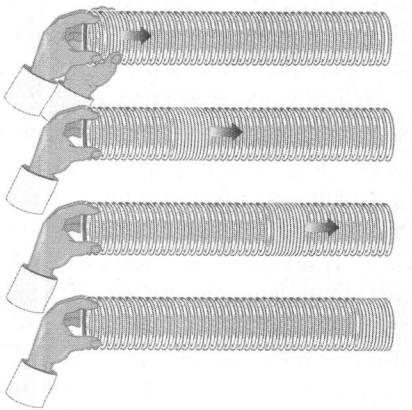

Figure 11.2 A longitudinal wave in a spring.

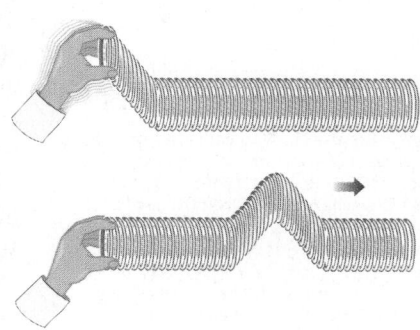

Figure 11.3 A transverse wave in a spring.

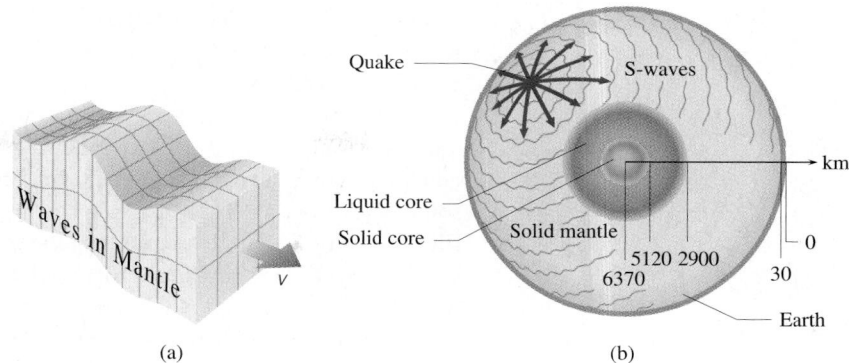

Figure 11.4 (*a*) Transverse seismic waves like this one are called shear waves, or S-waves. (*b*) Because of the Earth's molten core, which cannot support transverse waves, the S-waves from a quake on one side of the planet are picked up everywhere except diagonally across on the other side. Shear waves in the mantle travel at about 8 km/s.

spring. *When the sustaining medium is displaced perpendicular to the direction of propagation, the wave is transverse* (meaning "lying across"). Plucked guitar strings oscillate as transverse waves.

Transverse mechanical waves can be supported provided the medium resists shear, which is something fluids don't do very well. Accordingly, we can expect these waves will not be sustained within the body of a liquid or gas (Fig. 11.4). {For an animated look at these forms click on TRANSVERSE AND LONGITUDINAL WAVES under INTERACTIVE EXPLORATIONS in CHAPTER 11 on the CD.}

Waveforms

Waves in strings, wires, and ropes are of practical concern to piano tuners and people who worry about vibrating powerlines, but our study of them is more general. For us, they provide an especially simple visual model that can be applied to all forms of waves. Figure 11.5 depicts a transverse **wavepulse** set up in a taut rope. The outline, *profile*, or shape of the pulse, is determined by the motion of the driver (the hand), while the physical aspects of the medium (the tension and inertia of the rope) determine the speed of the wave. This is among the most commonplace of all forms of the wave; the sound of a shotgun, a grunt, or an ocean tidal wave are each wavepulses.

If the hand now oscillates up and down in a regular way, it can generate a disturbance like the **wavetrain** depicted in Fig. 11.6. That disturbance can be viewed as a steady sinu-

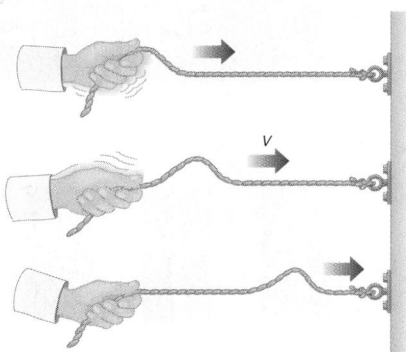

Figure 11.5 A wavepulse riding along a taut rope.

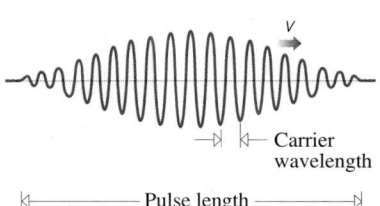

Figure 11.6 This wavetrain can be thought of as a constant-frequency carrier whose amplitude is modulated—that is, made to vary in time.

soidal oscillation called the **carrier**, which gradually varies in amplitude. A wavetrain, whatever the details of its shape, has a beginning and an end. A burst on a whistle is an acoustical wavetrain.

It is possible to sustain a vibration for a considerable time and thereby generate a very long wavetrain. Figure 11.7 shows the waveform produced via a microphone by a sustained note played on a saxophone. It could equally well have been created on a taut rope through some fancy jiggling. The wavetrain can be visualized as made up of a single repeated profile wave (Fig. 11.7*b*). However long the overall wavetrain is, it is finite—there was a time before the note was blown and there will be a time when it ends. That's the nature of real wavetrains as compared with mathematical ones. In many situations, the disturbance may be extremely long with an enormous number of repetitions. It then becomes much simpler mathematically to assume that the wavetrain is infinitely long. *An idealized disturbance composed of countless repetitions of the same profile is said to be* **periodic**.

To Find the Velocity of Waves

A periodic wave (whether it's a sinusoid, a saw tooth, or anything else) can be envisioned as the endless repetition of some basic disturbance (whose shape is called the *profile*). The **period** (T) of such a wave is the time it takes for a single profile to pass a point in space. It's the time for one cycle to occur; *the number of seconds per cycles*. Whatever the shape of the wave, it repeats itself after a time T, and it does that over and over again endlessly. The inverse of that ($1/T$) is the **frequency** f, the number of profiles passing per second, *the number of cycles per second*. The distance in space over which the wave executes one cycle of its basic repeated form is the **wavelength**, λ (Greek lowercase lambda): It's the length of the profile.

Imagine that you are at rest and a wavetrain on a string is progressing past you. The number of profile waves that sweep by per second is f, and the length of each is λ. In 1 s, the overall length of the disturbance that passes you is the product $f\lambda$. If, for example, each is 2-m long and they come at a rate of five per second, then in 1 s, 10 m of wave fly by. This is just what we mean by the speed of the wave ($v = |v|$)—the rate in m/s at which it advances. Said slightly differently, since a length of wave λ passes by in a time T, its speed must equal $\lambda/T = f\lambda$. We now have an expression for the speed of any progressive *periodic* wave, be it sound, water ripple, or light:

$$v = f\lambda \tag{11.1}$$

Incidentally, Newton derived this relationship in the *Principia* (1687) in a section called "To find the velocity of waves."

The **period** is the time it takes for a wave to go through one complete oscillation.

The **frequency** is the number of complete oscillations the wave makes in a second.

The **wavelength** is the spatial distance over which the wave makes one complete oscillation.

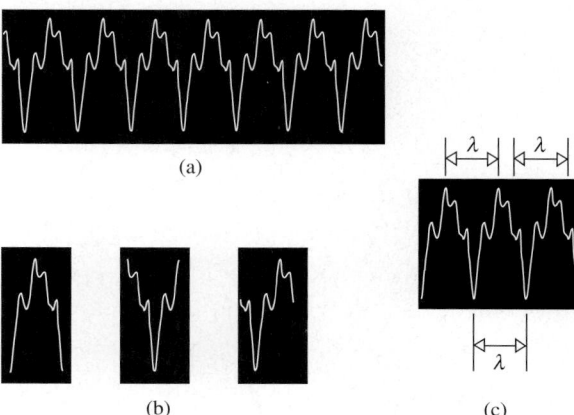

(a)

(b)   (c)

Figure 11.7 (*a*) The waveform generated by a saxophone. We can imagine any number of profile-elements (*b*) that, when repeated, create the waveform (*c*). The length over which the wave repeats itself is called the wavelength, λ.

Example 11.1 **[I]** A youngster in a boat watches waves on a lake that seem to be an endless succession of identical crests passing, with a half-second between them. If one wave takes 1.5 s to sweep straight down the length of her 4.5-m-long boat, what are the frequency, period, and wavelength of the waves?

Solution The words "waves" and "endless succession" suggest that this is a problem dealing with periodic waves. (1) TRANSLATION—A wave with a specified interval between crests travels a known distance in a known time; determine its (a) period, (b) frequency, (c) speed, and (d) wavelength. (2) GIVEN: The waves are periodic; there is $\frac{1}{2}$ s between crests; $L = 4.5$ m, and $t = 1.5$ s. FIND: T, f, v, and λ. (3) PROBLEM TYPE—Wave motion/periodic waves.

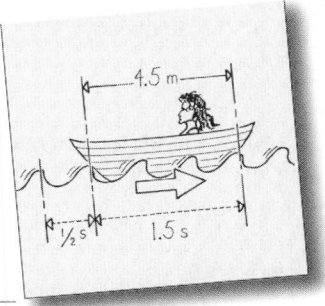

(4) PROCEDURE—Use the basic equation for periodic waves, $v = f\lambda$. (5) CALCULATION—(a) The time between crests *is* the period, so $\boxed{T = \frac{1}{2} \text{ s}}$. (b) Therefore

$$f = 1/T = \boxed{2.0 \text{ Hz}}$$

(c) As for the speed

$$v = \frac{L}{t} = \frac{4.5 \text{ m}}{1.5 \text{ s}} = \boxed{3.0 \text{ m/s}}$$

We now know T, f, and v and must determine λ. From Eq. (11.1)

$$\lambda = \frac{v}{f} = \frac{3.0 \text{ m/s}}{2.0 \text{ Hz}} = \boxed{1.5 \text{ m}}$$

Quick Check: $Tv = \lambda = (\frac{1}{2}\text{s})(3.0 \text{ m/s}) = 1.5$ m. $v = f\lambda = (2.0 \text{ Hz})(1.5 \text{ m}) = 3.0$ m/s.

[For more worked problems click on the WALK-THROUGHS *in* CHAPTER 11 *on the* CD.]

Harmonic Waves

The easiest of repeating disturbances to treat theoretically is the **harmonic wave**, which rises and falls sinusoidally without end. This is *the* fundamental waveform because, as we will see later, all real waves can be synthesized from overlapping harmonic waves.

Remembering the treatment of SHM, examine Fig. 11.8, which is a plot of

$$y = A \sin \frac{2\pi x}{\lambda} \tag{11.2}$$

The argument of the sine function $(2\pi x/\lambda)$ is the **phase** and it's unitless, as it must always be because the sine is the ratio of two quantities. Equation (11.2) describes the profile of a harmonic wave oscillating in the y-direction, traveling along the x-axis, frozen at $t = 0$. At any location x, the ratio x/λ is the number of wavelengths from the origin out to that point and, since there are 2π rad per wavelength, $2\pi x/\lambda$ is the number of radians out to the point at x. The profile repeats itself with a wavelength λ so that $y = 0$ when $x = 0, \lambda, 2\lambda, 3\lambda,\dots$. The *amplitude* is A, and it's always positive. The wave builds up and falls off between values of $+A$ and $-A$ in the y-direction. **The energy associated with a wave is proportional to the amplitude of the wave squared**, and that's true for all waves. The amplitude of a sound wave determines how loud it is, and the amplitude of a light wave determines how bright it is. Figure 11.9 on p. 377 shows a harmonic wave advancing one wavelength during a time interval of one period (T).

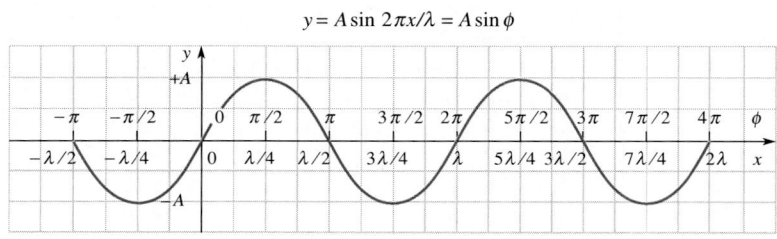

Figure 11.8 A harmonic function, which is to serve as the profile of a harmonic wave. One wavelength corresponds to a change in phase ϕ of 2π rad.

Example 11.2 **[II]** The profile of a harmonic wave, traveling at 1.2 m/s on a string, is given by

$$y = (0.02 \text{ m}) \sin (157 \text{ m}^{-1})x$$

Determine its amplitude, wavelength, frequency, and period.

Solution This problem deals with the mathematical representation of a sinusoidal wave. (1) TRANSLATION—The equation of the profile of a wave is known, as is its speed; determine its (a) amplitude, (b) wavelength, (c) frequency, and (d) period. (2) GIVEN: The profile, and $v = 1.2$ m/s. FIND: A, λ, f, and T. (3) PROBLEM TYPE—Wave motion/harmonic waves. (4) PROCEDURE—Compare the equation for the profile of a harmonic wave [Eq. (11.2)] with y as given. Use the basic equation for periodic waves, $v = f\lambda$. (5) CALCULATION—(a)

$$y = A \sin \frac{2\pi x}{\lambda} \qquad [11.2]$$

$$y = (0.02 \text{ m}) \sin (157 \text{ m}^{-1})x$$

it follows that $\boxed{A = 0.02 \text{ m}}$. (b) Furthermore, $2\pi/\lambda = 157$ m^{-1} and so $\lambda = 2\pi/(157 \text{ m}^{-1}) = \boxed{0.040\,0 \text{ m}}$. (c) The relationship between frequency and wavelength is fixed by $v = f\lambda$, and so

$$f = \frac{v}{\lambda} = \frac{1.2 \text{ m/s}}{0.040\,0 \text{ m}} = \boxed{30 \text{ Hz}}$$

(d) The period is the inverse of the frequency: $T = 1/f$ $= \boxed{0.033 \text{ s}}$.

Quick Check: $f/v = 1/\lambda$; hence, the phase $(2\pi/\lambda)x = (2\pi f/v)x = [2\pi(30 \text{ Hz})/(1.2 \text{ m/s})]x = (157 \text{ m}^{-1})x$.

11.2 Transverse Waves: Strings

The speed of a mechanical wave is determined by the inertial and elastic properties of the medium and not in any way by the motion of the source. The central concerns are how much mass is being accelerated and with how much force does the medium resist deformation. This can be shown for waves on strings using the calculus, but for our purposes it will suffice to just state the result. {For a derivation due to P. G. Tait in the 1800s, click on **WAVES ON STRINGS** under **FURTHER DISCUSSIONS** on the **CD**.}

Consider a pulse traveling with a speed v along a lightweight, flexible string; this might, for example, be a string of a guitar, violin, or piano. The string is kept taut and there's a constant tension F_T. The tension provides the force that pulls the string back to its equilibrium position; the greater the tension, the greater the acceleration, the faster the return, and the faster the wave can travel. At any moment the motion is occurring in a portion of the string, so the total mass of the string is irrelevant. What's important is the mass of a moving piece, and that depends on the **mass per unit length**, m/L. The more massive a portion of the string is, the more difficult it is to accelerate. For a wave of modest amplitude,

[transverse waves on a string]

$$v = \sqrt{\frac{F_\text{T}}{m/L}} \qquad (11.3)$$

Example 11.3 **[II]** A 2.0-m-long horizontal string having a mass of 40 g is slung over a light frictionless pulley, and its end is attached to a hanging 2.0-kg mass. Compute the speed of the wavepulse on the string. Ignore the weight of the overhanging length of rope.

Solution This is a problem about waves on a string. (1) TRANSLATION—A string of known mass and length is under a computable amount of tension; determine the speed of a wave on the string. (2) GIVEN: A string of $l = 2.0$ m, $m = 40$ g supporting a 2.0-kg load. FIND: v. (3) PROBLEM TYPE—Wave motion/on a

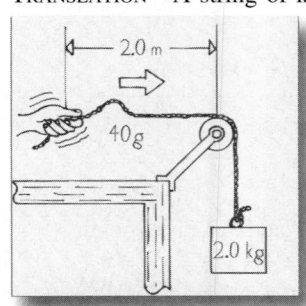

string. (4) PROCEDURE—Use the specific equation for v on a taut string [Eq. (11.3)]. (5) CALCULATION—We need F_T and m/L. The tension is just the load in newtons, namely, (2.0 kg)(9.81 m/s^2) = 19.62 N, while for the string $m/L = (0.040$ kg)/(2.0 m) = 0.020 kg/m. Therefore,

$$v = \sqrt{\frac{F_\text{T}}{m/L}} = \sqrt{\frac{19.62 \text{ N}}{0.020 \text{ kg/m}}} = \boxed{31 \text{ m/s}}$$

which is equivalent to 70 mi/h.

Quick Check: If the mass was 1 kg, $F_\text{T} = 9.8$ N and $v = 22$ m/s, and this quantity times $\sqrt{2} = 1.414$ is 31 m/s.

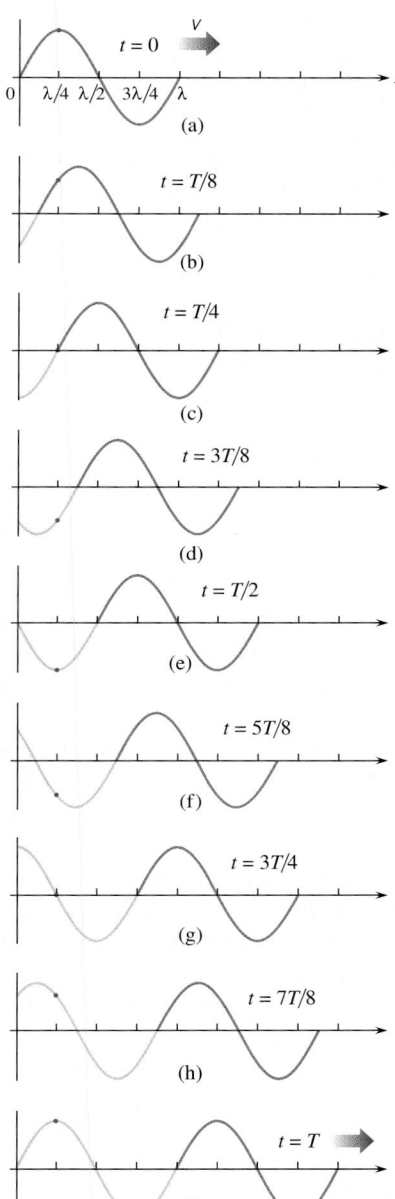

Figure 11.9 A harmonic wave moving along the *x*-axis during a time of one period. Note that any one point on the rope only moves vertically.

When m/L is large, there is a lot of inertia and the speed is low. When F_T is large, the string tends to spring back rapidly, and the speed is high.

Reflection, Absorption, and Transmission

Every variety of wave from sound to light can undergo reflection, absorption, and transmission on interacting with material media. Interesting things happen to waves at discontinuities in the media in which they travel.

In Fig. 11.10, one end of the rope is held stationary while energy is pumped in at the other end. The **reflected** wave ideally carries away all the original energy. Because it is inverted, it's said to be 180° out-of-phase with the incident wave. Alternatively, if the far end of the rope is free (Fig. 11.11), it will rise up as the pulse arrives until all the energy is stored elastically. The rope then snaps back down, producing a reflected wavepulse that is right side up. A similar effect occurs when an ocean wave slams into a breaker wall and climbs high above the crests behind it.

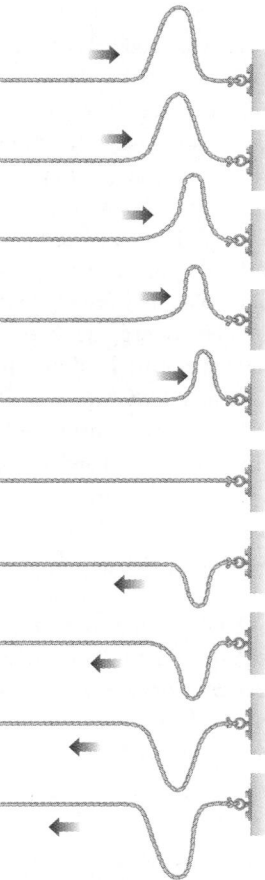

Figure 11.10 The reflection of a pulse on a rope with a fixed end point. As the pulse arrives, it exerts a vertical force on the fixed anchor point, which in turn exerts an equal and opposite force on the string. When the string tugs up, the anchor point tugs down. This downward force on the rope generates an upside-down reflected pulse traveling in the opposite direction.

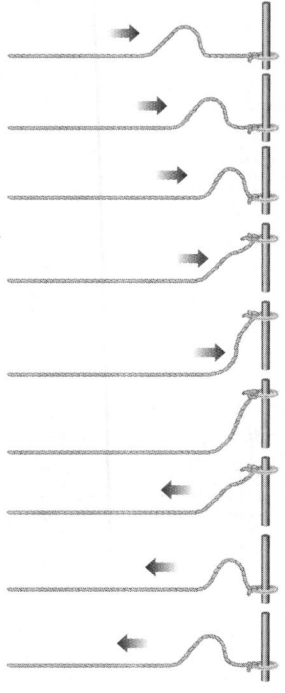

Figure 11.11 The reflection of a pulse on a rope with a free end point. That free end rises until all the energy of the end segment is stored elastically. It comes to rest at a maximum vertical displacement of twice the height of the crest. Carried up by its inertia, the end segment pulls upward on the rope, generating a reflected wavepulse that travels back toward the source, right side up and simply reversed.

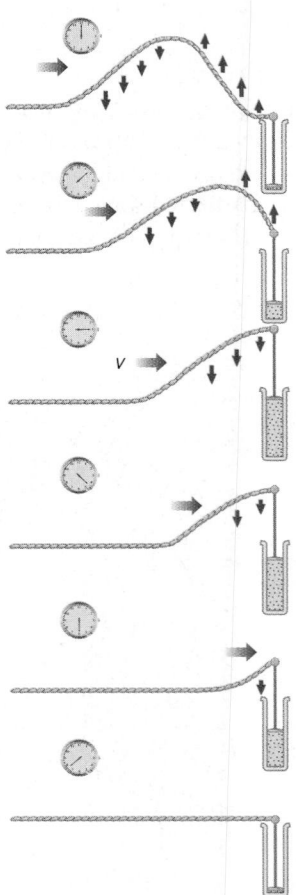

Figure 11.12 A wave in a string terminated on a dashpot. The vertical speed of the rope is proportional to the slope of the wave. The drag exerted by the dashpot is proportional to the vertical speed of the piston.

DASHPOTS & DRAG

Consider what would happen if we mounted a little dashpot damper to the end of the rope (Fig. 11.12). A dashpot is a fluid-filled cylinder sealed with a movable piston. The dashpot opposes motion, via viscous damping, with a force that varies as the speed of the point to which it is fixed (i.e., the speed of the piston). If the dashpot exerts just the right amount of drag on the rope, the segment immediately to the left of it experiences the same motion as if the rope continued beyond it indefinitely—the wave's energy flows via friction into the dashpot. In this ideal case, the wave will be totally absorbed. The muffler in a car serves a similar function converting large-amplitude sound waves into thermal energy.

If we draw off energy from the oscillating end point of a rope via friction, the reflected pulse has a proportionately diminished amplitude and we call the process **absorption**. It occurs most effectively when the rope encounters friction that is speed-dependent. To understand why, realize that any small segment of the rope only moves vertically. The vertical force exerted on any such segment due to the rope just in front of it is proportional to that segment's vertical speed. When a wave travels from left to right, at any point the segment on the left does work on the segment immediately to the right of it and energy flows along with the disturbance.

When a wave passes from one medium to another having different physical characteristics, there will be a redistribution of energy. Figure 11.13a shows a wavepulse initially traveling in a low-density rope impinging on the interface with a high-density rope. Like the fixed point of Fig. 11.10, the large inertia of the second medium at the junction retards the easy motion of the boundary. There is again an oppositely directed reaction force, and the reflected wave is phase-shifted by 180°. But the second medium is also displaced, and a portion of the incident energy appears as a **transmitted** wave. The time it takes to generate the reflected and transmitted waves must be identical since they share a common source. Yet the speeds of the pulses must be different because they have the same tension and different densities. The pulses must therefore have different lengths.

When the first medium is denser than the second (Fig. 11.13b), the situation resembles that of Fig. 11.11 with a free end point: no phase shift results and the transmitted wave has a greater length. The process of reflection and transmission at an interface between media occurs for all waves regardless of whether they are longitudinal or transverse, though the details differ.

Evidently, if the incident wave is periodic, the transmitted wave has the same frequency but a different speed and therefore a different wavelength: *the larger the density of the transmitting medium, the smaller the length of the wave*. The fact that the frequencies of the incident, reflected, and transmitted waves are normally the same is true for every sort of disturbance from sound to light.

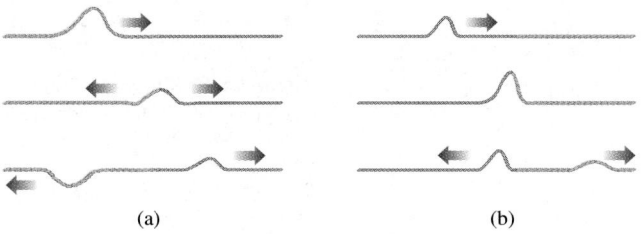

(a)   (b)

Figure 11.13 Reflection and transmission of a pulse at the junction of two media. The darker rope has a greater linear mass-density.

11.3 Compression Waves

Atoms of an elastic substance can be shifted by an applied force, and the resulting state of being compressed will travel through the medium as a *compression wave*. In solids, such a disturbance is called a *longitudinal elastic wave*. In fluids, which cannot support transverse disturbances, it's called an *acoustic wave*. An earthquake or a large truck rumbling down the street can shake a building, rattling everything in it with streams of compression waves.

Let's model a succession of atoms or molecules in an elastic medium (either solid or fluid of sufficient density) using a row of spheres separated by springs (Fig. 11.14). The particle at the far left is made to vibrate in SHM along the line by some source of energy. Wherever the springs are elongated, the density of particles is diminished, and that is a ***rarefaction***. Wherever the springs are compressed, the density is increased, and that is a ***condensation***. **The propagation of a compression wave takes place in the direction along which the particles of the medium oscillate, and it is marked by a series of alternate condensations and rarefactions**. There are tremendous numbers of atoms in an ordinary medium, and they behave as though the distribution were continuous; rarefactions gradually blend into condensations. Figure 11.15 shows that at any instant the particles within the condensations move forward in the propagation direction, while those in the rarefactions move backward. {For an animated study of compression disturbances click on TRANSVERSE AND LONGITUDINAL WAVES under INTERACTIVE EXPLORATIONS in CHAPTER 11 on the CD.}

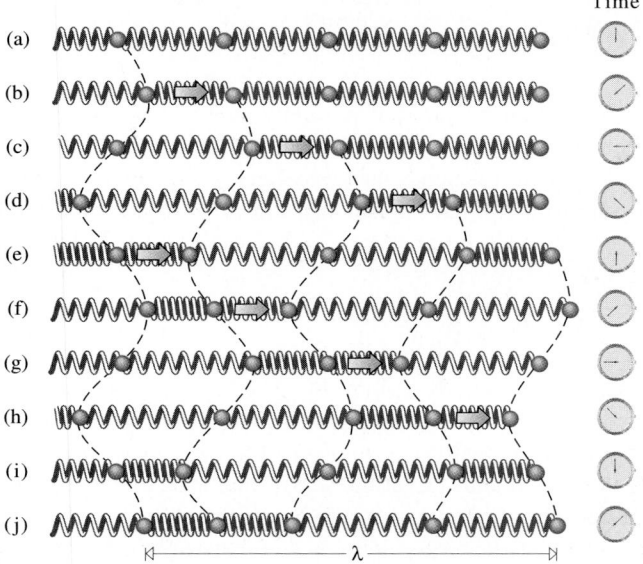

Time

Figure 11.14 A sequence of views of a line of masses attached by springs. In (*a*) the system is undisturbed. The first mass on the left oscillates in SHM. The resulting disturbance propagates to the right [(*b*) through (*j*)] as a series of compressions and elongations that repeats after a distance λ. Each mass oscillates sinusoidally in SHM.

The Speed of Compression Waves

We now consider the speed of a compression wave in a liquid. It can be anticipated that any formula for v would involve both the restoring force generated by the sustaining medium (via some measure of its elastic response) and its density. {For a complete derivation click on COMPRESSION WAVES under FURTHER DISCUSSIONS on the CD.}

Imagine a compression pulse of high pressure traveling through a liquid at rest. The liquid is at a pressure P, and the pulse pressure exceeds that by ΔP. The elastic properties of the medium are embodied in its ability to resist compression under pressure. Recall that the Bulk Modulus B (p. 341) is defined as

$$B = -\frac{\Delta P}{\Delta V/V_0}$$

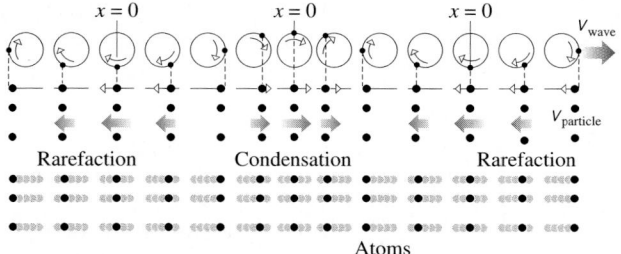

Figure 11.15 A compression wave passing to the right through a distribution of elastically interacting atoms. The peaks of the condensations and rarefactions occur at zero atomic displacements ($x = 0$). Note that the atoms in the condensations move in the direction of propagation of the wave. Start at the right and examine each reference circle in turn. Remember that each atom oscillates around its equilibrium position ($x = 0$) in SHM.

Figure 11.16 Earthquake compression waves are called primary waves, or P-waves, because they get to seismic stations before the S-waves (see Fig. 11.4 on p. 373). They predominate in the disturbances set up by underground nuclear explosions and are used to distinguish between natural quakes and weapons tests.

SPEED, SALINITY, & SONAR

The density and therefore the speed of compression waves in the body of the sea depend on the temperature, salinity, and depth, which is why Example 11.4 is careful not to have the waves propagating very deep. While the speed is 1.4 km/s at the surface, at a depth of 5 km it's about 1.5 km/s. Moreover, a rise in temperature of 1°C increases the speed by 3.7 m/s, while an increase of 1% in salinity adds about 1.2 m/s to it—both of which would be of great interest to a dolphin or a sonar operator in a submarine. By comparison in the rock mantle of the Earth, compression waves reach speeds of 13 km/s (Fig. 11.16).

The Bulk Modulus increases as the medium becomes increasingly rigid, and we can expect that v depends on B. The more mass there is in each unit volume, the slower the wave "should" travel, so we can anticipate that v depends on $1/\rho$, where ρ is the mass density. Also remember from Bernoulli's Equation that $\Delta P/\rho$, and therefore B/ρ has the same units as v^2. In any event,

[compression waves in a liquid]
$$v = \sqrt{\frac{B}{\rho}}$$
(11.4)

It takes a large increase in pressure to decrease the volume of a solid as compared to a liquid, and B is generally larger for dense media—it increases as the medium becomes increasingly rigid (Table 10.4), and we can expect that v depends on B. The more mass there is in each unit volume, the slower the wave "should" travel, so we can anticipate that v depends on $1/\rho$. Also remember from Bernoulli's Equation that $\Delta P/\rho$, and therefore B/ρ, has the same units as v^2. Still, since B is generally larger for dense media, v usually increases as ρ increases, which is not obvious from Eq. (11.4).

Example 11.4 **[I]** An explosion occurs not far beneath the surface of the Atlantic Ocean. Compute the speed of the resulting compression wave measured by instruments several meters below a ship.

Solution This one's about the speed of compression waves in water. (1) TRANSLATION—Determine the speed of a compression wave in seawater. (2) The source is not very deep; therefore, the density can be taken from Table 9.1. GIVEN: $\rho = 1.03 \times 10^3$ kg/m³ and from Table 10.4 $B = 2.1$ GPa. FIND: v. (3)

PROBLEM TYPE—Wave motion/in a liquid. (4) PROCEDURE—Use the specific equation for v in a liquid. (5) CALCULATION—Substituting into Eq. (11.4),

$$v = \sqrt{\frac{B}{\rho}} = \sqrt{\frac{2.1 \times 10^9 \text{ Pa}}{1.03 \times 10^3 \text{ kg/m}^3}} = \boxed{1.4 \times 10^3 \text{ m/s}}$$

Quick Check: $B = v^2\rho \approx (1400 \text{ m/s})^2(1000 \text{ kg/m}^3) \approx 2 \times 10^9$ kg/m·s².

Using Eq. (11.4) as a guide, we can "feel" our way through to a formula for the speed of a compression wave in a slender solid (as in Fig. 11.1). Young's Modulus provides a measure of the one-dimensional rigidity of a solid medium (i.e., rods, wires, etc.), and it can be expected that v depends on Y. Again, we can anticipate that v depends on $1/\rho$. Keep in mind too that Y/ρ has the same units as v^2. In any event, a complete analysis yields

[compression waves in a slender solid]
$$v = \sqrt{\frac{Y}{\rho}}$$
(11.5)

This expression works fine when the medium is narrow compared to λ and we don't have to worry about lateral changes. The general formula for waves in solids with large cross sections looks like Eq. (11.5) but has a few more terms under the square root sign. The speed of compression waves (e.g., sound) in solids is usually higher than in liquids, although for a soft (low Y), dense material like lead the speed is only 1.23 km/s. At the other extreme, for granite, which is quite rigid (high Y) and yet not very dense, the speed, 6.00 km/s, is relatively large.

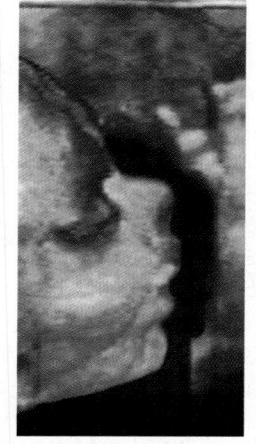

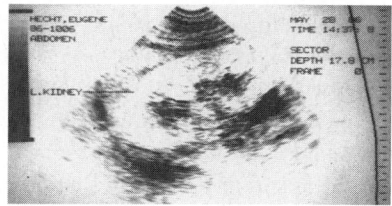

An ordinary low resolution sonogram of a left kidney (the white C-shaped form is facing down toward 5 o'clock). The speed of compression waves in the body averages $\approx 1.5 \times 10^3$ m/s, and so frequencies of 1 MHz to 10 MHz have wavelengths of 1.5 mm to 0.15 mm and can reveal fine details.

Ultrasound and Infrasound

Dolphins operate their underwater tracking system by radiating compression wavetrains, high-frequency chirps with carrier wavelengths (Fig. 11.6) of about 1.4 cm. This echo ranging, which is quite similar to sonar, determines the round-trip time for a pulse to be reflected back to the dolphin. *Waves reflect effectively off objects that are at least as large as about one wavelength*; here, the carrier wavelength is small enough to still "see" rather little fish. To produce emissions of that wavelength (since $v = f\lambda = 1.4 \times 10^3$ m/s), the frequency has to be up around 10^5 Hz. These waves are well beyond the human audible range (which is less than 20 kHz) and are therefore called **ultrasonic**.

Some autofocus cameras imitate bats, sending out ultrasonic pulses and using the round-trip time to determine target range. Ultrasonic techniques are routinely used in medicine in both diagnosis and treatment. Intense concentrations of energy can be directed into the body to destroy tumors and stones. In yet another application, low-energy ultrasonic waves are used to probe the body in a way that is widely believed to be far less hazardous than X-rays.

Infrasound corresponds to compression waves of subaudible frequencies (< 20 Hz). A good deal of this energy is present in the environment, especially in the vicinity of airplanes, elephants, thunderstorms, fast-moving cars, and overamplified rock bands.

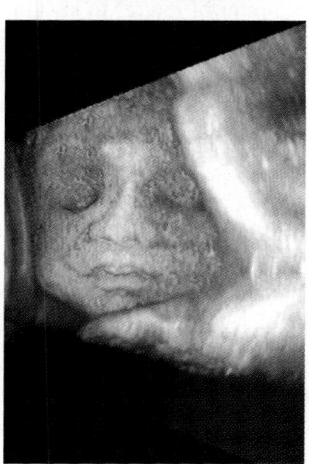

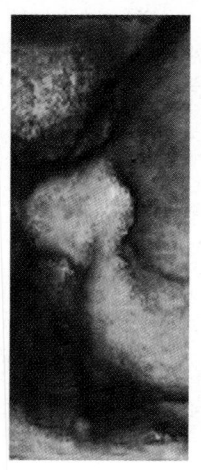

High-resolution sonograms of the head of a fetus in the womb.

A Wellington airplane under 210 m of water at the bottom of Loch Ness, revealed by side-scanning sonar.

Sound

The specific frequency bandwidth from 20 Hz to 20 kHz is the spectrum heard by the so-called "average" person, who is actually a statistical fiction. Anyone in our society who can hear across that entire range is something of a rarity. Even so, the frequency response of the human ear-brain system diminishes with age beyond about 20 years. Moreover, the exposure to loud sounds, as is common these days, desensitizes the ear at both frequency extremes.

11.4 Acoustics: Sound Waves

That sound is a wave seems to have first been suggested by the Roman architect Marcus Vitruvius Pollio. While considering the design of amphitheaters, he likened sound to water ripples moving through space "upward, wave after wave." His work marked the beginning of the study of the physics of sound—that is, acoustics (from the Greek *akoustikos*, relating to hearing). The recognition that a sound wave is longitudinal follows from the fact that it propagates in air, which, like all fluids, has no stiffness, cannot resist shear, and cannot sustain a transverse wave. Because a wave does not transport the medium, its speed can be extraordinarily large—a bolt of lightning a mile away produces a clap of thunder heard only a moment ($\approx$5 s) later.

Sound propagates in any medium that can respond elastically and thereby transmit vibrational energy. Air at ordinary density is certainly springy. (Squeeze an empty plastic soda bottle capped and uncapped, and the point will be made.) The fact that the region connecting the source and detector must contain an adequate amount of matter to sustain the disturbance was established experimentally (1672) by von Guericke. He produced a partial vacuum within a chamber containing a noise-making mechanism. As the air was removed, the sound became weaker until it vanished altogether. Sound does not propagate in vacuum. All the science fiction movies that depict roaring explosions in space have missed the point.

Consider a loudspeaker being driven by a sinusoidal signal, a *pure tone*. The flexible cone vibrates harmonically, pumping the air in front of it into a series of condensations and rarefactions (Fig. 11.17). The resulting pressure variations that constitute the sound wave are quite small. A very loud sound corresponds to a pressure change of less than about 10 Pa (10^{-5} atm), and quiet music exists at a gauge pressure of $< \pm0.002$ Pa. The speaker cone usually moves back and forth a fraction of a centimeter, and any given air molecule moves an equally small amount (Fig. 11.15). ***The distance the molecules move determines the amplitude of the wave and, therefore, its energy and the loudness of the sound.*** In the same way, a large drum head moves through a relatively large distance and makes a large sound.

The time it takes the speaker cone to pass through a cycle fixes the period and frequency of the sound. The distance in air between successive condensations (or rarefactions) is the wavelength (Table 11.1). The wavelength, too, is determined by how fast the cone

Table 11.1

Wavelengths in Air at 20 °C			
Source	**Frequency** (Hz)	**Wavelength** (m)	(ft)
Organ, lowest note	16	21	70
Piano, lowest note	28	12	40
Bass voice, low C	65	5.3	17
Middle C	262	1.3	4.3
Alto voice, high F	698	0.49	1.6
Soprano voice, high C	1047	0.33	1.1
Piano, highest note	4186	0.08	0.27

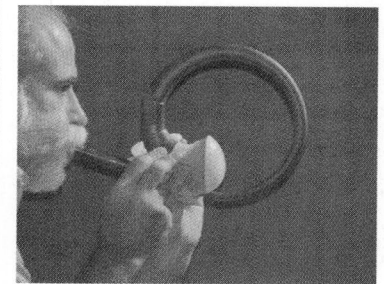

EXPLORING PHYSICS ON YOUR OWN

Sound Waves: Many people think that when sound blasts from a source, air is literally shot out from it, rushing away at the speed of sound. The truth is that the air stays pretty much where it is and just oscillates—the disturbance propagates out, not the air. To see that, get a horn, a toy one will do, and cover the end with a thin rubber sheet—I cut up a disposable rubber glove. Stretch the rubber over the end of the horn and fix it in place with a rubber band. Now blow a sustained note. The diaphragm will bulge out and stay out, while it oscillates with a tiny amplitude, all the time you're blowing the note. Amazingly, there'll be little or no diminution in the volume of the sound. Put your hand in front of the rubber and you can feel the air vibrating. How does the loudness relate to the bulge?

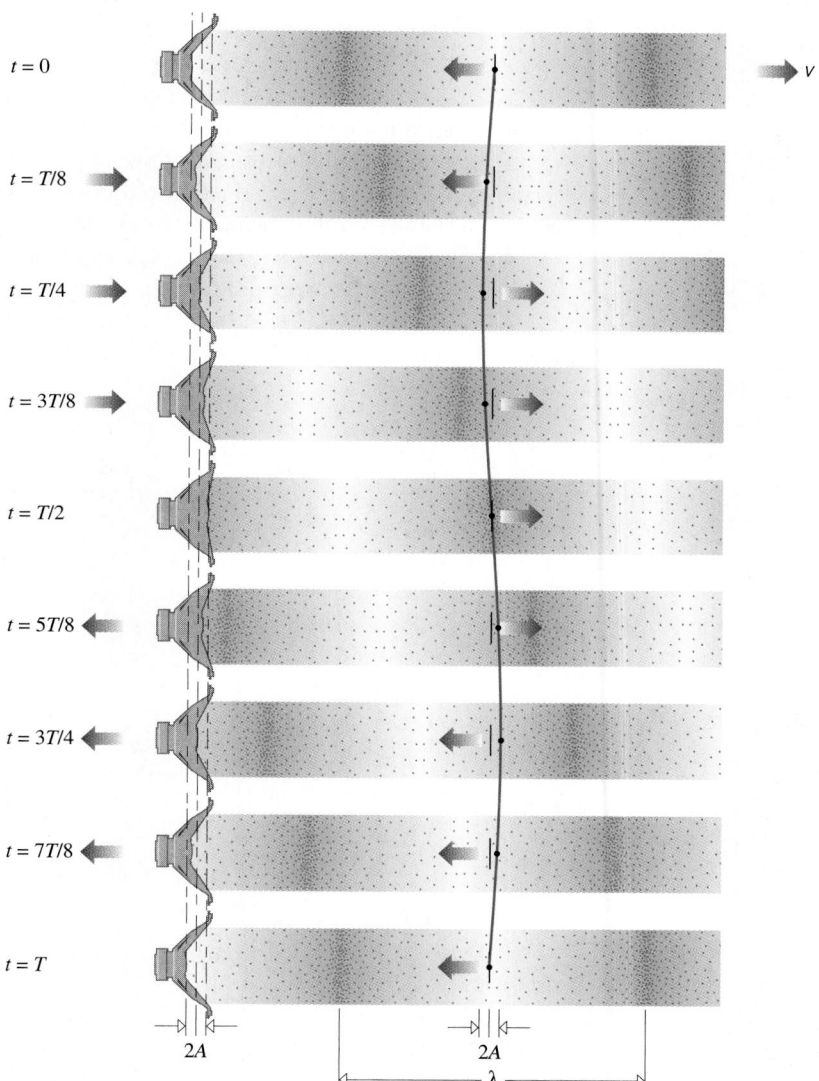

Figure 11.17 A sound wave coming from a loudspeaker. Note that the speaker cone sweeps through a distance of $2A$. Each atom oscillates through a distance of $2A$ as well. Compare that to the wavelength λ.

moves through a cycle. Energy streams into the air all the while the cone is in motion. A slow progression, even though it is through a small cone displacement (<1 cm), can launch a wave with a λ of as much as ≈ 17 m at 20 Hz.

Example 11.5 **[I]** By international agreement, most orchestras tune to a frequency of 440 Hz, which is called A440 (the A note above middle C). Given that the speed of sound in air at room temperature is 343.9 m/s, what is the wavelength of A440?

Solution The words "frequency," "speed," and "wavelength" bring to mind $v = f\lambda$. (1) TRANSLATION—Knowing the speed and frequency of a sound wave, determine its wavelength. (2) GIVEN: $f = 440$ Hz and $v = 343.9$ m/s. FIND: λ. (3) PROBLEM TYPE—Wave motion/sound. (4) PROCEDURE—The problem deals with frequency and wavelength, and that should always bring to mind the basic equation for v for all waves. (5) CALCULATION—$v = f\lambda$ and

$$\lambda = \frac{v}{f} = \frac{343.9 \text{ m/s}}{440 \text{ Hz}} = \boxed{0.782 \text{ m}}$$

Quick Check: For 100 Hz, which is 4.4 times smaller than f, $\lambda = 3.4$ m; hence, 4.4(0.78 m) = 3.4 m. It's a good thing to remember that sound waves extend from $\approx$20 mm to $\approx$20 m. Happily, 0.78 m is in the right range.

Unlike a transverse wave on a rope, compression waves are difficult to draw. Instead, we could create a curve with an axis along the propagation direction, whose magnitude at each point corresponds to the displacement of the particles of the medium (positive in the propagation direction, negative opposite to it). For a single-frequency compression wave, the resulting curve is a sinusoid. Alternatively, we could plot either the density or the gauge pressure within the medium, and these would also be sinusoidal (Fig. 11.18).

Since pressure can be measured directly with a microphone and readily displayed on an oscilloscope, plotting P_G versus t will be the preferred approach. As an example, Fig. 11.19 shows the sound generated by a tuning fork. By contrast, a struck wooden stick (Fig. 11.20) produces a short-lived wavetrain only a few meters long.

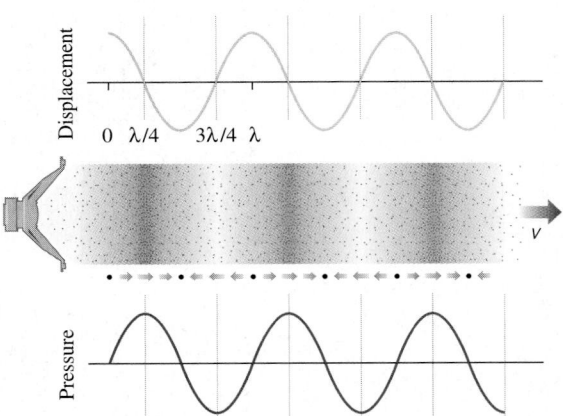

Figure 11.18 A sinusoidal sound wave. Where the pressure is maximum, above and below atmospheric, the displacement is zero. When the pressure is positive, as it is in condensations, atoms are shifted in the direction of motion of the wave.

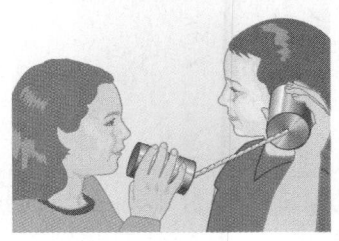

EXPLORING PHYSICS ON YOUR OWN

The Old String Phone: Punch a hole in the bottom of a topless tin can (or a paper cup), pass a string through it, and make a large knot at the end so that it will not come out the hole. Then do the same thing several meters away at the other end of the string with a second can (bottoms toward each other). With one can at your mouth and the other at your listener's ear, pull the string tight and talk into the open end—you've made a kid's "telephone." Your voice vibrates the bottom of the can, pumping energy into the string and sending compression waves down its length, ultimately setting the bottom of the receiver can vibrating just like a speaker cone.

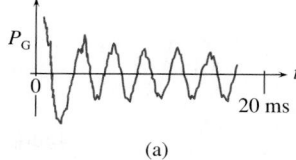

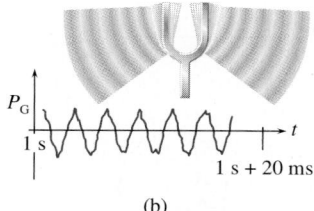

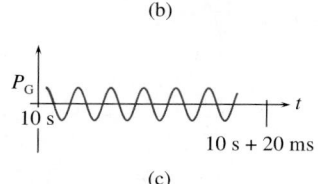

Figure 11.19 Oscilloscope traces corresponding to gauge pressure produced by a tuning fork: (*a*) after the fork is initially struck; (*b*) after oscillating about 1 s; (*c*) after oscillating about 10 s. As time goes on, the high-frequency, short-wavelength components vanish, leaving a smooth, nearly sinusoidal wave. Precision tuning forks can maintain their frequency with an accuracy of 1 part in 100 000.

Figure 11.20 Gauge pressure produced by striking a wooden stick with a metal rod, as shown on an oscilloscope. The short-lived vibration dies out in about 10 ms.

The Superposition of Waves

A fascinating characteristic shared by all waves is that two or more of them moving through the same region of space will superimpose and produce a combined effect. Waves maintain their integrity upon overlapping; they cohabit the medium without themselves being permanently changed. They do not interact with each other in the sense that they do not scatter one another as would two crossing beams of particles. *In the region where two or more waves overlap, the resultant is the algebraic sum of the various contributions at each point.* This is called the **Superposition Principle**, and it was introduced by Dr. Thomas Young. As we will come to see, it's among the most far-reaching and fundamental insights in wave theory.

Figure 11.21 shows the effect of superimposing two harmonic waves of the same frequency and amplitude. At every value of *x*, we simply add the corresponding heights of the two sine curves, taking measures above the axis as positive and below it as negative. *The sum of any number of harmonic waves of the same frequency traveling in the same direction is also a harmonic wave of that frequency.* When sine waves with different fre-

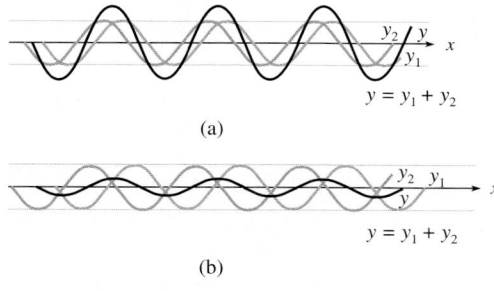

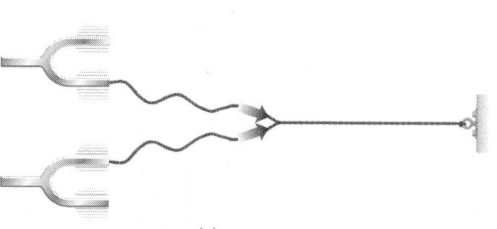

Figure 11.21 The superposition of waves y_1 and y_2, which combine to yield y. The two component waves have the same amplitude and differ in phase. In (*a*) the two are nearly in-phase, and the resultant is large. In (*b*) they are nearly out-of-phase by 180°, and the resultant is small. In (*c*) they are out-of-phase by exactly 180°, and the two waves cancel completely.

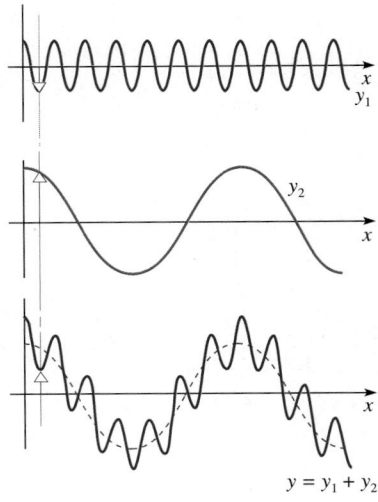

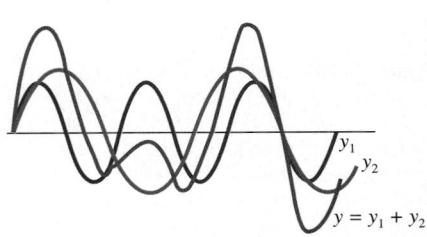

Figure 11.22 The superposition of waves y_1 and y_2. At each value of x, we add the magnitude of y_1 to that of y_2.

Figure 11.23 The superposition of two harmonic waves of different frequency. The resultant $y = y_1 + y_2$ is periodic, but it certainly is not harmonic.

quencies are added to one another, the composite disturbance is not harmonic (Figs. 11.22 and 11.23).

Fourier Analysis. A beautiful mathematical technique for synthesizing waveforms was devised in 1807 by the French physicist Jean Baptiste Joseph, Baron de Fourier. He proved that a periodic wave having a wavelength λ can be synthesized by a sum of harmonic waves whose wavelengths are λ, $\lambda/2$, $\lambda/4$, and so forth. In fact, any wave profile encountered in nature—pulse or periodic—can be envisioned as the result of overlapping sines and cosines. Any waveform can be decomposed into harmonic components, and this can be accomplished with sound, to some extent, by the human ear and with light by a prism.

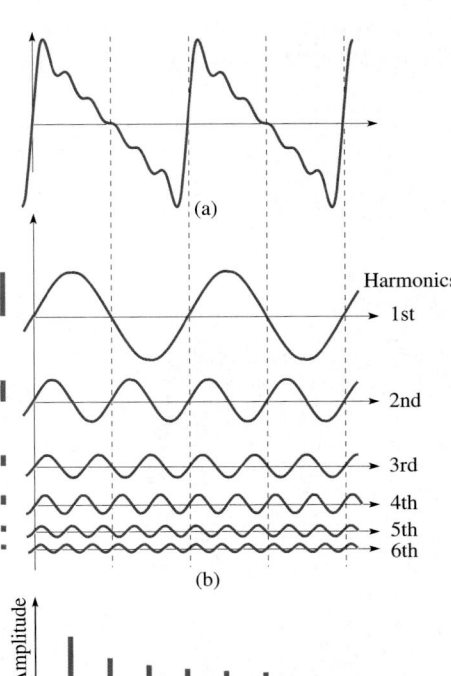

Figure 11.24 (a) The synthesis of a sawtooth wave. (b) Here, the first six harmonics combine to form a curve that clearly resembles a sawtooth. Another dozen terms will make it still better. (c) A plot of the amplitudes of the contributing harmonic terms plotted against frequency. This is called the *frequency spectrum*.

According to Fourier, a periodic function of frequency f can be created out of an appropriate sum of sinusoidal terms, the first of which will also have a frequency f. This is the **fundamental** or *first harmonic* ($f_1 = f$). The next term has a frequency of $f_2 = 2f$ and is called the *second harmonic*, and the next has a frequency of $f_3 = 3f$ and is the *third harmonic*, and so on. In the time the fundamental takes to complete one cycle ($1/f_1$), every higher harmonic will complete a whole number of cycles, and the entire set will be back where it started. Thus, the frequency of the combined waveform is the frequency of the fundamental, which was made equal to f, the frequency of the periodic wave being analyzed (Fig. 11.24). If the several pure tones in Fig. 11.25a are combined, they produce the acoustic wave in Fig. 11.25b. Electronically synthesized music and talking computer chips do exactly this sort of Fourier addition (Fig. 11.26).

Figure 11.25 (*a*) Six sinusoidal wave-forms. (*b*) The result of combining these waveforms. (*c*) The associated spectrum. Each vertical line represents the amplitude of the corresponding harmonic in (*a*). Note that $f_2 = 2f_1$, $f_3 = 3f_1$, $f_4 = 4f_1$, and so forth.

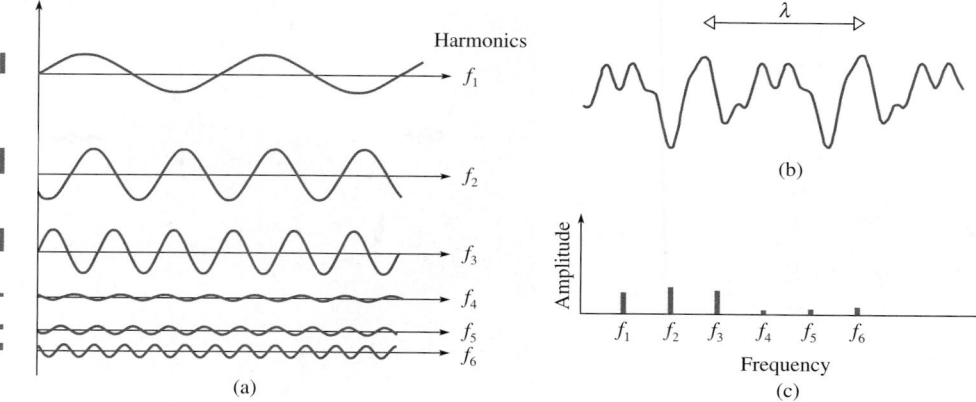

(a)

(b)

(c)

11.5 Wavefronts & Intensity

In two dimensions, the harmonic disturbance spreading out on the surface of a pool of water from a rhythmically tapping toe (Fig. 11.27) forms a pattern of expanding circular ripples. The lines connecting all the points where the water is rising or falling in-step form a series of concentric circles. At every point on any one of these circles, the ripple traveling out radially is at the same stage, or phase, of its essentially sinusoidal cycle, whether that's a crest or a trough or anything between. *The phase of the disturbance is constant at every point on a given circle, and this is a **circular wave**.* In general, ***each curve connecting all of the neighboring points of a wave that are in-phase is known as a* wavefront**.

A **wavefront** is a surface of constant phase.

Sound waves are three-dimensional. Imagine a *point source* radiating waves uniformly in all directions, as in Fig. 11.28 on p. 388. Here, **the surfaces of constant phase**, the wavefronts, are spherical. If we examine the compression wave from an exploding firecracker $\approx$10 m away, the portion of the wavefront intercepted by an ear or a microphone will be essentially spherical.

A sound wave represents a moving change in the distribution of atoms in a medium, and that change is associated with the flow of energy. The rate of transfer of energy is the *acoustic power*. A person speaking at a normal conversational level emits power at about 10^{-5} J/s and shouts at about 1 mJ/s. The energy of a mechanical wave is distributed over its wavefront. This raises a practical point: when a detector intersects a portion of a wave-

Figure 11.26 Each button on a Touchtone telephone produces a pair of nearly pure, widely separated tones. Push the 7-button, and you can hear the superposition of 852 Hz and 1209 Hz.

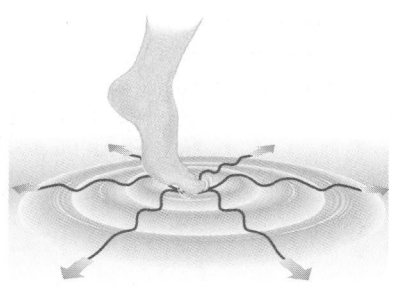

Figure 11.27 A toe wiggling in the water generates circular surface waves.

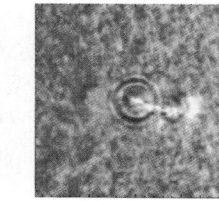
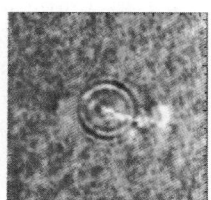
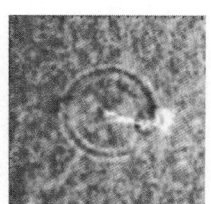

A solar flare on the Sun causes circular seismic ripples to flow across the surface.

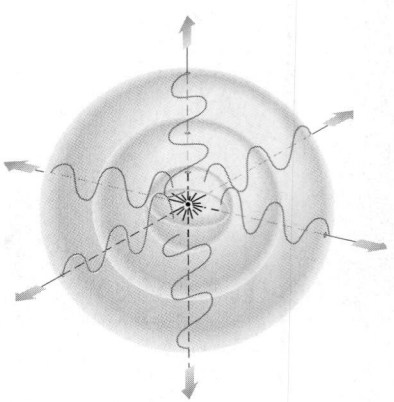

A steady drip of rainwater produced these circular waves.

Figure 11.28 A wave from a point source spreads out uniformly in all directions. Energy flows outward in the form of a spherical wave. As the area $(4\pi R^2)$ gets larger, the concentration of energy diminishes and the amplitude of the wave decreases as $1/R$.

The **intensity** of a wave is the average energy delivered per unit area per unit time.

The **Inverse Square Law**: For any uniformly radiating point source, the intensity of the emission drops off inversely with the square of the distance.

front, it records an amount of energy that depends on its own receiving area and on the time during which it is receptive. Both of these depend on the particular detector, but we all need to be able to carry out the same measurements regardless of the detectors. Thus it's not the total energy arriving on the particular instrument that we should measure but the energy per unit time, per unit area—the concentration of power. The **intensity** (I) *of a wave is defined as the average power divided by the perpendicular area across which it is transported*:

$$I = \frac{P_{av}}{A} \tag{11.6}$$

Intensity has the units of W/m^2. When a 2-m^2 area receives a total of 1 W of wave energy flowing perpendicularly onto it, the incident intensity is 0.5 W/m^2.

As a spherical wave expands outward, its area $(4\pi R^2)$ increases, and since the same power flows through ever-increasing areas, its concentration—the intensity—diminishes *inversely with the square of the radius*; this is the **Inverse Square Law**. It is the reason why you cannot read a book by starlight or hear what the players on the field are saying from up in the stands. Remember that the amplitude of a *spherical wave* must diminish with distance because the energy it carries spreads thinner and thinner as the wavefront gets larger.

By allowing a spherical wave to expand (Fig. 11.29) over a great distance, the wavefronts will get larger and flatter, ultimately resembling planes. The light entering a telescope from a star is usually indistinguishable from a perfectly flat wavefront.

Example 11.6 **[I]** An underwater explosion is detected 100 m away, where the intensity is recorded to be 1 GW/m^2. About 1 s later, the sound wave is recorded 1.5 km away from ground-zero. What will its intensity be at that distance? Ignore absorption losses.

Solution The problem talks about intensity at two different locations and that calls to mind the Inverse Square Law. (1) TRANSLATION—Knowing the intensity of a wave at a given

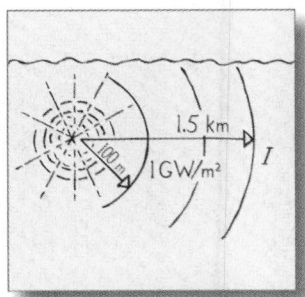

distance, determine its intensity at a farther distance. (2) Assuming that the compression wave is spherical, the distances provided represent radii. GIVEN: $R_1 = 100$ m, $I_1 = 1$ GW/m^2, $\Delta t = 1$ s, and $R_2 = 1.5$ km. FIND: I_2. (3)

PROBLEM TYPE—Wave motion/sound/the Inverse Square Law. (4) PROCEDURE—Whenever you have energy considerations involving spherical waves of different radii, think about the Inverse Square Law. The power flowing through the first sphere will equal the power through the second. (5) CALCULATION:

$$I_1(4\pi R_1^2) = I_2(4\pi R_2^2)$$

and

$$I_2 = \frac{I_1 R_1^2}{R_2^2} = \frac{(1 \times 10^9 \text{ W/m}^2)(100 \text{ m})^2}{(1.5 \times 10^3 \text{ m})^2}$$

$$I_2 = 4.4 \times 10^6 \text{ W/m}^2 = \boxed{4 \text{ MW/m}^2}$$

Quick Check: I goes as $1/R^2$ and 100 m is 15 times closer than 1.5 km; therefore, $I_1 = 15^2 I_2 \approx 225(4 \text{ MW/m}^2) \approx 0.9$ GW/m^2.

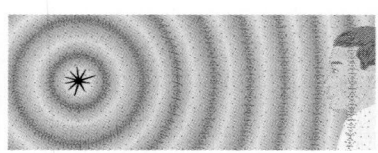

Figure 11.29 As a spherical wave expands, its radius increases and the wavefronts flatten. At very large distances, the wavefronts are quite flat and the disturbance resembles a plane wave.

11.6 The Speed of Sound in Air

By the seventeenth century, Galileo, Huygens, and others were already studying the finer points of acoustics, and the wave theory was widely accepted. In Paris, Father Mersenne (ca. 1636) used a crude echo technique to first measure the speed of sound. He found a value of about 1000 ft/s, which is not far off from the present value of 331.45 m/s (i.e., 1087.4 ft/s or 741 mi/h) in air at 0 °C and 1 atm (Table 11.2). *In a confined sample, the speed of sound increases with temperature*. The springiness of a gas (e.g., in an inflated ball) comes from the motion of its atoms and is measured via the pressure—the higher the temperature, the more violent the motion and the more the gas resists compression. At 20 °C and 1 atm sound travels at 343 m/s.

We have already formulated a theoretical expression for the speed of a compression wave in an elastic fluid [Eq. (11.4)], which certainly ought to tell us something about the speed of sound in air. Indeed, the equation was first derived by Newton, although he called the term represented by B the "elastic force." Sir Isaac argued that this elastic force equaled the pressure of the atmosphere and concluded that

$$v = \sqrt{\frac{P}{\rho}}$$

which gives sound a speed of only 979 ft/s. This equation works fine for liquids where the compressibility is small, but it's missing something when it comes to gases.

The discrepancy was explained in 1816 by Pierre Simon Laplace, who recognized that Newton's derivation required the temperature of the gas to be constant, and it wasn't. The regions that are significantly compressed increase in temperature, while those that are rarefied decrease in temperature (Fig. 14.13, p. 502). Because the waves are long and travel quickly, there is not enough time for heat to be conducted from one region to the other, and so the effect is not neutralized. Laplace showed that

[speed of sound in air] $$v = \sqrt{\frac{\gamma P}{\rho}} \qquad (11.7)$$

where γ is a constant that depends on the gas. For gases that are diatomic (i.e., two atoms form each molecule) such as hydrogen, oxygen, and (approximately) air, $\gamma = 1.4$. And this is in excellent agreement with observation (Table 11.2). {For a derivation click on **SPEED OF SOUND IN AIR** under **FURTHER DISCUSSIONS** on the **CD**.}

One thing to observe from Eq. (11.7) is that *the speed of sound does not depend on frequency*. This is confirmed every time you listen to music in a large hall or stadium and all the sounds stay together: the high frequencies reach your ears along with the low frequencies; otherwise, only the people in the front rows would hear anything recognizable.

The speed of sound in air does depend on temperature. If the pressure of some region of the atmosphere changed while the temperature remained fixed, the density would change in proportion [via Boyle's Law (p. 435)]. That change would leave P/ρ constant and v unaltered: the speed of sound in the thin air on a mountaintop is the same as it is at sea level, provided the temperature is the same. Changing the temperature will alter the density of

Table 11.2

The Speed of Sound

Material	Temperature (°C)	Speed (m/s)
Vulcanized rubber	0	54
Carbon disulfide	0	189
Air	0	331.45
Water vapor	0	401
Cork		500
Carbon tetrachloride	23	929
Helium	0	970
Chloroform	24	1001
Lead	20	1227
Hydrogen	0	1270
Mercury	≈25	1450
Fresh water	25	1493
Seawater	20	1513
Glycerin	22	1986
Platinum	20	2690
Brass		3500
Copper	20	3560
Brick		3652
Oak		3850
Aluminum		5104
Granite		6000

Example 11.7 [I] Use Eq. (11.7) to determine the speed of sound in air at STP.

Solution   This one's straightforward. (1) TRANSLATION— Compute the speed of sound in air at STP. (2) GIVEN: STP, from Table 9.1, $\rho = 1.29$ kg/m³, and $P = 1$ atm $= 1.013 \times 10^5$ Pa. FIND: The speed v. (3) PROBLEM TYPE—Wave motion/ sound/speed in air. (4) PROCEDURE—Use Eq. (11.7). (5)

CALCULATION:

$$v = \sqrt{\frac{1.4P}{\rho}} = \sqrt{\frac{1.4(1.013 \times 10^5 \text{ Pa})}{1.29 \text{ kg/m}^3}} = \boxed{332 \text{ m/s}}$$

Quick Check: This compares well with $v = 331.45$ m/s at STP.

Example 11.8 **[I]** What is the speed of sound at room temperature (20 °C) and normal atmospheric pressure?

Solution The speed of sound in air depends on the temperature. (1) TRANSLATION—Compute the speed of sound in air at 20 °C. (2) GIVEN: A temperature of 20 °C. FIND: *v*. (3) PROBLEM TYPE—Wave motion/sound/speed in air. (4) PROCEDURE—The speed increases from its 0 °C value of 331

m/s by 0.60 m/s for each degree increase. (5) CALCULATION—There is a 20° increase in temperature and so

$$v = 331 \text{ m/s} + 20(0.60 \text{ m/s}) = \boxed{343 \text{ m/s}}$$

Quick Check: The density of air at 20 °C can be approximated from Table 9.1 as 1.2 kg/m³; using Eq. (11.7), $v \approx 343.8$ m/s.

the air, but barometric pressure is determined by the weight of the column of air, which is independent of temperature. Since the density decreases as the temperature increases, the speed will increase with the square root of the temperature. *Over the usual range of temperatures encountered at sea level, the speed of sound changes by about ±0.60 m/s per change of ±1.0 °C.* {For more on why loud sounds travel slightly faster than quiet ones click on SOUND: AMPLITUDE & SPEED under FURTHER DISCUSSIONS on the CD.}

11.7 Hearing Sound

The human sound receiver, though not fully understood, is composed of three subsystems: the outer, middle, and inner ears (Fig. 11.30). Sound intercepted by the external ear is channeled via the auditory canal to the eardrum, which is driven into oscillation. This three-part structure is the *outer ear*. Besides aiding in determining the direction of sound, it provides a resonant cavity (p. 397) that amplifies (by about 100%) sounds in the range from ≈3 kHz to ≈4 kHz, making the entire system especially sensitive to this mid-frequency band. The ear is 1000 times more responsive at 1 kHz than at 100 Hz. Were the ear more sensitive to the low-frequency range, we would have to contend with a constant drone of internal body noises.

The *middle ear* links the eardrum, via three pivoted bones, to the oval window. This arrangement increases the force exerted on the flexible window into the inner ear. In addition, since the effective area of the eardrum is about 20 times larger than that of the oval window, there is another increase in pressure.

The *inner ear* is the transducer that converts a mechanical pressure input into an electrical nerve output. It begins at the oval window, which is the entrance to a fluid-filled coiled cavity known as the *cochlea*. The cochlea contains the *basilar membrane* on which is a structure of over 20 000 strands of hairlike receptor cells leading to nerve fibers. Hydraulic compression waves generated at the oval window induce vibrations in the basilar membrane, which flutters transversely along its length. At the front end of the cavity where the membrane is narrow and stiff, it vibrates at high frequencies; toward the rear where the membrane is slack, its resonant frequency is lower. As the membrane oscillates, the receptor cells in the different regions flutter and trigger impulses at a rate that depends on both the amplitude and frequency of the motion. The extent of the displacement of the basilar membrane determines the firing rate of the receptor cells, which, in turn, manifests itself in the perception of loudness.

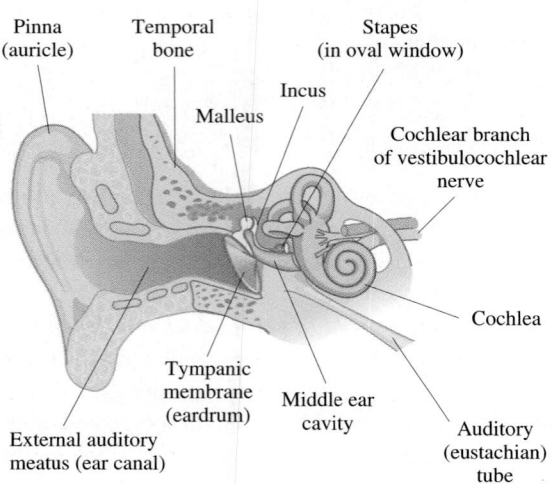

Pinna (auricle)
Temporal bone
Stapes (in oval window)
Incus
Malleus
Cochlear branch of vestibulocochlear nerve
Cochlea
Tympanic membrane (eardrum)
Middle ear cavity
Auditory (eustachian) tube
External auditory meatus (ear canal)

Figure 11.30 The human ear.

Pitch

Pitch is a human perception resulting from the sensing of acoustic energy. It is an ear-brain response mainly to the frequency of sound. Certainly we know what we mean by the term: a soprano voice is high-pitched, a bass voice is low-pitched. The pitch of a pure tone (a sinusoidal wave with a single fixed wavelength, as approximated by a tuning fork) more or less corresponds to frequency—the higher the frequency, the higher the pitch (Fig. 11.31). Nonetheless, the ear-brain system is far from simple, and there is no one-to-one relation-

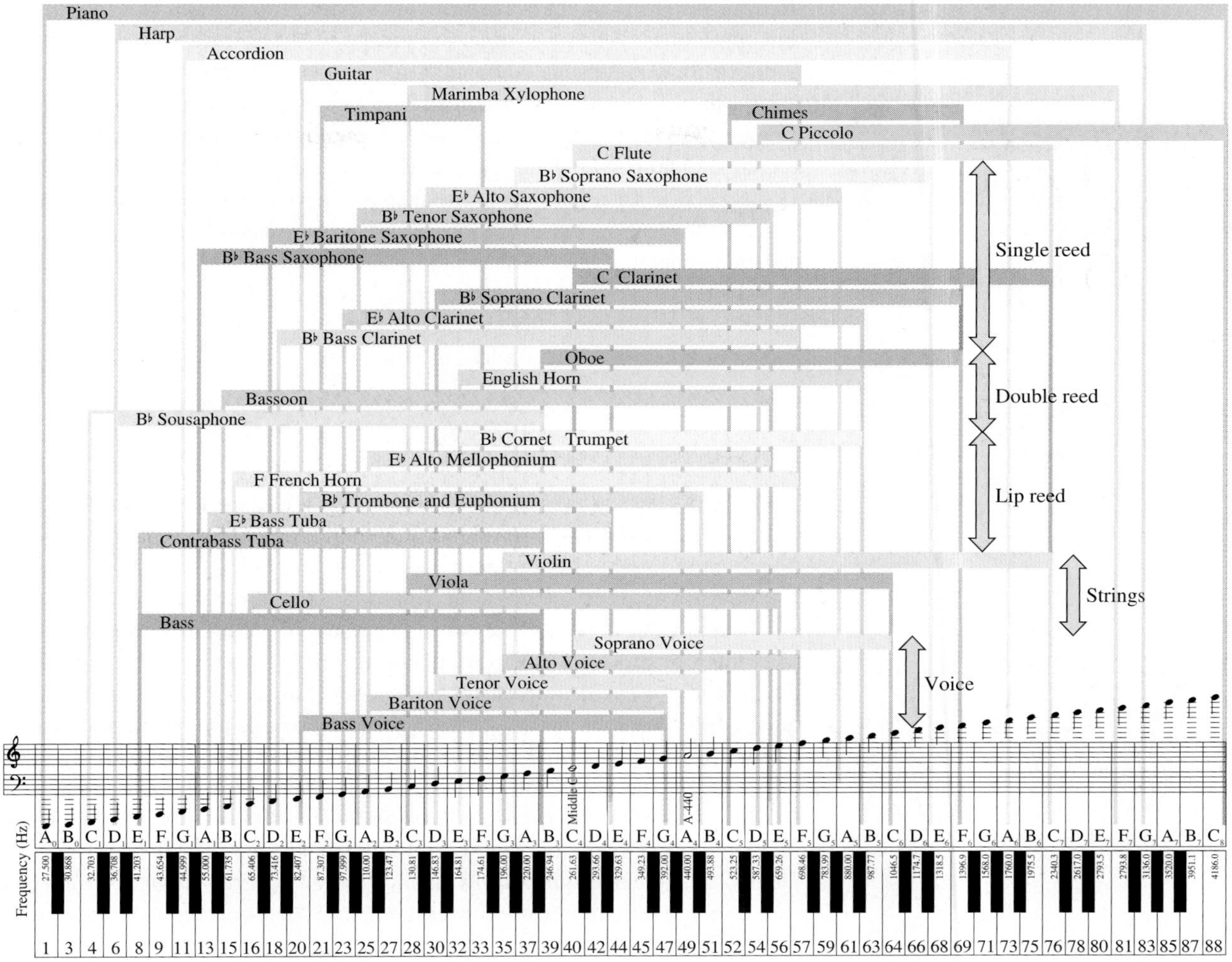

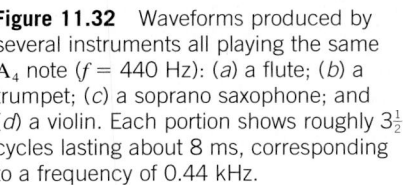

Figure 11.31   The frequency ranges of voices and instruments.

ship between frequency and pitch. For example, a high-frequency tone will be heard to increase in pitch as its intensity increases, while a low-frequency tone will cause the opposite response.

Timbre

Another auditory sensation that allows us to distinguish between sounds is *timbre*. Listening to a flute, a trumpet, a saxophone, or a tuning fork, each producing the same note (same pitch) at the same loudness, there would be little difficulty telling one from the other (Fig. 11.32). The catchall attribute that allows for this distinction is the timbre, which

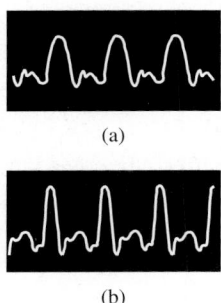

(a)

(b)

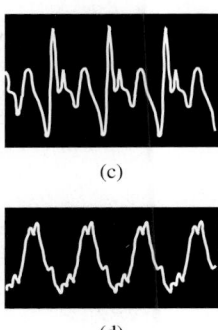

(c)

(d)

Figure 11.32   Waveforms produced by several instruments all playing the same A_4 note ($f = 440$ Hz): (*a*) a flute; (*b*) a trumpet; (*c*) a soprano saxophone; and (*d*) a violin. Each portion shows roughly $3\frac{1}{2}$ cycles lasting about 8 ms, corresponding to a frequency of 0.44 kHz.

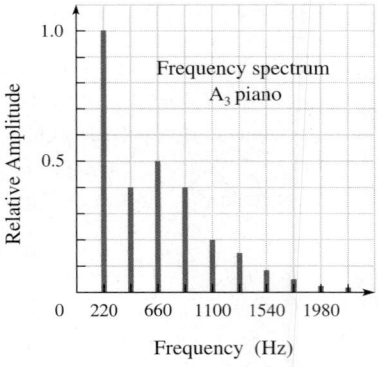

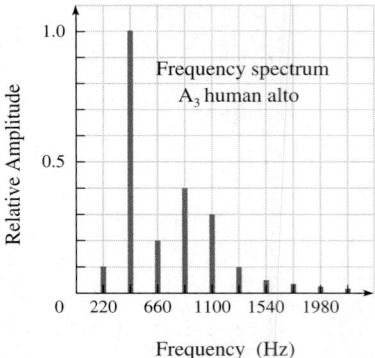

Figure 11.33 The frequency spectra for the A₃ (220 Hz) note sounded by a piano and an alto voice.

depends primarily on the waveform—that is, the frequencies present, their relative phases, and amplitudes. When the same tones made by these various instruments are picked up by a microphone and displayed on an oscilloscope, they are seen to be substantially different. Similarly, the same notes produced by a piano and a singer have distinctly different frequency spectra (Fig. 11.33), and the ear immediately distinguishes the vocalist from the accompaniment, even when the sounds are made at the same moment.

An actual tone will contain frequencies higher than the fundamental (or first harmonic), and these are called **overtones**. An overtone need not be a whole-number multiple of the fundamental; that is, it need not be a harmonic. It is the number of overtones (whether they are harmonics or not) and their relative amplitudes that more than anything else determines timbre. Probably the most important difference between a fine musical instrument and an ordinary one is its timbre, or tone color.

Loudness

The auditory sensation of loudness depends on the frequency spectrum, duration, and, most importantly, intensity of the sound. In general, the intensity of an elastic wave is proportional to the square of both the frequency and amplitude ($I \propto f^2 A^2$). The intensity of sound is proportional to the square of the amplitude (P_0) of the pressure wave ($I \propto P_0^2$).

The breadth of sound intensities that occur in nature, from the faint buzz of a mosquito to the roar of a volcano, is tremendous, and the human ear is remarkably sensitive to much of it. We can bear intensities in excess of around 1 W/m^2, which is so loud it can be quite uncomfortable. In fact, at this level, one begins to feel the sound as opposed to just hearing it. That *threshold of pain* corresponds to a gauge pressure change of only about 10^{-4} atm, or roughly 10 Pa. At the other extreme, the ear can just detect sounds at intensities as low as 10^{-12} W/m^2. That *threshold of hearing* is associated with a pressure change of about 10^{-10} atm, or 10^{-5} Pa. Lord Rayleigh showed that, at 1 kHz, this change is equivalent to air molecules vibrating with an amplitude of only about 10^{-6} mm (see Problem 80).

The ear functions over a pressure range that corresponds to a factor of a million (10^{-4} atm to 10^{-10} atm), and the intensity range—the square of that—extends a factor of 10^{12}. This is an amazing breadth of operation for any sensor—it's the equivalent of using the same device to measure both the diameter of an atom and the length of a football field. Figure 11.34 depicts the limits of sensitivity as a function of frequency.

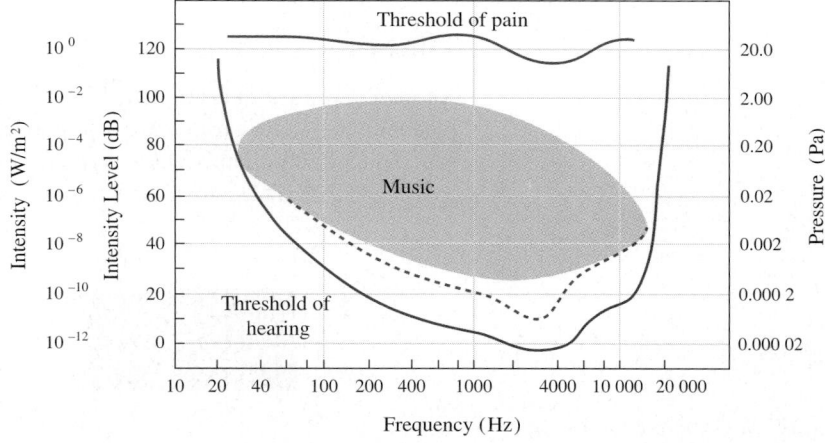

Figure 11.34 About 1% of the people in the United States can hear sound below the levels of the lower curve. Perhaps 50% can hear below the dashed curve. The shaded region corresponds to the levels of normal music.

We judge the relative loudness of a sound not by the difference in intensity between it and some reference but by their ratio. Doubling the intensity of a faint sound will produce a sensation of some increase in loudness. Doubling the intensity of an already loud sound will nevertheless produce the sensation of the same increase in loudness. In fact, **multiplying any intensity by 10 will generally be perceived as approximately doubling it in loudness**; 10^{-8} W/m^2 sounds twice as loud as 10^{-9} W/m^2, and 0.1 W/m^2 is heard to be twice as loud as 0.01 W/m^2. As we will see presently, that's equivalent to saying that the ear responds logarithmically.

> To double the **loudness** of a source, its intensity must be increased by a factor of ten.

11.8 Sound-Level

The **sound-level** (or *intensity-level*) of an acoustic wave is defined as *the number of factors of 10 that its intensity is above the threshold of hearing*: $I_0 = 1.0 \times 10^{-12}$ W/m^2. To honor Alexander Graham Bell, who did important research in acoustics, the unit of sound intensity-level is called the *bel*. The performance of audio amplifiers, tuners, loudspeakers, and so forth, is given in terms of the smaller unit, 1/10 of a bel, which is a **decibel** (abbreviated dB). The decibel, like the radian, is not a physical unit expressible in terms of meters, kilograms, or seconds. It is simply a reminder of the meaning of the number preceding it: 10 dB means something very different from just 10 or from 10 rad.

By definition, the logarithm to the base 10 of any number X equals the power to which 10 must be raised to equal X. In other words, if $X = 10^Y$, then the $\log_{10} X = Y$. Therefore, $1000 = 10^3$ and so $\log_{10} 1000 = 3$. If we form the ratio of the intensity of a wave I to the reference level I_0 (remembering how the ear responds to such ratios) and then take its logarithm, we get the number of factors of 10 that I is above I_0: this number is the intensity-level in bels, and 10 times that gives it in decibels. The **intensity-level** β in dB of any sound is

To take the **antilog** of β, raise 10 to the power of each side of the equation using the fact that

$$10^{\log_{10} x} = x$$

Accordingly, dividing both sides of Eq. (11.8) by 10,

$$\beta/10 = \log_{10} I/I_0$$

Then exponentiate both sides,

$$10^{\beta/10} = I/I_0$$

$$\beta = 10 \log_{10} \frac{I}{I_0} \tag{11.8}$$

For example, a sound wave with an intensity of 10^{-6} W/m^2 (about the level of normal conversation with a person 1 m away) has an intensity-level of

$$\beta = 10 \log_{10} \frac{10^{-6}\,\text{W/m}^2}{10^{-12}\,\text{W/m}^2} = 10 \log_{10} 10^6 = 10(6) = 60 \text{ dB}$$

Table 11.3

Intensity Ratios and Intensity-Level

Ratio of intensities	Intensity-level (dB)
100 000 000 000:1	110
1 000 000:1	60
10 000:1	40
100:1	20
10:1	10
4:1	6
2:1	3
1:1	0
1/2:1	−3
1/4:1	−6
1/10:1	−10
1/100:1	−20
1/1000:1	−30

The sound has an intensity of 1 million times the threshold intensity and therefore has a sound-level of 6 bel, or 60 dB—it's 6 factors of 10 or 10^6 above threshold (see Table 11.3).

Since $\log_{10} 1 = 0$, the intensity-level at the threshold of hearing is $\beta = 10 \log_{10} 1 = 0$ dB. The entire intensity range from 1.0 W/m^2 to 10^{-12} W/m^2 corresponds to 12 bels (12 multiplicative factors of 10) or 120 dB. A range of 1 to 1 million million is thus converted into a range of 0 dB to 120 dB (Table 11.4). *A young healthy ear can distinguish between sound-levels that differ by as little as 1 dB*; 2 dB or greater is more often the case, however.

Two logarithm identities will be useful in performing sound-level calculations {see **MATH REVIEW: PART A-3** on the CD 💿 }

Table 11.4

Approximate Intensity-Levels in dB

Source	Intensity-level (dB)	
Large rocket	≈ 180	
Jet engine	140	
Jet takeoff (30 m − 60 m)	≈ 125	intolerable
Rock concert (1.0 W/m²)	≈ 120	painful
Yell into the ear (20 cm)	120	(immediate danger)
Pneumatic hammer	110	
Passing subway train (10⁻² W/m²)	100	
Car without muffler	100	(damage after 2 hours)
Shout (1.5 m)	100	
Car horn, loud	95	very loud
Heavy truck (15 m)	90	
City street	80	(damage after 8 hours)
Hair dryer	80	
Loud music	80	
Freeway traffic	75	
Automobile interior	≈ 70	loud
Toilet flushing	≈ 67	
Noisy store (10⁻⁶ W/m²)	60	
Conversation, average (1 m)	60	
Office	50	moderate
Living room, city	40	
Bedroom (10⁻⁹ W/m²)	30	
Library	30	quiet
Broadcast studio (10⁻¹⁰ W/m²)	20	
Whisper	20	very quiet
Cat purring	15	
Rustling leaves	≈ 10	barely audible
Threshold (10⁻¹² W/m²)	0	

THE HUMAN VOICE & TALKING ELEVATORS

An adult male generates basic voicing vibrations at ≈ 80 Hz to ≈ 240 Hz for speech and up to ≈ 700 Hz for song. By comparison, the adult female voice range is ≈ 140 Hz to ≈ 500 Hz in speech and up to ≈ 1100 Hz in song. Given enough control over a wide spectrum of pure tones (≈ 80 Hz to ≈ 1100 Hz), we presumably could artificially duplicate all human vocal sounds—a process just beginning with "talking" alarm clocks, elevators, computers, and cameras.

$$\log_{10} \frac{A}{B} = \log_{10} A - \log_{10} B \tag{11.9}$$

and

$$\log_{10} AB = \log_{10} A + \log_{10} B \tag{11.10}$$

For example, suppose we increase the intensity of a sound by some multiplicative factor; by how much would the intensity-level change? In general, for a change from I_1 to I_2

$$\Delta\beta = \beta_2 - \beta_1 = 10 \log_{10}\left(\frac{I_2}{I_0}\right) - 10 \log_{10}\left(\frac{I_1}{I_0}\right)$$

and so from Eq. (11.9)

[in dB]
$$\Delta\beta = 10 \log_{10}\left(\frac{I_2}{I_1}\right) \tag{11.11}$$

An increase of 10 dB in **sound-level** corresponds to a sound that's twice as loud.

► **Increasing the intensity by a factor of 10 changes the sound-level by 1 bel or 10 dB;** increasing it by 100 changes β by 2 bels, or 20 dB, and so on. A 60-dB sound compared to a 20-dB sound is 40 dB (4 bels) greater, which means—from Eq. (11.11)—that $\log_{10}(I_2/I_1) = 4$ and the intensity of one is 10^4 times the other. Since every increase of 10 dB corresponds to a doubling in loudness, an 80-dB sound is louder than a 60-dB sound by (two 10-dB steps or) a factor of 4.

Example 11.9 **[II]** Two public address systems are being compared, and one is perceived to be 32 times louder than the other. What will be the difference in sound-levels between the two when measured by a dB-meter?

Solution Loudness and sound-level are related. (1) TRANSLATION—One source is 32 times louder than another; determine the difference in their sound-levels. (2) GIVEN: A factor of 32 in *loudness*. FIND: $\Delta\beta$. (3) PROBLEM TYPE—Wave motion/sound/sound-level. (4) PROCEDURE— We are given

the change in loudness, which correlates to a change in sound-level by way of the fact that a doubling in loudness is roughly a change of 10 dB. (5) CALCULATION—Since $32 = 2^5$, $\Delta\beta = 5(10\ \text{dB}) = \boxed{50\ \text{dB}}$.

Quick Check: $\Delta\beta = 10 \log_{10}(I_2/I_1) = 50$; hence, $\log_{10}(I_2/I_1) = 5$ and $(I_2/I_1) = 10^5$, the intensities differ by 5 factors of 10, each one doubling the loudness, which increases 2^5, or 32-fold.

Example 11.10 **[III]** Imagine that 10 identical violins are each about to play something at 70 dB. If they join in one at a time, what will be the sound-level as each contributes?

Solution This problem is about intensity and sound-level. (1) TRANSLATION—Determine the increasing sound-level of ten 70-dB sources added one at a time. (2) GIVEN: 10 sources at 70 dB each. FIND: The corresponding sound-levels. (3) PROBLEM TYPE—Sound/sound-level. (4) PROCEDURE—The intensity goes from I to $2I$ to $3I$ up to $10I$. We could solve the problem using Eq. (11.11), but let's go back to Eq. (11.8) to compute β and use Eq. (11.10). (5) CALCULATION—For two violins

$$10 \log_{10} \frac{2I}{I_0} = 10 \log_{10} 2 + 10 \log_{10} \frac{I}{I_0}$$

and since $\log_{10} 2 = 0.30$, the sound-level becomes $\boxed{3\ \text{dB} + 70\ \text{dB}}$. *Doubling the intensity produces a 3-dB sound-level increase.* With three instruments playing, the sound-level rises to $\boxed{75\ \text{dB}}$; with four, $\boxed{76\ \text{dB}}$; with five, $\boxed{77\ \text{dB}}$, and so on up to 10, when it reaches $\boxed{80\ \text{dB}}$. Ten violins produce an intensity 10 times that of one and therefore 1 bel (10 dB) greater, and that essentially corresponds to a doubling of the perceived loudness.

Quick Check: $\Delta\beta = 10 \log_{10}(I_2/I_1) = 10 \log_{10} 2 = 3.0$ dB and $\beta + \Delta\beta = 70$ dB $+$ 3 dB.

It follows that if we have a 12-W audio system and wish to replace it by one that is equally efficient but capable of being twice as loud, we will need to increase its power, not twofold but by a factor of 10, raising it to 120 W.

11.9 Sound Waves: Beats

Anything that will cause compression waves in the audible range in a material medium is a source of sound. Smack your hands together or tap a hammer and a sound will be made—it will not be sustained very long, but it certainly will be an acoustical wave (Fig. 11.20). Physically, **noise** refers to an unrelated jumble of disturbances (Fig. 11.35), a *nonperiodic*, randomly changing wave with an essentially continuous frequency spectrum. When the spectrum has a *broad bandwidth encompassing the entire audible range with equal intensity everywhere*, it's called *white noise*, and we hear it as a mix of wind, pouring water, and continuous radio static.

Suppose that the tapping of a hammer occurs at a slow, regular pace. A series of separate impulses will be heard, increasing in rate as the tapping increases, until at roughly around 20 impacts per second, the ear-brain will perceive the clatter as a continuous sustained hum. Galileo noticed that he could produce a tone by drawing a knife blade over a serrated edge, generating a rapid succession of taps. The periodic nature of the process was apparent to him, and he inferred that the pitch was proportional to the rapidity of the tapping.

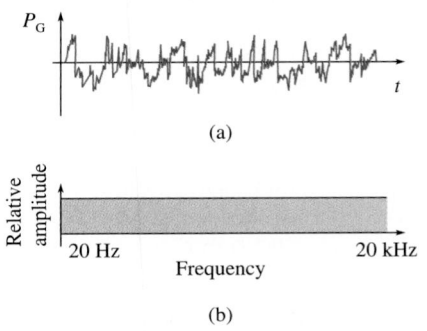

Figure 11.35 (a) Random aperiodic noise is called *white noise* when (b) its frequency spectrum has a constant amplitude all across the audible frequency range.

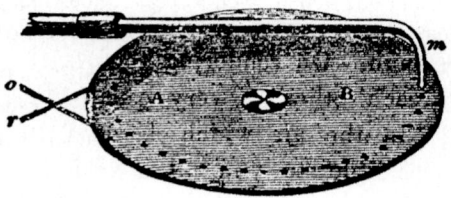

Figure 11.36 Professor Robison's siren. Puffs of air blowing through the holes in a rotating disk created a sound "equal in sweetness to a clear female voice."

Beats

With a standard source like a siren, we can determine the frequencies of other sound-generating devices, such as tuning forks and vibrating wires. The most common technique for the purpose (still used by traditional piano tuners) relies on the phenomenon of **beats**. Mersenne was the first to report the effect, and it was he who gave it that name. Describing the result of playing two organ pipes that differ slightly in pitch, he wrote that they "produce a throbbing or thrumming sound similar to that of beating a tambourine." The effect is a manifestation of interference: two or more waves overlap and produce a resultant wave.

As was seen earlier (Fig. 11.21), two waves of the same frequency superimpose to form a stable resultant wave. When they are in-phase, or nearly so, crests fall on crests, troughs on troughs, and the resultant is large. When they are 180° out-of-phase, or nearly so, crests fall on troughs, and the resultant is small or even nonexistent. Figure 11.37 depicts the overlapping and interference of two pure tones having slightly different frequencies, traveling in the same direction. The two waves, because they are not quite able to stay in-step, gradually (and periodically) go in- and out-of-phase; at one instant, rising and falling together, they reinforce one another to form a large amplitude sound. Then, one rising while the other is falling, they subtract to cancel each other altogether. The two waves of frequency f_1 and f_2 combine to produce a carrier wave having the average frequency $f_{av} = (f_1 + f_2)/2$. Typically f_1 and f_2 are relatively high, as is f_{av}. This high-frequency carrier is *amplitude modulated*, slowly rising and falling with the relatively low **beat frequency** of $(f_1 - f_2)$. What you hear is a high-pitch tone—the carrier—building and fading in intensity at the beat frequency. Keep in mind that you can only hear beats if both f_1 and f_2 are in the range of audibility. (A mathematical analysis of the effect is explored in Problem 135.)

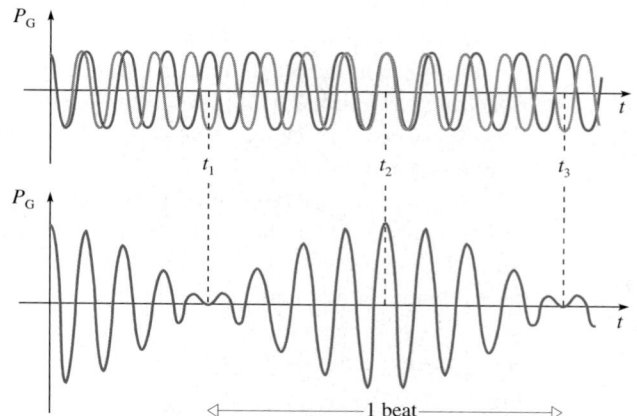

Figure 11.37 When two signals of different but nearly equal frequencies (f_1 and f_2) superimpose, they create a beat pattern, which is true of sound as well as light. The intensity rises and falls with a frequency equal to the difference ($f_1 - f_2$). Note that at t_1 and t_3 the signals are 180° out-of-phase, while at t_2 they are in-phase.

1 beat

Example 11.11 **[I]** A siren is used as a frequency reference during the fine adjustment of a middle-A tuning fork. The siren is slowly reduced in frequency and, together with the tuning fork, produces a quavering tone. This warbling also decreases in frequency until the siren reaches and holds at 440 Hz. At that point, the tone varies in loudness from maximum to minimum and back to maximum in $\frac{1}{4}$ s. What is the frequency of the fork?

Solution The word "warbling" suggests that we're dealing with beats. (1) TRANSLATION—A sound of known frequency combines with an unknown tone to produce beats having a given period; determine the unknown frequency. (2) The quavering tone tells us that this is a problem in beats. GIVEN: $f_1 =$ 440 Hz and the period of the beats is $\frac{1}{4}$ s. FIND: The frequency of the fork, f_2. (3) PROBLEM TYPE—Wave motion/sound beats. (4) PROCEDURE—The beat frequency is 1 over the beat period. (5) CALCULATION:

$$\frac{1}{\frac{1}{4}\,\text{s}} = (f_1 - f_2) = 440\,\text{Hz} - f_2$$

and

$$\boxed{f_2 = 436\,\text{Hz}}$$

Quick Check: $1/0.25\,\text{s} = 4\,\text{Hz}$ and since the siren was coming down, its frequency must exceed that of the tuning fork by 4 Hz.

A variety of vibrating objects can serve as acoustical sources that produce sustained sounds with discernible pitches. An unburdened honeybee fluttering its wings at about 440 Hz generates a buzz at what has come to be the musician's standard A_4 (middle-A) frequency. Strings and reeds, membranes, rods, bars, lips, air columns, and tongues can all be set vibrating, all making sounds, even those we call music, whatever that is.

11.10 Standing Waves

When a wavetrain of any kind is created in some real, finite medium (whether a string, a drum, or the Earth itself), it will propagate outward until it encounters an end or boundary. There, some fraction of the wave-energy will usually be reflected backward (p. 377), and if the original disturbance is sustained, the medium will quickly fill with waves traveling back and forth over one another. These disturbances will combine or interfere to form a steady-state distribution of energy known somewhat paradoxically as a **standing** or **stationary wave**. This situation is remarkably commonplace. It happens within every sort of musical instrument from the piano to the triangle (and, for that matter, within every kind of laser as well). It occurs in the head, throat, and mouth when we speak; in the ear canal when we listen; and in the shower stall when we sing. Standing waves exist when you ring a bell, strike a tuning fork, run water from the tap into an open sink, or blow across an empty soda bottle.

The phenomenon was first appreciated in 1821 by the brothers Ernst and Wilhelm Weber in Leipzig. They recognized that the effect (which they observed while pouring mercury) arose when "two wavetrains of the same wavelength and intensity but [traveling] in opposite sense encounter each other." Figure 11.38 shows how the process occurs for two harmonic waves passing over one another. What results is a fixed pattern of **nodes** where the resultant is zero and, midway between them, **antinodes** where the resultant is a maximum. Every point on each separate portion of the wave between successive nodes oscillates in-phase, with its own amplitude. There is no transporting of energy, and, in that sense, one might not even consider it a wave at all. Still, there are good mathematical reasons to call the sum of two waves a wave even if it isn't going anywhere; and so, along with the brothers Weber, we distinguish between progressive (traveling) waves and stationary (standing) waves. Here, *the wavelength is twice the node-to-node distance*.

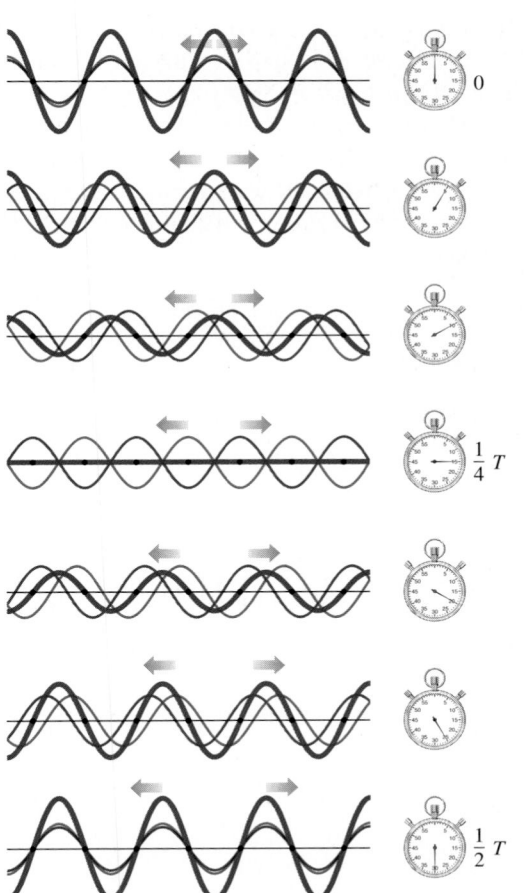

Figure 11.38 The creation of standing waves. Two waves of the same amplitude and wavelength traveling in opposite directions form a stationary disturbance that oscillates in place. The green curve has been made slightly smaller than the red one in the first and last parts, so that they can be better distinguished.

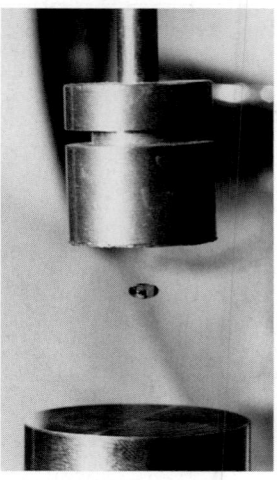

Ultrasonic levitation. A drop of water floating in a so-called energy well. Ultrasonic waves, one traveling up, the other down, create a standing-wave pattern. The suspended object is confined to a nodal region.

Standing Waves on Strings

One of the most common arrangements is the taut string with both ends held immovable, which is the basic setup for the piano and all stringed instruments (Fig. 11.39). Energy can be continuously pumped into such a system by attaching one end to a source, like a driven tuning fork or vibrator, which has such a small amplitude that it might well be considered a fixed point. Envision a harmonic wave with a tiny amplitude continuously introduced at the left, moving to the right, down the length of the string L at a speed v, as in Fig. 11.40. Presume that it will be reflected back to the left without any loss in amplitude but with a 180° phase shift (p. 377). This reflected wave will soon arrive back at the origin (i.e., at the vibrator), there to again be phase-shifted another 180° as it is reflected (from that effectively fixed end) off to the right. What conditions have to be met if this twice-reflected wave is to move off exactly in-step with the continuously flowing original disturbance? We want both waves traveling to the right to be in-phase, meaning that the round-trip time $2L/v$ must equal the period T since the two 180° phase shifts due to reflection add up to 360° and so essentially cancel. Inasmuch as $v = f\lambda$, $2L/v = T$ yields

$$L = \tfrac{1}{2}\lambda$$

and the round trip is one wavelength long. This being the case, each trip down and back will bring the reflected wave to the fork just in time to combine in step with the newly emitted wave. Each such trip will cause the wave in the system to grow slightly larger.

As can be seen in Fig. 11.40, if the length of the string is half a wavelength, its two ends will remain motionless as the center oscillates as an antinode. This synchronization of the waves, which depends on matching the vibrator's frequency with the size of the medium, is a resonance effect. Initially, the disturbance on the string will continue to build as energy is pumped in until the increasing frictional losses stabilize the system. At that point,

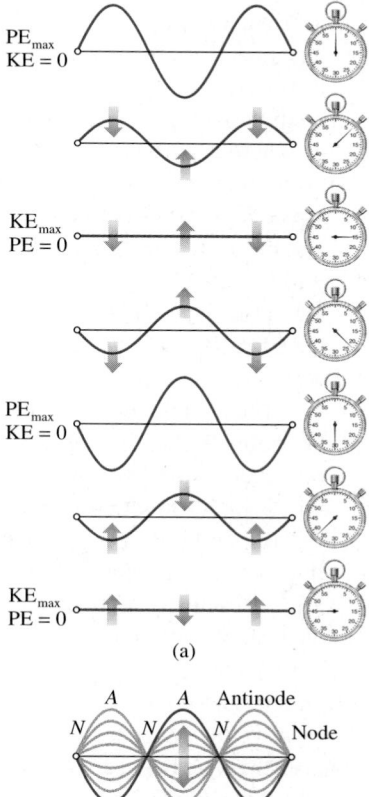

PE_{max}
$KE = 0$

KE_{max}
$PE = 0$

PE_{max}
$KE = 0$

KE_{max}
$PE = 0$

(a)

Figure 11.39 (a) A standing wave on a string that has both ends fixed. (b) represents a composite of all the configurations demonstrated in (a).

A A Antinode
N N N Node

(b)

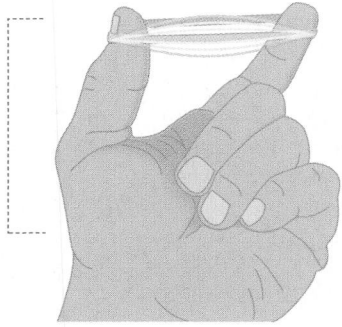

the energy provided by the vibrator at each oscillation equals the sum of the losses, and there is no more growth of the wave, whose amplitude is nonetheless much larger than that of the source. *The ability to amplify the input is an extremely important feature of standing-wave systems.*

Similarly, if the ends are fixed and the string is displaced, it will naturally oscillate in this configuration, or *mode*. Yet, unless every point on the string is carefully displaced along a smooth sine curve, there will be higher-order standing-wave modes present as well. These occur when the round-trip time corresponds to $2T$, $3T$, $4T$, and so forth, or NT where N is a whole number (Fig. 11.41). Then

[standing waves on a string]

$$L = \tfrac{1}{2}N\lambda \qquad (N = 1, 2, 3, ...) \qquad (11.12)$$

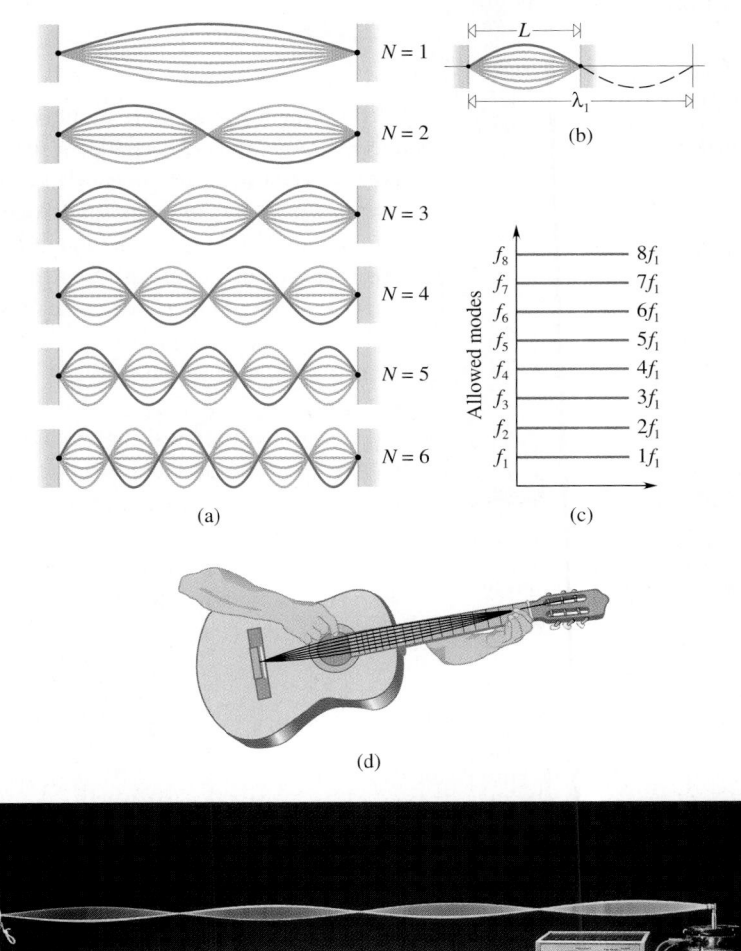

Figure 11.41 (*a*) Standing-wave modes with a node at both ends. (*b*) The wavelength of the $N = 1$ fundamental equals $2L$. (*c*) Allowed modes of oscillation. (*d*) The fundamental of a string on a guitar. Here the amplitude of the oscillation is greatly exaggerated and the other strings aren't shown.

Figure 11.40 A standing wave. A wave enters from the left (*a*) and is reflected (dashed) from the termination point on the right (*d*). Now there are two overlapping waves that combine to form a standing wave (red), oscillating (*g* through *k*) in place.

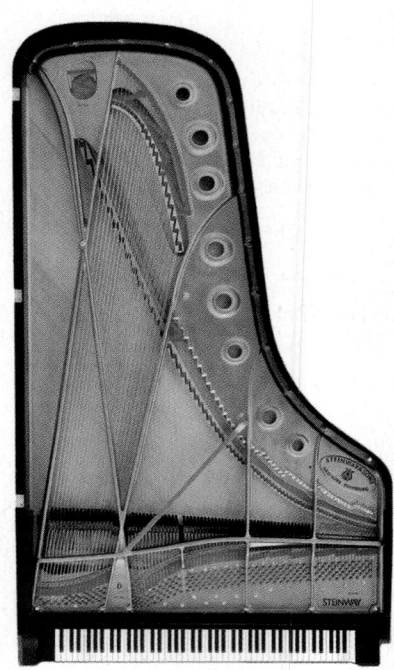

The low pitch strings on a piano are long (L is large), thick and heavy (m/L is large), and under relatively little tension (F_T is small).

TO VIBRATE & QUIVER

Galileo understood that an oscillating string causes "the air immediately surrounding the string to vibrate and quiver," thus producing a sound of the same frequency. A plucked or struck string actually vibrates in several modes at once and generates several tones (a fact observed by Mersenne). Strike a key on a piano, and the string will generate a strong fundamental along with a number of much quieter overtones (Fig. 11.33), which is what gives the instrument its characteristic tone color. Strings are usually bowed or plucked near their ends so as to produce a wide range of harmonics and thereby a rich sound.

and the wavelength of the Nth mode is $\lambda_N = 2L/N$. Accordingly, the pattern shown in Fig. 11.39 comes into being when $N = 3$.

We can find the resonant frequencies f_N of the system using $v = f\lambda$, along with Eq. (11.14), which becomes $L = \frac{1}{2}Nv/f_N$, and $f_N = Nv/2L$. Using Eq. (11.4) for the speed of a transverse wave on a string,

[standing waves on a string]
$$f_N = \frac{N}{2L}\sqrt{\frac{F_T}{m/L}} \qquad (N = 1, 2, 3, 4, ...) \qquad (11.13)$$

Each frequency is a whole-number multiple of the **fundamental**, or lowest resonant frequency: $f_N = Nf_1$; and each higher resonant frequency is a harmonic. The Nth harmonic will have N antinodes, so it is easy to recognize when you see it. Usually, each string of an instrument is played in its fundamental mode. In the case of a guitar, for example, the strings sound different even though they all have the same length because the tensions and linear mass-densities are different. By "fingering" a string (that is, by pressing it down onto a fret), thereby fixing a node at that point and effectively shortening the length of string that is able to vibrate, the player increases the fundamental frequency (Fig. 11.41d). Similarly, by increasing the tension on the vocal cords we can increase the frequency and change to a *falsetto* voice.

Because vibrating strings are not good at pushing large amounts of air, they are not, by themselves, very loud. In instruments, strings are therefore coupled to a large section of the body in such a way as to transmit vibrations to it—the sounding board in pianos or the sounding box in violins and guitars. An alternative is to couple the strings to a pick-up and amplify the sound electronically. {For an animated look at standing waves click on **NORMAL MODES OF A STRING** under **INTERACTIVE EXPLORATIONS** on the **CD**. This simulation will give you a fine sense of the interplay of the several parameters.}

STUDY GUIDE

When a string (guitar, piano, etc.) vibrates in some standing-wave pattern it creates a sound wave with the same frequency, but not the same wavelength.

Example 11.12 **[I]** What must be the tension in the E_5-string of a violin if it is to be tuned to 660 Hz? The length of the string from bridge to peg is 330 mm, and its mass-per-unit-length is 0.38 g/m.

Solution The frequency depends on the tension and density. (1) TRANSLATION—A string of known length and mass-per-unit-length is to oscillate, with its ends fixed, at a specified frequency; determine the required tension. (2) GIVEN: $L = 0.330$ m, $m/L = 0.38$ g/m, and $f_1 = 660$ Hz. FIND: F_T. (3) PROBLEM TYPE—Wave motion/standing waves/on a string. (4)

PROCEDURE—The violin is tuned to a fundamental of 660 Hz; hence, $N = 1$ in Eq. (11.13). (5) CALCULATION—Square both sides of Eq. (11.13) and solve for the tension:

$$F_T = (m/L)(2Lf_1)^2 = (0.38 \times 10^{-3} \text{ kg/m})4(0.330 \text{ m})^2(660 \text{ Hz})^2$$

and $\boxed{F_T = 72 \text{ N}}$ or about 16 lb.

Quick Check: $f_1 = \dfrac{1}{2L}\sqrt{\dfrac{72 \text{ N}}{0.38 \times 10^{-3} \text{ kg/m}}} = 660$ Hz.

A pail used to wash a floor contained a suspension of fine dirt particles in water. When placed in a curved sink, the pail gently rocked along a fixed axis, setting up standing waves and distributing the particles in ridges as they settled.

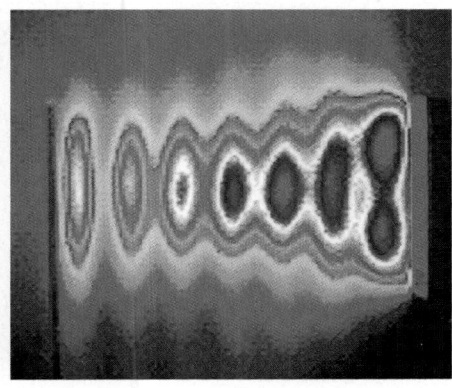

A two-dimensional standing-wave pattern formed between a source and a reflector. Electromagnetic waves from a 3.9-GHz antenna enter from the right. They reflect off a metal rod and travel back to the antenna. The pattern is made visible by absorbing the microwave radiation and recording the resulting temperature distribution with an IR camera.

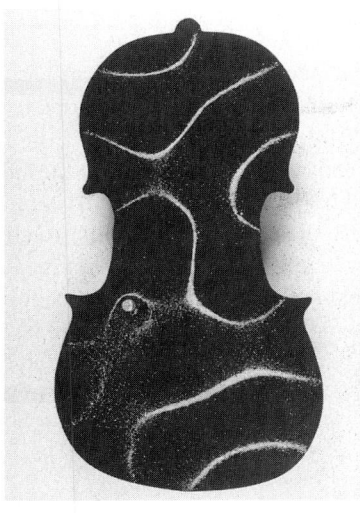

Standing waves, known as a Chladni pattern, on a vibrating (1868 Hz) violin back. Fine sand jiggles around until it lines up along the motionless nodal lines.

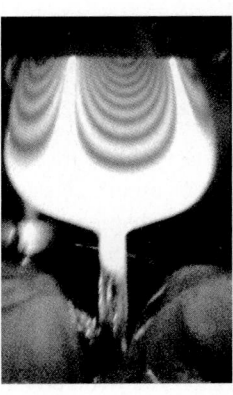

A wineglass oscillating in two of its several possible standing-wave modes. The bright region is the stationary portion of the glass. The photos were made using interferometric holography.

A standing-wave pattern on the side of a car due to vibrations caused by its running engine. The scale is in microns, where $1\ \mu\text{m} = 10^{-6}\,\text{m}$. The photo was made using a holographic technique.

Standing Waves in Air Columns

Standing waves can be sustained by the air within almost any shaped chamber from the vocal tract to the shower stall. Indeed, we speak and hear and make music via the acoustical resonance of air-filled chambers of one sort or another that amplify initial sounds. Most musical instruments driven by streams of air (organs, woodwinds, and brasses) utilize this same kind of standing-wave mechanism for increasing intensity.

There are many ways to excite a column of air into oscillation: most woodwinds use a small vibrating reed; the brasses rely on the vibration of the player's lips; and the flute and organ pipe are driven by a fluttering jet of air that pumps the column (Fig. 11.42). Similarly, blowing across the top of a soda bottle sets the air within it vibrating. Whatever the source, the air column initially oscillates with a wide range of frequencies, but only those corresponding to the standing-

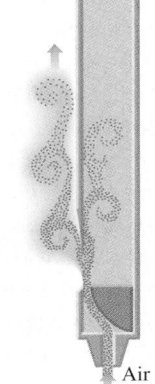

Figure 11.42 An organ pipe. The primary driving mechanism is a wavering, sheetlike jet of air from the flue-slit, which interacts with the upper lip and the air column in the pipe to maintain a steady oscillation.

Air

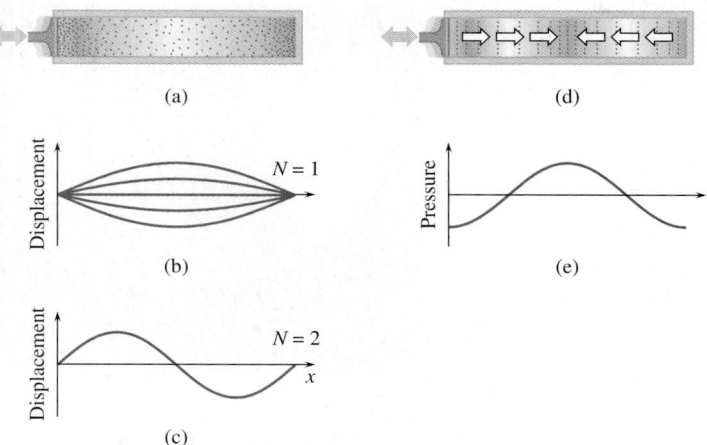

Figure 11.43 (*a*) A sealed pipe. The dark blue regions are at density or pressure antinodes. (*b*) The fundamental mode. (*c*) The second harmonic at the instant a positive peak is at $x = \frac{1}{4}\lambda_2$. (*d*) The direction of motion of the molecules at that moment. (*e*) The associated pressure, which is 90° out-of-phase with the displacement represented in (*c*).

wave modes of the particular chamber will be sustained and amplified. There are three basic configurations for tubular enclosures: closed at both ends, closed at one end, and open at both ends.

Figure 11.43 shows a chamber closed off at the right end by a wall and sealed at the left end with a slightly movable piston or diaphragm, something like the situation that exists when you hum with your throat or mouth closed. Because the ends are essentially fixed, they will always be the locations of displacement nodes—the air molecules cannot move beyond these points (and, as we saw on p. 384, that means these are also pressure antinodes). Small oscillations of the piston will excite the air column, and a standing-wave pattern will occur. Longitudinal sound waves do not shift phase on reflection from a fixed surface as do the transverse waves on a string—*a rarefaction is reflected from a wall as a rarefaction*. Even so, because there are two 180° shifts for a string with both ends fixed, its standing-wave displacement pattern will be the same as that of the sound wave in a closed chamber of length *L*—displacement nodes at both ends. Here again

[chamber—both ends closed] $$L = \tfrac{1}{2}N\lambda \qquad [11.12]$$

and the wavelength of the *N*th mode is $\lambda_N = 2L/N$. This is all fine, but if you want to really make some intense sounds, you must allow the energy to escape the instrument more efficiently: you must open your mouth and sing out, or open the end of the chamber.

The resonant chamber or cavity can be open either at one end or both, and there are organ pipes of each kind, just as there are musical instruments of each kind: the soda bottle and the trumpet (closed by the mouth) versus the flute and piccolo. The organ pipe open at one end can be thought of as half the closed chamber of Fig. 11.43. At the hole where it is open to the atmosphere, the gauge pressure is zero, and a pressure node (Fig. 11.44) always exists there (antinodes are drawn dark blue, nodes are pale blue). Unimpeded at that point, the molecules of air can move freely, and a displacement antinode exists there. This situation happens because *sound waves are phase-shifted by 180° when they reflect off an open end*. It is as if the compression wavefront travels more easily in the open air, stretching the bonds with the layers behind it and sending back a reflected rarefaction. *At an open end, a compression is reflected as a rarefaction and vice versa.* It should be mentioned that the displacement antinodes do not actually occur precisely at the open end, but this effect is a small one for narrow pipes, and we can overlook it here. For the singly open pipe

The pipes of an organ.

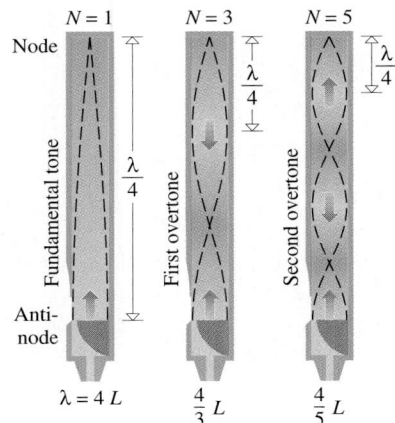

Figure 11.44 A few standing-wave modes for an organ pipe open at its lower end. The arrows show the direction in which air moves for half the cycle, whereupon the modes reverse. The dashed lines indicate the displacement nodes and antinodes. All brass instruments are closed at one end by the mouth of the player.

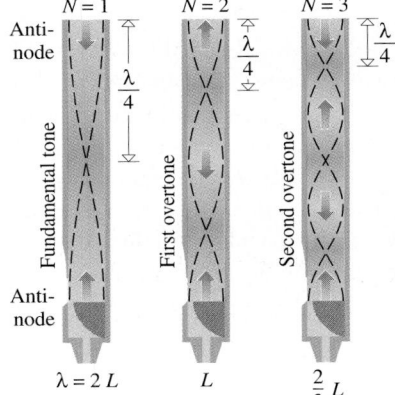

Figure 11.45 A few standing-wave modes for an organ pipe open at both ends. The arrows show the direction in which air moves for half the cycle, where-upon the modes reverse. This time, both ends of the tube are at atmospheric pressure ($P_G = 0$) and there are pressure nodes at both.

[chamber—one end open]
$$L = \tfrac{1}{4}N\lambda \qquad (11.14)$$

where N is an *odd integer*. Again, since $v = f\lambda$, it follows that $L = \tfrac{1}{4}Nv/f_N$ and the resonant frequencies are given by

[chamber—one end open]
$$f_N = \frac{Nv}{4L} \qquad (N = 1, 3, 5, ...) \qquad (11.15)$$

Here v is the speed of sound in air. Pressing on a valve of a trumpet opens a loop of tubing into the airstream, effectively adding to the length of the pipe and increasing the wavelength and decreasing the pitch.

The doubly open pipe of Fig. 11.45 must have displacement antinodes at both ends and so must have a node at its middle. Clearly, the pipe has to be half a wavelength long; that is

$$L = \tfrac{1}{2}N\lambda \qquad [11.12]$$

[chamber—both ends open]
$$f_N = \frac{Nv}{2L} \qquad (N = 1, 2, 3, ...) \qquad (11.16)$$

Example 11.13 **[I]** As we saw earlier, the auditory canal, which is essentially a narrow chamber closed at one end by the eardrum, is a standing-wave cavity. Knowing that it is about 2.7 cm long determines the fundamental frequency that will be amplified at room temperature.

Solution This one's about standing waves. (1) TRANSLATION—Knowing the length of an air cavity open at one end, determine the fundamental frequency. (2) GIVEN: $L = 2.7$ cm and $v = 343$ m/s. FIND: f_1. (3) PROBLEM TYPE—Wave motion/sound/standing waves/air column. (4) PROCEDURE—

The frequency of the standing-wave pattern is given by Eq. (11.15), and it could be used directly. Instead, find λ and use $v = f\lambda$. (5) CALCULATION—$L = \tfrac{1}{4}N\lambda$, with $N = 1$ and $\lambda = 4L$ $= 4(0.027$ m$) = 0.108$ m, and so

$$f_1 = \frac{v}{\lambda} = \frac{343 \text{ m/s}}{0.108 \text{ m}} = \boxed{3.2 \text{ kHz}}$$

Quick Check: 3 kHz is in the mid-frequency audible region where it should be (p. 390).

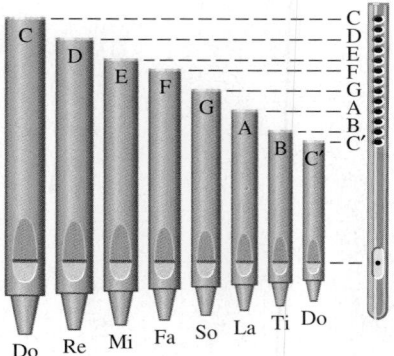

Figure 11.46 Organ pipes and a flute. The length of the resonant chamber determines the pitch.

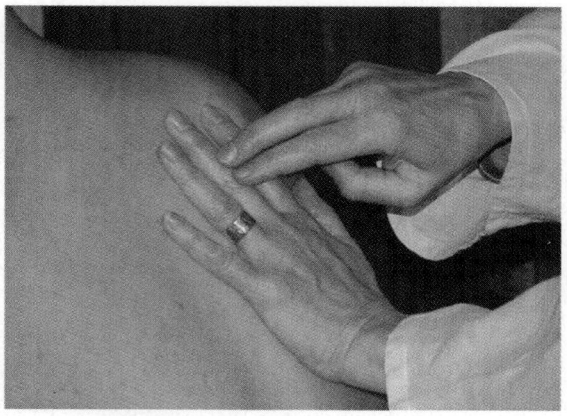

By tapping on the fingers, which are in contact with the back, compression waves are introduced into the chest. This percussion procedure, introduced by L. Auenbrugger in 1761, allows one to recognize various pathological conditions from the resulting sounds.

Except for the location of the nodes, this result is the same one we obtained for the closed pipe and the string fixed at both ends. The flute is an open pipe, open to the atmosphere both at the mouthpiece or embouchure and at the far end. Like many wind instruments, it contains a row of holes that can be covered by the fingers. Opening any one has the effect of cutting the tube off at that length and therefore shortening the wavelength and raising the pitch (Fig. 11.46). As a rule, blowing gently on a pipe (or soda bottle) primarily excites the fundamental. When air is blown much more forcefully across the opening of a flute, the column can resonate at the second harmonic, and the entire scale will be raised one octave corresponding to this doubling.

The metal reeds in a music box mechanism are displaced by rotating pins and set vibrating. Note the different lengths of the reeds that oscillate with one end fixed. The longer the reed, the longer the wavelength of the tone it produces.

11.11 The Doppler Effect   (Optional)

Every kind of wave moves through a homogeneous medium at a constant speed that depends only on the physical properties of the medium. That's true irrespective of any motion of the source—it launches the waves and away they go. Still, the perception of a

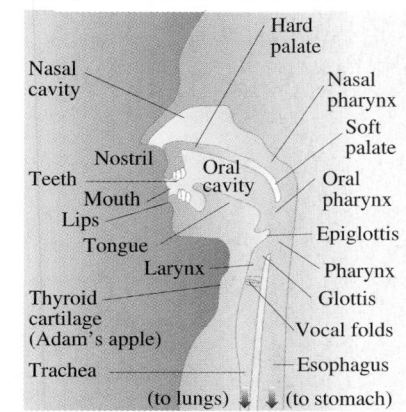

Hard palate
Nasal cavity
Nasal pharynx
Soft palate
Nostril
Oral cavity
Teeth
Mouth
Oral pharynx
Lips
Tongue
Epiglottis
Larynx
Pharynx
Thyroid cartilage (Adam's apple)
Glottis
Vocal folds
Trachea
Esophagus
(to lungs) (to stomach)

EXPLORING PHYSICS ON YOUR OWN

Humming & Resonant Cavities: The human vocal tract—larynx, pharynx, and nasal and oral cavities—can crudely be thought of as a singly open pipe with the vocal cords at the closed end and the mouth at the open end. It therefore sustains a series of standing-wave modes that amplify the sounds made primarily by the vocal cords. Unlike other musical instruments, this one changes its shape. When the fundamental generated by the vocal cords is matched by a vocal tract resonance at that frequency, the amplitude of the note will be a maximum. Such is the business of a skilled singer. With your mouth closed, first hum the lowest pitch you can, then some mid-range pitch, and finally the highest pitch. Notice how you instinctively close off a small region way up in your nose wherein you make the high-pitched (small λ) sounds. To get the lowest pitch (longest λ), you open the back of the throat ("push the sound into your chest") and resonate the largest cavity possible. Hold your neck in your hand and feel the difference in the vibrations.

STUDY GUIDE

The Doppler Effect treats the measured frequency of sound when the source and observer are in relative motion. There are many possible scenarios, and we'll examine several of them [Figs. 11.50 through 11.53]. Study the material and make sure you follow the basic logic as it unfolds. There'll be a number of equations developed [Eqs. (11.19) through (11.22)] each of which you should understand but not memorize; they'll all be summarized later in one final formula, Eq. (11.23). So, read the whole section, paying attention to the physics, and don't worry about retaining all the details. Then come back to Eq. (11.23) as a summary. It's the central equation for almost all problem solving.

wave's frequency and wavelength can be significantly altered by a relative motion between the observer and the source. Almost everyone has heard a shift in frequency when a car blowing its horn passes by. The pitch while it is approaching is somewhat higher than when at rest, and as it passes, the pitch drops. The result is a kind of eeeeoooo sound. This phenomenon is known as the **Doppler Effect** after the Austrian physicist Johann Doppler, who first worked out the analysis for sound in 1842. He also correctly suggested that the effect would apply to light. The predictions of the theory were soon confirmed (1845) by Ballot, who arranged an acoustical extravaganza that must have been a joy to behold. For two days, a locomotive, flatcar, and band of trumpeters performed—coming and going—for a group of musician observers.

Figure 11.47 indicates how a frequency shift arises for a source moving toward a stationary observer at a rate of half the speed of sound. The motion of the source essentially closes up the gap between successive wavefronts. The observer will measure a wavelength that is halved and a frequency that is doubled. In the same way, Fig. 11.48 depicts a stationary source radiating sound to the right and an observer moving left. The listener rushing toward the source overtakes the wavefronts, intercepting more of them per second.

Consider a point source moving with a speed v_s and an observer moving with a speed v_o, both traveling along the same line. Figure 11.49 is a multiple exposure showing the point source's location at five consecutive moments and the associated wavefronts. Using this figure as a reference, we **take the direction from the observer to the source as positive**. Let's start ($t = 0$) when the source was at point 0 whereupon it moved a distance $v_s t$ in a time t. Since its frequency is f_s, it has generated a number of waves equal to $f_s t$. Propagating at the **speed of sound** (v), the front emitted at $t = 0$ is now expanded out to a radius of vt, where it has encountered the observer who is moving to the right at v_o. Notice that the wavelength on either side of the source is different. On the left, where the observer is, λ equals the distance between the source and the observer ($vt + v_s t$) divided by the number of waves that span that space ($f_s t$), and so

The **speed of sound** is v.
The **speed of the observer** is v_o.
The **speed of the source** is v_s.
The **positive direction** is from observer to souce.

$$\lambda = \frac{v + v_s}{f_s} \qquad (11.17)$$

On the other side, the same number of waves is squeezed into a space of ($vt - v_s t$), and therefore λ in the forward direction equals ($v - v_s$)/f_s. Once we have the wavelength, we can find the frequency by dividing the relative speed by λ.

As far as the observer is concerned, she moves right at a rate v_o (with respect to the medium) as the sound travels toward her at v (with respect to the medium), and so she

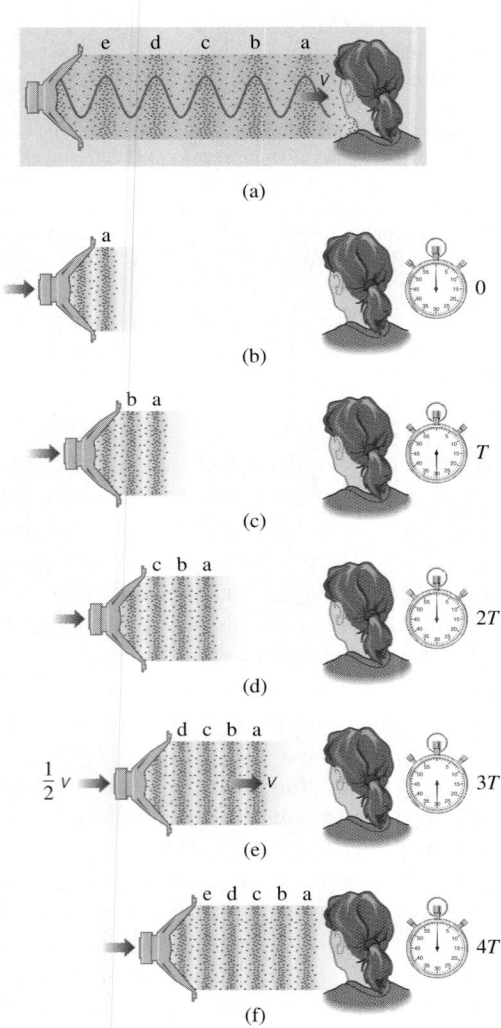

Figure 11.47 The Doppler Effect for a source moving toward a stationary observer at half the speed of sound (v). Compare the wavelength in (a), where the source is at rest, to the wavelength in (f), where it is moving at $\frac{1}{2}v$. Note how the source advances $\frac{1}{2}\lambda$ during the emission of one complete wave.

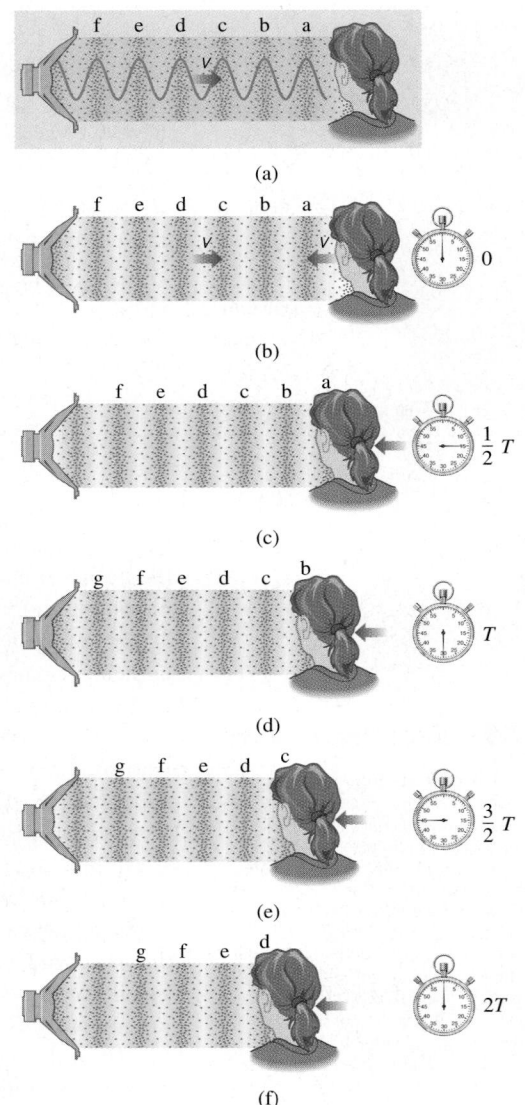

Figure 11.48 The Doppler Effect for an observer moving at speed v toward a stationary source of sound of speed v. Notice how the moving observer receives fronts a, b, c, and d in a time of two periods. Had she remained at rest as in (a), she would have received only two wavefronts a and b in that time.

sweeps through the wavefronts at a rate of ($v + v_{\mathrm{o}}$). Consequently, the observer will perceive a frequency of

$$f_{\mathrm{o}} = \frac{v + v_{\mathrm{o}}}{\lambda} \qquad (11.18)$$

Combining these last two expressions,

$$\left[\begin{array}{c}\text{observer approaching: source receding}\\ \text{same direction}\end{array}\right] \qquad f_{\mathrm{o}} = f_{\mathrm{s}} \frac{v + v_{\mathrm{o}}}{v + v_{\mathrm{s}}} \qquad (11.19)$$

STUDY GUIDE

The two + signs in Eq. (11.19) mean that both the observer and the source are moving in the positive direction.

This is the situation depicted in Fig. 11.50. Notice that v_{o} and v_{s} are both preceded by plus signs here and that both are in the positive direction. The motion of the source, heading

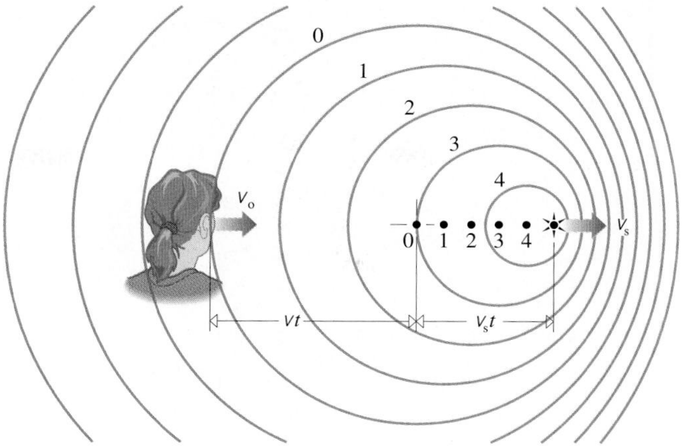

Figure 11.49 The source on the right is moving to the right at speed v_s, which is taken to the positive. The observer is also moving to the right, this time with a speed v_o which is taken as positive.

away from the observer, tends to increase the observed wavelength and decrease the observed frequency (behind itself). That means the denominator $(v + v_s)$ must be larger when v_s is not zero, whereas the motion of the observer, heading into the sound wave, tends to increase that frequency. This means the numerator $(v + v_o)$ must be larger when v_o is not zero.

When $v_o > v_s$ the observer begins to overtake the source ($\text{o}\longrightarrow \text{s}\rightarrow$) and they actually approach one another, $(v + v_o)/(v + v_s) > 1$. The wavefronts, which are closer together, arrive more frequently and $f_o > f_s$. When $v_s > v_o$ they recede ($\text{o}\rightarrow \text{s}\longrightarrow$) from one another, $(v + v_o)/(v + v_s) < 1$. The wavefronts, which are farther apart, arrive less frequently and $f_o < f_s$. (In short, motion toward—f shifts up; motion away—f shifts down.)

Now suppose the source and observer change places (Fig. 11.51). The motion of the source to the right causes an observed decrease in wavelength and an increase in frequency. That means the denominator in Eq. (11.19) must be $(v - v_s)$; the source moves in the negative direction. Furthermore, the observer is now moving away from the source, and the sound waves sweep over her at a speed $(v - v_o)$, which becomes the numerator in Eq, (11.19); the observer moves in the negative direction. Hence

$$\begin{bmatrix}\text{observer receding: source approaching}\\ \text{same direction}\end{bmatrix} \qquad f_o = f_s \frac{v - v_o}{v - v_s} \qquad (11.20)$$

When $v_o = 0$ the observed frequency must be larger, and when $v_s = 0$ the observed frequency must be smaller. Realize that the motion depicted in Fig. 11.51 is essentially just that of Fig. 11.50 with both directions of motion reversed. *Changing the directions of motion changes the signs in Eq. (11.19).*

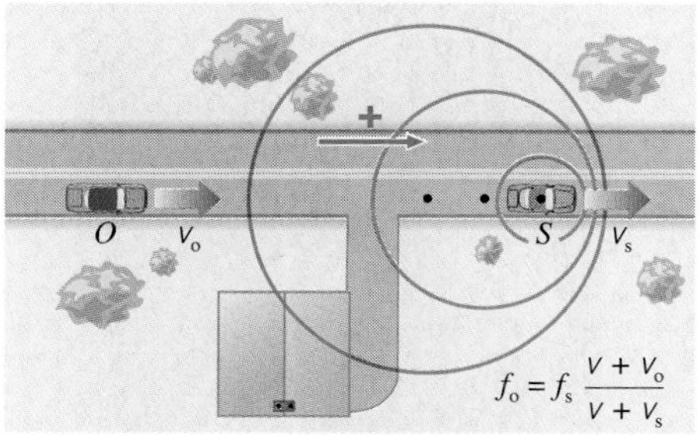

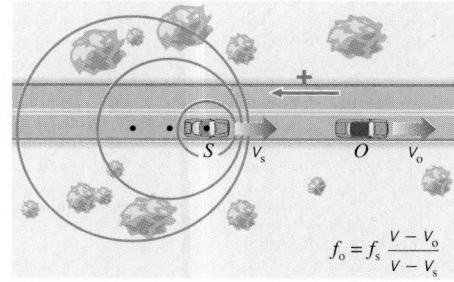

$$f_o = f_s \frac{v + v_o}{v + v_s}$$

$$f_o = f_s \frac{v - v_o}{v - v_s}$$

Figure 11.50 The car on the right is the source of a sound wave at a frequency f_s. The moving observer on the left hears the sound at a frequency of f_o.

Figure 11.51 The car on the left is the source of a sound wave at a frequency f_s. The moving observer on the right hears the sound at a frequency of f_o.

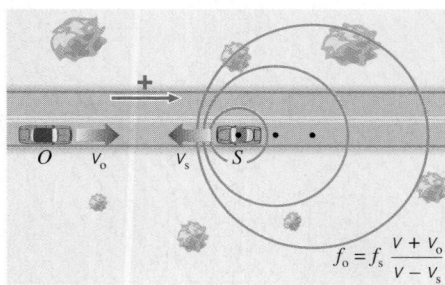

$$f_o = f_s \frac{v + v_o}{v - v_s}$$

Figure 11.52 The car on the right is the source of a sound wave at a frequency f_s. The moving observer on the left hears the sound at a frequency of f_o.

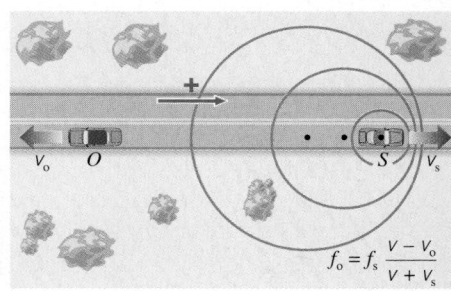

$$f_o = f_s \frac{v - v_o}{v + v_s}$$

Figure 11.53 The car on the right is the source of a sound wave at a frequency f_s. The moving observer on the left hears the sound at a frequency of f_o.

When the observer moves right and the source moves left (Fig. 11.52),

$$\left[\begin{array}{c}\text{approaching one another}\\ \text{opposite directions}\end{array}\right] \qquad \frac{f_o}{v + v_o} = \frac{f_s}{v - v_s} \qquad (11.21)$$

When the observer moves left and the source moves right (Fig. 11.53),

$$\left[\begin{array}{c}\text{receding from one another}\\ \text{opposite directions}\end{array}\right] \qquad \frac{f_o}{v - v_o} = \frac{f_s}{v + v_s} \qquad (11.22)$$

If the observer is at rest ($v_o = 0$) and the source is moving toward her (o ←s), keeping in mind Figs. 11.49, 11.51 and 11.52, we enter $-v_s$ into Eq. (11.19), which becomes equivalent to Eq. (11.21). Then $f_o = vf_s/(v - v_s)$, and since $v/(v - v_s) > 1$ it follows that $f_o > f_s$; the observer detects a shift up in frequency. As you might expect, it doesn't matter in which direction the source is traveling as long as it is approaching the observer: (o ←s) is identical to (s→ o).

If the source moves away from the motionless observer (o s→), we enter $+v_s$ into Eq. (11.19), which becomes equivalent to Eq. (11.22), since ($v_o = 0$). Then $f_o = vf_s/(v + v_s)$, and since $v/(v + v_s) < 1$ it follows that $f_o < f_s$; the observer detects a shift down in frequency. (In short, motion toward—f shifts up; motion away—f shifts down.)

When the source is at rest ($v_s = 0$) and the observer approaches it (o→ s), we enter $+v_o$ into Eq. (11.19), which becomes equivalent to Eq. (11.21). Then $f_o = f_s(v + v_o)/v$ and since $(v + v_o)/v > 1$ it follows that $f_o > f_s$. Similarly, if the observer moves away (←o s),

The transmitter and receiver of a Doppler-shifted radar system mounted on an unmarked police car.

DOPPLER, COPS, & THROMBOSIS

When waves are sent out from a stationary transmitter and reflect back off some moving target, that target functions as though it were a source of waves. The resulting Doppler shift (which is usually determined by beating the returning wave against a signal having the original frequency) provides the speed of the target (Fig 11.54). The police do this little trick with microwaves to clock passing cars on the highway, and the same approach is used to track both satellites and dangerous weather conditions. Likewise, Doppler-shifted ultrasonic waves (8 MHz) are being used to monitor the flow of blood, aiding in the diagnosis of conditions such as deep-vein thrombosis. Motion detectors (e.g., in security systems) also use Doppler-shifted ultrasound.

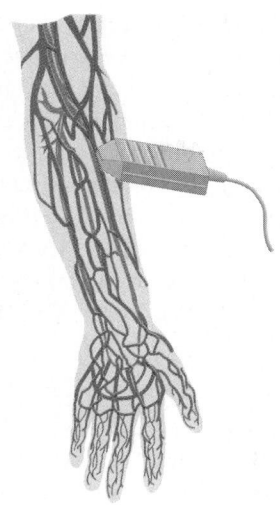

Figure 11.54 An ultrasonic source and receiver used to measure blood flow. Ultrasound reflected back from moving blood cells is Doppler shifted in accord with Eq. (11.23).

we enter $-v_o$ and $f_o < f_s$. (In short, **motion toward—f shifts up; motion away—f shifts down**.)

Once we understand Eqs. (11.19) through (11.22) they might all just as well be combined:

$$f_o = f_s \frac{v \pm v_o}{v \pm v_s}$$

(11.23)

Draw an arrow from the observer to the source; that's the positive direction. When the velocity of the source is in that direction, we use a plus sign in front of v_s in Eq. (11.23), and the same for v_o and the observer. When either velocity is in the opposite direction, the corresponding speed goes into Eq. (11.23) with a minus sign. In all cases, when the source and observer are approaching each other the frequency shifts up, and when they're separating it shifts down.

Example 11.14 **[II]** An automobile traveling at 20.0 m/s (i.e., 45 mi/h) blows a horn at a constant 600 Hz. Determine the frequency that will be perceived by a stationary observer both as the car (a) approaches and (b) recedes. Take the speed of sound to be 340 m/s.

Solution When you read about a moving source of sound, think Doppler. (1) TRANSLATION—A source of sound of a specified frequency is moving at a known speed first toward, then away from, a stationary observer; determine the perceived frequencies. (2) GIVEN: $v_s = 20.0$ m/s, $v = 340$ m/s, and $f_s = 600$ Hz. FIND: f_o. (3) PROBLEM TYPE—Wave motion/sound/Doppler Effect. (4) PROCEDURE—**The positive direction is from the observer to the source.** Set $v_o = 0$ and use Eq. (11.23). (a) Keeping in

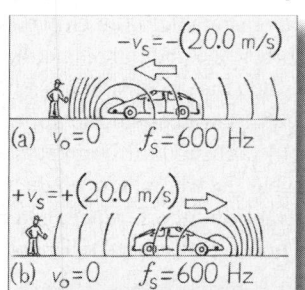

mind Fig. 11.51 or 11.52, with $v_o = 0$, the source is approaching in the negative direction and we use $-v_s$ in Eq. (11.23). (b) In Fig. 11.50 or 11.53, with $v_o = 0$, the source is receding in the positive direction and we use $+v_s$ in Eq. (11.23). (5) CALCULATION—As the car approaches the observer

$$f_o = \frac{vf_s}{v - v_s} = \frac{(340 \text{ m/s})(600 \text{ Hz})}{(340 \text{ m/s}) - (20.0 \text{ m/s})} = \boxed{638 \text{ Hz}}$$

As it recedes

$$f_o = \frac{vf_s}{v + v_s} = \frac{(340 \text{ m/s})(600 \text{ Hz})}{(340 \text{ m/s}) + (20.0 \text{ m/s})} = \boxed{567 \text{ Hz}}$$

Quick Check: $v/(v \pm v_s) = (340$ m/s$)/(340$ m/s ± 20 m/s)—namely, 1.063 and 0.944; 1.063(600 Hz) = 638 Hz; 0.944(600 Hz) = 567 Hz.

All this traveling around is taking place with respect to the stationary medium, and the amount of the frequency shift actually depends on who is doing the moving—it's not symmetrical. The relative motion of source to observer is important here, but so, too, is the motion of each with respect to the medium. *Whichever is moving relative to the medium will experience a wind blowing.* Thus, the first part of Example 11.14 finds that the frequency of an on-coming 600-Hz source (approaching at 20 m/s) as measured by a stationary observer is 638 Hz. If instead, the observer is moving at 20.0 m/s toward a stationary 600-Hz source, we get a different frequency of $f_o = (v + v_o)f_s/v = (340$ m/s $+ 20$ m/s$)(600$ Hz$)/(340$ m/s$) = 635$ Hz. {For an opportunity to study all of this via computer animations click on **THE DOPPLER EFFECT** under **INTERACTIVE EXPLORATIONS** on the **CD**.}

Example 11.15 **[III]** Derive a general expression for the Doppler Shift seen when a sound wave is bounced off an approaching target and returned to the stationary source, where it is viewed by an observer at rest.

Solution This one's a Doppler problem. (1) TRANSLATION—Sound of a specified frequency is reflected from a moving target; determine the frequency perceived by a stationary observer at the source. (2) GIVEN: $v_s = 0$, v, v_t, and f_s. FIND: f_o. (3) PROBLEM TYPE—Wave motion/sound/Doppler Effect. (4) PROCEDURE—Imitating Fig. 11.50, imagine the source (also the primary observer) on the right, and the target (intermediate observer) on the left. There are two Doppler Shifts involved in this process: the wave received by the moving target is shifted and then reemitted at that new frequency (f_t). But the target is moving as it reemits so there will be yet another shift as perceived by the primary observer. (5) CALCULATION—**The positive direction is from the observer to the source.** In the first phase of the analysis, the target (now an "observer" moving to the right w.r.t. the medium) receives the wave at a shifted frequency $f_o = f_t$ which can be calculated using Eq. (11.23). The source is at rest w.r.t. the medium ($v_s = 0$) and sends out a frequency f_s. The target observer is moving to the right, $v_o = v_t$, it's approaching the source (o→ s), and that information must be entered into Eq. (11.23) via a plus sign on the observer side;

$$\frac{f_o}{v + v_o} = \frac{f_s}{v \pm v_s}$$

Because $f_o = f_t$, $v_o = v_t$, and $v_s = 0$,

$$f_t = \frac{(v + v_t)f_s}{v}$$

The target is approaching the source and sees an up-shifted frequency, $f_t > f_s$. This same frequency is next reflected. In so doing, it becomes the source wave to be received by the primary observer, much like Fig. 11.51 with $v_o = 0$. This reflecting source is approaching the motionless observer (s→ o) and so v_s enters Eq. (11.23) as $-v_s$,

$$\frac{f_o}{v \pm v_o} = \frac{f_s}{v - v_s}$$

Because $f_s = f_t$, $v_s = v_t$ and $v_o = 0$,

$$f_o = \frac{vf_t}{v - v_t}$$

Notice that both of these shifts raise the frequency because the target is approaching. Substituting for f_t,

$$f_o = \frac{(v + v_t)f_s}{v - v_t} \qquad (11.24)$$

If the target was receding, the signs in front of v_t would have to be changed accordingly.

Quick Check: First, the units are right. Second, the observed frequency (f_o) is greater than the source frequency (f_s), which is correct. Third, as $v_t \to 0$, $f_o \to f_s$, which is as it should be.

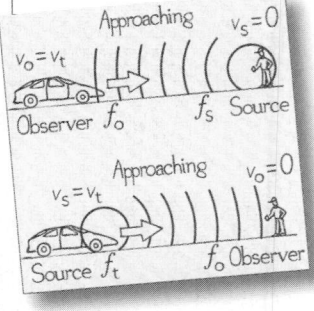

As applied to light, the Doppler Effect provides a key insight into the creation of the Universe. The light from the vast majority of galaxies is seen to be shifted toward the low-frequency end of the spectrum. This shift implies that if the Doppler Effect is the cause (and no other possible mechanism is known), then all these thousands of millions of galaxies must be moving away from us—a notion that can only be reasonable if the Universe is expanding so that every galaxy is moving away from every other one and ours is neither at the center nor flying off on its own. If that picture is correct, then the expansion must have begun back at some moment of creation. From today's recession rates and the apparent size of the Universe, we conclude that it came into being ≈14 thousand million years ago in a cosmic fireball known as the *Big Bang*. One can almost hear the trumpeters on the flatcar celebrating it all. {For a discussion of the reflection, refraction, interference, and diffraction of sound waves click on **PROPAGATION OF SOUND** under **FURTHER DISCUSSIONS** on the **CD**.}

Core Material & Study Guide

MECHANICAL WAVES

A progressive wave is a self-sustaining disturbance of a medium. A periodic wave propagates with a speed

$$v = f\lambda \qquad [11.1]$$

This is the most fundamental equation in the chapter. The profile (p. 373) of a harmonic wave is given by

$$y = A \sin \frac{2\pi}{\lambda} x \qquad [11.2]$$

For a general introduction to waves, reread Sections 11.1 (Wave Characteristics) and study Examples 11.1 and 11.2. Throughout this chapter examine the appropriate **WARM-UPS** and **WALK-THROUGHS** on the **CD**.

The speed of a transverse wave on a *stretched* string is

$$v = \sqrt{\frac{F_T}{m/L}} \qquad [11.3]$$

Study Section 11.2 (Transverse Waves: Strings) and make sure you understand Example 11.3.

For *compression waves in a liquid*

$$v = \sqrt{\frac{B}{\rho}} \qquad [11.4]$$

For *compression waves in a slender solid*

$$v = \sqrt{\frac{Y}{\rho}} \qquad [11.5]$$

Consult Section 11.3 (Compression Waves) for specifics on calculating wave speeds in liquids and solids, and then go over Example 11.4.

SOUND

Sound is *a compression wave in any material medium oscillating within the frequency range of 20 Hz to 20 kHz.* The **intensity** (*I*) of a sound wave is defined as *the average power radiating divided by the perpendicular area across which it is transported*:

$$I = \frac{P_{av}}{A} \qquad [11.6]$$

The speed of sound in a diatomic gas such as air is given by

$$v = \sqrt{\frac{1.4P}{\rho}}$$

Sound travels in air at 331.45 m/s at STP. Over the usual range of temperatures encountered at sea level, the speed of sound changes by about 0.60 m/s per change of 1.0° C. The general principles of sound are discussed in Sections 11.4 (Acoustics: Sound Waves), 11.5 (Wavefronts & Intensity), and 11.6 (The Speed of Sound in Air).

The **intensity-level** (or *sound-level*) of an acoustic wave is defined as *the number of factors of 10 that its intensity is above the threshold of hearing*: $I_0 = 1.0 \times 10^{-12}$ W/m². The intensity-level is measured in *decibels* (abbreviated dB) and is given by

$$\beta = 10 \log_{10} \frac{I}{I_0} \qquad [11.8]$$

For a change from I_1 to I_2

$$\Delta\beta = 10 \log_{10} \frac{I_2}{I_1} \qquad [11.11]$$

The above material is treated in Section 11.8 (Sound-Level). Given that you are not likely to be very familiar with logs, it's especially important to study Examples 11.9 and 11.10.

When two wavetrains of slightly different frequency overlap, they produce **beats** at the difference frequency (see Section 11.9). When two wavetrains of the same wavelength and amplitude traveling in opposite directions overlap, they create a **standing wave** (p. 397). For a system with zero displacement at both ends

$$L = \tfrac{1}{2}N\lambda \quad (N = 1, 2, 3, ...) \qquad [11.12]$$

and the wavelength of the Nth mode is $\lambda_N = 2L/N$. The resonant frequencies of a string fixed at both ends are

$$f_N = \frac{Nv}{2L} = \frac{N}{2L}\sqrt{\frac{F_T}{m/L}} \quad (N = 1, 2, 3, ...) \qquad [11.13]$$

For a pipe open at one end

$$L = \tfrac{1}{4}N\lambda \qquad [11.14]$$

and

[antinode at one end] $\qquad f_N = \frac{Nv}{4L} \qquad (N = 1, 3, 5, ...) \qquad [11.15]$

For a pipe open at both ends

$$L = \tfrac{1}{2}N\lambda \qquad [11.12]$$

and

[antinode at both ends] $\qquad f_N = \frac{Nv}{2L} \qquad (N = 1, 2, 3, ...) \qquad [11.16]$

If you wish to master the above material, study Section 11.10 (Standing Waves) and review Examples 11.12 and 11.13.

This pattern of particles at the bottom of a cup of tea reveals that there once was a standing wave in the liquid.

The Doppler Effect is a shift in frequency when the source and/or the observer are in motion (p. 405). Draw an arrow from the observer to the source; that's the positive direction and it allows us to select the appropriate signs in the expression relating the frequencies, namely

$$f_o = f_s \frac{v \pm v_o}{v \pm v_s} \qquad [11.23]$$

The Doppler-shifted frequency seen when a wave is bounced off an approaching target and returned to the stationary source where it is viewed by an observer at rest is

$$f_o = \frac{(v + v_t)f_s}{v - v_t}$$ [11.24]

The Doppler Effect is the subject of Section 11.11.

Study each of the Coordinated and Progressive Problems on pages 416 through 422 and try to do all the I-level problems.

Key Terms

progressive wave	infrasound
longitudinal wave	sound
transverse wave	pour tone
wavepulse	Superposition Principle
wavetrain	Fourier analysis
carrier	fundamental

profile	ultrasound
periodic wave	wavefront
period	acoustic power
frequency	intensity
wavelength	Inverse Square Law
harmonic wave	pitch
phase	timbre
amplitude	overtones
reflection	loudness
absorption	sound-level
transmission	intensity-level
compression waves	decibel
acoustic waves	beats
rarefaction	standing waves
condensation	Doppler Effect

Discussion Questions

1. Imagine two symmetrical pulses that are identical in every way, except that one is inverted. These pulses are moving toward each other on a rope in opposite directions. What happens to the energy of the rope at the instant of complete cancellation?

2. In the real world, a pulse sent down a very long, taut rope diminishes in amplitude and ultimately vanishes. Describe what's happening in terms of Conservation of Energy.

3. Imagine a long string of beads hung vertically from a high ceiling with a weight fixed to its lower end. The bottom is then rapidly displaced, generating a transverse pulse. Describe the motion of the disturbance, paying particular attention to how it might differ if the string were held taut horizontally.

4. Picture a long row of standing dominos, one a little behind the other so that knocking over the first will knock over the second and so on. Now imagine an idealized tube an inch or two in diameter extending from New York to California. Suppose this tube is filled with greased, essentially frictionless marbles and you push one more in at the East Coast end only to have one pop out at the West Coast end. What's happening, and how do these arrangements of dominos and marbles relate to this chapter?

5. The accompanying list is a brief accounting of the speeds of compression waves in various materials. Considering the physical characteristics of each of the materials, discuss the relative speeds and explain why they are reasonable in light of your knowledge of the process of wave propagation.

Material	Speed (km/s)
Clay	1
Sandstone	2
Limestone	4
Granite	5
Salt	6

6. Extracorporeal shock wave lithotripsy (ESWL) is a technique for pulverizing kidney stones within the body. The patient is placed in a bath of water. An intense electric spark within the tank below the patient rapidly vaporizes a small amount of water, producing a

shock wave that is reflected from a curved mirror that focuses it onto the stone. After a thousand or two blasts, a typical stone is destroyed. How is the shock wave formed and what kind of wave is it? For what is it used in a physical sense? Why isn't the skin destroyed as well as the stone?

7. Figure Q7 shows two different pulses heading toward each other. Draw several pictures in sequence, illustrating the result as they superimpose and separate.

Figure Q7

8. In December 1916, a major disaster befell a contingent of the Austrian army in the Alps fighting in World War I. Almost as soon as they began a cannon barrage, an avalanche of snow came roaring down, burying thousands of soldiers. What does that suggest about sound?

9. As an orchestra literally warms up, its sound changes. What do you think happens to the pitch of the strings and the wind instruments as a result of their being played for a while? Explain.

10. Suppose you inhale a little helium (too much will displace oxygen and you will get into trouble even though helium is otherwise nontoxic) and speak while exhaling it. The resulting voice will have a comic high-pitched sound. Why? What would happen to an organ pipe blown with carbon dioxide?

11. The ear has a protective overload mechanism called the *acoustic reflex*. A sound in excess of about 85 dB causes muscles attached to the eardrum and ossicles to engage, which provides a safety margin of about 20 dB or 30 dB (the equivalent of ear plugs). This reflex takes about 30 ms to 40 ms to cut in, and maximum effect only occurs after about 150 ms. What is good and bad about such a system in our modern world? Why do many hunters show loss of hearing (especially in the right ear)?

12. Suppose you sing or yell into an open piano with its dampers lifted off the strings. You are likely to hear the instrument "speak" back in response. Explain. Imagine that you have two identical tuning forks and you tap one and then place it near the other. What, if anything, will you hear if you now silence the first fork? "If one bows the bass string on a viola rather smartly," wrote Galileo, "and brings near it a goblet of fine, thin glass having the same tone as

that of the string, this goblet will vibrate and audibly resound." Explain.

13. As a rule, instruments such as the piano that utilize struck strings will arrange to have the striking take place at a point 1/7th of the way down the string. Considering the fact that the 7th harmonic is generally unpleasantly dissonant, explain why the strings are struck where they are.

14. Imagine two people turning a jump rope while a third hops up and down. Describe every aspect of the process using the ideas developed in this chapter.

15. The tuning fork, which was invented in 1712 by John Shore, Handel's trumpeter, can vibrate (Fig. Q15) briefly in a mode that produces its first overtone, or clang tone, at a frequency of 6.27 times the fundamental. If you wish to have the fork produce a nearly pure tone and so suppress the high-pitched clang tone, where should it be struck? Explain. Compare the tuning fork to the rod in the previous question.

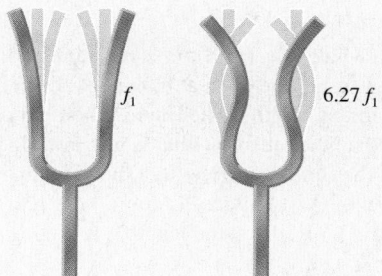

Figure Q15

16. Discuss two ways in which temperature changes will affect the pitch of an organ pipe.

17. A tuning fork is mounted to a hollow open-ended wooden sounding box that amplifies the sound. Explain how it manages to do that. What can you anticipate would be the difference between the sounding boxes for two such forks at, say, 1000 Hz and 50 Hz?

18. In light of Question 17, what can you say about the reasons for the difference in size between a violin (60-cm long) and a double bass (200-cm long)?

19. The long glass tube (Kundt's tube) in Fig. Q19 is sealed at the lower end with a movable plunger and at the upper end with a loosely fitted piston attached to a stationary rod (itself fixed at the middle). The tube contains an even layer of light, dry, very fine powder (lycopodium, precipitated silica, or cork dust). When the rod is stroked lengthwise with a rosined cloth or chamois, it gives out a high-pitched sound, and the powder flies up in violent agitation. When the tone is stopped, the powder settles back into heaps, and by adjusting the location of the plunger, these piles become well defined, uniform, and periodic. The device, which was invented by A. E. Kundt of Germany in 1876, can be used to measure the speed of sound (Problem 132). Discuss what is happening to the rod, to the air in the tube, and to the powder.

20. The behavior of sound waves can be studied using electronic frequency generators, microphones, and oscilloscopes. Suppose, then, that the speaker of Fig. Q20 sends out a single short (5-ms) compression pulse that travels into the doubly open tube, which is 20-m long. What signal will appear on the scope as provided by the pressure microphone at the middle of the pipe?

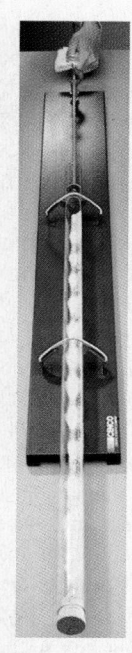

Figure Q19

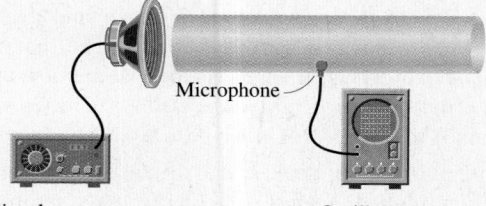

Figure Q20   Signal generator   Oscilloscope

21. When a string on, say, a violin is plucked, particularly toward one end, it initially takes on a nearly triangular shape rather than a sinusoidal one. Why does that result in a more "metallic" sound? Similarly, plucking the strings of a guitar with the fingers produces a more mellow sound than using a pick. Remember that the banjo is played with a pick and the harp with the fingers. Explain. Keep in mind that a signal with fine details, such as sharp corners, will require high-frequency Fourier components in order to synthesize it.

Multiple Choice Questions

1. A wave in a medium transfers   (a) only mass   (b) mass and energy   (c) neither mass nor energy   (d) energy   (e) none of these.

The next five questions pertain to Fig. MC2 which is a snapshot of a wave traveling to the right on a stretched string.

2. What is the wavelength of the wave?   (a) 0.30 m   (b) 0.60 m   (c) 1.20 m   (d) 2.40 m   (e) none of these.

3. What is the amplitude of the wave?   (a) 1.0 mm   (b) 2.0 mm   (c) 3.0 mm   (d) 4.0 mm   (e) none of these.

4. As the wave travels to the right, point-A on the string will   (a) move toward point-3   (b) move toward point-7   (c) move toward point-2   (d) move toward point-*1*   (e) none of these.

5. As the wave travels to the right, point-B on the string will   (a)

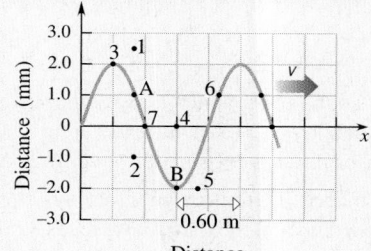

Figure MC2   Distance

move toward point-5   (b) move toward point-7   (c) move toward point-4   (d) move toward point-2   (e) none of these.

6. If the wave travels at 2.40 m/s, what is its frequency?   (a) 2.00 Hz   (b) 2.40 Hz   (c) 3.00 Hz   (d) 3.40 Hz   (e) none of these.

7. The speed of a wave on a stretched string _____ when the tension in it is doubled. (a) doubles (b) increases by a factor of 4 (c) increases by a factor of 1.414 (d) decreases by a factor of 2 (e) none of these.

8. Two strings of the same length are being used in a wave experiment. The first, which has twice the mass of the second, is stretched twice as taut as the second. The speed of the first's wave is (a) twice that of the second (b) the same as the second (c) greater than the second by $\sqrt{2}$ (d) less than the second by $\sqrt{2}$ (e) none of these.

9. A periodic wave passes an observer, who records that there is a time of 0.5 s between crests. (a) the frequency is 0.5 Hz (b) the speed is 0.5 m/s (c) the wavelength is 0.5 m (d) the period is 0.5 s (e) none of these.

10. A string of a certain linear mass-density is attached end-to-end to a rope whose linear mass-density is nine times greater. The combination is stretched taut, and a pulse is sent down the string with a speed v. Its speed in the rope is (a) $9v$ (b) v (c) $3v$ (d) $v/9$ (e) none of these.

11. For a harmonic wave of a certain type in a given medium, doubling the frequency has the effect of (a) halving the speed (b) halving the wavelength (c) doubling the amplitude (d) doubling the period (e) none of these.

12. A stretched rope is dipped in water and thereby made wet. Without changing the tension, a pulse is then sent down its length. Compared to when the rope was dry, we can now expect the speed of the pulse to have (a) decreased (b) increased slightly (c) remained the same (d) doubled (e) none of these.

13. Figure MC13 depicts the displacement of a particle in a medium plotted against time as a wave propagates through. The period of the wave is (a) 0.1 s (b) 0.2 s (c) 0.3 s (d) 0.4 s (e) none of these.

14. Figure MC13 depicts the displacement of a particle in a medium plotted against time as a wave propagates through. The frequency of the wave is (a) 2.5 Hz (b) 5.0 Hz (c) 3.0 Hz (d) 0.4 Hz (e) none of these.

Figure MC13

15. Quite generally, doubling the amplitude of a wave (a) doubles the frequency (b) halves the period (c) quadruples the energy (d) doubles the speed (e) none of these.

16. The speed of sound in air at 40 °C by comparison to the speed at 0 °C is (a) much lower (b) a little lower (c) the same (d) higher (e) none of these.

17. A dB-meter placed in front of an audio system reads 62 dB. The volume is turned up and the meter now reads 82 dB. Evidently, the sound got (a) twice as loud (b) half as loud (c) 20 times louder (d) 10 times louder (e) none of these.

18. A 60-dB sound-level corresponds to an intensity of (a) 10^{-10} W/m^2 (b) 10^6 W/m^2 (c) 10^{-12} W/m^2 (d) 10^{12} W/m^2 (e) none of these.

19. A sound that has an intensity-level of 100 dB is how many times more intense than a sound of 20 dB? (a) 5 (b) 8 (c) 1000 (d) 10^8 (e) none of these.

20. On a day when you are listening to music at a sound-level of 20 dB above the threshold of audibility, an average person would say that it was (a) very loud (b) of normal listening level (c) very soft (d) slightly loud (e) none of these.

21. A sound-level meter placed in front of the loudspeaker of a 60-W audio system reads 70 dB. All else being equal, when placed in front of a 120-W system, the meter will read (a) 120 dB (b) 140 dB (c) 63 dB (d) 73 dB (e) none of these.

22. In order to cause the loudness of a sound to appear half its original value, it must be (a) increased by 50 dB (b) decreased by 50 dB (c) decreased by 10 dB (d) decreased by 3 dB (e) none of these.

23. A doubly open organ pipe, in comparison to a singly open pipe of the same length, has a fundamental frequency that is (a) half as great (b) twice as great (c) three times larger (d) $1/\pi$ times smaller (e) none of these.

24. The flute and piccolo are in the same family of instruments, and each functions as a pipe open at both ends. Both have the same fingering but different lengths. The flute is about 66-cm long and is an octave lower than the piccolo—that is, the piccolo plays at twice the frequency of the flute. It follows that the piccolo, in comparison to the flute, is about (a) twice as long (b) half as long (c) eight times as long (d) the same length (e) none of these.

25. Figure MC25 shows the frequency spectrum of a flute playing a D_4 note (294 Hz). It's clear from the diagram that the flute is likely to be (a) made of metal (b) 66-cm long (c) open at both ends (d) cylindrical (e) none of these.

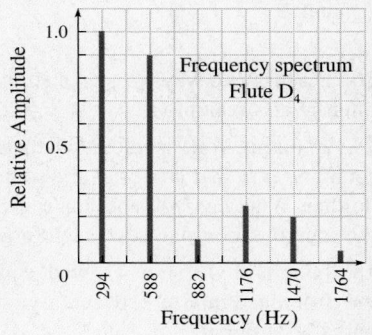

Figure MC25

26. What will happen to the fundamental frequency of an organ pipe open at both ends if a cap is placed over one end sealing it? The frequency will (a) go up 20% (b) be halved (c) stay the same (d) double (e) none of these.

27. If one stereo sounds 16 times louder than another, the difference in their sound-levels is about (a) 40 dB (b) 30 dB (c) 16 dB (d) 160 dB (e) none of these.

28. A clarinet has a single vibrating reed at one end coupled to a cylindrical resonance tube that is flared open at the other end. The coupling of reed and air column produces an interesting spectrum (Fig. MC28), even though there is a node at the reed. Typically, there is a strong first harmonic, little second, a strong third, some fourth, a strong fifth, and an appreciable sixth. It appears that the instrument functions as if it were (a) purely closed at both ends (b) purely open at one end (c) a mix of somewhat open at both ends and yet effectively closed at one (d) purely open at both ends (e) none of these.

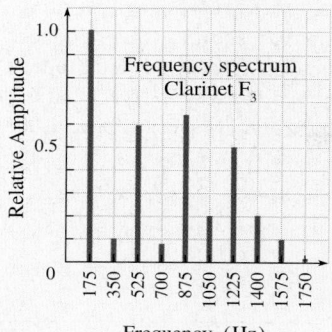

Figure MC28

Frequency spectrum Clarinet F_3

Relative Amplitude

Frequency (Hz)

29. If you stick a small lump of clay to the end of one or both tines of a tuning fork, you can expect the fundamental frequency to (a)

remain unchanged (b) go down (c) go up (d) vary from moment to moment (e) none of these.

30. Two tuning forks of frequency 380 Hz and 384 Hz are excited simultaneously. The resulting sound will waver in intensity with a frequency of (a) 382 Hz (b) 0 (c) 4 Hz (d) 2 Hz (e) none of these.

31. A vehicle heading down a highway at 80 km/h is being pursued by a police car that is right behind it, also traveling at 80 km/h. If the cop's siren puts out a 500-Hz wail, what frequency will the perpetrator hear? (a) < 500 Hz (b) > 500 Hz (c) 500 Hz (d) 80 Hz (e) none of these.

For more Multiple Choice Questions with answers click on WARM-UPS in CHAPTER 11 on the CD.

Suggestions on Problem Solving

1. The most basic equation in wave theory is $v = f\lambda$. It is likely to be encountered, in one way or another, in the solution of many of the problems. We have equations for the speeds of the different types of waves and, in conjunction with those, $v = f\lambda$ provides a means of determining either f or λ when the other is given.

2. Another fundamental relationship is $T = 1/f$. This is easy enough to remember, although the tendency at first is not to associate it with $v = f\lambda$, but of course, $v = \lambda/T$.

3. The speed of a transverse wave on a stretched string depends on the *linear* mass-density (mass/length). Be especially careful when this is given in units of g/m or g/cm, as it often is—convert to kg/m. All the other wave-speed equations containing density refer to the usual notion of mass/volume.

4. There are several facts associated with the notion of sound-level that are worth committing to memory. An intensity-level that is +10 dB higher or −10 dB lower than another will be heard to be twice as loud, or twice as soft, respectively. Doubling (or halving) either the acoustic power or intensity will increase (or decrease) the sound-level by 3 dB. A change of ≈1 dB is the smallest audible increment in loudness you are likely to be able to hear. Knowing this kind of thing helps to establish the physical meaningfulness of numerical answers, and that should always be a primary concern.

5. When a physical entity, a string or a bar, vibrates and makes a sound, it pumps back and forth and sets the air around it vibrating at the same rate, the same *frequency*. The corresponding wavelength of

the string depends on its physical conditions and the generated wavelength of the sound depends on its physical conditions, as manifested by the speed of the wave. The two wavelengths will generally *not* be the same. The natural emphasis is therefore on frequency and speed. **The energy of a wave is associated with its frequency** (p. 392).

6. Bear in mind that the standing-wave modes for all systems will have the same basic form; namely, the frequency will equal the speed over the wavelength. For a system bounded by two nodes (or two antinodes), the wavelength is some fraction of the fundamental $\lambda_1 = 2L$; that is, $2L/N$ where ($N = 1, 2, 3, \ldots$). Hence, $f_N = v/(2L/N)$. When the system is bounded at its ends by a node and an antinode, respectively, $\lambda_1 = 4L$ and only odd fractions of that (namely, $4L/N$ where $N = 1, 3, 5, \ldots$) are possible. Each different type of wave will have its own speed, but these frequency expressions hold for all of them. Try to think in terms of $f_N = v/\lambda_N$, where the form of λ_N can be determined from a mental picture of the particular mode pattern, such as Figs. 11.41, 11.44, or 11.45—that may be easier for you than memorizing the equations. Professionals store the basics and derive the rest as needed.

7. If you are having difficulties with the signs of any of the terms when doing Doppler Effect problems, remember that if the source and observer are approaching each other there must be a shift up in pitch and, hence, down in wavelength. If they are receding from one another, the shift is down in pitch and up in wavelength.

Problems ✦ Coordinated Problems ✦ Progressive Problems ✦ Solutions

STUDY GUIDE **1. Coordinated Problems:** The three problems within each magenta-colored grouping are solvable in similar ways. Note that the first of these always has a hint; moreover, its solution is provided in the back of the book. *Work out each of these sets; they'll strengthen technique and build confidence.* **2. Progressive Problems:** The problems introduced in blue unfold step-by-step carrying along the analysis in a more suggestive way than is customary. *Work out all of these; they'll guide you through the analytic process and help develop problem-solving skills.* **3. Worked-Out Solutions:** Studying worked-out solutions is an important part of learning how to solve problems. Accordingly, additional *solutions* to a number of model problems are given below. *Make sure you understand each of them before you go on to the next problem.* **4.** Also provided in the back of the

book are the *Answers* to all odd-numbered problems, as well as worked-out *solutions* to those with boldface numbers. Problem numbers in italic indicate that a solution appears in the Student Solutions Manual.

SECTION 11.1: WAVE CHARACTERISTICS

1. [I] Figure P1 shows two views—(a) optical and (b) acoustical of a computer circuit. The acoustic microscope has the highly desirable ability to see several micrometers below the surface. The

device focuses ultrasonic waves at a frequency of 3 GHz through a droplet of water onto the object. The speed of the waves in the water, which couples the microscope to the specimen, is 1.5 km/s. Compute the wavelength of the radiation and compare it to the wavelength of green light in air ≈ 500 nm.

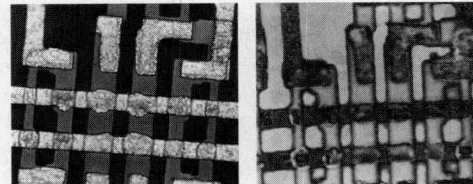

Figure P1

2. [I] A long metal rod is struck by a vibrating hammer in such a way that a compression wave with a wavelength of 4.3 m travels down its length at a speed of 3.5 km/s. What was the frequency of the vibration?

3. [I] An A note of 440 Hz is played on a violin submerged in a swimming pool at the wedding of two scuba divers. Given that the speed of compression waves in pure water is 1498 m/s, what is the wavelength of that tone?

4. [I] A wave on a string travels a 10-m length in 2.0 s. A harmonic disturbance of wavelength 0.50 m is then generated on the string. What is its frequency?

5. [I] Show that, for a periodic wave, $\omega = (2\pi/\lambda)v$.

> **SOLUTION:** $\omega = 2\pi f$, but here we have v and λ instead of f. That suggests using $v = \lambda f$ to replace f. Since $f = v/\lambda$, $\omega = (2\pi/\lambda)v$.

6. [I] Imagine a sinusoidal pressure wave for which the distance between successive maxima is 2.25 m. If exactly 10 such peaks pass a microphone each 1.00 s, what is the speed of the wave?

7. [I] A harmonic wave has a profile described by the expression $y = (1.2 \text{ cm}) \sin [2\pi x/(10.0 \text{ cm})]$. What are the values of the amplitude and wavelength of the wave? If the speed of the wave is 2.00 m/s, what is its frequency? [*Hint: Term by term compare this expression with Eq. (11.2), and remember that $v = f\lambda$.*]

8. [I] The speed of a sinusoidal wave is 4.00 m/s, and its profile is specified by the expression $y = (10 \text{ cm}) \sin [2\pi x/(0.031 \, 4 \text{ cm})]$. Determine its period.

9. [I] A harmonic wave has a profile given by $z = (2.5 \text{ cm}) \cos (0.079 \, 6 \text{ m}^{-1})y$. What are its amplitude, axis of travel, and wavelength?

10. [II] THIS PROBLEM EXPLORES THE GENERAL CHARACTERISTICS OF A PERIODIC WAVE. Figure P10 shows a wave traveling at 5.0 m/s on a beaded string. Any given bead vibrates up and down 3.0 times in each 6.0 s interval. (a) What is the amplitude of the wave? (b) What is the relationship between the frequency of the vertical oscillation of any given bead and the frequency of the wave itself? (c) What is the frequency of the wave? (d) Determine the wavelength of the wave.

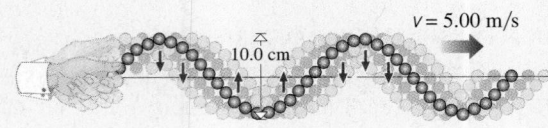

$v = 5.00$ m/s

10.0 cm

Figure P10

11. [II] THIS PROBLEM EXPLORES THE GENERAL CHARACTERISTICS OF A PERIODIC WAVE. Figure P11 shows a wave produced on a taut string by a 50-Hz generator. (a) What is the amplitude of the wave? (b) What is the frequency of the wave? (c) What is the wavelength? (d) Determine the speed of the wave. (e) What is the wave's period?

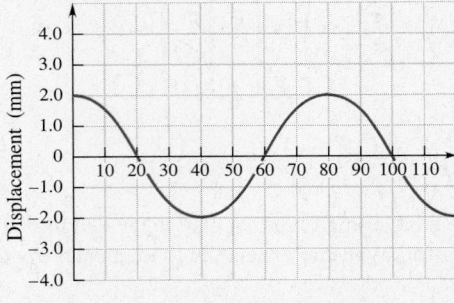

Figure P11   Distance (cm)

12. [II] Write an expression for the profile of a sinusoidal wave traveling in the x-direction at a speed of 10.0 m/s if it has a frequency of 20.0 Hz and an amplitude in the y-direction of 0.010 m.

13. [II] Imagine a sinusoidal wave traveling to the right on a stretched string. Suppose that the distance between successive positive peaks is 50.0 cm. There is a dot of red paint on the rope and at a given instant the displacement of that dot is zero, and it is moving downward. At that moment the displacement of the rope 12.5 cm to the right of the dot is +4.0 cm. What is the displacement, at that instant, of a point 60.0 cm to the right of the dot?

14. [II] A transverse harmonic wave on a beaded string has an amplitude of 2.5 cm and a wavelength of 160 cm. If at $t = 0$ a life-sized photo shows that the height of the wave at $x = 0$ is zero and at $x = 40$ cm it is +2.5 cm, then what is the string's displacement at $x = 10$ cm?

15. [II] A member of the Vespertilionidea family of bats typically emits a sequence of chirps, wavetrains lasting about 3.0 ms and having a carrier frequency that varies from 100 kHz to about 30 kHz. Generally, there is a time between individual chirps of 70 ms. Assuming the air speed of one of these wavetrains is 330 m/s, how far away can an object be and still be detected without being masked by the next outgoing chirp?

16. [II] A radar beam transmitted with a carrier frequency of 9.8 GHz is composed of a stream of nearly harmonic wavetrains each lasting 1.0 μs. Determine the number of wavelengths of the carrier that are in each pulse and show that this number is independent of v, provided the speed of the pulse equals the speed of the carrier wave.

17. [II] Figure P17 represents a rather idealized square water wave. Show that doubling its amplitude would increase its energy by a factor of four, thus again making the point that the energy of a wave is proportional to its amplitude squared.

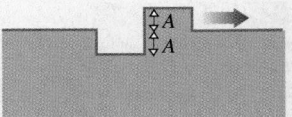

Figure P17

18. [II] The function

$$y = (2.0 \text{ m}) \sin \frac{2\pi}{\lambda} (x - vt)$$

describes the displacement of a harmonic wave traveling in the positive x-direction at a speed v. If the frequency of the wave is 2.0 Hz, what is the value of y at $x = 0$ when $t = 4.0$ s?

19. [II] Show mathematically that shifting the phase of a sinusoidal wave profile by $\frac{1}{2}\pi$ rad turns it into a cosine profile. Begin with

$$y = A \sin\left(\frac{2\pi}{\lambda} x \mp \tfrac{1}{2}\pi\right)$$

20. [III] Determine the profile of the wavefunction

$$y = A \sin 2\pi\left(\frac{x}{\lambda} - \frac{t}{T}\right)$$

at $t = 0$. Find the values of y at $x = 0$, $\lambda/8$, $\lambda/4$, $3\lambda/8$, $\lambda/2$, $5\lambda/8$, $3\lambda/4$, $7\lambda/8$, and λ. Make a plot of the profile. Now do the same thing for the wavefunction

$$y' = A \sin 2\pi\left(\frac{t}{T} - \frac{x}{\lambda}\right)$$

where T is its period and compare your results.

SECTION 11.2: TRANSVERSE WAVES: STRINGS

21. [I] During a lab experiment a long bronze wire is held taut by fixing one end, passing the other over a pulley, and attaching a known hanging mass to it. The wire is then tapped, creating a transverse wave whose speed is measured. If the hanging mass is now quadrupled, what, if anything, will happen to the speed of subsequent transverse waves?

22. [I] During a lab experiment a long silver wire is held taut by fixing one end, passing the other over a pulley, and attaching a known hanging mass to it. The wire is then tapped, creating a transverse wave whose speed is measured. If a new wire of the same length and quadruple the mass is now used, and all else is unchanged, what, if anything, will happen to the speed of subsequent transverse waves?

23. [I] The A string of a violin has a linear mass-density (i.e., a mass per unit length) of 0.59×10^{-3} kg/m and is stretched to a tension of 10 N. What is the speed of the transverse waves it can sustain? Incidentally, the four violin strings, when mounted and properly tuned, produce a net tension on the instrument of about 220 N (50 lb). [*Hint: Eq. (11.3) provides a relationship between speed, tension, and mass per unit length for a stretched string.*]

24. [I] A wire having a mass of 60.0 g is 6.00-m long. If transverse waves travel along it at 200 m/s, what is the tension in the wire?

25. [I] The end of a 100-cm-long wire is plucked, and a transverse wave travels along it at 180 m/s. If the tension in the wire is measured to be 480 N, what is its mass?

26. [I] A heavy rope is tied to a tree 30 m away, and it is then pulled taut with a force of 360 N. The rope is struck near its end with a stick, thus setting up a transverse wavepulse. If the rope has a linear mass-density of 0.06 kg/m, how long will it take for the pulse to make the round trip from end to tree to end?

27. [I] If the speed of a wavepulse set up on the longest string of a grand piano is 129 m/s and if the mass per unit length of the string is 64.9×10^{-3} kg/m, what is the tension? Typically, the net tension on the frame of a concert grand due to all the strings is around 2.7×10^5 N or 30 tons (60 000 lb).

28. [I] Figure P28 depicts two identical symmetrical transverse wavepulses, each traveling at a speed of 5.0 m/s heading toward one another. Draw the resultant disturbance 2.0 s later and again 2.0 s after that.

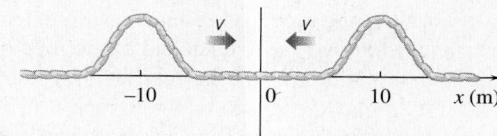

Figure P28

29. [I] Figure P29 depicts two identical symmetrical transverse wavepulses, each traveling at a speed of 5.0 m/s heading toward one another. Draw the resultant disturbance 2.0 s later and again 2.0 s after that.

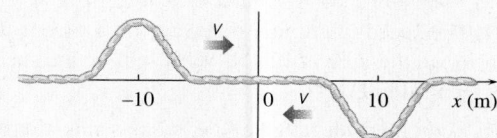

Figure P29

30. [II] Write an expression for the profile of a transverse harmonic wave of amplitude 10 cm and wavelength 1.2 m traveling along a rope in the positive z-direction and oscillating in the yz-plane. The height of the wave is zero at $t = 0$ and $z = 0$.

31. [II] The profile of a transverse harmonic wave on a long taut nylon thread is described in SI units by the function

$$y = 0.040 \sin 2\pi x$$

Given that the wave travels at a speed of 2.0 m/s, determine the maximum *transverse* acceleration of any point on the thread.

32. [II] The profile of a transverse harmonic wave on a long beaded string is described in SI units by the function

$$y = 0.020 \sin (6.28x)$$

Given that the wave has a speed of 5.0 m/s, determine the maximum *transverse* speed of any bead on the string.

33. [II] A string has a mass per unit length of 2.50 g/m and is put under a tension of 25.0 N as it is stretched taut along the z-axis. The free end is attached to a tuning fork that vibrates at 50.0 Hz, setting up a transverse wave on the string having an amplitude of 5.00 mm. Determine the speed, angular frequency, period, and wavelength of the disturbance.

34. [II] Using Problem 33, write an expression for the profile of the wave given that, at $t = 0$, the end of the rope ($x = 0$) is at $y = +5.00$ mm. The wave travels in the z-direction.

35. [II] THIS PROBLEM EXPLORES THE BEHAVIOR OF AN OSCILLATING STRING. One end of a long horizontal string is passed over a frictionless pulley, and a 3.0-N weight is hung from it. The other end is tied to a hook in the wall. A 150-Hz oscillator sets up a transverse harmonic traveling wave on the string that has a peak-to-positive-peak distance of 45.0 cm. (a) What is the frequency and wave-

length of the wave on the string? (b) Determine the speed of the wave. (c) What is the tension in the string? (d) Write an expression for the speed in terms of the tension. (e) If the oscillating horizontal portion of the string is 2.0 m long, what is its mass?

36. [II] THIS PROBLEM EXPLORES THE BEHAVIOR OF A TRANSVERSE WAVE ON A TAUT ROPE. A long rope having a mass per unit length of 0.025 kg/m is stretched horizontally. Each end of the rope passes over a light frictionless pulley. A 10.0-kg mass is then hung from each end. (a) Determine the tension in the rope. (b) If someone plucks the rope, write an expression for the speed of the resulting wave in terms of the tension. (c) Compute the speed of the wave. (d) If a point on the rope is displaced transversely in SHM at 3.00 Hz, what will be the resulting wavelength?

37. [II] A heavy nylon guitar string with a linear mass-density of 7.5 g/m is stretched to a tension of 80 N. What is the speed of the transverse wave that can be generated on the string? What tension will be required to double the speed?

38. [II] The speed of a transverse wave on a wire is 90 m/s when the tension in the wire is 120 N. What will be the speed if the tension is reduced to 80 N?

39. [II] A homemade telegraph system sends transverse pulses along a stretched string. It operates between two neighboring houses using 12 m of a string having a total weight of 0.20 N. What should be the tension in the string if the signals are to travel at least as fast as they would were the users simply to yell? Take the speed of sound in air to be 333 m/s.

40. [III] A transverse harmonic wave on a long beaded string is described in SI units by the function

$$y = 0.02 \sin (6.28x - 15t)$$

If we have a detector at $x = 0$, what will be the speed of the bead at that location at a time $t = 1.2$ s?

41. [III] A uniform rope of length L and mass m hangs freely straight down from a hook in the ceiling. Taking the bottommost point as $y = 0$, write an expression for the mass of a segment of arbitrary length Δy. What is the tension in the rope at a height Δy? Derive an expression for the speed of a transverse wave set up by wiggling the lower end. What is the maximum speed? Determine an expression for the speed of the wave if a mass M is hung on the bottom of the rope.

42. [III] The very end of a 14.5-m-long wire having a mass of 0.12 kg is clamped in a vise. The other end is tugged on, putting a tension of 50 N in the wire. That end is then driven transversely by a tuning fork oscillating at 440 Hz, and the wire vibrates with a maximum displacement of 0.15 mm. (a) Find the wavelength and period of the resulting disturbance. (b) Determine the maximum transverse speed of any point on the wire.

43. [III] Given the equation

$$y = A \sin 2\pi \left(\frac{t}{0.01} - \frac{x}{40} \right)$$

describing the displacement of a string of beads in SI units, write an expression for the motion of a bead at $x = 5.0$ m as a function of time. What is the value of y at $t = 0$ and $t = 0.01$ s? When is $y = 0$? Graph your results.

SECTION 11.3: COMPRESSION WAVES

44. [I] According to folklore, one can best listen for trains, out in the countryside, by putting an ear to the rails. Steel track has a density of 7.8×10^3 kg/m^3 and a Young's Modulus of 200 GPa. What is the speed of compression waves in the track (the speed of sound in air is ≈ 330 m/s)?

45. [I] The speed of a compression wave (e.g., sound) in a metal rod is measured to be 4319 m/s. Given that the rod has a density of 19.3×10^3 kg/m^3, determine its Young's Modulus.

46. [I] Show that the $\sqrt{B/\rho}$ has the units of speed.

47. [I] An underwater explosion occurs in a fresh water lake. What is the speed of the resulting compression wave if the Bulk Modulus of water is 2.0 GPa? [*Hint: When treating liquids the wave speed is determined by the Bulk Modulus and the density.*]

48. [I] At 20 °C mercury has a density of 13.55×10^3 kg/m^3. The measured speed of sound (at 1 atm) in mercury is 1451 m/s. Determine its Bulk Modulus.

49. [I] What is the speed of a compression wave in ethyl alcohol?

50. [I] Considering Problem 48, what is the wavelength of a 1000-Hz compression wave in mercury? Compare that to its wavelength in air where $v = 331.4$ m/s.

51. [II] A particular kind of glass has a Shear Modulus of 25 GPa, a Young's Modulus of 62 GPa, and a density 2.5×10^3 kg/m^3. A 100-m-long rod of this material is struck at one end with a hammer. How long will it take for the compression wave to run down the rod, hit the end, and be reflected so that it returns to where it started?

52. [II] Use the method of dimensional analysis to "derive" an expression for the speed of a compression wave in a solid rod. We guess that the answer depends on Young's Modulus and the density and write

$$v = KY^a \rho^b$$

where a and b are to be determined and K is a dimensionless constant (like 2π). Finish out the calculation by balancing the units.

53. [II] Laboratory measurements show that a quantity of a certain liquid decreases in volume by 9.52×10^{-3} percent when it is exposed to a pressure increase of 200 kPa. If the liquid has a density of 1.025×10^3 kg/m^3, what is the speed at which compression waves like sound traverse it?

54. [II] Imagine a uniform wire of density ρ, cross-sectional area A, and modulus Y. What must the tension on the wire be (in terms of A and Y) such that both transverse and longitudinal waves on the wire have the same speed? Incidentally, the tension will have to be unrealistically high—as a rule, longitudinal waves travel much faster than transverse ones.

SECTION 11.4: ACOUSTICS: SOUND WAVES

55. [I] In music, the standard A$_4$ note has a frequency of 440 Hz. What are its period and wavelength at room temperature?

56. [I] A person standing at one side of a playing field on a cold winter night emits a brief yell. The short acoustical wavetrain returns 1.00 s later as an echo having "bounced off" a distant dormitory. Approximately how far away was the building?

57. [I] A string vibrating at 1000 Hz produces a sound wave that travels at 344 m/s. How many wavelengths will correspond to 1 m?

58. [I] A low-frequency loudspeaker is called a woofer. At what frequency will a tone have a wavelength equal to the diameter of a 15-in. woofer? Take the speed of sound to be 344 m/s.

59. [I] A groove on a monophonic phonograph record wiggles laterally such that its amplitude and frequency more or less correspond to the sound that is recorded. If at a given moment the needle is moving through the groove at 0.50 m/s, what would be the wiggle-wavelength for a 1.5-kHz tone?

60. [I] The speed of sound in ether (25 °C, 1 atm) is 976 m/s. What is the wavelength in ether produced by a tuning fork oscillating at 1000 Hz?

61. [I] The contact time between the hammers and strings of a piano is one of the determining factors of the tone of the instrument; if it's too long, the higher overtones will be damped out. Though contact time varies across the keyboard, an average is about half of a period. To two significant figures, what is the contact time for middle-C at 261.6 Hz?

62. [II] Suppose that a tuning fork in air, where the speed of sound is 343 m/s, produces a tone having a wavelength of 0.779 5 m. The fork is immersed in acetone and then tapped. How long a wave will be created in the liquid given that acetone (20 °C, 1 atm) supports sound waves that travel at 1203 m/s?

63. [II] A bar of aluminum alloy 10-m long with a cross-sectional area of 1.0 cm^2 has a mass of 2.7 kg and a Young's Modulus of 7.0 $\times$ 10^{10} N/m^2. If the end is tapped at a rate of 100 Hz, how long will it take for the sound wave to reach the other end of the bar? What will be its wavelength?

64. [II] Galileo determined experimentally that the fundamental frequency of a string fixed at both ends was inversely proportional to both its diameter and to the square root of its density. Prove that this description is true theoretically.

65. [II] The speed of sound (20 °C, 1 atm) in pure water is 1482.3 m/s as compared to 1522.2 m/s in seawater. What is the ratio of the wavelengths of a 1000-Hz tone in fresh water to that in seawater?

SECTION 11.5: WAVEFRONTS & INTENSITY

SECTION 11.6: THE SPEED OF SOUND IN AIR

66. [I] A detector placed perpendicular to a streaming sound wave has an active area of 10.0 cm^2 and records energy incident at a rate of 25.0 μJ/s. What is the intensity of the wave?

67. [I] Envision a sound wave incident perpendicularly on a detector that records an incident acoustic intensity of 50.0 $\times$ 10^{-4} W/m^2. If the sensing aperture of the detector is 4.0 cm in radius, what is the amount of power entering it?

68. [I] The intensity of the acoustic wave from an underground explosion measured 5.0 km away is 1.6 $\times$ 10^4 W/m^2. What intensity will be recorded at a site 50 km away? Assume no losses.

69. [I] The speaker in a sound system has a diameter of 38 cm. If it pumps out sound uniformly over its entire surface with an intensity of 10.0 mW/m^2, how much power is radiated?

70. [I] Air is pumped into a chamber that is then sealed, and the speed of sound through the gas is measured. The temperature of the gas is subsequently increased, and a gauge shows that the inside pressure doubled. Compare the speed of sound before and after the temperature rise.

71. [I] A measured quantity of air is put into a chamber that is fitted with a movable piston so that the pressure is kept constant. The temperature of the gas is subsequently increased and its volume doubles. Compare the speed of sound before and after the temperature rise. This is similar to what happens in the atmosphere when it's heated by the Sun.

72. [I] At what temperature will the speed of sound in air at standard pressure equal 320 m/s?

73. [I] If a tuning fork puts out a tone at 440 Hz, what is its wavelength in air at 25 °C?

74. [I] By what percentage does the speed of sound change when the air temperature rises from 0 °C to 30 °C?

75. [II] A point source of sound waves emits a disturbance with a power of 50 W into a surrounding homogeneous medium. Determine the intensity of the radiation at a distance of 10 m from the source. How much energy arrives on a little detector with an area of 1.0 cm^2 held perpendicular to the flow each second? Assume no losses.

76. [II] As a sound wave travels along it causes an oscillatory displacement of the air molecules that has an amplitude s_0. If a detector of area A is placed perpendicular to the propagation direction it can be shown that the average incident power is given by

$$P_{av} = \frac{1}{2}\omega^2 s_0^2 \rho A v$$

Check that this has the correct units. Write an expression for the intensity of the sound wave.

77. [II] An ideal gas (p. 438) is described by the expression $PV = nRT$ where n is the number of kilomoles present in the sample, T is the temperature, and R is a constant equal to 8.314 $\times$ 10^3 J/kmol·K. Most gases, air included, behave as if they were ideal gases. If M is the mass per kilomole, assuming Eq. (11.7) applies, prove that

$$v = \sqrt{\gamma RT/M}$$

Note that when n is expressed in kilomoles, R must be in terms of kilomoles, as well.

78. [II] Both gaseous hydrogen and oxygen consist of diatomic molecules. What is the ratio of the speed of sound in hydrogen to the speed in oxygen at the same temperature and pressure? [*Hint: Use the results of the previous problem.*]

79. [III] If the displacement of the oscillating air molecules in a sound wave is written in terms of a cosine function of amplitude A, then (with a bit of calculus and some effort) the corresponding gauge pressure can be expressed as

$$P = \frac{2\pi}{\lambda} BA \sin\left(\frac{2\pi}{\lambda}x - \omega t\right)$$

(a) Show that both sides of this expression have the same units. (b) What can you say about the displacement as compared to the pressure? (c) Show that the maximum value of the change in pressure (P_0) from the no-wave ambient value, the so-called *pressure amplitude*, is given by

$$P_0 = \frac{2\pi}{\lambda} \rho v^2 A$$

(d) Discuss, in general terms, why it is reasonable to have B and A in the equation for P.

80. [III] The pressure amplitude associated with the loudest sounds tolerable by human beings is roughly 30 Pa. Such pressure waves vary ±30 Pa with respect to atmospheric pressure (10^5 Pa). Determine (using the results of Problem 79) the maximum displacement of the air molecules in such a disturbance having a frequency of 1000 Hz at human body temperature. Take the density of the air to be 1.22 kg/m³.

SECTION 11.8: SOUND-LEVEL

81. [I] A faint sound with an intensity of 10^{-9} W/m² is measured by a sound-level meter—what will the reading be in dB?

82. [I] One of the most important acoustical characteristics of a room is its *reverberation time*, the time it takes for a sound to decrease 60 dB. What does that mean as far as the sound's intensity is concerned? For a concert hall, the reverberation time is typically 1 to 3 s.

83. [I] A radio playing quietly produces a sound intensity of about 10^{-8} W/m². What is the corresponding sound-level?

84. [I] Someone playing a CD at 60 dB wants to make the music twice as loud. At what sound-level should it be played?

85. [I] Two (otherwise identical) audio systems at a demonstration are blasting away, with one putting out 10 times the acoustic power of the other. What is the difference in their sound-levels?

86. [I] Two people having a normal conversation are 1.0 m from a sound-level meter. Approximately what will the meter read?

87. [I] Two audio systems each produce 50 W of acoustic power at the location of a microphone. What is the difference in their sound-levels in dB at that point?

88. [I] Does a 0-dB sound-level mean there is no sound? Explain.

89. [I] The intensity of a sound is tripled. By how many decibels does it increase?

90. [I] THIS PROBLEM TREATS THE RELATIONSHIP BETWEEN INTENSITY AND SOUND-LEVEL. We want to compute the sound-level given that the intensity of a sound wave is 6.0 μW/m². (a) Write an expression for the sound-level in terms of I. (b) What is the usual reference intensity? (c) Compute the sound-level in dB.

91. [I] THIS PROBLEM TREATS THE RELATIONSHIP BETWEEN INTENSITY AND SOUND-LEVEL. A vacuum cleaner generates a sound-level of 52 dB, and we need to find the corresponding intensity. (a) Write an expression for the sound-level in terms of I. (b) What is the usual reference intensity? (c) Plug in all the numbers you know and then raise 10 to the power of each side of the equation. (d) Using the fact that $10^{\log_{10} x} = x$, compute the intensity.

92. [I] On occasion, it turns out that we know the sound-level of a sound and wish to find out the associated intensity. Show that

$$I = 10^{\beta/10} I_0$$

93. [I] The noise of traffic registers 77 dB on a meter. To what intensity does that correspond?

94. [I] A cannon produces a 90-dB sound-level at a certain distance from a detector. What will the device read when two such cannons at that same distance are fired?

95. [II] What is the sound-level 10 m away from a point source radiating 1.2 W of acoustic power?

SOLUTION: To find the sound-level we need the intensity. We have the power at the point source and so at a distance R away $I = P/4\pi R^2 =$

$$(1.2 \text{ W})/4\pi(10 \text{ m})^2 = 9.55 \times 10^{-4} \text{ W/m}^2.$$

$$\beta = 10 \log_{10} \frac{I}{I_0} = 10 \log_{10} \frac{9.55 \times 10^{-4} \text{ W/m}^2}{1.0 \times 10^{-12} \text{ W/m}^2} = 90 \text{ dB}$$

96. [II] THIS PROBLEM TREATS THE RELATIONSHIP BETWEEN ACOUSTIC POWER AND SOUND-LEVEL. A point source emits sound uniformly in all directions, at a power that we would like to determine. An observer standing 5.0 m from the source measures a sound-level of 40 dB. (a) Write an expression for the sound-level. (b) What is the usual reference intensity? (c) Determine the intensity 5.0 m from the source. (d) What is the power radiated at the source?

97. [II] THIS PROBLEM DEALS WITH THE RELATIONSHIP BETWEEN INTENSITY AND SOUND-LEVEL. A large engine has a decibel noise level of 120 dB above the threshold of hearing, and we want to find the intensity of the sound. (a) What physical quantity does 120 dB correspond to? (b) Write an expression for the sound-level in terms of intensity. (c) Determine the intensity of the sound. (d) If two such machines were operating right next to each other, what would be the intensity of the noise nearby? (e) What would be the sound-level?

98. [II] Two sounds have intensities of 10^{-2} W/m² and 10^{-10} W/m². What will be the difference in their sound-levels as read by a sound-level meter?

99. [II] A small but noisy printer produces an acoustic intensity of 56×10^{-5} W/m² at a point 5.0 m away. Approximately what value will a dB-meter read at that location and, again, at 20.0 m from the printer?

100. [II] With Problem 99 in mind, what will be the *change* in sound-level measured for any point source at 5.0 m and then at a 20.0-m distance?

101. [II] A small source emits nearly spherical waves with an acoustic power of 60 W. How far away must a sound-level meter be if it is to read 60 dB?

102. [II] The sound-level of a large engine is measured to be 130 dB at a distance of 10.0 m. Approximately what intensity will exist at a point 100 m away?

103. [II] The sound-level 2.0 m from a pneumatic chipper is 120 dB. Assuming it radiates uniformly in all directions, how far from it must you be in order for the level to drop 40 dB down to something more comfortable?

104. [II] How much acoustic power impinges on a 10-cm² detector when the intensity-level is 70 dB?

105. [II] Given that the standard reference pressure for sound is $P_0 = 2 \times 10^{-5}$ Pa, write an expression for the sound-level in terms of the pressure of the disturbance P. [*Hint: p. 392.*]

106. [II] Someone turns on a radio at 65.0 dB while vacuuming the floor at 80.0 dB. What will be the total sound-level in the room?

107. [III] If one audio system sounds six times louder than another, what is the difference in their sound-levels?

SECTION 11.9: SOUND WAVES: BEATS
SECTION 11.10: STANDING WAVES

108. [I] Two sound waves of angular frequencies 900.0 rad/s and 896.0 rad/s overlap. What is the resulting beat frequency?

109. [I] The sound from a tuning fork of 1000 Hz is beat against the unknown emission from a vibrating wire. If beats are heard at a

frequency of 4 Hz, what can be said about the frequency of the wire?

110. [I] With Problem 101 in mind, what can you say if a small piece of tape is fixed to the tuning fork and the beat frequency now increases?

111. [I] On tapping two tuning forks, an observer hears a succession of intensity maxima arriving at a rate of one every 0.99 s. What is the difference in frequency between the two forks?

112. [I] THIS PROBLEM EXPLORES THE PHENOMENON OF BEATS. Two tuning forks are struck and they produce sounds of frequency 440 Hz and 444 Hz that subsequently overlap. (a) Describe what an observer will hear? (b) Determine the frequency of the resulting wave? (c) Compute the number of beats per second?

113. [I] Two waves of wavelength λ, traveling in opposite directions give rise to the standing-wave pattern shown in Fig. P113. Determine λ.

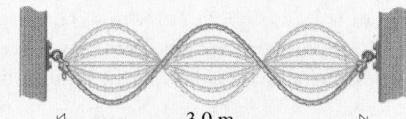

Figure P113 $\longmapsto$————3.0 m————$\longmapsto$

114. [I] THIS PROBLEM TREATS THE PHENOMENON OF STANDING WAVES. A 1.20-m long guitar string is held under tension and made to vibrate, with a central antinode, at middle C (262 Hz). (a) What is the frequency of that oscillation? (b) Draw a picture of the vibrating string. (c) Determine the wavelength of the fundamental oscillation of the string. (d) What is the frequency of the resulting sound? (e) What is the wavelength of the resulting sound wave, given that the speed of sound is 341 m/s?

115. [I] A standing-wave pattern consists of a series of nodes and antinodes. If the distance between successive antinodes is N meters, what are the wavelengths of the two waves that produce this pattern?

116. [I] A thin wire is stretched between two posts 50 cm apart. It is then bowed and thereby set into oscillation. What are the wavelengths of the fundamental and the first overtone of the system?

117. [I] A taut string is fixed at both ends, which are 0.50 m apart. It is then set into resonance at its sixth harmonic. Determine the wavelength of the oscillation and draw the standing-wave pattern.

118. [I] A string stretched between fixed posts is 250-cm long and oscillates in its fundamental mode at 100 Hz. Determine the speed of a transverse wave on the string.

119. [I] A piece of steel piano wire is held fixed at both ends under a tension of 100 N. The free length of wire is 1.00 m, and it has a mass of 2.5 g. What is its fundamental frequency?

120. [I] A narrow tube 1.00 m long is closed rigidly at one end and with a piston at the other. Given that the speed of sound is 335 m/s, what is the frequency of the tube's fundamental oscillatory mode?

121. [I] Show theoretically that for a string fixed at both ends, its standing-wave frequencies are given by

$$f_{\text{N}} = \tfrac{1}{2}N\sqrt{\frac{F_{\text{T}}}{Lm}}$$

where L is its length and m its mass.

122. [I] A trumpet is a bent tube roughly 140-cm long and closed at one end by the player's mouth. Determine the fundamental (which will be quite difficult to blow) and the first three overtones. Take the temperature to be 20°C.

123. [II] THIS PROBLEM DEALS WITH THE CONCEPT OF BEATS. A taut wire vibrates making a constant tone. When a 260-Hz tuning fork is struck, 6.0 beats/s result and when a 270-Hz tuning fork is struck, 4.0 beats/s result. (a) What possible frequencies of the wire would produce 6.0 beats/s? (b) What possible frequencies of the wire would produce 4.0 beats/s? (c) Determine the frequency at which the wire vibrates.

124. [II] THIS PROBLEM WILL HELP US UNDERSTAND STANDING WAVES. The generator in Fig. P124 sends a wave down the string that reflects back off the wall, which is 4.0 m away. These two oppositely directed waves, each traveling at 2.0 m/s, create the pattern shown. (a) How many nodes, and how many antinodes are there? (b) Looking at the diagram, what is the wavelength of the wave produced by the generator? (c) Determine the frequency of the generator.

Wave generator

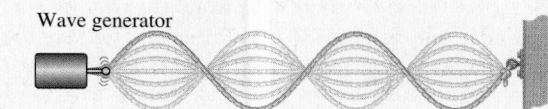

Figure P124

125. [II] A B-string from a guitar is held fixed at both ends under tension with a vibrating length of 33 cm. It oscillates at its fundamental frequency of 246 Hz. What are the wavelengths of the transverse wave on the string and the sound wave in the air at room temperature? Take the speed of sound in the air to be 344 m/s.

> **SOLUTION:** The wavelength on the string is twice the node-to-node distance of the fundamental: $\lambda = 2(33 \text{ cm}) = 66 \text{ cm}$. For the sound wave in air $\lambda = v/f = (344 \text{ m/s})/(246 \text{ Hz}) = 1.40 \text{ m}$

126. [II] Imagine a hypothetical piano with all strings made of the same material and all under the same tension. The piano extends from 27.5 Hz to 4186 Hz, which is over seven octaves—that is, seven doublings of frequency. If the highest note corresponds to a string 15-cm long, how long will the lowest string have to be? What can you conclude about this approach to piano design?

127. [II] Two identical piano strings are both tuned to 440 Hz. The tension in one is then increased by 1.00%, and both strings, which are kept at the same length, are activated so that they sound their fundamental frequencies. What will be the resulting beat frequency?

128. [II] A narrow glass tube 0.50-m long and sealed at its bottom end is held vertically just below a loudspeaker that is connected to an audio generator and amplifier. A tone with a gradually increasing frequency is fed into the tube, and a loud resonance is first observed at 170 Hz. What is the speed of sound in the room?

129. [II] An organ pipe that ordinarily sounds at 600 Hz at 0 °C is connected to a source of helium at that temperature. At what frequency will it now operate?

130. [II] A quartz tube open at both ends has a fundamental resonant frequency of 200 Hz at 0 °C. Neglecting any changes in length, by how much will the fundamental change when the tube is sounded in a chamber at 40 °C?

131. [II] A wire stretched between two posts and under a tension of 200 N oscillates at a fundamental frequency of 420 Hz. At what tension would it oscillate instead at 430 Hz?

132. [II] Referring to Kundt's tube (Fig. Q19), if v_a and v_r are the speeds of sound in the air and rod, respectively, and if L is the length of the rod and l the node-to-node distance, show that

$$v_r = \frac{v_a L}{l}$$

In other words, knowing the speed of sound in air, we can determine it in the material of the rod.

133. [II] A copper bar 1.00-m long is clamped at its middle and set vibrating. Given that $v = \sqrt{Y/\rho}$ and the bar has a density of $8.9 \times 10^3 \ kg/m^3$ and a Young's Modulus of $11 \times 10^{10} \ N/m^2$, what will its frequency be?

134. [II] A C-flute with all its holes covered plays a middle C (262 Hz) as its fundamental. Assuming room temperature (20°C) and overlooking end corrections, how long should the flute be from embouchure hole to end?

135. [III] Consider the expressions for two waves of the same amplitude with slightly different frequencies traveling in the same direction:

$$y_1 = A \sin(\omega_1 t) \qquad \text{and} \qquad y_2 = A \sin(\omega_2 t)$$

where we simplify things a little by just looking at them at the point $x = 0$. Now, supposing the two waves were to overlap, derive the following equation for the combined waveform:

$$y = 2A \cos\left[\tfrac{1}{2}(\omega_1 - \omega_2)t\right] \sin\left[\tfrac{1}{2}(\omega_1 + \omega_2)t\right]$$

and interpret each term in reference to Fig. 11.37.

136. [III] Consider a long, narrow rod of length L clamped at both ends. If the rod is rubbed with a rosined cloth, a compression wave will travel its length. Show that the expression for the resulting standing-wave modes is

$$f_N = \frac{N}{2L}\sqrt{\frac{Y}{\rho}}$$

where $N = 1, 2, 3, \ldots$. A novice playing with a violin is likely to move the bow lengthwise along a string. Explain why this movement will generally produce a shrill squeaking sound.

SECTION 11.11: THE DOPPLER EFFECT

137. [I] THIS PROBLEM EXAMINES THE RELATIONSHIP BETWEEN THE MOTION OF A SOUND SOURCE AND ITS PERCEIVED WAVELENGTH. A 310-Hz siren is on the roof of a car traveling at 31.0 m/s toward an observer who is at rest. If the speed of sound is 341 m/s, (a) what is the wavelength measured by the observer when the car is at rest? (b) What is the wavelength the observer measures when the car approaches? (c) What is the wavelength the observer measures when the car recedes from it?

138. [I] A hawk flying horizontally screeches at a central frequency of 900 Hz. A person standing still on the ground sees the bird flying away at 10.0 m/s. At what frequency does she hear the screech? Take the speed of sound to be 341 m/s.

139. [I] A police car, its siren blaring at 1000 Hz, is traveling at 20.00 m/s while chasing a garbage truck moving at 15.00 m/s a block in front of it. What apparent frequency will the garbage collectors hear? The speed of sound is 330.0 m/s.

140. [I] A train whistle is blown by an engineer who hears it sound at 650 Hz. If the train is heading toward a station at 20 m/s, what will the whistle sound like to a waiting commuter? Take the speed of sound to be 340 m/s.

141. [I] A car with its horn blaring at 500 Hz passed a woman standing on the street at 25 m/s. What frequency did she hear as the car receded into the distance? The speed of sound that day was 344 m/s.

142. [I] An ultrasonic wave at 80 000 Hz is emitted into a vein where the speed of sound is about 1.5 km/s. The wave reflects off the red blood cells moving toward the stationary receiver. If the frequency of the returning signal is 80 020 Hz, what is the speed of the blood flow?

143. [I] A man running toward the stage in a theater hears an A_4 note from a stationary tuning fork to have a frequency of 441 Hz instead of its more normal 440 Hz. About how fast is he going?

144. [II] A point source of sound on top of a police car emits a signal at 1000 Hz. If the car is traveling in a straight line at 30 m/s, what will be the wavelength perceived by people standing on the road both directly in front of and behind the car? Take the speed of sound to be 335 m/s. What is the wavelength as measured in the car?

145. [II] A bat flying straight toward a wall at 11 m/s emits a pulse at 60 kHz. What is the frequency of the reflected ultrasonic echo wave reaching a stationary observer standing under the bat. Take the speed of sound to be 341 m/s.

146. [II] THIS PROBLEM TREATS THE DOPPLER EFFECT WHEN SOUND IS REFLECTED BACK TO A MOVING SOURCE. A 310-Hz siren is on the roof of a police car traveling at 31.0 m/s directly toward a large wall. The speed of sound is 341 m/s. (a) As far as an observer at the wall is concerned, what is the positive direction? (b) What is the frequency (f_w) measured by that observer standing in front of the wall? Consider the sound reflected off the wall, and take the wall as a source at rest. (c) What is the positive direction now? (d) Determine the frequency of the reflected sound as measured by the approaching cop?

147. [II] An observer with a stationary source of sound sends out a signal directly toward an approaching target. Write an expression for the frequency of the beats heard by the observer.

148. [II] A sound wave at 1000 Hz is sent out from a stationary source toward a target approaching at 20.0 m/s. What will be the frequency of the returning signal? Take the speed of sound to be 340 m/s.

Chapter 12

Thermal Properties of Matter

*C*hapters 12, 13, and 14 are mostly about internal energy, the energy stored via the motion and interaction of the atoms of a system. The discipline is called *thermal physics*, and it deals, broadly speaking, with temperature (Chapter 12) and the transfer (Chapter 13) and transformation (Chapter 14) of energy. The study begins with our rather crude bodily appreciation of hot and cold; with the what-can-you-put-in-your-mouth-without-burning-it understanding of temperature. Our sensory perception of temperature, however unreliable, must correspond to some unseen aspect of the physical world. We will find that, *even though temperature may well be a basic quantity like mass and time, it can usually be associated with the concentration of thermal energy in a material system.* This chapter deals with a variety of temperature-dependent properties of bulk matter.

Temperature

Like so many other physical quantities, temperature was measured long before it was understood. Galileo appears to have invented (ca. 1592) the first device for indicating "degrees of hotness." He simply placed the end (Fig. 12.1) of an inverted narrow-necked flask, warmed in his hands, into a bowl of water (or wine). As it cooled, the liquid was drawn up, partially filling the neck. Air captured in the bulb at the top either expanded or contracted when subsequently heated or cooled, and the column fell or rose proportionately.

The medical applications of the thermometer were recognized almost immediately, and normal body temperature became a focus of interest. In 1631, J. Rey, a French physician, inverted Galileo's device, filling the bulb with water and leaving much of the stem with air in it. In that configuration, which is more like today's thermometer, the liquid's expansion operates the device. Within 70 years of Galileo's invention, sealed pock-

STUDY GUIDE

The chapter begins with a discussion of temperature (p. 423) and the three scales by which it's measured: Fahrenheit, Celsius, and most importantly, **Kelvin**. Make sure you understand their interrelationships. The chapter also treats the **thermal expansion** of solids and liquids (p. 429), and the behavior of gases (p. 435).

Figure 12.1 Galileo's thermoscope is not called a thermometer because the scale was arbitrary. The egg-sized globe at the top is the sensor. The gas within it expands or contracts, and the liquid level rises and falls.

et-sized thermometers containing either alcohol or mercury were in use. (Both freeze at much lower temperatures than water, and sealing the devices isolated them from atmospheric pressure changes.)

By exploring the behavior of matter with the aid of the thermometer, it soon became evident that there were a number of physical occurrences that happened at fixed temperatures. Boyle, Hooke, and Huygens independently recognized (1665) that that fact could provide a reliable reference point for any thermometer. Hooke suggested using the freezing point of water, and Huygens offered its boiling point. C. Renaldini (1694) used both the freezing and boiling points of water to standardize two widely spaced points on his thermometer (something Newton would do independently seven years later).

The scale we still use in the United States is a gift of the instrument maker G. D. Fahrenheit (1717). Not wanting to deal with negative values, he set the 0 mark at the coldest temperature he could produce, that of a mixture of water, ice, and sea salt. Following Newton's lead, Fahrenheit set the upper reference at normal human body temperature, which he took as 96 (probably because that number was divisible by 12 and easily halved and quartered). Upon dividing the distance between these two fixed points into 96 equal-sized divisions, or degrees, he could then extend the scale as high or low as he wished. That done, water froze at what turned out to be 32 °F (which is read *degrees Fahrenheit*) and boiled at 212 °F.

The scale that is still used in modern scientific work (although it is gradually being replaced) is associated with the name of a man who neither first proposed it nor ever actually produced it. Anders Celsius (1742) used the freezing and boiling of water for the reference points and then divided the distance between them into 100 equal parts. Strangely enough, Celsius set the temperature of freezing water at 100° and boiling at 0°. Some years later, with these two numbers more reasonably interchanged (Fig. 12.2), the arrangement came to be known as the *centigrade scale* (from the Latin *centum*, meaning 100, and *gradus*, degree). The Tenth International Conference on Weights and Measures (1954) changed the name to the *Celsius scale*.

Since the Fahrenheit scale goes from 32° to 212° (a total of 180 divisions) and the Celsius scale goes from 0° to 100°, each Fahrenheit degree is smaller than a Celsius degree. A change of 1 C° is equivalent to a larger change of (180/100) F°, or (9/5) F°. A change from freezing (0 °C) of a certain number of Celsius degrees, T_C, corresponds to a change from freezing (32 °F) of $(9/5)T_C$ Fahrenheit degrees. In short,

$$T_F = 32° + \tfrac{9}{5}T_C \qquad \text{or} \qquad T_C = \tfrac{5}{9}(T_F - 32°) \qquad (12.1)$$

Rather than memorizing these equations, it might be easier to make the transformations logically. For example, human body temperature is 98.6 °F, which is 66.6 Fahrenheit degrees above freezing—equivalent to (5/9)66.6 = 37.0 of the larger Celsius degrees above freezing (0 °C). Hence, body temperature is 37.0 °C.

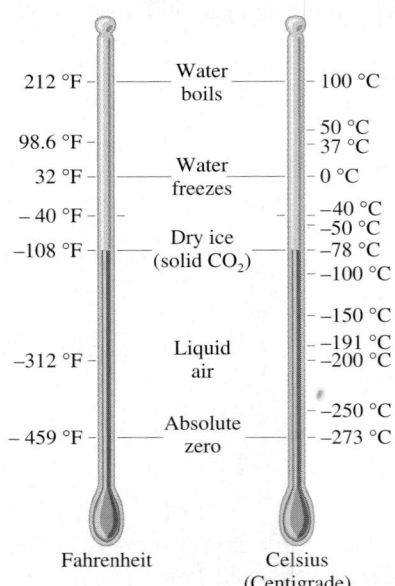

Figure 12.2 A comparison of the Fahrenheit and Celsius temperature scales.

At present we know of no purely mechanical quantity—that is, one expressible in terms of mass, length, and time only—which can be used, however inconveniently, in place of temperature. We are inclined to conclude that temperature probably is itself a basic concept.

A. G. WORTHING (1940)

Example 12.1 **[I]** Two thermometers, one marked in Fahrenheit and the other in Celsius, are placed in a bath. At what temperature will both thermometers read the same?

Solution This problem compares Fahrenheit and Celsius temperature scales. (1) TRANSLATION—Determine the temperature at which both the Fahrenheit and Celsius scales have the same numerical value. (2) GIVEN: $T_F = T_C$. FIND: T_F. (3) PROBLEM TYPE—Thermo/temperature scales. (4) PROCEDURE —Use either relationship in Eq. (12.1). (5) CALCULATION— Setting $T_F = T_C$, we have

$$T_F = 32° + \tfrac{9}{5}T_F$$

$$-\tfrac{4}{5}T_F = 32°$$

and

$$\boxed{T_F = T_C = -40°}$$

The two scales are the same at −40°.

Quick Check: $T_C = (5/9)(T_F - 32°) = (5/9)(-72°) = -40°$.

{For more worked problems click on **WALK-THROUGHS** *in* **CHAPTER 12** *on the CD.}*

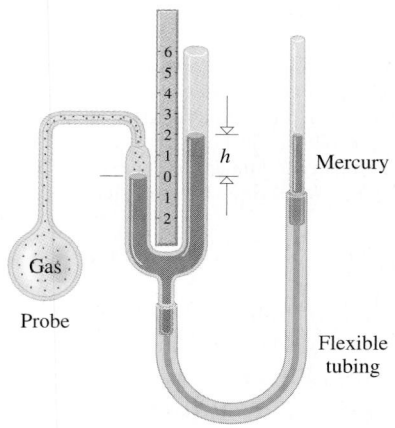

Figure 12.3 A constant-volume gas thermometer. The chamber at the left is the sensing bulb. The gas in the bulb changes volume when it experiences a temperature change. That causes the levels of mercury in the U-tube manometer to shift, and *h* gives us the pressure in the probe. Lifting the movable mercury column on the right shifts the other two mercury levels and allows the volume of gas to be kept constant.

As we will see, the manner in which any material—mercury, alcohol, water, glass, whatever—expands on heating is characteristic of that material. What this means practically is that thermometers that differ in their physical construction and yet have the same two fixed reference points will necessarily agree exactly only at those two points.

The range of temperatures we deal with is extensive, and there is now a whole arsenal of different kinds of thermometers, each with its own virtues. The familiar mercury-in-glass instrument is only useful between the points where mercury freezes and glass melts. Moreover, it can be used reliably only when its presence doesn't affect the temperature being determined. If you want to measure the temperature of a flea or a thermonuclear fireball, the old mercury-in-glass standby will not be of much help. Accordingly, there are electrical resistance thermometers, optical thermometers, thermocouples, and constant-volume gas thermometers, to name a few. Of these, the standard for accuracy and reproducibility is still the constant-volume gas instrument (Fig. 12.3), though it is large, slow, delicate, and inconvenient.

12.1 Thermodynamic Temperature & Absolute Zero

The Fahrenheit and Celsius scales have little or nothing to do with any fundamental aspect of temperature. After all, the freezing point of water at Earth's atmospheric pressure has no obvious relationship to any basic aspect of the Universe. Presumably, an alien scientist on a planet that has no water will devise a thermometer that measures temperature equally well. More to the point, is there a universal zero of temperature linked to the very essence of matter and energy, a zero that all scientists (human or otherwise) might discover? The answer is yes, and it's called the *absolute zero of temperature*.

Sometime around 1702, G. Amontons devised an improvement on Galileo's thermoscope that has evolved into the modern constant-volume gas thermometer (Fig. 12.3). Figure 12.4 shows the behavior of a typical gas as its temperature is lowered. The important feature is that the graph is a straight line—the pressure falls linearly until the gas liquifies and the process abruptly ends. Remarkably, all gases behave nearly the same way. Moreover, when the straight line for any gas is extended downward, it crosses the zero-pressure axis at a temperature of −273.15 °C. Similarly, if the pressure is kept constant, the volume of a gas will collapse linearly, again approaching the zero-volume axis at the same −273.15 °C (i.e., −459.7 °F), as in Fig. 12.5. This point is now believed to be the very limit of low temperature, **Absolute Zero**.

Amontons, studying the behavior of a number of gases, showed that each volume changed by the same percentage for a given change in temperature. Astutely, he recognized that a decrease in pressure down to zero would accompany a decrease in temperature down to some limiting value of coldness. Later, J. H. Lambert repeated these experiments with greater accuracy, concluding (ca. 1779) that there was indeed a temperature limit. "Now a degree of heat equal to zero," wrote Lambert, "is really what may be called absolute cold."

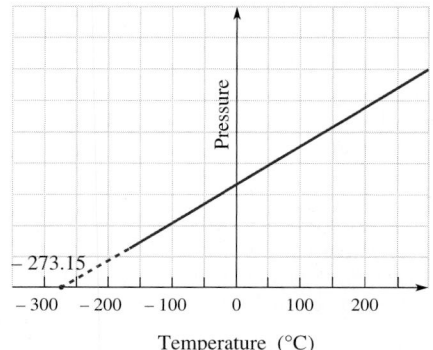

Figure 12.4 Pressure versus temperature for a gas at constant volume. As the gas is cooled, its atoms lose some of their thermal energy; they travel more slowly and collide with the chamber walls less frequently and less forcefully. As a result, the pressure in the chamber drops. All gases behave essentially the same way. Their *P* versus *T* graphs have different slopes, but they are all straight lines, and they all head toward −273.15 °C, even if they liquify before getting there.

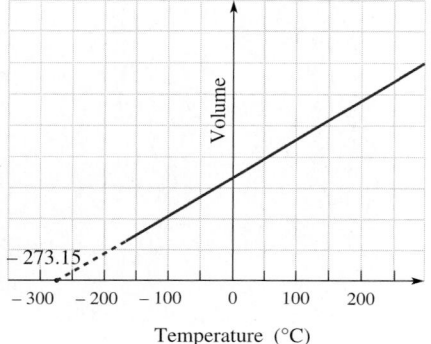

Figure 12.5 Volume versus temperature for a gas at constant pressure. As the temperature of a gas is lowered, its atoms are less energetic. To keep the pressure constant, the volume must decrease, which will make it easier for atoms to strike the chamber walls frequently, thereby sustaining the pressure. Although they liquify first, all gases shrink toward zero volume at $T = -273.15$ °C.

EXPLORING PHYSICS ON YOUR OWN

Defining Temperature Via Our Senses Is Naive: Take three bowls. In one put the hottest water your hands can tolerate comfortably; in another put ice cubes and cold water; fill the third with lukewarm water. Immerse both hands in the lukewarm bath for about a minute. Note what temperature you sense with each hand. Now place your right hand in the hot water and your left in the cold water for a minute. Remove them, quickly dry them and immediately immerse both back in the lukewarm bath. Surprise! What do you feel now? We'll come back for an explanation later (p. 476).

Absolute Zero certainly seems the place to begin a temperature scale. Still, the idea received little attention for well over half a century. Then, in 1848, William Thomson (later to become Lord Kelvin) formalized the concept theoretically. On the basis of very different thermodynamic considerations (p. 509), he proposed an **absolute temperature scale** that turned out to be in perfect agreement with the results from the constant-volume gas thermometer.

Today, the practical laboratory temperature reference is the *triple point* of water: the unique circumstance of temperature (0.01 °C) and pressure (610 Pa) where solid, liquid, and vapor coexist. Absolute Zero (−273.15 °C) and the triple point (0.01 °C) are separated by 273.16 Celsius degrees.

Temperature is now recognized as a fundamental physical quantity like mass, length, and time, so the old degree symbol is discarded. As with all other quantities, the **thermodynamic** or **absolute temperature** is given a unit, the *kelvin* (abbreviated K). The SI value of 1 K is defined as 1/273.16 of the temperature of the triple point of water (p. 441). Note that 1 kelvin is effectively the same size as 1 degree Celsius. For practical reasons (not the least of which are all the volumes of research data already in existence), the scheme is constructed to mesh with the Celsius system (Fig. 12.6). Using an arrow to mean "equivalent to," Absolute Zero is 0 K → −273.15 °C ; therefore 0 °C → 273.15 K , and room temperature is just about 293 K (which is read 293 kelvins). The numerical equivalent is summarized symbolically by

$$T \rightarrow T_C + 273.15 \qquad (12.2)$$

This expression, awkward as far as units are concerned, is more a statement of the logic of making numerical conversions than it is an equation—the 273.15 can be either degrees or kelvins.

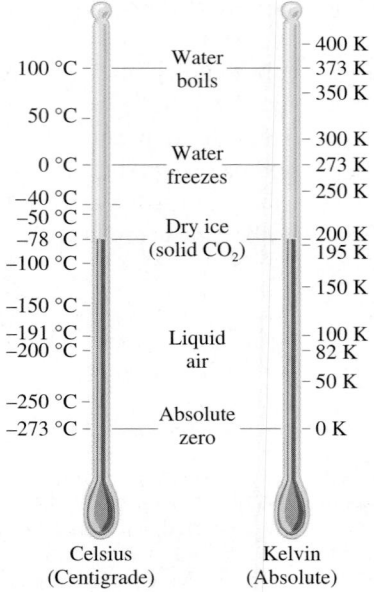

Celsius (Centigrade)		Kelvin (Absolute)
100 °C	Water boils	400 K / 373 K / 350 K
50 °C		300 K
0 °C	Water freezes	273 K / 250 K
−40 °C / −50 °C / −78 °C / −100 °C	Dry ice (solid CO$_2$)	200 K / 195 K
−150 °C		150 K
−191 °C / −200 °C	Liquid air	100 K / 82 K / 50 K
−250 °C / −273 °C	Absolute zero	0 K

Figure 12.6 A comparison of the Celsius and Kelvin scales. All scientific work is done in these two systems, with the Kelvin scale being preferred.

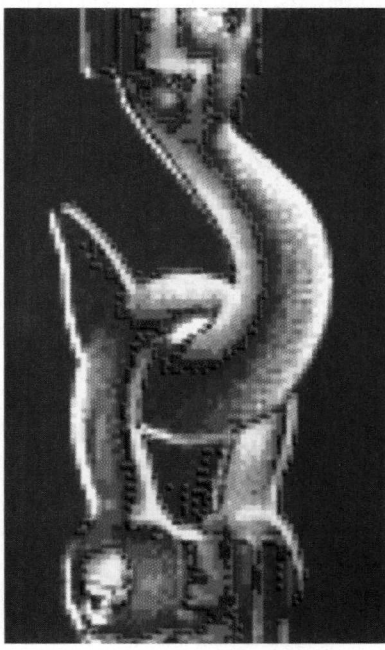

In a process called thermoelasticity, a solid undergoes minute changes in temperature when it's stretched or compressed. Here a hook is supporting a 700-lb load. The inner cooler orange region of the hook is under tension, while the outer warmer blue region is under compression. The dark areas experience no stress. The camera that took this picture had a temperature resolution of 1.0 mK.

Example 12.2 **[I]** Gold freezes at 1337.58 K. What is the equivalent Celsius temperature?

Solution   This problem clearly deals with Kelvin and Celsius temperatures. (1) TRANSLATION—Knowing a temperature in kelvins, determine the equivalent in Celsius.   (2) GIVEN: $T =$ 1337.58 K. FIND: T_C. (3) PROBLEM TYPE—Thermo/temperature scale conversion.   (4) PROCEDURE—Eq. (12.2) summarizes the relationship between thermodynamic temperature and Celsius. (5) CALCULATION—Subtracting 273.15 from both sides yields

$$T_C = (1337.58 - 273.15)\,°C = \boxed{1064.43\,°C}$$

Quick Check: This result is 1064.43 °C above 0 °C, which is $\approx 273\,°C$ above 0 K. Hence, it is ≈ 1337 K above 0 K.

The Range of Temperatures

The Universe sustains an incredible temperature range that's completely alien to our luke-warm intuition: we live out our delicate lives within a tiny band of hot and cold (see Table 12.1). The highest temperatures likely to exist at this moment are to be found deep within certain stars: $\approx 4 \times 10^9$ K seems to be the theoretical extreme. At a temperature only about 10 times higher, matter fragments into subatomic particles. A hydrogen bomb ignites at

Lord Kelvin, born William Thomson (1824–1907), was one of the most accomplished scientists of the nineteenth century.

Man has no Body distinct from his Soul;
for that called Body is a portion of Soul
discern'd by the five Senses, the chief
inlets of Soul in this age.
Energy is the only life and is from the Body;
and Reason is the bound or outward
circumference of Energy.
Energy is Eternal Delight.

WILLIAM BLAKE (1793)
THE MARRIAGE OF HEAVEN AND HELL

Table 12.1

Temperatures of Several Physical Phenomena

°C		°F
−273.2	Absolute Zero	−459.7
−269	Helium boils	−452
−196	Nitrogen boils	−320
−183	Oxygen boils	−297
−79	Dry ice (CO_2) freezes	−109
−39	Mercury freezes	−38
0	Water freezes	32
3.8	Heavy water freezes	38.9
≈20	Room temperature	≈68
31	Butter melts	88
≈37	Body temperature	≈98.6
≈54	Paraffin melts	≈129
78	Alcohol boils	172
100	Water boils	212
101.4	Heavy water boils	214.6
108	Saturated salt solution boils	226
232	Tin melts	449
327	Lead melts	621
445	Sulfur boils	833
657	Aluminum melts	1215
801	Salt (NaCl) melts	1473
961	Silver melts	1762
1063	Gold melts	1945
1083	Copper melts	1981
1000–1400	Glass melts	1830–2550
1300–1400	Steel melts	2370–2550
1530	Iron melts	2786
1620	Lead boils	2948
1774	Platinum melts	3225
1870	Bunsen burner	3398
2450	Iron boils	4442
≈3410	Tungsten melts	≈6170
3500	Oxyacetylene flame	6332
5500	Carbon arc	9932
6000	Surface of the Sun	10 832
6020	Iron welding arc	10 868

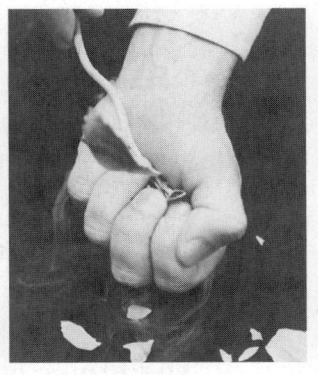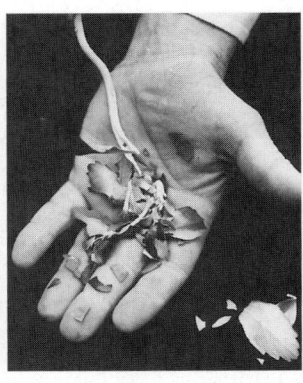

A sprig of pachysandra frozen in a bath of liquid nitrogen. Holding the plant in the liquid for a minute or so drops its temperature down to −196 °C (i.e., −320 °F). The leaves are then as brittle as glass and shatter into sharp fragments when squeezed. The sprig is so cold it is able to condense the water vapor in the air into streamers. The hand isn't harmed because the mass of the sprig is small, as is its heat capacity (p. 460).

roughly 40×10^6 K and blazes up to nearly 10 times that, whereas the innards of the Sun churn away at a mere 15×10^6 K. Atoms can form, if only momentarily, in the seething body of a star even at 10^5 K.

At temperatures lower by a factor of 10, matter exists in clouds of atoms, ions, electrons, and an occasional molecule. There is evidence for the presence of a variety of molecules in cool spots (< 5000 K) on the surface of the Sun. If we take a cloud of steam up to ≈3000 K, about one-quarter of the water molecules rupture into separate atoms of oxygen and hydrogen. The hottest thing you are likely to find around the house is a tungsten lightbulb filament, which operates at ≈2800 K (2500 °C). By comparison, lava is molten at ≈2000 K. Lead melts at 327 °C (621°F), paper burns at ≈230 °C (450 °F), and now we are down to the temperatures of baking apple pie and a hot cup of tea.

Ice melts at 273.15 K (0 °C), but if salt (NaCl) is added to icy slush, the temperature of the mix drops as low as 252 K (−21°C). That's the way people traditionally cool homemade ice cream during churning. Mercury freezes at 234.29 K (−38.87 °C). Adding four parts by weight of crystalline calcium chloride to three parts of water will produce a mix at 218 K (−55 °C). Dry ice, solid CO_2, exists at 194 K (−79 °C), and by adding ether or alcohol, the resulting rapid evaporation lowers the temperature further. By pumping on the mixture, carrying off the vapor, temperatures as low as ≈163 K (−110 °C) can be achieved.

The mainstay of the modern low-temperature laboratory is liquid nitrogen, a clear, colorless fluid that looks like water and boils away at about 77.4 K (−196 °C). It is easy to handle and relatively inexpensive (about the price of beer). Hydrogen gas can also be liquefied, but not until it is brought down to 20.4 K (−253 °C). The atoms within the molecules of any substance at that penetrating cold are still capable of motion, but they do not have enough energy to rearrange themselves into new compounds; all chemical and biological activity has long ceased. Still, the atoms vibrate, and a piece of frozen oxygen carries compression waves like any other solid.

The most exotic of all cryogenic materials is liquid helium. Clear, colorless, and incredibly cold, it boils in an open container at 4.25 K (−268.9 °C). Difficult to handle, it must be stored in a special thermos bottle surrounded by liquid nitrogen. It vaporizes easily, and, since liquid helium sells at the price of a decent champagne, that's to be avoided. By pumping away the evaporating helium, the remaining liquid is further

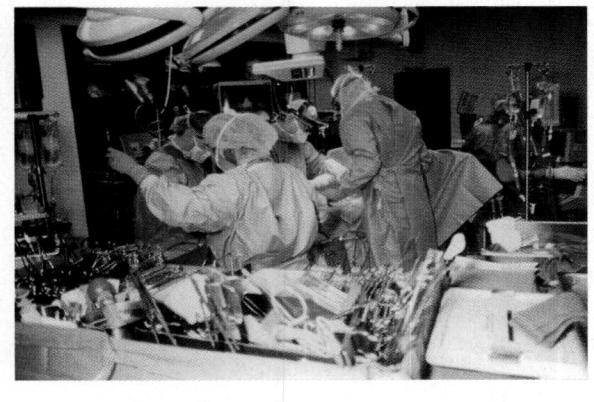

During controlled hypothermia a patient's body temperature is appreciably lowered. The procedure is often used in brain or heart surgery where normal blood circulation might be interrupted for long periods of time. An organ's need for oxygen decreases with temperature. Accordingly, cardiac operations are often performed in the range from 10 °C to 15 °C as compared to normal body temperature of 37 °C.

cooled, and the temperature can be lowered appreciably. At 2.18 K, helium undergoes a dramatic change, becoming not a solid but a so-called *superfluid*. In that state, it is the best-known conductor of heat, though it remains an electrical insulator.

Dropping the temperature of a system becomes increasingly difficult and more costly in terms of energy expended, the lower we go. In order to cool a sample much below 1 K, a technique called adiabatic demagnetization is generally used; with it, solid specimens have been taken down to about 0.003 K. To plunge still lower, an even more subtle method of nuclear adiabatic demagnetization is employed. In the quest for Absolute Zero, bulk matter has been cooled to below ≈0.000 000 02 K.

At these low temperatures, only a faint rustling of atoms persists. Yet, enough energy remains in the atoms to keep superfluid helium from solidifying, making it the only liquid that does not become a solid under ordinary pressure. To form crystalline helium, the atoms must be forced close together. Even below 1 K, where there is hardly any disruptive atomic jiggling left, it still takes about 30 times atmospheric pressure to solidify helium. ► *Absolute zero corresponds to a state of minimum but finite atomic motion*: at 0 K, a tiny **zero-point energy** associated with the vibration of atoms remains.

Even at an absolute temperature of zero, the atoms and molecules of a substance continue to vibrate with some minimum energy known as the **zero-point energy**.

Thermal Expansion

The realization that most materials expand when heated is an ancient one: centuries before the thermometer was invented, blacksmiths were putting red-hot iron rims on wooden wagon wheels (so they would cool and shrink tight). Nevertheless, it was the ability to measure temperature that made possible the quantitative study of **thermal expansion**.

12.2 Linear Expansion

For the moment, let's limit ourselves to long narrow solids so that one dimension predominates: bars, pipes, wires, girders, and so forth. Experimentally, it's found that the change in length (ΔL) of a solid bar of original length L_0 depends on the change in temperature (ΔT) it experiences when heated uniformly; that is, $\Delta L \propto \Delta T$. Heating the bar causes its atoms to vibrate more actively with greater amplitudes, and ***colliding with their neighbors, they put more distance between one another*** (Fig. 12.7). At any one temperature, each atom vibrates in three dimensions about a more or less fixed equilibrium point in the solid. As thermal energy is added to the material, the equilibrium positions of the rapidly oscillating atoms separate and the solid gradually expands.

The weaker the interatomic cohesive force is, the greater are the excursions of the atoms from their previous equilibrium separations, and the more the material expands as a

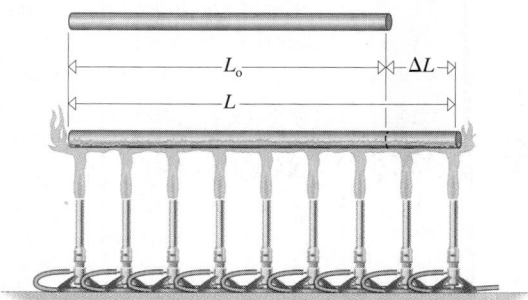

Figure 12.7 When heated the atoms in a narrow rod oscillate more vigorously about their equilibrium positions and the rod expands. Here that expansion is exaggerated to make the point.

Table 12.2

Approximate Values* of Coefficients of Linear Expansion

Material	Coefficient (α) (K^{-1})
Aluminum	25×10^{-6}
Brass (yellow)	18.9×10^{-6}
Brick	10×10^{-6}
Diamond	1×10^{-6}
Cement and concrete	$10-14 \times 10^{-6}$
Copper	16.6×10^{-6}
Glass (ordinary)	$9-12 \times 10^{-6}$
Glass (Pyrex)	3×10^{-6}
Glass (Vycor)	0.08×10^{-6}
Gold	13×10^{-6}
Granite	8×10^{-6}
Hard rubber	80×10^{-6}
Invar (64% Fe, 36% Ni)	1.54×10^{-6}
Iron (soft)	$9-12 \times 10^{-6}$
Lead	29×10^{-6}
Nylon (molded)	81×10^{-6}
Paraffin	130×10^{-6}
Platinum	8.9×10^{-6}
Porcelain	4×10^{-6}
Quartz (fused)	0.42×10^{-6}
Steel (structural)	12×10^{-6}
Steel (stainless)	17.3×10^{-6}

*At temperatures around 20 °C.

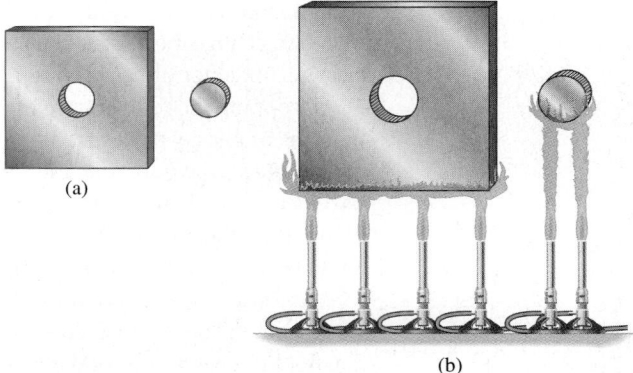

Figure 12.8 When an essentially two-dimensional object (a) is uniformly heated, its linear dimensions increase. (b) Accordingly, the plate gets larger, the hole in it gets larger, and the plug that fills that hole gets larger. Of course, if we imagine the plug pushed back into the hole, both would expand (or shrink) identically, as if the plate were whole.

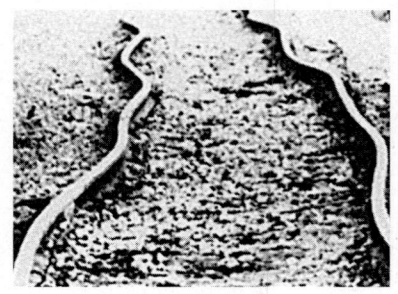

Years ago trains rattled along because fairly large spaces had to be left between the lengths of steel track. This was done so that the rails could expand in warm weather. The photo shows what would happen when that precaution wasn't taken. New low-expansion steel alloys have smoothed the ride considerably.

whole. Since even in the liquid state the atoms are not much farther apart, it can be expected that this increase in separation will be small. A typical metal expands about 7% when its temperature rises from near 0 K to its melting point. If for some small ΔT the atoms or molecules on average separate from one another by an additional 0.001%, we can expect the rod to increase in length by that same 0.001%. Thus, a long rod will increase more than a short one, but both will change by the same percentage. As a consequence, $\Delta L \propto L_0$ and we can anticipate that $\Delta L \propto L_0 \Delta T$. The change in length depends on both the original length and the change in temperature.

Once again, we have a proportionality that can be made into an equation by introducing a constant of proportionality, in this case α:

[linear expansion]
$$\Delta L = \alpha L_0 \Delta T \qquad (12.3)$$

Here, ΔT can be in either degrees Celsius or kelvins, since *the change in temperature is numerically the same for both*. At this point, we go into the laboratory and measure α (the **temperature coefficient of linear expansion**) and find that it has a specific value for every material tested (Table 12.2). The vast majority of materials have positive values of α; rubber under tension is a notable exception, and there are a number of ceramics (having strong atomic bonding) whose expansion coefficients at room temperature are nearly zero or even negative (Fig. 12.8).

It's important to realize the logic of this process because it's used over and over in the development of physics. We do not yet have a theory complete enough to write out exactly what ΔL equals on an atomic level. Instead, we determine on which macroscopic parameters it depends (L_0 and ΔT) and then lump the effects of everything else into an empirical constant specific to each material, which we assume takes into account all the atomic details. If we knew more physics, we could presumably write an equation that would show exactly why lead expands 53 times more than quartz.

We can say that, because of its relatively weak interatomic bonds, lead is soft (has a low Young's Modulus), melts at a relatively low temperature (327 °C), and has a high coefficient of expansion. The idea that the melting point and α are interrelated can be seen in Fig. 12.9. {For more on this subject click on **MELTING POINTS & EXPANSION COEFFICIENTS** under **FURTHER DISCUSSIONS** on the CD.}

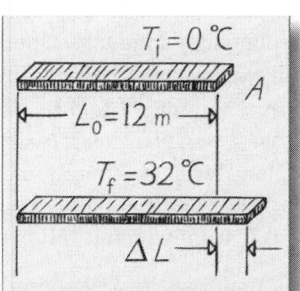

Figure 12.9 The higher the melting point of a material, the stronger the interatomic binding and the smaller the amount of expansion.

Example 12.3 **[II]** A structural steel beam holds up the scoreboard in an open-air stadium. If the beam is 12 m long when it's put into place on a winter day at 0 °C, (a) how much longer will it be in the summer at 32 °C? (b) Given that it has a cross-sectional area of 100 cm², how much force would it take to stretch the beam that much?

Solution The fact that we're given the material, its length, and temperature suggests at the start that this is probably a problem concerning linear thermal expansion. (1) TRANSLATION — A narrow (steel) object of known length experiences a specified temperature change; determine the change in its length. Knowing the cross-sectional area, determine the force needed to produce the same change in length. (2) GIVEN: (a) $L_0 = 12$ m, $T_i = 0$ °C, and $T_f = 32$ °C, and (b) $A = 100$ cm². FIND: (a) ΔL and (b) F. (3) PROBLEM TYPE—(a) Linear thermal expansion and (b) Young's Modulus. (4) PROCEDURE—Eq. (12.3), $\Delta L = \alpha L_0 \Delta T$, summarizes the relationship between

temperature change and length change. The force needed to produce the same ΔL is gotten via Y. (5) CALCULATION—(a) From Table 12.2, $\alpha = 12 \times 10^{-6}$ K^{-1}, and so

$$\Delta L = \alpha L_0 \Delta T = (12 \times 10^{-6}\ \text{K}^{-1})(12\ \text{m})(32\ \text{K}) = \boxed{4.6\ \text{mm}}$$

or almost $\frac{1}{2}$ cm. (b) To find the force, remember $Y = $ stress/strain. From Eq. (10.8) and Table 10.2

$$F = YA \frac{\Delta L}{L_0}$$

$$F = \frac{(200 \times 10^9\ \text{Pa})(100 \times 10^{-4}\ \text{m}^2)(4.6 \times 10^{-3}\ \text{m})}{12\ \text{m}}$$

and $\boxed{F = 77 \times 10^4\ \text{N}}$ or 17×10^4 lb. This is also the force exerted by the beam on whatever might be resisting its change in length. Evidently, provision will have to be made in the design of the structure to allow for thermal expansion and contraction.

Quick Check: Steel changes length by 0.001 2% per kelvin. Here, ΔT is 32 K, so the length change is 0.038 4% $L_0 = 4.6$ mm.

SMALL COMPLICATIONS

Equation (12.3) is a useful relationship, even if it's not a fundamental one. Provided we're careful about applying it, it works nicely. Something to watch out for is that α really depends on T, increasing slightly as T increases. The literature contains extensive tabulations of α for a variety of substances over various ranges of temperature. Moreover, materials that have their atoms arranged differently in different directions (anisotropic materials, e.g., crystalline quartz and bismuth) expand and contract differently in those different directions. Indeed, when heated, graphite expands along one axis and contracts along another. But we needn't worry about that, beyond realizing that none of this is as simple as it might appear.

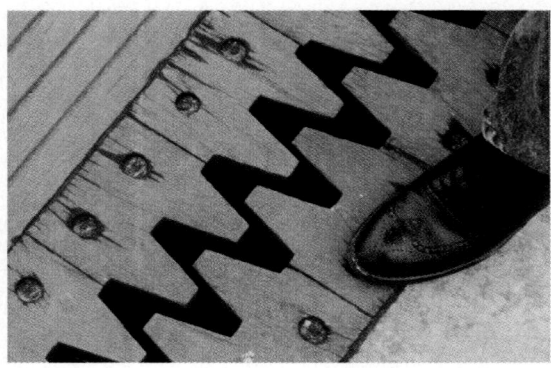

A common device that allows an elevated roadway to expand and contract without affecting traffic. This one is part of an old wooden bridge over the Seine River in Paris.

12.3 Volumetric Expansion: Solids & Liquids

Using the same procedure as above, we can create an expression for the change in volume of a substance (Fig. 12.10) that has undergone a change in temperature ΔT;

$$\Delta V = \beta V_0 \Delta T \qquad (12.4)$$

Here, V_0 is the original volume, and β is known formally as the **temperature coefficient of volume expansion** (see Table 12.3). Notice that the coefficients for liquids are about 50 times greater than for solids, which is to be expected. The weak intermolecular bonds inherent in that state account for the fact that liquids are both more compressible and more readily expanded thermally than are solids. As a rule, β *decreases as the thermodynamic temperature decreases, approaching zero as* 0 K *is approached*. For example, alcohol expands roughly 20% more per kelvin change near 373 K (100 °C) than at 273 K (0 °C). Fortunately, β for mercury, over the same range, varies by less than 0.01%, making it a superior thermometric substance.

Much of what was said before regarding α pertains here. The atoms again separate, but now that happens in three dimensions rather than one. Indeed, a comparison of Tables 12.2 and 12.3 suggests that $\beta \approx 3\alpha$ is generally the case for isotropic solids. That relationship between α and β can be understood by imagining a rectangular solid with sides a, b, and c. If this block is changed in temperature by ΔT, we can expect each side to change its length proportionately. Accordingly, it will go from an initial volume of (abc) to a final one of $(a + \alpha a\,\Delta T)(b + \alpha b\,\Delta T)(c + \alpha c\,\Delta T)$. Multiplying out this expression provides the new volume, but α is a very small number and we can drop any terms containing α^2 or α^3 because they will be still smaller. The results are a final volume approximately equal to $(abc + 3\alpha abc\,\Delta T)$ and a change in volume of $\Delta V \approx 3\alpha V_0\,\Delta T$, from which it follows that $\beta \approx 3\alpha$.

Table 12.3

Approximate Values* of Coefficients of Volumetric Expansion

Material	Coefficient (β) (K^{-1})
Solids	
Aluminum	72×10^{-6}
Asphalt	$\approx 600 \times 10^{-6}$
Brass (yellow)	56×10^{-6}
Cement and concrete	$\approx 36 \times 10^{-6}$
Glass (ordinary)	$\approx 26 \times 10^{-6}$
Glass (Pyrex)	9×10^{-6}
Invar	2.7×10^{-6}
Iron	36×10^{-6}
Lead	87×10^{-6}
Paraffin	590×10^{-6}
Porcelain	11×10^{-6}
Quartz (fused)	1.2×10^{-6}
Steel (structural)	36×10^{-6}
Liquids	
Acetone	1487×10^{-6}
Ethyl alcohol	1120×10^{-6}
Gasoline	950×10^{-6}
Glycerin	505×10^{-6}
Mercury	182×10^{-6}
Turpentine	973×10^{-6}
Water	207×10^{-6}

*At temperatures around 20 °C.

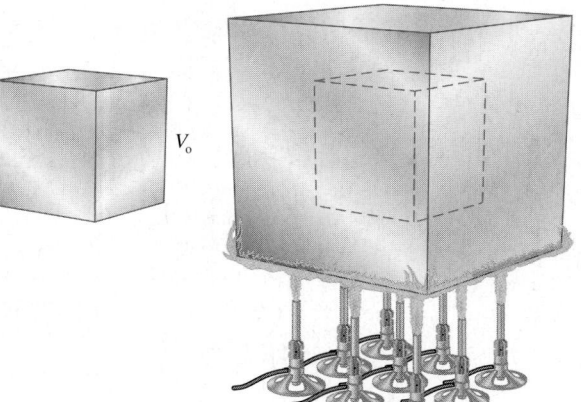

Figure 12.10 When a three-dimensional object is uniformly heated, its linear dimensions increase, and its volume increases by a proportionate amount ΔV. Here the change in volume is exaggerated to make the point.

A bimetallic strip bends when it's heated because the two metals expand by different amounts.

This thick-walled glass mug cracked when it was quickly filled with hot water. The inside and outside of the mug expanded at different rates.

BOWLING BALLS & THE BAY BRIDGE

When a bowling ball is heated, the finger holes get bigger, as does the cavity within a mercury-in-glass thermometer. That's why running hot water on a metal jar lid loosens it—the coefficient of expansion for steel is greater than for glass. A jar is larger when hot than cold and it holds more. In fact, because ordinary glass is a poor conductor of heat, putting a thick-walled glass vessel into an already hot oven is usually disastrous. The exposed faces of the vessel get hot while the region within the glass stays cool. Uneven expansion can crack the glass unless it's heated or cooled slowly. The advantage of Pyrex glass is that its expansion coefficient is three times smaller than that of ordinary glass. Spaces are left between slabs of concrete on streets and roads to keep them from buckling on hot days. Most exposed outdoor steel structures (overpasses, trestles, etc.) have expansion joints to provide for changes in length. During a particularly cold week in 1937, the Bay Bridge in San Francisco contracted a total of 4 ft, 5 in. Often, small bridges are fastened only at one end, while the other rests on rollers to permit expansion. It has even been necessary, when finishing a large bridge in summer, to pack the final spans in ice for several hours to shrink them so they fit in place. Dental fillings and caps can experience substantial temperature variations in the mouth and are made from materials that closely match the β for teeth.

Suppose that we had a solid, homogeneous object whose shape is of no particular interest. Picture anywhere inside of it a portion of the material enclosed within a smaller imaginary sphere. When the big object is heated, it expands uniformly, and, since no holes form inside of it, we can assume that the little imaginary sphere of material expands at the same rate. If the little sphere of matter is removed, the resulting cavity must expand at the same rate as the sphere itself. **When an object expands (or contracts), all its protrusions and cavities expand (or contract) proportionately.**

Thermal expansion is utilized to advantage in the *bimetallic* strip, where flat lengths of two metals with different values of α are bonded together. When heated, the strip bends toward the side with the lower coefficient. These strips are used to make thermometers and circuit breakers or to open and close electrical switches in thermostats.

Example 12.4 **[II]** A driver pulls into a gas station and casually says "fill it up." The attendant does just that, filling the steel tank to the very brim with 56 liters of gasoline at 10 °C. The trip home is a short one, and the heated garage is at 20 °C. How much gasoline will overflow onto the floor?

Solution This problem deals with the change in volume when there is a change in temperature. (1) TRANSLATION—A (steel) container of known volume is filled with a known liquid; determine the difference between the final volumes of the tank and liquid after they both undergo a specified temperature change. (2) GIVEN: $V_0 = 56$ liters, $\Delta T = +10$ K; the liquid is gasoline, and the tank is steel. FIND: The overflow. (3) PROBLEM TYPE—Thermo/volumetric thermal expansion. (4) PROCEDURE —Since we're dealing with expanding volumes, compare the change in volume $\Delta V = \beta V_0 \, \Delta T$ for tank and liquid. (5) CALCULATION—First, convert the volume from liters to meters cubed: 56 liters = $56 \times 10^3 \, \text{cm}^3 = 0.056 \, \text{m}^3$. The change in volume of the gasoline is

$$\Delta V_g = \beta_g V_0 \Delta T$$
$$\Delta V_g = (950 \times 10^{-6} \, \text{K}^{-1})(0.056 \, \text{m}^3)(+10 \, \text{K}) = +5.32 \times 10^{-4} \, \text{m}^3$$

But the tank's capacity also increases; the tank's 0.056-m³ cavity expands at the same rate as a solid steel mass in that shape. Hence, the tank's volume increases by

$$\Delta V_t = \beta_t V_0 \Delta T$$
$$\Delta V_t = (36 \times 10^{-6} \, \text{K}^{-1})(0.056 \, \text{m}^3)(+10 \, \text{K}) = +0.202 \times 10^{-4} \, \text{m}^3$$

The difference, $\Delta V_g - \Delta V_t = \boxed{0.51 \times 10^{-3} \, \text{m}^3}$ (i.e., 0.51 liter), overflows.

Quick Check: $\Delta V_g / \Delta V_t = \beta_g / \beta_t = (950 \times 10^{-6} \text{K}^{-1})/(36 \times 10^{-6} \text{K}^{-1}) = 26.4$; whereas $\Delta V_g / \Delta V_t = (+5.32 \times 10^{-4} \, \text{m}^3)/(+0.202 \times 10^{-4} \, \text{m}^3) = 26.3$.

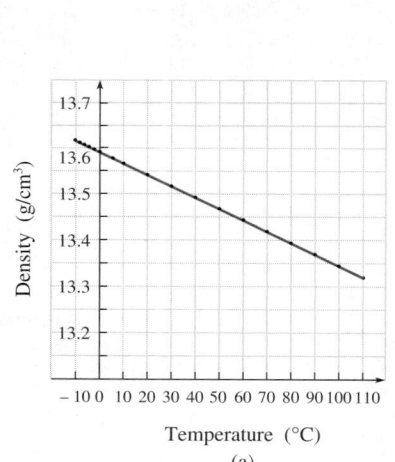

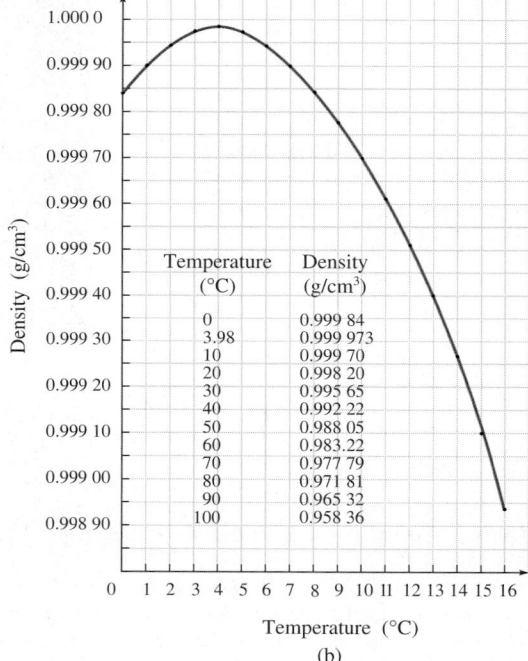

Temperature (°C)	Density (g/cm³)
0	0.999 84
3.98	0.999 973
10	0.999 70
20	0.998 20
30	0.995 65
40	0.992 22
50	0.988 05
60	0.983 22
70	0.977 79
80	0.971 81
90	0.965 32
100	0.958 36

Figure 12.11 (a) The density of mercury plotted against temperature in °C. Mercury is a more representative liquid than water. (b) The density of water is a maximum at 3.98 °C.

The bimetallic coil in a thermostat controlling the temperature in a room. As the coil expands, it tilts the vial of mercury, which breaks the electrical connection and shuts off the furnace.

The Expansion of Water

Water is especially important to us, and that makes its physical behavior of great interest. Like the majority of liquids, water contracts slightly as it is cooled (Fig. 12.11), becoming a trifle more dense. In a unique way, however, this contraction gradually diminishes until at 277.13 K (3.98 °C), it stops altogether. Between 373.15 K (100 °C) and 277.13 K (3.98 °C), water contracts about 4%, which is not very much, but still about 20 times more than would occur over that range for most solids. Water molecules can be envisioned as widely spaced, moving about fairly freely, and forming and unforming small groupings. They slowly move close together as the temperature descends. Below 277.13 K (3.98 °C), the ordered clusters of water molecules rapidly get larger. Since the molecules unite with only four neighbors, these groupings contain an appreciable amount of empty space. Thus, below 3.98 °C the volume of liquid begins to expand and the density decreases (Fig. 12.11b).

Water crystallizes into ice, which is an orderly structure that takes up even more space (Fig. 12.12). Accordingly, the volume increases abruptly—at 273 K (0 °C) the density of water is 999.8 kg/m³, whereas the density of ice is only 917 kg/m³. Lowering the temperature further causes the ice to contract ($\alpha = 52.7 \times 10^{-6}$ K^{-1} at ≈ 0 °C), and it continues to do so all the way down to ≈ 75 K. By comparison to most materials, the fact that solid water

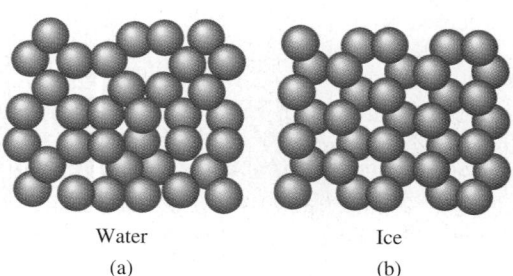

Water
(a)

Ice
(b)

Figure 12.12 (a) The molecules of H$_2$O form small orderly groupings in the liquid state. (b) When the solid forms, the molecules come together in a highly open, ordered structure.

EXPLORING PHYSICS ON YOUR OWN

The Expansion of Water Ice: Fill a small glass bottle to the top with water and stand it, unsealed, in the freezer. A cylindrical plug of ice will rise up out of the bottle as the water freezes and expands. If you used a bottle with a tight metal cap the water in the neck would likely freeze first (metal is a better conductor of heat than glass—p. 477) and plug the top of the bottle. With no way to further expand, the pressure in the liquid would rise as more ice is formed (water isn't very compressible—p. 341). The interatomic forces are tremendous, and ultimately the glass would burst. This is why water pipes break in cold weather. You can see the same thing (without the broken glass) by putting a can of soda in the freezer for a few hours (place it on a dish to avoid any mess). The last time I tried this a long blade of ice jutted out through a gash in the side of the can.

floats in liquid water is a rarity. There are a few other substances, such as bismuth, antimony, and cast iron, that behave the same way, though as a rule, solids are more dense and sink in their liquids.

The implications of this for life on Earth are crucial. At the onset of winter, a lake or stream is gradually cooled at its surface. The colder, denser water descends until finally even the surface layer reaches 3.98 °C. Thereafter the upper layer becomes less dense as the air temperature drops, and floats at the top of the lake. At 0 °C, ice forms at the surface and thereafter grows downward, but since it is a much poorer conductor of heat than water, its very presence slows further ice formation. The sheet of surface ice essentially seals off the lake from the winter cold, allowing life to go on beneath it. {To learn more about the fact that liquid water at a pressure of 1 atm can exist at temperatures well below 0 °C click on **SUPERCOOLING** under **FURTHER DISCUSSIONS** on the **CD.**} 💿

The Gas Laws

Because a gas expands to fill the container that confines it, gases do not undergo a simple temperature-dependent volume change. The change in volume of condensed matter is linked to both the amount of material present and the temperature variation. The same is true for a gas, although the relationships are more subtle. The amount of gas in a given container contributes to determining the pressure therein, and it is that pressure P along with the temperature T, the volume V of the container, and the mass m of the gas that are the observables.

12.4 The Laws of Boyle, Charles, & Gay-Lussac

There are three simple laws that work nicely as long as the molecules of the gas don't get too close to one another. After all, if the same equation is to describe both O_2 and CO_2, which are different in size and have different intermolecular cohesive forces, there will have to be lots of space between molecules so that these differences have no noticeable effect. Thus, we avoid high densities, high pressures, and low temperatures (near the point of liquefaction).

A typical gas at STP (273.15 K or 0 °C and 0.101 MPa or 1 atm) contains 2.7×10^{19} molecules per cubic centimeter. Each such particle, in the case of a monatomic gas such as helium or argon, has a diameter of roughly 0.2 nm. On average, they will be separated from one another by about 15 atomic-diameters. Imagining each little particle at the center of a cube 15 atomic-diameters on a side, the ratio of the empty volume to the actually occupied volume is about $15^3:1$, or 3375:1. Under such circumstances, each atom functions independently, and the gas as a whole behaves predictably.

In 1662 the Honourable Robert Boyle, the seventh son of the Earl of Cork, conducted a study of gases that culminated in the recognition of a simple interdependence between pressure and volume (Fig. 12.13). **Keeping the temperature constant, the volume of a**

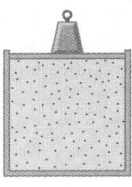
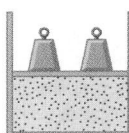
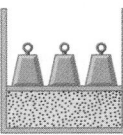

Figure 12.13 When the pressure exerted on a gas is doubled, the volume is halved when it is tripled, the volume is reduced to one-third.

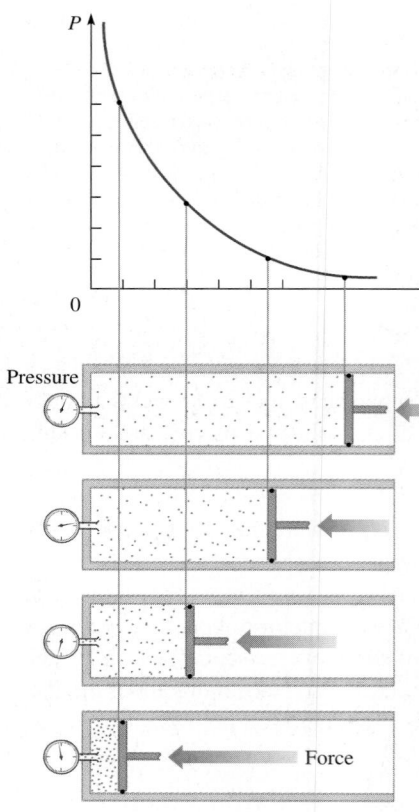

Figure 12.14 Here the volume of a gas is changed while the temperature is maintained constant. The product of P and V is constant and a plot of P versus V is a hyperbola: $PV = $ constant.

gas varies inversely with the absolute pressure, which is equivalent to saying that *absolute pressure times volume is constant.* **Boyle's Law** is then

[constant T and m]
$$PV = \text{constant} \qquad (12.5)$$

Given a gas-filled chamber that can expand and contract, the pressure within the gas varies inversely with V; double the volume and the pressure is halved, and vice versa. As we saw earlier in Fig. 1.14, a plot of pressure versus volume at constant temperature is a hyperbola (Fig. 12.14).

Consider a balloon floating upward in the air or a bubble rising in a liquid. As the external pressure drops, the internal counterpressure must also decrease. If PV is to be constant, the volume must increase; the balloon grows larger as the pressure decreases. Similarly, Fig. 12.15 shows how Boyle's Law applies to the way we breathe.

The imagery of a gas as a swarm of tiny bouncing, colliding atoms or molecules was first applied analytically by Daniel Bernoulli (1738), who succeeded in deriving Boyle's Law using the concept (p. 443). A molecule impacting on, and ricocheting off, a chamber wall changes its momentum in the process. That, in turn, corresponds to a force on the molecule and an equal-magnitude oppositely directed force on the wall. The ongoing bombardment of millions of molecules blurs into a seemingly continuous force and pressure. {For more details click on **BOYLE'S LAW** under **FURTHER DISCUSSIONS** on the **CD**.}

If we keep the pressure constant and vary the temperature of a gas, we find, as did Amontons, that $\Delta V \propto \Delta T$. Indeed, the behavior of most gases agrees with Eq. (12.4), $\Delta V = \beta V_0 \Delta T$, where the expansion coefficient is almost the same for all of them, namely, $\beta \approx 0.0037 \text{ K}^{-1} = 1/(273 \text{ K})$. Thus, if we plot V versus T at constant P, as in Fig. 12.5, the slope $\Delta V/\Delta T = \beta V$ is a constant and the graph a straight line. In other words, **when the pressure is kept constant, the volume of a given amount of any gas varies directly with the thermodynamic temperature**. This is known as **Charles's Law**, after Citizen Jacques Alexandre César Charles of Paris (ca. 1787)—the man who inadvertently made the first solo balloon ascent when his unhappy assistant accidentally let go while climbing out of the

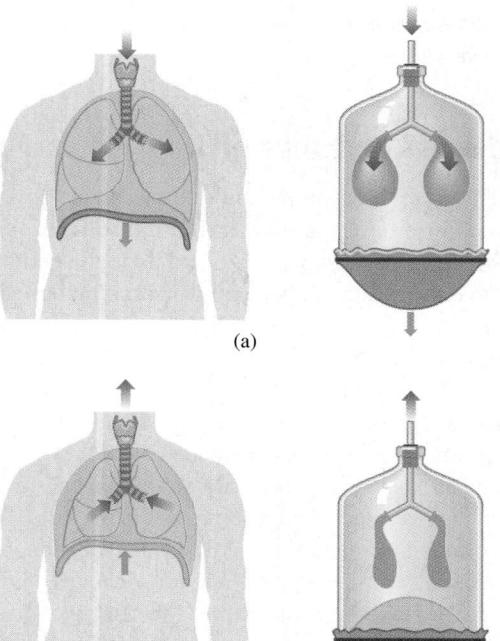

(a)

(b)

Figure 12.15 A bottomless jar surrounding two balloons and closed off with a rubber membrane can make a simple model of the lung system. (a) To inhale, the diaphragm pulls downward, enlarging the chest volume and thereby dropping the pressure ($PV = $ constant). The chest cavity is usually at a negative gauge pressure (to keep the lungs open against their surface tension), but now the lung pressure is negative and air flows in. (b) When the chest compresses and the diaphragm moves up, chest pressure rises, the gauge pressure in the lungs becomes positive, and air is expelled.

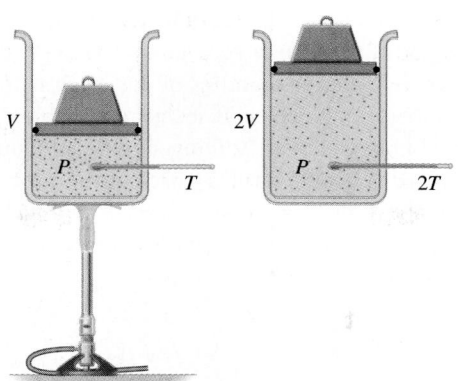

Figure 12.16 The weight on the piston keeps the pressure constant at P. Heating the gas, and thereby doubling T, must double V.

gondola. Putting it another way, $V \propto T$, or

[constant P and m]
$$\frac{V}{T} = \text{constant} \qquad (12.6)$$

All gases increase in volume when the temperature increases. Doubling the temperature of a gas at constant pressure doubles its volume (Fig. 12.16).

The third basic gas relationship, called **Gay-Lussac's Law** after Joseph Louis Gay-Lussac (1802), maintains that **when the volume is kept constant, the absolute pressure of a given amount of any gas varies directly with the thermodynamic temperature**; that is, $P \propto T$, or

[constant V and m]
$$\frac{P}{T} = \text{constant} \qquad (12.7)$$

The pressure in an automobile tire (V = constant) rises in the summer as T rises, and it's dangerous to throw aerosol cans in a fire because P increases so much that they explode.

Multiplying these three relationships (and taking the square root) yields

[constant m]
$$\frac{PV}{T} = \text{constant} \qquad (12.8)$$

for a fixed quantity of gas. Note that each of the constants in the last four equations is different.

Example 12.5 **[I]** A tank having a volume of 1.00 m³ is filled with air at 0 °C to 20.0 times atmospheric pressure. How much volume will that gas occupy at 1.00 atm and room temperature?

Solution This problem, which mentions temperature, pressure, and volume, is about a gas and therefore think "gas laws."
(1) TRANSLATION—A gas with a given volume, temperature, and pressure undergoes a known change in both pressure and temperature; determine its new volume. (2) GIVEN: V_i = 1.00 m³, T_i = 273.15 K, P_i = 20.0 atm, P_f = 1.00 atm, and T_f = 293.15 K. FIND: V_f. (3) PROBLEM TYPE—Thermo/gas laws. (4) PROCEDURE—All the quantities we're dealing with

$T_i = 0\,°C$ $T_f = 20\,°C$

$V_i = 1.00\,m^3$
$P_i = 20.0\,atm$ $V_f = ?$
$P_f = 1.00\,atm$

(V, P, and T) are related via Eq. (12.8). (5) CALCULATION—*You must use absolute temperatures*. Since the ratio in Eq. (12.8) is constant,

$$\frac{P_i V_i}{T_i} = \frac{P_f V_f}{T_f}$$

Therefore

$$V_f = \frac{P_i V_i T_f}{P_f T_i} = \frac{20(1.00\ \text{m}^3)(293.15\ \text{K})}{273.15\ \text{K}} = \boxed{21.5\ \text{m}^3}.$$

where because the two pressures divide we need not first convert their units. There is a slight increase in volume due to the increase in temperature, but the predominant increase is due to the 20-fold drop in pressure.

Quick Check: The rise in temperature increases the volume by a factor of $\approx 293/273 \approx 1.07$; the pressure change increases it by a factor of 20; hence, $1.07 \times 20 \approx 21.5$ and $V_f \approx 21.5(1\ \text{m}^3)$.

We still have one quantity to account for, and that's the mass of the gas. With all else unchanged, the volume of a gas must depend directly on the *amount* of gas present. Consider two identical bottles of gas, each with the same pressure, temperature, volume, and number of molecules. It seems reasonable that we could attach them together, remove the separating wall, and nothing significant would change on a macroscopic level—twice the number of gas molecules, twice the volume. Similarly, if V and T are fixed, increasing the number of molecules will increase the number of their collisions with the walls and, therefore, will increase the absolute pressure. Consequently, $PV \propto m$, or

[constant T]
$$\frac{PV}{m} = \text{constant} \qquad (12.9)$$

Example 12.6 **[II]** A 100-liter storage tank is slowly being filled with gas. At 5.00 times atmospheric pressure, the tank holds 0.60 kg of gas. If the temperature is kept constant, how much gas will be in the tank when the pressure is raised to 10.00 atm?

Solution This problem, which mentions pressure, mass, and temperature, is about a gas and therefore think "gas laws." (1) TRANSLATION—A known mass of gas has a given volume and pressure. More gas is added, producing a known change in pressure while keeping the volume and temperature fixed; determine the final mass of gas. (2) GIVEN: $V = 100$ liters = constant, $P_i = 5.00$ atm, $m_i = 0.60$ kg, and $P_f = 10.00$ atm. FIND: m_f. (3) PROBLEM TYPE—Thermo/gas laws. (4) PROCEDURE—All the

quantities we're dealing with (V, P, and m at constant T) are related via Eq. (12.9). (5) CALCULATION—It follows from Eq. (12.9) that the initial and final values of the ratio PV/m must be equal:

$$\frac{P_i V_i}{m_i} = \frac{P_f V_f}{m_f}$$

and $\quad m_f = \dfrac{P_f m_i}{P_i} = \dfrac{(1.013\ \text{MPa})(0.60\ \text{kg})}{0.506\ 6\ \text{MPa}} = \boxed{1.2\ \text{kg}}$

Quick Check: To double the pressure, double the mass: $\Delta m = 0.60$ kg. (Notice that it would have been easier to divide the two pressures before converting their units.)

12.5 The Ideal Gas Law

Instead of using m in the gas law, we can express the amount of gas in terms of either the **number of moles** (n) present (p. 284) or the **number of molecules** (N) where $m \propto n$ or $m \propto N$; that is, $PV \propto nT$ or $PV \propto NT$. The masses of the individual particles play a less central role than the total number of particles. We saw this earlier when we found that the same number of molecules of any gas at the same temperature and pressure occupies the same volume. It seems reasonable that if a single equation is to describe both H_2 and CO_2, we should deal with equal numbers of molecules, not equal masses of gas.

Introducing a constant of proportionality R, the expression $PV \propto nT$ becomes

$$PV = nRT \qquad (12.10)$$

> The quantitiy T in the several gas laws is the **thermodynamic** or **absolute temperature**.

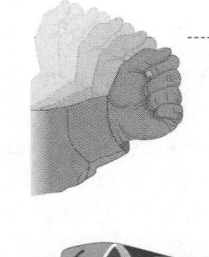

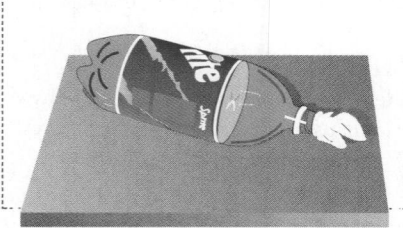

EXPLORING PHYSICS ON YOUR OWN

The Gas Laws & Pop Guns: Put an inflated balloon in the freezer. The gas molecules inside the balloon keep it puffed up by constantly bombarding the inner walls. Cooling the gas, removing kinetic energy from its molecules, lessens their bouncing, and the balloon collapses. Let it come back to room temperature (or better yet put it in warm sunlight) and it will reinflate itself. Take a warm partially filled balloon and squeeze it. What happens to the pressure in the gas (when the rubber stretches, it means the pressure has increased)? Have you ever sat on an air-filled cushion? Roll up a tissue and jam it in the opening of a 2-liter plastic soda bottle so that it fits tightly. Place the bottle horizontally on a table; point it away from people and anything breakable, and slam down on the middle of the bottle with your fist. The tissue will shoot out via Boyle's Law. Carefully try it with a cork—about a hundred years ago the cork-shooting compressed gas pop gun was a favorite toy.

Example 12.7 **[I]** What is the volume of 1.00 mole of any gas at STP? It's safe to assume that at STP the gas will behave like an ideal gas.

Solution This problem is about an ideal gas and therefore think "Ideal Gas Law." (1) TRANSLATION—Determine the volume of a specified amount of an ideal gas at a known temperature and pressure. (2) GIVEN: $n = 1.00$ mol, $T = 273.15$ K, and $P = 1.013 \times 10^5$ Pa. FIND: V. (3) PROBLEM TYPE—Thermo/Ideal Gas Law. (4) PROCEDURE—All the quantities we're dealing with (V, P, T, and n) are related via Eq. (12.10).

(5) CALCULATION—From the Ideal Gas Law

$$V = \frac{nRT}{P} = \frac{(1.00 \text{ mol})(8.315 \text{ J/mol·K})(273.15 \text{ K})}{1.013 \times 10^5 \text{ Pa}}$$

$$\boxed{V = 0.022\,4 \text{ m}^3}$$

Quick Check: This result is confirmed by the discussion on p. 284.

Remarkably, experiments reveal that R is nearly the same for all gases at low pressures: $R = 8.314\,51$ J/mol·K, and it's called the **Universal Gas Constant**. Still, we must limit the circumstances under which this formula is applied—it doesn't work well for real gases at high densities. A model gas, one for which this expression works perfectly, is known as an *ideal gas*, and Eq. (12.10) is the **Ideal Gas Law**.

Avogadro's number $N_A = 6.022 \times 10^{23}$ is the number of molecules per mole, and so the number of molecules N present in a sample of gas is just nN_A. Hence

$$PV = nRT = \frac{N}{N_A} RT$$

Accordingly, we define a new constant, called **Boltzmann's Constant**, $k_B = R/N_A = 1.380\,66 \times 10^{-23}$ J/K and thereby obtain the final form of the Ideal Gas Law

$$PV = Nk_B T \tag{12.11}$$

This expression may seem to be just another way of saying the same thing, but Boltzmann's Constant is actually one of the fundamental constants of Nature (p. 444).

Table 12.4

The masses of one gram mole of several gases

Material	Mass (kg)
Hydrogen	0.002
Helium	0.004
Nitrogen	0.028
Air (average)	0.029
Oxygen	0.032
Argon	0.040
Carbon dioxide	0.044
Ozone	0.048

Example 12.8 **[II]** (a) Estimate the number of air molecules there are in a room 4.00 m by 6.00 m by 3.00 m at standard pressure and room temperature (20 °C). (b) How many moles is that? (c) How much volume would the gas occupy if the temperature were 273 K (0 °C), all else constant? (d) Roughly how much does the air weigh? (See Table 12.4.) Assume the air is an ideal gas.

Solution This problem is about an ideal gas and therefore think "Ideal Gas Law." (1) TRANSLATION—A gas occupies a known volume at a known temperature and pressure; determine (a) the number of molecules present, (b) the number of moles present, (c) the volume of the gas at some other temperature, and (d) the weight of the gas. (2) GIVEN: $T = 293$ K (20 °C), $P = 0.101$ MPa, and $V = 4.00 \times 6.00 \times 3.00$ m^3. FIND: N. (3) PROBLEM TYPE—Thermo/Ideal Gas Law. (4) PROCEDURE—All the quantities we're dealing with (T, P, V, and N) are related via Eq. (12.11). (5) CALCULATION—(a) From the Ideal Gas Law

$$N = \frac{PV}{k_B T} = \frac{(0.101\,3 \times 10^6 \text{ Pa})(72.0 \text{ m}^3)}{(1.381 \times 10^{-23} \text{ J/K})(293 \text{ K})}$$

$$\boxed{N = 1.803 \times 10^{27} \text{ molecules}}$$

(b) To find the number of moles, divide N by Avogadro's number:

$$n = \frac{N}{N_A} = \frac{1.803 \times 10^{27}}{6.022 \times 10^{23}} = \boxed{2.99 \times 10^3 \text{ moles}}$$

(c) At STP, each mole occupies 22.414 liters; hence

$$V_{STP} = (22.414 \text{ liters/mol})(2994 \text{ mol})$$

$$V_{STP} = 67.1 \times 10^3 \text{ liters} = \boxed{67.1 \text{ m}^3}$$

as compared to 72 m^3 at room temperature. (d) From Table 12.4, each mole of air has a mass of 0.029 kg; hence, the total mass of air is

$$M = (0.029 \text{ kg/mol})(2994 \text{ mol}) = 86.8 \text{ kg}$$

which weighs $\boxed{0.85 \text{ kN}}$ or 1.9×10^2 lb.

Quick Check: Using Eq. (12.11) at STP, $V_{STP} = Nk_B(273 \text{ K})/P_A = (1.8 \times 10^{27})(1.381 \times 10^{-23} \text{ J/K})(273 \text{ K})/(0.101 \times 10^6 \text{ Pa}) = 67$ m^3.

12.6 Phase Diagrams (Optional)

Suppose we fill a cylinder (fitted with a tight piston as in Fig. 12.17) with a gas and study its behavior as the temperature, pressure, and volume are varied: under what conditions will the gas become a liquid or a solid? Figure 12.18 is a typical pressure versus temperature or *PT*-diagram, also known as a ***phase diagram*** because it summarizes the relationships between the solid, liquid, and gaseous states of a substance, in this case, water. Two different phases coexist at temperatures and pressures lying along each of the three boundary lines.

Point *C* is a special place called the *critical point*. Here, the molecules are close enough to one another that their mutual (electromagnetic) attraction, the cohesive force, is no longer negligible. At *C*, *liquid and gas coexist at the same density* and are indistinguishable. Above that critical temperature, no amount of pressure will liquefy the gas; it can become denser and denser, but there will never be a clear gas-liquid transition. Above 647 K (374 °C) water can exist only as a gas no matter what the pressure. The critical temperature is a measure of the strength of the intermolecular cohesive force. The relatively high value (647 K) for water (as compared, e.g., to 304 K for carbon dioxide) is a reflection of the sizable attraction between its polar molecules (p. 3).

Suppose the gas-filled cylinder (Fig.12.17) is held at a constant temperature not far below the critical temperature—the experiment begins at point-1 in Fig. 12.18. Increasing the pressure, by pushing in the piston, will cause the data to rise up along a straight line (constant *T*). Incidentally, it's customary to call the substance *vapor* rather than gas when it's below the critical temperature—***vapors can be condensed via compression, gases cannot***. Compressing the vapor further increases *P* and carries the system up from point-1 to point-2 to point-3, where the vapor is just about to condense. Still more compression results in continued liquefaction at a constant pressure, bringing the system first to point-4 and then, when all is liquid, to point-5. Beyond that it takes considerably more pressure to compress the liquid, bringing it to point-6.

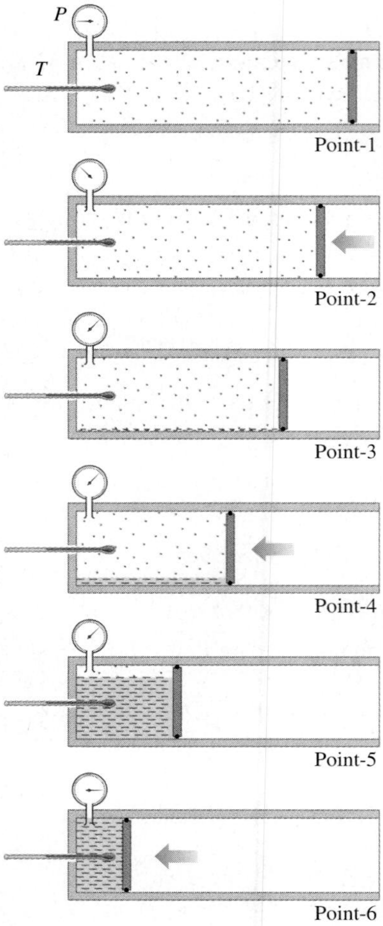

At point-4, the chamber contains both liquid and vapor, and the amount of each is stable. Some molecules that happen to be moving fast enough up near the surface will leave the liquid altogether, in the usual process of evaporation. Others in the vapor above it will crash into the liquid, becoming part of it. When the rates of evaporation and condensation are equal, the system remains in a state of equilibrium. If the concentration of vapor above the liquid is stable, it exerts, on average, a stable pressure on the liquid known as the **equilibrium vapor pressure**, or the *saturated vapor pressure*, or, when there is no ambiguity, just the vapor pressure. This pressure depends both on the nature of the liquid and on the temperature. The larger the cohesive force, the lower the vapor pressure. The higher the temperature, the higher the vapor pressure. At 293 K (20 °C), strongly interacting water (which is polar and rather special) has a vapor pressure of only 2.3 kPa. In comparison volatile, weakly bound chloroform has a vapor pressure of 21.3 kPa—it can keep evaporating much longer before equilibrium sets in (Fig. 12.19).

Figure 12.17 The isothermal compression of a real gas below its critical temperature. These are the six points on the isothermal shown in Fig. 12.18

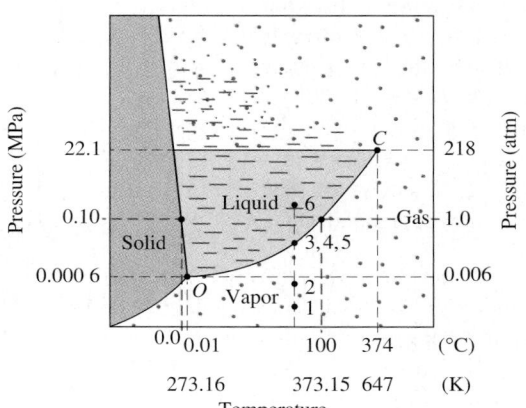

Figure 12.18 Phase diagram for water. If we take a block of ice at atmospheric pressure and slowly raise its temperature, it melts completely at 273 K (0 °C) and remains liquid until 373 K (100 °C), whereupon it completely vaporizes.

SPRAY CANS, SMELLY FEET, & POPCORN

Large molecules have lots of inertia and tend not to evaporate—all else equal, they have low vapor pressures. At room temperature fingernail polish remover (acetone, C_3H_6O) evaporates much more easily than rubbing alcohol (C_3H_8O), which is a larger (two more hydrogen) molecule. Paraffin (candle wax, $C_{30}H_{62}$) hardly evaporates at all, which is why it has little or no odor. By contrast, butyric acid ($C_4H_8O_2$), the principal offender of smelly feet, is a small high-vapor pressure, easily diffused molecule.

Aerosol spray cans (such as for hair lacquer or paint) contain a liquid propellant with a room-temperature vapor pressure well above atmospheric so that it can continue to sustain a high pressure on the surface of the product being expelled (p. 297). Because the vapor pressure of water exceeds 101 kPa beyond 100 °C, the tiny amount of water in a dried kernel of corn evaporates when heated, building up pressure until it explodes as popcorn.

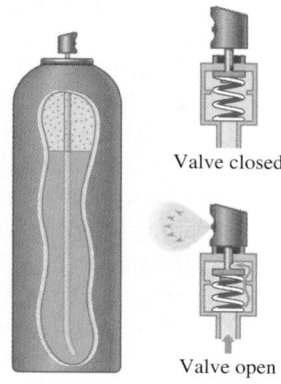

Figure 12.19 A liquid with an especially high vapor pressure is mixed in with the product to be sprayed. As a result, the pressure inside the can is always greater than atmospheric.

Valve closed

Valve open

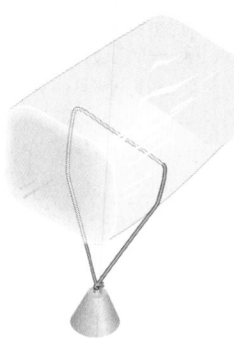

Figure 12.20 A wire that supports a load passes through a block of ice because the melting point drops due to the high pressure just under the wire. Above the wire, the water refreezes.

Hoarfrost.

The curve from O to C corresponds to the equilibrium vapor pressure and is, as was recognized by Dalton (1804), logarithmic. It's often called the ***boiling-point curve***. If the line from O to C is crossed, there will be a clear liquid-vapor phase transition. By contrast, if the system is carried (by altering P and T) from the region on the lower right where it is a gas (up around and over C), it gradually blends into a liquid, without there being any phase transition. That's what would be seen if we could descend into the hydrogen-helium atmosphere of the giant planet Jupiter. Under its own crushing pressure, the atmosphere gets thicker and thicker, imperceptibly merging into the liquid body of the planet.

For every pressure above that of point O, there is a temperature at which a pure solid will be in equilibrium with its liquid state. The locus of all such melting (or freezing) points is the line rising up from O on the left, which is the ***melting-point curve***. Notice that this solid-liquid transition line leans to the left. Raising the pressure by 1 atm lowers the melting temperature by about 0.007 2 °C (see Fig. 12.20).

At point O in Fig. 12.18, which is the **triple point**, all three phases coexist in equilibrium. Like liquids, solids also have a vapor pressure, though it may be extremely small. Raising the temperature of a solid at a fixed pressure below that of the triple point will cause it to pass directly from solid to vapor, a process known as *sublimation*. The line descending from O with decreasing temperature corresponds to the vapor pressure and is often called the ***sublimation curve***. The intermolecular bonds in dry ice are so weak that even at room temperature it will transform directly to the vapor state, as will camphor and naphthalene (mothballs). In this spirit, take a block of frozen coffee, ice cream, or even blood at a temperature below the triple point; lower the pressure and pump off the vapor as the solid is gently heated. The ice will sublime, leaving a light porous residue, which is said to be *freeze-dried*.

When the sublimation process is run the other way via a reduction in temperature, the vapor condenses out directly to the solid state. This condensation is what produces snow in the upper atmosphere and hoarfrost down at the surface of the Earth. The latter is the feathery crystalline deposit of ice that can be seen on the inside of a window pane in winter or lining the plastic window of a cake box in a freezer.

{For more discussion of the critical point and the liquefaction of gases, click on **REAL GASES: LIQUEFACTION** under **FURTHER DISCUSSIONS** on the **CD**.}

12.7 Kinetic Theory (Optional)

In ancient times, the Greek philosopher Plato observed that "heat and fire . . . are themselves begotten by impact and friction: but this is motion." That insight, that thermal phenomena arise from motion, was echoed almost 2000 years later by Francis Bacon, who maintained that "the very essence of heat ... is motion and nothing else." The picture of a gas as a tumult of "very minute corpuscles which are driven hither and thither" was used as early as

1738 by Bernoulli to explain Boyle's Law {click on **BOYLE'S LAW** under **FURTHER DISCUSSIONS** on the **CD**}. By the middle of the 1800s, this motion theory, now known as **Kinetic Theory**, was one of several competing formalisms concerning the nature of heat and temperature. Keep in mind that the reality of the atom was not established until the early twentieth century.

Kinetic Theory, as further developed mainly by James Clerk Maxwell and Ludwig Boltzmann, is a mathematical description of how a vast number of minute, rapidly moving particles can manifest itself macroscopically in the observed properties of bulk matter. We will examine only the basics of the theory and use it to gain some insights into the nature of gases. Following Maxwell, we model a gas as "an indefinite number of small, hard, and perfectly elastic spheres acting on one another only during impact." Clearly, we are talking about an ideal gas. To make the treatment still easier, we assume that the gas particles are essentially points having no extension that do not collide with one another. This is certainly not the case, nor is it a necessary condition. (The inclusion of interparticle collisions would bring us into the study of Statistical Mechanics.)

Figure 12.21 depicts a piston in a cylinder of length L and cross-sectional area A. The pressure exerted on the piston originates from the countless millions of impacts made on it by the gas molecules each and every second. Inasmuch as the system is in equilibrium, this is the pressure P everywhere within the cylinder. The motion of a molecule can be resolved into its x-, y-, and z-components (Fig. 12.21a). The collisions are perfectly elastic (conserving KE), and so each time a particle bounces off the piston, its velocity's y- and z-components, parallel to the face of the piston, remain unaltered; only the x-component changes, reversing itself from $+v_x$ to $-v_x$. Each impact produces a change in the momentum of a single particle of

$$\Delta p_x = 2mv_x$$

Each molecule moves along the length of the cylinder at a speed of v_x, which is constant (even though it may well be bouncing around from side to side in the process). Hence, it travels from the piston to the far face and back a distance of $2L$ in a time

$$\Delta t = \frac{2L}{v_x}$$

This is the time between collisions on the piston: the number of seconds per collision. The reciprocal of that value is the number of collisions per second, or the rate of collision:

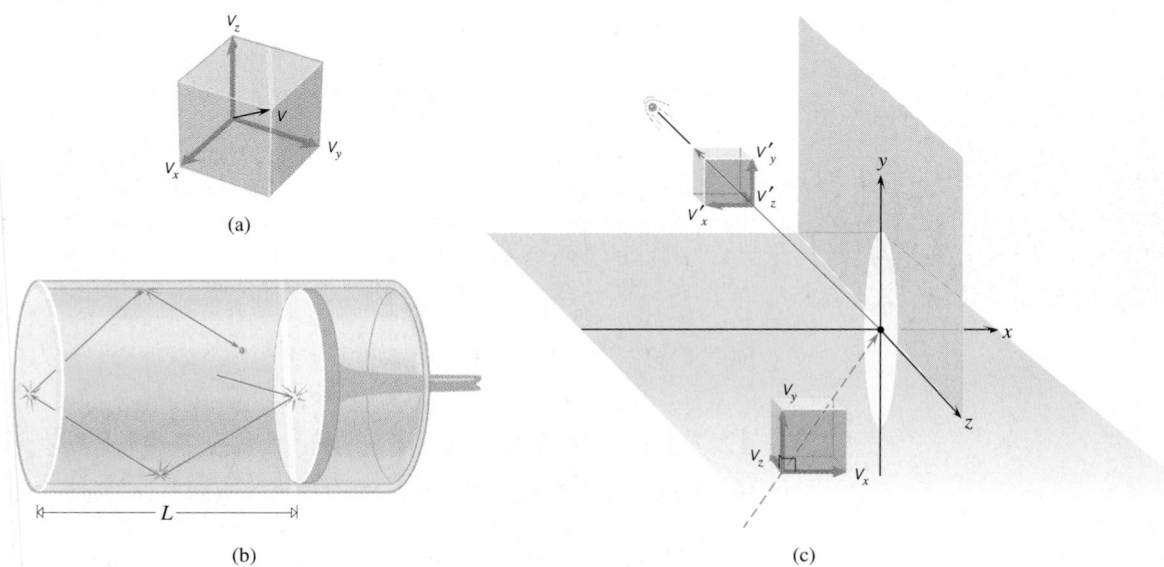

(a)

(b)

(c)

Figure 12.21   Atoms colliding with the face of a piston. Only the v_x component changes its direction on collision with the surface.

$1/\Delta t = v_x/2L$. Multiplying the number of collisions per second $(1/\Delta t)$ by the change of momentum per collision (Δp) gives us the *rate-of-change of momentum*:

$$\frac{\Delta p_x}{\Delta t} = \frac{2mv_x}{2L/v_x} = \frac{mv_x^2}{L}$$

From Newton's Second Law, this equals the average force exerted by that one particle on the piston. Assuming, just for a moment, that all N of the molecules in the cylinder behave the same way, the total average force on the piston due to the hail of particles is

$$F_{av} = N\frac{mv_x^2}{L}$$

By definition, the force per unit area is the pressure

$$P = \frac{Nmv_x^2}{LA}$$

This equation can be expressed a little more conveniently using the three-dimensional equivalent of the Pythagorean Theorem, which tells us that the total speed squared equals the sum of the squares of its components, that is,

$$v^2 = v_x^2 + v_y^2 + v_z^2$$

Since there are no external forces acting on the system, on average, we can expect that the gas particles will be moving in all directions with equal likelihood: $v_x^2 = v_y^2 = v_z^2$, and so

$$v_x^2 = \tfrac{1}{3}v^2$$

Ludwig Boltzmann (1844–1906). Despondent and disturbed by the idea that his life's work was a waste, Boltzmann killed himself in 1906. Within a few decades, the atom was established as a reality, and Boltzmann's brilliant achievement was recognized as seminal.

Accordingly
$$P = \frac{Nmv^2}{3LA} \qquad (12.12)$$

This would be the pressure if each molecule had the same speed, but in fact they don't. Not surprisingly, there is a temperature-dependent distribution of speeds, something Maxwell and Boltzmann each worked out theoretically. For us, it will suffice to recognize that fact, and we will use an average speed in the pressure equation in place of v^2. The sum of the squares of the speeds of each of the N particles divided by N is the average, or *mean*, of the speed squared; consequently

$$(v^2)_{av} = \frac{v_1^2 + v_2^2 + \cdots + v_N^2}{N}$$

and

$$P = \frac{Nm(v^2)_{av}}{3LA}$$

Noticing that the volume of the cylinder is $V = LA$, we get

$$P = \frac{Nm(v^2)_{av}}{3V}$$

Since Nm is the total mass of the gas, the density is $\rho = Nm/V$ and

$$P = \tfrac{1}{3}\rho(v^2)_{av} \qquad (12.13)$$

Moreover
$$PV = \tfrac{1}{3}Nm(v^2)_{av}$$

This equation is Boyle's Law. Remembering that the translational kinetic energy of an object is $\tfrac{1}{2}mv^2$, it follows that

$$PV = \tfrac{2}{3}N\,\text{KE}_{av} \qquad (12.14)$$

Pressure depends on the number of molecules per unit volume as well as on their average translational KE.

Comparing this situation with the Ideal Gas Law, Eq. (12.11), we see that

$$KE_{av} = \tfrac{3}{2}k_B T \tag{12.15}$$

Roughly speaking, the **temperature** of a substance is a measure of the average kinetic energy of its constituent particles (e.g., atoms, ions, molecules, and/or free electrons).

The temperature of an ideal gas is proportional to the average translational kinetic energy of its molecules. The microscopic average translational KE of the molecules manifests itself macroscopically in the form of temperature (Fig. 12.22).

Real gases, especially at low temperatures, do interact, and that must be treated via Quantum Mechanics and not by Newtonian physics. Although the relationship between translational KE and temperature is undeniably an important insight, one that can fruitfully guide our thinking, it has its shortcomings. Remember that a substance must have a zero-point energy (p. 429) even when $T = 0$ K, something not in accord with Eq. (12.15), which therefore cannot be the whole truth at low temperatures. All of this underscores the conclusion that *temperature is a fundamental quantity and, like length, time, and mass, is not expressible in terms of other quantities*.

The Speeds of Molecules in a Gas

It follows from Eq. (12.15) and the fact that $KE_{av} = \tfrac{1}{2}m(v^2)_{av}$ that

$$(v^2)_{av} = \frac{3k_B T}{m}$$

which is the mean of the speed squared. If we take the square root of it, we get something called the *root of the mean of the square* (*rms*, for short):

$$v_{rms} = \sqrt{(v^2)_{av}} = \sqrt{\frac{3k_B T}{m}} \tag{12.16}$$

This is not quite the average speed, but a statistical analysis shows that $v_{av} = 0.92\, v_{rms}$, and so Eq. (12.16) gives us a measure of the typical speed of a gas particle.

Example 12.9 **[II]** What is the *rms*-speed of a hydrogen molecule (H_2) at STP? Take the molar mass of hydrogen molecules to be 2.0 g/mol.

Solution The mention of *rms*-speed should call to mind "Kinetic Theory." (1) TRANSLATION—Knowing the mass per mole of a gas and its temperature and pressure, determine the *rms*-speed of its molecules. (2) GIVEN: $T = 273$ K, $P = 0.101$ MPa, and the molar mass equals 2.0 g/mol. FIND: v_{rms}. (3) PROBLEM TYPE—Kinetic Theory. (4) PROCEDURE—The only relationship we have for *rms*-speed is provided by Eq. (12.16). (5) CALCULATION—We first compute m, the mass of a molecule, which is the molar mass divided by the number of molecules per mole:

$$m = \frac{0.002\,0 \text{ kg/mol}}{6.022 \times 10^{23} \text{ molecules/mol}} = 3.32 \times 10^{-27} \text{ kg}$$

From Eq. (12.16)

$$v_{rms} = \sqrt{\frac{3k_B T}{m}} = \left[\frac{3(1.38 \times 10^{-23} \text{ J/K})(273 \text{ K})}{3.32 \times 10^{-27} \text{ kg}} \right]^{\frac{1}{2}}$$

and so $\boxed{v_{rms} = 1.8 \text{ km/s}}$ or about 6×10^3 ft/s, which is faster than a bullet.

Quick Check: m is correctly just about the mass of two protons. From Eq. (12.13) and Table 9.1 (p. 286) listing densities, $v_{rms}^2 \approx 3P/\rho \approx 3(1.01 \times 10^5 \text{ Pa})/(0.090 \text{ kg/m}^3)$; $v_{rms} \approx 1.8$ km/s.

We can expect sound waves to travel at speeds less than that of the individual molecules. The speed of sound in hydrogen (Table 11.2) is 1.27 km/s $= 0.7 v_{av}$, and that's reasonable.

Table 12.5

Molecular parameters at STP

Property	Hydrogen (H$_2$)	Oxygen (O$_2$)
Number of molecules per cm^3	2.7×10^{19}	2.7×10^{19}
Diameter of molecule	0.24 nm	0.32 nm
Mass per cm^3	8.99×10^{-8} kg	1.43×10^{-6} kg
Average speed	18.4×10^2 m/s	4.6×10^2 m/s
Collisions per second	10×10^9	4.6×10^9
Mass of molecule	3.3×10^{-27} kg	53×10^{-27} kg
Mean free path	1.8×10^{-7} m	1.0×10^{-7} m
Volume per gram	11.2×10^3 cm^3	700 cm^3

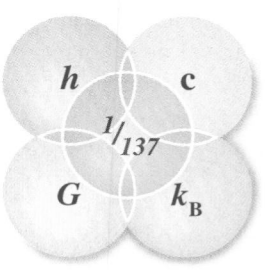

Figure 12.22 Boltzmann's constant bridges the domains of the macroscopic and microscopic. Along with the gravitational constant (G), the speed of light in vacuum (c), Planck's Constant (h), and the fine structure constant (1/137) it's one of the fundamental constants of Nature.

Our present understanding is that the molecules of a gas undergo a vast number of collisions with each other—about 5×10^9 per second for each molecule in the air around us, which is about 10^5 collisions for each centimeter of path traveled. The uninterrupted distance between collisions—the *mean free path*—is about 10^{-5} cm (Table 12.5), which is ≈500 atomic-diameters. That's why it takes time for the scent of a perfume to fill a room after the bottle is opened. The molecules, even though they move at great speeds, travel zigzag paths, only gradually diffusing outward. {To study this behavior via a great simulation, click on **THE KINETIC MODEL OF AN IDEAL GAS** under **INTERACTIVE EXPLORATIONS** on the **CD**.}

The effect of the tremendous number of intermolecular collisions does *not* even out the speeds of the particles but distributes them in a stable pattern over a broad range from zero up toward infinity. The corresponding curves represent the number of molecules in each interval of speed (Fig. 12.23). They peak at a speed that is likely to be possessed by more particles than any other, and that's roughly 13% below the average speed v_{av} and 23% below v_{rms}. Most of the particles are grouped around this peak speed; in fact, there will probably be no more than 1 out of about 10 000 molecules with speeds in excess of $3v_{av}$. Nonetheless, the immense number of particles present ensures that small but finite quantities will be traveling very slowly and very rapidly. Realize, too, that the curve represents a stable pattern even though, from moment to moment, collision to collision, the speed of any one molecule will change drastically. These are the *Maxwell–Boltzmann distributions*, each displaying the most probable spectrum of molecular speeds available to the system at a particular temperature. Though purely theoretical when proposed, the distribution received direct confirmation in the late 1920s and early 1930s via molecular beam experiments. {For a wonderful animated examination of the behavior of molecules click on **MAXWELL'S**

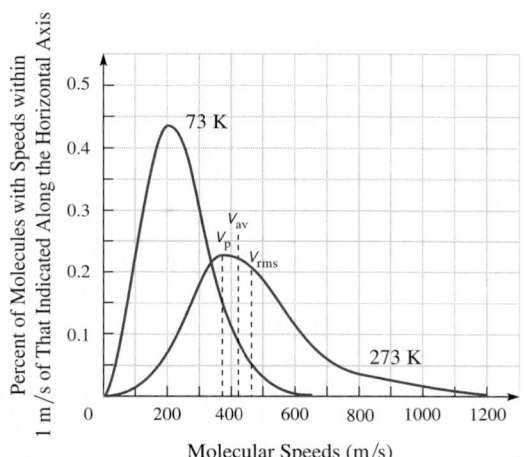

Figure 12.23 The Maxwell–Boltzmann speed distributions for molecular oxygen at $T = 73$ K and $T = 273$ K. The area under either curve between any two speeds, say, 400 m/s to 600 m/s, is the percentage of the total number of particles N in that range. The total area under either curve is 100%. Multiply the percentage in any speed range by the total number of molecules present, and you obtain the number with that range of speeds.

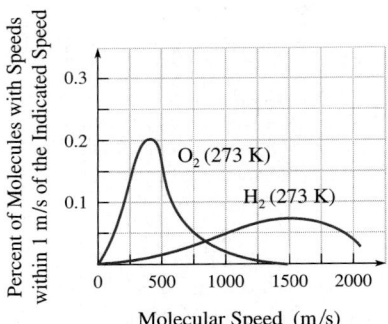

Figure 12.24 At the same temperatures, oxygen and hydrogen molecules have the same average kinetic energy; therefore, the less massive hydrogen molecules, on average, move faster.

DISTRIBUTION under INTERACTIVE EXPLORATIONS on the CD. By playing with this simulation for a while, you'll get a good understanding of what Fig. 12.23 really means.}

For a mix of two different gases held at the same temperature in a chamber—insofar as they behaved as ideal gases—Eq. (12.15) tells us that each would have the same average translational KE. The implication is that massive gas particles travel slower than light ones, and that's been confirmed experimentally (see Fig. 12.24). {For more on the subject, click on DIFFUSION under FURTHER DISCUSSIONS on the CD.}

Core Material & Study Guide

TEMPERATURE

Temperature is a fundamental quantity that describes an aspect of the physical state of a system. The two common temperature scales, Fahrenheit and Celsius, are related by

$$T_F = 32° + \tfrac{9}{5}T_C \quad \text{or} \quad T_C = \tfrac{5}{9}(T_F - 32°) \quad [12.1]$$

The **thermodynamic temperature** is measured in units of *kelvins* (K):

$$T \rightarrow T_C + 273.15 \quad [12.2]$$

and the Absolute Zero of temperature corresponds to 0 K (−273.15 °C). To learn about the various temperature scales read through Section 12.1 (Thermodynamic Temperature & Absolute Zero) and study Examples 12.1 and 12.2. This is basic, necessary material. **Make sure to go over all the CD WARM-UPS and WALK-THROUGHS for this chapter.**

THERMAL EXPANSION

Most solids expand (or contract) upon having their temperatures raised (or lowered) and

$$\Delta L = \alpha L_0 \Delta T \quad [12.3]$$

Here, ΔT can be in either °C or kelvins, and α is the *temperature coefficient of linear expansion* (p. 429). The change in volume of a substance that undergoes a change in temperature ΔT is

$$\Delta V = \beta V_0 \Delta T \quad [12.4]$$

V_0 is the original volume, and β is the *temperature coefficient of*

volume expansion (p. 432). The above material is treated in Sections 12.2 (Linear Expansion) and 12.3 (Volumetric Expansion: Solids & Liquids). Make sure you understand Examples 12.3 and 12.4—go over these solutions several times and read the Suggestions on Problem Solving *before* attacking the problem set.

THE GAS LAWS
Boyle's Law is

[constant T and m] $\qquad PV = \text{constant} \qquad [12.5]$

Charles's Law is

[constant P and m] $\qquad \dfrac{V}{T} = \text{constant} \qquad [12.6]$

Gay-Lussac's Law is

[constant V and m] $\qquad \dfrac{P}{T} = \text{constant} \qquad [12.7]$

These three relationships can be combined:

[constant m] $\qquad \dfrac{PV}{T} = \text{constant} \qquad [12.8]$

When the mass of gas can change but the temperature is held constant, $PV \propto m$, or

[constant T] $\qquad \dfrac{PV}{m} = \text{constant} \qquad [12.9]$

All of these relationships are treated in Section 12.4 (The Laws of Boyle, Charles, & Gay-Lussac). Study the accompanying Examples 12.5 and 12.6. The above equations lead to the more inclusive

Ideal Gas Law:

$$PV = nRT \qquad [12.10]$$

where $R = 8.314\,51$ J/mol·K is called the **Universal Gas Constant**. Incidentally, R is given here in gram moles. We define a new constant, called **Boltzmann's Constant**, $k_B = R/N_A = 1.380\,66 \times 10^{-23}$ J/K and thereby obtain

$$PV = Nk_B T \qquad [12.11]$$

The Kinetic Theory of gases leads to Boyle's Law, in the form

$$PV = \tfrac{2}{3} N\,KE_{av} \qquad [12.14]$$

where

$$KE_{av} = \tfrac{3}{2} k_B T \qquad [12.15]$$

The average translational KE of the molecules of an ideal gas is related to the temperature of that gas. Furthermore

$$v_{rms} = \sqrt{\frac{3k_B T}{m}} \qquad [12.16]$$

Section 12.7 (Kinetic Theory) treats the above material, and Example 12.9 deals with *rms*-speed.

Key Terms

temperature	Gay-Lussac's Law
Fahrenheit scale	Ideal Gas Law
Celsius scale	Universal Gas Constant
thermodynamic temperature	Boltzmann's Constant
absolute temperature scale	critical point
Absolute Zero	vapor
zero-point energy	equilibrium vapor pressure
temperature coefficient	triple point
of linear expansion	sublimation
temperature coefficient	KineticTheory
of volume expansion	mean free path
Boyle's Law	Maxwell-Boltzmann
Charles's Law	distribution

Discussion Questions

1. Explain the meaning of each of the Key Terms above.

2. List all the possible sources of errors you can think of that might limit the accuracy of a temperature reading made with a typical mercury-in-glass thermometer. Be as imaginative as you can and, as Descartes put it, "doubt of everything"; that is, take nothing for granted.

3. Plunge a sensitive mercury-in-glass thermometer into a bath of hot water. The temperature reading will first drop for a moment and then gradually rise. Explain.

4. Type metal, which is poured into molds to make printer's type, is an alloy of lead, antimony, and tin. Similarly, cast iron, which is an alloy of iron and carbon, is also used in making castings. What uncommon physical property do you think these materials might share with water? Explain.

5. A demonstration device known as a cast-iron bomb is a thick-walled sphere that is filled to the top with water and then sealed with a tightly fitting iron plug. The thing is then submerged in a pail of ice, water, and salt. What happens next? Explain. How does this same process, occurring naturally, affect rocks and mountains?

6. The two steel plates in Fig. Q6 have been attached to one another with a rivet, which was slipped into the hole running through the plates while red hot and "headed over" with a hammer. Why were the rivets put in while hot? (Incidentally, the usual approach was to throw the red-hot glowing rivets, almost the size of your fist, through the air from the fire, up to the installing team, which caught them in buckets.)

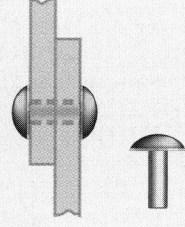
Figure Q6

7. Suppose that you are visited by an inventor who has found a liquid that has exactly the same β as glass. It is proposed that the wonder fluid be dyed red so that it can be seen easily and used to make inexpensive, but incredibly accurate thermometers. Would you invest in such a scheme? Explain the physical reasons behind your decision.

8. The old-fashioned way to break boulders is to heat them up for a long time in a big fire and then quickly douse them with cold water. It is a lot easier than hitting at them with sledge hammers. Explain.

9. Anyone who has ever worked in a chemistry lab knows that such places invariably house a multitude of beakers and test tubes. Why is it that such equipment is generally made of Pyrex (or some other special glass), and why are they so thin-walled and delicate? (You never see a good, thick, sturdy test tube.)

10. In light of Question 9, why are glass incandescent bulbs so thin? What might happen to a hot light bulb if a drop of cold water fell on it? *Don't try it!* (This can be a problem to sloppy house painters.) What does that suggest about outdoor lighting? How does this relate to traditional photographic flashbulbs?

11. Though it was no doubt observed countless times before, Michael Faraday was the first person to report that two pieces of ice can be frozen together as one by simply pressing them against each other. Explain. What does this phenomenon have to do with snowballs?

12. I remember days as a kid when there was snow everywhere, but the stuff just would not pack into snowballs no matter how we tried. Although we concluded that it was just "terrible snow," what might have been the actual cause of the problem?

13. Professor James Thomson, Lord Kelvin's older brother, was the

one who suggested that the three transition curves in the phase diagram (Fig. 12.18) actually meet at a single point—the triple point. Show logically that this must be the case (Fig. Q13).

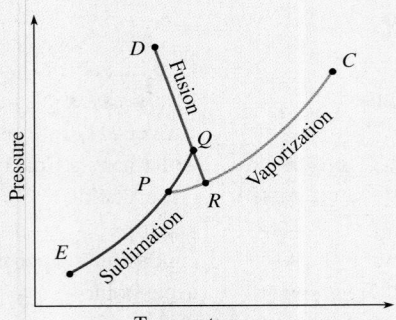

Figure Q13

14. The thick-walled sealed tube in Fig. Q14 shows liquid carbon dioxide in equilibrium with its vapor at room temperature and a pressure of about 6.2 MPa (i.e., 61 atm). Describe what will happen in the tube as the temperature is slowly raised to 31.2 °C. The critical temperature and pressure of CO_2 are 304.3 K and 7.38 MPa, respectively.

15. Under what circumstances, if any, can water freeze and boil at almost the same time? What would happen to a dish of water in an evacuated region where the vapor is efficiently drawn away?

Figure Q14

16. Can we expect the ordinary gases of the air (oxygen, hydrogen, nitrogen, etc.) to behave like an ideal gas? Explain.

17. Sir James H. Jeans, a physicist who made a number of important contributions to modern physics during the early part of the twentieth century, wrote the following in *The Growth of Physical Science*:

> In warmer air, the molecules must, of course, move more rapidly, in cooler air more slowly. In general the energy of motion per molecule is proportional to the temperature, measured from the absolute zero at which the energy of motion is nil.

Which equations derived in the chapter confirm his comment? What, if anything, might you take exception to?

18. Given two 1.0-m³ sealed evacuated chambers surrounded by the same constant temperature bath and suppose we put two moles of hydrogen in one of them and two moles of nitrogen in the other. What can be said about the pressures in the two chambers? Undeniably, the nitrogen molecules are much more massive than the hydrogen molecules. How can that difference be understood in light of your conclusion about the pressures? What, if anything, can be said about the speeds of the different molecules?

19. If we should ever be visited by some advanced extraterrestrial society, is it likely that they, too, would have arrived at the same notion of an absolute zero of temperature? How reasonable is it to expect that any thermometer they might pull out would have a scale we would be familiar with? If you answered affirmatively, which scale?

20. The air in an ordinary room heated by a stove of some sort will rise in temperature but may not increase its net amount of molecular translational energy. Explain. Be careful about the assumptions you make.

21. What phase is water in at (a) 1.0 atm and 102 °C (b) 2.0 atm and 50 °C (c) 0.006 atm and 5 °C (d) 0.001 atm and 1.0 °C (e) 0.000 1 atm and 0.01 °C?

Multiple Choice Questions

1. The Celsius equivalent of 90 °F is (a) 90° (b) 32° (c) 194° (d) 45° (e) none of these.

2. At the temperature known as absolute zero (a) all motion ceases (b) time ceases (c) translational atomic KE is zero (d) atomic motion is at a minimum (e) none of these.

3. The surface of a very hot star has a temperature of 2×10^5 °C. Very roughly, how many times hotter is that than room temperature (293 K)? (a) 10^3 (b) 10^5 (c) 10^{-3} (d) 10^{-5} (e) none of these.

4. What is the order of magnitude of the fractional change in the length of a metal bar when it is altered in temperature by 1 K? (a) 10^{-6} (b) 10^{-4} (c) 10^{-5} (d) 10^6 (e) none of these.

5. Water has its maximum density at a temperature of about (a) 0 °C (b) 32 °F (c) 4 °C (d) 273 K (e) none of these.

6. If the temperature of a sealed rigid chamber full of gas is increased, the pressure will (a) remain constant (b) decrease proportionately (c) increase proportionately (d) first decrease and then increase (e) none of these.

7. Which graph in Fig. MC7 best represents Boyle's Law? (a) *a* (b) *b* (c) *c* (d) *d* (e) none of these.

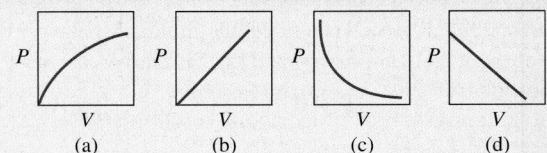

Figure MC7   (a)   (b)   (c)   (d)

8. The pressure in a chamber filled with gas arises from (a) the gravitational interaction between the atoms and molecules of the gas (b) collisions between the molecules of the gas itself (c) collisions between the molecules of the gas and the walls of the chamber (d) the electromagnetic force between the atoms and molecules of the gas (e) none of these.

9. Which graph in Fig. MC9 best represents the relationship between the volume and temperature of a low-density gas at constant pressure and mass? (a) *a* (b) *b* (c) *c* (d) *d* (e) none of these.

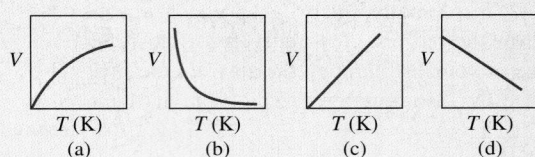

Figure MC9   (a)   (b)   (c)   (d)

10. Consider a gas that resembles an Ideal Gas. Which of the following is not true? (a) The net volume occupied by the molecules is much smaller than the volume of the gas (b) The intermolecular forces are strong (c) The molecules of the gas are in continuous rapid random motion (d) Compared to their sizes, the separations between the molecules are large (e) none of these.

11. Which graph in Fig. MC11 best represents the relationship between the pressure and temperature of a low-density gas at constant volume and mass? (a) a (b) b (c) c (d) d (e) none of these.

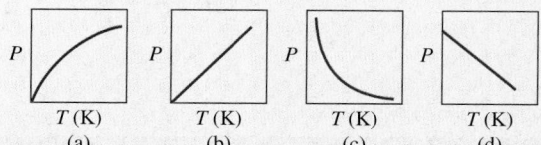

Figure MC11
       (a)   (b)   (c)   (d)

12. If the pressure in a sealed flexible chamber full of a fixed mass of gas is doubled, while the temperature is kept constant, the volume of the gas will (a) remain constant (b) be halved (c) quadruple (d) double (e) none of these.

13. What is the approximate mass of 11.11 liters of molecular hydrogen gas at STP? Each hydrogen atom has an average mass of 1.007 97 u. (a) 1.0 g (b) 1.0 kg (c) 11.1 g (d) 22.4 liters (e) none of these.

14. The density of carbon dioxide at STP is (a) 1.96 kg/m^3 (b) 44 g/liter (c) 44 kg/m^3 (d) 6.02×10^{23} g/liter (e) none of these.

15. Given that we have 17.0 g of ammonia gas (NH_3), the volume it occupies at STP is (a) 17 liters (b) 22.4 liters (c) 22.4×10^6 m^3 (d) 17 m^3 (e) none of these.

16. Which graph in Fig. MC16 best represents a plot of the average kinetic energy of the molecules of an Ideal Gas against the absolute temperature of that gas? (a) a (b) b (c) c (d) d (e) none of these.

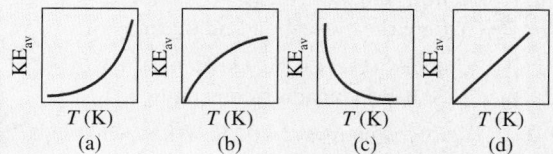

Figure MC16
       (a)   (b)   (c)   (d)

17. The vibrational energy of the atoms and molecules of a material is a minimum at a temperature of (a) 0 °C (b) 273 K (c) 0 °F (d) 0 K (e) none of these.

18. According to Kinetic Theory, the molecules of a gas at a given temperature (a) all move with a speed v_{rms} (b) all move at speeds in excess of v_{rms} (c) all move at speeds less than v_{rms} (d) move at speeds above, at, and below v_{rms} (e) none of these.

19. In a mixture of oxygen and hydrogen gas, we can expect, on average, that (a) the hydrogen molecules will be moving faster than the oxygen molecules (b) both kinds of molecules will be moving at the same speed (c) the oxygen molecules will be moving more rapidly than the hydrogen molecules (d) the KE of the hydrogen will exceed that of the oxygen (e) none of these.

20. A sample of gas is held at a constant pressure in a cylinder closed by a movable piston. If the volume is halved, how will the new *rms*-speed of the molecules compare with the original *rms*-speed? It will be (a) $\sqrt{2}$ times greater (b) the same (c) 2 times greater (d) 4 times greater (e) none of these.

21. A sample of gas is held at a constant pressure in a cylinder closed by a movable piston. If the volume is doubled, how will the new *rms*-speed of the molecules compare with the original *rms*-speed? It will be (a) $\sqrt{2}$ times greater (b) the same (c) 2 times greater (d) 4 times greater (e) none of these.

22. Molecular hydrogen has a mass per mole of 2 g/mol, while the equivalent value for molecular oxygen is 32 g/mol. That means that the ratio of the *rms*-speed of a hydrogen molecule to that of an oxygen molecule will be (a) 2:1 (b) 4:1 (c) 8:1 (d) $\sqrt{2}$:1 (e) none of these.

23. An atom of argon ($^{40}_{18}$A) in the air at 20 °C has an average translational kinetic energy of (a) 6.0×10^{-21} J (b) 6.0×10^{-21} J (c) 6.0×10^{-21} m/s (d) 21×10^{-6} J/s (e) none of these.

24. The molecules of a gas have a certain average translational kinetic energy at 10 °C. They will have twice that average value at (a) 20 °C (b) 283 K (c) 566 K (d) 14.1 °C (e) none of these.

25. According to Kinetic Theory, the molecules of a gas at a given temperature (a) all have the same speed (b) all have the same direction of motion (c) all have the same kinetic energy (d) all have the same momentum (e) none of these.

26. If the temperature of an ideal gas is increased from 100 K to 400 K, the *rms*-speed of the particles will change by a multiplicative factor of (a) 2 (b) $\frac{1}{2}$ (c) $\frac{1}{4}$ (d) $\sqrt{2}$ (e) none of these.

27. A flat metal disk has a small off-center hole drilled in it. When heated, (a) the disk will expand and so will the hole (b) the disk will expand, but the hole will remain unchanged (c) the disk will expand, but the hole will shrink (d) the disk will remain unchanged, but the hole will enlarge (e) none of these.

For more Multiple Choice Questions with answers click on WARM-UPS in CHAPTER 12 on the CD.

Suggestions on Problem Solving

1. When confronted by a problem on gases, note any physical parameters that are held constant. Use the Ideal Gas Law in one of its two forms to summarize all the gas laws. Thus, if T and m (i.e., n) are constant, the problem deals with variations in pressure and volume and $PV = $ constant; if V and m (i.e., n) are constant, the problem deals with variations in pressure and temperature and $P/T = $ constant, and so on. When given or asked for N, use $PV = Nk_BT$.

2. To determine the mass of any molecule, either consult the Periodic Table or Table 12.4. Atomic masses are usually given in atomic mass units (u), where 1.000 00 u $= 1.660\,54 \times 10^{-27}$ kg. The numbers in the Periodic Table are averages for naturally occurring mixes of isotopes (something we will come back to later), so there is no need to carry around all those figures.

3. In problems concerning variations in the size of an object, the change in temperature is central. Whether you use °C or K, ΔT will be the same; however, you cannot use Fahrenheit when α or β is

given in SI. **All the gas equations are formulated in terms of thermodynamic temperature, and using anything other than kelvins will generally produce wrong answers.**

4. The gas equations often lead to a symmetrical form like $P_iV_i/T_i = P_fV_f/T_f$, in which case the units cancel from both sides. It is then allowable to enter mixed units into the equation, but be very careful. Like it or not, there will always be a body of historical data in the literature that uses all the old, now shunned, units, and we should be able to deal with that, too. When one unit can be converted to another by a multiplicative constant, then the ratios of quantities in either set of units will be the same: P_i/P_f is unitless and the same whether determined in Pa or mm Hg. *But* when the units are converted one to the other by the addition of a numerical factor, this will not be true: T_i/T_f is *not* the same in degrees Celsius as in kelvins. (Just think about the chaos that would arise in the equations at 0 °C.) So, yes you can use liters on both sides, or atmospheres, but you must use kelvins for temperature.

5. **The gas equations are formulated in terms of absolute pressure. Using gauge pressure will produce wrong answers.**

6. Remember that $\Delta L/L_0 = \alpha\,\Delta T$, and so α contains no length unit. It is the fractional change per kelvin (or per °C, which is the equivalent), and hence, ΔL and L_0 can be entered in any unit you like.

Problems ✦ Coordinated Problems ✦ Progressive Problems ✦ Solutions

SECTION 12.1: THERMODYNAMIC TEMPERATURE & ABSOLUTE ZERO

1. [I] What is normal body temperature (98.6 °F) in degrees Celsius?

2. [I] A thermometer reads the temperature of a room to be 70.0 °F. How much is that in Celsius?

3. [I] Examine Fig. P3 and explain what it represents. Why is the curve a straight line? Use the figure to check your answers to the previous two questions.

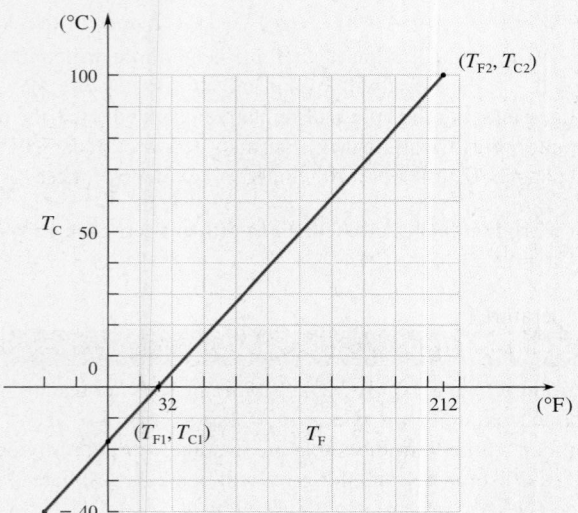

Figure P3

4. [I] What is the slope of the curve in Fig. P3?

5. [I] Determine the value of T_C at the point where the curve in Fig. P3 crosses the vertical axis (i.e., the y-intercept). In other words, find T_{C1}. Use the diagram and a little geometry. Now check your answer using Eq. (12.1).

6. [I] What are the coordinates of the leftmost point on the curve?

7. [I] The temperature of the Universe can be said to be 2.726 K. What is that in Celsius?

8. [I] A person who isn't feeling very well takes his temperature with a Celsius thermometer and finds a value of 39.0 °C. What is that in Fahrenheit?

9. [I] Convert a temperature reading of 58 K into Celsius.

10. [I] What is the ratio of the size of one Celsius degree to one kelvin unit?

11. [I] The visible region of the Sun, the photosphere, is at a temperature of 5.6×10^3 K. What is that temperature in Fahrenheit?

> **SOLUTION:** $T \rightarrow T_C + 273.15$; $T_C = T_K - 273.15 = 5327$ °C; $T_F = 32° + \frac{9}{5}T_C = 32° + \frac{9}{5}(5327\ °C) = 9.6 \times 10^3$ °F.

12. [I] Helium boils at -268.9 °C and melts at -272 °C (at 26 atm). What are the corresponding thermodynamic temperatures?

13. [I] A block of steel is heated from 0 °F to 100 °F. What is the change in its thermodynamic temperature?

14. [I] Hydrogen melts at 14.01 K. What is that in °C?

15. [I] Consider a temperature change of 42 Celsius degrees. What change in kelvins is that equivalent to?

16. [II] Will doubling the absolute temperature of a substance, for example in going from 100 K to 200 K, double its Celsius temperature? Explain. Will doubling the Celsius temperature of a substance, for example in going from 100 °C to 200 °C, double its Fahrenheit temperature? Explain.

17. [II] If the temperature at the center of a star is 10 MK, what is the equivalent in °C and °F?

18. [II] A uranium fuel rod in a reactor operates at about 4000 °F. What is that temperature in kelvins?

19. [II] The temperature of steam in a typical fossil-fuel plant is about 770 K. What would the temperature be in Fahrenheit degrees?

20. [II] For lead, what is the unitless ratio of its boiling point to its melting point computed in kelvins and in °C? See Table 12.1 or 13.6.

21. [II] Referring to Fig. P3, use geometry to prove that for any positive value of temperature (T_C) on the Celsius scale $T_C = (T_F - 32)[100/(212 - 32)]$. Compare this to Eq. (12.1).

22. [II] Referring to Fig. P3, show that $T_C = [T_F(T_{C2} - T_{C1})/T_{F2}] + T_{C1}$. Using this expression determine the Celsius equivalent of 68 °F.

SECTION 12.2: LINEAR EXPANSION

SECTION 12.3: VOLUMETRIC EXPANSION: SOLIDS & LIQUIDS

23. [I] A 1.00 m long Pyrex glass rod is heated so its temperature rises 80.0 °C. By how much will its length increase? [*Hint: You might want to use Table 12.2.*]

24. [I] A 10.0 m tall structural steel pole at 37.5 °C supports a streetlamp. If, overnight, the pole shrinks by 2.00 mm, by how much has its temperature changed?

> SOLUTION: $\Delta L = L_0 \alpha \Delta T$; 2.00×10^{-3} m $= (10.0$ m$)(12 \times 10^{-6})\Delta T$; $\Delta T = 16.7$ °C $= 17$ °C.

25. [I] By how much would a 1.00 m long aluminum rod at 20 °C increase in length if its temperature were raised 1.00 °C? [*Hint: Use Table 12.2 to get α for aluminum.*]

26. [I] By how much would a 1.00 m long brass rod shrink if its temperature were lowered 1.00 °C?

27. [I] By how much would a 10 m long aluminum bar change its length in the process of going from 30 °C to 50 °C?

28. [I] A steel girder 10-m long is installed in a structure on a day when the temperature is 5 °C. By how much will it increase in length when it warms up to 35 °C?

29. [I] The roadway of the Golden Gate Bridge is 1280-m long, and it is supported by a steel structure. If the temperature varies from 0 °C to 35 °C, how much does the length change?

30. [I] A stainless steel rod is heated from room temperature (68 °F) to 168 °F. By what percentage of its original length will its length increase?

31. [I] THIS PROBLEM EXPLORES THE PHENOMENON OF THERMAL EXPANSION. A sidewalk is to be made of poured concrete having a coefficient of linear expansion of 12×10^{-6} K^{-1}. Each slab is to be 2.4 m long by 2.4 m wide, and it's to be installed when the temperature is 10 °C (50 °F). We want to determine the appropriate space to leave between slabs if we can expect the temperature to rise, at most, to a 46 °C (115 °F). (a) What is the mathematical relationship between the changes in length (ΔL) and temperature (ΔT) of each slab? (b) What is the relationship between ΔL and the minimum space that must be left between slabs? (c) What is the numerical value of ΔT? (d) Compute the minimum space that must be left between slabs.

32. [I] A thin sheet of copper 50.00 cm by 20.00 cm at 30.00 °C is heated to 60.00 °C. What will be the new area of one face? (See Problem 33.)

33. [I] Redo Problem 32 using the equation derived in Problem 40. Compare the two results.

34. [I] A steel television tower is 150.00-m tall at 10 degrees below zero Fahrenheit. What is its height at 95 °F?

35. [I] THIS PROBLEM EXPLORES THE PHENOMENON OF THERMAL EXPANSION. A length of copper tubing is installed in a private house to carry hot water, and we want to calculate the change in the tube's cross-sectional area when it does. At 20.0 °C the tubing has an inner diameter of 2.00 cm. (a) What is the area of the opening in the pipe? (b) With hot water at 88.0 °C flowing through the pipe, what is the new diameter of its opening? (c) What will be the change in the pipe's cross-sectional area?

36. [II] THIS PROBLEM EXPLORES THE PHENOMENON OF LINEAR THERMAL EXPANSION. A metal rod is submerged in an ice and water mixture, and its length is measured and plotted in Fig. P36. It is then immersed in boiling water, and its length is measured and plotted again. (a) What are the Celsius temperatures of the two plotted points? (b) Connect the two points by a straight line. What assumption must be made in doing so? (c) What is the physical meaning of the slope of the line? (d) Determine the linear coefficient of thermal expansion of the metal. (e) What material is this likely to be? (f) What will be the length of the rod at 60.0 °C?

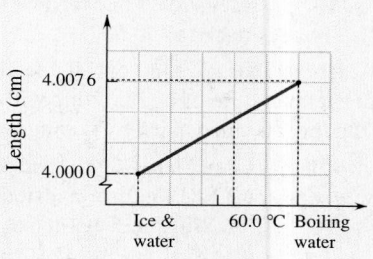

Figure P36

37. [II] THIS PROBLEM EXPLORES THE PHENOMENON OF THERMAL EXPANSION. A solid brass rod having a diameter of 4.000 cm is to be slid into a brass tube which has an inner diameter of 3.995 cm (both measured at room temperature, i.e., 20 °C). (a) How might we accomplish this feat? (b) What is the mathematical relationship between the changes in diameter and temperature of the tube? (c) What change in the temperature of the pipe will cause its inner diameter to equal the diameter of the rod? (d) To what temperature should the pipe be heated so the rod (still at 20 °C) fits inside it?

38. [II] THIS PROBLEM EXPLORES THE PHENOMENON OF THERMAL EXPANSION. A sample of aluminum has a density of 2.70×10^3 kg/m^3 at room temperature (20 °C). We want to determine its density at some elevated temperature. (a) What is the definition of the density? (b) What happens to the volume as the temperature changes? (c) If after the temperature is changed, the initial and final volumes and densities are V_0, ρ_0, and V, and ρ, respectively, explain why $V_0\rho_0 = V\rho$. (d) Write an expression for the density in terms of T. (e) What is the density of aluminum at 120 °C? (For the purpose of this calculation take the coefficient of volumetric expansion to be 72.0×10^{-6} K^{-1}.)

39. [II] A carpenter's steel measuring tape is 10.000-m long and 1.00-cm wide. Assuming that it was calibrated in the factory at room temperature (20 °C), what is its length on a summer day when the temperature is 96.8 °F? (For this steel, $\rho = 7860$ kg/m^3, $\alpha = 12.15 \times 10^{-6}$ K^{-1}, and $\beta = 36.46 \times 10^{-6}$ K^{-1}.) Comment on the practical aspects of your results.

40. [II] You have a sheet of material of area A_0 whose temperature is changed by an amount ΔT. Prove that the area will change by

$$\Delta A = 2\alpha A_0 \Delta T$$

41. [II] The mercury in a thermometer at 32 °F has a volume of 0.50 cm^3. What will be its volume at 212 °F?

42. [II] Imagine that we have a mercury-in-Pyrex thermometer. What is the *apparent* volume coefficient of expansion of mercury in Pyrex? That is, what is the effective coefficient, taking into consideration the expansion of the body of the device?

43. [II] A soft glass beaker with an inside diameter of 10.00 cm is filled to the very top with 800.0 cm³ of pure mercury at 95 °C. How much will the level change if the temperature is lowered to 0 °C? [*Hint: Remember that both the beaker and the mercury expand.*]

44. [II] A thermometer has a quartz body within which is sealed a total volume of 0.400 cm³ of mercury. The stem contains a cylindrical hole with a bore diameter of 0.10 mm. How far does the mercury column extend in the process of rising from 10 °C to 90 °C? Neglect any change in volume of the quartz.

45. [II] A bowl made of Pyrex is filled to the very brim with 100 cm³ of water at 10 °C. How much will overflow when the temperature of the filled bowl is raised to 50 °C?

46. [II] A sheet of brass has a 2.000-cm-diameter hole drilled through it while at a temperature of 20 °C. What will be the diameter of that hole if the sheet is surrounded by boiling water?

47. [II] A block of iron has an old label stuck to its side that gives its density as 7.85 g/cm³ (at 20 °C) and its coefficient of volume expansion as 36×10^{-6} °C⁻¹. What will its density be at 0 °C?

48. [II] When one cubic meter of water freezes at 0 °C, it becomes how many cubic meters of ice?

49. [III] A pendulum clock is made (at 20.00 °C) by supporting the *c.g.* of a bob at the very end of a 1.000 0-m-long aluminum rod. Determine its period. Now find its period on a winter's day when the heater goes out and the temperature drops to 0.000 °C. Did the clock run fast or slow that day?

50. [III] A 1-second pendulum clock is made at 20.000 °C by suspending a mass at its *c.g.* from a steel wire with a linear coefficient of expansion of $12.113\,5 \times 10^{-6}$ K⁻¹. What will its period be at 40.000 °C? Will it run fast or slow?

51. [III] Given that β is small, show that the final density of a solid specimen that has experienced a small temperature change of ΔT is

$$\rho \approx \rho_0(1 - \beta \Delta T)$$

52. [III] Show that the *thermal stress* (the stress due to an expansion or contraction of the object as a result of a temperature change) is given by

$$\text{Thermal stress} = Y\alpha\Delta T$$

53. [III] With Problem 52 in mind, suppose two concrete slabs are butted flat up against one another with their far ends fixed. Each is 10-m long, and the contact area is 1000 cm². If the concrete was poured on a day when the temperature was 5.0 °C, what will be the stress in them when the temperature is 42 °C? Will they fracture? Take $\alpha = 10 \times 10^{-6}$ K⁻¹ and $Y = 25$ GPa.

SECTION 12.4: THE LAWS OF BOYLE, CHARLES, & GAY-LUSSAC

54. [I] THIS PROBLEM EXPLORES THE BEHAVIOR OF A TENUOUS GAS. A fixed mass of gas, in a cylinder sealed by a movable piston, has a known absolute pressure (P_1), temperature (T_1), and volume (V_1). Suppose the temperature is kept constant. (a) What is the mathematical relationship between the pressure and volume? (b) If the piston is pushed in thereby decreasing the volume, what will happen to the pressure? (c) If the absolute pressure is tripled, what will be the final volume (V_2) in terms of the initial volume?

55. [I] THIS PROBLEM EXPLORES THE BEHAVIOR OF A TENUOUS GAS. A fixed mass of gas, in a cylinder sealed by a movable piston, has a constant absolute pressure (P), and a known temperature (T_1), and volume (V_1). (a) What is the relationship between the absolute temperature and volume? (b) Suppose the absolute temperature is doubled, on average will the molecules be moving around faster or slower? (c) Will the piston move in or out? (d) What will be the final volume (V_2) in terms of the initial volume?

56. [I] THIS PROBLEM EXPLORES THE BEHAVIOR OF A TENUOUS GAS. A fixed mass of gas, in a cylinder sealed by a movable piston, is held at a constant volume (V). The gas has a pressure (P_1) and an absolute temperature (T_1). (a) What is the mathematical relationship between the absolute temperature and absolute pressure of the gas? (b) Suppose the absolute temperature is halved, on average will the molecules be moving around faster or slower? (c) Will the piston move? in or out? (d) What will be the final pressure (P_2) in terms of the initial pressure?

57. [I] THIS PROBLEM EXPLORES THE BEHAVIOR OF A LOW-DENSITY GAS. A fixed mass of gas, in a cylinder sealed by a movable piston, has a volume (V_1), a pressure (P_1) and an absolute temperature (T_1). (a) What is the mathematical relationship between the absolute temperature, absolute pressure, and volume, for such a low-density gas? (b) Suppose that both the absolute temperature and absolute pressure are halved, on average will the molecules be moving around faster or slower? (c) Will the piston move? in or out? (d) What will be the final volume (V_2) in terms of the initial volume?

58. [I] A container of helium gas at STP is sealed and then raised to a temperature of 730 K. What will be its new pressure?

59. [I] The needle is removed from a hypodermic syringe, and that end is sealed. The air inside of it is slowly compressed (so that it remains at room temperature) to one-tenth its original volume. What is the final pressure (to two significant figures) in the syringe? [*Hint: We're dealing with pressure and volume changes at constant temperature.*]

60. [I] A bubble rises from the bottom of a tall open tank of water that is at a uniform temperature. Just before it bursts at the surface the bubble is three times its original volume. Determine the absolute pressure at the bottom of the tank.

61. [I] A 1.00-m³ tank is filled with nitrogen gas to a gauge pressure of 20.0 atm. It is then connected through a valve to a 9.00-m³ evacuated chamber. The valve is opened just a little, and the gas slowly flows until it finally stops on its own. Given that the temperature was constant throughout the transfer, what is the pressure in the final chamber?

62. [I] A container of gas is kept at a constant pressure (which is close to atmospheric) by supporting a glob of mercury as in Fig. P62. If its volume is initially 500×10^{-6} m³ at a temperature of 273 K, what will it be at 300 K?

63. [I] A quantity of helium gas at atmospheric pressure occupies the 300-cm³ volume of a cylinder fitted tightly

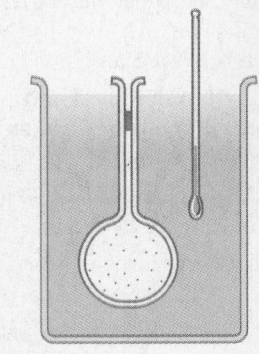

Figure P62

with a piston. If the temperature is initially 25.0 °C, what will be its final value when the gas occupies 200 cm³ at atmospheric pressure?

64. [I] A chamber sealed with a movable piston contains 5.50 liters of hydrogen at 28 °C and 81.3 kPa. What would its new volume be at STP?

65. [I] Given that we have 1200 cm³ of helium at 15 °C and 99 kPa, what will be its volume at STP?

66. [I] Determine the density of oxygen (O_2) at STP.

67. [I] What is the mass of a water molecule in kilograms? Assume the mass is simply the sum of the constituent masses. The masses of hydrogen and oxygen atoms are 1.008 u and 15.999 u, respectively.

68. [I] One mole of ammonia gas has a mass of 17.03 g. What is its density at STP to two significant figures?

69. [I] An expandable chamber contains 16.0 kg of molecular oxygen at STP. How many molecules are in it? How many moles?

70. [I] An expandable chamber contains 16.0 kg of molecular oxygen at STP. What's its volume?

71. [II] A large, aluminumized-Mylar™ balloon contains 1.000 m³ of helium when filled outdoors at 0 °C on a winter's day. What will its volume be at home at 20 °C?

> **SOLUTION:** The balloon is always at atmospheric pressure and so $V_i/T_i = V_f/T_f$ where $V_i = 1.000$ m³, $T_i = 273.15$ K, and $T_f = 293.15$ K. $V_f = T_f V_i/T_i = 1.1$ m³.

72. [II] Hydrogen gas at 273 K fills an expandable chamber to a volume of 30.0 liters at a pressure of 2.00 atm. The gas is then compressed to 15.0 liters at a pressure of 3.00 atm. What will its new temperature be?

73. [II] Show that for a given amount of gas

$$\frac{P_1}{T_1\rho_1} = \frac{P_2}{T_2\rho_2}$$

74. [II] Determine the density of hydrogen gas at STP. What is its density at 1.00 atm and at a temperature of 273 °C?

75. [II] THIS PROBLEM EXPLORES THE BEHAVIOR OF A LOW-DENSITY GAS. Figure P75 shows a constant-pressure gas thermometer immersed in a bath whose temperature is to be measured. The gas, which is the thermometric material, is sealed in at the left by a small drop of mercury blocking the open capillary tube, and at the right by a movable piston. The knob at the right moves the piston and is precisely calibrated to read the volume of the chamber (i.e., the volume of the gas, provided the mercury droplet is returned to its original location). The device is taken from a reference bath where T is known and V is measured, and placed into the environment whose temperature is to be determined. (a) What is the mathematical relationship between the absolute temperature and the volume of a gas, upon which this device is based? (b) How does the thing work? (c) Why is it called a constant-pressure thermometer? (d) What is the value of the pressure in the gas? (e) Suppose the chamber contains 1.50×10^3 cm³ of Ne gas at precisely 0 °C. What is the temperature of the new environment if the volume becomes 1.80×10^3 cm³?

76. [II] THIS PROBLEM EXPLORES THE BEHAVIOR OF A LOW-DENSITY GAS. A constant quantity of gas in a cylinder sealed by a movable piston occupies a volume of 590 cm³ at a gauge pressure of 1.9 atm. It is maintained at a fixed temperature of 298 K. (a) What is the

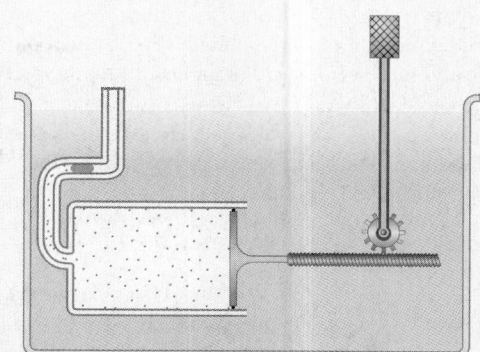

Figure P75

mathematical relationship between the absolute pressure and volume for a low-density gas of this sort? (b) Suppose the piston is pushed in, and the gas is compressed, will the pressure change, and if so, in what way? (c) If the new volume of the sample is 181 cm³, what will be the new gauge pressure (in atm) in the gas?

77. [II] A bubble rises a height h from the bottom of a tall open tank of liquid that is at a uniform temperature. Write an expression for the bubble's volume V_f just before it bursts at the surface in terms of its original volume V_i, and the uniform density of the liquid ρ.

78. [II] An automobile tire is pumped up to an absolute pressure of 33 lb/in², with air at a temperature of 40.0 °F. After driving for several hours, the temperature in the tire reaches 120 °F. Find the pressure in the tire at that point in SI units.

79. [II] THIS PROBLEM EXPLORES THE BEHAVIOR OF A LOW-DENSITY GAS. The device in Figure P79a consists of two glass tubes (the one on the right is sealed at the top and the one on the left is open), attached to each other by a rubber tube. Mercury is poured in and captures a sample of air in the tube at the right, whose cross-sectional area A, is known. By raising and lowering either side, the pressure and volume of the sample can be varied at will. We want to understand the resulting data graphed in Figure P79b. (a) Write an expression for the gauge pressure in the sample, in terms of the density of mercury (ρ_{Hg}). (b) Suppose that a mercury barometer in the lab has a column height H, write an expression for the absolute pressure in the sample in terms of H, h, g, and ρ_{Hg}. (c) Write an expression for the volume of the gas sample. (d) Explain the graph in Figure P79b.

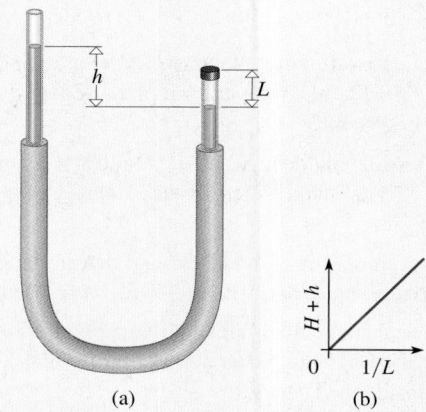

Figure P79   (a)   (b)

80. [II] THIS PROBLEM EXPLORES THE BEHAVIOR OF A LOW-DENSITY GAS. A fixed mass of gas, in a cylinder sealed by a movable piston, occupies a volume of 0.58 m³ at an absolute pressure of 49.8 kPa, and an absolute temperature of 398 K. (a) What is the mathematical relationship between the absolute temperature, absolute pressure, and volume, for such a low-density gas? (b) Suppose the absolute temperature is halved, and the pressure is dropped to 39.8 kPa, will the piston move, and if so, in what way? (c) What will be the new volume of the gas?

81. [III] A gas is being collected in a beaker over mercury. This is done by inverting a graduated beaker filled with mercury and then partially submerging it in a mercury-filled bowl. Gas is introduced into the beaker from below, bubbles up through the mercury, and displaces it. Suppose, at room temperature, 214 cm³ of gas is collected, leaving the surface of the mercury 3.90 cm above the level in the bowl. A barometer on the wall reads 101.05 kPa. If the gas is removed and slowly cooled to 0.00 °C at a pressure of 101.32 kPa, what will be its new volume?

82. [III] Hydrogen is being collected in an inverted beaker over water. The gas occupies a volume of 370 cm³, and the surface of the water inside the beaker is raised 5.00 cm while a barometer reads 759 mm. The temperature of the system is 17 °C. Given that water has a vapor pressure at that temperature of 1.92 kPa, find the pressure exerted by the hydrogen alone.

83. [III] Suppose 3.0 liters of oxygen at a pressure of 5.0 atm are added to 1.0 liter of nitrogen at 2.0 atm in a chamber with a volume of 4.0 liters. What is the pressure in the chamber if the temperature is kept constant throughout?

SECTION 12.5: THE IDEAL GAS LAW

84. [I] Three moles of a gas are to be stored at a temperature of 60 °C and a pressure of 93.0 kPa. How much volume will the gas occupy?

85. [I] If 1.40 moles of ammonia gas are to be put into a 10.0-liter container at a pressure of 202.6 kPa, at what temperature should the gas be maintained?

86. [I] A very good vacuum system can pump a chamber down to about 10^{-15} atm. Assuming a temperature of 273 K in the evacuated region, how many molecules of air are left in each cubic centimeter?

87. [I] With Problem 86 in mind, how many molecules are in 1.00 cm³ of air at STP?

SOLUTION: $PV = Nk_BT$; $N = PV/k_BT = (1.01 \times 10^5 \text{ Pa})(1.00 \times 10^{-6} \text{ m}^3)/(1.381 \times 10^{-23} \text{ J/K})(237.15) = 2.68 \times 10^{19}$.

88. [I] A spacesuit with a volume of 0.10 m³ contains air at a pressure of 0.01 MPa at a temperature of 20 °C. How many gas molecules are in the suit?

89. [I] Consider an ideal gas at 20 °C and a pressure of 2.00 MPa in a 1.00×10^{-2}-m³ tank. Determine the number of gram moles of gas present.

90. [I] Assume air is an ideal gas and determine how many molecules there are in exactly 1-m³ at 300 K and 1.00 atm.

91. [I] A sealed chamber with a volume of 5.00×10^{-4} m³ contains 0.050 mole of oxygen gas at a temperature of 100°C. What is the pressure inside the chamber?

92. [II] THIS PROBLEM EXPLORES THE BEHAVIOR OF AN IDEAL GAS. A tank having a capacity of 2.0 liters is maintained at room temperature (20 °C), and 4.0 moles of hydrogen gas are pumped into it. We want to determine the resulting pressure. (a) What is the mathematical relationship between the absolute pressure, the volume, the absolute temperature, and the number of moles of a low-density gas of this sort? (b) What is the absolute temperature of the gas? (c) What is the volume of the gas in m³? (d) What is the gauge pressure (in Pa) in the gas?

93. [II] THIS PROBLEM EXPLORES THE BEHAVIOR OF AN IDEAL GAS. A chamber contains 4.0 liters of a low-density molecular gas at a temperature of 74 °C, and a gauge pressure of 2.0 atm. We want to determine the number of gas molecules in the chamber. (a) What is the mathematical relationship between the absolute pressure, absolute temperature, and volume for a gas of this sort? (b) What is the absolute temperature of the gas? (c) What is the volume of the gas in m³? (d) What is the absolute pressure (in Pa) in the gas? (e) How many molecules are in the gas?

94. [II] Determine the average center-to-center distance between air molecules at 300 K and 1 atm. [Hint: See Problem 90.]

95. [II] Taking the normal lung capacity to be 500 cm³ and the pressure therein to be the equivalent of 761 mm Hg (which is still the way the medical texts are listing it), estimate the number of molecules per breath.

96. [II] If the density of nitrogen in a chamber at a pressure of 1.00 atm is found to be 1.245 kg/m³, what is the temperature of the gas?

97. [II] The volume of a sealed glass container filled with helium at 2.00 atm is 1000 cm³, and its temperature is 23.0 °C. What is the mass of the gas contained?

SOLUTION: $n = PV/RT = (2.0 \times 1.013 \times 10^5 \text{ Pa}) (1000 \times 10^{-6} \text{ m}^3)/(8.315 \text{ J/mol·K})(296.15 \text{ K}) = 0.082 3$ mole.

98. [II] Exactly 600 cubic centimeters of a gas are held in a constant-pressure chamber at 23.0 °C. What will be the temperature of the gas after the volume is increased to 750 cm³?

99. [II] At an altitude of about 12.5 km, the temperature of the Earth's atmosphere is roughly −55 °C and the pressure is around 19.4 kPa. How many kilograms of hydrogen gas (H_2) should be put in a balloon to fill it to 2000 m³ at that altitude?

100. [III] Prove that the volume coefficient of expansion (keeping P constant) of an ideal gas equals $1/T$.

SECTION 12.7: KINETIC THEORY

101. [I] Given five molecules with speeds of 1.0 m/s, 2.0 m/s, 3.0 m/s, 4.0 m/s, and 5.0 m/s, find their average speed.

102. [I] Given five molecules with speeds of 1.0 m/s, 2.0 m/s, 3.0 m/s, 4.0 m/s, and 5.0 m/s, find their rms-speed and compare it to the results of Problem 101.

103. [I] Given that the mass of an average oxygen molecule is $5.313\,6 \times 10^{-26}$ kg, what is its rms-speed at 293.15 K?

104. [I] What is the average translational kinetic energy of an oxygen molecule in the air at 0 °C?

105. [I] Determine the average translational kinetic energy of a molecule of a gas at room temperature (20 °C).

SOLUTION: $KE_{av} = \frac{3}{2}k_B T = \frac{3}{2}(1.381 \times 10^{-23} \text{ J/K})(293.15 \text{ K}) = 6.1 \times 10^{-21}$ J.

106. [I] Compute the *rms*-speed of carbon dioxide molecules (CO_2) when a quantity of gas is held in a chamber at 0.000 °C.

107. [I] This problem will help us better understand Kinetic Theory. Consider a few hundred nitrogen molecules moving around in an evacuated chamber, with an average KE of 7.0×10^{-22} J. We wish to determine the Celsius temperature of the gas. (a) What is the mathematical relationship between the average KE of the molecules, and the absolute temperature of a gas? (b) What would happen to the average KE if the absolute temperature of the gas was doubled? (c) What is the absolute temperature of the gas? (d) What is the Celsius temperature of the gas? (e) What would the temperature be if the gas was hydrogen, and the average KE was the same?

108. [II] This problem will help us better understand Kinetic Theory. For an object to escape from the surface of the Earth (p. 193) it must be moving at a speed of about 11 km/s. Given that the mass of a hydrogen molecule is 3.3×10^{-27} kg, we want to determine the approximate Fahrenheit temperature at which much of the hydrogen in the atmosphere would escape. (a) What is the mathematical relationship between the average KE of the molecules and the absolute temperature of a gas? (b) What is the KE_{esc} of a hydrogen molecule moving at the escape speed? (c) What is the absolute temperature of the gas when its average KE equals KE_{esc}? (d) What is the equivalent Celsius temperature of the gas? (e) What is the equivalent Fahrenheit temperature of the gas? (f) Why has the Earth's atmosphere lost most of its hydrogen in comparison to its other gases?

109. [II] What is the average amount of translational kinetic energy possessed in total by all the molecules of oxygen contained in 0.50 m^3 of gas at STP? [*Hint: Compute the average KE per molecule and then find out how many molecules there are.*]

110. [II] A total of 12.0×10^{23} molecules of a gas are stored at 100 °C in a glass sphere having a volume of 1000 cm^3. Find the total translational KE contained on average by all the molecules.

111. [II] A chamber is filled with 25.0 moles of a gas at 200 °C. What is the total amount of internal translational KE associated with the gas?

112. [II] Prove that the pressure produced by an ideal gas in a sealed chamber is proportional to both the number of molecules per unit volume and their average translational kinetic energy.

113. [II] Derive an expression for the *rms*-speed of an ideal gas in terms of its density and pressure.

114. [II] For a monatomic ideal gas like helium, where rotational KE is negligible because the atom is so small (and there's no interaction and no interatomic PE), show that the total energy of all the atoms—the **internal energy** of the gas—is given by

$$\text{Internal energy} = \frac{3}{2}PV$$

115. [II] What is the total amount of internal energy stored in a quantity of helium gas filling a bag the size of a typical room (3.5 m by 5.0 m by 6.0 m) at atmospheric pressure? [*Hint: See the previous problem.*]

116. [II] What is the ratio of the *rms*-speeds of nitrogen in the air at 0 °C and 100 °C?

117. [III] Considering the discussion on p. 444 and on the **CD** under **Diffusion**, what is the percentage difference between the *rms*-speeds at room temperature of the two different forms of uranium hexafluoride (UF_6)—namely, those containing either $^{238}_{92}U$ or $^{235}_{92}U$?

Chapter 13
Heat & Thermal Energy

*T*his chapter is about thermal energy, heat, and the associated behavior of bulk matter. We will attempt to answer such questions as: What is heat? What is the relationship between heat and temperature? What happens to bulk matter as thermal energy is added or removed? How is energy transferred from one place to another? The discourse is not just about boiling water and barbecues. As with all the creatures on this planet, thermal concerns are of vital personal interest at every moment of our lives. On a grander level, these ideas lay the groundwork for a more complete understanding of the concept of energy, and with that, a more complete understanding of the way change occurs throughout the Universe.

Thermal Energy

The atoms of a solid, constantly colliding with one another, forever vibrate every-which-way about their fixed equilibrium positions (Fig. 13.1). In a liquid, those positions shift and much of the long-range order of the solid state vanishes, but the oscillations persist. By comparison, the gas is a tumult of impacts and recoils.

Gas atoms are essentially free. There are no equilibrium positions, no oscillations; the motion is a perpetual high-speed zigzag dance punctuated by collisions. Using Kinetic Theory (p. 441)—the idea that matter is composed of moving submicroscopic particles—we can understand all the observable manifestations (such as temperature and heat) of this invisible atomic mayhem.

Caloric versus Kinetic

Kinetic Theory was widely accepted, in one form or another, throughout the seventeenth century. It was correctly believed by many that *heat* was a manifestation of atomic motion. Galileo had said as much and Plato, long before, appreciated the equivalence of motion and heat. But by the 1760s, a new and totally wrong idea had begun to take hold: heat was then supposed to be an imponderable, self-repellent, indestructible fluid called **caloric**. Ironically, the theory of caloric initially proved to be more practical than Kinetic Theory, which few if any at the time were able to successfully apply. Indeed, we still say things like "pour on the heat" and "soak up the heat" as if it were a liquid.

STUDY GUIDE

This chapter is about the thermal behavior of bulk matter. It begins with the basic definitions of thermal energy (p. 457), heat (p. 459), and temperature (p. 458). It then introduces the central idea of thermal equilibrium (p. 458), all of which is necessary to understanding everything that follows.

The remainder of the chapter deals with the way a sample changes temperature, melts, boils, and vaporizes.

As creatures that live within a narrow range of temperatures (we're often too hot or too cold), knowing a little Thermal Physics can be of great practical value.

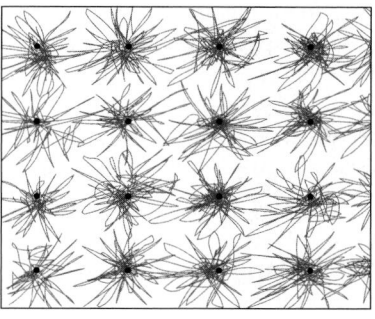 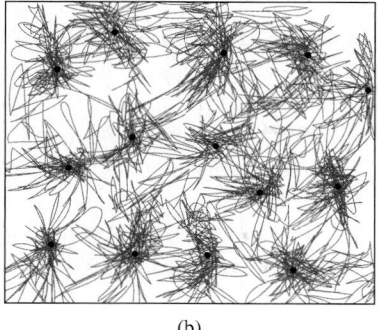 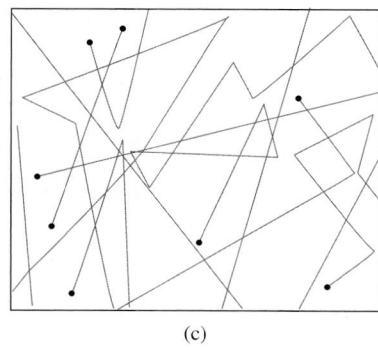

(a)   (b)   (c)

Figure 13.1   A schematic representation of atomic motion. (a) In the solid, each atom oscillates about a fixed lattice site and collides (interacting electromagnetically) with its immediate neighbors. (b) In the liquid, the equilibrium positions move and each atom vibrates over a larger region. (c) In the gas, the oscillations vanish and the motion is essentially unconstrained until two atoms collide.

Energy will remain in some sense the lord and giver of life, a reality transcending our mathematical descriptions. Its nature lies at the heart of the mystery of our existence as animate beings in an inanimate universe.

FREEMAN DYSON
(*Contemporary Physicist*)

In 1798, Benjamin Thompson (Count Rumford), having newly outfitted the arsenal at Munich, was watching the cannons being bored and realized that there was a tremendous amount of heat liberated in the process. He immediately arranged to have a brass gun-barrel blank, surrounded by several gallons of cold water, rotated against a dull steel borer. Gradually, the temperature rose until "the water *actually boiled*!" Heating by friction was certainly well known—aborigines and Boy Scouts still start fires that way, and people at the time even believed that forest fires began when the leaves on the trees rubbed too vigorously. Still, the central postulate of the heat-as-matter theory was that caloric was indestructible and uncreatable. And yet Rumford showed that heat could continue to be generated indefinitely as long as work was done turning the cannon barrel. It "appeared evidently to be *inexhaustible*" and could not therefore "possibly be a *material substance*." This was a death blow to caloric theory and a powerful affirmation of heat as motion.

Count Rumford. As a young man, he had seen the Boston Massacre, but that didn't keep him from becoming a major in the King's American Dragoons. When British General Gage marched on Lexington and Concord, starting the Revolutionary War, it was on the strength of secret intelligence provided in a letter written (in invisible ink) by Benjamin Thompson.

THE LIKEABLE ROGUE & THE FALL OF CALORIC

A lethal attack on the caloric theory was delivered in 1798 by one of the most fascinating characters ever to be involved in science. The American, Benjamin Thompson, a shrewd and elegant opportunist, was a professional soldier who saw little action yet rapidly rose to the rank of major general. Relying on an audacious blend of charm and cunning, he managed to be knighted by the King of England, became minister of war in Bavaria, and ultimately was made Count Rumford of the Holy Roman Empire. A tall, handsome, philandering, blue-eyed, likeable rogue, he fathered three children, of whom only one was legitimate. Thompson was a spy and a scoundrel, a quasi-beloved benefactor of the poor, a social reformer, a superb administrator (who even seems to have provided for a Bavarian brothel), and the founder of the British Royal Institution.

13.1 Heat and Temperature

In previous chapters, we concentrated on mechanical energy, the energy associated with an object that moved or interacted with other objects as a whole. In that sense, we've been dealing with organized, or *ordered energy*. On the other hand, it is possible to impart uncoordinated motions to the individual constituent atoms—disorganized random motion, not of the body as a whole, but within the body. **Thermal energy *is the net disordered* KE *possessed by a group of particles, usually the atoms of bulk matter.***

Thermal energy is the internal KE associated with the random motion of the submicroscopic particles (atoms, molecules, ions, and free electrons) that constitute a system.

If we throw an apple, it will not experience an increase in thermal energy, although it will have an increased amount of KE (with respect to us). A thermometer stuck in the apple will show no change due to the motion. All the apple's atoms will move together as a whole in an ordered fashion. By contrast, if the apple crashes into a wall, it deforms and some of that organized KE will invariably be converted into an increase in thermal energy. {For the origins of this insight, click on **1695-LEIBNIZ** under **FURTHER DISCUSSIONS** on the **CD.**}

Earlier, in our brief study of Kinetic Theory (p. 441), we found that for a gas, temperature was determined by the average translational kinetic energy of the molecules: $KE_{av} = \frac{3}{2}k_B T$. In a solid the molecules vibrate around their equilibrium positions, with very small amplitudes of about 3×10^{-11} m ($\approx 1/10$ of a molecular diameter at room temperature). There is a definite average location of each molecule (about which it vibrates), and therefore a long- range order. Liquids behave in a similar way except their molecules can also undergo tiny displacements as they vibrate; they drift around and there is little or no long-range order. In both solids and liquids at ordinary temperatures, the average vibrational energy of the molecules equals $3k_B T$, and we again have a relationship between average energy and T. Alas, this simple dependence gives way when the temperature descends below ≈ 300 K. There, quantum effects soon prevail and things get much more complicated. And so when we say that temperature is a measure of the average kinetic energy of the molecules of the system, there are some serious limitations to be kept in mind.

Temperature *(at values greater than 0 K) is a measure of the ability of moving particles, usually atoms, to directly impart thermal energy to a thermometer or any other object. It reflects not the net amount of random* **KE**, *but its average value: the concentration of thermal energy.* The temperature of a substance is independent of the total number of atoms present. The net amount of thermal energy in the Atlantic Ocean is huge, but the average random KE of its molecules is low and its temperature is low. By contrast, thermal energy depends on both the total number of atoms present and on their average random KE (which depends, in large measure, on temperature). The more atoms present and the more KE each possesses on average, the more thermal energy is present in the system. *Because the individual atomic motions are random, altering the thermal energy of a system has no effect on its net momentum.*

Temperature is essentially a measure of the average value of the internal KE associated with the random motion of the submicroscopic particles that constitute a system.

We now consider three basic ways in which the thermal energy of a body can be increased (or, if reversed, decreased). First, we can do work on the body: rub it, compress it, or in any way *distort it*; that is, physically displace some of its atoms. Blacksmiths sometimes kindled a fire using a soft iron rod made red hot by pounding on it with rapid hammer blows. **Work** *is the organized mechanical energy transferred into or out of a system by way of a force acting through a distance.* If a force is applied to overcome interatomic forces (e.g., friction), thereby displacing atoms, the energy introduced is rapidly distributed within the body through atomic collisions and much of it will appear as thermal energy. That's what happens when you warm your hands by rubbing them together. Increasing the total amount of thermal energy of an object (like your hand) increases the average random KE, which means an increase in temperature.

The second means of increasing thermal energy involves shining electromagnetic radiation (for example, photons of light, infrared, and ultraviolet) onto the body. **Radiant energy** (p. 473) is transferred directly to the absorbing atoms that, now agitated, collide with other atoms and mechanically redistribute the energy as thermal energy. That's what happens when you warm your face in the sunshine: radiant energy is absorbed, the average KE of the molecules rises, and your skin temperature goes up.

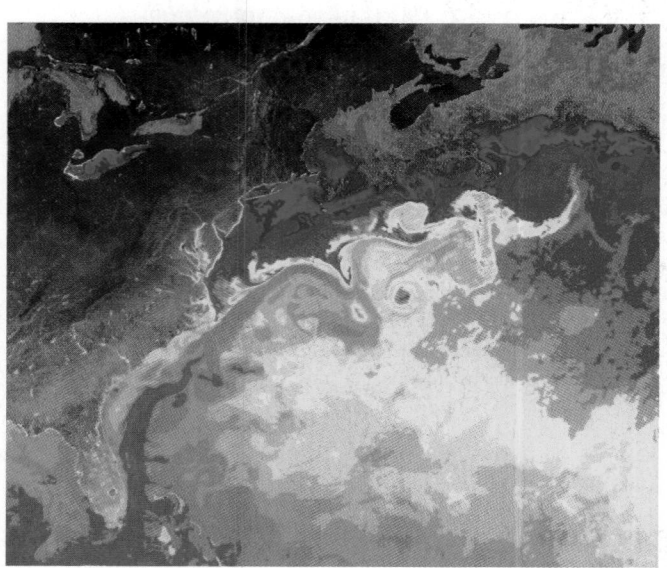

A temperature map of the surface of the East Coast of the United States produced by a space vehicle orbiting the planet. Florida is outlined by warm water (deep red) at the lower left. Long Island juts into the cool blue-green coastal waters of the Atlantic Ocean. The Great Lakes are at the upper left. Notice the warm vortices spun off by the Gulf Stream (at the center of the picture).

The weight of an automobile flattens the bottoms of its tires. And as a tire revolves a new region touches the ground and is flattened in turn. In effect, the tire is cyclicly distorted at a frequency dependent on the speed of the vehicle. Because the tire is far from being perfectly elastic, much of the work done on it (in changing its shape) ends up as thermal energy—the tire gets hot. Driving at high speed on a hot asphalt road can cause a poorly made tire to come apart, usually with catastrophic results. The tread on this tire ripped off, resulting in a violent blowout.

The third mechanism is the one we'll focus on in this chapter. We know from experience that the "hotness" of an object can be increased by placing it in contact with something at a higher temperature. A spoon at room temperature positioned in a flame will soon heat up. The high-temperature flame consists of particles randomly moving about with a high average KE. This violent storm of glowing gas bombards the atoms in the surface of the spoon, which have a much lower average KE. The lower temperature spoon-atoms take up energy on impact from the more energetic flame-atoms. The former are thus increased in their random motion, and thermal energy is thereby imparted to the spoon. Energetic spoon-atoms that are touching the flame strike quiescent spoon-atoms just behind them, and the jostling continues back until the entire object becomes hot—a process we ordinarily describe by saying that "heat flows" down the spoon. More formally, **heat** (*Q*) *is the thermal energy transferred, via atomic collisions, from a region of high temperature (where the average KE is high) to a region of lower temperature (where the average KE is low).*

Heat is the energy of motion transported by way of contact, from one group of randomly moving particles to another, exclusively as a result of a temperature difference. The energy carriers are most often atoms, molecules, ions, or free electrons.

A body contains or stores thermal energy, not heat. **Heat is thermal energy in transit**, and once transferred it is no longer called heat. (We blow wind into a balloon, but once inflated, what's stored is air.) As before, we are again more concerned with the change in energy of a system than with the net amount of energy present at any moment. An agency of change of thermal energy is heat, just as an agency of change of mechanical energy is work. Once transmitted, the abiding mechanical energy of a system is no longer called work—what is stored is KE or PE, not work. In general, *a system contains energy when it can undergo spontaneous change*.

13.2 Quantity of Heat

Joseph Black was the great eighteenth-century pioneer in thermal physics and probably the first to distinguish between "quantity of heat" and temperature. He saw temperature as reflecting the "intensity," or concentration, of what we would call thermal energy and he called "the matter of heat." A stew tastes salty because of the concentration of salt, and it is hot because of the concentration of thermal energy. The bigger the potful, the more salt must be sprinkled in and the more heat must be added. The ocean, cold as it is, contains far more thermal energy than the scalding stew, and no matter how salty the dish, the ocean will contain far more salt as well. The fact that its thermal energy is spread out thin, not concentrated, accounts for the ocean's low temperature, but the large amount of random KE is there nonetheless.

Sometime around 1760, Black introduced a simple scheme for quantifying heat that was operational and totally independent of what heat actually is—caloric or KE, it didn't really matter. One only needed to have a thermometer and a scale to weigh out some water. Following Black's lead, the **calorie** (cal) ultimately became the pre-SI metric unit of heat. It is defined as *the amount of heat that must be added to raise the temperature of 1 gram*

A **calorie** is the amount of heat that must be added to raise the temperature of 1 gram of water 1 degree Celsius.

of water 1 degree Celsius—from 14.5 °C to 15.5 °C. This particular temperature range is now specified because the heat required is very slightly different at other temperatures. As H. A. Rowland discovered (1878), it differs by less than 1% over the range from 0 °C to 100 °C, which, for our purposes, is negligible. There we are then: we can say exactly what a calorie is, even without knowing what heat is, just as we can say exactly what a kilogram is, without knowing what mass is.

By definition, an input of 1 cal will raise 1 g of water 1°C. We can assume (and experiment confirms) that 2 cal will raise the temperature of 1 g of water 2 °C, and so on. In other words, the resulting temperature change (a rise or fall) is proportional to the heat (in or out): $\Delta T \propto Q$. Similarly, 1 cal will raise the temperature of 2 g of water only $\frac{1}{2}$°C. That is, $\Delta T \propto 1/m$; the more water, the more heat is needed to raise the temperature, $\Delta T \propto Q/m$. Introducing a constant of proportionality c (to take care of the units), we have

$$Q = cm\Delta T = cm(T_f - T_i) \tag{13.1}$$

In the case of water, $c = 1$ cal/g·C°, exactly.

The calorie turns out to be inconveniently small, especially for weight watchers. So the world of dietetics (Table 13.1) uses the **kilocalorie** (kcal), sometimes cleverly called the Calorie (1 kcal = 1 Cal = 1000 cal).

Table 13.1

Energy Provided by Various Foods	
Food	**Kilocalories**
Martini	145
Peanut butter sandwich	330
Buttered popcorn (1 cup)	55
Vanilla ice cream ($\frac{1}{3}$cup)	145
Whiskey (1 shot)	105
Brownies (1)	140
Jelly doughnut	225
Dry red wine (1 glass)	75

DR. BLACK & MR. WATT

Four friends dominated the intellectual life of Scotland in the latter half of the 1700s. They were Adam Smith, the political economist; David Hume, the renowned philosopher; his personal physician, Dr. Joseph Black, professor of chemistry at Glasgow; and that university's resident scientific instrument maker, James (Jamy) Watt. Both Black and Watt were reserved advocates of the caloric theory, and that was reasonable enough; caloric was a practical contrivance, and these were practical men, more concerned with using heat than with hypothetical abstractions.

Example 13.1 **[I]** The equivalent of a glass of water (that is, 270 g of liquid) at 20.0 °C receives 1000 cal from a candle flame. Assuming that all the energy goes into the water without any losses, what is the final temperature of the liquid?

Solution The problem talks about the receipt of a number of calories, and at this point, that can only mean a temperature change. ***Specific heat should come to mind whenever there's a temperature change.*** (1) TRANSLATION—A known amount of heat is added to a known mass of a substance (i.e., water) at a specified initial temperature; determine the final temperature. (2) GIVEN: Q = 1000 cal, m = 270 g, and

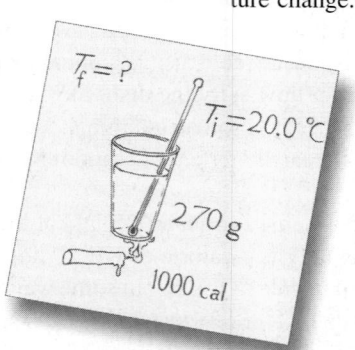

$T_i = 20.0$ °C. FIND: T_f. (3) PROBLEM TYPE—Thermo/heat/temperature change. (4) PROCEDURE—Eq. (13.1), $Q = cm\Delta T$, summarizes the relationship between temperature change and heat-in. (5) CALCULATION—Using the fact that $c = 1.00$ cal/g·C°

$$Q = 1000 \text{ cal} = cm(T_f - T_i)$$

$$Q = (1.00 \text{ cal/g·C°})(270 \text{ g})(T_f - 20.0 \text{ °C})$$

and

$$T_f = \frac{1000 \text{ cal}}{(1.00 \text{ cal/g·C°})(270 \text{ g})} + 20.0 \text{ °C} = \boxed{23.7 \text{ °C}}$$

Quick Check: 270 cal will raise 270 g by 1 °C; 1000 cal/270 cal = 3.7, hence the new temperature is 20.0 °C + 3.7 °C.

{For more worked problems click on **WALK-THROUGHS** *in* **CHAPTER 13** *on the CD.}*

BREAKFAST CRUNCHES & THE BTU

The fact that "calorie" is supposed to mean kilocalorie is often kept as a little secret (Table 13.2). When you read on the box that your favorite breakfast cereal "contains" 110 calories per serving, that's really 110 kilocalories; that amount of energy will be liberated upon the complete combustion of one serving of cereal. Some engineers and aspiring air-conditioner salesmen (mostly in the United States) hold fast to another fading anachronism, the Btu, or *British thermal unit*. One Btu is the amount of heat needed to raise 1 lb of water 1°F. The realization that heat is energy only came gradually in the mid-nineteenth century.

Table 13.2

Approximate Kilocalories Consumed per Hour

| Body weight (lb) | 100 | 150 | 200 | 250 |
Body mass (kg)	45	68	90	113
Sleeping it off	40	60	80	105
Sitting on the grass	65	95	130	165
Standing in line	70	100	140	170
Strolling through a park	130	195	260	320
Running from a mugger	290	440	580	730

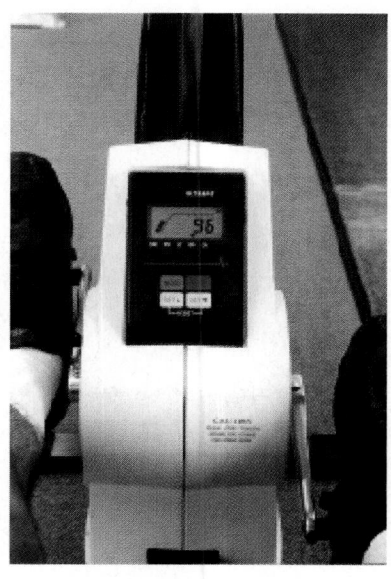

Modern exercise machines can tell you roughly how many calories you've "burned" as you sit there pedaling endlessly.

13.3 The Mechanical Equivalent of Heat

The German physician who took the next step was twenty-eight years old when his first essay appeared (1842). Julius Robert Mayer reinterpreted some long-forgotten results on expanding gases to determine a rough value for what came to be called the "mechanical equivalent of heat"—namely, the numerical relationship between work and heat.

As if the time was ripe for change, within one year of Mayer's paper there appeared in print the first report of the remarkable researches of James Prescott Joule. Unlike Mayer, Joule was a gifted experimenter who devised a whole range of apparatus using various means (e.g., electric currents, friction, and compression of gases) to generate "heat." His most famous was a paddle wheel device driven by falling weights (Fig. 13.2) where he

James Prescott Joule (1818–1889) was a well-to-do brewer's son from Manchester, England, and a pupil of John Dalton. He was greatly influenced by Rumford's research at the cannon factory.

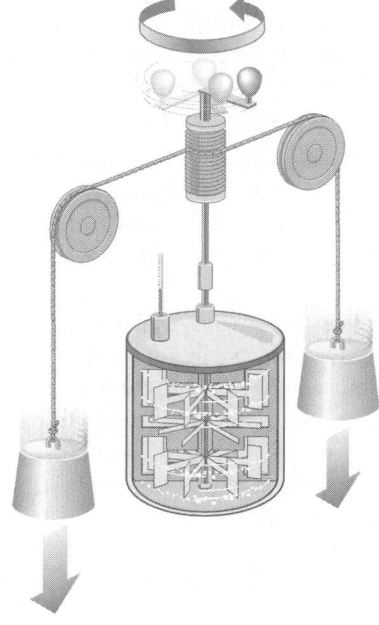

Figure 13.2 Joule's apparatus for determining the mechanical equivalent of heat. As the weights fall, the paddle turns, raising the temperature of the water. *Gravitational*-PE is converted into the KE of the paddle and the water. That KE is ultimately converted into thermal energy within the isolated chamber.

equated the work done by the weights to the thermal energy imparted to the water. Nowadays, we use SI units and say the **mechanical equivalent of heat** is

$$4.186 \text{ J} = 1 \text{ cal}$$

or

$$1 \text{ kcal} = 4186 \text{ J}$$

By international agreement, the calorie is no longer the recognized unit of heat. As with all forms of energy, the preferred unit is the *joule*. We can summarize much of what has been discussed so far with the statement that *the temperature of 1 kilogram of water can be increased (or decreased) by 1 kelvin through the addition (or removal) of 4186 joules of heat*.

Example 13.2 [I] The following statement is taken from Mayer's paper of 1842: "The warming of a given weight of water from 0 °C to 1 °C corresponds to the fall of an equal weight from the height of about 365 metres." Compare his value of the mechanical equivalent of heat with the modern value.

Solution The problem compares the change in *gravitational*-PE of an object to the equivalent amount of thermal energy. (1) TRANSLATION—A mass (of water) falls through a known distance; determine the change in its energy. (2) GIVEN: $h = 365$ m and $\Delta T = 1$ °C. FIND: Mayer's mechanical equivalent of heat. (3) PROBLEM TYPE—Thermo/heat/ PE_G/mechanical equivalent of heat. (4) PROCEDURE—By "a given weight," he means

any value of mg, so let's take $m = 1.00$ kg for convenience. We must find the corresponding energy associated with a fall of 365 m in SI units. (5) CALCULATION—The change in *gravitational*-PE is

$$PE_G = mgh = (1.00 \text{ kg})(9.81 \text{ m/s}^2)(365 \text{ m}) = 3.58 \text{ kJ}$$

That's the mechanical energy associated with the fall of a 1-kg mass (of water, or anything else), and it is supposed to equal the heat needed to raise the temperature of that mass (1 kg) of water 1 °C, which, from Eq. (13.1), is just 1000 cal = 1 kcal. According to Mayer, then, $\boxed{1 \text{ kcal} = 3.58 \text{ kJ}}$, which is about $\boxed{14\% \text{ too small}}$.

Just 1 kilocalorie per glass. Note that the Europeans use commas in their numbers where we use periods. By the way, 1 kcal is actually equal to 4.19 kJ, not 4.18 kJ.

....

The **Zeroth Law of Thermodynamics:** Two material objects in contact will spontaneously (i.e., all by themselves) reach the same uniform-temperature equilibrium state.

The Zeroth Law

Joseph Black began one of his lectures with an observation so central to any complete understanding of the behavior of heat that it has since come to be called the **Zeroth Law of Thermodynamics**:

> **We can perceive a tendency of heat to diffuse itself from any hotter body to the cooler ones around it, until the heat is distributed among them in such a manner that none of them is disposed to take any more from the rest. The heat is thus brought into a state of equilibrium…. The temperature of all of them [i.e., all the bodies] is the same.**

In other words, **when any two bodies at different temperatures come into contact, thermal energy flows from the higher to the lower until the temperatures of the two equalize and they are said to be in thermal equilibrium**. Thus, suppose you put a shoe, a cold can of coke, and a hot pizza, on a kitchen table and come back in a few hours. The table, the coke, the pizza, and the shoe will all be in equilibrium at the same temperature. *It doesn't matter what these things are made of*, and it doesn't matter what they feel like to the touch (p. 426), they'll all reach the same temperature.

Today, this equilibrium insight is used to introduce the idea that thermodynamic temperature is a unique variable describing the state of the system. After all, that's really the wonderful part of the process, the fact that the cold soda left on the table will ultimately reach the same temperature as the table (and anything else sitting on it) reveals something fundamental about Nature. When system-1 is placed in thermal contact with system-2 and no net energy flows from one to the other, the two exist in a special physical condition, or state; *they have the same temperature*. {For more on this subject click on **THE ZEROTH LAW** under **FURTHER DISCUSSIONS** on the CD.}

13.4 Specific Heat: Calorimetry (No Change in State)

We know that 4.2 kJ (i.e., 1 kcal) will raise the temperature of 1 kg of water 1 °C, but there is no reason to assume it will do the same thing to 1 kg of iron or peanut butter or pizza; still, that was the common wisdom before Dr. Black put things right. He found, quite to the contrary, that *every substance changed temperature by a characteristic amount when infused with a fixed quantity of heat*.

Black placed equal weights of different substances in identical vessels that were subsequently put on the same burner, or in front of the same fire, for the same length of time. The amount of heat transferred was presumably proportional to the time the sample was exposed to the source (Fig. 13.3). In this way, he confirmed that an equal mass of iron will get 9 times hotter than water in the same amount of time.

Another of his methods has evolved into the modern technique of calorimetry. It was fairly common knowledge in the community of scientists that if a given mass of cold water at, say, 5.0 °C is mixed with an equal amount of hot water at, say, 95 °C, the blend will come to an equilibrium temperature of 50 °C. The hot system loses a certain amount of heat to the cold system; the former drops 45 °C, the latter gains 45 °C—*equal changes, because equal amounts of the same material are being mixed*. We can see as much from Eq. (13.1): the *heat lost* by hot system-1 is

$$Q_1 = cm_1(T_{f1} - T_{i1})$$

where $T_{i1} > T_{f1}$ and $Q_1 < 0$. Similarly, the *heat gained* by cold system-2 is

$$Q_2 = cm_2(T_{f2} - T_{i2})$$

where $T_{f2} > T_{i2}$ and $Q_2 > 0$. But these are plainly equal in magnitude; that is, $-Q_1 = Q_2$; **heat-out is negative, heat-in is positive**. Moreover, $m_1 = m_2$, and so

$$-c(T_{f1} - 95\,°\text{C}) = c(T_{f2} - 5.0\,°\text{C})$$

At equilibrium, $T_{f1} = T_{f2} = T_f$ and the above equation yields $T_f = \frac{1}{2}(95\,°\text{C} + 5.0\,°\text{C}) = 50\,°\text{C}$ as expected. Alternatively, the temperature difference is 90 °C and, because the masses are equal, one drops 45 °C while the other rises 45 °C. The crucial points are that the final temperatures of the two samples are equal and that the heat lost by one is gained by the other (that's Conservation of Energy).

This approach of mixing different systems became important, especially after Fahrenheit performed much the same experiment using water and mercury. The hot mercury decreased in temperature more than the water rose in the process of their coming together at equilibrium. As Black interpreted the results, an equal mass of mercury required less heat to change its temperature 1°C than did water. He concluded that "quicksilver [mercury] has less *capacity for heat*" than does water. In time, the **specific heat capacity**

Figure 13.3 Imagine some water and an equal mass of iron at, say, 25 °C. If the same amount of heat is added to both, the temperature of the iron increases about 9 times as much as the water. When the water rises to 35 °C, the iron rises to 116 °C.

Table 13.3
Specific Heat Capacity for Some Materials*

Material	Specific heat capacity	
	kJ/kg·K	kcal/kg·K
Solids		
Aluminum	0.90	0.21
Clay (dry)	0.92	0.22
Copper	0.39	0.093
Glass	0.84	0.20
Gold	0.13	0.031
Human body (average)	3.47	0.83
Ice (water, −5 °C)	2.1	0.50
Iron	0.47	0.11
Polyamides (e.g., Nylon)	1.7	0.4
Polyethylenes	2.3	0.55
Polytetrafluorethylene		
(e.g., Teflon)	1.0	0.25
Lead	0.13	0.031
Marble	0.86	0.21
Platinum	0.14	0.032
Protein	1.7	0.4
Silver	0.23	0.056
Stainless steel (type 304)	0.50	0.12
Wood	1.8	0.42
Liquids		
Acetone	2.2	0.53
Alcohol (ethyl)	2.4	0.57
Ammonia	4.71	1.13
Mercury	0.14	0.033
Nitrogen (−200 °C)	1.98	0.474
Oxygen (−200 °C)	1.65	0.394
Sulfuric acid	1.4	0.34
Water	4.186	1.000
Gases		
Air (100 °C)	1.0	0.24
Argon	0.52	0.13
Carbon monoxide	1.0	0.25
Hydrogen	14.2	3.39
Methane	2.2	0.53
Steam (110 °C)	2.01	0.481

*At ≈ 20 °C.

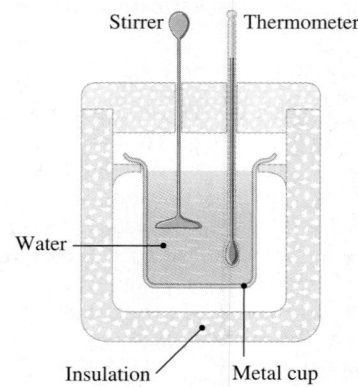

Stirrer   Thermometer

Water

Insulation   Metal cup

Figure 13.4 A water calorimeter.

(c) of a substance was defined to be ***the number of joules of heat that must be added to raise the temperature of 1 kg of the material 1 K*** (or 1°C). It follows from Eq. (13.1), $Q = cm\Delta T$, that each substance has its own value of c (Table 13.3). Nowadays the specific heat capacity is more often just called the **specific heat**. {For more on the origins of the idea click on **SPECIFIC HEAT CAPACITY** under **FURTHER DISCUSSIONS** on the **CD**.}

Since specific heat is the amount of heat, or better yet, thermal energy, (Q) that must be added or removed per kilogram to change the temperature of a sample one kelvin, it follows that

$$c = \frac{Q}{m\Delta T} \tag{13.2}$$

In all these thermal experiments, it is tacitly assumed that there is no significant heat loss. To make that a reality, we use a **calorimeter** as pictured in Fig. 13.4. It's simply a

Example 13.3 **[I]** A copper pot has a mass of 0.50 kg and is at 100 °C. How much thermal energy must be removed from it if its temperature is to be lowered to precisely 0 °C?

Solution Specific heat should come to mind whenever there's a temperature change. (1) TRANSLATION—A known mass (of copper) at a known initial temperature is to be cooled to a specified final temperature; determine the amount of thermal energy to be removed. (2) GIVEN: Cu, $m = 0.50$ kg, $T_i = 100$ °C, and $T_f = 0$ °C. FIND: Q. (3) PROBLEM TYPE—Thermodynamics/heat/specific heat. (4) PROCEDURE—The expression $Q = cm\Delta T$ relates heat in or out and the resulting temperature change. (5) CALCULATION—Here Q is the heat-out of the system, which is negative since $\Delta T = -100$ K.

$$Q = c_{Cu}m\Delta T = (390 \text{ J/kg·K})(0.50 \text{ kg})(-100 \text{ K}) = -19.5 \text{ kJ}$$

or to two significant figures $\boxed{-20 \text{ kJ}}$

Quick Check: $\Delta T = -100$ K. From Table 13.3, the specific heat of copper is 9.3% that of water; if the pot were made of water, we would have to remove $\frac{1}{2}100$ kcal; for copper we need only remove 9.3% of that, or 4.65 kcal = 19.5 kJ.

Example 13.4 **[II]** A quantity of what looks like lead shot, having a mass of 100 g, is poured into a test tube. The tube is then partially submerged in a beaker of boiling water and kept there until the shot reaches 100.0 °C. The hot shot, as it were, is then quickly transferred to a vessel containing 100 g of water at 20.00 °C. The contents are gently stirred until they come to an equilibrium temperature of 22.41 °C. What is the specific heat of the shot? The effect of the vessel is negligible.

Solution The idea of specific heat should come to mind whenever there is a temperature change. Moreover, this problem is about things at different temperatures coming together, and that should suggest heat-out = heat-in. (1) TRANSLATION—A known mass of a substance at a given temperature is added to a known mass of water at a different given temperature. The combination reaches a known final temperature; determine the specific heat of the substance. (2) GIVEN: $m_S = 100$ g, $m_w = 100$ g, $T_{iS} = 100.0$ °C, $T_{iw} = 20.00$ °C, and $T_{fS} = T_{fw} = T_f = 22.41$ °C.

FIND: c_S. (3) PROBLEM TYPE—Thermal energy/specific heat. (4) PROCEDURE—We can do this problem in several different ways. All rely on the facts that the thermal energy lost by the shot is gained by the water ($-Q_S = Q_w$) and that both are at the same final temperature. (5) From the definition, Eq. (13.2),

$$c_S = \frac{Q_S}{m_S\Delta T_S}$$

and we need Q_S and ΔT_S. To find Q_w consider that the water has changed temperature by (22.41 °C − 20.00 °C) = +2.41 °C = +2.41 K. There being 100 g of water present, each 1-degree change corresponds to 100(4.186 J); hence, the water gains $Q_w = +1008.8$ J and the shot loses $Q_S = -1008.8$ J. We know the shot changes temperature by $\Delta T_S = (22.41$ °C − 100 °C) = −77.59 °C; therefore,

$$c_S = \frac{-1008.8 \text{ J}}{(0.100 \text{ kg})(-77.59 \text{ K})} = \boxed{130 \text{ J/kg·K}}$$

Quick Check: From Eq. (13.1) and the fact that the heat-out equals the heat-in, $-Q_S = Q_w$; $-c_S m_S\Delta T_S = c_w m_w\Delta T_w$; hence, $c_S = -c_w\Delta T_w/\Delta T_S = -(4186$ J/kg·K)(22.41 °C − 20.00 °C)/(22.41 °C − 100 °C) and $c_S = 130$ J/kg·K.

---

well-insulated chamber whose vacuum or Styrofoam walls keep the contents from exchanging heat with the outside. A thin metal cup with a low specific heat and a small mass is used to hold a measured amount of water. The cup changes temperature easily and can store very little thermal energy in the process; it tends to remove very little from the system. That's a lot like the behavior of aluminum foil used in cooking—it can be handled, albeit carefully, right out of the oven in spite of its high temperature because the quantity of heat given off in coming to equilibrium with your fingers is necessarily small (mc is small). Typically, a sample whose specific heat is to be measured is raised to a known temperature

EXPLORING PHYSICS ON YOUR OWN

Heat Capacity: The heat capacity of a given object is defined as the amount of heat it takes to raise the temperature of the object 1 K. For a homogeneous body it equals the mass times the specific heat of the material: mc. A small mass of a low specific heat material (like copper) will require very little energy to reach a substantial temperature. A few inches of water gently boiling in a pot will be our constant-temperature (100 °C) heat reservoir. In turn, place into the bath a variety of objects—for example, a copper penny, a quarter, a glass marble (one you don't care much about because it's likely to crack, p. 433), a cork, a 2-in. ball of aluminum foil, and a 2-in. square sheet of the same foil. Let each sit in the water for a minute or so, then remove it with tongs or chopsticks, quickly dry it, and carefully handle it to get a sense of how hot it feels and how long it stays hot—be especially cautious until you learn what to expect. The glass marble will feel as though it had stored a lot more energy than the penny—it will stay hotter longer. Why? *Amazingly, the sheet of aluminum might actually seem cool!* What does it feel like while floating on the water? Clearly, it must have rapidly cooled (see p. 469). Explain. Now put a ceramic mug in the water. It will store a considerable amount of thermal energy, so handle it especially carefully.

Example 13.5 [I] A calorimeter consisting of a thin copper cup of mass 150 g containing 500 g of water is at a temperature of 20.0 °C. A 225-g sample of an unidentified material at 508 °C is lowered into the bath, and the device is sealed. After a few minutes, the system reaches a constant temperature of 40.0 °C. Determine the specific heat of the sample. Overlook any losses due to the thermometer, stirrer, or insulation.

Solution   Specific heat should come to mind whenever there's a temperature change. Moreover, this problem is about things at different temperatures coming together and that should suggest heat-out = heat-in. (1) TRANSLATION—Two known masses (copper and water) are in contact at a given initial temperature. A third substance of known mass at a different specified temperature is added. The combination reaches a known final temperature; determine the specific heat of the substance. (2) GIVEN: $m_c = 0.150$ kg, $m_w = 0.500$ kg, $T_{ic} = T_{iw} = 20.0$ °C, $m_s = 0.225$ kg, $T_{is} = 508$ °C, and $T_f = 40.0$ °C. FIND: c_s.   (3) PROBLEM TYPE—Heat/calorimetry.   (4) PROCEDURE—The

heat lost by the substance ($-Q_s$) equals the heat gained by the copper and water ($Q_w + Q_c$), and all are at the same final temperature.   (5) CALCULATION—From the definition, Eq. (13.1),

$$-Q_s = Q_w + Q_c$$

$$-m_s c_s \Delta T_s = m_w c_w \Delta T_w + m_c c_c \Delta T_c$$

Using Table 13.3,

$$-(0.225 \text{ kg})(c_s)(40 \text{ °C} - 508 \text{ °C}) = (0.500 \text{ kg}) \times$$
$$(4186 \text{ J/kg·K})(20 \text{ K}) + (0.150 \text{ kg})(390 \text{ J/kg·K})(20 \text{ K})$$

and $105.3 c_s = (41\,860 + 1170) \text{ J/kg·K}$

hence

$$\boxed{c_s = 409 \text{ J/kg·K}}$$

Quick Check: Using Table 13.3, [(0.15 kg)(0.093 kcal/kg·K) + (0.5 kg)(1.0 kcal/kg·K)](20 K) = 10.28 kcal = c_s(0.225 kg)(468 K); c_s = 0.0976 kcal/kg·K = 409 J/kg·K.

HOT-WATER BAGS & ONIONS

Water has an extraordinarily high specific heat, a fact that is crucial to all life on the planet. A quantity of water will change temperature only when a great deal of heat is added or removed; it heats up slowly and cools off slowly. The legendary hot-water bag is the ultimate testament. A potato or a lima bean, fished out of a hot stew, can be eaten with a little caution, but a juicy onion (mostly water) will have to be treated with respect. To lower its temperature from 100 °C in the pot to something you can deal with will require removing a lot more heat first. And if you don't bother, that considerable amount of heat will be deposited in your mouth as it descends down to equilibrium with your tongue. We are mostly water ourselves, and that high value of c is one factor that helps us keep from rapid temperature changes. Cities on the coast of large bodies of water get that same benefit. It is also the reason water is used to carry thermal energy in radiators that heat houses and in cooling systems that maintain automobile engines.

and then immersed in the water. The change in temperature of the water allows the required specific heat to be computed.

The Calorie content, or better still the *energy content*, of your breakfast crunchies can be found with a modified device known as a *bomb calorimeter*. A weighed sample of dried material is placed into a sealed chamber containing a heating coil and an overpressure of oxygen. That chamber is then positioned in the water bath of an otherwise traditional calorimeter. An electric current heats the coil (with a known amount of energy), and the crunchies burn to dust. The same scheme can be used to determine the energy content of fuels, the so-called **heats of combustion** (see Tables 13.4 and 13.5).

{For a review of the underlying atomic processes and the Law of Dulong and Petit click on ATOMIC THEORY & SPECIFIC HEAT under FURTHER DISCUSSIONS on the CD.}

Table 13.4 Approximate Heats of Combustion	
Material	**Heat of combustion** (MJ/kg)
Alcohol, methyl fuel	22
(denatured)	27
Anthracite (hard coal)	33
Bituminous coal	30
Bread	10
Butter	33
Carbohydrates	17
Charcoal	28
Diesel oil	45
Dung	17
Fats	38
Gasoline	48
Methane	56
Oil, furnace	44
Propane	50
Proteins	17
TNT	5
Wood	15

Table 13.5 Approximate Energy Content of Various Materials	
Source	**Energy** (J)
Ton of coal	3×10^{10}
Barrel of oil	6.3×10^{9}
Ton of TNT	4.2×10^{9}
Gallon (U.S.) of oil	1.5×10^{8}
Gallon (U.S.) of gasoline	1.3×10^{8}
Kilogram of fat	3.8×10^{7}
Pound of coal	1.5×10^{7}
Pound of wood	6.8×10^{6}
Cubic foot (1 atm, 16 °C) of natural gas	1.1×10^{6}

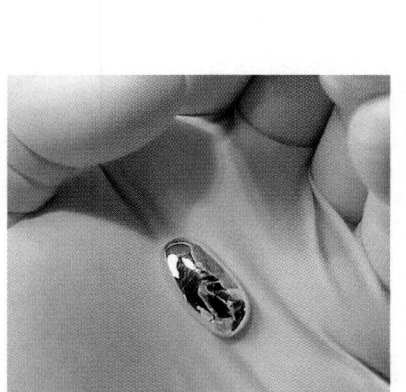

The element gallium has a heat of fusion of 80 kJ/kg and a melting temperature of 29.8 °C (i.e., 85.6 °F)—it melts in the hand.

At the **melting point** the atoms of a crystalline substance are moving so vigorously that any additional energy goes into "breaking" the intermolecular bonds that hold the solid together. The internal energy of the system increases, although the temperature remains constant throughout the melting process.

Change of State

The transport of heat into or out of a substance can alter it in a number of ways. Of course, it can change the temperature of the sample, but it can also change the physical structure, or *phase*, of the sample.

13.5 Melting and Freezing

When a solid is transformed into a liquid by the addition of thermal energy, the process is known as ***melting***, though at one time it was called ***fusion*** (as when you fuse or melt two objects together). The common wisdom in the mid-1700s was that once at the melting point, the addition of heat would cause a rise in temperature accompanied by the melting of the body. To the contrary, Black was convinced that when a body melts, "a large quantity of heat enters into it … without making it apparently warmer…. This heat must be added in order to give it the form of a liquid …." He established experimentally that ***when a solid is warmed to its melting point, the slow continued addition of heat results in the material's gradual and complete liquefaction at that fixed temperature***. Only then does the temperature again begin to rise.

The heat added to a system while it melts apparently has no effect on the temperature, and so Black called it *latent* (meaning unrevealed) *heat*. Imagine that we take two thin glass bottles, put a kilogram of pure water in each, and cool them both to as close to 0 °C as possible, leaving one liquid and freezing the other. Now bring them both into a warm room and observe the time it takes for the water to rise 1°C (on gaining 4.2 kJ, or 1 kcal). Next, observe how long it takes the ice to completely melt. What you will find, as Black did, is that the melting takes about 80 times longer.

Experiments have established that 1 kg of ice at 0 °C will be transformed into 1 kg of liquid water at 0 °C when 334 kJ (80 kcal) of heat are added. The amount of heat that must be added to a kilogram of any solid at its melting temperature in order to liquefy it (or, alternatively, the same amount of heat that must be removed from a kilogram of liquid at its freezing temperature in order to solidify it) is the **latent heat of fusion**, or just the **heat of fusion**. It's an unfortunate term because it is energy per unit mass (J/kg) and not energy. Be

Table 13.6
Approximate Heats of Fusion and Vaporization

Material	Melting point (°C)	Heat of fusion (kJ/kg)	Boiling point (°C)	Heat of vaporization (kJ/kg)
Antimony	630.5	165	1380	561
Alcohol, ethyl	−114	104	78	854
Copper	1083	205	2336	5069
Gold	1063	66.6	2600	1578
Helium			−268.93	21
Hydrogen	−259.31	58.6	−252.89	452
Lead	327.4	22.9	1620	871
Mercury	−38.87	11.8	356.58	296
Nitrogen	−209.86	25.5	−195.81	199
Oxygen	−218.4	13.8	−182.86	213
Silver	960.8	109	1950	2336
Water	0.0	333.7	100.0	2259

that as it may, the heat of fusion (L_f) for water is

[water at 0 °C]
$$L_f = 334 \text{ kJ/kg} \approx 80 \text{ kcal/kg}$$

Table 13.6 lists the heats of fusion for a representative selection of substances at a pressure of 1 atm. As we will see presently, L_f depends on pressure. It follows that to change the phase of a mass m already at its melting point requires the addition (or removal) of an amount of heat, or better yet, thermal energy, Q such that

$$Q = \pm mL_f \tag{13.3}$$

where the positive sign corresponds to heat-in (melting) and the negative sign to heat-out (solidification).

Example 13.6 **[II]** A container holding 0.250 kg of water at 20.0 °C is placed in the freezer compartment of a refrigerator. How much energy must be removed from the water to turn it into ice at 0 °C?

Solution Specific heat should come to mind whenever there's a temperature change. Here we also have a change in state from water to ice. (1) TRANSLATION—A known mass of liquid (water) at a given initial temperature is to be frozen (at 0 °C); determine the amount of energy that must be removed. (2) GIVEN: Water, $m_w = 0.250$ kg, $T_{iw} = 20.0$ °C, and $T_{fw} = 0$ °C. FIND: Q_{out}. (3) PROBLEM TYPE Thermodynamics/heat/heat of fusion. (4) PROCEDURE—The heat-out (i.e., the heat that must be removed) must be enough to drop the water at 20.0 °C down to 0 °C and then transform it to ice (also at 0 °C). Both processes corre-

spond to negative amounts of heat. (5) CALCULATION—Using Eq. (13.1) for the temperature change and Eq. (13.3) for the freezing,

$$Q_{out} = c_w m_w (T_{fw} - T_{iw}) + (-m_w L_f)$$

Substituting in the numbers,

$$Q_{out} = (4.186 \text{ kJ/kg·K})(0.250 \text{ kg})(0 \text{ °C} - 20.0 \text{ °C})$$
$$- (0.250 \text{ kg})(334 \text{ kJ/kg})$$

$$Q_{out} = -20.9 \text{ kJ} - 83.5 \text{ kJ} = \boxed{-1.04 \times 10^5 \text{ J}}$$

Quick Check: To drop the temperature 20 K, we must remove $(\frac{1}{4} \text{ kg})(20 \text{ K})(1 \text{ kcal/kg·K}) = 5$ kcal. To freeze it, another $(80 \text{ kcal/kg})(\frac{1}{4} \text{ kg}) = 20$ kcal must be removed, for a total of 25 kcal = 105 kJ.

From an atomic perspective, it's apparent that the liquid state has more internal energy than the solid state at the same temperature; after all, heat is transferred into the sample as it melts. Still, no corresponding rise occurs in temperature, and that's true for monatomic systems, like the metals, as well as molecular systems, like water. It would seem that the energy input must be increasing the PE, which has no effect on temperature. The energy associated with the heat of fusion liberates the atoms or molecules from the rigid bonding of the solid state, allowing them to move apart into a less ordered, more fluid configuration.

COLD BEERS & FRUIT CELLARS

Since heat must flow into a body when it melts, melting is a cooling process in the sense that it removes thermal energy from the immediate environment. A block of ice in a picnic cooler slowly melts, drawing off a lot of thermal energy (80 kcal/kg) from the surrounding warm cans of beer. After a while, the beer is cold and the ice is melted. The same is true for ice cubes in a glass of warm soda—that's why the ice is added, and that's why it always melts. In reverse, freezing can be thought of as a warming process: one that exhausts thermal energy into the immediate environment. Years ago, people used to protect their fruit cellars and greenhouses from freezing by storing large vats of water in them. As the water began to freeze, it (in effect) liberated thermal energy, making it that much harder for the cold air to further cool the room.

13.6 Vaporization

The transformation of a liquid or solid into a gas is known as **vaporization**. The resulting product is called a *vapor* because it deviates radically from the behavior predicted by Boyle's Law, the behavior of an ideal gas (p. 438). The change from a liquid to a gas that occurs continuously at a free surface of the liquid (i.e., one not in contact with a solid or other liquid) is known as **evaporation**.

A liquid is composed of a large number of swarming particles, atoms, or molecules that move around with a distribution of kinetic energies much like that of a gas. Even at low temperatures, a few of those traveling in the vicinity of the surface move up toward it with speeds great enough to overcome the cohesive forces and escape the liquid altogether (Fig. 13.5). Some molecules will return to the liquid as new ones leave it. As long as the vapor pressure above the liquid is less than the saturated value for that temperature, evaporation continues. And this is essentially independent of the presence of other gases (such as air). If the vapor pressure above the liquid is made to exceed the saturated value (e.g., by pressing down on the vapor with a piston), there will be condensation. Similarly, a drinking glass filled with ice cubes can cool the surrounding moist air, thereby lowering its saturated vapor pressure for water enough to allow condensation onto the cold surface.

The rate of evaporation can be increased by altering the physical condition of the liquid in two ways: First, by increasing the surface area (a wet shirt dries slowly if it's balled up, and brandy evaporates from a wide snifter a lot faster than from an uncorked bottle). Second, increasing the temperature of the liquid accelerates evaporation because at greater temperatures the molecules have higher average kinetic energies and more likelihood of escaping the surface. Warming the clothes in a dryer, heating the back window of your car

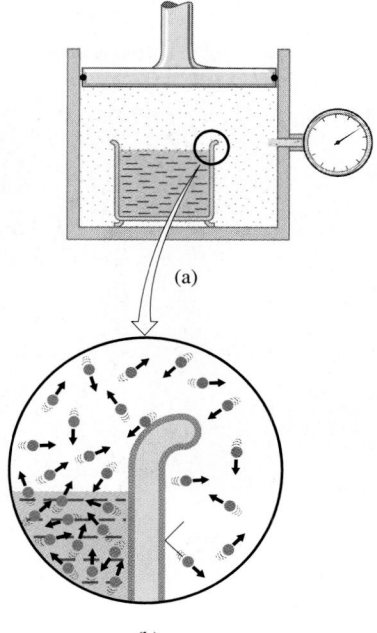

(a)

(b)

Figure 13.5 (*a*) A liquid in equilibrium with its vapor at a constant pressure. (*b*) On the average, as many molecules leave the surface from below as enter it from above.

to drive off the condensation, or holding the bowl of the brandy snifter in a warm hand are examples of this effect.

Just as with melting, energy must be supplied to the molecules being liberated from the cohesion of the liquid state. Vaporization is an even more drastic physical change than melting: the molecules are both torn free and appreciably separated, a process requiring the input of a substantial amount of energy. In fact, the energy per molecule needed to vaporize a substance is often as much as ten to twenty times greater than the energy to melt it. As a rule, the vapor contains far more energy than an equal mass of liquid at the same temperature. Accordingly, **the heat of vaporization (L_v) is the amount of thermal energy required to evaporate 1 kilogram of a liquid at a constant temperature**, or if that energy is removed from a kilogram of vapor, to condense it under the same conditions. As we will see presently, this temperature is usually the boiling point, but it need not be. Moreover, the heat of vaporization generally decreases as T increases (Fig. 13.6). For example, L_v for water at 0 °C is 2.49×10^3 kJ/kg, whereas at skin temperature, $\approx$33 °C, it's down somewhat to 2.42×10^3 kJ/kg. For water at 100 °C

[water at 100 °C] $$L_v = 2259 \text{ kJ/kg} \approx 540 \text{ kcal/kg}$$

Since the density of the liquid is greater than the density of the vapor, it's evident that energy is being supplied to each escaping molecule to overcome the attraction of the other molecules. Yet this density difference decreases as the sample approaches its critical temperature (p. 440). There, the two states are indistinguishable, and the heat of vaporization is zero.

As Table 13.6 shows, each substance has a characteristic heat of vaporization. Thus, the amount of heat Q that must be supplied (or removed) in order to vaporize (or condense) a mass m of liquid (or vapor) at a specific constant temperature is given by

$$Q = \pm m L_v \tag{13.4}$$

With this equation, we could compute how much energy had to be supplied to your wet bathing suit the last time you impatiently waited around for it to dry. Clearly, one of the synthetic fabrics that does not absorb much water in the first place (so that m is small)

Figure 13.6 A graph of temperature versus thermal energy added to, or removed from, 1 kg of H_2O. The sample is raised from −10 °C to above 100 °C and passes from ice to ice plus water, to water, to water plus steam, to steam.

Example 13.7 **[II]** At what rate would you have to perspire to rid your body of the 75 W of thermal power you generate metabolically just sitting around resting? Assume a skin temperature of 33 °C. As we will see, this is not our only cooling technique, and we generally don't have to sweat that much. Still, an athlete may lose several liters of water to perspiration during a good workout. What happens when you dry off the perspiration with a towel?

Solution This problem is about evaporation and the cooling that results from it. (1) TRANSLATION—The vaporization of water at a known temperature is to provide a specified rate of cooling (in watts); determine the needed evaporation rate (in kg/s). (2) GIVEN: A power output of 75 W and $T = 33$ °C. FIND: The evaporation rate. (3) PROBLEM TYPE—Heat of vaporization. (4) PROCEDURE—Use L_v for water at $T = 33$ °C. (5) CALCULATION—First, 75 W = 75 J/s. From the above

remarks, the heat of vaporization of water at skin temperature is 2.42×10^3 kJ/kg. You must lose energy at a rate of

$$75 \text{ J/s} = \frac{Q}{t} = \frac{mL_v}{1 \text{ s}}$$

and

$$\frac{m}{1 \text{ s}} = \frac{75 \text{ J/s}}{L_v} = \frac{75 \text{ J/s}}{2.42 \times 10^6 \text{ J/kg}} = 3.1 \times 10^{-5} \text{ kg/s}$$

If this were your only available cooling mechanism, you would sweat away 3.1×10^{-2} g/s, or about $\boxed{0.11 \text{ kilogram per hour}}$. Normally we sweat about $\frac{1}{2}$ liter per day, or about the equivalent of 1.2 MJ. Toweling perspiration off the skin or letting it drip off is of no use—it has to evaporate to do any cooling.

Quick Check: The power equals L_v times the mass evaporated per second; $(2.4 \times 10^3 \text{ kJ/kg})(3.1 \times 10^{-5} \text{ kg/s}) = 74$ W.

A bubble chamber consists of a tank filled with a liquefied gas, in this particular case a mix of neon and hydrogen. The liquid is held under elevated pressure at a temperature just below its boiling point. The pressure is then lowered so rapidly that for a moment the liquid is in a state where it normally would be boiling but isn't. This "superheated" liquid is extremely unstable. When an energetic charged particle passes through a medium, it ionizes some of the atom it collides with. In this case it causes enough of a disturbance for the superheated liquid to "boil" locally, forming tiny vapor bubbles around the ionized atoms, thereby creating a visible track.

requires the least amount of thermal energy and dries the most quickly, especially if it's a warm, dry day.

Since a rather substantial amount of thermal energy must be supplied to the system (often drawn from the immediate environment), vaporization can be considered a cooling process: the molecules that escape are the energetic ones, leaving behind their lower-average-KE brethren in the liquid, which therefore must have a reduced temperature (unless the system is being supplied heat from outside). That's in part why we perspire and why dogs pant. It's why you might lick a finger to test the wind or find your unattended bath water colder than the room even though it started out warm a few hours before.

The primary mechanism that causes hot, moist food to cool is evaporation. That's why you blow on a hot cup of soup—without air currents to carry away the vapor, it builds up above the soup and becomes saturated. That build-up decreases the evaporation rate and slows down the cooling process. To increase evaporation, you blow away the vapors every now and then. The hot food at a barbecue always seems to cool too fast because of the active air currents outdoors.

13.7 Boiling

Evaporation takes place at a liquid's free surface at any temperature, but under special circumstances it can also occur throughout the body of the liquid. Since vapor generated at any point within the liquid has a much lower density than the surrounding medium, it pushes the liquid in the immediate vicinity out of the way, creating (under hydrostatic pressure) a little gaseous sphere. The formation of such a bubble of vapor signals the beginning of the familiar process of **boiling**. To be precise, any liquid (water, olive oil, or molten gold) will boil at a specific temperature where its saturated vapor pressure equals the surrounding pressure. After all, to exist for an appreciable time, a bubble must be essentially in equilibrium with the surrounding liquid. That means the pressure inside of it—the vapor pressure—must equal the ambient hydrostatic pressure, and that's usually ≈1 atm. The vapor pressure of water, which is 0.006 atm at 0 °C and rises to 0.2 atm at 60 °C, equals 0.1 MPa (1 atm) at 100 °C. That should be no surprise; the Celsius scale was fixed via the temperature of boiling water at atmospheric pressure, so we have just gone full circle.

When a pot of water has been heated for a while, bubbles begin to form around the walls and bottom of the vessel. The nucleation sites are centered on minute pockets of air or particles of dust. Some of the bubbles are air, driven out of solution by the rise in tempera-

Example 13.8 **[II]** How much heat must be added to a 1.0-kg mass of water ice at $-10\ °C$ and atmospheric pressure, in order to transform it into superheated steam at $110\ °C$? Compare the energy associated with each stage of the process and confirm Fig. 13.7.

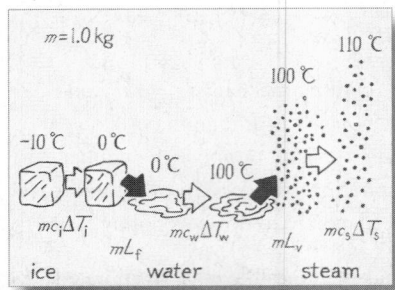

Solution This problem is about raising the temperature and changing the state of a substance, and of course, specific heat should come to mind whenever there's a temperature change. (1) TRANSLATION—A known quantity of ice at a known initial temperature is to be transformed into steam at a specified final temperature; determine the needed input of heat. (2) GIVEN: $m = 1.0$ kg, initial temperature $-10\ °C$, and final temperature $110\ °C$. FIND: Q_{in}. (3) PROBLEM TYPE—Heat/temperature change/change of state: ice → water → steam. (4) PROCEDURE—We have to raise the temperature of the ice (ΔT_i) to $0\ °C$, melt it, raise the temperature of the water (ΔT_w) to $100\ °C$, vaporize it, and raise the temperature of the steam (ΔT_s) to $110\ °C$. Remember that each state has its own value of specific heat. (5) CALCULATION—Using $L_f = 334$ kJ/kg, $L_v = 2.26 \times 10^3$ kJ/kg, $c_w = 4.2$ kJ/kg·K, $c_i = 2.1$

kJ/kg·K, $c_s = 2.0$ kJ/kg·K, and the relationship

$$Q_{in} = mc_i\Delta T_i + mL_f + mc_w\Delta T_w + mL_v + mc_s\Delta T_s$$

$$Q_{in} = (1.0\ \text{kg})(2.1\ \text{kJ/kg·K})[0\ °C - (-10\ °C)]$$
$$+ (1.0\ \text{kg})(334\ \text{kJ/kg})$$
$$+ (1.0\ \text{kg})(4.2\ \text{kJ/kg·K})[100\ °C - 0\ °C]$$
$$+ (1.0\ \text{kg})(2.26 \times 10^3\ \text{kJ/kg})$$
$$+ (1.0\ \text{kg})(2.0\ \text{kJ/kg·K})[110\ °C - 100\ °C]$$

$$Q_{in} = 21\ \text{kJ} + 334\ \text{kJ} + 420\ \text{kJ} + 2260\ \text{kJ} + 20\ \text{kJ}$$

$$Q_{in} = 3055\ \text{kJ}$$

or to two significant figures $\boxed{Q_{in} = 3.1 \times 10^3\ \text{kJ}}$. Notice that most of this energy was required to make the steam.

Quick Check: Because the calorie was defined via the behavior of water, problems that deal exclusively with water are easier to treat in the old units:

$$Q_{in} = (1\ \text{kg})(0.50\ \text{kcal/kg·K})(+10\ \text{K})$$
$$+ (1\ \text{kg})(80\ \text{kcal/kg})$$
$$+ (1\ \text{kg})(1.00\ \text{kcal/kg·K})(+100\ \text{K})$$
$$+ (1\ \text{kg})(540\ \text{kcal/kg})$$
$$+ (1\ \text{kg})(0.48\ \text{kcal/kg·K})(+10\ \text{K})$$

$$Q_{in} = 5.0\ \text{kcal} + 80\ \text{kcal} + 100\ \text{kcal} + 540\ \text{kcal} + 4.8\ \text{kcal}$$
$$Q_{in} = 729.8\ \text{kcal} = 3.1 \times 10^3\ \text{kJ}$$

ture. Once created at the hot bottom, air bubbles remain intact until they reach the surface. By contrast, bubbles of water vapor rapidly form, rise a little, and disappear. When they enter the colder water above, they are cooled and suffer a decrease in vapor pressure, collapsing, condensing, and vanishing before reaching the surface. This compression can occur violently and is often noticeably noisy. Only when the upper regions of the water reach $100\ °C$ will true boiling take place throughout the liquid, and then bubbles reach and burst at the surface. That bursting process liberates a great deal of vapor and with it a great deal of energy—the heat of vaporization of water at $100\ °C$ is 2.259×10^3 kJ/kg (540 kcal/kg). If heat is continuously fed into the water, it will continue to boil, and raising the flame under the pot will only increase the rate of boiling without changing the temperature of the water. More heat-in generates more vapor, which at temperatures $\geq 100\ °C$ is known as **steam**.

Keep in mind that the bubbles only rise because of their buoyancy, which exists because of the hydrostatic pressure, which exists because of gravity. Out in space water boils very differently—the bubbles tend to coalesce at the bottom of the vessel.

Each kilogram of steam at $100\ °C$ can conveniently be used to carry that 2.3 MJ from one place to another, from a basement boiler to a third floor radiator. On condensing in the radiator, the steam liberates its 2.3×10^3 kJ/kg to the environment. Of course, steam can be raised to temperatures well in excess of the 1-atm boiling point, the ultimate limitation being the breaking up of the water molecules. In most modern applications, steam above $500\ °C$ is used. This ***superheated steam*** (as it's called above $100\ °C$) whirls the great turbines that drive the majority of large ships and electric generators.

COOKING UNDER PRESSURE

In the pressure cooker (invented by Denis Papin in the late 1600s) the liberated vapor builds up pressure inside a heated sealed vessel. The domestic pressure cooker usually operates at a gauge pressure of 1 atm and a boiling-water temperature of $121\ °C$ ($250\ °F$). The chemical reactions of cooking roughly double their rate with every $10\ °C$ increase beyond $100\ °C$. The $21°C$ rise is significant, and the pot will cook things much more quickly than the traditional open-to-the-atmosphere culinary approach. The hospital autoclave is a large-scale pressure cooker for sterilizing instruments.

The boiling point of a substance depends on the external pressure. Lower the surface pressure, and bubbles can form at a lower vapor pressure, which corresponds to a lower temperature (they are squeezed less and form easier). A pot of water on a campfire atop Mont Blanc (5 km) will boil at only 83 °C, while on Mt. Everest (9 km) making a decent cup of tea on an open fire would be impossible (the pressure is ≈37 kPa; that is, 0.4 atm, and the boiling point is ≈74 °C). This effect has been used commercially to boil off water from delicate foodstuffs like milk and syrups (in the manufacture of sugar) without cooking them. They're boiled in a low-pressure vessel where the vapor is continuously pumped off. If we put a small sample of water in a vacuum system and vigorously pump off the vapor, dropping the saturated pressure, the liquid will boil at room temperature. This being a cooling process, the sample will rapidly decrease in temperature and freeze into ice.

The Transfer of Thermal Energy

Put a hot cup of coffee on a table in your kitchen and come back in an hour and it will certainly be cooler, just as a cold bottle of soda left out will get warmer. Objects transfer energy to and from the environment in three basic ways: radiation, convection, and conduction.

13.8 Radiation

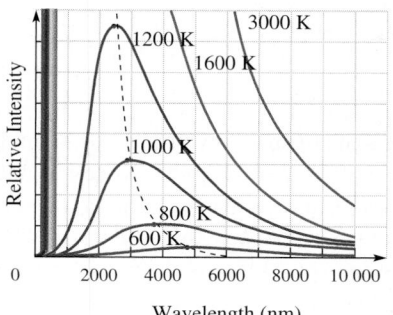

Figure 13.7 Thermal radiation: The electromagnetic energy radiated by a body at various temperatures as a function of wavelength. The narrow band from ≈380 nm to ≈760 nm is the visible region of the spectrum.

All bodies radiate electromagnetic energy as a natural result of their constituent atoms randomly oscillating. This **thermal radiation** is characterized by a broad, continuous range of wavelengths (Fig. 13.7), which arises from interactions among the atoms of solids, liquids, and dense gases. At this very instant, because of the warmth of your body, you are emitting copious amounts of invisible infrared (IR) electromagnetic radiation (which can easily be observed by IR cameras). The reverse process of absorption becomes evident whenever you put your face in the sunshine (or near an open furnace) and feel a rush of warmth as the radiant energy, especially the IR, is converted to thermal energy in your skin.

As long ago as 1791, P. Prevost maintained that matter at a constant temperature was in equilibrium, absorbing as much radiant energy as it emitted. ***The amount of thermal radiation emitted by a body depends on its surface condition (color, texture, exposed area, etc.) and on its temperature.*** The higher the temperature, the more energy it radiates per second. On the other hand, ***the amount of thermal radiation absorbed by a body depends on its surface condition and on the nature of the incident radiant energy (wavelength, intensity, etc.), which, in turn, depends on the temperature of the source***.

All else equal, black rough materials emit radiant energy effusively, whereas polished metals such as silver and copper are (20 or 30 times) less given to radiate, and white surfaces are usually between the two. As a rule, **a good emitter is a good absorber**, and vice versa (it would follow that a poor absorber is a good reflector or a good transmitter—if it cannot absorb the energy that lands on it, it has to get rid of it some other way). Were it not so, we could put our great little absorber out in the sunshine and it would get hotter and hotter indefinitely. As is, an object bathed in radiant energy from a higher temperature source (but otherwise isolated from the environment) will absorb energy and rise in temperature. It will increase its initially (low T) low rate of emission until, at an appropriately elevated temperature, emission precisely balances absorption. At that fixed equilibrium temperature, the body will thereafter continue to radiate as much as it absorbs. That's what happens when you park your car in the sunshine: if it's black, it heats up a lot and quickly; if white, it heats up a little and slowly. In neither case does the temperature keep rising until the car melts. The more surface exposed to the radiation, the more energy will be absorbed and the more rapidly the temperature will rise—that's why we sit in the shade when it's hot and why, if there is no shade, a camel will sit facing the Sun rather than expose the length of its body to the Sun's rays.

Much of the Hubble space telescope is wrapped in a highly reflecting metalized plastic covering. It reflects radiant energy from the Sun and helps maintain an appropriate range of operating temperatures.

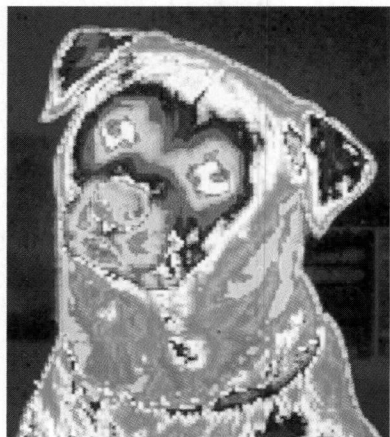

This photo of a dog was made using only the infrared radiant energy emitted by the animal. Such images are essentially temperature mappings.

SKIN COLOR & RADIATION

We humans radiate rather well—light-skinned folks emit at about 60% of the maximum rate for an ideal (black) body at that temperature, while dark-skinned people approach about 80%. The difference is not unreasonable given that lighter skins probably evolved in cooler climates. Sitting quietly reading a book, you lose up to 50% of your exhausted thermal energy to the room by way of radiation. The surface temperature of the Sun (6000 K) is a lot higher than your surface temperature (310 K). As we'll see later, the amount of energy a body radiates goes as T^4, and since the Sun's temperature is ≈ 20 times greater, standing outside, it radiates 20^4 or $\approx 160\,000$ times more energy toward you than you radiate toward it. At night the reverse is true; the dark sky has an ambient temperature of only ≈ 3 K and you radiate far more (≈ 100 W) to it than it sends down to you. If you're cold outside at night, stand under something (e.g., a tree, or a marquee)—it will be a lot warmer than 3 K.

In sunshine, the dark-colored leaf absorbs a good deal of radiant energy and its temperature rises. As a result, the snow beneath it melts and the leaf sinks down well below the snow's surface.

If it were possible to put our hot cup of coffee off alone in an empty universe, it would continue to radiate away energy, slowly approaching Absolute Zero. But that process is not what happens in our far-from-empty Universe. Here, any object is actually immersed in a local sea of electromagnetic radiation, and it continuously exchanges radiant energy with that environment. The rate at which it radiates is determined by its thermodynamic temperature; the rate at which it absorbs is determined by the character of the incident radiation and therefore by the effective temperature of the surroundings. That holds true even in the distant reaches of space where the ambient radiation has a thermal spectrum consistent with a temperature of about 3 K (indicating to physicists that the radiation was once in equilibrium with matter in a denser, hotter, earlier time, presumably near the beginning of the Universe). Space is awash with randomly fluctuating electromagnetic fields that correspond to thermal radiation.

Putting it rather simplistically, when an object (like our coffee) is hotter than the local radiation environment, it radiates more than it absorbs and tends to cool down to the surrounding temperature. Your kitchen is filled with a spectrum of radiant energy that is typical of a room-temperature thermal source, and the hot coffee loses energy to it. On the contrary, our cold bottle of soda still radiates, but much less than it absorbs from the radiation environment of the kitchen, and so it slowly warms up. This mechanism has nothing to do with any contact the bottle might have with the table or the air. It would not be affected if the bottle were suspended inside of a vacuum. Radiant energy passes through vacuum unimpeded, which is obvious in the case of solar radiation.

Someone who lives on a 2400 kcal daily diet, and whose weight is stable, is operating on average at 120 J/s. In other words, the organs and tissues of the body generate thermal energy at an average rate of about 120 W. To maintain a constant body temperature that person would have to lose energy at that same rate. That's accomplished primarily through radiation, convection, evaporation of sweat, and breathing. The net amount of energy radi-

An ordinary hot water pipe can be turned into a moderately efficient radiator by covering it with metal vanes. Both radiation and convection increase with surface area and that's just what the vanes accomplish. They are a dark color to increase radiation; they are metal to conduct heat out from the pipe, and they are separated to allow air to circulate between them.

ated depends on the temperature of the surroundings and the exposed surface area. A nude human body, with a skin temperature of 34 °C (93.2 °F) even in a warm room at 25 °C (77 °F) will radiate, at a rate of about 63 W. Standing outside on a clear night you can get quite cold radiating out into space. (For a more complete treatment of radiation see Sect. 28.1.)

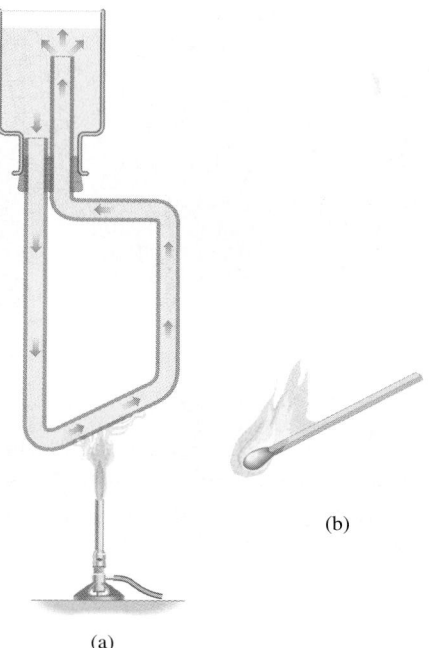

(a)

(b)

Figure 13.8 (*a*) Convection currents. Heated liquid rises, cooler liquid descends. The effect is due to a decrease in density as *T* increases and is driven by gravity—no gravity, no free convection. (*b*) With a tilting match, convection currents carry thermal energy up to the unburned wood, which fuels the process.

13.9 Convection

When a region of a fluid is heated (e.g., the air around a candle flame), its density decreases via thermal expansion and it rises. Displacing cooler portions in its path, the resulting fluid current constitutes a flow of thermal energy, up and away from the source. Driven by gravity, this process is known as *free* or **natural convection** (Fig. 13.8). It carries off thermal energy at a rate in proportion to the temperature difference between the body and the surrounding fluid (ΔT). Reasonably enough, that rate (Q/t) is also proportional to the exposed area (A). In other words, the rate at which thermal energy is carried off by convection is $Q/t \propto A\Delta T$. For example, a nude human body having an area of 1.2 m² in air at 25 °C (77 °F) loses about 29 W through convection.

The household radiator warms a space via convection currents that swirl up and around the room. A hot cup of coffee will produce the same effect in the air on a smaller scale. During the day on the coast, the land (heated more quickly by the Sun because of its low specific heat) heats the air above it. The warm air circulates up and around to create a cool-

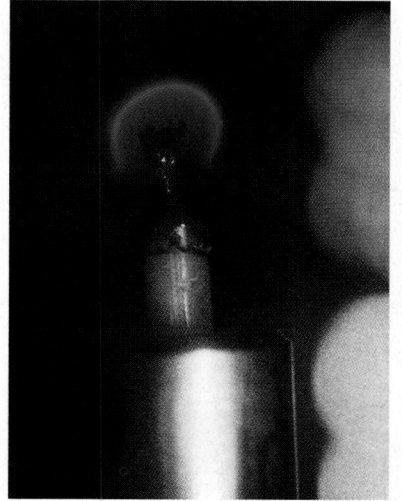

A candle burning in the Microgravity Laboratory of the *Space Shuttle*. While in Earth-orbit, convection currents are far less pronounced than they are under normal gravitational conditions. Ordinarily, hot air, being less dense than cool air, rises. Convection currents feed oxygen to the flame, which also rises.

NAKED PEOPLE & STORM WINDOWS

Another reason to wear clothing (if you need one) is to prevent convection currents from carrying off too much thermal energy from your body on cold days. After all, we are tropical animals who could not survive naked very much below 20 °C. Stopping convection currents is also a good reason to insulate water heaters, put up storm windows, and sleep under blankets when it's cold. The blanket, unless it's electric, does not supply heat; it just cuts down on the loss of the thermal energy you generate—covering a shut-off cold automobile engine with a blanket will not warm it up.

Figure 13.9 Convection currents coming off the sea.

An eagle soaring on updrafts of warm air corresponding to convection currents.

ing breeze off the sea (Fig. 13.9). At night, when the sea tends to be warmer than the land, the air current reverses. Large birds such as eagles and vultures (as well as sailplane pilots) seek out warm air updrafts. They conserve their energy by riding convection currents, hovering effortlessly or gaining altitude in the uprush of these so-called thermals. On a larger scale, the trade winds are caused by atmospheric convection currents rising up from the tropics. Ocean currents, like the Gulf Stream, are also the result of free convection.

13.10 Conduction

We began this chapter with a discussion of what heat is, and now we end with a few remarks about how it is conducted—namely, by atomic collisions. At the center of the concept is the idea of a temperature difference: ΔT. Heat is only conducted from a high to a low temperature. The measure of the ability of a substance to transmit heat is known as its **thermal conductivity** (k_T), and it depends on the atomic structure of the material. Metals are roughly 400 times better conductors of heat than other solids, which are a trifle better than most liquids, which, in turn, are about 10 times better than the gases (Table 13.7).

It is usually true that a good conductor of heat is a good conductor of electricity, a parallel that is no coincidence. A metal can be envisioned as a lattice of positive ions immersed in a tenuous sea of essentially free electrons. Both heat and electricity are transported largely by the motion of these unbound electrons, which behave like a highly mobile fluid with-

Nowadays buildings are wrapped in insulation (the pale pink under the bricks) to cut down on thermal conduction.

Table 13.7

Approximate Values* of Thermal Conductivities

Material	Thermal conductivity, k_T (W/m·K)	Material	Thermal conductivity, k_T (W/m·K)
Metals		Linen	0.088
Aluminum	210	Paper	0.13
Brass (yellow)	85	Paraffin	0.25
Copper	386	Plaster of Paris	0.29
Gold	293	Polyamides (e.g., Nylon)	0.22−0.24
Iron	73	Polyethylenes	0.3
Lead	35	Polytetrafluroethylene	
Platinum	70	(e.g., Teflon)	0.25
Silver	406	Porcelain	1.1
Steel	≈46	Rubber, soft	0.14
		Sand, dry	0.39
Other solids		Silk	0.04
Asbestos	0.16	Snow, compact	0.21
Brick, common red	0.63	Soil, dry	0.14
Cardboard	0.21	Wood, fir, parallel to grain	0.13
Cement	0.30		
Chalk	0.84	*Liquids*	
Concrete and cement mortar	1.8	Acetone	0.20
cinder block	0.7	Benzene	0.16
Down	0.02	Alcohol, ethyl	0.17
Earth's crust	1.7	Mercury	8.7
Felt	0.036	Oil engine	0.15
Flannel	0.096	Vaseline	0.18
Glass	0.7−0.97	Water	0.58
fiberglass	0.04		
Granite	2.1	*Gases*	
Human tissue (no blood)	0.21	Air	0.026
fat	0.17	Carbon dioxide	0.017
Ice	2.2	Nitrogen	0.026
Leather	0.18	Oxygen	0.027

*Near room temperature.

in the metal. In poor conductors of electricity, the transport of heat is primarily by the slower mode of atomic collisions. Thus, amorphous nonmetallic solids (e.g., glass and plastic) are only modest thermal conductors. But the ability to conduct energy is enhanced a bit when the substance is crystalline and the lattice of atoms can also vibrate as a continuous structure (as it does when it carries a sound wave). Whether it is via the random motion of

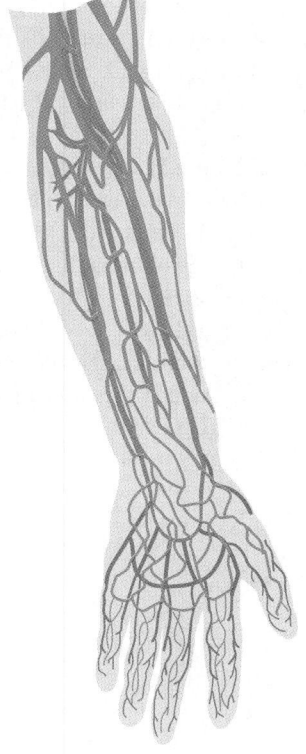

Figure 13.10 Because the veins and arteries are close together, warm arterial blood on the way down the arm loses thermal energy to venous blood returning to the body's core, making the hands cold but the core warm.

BLOOD, BLUSHING, & HEAT

Humans are wrapped in a protective layer of skin and fat that's a relatively poor thermal conductor. In a warm environment, the direct transmission of excess heat from inside the body to the skin's surface, where it can be expelled, is impractical. Instead, thermal energy is first conducted into the blood, which is then circulated up to the body's surface via a network of capillaries—that's what blushing is all about. At the other extreme, when it's cold and thermal energy must be retained, the capillaries near the surface contract, cutting off the flow of blood. *Countercurrent exchange* sustains the body's core temperature by cooling the blood to the arms and legs, thereby lowering their temperature and ability to lose thermal energy. The arteries and veins are close together in the limbs (Fig. 13.10) and arterial blood (carrying nutrients and oxygen) loses heat to the venous blood (returning to the heart) before it gets out to the extremities. Your hands and feet get cold to keep your vital organs warm—better to sacrifice a finger or two to frostbite than lose a liver.

This material was developed for the heat-shield tiles of the *Space Shuttle*. It has both an extremely low thermal conductivity and small specific heat capacity—so much so, that a cube of it can be held at its edges even when red hot.

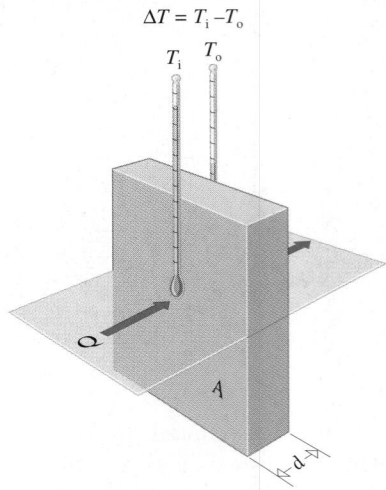

Figure 13.11 The rate (Q/t) of transport of heat is proportional to the temperature gradient ($\Delta T/d$) and the area (A).

atoms or molecules or electrons, *a temperature difference propels heat* much as a gravity difference propels a river.

The transfer of energy always occurs from the action of one of the fundamental Four Forces. In the case of thermal conduction, the force is electromagnetic. A group of atoms with a high average thermal energy transfers random KE to an adjacent group of atoms having a lower average thermal energy, and we observe the process as heat conduction across a temperature difference.

The molecules of a gas are widely separated, which makes them much less able to communicate thermal energy via collisions. Accordingly, air is a poor thermal conductor, and you can hold your hand beside a candle flame (as opposed to above it in the convection current) and barely feel it. Anything that will trap a layer of air will function as a thermal insulator. That's what a storm window does and why wearing many layers of clothing works nicely in the winter. Your body generates thermal energy, and the clothing, along with the air it captures, keeps it from escaping. In part, the containment of air is also what makes fibrous materials such as hair, wool, fur, and fiberglass such effective thermal insulators. Trapped layers of motionless or "dead" air make excellent insulation. Because the thermal conductivity of air is low, our bodies warm only a thin layer of it, and if it stays put we don't lose much more heat to it via further conduction. All of that changes if there's a breeze and cool air is circulated over the skin. That increase in energy loss is called *wind chill,* and it's why you blow on hot coffee. It's also why you might wear a thin nylon wind breaker. A swim in a cold ocean will quickly make the point that liquids are better conductors than gases.

The steady-state amount of heat Q that flows per unit time Q/t through a sample has been studied extensively, both theoretically and in the laboratory. Because it's actually a complicated problem, we will consider only the simplest geometry and use the classic approach of compressing all the subtle atomic details into a single material constant. The amount of heat that is conducted per second through a uniform slab or rod of material (Fig. 13.11) is found to be proportional to the *fixed* temperature difference ΔT across its two faces, $Q/t \propto \Delta T$. We expect that our sample slapped against a blazing furnace wall would transmit more heat to your hand than if it were pressed against a lukewarm radiator. Moreover, it is reasonable to assume that the amount of molecular agitation that can be propagated down the sample should depend directly on its cross-sectional area A: $Q/t \propto A$.

If the sample has a thickness or length d, the amount of heat that will traverse it per second in steady-state flow is found to be inversely proportional to d. In other words, Q/t is proportional to the **temperature gradient** $\Delta T/d$ (which, if we were looking at the rate at which balls were rolling down a hill, would be the equivalent of how steep the hill was, and not just how high it was). Remember the end-face temperatures (Fig. 13.11) are held constant, and the closer those faces are, the greater the gradient and the more rapidly the collision energy is moved along. In summary, then, $Q/t \propto A(\Delta T/d)$, and if we introduce a material constant of proportionality, namely, the thermal conductivity (k_T), then

$$\frac{Q}{t} = k_T A \frac{\Delta T}{d}$$ (13.5)

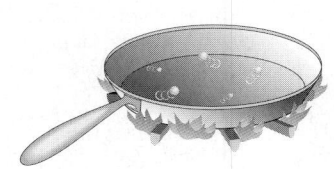

EXPLORING PHYSICS ON YOUR OWN

Floating Drops: Take an ordinary pot or pan, place it on a stove, and heat it up. Before it gets very hot, pour in a few drops of water and observe what happens. The water sizzles off, quickly evaporating completely. Let the pan get still hotter until, when a few drops are added, the water sizzles wildly, but much of it quickly beads up and remains. Drops will then seemingly hover above the bottom of the pan and sail around as if floating weightlessly. They can continue to exist like this for a remarkably long time without getting noticeably smaller. Now shut the flame off and watch the beads as the pan cools. Explain what's happening in terms of evaporation and thermal conductivity.

Example 13.9 **[I]** A pane of window glass is 0.90 m wide by 1.5 m high and 4.0 mm thick. It's a cold blustery winter's day in the Midwest. The temperature of the inside face of the window is 10 °C, and the outside face is at −9.0 °C. How much thermal power is being transported through the windows (k_T = 0.84 W/m·K)?

Solution This problem is about heat conduction and that means Eq. (13.5). (1) TRANSLATION—A flat object of known dimensions and material (glass) has a specified temperature difference across its two faces; determine the rate at which heat is conducted through it. (2) GIVEN: The pane is 0.90 m by 1.5 m, d = 4.0 mm, ΔT = 10 °C − (−9.0 °C), and k_T = 0.84 W/m·K. FIND: Q/t. (3) PROBLEM TYPE—Thermo/heat/thermal conduction. (4) PROCEDURE—The rate at which heat flows is provided by Eq. (13.5). (5) CALCULATION—First we need the area A = 0.90 m × 1.5 m = 1.35 m², then

$$\frac{Q}{t} = \frac{k_T A \Delta T}{d} = \frac{(0.84 \text{ W/m·K})(1.35 \text{ m}^2)(19 \text{ K})}{4 \times 10^{-3} \text{ m}} = 5386.5 \text{ W}$$

and to two significant figures $\boxed{Q/t = 5.4 \text{ kW}}$. Notice that the inside face of the window is a lot colder than room temperature. That's the result of the blanketlike quality of a thin layer of stationary air that usually forms against the window. Even more power would be wasted if the inside air were moving across the inner face of the window, keeping its temperature closer to that of the room. This underscores the virtues of having closed curtains or window shades (in addition to storm windows), especially at night when little radiation enters via the windows.

Quick Check: $\Delta T = Qd/tk_T A$ = (5386.5 W)(0.004 0 m)/ (0.84 W/m·K)(1.35 m²) = 19 K.

where k_T has the units of J/s·m·K or W/m·K. This expression is usually called **Fourier's Conduction Law**, after J. B. J. Fourier (1807). It provides the rate of flow of heat across an area A in a perpendicular direction in terms of the temperature gradient ($\Delta T/d$) in that direction. For a given ΔT, the greater d is, the lower Q/t is (the thicker your clothing is on a winter's day, the lower the rate of heat loss). When the human body is exposed to cold, the surface blood vessels contract, and that contraction goes on deeper and deeper, providing a thicker (larger d) layer of low k_T insulation as needed.

In the steady state, the amount of heat flowing into a conducting system traverses it and emerges; there is no loss or gain of energy, no storage. What that suggests (for Q/t = constant) is that regions of low conductivity correspond to large temperature gradients and often to large temperature changes. A thin layer of insulating material generally experiences a large change in T across it—that's what a blanket does.

When you awake on a chilly morning, miss the rug and step onto the "cold" bathroom floor, it's a lesson in conductivity. The floor and the rug must have the same temperature—they have been in equilibrium all night. But they certainly don't feel equally warm. The floor tiles have about 10 times the conductivity the rug has and will draw a great deal of heat per second from your warmer feet. That loss of heat is what produces the perception of coldness more so than the temperature difference—we are not very good thermometers.

Streaks of snow remain on the roof where the shingles stay cold, above the overhang and the thick roof beams. This roof is obviously uninsulated; heat flows up from the warm room below, passing out between the beams.

THERMAL ENERGY

Thermal energy is the disordered KE of any group of particles, usually the atoms constituting bulk matter. **Temperature** is essentially a measure of the concentration of thermal energy (p. 457). **Heat** (Q) is the thermal energy transferred, via atomic collisions, from a region of high temperature to a region of lower temperature (p. 459). These definitions are in Section 13.1 (Heat and Temperature)—they are basic and should be mastered before continuing.

In Section 13.2 (Quantity of Heat) we learn how to quantify heat. The amount of heat that must be added to raise the temperature of 1 gram of water 1 degree Celsius—from 14.5°C to 15.5°C— is a calorie. And then in Section 13.3 (The Mechanical Equivalent of Heat) we find that heat is energy and

$$4.186 \text{ J} = 1 \text{ cal} \qquad \text{or} \qquad 1 \text{ kcal} = 4186 \text{ J}$$

Study Example 13.2.

The **Zeroth Law of Thermodynamics** maintains that *when two systems are at the same temperature as a third*, *they are all three at the same temperature as each other* (p. 462)). Two systems are in thermal equilibrium only when their temperatures are equal.

In general, the **specific heat** (c) of a substance is the number of joules of heat that must be added to raise the temperature of 1 kg of the material 1 K; thus

$$Q = cm\,\Delta T = cm(T_f - T_i) \qquad [13.1]$$

where the SI units of c are J/kg·K. Water stands out for its high specific heat capacity: as liquid (4.2 kJ/kg·K, or 1.0 kcal/kg·K), as ice (2.1 kJ/kg·K, or 0.50 kcal/kg·K), and as steam (2.0 kJ/kg·K, or 0.48 kcal/kg·K). There are very few equations in this chapter and each is important. This material is treated in Section 13.4 (Specific Heat: Calorimetry), and you should understand Examples 13.3 − 13.5.

CHANGE OF STATE

The amount of heat that must be added to a kilogram of any solid at its melting temperature in order to liquefy it is the **heat of fusion** (p. 467). For water $L_f = 334$ kJ/kg ≈ 80 kcal/kg. Changing the phase of a mass m requires the addition (or removal) of an amount of heat Q:

$$Q = \pm mL_f \qquad [13.3]$$

The **heat of vaporization** (L_v) is the amount of thermal energy required to evaporate 1 kilogram of a liquid at a constant temperature, thus

$$Q = \pm mL_v \qquad [13.4]$$

The heat of vaporization of water at 100°C is 2.259×10^3 kJ/kg $\approx$ 540 kcal/kg. These changes of state are treated in Sections 13.5 (Melting and Freezing) and 13.6 (Vaporization). The student should study Examples 13.6 to 13.8, where it all comes together.

THE TRANSFER OF THERMAL ENERGY

Objects transfer thermal energy to and from the environment in three basic ways: **radiation**, **convection**, and **conduction**. The amount of heat conducted per second through a uniform slab or rod of material is

$$\frac{Q}{t} = k_T A \frac{\Delta T}{d} \qquad [13.5]$$

where k_T is the **thermal conductivity**. Example 13.9 in Section 13.10 (Conduction) illustrates how to apply this equation.

Before attempting any of the problems, reread the appropriate sections and then study the Suggestions on Problem Solving.

Read all the Suggestions on Problem Solving and go over the Examples in the text. **Look at the CD WARM-UPS, then study the WALK-THROUGH EXAMPLES.** Do the odd-number Multiple Choice Questions in the textbook and check your answers. You should then be ready to try the problems. Do all the odd I-level Problems first. The complete solutions for many of them (those with boldfaced numbers) are provided. Once you feel confident, go on to the II-level Problems.

Key Terms

thermal energy	heat of fusion
temperature	vaporization
radiant energy	evaporation
heat	heat of combustion
calorie	heat of vaporization
kilocalorie	boiling
mechanical equivalent of heat	steam
Zeroth Law	superheated steam
specific heat	thermal radiation
heat-out	convection
heat-in	conduction
calorimeter	thermal conductivity
heat of combustion	temperature gradient
melting	Fourier's Conduction Law

Discussion Questions

1. Discuss the physics involved in making a soft piece of steel red hot by hammering it. Will that work as well for hard steel?

2. Imagine that you have put a hot pot of soup in the refrigerator and Fig. Q2 is a plot of its subsequent temperature versus time. Assuming that radiation is negligible, explain the main features of the curve.

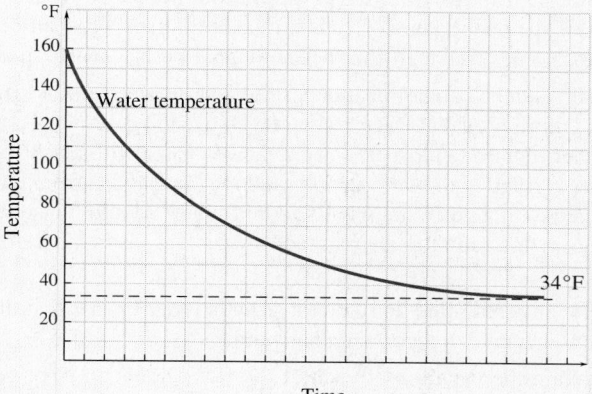

Figure Q2

3. The rate of energy consumed by the human body in the process of doing something, divided by the area of the body, is found to be fairly constant for different weights and body types; it's called the *metabolic rate* (usually expressed in kcal/m²·h). Measurements show that the metabolic rates associated with bicycling and shivering are about equal at 250 kcal/m²·h. Considering that muscles are only about 20% efficient, the rest being lost as thermal energy, compare these activities from an energy perspective. What function does shivering serve?

4. Imagine that we place a wooden rod and a brass rod, which are otherwise identical, in a freezer. Which will cool faster? Compare what they would feel like after being in the freezer for a few hours. Explain.

5. Why are steel saucepans and tea kettles sometimes covered with copper on their outside bottoms? Why are high-quality saucepans thick and tea kettles thin-walled?

6. EXPLORING PHYSICS ON YOUR OWN: Wrap an ice cube in aluminum foil (bright side out) and another in black cloth and place them both in the sunshine. What will probably happen and why?

7. Having evolved an expression for the rate of heat transfer via conduction, see if you can do the same thing for convection. Explain each assumption.

8. Ordinary tungsten incandescent light bulbs are evacuated. Explain why, and discuss the effect it has on the amount of power consumed and the amount of heat wasted. [*Hint*: Why do light bulbs often get blacker inside the longer they are used?]

9. Is it likely that you will feel more or less comfortable in a low-humidity environment at 80 °F than in a high-humidity one at 70 °F? Explain. Why do many people become "flushed" when they get overheated?

10. Suppose you had to remove a large hot pot from the stove and to protect your hands you could use either of two cloth towels that are identical, except one is dry and the other is thoroughly wet with cool water. Which would you use and why?

11. EXPLORING PHYSICS ON YOUR OWN: Place some water in a metal ice cube tray in a freezer (or outside if it's below 0 °C) and periodically observe what's happening. Where does the ice first begin to grow? Why is there often a region of water at the center of the cube? Now try the same thing with two trays, one with warm water and one with cool. Explain how it might be possible for the warm water to freeze first.

12. Most modern automobile cooling systems operate at a gauge pressure of 1 atm. The cooling fluid is usually a 50% mixture of water and ethylene glycol. What purpose does the ethylene glycol serve as regards the cooling process? Why is the system pressurized?

13. Figure Q13 depicts an experiment on conductivity performed by Forbes in 1864. Two metal containers at fixed temperatures of 100 °C and 0 °C are connected to each other by a metal rod. In the first case, the rod is exposed; in the second, it's insulated. Steady-state temperature readings are made all along the rod and the values plotted as the two *T*-versus-*d* curves in the diagram. Interpret the significance of the arrows and explain the curves. What is the meaning of the slope at each point on each curve? Does what is happening agree with the picture of heat as an incompressible liquid?

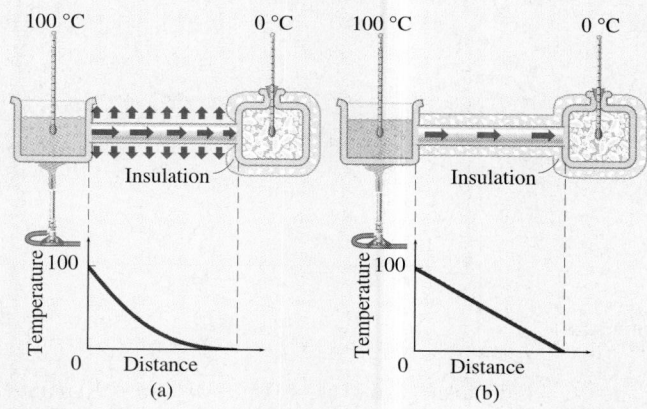

Figure Q13

14. Given a wet shirt hanging on a clothesline, what will determine the rate at which the water evaporates from it? Why will it dry faster in the sunshine?

15. Figure Q15 shows a demonstration devised by the English physicist John Tyndall. Four 1-kg cylinders made respectively of aluminum, iron, copper, and lead, each with the same cross-sectional area, are heated to 100 °C in boiling water. They are then all

placed upright upon a block of paraffin. Describe what happens and why.

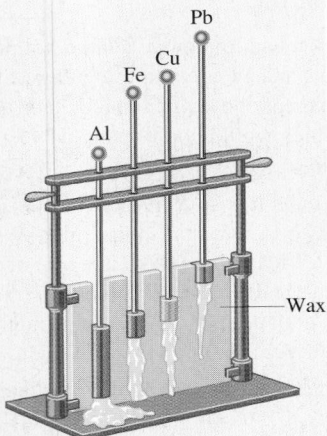

Figure Q15

16. How do storm windows work? Should they be sealed tight or left open a little? It's a hot, sunny day and you have the central air conditioning running full blast. Should you lower the blinds and draw the curtains?

17. If your goal is to keep warm in a very cold climate, should you wear a fur coat with the hair inside or out? Explain. If your goal is to keep cool in a hot environment where drinking water is plentiful, what should you wear outdoors? What if water is scarce?

18. We saw earlier that when a stream of air flows over a body, there is usually a thin layer in contact with the surface that is at rest. This same effect should exist when a flame carries hot gas over the bottom of a tea kettle. Explain how it is possible to boil water held in a paper milk container that is placed on an open flame. Why should you avoid turbulence in the hot air?

19. Imagine that you are floating around in space one or two hundred thousand miles from Earth with a thermometer. What will determine the temperature of your suit (assuming no active air conditioning), and what will the thermometer floating nearby read?

20. Leslie's cube is a hollow copper box whose outer surfaces are finished differently. One face is smooth, shiny, and gold-plated; one is roughened with sandpaper; and another is coated with black soot. The box is filled with boiling water, and the emanations from it are examined with a detector. What will be found? Should household heating system radiators be painted shiny silver, as is most often done?

21. Knowing that birds don't sweat, why is it reasonable that they have a higher body temperature than humans?

22. Explain the meaning of each of the Key Terms on page 480.

Multiple Choice Questions

The first eight questions below refer to Fig. MC1 which is a plot of temperature versus the input of thermal energy for a 0.5 kg sample of a strange blue metal taken from the fictitious planet Mongo. It is observed that at T = 0 °C the material is solid and that it does not sublimate.

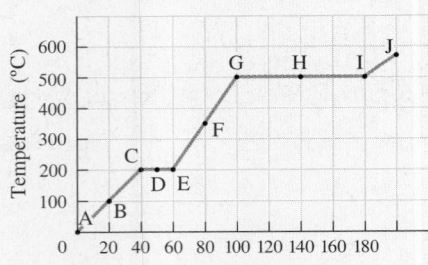

Figure MC1

1. The metal is completely solid at   (a) only point-*B*   (b) point-*D* and point-*H*   (c) only point-*F*   (d) point-*E*   (e) none of these.

2. The metal begins to vaporize at   (a) point-*G*   (b) point-*C*   (c) point-*E*   (d) point-*I*   (e) none of these.

3. At what temperature does the material melt?   (a) 300 °C   (b) 500 °C   (c) 200 °C   (d) 600 °C   (e) none of these.

4. At what temperature does the material boil?   (a) 300 °C   (b) 500 °C   (c) 200 °C   (d) 600 °C   (e) none of these.

5. At what minimum temperature is the material completely vaporized?   (a) 300 °C   (b) 500 °C   (c) 200 °C   (d) 600 °C   (e) none of these.

6. How much energy must be removed from the sample, keeping it in the liquid state, to lower its temperature from 500 °C to 200 °C?   (a) 100 kJ   (b) 80 kJ   (c) 60 kJ   (d) 40 kJ   (e) none of these.

7. Roughly how much thermal energy must be added to the sample, to raise its temperature from 199.999 °C to 500.001 °C?   (a) 140 kJ   (b) 120 kJ   (c) 180 kJ   (d) 220 kJ   (e) none of these.

8. Roughly what is the specific heat (in units of kJ/kg·K) of the sample in its liquid state?   (a) 2   (b) 0.08   (c) 0.13   (d) 0.27   (e) none of these.

9. Two bodies that are not initially in thermal equilibrium are placed in intimate contact. After a while the   (a) temperature of the cooler one will rise the same number of kelvins as the temperature of the hotter one drops   (b) amount of thermal energy contained by both bodies will be equal   (c) specific heats of both bodies will be equal   (d) thermal conductivity of each body will be the same   (e) none of these.

10. An open beaker of pure water is gently boiling at atmospheric pressure. A thermometer held deep in the water will likely read a temperature   (a) equal to 100 °C   (b) a little less than 100 °C   (c) a little greater than 100 °C   (d) equal to 212 °C   (e) none of these.

11. Two blocks of aluminum, one having a mass of 1.0-kg, the other having a mass of 2.0-kg, are in thermal equilibrium with a third block of brass at 100 °C. The two aluminum blocks are at temperatures, respectively, of   (a) 100 °C and 50 °C   (b) 50 °C and 100 °C   (c) 100 °C and 100 °C   (d) 200 °C and 100 °C   (e) none of these.

12. Referring to Question 3, the ratio of the net thermal energy of the 1.0-kg aluminum block to that of the 2.0-kg block is   (a) $\frac{1}{2}$ (b) 4   (c) 2   (d) $\frac{1}{4}$ (e) none of these.

13. An open pot of water is boiling on a gas stove when someone raises the flame. The result will be   (a) a substantial increase in the temperature of the water   (b) a tiny decrease in the rate of evapo-

ration   (c) an increase in the rate of boiling   (d) an appreciable increase in both the rate of boiling and in the temperature of the water   (e) none of these.

14. Figure MC14 shows a glass flask in which water was boiled until the steam drove out much of the air. Thereafter it was tightly corked and inverted. Cold water was then sprinkled over the spherical bottom, and the water inside   (a) froze because of the evaporation   (b) boiled because of the condensation of steam   (c) froze because of conduction   (d) boiled because of convection of thermal energy   (e) none of these.

Figure MC14

15. Everyone who has ever walked barefoot on a beach in summer has noticed how fast the dry sand gets hot in the morning. That's because sand has a   (a) light color   (b) fairly low specific heat   (c) high thermal conductivity   (d) great deal of convection   (e) none of these.

16. Figure MC16 shows a test tube containing water and some ice (held down by a weight). As shown, the water at the top is boiling. Is that possible and, if so, why? (a) yes, because water has a high heat of vaporization   (b) no, because water has a low specific heat capacity   (c) yes, because water has a low thermal conductivity   (d) no, because water has a low heat of fusion   (e) none of these.

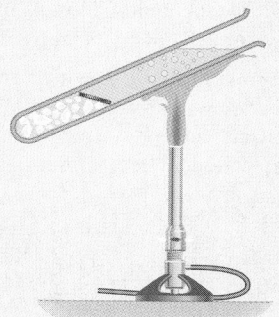

Figure MC16

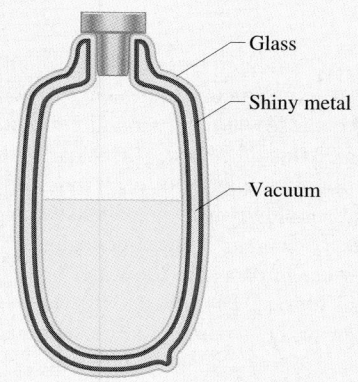

Glass

Shiny metal

Vacuum

Figure MC17

17. Figure MC17 depicts a Dewar flask, or thermos bottle. The space between the two glass walls is evacuated (via a tube that is then sealed). This is done to   (a) decrease conduction   (b) decrease the specific heat capacity   (c) decrease radiation   (d) decrease weight   (e) none of these.

18. Referring to the Dewar flask of Fig. MC17, why is the outer surface of the inner wall covered with a shiny layer of metal?   (a) to decrease conductivity   (b) to decrease radiation to and from the inner chamber   (c) to increase convection currents in the glass   (d) to increase the loss of coldness   (e) none of these.

19. Given that a thermos bottle (Fig. MC17) contains a cold drink, what function does the metal film on the inner surface of the outer wall serve?   (a) it keeps heat from being conducted inward beyond that point   (b) it looks nice shiny   (c) it reflects radiant energy incident from outside   (d) it cools the inner chamber by emitting radiation   (e) none of these.

20. Smoke is drawn out of a fireplace by an updraft that is initiated by wind blowing across the top of the chimney. To help matters along, it is recommended that a burning piece of newspaper be placed up the exhaust hole in the fireplace for a minute or two before lighting the fire. This is done in order to   (a) increase the conductivity of the air   (b) increase the updraft via convection currents   (c) decrease the specific heat of the air   (d) warm up the air so it burns better   (e) none of these.

21. Suppose you pour a hot cup of coffee and the phone rings so you can't drink it. To keep it hot as long as possible   (a) add the cool milk and sugar immediately   (b) don't add the cool milk until you're ready to drink it   (c) stir it once gently without the milk, but don't add the sugar   (d) don't add the milk and use a black mug if possible   (e) none of these.

22. Which of the following materials has the highest thermal conductivity?   (a) wood   (b) water   (c) air   (d) gold   (e) none of these.

23. The purpose of a cover on a soup tureen is to   (a) increase conductivity   (b) decrease convection   (c) increase radiation   (d) decrease conduction   (e) none of these.

24. The thermal effect of dissolving sugar in a cup of hot tea   (a) is to lower the temperature   (b) is to raise the temperature   (c) is to leave the temperature unchanged   (d) cannot be predicted   (e) none of these.

25. Metal pots are often made shiny on the outside especially on the top and side, and that makes sense thermally because   (a) this conducts heat better   (b) this radiates less energy out from the pot   (c) this lowers the loss to conduction   (d) this appreciably decreases convection losses   (e) none of these.

26. Fanning yourself on a hot day (say, 100°F, or 38°C) cools you by   (a) increasing conductivity   (b) decreasing the mean free path of the air   (c) increasing the radiation rate of the skin   (d) increasing the evaporation of perspiration   (e) none of these.

27. Franklin's experiment (Fig. MC27) consists of a glass bulb connected to a wide, initially open tube. The whole thing is heated until steam drives the air from the cylinder, whereupon it is sealed.

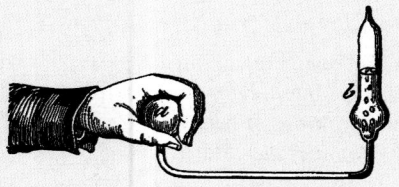

Figure MC27

Then, by holding the bulb in the hand, the water is warmed and boils in the tube because   (a) the temperature can quickly reach 100 °C   (b) the water is a good conductor and the heat travels to the cylinder where it is concentrated   (c) the vapor pressure in the cylinder is quite low   (d) radiation is trapped in the connecting tube and brought to bear on the cylinder   (e) none of these.

28. Contemporary radiators of all sorts consist of the heated body—a pipe or transistor, for example—to which are affixed a series of spaced flat vanes, often painted black. The function of the vanes is to   (a) increase the convection currents and so decrease conduction   (b) increase the heated surface area to increase radiation and convection   (c) increase the ability to block air currents thereby increasing conduction   (d) conduct more heat to the stagnant air layer and so increase radiation   (e) none of these.

For more Multiple Choice Questions with answers click on WARM-UPS in CHAPTER 13 on the CD.

Suggestions on Problem Solving

1. Common practice is to represent zero degrees Celsius as 0 °C with no reference to significant figures. *Take 0°C to have the same number of significant figures as the other temperatures given.*

2. Because we are generally dealing with temperature changes, we need not transform the data from C° to K. When using the equation $Q = cm\Delta T$, you can express the temperatures in Celsius degrees or kelvins, whichever is more convenient. Still, if you use Celsius and calculate T_f, the answer will be in Celsius.

3. Be careful with calculations when mixing various substances in *different states* (for example, water and ice, or steam and water) in a calorimeter. Whenever there is a change of state, it is entirely possible that not all of the substance will actually make the transition. The governing equation $-Q_{out} = Q_{in}$ is correct, but saying that $Q_{in} = mL_f + cm\Delta T$ assumes that the entire mass m has melted, and that need not be the case. We must compute the thermal energy that would be made available on descending to the transition temperature and then check to see if that is enough to accomplish the change of state for the whole mass m. If it isn't, only part of the ice melts (or part of the steam condenses, or whatever).

4. Another common error is to use L_v when you should be using L_f, and vice versa. Double check that you have made the correct substitution before carrying out the calculation.

5. The analyses of many of the mixing problems begin with $-Q_{out} = Q_{in}$. The minus sign is there because the object providing the thermal energy decreases in temperature: $\Delta T = (T_f - T_i)$ is negative.

6. In the old cgs system of units, the specific heat capacity of water is 1 cal/g·C°, which in mks is equivalent to 1 kcal/kg·C°. *Watch out for the units*; those are kilocalories, kilojoules, and kilograms.

Problems ✦ Coordinated Problems ✦ Progressive Problems ✦ Solutions

STUDY GUIDE **1. Coordinated Problems:** The three problems within each magenta-colored grouping are solvable in similar ways. Note that the first of these always has a hint; moreover, its solution is provided in the back of the book. *Work out each of these sets; they'll strengthen technique and build confidence.* **2. Progressive Problems:** The problems introduced in blue unfold step-by-step carrying along the analysis in a more suggestive way than is customary. *Work out all of these; they'll guide you through the analytic process and help develop problem-solving skills.* **3. Worked-Out Solutions:** Studying worked-out solutions is an important part of learning how to solve problems. Accordingly, additional *solutions* to a number of model problems are given below. *Make sure you understand each of them before you go on to the next problem.* **4.** Also provided in the back of the book are the *Answers* to all odd-numbered problems, as well as worked-out *solutions* to those with boldface numbers. Problem numbers in italic indicate that a solution appears in the Student Solutions Manual.

SECTION 13.2: QUANTITY OF HEAT

SECTION 13.3: THE MECHANICAL EQUIVALENT OF HEAT

1. [I] A cup containing 100 g of pure water is placed over a flame until the liquid rises from room temperature (20 °C) to 40 °C. How many calories of heat were transmitted to the liquid? [*Hint: Use the definition of the calorie.*]

2. [I] Heat in the amount of 100 kcal enters 10.0 kg of water at 23.0 °C. What will be the resulting temperature change?

3. [I] Ten grams of water at 20.0 °C are placed in a freezer. How many calories will have to be removed in order to lower the temperature of the liquid to 0°C?

4. [I] One thousand cubic centimeters of pure water at 100 °C loses 1.00 kcal to its immediate environment. By how much will its temperature change?

5. [I] The cooling system of an engine contains 30 liters of water. If, during a brief run, the water carries off 500 kcal, by how much will its temperature rise?

6. [I] How much thermal energy must be added to 5.0 kg of pure water at 0 °C to raise the temperature to 37 °C?

7. [I] If 15 kcal of heat are added to a system, how many joules of energy are added?

8. [I] A wrapped sugar cube is labeled to produce 92 kJ. How many kilocalories is that?

9. [I] One hundred grams of water at 100 °C are added to 100 g of water at 20.0 °C. What will be the final temperature of the mix?

10. [I] In Fig. P10 which graph best describes the variation in temperature as an amount of thermal energy Q is added to a fixed quantity of water? Assume there are no phase changes. Explain your selection. Where appropriate, discuss what's wrong with each diagram.

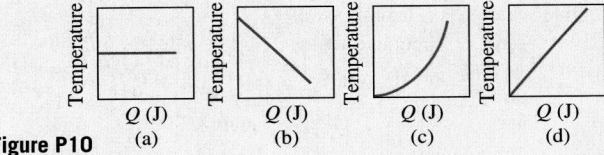

Figure P10

11. [I] THIS PROBLEM EXPLORES THE MECHANICAL EQUIVALENT OF HEAT. In one of his most famous experiments Joule (Fig. 13.2) simultaneously dropped two 14-kg masses, letting each fall through a distance of 2.0 m, and he repeated that 20 times. (a) What was the change in the gravitational potential energy of each mass on falling the 2.0 m? (b) As the masses slowly fell, the paddle wheel churned up the water in an insulated container. Assuming negligible loses, how many joules in total were imparted to the water? (c) How much thermal energy, in kilocalories, was added to the water? (d) The device contained 7.0 kg of water. Assuming all the energy went into the water, what was the resulting change its the temperature?

12. [I] The human brain operates with an average power consumption of about 20 W. What is that in kcal/h?

13. [I] Alcohol has a rather high caloric value of about 7 kcal/g, which is completely converted into energy in the body. Because it contains no vitamins or minerals, people who drink to excess often have nutritional problems. How much energy is liberated by the body (as thermal energy) via one shot or drink, about 30 g?

14. [I] A person just standing around consumes energy at a rate of about 100 kcal/h. How many watts is that?

15. [I] The Frenchman Gustave Hirn, in the 1800s, was the first per-

son to make a quantitative study of the thermal energy involved in collisions. He crushed a lead ball with a steel cylinder of known mass dropping through a known height and quickly transferred it to a calorimeter. Hirn essentially found that dropping a 423-g mass a distance of 1.0 m produced 1.0 cal. Explain this result and comment on its accuracy.

16. [II] THIS PROBLEM EXPLORES THE TEMPERATURE CHANGE WHEN DIFFERENT QUANTITIES OF WATER ARE MIXING. A sample of water at an initial high Celsius temperature T_{Hi} has a mass m_H. It's to be mixed with a sample of water at an initial low Celsius temperature T_{Li} having a mass m_L. Once combined, a new final Celsius temperature T_f is reached. (a) Write an expression for the heat given up by the high-temperature sample. (b) Write an expression for the heat gained by the low-temperature sample. (c) Taking the heat that flows out of a sample to be negative, and the heat that flows into a sample to be positive, show that

$$T_f = \frac{m_H T_{Hi} + m_L T_{Li}}{m_L + m_H}$$

17. [II] A quantity of cold water at 6.90 °C weighing 20.0 N is mixed with a sample of warm water at 36.9 °C weighing 40.0 N, what will be the final temperature of the mix?

18. [II] A 1000-kg car traveling at 60 km/h is brought to a stop by its braking system. How much thermal energy is evolved in the brakes? Friction with the ground does no work unless there is slipping.

19. [II] Water at 20 °C flows through a 4.0-kW heater at a rate of 1.5 liter/min. Assuming no losses, at what temperature will it emerge? [*Hint: Begin with Eq. (13.1) and divide both sides by time. Watch out for units, especially those of c.*]

20. [II] A refrigeration system removes 10.0 W of thermal energy from the water flowing through it at a rate of 60.0 cm³/min. By how much will the temperature change each second?

21. [II] A constant stream of water flowing at 1.00 kg/s cools an engine. The water enters the engine at 15.0 °C and leaves at 85.0 °C. How much thermal energy is removed each minute? Give your answer in joules.

22. [II] One gram of carbohydrate burned in a calorimeter liberates 4.10 Cal, and it is estimated that in the body about 98% of that energy finds its way to the cells. If a woman consumes 150 Cal/h jogging, how long must she exercise to burn off 100 g of carbohydrate?

23. [II] Suppose we could convert 1.00 kcal completely into work, thereby raising a 1.00-kg mass in a uniform gravitational field ($g = 9.807$ m/s²). How high would it get?

> **SOLUTION:** First we need to determine the energy available in joules. That energy goes into *gravitational*-PE where $PE_G = mgh$. Remembering that 1.00 kcal = 4.186 kJ, this much *gravitational*-PE is equivalent to
>
> $$PE_G = mgh = 4.186 \text{ kJ} = (1.00 \text{ kg})(9.807 \text{ m/s}^2)h$$
>
> and $h = (4.186 \text{ kJ})/(9.807 \text{ m/s}^2) = 0.427 \text{ km}$

24. [II] A person who has a surface area of 1.65 m² consumes energy at about 70 kcal/h while resting. Determine the corresponding metabolic rate (see Discussion Question 3) and compute the approximate amount of thermal energy that will be generated in 2 h.

25. [II] A 50-W immersion heater is placed in a beaker containing 1.0 kg of water at 20 °C. How long will it take to raise the temperature to 100 °C? Neglect any heat losses to the air or beaker.

SECTION 13.4: SPECIFIC HEAT: CALORIMETRY (NO CHANGE IN STATE)

26. [I] How much energy does it take to raise the temperature of 500 g of mercury from −19 °C to 61 °C? Give the answer in both joules and kilocalories.

27. [I] How much thermal energy must be removed from 30 g of tin with a specific heat of 0.060 kcal/kg·K to drop its temperature from 373 K to 283 K? [*Hint: Use the definition of Q and the specific heat of tin.*]

28. [I] How much heat does it take to raise the temperature of 0.40 kg of aluminum from 50 °C to 60 °C given that it has a specific heat of 0.217?

29. [I] If a bar of pure copper is found to absorb 16 kJ in the process of having its temperature raised from 293 K to 353 K, what is the mass of the metal?

30. [I] A 1.00-kg block of ice at −10.0 °C is to be raised to 0.00 °C. How much heat would be necessary? Give your answer in joules.

31. [I] How much energy will be given out upon the complete combustion of 1 gallon of gasoline? One gallon equals 3785 cm³, and the density is 0.68×10^3 kg/m³.

32. [I] Two beakers of water, one at 15 °C and the other at 95 °C, each contain 1.52 kg of the liquid. If the water is combined, what will its final temperature be, assuming no heat losses from the liquid to the environment?

33. [I] THIS PROBLEM EXPLORES HOW TEMPERATURE CHANGES WITH THE INPUT OF POWER. One way to produce thermal energy in underdeveloped countries that are poor in combustable fuels is to collect sunlight. Suppose a 1.5-m diameter aluminum foil reflector concentrates 510 W of radiant energy onto half a liter of water. (a) Assuming all the energy is imparted to the water, how many joules flow in per second? (b) How much heat is needed to raise the temperature of the water 1.0 °C? (c) How long will it take to raise the temperature of the water 1.0 °C? (d) How long will it take to raise the temperature of the water from 20 °C to 100 °C? Assume any energy losses are negligible.

34. [I] In the winter of 2001 a 13-month-old Canadian girl wandered outside in −20 °C weather wearing only her diaper. When she was found, virtually frozen in the snow, her core body temperature had fallen from its normal value of 37 °C to 15 °C. Despite the fact that the 9-kg toddler had no circulation for some time, she survived without brain damage. How much heat did she lose to the frigid environment? Take the specific heat of the human body to be 3.47 kJ/kg·K.

35. [I] THIS PROBLEM EXPLORES HOW TEMPERATURE CHANGES WITH THE ADDITION OF HEAT. A 1.82-kg copper teapot contains 1.90 kg of water and both are at 20 °C. We want to find out how much heat it will take to raise the temperature of the pot and its contents up to 100 °C. (a) What are the values of the specific heats of both the water and the copper? (b) How much heat (in joules) is need to bring the water to 100 °C? (c) How much heat (in joules) is needed to bring the pot to 100 °C? (d) How many kilocalories must be provided by the stove?

36. [I] THIS PROBLEM EXPLORES THE NOTION OF HEAT CONTENT. For over a thousand years it was common to defend a castle by pouring

hot oil (with a specific heat of ≈ 2.1 kJ/kg·K and a boiling point of ≈ 300 °C) down on the heads of the unrelenting invaders. We want to explore the military advantages of oil over water. Assuming the liquid is as hot as possible, and skin temperature is 34 °C, (a) what is the temperature change that would occur for 1.0 kg of each liquid on encountering some unfortunate solder? (b) How much thermal energy would, at most, be transferred per kilogram of each liquid to that poor guy below? (c) So was it worth while to go through all the effort of using oil? Explain. (d) Which property was more important, a high specific heat or a high boiling point?

37. [I] It is found experimentally that by adding 100 g of iron, with a specific heat capacity of 0.113 kcal/kg·K at 80 °C, to a quantity of water at 25 °C, the new equilibrium temperature is 30 °C. What is that amount of water?

38. [I] Suppose we combine 100 g of aluminum with a specific heat capacity of 920 J/kg·K at a temperature of 495 °C with 99.9 g of water at 0.010 °C. What will be the final temperature?

39. [I] A quantity of aluminum shot ($c = 910$ J/kg·K) at 473 K is mixed with 4.95 kg of water at room temperature, and in a little while, the whole thing comes to equilibrium at 300 K. How much aluminum was used?

40. [I] A 340-g glass mug at 20.0 °C is filled with 250 milliliters of water at 96.0 °C. Assuming no losses to the external environment, what is the final temperature of the mug?

41. [II] According to the *Shooter's Bible*, a 357-magnum bullet has a kinetic energy of 540 ft·lb at 50 yards from the pistol. If it strikes and comes to rest inside a 1.0-kg block of wood (with a specific heat capacity of 1700 J/kg·K), by how much would the block's temperature change? The bullet is lead and weighs 158 grains (1.0 grain is equivalent to 0.064 8 g).

42. [II] In a time of 10.0 s, 20.0 kg of steam enter an engine at 120 °C and leave it at 100 °C. What is the rate (in watts) at which energy is transferred to the engine?

43. [II] The 200.5-g copper cup in a calorimeter contains 118.1 g of olive oil with a specific heat of 0.471 kcal/kg·K at room temperature (20.0 °C). An 84.0-g block of glass at 300 °C is dropped into the oil. Determine the final temperature of the cup, oil, and glass. [*Hint:* Remember that $Q_{in} = -Q_{out}$ and use the definition of Q.]

44. [II] Suppose we have 0.30 kg of unknown material. It is heated to 371 K and placed into 0.60 kg of room-temperature (20 °C) water. The water is in a copper cup of mass 0.14 kg surrounded by insulating material, and the final equilibrium temperature is 295 K. Compute the specific heat.

45. [II] A 20-g dried sample of food is placed in an aluminum bomb calorimeter and burned completely. The device consists of an aluminum chamber (0.50 kg) containing the sample, a surrounding water bath (2.00 kg), and an aluminum cup (0.60 kg) holding the water. A thermometer in the water shows that the calorimeter changes temperature from 20 °C to 32 °C; how many kilocalories did the food make available? Neglect the mass of the ash.

46. [II] Let's model the human body as being composed of 75% water and 25% protein. Given that the specific heat of protein is 0.4 cal/g·C°, approximate the specific heat capacity of the body in units of kJ/kg·K.

47. [II] Due to the presence in the body of both minerals (in the

bone) and fat, its specific heat is closer to 0.83 cal/g·C° than the value computed in Problem 46. Given that a 150-lb human at a temperature of 98.6 °F must be dropped to a hypothermic value of 93 °F, how much heat must be removed?

48. [II] A 1-kg sample of a material is surrounded by insulation material as in Fig. P48. An electric heater is inserted into a hole in the sample, as is a thermometer. Given that 15 W are supplied by the heater for 16.667 min and the thermometer changes temperature by 16.67 K, what is the specific heat of the sample? Ignore any losses.

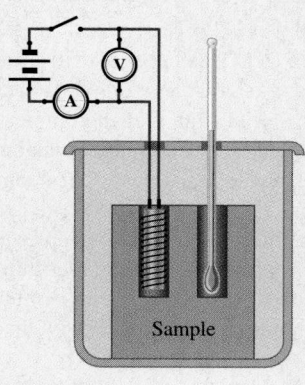

Figure P48

49. [II] The specific heat of a 0.210-kg sample is to be determined. It's first heated to 373 K and then placed into a bath of 0.082 kg of water at 288 K, which is held in a 0.121-kg copper cup at that same temperature. If the final temperature of the system is 310 K, what is the specific heat capacity of the sample?

SOLUTION: When the hot sample(of mass m_x and specific heat c_x) is immersed, the heat-out of it (Q_o) equals the heat-in to the calorimeter (Q_i):

$$Q_o = -c_x m_x \Delta T_x = Q_i = c_w m_w \Delta T_w + c_c m_c \Delta T_c$$
$$-c_x (0.210 \text{ kg})(310 \text{ K} - 373 \text{ K}) = (4.186 \text{ kJ/kg·K})(0.082 \text{ kg})(310 \text{ K} - 288 \text{ K}) + (0.39 \text{ kJ/kg·K})(0.121 \text{ kg})(310 \text{ K} - 288 \text{ K})$$
$$c_x (13.230 \text{ kg·K}) = 7.551\,5 \text{ kJ} + 1.038\,2 \text{ kJ}$$
$$c_x = 0.65 \text{ kJ/kg·K}$$

50. [II] A handful (405 g) of lead shot is removed from boiling water and dropped into a 99.0-g glass beaker containing 198 g of water at 20 °C. If the shot and glass have specific heat capacities of 0.031 kcal/kg·K and 0.20 kcal/kg·K, respectively, find the equilibrium temperature of the system.

51. [II] A 70-kg person engaged in moderate physical activity produces thermal energy internally at a rate of 200 kcal/h. If all the cooling mechanisms become inoperable so there is no dissipation of this energy, how long will it take before the person collapses with a body temperature of 43 °C (109 °F)? (A body temperature of about 44 °C begins to cause irreversible protein damage.) [*Hint:* The specific heat of the body is about 3.5 kJ/kg·K.]

52. [II] How much excess heat is produced when a 65-kg person with the flu experiences a rise in temperature from 98.6 °F to 102 °F? Take the average specific heat of the human body to be 0.83 cal/g·C°.

53. [II] How much hard coal must be burned to raise the temperature of 1.00 kg of water from 0 °C to 100 °C?

54. [III] A house is heated electrically with a system that is 100% efficient; all the power consumed is converted into thermal energy via baseboard heaters. Suppose that it takes 2500 kW·h of energy per month to heat the place. How much wood would be needed to do the same job? Assume the wood stove, which must be exhausted to the outside, is 35% efficient. Given that a cord of wood is a stack of roughly about 1000 kg, how many cords would have to be burned each month?

55. [III] Figure P55 shows a device for measuring the specific heat of a solid rod of material by mechanical means. The rod is rotated, and the rope wrapped around it slides on its surface, producing a frictional drag. The crank is turned at a rate that takes all the tension off the rubber band holding the rope's far end. A thermometer is set in a hole in the

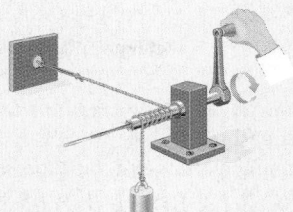

Figure P55

specimen that is separated from the rest of the system by thermal insulators. Explain how the apparatus works and how it is used to find c. If the load is 6.00 kg, the mass of the sample 250 g, its diameter 3.00 cm, the number of rotations 240, and the corresponding temperature increase of the rod is 12.0 K, determine its specific heat.

56. [III] In 1899, Callendar and Barnes devised a new method for measuring the specific heat of water as a function of temperature. It utilized a scheme (Fig. P56) in which only constant temperatures occurred that could be measured slowly and with great accuracy using platinum resistance thermometers. Water was made to flow at a constant rate through a tube that was thermally insulated by a vacuum jacket. The tube contained an electrical heating coil that generated a constant known amount of thermal power (P). In steady-state operation, the heat generated electrically is carried off by the water that enters the tube at a temperature T_i and leaves at T_o. Write an expression for the specific heat of water in terms of the experimental variables. Explain your reasoning.

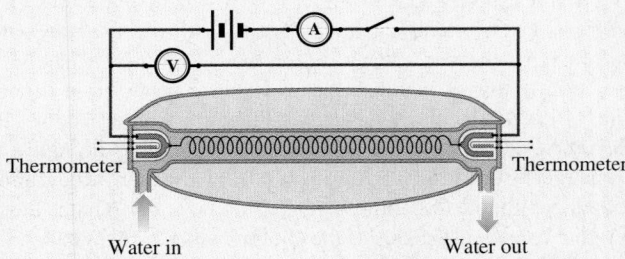

Thermometer   Thermometer

Water in   Water out

Figure P56

57. [III] How does the system in Problem 56 lose energy and thus introduce error? Suppose the power loss is P_l. Repeat the previous experiment at a new flow rate such that a mass of water m' passes through in time t, but also change the power input to P' so that T_o is the same. Show that the thermal losses can be eliminated from the equations, whereupon

$$c_w = \frac{(P - P')t}{(m - m')\,\Delta T}$$

Explain your reasoning.

58. [III] Marsh gas [that is, methane (CH_4)] has a heat of combustion of 212 kcal/mol. Show that this value agrees with Table 13.4. If 10 liters of it at STP are burned completely and

$$CH_4 + 2O_2 \rightarrow CO_2 + 2H_2O$$

how much energy is released in the process? What is the net energy released per molecule burned? The formation of methane gives off an amount of energy of $(-)17.9$ kcal/mol, whereas the formation of carbon dioxide and water releases $(-)94.0$ kcal/mol and $(-)68.3$ kcal/mol, respectively. The heat of formation of the elements in

their natural states is zero. From this fact, determine the heat of combustion of methane.

SECTION 13.5: MELTING AND FREEZING
SECTION 13.6: VAPORIZATION

59. [I] At what temperature (at a pressure of 1 atm) does the metal antimony exist as both a liquid and a vapor?

60. [I] What is the total temperature range over which liquid mercury exists at ordinary pressure?

61. [I] How much heat must be added to a 1000-cm^3 block of ice at 0 °C in order to melt it?

62. [I] How much heat must be removed from 1000 cm^3 of water at 0 °C in order to freeze it?

63. [I] Determine the thermal energy that must be added to a 200-g block of copper at 1356 K to melt it without raising its temperature. [See Table 13.6.]

64. [I] A layer of ice 2.0-mm thick covers the 0.75-m^2 windshield of a car. How much power must be supplied by a heater if the ice is to be melted in 2 min and the air temperature is 0 °C? Assume no losses.

65. [I] How much ice at 0 °C must be added to 0.50 kg of water at 20 °C if it is all to melt, leaving only water at 0 °C? (Do the problem using SI units and see Problem 66.)

66. [I] Redo Problem 65 using kilocalories rather than joules and compare your answers.

67. [I] Silver vapor at atmospheric pressure condenses at a temperature of 2466 K. How much thermal energy must be removed from 1.5 kg of silver vapor if it's to form a liquid at that temperature? [*Hint: You might need Table 13.6.*]

68. [I] Determine the thermal energy that must be added to 200 g of molten copper at 2336 °C to completely vaporize it.

69. [I] What is the final state of the system consisting of 500.0 g of liquid mercury at its boiling temperature given that 16.244 kcal is added and there are no losses?

70. [I] Roughly how much thermal energy is needed to melt a 3.00-kg block of silver originally at 950.8 °C? Where does most of that energy go?

SOLUTION: The silver has to be raised to its melting temperature (which from Table 13.6 is 960.8 °C) and then provided with 109 kilojoules for each kilogram that's to be melted. Thus

$$Q = c_s\,m\Delta T + mL_f$$
$$Q = (0.23 \text{ kJ/kg}\cdot\text{°C})(3.00 \text{ kg})(10.0 \text{°C}) + (3.00 \text{ kg})(109 \text{ kJ/kg})$$
$$Q = 334 \text{ kJ and most of it goes into melting the silver.}$$

71. [I] A substance is placed in a heating chamber, and thermal energy flows into it at a fixed rate. Given that the sample changes state in the process, which of the accompanying diagrams (Fig. P71) might best apply? Explain your answer indicating what's wrong with the other selections.

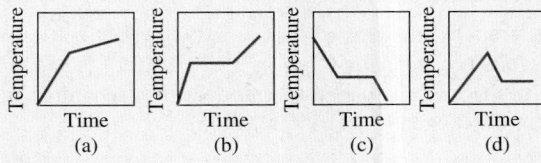

Figure P71

72. [I] A cup of water, 0.25 kg, at room temperature is set on the stove in an open pot to make some tea. The phone rings and you get distracted only to return to the pot as the last drops of water boil off completely. How much heat was transferred to the water?

73. [I] A 0.50-kg quantity of mercury, with an average specific heat capacity of 138 J/kg·K, at 240 K is to be brought to its boiling point and maintained at that temperature until completely vaporized. How much thermal energy must be provided?

74. [I] Compare the amount of heat it takes to raise an arbitrary amount of water from 0 °C up to 100 °C with the amount of heat needed to vaporize that water.

75. [I] Large air conditioners are often rated in tons. A 1-ton unit has a cooling capacity equivalent to changing 2000 lb of water per day at 0 °C into ice at 0 °C. Determine (to two significant figures) the thermal power removed from a building by a 1-ton air conditioner.

76. [I] Imagine that you have 2.7 g of ice at −12 °C and you want to turn it into steam at 134 °C, how much heat must you provide?

77. [II] THIS PROBLEM EXPLORES THE TRANSFER OF ENERGY FROM A SPRING TO A BATH. An insulated tank contains water and crushed ice. A vertical spring having an elastic constant of 500 N/m supporting a 0.5-kg brass mass is suspended in the bath and the whole system is in equilibrium. (a) What is the temperature of the bath, spring, and mass? (b) The spring is stretched 25.0 cm and released. How much elastic energy is imparted to the spring? (c) After the mass stops oscillating and equilibrium is reached, how much thermal energy has been imparted to the ice? Explain your answer. (d) How much ice will be found to have melted?

78. [II] A 0.100-kg lump of ice is floating in 900 cm³ of water in a cup at 0 °C. How much thermal energy must be added in order to vaporize all the water in the cup? Assume there are no losses. Where does most of the energy go?

79. [II] Suppose we combine 100 g of aluminum with a specific heat of 920 J/kg·K at a temperature of 495 °C with 99.9 g of ice at 0.00°C. What will be the final temperature?

80. [II] How much heat must be removed from 1500 g of steam at 423 K to change it into water at 303 K?

81. [II] A 1.00-cm-thick sheet of ice with an area of 2.00 m² covers a piece of sidewalk. Since it's a nice clear day, we simply sprinkle a thin layer of black soot on the ice exposed to the Sun's rays. The idea is to melt the stuff rather than going out to break it up with a chopper. If the air temperature is 0 °C and the solar energy per unit area per unit time impinging perpendicularly on the ice is 350 W/m², how long will it take before it melts? Assume 100% absorption and no losses.

82. [II] A 500-g lead mass is heated to 150 °C and placed on a block of ice at 0 °C. How much, if any, ice will melt?

83. [II] Knowing that the heat of vaporization of water at STP is 2492 kJ/kg, approximate the amount of energy required by one molecule to escape from the liquid.

84. [II] Compute the heats of fusion and vaporization of water in electron volts per molecule (1 eV = 1.602 × 10⁻¹⁹ J) and compare those with the energy needed to tear apart a water molecule into oxygen and hydrogen (2.96 eV/molecule).

85. [II] How much steam at 100 °C must be fed into 1.00 kg of water at 10 °C to bring it up to 50 °C?

86. [II] A lead bullet is heated up to 600 K and then fired at a stone wall, where it comes to rest in a completely molten state. Assuming no energy losses, compute its impact speed.

87. [II] A 10.0-g lead bullet at 23.0 °C slams into a stone wall and, squashing, comes to rest. Assuming no loss of energy to the environment, how fast must the bullet be traveling if it is to totally melt?

88. [II] A 10-kg block of ice at 0 °C falls, impacting in a way that brings it to rest with no loss of thermal energy. From what height should it drop if 0.010% of the ice is to melt?

89. [II] How many grams of steam at 110 °C must be added to 1 kg of ice at 0 °C to melt it without raising its temperature? Do this one in both kilocalories and joules and check your answer.

90. [II] THIS PROBLEM EXPLORES THE TRANSFER OF ENERGY VIA CONDENSATION. A 51.0-g aluminum calorimeter cup holds 252 g of water in equilibrium at 23.0 °C. We are going to bubble steam at 100 °C into the water raising its equilibrium temperature to 53.0 °C. (a) If m_s is the mass of steam introduced (all of which is condensed), write an expression for the amount of thermal energy provided (heat-out) when the steam condenses. (b) Write an expression for the amount of thermal energy provided (heat-out) when the "steam" drops to 53.0 °C. (c) Write an expression for the amount of thermal energy absorbed (heat-in) when the water rises to 53.0 °C. (d) Write an expression for the amount of thermal energy absorbed (heat-in) when the cup rises to 53.0 °C. (e) Assuming no losses, set the heat-out equal to the heat-in and solve for the amount of steam, m_s.

91. [II] THIS PROBLEM EXPLORES THE TRANSFER OF ENERGY VIA EVAPORATION. Power plants are often built next to rivers so they can get rid of excess thermal energy. When there isn't a river handy, they use cooling towers where large amounts of water are evaporated. The 1143-MW Callaway Nuclear Plant in Missouri cools about 585 000 gallons of water per minute (circulating through a heat exchanger) from 125.0 °F to 95.0 °F. (a) How many kilograms are cooled per minute (1.0 gallon = 3.785 liters)? (b) What is the temperature change in Celsius of the cooling water? (c) What's the net amount of thermal energy that must be removed from that water per minute? (d) Given that at that temperature the heat of vaporization is 2.38 MJ/kg, at what rate (in kg/min) should the plant evaporate water? (e) How much is that in gallons per minute?

Figure P91

92. [II] An unheated garage in New Jersey in the winter is maintained inside at 0 °C by the freezing of a tank of 1000 kg of water,

despite the fact that it's colder than that outside. The garage loses heat at a rate of 10^4 cal/min. How long will it be able to stay at 0 °C?

93. [II] A calorimeter contains 398 g of water in a 102-g copper cup at 5.1 °C. To this is added 40.5 g of crushed ice at −7.6 °C. What is the final temperature of the system?

94. [II] As little as 60 years ago, most people kept food in iceboxes that were simply well-insulated cabinets. A large block of ice, delivered roughly once a week, was placed inside above a drip pan. The remaining space was packed with food, the stuff to be colder kept nearer the ice. How does the system work? What's the equilibrium temperature inside the icebox? How much heat would a typical 100-lb block remove from the food before it melted?

95. [II] A block of ice at 0 °C is placed on a balance, and an equal mass of water at 80 °C is measured off. The water and ice are then combined in an isolated chamber. Determine the equilibrium temperature of the system and describe in detail what it consists of.

96. [II] How much wood would have to burn completely to provide enough energy (assuming no losses) to vaporize 1.00 kg of water boiling away at 100 °C?

97. [III] How much thermal energy must be added to 0.900 kg of water to change its temperature from 5.0 °C to 100 °C? If that same quantity of energy is added to 1.20 kg of silver at 5.0 °C, what will its final temperature be?

SECTION 13.10: CONDUCTION

98. [I] An iron pot filled with boiling water is on top of a stove. The bottom has a surface area of 200 cm² and a thickness of 3.00 mm. If the bottom has a temperature of 500 °C, how much heat flows through it per second?

99. [I] The 2.0-mm-thick glass pane in a window has an area of 0.250 m² and a thermal conductivity of 0.80 W/m·K. If the outside temperature is −18 °C and the inside temperature is 21 °C, how much heat is conducted each second?

100. [I] Body tissue without blood flowing through it has a thermal conductivity of about 18 Cal·cm/m²·h·C°. It's often given in those units in the life sciences because the surface area of the body is generally determined in m² and the thickness of the tissue in cm. Convert the conductivity to SI units.

101. [I] The following provides some idea of the inadequacy of conduction as a mode of heat transfer for the human body (see Problem 100). Suppose a person has a surface area of 1.4 m² and an average tissue thickness of 2.0 cm between what we might consider the inner and outer body. If the skin temperature is 33 °C and the inner body temperature is 37 °C, how much heat will be conducted to the surface per hour? Compare that to the fact that about 100 kcal/h are developed while just standing around staying alive.

102. [I] Determine the amount of heat that is conducted through a 4.0-m² portion of a brick wall 15-cm thick in the course of 1.0 h if the inside temperature is 20 °C and the outside temperature is 0 °C.

103. [II] What thickness of brick, with a thermal conductivity of 0.60 W/m·K, will conduct heat at the same rate as a 10-cm layer of dead air under the same conditions?

104. [II] A wall is made up of a 10-cm layer of common red brick in contact with a 20-cm layer of cinder block concrete. If the inside concrete face has a temperature of 20 °C and the outside brick face is at 0 °C, how much heat flows through a 1.0-m² area each hour? What is the power loss?

105. [III] A double-glazed window consists of two sheets of glass (each 3.0-mm thick) separated by a 5.0-mm layer of air. Determine the ratio of the thermal conductivity per unit area to that of a single 3.0-mm pane under the same conditions.

Chapter 14
Thermodynamics

*N*ow that we know a little about the basics of thermal energy, temperature, and heat, we can turn our attention to **Thermodynamics**, *the study of thermal energy, its transfer, transformation, degradation*, and *dispersal*. Although it was conceived (1824) out of narrow pragmatic concerns about steam engines, Thermodynamics has evolved into a grand discourse on energy and change. Energy is a measure of change (the change in position, speed, mass, temperature, etc.), and change informs time. At base, Thermodynamics is the study of change, the study of the unfolding of events, the progression of the Universe.

Thermodynamics treats the thermal behavior of matter, and when we consider some specific entity or group of entities, it's called a *system*: It might be a bottle of gas, a jet engine, or a yard full of chickens. Whatever else there is beyond the system (i.e., the rest of the Universe) is called the *surroundings*. A system might be connected to its surroundings in any number of ways: for example, it might allow heat, or electromagnetic radiation, to cross its boundaries, or it might even emit acoustical energy.

On the other hand, we can imagine a system to be completely insulated from the rest of the environment—in which case it is *isolated*. The ultimate isolated system is the entire Universe, there being no surroundings for it to interact with. Each of us can be considered a system in rather elaborate contact with our surroundings, transferring heat to and from it, exhausting and intaking gases, radiating and absorbing electromagnetic energy, ingesting and excreting material, and so forth. Living creatures are just as much the subject of Thermodynamics as are steam engines.

The First Law of Thermodynamics

Perhaps the first person to grasp the nature of heat and its relationship to, and unity with, all the various forms of energy was Dr. J. Mayer (p. 461). His work contained the broad notion of Conservation of Energy, but it generated little interest beyond ridicule. Before the decade passed, Mayer was overwhelmed by a deep depression fed by derision and made more intense with the death of two of his children. Lost in despair, he leapt from a second-story window (1849) in a painfully unsuccessful attempt at suicide. Within two years, he was a straitjacketed inmate in an insane asylum.

Though Mayer was first (1842) to recognize that all the various forms of energy were one, Hermann von Helmholtz, a renowned German physiologist (1847), independently for-

This chapter is about Thermodynamics, but it's also about a very practical class of machines—heat engines. We'll talk about lots of abstract ideas like internal energy (p. 492), reversibility (p. 500), efficiency (p. 505), and entropy (p. 513), but just beyond our introductory discourse is the roaring world of steam turbines producing electricity, gasoline motors powering countless automobiles, and blasting jet engines propelling airplanes. All of these are relatively new inventions; there were no practical heat engines even three hundred years ago.

The laws of Thermodynamics "control, in the last resort, the rise and fall of political systems, the freedom or bondage of nations, the movements of commerce and industry, the origins of wealth and poverty, and the general physical welfare of the race."

FREDRICK SODDY (1877–1956)
NOBEL PRIZE-WINNING CHEMIST

mulated the idea of Conservation of Energy, as did Joule. In 1847, Joule gave a short uninvited talk at a scientific meeting. He was cautioned to be brief, and discussion was neither expected nor encouraged, but a young man in the audience began to raise some interesting questions and soon there was a lively discussion that ultimately caused a sensation in the scientific community. That young Scotsman who was so moved by what he had heard was Professor William Thomson, the future Lord Kelvin.

14.1 Conservation of Energy

The most complete conception of the Law of Conservation of Energy includes *all* kinds of energy and is known as the **First Law of Thermodynamics**:

> **Energy can neither be created nor destroyed, but only transferred from one system to another and transformed from one form to another.**

When a 1-N apple falls 10 m from a tree into your hand, the original 10 J of *gravitational*-PE is transformed into 10 J of KE at the instant before it lands (minus a tiny bit transferred to the air via friction). When the apple comes to rest (neglecting a minute amount of energy lost via sound as it thumps down), the 10 J are shared as thermal energy between hand and apple, and so both increase in temperature slightly. A meteorite with much more KE may even melt on impact, but all the energy present at any moment is constant, whatever its form.

If we do 10 J of work compressing a spring, and then tie it up and lower it into an insulated acid bath, what happens to the *elastic*-PE as the spring dissolves? The energy stored via the interatomic forces must be liberated when the spring vanishes. We can expect the temperature of the bath to rise slightly more when dissolving the compressed spring than when dissolving a relaxed one.

To slow down a spaceship reentering the atmosphere, some of its KE is deliberately converted into thermal energy in the air and heat shield. For that matter, every time you activate the brakes on a moving car, you are doing the same thing: converting KE—yours and the car's—to random thermal energy via friction and the First Law. As Mayer suggested, friction that occurs during the tidal action of the oceans, the back and forth rubbing against the ground, heats the waters a bit. And that thermal energy comes from the rotational KE of the planet—the Earth experiences drag and slows down.

THE TRANSFER & TRANSFORMATION OF ENERGY

Whatever energy is, it is associated with the change imparted to a system as the result of interactions. If a positive amount of work is done *on* a sample of matter (on the worked), simultaneously an equal amount of work will be done *by* that sample of matter (on the worker). In that sense, change is transferable, and insofar as it is transferred, energy (ΔE) is transferred. *The transformation of energy is mediated by matter and occurs in the region of that matter.* And this goes for light as well since it is, as de Broglie put it, "the most refined form of matter" (p. 3). Energy is a property of matter; it does not have a separate existence. It does not spontaneously vanish at one place in the Universe and simultaneously appear at another, even though that would not explicitly violate the First Law. *The transformation of energy occurs through the action of the Four Forces.* The First Law, fleshed out with the ideas of heat, work, *gravitational*-PE, *electrical*-PE, KE, and so on, allows us to build an understanding of the operation of the natural world (Fig. 14.1).

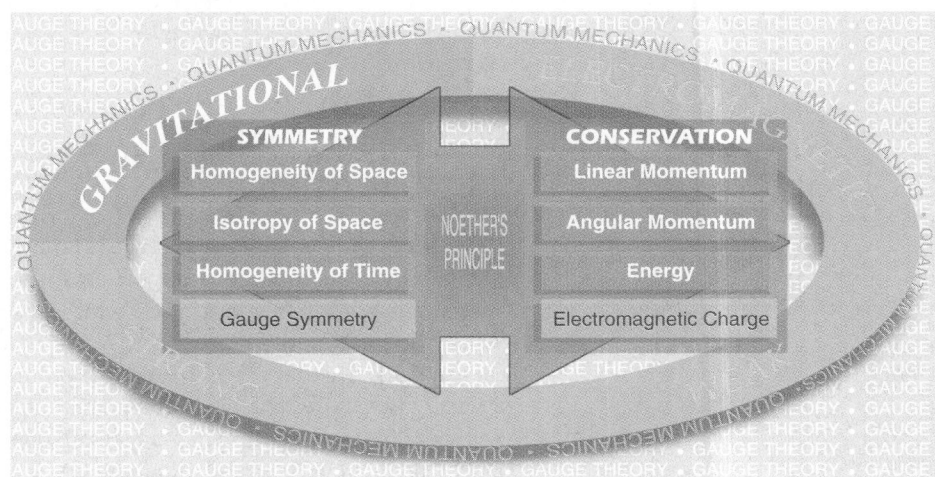

Figure 14.1 The First Law of Thermodynamics established Conservation of Energy as a central principle in physics. Now thermal energy joins other forms of energy, broadening the concept of conservation.

Work, Heat, and Internal Energy

> The **internal energy** (U) of an ideal gas is all translational KE.

Consider an ideal gas within a container at rest. Its constituent particles are effectively non-interacting (PE = 0) point-masses. These are endowed with neither rotational nor vibrational KE, and the gas's total energy, its **internal energy** U, is purely random translational KE. That being the case, it follows from Kinetic Theory that *the internal energy of the ideal gas depends only on its temperature* [Eq. (12.15), $KE_{av} = \frac{3}{2}k_B T$]. We will be using the ideal gas as a guide to the behavior of real gases, which usually respond in more or less the same way. More generally, when the atoms or molecules of a system interact, as they do in a dense gas, liquid, or solid, potential energy is stored via that interaction. Then the internal energy comprises both the KE and PE of the molecules.

> A system composed of atoms or molecules contains energy within its confines known as **internal energy**. In general, this is a combination of PE, via any interactions (principally electromagnetic), and KE, via the various motions.

Both heat (Q) and work (W) represent the transfer of energy. That raises the issue of the signs we might assign to Q and W, inasmuch as the transfer can be into or out of the system. It doesn't matter which convention we choose, as long as energy flowing out is opposite in sign to energy flowing in. Earlier, in mechanics, we defined the work done *on* a system (work-in) to be positive, and we'll stick with that to be consistent. **Work done *on* a thermodynamic system *by* the environment (work-in) is positive, while work done *by* the system *on* the environment (work-out) is negative.** Using this convention, energy-in is positive, whether it's heat or work.

> **Work-in** is positive. **Heat-in** is positive.

We can increase the internal energy ($\Delta U > 0$) of a system either by supplying it with heat ($Q > 0$) or by doing work on it ($W > 0$). The corresponding statement of Conservation of Energy, embracing the concept of heat,

> The **First Law**: The increase (*or decrease*) in the internal energy of a system equals the energy transferred in (*or out*) via heat, plus the energy transferred in (*or out*) via work.

$$\Delta U = Q + W \qquad (14.1)$$

is known as the **First Law of Thermodynamics**. Each of these quantities can be positive or negative; whatever increases U is positive, whatever decreases U is negative. Thus, if 10 J of heat flow *out* of the system $Q = -10$ J, and if 10 J of work are simultaneously done *by* the system $W = -10$ J. In all, that system loses 20 J and $\Delta U = -20$ J; the molecules of the system give up 20 J of internal energy to the environment.

Example 14.1 **[I]** A system in contact with a furnace receives 5000 J of heat while doing 2700 J of work on its surroundings. What, if any, is the change in the internal energy of the system?

Solution The mention of both "heat" and "work" should bring to mind the First Law of Thermodynamics. (1) TRANSLATION—A known quantity of heat is added to a system that does a known amount of work; determine the internal energy change. (2) GIVEN: $Q = +5000$ J = heat-in, $W = -2700$ J = work-out.

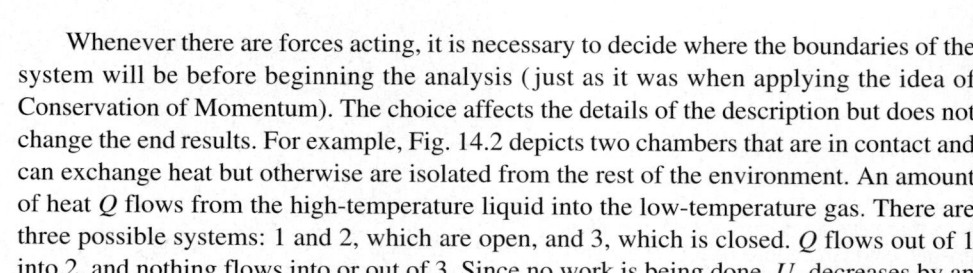

FIND: ΔU. (3) PROBLEM TYPE—Thermodynamics/First Law. (4) PROCEDURE—The relationship between the quantities Q, W, and ΔU is given by the First Law. (5) CALCULATION—Using Eq. (14.1), the change in internal energy is

$$\Delta U = Q + W = (+5000 \text{ J}) + (-2700 \text{ J}) = \boxed{2300 \text{ J}}$$

Quick Check: The system gains energy, and that's reasonable. Moreover, $Q = \Delta U - W = 2300 \text{ J} - (-2700 \text{ J}) = 5000 \text{ J}$.

[For more worked problems click on **WALK-THROUGHS** *in* **CHAPTER 14** *on the* **CD**.*]*

Whenever there are forces acting, it is necessary to decide where the boundaries of the system will be before beginning the analysis (just as it was when applying the idea of Conservation of Momentum). The choice affects the details of the description but does not change the end results. For example, Fig. 14.2 depicts two chambers that are in contact and can exchange heat but otherwise are isolated from the rest of the environment. An amount of heat Q flows from the high-temperature liquid into the low-temperature gas. There are three possible systems: 1 and 2, which are open, and 3, which is closed. Q flows out of 1 into 2, and nothing flows into or out of 3. Since no work is being done, U_1 decreases by an

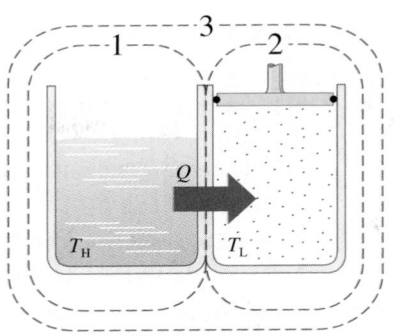

Figure 14.2 Because of the action-reaction nature of interactions, whenever forces act, it is necessary to establish the boundaries of the system being analyzed.

amount equal to Q, U_2 increases by an amount equal to Q, and $\Delta U_3 = 0$.

By definition, an isolated system is thermally insulated ($Q = 0$) and can neither do work on the surroundings nor have work done upon it ($W = 0$). It follows, then, that $\Delta U = 0$. Another statement of the First Law is then

The internal energy of an isolated system is constant (even though that energy may be transformed from one type to another).

This applies to all isolated systems, including the Universe itself. That's a presumptuous statement to make considering our meager role in the Cosmos, but to date, the First Law has never failed, so we keep saying it. For an isolated system, the internal energy is constant regardless of what processes go on within it—be they chemical, nuclear, or biological.

Perpetual Motion Machines

The internal energy of a finite system is finite. Clearly, then, if work is to be extracted from it, an equivalent amount of energy must be supplied, or the internal energy will diminish and ultimately no longer be available to drive the system. A *perpetual-motion machine* is one that puts out more work than its input of energy and can presumably continue to do so forever. We call such a mythical device a *perpetual-motion machine of the first kind,* and it most certainly cannot exist because it violates the First Law. Despite the fact that people are still attempting to patent such machines, they cannot possibly work. Recently, a Los Angeles jury acquitted two men of 50 counts of fraud, theft, and conspiracy. They had raised several hundred thousand dollars from backers of a compressed air turbine that was supposedly a perpetual-motion machine. According to the deputy district attorney, "The jurors told me they believed this machine produces more energy than goes into it."

All of this is not to say that the First Law precludes the existence of isolated systems that could, for example, spin forever or bounce forever—provided they do not output work or any other form of energy, they are not forbidden by the First Law.

NONISOLATED SYSTEMS: THE EARTH & YOU

It's possible for the internal energy of a nonisolated system to be constant as long as energy-in equals energy-out. This circumstance applies to the Earth as a whole, but to deal with it we have to add a radiation term to Eq. (14.2). Take the system to be everything within a great sphere enclosing the atmosphere. Practically speaking, the amount of work done on or by the Earth is negligible ($W = 0$), and since it is thermally insulated, being surrounded by vacuum, $Q = 0$. Still, energy pours in in the form of electromagnetic radiation (mostly from the Sun and mostly visible) at a rate of about 1.7×10^{17} J/s. The Earth reflects some of this and, furthermore, radiates energy on its own (mostly infrared). The result is that it emits just about at a rate of $(-)1.7 \times 10^{17}$ J/s. Thus, the total internal energy of the Earth (i.e., everything on it and everything in it) is more or less constant. Indeed, as long as you are not changing your weight, much the same thing can be said about you—energy goes in, changes form, and goes out.

14.2 Thermal Processes & Work

> **A system is in equilibrium when there is no tendency for it to undergo spontaneous macroscopic change.** Despite the fact that its atoms are moving around randomly, it manifests constant measurable macroscopic *physical properties*, such as pressure, volume, and temperature. A configuration of the system, corresponding to a specific set of values of these physical properties (P, V, T,…), is called a *state*. As long as they are constant, the system is said to be in a particular state, even though its constituent atoms are continuously moving. Accordingly, an ideal gas in equilibrium can be described by Eq. (12.10), $PV = nRT$, which is its *equation of state*. A *process* occurs when the system changes from one state (one set of values of its physical properties) to another state. The system returns to its original state when all of its macroscopic physical properties resume their original values.

When a process is taking place, there is some observable alteration of the system: something measurable (for example, P, V, or T) changes. A system can be varied in a

A macroscopic system is in **equilibrium** when its directly observable physical characteristics (temperature, pressure, etc.) do not change spontaneously.

variety of ways, but there are four basic modes that are especially simple to deal with (Fig. 14.3):

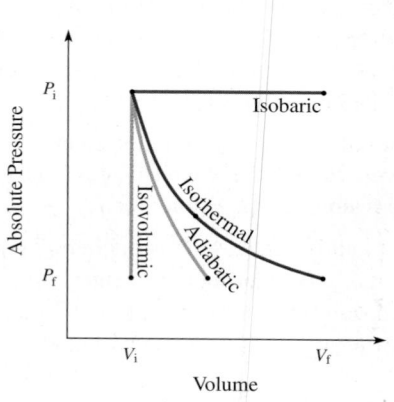

Figure 14.3 Plots of the four basic thermal processes. Here they all start at the same pressure and volume so you can compare them more readily.

isothermal   A process is isothermal when the temperature of the system is constant. The word comes from the Greek *isos*, equal, and *therma*, heat. Practically, an isothermal process is maintained by adding or removing heat from the system.

isobaric   A process is isobaric when the pressure of the system is constant. The word comes from the Greek *baros*, weight.

adiabatic   A process is adiabatic when no heat is transferred to or from the system. The word comes from the Greek *adiabatos*, not passable. An adiabatic process can be achieved in practice by either thermally insulating the system or performing the process so rapidly that there is no opportunity for heat to flow.

isovolumic   A process is isovolumic (or isovolumetric, or isochoric) when the volume of the system is constant.

The thermal processes involving vapors and gases are particularly important: they play a central role, for example, in the operation of the electric power–producing steam engine, the internal combustion automobile engine, and the jet propulsion airplane engine.

Work and the First Law

Let's now express the work W in Eq. (14.1) in terms of directly measured variables. Imagine a system that expands against an external pressure, changing its volume and doing work. The primary features are the existence of an applied force and the volume change. What that system *is* really doesn't matter much—it could be a solid, liquid, or gas. To keep things simple, let the system consist of a cylinder of gas sealed with a weightless, frictionless piston of area A. Some influence in the surroundings pushes down on the piston with a constant force F, as shown in Fig. 14.4. The gas and piston are in equilibrium, and the downward force F is exactly opposed by an upward force produced by the gas—namely, PA. The gas is now somehow made to expand very slowly (quasi-statically) so that it always remains essentially in equilibrium with the piston (here, that means P is constant)—the process is isobaric. This could be accomplished by slowly providing heat to the gas, which will increase the average speed of the molecules, thereby sustaining P, even as V increases (p. 443). If the displacement of the piston is Δl upward, the work done *on the surroundings (the atmosphere) by the expanding gas* is a negative quantity

[constant pressure] $$W = -F\Delta l = -PA\,\Delta l \tag{14.2}$$

Inasmuch as the change in volume is $V_f - V_i = \Delta V = A\Delta l$, **the work done in an isobaric**

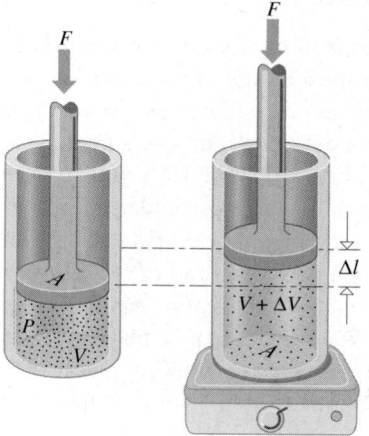

Figure 14.4 A gas sealed in a cylinder by a weightless, frictionless piston. The constant downward-applied force F equals PA, and when the piston is displaced, downward work is done on the gas.

process becomes

[constant pressure]
$$W = -P\Delta V \qquad (14.3)$$

When a system expands, ΔV is positive, and if work is done it's done *by* the system and W is negative. When the system contracts, ΔV is negative, work is done *on* the system, and W is positive.

Example 14.2 **[II]** A cylinder closed off with a movable piston contains 10.0 g of steam at 100 °C. The system is heated, and its temperature rises 10.0 °C as the steam expands 30.0 × 10^{-6} m³ at a constant pressure of 0.400 MPa. Determine (a) the work done by the steam and (b) the change in its internal energy. Take the specific heat capacity of steam to be 2.02 kJ/kg·K.

Solution The steam expands and does work on its environment; it loses energy in the process. But it's also heated and gains some energy. (1) TRANSLATION—A known mass of steam at a given temperature is heated. Its temperature rises by a known value, and it increases in volume by a specified amount

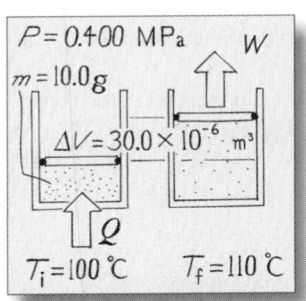

$P = 0.400$ MPa W
$m = 10.0$ g
$\Delta V = 30.0 \times 10^{-6}$ m³
Q
$T_i = 100$ °C $T_f = 110$ °C

at a constant known pressure; determine (a) the work done and (b) the internal energy change. (2) GIVEN: $m = 10.0$ g (of water), $T_i = 100$ °C, $\Delta T =$

10.0 K, $\Delta V = 30.0 \times 10^{-6}$ m³, $P =$ constant $= 0.400$ MPa, and $c_s = 2.02$ kJ/kg·K. FIND: ΔU. (3) PROBLEM TYPE—Thermo/work/First Law. (4) PROCEDURE—(a) The process is isobaric, and so W follows from Eq. (14.3). (b) The relationship between Q, W, and ΔU is given by the First Law. (5) CALCULATION—(a) Using Eq. (14.3)

$$W = -P\Delta V = -(0.400 \times 10^6 \text{ Pa})(30.0 \times 10^{-6} \text{m}^3) = \boxed{-12.0 \text{ J}}$$

(b) Since $\Delta U = Q + W$, we will need Q, the heat-in. $Q = c_s m \Delta T$, and so $\Delta U = c_s m \Delta T + W$; as a result,

$$\Delta U = (2.02 \times 10^3 \text{ J/kg·K})(0.0100 \text{ kg})(10.0 \text{ K}) - 12.0 \text{ J}$$
$$\Delta U = 202 \text{ J} - 12.0 \text{ J} = \boxed{190 \text{ J}}$$

Quick Check: Let's first calculate Q and from that ΔT: $Q = c_s m \Delta T = \Delta U + W = 190 \text{ J} + 12.0 \text{ J}; \Delta T = Q/c_s m = (202 \text{ J})/c_s m = 10 \text{ K}$.

Example 14.3 **[III]** An aluminum cube 20 cm on a side is heated from 50 °C to 150 °C in a chamber at atmospheric pressure. Determine the work done by the cube and the change in its internal energy. If the same process were carried out in vacuum, what would the change in internal energy be then?

Solution The cube expands and does work on its environment; it loses energy in the process. But it's also heated and gains some energy. (1) TRANSLATION—A known volume of a substance (aluminum) is heated and goes from a known initial to a known final temperature at a specified pressure; determine (a) the work done, (b) the internal energy change, and (c) the internal energy change when the pressure is zero. (2) GIVEN: Aluminum cube, $L = 20$ cm, $T_i = 50$ °C, and $T_f = 150$ °C. FIND: W, and ΔU when $P = 1$ atm and when $P = 0$. (3) PROBLEM TYPE—Thermo/work/First Law. (4) PROCEDURE—(a) This isobaric process involves ΔU and W, and that immediately suggests $\Delta U = Q + W$: the block expands against the hydrostatic pressure of the atmosphere and does work on it. We have to determine Q and W to find ΔU via the First Law. (5) CALCULATION—(a) To get $W = -P\Delta V$, we need ΔV resulting from ΔT, and that follows from Eq. (12.4) and Table 12.3:

$$\Delta V = \beta V_0 \Delta T$$
$$\Delta V = (72 \times 10^{-6} \text{ K}^{-1})(8.0 \times 10^{-3} \text{ m}^3)(100 \text{ K})$$

$$\Delta V = 5.76 \times 10^{-5} \text{ m}^3$$

and so

$$W = -P\Delta V = -(0.101\ 3 \text{ MPa})(5.76 \times 10^{-5} \text{ m}^3) = \boxed{-5.8 \text{ J}}$$

Now, to find Q, we use $Q = c_a m \Delta T$, but we will need m first. Since $m = \rho V$ and from Table 9.1 (p. 286), it follows that

$$m = (2.7 \times 10^3 \text{ kg/m}^3)(8.0 \times 10^{-3} \text{ m}^3) = 21.6 \text{ kg}$$

Using Table 13.3:

$$Q = c_a m \Delta T = (0.90 \text{ kJ/kg·K})(21.6 \text{ kg})(100 \text{ K}) = 1.94 \text{ MJ}$$

Therefore,

$$\Delta U = Q - P\Delta V = 1.94 \text{ MJ} - 5.8 \text{ J} = \boxed{1.9 \text{ MJ}}$$

The amount of work done is negligible by comparison to Q, so it doesn't matter whether or not this process is carried out in vacuum (where $W = 0$); the internal energy change will still be 1.9 MJ.

Quick Check: Using rounded-off numbers, $\Delta V \approx (70 \times 10^{-6} \times 10 \times 10^{-3} \times 100)$ m³ $\approx 7 \times 10^{-5}$ m³, which, because we rounded up, is high; $W \approx -0.1 \text{ MPa} \times 7 \times 10^{-5}$ m³ ≈ -7 J; $Q \approx (1 \times 10^3 \times 20 \times 100)$ J ≈ 2 MJ.

The pressure-versus-volume or *PV*-diagram for an arbitrarily expanding gas is shown in Fig. 14.5*a*. Notice that **the work done is the area under the *PV*-curve**, whatever its shape happens to be. **For an expansion, the area beneath the curve, the work done *by* the gas, must be taken to be negative.** By contrast, the gas depicted in Fig. 14.5*b* undergoes compression, and so **the area beneath the curve, the work done *on* the gas, must be taken to be positive**.

Example 14.4 **[I]** A cylinder containing a gas is sealed with a nearly frictionless piston at a pressure of 0.20 MPa. The cylinder is placed in contact with a source of heat, and the gas very slowly expands, moving the piston, whose area is 1000 cm², a distance of 5.0 cm. Assuming the process is isobaric, and given that 300 J of heat enters the system, determine its change in internal energy.

Solution The expanding gas does work and is simultaneously heated, all of which should call to mind the First Law of Thermo. (1) TRANSLATION—A gas at a known constant pressure is heated and expands a known amount (against a piston) upon the addition of a specified quantity of heat; determine the internal energy change. (2) GIVEN: $P = 0.20$ MPa = constant, $A = 1000$ cm², $\Delta l = 5.0$ cm, and $Q = 300$ J. FIND: ΔU. (3) PROBLEM TYPE—Thermo/First Law. (4) PROCEDURE—This isobaric process involves ΔU and W, which immediately suggests $\Delta U = Q + W$: the block expands against the piston and does work on it. We have to determine Q and W to find ΔU

via the First Law. (5) CALCULATION—Because this process is isobaric, and it's an expansion, $W = -P\Delta V$

$$\Delta U = Q - P\Delta V \qquad (14.4)$$

We need $\Delta V = (1000$ cm²$)(5.0$ cm$) = 5000$ cm³ $= 5.0 \times 10^{-3}$ m³. Then

$$\Delta U = (300 \text{ J}) - (0.20 \times 10^6 \text{ Pa})(5.0 \times 10^{-3} \text{ m}^3)$$

and

$$\boxed{\Delta U = -0.70 \text{ kJ}}$$

The gas does much of the work of expansion at the expense of internal energy to the extent of a 700-J reduction in *U*.

Quick Check: Approximating the numbers we'll recompute Q using the above determined value of ΔU: $\Delta V = 5 \times 10^3$ cm³ $\approx 10^{-2}$ m³; $P\Delta V \approx 10^5$ Pa $\times 10^{-2}$ m³ $\approx 10^3$ J; $Q = \Delta U + P\Delta V \approx -700$ J $+ 10^3$ J ≈ 300 J.

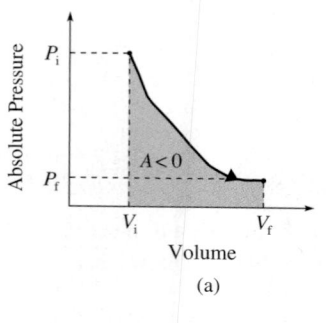

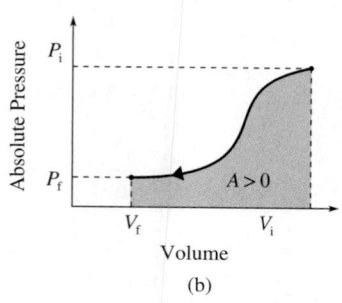

Figure 14.5 (a) During an arbitrary expansion of a gas, the area under the curve is the work done *by* the gas. This area must be taken to be negative $(A < 0)$. (b) During an arbitrary contraction of a gas, the area under the curve is the work done *on* the gas. This area must be taken to be positive $(A > 0)$.

The *PV*-diagram for an isobaric expansion of a gas is shown in Fig. 14.6*a*. From an initial state-*I*, the system was transformed at a constant pressure to a different final state-*F*. If the pressure of the gas is now decreased (Fig. 14.6*b*) while maintaining a constant volume $\Delta V = 0$, the system arrives at a new state-*F* with no additional work done (the area under the curve is unchanged). In the case of an ideal gas, we can imagine the various processes of constant *P*, *V*, or *T* as occurring on the surface of a three-dimensional *PVT*-diagram (Fig. 14.6). The points *A*, *B*, and *C* in Fig. 14.6*b* correspond to those in Fig. 14.6*c*, where the system goes from *A* to *B* to *C*.

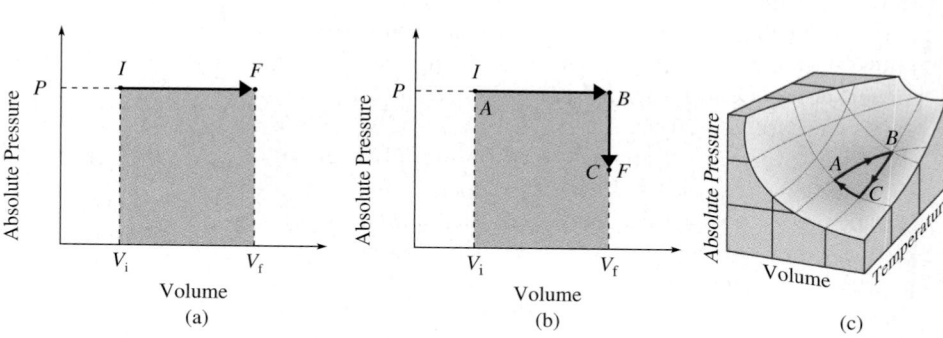

Figure 14.6 (a) The volume of a gas is changed while the pressure is kept constant. The work done by the gas in expanding is $-P\Delta V$, which is equal to the area under the curve. (b) No work is done in decreasing the pressure, carrying the system from *B* to *C* and a final state-*F*. The work done by the gas in going from *I* to *F* is equal to the area under the curve. (c) *PVT*-diagram for an ideal gas.

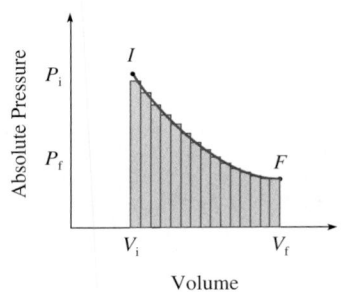

Figure 14.7 The area under a *PV*-curve equals the work done on or by the gas. Here, that area is divided into a large number of small rectangles so that it can be computed.

If the pressure varies continuously as the volume changes (Fig. 14.7), the area under the *PV*-curve will still correspond to the work done. That area can be found by dividing it up into small rectangular segments (as in Fig. 6.9, p. 178), each terminated by the average value of the pressure over that interval. The work done during the *i*th constant-*P* segment ΔW_i is given by Eq. (14.3) as $-P_i \Delta V_i$, and the sum of all such contributions, as they get narrower and their number approaches infinity, approaches the total work exactly. The only proviso is that the system be in equilibrium everywhere. If it isn't, $-PA$ will not equal F over some of those vanishingly narrow area segments. We'll return to this issue presently.

Depending on what we do to it, the system can be transformed from its initial condition (state-*I*) to its final condition (state-*F*) by various routes, each corresponding to a different amount of work done (Fig. 14.8). Clearly, first dropping *P* at constant *V* and then changing the volume encompasses less area (and so does less work) than initially keeping *P* constant. ***The work done on or by a system depends on the manner in which it is transformed from its initial to its final state—work is not a state variable, it does not reflect the state of the system.*** Remember that heat may also be entering or leaving the system and that the final state is determined by a balance of both these contributions. The path taken in the *PV*-diagram must be fully specified if the work is to be determined. Still, whatever path is taken to get there, the final state is the same—its atoms presumably should not remember the sequence of events that brought them there. Consequently, we can expect that U_f will depend on P_f and V_f but will be the same for all paths leading to that final state. Experiments bear out this conclusion: ΔU is always found to be path-independent. **The internal energy of a system is a unique measure of the state of the system.** Although *U* is path-independent, *W* is not and it follows from the First Law that *Q* is not. Both *W* and *Q* depend on the history of the system and combine to sustain *U* so that it is independent of that history.

{It's highly recommended that you now click on ISOBARIC PROCESSES under INTERACTIVE EXPLORATIONS on the CD. This simulation will allow you to study the atomic behavior of a gas at constant pressure. Spend some time playing with this piece; it's a wonderful opportunity to see how everything fits together.}

14.3 Isothermal Change of an Ideal Gas

Suppose that the device in Fig. 14.9*a* contains an ideal gas that is to expand isothermally. Accordingly, imagine that the cylinder is made of a highly conducting material, is thin-walled, and is surrounded by a large constant temperature bath (a ***thermal reservoir***). Now, if the gas is allowed to expand very slowly, raising the piston, heat will be able to enter the system at a rate such that the gas remains at a constant temperature. Quite generally, an expanding gas does negative work on the movable wall of its container, the piston. When a molecule collides with a wall that is moving away and rebounds off, it has a final speed that is less than its initial precollision speed (Fig. 14.9*b*). The decrease in molecular KE equals the work done on the piston. That work is therefore done at the expense of internal energy.

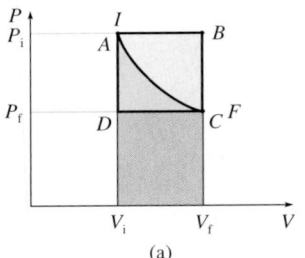
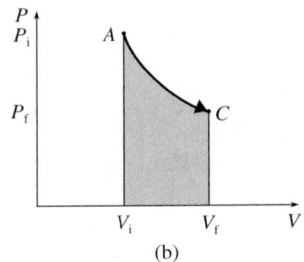
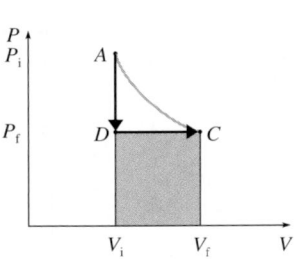
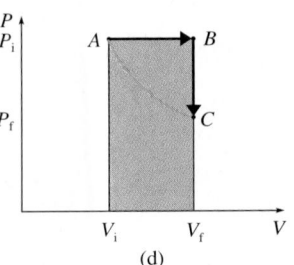

      (a)   (b)   (c)   (d)

Figure 14.8 A system (*a*) may go from an initial state-*I* to a final state-*F* along several different routes, each corresponding to doing a different amount of work. Parts (*b*), (*c*), and (*d*) depict different procedures, areas, and amounts of work, respectively.

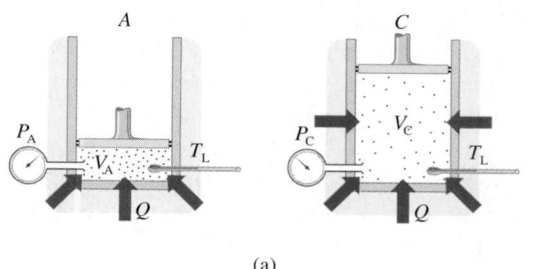

Figure 14.9 (*a*) A gas expanded isothermally (see Fig. 14.10) from state *A* to state *C*. As heat enters, the piston rises and pressure drops. (*b*) When molecules bounce off a receding piston, they lose momentum. (*c*) When they bounce off an advancing piston, they gain momentum.

(a)

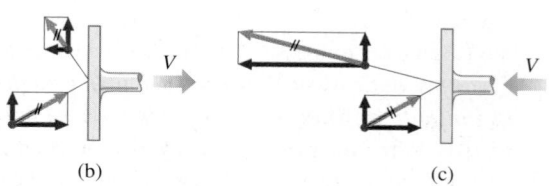

(b)   (c)

Thus, ***unless heat is supplied to the gas, its temperature will drop***. Here, the cylinder is in contact with a thermal reservoir, and any tendency for the temperature to change is negated by a flow of heat across the boundary.

During an **isothermal** (constant *T*) change, $\Delta U = 0$ for an **ideal gas**.

In an isothermal expansion of an ideal gas, since *U* depends only on the translational KE, which depends on *T*, which is constant, it follows that $\Delta U = 0$ and $Q = -W$: the work done by the gas is provided by the heat-in from the reservoir. In effect, the gas serves as a conduit that carries energy from reservoir to piston. On the other hand, when a gas is compressed, the molecules bounce off an approaching wall (Fig. 14.9*c*), picking up speed like a tennis ball colliding with an advancing racket. The KE of the gas increases; the piston does work on the gas; and unless heat can escape the system, its temperature will rise. Figure 14.10*a* depicts the behavior of an ideal gas being moved conceptually (at a constant temperature) along an isothermal curve (also called an **isotherm**) from point *A* to point *C*—the same path is pictured in Fig. 14.6 as well. Notice how going from *A* to *B* entails raising the gas from temperature T_L to a higher temperature T_H, crossing several isotherms along the way. It's helpful when interpreting *PV*-diagrams to envision a grid of isotherms ranging from low to high temperatures. Here, since $P = nRT/V = $ (constant)$/V$, the isotherms are hyperbolas.

Consider the work done by the isothermal expansion of the ideal gas. The area under the curve in Fig. 14.10*b* is gotten by summing all of the tiny ΔW area elements (or, via calculus, by integration). Here, $\Delta W = -P\Delta V$, but if the process is done slowly so that the system never deviates appreciably from equilibrium, $P = $ (constant)$/V$ and the pressure

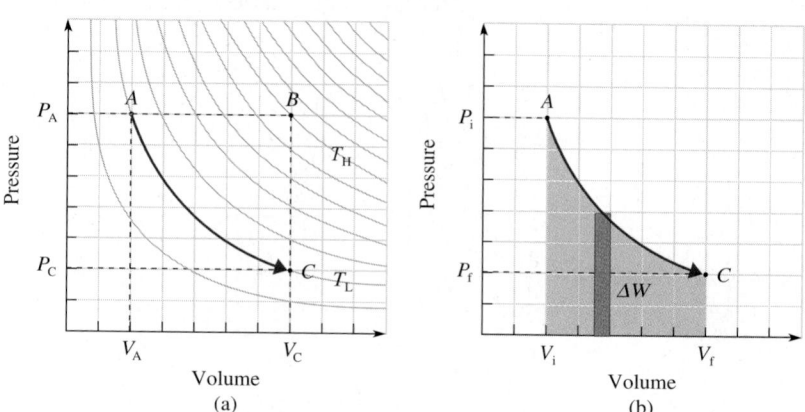

Figure 14.10 (*a*) An isothermal expansion of a gas. (*b*) The area under the *PV*-diagram equals work done.

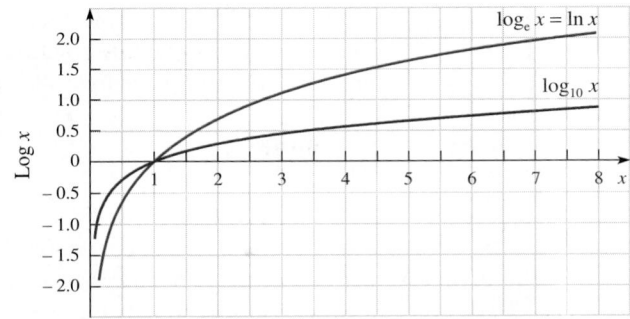

Figure 14.11   The function ln x equals $-\infty$ at $x = 0$, 0 at $x = 1$, and continues to rise slowly as x increases.

can be determined in terms of the volume at each element. Thus, $\Delta W = -(\text{constant})\Delta V/V$; more work is done (i.e., the area element is taller) initially, when the volume is small and the pressure high rather than later, when the volume is large and the pressure low. This particular functional dependence is extremely common in the real world: ΔW is proportional to both ΔV and to $1/V$. The situation where the change in something depends both on the change in another quantity and, inversely, on how much of that other quantity exists at each step is encountered frequently in physics. That kind of dependence is called *logarithmic* {see **MATH REVIEW: PART A-3** on the **CD** 💿 }, and we define the natural logarithm (abbreviated ln) by way of it. We encountered the $\log_{10}$ when we dealt with sound intensity. This *natural* logarithm (so called because it occurs in so many natural processes) is the log to a special base; rather than 10, it's the number known as e equal to 2.718.... In other words, $\ln x = \log_e x = 2.30 \log_{10} x$, as shown in Fig. 14.11, and the corresponding values of ln x can be determined using a calculator.

The area under the isotherm varies logarithmically with volume—that is, W varies logarithmically with V. ***The work done on or by an ideal gas during an isothermal volume change from V_i to V_f is***

[isothermal ΔV process]
$$W = -nRT \ln \frac{V_f}{V_i}$$
(14.5)

where $W = Q$ and, of course, $P_iV_i = P_fV_f = nRT$. Notice that when $V_f > V_i$, as when the volume expands, the logarithm is positive, $W < 0$, and the system does work on the environment. At least for the purpose of analysis, many processes are essentially isothermal and, as a rule, they occur slowly—such as a very small leak in a tank of compressed gas at room temperature, or an open pot of water boiling away on a low flame.

CALCULATOR TIPS

To find the natural log of a number, first locate the $\boxed{\text{ln}}$ key. Using a single-line (old-fashioned) calculator (without parens) to determine, for example, $\ln \frac{1}{3}$ enter

$\boxed{1}\ \boxed{\div}\ \boxed{3}\ \boxed{=}\ \boxed{\text{ln}}$

Whereas on a two-line calculator you'll enter

$\boxed{\text{ln}}\ \boxed{(}\ \boxed{1}\ \boxed{\div}\ \boxed{3}\ \boxed{)}\ \boxed{=}$

and get

$-1.09861229\ ^{00}$

Example 14.5 **[I]** Determine the amount of work needed to compress 4.0 g of oxygen at STP down to one-third its original volume, keeping the temperature constant. Assume it behaves as an ideal gas.

Solution   This is a constant-temperature (isothermal) compression. (1) TRANSLATION—A known mass of gas at STP is isothermally compressed a specified amount; determine the work done on the gas. (2) GIVEN: Oxygen at STP, $m = 4.0$ g, and $V_f = V_i/3$. FIND: W. (3) PROBLEM TYPE—Thermo/isothermal compression. (4) PROCEDURE—The work for an isothermal volume change follows from Eq. (14.5), and so we need n, the number of moles. (5) CALCULATION—The atomic

mass of O_2 is 2(16) = 32; as a result, there are 32 g/mol and $n = (4.0\text{ g})/(32\text{ g/mol}) = 0.125$ mol. It follows that

$$W = -nRT \ln \frac{V_f}{V_i} = -(0.125\text{ mol})(8.314\text{ J/mol·K})(273\text{ K})(\ln \tfrac{1}{3})$$

and $\boxed{W = +0.31\text{ kJ}}$. The plus sign shows that work was done on the gas and that, since the temperature did not change, the gas had to lose the same amount of energy as heat-out.

Quick Check: Redoing the calculation using simplified numbers, we get 4 g $= \frac{1}{8}$ mol; $nRT \approx \frac{1}{8}(8\text{ J/mol·K})(273\text{ K}) \approx 273$ J/mol; $\ln \frac{1}{3} \approx -1$; $W \approx +0.3$ kJ.

{Click on I<small>SOTHERMAL</small> P<small>ROCESSES</small> under I<small>NTERACTIVE</small> E<small>XPLORATIONS</small> on the **CD**. This simulation will allow you to study the atomic behavior of a gas at constant temperature. Play with this piece for a while; it will definitely help clarify the ideas.}

Reversibility

If the isothermal process that carried the gas from point *A* to point *C* in Fig. 14.10 can be retraced so that the system returns from P_f, V_f, and *T* to P_i, V_i, and *T* along the same curve, then we say that that process was **reversible**. Another way to put it is that the system must always be infinitesimally close to thermodynamic equilibrium even as it passes from one state to the next.

In truth, reversible processes are idealizations: they can be approximated but never actually attained. For an isothermal process such as that in Fig. 14.9*a* to be reversible, the piston must be weightless and frictionless. Moreover, the volume change would have to be done very slowly, keeping the gas in equilibrium and not generating any organized motion (such as eddies or turbulence) that would take up energy. *Any process that entails friction losses is irreversible.* Consequently, real processes can, at best, only resemble being reversible, but even so, such behavior is of great practical interest.

Push a book across the surface of a table and you irreversibly convert mechanical energy into thermal energy via friction. Open the valve on a tank of compressed gas and let the contents blow out into a vacuum chamber and the resulting rapid free expansion is irreversible. The system does not remain in equilibrium during the process, and there are different values of pressure at different points in the chamber. Thus we could know the pressure and volume at the start and at the end, but could not even draw a *PV*-diagram of the entire process. Despite the natural prevalence of irreversibility, we'll limit our discussion to the far more manageable terrain of essentially reversible processes.

> A process carried out by a system is **reversible** if the system and its environment change in such a way that both can be returned to their original condition; both are completely restorable in the sense that the measurable state variables (*P, V, T,* and *m*) for both can be made to return to their original values without otherwise affecting the Universe.

REVERSIBILITY & MINUTE CHANGE

Imagine a vertical, thermally insulated cylinder containing some gas sealed in with an upright frictionless piston on which is piled a mass of fine sand. The sand provides the downward force that's balanced by the trapped gas, leaving the piston motionless in equilibrium. If one minuscule sand grain is added to the pile, the piston will descend through a tiny displacement, in an essentially quasi-static fashion. One grain at a time, the compression can continue until we choose to stop at any moment. If sand is now slowly removed, the process will be reversed into an expansion that carries it through the same states it has already experienced. The fact that a process can be stopped anywhere along the way and turned around by a minute change in the operating conditions is the hallmark of reversibility.

14.4 Adiabatic Change of an Ideal Gas

Imagine an *adiabatic* process occurring in a system, that is, one where no heat crosses the boundary. This can be accomplished, with varying degrees of success, by thermally insulating the system. Alternatively, processes that occur suddenly tend to be adiabatic because heat takes a fair amount of time to flow. Thus, sound waves in air are adiabatic since there isn't enough time for heat to flow from the compressions to the rarefactions (see p. 389). Adiabatic processes are important in many situations, not the least of which is the operation of the internal combustion engine.

Consider an ideal gas, sealed in a cylinder closed off with a piston, all of which is thermally insulated so that **no heat can enter or leave**. The process about to be performed will be adiabatic, but it may or may not be reversible—that depends on how it's carried out. Let's stay with a reversible adiabatic process and do everything nice and slowly so as to maintain equilibrium. Now, compressing the gas (*W* > 0), it follows that *U* must increase as the work done on the gas goes directly into internal energy($\Delta U = W$) and the temperature increases. Accordingly, *PV*, which equals *nRT*, is no longer constant. The curve corresponding to this process in a *PV*-diagram is called an ***adiabat*** (Fig. 14.12*a*) and is evidently everywhere steeper than an isotherm. That's the case because as the volume is decreased, the temperature rises, sustaining a higher pressure than would be if *T* were constant. On the other hand, as the gas expands adiabatically, it cools, sustaining a lower pressure at any value of *V* than would the isothermal process.

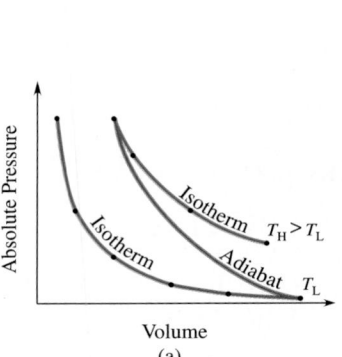

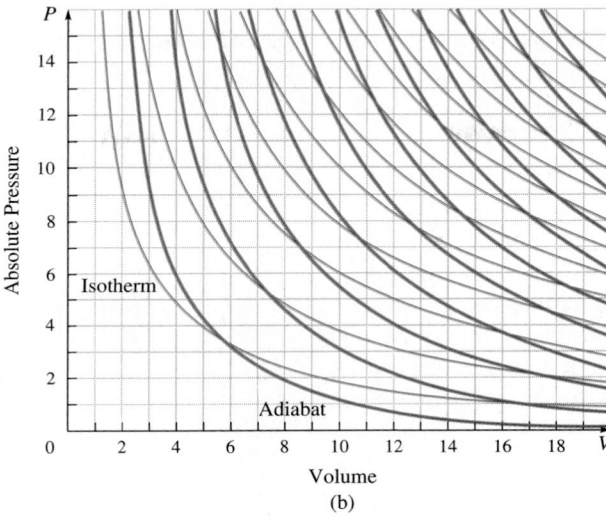

Figure 14.12 (a) The *PV*-diagram for an ideal gas; isothermals and adiabats. Each isothermal corresponds to a specific temperature. (b) As we move along the set of curves, going up and to the right, the temperature rises. Whenever you see a *PV*-diagram it's useful to think of such an array of isothermals forming a temperature grid.

For an isothermal process, *PV* = *constant* by way of Boyle's Law. In an adiabatic process (*T* ≠ constant), *P* drops more quickly with increasing *V*, and it is found experimentally that

[reversible adiabatic]
$$PV^{\gamma} = \text{constant} \qquad (14.6)$$

where γ is a constant that equals 1.67 for monatomic gases, ≈1.4 for diatomic gases, and ≈1.3 for polyatomic gases. It can be shown theoretically, though we shall not, that γ is actually the ratio of the specific heat of the gas at constant pressure to that at constant volume. Another way to write Eq. (14.6) is

[ideal gas]
$$P_i V_i^{\gamma} = P_f V_f^{\gamma} \qquad (14.7)$$

and, since the gas is still ideal, the resulting temperature can be determined from

$$\frac{P_i V_i}{T_i} = \frac{P_f V_f}{T_f} \qquad (14.8)$$

When the cap is removed from a beer bottle, gas in the neck expands rapidly and adiabatically. It does work on the air and loses thermal energy, overcoming its own internal interactions. As a result, the temperature in the neck of the bottle can drop from ≈41 °F to ≈−31 °F. The sudden change causes water vapor in the neck to condense into a cloud of tiny droplets.

Example 14.6 **[II]** Argon gas (which, like the other Noble Gases, is monatomic) is very slowly compressed adiabatically in a well-insulated cylinder to half its original volume of 0.100 m³. If it is at atmospheric pressure and 27.0 °C to start, what will be the final pressure and temperature of the gas?

Solution This problem treats an adiabatic compression, and that should call to mind $PV^\gamma = $ constant. (1) TRANSLATION—A known volume of monatomic gas at a known pressure and temperature is adiabatically compressed a specified amount; determine the resulting temperature and pressure. (2) GIVEN: An adiabatic process, $V_i = 0.100$ m³, $P_i = 1$ atm, $V_f = \frac{1}{2}V_i$, and $T_i = 27.0$ °C. FIND: P_f and T_f. (3) PROBLEM TYPE—Thermo/ adiabatic compression. (4) PROCEDURE—The process is adiabatic and involves P, V, and T, so we begin with Eq. (14.7). (5)

CALCULATION—The gas is monatomic; $\gamma = 1.67$, and

$$P_f = P_i \left(\frac{V_i}{V_f}\right)^\gamma = (0.101\,3\ \text{MPa})(2)^{1.67}$$

$$P_f = (0.101\,3\ \text{MPa})(3.182) = \boxed{0.322\ \text{MPa}}$$

To find the temperature, use Eq. (14.10), yielding

$$T_f = T_i \left(\frac{P_f}{P_i}\right)\left(\frac{V_f}{V_i}\right) = (300\ \text{K})(3.182)\left(\tfrac{1}{2}\right) = \boxed{477\ \text{K}}$$

Quick Check: Using Eq. (14.7), $P_i V_i^\gamma = (0.1\ \text{MPa})(0.1\ \text{m}^3)^{1.67} = 2 \times 10^3$, whereas $P_f V_f^\gamma = (0.3\ \text{MPa})(0.05\ \text{m}^3)^{1.67} = 2 \times 10^3$.

A common instance of an adiabatic process is the rapid compression of air with a hand pump on the downstroke. You do work on the gas (which is positive) and $Q = 0$; energy flows into the gas and it rises in temperature ($\Delta U = W > 0$), heating the pump quite noticeably (Fig. 14.13). An example of the reverse process, where the system does work, is the discharge of a carbon dioxide fire extinguisher. This happens so fast, it's again adiabatic; now energy flows out of the gas and the temperature drops. Initially at room temperature, the CO_2 blasts out into a region of atmospheric pressure and does (negative) work on the surrounding air. Since $\Delta U = W < 0$, the temperature of the gas drops, often to the point of forming dry ice crystals. Meanwhile, the canister is cooler, too, because the CO_2 remaining in it consists of the slower molecules (lower KE) left behind. The same sort of thing happens when you fill a spare tire from a can of compressed gas—the can gets cold.

{It's highly recommended that you now click on **ADIABATIC PROCESSES** under **INTERACTIVE EXPLORATIONS** in **CHAPTER 14** on the **CD**. This simulation will allow you to study the behavior of a gas when no heat enters or leaves the system. Spend some time playing with this piece; doing that will clarify the ideas and bring insights a traditional textbook cannot.}

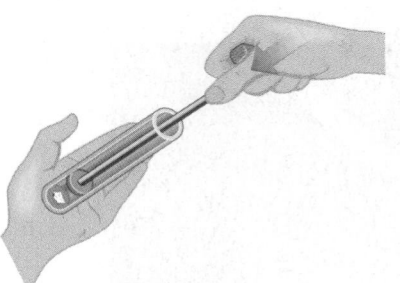

Figure 14.13 The fire syringe is a small cylinder with a tight-fitting piston. When the piston is rapidly driven in, the gas is compressed, and V decreases while both P and T increase. A small piece of tissue paper will burst into flames within the syringe.

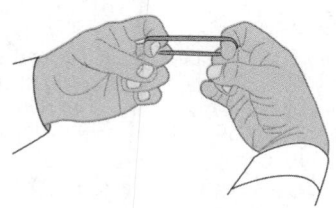

EXPLORING PHYSICS ON YOUR OWN

Work, Rubber Bands, & Internal Energy: Stretch a broad rubber band, doing work on it in one quick adiabatic jerk. The internal energy of the rubber band increases, and not all of it is in the form of PE; there's plenty of random KE imparted to the atoms as the long twisted molecules are pulled into alignment. As a result, the band's temperature will rise appreciably—you can easily detect the increase with your lips. Hold that fat rubber band stretched for a while until it reaches room temperature and then let it quickly contract, and it will cool appreciably. The elastic force does work as the band contracts, decreasing the internal energy and dropping the temperature.

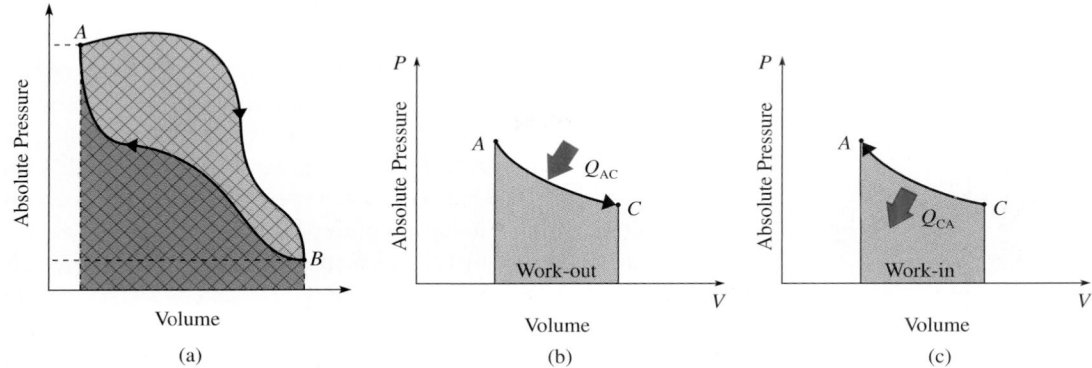

Figure 14.14 (*a*) The work done *by* the gas during an arbitrary expansion from *A* to *B* corresponds to the cross-hatched area under the curve. The work done *on* the gas during an arbitrary contraction from *B* to *A* corresponds to the green area beneath the curve. And the net work done going from *A* to *B* and back to *A* is equal to the blue area. (*b*) The *PV*-diagram for an isothermal expansion during which a quantity of heat Q_{AC} enters the system as it goes from *A* to *C*. (*c*) During the isothermal compression back to *A*, the same amount of heat Q_{AC} is ejected.

Cyclic Processes: Engines and Refrigerators

Let's suppose that we are working with reversible processes, and, furthermore, once having carried out several of them, we wish to return the gas to its original state ($\Delta U = 0$). The *PV*-diagram would then represent a *cycle* and would appear as a closed figure that could be swept through without change, over and over again. Figure 14.14*a* shows a hypothetical cyclic process. During the expansion from *A* to *B,* there is a large amount of negative work done W_{AB} (the cross-hatched area). On returning to *A* via compression, a smaller amount of positive work is done, W_{BA} (the green area). Since $|W_{AB}| > |W_{BA}|$, the net amount of work done by the gas on the environment (the blue area) is negative, so we can conclude that an equal net amount of heat entered the system.

Let's now be more specific. The simplest program is to put an ideal gas in a cylinder closed by a piston (Fig. 14.3), surround it with a constant temperature bath, and expand it along an isothermal, as in Fig. 14.14*b*. An amount of heat Q_{AC} flows in from the bath as the system expands from *A* to *C*: $W_{AC} < 0$, and because $\Delta T = 0$, $\Delta U = 0$ for an ideal gas (p. 497); work-out plus heat-in equals zero and $Q_{AC} > 0$. The work done by the gas is represented by the area under the curve, which, because it's an expansion, must be taken to be negative. If the system returns to point *A* via the same isothermal in reverse, as in Fig. 14.14*c*, positive work is done on the gas and heat flows out of it, where $W_{AC} = -W_{CA}$ and $Q_{AC} = -Q_{CA}$, and the system is energetically back to its starting point. Notice that if we add the positive work done on the gas during the compression leg to the negative work done during the expansion leg, the result is the net work done over the cycle, or in this special case, zero. That result is again represented by the *area enclosed by the closed curve*—zero.

Figure 14.15 shows a cycle where the net work done by the gas (the area enclosed) is nonzero. In going from *A* to *B*, work is done by the gas as the volume increases ($V_B > V_A$). Its temperature rises (the gas moves from point *A* on one isotherm to point *B* on another higher-temperature isotherm), so *U* increases as heat enters the system. In going from *B* to *C*, no work is done and the temperature drops, so *U* decreases and heat leaves the system. In going from *C* back to *A* along an isotherm the gas is compressed ($V_C > V_A$), and work is done on it. Because it's an isotherm $\Delta U = 0$, and since $\Delta U = Q + W$ the positive work done on the gas must be accompanied by an equal amount of negative heat-out. The total cycle (*A* to *A*) draws heat in and does an equivalent amount of work on the environment.

This cycle suggests the operation of an engine. ***An engine is a device that converts some form of energy (electrical, gravitational, chemical, thermal, etc.) into work.*** In the broadest sense, both a cannon and a solid-fuel rocket motor are engines. But these are

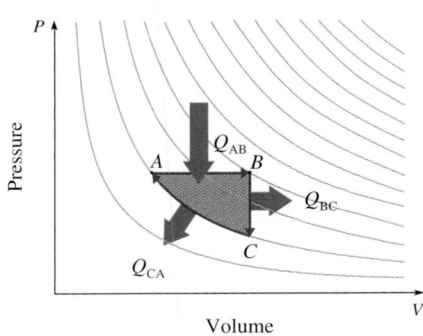

Figure 14.15 The blue area corresponds to a net amount of negative work done *by* the system. The work done *on* the system during the compression from *C* to *A* is positive.

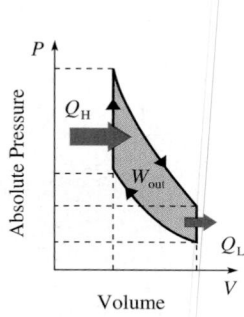

Figure 14.16 The *PV*-diagram for the Otto cycle. Heat flows in at a high temperature (Q_H), and heat flows out (Q_L) at a low temperature. The difference is equal to the net work-out (W_{out}). This is an idealized representation of the processes associated with the gasoline engine introduced by Nikolays Otto in 1876.

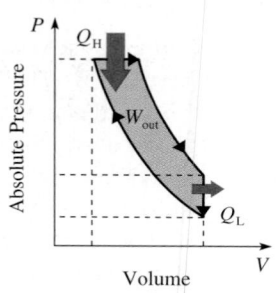

Figure 14.17 The *PV*-diagram for the Diesel cycle. Heat flows in at a high temperature (Q_H), and heat flows out (Q_L) at a low temperature. The difference is equal to the net work-out (W_{out}). This is an idealized representation of the processes associated with the internal combustion engine introduced by Rudolf Diesel in 1892.

essentially one-shot affairs, and we will do better to limit the term *engine* to cyclic devices that can continue to operate for as long as we provide energy (fuel) to them. *A heat engine is a cyclic device that converts thermal energy into work-out.*

There are lots of potentially useful cyclic processes and a number of practical work-producing thermal engines that are based on them. The gasoline motor in the family automobile is probably the most familiar thermal engine. It's based on the Otto cycle depicted in Fig. 14.16, which consists of two adiabats and two isovolumics. Many of the trucks being driven around today are powered by diesel engines. These are based on the Diesel cycle, shown in Fig. 14.17, which consists of two adiabats bounded by an isobar and an isovolumic (see Discussion Question 1). Diesel engines are generally heavier but more efficient than gasoline engines. They can be difficult to start, which is why truckers often leave their vehicles running while parked. {For an analysis of the modern gasoline engine click on **INTERNAL COMBUSTION** under **FURTHER DISCUSSIONS** on the **CD**.}

14.5 The Carnot Cycle

When Nicholas Léonard Sadi Carnot (pronounced car-no) published his masterwork *Reflections on the Motive Power of Heat* in 1824, he was twenty-eight years old. Carnot's father, Lazare, had been minister of war to Napoleon Bonaparte, and Sadi had fought in the outskirts of Paris in 1814. Recognizing that military power came with industrial mastery, Carnot turned his special ingenuity to bear on the great symbol of the age, the fire-chariot of the Industrial Revolution: the steam engine.

Carnot began his analysis with something more manageable: a simple reversible cycle that corresponds to a purely hypothetical device. Remarkably, the conclusions he drew were applicable, in one way or another, to all heat engines from the gas turbine to the power system that propels the family car. The Carnot engine (Fig. 14.18) is just a gas-filled cylinder and piston that can alternately be brought into contact with, or isolated from, either a high-temperature reservoir (steam) that serves as a source of heat or a low-temperature reservoir (cooling water) that serves as a sink into which heat is exhausted. The Carnot cycle is a four-stage reversible sequence (Fig. 14.19) consisting of: (1) an isothermal expansion (from *A* to *B*) with the reception of *heat-in* (Q_H) *at a constant high temperature*; followed by (2) an adiabatic expansion (from *B* to *C*); followed by (3) an isothermal compression (from *C* to *D*) with the rejection of *heat-out* (Q_L) *at a low temperature*; followed by (4) an adiabatic compression (from *D* to *A*). The net flow of heat to the system is ($|Q_H| - |Q_L|$).

Nicolas Léonard Sadi Carnot (1796–1832).

Figure 14.20 shows the corresponding stages of the engine. The *A-B-C* portion, the expansion, is the *power stroke*, during which positive work is done. For the first part of it (*A-B*), the work is provided for by an input of heat. During the second part (*B-C*), the work is provided for by draining some of the internal energy of the gas—that's why its temperature drops. The piston may transmit work to the environment by way of a crankshaft. And it would be the inertia of the rotating crankshaft that drives the system back from *C* to *D* to *A*, doing work on the gas and returning it to its original condition.

There are several crucial points here: by using both a high-temperature source (e.g., steam) and a low-temperature sink (e.g., cooling water), the cycle rides out and back on isothermals but now encloses a nonzero area, unlike Fig. 14.14; less heat is exhausted into the sink than entered from the source. Since $\Delta U = 0$, the net heat-in provides the energy for

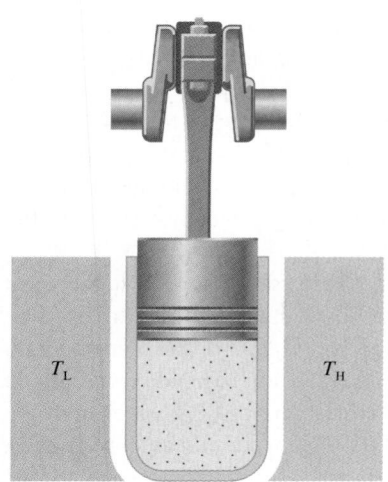

Figure 14.18 A Carnot engine consists of a cylinder and piston filled with gas. The device is able to be brought into contact with either a high-temperature or a low-temperature reservoir.

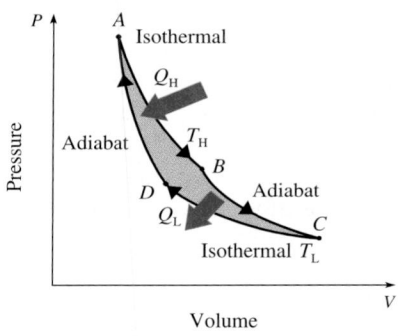

Figure 14.19 The four-stage reversible sequence representing a Carnot cycle.

the work-out, and the physics of that relationship is more apparent if we write it in terms of absolute values so we don't have to worry about signs: $|Q_H| - |Q_L| = |W_o|$. The lower the sink temperature, the more work would be done, and the less heat would be exhausted.

{Click on **THE CARNOT CYCLE** under **INTERACTIVE EXPLORATIONS** on the **CD**. Spend some time working with this simulation; its interactive nature will provide insights a traditional textbook cannot.}

Efficiency

The Carnot engine is important primarily because it represents a class of idealized devices (reversible engines) that can be shown to be the most efficient possible. Thus, it sets an upper limit on the attainable efficiency of any real heat engine, no matter how cleverly contrived, and that's a marvelous practical tool to have.

Most generally, the energy **efficiency** (*e*) of a process is the positive quantity defined as

$$\text{Efficiency} = \frac{\text{useful energy-out}}{\text{energy-in}} \qquad (14.9)$$

which is the ratio of what you get out (that is of value) to what you put in (to accomplish the process). With heat engines, the useful energy output is work, and the energy input is heat; over one cycle

[actual engine] $\qquad \text{Efficiency} = \dfrac{|\text{work-out}|}{|\text{heat-in}|} = \dfrac{|W_o|}{|Q_i|}$

Numerical values obtained here should be thought of as percentages: if the engine takes in 4000 J of heat and outputs 2000 J of work, its efficiency is 50%. This expression is applicable to all heat engines, real or imagined. For a somewhat idealized cyclic heat engine, which is both frictionless and otherwise lossless (e.g., no radiation), the First Law requires that $|W_o| = |Q_H(\text{in})| - |Q_L(\text{out})|$, and so starting with $e = |W_o|/|Q_i|$

[ideal engine] $\qquad e = \dfrac{|Q_H| - |Q_L|}{|Q_H|} = 1 - \dfrac{|Q_L|}{|Q_H|} \qquad (14.10)$

Because all engines, no matter how well built, will lose some energy via friction, this method of determining efficiency will generally produce a value that's higher than the actual efficiency given by Eq. (14.9). If besides exhausting energy, $|Q_L|$, to the low-temperature reservoir, there is a loss of energy via friction, $|W_f|$, the numerator in Eq. (14.10) would just become $|Q_H| - |Q_L| - |W_f|$.

Figure 14.20 The various stages of the Carnot cycle.

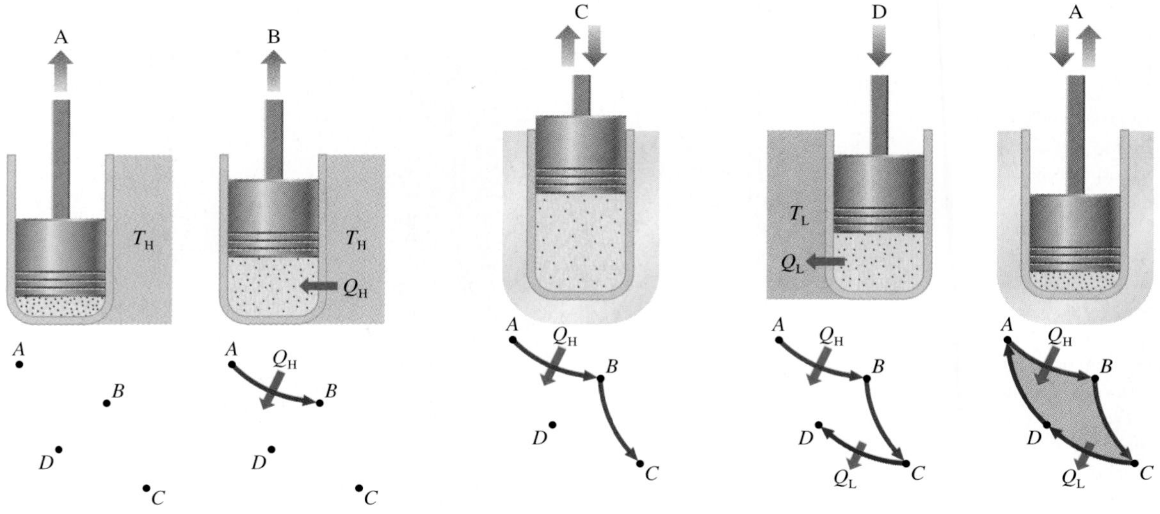

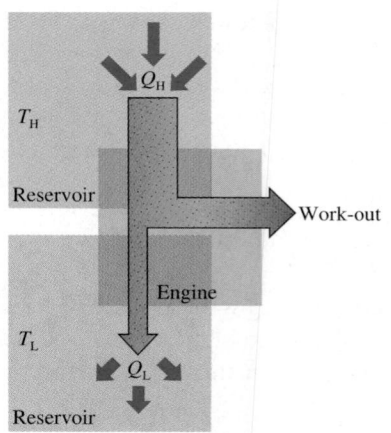

Figure 14.21 A schematic representation of a heat engine operating between high- and low-temperature reservoirs. Q_H flows in, Q_L flows out, and the engine does work. Energy-in equals energy-out.

Figure 14.21 schematically depicts a heat engine operating between a source and sink at high and low temperatures, respectively. Note that the efficiency increases as $|Q_L|$ decreases, becoming 1 (that is, 100%) only when there is no exhaust heat, a condition that has never been achieved (and one that is expressly forbidden by the Second Law of Thermodynamics, p. 510). Actual engines dissipate energy via friction and lose appreciable amounts of energy to the environment by convection, conduction, and radiation. That's why the hood of a car gets hot after a few minutes of operation, and the hotter it gets, the more wasted energy there is (Table 14.1).

The form of Eq. (14.10) in terms of heat is really not very convenient; it would be far simpler to measure if it were expressed in temperatures instead, and Carnot did just that. "According to established principles at the present time," he wrote, "we can compare with sufficient accuracy the motive power of heat to that of a waterfall." The idea was simple enough: consider a waterfall and an appropriate ideal harness-the-rushing-water-machine located somewhere in the cataract, as shown in Fig. 14.22. The maximum amount of energy available that can be converted to work by the machine is ΔPE, and that's proportional to the height the water drops. At most, with the machine at the very bottom, the available energy is proportional to h_H. With the machine anywhere else it's proportional to $(h_H - h_L)$. The fractional amount of energy available is then $(h_H - h_L)/h_H$, where any other parameters, like flow rate, cancel when the ratio is formed. Carnot wrongly believed heat was a liquid, and so with caloric flowing like water, he argued by analogy that the theoretical efficiency of an ideal heat engine should be $(T_H - T_L)/T_H$. Notice how the discussion has built into it a zero-temperature reference level—the denominator is the maximum change in temperature $(T_H - 0)$. Heat is naturally propelled down a temperature gradient just as water is naturally propelled down a gravity gradient. Moreover, $Q = cm\Delta T$ and $W = gm\Delta h$. Delightfully, even though there is no such heat fluid, Carnot's answer is correct. Before he died at age thirty-six in the cholera epidemic that swept through Paris in 1832, Carnot (his unpublished papers reveal) had already rejected caloric and even anticipated the First Law.

The preceding temperature statement of the efficiency of a Carnot engine (e_c) can be established more rigorously, but we would gain little from the effort. Let's instead focus on understanding the concept, represented by

[ideal engine]

$$e_c = 1 - \frac{T_L}{T_H} = \frac{T_H - T_L}{T_H}$$

(14.11)

Table 14.1

Approximate Efficiencies of Various Devices	
Device	**Efficiency (%)**
Electric generator	70−99
Electric motor	50−93
Dry cell battery	90
Domestic gas furnace	70−85
Storage battery	72
Hydrogen-oxygen fuel cell	60
Liquid fuel rocket	47
Steam turbine	35−46
Fossil-fuel power plant	30−40
Nuclear power plant	30−35
Nuclear reactor	39
Aircraft gas turbine engine	36
Solid-state laser	30
Internal combustion gasoline engine	20−30
Gallium arsenide solar cells	>20
Fluorescent lamp	20
Silicon solar cell	12−16
Steam locomotive	8
Incandescent lamp	5
Watt's steam engine	1

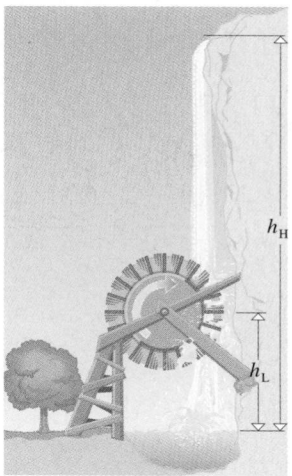

Figure 14.22 A waterfall analogy for the heat engine.

The maximum possible efficiency for any heat engine operating between two temperatures T_H and T_L is the **Carnot efficiency**, e_c.

Keep in mind that this expression is independent of the engine's working fluid—it could be a hot gas in an internal combustion engine, or steam in a steam turbine. Remember, too, that **these are absolute temperatures**. The lengths of the isotherms and adiabats for different Carnot cycles can be different, but all Carnot engines operating between the same two temperatures have the same efficiency. Incidentally, Eq. (14.11) along with Carnot's other work only came to prominence fifteen years after his death when Kelvin brought it to the attention of the scientific world.

As we shall see presently, the **Carnot engine has the highest allowable efficiency for a heat engine operating between any two fixed temperatures**. Nonetheless, that efficiency can only equal 1 (that is, 100%) when the cold-sink temperature is 0 K. In other words, *even the Carnot engine operates at less than 100% efficiency*. Still, e_c tells us the best we can hope for while operating between T_H and T_L. One way to ensure that the limiting value of Eq. (14.11) is as large as possible is to make T_L as low as possible, but there are practical restraints on that, too. In most real situations, the cold-sink is the open air or cool water from a river, so there isn't very much latitude on the low-temperature end. Of course, you could cool the low-sink with some sort of refrigerator, but that will dissipate additional amounts of energy and negatively affect the overall efficiency of the system. On the other hand, increasing T_H to increase e_c is far more practical. That's why power plants operate with superheated steam and why modern jet engines and auto engines alike are being designed to run at increasingly higher temperatures.

Carnot had discussed the possibilities of powering an engine by igniting an inflammable gas in a cylinder, but it was J. Lenoir (1859) who produced the first workable internal combustion engine. A traveling salesman and mechanic named Nikolaus Otto happened on a newspaper account of Lenoir's work (1876) and within a year built and patented the first four-stroke internal combustion engine. The automobile, airplane, and countless other noisy progeny from the lawnmower to the motorcycle and chainsaw all owe their existence to Otto and the founder of Thermodynamics, Sadi Carnot.

Example 14.7 **[I]** Determine the maximum possible efficiency for a steam engine operating between 200 °C and 27.0 °C.

Solution Mention of the efficiency of the best possible heat engine calls to mind Carnot. (1) TRANSLATION—A heat engine operates between two specified temperatures; determine its maximum efficiency. (2) GIVEN: $T_H \rightarrow 200$ °C and $T_L \rightarrow 27.0$ °C. FIND: e_c. (3) PROBLEM TYPE—Thermo/Carnot efficiency. (4) PROCEDURE —We need only substitute into the defining equation, Eq. (14.11). (5) CALCULATION—First con-

vert the temperatures to kelvins:

$$e_c = 1 - \frac{300 \text{ K}}{473 \text{ K}} = 1 - 0.634 = 0.365 = \boxed{36.5\%}$$

In practice, losses will cut that down by about one-third, but this efficiency is the very best we can hope for operating between those two temperatures.

Quick Check: $(T_H - T_L)/T_H = (473 \text{ K} - 300 \text{ K})/(473 \text{ K})$ $= 37\%$.

14.6 Refrigeration Machines & Heat Pumps

A heat engine uses thermal energy flowing from a high-temperature reservoir to a low-temperature reservoir to produce work. A *refrigeration machine* does the reverse—it uses work to transfer thermal energy from a low-temperature reservoir to a high-temperature reservoir. When the region to be cooled is a food chest, the device, an ordinary household refrigerator, simply expels the extracted heat into the room. That's fine in the winter, but it's admittedly a little dumb in the summer. On the other hand, to cool the room itself, the refrigeration machine, known as an air conditioner, simply dumps the hot exhaust outside. If we turn the air conditioner around and cool the outdoors, regardless of its temperature, blowing the exhaust energy into the house, we have a heating system: such a reversible device is called a *heat pump*.

POWER PLANTS & AUTO ENGINES

A modern fossil-fuel electric-generating power plant uses superheated steam to provide a high-temperature source at perhaps 500 °C (≈800 K). High-pressure steam violently expands into the turbine, impacting upon and driving the spinning blades (Fig. 14.23). The steam is expelled into a cooled condenser at ≈373 K. The maximum theoretical efficiency follows from Eq. (14.11) and is 53%, although thermal losses generally reduce that to about 40% in modern facilities (much of the untransformed energy goes up the smokestack).

The automobile engine functions at approximately 3300 K, exhausting its spent fuel at about 1400 K. That gives it a theoretical efficiency of roughly 58%. Even though the exhaust gas escapes into the surrounding cool air, it leaves the engine, where it can do work, at a much higher temperature T_L. In reality, most of the fuel energy fed into the system is lost by way of thermal energy escaping to the air around the engine and out with the still hot exhaust gas.

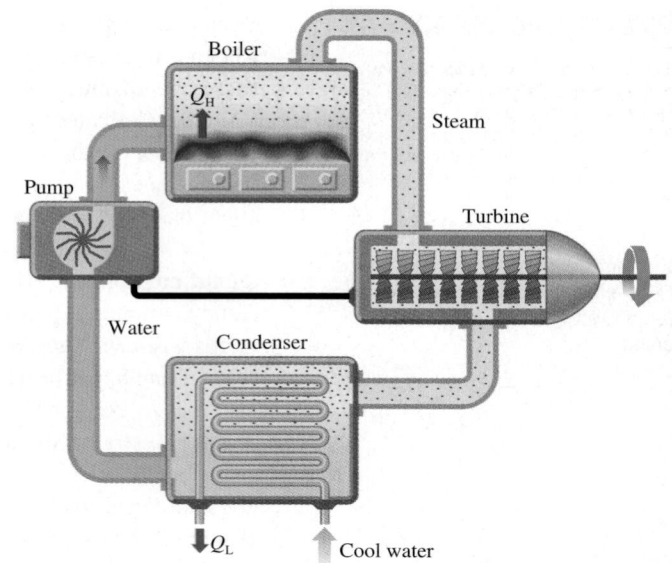

Figure 14.23 A modern fossil-fuel electric-generating power plant. Steam drives the turbine, transferring energy to it. The turbine powers a high-voltage electric generator. A large pressure difference is maintained across the turbine by condensing the steam. The steam turns back to water in the condenser (the cold-sink). Water of extremely high purity is recirculated in the sealed system.

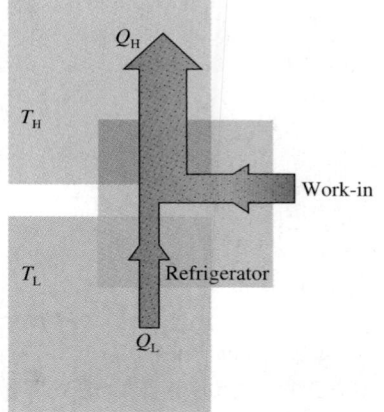

Figure 14.24 A schematic representation of a refrigeration machine. It takes heat Q_L from a low-temperature region and dumps it in a high-temperature region, at the cost of a certain amount of work-in.

Figure 14.24 schematically depicts a continuously operating refrigeration machine, which is simply an engine run in reverse—Fig. 14.25 is probably a more familiar version. If it were a Carnot refrigerator, the PV-diagram would be identical to Fig. 14.19 but run through counterclockwise. From the First Law it follows that |heat-out| = |heat-in| + |work-in| or

$$|Q_H(\text{out})| = |Q_L(\text{in})| + |W_i| \qquad (14.12)$$

The effectiveness of a refrigeration machine cannot be measured by way of the efficiency, Eq. (14.10), because there is no work-out. Consequently, we introduce the ***coefficient of performance*** (η), which is defined as the ratio of the amount of heat removed from the cold reservoir (the heat-in to the system) to the amount of work done on the system, represented symbolically as

$$\eta = \frac{|Q_L|}{|W_i|} = \frac{|Q_L|}{|Q_H| - |Q_L|} \qquad (14.13)$$

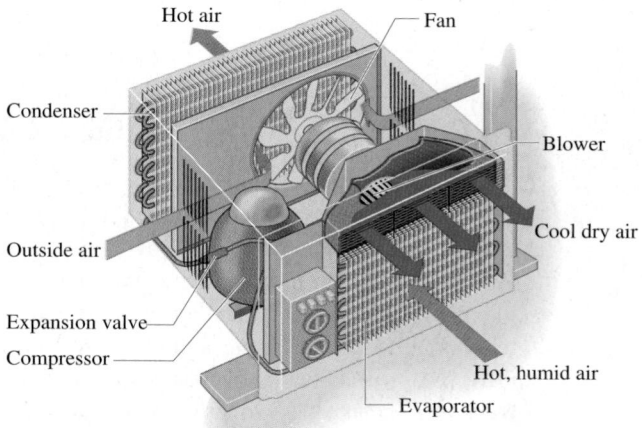

Figure 14.25 A window-mounted air conditioner. Hot, humid room air is drawn over the cold evaporation coils, where it is cooled. Water vapor condenses, runs off, and collects in the base pan, dripping away outside.

The larger η is, the more effective is the refrigeration machine, and values of about 5 are typical.

The best possible coefficient will correspond to a Carnot engine running in reverse. Accordingly, set Eq. (14.10) equal to Eq. (14.11), whereupon

$$e_c = 1 - \frac{|Q_L|}{|Q_H|} = 1 - \frac{T_L}{T_H}$$

and so

$$\frac{|Q_L|}{|Q_H|} = \frac{T_L}{T_H} \tag{14.14}$$

The *coefficient of performance for an ideal system* in terms of the operating temperatures can now be determined from Eq. (14.13) using Eq. (14.14); accordingly

[ideal system]
$$\eta_c = \frac{T_L}{T_H - T_L} \tag{14.15}$$

{For more on how an actual refrigerator works click on **REFRIGERATION MACHINES** under **FURTHER DISCUSSIONS** on the **CD**.}

Example 14.8 **[II]** Determine the best possible coefficient of performance that can be achieved for an air conditioner maintaining a room at 21 °C (70 °F) when the outside temperature is 36 °C (97 °F). Suppose 5.0 MJ of heat leaks into the room in an hour. The machine exhausts the waste heat that it generates in operating, outside the room via a cooling fan. How much work would that air conditioner have to do per hour to maintain the temperature of the room? How much heat in total is exhausted outside per hour?

Solution The phrase "the best possible coefficient of performance," should immediately call to mind η_c. (1) TRANSLATION —A refrigeration machine operating between two specified temperatures must remove a known amount of heat leaking in; determine (a) its maximum coefficient of performance, (b) the work per hour it must do, and (c) the heat per hour it must exhaust. (2) GIVEN: $T_L \rightarrow 21\,°C$, $T_H \rightarrow 36\,°C$, and $Q(\text{in}) = 5.0$ MJ. FIND: η_c, W_i, and Q_H. (3) PROBLEM TYPE Thermodynamics/refrigeration/coefficient of performance. (4) PROCEDURE—Remember that $\Delta U = 0$, and so ΔQ equals the work. (5) CALCULATION —(a) Using the definition, Eq. (14.15),

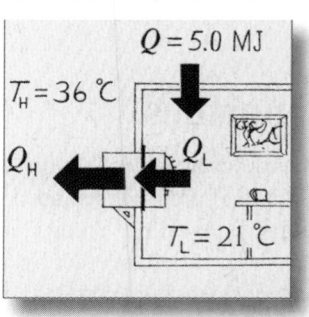

$Q = 5.0$ MJ
$T_H = 36\,°C$
Q_H Q_L
$T_L = 21\,°C$

$$\eta_c = \frac{T_L}{T_H - T_L} = \frac{294\ K}{(309\ K) - (294\ K)} = 19.6$$

(b) The hourly heat leak, $Q(\text{in})$, enters while the system is being maintained at T_L and equals Q_L. The work that must be done each hour to remove Q_L joules of thermal energy follows from the definition of η, Eq. (14.13):

$$W_i = \frac{Q_L}{\eta_c} = \frac{5.0\ MJ}{19.6} = \boxed{0.26\ MJ}$$

(c) To remove 5.0 MJ, this machine does only 0.26 MJ of work. In the process, that 0.26 MJ is converted to thermal energy, which must also be exhausted outside. From Eq. (14.12), we have

$$Q_H = Q_L + W_i = 5.0\ MJ + 0.26\ MJ = \boxed{5.3\ MJ}$$

This is the total amount of heat exhausted to the outside.

Quick Check: For every 19.6 J it removes, the machine does 1 J of work and produces 1 J of waste thermal energy. When it does 0.26 MJ of work, it removes 5 MJ of thermal energy from the room and produces an additional 0.26 MJ itself.

Absolute Temperature

A Carnot engine takes in a certain amount of heat (Q_H) at a high temperature (T_H) and exhausts a lesser amount (Q_L) at a lower temperature (T_L), the energy difference going into work-out. The amount of heat exhausted, and therefore essentially wasted, is directly proportional to T_L: the lower the sink temperature, the less the value of Q_L, the more work is done, and the higher the efficiency. That much has been considered already:

$$\frac{|Q_L|}{|Q_H|} = \frac{T_L}{T_H} \qquad [14.14]$$

Kelvin proposed that this ratio be used to define an absolute thermodynamic temperature scale (p. 425) that utilized the Carnot engine as a thermometer. This would have the virtue of eliminating the dependence on the working substance that plagues all ordinary thermometers. Kelvin assumed the validity of Eq. (14.14) and used it to define the ratio of two thermodynamic temperatures. Inasmuch as Eq. (14.14) is known to be satisfied when an ideal gas is used in a gas thermometer (since we used the ideal gas to derive the equation in the first place), the gas thermometer must agree with the Kelvin thermometer.

Suppose, then, that we have a Carnot cycle operating between the temperatures of boiling and freezing water. Now divide the cycle (Fig. 14.26) into 100 equal-area little Carnot subcycles, each doing the same work ΔW and each residing between isotherms 1 K apart. As we progress down from one cycle to the next, more and more net work is being done, and less and less heat is exhausted to the next little engine. It follows that if we continue to add more of these little 1-K cycles below the ice-point temperature, sooner or later we will reach a point where the heat-in to one of them all goes into the work-out (ΔW), and no heat is exhausted to the last low-temperature sink, which must be the Absolute Zero of temperature: 0 K. At that point the total amount of heat-in, Q_H, is transformed entirely into work-out, $\sum \Delta W$. Presumably there could be no temperature lower than 0 K, since that would imply an efficiency greater than 1, via Eq. (14.11), and more work-out than heat-in, which would violate the First Law.

As we will see later, the notion that all the heat flowing into the last little engine is converted to work violates the Second Law of Thermodynamics, and that suggests that it may be impossible to actually bring bulk matter to a temperature of Absolute Zero.

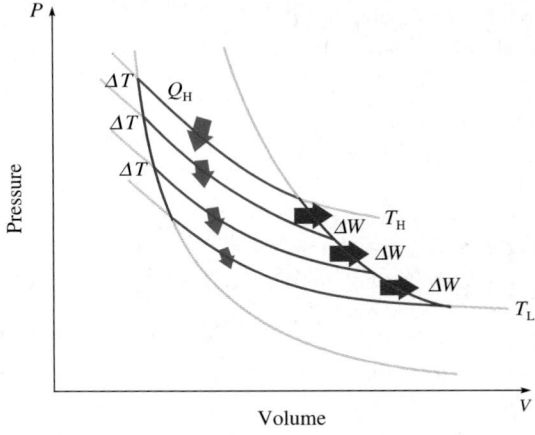

Figure 14.26 A Carnot cycle divided into a number of little Carnot cycles. Each one expels less heat than the one before.

The Second Law of Thermodynamics

The Second Law of Thermodynamics was introduced into physics by way of a rather mundane statement about heat flow. But it evolved into a principle of great scope and significance, as far-reaching as the First Law, and perhaps even more revealing. Things happen via change, and energy is the measure of that change. Everything is tied together by a web of interactions, and change tends to spread out—one entity influencing another—and so energy tends to disperse. Strike a match and the smoke, sound, thermal energy, and light will all disperse. The Second Law is about the dispersal of energy, the unfolding of events. Time itself is given direction by the Second Law.

The germ of the idea was already there when scientists realized from countless observations that:

> **Heat flows naturally from a region at high temperature to a region at low temperature. By itself, heat will not flow from a cold to a hot body.**

And this is one of several equivalent but different statements of the Second Law.

Once the concept of internal energy was formulated, this observation was enriched with the insight that the spontaneous flow of heat from one object to another is independent of the amount of internal energy in either entity. Heat is propelled by a temperature gradient—no gradient, no heat. From Carnot's work, we see that a cyclic heat engine must operate between high- and low-temperature reservoirs. If both temperatures are the same, its efficiency is zero and no heat enters the engine, across which there must be a gradient.

Furthermore, with any low-sink temperature other than zero, even for an ideal engine, there will be some unavoidable loss of heat Q_L to the reservoir: the efficiency must be less than 100%. That much follows from Eq. (14.14), if $T_L \neq 0$, $Q_L \neq 0$. There is an energy equivalence between heat and work, but there is also a *dissymmetry*, a difference on the microscopic level, that has its roots in the difference between order and disorder. We can convert all the ordered energy we wish *entirely* into thermal energy—the brakes on your car do it, and Rumford's cannon borer (p. 457) did it. Nature allows the continued spontaneous transformation of work entirely into thermal energy, but it does not allow the reverse. The transformation of disordered heat into ordered work must be paid for, as it were, by an additional loss of heat Q_L (Fig. 14.20). The formal wording of the Second Law in terms of the above ideas is known as the ***Kelvin–Planck statement***:

> **No ongoing process is possible for which the sole result is the removal of heat from a source and its complete transformation into work.**

In other words, ***it is quite impossible to produce an engine that will generate work by extracting heat from a reservoir without expelling some waste heat to a lower-temperature sink*** (Fig. 14.27). An anti-Kelvin device, which could continuously do work extracting thermal energy from a single reservoir with no concern about temperature gradients, is a **perpetual motion machine of the second kind**. Although such a device would not violate the First Law, it would violate the Second Law. It could not create energy or, for that matter, run forever, but it could do the next best thing—namely, draw off energy at no cost from the vast low-temperature sources of atmosphere and ocean. No such machine capable of solely producing ordered work out of disordered heat is possible.

The Clausius Formulation

It's commonplace for hot objects to cool spontaneously, all by themselves, losing heat to a colder environment, but we presumably never observe cold objects spontaneously getting colder by losing heat to a warmer environment. And that's true even when heat flows from a warm body with a tiny amount of internal energy (a cup of tea) into a cold body with an immense amount of internal energy (the atmosphere). In the same way, an object can start at the top of a hill and spontaneously roll down it, but never does one begin at the bottom and roll uphill unaided by any external agent. There is a specific way in which all processes (not just those involving heat) naturally unfold, and the Second Law, at its heart, is about just that: the natural direction in which energy flows as it irreversibly redistributes itself.

Figure 14.27 A schematic representation of an impossible anti-Kelvin device. It converts heat-in to work with no wasted heat-out.

Rudolf Clausius (1822–1888).

Rudolf Clausius proposed an alternative statement of the Second Law that at first glance seems quite different from Kelvin's:

> **No ongoing process is possible in which the sole result is the transfer of a given amount of thermal energy from a body at low temperature to a body at a higher temperature.**

The Clausius statement of the Second Law: It is impossible to construct a device that will cause heat to flow in an ongoing way from a low temperature to a high temperature without producing some other effect (e.g., waste heat) in the process.

Refrigerators are not forbidden, provided they produce some other result in addition to the transfer of thermal energy. The above statement may imply that an input of work is necessary to transport heat "uphill," but it doesn't say so explicitly. The Kelvin–Planck statement talks about the conversion of heat to work. The Clausius statement talks about the natural direction of the flow of heat. Both suggest a hidden asymmetry in the manner in which processes spontaneously unfold.

To prove that these two perspectives are in fact equivalent, refer to Fig. 14.28, which depicts an ordinary reversible engine being driven in reverse by a hypothetical anti-Kelvin engine (one that violates Kelvin's statement). The latter converts an amount of heat (Q_{HK}) directly into work, which drives the ordinary engine as a refrigerator. It, in turn, exhausts heat (Q_{HO}), some of which can be imagined circulating back to the anti-Kelvin device. The net effect is that the combined system has the sole result of taking heat from the low-temperature reservoir and dumping it into the high-temperature reservoir, in direct conflict with the Clausius statement. Thus, if an anti-Kelvin device were possible, it could be used to make an anti-Clausius device. Similarly, the ordinary engine on the left in Fig. 14.29 is effectively being supplied heat by a hypothetical anti-Clausius engine that carries heat from the low to the high reservoir. The ordinary engine draws off heat (Q_{HO}), expels heat (Q_L), and performs work. The net effect of the combined system is solely to convert an amount of heat Q_H into work in direct conflict with the Kelvin statement. Thus, if an anti-Clausius engine were possible, it could be used to make an anti-Kelvin engine. We must conclude that if either statement is false, the other is false—they might be wrong, but they must be equivalent.

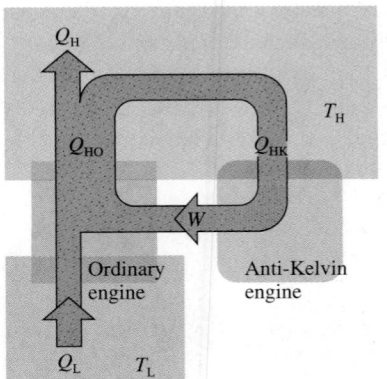

Figure 14.28 If an anti-Kelvin engine could exist, it could be attached to an ordinary engine, and the combination would violate the Clausius statement of the Second Law. It could take in Q_L and dump it as Q_H and do nothing else.

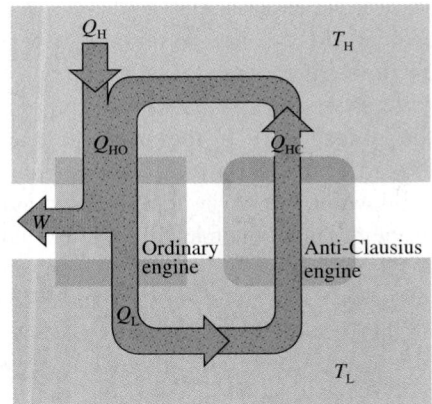

Figure 14.29 If an anti-Clausius engine could exist, it could be attached to an ordinary engine, and the combination could violate the Kelvin statement of the Second Law. It could remove heat Q_H from a source, converting it entirely into work.

14.7 Entropy

In 1865 Clausius introduced a new concept, *entropy*, in order to distinguish between the ideas of *conservation* and *reversibility*. When a match burns, energy is conserved, but that's not the whole story of the change. The match will never spontaneously reform itself from the ash and smoke—the process is irreversible and the energy naturally disperses. This is very unlike the *idealized* occurrences of (lossless) Newtonian dynamics that are reversible, and conservation tells us all we need to know. Dynamics is independent of the direction of time: a lossless bouncing ball bounces forever and is independent of whether time runs forward or backward—the ball falls, climbs, and falls again or climbs, falls, and climbs again, either way. By contrast, heat conduction and friction are irreversible changes; indeed, without irreversible processes, life as we live it would be impossible.

When an engine operates, converting heat-in into work-out, some fraction of the input energy is irreversibly lost as heat-out. In effect, the engine takes in heat and puts out two distinct channels of energy: highly useful ordered energy (work) and not-so-useful disordered energy (heat). The expelled waste thermal energy is at a low temperature and will thereafter be quite difficult to extract and use again. To be sure, the more efficient the engine, the lower the sink temperature is likely to be, and the harder it will be to get at that exhausted thermal energy in the future. It's like water pouring over the falls in Fig. 14.22; once it has descended down to sea level, it is a lot less useful. In a real sense, ***the quality of thermal energy (i.e., its ability to do work on a macroscopic scale*) is dependent on its temperature***. When an engine operates, it does not use up energy—energy is conserved—it transforms energy, and it does that at the cost of degrading a certain amount of high-quality, high-temperature energy into low-quality, low-temperature energy. Energy doesn't vanish, it is dissipated (i.e., unconcentrated).

Suppose there is an ideal gas in a chamber and it's released through a valve into an insulated evacuated vessel. The gas spontaneously expands, filling the vessel; but there is no change in internal energy because no work is done. The pressure and volume both change; the temperature remains constant. The system in its initial state transforms, all by itself, to a new final state where the energy is the same. And yet, if we want to return the gas to the original chamber, we will have to do work on it—the gas will not spontaneously squeeze back through the valve. Once the valve is opened, the gas is no longer in equilibrium, and the spontaneous rush toward equilibrium is irreversible. The system is directed by a hidden imperative (other than maintaining the total energy) to seek a new dispersed state from which it will not return on its own.

In the process of spontaneously expanding, the gas loses some of its ability to do work (P is lower), though its energy is unchanged. The amount of work it takes to recompress the gas would seem to be equivalent to the ability to do work it lost. In any event, the final state of the gas is profoundly different from the initial state. What we want now is a new property of the system that will reflect that difference; a property that will allow us to know if spontaneous change will take place between any two states, a property that governs the direction of natural change in all systems from galaxies to embryos. Clausius proposed that there was such a measure, and he gave it the symbol S and the name **entropy**.

Experience suggests that the entropy of a system be defined such that it increases as the system degrades. In other words, natural changes (i.e., spontaneous ones) carry an isolated system toward disorder (disorganization) and are accompanied by an increase in entropy. Unnatural changes, which must be forced on a nonisolated system by an outside agency, can bring it more order and decrease *its* entropy. **Reversible changes are entropy neutral**;

ORDER, DISORDER, & A DRIVE AROUND TOWN

The family car is an example of how highly ordered available chemical energy (in the fuel) is converted into highly disordered, low-temperature unavailable thermal energy. Drive for 20 miles and you will convert a gallon of gasoline into a ribbon of heated air. Then try to drive back by reclaiming those 1.3×10^8 (now essentially useless) joules that are still there warming the sky.

---

*The familiar definition of energy as "the ability to do work" (and that presumably means work on a macroscopic scale, an organized phenomenon) is inconsistent with the idea that energy is both conserved (First Law) and degraded (Second Law).

they leave S for an isolated system unaltered. Since entropy has something to do with degradation and disorder, it's not surprising that a reversible process carried out by an isolated system (which must be able to return to exactly the state from which it started) should have $\Delta S = 0$. The entropy equivalent of the Second Law, which is often called the *Entropy Principle*, is as follows:

> **When an isolated system undergoes a change, passing from one state to another, it will do so in such a way that its entropy will increase, or at best remain the same.**

This imperative gives direction to the natural unfolding of events. Remember that entropy describes the state of a system. It is a property, like internal energy and temperature, of engines and ice cubes and cats.

Whenever a process occurs in an isolated system, its entropy increases, or at best ideally remains the same. If the system is not isolated, then its entropy can be made to decrease by impressing order upon it from outside via external agencies. In that case (according to the Entropy Principle), the entropy of the surroundings will have to increase even more; so the entropy of the ultimate isolated system, the entire Universe, nonetheless increases.

> **Every process increases (or at best ideally leaves unchanged) the entropy of the Universe.**

The essential measure of the difference between reversible and irreversible processes is the change in entropy of the Universe. The essential measure of the difference between what occurs naturally and what does not occur naturally is the change in entropy of the Universe.

The Second Law and Entropy

We can get some insight into the behavior of entropy by using the Second Law as a guide. The Kelvin–Planck statement of the Second Law is subsumed into the entropy version provided that *the entropy of a system increases when it receives heat, decreases when it loses heat, and remains unchanged by work done in the absence of friction*: $S \propto Q$. In that context, work is an organized mechanical change in energy and, as such, does not affect the microscopic disorder of a system. (By contrast, any mechanism—such as friction—that disperses energy into the agitation of the system's particles will increase entropy.) Consider an isolated system composed of a reservoir of thermal energy, an anti-Kelvin engine, and its immediate surroundings. Running the engine off the reservoir will have only one effect: the entropy of the reservoir will decrease, and therefore the engine is forbidden.

Similarly, the Clausius statement will be subsumed if we assume that *the higher the temperature of the heat transferred, the less entropy change results*: $S \propto 1/T$. Imagine an isolated system consisting of two objects, one hot, the other cold, each with the same net internal energy. Bring them into contact—heat spontaneously flows out of the hot one into the cold one, and they both reach a new equilibrium temperature. The same amount of heat is lost by one as is gained by the other. But the high-temperature heat is "high quality," and its loss only slightly decreases the entropy of the high-temperature object. Whereas the heat delivered at a low temperature is of "low quality" and increases the entropy of that body appreciably, the net effect of this spontaneous transfer of heat is properly an increase in the entropy of the system and, hence, the Universe.

In this way, we know that entropy is directly proportional to the heat supplied and inversely proportional to the temperature at which it's supplied. Moreover, for a Carnot cycle,

$$\frac{|Q_L|}{|Q_H|} = \frac{T_L}{T_H} \tag{14.14}$$

and

$$\frac{|Q_H|}{T_H} - \frac{|Q_L|}{T_L} = 0$$

Remember that heat-in (Q_H) is positive and heat-out (Q_L) is negative, so we can drop the absolute value signs and rewrite this equation as

$$\frac{Q_H}{T_H} + \frac{Q_L}{T_L} = 0 \qquad (14.16)$$

We interpret this expression to represent the zero net change in entropy experienced by a Carnot engine over its reversible cycle. Entropy "flows" in from the high-temperature source, "flows" out to the low-temperature sink, and the engine returns to its original state. Accordingly, Clausius defined the change in the entropy of a system (ΔS) experienced during a ***reversible process*** as the ratio of the heat-in to the absolute temperature at which it enters:

STUDY GUIDE

Remember that the expression $\Delta S = Q/T$ can only be applied to finite values of ΔS provided T is constant.

[reversible isothermal process]

$$\Delta S = \frac{Q}{T} \qquad (14.17)$$

When heat enters a system, its entropy increases; when heat leaves, its entropy decreases. Observe that the entropy of a system will increase by the same 1 unit when 1000 J enters at 1000 K or, alternatively, when 0.001 J enters at 0.001 K. ***To use Eq. (14.17) for finite entropy changes (ΔS), the temperature has to be constant.*** We could generalize Eq. (14.17) using derivatives ($dS = dQ/T$) and talk about adding up infinitesimal reversible processes via integration. Without the ability to apply calculus to the analysis, we'll just have to be satisfied with a rather limited range of application of Eq. (14.17), but it's good to know that there's lots more to the subject than one can find in these pages.

Example 14.9 **[I]** By how much does its entropy change when a 10-kg block of ice at 0 °C is completely melted into water at 0 °C?

Solution This problem is obviously about entropy, and we know how to deal with entropy provided T = constant. (1) TRANSLATION—A known mass of a substance (ice) melts at a specified temperature; determine the resulting entropy change. (2) GIVEN: m = 10 kg, and ice melting at 0 °C. FIND: ΔS. (3) PROBLEM TYPE—Thermo/entropy. (4) PROCEDURE—To find the entropy for this isothermal process, we use the defining expression, Eq. (14.17). (5) CALCULATION—Here T is absolute, namely, T = 273 K. First determine Q for the melting. From $Q = mL_f = (10 \text{ kg})(334 \text{ kJ/kg}) = 3.34 \text{ MJ}$, it follows that

$$\Delta S = \frac{Q}{T} = \frac{3.34 \text{ MJ}}{273 \text{ K}} = \boxed{12 \text{ kJ/K}}$$

Quick Check: $Q = T\Delta S = (273 \text{ K})(12 \text{ kJ/K}) = 3.3 \text{ MJ}$. We can expect an increase in entropy corresponding to the increase in molecular disorder as the crystalline solid gives way to the liquid.

THE CARNOT ENGINE & ENTROPY

Suppose the isolated system is a Carnot engine with the attendant source and sink (Fig. 14.18). In isothermally expanding from A to B, the source loses entropy, $\Delta S_H = (Q_H/T_H)$, and the engine gains the same amount—the net change is zero. The expansion to C is adiabatic ($Q = 0$) and therefore **isentropic** ($\Delta S = 0$). Over the isothermal compression to D, the engine loses entropy, $\Delta S_L = (Q_L/T_L)$, and the sink gains the same amount. The final adiabatic compression back to A is isentropic ($\Delta S = 0$), and it closes the cycle. Since $\Delta S_H = \Delta S_L$, from Eq. (14.16), the *net change in entropy of the entire system during this reversible cycle is indeed zero.* The Carnot engine is just efficient enough to exhaust the smallest allowable amount of waste heat. Had it been still more efficient, expelling less low-temperature heat, the entropy change would have been negative and that's forbidden by the Second Law. ***All real engines will exhaust more heat and produce a net increase in entropy.***

Of course, entropy increases in irreversible processes even though we don't have a formula that spells out exactly how. Still, ***we can get around that and use Eq. (14.17) to calculate the entropy change using a hypothetical reversible process between the same initial and final states***. Since the entropy change only depends on these two states, ΔS for the reversible process will be the same as for the irreversible process. For example, imagine that an ideal gas irreversibly undergoes a free expansion to a new state. Now suppose that the gas is returned to its initial state via a quasi-static isothermal compression, which decreases its entropy by an amount ΔS, given by Eq. (14.17). Since the system is back where it started, ΔS must be the increase in entropy it experienced during the irreversible expansion.

EXPLORING PHYSICS ON YOUR OWN

Heat Engines, Soda Cans, & Plastic Bags: Put about a half inch of water in an empty soda can and heat it until you can hear boiling and steam comes out the opening. With gloved hands, quickly turn the can over and plunge it into a bowl of cold water. The can will be immediately crushed. What does work on the can? Imagine this as part of the cycle of a soda-can heat engine; what are the high-and low-temperature reservoirs? Does heat flow into and out of the can? What would happen if you upended the hot can into a bowl of equally hot water? What does that say thermodynamically? Now fill a zip-lock plastic bag with air. Fold over the top edge and tape it closed so it's really air tight. Put it into a freezer for an hour and it will be somewhat collapsed when you take it out. Put the cold bag on a table and quickly place an object (I used a small book) on the bag. The bag will expand and do work raising the load. The freezer, the bag, and the room can be thought of as a rather poor heat engine (you have to put the bag back in the refrigerator for each cycle) operating between high- and low-temperature reservoirs. Explain.

Example 14.10   [II] A very large hot object at 573 K is put in contact with a very large cool object at 273 K, and 20.0 kJ of heat flow irreversibly from one to the other. If neither object changes temperature appreciably, by how much is the entropy of the Universe changed?

Solution   This problem is obviously about entropy, and we know how to deal with entropy provided $T =$ constant.   (1) TRANSLATION—An object at a known temperature experiences a specified temperature change as a result of the transfer of a known amount of energy; determine the resulting entropy change.   (2) GIVEN: $T_H = 573$ K, $T_L = 273$ K, and the heat $Q = 20.0$ kJ. FIND: ΔS.   (3) PROBLEM TYPE—Thermo/entropy. (4) PROCEDURE—The entropy change of the Universe equals the net entropy change of the high- and low-temperature objects. (5) CALCULATION—Use the defining expression, Eq. (14.17),

$$\Delta S = \Delta S_L + \Delta S_H = \frac{Q}{T_L} + \frac{Q}{T_H}$$

Remembering that heat-in is positive

$$\Delta S = \frac{+20.0 \times 10^3 \text{ J}}{273 \text{ K}} + \frac{-20.0 \times 10^3 \text{ J}}{573 \text{ K}}$$

$$\Delta S = 73.26 \text{ J/K} - 34.90 \text{ J/K}$$

and to three figures $\boxed{\Delta S = +38.4 \text{ J/K}}$.

Quick Check: $|\Delta S_H| < |\Delta S_L|$ is proper since high-temperature heat has less entropy than low-temperature heat. Moreover, $T_H/T_L = |\Delta S_L|/|\Delta S_H| = (573 \text{ K})/(273 \text{ K}) \approx 2.1 \approx (73.3 \text{ J/K})/(34.9 \text{ J/K})$.

Order and Disorder

It was not until 1878 that Boltzmann reformulated entropy, giving it a new fundamental clarity. He recognized that entropy was a measure of the disorder of the Universe on an atomic level. Once lit, the neatly arranged molecular pattern of a match spontaneously transforms into the mayhem of gas, smoke, and flame. The disorder of the Universe increases, entropy increases; energy once concentrated is soon spread out with an evenness that is the essence of thermodynamic disorder. ***Maximum disorder (disorganization) corresponds to a random tumult, an undifferentiated uniformity.*** The easily differentiated configuration of a lit firecracker represents a high degree of order. The uniform cloud of dust and smoke after it explodes corresponds to disorder.

We never measure entropy, but only entropy change, ΔS, which accompanies physical change. And a system naturally changes (energy naturally flows) so as to increase universal entropy ($\Delta S \geq 0$). **The quantity ΔS measures the degree of disorder associated with any change (ΔE) in a system.** *An isolated system in a maximum entropy configuration cannot of itself change macroscopically; it is in equilibrium.* Every physical and chemical process in nature occurs in such a way that the net entropy of all the matter involved increases—the disorder of the Universe inexorably moves toward a maximum.

Mixtures of any kind are more disordered than were their separate constituents. Two gases introduced into a chamber will "always" mix, even though there is no energy incentive to do so. Colored grains of sand shaken in a bag will "always" mix. Yet neither the gas-

THE UNIFORMITY OF DISORDER

We tend, at first, to think of disorder as a scene with everything tossed about, a shoe here, a sock there. But that's only crude disorder. Splendid disorder requires a fine, dispersed dust—no shoes, no socks, but a uniform cloud of agitated atoms. A bull in a china shop will spontaneously powder the pretty cups, crush the delicate vases, and increase the uncompromising entropy of the Universe. The disorder of a distribution of particles is reflected by both their separations and their speeds. If all the particles in a perfect crystal were at rest, that would correspond to its state of maximum order, only approached at 0 K.

es nor the sands are likely to spontaneously unmix, to neatly separate, to decrease in entropy. For that matter, you can shake the bag of sand for centuries, and though it is possible, it is most improbable that the different colored specks will ever separate perfectly again. True, you could reach into the system (then no longer isolated) and sort out the sand—you could reestablish order and decrease entropy. But you would spend more order in the effort; you would disorder your breakfast more than you would order the sand—the Universe would lose again. It must lose again.

Wherever entropy is decreasing—in a factory where someone is stringing beads, among the ice cubes in a freezer, or with the cells forming a living embryo—order is rising out of disorder. But whenever that occurs locally, there will always be a hidden harvest of mayhem that overbalances the scales on the side of universal disorder. The freezer will make its ice cubes, but the system isn't isolated. The cost will be in electrical energy converted to heat and in the end, $\Delta S > 0$.

THE ARROW OF TIME

Of all the laws we have studied, only the Second gives a direction to the unfolding of events, to the progression of processes. It points the arrow of time in the "forward" direction—toward maximum disorder. Time stops when all events stop, when nothing happens, and the Second Law provides the direction to processes and to time. Have you ever seen the gases come back, the smoke descend, the light return, the heat pour in, the flame collapse, and a match reform, unburnt and whole? Will the gray hairs spontaneously darken, the skin grow tight and supple, the sags unsag? Not likely! And yet energy would be conserved. If everything reversed and processes ran toward order, would time be going backward? All by itself the game runs downhill, and it is the Second Law that shows the way.

Entropy as Probability

Boltzmann found the mathematical fingerprint of disorder in a statistical relationship describing the likelihood that a particular molecular arrangement would occur. Were we to take all the separate molecules that make up a match and put them in a box and shake them, the chance of finding them all bound together, neatly arranged as a match, is incredibly small. It is faintly possible but, given the astounding number of possible combinations that do not lead to a match, exceedingly unlikely. Unwrap a deck of playing cards and observe the arrangement. That ordered state is just one out of 80 million, million, million, million, million, million, million, million, million, million possible arrangements of those 52 cards. Order is far less likely than disorder. Once the molecules in an egg are scrambled, the chance that all of them will simultaneously find their way back to so unlikely an arrangement as the original ordered ovum is slim indeed. Every process irreversibly scrambles a bit of the cosmic egg.

A system of continually agitated molecules moves from one state to the next because the new configuration is more disordered and therefore statistically more probable. Throw four coins on the table and it is most likely that two will be heads and two tails. Each coin has two ways of landing, and there are four coins. Accordingly, there are a total of 2^4 or 16 different ways the four coins can land (16 *microstates*), as shown in Table 14.2. Of these, 6 equivalent microstates correspond to 2-heads-2-tails (i.e., 1/2-heads). That's 6/16, or 38%, which is the most likely state to occur. By comparison, there is only one way (1 microstate) where all the coins will be heads, and the probability of that highly ordered state happening

Table 14.2

Possible Arrangements of Four Coins

Number of heads	Equivalent microstates	Number of microstates
0	TTTT	1
1	HTTT, THTT, TTHT, TTTH	4
2	HHTT, THHT, TTHH, HTTH, THTH, HTHT	6
3	HHHT, THHH, HTHH, HHTH	4
4	HHHH	1
	Total number of microstates	16

Entropy.

is only 1/16, or 6%. Clearly, the probability of finding the system in some particular state, say 2-heads-2-tails, is proportional to the number of possible equivalent microstates ($\mathcal{W}$) composing that state—in this case, 6.

A crystal is a regular structure in which, especially at low temperatures, there is very little variation in the arrangement of its atoms. Each such configuration is a microstate of the system, and there are relatively few of them possible. The crystal is highly ordered and possesses little entropy. Still, as its internal energy increases, the number of ways that energy can be distributed among the constituent particles increases, which increases the number of available microstates and, hence, the entropy. In contrast, a mass of gas has a tremendous number of possible equivalent microstates that will produce the same macrostate with the same values of temperature, volume, and pressure. The gas is highly disordered and possesses a large amount of entropy.

If there were 100 coins in our system, there would still be only 1 microstate corresponding to all-heads, but now there would be a total of 2^{100} or $\approx 10^{30}$ microstates. That makes the chance of throwing all-heads about 1 out of 10^{30}. Since there are roughly 32×10^6 seconds per year, you can throw down the coins once every second for 3×10^{22} years, and you are not likely to hit all-heads more than once, though it certainly could happen on the first shot. By comparison, there are about 10^{29} ways to get 1/2-heads and that's by far the most likely state, occurring 10% of the time. About 90% of all throws will result in between 45 and 55 heads. Sequential probabilities build by multiplication. The chance of getting a head with 1 coin is 1/2 per throw (50%), while the chance of following that with another head is $1/2 \times 1/2$ and so on. The probability of throwing 4 heads in a row is $1/2 \times 1/2 \times 1/2 \times 1/2 = 1/16$, just as it was when we threw them all at once, which is reasonable since they don't talk to each other as they fall. By contrast, sequential processes would progressively *add* entropy changes. A system usually increases in entropy every time something happens to it.

The more microstates a state has, the more likely it is that that state will be observed, and the more disordered it must be. Hence, we can guess that $S \propto \mathcal{W}$, but not directly, because one is additive and the other multiplicative. This suggests using a function like the logarithm to overcome the difference, because $\ln ab = \ln a + \ln b$; taking the logarithm turns products into sums.

Boltzmann (1877) proposed that the entropy could be explicitly connected to the notion of disorder by defining it in a new way, which was nonetheless consistent with Clausius's definition:

$$S = k_B \ln \mathcal{W} \qquad (14.18)$$

where the k_B is Boltzmann's Constant (see Fig. 14.30 and p. 444). $\mathcal{W}$ is unitless, so the units of S are the units of k_B, namely, J/K. Keep in mind that the logarithm (Fig. 14.11) rises rapidly at low values and levels off at high values. That would reflect the greater disorder-

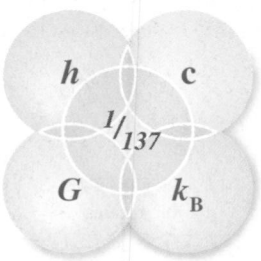

Figure 14.30 Fundamental constants. The Boltzmann Constant connects the everyday world with the hidden atomic tumult that lies beneath the observable surface.

ing effect of heat-in at lower temperatures. Even though fewer microstates are available at low temperatures, S is more sensitive to changes in internal energy at small T.

The probabilistic version of the Second Law does not forbid a state from spontaneously moving to a lower entropy; rather, it maintains that such a case would be most unlikely. If this statistical formulation is true to life, and it certainly seems to be, anything is possible—any process that can occur via the First Law may occur, if only remotely, via the Second. It is possible for the egg to unscramble, but it's unthinkably improbable.

To live at all is to churn through spontaneous processes, to be out of equilibrium, to pour forth entropy (while usually exhausting just about as much energy as we take in). Life feeds on order, and even as it builds, it disperses.

Core Material & Study Guide

THE FIRST LAW OF THERMODYNAMICS

Thermodynamics (or thermo, as it's affectionately called) is the study of thermal energy—its relationship to heat, mechanical energy, and work, and the conversion of one to the other.

The **First Law of Thermodynamics** is

$$\Delta U = Q + W \qquad [14.1]$$

An equivalent statement of the First Law is

> **The internal energy of an isolated system is constant (even though it may be transformed from one type of energy to another).**

These ideas are developed in Section 14.1 (Conservation of Energy); reread it and study Example 14.1. This is very basic important stuff. As always, **look at the WALK-THROUGH EXAMPLES on the CD.** Before you attempt any problems read the Suggestions on Problem Solving. Remember to work out the coordinated groups of three similar problems (colored magenta).

A system is in **equilibrium** when there is no tendency for it to undergo spontaneous change (on a macroscopic level). If a system is always infinitesimally close to thermodynamic equilibrium, even as it passes from one state to the next, the process that is occurring is said to be **reversible** (p. 500). The work done *by* the gas in an isobaric process is

[isobaric] $\qquad W = -P\Delta V \qquad [14.3]$

This equation, along with Eq. (14.1), deals with an important class of problems represented by Examples 14.2 to 14.4 in Section 14.2 (Thermal Processes & Work). Almost every intro course will include this material.

The work done on or by an ideal gas during an isothermal volume change from V_i to V_f is

[isothermal] $\qquad W = -nRT \ln \dfrac{V_f}{V_i} \qquad [14.5]$

Study Section 14.3 (Isothermal Change of an Ideal Gas) and examine Example 14.5 to see how to apply the ideas analytically.

For a reversible adiabatic process

[adiabatic] $\qquad PV^\gamma = \text{constant} \qquad [14.6]$

where γ is a constant that equals 1.67 for monatomic gases, $\approx$1.4 for diatomic gases, and $\approx$1.3 for polyatomic gases. Reread Section 14.4 (Adiabatic Change of an Ideal Gas), and when you are comfortable with it study Example 14.6, which is a typical application of the ideas.

CYCLIC PROCESSES: ENGINES AND REFRIGERATORS

A *heat engine* is a cyclic device that converts thermal energy into work-out. A **Carnot engine** is a reversible device operating in a cycle composed of two adiabats and two isothermals. The efficiency of an engine is

[actual engine] $\quad \text{Efficiency} = \dfrac{|\text{work-out}|}{|\text{heat-in}|} = \dfrac{|W_o|}{|Q_i|}$

[ideal engine] $\quad e = \dfrac{|Q_H| - |Q_L|}{|Q_H|} = 1 - \dfrac{|Q_L|}{|Q_H|} \qquad [14.10]$

The efficiency of a Carnot engine is

[ideal engine] $\quad e_c = 1 - \dfrac{T_L}{T_H} = \dfrac{T_H - T_L}{T_H} \qquad [14.11]$

A *refrigeration machine* uses work to transfer thermal energy from a low-temperature reservoir to a high-temperature reservoir. The *coefficient of performance* (η) is defined as the ratio of the amount of heat removed from the cold reservoir (the heat-in to the system) to the amount of work done on the system; consequently,

$$\eta = \dfrac{|Q_L|}{|W_i|} = \dfrac{|Q_L|}{|Q_H| - |Q_L|} \qquad [14.13]$$

For a **Carnot engine**

$$\dfrac{|Q_L|}{|Q_H|} = \dfrac{T_L}{T_H} \qquad [14.14]$$

The coefficient of performance for an ideal system is

[ideal system] $\quad \eta_c = \dfrac{T_L}{T_H - T_L} \qquad [14.15]$

Sections 14.5 (The Carnot Cycle) and 14.6 (Refrigeration Machines & Heat Pumps), in addition to being of some practical interest, lay the ground work for the Second Law of Thermo, so make sure you study these discussions and can apply the ideas (Examples 14.7 and 14.8).

THE SECOND LAW OF THERMODYNAMICS

The **Kelvin–Planck statement** of the Second Law is

> **No ongoing process is possible for which the *sole result* is the removal of heat from a source and its complete transformation into work.**

The **Clausius statement** of the Second Law is

> **No ongoing process is possible in which the *sole result* is the transfer of thermal energy from a body at low temperature to a body at a higher temperature.**

A Carnot engine has the same efficiency as any other reversible engine operating between the same two temperatures. No engine can exceed the efficiency of a reversible engine.

Clausius proposed that the measure of the disorder of a system is the **entropy** S. The Second Law in terms of entropy is

> **When an isolated system undergoes a change, passing from one state to another, it will do so in such a way that its entropy will increase, or at best remain the same.**

Clausius defined the change in entropy of a system (ΔS) experienced during a *reversible* process as

$$\Delta S = \frac{Q}{T} \qquad [14.17]$$

A system with maximum entropy is in equilibrium. Boltzmann proposed that entropy could be explicitly connected to the notion of disorder by defining it as

$$S = k_B \ln \mathcal{W} \qquad [14.18]$$

where $\mathcal{W}$ is the number of microstates. This material is treated in Section 14.7 (Entropy). Examples 14.9 and 14.10 are typical of the analysis that can be done on this level.

Key Terms

internal energy	Carnot cycle
First Law of Thermo	efficiency
perpetual motion	refrigeration machine
equilibrium	coefficient of performance
isothermal process	Second Law of Thermo
isobaric process	Kelvin-Planck statement
adiabatic process	Clausius statement
isovolumic process	entropy
reversibility	Entropy Principle
adiabat	disorder
isotherm	microstate
heat engine	

Discussion Questions

1. Figure Q1 is the *PV*-diagram for a diesel engine. It uses the internal combustion of a gaseous fuel in a cylinder-and-piston arrangement. Here, the fuel is sprayed into the cylinder at *C*. Explain each part of the cycle. Why do you think only air is compressed from *B* to *C* with the fuel injected only at *C*? Is work done? When does heat enter and leave the system? What kind of processes are occurring? At what point does the cycle end? Where in the cycle does the exhaust valve open to eject spent fuel? Use the gasoline engine as a guide since it's fairly similar to the diesel. {For an analysis of the modern gasoline engine click on **INTERNAL COMBUSTION** under **FURTHER DISCUSSIONS** on the **CD**.} 💿

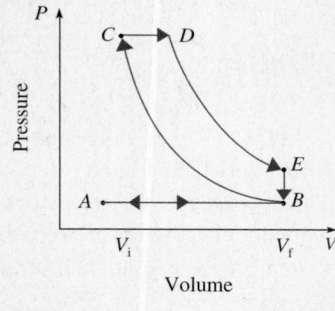

Figure Q1

2. What if anything do you work against in the process of blowing up an ordinary balloon? Does it take the same amount of effort to blow one up (assuming you get around the technical details) in space as on Earth? How does this illuminate the fact that when a gas expands freely into an evacuated chamber its temperature stays fixed, but if it expands into the air its temperature drops?

3. Figure Q3 depicts a reciprocating steam engine. Explain how it works.

4. The so-called *Third Law of Thermodynamics*, which to some physicists does not seem as secure as the other two, states that:

It is impossible to bring a body to a temperature of Absolute Zero in a finite number of steps.

In other words, it is widely believed that Absolute Zero is unattainable. Discuss what this means and determine whether or not it fits in with the Second Law.

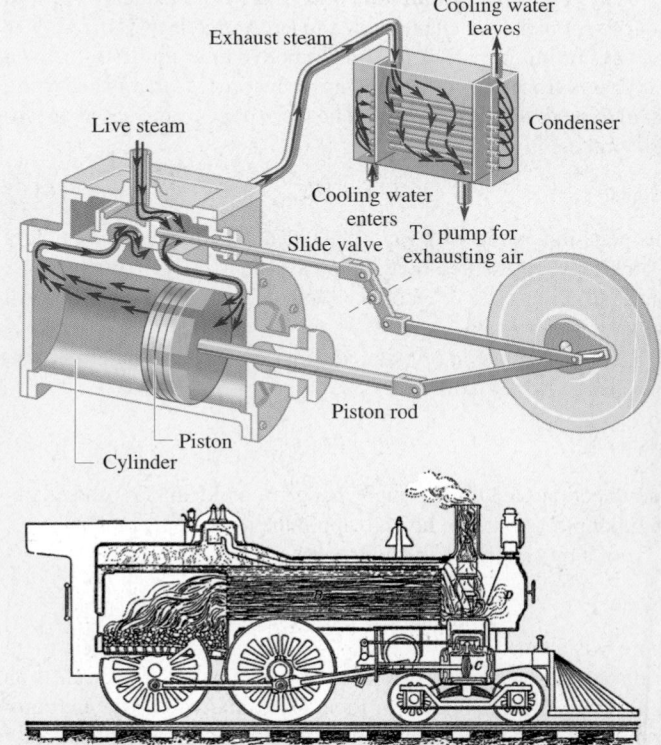

Figure Q3

5. Discuss the thermodynamics of a possible mechanism for the formation of clouds as a result of rising columns of moist, warm air coming from near the ground.

6. You are offered the opportunity to "buy in on the ground floor" of a company selling a new machine. It is a container designed to keep drinks at a fixed temperature indefinitely. One loads it with hot liquid; the heat that inadvertently leaks off is cleverly converted into work that is fed to a stirrer that continually churns the liquid putting the thermal energy back and maintaining the temperature. Do you buy? Explain. Discuss the process from an entropy perspective.

7. What entity is the ultimate source of the energy stored in coal, wood, and oil? Explain. Does this arrangement violate the entropy version of the Second Law?

8. Suppose someone is placed in a totally isolated room with enough food for one day. What will inevitably happen to him? Explain your answer in terms of both the First and Second Laws.

9. What is the ultimate source of the energy extracted via a hydro-electric power plant? Explain.

10. A red-hot coal is hung on an insulating fiber inside a room-temperature metal box that is sealed, evacuated, and completely isolated from the rest of the environment with massive amounts of insulation. Discuss the situation from an entropy perspective. What happens to the coal and the box?

11. Imagine a thermally insulated cylinder-and-piston containing an ideal gas at a certain initial volume *V*. Suppose the gas, pushing on the piston, is allowed to expand to a volume 2*V*. Discuss, in thermodynamic terms, every aspect of the process. What kind of process is it? What happens to the internal energy of the gas? What happens to *P*, *V*, and *T*? Is work done? By what, on what? Does the entropy change?

12. Suppose you fly in a jet plane from New York to London. What is the net effect on the energy and entropy of the planet?

13. Suppose an automobile engine burns its hydrocarbon fuel at about 2400 K while the outside air temperature is 300 K. Explain why the actual efficiency is far less than the corresponding Carnot efficiency between those two temperatures.

14. What would you say was the overall effect of the rise of technology on the entropy of the Universe?

15. A large room may have some 10^{28} air molecules filling it. Is it possible for these molecules to suddenly all end up in one corner of the room, leaving us gasping for breath? Do we have to worry about that happening and if not, why not? How would matters change if there were only 10 molecules in the room?

16. The gas in the combustion chamber of a rocket is in violent random motion at high temperature. It leaves the engine through a nozzle at high speeds. Describe what's happening in terms of energy. Considering that the exhaust emerges in a well-defined direction, what is the temperature of the exhaust relative to that of the combustion chamber?

Multiple Choice Questions

1. A cylinder-and-piston contains an ideal gas whose volume is to be halved. The work done on the gas will be   (a) greater if the compression is isothermal rather than adiabatic   (b) greater if the compression is adiabatic rather than isothermal   (c) the same if it's either adiabatic or isothermal   (d) equal to the change in internal energy in all cases   (e) none of these.

2. A cylinder-and-piston contains a gas that is made to expand quasi-statically from *A* to *B* at a constant temperature as shown in Fig. MC2. The process is   (a) adiabatic and heat leaves the gas   (b) isothermal and heat enters the gas   (c) adiabatic and heat enters the gas   (d) isothermal and heat leaves the gas   (e) none of these.

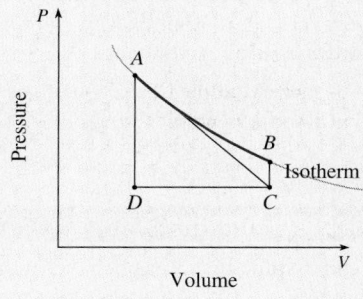

Figure MC2

3. Referring to the previous question and Fig. MC2, when the gas is carried from *B* to *C*, the process is   (a) isobaric and no heat is transferred   (b) isovolumic and no heat is transferred   (c) adiabatic and no heat is transferred   (d) isovolumic and heat leaves the gas   (e) none of these.

4. Referring to the previous question and Fig. MC2, when the gas is carried from *C* to *D*, the process is   (a) isothermal and no heat is transferred   (b) isovolumic and no heat is transferred   (c) adiabatic and heat enters the gas   (d) isovolumic and heat leaves the gas   (e) none of these.

5. Referring to the previous question and Fig. MC2, when the gas is carried from *D* to *A*, the process is   (a) isothermal and no heat is transferred   (b) isovolumic and no heat is transferred   (c) adiabatic and heat enters the gas   (d) isovolumic and heat enters the gas   (e) none of these.

6. When Joule wrote "the grand agents of nature are, by the Creator's fiat, indestructible" (1843), he was talking about   (a) the F.B.I.   (b) the Second Law   (c) an automobile   (d) the First Law   (e) none of these.

7. An ideal gas is contained in a cylinder-and-piston that is 0.5-m long, and 50 J of heat is added while the piston is held in place. The work done by the gas is   (a) 50 J   (b) 0   (c) 25 J   (d) 250 J   (e) none of these.

8. If a system does work adiabatically,   (a) the temperature stays the same   (b) the internal energy stays the same   (c) the internal energy decreases   (d) the internal energy increases   (e) none of these.

9. Fig. MC2 shows the *PV*-diagram for a gas. In reference to the work done by the gas in going from state *A* to state *C*   (a) the route *A-C* corresponds to a minimum   (b) the route *A-D-C* is equivalent to route *A-B-C*   (c) all the routes are equivalent   (d) route *A-D-C* corresponds to a maximum   (e) none of these.

10. When this book is given a push across a desktop, it comes to rest losing its initial KE in a(n)   (a) reversible process   (b) irreversible process   (c) process that is isothermal   (d) process that is isentropic   (e) none of these.

The next six questions refer to Fig. MC11, which is a plot of pressure versus volume as an ideal gas undergoes several different processes.

11. In going from point-*A* to point-*B*   (a) work is done by the gas and heat flows out of it   (b) work is done by the gas and heat flows into it   (c) work is done on the gas and heat flows out of it   (d)

work is done on the gas and heat flows into it   (e) none of these.

12. Which of the following is true?   (a) $T_A > T_B$ (b) $T_A < T_C$ (c) $T_C > T_B$ (d) $T_D > T_B$ (e) none of these.

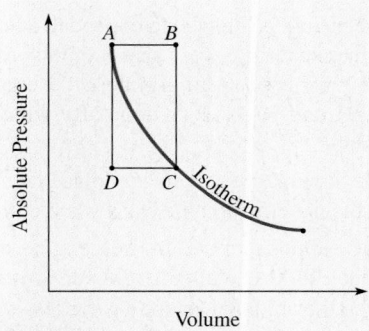

Figure MC11

13. In going from point-B to point-C   (a) work is done by the gas and heat flows out of it   (b) no work is done by or on the gas, but heat flows into it   (c) no work is done by or on the gas, but heat flows out of it   (d) no work is done by or the gas, and no heat flows into or out of it   (e) none of these.

14. Comparing internal energies, which of the following is true?   (a) $U_A > U_B$ (b) $U_A < U_C$ (c) $U_D > U_B$ (d) $U_B > U_C$ (e) none of these.

15. Comparing internal energies, which of the following is true?   (a) $U_A = U_B$ (b) $U_A = U_C$ (c) $U_D = U_B$ (d) $U_B = U_C$ (e) none of these.

16. In going from point-A to point-C to point-D, and back to point-A   (a) a net amount of work is done on the gas, but its internal energy is unchanged   (b) no net work is done on the gas, but its internal energy increases   (c) a net amount of work is done by the gas, but its internal energy is unchanged   (d) no net work is done on the gas, but its internal energy decreases   (e) none of these.

17. A gas is compressed adiabatically by a force of 500 N acting through a distance of 5.0 cm. The net change in its internal energy is   (a) +500 J   (b) +25 J   (c) −500 J   (d) −25 J   (e) none of these.

18. In an adiabatic operation, the value of γ for mercury vapor is likely to be about   (a) 1.67   (b) 1.4   (c) 1.3   (d) 1   (e) none of these.

19. The statement that heat flows spontaneously only from a high-temperature object to a low-temperature object is   (a) not always true   (b) the First Law of Thermodynamics   (c) the Second Law of Thermodynamics   (d) only true for isothermal processes   (e) none of these.

20. An ice-filled glass of soda is so cold that water from the air condenses out on it in beads. As a result, the entropy of that water must have   (a) increased   (b) remained unchanged   (c) first increased then decreased   (d) decreased   (e) none of these.

21. As it regards the ordinary processes of everyday life, it can be said that the Second Law of Thermodynamics   (a) distinguishes the possible from the impossible   (b) is quite inapplicable   (c) demands that energy be conserved, no matter what   (d) requires that the sum of the heat-in and the heat-out be zero   (e) none of these.

22. The addition of heat to a system   (a) decreases its order and so decreases its entropy   (b) decreases its order and its internal energy   (c) decreases its order and increases its entropy   (d) increases its disorder and decreases its entropy   (e) none of these.

23. The disordering effect of heat   (a) increases with increasing temperature   (b) decreases with increasing entropy   (c) is constant   (d) increases with decreasing temperature   (e) none of these.

24. The area under a process curve on a TS-diagram is the   (a) heat flowing into or out of the system   (b) the work done on or by the system   (c) the change in pressure of the system   (d) the change in entropy of the system   (e) none of these.

25. The entropy associated with a particular state of a system is   (a) independent of probability of any sort   (b) dependent on the logarithm of the probability of the state occurring   (c) inversely proportional to the number of microstates associated with that state   (d) independent of the number of microstates associated with that state   (e) none of these.

26. The statement that processes take place in such a way that the entropy of the Universe tends toward a maximum should be interpreted as saying   (a) that the Universe is moving inextricably toward equilibrium   (b) that nothing much is happening in the Universe   (c) that energy is approaching a minimum   (d) that order is rising out of chaos   (e) none of these.

For more Multiple Choice Questions with answers click on WARM-UPS in CHAPTER 14 on the CD. 💿

Suggestions on Problem Solving

1. Keep in mind that all the calculations involving temperature in this chapter use kelvins. Look out for data given in °F; convert to degrees Celsius and then to kelvins early in the analysis.

2. Whether Q_H and Q_L are flowing into or out of the device in a particular problem can be confused, so it's a good idea to draw an energy flow diagram, like Fig. 14.21. This is especially true when dealing with refrigerators.

3. When an engine is reversible, the values of Q_L, Q_H, and W must remain the same in either mode of operation. Of course, when it's run as an engine, Q_H is the heat-in whereas, as a refrigerator, Q_H is the heat-out, but numerically the values stay the same. In other words, the energy channels in schematic diagrams like Fig. 14.29 don't change width on reversing a reversible engine; they do change if the engine is irreversible.

4. Adiabatic processes utilizing gases require values for γ, and most often air is involved, for which $\gamma = 1.4$. If not explicitly stated, you can assume that a gas behaves adiabatically if the process occurs rapidly.

5. The heat leaving an object is negative, and so the change in entropy that it experiences is also negative. Keep careful track of the signs throughout entropy calculations. If the temperature drops $\Delta T < 0$, $Q = cm\Delta T < 0$ and $\Delta S = Q/T$.

6. It often happens that a quantity you might think would be needed explicitly to solve a problem cancels out of the equations. That sort of thing happens especially with before-and-after volumes and pressures of gases. Moreover, there are situations involving the gas laws where it's useful to work on a per-unit-volume basis. Again, because the equations for gases often have before-and-after pressures and volumes (that is, P and V on both sides), you can use non-SI units such as atm or cm³. That will save the time to convert, but it is risky if you are not careful. It's better to play it safe and convert everything to SI right away. To illustrate the point, a few of the solutions given in the back of the book use mixed units.

Problems ✦ Coordinated Problems ✦ Progressive Problems ✦ Solutions

STUDY GUIDE **1. Coordinated Problems:** The three problems within each magenta-colored grouping are solvable in similar ways. Note that the first of these always has a hint; moreover, its solution is provided in the back of the book. *Work out each of these sets; they'll strengthen technique and build confidence.* **2. Progressive Problems:** The problems introduced in blue unfold step-by-step carrying along the analysis in a more suggestive way than is customary. *Work out all of these; they'll guide you through the analytic process and help develop problem-solving skills.* **3. Worked-Out Solutions:** Studying worked-out solutions is an important part of learning how to solve problems. Accordingly, additional *solutions* to a number of model problems are given below. *Make sure you understand each of them before you go on to the next problem.* **4.** Also provided in the back of the book are the *Answers* to all odd-numbered problems, as well as worked-out *solutions* to those with boldface numbers. Problem numbers in italic indicate that a solution appears in the Student Solutions Manual.

SECTION 14.1: CONSERVATION OF ENERGY

SECTION 14.2: THERMAL PROCESSES & WORK

1. [I] An amount of heat equal to 500 J flows into a system that does 250 J of work on its surroundings. What is the change in the internal energy of the system? [*Hint: The First Law gives you the answer provided you're careful with the signs.*]

2. [I] How much work is done by a system if 1000 J of heat enter it while its internal energy increases by 250 J?

3. [I] A system has 150 J of work performed upon it while its internal energy is made to decrease by 300 J. Determine the heat transferred to or from the system.

4. [I] What is the change in the internal energy of a system if it loses 600 J of heat while doing 400 J of work?

5. [I] A 100-g sample of gas is contained in a sealed insulated tank. Heat is slowly transferred to the gas (via an electric heater), and its internal energy increases by 500 J. Determine the amount of heat added to the gas.

6. [I] THIS PROBLEM EXPLORES THE FIRST LAW OF THERMO. A 75.1-kg sample of pure water at 14.9 °C is placed in contact with a heat reservoir which is at 24.9 °C. The water subsequently reaches equilibrium with the reservoir, and we want to find its change in internal energy. (a) What's the final temperature of the water? (b) What causes the water to change temperature? (c) Knowing that its specific heat is 4.186 kJ/kg·K, how much heat is required to change the temperature of the water? (d) Assuming no work is done by the liquid, where does the energy associated with the heat-in end up? (e) Determine the change in internal energy of the water.

7. [I] THIS PROBLEM EXPLORES THE FIRST LAW OF THERMO. A vertical cylinder sealed with a movable piston having an cross-sectional area of 0.20 m² is filled with a tenuous gas. The cylinder is placed over a burner and slowly heated. The weight of the piston maintains a constant pressure in the gas of 16.0 kPa. (a) What happens to the volume of the gas as heat is applied? Explain. (b) If the piston travels 8.00 cm, what is the resulting change in volume? (c) How much work was done on or by the gas? Which was it? Measurements show that 400 J of heat was supplied to the gas. (d) What, if any, was the change in the internal energy of the gas?

8. [I] A sealed steel tank was filled with a tenuous gas and then placed over a block of ice and slowly cooled. Measurements show that 400 J of heat were removed from the gas. (a) How much work was done on or by the gas? Explain. (b) What, if any, was the change in the internal energy of the gas?

SOLUTION: Here V is constant and so no work is done on or by the gas. Moreover, if heat flows out of the gas, T decreases and P decreases. The

First Law becomes $\Delta U = Q + W = Q + 0$ and $\Delta U = -400$ J, the internal energy decreases.

9. [I] As the result of an isovolumic process, an ideal gas absorbs 5.0 kJ of heat. The chamber containing the gas is not well insulated, and it loses 1.0 kJ. (a) What, if anything happened to the pressure of the gas? (b) What, if any, is the change in the internal energy of the system?

SOLUTION: (a) In an isovolumic process V is constant, and so no work is done on or by the gas. Moreover, if heat flows into the gas, T increases and P increases. (b) The First Law becomes $\Delta U = Q + W = Q + 0$, where $Q = 5.0$ kJ $- 1.0$ kJ and so $\Delta U = 4.0$ kJ, the internal energy increases.

10. [I] An ideal gas is compressed to one-half its original volume by a piston moving in a cylinder. During the process, 500 J of heat leave the gas, but its temperature remains fixed. How much work is done on or by the gas?

11. [I] An amount of heat equal to 100 cal is transferred to a system while 100.4 N·m of work is done on it. What is the net change in its internal energy?

12. [I] When a chunk of ice melts, it changes from a density of 920 kg/m³ to 1000 kg/m³, and work is done on it by the atmosphere. Compare the work done per kilogram on the ice to the heat-in per kilogram.

13. [I] An ideal gas in a cylinder-and-piston is made to follow the cycle shown in Fig. P13 from A to B to C to A. What is the work done in terms of the initial volume and pressure?

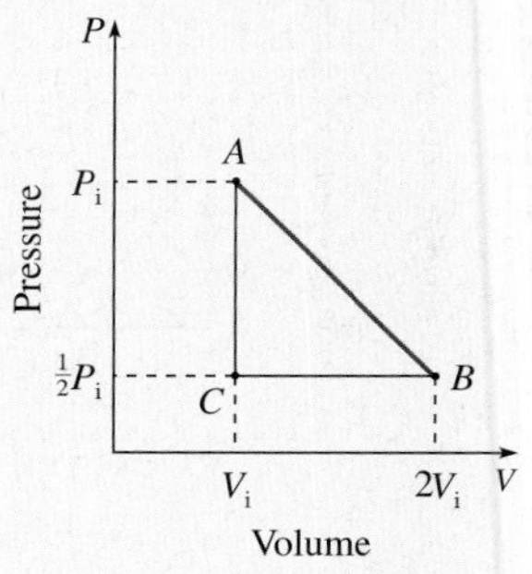

Figure P13

14. [I] A quantity of gas in a cylinder sealed with a piston absorbs 1.00 kcal of heat as it expands by 15 liters at constant pressure. Determine that pressure if the internal energy is found to increase 286 J.

15. [I] A cylinder sealed with a piston contains 20.0 m³ of pure oxygen at atmospheric pressure. The gas is then forced into a tiny high-pressure container of negligible volume. What is the amount of work the gas can do if it is subsequently released into a large empty, lightweight plastic bag? What happens to the temperature of the gas upon entering the bag? What would happen to the temperature of the gas if it were released into the air?

16. [I] A cylinder-and-piston containing an ideal gas is placed in contact with a thermal reservoir. The volume of the gas is very slowly changed from 50 liters to 10 liters as 40 J of work is done on it by an external agency. Determine the change in the internal energy of the gas and the amount of heat flowing into or out of the system.

17. [I] A gas experiences a change from state I to state F as indicated in Fig. P17 on p. 524. Determine the work done.

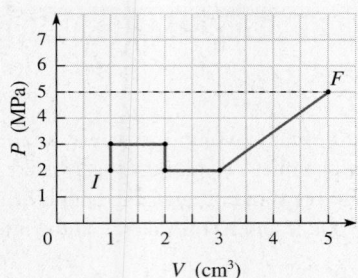

Figure P17

18. [I] A 15-cm diameter brass sphere is heated from 0 °C to 280 °C. How much work does it do on the atmosphere in the process?

19. [I] An ideal gas in a cylinder-and-piston is made to follow the cycle shown in Fig. P19 from A to B to C to D. What is the work done in terms of the initial volume and pressure?

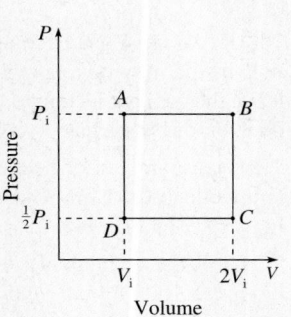

Figure P19

20. [II] THIS PROBLEM DEALS WITH CONSERVATION OF ENERGY AND THE *PV*-DIAGRAM. The absolute value of the work done on, or by, the gas while following path-1 from A to B in Fig. P20 is 458 J. (a) How do you know whether this is work done *on* or *by* the gas? Explain. (b) The gas simultaneously experiences a decrease in its internal energy of 178 J. What concept relates the work and this change in internal energy? (c) How much heat flows *into* or *out of* the gas, and which is it? (c) Now suppose the gas is processed along path-2, and 50.0 J of heat enters it. (d) What is its change in internal energy? Explain. (e) How much work is done *on* or *by* the gas, and which is it? Explain.

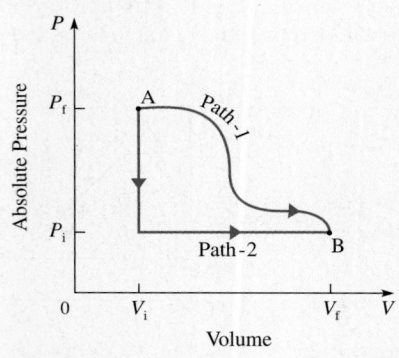

Figure P20

21. [II] THIS PROBLEM DEALS WITH CONSERVATION OF ENERGY AND THE *PV*-DIAGRAM. The ideal gas whose history is depicted in Fig. P21 has an internal energy of 25.0 kJ at point-*D*. (a) How much work is done *on* or *by* the gas in going from point-*A* to point-*D*? Explain. (b) How might the gas be made to go from point-*A* to point-*D*? (c) If 4.5 kJ of heat must be added to the gas to go from point-*D* to point-*A*, what is its internal

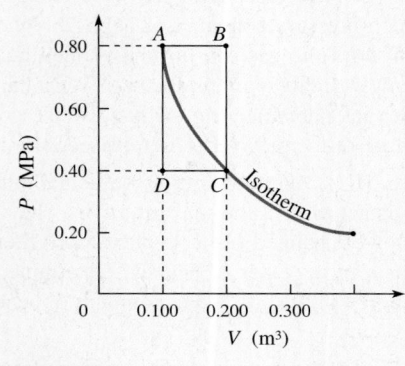

Figure P21

energy at *A*? (d) Compare the temperature at *A* with that at *B*, that is, which is higher? Explain. (e) Compare the internal energy at *A* with that at *B*, that is, which is higher? Explain. (f) How much work is done *on* or *by* the gas in going from point-*A* to point-*B*? Explain. (g) What is the internal energy of the gas at *C*? Explain.

22. [II] An ideal gas is compressed from 20 liters to 2.0 liters at a constant pressure of 3.0 atm by drawing heat from it so that its temperature drops. At that point, its volume is held constant, and heat is added to it so that the pressure rises and it returns to its original temperature. Draw a *PV*-diagram. Compute the net work done on or by the gas. Determine the net heat flow.

23. [II] A piston with a cross-sectional area of 0.50 m² seals a gas within a cylinder at a pressure of 0.30 MPa. Ten kilojoules of heat is very gradually added to the cylinder, and the piston rises 6.5 cm as the gas expands isobarically. What work is done by the gas? What is the change in its internal energy? [*Hint: The pressure is constant, and so the work done depends on the volume change. Assume the gas is ideal.*]

24. [II] A gas in a thin plastic bag at atmospheric pressure receives 10 000 J of heat and, in the process, puffs up the bag, increasing its volume by 1.00 m³. By how much is the internal energy of the gas altered?

25. [II] A cylinder sealed with a piston of negligible weight (having a face-area of 1.00 m²) contains helium at atmospheric pressure. Thirty thousand joules of heat are removed from the gas and the piston is simultaneously, isobarically moved inward 0.25 m. What, if any, is the change in the internal energy of the gas?

26. [II] A quantity of 5.0 kg of water in a vacuum chamber is supplied with heat, and its temperature increases by 10 °C. What is the change in its internal energy?

27. [II] A large thermally insulated cylinder is sealed with a piston. Steam at 100 °C enters the cylinder and pushes the piston outward against a pressure of 99 kPa, increasing the occupied volume by 4950 cm³. How much water condenses inside the cylinder?

28. [II] A 2.0-mole sample of a gas at 0 °C, which behaves like an ideal gas, is compressed to half its original volume isobarically. How much work must be done on the gas?

29. [III] When water is vaporized, it expands and does work in the process. What percent of the heat of vaporization at 100 °C is expended on expanding the liquid water into vapor? The density of steam at atmospheric pressure is 0.598 kg/m³.

30. [III] A 1.0-m² panel of solar cells absorbs 35% of the total radiation (of intensity 0.83 kW/m²) impinging on it from the Sun. It generates electrical energy, which is fed to a motor that raises a load, doing work at a rate of 150 J each second. Assuming no other losses, determine the rate at which the internal energy of the system (panel and motor) is changing.

SECTION 14.3: ISOTHERMAL CHANGE OF AN IDEAL GAS
SECTION 14.4: ADIABATIC CHANGE OF AN IDEAL GAS

31. [I] THIS PROBLEM WILL HELP US LEARN ABOUT THE FIRST LAW OF THERMO. A low-density gas at 61.0 °C undergoes an isothermal compression. It takes 10.5 kJ of work to accomplish the compression. (a) What is the resulting change in the temperature of the gas? (b) What is the resulting change in the internal energy of the gas? (c) Was any heat added or removed from the gas during the compression? Explain your answer. (d) If heat was transferred into or out of the gas, how much was it?

32. [I] A system, which is very well insulated, undergoes an adiabatic process that delivers 10.0 kJ of work into the environment. What, if any, is the change in the internal energy of the system?

33. [I] Consider a cylinder sealed with a piston containing 6000 cm^3 of an ideal gas at a pressure of 10.0 atm. A quantity of heat is slowly introduced, and the gas expands at a constant temperature to 12 000 cm^3. How much work will the gas do on the piston? [*Hint: This is an isothermal expansion.*]

34. [I] Heat is added to 4.00 m^3 of helium gas in an expandable chamber that thereupon increases its volume by 2.00 m^3. If in the process 2.00 kJ of work are done by the gas, what was its original pressure?

35. [I] An ideal gas in a piston-sealed cylinder has a pressure of 0.101 MPa and a volume of 10.0 liters. The piston is slowly forced inward, the volume is halved, and all the while the temperature is kept constant. How much work is done on the gas?

36. [I] A balloon contains 5.00 m^3 of argon gas. It is slowly heated at a constant temperature so that its volume triples. If the gas does 1.00 kJ of work in the expansion, what was the initial pressure in the balloon?

37. [I] The ideal gas in a flexible container is cooled while the container is made to shrink, thereby maintaining a constant temperature. The initial pressure and volume were 0.300 MPa and 3.00 m^3, respectively. Given that the final pressure was 0.900 MPa, how much work was done on the gas?

38. [I] An ideal gas contained by a cylinder-and-piston is made to double its volume isothermally, at 27 °C. If the chamber contains 5.0 moles of gas, how much work is done in the process?

39. [I] THIS PROBLEM EXPLORES THE ISOTHERMAL BEHAVIOR OF A GAS. Take a tire pump having a 0.60-liter capacity and seal off the output nozzle. The temperature of the air in the pump is 293 K, and it's kept at that value as the gas is slowly compressed to half its original value. Given that the gas is essentially ideal and that its initial absolute pressure is 1.00 atm, (a) what is the value of the ratio of the final volume to the initial volume? (b) What is the value of the ratio of the initial pressure to the final pressure? (c) What is the value of the final pressure? (d) Is work done on or by the gas during the compression? (e) Write an expression for the work done in terms of P_i and V_i. (f) How much work was done?

40. [II] THIS PROBLEM EXPLORES THE BEHAVIOR OF A GAS FROM AN ENERGY PERSPECTIVE. Referring to Fig. P40, which shows the process history of an ideal gas, (a) how much work is done in carrying the gas from point-*A* to point-*B*? (b) Is it done on, or by, the gas? (c) Does the internal energy increase, decrease, or stay the same in that process? Explain. (d) Does heat flow in, out, or neither? Explain. (e) How much work is done in carrying the gas from point-*B* to point-*C*? (f) Does the internal energy increase, decrease, or stay the same? Explain. (g) Does heat flow in, out, or neither in that process? Explain. (h)

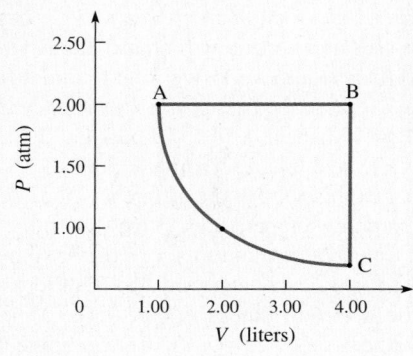

Figure P40

How much work is done in carrying the gas from point-*C* to point-*A*? This should be substantially less than W_{BA}; is it? (i) Does the internal energy increase, decrease, or stay the same? Explain. (j) Does heat flow in, out, or neither? Explain. (k) How much, if any, heat flows?

41. [II] A 2.0 liter quantity of an ideal gas contained in a cylinder-and-piston assembly is placed in intimate contact with a large thermal reservoir at 283.1 K. The gas, and there is 0.119 mol of it, expands and does 28.0 J of work. What is the final volume of the gas?

> **SOLUTION:** This is an isothermal process performed on an ideal gas for which
>
> $$W = -nRT \ln \frac{V_f}{V_i} \quad \text{and so} \quad \ln \frac{V_f}{V_i} = -W/nRT$$
>
> $$\ln \frac{V_f}{V_i} = -\frac{-28.0 \text{ J}}{(0.119 \text{ mol})(8.315 \text{ J/mol·K})(283.1 \text{ K})} = 0.100$$
>
> Now exponentiate both sides; that is, raise e to the power of each side of the equation in order to undo the logarithm.
>
> $$\exp\left[\ln \frac{V_f}{V_i}\right] = \exp(0.100) \quad \text{and} \quad \frac{V_f}{V_i} = e^{0.100}$$
>
> and $V_f = 1.105 V_i = (1.105)(2.0 \text{ liters}) = 2.2 \text{ liters}$

42. [II] A 1.29-liter sample of ideal gas, composed of 0.119 mol, was made to fill an expandable container at 10.0 °C. It was then gently heated, even as its temperature was maintained constant. In the process the gas expanded and did 13.9 J of work. What is the final pressure of the gas?

43. [II] An ideal gas contained in a cylinder-and-piston assembly undergoes a reversible isothermal expansion. Derive an expression for the work done by the gas in terms of the number of moles present, the temperature, the gas constant, and the initial and final pressures.

44. [II] With the results of Problem 43 in mind, compute the amount of work done by 2.00 moles of an ideal gas in an isothermal expansion at 400 K from an initial pressure of 12 atm down to 1.5 atm.

45. [II] Figure P45 shows two reversible cycles *A-B-C-A* and *C-D-E-C* carried out on an ideal gas (the curves are isothermals). Prove that the work done is the same for each cycle.

46. [II] An ideal diatomic gas at STP is adiabatically compressed to half its original volume. Determine its final temperature and pressure.

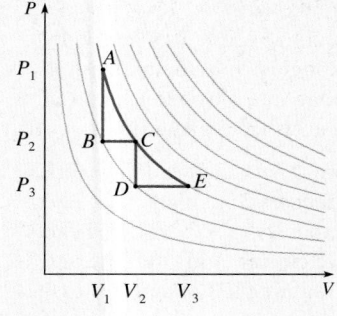

Figure P45

47. [II] A cylinder-and-piston with a volume of 22.4 liters contains 1.00 mole of diatomic hydrogen at a pressure of 1.00 atm. The gas is allowed to expand adiabatically to twice its original volume, whereupon it is compressed back to its initial volume isothermally. (a) What is the lowest pressure the system attains? (b) What is the lowest temperature the system attains? Draw a *PV*-diagram showing everything that's happening.

48. [II] Considering Problem 47, how much work is done on isothermally returning the system to its original volume? What is the final temperature of the system? What is the final pressure?

49. [II] A bicycle tire filled with air to a pressure of 4.5 atm is rapidly emptied into a large bag at atmospheric pressure. If the air in the tire was initially at 27 °C, what will be its final temperature in the bag? Assume ideal gas behavior.

50. [II] Helium gas at 150 kPa is pumped into a 40.0-cm³ cylinder sealed with a piston. The piston is then rapidly pushed in, adiabatically compressing the gas to 1.00 cm³. Given that the gas was originally at 10.0°C, determine its new temperature.

51. [II] Please show that for an adiabatic volume change of an ideal gas

$$\frac{T_i}{T_f} = \left(\frac{V_f}{V_i}\right)^{\gamma - 1}$$

52. [III] A diesel engine adiabatically compresses an air-fuel mixture from its initial volume at atmospheric pressure and a temperature of 25.0°C to 1/20 that volume. What is its final temperature? (Take $\gamma = 1.35$.)

53. [III] The compression ratio (V_i / V_f) of a gasoline engine is 7.5. Determine the ratio of the temperature of the gas before and after the compression. (Take $\gamma = 1.35$.) Explain your results.

SECTION 14.5: THE CARNOT CYCLE

SECTION 14.6: REFRIGERATION MACHINES & HEAT PUMPS

54. [I] Study the thermodynamic cycle shown in Fig. P54. Given that the path from $A \rightarrow B$ is an isotherm, that from $B \rightarrow C$ is an isochor, that from $C \rightarrow D$ is an adiabat, that from $D \rightarrow E$ is an isobar, and that from $E \rightarrow A$ is an adiabat. For each process, does heat flow in or out? How does the temperature change during each process? Is work done on or by the system during each process? Explain each of your conclusions.

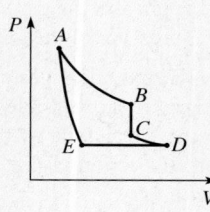

Figure P54

55. [I] In Fig. P55 each box in the grid is 1.0×10^5 Pa high by 1.0×10^{-5} m³ wide. Roughly, how much work is done on or by the system in one cycle? Explain.

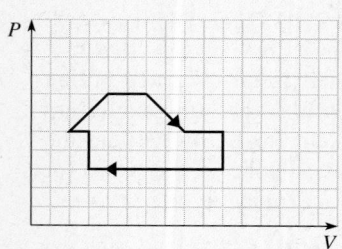

Figure P55

56. [I] An engine doing work receives 100 kJ of heat and exhausts 75 kJ of heat. Determine its efficiency, assuming it's otherwise lossless.

57. [I] A person consumes food at 98.6 °F and exhausts waste heat at 20 °C. What is the maximum theoretical efficiency for a heat engine operating at the same temperatures?

58. [I] What is the maximum theoretical efficiency of a steam engine operating between 100 °C and 400 °C?

59. [I] The temperature of the input steam to a steam engine is 180 °C, and the exhaust water leaves at 80.0 °C. What is the best efficiency we can hope for for such a device?

60. [I] To improve the theoretical efficiency of a heat engine, would it be better to lower T_L by 10 K or raise T_H by 10 K? Explain.

61. [I] THIS PROBLEM WILL HELP US BETTER UNDERSTAND THE BEHAVIOR OF A THERMAL ENGINE. During each cycle a well-insulated heat engine draws in 2.01 kJ from a high-temperature reservoir and expels 1.69 kJ to a low-temperature reservoir. We want to determine the engine's efficiency. Assuming no losses, (a) what is the change in the internal energy of the engine at the end of each cycle? Explain. (b) How much net heat is transferred into or out of the engine per cycle? (c) How much work is done by, or on, the engine per cycle? And is it *work-in* or *work-out*? (d) What is the efficiency of the engine?

62. [I] THIS PROBLEM EXPLORES THE BEHAVIOR OF A THERMAL ENGINE. A person is designing an engine that is supposed to burn hydrogen gas and run between 700 K and room temperature (20 °C) (a) What is its operating temperature range in kelvin? (b) What is the maximum theoretical efficiency of our engine? (c) The designer, using some very clever tricks, thinks the engine can do 1.00 kJ of work for every 1.66 kJ of heat entering it. Is she right? Explain.

63. [I] An otherwise lossless thermodynamic engine absorbs 1.99 kJ of heat from a high-temperature reservoir and discharges 1.50 kJ to a low-temperature reservoir during each cycle. What is its efficiency?

SOLUTION: The engine is lossless and so

$$e = \frac{|Q_H| - |Q_L|}{|Q_H|} = \frac{1.99 \text{ kJ} - 1.50 \text{ kJ}}{1.99 \text{ kJ}} = 24.6\%$$

64. [I] Suppose the engine in the previous problem loses an additional 0.10 kJ of energy during each cycle in overcoming friction, what's its efficiency now? How much work can it deliver?

65. [I] A heat engine operating between 200 °C and 100 °C has an efficiency that is only 20.0% of what is ideally possible. What's its efficiency?

66. [I] In a given amount of time, a refrigerator extracts 75 kJ from a cool chamber while exhausting 100 kJ. Determine its coefficient of performance.

67. [I] A 1.0-ton air conditioner has the capacity to freeze 2000 lb of ice at 0 °C in a day. What is the equivalent rate of transport of heat in watts?

68. [I] What is the coefficient of performance of a Carnot refrigerator operating between 0 °C and 85 °C?

69. [I] A refrigerator is to remove heat from a chamber at −3.0 °C and vent it out to a room at 27 °C. Ideally how many joules of work will be required per joule of heat removed?

70. [II] THIS PROBLEM DEALS WITH A THERMAL ENGINE. An ideal gas is used as the working fluid in a thermal process that essentially converts heat into work. We want to find its actual efficiency and compare that to the best possible efficiency attainable. The gas absorbs 4980 J of heat at 505 °C, puts out 2012 J of work, and exhausts waste heat to a reservoir at −30.0 °C. (a) What is the mathematical expression for the efficiency of a Carnot engine in terms of its operating temperatures? (b) Determine the best possible efficiency our engine could have. (c) Assuming the engine is lossless, what physical quantity equals the difference between the heat-in and the heat-out, that is $|Q_H| - |Q_L|$? (d) What is the numerical value of $|Q_H| - |Q_L|$? (e) Compute the value of the exhausted heat $|Q_L|$, assuming the engine to be otherwise lossless. (f) Calculate the actual efficiency of the engine.

71. [II] THIS PROBLEM EXPLORES THE BEHAVIOR OF A THERMAL ENGINE. Suppose we have a thermal engine operating between 301 °C and 149 °C, and want to increase its efficiency. (a) What is the mathematical expression for the efficiency of a Carnot engine in terms of its operating temperatures? (b) What is the maximum theoretical efficiency of our engine? (c) How can we increase that efficiency without changing the low-temperature thermal sink? (d) How might we increase the maximum theoretical efficiency by 35.0%?

72. [II] A reversible heat engine operates between 325 °C and 25 °C, taking in 610 kcal of heat per cycle. Determine the efficiency of the system, the work done per cycle, and the heat exhausted per cycle. If the engine runs at 5.0 cycles per second, what is the power delivered by it?

73. [II] It is claimed that a heat engine has been built that takes in 102 kJ of thermal energy and produces 26 kJ of useful work. It supposedly operates between 144 °F and 5°F. Is such a device physically possible? What's the least amount of heat an engine, operating between those same temperatures, would require to do 100 kJ of work? [*Hint: Be careful about the temperature scale. Check out the Carnot efficiency.*]

74. [II] A heat engine is being designed to operate between 200 °C and 20.0 °C. At best we should be prepared to supply an amount of heat Q to it in order to get out 1000 J of work. What is the value of Q?

75. [II] We have a furnace that can supply 100 MJ/h of heat at 300 °C to an engine exhausting waste at 20.0 °C. Under the best of circumstances, how much work could we expect to get out of the system per hour?

76. [II] Suppose that in the course of a day you consume one peanut butter sandwich (330 kcal), a jelly doughnut (225 kcal), 1.0 kg of milk (650 cal/g), and two martinis (145 kcal each). You spend a quiet day averaging an output of about 60 watts of useful work for 6 hours, above and beyond the energy consumed just staying alive. What's your efficiency?

77. [II] THIS PROBLEM TREATS THE BEHAVIOR OF A THERMAL ENGINE. A heat engine is to put out 7.20 kJ of work for every 10.00 kJ of heat supplied to it. Given that its maximum operating temperature is 900 K, we want to explore its operation. (a) What's its actual efficiency? (b) What percentage of the energy supplied is "lost"? (c) Where might some of that energy have gone? (d) Suppose the engine exhausts 2.00 kJ to a low-temperature reservoir. How much energy is "lost" by all other means including friction?

78. [II] THIS PROBLEM WILL HELP US UNDERSTAND THE BEHAVIOR OF A THERMAL ENGINE. An ideal heat engine operates in a 1.49 s long cycle bounded by two isotherms and two adiabats. The upper and lower reservoir temperatures are 1000 K and 250 K, respectively. (a) What is the mathematical expression for the efficiency of a Carnot engine in terms of its operating temperatures? (b) Determine the anticipated efficiency of our engine. (c) If the engine takes in 24.5 kJ of heat during each cycle, how much work can it perform per cycle? (d) What's the maximum amount of mechanical power the engine can deliver? (e) How much heat does the engine exhaust to the low-temperature reservoir per cycle?

79. [II] A Carnot engine operates at an efficiency of 42.2% with a high-temperature reservoir at 473 K. If the efficiency is to be raised to 50% using a new high-temperature reservoir, what will its temperature have to be?

80. [II] The design of a heat engine calls for it to operate between a source of heat at 67.0°C and a heat sink at 17.0°C. The engine is to function at 90.0% of the Carnot efficiency. How much work can we expect to get out for each kilojoule of heat provided?

81. [II] During each cycle of operation of a heat engine, it takes in 1200 J of heat from a hot reservoir and exhausts 450 J of heat to a cool reservoir. What is its efficiency? The engine is now run backward between the same reservoirs and, while delivering 1200 J per cycle to the high-temperature reservoir, it is driven by 1000 J of applied work. Determine whether this engine is reversible or irreversible in a thermodynamic sense.

82. [II] The air conditioner in my bedroom has a little plaque on it that says 18 000 Btu/h, 230 V, 9.0 amps. That means that it consumes electrical power at a rate of (230 V)(9.0 amps) = 2070 W while removing 18 000 Btu each hour from the room. Given that 1 Btu = 1.055×10^3 J, determine the coefficient of performance of the machine.

83. [II] An old-fashioned commercial 1-ton refrigerator removes heat from water in a chamber at 32 °F at such a rate as to produce 1.0 ton (2000 lb) of ice (at 32 °F) per day. The heat is exhausted to the environment at 88 °F. At its very best, how much power would be needed to drive this system?

84. [II] We have designed a 150-W electrical refrigeration system that removes the 360 J/s of heat leaking into a test chamber and expels that thermal energy (the so-called *heat load*) into the room (which is at 23 °C). Determine the lowest temperature we can expect to reach in the chamber.

85. [II] A heat pump, which removes thermal energy from the outdoors and exhausts it inside to heat the house, has a coefficient of performance of 5.5. What is the ratio of the heat supplied to the house to the work done? How much energy do you pay for as compared to the amount you get pumped into the house? How does this compare with heating with electricity directly?

86. [II] A refrigerator maintains food at 40 °F when the room temperature is 80 °F. Ideally, how much heat will be removed from the food per joule of work done by the compressor on the refrigerant?

87. [II] You are designing an electric refrigerator to remove 360 J of heat per minute from a cold chamber at −4.0 °C and exhaust that energy into a room at 27 °C. At the very best, how much electrical power will it require?

SECTION 14.7: ENTROPY

88. [I] A room containing several heat sources (such as people, lights, and a TV) is at a constant 22 °C maintained by an air conditioner that pumps heat out. How much entropy is removed for each 5000 J of heat removed?

89. [I] Imagine a quantity of boiling water at a constant pressure. What is the resulting change in the entropy of the H_2O molecules when the system receives 1.00 kJ of heat? [*Hint: Boiling takes place at a constant temperature.*]

90. [I] A block of metal is placed in contact with a furnace wall that is at a fixed temperature of 200 °C. During a measured time interval the block's temperature rises, indicating that it has received 1.20 kJ of heat. How much entropy has the furnace lost in the process?

91. [I] Five kilojoules of heat enter a very large chamber that remains at a constant temperature of −10.0 °C. What is the resulting change in the entropy of the chamber?

92. [I] THIS PROBLEM WILL HELP US BETTER UNDERSTAND THE CONCEPT OF ENTROPY. A house is kept at 23 °C while the temperature outside is 0.0 °C. The house loses heat at a rate of 8000 W. We want to study the entropy transferred to the environment. (a) How much thermal energy is lost by the house every minute? (b) Write a mathematical expression for the entropy change per minute in terms of the net heat that flowed in or out. (c) What is the increase in the entropy of the environment every minute?

93. [I] THIS PROBLEM WILL HELP US BETTER UNDERSTAND THE CONCEPT OF ENTROPY. Four kilograms of ice at 0.000 °C are slowly and completely melted until there exists 4.00 liters of water at 0.000 °C. We wish to determine the net change in entropy that occurs. (a) Write a mathematical expression for entropy change at a constant temperature. (b) How much heat is needed to melt 4.00 kg of ice at 0.000 °C? (c) At what kelvin temperature does the melting occur? (d) What is the change in entropy of the ice-water system?

94. [I] By how much does its entropy change when 10 g of water at 0 °C is completely frozen into ice at 0 °C? Make it clear whether your answer is an increase or decrease in entropy.

95. [I] Draw a T-versus-S diagram for the Carnot cycle and label it to match Fig. 14.19.

96. [I] A thick metal bar, insulated everywhere but on its ends, is placed between two thermal reservoirs at 600 K and 300 K, respectively, such that heat flows and a stable temperature distribution exists along the bar. If in a certain time interval 10.0 kcal of heat traverses the bar, what is the attendant change in entropy of (a) both reservoirs, (b) the bar, (c) the Universe?

97. [I] A 500-g piece of copper melts at 1083 °C. Determine its change in entropy in the process.

98. [I] By how much does the entropy of 1.00 kg of water change when it is completely converted to steam?

99. [I] With Problem 98 in mind, compare the changes in entropy that result when 1.00 kg of water ice is melted at 0 °C with that when the liquid is vaporized at 100 °C. Account for the great difference in these two values.

100. [II] A 100-g chunk of ice at 0 °C is placed on a very large marble table, which is at 27 °C. What, if any, is the change in entropy of the ice-table system when the ice melts completely?

101. [II] A 20-kg sample of pure water at 40.0 °C is mixed with a 20-kg sample of pure water at 32.0 °C. Using the average temperatures of each sample, determine the approximate net change in entropy of the total quantity of water.

102. [II] THIS PROBLEM WILL HELP US BETTER UNDERSTAND THE CONCEPT OF ENTROPY. A Carnot engine draws 1000 J of heat from a reservoir at 500 K. It does a certain amount of work, while exhausting heat to a reservoir at 300 K. (a) Determine the efficiency of the engine. (b) How much work is done by the engine in a single cycle? (c) How much heat flows *out* of the high-temperature reservoir per cycle? (d) What is the change in the entropy of the high-temperature reservoir per cycle? (e) Does that represent an increase or decrease in entropy? (f) How much heat was exhausted to the low-temperature reservoir per cycle? (g) What is the

change in the entropy of the low-temperature reservoir per cycle? (h) What is the change in the entropy of the engine itself per cycle? Explain. (i) What is the change in the entropy of the entire system per cycle?

103. [II] THIS PROBLEM WILL HELP US BETTER UNDERSTAND THE CONCEPT OF ENTROPY. An ideal gas undergoes a quasi-static isothermal expansion during which its volume doubles. We want to study its entropy change. (a) What temperature change does the gas experience? (b) If P_i is its initial absolute pressure, what will be its final pressure? (c) Determine the change in internal energy. Explain. (d) What is the relationship between the work done and the heat-in? (e) Write an expression in terms of T, for the work done. (f) Show that $\Delta S = 0.693nR$.

104. [II] An ideal gas expands isothermally from V_i to V_f. Prove that the associated entropy change is given by

$$\Delta S = nR \ln \frac{V_f}{V_i}$$

105. [II] Using the data of Fig. P105, which is a TS-diagram for a thermal engine, determine Q_H, Q_L, the net work done by the system, and its efficiency.

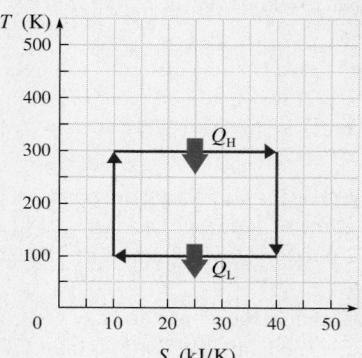

Figure P105

106. [II] A Carnot refrigerator is driven by 2.00 J of mechanical work each second, from an electric motor. It, in turn, removes heat from a cold region (0 °C), expelling it into the room (27 °C). What is the change in entropy of the cold region?

107. [II] Consider an ideal gas in a cylinder sealed with a frictionless piston. The gas is isothermally compressed at 22.0 °C as 800 J of work is done on it. Determine the resulting change in entropy of the gas.

108. [III] Derive an expression for the entropy change experienced by a gas during a free expansion from a volume V_i to a volume V_f.

109. [III] Two identical thermally insulated chambers are connected by a small valve. One chamber contains 1.0 mole of an ideal diatomic gas; the other chamber is evacuated. The valve is then opened, and the gas rushes through the valve so that it occupies both chambers. What is the resulting change in entropy?

110. [III] A 200-kg boulder rolls, slides, and slips down the side of a mountain, descending 200 m vertically before coming back to rest. It's a nice cool winter's day, and the temperature of the forest is 7.0 °C. What is the change in entropy of the Universe as a result of the tumble?

Chapter 15
Electrostatics: Forces

*T*hus far, we have concentrated on the gravitational interaction and the associated property of matter called *mass*. The electromagnetic interaction was always there in our considerations but in the guise of such macroscopic notions as friction, cohesion, elasticity, and so forth. Now we focus on it and on the associated characteristic of matter called *charge*. The electromagnetic interaction is responsible for holding the subatomic particles together as atoms, for holding the atoms together as molecules (p. 3), for holding the molecules together as objects, and indeed, for holding your nose onto your face. All biological processes are governed by the interaction of charges: seeing, feeling, moving, thinking, living—all of it.

The story begins in ancient times with amber, a yellow-brown material (fossilized pine-tree resin) used for jewelry for thousands of years. Probably while polishing amber, the Greeks noticed that it had an extraordinary quality: rubbed with cloth or fur, it attracts small bits of lightweight matter—tufts of lint and hair, for example. Even Plato wrote about the "marvels concerning the attraction of amber."

The Greek word for amber is *elektron*, and by the mid-seventeenth century it was already being suggested that a substance, activated by rubbing, possessed some sort of "amber stuff," or *electricity*. And people came to refer to the *charge* of electricity imparted to an object in the same sense as an "amount" (much like a charge of gunpowder).

When rubbed with woolen cloth, a chunk of amber or a plastic comb can pick up a small piece of Styrofoam™ broken from a coffee cup. From our understanding of the laws of mechanics we can conclude that there is yet another force, the *electric force*. Something must be causing it, and so we call the physical attribute responsible for that interaction **electric charge** or, more generally, ***electromagnetic charge***. (Electric and magnetic phenomena are one and the same—both are manifestations of charge.)

The subject of electromagnetism is usually arranged in subdivisions, one of which is **electrostatics**, *the study of charges at rest*.

Electromagnetic Charge

Charge gives rise to electric force (*or more precisely, to electromagnetic force*), and we are only now beginning to figure out how it manages to do that. Charge is fundamental and cannot be described in terms of simpler, more basic concepts. We know it by what it does, not by what it is—if you like, it is what it does, and that's that.

STUDY GUIDE

This chapter begins our treatment of the Electromagnetic Interaction. It contains dozens of new concepts (charge, permittivity, dielectric constant, etc.) along with two grand ideas: Coulomb's Law (p. 538) and the electric field (p. 544). All of the problems in the back of the chapter will deal with one or the other of these two primary notions. But you can't begin to learn about them without first knowing something about charge.

The first part of this chapter (p. 529), Electromagnetic Charge, provides the needed qualitative background material. With that under your belt we can quantify things with Coulomb's Law (Sect. 15.3). And that brings us to problem solving. As ever, the Devil is in the details. Accordingly, all the little technical details you will need to master are carefully elaborated in the Examples. Study them!!!! Without reading the text and working through every step of each Example it's not likely you'll be very successful in doing the problems at the end of the chapter (p. 565).

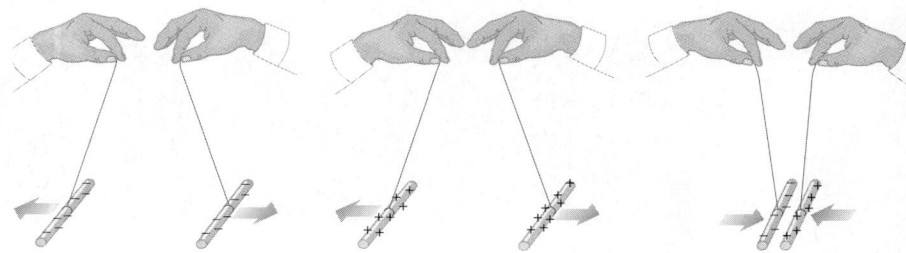

Figure 15.2 Attraction and repulsion of charges. Like charges repel; unlike charges attract.

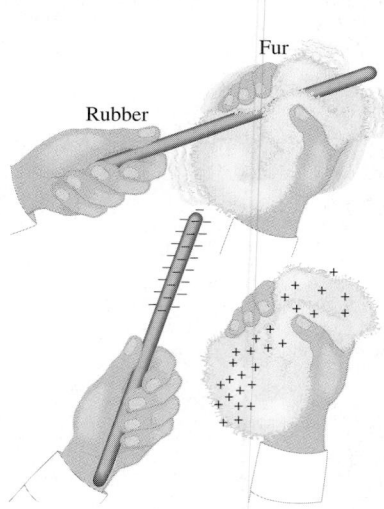

Figure 15.1 Charging by contact. Electrons are transferred from fur to rod. The hard-rubber rod becomes negatively charged, the fur becomes positively charged.

If we vigorously stroke a plastic pen on a woolen glove and hold the two apart but close, fibers on the glove will stand on end, straining to reach up to the pen, making the attraction obvious. It was a French botanist, C. Dufay, who first studied the *repulsive* interactions of electricity. He found that objects of the same material electrified in the same way repelled one another. Two pieces of glass rubbed with silk will repel each other, just as two chunks of amber rubbed with fur will (Fig. 15.1). Yet the charged glass will attract both the silk and the charged amber, and vice versa. Sometime around 1734, Dufay concluded "that there are two distinct Electricities"—two kinds of electric charge—and he was right. Summarized (Fig. 15.2) in contemporary terms:

Like charges repel; unlike charges attract.

15.1 Positive & Negative Charge

Today we follow Benjamin Franklin, arbitrarily calling the two kinds of charge *positive* and *negative*. Ordinarily we deal with the electrical behavior of solids (plastic combs, nylon sweaters, TV screens, etc.), and there the positive charges are locked up in the nuclei of the essentially stationary atoms, while some negative charges are free to be transferred. There are two kinds of charge, but in the vast majority of cases, only the negative is mobile.

An object that contains the same amount of positive as negative charge in close proximity attracts and repels an external charge equally and thereby cancels its own ability to exert a net force. It behaves as though it had no charge at all and is electrically **neutral**. That ability to combine charges to produce a null response allows us to create a kind of electrical algebra: 10 units of positive charge and 10 units of negative charge add up to a net of zero units of charge, in the sense that together they produce no observable external electric force. Any macroscopic object, yourself included, contains a vast number of minute individual charges, but on the whole it is usually neutral. That's why you're not noticeably pulled on by the electric force as you are by the gravitational force.

Charge Is Quantized and Conserved

An *electron is negatively charged*; it repels other electrons and it attracts protons. The latter are positively charged. There is no way to discharge an electron, no way to peel off its charge and make it naked, neutral. If the electron is truly a fundamental particle, and most physicists believe it is, charge is an inseparable aspect of the thing in itself.

The amount of the charge of the electron, q_e, has been determined experimentally. Moreover, all electric charge comes in whole number multiples of that basic amount. Whether positive or negative, **charge is quantized**; it appears in certain specific amounts—every subatomic entity that has been observed to date has had a charge of either 0, $\pm q_e$, or $\pm 2 q_e$. Since matter on the subatomic scale comes in specific lumps (only a small number of fundamental particles exist), it's not surprising that charge comes in lumps as well.

A neutral "chargeless" object is electrified by either losing or gaining charge; the glass rod loses electrons, the silk used to rub it gains them. Unlike mass (which can be converted

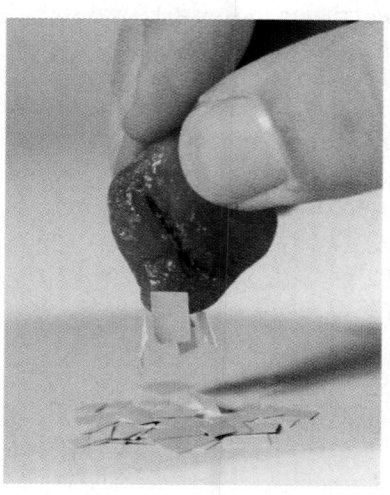

A chunk of amber rubbed on fur picks up small pieces of colored paper.

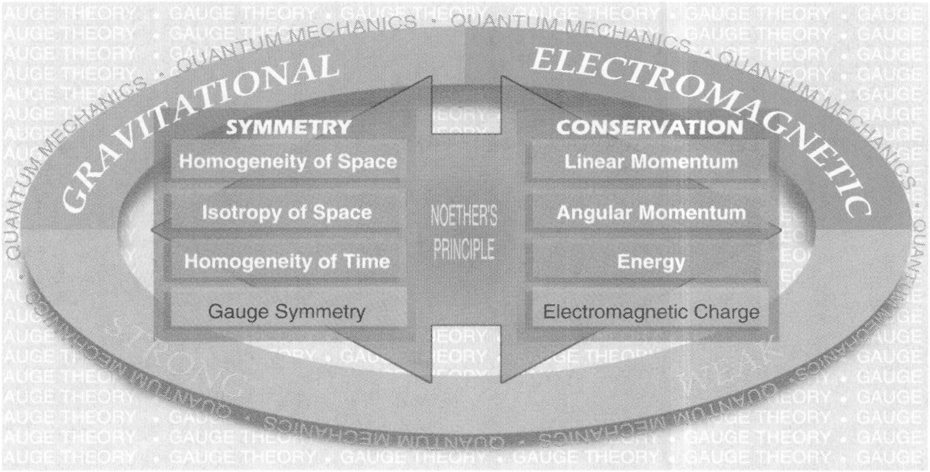

Figure 15.3 The electromagnetic force is the interaction of charge, just as the gravitational force is the interaction of mass. Charge is conserved—a concept we'll come back to later (p. 582).

> Electricity is of two kinds, positive and negative. The difference is, I presume, that one comes a little more expensive, but is more durable; the other is a cheaper thing, but the moths get into it.
>
> STEPHEN LEACOCK (1869–1944)
> CANADIAN HUMORIST

into energy, Ch. 26), electromagnetic charge is scrupulously conserved. ***The total charge (the difference between the amount of positive and negative charge) within an isolated system is always constant.*** That's the **Law of Conservation of Charge** (Fig. 15.3). Whenever a positive charge appears, somewhere in the vicinity there will appear an equal negative charge. As far as we can tell, the net amount of charge in the Universe is constant, which does not mean charge cannot be created; it just means that, if it is, as much positive as negative will be produced. This constant net charge of the Universe may well equal zero (that's a pretty thought), but clearly no one knows.

> **Conservation of Charge:** The net charge (the amount of positive charge minus the amount of negative charge) in an isolated system remains constant.

Charging by Rubbing

We are surrounded by charges that form a subtle electrostatic environment. Everyone has probably felt the sparks that often punctuate a short shuffle across a carpet on a dry winter's day. Most of us have stroked plastic foodwrap so that it adheres to some container, rubbed a balloon on a shirt to stick it on a wall, used a copy machine, or fussed with clinging clothes. All sorts of things get charged by rubbing, form water droplets in the clouds to anthrax spores blown into the air by terrorists—electrostatic effects are everywhere.

Atoms are neutral; they possess as many negative electrons as positive protons. But the outer electrons are the least strongly bound and are most easily shed. Different materials have different affinities for electrons. When two substances are put in contact, one of them may give up some of its loose electrons while the other draws them into itself. When a sheet of plastic is pressed down onto a metal plate, electrons will be transferred from the donor plastic to the grabber metal. The plastic, having lost electrons, now contains a number of immobile positive ions (atoms missing negative charge) on its surface and, as a whole, has become charged. The positive plastic attracts the negative metal, and the two cling to one another.

In much the same way, when a hard-rubber rod is stroked with a piece of fur, the rod draws off electrons, becoming negatively charged, and the donor fur assumes an equal, positive charge. *The rubbing seems to do little beyond increasing the area brought into intimate contact.* Moreover, there are degrees of grabbers, and a substance that can snatch electrons away

PROTONS, QUARKS, & FRACTIONAL CHARGE

It is curious that the proton has a charge $(+q_e)$ equal in size to that of the electron $(-q_e)$, since these particles are otherwise quite different. Modern experiments have shown that the magnitudes of the charges of the electron and proton are so nearly alike that if their ratio does differ from 1 at all, it does so by less than 10^{-20}. That's fascinating, especially since the proton seems to have a complex structure. Contemporary theory maintains that most heavy subatomic particles are actually composite systems (Ch. 31) made up of several fundamental entities called *quarks*. These, in turn, are supposed to have charges of $\pm\frac{1}{3}q_e$ and $\pm\frac{2}{3}q_e$. It is believed, however, that quarks cannot ordinarily exist in the free state (i.e., independently), and so the observable unit of charge remains q_e.

A new and extremely sensitive technique for revealing fingerprints has been developed at Los Alamos National Laboratory. It makes use of the fact that unlike charges attract. Negatively charged gold particles adhere to the positively charged proteins, which are always left behind when fingers touch a surface.

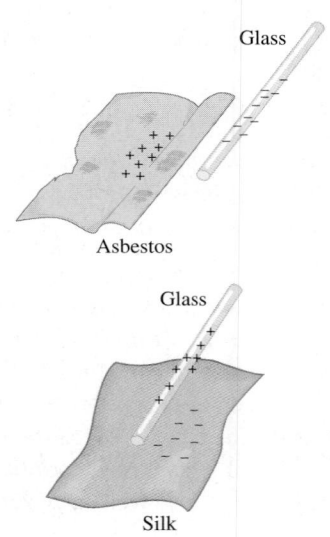

Figure 15.4 Electrons are transferred in contact from the asbestos to the glass and from the glass to the silk.

from one material may well find itself serving as donor to a still more potent grabber. Glass rubbed with asbestos will draw off electrons from that fibrous material, becoming negative; but if stroked with some persistence against silk or flannel, the glass will emerge positively charged, having lost electrons (Fig. 15.4). We can bring some order to the whole business by arbitrarily defining glass-rubbed-on-silk to be *positive*, and then any charged object that is attracted to it is *negative*.

By comparing the behavior of various materials, a listing (Table 15.1) known as the *triboelectric* (*tribo*, meaning friction) *sequence* has been formulated. The materials toward the top of the list tend to lose electrons easily; those near the bottom tend to gain them effectively. The farther apart on the list the two are, the more intense the resulting electrification, which is why rabbit fur and hard rubber are still the mainstays of electrostatic demonstrations.

The Transfer of Positive Charge

A negatively charged object contains an excess of electrons that repel each other. When such an object is placed in contact with a neutral body, some of those electrons are forced over onto that body, charging it negatively. Similarly (Fig. 15.5), a positively charged body has a deficiency of electrons or, equivalently, an excess of positive ions. When placed in contact with a neutral body, it attracts and draws off electrons, becoming less positive itself, while causing the now electron-diminished body to become positive. *Only electrons are transferred, but the system behaves exactly as though positive charge flowed from the positively charged object to the neutral one.* That being the case, it's common to talk about the transfer of positive charge even though no positive charge actually moves.

15.2 Insulators & Conductors

If you rub the middle of a plastic comb with a piece of wool, that region will have the ability to pick up tissue tufts, but the comb's ends will not—they remain neutral, even though the middle is electrified. This behavior represents a class of materials (such as wood, plastic, glass, air, hair, cloth, leather—and dry elephant), all of which are variously known as insulators, nonconductors, or dielectrics. Charges in a nonconductor have limited mobility and will only move when their mutual repulsion is great enough to overcome the tendency

Table 15.1

The Triboelectric Sequence	
Asbestos	On contact between
Fur (rabbit)	any two substances
Glass	shown in the col-
Mica	umn, the one
Wool	appearing above
Quartz	becomes positively
Fur (cat)	charged and the one
Lead	below becomes
Silk	negatively charged.
Human skin,	
aluminum	
Cotton	
Wood	
Amber	
Copper, brass	
Rubber	
Sulfur	
Celluloid	
India rubber	

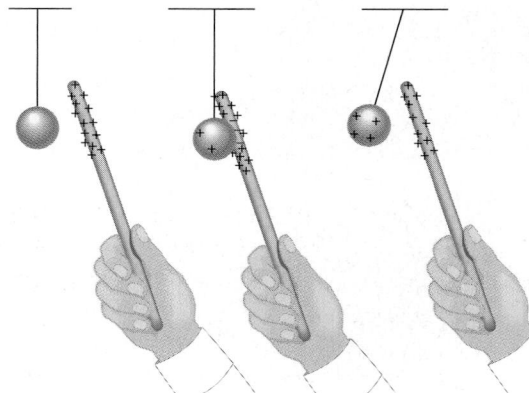

Figure 15.5 Charging a sphere of pith positively by direct transfer of electrons from ball to rod. This leaves the ball positively charged, and it is immediately repelled from the rod.

PITH BALLS & PACHYDERMS

The arrangement in Fig. 15.2 suggests a straightforward scheme for determining the polarity of any distribution of charge. Suppose we want to find the sign of a charged elephant and don't have a modern electrometer handy. Just suspend two identical, lightweight neutral objects—two Ping-Pong balls, say, or two pith balls.* Now touch a rod of glass-rubbed-on-silk to one of the pith balls, thereby charging it positively. Next, back the electrified pachyderm into the second pith ball, which will take on some of the unknown elephant-charge. If the two pith balls then attract, the elephant *may* be negative; if they repel, it surely is positive (see Discussion Question 15).

Electrical action-reaction before the Principia. It is commonly believed, that Amber attracts the little Bodies to itself; but the Action is indeed mutual, not more properly belonging to the Amber, than to the Bodies moved, by which it also itself is attracted;…"

MAGALOTTI (1665)
FLORENTINE ACADEMY

to be held in place by the host atoms. *When an insulator receives a charge, it retains that charge, confining it within the localized region in which it was introduced.**

A **conductor** *allows charge introduced anywhere within it to flow freely and redistribute*. Metals (e.g., copper, gold, and aluminum) are among the best conductors at ordinary temperatures. Incidentally, the fact that there are both good and bad conductors of electricity was first realized in 1729 by Stephen Gray, who carried out a series of brilliant experiments, despite the severe handicap of being a pensioner in a London poorhouse.

The distinction between conductors and nonconductors (and it's not always a clear one) arises from the relative mobility of charge within the material. The atoms of metals hold their own outermost electrons weakly, and so a bulk sample contains a tremendous number of free electrons, roughly one per atom. Any added electrons join that sea, which moves among the unmoving positive ions. Alternatively, the atoms of a nonconductor hold fast to their own electrons and will even latch on to excess ones introduced on them (Fig. 15.6). Despite this, *no material is a perfect insulator*; all allow some redistribution of charge. Human skin is a respectable nonconductor compared to copper, although it's a fairly decent conductor when compared to glass. Pure water is a modest insulator, but a pinch of some dissolved impurity such as table salt will provide enough ions to turn it into a good conductor (*ion* means *goer* or *traveler*).

Air is a good insulator, particularly dry air, even though it contains some 300 ions per cubic centimeter. Nevertheless, if enough negative charge builds up on an object, electrons under the influence of their mutual repulsion can be propelled into the surrounding gas. The air will have some of its own electrons ripped off, becoming ionized and creating a tempo-

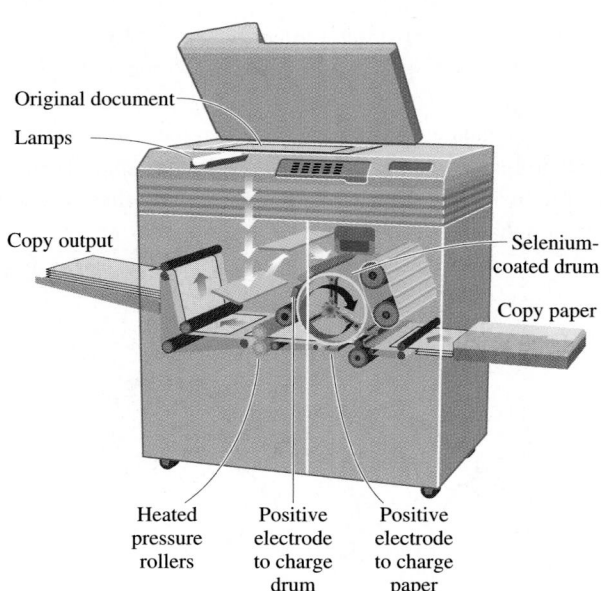

Original document
Lamps
Copy output
Selenium-coated drum
Copy paper
Heated pressure rollers
Positive electrode to charge drum
Positive electrode to charge paper

Figure 15.6 The drum is the centerpiece of both the electrostatic (photo) copier and the laser printer. It's an aluminum cylinder with a coating of photoconducting selenium, which is an insulator in the dark and a conductor in the light. The entire surface is first positively charged, then the image of the document is projected on the drum. Wherever light falls, the charge is conducted away. The dark regions remain positive and attract negatively charged toner powder. A sheet of highly positively charged paper then picks up the toner, to which it is fixed by heated rollers.

*Pith is the pulpy, soft stuff in the center of certain dried plant stems. It was the Styrofoam™ of generations past, used in work with electricity for centuries and still to be found in physics storage rooms around the world.

*The great bane of electrostatic demonstrations is dampness. Warm the objects being charged; work in a dry place; and keep metal, glass, and plastics clean by occasionally wiping them with alcohol. Airborne water molecules are electrically polarized; the hydrogen "Mickey Mouse ears" are positive and tend to attract electrons and so inevitably discharge apparatus.

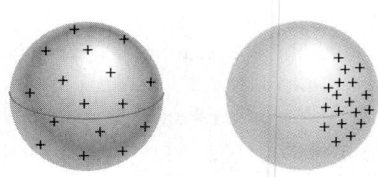

Figure 15.7 The distribution of charge placed on the surfaces of a conductor and a nonconductor.

Figure 15.8 Charge tends to bunch up on the pointed regions of a conductor.

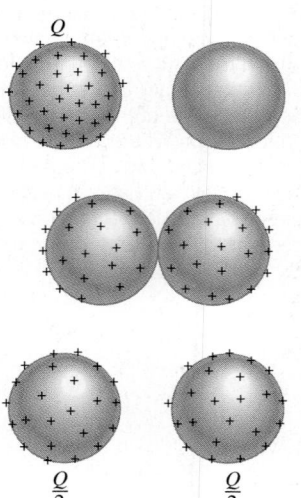

Figure 15.9 The distribution of charge on the surfaces of identical conductors. The charge divides evenly on the two spheres.

rary conductive pathway along which the bulk of the charge then flows. Collisions with the gas increase its temperature and cause some of the atoms to emit light. The result is the familiar glowing trail known as a **spark**.

Charged Conductors

A cluster of identical charges introduced on a conductor experiences a mutual repulsion that sends them all scurrying. Constantly pushed apart, they move until they can separate no farther and are as distant from one another as possible (Fig. 15.7). Charges tossed onto a metal sphere will very quickly stream around until they are *uniformly* distributed and at rest on the outer surface. **No matter what the shape of the conductor, excess charge always resides on its outer surface.** Charges simply push each other to the very extremities of the object. With a nonspherical conductor the charge distribution will be nonuniform, bunching up somewhat in the remote regions (Fig. 15.8). Each charge is impelled to get as far away from as many others as possible, even if that means getting closer to a few charges in the process. That's why charge tends to concentrate on the sharp protrusions of a conductor, a fact well known to people who worry about sparks.

Another manifestation of the free flow of electrons in a conductor is the manner in which charge is transferred. Envision a negative conductor made to touch an uncharged metal body. Electrons are propelled onto the neutral body by their mutual repulsion, which depends on how densely packed they were to begin with. The charge flows much as a fluid flows from a filled chamber into a connecting empty chamber. This gravity-powered flow continues until the liquid levels are the same, the pressures equalize, and equilibrium is reached. As we will see (p. 581), a very similar balance determines the amount of charge transported. Evidently, if a total excess charge Q is placed on one of two identical metal spheres (Fig. 15.9) and those spheres are brought into contact and then separated, a charge of $\frac{1}{2}Q$ will end up on each of them.

In 1786 Rev. A. Bennet introduced a device that, until the modern era of electronics, was the premier electrostatic indicator. The **gold-leaf electroscope** is basically two extremely thin metal leaves hanging parallel to each other from a conducting wire, all surrounded by a protective glass enclosure (Fig. 15.10). A charge introduced onto the metal knob on top spreads out; some travels down the wire and divides equally onto each of the leaves. Repelling one another, the leaves spring apart at an angle proportional to the net charge. And there they stay, leaves apart, until the charge is deliberately removed or until it leaks off into the air.

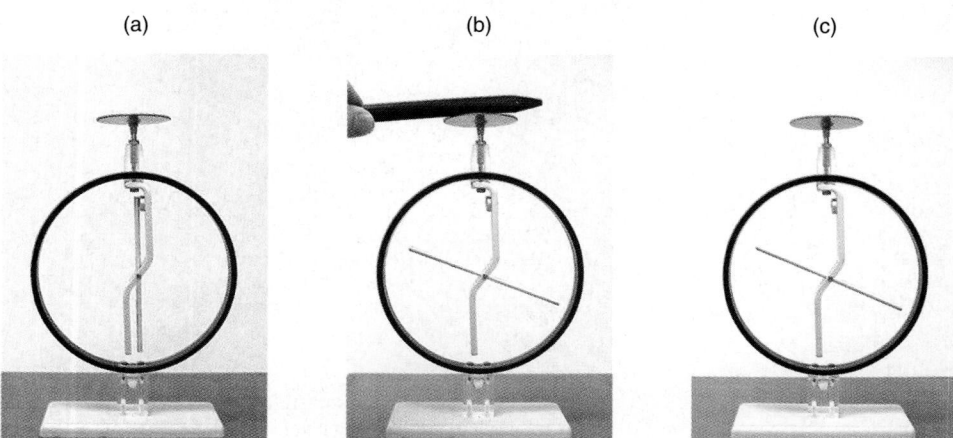

A rugged modern demonstration electroscope. As shown in (a), the device is uncharged and the rod (which is pivoted slightly off center) is in equilibrium. (b) Negative charge is being transferred from the hard rubber wand (which was rubbed with fur) to the electroscope. (c) Charge has been deposited on the device, and the pivoted rod has swung away (via electrostatic repulsion) from the fixed vertical bar, both of which are now negative. It will stay like this until the charge (i.e., electrons) leaks off into the moist air.

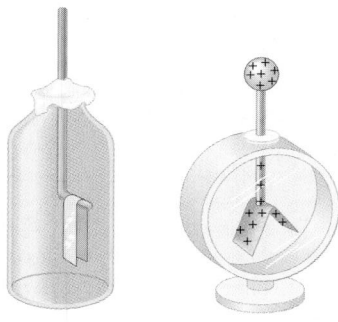

Figure 15.10 Electroscopes home-made and otherwise. No longer used in the laboratory, the electroscope is now primarily a teaching device. To make one, remove the thin aluminum foil from the wrapper on a stick of gum. Hang it on a thick wire, lower it into a bottle, and seal the bottle with wax or clay.

Electrons swarm in clouds around the nuclei of atoms and are incredibly numerous. The most weakly bound of these electrons can leave their host atoms and be moved around giving rise to the effects we will study in this chapter. Pictured here (in red) are the tracks of electrons knocked out of their atoms by (trackless) photons in a process called Compton scattering (Sect. 28.4).

To **ground** an electrical device literally means to connect it to the ground (to the moist soil) in such a way that charge can be transferred to the Earth.

Any conductor in good contact with the Earth (like the water pipes in a building) will carry off all the charge one could possibly want to discard, depositing it in the moist ground. That's precisely what is meant by the phrase to **ground** something (or in Britain, to "earth" it). Incidentally, the pictorial symbol for ground is ⏚.

Electrostatic Induction

It's not necessary for a charged object to physically touch an electroscope in order that the leaves respond to its presence (Fig. 15.11). A negatively charged object, such as a hard-rubber wand stroked with fur, located anywhere near the top of the electroscope will repel free electrons within the conducting knob and support wire. These will be forced down into the leaves, which will spring apart and stay that way as long as the charged object remains nearby to maintain the imbalance. Figure 15.12 is a closeup of what happens in a metal under these conditions. The nearer the wand comes, the greater the repulsion, and the more electrons enter the leaves, which makes the leaves more negative, and they spread farther apart. When the wand is removed, the displaced electrons immediately flow back, being

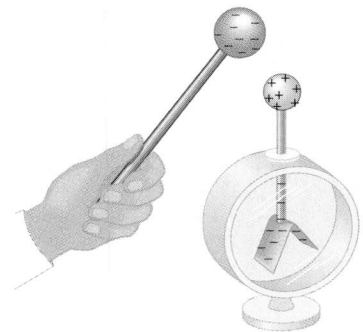

Figure 15.11 Inducing a charge on an electroscope. The negative rod repels electrons down into the leaves. The two equally charged leaves repel each other and move apart.

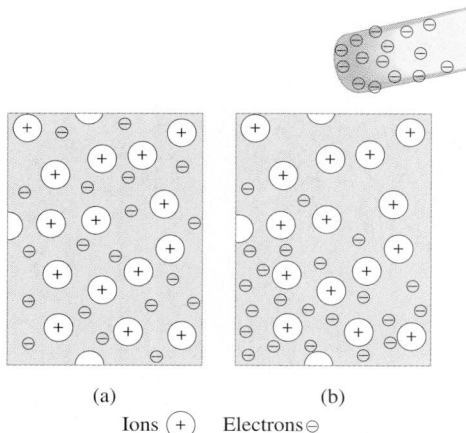

(a)   (b)

Ions ⊕   Electrons ⊖

Figure 15.12 (a) The conductor is neutral, and its electrons are uniformly distributed. (b) A charged rod repels electrons downward. The top of the conductor becomes positive, the bottom negative.

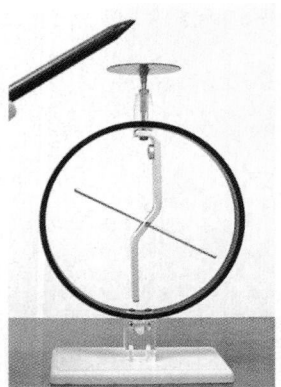

Charging an electroscope by induction. The negatively charged wand repels electrons down into both the pivoted rod and the fixed vertical bar. These repel each other and the rod swings away from the vertical. When the charged wand is removed, the rod swings back to its vertical equilibrium position. The electroscope as a whole is always neutral.

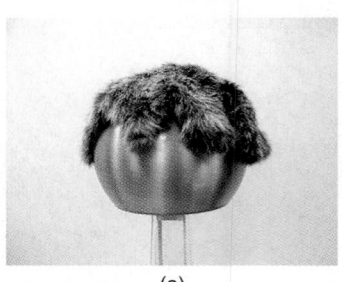

(a)

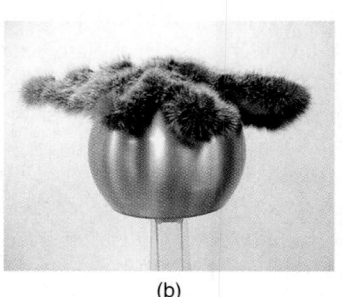

(b)

(a) Uncharged rabbit fur on an uncharged electrostatic generator. (b) Once charged, the fur repels the similarly charged generator and lifts up and off. Individual hairs repel each other and stand on end. (See Fig. Q14 on p. 563.)

mutually repelled by their brethren as well as attracted by the fixed positive ions of the metal (knob, wire, etc.). The leaves hang vertically, and the electroscope, which was neutral in total, reverts to its normally unsegregated charge distribution.

Instead of being transferred to the electroscope, the negative charge on the wand has **induced** a negative charge on the leaves. We could have equally well imagined that the negative wand attracted positive charges toward it, although that is not what happened. Alternatively, a positively charged wand would have attracted electrons upward and thus induced a positive charge on the leaves.

Suppose that the knob is now grounded and the game is repeated, bringing a negatively charged ball nearby (Fig. 15.13). Under the influence of the ball, mobile electrons in the metal (knob, support wire, and leaves), interacting in mutual repulsion, get a lot farther away from each other and the ball by flowing into ground. The scope becomes positively charged, and the leaves stand apart. Moving the ball away allows electrons to return up from ground, and the system relaxes back to its original neutral condition.

Now bring the ball back and hold it close. The scope is again made positive, the leaves again part, but this time we go one step further and disconnect the ground lead, isolating the device. That done, the scope is shy a clutch of electrons and is more or less permanently positively charged with the leaves standing apart. Again, charge was induced—the electroscope was never touched by the ball.

By contrast, electrons within a dielectric, such as a tuft of tissue paper, are far less mobile and ordinarily remain atom-bound. When a negatively charged rod is brought near such a nonconductor, it repels the electron clouds surrounding the atoms and in effect distorts or *polarizes* them (Fig. 15.14). Each nucleus, now slightly exposed, represents the positive end of the elongated atom, and the electron cloud bunched up on the opposite side constitutes the negative end. The result is that one side of the dielectric is electrified positively, the opposite side negatively. A charge has been induced, and it shows up on the surfaces of the specimen.

Figure 15.13 By letting electrons flow off to ground, the wire allows the electroscope to become permanently charged.

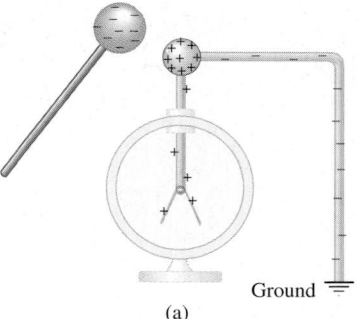

Ground

(a)

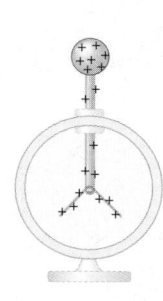

(b)

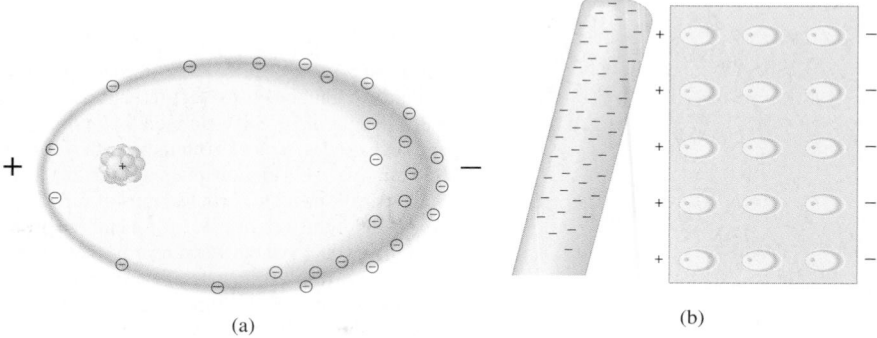

Figure 15.14 (a) A polarized atom. Electrons shifted to the right expose the nucleus. In this polarized state, the right end is negative and the left end is positive. (b) An electrically polarized dielectric.

Figure 15.15 A negatively charged comb picking up bits of tissue paper.

This is just what happens when you negatively charge a comb with a piece of wool or your clean hair and then use it to pick up bits of lint or tissue paper (Fig. 15.15). The region of the tissue closest to the comb (where the force is strongest) takes on a positive charge; the region farthest away becomes negative. That's why, on a dry day, a piece of tissue will stick to your TV screen and why that screen is always covered with dust. It's also why a charged balloon sticks to a wall: it induces an opposite surface charge on the wall, and the two cling together until the excess electrons leak off the balloon.

The Electric Force

In the mid-eighteenth century, the question of the nature of the force acting between charges was a major scientific issue. Naturally, speculation had a decidedly Newtonian bent, relying heavily on discoveries regarding gravity. As early as 1760, Daniel Bernoulli confirmed that his own rather crude experiments were at least consistent with an inverse-square law for electrical attraction and repulsion.

Benjamin Franklin observed that pith balls lowered inside a metal cup appeared to be unaffected by whether the cup was charged or not. We know that excess charge repels itself to the outer extremities of the cup, leaving the inner surface chargeless. That fact is easy to confirm with a *proof-plane*, a small metal disk on an insulating handle (Fig. 15.16). If the proof-plane is touched to the outside of the electrified cup, an electroscope shows that the proof-plane has become charged. Yet touching the inside wall of the cup leaves the proof-plane neutral—***there is no charge on the inside wall of a hollow electrified conductor***. That's clear, but it doesn't explain why there is no electric force inside the hollow—the pith ball is not attracted to a nearby wall even though there are charges just on the other side of

There will be no charge on the inside surface of a hollow conductor (provided the conductor does not surround any additional charge).

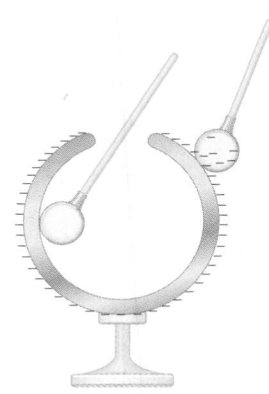

Figure 15.16 Proof-planes inside and outside a hollow charged conductor.

This old AM radio represents so much practical physics that we'll come back to it several times throughout our study of electromagnetic theory.

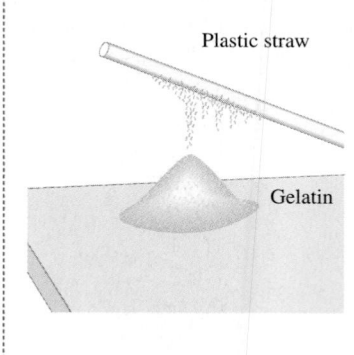

Plastic straw

Gelatin

that wall. Franklin communicated these findings to his English friend Joseph Priestley, the discoverer of oxygen, who used them to argue that, like gravity, electric force varied as $1/r^2$. {For more of this history click on THE ELECTRIC FORCE under FURTHER DISCUSSIONS on the CD.}

So it was, that by the time Charles Augustin de Coulomb published his definitive study (1785), it was already widely suspected that the electrical force between two *point-charges* varied inversely with the square of their separation.

15.3 Coulomb's Law

Coulomb's apparatus (Fig. 15.17) consisted of a lightweight insulating rod hung horizontally from its middle on a long, fine silver wire. On one end of the rod was a small pith ball covered with gold leaf; on the other, counterbalancing it, was a paper disk. Positioned behind the suspended ball was another identical one fixed in place so that the two were initially in contact. They were then both electrified simultaneously upon being touched by a charged rod. Immediately, the suspended ball swung away from the fixed one, twisting the silver wire and coming to rest some distance away (Fig. 15.18).

Coulomb had already determined how much force it took to twist the wire through any angle. Now with the movable sphere pushed aside, he twisted the suspension wire by hand, forcing that sphere back toward its original position near the fixed sphere, against the balls' mutual repulsion. While doing that, he measured the separations and determined the forces holding the two conducting spheres at each location. His results were clear: "The repulsive force between two small spheres charged with the same type of electricity is inversely proportional to the square of the distance between the centers of the two spheres." The electric

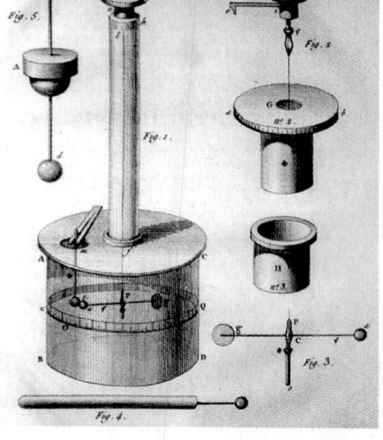

Figure 15.17 C. A. Coulomb's torsion balance (from *Histoire et Mémoires de L'Académie Royale des Sciences,* 1785). A light horizontal rod has a small sphere (*a*) on its left end and a disk counterweight (*g*) on its right end. An identical sphere (*t*) is inserted at the left through a hole (*m*).

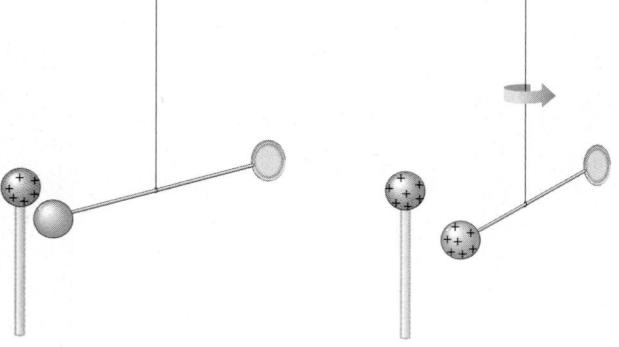

Figure 15.18 A detail of Coulomb's device. Once charged, the two spheres repel and move apart. That behavior twists the wire until the force it exerts matches the electrostatic repulsion.

COULOMB'S CHARGED SPHERES

Provided the two interacting spheres are small compared to their separations, we can expect the charge distributions on each of them to remain fairly, though not precisely, uniform. The presence of either sphere causes a shift of charge on the other sphere (that's part of the mechanism of interaction), just as a rock sags when a butterfly lands on it. When the spheres are very small, the repulsive forces between charges on either one of them will be powerful enough to keep the distribution almost uniform, especially if the other disturbing sphere is far away. With this restriction in mind, *we assume uniform charge distributions on each of the interacting spheres.*

Newton showed that a particle of mass, located outside a thin spherical shell, will be drawn gravitationally toward the sphere's center as if all the shell's mass is concentrated at that point {**Click on The Gravity of a Sphere on the CD**} . In an analogous way, we can conclude that *a uniformly charged conducting sphere produces the same electric force as if all its charge is concentrated at a single point at its center.* The force law can be assumed to be precise for two point-charges separated by a distance r (and nearly so for two small charged spheres).

force (F_E) of interaction between the spheres (and, by Newton's Third Law, the forces are equal in magnitude and opposite in direction even if the charges are unequal) is proportional to $1/r^2$, where r is the center-to-center distance

$$F_E \propto \frac{1}{r^2}$$

as is the gravitational interaction. Modern experiments confirm that if the power to which r is raised differs from 2, it must deviate by *less* than 3×10^{-16}.

Coulomb ingeniously determined how the force law varied with the magnitude of the two interacting charges q_1 and q_2. By touching the charged conducting sphere with an identical uncharged conducting sphere (Fig. 15.9), he was able to exactly halve the charge on it, even without knowing how much charge there was. In the same way, he could neatly reduce it to a quarter, an eighth, and so on, of its initial value simply by repeating the touch-and-halve procedure with two identical spheres. He found for the two interacting spheres that halving the charge on one halved the force, while halving the charge on both quartered it. Coulomb concluded that the force of interaction was proportional to the product of the charges $q_1 q_2$ and so

$$F_E \propto \frac{q_1 q_2}{r^2}$$

The similarity with the gravity formula is obvious and striking, though quite unexplained theoretically even to this day.

The above relationship works well provided the charge distribution on each sphere is uniform. That can be accomplished in practice if the two conducting spheres are tiny. By contrast, suppose the spheres are large compared to their separation and both are identically charged with several billion extra electrons. The electrons on one sphere will repel those on the other. Under the influence of that repulsion, and being free to move on a conductor, the electrons will end up being more concentrated on the far sides of the two spheres, thereby making the above equation essentially useless. So the idea is to work with tiny spheres where the added electrons start out so crowded together that they cannot shift around very much when the two spheres approach each other. The above inverse-square law only applies exactly if the two spheres are infinitesimally small, or equivalently if q_1 and q_2 are point-charges. Accordingly, we'll remind ourselves of that by using the notation $q_\bullet$ to represent such a point-charge, one that's infinitesimal in spatial size, no matter what the value of its charge is.

The right side of the above relationship could be made identical to the left side (not just proportional to it) by defining the unit of charge appropriately: one unit of charge on each sphere would produce one unit of force when the separation was one unit of length. That definition was formulated in the early days in the system that used *centimeters*, *grams*, and *seconds*. There, the unit of force was the *dyne* (1 dyne = 10^{-5} N) and the unit of charge was the now long-forgotten *franklin*. But the force law is really not a very good way to measure charge—the distributions on the spheres are not perfectly uniform. So nowadays the unit of charge is defined via the more precisely measured interactions of electrical currents (p. 696). Accordingly, there is no reason for the right side of the above proportionality to equal the left side; in fact, there is no reason even to expect that the units will be equivalent. If charge is defined in some other way, the force law must contain a constant of proportionality (k) that also has units. As we'll see, this constant is determined by the material the charges are embedded in and has a different value for each such substance.

We begin with the simplest case where the charges are in vacuum; that is, they are not surrounded by ordinary atomic matter (such as, water, oil, etc.). In that case the constant k

Charles Coulomb (1736–1806).

Coulomb's Law: The force between two point-charges is directly proportional to the product of the charges and inversely proportional to the square of the distance separating them.

The sign of the force indicates that it is either (−) attractive or (+) repulsive.

is written as k_0 to remind us that we're dealing with the specific medium of vacuum. The electric force between two point-charges, $q_{\cdot 1}$ and $q_{\cdot 2}$, separated by a distance r in vacuum is then

[force between two point-charges]

$$F_E = k_0 \frac{q_{\cdot 1} q_{\cdot 2}}{r^2}$$

(15.1)

The force acts along the line connecting the charges; like charges repel, whereas unlike charges attract. This is a modern formulation of **Coulomb's Law**. It applies precisely to uniformly charged spheres and less precisely to everything from charged bugs to oil droplets and dust motes. The internationally accepted (SI) unit of charge is the *coulomb* (C) and

$$k_0 = 8.987\,551\,79 \times 10^9 \text{ N·m}^2/\text{C}^2 \approx 9.0 \times 10^9 \text{ N·m}^2/\text{C}^2$$

Suppose the charges are in some medium other than vacuum. Atoms are themselves tiny systems of charge (p. 4), and their presence in the surrounding vicinity of $q_{\cdot 1}$ and $q_{\cdot 2}$ influences what's going on. For example, the value of the constant (k) for water is measured to be about $\frac{1}{80}$ that for vacuum, making the forces between charges embedded in water that much smaller. Because k for air equals $0.999k_0$ it's common practice to just use k_0 for air as well.

Imagine the very unlikely situation of two small spheres, each carrying 1 coulomb of charge. If these are separated in vacuum by 1 meter, each will experience a force of about 9×10^9 N. That's a tremendous force (equal to the weight of more than 2400 jumbo jets), and it arises because a coulomb is a tremendous amount of charge. Since the best modern experimental value of the ***charge of the electron*** is $-1.602\,177\,33 \times 10^{-19}$ C (within an error of $\pm 0.000\,000\,49 \times 10^{-19}$ C), 1 coulomb corresponds to a charge of more than six million million million electrons.

It has been confirmed experimentally that when several point-charges are present, each exerts a force given by Eq. (15.1) on every other point-charge. **The interaction between any two charges is independent of the presence of all other charges.** The electrical force is a vector quantity; therefore, *the net force on any one charge is the vector sum of all the forces exerted on it due to each of the other charges interacting with it independently*. Only in the special case where the charges are arrayed along a straight line will the force vectors be colinear and can they be added or subtracted as scalars (provided we keep track of their directions). {For an animated study of Coulomb's Law, click on **INTERACTION OF ELECTRIC CHARGES** under **INTERACTIVE EXPLORATIONS** in **CHAPTER 15** on the **CD**.}

In theory, if we know the details of some charge distribution, no matter how complicated, we can compute the net force it exerts on a single external charge $q_{\cdot}$. Practically, the summation of all the individual Coulomb interactions can be accomplished either via calculus (approximating the situation as if the charge were distributed continuously) or, more directly, using a computer to carry out the large number of separate applications of Eq. (15.1). Being more concerned with the underlying ideas, our analysis is limited to the simplest cases of just a few charges.

Storing even 1 coulomb is a formidable task, and yet the Earth appears to be carrying a horrendous charge of roughly $-400\,000$ C. In fact, in storm-free regions, the ground leaks about 1500 C every second to the atmosphere. That flow is returned by lightning bolts (up to 20 C each). This ability of the Earth to store vast quantities of charge (because of its great size) makes it an ideal dumping ground for our excess charges.

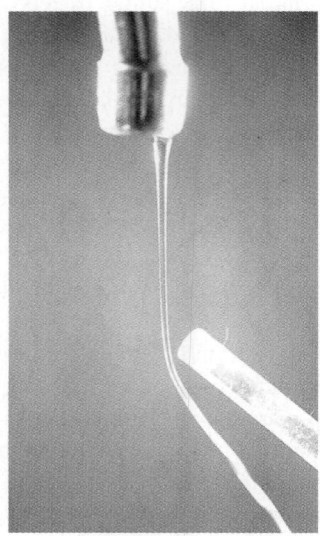

Water is composed of polarized molecules. When a charged rod is brought near the stream, these molecules align themselves so that they are drawn toward the rod. Do you think nonpolar liquids such as gasoline will show the same effect?

ON THE CONSTANT k_0

Until 1983, k_0 was determined by measurements, much as is the constant G. It turns out that numerically k_0 is precisely equal to 10^{-7} times the speed of light in vacuum squared (c^2) and, since c is now fixed by definition (p. 9), so too is k_0. As we'll see, it's no accident that k_0 and c are so intimately related. Notice that the units of k are whatever it takes to cancel out all the units on the right of Eq. (15.1) and still leave N, the unit of force.

EXPLORING PHYSICS ON YOUR OWN

More Electrostatics: For this demonstration you'll need a plastic wand, something to rub it with, several quarter-inch chunks of Styrofoam™ packing material (or even better the small spheres of Styrofoam™ used to stuff toy animals), and a plastic soda bottle. Put the Styrofoam™ in the clean dry soda bottle whose label you've removed. Rub one side of the bottle with wool, plastic, or fur. (I had great success using a soft plastic eyeglass case, of all things.) If you don't have any fur, try a paint brush. Holding the bottle horizontally, the spheres will stick to, and hang from, the region you've charged. Why? Notice how localized the effect is. Bring a finger nearby or touch the bottle at the place where the spheres are attached and see what happens. What do the hanging spheres do when you exhale moist air over the charged region? With the bottle vertical, bring a heavily charged wand (I used a PVC tube rubbed with a plastic bag) near its bottom, and the little spheres will leap into the air. You might even get a few of them to float momentarily.

THE COULOMB & YOU

By human standards, the coulomb is definitely a great deal of charge. We are accustomed to taking our charge in far smaller doses, usually in the form of sparks that carry much less than a microcoulomb (1 μC = 10^{-6} C). Usually, rubbing will build up charge on an ordinary-sized object with a density of up to 10 nanocoulombs per square centimeter (1 nC = 10^{-9} C), which is a practical limit beyond which there will tend to be discharging into the air.

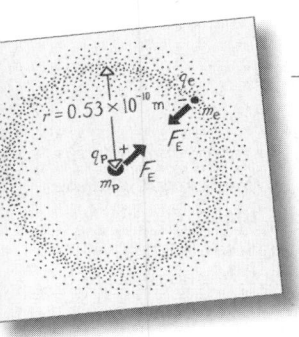

Example 15.1 [II] A hydrogen atom consists of an electron of mass $9.109\,4 \times 10^{-31}$ kg, moving about a proton of mass $1.672\,6 \times 10^{-27}$ kg at an average distance of 0.53×10^{-10} m. Determine the ratio of the electric and gravitational forces acting between the two particles.

Solution   This problem requires the application of Coulomb's Law along with the Law of Universal Gravitation to the hydrogen atom. (1) TRANSLATION—Two particles of known mass and charge are separated by a known distance; determine the ratio of their electric and gravitational forces.   (2) GIVEN: $\pm q_e = \pm 1.60 \times 10^{-19}$ C, $m_e = 9.109\,4 \times 10^{-31}$ kg, $m_p = 1.672\,6 \times 10^{-27}$ kg, and $r = 0.53 \times 10^{-10}$ m. FIND: F_E/F_G.   (3) PROBLEM TYPE—Electrostatics /Coulomb's Law.   (4) PROCEDURE—The electric force is calculated directly from Coulomb's Law.   (5) CALCULATION—Noting that the two charges have opposite signs,

$$F_E = k_0 \frac{q_e q_p}{r^2} = (8.99 \times 10^9 \text{ N·m}^2/\text{C}^2) \times$$

$$\frac{(-1.60 \times 10^{-19} \text{ C})(+1.60 \times 10^{-19} \text{ C})}{(0.53 \times 10^{-10} \text{ m})^2}$$

and $F_E = -8.2 \times 10^{-8}$ N

The minus sign here simply indicates that the force is attractive and that usage is widely adhered to in electrostatics. The sign of the force tells us only if it is attractive (−) or repulsive (+). The force on the proton due to the electron is equal in magnitude and opposite in direction to the force on the electron due to the proton, but each equals -8.2×10^{-8} N. Had we introduced this convention earlier, the Law of Universal Gravitation [Eq. (5.5)] would have had a minus sign in front of it since it's always attractive and the masses are always positive;

$$F_G = -G \frac{m_e m_p}{r^2} = -(6.67 \times 10^{-11} \text{ N·m}^2/\text{kg}^2) \times$$

$$\frac{(9.11 \times 10^{-31} \text{ kg})(1.67 \times 10^{-27} \text{ kg})}{(0.53 \times 10^{-10} \text{ m})^2}$$

and $F_G = -3.6 \times 10^{-47}$ N

Accordingly, $\boxed{F_E/F_G = 2.3 \times 10^{39}}$; the electrical force is far stronger than the gravitational force.

Quick Check: $F_E/F_G = k_0 q_e q_p/G m_e m_p$, and that equals 2.3×10^{39}.

{For more worked problems click on **WALK-THROUGHS** *in* **CHAPTER 15** *on the CD.}*

Coulomb's Law looks just like the Law of Universal Gravitation, but it will be used here in a more advanced way. In the past, we only determined the gravitational force between two masses, and that was done as a scalar computation. In the case of electrostatic force, it's traditional to calculate the force on a charge due to several other charges, and that requires adding the forces vectorially.

Moreover, the electrostatic force can be attractive or repulsive. So we have to be especially careful with signs. Accordingly, study Examples 15.2 and 15.3, going through every step. Example 15.2 treats the simple case of colinear forces, and Example 15.3 deals with the more involved situation of forces at arbitrary angles.

There are several pieces of business that must be understood if Coulomb's Law is to be applied correctly, and they're all there in these examples. The same challenges will arise later when dealing with electric fields, so don't go on to the next section before mastering Coulomb's Law!

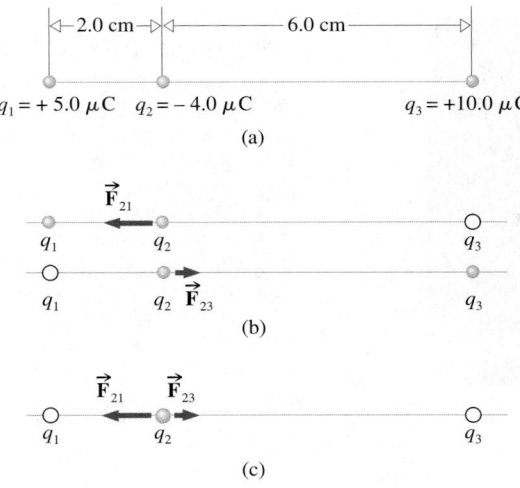

Figure 15.19 Opposite charges attract; like charges repel. Notice how strongly the distance $(1/r^2)$ affects the force: $F_{23} < F_{21}$, even though $q_3 > q_1$.

Example 15.2 **[II]** Figure 15.19a shows three tiny uniformly charged spheres. Determine the net force on the middle sphere due to the other two.

Solution This problem is designed to help you learn how to apply Coulomb's Law in a situation where the forces are colinear. (1) TRANSLATION—Determine the net force exerted on a known charge by two other known charges at specified displacements from it. (2) GIVEN: $q_1 = +5.0\ \mu C$, $q_2 = -4.0\ \mu C$, $q_3 = +10.0\ \mu C$, $r_{21} = 2.0$ cm, and $r_{23} = 6.0$ cm. FIND: F_2. (3) PROBLEM TYPE—Electrostatics/Coulomb's Law/vectors. (4) PROCEDURE—The electric force is calculated from Coulomb's Law, treating each of the several interactions as vectors. (5) CALCULATION—Watch out for units—we want everything in SI. The net force on charge-2 is the vector sum of the force exerted on 2 by 1, namely, $\vec{F}_{21}$; and the force exerted on 2 by 3, namely, $\vec{F}_{23}$:

$$\vec{F}_2 = \vec{F}_{21} + \vec{F}_{23}$$

Let's find $\vec{F}_2$ by just keeping track of the signs using a diagram. **Always sketch the problem showing the directions of the forces**, as is done in Fig. 15.19b. Because the charges have opposite polarities, both forces are attractive and act in opposite directions. Coulomb's Law provides the numerical values of those forces:

$$F_{21} = k_0 \frac{q_2 q_1}{r_{21}^2} = (8.99 \times 10^9\ \text{N·m}^2/\text{C}^2) \times$$

$$\frac{(-4.0 \times 10^{-6}\ \text{C})(+5.0 \times 10^{-6}\ \text{C})}{(2.0 \times 10^{-2}\ \text{m})^2} = -450\ \text{N}$$

$$F_{23} = k_0 \frac{q_2 q_3}{r_{23}^2} = (8.99 \times 10^9\ \text{N·m}^2/\text{C}^2) \times$$

$$\frac{(-4.0 \times 10^{-6}\ \text{C})(+10.0 \times 10^{-6}\ \text{C})}{(6.0 \times 10^{-2}\ \text{m})^2} = -100\ \text{N}$$

These minus signs (attraction) have been integrated into the solution via the directions of the forces in the diagram and are no longer of any concern—**forget about them!** The two vectors in Fig. 15.19c must now be added together. As ever, when treating colinear vectors, we take a direction (to the right) to be positive, whereupon

$$F_2 = F_{21} + F_{23} = (-450\ \text{N}) + (+100\ \text{N}) = -350\ \text{N}$$

The resulting force is

$$\boxed{\vec{F}_2 = 3.5 \times 10^2\ \text{N–TO THE LEFT}}$$

Quick Check: Comparing F_{23} and F_{21} using Coulomb's Law, $F_{23} = (2/3^2)F_{21}$; hence, $F_2 = (7/9)F_{21} = (7/9)(450\ \text{N}) = 350\ \text{N}$.

Example 15.3 **[III]** Figure 15.20*a* depicts three small charged spheres at the vertices of a 3-4-5 right triangle. Calculate the force (magnitude and direction) exerted on q_3 by the other two charges.

Solution   This problem is designed to help you learn how to apply Coulomb's Law in a situation where the forces are not colinear.   (1) TRANSLATION—Determine the net force exerted on a known charge by two other known charges at specified displacements from it.   (2) GIVEN: $q_1 = +50\ \mu\text{C}$, $q_2 = -80\ \mu\text{C}$, $q_3 = +10\ \mu\text{C}$, $r_{12} = 50$ cm, $r_{31} = 30$ cm, and $r_{32} = 40$ cm. FIND: F_3.   (3) PROBLEM TYPE—Electrostatics/Coulomb's Law/vectors.   (4) PROCEDURE—The electric force is calculated from Coulomb's Law, treating each of the several interactions as vectors. **Always sketch the problem showing the directions of the forces.**   (5) CALCULATION—To start, determine if the forces are attractive or repulsive and then draw them acting center-to-center on q_3, as in Fig. 15.20*b*. Since these two force vectors are not colinear, we must, as usual, resolve them into perpendicular components and sum those individually. Figure 15.20*c* provides the appropriate geometry. But first we need to compute the scalar values of the two force vectors

$$F_{31} = k_0 \frac{q_3 q_1}{r_{31}^2} = (8.99 \times 10^9\ \text{N·m}^2/\text{C}^2) \times$$

$$\frac{(+10 \times 10^{-6}\ \text{C})(+50 \times 10^{-6}\ \text{C})}{(30 \times 10^{-2}\ \text{m})^2} = +50\ \text{N}$$

which is positive because like charges repel. Furthermore

$$F_{32} = k_0 \frac{q_3 q_2}{r_{32}^2} = (8.99 \times 10^9\ \text{N·m}^2/\text{C}^2) \times$$

$$\frac{(+10 \times 10^{-6}\ \text{C})(-80 \times 10^{-6}\ \text{C})}{(40 \times 10^{-2}\ \text{m})^2} = -45\ \text{N}$$

which is negative because unlike charges attract. From Fig. 15.20*c*, forgetting the attractive-minus sign, we have

$$F_{3x} = F_{32} \cos 36.9° + F_{31} \cos 53.1°$$

$$F_{3x} = (45\ \text{N})(0.800) + (50\ \text{N})(0.600)$$

and

$$F_{3x} = 36\ \text{N} + 30\ \text{N} = 66\ \text{N}$$

whereas

$$F_{3y} = -F_{32} \sin 36.9° + F_{31} \sin 53.1°$$

$$F_{3y} = -(45\ \text{N})(0.600) + (50\ \text{N})(0.800)$$

and

$$F_{3y} = -27\ \text{N} + 40\ \text{N} = +13\ \text{N}$$

The magnitude of the net force is

$$F_3 = \sqrt{F_{3x}^2 + F_{3y}^2} = \sqrt{(66\ \text{N})^2 + (13\ \text{N})^2} = \boxed{67\ \text{N}}$$

and its direction, as in Fig. 15.20*d*, is

$$\theta = \tan^{-1} \frac{F_{3y}}{F_{3x}} = \tan^{-1}\left(\frac{13\ \text{N}}{66\ \text{N}}\right) = \boxed{11°}$$

Quick Check: There is no different easy way to arrive at the above results, but since the force between two $\pm 10\ \mu\text{C}$ charges 10 cm apart is $\approx \pm 100$ N, these values are at least the right order-of-magnitude. Repelled by q_1 and attracted to q_2, the force on q_3 must be in either the first or fourth quadrants. The effect of q_1 is greater than q_2 because the force drops off as $1/r^2$.

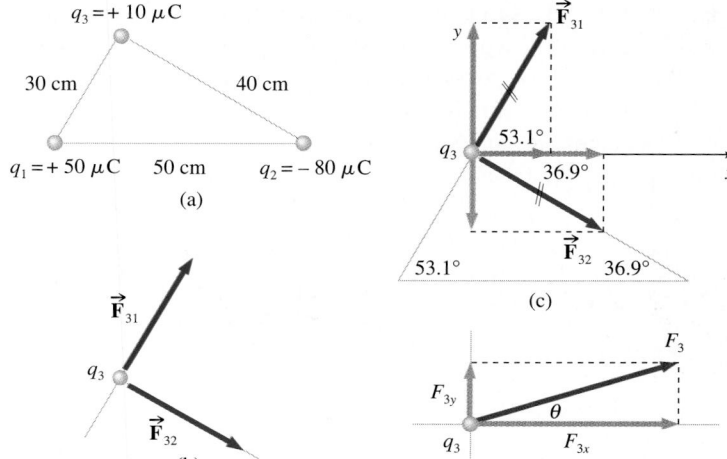

Figure 15.20   The force $\vec{F}_3$ on a charge q_3 due to two other charges q_1 and q_2.

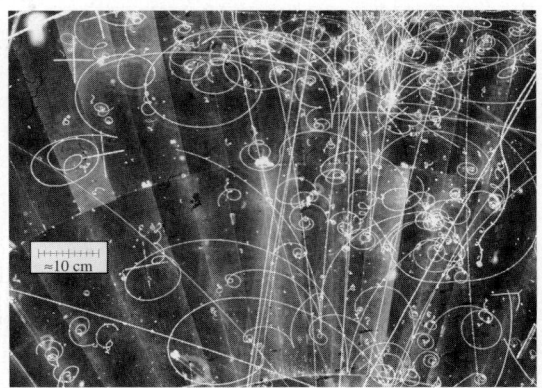

The medium in this familiar bubble chamber photo is roughly a 2:1 mixture of liquid neon and hydrogen. When an electron (clockwise spirals) or a positron (counterclockwise spirals) electrically interacts with the positive nuclei of the medium, it experiences an acceleration and radiates away energy. The stronger the Coulomb force, the greater the acceleration and the more rapidly the electron (or positron) spirals inward. The effect is enhanced by the small mass of the electron (or positron), which therefore accelerates much more rapidly than do other particles. Another contributing factor is the neon whose nucleus contains 10 protons as compared to the 1 proton in a hydrogen nucleus. That results in a Coulomb force on the traveling electron which is much greater than would be the case in a hydrogen bubble chamber, where the spirals are not nearly so tight.

The Electric Field

Here we are once more, this time wand in hand lifting lint and pulling pith and yet not actually touching either—action-at-a-distance again, the fundamental puzzle of affecting motion without apparent contact. Physicists are still struggling to learn how one speck of matter reaches across the "void" to influence another. Today, the most prevalent theoretical picture is based on the concept of the *field*, which was introduced to help visualize the distribution of forces in the space surrounding a material object (p.162). Inevitably, it itself became the medium for an explanation of action-at-a-distance.

A **field of force** *exists in a region of space when an appropriate object placed at any point therein experiences a force.* We can imagine a gravitational force field surrounding any object of mass M, and in the same way we can imagine an electric force field surrounding an object of charge Q.

Often we'll have a *primary charge distribution* whose influence, in the surrounding space, we wish to study. Accordingly, suppose for the moment that we have a small sphere carrying a uniform positive charge. As a probe, we use a tiny charged pith ball resembling a point-charge (Fig. 15.21). By tradition, our detector (called a **test-charge**) is always positive. Accordingly, it will be repelled from the charged sphere no matter where it is positioned. At every point in the region surrounding the sphere (or primary charge distribution), the detector (which is a positive point-charge q_o) will experience a force of a specific

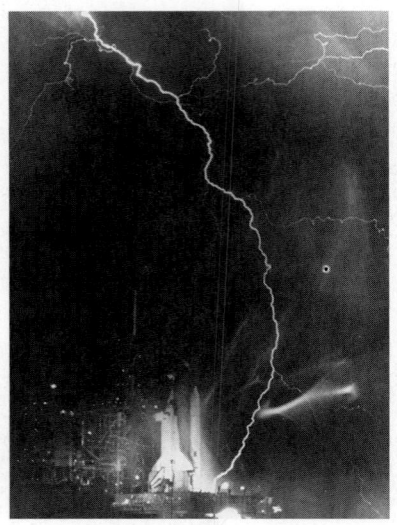

Here, a bolt of lightning just misses a *Space Shuttle*. Years earlier, a similar bolt destroyed an unmanned rocket.

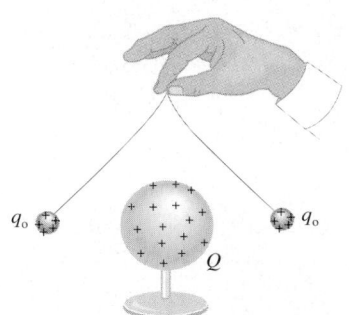

Figure 15.21 A primary charge Q affects a test-charge q_o, repelling it radially.

Figure 15.22 The vector force field surrounding a small positive primary charge. The force is large near the charge and small far from it. At every point in the surrounding space we can imagine that there is a force vector.

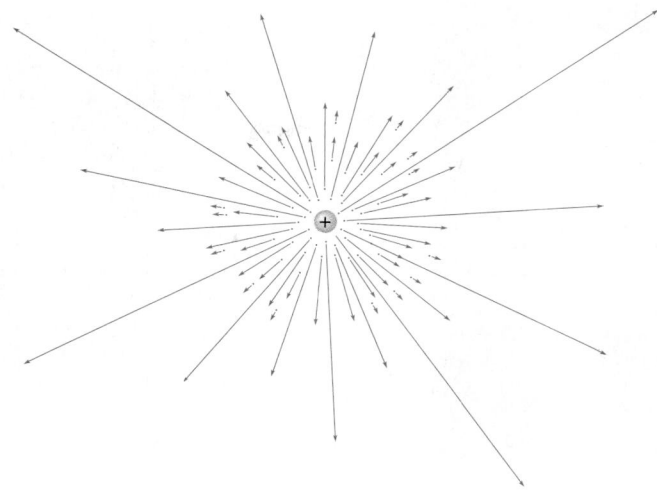

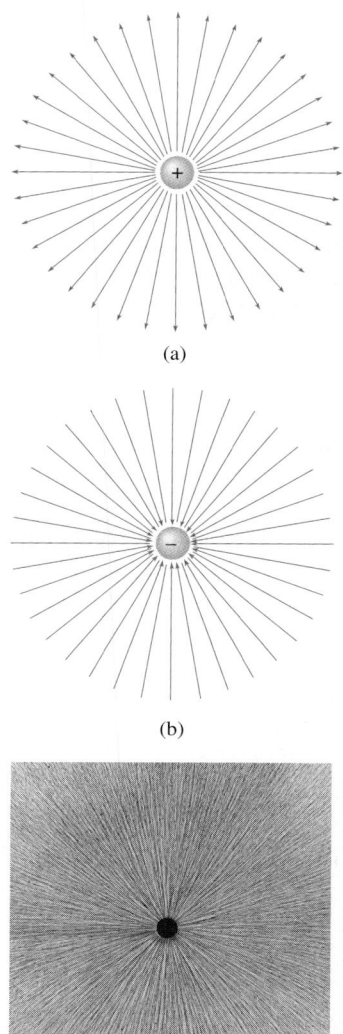

(a)

(b)

(c)

Figure 15.23 (a) Lines-of-force emanating from a positive point-charge. (b) Lines-of-force converging to a negative point-charge. (c) Fine rayon fibers suspended in oil tend to align themselves when in the vicinity of a charged object. The fiber patterns can be thought of as revealing the lines-of-force.

strength and direction, and so at every point we assign a corresponding force vector (Fig. 15.22). It should be noted that if we actually drew a vector at *every* point, the picture would be solid black and useless; the illustration is a compromise. That, then, is the vector force field associated with this particular charged object. It happens to be radial and uniformly outward in all directions because the object is a positive uniformly charged sphere.

Michael Faraday was the first to introduce a visual representation of the electric force field. His scheme, using *lines-of-force*, is a more convenient alternative to our field of force vectors. Faraday's version is equivalent to merging successive vectors into a continuous line that's tangent to all of them. ***The force experienced by a positive test-charge at any point in space is in the direction tangent to the line-of-force at that point.*** Notice that the lines-of-force "flow" radially outward from a positive point-charge, as they do for a uniformly charged sphere (Fig. 15.23).

The lines diverge, getting farther apart as they extend out. The same number of lines pass out through a small imaginary surrounding sphere centered on the charge as pass through a larger concentric sphere whose surface is more distant (Fig. 15.24). As the area ($A = 4\pi r^2$) of the encompassing sphere increases, the density of lines decreases in proportion to $1/A$ or, equivalently, to $1/r^2$. Happily, Coulomb's Law tells us that the force exerted by a point-charge $q_{\bullet}$ on a test-charge q_{o} also drops off as $1/r^2$. It follows that the density or concentration of the lines-of-force corresponds to the strength of the force field. Since this is true for a single point-charge, it's true for the superposition of the force fields of many point-charges. ***The more lines drawn in a region—that is, the denser the concentration of lines—the stronger the field they represent; the farther apart the lines, the weaker the field.***

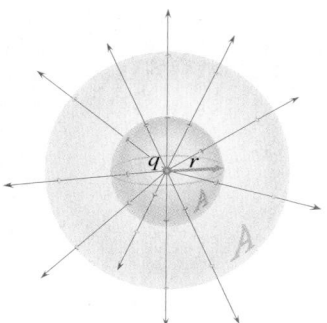

Figure 15.24 The lines-of-force stream through surrounding spheres. The area (A) of a sphere is $4\pi r^2$ and, as r increases the density of lines, the number per unit area decreases as $1/r^2$, which is in accord with Coulomb's Law.

Michael Faraday (1791–1867) was one of 10 children of a blacksmith in London. As a youngster, with little formal education, Michael was apprenticed to a bookbinder. But he longed to enter "the service of Science" and ultimately became one of the greatest experimentalists of all time.

THE BOOKBINDER & THE CHEMIST

To appreciate the central creation in all of this—the electromagnetic field—we go back more than 100 years to the man who gave it form, the consummate experimentalist Michael Faraday. In 1812, at the age of 21, Faraday was given free tickets to the evening lectures on chemistry delivered by Sir Humphry Davy at the Royal Institution in London. The young man was so dazzled by the experience that he sent Davy a leather-bound copy of his meticulous notes, accompanied by a request for a job as an assistant. A while later Davy fired his lab man for brawling and, remembering Faraday's flattering gesture, offered him the job of bottle washer. The bright young man accepted and quickly rose from lackey to protégé to rival.

15.4 Definition of the $\vec{\mathbf{E}}$-field

One obvious disadvantage of concentrating on force is that its magnitude at every point in space depends not only on the primary charge distribution, but also on the size of the test-charge q_o. What we really want is a map showing the field of a primary body independent of the detector, a map that could be used to compute the force at every point in space when any size charge is placed there. Regardless of its source, we define the **electric field** ($\vec{\mathbf{E}}$) *at a point in space to be the electric force experienced by a positive test-charge at that point divided by that charge*

[definition of $\vec{\mathbf{E}}$] $$\vec{\mathbf{E}} = \frac{\vec{\mathbf{F}}}{q_o}$$ (15.2)

Electric field has the SI units of newtons per coulomb (N/C). The tangent to a line-of-force at any point is the direction of the *E*-field. The custom is to draw the same pattern of lines but to call them *electric field lines*. Only in the special case when the *E*-field is uniform (and the field lines are therefore parallel) is $\vec{\mathbf{E}}$ a constant. More commonly $\vec{\mathbf{E}}$ varies from point to point in space depending on the charge distribution.

Conversely, knowing $\vec{\mathbf{E}}$ at any location in space (whatever the source), we can calculate the force $\vec{\mathbf{F}}$ that would arise on any point-charge q placed at that location;

[force on a point-charge q] $$\vec{\mathbf{F}} = q\,\vec{\mathbf{E}}$$ (15.3)

I can hardly imagine anyone who knows the agreement between observation and calculation, based on action at a distance, to hesitate an instant between this simple and precise action on the one hand and anything so vague and varying as lines of force on the other.

SIR GEORGE B. AIRY (1801–1892)
*BRITISH ASTRONOMER AND
MATHEMATICIAN*

FARADAY & THE FIELD

To Faraday, these intricate patterns of lines came to stand for an invisible physical reality. For him, the field pervading space became an entity that reached from puller to pulled. An electrified object sends out its field into space, and the pith ball detector immersed within that web interacts with the field in proportion to its own charge. Charge-field-charge—that concept was Faraday's answer to the magic of action-at-a-distance. The field was not only an illustrative device; it was a reality (though no less mysterious than the ghostly action-at-a-distance it now replaced). After all, what was the field made of, if anything?

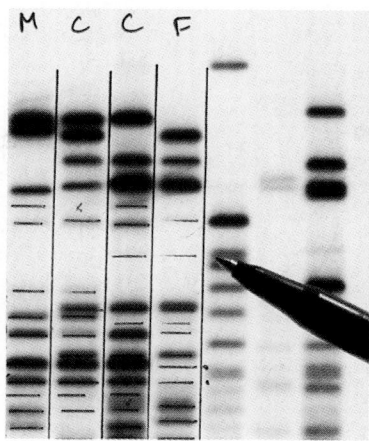

Many large so-called macromolecules are charged. These can be made to move through an appropriate medium by applying an electric field, a process called electrophoresis. The mobility of each such molecule depends on its size, shape and mass. Accordingly, blends of different molecules can be made to separate as they migrate under the influence of an E-field. Shown here are columns composed of bands of DNA produced by electrophoresis. The four on the left represent a family (Father F, mother M, and two children C.) The distribution of bands is a "finger print" unique to each person. Notice how both children share bands with each parent, proving that they are indeed related.

Notice that $\vec{F}$ and $\vec{E}$ point in the same direction when $q_{\cdot}$ is positive. Equation (15.3) works beautifully for subatomic particles like the electrons streaming down the picture tube in your TV set, but it can also be applied approximately to charged pith balls, bugs, paint droplets, and so forth.

Example 15.4 **[I]** At a particular moment the electric field at a point 30 cm above an electric blanket is 250 N/C, straight up. Compute the force that acts on an electron at that location, at that moment.

Solution Here we are given an electric field and asked to find a force, and that should call to mind the defining equation for E. (1) TRANSLATION—A known electric field exists at a point in space; determine the force on an electron at that location. (2) GIVEN: $q_{e} = -1.602 \times 10^{-19}$ C and $E = 250$ N/C. FIND: $\vec{F}$. (3) PROBLEM TYPE—Electrostatics/electric field. (4) PROCEDURE—The electric field and the corresponding force on

a point-charge are related by $F = q_{\cdot}E$. (5) CALCULATION—The point-charge is an electron. Taking up as positive,

$$F = q_{e}E = (-1.602 \times 10^{-19} \, \text{C})(250 \, \text{N/C}) = -4.01 \times 10^{-17} \, \text{N}$$

The minus sign arises from the charge and tells us that the force on the electron is in the opposite direction to the E-field, which is upward. In other words,

$$\boxed{\vec{F} = 4.01 \times 10^{-17} \, \text{N} - \text{DOWNWARD}}$$

Quick Check: $E = F/q = (-4 \times 10^{-17} \, \text{N})/(-1.6 \times 10^{-19} \, \text{C})$ $= 2.5 \times 10^2$ N/C.

The Field of a Point-Charge

Figure 15.23a shows the direction in which a positive test-charge $q_{\circ}$ would experience a force and so tend to accelerate. It is also a mapping of the electric field of the primary point-charge $+q_{\cdot}$. Alternatively, when the primary charge is negative, the field lines converge inward toward it, as in Fig. 15.23b. The special thing about a point-charge source is that we can determine a formula for its E-field using Eq. (15.2). Thus, substituting for F from Coulomb's Law, we obtain

The **electric field of a point-charge**, $E = k_0 q_{\cdot}/r^2$, immediately follows from the definition of the field, $E = F/q_{\circ}$ and Coulomb's Law (and needn't be memorized).

$$E = \frac{F}{q_{\circ}} = k_0 \frac{q_{\cdot}q_{\circ}}{r^2} \frac{1}{q_{\circ}}$$

and the scalar value of the electric field of a point-charge $q_{\cdot}$ is

[field of a point-charge in vacuum]

$$E = k_0 \frac{q_{\cdot}}{r^2} \qquad (15.4)$$

This result is extremely important: it allows us to calculate the E-field anywhere in space due to a known distribution of point-charges (electrons, protons, whatever). If there are just

E-FIELDS: FAIR WEATHER AND FOUL

Prior to the introduction of electrical technology in the nineteenth century, the strongest electric field most humans were likely to encounter was the static atmospheric field of about 120 N/C to 150 N/C in fair weather and up to 10 000 N/C during thunderstorms (Table 15.2).

Table 15.2
Electric Fields

Source	Field strength (N/C)
Background radiation in space	3×10^{-6}
In-house wires	10^{-2}
Radio waves	$\approx 10^{-1}$
Outside an electrified building	$\approx 10^{-1}$
Center of typical living room	≈ 3
In a fluorescent tube	10
30 cm from electric clock	15
30 cm from stereo	90
Laserbeam (low power)	10^{2}
Atmosphere (fair weather)	≈ 150
30 cm from electric blanket	250
Built up by splashing water in a shower	800
Sunlight (average)	10^{3}
Atmosphere (thunderstorm)	10^{4}
Van de Graaff accelerator	2×10^{6}
Breakdown of air	3×10^{6}
X-ray tube	5×10^{6}
At cell membrane	10^{7}
Created by pulsed laser system	5.7×10^{11}
At electron in hydrogen atom	6×10^{11}
Surface of a pulsar	$\approx 10^{14}$
Surface of uranium nucleus	2×10^{21}

The conducting strips on an airplane's wing tips allow charge, built up on the plane as it moves through the air, to leak off.

a few such charges, the calculation can be done directly by hand. Each charge contributes a field whose strength at any point is given by Eq. (15.4) and whose direction is radially inward (when $q_\bullet$ is negative) or outward (when $q_\bullet$ is positive), as in Fig. 15.25. These contributions add vectorially to produce the net $\vec{E}$.

The electrically responsive cells in a shark allow it to detect the weak electric fields created by the operation of the muscles of its prey. Sharks can sense fields of as little as 10^{-6} N/C.

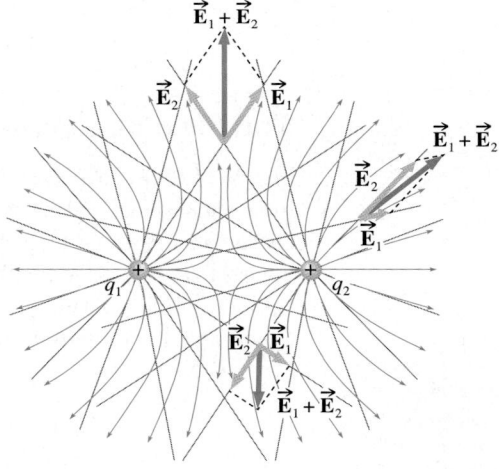

Figure 15.25 The two fields $\vec{E}_1$ and $\vec{E}_2$ superimpose, coexisting without interacting. The net field $\vec{E}$ is the vector sum of the two. In this diagram, at each point in space, $\vec{E} = \vec{E}_1 + \vec{E}_2$. Lines of $\vec{E}$ never intersect one another.

Example 15.5 **[II]** Figure 15.26 shows two point-charges (or tiny charged spheres) each of $+10$ nC separated in air by a distance of 8.0 m. Compute the electric field at points A, B, and C.

Solution Given one or more charges you can find the resulting E-field at any point in space. (1) TRANSLATION—Determine the electric field at several locations in space at specified distances from a known distribution of point-charges. (2) GIVEN: $q_1 = q_2 = +10 \times 10^{-9}$ C. FIND: $\vec{\mathbf{E}}$ at points A, B, and C. (3) PROBLEM TYPE—Electrostatics/electric field. (4) PROCEDURE —The electric field of a point-charge is $E = kq_./r^2$. (5) CALCULATION—Starting at A, make a sketch of the layout (Fig. 15.26b) and then draw in vectors for the field $\vec{\mathbf{E}}_1$ due to q_1 and $\vec{\mathbf{E}}_2$ due to q_2. To do that, imagine a positive test-charge at A. The force on it due to q_1 acts along the center-to-center line, is repulsive, and so points to the right. That means the field $\vec{\mathbf{E}}_1$ at A is to the right along the axis. Similarly, the force due to q_2 on our imaginary test-charge is to the left, as is $\vec{\mathbf{E}}_2$. Next, calculate E_1 and E_2 via Eq. (15.4) and add them vectorially. Happily we are spared that last effort because $E_1 = E_2$, and so the two cancel and the field at A is $\boxed{\vec{\mathbf{E}}_A = 0}$. A test-charge exactly at A would just sit there force-free in unstable equilibrium.

At point B, the two E-fields act as drawn in Fig. 15.26c and we must find their components. First, let's calculate E_1 and E_2. Since the charges and distances happen to be the same, the mag-

nitudes of the two contributing fields are equal; consequently

$$E_1 = E_2 = k_0 \frac{q_.}{r^2}$$

$$E_1 = E_2 = (8.99 \times 10^9 \text{ N·m}^2/\text{C}^2) \frac{(+10 \times 10^{-9} \text{ C})}{(4.0/\sin 45°)^2} = 2.81 \text{ N/C}$$

Now for the vector components. From the geometry, the fields are seen to act at 45°. This time, because the charges are equal and point B is on the midline, the horizontal field components are equal and opposite and cancel. Only the vertical components contribute

$$E_B = E_1 \sin 45° + E_2 \sin 45° = 2(2.81 \text{ N/C})(0.707)$$

and $\boxed{E_B = 4.0 \text{ N/C}}$

directed straight up in the positive y-direction. The same value applies to point C, where $\boxed{E_C = 4.0 \text{ N/C}}$ directed downward in the negative y-direction.

Quick Check: As we will see presently, these results agree with the diagram (Fig. 15.27) for the field of two equal charges. The field of a 1-nC positive charge at 1 m is ≈ 9 N/C, and these answers are the right order-of-magnitude.

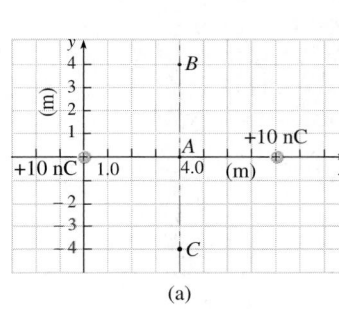

(a)

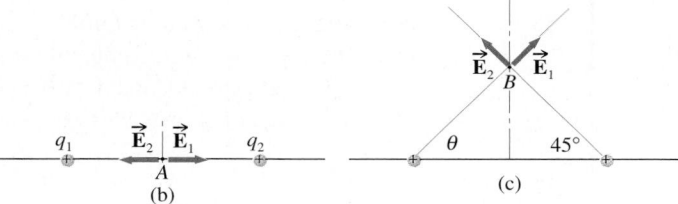

(b) (c)

Figure 15.26 The electric fields of two charges q_1 and q_2 at points A, B, and C.

Charge Immersed in a Dielectric Medium

An English electrical engineer, Oliver Heaviside, pointed out sometime around 1892 that many of the equations for the E-fields of different charge distributions contained a factor of 4π. He suggested that the constant k_0 in Coulomb's Law be written in terms of $1/4\pi$ and some new constant ε_0, so that

$$k_0 = \frac{1}{4\pi\varepsilon_0} \tag{15.5}$$

This would have the effect of causing 4π to appear in expressions for the E-field only in situations where the charge distribution was spherically symmetric and 2π to appear when it was axially (or cylindrically) symmetric. Since this was a purely formal device, the system of units that resulted was said to be *rationalized*. The SI system embraced the practice, and

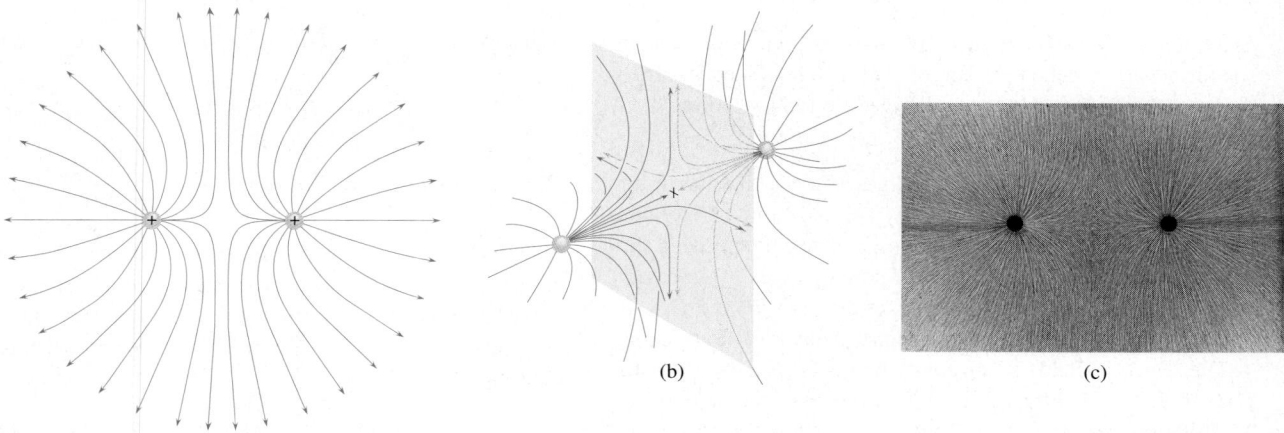

(b)

(c)

(a)

Figure 15.27 (a) The electric field lines due to two positive charges of equal value. (b) The E-field is three-dimensional. (c) Rayon fibers suspended in oil align with electric field lines. Incidentally, when the charges are separated by a distance $x = 2a$ the field has a maximum at $y = a/\sqrt{2}$.

in 1960 it became official. The constant ε is called the permittivity (from the Latin *permittere*, to let go through). In the case of vacuum, $k_0 = 1/4\pi\varepsilon_0$, where ε_0 is referred to as the **permittivity of free space**. It follows from the value of k_0 that

$$\varepsilon_0 = 8.854\,187\,8 \times 10^{-12}\ \mathrm{C^2/N\cdot m^2}$$

When the charges we are dealing with are immersed in a substantial medium, we replace k_0 by $k = 1/4\pi\varepsilon$; Table 15.3 lists the permittivities of several common materials. It also provides the corresponding **relative permittivity** $(\varepsilon/\varepsilon_0)$ for each. Often called the **dielectric constant** (K_e), this ratio is unitless and applies to all systems of units. In the early days when there were several competing systems, tabulating ratios was often the only sensible way to present experimental data. The users simply multiplied K_e by the appropriate value of ε_0 in the system of units they happened to be working with.

Table 15.3
The Permittivity (ε) and Relative Permittivity*
$(\varepsilon/\varepsilon_0)$ of Some Common Substances

Substance	Permittivity $(\mathrm{C^2/N\cdot m^2})$	Relative Permittivity $(\varepsilon/\varepsilon_0)$
Vacuum	8.85×10^{-12}	1.000 00
Air	8.85×10^{-12}	1.000 54
Body tissue	71×10^{-12}	8
Glass	$44 \times 10^{-12} - 89 \times 10^{-12}$	5−10
Mica	$27 \times 10^{-12} - 53 \times 10^{-12}$	3−6
Nylon	31×10^{-12}	3.5
Paper	$18 \times 10^{-12} - 35 \times 10^{-12}$	2−4
Polyethylene	20×10^{-12}	2.3
Polystyrene	23×10^{-12}	2.6
Rubber	$18 \times 10^{-12} - 27 \times 10^{-12}$	2−3
Silicone oil	$19 \times 10^{-12} - 25 \times 10^{-12}$	2.2−2.8
Sodium chloride	50×10^{-12}	5.6
Teflon	19×10^{-12}	2.1
Ethanol (25 °C)	2.2×10^{-10}	24.3
Methanol (20 °C)	3.0×10^{-10}	33.6
Water (20 °C)	7.1×10^{-10}	80

*Also called the *dielectric constant*.

Example 15.6 **[I]** A point-charge of 10 μC is surrounded by water with a dielectric constant of 80. Calculate the magnitude of the electric field 20 cm away.

Solution Whenever there's a medium other than vacuum we use k rather than k_0. (1) TRANSLATION—A known point-charge is immersed in a known medium; determine the E-field at a specified distance. (2) GIVEN: $q_{\bullet} = 10$ μC, $r = 0.20$ m, and $K_e = 80$. FIND: E. (3) PROBLEM TYPE—Electrostatics/electric field/dielectric constant. (4) PROCEDURE—Begin with the electric field of a point-charge $E = kq_{\bullet}/r^2$ and use the permittivity of the medium. (5) CALCULATION—Since we are given

the dielectric constant, $\varepsilon = K_e \varepsilon_0$. For a point-charge

$$ E = k\frac{q_{\bullet}}{r^2} = \frac{1}{4\pi\varepsilon}\frac{q_{\bullet}}{r^2} = \frac{1}{4\pi K_e \varepsilon_0}\frac{q_{\bullet}}{r^2} $$

and

$$ E = \frac{10 \times 10^{-6}\text{ C}}{4\pi(80)(8.85 \times 10^{-12}\text{ C}^2/\text{N}\cdot\text{m}^2)r^2} = \boxed{28\text{ kN/C}} $$

Quick Check: In vacuum $E \approx (9 \times 10^4/0.04)$ N/C $\approx 2.3 \times 10^6$ N/C. With water, it will be a factor of 80 smaller; that is, $E \approx 28$ kN/C.

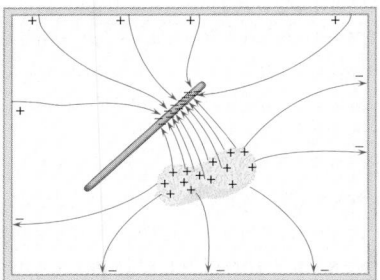

Figure 15.28 A hard-rubber rod rubbed on fur. Many of the field lines begin or end on the walls. *The net enclosed charge is zero.*

Field Lines

Because electric field lines diverge away from a positive point-charge and converge in toward a negative point-charge, we say the former is the *source* of the field and the latter the *sink*. **In electrostatic situations, field lines always begin on positive charge and end on negative charge.** Both ends of the field lines are not always pictured, but they terminate somewhere on a charge (Fig. 15.28), even if that charge is on the walls of the room or beyond.

Figure 15.27 shows the E-field of two equal positive point-charges (each $q_{\bullet}$). The pattern is three-dimensional (Fig 15.27b), and it actually corresponds to the picture we would get if we rotated the drawing (Fig. 15.27a) about the line connecting the charges. As was found analytically (Fig. 15.26), the field is zero at the very center. Very far from the pair (at a distance much greater than their separation), the field will appear as if it were due to a single positive charge ($2q_{\bullet}$).

Lines of the net field never cross. If they did, the field would have two different values at that point, which is silly since two superimposed fields combine to yield a single net field. A positive test-charge at that point must experience a single net force, and that is the direction of $\vec{\mathbf{E}}$.

Figure 15.29 shows positive and negative charges of the same magnitude, a configuration that is special enough to have its own name—it's called a **dipole**. Because the charges

Figure 15.29 (a) A dipole field set up by two equal opposite charges. (b) Rayon fibers suspended in oil reveal the pattern of electric field lines. (c) The dipole field seen from far away, showing how it rapidly weakens. (d) Water is a polar molecule, which means that one end is positive and the other is negative. It behaves like a dipole. (e) The Elephant Gnathonemus produces a dipole electric field and detects nearby objects by their effects on that field.

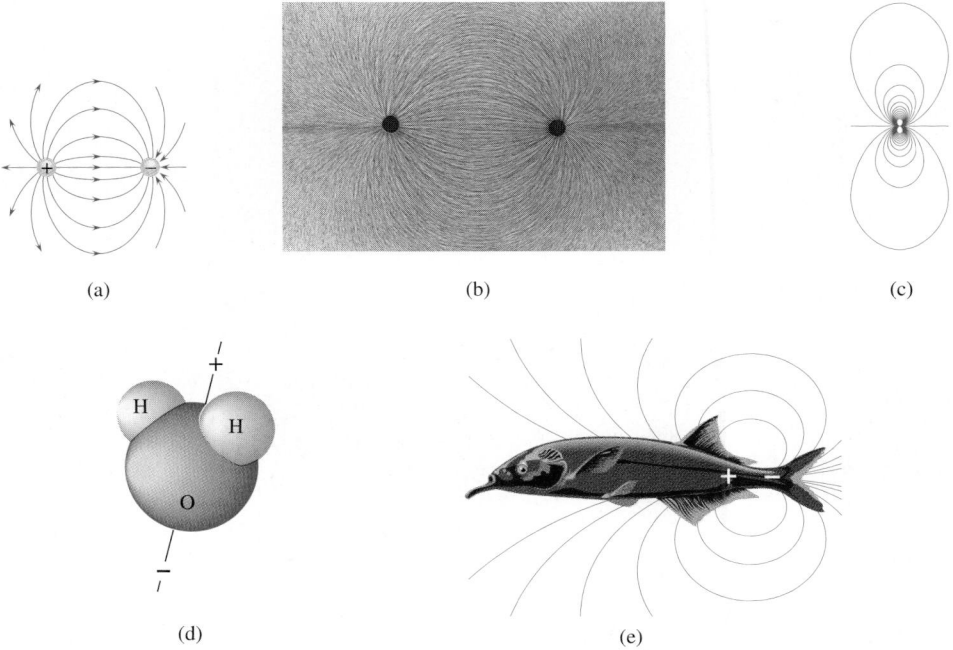

(a)

(b)

(c)

(d)

(e)

have opposite signs and equal magnitudes, the field is not zero anywhere nearby (as it was in Fig. 15.27). Still, at greater and greater distances, the lines get increasingly sparse, and the field (which drops off as $1/r^3$) rapidly decreases. It must, since as seen from very far away the charges will essentially coalesce and cancel (Fig. 15.29c). {For a lovely simulation of the E-fields of one and two point-charges, click on **ELECTRIC FIELDS OF POINT-CHARGES** under **INTERACTIVE EXPLORATIONS** in **CHAPTER 15** on the **CD**.}

Whenever a physical system has an asymmetry in its charge distribution, it's likely to produce an E-field that in some way resembles a dipole. The dipole is therefore important in understanding a wide range of electric systems from molecules (Fig. 15.29d) to fish (Fig. 15.29e). In fact, *there is no way to understand chemistry or molecular biology without understanding electrostatics*. For example, each hydrogen atom in a water molecule shares its electron with the oxygen atom. That leaves the single nuclear proton in each H atom somewhat exposed, with the result that the hydrogen end of the molecule is electrically positive. The oxygen end, now harboring two additional electrons, is negative. The water molecule, which is neutral as a whole, is electrically **polarized** and behaves as a dipole. As a result, water molecules attract themselves and other polarized molecules. That's what makes water such a wonderful solvent. Water readily dissolves a variety of compounds, of which salt and sugar are perhaps the most familiar. As a rule, polar molecules electrostatically interact strongly with polar compounds (e.g., sugar).

When table salt, NaCl, is formed, an electron is transferred from a sodium atom (which becomes an Na$^+$ ion) to the chlorine atom (which becomes a Cl$^-$ ion). The ions electrostatically attract each other to form a stable structure (Fig. 15.30a). Water desolves a chunk of salt crystal by nuzzling in and surrounding each ion. As can be seen in Fig. 15.30b the positive hydrogen ends of the water molecules attract and tend to face the negative chlorine ions, while the negative oxygen ends electrostatically attract and rotate around to face the positive sodium ions. It's in part because the water molecule is a dipole that it's the central agent in all the biochemical processes that take place on this planet.

The dipole with its equal-magnitude charges represents an especially symmetrical situation. If we revolve Fig. 15.29a about its central axis, we'll get an accurate three-dimensional picture (similar in kind to Fig. 15.27b) in which the density of the lines corresponds to the strength of the field. Unfortunately, that's not always the case. *The electric field is a three-dimensional concept, and in general, when we try to represent it in two dimensions, the field lines clump up and there's no longer an exact correlation between field strength and line density.*

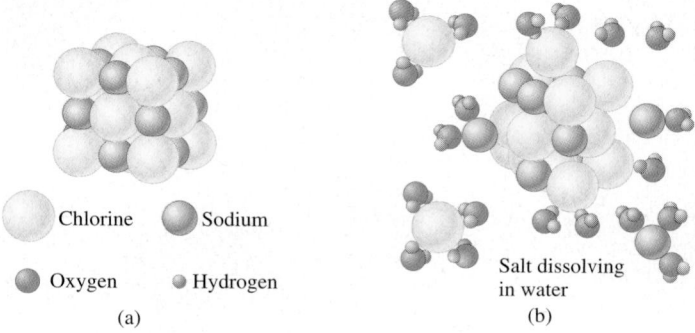

Chlorine   Sodium

Oxygen   Hydrogen

(a)

Salt dissolving in water

(b)

Figure 15.30 (a) An ionic crystal of table salt. (b) Water dissolving a crystal of NaCl. Notice how the positive hydrogen ends of the water molecules attract the negative chlorine ions.

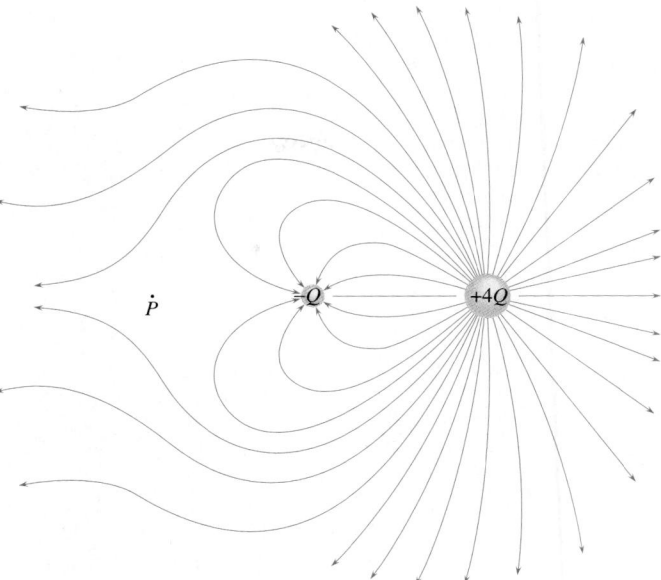

Figure 15.31   A rough sketch of the electric field of two charges $+4Q$ and $-Q$. Notice that the field is zero at point P.

When the charges are unequal, as in Fig. 15.31 where one is $+4Q$ and the other $-Q$, the field configuration becomes a bit more complex, even going to zero at point P. Here the lines tell us the direction of the E-field, but there is no longer a consistent correlation between line density and field strength. {For more on this shortcoming click on **Electric Field Lines** under **Further Discussions** on the **CD**.}

Figure 15.32 takes a different graphical approach, one that's useful in visualizing complicated fields, even though it's not very practical for notebook diagrams. In this representation the density of lines is of no significance; instead we use color to indicate the relative field strength. The field is strongest where the lines are red, and it falls off, fading to orange,

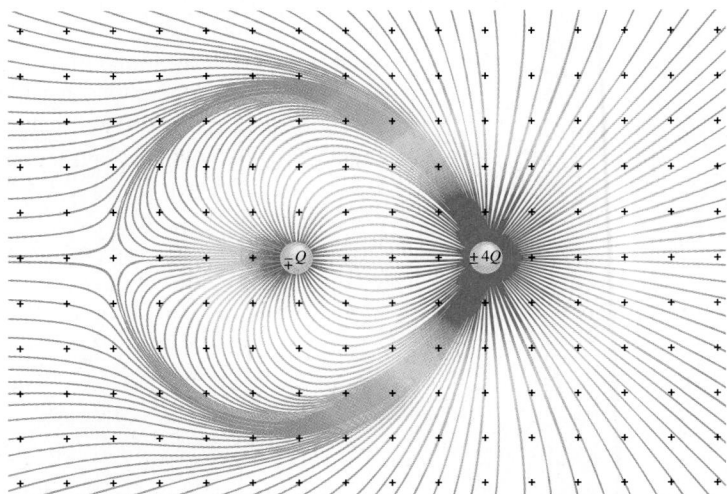

Figure 15.32   The electric field surrounding two charges—the one on the right is $\pm 4Q$, the one on the left is $\mp Q$. Here the density of the lines is arbitrary. The strength of the field is indicated by the color of the field lines; red represents the strongest field, orange is weaker, yellow is still weaker, and down to blue which marks the faintest field.

Figure 15.33 The same charges depicted in Fig. 15.32 are shown here viewed from farther away. Notice how the field is already beginning to resemble that of a sphere of charge $\pm 3Q$. The outer portion of the yellow region is already nearly circular. From far away (far compared to the separation between charges) any charge distribution will look like a point-charge with a value equal to the net charge.

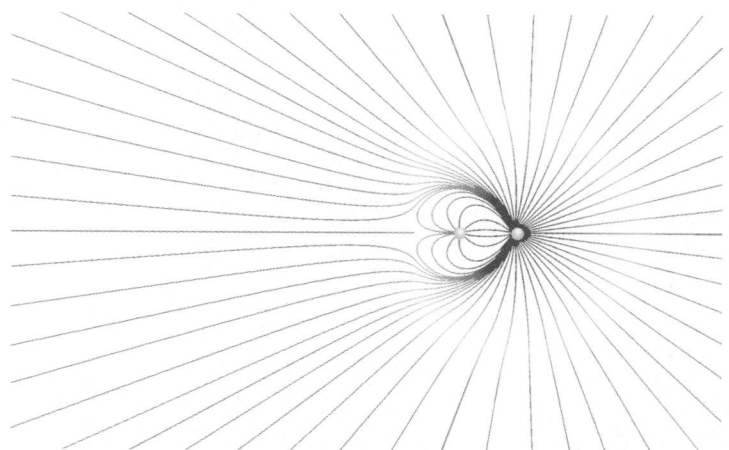

yellow, green, and finally blue. At great distances from the two unlike charges (in Fig. 15.32) the field must resemble the field of a single net charge of $+3Q$; Fig. 15.33 shows how the lines straighten and all seem to project back toward a single point centered on the $+4Q$ charge.

Imagine two large, parallel horizontal metal plates each equally charged—the upper one positive, the lower negative (Fig. 15.34). The charge distributes itself more or less uniformly over the inner faces of the plates; the mutual attraction draws most of the opposite charges as close together as possible (leaving just a few stray charges on the outer surfaces). A positive test-charge placed anywhere in the gap (away from the ends) would be repelled from the top plate, attracted to the bottom plate, and down the charge would go. The E-field between the plates is down and *uniform*, the field lines are parallel, and E is constant. This is the most convenient way to produce a uniform field, and so it's used in many devices, including cathode-ray tubes. {To see how a charged particle moves through such a field, open the **CD**, go to **INTERACTIVE EXPLORATIONS**, and click on **MOTION OF A CHARGED PARTICLE IN AN ELECTRIC FIELD**.}

Conductors and Field Lines

Imagine a neutral conductor, solid or hollow, in a field-free region of space. Now suppose we add some electrons to it, either on the surface or inside. The mutual repulsion between free charges causes them to redistribute. The charges are initially bunched up, there is a transient E-field; and very quickly they move apart, coming to rest on the outer surface. The process generally only takes a fraction of a second, depending on the physical details of the

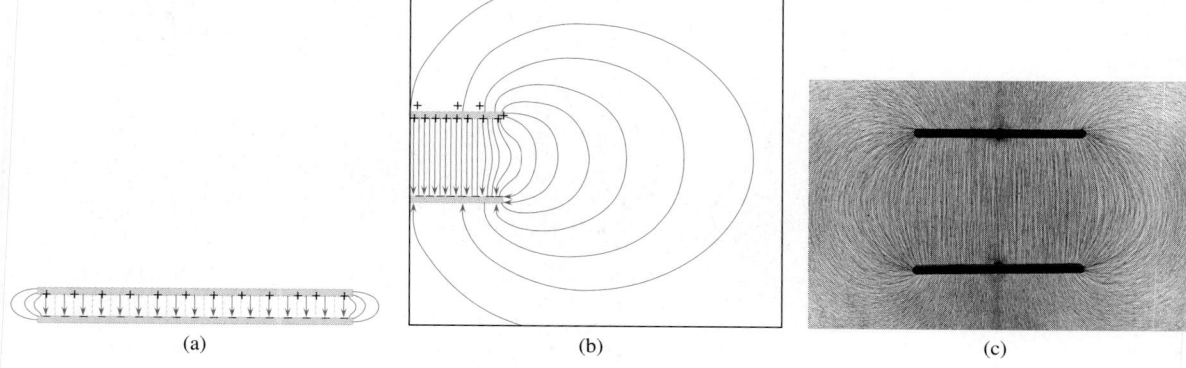

(a) (b) (c)

Figure 15.34 (a) The E-field of a parallel-plate capacitor. (b) A closeup of edge effects. (c) The field as illustrated by rayon fibers suspended in oil.

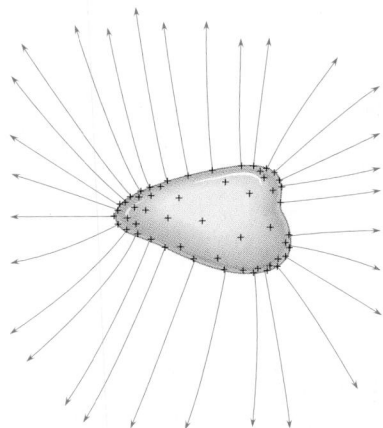

Figure 15.35 The electric field of a charged conductor is everywhere perpendicular to the surface. Charge concentrates on regions with a small radius of curvature.

A tube in the old radio pictured on p. 537. It's surrounded by a metal shield that reduces interference caused by stray electric fields. Transistors and sensitive integrated circuits are commonly shielded by encasing them in metal.

The external *E*-field is perpendicular to a conductor. Moreover, there is no field inside a conductor unless it surrounds a net charge.

conductor. Once settled, the distribution is such that each free charge experiences a zero net force. If it did not, the charge would accelerate until it could move no more, and equilibrium would ultimately be established. If none of the charges experiences a net electric force, none of them is in an electric field. Since field lines either begin or end on charge, the field could only extend inside the conductor if there were a remaining excess of free charges there and that's impossible. The electrostatic field inside a charged conductor, anywhere beneath the surface, is zero (provided it does not encompass a space in which there is an isolated charge). Faraday dramatically proved the point by constructing a room within a room, covering the inner enclosure with tinfoil. He sat inside this *Faraday cage*, as it has come to be called, with an electroscope at hand, while the entire structure was charged by an electrostatic generator. No field could be detected inside, even while sparks were flying outside.

As for the electrostatic field external to the body of a conductor—it must always be perpendicular to the surface. It doesn't matter whether we are talking about a charged object (Fig. 15.35) and its own field or a neutral conductor in an external field (Fig. 15.36). If the field were not perpendicular, it would have a component parallel to the surface. There would be forces on the free surface charges, and they would move. Static equilibrium, which is ordinarily observed to set in quickly, can only be reestablished when there are no tangential field components.

Unlike gravity, we can shield against electric fields simply by surrounding the region to be isolated with a closed conductor. This is true as well in the case of nonstatic fields because the electrons can redistribute themselves very quickly. Electronic components are

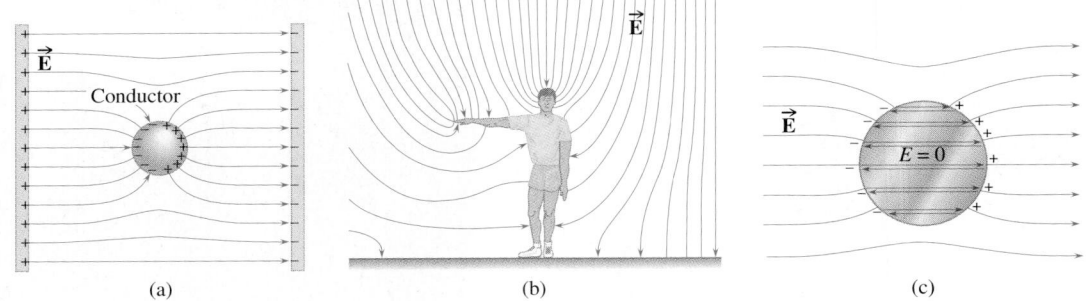

(a)   (b)   (c)

Figure 15.36 Neutral conductors immersed in an applied *E*-field. (a) A conducting sphere in the uniform field of a parallel-plate capacitor. (b) A person in the *E*-field of the out-of-doors environment. The presence of a conductor distorts the field. (c) The external field polarizes the conductor, making one side positive and the other negative, which creates a self-field that cancels the applied field and leaves a zero internal electric field.

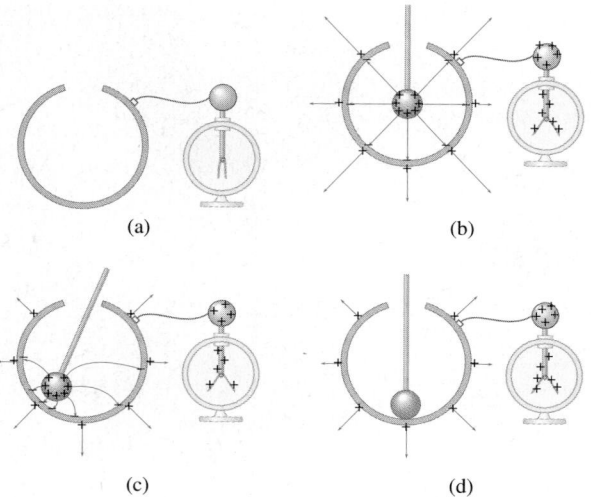

(a)   (b)

(c)   (d)

Figure 15.37 The Faraday ice-pail experiment. A charge held inside the conductor causes an equal and opposite charge to move to the outer and inner surfaces.

often encased in metal cans, and wires (for example, the leads for your stereo amplifier and cable TV) are surrounded by braided copper sheathing to keep out stray electric fields. If a portable radio is placed in a closed metal pot, the *E*-field signal will not be able to reach the antenna. That's why a car radio's reception is so poor in a metal-encased tunnel or on a steel bridge.

The electrostatic field within the material of a conductor is always zero, but it is possible to have a field in the hollow of a conductor if we put an isolated charge inside it. Imagine that we introduce a positively charged ball into a nearly closed conductor attached to an electroscope (Fig. 15.37), just as Faraday did in 1843 using a small ice-pail. Once inside, it doesn't matter where the ball is, the outside will be charged positively, and the leaves of the electroscope will stand apart at an unchanging angle. If the ball touches the inside surface, it discharges, becoming neutral, but the leaves do not move at all—the same charge was induced on the outside as was carried by the ball.

The man in the Faraday cage at the left of center sits safely as sparks from a large Van de Graaff generator flash all around him.

Notice that the net charge inside the conductor is zero both before and after the ball touched the inner surface.

15.5 Gauss's Law   (Optional)

The *E*-field of any extended charge distribution can be found directly using Coulomb's Law, but the process is often difficult. Fortunately, there is another calculational approach that's relatively simple, in situations where the charge distribution is symmetrical. It's due to K. F. Gauss, the great nineteenth-century German mathematician and astronomer. Gauss's Law, which derives from the ideas of fluid dynamics, is a relationship between the "amount" of the electric field passing through a closed area and the enclosed sources and sinks of the field, namely, the charge.

To begin, we have to quantify this notion of "amount" of field. Figure 15.38 shows the field "streaming" from a positive point-charge as if it were water from a nozzle. The analogy naturally suggests that we take the "amount" of field passing perpendicularly $(E_{\perp 1})$ through area section ΔA_1 to equal the "amount" passing perpendicularly $(E_{\perp 2})$ through area section ΔA_2. The strength of the field at r_1 is greater than the strength at r_2, but the product $\Delta A_1 E_{\perp 1}$ is identical to $\Delta A_2 E_{\perp 2}$; the area increases as r^2 and the field falls off as $1/r^2$. ***This product of the area and the strength of the field passing perpendicularly through it is the flux of the electric field.*** (It should remind you of the Continuity Equation in fluid dynamics, p. 305).

Now imagine a closed surface in some region where there is a field; the net flux of the field emerging from the surface is the difference between the flux-out (which we'll take as positive) and the flux-in (which we'll take as negative). Usually that's zero, and we can see as much in Fig. 15.38 where the same flux enters and leaves the wedge-shaped volume bounded, top and bottom, by ΔA_2 and ΔA_1; the sides of the wedge are parallel to the field, and no flux enters or leaves through the sides. But if the surface encloses a source of the field, a total positive charge, then the flux is not zero. Envision a sphere surrounding a

> The **flux** of some field is the amount of that field passing perpendicularly through the area, multiplied by that area.

Figure 15.38   The electric field passing perpendicularly through area segment ΔA_1 passes perpendicularly through segment ΔA_2 such that $\Delta A_1 E_{\perp 1} = \Delta A_2 E_{\perp 2}$.

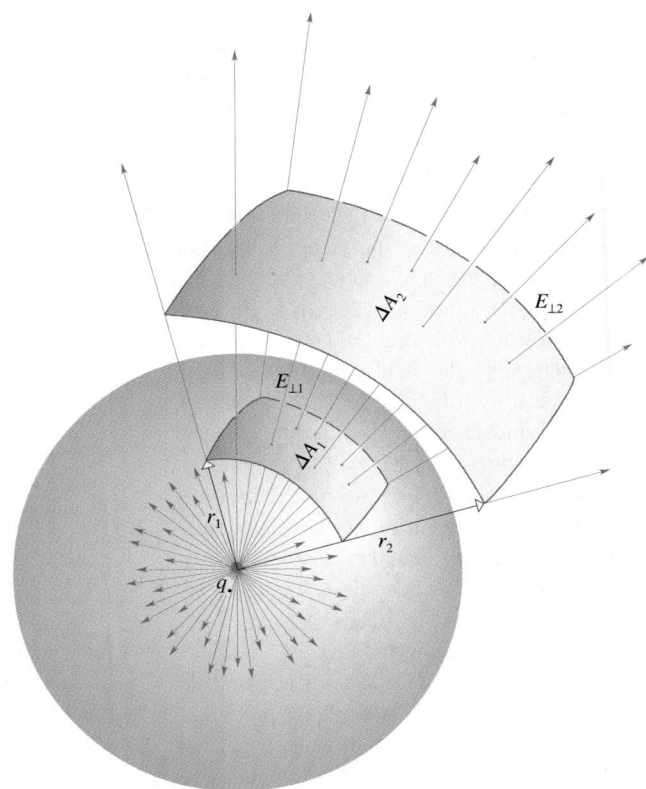

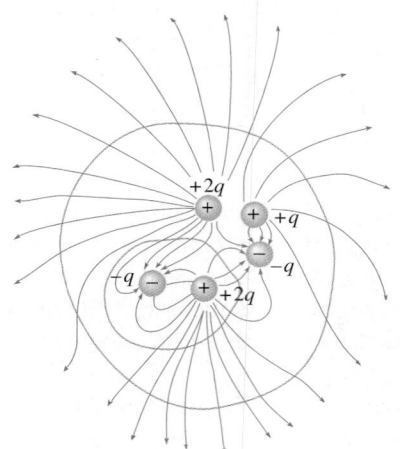

Figure 15.39 A schematic representation of the *E*-field of a distribution of charge. From far away it should look like the field of $+3q$. Given that there are eight lines per $\pm q$, surround any region by a closed area and subtract the number of lines in from those out..

point-charge q (as in Fig. 15.38); a net flux then passes out of the surface. What Gauss showed was that

> **for any closed surface the net flux of the electric field passing through that surface, whatever its shape, is proportional to the total enclosed charge.**

If there are an equal number of positive and negative charges inside the surface, the sum of the charges equals zero ($\sum q = 0$); the field begins on the positive ones, ends on the negative ones, and no net electric flux emerges (Fig. 15.39).

Typically, we'll need to calculate the field of some known charge distribution. To do that we'll surround it with a cleverly chosen closed surface and compute the net emerging flux. Often a surface such as a cylinder, or the wedge in Fig. 15.38, can be conveniently thought of as composed of several area sections ($\Delta A_1, \Delta A_2$, etc.), and we want the sum of the flux through all of them, namely, $\sum E_\perp \Delta A$. Stated symbolically, **Gauss's Law** becomes

[Gauss's Law in vacuum]

$$\sum E_\perp \Delta A = \frac{1}{\varepsilon_0} \sum q \tag{15.6}$$

where the charge distribution is surrounded by vacuum. {For a derivation click on **Gauss's Law** on the CD.} 💿

Because of the complexity of carrying out the calculation on the left side of Eq. (15.6), we will limit the discussion to situations where the imaginary closed surface, the so-called *Gaussian surface*, leads to simple results. Not being able to use calculus, we will only deal with symmetrical charge distributions. In such cases, *we can construct a closed surface out of faces that are either parallel to the field* (in which case $E_\perp = 0$) *or perpendicular to the field* (in which case $E_\perp = E$), *which is constant over the surface selected*. Under these circumstances, we'll be able to stick with simple Gaussian surfaces, such as spheres and cylinders.

Example 15.7 **[II]** A long, straight wire of length *L*, in air, carries a uniform positive charge *Q*. Use Gauss's Law to find the electric field at a perpendicular distance *r* from the wire at a point that is far from the conductor's ends. To operate in a field region free of the complications of end effects, we limit the analysis to the domain where $L \gg r$.

Solution Whenever you have a Gauss's Law problem immediately start thinking about the symmetry of the situation. (1) Translation—Determine the *E*-field in the vicinity of a uniformly charged straight wire of known length and charge. (2) Given: Total charge *Q*, distance *r*, and wire length *L*. Find: *E*. (3) Problem Type—Electrostatics/electric field/Gauss's Law. (4) Procedure—First, determine the field configuration with an eye to finding its symmetries. From the facts that the charge is uniform and the field lines must begin on positive charges and be perpendicular to the conductor's surface, we can assume that *the field is radially outward* (Fig. 15.40). Because the wire is very long, the field in the middle region cannot have a component toward either end—either end is equally remote, neither is special; **the field must be radial**. Considerations like these are the first step in the application of Gauss's Law. (5) Calculation—Given the cylindrical symmetry of the field, surround the wire with a cylindrical Gaussian surface of arbitrary radius *R* and length *l*, closed with flat end caps of area A_1 and A_3. Apply Eq. (15.6) over the closed imaginary surface made up of A_1, A_2, and A_3, which will involve the field at a dis-

tance of *r* from the wire—just what we are after. Processing the left side of the equation yields

$$\sum E_\perp \Delta A = E_{\perp 1}A_1 + E_{\perp 2}A_2 + E_{\perp 3}A_3 = \frac{1}{\varepsilon_0}\sum q$$

But $E_{\perp 1}$ and $E_{\perp 3}$ are both zero because the field is parallel to the ends. Furthermore, the entire field passes perpendicularly through the curved surface and also has a constant value over that surface; that is, $E = E_{\perp 2}$, and so

$$EA = E(2\pi rl) = \frac{1}{\varepsilon_0}\sum q$$

where $A_2 = 2\pi rl$. Now for the right side: If the total wire of length *L* carries a charge *Q*, then the **charge per unit length** (λ) is Q/L and the charge inside the Gaussian surface ($\sum q$), which we made to be of length *l*, is λl. Consequently

$$E(2\pi rl) = \frac{1}{\varepsilon_0}\lambda l$$

and

$$\boxed{E = \frac{\lambda}{2\pi r\varepsilon_0}} \tag{15.7}$$

Since *r* is unspecified, we have a formula for *E* everywhere beyond the wire where the fringing at the ends is negligible.

Quick Check: This result should have the units of F/q. Since λ is charge over length, it follows from Coulomb's Law that the units are all right. We can expect that at $r = 0$, $E = \infty$, and at $r = \infty$, $E = 0$, and both are the case. Moreover, the 2π is there because of the cylindrical symmetry.

Figure 15.40 (a) The field of a line charge. (b) The corresponding Gaussian surface..

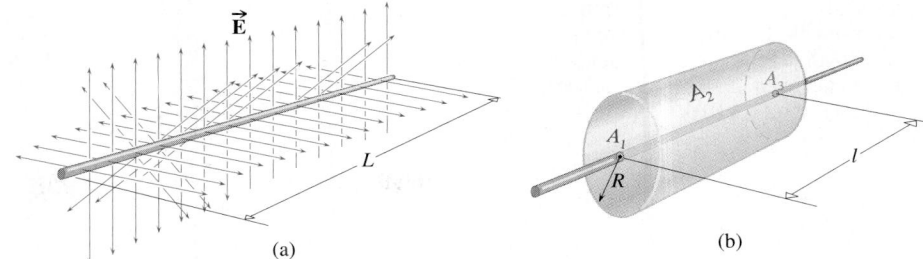

(a)   (b)

The expression determined in Example 15.7 is independent of the Gaussian surface used to derive it. That's reasonable since there are an infinite number of possible surfaces that can be used (most of them with great difficulty). Thus l, which related only to the Gaussian surface, cancels out of the analysis. The need sometimes to surround only a portion of the entire charge distribution within a surface will require that we introduce *charge densities*

[linear charge density]
$$\lambda = \frac{Q}{L}$$

[surface charge density]
$$\sigma = \frac{Q}{A}$$

[volume charge density]
$$\rho = \frac{Q}{V}$$

Consider the *very large flat sheet of charge* shown in Fig. 15.41 that is embedded in a medium that has a permittivity ε_0. Its E-field (not far from the middle), by symmetry, must be perpendicular, outward, and *uniform*. Thus, a Gaussian surface in the shape of a cylinder with end faces of area $A = A_1 = A_2$ encompasses a charge of σA. As a result

$$\sum E_\perp \Delta A = E_{\perp 1} A_1 + E_{\perp 2} A_2 + E_{\perp 3} A_3 = \frac{1}{\varepsilon_0} \sum q$$

and since $E = E_{\perp 1} = E_{\perp 3}$ and $E_{\perp 2} = 0$,

$$EA + EA = \frac{\sigma A}{\varepsilon_0}$$

Finally, *the E-field of a large sheet of charge* is

[large sheet of charge]
$$E = \frac{\sigma}{2\varepsilon_0} \tag{15.8}$$

This is an interesting equation, which we'll promptly apply to the practical case of the parallel plate capacitor. But before we do, let's take a moment to appreciate the fact that **the $\vec{E}$-field is constant**; it's independent of the distance from the sheet. That's an extraordinary result, and it's not immediately obvious. To see how it happens, consider a point P initially close to the sheet, somewhere near its middle. Call the perpendicular line from P to the plane of charge the central axis. All the charges far from this axis contribute fields at P that are nearly parallel to the sheet. Each has a large field component parallel to the sheet and a small axial one perpendicular to it. But for every charge above P, there's one below, and for every charge to the right, there's one to the left. And so all the large field components cancel. That must be true, since by symmetry the net field could not point in any direction

Johann Karl Friedrich Gauss

Figure 15.41 (a) A large sheet of charge. (b) The Gaussian surface (a cylinder perpendicular to the sheet) used to calculate $\vec{E}$.

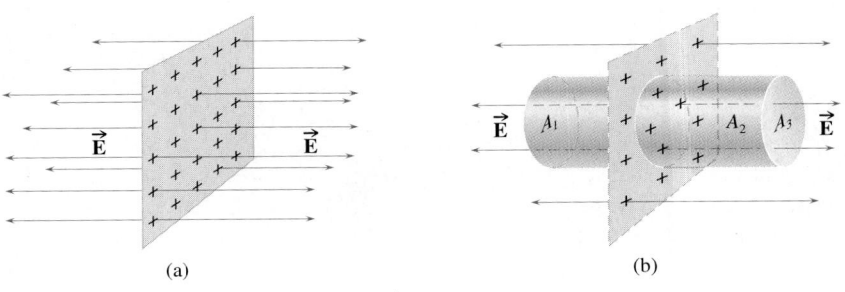

(a)   (b)

Figure 15.42 The *E*-field between two charged sheets. The ability of a parallel-plate configuration to produce a uniform field region is used in all sorts of devices; for example, see Fig. 15.43.

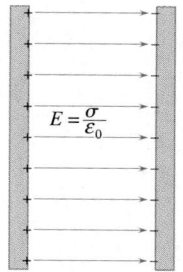

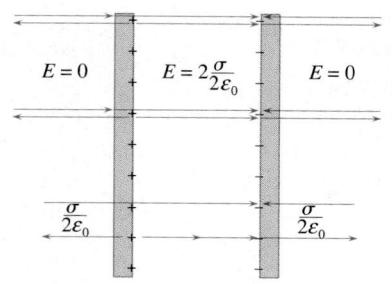

QUANTUM FIELD THEORY

A charged particle interacts with other charged particles. It creates a web of interaction around itself that extends out into space. That imagery leads to the concept of the electric field, which is a representation of the way the electromagnetic interaction reveals itself on a macroscopic level. Classically, we say that one charge sets up an *E*-field in space and another charge immersed in that field interacts directly with it, and vice versa. The picture is straightforward, but many questions come to mind. Does the *E*-field have a physical presence? Is there some continuous field-stuff filling space? Is anything actually flowing? How does the field produce a force on a charge? Does it take time to exert its influence?

As early as 1905, Einstein already considered the classical equations of Electromagnetic Theory to be descriptions of the average values of the quantities being considered. Classical theory beautifully accounted for everything measured, but it was oblivious to the exceedingly fine granular structure of the phenomenon. Using thermodynamic arguments, Einstein proposed that electric and magnetic fields were quantized, that they are particulate rather than continuous.

Today, we are guided by Quantum Mechanics, a highly mathematical theory that provides tremendous computational and predictive power but is nonetheless disconcertingly abstract. Contemporary physics holds that all fields are quantized; that each of the fundamental Four Forces is mediated by a special kind of field particle. These *messenger particles* are continuously absorbed and emitted by the interacting material particles (electrons, protons, etc.). It is this ongoing exchange that *is* the interaction. The mediating particle of the electric field is the **virtual photon**. This massless messenger travels at the speed of light and transports momentum and energy. When two electrons repel one another, or an electron and proton attract, it is by emitting and absorbing virtual photons and thereby transferring momentum from one to the other, that transfer being a measure of the action of force. {For more on the subject click on VIRTUAL PHOTONS under FURTHER DISCUSSIONS on the CD.}

other than perpendicular. Accordingly, only the small perpendicular components along the central axis do not cancel. These add to the stronger nearly perpendicular fields due to charges in the vicinity of the central axis. When *P* moves away from the sheet, the fields due to these latter charges close to the central axis drop off; as they do, the fields produced by the more off-axis charges make smaller and smaller angles with the axis, and now contribute more effectively to the perpendicular field. Thus, as the fields due to charges near the central axis diminish with distance, the number of charges contributing effectively will increase as *P* moves away. For an infinitely large, charged sheet, the result is a constant perpendicular *E*-field.

Figure 15.42 pictures a pair of parallel, equally and oppositely charged metal plates. The two fairly uniform sheets of opposite (excess) charge mutually attract and hold each other in place on the inner faces of the plates. Except for fringing at the edges (see Fig. 15.34), the net field everywhere should be the vector sum of the two overlapping *uniform* fields set up by the two opposite sheets of charge. Outside this *parallel plate capacitor*, the two uniform fields are oppositely directed and cancel—ideally, there is no *E*-field outside the plates (Fig. 15.43). Inside, the two fields, each given by Eq. (15.8), are in the same direction and add: $E = 2(\sigma/2\varepsilon_0)$ and *the field between the plates of*

a parallel plate capacitor is

[parallel plate air-filled capacitor]
$$E = \frac{\sigma}{\varepsilon_0}$$
(15.9)

It's left as Problem 65 to show that this same result follows directly from Gauss's Law.

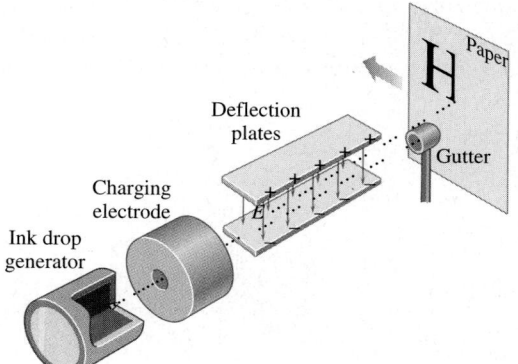

Figure 15.43 One type of ink jet printer fires charged droplets of ink at the paper. The droplets pass through a set of parallel plates, which carries a charge proportional to the control signal. The resulting *E*-field steers the beam of ink vertically as the paper moves by.

ELECTROMAGNETIC CHARGE

Charge is the property of matter that gives rise to electric force. *Like charges repel; unlike charges attract.* Charge is conserved. The basic ideas about the behavior of charge are discussed in Sections 15.1 (Positive & Negative Charge) and 15.2 (Insulators & Conductors)—reread this material (pp. 530–532) before going on.

THE ELECTRIC FORCE

The electrostatic force between two point-charges, or two uniformly charged spheres, is

[point-charges in vacuum] $$F_{\text{E}} = k_0 \frac{q_{\cdot 1} q_{\cdot 2}}{r^2} \qquad [15.1]$$

which is **Coulomb's Law.** The unit of charge is the *coulomb* (C), and k is a constant specific to the medium in which the charges are embedded. In the special case of vacuum, $k_0 = 8.987\,551\,79 \times 10^9$ N·m^2/C$^2 \approx 9.0 \times 10^9$ N·m^2/C^2. The *charge of the electron* is $-1.602\,177\,33 \times 10^{-19}$ C. The net force on any one charge is the vector sum of all the forces exerted on it due to each of the other charges interacting with it independently (p. 542). Study Section 15.3 (Coulomb's Law); its role in problem solving is central. Make sure you understand Examples 15.1 to 15.3. Go through the **WARM-UPS** and study the **WALK-THROUGHS** on the **CD**.

THE ELECTRIC FIELD

The electric field ($\vec{\mathbf{E}}$) *at any point in space is the electric force experienced by a positive test-charge* ($q_\circ$) *at that point, divided by that charge:*

$$\vec{\mathbf{E}} = \frac{\vec{\mathbf{F}}}{q_\circ} \qquad [15.2]$$

in units of newtons per coulomb (N/C). The field is represented by field lines that begin and end on charges. The denser the concentration of lines, the greater the field. The force that would arise on any point-charge $q_\cdot$ placed in the field is

$$\vec{\mathbf{F}} = q_\cdot \vec{\mathbf{E}} \qquad [15.3]$$

The *electric field of a point-charge* $q_\cdot$ is

[point-charge in vacuum] $$E = k_0 \frac{q_\cdot}{r^2} \qquad [15.4]$$

In terms of the **permittivity** (ε) of the surrounding medium

$$k_0 = \frac{1}{4\pi\varepsilon_0} \qquad [15.5]$$

The *permittivity of free space* is a constant defined as $\varepsilon_0 = 8.854\,187\,8 \times 10^{-12}$ C^2/N·m^2. This material is discussed in Sect. 15.4 (Definition of the $\vec{\mathbf{E}}$-field). Coulomb's Law and Eq. (15.2) are the basic defining statements of this entire chapter. Make sure you understand Examples 15.4 to 15.6.

The field inside a charged conductor in the steady state is zero. External field lines must be perpendicular to the surface of any con-

ductor. **Gauss's Law** (p. 557) is

$$\sum E_\perp \Delta A = \frac{1}{\varepsilon_0} \sum q \qquad [15.6]$$

{For a complete derivation click on **GAUSS'S LAW** on the **CD**.} The electric field of a long, straight charged wire in vacuum is

$$E = \frac{\lambda}{2\pi r \varepsilon_0} \qquad [15.7]$$

The $1/2\pi r$ tells us that we have cylindrical symmetry. We define the *linear charge density* as $\lambda = Q/L$, the *surface charge density* as $\sigma = Q/A$, and the *volume charge density* as $\rho = Q/V$. The *E*-field, in air, of a large sheet of charge is

[large sheet of charge] $$E = \frac{\sigma}{2\varepsilon_0} \qquad [15.8]$$

The field between the plates of an air-filled parallel-plate capacitor is

[parallel plates air-filled] $$E = \frac{\sigma}{\varepsilon_0} \qquad [15.9]$$

These important results are to be found in Section 15.5 (Gauss's Law); reexamine Example 15.7.

Before attempting to do the problems, review the appropriate section and any previous material that you feel keeps you from mastering that section; study the Examples in that section; study the **WARM-UPS** and **WALK-THROUGHS** on the **CD** that apply; read all the Suggestions on Problem Solving. Work out the I-level Coordinated Problems that are grouped in threes, that is, the ones in magenta. Do the I-level Progressive Problems. Then do as many of the remaining regular I-level problems as you can. At that point if you're not exhausted, start over with the II-level problems.

Key Terms

charge	test-charge
electron	line-of-force
Conservation of Charge	electric field
conductor	dielectric
insulator	permittivity
spark	relative permittivity
electroscope	dielectric constant
induced charge	source
electrically polarized	sink
proof-plane	dipole
point-charge	Faraday cage
Coulomb's Law	Gauss's Law
coulomb	flux
charge of the electron	Gaussian surface
field of force	

1. A common demonstration is to toss some small fragments of paper onto a highly charged conducting sphere. The paper initially sticks to the sphere and after a little while pops off as if shot into the air. In 1676, Newton sent a note to the Royal Society describing a similar experiment where he charged a glass disk and held it horizontally above "some little fragments of paper." These he observed "sometimes leaping up to the glass, and resting there awhile; then leaping down and again resting." Explain what was happening.

2. Imagine a spherical dielectric shell—one made of glass, for example. A charge of $+Q$ is carefully distributed uniformly over the entire outer surface. Is there an electric field anywhere inside the shell? Explain.

3. Figure Q3 shows a dipole in a room (which is represented by a large grounded conductor). Make a rough sketch of its E-field and explain your reasoning.

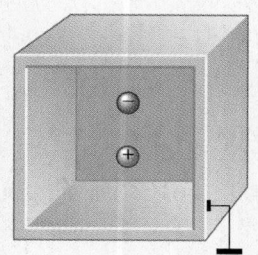

4. Static electric buildup can be troublesome as, for example, in a photographic processing laboratory where film easily becomes charged (usually positively), attracting dust, and even creating sparks. To control the effect, there are static eliminators, one variety of which uses radioactive polonium-210. This material pours out a constant stream of positively charged alpha particles. How would that help?

Figure Q3

5. EXPLORING PHYSICS ON YOUR OWN: William Gilbert in the sixteenth century devised a simple instrument to detect electrification, which he called a *versorium*. To make one, just balance a length of any material "lightly pivoted on a needle." A piece of a drinking straw stuck on a tack set in clay will work nicely once it's made free to rotate (Fig. Q5). Or fold a narrow strip of paper into a long upside-down V-shape and balance it horizontally on a tack. Now rub something like a comb to electrify it. Use the versorium to see if you have succeeded—that is, wave the comb near the detector. Put your versorium near a TV tube and see if it can detect when you turn the electron beam on. What happens and why?

Figure Q5

6. What, if anything, might happen to the leaves of a charged electroscope if you waved a lit match around near it? Explain.

7. Stephen Gray attached a lead ball to the end of a glass tube with a 3-ft length of moistened "parcel string." Under the ball he placed some fragments of brass leaf (very thin foil). When he rubbed the tube, the leaf was attracted up to the ball. Explain what was happening. (By the way, Gray later redid the experiment using 80 feet of string, and "the Tube being rubbed, the Ball attracted the Leaf-Brass.")

8. EXPLORING PHYSICS ON YOUR OWN: Why do long strips of ordinary plastic tape freshly pulled from the roll usually crumple and stick to themselves and everything nearby? Go in a dark closet and pull off a length of transparent plastic tape—sparks will blaze brilliantly at the point of contact with the roll.

9. An electric typewriter on a desk not far from a TV set disturbs the picture, generating colored bands across the screen. The roof antenna is attached to the set via an ordinary flat twin-lead TV wire. What would you suggest as a possible solution to the problem and why?

10. Figure Q10 shows a hollow charged conductor with two proof-planes touching each other inside of it. The scheme is used to determine that no E-field is inside a charged conductor. Explain how that might be accomplished.

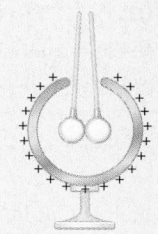

11. A neutral, banana-shaped solid conductor is grounded to the plumbing via a wire. A negatively charged sphere is brought near it, the ground lead is disconnected, and the sphere removed. Describe the resulting E-field and surface charge density on the metal banana, if any.

Figure Q10

12. Why are electrostatic phenomena, like sparks when you take off a sweater in the dark, more common in winter than in summer?

13. During the time Newton was president of the Royal Society, its curator of experiments was Francis Hauksbee. The following (Fig. Q13) is a demonstration Hauksbee performed that seems a very early hint of Faraday's lines-of-force. Onto a semicircular wire frame, Hauksbee fastened "several pieces of Woollen Thread ... so as to hang down at pretty nearly equal distances." At the center, he placed a glass "Tube," which was subsequently charged by rubbing. The threads immediately swung about so that they pointed directly at the center of the glass. Explain what was happening and discuss the implications regarding field lines.

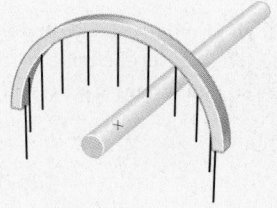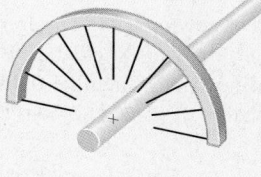

Figure Q13

14. Figure Q14 shows a variation of the most widely used electrostatic generator, a device that carries the name of its inventor, Robert J. Van de Graaff (1902–1967). The two pulleys are covered with different materials so that when they are contacted by the motor-driven belt, the belt acquires a negative charge from the bottom pulley and a positive charge from the top one. In some research models, electrons are literally sprayed on the belt at the base. Figure out how the generator works and explain the crucial role of the hollow conductor dome. What, if anything, limits the amount of charge that can be built up on the dome? How would that compare to the case where the belt delivered the charge to the outside surface? (See the photos on p. 536.)

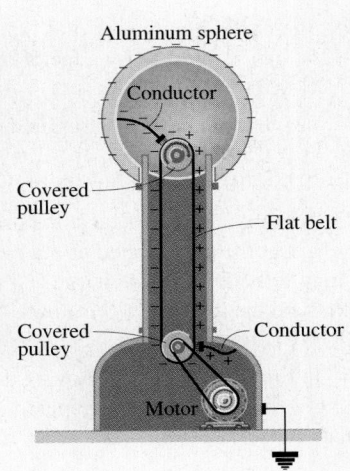

Aluminum sphere

Conductor

Covered pulley

Flat belt

Covered pulley

Conductor

Motor

Fgure Q14

Fgure Q14

15. Suppose that you approach an elephant with two known oppositely charged pith balls in order to determine if the beast is itself charged. The first ball (+) is attracted, the second (−) repelled. What does that tell you? Which ball provides the unambiguous evidence? Explain.

16. Explain the meaning of each of the Key Terms on page 561.

Multiple Choice Questions

The next six questions refer to the gold-leaf electroscopes in Fig. MC1. Each figure contains a charged nonconducting wand.

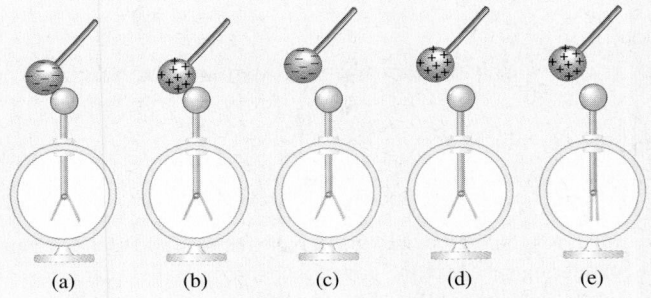

(a)   (b)   (c)   (d)   (e)

Figure MC1

1. In Fig. MC1*a* (a) the spherical knob is positive, the leaves are negative (b) the knob is negative, the leaves are positive (c) the knob is neutral, the leaves are negative (d) the knob is neutral, the leaves are positive (e) none of these.

2. In Fig. MC1*b* (a) the spherical knob is positive, the leaves are negative (b) the knob is positive, the leaves are positive (c) the knob is neutral, the leaves are negative (d) the knob is positive, the leaves are neutral (e) none of these.

3. In Fig. MC1*c* (a) the spherical knob is positive, the leaves are negative (b) the knob is positive, the leaves are positive (c) the knob is neutral, the leaves are negative (d) the knob is positive, the leaves are neutral (e) none of these.

4. In Fig. MC1*d* (a) the spherical knob is positive, the leaves are negative (b) the knob is positive, the leaves are positive (c) the knob is neutral, the leaves are negative (d) the knob is negative, the leaves are neutral (e) none of these.

5. In Fig. MC1*e* (a) the spherical knob is positive, the leaves are negative (b) the knob is negative, the leaves were positive (c)

the knob is negative, the leaves were negative (d) the knob is negative, the leaves were neutral (e) none of these.

6. Suppose all the charged rods in Fig. MC1 are now removed, we can expect that (a) the leaves in *a*, *b*, and *c* will hang vertically (b) the leaves in *a*, *c*, and *d* will hang vertically (c) the leaves in *c*, *d*, and *e* will hang vertically (d) the leaves in *a*, *b*, and *e* will not hang vertically (e) none of these.

7. The electrostatic force between a negative electron and a neutral neutron is (a) negative and attractive (b) positive and repulsive (c) zero (d) sometimes attractive and sometimes repulsive (e) none of these.

8. By comparison with the force of gravity, the electrical attraction between an electron and a proton (a) is just about the same size (b) is very much stronger (c) is very much weaker (d) cannot be compared (e) none of these.

9. When rubbed with a piece of wool, sulfur and glass will become charged (a) positively and negatively, respectively (b) negatively and positively, respectively (c) both positively (d) both negatively (e) none of these.

10. When the center-to-center separation between two small charged spheres is doubled, the electric force between them (a) is halved (b) doubles (c) is quartered (d) is quadrupled (e) none of these.

11. The SI units of electric flux are (a) N/C^2 (b) $N \cdot m/C$ (c) $N \cdot m^2/C$ (d) $C/N \cdot m$ (e) none of these.

12. Some trucks (especially those carrying combustible fuels) have conducting tires while others use straplike bands attached to their bottoms that drag along the ground in an attempt to (a) discharge static electricity (b) get rid of moisture (c) build up negative charge (d) protect them in case they get hit by lightning (e) none of these.

13. Suppose we have three identical conducting spheres and one of them carries a charge of Q. If they are all brought into contact and

then separated   (a) they will each have a charge of $Q/3$ (b) they will each have a charge of Q (c) only one will be charged with Q (d) they will all be discharged   (e) none of these.

14. A metal sphere is grounded through a switch, and a positively charged balloon is brought near it. The switch is opened and the balloon taken away. The sphere is now   (a) neutral   (b) negatively charged   (c) positively charged   (d) charged, but we cannot know its polarity   (e) none of these.

15. A metal sphere is grounded through a switch, and a positively charged balloon is brought near it. The balloon is then taken away, and the switch is opened. The sphere is now   (a) neutral   (b) negatively charged   (c) positively charged   (d) charged, but we cannot know its polarity   (e) none of these.

16. If the charge on each of two identical tiny spheres is doubled while their separation is also doubled, their force of interaction will (a) double   (b) become halved   (c) be quartered   (d) stay unchanged   (e) none of these.

17. If you are stretched out in a bathtub full of water, you are likely (a) not to be grounded because the water is in a tub   (b) to be grounded if the tub has metal feet   (c) not to be grounded because you are inside a conductor   (d) to be grounded because the water connects you electrically to the pipes   (e) none of these.

18. The E-field of a point-charge 4.0 m away is measured to be 100 N/C. The field 2.0 m away from that charge is   (a) 400 N/C   (b) 50 N/C   (c) 100 N/C   (d) 800 N/C   (e) none of these.

19. A nonconductor is charged and then brought near a conductor. Consequently   (a) the two electrostatically repel each other   (b) the two electrostatically attract each other   (c) only the nonconductor is repelled   (d) there is no electrostatic interaction at all (e) none of these.

20. Three charges ($q_1 = 1$ nC, $q_2 = 2$ nC , and $q_3 = 5$ nC) are separated in space (by distances of $r_{12} = 1$ m, $r_{23} = 2$ m, and $r_{13} = 3$ m). The ratio of the magnitudes of the forces on q_3 due to q_1 and due to q_2 is   (a) 2/9   (b) 5/2   (c) 5/9   (d) 9/2   (e) none of these.

21. Two charges in vacuum attract each other. The same two charges, with the same separation, are now immersed in ethanol, which has a relative permittivity ($\varepsilon/\varepsilon_0$) of 25. The interaction is now

(a) increased by a factor of $\sqrt{25}$ (b) decreased by a factor of $\sqrt{25}$ (c) decreased by a factor of 25   (d) increased by a factor of 25   (e) none of these.

22. An electron in an electric field of 100 N/C experiences a force of   (a) 1.6×10^{-10} N   (b) 1.6×10^{-21} N   (c) 3.2×10^{-17} N (d) 1.6×10^{-17} N   (e) none of these.

23. A charge $+q$ is placed at the origin of a coordinate system, and a charge of $+Q$ is located at $+a$ on the x-axis. The force on $+Q$ is found to be $\vec{F}$. A third charge $-q$ is now placed at $+2a$ on the x-axis, and the force on $+Q$ is now   (a) zero   (b) $\vec{F}$ (c) $\frac{1}{2}F$ (d) $2\vec{F}$ (e) none of these.

24. In Fig. MC24, which drawing depicts the net field of a tiny positively charged conducting sphere in the gap of a charged parallel-plate capacitor?   (a)   (b)   (c)   (d)   (e) none of these.

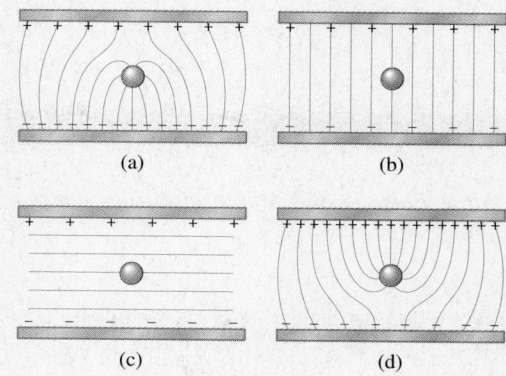

(a)   (b)

(c)   (d)

Figure MC24

25. In Fig. MC24, which drawing depicts the net field of a tiny neutral conducting sphere in the gap of a charged parallel-plate capacitor?   (a)   (b)   (c)   (d)   (e) none of these.

For more Multiple Choice Questions with answers click on WARM-UPS in CHAPTER 15 on the CD.

Suggestions on Problem Solving

1. Remember that the signs arising from the application of Coulomb's Law just relate to whether the force is attractive or repulsive. The forces between two point-charges are equal in magnitude, opposite in direction, and act along their center-to-center line. **Draw a diagram, sketch in the directions of the forces, and then forget those signs.** As ever, take to the right and up as the positive directions.

2. Given several point-charges, you may need to find the net force acting *on one of them*. Draw the direction of each force acting on the charge. Next, determine the magnitude of each force via Coulomb's Law. Resolve each force vector into perpendicular components and find the resultant of all the components in the usual way—that is, the net force.

3. The electric field *at a point* in space due to several point-charges is calculated much as is the force but using Eq. (15.4) instead of Coulomb's Law. Imagine a positive test-charge at that point and

draw in the field vectors accordingly. Be especially careful about signs: **the force on a positive charge in an $\vec{E}$-field is in the direction of the field; the force on a negative charge is opposite to the field.**

4. When applying Gauss's Law, first find the direction of the E-field so that you can take advantage of its symmetries. Remember that there is no field inside a conductor and embedding all or part of the Gaussian surface inside the conductor puts it in a zero-field region. The fields of all of the easy configurations have long since been figured out using Gauss's Law. Once you have learned how to make the standard dozen-or-so calculations, you have pretty much mastered the thing. *Most professionals do not remember the fields due to complex configurations of charges (e.g., inside a charged coaxial cable); they just quickly derive it when needed, using Gauss's Law, and that's the best reason to learn the method.*

Problems ✛ **Coordinated Problems** ✛ **Progressive Problems** ✛ **Solutions**

STUDY GUIDE **1. Coordinated Problems:** The three problems within each magenta-colored grouping are solvable in similar ways. Note that the first of these always has a hint; moreover, its solution is provided in the back of the book. *Work out each of these sets; they'll strengthen technique and build confidence.* **2. Progressive Problems:** The problems introduced in blue unfold step-by-step carrying along the analysis in a more suggestive way than is customary. *Work out all of these; they'll guide you through the analytic process and help develop problem-solving skills.* **3. Worked-Out Solutions:** Studying worked-out solutions is an important part of learning how to solve problems. Accordingly, additional *solutions* to a number of model problems are given below. *Make sure you understand each of them before you go on to the next problem.* **4.** Also provided in the back of the book are the *Answers* to all odd-numbered problems, as well as worked-out *solutions* to those with boldface numbers. Problem numbers in italic indicate that a solution appears in the Student Solutions Manual.

SECTION 15.1: POSITIVE & NEGATIVE CHARGE
SECTION 15.2: INSULATORS & CONDUCTORS
SECTION 15.3: COULOMB'S LAW

1. [I] By roughly how much does the mass of a copper object change when, upon being stroked with a piece of woolen cloth, it acquires an excess charge of $-1.0\ \mu C$?

2. [I] How many electrons are needed to produce a charge of -1.0 C?

3. [I] THIS PROBLEM EXPLORES THE BEHAVIOR OF CHARGE. Two identical very small conducting spheres carrying charges of $+6.0\ \mu C$ and $-2.0\ \mu C$ come into contact for a while and are then separated. (a) What happens, in general, when two charged conducting spheres touch? (b) What's the net charge on our specific two spheres once they are brought into contact? (c) If a net charge Q was on two identical spheres when they were touching, what charge would be on each sphere after separation? (d) How much charge is on each of our original spheres after they're parted?

4. [I] THIS PROBLEM EXPLORES THE BEHAVIOR OF CHARGE. Imagine that we have two tiny charged dust particles. (a) Under what circumstances, if any, would the force on either one due to the other become zero? (b) What happens to the force on each speck of dust as their center-to-center separation (r) decreases? (c) How does the force on either charged particle vary with r? (d) Considering Fig. P4, which sketch best represents the force on each particle versus their center-to-center separation?

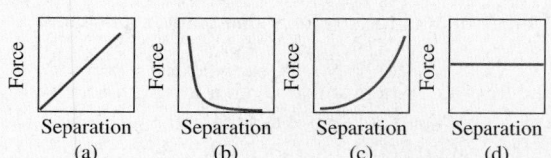

Figure P4

5. [I] THIS PROBLEM EXPLORES THE FORCE EXERTED BY CHARGE. A cloud of hairspray becomes charged as it leaves the nozzle of a spray can. Two droplets that happen to be 2.00 mm apart, each carry a charge of -16.02×10^{-19} C. (a) Is the force between them attractive or repulsive? Explain. (b) How many electrons has each sphere picked up? (c) What law describes the electrostatic force between the droplets. (d) Determine the value of the electrostatic force between the droplets.

6. [I] While flying around, two little bugs pick up charges of $+3.0 \times 10^{-14}$ C and $+4.0 \times 10^{-14}$ C, respectively. What will be the approximate electrostatic force between them when they are 2.00 cm apart?

> **SOLUTION:** If you want the electrostatic force between two charges, you'll need Coulomb's Law: $F_E = k_0 q_{\cdot 1} q_{\cdot 2}/r^2$
>
> $$F_E = (8.99 \times 10^9\ \text{N·m}^2/\text{C}^2)\frac{(3.0 \times 10^{-14}\ \text{C})(4.0 \times 10^{-14}\ \text{C})}{(2.00 \times 10^{-2}\ \text{m})^2}$$
>
> $$F_E = 2.7 \times 10^{-14}\ \text{N}$$
>
> The charges are both positive, and the force is repulsive.

7. [I] Two tiny gold-covered pith balls carrying the same charge are 1.0 m apart in vacuum and experience an electrical repulsion of 1.0 N. What is the charge on each of them?

8. [I] A very small conducting sphere in air carries a charge of 5.0 picocoulombs and is 0.20 m from another such sphere carrying a charge Q. If each sphere experiences a mutual electrical repulsion of 2.0 μN, find Q.

9. [I] Two protons are fired directly at each other in a vacuum chamber. What is the force each experiences at the instant they are 1.0×10^{-14} m apart? [*Hint: The particles are identically charged and will experience a Coulomb repulsion.*]

10. [I] Two point-charges of $+0.50\ \mu C$ are 0.10 m apart. Determine the electric force they each experience in air.

11. [I] Two equally charged small spheres repel each other with an electric force of 1.0 N when 0.50 m apart, center-to-center, in air. What is the charge on each sphere?

12. [I] Two tiny spores in vacuum carrying charges of $+1.00\ \mu C$ and $-1.00\ \mu C$ are 1.00 m apart, center-to-center. What is the electrostatic force each experiences?

13. [I] How far apart must an electron and a proton be in vacuum if they are to experience an attractive force of 1.00 N? Assume each is a point-charge.

14. [I] Compute the gravitational attraction between two electrons separated by 1.0 mm in vacuum, and compare that with the electrical repulsion they experience.

15. [I] An equilateral triangle with sides of 2.0 m is inscribed within a circle. A tiny charged sphere carrying ($+10\ \mu C$) is then fixed at each vertex, and one of $-25\ \mu C$ is placed at the center of the circle. What is the net force acting on that central charge (magnitude and direction)?

16. [I] Two charged spheres each containing a quantity of excess electrons equal in number to Avogadro's number are separated by 1000 km in vacuum. Compute to two significant figures the electrical interaction between the spheres.

17. [I] Three small spheres immersed in air, each carrying a charge of $+25$ nC, are located at the vertices of a right isosceles triangle with a hypotenuse of 1.414 m, lying along the x-axis. Find the net electric force on the charge opposite the hypotenuse, located above it on the positive y-axis.

18. [I] A narrow conducting ring in vacuum, having a diameter of 10.0 cm, is given a charge of $+20.0$ nC; determine the electrostatic force that would act on a tiny sphere carrying a charge of $+1.00$ nC placed directly at the center.

19. [I] A square nonconducting framework is set up in space with a tiny metal sphere mounted at each corner. The spheres at either end of the 45° diagonal are charged equally with +45 nC each, while the other two spheres are both given charges of −45 nC. What is the net force on a charge of 10 nC at the very center of the square?

20. [II] How many electrons are there in a tablespoon (15 cm³) of water? What is the net charge of all of these electrons?

21. [II] Three small negatively charged metal spheres in vacuum are fixed on a horizontal straight line, the x-axis. One (−12.5 μC) is at the origin, another (−5.0 μC) is at $x = 2.0$ m , and the third (−10.0 μC) is 1.0 m beyond that at $x = 3.0$ m. Compute the net electric force on the last sphere due to the other two. [*Hint: Draw a picture. Compute the effect of each charge on the target charge; draw in the corresponding force vectors and then add them.*]

22. [II] Imagine a vertical line marked off in 1.00-cm intervals and suppose you have three tiny identical silver spheres. While they are all in contact, you transfer 30.0 μC to the group and then fix them, one at a time, to points $y = 1, 2,$ and 3. What is the electrostatic force on the last sphere? Take them to be surrounded by air.

23. [II] At a given instant three electrons, in vacuum, happen to lie on a straight line. The first and second are separated by 1.00×10^{-6} m, while the second and third are 2.00×10^{-6} m apart. What is the net electrostatic force on the middle electron, the second one?

24. [II] Two charges of +4.0 nC and −1.0 nC are fixed to a baseline at a separation of 1.0 m. Where on the baseline should a third charge of +2.0 nC be placed if it is to experience zero net electric force?

25. [II] Two very small identical conducting spheres carrying charges of +40.0 nC and −80.0 nC are 1.00 cm apart, center-to-center, in vacuum. Determine the force each experiences. They are then brought into contact and subsequently again separated to 1.00 cm. Determine the force each now experiences.

26. [II] Point-charges of +1.0 nC, +3.0 nC, and −3.0 nC are located, one each, at the three corners of an equilateral triangle of side-length 30 cm. Find the net electrostatic force exerted on the 1.0-nC charge. Assume the surrounding medium is vacuum.

27. [II] Figure P27 shows four point-charges fixed at the corners of a rectangle, in vacuum. Please compute the net electrostatic force acting on the 100-μC charge.

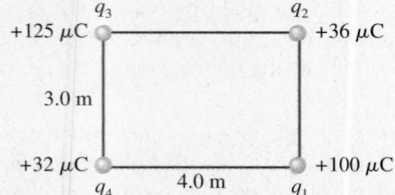

Figure P27

28. [II] Three very small charged spheres are located in a plane in air as follows: $q_1 = +15$ μC at point (0, 0), $q_2 = −20$ μC at point (2, 0), and $q_3 = +10$ μC at point (2, 2). Find the net force acting on the last charge.

29. [II] Redo Problem 27 where now q_4 is −32 μC.

30. [II] A cubic framework is inscribed within a sphere of 1.0-m radius. Tiny conducting spheres are then fixed at each corner and subsequently charged. There are four diagonals, and the pairs of

spheres on the ends of each diagonal are equally charged with +10 nC, −10 nC, +20 nC, and −20 nC, respectively. What would be the resulting electric force on a +100-nC point-charge located at the center of the sphere? The surrounding medium is air.

31. [II] Two 2.0-g pith balls hang in air on essentially weightless cotton threads 50-cm long from a common point of support. The balls are then equally charged and they spring apart, each making an angle of 10° with the vertical. Find the magnitude of the charge on each ball.

32. [III] Two charges $+q$ and $−q$ reside in vacuum on the y-axis at locations of $-\frac{1}{2}d$ and $+\frac{1}{2}d$, respectively. Determine the force on a third charge $+Q$ located at a distance of $+x$ from the origin on the x-axis.

33. [III] Three free charges (two of which are $+Q$ and $+2Q$, separated by a distance d) are in equilibrium. Find the size, polarity, and location of the third charge.

SECTION 15.4: DEFINITION OF THE $\vec{\mathbf{E}}$-FIELD

34. [I] A person becomes charged after walking across a woolen rug. A tiny piece of lint floating nearby carries a charge of -16.0×10^{-19} C and interacts with the person's E-field. If the lint experiences an electrostatic force of 3.0×10^{-14} N, what is the magnitude of the person's electric field at that location?

> **SOLUTION:** We're asked to find the magnitude of the electric field, so we begin with the scalar version of its definition:
>
> $$E = F_E / q_o = (3.0 \times 10^{-14} \text{ N})/(-16.0 \times 10^{-19} \text{ C})$$
> $$|E| = 1.9 \times 10^4 \text{ N/C}$$

35. [I] A minute particle of toner carrying a charge of 5.0×10^{-15} C floats toward a sheet of paper in a copy machine. If the particle experiences an electrostatic force of 25.0×10^{-25} N, what is the value of the electric field at that location?

36. [I] Determine the electric force acting on an electron placed in a uniform north-to-south E-field of 8.0×10^4 N/C in vacuum.

37. [I] A test-charge of +5.0 nC placed at the origin of a coordinate system experiences a force of 4.0×10^{-6} N in the positive y-direction. What is the electric field at that location? Assume the medium is vacuum.

38. [I] A +10-μC test-charge at some point beyond a charged sphere experiences an attractive force of 40 μN. Please compute the value of the E-field of the sphere at that point in a vacuum.

39. [I] A very small conducting sphere carrying a charge of −20 nC is attached to a force gauge and lowered into a uniform electric field. A force of 2.0 nN due east keeps the sphere in equilibrium. Describe the E-field, assuming air is the medium. [*Hint: Use the defining expression for the electric field. Be especially careful about the sign of the charge and the direction of $\vec{\mathbf{F}}$.*]

40. [I] A region of space is permeated by an electric field. When a tiny gold sphere carrying a charge of +50.0 μC is placed at a particular point in the field, it experiences an electrostatic force of 100.0×10^{-6} N due south. Write an expression for the electric field vector at the location of the sphere.

41. [I] A −40.0-nC test charge is placed in a 2000.0-N/C electric field pointing due west in a region of space. Describe the electrostatic force the charge will experience.

42. [I] A small positively charged object is placed, at rest, in a uniform electric field in vacuum. Write an equation giving its speed after a time t in terms of its mass m and charge q.

43. [I] Determine the magnitude and direction of an E-field if an electron placed in it, in vacuum, is to experience a force that will exactly cancel its weight at the Earth's surface.

44. [I] What is the magnitude and direction of the electric field of an electron at a point 1.0 m away in vacuum?

45. [I] THIS PROBLEM WILL HELP US UNDERSTAND THE ELECTRIC FIELD OF A POINT-CHARGE. Suppose the electric field at a given location beyond a point-charge in vacuum is $\vec{E}$. (a) What is the new electric field at the same location in terms of $\vec{E}$, if the charge is doubled? (b) What is the new electric field if now the original charge is maintained, but the distance is doubled? (c) What is the new electric field in terms of $\vec{E}$ if both the original charge and the distance are doubled?

46. [I] THIS PROBLEM EXPLORES THE ELECTRIC FIELD ARISING FROM SEVERAL POINT-CHARGES. Figure P46 depicts two uniformly charged minute spheres in vacuum. If each sphere carries a charge of $+2.00\ \mu C$, (a) what is the value of the electric field at point-A due to Q_1? (b) What's the direction of that field? (c) What is the value of the electric field at point-A due to Q_2? (d) What's the direction of that field? (e) What is the value of the electric field at point-A due to both Q_1 and Q_2? (f) What's the direction of that net field?

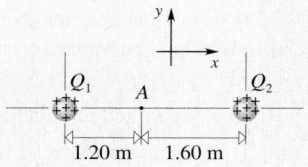

Figure P46

47. [I] Consider a hydrogen atom to be a central proton around which an electron circulates at a distance of 5.3×10^{-11} m. Find the electric field at the electron due to the proton.

48. [I] Two point-charges of $+10$ nC and -20 nC lie on the x-axis at points $x = 0$ and $x = +10$ m, respectively. Find the electric field on the axis at point $x = +5.0$ m. Assume vacuum.

49. [I] Positive point-charges of $+20\ \mu C$ are fixed at two of the vertices of an equilateral triangle with sides of 2.0 m, located in vacuum. Determine the magnitude of the E-field at the third vertex.

50. [I] Redo Problem 49, this time with charges of $+20\ \mu C$ and $-20\ \mu C$ at either end of the baseline (creating the field at the remaining vertex).

51. [I] Two point-charges of $+50$ nC each are separated in air by 1.414 m. What is the value of the net electric field they produce at a point that is 1.0 m away from both of them?

52. [II] Two tiny gold spheres lie on the x-axis at $x = 0$ and $x = 1.00$ cm. The one at the origin carries a charge of $+40.0$ nC, whereas the other has a charge of -80.0 nC. What is the value of the net electric field they produce at a point that is 2.00 m away from each of them?

53. [II] We can produce a uniform $\vec{E}$-field by charging two parallel plates. Suppose such a uniform field of 30.00 N/C due north is overlapped with a uniform field of 40.0 N/C due east; determine the net $\vec{E}$-field in the region of overlap.

54. [II] A uniform field of 1.200 N/C pointing due north exists in a region of space. A point-charge of 6.00×10^{-12} C is then placed in that original field. What is the magnitude of the net field now at a point 20.0 cm due east of the charge?

55. [II] Charged parallel plates are used to set up a uniform E-field of 1.20 N/C pointing due west in a region of vacuum. A tiny silver sphere exists in the space, and it is desired to charge the sphere such that the net field so produced at a point 20.0 cm due south of it will be 1.805 N/C−AT 48.33° SOUTH OF WEST; what charge should the sphere receive?

56. [II] Three point-charges are placed at the corners of an isosceles triangle. At the left and right, end points of the base are $+1.0\ \mu C$ and $+1.0\ \mu C$, respectively, and at the vertex $+3.0\ \mu C$. The base of the triangle is 40-cm long, and the altitude is 30-cm high. Find the E-field at the midpoint of the baseline.

57. [II] An electron is placed in a uniform electric field of 1.5×10^4 N/C. Please determine its acceleration.

58. [II] Two point-charges of $+10$ nC and -20 nC lie on the x-axis at points $x = 0$ and $x = +10$ m, respectively. Find a point where the net electric field is zero, if such a point exists.

59. [II] A section of an advertising sign consists of a long tube filled with neon gas having electrodes inside at both ends. A uniform electric field of 20 kN/C is set up between the electrodes, and neon ions accelerate along the length of the tube. Given that the ions each have a mass of 3.35×10^{-26} kg and are singly ionized, determine their acceleration.

60. [II] THIS PROBLEM DEALS WITH THE ELECTRIC FIELD AND ITS ACTION ON CHARGE. A uniform downward electric field of 6.0×10^4 N/C exists in a region of space. A tiny charged chunk of Styrofoam of mass 0.010 g floats motionlessly in the field. (a) In what direction is the electrostatic force on the Styrofoam? (b) What is the polarity of the charge on the Styrofoam? (c) What two forces act on the chunk? (d) Draw a free-body diagram of the particle. (e) Compute the charge on the piece of Styrofoam. Ignore air friction.

61. [II] THIS PROBLEM EXPLORES THE ELECTRIC FIELD AND ITS ACTION ON CHARGE. Figure P61 shows a small charged 1.00-g sphere in equilibrium at the end of an essentially weightless fiber. Knowing that the E-field is uniform and has a value of 200 N/C, we want to determine the charge on the sphere. (a) What is the sign of the charge on the sphere? Explain. (b) What three forces act on the sphere? (c) Draw a free-body diagram of the sphere. (d) Prove that $mg = F_T \cos 12.0°$. (e) Determine the value of the tension in the fiber. (f) Prove that $F_E = F_T \sin 12.0°$. (g) Determine the charge on the sphere.

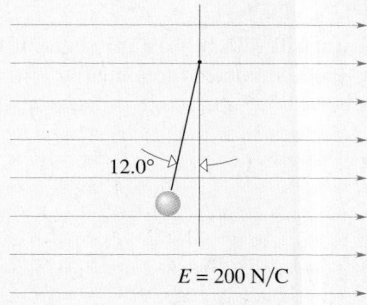

Figure P61

SECTION 15.5: GAUSS'S LAW

62. [I] Considering Fig. P62, which part best represents the magnitude of the electric field arising from a uniformly charged long narrow rod versus the perpendicular distance from its middle? Explain.

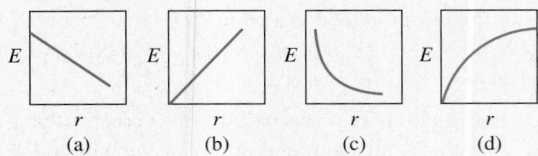

Figure P62

63. [I] THIS PROBLEM EXAMINES THE ELECTRIC FIELD ARISING FROM A UNIFORM SHEET OF CHARGE. Positive charge is sprayed uniformly onto a nonconducting flat horizontal sheet imparting a charge density of 4.0 $\mu C/m^2$. The sheet is then placed in a vacuum chamber, and a tiny charged 0.020-g oil droplet is made to float at rest 1.00 cm above it. (a) In what direction is the electrostatic force on the droplet? (b) What is the polarity of the charge on the droplet? (c) What two forces act on the droplet? (d) Draw a free-body diagram of the droplet. (e) What is the expression for the E-field above the sheet as already derived using Gauss's Law? (f) Compute the charge on the droplet.

64. [I] Determine a formula for the electric field of a point-charge Q in vacuum using Gauss's Law.

65. [I] Using Gauss's Law, determine the electric field in the air gap of a charged parallel-plate capacitor. [*Hint: Use a cylindrical Gaussian surface with one endface embedded in the metal of one of the plates.*]

66. [I] Two flat metal plates of area 2.0 m^2 are placed parallel to each other and both are then charged, one with $+10$ μC and the other with -10 μC. Determine the electric field in the air gap anywhere far from the edges.

67. [I] Write an expression for the external electric field far from the ends of a uniformly charged narrow rod embedded in a medium having a dielectric constant K_e. The expression should be in terms of R, the radial distance away from the rod; L, the length of the rod; Q, the charge on the rod, and K_e.

68. [I] What is the strength in air of the electric field at a radial distance of 1.00-cm from the middle region of a 10.0-m long straight wire carrying a uniform charge of 400.0 μC?

69. [I] In vacuum, there is a vertical (10 m × 10 m) sheet of positive charge corresponding to the xy-plane. If there are 2.2 C of charge uniformly distributed on the sheet, determine the electric field in the space on either side beyond it.

70. [I] There is a vertical (20 m × 10 m) sheet of charge corresponding to the yz-plane immersed in a medium having a relative permittivity of 20. If -4.0 mC of charge is uniformly distributed on the sheet, determine the electric field in the space on either side beyond it.

71. [I] A pair of flat, horizontal parallel aluminum plates 100 cm × 50 cm has a 1.0-mm vacuum gap between them. How should they be charged if there is to be a uniform upward E-field of 1000 N/C in the gap?

72. [I] Two flat metal plates of area 2.0 m^2 are placed parallel to each other and both are then charged, one with $+10$ μC and the other with -10 μC. Determine the electric field inside anywhere far from the edges when the gap is completely filled with mica (use a permittivity of 41×10^{-12} $C^2/N \cdot m^2$). How does the field now compare to the case where the gap was empty?

73. [II] A large thin metal spherical shell of radius R has a small hole through which is introduced a quantity Q of charge that's placed on the inner surface. Write an equation for the electric field inside and outside the sphere. What is its direction? Assume the sphere is in vacuum.

74. [II] Use Gauss's Law to prove that there can be no net charge within a hollow conductor.

75. [II] Use Gauss's Law to find a formula for the electric field very close to any charged conducting surface in vacuum.

76. [II] With the previous problem in mind, compute the surface charge density on a conducting surface in the immediate vicinity of a point (in vacuum) where the electric field due to the charge is 5.0 $\times 10^5$ N/C.

77. [II] Use Gauss's Law to determine a formula for the electric field outside (and far from the ends) of a long, uniformly charged cylinder of radius R with a positive surface charge density of σ.

78. [II] Imagine a long open-ended metal pipe of length L, inner radius R_i, and outer radius R_o. Write expressions for the electric field inside and outside the pipe some time after the inner surface is sprayed with a charge Q. State your assumptions.

79. [II] Consider an idealized uniform spherical shell of charge $+Q$ surrounded at a distance by a concentric idealized shell of uniform charge $-Q$. Determine the E-fields between and beyond the shells. Assume vacuum.

80. [II] Consider the case of a small metal sphere suspended at the center of the cavity within a large hollow metal sphere. The central conductor is electrified with a charge of $+Q$. Write expressions for the E-field both inside the gap and outside the large sphere. Assume vacuum everywhere.

81. [III] A uniform ball of charge (Q) has a radius R. Determine the electric field inside the ball.

82. [III] With Problem 81 in mind, determine a formula for the field outside the sphere and draw a curve of E versus r.

83. [III] If a proton is considered a uniform ball of charge of radius 1.0×10^{-15} m, what is the E-field just beyond its surface?

Chapter 16
Electrostatics: Energy

*A*n understanding of electrostatics is necessary before we study a number of other important topics, from electric currents to atomic theory. This chapter continues the development, elaborating the primary idea of *electric potential*, the concept that corresponds in informal language to the notion of *voltage*. Of course, everybody knows what voltage is—batteries come in 1.5- and 9-volt varieties, and in the United States household wall outlets deliver electricity at a voltage of 110 volts—but what does that really mean?

Electric Potential

Chapter 15 dealt with electric *force* and its more convenient equivalent, electric field. This chapter focuses on *energy* as it relates to electrostatics. Both $\vec{F}$ and $\vec{E}$ are vector quantities and sometimes are a little complicated to deal with. Energy is a scalar, and that will make the analysis a lot simpler.

16.1 Electrical-PE & Potential

It takes a force to raise a mass in the Earth's gravitational field (Fig. 16.1*a*). We do work on the object to overcome the downward force field. And the mass goes up from an initial value of *gravitational*-PE to a higher one. The Coulomb force has the same mathematical form as the gravitational force, and so it, too, must be conservative (p. 186); potential energy is associated with each of these interactions. When a charge is made to move *against* the influence of an electric field, as for instance in Fig. 16.1*b*, it will experience a change in its *electrical*-**PE**. Such motion requires the application of an external force and the expenditure of energy by the agency providing that force (Fig. 16.2).

The opposite occurs when the field does positive work on the charge and propels it to a new lower-energy location. If an electron is released in the vicinity of a positively electrified object, it is drawn toward the plus charges; work is done *by the field* on the electron. As it "descends" in the field, the electron loses ***electrical potential energy*** (PE_E) while it gains speed, and its KE increases equivalently. The electron "falls" to the positive object (or away from a negative one), just as a proton "falls" to a negative object (or away from a positive one), just as a rock falls to the Earth.

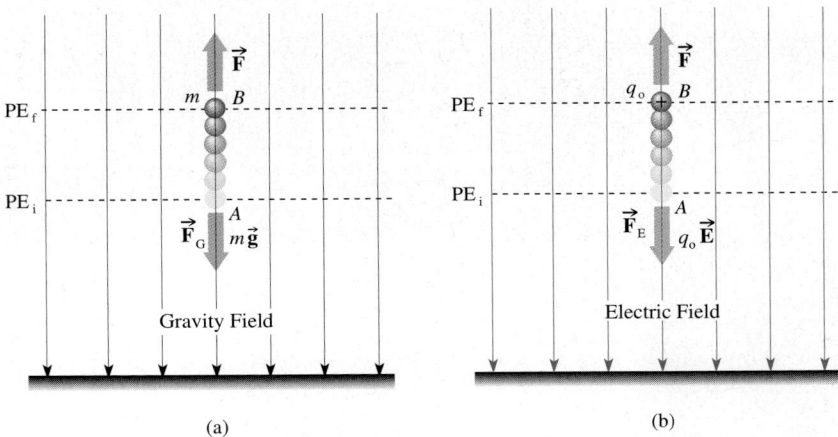

(a)   (b)

Figure 16.1 (a) An object, with a mass m, being raised in a uniform downward gravitational field by a force $\vec{\mathbf{F}}$, experiences a downward gravitational force $\vec{\mathbf{F}}_G = m\vec{\mathbf{g}}$. (b) An object, with a charge q_o, being raised in a uniform downward electric field by a force $\vec{\mathbf{F}}$, experiences a downward electric force $\vec{\mathbf{F}}_E = q_o\vec{\mathbf{E}}$.

A piano held in midair next to a pea has more *gravitational*-PE with respect to the Earth than does its little green companion because it has more mass. Similarly, a highly positively charged pith ball immersed in an *E*-field will experience a greater force and have a larger *electrical*-PE than will a single proton at that point in the field because of its greater charge.

Potential

Suppose we have a charged body of some sort. We'll refer to it as the primary charge because we want to know about *its* influence on the surrounding scene. In Chapter 15, we initially pictured the force field of a primary charge. But that was dependent on the test-charge used to make the survey, so we then eliminated that dependence by dividing out the test-charge (q_o). The result was the $\vec{\mathbf{E}}$-field, which depends only on the primary charge and tells us all about the distribution of force (acting on *any* charge placed in the field) in the surrounding space.

What we want now is an energy measure that is again independent of the test-charge being used for the survey. Accordingly, we first find the *electrical*-PE of a test-charge at all points in space within the field of the primary charge and divide each such value by q_o. This

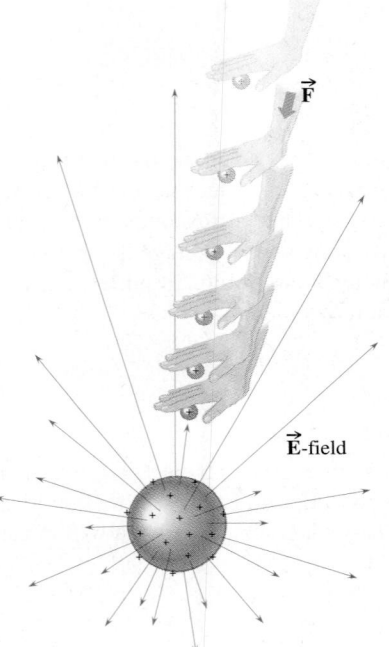

Figure 16.2 Work must be done in moving a charge against an electric field. Here a positive charge is forced toward a positively charged sphere. It's being pushed against the *E*-field.

Smoke particles are removed before they leave the stacks at this plant. The gray structures at the bases of the smoke-stacks are the electrostatic precipitators. (See Fig. Q21.)

The **electrical potential** at any point in an *E*-field is the potential energy of a test charge at that location divided by the value of that test charge.

associates a number (positive or negative) with every point in space and yields a map of the *potential energy per unit charge*. That scalar quantity is known as the **electric potential** (*V*) or just the **potential** for short:

[PE$_E$ per unit charge]   Electric potential $= \dfrac{\textit{electrical}\text{-PE}}{\text{charge}}$ (16.1)

The units of potential are joules per coulomb (J/C), and, in honor of Alessandro Volta, 1 J/C is defined to be 1 volt (1 V). It's common to speak of the electric potential as the *voltage*.

Example 16.1 **[I]** The most commonly used unit of electric field is the volt per meter. Show that this is consistent with the units introduced thus far.

Solution Whenever there's a question about the equivalence of units, go back to the defining equations. (1) TRANSLATION— Show the equivalence between N/C and V/m. (2) GIVEN: N/C. FIND: Show that 1 N/C = 1 V/m. (3) PROBLEM TYPE— Electrostatics / unit conversion. (4) PROCEDURE—The brute force way is simply to convert everything into the most basic form possible in terms of meters, kilograms, seconds, and coulombs. (5) CALCULATION: Beginning with 1 N/C, it follows from $F = ma$ that 1 N = 1 kg·m/s², and so

$$1\,\frac{\text{N}}{\text{C}} = \frac{1\ \text{kg·m/s}^2}{\text{C}}$$

Now to convert 1 V/m into m, kg, s, we use the definition $\Delta V = \Delta\text{PE}_E/q_\text{o}$ to get 1 V = 1 J/C. Recall (from the definition of work) that 1 J = 1 N·m and therefore

$$1\,\frac{\text{V}}{\text{m}} = 1\,\frac{\text{J/C}}{\text{m}} = 1\,\frac{\text{N·m/C}}{\text{m}} = 1\,\frac{\text{N}}{\text{C}} = \frac{1\ \text{kg·m/s}^2}{\text{C}}$$

and the two are, indeed, identical.

Quick Check: We could have stopped one step back and not made the first and last conversions, but we didn't know it was going to work out that way. Alternatively, multiply the top and bottom of 1 N/C by m: 1 N·m/C·m = 1 (J/C)/m = 1 V/m.

{For more worked problems click on **WALK-THROUGHS** *in* **CHAPTER 16** *on the CD.}*

Potential Difference

Imagine an electric field in space (e.g., the one depicted in Fig. 16.1*b*). A positive charge is transported from point *A* to point *B* as a result of an applied force $\vec{\mathbf{F}}$. Here the force pulls the charge against the field. It does a positive amount of work $W(A{\rightarrow}B)$ and increases the charge's potential energy. Were it to be released at *B*, the positive charge would spontaneously sail back in the direction from which it came (much as a mass rolls down a mountain) and if the field were uniform, it would return to *A*. {For an explanation of why free charges don't generally move along field lines click on **MOTION IN AN *E*-FIELD** under **FURTHER DISCUSSIONS** on the CD.} Clearly, the PE$_E$ at *B* is greater than at *A*. Since the potential *V* is the potential energy per unit charge, the potential at *B*, namely, V_B, is greater than the potential at *A*, namely V_A.

The quantity ΔV is referred to as the **difference in potential** (or the **potential difference**) between the initial (*A*) and final (*B*) points in the field and is often written as

$$\Delta V = V_B - V_A \qquad (16.2)$$

The potential difference between any two points A and B numerically equals the work done per unit charge, against the field in moving a positive test charge q_o from A to B with no acceleration:

[positive charge] $V_B - V_A = \dfrac{W(A{\rightarrow}B)}{q_\text{o}}$ (16.3)

(see Table 16.1). In general, the work done can be greater than zero, less than zero, or zero; the same is true for ΔV. By definition

$$\Delta V = \frac{\Delta\text{PE}_E}{q_\text{o}} \qquad (16.4)$$

Table 16.1

Common Potential Differences	
Biochemical	1 mV − 100 mV
Dry cell	1.5 V
Automobile battery	12 V
Household electricity	
U.S.A.	110 V − 120 V
Much of Europe	
and Asia	240 V − 250 V
Voltage induced in	
600-mile Alaska pipeline	
by solar storm	1000 V
Power-plant generator	24 000 V
Transmission lines	
Local	4 400 V
Cross country	120 000 V
Extra high voltage	500 kV − 1 MV
Van de Graaff	
generator (1940)	4.5 MV
Folded tandem generator	25 MV
Lightning	10^8 V − 10^9 V

Example 16.2 **[I]** A small sphere carrying a positive charge of 10.0 μC is moved against an E-field through a potential difference of +12.0 V. How much work was done by the applied force in raising the potential of the sphere?

Solution

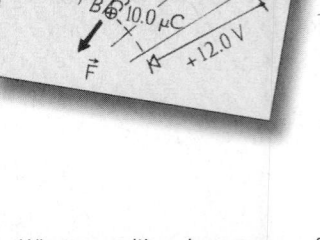

This example is about potential, and its definition in terms of work. (1) TRANSLATION—A known charge is moved through a specified potential difference; determine the work done. (2) GIVEN: q_o = 10.0 μC and ΔV = +12.0 V. FIND: W. (3) PROBLEM TYPE—Electrostatics/potential difference. (4) PROCEDURE—The defining relationship between ΔV, q_o, and W is Eq. (16.3). (5) CALCULATION—$(V_B - V_A) = W(A{\to}B)/q_o$ and so

$$W = \Delta V q_o = (+12.0\ \text{V})(10.0 \times 10^{-6}\ \text{C}) = \boxed{120\ \mu\text{J}}$$

Quick Check: Use the above answer to calculate one of the givens; $q_o = W/\Delta V = (120\ \mu\text{J})/(12\ \text{V}) = 10\ \mu\text{C}$.

Thus, for example, if +1.0 C of charge moves within an electric field from a place where the potential is higher to one where it is lower, and in the process drops in voltage by ΔV = 1.0 V, there will be a corresponding reduction in the charge's PE_E of 1.0 J.

Remember that energy is a relative quantity. There is no signpost floating in space that says, "This is the absolute zero of PE." Only changes or differences in *electrical*-PE can be determined, and therefore *only differences in potential are important* (Table 16.1). A positive charge spontaneously descends from a high potential (V_A) to a low potential (V_B), losing PE_E and gaining KE. Thus, the *change in potential ($\Delta V < 0$) equals the negative of the work done per unit charge by the field on the positive charge.* **Work done against the field always increases PE; work done by the field always decreases the PE.** As a rule, the change in PE equals the negative of the work done *by* a conservative force field.

> When a positive charge moves from a high potential to a low potential, it drops in potential energy. When released, such a charge will spontaneously travel in the field in such a way as to descend in potential.

Potential was defined in terms of positive charge and the idea seems counterintuitive when applied to negative charge. Naturally enough, a negative charge released in an E-field moves spontaneously, experiencing a drop in *electrical*-PE. But in doing so it travels from a low potential to a higher potential ($\Delta V > 0$). That's evident from Eq. (16.4), where ΔPE_E is negative and q_o is negative, so that ΔV is positive. When a negative charge moves from a point of high potential in a field to a point of low potential, it *increases* its PE_E.

The work done on a charge in moving it from point A to point B is independent of the path taken, just as the work done in climbing a mountain is independent of the path. If it were not, we could make a perpetual-motion machine, going from A to B along a small-change-in-potential path and coming back along a larger one, thereby gaining a bit of energy on each cycle.

When a particle (Fig. 16.3) with a charge $+q_e$ moves in a field, dropping 1 volt in potential, it will thereby decrease its *electrical*-PE by $\Delta PE_E = q_e \Delta V = 1.6 \times 10^{-19}$ J. Because many phenomena on the atomic level involve energies of about this amount, a new measure of energy was introduced called the **electron volt** (eV):

> An **electron volt** is the amount of energy corresponding to an electron falling through a potential difference of one volt.

$$1\ \text{eV} = 1.6 \times 10^{-19}\ \text{J}$$

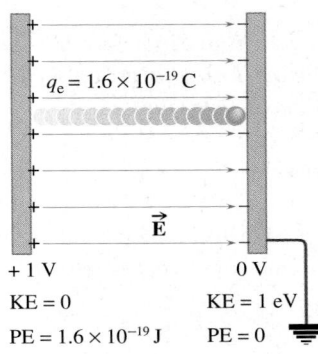

$q_e = 1.6 \times 10^{-19}$ C

$\vec{E}$

+1 V   0 V
KE = 0   KE = 1 eV
PE = 1.6×10^{-19} J   PE = 0

Figure 16.3 A proton falling through a potential difference of 1 volt. The proton follows the E-field, being pushed along by the field, and loses an amount of potential energy equal to 1 unit of charge times 1 volt, or 1 eV.

Example 16.3 **[II]** Figure 16.4 depicts the basic elements of a cathode-ray tube, the kind found in computer terminals and oscilloscopes. Electrons are "boiled" off a heated cathode and emerge through a pinhole, being drawn toward the first anode, which is at a relatively small positive potential above the cathode. A second anode that is 8000 V to 20 000 V above the cathode (depending on the design of the tube) accelerates the beam up to speed. Determine the change in the potential energy of each electron upon traversing the gun, given that the second anode has a voltage of 20 kV above the cathode. Assuming that an electron has negligible motion at the cathode, find its final speed.

Solution This problem is about potential, and its definition in terms of potential energy. (1) TRANSLATION—A known charge (with a known mass) moves through a specified potential difference; determine the change in potential energy and the final speed. (2) GIVEN: $v_i = 0$ and $\Delta V = 20$ kV. FIND: ΔPE_E and v_f. (3) PROBLEM TYPE—Electrostatics / potential difference / energy. (4) PROCEDURE—The change in PE_E goes into KE from which we can find v_f. (5) CALCULATION—Any one electron traverses a potential difference of 20×10^3 V and drops in potential energy by

$$\Delta PE_E = q_e \Delta V$$

This result is true regardless of the details of the field existing within the gun. Remember that a negative charge loses PE_E as it *increases* in potential (however weird that may sound)

$$\Delta PE_E = (-1.6 \times 10^{-19}\,\text{C})(20\,000\,\text{V}) = \boxed{-3.2 \times 10^{-15}\,\text{J}}$$

The electron's energy is conserved; therefore,

$$KE_i + PE_i = KE_f + PE_f$$

and since $KE_i = 0$

$$KE_f = PE_i - PE_f = -\Delta PE$$

and

$$\tfrac{1}{2} m_e v_f^2 = -q_e \Delta V$$

Consequently

$$v_f = \left[\frac{-2q_e\Delta V}{m_e}\right]^{\frac{1}{2}} = \left[\frac{-2(-3.2 \times 10^{-15}\,\text{J})}{9.1 \times 10^{-31}\,\text{kg}}\right]^{\frac{1}{2}}$$

$$\boxed{v_f = 8.4 \times 10^7\,\text{m/s}}$$

Quick Check: The high voltage and tiny mass reasonably result in a tremendous speed of $\approx \frac{1}{4}c$. Note that the electron experiences an energy drop of 20 keV = 3.2×10^{-15} J = ΔKE.

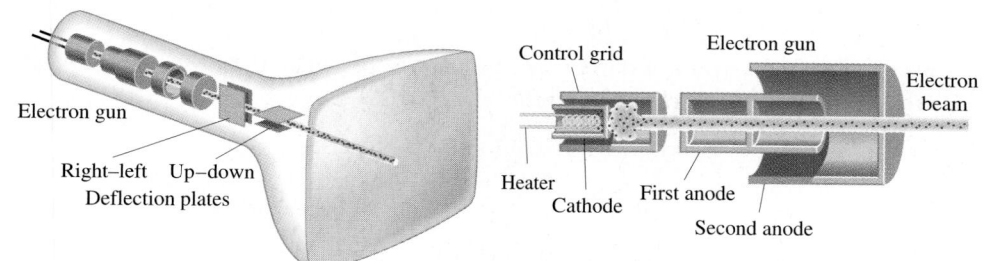

Electron gun

Right–left Up–down
Deflection plates

Control grid

Electron gun

Electron beam

Heater
Cathode First anode

Second anode

Figure 16.4 A cathode-ray tube (see Fig. Q19). A beam of electrons passes between two perpendicular sets of plates. These are appropriately charged, and the beam is deflected to any desired point on the face of the tube.

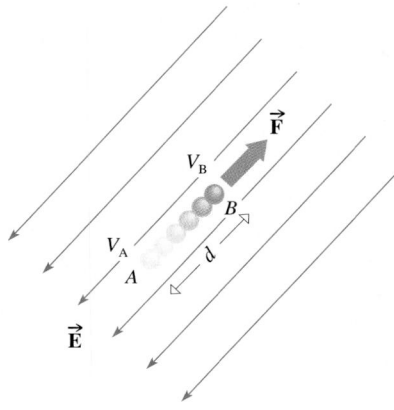

Figure 16.5 A positive charge moved against a uniform E-field by an applied force $\vec{F}$. The charge is transferred from point A to point B a distance d.

Potential in a Uniform Field

Let's now derive an expression for the potential difference between two points in a **uniform electric field**. The virtues of such a field are that its lines are parallel and straight as shown in Fig. 16.5. There an applied force F carries a positive charge q_0 from rest at A to rest at B a distance d against the field. The work done is $W(A{\to}B) = Fd$ where the applied force, in order to just overcome the field, must equal the electric force due to the field, namely, $F = q_0 E$. Therefore,

$$W(A{\to}B) = q_0 Ed$$

But the work done in moving against the field stores energy in the amount $W(A{\to}B) = \Delta PE_E$ and therefore

$$\Delta PE_E = q_0 Ed$$

By definition

$$V_B - V_A = \Delta PE_E / q_0$$

The man straddling the 138 000-V power line landed there in a parasail accident. The brief surge of charge that brought him up to 138 kV burned his hands and feet, but otherwise he was fine.

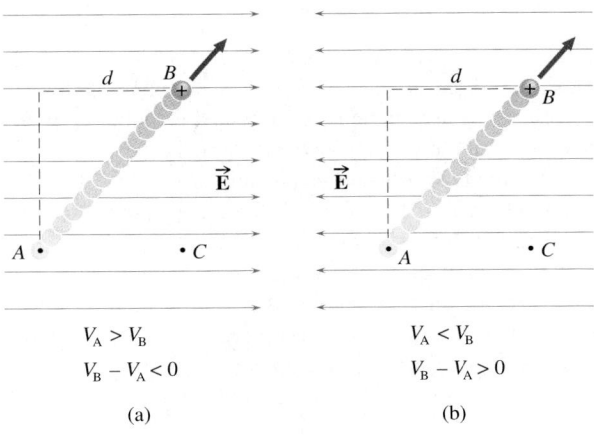

$$V_A > V_B$$
$$V_B - V_A < 0$$
(a)

$$V_A < V_B$$
$$V_B - V_A > 0$$
(b)

Figure 16.6 A positive charge moving diagonally across a uniform electric field. In (a) there is a component of displacement parallel to the field, and the potential drops in going from A to B. In (b) there is a component of displacement against the field, and the potential rises in going from A to B.

In general, for displacements either parallel or antiparallel to a uniform field

[uniform field]
$$V_B - V_A = \pm Ed \qquad (16.5)$$

The potential difference is $+$ (a voltage rise) when the displacement is opposite to the field and $-$ (a voltage drop) when it is parallel to the field. For a charge displaced perpendicular to the field, no electric force acts on it and no work is done by or against the E-field. That means that *in going from A to B along any path the potential difference only depends on the displacement parallel or antiparallel to the field*.

Observe in Fig. 16.6a that

$$V_C - V_A = -Ed$$

and so

$$V_B - V_A = V_C - V_A$$

Points B and C are at the same potential. Another way to see this is to realize that if the points B and C are such that there is a path from one to the other that is everywhere perpendicular to the field, there is no component of $\vec{E}$ parallel to the displacement and therefore no component of $\vec{F}_E$ parallel to the displacement, and consequently $W(A \rightarrow B) = 0$. No work is done by or against the field in moving from B to C, which is the case gravitationally when a mass is moved horizontally on the Earth's surface. Since $\Delta W = q\Delta V$, $\Delta V = 0$, and the two points are at the same potential.

The simplest situation, that of moving a positive charge along (or opposite to) a uniform E-field, is depicted in Fig. 16.7. Then $V_B - V_A = +Ed$, while $V_A - V_B = -Ed$. There are a number of devices—such as batteries, generators, fuel cells, solar cells, and thermoelectric cells—that can provide electrical energy and sustain a fairly constant potential difference across their two output terminals. In Fig. 16.7, **the positive terminal of the battery is 12 V higher than the negative terminal**. These terminals are connected to the parallel plates by excellent conductors, along which it is assumed there will be no appreciable loss in electrical energy and no measurable decrease in potential. That's a little unrealistic, but we will get back to that point in Chapter 17 when resistance is considered. This battery will maintain a constant 12-V difference across the parallel plates and sustain a constant field in the gap. {For more click on **POTENTIAL** under **FURTHER DISCUSSIONS** on the CD.}

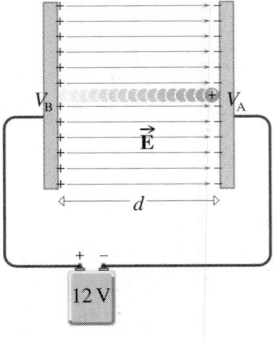

Figure 16.7 A battery connected across two parallel metal plates. There is a uniform E-field between the plates, and a positive charge experiences a drop in potential $V_A - V_B = -Ed$ upon traveling from the left plate to the right plate.

Example 16.4 **[I]** If the separation of the plates in Fig. 16.7 is 2.0 mm, determine the magnitude of the electric field in the air gap.

Solution You should immediately be thinking about the *E*-field inside a parallel-plate capacitor. (1) TRANSLATION—A known displacement in a uniform *E*-field corresponds to a specified voltage difference; determine the field. (2) GIVEN: $d = 2.0$ mm and $V = 12$ V. FIND: *E*. (3) PROBLEM TYPE—Electrostatics/potential difference/uniform field/parallel plate capacitor. (4) PROCEDURE—The field is uniform; we have *d* and *V* and want *E*. That suggests Eq. (16.5). (5)

CALCULATION—The plus side is at a higher potential than the minus side: here $V_B > V_A$ by 12 V and so

$$V_B - V_A = +Ed \qquad [16.5]$$

$$12 \text{ V} = +E(2.0 \times 10^{-3}\text{m})$$

Therefore

$$\boxed{E = 6.0 \text{ kV/m}}$$

Quick Check: This is a fair-sized field (Table 15.2), but that's reasonable since the spacing over which the voltage drops 12 V is only 2 mm. That corresponds to a drop of 6 V/mm, or 6 kV/m.

16.2 Potential of a Point-Charge

Earlier, we calculated the *E*-field of a point-charge and saw how that could be used to determine the fields for all sorts of charge distributions. The same is true for the potential; that is, once we have the potential of a point-charge, we can use it to determine the potentials of other more complicated systems. To that end, return to p. 184 where we discussed the *gravitational*-PE of a test-mass *m* in the field of a spherical mass *M*. Because the gravitational field of a point-mass and the electric field of a point-charge both vary inversely with r^2, finding the corresponding change in PE is a bit complicated. The change in PE is the work done, which is the force times the displacement, but the force is now a variable. Still, the mathematical problem was solved {for the analysis click on **GRAVITATIONAL-PE** under **FURTHER DISCUSSIONS** on the CD 🔘 } and it led to

$$\Delta\text{PE}_\text{G} = GmM\left(\frac{1}{R} - \frac{1}{r}\right)$$

This is the increase in *gravitational*-PE that occurs when a mass *m* is moved from *R out* to *r against* the *attractive* $1/r^2$-gravitational force produced by a sphere of mass *M*. It follows that

$$[r_A \text{ to } r_B] \qquad \Delta\text{PE}_\text{E} = k_0 q_0 q_. \left(\frac{1}{r_B} - \frac{1}{r_A}\right)$$

This gives us the change in *electrical*-PE that occurs when a positive test-charge q_0 is moved from r_A *inward* to r_B ***against*** the *repulsive* $1/r^2$-electric field produced by a point-charge (or tiny sphere) of positive charge $q_.$, as shown in Fig. 16.8. In that case $r_A > r_B$ and $\Delta\text{PE}_\text{E} > 0$. Similarly, when a test-charge q_0 is moved from r_A out to r_B by the repulsive $1/r^2$-electric field, $r_A < r_B$ and $\Delta\text{PE}_\text{E} < 0$. Inasmuch as

$$\Delta V = \Delta\text{PE}_\text{E}/q_0$$

$$[r_A \text{ to } r_B] \qquad V_B - V_A = k_0 q_. \left(\frac{1}{r_B} - \frac{1}{r_A}\right) \qquad (16.6)$$

This result is the sought-after potential difference encountered in moving from point-*A* to point-*B* in the electric field of a tiny charged sphere or, equivalently, a point-charge $q_.$, which is either positive or negative.

Again, as we saw in the case of gravity, it is often useful to take the zero of potential, which is totally arbitrary, at infinity (where the force is zero). To that end, let $r_A \to \infty$,

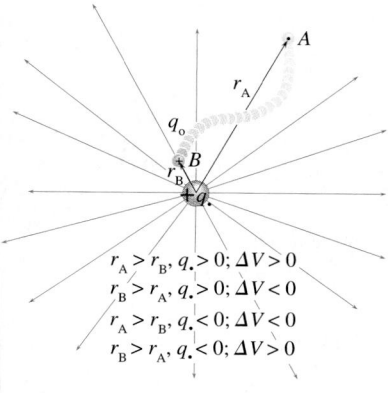

$$r_A > r_B, q_. > 0; \Delta V > 0$$
$$r_B > r_A, q_. > 0; \Delta V < 0$$
$$r_A > r_B, q_. < 0; \Delta V < 0$$
$$r_B > r_A, q_. < 0; \Delta V > 0$$

Figure 16.8 Going from *A* to *B*. Here, a positive charge q_0 is moved inward toward *Q*, going from a distance r_A to a distance r_B. The potential difference $(V_B - V_A)$ is positive because q_0 is pushed downward against the field. Its PE$_\text{E}$ increases and, if released, q_0 will be pushed out and away.

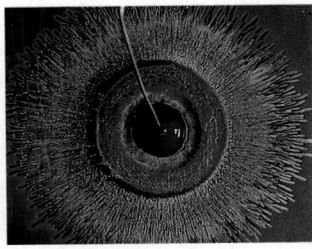 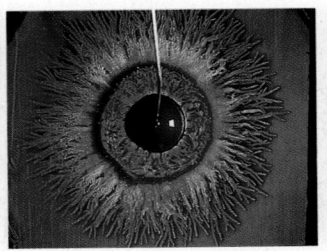 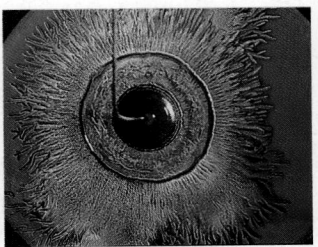

Three samples of blood in a radial electric field. An electrosensitive liquid crystal is spread over a microscope slide. A drop of blood is placed at the center and grounded via a vertical wire. The liquid crystal colors correspond to different voltages (blue ≈150 V, green ≈120 V, yellow ≈90 V, and red ≈50 V) dropping to zero at the center. (a) normal blood, (b) blood from osteosarcoma patient, and (c) blood from leukemia patient.

whereupon $1/r_A \to 0$ and

[∞ to r_B]

$$V_B - 0 = k_0 q \cdot \left(\frac{1}{r_B} - 0 \right)$$

represents the electric potential at the finite point B with respect to zero at infinity. To simplify matters, we might as well drop the subscript B since the equation applies to any point in the region beyond the charged sphere. Accordingly,

[$V = 0$ at ∞]

$$V = \frac{k_0 q \cdot}{r} \qquad (16.7)$$

> The **potential** at some radial distance r from a point-charge q varies directly with q and inversely with r.

> In theoretical work, the **potential at infinity** is usually taken to be zero.

▶ This is the electric potential at a distance r from an isolated point-charge q measured with respect to the zero at ∞. Alternatively, we can say that this is *the work done per unit charge against or by the E-field of a point-charge q in bringing a test-charge from infinity to a distance r from q*. The potential at a point can be positive, negative, or zero depending on the source q, and, of course, it is a directionless quantity, a scalar. Figure 16.9 shows the PE_E of a point-charge placed in the $\vec{E}$-field of a point-charge; it's significant because it corresponds to the electron-proton interaction of a hydrogen atom.

Potential varies continuously as a function of position. Discontinuities appear only at charges, where $V = \infty$. Therefore, nearby a positive (or negative) charge V is large and positive (or negative). Since, as a rule, we take $V = 0$ at ∞, it must approach zero with distance from the charge. Even so, in the neighborhood of several charges V can get rather complicated (p. 580).

Example 16.5 **[I]** What is the potential difference encountered in going from close in (point-B) to far out (point-A) in Fig. 16.8, if the tiny sphere carries a charge of $Q = +10\ \mu C$, $r_A = 20$ cm, and $r_B = 10$ cm?

Solution This problem is about the potential of a charged sphere. (1) TRANSLATION—Determine the potential difference between two specified points in the field of a known point-charge. (2) GIVEN: $Q = +10\ \mu C$, $r_A = 20$ cm, and $r_B = 10$ cm. FIND: ΔV. (3) PROBLEM TYPE—Electrostatics/potential difference/point-charge. (4) PROCEDURE—The potential difference due to a point-charge is given by Eq. (16.6), and since $Q > 0$ and $r_A > r_B$, we can expect ΔV to be a voltage drop. (5) CALCULATION—Going from point-B to point-A, a positive test-

charge moves out along the field and so

$$V_A - V_B = -(V_B - V_A) = -k_0 Q \left(\frac{1}{r_B} - \frac{1}{r_A} \right)$$

$$V_A - V_B = -(8.99 \times 10^9\ \text{N} \cdot \text{m}^2/\text{C}^2)(+10 \times 10^{-6}\ \text{C})$$
$$\times (10\ \text{m}^{-1} - 5.0\ \text{m}^{-1})$$

and

$$\boxed{V_A - V_B = -0.45\ \text{MV}}$$

Quick Check: From Eq. (16.7), the potentials at distances of r_B and r_A are $+0.90$ MV and $+0.45$ MV, respectively. Hence, going from point-B to point-A, there is a change in potential of -0.45 MV.

Figure 16.9 Different ways to represent potential. (a) Field lines and equipotentials for a small positive charge. (b) The equipotential surfaces are concentric spheres centered on the charge. (c) A continuous color representation of the potential. The different bands of color mark the gradual diminution of the potential as we move out from the charge. (d) Here the height of the peak corresponds to the size of the potential above the zero base level.

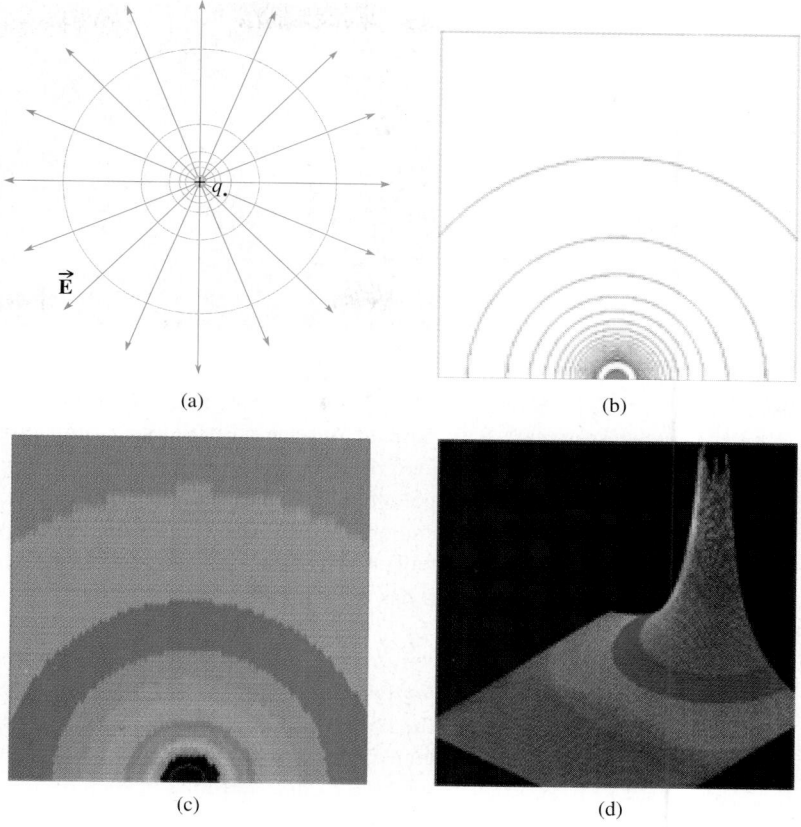

(a)   (b)

(c)   (d)

Equipotentials

Return to Fig. 16.6 and observe that we can move a positive test-charge from point-*B* to point-*C* so that the displacement is everywhere perpendicular to the uniform *E*-field and no work is done. The potential does not change in the process, and we say that *the line from B to C, which is everywhere perpendicular to the field, is an equipotential line*—a line along which the potential energy of a test-charge remains unchanged.

Envision a positive point-charge or, equivalently, a tiny charged conducting sphere (Fig. 16.9). The potential at any fixed distance beyond the charge is given by

The **potential** at a distance *r*, due to a spherically symmetrical distribution of charge *q* (having a radius *R*) is equal to the potential of a point-charge provided *r* > *R*.

[point-charge] $$V = \frac{k_0 q}{r}$$ [16.7]

which is constant in all directions. In other words, the **equipotential surfaces** of a spherically symmetric charge distribution are a series of concentric nested spheres, everywhere perpendicular to the *E*-field—this is a crucial point. Each one of the ever-larger sequence of circles (Fig. 16.9*b*) represents a change (here a drop) of a specific fixed voltage as we move away from the positive charge (the precise value of this change depends on the charge). The fact that there are many lines crowded together near the charge shows that the potential falls off quickly in that region. Another way to see this is to plot a map where potential corresponds to color, dropping from its highest value at red to orange, yellow, green, across the spectrum to the lowest voltage at purple (Fig. 16.9*c*). Remembering that we are talking about a spherically symmetric potential, it's helpful to represent it via a so-called wire-grid diagram (Fig. 16.9*d*) that depicts the value of *V* plotted above (+) and below (−) the horizontal plane.

These ideas should bring to mind the concept of a conductor. Since we found that the electric field in the vicinity of the surface is always perpendicular to a conductor regardless of its shape, *the conductor's surface must also be an equipotential*. That's true whether the conductor is charged or not. Another way to appreciate this is to realize that there is no

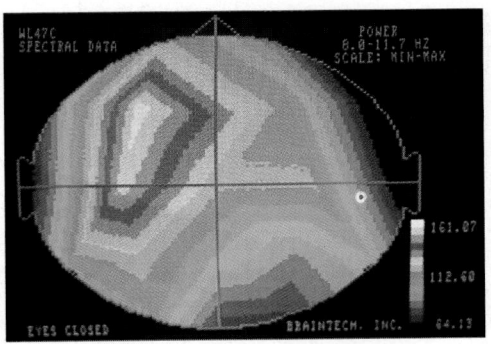

A map of the equipotentials in the brain of a person with epilepsy. This picture was made about 0.1 s after the person received a stimulus.

E-field within the substance of a conductor in electrostatic equilibrium. A test-charge transported from one point within a conductor to another will not have to be moved against an electric force, will not have work done on it, and will therefore experience no change in potential. By analogy, although a ball will roll down a hill by itself, if it were placed on the smooth surface of an idealized spherical planet, it would not spontaneously move at all. The ball, by itself, can only drop to a lower *gravitational*-PE, and yet everywhere on the surface it has the same PE.

The entire body of a conductor, devoid as it is of any *E*-field, must be an **equipotential volume**. Indeed, provided there are no encompassed isolated charges (as in Fig. 15.37), *the total region contained within a conductor, hollow or not, is at the same potential*. Because the Earth itself is a conductor, its surface is an equipotential and, by custom, it is often tak-

en to be the zero of potential. That's just what is done in the electrical system in your home, where the "hot" terminal in each of the outlets oscillates between ±120 V with respect to the zero-potential terminal. The latter is physically connected to ground (p. 535).

Figure 16.10 shows both the *E*-field lines and the associated equipotentials for several configurations. In each case, the equipotentials are drawn successively at fixed-voltage intervals, Δ*V*. When the field is uniform, as in Fig. 16.11, the equipotentials are evenly spaced. The inclusion pictorially of equipotentials beautifully complements the field-line diagrams. A drawing of field lines immediately reveals the direction in which a charge will experience a force when placed anywhere in the region, and that corresponds to the direction in which it will accelerate as well. Moreover, if the field is uniform, the charge will move along a field line. A glance at a diagram containing equipotentials reveals the energy change that will occur when a charge moves from one point to another along *any* path.

Out of doors, in the highly charged environment of a thunderstorm, the hair on a person's head can become charged by induction. Once charged, the individual strands repel and separate from each other like the leaves of an electroscope (see photo on p. 536). If you ever feel your skin tingle or hair stand on end, squat as low to the ground as you can with your hands on your knees and your head down. As they smiled for this photo, these two brothers didn't recognize the tremendous danger they were in. A moment later they and their sister were killed by a bolt of lightning.

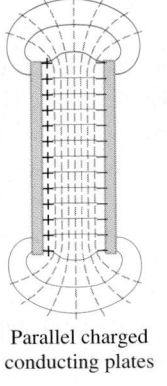

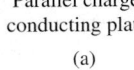

Parallel charged conducting plates

(a)

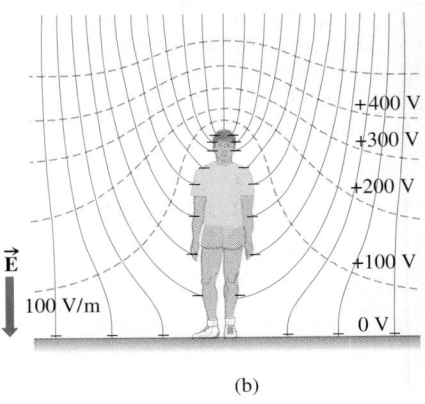

+400 V
+300 V
+200 V
+100 V
0 V

$\vec{E}$
100 V/m

(b)

Figure 16.10 Electric fields and equipotentials. (a) A pair of charged conducting parallel plates. The person in (b) is grounded and so is at zero potential.

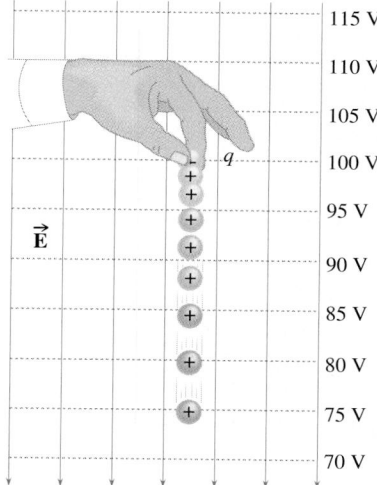

Figure 16.11 After falling through a potential difference of 25 V, the charge has lost an amount of potential energy (25 V)q equal to its gain in kinetic energy.

Together, the field lines and equipotentials provide a complete picture of the influence of the primary charge distribution everywhere in space. {Go to the **CD** and under **INTERACTIVE EXPLORATIONS** click on **MOTION OF A CHARGED PARTICLE IN AN ELECTRIC FIELD**.}

16.3 The Potential of Several Charges

Potential is a scalar quantity. If several charges are present, the sum of their superimposed potentials anywhere in the surrounding space is equal to the algebraic sum of the individual contributions. We can make use of this fact provided we stick with point-charges because we already have, via Eq. (16.7), the potential function for a point-charge. *If there are two or more point-charges, the net potential at any location will be the scalar sum of the potentials at that location due to each charge.* It's possible that at some point in space a positive potential due to one charge can be canceled by a negative potential due to another charge. If we imagine a positive test-charge brought to that zero-potential point from infinity, in the process it will be repelled by one charge as much as it is attracted by the other, and no net work will be done by or against the field. **There can be a net $\vec{E}$-field at a point, even though the potential at that point is zero** (*e.g., at the very center of Fig. 15.29a*). Conversely, **there can be a nonzero potential at a point where the net field is zero** (*e.g., at the very center of Fig. 15.27a p. 550*).

Figure 16.12a shows the fields and potentials for two equal charges. When both charges are positive, the field lines emerge from the system, and as we move out along them, the positive potential drops off, approaching zero at infinity (Fig. 16.12b and c). At the very center, there is a finite positive potential (Fig. 16.12d). Note how the equipotentials combine into a single surface that becomes increasingly more like a sphere; from very far away, the system looks like a single positive charge.

By contrast, in the case of the dipole (Fig. 16.13a), the field extends out the positive side and in the negative side. The equipotentials surrounding the positive charge are posi-

Figure 16.12 (a) Two equal positive point-charges seen from a distance of several times their separation. Shown are the field lines and equipotentials in a plane containing the two charges. (b) Here we see the equipotentials in a bit more detail, closer in. (c) The same information about the potential, this time plotted using regions of color. (The saw-tooth edges are a computer artifact.) (d) A wire-frame plot of the potential existing in the plane depicted in (b). Remember that in three dimensions the equipotential surfaces near the charges are (positive) spheres. A little farther away, the surfaces encompassing both charges begin as elongated peanut-shell shapes. These become concentric spheres far from the charges.

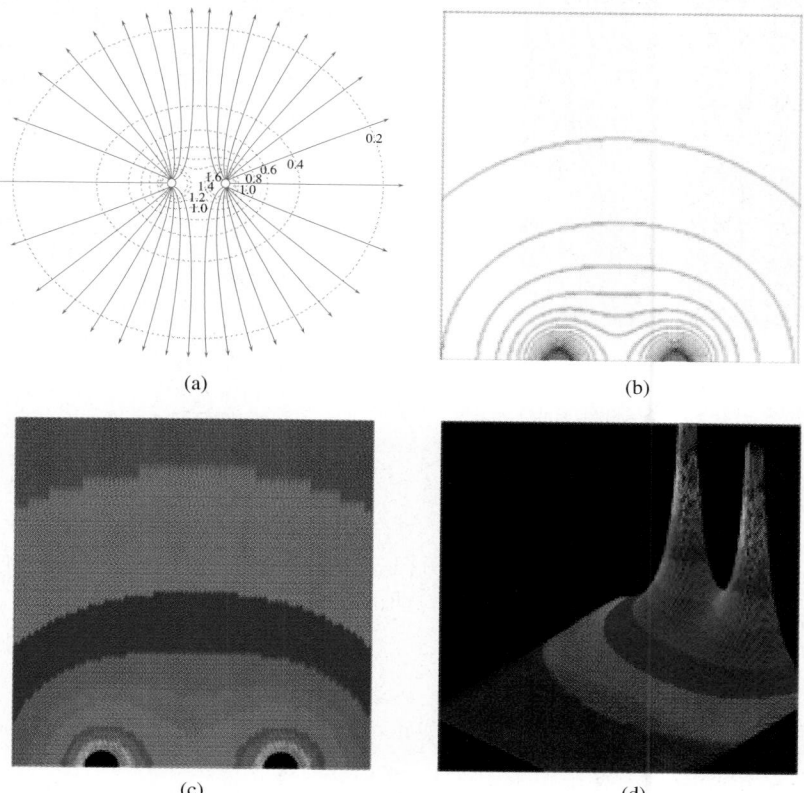

(a)

(b)

(c)

(d)

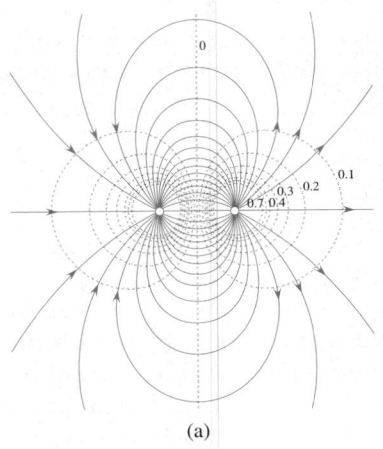

(a)

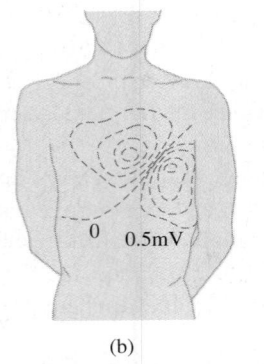

(b)

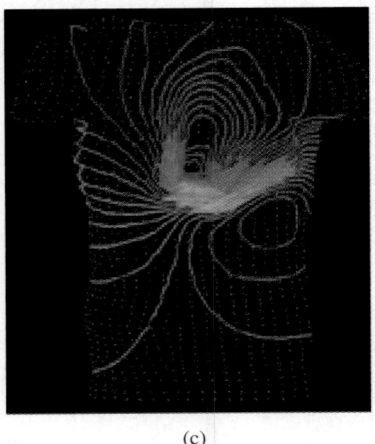

(c)

Figure 16.13 (a) Equipotentials and field lines for two opposite charges of equal magnitude (a dipole). (b) Equipotentials (in millivolts) at some instant across the chest due to heart activity. (c) An actual measurement of those equipotentials (blue, most negative, to red, most positive) on the surface of the skin.

A complex distribution of charges results in a complex pattern of field lines and equipotentials. See if you can figure out what's happening here.

tive, and those surrounding the negative charge are negative. There is a zero-potential plane down the middle to which the field is everywhere perpendicular. A test-charge can be brought from infinity by traveling within the plane and therefore along a path perpendicular to the field. No work is done, and no change from zero occurs in the potential. Of course, since the *E*-field is conservative, we can actually come from infinity to any point on the plane via any path, and the change in potential will still be zero. Only when the charges are equal and opposite will the $V = 0$ equipotential be a plane extending out to infinity. Otherwise, if there are two unequal charges, it closes around the smaller one—watch for that in the diagrams to follow.

{It's recommended that you now click on **ELECTRIC FIELDS OF POINT-CHARGES** under **INTERACTIVE EXPLORATIONS** in **CHAPTER 16** on the **CD**. This simulation will allow you to study the potentials of a few simple point-charge configurations. Do spend some time working with this piece; it's a great way to see how everything we've talked about fits together.}

ELECTROCARDIOGRAPHY

In the course of its normal functioning, a muscle produces electrical potential differences. In the case of the heart, these potential differences are so great that they result in measurable voltages at the skin (of up to ≈1 mV) that are easily monitored via external electrodes in a process called electrocardiography (ECG). The proper operation of the heart depends on the generation and propagation of voltage pulses that are characteristic of the various phases of the pumping cycle, and the ECG monitors those processes. In the basic arrangement, wires are attached to the right (RA) and left (LA) arms and the left leg (LL). The potential differences between RA and LA, RA and LL, LA and LL are all recorded. The typical output is a trace of time versus voltage difference, and its main features are labeled P, Q, R, S, and T. The P region reveals the activity of the atria, while the QRS complex shows the functioning of the ventricles. The heart relaxes, and the T portion marks the preparation of the ventricles for the next cycle.

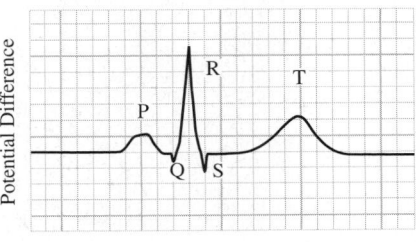

One complete cardiac cycle of a normal heart recorded by an electro-cardiograph.

Example 16.6 **[II]** Figure 15.26, p. 549, shows two charges each of $+10$ nC at $(0, 0)$ and $(8.0, 0)$. Determine the net potential at point $(4.0, 0)$. Then let one of the charges be -10 nC and recompute the potential at that same point.

Solution This example is about the potential produced by point-charges. (1) TRANSLATION—Two known charges are located at known positions; determine the potential at a specified point in space. (2) GIVEN: $Q_1 = Q_2 = \pm 10$ nC at $(0, 0)$ and $(8.0, 0)$, respectively. FIND: V at point A, $(4.0, 0)$. (3) PROBLEM TYPE—Electrostatics/potential/point-charges. (4) PROCEDURE—The potentials of every point-charge [as given by Eq. (16.7)] add algebraically. (5) CALCULATION—For each charge, there is a contribution given by $V = k_0 Q/r$, and so at point A

$$V_A = \frac{k_0 Q_1}{r_1} + \frac{k Q_2}{r_2}$$

$$V_A = \frac{(8.99 \times 10^9 \text{ N} \cdot \text{m}^2/\text{C}^2)(+10 \times 10^{-9} \text{ C})}{4.0 \text{ m}}$$

$$+ \frac{(8.99 \times 10^9 \text{ N} \cdot \text{m}^2/\text{C}^2)(+10 \times 10^{-9} \text{ C})}{4.0 \text{ m}}$$

$$\boxed{V_A = 45 \text{ V}}$$

Although the E-field is zero at that point (see Fig. 15.27), the potential is not; it is positive. Work must be done against the fields of both charges to haul a test-charge in from infinity. With either charge negative, we have a dipole and

$$V_A = \frac{(8.99 \times 10^9 \text{ N} \cdot \text{m}^2/\text{C}^2)(+10 \times 10^{-9} \text{ C})}{4.0 \text{ m}}$$

$$+ \frac{(8.99 \times 10^9 \text{ N} \cdot \text{m}^2/\text{C}^2)(-10 \times 10^{-9} \text{ C})}{4.0 \text{ m}}$$

$$\boxed{V_A = 0 \text{ V}}$$

Quick Check: In the field of a dipole, as a test-charge is moved in from infinity, the positive charge repels it, and work must be done to overcome that repulsion. At the same time, the negative charge attracts the test-charge, doing negative work on it. The result of the push-pull is that no net work needs to be done on the test-charge to bring it to A, and the potential there is the same as it was at the start of the journey at infinity— namely, zero.

Potential and the Distribution of Charge

In 1672, Otto von Guericke devised a machine that greatly reduced the effort it took to build up charge by rubbing. He produced a large sphere of sulfur (because that material was easily electrified) and mounted it on a crank-driven mechanism that whirled it around so it could be stroked as it spun. Spinning within his cupped hand, it built up sizable quantities of "electric virtue" that sparked away impressively. Before long, all sorts of revolving rubbing machines were devised, and everything in sight from milk buckets to chickens was being charged by these *electrostatic generators*.

Imagine von Guericke's sulfur sphere. As it becomes increasingly more charged, it takes increasingly more work to charge it further. There is a relationship between the charge, the geometry, and the work needed to bring yet another electron to the sphere. With a given charge Q distributed over a *large* sphere, the net force exerted on a new incoming electron (since the Coulomb force drops off as $1/r^2$) is smaller than if that same Q was on a tiny sphere. After all, on a large sphere, the individual charges that constitute Q are far apart from one another and from the new charge. **The potential of a charged sphere increases, both as its radius decreases and as its net charge increases.** The more tightly packed the surface charge, the more externally supplied work must be done against it, and the higher the potential. We already know that for a charged sphere of radius R, the voltage at its surface is

[a charged sphere] $$V = \frac{k_0 Q}{R}$$ [16.7]

Furthermore, since $Q = 4\pi R^2 \sigma$, where σ is the surface charge density, the voltage at its surface is

[a charged sphere] $$V = 4\pi k_0 R \sigma$$ (16.8)

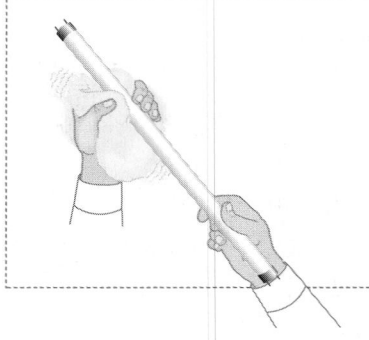

Because a charged conducting sphere has no internal *E*-field, it corresponds to an equipotential volume, and so Eq. (16.7) is also the potential at its center. If we have two different-size spheres, each with the same σ, the larger would hold more charge, more work would be needed to charge it, and so its potential would be higher.

An object like von Guericke's sphere cannot be charged endlessly; neither the potential nor σ keeps going up as the thing is rubbed indefinitely. The sphere, which initially strongly drew electrons from the hand, becomes highly negatively charged. The charges are increasingly crowded together until a point is reached, depending on the size of the sphere, where they strongly repel and block any further arrival of electrons. The sphere reaches a maximum potential. In contrast, the Van de Graaff generator avoids that physical limitation by having charge introduced inside a conductor, where there is no repelling field (Fig. Q14, p. 562). Its maximum potential can be so great that it is ultimately limited by the surrounding air. The very high electric field at the surface of the generator will ionize the air, causing it to become a conductor—a phenomenon known as **dielectric breakdown**. Air breaks down, ionizing and supporting sparks when the voltage across it is $\approx$30 kilovolts per centimeter. A little spark one-eighth inch long, the kind that comes from rug shuffling, corresponds to a potential difference of almost 10 kV.

Suppose we run an electrostatic generator at a sustained potential V and wish to transfer some charge from it to a neutral conducting object. Bringing the two in contact allows charge to flow from the generator to the conductor. *That flow of charge continues until the object reaches the same potential as the generator.* Similarly, suppose a neutral conductor is brought in contact with a charged conductor. Charge will flow to the formerly neutral body *until both reach the same potential*.

When there is a potential difference between two bodies, ΔV equals the work that must be done per unit of positive charge, to transfer charge from one body to the other. If the two are brought into contact, charge will spontaneously flow "downhill," and the potential will rapidly equalize at a value that may be positive, negative, or zero.

Power lines in rural areas are often operated at several hundred thousand volts. The voltage drop per meter from the line down to the zero of ground can be quite high. For this farmer, it's enough to light the fluorescent bulbs he's holding in his hands.

16.4 Conservation of Charge

Conservation of Charge had been known for over a century before physicists, guided by Noether's Principle (the notion that for every conserved quantity there is a symmetry), began to search for a corresponding symmetry associated with classical Electromagnetic Theory. What they found was an invariance arising from the arbitrariness of the electric and magnetic (which we are not going to discuss) potentials. That mathematical behavior is called *gauge symmetry*, where the word "gauge" relates to the way physical quantities are measured. Many physicists now believe that this symmetry is the fundamental characteristic shared by all correct theories (Fig. 16.14).

Conservation of Charge arises from the fact that the proper formulation of electromagnetism is gauge symmetric. That sounds complicated, and we will not attempt to explain gauge symmetry until Chapter 31, but we can get a sense of the central idea very simply. Let's look at just the electric-potential part of the symmetry. There is no observable effect that depends on an absolute value of electric potential—the zero to which all measurements

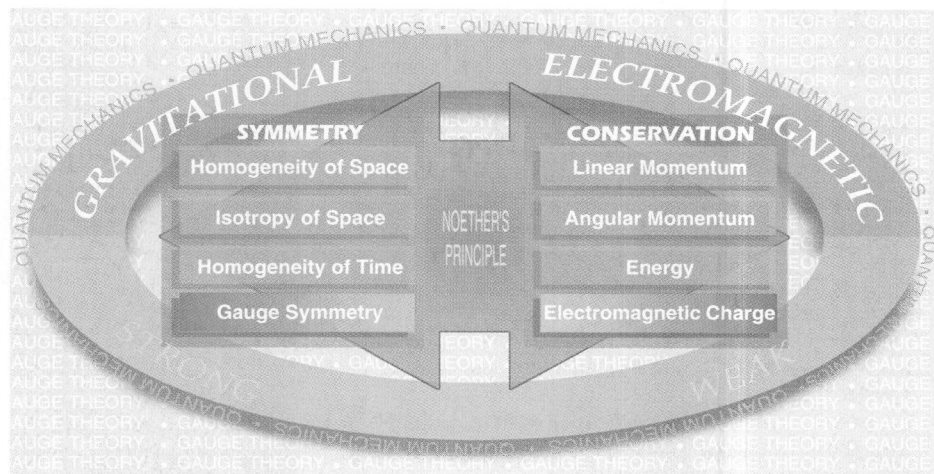

Figure 16.14 Like linear momentum, angular momentum, and energy, charge is conserved.

Subatomic events played out in a liquid-hydrogen bubble chamber. A gamma-ray photon (no track) descends from the top of the picture. It strikes an atom and knocks out an electron (long green track). The remainder of the photon's energy goes into creating an electron positron pair. Because of the applied magnetic field, these two low-energy particles move along tight spirals. Another invisible photon creates a second electron-positron pair further down in the picture. This time, all of the photon's energy goes into the pair. Each particle has more KE and therefore moves along a path that is nearly straight. Note how charge is conserved.

are referenced, or gauged, is arbitrary. We will show that the invariance of phenomena with respect to the zero of electric potential is consistent with Conservation of Charge.

Imagine a laboratory immersed in a uniform E-field (Fig. 16.15). We, outside the lab, fix the arbitrary zero-potential anywhere we like, and then bring a positive charge up to the lab from that $V = 0$ level. We do a precise amount of work on the test-charge and it has a corresponding precise ΔPE_E, as far as we are concerned. Yet no experiment performed *inside* the laboratory can measure the PE_E of the charge *with respect to the chosen $V = 0$ level*. In fact, the laboratory can be translated up, down, or sidewise in the uniform field, and the experimenter inside at the new location will not observe any change—that's due to the arbitrariness of the potential.

What would happen if it were possible for a charge to vanish while all the remaining laws of physics—such as Conservation of Energy and Conservation of Momentum—were precisely in effect? The energy possessed by the charge would have to be liberated to the system in the lab in some way if energy were to be conserved. (In whatever manner that energy is transferred, momentum must also be conserved.) The experimenter in the lab and we outside are at rest with respect to each other and must see the same amount of liberated energy. That's paradoxical—the amount of potential energy the charge possesses is arbitrary, and yet the amount of energy that would be liberated if the charge were to vanish is not. Furthermore, by measuring the energy given out, observers in the laboratory could presumably determine the charge's absolute potential, which is impossible: the $V = 0$ level can be anywhere. Hence, **a single charge cannot vanish**. On the other hand, if an equal pair of positive and negative charges were brought to the lab, *no* net work would be done against

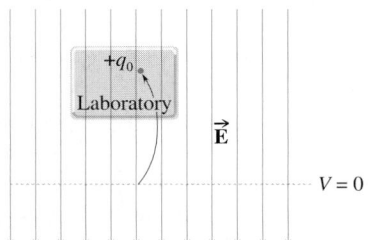

Figure 16.15 A charge in a uniform E-field is raised from an arbitrary zero-V level up to a laboratory. No experiment can be performed in the lab that will reveal the original level from which the charge was raised.

When charge is either created or destroyed, equal amounts of positive and negative charge are always involved—**charge is conserved**. Here we see the creation of several electron-positron pairs (red and blue, respectively) through the action of (trackless) photons. In the middle left, a fast-moving 200-MeV positron (long blue track which is half of an electron-positron pair created in the lower part of the picture) stops abruptly when it collides with an electron within one of the atoms in the liquid. Both charged particles are annihilated and an energetic photon is created in the process. (There is probably also another very low-energy photon produced, but it goes unnoticed.) The photon moving to the left invisibly carries away most of the energy of the two particles. It travels with a momentum more-or-less equal to that of the positron. About 10 cm away, the photon transforms, creating another electron-positron pair (upper left). These spiral more tightly than the original pair because they have less momentum to start with.

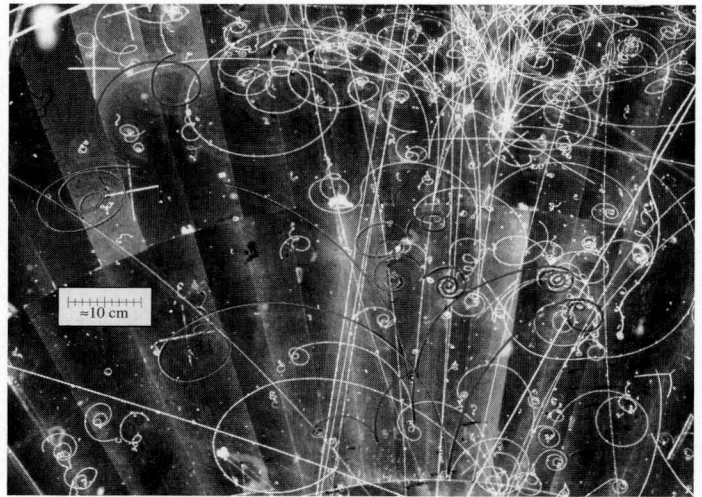

the field. They certainly could annihilate each other, or some other pair, still conserving charge, energy, and momentum. *In this Universe, energy is relative and charge must be conserved* (see the photo at the left).

Capacitance

In the early days, rubbing machines could develop prodigious voltages but only tiny trickles of charge—the pressing problem was to store the charge, so that it could be built up and then dumped in a single powerful blast. It was soon recognized that when a conductor was electrified, the size of the conductor determined the amount of charge it could store. In the beginning, bars of metal were used, gun barrels and the like, and sometimes even people themselves, but more ambitious practitioners suspended massive cannons, charging them to tremendous levels. Any such charge-storing device was dubbed a *condenser* by Volta, but that term has now been replaced by the word **capacitor**.

16.5 The Capacitor

It was Volta who introduced the expression "electrical capacity" in analogy with the concept of heat capacity. At a given potential (V), the amount of charge (Q) that can be stored by a body depends on its physical characteristics, all of which we lump together under the name **capacitance** (C). The more charge, the greater the capacitance; while the less voltage that is needed to accomplish the feat, the greater the capacitance. In other words, C must vary directly with Q and inversely with V:

$$C = \frac{Q}{V} \tag{16.9}$$

The unit of capacitance is coulombs per volt, and to honor Faraday, it is called a *farad* (F): 1 farad = 1 F = 1 C/V. *Keep in mind that capacitance is always a positive quantity.*

One farad is a rather large capacitance; microfarad (1 μF = 10^{-6} F) and picofarad (1 pF = 10^{-12} F) capacitors smaller than the size of a grain of rice are commonly used in radios and TV sets. But fairly hefty capacitors (about the size of a soup can) can still be found in air conditioners where large amounts of charge are dumped into the compressor motor to get it going. A new variety of small activated carbon–sulfuric acid capacitors are now available; a 1-F model fits in the palm of your hand.

THE GREAT MAGNETICO-ELECTRICO BED

Sometime around 1706, Hauksbee replaced von Guericke's sulfur ball with a large glass globe, thereby creating a friction electric machine that was to become popular throughout Europe. By mid-century the science of electricity was drawing eager crowds who came to see the wonders performed by itinerant lecturers with cartloads of mysterious paraphernalia. In London, one of Hauksbee's Influence Machines was even installed at the so-called Temple of Health, that it might provide a therapeutic environment around the "magnetico-electrico" bed, ostensibly to aid in matters connubial.

To find the capacitance of a metal sphere of radius R, suppose it's carrying a charge Q. Its potential is $V = k_0 Q/R = Q/4\pi\varepsilon_0 R$ and so

Capacitance is a measure of the ability of a device to store charge.

[capacitance of a sphere]

$$C = \frac{Q}{V} = 4\pi\varepsilon_0 R \qquad (16.10)$$

The larger the sphere, the greater the capacitance; but $(4\pi\varepsilon_0)$ is very small $(\approx 10^{-10})$, and even a large sphere will have only a modest capacitance. That's true as well for other shapes that would be much harder to calculate directly, such as your body.

Example 16.7 **[I]** Determine the capacitance of an isolated metal sphere 50 cm in diameter and immersed in vacuum.

Solution This problem deals with the capacitance of a sphere. (1) TRANSLATION—An isolated conducting sphere of known diameter is in vacuum; determine its capacitance. (2) GIVEN: Radius $R = 25$ cm and $\varepsilon = \varepsilon_0$. FIND: C. (3) PROBLEM TYPE—Electrostatics/capacitance/sphere. (4) PROCEDURE—The capacitance of a conducting sphere follows from the defin-

ition $C = Q/V = 4\pi\varepsilon_0 R$. (5) CALCULATION:

$$C = 4\pi\varepsilon_0 R = \frac{R}{k_0} = \frac{0.25 \text{ m}}{(8.99 \times 10^9 \text{ N·m}^2/\text{C}^2)} = \boxed{28 \text{ pF}}$$

Quick Check: This sphere has an area of ≈ 0.8 m^2 as compared to the 2-m^2 area of a person whose capacitance is ≈ 50 pF.

Alessandro Giuseppe Antonio Anastasio Count Volta (1745–1827).

The breakthrough in storing charge was made almost by accident in 1745 by G. von Kleist and again independently by P. van Musschenbroek. Both of these European gentlemen were experimenting with electricity and happened to insert the conductor being charged into a hand-held jar; they were probably attempting to collect "electric fluid." Van Musschenbroek dangled a brass wire, attached to a gun barrel that was being charged, into a flask "partly filled with water." It discharged, and all at once his body convulsed "as if it had been struck by lightning;… I thought it was all up with me," he recounted.

THE CAPACITANCE OF A COW

You are a capacitor that can store enough charge at a high enough voltage to produce observable sparks. In fact, the capacitance of a human (measured while standing on 5 cm of insulation) is roughly from 100 pF to 110 pF (for people 68 kg to 105 kg, respectively). Compare that to a cow, which is a lot larger and comes in a fairly standard 200-pF model. These values reflect a good deal of interaction with the Earth and, as we will see shortly, the presence of another conductor will appreciably increase the capacitance of any object. Thus, an isolated person, several meters above the Earth, has a capacitance of only about 50 pF.

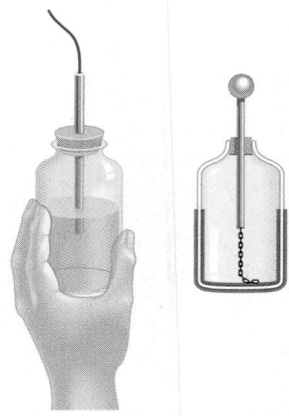

Figure 16.16 A Leyden jar. The improved version is lined with metal foil. A small chain connects the central wire with the inner metal surface.

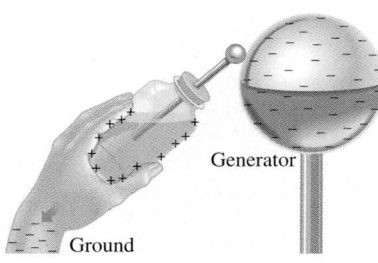

Figure 16.17 Charging a Leyden jar with the help of a generator and a grounded experimenter.

They had unknowingly constructed a device (Fig. 16.16) in which a conductor (the wire) was separated from another grounded conductor (a sweaty hand) by an insulating medium (the glass). Ordinarily, the isolated conductor being charged would rapidly reach the potential of the generator and thereafter repel any further charge. The new arrangement of conductor-insulator-conductor forestalled that cutoff. The charge put on one conductor, the central wire, induced an equal and opposite charge on the other conductor, the moist hand. That induced charge, having the opposite polarity and being relatively nearby, acted to reduce the wire's repulsion of additional charge. The result was a considerable increase in the charge stored before the device, which came to be called a *Leyden jar*, reached the potential of the generator (Fig. 16.17).

The Parallel-Plate Capacitor

The renowned tamer of lightning and writer of racy prose, Ben Franklin, had his Leyden jars, too, but he went one step further. Franklin was among the first to use a new and more convenient configuration consisting of flat metal plates separated by sheets of window glass. That simple arrangement allowed for a dramatic increase in the size of the conductor-insulator-conductor sandwich. Franklin's flattened Leyden jar is a **parallel-plate capacitor**.

At the end of the eighteenth century, Volta carried out a series of measurements of the potentials of large charged objects. He attached the test object via a conductor to a grounded electroscope. The leaves would then spring apart in proportion to the charge impressed on them. That, in turn, is proportional to the potential of the scope, which equals the potential of the object. In other words, the angle of the leaves of the electroscope is effectively proportional to the difference in potential between them and the grounded case. Suppose the test object is a positively charged flat plate as in Fig. 16.18. If an identical grounded plate is brought nearby, the leaves gradually descend as it approaches. In effect, the negative charge induced on the new plate will draw up some of the positive charge from the leaves, holding it fixed on the near side of the positive plate. But that means that introducing the second (now negatively charged) plate drops the potential of the first plate, which can be restored to its original value by further increasing the charge. *The second plate considerably enhances the ability of the parallel-plate capacitor to store charge at a given voltage.*

Figure 16.18 The operation of the capacitor as a charge storer. (a) The charges on a single plate experience a repulsive force, which contributes to establishing the potential. In (b), we bring a grounded neutral plate close to the positive plate. A group of negative charges are drawn onto this second plate, essentially neutralizing much of the positive charge and reducing the potential.

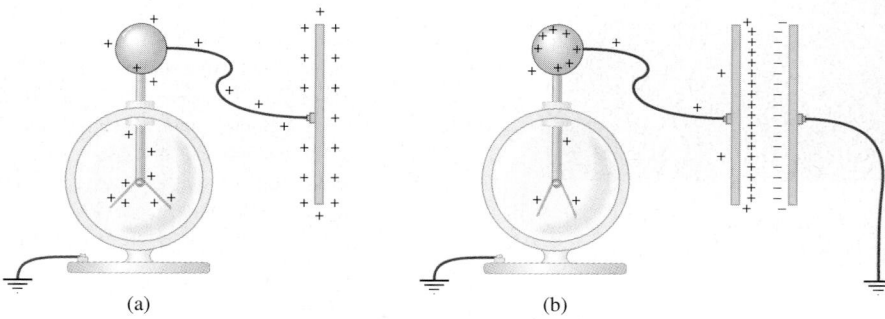

(a)   (b)

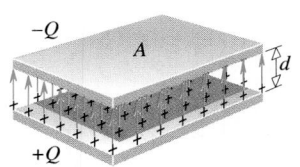

Figure 16.19 A charged parallel-plate capacitor. Each plate has an area A and stores a charge Q.

Internet

For an interactive simulation exploring Eq. (16.11) go to
http://www.micro.magnet.fsu.edu/electromag/java/
and click on *Factors Affecting Capacitance*.

We now determine the capacitance of the parallel-plate capacitor (Fig. 16.19) as a function of its physical characteristics. If the plates carry opposite charges of $\pm Q$ and have a difference of potential ΔV between them, then $C = Q/\Delta V$. We can calculate this potential difference in terms of the E-field between the plates starting with

$$\Delta V = Ed$$

from Eq. (16.5), where d is the plate separation. Recall that

$$E = \frac{\sigma}{\varepsilon} = \frac{Q}{A\varepsilon} \qquad [15.9]$$

wherein A is the area of *each* plate and $\sigma = Q/A$. It follows that

$$\Delta V = Ed = \frac{Qd}{A\varepsilon}$$

and

$$C = \frac{Q}{\Delta V} = \frac{Q}{Qd/A\varepsilon}$$

which simplifies to

[parallel plates]
$$C = \frac{\varepsilon A}{d} \qquad (16.11)$$

To produce as large a capacitance as possible, we must make A large and d small and use a material in the gap with a large permittivity (Table 15.3).

Example 16.8 **[I]** Determine the size of a 1.00-F parallel-plate capacitor if the plates are square and separated by 1.00 mm of air. How would things change if the gap were filled with a sheet of glass having a relative permittivity of 10.0?

Solution Here we're dealing with a parallel-plate capacitor. (1) TRANSLATION—Two parallel square plates with an air gap of known thickness have a known capacitance; determine the plate area. (2) GIVEN: $C = 1.00$ F, $d = 1.00$ mm, $\varepsilon =$ both ε_0 and $10.0\varepsilon_0$, and $A = L \times L$. FIND: L. (3) PROBLEM TYPE—Electrostatics/capacitance/parallel plates. (4) PROCEDURE—This problem involves the physical characteristics of a parallel-plate capacitor and one equation should come to mind immediately: $C = Q/V = \varepsilon A/d$. (5) CALCULATION—Since the gap is filled with air

$$C = \frac{\varepsilon_0 A}{d}$$

and

$$A = L^2 = \frac{dC}{\varepsilon_0} = \frac{(1.00 \times 10^{-3}\text{ m})(1.00\text{ F})}{(8.85 \times 10^{-12}\text{ C}^2/\text{N}\cdot\text{m}^2)} = 0.113 \times 10^9\text{ m}^2$$

and so $\boxed{L = 10.6\text{ km}}$. The plates are gigantic, about 6.6 miles on a side. When glass replaces air, ε replaces ε_0; and since $\varepsilon = 10.0\varepsilon_0$, the area is 10.0 times smaller and $L = (10.6\text{ km})/\sqrt{10.0} = 3.35$ km.

Quick Check: $C \approx (10^{-11}\text{ C}^2/\text{N}\cdot\text{m}^2)(10^8\text{ m}^2)/(10^{-3}\text{ m}) \approx 1$ F.

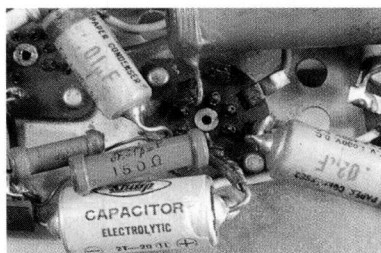

Capacitors in the circuit of the AM radio shown on p. 537. Condenser is the old-fashioned word for capacitor.

A computer keyboard. When a key is depressed, the spacing of the plates in an air capacitor changes, thereby changing the capacitance and registering the keystroke.

Figure 16.20 The effect of the dielectric in a parallel-plate capacitor. It becomes polarized and thereby reduces the internal field. (a) Polarization of the dielectric. (b) The E-field due to the polarized dielectric is opposite the E-field due to the charge on the plates. (c) The internal E-field is reduced, and more charge can be stored at a given potential difference.

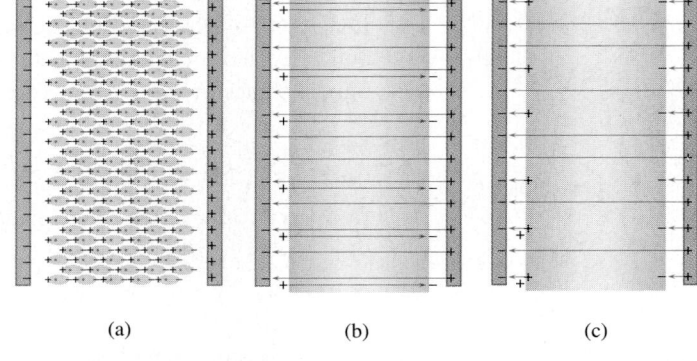

(a)   (b)   (c)

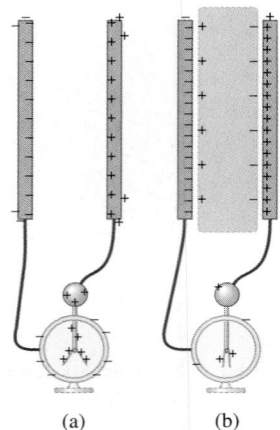

(a)   (b)

Figure 16.21 (a) A charged capacitor attached to an electroscope that indicates the potential difference. (b) Inserting a dielectric essentially neutralizes some of the charge on the plates and lowers the potential difference.

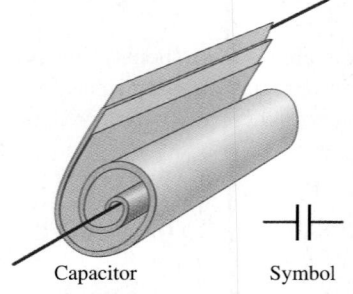

Capacitor   Symbol

Figure 16.22 When the metal foil, dielectric, metal foil sandwich is rolled up and sealed, it forms a parallel-plate capacitor with a fixed capacitance.

A dielectric has a larger value of permittivity than vacuum and therefore a smaller internal field ($E = \sigma/\varepsilon$), as is evident in Fig. 16.20. The dielectric becomes polarized in the external field of the charged plates and takes on a surface charge. As a result, a small internal *self-field* opposes the applied field within the dielectric. The effect is a weaker net field in the gap and a lower voltage ($\Delta V = +Ed$) across it. Slipping a sheet of insulation into a charged capacitor (Fig. 16.21) effectively cancels some of the charge on the plates via the opposite polarity induced charge on the surfaces of the dielectric.

Modern parallel-plate capacitors come in different forms. The most common is a sandwich of metal foil (aluminum), dielectric (waxed paper, mylar, etc.), and metal foil rolled up into a tight little cylinder and sealed (Fig. 16.22). It's represented symbolically in diagrams by two parallel lines of equal length: ⊣⊢.

16.6 Capacitors in Combination

Inasmuch as the early interest in capacitors was purely for charge storage, it's not surprising that Leyden jars were wired to one another to increase that ability. To the same end, Franklin connected almost a dozen parallel-plate capacitors together to produce great blasts of electricity.

The Parallel Circuit

Franklin, like others, used two basic wiring schemes to create two different results. Figure 16.23 shows the so-called **parallel** arrangement of three capacitors, though any number of them could be connected that way. The distinguishing characteristic of the parallel scheme is that one terminal (or plate) from each capacitor is connected to the same common wire, and all the remaining plates are connected to a second common wire. All the top plates are attached and must be at the same potential, and all the bottom plates are attached and must be at the same potential. **When a wire is used to connect two points together, those two points, and every point along the wire, are assumed to be at the same potential.** The result here is essentially one large capacitor (whose two plates are each segmented into three connected parts).

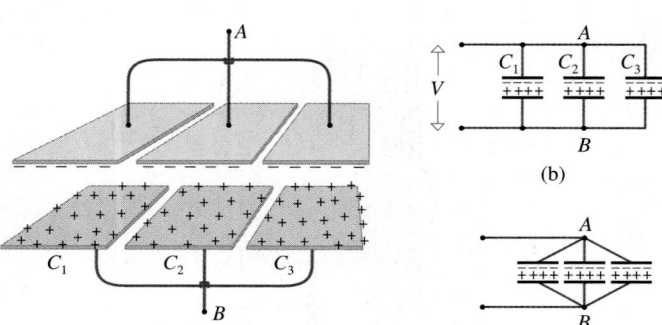

(a)   (b)   (c)

Figure 16.23 Three capacitors in parallel represented in several effectively identical ways.(a) The equivalent capacitance is $C = C_1 + C_2 + C_3$. The circuits in (a), (b), and (c) are all electrically the same.

This modern capacitor has a remarkably high capacitance of 1 farad, despite its small size.

A capacitor is a circuit element having two terminals: call them *A* and *B*. Two or more capacitors are wired in **parallel** when all of the *A*-terminals are connected to one another and all of the *B*-terminals are connected to one another.

Figure 16.24 shows a capacitor attached across a battery (p. 609), which is represented by the standard symbol ⊣⊢. The longer lighter line corresponds to the higher potential and is labeled +. It is *V* volts higher than the other terminal, labeled −. Negative charge flows out of the − terminal of the battery in the form of electrons. These charges gradually build up on the negative low-potential plate of the capacitor. They, in turn, repel an equal number of electrons off the other plate, leaving it positive, and this charge circulates back to the battery. The path is called a **circuit** and electrons continue to flow, as if around the circuit, for the brief time until the capacitor reaches the same potential difference as the battery. At that point, any further charge is repelled by the capacitor as forcefully as it is propelled by the battery. There is a field traversing the gap, and the voltage difference across the capacitor is given by $\Delta V = Ed$. This difference increases during the charging until it finally equals the battery's voltage, at which time the charging ceases.

It is customary to write the potential difference (ΔV) across any circuit element (e.g., capacitors, batteries, and resistors) simply as *V*. In Fig. 16.23, where the capacitors are in parallel, the voltage across the combination *V* must equal the voltage across each one:

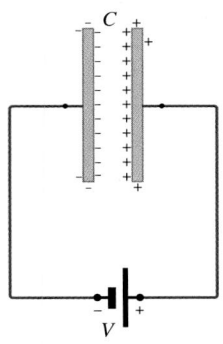

Figure 16.24 A capacitor across a battery. The capacitor's plates become charged as shown, and there is a voltage *V* across them.

[capacitors in parallel]

$$V = V_1 = V_2 = V_3$$

The three capacitors will reach exactly the same state if they are charged to voltage *V* all as a single unit or if each is charged separately and then attached together. The net charge stored (*Q*) is the sum of the individual amounts stored on each capacitor, and so

[capacitors in parallel]

$$Q = Q_1 + Q_2 + Q_3$$

Hence,

$$CV = C_1V_1 + C_2V_2 + C_3V_3$$

[capacitors in parallel]

$$C = C_1 + C_2 + C_3 \qquad (16.12)$$

where we just add on more terms if there are more capacitors. The **equivalent capacitance of several capacitors in parallel is the sum of all the individual capacitances**. A single

An air capacitor. This variable capacitor is from the tuning circuit of the radio shown on p. 537. Turning the knob of the radio causes a set of movable metal plates to slide between a set of stationary plates, thereby changing the capacitance and the station.

capacitor C, given by Eq. (16.12), would be electrically indistinguishable from the parallel array of *smaller* capacitors C_1, C_2, and C_3.

There are many kinds of circuit elements, and they can all be in parallel with each other provided both terminals of one are connected to both terminals of another. *A place where three or more leads come together is called a* **node**. The two regions in Fig. 16.23 where the three terminals are connected, top and bottom, constitute two nodes (A and B). *When elements are in parallel, any number of leads or circuit branches can converge at a node.*

Example 16.9 **[I]** Figure 16.25 shows two capacitors attached to a 12-V battery. Determine the equivalent capacitance and the charge it would carry. What is the charge on each of the capacitors in the figure?

Solution Here we examine the behavior of several parallel-plate capacitors wired together in a special way. (1) TRANSLATION—Two known capacitors and a known battery form a circuit; determine the equivalent capacitance, its charge, and the charge on each capacitor. (2) GIVEN: $C_1 = 20$ μF, $C_2 = 30$ μF, and $V = 12$ V. FIND: C, Q, Q_1, and Q_2. (3) PROBLEM TYPE—Electrostatics / capacitors / circuit. (4) PROCEDURE—First, notice that the top terminals of both capacitors are connected, as are the bottom terminals—the capacitors are in parallel. You might also want to redraw the diagram, as is done in part (*b*). (5) CALCULATION—Since the potential across each capacitor is 12 V

$$V = \frac{Q_1}{C_1} = \frac{Q_2}{C_2}$$

$$Q_1 = (12 \text{ V})C_1 \qquad \text{while} \qquad Q_2 = (12 \text{ V})C_2$$

$$Q_1 = (12 \text{ V})(20 \times 10^{-6} \text{ F}) = \boxed{2.4 \times 10^{-4} \text{ C}}$$

and $\quad Q_2 = (12 \text{ V})(30 \times 10^{-6} \text{ F}) = \boxed{3.6 \times 10^{-4} \text{ C}}$

Since the capacitors are in parallel

$$C = C_1 + C_2 = (20 \text{ } \mu\text{F}) + (30 \text{ } \mu\text{F}) = \boxed{50 \text{ } \mu\text{F}}$$

The charge on the equivalent capacitor is determined from $C = Q/V$; namely,

$$Q = CV = (50 \times 10^{-6} \text{ F})(12 \text{ V}) = \boxed{6.0 \times 10^{-4} \text{ C}}$$

Quick Check: For parallel capacitors we must have $Q = Q_1 + Q_2 = 2.4 \times 10^{-4} \text{ C} + 3.6 \times 10^{-4} \text{ C} = 6.0 \times 10^{-4} \text{ C}$.

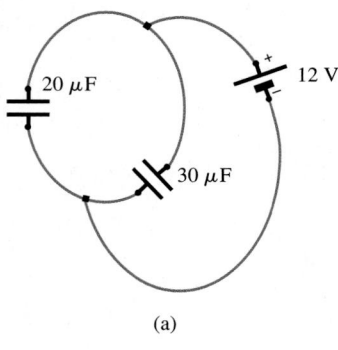

(a)

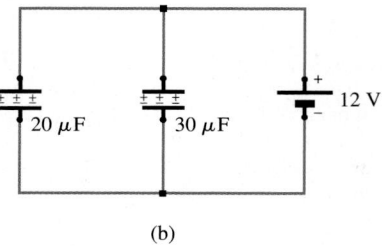

(b)

Figure 16.25 (a) A circuit of two capacitors and a battery. (b) A redrawn equivalent.

The Series Circuit

Two circuit elements are in **series** when terminal-*A* of one is connected to terminal-*B* of the other and no additional elements in the circuit are attached to that junction.

Another basic way to connect circuit elements is known as **series**, and it's illustrated in Fig. 16.26. Here *one and only one terminal of a circuit element is connected to one and only one terminal of an adjacent circuit element.* **Two elements will not be in series if any other branch in the circuit connects to the point at which the two are attached.** Imagine that the three series capacitors are put across a battery of voltage V, as in Fig. 16.26*c*. Electrons travel from the battery to the negative plate of C_3, giving it a charge of $-Q$. These, in turn, repel an equal quantity of charge ($-Q$) from the positive plate of C_3 to the negative plate of C_2. As the negative plate of C_2 charges up to $-Q$, it repels an equal number of electrons from its positive plate to the negative plate of C_1. An amount of elec-

When it comes to problem solving, it's important to remember the following:

When capacitors (with different capacitances) are wired in **parallel** across a voltage source, the voltage across each capacitor will be the same, although the charge on each will be different. *The sum of the charges on each capacitor will add up to equal the charge on the equivalent capacitance.*

When capacitors (with different capacitances) are wired in **series** across a voltage source, the voltage across each capacitor will be different, although the charge on each will be the same. *The charge on each capacitor will equal the charge on the equivalent capacitance.* The sum of the voltages on each capacitor will add up to equal the voltage on the equivalent capacitance.

Note that if one terminal of a fourth capacitor was attached to either *A*, *B*, *C*, or *D* in Fig. 16.26c it would not become charged, and its presence would have no effect at all on the circuit. The original three capacitors would still be in series. *A circuit element with only one terminal attached to the circuit is not in the circuit* and it might just as well be removed.

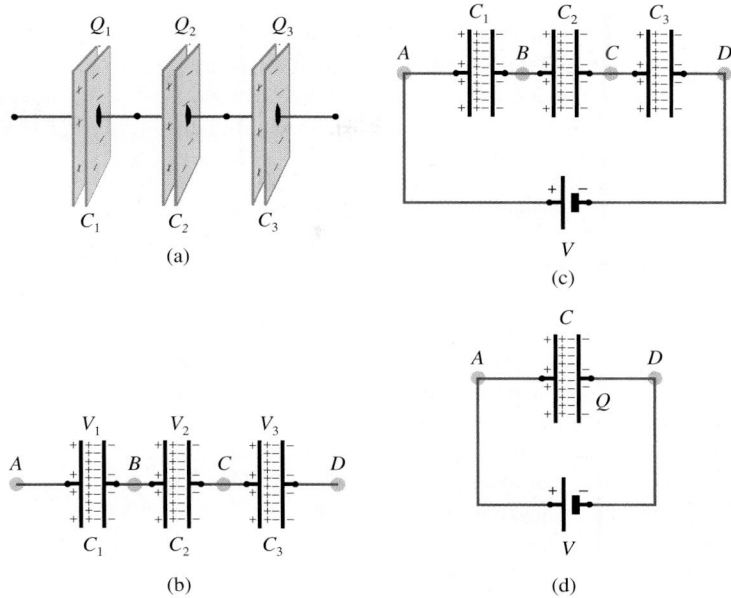

Figure 16.26 (a) Three capacitors in series and (b) their graphic representation. (c) The same series capacitors across a battery. (d) The equivalent circuit wherein $\frac{1}{C} = \frac{1}{C_1} + \frac{1}{C_2} + \frac{1}{C_3}$.

trons equivalent to $-Q$ is repelled back to the positive terminal of the battery. Hence,

[capacitors in series]
$$Q = Q_1 = Q_2 = Q_3$$

Notice that the net charge stored by the three capacitors is effectively $+Q$ on the leftmost plate and $-Q$ on the rightmost plate; the rest of the charge on the remaining plates ($-Q$ and $+Q$; $-Q$ and $+Q$, respectively) cancels out—only Q's worth of electrons went in on the right and out on the left. Thus, the equivalent capacitor will also have a charge Q (which is *not* the sum of the charges on the individual capacitors).

The equivalent capacitor will be across the battery, just as is the series string of C_1, C_2, and C_3. Hence, the voltage across C is V. If we go from point A on the left to point D on the right, we will, by necessity, drop in potential by an amount V; the battery is connected to points A and D. Yet, in going from point A to point B, we drop V_1; in going from B to C, we drop an additional V_2; in going from C to D, we drop another amount V_3; consequently

[capacitors in series]
$$V = V_1 + V_2 + V_3$$

Using the definition $V = Q/C$,

$$\frac{Q}{C} = \frac{Q_1}{C_1} + \frac{Q_2}{C_2} + \frac{Q_3}{C_3}$$

But all the charges are equal and therefore

[capacitors in series]
$$\frac{1}{C} = \frac{1}{C_1} + \frac{1}{C_2} + \frac{1}{C_3} \qquad (16.13)$$

The reciprocal function on the calculator is wonderfully helpful in computing Eq. (16.13) directly. For example, to compute C where $1/C = 1/4 + 1/3$, hit the following sequence of keys on your calculator:

| 4 | 1/x | + | 3 | 1/x | = | 1/x |

to get 1.7. The nice thing is that you can effortlessly apply this technique to a string of as many capacitors as you like.

and we just keep adding on terms if there are more capacitors. Notice that C_1 is in series with C_3 even though there is an intervening element. **Two elements in series with a third** (C_1 and C_2; C_3 and C_2) **are in series with each other** (C_1 and C_3). {For an introduction to the practical issue of leakage click on **REAL CAPACITORS** under **FURTHER DISCUSSIONS** on the **CD**.}

Example 16.10 **[II]** The circuit shown in Fig. 16.27a consists of a 12-V battery and three capacitors. Determine both the voltage across and charge on each capacitor after the switch S is closed and electrostatic equilibrium is established. Find the equivalent capacitance of the network.

Solution Here we're dealing with several parallel-plate capacitors wired together in a specific way. (1) TRANSLATION— Three known capacitors and a known battery form a circuit; determine the voltage and charge on each capacitor and the equivalent capacitance. (2) GIVEN: $C_1 = 2.0\ \mu$F, $C_2 = 2.0\ \mu$F, $C_3 = 5.0\ \mu$F, and $V = 12$ V. FIND: C, V_1, V_2, V_3, Q_1, Q_2, and Q_3. (3) PROBLEM TYPE—Electrostatics/capacitors/circuit. (4) PROCEDURE—First, redraw the circuit as in Fig. 16.27b to make things a bit clearer. Then simplify using series and parallel combinations. (5) CALCULATION—The two 2.0-μF capacitors are in series, and their equivalent capacitance C is given by

$$\frac{1}{C} = \frac{1}{C_1} + \frac{1}{C_2}$$

which is easy to compute with a calculator. However, if you would rather do it in your head, remember the equivalent form (which is left as a problem to prove)

$$C = \frac{C_1 C_2}{C_1 + C_2} = \frac{(2.0\ \mu\text{F})(2.0\ \mu\text{F})}{2.0\ \mu\text{F} + 2.0\ \mu\text{F}} = 1.0\ \mu\text{F}$$

As shown in Fig. 16.27c, this 1.0-μF equivalent of the series pair is itself in parallel with the 5.0-μF capacitor, and so they can be combined into a single capacitor via Eq. (16.15); namely,

$$C = 5.0\ \mu\text{F} + 1.0\ \mu\text{F} = \boxed{6.0\ \mu\text{F}}$$

which is the equivalent capacitance of the whole network. Now, working back from the simplified diagram of Fig. 16.27c, $V = 12$ V,

$$Q_3 = C_3 V_3 = (5.0\ \mu\text{F})(12\text{ V}) = \boxed{60\ \mu\text{C}}$$

There are 12 V across the combination of the two 2.0-μF capacitors, and, hence, there must be a potential difference of 6.0 V across each one. Therefore,

$$Q_1 = Q_2 = (2.0\ \mu\text{F})(6.0\text{ V}) = \boxed{12\ \mu\text{C}}$$

Quick Check: The charge on the equivalent capacitor is $Q = CV = 72\ \mu$C. This must equal the net charge on the two series capacitors (that is, the charge on either one—namely, 12 μC) plus $Q_3 = 60\ \mu$C, and, happily, it does.

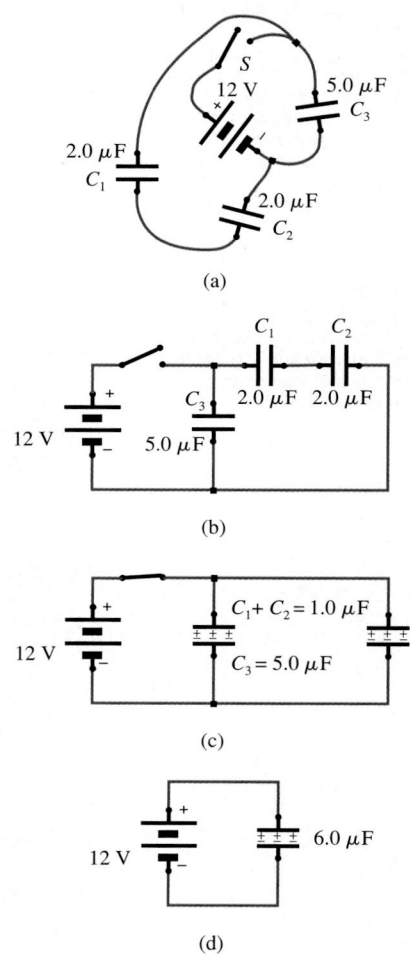

Figure 16.27 (a) A capacitive circuit and (b) a redrawn version. (c) A simplified configuration and (d) the simplest equivalent circuit.

16.7 Energy in Capacitors

Charging a parallel-plate capacitor requires that work must be done on the charges to bring them from wherever they are to the plates. Each electron is forced over to the plate against the repulsive action of all the other electrons that preceded it. Suppose that we wish to electrify a capacitor with a total charge Q to a potential difference V, starting from a potential difference of zero. The end result is two oppositely charged plates in close proximity. The amount of energy stored in the process is independent of the details of how the capacitor got charged, just as the amount of energy stored via a boulder on a mountaintop is independent of how it got up there. Accordingly, let's work with the simplest scenario. Assume that we carry an amount of charge $+Q$ from one plate across the gap to the other, thereby leaving the first plate charged with $-Q$ (Fig. 16.28).

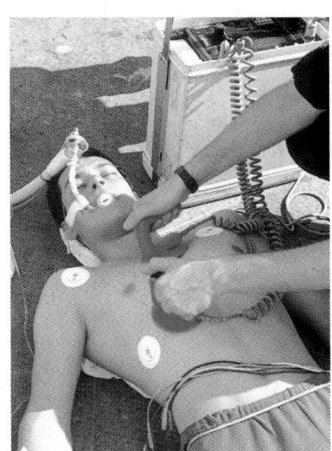

Sending a pulse of charge across the chest can start the heart beating with a steady rhythm. The necessary energy (about 360 J) is first built up and stored in a capacitor.

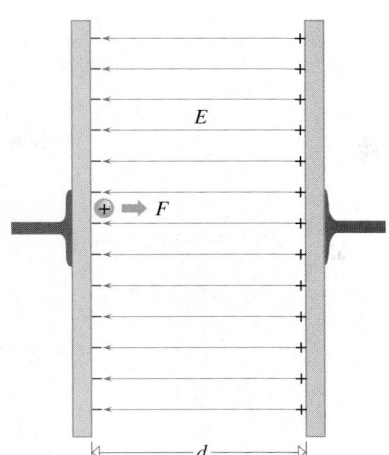

Figure 16.28 Carrying a charge through a distance d against a field requires an external force and the doing of work.

In the beginning, when the voltage is near zero and each new charge is only repelled by the few previous arrivers, the work done is small. As the charge on the plates increases, the voltage increases, the repulsive force increases, and the work expended increases; this process goes on until a potential difference of V is reached. Once again, we are faced with determining the work done in a process where the force is changing, but this time things are simple. We can treat this sort of problem as if all the charge were transported at once against an average potential difference.

Because $V = Q/C$, the variation in voltage as charge is built up is linear, going from 0 to V. Hence, the average potential difference is just the sum of the initial 0 and final V divided by two (recall the Mean-Speed Theorem, p. 58): $V_{av} = (0 + V)/2 = \frac{1}{2}V$. The work done in charging the capacitor, $W = QV_{av}$, is

$$W = Q\tfrac{1}{2}V$$

and if we think of this energy as being stored as *electrical*-PE, we have

$$PE_E = \tfrac{1}{2}QV \qquad (16.14)$$

Equivalently

$$PE_E = \tfrac{1}{2}CV^2 = \tfrac{1}{2}Q^2/C \qquad (16.15)$$

If we want to know how PE_E varies with, say, Q, we must pick the expression that contains Q and no other quantity depending on Q; thus, $PE_E = \frac{1}{2}Q^2/C$ tells us that doubling Q quadruples PE_E.

Example 16.11 **[I]** How much energy is stored in each of the capacitors of Fig. 16.27 in the process of charging them?

Solution This example treats the energy of a charged capacitor. (1) TRANSLATION—Three known capacitors and a known battery form a circuit; determine the energy stored in each capacitor. (2) GIVEN: $C_1 = 2.0\ \mu F$, $C_2 = 2.0\ \mu F$, $C_3 = 5.0\ \mu F$, and $V = 12$ V. FIND: The potential energy stored in each capacitor. (3) PROBLEM TYPE—Electrostatics/capacitors/circuit/energy. (4) PROCEDURE—We have the results of Example 16.10—namely, $Q_1 = Q_2 = 12\ \mu C$ and $Q_3 = 60\ \mu C$—and can find PE_E using C and V, or Q and V, or Q and C. (5) CALCULATION—Using Eq. (16.14),

Capacitor-1 $PE_E = \frac{1}{2}QV = \frac{1}{2}(12\ \mu C)(6.0\ V) = \boxed{36\ \mu J}$

Capacitor-2 $PE_E = \frac{1}{2}QV = \frac{1}{2}(12\ \mu C)(6.0\ V) = \boxed{36\ \mu J}$

Capacitor-3 $PE_E = \frac{1}{2}QV = \frac{1}{2}(60\ \mu C)(12\ V) = \boxed{0.36\ mJ}$

for a grand total of 0.43 mJ.

Quick Check: If the equivalent capacitor is truly equivalent to all the capacitors in the system, it should store the same amount of energy. Using Eq. (16.15), it follows that

$$PE_E = \tfrac{1}{2}CV^2 = \tfrac{1}{2}(6.0\ \mu F)(12\ V)^2 = 0.43\ mJ$$

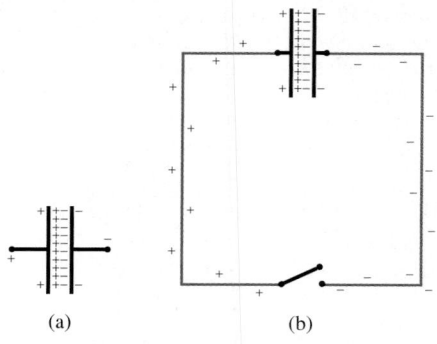

(a)   (b)

Figure 16.29 (a) A charged capacitor. (b) Closing the switch would allow charge to flow and cancel, leaving a net charge of zero.

One good way to provide a tremendous blast of electrical energy is to store charge in a giant capacitor. This capacitor bank at the Lawrence Livermore National Laboratory supplies energy to the Nova laser, the most powerful in the world.

Figure 16.29*a* shows a charged capacitor and therefore one with a potential difference across its plates. Each electron, given the opportunity, would spontaneously descend in potential energy. When a wire—a so-called **short circuit**—is connected across the capacitor, electrons immediately move from the − plate to the + plate, canceling the charge and reducing the potential difference across the capacitor to zero. {For more on the energy stored in the field click on **ENERGY OF THE ELECTROMAGNETIC FIELD** under **FURTHER DISCUSSIONS** on the CD.} 🔘

Example 16.12 **[III]** The two capacitors C_1 and C_2 in Fig. 16.30 were first put across different batteries so that they took on voltages of $V_1 = 12$ V and $V_2 = 6.0$ V, respectively. They were then attached as shown. Compute the charge and energy stored in each capacitor once the switch is closed and electrostatic equilibrium restored.

Solution This problem treats the redistribution of charge among capacitors. (1) TRANSLATION—Two known capacitors are charged to different known initial voltages. They are then attached to one another; determine the final charge and energy stored in each capacitor. (2) GIVEN: $C_1 = 4.0$ μF, $C_2 = 2.0$ μF, $V_{i1} = 12$ V, and $V_{i2} = 6.0$ V. FIND: The potential energy stored and the charge on each capacitor once the switch is closed. (3) PROLEM TYPE—Electrostatics/capacitors/circuit/energy. (4) PROCEDURE—A wire connects the two positive plates even before the switch is closed, but no redistribution of charge can occur because the charges are effectively bound in place on each capacitor by the oppositely charged opposing plate. Even after the switch is closed, electrons on the negative plates cannot get to the positive plates; and although the charge will redistribute itself, the net amount $Q_{i1} + Q_{i2}$ is apparently fixed. (5) CALCULATION—Let's find the two final charge distributions. Two unknowns will require two equations, so begin by determining the net charge. Before they were attached to one another

$$Q_{i1} = C_1 V_{i1} = (4.0\ \mu\text{F})(12\ \text{V}) = 48\ \mu\text{C}$$

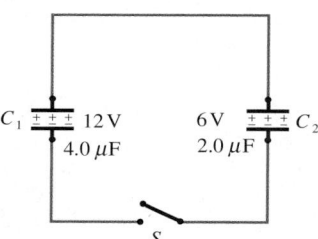

Figure 16.30

and

$$Q_{i2} = C_2 V_{i2} = (2.0\ \mu\text{F})(6.0\ \text{V}) = 12\ \mu\text{C}$$

The net initial charge is

$$Q_{i1} + Q_{i2} = 60\ \mu\text{C}$$

Once connected, the charge redistributes itself until any potential differences between the two positive plates and between the two negative plates vanish. Each capacitor has the same final voltage V_f across it (they're in parallel), and

$$V_f = \frac{Q_{f1}}{C_1} = \frac{Q_{f2}}{C_2}$$

whereupon

$$Q_{f1} = \frac{Q_{f2}(4.0\ \mu\text{F})}{2.0\ \mu\text{F}} = 2Q_{f2}$$

But the total charge is 60 μC; therefore $\boxed{Q_{f1} = 40\ \mu\text{C}}$ and

continued

$\boxed{Q_{f2} = 20\ \mu C}$. To find the potential energy use Eq. (16.15), and thus for C_1

$$PE_{E1} = \frac{\frac{1}{2}Q_{f1}^2}{C_1} = \frac{\frac{1}{2}(40\ \mu C)^2}{4.0\ \mu F} = \boxed{0.20\ \text{mJ}}$$

and for C_2

$$PE_{E2} = \frac{\frac{1}{2}Q_{f2}^2}{C_2} = \frac{\frac{1}{2}(20\ \mu C)^2}{2.0\ \mu F} = \boxed{0.10\ \text{mJ}}$$

Quick Check: The larger capacitor, which has twice the capacitance of the other, stores twice the charge and twice as much energy in reaching the same voltage.

Core Material & Study Guide

ELECTRIC POTENTIAL

The **electric potential** V, or just the *potential*, is defined as

$$\text{Electric potential} = \frac{electrical\text{-}PE}{charge} \qquad [16.1]$$

where $1\ \text{J/C} = 1\ \text{V}$. *The potential difference ΔV between two points A and B is the work done against the field in moving a unit positive charge from A to B:*

$$\Delta V = V_B - V_A = \frac{W(A \rightarrow B)}{q_o} \qquad [16.3]$$

and

$$\Delta V = \frac{\Delta PE_E}{q_o} \qquad [16.4]$$

In a *uniform electric field*, the potential difference is

[uniform *E*-field] $$V_B - V_A = \pm Ed \qquad [16.5]$$

The potential difference is + when the displacement has a component that is opposite to the field and − when it has a component parallel to the field. Review Section 16.1 (*Electrical*-PE & Potential) and study Examples 16.1 to 16.4. Equations (16.3) and (16.4) are equivalent general definitions of potential difference, whereas Eq. (16.5) is a simple special case. **Look at the CD Walk-Through Examples.**

When a positive test charge is moved from r_A to r_B against the $1/r^2$-electric field of a *sphere* of positive charge q, it experiences a change in potential

[r_A to r_B] $$V_B - V_A = k_o q \cdot \left(\frac{1}{r_B} - \frac{1}{r_A}\right) \qquad [16.6]$$

The potential at a distance r from a positive point-charge, measured with respect to the zero at ∞, is given by

[$V = 0$ at ∞] $$V = \frac{k_o q \cdot}{r} \qquad [16.7]$$

These two equations summarize Sections 16.2 (Potential of a Point-Charge) and 16.3 (The Potential of Several Charges). Potential is a scalar quantity, and the potential of each contributing point-charge adds algebraically to that of all the others. Make sure you understand Examples 16.5 and 16.6; they're typical.

CAPACITANCE

Capacitance (C) is a measure of *the capacity to store charge*:

$$C = \frac{Q}{V} \qquad [16.9]$$

the units of which are coulombs per volt, where 1 farad $= 1\ \text{F} = 1\ \text{C/V}$. For an isolated sphere in vacuum

[capacitance of a sphere] $$C = \frac{Q}{V} = 4\pi\varepsilon_0 R \qquad [16.10]$$

For a parallel-plate capacitor

$$C = \varepsilon A/d \qquad [16.11]$$

These ideas are treated in Section 16.5 (The Capacitor), which introduces the concept and describes spherical and parallel-plate capacitors. Study Examples 16.7 and 16.8, which deal with single capacitors of these two types.

When two or more capacitors are connected, the equivalent capacitance for them in **parallel** is

[parallel] $$C = C_1 + C_2 + C_3 + \cdots \qquad [16.12]$$

and in **series** is

[series] $$\frac{1}{C} = \frac{1}{C_1} + \frac{1}{C_2} + \frac{1}{C_3} + \cdots \qquad [16.13]$$

Groupings of capacitors in series and parallel are treated in Section 16.6 (Capacitors in Combination). Make sure you know the difference between series and parallel and then re-examine Examples 16.9 and 16.10; *draw lots of intermediate diagrams as you simplify each circuit*.

The energy stored in the process of charging a capacitor is

$$PE_E = \tfrac{1}{2}QV \qquad [16.14]$$

$$PE_E = \tfrac{1}{2}CV^2 = \tfrac{1}{2}Q^2/C \qquad [16.15]$$

Reread Section 16.7 (Energy in Capacitors) and review Examples 16.11 and 16.12. Any problem that talks about capacitors and energy probably involves these equations.

Key Terms

Elecrtical-PE	capacitor
potential	Leyden jar
potential difference	parallel-plate capacitor
electron volt	parallel circuit
equipotential	equivalent capacitance
electrostatic generator	series circuit
dielectric breakdown	energy of a capacitor
conservation of charge	

Discussion Questions

1. A negative charge in an electric field moves from a point where the potential is zero to a point where it is -100 V. Discuss the energy change and the work done.

2. Imagine a hollow, spherical, positively charged conductor of radius R that is far away from any other bodies. Draw a graph of its potential V as a function of r out from its center. Discuss your results.

3. With Question 2 in mind, suppose we place a neutral hollow conducting sphere in the vicinity of the charged sphere. Draw a rough graph of the potential along the center line of both spheres, and compare it to the previous potential without the neutral conductor. Discuss your results.

4. With Question 3 in mind, suppose we now ground the neutral conductor. Describe the potential along the center line. Discuss your results if they are any different from those of Question 3.

5. Using equipment like that depicted in Fig. 15.37, Faraday showed that when a neutral conductor touched the inside of a hollow charged conductor, it remained neutral. Use that result to justify the conclusion that *provided there is no external field, a neutral conductor assumes the potential of the region of space in which it is introduced.* What happens to the potential of the neutral conductor (the little sphere) if the charge on the surrounding body in Fig. 15.37 is increased?

6. Describe the potential of a negative point-charge as a function of distance r. Explain your thinking.

7. A small, neutral spherical conductor is placed between the plates of a large, charged parallel-plate capacitor. Describe the E-field and the equipotentials in the gap.

8. A positive point-charge is located a short distance above a large conducting horizontal plane. Describe the field lines and equipotentials. Compare your results with that of an electric dipole. Explain your conclusions.

9. Imagine a square with point-charges at each corner. At the ends of one diagonal, the charges are $+q_*$; at the ends of the other they are $-q_*$. This arrangement is called an *electric quadrupole*. Make a rough sketch of the field lines and equipotentials and explain your reasoning.

10. Figure Q10 shows a positive point-charge located at the corner of two intersecting conducting planes. Describe the E-field and the equipotentials. Compare your answer to that of the previous question. Explain your observations.

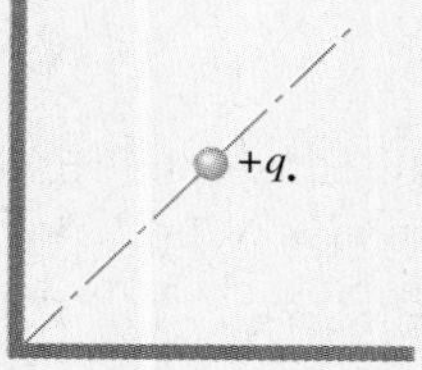

11. Is it correct to maintain that *when a positive charge is deposited on a body it raises the potential of that body and, moreover, it raises the potential of the entire vicinity, including that of any other bodies in the vicinity*? Contrarily, *does the introduction of a negative charge lower the potential of every point in the neighboring region*? Explain your answer in detail. How might you prove your conclusion experimentally?

12. Is it correct to maintain that, in general, *the potential of a conductor, which certainly depends on its own charge, also depends on the distribution of charge everywhere else nearby*? Under what cir-

cumstances, if any, does the potential of a charged conductor only depend on its charge, structure, and the medium it is in?

13. We know that points infinitely far from all charge are at zero potential; yet, as a practical matter, we take the Earth, which may well be charged, to be at zero-potential. How can this inconsistency be resolved?

14. Water has a rather large dielectric constant $(\varepsilon/\varepsilon_0)$. How might that characteristic contribute to its ability to keep substances such as table salt, once dissolved, in solution?

15. Given a region that is an equipotential volume, what can you say about the possibility that there is a net charge within it?

16. Given that a particular equipotential volume (different from its surroundings) is bounded by an equipotential surface, what can you say about the charge, if any, on that surface?

17. It is sometimes desirable, especially in high-voltage applications, to replace a capacitor by an equivalent string of several capacitors in series. Discuss the possible reasons for this. How do the sizes of the several series capacitors compare to the original single one?

18. Envision a parallel-plate capacitor across the terminals of a battery. When a dielectric is inserted between the plates, the capacitance increases, as does the charge, and more energy is stored by the device. Where is that additional energy stored?

19. Figure Q19 is a profile view showing the equipotentials in the space between the control grid and the first anode of an electron gun (Fig. 16.4). The holes in the ends of the metal surfaces of the grid and anode cause a distortion of the equipotentials, and the arrangement serves as an *electron lens*. The beam is brought to a focus at the "crossover" point P. Explain how this process happens. (Incidentally, another such lens in the second anode focuses the beam onto the screen.)

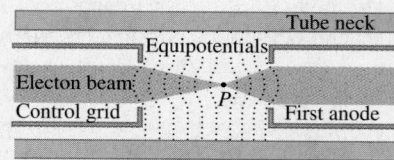

Figure Q19

20. The two identical capacitors in Fig. Q20 are charged at different voltages such that $Q_1 > Q_2$. What will be the charge on each capacitor after the two switches are closed? Explain your answer completely.

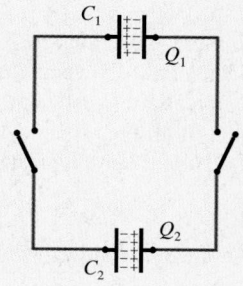

Figure Q20

Figure Q10

21. The Cottrell precipitator is shown mounted to a chimney in Fig. Q21. How do you think it works to remove 99% of the ash and dust that passes through it?

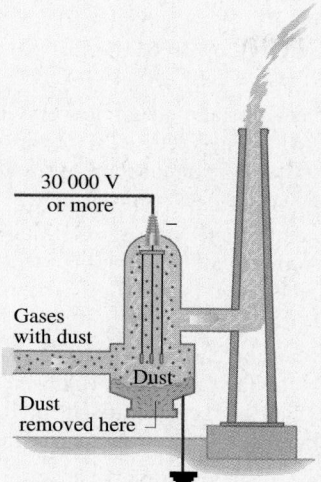

Figure Q21

Multiple Choice Questions

The first six questions refer to Fig. MC1 which shows two flat parallel metal plates separated by a distance of 10.0 cm. The plates are connected to a constant 100-V source.

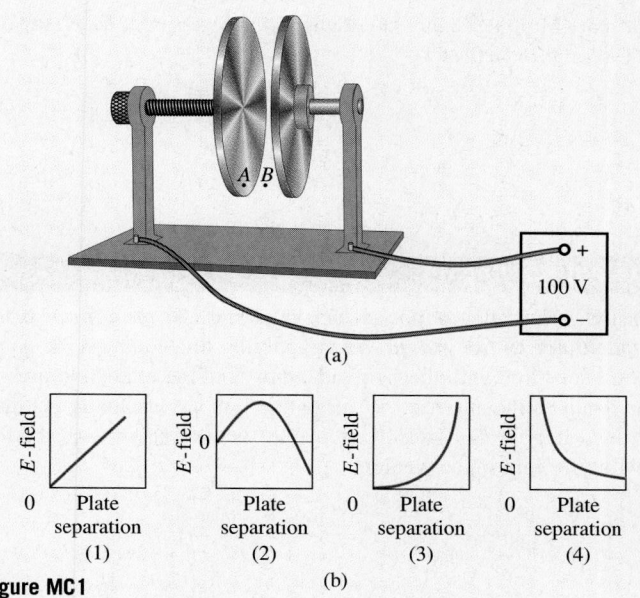

(a)

(1)   (2)   (3)   (4)

Figure MC1

(b)

1. The direction of the E-field is from   (a) horizontal, left to right   (b) horizontal, right to left   (c) vertical, top to bottom   (d) vertical, bottom to top   (e) none of these.

2. What is the value of the electric field between the plates?   (a) 1.00 N/C   (b) not enough information to know   (c) 10.0 C/N   (d) 1.00×10^3 C/N   (e) none of these.

3. Which graph in Fig. MC1b best represents the E-field between the plates as the separation increases?   (a) 1   (b) 2   (c) 3   (d) 4   (e) none of these.

4. If the voltage across the plates is doubled while their separation is halved, the E-field will   (a) quadruple   (b) be halved   (c) be quartered   (d) double   (e) none of these.

5. The voltage at point A is   (a) equal to the voltage at point B (b) greater than the voltage at point B (c) less than the voltage at point B (d) know nothing about it   (e) none of these.

6. If a negative point-charge is brought from point A to point B its electrostatic potential energy will   (a) remain unchanged   (b) equal zero   (c) increase   (d) decrease   (e) none of these.

7. The quantity 1 C·V is equivalent to   (a) 1 V/m   (b) 1 N·m   (c) 1 C/N   (d) 1 V/N   (e) none of these.

8. In the case of a nonconductor   (a) its surface must be at a single potential   (b) its entire volume, except for the surface, is at a constant potential   (c) different regions may well be at different potentials   (d) the potential must be zero everywhere within it   (e) none of these.

9. Electric field lines always point toward   (a) ground   (b) a region of higher potential   (c) a region of lower potential   (d) positive charge   (e) none of these.

10. The potential as we get closer and closer to a point-charge   (a) approaches $\pm\infty$ (b) is zero   (c) is indeterminate   (d) is exceedingly small but not zero   (e) none of these.

11. Given a group of nearby charges whose net value is nonzero, the equipotential surface at a very great distance is   (a) nearly a plane   (b) nearly a sphere   (c) quite indeterminate   (d) nearly a cylinder   (e) none of these.

12. Any closed equipotential surface that does not surround a net charge must   (a) be at zero-potential   (b) be a sphere   (c) enclose an equipotential volume   (d) be infinitely small   (e) none of these.

13. Suppose we examine a charged metal cup with a device that measures potential with respect to ground. What will happen as the probe from the device that touches the cup changes from contact with the outside to contact with the inside?   (a) the reading will ascend   (b) the reading will descend partway   (c) the reading will go to zero   (d) nothing   (e) none of these.

14. A volume of space is found to have a constant potential everywhere within it. It follows that in that region   (a) the *E*-field is zero   (b) the potential is zero   (c) the *E*-field is finite and uniform   (d) the potential gradient is a nonzero constant   (e) none of these.

15. When two charged metal objects are connected to each other by a conducting wire, the one that gains electrons or, equivalently, loses positive charge is said, in comparison to the other object, to have had   (a) a greater electrical potential energy   (b) a lower capacitance   (c) a lower dielectric constant   (d) a higher potential   (e) none of these.

16. Electrical potential determines the flow of positive charge just as   (a) pressure determines the flow of fluid, and temperature the flow of thermal energy   (b) kinetic energy determines the flow of matter, and charge the flow of electricity   (c) temperature determines the flow of entropy, and entropy the flow of heat   (d) power determines the flow of fluid, and efficiency the flow of work   (e) none of these.

17. Generally, when any conductor is connected to ground   (a) nothing happens   (b) charge flows so that the conductor takes on a potential above zero   (c) charge flows so that the conductor takes on a potential below zero   (d) charge flows so that the conductor takes on the potential of the ground   (e) none of these.

18. The movement of charge in an electric field from one point to another at a constant speed without the expenditure of work, by or against the field   (a) is impossible   (b) can only occur along a field line   (c) can only occur along an equipotential   (d) can only occur in a uniform field   (e) none of these.

19. A body that when grounded takes on electrons is said to have originally had   (a) a negative potential   (b) zero-potential   (c) a positive potential   (d) an original net negative charge   (e) none of these.

20. A neutron somehow picks up 10 eV, which is equivalent to it increasing its   (a) charge by 10 C   (b) electrical potential by 10 V   (c) energy by 16×10^{-19} J   (d) capacitance by 10 μF   (e) none of these.

21. The electrostatic potential everywhere inside a hollow conductor is   (a) always zero   (b) never positive   (c) always a nonzero constant   (d) constant, provided there are no enclosed isolated charges   (e) none of these.

22. The capacitance of a parallel-plate capacitor is   (a) independent of the plate separation   (b) dependent on the charge   (c) dependent on the voltage   (d) independent of the plate area   (e) none of these.

23. We can increase the capacitance of a parallel-plate capacitor by   (a) cooling the plates   (b) bringing the plates closer together   (c) decreasing the permittivity of the medium in the gap   (d) increasing the voltage   (e) none of these.

24. If the voltage across a capacitor is doubled, the amount of energy it can store   (a) doubles   (b) is halved   (c) is quadrupled   (d) is unaffected   (e) none of these.

25. If the charge on a capacitor is halved, its stored energy   (a) is halved   (b) is quartered   (c) is unchanged   (d) is doubled   (e) none of these.

For more Multiple Choice Questions with answers click on WARM-UPS in CHAPTER 16 on the CD.

Suggestions on Problem Solving

1. When calculating the PE_E of a charge, be careful with signs. An electron at a positive potential has a negative potential energy. Similarly, when a charge "falls" through a potential difference, thereby losing potential energy, the corresponding change in KE is positive. A negative charge "falls" (in the opposite direction to the *E*-field) to a *higher potential*; a positive charge "falls" (in the same direction as the *E*-field) to a lower potential.

2. If a parallel-plate capacitor has charges on its two plates of $+Q$ and $-Q$, the charge on the capacitor is Q, and that value is indicated in the defining equation $C = Q/V$. Similarly, the area *A* is essentially the area of overlap, generally the area of *either* plate.

3. When looking at equations, read *C* as capacitance and *Q* as charge. But C is a coulomb, which is the unit of charge. Don't confuse them.

4. Keep the following quantities handy: $\varepsilon_0 = 8.85 \times 10^{-12}$ $C^2/N \cdot m^2$, $1/\varepsilon_0 = 1.129 \times 10^{11}$ $N \cdot m^2/C^2$, $4\pi\varepsilon_0 = 1.113 \times 10^{-10}$ $C^2/N \cdot m^2$, and of course $1/4\pi\varepsilon_0 = k_0 = 8.988 \times 10^9 N \cdot m^2/C^2$.

5. Remember that Eq. (16.13) yields an equivalent capacitance for a series string of capacitors, *which is always less than the smallest capacitance in the group*. When applying this equation, be wary. Don't substitute into the right side, carry out the math, and present the result as the answer, forgetting that you have actually computed $1/C$ and not *C*. For only two capacitors you might find the following form more convenient:

$$C = \frac{C_1 C_2}{C_1 + C_2}$$

In any event, you could use it for a quick check to see that the magnitude of your answer is correct.

6. To remember the set of equations for a point-charge $q_.$, memorize only Coulomb's Law: force is $k_0 q q_./r^2$. By definition, divide that quantity by charge to get the field, $k_0 q_./r^2$; in effect, multiply force by distance to get work and PE, $k_0 q q_./r$; by definition, divide that quantity by *q* to get potential, $k_0 q_./r$. Although simplistic and not rigorous, this scheme is helpful as a memory device.

| **Problems** ✦ | **Coordinated Problems** ✦ | **Progressive Problems** ✦ | **Solutions** |

1. Coordinated Problems: The three problems within each magenta-colored grouping are solvable in similar ways. Note that the first of these always has a hint; moreover, its solution is provided in the back of the book. *Work out each of these sets; they'll strengthen technique and build confidence.* **2. Progressive Problems:** The problems introduced in blue unfold step-by-step carrying along the analysis in a more suggestive way than is customary. *Work out all of these; they'll guide you through the analytic process and help develop problem-solving skills.* **3. Worked-Out Solutions:** Studying worked-out solutions is an important part of learning how to solve problems. Accordingly, additional *solutions* to a number of model problems are given below. *Make sure you understand each of them before you go on to the next problem.* **4.** Also provided in the back of the book are the *Answers* to all odd-numbered problems, as well as worked-out *solutions* to those with boldface numbers. Problem numbers in italic indicate that a solution appears in the Student Solutions Manual.

SECTION 16.1: *ELECTRICAL*-PE & POTENTIAL

1. [I] THIS PROBLEM EXPLORES THE CONCEPTS OF POTENTIAL, WORK, AND ENERGY. We want to determine the work that must be delivered to a point-charge of +30.0 microcoulombs in order to move it from a location at 2.0 V to a location at 8.0 V. (a) How much of a potential difference does the charge experience? (b) Does its *electrical*-PE increase or decrease? (c) What is the change in the magnitude of its *electrical*-PE? (d) How much work was done in bringing the charge to its new location? (e) Was that work done on the point-charge by the field or by some external applied force?

2. [I] A small 12.0-V electric heating element provides 60 kJ of thermal energy in a certain amount of time. That energy is delivered electrically via the charge. How much charge must flow through the device during that interval?

> SOLUTION: The potential difference is 12.0 V and that's a voltage drop. The charge delivers, or "loses," electrical energy in the amount of 60 kJ. $\Delta V = \Delta PE_E/q_o$ so that $\Delta PE_E = (12.0 \text{ V})q_o = 60$ kJ and $q_o = 5.0$ kC.

3. [I] A +20.0-nC charge in an *E*-field is moved by an applied force. If 60.0 nJ of work are done on the charge by the force, what is the change in potential experienced by the charge? [*Hint: Use the definition of potential difference.*]

4. [I] A tiny positive charge of 50.0 μC is in a nonuniform electric field. An external force carries the charge from its original location, where the potential is 200.0 V, to a new location expending 100.0 mJ of work in the process. What is the potential of this new location?

5. [I] There is an *E*-field in a region of space, and a +50.0-nC charge is placed at a point where the potential is 500 V. When released the charge moves, while the field does 200.0 μJ of work on it. What is the potential of the final location of the charge?

6. [I] While carrying a charge of +30.0 nC, a small pith ball is released in an electric field. It subsequently moves and drops 150 V in the process. How much work was done on the charge by the field?

7. [I] THIS PROBLEM EXPLORES THE CONCEPTS OF POTENTIAL, WORK, AND ENERGY. A minute metal sphere has 1000 electrons removed from its surface. It is then placed at a point in an electric field where the potential is +2.00 V. (a) What is the charge on the sphere? (b) What is its electrostatic-PE at that location? (c) The sphere is subsequently moved to a new location where the potential is +12.00 V. Does it experience a decrease or an increase in potential,

and by what amount? (d) Does the sphere gain or lose energy as a result of being moved, and how much? (e) Does the sphere have work done on it by the field, or must some external agency be acting on it? Explain. (f) How much work must be done in moving the sphere against the *E*-field?

8. [I] THIS PROBLEM WILL HELP US BETTER UNDERSTAND THE CONCEPTS OF POTENTIAL, WORK, AND ENERGY. A small droplet carrying −30.0 nC of charge is moved in an electric field from a location (point-*A*) where the potential is +1.00 V to a place (point-*B*) where the potential is −14.00 V. We want to determine what happened to it from an energy perspective. (a) Draw a picture of what's going on. (b) Is point-*A* at a higher or lower potential than point-*B*? (c) Does the droplet experience a decrease or an increase in potential, and by what amount? (d) Does the droplet gain or lose energy, and how much? (e) Does the droplet have work done on it by the field, or must some external agency be acting on it? Explain.

9. [I] A tiny sphere carrying a charge of −25.0 nC is moved 100 cm in a uniform electric field with no acceleration. It goes from a location at a potential of zero to a point where the potential is 100 V. How much work is done on it by the applied force? What is the significance of the sign of ΔW?

10. [I] Imagine a uniform *E*-field. A charge of −1.00 μC is moved parallel to the field by a force acting in the same direction as the field. If 500 nJ of work are done by the force and if the charge was initially at 1.00 V, what will be its final potential?

11. [I] The *E*-field in a region of space is uniform and equal to 10.0 V/m. What voltage difference will a 20.0-μC charge experience when displaced 10.0 cm in the direction of the field?

12. [I] A stream of singly ionized gold atoms pouring from a small oven impinges on a metal target. The target is attached to the positive terminal of a 12-V battery, and the oven is attached to the negative terminal. How much kinetic energy do the ions pick up in the process of crossing over to the target?

13. [I] What voltage should be put across a pair of parallel metal plates 10.0 cm apart if the field between them is to be 1.00 V/m?

14. [I] Two charged parallel metal plates, inside the evacuated cathode-ray tube of a radar system, are separated by 1.00 cm and have a potential difference of 25.0 V. What is the value of the electric field in the gap?

15. [I] An electron is to be accelerated from rest at a grounded cathode to a metal plate at +500 V. Express in electron volts how much kinetic energy it will gain. How much electrical potential energy will it lose, if any?

16. [I] It is fairly easy to strip the two electrons off a helium atom, leaving a bare nucleus of two neutrons and two protons. That so-called *alpha particle* is to be accelerated, essentially from rest, up to a KE of 100 keV by having it "fall" through a potential difference. What is the necessary voltage difference?

17. [II] An electron initially at rest in an X-ray tube crosses a potential difference of 30 kV and crashes into a target that then emits radiation. Compute the KE of the electron (in joules) at impact. Determine its maximum speed.

18. [II] THIS PROBLEM TREATS THE FIELD BETWEEN CHARGED PARAL-

LEL PLATES. A small conducting ball carries a charge of −44.0 nC. It is placed between two large horizontal metal plates that are 40 mm apart. A 30-V battery is attached across the two plates (as in Fig. 16.24), and we want to find the electrostatic force on the ball. (a) Write an expression relating the voltage across the plates and the E-field between them. (b) Compute that E-field. (c) Determine the electrostatic force on the ball. (d) If the bottom plate is at a higher potential than the upper plate, in what direction is the force? Explain.

19. [II] THIS PROBLEM EXPLORES THE CONCEPTS OF POTENTIAL, WORK, AND ENERGY. A uniform electric field of 12.9 N/C exists in a region of space. A small gold-covered pith ball carrying a charge of +20.0 μC is pulled against the field (i.e., in the opposite direction to it) a distance of 15.0 cm. (a) What is the force on the charge? (b) How much work was done on the charge by the externally applied force? (c) What is the change in the electrostatic potential energy of the ball? (d) What is the change in the potential experienced by the ball? (e) If the ball is at a final position where the voltage is 100 V, at what voltage did it start?

20. [II] A proton is released from rest in a uniform electric field of 500 V/m. How fast will it be moving after traveling 40 cm in and parallel to the field?

21. [II] Two horizontal parallel metal plates 10.0 cm apart in a vacuum chamber are to be used to suspend an electron in "midair." What voltage must be put across the plates?

22. [II] Two parallel copper plates each of area A carry charges of +Q and −Q. The gap between them is filled with a material having a dielectric constant K_e. Show that if a positive charge q is moved a distance d along but in the opposite direction to a field line, an amount of work will be done on it given by the expression

$$W = qQd/AK_e\varepsilon_0$$

23. [II] A region of space is traversed by a uniform electric field of strength 10.0 V/m directed due east. A small metal sphere on which has been deposited a charge of +50.0 nC is in the field. The sphere is moved (at a constant speed under the action of an external force) 10.0 cm north, 50.0 cm east, 20.0 cm south, and 50.0 cm east. How much work was done in total on the sphere by the applied force? What was its net change in potential?

24. [II] A uniform electric field in air has a strength of 100 V/m and points due north. An applied force causes an electron to travel without accelerating through the field a distance of 50.0 cm in a direction 36.9° north of west. What is the change in potential that it experiences? How much work was done on it by the applied force?

25. [II] The fluid within a living cell is rich in potassium chloride, while the fluid outside it predominantly contains sodium chloride. The membrane of a resting cell is far more permeable to ions of potassium than sodium, and so there is a transport out of positive ions, leaving the cell interior negative. The result is a voltage of about −85 mV across the membrane, called the *resting potential*. The membrane (about 50 atom-layers) is roughly 8.0-nm thick. Assuming the E-field across the cell membrane is constant, determine its magnitude.

SECTION 16.2: POTENTIAL OF A POINT-CHARGE

26. [I] A 2.00-mm-diameter conducting sphere in vacuum is charged with +30.0 nC. Determine the potential at a distance of 2.00 m from its center. [*Hint: Since 2.00 m >> 2.00 mm, this is essentially a point-charge.*]

27. [I] Figure P27 shows two hollow concentric metal spheres of radii ρ and R. If the inner one is charged with +Q and the outer surface is grounded, what is the potential at any point P outside the larger sphere? Explain your thinking.

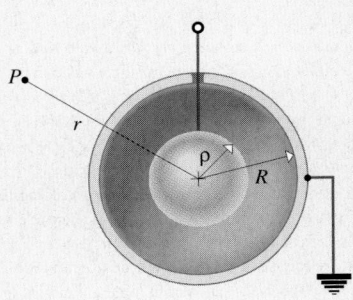

Figure P27

28. [I] At what distance from a 100-μC point-charge in air will the potential be 100 V?

29. [I] We wish to place a small charged sphere at a distance of 10.0 m from a point in space such that the voltage at that point is 1.00 V. How much charge should be on the sphere?

30. [I] A 10.0-cm-diameter solid gold sphere carries a charge of +0.100 μC. What is the potential 10.0 m away in the surrounding air?

31. [I] High atop a nonconducting rod is a small sphere carrying a uniform charge of +25.0 μC. Two detectors are located, respectively, 1.00 m due south and 20.0 m due east. What is the potential difference between the two locations? Which is at the higher voltage? [*Hint: We are being asked to find the potential at two locations in space due to a point-charge.*]

32. [I] Out in space there is a 1.0-cm copper sphere uniformly charged with −100.0 nC. Two small probes are located at point A (2.00 m due south) and point B (10.0 m north of west). What is the potential at B measured with respect to that at A?

33. [I] Imagine that a small metal sphere left by some alien civilization is floating in space. A probe is launched directly at the sphere in order to study it. When the probe is 1000 m from its target, it detects a voltage of 110.0 V; and when it's at 100.0 m, it records 10.0 V. What, if any, is the charge on the sphere?

34. [I] While surrounded by vacuum, a minute pith ball is charged with +50.0 μC. What is the potential 5.0 m away from it?

35. [II] A metal target sphere of 20-cm diameter suspended out in space is given a charge of +1.00 nC. How much work must be done on a proton in taking it from very far away (essentially infinity) to the surface of the sphere?

36. [II] A hydrogen atom consists of a proton around which circulates an electron at an average distance of 0.053 nm. Determine the potential at that distance due to the proton and find the potential energy of the electron (in joules).

37. [II] With Problem 35 in mind, through what voltage must the proton be accelerated from rest (by some sort of space weapon) if it is to arrive at the sphere at a speed of 8.5×10^4 m/s? Neglect any gravitational effects.

38. [II] Refer to Fig. P27. If the inner sphere is charged with +Q and the outer surface is no longer grounded, how does the charge distribute itself and what is the potential in the region beyond the spheres in terms of ρ, R, and Q? Explain your answer. Write an expression for the field beyond the spheres.

39. [II] The radius of a steel ball is 1.00 mm, and it has a surface

charge density of 2.00 C/m². It is embedded in a medium that has a dielectric constant of 2.0. What is the potential 2.00 m from the center of the sphere?

40. [II] In the middle of a large tank of ethanol (at 25 °C) is suspended a small hollow copper sphere. If +40.0 μC of charge are deposited on the inside surface of the sphere what will be the electric potential 2.6 m from its center?

41. [II] A point-charge of −100 nC is surrounded by oil having a permittivity of 20×10^{-12} C²/N·m². How much work would be done on a proton in transporting it from a radial distance of 0.50 m to 1.50 m?

42. [III] If the inner sphere in Fig. P27 is charged with $+Q$ and the outer surface is grounded, show that

$$\Delta V = kQ\left(\frac{1}{\rho} - \frac{1}{R}\right)$$

is the expression for the potential difference across the gap.

SECTION 16.3: THE POTENTIAL OF SEVERAL CHARGES

43. [I] What is the electric potential at a point 1.00 m from each of two +30.0-nC point-charges? [Hint: Potential is a scalar quantity.]

44. [I] A point-charge of +50.0 nC is 0.50 m from a point-charge of −50.0 nC. What is the potential at a point equidistant from the two charges?

45. [I] Two small gold beads carry charges of +30.0 nC and −40.0 nC, respectively. What is the potential at a point 2.00 m from both spheres?

46. [I] Two steel ball bearings 1.00 m apart in air carry charges of +40.0 μC and −20.0 μC. What is the potential at a point 1.00 m from the negative charge and 2.00 m from the positive charge?

47. [I] Two small spheres carrying charges of +30.0 μC and −50 μC are 100 cm apart in air. What is the potential at a point on the center-to-center line midway between them?

48. [I] THIS PROBLEM EXPLORES THE POTENTIAL OF A COLLECTION OF CHARGES. A tiny conducting sphere carrying a charge of either +60 μC or −60 μC is fixed at each of the four corners of a square 20 cm on a side. Going around the square from corner to corner, the charges have alternating polarities of +−+−. (a) Show that the diagonal of the square has a length of $20\sqrt{2}$ cm. (b) What is the magnitude of the potential due to each charge at the center of the square? (c) What is the net potential at the center? Explain. (d) What is the electric field at the center? Explain. Assume the medium is vacuum.

49. [I] A large hollow metal sphere with a radius of 1.00 m carries a charge of +2.00 μC. What is the potential at its surface? What is the voltage at its center?

> SOLUTION: The entire inside of the sphere is an equipotential volume which has the same voltage as the surface. The potential at the surface is given by $V = k_0Q/R = (8.99 \times 10^9 \text{ N·m}^2/\text{C}^2)(+2.00 \times 10^{-6} \text{ C})/ (1.00 \text{ m}) = 18.0 \text{ kV}$.

50. [I] THIS PROBLEM DEALS WITH THE POTENTIAL OF A SPHERE. Consider a 30.0-cm diameter hollow metal sphere in air. It's momentarily touched to a running Van de Graaff generator (Fig. Q14, p. 563) operating at −30.0 kV. The sphere becomes charged, and we want to determine the magnitude of that charge.

(a) What is the radius of the sphere in meters? (b) What is the voltage of the sphere? (c) What is the charge on the sphere?

51. [I] THIS PROBLEM DEALS WITH THE POTENTIAL OF A SPHERE. A solid aluminum sphere has a radius of 8.99 cm. It is brought into contact with a charged hollow copper sphere, and both come to equilibrium at 8.00 kV. We want to determine the resulting surface charge density on the solid sphere. (a) What is the radius of the solid sphere in meters? (b) What is the voltage of the aluminum sphere once the two spheres are separated? (c) What is the net charge on the aluminum sphere? (d) What is the surface charge density on that sphere?

52. [II] A +60.0-nC point-charge is at each of the vertices of an equilateral triangle. What is the potential at a point 25.0 cm from each of the charges?

53. [II] Two very small metal spheres carrying charges of +10 μC and −25 μC are located at coordinates (0, 0) and (3.0 m, 0), respectively, in air. How much work would have to be done to bring a third sphere with a charge of −10 μC from very far away to the point (0, 4.0 m)?

54. [II] Tiny conducting spheres carrying charges of +60 μC are fixed at each of the four corners of a square 20 cm on a side. What is the potential at the very center? What is the electric field at the center? Assume the medium is vacuum.

55. [II] Three 80.0-nC point-charges form a triangle. What is the potential 1.00 m from two of them and 2.00 m from the third?

56. [II] Point-charges of +300 nC, −700 nC, +500 nC, and −100 nC are located in sequence at the corners of a square 40 cm on a side. Determine the potential at the center of the square.

57. [II] THIS PROBLEM EXAMINES THE POTENTIAL OF SEVERAL CHARGES. Two identical very small silver balls in air are touching each other when a charge of +2.00 μC is deposited on one of them. They are then separated, one being fixed at +1.00 m, the other at −1.00 m on the x-axis. A third small gold ball carrying +4.00 μC is brought from across the street to a point-A +2.00 m up on the y-axis. And we want to find out how much work it took to bring over the third ball. (a) How much charge ultimately resides on each of the silver balls? (b) What is the distance from either silver ball to point-A? (c) What is the potential at point-A due to either silver ball? (d) What is the potential at point-A due to both silver balls? (e) What is the electrostatic potential energy of the gold ball at point-A?

58. [II] THIS PROBLEM EXAMINES POTENTIAL AND THE WAY CHARGE DISTRIBUTES ITSELF AMONG CONDUCTORS. Two identical hollow gold spheres 2.00 cm in diameter having a center-to-center separation of 1.00 m are connected to each other by a very fine gold wire. A charge of +200 nC is sprayed onto one of the spheres, and we want to determine the potential at its center once the system settles down. (a) Assuming that a negligible amount of charge resides on the wire, how much charge is on each sphere once equilibrium is established? (b) Would your answer be different if the spheres were not the same size? Explain. (c) Compare the potential of each sphere. (d) Imagining that one of the spheres is removed, what is the voltage at the surface of the remaining sphere? (e) What is the voltage at the center of the remaining sphere? (f) Now suppose both spheres are present, what is the net voltage at the center of each sphere?

59. [II] THIS PROBLEM DEALS WITH THE POTENTIAL OF A SPHERE.

Recall the Van de Graaff generator (Fig. Q14, p. 563). Suppose the top of it consists of a large aluminum sphere having a radius of 20.0 cm, which carries an initial charge $Q_{Li} = -1.00$ μC. With the motor shut off, a small uncharged 10.0-cm diameter metal sphere is momentarily touched to the dome of the generator. We want to find out how much charge ends up on both the small (Q_{Sf}) and large (Q_{Lf}) spheres. (a) What is the radius of each sphere in meters? (b) What is the initial voltage of the generator; that is, what's the initial potential of the large sphere? (c) What can you say about the potential of each sphere immediately after they're separated? (d) Using the fact that the spheres' voltages are equal, prove that $Q_{Lf}/(0.200$ m$) = Q_{Sf}/(0.050$ m$)$. (e) Using Conservation of Charge along with the previous equation, determine Q_{Sf} and Q_{Lf}.

60. [III] THIS PROBLEM DEALS WITH THE POTENTIAL OF SEVERAL CHARGED SPHERES. Three hollow gold balls 20.0 cm in diameter lie along the x-axis in air with their centers at locations $x_1 = -1.00$ m, $x_2 = 0$, and $x_3 = +1.00$ m. Very fine gold wires connect the first ball to the second and the second to the third. A charge of q is sprayed onto one of the balls. We want to determine how that charge finally redistributes itself. (a) One might assume that an equal charge ends up on each ball; why is that not the case? (b) Assuming that a negligible amount of charge resides on the wires, what can you say about the charges $(q_1$ and $q_3)$ on the two end balls? (c) Show that the potential at $x_2 = 0$ is $V_2 = 2k_0q_1/(1.00$ m$) + k_0q_2/(0.100$ m$)$ where q_2 is the charge on the middle ball. (d) Write an expression for V_1, the potential at the centers of either end ball. (e) Explain why $V_1 = V_2$. (f) Using this and Conservation of Charge, solve for q_1, q_2, and q_3 in terms of q.

61. [III] A total of eight tiny conducting spheres, each carrying a charge of -100 nC, are placed one each at the corners of a cube 1.0 m on a side. Find the potential at the center. What is the electric field at the center?

62. [III] Two very small conducting spheres surrounded by transformer oil are charged with $+50.0$ μC and -40.0 μC, respectively. Determine the point on the line connecting them where the potential is zero, if indeed such a point exists. The center-to-center separation is 1.00 m.

SECTION 16.5: THE CAPACITOR

63. [I] A 100-pF capacitor is charged by putting it across a 1.5-V battery. What is the charge on its plates?

64. [I] A 48.0-μF capacitor, with an impregnated paper dielectric, is placed across the terminals of a 12-V battery. How much charge flows from the battery to the capacitor?

65. [I] Cathode-ray tubes used in computers and TV sets are coated inside and out with a conducting graphite paint called *Aquadag*. The inside layer is connected to the second anode (see Fig. 16.4) and helps to accelerate the beam and keep it narrow. The outer layer is connected to the chassis, which is ground. This arrangement creates an Aquadag-glass-Aquadag capacitor (which is connected across the power supply providing the high voltage for the anode and tends to smooth out voltage variations). The capacitance is generally only about 500 pF, but all such tubes always carry a warning that even though the set has been turned off, the CRT must be discharged, otherwise you risk a dangerous shock. Explain why this warning is necessary and compute the charge for a tube voltage of 20 kV.

66. [I] Estimate the capacitance of the Earth. Its radius is 6371 km. Give the answer to two significant figures.

67. [I] What is the radius of a conducting sphere in air if it has a capacitance of 10 pF (or as it used to be called, $\mu\mu$F)?

68. [I] As a very rough estimate, assume you have about the same capacitance as a conducting sphere 3/4 m in diameter. How much charge can you store at a potential of 100 V?

69. [I] A 10.0-μF parallel-plate air capacitor is charged so that it carries -200 μC on one plate and $+200$ μC on the other. What is the potential difference across it?

70. [I] What is the capacitance of two parallel metal plates each with an area of 100 cm^2 separated by 1.0 mm of air?

71. [I] What is the capacitance of two 600-cm^2 copper sheets separated in air by 0.50 mm?

72. [I] Thus far, we have used units for ε that relate back to Coulomb's Law. Accordingly, prove that 1 C^2/N·m^2 is equivalent to 1 F/m.

73. [I] Two metal balls hanging from insulating threads are separated by 1.0 cm. Each carries a charge of 5.0 μC, although one is positive and the other is negative. If there is a potential difference of 200 V between them, what is the capacitance of the pair?

74. [I] Show that capacitance has the dimensions of $[Q^2T^2]/[ML^2]$. ($Q \rightarrow$charge, $T \rightarrow$time, $M \rightarrow$mass, and $L \rightarrow$length.)

75. [I] A parallel-plate capacitor immersed in transformer oil carries a charge of $+20$ μC on one plate and -20 μC on the other when there is a voltage of 4.0 V across it. What is its capacitance?

76. [I] THIS PROBLEM EXAMINES THE CAPACITANCE OF PARALLEL PLATES. Two flat circular metal plates, each of area A, are separated by a distance d. The device, which has a capacitance C_0, has its two plates connected to the two terminals of a 6.0-V battery. The capacitor receives a charge q and is then disconnected from the battery. (a) What is the voltage across the plates? (b) The two plates are pulled apart to a separation of $2d$. What is the new capacitance in terms of C_0? (c) What is the charge on the capacitor in terms of q? (d) What is the new voltage across the capacitor? (e) In light of your previous answer, is energy supplied to the capacitor, and if so how?

77. [I] THIS PROBLEM EXAMINES THE CAPACITANCE OF PARALLEL PLATES. Two flat circular metal plates, each of area A, are separated by a distance d. The device, which has a capacitance C_0, has its two plates connected to the two terminals of a 12.0-V battery. The capacitor receives a charge q and is then disconnected from the battery. (a) What is the voltage across the plates? (b) A sheet of glass having a relative permittivity of 10 is slid between the plates. Write an expression for the electric field now, in terms of the field E_0 which existed before when air occupied the gap. (c) What is the new capacitance in terms of C_0? (d) What is the new voltage across the capacitor?

78. [II] Two sheets of aluminum each of area 200 cm^2 are separated in air by 2.0 mm. Determine the charge it can store when attached to a 12.0-V battery.

79. [II] What is the capacitance of two parallel metal plates each with an area of 100 cm^2 separated by 1.0 mm if the gap is filled with glass having a dielectric constant of 10?

80. [II] The plates of a capacitor consist of two 2.00-m^2 sheets of copper separated by a 0.10-mm-thick sheet of teflon. How much

charge will have to be put on the plates to produce a voltage difference of 20.0 V across the plates?

81. [II] A little flat ceramic capacitor consists of two circular plates 0.50 cm in diameter separated by a dielectric 1/3-mm thick with a relative permittivity of 4.8. Compute its capacitance.

82. [II] The plates of a parallel-plate capacitor separated by a 0.50-mm-thick sheet of nylon carry a surface charge density of 100.0 nC/m^2. What voltage difference appears across the plates?

83. [II] Figure P83 depicts a neuron, or nerve cell, and shows the long signal-carrying axon (which, from spine to fingers, can be more than a meter in length). As discussed in Problem 25, the axon membrane is usually positive on the outside and negative on the inside. The dielectric constant of the membrane has been measured to be about 7. Given that the membrane wall is a mere 6.0-nm thick and that the axon radius is 5 μm, determine its capacitance per unit area. Explain any assumptions you make.

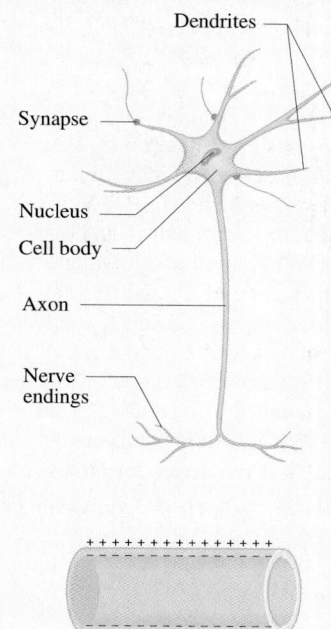

Figure P83

84. [II] A Leyden jar, like the one shown in Fig. 16.16, has a 22-cm-diameter base, and metal foil (inside and out) that goes up to a height of 35.5 cm. The inside metal is attached to the central conductor by a wire. The glass has a thickness of 2.25 mm and a dielectric constant of 7.2. Estimate its capacitance.

85. [II] With Problem 83 in mind, determine the surface charge density of the axon. In the resting state, when there is no signal being transmitted, the potential difference across the membrane is about 70 mV.

86. [II] With the previous problem in mind, determine the capacitance per unit length of the axon.

87. [III] Fig. P27 shows two hollow metal concentric spheres of radii ρ and R. If the inner one is charged with $+Q$ and the outer surface is grounded, show that the capacitance is

$$C = \frac{\rho R}{k(R - \rho)}$$

Moreover, if the gap (d) is very small and A is the area of either inside surface ($4\pi\rho^2 \approx 4\pi R^2$), show that

$$C \approx \frac{\varepsilon A}{d}$$

SECTION 16.6: CAPACITORS IN COMBINATION

88. [I] Two capacitors of 175 μF and 200 μF are wired in parallel. What's their equivalent capacitance?

89. [I] A 3.0-nF capacitor is wired in series with a 6.0-nF capaci-

tor. What's their equivalent capacitance? [Hint: Capacitors in series add as one over the capacitance.]

90. [I] Three 3.0-μF capacitors are arranged in series and put in a box so that only the first and last terminals jut out. What capacitance would be measured across those two terminals?

91. [I] Two 6.0-μF capacitors are wired together in parallel, and that combination is put in series with a 4.0-μF capacitor. What is the total capacitance of the three elements?

92. [I] Imagine four 1.0-μF capacitors forming two groups, each consisting of two capacitors in parallel. These two pairs are then wired in series. What is the equivalent capacitance of all four?

93. [I] Given three capacitors of values 60 pF, 30 pF, and 20 pF, determine the net capacitance when they are placed successively in series and then in parallel.

94. [I] What is the equivalent capacitance of the circuit between the terminals-A and B indicated in Fig. P94?

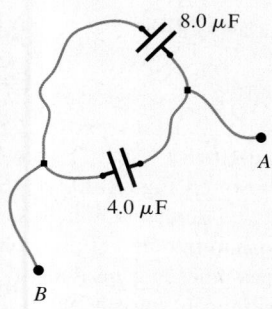

Figure P94

95. [I] Suppose we make an alternating stack of aluminum foil and paper (numbering each sheet of metal successively): 101 sheets of foil, 100 of paper, each 12 cm by 50 cm. The paper has a thickness of 0.22 mm and a relative permittivity of 4.1. If all the odd-numbered foil sheets are connected to a common wire and all the even-numbered sheets are connected to a second common wire, what will be the capacitance measured between the two wires? [Hint: There's one sheet of dielectric per capacitor.]

96. [I] What is the equivalent capacitance of the circuit between the terminals-A and B indicated in Fig. P96?

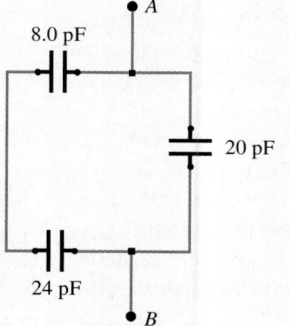

Figure P96

97. [I] What is the equivalent capacitance of the circuit between the terminals-A and B indicated in Fig. P97?

98. [I] What is the equivalent capacitance of the circuit between the terminals-A and B, and again between A and C, as indicated in Fig. P98?

99. [I] What is the equiva-

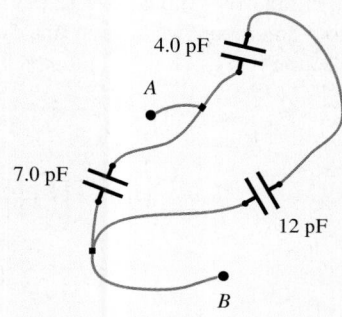

Figure P97

lent capacitance of the circuit between terminals-*B* and *C* in Fig. P98?

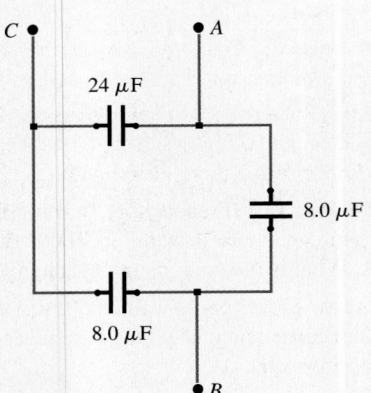

Figure P98

100. [II] THIS PROBLEM DEALS WITH SEVERAL CAPACITORS WIRED TOGETHER. Two capacitors C_1 and C_2 are wired together as shown in Fig. P100. A 16.0-V source is placed across terminals-*A* and *B*, and we wish to find the resulting charges and voltages on each capacitor. (a) What, if anything, can you say about the arrangement of the circuit? (b) What is the equivalent capacitance of the circuit between terminals-*A* and *B*? (c) Compute the charge on the equivalent capacitor. (d) What is the charge on C_1 and on C_2? (e) Determine the voltages across C_1 and C_2.

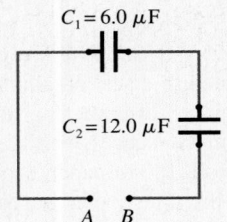

Figure P100

101. [II] THIS PROBLEM DEALS WITH SEVERAL CAPACITORS WIRED TOGETHER. In the circuit shown in Fig. P101 $C_1 = 8.0$ μF and $C_2 = 4.0$ μF. A 24.0-V source is placed across terminals-*A* and *B*, and we wish to find the charges and voltages on each capacitor. (a) What, if anything, can you say about the arrangement of the circuit? (b) What is the equivalent capacitance of the circuit between terminals-*A* and *B*? (c) Determine the voltages across C_1 and C_2. (d) Compute the charge on the equivalent capacitor. (e) What is the charge on C_1 and on C_2? (f) Compare your last two answers.

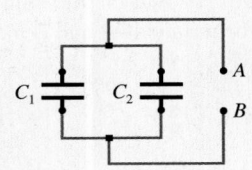

Figure P101

102. [II] In the circuit shown in Fig. P102 $C_1 = 6.0$ μF and $C_2 = 6.0$ μF and $C_3 = 3.0$ μF. A 12.0-V source is placed across terminals-*A* and *B*, and we wish to find the charges and voltages on each capacitor.

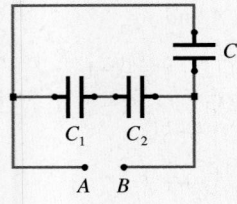

Figure P102

103. [II] THIS PROBLEM DEALS WITH SEVERAL CAPACITORS WIRED TOGETHER. In the circuit shown in Fig. P103 $C_1 = 12.0$ μF, $C_2 = 24.0$ μF, $C_3 = 6.0$ μF, $C_4 = 12.0$ μF, $C_5 = 12.0$ μF, and $C_6 = 12.0$ μF. We wish to determine the charge stored by the circuit when a voltage of 12.0 V is put across terminals-*A* and *B*. (a) What is the equivalent capacitance of C_1 and C_2, call it C_7? (b) What is the equivalent capacitance of C_3 and C_4, call it C_8? (c) What is the equivalent capacitance of C_5 and C_6, call it C_9? (d) What is the equivalent capacitance of C_7 and C_8, call it C_{10}? (e) What is the equivalent capacitance of C_9 and C_{10}, call it C, the equivalent capacitance of the circuit? (f) Determine the net charge stored when a voltage of 12.0 V is put across terminals-*A* and *B*.

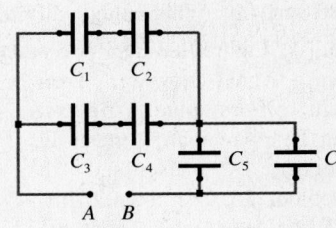

Figure P103

104. [II] What is the equivalent capacitance of the grouping shown in Fig. P104?

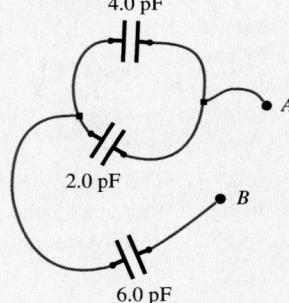

Figure P104

105. [II] What is the equivalent capacitance of the circuit between terminals-*A* and *B* in Fig. P105?

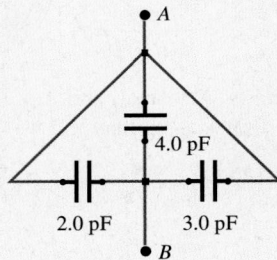

Figure P105

106. [II] What is the equivalent capacitance of the circuit between terminals-*A* and *B* in Fig. P106?

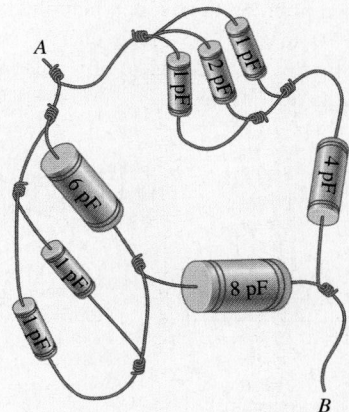

Figure P106

107. [II] What is the equivalent capacitance of the circuit between terminals-A and B in Fig. P107? The capacitors are 6.0 μF, 6.0 μF, 5.0 μF, 5.0 μF, and 7.5 μF.

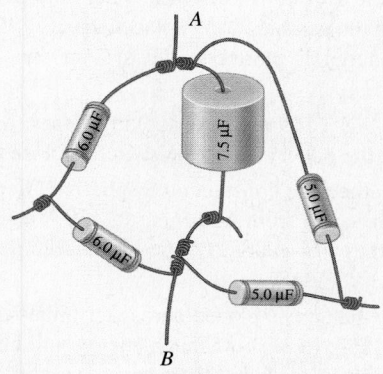

Figure P107

108. [II] What is the equivalent capacitance of the circuit between terminals-A and B in Fig. P108 if each of the capacitors is 2.0 pF?

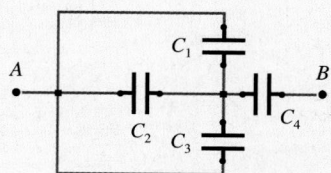

Figure P108

109. [II] What is the voltage across the 9.0-pF capacitor in the circuit shown in Fig. P109 after the switch is closed? Determine the equivalent capacitance across the battery. How much charge is drawn from the battery?

110. [II] Determine the voltage across the 4.0-μF capacitor after the switch is closed in the circuit shown in Fig. P110.

111. [II] Two capacitors of 100 μF and 50 μF are separately charged to 250 μC and 100 μC, respectively. They are then attached in parallel so the + plate of one goes to the − plate of the other, and vice versa. Determine the final voltage across the two.

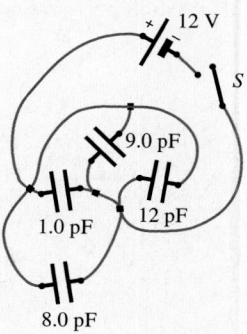

Figure P109

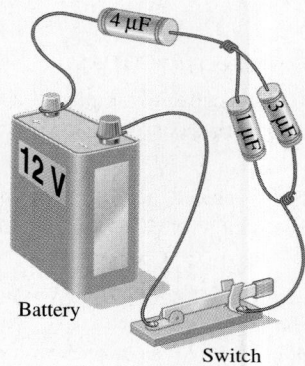

Figure P110

112. [II] Figure P112 shows four metal plates with the outer pair and inner pair wired together. Given that there is air in the gaps, which are $\frac{1}{2}$-mm wide, and that the area of each plate is 0.01 m², what is the capacitance of the device?

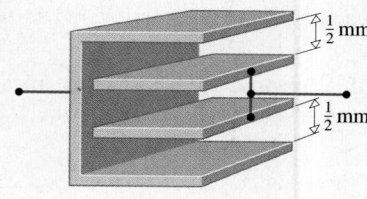

Figure P112

113. [III] Figure P113 shows three plates each 50 cm by 60 cm and separated by 1.0 mm. The plates are immersed in transformer oil with a dielectric constant of 4.5. Find the total capacitance of the system.

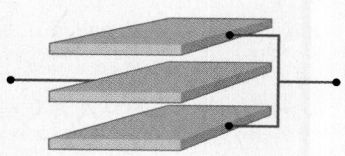

Figure P113

114. [III] What is the difference in potential between points A-D, B-D, and C-D in Fig. P114?

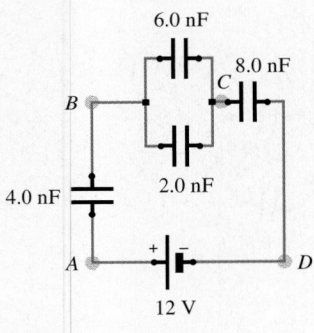

Figure P114

SECTION 16.7: ENERGY IN CAPACITORS

115. [I] When a 20.0-μF capacitor is placed across the terminals of a 12.0-V battery it becomes charged. How much energy is stored in the process?

116. [I] A capacitor is placed across the terminals of a 9.0-V battery and receives a charge of 20.0 μC. How much energy does the capacitor store?

117. [I] After receiving a charge of 40.0 μC, an 80.0-μF capacitor is separated from a battery. How much energy has been imparted to the capacitor? [*Hint: We know the charge and the capacitance, and need the energy stored.*]

118. [I] If 2.00 J are to be stored in a 10.0-μF capacitor, what voltage must be applied to it?

119. [I] A capacitor stores 3.00 mJ of energy when placed across a voltage source of 100.0 V. How much charge does the capacitor receive?

120. [I] Two capacitors are wired in parallel, and the combination is placed across the terminals of a battery. Prove that the sum of the energy stored in each equals the energy stored in the equivalent capacitor.

121. [I] A 20-pF capacitor with a mica insulating sheet $\frac{1}{2}$-mm thick is put across a 12-V battery. How much energy will it store?

122. [I] A bolt of lightning corresponding to about 20 C descends through a potential difference of upwards of 150 MV. How much energy is involved? Incidentally, at any one time there might be 2000 thunderstorms rolling over the Earth with 100 lightning bolts flashing every second!

123. [II] When a pulse is propagated down an axon, there is a shift of ions across a segment of membrane that causes a reversal of polarity and an overall change in potential of about 100 mV (see Fig. P123). This voltage spike (and the associated repolarization of the membrane) propagates along from one region to the next and constitutes the signal, just as a flame propagates down a length of fuse. After a few milliseconds, this so-called *action potential*

pulse passes any given point on the axon, which then returns to its resting potential (-70 mV). How much energy is required to recharge a 1-m length of axon in the wake of a pulse so that it will be ready to transmit the next pulse?

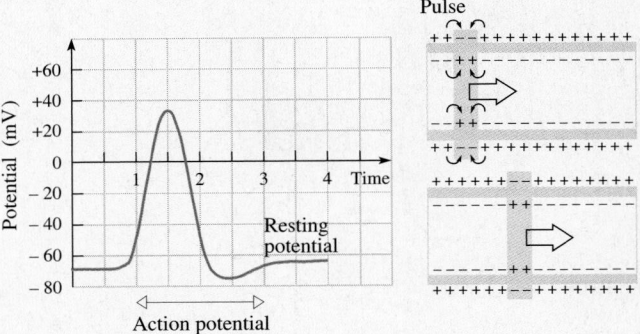

Figure P123

124. [II] Two capacitors (one 10.0 μF, the other 20.0 μF) are connected in series, and the combination is placed across a 12.0-V battery. How much energy is stored in the system once it's fully charged?

125. [II] Return to Fig. P109 and, overlooking losses, find the net energy provided by the battery when the switch is closed.

126. [II] In fair weather, the constant atmospheric E-field near the surface of the Earth varies from 120 V/m to 150 V/m. Compute the corresponding range of energies stored per cubic meter in the field.

127. [III] If a 12-V battery is placed across terminals-A and B in the circuit shown in Fig. P127, how much energy will be stored by the capacitive network? Notice that there are two places where one wire goes over another without touching it.

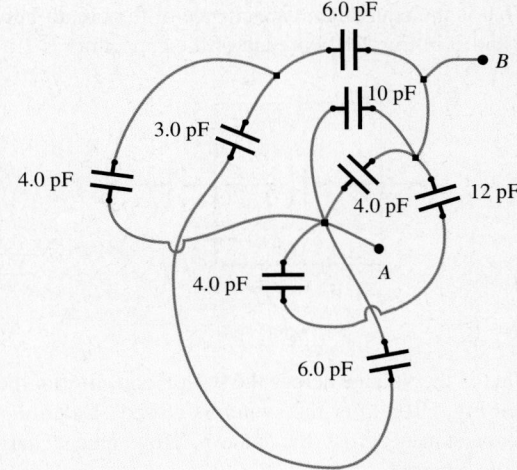

Figure P127

Silver
Cardboard
Zinc

Silver
Cardboard
Zinc

E L E C T R O M A G N E T I S M

Chapter 17
Direct Current

*I*f there is a life's blood in our technology, it's most assuredly electricity—electricity coursing along wire veins, delivering power and information, trickling through a metal nervous system just as it trickles through our own. **The ordered flow of charge is called electric current,** whether we're talking about electrons propelled down a wire by a battery or protons hurled through space by an exploding star. And currents carry energy. Much of the energy we "consume" is delivered by electricity conveniently on tap at wall outlets everywhere: refrigerators, VCRs, computers, heart-lung machines, all plug into the electric stream and draw energy from it.

Flowing Electricity

This chapter is about electric currents, the batteries that sustain them, and the resistances that impede them. When you think about "electricity" it's likely you are thinking about charges in motion, and that's current.

17.1 Electric Current

To quantify *electric current* (I), envision a stream of positive charge (Fig. 17.1). Now picture an imaginary reference plane cutting across the flow at some point of observation and determine the net charge Δq traversing the plane in a time Δt. The ratio $\Delta q/\Delta t$ is the average rate at which charge passes the point of observation during that interval, the ***average current***. Often, the flow is constant, in which case

Current is the flow of charge; the amount of charge moving past an observer each second.

[constant current]

$$I = \frac{\Delta q}{\Delta t}$$ (17.1)

Figure 17.1 A beam of positive particles of cross-sectional area A constituting a current I. During each interval of time Δt, an amount of charge Δq passes through the plane, such that $I = \Delta q/\Delta t$.

A beam of electrons (60 kiloamps, 3 MeV) deflected by both air scattering and a magnetic field. This is as much an electrical current as any that ever negotiated a toaster.

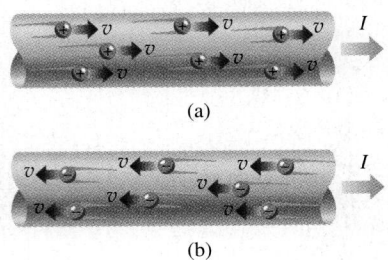

(a)

(b)

Figure 17.2   (a) The direction of an electric current is the direction of flow of positive charge. (b) Thus, it is opposite to the flow of negative charge.

The SI unit of electric current is the *ampere* (A) or simply the *amp*, where 1 amp corresponds to a flow of 1 coulomb of charge per second;

$$1 \text{ A} = 1 \text{ C/s}$$

Minute currents in the microamp range (10^{-6} A = 1 μA) are commonplace, even within the human body. Microamp currents are generated in bone and connective tissue during exercise and seem to play a vital role in sustaining the health of these structures.

Normally, a current in a metal wire is a stream of free electrons. But the **mobile charge carriers** constituting a current can be positive, negative, or both. The latter may be the case, for example, in a semiconductor or a plasma.

The mobile charge carriers in Fig. 17.1 happen to be positive—just the sort of picture Ben Franklin had in mind for a current. Because of him, *the direction of flow of positive charge is traditionally taken to be the direction of current, regardless of the actual sign of the participating carriers*. Since electrons are the carriers in ordinary wires, this custom can be a little awkward at times, though it's easy enough to live with. *A flow of negative carriers to the left is equivalent to an equal flow of positive carriers to the right* (Fig. 17.2).

Example 17.1 [I] A constant downward electron beam transports 3.20 μC of negative charge in 200 ms across the vacuum chamber of an electron microscope. Determine the beam current and the number of electrons traversing the chamber per second.

Solution   The first sentence talks about the passage of charge in a given amount of time, and that should bring to mind the definition of current.   (1) TRANSLATION—A known quantity of charge flows during a specified time; determine the current and number of electrons passing by per second.   (2) GIVEN: $\Delta q = 3.20 \times 10^{-6}$ C and $\Delta t = 200 \times 10^{-3}$ s. FIND: I and the number of electrons per second.   (3) PROBLEM TYPE—Electric current. (4) PROCEDURE—Use the definition of current and don't worry about signs.   (5) CALCULATION—From Eq. (17.1)

$$I = \frac{\Delta q}{\Delta t} = \frac{3.20 \times 10^{-6} \text{C}}{200 \times 10^{-3} \text{ s}} = \boxed{16.0 \ \mu\text{A}}$$

The current is *upward* and equal to 1.60×10^{-5} C/s. The current transfers -1.60×10^{-5} C/s, and so the number of electrons transported per second, each with a charge of -1.60×10^{-19} C, is

$$\frac{-1.60 \times 10^{-5} \text{C/s}}{-1.60 \times 10^{-19} \text{ C/electrons}} = \boxed{1.00 \times 10^{14} \text{ electrons/s}}$$

Quick Check: 16 μA flowing for 200 ms transports 3.2 μC. Since 1 A corresponds to (1 A)/(1.6 $\times 10^{-19}$ C) = 6.2 $\times 10^{18}$ electrons/s, 16 μA is about 10 $\times 10^{13}$ electrons/s.

{For more worked problems click on WALK-THROUGHS *in* CHAPTER 17 *on the* CD.*}*

A current (either a flow of electricity in a wire or water in a pipe) is usually impeded in some way by the environment through which it progresses. That *resistance* inevitably results in the expenditure of energy from the flow. To be sustained, a current must be driven by an external source of energy. In the case of electricity, a nonelectrostatic source must continuously supply energy to the mobile charge carriers, pushing them along. It's not obvious how to perform such a process; indeed, sustained currents were unknown up until the end of the 1700s. It was only then that a strange series of events led to the development of the electric battery, thereby plunging the world into the Age of Electricity.

Volta demonstrating his "pile" to Napoleon in a painting by a contemporary, A. E. Fragonard. Some years later, Volta gave Faraday a battery of this sort.

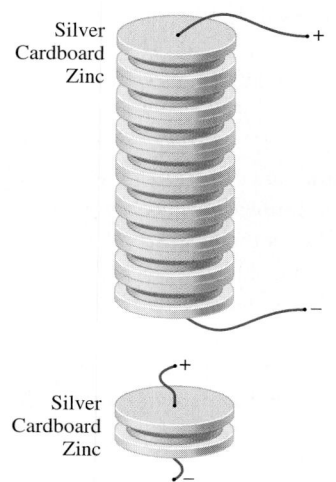

Figure 17.3 The voltaic pile, in which the elementary cell was a zinc-cardboard-silver sandwich. Volta wrongly thought that it was the zinc-silver contact that provided the effect.

The Battery

In 1780, using frogs, Luigi Galvani, a noted Italian anatomist, made the first of several accidental discoveries that would arouse a storm of controversy. Each of his little green participants was "humanely" terminated, being impaled through the spinal cord on a sharp bronze hook. Thus prepared, the frogs waited their turns hanging on an iron trellis in the garden. To his surprise, Galvani noticed that every now and then the disembodied legs would spastically convulse for no apparent reason. He concluded that he had discovered "in the animal itself" a natural source of electricity.

Alessandro Giuseppe Antonio Anastasio Volta, already renowned as an "electrician," soon became Galvani's staunchest critic. It seemed to Volta that the source of the electricity was not the animal, but the two different metals brought in contact. The frog was simply a current meter. Recognizing that his tongue was a highly sensitive and convenient muscle, Volta became his own guinea pig (or, in this case, frog). He placed different pairs of metals on his tongue and brought them into contact. Expecting to feel a contractive spasm, he was surprised when the arrangement produced a metallic taste that lasted as long as the two metals were in contact.* This implied the incredible notion that "the flow of electricity from one place to another is continuing without interruption." He had created a source of sustained current.

Volta soon produced two devices that were the forerunners of the modern electric battery. He replaced the hook, frog, and trellis with something much more convenient. The **voltaic pile** was a stack of small disks of zinc, brine-soaked cardboard, and silver, layered in order: zinc-cardboard-silver, and so on (Fig. 17.3). Each zinc-cardboard-silver sandwich formed a unit (much like the aluminum, wet tongue, and silver) called a *cell*. These were repeated about 20 times (with the cells effectively in series) to build up the voltage. The result was a device that produced an appreciable continuous current. The other variation on the theme was the *voltaic wet cell*. It consisted of a drinking glass filled with brine or a dilute acid into which two metal strips were immersed, one of zinc, the other of copper. By putting several of these cells in series (Fig. 17.4), Volta created the world's first *electric battery*. **A battery is two or more cells electrically attached to one another.**

How Batteries Work

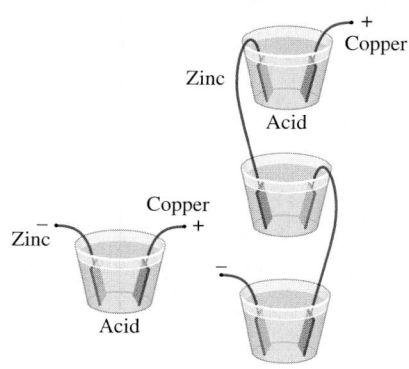

Figure 17.4 A single voltaic wet cell and three wired in series to form a battery.

Almost any two different solid conductors immersed in a variety of active solutions, known as *electrolytes*, function more or less as a battery. Chemical energy stored in the interatomic bonds (typically at less than about 3 eV) is converted into *electrical*-PE as the solution,

---

* Volta was not the first to perform this little experiment, nor did he ever fully understand it.

But when I took the frog into a closed room, laid it on an iron plate, and began to press the hook that was fixed in its spinal cord against the plate, lo and behold, the same contractions and the same kicks!

GALVANI

THE MARTYRED FROG

By the late 1700s, many researchers working with Leyden jars (p. 586) had experienced muscle spasms due to accidental shocks. Fascinated by the relationship between electricity and life, they began to study the electrical excitation of muscles in animals. The frog, whose muscular legs were a popular delicacy, became a logical martyr to the cause. In a short time, dissected frog's legs were twitching and convulsing on command from one end of Europe to the other. Even Faraday kept a froggery in the basement of the Royal Institution.

and one or both of the conducting plates, the *electrodes*, become involved in the chemical reaction. In the voltaic wet cell, the acid attacks the copper (Cu) and some of its positive Cu^{++} ions go into solution, leaving behind a negatively charged plate. Similarly, zinc (Zn) ions go into solution, too, but zinc is more soluble than copper and the zinc plate becomes even more negative—with even more excess electrons. The copper is slightly lower in potential than the acid; the zinc is a lot lower than the acid. The result is a difference in potential between the electrodes, with the copper higher and so positive, and the zinc lower and negative (Fig. 17.5).

The *electromotive series* (Table 17.1) is a list of metals in decreasing order of the tendency of each to become ionized by losing an electron. If we construct a simple cell with two metal electrodes immersed in a uniform electrolyte, the cell's voltage can be approximated from the values in the table. Accordingly, the potential of copper is +0.34 V, whereas zinc is −0.76 V (each with respect to hydrogen as the reference). Thus, copper is (0.34 V) − (−0.76 V) = +1.1 V higher than zinc, and this value is the largest voltage we can expect to produce across the terminals of a simple copper-zinc cell.

A potential difference that can be used to supply energy and thereby sustain a current in an external circuit is called an **electromotive force**, *or* **emf** (pronounced *ee em ef*), although that's a misnomer—it's not force at all. Practically, **the emf is the voltage measured across the terminals of a source when no current is being drawn from or delivered to it.**

Table 17.1

Electromotive Series for Several Metals*

Substance	Electrode potential (V)
Lithium	−3.0
Potassium	−2.9
Sodium	−2.7
Aluminum	−1.7
Zinc	−0.76
Iron	−0.44
Tin	−0.14
Lead	−0.13
Copper	+0.34
Mercury	+0.80
Silver	+0.80
Platinum	+1.2
Gold	≈+1.3

*These values correspond to the potentials required to ionize the corresponding metal atoms. The signs are a matter of convention referenced to hydrogen (and measured at 25 °C).

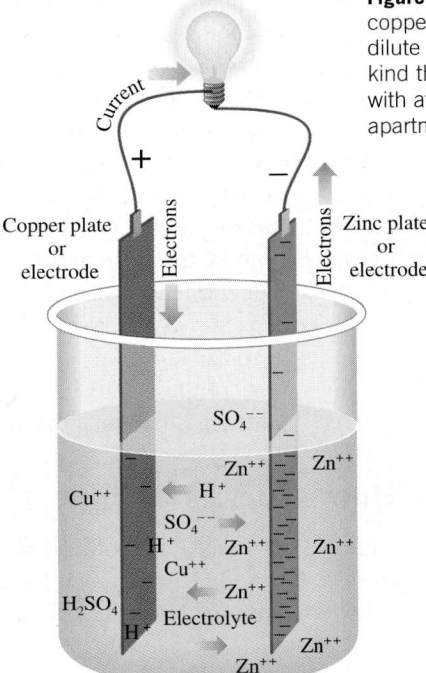

Figure 17.5 A voltaic wet cell composed of a copper and zinc plate immersed in a solution of dilute sulfuric acid (H_2SO_4). This device is the kind the young poet Percy Shelley was playing with at Oxford, burning holes in the carpet of his apartment with the splashed acid.

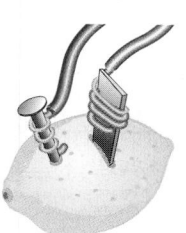

EXPLORING PHYSICS ON YOUR OWN

Making a Battery Put some aluminum foil and a piece of sterling silver on your wet tongue and let them touch each other—what do you taste? Try some other common metals. Alternatively, a strip of copper (some wire or a penny) and a strip of zinc (or a galvanized nail) immersed in salt water easily puts out ≈0.7 mA at ≈0.6 V. Add some vinegar and the emf will go up to perhaps 0.9 V. A small speaker from an old radio makes a fine current detector as does a homemade galvanometer (Fig. MC20 on p. 701). Attach one wire, or *lead*, from the cell to the speaker and then tap the other lead from the cell on the remaining terminal. The speaker will clatter with each touch. Stick a carbon rod (a pencil lead will do) into a lemon and surround it with a bunch of galvanized nails touching each other. The resulting cell will deliver ≈0.5 mA at a voltage of roughly 0.7 V. Replace the carbon with something as ordinary as a paper clip, and the lemon cell will still generate an emf, though down now to ≈0.2 V. Try a grapefruit, potato, or even some sauerkraut. Put several lemon cells in series and see if you can light a small flashlight bulb.

Figure 17.6 (a) The basic dry cell, essentially as Leclanché developed it in 1868. The modern version has a so-called *depolarizing* region of manganese dioxide around the carbon rod. That region controls the buildup of hydrogen ions that would otherwise reduce the emf. (b) A mercury cell used in calculators and watches.

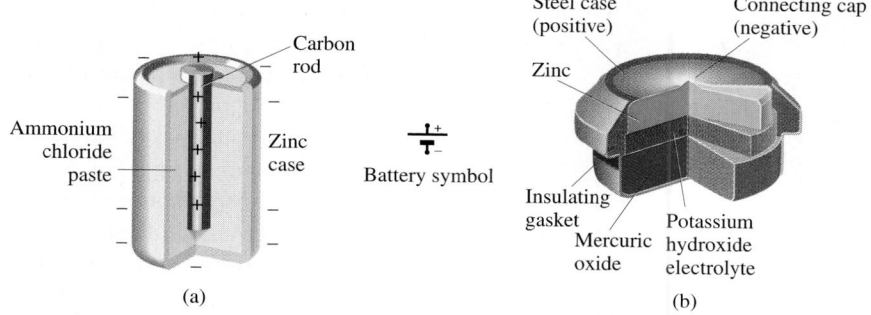

(a)

(b)

A given kind of cell will generate a voltage difference that is determined by its chemical makeup, independent of size. The size determines the total current a cell can deliver, not the voltage; the greater the quantity of each substance chemically reacting, the more charge is liberated. An ordinary flashlight "battery," a *dry cell* (Fig. 17.6a), has an emf of 1.5 V. A *mercury cell*, one of those little button-sized calculator, watch, and hearing-aid "batteries," has an emf of about 1.4 V; while the *lead storage cell* of auto fame is a 2-V device. One of its great virtues is that it can be recharged by the generator in the car. The *nickel-cadmium cell* used in rechargeable computer battery packs has an emf of 1.2 V.

Cells in Series and Parallel

To boost the potential difference, cells are often connected in series. The crucial point is that *the voltage across the series-connected battery is the sum of the voltages across each constituent cell*. Refer to Fig. 17.7a. Point *B* is 1.5 V higher than *A*, and point *D* is 4.5 V higher than *A*. This sort of series stacking is just what you are doing when you load two, three, or four D-cells, top (+) to bottom (−), into a flashlight or a portable radio in order to

An old 45-V battery made up of thirty 1.5-V dry cells in series.

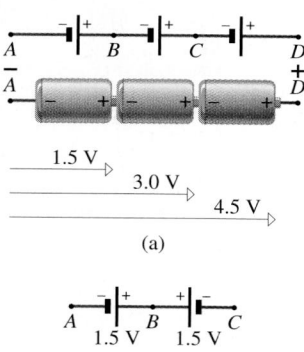

Figure 17.7 Cells in series. As connected in (a), the voltages add up, and point *D* is 4.5 V above point *A*. As connected in (b), the voltages subtract and *A* and *C* are at the same potential.

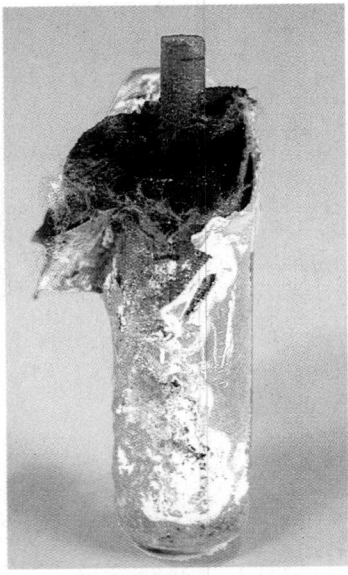

A little dry cell torn open to show the central carbon rod. The zinc casing is quite corroded.

Circuit elements can draw off energy from a current, but as a rule **the same amount of current that enters an element leaves that element**.

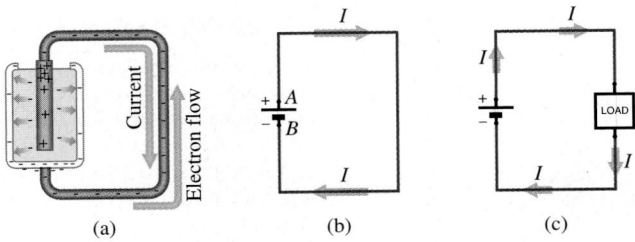

Figure 17.8 (a) (b) The flow of charge through a dry cell. This diagram represents a short circuit, which will quickly destroy a cell. (c) A more practical arrangement has the battery send current through a load, such as a motor or a light bulb.

get the 3.0 V, 4.5 V, or 6.0 V needed to operate the device. It's also the way the cells are wired in a car battery to yield 12 V.

If a wire is connected across the terminals of a cell (as in Fig. 17.8*a*), a current will move around the closed conducting path. In practice, this is a bad idea since a wire having little resistance corresponds to a ***short circuit*** that will drain the cell. It could also be dangerous if enough current flows to melt the wire. The battery provides power to a ***load***—for example, a motor, radio, or light bulb (Fig. 17.8*c*). **A steady-state current can only exist in a closed circuit, and the same current flows in and out of the load.** *Current is never used up by a circuit element.* Because the polarity of the cell is fixed, ***the direction of the flow of charge is constant***, and we say that *I* is a **direct current** (dc). If the same thing is done with a battery comprising several cells in series (Fig. 17.9), the same amount of current will pass through each cell. *In series the voltages add, while the current remains unaltered as it passes in and out of each element* (reminiscent of the way series capacitors behave where the voltages add and the net charge is the charge on any one). By contrast, *cells attached in parallel* (Fig. 17.10*a*) *form a battery whose voltage is the same as the individual voltages but whose current capacity is the sum of the individual current outputs* (again matching the behavior of capacitors). If you want a lot of current at a low voltage, put the cells in parallel; if you want both a large current and a large voltage, stack the cells in parallel and then put the stacks in series.

Battery manufacturers provide (though not often right on the battery where it ought to be) a crude measure of the current capacity in the form of an **amp-hour rating**. A little 1.5-V AA-cell, the common penlight finger-sized "battery," is rated at about 0.6 amp-hour, whereas the larger 1.5-V D-cell can deliver as much as 3 amp-hours. Presumably, one can draw a steady 0.3 amp for 10 hours or 0.1 amp for 30 hours before discharging a D-cell. It will never be able to put out 300 amps for 0.01 hour, but you get the point. {For more information click on **BATTERIES** under **FURTHER DISCUSSIONS** on the **CD**.}

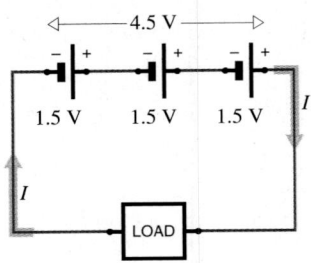

Figure 17.9 When cells in series form a battery, the same current *I* passes through each cell: *I* circulates around the circuit and is undiminished as it enters and leaves each element.

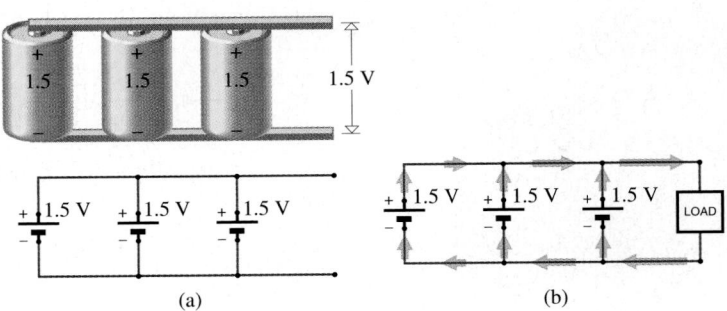

Figure 17.10 (a) Three 1.5-V cells in parallel. The voltage across the battery is the voltage across each cell. (b) Currents in the various segments of a parallel array of cells with a load across the terminals. The current through the load is the sum of the currents provided by each cell.

Example 17.2 **[I]** Design a battery to be constructed of D-cells (each rated at 3 amp-hours) that will provide a maximum operating current of 5.0 A with an emf of 4.5 V. One restriction is that no cell be required to supply in excess of 1.0 A. What can you expect will be the lifetime of your battery while delivering maximum current?

Solution   This problem is about combining cells in series and parallel. (1) TRANSLATION—Design a battery to have a specified maximum current and emf, and be made of cells of known voltage and amp-hour rating. Knowing the maximum allowed cell current, determine the battery's lifetime at maximum current. (2) GIVEN: For the battery $I_b(max) = 5.0$ A and $V_b = 4.5$ V; for each cell $V_c = 1.5$ V and $I_c(max) = 1.0$ A. FIND: The appropriate arrangement of cells and the battery lifetime. (3) PROBLEM TYPE—Electric current/batteries in series & parallel. (4) PROCEDURE—Currents add for cells in parallel and voltages add for cells in series. (5) CALCULATION—First, let's deal with the current:

The 1.0-A restriction per cell demands we put five cells in parallel to form a 5.0-A unit, as in Fig. 17.11a. The maximum current through each cell would then be 1.0 A as the unit provides the required 5.0 A with an emf of 1.5 V.

To bring the voltage up to 4.5 V, we put three such units in series, as in Fig. 17.11b. With a 3-amp-hour rating, operating at a maximum of 1.0 A, we anticipate a lifetime of $\boxed{3\ \text{h}}$ for each cell and the battery as a whole.

Quick Check: The 5.0-A current entering an adjacent unit splits up into five separate 1.0-A currents, one of which passes through each cell. These reunite as a 5.0-A stream upon leaving the unit.

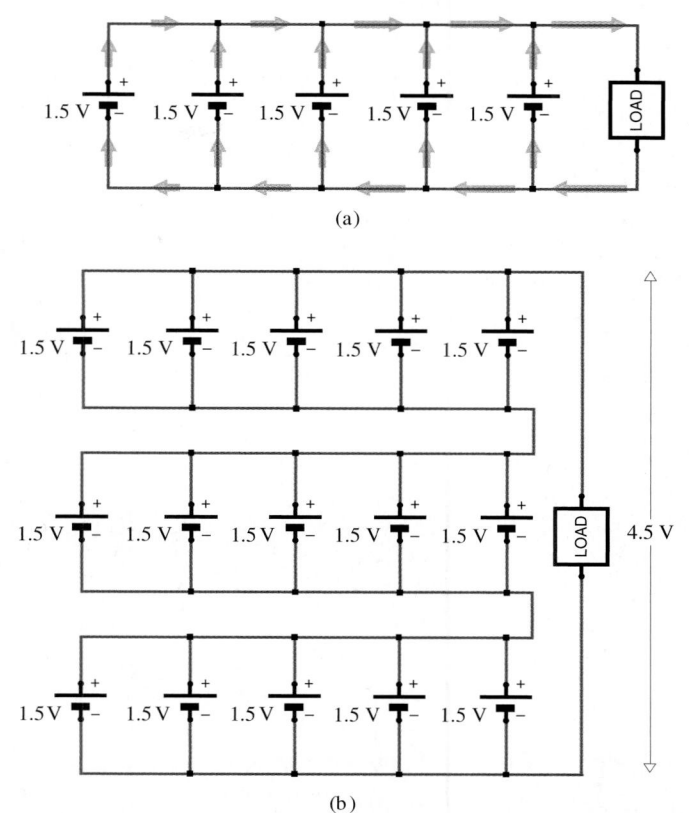

(a)

(b)

Figure 17.11   A battery made up of D-cells each rated at 3 amp-hours. (a) The voltage across this battery is the voltage across each of its cells—namely, 1.5 V. Thus, the voltage across the load is 1.5 V. (b) By putting three strings of cells in series, the voltage across the load is now 4.5 V.

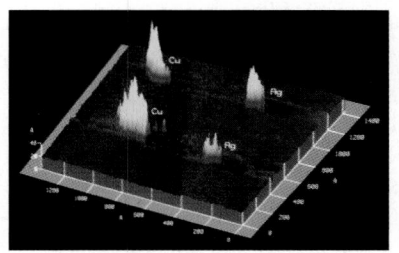

The world's smallest battery. Measuring 70 nm long and made of fewer than 5×10^5 atoms, it is 1/100 the diameter of a red blood cell. It consists of tiny pillars of copper and silver deposited on a graphite surface. When immersed in a copper sulfate solution, it produces an emf of 20 mV for about 45 minutes.

Electric Fields and the Drift Velocity

The electric field produced by an isolated battery rises out of the positive terminal and returns to the negative terminal, looking more or less like the field of a dipole (Fig. 17.12). A negative mobile charge carrier in the vicinity of the anode would be drawn toward it, and one near the cathode would be repelled. Now suppose we attach a few meters of ordinary copper hookup wire across the terminals of the battery. Electrons, forced away from the negative terminal, stream into the wire, repelling each other and spreading out to the wire's

Figure 17.12   The electric field of a dry cell. The field lines are distributed symmetrically around the central axis.

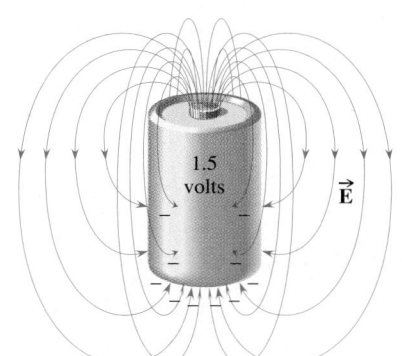

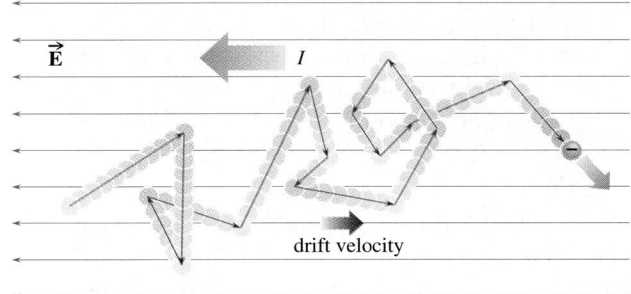

Figure 17.13 The zigzag path of an electron colliding with impurities and imperfections. It has an overall drift velocity, which (because of its negative charge) is opposite to the *E*-field.

surface as they progress along its length. The surface becomes nonuniformly charged, although there is no net charge on the wire as a whole. The copper resists the flow of charge, and so there must be a sustained driving force (i.e., an *E*-field within the wire maintained by a battery, generator, or power supply). Once the surface charges accumulate in a region, they naturally inhibit any further lateral flow of charge and the bulk of the current progresses down the length of the wire. A surface charge rapidly builds up that creates a uniform field along the length of the wire (and a field outside it as well).

Once a uniform axial *E*-field is established, a constant current progresses around the closed circuit. That current may be imagined as continuously emerging from the positive terminal of the battery, moving along the wire (where it loses energy due to the resistance), and coming back into the battery (where it's pumped up in energy), only to be sent out and around again. In the steady state there is current everywhere in the closed circuit and no further buildup of charge anywhere; **current neither bunches up nor gets used up**. Open the circuit, the surface charges redistribute, the field vanishes, and the current ceases; an *E*-field cannot be maintained inside a conductor under electrostatic conditions.

Figure 17.13 depicts the zigzag path of a typical electron as it makes its way along an ordinary metal conductor. Pushed by an externally sustained *E*-field, the electron advances but quickly scatters every-which-way off thermally vibrating metal ions, impurities (i.e., foreign atoms), and imperfections (i.e., lattice flaws and crystal grain boundaries). In copper at room temperature, the electron travels at about 10^6 m/s, collides roughly every 10^{-14} s, and gets little farther than 10^{-8} m (or several hundred atom lengths) between impacts. Yet the field relentlessly forces it to move axially along the wire in the opposite direction to $\vec{E}$ with an average drift speed. Electrons at room temperature make progress down a wire very slowly, typically at a mere 1 mm/s or so. {For more analysis click on **ELECTRON DRIFT VELOCITY** under **FURTHER DISCUSSIONS** on the **CD**.} 🖱️

IF THE DRIFT SPEED IS SO SMALL, HOW DOES A TELEPHONE TRANSMIT ELECTRICAL SIGNALS AT CLOSE TO THE SPEED OF LIGHT?

The answer is simple: The electron we push on in New York is not the one that tickles the phone in San Francisco. The starting electron might take 16 minutes to travel the first meter of the journey; it may not even be out the door before the message is over! The situation is like a long pipe filled with water. You push inward at this end, and a pulse of water carrying energy spurts out the other end. A disturbance of the water rapidly propagates down the pipe even though any given sample of liquid hardly moves. It's the electric field that can be thought of as traveling down the wire at near the speed of light, carrying the signal, setting the electrons in motion before it. When we buy electrical energy, we don't buy electrons; there are plenty of those in the wires we already have—we buy additional amounts of electron motion. *The power company pushes around our electrons and sends us a bill for how much work they did in the process.*

Resistance

There is a great diversity in the ability of materials to conduct electricity, and Georg Simon Ohm set himself the task of finding order in the seeming chaos. His experiments had to be modest in scale. Ohm was a high school teacher in Cologne and never far from poverty. He

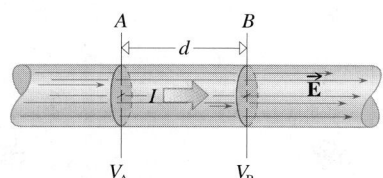

Figure 17.14 Moving along the uniform E-field from A to B a distance d, there is a voltage drop $V = -Ed$.

Georg Simon Ohm (1787–1854).

A versatile little multimeter set to operate as an ohmmeter. Here it's being used to measure the resistance of slightly damp skin. The reading is 1.04 MΩ, but it drifts around.

had been inspired to the particular challenge by the recent publications of Fourier, whose work established that the rate of flow of heat along a conducting rod was proportional to the temperature difference between its ends. Ohm wondered whether the rate of flow of charge along a conducting rod was likewise proportional to the voltage difference between its ends. Remember that in the steady state there is a uniform E-field along the current-carrying rod and so $V_B - V_A = \pm Ed$. As we move a distance d from point A to point B in the direction of $\vec{E}$, that is, in the direction of I in Fig. 17.14, there will be a voltage drop, $V = -Ed$. If we attach a length L of almost any kind of hookup wire across a battery, there will be a total voltage drop of $-EL$ along the wire; that, in turn, is determined by the voltage the battery sustains across its terminals.

17.2 Ohm's Law

Suppose we take a sample of metal wire and attach it successively to the terminals of different batteries, thereby applying different known voltages across it. In each case, we could measure the resulting current passing through the sample with a device known as an *ammeter*. (Ohm used a kind of magnetic torsion balance to accomplish the same thing.) What we would find is that the current the battery could force through a specimen depends linearly on the applied voltage, $I \propto V$; doubling the voltage doubles the current. And this relationship is true for a variety of different conducting materials—such as gold, copper, and brass—though each behaves in a characteristic way. Ohm suggested that every sample manifested a **resistance** (R) to the flow of charge. The greater the resistance (symbolized diagrammatically by -W-), the less current any battery could push through it. The current in the circuit, the current through the sample, varies directly with the applied V and inversely with R. In 1826, Ohm published his results:

$$I = \frac{V}{R} \tag{17.2}$$

or
$$V = IR \tag{17.3}$$

Ohm's Law. Because an ordinary specimen has resistance, there must be a voltage across it if a current is to be propelled through it. For most conductors *that voltage difference equals the product of the resistance of the specimen times the current being forced through it.*

Ohm's Law, as this relationship is called, deals with a rather limited, rather special set of circumstances, and yet it is of tremendous practical value. It applies to conductors (at a constant temperature), notably the common metals and several nonmetallic conductors as well. A plot of V versus I (Fig. 17.15) is a straight line *passing through the origin* with a slope of R. Put slightly differently, R is independent of I and V for these important materi-

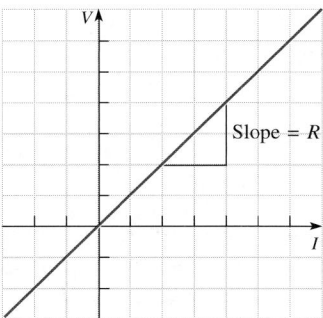

Figure 17.15 An ohmic circuit element has a linear V versus I curve. The slope of the line equals R.

Internet

For a nice interactive look at Ohm's Law go to

http://www.micro.magnet.fsu.edu/electromag/java/ohmslaw/

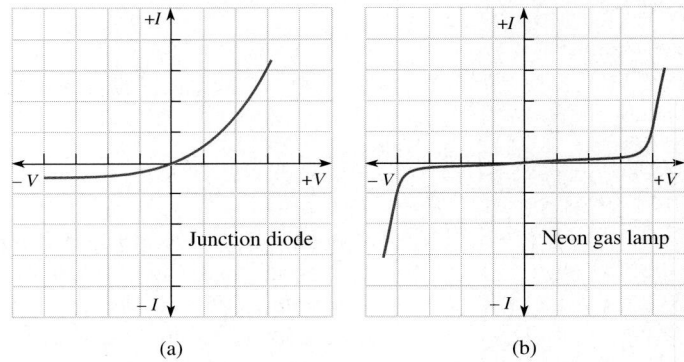

Figure 17.16 Many electrical devices, like the two shown here, are nonohmic.

als. Moreover, for a circuit element to be *ohmic*, reversing the potential difference across it must simply reverse the current through it. A cup of copper sulphate solution with copper electrodes in it obeys Ohm's Law and is an ohmic conductor, though not one you are likely to find in a typical circuit. There are many materials and devices that are *nonohmic*—an ionized gas is one (Fig. 17.16*b*). So Ohm's "Law" is a useful practical statement that applies to an important class of materials, but it's nothing like the grand fundamental pronouncements of Coulomb's Law or the Law of Universal Gravitation.

To honor Ohm—recognition for his work was depressingly late in coming—the unit of resistance was named the *ohm* and symbolized by the Greek capital letter omega, Ω. Ohm's Law serves to define resistance as $R = V/I$, and so 1 ohm = 1 volt per ampere.

All kinds of electrical devices, including the very wires that connect them, have resistance. Even so, there are specific circuit elements called **resistors** whose primary function is to introduce a certain known resistance and, in so doing, control the currents and voltages

Example 17.3 **[I]** A small ohmic light bulb is placed in series with two D-cells, as shown in Fig. 17.17. The ammeter in series with the bulb reads the current in the circuit (0.50 A) without introducing any appreciable voltage drop across its own terminals. The voltmeter attached to the terminals of the bulb reads the voltage across it (3.0 V) without introducing any appreciable change in the current through the bulb. What is the resistance of the bulb?

Solution The mention of "resistance" should always bring to mind Ohm's Law. (1) TRANSLATION—The current through and voltage across a resistive element are both known; determine its resistance. (2) GIVEN: At the bulb, $V = 3.0$ V and $I = 0.50$ A. FIND: R. (3) PROBLEM TYPE—Electric current/resistance. (4) PROCEDURE—The ideas of current, voltage, and resistance are related by Ohm's Law. Note that the total voltage across the bulb is the net voltage produced by the batteries, 1.5 V + 1.5 V. (5) CALCULATION—Solving Ohm's Law for R yields

$$R = \frac{V}{I} = \frac{3.0 \text{ V}}{0.50 \text{ A}} = \boxed{6.0 \ \Omega}$$

Quick Check: $V = IR = (0.50 \text{ A})(6.0 \ \Omega) = 3.0$ V. The low resistance draws a sizable current even at a low voltage.

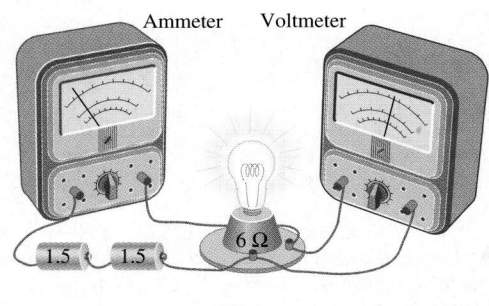

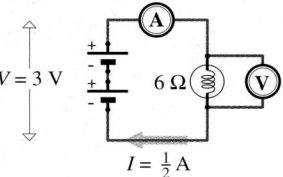

Figure 17.17 A circuit consisting of two 1.5-V cells and a lamp, all in series. The ammeter -Ⓐ- is placed in the arm of the circuit through which the desired current passes. The voltmeter -Ⓥ- is placed across the two points where a potential difference is to be determined.

Learning the resistor color code may be optional, but if you've got a lab on circuits you'll definitely have to understand it. When you work with color-coded resistors, remember that the third band is the number of zeros that must follow the first two numbers. Thus, red (2)-brown (1)-blue (6) is 21 000 000. And yellow (4)-violet (7)-brown (1) is 470. Be especially careful when the third band is black (0); grey (8)-blue (6)-black (0) is 86, NOT 860.

Internet

If you would like to play with an interactive presentation of the resistor color code go to

http://www.micro.magnet.fsu.edu/electromag/java/resistor/

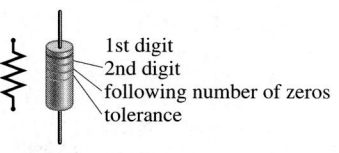

1st digit
2nd digit
following number of zeros
tolerance

(a)

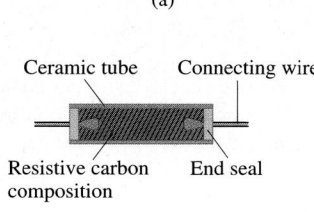

Ceramic tube   Connecting wire

Resistive carbon   End seal
composition

(b)

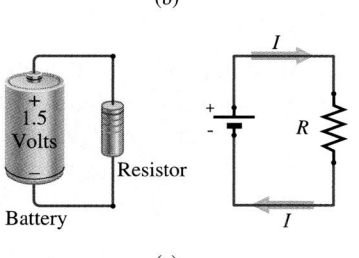

Battery   Resistor

(c)

Figure 17.18 (a) Although resistors come in many forms, the most common is the little striped brown cylinder. These are carbon composition resistors, as shown in (b). The stripes are a color code indicating the resistance. (c) shows the schematic representation.

Color Code

Black	0
Brown	1
Red	2
Orange	3
Yellow	4
Green	5
Blue	6
Violet	7
Gray	8
White	9
Gold	± 5%
Silver	± 10%

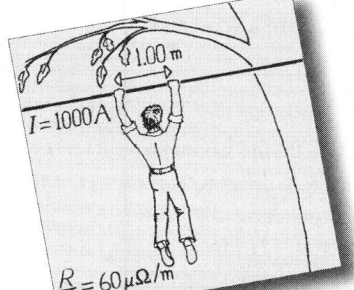

in a circuit. They range from fractions of an ohm to millions of ohms (megohms, MΩ). Used in almost all electronic devices from radios to computers, they regulate the flow of charge (Fig. 17.18).

Example 17.4 **[I]** Suppose someone falling out of a tree grabs an overhead power line. The wire has a resistance of 60 microhms per meter and is carrying a dc current of 1000 amps. With hands a meter apart, what is the voltage across him? Will the unfortunate soul get much of a shock?

Solution This problem talks about "resistance," "current," and "voltage," and that means Ohm's Law. (1) TRANSLATION—Determine the voltage drop across a specified length of a specimen of known resistance per unit length, carrying a known current. (2) GIVEN: $R/L = 60\ \mu\Omega/\text{m}$ and $I = 1000$ A. FIND: V across 1.0 m. (3) PROBLEM TYPE—Electric current/resistance. (4) PROCEDURE—The ideas of current, voltage, and resistance are related by Ohm's Law. (5) CALCULATION—The

voltage drop across 1.0 m of the line is

$$V = IR = (1000\ \text{A})(60\ \mu\Omega/\text{m} \times 1.0\ \text{m})$$

and
$$V = 0.060\ \text{V} = \boxed{60\ \text{mV}}$$

You know that you can handle the terminals of a 1.5-V dry cell without feeling any electricity, so 60 mV is much too small a voltage to push a detectable amount of current through a human body (see Sect. 21.6).

Quick Check: $V = IR = (10^3\ \text{A})(6 \times 10^{-5}\ \Omega) = 6 \times 10^{-2}$ V.

PLASTIC CONDUCTORS

The plastic polyacetylene is ordinarily a semiconductor, but when doped with iodine, it becomes a conductor. (A material is said to be *doped* when small amounts of a foreign substance are introduced into it.) An iodine atom removes an electron from a carbon atom in the polymer chain, leaving behind a "hole" that behaves like a positively charged particle. The holes advance in the direction of the E-field as if they were a flow of positive charge, a current. This metallic-looking plastic is, ounce-for-ounce, twice as conductive as copper, and, though still in the developmental stage, it promises a new age of inexpensive plastic electronic devices.

17.3 Resistivity

The resistance of a piece of wire, or anything else for that matter, is specific to that sample—it tells us nothing about the material of which the wire is made. Alternatively, it would be nice to have some measure of how each kind of material behaves independent of geometry. We did the same sort of thing when we expressed the spring constant (which is specific to each spring) in terms of Young's Modulus (p. 340), which is characteristic of the material making up the spring. Here, we follow Ohm's lead and, for the very simplest geometry, guess at what he found experimentally. How does the resistance of a rod vary with its shape and composition? The analogy with water flowing through pipes suggests that the resistance is directly proportional to the length of the conductor (L)—the longer the rod, the more scattering the electrons will experience in traversing it. Similarly, the narrower the pipe, the more the resistance; so we can anticipate that electrical resistance will also vary inversely with the cross-sectional area, $R \propto 1/A$. The amount of current should depend on E, and E in turn depends on V; the greater the voltage, the more the current. Hence, $I \propto AV$; and so from Ohm's Law, $R \propto 1/A$. In any event, Ohm found experimentally that $R \propto L/A$ and introduced a material-dependent constant of proportionality ρ, the **resistivity**. To make the statement an equation,

$$R = \rho \frac{L}{A} \qquad (17.4)$$

A long extension cord (large L) should have heavy-gauge wire (large A) to keep R small. Keeping in mind that $R = V/I$, this result is similar to Eq. (13.5), which describes the heat current (Q/t) driven by a temperature difference (ΔT): the ratio of driving influence to resulting thermal current is proportional to the sample's length over its cross-sectional area.

Table 17.2 lists the resistivities, in units of ohm-meters ($\Omega \cdot m$), for a number of important materials. Substances with resistivities of less than about 10^{-5} $\Omega \cdot m$,

Table 17.2

Resistivities*

Substance	Resistivity (ρ) ($\Omega \cdot m$)
Aluminum	2.8×10^{-8}
Brass	$\approx 8 \times 10^{-8}$
Constantan (60% Cu, 40% Ni)	$\approx 44 \times 10^{-8}$
Copper	1.7×10^{-8}
Iron	$\approx 10 \times 10^{-8}$
Manganin ($\approx$84% Cu, $\approx$12% Mn, $\approx$4% Ni)	44×10^{-8}
Mercury	96×10^{-8}
Nichrome ($\approx$59% Ni, $\approx$23% Cu, $\approx$16% Cr)	100×10^{-8}
Platinum	10×10^{-8}
Silver	1.6×10^{-8}
Tungsten	5.5×10^{-8}
Carbon	3.5×10^{-5}
Germanium	0.46
Silicon	$100-1000$
Glass	$10^{10}-10^{14}$
Neoprene	10^{9}
Polyethylene	$10^{8}-10^{9}$
Polystyrene	$10^{7}-10^{11}$
Porcelain	$10^{10}-10^{12}$
Teflon	10^{14}
Sodium chloride (saturated solution)	0.044
Blood	1.5
Fat	25

*Values determined at or near 20 °C.

A close-up of the circuitry in the radio pictured on p. 537. The gray cylinder is a 1.5-kΩ wire-wound resistor. Such resistors are made by wrapping wire of some exotic alloy such as nichrome, manganin, or constantan around an insulating core. They are more precise and have better temperature stability than the less expensive carbon resistors.

such as silver and copper, are called **conductors**. **Insulators** such as glass, rubber, and teflon typically have resistivities greater than about 10^5 $\Omega \cdot m$. Between 10^{-5} and 10^5 $\Omega \cdot m$ are the so-called **semiconductors**, like silicon and germanium.

Example 17.5 **[I]** A length of nichrome ribbon with a rectangular cross section of 0.25 mm × 1.0 mm is to be used as the heating element in a toaster. How long should it be if it's to have a total resistance of 1.5 Ω at room temperature?

Solution When a resistance problem mentions the material and geometry of a conductor, you can expect the resistivity to be involved. (1) TRANSLATION—Determine the length of a specimen of known resistivity and cross section if it is to have a specified resistance. (2) GIVEN: Cross section 0.25 mm × 1.0 mm, nichrome ribbon, resistance of $R = 1.5$ Ω. FIND: L. (3) PROBLEM TYPE—Electric current/resistance/resistivity. (4) PROCEDURE—There is only one relationship that allows us

to calculate the resistance of a specimen in terms of its physical characteristics: $R = \rho L/A$. (5) CALCULATION—The cross-sectional area is

$$A = (0.25 \times 10^{-3})(1.0 \times 10^{-3}) = 0.25 \times 10^{-6} \text{ m}^2$$

Using Table 17.2, we have

$$L = \frac{RA}{\rho} = \frac{(1.5 \ \Omega)(0.25 \times 10^{-6} \text{ m}^2)}{100 \times 10^{-8} \ \Omega \cdot m} = \boxed{0.38 \text{ m}}$$

Quick Check: The resistance of this wire per meter ($L = 1$ m) is $\rho 1/A = 4$ Ω/m. Using ratios, $1.5/4 = L/1$ and $L = 0.38$ m.

The Temperature Dependence of Resistivity

When the temperature of a conductor increases, the corresponding increase in the random vibrations of its atoms and ions increases the scattering of electrons, impeding their progress and elevating the resistivity of the material. For example, during the fraction of a second while the tungsten filament in a light bulb rises roughly 2000 °C as it becomes incandescent, its resistance increases by a factor of about 10. Experiments show that ρ usually varies *almost* linearly with modest changes in temperature (ΔT), as shown in Fig. 17.19. Accordingly, we write an expression for ρ at any temperature relating it to a known value (ρ_0) at some reference temperature:

$$\rho \approx \rho_0(1 + \alpha_0 \Delta T) \tag{17.5}$$

Table 17.3

Temperature Coefficients of Resistivity*

Substance	α_0 (K^{-1})
Aluminum	0.003 9
Brass	0.002
Constantan (60% Cu, 40% Ni)	0.000 002
Copper	0.003 93
Iron	0.005 0
Manganin ($\approx$84% Cu, $\approx$12% Mn, $\approx$4% Ni)	0.000 000
Mercury	0.000 89
Nichrome ($\approx$59% Ni, $\approx$23% Cu, $\approx$16% Cr)	0.000 4
Platinum	0.003 927
Silver	0.003 8
Tin	0.004 2
Tungsten	0.004 5
Carbon	$-0.000 5$
Germanium	-0.05
Silicon	-0.075
Sodium chloride (saturated solution)	-0.005

*Values determined at or near 20 °C.

Values of α_0, the ***temperature coefficient of resistivity***, are given in Table 17.3, and ΔT is the difference in temperature (either in °C or K) from the reference value. Because α_0 also varies somewhat with temperature, it has been subscripted to indicate that it, too, must be correlated to the reference temperature. For precise work, there are handbooks of physical data that supply values of both ρ_0 and α_0 at different temperatures. The approximation sign in Eq. (17.5) is a reminder that the formula holds best in the vicinity of the reference temperature at which α_0 and ρ_0 are given. Notice that for most pure metals $\alpha_0 \approx 1/273$ K^{-1} (see Problem 72).

The temperature coefficients of the semiconductors in Table 17.3 are negative; they become less resistive when the temperature increases. As with conductors, increasing T increases the scattering of charge carriers. The density of charge carriers in a semiconductor increases strongly with T—carriers that were

The tungsten helical filament of an ordinary incandescent light bulb. It has a high resistance and becomes white hot at 110 V.

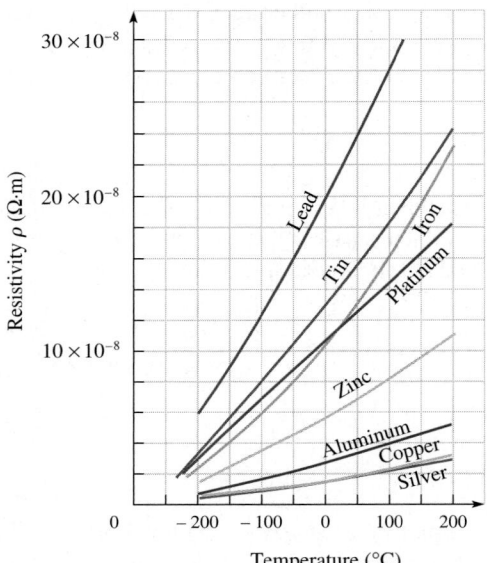

Figure 17.19 The resistivities of several metals over a range of temperatures. All rise with *T*, and all are nearly linear.

Resistivity ρ ($\Omega \cdot$m)

30×10^{-8}

20×10^{-8}

10×10^{-8}

Lead

Tin

Iron

Platinum

Zinc

Aluminum

Copper

Silver

Temperature (°C)

0   −200   −100   0   100   200

initially not free to move can be set loose after absorbing thermal energy. This has the effect of increasing *I* for a given *V* and, therefore decreasing *R*, via Eq. (17.3).

Example 17.6 **[II]** One of the most useful devices for measuring temperature is the platinum-resistance thermometer. Typically, about 2.0 m of pure platinum wire 0.1 mm in diameter is formed into a coil with a resistance at 0 °C of 25.5 Ω. Taking the temperature coefficient of resistivity to be 0.003 927 K^{-1}, determine the change in resistance corresponding to a 1.00 °C change in temperature. At what temperature is the thermometer if its resistance is 35.5 Ω?

Solution When a resistance problem mentions the material and geometry of a conductor, you can expect the resistivity to be involved. (1) TRANSLATION—A wire of known length, diameter, temperature coefficient of resistivity, and temperature has a specified resistance; determine its change in resistance per °C and its temperature corresponding to a specified resistance. (2) GIVEN: At 0 °C R_0 = 25.5 Ω, and α_0 = 0.003 927 K^{-1}. FIND: ΔR corresponding to ΔT = 1.00 °C and *T* corresponding to *R* = 35.5 Ω. (3) PROBLEM TYPE—Electric current/resistance/resistivity/temperature. (4) PROCEDURE—We know how to deal with the temperature dependence of the resistivity and from that can find the corresponding dependence of resis-

tance on temperature. (5) CALCULATION—Since

$$\rho \approx \rho_0(1 + \alpha_0 \Delta T) \qquad [17.5]$$

it follows from $R = \rho L/A$ that

$$R \approx R_0(1 + \alpha_0 \Delta T) \qquad (17.6)$$

We want the change in resistance $(R - R_0)$ that occurs when $\Delta T = \pm 1$, which can either be in °C or K

$$R_0 \alpha_0 \Delta T = R - R_0$$

$$(25.5\ \Omega)(0.003\ 927\ K^{-1})(1.00\ K) = 0.100\ \Omega$$

The thermometer changes resistance by $\boxed{0.100\ \Omega\ \text{per}\ 1\ K}$ or 1 °C. A change of +10.0 Ω from its 0 °C reading means a temperature of $\boxed{+100\ °C}$.

Quick Check: For a change of ΔT = 100 K, $R - R_0$ = $(25.5\ \Omega)(0.003\ 927\ K^{-1})(100\ K)$ = 10.0 Ω.

Superconductivity

Resistance, although helpful in controlling currents, is the bane of electrical technology; it limits the operation of almost everything. If we could get rid of it, we could put nuclear power plants safely away from populated areas, cut the cost of electricity, float cars frictionlessly on magnetic fields, make tiny powerful motors, and improve computers. We could revolutionize the entire technology. Today, the goal of eliminating resistance seems just over the horizon in the realm of superconductivity.

Table 17.4
Critical Temperature of Some Superconducting Elements

Element	T_c (K)
Aluminum	1.175
Beryllium	0.026
Cadmium	0.52
Gallium	1.083 3
Indium	3.405
Lead	7.23
Mercury (α)	4.154
Molybdenum	0.916
Niobium	9.25
Osmium	0.655
Protactinium	1.4
Tantalum	4.47
Tin	3.721
Titanium	0.39
Tungsten	0.015 4

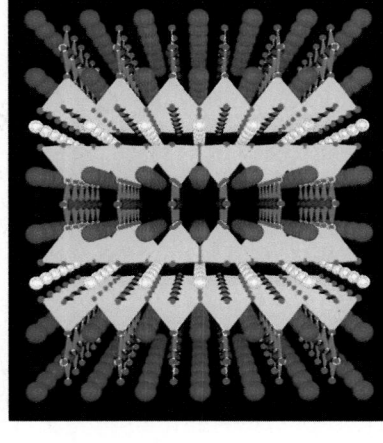

A computer model of a high-temperature, superconducting compound. The atoms are yttrium (gray), barium (green), copper (blue), and oxygen (red).

Soon after H. Kamerlingh Onnes succeeded in liquefying helium in 1908, he began a study of the temperature dependence of the dc resistance of metals. The use of liquid helium as a refrigerant allowed him to routinely operate down to about 1 K (-458 °F). Onnes, working first with platinum, found that its resistance dropped as the temperature descended, though it leveled off at a fairly constant value below roughly 4 K. He next selected mercury to examine because it could be obtained at ultrahigh purity, and he wrongly believed that the resistance of any pure metal would vanish as it approached 0 K—something certainly suggested by Fig. 17.19. The resistance of mercury did slowly decrease with T, but at 4.2 K it inexplicably plunged to an unmeasurably small value. "Mercury has passed into a new state," wrote Onnes, "which on account of its extraordinary electrical properties may be called the superconducting state."

This total absence of dc resistivity below a **critical temperature** (T_c) is known as **superconductivity**. Onnes was lucky in selecting mercury: only 27 elements become superconducting under ordinary pressure, and many of those do so at critical temperatures well below 4.2 K (see Table 17.4). Incidentally, platinum does not make the transition to superconductivity, nor do the other good conductors like copper, silver, and gold. Still, well over a thousand alloys and compounds undergo this remarkable transformation.

Although a substance in its normal state has resistance and so must have an emf across it and an E-field within it to sustain a current, no such electric field exists in a superconductor. A current once initiated will continue on its own in a closed superconducting loop, perhaps indefinitely. One such experiment ran for over $2\frac{1}{2}$ years with no observed diminution in the circulating supercurrent. By contrast, a current circulating in a normal resistive material without any driving force would last less than a second. Careful measurements indicate that the decay time for a supercurrent is at least 10^5 years, implying that if there is any resistance at all in the superconducting state, it's at least 10^{-12} times that of the normal state. {For more on the theory of the phenomenon click on **Superconductivity** under **Further Discussions** on the **CD**.}

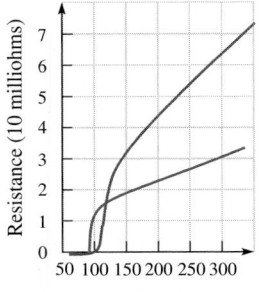

Temperature (K)

Figure 17.20 Resistance versus temperature of a thallium compound (blue) and a europium compound (red). Both become superconducting below around 100 K. These are typical of high-temperature superconductors. (Adapted from *Physics Today*, April 1988.)

ROOM TEMPERATURE SUPERCONDUCTIVITY

A breakthrough occurred in 1986, when it was discovered that an exotic ceramic compound of barium, lanthanum, copper, and oxygen had an unprecedentedly high critical temperature of 35 K. The race was on. By the beginning of 1987, physicists had prepared a ceramic (substituting yttrium for lanthanum) for which T_c was 98 K (-283 °F), well above the temperature of liquid nitrogen (77 K). And by early 1988, a thallium compound (Tl-Ca-Ba-Cu-O) with a rather balmy critical temperature of 125 K had been produced (Fig. 17.20). Though still far from room temperature, values of T_c continue to creep upward, and a variety of new high-temperature superconducting devices now exists.

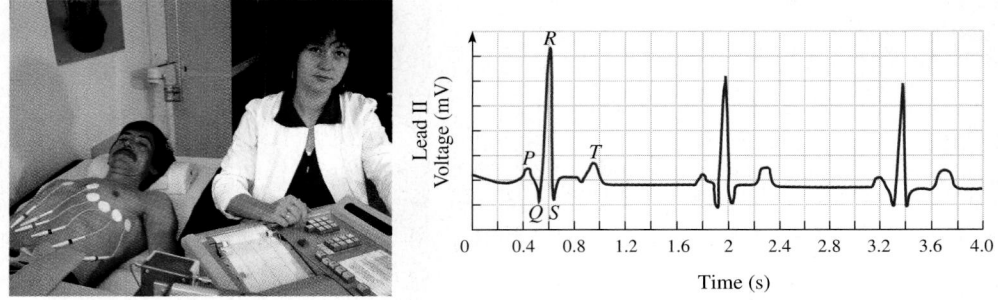

The electrocardiograph records potential differences at the body's surface due to the electrical activity of the heart. The voltage shown here is between the left leg and the right arm. The region labeled *P* arises when the atria contract. The contraction of the ventricles corresponds to *QRS,* and *T* marks the preparation of the ventricles for the next contraction.

17.4 Voltage Drops and Rises

Figure 17.21*a* depicts an idealized battery (having no internal resistance of its own) connected to a resistor by two ideal zero-resistance leads (generally, heavy-gauge copper wire will do nicely, but you can imagine these to be superconductors). The voltage across the battery (V) is its emf; and since the hookup wires have no resistance, there is no drop in potential along them—points A and A' are at the same voltage, as are points B and B'. In this simple circuit, the voltage across the terminals of the battery equals its emf, and equals the voltage across the resistor.

In the steady state, charges introduced from the battery into that idealized wire coast along freely ($E = 0$) until they encounter the resistor. If we suppose that the mobile charge carriers are positive, they effortlessly traverse the ideal wire and arrive at the resistor (at point A'). Impeded in its progress, positive charge accumulates in the vicinity of that point, repelling positive charges out and away from the other end of the resistor, which becomes negatively charged. An E-field now exists within the resistor that has an axial component along its length. With the potential across R equal to its maximum value, namely that of the battery (V), the E-field drives a steady-state current I through the resistor. From point B' the charges coast at a constant potential back to the battery. Thus, there are surface charges distributed around the circuit with concentrations ($+$) at A and A' and ($-$) at B and B'.

Note that I leaves the high potential (positive) side of the battery and enters the high potential (positive) side of the resistor. In passing through the resistor, the mobile charge carriers are scattered somewhat, imparting random KE (i.e., thermal energy) to the atoms of the resistive medium. As a result, energy is transferred from the current to the resistor— each positive charge q drops in potential V and thus in potential energy qV, on traversing the resistor. The carriers return to the negative terminal of the battery, which raises their potential by V, pumping them up in energy qV, and sending them out again. The battery supplies electrical energy, and the resistor dissipates that exact amount as thermal energy (in accord with the Law of Conservation of Energy). Current will circulate until the battery

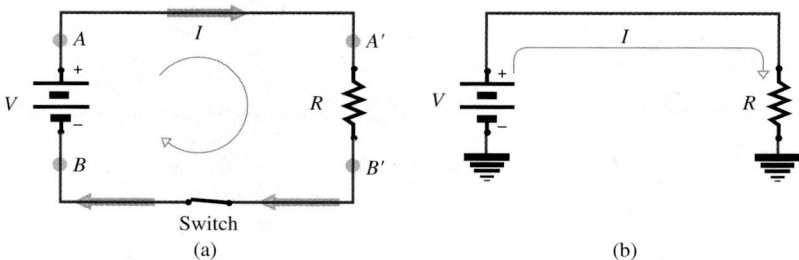

Figure 17.21 (a) Points -*A and* *A'* are at the same potential and so too are *B* and *B'*. (b) This circuit is essentially identical to the circuit in (a).

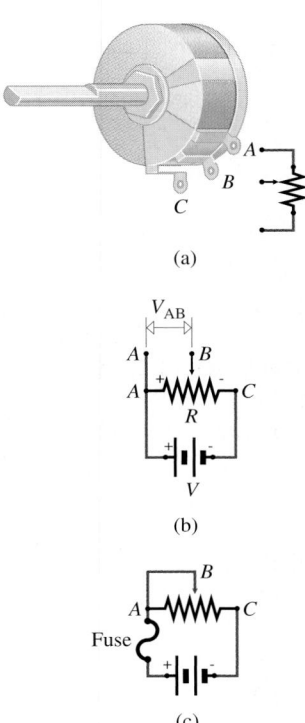

Figure 17.22 A potentiometer is a variable resistor. It is commonly found in radios, stereos, amplifiers, and tape decks.

supplies all the energy it can and inevitably runs down. **The current is the transporter of energy.**

No point in the circuit is grounded, and so no point has a potential referenced to the zero of ground. The circuit is said to *float*, and what is important is the voltage of one point with respect to another: **voltage differences are the practical quantities**. In practice, the metal chassis that houses a circuit (or a large conducting region on a printed circuit board) is often used as a common terminal known as the **circuit ground,** which need not be attached to Earth ground. Indeed, in most automobiles the entire metal body of the vehicle is the circuit ground, and it's isolated from the Earth by the rubber tires. All the points constituting ground are by definition at 0 V with respect to other points on the circuit. Figure 17.21*b* shows the way this is represented and how the current simply flows around the closed circuit through ground.

Figure 17.22*a* depicts a variable resistor used to divide a voltage, namely, the voltage across terminals *A* and *C*. When serving that function it's called a *potentiometer*, a pot for short. This device is usually a long coil of metal wire (its end points being *A* and *C*) with a central slide pressing against it, making contact with the wire (at a variable point *B*). When placed across an ideal battery (Fig. 17.22*b*), the emf (*V*) equals V_{AC}, and V_{AB} is any desired fraction thereof. This sort of *voltage divider* is very useful. When you turn the knob on the volume control of a radio or stereo, you're moving the slide on just such a pot.

Alternatively, in Fig. 17.22*c* as the slider (*B*) is moved toward *C,* more and more resistance is bypassed or "shorted" by the wire from *A* to *B*. In other words, current flows from *A,* up and over to *B,* to *C,* bypassing much of the resistance of the device. The closer *B* comes to *C* the less resistance appears across the battery, and the more current will flow in the circuit. A variable resistor used like this to control current is called a ***rheostat***. Notice that if *B* touches *C* the battery will be "short circuited" and rapidly drained and often destroyed. Moreover, if the battery is large and able to supply a lot of current, shorting it can be a dangerous thing to do. To avoid such possibilities a *fuse* has been installed that will harmlessly blow if the current in the circuit becomes too large.

17.5 Energy and Power

A current in a circuit is like a moving fluid capable of transporting energy from the source of emf (such as a battery, generator, or solar cell) to some device (a toaster or a TV), where it can subsequently be utilized. The transport of energy by a current is one of the fundamental features of electricity.

Suppose that a small amount of charge Δq traverses a circuit element and moves through a constant potential difference of *V*. It will, in doing so, change its electrical potential energy (ΔPE_E) by an amount $\Delta q V$. The rate at which this happens, the time rate of transfer of energy, is

$$\frac{\Delta PE_E}{\Delta t} = \frac{\Delta q}{\Delta t} V$$

which, by definition (p. 195), is the **power**, P (either delivered or consumed), over a time Δt. Since $\Delta q/\Delta t$ is the current, the power at any instant is

[electrical power in general] $$P = IV$$ (17.7)

The unit of electrical power is an ampere-volt, where

$$1 \text{ A·V} = (1 \text{ C/s})(1 \text{ J/C}) = 1 \text{ J/s} = 1 \text{ W}$$

In Fig. 17.21, the power delivered by the battery to the current *I* in raising its potential *V* is *IV*. Similarly, the power delivered thermally to the resistor by the current *I* while dropping in potential by *V* is also *IV*. If several batteries are connected in a circuit, it is possible that a current may *enter* one of them at its positive terminal. Such a current would emerge

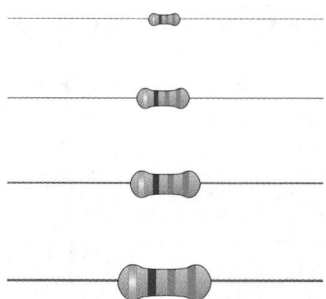

Figure 17.23 Film resistors are made by depositing a thin layer of carbon (carbon film) or nickel chromium (metal film) on a ceramic core. These are high precision circuit elements. Here we see a selection of them (starting at the top) with power ratings of $\frac{1}{8}$W, $\frac{1}{4}$W, $\frac{1}{2}$W, and 1 W.

after a drop in potential—that is, after depositing energy in that battery. It is by supplying energy to a battery in this fashion that it can be recharged.

Using Ohm's Law, $V = IR$, **the power dissipated in a resistance R** can be expressed as $P = IV = I(IR)$ and

[electrical power: resistance]

$$P = I^2R$$

(17.8)

Similarly $P = IV = (V/R)V$ and

[electrical power: resistance]

$$P = \frac{V^2}{R}$$

(17.9)

J. P. Joule (1841) first showed experimentally that the "heating-power" of an electric current through a resistance had the form of Eq. (17.7). To recognize his accomplishment, we speak of the thermal energy produced in this way as **joule heat**. Commercial resistors generally carry a power rating that corresponds to the maximum amount of power the device can dissipate without being damaged by joule heating. As a rule, for a given type of resistor, the larger the resistor, the larger the surface area, and the more power it can dissipate (Fig. 17.23).

Example 17.7 **[II]** The circuit in Fig. 17.24, containing a battery of unknown voltage, carries a current of 5.0 A flowing as shown. Determine the power either dissipated or provided by each circuit element and the net power dissipated.

Solution When you read the word "power" in an electric current problem, two equations should immediately come to mind: $P = IV$ and $P = I^2R$. (1) TRANSLATION—Two batteries (one specified, one not specified) and a known resistor form a circuit carrying a known current; determine the power associated with each element. (2) GIVEN: A 12-V battery, an unknown battery, $I = 5.0$ A flowing clockwise, and $R = 10\ \Omega$. FIND: Power for each. (3) PROBLEM TYPE—Electric current/power. (4) PROCEDURE —Determine which elements are supplying power and which are dissipating it. Inasmuch as we have the current, use $P = IV$ for the batteries and, knowing R, use $P = I^2R$ for the resistor. (5) CALCULATION—The same 5.0-A current passes through both the 12-V battery and the resistor in going from A to B. The power *delivered to* (since I enters on the $+$ side) the 12-V battery is

$$P = IV = (5.0\ \text{A})(12\ \text{V}) = \boxed{60\ \text{W}}$$

It's receiving energy from the current. **Resistors only dissipate**

power, and here that's in the amount of

$$P = I^2R = (5.0\ \text{A})^2(10\ \Omega) = \boxed{0.25\ \text{kW}}$$

The net power dissipated is therefore 0.060 kW + 0.25 kW = $\boxed{0.31\ \text{kW}}$, and that's provided by the second battery—current leaves its $+$ terminal.

Quick Check: The voltage of the second battery is $V_{AB} = 12\ \text{V} + IR = 12\ \text{V} + 50\ \text{V} = 62\ \text{V}$. The power it delivers is therefore $P = IV = (5.0\ \text{A})(62\ \text{V}) = 0.31$ kW, which is the power dissipated.

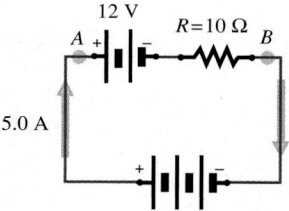

Figure 17.24 Here each circuit element either provides or dissipates power.

EXPLORING PHYSICS ON YOUR OWN

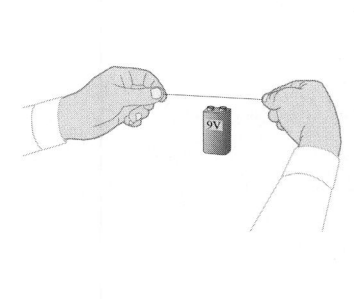

Joule Heating & the Fuse This experiment is tremendously exciting, but it requires care and attention to good safety practices, so read it through before starting. You'll need an ordinary 9-V battery and a very thin conductor with a high resistance (use a small-diameter strand of steel wool). Stretch a single filament of steel across the terminals, and within seconds a region of the fiber between the terminals will glow red hot and burst into a shower of molten beads. This is the way a fuse works to limit the current in a circuit. Why did the wire melt? Of course, it's these molten beads that one has to be concerned about. Accordingly, use only a long filament at least three times the length between the terminals; hold it at its ends and position the battery near the middle so that you have plenty of time to drop the two pieces if they keep burning. Before trying this, remember to **wear safety goggles**, work on an appropriate surface (e.g., inside a sink), away from anything flammable, and take all necessary precautions. Alternatively, convince your instructor to do it in class.

Example 17.8 [II] What is the equivalent in joules of a kilowatt-hour? Calculate the price of electrical energy in dollars per kilowatt-hour as supplied by a 50¢ D-cell having a 3.0 amp-hour rating. Compare that with the ≈10¢ per kW·h for electricity on tap from a wall outlet.

Solution A kilowatt is a quantity of power, and that should call to mind $P = IV$. (1) TRANSLATION—Knowing the cost and amp-hour rating of a cell, determine its price per kilowatt-hour. (2) GIVEN: 50¢ for 3.0 amp-hours. FIND: Price per kW·h. (3) PROBLEM TYPE—Electric current/power. (4) PROCEDURE—Determine how much energy a kilowatt-hour corresponds to and how much energy 3.0 amp-hours delivered at 1.5 V corre-

sponds to. (5) CALCULATION—First, 1.0 kW equals 1000 J/s; hence, 1.0 kW·h = (1000 J·h/s)(3600 s/h), or $\boxed{3.6 \text{ MJ}}$. The cell delivers 3.0 amp-hours at 1.5 V or, equivalently, 3.0 A at 1.5 V for 1.0 h. Since $P = IV$, that's an amount of energy of

$$(3.0 \text{ A})(1.5 \text{ V})(1.0 \text{ h}) = 4.5 \text{ W·h}$$

or 4.5×10^{-3} kW·h. At 50¢ per cell, that's (50¢)/(4.5 × 10^{-3} kW·h) or (not worrying about significant figures) $\boxed{\$111 \text{ per kW·h}}$.

Quick Check: The energy of the cell is 4.5 W·h, which, divided by the operating voltage, yields the number of amp-hours: 3.

Similarly, for a given type of cell, the bigger it is, the higher the amp-hour rating, the more energy it stores. That's why portable devices that need a lot of energy, such as heavy-duty flashlights or large speakers, operate on D-cells rather than A-cells, even though both have the same emf of 1.5 V.

Core Material & Study Guide

FLOWING ELECTRICITY

The ordered flow of charge is called **electric current** (I):

$$I = \frac{\Delta q}{\Delta t} \qquad [17.1]$$

The SI unit of current is the *ampere*: 1 A = 1 C/s. A nonelectrostatic potential difference that sustains a current in an external circuit is called an **electromotive force**, or emf (p. 610). Reread Section 17.1 (Electric Current) and go over Example 17.1. Get a rough idea of what a battery is (p. 609) and how to hook up cells in series and parallel (p. 611). **Look at the CD WARM-UPS, then study the WALK-THROUGH EXAMPLES.**

RESISTANCE

The current passing through a resistor varies directly with the applied V and inversely with R:

$$V = IR \qquad [17.3]$$

which is **Ohm's Law**, and it applies to a limited range of materials. The unit of resistance is the *ohm* (Ω). There are a variety of problems that deal with Ohm's Law here and in the next chapter as well, so this material must be mastered now. Study Section 17.2 (Ohm's Law) and review Examples 17.3 and 17.4.

The resistance of a wire can be expressed in terms of its length L, cross-sectional area A, and a material-dependent constant of proportionality ρ, the **resistivity**, as

$$R = \rho \frac{L}{A} \qquad [17.4]$$

Resistivity is temperature-dependent (p. 619); accordingly

$$\rho \approx \rho_0(1 + \alpha_0 \Delta T) \qquad [17.5]$$

α_0 is called the ***temperature coefficient of resistivity***. The total absence of dc resistivity below some **critical temperature** (T_c) is known as **superconductivity** (p. 620). The preceding material is considered in Section 17.3 (Resistivity) and forms the basis for a number of problems. Example 17.5 is typical of problems relating resistance to the physical characteristics of the sample, whereas Example 17.6 deals with temperature dependence.

The power either provided or dissipated by a circuit element is

$$P = IV \qquad [17.7]$$

The unit of electrical power is the *watt*. In particular, the power *dissipated* in a resistance R is

$$P = I^2 R \qquad [17.8]$$

or

$$P = \frac{V^2}{R} \qquad [17.9]$$

Figure 17.25 summarizes the diagrammatic representations of the circuit elements introduced thus far.

Key Terms

electric current	semiconductor
ampere	insulator
mobile charge carrier	conductor
battery	temperature coefficient of resistivity

voltaic pile	superconductivity
direct current	critical temperature
amp-hour rating	potentiometer
resistance	rheostat
Ohm's Law	fuse
ammeter	power dissipated
ohmic	power delivered
resistivity	joule heat

∿	Fuse	⌒	Circuit breaker
╢╟	Capacitor	Ⓥ	Voltmeter
╫╟	Variable capacitor	Ⓐ	Ammeter
⌁W⌁	Resistor	╫	dc cell or voltage source
⌁W⌁	Variable resistor	╫╫	Battery or dc source
⌁WWW⌁	Potentiometer	╫╫	Variable dc voltage source

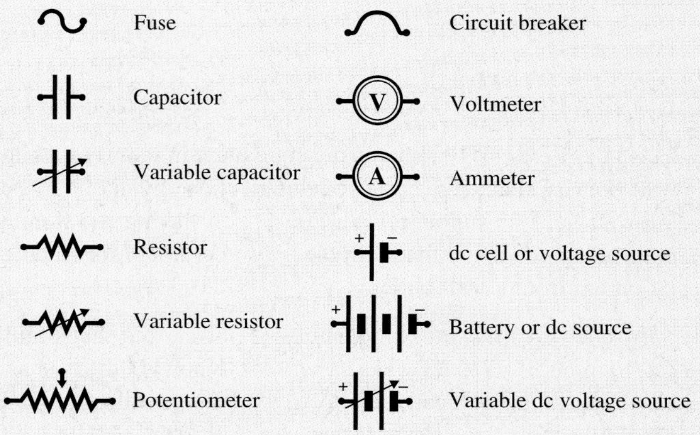

Figure 17.25

Discussion Questions

1. Explain the meaning of each of the Key Terms above.

2. The accompanying photo pictures several carbon composition resistors. Going from the larger to the smaller the color bands are (yellow, violet, brown, gold), (red, red, brown, silver), (orange, orange, brown, silver), (brown, black, yellow, silver), (red, red, black, silver), and (gray, red, red, silver). What do these markings tell us? What are the values of the resistors? Why are the resistors different sizes?

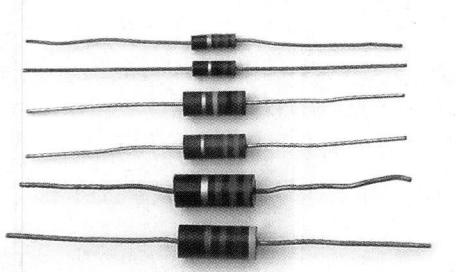

Figure Q2

3. Ohm once remarked that the laws of electricity "are so similar to those given for the propagation of heat... that even if there existed no other reasons, we might with perfect justice draw the conclusion that there exists an intimate connection between these natural phenomena." What was he alluding to? Discuss the nature of electrical and thermal conductivity for metals.

4. Figure Q4 shows several 9-V batteries opened up so that we can look inside. Explain what you see.

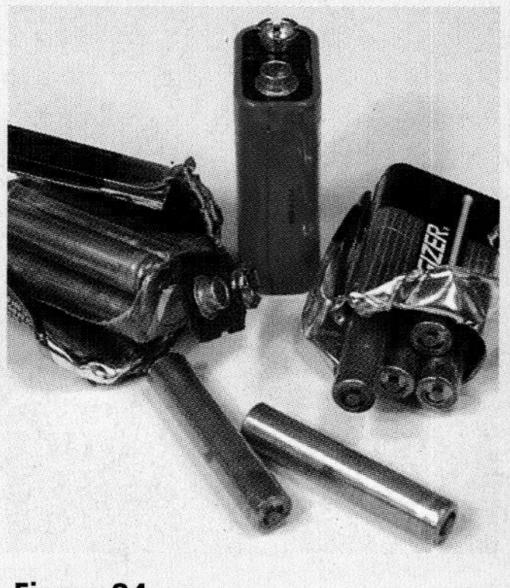

5. Suppose you wish to set up a circuit with a battery and some resistors so that you can measure currents and voltages and confirm the ideas of this chapter. How should you select the hookup wire, or doesn't it matter?

Figure Q4

6. What happens when you turn the key in the ignition of your automobile? Should you start your car on a cold rainy night with the wipers, lights, heater, and defroster all on? Explain.

7. Figure Q7 shows a 1.5-V dry cell attached to a 1.5-V flashlight bulb via two single-pole double-throw knife switches. Each switch attaches the central post (or pole) to either the right or left terminal on the device—the blade is only thrown to a horizontal position. You probably have an arrangement of this kind in your house, especially if it has a long flight of steps that are lighted. What's it for? Make a simplified drawing of the circuit and discuss its operation. As shown, is the light on? What happens when either switch A or switch B is thrown?

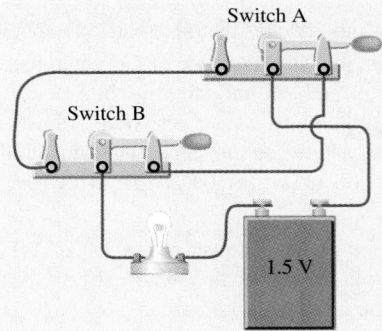

Figure Q7

8. Suppose you parked your car and left the lights on all night. Why would you have trouble starting it the next day? Would it be more bothersome to start on a really cold winter's day? After several tries, the engine just makes some disheartening clicking noises and you've had it—what has happened? A friend comes to your aid and "jump starts" your car. How is that done and what does it accomplish?

9. With Question 8 in mind, it's assumed that once the engine is running, all will return to normal. Explain. In other words, how does your engine keep running once the jumper cables are removed, given that the battery is still exhausted? Is it then a good idea to let the engine run or drive for a while without turning on any unnecessary electrical devices? Why? What's happening in the electrical system while you drive the car with the radio playing?

10. Figure Q10 shows an ordinary metal-body flashlight. Explain how it works and discuss the significance of the arrangement of the dry cells.

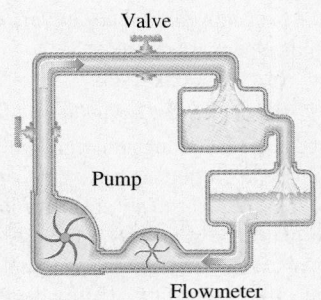

Figure Q10

11. Describe Volta's tin-tongue-silver cell. Why don't you taste anything until the two metals touch? Volta thought that the current of "electric fluid" in his pile arose from a "contact force" between the two metals. It was not until about 25 years later that the chemical action taking place was recognized (by C. A. Becquerel, A. De La Rive, G. F. Parrot, and others).

12. Figure Q12 shows a vertical arrangement of pipes with water flowing around the circuit at a rate that keeps the levels constant. This system is a liquid analogy to an electrical circuit. Draw the corresponding electrical diagram and discuss the relationship between the various parameters. Consider the system from an energy perspective. What does the pump "pump" besides water? No wonder nineteenth-century scientists liked the notion of "electric fluid." Have you ever heard a modern electrician mutter things like "turn on the juice"?

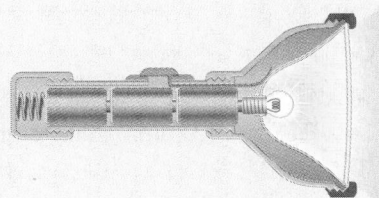

Figure Q12

13. Discuss everything of significance you know about the dry cell—that is, the ordinary flashlight "battery." Would it still work if it were actually dry? Ask a few elderly people if they remember how leaky old-fashioned D-cells were when they were kids—explain. Why do expensive electronic devices recommend that they not be stored with the batteries in them?

14. Question 12 deals with a fluid system using gravity. Figure Q14 represents a horizontal variation that is independent of gravity. Relate the liquid and electrical parameters. Considering fluid friction, talk about the influence of the narrow-bore tube.

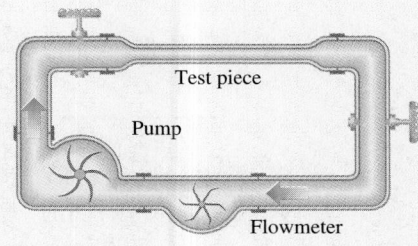

Figure Q14

15. When an ordinary tungsten light bulb burns out, it's likely to flash bright blue-white for an instant. Explain what is happening. Suggest a possible reason these lamps usually burn out soon after being turned on.

16. I once saw someone accidentally drop a wrench across the terminals of a 12-V car battery. There was a tremendous sparking flash, and molten beads of metal flew all over the place. Explain. Why did the wrench only melt where it touched the terminals of the battery?

17. When lightning hits the ground, it often spreads out radially as it penetrates downward. Thus, it is possible for a large four-legged animal standing with its bodyline radial with the bolt to be electrocuted. Explain. Considering ground currents, what is the best posture for a human out in a thunderstorm—lying down, squatting, or standing up? Read the newspaper article (Fig. Q17).

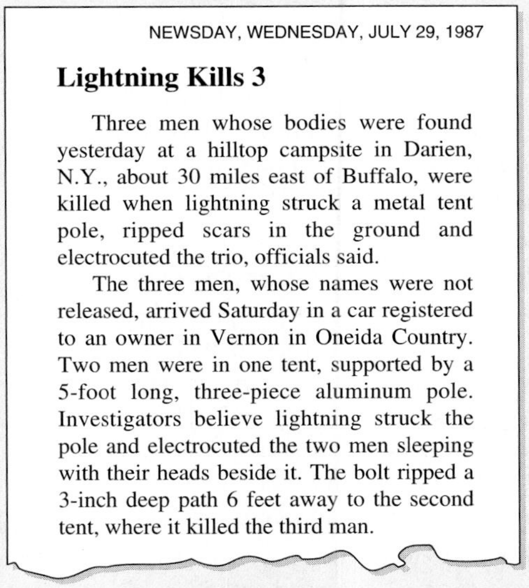

NEWSDAY, WEDNESDAY, JULY 29, 1987

Lightning Kills 3

Three men whose bodies were found yesterday at a hilltop campsite in Darien, N.Y., about 30 miles east of Buffalo, were killed when lightning struck a metal tent pole, ripped scars in the ground and electrocuted the trio, officials said.

The three men, whose names were not released, arrived Saturday in a car registered to an owner in Vernon in Oneida Country. Two men were in one tent, supported by a 5-foot long, three-piece aluminum pole. Investigators believe lightning struck the pole and electrocuted the two men sleeping with their heads beside it. The bolt ripped a 3-inch deep path 6 feet away to the second tent, where it killed the third man.

Figure Q17

18. High-voltage cables are generally uninsulated, and a bird standing on one of them may have its feathers puffed up. Why? Why aren't our little feathered friends electrocuted on their perches? Could a chicken walk safely on the third rail of an electrified railroad?

Multiple Choice Questions

The first four questions below refer to Fig. MC1, which shows three light-bulb circuits.

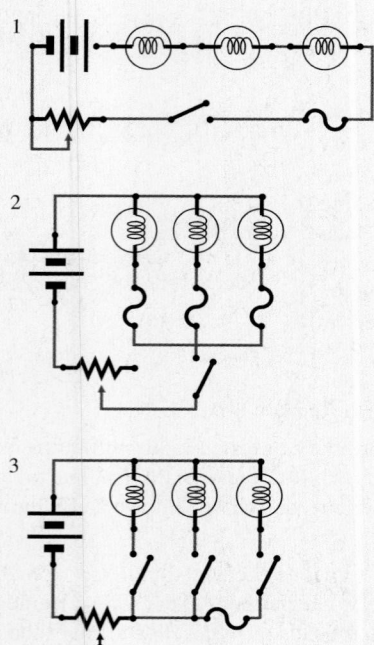

Figure MC1

1. Which of the three circuits allow the brightness of the lamps that are on to be varied? (a) 1, 2, and 3 (b) 2 and 3 (c) 1 and 3 (d) 1 and 2 (e) none of these.

2. Which of the three circuits is protected against each bulb being short circuited? (a) 1, 2, and 3 (b) 2 (c) 1 and 2 (d) 1 and 3 (e) none of these.

3. Which of the three circuits allow the bulbs to be lit individually? (a) 1 and 3 (b) 2 and 3 (c) 1 and 2 (d) 3 (e) none of these.

4. Which of the three circuits turns all the bulbs on at once? (a) 1 and 3 (b) 2 and 3 (c) 1 and 2 (d) 3 (e) none of these.

5. If 10 amperes circulate in a closed circuit, how much charge passes any point therein in 2 s? (a) 10 C (b) 5 C (c) 20 C (d) 200 C (e) none of these.

6. The difference in potential between the electrodes of a voltaic cell when there is no current being drawn is (a) zero (b) 1.5 V (c) its emf (d) the power (e) none of these.

7. The emf of a voltaic cell (a) is independent of the chemical interactions taking place within it (b) is dependent on the size of the cell (c) is dependent on its amp-hour rating (d) is independent of the plate size (e) none of these.

8. Usually, the larger a voltaic cell is, the more (a) voltage it can supply (b) current it can supply (c) potential it can develop (d) emf it can sustain (e) none of these.

9. The moving nonconducting belt of a Van de Graaff generator carries 10 μC of charge up to the metal sphere (Fig. Q14, Chapter 15) each second. The steady-state potential difference between the sphere and the grounded source of charge is 3 $\times$ 10^6 V. What, if any, is the emf of the generator? (a) 10 μV (b) 3 μV (c) 3 MV (d) 30 μV (e) none of these.

10. A statue of a chicken made of pure gold has a resistance of 0.10 mΩ between the beak and the tail. An exact duplicate of the piece is made in an alloy that has a resistivity 10 times greater than gold. The resistance between the same two points will now be (a) 0.10 mΩ (b) 1.00 mΩ (c) 0.01 mΩ (d) 10.0 mΩ (e) none of these.

11. A metal wire has a resistance of 1.0 Ω. What will be the resistance of a wire made of the same material but twice as long and with half the cross-sectional area? (a) 0.40 Ω (b) 2.00 Ω (c) 0.02 Ω (d) 40.0 Ω (e) none of these.

12. The resistance of a superconducting material drops, essentially to zero, rather suddenly when (a) the sample is heated above T_c (b) the sample is exposed to a magnetic field (c) the sample is cooled below T_c (d) the sample experiences a current I_c (e) none of these.

13. Referring to Fig. MC13, the voltage at point B is (a) +12 V (b) −12 V (c) 0 V (d) −6 V (e) none of these.

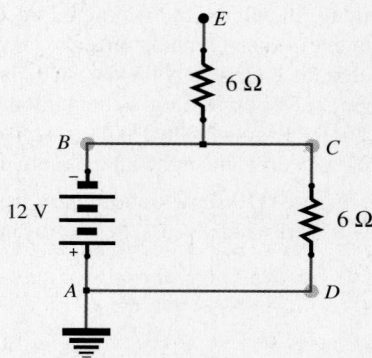

Figure MC13

14. Referring to Fig. MC13, the voltage at point D is (a) +12 V (b) −12 V (c) 0 V (d) −6 V (e) none of these.

15. Referring to Fig. MC13, the voltage at point C is (a) +12 V (b) −12 V (c) 0 V (d) −6 V (e) none of these.

16. Referring to Fig. MC13, the voltage at point E is (a) +12 V (b) −12 V (c) 0 V (d) −6 V (e) none of these.

17. An 8.00-Ω speaker is connected across the output terminals of a power amplifier that delivers 64.0 W to it. The current supplied to the speaker is (a) 2.83 A (b) 8.00 A (c) 2.00 A (d) 64.0 A (e) none of these.

18. Figure MC18 shows the voltage across a tungsten light bulb filament plotted against the current through it. The curve bends upward because (a) the filament is getting hot and running out of electricity (b) it takes more voltage to propel the same current than it would if the device were ohmic because the resistance increases (c) the resistance goes down, so we get more

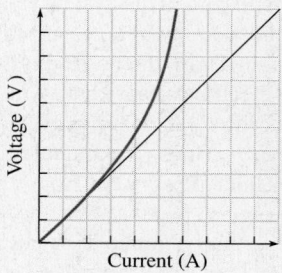

Figure MC18

voltage for a given current   (d) the resistance stays constant, but the current decreases   (e) none of these.

19. A copper wire has a resistance of 10 Ω. What will be its new resistance if the wire is shortened by cutting it in half?   (a) 20 Ω   (b) 10 Ω   (c) 5 Ω   (d) 1 Ω   (e) none of these.

20. When the temperature of a length of aluminum wire is lowered, itsresistance   (a) increases slightly   (b) decreases correspondingly   (c) stays the same   (d) increases as ΔT (e) none of these.

21. If a potential difference of 12 V across a resistor results in a current of $\frac{1}{2}$ A, how much power is dissipated?   (a) 48 W   (b) 12 W   (c) 4 W   (d) 6 W   (e) none of these.

22. Imagine a length of ordinary insulated hookup wire. The wire's resistance is not dependent on   (a) the conductor's length   (b) the conductor's radius   (c) the material making up the insulator   (d) the material making up the conductor   (e) none of these.

23. The units of $\Omega \cdot A^2$ correspond to   (a) current   (b) energy   (c) power   (d) voltage   (e) none of these.

24. One microvolt (1 μV) corresponds to   (a) 10^6 V   (b) 10^{-6} V   (c) 1000 V   (d) 1/1000 V   (e) none of these.

25. A resistor operating at 100 V generates joule heat at a rate of 20 W. When placed across a 50-V source, it will draw   (a) 0.10 A   (b) 5 A   (c) 20 A   (d) 4.0 A   (e) none of these.

For more Multiple Choice Questions with answers click on WARM-UPS in CHAPTER 17 on the CD. ⊙

Suggestions on Problem Solving

1. An important type of derivation involves determining the current, given a flow of charge or vice versa. For example, a detector measures N particles, each carrying a charge q, arriving *per second per unit of its surface area*—what current strikes the instrument if its area is A? The basic definition is $I = \Delta q / \Delta t$, but to use this equation you must first find Δq, the amount of charge that strikes the detector during the time Δt. We are given N, and so Nq is the amount of charge arriving *per second* per unit of surface area. Consequently, NqA is the total amount of charge arriving on the detector per second and is equal to I. We obtained I without finding Δq explicitly because N had the time built into it to begin with. This kind of analysis of flow is very important in physics—we've seen it before and we'll see it again

2. Remember that all parts of an ideal hookup wire in a circuit are at the same potential. Drops or rises in potential occur only across circuit elements (resistors, batteries, etc.); and in the case of a resistor, that means only when a current is passing through it (take a look at Multiple Choice Question 16). **A steady-state current cannot exist in a length of conductor unless it forms part of a closed path.**

3. Whenever you have a circuit diagram with a battery in it, place $+$ and $-$ signs at the appropriate terminals. Current, being imagined as composed of positive charge, always emerges from the positive terminal (provided there are no other batteries bucking it) and reenters at the negative one. In a simple circuit where you know the direction of the current, follow it around, labeling all the resistors with $+$ signs where current enters and $-$ signs where it leaves. This will show the voltage drops and rises, something that will be explored further in Chapter 18.

4. The voltage of any point in a circuit is only known relative to some other point in that circuit. Thus, one side of a battery might be 12 V higher than the other side (though it could be 10000 V higher than your nose). For example, the entire chassis of an automobile serves as a common conductor called "ground" in the trade, though it isn't usually grounded and so actually floats—the $+$ terminal of the battery is 12 V higher than the engine block. Often a point in a circuit is grounded, as is done with stereo systems and computers. Its potential is then taken as zero, and all voltage drops or rises are referenced with respect to it.

Problems ✚ Coordinated Problems ✚ Progressive Problems ✚ Solutions

STUDY GUIDE **1. Coordinated Problems:** The three problems within each magenta-colored grouping are solvable in similar ways. Note that the first of these always has a hint; moreover, its solution is provided in the back of the book. *Work out each of these sets; they'll strengthen technique and build confidence.* **2. Progressive Problems:** The problems introduced in blue unfold step-by-step carrying along the analysis in a more suggestive way than is customary. *Work out all of these; they'll guide you through the analytic process and help develop problem-solving skills.* **3. Worked-Out Solutions:** Studying worked-out solutions is an important part of learning how to solve problems. Accordingly, additional *solutions* to a number of model problems are given below. *Make sure you understand each of them before you go on to the next problem.* **4.** Also provided in the back of the book are the **Answers** to all odd-numbered problems, as well as worked-out *solutions* to those with boldface numbers. Problem numbers in italic indicate that a solution appears in the Student Solutions Manual.

SECTION 17.1: ELECTRIC CURRENT

1. [I] The moving nonconducting belt of a Van de Graaff generator carries 10 μC of charge up to the metal sphere (Fig. Q14, Chapter 15) each second. Determine the corresponding current.

2. [I] THIS PROBLEM EXPLORES THE BASIC NATURE OF CURRENT. A small motor is attached to a 12-V battery, and 0.48 millicoulombs of charge are supplied during a period of 10 milliseconds. (a) How much charge in coulombs are we talking about? (b) How much time in seconds did the charge transfer take? (c) What is the definition of current? (d) How much current flows through the motor?

3. [I] A 1.5-V D-cell has a rating of 3.0 amp-hours. How many coulombs of charge can it provide before running down?

> **SOLUTION:** The cell can provide 3 A for 1 h, where 1 A = 1 C/s. That's 3 C/s for 3600 s or (3 C/s)(3600 s) = 10800 C = 0.011 MC.

4. [I] A beam of positrons carries 1.4 C past a point in space in 2.0 s. To what current does that correspond?

5. [I] If a quantity of singly ionized sodium ions (Na^+) equal to Avogadro's number streams past a point in 1000 s, what is the current?

6. [I] The photo on p. 608 shows an electron beam representing a current of 60 kA. How many electrons flow out of the device per second?

7. [I] The starter motor in an automobile is a small but powerful electrical device that "turns over" the main gasoline engine, moving it through a cycle to get it started. Typically, it will draw about 180 A from the battery for perhaps 2.0 s. How much charge flows through the circuit?

8. [I] A particle accelerator contains two beams flowing side-by-side in opposite directions; the beams will ultimately be made to collide. One is a stream of protons (each with a charge of $+1.60 \times 10^{-19}$ C), the other a stream of antiprotons (each with a charge of -1.60×10^{-19} C). Given that either beam can deliver 1.0×10^{14} particles per second to the colliding region, what is the net current in the machine as the two streams race past each other?

9. [I] A synchrotron accelerates protons up to nearly the speed of light, imparting energies to them of 500 MeV. If the beam current is 1.0 mA, how many protons will hit a target in 0.10 s?

10. [I] With the previous problem in mind, how many protons are there in each 1.0-cm-long segment of the beam? Assume a uniform particle density throughout the beam.

11. [I] A wire is connected across the terminals of a battery for precisely 60.0 s during which time a constant current of 2.00 A circulates around the loop. What was the net charge that flowed past any point on the wire?

12. [I] A portable tape recorder is powered by six 1.5-V AA-cells in series. What is its operating voltage?

13. [I] A torpedo, a giant saltwater ray that can develop a voltage of 220 V, is covered with cells known as electroplaques, each of which produces a potential difference of about 0.15 V. Several thousand rows (each made up of a series-connected array of cells) are then connected in parallel to build up a sizable current. How many cells would you guess form each row? Why do freshwater electric fish in general develop higher voltages than saltwater ones?

14. [I] A NiCd-cell has a voltage of 1.2 V and an amp-hour rating of 34. It is a sealed storage cell with a nickel anode and a cadmium cathode immersed in an alkaline electrolyte. How long can it operate when providing 2.0 A to some load?

15. [I] A battery is rated at 10 amp-hours. How much charge does that correspond to?

16. [I] The positive terminal of a 1.5-V dry cell is attached to ground via a length of heavy hookup wire. The negative terminal is then connected, via the same kind of wire, to a neutral brass ball. What is the potential of the ball? Describe its state of charge. What is the potential of the central carbon electrode?

17. [I] THIS PROBLEM EXPLORES THE VOLTAGES IN A CIRCUIT. Consider Fig. P17. (a) What is the voltage of point-A with respect to ground? (b) What is the voltage of point-D with respect to ground? (c) What is the voltage of point-B with respect to ground? (d) What, if anything, is the value of the voltage (V) across the load? (e) Which point B or C, is at a higher potential? (f) Is there a current in the circuit, and if so, in which direction does it flow?

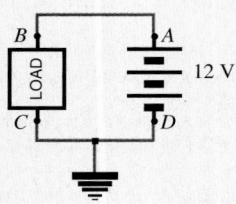

Figure P17

18. [II] Consider a 1.0-A current flowing in a long wire. Roughly 10^{22} electrons move in each meter-long segment of the wire. (a) How much charge moves along each meter of wire? (b) What is the average speed of the electrons?

> **SOLUTION:** The definition of current is $I = Q/t$ where a charge Q is transported in a time t. In that time t, traveling at an average speed v_{av}, charge moves a distance l where $l = v_{av} t$. (a) Recall that if 10^{22} electrons move, a charge of $q_e 10^{22}$ moves. (b) Hence $I = Q/t = Qv_{av}/l$ and the moving charge per meter $Q/l = q_e 10^{22} = 1.6 \times 10^3$ C. $v_{av} = (1.0$ A$)/(1.6 \times 10^3$ C$) = 6 \times 10^{-4}$ m/s.

19. [II] A pure gold wire with a 1.00-mm $\times$ 1.00-mm cross section carries a flow of electrons having a current density (I/A) equal to 1.0 MA/m². How long will it take for an amount of electrons equal to Avogadro's number to pass a point on the wire?

20. [II] An ion generator used to clean room air puts out a stream of negatively charged molecules that attach themselves to airborne pollutants, which are then collected electrostatically. Oxygen molecules tend to pick up electrons, thereby becoming negative oxygen ions. According to one company's literature, at 1.0 m from a particular generator, a detector would record the arrival of "168 million ions/sec./cm²." Assuming that each ion is singly charged, what current impinges on a 10.0-cm² target at 1.0 m from the device?

21. [II] Figure P21 is a diagram of a portion of the electrical system of an automobile. List which switches—$A, B, C, D, E,$ and F—must be closed in order to (a) blow the horn; (b) turn on the headlights; (c) turn on the tail lights; (d) turn on only the parking lights; (e) activate the inside dome light. When do the side marker lights go on? *Note that the whole body of the car, including the engine block, is generally wired together as a common "ground."*

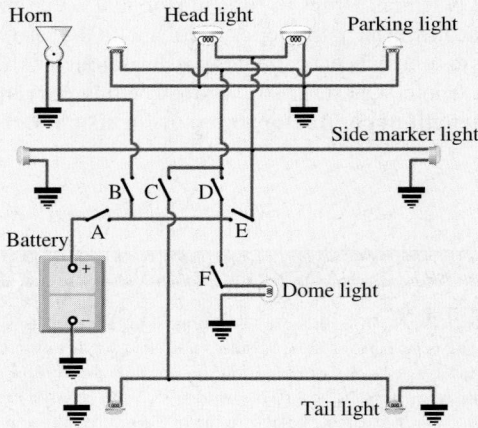

Figure P21

22. [II] A particular model of automobile battery, when fully charged, can deliver roughly 4.0×10^5 C before becoming completely run down. What is the amp-hour rating of such a battery?

23. [II] We wish to make up a panel of silicon solar cells that will provide at least 440 mA at 9.0 V when placed in an appropriately illuminated region. Given that each cell has an output of 22 mA at 0.45 V, design the panel. [*Hint: First deal with the voltage.*]

24. [II] A small cell has an emf of 0.80 V and can supply 10.0 mA under normal operating conditions. How should a battery be con-

structed if it is to have an emf of 12.0 V and an operating current of 30.0 mA? How many cells will be needed?

25. [II] Suppose you have two different kinds of cells: four large 1.5-V cells that can operate at 4.0 A and ten small 1.5-V cells that can operate at 1.0 A. Design a 4.5-V battery that can supply 8.0 A and, moreover, use all the large cells. How many of the small size will you need?

26. [II] Show that the battery in Fig 17.11*b* is equivalent to having five rows in parallel each with three cells in series.

27. [III] In the process of recharging a rundown automobile battery, it's attached to an electronic 12-V charger. As soon as the device is turned on, an ammeter shows that the battery draws 7.0 A; but as it revives, the current slowly drops until after 6.0 hours it's down to 3.0 A. Assuming the current decreased linearly with time, how much charge passed through the battery?

SECTION 17.2: OHM'S LAW

28. [I] A 6.0-Ω resistor is placed across the terminals of a 12.0-V battery. How much current flows through the resistor?

29. [I] Consider an ordinary light bulb which has a resistance of 100 Ω when illuminated. What is the value of the current drawn by the bulb while 110 V are across it? When measured fresh out of the box the bulb had a resistance of around 20 Ω; explain this difference.

> **SOLUTION:** We have a 100-Ω resistance with 110 V across it. Using Ohm's Law, $V = IR$, $I = (110 \text{ V})/(100 \text{ Ω}) = 1.10$ A. The resistance increases with temperature.

30. [I] A piece of wire at a constant temperature is placed across a variable voltage source, and the current through it is measured. If the wire is ohmic, which part of Fig. P30 best represents its behavior as the voltage is increased?

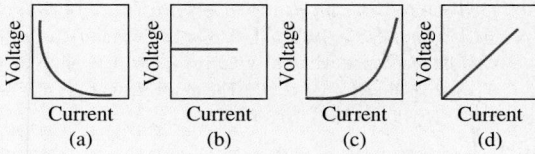

Figure P30

31. [I] THIS PROBLEM EXPLORES THE VOLTAGES AND CURRENT IN A CIRCUIT. Consider Fig. P31. (a) What is the voltage of point-*C* with respect to ground? (b) What is the voltage of point-*D* with respect to ground? (c) What is the voltage of point-*A* with respect to ground? (d) What, if anything, is the value of the voltage (*V*) across the resistor *R*? (e) Which point, *A* or *B*, is at a higher potential? (f) Write an expression relating *R* and *V*. (g) What is the value of the resistance *R*?

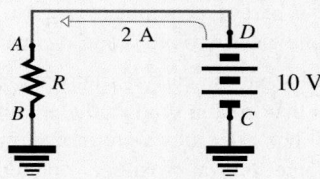

Figure P31

32. [I] Determine the current, if there is any, in the circuit in Fig. P32.

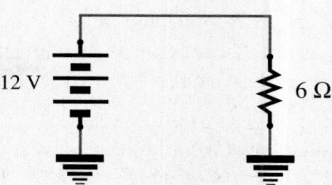

Figure P32

33. [I] A length of wire with a resistance of 150 Ω is placed across the terminals of a battery comprised of six 1.5-V cells in series. How much current will it draw from the battery? [*Hint: The total voltage of the battery is across the wire.*]

34. [I] Three 1.5-V cells are connected in series, and the combination is attached across a resistor. If 3.0 A flows through the resistor, what is its resistance?

35. [I] What voltage will be measured across a 1000-Ω resistor in a circuit if we determine that there is a current of 2.50 mA flowing through it?

36. [I] The two electric probes of an ohmmeter touching damp skin measure a resistance of 120 kΩ. If a voltage of 200 V is now put across these probes, what current will flow through that region of the skin?

37. [I] THIS PROBLEM EXPLORES RESISTANCE IN A CIRCUIT. Considering Fig. P37 where the colors on the resistor are green-blue-black-silver, (a) what is its resistance? Suppose that the red and yellow leads clipped to the resistor are attached to the two terminals of a 1.5-V cell. (b) What is the voltage across the the resistor? (c) Write a mathematical expression for the relationship between that voltage, the resistance, and the resulting current. (d) Determine the current flowing through the resistor.

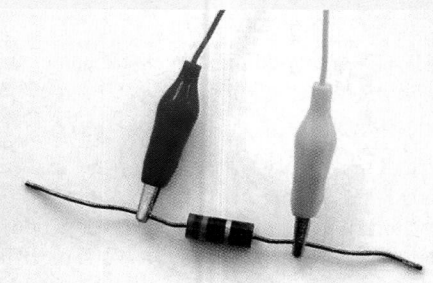

Figure P37

38. [I] A small high-torque variable-speed dc motor requires an input of 10 mA at 3.0 V. Determine the electrical resistance of the motor.

39. [I] A 100-V electric heater draws 10 A. What is its resistance?

40. [I] If electric contacts are placed on the scalp, time-varying differences in potential will be observed. These can be recorded by an electroencephalograph. Voltage differences of ≈0.5 mV will appear across resistances of ≈10 kΩ. What size currents are involved?

41. [I] A wooden stick in contact with the metal sphere of a 100-kV Van de Graaff generator carries a current of 2.0 μA down to ground. Calculate the stick's resistance.

42. [I] If the current in a 10-Ω resistor is 500 mA, what is the voltage across its terminals?

43. [II] Six 1.5-V cells are connected in two strings, each of which has three cells in series. These two are then attached to each other in parallel and put across a 90.0-Ω resistor. How much current flows through the resistor?

44. [II] Figure P44 shows a variable resistance (total 100 Ω) across which is a voltage drop of 12 V supplied by a battery. What must be the resistance of that portion of the resistor between A and B (namely, R_{AB}) if the voltage V_{AB} is to be (a) 12 V; (b) 6.0 V; (c) 3.0 V?

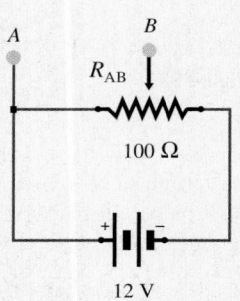

Figure P44

45. [II] Two 1.0-MΩ resistors in series form a closed circuit with a 200-V source. What is the voltage drop across each resistor? What is the current through each resistor?

46. [II] A variable slidewire resistor with a total of 50 Ω is made by wrapping a single layer of 200 turns of varnished copper wire around an insulating cylindrical core. When it carries a current of 3.0 A, what is the voltage drop across each turn of wire?

47. [II] Three identical resistors are wired one-to-the-next in series, and the string of them is then placed across the two terminals of a 300-V power source, thereby forming a closed circuit. If 4.0 A flows around the circuit, what is the voltage drop across each resistor? What is the resistance of each resistor?

48. [II] Two 12-V batteries are connected with the positive terminal of one attached to the negative terminal of the other. The two remaining terminals are then each attached to one of the terminals of a 600-Ω resistor, thereby forming a closed circuit. How much current passes through each battery?

SECTION 17.3: RESISTIVITY

49. [I] Given a solid cube of metal, under what circumstances, if any, will $R = \rho$?

50. [I] Prove that the correct units for resistivity are Ω·m.

51. [I] THIS PROBLEM EXAMINES THE IDEAS OF RESISTIVITY AND RESISTANCE. Earlier we looked at the way a voltage spike propagates along a nerve cell (see Fig. P123, p. 606) . There is something called the *cable model* of the nerve axon. It assumes currents are parallel to the axis of the axon, which is taken to be a fluid cylinder. Axoplasm fills the axon and has a resistivity of ≈0.5 Ω·m. The axon (see Fig. P83, p. 603) is a thin, ≈10 mm long, cylinder having a diameter of ≈10 μm. From that we want to compute the axon's resistance per unit length. (a) What physical characteristics of the axon determine its resistance? (b) What are its dimensions in meters? (c) What is the cross-sectional area of the axon in meters squared? (d) Now that we have everything in proper units, approximate the resistance of the axon. (e) Why is the resistance so high? (f) What is the resistance per unit length of the axon?

52. [I] A 1.0-m-long wire of pure silver at 20 °C is to have a resistance of 0.10 Ω. What should be its diameter?

53. [I] Considering a length of metal wire, show that Ohm's Law can be written as

$$I = \frac{E}{\rho/A}$$

54. [I] A nichrome wire with a cross-sectional area of 1.5×10^{-6} m² is to be used in a heater. If the design calls for a 3.0-Ω coil, what length of wire will be needed?

55. [I] A 5.0-m-long wire has a cross-sectional area of 2.0 mm² and a resistance of 40 mΩ. What is the resistivity of the material constituting the wire?

56. [I] According to the American Wire Gauge system, No. 0000 wire, which is the heaviest, has a diameter of 11.7 mm. What would be the resistance of 100 m of copper AWG No. 0000 at 20 °C?

57. [I] Copper telegraph wire has a resistance of about 10 Ω per mile. What's its diameter in millimeters?

58. [I] A wire of pure gold is drawn through a die so that it is stretched out to twice its original length. Given that its volume is unchanged in the process and its new cross-sectional area is constant, compare the new with the original resistance.

59. [I] A narrow rod of pure iron has a resistance of 0.10 Ω at 20 °C. What is its resistance at 50 °C? [*Hint: Reexamine Example 17.6.*]

60. [I] A carbon rod used to generate the bright light in a movie-theater projector has a resistance of 110 Ω at 20.0 °C. What will be its resistance at 520 °C? (Incidentally, the hottest point on a functioning carbon arc, and the point of greatest luminosity, is typically at about 3500 °C.)

61. [I] A narrow rod of manganin has a resistance of 0.10 Ω at 20 °C. What is its resistance at 50 °C?

62. [II] At room temperature the resistance of a length of aluminum rod is 250 mΩ. (a) If the rod has a cross-sectional area of 0.025 cm², how long is it? (b) How much current will it draw when 9.0 V are put across its ends? (c) How much current would it draw if the rod was shortened by 20%?

> SOLUTION: (a) We want to find the length, and that's a physical characteristic of the rod. That calls to mind $R = \rho L/A$; using Table 17.2, $L = RA/\rho = (0.250\ \Omega)(0.025 \times 10^{-4}\ \text{m}^2)/(2.8 \times 10^{-8}\ \Omega\cdot\text{m}) = 22$ m. (b) We have a 0.250-Ω resistance with 9.0 V across it. Using Ohm's Law, $V = IR$, $I = (9.0\ \text{V})/(0.250\ \Omega) = 36$ A. (c) The new resistance is $R' = 80\%R$; therefore $I' = I/0.80 = (36\ \text{A})/0.80 = 45$ A.

63. [II] THIS PROBLEM EXAMINES THE IDEAS OF RESISTIVITY AND RESISTANCE. A piece of copper wire 210 cm long has a cross-sectional area of 1.00×10^{-4} m². We want to know how much current it will draw from a 12-V battery. (a) First put everything in proper units. (b) What physical characteristics of the wire determine its resistance? (c) Write an expression for the resistance of the wire. (d) Compute the resistance of the wire. (e) If we attach each end of the wire to one of the terminals of a 12-V battery, what potential difference will exist across the length of the wire? (f) How much current will pass through the wire? (g) Is shorting a battery a good idea?

64. [II] A length of copper wire 1.0-m long with a cross-sectional area of 1.0 mm² is part of a circuit carrying 1.0 A. What is the value of the steady-state electric field within the wire?

65. [II] It is said that the carbon filaments in the early incandescent light bulbs that Edison produced lost more than two-thirds of their resistance soon after they were turned on. Explain this occurrence and compute their approximate operating temperature. The latter was actually about 1900 °C.

66. [II] Imagine that you are going to connect a remote speaker to your stereo system. The speaker has a resistance of 4.0 Ω, and so

you want to use hookup wire whose total resistance is small by comparison, say, 0.25 Ω. If the speaker is to be 15 m away, what diameter copper wire should be used? [*Hint: You'll need two leads.*]

67. [II] A platinum resistance thermometer made up of a coil of wire with a resistance of 10 Ω at 20 °C is placed in a chamber at 420 °C. What will be its new resistance if α_0 is fairly constant at 0.003 9 K^{-1}?

68. [II] Show that if a conductor has its temperature changed, it will experience a fractional change in its resistivity given by

$$\frac{\Delta\rho}{\rho_0} = \alpha_0 \Delta T$$

What assumption must be made here?

69. [II] An iron wire at 20 °C is heated until its resistance doubles. At what temperature will that occur? Assume the temperature coefficient is constant over that temperature range.

70. [II] When currents are transported at very high voltages, it's desirable to keep the electric fields surrounding the conductors down to levels that will not break down the surrounding air, thereby controlling sparking. Accordingly, conductors at power plants and high-voltage laboratories are often large-diameter pipes. What is the resistance per meter of a copper pipe 2.5-cm thick with an inside diameter of 15.0 cm?

71. [III] A modern incandescent lamp has a tungsten filament with a melting point of 3400 °C. Ordinarily, with the bulb evacuated, it's operated at about 2200 °C. The efficiency can be improved almost threefold by raising the temperature to 2800 °C, but that causes the tungsten to evaporate and shortens the life appreciably. The alternative (an idea introduced by Langmuir) is to fill the bulb with nitrogen or argon to suppress evaporation of the tungsten. What is the fractional change in the resistance of the filament when raised from 2200 °C to 2800 °C? Use the value of α_0 found in Table 17.3. Given the following values of α_0 for tungsten (from the *Handbook of Chemistry and Physics*)—0.004 5 at 18 °C, 0.005 7 at 500 °C, 0.008 9 at 1000 °C— how good was the calculation you just made? Discuss your answer and make a very rough estimate of α_0 at 2500 °C. Now estimate the fractional change in resistance using this new value.

72. [III] Figure 17.19 shows how the resistivities of various pure metals each seem to be heading toward zero at some Celsius temperature $-T'$. Make a plot of R versus T (in °C), assuming it to be a straight line. Is that reasonable? Taking $T = 0$ °C as the reference (that is, the line crosses the R-axis at $R = R_0$ and $T = 0$), show that for any point (T, R) on the line

$$R = R_0 \frac{T' + T}{T'}$$

and that $\alpha_0 = 1/T'$. How does this result compare with the values of resistivity given in Table 17.3? Notice how most values approximate $1/273 = 3.7 \times 10^{-3}$.

SECTION 17.5: ENERGY AND POWER

73. [I] A portable tape recorder has a plate on its underside indicating that it uses 1 W at 9 V dc. What net current does it draw from its battery?

74. [I] Referring to Problem 73, what is the net resistance of the tape recorder?

75. [I] A high-torque dc motor designed to drive cassette decks has a no-load speed of 7400 rpm at 9 V with a torque of 9.6 in.·oz and a no-load current of 16 mA. How much power will it draw from a 9-V battery when turning freely? [*Hint: At what voltage is it operating and how much current does it draw?*]

76. [I] A tiny motor mounted on a toy car operates across two D-cells in series. If it draws 0.15 A, how much power does it drain from the battery pack?

77. [I] Six D-cells in series power a tape recorder. If the device dissipates 15 W, how much current does it require?

78. [I] What is the maximum current that should be passed through a 100-Ω, 10-W resistor?

79. [I] The little speaker in a portable radio is labeled 8 Ω, 0.2 W. To what current does that correspond?

80. [I] A windmill with 6-ft propellers generates dc at 12 V. In a strong wind it will produce up to 200 W of electrical power, which is fed to a 230-amp-hour battery. What's the maximum current the windmill will deliver?

81. [I] An automobile starter motor will draw about 180 A from the 12-V battery of a car for perhaps 2.0 s in the process of starting the gasoline engine. How much power does it use?

82. [I] A light bulb operating at 110 V draws 1.82 A. Determine its resistance and the amount of power it draws.

83. [II] THIS PROBLEM EXAMINES THE CONCEPT OF ELECTRIC POWER. Operating at 110 V a light bulb draws 0.50 A. The bulb converts electrical energy into thermal energy and electromagnetic radiant energy. (a) How much charge passes through the bulb in 10 minutes? (b) Making use of part (a), how much energy does that charge deliver in the process of descending 110 V? (c) Using these results determine the power rating of the light bulb. (d) Check your answer using $P = IV$. (e) What is the resistance of the bulb?

84. [II] THIS PROBLEM DEALS WITH THE CONCEPT OF ELECTRIC POWER. The resistor in Fig. P84 has a power rating of 2.0 W, and we want to determine the limitations that places on voltage and current in the circuit. (a) In terms of the voltage of the battery, what is the potential difference across the resistor? (b) Write an expression for the power dissipated by the resistor in terms of the voltage across it. (c) What is the maximum voltage the battery can have if the resistor is not to be damaged? (d) Write an expression for the power dissipated by the resistor in terms of the current through it. (e) What is the maximum current the resistor can handle?

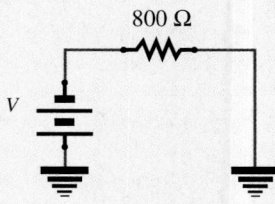

800 Ω

V

Figure P84

85. [II] A powerhouse near a waterfall has a large dc generator that produces electricity for a factory 0.50 mile away. The energy is transmitted over two cables each with a resistance of 0.25 Ω/mile. Given that the factory requires 45 kW at a voltage of 110 V to run its equipment, what must be the output of the powerhouse?

86. [II] A 0.8-in.-square silicon solar cell delivers 90 mA at 0.45 V when illuminated by sunlight (at 100 mW/cm^2). Suppose 10 such cells are connected in series; how much power could the panel deliver?

87. [II] For greater flexibility, electrical wires often consist of several fine strands twisted together rather than being constructed of one thick lead. A copper wire is made up of 10 fine fibers, each with a resistance of 2.0 mΩ. When placed across a voltage difference, a total current of 0.12 A traverses the wire. (a) What is the net resistance of the length of wire? (b) What voltage exists across it? (c) How much power is dissipated by each strand? [*Hint: How much current does each strand carry?*]

88. [II] A credit card–sized calculator uses two tiny 1.5-V cells in series that provide a normal operating power of 0.000 18 W. Determine the current passing through each cell when the device is in use.

89. [II] An old-fashioned trolley car draws 12 A from an overhead wire at +500 V (the rails are grounded). What power is delivered to the motor? If the motor is 86% efficient, what power does it develop in propelling the car?

90. [II] When put across the terminals of two D-cells in series, a small flashlight bulb draws 330 mA. How much power does it consume? How much energy does it take from the cells in 1.0 minute of operation?

91. [II] The moving nonconducting belt of a Van de Graaff generator carries 10 μC of charge up to the metal sphere (Fig. Q14, Chapter 15) each second. The steady-state potential difference between the sphere and the grounded source of charge is 3 MV. What minimum power must be supplied to the generator to sustain its operation?

92. [II] A stereo tuner-amplifier with a maximum power output of 50 W per channel (that is, 50 W to each of two speakers) has its right channel connected as shown in Fig. P92. Since the speaker

will be destroyed if it receives more than 36 W, a fuse is installed to limit the current entering it. What should be the rated maximum current of the fuse?

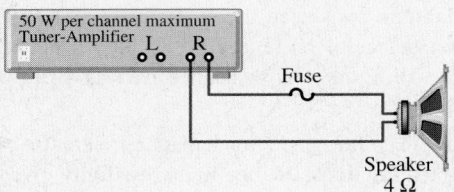

Figure P92

93. [II] When turned on, a flashlight bulb draws 1/3 A at 3.9 V. How much power does it require? Incidentally, more than 95% of that power appears in the form of thermal energy and not light. What is the resistance of the filament in the lamp?

94. [II] How much current does an ideal 1.00-hp dc electric motor draw when operating off a portable 100-V generator?

95. [II] A NiCd battery has an amp-hour rating of 10 and consists of five cells connected in series. What is the maximum power the battery will deliver if it is to operate for 5.0 h?

96. [III] A silicon solar cell 5 mm × 4 mm produces a current of 5 mA at 0.45 V under sunlight illumination of 100 mW/cm^2. Determine its efficiency.

97. [III] Determine the power rating in kilowatts of an electric heater that will raise the temperature of 10 liters of water from 25 °C to 85 °C in 15 minutes, assuming no loss of thermal energy. Given that the heater coil has a resistance of 10 Ω, how much current does it draw?

Chapter 18
Circuits

Now that we are familiar with the basics of dc—batteries, resistors, capacitors, and Ohm's Law—we can apply this knowledge to the treatment of circuits, where several elements are attached together. For example, we can study the electrical system of an automobile or design an experimental setup to measure the response of pigeons to the sight of popcorn. Today, virtually every field of science uses electric circuits in its research.

A variety of sensors, or *input transducers*, convert nonelectrical signals into electrical ones—the microphone is a familiar example. Any up-to-date hospital intensive care unit can electronically monitor a patient's body temperature, blood pressure, and respiration, measuring infusions and drainage while recording electrocardiograms, all with transducers. One kind of strain gauge is made from fine alloy wire that changes its resistance as it's stretched and compressed. Such gauges are widely used to monitor mechanical apparatus and in medical studies of bones, joints, and the behavior of muscles.

Circuit Principles

A circuit, in simple form, is a few elements (e.g., a resistor and a battery) joined together to make at least one closed current path (usually with some purpose in mind beyond merely confounding students). Often a system converts electrical energy into light via an incandescent lamp or into thermal energy via a toaster. There might be an input transducer, a microphone, at one end and an output transducer, a speaker, at the other end. In any event, we are generally interested in determining a variety of circuit parameters, such as voltage drops, currents, power dissipated, and so on—you can't send a 100-W signal out to a 20-W speaker, at least not twice. The only limitation in this chapter is that the currents and voltages are dc—they may rise and fall, but they do not fall below zero; they do not reverse direction, the sources do not change polarity.

18.1 Sources and Internal Resistance

So far, the battery has been taken to be an ideal constant-voltage source, assuming that its terminal voltage remains fixed regardless of the resistive load across it or the length of time it provides power. Alas, that assumption is generally not the case. Although mercury

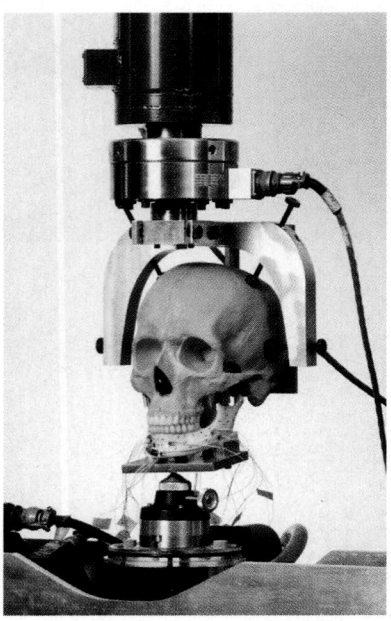

A testing device applies axial loading to the jaw. Here the mandible was fractured and repaired with a surgical plate assembly. Strain gauges around the jaw convert the resulting strain distribution into electrical signals.

cells approximate this behavior, they are small and usually supply very little current. Most other batteries show a marked decrease in terminal voltage as the current supplied by them increases (Fig. 18.1). Start your car with the headlights on. As the starter motor draws hundreds of amps, the battery's terminal voltage drops and the lights become noticeably dimmer.

When a voltaic cell provides current to an external circuit, there is a transport of charge from one electrode to the other across the electrolyte, and that does not happen in an unimpeded way. The cell resists current traversing it in either direction. This **internal resistance** (r) is an inseparable aspect of any real battery, cell, or other power source. Insofar as the plot in Fig. 18.1a is linear, r is ohmic, and the battery can be represented most simply by a resistor in series with an ideal emf ($\mathscr{E}$), as in Fig. 18.2. The + and − terminals of the source are the points A and B, respectively, and it doesn't matter on which side the internal resistor is imagined as long as it is between those terminals.

When the battery supplies current to an external circuit, as in Fig. 18.3a, I leaves the positive terminal A, traverses the circuit, returns to the battery at terminal B, and passes internally to A. (*Label the resistor with + and − signs, indicating the ends at which current enters and leaves, respectively.*) If we trace from B to A, there will be a voltage drop

> As soon as you know the direction of **current** flow in a circuit, mark all resistors with a + on the side current enters and a − on the side it leaves. There is always a voltage drop (+ → −) across a resistor associated with a loss of energy as current traverses it.

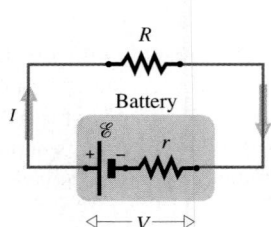

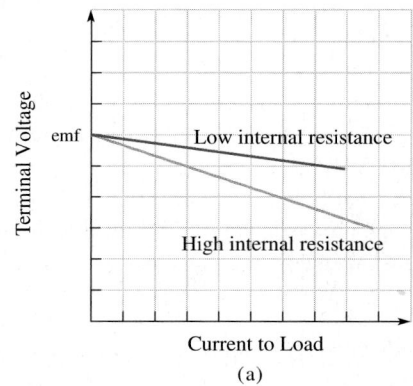

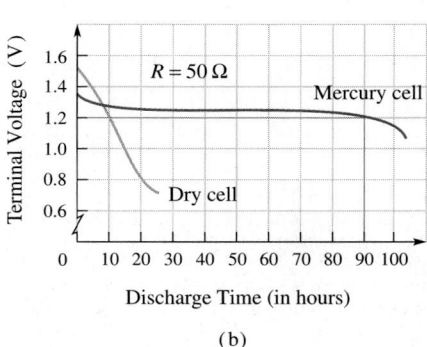

Figure 18.1 The terminal voltage of a battery will change as more and more current is drawn from it. (a) The effect is small with a fresh battery. (b) Most cells suffer a drop in terminal voltage when they are in prolonged use.

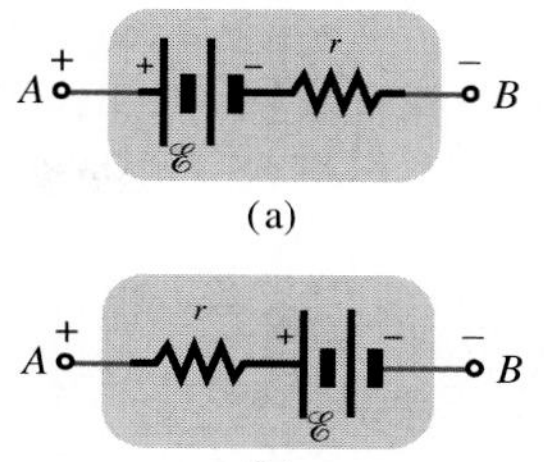

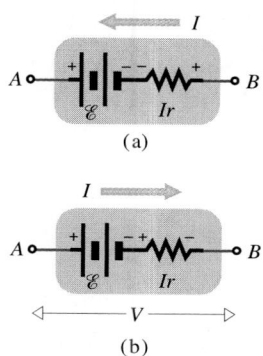

Figure 18.2 A real battery can be represented by an emf in series with an internal resistance r. The order doesn't matter since (a) and (b) are equivalent. The voltage difference between A and B is the terminal voltage of the battery.

Figure 18.3 (a) When the battery supplies current to an external circuit (not shown here), there is a voltage drop across r, and $V < \mathscr{E}$. (b) When the battery receives current from an external source (not shown here), $V > \mathscr{E}$.

($-$) across the internal resistor of $-Ir$ followed by a rise ($+$) equal to the emf, and so with $V = V_A - V_B$ we have

[power out]
$$V = \mathscr{E} - Ir \qquad (18.1)$$

The **terminal voltage** V, *the potential difference measured by an ideal voltmeter across the battery*, is less than the emf when the current goes from B to A and the battery provides power.

When the external circuit supplies current to the battery, as in Fig. 18.3b, I enters at the positive terminal A and passes internally to terminal B. The battery is having power supplied to it—it's being charged. Tracing from B to A, we move first from $-$ to $+$ across the internal resistor, which corresponds to a voltage rise ($+$) of $+Ir$, followed by another rise ($+$) equal to the emf. Hence

[power in]
$$V = \mathscr{E} + Ir \qquad (18.2)$$

and the terminal voltage exceeds the emf.

Example 18.1 **[I]** A voltage source with an emf of 12 V and an internal resistance of 0.40 Ω supplies 1.5 A to an external load. If a high-quality voltmeter is placed across the terminals of the source, what will it read? What will it read if a switch in series with the source open-circuits the system?

Solution Any mention of the "internal resistance" of a source should bring to mind Eq. (18.1). (1) TRANSLATION—A source with a known emf and internal resistance provides a specified current; determine its terminal voltage with both this and zero current. (2) GIVEN: $\mathscr{E} = 12$ V, $r = 0.40$ Ω, and $I = 1.5$ A. FIND: V when $I = 1.5$ A and when $I = 0$. (3) PROBLEM TYPE—Electric current/internal current/internal resistance. (4) PROCEDURE—Determine whether the battery is supplying

energy or receiving it. In this case it's supplying, so draw a circuit diagram with the current leaving the positive terminal as in Fig. 18.3a. The terminal voltage (when $I \neq 0$) is given by Eq. (18.1). (5) CALCULATION:

$$V = \mathscr{E} - Ir = 12 \text{ V} - (1.5 \text{ A})(0.40 \text{ Ω}) = 11.4 \text{ V}$$

or, to two significant figures, $\boxed{V = 11 \text{ V}}$. Opening the switch causes I to become zero, whereupon $\boxed{V = \mathscr{E} = 12 \text{ V}}$.

Quick Check: First see if $V < \mathscr{E}$ as it should be. The drop across r is $Ir = (1.5 \text{ A})(0.40 \text{ Ω}) \approx 1$ V. $V \approx 12 \text{ V} - 1 \text{ V} \approx 11$ V. *[For more worked problems click on* **WALK-THROUGHS** *in* **CHAPTER 18** *on the* **CD.***]*

The internal resistance of a fresh battery is small: $\approx 0.05\ \Omega$ or so for a D-cell and only $\approx 2\ m\ \Omega$ for a 12-V lead storage battery. As a battery ages, its internal resistance increases while its emf gradually decreases. Still, it's possible to have an old 1.5-V dry cell with an internal resistance as high as several ohms and an emf that is only a few percent less than normal. As a consequence, using a meter to measure the terminal voltage of a worn D-cell with no load on it ($I = 0$) may yield a value near its normal emf. Clearly, that's an unreliable indication of the condition of the cell. Measuring V while there is a load on the cell (e.g., while there is a flashlight bulb across it) is a better test. For most applications, if V does not exceed at least 1.1 V, the cell should be replaced.

Example 18.2 **[I]** A 5.9-Ω load resistor is placed across the terminals of a battery that has an emf of 12 V and an internal resistance of 0.10 Ω. Write a general expression for the current in the circuit and then compute its specific value. What's the terminal voltage of the battery?

Solution Any mention of the "internal resistance" of a source should bring to mind Eq. (18.1). (1) TRANSLATION—A source with a known emf and internal resistance is attached across a specified load; determine the current and the terminal voltage. (2) GIVEN: $R = 5.9\ \Omega$, $r = 0.10\ \Omega$, and $\mathscr{E} = 12$ V. FIND: I in general and in particular, and V. (3) PROBLEM TYPE—Electric current/circuit/internal resistance. (4) PROCEDURE—First draw a circuit diagram (Fig. 18.4). Current leaves the + side of the battery. We can find I, using Ohm's Law, if we know the voltage across R, namely, the terminal voltage. So, the first thing is to find V. From Eq. (18.1), $V = \mathscr{E} - Ir$, which in turn

equals IR. (5) CALCULATION:

$$V = IR = \mathscr{E} - Ir$$

and

$$I(R + r) = \mathscr{E}$$

so

$$I = \frac{\mathscr{E}}{R + r} \qquad (18.3)$$

Specifically,

$$I = \frac{12\ V}{5.9\ \Omega + 0.10\ \Omega} = \boxed{2.0\ A}$$

As for the terminal voltage

$$V = \mathscr{E} - Ir = 12\ V - (2.0\ A)(0.10\ \Omega) = 11.8\ V$$

or to two significant figures $\boxed{V = 12\ V}$.

Quick Check: $V = IR = (2.0\ A)(5.9\ \Omega) = 11.8\ V$.

If an ammeter and voltmeter are added to the circuit of Fig. 18.4, as indicated in Fig. 18.5, it becomes an easy matter to determine r. Note that **the ammeter is always placed in the branch of the circuit through which the required current passes**, whereas **the voltmeter is always placed across the two points whose potential difference is to be measured**.

Before we move on, let's standardize some of the terminology: a **branch** is one or more circuit elements (in series) carrying a single current. A **node** is a place (often a single point but not necessarily) where three or more branches come together. Nodes are interconnected by branches, and branches begin and end on nodes. A circuit that has no nodes (Fig. 18.4) can be thought of as a single branch that closes on itself. Any closed current path

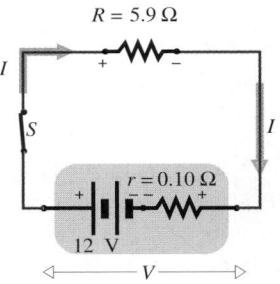

$R = 5.9\ \Omega$

Figure 18.4 A battery with an emf of 12 V and internal resistance r supplies a current I to the resistor R when the switch S is closed. The current (a flow of positive charge) leaves the positive terminal of the battery and returns, with less energy, to the negative terminal.

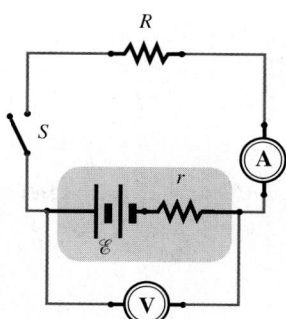

Figure 18.5 Using high-quality meters, we can accurately determine r. Here, an ammeter (A) is placed in series with the battery, and a voltmeter (V) is placed across it.

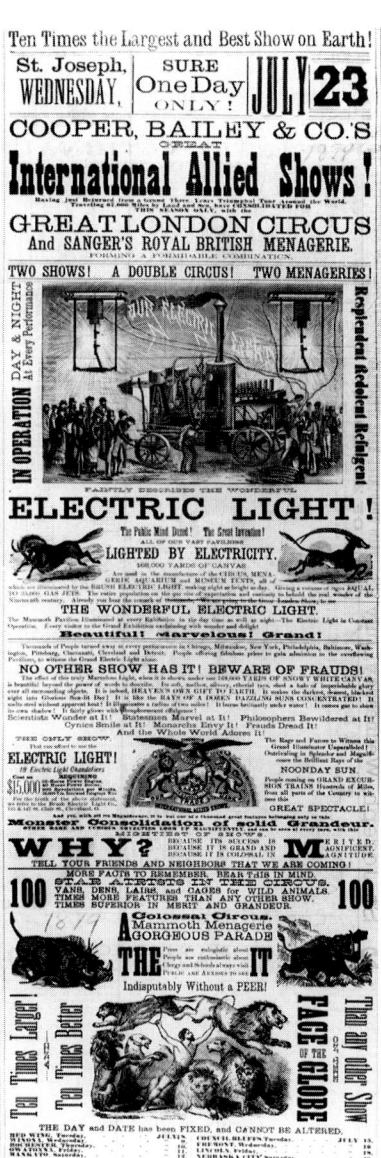

A circus poster from 1879. A steam engine powered a generator that supplied current to the arc lamps that lit this circus "day and night."

is a **loop**. The circuit in Fig. 18.4 is a single loop, and the ammeter is positioned in that loop. The two points where the voltmeter attaches to the circuit (Fig. 18.5) are nodes, and the voltmeter constitutes one of three branches.

18.2 Resistors in Series and Parallel

Suppose we have a circuit with several branches and lots of resistors and sources and we want to compute the voltage across, and the current through, each element. The computation is likely to be difficult, but two special configurations are particularly easy to deal with. When the resistors are in *series* or in *parallel* or a combination of both, they can be replaced by a single *equivalent resistor* (as was done earlier with capacitors, p. 584). Then Ohm's Law can be applied and the problem solved step by step. How do we arrive at this **equivalent resistance** (R_e)?

Figure 18.6a shows an ideal voltage source in series with two resistors R_1 and R_2. *The same current I enters and leaves each element in series.* The difference in potential between points A and B is the voltage of the source V. If we go from B to A across the battery, we encounter a voltage rise of $+V$. Similarly, tracing from B to A across the resistors, we encounter a rise of V_2 followed by another rise of V_1, which must also equal V; thus

$$V = V_1 + V_2$$

and using Ohm's Law

$$V = IR_1 + IR_2$$

We want an equivalent resistance (R_e) that will replace the load, as indicated in Fig. 18.6b, so that the same current I circulates. Hence

$$V = IR_e$$

and so

$$IR_e = IR_1 + IR_2$$

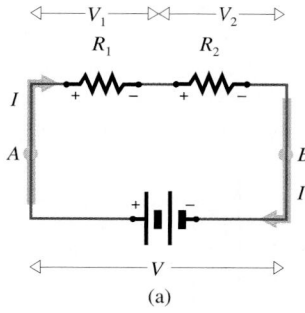

(a)

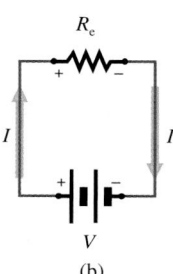

(b)

Figure 18.6 (a) An ideal voltage source in series with two resistors. (b) The equivalent circuit where $R_e = R_1 + R_2$.

One of the innovations that makes modern miniature electronics possible is the printed circuit. Here all of the various components are wired together by narrow metal strips that are predeposited on an insulating circuit board. As long as the currents are small, the strips can be quite thin. The black rectangles are integrated circuits that plug into the board. If you look carefully, you can see rows of resistors at the bottom and right, and a few flat mustard-colored capacitors at the lower left.

Consequently

[in series]
$$R_e = R_1 + R_2 \qquad (18.4)$$

The equivalent resistance is simply the sum of the series resistances, no matter how many there happen to be. To see how this works, suppose $R_1 = 5.0\ \Omega$, $R_2 = 1.0\ \Omega$, and $V = 12$ V. From Eq. (18.4), $R_e = 6.0\ \Omega$, and using Ohm's Law for Fig. 18.6*b*, $I = V/R_e = 2.0$ A. Returning to the original circuit of Fig. 18.6*a*, $V_1 = IR_1 = 10$ V and $V_2 = IR_2 = 2.0$ V and, as expected, $V_1 + V_2 = 12$ V. The analysis of the circuit in Fig. 18.4 can now be reinterpreted in light of the above discussion; $R_e = R + r$, and since $I = V/R_e$, $I = V/(R + r)$, which is Eq. (18.3).

Figure 18.7*a* shows two resistors in parallel with a constant-voltage dc source, though there could equally well have been ten of them. The (as yet undetermined) current *I* supplied by the source comes to node-*A* and splits into two branch currents I_1 and I_2, which are (as we will find presently) inversely proportional to the resistances of the two branches. These currents recombine into *I* at node-*B*:

$$I = I_1 + I_2$$

The voltage across each resistor is *V*, and therefore Ohm's Law gives

$$I = \frac{V}{R_1} + \frac{V}{R_2}$$

But in the equivalent circuit of Fig. 18.7*b*, $I = V/R_e$ and so

$$\frac{V}{R_e} = \frac{V}{R_1} + \frac{V}{R_2}$$

Canceling *V* yields

[in parallel]
$$\frac{1}{R_e} = \frac{1}{R_1} + \frac{1}{R_2} \qquad (18.5)$$

and the summation would just continue were there additional resistors in the circuit.

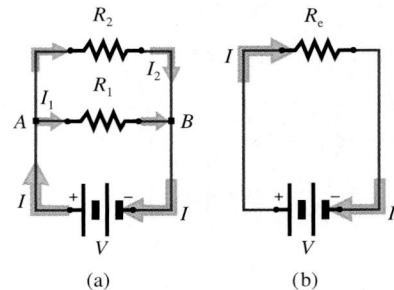

Figure 18.7 (a) An ideal voltage source across two resistors in parallel. Notice how the current *I* splits at node *A* and recombines at *B*. (b) The equivalent circuit where in $1/R_e = 1/R_1 + 1/R_2$.

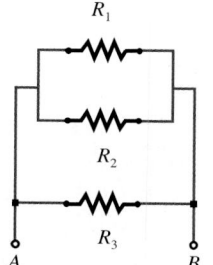 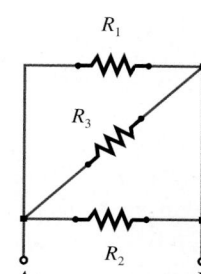 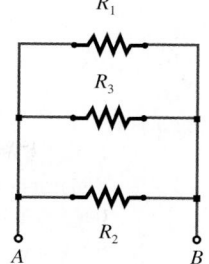 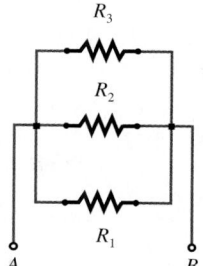 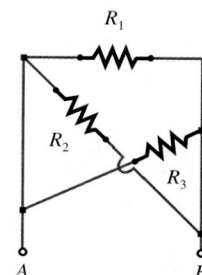

Figure 18.8 Each of these circuits has the same equivalent resistance between A and B. In each version, R_1, R_2, and R_3 are in parallel.

Equation (18.5) for two resistors can be rewritten as

[two resistors in parallel]
$$R_e = \frac{R_1 R_2}{R_1 + R_2} \qquad (18.6)$$

which makes it apparent that when either resistor is much larger than the other, say, $R_1 >> R_2$, then $R_1 + R_2 \approx R_1$ and $R_e \approx R_2$; the equivalent resistance approximates the *smaller* of the two. **The equivalent parallel resistance is always less than any of the contributing resistances.** All of the circuits in Fig. 18.8 are electrically the same, and all have the same equivalent resistance between A and B. Notice that when there are two resistors in parallel and $R_1 = R_2 = R$, $R_e = \frac{1}{2}R$, and similarly (see Problem 25) when we have N identical resistors in parallel

[N identical resistors in parallel]
$$R_e = \frac{R}{N} \qquad (18.7)$$

Example 18.3   [II] Figure 18.9a represents three light bulbs with resistances of 2.0 Ω, 4.0 Ω, and 8.0 Ω attached across a source with an emf of 6.0 V and an internal resistance of 1.0 Ω. Find the current through each bulb.

Solution   To solve this one we'll need the equivalent resistance of the circuit. (1) TRANSLATION—Three known resistors are wired across a battery with a known emf and internal resistance; determine the current through each resistor. (2) GIVEN: $R_1 = 2.0$ Ω, $R_2 = 4.0$ Ω, $R_3 = 8.0$ Ω, $r = 1.0$ Ω, and $\mathscr{E} = 6.0$ V. FIND: I_1, I_2, and I_3. (3) PROBLEM TYPE—Electric current / circuit / internal resistance. (4) PROCEDURE—Draw a diagram at each step of the simplification. As a rule (unless overridden by some other more powerful source), the current leaves the + side of the battery. The three bulbs are in parallel, and that resistance is in series with r. We can find I using Ohm's Law because the voltage across R_e is the emf. Then work backward. (5) CALCULATION—For the three bulbs

$$\frac{1}{R} = \frac{1}{2.0 \ \Omega} + \frac{1}{4.0 \ \Omega} + \frac{1}{8.0 \ \Omega} = 0.875 \ \Omega^{-1}$$

and $R = 1.14$ Ω, which is the equivalent resistance of the load. R and r are in series, and the equivalent resistance of the entire circuit (R_e) is $(R + r) = 2.14$ Ω. With R_e across the emf in Fig. 18.9c, the current provided by the source (via Ohm's Law) is

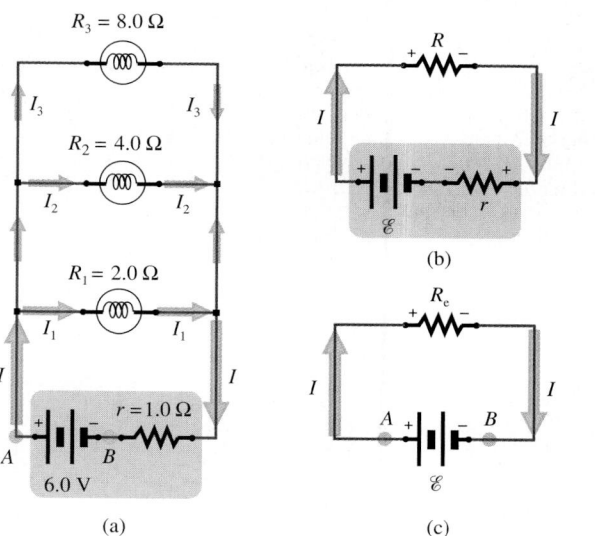

Figure 18.9 (a) Three light bulbs in parallel with a real battery. (b) The bulbs have resistances that are in parallel such that $1/R = 1/R_1 + 1/R_2 + 1/R_3$. (c) The equivalent resistance "seen" by the emf between points A and B is R_e.

$$I = \frac{\mathscr{E}}{R_e} = \frac{6.0 \ \text{V}}{2.14 \ \Omega} = 2.8 \ \text{A}$$

continued

Going back to Fig. 18.9*b*, we can find the terminal voltage across *R* and then each branch current. The terminal voltage follows from Eq. (18.1); namely, a drop of $-Ir = -2.8$ V followed by a rise of $\mathscr{E} = +6.0$ V for a total of $V = +3.2$ V. The branch currents are then

$$I_1 = \frac{V}{R_1} = \frac{3.2 \text{ V}}{2.0 \text{ }\Omega} = \boxed{1.6 \text{ A}}$$

$$I_2 = \frac{V}{R_2} = \frac{3.2 \text{ V}}{4.0 \text{ }\Omega} = \boxed{0.80 \text{ A}}$$

and

$$I_3 = \frac{V}{R_3} = \frac{3.2 \text{ V}}{8.0 \text{ }\Omega} = \boxed{0.40 \text{ A}}$$

Quick Check: Note that *R* is less than any of the constituent resistors, which is as it should be. $I = I_1 + I_2 + I_3 = 1.6$ A + 0.80 A + 0.40 A = 2.8 A, which is appropriate. The smaller the resistance of the branch, the more current it carries.

The analysis of Example 18.3 shows that the current in each branch in a parallel circuit depends inversely on the resistance; the larger the resistance, the less current flows through that branch. We used Ohm's Law and the known voltage to find the individual branch currents and the same can be done for the simpler circuit in Fig. 18.10. If, however, we know *I* but not *V*, it can be shown (see Problem 48) that

$$I_1 = \frac{R_2}{R_1 + R_2} I \tag{18.8}$$

and

$$I_2 = \frac{R_1}{R_1 + R_2} I \tag{18.9}$$

Figure 18.11*a* depicts a cluster of attached resistors. There is no source in the circuit and so no current, but we can still think about adding one, say, across points *A* and *D*, and consequently still be concerned with the equivalent resistance of the network. The node points are labeled to keep track of the process as the circuit is successively contracted. Resistors R_2 and R_3 are in parallel. The result of adding those (viz., 3 Ω) is in series with R_1 and R_4, giving a total of 10 Ω. This resistance is in parallel with R_5, since both resistors are attached at *A* and *D*, yielding $R_e = 5.0$ Ω between points *A* and *D*. This means that the

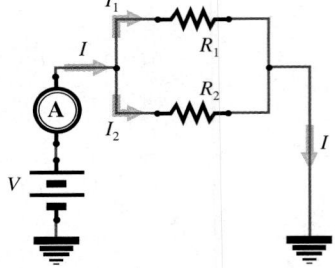

Figure 18.10 Current *I* splits into the two branch currents I_1 and I_2 which are given by Eqs. (18.8) and (18.9). On passing through the resistors, I_1 and I_2 recombine into *I*, which travels through ground on the way back to the battery.

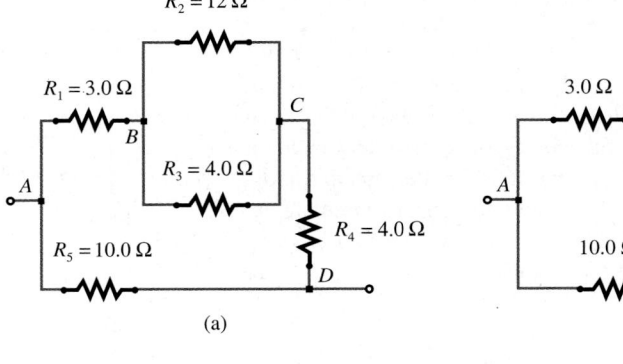

(a)

(b)

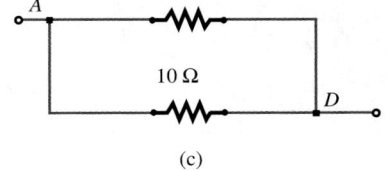

(c)

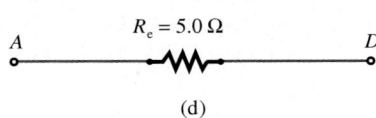

(d)

Figure 18.11 A cluster of resistors and the successive combinations and simplifications leading to a single equivalent resistance R_e between points *A* and *D*.

As a rule, two resistors are in **series** when one and only one terminal of one resistor is connected to one and only one terminal of the other resistor. If that point of connection becomes a node by being attached to another current-carrying branch, the two resistors are no longer in series. By contrast, if a third resistor is attached to that point of connection (making a Y shape) and it just floats out there without carrying a current, the original two resistors are still in series (see, e.g., Problem 43). Note that a voltmeter can be placed across any resistor without changing its series status. Furthermore, if a source (e.g., a battery) is placed in the branch between two resistors that are in series, they remain in series even though they are no longer directly connected to one another.

Two resistors are in **parallel** when both terminals of one resistor are connected to both terminals of the other resistor. Unlike the series connection, there may or may not be other current-carrying branches attached to either terminal of either of these two resistors.

Two **resistors** in series with a third resistor are in series with each other.

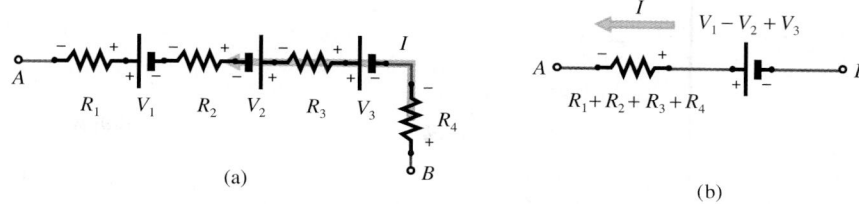

Figure 18.12 A series of sources separated by resistors is equivalent to a single source having the net voltage and a single resistor having the combined resistance.

whole cluster behaves in precisely the same way as would a single 5-Ω load. Put a battery across A and D, and the same current will be delivered to either configuration. This value is *not* necessarily the equivalent resistance between A and C, or A and B (Problems 32, 33, and 34)—each of those is quite distinct.

The resistors in Fig. 18.6a can be slid around so that the battery is between them and nothing changes. ***The same current passes through every resistor in a given branch, regardless of the presence of sources in that branch, and the resistors are in series even though they are not directly connected to one another.*** In the process of simplifying a circuit, sources and resistors in series can be interchanged, provided they are all returned after the currents are computed (Fig. 18.12). {Many of the problems in this chapter can be checked using the **INTERACTIVE EXPLORATION** called **DIRECT CURRENT CIRCUITS** found in **CHAPTER 18** on the **CD.**}

Fifty years ago a typical commercial device like a radio (see photo on p. 537) or TV was a jumble of wires and circuit elements mounted on a metal box, but things began to get a lot more organized and compact with the introduction of the printed circuit (ca. 1941). Today you're much more likely to encounter neat rows of circuit elements interconnected by thin ribbons of conducting metal "printed" onto a plastic board. Accordingly, it's appropriate for us to learn a little about how to work with printed circuits, and Fig. 18.13 is a nice place to start.

This is a protoboard used to wire up temporary circuits without having to solder the leads. The columns are each numbered, whereas the rows have letter designations. Labeled *A-E,* the five socket-hole groupings in any numbered column are all electrically attached to one another. The same is true for the strips of holes *F-J,* which are isolated from holes *A-E.*

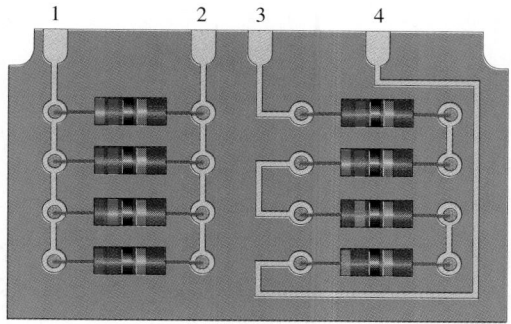

Figure 18.13 A printed circuit board. Between terminals-1 and 2, the resistors are all in parallel. Between terminals-3 and 4, the resistors are in series.

Example 18.4 **[III]** Consider the circuit in Fig 18.14a (a) If possible, simplify it and determine an equivalent resistance between C and G. (b) What current is provided by the source? (c) What is the voltage across points G and E?

Solution To solve this one we'll need the equivalent resistance of the circuit. (1) TRANSLATION—A circuit is specified with known resistors and a battery; determine (a) the equivalent resistance, (b) the current provided by the battery, and (c) the voltage across points G and E. (2) GIVEN: Nine resistors, $R = 1.0$ kΩ each, and $V = 12$ V. FIND: R_e, I, and V between G and E. (3) PROBLEM TYPE—Electric current/circuit. (4) PROCEDURE—This circuit can be made to look a lot more manageable by redrawing it. (5) CALCULATION—Imagine the wires to be flexible and lift up the inside square, with the resistor and source attached, and place it outside E–F–G–H, as in Fig. 18.14b. Branches A–B–C and A–D–C are in parallel, as are E–F–G and E–H–G, and each has a resistance of 1.0 kΩ + 1.0 kΩ = 2.0 kΩ (Fig. 18.14c). The resistance of each square E–F–G–H and A–B–C–D reduces to

$$\frac{1}{2.0 \text{ k}\Omega} + \frac{1}{2.0 \text{ k}\Omega} = \frac{1}{R}$$

and $R = 1.0$ kΩ. The three 1.0-kΩ resistors in Fig.18.14d are in series with the source. (a) The equivalent resistance in Fig. 18.14e is $\boxed{3.0 \text{ k}\Omega}$. (b) Since $V = IR_e$,

$$I = \frac{V}{R_e} = \frac{12 \text{ V}}{3.0 \text{ k}\Omega} = \boxed{4.0 \text{ mA}}$$

(c) A current of 4.0 mA leaves the battery and splits at C. Because the two branches C–D–A and C–B–A have the same resistance, the current divides into two equal streams of 2.0 mA each. The voltage drop in going from C to A is given by $V_{AC} = IR = (2.0 \text{ mA})(1.0 \text{ k}\Omega + 1.0 \text{ k}\Omega) = 4.0$ V. In going from A to E, there is another drop of $V_{EA} = IR = (4.0 \text{ mA})(1.0 \text{ k}\Omega) = 4.0$ V. C is 12 V above G, A is 8.0 V above G, and E is $\boxed{4.0 \text{ V}}$ above G.

Quick Check: 4.0 mA leaves A, enters E, and splits so that 2.0 mA goes through both branches E–F–G and E–H–G. The voltage drop from E along either branch to G is (2.0 mA)(2.0 kΩ) = 4.0 V.

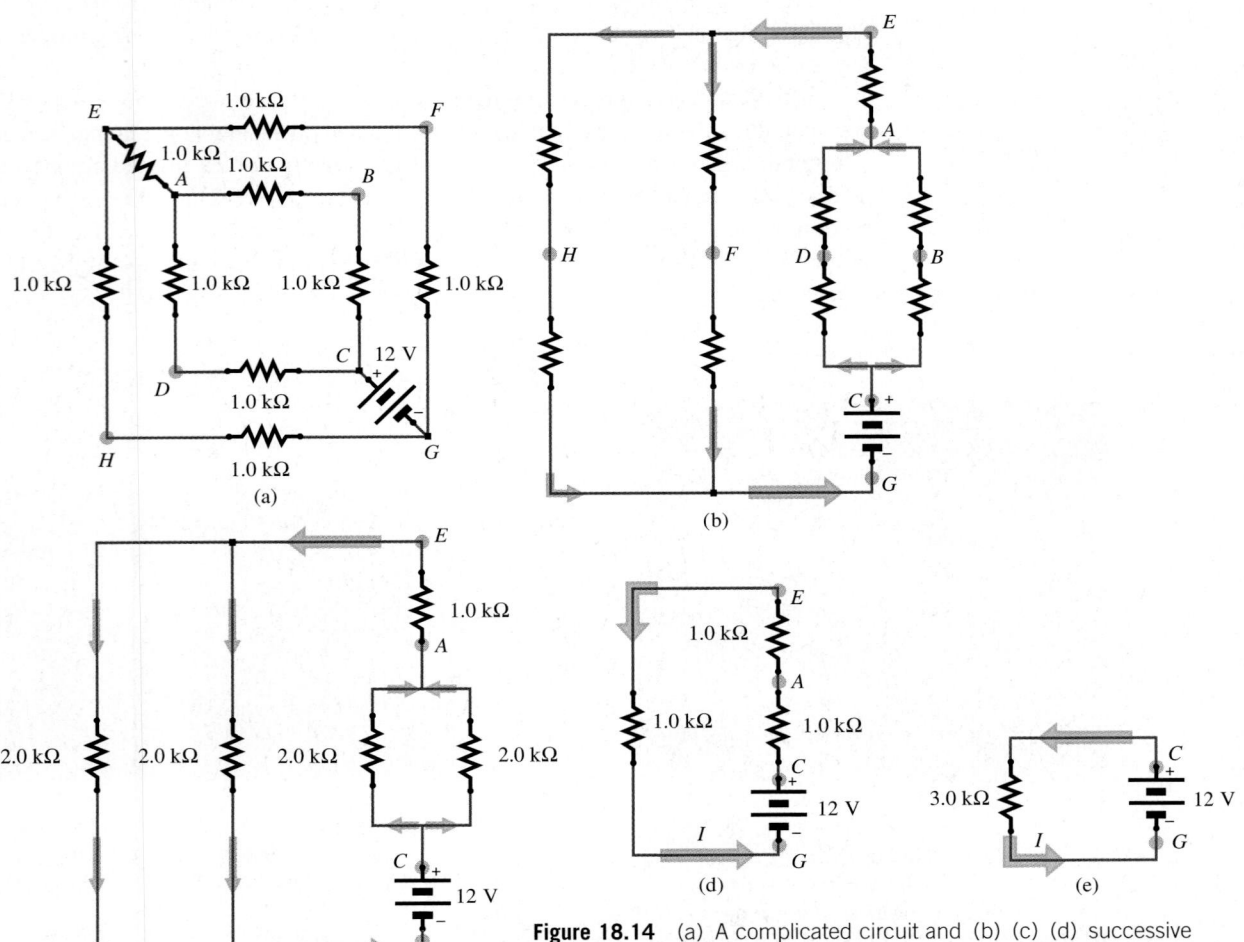

Figure 18.14 (a) A complicated circuit and (b) (c) (d) successive stages of simplification. (e) The equivalent resistance between points C and G is 3.0 kΩ, and that's across the 12-V source.

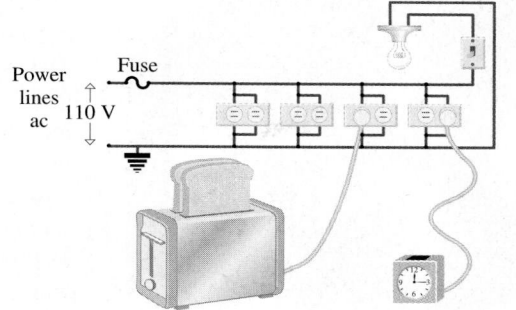

Figure 18.15 Household wiring, using alternating current, puts all appliances and lights in parallel across ≈110 V.

Often power is provided to a number of devices by putting them each in parallel across the source. That's what you do when you "plug in" a toaster (p. 764) or a TV set—it is placed across the two leads that supply electrical energy to the home (Fig. 18.15). All household appliances are in parallel—in that way each can be designed to operate at a specific voltage (≈120 V in the United States, ≈220 V most everywhere else). That voltage can be provided to each appliance independent of whatever else is drawing power in the circuit, which would not be the case were they in series. Moreover, if any one device open-circuits (as when a light bulb burns out), the rest are unaffected. The same is true if you have several telephone extensions—they are all in parallel, supplied with the same low-voltage dc. {To explore the transfer of power from a source to a load click on **The Maximum Power Transfer Theorem** under **Further Discussions** on the **CD**.}

18.3 Ammeters and Voltmeters (Optional)

Electrical meters of all kinds come in two varieties, analog and digital. An analog meter usually has a pointer that moves across a scale, whereas a digital meter provides direct numerical values of whatever is being measured. An analog meter is constructed around a moving coil galvanometer (p. 693). The latter is a coil of wire with a resistance r of perhaps 200 Ω or less, having a pointer fixed to it, placed between the poles of a magnet. Passing a current through the coil causes it to twist proportionately and the pointer swings around. The digital meter is an electronic device that takes a reading from the circuit, which it amplifies, digitizes, and displays. Digital meters tend to be more accurate, sturdier, and more expensive than analog meters, but they are both used in the same way.

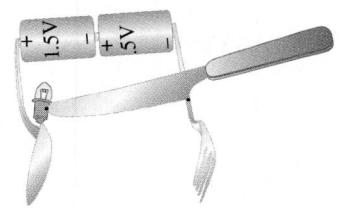

EXPLORING PHYSICS ON YOUR OWN

Simple Circuits: For this experiment you'll need to start with an ordinary two D-cell flashlight and a piece of wire about one-foot long. Take the flashlight apart, removing the bulb and the two D-cells. If you don't have any insulated copper wire, you can use a wire coat hanger (scrape the paint off the ends) or even a metal fork or two. Put 3.0 V across the bulb and it will light brightly. Test the conductivity of various items (such as aluminum foil, flesh, paper, water, saltwater, nails) by placing them in series with the bulb and noting its brightness. Now get a few more flashlight bulbs and batteries and some wire and explore series and parallel hookups. The brightness of the lamp corresponds to the power it's dissipating, which depends on the current flowing through it. Get a clear 110-V light bulb and see if you can make it glow with a 9-V battery.

Looking under the hood of an electric car. The two big boxes in the middle are fuel cells. The battery is at the lower right.

Figure 18.16*a* represents the basic analog detector, a galvanometer comprising a resistor in series with an ideal rotating coil. The input that causes a maximum deflection of the pointer is called the *full-scale current*, and for an inexpensive device, that's about 1 mA. To use the galvanometer as an **ammeter**, a current meter, we must direct a small fixed fraction of the main stream of current through the galvanometer. That's accomplished by placing a small *shunt* resistor ($R_s \ll r$) across the galvanometer (Fig. 18.16*b*). The resulting instrument is an ammeter, and it is always positioned within the branch where the current I is to be measured. The branch current I entering the ammeter splits with part I_g going through the galvanometer and most of it, I_s, being shunted through the small resistor R_s. The specific value of R_s depends on the full-scale current sensitivity of the galvanometer. Thus, if the coil has a full-scale current of $I_g = 1.0$ mA and we want the ammeter to read $I = 1.0$ A at full scale, then, since $I = I_g + I_s$, it follows that $I_s = 0.999$ A. Because the voltage drops across R_s and $r = 200$ Ω are equal,

$$I_g r = I_s R_s$$

hence
$$R_s = \frac{I_g r}{I_s} = \frac{(1.0 \text{ mA})(200 \text{ Ω})}{0.999 \text{ A}} = 0.20 \text{ Ω}$$

Notice that the analog dc ammeter has a specific polarity; the current must enter at the + terminal or the meter may be damaged. One nice feature of the digital ammeter is that we usually needn't worry about polarity.

A classic **voltmeter**, Fig. 18.16*c*, is a galvanometer in series with a resistance R. The voltmeter is always placed *across* the two points whose potential difference is to be measured. To ensure that little current passes through the coil, R is very large (usually in the high kilohm to megohm range). For example, suppose the voltmeter is to measure from 0 to 100 V, using our cheap galvanometer with a full-scale current of $I_g = 1.0$ mA and $r = 200$ Ω. That current must exist when the voltage is a maximum of 100 V; full scale, then, since $I = I_g + I_s$, it follows that $I_s = 0.999$ A. Because the voltage drops across R_s and $r = 200$ Ω are equal,

$$I_g r = I_s R_s$$

hence
$$R_s = \frac{I_g r}{I_s} = \frac{(1.0 \text{ mA})(200 \text{ Ω})}{0.999 \text{ A}} = 0.20 \text{ Ω}$$

and $R = 10 \times 10^4$ Ω. This meter is not a very good one because R is not really large enough. If it were placed across a several-kilohm resistor in a circuit, it would appreciably affect the

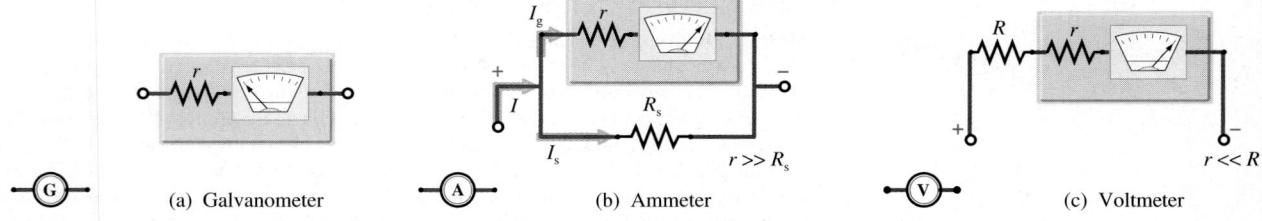

(a) Galvanometer   (b) Ammeter   (c) Voltmeter

Figure 18.16 (a) A galvanometer having an internal resistance r. (b) When the galvanometer is placed in parallel with a low-resistance shunt R_s, we have an ammeter. (c) When the galvanometer is in series with a high-resistance R, we have a voltmeter.

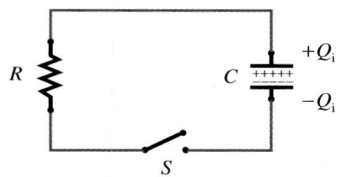

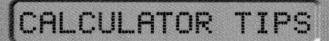

Figure 18.17 A charged capacitor *C* in series with a resistor and an open switch *S*. The charge on each plate is $\pm Q_i$.

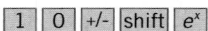
current and produce an erroneous reading. By contrast, a fine electronic voltmeter might have a resistance of 10^8 Ω or more. Notice that the dc analog voltmeter also has a specific polarity and must be connected with its + terminal at the higher potential.

18.4 *R-C* Circuits

Any circuit containing a capacitor also has some resistance, even if it's only due to the wiring. Such **R-C circuits** are both commonplace and important—especially if you happen to be wearing a cardiac pacemaker. And that points to one of the most interesting features of *R-C* circuits. The transient states of the resistive circuits considered thus far are extremely brief, and so voltages and currents are taken to be constant at their steady-state values. By contrast, *R-C* circuits have time-dependent voltages and currents. That makes them very useful for creating a variety of circuits that produce time-varying signals (in a pacemaker, for example).

Imagine a capacitor with its plates charged at $\pm Q_i$ as shown in Fig. 18.17. There is an initial voltage $V_i = Q_i/C$ across the capacitor, but no charge flows while the switch *S* is open. At $t = 0$, the switch is closed, the voltage across the resistor is immediately V_i, and the charge on the capacitor begins to redistribute itself under the influence of the Coulomb force. For an instant there is an initial current through the circuit of

$$I_i = \frac{V_i}{R} = \frac{Q_i}{RC}$$

But as the charge on the capacitor decreases ($\Delta Q < 0$), the voltage decreases, and that decreases the current: $I = -\Delta Q / \Delta t$. At any moment, the voltage across the resistor IR equals the voltage across the capacitor Q/C, such that

$$IR = -\frac{\Delta Q}{\Delta t} R = \frac{Q}{C}$$

and

$$\frac{\Delta Q}{\Delta t} = -\frac{1}{RC} Q \qquad (18.10)$$

which is a rather special and yet wonderfully common situation. **Whenever a physical quantity changes ($\Delta Q/\Delta t$) at a rate that depends on the quantity itself (Q), the quantity varies exponentially** {see **MATH REVIEW: PART A-3** on the **CD** }. If we plot Q versus t (Fig. 18.18*a*), *the slope of the curve is proportional to the value of the curve:* when

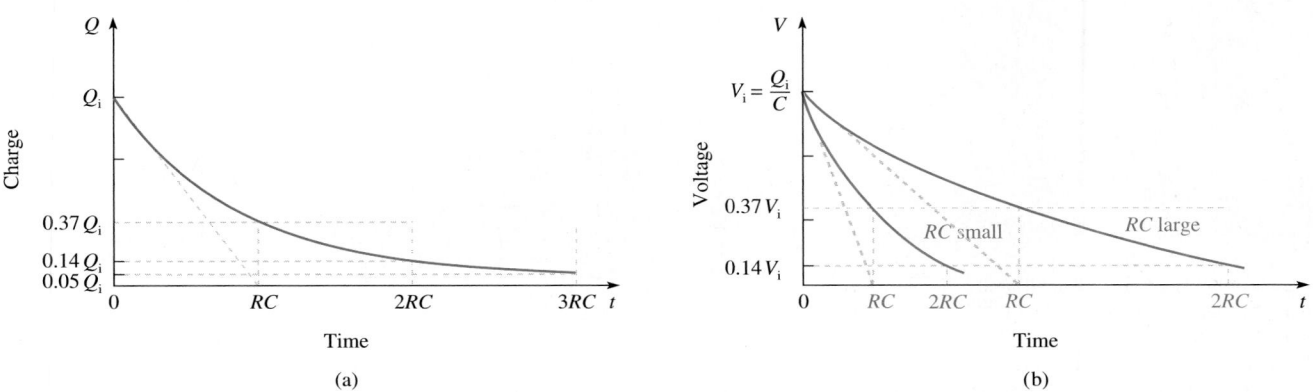

Figure 18.18 Curves for the charge and voltage of the capacitor in Fig. 18.17 once *S* is closed. (a) After each interval of time equal to *RC*, the curve of *Q* versus *t* drops to 36.787 9% of the previous value. (b) The voltage also decays exponentially. The smaller the *RC* is, the faster the curves decay.

the charge is large, the slope ($\Delta Q/\Delta t$) is negative and large; when Q is small, the slope is small. We will examine the exact shape of the e^x curve later, in Section 30.8. For the moment, it will suffice just to see what the solution to Eq. (18.7) looks like, namely,

$$Q = Q_i e^{-t/RC} \tag{18.11}$$

Since $e^0 = 1$, at $t = 0$, $Q = Q_i$, which makes sense. At $t = \infty$, $e^{-\infty} = 0$ and $Q = 0$, which also makes sense; the capacitor is discharged. Notice from Eq. (18.10) that at $t = 0$, the slope is $-Q_i/RC$. The initial slope intersects the time axis at $t = RC$. At that time, $Q = Q_i e^{-1} = 0.37 Q_i$. The quantity RC is called the **time constant** because it influences the temporal behavior of the circuit. That's evident in the plots of the voltage across the capacitor in Fig. 18.18*b* for two different values of RC. The voltage drops 63% of the way down toward zero in one time constant, and the smaller RC is, the faster that drop happens. The smaller R is, the less the resistor retards the flow of charge; and the smaller C is, the faster the capacitor discharges.

> In an *R-C* circuit the **time constant** equals *RC*.

The circuit in Fig. 18.19 shows an uncharged capacitor in series with a resistor and an ideal battery. When the switch is closed, charge flows and a time-dependent current begins to circulate. At that moment, the charge on the capacitor Q is essentially zero as is the voltage drop across it; hence,

$$\mathcal{E} = I_i R$$

At any instant thereafter

$$\mathcal{E} = IR + Q/C$$

Since $\mathcal{E}$ is constant, as the charge builds up (Fig. 18.20*a*), the current correspondingly dies off (Fig. 18.20*b*). As the current approaches zero, $\mathcal{E}$ approaches Q/C, and the maximum charge on the capacitor approaches $C\mathcal{E}$. The current never actually reaches zero; after 20 time constants ($t = 20RC$), it's down to $I = I_i e^{-20} = 2 \times 10^{-9} I_i$, which isn't much.

R-C circuits are often used as timers to control the periodic activation of other devices. For example, the intermittent action of an automobile windshield wiper is controlled by an *R-C* circuit. A cardiac pacemaker circuit with a time constant RC of 0.8 s can be used to reach a certain triggering voltage at a frequency of $1/(0.8 \text{ s}) = 1.25$ times per second, thereupon activating a signal to the heart at a rate of 75 pulses per minute. {At this point you should activate the **CD**, go to **INTERACTIVE EXPLORATIONS** in **CHAPTER 18** and click on **THE *R-C* CIRCUIT**. Working this simulation is a great way to master all the subtleties.}

An electronic photographic flash unit. The rechargeable battery is a stack of four cells at the upper left. The large cylinder is the capacitor that stores charge and energy. To fire the flash, the capacitor dumps its charge, via the red and blue leads, into the lamp. After a short wait, determined by *RC*, it's recharged and ready to go again.

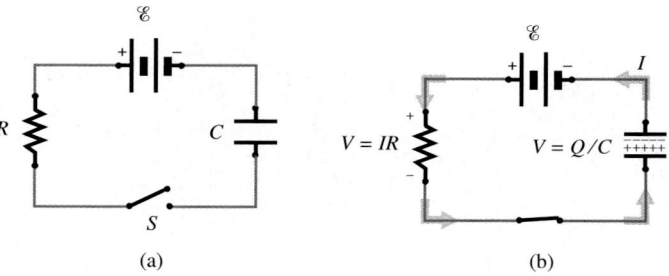

(a) (b)

Figure 18.19 (a) When the switch in the circuit is closed, a transient current appears. (b) Positive charge goes out of the battery, across the resistor, and builds up on one plate of the capacitor. An equal amount of charge is repelled off the other plate and returns to the negative terminal of the battery.

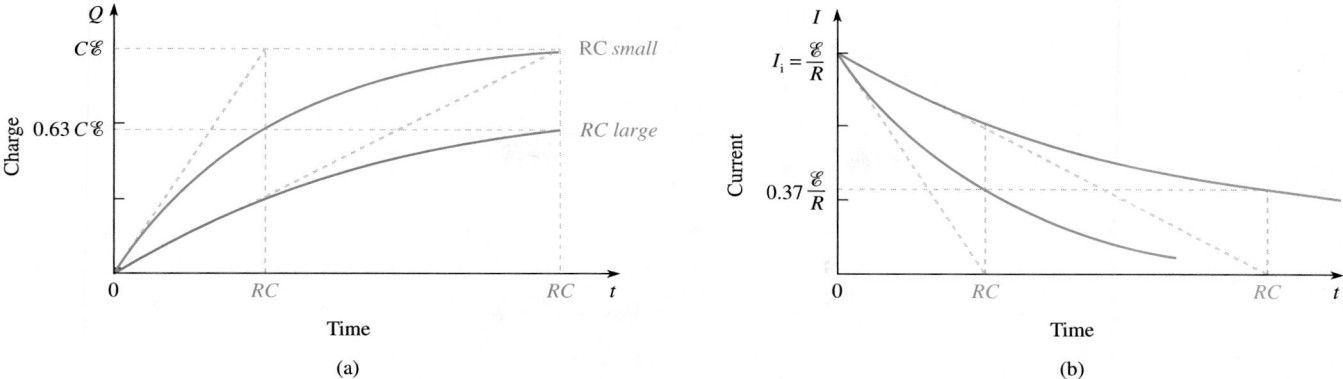

Figure 18.20 (a) As the charge on the capacitor in Fig. 18.17 builds up, (b) the current in the circuit dies away. The smaller the *RC* is, the faster the curve (a) rises or (b) decays.

Network Analysis   (Optional)

The first approach to analyzing any circuit is to reduce it to a manageable equivalent scheme, but there are many configurations that can't be simplified. Figure 18.21 shows a representative selection wherein the elements are neither in series nor in parallel. Complicated circuits are often called *networks*—Fig. 18.22 is an example of the genre. But before we get into the systematic approach to the solution of such circuits, it's appropriate to look at some special cases that are both important and remarkably simple. For example, although the circuit in Fig. 18.22 appears to be a small horror, some aspects of it are easily solved. What is the current through, and the voltage drop across, the 6.0-Ω resistor (R_1)?

A cardiac pacemaker uses an implanted *R-C* circuit to supply periodic voltage pulses to the heart.

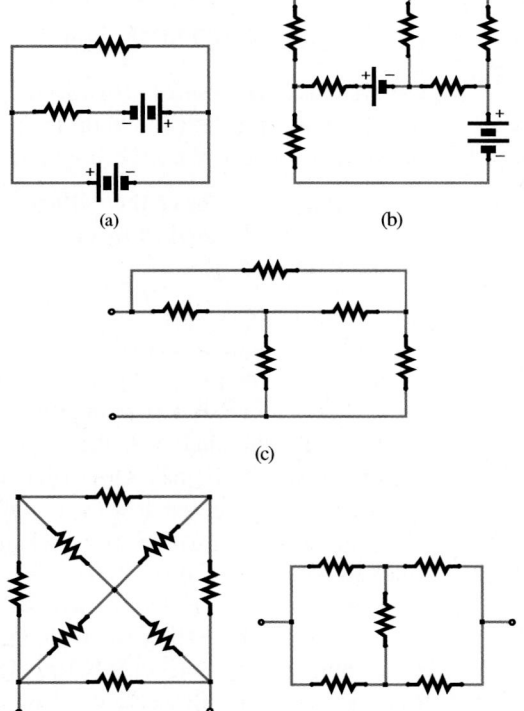

Figure 18.21 Several circuits in which the elements are neither in series nor in parallel.

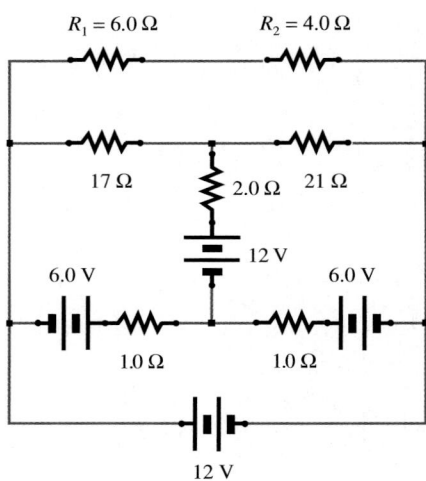

Figure 18.22 We can easily calculate the current through R_1 since the voltage across the upper branch is 12 V.

Happily, the voltage drop across the whole uppermost branch is apparent—it's 12 V because the two resistors (6.0 Ω and 4.0 Ω in series) are directly across the terminals of the 12-V source. All the complex circuitry in the middle has no effect on the voltage across the upper (10-Ω) branch. Thus, whatever the current may be, the drop across R_1 is 6/10 of 12 V, or 7.2 V. The current is then $I = V_1/R_1 = (7.2 \text{ V})/(6.0 \text{ } \Omega) = 1.2$ A or, alternatively, $I = V/(R_1 + R_2) = 1.2$ A.

Figure 18.23 depicts another rather formidable network, but there's something special here, too. What is the voltage drop across, and the current through, the 6.0-Ω resistor? Notice that when the current reaches node A, it "sees" two identical paths and therefore splits into equal parts. The voltage drop across R_1 is equal to the drop across R_3, and so points C and D are at the same potential. There is no voltage difference across R_5 and no current through it; R_5 has no effect on this so-called *bridge circuit*! It's as if the branch from C to D were removed.

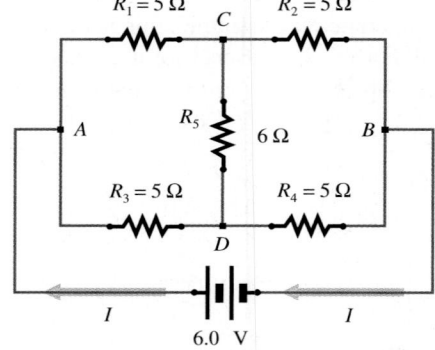

Figure 18.23 By symmetry, the voltage at C equals the voltage at D, so $V_{CD} = 0$. Therefore, no current passes through R_5, and R_1 and R_2 are in series, as are R_3 and R_4.

18.5 Kirchhoff's Rules

The systematic analysis of networks proceeds from two basic rules set forth by the German physicist Gustav Robert Kirchhoff in the late 1800s. There's really nothing new about these ideas; what *is* new is the application. Kirchhoff's first rule is

> **The algebraic sum of the voltage rises and drops encountered in going around any closed path formed by any portion of a circuit must be zero.**

This **Loop Rule** maintains that if we start at some arbitrary point in a circuit at a given potential and follow any path that returns to that same point, there can be no net change in potential.

In exactly the same way, there is no change in *gravitational*-PE if you return to the same spot after meandering up and down a mountainside. Thus, the sum of the voltage rises (taken as positive) and the voltage drops (taken as negative) must equal zero. For a charge traversing a loop and returning to its starting point, the net energy gained must equal the net energy lost. The Loop Rule is a restatement of the Law of Conservation of Energy.

Gustav Robert Kirchhoff (1824–1887).

Figure 18.24 The analysis of a circuit using the Loop Rule.

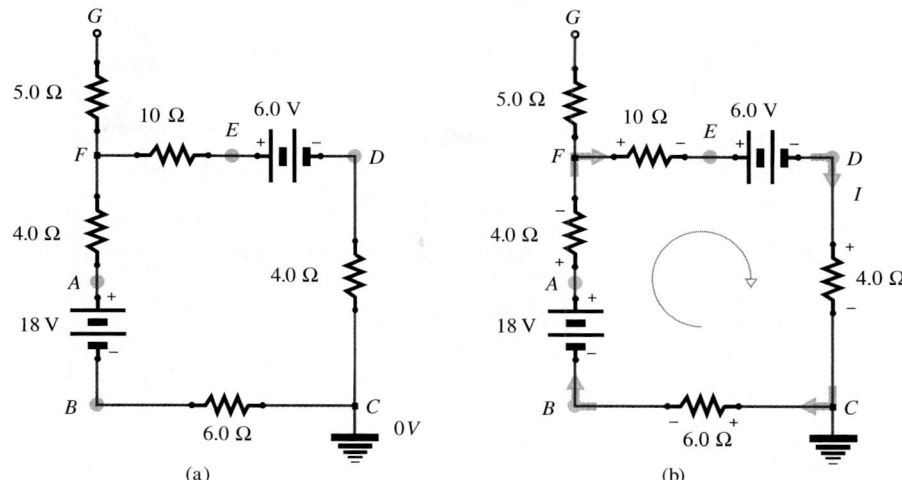

(a)   (b)

To apply the Loop Rule, let's determine the potentials at points A, B, C, D, E, and F in Fig. 18.24. This circuit has only one loop, and therefore only one branch, and one unknown branch current, I. Kirchhoff's Rules must be applied (thereby producing an equation) as many times as there are unknowns in the system.

The next step is to guess at the directions of the branch currents and draw them in. Assume that the larger battery dominates and the current emerges from its positive terminal. But not to worry, if ever the guess is wrong, the solution for I will simply come out negative, indicating that it's really in the opposite direction. Now label all the rises and drops with $+$ and $-$ signs depending on the direction of the current. **The signs of the emfs are independent of the current.** At this point, we are ready to apply the Loop Rule. Start at any point, say A, and go around the loop in either direction, writing the loop equation as you go. Let's do it clockwise: from A to F, there's a drop of $-4.0I$; from F to E, a drop of $-10I$; from E to D, a drop of -6.0 V; from D to C, a drop of $-4.0I$; from C to B, a drop of $-6.0I$; and from B to A, a rise of $+18$ V. Thus

$$-4.0I - 10I - 6.0 - 4.0I - 6.0I + 18 = 0$$

and

$$-24I + 12 = 0$$

therefore

$$I = \tfrac{1}{2} \text{ A}$$

The current is positive, and all the drops are correct as indicated. Since C is grounded and at a potential of zero, B is at -3.0 V. That is, to go from B to C, we rise ($\tfrac{1}{2}$ A)(6.0 Ω): A is at $+15$ V, F is at $+13$ V, E is $+8.0$ V, and D is $+2.0$ V. Notice that the potential at any point—for example, E—is indeed the same whether we go from C to D to E or from C to B to A to F to E.

Kirchhoff's second rule is known as the **Node Rule**:

> **The sum of all the currents entering any node in a circuit must equal the sum of all the currents leaving that node.**

In the steady state, there is neither an accumulation nor a diminution of charge anywhere in the circuit. Hence, at a node, the net amount of charge flowing in must equal the net amount flowing out—that's the Principle of Conservation of Charge. Figure 18.25 illustrates the point: $I_1 + I_2 = I_3 + I_4$.

To see how all of this comes together in practice, let's apply Kirchhoff's Rules to the network depicted in Fig. 18.26a, determining the current through every element. First, notice that the circuit has only two nodes—C and E—and therefore there are three branch-

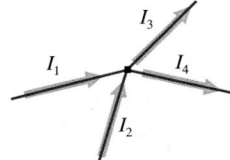

Figure 18.25 The Node Rule states that the sum of the currents entering a node must equal the sum of the currents leaving the node.

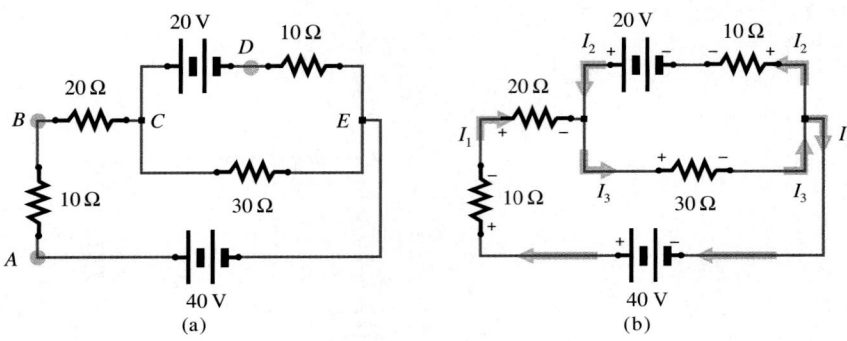

Figure 18.26 A network to be analyzed using Kirchhoff's Rules. Each branch has a current assigned to it.

es: *C–B–A–E, C–D–E,* and *C–E*. Assign a current arbitrarily to each branch, as shown in Fig. 18.24*b*, and label with signs all the voltage drops and rises.

Three branches means three unknown currents, which requires three equations. But when it comes to picking those three equations, we have to be cautious. There are three loops: the lower one *A–B–C–E–A,* the upper one *A–B–C–D–E–A,* and the little one *C–D–E–C,* as well as two nodes *C* and *E*. Thus, there are five possible equations, *but not any three will do*—we need three independent equations. To that end, ***use a mix of loop and node equations and always avoid using all the equations of either kind***. For instance, here the two node equations are

[at *C*] $$I_1 + I_2 = I_3$$

[at *E*] $$I_3 = I_1 + I_2$$

These are identical and not independent—only one of them can be used. *If two nodes involve the same set of currents, they will not produce independent equations* (and you should not waste time playing with them both). The three loop equations are also not independent. Consequently, we utilize any two loop equations and either node equation.

Using loop *A–B–C–E–A* and going around clockwise starting at *A*, we have

$$-10I_1 - 20I_1 - 30I_3 + 40 = 0$$

or $$-30I_1 - 30I_3 + 40 = 0 \qquad (i)$$

Similarly, going clockwise around loop *C–D–E–C* starting at *C* yields

$$-20 + 10I_2 + 30I_3 = 0 \qquad (ii)$$

These two equations can be solved simultaneously if we use the node equation to eliminate one of the variables, say, $I_2 = I_3 - I_1$. Hence, Eq. (ii) becomes

$$-20 + 10(I_3 - I_1) + 30I_3 = 0$$

and $$-10I_1 + 40I_3 - 20 = 0 \qquad (iii)$$

Multiplying this by 3 and subtracting it from Eq. (i) yields

$$-150I_3 + 100 = 0$$

and $I_3 = 2/3$ A. It follows from Eq. (i) that $I_1 = 2/3$ A. And, finally, something of a surprise: $I_2 = 0$. As a *quick check*, we see that the voltage difference in going from *C* to *E* is -20 V via either branch; moreover, the sum of the drops and rises is indeed zero around each loop.

As a last point, observe that loop *A–B–C–D–E–A* produces (going clockwise from *A*) the equation

$$-10I_1 - 20I_1 - 20 + 10I_2 + 40 = 0$$

which is

$$-30I_1 + 10I_2 + 20 = 0 \qquad \text{(iv)}$$

If we subtract Eq. (i) from this result, we get Eq. (ii) and vice versa, proving that we could not solve the problem with just those three—they're not independent.

If there are *N* nodes, use *N*−1 node equations along with as many loop equations as is necessary to form a total number of equations equal to the number of unknowns. Every element of the circuit must appear in at least one loop equation.

Example 18.5 **[III]** Determine the power delivered to the circuit by the 12-V source in Fig. 18.27*a*.

Solution The first thing to do is to simplify the circuit as much as possible. (1) TRANSLATION—A circuit is specified with known resistors and batteries; determine the power supplied by the battery. (2) GIVEN: The circuit diagram. FIND: P, the power delivered by 12-V source. (3) PROBLEM TYPE—Electric current/circuit/network analysis. (4) PROCEDURE—The presence of several batteries often leads to a circuit that cannot be simplified down to a single loop. This circuit can be made to look a lot more manageable by redrawing it. What's needed is the current through the 12-V source: P = *IV*. Let's first simplify the circuit as much as possible—Figs. 18.27*b*, *c*, and *d* show the series and parallel combinations. (5) CALCULATION— There are two nodes in the final circuit configuration, *A* and *B*, and three branches, *B–F–A*, *B–H–A*, and *B–C–A*. Hence, we assign three branch currents. The same currents are involved at *A* and *B*, so only one node equation will be used; accordingly

$$I_1 + I_3 = I_2 \qquad \text{(i)}$$

Starting at *B* and going clockwise around loop *B–F–A–H–B*, the Loop Rule yields

$$+4.0I_2 - 12 + 6.0I_1 = 0 \qquad \text{(ii)}$$

Starting at *B* and going clockwise around loop *B–H–A–C–B*, the Loop Rule yields

$$-6.0I_1 + 12 + 4.0I_3 - 6.0 = 0$$

or

$$-6.0I_1 + 4.0I_3 + 6.0 = 0 \qquad \text{(iii)}$$

We just want I_1, so use Eq. (i) to get rid of I_2 in Eq. (ii), which then becomes

$$10I_1 + 4.0I_3 - 12 = 0 \qquad \text{(iv)}$$

Subtracting Eq. (iv) from Eq. (iii) results in

$$-16I_1 + 18 = 0$$

and $I_1 = +1.125$ A. Current passes through the source from − to +; hence, power *is* delivered by the source. Consequently

$$P = IV = (1.125 \text{ A})(12 \text{ V}) = 13.5 \text{ W}$$

or to two figures $\boxed{P = 14 \text{ W}}$.

Quick Check: From Eq. (ii), $I_2 = 1.31$ A and the potential difference from *B* to *A* via *F* is +5.24 V, which must equal the rise from *B* to *A* via *H*; namely, $-(6.0 \text{ } \Omega)(1.125 \text{ A}) + 12 \text{ V} = +5.25$ V.

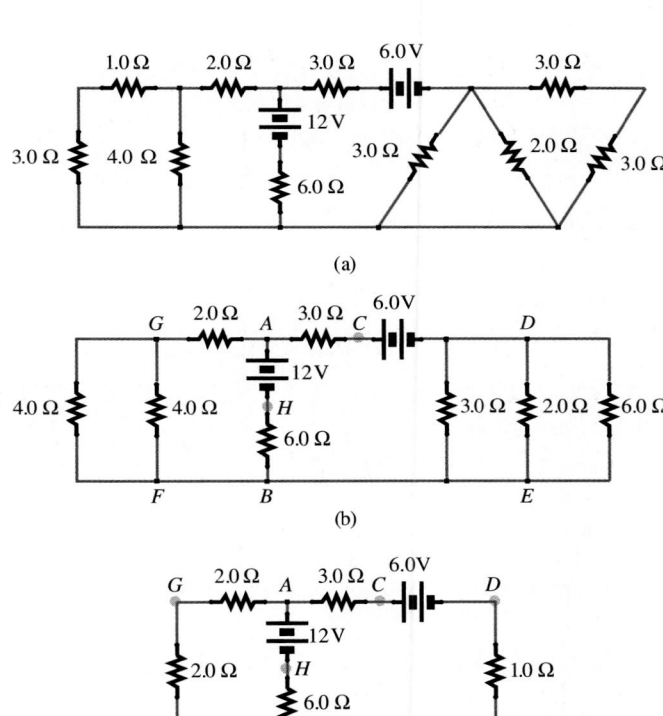

(a)

(b)

(c)

(d)

Figure 18.27 A circuit to be analyzed. It is first simplified as much as possible.

CIRCUIT PRINCIPLES

The voltage (V) across the terminals of a source of emf ($\mathcal{E}$) having an internal resistance (r) is

[power out] $$V = \mathcal{E} - Ir \qquad [18.1]$$

when the source provides power and

[power in] $$V = \mathcal{E} + Ir \qquad [18.2]$$

when power is provided to it (p. 637). The idea that $\mathcal{E} \neq V$ is discussed in Section 18.1 (Sources and Internal Resistance), and typical problems are examined in Examples 18.1 and 18.2. This is a minor issue, but it can have practical significance and so is generally treated. Study the **CD Warm-Ups** and **Walk-Throughs**. 💿

The **equivalent resistance** of two or more resistors in series is

[series] $$R_e = R_1 + R_2 + \cdots \qquad [18.4]$$

The equivalent resistance of two or more resistors in parallel is

[parallel] $$\frac{1}{R_e} = \frac{1}{R_1} + \frac{1}{R_2} + \cdots \qquad [18.5]$$

For two resistors, this equation can be expressed as

[two resistors in parallel] $$R_e = \frac{R_1 R_2}{R_1 + R_2} \qquad [18.6]$$

[N identical resistors in parallel] $$R_e = \frac{R}{N} \qquad [18.7]$$

When current divides into two parallel branches

$$I_1 = \frac{R_2}{R_1 + R_2} I \qquad [18.8]$$

and $$I_2 = \frac{R_1}{R_1 + R_2} I \qquad [18.9]$$

These ideas, which are considered in Section 18.2 (Resistors in Series and Parallel), are among the most important in the chapter. Study Examples 18.3 and 18.4, but keep in mind that this last prob-

lem is of the III-variety—it ain't easy. Do as many I-level problems as you can.

When a capacitor is discharged through a resistance R, the process takes place gradually according to the expression

$$Q = Q_i e^{-t/RC} \qquad [18.11]$$

and the decay is exponential. And similarly the current through a capacitor drops off exponentially. This behavior is treated in Section 18.4 (R-C Circuits), and there is a selection of relevant problems in the back of the chapter.

NETWORK ANALYSIS

Kirchhoff's **Loop Rule** is *the algebraic sum of the voltage rises and drops encountered in going around any closed path formed by any portion of a circuit must be zero.* Kirchhoff's **Node Rule** is *the sum of all the currents entering any node in a circuit must equal the sum of all the currents leaving that node.* Together these rules provide a powerful technique for analyzing circuits that can be found in Section 18.5 (Kirchhoff's Rules). If it's your intention to master this material, reread that section, go over the analysis on p. 649, and study Example 18.5 before trying the problems.

Key Terms

internal resistance	voltmeter
terminal voltage	shunt
node	R-C circuit
loop	time constant
equivalent resistance	network
resistor in series	Kirchhoff's Rules branch
resistor in parallel	Loop Rule
ammeter	Node Rule

1. Figure 18.5 depicts a battery with an internal resistance of r in series with a load R. What experimental procedure might be used to determine r?

2. What is meant by a complete or closed circuit, and what is the relationship between such a circuit and a steady-state direct current? When the switch in Fig. 18.17 is closed, a transient current circulates in the circuit. Explain how this could happen in spite of the gap in the capacitor that stops charge from literally crossing it.

3. Figure Q3 is a reproduction of a circuit taken from a simulated MCAT exam created by a well-known company that prepares students for that test. The question asks for the current I, and the answer given is "9 amps toward G." What's wrong with that and why do you think they give 9 A as the answer?

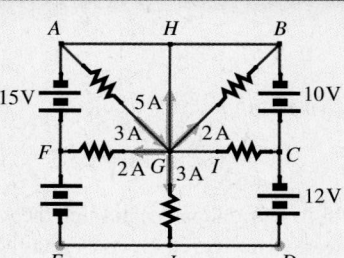

Figure Q3

4. In light of Question 2, discuss the statement: *In the steady state, a capacitor acts as an open circuit to dc.*

5. The circuit of Fig. Q5 contains a gas-filled lamp—the sort of flashing light often used at construction sites. At low voltages, the

lamp has nearly infinite resistance, but at a certain breakdown voltage, which is less than the terminal voltage of the battery, it becomes a very good conductor. A blast of charge through the lamp causes it to flash brightly. What happens once the switch is closed?

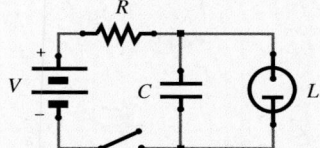

Figure Q5

6. Suppose we have a fluid voltaic cell across which we place a load resistor. Next we measure the current from, and voltage across, the source with an ammeter and a voltmeter, respectively. It can be observed that when the plates are moved apart, the current diminishes. Moreover, when the plates are partly lifted out of the electrolyte, the current again drops, although in both cases the voltage remains unchanged. What's happening to the internal resistance of the cell? Relate your conclusions to the fluid-flow analogy of current.

7. *R-C* circuits are often used to change the shape of a signal. The arrangement of Fig. Q7 shows a network being fed a square-wave signal comprising a series of rectangular dc voltage pulses rising to $+V$ and falling to 0. Part (b) shows the corresponding voltage across the capacitor. Explain its shape. What can you say about the value of RC as compared to the width of each pulse? Draw a curve of the output signal as seen across the resistor and explain its features. Write a general expression relating V, V_R, and V_C.

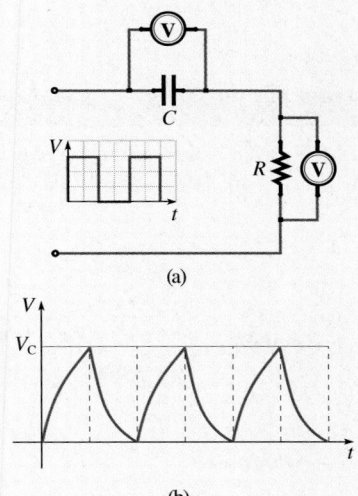

Figure Q7

8. Explain how a flashlight bulb works. Draw a picture showing the relationship between a bulb's filament and its two terminals. Why does the brightness depend on the current through the bulb? What effect would increasing the voltage across a bulb have on its brightness? Incidentally, light bulbs are not ohmic.

9. Figure Q9 depicts several voltage-versus-time curves. These might be the outputs of a generator or transducer, for example. Which of them corresponds to dc? Explain.

10. Draw a flashlight bulb and show where the current usually enters and leaves. Draw a D-cell and label its two terminals. Now

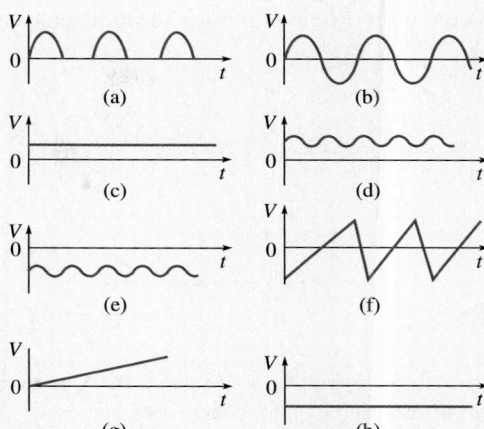

Figure Q9

draw a diagram showing how to connect the bulb and cell with two lengths of wire. How might you connect them with one wire? Show the path of the current in each case.

11. Figure Q11 shows three arrangements of identical flashlight bulbs along with an ideal 6-V source. Compare the brightness of all the bulbs, listing them in descending order. Explain.

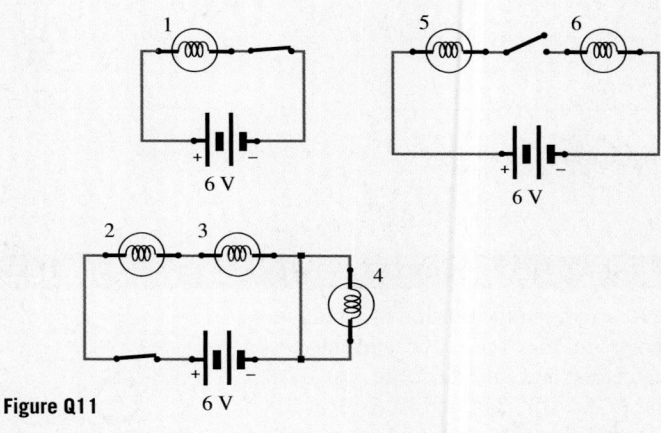

Figure Q11

12. Compare the brightness of all the bulbs in Fig. Q12, listing them in descending order. The battery is ideal. Explain.

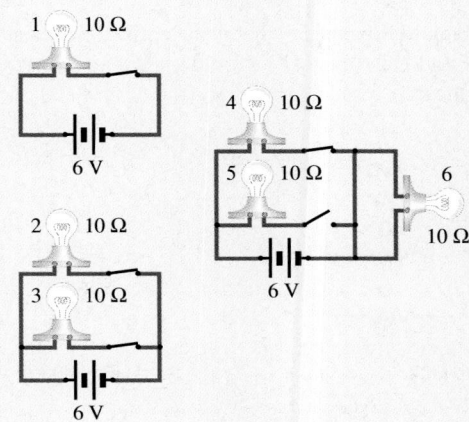

Figure Q12

13. Compare the brightness of each identical bulb in Fig. Q13. Explain.

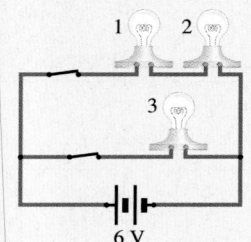

Figure Q13

14. All the bulbs in Fig. Q14 are identical. Discuss the relative brightness of each and explain your answer.

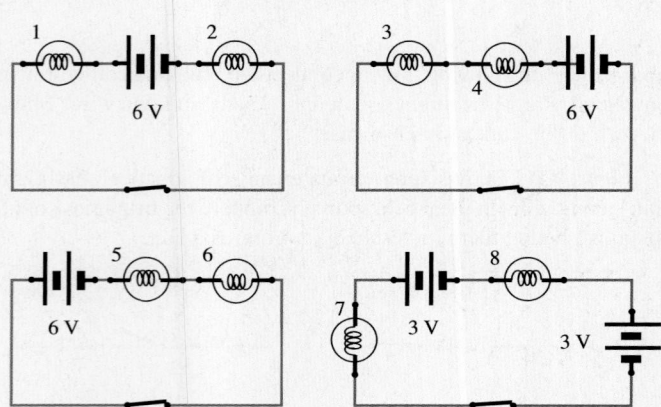

Figure Q14

15. If all the resistors in Fig. Q15 are identical, rate them in order of the amount of current passing through each one. Explain.

16. Explain the meaning of each of the Key Terms on page 654.

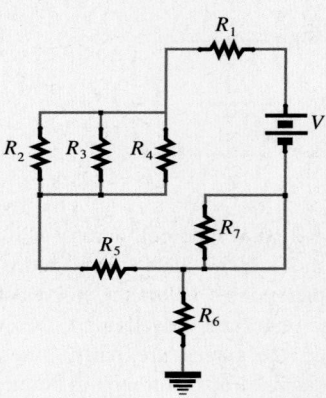

Figure Q15

Multiple Choice Questions

1. Referring to the portion of a circuit shown in Fig. MC1, the currents in ammeter-1 and ammeter-2 are (a) 8 A and 12 A (b) 20 A and 10 A (c) 20 A and 4 A (d) 4 A and 12 A (e) none of these.

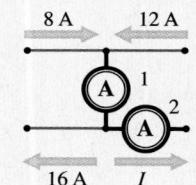

Figure MC1

2. Referring to Fig. MC2, what is the equivalent resistance between points A and B? (a) 27 Ω (b) 9 Ω (c) 6 Ω (d) 3 Ω (e) none of these.

3. Referring to Fig. MC3, what is the equivalent resistance between points A and D? (a) 6 Ω (b) 9 Ω (c) 12 Ω (d) 18 Ω (e) none of these.

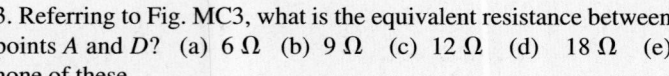

Figure MC2

Figure MC3

4. Referring to Fig. MC4, what is the equivalent resistance between points A and D? (a) 64 Ω (b) 10 Ω (c) 12 Ω (d) 70 Ω (e) none of these.

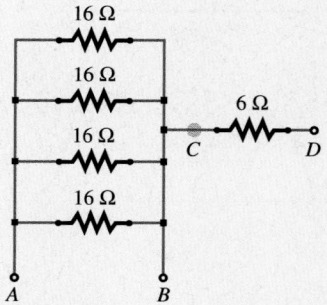

Figure MC4

5. The equivalent resistance of a 4-Ω resistor in parallel with a 12-Ω resistor is (a) 48 Ω (b) 8 Ω (c) 16 Ω (d) 3 Ω (e) none of these.

6. An ideal source of emf is connected in series with a resistor in a closed loop, and a current of 0.50 A circulates. An additional 6.0-Ω resistor is added in series, and the current drops to 0.30 A. The original resistance was (a) 9.0 Ω (b) 6.0 Ω (c) 15 Ω (d) 3.0 Ω (e) none of these.

7. In the previous question, what was the emf? (a) 9.0 V (b) 12 V (c) 15 V (d) 4.5 V (e) none of these.

8. A 24-W, 12-V ohmic lamp is placed in a circuit with 6 V across it, at which time it draws a current of (a) 2 A (b) 1 A (c) 0 (d) $\frac{1}{2}$ A (e) none of these.

9. The device in Fig. MC9 consists of a rotating switch, five different resistors, and a galvanometer. What is the apparatus? (a) an ohmmeter (b) a multirange voltmeter (c) a multirange ammeter (d) a multirange wattmeter (e) none of these.

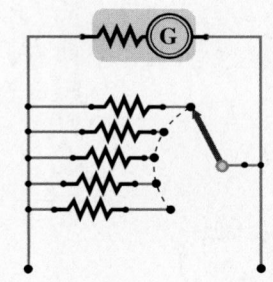

Figure MC9

10. The circuit in Fig. MC10 contains four identical light bulbs in series with an ideal battery. A wire is then clipped to points B and E and (a) lamps 3 and 4 become brighter (b) lamp 1 gets dimmer (c) lamps 2, 3, and 4 remain unaffected (d) lamp 1 brightens (e) none of these.

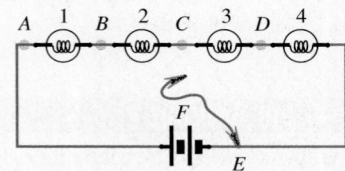

Figure MC10

11. In Fig. MC10, a wire is clipped to points B and E, and (a) lamps 2, 3, and 4 become brighter (b) lamps 1 and 2 get dimmer (c) lamps 2, 3, and 4 go out altogether (d) lamp 1 goes out (e) none of these.

12. In Fig. MC10, a wire is clipped to points F and E and (a) lamps 1, 2, 3, and 4 become brighter (b) lamp 1 gets slightly dimmer (c) lamps 2, 3, and 4 remain unaffected (d) lamps 1, 2, 3, and 4 get slightly dimmer (e) none of these.

13. In Fig. MC10, a wire is clipped to points C and E and (a) more current is drawn from the battery (b) less current is drawn from the battery (c) the same current is drawn from the battery (d) not enough information is given to tell what will happen to the current (e) none of these.

14. In Fig. MC10, a wire is clipped to points C and E and (a) the voltage across lamps 1 and 2 increases (b) the voltage across lamps 3 and 4 increases (c) the voltage across only lamp 1 increases (d) the voltage across lamp 2 decreases (e) none of these.

15. In Fig. MC10, a wire is clipped to points C and E, and (a) the power delivered by the battery remains unchanged (b) the power delivered by the battery increases (c) the power dissipated by the circuit decreases (d) the power dissipated by the circuit is halved (e) none of these.

16. Suppose the four lamps in Fig. MC10 are rewired so that they are in parallel across the battery: (a) each lamp will put out the same amount of light as before (b) each lamp will put out more light than before (c) each lamp will put out less light than before (d) there is not enough information to tell what will happen (e) none of these.

17. The two circuits (i) and (ii) shown in Fig. MC17 are each com-posed of identical resistors connected to a battery. The resistors in (i) and the resistors in (ii), respectively, are hooked up in (a) series and series (b) parallel and parallel (c) series and parallel (d) parallel and series (e) none of these.

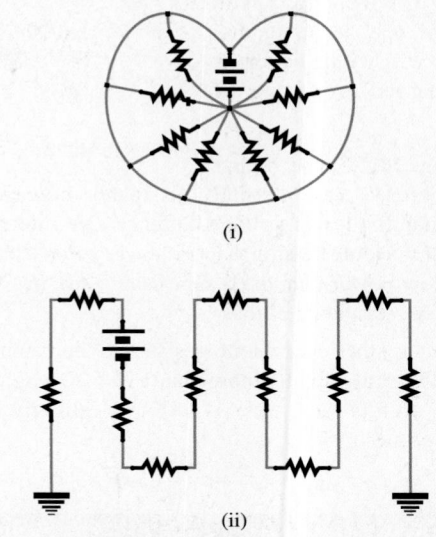

(i)

Figure MC17 (ii)

18. A battery of emf 12 V and internal resistance 0.2 Ω is placed across a variable resistor. The resistor is adjusted until it dis-sipates a maximum amount of power, at which point the current through it is (a) 12 A (b) 0.4 A (c) 30 A (d) 60 Ω (e) none of these.

19. A battery attached to a load supplies 2 A with a terminal voltage of 12.0 V. If the battery dissipates 0.4 W, its emf is (a) 11.8 V (b) 12 V (c) 12.2 V (d) 12.4 V (e) none of these.

20. Two resistors of 5 Ω and 20 Ω are connected in parallel across an ideal source of 20 V. The current supplied by the source is (a) 4 A (b) 5 A (c) 20 A (d) 1 A (e) none of these.

21. Figure MC21 shows a portion of a network. The current I is (a) +2 A (b) +4 A (c) not enough informa-tion to tell (d) −4 A (e) none of these.

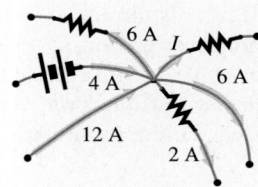

Figure MC21

22. What is the current in any one of the 4-Ω resistors in the circuit of Fig. MC22? (a) 12 A (b) 1.2 A (c) 3 mA (d) 0 (e) none of these.

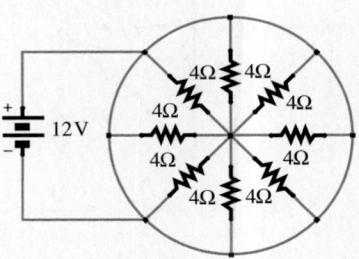

Figure MC22

23. In Fig. MC23, (a) the battery with an emf of 12 V is being charged (b) the battery with an emf of 6 V is being charged (c) the battery with an emf of 6 V is being discharged (d) neither battery is being charged (e) none of these.

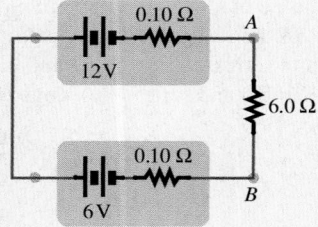

Figure MC23

24. In Fig. MC23, the higher voltage battery (a) internally dissipates more power via joule heating than the lower voltage battery (b) internally dissipates less power via joule heating than the lower voltage battery (c) dissipates more power than the 6-Ω resistor (d) has a terminal voltage of 14 V (e) none of these.

25. What is the equivalent resistance between A and B in Fig. MC25? (a) slightly more than 1 Ω (b) slightly more than 1 kΩ (c) slightly more than 17 kΩ (d) slightly less than 1 kΩ (e) none of these.

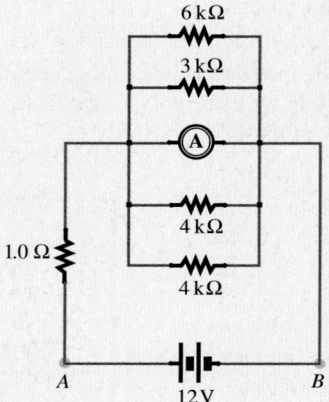

Figure MC25

For more Multiple Choice Questions with answers click on WARM-UPS in CHAPTER 18 on the CD.

Suggestions on Problem Solving

1. A common error is to use Eq. (18.5) for the equivalent resistance of a parallel string of resistors, computing and adding $1/R_1$ and $1/R_2$ and then presenting that as R_e rather than $1/R_e$; look out for this one. Note, too, that the reciprocal of the sum is *not* equal to the sum of the reciprocals: $1/(R_1 + R_2) \neq 1/R_1 + 1/R_2$. Equation (18.6), $R_e = (R_1R_2)/(R_1 + R_2)$, is very useful for getting a quick sense of parallel combinations without a calculator. The expression $R_1R_2R_3/(R_1 + R_2 + R_3)$ is *not* the correct formula for three resistors in parallel!

2. When analyzing a circuit, study it carefully. Work with a drawing of the network. If the original version is not neatly diagrammed, redraw it before proceeding. Is there anything extraordinary about the circuit (as, for example, in Fig. Q3)? Are there any obvious shortcuts to finding the required quantities (as, for example, in Fig. 18.22)? Is the circuit symmetrical in such a way that some sort of bridge is present so that elements might be redundant (as, for example, in Fig. 18.23)? Remember, *if no current traverses a particular resistor, it can be removed from the circuit*. All right, then, what current passes through the 8-Ω resistor in Fig. 18.28? Zero!

All the sources buck each other—there's no current anywhere. Try Fig. 18.29. What's the current in the 6-Ω resistor? By inspection, it must be 2 A to the left. Both ends of the bottom branch are grounded, so the net potential difference between E and A is zero. Going from A to E, there must be a drop of 12 V across the resistor since there is a rise of 12 V across the battery. In Fig. 18.29, what is the current in the 9-Ω resistor? Again, it's zero because there is no voltage across E–A. All the current from the battery ($I = 2$ A) goes from E into the zero resistance path back to A via ground. When ground connections are indicated, it should be assumed that all such leads are wired to a common line even if not shown.

Referring to Fig. 18.30, symmetry demands that current only circulate in the outer elements. Points A and B are at the same potential because the circuit is symmetrical. Therefore, no current can go across the resistors in that branch. The current through both batteries is 2 A.

3. The first step in an analysis is to simplify the circuit by combining series and parallel groupings wherever possible. Redraw the circuit whenever needed, keeping track of what you have done so

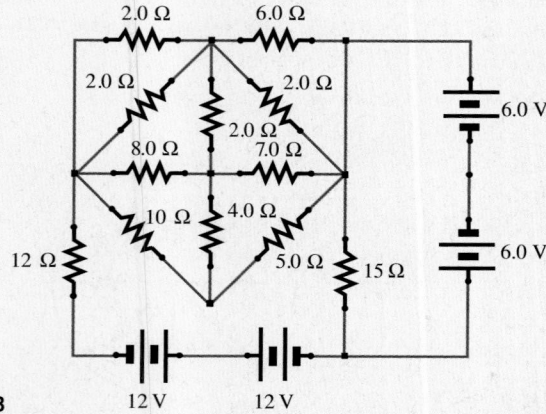

Figure 18.28

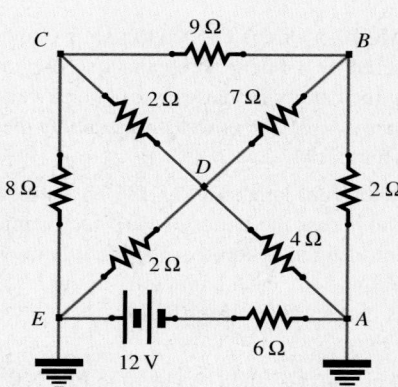

Figure 18.29

that you can work your way back to the original. (a) If the system reduces to one source across an equivalent resistance, use Ohm's Law to find what is unknown and work backward reconstructing the original circuit step by step. (b) If the circuit cannot be further simplified, apply Kirchhoff's Rules to what you have. (c) Line up the several variables, essentially making columns of them so that you can see what needs to be done at a glance. For example,

$$-3I_2 - 2I_3 + 12 = 0 \qquad \text{(i)}$$

$$+2I_3 - 3I_4 - 6 = 0 \qquad \text{(ii)}$$

$$-3I_2 - 3I_4 + 6 = 0 \qquad \text{(iii)}$$

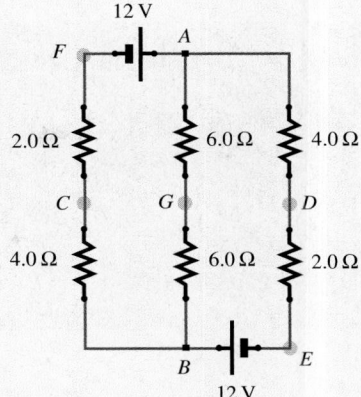

Figure 18.30

Problems ✦ Coordinated Problems ✦ Progressive Problems ✦ Solutions

STUDY GUIDE **1. Coordinated Problems:** The three problems within each magenta-colored grouping are solvable in similar ways. Note that the first of these always has a hint; moreover, its solution is provided in the back of the book. *Work out each of these sets; they'll strengthen technique and build confidence.* **2. Progressive Problems:** The problems introduced in blue unfold step-by-step carrying along the analysis in a more suggestive way than is customary. *Work out all of these; they'll guide you through the analytic process and help develop problem-solving skills.* **3. Worked-Out Solutions:** Studying worked-out solutions is an important part of learning how to solve problems. Accordingly, additional *solutions* to a number of model problems are given below. *Make sure you understand each of them before you go on to the next problem.* **4.** Also provided in the back of the book are the *Answers* to all odd-numbered problems, as well as worked-out *solutions* to those with boldface numbers. Problem numbers in italic indicate that a solution appears in the Student Solutions Manual.

SECTION 18.1: SOURCES AND INTERNAL RESISTANCE

1. [I] An old cell with an emf of 1.5 V has an internal resistance of 1.0 Ω. How much current will initially flow if its terminals are short-circuited?

2. [I] A dc source with an internal resistance of 0.10 Ω is connected across a length of nichrome wire having a resistance of 20 Ω. If a voltmeter across the nichrome indicates a drop of 10 V, what is the emf of the source?

3. [I] A D-cell with an emf of 1.50 V and an internal resistance of 0.05 Ω is placed in a circuit with several resistors. The cell provides 0.50 A to the circuit. What is the terminal voltage of the cell? [*Hint: How much current flows through the cell?*]

4. [I] A battery with a terminal voltage of 11.95 V when delivering 0.50 A has an internal resistance of 0.10 Ω. What is its emf?

5. [I] The battery in a circuit has an emf of 9.0 V. It is attached to a resistor and an ammeter that shows a current of 0.10 A. If a voltmeter across the battery's terminals reads 8.9 V, what is its internal resistance?

6. [I] A dc source with a terminal voltage of 100 V internally dissipates 40 W as it delivers 2.0 A. What is its emf?

7. [I] The terminal voltage of a well-used battery is 8.85 V, and it has an internal resistance of 0.30 Ω. How much power will it dissipate when supplying 0.50 A to an external circuit?

8. [II] Determine the terminal voltage of a battery having an emf of 12.0 V and an internal resistance of 0.20 Ω, knowing that when placed across its terminals, a 40.0-Ω resistor dissipates 10.0 W.

9. [II] What is the total power dissipated when a 10.0-Ω resistor is placed across a battery having an emf of 9.0 V and an internal resistance of 0.20 Ω. [*Hint: The current through the 10.0-Ω resistor is 0.882 A.*] What is the terminal voltage of the battery?

10. [II] Imagine that *N* is the number of identical cells in series, each having an internal resistance *r* and emf *𝓔*. Derive an expression for the current through a load resistor *R* placed across the terminals of the battery.

11. [II] Determine the terminal voltage of a battery after being fully recharged by an external power supply that continues to provide 0.20 W to it. The battery has an emf of 12.0 V and an internal resistance of 0.20 Ω.

12. [II] In order to be recharged, a battery is attached to an external power source. The terminal voltage of the battery is 9.14 V, and its internal resistance is 0.10 Ω. Given that it dissipates 0.20 W, what is its emf?

13. [II] A battery provides 80.0 W of power to a resistive load that has a resistance of 20.0 Ω. A voltmeter across the battery reads 5.8 V when the load is in place and 6.0 V when it is removed. Determine the internal resistance of the battery.

14. [II] The open-circuit (or zero-current) voltage across the terminals of a battery is measured, via a voltmeter, to be 12.0 V. The battery is then attached to a load that draws 1.0 A and the voltmeter now reads 11.9 V. What is the internal resistance of the battery? How much power does the battery dissipate? How much power does the battery supply to the load?

SECTION 18.2: RESISTORS IN SERIES AND PARALLEL

15. [I] Consider the circuit board in Fig. P15 where the large conducting area is ground. (a) What would an ohmmeter read across terminals-1 and 2? (b) What would an ohmmeter read across terminals-2 and 3? (c) What would an ohmmeter read across terminals-3 and 4? (d) What is the resistance between terminals-4 and 5? (e) What would an ohmmeter read across terminals-1 and 5?

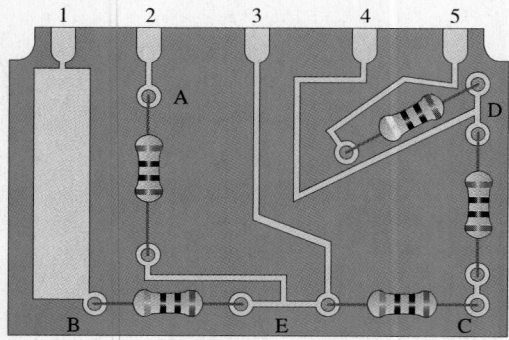

Figure P15

16. [I] THIS PROBLEM WILL HELP US DEAL WITH CIRCUIT DIAGRAMS AND RESISTANCE. With the switch S open in the circuit of Fig. P16, we put an ohmmeter between points-A and C. (a) Does the branch from A to B to C have any effect on the measurement? Explain. (b) What is the value of the resistance the ohmmeter will measure between points-A and C? (c) What will the ohmmeter read if we touch one of its probes to A and the other to ground? Explain. (d) What will the ohmmeter read if we touch one of its probes to D and the other to ground? (e) What will the ohmmeter read if we touch one of its probes to E and the other to ground?

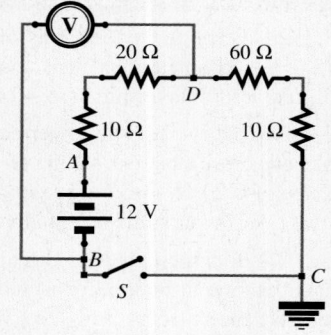

Figure P16

17. [I] What is the current in the circuit of Fig. P17 before and after the wire is clipped on at C and B? What happens to the 12-Ω bulb after the wire is attached? The battery has a negligible internal resistance.

18. [I] Two 30.0-Ω resistors are attached in series and then placed across the terminals of a 12.0 -V battery having a negligible internal resistance. How much current flows through the circuit?

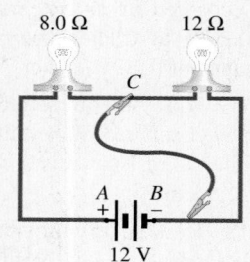

Figure P17

19. [I] Each lamp in Fig. P19 has a resistance of 20 Ω. How much current is drawn from the battery before and after the switch is closed? How does the power supplied to the circuit change when the switch is closed? The battery has a negligible internal resistance.

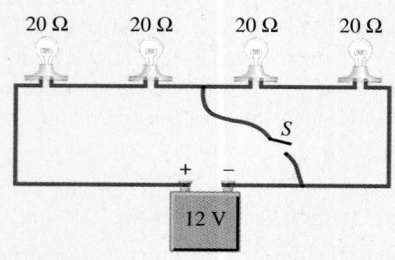

Figure P19

20. [I] Three 3.0-Ω resistors are connected in parallel and the combination is placed across a battery that has a terminal voltage of 9.0 V. How much current flows through the battery?

21. [I] Two 2.0-Ω resistors are in parallel. What is their equivalent resistance?

22. [I] Three resistors with values of 2.0 Ω, 3.0 Ω, and 6.0 Ω are connected in parallel. What is the equivalent resistance?

23. [I] A portable generator in a field hospital produces dc at 100 V. Five 100-W lamps are attached in parallel, and the string placed across the terminals of the generator. How much current must the source provide for the lamps to operate as designed?

24. [I] Three resistors with values of 1/2 Ω, 1/3 Ω, and 1/6 Ω are connected in parallel. What is the equivalent resistance?

25. [I] Show that if there are a number N of identical resistors in parallel, each with a value R, then $R_e = R/N$.

26. [I] THIS PROBLEM DEALS WITH AN ARRAY OF RESISTORS. Each resistor in Fig. P26 is color coded orange-black-black-silver. (a) What is the resistance of each resistor? (b) With the switch S open, we put an ohmmeter between points-D and C and measure the resistance. Does the branch from D to F to C have any effect on the measurement? Explain. (c) What can you say about points A, B, C, and E? (d) What is the value of the resistance the ohmmeter will measure between points-D and C? (e) What will the ohmmeter read if we touch one of its probes to D and the other to ground? Explain.

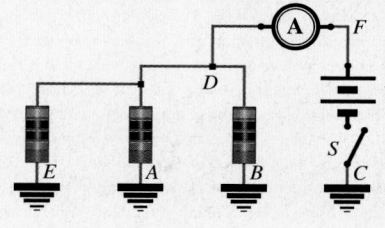

Figure P26

27. [I] Suppose each resistor in Fig.P26 was color coded orange-black-orange-silver. With the switch closed, how much voltage will be needed to produce a current in the ammeter of 120 μA? In what direction will it flow?

SOLUTION: The resistance of each resistor is orange (3), black (0), orange (3) or 30 kΩ. Three of them in parallel is equivalent to 10 kΩ. $V = IR$ and so $V = (120 \times 10^{-6}$ A$)(10 \times 10^3$ $\Omega) = 1.2$ V. Current flows from F to D, out of the plus terminal of the source.

28. [I] If in Fig. P28 the ideal ammeter reads 3.20 A, what will the ideal voltmeter read?

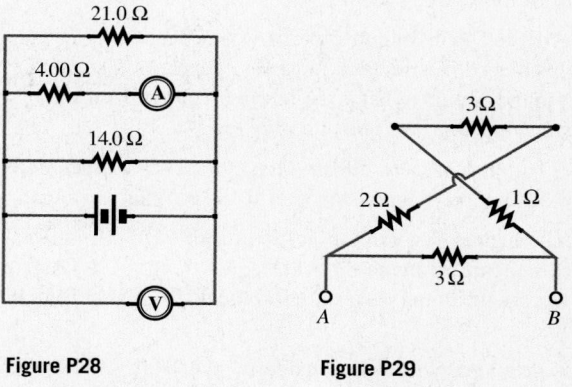

Figure P28   **Figure P29**

29. [I] What is the equivalent resistance of the circuit in Fig. P29 between terminals A and B? Note that the wires cross but do not make contact at the center.

30. [I] Determine the equivalent resistance of the circuit between points A and B in Fig. P30.

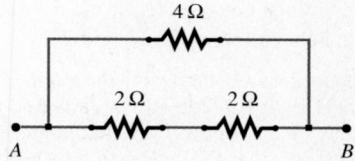

Figure P30

31. [I] Determine the equivalent resistance of the circuit between points *A* and *B* in Fig. P31.

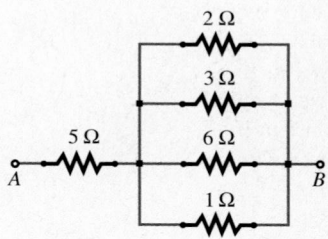

Figure P31

32. [I] Determine the equivalent resistance of the circuit between points *A* and *B* in Fig. P32.

33. [I] Determine the equivalent resistance of the circuit between points *A* and *C* in Fig. P32.

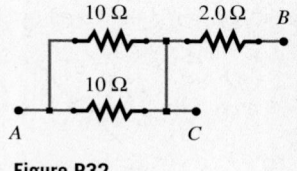

Figure P32

34. [I] Determine the equivalent resistance of the circuit between points *B* and *C* in Fig. P32.

35. [I] Determine the equivalent resistance of the circuit between points *A* and *B* in Fig. P35.

36. [I] Determine the equivalent resistance of the circuit between points *C* and *B* in Fig. P35.

37. [I] Figure P37 shows a portion of a circuit, the rest of which contains both sources and resistors. If the ammeter reads 9 A, what current passes through each resistor shown?

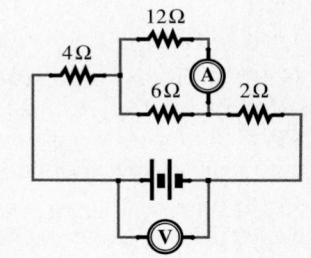

Figure P35

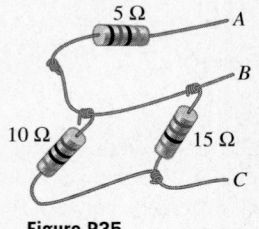

Figure P37

38. [I] Given that the ammeter in Fig. P38 reads 1.0 A, what is the emf of the ideal dc source as indicated by the voltmeter?

39. [I] How much current passes through each lamp in Fig. P39?

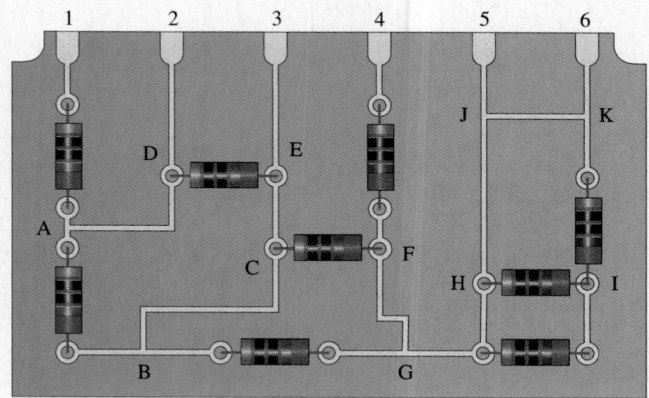

Figure P38

The battery has negligible internal resistance.

40. [I] A 10-Ω resistor is attached at one end to the + terminal of a 20-V dc source whose − terminal is grounded. The resistor's other end is attached to the + terminal of a 40-V dc source whose − terminal is grounded. What current traverses the resistor?

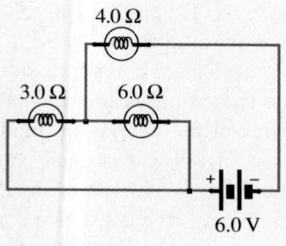

Figure P39

41. [I] A 20-V lamp designed to dissipate 80 W is placed in series with a resistor *R* and a 60-V dc source. What value should *R* be for the lamp to operate properly?

42. [I] What is the equivalent resistance between points *D* and *B* in the circuit shown in Fig. P42?

43. [II] THIS PROBLEM WORKS WITH A PRINTED CIRCUIT BOARD. Consider the circuit board in Fig. P43 where all the resistors are color coded red-black-black-silver. (a) What is the value of each resistor? We want to find out what an ohmmeter would read across terminals 5 and 6. (b) Redraw the circuit between these terminals simplifying what the ohmmeter would "see." (c) What path(s) would current take in going from 5 to 6? (d) What is the crucial role played by the wire from *J* to *K*? (e) What is the resistance between terminals-5 and 6.

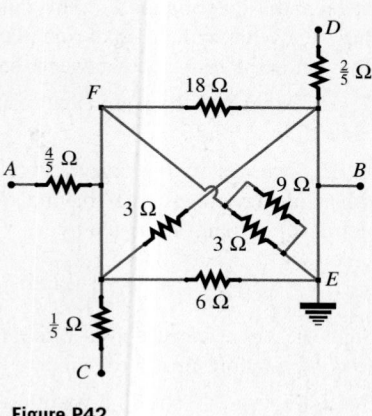

Figure P42

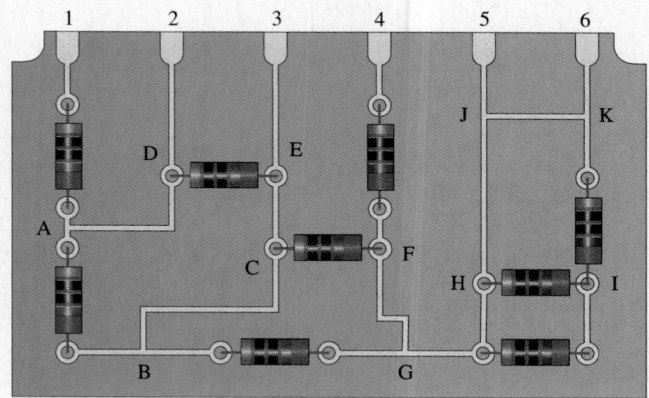

Figure P43

44. [II] THIS PROBLEM WORKS WITH A PRINTED CIRCUIT BOARD. Consider the circuit board in Fig. P43, but this time suppose all the resistors were color coded red-black-brown-silver. (a) What is the value of each resistor? We want to find out what an ohmmeter would read across terminals-1 and 2. (b) Redraw the circuit between these terminals simplifying what the ohmmeter would "see." (c) What path(s) would current take in going from 1 to 2? (d) What is the crucial role played by the wire from *D* to *A*? (e) What is the resistance between terminals-1 and 2?

45. [II] THIS PROBLEM WORKS WITH A PRINTED CIRCUIT BOARD. Consider the circuit board in Fig. P43, but this time suppose all the resistors were color coded orange-black-brown-silver. (a) What is the value of each resistor? We want to find out what an ohmmeter would read across terminals-2 and 3. (b) Redraw the circuit between these terminals simplifying what the ohmmeter would "see." (c) What is the crucial role played by the wire from E to B? (d) What is the resistance between terminals-2 and 3?

46. [II] THIS PROBLEM WORKS WITH A PRINTED CIRCUIT BOARD. Consider the circuit board in Fig. P43, but this time suppose all the resistors were color coded brown-black-yellow-silver. (a) What is the value of each resistor? We want to find out what an ohmmeter would read across terminals-3 and 4. (b) Redraw the circuit between these terminals simplifying what the ohmmeter would "see." (c) What is the crucial role played by the wire from E to B? (d) What is the resistance between terminals-3 and 4?

47. [II] Determine the equivalent resistance of the circuit between points A and B in Fig. P42.

48. [II] Two resistors R_1 and R_2 are in parallel with each other and with an ideal source ($r = 0$) having a terminal voltage V. Show that the branch currents are given by

$$I_1 = I\left(\frac{R_2}{R_1 + R_2}\right) \quad \text{and} \quad I_2 = I\left(\frac{R_1}{R_1 + R_2}\right)$$

where the larger current goes through the smaller resistor. (These are good relationships to remember.)

49. [II] Figure P49 shows a portion of a circuit, the rest of which contains both sources and resistors. If the ammeter reads 9 A, what current passes through each resistor in the diagram?

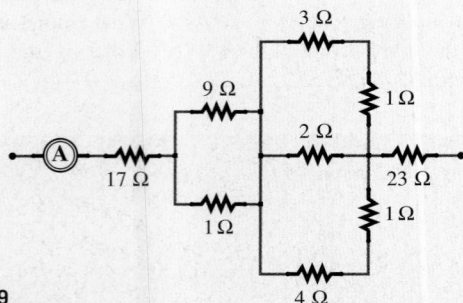

Figure P49

50. [II] Find the equivalent resistance between points C and E in Fig. P42.

51. [II] THIS PROBLEM DEALS WITH VOLTAGE AND CURRENT IN A DC CIRCUIT. Referring to Fig. P16 and Problem 16, we want to find the voltmeter reading when the switch in the circuit is closed, but we first have to find the current. (a) What is the equivalent resistance of the circuit? (b) What is the voltage across this equivalent resistance? (c) What is the direction of current around the circuit. (d) How much current flows? (e) What is the voltage drop across each resistor? (f) What is the sum of the voltage rises and drops between B and D? (g) What value will the voltmeter read? (h) Which voltmeter terminal has a higher potential?

52. [II] Consider the circuit board in Fig. P15 where the large printed area is ground. (a) If 75 V is put across terminals-1 and 5 (with 5 being +), what would a voltmeter read across terminals-1 and 4?

(b) What is then the voltage between terminals-1 and 3? (c) What, if any, is the voltage between terminals-1 and 2?

SOLUTION: The equivalent resistance between terminals-1 and 5 is 40 Ω + 20 Ω + 60 Ω + 30 Ω = 150 Ω. $V = IR$ and so a current of $I = (75 \text{ V})/(150 \text{ Ω}) = 0.50$ A flows. (a) That produces a voltage drop across the 40-Ω resistor of $V = IR = (0.50 \text{ A})(40 \text{ Ω}) = 20$ V and so the voltage at D is 75 V − 20 V = 55 V above ground. (b) The voltage at A equals the voltage at E, which is 75 V − 20 V − 10 V − 30 V= 15 V above ground. (c) There's no current in the branch from 2 to E so there's no voltage drop between 2 and E. The drop across the 30-Ω resistor is 15 V and so E is 15 V above ground.

53. [II] What is the equivalent resistance between points A and B of the circuit shown in Fig. P53? How much power would be dissipated by this circuit if a constant 20-V dc source with a 0.10-Ω internal resistance were placed across A and B?

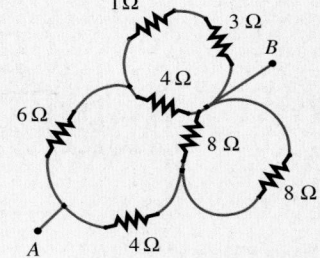

Figure P53

54. [II] How much current passes through each of the 5.0-Ω resistors in Fig. P54? How much power is delivered by the dc source?

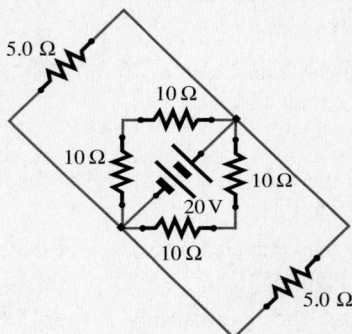

Figure P54

55. [II] Given the circuit in Fig. P55, calculate the current in each resistor. What power is delivered by the battery? What is the potential difference between A and C?

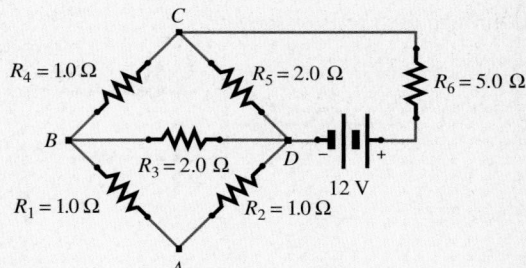

Figure P55

56. [II] The battery in Fig. P56 has an emf of 9.0 V and an internal resistance of 0.50 Ω. (a) What current does it supply? (b) What is the total power dissipated by the entire circuit? (c) What is the terminal voltage of the battery?

57. [II] How much power is dissipated by the automobile circuit in Fig. P57 when switches A, B, C, and D are all closed?

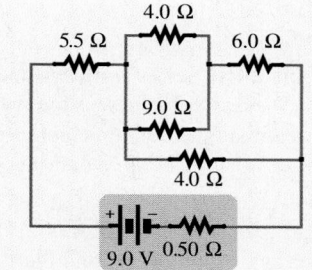

Figure P56

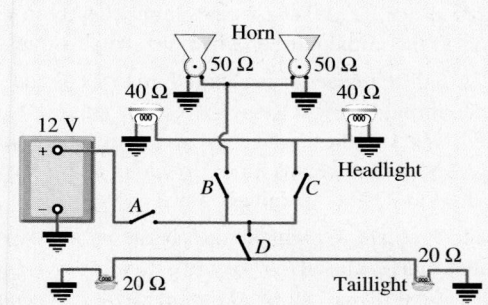

Figure P57

58. [II] Find the current supplied by the source in Fig. P58. The resistors are mounted around a cylindrical form.

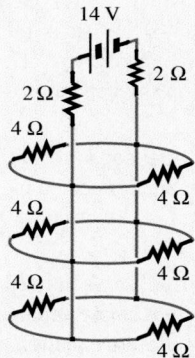

Figure P58

59. [II] What is the current provided by the battery in Fig. P59, given that its internal resistance is 0.50 Ω? What is its terminal voltage?

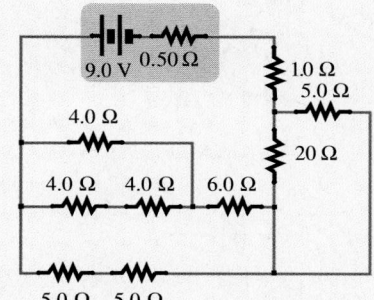

Figure P59

60. [II] THIS PROBLEM IS CONCERNED WITH A CIRCUIT CONSTRUCTED ON A PROTOBOARD. The circuit shown in Fig. P60 is mounted on a protoboard (go back and reread the caption for the photo on p. 643). There are four resistors color coded brown-red-black-silver, and one on the far left color coded brown-gray-black-silver . Power will be supplied through the red and green clips, and a voltmeter reading will be made using the black and yellow clips. We want to determine the source voltage when the voltmeter reads 6.0 V.   (a) What can you say about the way the resistors are wired?   (b) With no source attached, how much resistance will an ohmmeter measure across the red and green clips?   (c) How much current flows through the 12-Ω resistor when the voltmeter across it reads 6.0 V? (d) What is the voltage across the equivalent resistance?   (e) What is the voltage across the power supply?   (f) What is the voltage across either of the two vertical resistors at the right?

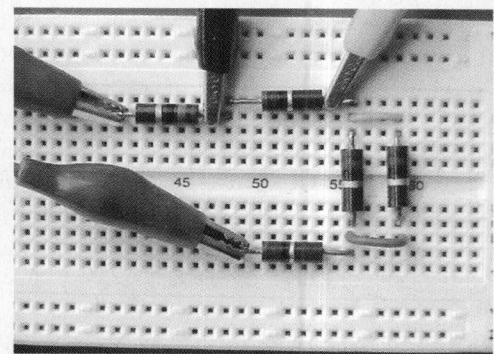

Figure P60

61. [III] THIS PROBLEM DEALS WITH RESISTORS ON A PRINTED CIRCUIT BOARD. All the resistors in the circuit shown in Fig. P61 are color coded brown-red-red-silver. We want to put in a current via one set of terminals and read out a voltage across another set of terminals. What is the resistance between terminals   (a) 1 and 2?   (b) 1 and 5?   (c) 3 and 4?   (d) 1 and 4?   (e) 6 and 7?   (f) 5 and 7?   (g) If 26 V is put across terminals-1 and 2 how much current will flow from 1 to 2?   (h) What voltage will then appear across 1 and 7? (i) across 1 and 5?

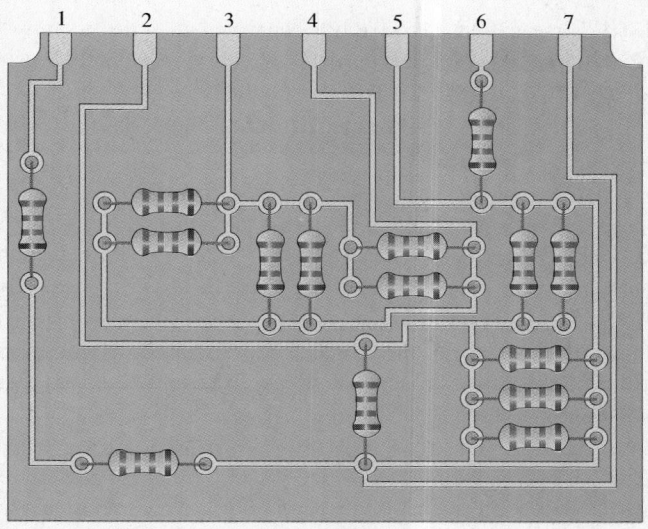

Figure P61

62. [III] Imagine a wire cube with identical resistors R in each arm (Fig. P62). What is the equivalent resistance between points A and B?

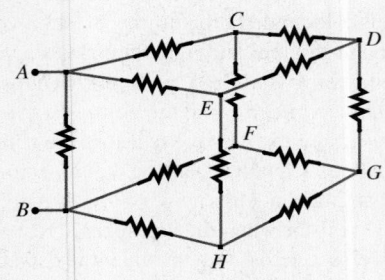

Figure P62

SECTION 18.3: AMMETERS AND VOLTMETERS

63. [I] In an ammeter the coil has a resistance of 100 Ω and the shunt resistance is 0.20 Ω. If 2.0 A pass through the shunt, how much current enters the coil?

64. [I] If the total resistance of a voltmeter is 6.0 kΩ and it contains a 50-Ω coil movement, describe the other circuit element in the meter.

65. [I] An ammeter movement consists of a 50.0-Ω coil. If 0.10% of the current entering the meter is to pass through the coil, how big must the shunt resistor be?

66. [II] Design a voltmeter using a galvanometer with a coil resistance of 100 Ω and a full-scale current of 1.00 mA that will measure 100 V full scale.

67. [II] Design a shunt such that a galvanometer with a coil resistance of 100 Ω and a full-scale current of 1.00 mA can be used as an ammeter to measure up to 1.00 A.

68. [II] The coil of a galvanometer has a resistance of 20 Ω, and it deflects full scale when a current of 0.50 mA passes through it. By shunting the coil with a 2.0-mΩ resistor, it becomes an ammeter. What full-scale current will it now read?

SECTION 18.4: *R-C* CIRCUITS

69. [I] Three circuit elements are connected in series to form a closed loop; a 100-μF capacitor, a 12-V battery, and a 50-Ω resistor. What is the time constant of the circuit?

70. [I] Show that the time constant of an *R-C* circuit has the correct units.

71. [I] An electronic flash fires a blast of energy from a 800-μF capacitor into a xenon lamp. It recharges through a series resistor of 5.0 kΩ. How long will it take to recharge 63% of its maximum charge?

72. [I] A charged 600-μF capacitor is in series with an open switch and a 4.0-kΩ resistor. If the switch is closed, how long will it take for the charge on the capacitor to decay down to 37% of its original value?

73. [I] An uncharged 1.0-μF capacitor is in series, through a switch, with a 2.0-MΩ resistor and a 12.0-V battery (with negligible internal resistance). The switch is closed at $t = 0$ and a current I_i immediately appears. Determine I_i. How long will it take for the current

in the circuit to drop to $0.37I_i$? [*Hint: What is the value of the current after one time constant?*]

74. [I] A resistor is placed in series with an uncharged 2.0-μF capacitor, and a 12.0-V battery is put across the two. If the current that immediately flows around the circuit is measured to be 12 μA, determine the resistance. What is the time constant of the circuit?

75. [I] A 10.0-V dc power supply is wired across the series combination of an uncharged 0.20-μF capacitor and a 1.0-MΩ resistor. What are the initial current in the circuit and the current after one time constant has passed?

76. [I] A 3.0-μF capacitor is put across a 12-V ideal battery. After an hour, it is disconnected and put in series, through a switch, with a 200-Ω resistor. (a) What is the initial charge on the capacitor? (b) What is the initial current when the switch is closed?

77. [II] A 6.0-μF capacitor is charged up to 12 V and subsequently connected through a switch to a 100-Ω resistor. At $t = 0$, the switch is closed. What is the initial current through the circuit? Draw a rough plot of current versus time. How long does it take for the current to drop to 37% of its initial value?

78. [II] In Problem 77, what is the charge on the capacitor 6.0 ms after the switch is closed?

79. [II] A 12-V battery with negligible internal resistance is placed across a series combination of a 1.0-MΩ resistor and a 12.0-μF capacitor for 10 hours. The battery is then removed and the circuit closed at $t = 0$. What is the current through the resistor at $t = 24$ s?

80. [II] Three circuit elements are connected in series to form a closed loop: an uncharged 20.0-μF capacitor, a 100-V dc power supply, and a 10.0-MΩ resistor. What is the maximum charge that will be stored in the capacitor? What is the initial charge on the capacitor and current through the resistor? Determine the current in the circuit and the charge on the capacitor after an interval of one time constant.

81. [II] An uncharged 10.0-μF capacitor, a 80.0-V dc power supply, and a 20.0-MΩ resistor are connected in series to form a closed loop. Determine the current in the circuit 400 s after the circuit is closed.

82. [III] An uncharged 10.0-μF capacitor, a 120.0-V dc power supply, and a 40.0-MΩ resistor are connected in series to form a closed loop. Using Fig. 18.20a write an equation for the charge on the capacitor as a function of time. Determine that charge 4.00 s after the circuit is closed.

SECTION 18.5: KIRCHHOFF'S RULES

83. [I] Use Kirchhoff's Loop Rule to solve for the currents in branches A–D–C and A–B–C of the circuit in Fig. P83. Then use the Node Rule to find the current in branch A–C.

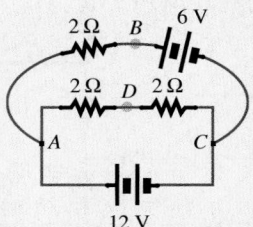

Figure P83

84. [I] Find the current through each element of the circuit in Fig. P72.

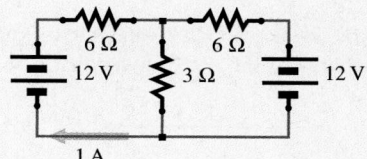

Figure P84

85. [I] Apply Kirchhoff's Rules to the circuit of Fig. P73, and solve for the three branch currents. Next, simplify the network, determine all the currents, and check your answers.

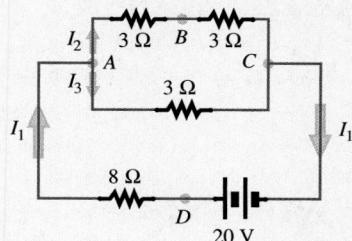

Figure P85

86. [I] Solve for the unknown source voltage and the power delivered by the 12-V battery in Fig. P86.

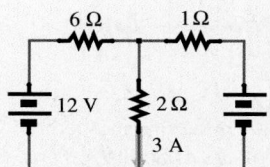

Figure P86

87. [I] The transistor circuit of Fig. P87 is to be checked for proper operation. According to specifications, the collector voltage (point C) should be at a constant +6 V with respect to ground. If that's the case, what should be the voltage measured at point D? [*Hint: Read Discussion Question 4 to learn that a capacitor is an open circuit to dc.*]

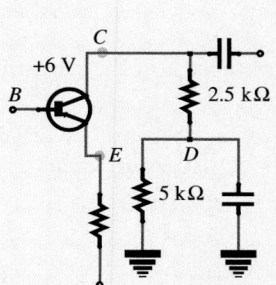

Figure P87

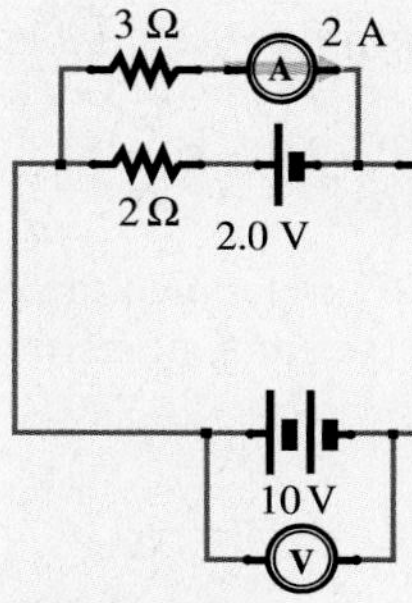

Figure P88

88. [I] If the ammeter and voltmeter in Fig. P88 read 2.0 A and 10 V, respectively, what current passes through the 1.0-Ω resistor?

89. [I] The ammeter in Fig. P89 reads 2.0 A; what will the voltmeter read?

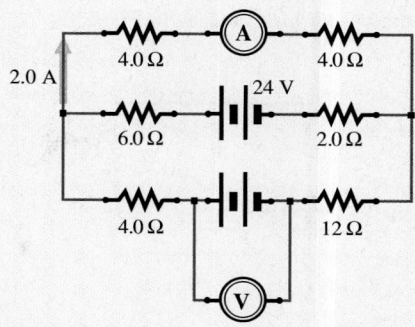

Figure P89

90. [I] Referring to the bridge circuit of Fig. P90, if $I = 6$ A, $I_2 = 4$ A, and $I_3 = 0$, find I_1, I_4, and I_5.

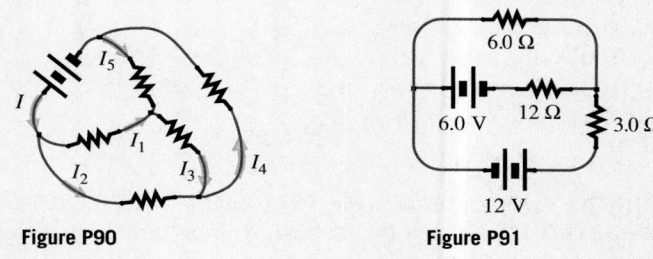

Figure P90   **Figure P91**

91. [I] Determine the current in each branch of the circuit of Fig. P91.

92. [I] Find the current in each resistor of the circuit in Fig. P92 using Kirchhoff's Rules. Then simplify the circuit and compare your results.

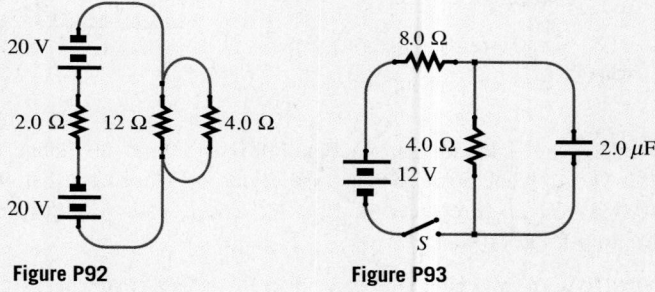

Figure P92   **Figure P93**

93. [I] The switch in the circuit of Fig. P93 is closed, and a steady state is established. What is the charge on the capacitor?

94. [I] Figure P94 shows a 200-V dc generator (the pictorial symbol used is another fairly common one) supplying 100 A to a load via a two-lead cable having a resistance of 0.20 Ω per length of conductor. What is the voltage across the load?

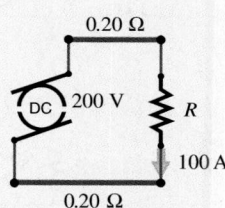

Figure P94

95. [II] Find the values of R_1 and $\mathscr{E}_1$ in Fig. P95.

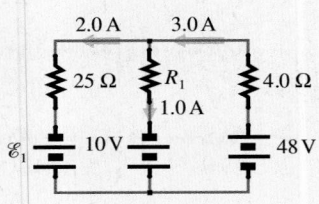

Figure P95

96. [II] The variable resistor in Fig. P96 is adjusted to 20 Ω, whereupon the ammeter (which has a negligible internal resistance) reads zero. Use Kirchhoff's Rules to determine the power provided by the sources. What is the voltage at points B, C, and D?

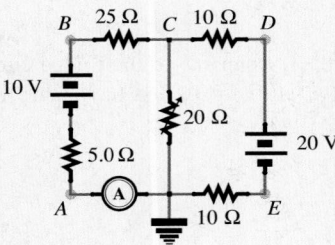

Figure P96

97. [II] The variable resistor in Fig. P97 is adjusted until the ammeters read (#1) 120 mA and (#2) 80 mA, with the directions of the currents as shown. Find the value of R.

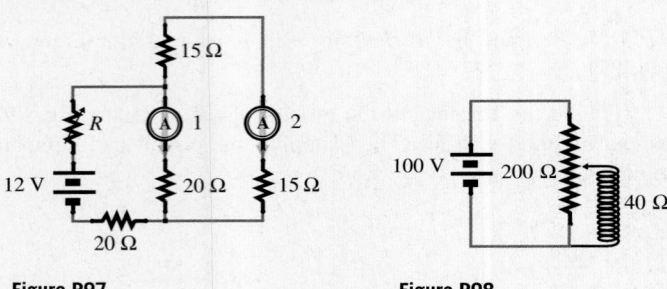

Figure P97 **Figure P98**

98. [II] The potentiometer in Fig. P98 has a total resistance of 200 Ω. At what position must the slider be placed so that the 40.0-Ω coil of wire receives 1.00 A? Check your results using Kirchhoff's Loop Rule.

99. [II] With only switch S_1 closed in Fig. P99, (a) what is the steady-state reading of the voltmeter? (b) what is the charge on the 3.0-μF capacitor?

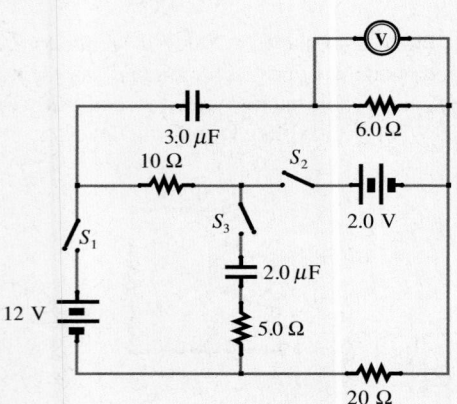

Figure P99

100. [II] With Problem 99 in mind, how much power does the 12-V battery supply in the steady state a few minutes after all the switches are closed? What's the charge on the 2.0-μF capacitor?

101. [II] Find the values of R, V, and all the unknown branch currents in the network of Fig. P101, given that $I_3 = 1.0$ A.

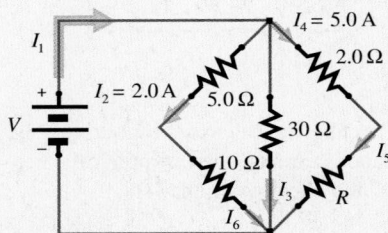

Figure P101

102. [II] Solve for the currents in each branch of the circuit in Fig. P102.

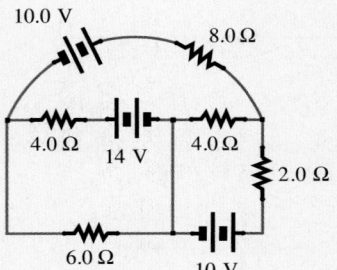

Figure P102

103. [II] Given that 5.0 A passes along the branch from C to B in Fig. P103, what is the voltage of points A, D, E, F, and G?

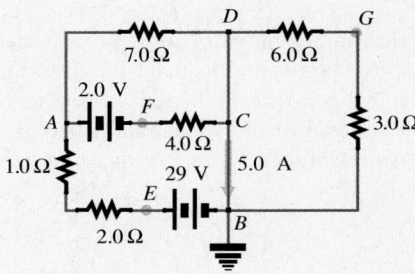

Figure P103

104. [III] Solve for the currents in each branch of the circuit in Fig. P104.

105. [III] Determine the equivalent resistance between the terminals of the group of resistors shown in Fig. P105 using Kirchhoff's Rules.

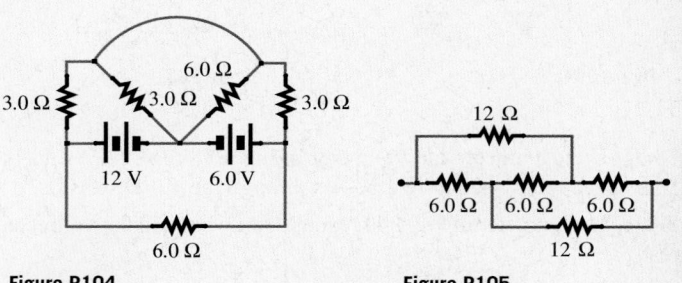

Figure P104 **Figure P105**

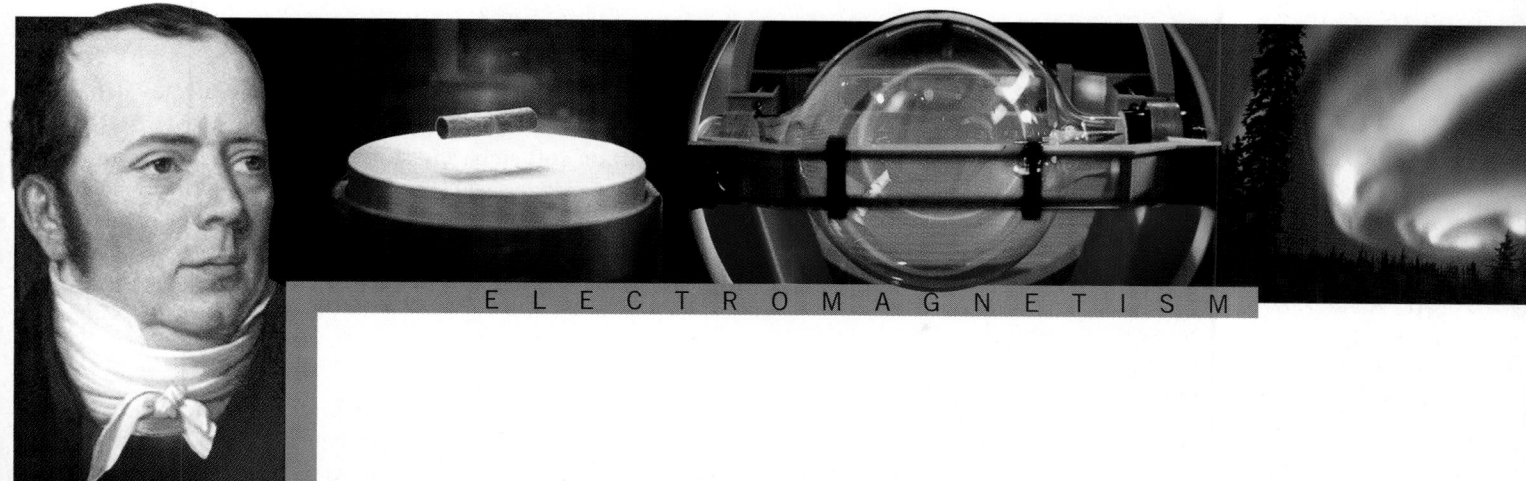

Chapter 19
Magnetism

*O*ur understanding of magnetism has evolved into a sophisticated picture of whirling electrons, of fields and currents. Yet the modern era of magnetism began less than two hundred years ago, and it's not surprising that the theory remains incomplete—for example, we still can't determine theoretically the magnetic characteristics of the proton or the neutron.

The strange power of natural magnets (originally called **lodestones**) to cling to iron tools was discovered in ancient times. The lodestone is an oxide of iron (Fe_3O_4) known as magnetite. That rich iron ore, some of it permanently magnetized, occurs in many parts of the world. Indeed, the word *magnet* comes from the Greek (*magnes*), which probably derives from the ancient colony of Magnesia, where the ore was mined 2500 years ago.

After 1820 there was a surge of activity and innovation that followed the discovery by Oersted (p. 677) that electric currents give rise to magnetic force. For centuries, electricity and magnetism had been taken as two distinct powers, and now they were connected. **Charges generate electric fields. Charges in motion, in addition, generate magnetic fields.** The two fields are different manifestations of a single phenomenon—*electromagnetism*.

STUDY GUIDE

Magnetism is one of those phenomena that we're all familiar with on some level. Everyone knows what a compass is, more or less, and who among us has never seen a refrigerator magnet (perhaps even sporting a plastic pineapple)? Be that as it may, magnetism is so basic that it appears everywhere; it's associated with every form of matter, including light. There is no place in the Universe that is free of magnetism, and that's true from the subatomic domain to the vast reaches of intergalactic space.

We'll first study magnetism associated with materials, so-called permanent magnets (p. 667), and with that learn the vocabulary of the discourse. Once we have some idea of what a magnetic field is, we'll explore how macroscopic currents produce such fields (p. 677). Finally, we'll study how charged particles are affected by magnetism. Every time you turn on a TV set with a picture tube you're shooting a beam of electrons through a magnetic field.

Magnets and the Magnetic Field

The earliest scholar to study the lodestone was probably Thales (ca. 590 B.C.E.). Almost 150 years later, the philosopher Socrates dangled soft iron rings clinging to one another beneath a lodestone. That's the same game most of us have played with a magnet and a box of paper clips. Just as a charged comb can induce charge on scraps of paper, a magnet can magnetize nearby pieces of iron.

19.1 Permanent Magnets

There are many variations on what passes for a chunk of iron, and they all behave differently magnetically. Cast iron is a hard, brittle material rich in carbon (from 2% to about 7%), whereas pure iron is rather soft. Between these two extremes is the carbon-iron alloy *steel*. Soft iron retains its magnetized condition only as long as the inducer (the lodestone)

THE ANCIENT COMPASS

Chinese legend has it that Emperor Hwang-ti (ca. 2600 B.C.E.) was guided in battle through dense fog by a small pivoting figure that always pointed south, a lodestone embedded in its outstretched arm. By about 1100 A.D., the magnetic compass had come to the West, and the lodestone's ability to align itself with the north-south axis of the "Universe" was awesome. By the beginning of the second century A.D., the Chinese were versed in the art of making permanent magnets. A manuscript from the period suggests stroking an iron rod or needle from end to end along a lodestone, repeatedly, and *always in the same direction*.

Figure 19.1 Several small steel needles attracted to the surface of a spherical magnet. Only at the poles do the needles stand straight up. A compass pointer pivoted so that it can swing vertically shows the same behavior in the Earth's magnetic field.

On the Lodestone. The stone not only attracts iron rings but also imparts to them a similar power whereby they attract other rings.

PLATO recounting the words of SOCRATES
(CA. 400 B.C.E.)

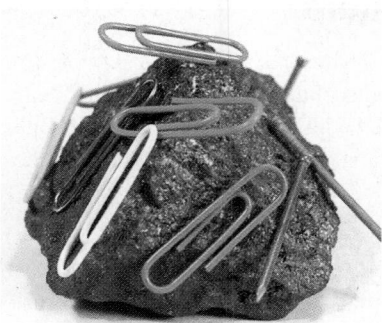

A lodestone with a few steel nails and paper clips clinging to it. It is a chunk of magnetite, a common iron ore, magnetized by the Earth's magnetic field.

is kept nearby. When removed, the specimen quickly demagnetizes, just as a charged piece of paper quickly depolarizes. Low-carbon soft steel, the stuff of paper clips and common nails, demagnetizes a bit more gradually. In contrast, a piece of hard steel, once magnetized, retains much of its power, and we speak of it as a **permanent magnet**, although that's an exaggeration.

Poles

On a summer's day in 1269 a French engineer, Peter de Maricourt—alias Peter Peregrine, Peter the Pilgrim—sat down to write a long letter to a friend. (Peter was whiling away time in the trenches with an army that was laying siege to a city in Italy.) In the letter, he described his researches on magnetism: Military engineers were concerned about the compass, often being required to construct long tunnels leading under fortress walls.

Peregrine was the first to introduce the concept of the **magnetic pole**. He had several lodestones ground into the shape of a sphere to resemble the Earth. A steel needle placed anywhere on the surface of one of these spherical magnets aligned itself in a particular way (Fig. 19.1). By drawing lines on the stone in the directions assumed by the needle, he determined that they all crossed at two opposing points, just as "all the meridian circles of the Earth meet in the two opposite poles of the world." If a piece of a needle were placed in contact with the lodestone, it would stand straight upright at, and only at, the poles.

Most magnets have two poles, where the force is clearly strongest. A straight bar magnet is the simplest two-pole configuration (known as a **dipole**), and Peregrine made iron bar magnets as well. Beyond that, it is possible for a magnet to have any number of poles, odd or even, provided it's two or more. Some modern flexible magnets (the narrow brown plastic ones) are made in long strips with hundreds of poles.

Peregrine next put a spherical magnet in a wooden bowl and set it afloat in a large vessel of water. As soon as it was released, it spun around, bowl and all, the *north-seeking* or **north pole** always pointing northward, the *south-seeking* or **south pole** always pointing southward. Of course, this was a compass, not all that different from the crude floating-needle instruments already in widespread use. Holding another magnet, whose poles were determined and marked, he approached the floating stone. When the north pole of one was brought near the south pole of the other, the little boat lunged toward the hand-held stone; but when either two north poles or two south poles were positioned near each other, the boat was pushed away. Peregrine had discovered the basic mutual interaction of all magnets: **like magnetic poles repel and unlike magnetic poles attract** (Fig. 19.2).

Naturally, Peregrine tried to isolate a single **monopole**, a piece of magnet that was simply and only north polar or south polar. And what more obvious way to do that than to split a magnet in two (Fig. 19.3)? Surprise! No matter how we break a magnet, the fragments are always bipolar—*the monopole cannot be isolated*. It's as if a magnet were composed of a

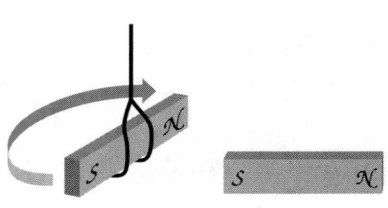
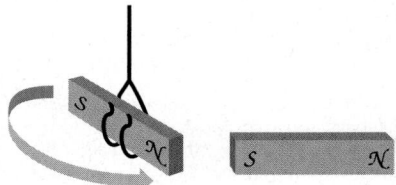

Figure 19.2 The attraction of unlike poles and the repulsion of like poles can be observed easily by suspending one bar magnet on a thread. The suspended magnet will swing away from the like pole of a second, stationary magnet and swing toward the stationary magnet's unlike pole.

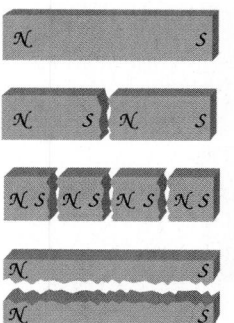

Figure 19.3 The fragments of a bar magnet always have two poles.

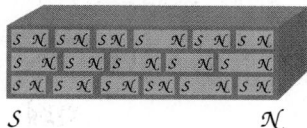 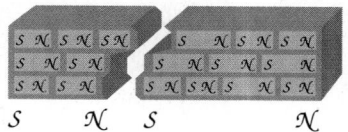

Figure 19.4 A magnet behaves as if it were composed of tiny bipolar units, tiny bar magnets, or dipoles.

succession of microscopic bar magnets with opposite poles touching and neutralizing each other everywhere but at the ends. When the magnet is broken, the appropriate poles appear (Fig. 19.4). That behavior will be understood only after we learn that **the electron itself is the fundamental dipole magnet**. {For more on the hypothetical monopole particle and the Big Bang click on **MODERN MONOPOLES** under **FURTHER DISCUSSIONS** on the **CD**.}

A magnetic top (south pole down) floating above a hollow square magnet (south pole up) hidden in the black base. Because it's spinning, it's gyroscopically stable and will hover in midair for a remarkably long time.

WILLIAM GILBERT & THE MAGNETIC FIELD

In 1600 William Gilbert (p. 147), then physician to Queen Elizabeth I, compared the Earth to a large spherical lodestone, maintaining that the compass needle was drawn to the planet's magnetic pole, not to the heavens as everyone else had thought. It was simply a matter of one magnet pulling on another. Dr. Gilbert probed the region surrounding a magnet with a little compass and concluded that "Rays of magnetick virtue spread out in every direction in an orbe." That statement seems almost identical in spirit to the nineteenth-century vision of Faraday's *lines-of-force* of the magnetic field. One need only connect the little compass arrows with smooth arcs to transform the imagery from an "orbe of virtue" into a "magnetic field." Less than half a century after Gilbert, Descartes carried the mapping process one step further. He sprinkled iron filings around a magnet, and they aligned themselves like minute compass needles to form curved continuous filaments, which even more potently *suggest* lines-of-force, lines of *magnetic field* (Fig. 19.5).

I have seen Samothracian iron rings even leap up and at the same time iron filings move in a frenzy inside brass bowls, when this Magnesian mineral was placed beneath.

LUCRETIUS
(*FIRST CENTURY B.C.E.*)

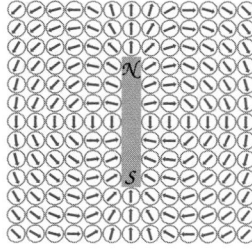 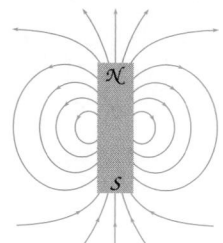 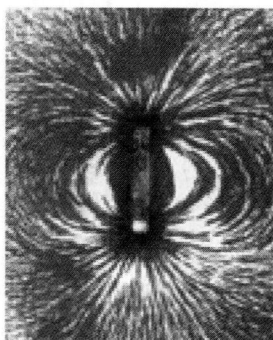

Figure 19.5 The force field around a bar magnet, as revealed by an array of small compasses. This drawing shows what is happening in only one plane. The field is three-dimensional. The photo shows iron filings lining up in the vicinity of a small bar magnet. By tradition, we say the field lines emerge from the north pole, curve around, and enter the south pole. But by just looking at this photo, we can't tell which pole is which.

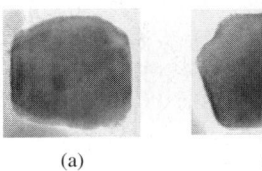

(a)   (b)

(a) A magnetite crystal found in a bacterium. (b) Similar crystals, this one about a millionth of an inch long, have been found in the human brain.

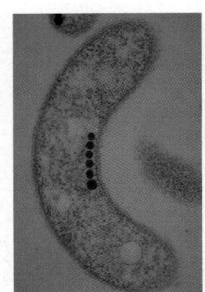

A magnetotactic bacterium. The dark line of dots is a chain of magnetite crystals that functions as a compass needle.

19.2 The Magnetic Field

Again we are faced with the mystery of action-at-a-distance, and again our answer is to assume that every magnet establishes, in the space surrounding it, a **magnetic field** (B). As before, we say a field exists in a region of space when an appropriate object placed at any point therein experiences a force. The fields span the space and communicate the interaction—the fields mediate the Third Law's action-reaction. Remember that we probed the electric field with a test-charge; now we map the magnetic field, not with a monopole, but with the next best thing, a dipole, a tiny test-compass.

A compass needle, able to turn freely, placed near the south pole of a bar magnet simultaneously has its own south pole repelled and its north pole attracted (Fig. 19.6a). It experiences a net torque and twists around into a new equilibrium orientation such that the torque vanishes (Fig. 19.6d). That's why a compass needle aligns itself with the local field. The strength of a magnetic field (B) at every point in space can be determined from the torque tending to realign the test-compass. The SI unit for B is the *tesla* (after Nikola Tesla), abbreviated T (Table 19.1). The compass is a tiny bar magnet that will settle tangent to the field, and *we arbitrarily take the arrow from its south pole to its north pole as the direction of the field in which it is immersed*.

Once again the concentration of field lines, the number per unit cross-sectional area, is proportional to the strength of the field. The field lines in Fig. 19.7 are more densely

> The **direction of a magnetic field** is the direction in which a tiny compass needle will point at that location.

> The **unit of magnetic field** is the tesla (T). The Earth's field is around 0.5×10^{-4} T. The most powerful permanent magnets (samarium-cobalt or neodymium-iron-boron) have fields of ≈ 0.3 T to ≈ 0.4 T.

> The magnetic field of a **compass needle** runs along its length from its south to its north pole; that's the direction of the pointer. When placed in an external B-field, the compass needle swings so that its field is aligned with the external field.

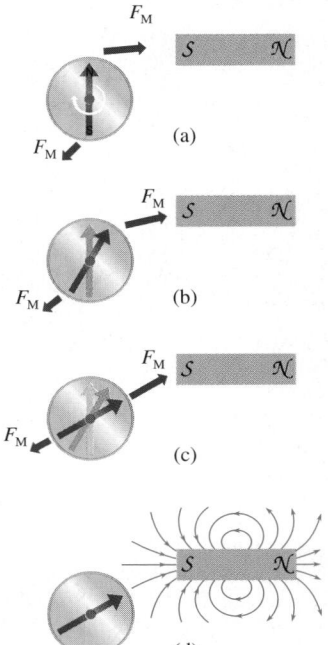

Figure 19.6 A compass needle mounted so that it can turn freely experiences a torque in the B-field of a magnet and swings around until that torque vanishes.

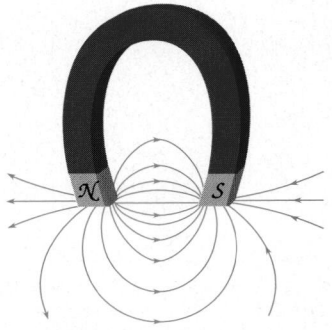

Figure 19.7 The horseshoe magnet concentrates its field in the immediate region between its poles.

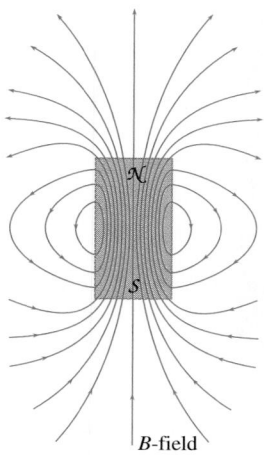

(a)   (b)

If we suppose that each iron filing is a compass needle, then the pattern they form reveals the magnet's *B*-field lines. In (a), the two magnets have like poles facing each other. In (b), unlike poles face each other. Of course, these views are in only one plane; the field is three-dimensional.

Figure 19.8 The magnetic field inside and outside of a permanent magnet. Electric field lines begin on positive charges and end on negative charges. Here there are no magnetic charges (monopoles), so magnetic field lines are closed. They must turn back on themselves. The field in three dimensions is like a system of concentric misshapen doughnuts.

concentrated near the poles, where *B* is largest. By convention, **the field extends out from the north pole and into the south pole**. The north end of the test-compass is attracted to the south pole of the magnet responsible for *B* (and the south end is attracted to its north pole). As usual, *field lines never intersect, nor should they begin or end on anything but sources and sinks:* if there were monopoles clustered at both ends of a bar magnet, the field lines would begin and end there. But there are not, and **the lines of *B* form closed loops** (Fig. 19.8).

A magnet and a box of paper clips. Each clip is magnetized, becoming a small temporary magnet. The clips pack in densely where the field is strong. They form bridges that arch from pole to pole, crudely suggesting a pattern of field lines.

Table 19.1
Magnetic Fields

Source	Field (T)
Nucleus (at surface)	10^{12}
Neutron star (at surface)	$\approx 10^8$
Highest yet attained in laboratory	
explosive compression ($\approx 10^{-6}$ s)	1.5×10^3
pulsed coils ($\approx 10^{-3}$ s)	100
constant (dc, superconducting, 1993, MIT)	37.2
constant (dc, room temperature)	23.5
No acute effects on bacteria, mice, or fruit flies	14
Large laboratory electromagnet	5
Sunspot (within)	0.3
Human exposure limit	
(full body, dc, for minutes)	≈ 0.2
Small ceramic magnet (nearby)	≈ 0.02
Small bar magnet (near pole)	10^{-2}
Sun (at surface)	10^{-2}
Jupiter (at poles)	8×10^{-4}
Hair dryer (60 Hz, nearby)	$1 \times 10^{-7} - 0.7 \times 10^{-4}$
Can opener (60 Hz, nearby)	$0.5 \times 10^{-4} - 1.5 \times 10^{-4}$
Earth (dc, at surface)	0.5×10^{-4}
Transmission line	
(maximum under, 765 kV, 4 kA)	$\approx 0.5 \times 10^{-4}$
Blender (60 Hz, nearby)	$0.3 \times 10^{-5} - 1.0 \times 10^{-5}$
Sunlight (rms)	3×10^{-6}
Refrigerator (60 Hz, nearby)	$\approx 2 \times 10^{-6}$
Color TV (60 Hz, nearby)	$\approx 1 \times 10^{-6}$
Toaster (60 Hz, nearby)	$5 \times 10^{-7} - 2 \times 10^{-6}$
Mercury (at surface of planet)	2×10^{-7}
AM radio wave	10^{-9}
Human body (produced by)	$\approx 3 \times 10^{-10}$
Interstellar space	$\approx 10^{-10}$
Earth (50−60 Hz, at surface)	10^{-12}
Shielded region (smallest value measured)	1.6×10^{-14}

William Gilbert, English physician and
physicist (1544–1603).

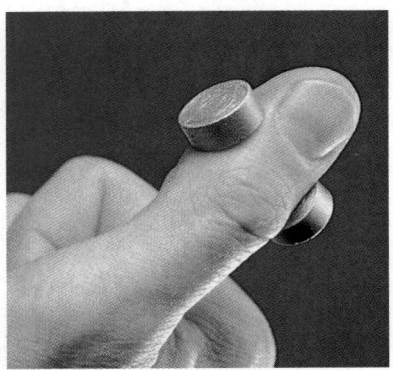

These two extremely powerful magnets are made of a new (1985) alloy of 14 parts
iron, 2 parts neodymium, and 1 part boron. The alloy is usually powdered, the
grains aligned in a strong *B*-field, and then the material is heated until it fuses.

The Earth's Magnetism

William Gilbert was right, four hundred years ago, when he suggested that the Earth
behaves as if it contained, as its main field source, a dipole. It acts as if a relatively short bar
magnet were embedded at its center (Fig. 19.9), tilted over at about 11.5° along the 70° W
meridian. But matters are much more complicated than that.

The core of the planet is too hot to be totally solid and even
too hot (≈ 2500 K) for magnetic materials to remain magnetic—we are not dealing with a buried bar magnet. Besides that,
the field changes in time, both in size and direction. There is
also compelling evidence that the field has reversed itself some
300 times in the past 170 million years and may have done so
just 30 000 years ago.

As a further complication, the *dip poles*, where the field
points straight up and down, not only wander, but are 500
miles from the *geomagnetic poles* (the poles of Gilbert's magnet toward which a compass points). Moreover, neither the
geomagnetic nor the dip poles are at the geographic poles, as fixed by the planet's spin
axis. The final irony is linguistic: the Earth's northern magnetic pole is actually a *south*
pole, and the southern magnetic pole is—you guessed it—a *north* pole. That's why the
north-seeking pole of a compass needle points north—it's attracted toward the Earth's
south magnetic pole.

THE COMPASS DOESN'T ALWAYS POINT NORTH

The Chinese of the eleventh century were already aware that the compass needle does not align itself everywhere along a true north-south
direction. That knowledge came to Europe at least as early as 1436, but
it's still a surprise to some to learn that a magnetic compass rarely points
true north. While a compass in Chicago does align itself almost due
north, one in Los Angeles tilts about 15° to the east, and one in New
York leans almost 15° west of north.

Figure 19.9 (a) The Earth's magnetic influence resembles that of a tilted bar magnet. A
compass needle aligns itself with the field and
points roughly toward the north geographic
pole, which is not far from the Earth's south
magnetic pole. The field extends thousands of
kilometers out into space and is rotationally
symmetrical around the axis of the hypothetical
bar magnet. (b) The declination of a compass
from true north. The red lines show the directions in which a compass needle will align. In
New York City a compass points about 10°
west of north.

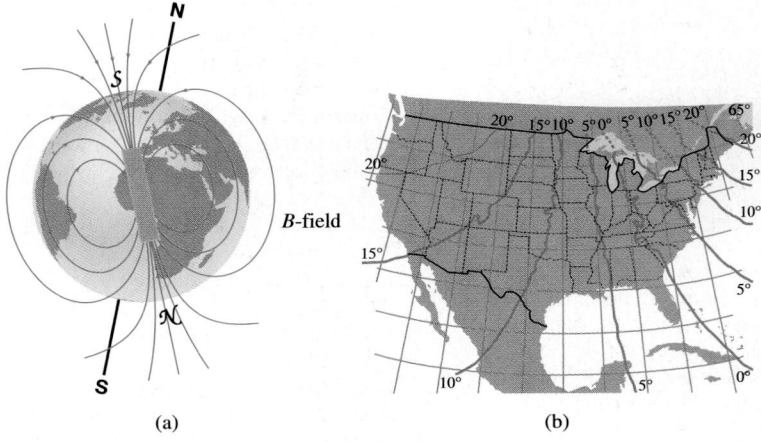

(a)   (b)

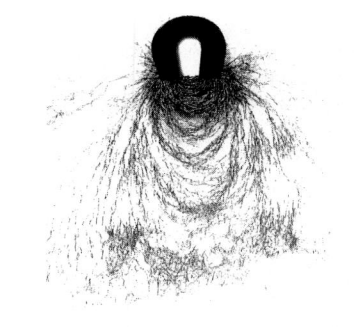

Magnetism on an Atomic Level

As we shall see, charge in motion produces magnetic force, and in particular a current moving in a circular path is magnetically identical to a dipole (p. 681). Moreover, electrons behave in ways that suggest they are perpetually spinning.* It follows that ***the electron itself corresponds to a circulating charge and is the ultimate subatomic dipole magnet*** (Fig. 19.10). The observed magnetic response of electrons is consistent with each having a purely dipole field down to a radius of 10^{-12} m. The orbitlike motion of an electron about the nucleus of an atom also constitutes a current and produces an additional magnetic field. Atomic nuclei can generate dipole fields too, but these are typically a thousand times weaker. Together, the two electron mechanisms (spin and orbit) account for the magnetic behavior of all the various forms of bulk matter.

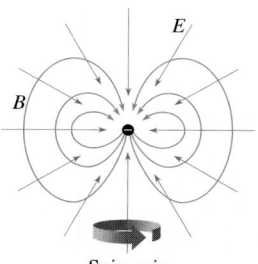

Figure 19.10   The electric and magnetic fields of an electron. The *E*-field is radially inward. The *B*-field forms a system of concentric doughnut shapes about the spin axis. The magnetic field is that of a minute dipole.

Usually, electrons in atoms come in oppositely moving pairs and their magnetic fields cancel—the vast majority of atoms and molecules have no net magnetic field. Still, no material exists that will not respond magnetically to an applied *B*-field. Most substances (water, glass, copper, lead, salt, rubber, diamond, wood, and so on), the majority of gases (i.e., nitrogen, carbon dioxide, and hydrogen), and millions of organic compounds (e.g., plastics) are very weakly magnetic in what might at first seem a startling way. They are *repelled* by the pole of a strong magnet. Faraday was the first to observe this peculiar phenomenon (1845), and he called it **diamagnetism**. It's still amusing to swing a small piece of glass dangling on a thread into a field almost 100 000 times stronger than the Earth's,

only to have the glass pop right out and stay out at some gravity-defying angle. Humans are mostly water and should be able to feel this repulsion. But that requires a tremendously strong *B*-field (you feel nothing even at 4 T, which is quite large), and very few people have actually experienced the repulsion.

Diamagnetism is associated with the orbital motions of atomic electrons. Turning on a *B*-field changes their angular momenta, and that added motion produces a field that opposes the applied field (p. 714). Accordingly, diamagnetism is present in all substances, although it is observable only when not swamped out by other much stronger effects.

When an atom has an odd number of electrons or a structure in which not all its electrons are paired, the atom will have a net magnetic dipole field of its own. In a substance made up of countless such atoms, these dipoles *en masse* produce only a feeble response

---

*Although physicists commonly talk about spinning subatomic particles in the same way they talk about spinning basketballs, in their heart of hearts they know better. Quantum Mechanics has made it clear that the old notion of spin needs a modern interpretation. By spin, we mean a fundamental quality (like mass and charge, whatever they are) that is associated with the manifestation of angular momentum, but not necessarily with the existence of spinning, with turning round and around. The latter picture produces relativistic inconsistencies and must be rejected. Thus, an electron has spin, though it is not spinning in the usual sense of the word. Whatever it's doing, we are confident that an electron has an intrinsic angular momentum and an intrinsic magnetic dipole field.

The little floating cylinder is a powerful neodymium-iron-boron magnet. There's a large electromagnet above it, pulling it up. The fingers—composed of diamagnetic water, proteins, and organic molecules—provide a weak repulsive force that nonetheless stabilizes the little permanent magnet so that it neither falls nor is drawn upward.

A live frog levitating in the hollow core of a superconducting electromagnet. The blend of water, proteins, and organic molecules that constitute the frog are diamagnetic and are therefore pushed out of the high-strength region of an inhomogeneous *B*-field.

because the ordinary thermal agitation of the atoms keeps them disoriented. Substances of this kind are called **paramagnetic** and they include the elements aluminum, oxygen, sodium, platinum, and uranium, among others. If placed near a powerful magnet, they will be drawn in toward a pole, but only weakly.

The last of the three major magnetic classes of substances is known as **ferromagnetic**. This group, to which newly concocted materials are always being added, includes magnetite, a troop of alloys such as steel and Alnico, and a number of elements. Iron, cobalt, and nickel have been known for centuries to be "magnetic" at room temperature, and there are a half-dozen other elements that become ferromagnetic at low temperatures. On average, ferromagnetic materials have a greater number of unpaired, spin dipoles per atom. But more importantly, *these dipoles enter into large-scale cooperative alignments*. In other words, the uncompensated spin dipoles of each atom interact strongly with the dipoles of adjacent atoms, locking together in a parallel orientation that tends to persist even at room temperature.* Such substances are strongly attracted to the poles of a magnet and are themselves easily magnetized.

Magnetic Domains

Ferromagnetic substances are composed of very many microscopic **domains**—islands of order—throughout each of which tremendous numbers of atomic spin dipoles are aligned parallel to one another. Each domain is a tiny ($\approx 5 \times 10^{-5}$ m across) magnet that can be viewed under a microscope. In an unmagnetized specimen, the orientations of these domains are random and their fields cancel (Fig. 19.11*a*). That configuration is a compromise between total order of the spin dipoles and total disorder and exists because it is energetically the most economical.

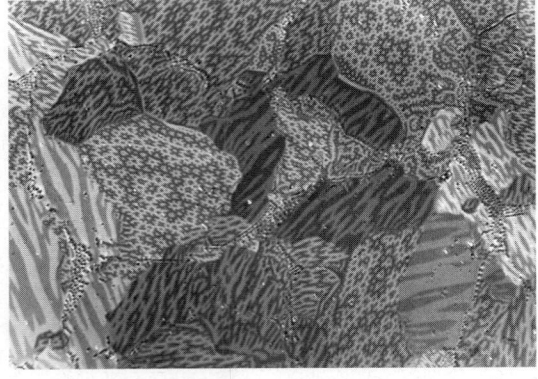

The magnetic-domain structure of a polycrystalline sample of cobalt samarium. Each little domain has its own random magnetic field orientation. The sample as a whole is said to be unmagnetized.

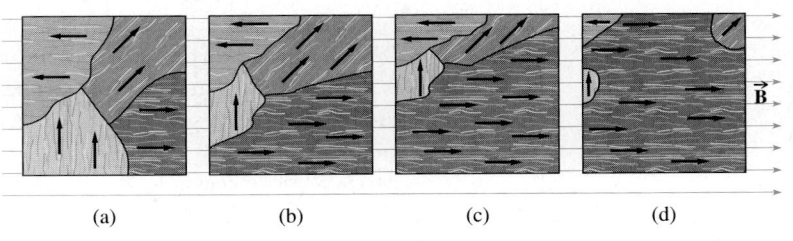

(a)   (b)   (c)   (d)

Figure 19.11 The growth of domains aligned in the direction of an applied magnetic field. As the applied *B*-field increases, the pattern goes from (a) to (b) to (c) to (d). Regions parallel to the field grow at the expense of those not parallel.

*That powerful coupling is a quantum-mechanical effect, first explained by Werner Heisenberg in 1928.

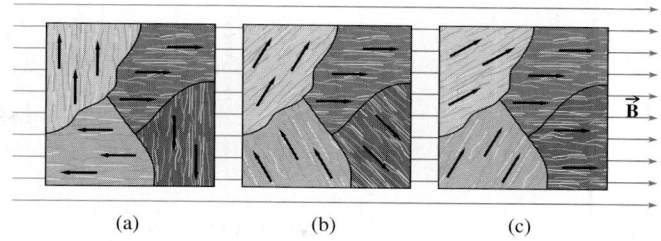

Figure 19.13 The reorientation of domains, bringing about an alignment with the applied field. As the applied magnetic field increases from (a) to (b) to (c), the electrons in the domains rotate so that their fields align with the applied field. The domains do not change shape as they did in Fig. 19.11. This realignment mechanism occurs in materials that become permanently magnetized.

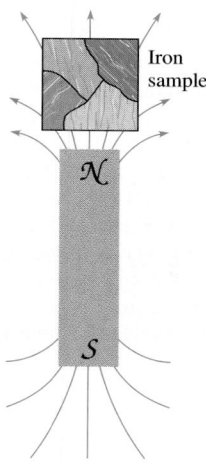

Figure 19.12 An iron sample placed in the magnetic field of a bar magnet. The domains of the sample are influenced, and they align with the applied field. The sample becomes magnetized and is drawn toward the bar magnet.

When a ferromagnetic sample is brought into a magnetic field, its domain structure can be drastically altered in two basic ways. The low-energy process (which occurs even with fairly weak fields) is one in which *the many domains that happen to be already aligned with the applied field grow at the expense of the domains that are misaligned at the start and subsequently shrink*. This is what occurs when soft iron is placed in a field and becomes magnetized (Fig. 19.12). This state with aligned domains and a resulting induced magnetic field (in which there is a good deal of energy) is unstable. Without support from outside, the induced field collapses and the iron spontaneously demagnetizes, rearranging domains.

The other magnetizing process, which requires a higher applied *B*-field, results in the more or less *irreversible reorientation of the domains* (Fig. 19.13). All the electron dipoles coupled together within each domain can literally be rotated into alignment with the applied field, just like a compass needle. This mechanism prevails in substances that become permanently magnetized. Domains in materials with irregular internal structures, such as steel, cannot easily change shape. The domains are pulled around instead, and having once been forced to do so against a kind of internal friction, they tend to stay put. When a steel knife blade (some stainless steel is "nonmagnetic") is stroked with a strong magnet, the effect is to rotate the domains into that direction, to "comb" them into alignment.

Permanent magnets and ferromagnetic materials are used in VCRs, TVs, stereo headsets, automobiles, speakers, tape decks, motors, and telephones. They float in space in a thousand different satellites, are on the backs of millions of credit cards, and in the inks on countless dollar bills and personal checks. (Try holding a dollar bill near a powerful magnet.)

Anything that reorients the domains of a piece of magnetized steel diminishes its supposedly permanent field. Banging on a magnet with a hammer will do just that, by knocking regions out of alignment—providing the domains with enough energy to unpin

An image of George Washington produced by scanning a one-dollar bill using an extremely sensitive magnetic detector. The superconducting quantum interference device (SQUID) detects minute magnetic field variations caused by particles in the ink on the bill. (See the photo on p. 678.)

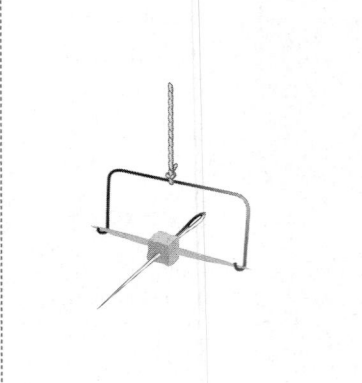

Having made many and divers compasses ... I found continuallie that after I had touched the yrons with the stone, that presentlie the north point thereof wouilde bend or decline downwards under the horizon in some quantitie.

ROBERT NORMAN
The Newe Attractive (1581)

themselves. If the magnet's temperature is made high enough, the vibrating atoms will jostle themselves out of alignment and disrupt the long-range order. Pierre Curie realized that there is a limiting temperature, now called the **Curie temperature**, for each substance beyond which ferromagnetism vanishes and the material becomes paramagnetic. In 1894, Curie found that this critical temperature for iron is 770 °C, though it had been known for a long time that red-hot iron has no magnetic power. The Curie temperature for magnetite is roughly 575 °C; for the other important iron ore, hematite, it's about 675 °C. Interestingly, in all three cases, the Curie temperature is well below the melting point.

Permeability

The presence of a dielectric in an applied electric field has the effect of producing a weaker net internal *E*-field (p. 588). Similarly, when a diamagnetic material is placed in an applied magnetic field, it decreases the net *B*-field within the medium, but the effect is very small. On the other hand, the presence of a paramagnetic material will very slightly enhance the field within the medium. By contrast, upon being immersed in a *B*-field, a ferromagnetic material becomes strongly magnetized. That, in turn, contributes an additional field component that adds to the original field and tends to cause the new net field to follow the contours of the metal (Fig. 19.14*a*). Lord Kelvin called this property **permeability**, observing that iron is hundreds of times more permeable than air. Special alloys that are millions of times more permeable are used to shield things magnetically such as delicate wristwatches and color TV picture tubes.

Just the opposite effect occurs in superconductors, which are both perfectly conducting and perfectly diamagnetic. Below a certain applied field strength, a superconductor will maintain itself in a state where, internally, $\vec{\mathbf{B}} = 0$. When the sample (immersed in a magnetic field) is cooled below its critical temperature, it becomes superconducting. As in Fig. 19.14*c*, it then almost completely expels the *B*-field from its interior and is therefore said to

Figure 19.14 (a) When a material with a high permeability is placed in a *B*-field, most of the field lines pass through the material, effectively shielding the region it surrounds. (b) By contrast, a substance (such as lead) in its nonsuperconducting or normal state allows the *B*-field to pass through it. (c) When the temperature drops and the material becomes superconducting, it completely expels the applied *B*-field via the Meissner Effect.

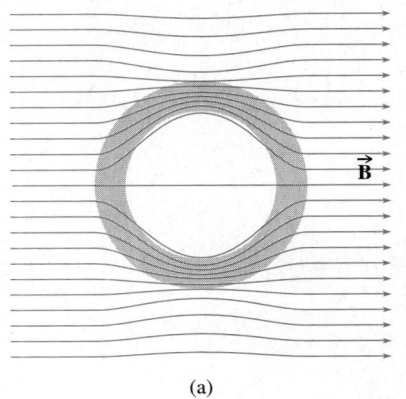

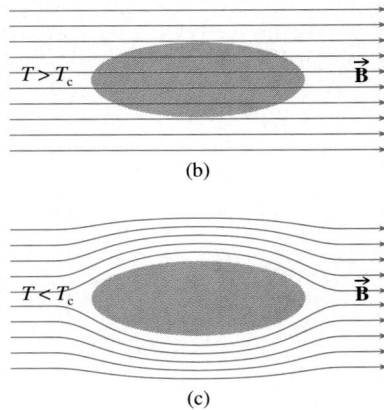

(a)

(b)

(c)

be *perfectly diamagnetic*. Only if the applied field is made to exceed the so-called *critical field* will it again penetrate the body of the specimen. The diamagnetic behavior of superconductors was discovered experimentally in 1933 and is known as the **Meissner Effect**.

Electrodynamics

A charged particle, whether at rest or in motion, has an electric field $\vec{\mathbf{E}}$, which is not much more than saying that charges interact electrically. Now, suppose that the particle has no intrinsic magnetism of its own. Nonetheless, when such a charge moves in space, it exerts magnetic forces and possesses a magnetic field. This *magnetism arises out of motion, and motion is relative*. A beam of protons is a current I and as such exerts both electric and magnetic forces. And yet, if we run along with the flow, essentially causing I to become zero, the magnetic force vanishes. It must vanish, because there is no longer a source of the B-field.

The Special Theory of Relativity provides the realization that current-generated magnetism is a facet of electricity—a modification of the electrical interaction arising from relative motion appears as the magnetic force. Electricity and magnetism are the two sides of a single phenomenon—*electromagnetism*—that looks different to observers in relative motion.

The study of the electromagnetic interaction is known as **electrodynamics**, a word prophetically coined by Ampère, who initiated the unification over 150 years ago when he linked the source of magnetism to currents.

19.3 Currents and Fields

On July 21, 1820, Hans Oersted, professor of physics at Copenhagen University, delivered a lecture on electricity to some advanced students. By chance, a wire leading to a voltaic pile was nearly parallel to and above a compass that happened to be on the table along with other paraphernalia. When the circuit was closed, the needle swung around almost perpendicular to the current-carrying wire as if gripped by a powerful magnet (see photo).

The news of Oersted's discovery reached Paris on September 4, when Dominique F. J. Arago reported it to a skeptical gathering of the Paris Academy of Sciences. A young professor, André Marie Ampère, attended the talk, and within two weeks he completed a series of experiments of his own. Ampère showed that the magnetic force experienced by a compass needle in the vicinity of a current-carrying straight wire acted at right angles to the current along a series of concentric circles (Fig. 19.15). *Point the thumb of the right hand in the conventional direction of the current and the direction in which a compass will point,*

Hans Christian Oersted (1777–1851). Besides his work in electricity and magnetism, Oersted was the first to prepare pure metallic aluminum (1825).

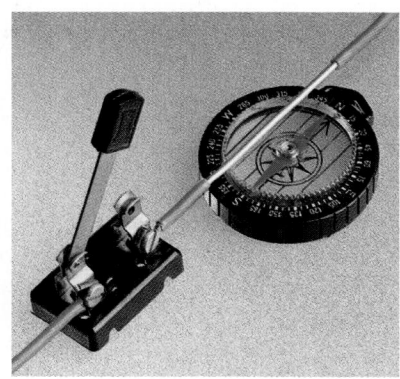

Oersted's demonstration. (a) With no current in the wire, the compass needle points north. (b) When a current exists, the needle swings so that it almost aligns with the new field created by the current. The Earth's field causes a small northerly deflection of the needle.

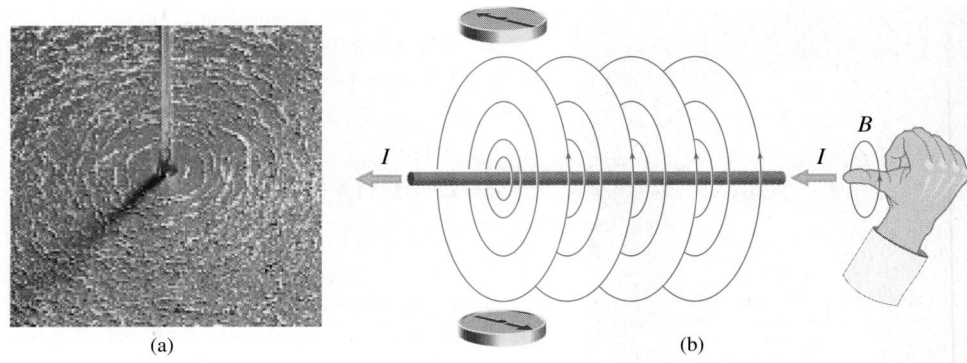

(a)   (b)

Figure 19.15 (a) The circular magnetic field surrounding a current-carrying wire, as revealed in an iron-filing pattern. (b) When the thumb on the right hand points in the direction of the current (*I*), the fingers curl in the direction of the circular magnetic field (*B*).

The **Right-Hand Current Rule** maintains that if you point the thumb of your right hand in the direction of any current (of positive charge) your fingers will curl in the direction of the resulting magnetic field.

the direction of the B-field, is given by the direction of the fingers curling around the wire at that location. (Call this the **Right-Hand-Current Rule**.) That rather surprising perpendicular character of the force explains why it took so long to discover.

The Current-Carrying Wire

A straight current-carrying wire generates a circular or, more accurately, a cylindrical magnetic field in the space surrounding it. *That field is constant at any given perpendicular distance from the wire and gets weaker as that distance increases.* Experiments, principally by Jean Biot and Félix Savart (1820), established that the *B*-field near a long straight wire in air is directly proportional to the current *I* and inversely proportional to the perpendicular distance *r* from the wire: $B \propto I/r$. The next logical step is to introduce a constant of proportionality that balances the units, yielding teslas on both sides, and thereby produce an equality. Since the SI system is "rationalized," the constant is defined so that $1/2\pi$ shows up in the equations for fields when there is axial symmetry, as there is here. Furthermore, it was found that the *B*-field depends on the magnetic behavior of the medium in which the wire is immersed. We follow tradition and introduce the constant $\mu/2\pi$. The Greek letter mu (μ) represents the medium-dependent constant that, naturally enough, is called the **permeability**.*

The magnetic field at any point outside a long straight current-carrying wire is then

[long straight wire]
$$B = \frac{\mu}{2\pi}\frac{I}{r} \tag{19.1}$$

In vacuum, the value of the permeability is, by definition,

$$\mu_0 = 4\pi \times 10^{-7} \text{ T·m/A}$$

An expression related to Eq. (19.1) will be used later to define the unit of current, and that's where this value of μ_0 comes from. Equation (19.1) works fine provided μ is constant, which it is for diamagnetic and paramagnetic media. It isn't constant for ferromagnetic media, where μ changes as the *B*-field changes (Fig. 19.16), and Eq. (19.1) should then be

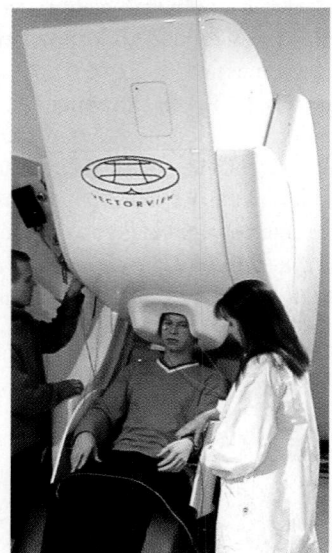

Today, superconducting quantum interference devices (SQUIDs) are being used to measure the minute ($\approx 10^{-13}$ T) magnetic fields generated by currents in the brain and heart. The magnetoencephalograph can locate the source of nerve signals in the brain to within a few millimeters. (See photo on p. 675.)

---

*The constant $\mu/2\pi$ happens to have the μ on top because the permeability was originally defined so that Coulomb's Law for magnetic charge (no longer of any interest) would have a factor of $1/4\pi\mu$ in front to match the $1/4\pi\varepsilon$ in front of Coulomb's Law for electrical charge.

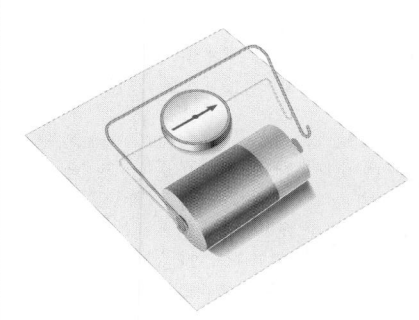

EXPLORING PHYSICS ON YOUR OWN

The Field of a Current-Carrying Wire: Oersted's experiment is very easy to duplicate. Just align a compass due north (the floating needle you made earlier—p. 676—will work nicely). Stretch a foot or so of any ordinary insulated wire from north to south directly above and close to the compass; the needle will be unaffected. Attach one end of the wire to the negative terminal of a D-cell and tape the other end of the wire to the positive terminal—this is essentially a "short" across the battery, so you don't want to leave it connected for long. The compass will immediately swing to a new position and then return north-south. If you want to leave the current on longer, either use a greater length of wire or put a flashlight bulb in series so as to increase the resistance and drop the current. See if you can swing the needle around through 180°.

used with care. For ferromagnetic media, μ is usually much larger than μ_0. Typically, in media that are likely to be of concern to us such as air or water, $\mu \approx \mu_0$. For diamagnetic media $\mu/\mu_0 < 1$, whereas for paramagnetic media $\mu/\mu_0 > 1$; but in both cases $\mu/\mu_0 \approx 1$. The differences are small. In air $\mu_{\text{air}}/\mu_0 = 1 + 3.6 \times 10^{-7}$, whereas $\mu_{\text{water}}/\mu_0 = 1 - 0.88 \times 10^{-5}$. Thus, we can rewrite Eq. (19.1), **the magnetic field of a long straight wire in vacuum,** as

To help you remember Eq. (19.1), or (19.2), keep in mind that B depends directly on both I, and on the surrounding medium via μ. Thus $B \propto \mu I$. The $1/2\pi r$ arises because of the cylindrical symmetry, just as it did earlier in Eq. (15.7) on p. 558.

[long straight wire]

$$B = \frac{\mu_0}{2\pi}\frac{I}{r}$$

(19.2)

This provides the strength of the B-field (Fig. 19.17) of any line of current—a straight beam of protons transporting I coulombs per second through space will produce a B-field given by Eq. (19.2). The B-field in a subway, 1.5 m from the third rail carrying 5000 A (at 600 V), is $\approx 7 \times 10^{-4}$ T and can be easily detected with a compass. {For an animated look at these ideas click on **MAGNETIC FIELD OF A STRAIGHT CONDUCTOR** under **INTERACTIVE EXPLORATIONS** on the **CD**.}

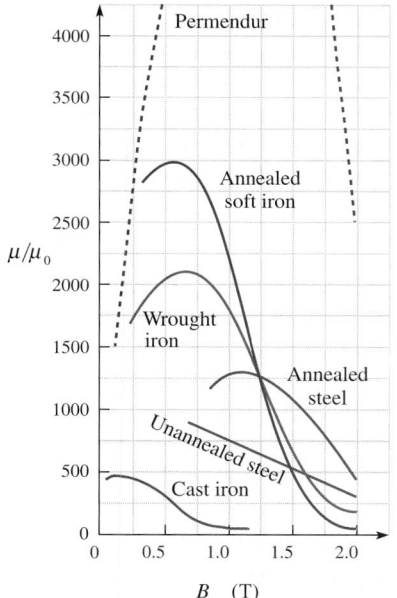

Figure 19.16 The permeability of magnetic materials is not constant but instead changes as the field affects the material. Notice how the permeability initially rises, peaks, and then falls off as the material reaches saturation. Permendur, which is half iron and half cobalt, keeps a high μ even at large values of B.

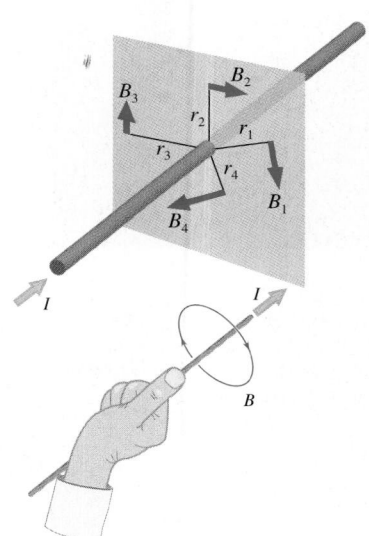

Figure 19.17 The B-field surrounding a current-carrying straight wire lies in a plane perpendicular to the wire. At any point on that plane, the B-field is perpendicular to the line from the wire to the point and decreases inversely with the distance from the wire to the point.

Example 19.1 **[I]** The overhead power cable for a street trolley is strung horizontally 10 m above the ground. A long straight section of it carries 100 amps dc due west. Describe the magnetic field produced by the current, and determine its value at ground level just under the wire. Compare that to the strength of the Earth's field.

Solution We've got a straight current-carrying wire and that produces a known *B*-field. (1) TRANSLATION—Determine the *B*-field at a specified distance from a straight wire carrying a known current. (2) GIVEN: $r = 10$ m and $I = 100$ A. FIND: *B*. (3) PROBLEM TYPE—Magnetic field/current/straight wire. (4) PROCEDURE—First, draw a diagram. Figure 19.15*b* will suffice, with the current assumed heading west. At ground level (at a point beneath the westerly current), the Right-Hand-Current

Rule tells us that $\vec{B}$ points due south. We already have an expression for the field of a long straight current-carrying wire in terms of *I* and *r*. (5) CALCULATION—Using Eq. (19.2) and $\mu_0 = 4\pi \times 10^{-7}$ T·m/A,

$$B = \frac{(4\pi \times 10^{-7} \text{ T·m/A})(100 \text{ A})}{2\pi(10 \text{ m})} = \boxed{2.0 \times 10^{-6} \text{ T}}$$

which, from Table 19.1, is only 4% of the Earth's field.

Quick Check: $\mu_0 = 1.256\ 6 \times 10^{-6}$ T·m/A, $B \approx (10^{-6}$ T·m/A$)(10^2$ A$)/60$ m $\approx 2 \times 10^{-6}$ T.

{For more worked problems click on **WALK-THROUGHS** *in* **CHAPTER** *19 on the* **CD.***}*

The Current Loop

During those first weeks of excitement, Ampère had another lovely idea. Since a straight current-carrying wire is surrounded by concentric rings of magnetic force, bending the conductor into a loop should concentrate that force (as does bending a bar magnet into a horseshoe magnet). We can imagine the lines of *B* that extend far out into space on one side of a straight wire being carried around and crowded within the small region, then encompassed by the loop (Fig. 19.18). The field inside the loop is much stronger than the field outside. Again the Right-Hand-Current Rule provides the direction of *B*. The field will effectively vanish if the loop is squashed such that the current doubles back on itself—oppositely directed fields then overlap and cancel. The current-circle brings to mind the picture of a dipole field. Recognizing this similarity, Ampère boldly suggested that **currents are the basic cause of magnetism** (and to the degree that the intrinsic magnetism of an electron is due to the motion of its charge, he was right).

Biot and Savart determined experimentally that the field at the very center of a current loop points axially outward (Fig. 19.19) along the *z*-axis. They found *B* to be directly proportional to *I* and inversely proportional to the radius *R* of the loop:

A. M. Ampère (1775–1836). Legend has it that Ampère was the classic absent-minded professor, who once even forgot to attend a dinner with the Emperor Napoleon.

[circular loop at center]

$$B_z = \frac{\mu_0 I}{2R} \tag{19.3}$$

In this case, the field's cross section over a plane perpendicular to the central *z*-axis is different at different values of *z*, and so the factor of 2π is not present.

Stacking several loops in parallel results in the overlapping of their individual fields, providing a proportionately increased net effect. Each loop simply adds (vectorially) its field to the fields of all the others. A tight short coil (Fig. 19.19) composed of *N* closely wrapped turns of wire each carrying a current *I* has a field at its center of

[circular coil at center]

$$B_z = N \frac{\mu_0 I}{2R} \tag{19.4}$$

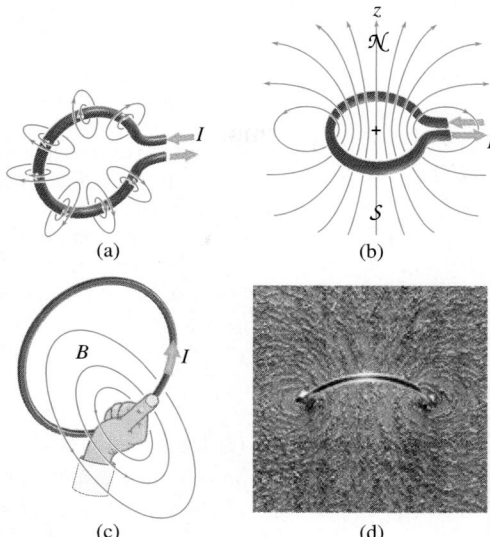

(a)   (b)

(c)   (d)

Figure 19.18 (a) Each segment of a current-carrying loop is surrounded by a circular *B*-field. (b) These combine to produce a dipole field very much like the field of a bar magnet. Remember that the field is 3-dimensional and more or less axially symmetrical around the central *z*-axis. (c) The Right-Hand-Current Rule gives the direction of $\vec{\mathbf{B}}$. (d) The field pattern, as revealed with iron filings, in a plane perpendicular to the loop.

This sort of coil, with a negligible length compared to its diameter, resembles a stubby disk magnet. If the coil is delicately pivoted so that it can rotate freely about a diameter, it will swing into alignment with an applied field just as a compass would. Ampère observed as much in 1820. This twisting behavior is the basis of both the moving coil galvanometer and the electric motor (p. 694).

The Solenoid

Carrying the coil concept one step further, Ampère wound wire into a long helix (Fig. 19.20) or *solenoid* (from the Greek *solen* meaning "tube") and found that, with a current passing through it, it acted like a bar magnet. Within the space encompassed by a long, nar-

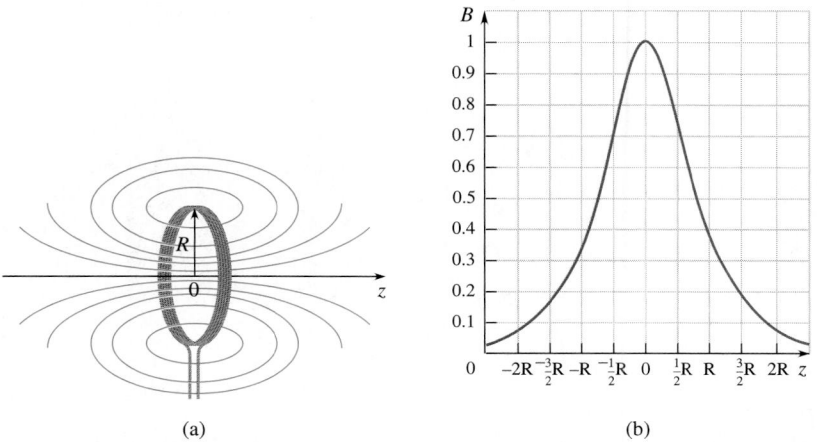

(a)   (b)

Figure 19.19 (a) A narrow circular current-carrying coil. (b) The *B*-field measured along the central *z*-axis. The curve of *B* is fairly straight around $z = R/2$, where there is a turning point (see Discussion Question 18).

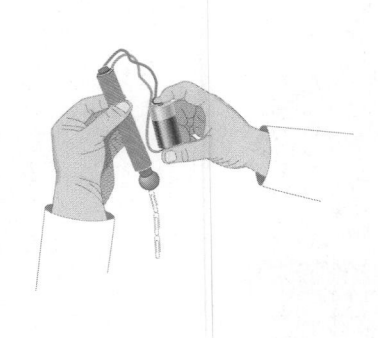

EXPLORING PHYSICS ON YOUR OWN

Make an Electromagnet: Start with a slug of iron or soft steel up to about 4 or 5 in. long and 1/4 in. in diameter: a large bolt or very big nail—I used a door hinge pin. Wrap and tape a layer of cardboard around it so that the rod can be removed later. Now wrap about 150 turns of insulated wire (#22 or heavier) around the sleeve in three or four layers. With this many turns we can get away with using a single D-cell—you might try fewer turns and a big 6-V battery. Put your disconnected electromagnet about a foot from a compass (p. 676) and there'll be little or no effect. But if you touch its leads to the terminals of a D-cell, the needle will jump. To determine the strength of your electromagnet count the number of paper clips it can lift in a chain—drop them by interrupting the current. Now remove the steel core and try it again. What does the core do? Why did we use a material with a high permeability? What happens when the core is brass or wood or plastic? Determine the polarity of your solenoid using a compass. Reverse the current leads and repeat the last observation.

row (at least 10 times longer than it is wide), tightly wound solenoid, the *B*-field is strong and quite uniform, especially in the middle and around the central *z*-axis. The solenoid is one of the most useful magnetic devices—a typical home has dozens of them operating bells, chimes, and speakers. The solenoid is the central component of the relay that mechanically controls equipment such as washing machines, dishwashers, clothes driers, and furnaces. As a circuit element, the solenoid is in radios, TVs, and computers.

A solenoid is helical and not quite the same as stacking a bunch of separate flat loops; here, the current progresses from one end of the coil to the other and that adds a small additional contribution to *B* (p. 686). On the other hand, if the solenoid is wound with overlapping turns, an even number of layers will bring the end wire back to the beginning and cancel the longitudinal current.

It's reasonable to assume that the field inside a solenoid increases directly with *I*. Moreover, the field should increase with the number of current loops contributing, but here there's the additional concern of packing the loops as close together as possible. Carrying the same current, 100 turns spread over a length of a meter will produce a much weaker

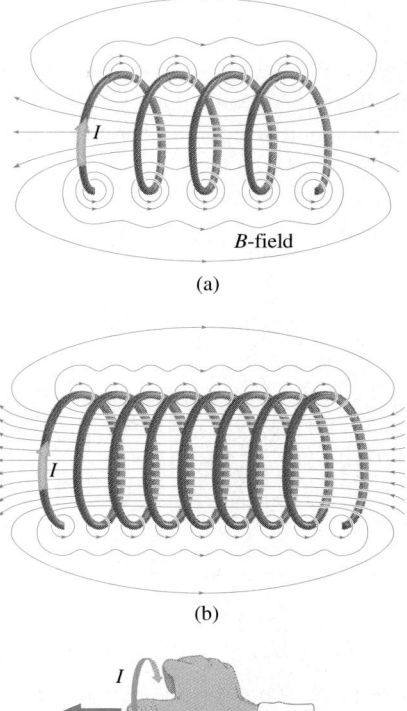

Figure 19.20 The solenoid. (a) The magnetic field of a loosely wound current-carrying coil. (b) When the coil is wound tighter and there are more loops, the field inside becomes larger and more uniform. (c) The Right-Hand-Current Rule provides the direction of $\vec{B}$. Alternatively, there's a Right-Hand-Solenoid Rule: when the fingers of the right-hand curl around a solenoid in the direction of the current, the thumb points in the direction of the $\vec{B}$-field.

B-field

(a)

(b)

(c)

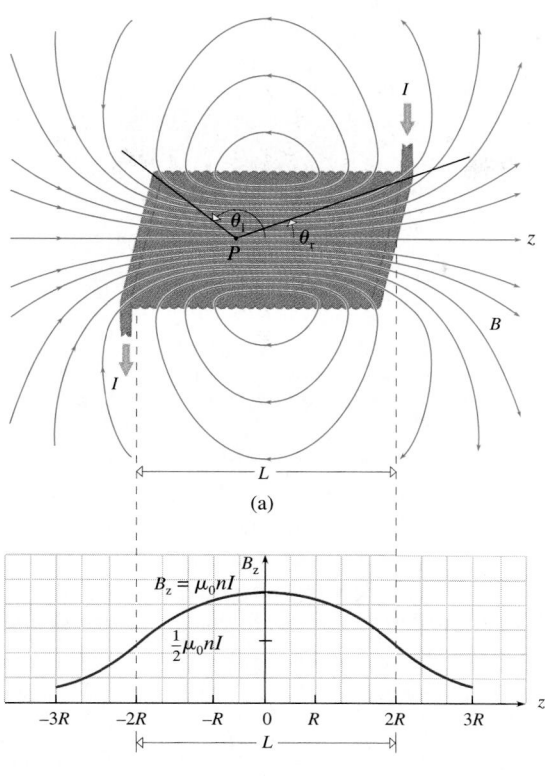

Figure 19.21   (a) The *B*-field of a finite solenoid carrying a current *I*. (b) The value of the field at any point *P* on the central *z*-axis is called B_z, and it varies with θ_l and θ_r. At the very center of the coil ($z = 0$), the field is a maximum equal to $\mu_0 nI$.

(a)

(b)

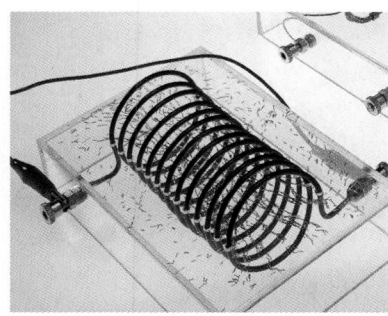

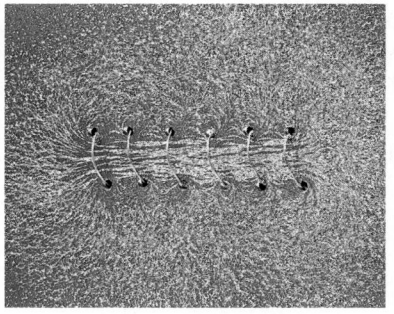

The magnetic field lines produced by a current-carrying coil. The pattern is again formed using iron filings.

inside field than 100 turns distributed over a centimeter. What's needed is a measure of the density of the winding, namely, the ***number of turns per unit length*** (*n*) of solenoid. If *L* is the length of the coil and *N* the total number of turns, $n = N/L$. Then $B \propto nI$, and it's found experimentally (Fig.19.21) that in air

[solenoid at center; axial field] $B_z \approx \mu_0 nI$ (19.5)

The field at any axial point *P* is actually given by

$$B_z = \tfrac{1}{2}\mu_0 nI(\cos \theta_r - \cos \theta_l)$$

where θ_r and θ_l are the angles made with the right and left edges of the coil. As $\theta_r \to 0$ and $\theta_l \to 180°$, as they would for an infinitely long coil, the field approaches that of Eq. (19.5).

THE ELECTROMAGNET & MRS. HENRY'S PETTICOATS

Sometime around 1825, W. Sturgeon wrapped 18 turns of bare wire around a varnished iron bar and sent a current through the coil; in so doing, he created the first powerful ***electromagnet***. The field set up by the current aligned the domains within the iron to produce a combined magnetic field of unprecedented strength (Fig. 19.22). The American physicist Joseph Henry heard about the feat and set about to better it. Legend has it that he tore apart his wife's petticoats so that he could insulate his wires with the silk stripping. By using many turns of insulated wire, Henry enhanced the field while keeping *I* relatively low. In 1831, he produced a modest-sized device powered by an ordinary battery that could lift more than a ton of iron. When the current was interrupted, the soft iron core almost completely demagnetized spontaneously, and the load was released.

Joseph Henry (1797–1878). To honor him for his many original contributions, the International Electrical Congress of 1893 named the unit of inductance the henry.

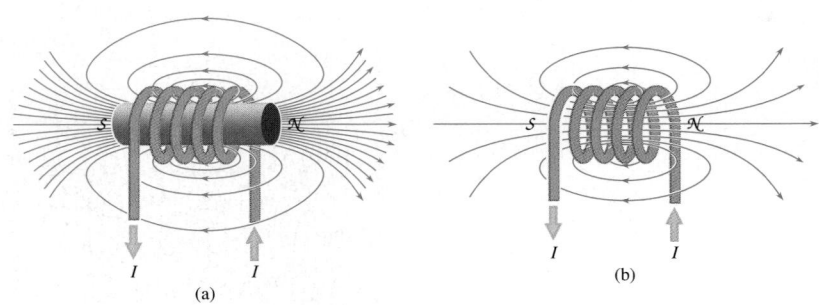

Figure 19.22 The field of a current-carrying coil (a) with and (b) without a ferromagnetic core. The B-field set up by the current in the coil magnetizes the core, aligning its domains, and that increases the field even more.

For a coil 10 times longer than its diameter, Eq. (19.5) yields results that are only about 0.5% too large, which is good enough for our purposes.

If such a coil is imagined cut across the middle into two equal-length solenoids carrying the same current as the original, we can expect that the fields at their ends (which add up to $\mu_0 nI$ in the middle of the uncut coil) should be

[solenoid at ends] $$B_z \approx \tfrac{1}{2}\mu_0 nI$$

That suggests that about half the field "leaks out" between the windings of a solenoid.

{To read about a wonderful experimental confirmation of Ampère's Hypothesis by none other than A. Einstein click on **AMPÈRE'S CURRENTS** under **FURTHER DISCUSSIONS** on the **CD**.} 💿

Example 19.2 **[I]** A 20-cm-long solenoid with a 2.0-cm inside diameter is tightly wound on a hollow quartz cylinder. There are several layers with a total of 20×10^3 turns per meter of a niobium-tin wire. The device is cooled below its critical temperature and becomes superconducting. Since the wire is then without resistance, it can easily carry 30 A and not develop any I^2R losses. Compute the approximate field inside the solenoid near the middle. What is its value at either end?

Solution We've got a current-carrying solenoid and that produces a known B-field. (1) TRANSLATION—Determine the B-field near the center of a solenoid of known length, diameter, and number of turns per meter, carrying a specified current. (2) GIVEN: $n = 20 \times 10^3$ m^{-1}, $I = 30$ A, and $D = 2.0$ cm. FIND:

B_z. (3) PROBLEM TYPE—Magnetic field/current/solenoid. (4) PROCEDURE—The solenoid is long and narrow and will obey the approximations that led to Eq. (19.5). (5) CALCULATION—Using $\mu_0 = 1.257 \times 10^{-6}$ T·m/A,

$$B_z \approx \mu_0 nI = (1.257 \times 10^{-6}\ \text{T·m/A})(20 \times 10^3\ \text{m}^{-1})(30\ \text{A})$$

and $$\boxed{B_z \approx 0.75\ \text{T}}$$

which is a formidable field, over 10^4 times that of the Earth. The field at either end is about half this, $\boxed{0.38\ \text{T}}$.

Quick Check: Round off all the numbers and recalculate; $B_z \approx \mu_0 nI \approx (10^{-6}\ \text{T·m/A})(2 \times 10^4\ \text{m}^{-1})(3 \times 10^1\ \text{A}) \approx 0.6$ T.

Talking about Ampère. He further deduced from this analogy the consequence that the attractive and repulsive properties of magnets depend on electric currents which circulate about the molecules of iron and steel.

D. F. J. ARAGO (1820)
FRENCH PHYSICIST

19.4 Ampère's Law (Optional)

There are two equivalent general schemes for calculating magnetic fields due to currents: one devised by Laplace based on the data of Biot and Savart and the other by Ampère. We will study **Ampère's Law**, which is simpler to use, provided the geometry is simple. We cannot *derive* the law from basics (i.e., from the properties of the electron)—physicists simply don't know enough yet to do that. It can, however, be *deduced* from experimental observations. {For an old-fashioned, but elegant derivation click on **AMPÈRE'S LAW** under **FURTHER DISCUSSIONS** on the **CD**.} 💿

Steel nails align themselves to reveal the *B*-field configuration surrounding the opposite poles of two powerful superconducting magnets. Frost around the necks of the stainless steel Dewar flasks is a clue to the fact that they're operating at liquid helium temperature (around 4.2 K).

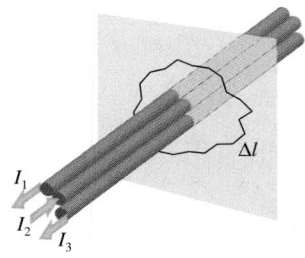

Figure 19.23 The net current, $\sum I$, flowing through the closed path is the difference between the currents flowing in one direction and the currents flowing in the opposite direction.

Ampère's Law is usually applied using calculus, but in situations where the geometry is simple and the configuration of the field is known, it can be exploited with a little algebra and a bit of trigonometry. Consider Fig. 19.23, which depicts several wires carrying currents. We know that there will be some resulting *B*-field in the space surrounding the wires and it will depend on those currents. Imagine an arbitrary path (within the *B*-field) that encloses the current distribution. In Fig. 19.23 that path, known as an Ampèrian loop, is made up of a bunch of tiny connected segments of length Δl. Whatever configuration the field has it will have some value (which might be zero) at the location of each path segment. If the closed path is large and at a great distance, *B* will be far from the currents and small in the region of that loop. Similarly, if the path is close in and small, *B* will be near the currents and large. This reciprocity suggests that we use the product of the field and the path-length. More precisely, we need the component of the field parallel ($B_\parallel$) to each tiny segment (Δl). It turns out that the quantity $B_\parallel \Delta l$ for each and every tiny segment, summed once around the loop ($\sum B_\parallel \Delta l$), is indeed constant for *all* closed paths. Not surprisingly, that constant is proportional to the net current ($\sum I$) enclosed by the loop: the more current, the stronger the field and the greater the value of ($\sum B_\parallel \Delta l$). In other words, $\sum B_\parallel \Delta l \propto \sum I$, independent of path. (Keep in mind that when we move a length Δl, in the process of going around a loop, in the direction of $B_\parallel$ the contribution $B_\parallel \Delta l$ is positive, whereas if we move a length Δl in a direction opposite to $B_\parallel$ that contribution $B_\parallel \Delta l$ is negative.) Summarizing all of this, **Ampère's Law** states that

[over any closed path]
$$\sum B_\parallel \Delta l = \mu_0 \sum I \qquad (19.6)$$

To apply the law consider the simplest case of a straight wire carrying a current *I*. *Wherever possible, we want B either perpendicular ($B_\parallel = 0$) or parallel ($B_\parallel = B$) to the chosen encompassing path.* The field here (Fig. 19.24) is known to be circular, so we select a circular Ampèrian path of radius *r*, around which $B_\parallel = B = constant$. Consequently,

$$\sum B_\parallel \Delta l = \sum B \Delta l$$

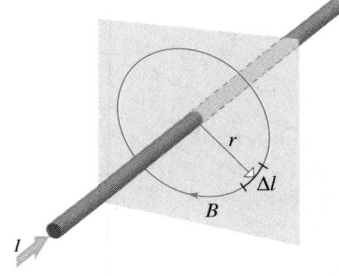

Figure 19.24 Using Ampère's Law. The circle of radius *r* surrounding the current-carrying wire is the Ampèrian path we have chosen. The *B*-field is everywhere parallel to the path elements Δl.

The speaker of the radio shown on p. 537. You can see the permanent magnetic housing at the rear of the speaker. The paper cone is attached to a current-carrying coil that causes the cone to vibrate. (See Fig. Q9, p. 698.)

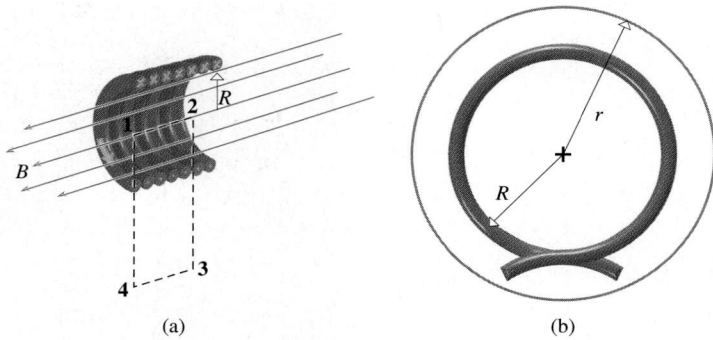

Figure 19.25 Using Ampère's Law. A helical coil carrying a current (out of the page and to the right at the bottom and into it and to the left at the top). (a) To find the field inside the solenoid, use path 1–2–3–4. (b) To find the field outside, use a circle of radius r.

and because B is constant, it can be factored out of the summation:

$$\sum B_{\parallel} \Delta l = B \sum \Delta l$$

Furthermore, $\sum \Delta l$ around a circular path equals the circumference, $2\pi r$, and so $\sum B_{\parallel} \Delta l = B2\pi r$. Because the net current is just I,

$$B2\pi r = \mu_0 \sum I$$

and

$$B = \frac{\mu_0 I}{2\pi r} \qquad [19.2]$$

As another and last example, let's find the field inside a very long solenoid (Fig. 19.25). Its asymmetry requires a bit of tricky maneuvering. As an Ampèrian loop that follows the B-field, with segments either parallel or perpendicular to it, we construct the path 1–2–3–4–1. The summation $\sum B_{\parallel} \Delta l$ is carried out over the four straight segments. Since the field lines close on themselves and since the system is symmetrical, there cannot be a radial component of the field. Thus, B must be perpendicular to segments 2–3 and 4–1, and they make no contribution. Even if there were a radial field, in going around the loop, the contribution along 2–3 would be equal and opposite to that from 4–1, and they would cancel. Now for segment 3–4: the field outside an infinitely long solenoid is zero, and the field outside a finite but long solenoid must be small, axial, and drop off with r. In any event, we take 3–4 so far from the solenoid that the field there is negligible and that segment makes no appreciable contribution. What remains is segment 1–2, over whose length L the field is parallel (that is, $B_{\parallel} = B_z$). Thus, if the number of turns of wire encompassed by the path is N, the net current is NI, and Ampère's Law yields

$$B_z L = \mu_0 N I$$

but N/L is the number of turns per unit length n, so

$$B_z = \mu_0 n I \qquad [19.5]$$

Note that the field inside is independent of its distance from the central z-axis and must be uniform. It's also independent of the cross-sectional shape of the coil.

If the solenoid is wound in a single helical layer, charge is transported from one end to the other, and there is a net current I in the z-direction that produces a circular external field. Applying Ampère's Law using the path shown in Fig. 19.25b, we find that the field has a component B_{ϕ} around the solenoid, given by

[solenoid; outside] $$B_{\phi} = \mu_0 I / 2\pi r$$

This is the field of a straight wire, and it's very small compared to the axial field inside the solenoid, where n might be 10^4 turns per meter. (See Discussion Question 11.)

Magnetic Force

Oersted demonstrated that a current exerted a force on a compass needle. To establish that this was a purely magnetic interaction, Ampère did away with the iron needle altogether. He passed a current through two parallel wires, one of which was suspended so that it could swing in response to the B-field of the other, and swing it did. Inasmuch as currents exert forces on magnets, it follows from Newton's Third Law that magnets ought to exert forces on currents. As we will see, Ampère's two-wire experiment proves the point, as do a number of other elegant arrangements, including the electric motor (p. 694) and generator (p. 720). We know now that these interactions, which are describable on a macroscopic level in terms of currents, are fundamentally due to a magnetic force experienced by mobile charge carriers.

19.5 The Force on a Moving Charge

Nowadays it's easy enough to send a beam of charged particles (e.g., in a cathode-ray tube) through a known field and observe the effects firsthand. Several conclusions are forthcoming: a point-charge q_+ moving through a magnetic field ($\vec{B}$) with a velocity $\vec{v}$ experiences a force $\vec{F}_M$, which, reasonably enough, is proportional to q_+, v, and B; that is, $F_M \propto q_+ vB$—thus, no relative motion ($v = 0$), no magnetic force. Further, the ***two vectors $\vec{v}$ and $\vec{B}$ determine a plane, and the force is perpendicular to that plane***, as shown in Fig. 19.26. ***If a particle with an opposite charge is introduced, the force reverses***—the sign of q affects the sign of F_M.

The magnitude of the force depends on the angle θ between $\vec{v}$ and $\vec{B}$: in particular, *when $\theta = 0$ or $180°$ and the particle is moving along or opposite to the field, the force is zero. When $\theta = 90°$ or $270°$ and the particle is moving perpendicular to the field, the force*

Figure 19.26 Charged particles, (a) positive and (b) negative, moving in a magnetic field. In each case, the resulting magnetic force $\vec{F}_M$ is perpendicular to the plane of $\vec{v}$ and $\vec{B}$.

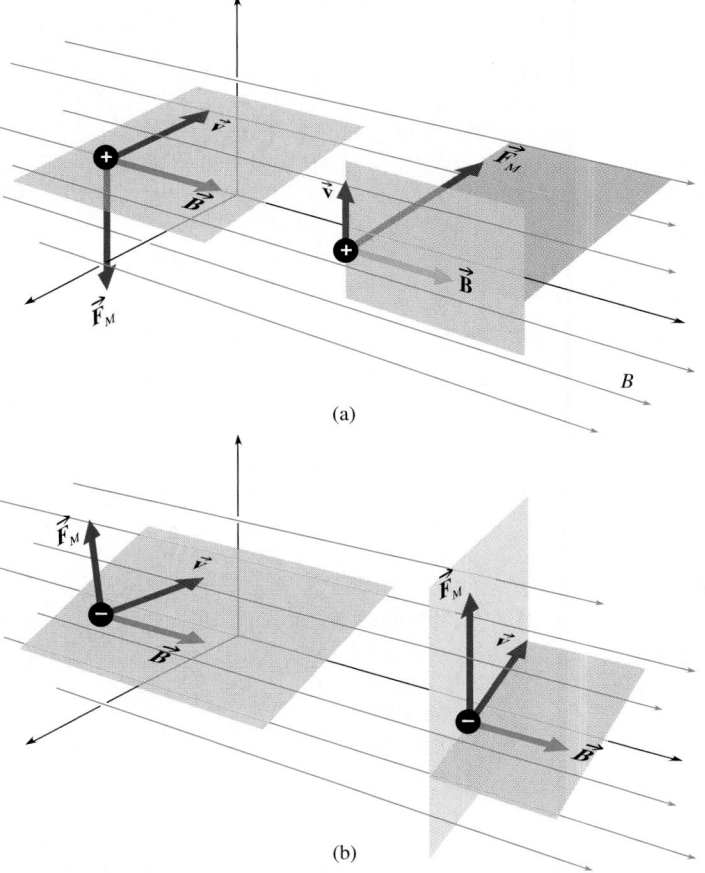

(a)

(b)

Figure 19.27 Two ways to determine the directional relationship between $\vec{v}$, $\vec{B}$, and $\vec{F}_M$. (a) Point the fingers of the right hand in the direction of $\vec{v}$. Close them (through the smallest angle) toward $\vec{B}$, and the extended thumb points in the direction of $\vec{F}_M$. (b) A positive charge moving with a velocity perpendicular to a magnetic field experiences a force perpendicular to both.

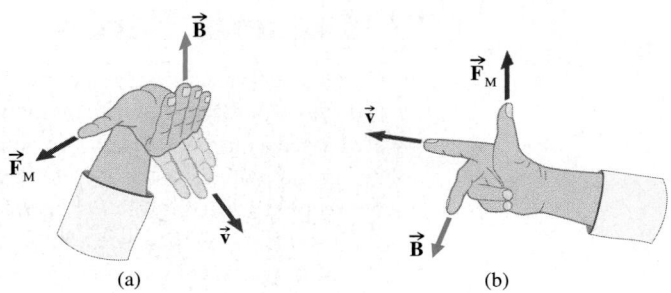

(a)   (b)

is a maximum ($q_.vB$). In other words, $F_M \propto \sin\theta$, and putting it all together

$$F_M = q_.vB \sin\theta \qquad (19.7)$$

There is no constant of proportionality here because the equation is basically the one that will be used to define B. The zero-force line can be defined as the direction of the magnetic field. From the formula, a 1-N force will be exerted on a 1-C charge moving at 1 m/s at 90° to a 1-T magnetic field.

Not surprisingly, the directions of the force, the field, and the velocity are related by a right-hand rule. If you put the fingers of your *right* hand in the direction of $\vec{v}$ and curl them through the smallest angle to $\vec{B}$, your thumb will point in the direction of $\vec{F}_M$, as in Fig. 19.27a. This relationship between vectors is called a *cross product*, and we say that $\vec{F}_M$ is in the direction of $\vec{v}$ cross $\vec{B}$, usually written as $\vec{v} \times \vec{B}$. This is the most versatile direction rule, but there are others (see e.g., Fig. 19.27b). {To review the vector product click on **CROSS PRODUCT** under **FURTHER DISCUSSIONS** on the **CD**.}

The direction of the **magnetic force** $\vec{F}_M$ is in the direction of $\vec{v}$ cross $\vec{B}$.

Example 19.3 **[I]** A conventional water-cooled electromagnet produces a 3.0-T uniform magnetic field in the 4-in. gap between its flat pole pieces. The field is aligned horizontally pointing due north. A proton is fired into the field region at a speed of 5.0×10^6 m/s. It enters traveling in a vertical north-south plane, heading north and downward at 30° below the horizontal. Compute the force vector acting on the proton at the moment it enters the field.

Solution Here a charged particle is moving in a B-field and that should call to mind $\vec{v} \times \vec{B}$. (1) TRANSLATION—A known particle is traveling with a specified velocity through a known magnetic field; determine the force it experiences. (2) GIVEN: A proton with $v = 5.0 \times 10^6$ m/s, at 30° below the horizontal in the northerly direction, and $B = 3.0$ T, north. FIND: $\vec{F}_M$. (3) PROBLEM TYPE—Magnetic field/force on a moving charge. (4) PROCEDURE—First, make a drawing—Fig. 19.28. The proton has a *positive* charge of $+1.60 \times 10^{-19}$ C and so $\vec{v} \times \vec{B}$ is due east, $\vec{F}_M$ is due east. The basic force-on-a-moving-charge relationship is $F_M = q_.vB \sin\theta$. (5) CALCULATION—The angle between $\vec{v}$ and $\vec{B}$ is $\theta = 30°$ and so with $q_. = q_e$

$$F_M = q_e vB \sin\theta$$

$$F_M = (+1.6 \times 10^{-19} \text{ C})(5.0 \times 10^6 \text{ m/s})(3.0 \text{ T})(\sin 30°)$$

and

$$\boxed{F_M = 1.2 \times 10^{-12} \text{ N}}$$

Quick Check: $F_M = q_e vB \sin\theta \approx (10^{-19} \text{ C})(10^7 \text{ m/s}) \times (1 \text{ T})(1) \approx 10^{-12}$ N.

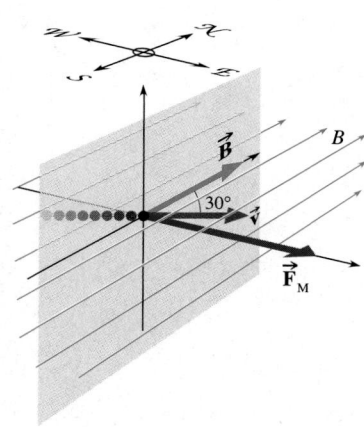

Figure 19.28

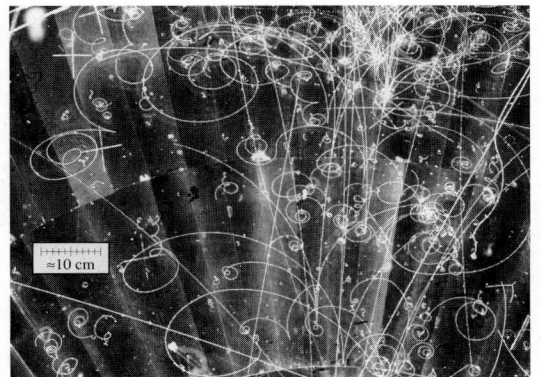

A bubble-chamber photo showing the paths of charged particles moving across a downward 3-T magnetic field. The curvature of a track can be measured and from that the momentum of the particle can be determined since, from Eq.(19.8) below, $p = q.BR$. The smaller the momentm, the smaller is R; that's why the very lightweight electrons and positrons spiral so tightly. The nearly straight tracks are made by massive high-momentum particles (with large values of R). Look along their paths and you'll see a number of tiny (low-energy) spirals coming off them; these are electrons knocked out of their hydrogen or neon atoms by the passing particle.

A TV special—"Monty Hall meets the magnet." The B-field of this fairly strong horseshoe magnet exerts forces on the electron beam and distorts the TV picture. This demonstration can be done with a black-and-white set with no risk of harm to it, but it's not advisable to try it on a color picture tube.

Because $\vec{\mathbf{F}}_M$ is perpendicular to $\vec{\mathbf{v}}$, it's purely a *deflecting* force; it changes the direction of $\vec{\mathbf{v}}$ without altering v. Since there can never be a component of magnetic force along the motion, there will be no tangential acceleration. **No work will be done on a moving charge by a $\vec{\mathbf{B}}$-field, and no change in its energy can occur in the process.**

The Trajectory of a Free Particle

Imagine a positively charged particle q entering perpendicularly into a uniform magnetic field (Fig. 19.29). Because the magnetic force is always perpendicular to the velocity, an otherwise free particle will experience a centripetal acceleration that is also always perpendicular to the motion, that is, radial. The particle will be forced to move along a circular arc. If the field is strong enough and the particle stays in it long enough, and moreover doesn't lose any energy, it will swing out a complete circle. In reality, free accelerating charges radiate electromagnetic energy and spiral inward.

Here, a positive particle is carried into a counterclockwise trajectory; had the particle been negative, it would have simply swung the other way into a clockwise circle, as in the photo on p. 583. We know that an object will move in a circle of radius R provided there is a centripetal force on it whose magnitude is

$$F_C = \frac{mv^2}{R} \qquad [5.2]$$

Here, with $\theta = 90°$

$$q.vB = \frac{mv^2}{R}$$

and

$$R = \frac{mv}{q.B} \qquad (19.8)$$

Figure 19.29 (a) A positive particle traveling through a uniform B-field experiences a force that causes it to move in a curved path. (b) The force is always perpendicular to the velocity, so the particle will ideally move in a circle. (It actually radiates, loses energy, and spirals inward.)

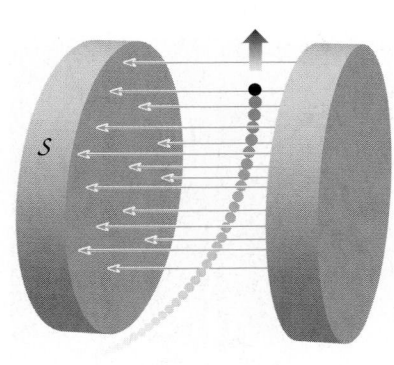

(a)

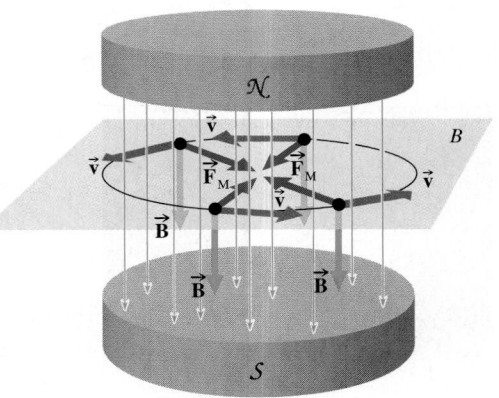

(b)

(a) A beam of electrons in an old cathode-ray tube. (b) Here I displaced the beam downward using a horseshoe magnet. (c) A beam of electrons bent into a circular orbit by the magnetic field of a set of large Helmholtz coils. The small amount of gas in the tube glows when it's ionized by collisions with the electrons, making the trajectories of the electrons visible as a pale purple circle.

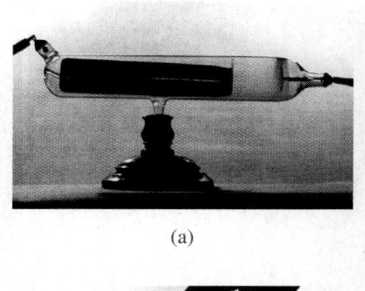

(a)

(b)

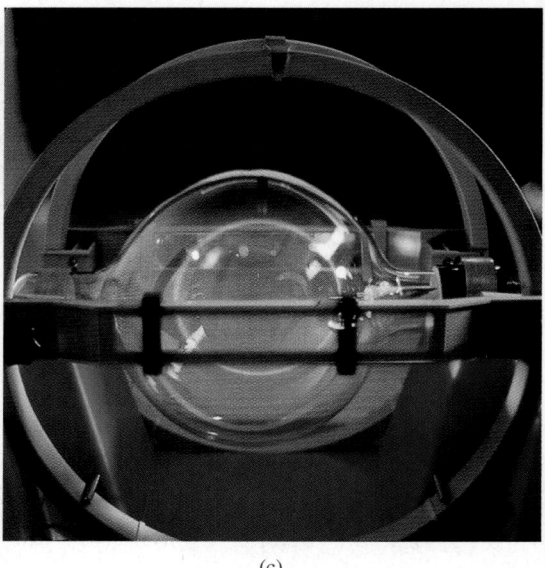

(c)

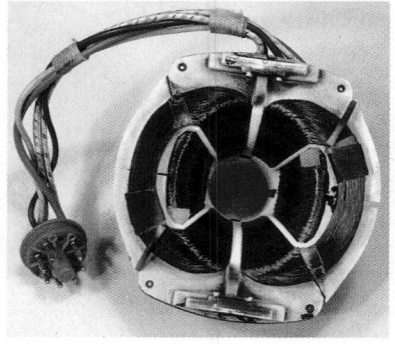

The *B*-field coils, or yoke, which surround the neck of a TV picture tube. They steer the electron beam so that it sweeps across the screen.

For a given q_+, the radius R of the path is determined by the momentum mv and the field. If we somehow impart energy to the particle (for example, via an applied E-field), thereby increasing its momentum, and if at the same time we increase B proportionately, the radius of the orbit can be kept fixed. This is the central feature of all modern ring-shaped particle accelerators. To get the largest possible $mv = q_+BR$, we need both the largest available field and orbital radius. Today, the approach is to construct a great doughnut-shaped hollow chamber miles in diameter, pump out all the air within, and surround it with powerful superconducting electromagnets, so that B can be as large as possible. A beam of, say, protons is fired in tangentially and, by adjusting mv and B for the fixed R of the chamber, the particles are brought up to tremendous energies (Ch. 31). {To play with these ideas, click on **CHARGED PARTICLES IN A MAGNETIC FIELD** under **INTERACTIVE EXPLORATIONS** in **CHAPTER 19** on the **CD**. After you finish with that piece, click on **VELOCITY SELECTOR**, at the same location, to see how both the electric and magnetic fields can be used to control a stream of charged particles.}

On a grand scale, the magnetic field acts on charged particles as a kind of cosmic accelerator provoking and guiding a range of violent occurrences from X-ray emissions and stellar flares to the blazing northern and southern lights—the auroras—in the Earth's atmosphere. Charged particles from the Sun stream down and are trapped by the planet's magnetic field. They collide with the molecules of the atmosphere, which then give off an eerie light. {Click on **AURORAS & RADIATION BELTS** under **FURTHER DISCUSSIONS** on the **CD**.}

STEERING A TV'S ELECTRON BEAM

The *deflection* yoke on the neck of a TV picture tube consists of a pair of coils that create a set of crossed magnetic fields, one for vertical deflection and one for horizontal deflection. The current through each coil creates a variable magnetic field that steers the electron beam across the face of the tube.

The aurora borealis seen from the surface of the Earth.

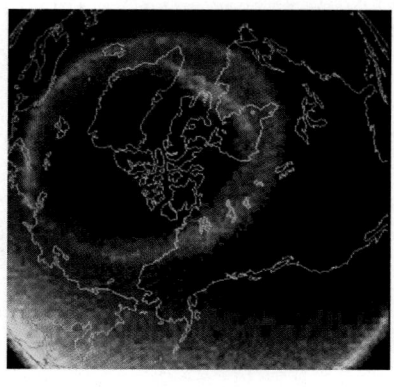

A high-altitude view of the aurora taken by a satellite about three Earth-radii away from the planet. A computer graphic shows the location of land masses. The U.S.A. is at the right of center.

A closeup image of the surface of the Sun taken by NASA's Transition Region and Coronal Explorer (TRACE) spacecraft. These coronal loops are immense coils of hot ionized gas that reveal the magnetic field lines in the outer atmosphere of the Sun.

19.6 Forces on Wires

Free charges experience forces when they traverse a magnetic field, something that would happen whether they were in a beam in vacuum or traveling as an ordinary current down a copper wire. When constrained to move within a conductor, the charges impart an average force on the conductor. Consider a quantity of charge Δq streaming along a wire in a $\vec{\mathbf{B}}$-field such that in a time interval Δt the charge travels a length of wire l. The rate of flow of charge ($\Delta q/\Delta t$) is the constant current in the wire, I. From Eq. (19.7) the force exerted on the charge Δq is $F_M = \Delta q \, vB \sin \theta$. Multiply the top and bottom of the expression on the right by Δt to arrive at $F_M = (\Delta q/\Delta t)(v\Delta t) B \sin \theta$, where $\Delta q/\Delta t = I$ and $v\Delta t = l$. Therefore, a straight wire segment of length l carrying a current I and making an angle θ with a fixed $\vec{\mathbf{B}}$-field will experience a force

[current-carrying wire]

$$F_M = IlB \sin \theta \qquad (19.9)$$

The direction of the force on the wire is the same as the direction of the force on the individual mobile carriers (always taken as positive charges). In other words, the direction of the force is given by the direction of $\vec{\mathbf{v}} \times \vec{\mathbf{B}}$, as shown in Fig. 19.30 (or if you like, index finger in the direction of $\vec{\mathbf{v}}$, middle finger in the direction of $\vec{\mathbf{B}}$, and thumb in the direction of $\vec{\mathbf{F}}_M$). Ampère used the arrangement of Fig. 19.31 to establish that no matter how he applied a field, the force on the current-carrying wire segment was always perpendicular to it. {For an alternative derivation of Eq. (19.9) click on **FORCES ON WIRES** under **FURTHER DISCUSSIONS** on the CD.}

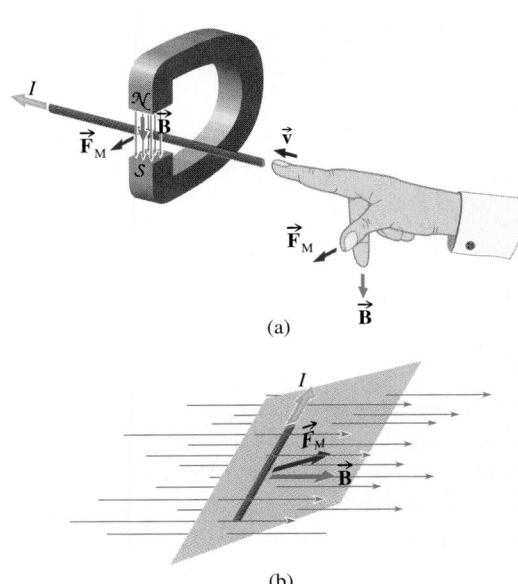

(a)

(b)

Figure 19.30 (a) The force on a current-carrying wire in a magnetic field. The current is in the direction of $\vec{\mathbf{v}}$. (b) Positive charges moving up in the direction of I and so $\vec{\mathbf{v}}$ is upward along the wire. $\vec{\mathbf{F}}_M$ is then in the direction of $\vec{\mathbf{v}} \times \vec{\mathbf{B}}$.

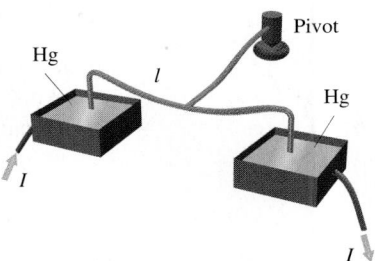

Figure 19.31 An arrangement used by Ampère to prove that F_M is always perpendicular to I. A segment of wire was set so as to freely rotate about a pivot. No matter what the direction of the applied magnetic field was, the wire never rotated because there was never a force tangent to the wire.

Example 19.4 **[II]** A flat, horizontal rectangular loop of wire is positioned, as shown in Fig. 19.32a, in a 0.10-T uniform vertical magnetic field. The sides of the rectangle are F–C equal to 30 cm and C–D equal to 20 cm. Determine the total force acting on the loop when it carries a current of 1.0 A.

Solution We've got a current-carrying loop in a B-field and that, having read the section, should suggest $F_M = IlB \sin\theta$. (1) TRANSLATION—A loop of known dimensions, carrying a specified current, is situated in a known magnetic field; determine the force it experiences. (2) GIVEN: $B = 0.10$ T, F–$C = 30$ cm, C–$D = 20$ cm, and $I = 1.0$ A. FIND: $\vec{F}_M$. (3) PROBLEM TYPE— Magnetic field/force on current-carrying wire. (4) PROCEDURE—First, make a drawing—Fig. 19.32b. Current travels from the positive terminal of the battery clockwise around the circuit. The directions of the forces on each segment are

arrived at via $\vec{v} \times \vec{B}$ and are indicated in the diagram. Because the forces on segments F–C and D–E are equal and opposite, they cancel. The total force acting on the loop is the force acting on segment C–D. (5) CALCULATION—Using Eq. (19.13),

$$F_M = IlB \sin\theta = (1.0\ \text{A})(0.20\ \text{m})(0.10\ \text{T})(\sin 90°)$$

and

$$\boxed{F_M = 0.020\ \text{N}}$$

Quick Check: A 1-A current in a 1-m wire at 90° to a 1-T field experiences a force of 1 N. Here, the length is 0.2 of that value and the field is 0.1 of it, so we can expect a force 0.02 times smaller.

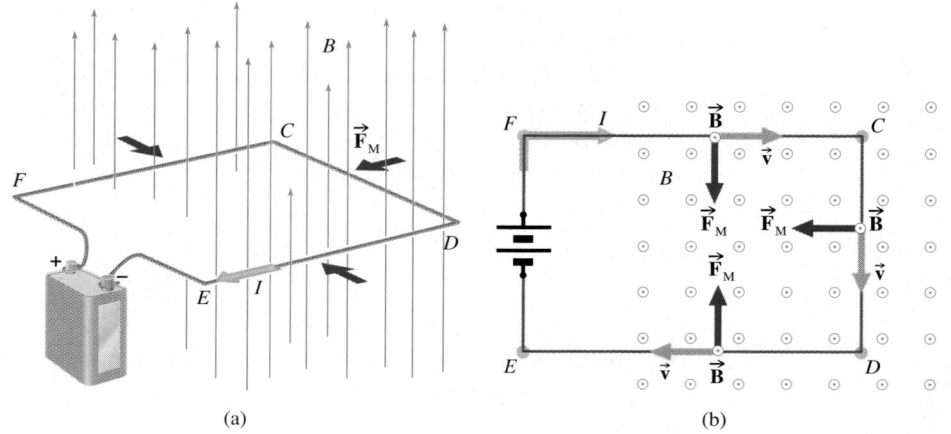

(a)   (b)

Figure 19.32 (a) A rectangular loop of wire in an idealized, uniform vertical field. (b) The forces on the wire segments are inward everywhere since the B-field is up out of the page.

The Torque on a Current Loop

Imagine a lightweight current-carrying coil in the shape of a rectangle (Fig. 19.33) supported in a vertical plane so that it turns easily about a vertical axis. The lengths of its horizontal and vertical sides are l_h and l_v, respectively. When placed in a uniform horizontal B-field, the forces on the top and bottom wires are oppositely directed and act parallel to the axis of rotation and so have no effect on the allowed motion of the coil. But the forces on the vertical segments act at a distance from that axis and produce a torque that tends to twist the coil so that it becomes perpendicular to the field. The coil is equivalent to a little dipole with the z-axis (Right-Hand-Current Rule) pointing from the south to the north pole. Like a compass, it tends to swing around such that the z-axis aligns with the B-field and $\phi = 0$.

The forces on the two vertical segments are equal and constant, independent of the angle between the z-axis and $\vec{B}$ because they are always perpendicular to the field. For each

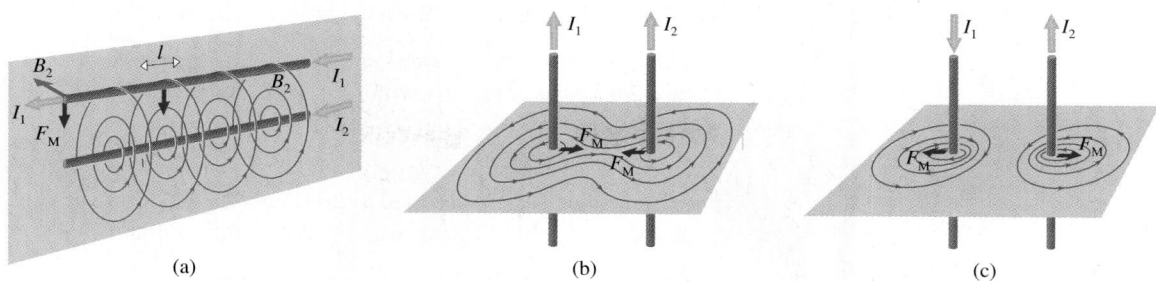

Figure 19.36 (a) Two parallel wires carrying current. Each wire is in the B-field of the other. (b) When the currents are parallel, they are forced toward one another. (c) When the currents are antiparallel, they are forced apart.

Two Parallel Wires—the Ampère

Ampère passed a current through two long parallel wires and found that they experienced equal and opposite forces. The idea was to demonstrate that a current-carrying wire interacts with the B-field in which it is immersed, regardless of whether that field is due to a permanent magnet or another current.

Figure 19.36*a* shows how the field of I_2 extends over to I_1, a distance d away, where it has a value given by Eq. (19.2) of

$$B_2 = \frac{\mu_0 I_2}{2\pi d}$$

and points directly to the left. If l is some length of the upper wire, $F_M = IlB \sin \theta$ provides the force on it where here $\theta = 90°$. Thus, the force per unit length on wire-1 due to the field of wire-2 is

[parallel current-carrying wires] $$\frac{F_M}{l} = I_1 B_2 = \frac{\mu_0 I_1 I_2}{2\pi d} \qquad (19.11)$$

The same force acts on wire-2 due to the field of wire-1 (the equation is symmetrical in the currents), except that B_1 is to the right and the force (in the direction of $\vec{v} \times \vec{B}$) is upward in the plane. Newton's Third Law makes the same prediction. The wires attract each other. Reversing either one of the currents reverses its field and the forces both reverse, becoming outwardly directed and therefore repulsive.

Example 19.5 **[I]** What is the force per unit length experienced by each of two extremely long parallel wires carrying equal 1.0-A currents in opposite directions while separated by a distance of 1 m in vacuum?

Solution We derived Eq. (19.11) for the force between two parallel currents, and it would seem now's the time to use it. (1) TRANSLATION—Two long parallel wires separated by a known distance carry a specified current in opposite directions; determine the force per unit length each experiences. (2) GIVEN: $I_1 = I_2 = 1.0$ A and $d = 1$ m. FIND: $\vec{F}_M$. (3) PROBLEM TYPE— Magnetic field/force on two current-carrying wires. (4)

PROCEDURE—First, make a drawing—Fig. 19.36*c*. The defining equation is Eq. (19.17). (5) CALCULATION:

$$\frac{F_M}{l} = \frac{\mu_0 I_1 I_2}{2\pi d} = \frac{(4\pi \times 10^{-7} \text{ T·m/A})(1.0 \text{ A})(1.0 \text{ A})}{2\pi (1 \text{ m})}$$

and so

$$\boxed{F_M/l = 2 \times 10^{-7} \text{ N/m}, \quad \text{repulsion}}$$

Quick Check: $B_2 = \mu_0 I_2/2\pi r = (2 \times 10^{-7} \text{ T}); \ F_M/l = I_1 B_2 = 2 \times 10^{-7} \text{ N/m}.$

The ideas in Example 19.5 form the basis for the SI definition of the ampere:

> *One ampere is the constant current that, if present in two infinitely long parallel straight wires one meter apart in vacuum, produces a force of exactly 2×10^{-7} newtons per meter of length.*

Practically, we can't use infinitely long wires, but very precise measurements can be made with a delicate instrument called a *current balance*. It turns out to be much more reasonable to define the ampere first and use it to define the coulomb:

> *One coulomb is that amount of charge transported in one second through any cross section of a wire carrying a current of one ampere.*

Along with the kilogram, the meter, and the second, the ampere is considered a basic unit. {For an animated look at these ideas click on FORCES ON PARALLEL CONDUCTORS under INTERACTIVE EXPLORATIONS in CHAPTER 19 on the CD.}

Core Material & Study Guide

MAGNETS AND THE MAGNETIC FIELD

Like magnetic poles repel and unlike magnetic poles attract. The SI unit for B, the magnetic field, is the *tesla* (T). The field of a magnet extends out from the north pole and into the south pole. The field lines form closed loops (p. 672). The basics of magnetism are spelled out in Sections 19.1 (Permanent Magnets) and 19.2 (The Magnetic Field). There are few problems specifically for this material, but it should be read carefully before going on.

ELECTRODYNAMICS

A straight current-carrying wire generates a circular magnetic field in the space surrounding it given by

[straight long wire] $$B = \frac{\mu_0}{2\pi} \frac{I}{r}$$ [19.2]

where μ is the permeability and for vacuum $\mu_0 = 4\pi \times 10^{-7}$ T·m/A. For a current loop of radius R

[circular loop, center] $$B_z = \frac{\mu_0 I}{2R}$$ [19.3]

If there are N turns of wire

[circular coil, center] $$B_z = N \frac{\mu_0 I}{2R}$$ [19.4]

For a long coil with n turns per unit length

[solenoid, center] $$B_z \approx \mu_0 nI$$ [19.5]

[solenoid, ends] $$B_z \approx \tfrac{1}{2}\mu_0 nI$$

All of these equations are discussed in Section 19.3 (Currents and Fields). Most of the problems in this chapter will require applying these ideas to a variety of situations; review Examples 19.1 and 19.2.

Ampère's Law

$$\sum B_\parallel \Delta l = \mu_0 \sum I$$ [19.6]

provides a means of computing magnetic fields (p. 685). If you wish to master this material, reread Section 19.4 (Ampère's Law).

MAGNETIC FORCE

The magnitude of the force on a charge moving in a B-field depends on the angle (θ) between $\vec{v}$ and $\vec{\mathbf{B}}$:

$$F_{\mathrm{M}} = q.vB \sin\theta$$ [19.7]

where the direction of $\vec{\mathbf{F}}_{\mathrm{M}}$ is the direction of $\vec{v} \times \vec{\mathbf{B}}$ (p. 688). Reread Section 19.5 (The Force on a Moving Charge) and study Example 19.3.

The force on a current-carrying wire of length l is

[force on a moving q.] $$F_{\mathrm{M}} = IlB \sin\theta$$ [19.9]

The torque on a coil of N turns with an area A, making an angle ϕ with the field, is

[current loop] $$\tau = NIAB \sin\phi$$ [19.10]

The force per unit length exerted on each other by two current-carrying straight wires is

[parallel wires] $$\frac{F_{\mathrm{M}}}{l} = I_1 B_2 = \frac{\mu_0 I_1 I_2}{2\pi d}$$ [19.11]

One ampere is the constant current that, if present in two infinitely long parallel straight wires one meter apart in vacuum, will produce a force of exactly 2×10^{-7} newtons per meter of length. All of this material, which is the source of innumerable problems, is discussed in Section 19.6 (Forces on Wires). Examples 19.4 and 19.5 are typical.

Key Terms

lodestone	permeability
permanent magnet	Meissner Effect
magnetic pole	electrodynamics
monopole	Right-Hand-Current Rule
magnetic field	solenoid
diamagnetism	Ampère's Law
paramagnetism	cross product
ferromagnetism	magnetic dipole moment
domains	commutator
Curie temperature	brushes

Discussion Questions

1. Assuming that space is isotropic, discuss the symmetry of the electric field of a point-charge, both at rest and in uniform motion. What can you expect for the symmetry of the magnetic field of a charge moving with a constant velocity (i.e., a current)?

2. Why is a chunk of iron attracted to either pole of a magnet? What does this tell you about the reliability of attraction as a test of whether something is permanently magnetized or not? A classic puzzle involves two seemingly identical rods, one steel and magnetized (with poles at its ends), the other soft iron and not magnetized. How can you tell which rod is which, using nothing else and not bending or breaking either rod?

3. James Clerk Maxwell confirmed experimentally that the *B*-field of a long straight current-carrying wire drops off inversely with distance. His apparatus is pictured in Fig. Q3 and consists of a lightweight disk, free to rotate, on which rest four bar magnets. No matter how large the current through the central wire, there was never any rotation of the disk. Given that the poles are at distances of R_S and R_N from the wire, use the idea of fictitious magnetic charges and the torques they would experience to explain the experiment.

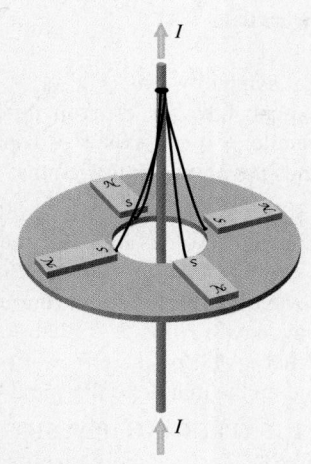

Figure Q3

4. An electric door bell is basically an automatically interrupted electromagnet. Referring to Fig. Q4, describe how it works.

5. Figure Q5 shows a horseshoe magnet with and without a piece of iron, called a *keeper*, across its poles. Explain what happens to the field. To understand the virtues of a keeper, realize that an ordinary bar magnet produces a field that extends externally from N to S and then continues through the magnet to form closed loops. But we can think of the poles at the bar's ends setting up an opposing internal field (N to S), which tends to demagnetize the magnet. The keeper essentially cancels these bare poles via induction. Figure Q5*c* shows the arrangement (with two soft-steel keepers) usually used to package a pair of bar magnets—explain how it works.

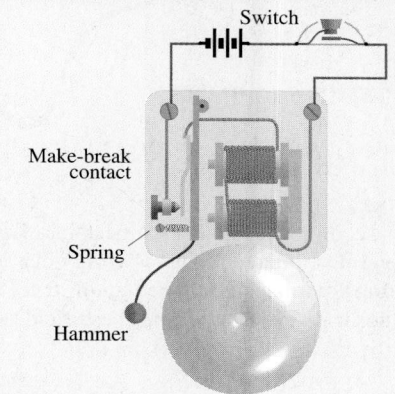

Figure Q4

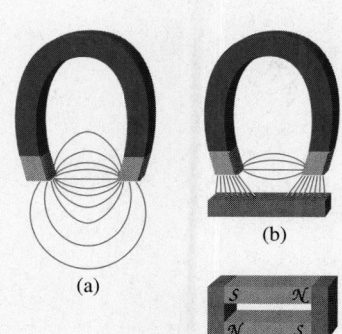

Figure Q5

6. Suppose you placed a candle flame between the poles of a powerful electromagnet so that it was in a nonuniform high-field region. What, if anything, would happen to the flame? Why? What would happen to a soap bubble filled with smoke?

7. A good way to demagnetize the heads of a tape recorder (or anything else) is to send an alternating current—one that reverses direction periodically—through a nearby coil, which is then slowly moved away from the heads, decreasing *B*. Discuss how this process will demagnetize an object.

8. Figure Q8 is a schematic diagram of a setup for tape-recording the output of a microphone. Explain how it works.

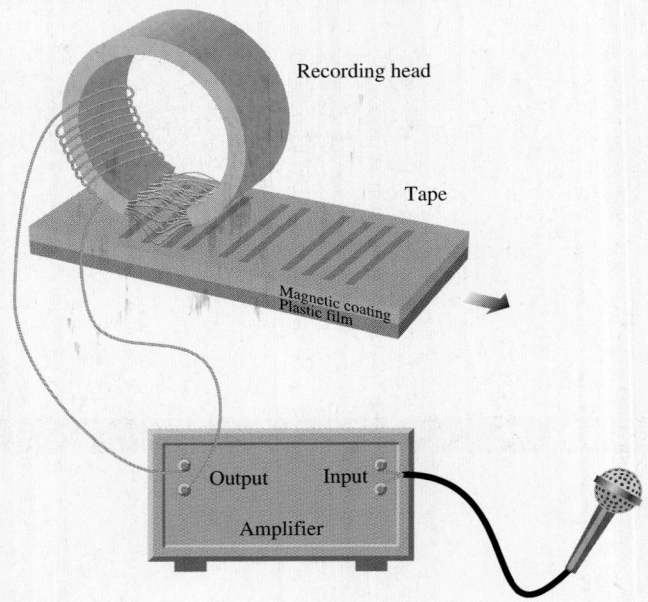

Figure Q8

9. A speaker consists of a coil fixed to the back of a flexible cone, as in Fig. Q9. The coil is mounted in an assembly that is essentially a hollow cylindrical permanent magnet (south pole) and a solid cylindrical core (north pole). The permanent B-field is radial out

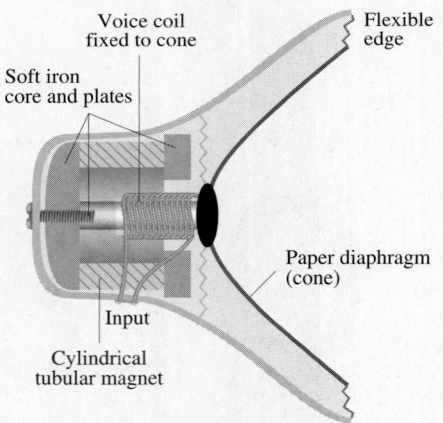

Figure Q9

from the center, and the coil wire is everywhere perpendicular to it. Describe how the device works.

10. A lovely way to shield against the Earth's magnetic field was used in a monopole experiment at Stanford University. A deflated lead-foil balloon is cooled below its critical temperature. Inflating the balloon then provides a field-free region inside of its hollow. How does that work?

11. Figure Q11 shows a straight section of wire carrying a downward current surrounded by a spiral of wire carrying the same current but upward. What will happen to these wires (they are pivoted and free to rotate as a unit) when a current-carrying coil or a bar magnet is brought nearby? Discuss the situation from the perspec-

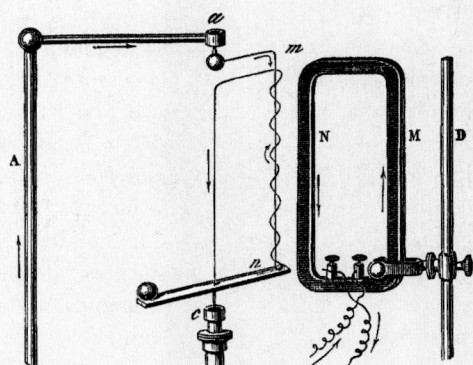

Figure Q11

tive of Ampère's Law. How does it relate to the field outside of a solenoid as considered in Fig. 19.25*b*?

12. It's been known since the late nineteenth century that a short coil can act like a lens, deflecting charged particles toward the cen-

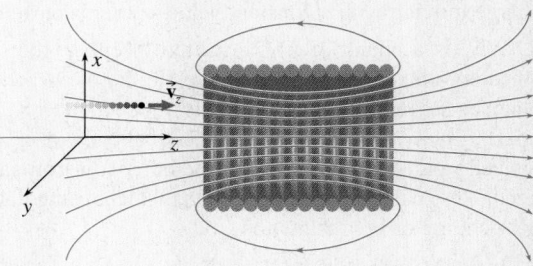

Figure Q12

tral symmetry axis (Fig. Q12). Accordingly, explain how the electron initially traveling parallel to the *z*-axis, but displaced from it, ends up moving in toward that axis.

13. A very flexible helical coil is suspended (Fig. Q13) so that its lower end just dips into a cup of mercury. What will happen when a sizable current is sent through the coil? Incidentally, this is called *Roget's Spiral*. What would be the effect of putting an iron rod up the middle of the spiral?

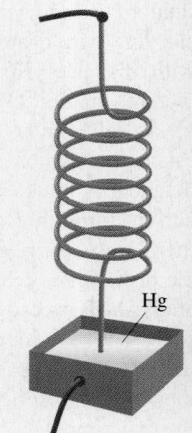

Figure Q13

14. **EXPLORING PHYSICS ON YOUR OWN:** Most radiators, steel wastepaper baskets, and metal garbage cans in the world are magnetized. Explain how this might happen naturally, and figure out the likely polarity. Use a compass to confirm your conclusions. Steel ships are also magnetized, which prompted someone to wrap ac current-carrying coils around ships during World War II in order to foil magnetic mines. W. Gilbert makes the following observation (in his book of 1600) concerning a "glowing mass of iron": "Let the smith be standing with his face to the north, his back to the south. Let him always, while he is striking the iron direct the same point of it toward the north and let him lay down that end toward the north [during cooling]." What will happen to the iron?

15. In 1821, Faraday devised a primitive motor he called a *rotator*. Actually, he produced two versions of it in a single unit (Fig. Q15).

One had a pivoted wire carrying a sizable current swinging around a vertical, fixed bar magnet immersed in mercury. The other had a pivoted magnet rotating around a fixed current-carrying wire, also in mercury. Explain how they worked.

Figure Q15

16. Describe what you think will happen to a long glass tube filled with very fine iron filings when it is placed in a strong magnetic field, shaken while still in that field, and then gently removed. What effect will subsequently shaking it have? Compare this with a solid bar of iron from the perspective of domains.

17. Figure Q17 shows a small cylindrical permanent magnet floating above a superconducting tin disk bathed in liquid helium at ≈1.2 K. The magnet was placed on the disk, and the latter was cooled below its transition temperature, at which point the magnet spontaneously jumped into the air. Explain what happened. This same kind of magnetic levitation is being applied via high-temperature superconductors to produce frictionless magnetic bearings for gyroscopes, computer disk drives, and the like.

Figure Q17

18. Since the field of a narrow coil changes linearly with z at a distance of around $R/2$, as shown in Fig. 19.19b, Helmholtz put two such coils (each with N turns) parallel to one another a distance R apart. The result is a fairly uniform B-field over the large central region given by

$$B_z \approx \frac{0.72\mu_0 NI}{R}$$

and depicted in Fig. Q18. Discuss the advantages of these Helmholtz coils as a provider of uniform field (see the photo on the upper right of p. 690). In what direction does the current progress in the coils in the diagram? How much current would it take to provide a region in which the Earth's field was canceled for coils of 1.0-m diameter with 200 turns each?

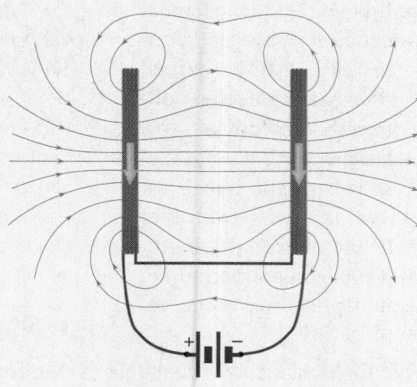

Figure Q18

19. Explain the meaning of each of the Key Terms on page 697.

Multiple Choice Questions

1. The field-line pattern around the two bars in Fig. MC1 shows that (a) neither bar is a permanent magnet (b) both bars must be permanent magnets with like poles adjacent to each other (c) either both bars are permanent magnets with like poles adjacent, or one is permanent and one is a soft iron bar (d) both must be identical permanently magnetized bars with opposite poles adjacent to each other (e) none of these.

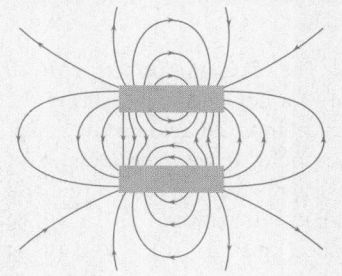

Figure MC1

2. The field-line pattern around the two bars in Fig. MC2 shows that (a) neither bar is a permanent magnet (b) both bars must be permanent magnets with like poles adjacent to each other (c) either

both bars are permanent magnets with like poles adjacent, or one is permanent and one is a soft iron bar (d) both bars must be permanently magnetized with opposite poles adjacent to each other (e) none of these.

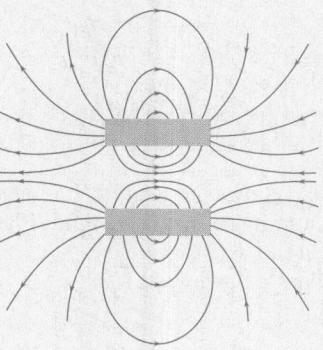

Figure MC2

3. Iron filings are sprinkled around the two bars in Fig. MC3 making a pattern showing that (a) neither bar is a permanent magnet (b) both bars must be permanent magnets with like poles adjacent to each other (c) the top bar is soft iron, the bottom is a permanent magnet (d) both must be identical permanent bar magnets with opposite poles adjacent to each other (e) none of these.

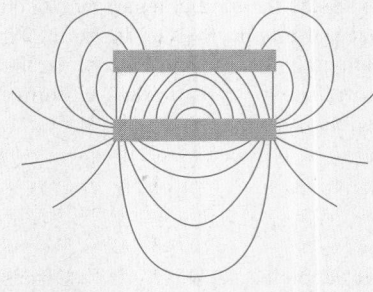

Figure MC3

4. A long, magnetized needle is floated vertically with its north pole up. It is held in place, and a bar magnet is brought near, as shown in Fig. MC4. When released, the floating needle will (a) stay exactly where it is (b) rush toward the north pole of the magnet in a straight line (c) swing in an arc away from the north pole and over to the south pole (d) move in a straight line to the south pole (e) none of these.

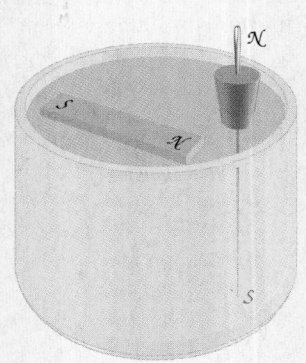

Figure MC4

5. The net force on a magnetic dipole in a uniform magnetic field is (a) toward the north pole (b) toward the south pole (c) zero (d) not enough information given (e) none of these.

6. At the instant shown in Fig. MC6 (assuming no interaction between them), which of the little magnets experiences a net downward magnetic force? (a) 1, 2, and 5 (b) 1 and 5 (c) 2 and 4 (d) 1, 3, 5, and 6 (e) none of these.

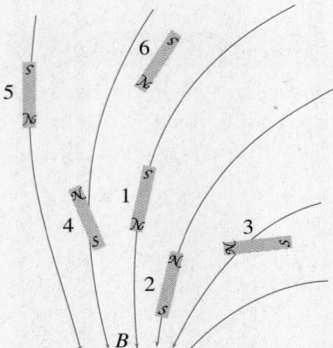

Figure MC6

7. Figure MC7a shows the B-field of a long current-carrying wire immersed in a uniform magnetic field. Figure MC7b depicts the resultant field. The wire experiences (a) zero force (b) a force to the right (c) a downward force (d) a force to the left (e) none of these.

8. Referring to Fig. MC8 (and taking "up" as out of the plane) (a) particle 1 experiences an upward force while particle 3 experiences

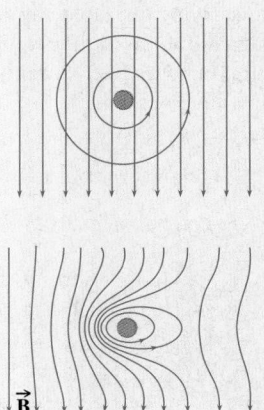

Figure MC7

a downward force (b) particle 1 experiences an upward force while particle 4 experiences a force due north (c) particle 3 experiences an upward force while particle 2 experiences no force at all (d) particle 4 experiences an upward force while particle 2 experiences a downward force (e) none of these.

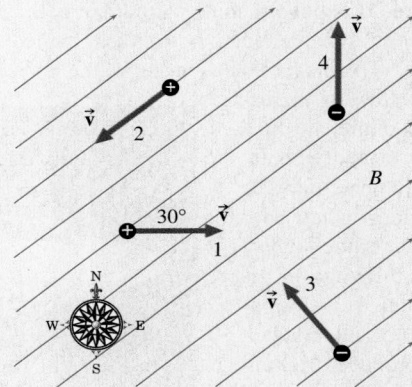

Figure MC8

9. A charge can move through a magnetic field and not experience a force by (a) moving quickly (b) traveling parallel to $\vec{B}$ (c) moving perpendicular to $\vec{B}$ (d) traveling very slowly (e) none of these.

10. The diagram of Fig. MC10 shows a device for measuring the vertical force on a sample in a magnetic field. If we hang a chunk of fresh apple from its end (a) it will be drawn downward (b) it will be pushed upward (c) it will swing to the left (d) nothing will happen (e) none of these.

11. If the force on a current element is imagined as arising from a flow of positive charge carriers, how will it change if the same current is considered to be due to an oppositely directed flow of negative carriers? (a) it will be unchanged (b) it

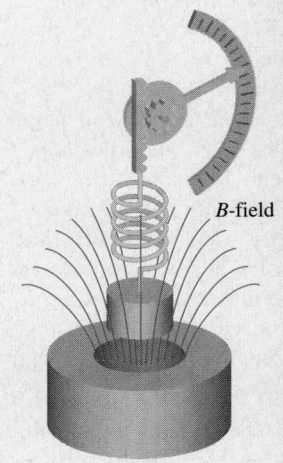

Figure MC10

will reverse direction   (c) it will be perpendicular   (d) it will be at 45° in the first quadrant   (e) none of these.

12. A straight length of current-carrying wire is in a uniform magnetic field. If the wire does not experience a force,   (a) everything is as it should be   (b) it must be parallel to $\vec{B}$ (c) we have an impossible situation   (d) the wire must be perpendicular to $\vec{B}$ (e) none of these.

The next five questions refer to Fig. MC13, which shows a solenoid in series with a battery.

13. Where is the north pole of the solenoid? Near (a) C (b) A (c) D (d) B (e) none of these.

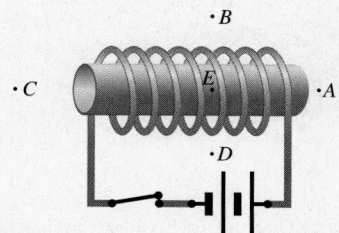

Figure MC13

14. Where, outside of the coil, is the B-field strongest? (a) C (b) A (c) D (d) B (e) none of these.

15. If a small compass is placed at point-D, in what direction will its needle point? (a) any direction since there is no field at D (b) straight up (c) to the right (d) to the left (e) none of these.

16. If the north pole of a bar magnet is brought to point-A (a) the bar magnet will be repelled (b) the solenoid will be repelled (c) the bar magnet will be attracted but the coil will be repelled (d) the bar magnet will be repelled but the coil will be attracted (e) none of these.

17. The B-field at A is roughly (a) twice the field at E (b) $\sqrt{2}$ the field at E (c) equal to the field at E (d) half the field at E (e) none of these.

18. Which of the diagrams in Fig. MC18 best represents the magnetic field of the solenoid? (a) 1 (b) 2 (c) 3 (d) 4 (e) none of these.

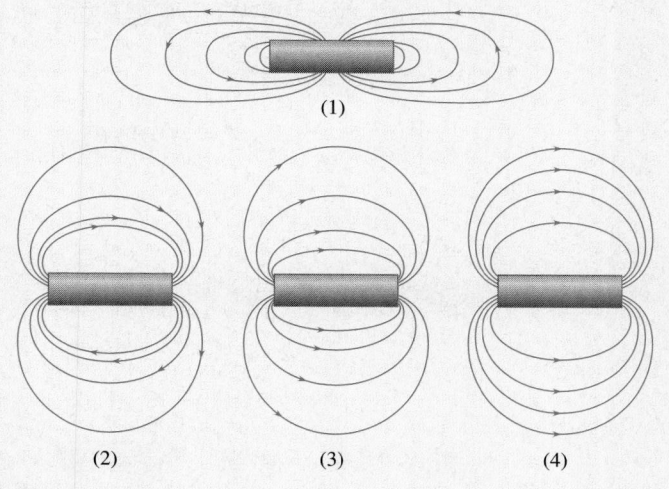

Figure MC18

19. A circular flat coil of N turns and enclosed area A, carrying a current I, has its symmetry z-axis parallel to a uniform B-field in which it is immersed. The torque on the coil is   (a) zero   (b) $NIBA$ (c) NBA (d) IBA (e) none of these.

20. Figure MC20 shows a compass with several turns of wire wrapped around it in the north-south direction. Such a device can best be used in a circuit to   (a) measure power   (b) indicate the presence of resistance   (c) measure a voltage difference   (d) indicate the presence of current   (e) none of these.

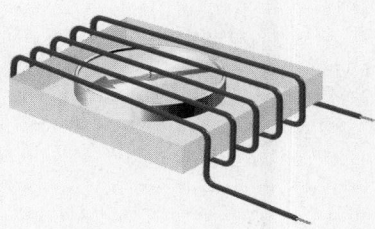

Figure MC20

21. Just wrap 150 turns of a heavy (#22) insulated wire around an iron rod (for example, a door hinge pin) and attach the leads to a 1.5-V D-cell via a switch and you have   (a) a radio   (b) an electromagnet   (c) a galvanometer   (d) an ammeter   (e) none of these.

22. The American physicist Henry Rowland (1876) placed some charges (in a fixed location) on a nonconducting disk, which he then rotated at high speed about its central axis near a delicate compass. (a) the current created a B-field that deflected the compass   (b) there was no E-field, so nothing happened to the compass   (c) the charges attracted the compass, which moved toward the disk   (d) the charges repelled the compass via Coulomb's Law   (e) none of these.

23. Which way will the bar move in Fig. MC23 once the current is established?   (a) straight up   (b) to the left   (c) to the right   (d) it will not move at all   (e) none of these.

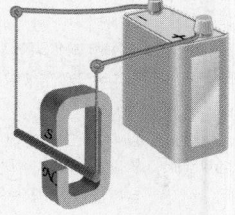

Figure MC23

24. The horseshoe-shaped iron bar in Fig. MC24 is wrapped with a wire, and once the dc current is turned on   (a) it becomes demagnetized   (b) the right end becomes a south pole and the left a north pole   (c) the left end becomes a south pole and the right a north pole   (d) both ends become north poles and the bottom becomes a south pole   (e) none of these.

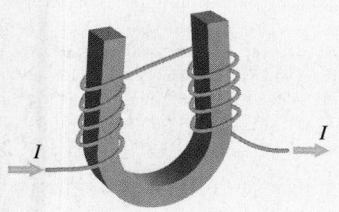

Figure MC24

25. If the current through a long solenoid is doubled while the coil's length is also doubled, keeping the total number of turns constant, the magnetic field at a point inside near the axis is   (a) four times larger   (b) half the original size   (c) unchanged   (d) one-quarter the size   (e) none of these.

26. The two (noninteracting) samples (1 and 2), which have swung into alignment in the B-field, as shown in Fig. MC26, may be, respectively,   (a) ferromagnetic and paramagnetic   (b) diamagnetic and ferromagnetic   (c) paramagnetic and diamagnetic   (d) paramagnetic and ferromagnetic   (e) none of these.

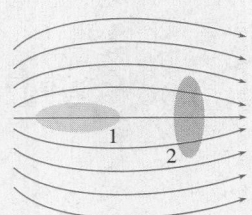

Figure MC26

27. The movable loop in Fig. MC27 and the coil carry currents in the directions indicated. The loop will   (a) looking down from above, rotate counterclockwise   (b) experience no force   (c)

looking down from above, rotate clockwise   (d) experience an upward force   (e) none of these.

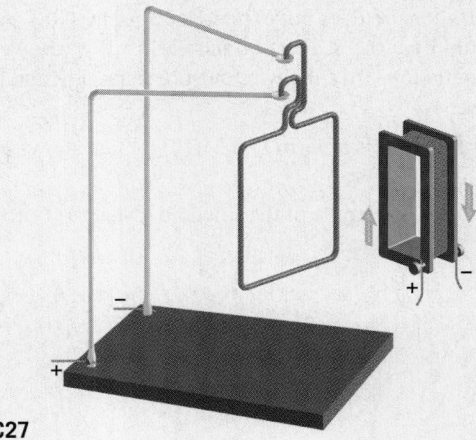

Figure MC27

For more Multiple Choice Questions with answers click on WARM-UPS in CHAPTER 19 on the CD.

Suggestions on Problem Solving

1. This chapter has several equations that depend directly on q and produce reversed results when the sign of the charge changes. As soon as you see you are dealing with negative charges, **be extra careful of directions**.

2. It's helpful to remember that $\mu_0/4\pi = 10^{-7}$ N/A^2 and so $\mu_0/2\pi = 2.001 \times 10^{-7}$ N/A^2, where $\mu_0 = 1.257 \times 10^{-6}$ N/A^2.

3. Physicists don't usually memorize all the equations for the fields arising from the various current configurations. Instead, many just remember Ampère's Law and derive what they need. When applying Ampère's Law to a new situation, first establish the direction of

the B-field. That's usually done via symmetry arguments and the realization that the field lines are closed. Since there are no monopoles in magnets, B-field lines are not going to be radial, as they were with the E-fields of a small spherical charge or a line charge. Select an Ampèrian loop that is everywhere perpendicular or parallel to $\vec{B}$. Any lengths or distances that are specific to your Ampèrian path and not arbitrary must cancel out of the final equation for B.

Problems ✦ Coordinated Problems ✦ Progressive Problems ✦ Solutions

STUDY GUIDE   **1. Coordinated Problems:** The three problems within each magenta-colored grouping are solvable in similar ways. Note that the first of these always has a hint; moreover, its solution is provided in the back of the book. *Work out each of these sets; they'll strengthen technique and build confidence.* **2. Progressive Problems:** The problems introduced in blue unfold step-by-step carrying along the analysis in a more suggestive way than is customary. *Work out all of these; they'll guide you through the analytic process and help develop problem-solving skills.* **3. Worked-Out Solutions:** Studying worked-out solutions is an important part of learning how to solve problems. Accordingly, additional *solutions* to a number of model problems are given below. *Make sure you understand each of them before you go on to the next problem.* **4.** Also provided in the back of the book are the *Answers* to all odd-numbered problems, as well as worked-out *solutions* to those with boldface numbers. Problem numbers in italic indicate that a solution appears in the Student Solutions Manual.

SECTION 19.2: THE MAGNETIC FIELD
SECTION 19.3: CURRENTS AND FIELDS

1. [I] A small bar of annealed soft iron is placed between two magnetic poles as shown in Fig. P1. Which of the rough sketches best represents the resulting B-field? Explain your answer and state what's wrong with each incorrect drawing.

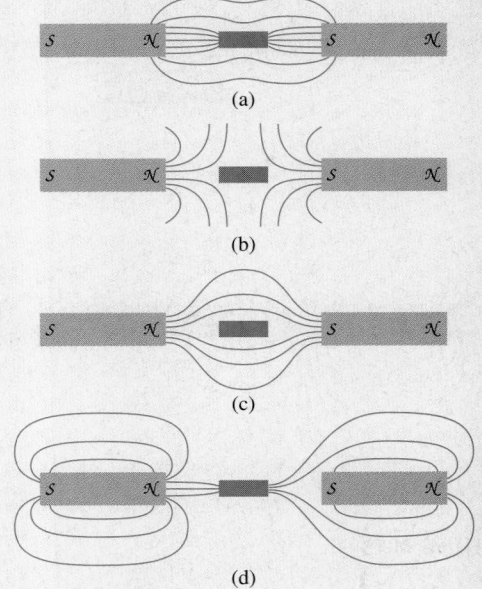

Figure P1

(a)

(b)

(c)

(d)

2. [I] Two bar magnets cling to a soft iron keeper as shown in Fig. P2. Which of the rough sketches best represents the resulting *B*-field? Explain your answer and state what's wrong with each incorrect drawing.

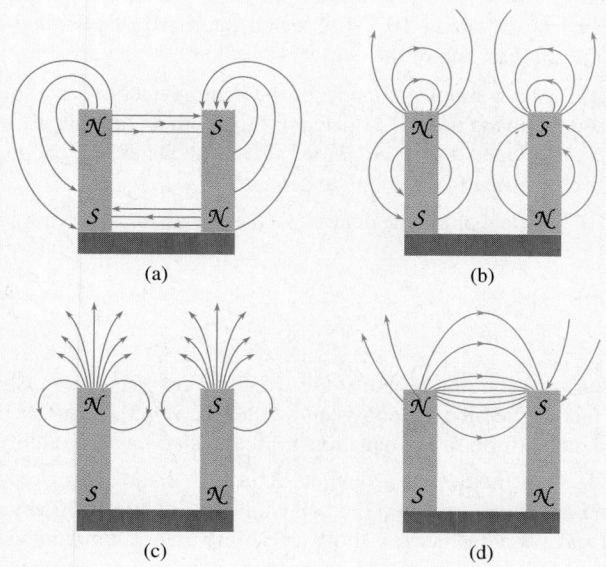

(a)

(b)

(c)

(d)

Figure P2

3. [I] Two mutually perpendicular horizontal uniform magnetic fields, B_1 in a northerly direction and B_2 in an easterly direction, overlap in a region of space. Knowing that the fields add vectorially, show that a small compass needle will align itself at an angle θ north of east such that

$$B_1/B_2 = \tan \theta$$

4. [I] The most common pre-SI unit of magnetic field, one still to be found in use, is the *gauss* (1 tesla = 10^4 gauss). The magnetic field in intergalactic space is around 10^{-6} gauss, as compared with the 8-kilogauss field of a powerful samarium-cobalt permanent magnet. Express these quantities in teslas.

5. [I] Determine the *B*-field 50 cm from a long narrow straight wire carrying a current of 10 A immersed in air. [*Hint: The field is directly proportional to I, inversely proportional to r, and cylindrically symmetric; the medium has a permeability of μ_0.*]

6. [I] A magnetic field of 6.0 μT is to be produced 10 cm from a single long straight wire in air. How much current must it carry?

7. [I] The return stroke of a bolt of lightning typically carries a peak current of 20 kA up from the ground. What is the maximum magnetic field associated with the bolt 1.0 m away?

8. [I] A long straight wire carrying a current of 4.00 A is suspended in air and the field around it is measured with a magnetometer. It is found that at some point *P* the field is 0.660×10^{-5} T. How far from the wire is *P*?

9. [I] A straight wire 1.00 m long having a resistance of 1.2 Ω is attached to a 12-V battery. What is the magnitude of its *B*-field 2.0 cm away in air?

10. [I] A long, somewhat wiggly copper wire carries a current of 20 A. What is the *B*-field at a distance of 50 cm, given that this dimen-

sion is much greater than those of any of the bends in the wire? (See Discussion Question 11.)

11. [I] A long straight wire carries a current of 6.0 A. At what perpendicular distance from the middle of the wire will the magnetic field have a magnitude of 1.2 gauss? (See Problem 4.)

SOLUTION: 1.2 gauss = 1.2×10^{-4} T. $B = \mu_0 I/2\pi r$ and $r = \mu_0 I/2\pi B =$ $2 \times 10^{-7}(6.0$ A$)/(1.2 \times 10^{-4}$ T$) = 0.010$ m.

12. [I] THIS PROBLEM DEALS WITH THE MAGNETIC FIELD DUE TO A CURRENT-CARRYING COIL. Figure P12 shows a tight two-turn coil in air having a resistance of 6.0 Ω connected to a 18-V battery. The loop has a radius of 1.5 cm. (a) How much current will flow through the coil when the switch is closed? (b) In what direction will it flow, clockwise or counterclockwise? (c) In what direction is the *B*-field (generated by the current) at the center of the coil? (d) How big is the *B*-field produced by each turn at its center? (e) How big is the net *B*-field at the coil's center?

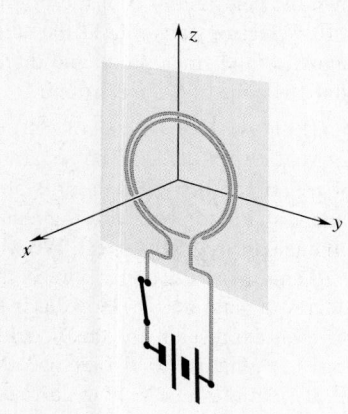

Figure P12

13. [I] A single flat loop of superconducting wire 50 cm in diameter carries a current of 25 A. What is the magnitude of the magnetic field at its center?

14. [I] A narrow flat circular coil with a diameter of 20 cm consists of 100 turns of wire. What is the magnetic field at its center when it carries a current of 5.0 A?

15. [I] A beam of protons is made to travel in a nearly circular orbit by a perpendicularly applied external *B*-field. Determine the magnetic field at the center of a 20-cm-diameter orbit produced by a 0.10-mA proton beam.

16. [I] A flat circular coil having a diameter of 25 cm is to produce a *B*-field at its center of 1.00 mT. If it has 100 turns, how much current must be provided to it?

17. [I] We wish to make a hollow solenoid 10 cm long and 1.5 cm in diameter having 200 turns of wire. How much current must we send through it to produce a field of roughly 0.50 mT inside the coil?

18. [I] An air-filled solenoid is to be 80 cm long and carry a current of 20 A. How many turns of a superconducting wire should it have if the field inside it near the middle is to be roughly 2.0 T?

19. [I] An air-core solenoid has 100 turns per centimeter and a resistance of 60 Ω. Determine the magnetic field inside it near its middle when it is connected across a 12-V battery. [*Hint: The field depends on the current in the coil.*]

20. [I] A thousand turns of fine wire are wrapped around a thin hollow cardboard core, making a 100-Ω coil 20.0 cm long and 2.00 cm in diameter. What *B*-field will exist inside the coil at its center when placed across a 20.0-V dc source?

21. [I] We wish to pass 200 mA through a 25.0-cm-long solenoid

in order to generate a field of 40.0 mT at its air-filled center. How many turns of wire must it have?

22. [I] A 5.0 cm-diameter solenoid 50 cm long is made by wrapping four layers each of 1000 turns of wire on a thin hollow plastic core. Calculate the approximate B-field generated near the coil's ends when it carries a current of 1.5 A.

23. [II] Suppose we remove the magnetic deflection yoke from the neck of a TV picture tube so that the electron beam travels straight down the central axis. For the brightness level given, there are 6.0 $\times 10^{12}$ electrons arriving at the screen per second. Determine the magnetic field (magnitude and direction) caused by the beam at a radial distance of 1.5 cm from it.

24. [II] THIS PROBLEM EXPLORES THE MAGNETIC FIELD OF TWO CUR-RENT-CARRYING WIRES. The separation between two long straight horizontal parallel wires in air is 50 cm. The wires lie in a vertical plane with the top one carrying a current of 6.0 A east and the bottom one carrying 6.0 A west. We want the magnetic field at a point P, 20 cm below the upper wire in the plane containing both wires. (a) Draw a diagram. (b) How far is P from each wire in meters? (c) Write an expression for the B-field of a current-carrying straight wire. (d) What is the $\vec{B}$-field at P due to each current? (e) What is the $\vec{B}$-field at P due to both currents?

25. [II] THIS PROBLEM DEALS WITH THE MAGNETIC FIELD OF TWO CUR-RENT-CARRYING WIRES. Two long straight horizontal parallel wires in air lie in a vertical east-west plane. The wires are 196 cm apart, and the top one carries a current to the west of 1.1 A. The bottom one carries a current of 2.2 A also to the west. We want the magnetic field at P, midway between the wires in the plane containing them. (a) Draw a diagram. (b) How far is P from each wire in meters? (c) Write an expression for the B-field of a current-carry-ing straight wire. (d) What is the $\vec{B}$-field at P due to each current? (e) What is the $\vec{B}$-field at P due to both currents?

26. [II] Two long horizontal straight parallel wires are 28.28 cm apart, and each carries a current of 2.0 A in the same direction, namely, due south. What is the B-field at a point that is a perpendicular distance of 20 cm from both wires?

27. [II] A superconducting niobium wire will return to the normal state when the magnetic field at its surface exceeds 0.100 T. If the wire has a diameter of 2.00 mm, what is the *critical current*; that is, what maximum current can it carry without quenching the super-conducting state?

28. [II] Two long cur-rent-carrying wires are depicted in Fig. P28. What is the value of the B-field at point P, 1.0 m away from the crossing point?

29. [II] Considering Problem 26, what would be the magni-tude of B at P if one of the currents was reversed?

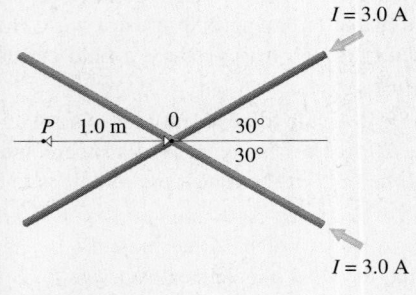

Figure P28

30. [II] Show that the dimensions of $\varepsilon_0\mu_0$ correspond to those of 1 over speed squared: $[T^2]/[L^2]$. This point will come up again later when we see that $\varepsilon_0\mu_0 = 1/c^2$. Indeed, since the speed of light is

now defined to be exactly 299 792 458 m/s, both ε_0 and μ_0 are also exact.

31. [II] A long straight vertical wire carries a vertically upward cur-rent of 15 A. A small horizontal compass is placed 10 cm away due north of the wire. At that point, the Earth's B-field has a horizontal component of 0.50 $\times 10^{-4}$ T directed due north. Determine the equilibrium direction of the compass needle.

32. [II] A rather simplistic model of the hydrogen atom has a single electron revolving around a nuclear proton with an orbital radius of 0.53 $\times 10^{-10}$ m at a speed of 2.19 $\times 10^6$ m/s. Determine the mag-netic field at the proton due to the electron.

33. [II] The field along the central z-axis of a current-carrying loop is

$$B_z = \frac{\mu_0 I R^2}{2(R^2 + z^2)^{\frac{3}{2}}}$$

where $z = 0$ is at the center of the circular loop of radius R. Show that this expression is consistent with Eq. (19.3). What is the approximate form of the equation at great distances away along z?

34. [II] A solenoid 30 cm long and 2.0 cm in diameter is wrapped around an aluminum core. The coil contains 500 turns of fine insu-lated wire carrying 5.0 A. If the permeability of the aluminum in the field is 1.257 $\times 10^{-6}$ T·m/A, what is the value of B produced at the center of the coil?

35. [II] Consider the Earth (radius 6.4 $\times 10^3$ km) from the perspec-tive of the Dynamo Theory, which maintains that the B-field is due to currents in the spinning planet. Given that its liquid iron outer core has a mean radius of about 2.3 $\times 10^3$ km and that the field at either pole is about 0.6 $\times 10^{-4}$ T, approximate the net current that would produce the dipole magnetic field. (Make your model as sim-ple as possible, but discuss its shortcomings.) Draw a rough sketch of the planet and the currents. In what direction does I progress? (Take a look at Problem 33 and Fig. 19.19.)

36. [II] Consider the solenoid of Problem 34, this time with a sili-con-iron core having a relative permeability of 600 times that of free space when immersed in the field. Determine the approximate field at the center of the coil.

SECTION 19.4: AMPERE'S LAW

37. [I] A uniform magnetic field exists in a region of space. Use Ampère's Law to show that there can be no net electric current in that region.

38. [I] Return to Fig. 19.23 and derive an expression for the B-field far from the bundle of current-carrying wires that are surrounded by air.

39. [I] With the previous problem in mind, suppose that in Fig. 19.23, $I_1 = 1.0$ A, $I_2 = 3.0$ A, and $I_3 = 2.0$ A. What is the $\vec{B}$-field at a radial distance from the wire of 1.20 m?

40. [I] Imagine a long thin-walled gold tube of radius R immersed in air. Suppose it's put across the terminals of a battery so that a current I flows along it from one end to the other. What is the $\vec{B}$-field in the air-filled region within the tube? Assume the current is uniformly distributed within the wall.

41. [I] A long, fairly straight length of ordinary 1-in. diameter cop-per water pipe is used to carry 50.0 A of current across a laborato-ry. Determine the resulting B-field at a radial distance of 1.00 m

from the pipe, somewhere near the middle. [*Hint: Use an Ampèrian loop concentric with the axis of the wire.*]

42. [I] A coaxial cable (like those used in cable TV installations and with VCRs) consists of a central conducting wire embedded in a cylindrical insulating core, which is then surrounded by another outer conducting cylinder usually made of wrapped aluminum foil or copper braid. Suppose such a cable carries a current of 2.00 A flowing in one direction along the wire and 2.00 A flowing in the opposite direction along the foil. Use Ampère's Law to find the magnitude of the magnetic field at a radial distance $r = 1.5$ mm that is between conductors.

43. [I] Use Ampère's Law to find the magnitude of the magnetic field outside the coaxial cable in the previous problem.

44. [II] A long solid copper rod of radius R carries a current I that we shall take to be uniform over its cross section. Use Ampère's Law to derive an expression for the magnetic field inside the wire at a distance $r < R$. [*Hint: The field is cylindrical inside and out.*]

45. [II] Figure P45 shows a toroidal coil of N turns carrying a current I. Determine the B-field inside the coil. How does it depend on position? What is the relationship between the B-field inside the torus and that of a single straight wire along the axis of symmetry of the coil carrying a current of NI?

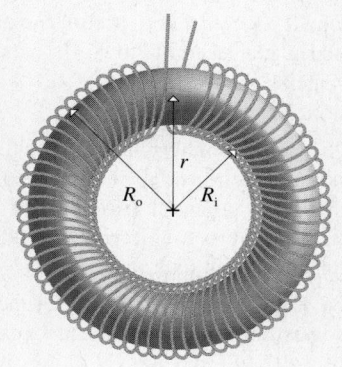

Figure P45

46. [II] In 1962, scientists at the Geophysical Observatory in Tulsa, Oklahoma, measured a magnetic field of 1.5×10^{-8} T being generated by a large tornado about 9.0 km away. Use Ampère's Law to approximate the current swirling downward within the funnel.

47. [II] With Problem 45 in mind, use Ampère's Law to prove that the B-field lies entirely within the confines of the torus.

48. [II] Why is it that a long solenoid can be regarded as a toroidal coil with an infinite radius?

49. [II] Figure P49 shows a portion of an infinite conducting sheet carrying a current (traveling in the z-direction) per unit of width (along the x-axis) given by i amps per meter. Imagine the sheet as if it were made up of an infinite number of adjacent parallel wires. Use symmetry arguments and the Right-Hand-Current Rule to establish that $\vec{B}$ is a constant everywhere (independent of distance) above and below the sheet. Further, show that B points in the negative x-direction above the sheet and in the positive x-direction below the sheet. Using Ampère's Law, show that

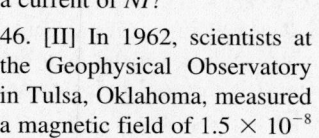

$$B = \tfrac{1}{2}\mu_0 i$$

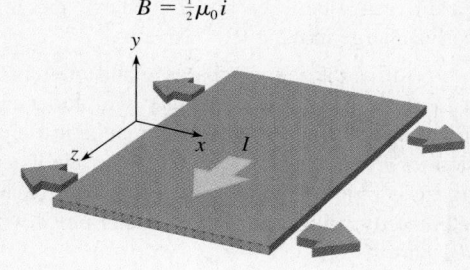

Figure P49

50. [III] A long straight hollow cylindrical conductor made of copper (for which $\mu \approx \mu_0$) with an inner radius of R_i and an outer radius of R_0 carries a current I that is distributed uniformly over its cross section and propagates along its very long length. Given that it is immersed in air, determine the magnetic fields in the inner and outer air-filled regions as well as in the conductor. Draw a graph of B against radial distance from the central axis.

51. [III] Imagine two horizontal parallel infinite planes, one above the other. If these are each thin conductors carrying equal currents, the top one traveling west to east, the bottom one east to west, find the B-field between them if the current per unit width is i. Take a look at Problem 43. What is the field outside the sheets?

SECTION 19.5: THE FORCE ON A MOVING CHARGE

52. [I] A small ball of plastic carries a positive charge and is shot at high speed along the x-axis in the horizontal x–y plane. It passes through a uniform horizontal B-field, which makes an angle with the x-axis of $+23.5°$. In what direction is the magnetic force it will experience?

53. [I] A negatively charged projectile is fired upward in the vertical z–y plane at an angle of $+20°$ to the y-axis. It travels through a uniform magnetic field directed downward in the negative z-direction. What is the direction of the resulting magnetic force? [*Hint: Remember that this is a negative particle.*]

54. [I] A proton travels downward in the vertical x–z plane at an angle of $-40°$ with respect to the x-axis. It passes through a uniform horizontal B-field directed in the $+y$-direction. In what direction is the resulting magnetic force?

55. [I] An electron travels in the x-z plane through a uniform magnetic field in the negative y-direction. The electron is moving at an angle of $+60°$ above the $+x$-axis. Determine the direction in which it experiences a magnetic force.

56. [I] Pulses of electrons are fired at increasing speeds at a fixed angle of $20°$ across a uniform magnetic field. Which of the curves in Fig. P56 best represents the magnetic force on an electron as a function of its speed?

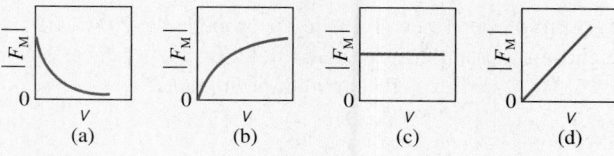

Figure P56

57. [I] A proton, sailing along inside a synchrotron at 2.99×10^7 m/s, experiences a perpendicular B-field of 3.8 T. What is the magnitude of the magnetic force it will experience?

SOLUTION: Because $\theta = 90°$ the expression $F_M = q_e vB \sin\theta$ becomes
$F_M = q_e vB = (1.60 \times 10^{-19}\text{ C})(2.99 \times 10^7\text{ m/s})(3.8\text{ T}) = 1.8 \times 10^{-11}\text{ N}$.

58. [I] A negatively charged particle traveling in the positive y-direction enters a region of space pervaded by a 1.0-T uniform magnetic field directed in the negative y-direction. Determine the direction of the magnetic force acting on the particle.

59. [I] THIS PROBLEM WILL HELP US BETTER UNDERSTAND THE FORCE ON A CHARGE MOVING THROUGH A MAGNETIC FIELD. There is a *B*-field and a charge moving across it, and we want to study the force on that charge. To be able to deal with all the directions, imagine an ordinary *xyz* coordinate system with the positive *x*-axis pointing east and the positive *y*-axis pointing north. (a) Suppose there is a substantial *B*-field of 2.99×10^4 gauss pointing north. How much is that in teslas? (b) A proton moving perpendicular to the *B*-field experiences a force of 2.10×10^{-12} N due east. What is the charge of the proton (is it positive or negative)? (c) Write a general expression relating F_M, q_e, v, B, and θ. (d) With Fig. P59 in mind, determine the particle's velocity (direction and magnitude).

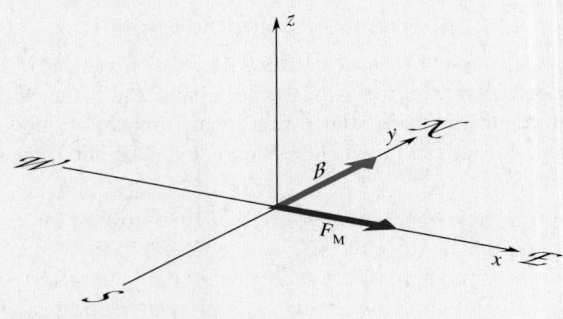

Figure P59

60. [I] THIS PROBLEM EXPLORES THE FORCE ON A CHARGE MOVING THROUGH A MAGNETIC FIELD. A uniform horizontal 3.80-T magnetic field points northeast (i.e., 45° north of east). An electron enters the field heading due north at a speed of 2.10×10^6 m/s. (a) Draw a diagram. (b) What is the charge of the electron (is it positive or negative)? (c) In what direction would the force act on a positive charge? (d) In what direction would the force act on a negative charge? (e) Write a general expression relating F_M, q_e, v, B, and θ. (f) Determine the force on the particle (direction and magnitude).

61. [I] Show that the units T·m/A and N/A² for μ are equivalent.

62. [I] Determine the magnetic force acting on a tiny sphere carrying a charge of 2.00 nC and traveling at a speed of 500 m/s perpendicular to a 2.00-T *B*-field.

63. [I] A proton in a research facility is propelled at 2.0×10^6 m/s perpendicular to a uniform magnetic field. If it experiences a force of 4.0×10^{-13} N, what is the strength of the field?

64. [I] Show that 1 T = 1 kg/s·C.

65. [I] Show that 1 T = 1 V·s/m².

66. [I] Show that 1 T = 1 N/A·m.

67. [I] An electron of mass 9.11×10^{-31} kg travels in a circular orbit within a large evacuated chamber. The orbit has a 2.0-mm radius and is perpendicular to a *B*-field of 0.050 T. What's the electron's speed?

68. [I] An electron moving at 2.5×10^6 m/s within a cathode-ray tube (CRT) makes an angle of 45° with a uniform 1.2-T magnetic field. What is the magnitude of the force it will experience?

69. [I] A beam of protons of various speeds enters a region where it encounters perpendicular electric and magnetic fields both at right angles to $\vec{v}$. Show that only particles for which $v = E/B$ will pass through undeflected. Accordingly, such an arrangement is called a *velocity selector*.

70. [I] A proton moving at 3.00×10^6 m/s through a uniform magnetic field experiences a maximum force of 5.2×10^{-12} N, straight upward when it's traveling due east. Determine $\vec{\textbf{B}}$.

71. [II] An electron having a speed of 5.0×10^6 m/s along the *x*-axis enters a region where there is a uniform 5.0-T *B*-field making an angle of 60° to the *x*-axis. Determine the magnitude of its acceleration. [*Hint:* $F_M = ma$.]

72. [II] Consider a tiny 1.00×10^{-6} kg sphere carrying a charge of 2.50 nC. The sphere is moving at 3.00×10^6 m/s in the positive *x*-direction. Given that it enters a region occupied by a uniform 0.300-T field making an angle of 30.0° to the *x*-axis, determine the magnitude of the sphere's acceleration.

73. [II] The magnetic field in a region is uniform, has a value of 0.500 T, and points in the positive *x*-direction. A small sphere with a mass of 12×10^{-5} kg and a charge of 30.0 μC is fired into that region at a speed of 950 m/s. If it enters traveling at 45.0° to the +*x*-axis, what is the size of the magnetic force it will experience?

74. [II] A cosmic ray proton traverses a uniform *B*-field perpendicularly at a speed of 2.0×10^7 m/s in the positive *x*-direction, experiencing an acceleration of 4.0×10^{12} m/s² in the positive *y*-direction. Determine $\vec{\textbf{B}}$.

75. [II] A doubly charged positive helium ion, one missing two orbital electrons, has a mass of 6.7×10^{-27} kg and is accelerated through a potential difference of 10 kV. It then enters a region in which there is a uniform perpendicular *B*-field of 1.50 T. What is its subsequent path in the field?

76. [II] A tiny plastic sphere of mass 0.10 mg is shot horizontally at a speed of 200 m/s in the *x–y* plane. Given that it carries a charge of -10 μC and travels along the positive *y*-axis, determine the nature of the smallest uniform magnetic field that will keep the particle moving horizontally despite the Earth's gravity.

77. [II] Write an expression for the momentum of a particle of charge q and mass m moving in a circular orbit of radius R in a uniform magnetic field B.

78. [II] A proton is sailing through the outer region of the Sun at a speed of 0.15c. It traverses a locally uniform magnetic field of 0.12 T at an angle of 25°. What is the radius of its helical orbit? [*Hint:* $v_\parallel$ and $v_\perp$ can be considered separately.]

79. [II] The American physicist E. H. Hall discovered (1879) that when a current travels along a conducting plate of width l, which is perpendicular to a magnetic field, a potential difference V appears across the plate as shown in Fig. P79. Prove that

$$V = vBl$$

The Hall probe makes a very convenient magnetometer. Discuss the difference you might expect if the probe is made of copper in one case, where the charge carriers are negative, and germanium, where the charge carriers are positive "holes." The Hall effect reveals a difference between positive charge moving to the right and negative charge moving left. Explain.

80. [II] A positive particle of charge q and mass m is accelerated through a potential V and subsequently bent into a circular orbit of radius R by a uniform perpendicular magnetic field B. Write an expression for R^2 in terms of V, B, and m/q; the latter is called the charge-to-mass ratio. A modern mass spectrometer allows R to be measured directly, and from that m is determined with a precision of about 1 part in 10^7.

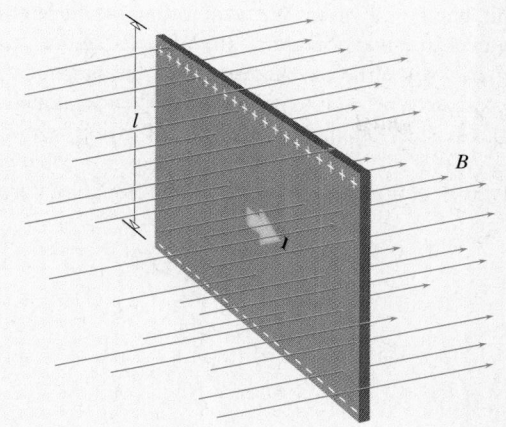

Figure P79

81. [II] With Problem 79 in mind, suppose we want to use the Hall effect to measure the blood's flow rate (v) in an artery. If we apply a transverse magnetic field, it will cause the positive and negative ions in the blood to separate, thereby producing a voltage across the artery. Given that the artery has an inner diameter of 4 mm, if a uniform 0.50-mT field produces a voltage difference of 1.0 μV, what is the flow rate?

82. [III] Suppose we have two tiny spheres 1.00 m apart, each carrying 1.00 C. If they now move along parallel straight paths (1.00 m apart) at 1.00 m/s, compare the electric force to the magnetic force they experience.

83. [III] A positively charged particle is moving in a circle under the influence of a perpendicular magnetic field. Write an expression for the frequency of its orbital motion and show that it is independent of both radius and speed. Overlook the fact that it radiates.

SECTION 19.6: FORCES ON WIRES

84. [I] The two poles of a large electromagnet create an upwardly directed vertical magnetic field. A straight length of wire running east-west passes between the poles. If a current is sent through the wire heading west, what is the direction of the magnetic force on the wire?

85. [I] A straight wire carrying an upwardly directed current is in the x–z plane making an angle of +20° with the horizontal x-axis. The wire is immersed in a uniform B-field in the positive z-direction. What is the direction of the force on the wire?

86. [I] THIS PROBLEM EXPLORES THE FORCE ON A CURRENT-CARRYING WIRE IN A MAGNETIC FIELD. A uniform horizontal 3.80-mT magnetic field points east. A straight 1.5 m long piece of wire is positioned so that it extends north-south in the field. The north end is attached to the positive terminal of a battery and the south end to the negative terminal. A current of 12.0 A flows in the wire.   (a) Draw a diagram. (b) In what direction does the current flow? (c) In what direction do the imaginary positive mobile charge carriers move? (d) What is the direction of $\vec{v} \times \vec{B}$ on those positive charges? (e) In what direction is the magnetic force on the wire? (f) Write a general expression relating F_M, I, l, B, and θ. (g) What is the value of θ in this situation? (h) Determine the force on the wire.

87 [I] The wire in the previous problem is rotated through an angle of 20.0° counterclockwise about a vertical axis. What is the magni-

tude of the magnetic force that now acts on the wire?

SOLUTION: θ is the angle between $\vec{v}$ and $\vec{B}$ and that's 90° − 20.0° = 70.0°. Hence $F_M = IlB \sin\theta = (68 \times 10^{-3}\,\text{N}) \sin 70.0° = 64 \times 10^{-3}\,\text{N}$.

88. [I] A 10-m-long wire is strung more or less horizontally in an east-west direction, and a current of 10 A is sent from west to east along it. Assuming the Earth's B-field at that location has a magnitude of 0.5×10^{-4} T and is horizontal and due north, find the magnitude of the force on the wire.

89. [I] A 1.50-m-long straight wire experiences a maximum force of 2.00 N when in a uniform 1.333-T B-field. What current must be passing through it? [Hint: Maximum force occurs when $\theta = 90°$.]

90. [I] When 2.00 A passes along a 3.00-m length of straight wire that is entirely immersed in a uniform B-field, it experiences a maximum force of 6.00×10^{-3} N. How strong is the field?

91. [I] A constant 3.50-A current moves along a straight wire that is perpendicular to a 2.00-T magnetic field. If the wire is 0.50 m long, how much magnetic force will it experience?

92. [I] A straight wire is positioned in a uniform magnetic field so that the force on it—4.00 N—is a maximum. If the wire is 20 cm long and carries 10.0 A, what is the magnitude of the B-field?

93. [I] A straight current-carrying ($I = 6.0$ A) wire makes an angle of 31.2° with a 0.01-T uniform B-field. What is the magnitude of the force exerted on a 1.0-cm length of the wire?

94. [I] A narrow flat coil wound on a square frame has 200 turns and sides of 20 cm. It carries a current of 1.25 A and is positioned in a 0.50-T magnetic field. What is the maximum torque that can be exerted by the field on the coil?

95. [I] A circular flat coil of wire encompassing an area of 1.3×10^{-3} m^2 has 20 turns and carries a current of 1.5 A. If its rotational symmetry axis makes an angle of 32° with a B-field of 0.90 T, what is the torque acting on it?

96. [I] Two parallel wires 50.0 m long are 25.0 cm apart and each carries a current of 10 A in the same direction. What is the force (magnitude and direction) between them?

97. [II] A single power line 50.0 m long is stretched more or less horizontally at an angle of 20° east of north at a location where the Earth's field is 0.50×10^{-4} T at 5.0° west of north. What is the total force on the wire when it carries 1.0 kA?

98. [II] A flat rectangular coil of 10 turns is suspended vertically from one arm of a beam balance that is brought into horizontal alignment with a few weights on the other pan. The bottom section of the coil, which is 10 cm long, is next exposed to a uniform perpendicular magnetic field of 0.60 T. A current of 300 mA is sent through the coil, which is pulled downward. How much weight must now be added to the other pan to restore balance?

99. [II] In a moving-coil galvanometer, the coil, which has N turns, surrounds a steel core so that the B-field always makes an angle of $\phi \approx 90°$ with the symmetry axis (the z-axis). If the current through the coil is reduced by 25%, how does the torque on it change?

100. [II] A single circular loop of wire 10 cm in radius is hung on a fine silver ribbon so that the plane of the loop is parallel to the magnetic field in a uniform region between the pole pieces of a large electromagnet. If the torque on the coil is 0.100 N·m when 5.0 A passes through the coil, what is the magnitude of the B-field?

101. [II] A long wire is stretched out horizontally to a 1.0-m-diameter tree, around it and back. Both ends are connected to the terminals of a battery and 10 A is drawn by the wire. If both lengths are parallel and the Earth's field is negligible, what force exists between each meter of the wires?

102. [II] THIS PROBLEM TREATS THE FORCE ON A CURRENT-CARRYING WIRE IN A MAGNETIC FIELD. Reread Discussion Question 9 and examine Fig. Q9 which shows a loudspeaker. A current is sent through the coil and it moves back and forth accordingly, pulling and pushing the paper cone. (a) Since **the B-field is radial** out from the center, what is the direction of the force on each segment of wire in the coil? Note that the wire is everywhere perpendicular to the field. (b) If the diameter of the coil is 6.2 cm, roughly how long is the current-carrying wire of the coil? (c) Suppose the B-field is 0.59 T and the coil has 210 turns. What is the net force on the coil when it carries a signal of 4.0 mA?

103. [III] THIS PROBLEM EXPLORES THE REPULSIVE FORCE ON TWO CURRENT-CARRYING WIRES IN A MAGNETIC FIELD. Figure P103 shows the ends of two straight wires each carrying a current I, one into and the other out of the plane of the page. The wires have a mass per unit length of 25 g/m. We want to find the current in each wire. (a) Draw a free-body diagram. (b) What is the mass per unit length in kg/m? (c) Write an expression for the magnetic force each wire experiences per unit length of wire when separated by a distance d. (d) Considering the tension in the string, show that $(F_M/l) = (F_W/l) \tan 2.0°$. (e) Prove that the separation between the wires is given by $d = 2(0.50 \text{ m}) \sin 2.0°$. (f) Equating parts b and c, solve for I.

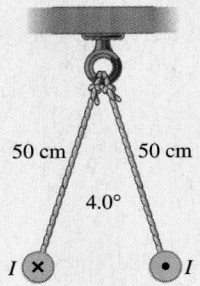

50 cm   50 cm

4.0°

I I

Figure P103

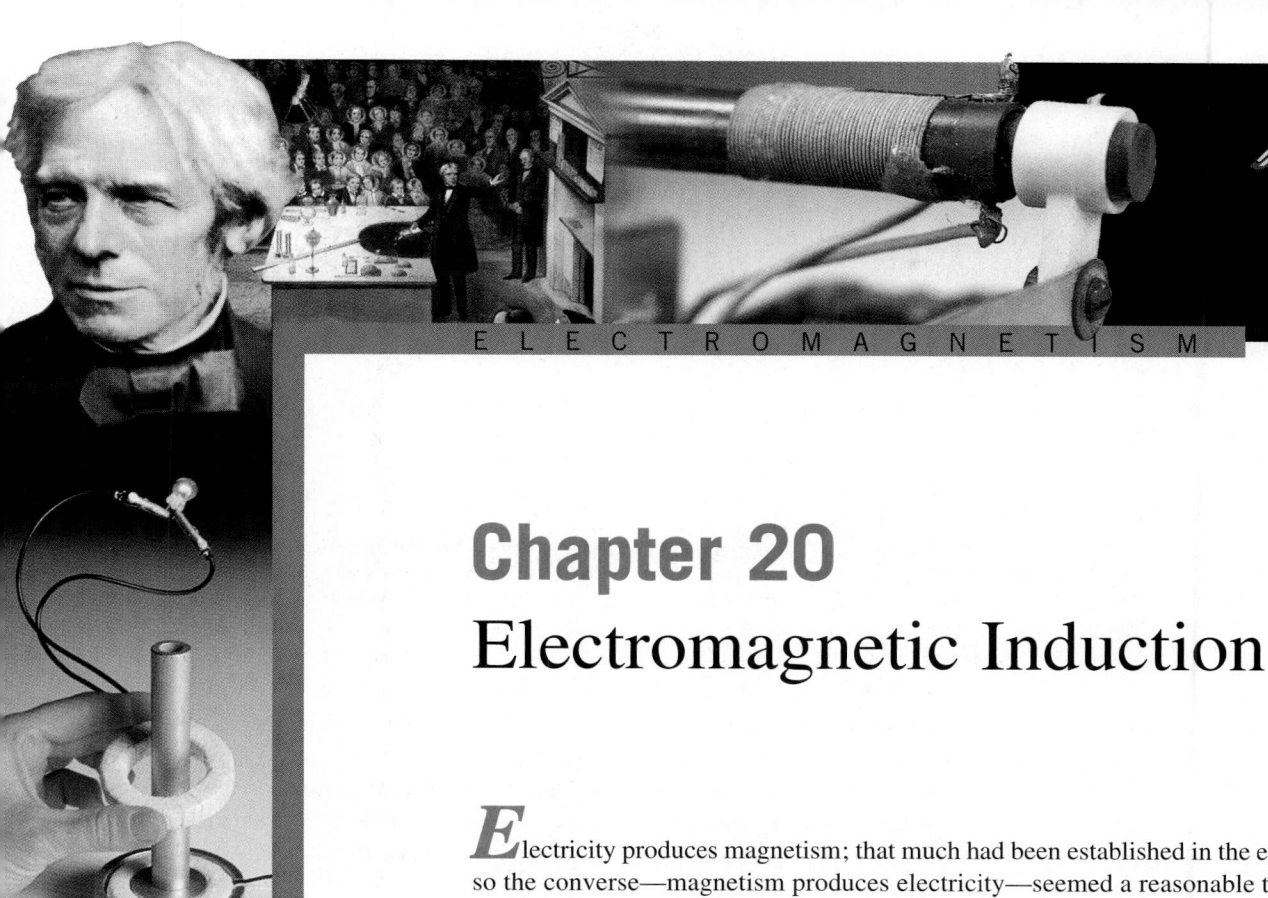

Chapter 20
Electromagnetic Induction

*E*lectricity produces magnetism; that much had been established in the early 1820s, and so the converse—magnetism produces electricity—seemed a reasonable thing to expect. And yet, the best researchers of the day could only come up with results that were ambiguous and unconvincing. Still, the agenda was obvious enough: a charge can electrify a nearby object by induction, and a magnet can magnetize a nearby piece of iron by induction; it was only reasonable to expect that a current should induce a current in a nearby conductor.

Since a steady current generates a steady magnetic field, should not a steady magnetic field generate a steady current? However logical that was, it was wrong. A steady magnetic field does not impart energy to free charges; it does no work on them, and yet for a current to exist it must get energy from somewhere. A changing magnetic field is something very different. It can impart energy to charges, and it can produce currents.

Electromagnetically Induced emf

In 1821, Ampère conducted an experiment during which he observed the momentary effects of what has come to be known as **electromagnetic induction**, but that was not the main thrust of his research and he didn't appreciate what was at hand. "Convert magnetism into electricity" was the brief remark Faraday jotted in his notebook in 1822, a challenge he set himself with an easy confidence that made it seem so attainable. Two years later, Arago by chance observed that the needle of a fine compass tended, after it was jolted, to oscillate for a shorter-than-normal time when it was in a housing with a copper bottom (p. 724). Arago then performed the inverse experiment and rotated a copper disk beneath a compass needle and found, inexplicably, that the needle followed the disk around.

After several years doing other research, Faraday returned to the problem of electromagnetic induction in 1831. His first apparatus made use of two coils mounted on a wooden spool. One, called the *primary*, was attached to a battery and a switch; the other, the *secondary*, was attached to a galvanometer (Fig. 20.1*a*). Initially, the results were totally negative until he installed a much more powerful battery. Even then, the effects were faint and transient, and yet their mere existence was enough for him. Faraday found that the galvanometer deflected in one direction just for a moment whenever the switch was closed, returning to zero almost immediately, despite the constant current still in the primary.

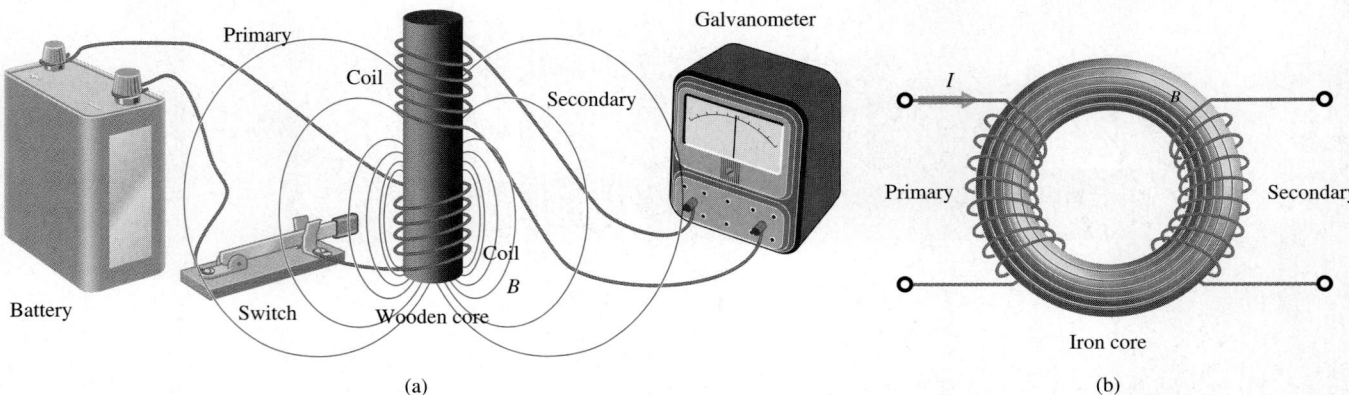

Figure 20.1 (a) A current in one coil produces a time-varying magnetic field that couples to the other coil, inducing a transient current in it. (b) An iron core becomes magnetized by the field of the primary current, increasing the field and improving the coupling to the secondary current.

Whenever the switch was opened, interrupting the primary current, the galvanometer in the secondary circuit momentarily swung in the opposite direction and then promptly returned to zero.

Using a ferromagnetic core to concentrate the "magnetic force," Faraday wound two coils around opposing sections of a soft iron ring (Fig. 20.1*b*). Now the effect was unmistakable—*a changing magnetic field generated a current*. Indeed, as he would continue to discover, *change* was the essential aspect of electromagnetic induction. Within a few weeks, the modest, untutored genius had explored the phenomenon and was ready to present his first paper on the subject.

Mr. Faraday (he declined knighthood) went on to use his new knowledge to build an electric generator (p. 721), one of the single most far-reaching accomplishments of the era. Later, when asked by Prime Minister Gladstone if his research had any practical value, Faraday quipped, "Why, sir, you will soon be able to tax it." Electromagnetic induction is now at the very heart of the generation, delivery, and utilization of electrical energy—and they do indeed tax it.

A LACK OF URGENCY

Joseph Henry in New York didn't hear of Faraday's achievement until April, and when he did, the news was a bitter disappointment. He had performed much the same experiment as Faraday well over a year before (1830), though the results went unpublished. Like Faraday, Henry appreciated the implications of the discovery even if he felt no urgency to publish them. When he finally dashed off an article in response to the news from Europe, the experiment he described was somewhat more sophisticated than that performed in London, and it even contained an important feature Faraday had missed (p. 725).

20.1 Faraday's Induction Law

Figure 20.2 depicts a modern version of an experiment performed by Faraday to explore the dependence of electromagnetic induction on variations in B. The oscilloscope displays, as a function of time, the voltage—otherwise known as the **induced emf** ($\mathscr{E}$)—across the coil's terminals. The exact shape of the resulting curve is not important; it depends on the details of the motion of the magnet. Observe that thrusting a south pole toward the coil pro-

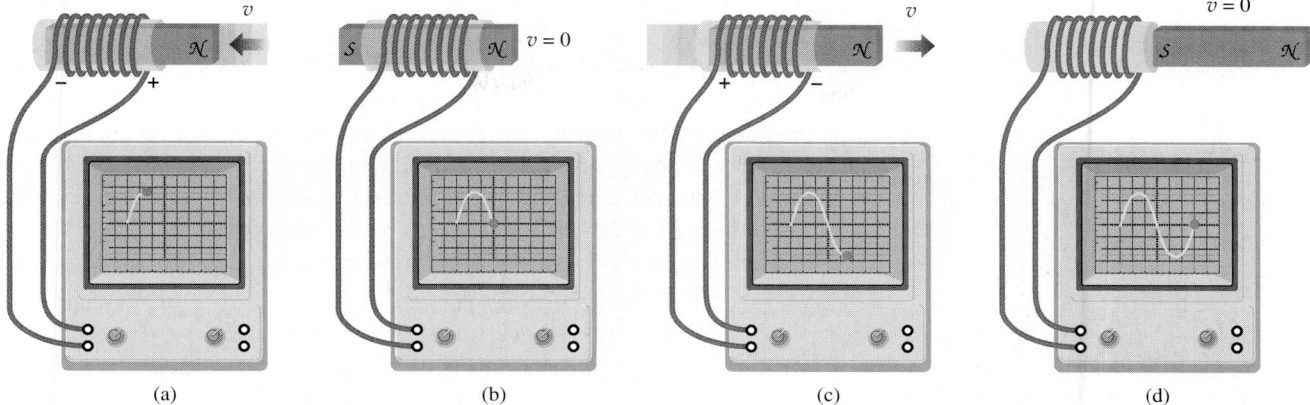

(a)   (b)   (c)   (d)

Figure 20.2 (a) As the S-pole approaches the right end of the coil, a current circulates from left to right through the coil and the right end of the coil becomes an S-pole. The right side of the coil becomes positive, the left side negative, and it acts like a battery, sending current clockwise through the external circuit. (b) When the magnet is stationary, the emf is zero. (c) Removing the magnet induces a north pole at the right end of the coil, which attracts the receding S-pole of the magnet. (d) No motion, no flux change, and $\mathscr{E} = 0$.

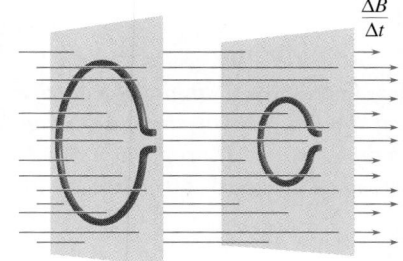

Figure 20.3 A time-varying magnetic field passes through both loops. It induces a larger emf in the larger loop, which embraces a greater perpendicular area.

duces a positive emf and yanking it away produces a negative emf (for this particular winding). Approaching the coil with a north pole reverses the polarity of the induced emf. Furthermore, the amplitude of the emf depends on how rapidly the magnet is moved; when $v = 0$, the emf $= 0$. In this arrangement, *the induced emf depends on the rate-of-change of B through the coil and not on B itself*. A weak magnet moved rapidly can induce a greater emf than a strong magnet moved slowly.

When the same changing B-field passes through two different wire loops, as in Fig. 20.3, the induced emf is larger across the terminals of the larger loop. In other words, here where the B-field is changing, ***the induced emf is proportional to the area A of the loop penetrated perpendicularly by the field***. If the loop is successively tilted over, as in Fig. 20.4, the area presented perpendicularly to the field ($A_\perp$) varies as $A \cos \theta$ and, when

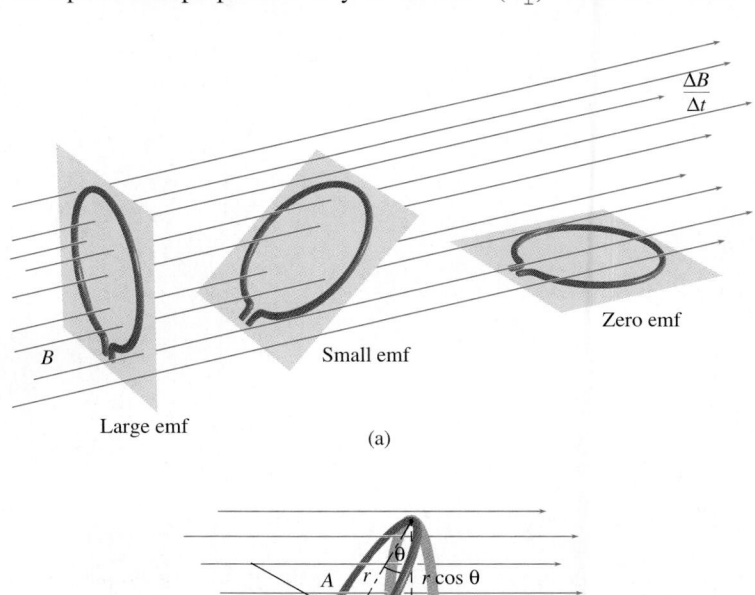

Large emf   Small emf   Zero emf

(a)

Figure 20.4 (a) The induced emf is proportional to the perpendicular area intercepted by the field. (b) The area of the ring ($A = \pi r^2$) projects as an ellipse. Vertically, r projects to $r \cos \theta$; horizontally, r projects to r. The area of the ellipse is $A_\perp = \pi(r \cos \theta)r = A \cos \theta$.

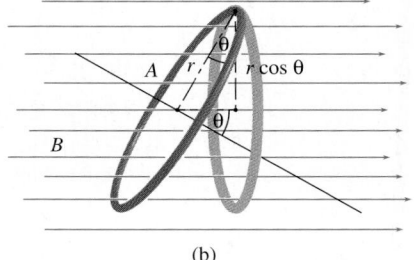

(b)

$\theta = 90°$, the induced emf is zero because no amount of B-field then penetrates the loop: when $\Delta B / \Delta t \neq 0$, emf $\propto A_\perp$. The converse also holds: ***when the field is constant, the induced emf is proportional to the rate-of-change of the perpendicular area penetrated.*** If a coil is twisted or rotated or even squashed (Fig. 20.5) while in a constant B-field so that the perpendicular area initially penetrated is altered, there will be an induced emf $\propto \Delta A_\perp / \Delta t$ and it will be proportional to B. In summary, when $A_\perp$ = constant, emf $\propto A_\perp \Delta B / \Delta t$ and, when B = constant, emf $\propto B \Delta A_\perp / \Delta t$ {see **MATH REVIEW: PART A-6** on the **CD** 💿 }.

All of this suggests that the emf depends on the rate-of-change of both $A_\perp$ and B, that is, on the rate of change of their product. This brings to mind the notion of the flux of the field (p. 557)—the product of field and area where the penetration is perpendicular. Accordingly, we define the **flux of the magnetic field** as

> **Flux** is the amount of field passing perpendicularly through an area multiplied by that area.

$$\Phi_M = B_\perp A = BA_\perp = BA \cos \theta \qquad (20.1)$$

The units of magnetic flux are tesla-meter², which is called a **weber** (Wb) to honor one of the early workers in magnetism, Wilhelm Weber: 1 Wb = 1 T·m². Clearly, B must be measured in Wb/m², which is the old unit that was replaced by the tesla: 1 Wb/m² = 1 T. It's common in the literature to find B referred to as **flux density** (flux per unit area).

The results of the preceding observations for a one-turn loop of wire can be stated as

$$\mathscr{E}_{av} = -\frac{\Delta \Phi_M}{\Delta t} \qquad (20.2)$$

where the significance of the minus sign will be considered presently. *A single loop of wire will experience an induced voltage (in volts) that equals the time rate-of-change of magnetic flux through it at any given instant (in SI units the constant of proportionality is unity).* If there are N turns of wire in a coil, each has an induced voltage that is in series with the others so that the net **induced emf** is

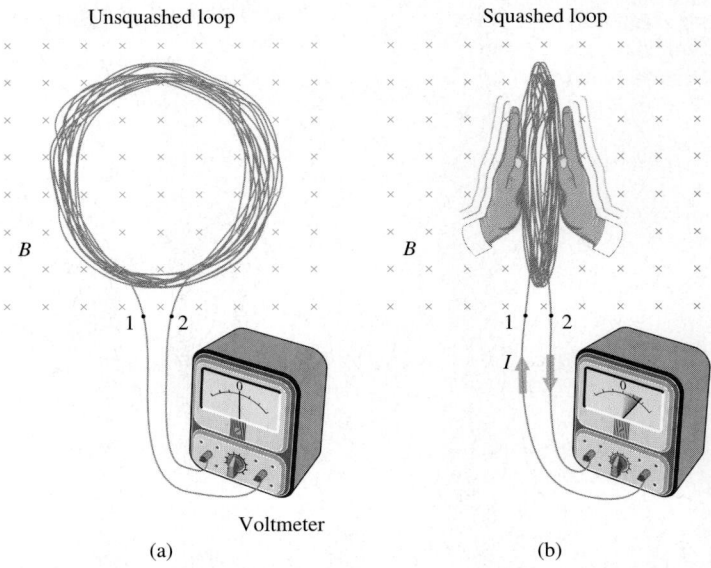

Figure 20.5 (a) A downward B-field perpendicular to an insulated wire loop. The coil starts at point 1, winds clockwise through a number of turns, and ends at point 2. (b) Squashing the loop reduces $A_\perp$, changes the flux, and, therefore, induces an emf. The voltmeter will show a momentary jump in voltage, after which it will return to zero.

Example 20.1 **[II]** A circular flat coil of 200 turns of wire encloses an area of 100 cm². The coil is immersed in a uniform perpendicular magnetic field of 0.50 T that penetrates the entire area. If the field is shut off so that it drops to zero in 200 ms, what is the average induced emf? Given that the coil has a resistance of 25 Ω, what current will be induced in it?

Solution   We've got a coil in a changing *B*-field, and that should remind you of Faraday's Law. (1) TRANSLATION—A coil with a known number of turns and area is exposed to a *B*-field changing in a specified way; determine the induced emf. Knowing the coil's resistance, determine the induced current. (2) GIVEN: $N = 200$, $A = 100$ cm², $B_i = 0.50$ T, $B_f = 0$, $\Delta t = 200$ ms, and $R = 25$ Ω. FIND: The induced emf and current *I*. (3) PROBLEM TYPE—Electromagnetic induction/Faraday's Induction Law/Ohm's Law. (4) PROCEDURE—The time rate-of-change of the flux equals the emf. We need to find the flux and its change per second. (5) CALCULATION—The *B*-field links the entire area perpendicularly; hence, the initial flux is

$$\Phi_M = BA = (0.50 \text{ T})(0.010\,0 \text{ m}^2) = 0.005\,0 \text{ T·m}^2$$

and so $\Delta\Phi_M = -0.005\,0$ T·m² since the final flux is zero. It follows from Eq. (20.3) that

$$\mathcal{E} = -N\frac{\Delta\Phi_M}{\Delta t} = -200\frac{(-0.005\,0 \text{ T·m}^2)}{0.200 \text{ s}} = \boxed{5.0 \text{ V}}$$

(Don't worry about the signs here; we'll deal with that part presently.) From Ohm's Law,

$$I = \frac{V}{R} = \frac{(5.0 \text{ V})}{(25 \text{ Ω})} = \boxed{0.20 \text{ A}}$$

Quick Check: Remember that a change of flux of 1 T·m² through 1 turn in 1 s produces 1 V. Here, $N\Delta\Phi_M = (200)(-0.005\,0$ T·m²), which is the change in the total flux linking the coil, equals one; so 5.0 V is reasonable.

{For more worked problems click on WALK-THROUGHS in CHAPTER 20 on the CD.} 🔘

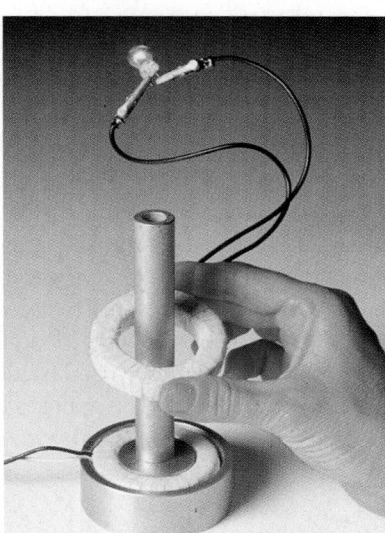

A time-varying current in the lower coil produces a time-varying magnetic flux that passes through the upper coil. That, in turn, induces a current in the upper coil, and the bulb lights.

The **induced emf** in a wire loop equals the negative of the time rate-of-change of the magnetic flux through that loop.

$$\mathcal{E}_{av} = -N\frac{\Delta\Phi_M}{\Delta t} \qquad (20.3)$$

which is one of the *fundamental equations of electromagnetism*. It is a summary of observations and not derivable from any previous formulas. The equation is generally known as **Faraday's Induction Law**, though Faraday himself never wrote it.

Equation (20.3) makes the point that the emf depends on the rate-of-change of flux linking all the turns of wire, and it's useful to formalize the notion by speaking about the **flux linkage** in the coil—namely, $N\Phi_M$. Thus, **the induced emf is the negative of the time rate-of-change of the flux linkage**. {For a delightful simulation click on **FARADAY'S INDUCTION LAW** under **INTERACTIVE EXPLORATIONS** on the **CD**.} 🔘

Lenz's Law

The negative sign in Eq. (20.3) relates the polarity of the induced emf to the flux change, which can occur in a variety of ways: the field can be increased, decreased, or moved; the loop area can be turned, squashed, or yanked out of the field, and yet there is always a consistent, reproducible sense in which the emf appears. It was a physicist working in Russia—Heinrich Friedrich Emil Lenz (pronounced *lents*)—who first (1834) published an elegant statement of the phenomenon that has come to be known as **Lenz's Law**:

> **The induced emf will produce a current that always acts to oppose the change that originally caused it.**

In Fig. 20.6a, a south pole approaches a coil, inducing an emf across its terminals. The emf will be such as to cause an induced current, which in turn creates an induced magnetic field (B_1). This field opposes the *change* of flux linking the coil. The instigating effect is an *increasing B-field directed to the right* (toward the approaching magnet's south pole). The opposing induced field, tending to cancel an increasing field to the right, will be to the left; and the induced current in the coil circulates appropriately. **The polarity of the emf indicates the direction in which it drives current in the external circuit** and would be measured by a voltmeter across the terminals of the coil. In other words, the approach of a south pole induces a current that causes the coil to manifest a south pole at its near end, which opposes the advance of the magnet and thereby the change of the field. If the magnet is withdrawn, as in Fig. 20.6b, the change—*a decreasing field to the right*—is opposed by an induced field to the right. The induced current direction is now reversed, as is the emf. A north pole is induced at the coil's right end, which attracts the magnet, tending to oppose its motion away.

Change is at the heart of the phenomenon of induction, just as it underlies the concept of energy—change is a manifestation of energy. Not surprisingly, Lenz's Law can be appreciated as a result of the Law of Conservation of Energy. For example, when an external agent moves the magnet in Fig. 20.6a, it must overcome the counterforce immediately exerted by the coil. The work done in the process provides the electrical energy needed to build up and sustain the induced current. The magnetic field is the intermediary between the external mover pushing the magnet and the resulting current. Were that not the case, the induced current would be created at no cost of mechanical work, and energy would not be conserved. No matter how the flux changes, work must be done on the system and the induced current, which is a manifestation of that work, must oppose the change—it doesn't necessarily stop it but always opposes it. *The work done by the magnet on the current* (W_{mc}) *must match the work done by the current on the magnet* (W_{cm}), *and that work is negative:*
$$W_{mc} = -W_{cm}.$$

STUDY GUIDE

When you have a time-varying B-field passing through a coil, ask yourself if the field is increasing or decreasing. Then determine the direction of the induced field B_1; it opposes that increase or decrease. Knowing the direction of B_1, you can use the Right-Hand-Current Rule (p. 678) to find the direction of the induced current I. To determine the polarity of the induced voltage on the coil, image a resistor across its terminals. The current I enters the resistor at the +, or higher-voltage, side and leaves at the −, or lower-voltage, side.

Figure 20.6 A coil with an air core, an ammeter, and a bar magnet. (a) The approach of an S-pole induces an emf (+ on the right, − on the left) that causes a current to circulate clockwise in the external circuit. (b) Removing the S-pole induces a counter-clockwise current and the reversal of the emf, driving it. Note the polarity of the connections on each meter and the sign of its reading. For simplicity the B-fields are drawn only in the plane of the page.

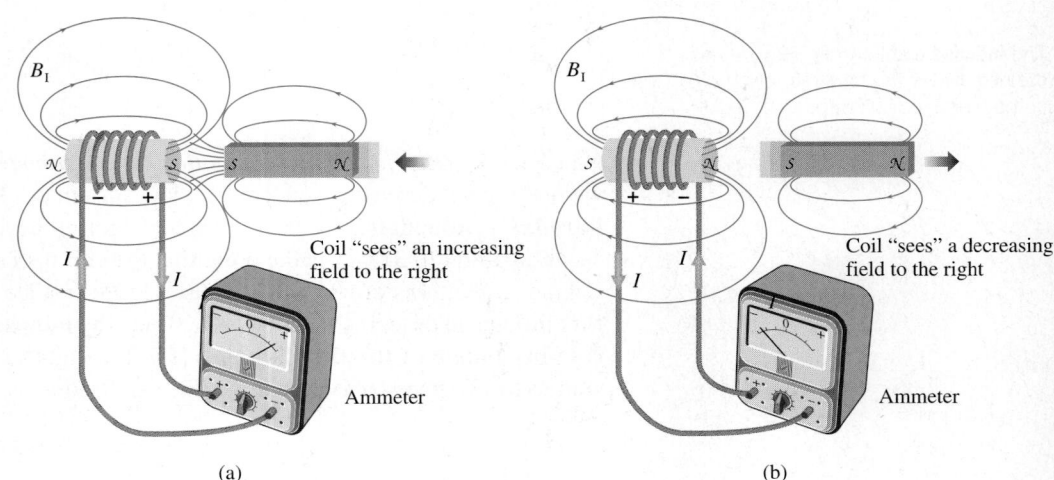

(a)   (b)

Example 20.2 **[II]** The loop in Fig. 20.5*a* has an area of 0.25 m² and is immersed in a uniform perpendicular 0.40-T downward magnetic field. If the coil has 200 turns and a resistance of 5.0 Ω and is squashed to a zero area in 100 ms, what average current will be induced in it during the collapse? Discuss the direction of the current and the polarity of the emf. What happens to the current?

Solution We've got a coil and a changing magnetic flux, and should call to mind Faraday's Law.   (1) TRANSLATION—A coil with a known number of turns and area, exposed to a known *B*-field, is collapsed in a specified time. Knowing the coil's resistance, determine the induced current.   (2) GIVEN: $A_i = 0.25$ m², $A_f = 0$, $B = 0.40$ T, $N = 200$, $R = 5.0$ Ω, and $\Delta t = 0.100$ s. FIND: *I*.   (3) PROBLEM TYPE Electromagnetic induction/ Faraday's Law/Lenz's Law/Ohm's Law.   (4) PROCEDURE— The time rate-of-change of the flux equals the emf. We need to find the flux and its change per second.   (5) CALCULATION— The initial flux is

$$\Phi_M = BA = (0.40 \text{ T})(0.25 \text{ m}^2) = 0.100 \text{ T·m}^2$$

and so $\Phi_M = -0.100$ T·m² since the final flux linking the loop is zero. The average emf is

$$\mathscr{E}_{av} = -N\frac{\Delta\Phi_M}{\Delta t} = -200\frac{(-0.100 \text{ T·m}^2)}{0.100 \text{ s}} = 2.0 \times 10^2 \text{ V}$$

From Ohm's Law, the average induced current is

$$I = \frac{V}{R} = \frac{(200 \text{ V})}{(5.0 \text{ Ω})} = \boxed{40 \text{ A}}$$

The current circulates clockwise around the loop so as to induce a *B*-field within its confines, which will tend to counter the decrease in downward flux. Terminal-2 is positive with respect to terminal-1.

Quick Check: The sides of the coil as it is crushed will experience a force via Eq. (19.13)—a force on a current-carrying wire in a *B*-field. That force will everywhere be in the direction of $\vec{\mathbf{v}} \times \vec{\mathbf{B}}$, where $\vec{\mathbf{v}}$ is in the direction of motion of the positive current. For a clockwise current, the force is to the left on the left side and to the right on the right side. The induced force will oppose the external force that is crushing the coil. Because of resistance, the energy associated with the current is dissipated as joule heat, and the current stops as the emf → 0; it would continue to circulate if the coil were superconducting.

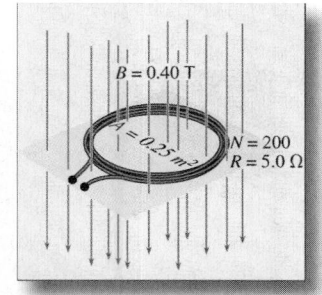

In one version of the magnetically levitated train (Fig. 20.7), there are superconducting electromagnets in the bottom of the car. These pass above two rows of closed coils mounted in the guideway, and as they do, currents are induced in the coils creating an opposing field (via Lenz's Law) that lifts the vehicle six inches into the air. It is propelled by successively energizing the coils in the sides of the guideway.

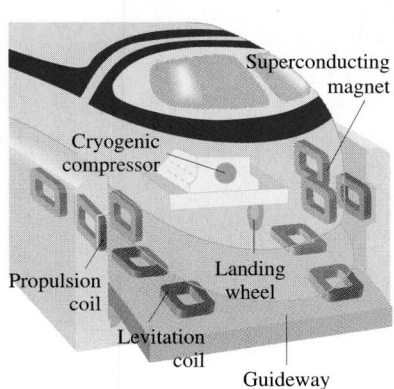

Figure 20.7   A high-speed maglev. Superconducting coils on the bottom of the car induce opposing *B*-fields in aluminum coils in the guideway that lift the car.

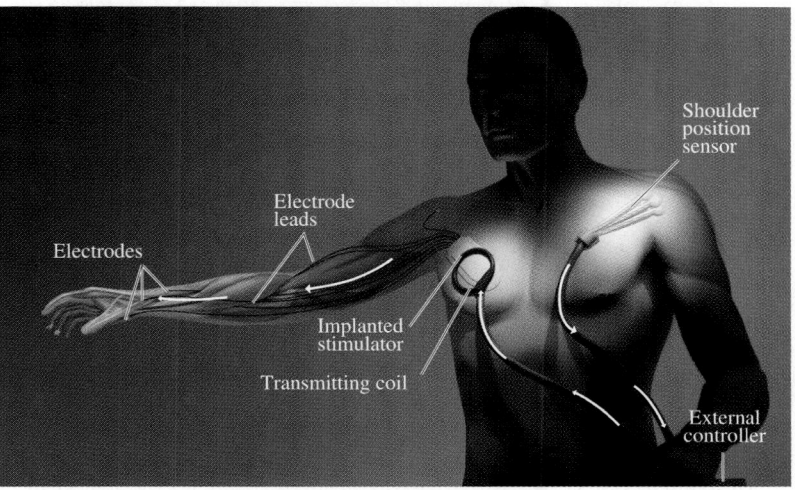

A device which allows people with a spinal cord injury to move paralyzed muscles. Here someone who retains shoulder mobility can activate a detector that sends signals to arm and hand muscles. An external controller feeds currents to a coil placed on the chest. Implanted under the skin beneath this transmitting coil is a pick-up coil in which there is an induced signal. That signal, in the form of tiny currents, activates the muscles of the arm and hand.

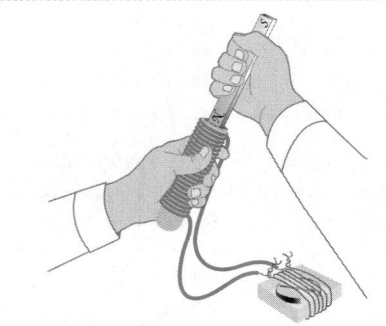

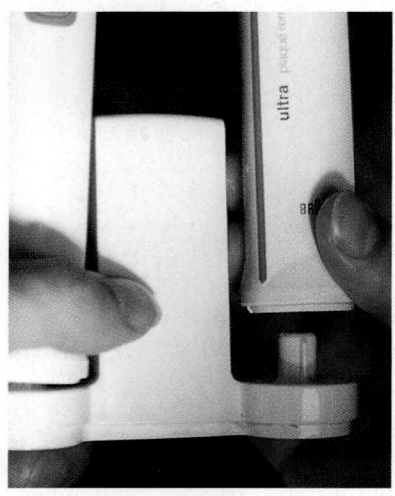

These two water-tight electric tooth-brushes are powered by rechargeable batteries. There is a coil sealed in the bottom of each handpiece which sits atop a ferromagnetic core that protrudes up from another coil sealed in the base. Just as in the photo on p. 713, an ac current in the base (which is plugged into a wall outlet) induces an ac current in the handpiece. That current is converted to dc (p. 768) and used to recharge the battery.

20.2 Motional emf

Suppose a length of ordinary wire is moved perpendicularly across a uniform magnetic field at a speed v (Fig. 20.8). The mobile charge carriers within it are dragged along to the right with the wire at the same speed. Moving through the field, they experience a force (in the direction of $\vec{\mathbf{v}} \times \vec{\mathbf{B}}$) equal to

$$[\theta = 90°] \qquad F_M = q_.vB \sin \theta = q_.vB \qquad [19.7]$$

parallel to the wire. In this case, the carriers are electrons and they flow downward. Accordingly, there is an induced emf across the wire—the top is at a higher potential than the bottom. As *within a battery*, negative charge is propelled from the + terminal to the − terminal.

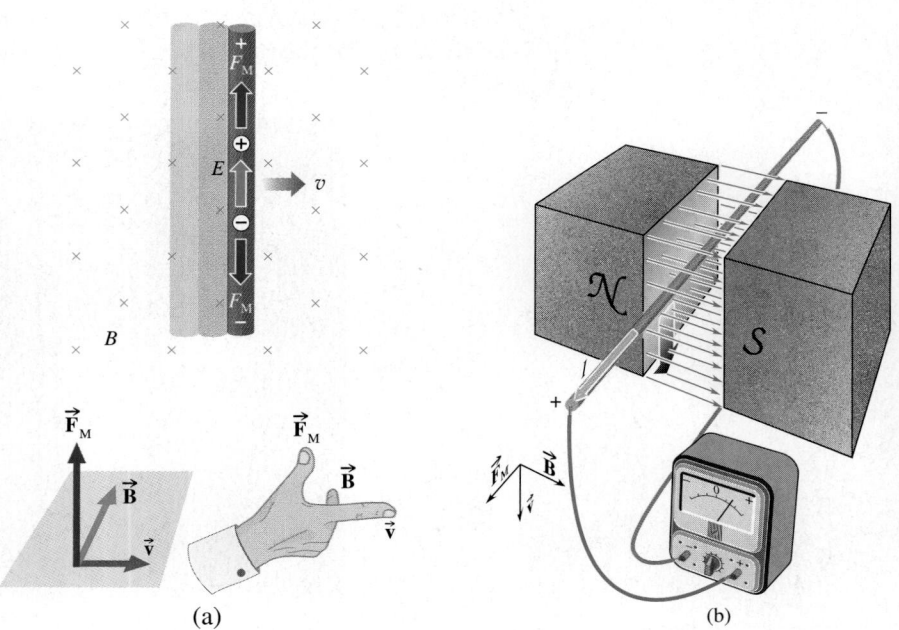

(a) (b)

Figure 20.8 (a) A positive charge experiences an upward force in the direction of $\vec{\mathbf{v}} \times \vec{\mathbf{B}}$ as the wire cuts across a B-field. (b) Here as the wire falls across a perpendicular B-field, electrons are forced to the rear. The front end becomes positive (resulting in a higher potential), just as if positive mobile charge carriers had been pushed forward. Current leaves the + end of the wire, enters the ammeter, and returns to the − end of the wire.

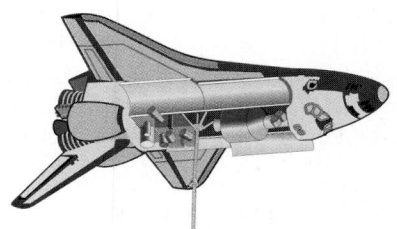

Once a charge begins to move along the wire with some scalar drift velocity v_d, it will experience yet another force ($F_\perp = q_s v_d B$) due to this new motion in the B-field. The charge is then also propelled perpendicularly across the wire, but that lateral motion is restrained by the boundaries of the conductor. In accord with Lenz's Law, the net transverse force on all the mobile charges is opposite to the velocity of the wire and must be overcome by an externally applied force if the motion of the wire is to be sustained. *The external agency that moves the wire supplies energy to the system* (by doing work on the wire against the net transverse force). That energy is imparted to the induced current (Fig. 20.9).

The motional emf (the change in potential) equals the work done on a positive test-charge in bringing it from one end of the rod to the other (i.e., the change in its potential energy) per unit charge (p. 571). The force $q_s vB$ acting parallel to the wire's length l does work on a charge in the amount of $q_s vBl$, and so the induced emf is

$$[\vec{v} \perp \vec{\mathbf{B}}] \qquad\qquad \mathscr{E} = vBl \qquad\qquad (20.4)$$

Here the wire moves perpendicular to the B-field, the angle θ between $\vec{v}$ and $\vec{\mathbf{B}}$ is 90°, and sin 90° = 1 in Eq. (19.7). It need not be so, in which case $v_\perp = v \sin \theta$ instead of v must appear in Eq. (20.4).

The downward transit of electrons makes the bottom end of the wire negative, leaving behind a positive upper end. This charge buildup continues until the repulsion thus produced on any subsequent approaching charge matches the driving force $q_s vB$, and the current stops. Simply put, the bottom end gets so negatively charged that the induced **motional emf** cannot push any more electrons down to it. If the motion of the wire stopped, the emf would go to zero and electrons would flow back up or, if you like, the hypothetical mobile positive charges would descend from the higher potential and move downward.

Because there is no closed circuit, there cannot be a steady current produced by this wire generator. Instead, at equilibrium, the separated charges will have created an electric field E that opposes any further motion of charge—the induced current is transient. As we saw earlier (p. 575), since emf = El, it follows that the electric field in the wire, which exactly counters the motional emf, is

$$E = vB \qquad\qquad (20.5)$$

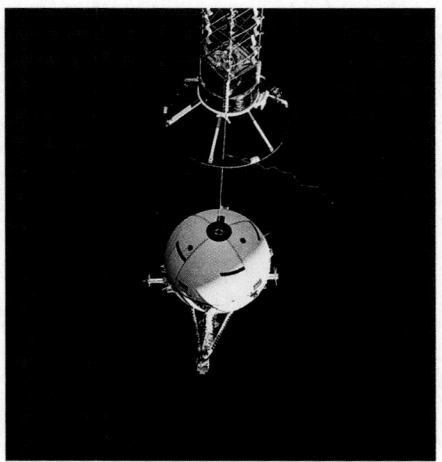

Figure 20.9 The Tethered Satellite System flew aboard a shuttle in February of 1996. The satellite, lowered at the end of a 19.7 km cable, swept through the Earth's B-field at 8 km/s. The system produced a peak current of 1 A and a voltage of 3500 V. As electrical energy is removed from the system, the shuttle's orbit decays.

Example 20.3 **[I]** A 1.0-meter-long wire held in a horizontal east-west orientation is dropped at a place where the Earth's magnetic field is 2.0×10^{-5} T, due north. Determine the induced emf 4.0 s after release.

Solution This problem deals with a wire moving across a *B*-field, and that always means an induced emf. (1) TRANSLA-TION—A wire of known length and orientation falls through a known *B*-field; determine the induced emf at a specified time. (2) GIVEN: $l = 1.0$ m, $t = 4.0$ s, and $B = 2.0 \times 10^{-5}$ T. FIND:

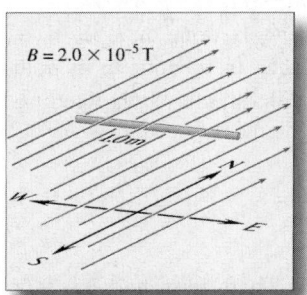

$B = 2.0 \times 10^{-5}$ T

The resulting emf. (3) PROBLEM TYPE—Electromagnetic induction/motional emf. (4) PROCEDURE—Because a wire is moving through a magnetic field and we want the emf, this problem involves the expression $\mathscr{E} = vBl$. We have *B* and *l* and need *v*. (5) CALCULATION—First compute the speed at $t = 4.0$ s;

$$v = v_0 + gt = 0 + (9.81 \text{ m/s}^2)(4.0 \text{ s}) = 39.2 \text{ m/s}$$

and using that,

$$\mathscr{E} = vBl = (39.2 \text{ m/s})(2.0 \times 10^{-5} \text{ T})(1.0 \text{ m}) = \boxed{0.78 \text{ mV}}$$

Quick Check: $v \approx (10 \text{ m/s}^2)(4 \text{ s}) \approx 40 \text{ m/s}$; $\mathscr{E} = vBl \approx 80 \times 10^{-5}$ V.

Cutting Field Lines

With the moving wire in Fig. 20.8, there is no closed loop and so no flux change; the induced emf arises from the motion of the charges across the field. We might expect that if the wire were kept stationary and the field moved to the left, the same relative motion would exist and the same emf would be induced; indeed, that's what happens experimentally. With this in mind, Faraday provided a useful alternative model in the spirit of Eq. (20.4). He envisioned the emf arising when the wire cut across magnetic field lines. Remember that the strength of a field *B* can be related to the number of field lines per unit area. Thus, through any perpendicular area *A*, the number of field lines is defined to be *BA*. When the wire in Fig. 20.8 moves at *v*, it sweeps across an area equal to *vl* per second. The number of field lines "cut" per second is therefore *vlB*, which, according to Eq. (20.4), equals the emf: *the induced emf equals the number of field lines cut per second by a conductor.*

Figure 20.10 shows the field of a permanent magnet being cut by a revolving wire. The field is rotationally symmetrical, and there is no flux through the loop bounded by the device; yet if the crank is turned at a steady rate, a steady emf will be induced and a steady current will circulate. Both the emf and the current will vary linearly with the rotation rate.

Figure 20.10 Consider the loop formed by the rotating U-shaped rod and the hookup wires (which lie in a plane containing the magnet). The *B*-field is rotationally symmetrical and does not penetrate the loop. Hence, there is no flux change as the rod rotates, and yet there is an emf. The moving wire cuts the field. There is a relative motion between the electrons in the wire and the field, and there is an induced emf. Note the polarity of the connections to the meter and the sign of its reading.

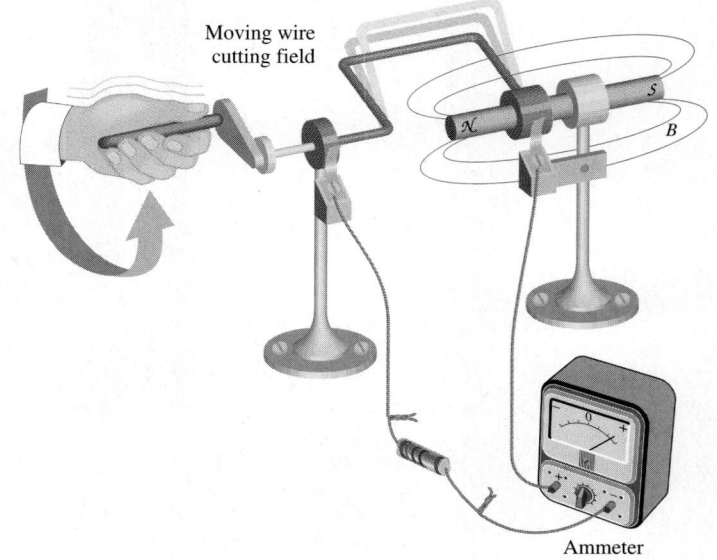

Moving wire cutting field

Ammeter

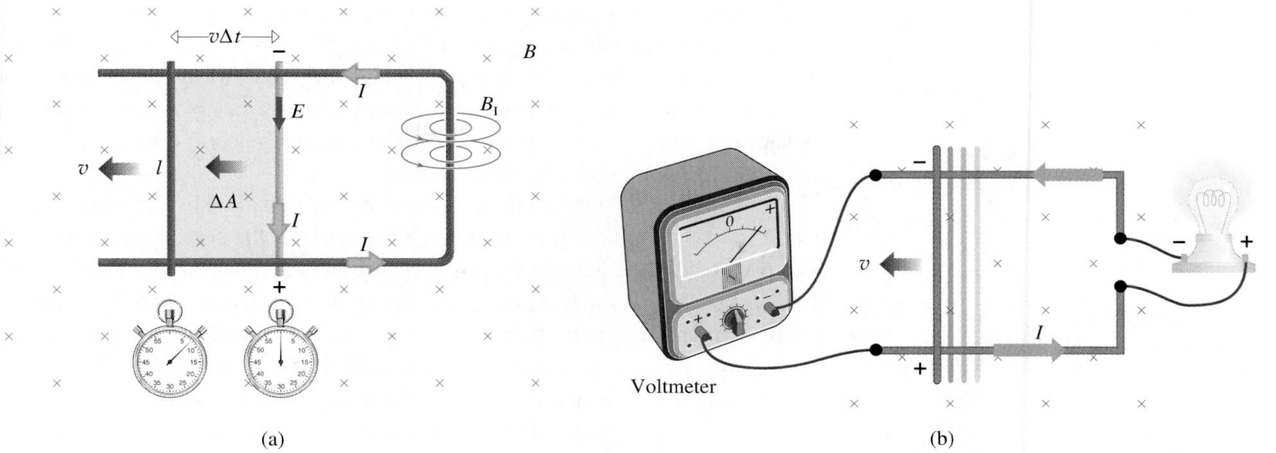

Figure 20.11 A straight wire sliding along, and in contact with, a stationary U-shaped conductor. (a) As the wire moves to the left, there is a time rate-of-change of the flux equal to $B\Delta A/\Delta t = B(lv\,\Delta t)/\Delta t = Bvl$. Notice how the induced field B_I *within* the loop opposes the increase of flux. (b) Current circulates counterclockwise as if the load were placed across a battery with $+$ and $-$ terminals.

Now suppose the moving wire of Fig. 20.8 is made to be part of a closed loop by placing it on a conducting U-shaped track (Fig. 20.11) having an appreciable resistance. Closing the loop will allow a current I to circulate, thereby extracting energy from the arrangement, if only as joule heat. The moving wire again produces an emf $= vBl$ much like a battery, and the resulting current provides energy at a rate of $P = I\mathscr{E}$; we have a dc generator. Notice that if the rod moves at a constant speed v, in a time Δt, it travels a distance $v\,\Delta t$ and sweeps out an area $\Delta A = v\,\Delta t\,l$. It therefore cuts a number of field lines equal to $B\,\Delta A = Bv\,\Delta t\,l$, and it does so at a rate of $B\,\Delta A/\Delta t = Bvl$, which indeed equals the emf! From Lenz's Law, the induced current progresses counterclockwise and thereby generates a force to the right opposing the applied force that is moving the wire.

Alternatively, we can say that the flux through the loop increases at a rate $B\Delta A/\Delta t = Bv\,\Delta t\,l/\Delta t = Bvl =$ emf, which is the same result. Again, Lenz's Law maintains that the induced current will generate a B-field (out of the plane, inside the loop) that tends to decrease the downward flux increase. Inasmuch as the flux through a perpendicular area is BA and the number of field lines through that area is BA, it's not surprising that these two interpretations of the experimental results are *nearly* equivalent. What is remarkable is that there appear to be two distinct notions [Eqs. (20.3) and (20.4)] associated with induced emfs. Still, *if there is a relative transverse motion between the B-field lines and a charge (so that field lines are crossed), the charge will experience a force*, and that seems to be at the heart of the phenomenon.

> What led me more or less directly to the special theory of relativity was the conviction that the electromotive force acting on a body in motion in a magnetic field was nothing else but an electric field.
>
> ALBERT EINSTEIN

Induced Magnetic and Electric Fields

A stationary charge has a constant E-field at every surrounding point in space, whereas a moving charge—a current—produces a time-varying E-field. But a current is also surrounded by a B-field. This suggests a fresh way of looking at things from a field perspective; namely, **a time-varying E-field produces a B-field**. Let's now examine the reciprocal idea that a time-varying B-field is always accompanied by an E-field.

Picture a long solenoid of radius R carrying a current (Fig. 20.12). To probe for an electric field, we place within the solenoid a small coaxial coil of radius r having one or more turns, a so-called *search coil* (reminiscent of the test-charge). If the primary current is now increased, the flux through the search coil will increase, an emf will be induced across its terminals, and an induced E-field will exist along it. Alternatively, we can envision the increasing current creating additional B-field lines that

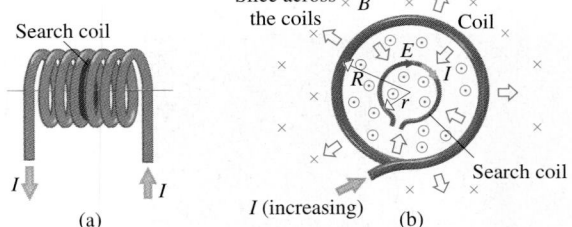

Figure 20.12 (a) A current-carrying coil. As I increases, B increases and, inside the coil, field lines move in toward the center. (b) A search coil within the solenoid experiences an E-field.

arise out of the wires and move inward toward the center of the primary (decreasing in density away from the wire). These lines cut the search coil, causing a positive charge therein to have an outward relative motion with respect to the field and thus be moved (in the direction of $\vec{v} \times \vec{B}$) to flow as a clockwise current, as required by Lenz's Law (Fig. 20.12b). At any point in the conductor of the search coil, a positive charge would experience a tangential force and a tangential electric field.

Since the wire loop of the search coil doesn't physically contribute to the E-field, it's useful to reconsider the preceding situation without the search coil in place. Experiments show that a charged particle will be accelerated in a uniformly varying B-field precisely as if there were a circular E-field present. In general, **an induced electric field accompanies a time-varying magnetic field**. Since there are no sources or sinks in the form of charges, *the induced E-field lines must close on themselves*. The search coil simply allowed us to detect the presence of an independent E-field associated with the time-varying B-field. This is exactly what the circular UHF loop antenna you might have on top of your TV set is supposed to do.

> **A time-varying B-field** leads to an emf via Eq. (20.3), and an emf is equivalent to the presence of an E-field.

Generators

Michael Faraday devised the world's first electric generator in 1831 (Fig. 20.13). It is a seemingly simple thing: a copper disk hand-cranked to rotate between the poles of a permanent magnet. Charges in the metal, set in motion across the field as the disk turns, experience a radial force acting along the conductor. A conventional current of positive charges moves outward in the direction of $\vec{v} \times \vec{B}$. Work done on the mobile charges raises them in potential such that the rim terminal is positive and the axis terminal is negative. As long as the rotation rate is kept constant (which is easy to do), the emf is quite steady, and a continuous current is supplied to a load by this dc generator, or **dynamo**. The emf produced was small and the device impractical (modern versions have been used in electroplating where precise dc is desirable); still, its effect on the world was probably as far-reaching as any other single invention in all of history.

THE GENERATION OF ELECTRICITY

The people of the world use electrical energy at a tremendous rate of about 10^{13} W. Almost all the electric current supplied commercially is generated with induction machines that produce a relative rotation between coils of wire and magnetic fields. The external power to rotate either the coils or the field is usually supplied by steam turbines (steam blasting against fanlike blades that are forced into rotation). The steam generally comes either from burning fossil fuels or from nuclear reactors.

Faraday delivering one of his famous Christmas lectures (1855). On the left in the front row is the Prince Consort between his two princely charges. This room at the Royal Institution in London still exists, and they still give demonstrations there.

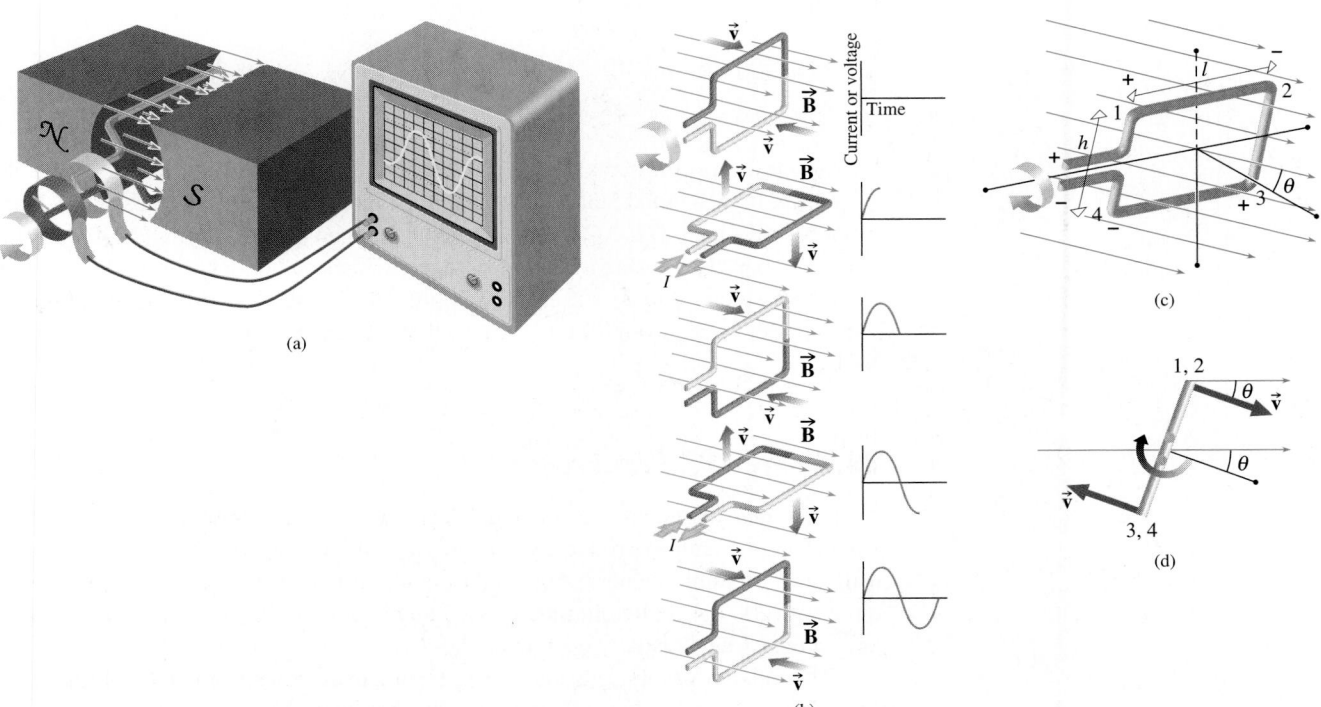

Figure 20.13   The first generator. Positive charge carriers moving in the direction of $\vec{v}$ experience a force in the direction of $\vec{v} \times \vec{B}$ that is radially outward along the disk. As long as the disk is turned in one direction, this device is a dc generator, also called a *dynamo*.

20.3 The AC Generator

A simplified version of an alternating current generator is shown in Fig. 20.14. Usually a coil of several turns of wire wound on an iron armature is rotated in the constant field of a magnet. Brushes rubbing against two slip rings attached to the ends of the coil carry off the induced current. As the coil rotates, its two long parallel sides (1–2 and 3–4), each of length l, move through the field in opposite directions. The emf induced across side 1–2, moving at $v_\perp = v \sin \theta$ with respect to B, follows from Eq. (20.4):

$$\mathcal{E} = v_\perp Bl = Blv \sin \theta$$

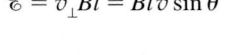

Figure 20.14   (a) A simple ac generator consisting of a loop rotating in a magnetic field. (b) Various stages in the rotation cycle, along with graphs of the corresponding generator outputs. (c) Detail of the conducting loop tilted at an angle θ. (d) The same loop shown in profile.

A little hand-cranked generator.

ALTERNATORS

The simple generator of Fig. 20.14, with a few horseshoe magnets providing the field, was once commonplace (you've probably seen someone in an old movie cranking a telephone or perhaps a detonator just prior to setting off explosives). When large voltages (several kilovolts) and currents (of 50 amps and more) are involved, the slip rings and brushes, which tend to spark and deteriorate, become too troublesome. Today, this sort of *revolving-armature* machine (the armature is the component in which the emf is induced) is in decline. The difficulties with it can be avoided by keeping the coil that carries the induced current stationary. Instead of rotating the armature, electromagnets are mounted on a turning shaft to produce a *revolving-field* machine. The current creating the revolving *B*-field is comparatively small, and slip rings and brushes can easily handle the task. This kind of *alternator* powers the electrical system in most automobiles.

At the moment pictured in Fig. 20.14c, $\vec{v} \times \vec{B}$ points from 2 to 1, so point 2 is at a lower potential than point 1 (work is done in bringing positive charges over to point 1). The same emf appears across the length 4–3 with point 4 lower than point 3. Since no emf is induced along the lengths of 1–4 and 2–3, the net emf around the single loop is $\mathcal{E} = 2(Blv \sin \theta)$ with point 4 negative (lower) and point 1 positive (higher): a conventional current will circulate from 4 to 1 driven by induction and then out (from the + terminal, near 1) into the external circuit, around and back into the loop (at the − terminal, near 4). When there are *N* turns of wire forming a coil, the emf of each loop is in series with the next, yielding a net emf of

$$\mathcal{E} = 2NBlv \sin \theta \tag{20.6}$$

When θ exceeds 180°, the polarity reverses. Since both θ and v can be written in terms of ω, we use the fact that when the angular speed ω is constant, $\theta = \omega t$. Moreover, $v = r\omega$, where here $r = \frac{1}{2}h$. Keeping in mind that lh is the area A, $lv = l(\frac{1}{2}h\omega) = \frac{1}{2}A\omega$ and the emf becomes

[emf of a generator]
$$\mathcal{E} = NAB\omega \sin \omega t \tag{20.7}$$

Regardless of the actual shape of the coil, the emf will be alternating with a frequency $f = \omega/2\pi$. In the United States and Canada f is typically 60 Hz, whereas in much of the rest of the world f is 50 Hz. AC voltages can be changed quite easily, and so you can plug an American hairdryer (using a converter) into a British wall outlet—the dryer isn't noticeably affected by the frequency difference. On the other hand, you will have a lot more trouble with something requiring timing signals, like playing an American video tape on a French VCR.

20.4 The DC Generator

There are many applications where dc electrical power is required, even though it's somewhat more difficult to produce. For example, although small motors (for things like hand drills and vacuum cleaners) run nicely on ac, large electrical motors generally don't do quite so well. The really big motors used on electric railway systems (such as trolley cars and subway trains) usually operate on dc.

The polarity reversals of the voltage from an ac generator can be eliminated using a split-ring *commutator* (Fig. 20.15), just like that on a simple dc motor. Thus, the negative half of the ac signal is reversed and made positive. The result is a bumpy direct current that rises and falls but never goes negative (Fig. 20.16). The single-coil arrangement produces

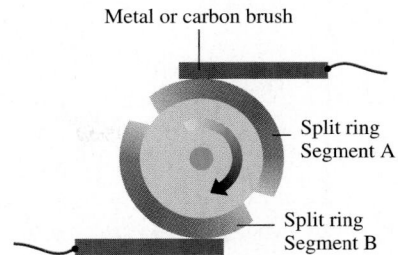

Metal or carbon brush

Split ring
Segment A

Split ring
Segment B

Figure 20.15 A two-segment commutator. At the moment shown, the top brush is touching segment *A* and the bottom one, segment B. As the ring turns, segment *A* moves into contact with the bottom brush. Reversing the connection reverses the signal.

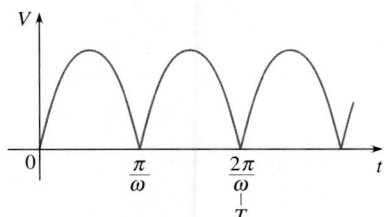

Figure 20.16 The pulsating dc from a single-coil generator with a two-segment commutator. Just as the sinusoidal voltage is about to go negative, the commutator flips it.

an emf that is zero twice during each revolution. Moreover, the voltage is relatively small for much of the cycle. By contrast, if the armature comprises two perpendicular coils (Fig. 20.17) and the commutator has four segments, so that the brushes are always in contact with the coil having the greatest emf, the output voltage across the terminals will be appreciably smoother, and the zeros will no longer occur. With 90° between coils, there will be a relative shift in the induced voltages of 90°, which is equivalent to a shift in time of $\pi/2\omega$, as shown in Fig. 20.17b. With three coils 60° apart, three voltages each shifted in time by $\pi/3\omega$ will be induced, yielding a final voltage equal to the envelope of the three curves, as in Fig. 20.18. Still more coils can be added, and the output voltage can thereby be made almost constant with only a slight ripple.

Example 20.4 **[I]** A simple single-coil dc generator rotates at a constant frequency of 60 Hz in a 0.40-T magnetic field. Given that the coil has 10 turns and encompasses an area of 1200 cm², what will be its maximum emf?

Solution A coil is rotating in a *B*-field with a specified frequency, and that ought to bring to mind visions of a sinusoidal emf. (1) TRANSLATION—A generator coil of known area and number of turns revolves at a specified frequency in a known *B*-field; determine the maximum induced emf. (2) GIVEN: $f =$ 60 Hz, $A = 1200$ cm², $B = 0.40$ T, and $N = 10$. FIND: $\mathscr{E}_m$. (3) PROBLEM TYPE—Electromagnetic induction/motional emf/ generator. (4) PROCEDURE—The basic formula for the emf of

an ac generator is Eq. (20.7), namely, $\mathscr{E} = NAB\omega \sin \omega t$. (5) CALCULATION—The maximum emf ($\mathscr{E}_m$) is the amplitude of the oscillating voltage given by

$$\mathscr{E} = NAB\omega \sin \omega t \qquad [20.7]$$

namely $\qquad \mathscr{E}_m = NAB\omega$

Here $\omega = 2\pi f = 2\pi(60 \text{ Hz}) = 376.99$ rad/s and

$\mathscr{E}_m = (10)(1200 \times 10^{-4} \text{ m}^2)(0.40 \text{ T})(376.99 \text{ rad/s}) = \boxed{0.18 \text{ kV}}$

Quick Check: $\omega \approx 6(60) \approx 360$ rad/s; $\mathscr{E}_m \approx 10 \times (0.12$ m²)(0.4 T)(360 rad/s) ≈ 0.17 kV.

Figure 20.17 (a) A two-coil dc generator. Compare this device with the one depicted in Fig. 20.14 and note how the brushes are in contact with one of the loops at maximum voltage. (b) Emf of the two-coil dc generator. Notice how this emf is equivalent to two voltages like the one shown in Fig. 20.16, shifted by $\pi/2\omega$ = $\frac{1}{4}$ *T* = 90° with respect to one another. A single coil contributes for $\frac{1}{4}$ of a cycle until the brushes shift and the other coil kicks in. It then contributes for the next $\frac{1}{4}$ cycle, and so on. At any instant only one coil is providing the voltage.

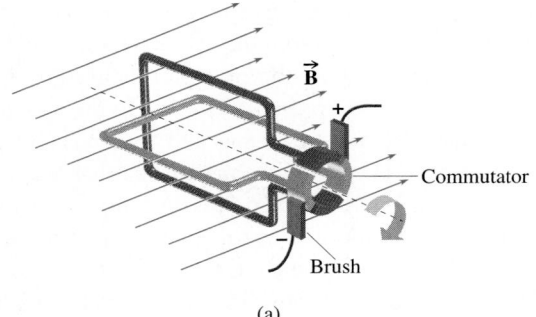

Commutator

Brush

(a)

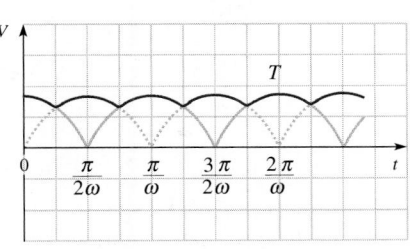

(b)

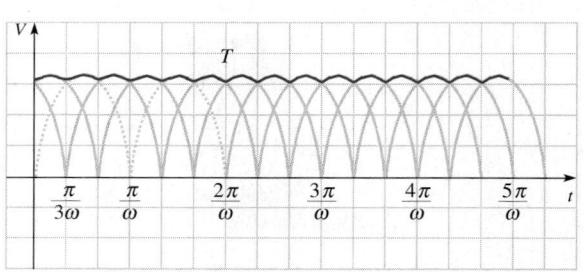

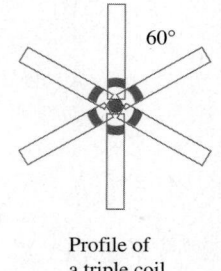

Profile of
a triple coil

Figure 20.18 The emf of a three-coil dc generator. Each component signal is shifted by $\pi/3\omega = \frac{1}{6} T = 60°$, where T is the period.

It's curious that the similarity between the dynamo and the dc motor was not recognized until the mid-nineteenth century and not really appreciated until a remarkable accident occurred in 1873. Not only do they look alike, they are actually identical: the action of the dynamo is just the reverse of the action of the dc motor. The dynamo converts mechanical energy into electrical energy—the armature is turned mechanically from the outside, and out comes a current. The motor converts electrical energy into mechanical energy—in comes a current from the outside, and around goes the armature.

The large hall at the 1873 Vienna Exposition was filled with modern gadgets. One of the Gramme dynamos driven by a steam engine was pouring forth electrical power when a workman unwittingly connected the output leads from another dynamo to the energized circuit. In an instant, the second device began to spark and whine and come alive, whirling around at great speed. The dynamo had become a motor. That impromptu mating of machines gave birth to the future as we have come to know it.

Eddy Currents

When Arago rotated a copper disk beneath a compass and found that the needle soon began revolving around with the disk, he was actually observing the effects of electromagnetic induction. The rotating disk "sees" a nonuniform time-varying magnetic field. An emf is induced, and loops of current are set up in the disk. These currents generate their own counter B-field that opposes, via Lenz's Law, the cause of the induction—the needle is thereby made to rotate such that its relative motion with respect to the disk tends to vanish.

If an extended conductor translates with respect to a B-field, which is not uniform over the entire conductor, or if different portions of the conductor move at different velocities (i.e., it rotates) with respect to the field, currents will be induced within it that circulate in closed paths. Instead of the transient current in the uniform-field situation of Fig. 20.8, a closed current loop can now form. For example, with Faraday's disk (open circuited, as shown in Fig. 20.19) the induced current loops, known as **eddy currents**, set up two dipole field regions, each north on one face of the disk and south on the other. The leading dipole, emerging from the external B-field on the left, attracts the poles of the magnet, thus tending to oppose the rotation. The trailing eddy-current dipole, moving into the B-field of the magnet, presents its north to the magnet's north and its south to the magnet's south, thus repelling the magnet and again resisting the motion. Consequently, the eddy currents produce a Lenz's Law drag on the rotating disk. Eddy currents are used to great advantage in heating induction furnaces, in damping the oscillations of delicate devices, and in the operation of a variety of instruments, such as the speedometer (p. 28) and the metal detector now familiar to air travelers and beachcombers alike. In the latter, a time-varying magnetic field induces currents in a metal object, and the resulting induced B-field reveals its presence to a pickup coil. {For more on the subject click on **EDDY CURRENTS** under **FURTHER DISCUSSIONS** on the CD.}

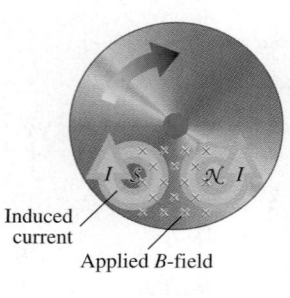

Induced
current

Applied B-field

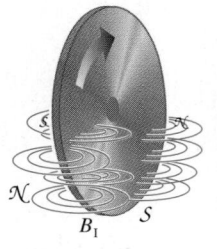

B_I

Figure 20.19 A copper disk rotating across a perpendicular magnetic field. Swirls of eddy currents circulate, opposing the motion. The induced magnetic field B_I opposes the turning of the disk by interacting with the applied field.

Self-Induction

Whenever a voltage from some external source is placed across the terminals of a coil, the resulting current will produce a magnetic field. But this phenomenon raises an interesting point first considered by Joseph Henry. A coil laced with an increasing field must experience an induced emf—what does it matter that the coil itself is involved in creating the field? Field lines stream out from the center of the wire of the coil, "cutting" it just as they would cut a search coil immediately adjacent to it. Mobile charges within the wire experience a time-varying B-field, a perpendicular force, and a resulting induced emf along the length of the conductor. Or, if you prefer, the flux linking the coil will be changing in time, and Eq. (20.3) therefore requires that there be an induced emf. With either interpretation, it follows from Lenz's Law that the induced emf must oppose the cause of itself, and so it is called a **back-emf**. Because of this opposition, the steady state is not reached instantaneously; the current builds gradually, as does its associated B-field. This process of **self-induction** retards the increase or decrease of current in a coil (and to a lesser extent in other circuit elements, hookup wires, transmission lines, and so on). As such, it is an extremely important aspect of almost all real ac devices from stereos to satellites.

20.5 Inductance

To quantify the self-induction of a current-carrying coil, we realize that the flux linkage $N\Phi_M$ is proportional to the current I producing it:

$$N\Phi_M = LI \tag{20.8}$$

The **inductance** of any circuit element (including the wiring) is a measure of its opposition to the change of current in that element.

The constant of proportionality L is called the **self-inductance**, or just the *inductance*, for short. In a sense, it is the electrical equivalent of inertia, a resistance to change. The SI unit of inductance is the **henry** (H); a flux linkage of 1 weber is established in a coil by a current of 1 ampere circulating therein when the coil's inductance is 1 henry. As we have seen so many times before, the constant of proportionality L is a composite of several of the system's physical characteristics. Here, it depends on the size and shape of the coil and on the surrounding medium. The flux linkage does not vary linearly with I if the permeability of the medium is not constant—that is, if it depends on B and therefore on I. Consequently, if there is a ferromagnetic material within the coil, or even nearby, L becomes a function of I, and though Eq. (20.8) still holds, it is inherently more complicated than it would be were L constant.

In practice, when the precise value of an inductance is needed, it's usually measured; but if the geometry is simple enough, it can be approximated theoretically. As an example, to compute the inductance of a long hollow coil of length l with n turns per unit length ($n = N/l$) and a cross-sectional area A, we recall that

$$B_z \approx \mu_0 nI \tag{19.5}$$

Assuming B to be constant across the region encompassed by the solenoid (although we know the field actually decreases in toward the center), we should get a useful expression, even if it's likely to be slightly too large. Using Eq. (20.8) the inductance of a long air-core solenoid becomes

[long narrow coil]
$$L = \frac{N\Phi_M}{I} = \frac{NBA}{I} \approx \frac{\mu_0 N^2 A}{l} \tag{20.9}$$

If some material with a permeability μ is made to fill the hollow, we simply replace μ_0 by μ. There are commercial inductors that allow their inductances to be adjusted by sliding a ferrite slug partway into the coil. Ferrites (oxides of magnesium, manganese, zinc, or nickel) are especially useful in suppressing core eddy currents at the high operating frequencies of radio and television circuits.

Example 20.5   [I] A 3.0-cm-long solenoid with a cross-sectional area of 0.50 cm² and comprising 300 turns of fine copper wire in a single layer is to be used as the antenna for a radio. The magnetic field component of the incoming electromagnetic signal will oscillate within the coil, induce an emf that will then be processed by the rest of the radio, and out will come music. (a) Determine the coil's inductance when the core is air-filled. (b) Approximate the inductance when a ferrite core is used instead, given that its relative permeability at the anticipated current level is 400.

Solution   We've got a coil and need its inductance, and that should bring to mind the defining equation (20.8) for inductance, as well as the only specific equation for L that we have (20.9). (1) TRANSLATION—Determine the inductance of a solenoid of known geometry and number of turns, with and without a specified

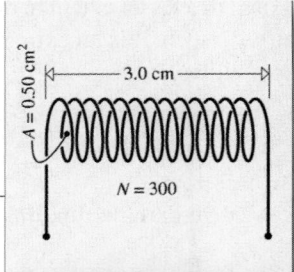

core.   (2) GIVEN: $l = 0.030$ m, $A = 0.50 \times 10^{-4}$ m², $N = 300$, and $\mu/\mu_0 = 400$. FIND: L.   (3) PROBLEM TYPE—Inductance. (4) PROCEDURE—The basic equation for the inductance of a solenoid is Eq. (20.9): $L \approx \mu_0 N^2 A/l$.   (5) CALCULATION:

$$L \approx \frac{\mu_0 N^2 A}{l}$$

$$L \approx \frac{(1.26 \times 10^{-6}\,\text{T·m/A})(300)^2(0.50 \times 10^{-4}\,\text{m}^2)}{(0.030\,\text{m})}$$

$$L \approx 1.89 \times 10^{-4}\,\text{H}$$

and to two significant figures $\boxed{L \approx 0.19\,\text{mH}}$. (b) Given $\mu = 400\mu_0$, L is increased by a factor of 400 and $\boxed{L \approx 76\,\text{mH}}$.

Quick Check: $N^2 A \approx 9 \times 10^4 (0.5 \times 10^{-4}) \approx 4.5$; $N^2 A/l \approx 150$; $L \approx \mu_0 150 \approx (1.25 \times 10^{-6})(150)\,\text{H} \approx 190\,\mu\text{H}$.

The Back-Emf

Faraday's Induction Law can be reformed in terms of the inductance, thereby providing an expression for the back-emf. From Eqs. (20.3) and (20.9) the average self-induced emf is

$$\mathscr{E} = -N\frac{\Delta\Phi_M}{\Delta t} = -\frac{\Delta(LI)}{\Delta t}$$

Provided that the inductance is constant (see **APPENDIX** A-6), $\Delta(LI) = L\,\Delta I$ and

$$\mathscr{E} = -L\frac{\Delta I}{\Delta t} \qquad\qquad (20.10)$$

The average induced emf is proportional to the time rate-of-change of the current in the coil. *The positive direction is that of I.* An inductance of 1 H will induce a back-emf of 1 V when the current through it changes at a rate of 1 A/s. This expression also tells us that 1 V = 1 H·A/s, and so 1 H = 1 V·s/A, or 1 H = 1 Ω·s. *If the coil itself has negligible resistance, there will not be an appreciable voltage drop across it due to the current traversing it and the back-emf will be the voltage measured across its terminals.*

The antenna of the old radio pictured on p. 537. The antenna consists of a coil wrapped around a ferrite core (the gray, cylindrical bar). The time-varying magnetic field component of an incoming electromagnetic wave induces a signal, an emf, in the coil.

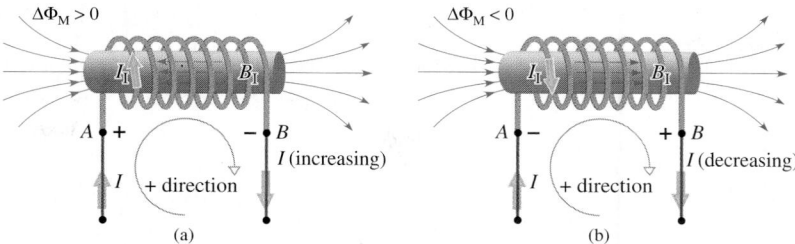

Figure 20.20 (a) I is increasing, an induced current I_I opposes that increase, and a negative back-emf (a voltage drop) appears across the terminals. (b) I is decreasing, I_I opposes that decrease, and a positive back-emf (a voltage rise) appears across the terminals.

Suppose an externally supplied current is made to *increase* through a coil passing from its terminals A to B (Fig. 20.20a). In response to the changing flux, an induced current (I_I) then moves from B to A, opposing ΔI and reducing I in the circuit, which in effect raises the potential of point-A, making it positive with respect to B. This back-emf is in the opposite direction to I and is negative—it's a voltage drop. The coil's inductance impedes the original *changing* current, thereby producing a voltage drop across its terminals (as if it now had some kind of "resistance" even though $IR = 0$). While the current is increasing, there will be a voltage across the terminals as shown. If the current begins decreasing, an induced current will pass from A to B to oppose the decrease. Point B will be positive with respect to point A, and the back-emf will reverse, becoming positive (Fig. 20.20b).

An **inductor** is a device designed expressly to introduce inductance into a circuit, something that is done for a variety of reasons. For instance, an inductor (or *choke*) opposes a changing current and will therefore impede the progress of alternating currents while passing, with very little opposition, steady currents. Not surprisingly, this impedance (p. 756) to ac increases with both the frequency of the current and the inductance of the inductor. Accordingly, inductors are often used to separate or filter out ac from dc. Typically, an inductor consists of a coil of wire wound on a hollow cylinder that may contain air or some kind of ferromagnetic core. Values usually range from microhenries (μH), used at high frequencies as in radio and television, to several henries, used, for instance, in low-frequency (60 Hz) choke-filtered power supplies. The circuit symbols for an inductor with and without a ferromagnetic core are ‒₀₀₀₀‒ and ‒₀₀₀₀‒ , respectively.

This adjustable ferrite-core inductor is part of the tuning circuit of the AM radio on p. 537. We'll come back to what it does when we consider resonant circuits (p. 758).

Example 20.6 **[I]** The current in a 50-μH coil (for which R is negligible) goes from 0 to 2.0 A in 0.10 s. Determine the average self-induced emf measured across its terminals.

Solution The emf of a coil depends on the time rate-of-change of the current. (1) TRANSLATION—A known inductance experiences a known current change in a specified time; determine the induced emf. (2) GIVEN: $L = 50$ μH, $I_i = 0$, $I_f = 2.0$ A, and $\Delta t = 0.10$ s. FIND: $\mathscr{E}$. (3) PROBLEM TYPE—Inductance/back-emf. (4) PROCEDURE—There is only one relationship that gives the self-induced emf of a coil in terms of

the changing current. (5) CALCULATION—From Eq. (20.10)

$$\mathscr{E} = -L\frac{\Delta I}{\Delta t} = -(50 \times 10^{-6} \text{ H})\frac{2.0 \text{ A}}{0.10 \text{ s}} = \boxed{-1.0 \text{ mV}}$$

The negative sign tells us that the back-emf opposes the increase in current. The polarity shown in Fig. 20.20 already took the sign into consideration.

Quick Check: The current is changing at a moderate rate of 20 A/s, but the inductance is only 50 microhenries, so we can expect an emf in the millivolt range.

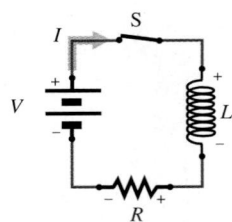

Figure 20.21 An *R-L* circuit with the switch just closed.

20.6 The *R-L* Circuit: Transients

Figure 20.21 shows a battery in series with an inductor *L* and a resistor *R*. The inductor is imagined to have zero resistance, although we could have just as well lumped its more realistic nonzero resistance into *R*. As soon as the switch is closed, an increasing current *I* just begins to circulate, but it's opposed by the back-emf, so it builds only gradually. From Kirchhoff's Loop Rule (p. 650), at any instant

$$V = L\frac{\Delta I}{\Delta t} + IR \qquad (20.11)$$

Remember that V, R, and L are fixed but I varies from moment to moment. Inasmuch as the battery voltage *V* is constant, it follows from the equation that as the current increases, it must increase ever more slowly. This occurs because as *IR* increases, the back-emf decreases and $\Delta I/\Delta t$ decreases. That's shown in the exponential *I*-versus-*t* curves of Fig. 20.22*a*. The slopes do indeed decrease with time. The inductor causes the current to rise slowly, reaching a maximum value of *V/R* as $\Delta I/\Delta t \to 0$, as $t \to \infty$.

Since

$$L\frac{\Delta I}{\Delta t} = \text{emf} = V - IR$$

at the instant the switch is closed ($t = 0$), the current begins to rise from its value of zero and, hence, the emf initially equals *V*; that is, $L(\Delta I/\Delta t) = V$. Thus, the initial slope of the curve at $t = 0$ is

[initial slope]
$$\frac{\Delta I}{\Delta t} = \frac{V}{L}$$

The **time constant** of an *R-L* circuit is *L/R*.

Had the current continued to increase at this initial rate of increase, it would have reached its maximum value (*V/R*) at a time *L/R* (such that the rise *V/R*, over the run *L/R*, equals the slope *V/L*). The quantity *L/R* is the **time constant**. After an interval of one time constant, the current reaches an amount of 0.632 of its final value (*V/R*). After an interval five time constants long, the current is within 1% of its steady-state value (Table 20.1).

Table 20.1

The Exponential Buildup of Current in an *R-L* Circuit

Number of time Constants	Percent of Final Current*
1	63.2
2	86.5
3	95.0
4	98.2
5	99.3
6	99.8
7	99.9

*$I = (V/R)(1-e^{-t/\tau})$ where $\tau = L/R$.

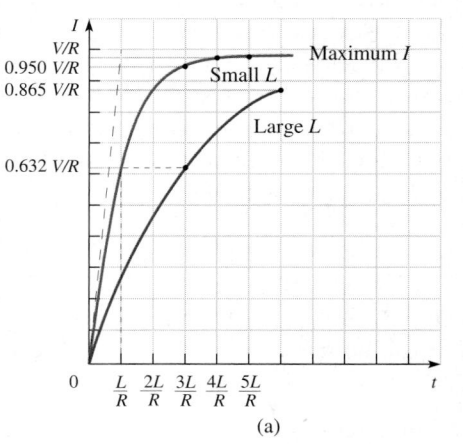

(a)

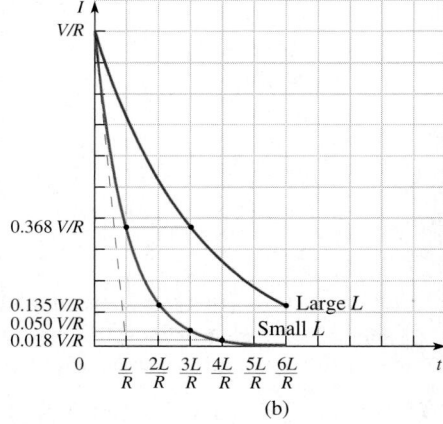

(b)

Figure 20.22 (a) The graph shows *I* versus *t* for an *R-L* circuit with a time constant of *L/R*. For comparison, another curve is included where the time constant is three times as large. Note that the curves rise to 63.2% of maximum value after *L/R*, and 63.2% of the remaining 36.8% after 2*L/R*, and 63.2% of the remaining 13.6% after 3*L/R*, and so on. (b) When the switch in Fig. 20.21 is opened and the current decays, the current drops 63.2% of the remaining amount during each successive one-time-constant interval. Thus, the current falls to 36.8% *V/R* after *L/R* and then 63.2% of 36.8% *V/R* to 13.6% *V/R* at 2*L/R*, then 63.2% of 13.6% *V/R* to 5.1% *V/R* at 3*L/R*, and so on.

Example 20.7 **[I]** Determine the time constants for a series *R-L* circuit (like that in Fig. 20.21) consisting in one case of a 100-Ω resistor and a 10-H inductor, and in the other case of the same resistor with a 1.0-H inductor. Compare the two resulting currents.

Solution The time constant for an *R-L* circuit is just L/R. (1) TRANSLATION—A known inductance is in series with a known resistance; determine the time constant. (2) GIVEN:

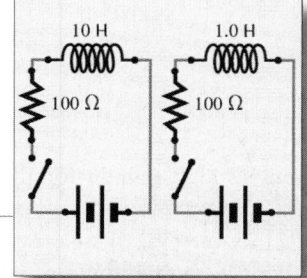

$L = 10$ H, 1.0 H, and $R = 100$ Ω. FIND: The time constants. (3) PROBLEM TYPE—Inductance/*R-L* circuit/time constant. (4) PROCEDURE—For an *R-L*

circuit the time constant is L/R. (5) CALCULATION:

$$\frac{L}{R} = \frac{10\,\text{H}}{100\,\Omega} = \boxed{0.10\,\text{s}}$$

as compared to

$$\frac{L}{R} = \frac{1.0\,\text{H}}{100\,\Omega} = \boxed{10\,\text{ms}}$$

With the larger inductor, we will have to wait 0.10 s for the current to reach 63% of its maximum value, as compared to only 10 ms for the smaller one.

Quick Check: The inductances differ by a factor of 10, so the time constants L/R must also: $(0.10\,\text{s})/(10\,\text{ms}) = 10$.

Notice that when *R* is very large, most of the voltage drop will be across the resistor, the back-emf will be small, and the delay it causes in the current buildup will be small. On the other hand, the larger the inductance, the greater the back-emf, the less tilted the *I-t* curve is initially, the larger the time constant, and the longer it takes for the current to reach maximum [which brings to mind the *R-C* circuit (p. 647), where the time constant is *RC*].

Opening the switch once the steady state has been established will cause the current to gradually decay, being sustained by the back-emf (Fig. 20.22*b*). Have you ever pulled the plug on an appliance that was still turned on and seen a spark at the ends of the prongs? If a lamp with a large resistance, compared to that of the coil, is positioned as in Fig. 20.23 and the switch is closed, it will only glow faintly in the steady state: most of the current will then be passing through the inductor since the back-emf is zero and the resistance low. Upon opening the switch, the lamp will not immediately go out but instead burn even more brightly for a moment and then gradually diminish in intensity. The back-emf at the instant *S* is opened can be greater than the steady-state voltage across the lamp—that condition will not last long, but it is what makes the lamp flare up initially. The larger the *L*, the larger the time constant, and the longer the lamp glows on.

{It's recommended at this point that you click on THE *R-L* CIRCUIT under INTERACTIVE EXPLORATIONS on the CD. This simulation will allow you to study the behavior of the circuit in Fig. 20.21 and to see how each parameter affects the rising current curve of Fig. 20.22.}

Table 20.2

The Exponential Decay of Current in an *R-L* Circuit

Number of time Constants	Percent of Final Current*
1	36.8
2	13.5
3	4.98
4	1.83
5	0.67
6	0.25
7	0.09

*$I = (V/R)e^{-t/\tau}$ where $\tau = L/R$.

WHERE IS THE ENERGY COMING FROM?

One might well ask at this point, "Where is the energy coming from that powers the lamp when the switch in Fig. 20.23 is opened?" Clearly, it's not from the battery. Classical theory maintains that the *B*-field must possess energy that it imparts to the coil in the form of an induced current, when the steady-state current, which sustained the field, is disrupted. The model suggests that when the switch is opened the field is radiated away; it sweeps out through the coil, inducing a current in it. In this way the field returns energy to the circuit. The realization that the field carries energy (and momentum), which are properties of matter, raises a number of interesting questions that will have to be dealt with later. For example, how exactly is energy transferred from the field to a charged particle?

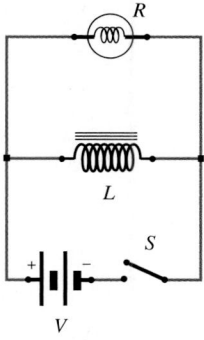

Figure 20.23 An inductor, a battery, and a lamp. The resistance of the lamp (*R*) is much greater than that of the coil, and the lamp gets little current when *S* is closed. Remarkably, it glows brightly for a moment after the switch is opened.

20.7 Energy in the Magnetic Field   (Optional)

In Chapter 16, we talked about the energy stored in a charged capacitor in the process of building up the electric field that spans the gap between its plates:

$$PE_E = \tfrac{1}{2}CV^2 = \tfrac{1}{2}\varepsilon_0(Ad)E^2 \qquad [16.15]$$

The field in the gap can be imagined as retaining this energy uniformly within the space of volume Ad, which it occupies. If we introduce the concept of *energy per unit volume of the electric field* (u_E), the preceding expression becomes

[energy density—electric]
$$u_E = \tfrac{1}{2}\varepsilon_0 E^2 \qquad (20.12)$$

This is the **energy density** of the electric field; wherever there is an E-field in space, this will be the energy per unit volume associated with it.

The analogous situation exists when we establish a current through an inductor, thereby building up a magnetic field. In Fig. 20.21, the battery must do work W against the back-emf if it is to send a current through the inductor. During a tiny interval of time Δt, the battery, which provides power to the inductor at a rate P, does a small amount of work

$$\Delta W = P\Delta t = I\mathcal{E}\Delta t = IL\frac{\Delta I}{\Delta t}\Delta t = IL\,\Delta I$$

where I is the current at any instant. We want the total amount of work done in increasing the current from $I = 0$ to $I = I_f$:

$$\sum\Delta W = \sum IL\,\Delta I$$

The right side is the area under a plot of IL versus I, as shown in Fig. 20.24. Since L is constant, this plot is a straight line and the total area (W) under it between 0 and I_f is the triangular area of height I_fL and base I_f; namely, $W = \tfrac{1}{2}LI_f^2$. This is the work done on the inductor, and it's also the magnetic potential energy PE_M stored. Remembering that we are considering the final current, we can drop the subscript f for simplicity. The energy stored in an inductor carrying a current I is then

$$PE_M = \tfrac{1}{2}LI^2 \qquad (20.13)$$

where we *assume* it's associated with the B-field. Thus, with a long solenoid of length l, where from Eq. (20.9) $L = NBA/I$, the field existing inside is $B \approx \mu_0 NI/l$, and so

$$PE_M = \tfrac{1}{2}LI^2 = \tfrac{1}{2}NBAI = \tfrac{1}{2}(Al)\frac{B^2}{\mu_0} \qquad (20.14)$$

Representing the *energy per unit volume of the magnetic field* as u_M and realizing that Al is the volume occupied by the B-field, we have

[energy density-magnetic]
$$u_M = \tfrac{1}{2}\frac{B^2}{\mu_0} \qquad (20.15)$$

Although this formula was derived for a long solenoid, it turns out to be generally valid. Apparently, it is possible to store electromagnetic energy in electric and magnetic fields in vacuum. As we shall see in Chapter 22, it is even possible to transport this energy across free space—that's what light does.

THE CLASSICAL ELECTROMAGNETIC FIELD

So far, we have dealt with a variety of subtle electromagnetic phenomena in terms of classical field theory. It allows us to understand everything we have looked at in terms of electric and magnetic fields using rather picturesque nineteenth-century images of flux and field lines. Indeed, these fields seem to become more substantial with each extension of the analysis. **The classical field appears as a continuous something that can store, transfer, and transport energy** (although it's not obvious exactly how it does that). This picture will have to be profoundly modified when we confront the modern concept that fields are quantized and radiant energy exists only in minute discrete bursts. Usually, however, when observed on a macroscopic scale, quantum fields average out to, and become indistinguishable from, classical fields. Though the classical electromagnetic field is an indispensable conceptual tool, as we'll see, the model can no longer be taken quite so literally.

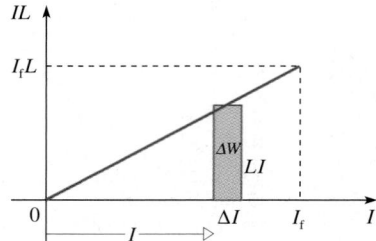

Figure 20.24 A plot of IL versus I. The area under the curve is the work done in building up a current of I_f through an inductor L.

Core Material & Study Guide

ELECTROMAGNETICALLY INDUCED EMF

The **flux of the magnetic field** is

$$\Phi_M = B_\perp A = BA_\perp = BA \cos \theta \qquad [20.1]$$

with units of **webers**: $1 \text{ Wb} = 1 \text{ T·m}^2$. As a result of flux changes, the average induced emf is

$$\mathcal{E} = -N \frac{\Delta \Phi_M}{\Delta t} \qquad [20.3]$$

which is **Faraday's Induction Law** (p. 712). **Lenz's Law** is

> *The induced emf will produce a current that always acts to oppose the change that originally caused it.*

This consequence of energy conservation goes hand in hand with Faraday's Law—to learn about both study Section 20.1 (Faraday's Induction Law) and examine Examples 20.1 and 20.2. These are important basic ideas. {Make sure to work out the WARM-UPS in CHAPTER 20 on the CD. That will ensure that you've got the basics. Then study the WALK-THROUGHS}.

When a wire of length l moves at a speed v perpendicularly through a B-field, the induced **motional emf** (p. 716) is

$$\mathcal{E} = vBl \qquad [20.4]$$

The fact that there is an emf, and therefore an E-field, whenever a charge and a B-field are in relative transverse motion, is fundamental. Study Section 20.2 (Motional emf).

GENERATORS

An ac generator revolving at a rate ω produces an emf given by

$$\mathcal{E} = NAB\omega \sin \omega t \qquad [20.7]$$

The practical application of induction is found in Sections 20.3 (The AC Generator) and 20.4 (The DC Generator); Example 20.4 is representative.

SELF-INDUCTION

The net flux through a current-carrying coil—the flux linkage $N\Phi_M$—is proportional to the current I producing it:

$$N\Phi_M = LI \qquad [20.8]$$

The constant of proportionality, L, is the **self-inductance**, or just the *inductance*. The self-inductance of a long air-core solenoid is

$$L \approx \frac{\mu_0 N^2 A}{l} \qquad [20.9]$$

The induced, or **back-emf**, in a coil is

$$\mathcal{E} = -L \frac{\Delta I}{\Delta t} \qquad [20.10]$$

Study Section 20.5 (Inductance) and review Examples 20.5 and 20.6.

When a coil is placed in series with a battery of voltage V and a resistor R, we have an R-L circuit where

$$V = L \frac{\Delta I}{\Delta t} + IR \qquad [20.11]$$

Reread Section 20.6 (The R-L Circuit: Transients), study Fig. 20.22, and make sure you understand Example 20.7.

The energy stored in an inductor carrying a current I is

$$PE_M = \tfrac{1}{2} L I^2 \qquad [20.13]$$

The *energy per unit volume of the magnetic field* is

$$u_M = \tfrac{1}{2} \frac{B^2}{\mu_0} \qquad [20.15]$$

Section 20.7 (Energy in the Magnetic Field) deals with these ideas and should come to mind when treating any problem in this chapter that talks about energy.

In case you've forgotten, read all the Suggestions on Problem Solving and go over the Examples in the text. **Look at the CD WARM-UPS, then study the WALK-THROUGH EXAMPLES.** Do the odd-number Multiple Choice Questions in the textbook and check your answers. You should then be ready to try the problems. Do all the odd I-level Problems first. The complete solutions for many of them (those with boldfaced numbers) are provided. Once you feel confident, go on to the level-II Problems.

Key Terms

electromagnetic induction	commutator
induced emf	eddy current
magnetic flux	back-emf
flux density	self-induction
Faraday's Induction Law	inductance
Lenz's Law	henry
motional emf	inductor
search coil	choke
dynamo	R-L circuit
ac generator	time constant
dc generator	energy density

Discussion Questions

1. In 1825, long before the successful work by Henry and Faraday on induction, Jean Daniel Colladon attempted to observe the effect by attaching a helical coil to a sensitive galvanometer. To shield the delicate instrument from the direct influence of the moving magnet, which he planned to wave around near the coil, Colladon wired the galvanometer to the rest of the circuit via two long leads so that it could be safely located in an adjacent room. Because he had no assistant, whenever he moved the magnet he had to walk over to the galvanometer to observe its response. Since the name of Colladon is never included with that of Faraday and Henry, what do you think he saw and what did he not see? Explain.

2. During a major period of solar activity, magnetic storms of such severity occur that voltages of upwards of 1 kV can appear across the length of the Alaska pipeline from Prudhoe Bay to Valdez. Explain how this happens. (In 1956, a particularly strong magnetic storm severely affected the first transatlantic voice cable. Such effects were a troublesome problem for early telegraph operators.)

3. Using rechargeable batteries, it is possible to power small appliances such as electric toothbrushes with watertight sealed units. Rather than having the customary two exposed terminals that plug directly into a dc power supply, these devices are simply positioned, handle down, in a well within a holder that is continuously attached to ac. During the long periods of nonuse, both the holder well and the handle become quite warm. As the toothbrush is slightly lifted from the base, you can feel a vibration (at what seems to be 60 Hz). How might such a system work? Does any "electricity" actually pass from the base to the batteries? A similar arrangement could be used to power an artificial heart by passing electromagnetic energy into the chest without the need for wires through the skin. Discuss how this might be done.

4. Imagine a superconductor in its normal state in the shape of a ring immersed in a magnetic field parallel to the central axis (that is, perpendicular to the plane of the ring). Suppose the ring is cooled and becomes superconducting. (a) Describe what happens to the field. (b) If the ring is pulled perpendicularly out of the field, what will happen to the flux in the hole in the "doughnut"? (c) Account for the energy associated with the work done on the ring (if any) in yanking it from the field region.

5. Imagine a superconducting ring supported horizontally so that it can be approached from below by a bar magnet. Describe and explain what will happen as the magnet (north pole upward) is brought near the ring. Since a superconductor has zero resistance, an induced E-field would result in an infinite current. How, then, must a superconductor behave in the presence of a changing B-field in order that this impossibility not occur?

6. A promising future source of energy is the controlled fusion reaction, the same phenomenon that powers the stars. Deuterium and tritium atoms are ionized. The resulting plasma of positive nuclei and electrons is confined to a ring-shaped region within a vacuum chamber (Fig. Q6) by magnetic fields. The plasma is raised up to temperatures in excess of 100 million K, whereupon the nuclei

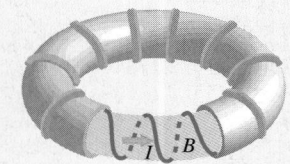

Figure Q6

undergo fusion and liberate great amounts of energy. The confinement field, which actually spirals around the toroid, is provided in part by external current-carrying coils wrapped around the chamber. A crucial component of this confinement field is generated by toroidal currents circulating in the plasma itself. Explain how such a toroidally directed current (along the axis of the chamber) could be induced in the plasma. Describe the kind of current required and trace the transfer of energy into the plasma.

7. Imagine a hand-cranked generator in parallel with a 50-W light bulb and a 100-W light bulb. Suppose each bulb can be switched out of the circuit. Compare the amount of effort it would take to light each bulb steadily with the effort needed to crank the generator in a sustained fashion without a load. Explain what causes the difference if there is one.

8. The switch in the circuit shown in Fig. Q8 is closed, and after a long wait, the resistor is set so that the two lamps are equally bright, at which point the switch is again opened. After another long interlude, the switch is closed. Describe what subsequently happens to the two lamps.

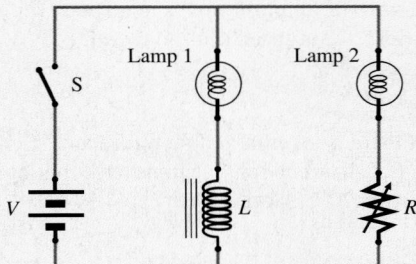

Figure Q8

9. Figure Q9 shows a 70-kV, 60-Hz power line in a remote area of the countryside. A shifty local resident has erected a large open loop just below the line with the intention of drawing off power. Is this possible and, if so, how would it be transferred? Where would the energy stolen come from? Would anyone be able to detect the loss? Might it be possible to bug a telephone using the same approach? Explain.

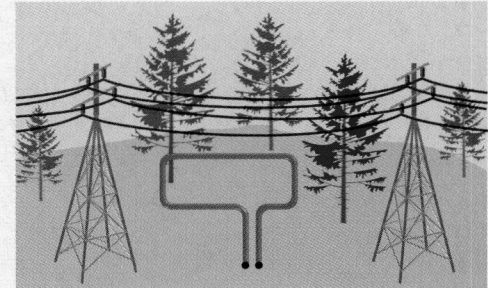

Figure Q9

10. Consider several inductors (L_1, L_2, and L_3) connected alternatively in series and then in parallel. What do you think will be the equivalent inductance in each case? Explain your reasoning.

11. A circuit contains an air-core coil of inductance L and resistance R in series with a power supply. Discuss and compare the amounts of energy stored with and without an iron core in place. Compare the final currents established. Where does the difference in energy come from?

12. Figure Q12 illustrates the construction of a so-called variable reluctance microphone. A diaphragm is attached to a light flexible rod made of ferromagnetic material that, in turn, is fixed via a permanent magnet to a C-shaped structure also made of magnetic material. How does it work? Comment on the way the two coils are wound.

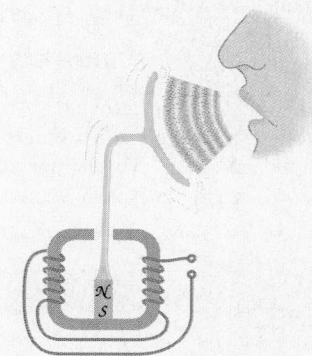

Figure Q12

13. Motors and generators are the same device, the distinction appearing in the operation rather than in the construction—mechanical power in and electrical out and you have a generator; electrical in and mechanical out, and you have a motor. It's reasonable, therefore, to ask, "Is a motor in some way a generator while it's turning?" Explain. If there is a generated back-current, how might it depend on the load? What do you think limits the speed of a free-turning motor? Describe the operation of the motor with a load attached. Why do many motors have open slots on their sides and little internal fan blades attached to their shafts? If you jam a motor—bind a drill or a blender so it's receiving current but not turning—it won't be long before you smell burning insulation. What's happening, and why?

14. Figure Q14 depicts a strip of recorded magnetic tape (which is like a succession of little magnets) passing under a playback head. The latter is a ferromagnetic C-shaped structure with a coil wrapped around it. How does the playback head read the tape? Comment on the relationship between the fineness of the read and write heads, the speed of the tape, and the density of information.

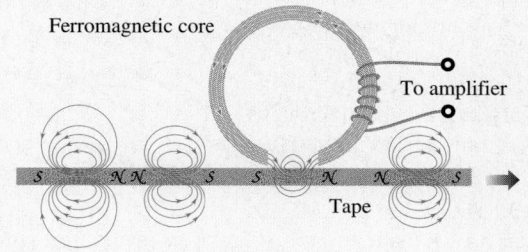

Ferromagnetic core
To amplifier
Tape

Figure Q14

15. The circuits in Fig. Q15 are adapted from an article by R. H. Romer entitled "What do 'voltmeters' measure?: Faraday's law in a multiply connected region" (*Am. J. Phys.*, **50**, no. 12, Dec. 1982, 1089). Part (a) illustrates the nonconservative nature of the induced E-field, in this instance surrounding a long solenoid (perpendicular to the page) carrying a time-varying current. The induced current, I_i, passes through both resistors. The meters (which draw negligible current) show different readings and even different polarities. Note that no flux links either circuit 1–3–4–2–10–9–1 or 1–8–7–2–5–6–1, and these must obey Kirchhoff's Loop Rule (p. 650). Each voltmeter actually reads the work done per unit charge in moving charge through the meter itself. Discuss what's happening. What does the meter on the left in part (b) read and why?

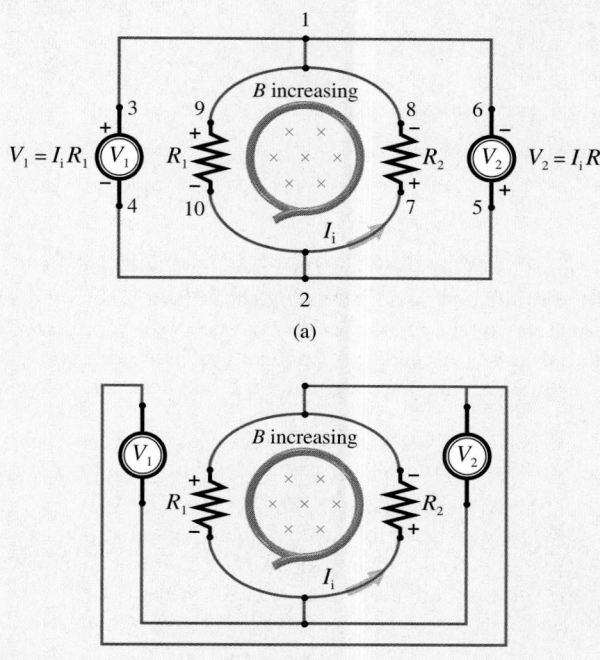

(a)

Figure Q15 (b)

16. Figure Q16 shows an electron orbiting between the poles of an electromagnet in a device called a betatron. The field is gradually being increased. How does the machine accelerate the electron? What keeps it in orbit?

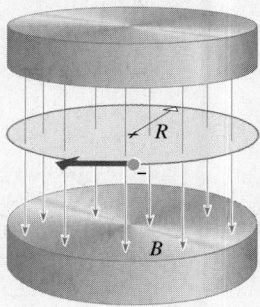

Figure Q16

17. Explain the meaning of each of the Key Terms on page 731.

Multiple Choice Questions

1. A single horizontal loop of wire is moving in a horizontal plane at a constant speed *across* a uniform vertical magnetic field that "fills" and surrounds the loop. The emf induced across its terminals will be   (a) time varying   (b) constant   (c) negative   (d) positive   (e) none of these.

2. Figure MC2, looking down, shows a copper wire moving across a horizontal magnetic field. There will be   (a) a negative voltage induced across its ends   (b) a positive voltage induced across its ends   (c) no voltage induced across its ends   (d) a time-varying voltage induced across its ends   (e) none of these.

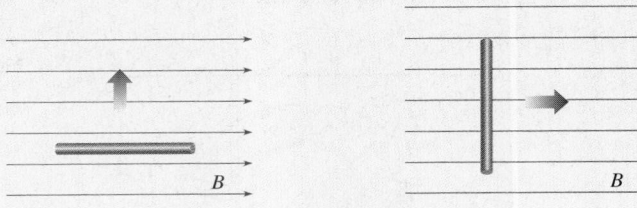

Figure MC2   **Figure MC3**

3. Figure MC3, looking down, shows a horizontal length of copper wire moving along a horizontal magnetic field. There will be   (a) a negative voltage induced across its ends   (b) a positive voltage induced across its ends   (c) no voltage induced across its ends   (d) a time-varying voltage induced across its ends   (e) none of these.

The next four questions refer to Fig. MC4, which shows a horizontal wire capable of moving through a horizontal B-field.

4. If the wire is suddenly lowered, the potential at point-C will be   (a) zero   (b) higher than at point-D   (c) lower than at point-D   (d) the same as at point-D   (e) none of these.

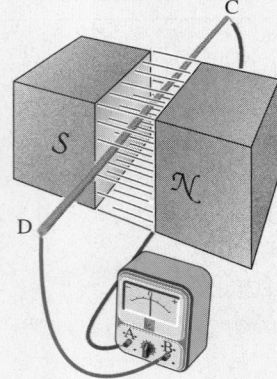

Figure MC4

5. If the wire is suddenly raised, the potential at point-C will be   (a) higher than at point-D, and B should be the + terminal on the voltmeter   (b) lower than at point-D, and B should be the + terminal on the voltmeter   (c) lower than at point-D, and A should be the + terminal on the voltmeter   (d) lower than at point-D, and B should be the + terminal on the voltmeter   (e) none of these.

6. If the two magnets are suddenly raised leaving the wire where it is, the potential at point-D will be   (a) zero   (b) lower than at point-C, and B should be the + terminal on the voltmeter   (c) lower than at point-C, and A should be the + terminal on the voltmeter   (d) higher than at point-C, and B should be the + terminal on the voltmeter   (e) none of these.

7. If the wire is moved horizontally toward the north pole the induced voltage will be   (a) zero   (b) lower at point-C than at point-D   (c) higher at point-C than at point-D   (d) higher at point-B than at point-A   (e) none of these.

8. A closed loop moves at a constant speed parallel to a long straight current-carrying wire, as in Fig. MC8.   (a) the induced current in the loop will progress clockwise   (b) there will be no induced current in the loop   (c) the induced current in the loop will progress counterclockwise   (d) the induced current in the loop will vary with the speed at which the loop moves   (e) none of these.

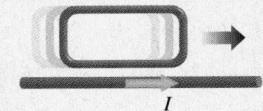

Figure MC8

9. The bar magnet in Fig. MC9 is moving at a constant speed toward the coil. The voltage measured across points *A* and *B* is   (a) higher at *B* than *A* and increasing   (b) higher at *A* than *B* and increasing   (c) zero   (d) higher at *A* than *B* and decreasing   (e) none of these.

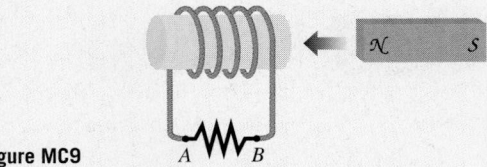

Figure MC9

10. The more rapidly a magnet approaches a coil (as in Fig. MC9), the   (a) lower the current in the coil   (b) greater the resistance of the coil   (c) greater the induced voltage across the coil   (d) more it is attracted   (e) none of these.

11. The wire loop of area 0.050 m^2 in Fig. MC11 is in a uniform downward B-field, which is increasing at 0.010 mT/s. The induced emf is such that the potential of   (a) *C* is higher than *A* by 0.50 μV   (b) *A* is higher than *C* by 0.50 μV   (c) *A* is the same as that of *C*   (d) *C* is higher than *A* by 0.50 mV   (e) none of these.

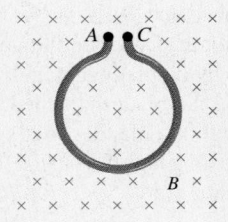

Figure MC11

12. The two coils in Fig. MC12 are wrapped on an iron bar. When the switch is closed   (a) a current momentarily passes through *R* from right to left   (b) a constant current circulates through *R* from right to left   (c) a current momentarily passes through *R* from left to right   (d) a constant current circulates through *R* from left to right   (e) none of these.

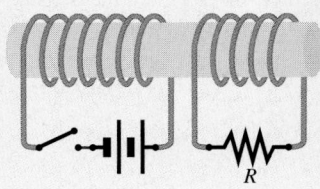

Figure MC12

13. The coil in Fig. MC13 has a core made up of a stack of insulated iron wires. The reason for using such a configuration is to   (a) generate as much thermal energy as possible   (b) be able to make it inexpensively   (c) reduce eddy current losses   (d) make it strong to resist magnetic bending   (e) none of these.

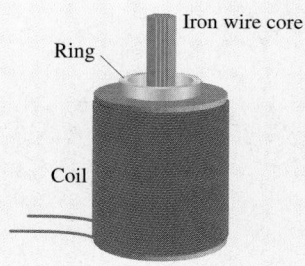

Figure MC13

14. When a loose metal ring is placed on the coil in Fig. MC13 and the latter is suddenly fed a large current, the ring will (a) pop into the air (b) initially vibrate and then stop (c) initially remain stationary but get very hot (d) have nothing happen to it (e) none of these.

15. The solenoid and battery of Fig. MC15 are moving at a constant speed toward the coil on the left. The voltage measured across points 1 and 2 is (a) zero (b) higher at 1 than 2 and increasing (c) higher at 2 than 1 and increasing (d) higher at 1 than 2 and decreasing (e) none of these.

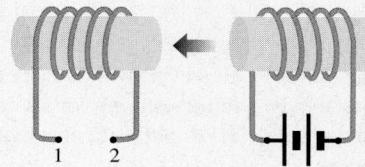

Figure MC15   1   2

16. We wish to produce a clockwise (looking down) current in the loop on the right in Fig. MC16. The variable resistor should be (a) left as is (b) decreased (c) increased (d) reversed (e) none of these.

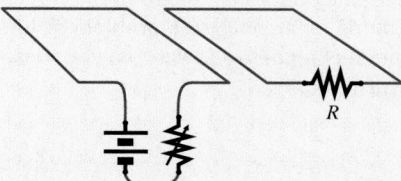

Figure MC16

17. The copper ring in Fig. MC17 is in a uniformly increasing magnetic field. The induced electric field within it is (a) clockwise and constant (b) counterclockwise and constant (c) clockwise and changing (d) counterclockwise and changing (e) none of these.

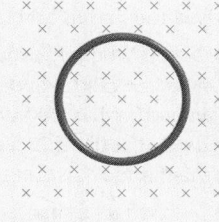

Figure MC17

18. Figure MC18 shows an aluminum ring and the current induced in it by the nearby magnet that is free to move along its central axis. (a) the magnet must be stationary (b) the magnet must be moving to the right (c) the magnet must be moving to the left (d) not enough information to say anything about the magnet (e) none of these.

19. A solenoid is physically altered by doubling the number of turns it has while halving the current through it, leaving everything else unchanged. (a) its self-inductance stays the same (b) its self-inductance doubles (c) its self-inductance is halved (d) its self-inductance is four times greater (e) none of these.

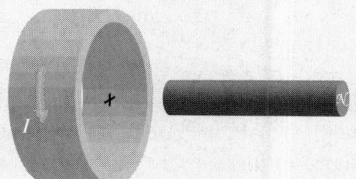

Figure MC18

20. The coil in Fig. MC20 is rotating at a constant rate about an axis perpendicular to the field. The induced voltage across its terminals will (a) always be zero when $\theta = 0$ (b) always be zero when $\theta = 90°$ (c) sometimes be zero when $\theta = 90°$ (d) never be zero (e) none of these.

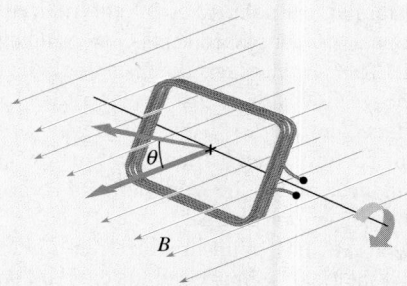

Figure MC20

21. A small light bulb is in series with an air-core coil and a dc power supply such that the lamp glows brightly. An iron core is then inserted into the coil, and an hour later the lamp (a) is brighter (b) goes out completely (c) grows dimmer (d) is unaffected (e) none of these.

22. A small light bulb is in series with an air-core coil and an ac power supply such that the lamp glows brightly. An iron core is then inserted into the coil, and the lamp (a) grows brighter (b) goes out completely (c) grows dimmer (d) is unaffected (e) none of these.

23. An electric locomotive going uphill draws power from the feeder lines. In its most efficient mode of operation, when it goes downhill, it (a) must draw even more power (b) can generate and return power to the lines (c) will neither draw nor produce power (d) draws the same amount of power (e) none of these.

24. Figure MC24 shows an end-view looking down onto a long narrow solenoid carrying an increasing clockwise current. Two identical light bulbs are connected in a circuit that encircles the solenoid. (a) the bulb on the right glows, the one on the left does not (b) the bulb on the left glows, the one on the right does not (c) both bulbs glow equally (d) neither bulb glows (e) none of these.

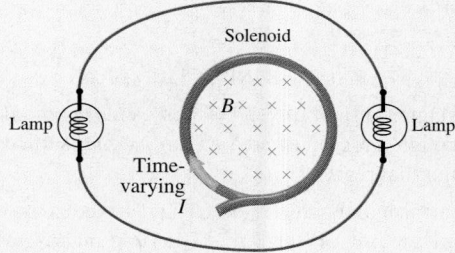

Figure MC24

25. Figure MC25 shows an end-view looking down onto a long narrow solenoid carrying an increasing clockwise current. Two

identical light bulbs are connected in a circuit that encircles the solenoid. A wire is then attached as shown. (a) the bulb on the right glows more brightly, the one on the left does not light at all (b) the bulb on the left glows more brightly, the one on the right does not light at all   (c) both bulbs glow equally   (d) neither bulb glows   (e) none of these.

For more **Multiple Choice Questions** with answers click on WARM-UPS in CHAPTER 20 on the CD. 💿

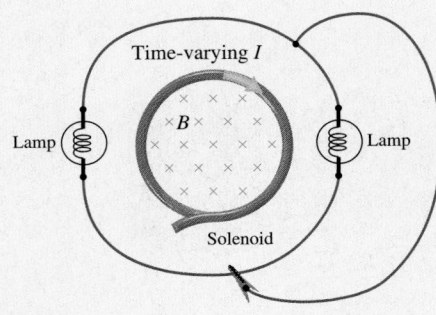

Figure MC25

Suggestions on Problem Solving

1. When determining the polarity of an emf induced in a coil, it's useful to imagine a resistor placed across the terminals, forming an external circuit. The current enters the resistor on the (+) high side and exits on the (−) low side, as in Figs.20.6, 20.11b, 20.14c, and 20.20. The induced current *inside* the coil travels from (−) to (+) just as it does in a battery, which often seems to be a sticky point for students. The induced emf is the work done, in accord with Lenz's Law, on the charge, per unit charge.

2. Be aware that $\Delta\vec{B}$ may point in a different direction from the $\vec{B}$-field. Thus, if the field is directed to the east and decreasing, $\Delta\vec{B}$ points west. If there is an induced current, its induced field will oppose this change and so must point east.

3. Imagine an electrostatic E-field and suppose that we carry a charge from point 1 to point 2. Work will be done and an emf will exist between 1 and 2, an emf that is independent of the path taken. The electrostatic E-field is conservative, but the E-field arising from a changing magnetic flux is *not* conservative. The emf does depend on the path taken between 1 and 2 since different amounts of work may well be needed. Given a time-varying B-field in some localized region of space, it's often possible to go around a closed loop so that the path either encompasses the varying flux totally or in part, or even avoids it, thereby leading to different path-dependent values of the emf. The work done in moving a charge around a closed path in an electrostatic E-field is zero—you come back to the same energy with which you started. By contrast, in the case of an induced E-field, if we move a positive charge around a loop, work may be done continuously. *The concept of potential is ambiguous with induced E-fields* (see Discussion Question 15).

4. The sign in Eq. (20.10) for the back-emf reminds us that the emf opposes that which causes it. If the current is building at a rate of, say, 2.0 A/s, then $\Delta I/\Delta t$ is positive and the emf is negative (meaning it opposes the buildup). However, without a clear picture of the system, the minus sign by itself doesn't provide a complete picture. In such cases, it's often easier to put aside the sign, remembering that the emf opposes its cause. For example, given that a certain back-emf is measured to be 10 V while the current buildup is 2.0 A/s, find the inductance. Since this situation represents an increase of current, the emf must be put into Eq. (20.10) as −10 V or you will get a negative L, which is impossible. Had the current been decreasing, it would be entered as −2.0 A/s and the emf would then be positive, again yielding a positive L. It's common to just give the numerical value of the emf in a problem (without talking about which terminal is higher or lower), so you must be careful not to come up with a negative L.

Problems ✦ Coordinated Problems ✦ Progressive Problems ✦ Solutions

STUDY GUIDE   **1. Coordinated Problems:** The three problems within each magenta-colored grouping are solvable in similar ways. Note that the first of these always has a hint; moreover, its solution is provided in the back of the book. *Work out each of these sets; they'll strengthen technique and build confidence.* **2. Progressive Problems:** The problems introduced in blue unfold step-by-step carrying along the analysis in a more suggestive way than is customary. *Work out all of these; they'll guide you through the analytic process and help develop problem-solving skills.* **3. Worked-Out Solutions:** Studying worked-out solutions is an important part of learning how to solve problems. Accordingly, additional *solutions* to a number of model problems are given below. *Make sure you understand each of them before you go on to the next problem.* **4.** Also provided in the back of the book are the *Answers* to all odd-numbered problems, as well as worked-out *solutions* to those with boldface numbers. Problem numbers in italic indicate that a solution appears in the Student Solutions Manual.

SECTION 20.1: FARADAY'S INDUCTION LAW

1. [I] A magnetic field of 1.2 mT passes perpendicularly and uniformly through a region of area 25 cm². What is the corresponding flux through that region?

2. [I] A uniform magnetic field of 100 mT passes through a loop of wire having an area of 0.020 m². The field makes an angle of 30° with the perpendicular to the plane of the loop. What is the magnetic flux through the loop?

3. [I] A flux of 6.0 mWb passes uniformly along an iron bar of cross-sectional area 50 cm². Compute the flux density in the bar.

4. [I] A coil wrapped around a portion of an iron ring generates a flux that passes through another coil also wrapped around the ring. This secondary has 100 turns and is penetrated by a flux of 0.016 Wb. Given that the ring has a cross-sectional area of 8.0 cm², what is the magnitude of the magnetic field in the toroid?

5. [I] A changing magnetic field has an initial value $\vec{B}_i$, which points due south and has a strength of 0.5 T. If the final value of the field, $\vec{B}_f$, is 0.6 T pointing due south, what is the magnitude and direction of the change in the field, $\Delta\vec{B}$? [*Hint:* $\Delta\vec{B} = \vec{B}_f - \vec{B}_i$.]

6. [I] A changing magnetic field has an initial value $\vec{B}_i$, which points downward and has a strength of 0.6 T. If the final value of the field, $\vec{B}_f$, is 0.5 T pointing upward, what is the magnitude and direction of the change in the field, $\Delta\vec{B}$?

7. [I] A changing magnetic field has an initial value $\vec{B}_i$, which points east and has a strength of 0.01 T. If the final value of the field, $\vec{B}_f$, is zero, what is the magnitude and direction of the change in the field, $\Delta\vec{B}$?

8. [I] A uniform magnetic field of 100 mT passes perpendicularly through a loop of wire having an area of 0.020 m². In 0.002 0 s the loop is rotated so that it is parallel to the field. What is the average time rate-of-change of flux through the loop?

9. [I] A single loop of wire with an encompassed area of 0.25 m² is perpendicular to a 0.40-T uniform magnetic field. The loop is yanked from the field in 200 ms; what is the induced emf?

10. [I] A 100-turn coil of wire is removed laterally from an axial B-field in 20 ms, and an emf of 1.0 V is induced across it in the process. How big was the field if the cross-sectional area of the coil is 4.0 cm²?

11. [I] A magnet is moved close to a coil of 150 turns in which it introduces a flux change of 30 mWb. If a voltage of 60 V appears across its terminals and if the resistance of the coil is negligible, how much time elapsed during the flux change?

12. [I] A coil of 100 turns is penetrated by a flux of 0.050 Wb in 0.020 s. Determine the numerical value of the average induced emf.

13. [I] The flux passing through a coil of 200 turns changes uniformly from a constant 0.010 Wb to 0.070 Wb in a time of 1.5 s. Find the numerical value of the emf induced during that interval.

14. [I] A 10-turn coil of wire of area 0.25 m² is in a perpendicular magnetic field that changes in time as illustrated in Fig. P14. Determine the induced voltage across the loop as a function of time.

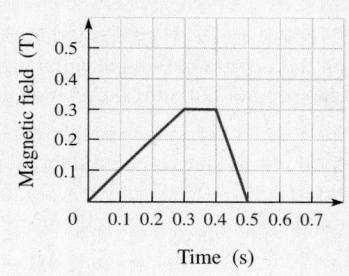

Figure P14

15. [I] A 100-turn rectangular coil 11.0 cm by 19.9 cm is in a uniform perpendicular 82.0-mT B-field. (a) If it takes 45.0 ms to yank the coil completely out of the field without changing its orientation, what average emf will be induced? (b) If the coil has a resistance of 2.0 Ω, how much current flows in it while it's exiting the field?

SOLUTION: (a) $\Phi_M = BA = (82.0 \times 10^{-3} \text{ T})(11.0 \times 10^{-2} \text{ m} \times 19.9 \times 10^{-2} \text{ m}) = 1.79 \times 10^{-3}$ T·m² and $\mathscr{E} = -N \Delta\Phi_M/\Delta t = 100(1.79 \times 10^{-3} \text{ T·m}^2)/(45.0 \times 10^{-3} \text{ s}) = 4.0$ V. (b) $V = IR, I = 2.0$ A.

16. [II] THIS PROBLEM EXPLORES FARADAY'S INDUCTION LAW. A tight ten-turn wire coil is held horizontally. The wire starts at terminal-A in Fig. P16 and goes around counterclockwise, ending at terminal-B. The cross-sectional area of the coil is 10.0 cm², and its resistance is 1.0 Ω. The vertical B-field, which is initially 1.75 T, is decreased at 0.10 T/s (a) Write an expression for the emf induced in the coil in terms of the magnetic flux. (b) Show that here $\mathscr{E} = -NA\,\Delta B/\Delta t$. Explain. (c) What is the area in m²? (d) What is the value of $\Delta B/\Delta t$? (e) Determine the magnitude of the induced emf. (f) What is the value of the induced

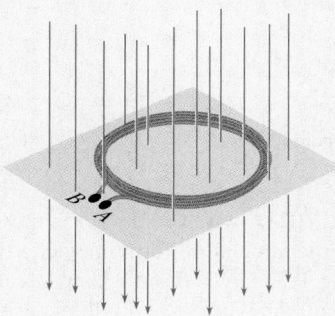

Figure P16

current? (g) What is the direction of the induced current? (h) Which terminal is at a higher potential (i.e, which should be marked +)?

17. [II] A specimen is to be exposed to a controllable magnetic field and is therefore positioned inside a 0.55-m-long narrow air-core solenoid of cross-sectional area 2.0×10^{-4} m² comprising 10 turns per cm. To monitor the field, a small search coil (connected to a voltmeter) is wrapped around the outside of the solenoid at its middle. When the current in the solenoid is increased from zero to 4.9 A in 5.00 ms, what will be the emf across the search coil given that it consists of 240 turns of fine wire?

18. [II] A flux of 8.0 mWb generated by a large electromagnet links a coil of 100 turns placed between its poles. Suppose the current in the electromagnet is gradually and continuously reduced to zero and then reversed so that a flux of 8.0 mWb is uniformly reestablished in the opposite direction in a time of 200 ms? What is the emf induced in the coil?

19. [II] A single loop of copper wire in the shape of a square 4.0 cm on each side is lying flat on a horizontal table. A large electromagnet is positioned with its north pole above and to the left a little so that the uniform magnetic field is downward onto the loop, making an angle of 30° with the vertical. Compute the average induced emf across the loop as the field varies linearly from 0 to its final value of 0.500 T in 200 ms. What is the direction of the induced current?

20. [II] A flat circular coil of 10 turns and diameter 10 cm (shown in Fig. P20) is located in a uniform magnetic field of 0.20 T that passes through the coil area at 45°. What will be the voltage, on average, across the terminals if the plane of the coil is smoothly rotated (through 45°) so that it's parallel to the field in a time of 0.10 s? Indicate which terminal has the higher potential. What will the voltage be when the coil comes to a stop?

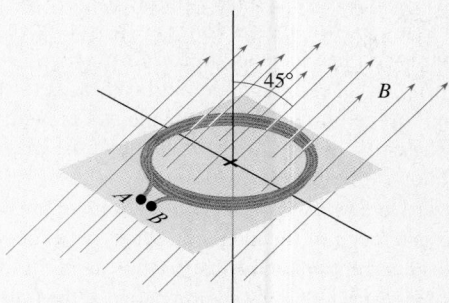

Figure P20

21. [II] Imagine a vertical magnetic field decreasing at a constant rate of 20 mT/s. A flat circular coil of 220 turns with a 20-cm diameter has a 30-μF capacitor across its terminals. If the coil is placed perpendicular to the field, what is the steady-state charge on the capacitor?

22. [II] Imagine a flat circular coil of N turns and resistance R connected to a ballistic galvanometer (that is, one with negligible damping). The flux through the coil (Φ_M) is changed and an induced charge ΔQ flows through the galvanometer, whose pointer is initially displaced by an amount $\theta \propto \Delta Q$, known as the *throw*. Show that

$$\Delta Q = \frac{N\Delta\Phi_M}{R}$$

23. [II] With Problem 22 in mind, suppose a small search coil is

attached to a ballistic galvanometer for which $\Delta Q = K\theta$, where K is a measured constant and θ is the initial deflection. The coil is first placed perpendicularly in a uniform B-field that is to be measured. It is then rapidly yanked out, causing the galvanometer to swing. Write an expression for B in terms of θ, R, K, N, and A, the area of the search coil.

24. [II] A flat circular coil of 20 turns with an area of $5.0 \times 10^{-2} \text{ m}^2$ is located in a uniform magnetic field of 10 mT. Initially, the magnetic flux passes through the coil perpendicularly. The coil is then rotated, in 150 ms, so that its central axis makes an angle of 50° with $\vec{B}$. Determine the average induced emf resulting from the rotation.

25. [II] A small search coil (of area 0.50 cm², resistance 0.50 Ω, and having 12 turns) is attached to a ballistic galvanometer. The coil is placed inside, at the center of, and coaxial with, a large solenoid. When the current in the solenoid is rapidly reversed, the galvanometer deflects, indicating a pulse of 2.0 μC of charge. Determine the initial B-field of the solenoid. [*Hint: Remember Problems 20 and 21.*]

26. [II] The current in a long solenoid of 210 turns generates a flux within it of 10 mWb. If that current is gradually reversed in a time of 200 ms, what will be the induced emf?

SECTION 20.2: MOTIONAL EMF

27. [I] An automobile traveling at 20.0 m/s up a constant slope has axles that are 1.8 m long. The Earth's field at that location is 0.40×10^{-4} T, and it's perpendicular to the road. What is the induced potential difference across one of these axles?

> **SOLUTION:** The axel moves across the B-field, and that makes this a problem about motional emf. $\mathscr{E} = vBl = (20.0 \text{ m/s})(0.40 \times 10^{-4} \text{ T}) \times (1.8 \text{ m}) = 1.4 \text{ mV}$.

28. [I] THIS PROBLEM WILL HELP US BETTER UNDERSTAND MOTIONAL EMF AS IT REVIEWS SOME ASPECTS OF ELECTROMAGNETISM. A current-carrying wire wrapped around an iron form (Fig. P28) produces a fairly uniform 500 gauss magnetic field. We want to study the voltage induced on the horizontal wire when it's moved. (a) From the current in the coils, determine the polarity of the magnet. (b) What is the direction of the B-field between the poles?   (c) If the wire is pulled upward at 5.00 m/s, in what direction will an electron in the wire move as a result of the magnetic force?   (d) In what direction is the induced current in the moving wire?   (e) If the B-field can be assumed to be restricted to the region between the poles, what is the voltage across the moving wire?

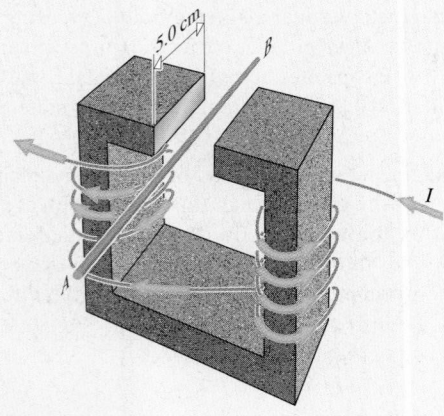

Figure P28

29. [I] A Boeing 747 jumbo jet with a wingspan of 60 m is flying in an area where the Earth's B-field is 0.050 mT. Given that the plane traverses the field perpendicularly at a speed of 200 m/s, what is the potential difference induced between its wing tips? [*Hint: The wings can be thought of as a single long conductor sweeping across the field.*]

30. [I] The magnetic field of a large electromagnet is a uniform 0.140 T directed east to west. A 10.0-cm length of straight copper wire is held horizontally in the field perpendicular to it and then moved downward at 0.50 m/s. What voltage will appear across the wire? Which end is positive?

31. [I] A 1.00-m-long wire held stationary in the east-west direction is in a uniform 0.200 T north-to-south field of a large magnet. The magnet is lowered vertically at a rate of 0.60 m/s. What voltage, if any, will appear across the wire, and which end will be positive?

32. [I] A straight 30-cm length of copper wire is moved at a constant speed of 0.50 m/s perpendicularly across a uniform magnetic field of 1.5 T. Determine the emf appearing across its ends.

33. [I] How could you measure the potential difference in Problem 29? If we attached a small light bulb across the wing tips, would it light?

34. [I] An emf of 0.45 V is induced in a straight conductor having a length of 20 cm moving at right angles to a magnetic field at a constant speed of 600 cm/s. Calculate the value of the field.

35. [I] A 100-cm-long straight copper rod is initially held horizontally pointing due east and then released. It falls across the Earth's field of 0.5×10^{-4} T, which is pointing north and 50° below the horizontal. What is the induced emf in the rod when it reaches a speed of 2.8 m/s?

36. [II] The vertical magnetic field of a large set of coils is uniform and equal to 0.045 T. A 20.0-cm-long straight copper tube is moved upward through the field at 25.0 cm/s cutting across it along a straight line that makes an angle of 60.0° with the field (i.e., the angle between $\vec{v}$ and $\vec{B}$ is 60.0°). Determine the induced voltage in the tube.

37. [II] The rectangular loop in Fig. P37 has a resistance R and is moving left with a constant speed v into a region of uniform magnetic field. The field is a little unrealistically assumed to be constant, dropping immediately to zero beyond its rectangular boundaries. Describe the induced current as a function of time.

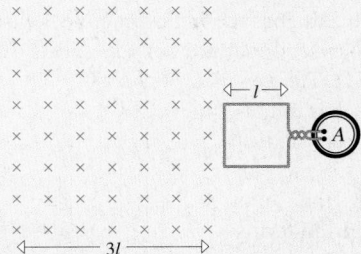

Figure P37

38. [II] A wire of length l is moved at a constant speed across a perpendicular uniform magnetic field B by a force F, as shown in Fig. P38. Given that the lamp has a resistance R while the resistance of all the wires is negligible, derive an expression for the force F assuming zero friction.

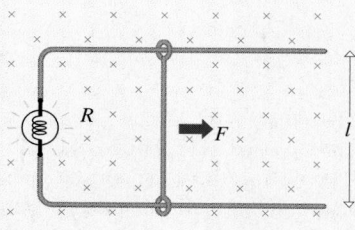

Figure P38

39. [II] With Problem 38 in mind, write an expression for the power supplied to the moving wire by the external agency. Show that this power is equal to I^2R dissipated in the lamp.

40. [II] A straight 1.00-m-long copper rod is moved perpendicularly across a uniform 0.11-T magnetic field. If a constant voltage of 0.50 V appears across its end, how much time will it take to traverse the 0.50-m-long field region?

41. [II] The lamp in Fig. P38 is rated at 5.0 W and draws 200 mA. The magnetic field has a value of 0.400 T. If the wire can be moved at a constant speed of 2.00 m/s, how long would it have to be to properly operate the lamp? [*Hint: Remember that* P = IV.]

42. [II] If instead of the lamp in Fig. P38 there is a little 2.0-W, 10.0-Ω motor, how fast must the wire move if it is 0.50 m long and the field is 0.20 T? Determine the force that must be exerted on the wire assuming zero friction.

43. [II] The lamp in Fig. P38 is removed and replaced by a 5.00-Ω resistor. The wire is 0.500 m long and can be moved at a constant speed of 1.20 m/s. If the resistor is to dissipate 10.0 W, how strong should the field be?

44. [III] Imagine two stationary (zero-resistance) vertical conductors with a horizontal wire crossbar (of length l) capable of moving downward freely while staying in contact (Fig. P44). The crossbar (of mass m) is dropped and it falls perpendicular to the uniform field B. Describe its motion completely, writing an equation for the acceleration and the maximum speed, if there is one.

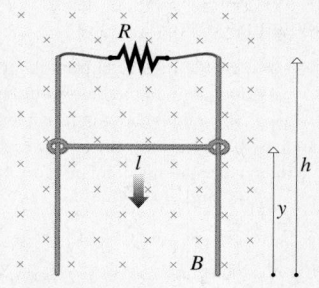

Figure P44

SECTION 20.3: THE AC GENERATOR

SECTION 20.4: THE DC GENERATOR

45. [I] A generator, which consists of a flat coil, is rotated about its central axis (Fig. 20.14*a*) so that the longer portion of it moves perpendicular to a 0.80-T magnetic field. The coil has an active length of 50 cm (i.e., there is a total of 50 cm of wire cutting the field normally). What is the emf when the machine turns at a rate such that each conductor moves at 4.0 m/s?

46. [I] We wish to design an ac generator using an armature on which is wound a flat tight rectangular coil (8.00 cm × 20.0 cm) of 150 turns. The generator output is to be sinusoidal with a maximum voltage of 20.0 V and a frequency of 50.0 Hz. How strong should the B-field be, and at what angular speed should the coil turn?

47. [I] A 25-turn coil is made to rotate in a uniform 0.35-T magnetic field. Each turn has two 10-cm wire lengths that cut the field at varying angles as the coil rotates. What is the emf when the wire is moving with a speed of 22 m/s at 90° to the field?

48. [I] In the United States, the national power grid supplies ac to ordinary consumers with a maximum voltage of 170 V at 60 Hz (the standard 120-V "average" household service). Given that a generator has a 100-turn coil of area 0.50 m², what strength magnetic field should be used?

49. [I] An ac generator produces an emf that (in SI units) has the form $\mathscr{E} = 100 \sin 376.99t$. What is the maximum voltage? What is the frequency of the output?

50. [I] The magnetic field on the platform of an electric train (e.g., the Washington-New York Amtrak line) is typically about 650×10^{-7} T. Suppose you set up a 1.00-m² coil having 100 turns of fine copper wire and rotated it at 100 Hz. What maximum emf can you expect to reach?

51. [II] We wish to make a dc generator that puts out a constant current. Accordingly, a flat copper disk with a radius of 25 cm is placed in a uniform magnetic field so that its axis of rotation is parallel to B. A 20-Ω resistor is connected between the axis and the rim of the disk. The disk is rotated at a rate of 360 rpm and a dc current of 1.25 mA passes through the resistor. Find B. [*Hint: Consider a radial strip of conductor and show that the rate at which it sweeps out area is $\frac{1}{2} r^2 \omega$. Also, look at Problem 53.*]

52. [II] A simple generator, which consists of a flat rectangular coil (Fig. 20.14) with a turning radius of 10 cm, is rotated about its central axis so that it moves perpendicularly across a 0.60-T magnetic field. The coil has an active length of 30 cm (that is, there is a total of 30 cm of wire cutting the field). What is the emf when the machine turns at 10 rev/s?

53. [II] A helicopter hovering in the air has its 5.0-m rotor blades revolving at 240 rpm in a plane that intersects the local B-field of the Earth (0.050 mT) at an angle (with the normal to the plane) of 40°. Determine the induced emf between the hub and the tip of each rotor blade. [*Hint: Look at the hint for Problem 51.*]

54. [II] A 25-turn flat coil is made to rotate in a uniform 0.35-T magnetic field. Each turn has two 10-cm wire lengths that cut the field at varying angles as the coil rotates. What is the emf when the wire is moving with a speed of 22 m/s at 30° with respect to the field?

55. [II] A simple ac generator consists of a coil, each turn of which has two 20-cm lengths of wire that cut the uniform 0.50-T B-field as they revolve at a radius of 5.0 cm from the central axis. Write an expression for the emf as a function of time given that the coil has 200 turns and revolves at 6000 rpm.

56. [III] Derive the expression for the emf of a rotating coil

$$\mathscr{E} = NAB\omega \sin \omega t \qquad [20.7]$$

by considering the work done by a torque acting through an angle.

SECTION 20.5: INDUCTANCE

57. [I] THIS PROBLEM EXPLORES THE UNITS OF INDUCTANCE AND PERMEABILITY. We want to show that permeability has the units of H/m. (a) What are the units of permeability as first introduced via Eq. (19.1)?   (b) What are the units of magnetic flux, expressed in terms of teslas (T) and meters (m)?   (c) Using the defining expression, Eq. (20.8), for inductance, express henries (H) in terms of teslas, meters, and amps (A). (d) Using the expression for the inductance of a coil, Eq. (20.9), show that the units of permeability are H/m.

58. [I] THIS PROBLEM IS ABOUT INDUCTANCE. A coil has a diameter of 0.50 cm, a length of 1.50 cm, and is formed of 349 turns of wire. (a) Write an expression for its self inductance. (b) Determine the coil's cross-sectional area in m^2. (c) If the core has a permeability of 0.30×10^{-3} H/m, compute the self-inductance.

59. [I] A long solenoid of 500 turns carrying a current of 3.8 A produces within itself a uniform magnetic flux of 2.0 mWb. Compute the self-inductance of the coil. [*Hint: What's the definition of self-inductance?*]

60. [I] If a current of 2.0 A in a coil produces a flux linkage of 6.0 mWb, what is the self-inductance?

61. [I] The self-inductance of a coil is 4.00 mH. If it carries 1.00 A through its 50 turns, what flux will be generated within it?

62. [I] An air-core coil having a self-inductance of 3.0 mH carries a current that creates a magnetic field within it of magnitude B. The coil is slid onto a ferromagnetic rod that has a relative permeability of 2000 when in a field of magnitude B. What will be the new inductance with the core in place?

63. [I] An air-core coil with 500 turns has an inductance of 200 mH. If a current of 2.0 A passes through the coil, what will be the flux within it at that moment?

64. [I] Show that the unit of μ_0 is H/m.

65. [I] What is the number of turns of a 2.5-H coil if, when 1.80 A is being carried, the flux linking it is 1.80 mWb?

66. [I] Determine the inductance of a 0.50-m-long solenoid having 200 turns wound on a hollow core 4.0 cm in diameter.

67. [I] A 40-cm-long air-core solenoid with a cross-sectional area of 4.5 cm^2 is to be constructed for a radio circuit so that it has an inductance of about 1.0 mH. How many turns of wire should it have?

68. [I] A coil with an inductance of 10 H has a current within it that is increasing at a rate of 4.0 A/s. What is the back-emf induced across the coil?

69. [I] If an emf of 10 V is induced across a coil when the current within it is changing at a constant rate of 2.0 A/s, what is its inductance?

70. [I] A current in a 5.0-H coil drops from 2.5 A to zero uniformly in a time of 10 ms. What is the induced emf?

71. [I] A current in a coil rises from zero to 4.0 A in 0.020 s. If the back-emf is measured to be 500 V, what is the inductance of the coil?

72. [II] We wish to make an inductor that will be used in an experiment and must be formed around an iron rod for which $\mu = 1.5 \times 10^3\mu_0$. The rod has a diameter of 2.0 cm and is 27 cm long. Given that the inductance is to be 300 mH, approximately, how many turns of wire will be needed? Why should we stress the word *approximately* here?

73. [II] A 5.0-A current through a 1.5-H coil is reversed, and in the process an average emf of 100 V is induced. Compute the time it took to reverse the current. [*Hint: The current goes from +5.0 A to −5.0 A.*]

74. [II] Consider a 2.0-mH coil carrying a current of 2.0 A clockwise. If the current drops to zero and rises in the opposite direction (counterclockwise) to 1.0 A in 500 μs, what average emf will appear across the coil?

75. [II] A switch is closed, and a power supply sends a current through a 2.0-mH coil. In 2.5 ms the current goes from a value of 250 mA in one direction to I in the opposite direction. If an average emf of 5.0 V appears across the coil, what is the value of I?

76. [II] Write an expression for the self-inductance of an air-core toroidal coil of N turns, with a mean radius of b and a cross-sectional area A.

77. [II] Inductance transducers, mounted in catheters (narrow tubes inserted into the body), have been used to measure blood pressure. Blood presses against a diaphragm (a few millimeters in diameter), which bends a proportionate distance backward. On the back side of the diaphragm is a small ferromagnetic shaft (of permeability μ) that is thereby displaced, entering into an air-core coil of length l (a distance d). The resulting change in inductance ΔL is measured electrically and calibrated to correspond to the blood pressure. Show that

$$\Delta L \approx \frac{d(\mu - \mu_0)N^2A}{l^2}$$

SECTION 20.6: THE *R-L* CIRCUIT: TRANSIENTS

78. [I] The coil in Fig. P78 has a negligible resistance. (a) What's the reading of the milliammeter an instant after the switch is closed? (b) If the source has a constant voltage of 20 V, $L = 20 \times 10^{-3}$ H, and $R = 2.0 \times 10^3$ Ω, what is the steady-state current attained after the switch is closed?

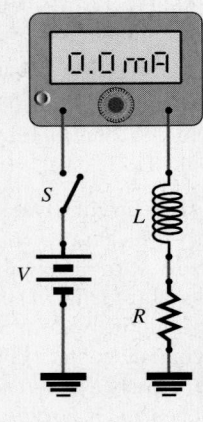

Figure P78

> **SOLUTION:** (a) The current in an *R-L* circuit cannot change immediately so the reading is zero. (b) The steady-state current is established after a long time, at which point the inductance has no effect on the current. Hence $V = IR$ and $I = (20 \text{ V})/(2.0 \times 10^3 \ \Omega) = $ 10 mA.

79. [I] THIS PROBLEM EXPLORES THE BEHAVIOR OF AN *R-L* CIRCUIT. A 1.0-H coil with a negligible resistance is in series with a resistor (R), a switch, an ammeter, and a 12-V battery. The switch is closed, and the ammeter reading slowly rises. The current finally becomes constant after 2.5 s. We want to find the value of R.   (a) Assuming that 2.5 s corresponds to about five time constants, what is the duration of one time constant?   (b) Write an expression for the time constant of the circuit.   (c) Find the value of R?   (d) What will the ammeter read after ten time constants?   (e) If a voltmeter is put across the inductor at that time, what value of voltage will it read?

80. [I] A 10.0-H inductor is in series with a 12-V battery, a 6.0-Ω resistor, and a switch. What is the steady-state current attained after the switch is closed?

81. [I] In Fig. 20.22 the current reaches a value of $0.63V/R$ in 1 time constant. Show that $1 - (1/e) = 0.632$, where $e = 2.718$ 3.

82. [I] A 2.5-H coil with negligible resistance is in series with a 0.80-Ω resistor, an open switch, and a 12.0-V battery. The switch is closed and the current in the circuit is measured. What will it be

immediately after the switch is closed? What will it be several hours later?

83. [I] When a switch is closed, a voltage of 20.0 V appears across a 10.0-mH coil having a resistance of 5.00 Ω. Determine the rate at which current begins to flow in the circuit. [*Hint: Initially* $I = 0$, *but* $\Delta I/\Delta t \neq 0$.]

84. [I] A coil having a negligible resistance of its own is put in series with a dc power supply set at 20.0 V, an open switch, an ammeter, and a 25-Ω resistor. If immediately after the switch is closed the rate at which the current begins to build up is 100 A/s, what is the inductance of the coil?

85. [I] A coil having a resistance of 20.0 Ω is put in series with a dc power supply set at 60 V, an open switch, a 10.0-Ω resistor, and an ammeter. What is the self-inductance of the coil if immediately after the switch is closed, the rate at which current in the circuit begins to increase is 200 A/s?

86. [I] A 0.400-H coil having a negligible resistance of its own is put in series with a dc power supply set at 100.0 V, an open switch, an ammeter, and a 25-Ω resistor. What is the current in the circuit at the moment that the rate-of-change of the current is 200 A/s?

87. [II] An experimental setup is comprised of an air-core solenoid (having an inductance of 300 mH and negligible resistance) connected in series to a sample of bone (of 25 Ω), an ammeter, and a 150-V dc source via a switch. The intent is to provide a gradually building current to the sample. (a) What will be the steady-state current? (b) Determine the time constant of the system. (c) Determine the current after such a time has elapsed.

88. [II] THIS PROBLEM EXPLORES THE BEHAVIOR OF AN *R-L* CIRCUIT. The coil in Fig. P78 has a negligible resistance. (a) What's the reading of the milliammeter an instant after the switch is closed? (b) Given that $L = 10$ mH, and $R = 1.0$ kΩ, first, express these values in henrys and ohms. (c) Determine the time constant for the circuit. (d) If the source has a constant voltage of 10 V, what is the steady-state current attained after the switch is closed? (e) What's the steady-state voltage across the inductor? (f) The switch is closed and after a time *t* the milliammeter reads 9.9 mA, how many time constants have gone by? (g) Determine *t* at that moment.

89. [II] THIS PROBLEM EXPLORES THE BEHAVIOR OF AN *R-L* CIRCUIT. The coil in Fig. P89 has a negligible resistance. (a) What's the reading of the milliammeter an instant after the switch S_1 is closed? (b) What's the reading of the milliammeter a very long time after the switch S_1 is closed? (c) Determine the time constant for the circuit. (d) If, after a long time, S_1 is opened and S_2 is closed what current will the milliammeter initially read, and will it be a positive or a negative current? Explain.

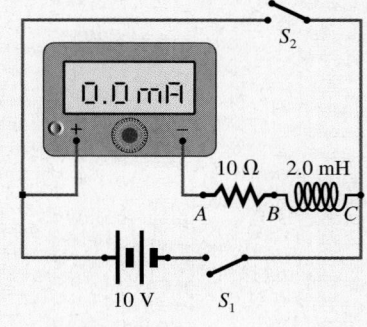

Figure P89

(e) How many time constants must we wait before the milliammeter reads about 140 mA? (f) How many seconds is that?

90. [II] An inductor ($L = 2.00$ H and $R = 4.0$ Ω) is connected in series with an ammeter (of negligible resistance) and a varying voltage source. Figure P90 is a plot of the current in the circuit versus time. Draw a corresponding plot of the voltage across the inductor versus time.

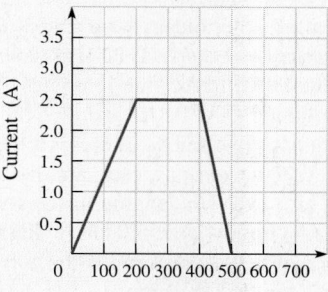

Figure P90

Time (ms)

91. [II] A 24-H iron-core inductor of negligible resistance is placed in series with a 12-V battery, a 6.0-Ω resistor, and a switch. Determine: (a) the steady-state current; (b) the time constant; and (c) roughly how long it would take for the current to reach within 1% of its maximum value.

92. [II] A 24-H iron-core inductor of negligible resistance is placed in series with a 12-V battery, a 6.0-Ω resistor, and a switch. Once the switch is closed, the current in the circuit is

$$I = \frac{V}{R}\left\{1 - \exp\left[\frac{-t}{L/R}\right]\right\}$$

Compute the current at a time of 2.0 s after the switch is closed.

93. [III] Derive an expression for the time rate-of-change of the current in an *R-L* circuit. Interpret your result in light of Fig. 20.22. Suppose a 200-Ω resistor is placed in series with a 50-mH inductor, a switch, and a 120-V dc power supply. Compute the rate-of-change of the current when $t = 0$, $t = L/R$, and $I = 1.0$ A.

94. [III] An air-core solenoid is attached to a 12-V battery via a switch. The switch is closed, and after a while the current reaches a constant level of 2.0 A. If the current changes at a rate of 12 A/s when $I = 1.0$ A, what is the resistance and inductance of the coil?

SECTION 20.7: ENERGY IN THE MAGNETIC FIELD

95. [I] A 40.0-mH coil is placed in a circuit, and after a considerable time the current in the circuit is 2.0 A. How much energy is stored in the inductor? [*Hint: The energy is proportional to the self-inductance and to the current squared.*]

96. [I] How much energy is stored in a 200-mH inductor when a current of 10 A circulates through it?

97. [I] If 10 J of energy is to be stored in a 500-mH inductor, how much current must be passing through it?

98. [I] A 40.0-mH inductor with an internal resistance of 10.0 Ω is placed across a dc power supply set at 80.0 V. How much energy is stored in the inductor once the steady state is reached?

99. [I] What is the inductance of a coil if it stores 15 J of energy when the current through it is 5.0 A?

100. [I] What is the total energy density in a region where the Earth's surface magnetic field is 0.50×10^{-4} T and its surface electric field is 100 V/m? Notice the difference in the energies stored in the two fields.

101. [II] A long narrow solenoid having a 2.0-cm radius is made of copper wire wrapped with two turns per millimeter on a plastic core. How much energy is stored per unit length of coil when 5.0 A circulates through it?

102. [II] The so-called solenoid in your car is mounted on the starter motor. When the key is turned, 12 V are applied to the solenoid, which then carries a current that generates a magnetic field. The ferromagnetic plunger (Fig. P102) is drawn into the coil, thereby moving a rod that does two things: it closes a heavy-duty switch sending about 100 A to the starter motor, and it engages a gear so that the motor can "turn over" the engine. If the solenoid (which is 200 turns wrapped on a hollow core 5.0 cm in diameter and 9.5 cm long) has a resistance of 1.20 Ω, determine the approximate energy density in the coil, given that the plunger fills the space and has a relative permeability of 500 at that field.

103. [II] Determine the approximate amount of energy stored in the Earth's magnetic field in the first 200 km above the planet's surface. Even though the dipole field drops off as $1/r^3$, this distance is comparatively small, so assume $B = 0.4 \times 10^{-4}$ T throughout. (That

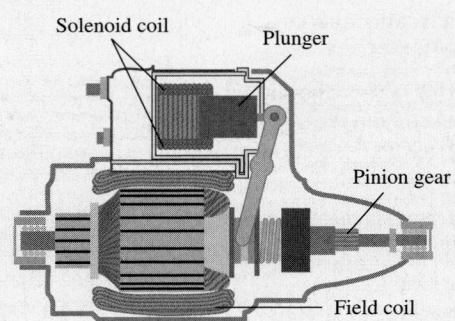

Figure P102

value averages-in the variation of B between the equator and pole.) The mean radius of the Earth is 6371.23 km. Compare your answer to the energy in a gallon of gasoline, $\approx 10^8$ J.

104. [II] An inductor in the motor circuit of a robot consists of an air-core solenoid of 100 turns with an inductance of 200 mH. If the coil carries a current of 5.0 A, what is the flux linkage? How much energy does the coil store?

Chapter 21
AC and Electronics

*H*owever rudimentary, batteries were the only practical source of sustained electric current throughout the early 1800s. By the 1870s, the dynamo (p. 720) had made the transition from laboratory curiosity to practical workhorse, and dc was still the preferred form of electricity. As yet undeveloped, alternating current (ac) was widely viewed as inherently inferior.

Alternating Current

Instead of maintaining a fixed polarity, each terminal of an ac generator, though always opposite to the other, alternates between $+$ and $-$. The electrons that constitute a typical alternating current move first forward then backward, oscillating essentially in place at some given number of cycles per second corresponding to the generator frequency. Remember that the electrons in a dc current drift quite slowly; what moves at nearly the speed of light is the disturbance of the electrons. An alternating current transports energy in the same way as does a direct current—namely, in the form of the organized kinetic energy of mobile-charge carriers.

When the practical high-resistance, low-current incandescent lamp was introduced by Edison around 1880, each installation came with its own 110-volt generator, and it was dc. Most other lamps of the era were low-resistance devices that had to be put in series so that the same high current could pass through each one ($P = I^2R$ and for a required P, a small R means a large I). No one lamp could be shut off without open-circuiting the system. Edison's carbon-filament lamps had a high resistance and could be put in parallel, where the feeder current would be divided into many small branch currents, each powering one bulb that could be turned on and off at will (p. 645)—which, of course, is the way your home is wired.

When central generating stations came into being and especially when they were located at remote energy sources like Niagara Falls (1895), the electrical power produced had to be transmitted over long distances. But the very wires that carry electricity have some resistance, and that poses a major problem. A medium-sized city might easily require ≈ 10 MW of power ($P = IV$). If that amount is to be provided at a modest 100 V or so, then 100 000 A will have to be supplied. There's the difficulty: the joule heating ($P = I^2R$) in the deliv-

STUDY GUIDE

Thus far, we've learned a substantial amount about dc circuits, and can deal with voltages and currents and Ohm's Law. Now we apply all of that to the practical world of alternating current. Because it's easier to transport economically, we use ac in our homes, businesses, and factories. Moreover, almost all of our electronic toys, from the DVD player to the TV, are basically ac devices. So there's a practical imperative for learning something about ac.

Nikola Tesla (1856–1943) was one of the leading figures in the development of alternating current. The spectacular display crackling around Tesla was produced by high-frequency (20 000 Hz) high-voltage ac. The handwritten inscription reads "To my illustrious friend Sir William Crookes of whom I always think and whose kind letters I never answer! June 17, 1901."

ery wires varies as I^2, not just as I. A two-wire line of one-quarter-inch-diameter copper wire has a resistance of $\approx 1.7\ \Omega$/mile. Carrying 10^5 A, the joule heating losses are $\approx 1.7 \times 10^{10}$ W/mile. For each mile, that's $\approx 1.7 \times 10^7$ kW and every hour the line loses 1.7×10^7 kW·h of energy. At a cost of roughly 10¢ per kW·h, sending 10^5 A down the line wastes 1.7 million dollars per hour per mile!

THE TITANIC STRUGGLE BETWEEN AC AND DC

Meanwhile, ac was quietly being transformed by a brilliant, wildly eccentric, young engineer named Nikola Tesla, who had briefly been associated with (and subsequently came to despise) Mr. Edison. With Tesla's invention of a practical ac induction motor, alternating current became far more appealing, particularly to shrewd visionaries like George Westinghouse who hired Tesla and bought the rights to his motor. Slowly, as the market expanded, there developed a titanic struggle for control of the industry between the hustlers of *high-voltage ac* (primarily Westinghouse and General Electric, the J. P. Morgan combine to which Edison would ultimately sell out) and those of *low-voltage dc* (led rather unscrupulously by Edison).

STUDY GUIDE

A current is **dc** when it flows in one direction along a conductor; it need not be constant.

A current is **ac** when it reverses the direction in which it flows along a conductor; it cannot be constant.

There was no economically feasible way out but to lower the current. Clearly, if the voltage were raised to 100 000 V, the same power could be efficiently delivered by 100 A! Thus, raising the voltage by a factor of 10^3 allows for the lowering of the current by 10^3, and that drops the associated power lost over the same lines by a factor of 10^6. Since there already existed a very simple way to raise and lower the voltage of ac (via transformers, p. 774) but no comparable means for dc (until recently), the contest ultimately went to the high-voltage-ac people.

Electrical power is usually generated at around 22 kV and boosted by a step-up transformer to about 500 kV, which is then transmitted long distances via high-tension lines (Fig. 21.1). Using a transformer, the voltage is then reduced to perhaps 60 kV for heavy industrial consumers—usually the more power required, the higher the voltage at which it's delivered. Substations in each area step the voltage down further to about 4 kV for distribution to local communities. This voltage is what's carried by those endless ugly overhead wires that crisscross most suburban towns. Thereafter, the voltage is finally reduced by small pole-mounted transformers, each of which feeds a cluster of buildings. **Line voltage** in the United States and Canada is ordinarily 110 V, 115 V, or 120 V and can be anywhere in that range, varying from time to time.

In the early days, there were a number of different frequencies supplied (at first 125 Hz and 133 Hz were common, and later 25 Hz, 35 Hz, 50 Hz, and 60 Hz became popular), depending on the producer. Lower frequencies were more suitable for use with the induction motor, but they still had to be high enough to keep incandescent lamps from flickering. Today, most of Europe and Asia uses 50 Hz (220 V), whereas the United States and Canada have adopted 60 Hz (110 V) as standard.

Alternating current was initially supplied (as was dc) using conventional two-wire lines in what is called a *single-phase* system—the voltage rises and falls sinusoidally (p. 721), as does the current. Under such circumstances, power is inefficiently available in pulses. Tesla was among the few to realize that by simultaneously generating several sinu-

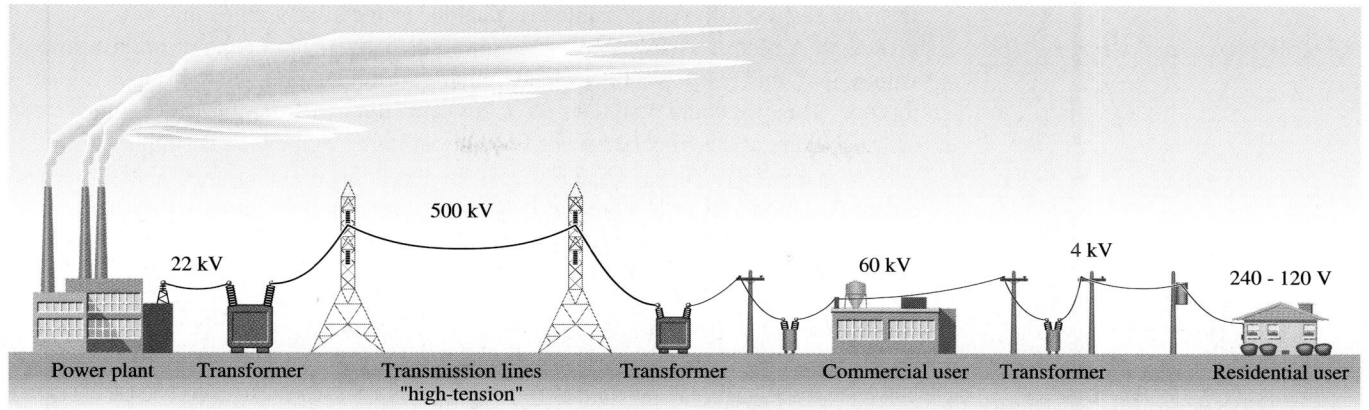

Figure 21.1 The transmission of ac power. Transformers are used to increase the voltage so power can be efficiently sent long distances via "high-tension" wires. The voltage is then stepped down for commercial and residential use. For simplicity, the ground and other "hot" leads are not shown.

An ac shaded-pole induction motor of the kind that powers devices such as blenders and can openers. The field coil sets up an oscillating *B*-field in the stator. Currents induced in the copper wire "shades" retard the changing *B*-field in those regions, producing a primitive rotating field that induces currents in the rotor. These currents result in a torque that sets the rotor turning.

soidal voltages, each shifted in phase with respect to the other, a transmission line could carry more power and actually does so at a constant, continuous rate. Using three wires instead of two and sending out three (120°-shifted) sinusoidal voltages between them (Fig. 21.2), Tesla was able to triple the average amount of power delivered by the line. Equally as important, this kind of *polyphase* ac was necessary to drive the big new induction motors that required no brushes or commutators and were extremely reliable. Under the irresistible pressure of these benefits, the entire ac power system was overhauled and the transmission of *three-phase* ac became almost universal. Today, although industrial users are often supplied directly with three-phase ac, the typical domestic consumer is usually provided with two single-phase 120-V ac lines (p. 763).

21.1 AC and Resistance

As we saw earlier [Eq. (20.7)], the emf produced by an ac generator can be represented by a sine function of angular frequency $\omega = 2\pi f = 2\pi/T$. In addition to the wall outlet, there are electronic devices known as *oscillators* that also provide a harmonically varying potential difference. They're often used in the laboratory, though there are many gadgets that have oscillators built into them—the modern radio receiver is one. Whatever the source, the terminal voltage then has the form

$$v = V_m \sin \omega t = V_m \sin 2\pi ft \qquad (21.1)$$

where *v* is the **instantaneous voltage**. The lowercase letter *v* is used here to distinguish it

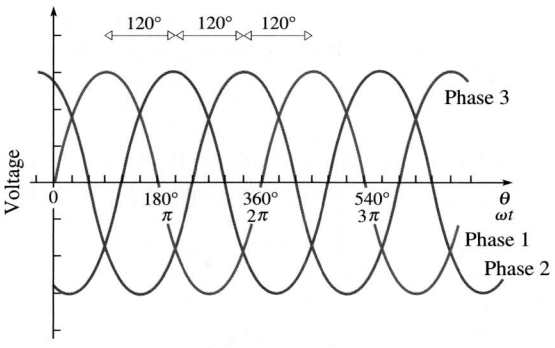

Figure 21.2 The emf produced by a three-phase ac generator. Each signal is shifted by 120°.

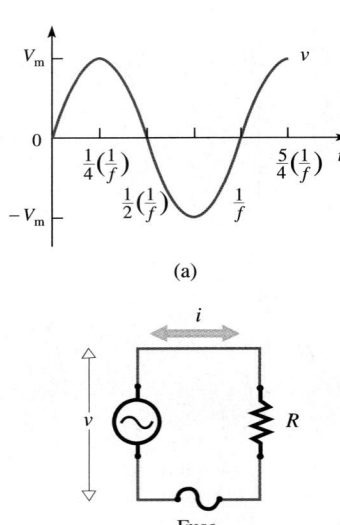

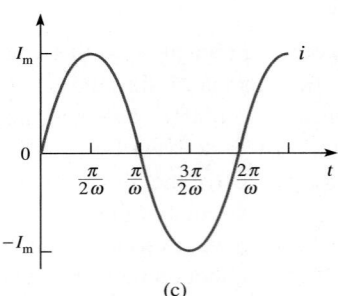

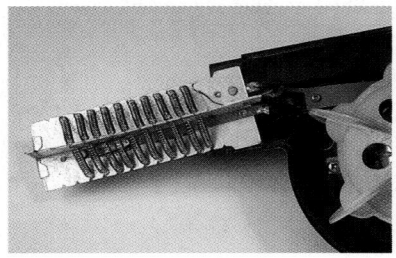

Figure 21.3 A sinusoidal voltage (a) applied to a resistive load (b) results in a sinusoidal in-phase current (c).

The heating coil in a hair dryer. Current passing through the coil causes it to become red hot ($P_{av} = I^2R$) via joule heating. The white plastic fan at the right blows air over the coil and out the nozzle.

from the several different possible practical measures of ac voltage that will be considered presently. A plot of v (Fig. 21.3) rises and falls between two peak or **maximum voltage** values, $+V_m$ and $-V_m$, which is what would be seen if we plugged the source into an oscilloscope. Then we could read the maximum value directly from the scale on the screen just as we could read the so-called **peak-to-peak voltage** (V_{pp}), which is simply $2V_m$.

With a purely resistive load R across the terminals of an ac source, a current will circulate that increases and decreases, reversing direction over and over again, precisely in step with the oscillating voltage—the applied emf directly drives the **instantaneous current** i. Ohm's Law takes the form

$$i = \frac{v}{R} = \frac{V_m}{R}\sin \omega t$$

and with the **maximum current** given by $I_m = V_m/R$ becomes

$$i = I_m \sin \omega t = I_m \sin 2\pi f t \qquad (21.2)$$

Both the voltage and the current are zero when $2\pi f t$ equals either 0, π, 2π, etc., since $\sin 0 = \sin \pi = \sin 2\pi = 0$. This occurs when $t = 0$, $t = 1/2f = \pi/\omega$, or $t = 1/f = 2\pi/\omega$, and so on, as shown in Fig. 21.3. For 60-Hz ac, the zeros happen at $t = 0$, $1/120$ s, $1/60$ s, and so on. The current reverses itself every $1/120$ s, and the period is $T = 1/f = 1/60$ s.

The early workers with electricity were faced with a practical problem when it came to measuring ac; they couldn't have the pointer on an ammeter or a voltmeter flutter back and forth at 60 Hz. Indeed, what does one even mean by the statement "the ac voltage* is 120 V?" Engineers were accustomed to dealing with dc, where it makes sense to speak of a 10-A current, but how does one describe the strength of an alternating current? It was decided to use the *ampere* as the unit of alternating current and define it in such a way as to be, insofar as possible, equivalent to the direct-current ampere. There were three possible methodologies that could be used: chemical, magnetic, or thermal. But only the thermal effect of a current varies as i^2 and is independent of the direction, or sign, of the current. Since both ac and dc produce joule heating, *the alternating-current ampere was defined as the quantity of current that produces the same amount of joule heating in a resistor as does a direct-current ampere during the same interval of time.*

Nowadays, ac ammeters are calibrated to read this **effective current** (I_{eff}): 10 amps ac effective will generate the same amount of thermal energy as 10 amps dc. So common is the usage, it is automatically assumed that a device rated at simply 10 amps ac refers to 10 A effective. And the same is true for ac voltmeters, which are, as a rule, calibrated in **effective voltage** (V_{eff}). Moreover, since we are likely to deal only in effective values, the subscripts are often dropped, and it is understood that the I and V mean effective current and voltage, respectively.

The effective values can be related to the maximum values by looking at the power; I_{eff} is defined by its ability to cause heating in a resistor, which is equivalent to its ability to transfer an average level of power. The **instantaneous power** p dissipated in a resistor is

$$p = i^2 R$$

What we actually measure via a calorimeter is the effect of such power averaged over many cycles. Inasmuch as the resistance R is constant

$$P_{av} = [i^2]_{av} R \qquad (21.3)$$

and this result by definition is to equal I^2R or, if you like, $I_{eff}^2 R$. The average rate at which thermal energy is developed in a resistor by an alternating current is

[average power]
$$P_{av} = I_{eff}^2 R \qquad (21.4)$$

*Despite the obvious awkwardness, such terms as ac current and ac voltage are now part of the jargon. In such cases, the ac should be read as ay cee and not alternating current.

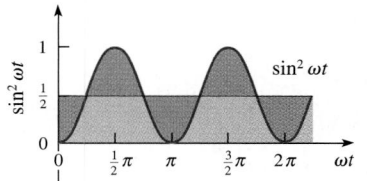

Figure 21.4 The area under the curve $\sin^2\omega t$ divided by the extent of the curve (ωt) equals the average value; $[\sin^2\omega t]_{av} = 1/2$.

Comparing these two equations, it follows that

$$I = I_{eff} = \sqrt{[i^2]_{av}}$$

The effective current equals the root of the mean of the square of the instantaneous current or, for short, the **rms current**. It's not uncommon in the literature to find the symbol I_{rms} used instead of I or I_{eff}. Had we started with $p = v^2/R$, we would have come upon a precisely analogous definition for the **rms voltage**.

Returning to Eq. (21.3) and substituting Eq. (21.2) into it yields

$$P_{av} = [(I_m \sin \omega t)^2]_{av} R = [I_m^2 \sin^2 \omega t]_{av} R$$

Because I_m^2 is constant,

$$P_{av} = I_m^2 [\sin^2 \omega t]_{av} R \qquad (21.5)$$

The average of any time-varying function taken over some interval is equal to the area under the curve divided by the length of the interval. In Fig. 21.4, the area under the curve equals the area under the constant $\frac{1}{2}$-line since the tops of the peaks can be imagined cut off and moved so that they fill the troughs. For any interval ωt, of several periods, the area $(\frac{1}{2} \times \omega t)$ divided by the duration (ωt) is just $\frac{1}{2}$: $[\sin^2 \omega t]_{av} = \frac{1}{2}$. Another way to see this is to start with the identity $\sin^2 \omega t + \cos^2 \omega t = 1$. Realizing that $\sin^2 \omega t$ and $\cos^2 \omega t$ are identical except for a 90° phase shift, they must both have the same average value over an interval much greater than one period. But if we take the average of both sides of the identity, it follows that $[\sin^2 \omega t]_{av} = [\cos^2 \omega t]_{av} = \frac{1}{2}$. From Eq. (21.5)

$$P_{av} = \tfrac{1}{2} I_m^2 R \qquad (21.6)$$

so, not surprisingly, since the instantaneous power $(I_m \sin^2 \omega t)$ oscillates, the average power is half the maximum power: $P_{av} = \frac{1}{2} P_m$. More to the point, since $P_{av} = I_{eff}^2 R$, we have the sought-after expression for the effective current in terms of the maximum current, namely,

$$I = I_{eff} = \frac{I_m}{\sqrt{2}} \qquad (21.7)$$

Similarly

$$V = V_{eff} = \frac{V_m}{\sqrt{2}} \qquad (21.8)$$

(**Memorize these last two relationships.**) The effective current or voltage is just 0.707 times the corresponding maximum value (Fig. 21.5). A wall outlet providing 120 volts effective is actually supplying a sinusoidal emf with a maximum terminal voltage of $V_m = \sqrt{2}(120 \text{ V}) = 170 \text{ V}$, which is a good reason to be even more careful with 120-V ac than 120-V dc.

From Ohm's Law, which is valid at any instant, $V_m = I_m R$ and

$$V_{eff} = I_{eff} R \quad \text{or} \quad V = IR \qquad (21.9)$$

Moreover

[resistive load] $$P_{av} = I_{eff}^2 R = I_{eff} V_{eff} \qquad (21.10)$$

which is nice and simple.

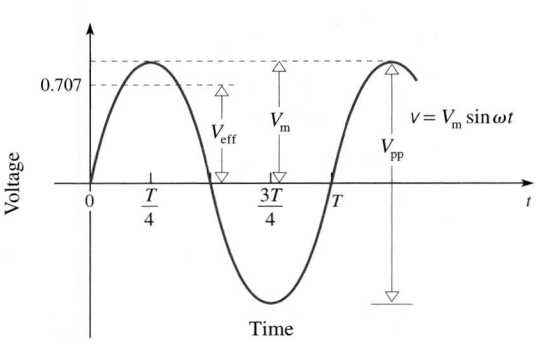

Figure 21.5 A sinusoidal voltage and the corresponding effective, maximum, and peak-to-peak values.

Example 21.1 **[II]** According to the little metal plate on a hair dryer, it's rated at 120 V and 1200 W. Assuming the load to be purely resistive (it really isn't because of the motor), how much current does it draw and what is its resistance? What is the maximum value of the current in the dryer?

Solution That "1000 W" is average ac power, which makes this a power problem concerning voltage, current, and resistance. (1) TRANSLATION—An ac device has a known (effective) voltage and average power rating; determine the (effective) current, resistance, and maximum current. (2) GIVEN: $V = 120$ V and $P_{av} = 1200$ W. FIND: I, I_m, and R. (3) PROBLEM TYPE—ac circuit/power. (4) PROCEDURE—From the definition of ac average power we can find I knowing V. (5) CALCULATION—The effective

current is

$$ I = \frac{P_{av}}{V} = \frac{1200\ \text{W}}{120\ \text{V}} = \boxed{10.0\ \text{A}} $$

The maximum current then follows from $I = I_{eff} = \dfrac{I_m}{\sqrt{2}}$:

$$ I_m = 1.4141 = \boxed{14.1\ \text{A}} $$

From Ohm's Law

$$ R = \frac{V}{I} = \frac{120\ \text{V}}{10.0\ \text{A}} = \boxed{12.0\ \Omega} $$

Quick Check: $P_{av} = I^2R = (10.0\ \text{A})^2(12\ \Omega) = 1200$ W.

[For more worked problems click on WALK-THROUGHS *in* CHAPTER 21 *on the CD.]*

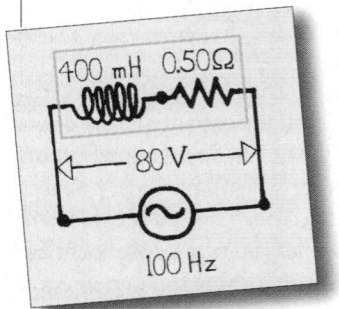

21.2 AC and Inductance

The distinctive aspects of ac become evident when there are inductors or capacitors in the circuit. It's then that the currents through, and the voltages across, these elements are *not in-phase*. As a result of these phase differences, all sorts of interesting things happen. For example, Ohm's Law doesn't apply in the same simple way it did with dc: in general, the current in an ac circuit (which is not purely resistive) is not equal to the applied voltage divided by the resistance of the circuit. Moreover, the algebraic sum of the ac voltage drops across each element of a series circuit (as measured by individual voltmeters) may not equal the applied voltage (as it does in a dc circuit). Energy is still conserved and Kirchhoff's Rules apply, but for sinusoidal voltages (and currents) that are out-of-phase we can't simply add their effective values algebraically.

To study the ac behavior of an inductor, consider the circuit shown in Fig. 21.6a. It consists of an inductor L, having negligible resistance, placed across the terminals of an ac source. At the instant shown in Fig. 21.6b, an increasing clockwise current exists, creating an increasing flux in the coil. Remember (p. 726) that in response, an induced current i_I will appear in the windings, and it will oppose the change in the flux via Lenz's Law. The induced current at that moment emerges from the inductor at point A, and there is a potential difference across the terminals of the inductor, which is $+$ at point A and $-$ at point B. The inductor then behaves much like a battery being charged. Because the current from the source is sinusoidal, there will be a continuously changing flux and a sustained harmonic back-emf. In the steady state, the back-emf is 180° out-of-phase with the emf of the source. In Fig. 21.6, trace around the circuit and, when the back-emf is a drop, the source voltage will be a rise. In the unattainable case where $R = 0$, the two are numerically equal. At any instant, the sum of the voltage rises and drops around the loop is then zero. *The voltage measured across the terminals of the inductor* ($v = L\,\Delta i/\Delta t$) *equals*

$$ v = V_m \sin \omega t $$

the voltage of the source.

We would like to deduce an expression for the instantaneous current, and that could be done directly using calculus since *the voltage depends on the time rate-of-change of the current.* Alternatively, we can determine the form of i, by analogy with the harmonic oscillator for which the position [Eq. (10.13)] and speed [Eq. (10.14)] are also harmonic functions and the speed is the time rate-of-change of position. When the position is $x = A\cos\omega t$, the corresponding speed is $v_x = -A\omega \sin \omega t$. Position x and speed v_x are related in

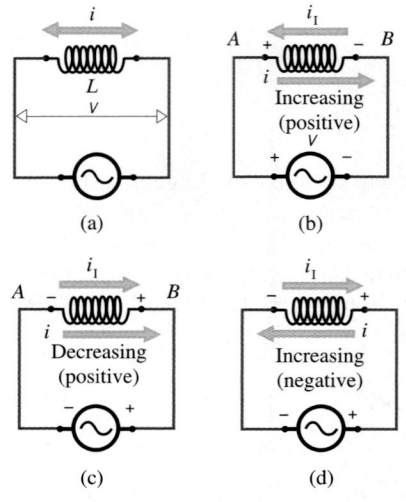

Figure 21.6 (a) An inductor placed across the terminals of an ac source. (b) Note that when i is positive (clockwise) and increasing, the terminal voltage v is positive. (c) When i is positive and decreasing, v is negative, and (d) when i is negative and increasing, v is negative.

the same way as are current i and voltage v: actually, as i and v/L. Comparing voltage/inductance with speed, we see that V_m/L corresponds to $-A\omega$ and so A corresponds to $-V_m/L\omega$. The expression for x suggests that

$$i = -\frac{V_m}{\omega L}\cos\omega t = -I_m\cos\omega t \qquad (21.11)$$

Figure 21.7 is a plot of this instantaneous current through, and the voltage across, the inductor. Notice how the current peaks at a later time than the voltage: **the current lags the voltage (or the voltage leads the current) by one-quarter cycle (90°)**. The instantaneous voltage depends on the product of the inductance and the *rate-of-change* of the current. The voltage is zero when the current curve levels off (i.e., when its slope is zero) at its maximum values. The voltage is maximum when the current curve is steepest (i.e., when its slope is maximum) at points where it crosses the time axis. *The higher the frequency, the tighter the curves are and the greater the slopes at these points.*

A real inductor inhibits current both because of its resistance and because of its inductance (via the associated back-emf). We can speak of the total effect as representing an ***impedance***, part of which, the ***reactance***, is due purely to the inductive behavior exclusive of any ***resistance***. It follows from Eq. (21.11) that

$$I_m = \frac{V_m}{\omega L} \qquad (21.12)$$

and, since $I = 0.707I_m$ and $V = 0.707V_m$,

$$V = \omega LI \qquad (21.13)$$

This result is so much like Ohm's Law, with the ωL term impeding the current, that we introduce a new quantity called the **inductive reactance** X_L, defining it as

[inductive reactance]
$$X_L = \omega L = 2\pi fL \qquad (21.14)$$

whereupon
$$V = IX_L \qquad (21.15)$$

Consequently, 1 volt per 1 amp corresponds to an inductive reactance of 1 ohm.

In this ideal situation, there is no resistance and the reactance equals the impedance. The greater the inductance, the greater the back-emf and the more the inductor inhibits or "chokes" the current. Note that in the case of dc ($\omega = 0$), there is no back-emf—the reactance is zero and there is no inhibition of current. An inductor with a large L and small R is usually called a *choke coil* because of its ability to control ac currents without wasting near-

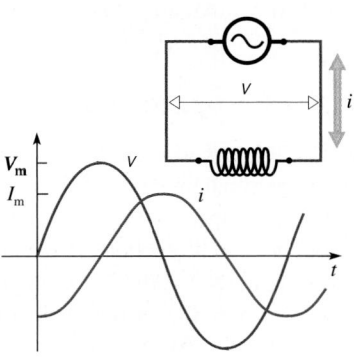

Figure 21.7 An inductor effectively holds back the current so that i (the current through it) lags v (the voltage applied by the source). (See Fig. 21.6.) In other words, graphed against t, the peak of the current occurs at a later time, farther along to the right, than does the voltage whose peak occurs earlier. **In an inductive ac circuit, voltage leads current.**

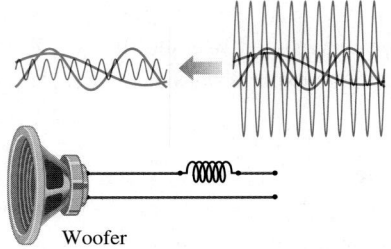

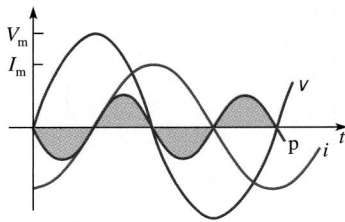

Figure 21.8 The inductor's impedance increases with frequency, so it passes low-frequency signals and blocks high-frequency signals.

Figure 21.9 The power drawn by a pure inductance averages to zero. Note how p = 0 whenever either v or i is zero. The relative heights of the curves are unimportant.

ly as much power as would a resistor. Because the inductive reactance increases with frequency, a choke can be used to suppress the high end of a mix of frequencies. For example, in a loudspeaker system there are usually at least two drivers, a small high-frequency tweeter and a large low-frequency woofer. By putting an inductor in series with the woofer (Fig. 21.8), high-frequency currents (representing high-pitched sound) are suppressed, allowing the woofer to respond to the signal range it was designed to handle best.

Example 21.2 **[II]** A radio circuit contains a 400-mH inductor with a resistance of 0.50 Ω. This is supplied (across its terminals) with an 80-V-effective ac signal of 100 Hz. Determine both the reactance of the coil and its *rms* current.

Solution We have the voltage and need the current, which makes this an inductive reactance problem. (1) TRANSLATION —An ac device with a known inductance and resistance has a specified (effective) voltage at a given frequency across it; determine the reactance and *rms* current. (2) GIVEN: $L = 0.400$ H, $R = 0.50$ Ω, $V = 80$ V, and $f = 100$ Hz. FIND: X_L and I. (3) PROBLEM TYPE—ac circuit/inductive reactance. (4) PROCEDURE —Use the definition of inductive reactance to determine X_L, and then I follows from the ac version of Ohm's Law. (5) CALCULATION:

$$X_L = \omega L = 2\pi f L = 2\pi(100 \text{ Hz})(0.400 \text{ H}) = \boxed{251 \ \Omega}$$

By comparison, R is negligible, and we can treat the coil as if it were a pure inductance. Hence

$$I = \frac{V}{X_L} = \frac{80 \text{ V}}{251 \ \Omega} = \boxed{0.32 \text{ A}}$$

Quick Check: $X_L = 2\pi f L \approx (6)10^2(4 \times 10^{-1}) \ \Omega \approx 0.24$ kΩ; $V = I X_L \approx (0.32 \text{ A})(\frac{1}{4} \text{ k}\Omega) \approx 0.08$ kV.

Figure 21.9 is a plot of the instantaneous power (p = iv) associated with the ideal inductor. Where i and v are both positive or both negative, p is positive; and where either one (i or v) is negative, p is negative. Thus, the area under the curve (subtracting what is below the axis from what is above it) per cycle is zero. Energy is stored in the alternating magnetic field, but *the average power supplied per cycle in a purely inductive circuit is zero*. Power is alternately absorbed from the generator (positive portion of the cycle) and then returned to it (negative portion of the cycle).

21.3 AC and Capacitance

Imagine a capacitor C placed across the terminals of a battery. We are already familiar with what happens in the transient state: charge piles up on the plates until the mutual repulsion between the charges stops any more from arriving. A *reverse potential* appears (via $v = q/C$) that increases until it reaches the battery potential, at which point the current stops. Because C is constant, $\Delta v = \Delta q/C$; a change in the charge is associated with a change in the voltage. By definition, the current is the time rate-of-change of q, which corresponds to a time rate-of-change of v: no $\Delta v/\Delta t$, no steady state i.

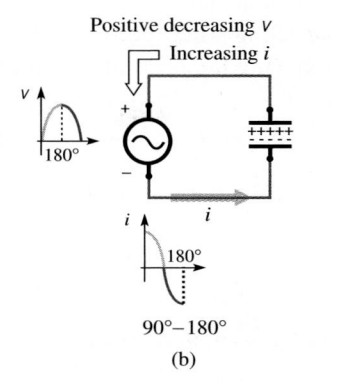

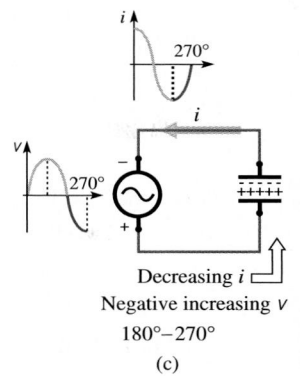

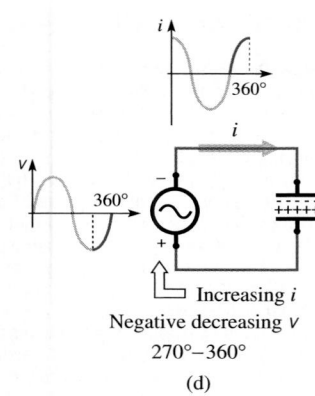

Figure 21.10   The voltage and current in an ac capacitive circuit. The positive direction is taken to be clockwise.

Now suppose the battery is replaced by an ac source (Fig. 21.10). We want to compare v across the capacitor to i passing through it. As the sinusoidal voltage of the generator rises positively (0° to 90° in the ac cycle), we can imagine positive charges traveling to the upper plate and an equal number repelled away from the lower plate. A large positive current immediately circulates, because at $t = 0$ there is no charge on the plates and no reverse potential to inhibit it. As the impressed voltage rises, the charge on the plates increases in step with it, and the reverse potential increases too, making it harder for more charge to be deposited—the clockwise current dies off.

At the moment the voltage peaks (90°) so that $\Delta v / \Delta t = 0$, the current is zero. But then the voltage begins to decrease (90° to 180° in the ac cycle), whereupon the capacitor starts to discharge: $\Delta v / \Delta t \neq 0$ and $i \neq 0$. Current appears in the negative counterclockwise direction. After the voltage passes through zero, it reverses and increases (180° to 270° in the ac cycle), increasing the charge on the plates in the reverse direction. That again decreases the counterclockwise current until it becomes zero, when the voltage peaks in the negative direction (270°). The negative applied voltage then decreases, and positive charge leaves the capacitor flowing clockwise and constituting a positive current (270° to 360° in the ac cycle). Apparently, **the instantaneous current in a capacitor leads the instantaneous voltage across it by one-quarter of a cycle** (90°).

A **capacitor** resists the change in voltage across it.

We can describe this behavior analytically in the following way. Since $v = q/C$ and $v = V_m \sin \omega t$, it follows that

$$q = C V_m \sin \omega t \qquad (21.16)$$

The charge and voltage are in-phase as shown in Fig. 21.11. Now to find the instantaneous current i that is the time rate-of-change of the charge. Again, using the harmonic oscillator, Eq. (21.16) suggests that we begin with the sinusoidal speed of the oscillator $v_x =$

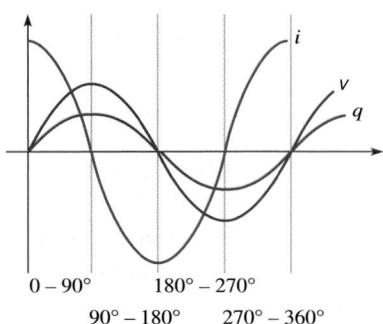

Figure 21.11   The current, charge, and voltage for an ac capacitive circuit. Graphed against t, the peak of the voltage occurs at a later time, farther along to the right, than does the current whose peak occurs earlier. **In a capacitive ac circuit, current leads voltage.**

STUDY GUIDE

A capacitor resists the buildup of charge on its plates and thereby resists the change in voltage. Still, the higher the frequency, the less the capacitor can react; there's less time for the charge to either build up or diminish. That's why X_C is inversely proportional to ω—the higher the frequency of the driving voltage, the less impeding the capacitor is to current flow.

Notice that as $\omega \to 0$, which corresponds to dc, $X_C \to \infty$; the ideal capacitor is open-circuited when $\omega = 0$. A capacitor blocks dc.

As far as the analysis of current and voltage in a circuit is concerned, **capacitive reactance** (X_C) behaves in a similar way to resistance. There are two important differences, however: ideal capacitors do not dissipate power (there's no joule heating), and (as we'll soon see) they affect the phases of the oscillating voltage and current.

For a purely capacitive ac circuit ($R = 0$, $L = 0$) Ohm's Law applies, provided the resistance is replaced by the **capacitive reactance**.

$-A\omega \sin \omega t$ and examine its time rate-of-change; namely (p. 348), the instantaneous acceleration $a_x = -A\omega^2 \cos \omega t$. Comparing speed and charge, we see that $-A\omega$ corresponds to CV_m. With acceleration having the form of the current, where $-A\omega^2 = (-A\omega)\omega$ corresponds to $CV_m \omega$, we get

$$i = \omega CV_m \cos \omega t = I_m \cos \omega t \qquad (21.17)$$

This time, $\omega CV_m = I_m$ and so the *rms* values share the same relationship—namely, $\omega CV = I$. It follows that

$$V = \frac{1}{\omega C} I \qquad (21.18)$$

and Ohm's Law suggests that we introduce a **capacitive reactance** X_C, defined as

[capacitive reactance]
$$X_C = \frac{1}{\omega C} = \frac{1}{2\pi f C} \qquad (21.19)$$

whereupon

$$V = X_C I$$

The reactance, in ohms (Problem 52), is a measure of the capacitor's opposition to the passage of alternating current.

When $\omega = 0$, the reactance is infinite, and no finite dc voltage will produce a current through an ideal capacitor—**capacitors block dc**. As the frequency goes up, the capacitive reactance goes down, in contrast to the inductive reactance; a given source voltage results in a higher and higher current as ω is increased. What stops the current is the peaking of the voltage arising from the buildup of charge on the plates. At a high frequency, the charge has little opportunity to pile up and thereby decrease the current before the applied voltage is reversed. Alternatively, increasing ω increases the slope of the v curve, which means a larger value for the time rate-of-change of q, and so a larger current. If the capacitor is getting charged and discharged more rapidly, the current must increase. Raising the capacitance lowers the reactance; a large capacitor can store more charge before the voltage across its plates rises to appreciably diminish the current. Alternatively, since $\Delta q / \Delta t = C \Delta v / \Delta t$, the larger is C, the greater the current.

Figure 21.12 indicates how a capacitor in series with a high-frequency tweeter in a speaker system will suppress the low frequencies in the input signal ($X_C \propto 1/\omega$). It will pass on, with little attenuation, the high-frequency components that the tweeter was designed to convert into sound.

Example 21.3 **[II]** A 50-μF capacitor is connected across the terminals of an oscillator set to have a sinusoidal output at 50 Hz with a maximum voltage of 100 V. Determine the effective current in the circuit. How would this change if the frequency were raised to 5.0 kHz?

Solution We have the voltage and need the current, which makes this a capacitive reactance problem. (1) TRANSLATION—An ac source of known frequency and maximum voltage has a known capacitance across its terminals; determine the current. (2) GIVEN: $C = 50$ μF, $V_m = 100$ V, and $f = 50$ Hz. FIND: I. (3) PROBLEM TYPE—ac circuit/capacitive reactance. (4) PROCEDURE—Use the definition of capacitive reactance to determine X_C, and then I follows from the ac version of Ohm's Law. (5) CALCULATION—Knowing the maximum voltage, we

can find the effective voltage and, from that and the reactance, we can find I. So, compute X_C:

$$X_C = \frac{1}{\omega C} = \frac{1}{2\pi f C} = \frac{1}{2\pi (50 \text{ Hz})(50 \text{ } \mu\text{F})} = 63.7 \text{ } \Omega$$

Now $V = 0.707 V_m = 70.7$ V and therefore

$$I = \frac{V}{X_C} = \frac{70.7 \text{ V}}{63.7 \text{ } \Omega} = \boxed{1.1 \text{ A}}$$

At 5.0 kHz, the reactance would be lower by a factor of 100 and the current raised by a factor of 100.

Quick Check: $V = IX_C = I/\omega C = I/2\pi f C \approx (1 \text{ A})/6(50 \text{ Hz})(50 \times 10^{-6} \text{ F}) \approx (1/0.015) \text{ V} \approx 0.07$ kV.

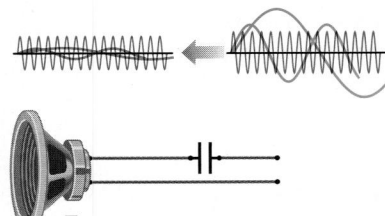

Figure 21.12 The capacitor represses low frequencies and passes high frequencies.

As with the inductor, because the instantaneous current and voltage are out-of-phase by 90°, no average power will be dissipated by an ideal capacitor. Energy stored in the electric field between the plates is returned to the source. ***Only resistance will dissipate power in an ac circuit, converting electrical energy into thermal energy.***

R-L-C AC Networks (Optional)

Electronic devices such as radios, televisions, stereos, and so on, utilize inductors, capacitors, and resistors to process electrical signals—currents and voltages. Arrangements of these elements can be used to reshape a signal, to filter out or perhaps accentuate certain frequencies, or to remove any dc that might be present, and so on.

21.4 *R-L-C* Circuits

Suppose we put an inductor, a capacitor, and a resistor in series across the terminals of an oscillator with an instantaneous voltage v, as shown in Figure 21.13. Now represent the current by either a sine or cosine function. Let it be

$$i = I_m \sin \omega t \tag{21.20}$$

This same current exists in each element of the circuit. And since the instantaneous voltage across the resistor (v_R) is in-phase with the current, we have

$$v_R = I_m R \sin \omega t \tag{21.21}$$

By comparison, the instantaneous voltage across the capacitor (v_C) lags the current by 90°, or $\pi/2$ radians; the sine function describing the voltage reaches the same value as $\sin \omega t$ at a later time. Using $V = I/\omega C$,

$$v_C = \frac{I_m}{\omega C} \sin\left(\omega t - \frac{\pi}{2}\right) \tag{21.22}$$

Comparing v_R and v_C, note that $\sin \omega t = 0$ at $t = 0$, whereas $\sin(\omega t - \pi/2) = 0$ *later*, when $t = \pi/2\omega$. On the other hand, the instantaneous voltage across the inductor leads the current by 90°, or $\pi/2$ radians, as

$$v_L = I_m \omega L \sin\left(\omega t + \frac{\pi}{2}\right) \tag{21.23}$$

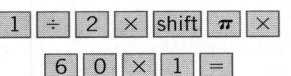
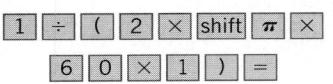
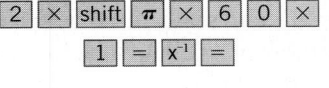
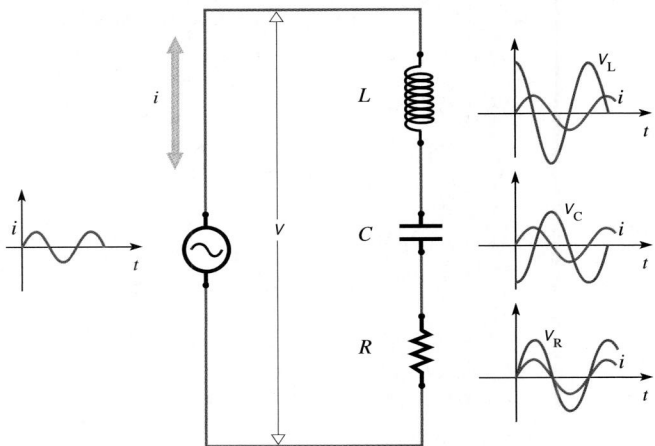

Figure 21.13 The same current passes through each element in a series circuit. The voltages across each are usually not in-phase and $v = v_L + v_C + v_R$. Notice that for the inductor, **voltage leads current**; for the capacitor, **current leads voltage**; and for the resistor, **voltage and current are in-phase**.

Thus, $\sin \omega t = 0$ at $t = 0$, whereas $\sin (\omega t + \pi/2) = 0$ *earlier*, when $t = -\pi/2\omega$.

In order for the applied voltage v to drive the current i through the circuit, it must equal the sum of the instantaneous voltages

$$v = v_R + v_C + v_L \tag{21.24}$$

If we added the three sinusoids via algebra and trigonometry (and the mathematics is laborious), we would end up with a resultant sinusoid

$$v = V_m \sin (\omega t + \theta) \tag{21.25}$$

Interestingly, the sum of any number of sine functions of the same frequency is a sine function of that same frequency (Fig. 11.21, p. 385). The complete analysis yields expressions for V_m and θ:

$$V_m = \sqrt{(I_m R)^2 + \left(I_m \omega L - \frac{I_m}{\omega C}\right)^2} \tag{21.26}$$

and

$$\theta = \tan^{-1} \frac{\omega L - 1/\omega C}{R} \tag{21.27}$$

which, when computed and substituted into Eq. (21.25), gives the potential difference across all three elements. We have not provided the details of that calculation because there is a much simpler practical scheme for arriving at the same results, a method called **phasor addition**.

Figure 10.19 (p. 346) illustrated how a line rotating at a rate of ωt can be projected onto either axis in order to generate harmonic functions. Suppose, then, that we draw an arrow of length V_m and have it rotate counterclockwise at a rate ω, where at any instant it makes an angle ωt. The corresponding sine and cosine components are shown in Fig. 21.14. This revolving "arrow," which looks like a vector and has some of the properties of a vector, is formally a different beast called a *phasor* (designated by boldface type). The important thing is that *phasors add like vectors*, and, in so doing, we in effect add their components—namely, the sinusoids. Combine the phasors and we combine the sinusoids, which is what we want, and that is easily done graphically.

Since the current and the voltage across the resistor are in-phase, we draw these two phasors one on top of the other in Fig. 21.15. Their lengths are I_m and $V_{Rm} = I_m R$, respectively, and their y-components are the sine functions of Eqs. (21.20) and (21.21). Because all of the phasors will be referenced to the current phasor, it's customary to simplify things a little by working at $t = 0$ (and remembering to put ωt in the expression for the phase when needed), which has the effect of placing the current phasor on the x-axis. Henceforth, all

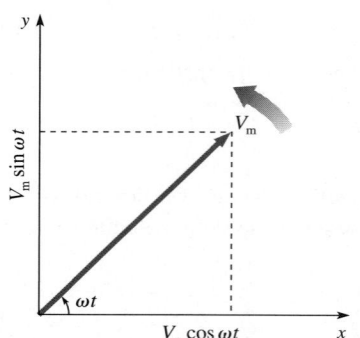

Figure 21.14 As the arrow rotates, the x- and y-components oscillate harmonically.

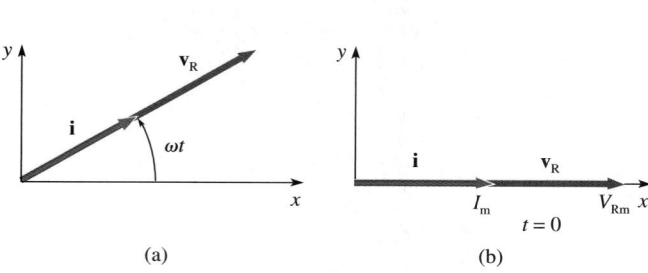

Figure 21.15 The phasors **i** and v_R are in-phase; that is, the ac current is in-phase with the voltage across the resistor.

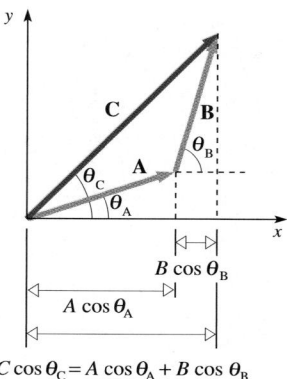

$$C \cos \theta_C = A \cos \theta_A + B \cos \theta_B$$

Figure 21.16 The addition of phasors **A** and **B**. Phasors add like vectors, and their projections onto the x-axis add like vector components.

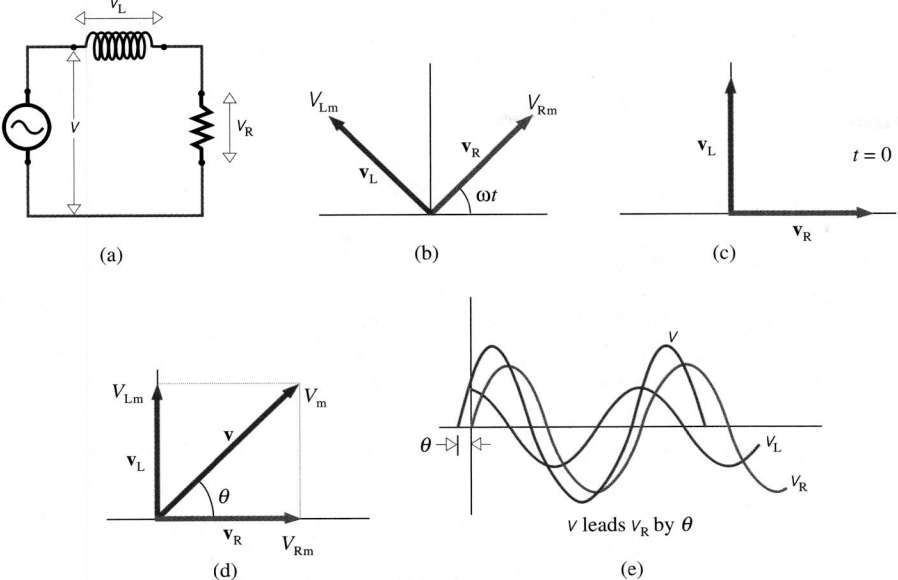

Figure 21.17 (a) An ac circuit containing inductance and resistance. (b) The voltage phasors actually rotate at a rate ω. (c) Rotation is eliminated when $t = 0$. (d) The amplitude of **v** is V_m. (e) v_L and v_R are 90° out-of-phase.

phasors will be drawn with respect to the x-axis. If its phase is $(\omega t + \alpha)$, it is tilted α radians above the x-axis; if its phase is $(\omega t - \beta)$, it is tilted β radians below the x-axis. Figure 21.16 shows how two phasors (which we shall distinguish using boldface type) **A** and **B** are added tip-to-tail like vectors. The projection on the x-axis is then the sum of the individual cosine functions, and the projection on the y-axis is the sum of the sine functions. ***Our phase shifts will only be ±90°.***

Consider a circuit with just an inductor and a resistor (Fig. 21.17). Because the voltage across the inductor leads the current by 90°, $\mathbf{v}_L$ leads $\mathbf{v}_R$ by 90°, and the two phasors (with lengths $V_{Lm} = I_m \omega L$ and V_{Rm}) are drawn as shown in Fig. 21.17b. For simplicity, this configuration is redrawn at $t = 0$ in Fig. 21.17c. If we put these instantaneous voltages across the terminals of a dual-beam oscilloscope, we'll see two sinusoids shifted 90° with respect to each other. The two phasors are next added "vectorially" to produce a resultant phasor **v** in Fig. 21.17d. The length of **v** corresponds to the maximum voltage (V_m), which will be measured across the combination of R and L. The angle θ (positive if it's a lead, negative if a lag) is the phase angle of the resultant voltage sinusoid with respect to the current through the circuit. And that's it; Eq. (21.25) provides the expression for v once we compute V_m and θ from the diagram. Figure 21.17e shows the individual instantaneous voltages that can be added point-by-point to get v. Notice how v is indeed shifted by some angle (namely, $\theta < 90°$) with respect to v_R. The larger is L, the larger is θ. Similarly, if R is made larger, θ becomes smaller.

Return to the *R-L-C* series circuit of Fig. 21.13. The voltage across the capacitor lags v_R by 90° and its phasor, of length $V_{Cm} = I_m/\omega C$, is added in along with $\mathbf{v}_R$ and $\mathbf{v}_L$ in Fig. 21.18. The two opposing phasors on the y-axis add to yield a single phasor of length $|V_{Lm} - V_{Cm}|$, which may or may not point in the positive y-direction, depending on which is larger in a particular situation. Finally, using the Pythagorean Theorem to find the length of the resultant phasor (Fig. 21.18c), we obtain

$$V_m = \sqrt{(V_{Rm})^2 + (V_{Lm} - V_{Cm})^2} \tag{21.28}$$

which is identical to Eq. (21.26). Moreover

$$\tan \theta = \frac{(V_{Lm} - V_{Cm})}{V_{Rm}} \tag{21.29}$$

which is equivalent to Eq. (21.27). Inasmuch as the same expression holds for the effective

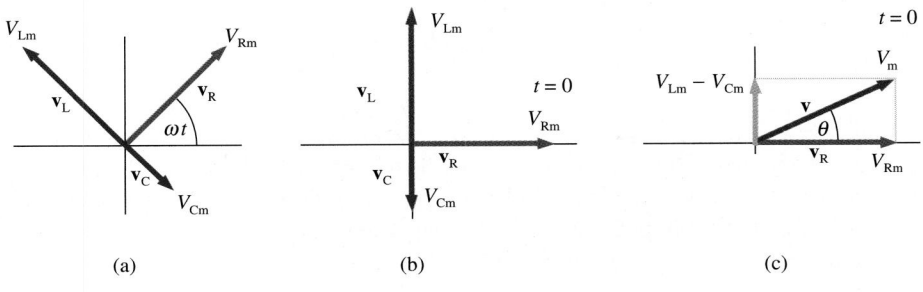

Figure 21.18 (a) The phasor diagram for a circuit containing resistance, inductance, and capacitance. (b) The magnitudes of the voltage phasors at $t = 0$. (c) The magnitude of the voltage across all the elements R, L, and C is V_m.

Example 21.4 **[II]** A series circuit contains a 240-Ω resistor, a 3.80-μF capacitor, and a 550-mH inductor. It's placed across the terminals of an ac generator set to 100 Hz. If an ammeter in the circuit reads 250 mA effective, what is the maximum voltage of the generator?

Solution This is a series *R-L-C* circuit. (1) TRANSLATION— A known resistor, capacitor, and inductor are in series across a source with a known frequency. If the effective current is known, determine the maximum voltage. (2) GIVEN: $R = 240$ Ω, $C = 3.80$ μF, $L = 550$ mH, $f = 100$ Hz, and $I = 250$ mA.

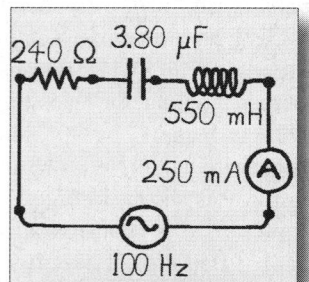

FIND: V_m. (3) PROBLEM TYPE—ac circuit/*R-L-C* series. (4) PROCEDURE— We have an expression for the maximum voltage, namely, Eq. (21.26). (5) CALCULATION—First determine I_m:

$$I_m = 1.414I = 353.5 \text{ mA}$$

Hence, $I_m R = 84.84$ V, $I_m \omega L = 122.2$ V, and $I_m/\omega C = 148.1$ V. And so

$$V_m = \sqrt{(I_m R)^2 + (I_m \omega L - I_m/\omega C)^2} \qquad [21.26]$$

hence $V_m = \sqrt{(84.84 \text{ V})^2 + (-25.9 \text{ V})^2} = \boxed{88.7 \text{ V}}$

Notice that *the voltages across the capacitor* (148 V) *and inductor* (122 V) *are higher than the voltage across all three components taken together* (88.7 V).

Quick Check: The reactances are $X_L = \omega L = 346$ Ω and $X_C = 1/\omega C = 419$ Ω, which are comparable. We can therefore expect the voltage drops V_{Lm} and V_{Cm} to be roughly equal, yielding only a small vertical phasor of magnitude $V_{Cm} - V_{Lm}$. The net voltage should be a little larger than $V_{Rm} = 84.8$ V, and it is.

As was the case with dc analysis, when solving ac circuit problems the goal is often to find the equivalent impedance and then use Ohm's Law.

The quantities X_C and X_L, which depend on frequency, usually need to be determined. (1) **Compute those individual reactances**.

Next, X_C and X_L must be combined, not algebraically, but by adding them as time-independent phasors (see Fig. 21.19). (2) Thus, **compute the combined reactance** $X = (X_L - X_C)$. This has the form it does because the phasors are always colinear and oppositely directed (180° out-of-phase).

(3) Then **compute the impedance of the entire circuit** using the Pythagorean Theorem: $Z = (R^2 + X^2)^{1/2}$. Notice that the sign of X is of no significance here.

(4) Knowing either the voltage or the current, **use the ac version of Ohm's Law** ($V = IZ$) **to compute** *I* **or** *V*.

As we'll see, this overall scheme works for both series and parallel circuits.

voltages, Eq. (21.28) tells us that the sum of the separate voltage readings made with a meter across each element will generally exceed the voltage across the source.

Going back to Eq. (21.26), factor out the current and divide both sides by $\sqrt{2}$ to get effective values, whereupon

$$V = I\sqrt{R^2 + \left(\omega L - \frac{1}{\omega C}\right)^2} = I\sqrt{R^2 + (X_L - X_C)^2}$$

The quantity $(X_L - X_C)$ is the **reactance** of the circuit; it's a measure of the net nonresistive influence impeding the current, and it's denoted by X:

$$X = (X_L - X_C) \qquad (21.30)$$

It follows that $V = I\sqrt{R^2 + X^2}$

Ohm's Law again suggests that the measure of a circuit's entire ability to restrain ac current (inductively, capacitively, and resistively) can be defined by its **impedance** (Z), where

[series circuit] $$Z = \sqrt{R^2 + X^2} \qquad (21.31)$$

given in ohms. Realize that when a circuit contains only a resistor, $Z = R$; only a capacitor, $Z = X_C$; or only an inductor, $Z = X_L$.

Ohm's Law survives into ac provided that it's written as

For any ac circuit Ohm's Law applies, provided the resistance is replaced by the **impedance**.

[ac Ohm's Law] $$V = IZ \qquad (21.32)$$

Notice how the impedance can also be thought of as a kind of vector quantity (one that isn't time-varying) in the sense that Z is equal to the magnitude of the resultant of adding X and R as if they were vectors (i.e., time-independent phasors). The associated diagram (Fig. 21.19) is often called the *impedance triangle*, and we see immediately that tan $\theta = X/R$, which is equivalent to Eqs. (21.27) and (21.29).

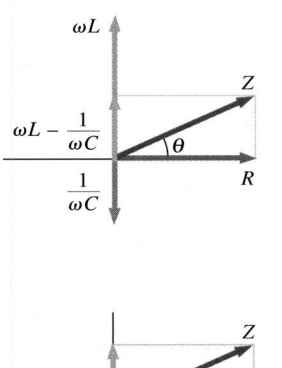

Figure 21.19 The impedance triangle. Reactances and resistance can be added as if they were vectors. The resultant is the impedance (Z) of the circuit. The "vector" $\mathbf{X}_C$ always points down, whereas $\mathbf{X}_L$ always points up. The "vector" $\mathbf{X}$ is the sum of the two, and so $X = X_L - X_C$. When $X_L > X_C$, θ is above the x-axis and positive. When $X_C > X_L$, X is negative, $\mathbf{X}$ points down, and θ is negative; it is beneath the "vector" $\mathbf{R}$, which always points in the positive x-direction. These vector-like quantities are actually phasors.

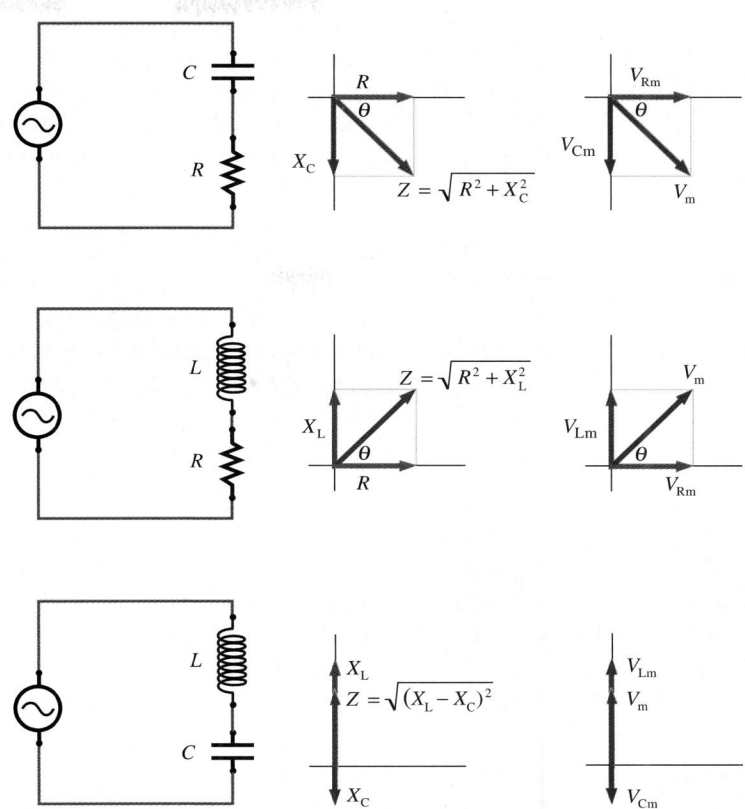

Figure 21.20 A summary of the behavior of various ac series circuits.

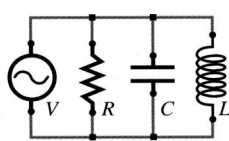

Figure 21.21 A parallel *R-L-C* circuit. As we'll see in Problem 92, the same techniques for analyzing series circuits apply to parallel circuits.

Figure 21.20 summarizes some of the results for two-element series circuits. Of course, a circuit may contain several resistors, capacitors, or inductors. We already know how to add any number of resistors (p. 639) and capacitors (p. 590) in series; suffice it to say without proof that inductors add as do resistors (see Discussion Question 10 in Chapter 20). To analyze an ac series circuit with many components, we first combine all of the same kinds of elements, whereupon all the preceding equations apply, with R, L, and C being the resultant values. Figure 21.21 shows a parallel *R-L-C* circuit which can be analyzed using the same methods you've just learned, and we'll do as much in Problem 92.

Power

On average, the power drawn by an *R-L-C* circuit is dissipated by the resistor: $P_{av} = I^2R$. We'd like to express P_{av} in terms of Z, and that can easily be done via the impedance triangle in Fig. 20.19. Thus $R = Z \cos \theta$,

$$P_{av} = I^2R = I^2Z \cos \theta$$

and since $V = IZ$,

$$P_{av} = IV \cos \theta \tag{21.33}$$

We must distinguish between the **real** or **dissipated power** ($P_{av} = IV \cos \theta$) and the **apparent power** (IV).

This expression is known as the **real**, or *average*, or **dissipated power**. It differs from the corresponding dc equation by the term $\cos \theta$, which is called the **power factor** of the circuit. The power factor is a measure of the relative influence of the resistance in dissipating power. With a purely resistive circuit, $Z = R$, $\cos \theta = 1$, and the real power is IV as expected. For either a purely inductive or capacitive circuit, the phase angle is either $\pm 90°$, and no real power is transferred from the source. A typical circuit has a power factor of less than 1 (or, since the power factor is often given as a percentage, less than 100%). For example, in a series ac circuit containing a resistor and an inductor for which the resistance equals the inductive reactance, it follows from the impedance triangle that $\theta = 45°$ and $\cos \theta = 0.707$. Small single-phase motors have power factors of around 0.6 to 0.8.

The product IV is called the **apparent power** and is measured in volt-amps (V·A) to distinguish it from real power. Practically, an amount of power equal to IV must be supplied even if a portion of that, $(1 - \cos \theta)IV$, is stored in the fields and returned to the source. A piece of electrical equipment that has an appreciable reactance, such as a transformer, fluorescent lamp, or motor, has a power factor much less than 100% and must, in order to operate properly, be supplied more power than it consumes. With a power factor of 80%, a motor that consumes 800 W must be supplied with 1000 V·A in order to operate. A large commercial user pays for the power that has to be provided, even if some of it is returned. To further examine some practical implications of the power factor, look at Discussion Question 15.

Example 21.5 **[II]** An oscillator set for 500 Hz puts out a sinusoidal voltage of 100 V effective. A 24.0-Ω resistor, a 10.0-μF capacitor, and a 50.0-mH inductor in series are wired across the terminals of the oscillator. (a) What will an ammeter in the circuit read? (b) What will a voltmeter read across each element? (c) What is the real power dissipated in the circuit?

Solution This is a series R-L-C circuit, and we follow the logic outlined in the Study Guide on p. 756. (1) Translation —A known resistor, capacitor, and inductor are in series across a source with a known frequency and effective voltage. Determine (a) the current, (b) the voltage across each element, and (c) the real power dissipated. (2) Given: $R = 24.0\ \Omega$, $C = 10.0\ \mu$F, $L = 50.0$ mH, $f = 500$ Hz, and $V = 100$ V. Find: I, V_L, V_C, V_R, and P_{av}. (3) Problem Type—ac circuit/R-L-C series. (4) Procedure—(a) Having V, to find the current via Ohm's Law we'll need the impedance. (c) The real power is given by $P_{av} = IV \cos \theta$. (5) Calculation—First determine the reactances:

$$X_L = \omega L = 2\pi(500\ \text{Hz})(50.0 \times 10^{-3}\ \text{H}) = 157.1\ \Omega$$

$$X_C = \frac{1}{\omega C} = \frac{1}{2\pi(500\ \text{Hz})(10.0 \times 10^{-6}\ \text{F})} = 31.8\ \Omega$$

Then compute the circuit's impedance:

$$Z = \sqrt{(24.0\ \Omega)^2 + (125.3\ \Omega)^2} = 127.5\ \Omega$$

Next, knowing the source voltage, use the ac version of Ohm's Law to find the current:

$$I = \frac{V}{Z} = \frac{100\ \text{V}}{127.5\ \Omega} = \boxed{784\ \text{mA}}$$

(b) Across each element, a voltmeter will read

$$V_R = IR = (784\ \text{mA})(24.0\ \Omega) = \boxed{18.8\ \text{V}}$$

$$V_L = IX_L = (784\ \text{mA})(157.1\ \Omega) = \boxed{123\ \text{V}}$$

$$V_C = IX_C = (784\ \text{mA})(31.8\ \Omega) = \boxed{24.9\ \text{V}}$$

(c) To determine the power, we must first compute the power factor:

$$\cos \theta = \frac{R}{Z} = \frac{24.0\ \Omega}{127.5\ \Omega} = 0.188$$

$$P_{av} = IV \cos \theta = (0.784\ \text{A})(100\ \text{V})(0.188) = \boxed{14.7\ \text{W}}$$

Quick Check: From the fact that $\cos \theta = 0.188$, $\theta = 79.2°$ and, using Eq. (21.29), $\tan \theta = (123\ \text{V} - 24.9\ \text{V})/(18.8\ \text{V}) = 5.22$ and $\theta = 79.2°$.

Series Resonance

An ac series circuit can function in a remarkable way at a specific frequency, $\omega_0 = 2\pi f_0$, known as its **resonant frequency**. The phenomenon is the electrical equivalent of the mechanical concept of resonance considered earlier (p. 354). Figure 21.22 depicts the frequency-dependent behavior of R, X_L, and X_C; what we see is that, at the resonant frequency, the capacitive and inductive reactances are equal. Inasmuch as

$$Z = \sqrt{R^2 + (X_L - X_C)^2} \qquad (21.34)$$

at resonance $\qquad\qquad\qquad Z = R$

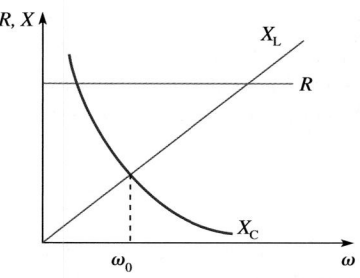

Figure 21.22 At the resonant frequency (ω_0), the inductive and capacitive reactances are equal.

The condition for resonance exists when

$$\omega L = \frac{1}{\omega C}$$

and that occurs at $\omega_0 = 2\pi f_0$, whereupon

[resonance]
$$f_0 = \frac{1}{2\pi\sqrt{LC}} \qquad (21.35)$$

At resonance $\theta = 0$, $Z = R$, and $P_{av} = IV$. Because the impedance is then a minimum, for a given voltage V, the current I is a maximum: $V = IR$. In other words, suppose that a wide range of frequencies is fed into an R-L-C circuit. If we adjust or *tune* L or C or both so that Eq. (21.35) holds at a particular frequency, say 1 kHz, then the circuit will have a peak current at 1 kHz—all other currents at other frequencies will be considerably less. Figure 21.23 indicates how the current curves are affected by the resistance of the circuit. {It's advisable now to open the **CD**, go to **INTERACTIVE EXPLORATIONS**, and click on **THE R-L-C SERIES CIRCUIT**. Study this simulation until you understand everything it deals with.}

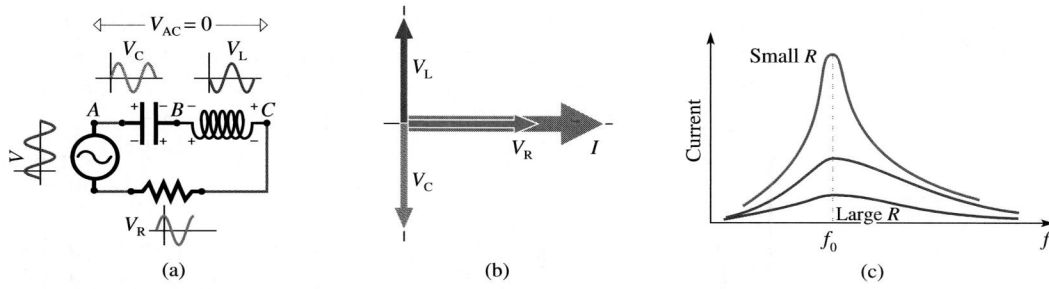

Figure 21.23 (a) A resonant circuit. Because the inductive and capacitive reactances are equal, the effective voltages across the inductor and capacitor are equal. But those voltages are 180° out-of-phase, and so even as the instantaneous voltages rise and fall sinusoidally, they always cancel. In other words, there's a voltage across the capacitor ($V_{AB} = V_C$) and across the inductor ($V_{BC} = V_L$), and yet the voltage across both (V_{AC}) is zero. Notice how the phases of the little voltage sine curves are shifted. (b) Because of those phase shifts, the phasors for the voltage across the inductor (pointing up), and the capacitor (pointing down) cancel each other. (c) At resonance, the current in a series R-L-C circuit reaches a maximum. The smaller the resistance, the sharper the peak, and the narrower the range of frequencies selected by the circuit.

Example 21.6 [II] A series circuit contains a 50.0-Ω resistor adjacent to a 200-mH inductor attached to a 0.050-μF capacitor, all connected across an ac generator with a terminal sinusoidal voltage of 150 V effective. (a) What is the resonant frequency? (b) What voltages will be measured by voltmeters across each element at resonance? (c) What is the voltage across the series combination of the inductor and capacitor?

Solution This is a series *resonant* circuit, and we follow the logic outlined in the Study Guide on p. 756. (1) TRANSLATION—A known resistor, capacitor, and inductor are in series across a source with a known frequency and effective voltage. Determine (a) the resonant frequency, (b) the voltage across each element, and (c) the net voltage across the inductor and capacitor. (2) GIVEN: $R = 50.0$ Ω, $C = 0.050$ μF, $L = 200$ mH, and $V = 150$ V. FIND: f_0, V_L, V_R, and V_C. (3) PROBLEM

TYPE—ac circuit/R-L-C series/resonance. (4) PROCEDURE—(a) We have an expression for the resonant frequency: $f_0 = 1/2\pi\sqrt{LC}$. (b) To get the voltages across each element via Ohm's Law we need the current and the reactances. (5) CALCULATION—First the resonant frequency:

$$f_0 = \frac{1}{2\pi\sqrt{LC}} = \boxed{1.59 \text{ kHz}}$$

and $\omega_0 = 10.0$ krad/s. To find the voltages, we need the reactances at ω_0, and the current:

$$X_L = \omega_0 L = (10.0 \text{ krad/s})(0.200 \text{ H}) = 2.00 \text{ k}\Omega$$

$$X_C = \frac{1}{\omega_0 C} = \frac{1}{(10.0 \text{ krad/s})(0.050 \times 10^{-6} \text{ F})} = 2.00 \text{ k}\Omega$$

continued

and at resonance, $Z = R$. Consequently

$$I = V/R = (150 \text{ V})/(50.0 \ \Omega) = 3.00 \text{ A}$$

and therefore

$$V_R = IR = (3.00 \text{ A})(50.0 \ \Omega) = \boxed{150 \text{ V}}$$

$$V_L = IX_L = (3.00 \text{ A})(2.00 \text{ k}\Omega) = \boxed{6.00 \text{ kV}}$$

$$V_C = IX_C = (3.00 \text{ A})(2.00 \text{ k}\Omega) = \boxed{6.00 \text{ kV}}$$

Notice that although there is 6.00 kV across the inductor and 6.00 kV across the capacitor, the corresponding instantaneous voltages are 180° out-of-phase—*the voltage across the combination is zero!*

Quick Check: The fact that the reactances are equal at resonance is a good indication that we haven't messed up the numbers.

AM Radio

A radio broadcasting system essentially converts sound (20 Hz–20 kHz) into electromagnetic waves (p. 784) that travel a lot faster and farther. We could transmit such waves at the same frequencies as the information, the sound, but that would require an antenna of tremendous size and is totally impractical. The solution is to use a convenient high-frequency radiowave (the electromagnetic **carrier wave**) and impress the information on it. In AM, or *amplitude modulation*, the carrier's amplitude is made to vary with the information. The high-frequency signal of Fig. 21.24 carries all the music or talk in the form of relatively low-frequency changes in height, or strength, of the signal. The valuable information is the envelope of the signal and the carrier itself will ultimately be discarded by the receiver.

Figure 21.25 is a rudimentary AM radio receiver. The antenna picks up a tumult of signals composed of the transmissions from all of the stations reaching it. That hodgepodge is available to the tuning circuit by way of the coupling between L_1 and L_2. When you turn the tuning knob on a radio, you are adjusting the capacitor C_1 to resonate the input circuit at the frequency of, say, WCBS. Only that frequency and its immediate surroundings will then be passed to the next stage, the crystal diode (p. 767). This is a one-way gate that chops off the negative portion of the signal. The resulting positive voltage is applied to a filter formed by C_2 and R. When the diode drops the current to zero, the capacitor discharges through R, but

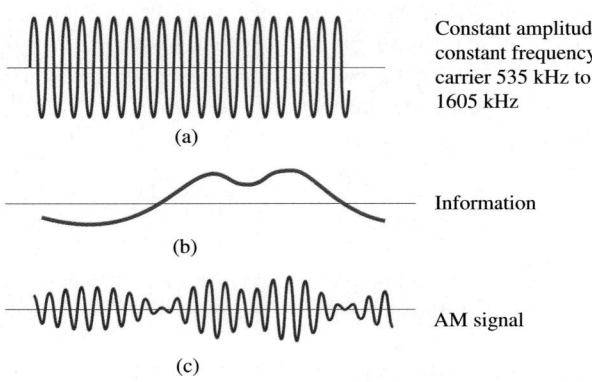

Constant amplitude, constant frequency carrier 535 kHz to 1605 kHz

Information

AM signal

Figure 21.24 An amplitude modulated (AM) signal. (a) A constant amplitude, constant frequency carrier is made to carry information, as in (b), by modulating its amplitude, as in (c).

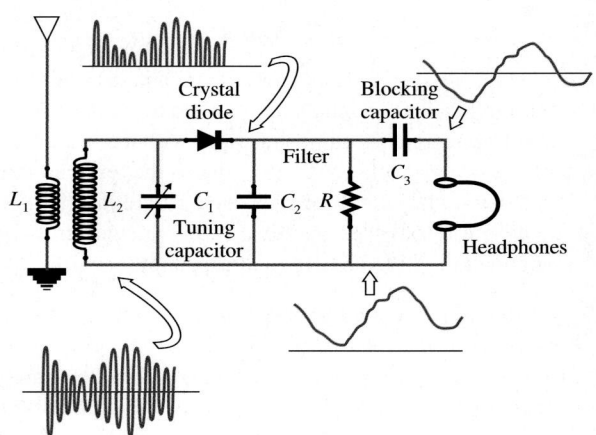

Figure 21.25 An early radio receiver showing a signal at various stages of being processed. The two coupled coils L_1 and L_2 constitute a transformer, a device we'll deal with next.

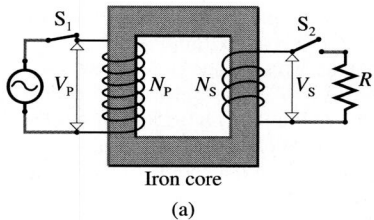

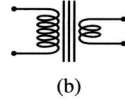

Figure 21.26 An iron-core transformer. (a) The input side is the primary; the output side is the secondary. (b) The symbol for an iron-core transformer.

the time constant of this *RC* circuit is large compared to the period of the carrier; the discharge is slow, and voltage across the filter hardly decreases before the diode passes current again and C_2 recharges. The result is a slightly wiggly, low-frequency voltage across *R* that otherwise corresponds in shape to the envelope of the carrier. The original information (Fig. 21.24*b*) oscillated above and below the axis, whereas this signal is only positive—it's been displaced by a constant positive voltage. That dc component is eliminated by a blocking capacitor (C_3). The resulting slowly oscillating voltage corresponds almost exactly to the oscillating sound wave. When it's fed into headphones, out comes a sound wave identical to the one that was heard in the studio at WCBS.

21.5 The Transformer

In 1831, Faraday discovered the principle of electromagnetic induction that underlies the transformer, but it took about 50 years before the latter became a practical instrument. In broad terms, the **transformer** *is an induction device used to convert energy in the form of a large time-varying current at a low voltage into nearly the same amount of energy in the form of a small time-varying current at a high voltage (or vice versa)*.

Imagine two coils wrapped around an iron core, as shown in Fig. 21.26. An ac power source is applied to the *primary* winding, and the switch is initially left open in the circuit of the *secondary* winding. The time-varying primary current I_p creates a time-varying magnetic flux that circulates through the secondary coil. Because of its high permeability, the iron core enhances the flux generated by the primary current by a factor of about 10 000. It provides an easy path for the field, which is very nearly totally constrained to pass within its volume, effectively coupling the two coils.

It might seem that closing the switch S_1 would essentially short-circuit the source, causing a tremendous input current; after all, the primary winding has a low resistance. Indeed, that's exactly what would happen if we put a constant dc source across the primary. With an ac source, the sizable self-inductance *L*, due largely to the core, will result in a substantial induced back-emf via Eq. (20.10). This back-emf will oppose the applied voltage, keeping the primary current I_p very small. Since the load on the source is essentially purely inductive, almost no energy is drawn from the source and, with the secondary open-circuited, no energy is transferred.

Ideally, the same time-varying flux passes through all the turns of both coils and so the induced emf on any one turn is the same as that on any other. Hence, the total induced emf on the primary is proportional to the total number of its turns (N_p) just as the total induced emf on the secondary is proportional to its number of turns (N_s). Assuming the windings have negligible resistance and therefore sustain no *IR* voltage drops, the induced emf across the primary will be numerically equal to the terminal voltage across it (V_p). Similarly, the emf induced across the secondary will equal its terminal voltage (V_s). It follows, then, that the ratio of the effective voltages will equal the ratio of the numbers of turns:

$$\frac{V_p}{V_s} = \frac{N_p}{N_s}$$

(21.36)

A speaker and output transformer from the AM radio shown on p. 537. The transformer allows the output amplifier of the radio to efficiently supply power to the speaker.

The coil with the higher number of turns corresponds to the higher voltage, which could be either the primary or the secondary, depending on how we wire it. *The ratio of the number of turns in the higher voltage winding to the number of turns in the lower voltage winding is called the* **turn ratio**. For example, a transformer with a 10:1 turn ratio has 10 times the number of turns on one coil as on the other.

For any given primary voltage, we need only select an appropriate turn ratio in order to produce any desired secondary voltage. When $V_p > V_s$, it's called a **step-down transformer**; and when $V_p < V_s$, it's a **step-up transformer**. The spark coil in an automobile is a commonplace example of the step-up variety. The 12-V dc from the battery is chopped into pulses by a switching device. The "coil" is a transformer that boosts this pulsating

A pole-mounted transformer used to drop the voltage from the supply cables. Three leads provide two separate 110-V lines for domestic service. One of the three wires from the transformer is common to both lines. It attaches to a horizontal, uninsulated support cable, around which the other two wires are wrapped. These two wires go off on the right and left to individual houses.

12-V input up to 20 kV, which activates the spark plugs that ignite the gasoline in each of the cylinders.

When a negligible amount of energy is lost via I^2R, the power-in equals the power-out:

$$I_p V_p = I_s V_s \tag{21.37}$$

When the voltage goes up the current goes down, and vice versa; the transformer must conserve energy. {For more on the topic click on **TRANSFORMERS** under **FURTHER DISCUSSIONS** on the CD.}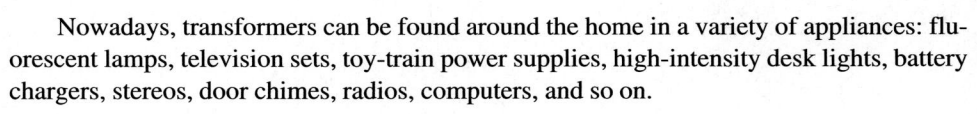

Example 21.7 **[I]** The ac adapter for a pocket calculator contains a transformer, two diodes, and a capacitor (p. 769). The transformer takes the 120-V ac at the wall down to 15.0-V ac, which the rest of the circuit converts to dc. If the transformer's secondary consists of exactly 50 turns, how many turns must the primary have? What is the turn ratio?

Solution This is a transformer problem, and that should immediately call to mind the equation for both the turn ratio and the power. (1) TRANSLATION—A transformer has a specified number of turns on its secondary and converts one known voltage into another; determine (a) the number of turns on the primary and (b) the turn ratio. (2) GIVEN: $V_p = 120$ V, $V_s = 15.0$ V, and $N_s = 50$. FIND: N_p and N_p/N_s. (3) PROBLEM TYPE—ac circuit/transformer. (4) PROCEDURE—We have the primary and secondary voltages as well as the number of turns on the secondary. We need the number of turns on the primary, and that suggests $V_p/V_s = N_p/N_s$. (5) CALCULATION—First the number of primary turns:

$$N_p = \frac{N_s V_p}{V_s} = \frac{50(120 \text{ V})}{15.0 \text{ V}} = \boxed{400 \text{ turns}}$$

This is a step-down transformer, and so the turn ratio is

$$\frac{N_p}{N_s} = \frac{400}{50} = \boxed{8}$$

Quick Check: The transformer is of the step-down variety, and we expect $N_p > N_s$. Further, the ratio of the voltages (120 V)/(15 V) does equal the turn ratio.

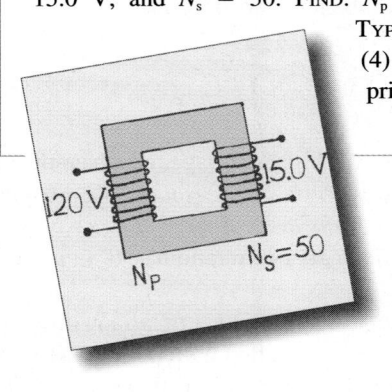

Nowadays, transformers can be found around the home in a variety of appliances: fluorescent lamps, television sets, toy-train power supplies, high-intensity desk lights, battery chargers, stereos, door chimes, radios, computers, and so on.

21.6 Domestic Circuits and Hazards

The ordinary consumer is provided—via three wires down from the pole (Fig. 21.27)—with a pair of single-phase 120-V lines. Two of these wires are "hot," and the third is neutral, grounded back at the transformer. Inside the building, the neutral is usually attached to

The electrical lines enter a house through an energy meter, which records the number of kilowatt hours delivered.

Figure 21.27 Domestic ac service.

Table 21.1
Electical Power Consumption

Appliance	Typical wattage (W)
Range	12 000–16 000
Clothes drier	5000–8000
Oven	4000–8000
Hair dryer	1000–1300
Dishwasher	1200–1500
Furnace blower	1200
Iron	1100
Toaster	1100
Waste disposal	1000
Oil burner	800
Refrigerator (big, double-door)	800
Vacuum cleaner	600
Washing machine	550
Blender	400
Fan	200
TV	100
Typewriter	90
Humidifier	40
Clock	4
Motors:	
1 hp	1500
1/2 hp	1000
1/4 hp	700
1/6 hp	450

the entering water pipe and all electrical metal conduits, receptacle boxes, wall brackets, and so on, are connected to it. The two hot leads (one red and one black), along with the neutral, go through a meter to a breaker or fuse box. The instantaneous voltages between red and ground (120 V) and black and ground (120 V) are 180° out-of-phase and can be combined to yield 240 V effective (between red and black), for use with heavy-duty air conditioners, water heaters, etc. Emerging from the breaker box are several parallel pairs of leads (black and white, hot and neutral, respectively) that carry power to the rest of the house. This arrangement is what's known as 240-V ac service, and it's fairly standard in North America. When you plug your TV into the wall, you are attaching one of the leads in the power cord to a black (hot) wire in the outlet and the other lead to a white (neutral) wire, across the two of which there is 120-V-effective, single-phase ac.

A typical home is wired internally with several separate lines leading to the main, each attached in series with a fuse or circuit breaker of its own. The idea is to limit the amount of current that can be drawn by any single line—too much current and the wires in the wall can get dangerously hot (via I^2R). Household wiring is typically rated at 15 or 20 amps effective and fused accordingly. If a line contains too large a fuse and is made to carry too large a current by plugging in too many appliances, there will be an appreciable voltage drop in the wiring itself. The temperature of the line will rise, and its resistance will therefore also rise. The terminal voltages at the outlets may then be appreciably less than 120 V, and lights will dim, TV pictures will shrink, and you will be risking burning the place down. If a line is rated at 20 A, it will provide (120 V)(20 A) = 2400 W, which is just enough power to run a dishwasher, simultaneously make a slice of toast, and allow you to watch it all happen under a 100-W light bulb (Table 21.1).

The newest three-wire system includes an additional ground lead connected to the metal housing of the appliance. The whole steel body of a washing machine or refrigerator is attached to ground via the round prong on the three-prong plug (Fig. 21.28). Experience has shown that occasionally the hot wire in an electric device can be exposed and the whole appliance becomes "hot." For example, suppose you move your washing machine to do some cleaning and then while pushing it back against the wall, a sharp corner of the cabinet cuts into the power cord and touches the hot lead. In the old two-wire system, the entire machine would sit there at 120 V waiting for someone (standing bare foot on a damp basement floor) to touch it and get a nasty shock. In the new system, as soon as the hot lead makes contact with the grounded case, the current is shorted out and the fuse blows.

Armored cable-2 leads

Plastic-sheathed cable-3 leads

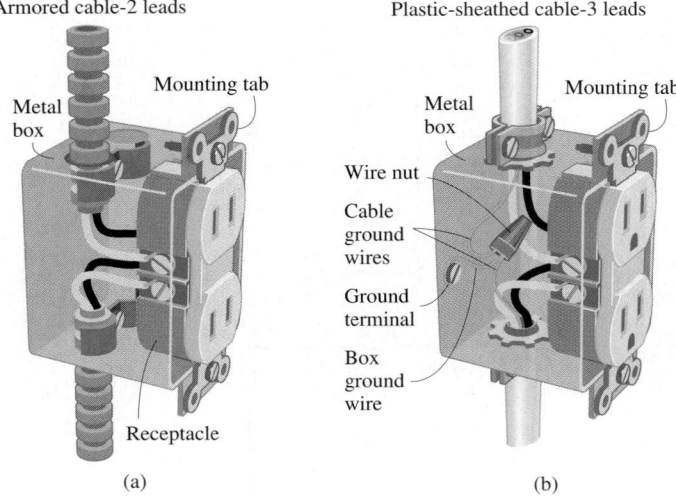

(a)

(b)

Figure 21.28 (a) An old-style two-prong outlet. (b) A modern three-prong outlet. The round third prong opening is internally connected to the metal mounting tab.

Usually, we can feel a 1-mA current, and up to about 5 mA, however unpleasant, is as a rule harmless. When a substantial amount of charge flows (>10 mA) through muscles, it causes wrenching spasms. If those muscles are in the hand, the effect may be little more than tiny burns and a deep ache. At above 15 mA or so, one loses voluntary muscle control—rather awkward if you're holding on to a "hot" wire and can't let go. Up to roughly 50 mA, currents will cause considerable pain, but probably no massive malfunction of any crucial body process. More sizable currents across the torso can paralyze the respiratory system and disrupt the steady pumping of the heart. A current of approximately 100 mA sustained for a second or more through the heart will cause it to go into a lethal condition of ventricular fibrillation (irregular beating). The idea is to avoid letting current pass through the body in general and certainly to keep it away from the heart. Electricians working with high voltages (120 V can be lethal, 240 V demands great care) will often position one arm well away from the circuit; they risk blasting the fingers of one hand but avoid creating a hand-to-hand pathway across the chest.

The resistance of the body is determined to a large extent by the contact resistance with the outer layer of the skin. The wet human stuff within each of us is rich in ions and is a fairly good conductor. Therefore, depending on the condition of the skin, the area of contact, and the intimacy of that contact, the resistance (hand-to-hand or head-to-foot) may vary from perhaps 100 kΩ to over 1.5 MΩ dry, and possibly 100 times less, wet. If we assume a body resistance of 100 kΩ and an outlet voltage of $V = 120$ V $= IR$, it follows that $I = 1.2$ mA—far from problematic. If you are dripping wet and $R = 1$ kΩ, $I = 120$ mA and you are in big trouble. Be exceedingly careful with 120-V ac and don't even consider puttering around with 240-V ac. One is not likely to be able to let go of a high-voltage line, and the resulting burns will quickly lower the skin resistance, along with the chance of survival.

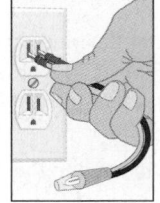
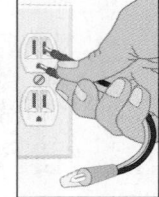
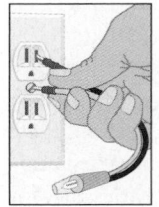
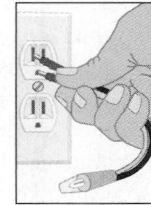

EXPLORING PHYSICS ON YOUR OWN

AC Home Wiring: The drawings on the left show a little neon discharge lamp that glows brightly when 110 V is across its terminals (with no concern for polarity). It's called a line voltage checker and sells for less than three dollars at any hardware store. It's a great gadget and every home should have one. Explain what should be observed in each part of the diagram if all is functioning normally. Try it with your tester. The shorter slot (here on the right) is supposed to be wired "hot" in modern receptacles. But in old houses that's not always the case—check a few outlets. What would happen if one lead of the tester was inserted in the right-hand slot and the other touched the faucet on the kitchen sink? Needless to say, be careful not to let any part of your body touch an exposed "hot" conductor.

Today almost every complex activity we engage in is supported by electronic devices. Here technicians are monitoring an angioplasty at a Minneapolis hospital.

Electronics   (Optional)

Over the past three decades, electronics has become dominated by solid-state components, principally semiconductor diodes and transistors. Each of us is likely to use dozens of integrated circuits (IC) in the course of an ordinary day; they are in computers, calculators, telephones, answering machines, cameras, cars, microwave ovens, and so on. An IC microprocessor "chip" (perhaps $\frac{1}{4}$ in. by $\frac{1}{4}$ in.) in the realm of what's called "very large-scale integration" might contain 450 000 minute transistors, along with a multitude of diodes, resistors, and capacitors. A digital watch requires about 5000 transistors; a little pocket calculator utilizes roughly 20 000 transistors; a computer, whose equivalent might once have filled a room with vacuum tubes, now contains a tiny 3 or 4 million-transistor chip and comfortably fits on a person's lap. The Pentium 2 (1997) has 7.5 million transistors, and new techniques promise microprocessors with about 1 billion transistors.

21.7 Semiconductors

An isolated atom can exist in any one of a number of distinct energy levels; its electron cloud can only have certain configurations. The atom typically has many electrons (for example, silicon has 14) distributed in closed shells about the nucleus. Only the outer group—the *valence electrons*—are involved in the behavior we are concerned with here. When these outer electrons (in silicon, there are 4) are in their lowest energy configuration, the atom as a whole is in its lowest energy state (the one it usually occupies), called the *ground state*. When two atoms are near each other, their interaction causes a slight shift in the allowed levels. When there is a tremendous number of interacting atoms, as in a solid, the shifted levels are so numerous and so close together that they form energy bands. The ground state is called the **valence band**, and the electrons therein are usually held to their respective atoms.

   Given enough energy, one of the outer electrons can be ripped from its atom, to move relatively freely through the lattice of atoms. That electron is then said to be in the more energetic **conduction band**. Figure 21.29 shows how, in an insulator, the almost totally empty conduction band is separated from the occupied valence band by a sizable gap (10 eV or greater) from which electrons are prohibited. Very few electrons can pick up enough energy thermally to be propelled across the gap. Thus, very few move freely in the insulator—hence, the high resistivity. By comparison, in a conductor the two bands overlap, and no clear distinction between the valence and conduction bands exists. Valence electrons are

A 4-megabit integrated circuit (IC) memory chip.

Figure 21.29 A representation of the band structure of solids. (a) The valence band in an insulator is occupied with charge carriers (electrons), but it's separated from the conduction band by a gap, so it is a poor conductor. (b) In the conductor, the two bands overlap and electrons flow easily. (c) A semiconductor has a gap, but it's small.

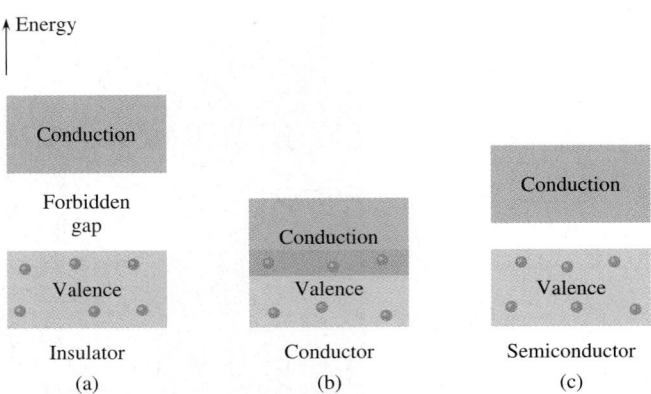

free to wander from atom to atom, and the resistance is low. An *elemental* (or *intrinsic*) semiconductor such as silicon or germanium has a small forbidden energy gap (in silicon it's only about 1.1 eV). Still, at room temperature (300 K), the average KE of the electrons ($k_B T \approx 4.1 \times 10^{-21}$ J) is about 0.026 eV, so very few high-energy electrons can jump the gap—the conduction band will be nearly empty.

Doping

Devices such as transistors and diodes are fabricated using *impurity* semiconductors prepared by adding minute quantities of foreign atoms (just a few parts per million) to an intrinsic semiconductor. The process is known as **doping**, and it affects the availability of mobile-charge carriers, producing two distinct kinds of systems. Figure 21.30 shows how the electrons of a silicon single crystal are shared (in so-called covalent bonds) between the four nearest neighbor atoms. When such a crystal is doped with a five-valence-electron atom (Fig. 21.31) like arsenic, the fifth electron is not locked in place—it does not fit and can move around freely within the crystal. Such an electron resides in an energy level just below the conduction band, into which it can easily be made to jump. Because these mobile-charge carriers are negative, the system is referred to as an **n-type** semiconductor.

If the silicon is doped with a three-valence-electron atom like gallium (Fig. 21.32), there will be a deficiency of one electron; in effect, there will be a **hole** in the negative distribution of electrons. An outer electron from a nearby silicon atom can drop out of its cloud and fill the hole, but this will leave a new hole in the place where the electron originally was. The hole moves about like a bubble in a cup of water—it is the absence of negativity and so behaves as if it were a positive mobile-charge carrier. Because the carriers are positive, this system is known as a **p-type** semiconductor. The presence of the gallium gives rise to a number of empty levels just above the valence band. Electrons can jump into these levels, leaving behind holes in the valence band that can then move around in response to an applied electric field, thus constituting a current.

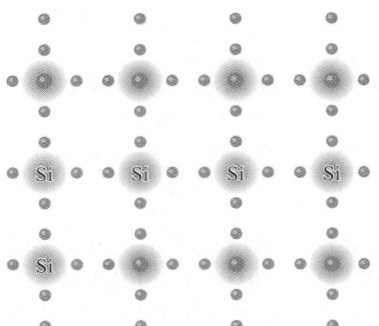

Figure 21.30 Silicon atoms sharing electrons in a crystal.

Figure 21.31 Silicon doped with arsenic. The impurity atoms result in filled levels just below the conduction band. Electrons from these levels can reach the conduction band relatively easily. These are the mobile-charge carriers.

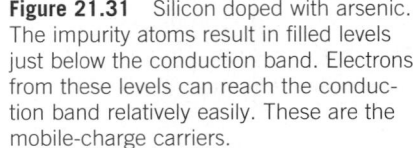
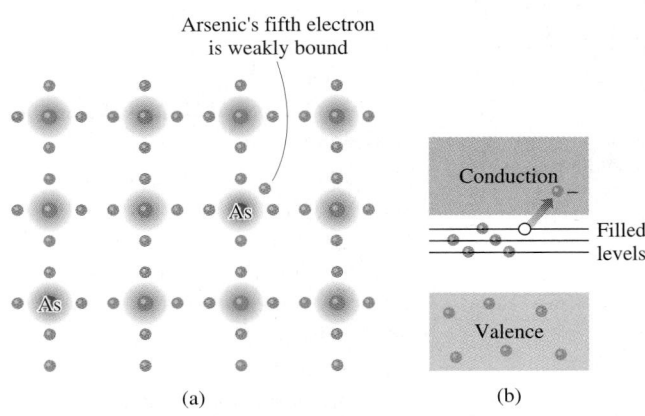

Figure 21.32 Silicon doped with gallium. The impurity atoms result in empty levels just above the valence band. Electrons from the valence band can reach these levels relatively easily. The holes left behind in the valence band are the mobile-charge carriers.

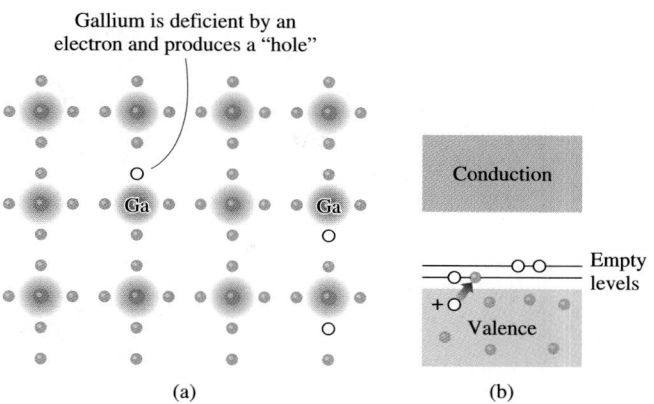

(a)   (b)

The *pn*-Junction and Diodes

One of the most practical arrangements of semiconductors is the ***pn*-junction**, the interface formed by joining a *p*-type to an *n*-type semiconductor. In practice, a polished slice of single crystalline *p*-type silicon is heated to around 1000°C and exposed to a vapor of arsenic or phosphorus, which diffuses into the surface. The uppermost layer is transformed into *n*-type silicon, which is then coated with a protective insulating layer of silicon dioxide. Because the single crystal structure is continuous across the junction, electrons from the *n*-type region can diffuse across to the *p*-type region, where they fill an equivalent number of holes. The *n*-type region is left positive (because of a deficiency of electrons) and the *p*-type region is negative (because of an excess of electrons). The resulting internal potential difference cuts off any further transfer of charge, leaving a central region depleted of carriers. This **depletion layer** is essentially an insulator, and the junction then resembles a charged capacitor (Fig. 21.33).

To see how such a *pn*-junction can function as a **diode** (that is, as a one-way gate passing current in one direction and blocking it in the other), examine Fig. 21.34. Here, an external potential difference is applied across the junction diode such that the positive terminal is attached to the *p*-side and the negative to the *n*-side. This opposes the internal potential difference, and the diode is **forward biased**. The positive terminal repels holes into the junction, where they are met by electrons repelled by the negative terminal—at first a tiny current traverses the diode. The depletion layer shrinks as the voltage is raised to about 650 mV (or about 300 mV for germanium), at which point the layer vanishes and the amount of current increases abruptly as carriers flow freely across the diode.

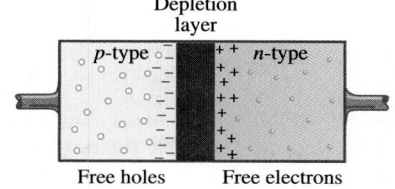

Figure 21.33 Schematic representation of a junction diode.

Figure 21.34 Current versus voltage for a junction diode. Observe the different voltage scales. A forward-biased junction diode conducts. A reverse-biased diode does not conduct. The symbol for the diode is an arrow in the direction it conducts a conventional current.

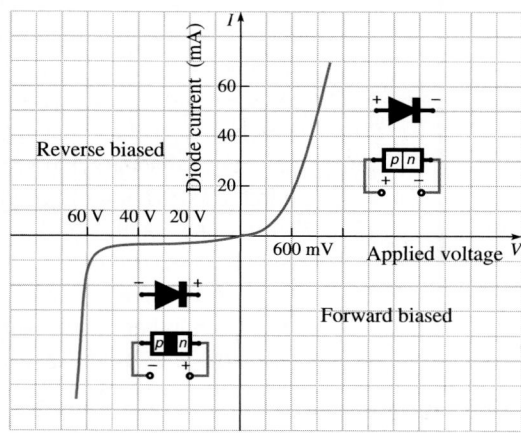

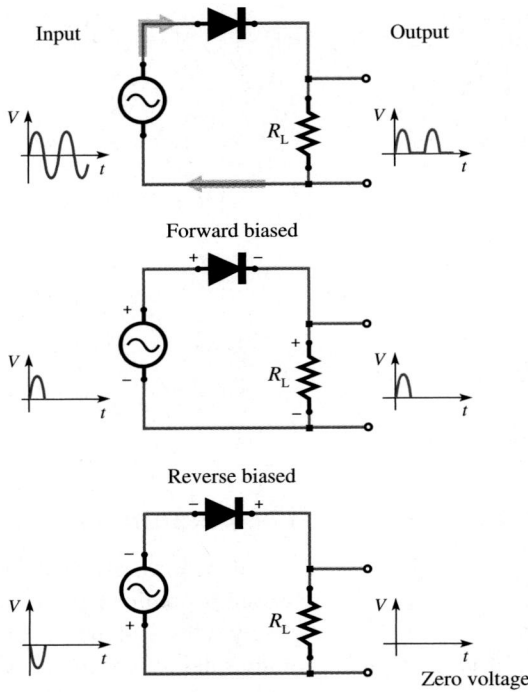

Figure 21.35 Rectification using a diode. Only when it's forward biased does the diode conduct. Think of it as an active switch that opens and closes depending on the polarity of the signal.

By contrast, when the diode is *reverse biased*, electrons are attracted away from the junction toward the positive terminal just as holes are attracted away toward the negative terminal. The depletion layer broadens, and only an exceedingly small current can traverse the diode.

Although we can take the opportunity to study only the diode, there are a number of other important applications of the *pn*-junction. These include the photovoltaic (solar) cell, the light-emitting diode (LEDs are the bright little red lights in VCR, camera, and stereo displays), and the *diode laser*.

Rectification

The process of converting ac to dc is called **rectification**, and the junction diode is a popular rectifier. The vast majority of electronic devices require some dc, and if the power source is 60 Hz ac, a rectifier will be needed. As arranged in Fig. 21.35, just the positive portion of the signal is passed by the diode, which will allow only a clockwise current. The output voltage across the *load resistor* R_L is dc, but hardly constant. To smooth out the voltage, we add a capacitor to make an *RC*-filter (Fig. 21.36). The voltage across C rises positively, and it becomes charged. As the voltage starts to drop from its maximum, the capacitor discharges into the load resistor (it can't send a current backward through the diode). With a long-time constant (*RC*), that discharge will be so slow that the next positive voltage peak will interrupt it and recharge the capacitor. The output, which still isn't flat, is said to have a *ripple* and, for obvious reasons, the circuit is known as a **half-wave rectifier**. A **full-wave rectifier** makes use of both the positive and negative peaks, producing a much smoother output with less ripple. One version, which you're likely to find if you cut open the ac adapter for your calculator or portable CD player, uses a center-tapped transformer and two diodes (Fig. 21.37).

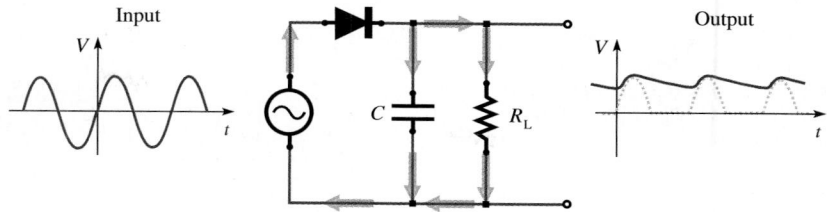

Figure 21.36 Rectification and filtering. The diode passes on only the positive peaks, and the *RC*-filter smooths them out.

Transistors

The transistor is effectively two diodes, back-to-back. Usually made of silicon, it is formed in a three-tiered sandwich of either *pnp* or *npn* doped regions. These *epitaxial* layers (from the Greek *epi* meaning *upon* and *taxis* meaning *arranged*) are grown in place so as to preserve the single-crystal structure (Fig. 21.38). One layer, called the **emitter** (labeled *E*), is highly doped. It therefore has a low resistance and is rich in mobile-charge carriers. The middle, very thin layer is the lightly doped *base* (labeled *B*). The remaining layer is the **collector** (labeled *C*), and it, too, is lightly doped. Reasonably enough, **the emitter emits mobile-charge carriers to the collector**. In a *pnp* transistor, the principal carriers are positive holes, and the arrow in the symbolic representation of the transistor is in the direction of traditional current. In an *npn* transistor, the principal carriers are negative electrons, and a positive current is imagined to pass from collector to emitter while the actual electron flow goes from emitter to collector. In either case, within the transistor symbol the emitter-base arrow always points from a *p*-type to an *n*-type region.

The transistor is usually placed in series with a dc source able to provide an appreciable current. It then serves as an electrical control valve, opening fully or partially, and

The full-wave rectifier in an ac-to-dc adapter. You can see the transformer (attached to the plug prongs via two yellow wires), two little (black) diodes, and a (blue) filter capacitor. This is one of those black-box plugs that supplies power to calculators, rechargeable flashlights, portable tape recorders, CD players, etc.

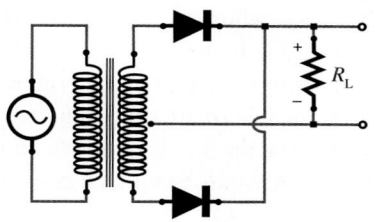

Figure 21.37 A full-wave rectifier. Note that the direction of the current through the load resistor is the same regardless of the polarity of the input signal. The output is unfiltered, but a capacitor across the output would smooth the signal appreciably.

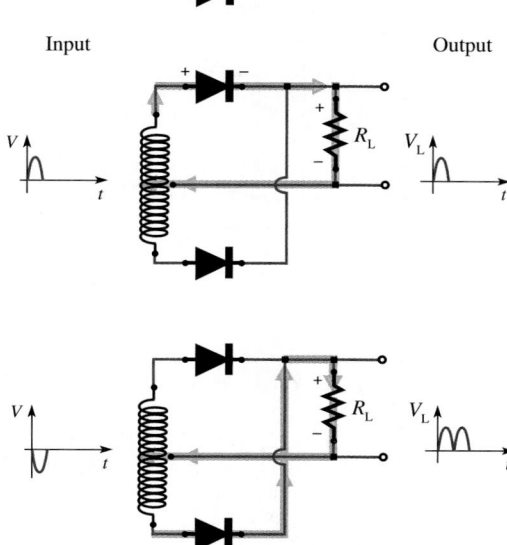

Figure 21.38 (a) The transistor is formed of doped layers of *p*- and *n*-type semiconductors. (b) The mobile-charge carriers in the *pnp* device are holes. (c) Electrons are the mobile-charge carriers in the *npn* transistor.

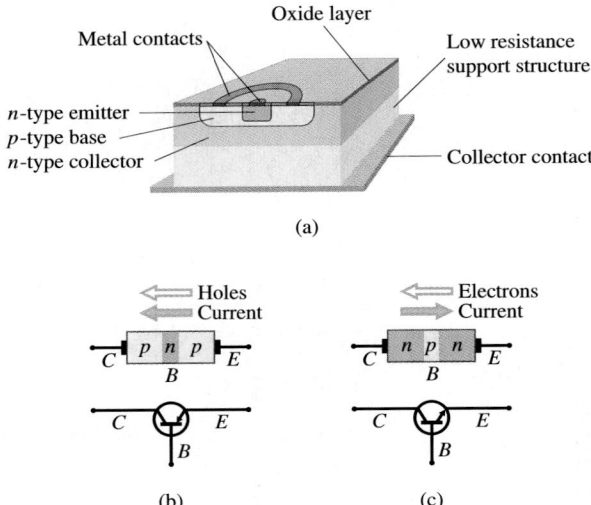

(a)

(b)   (c)

An integrated circuit.

allowing a proportionately large current to pass, or closing and cutting off the current altogether. This it does, in effect, by varying the emitter-collector resistance (indeed, the name *transistor* comes from combining the words *transfer* and *resistor*). The electrical control of the valve arises by way of a tiny current through the base—the base or signal current. Variations in this very small input swing the valve open proportionately and thus control the very much larger emitter-collector current. In other words, the tiny input signal current is *amplified* in the form of an identically shaped, but much larger, output current.

Figure 21.39 shows both a *pnp* and an *npn* transistor, each with a battery supplying a voltage between collector and emitter. Consider either transistor—say, the *npn*. It can be imagined as two *np* diodes back-to-back (*np-pn*)—two junctions formed, one on each surface of the base and separated by the layer's small ($\approx$10-μm) thickness. Recall that when the *p*-region is positive with respect to the *n*-region, the junction is forward biased, as in Fig. 21.34. Thus, while the *E-B* junction is forward biased by the battery, the *C-B* junction is reverse biased (and the same is true for the *pnp* transistor). Immediately after the switch is closed, if the forward bias exceeds about 650 mV for silicon (and $\approx$300 mV for germanium), electrons will easily flow from the emitter into the base. The majority of these will continue on, cross the thin base, move into the collector and out into the circuit, driven by the battery. But this current will not continue long. Holes in the base will soon be depleted,

Figure 21.39 (a) A *pnp* transistor and (b) an *npn* transistor, each with a voltage between collector and emitter. (c) The device functions as if it were two *np* diodes oppositely biased. (d) In the *npn* transistor, electrons flow from the emitter (*n*) to the collector. In the *pnp* transistor, holes flow from the emitter (*p*) to the collector.

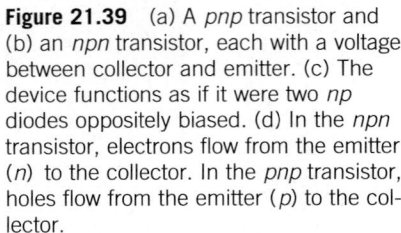

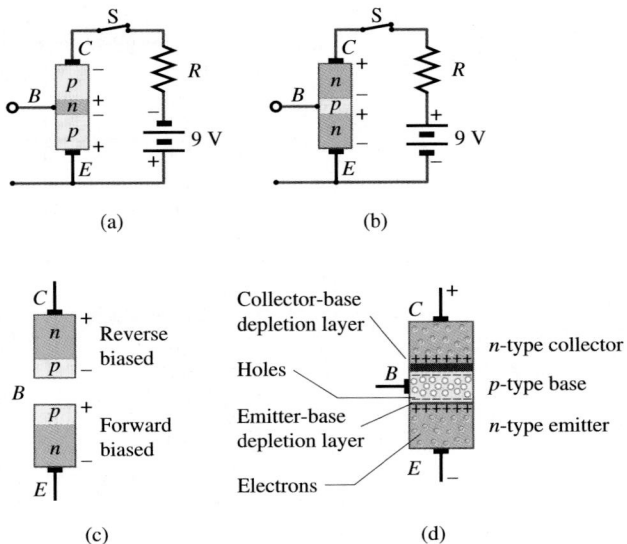

(a)   (b)

(c)   (d)

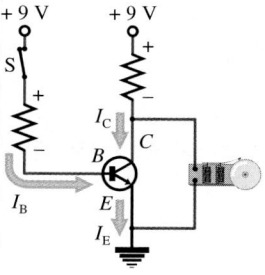

Figure 21.40 An *npn* transistor used as a switch, perhaps in a burglar alarm. Opening the switch *S* rings the bell. Here $I_E = I_B + I_C$. As long as current enters the base, the collector-emitter current is uninhibited. Interrupting I_B will cause the transistor to essentially open the circuit. With the path to ground blocked, current I_C goes through the bell instead, and it rings.

and that will change things drastically. The holes migrate across the *E-B* junction toward the negative terminal of the battery, and they can also be lost by recombining with electrons flowing toward the base from the emitter. The total effect is to cause the appearance of a net negative charge on the base that, opposing the electron flow from the emitter, will soon stop that current almost entirely (Fig. 21.39*d*). A relatively small charge inhibits the progress of a good deal of current that could be provided by the battery. The transistor is then like an opened switch with a nearly infinite resistance, which is identically the case with the *pnp* transistor, except there the principal carriers flowing from *E* to *C* are holes.

This blocked situation can be reduced or eliminated by injecting holes into the base of the *npn* transistor or, equivalently, drawing electrons out of it. By sending a small positive current into the base, a proportionately large positive current will pass from *C* to *E* and around the external circuit (Fig. 21.40). Again, the same thing will happen with a *pnp* transistor, provided electrons are injected into (or holes removed from) its base. *The presence of a small base current controls the flow of charge from emitter to collector.*

Amplifiers

Most transducers, used either in the home or laboratory, rarely generate enough electrical output to directly power an appropriate indicating device. Some high-quality microphones produce voltages of only a few tenths of a microvolt or less. The ac signals from a radio antenna or a tape recorder pick-up head are all a few millivolts and much too weak to be fed directly to a loudspeaker, especially if you want to hear a thundering performance. Such signals are usually amplified several times in successive stages before sending them out to the speakers or some other *load*. The design of an amplifier (often with one or two transistors) depends on the type of input and load, and there is an endless variety of possibilities.

Figure 21.41 depicts one of the most useful basic amplifier configurations, the *common emitter*. Its name comes from the fact that the emitter is common to both the base and collector circuits. An alternating signal is fed to the base on top of a constant dc bias voltage of the battery (perhaps 1.5 V). That way, the input into the base is never negative, and the base always receives the necessary influx of positive charge. Without the bias, the transistor would clip off the negative portion of the signal.

A small change in base or *input current* ΔI_B (typically several microamps) results in a large change in the collector or *output current* ΔI_C (perhaps a few milliamps), and we define the ratio $\Delta I_C / \Delta I_B$ as the *current gain* of the amplifier. And there is a corresponding definition for the *voltage gain*. According to the manufacturer, a general-purpose *npn* transistor like the 2N3904 has a maximum current gain of 400.

Although individual transistors are still being used to fabricate amplifiers and countless other devices, especially in small production runs, the integrated circuit is dominant in many areas of electronics. A *monolithic IC* is formed on a single crystal wafer, usually silicon, but there are other competing semiconductors as well. Using diffusion techniques (the vapor deposition of layers of materials and etching), hundreds of thousands of components

Figure 21.41 The common-emitter amplifier. The signal V_S is superimposed on a bias voltage V_B so that the input V_{BE} is always positive. The amplified output is taken off the load resistor. This kind of amplifier has a large current gain (50–250), a high-voltage gain, and a medium-input resistance (≈ 2 kΩ).

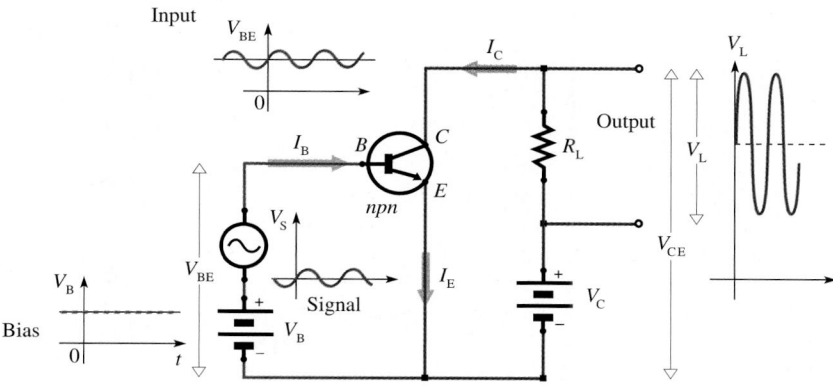

can be formed and "wired" together. A microscopic monolithic resistor is just a tiny sheet of isolated semiconductor. A monolithic capacitor is a little reverse-biased *pn*-junction. Monolithic transistors are usually *npn*, formed by successive diffusions. Although several hundred thousand transistors and all their accompanying resistors and capacitors can be made to fit on a dime, the basic operation of such a technological wonder is not any different from the kinds of circuits we have discussed.

Core Material & Study Guide

ALTERNATING CURRENT

A sinusoidal ac source has a terminal voltage of the form

$$v = V_m \sin \omega t = V_m \sin 2\pi f t \qquad [21.1]$$

and with a load resistor R across its terminals

$$i = I_m \sin \omega t = I_m \sin 2\pi f t \qquad [21.2]$$

The **effective** (or *rms*) **current** is given by (p. 747)

$$I = I_{eff} = \frac{I_m}{\sqrt{2}} \qquad [21.7]$$

and the **effective** (or *rms*) **voltage** is

$$V = V_{eff} = \frac{V_m}{\sqrt{2}} \qquad [21.8]$$

Thus, the average power dissipated is

[resistive load] $\qquad P_{av} = I_{eff}^2 R = I_{eff} V_{eff} \qquad [21.10]$

Reread Section 21.1 (AC and Resistance) and study Example 21.1. This is material basic to ac circuit analysis. Don't go on until you have mastered it. Try the I-level problems and **look at the WARM-UPS and WALK-THROUGH EXAMPLES on the CD.**

With a purely inductive load, *the current lags the voltage* (or the voltage leads the current) *by one-quarter cycle* (90°). The **inductive reactance** X_L is defined as

$$X_L = 2\pi f L = \omega L \qquad [21.14]$$

whereupon $\qquad V = IX_L \qquad [21.15]$

The average power supplied per cycle in a purely inductive circuit is zero (p. 750). Sections 21.1, 21.2 (AC and Inductance), and 21.3 (AC and Capacitance) lay out the fundamentals needed to apply Ohm's Law to ac. Make sure you can do problems similar to Example 21.2.

The instantaneous current in a capacitor leads the instantaneous voltage across it by one-quarter of a cycle (90°). The **capac-**itive reactance X_C is defined as

$$X_C = \frac{1}{2\pi f C} = \frac{1}{\omega C} \qquad [21.19]$$

whereupon $\qquad V = IX_C$

Only resistance will dissipate power in an ac circuit converting electrical energy into thermal energy (p. 753). Section 21.3 and the previous two sections provide the groundwork for the analysis to follow. Before proceeding, be sure to study Example 21.3.

R-L-C AC NETWORKS

In a series *R-L-C* circuit

$$V_m = \sqrt{(I_m R)^2 + \left(I_m \omega L - \frac{I_m}{\omega C}\right)^2} \qquad [21.26]$$

and $\qquad \theta = \tan^{-1} \frac{\omega L - 1/\omega C}{R} \qquad [21.27]$

The quantity $(X_L - X_C)$ is called the **reactance** of the circuit:

$$X = (X_L - X_C) \qquad [21.30]$$

It follows that $\qquad V = I\sqrt{R^2 + X^2}$

and the **impedance** is

$$Z = \sqrt{R^2 + X^2} \qquad [21.31]$$

in ohms. Ohm's Law for ac is then

$$V = IZ \qquad [21.32]$$

The average power dissipated is

$$P_{av} = IV \cos \theta \qquad [21.33]$$

where the product IV is called the *apparent power* (p. 757).

At **resonance**, where the capacitive and inductive reactances are equal (p. 759), we have

[resonance]
$$f_0 = \frac{1}{2\pi \sqrt{LC}}$$
[21.35]

$\theta = 0$, $Z = R$, and $P_{av} = IV$. A large variety of problems deal with ac circuits; to master them read Section 21.4 (*R-L-C* Circuits) and make sure you understand Examples 21.4 to 21.6.

For a transformer, the ratio of the effective voltages will equal the ratio of the numbers of turns, expressed as

$$\frac{V_p}{V_s} = \frac{N_p}{N_s}$$
[21.36]

Moreover
$$I_p V_p = I_s V_s$$
[21.37]

The transformer is an important device, and the preceding two equations provide the basis for understanding it— reexamine Section 21.5 (The Transformer) and go over Example 21.7.

ELECTRONICS

A *pn-junction* is the interface formed by joining a *p*-type to an *n*-type semiconductor. A *pn*-junction can function as a **diode**. The process of converting ac to dc is called *rectification,* and the junction diode is a popular rectifier (p. 767). The junction transistor, usually made of silicon, is formed in a three-tiered sandwich of either *pnp* or *npn* doped regions. It's a kind of electrical switch that can be used to control a large current and, in that sense, to amplify a signal.

Key Terms

alternating current	real power
line voltage	dissipated power
single-phase	power factor
polyphase	apparent power
instantaneous voltage	resonant frequency
maximum voltage	carrier wave
peak-to-peak voltage	amplitude modulation
instantaneous current	transformer
maximum current	primary
effective current	secondary
effective voltage	turn ratio
instantaneous power	valence band
rms current	conduction band
rms voltage	doping
average power	*n*-type
reactance	hole
inductive reactance	*p*-type
choke coil	*pn*-junction
capacitive reactance	diode
phasor addition	forward biased
impedance	rectification
impedance triangle	transistor

Discussion Questions

1. Heavy wire connects a light bulb to a large air-core coil of several hundred turns (Fig. Q1). The coil goes to a pair of fuses, which connect to a double-pole single-throw switch, and finally the leads plug into the power line via a wall outlet. Both leads contain fuses just to make sure the hot line, which is often unknown, is fused. When the switch is closed, will the lamp light? What, if anything, will happen to the light level after the iron core is inserted?

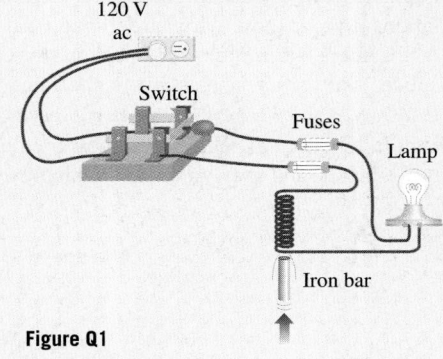

Figure Q1

2. Figure Q2 shows a galvanometer in a circuit with four diodes. Explain how this arrangement works as an ac voltmeter.

3. The transmission of elec-

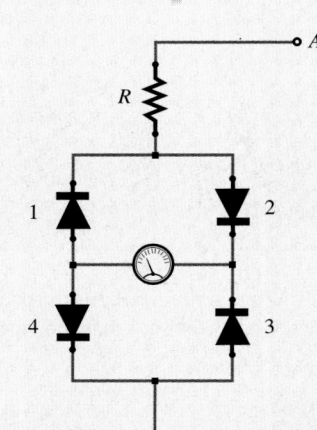

Figure Q2

trical power via high-voltage dc is becoming increasingly attractive. List a few reasons why this might be the case.

4. Does the input to a transformer have to be dc for there to be an output?

5. The circuit in Fig. Q5 is being studied, using an oscilloscope. The scope is essentially a voltage meter. Here, it's set to scan horizontally and at the same time be deflected vertically by the voltage signal coming in across its two terminals (see Problem 16). Sketch the curves that would result with the scope's probes across points *A* and *B*, *B* and *C*, and *A* and *C*. Discuss.

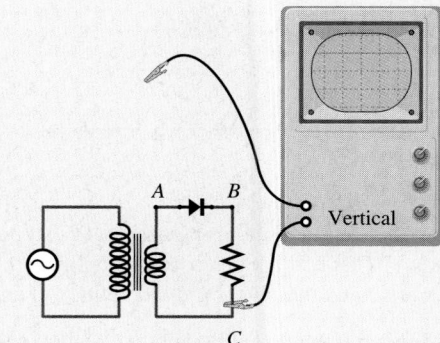

Figure Q5

6. Imagine that you have a faulty wall outlet and wish to replace it. Why would it be wise to shut off the voltage to the outlet? You go

to the fusebox and unscrew the fuse and return to the receptacle. Need you be concerned about touching the cover plate, which you want to unscrew and remove? What can you conclude if the tester held as in Fig. Q6 lights up? What do you do next?

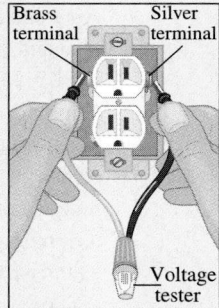

Figure Q6

7. Imagine that a fuse blows (or a breaker pops) and the lights in a room go out. You remove the toaster and shut off the hair dryer and then screw in a new fuse, but the lights don't go back on. How can you check the installation of the fuse with a neon tester?

8. Suppose a 240-V air conditioner doesn't seem to be working (at least it doesn't do anything when the on-switch is turned). A check of the outlet (Fig. Q8) reveals that a neon tester lights brightly between terminals B and C but not between A and B or A and C. What's wrong?

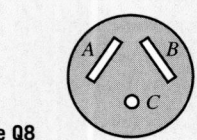

Figure Q8

9. Figure Q9 shows an R-C combination being fed by an ac signal, with the voltage across the capacitor as the output. It is left for Problem 85 to show that the ratio of the output to the input voltage is

$$\frac{V_o}{V_i} = \frac{1}{\sqrt{1 + (2\pi fRC)^2}}$$

What happens to this ratio when $f \to 0$? When $f \to \infty$? Why is it called a low-pass filter?

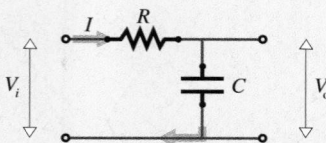

Figure Q9

10. Figure Q10 shows a three-prong adapter. What purpose does it serve and how? What will happen if you just plug in the adapter and don't bother attaching the plug's ground contact to anything?

11. Figure Q11 shows an R-C combination being fed by an ac signal, with the voltage across the resistor as the output. It is left

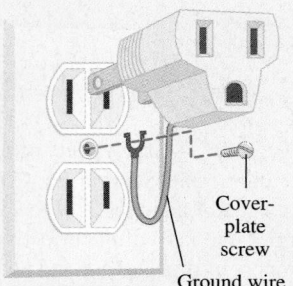

Figure Q10

for Problem 84 to show that the ratio of the output to the input voltage is

$$\frac{V_o}{V_i} = \frac{1}{\sqrt{1 + 1/(2\pi fRC)^2}}$$

What happens to this ratio when $f \to 0$? when $f \to \infty$? Why is it called a high-pass filter?

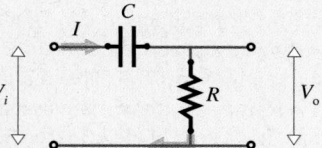

Figure Q11

12. The circuit shown in Fig. Q12 represents the wiring arrangement for a switch and a light fixture. As is customary in such installations, two or more wires are joined (that is, twisted together) by "wire nuts"—those are the small plastic caps wherever there is a splice. The cable is plastic-sheathed and contains two insulated leads and a third bare ground wire. Describe the circuit and draw a simplified diagram. Where does power enter? What does the switch do? What do the gray wires do? Which is the hot lead?

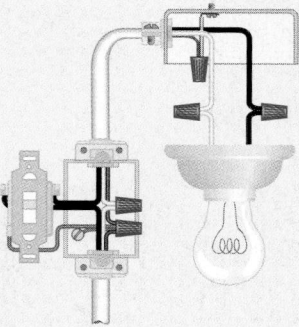

Figure Q12

13. What is the device pictured in the circuit diagram in Fig. Q13? How does it work—that is, what is the function of each portion of the circuit?

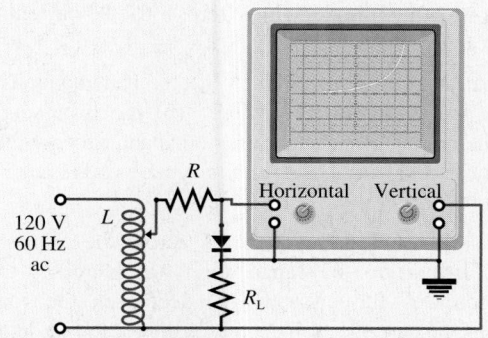

Figure Q13

Figure Q14

14. The laboratory setup in Fig. Q14 is used for testing diodes. The oscilloscope is essentially a voltage-measuring instrument, and that determines how it is used in the circuit. Voltages fed across the horizontal and vertical terminals cause the beam to be deflected proportionately along those directions. Explain what's happening here. What is being measured on the scope?

15. Imagine two circuits, one with a source, an ammeter, and a load having a low power factor, and the other with the same source and ammeter, but with a load having a high power factor. Suppose the loads are operated so that they dissipate the same amount of power. Which draws more current? Which power supply must have the high V·A rating? Which circuit might require leads with a larger wire gauge?

Multiple Choice Questions

1. In a sinusoidal-ac circuit, the *rms* current is   (a) $1.414I_m$ (b) $I_m/1.414$ (c) $I_m/0.707$ (d) $I/0.707$ (e) none of these.

2. In a sinusoidal-ac circuit, the peak-to-peak voltage equals   (a) 2 V   (b) 2(0.707) V   (c) 2(1.414) V   (d) 1.414 V/2   (e) none of these.

3. If we double the frequency in a sinusoidal-ac circuit, the effect on a capacitor in that circuit is to   (a) double its reactance   (b) increase its reactance by a factor of four   (c) leave its reactance unchanged   (d) halve its reactance   (e) none of these.

4. When the instantaneous voltage and current in an ac circuit are in-phase, we know that   (a) the total reactance is zero   (b) the capacitive reactance is zero   (c) the inductive reactance is zero   (d) the resistance is zero   (e) none of these.

5. In an *R-L* series ac circuit, increasing *L* and leaving everything else fixed has the effect of   (a) lowering the impedance   (b) lowering the reactance   (c) increasing the capacitive reactance   (d) increasing the power factor   (e) none of these.

6. In an *R-L* series ac circuit, increasing *R* and leaving everything else fixed has the effect of   (a) lowering the impedance   (b) lowering the reactance   (c) increasing the capacitive reactance   (d) increasing the power factor   (e) none of these.

The next four questions refer to Fig. MC7, which represents a series circuit composed of ideal elements.

7. This is a phasor diagram, at some arbitrary time *t*, for   (a) an *R-L* ac circuit   (b) an *R-C* ac circuit   (c) a purely resistive ac series circuit   (d) an *R-L-C* ac circuit   (e) none of these.

8. Here the inductive reactance   (a) is smaller than the capacitive

reactance   (b) equals the capacitive reactance   (c) is greater than the capacitive reactance   (d) is smaller than the resistance   (e) none of these.

9. The maximum voltage across the resistance is   (a) smaller than the maximum voltage measured across both the inductor and the capacitor   (b) equal to the maximum voltage measured across the inductor   (c) greater than the maximum voltage measured across the capacitor   (d) equal to the maximum voltage measured across the capacitor   (e) none of these.

10. The diagram tells us that the   (a) voltage across the inductor lags the voltage across the resistor   (b) voltage across the inductor leads the current through the resistor   (c) voltage across the capacitor leads the voltage across the inductor   (d) voltage across the inductor lags the voltage across the resistor   (e) none of these.

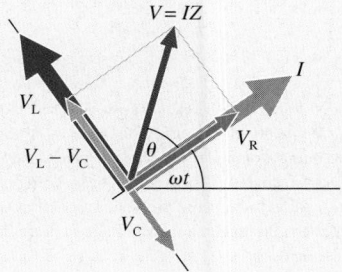

Figure MC7

11. In an *R-L-C* circuit, when the inductive and capacitive reactances are equal,   (a) the resistance is zero   (b) the phase factor

is zero   (c) the phase angle is zero   (d) the current is zero   (e) none of these.

12. In an *R-L* ac series circuit   (a) the instantaneous current and voltage are in-phase everywhere   (b) the instantaneous current leads the voltage across *L*   (c) the instantaneous voltage across *L* leads the current   (d) the instantaneous voltage across *L* lags the current by 90°   (e) none of these.

13. A variable capacitor is wired across the terminals of an ac source. Increasing the capacitance   (a) increases the reactance and decreases the current   (b) decreases the reactance and increases the current   (c) decreases the reactance, leaving the current unchanged   (d) decreases the reactance and decreases the current   (e) none of these.

14. In an ac series *L-C* circuit where the resistance is zero, we can expect that the phase angle will   (a) always be zero   (b) never be zero   (c) be + or −90°   (d) be 180°   (e) none of these.

15. Measuring across the two terminals that span an *R-C* ac series circuit, we find that   (a) the instantaneous current leads the voltage by 90°   (b) the instantaneous current lags the voltage by 90°

(c) the instantaneous current and voltage are in-phase   (d) the instantaneous power leads the resistance   (e) none of these.

16. A capacitor with a reactance of 100 Ω is in series with a 100-Ω resistor. The ac voltage across the two   (a) lags the current by 90°   (b) lags the current by 45°   (c) leads the current by 90°   (d) leads the current by 45°   (e) none of these.

17. A pure 1.0-H inductor is in series with a 0.20-μF capacitor across a 110-V, 60-Hz wall outlet. The resulting power factor is   (a) zero   (b) infinite   (c) 1.414   (d) 0.707   (e) none of these.

18. In an ac circuit, the power factor equals the   (a) apparent power   (b) real power   (c) real divided by the apparent power   (d) apparent divided by the real power   (e) none of these.

19. True or average power equals apparent power when   (a) the voltage is equal to the current   (b) the voltage is in-phase with the current   (c) the phase factor equals 0.707   (d) the current is very large   (e) none of these.

For more Multiple Choice Questions with answers click on WARM-UPS in CHAPTER 21 on the CD.

Suggestions on Problem Solving

1. When converting from effective values to maximum values, or vice versa ($V = 0.707V_m$ or $V_m = 1.414V$), keep in mind that you are simply multiplying by either $1/\sqrt{2}$ or $\sqrt{2}$. Which of these to use is easy to remember: if you are computing V_m, it's larger than *V*, and you multiply by 1.414 to get a larger number.

2. Remember that ω is in radians per second. Don't mix up degrees and radians when dealing with the phase angle. Accordingly, sin ωt must be computed using the radian setting on your calculator. For example, when $f = 60$ Hz and $t = 0.01$ s, sin $2\pi ft =$ sin 3.77 rad $= -0.59 =$ sin 216°.

3. In effect, an inductor holds back, or chokes, ac current, allowing voltage to develop, and a capacitor holds back ac voltage, allowing

current to develop. In series circuits, the phasor diagrams are for voltages (the current is the same for each element). In series, *whenever the phase angle is positive, voltage leads current*; the applied voltage leads the circulating current. *Whenever the phase angle is negative, voltage lags current*.

4. Many of the problems you will confront in this chapter can be checked by recomputing the results using a different approach. That is especially true in series circuits where the impedance triangle usually provides a complementary perspective.

Problems ✚ Coordinated Problems ✚ Progressive Problems ✚ Solutions

STUDY GUIDE   **1. Coordinated Problems:** The three problems within each magenta-colored grouping are solvable in similar ways. Note that the first of these always has a hint; moreover, its solution is provided in the back of the book. *Work out each of these sets; they'll strengthen technique and build confidence.* **2. Progressive Problems:** The problems introduced in blue unfold step-by-step carrying along the analysis in a more suggestive way than is customary. *Work out all of these; they'll guide you through the analytic process and help develop problem-solving skills.* **3. Worked-Out Solutions:** Studying worked-out solutions is an important part of learning how to solve problems. Accordingly, additional *solutions* to a number of model problems are given below. *Make sure you understand each of them before you go on to the next problem.* **4.** Also provided in the back of the book are the *Answers* to all odd-numbered problems, as well as worked-out *solutions* to those with boldface numbers. Problem numbers in italic indicate that a solution appears in the Student Solutions Manual.

SECTION 21.1: AC AND RESISTANCE

1. [I] The maximum output voltage of an ac generator is +120 V at a time equal to $\frac{1}{4}$ cycle. What will be the instantaneous terminal voltage   (a) $\frac{1}{4}$ cycle later and   (b) $\frac{1}{8}$ cycle later (at $\frac{3}{8}$ of a cycle)?

2. [I] A sinusoidal oscillator puts out an instantaneous voltage represented by $v = (75$ V) sin (376.99 rad/s)t. What is its frequency and maximum voltage?

3. [I] Write an expression for the instantaneous current delivered by an ac generator supplying 10 A effective, at 50 Hz.

4. [I] The maximum potential difference across the terminals of a 20.0-Hz sinusoidal-ac source is +50.0 V (occurring at $t = \frac{1}{4}$ cycle). If at $t = 0$, $v = 0$, find v at $t = 2.00$ ms.

5. [I] If the *rms* voltage read across a resistor by a voltmeter in a sinusoidal-ac circuit is 100 V, what is the maximum voltage?

6. [I] The effective current passing through a resistor *R* in a sinusoidal-ac circuit is 2.00 A. What is the maximum voltage drop across the resistor if $R = 100$ Ω?

7. [I] Determine the current drawn by a lit 100-W light bulb plugged into a 120-V wall outlet.

8. [I] THIS PROBLEM EXPLORES VOLTAGES AND CURRENTS IN A RESISTIVE AC CIRCUIT. A 10.0-Ω resistor and an ammeter are placed across the terminals of a power supply. (a) If a voltmeter across the resistor reads 25.0 V, what will the ammeter read? (b) What is the *rms* current in the circuit? (c) What maximum voltage will appear across the terminals of the source? (d) If an oscilloscope were used to measure current, what maximum current would it display? In other words, what would be the amplitude of the sinusoidal current curve?

9. [I] THIS PROBLEM DEALS WITH RESISTORS IN AN AC CIRCUIT. Two resistors, $R_1 = 1.00$ kΩ and $R_2 = 0.580$ kΩ, are in series with an ac ammeter across a 110-V (*rms*) ac power supply. We want to study the voltages and current in the circuit. (a) Draw a circuit diagram. (b) What is the equivalent resistance of the circuit? (c) Using Ohm's Law, find the *rms* current measured by the ammeter. (d) What are the voltage drops across each resistor? (e) Show that the sum of the voltage drops equals the source voltage.

10. [I] The instantaneous voltage measured across an ac source is $v = 200 \sin 2\pi 70t$. What is its effective voltage and frequency?

11. [I] A resistor is in series with an ac generator. An ammeter in series with the resistor reads 1.50 A, and a voltmeter across the resistor reads 75.0 V. What average power is being supplied by the source?

12. [I] If the source in Problem 10 is placed across a 200-Ω resistor, what current will be measured by an ac ammeter in series with the resistor?

13. [I] A 1.0-Ω resistor is attached across an ac generator. An oscilloscope shows that the sinusoidal current in the circuit has a maximum value of 0.50 A. What average power does the resistor dissipate?

14. [I] A 20.0-Ω resistor is placed across the terminals of a 60-Hz ac sinusoidal generator. If on average the resistor dissipates 200 W, what is the effective current flowing through it?

15. [II] If at a time of 2.00 ms after the start of a cycle ($v = 0$, at $t = 0$) a sinusoidal ac generator has an output of 95.1% of its maximum voltage, at what frequency is it operating? [*Hint: Write an expression for V as a function of t.*]

16. [II] A sinusoidal 60-Hz ac generator oscillating between ±100 V produces a voltage of 50 V at a given moment. At what value of t after the start of a cycle did this occur if at $t = 0$ the terminal voltage was zero?

17. [II] The current in a circuit, which varies from +2.00 A to −2.00 A, is provided by a 100-Hz ac sinusoidal generator ($i = 0$, at $t = 0$). What is the current at $t = 0.500$ ms after the start of a cycle?

18. [II] An ac generator and a load resistor are in series. An oscilloscope is connected across the resistor. The vertical deflection on the scope is caused by the voltage input and corresponds to a setting of 150 mV/cm. The horizontal axis is set for scan and is swept internally by the scope, producing a sinusoidal image on the screen. If the peak-to-peak displacement of the signal is 4.0 cm, what is the *rms* voltage across the resistor?

19. [II] The circuit in Fig. P19 needs your help. Every time the switch is closed the fuse blows. What could be wrong? [*Hint: One of the components is malfunctioning.*]

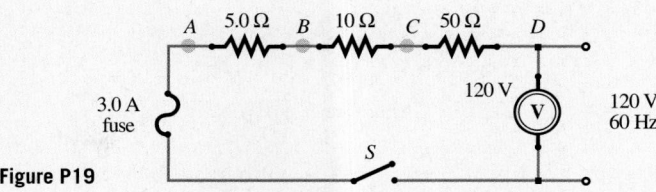

Figure P19

20. [II] Referring to Fig. P20, what will the two meters in the circuit read?

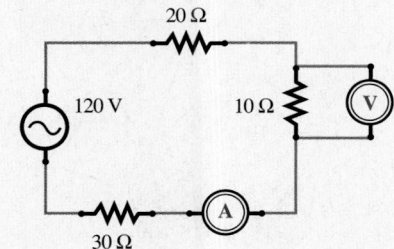

Figure P20

21. [II] Referring to Fig. P21, what will the two meters in the circuit read?

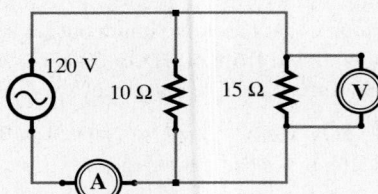

Figure P21

22. [II] A 1200-W hair dryer is plugged into a 120-V line. What's its resistance, and how much current does it draw?

23. [II] Figure P23 (on p. 778) shows a laboratory setup for studying the behavior of a resistive load (R) in an ac circuit. Here, a transformer inputs the 120 V from the wall outlet and supplies 30 V at 60 Hz, and the milliammeter reads 500 mA effective. Assuming the losses in the wiring to be negligible, what average power does the resistor dissipate? How would that change if the transformer were removed and the circuit plugged directly into the wall outlet?

24. [III] A 100-Ω resistor is wired across the terminals of an oscillator, which sends a current through it given by

$$i = 2.40 \cos 180t$$

in SI units. (a) Determine the period of the current oscillation. (b) Find the effective voltage of the source. (c) Write an expression for the instantaneous voltage across the load.

25. [III] Write an expression for the maximum instantaneous power dissipated in a sinusoidal-ac circuit consisting of a source and a resistor. How does this expression compare with the average power?

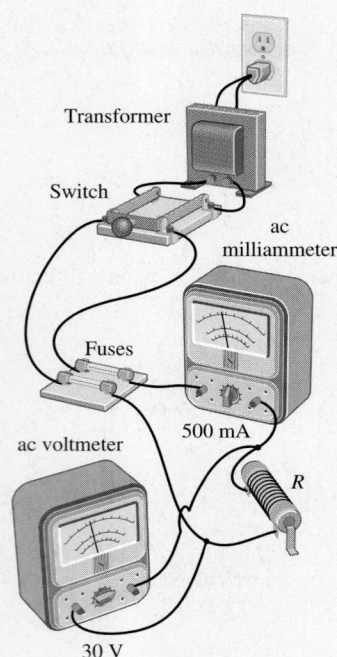

Figure P23

SECTION 21.2: AC AND INDUCTANCE

26.[I] A 100-mH inductor is put across the terminals of a variable frequency ac source that has a maximum output voltage of 36 V. If the frequency is varied from 60 Hz to 60 kHz, what is the corresponding range of the inductive reactance?

> **SOLUTION:** At 60 Hz, $X_L = 2\pi fL = 2\pi(60 \text{ Hz})(100 \times 10^{-3} \text{ H}) = 38 \ \Omega$; At 60 kHz, $X_L = 2\pi fL = 38 \text{ k}\Omega$.

27. [I] THIS PROBLEM EXPLORES INDUCTIVE REACTANCE IN AN AC CIRCUIT. A 100-Hz ac power supply has a maximum terminal voltage of 200 V. A 68-mH coil (with negligible resistance), an ammeter, and a switch, all in series, are put across the source. We want to determine the current reading of the ammeter when the switch is closed and that means, find the impedance (i.e., the reactance) and use Ohm's Law. (a) Draw a diagram. (b) Write an expression for the inductive reactance of the coil. (c) Compute the inductive reactance. (d) Write the appropriate expression corresponding to Ohm's Law. (e) Compute the maximum current in the circuit.

28. [I] An inductor is attached across the terminals of a variable frequency ac generator. If the frequency is halved, what happens to the inductive reactance?

29. [I] A 0.15-H coil with a negligible resistance has a reactance of 10 Ω when wired into a sinusoidal-ac circuit. Determine the frequency of the source. [*Hint: Inductive reactance is frequency dependent.*]

30. [I] What is the inductive reactance of a 0.200-H coil in a 100-Hz ac circuit?

31. [I] We are to design a coil having an inductive reactance of 0.50 kΩ to be used in a 60.0-Hz ac circuit. What should be its inductance?

32. [I] Determine the inductive reactance of a 10.0-mH coil connected to a 100-Hz ac generator.

33. [I] A high-quality 250-mH inductor (with negligible resistance) is attached across the terminals of a 60-Hz generator having an *rms* output of 125 V. Determine the reactance of the coil.

34. [I] What is the inductance of a coil having negligible resistance if at an angular frequency of 628.3 rad/s its reactance is 200 Ω?

35. [II] Prove that the ohm is the SI unit of inductive reactance.

36. [II] Determine the current that would be read by an ammeter in series with a 410-mH inductor of negligible resistance connected to a wall outlet at 120 V, 60 Hz.

37. [II] A high-quality coil with an inductive reactance of 100 Ω and negligible resistance is placed across the terminals of a 150-V ac generator. Determine the maximum current through the coil.

38. [II] A 20.0-mH inductor is connected across a variable frequency ac source. If the terminal voltage is 45 V, what is the current at 100 Hz and 100 kHz?

39. [II] The current supplied to a 100-mH inductor connected across the terminals of a sinusoidal oscillator is 350 mA. When operating at 100 Hz, what's the effective voltage of the oscillator?

SECTION 21.3: AC AND CAPACITANCE

40. [I] A 20-μF capacitor is put across the terminals of a variable frequency ac source that has a maximum output voltage of 36 V. If the frequency is varied from 60 Hz to 60 kHz, what is the corresponding range of the capacitive reactance?

> **SOLUTION:** At 60 Hz, $X_C = \dfrac{1}{2\pi fC} = \dfrac{1}{2\pi(60 \text{ Hz})(20 \times 10^{-6} \text{ F})} = 0.13 \text{ k}\Omega$; At 60 kHz, $X_C = \dfrac{1}{2\pi fC} = 0.13 \ \Omega$.

41. [I] THIS PROBLEM DEALS WITH CAPACITIVE REACTANCE. A 12.0-pF capacitor in series with an ac ammeter is put across the terminals of an electronic oscillator set at 100 kHz. When the oscillator is switched on, an ac voltmeter across it reads 6.0 V. We want to anticipate the current in the capacitor. (a) What is the capacitance in farads? (b) What is the frequency in hertz? (c) What is the capacitive reactance of the circuit? (d) Write the appropriate expression corresponding to Ohm's Law. (e) What is the value read by the ammeter?

42. [I] An 80-μF capacitor C is placed across the terminals of a 60-Hz sinusoidal-ac oscillator. What is the capacitive reactance of C?

43. [I] If a capacitor in series with a sinusoidal oscillator has a reactance of 200 Ω when $f = 50.0$ Hz, what reactance will it have when the frequency is raised to 5.00 kHz? [*Hint: The capacitive reactance is a function of frequency.*]

44. [I] What is the capacitance of the capacitor in the previous problem?

45. [I] A 2.0-μF capacitor is connected across a sinusoidal 60.0-Hz oscillator. What is its capacitive reactance?

46. [I] Determine the current that will be drawn by a 45.0-μF capacitor connected across a 240-V, 50.0-Hz source of sinusoidal-ac.

47. [I] A 60-Hz ac generator has a terminal voltage of 120 V. What size capacitor should be placed in series with it so that a current of 1.00 A circulates?

48. [I] Figure P48 depicts a simple circuit for studying the ac behavior of capacitors. What current will the milliammeter read? If either capacitor is disconnected, what will happen to the current?

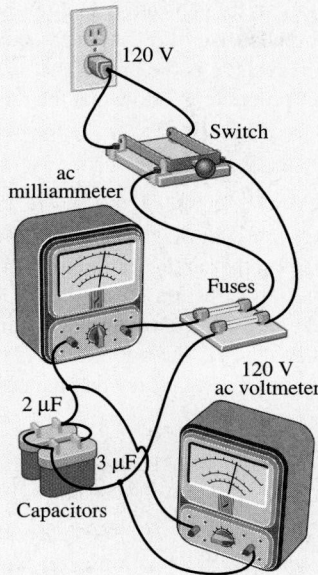

Figure P48

49. [II] The instantaneous voltage across a sinusoidal 100-Hz oscillator varies from +200 V to −200 V. If a 500-nF capacitor is connected across the source, what current will flow in the circuit?

50. [II] A sinusoidal oscillator with a terminal voltage of 125 V at a frequency of 55 Hz is put in series with a capacitor of 100 μF. What effective current will be provided by the oscillator?

51. [II] Two 40.0-μF capacitors in parallel are connected across a sinusoidal ac generator operating at 60.0 Hz. If the effective voltage of the generator is 120 V, how much current passes through each capacitor?

52. [II] Prove that the ohm is the SI unit of capacitive reactance.

53. [II] Figure P53 shows a 60-Hz ac laboratory setup for troubleshooting the 20-μF capacitor, which has been around for a while and needs to be checked before being used. (a) What can you say about it? Now, suppose we put another suspicious 10-μF capacitor across the first and find that the fuse blows every time the switch is closed. (b) What would you then conclude?

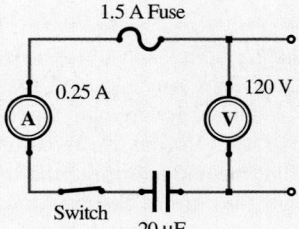

Figure P53

54. [II] Two capacitors of capacitance 6.0 nF and 3.0 nF are in series across the terminals of a 100-Hz sinusoidal oscillator with an output voltage that oscillates between +40.0 V and −40.0 V. Determine the *rms* current in the circuit.

SECTION 21.4: *R-L-C* CIRCUITS

55. [I] A coil has a resistance of 6.10 Ω and a reactance of 8.05 Ω. What is its impedance?

56. [I] A 6.0-μF capacitor in series with a 500-Ω resistor are both placed across a 120-V, 60-Hz ac source. Draw the impedance triangle for the circuit. Check your answer by computing Z.

57. [I] What is the impedance of a 10.0-nF capacitor operating in a 100-Hz ac circuit?

58. [I] An inductor has a reactance of 8.0 Ω and a resistance of 6.0 Ω. When connected across a 120-V ac outlet, how much current will it draw?

59. [I] A capacitor with a reactance of 160 Ω is in series with a 150-Ω resistor and an ac source. What is the net impedance of the load?

60. [I] A coil has an inductance of 32.0 mH and a resistance of 30.5 Ω. What is its impedance when placed across a 240-V, 200-Hz sinusoidal oscillator?

61. [I] If the total reactance of an *R-L-C* circuit is 1200 Ω and the total resistance is 500 Ω, what is the impedance?

62. [I] In a series *R-L-C* circuit, the inductive reactance is 1800 Ω, the capacitive reactance is 2600 Ω, and the resistance is 1000 Ω. What is the impedance?

63. [I] An *R-L-C* ac series circuit has a value of $(X_L - X_C) = +200$ Ω and a resistance of 400 Ω; what is the phase angle, and does the current lead or lag the applied voltage?

64. [I] A 2.0-kΩ resistor and a 2.0-μF capacitor are connected in series across a 60-Hz ac generator. What is the impedance of the circuit?

65. [I] A series circuit contains a capacitor, a resistor, and a 40-V, 240-Hz sinusoidal-ac source. If the net impedance of the load (resistor and capacitor) is 160 Ω, what value of current will be measured by an ammeter in the circuit?

66. [I] A coil with a reactance of 150 Ω and a resistance of 25 Ω is placed across a 120-V ac harmonic source. Determine the phase angle between the instantaneous voltage across the coil and the current.

67. [I] A capacitor with a reactance of 0.20 kΩ is placed in series with a 100-Ω resistor and the combination is attached to the terminals of an ac source. What is the phase angle between the instantaneous current and the instantaneous voltage across the combination?

68. [I] In the previous problem, what is the *rms* current if the source has a terminal voltage of 0.12 kV?

69. [I] A coil has a reactance of 1.55 kΩ and draws 26 mA from an ac generator with a terminal voltage of 60 V effective. What is its resistance?

70. [I] In an *L-C* series circuit with negligible resistance, the voltage drops across the inductor and capacitor are 100 V and 120 V, respectively. What is the phase angle?

71. [I] A circuit consists of a capacitor with a reactance of 2500 Ω in series with an inductor of negligible resistance and a reactance of 2000 Ω. What is the impedance of the circuit?

72. [I] A choke coil (of negligible resistance) with a reactance of 2.5 kΩ and a 2.5-kΩ resistor are in series across a 120-V ac generator. What average power is dissipated in the circuit?

73. [I] An R-L-C series circuit contains a 500-Ω resistor, a 5.0-H choke coil, and a capacitor. What value of capacitance will cause the circuit to resonate at 1000 Hz? What can you say about the practicality of your result?

74. [I] An R-L-C series circuit contains a 1.00-μF capacitor, a 5.00-mH coil, and a 100-Ω resistor. What is its resonant frequency?

75. [I] A little radio has a tuning circuit consisting of a variable capacitor and an antenna coil. If the circuit has a maximum current when $C = 300$ pF and $f = 40.0$ kHz, what is the inductance of the coil?

76. [I] This problem explores inductive and capacitive reactance. Consider a 60-mH inductor and a 25-μF capacitor. We want to determine at what frequency they will have the same reactance. (a) Write an expression for the inductive reactance. (b) Write an expression for the capacitive reactance. (c) Show that these reactances are equal at the angular frequency $\omega = 1/\sqrt{LC}$. (d) Determine the numerical value of the frequency f for which the reactances are equal.

77. [I] A fluorescent lighting device in Europe is powered by 50 Hz ac. Such lamps are generally connected to ballast coils to limit the current they draw. Here the fixture contains a large inductor (6.5 H) and a series capacitor whose purpose is to make the power factor equal one, so that commercial users don't have to pay for power not actually dissipated. What should that capacitance be?

> SOLUTION: The power factor is to be 1.0, hence $\cos \theta = 1$, and $\theta = 0$, which happens when $X_L = 2\pi fL = X_C = 1/2\pi fC$. Hence, $C = 1/4\pi^2 f^2 L = 1/4\pi^2 (50 \text{ Hz})^2 (6.5 \text{ H}) = 1.6 \ \mu$F.

78. [II] Suppose we want to find the inductance of a coil experimentally. We use an ohmmeter to determine that its resistance is 50 Ω. Next, we attach the coil to an oscillator, which supplies 300 mA to it at 20.0 V and 1.20 kHz. Determine L.

79. [II] A 30-mH coil with a resistance of 30 Ω is supplied by a 240-V, 200-Hz sinusoidal-ac source. Calculate the effective current it draws and the phase angle between the instantaneous current and the supply voltage.

80. [II] A 5.2-H choke coil with a resistance of 20 Ω is in series with a 1200-Ω resistor. The circuit is plugged into a transformer that puts out 60-V ac at 60 Hz. What voltage drop will appear across the resistor? Across the inductor?

81. [II] A 400-mH inductor with negligible resistance and a 185-Ω resistor are in series across a 60.0-Hz sinusoidal-ac source. If the voltage drop across the resistor is 370 V, what is the voltage across the inductor? What is the voltage of the source?

82. [II] Imagine yourself in a lab. On the table is an inductor rated at 1.5 H, supposedly with a resistance of 50 Ω. It's in series with a 0.5-A fuse. Every time the circuit is plugged into the 110-V, 60-Hz wall outlet the fuse blows. (a) Determine the current that should be in the circuit if all were well. (b) What would you guess was wrong? How might you confirm that?

83. [II] In an experimental setup, a capacitor and a 20-Ω resistor are in series across a 120-V, 60-Hz sinusoidal-ac source. An ammeter indicates that the circuit carries an *rms* current of 1.20 A. Determine the capacitance.

84. [II] Figure Q11 shows an R-C combination known as a high-pass filter (see Discussion Question 11). It's being fed an ac signal composed of a broad range of different frequency information. The voltage across the resistor is the output that is passed on to the next circuit for further processing. Show that the ratio of the output to the input voltage is

$$\frac{V_o}{V_i} = \frac{1}{\sqrt{1 + 1/(2\pi fRC)^2}}$$

85. [II] Figure Q9 shows an R-C combination known as a low-pass filter (see Discussion Question 9). It's being fed an ac signal composed of a broad range of different frequency information. The voltage across the capacitor is the output that is passed on to the next circuit for further processing. Show that the ratio of the output to the input voltage is

$$\frac{V_o}{V_i} = \frac{1}{\sqrt{1 + (2\pi fRC)^2}}$$

86. [II] A 9.00-μF capacitor, a 0.300-kΩ resistor, and a 300-mH inductor, all in series with an ac ammeter, are attached across the terminals of a 63.7-Hz ac source that has a maximum output voltage of 40 V. Draw the impedance triangle, and determine the phase angle between the terminal voltage and the current read by the ammeter.

> SOLUTION: $X_C = \dfrac{1}{2\pi fC} = \dfrac{1}{2\pi (63.7 \text{ Hz})(9.00 \times 10^{-6} \text{ F})} = 277.6$
>
> Ω; whereas $X_L = 2\pi fL = 2\pi (63.7 \text{ Hz})(0.300 \text{ H}) = 120.1 \ \Omega$. From Fig. P86, $\tan \theta = (X_L - X_C)/R = -157.5/300 = -0.525$ and so the phase angle is $\theta = -27.7°$.

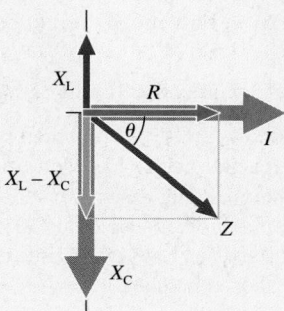

Figure P86

87. [II] This problem explores a series R-L-C circuit. A 210-Ω resistor in series with a 2.9-μF capacitor and a 390-mH inductor are put across a 100-Hz ac power supply with a peak-to-peak terminal voltage of 40.0 V. We want to find the maximum voltages across each element, and that means determine the impedance and use Ohm's Law in ac to get the current. See the Study Guide on p. 756. (a) Draw a diagram. (b) What's the inductance in henrys? Compute the reactances. Keep in mind the difference between f and ω. (c) What is the net reactance of the circuit? (d) Compute the total impedance. (e) What's the maximum source voltage? (f) Find the maximum current in the circuit. (g) What is the maximum

voltage across each element? (h) Does the sum of these voltages equal the maximum source voltage? Explain.

88. [II] THIS PROBLEM WILL HELP US UNDERSTAND THE SERIES *R-L-C* CIRCUIT. Consider the previous problem. We now want to determine the power dissipated by the circuit. (a) Draw the phasor diagram. Be careful here; the capacitive reactance is larger than the inductive reactance. (b) Use the diagram to compute the tan θ. (c) What is the value of the phase angle between the source voltage and the current? (d) Compute the power factor. (e) What is the effective voltage of the source? (f) What is the effective current? (g) Find the power dissipated in the circuit.

89. [II] THIS PROBLEM DEALS WITH POWER IN AN *R-L-C* CIRCUIT. An inductor, a capacitor, and a 20.0-Ω resistor are in series across an ac supply that provides a maximum voltage of 100 V. The total impedance of the *R-L-C* circuit is measured to be 60.0 Ω. We want to determine the power dissipated. (a) Remembering that we must use the effective values of current and voltage to compute average power, what is the value that would be measured by a voltmeter across the power supply? (b) Using the ac version of Ohm's Law, what is the effective current in the circuit? (c) Since the only element dissipating power is the resistor, how much power is going into joule heating?

90. [II] A series *R-L-C* circuit is connected across a 10.0-kHz source. The 1.2-H inductor and 1000-Ω resistor are fixed whereas the capacitor is variable. (a) At what capacitance will the current in the resistor be a maximum? (b) What will be the voltage drop across the resistor if the source voltage is 50 V? (c) What average power will be dissipated?

91. [II] Draw the impedance triangle for the circuit in Fig. P91 and determine the phase angle and impedance. Check your answers by direct computation. How much current is circulating? What is the resonant frequency of the circuit; and if it were driven at that frequency, how much current would be present?

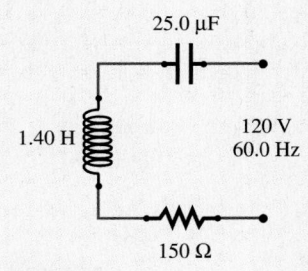

25.0 µF

1.40 H

120 V
60.0 Hz

150 Ω

Figure P91

92. [III] THIS PROBLEM WILL HELP US UNDERSTAND THE PARALLEL *R-L* CIRCUIT. Consider the circuit shown in Fig. P92*a*. In the past, to find the parallel combination of two resistors, we developed the *product-over-the sum rule*. We are going to use that same ideal to find the impedance of the *R-L* parallel circuit, but this time the sum must be the phasor sum, namely, $\sqrt{R^2 + X_L^2}$. (a) Explain Fig. P92*b* and then use it to show that $I^2 = I_R^2 + I_L^2$. (b) Use the fact that the voltages across each element in parallel are equal, to express I_R and I_L in terms of I, R, and X_L. (c) With the results of part-*b* in part-*a* prove that

$$Z = \frac{RX_L}{\sqrt{R^2 + X_L^2}}$$

(d) Show that

$$\theta = \tan^{-1}\frac{R}{X_L}$$

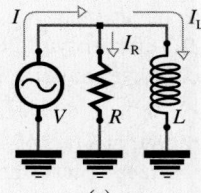
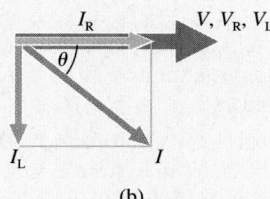

Figure P92 (a) (b)

93. [III] THIS PROBLEM WILL HELP US FURTHER UNDERSTAND THE PARALLEL *R-L* CIRCUIT. With the results of the previous problem in mind, suppose $V = 100$ V at 50.0 Hz, $R = 100$ Ω, and $X_L = 50.0$ Ω. (a) Determine the total impedance of the load. (b) What is the value of the phase angle? (c) Compute the value of the inductance. (d) How much current passes through the resistor? (e) How much current passes through the inductor? (f) How much current is supplied by the source?

SECTION 21.5: THE TRANSFORMER

94. [I] We want to construct a step-down transformer that will input 110 V and output as close to 12 V as possible. To accomplish this, we wrap 110 turns on the primary side of a ring-shaped iron core. How many turns should be put on the secondary assuming 100% efficiency.

95. [I] The input line to a transformer has an ac voltage of 1100 V. If the primary has 100 turns, how many turns should the secondary have if the output voltage is to be 550 V?

96. [I] A transformer has a 500-turn primary and a 2500-turn secondary. If the primary voltage is 120-V ac, what is the secondary voltage?

97. [I] A typical neon sign operates at a voltage of about 11.5 kV. If a transformer is to raise the 115-V line voltage up to that level, what should be its turn ratio? [*Hint: The turn ratio is defined on p. 761.*]

98. [I] A transformer drops a 120-V line voltage to 12 V, which is then used to operate a door bell. What must be the turn ratio of the transformer?

99. [I] If the turn ratio of a step-down transformer is 20:1 and the output is to be 6.0 V, what must be the input voltage?

100. [I] The primary of a transformer has 600 turns and a voltage across it of 240 V. How many turns should the secondary have if the output voltage is to be 60.0 V?

101. [I] A load resistor is placed across the secondary terminals of a step-down transformer with a turn ratio of 4:1. If the primary voltage and current are 120 V and 1.0 A, respectively, what will be the secondary voltage and current? Assume 100% efficiency.

102. [I] The primary side of a transformer carries 2.00 A at 120 V. What is the secondary current if the voltage across the secondary is 1200 V?

103. [II] A transformer is used to power a 6.0-V door chime. The 240-turn primary, which has an inductance of 3.0 H and negligible resistance, is connected across the standard 120-V, 60-Hz ac. (a) What turn-ratio should the transformer have? (b) With the secondary open-circuited, determine the primary current. [*Hint: Remember the ac version of Ohm's Law.*]

104. [II] A step-down transformer has an input voltage 1000 V and a turn ratio of 20:1. Assuming 100% efficiency, if the current in the output circuit is 40 A, how much power is delivered to the secondary?

105. [II] A transformer delivers power at a rate of 45 kW. If it loses 300 W in hysteresis and eddy current effects (the so-called iron loss) and 500 W in joule heat (the so-called copper loss), what is its efficiency?

106. [III] A transformer is supplied 8.20 kW by an ac source across its primary. The terminal voltage at the secondary is 220 V, and the load is such that there is a power factor of 82%. Determine the current in the secondary circuit.

107. [III] To transfer a maximum amount of power from a source to a load, the two should have the same impedance but often don't (audio amplifiers have a high impedance, speakers have a low impedance). A transformer can be used to match impedances (see photo on p. 761). Let Z_s be the load attached to the secondary and V_p and I_p be the voltage and current in the primary when the transformer and load are connected to the source. The source then sees an effective load Z_p; that is, it behaves as though Z_p were attached across its terminals (rather than the transformer and Z_s). Show that

$$Z_p = Z_s \left(\frac{N_p}{N_s} \right)^2$$

What turn ratio should a transformer have if it is to match a 3.2-kΩ amplifier to an 8.0-Ω speaker?

Chapter 22
Radiant Energy: Light

"What is light?" is one of those superbly simple questions we humans have pondered, and struggled with, for well over 2000 years. We are not without answers, marvelous answers that reflect the highest accomplishment of human ingenuity and creativity, but our understanding even now is far from complete, far from satisfactory. This chapter, and the next three as well, are about electromagnetic radiant energy, of which light is only that small part we happen to see. Thus, we begin the study of *Optics*, the study of the behavior of radiant energy.

The Nature of Light

Among the jumble of ideas that came from the ancient Greeks were the seeds of two significant lines of thought. The emission theory pictured light as a torrent of exceedingly minute, high-speed *particles*. The competing conception was advanced by Aristotle.

To the four elements of nature (fire, air, earth, and water), he added a fifth—the *aether*.* There was no such thing as empty space. All the void was filled with aether, and human vision "arises from a movement, produced by the body we perceive, in the interposed medium." Light was *aethereal motion*.

22.1 Waves and Particles

In the 1660s, Robert Hooke studied the color patterns associated with thin transparent films, such as soap bubbles (p. 905). The repetitive form of those patterns led him to a rudimentary *wave theory* in which light consisted of very rapid vibrations of the aether propagating

STUDY GUIDE

This chapter is an introduction to electromagnetic radiation. That word "radiation" tends often to be misunderstood and so it's a pity that we don't just call it "electromagnetic stuff" or better still, "electromagnetic matter." Instead, the phrase "electromagnetic energy" has long been widely used, and that's even more misleading.

There are three especially important equations in this chapter: $v = f\lambda$ (p. 787), $I = \frac{1}{2} c\varepsilon_0 E_0^2$ (p. 791), and $E = hf$ (p. 794). Make sure that you understand each of them. And then try and get an overall sense of the composition of the electromagnetic spectrum (p. 796).

*Though no longer accepted, the notion of a material aether has played a tremendously important role in the development of physics. This divine, invisible, elastic "goop" would ultimately become the transmitter of all sorts of influences from gravity and electricity to heat and "life force." The word itself was first used by the poet Hesiod (700 B.C.E.) to refer to the uppermost tenuous reaches of the Earth's atmosphere.

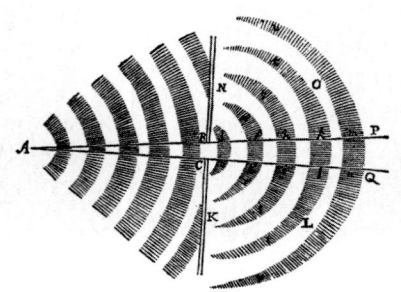

Figure 22.1 Taken from the *Principia*, this is Newton's picture of the diffraction of a wave as it passes through a hole. It's a fairly good representation, provided the hole is smaller than the wavelength.

Augustin Jean Fresnel (1788–1827).

Light. Caloric is the name assigned by the new nomenclature to the element of heat. It is the most subtile of all bodies, and exists in nature in very great quantities....Light is occasioned by the transmission of caloric, or the matter of heat, which travels through the air, in one second of time, about 170,000 miles....It is heat projected. The identity is the same; and consists of infinitely small particles thrown off in every direction from a luminous body.

H. G. SPAFFORD
*GENERAL GEOGRAPHY AND RUDIMENTS
OF USEFUL KNOWLEDGE (1809)*

A time-varying *B*-field is always accompanied by an *E*-field.

at tremendous speed. The fastest naturally occurring thing known at the time was sound, and light was even faster. So it seemed reasonable to suppose that only a wave, which is a *self-sustaining disturbance of a medium without the transport of matter*, could travel at such tremendous speed.

Although Newton seems earnestly to have tried to stay out of the wave-versus-particle debate, he nonetheless ultimately favored the corpuscular picture. His rejection of the wave theory came from his inability to explain the observed straight-line propagation of light. To Newton's mind, a wave should spread out markedly as it passes through an aperture (Fig. 22.1), filling almost the whole region beyond. It couldn't possibly produce a narrow beam, and yet such beams could easily be made. Unhappily, neither he nor anyone else of that era knew very much about waves.

Aware of Hooke's work with thin films, Newton proposed a remarkably prophetic solution. Light has a dual nature: it is a stream of material particles, but these are capable of setting up vibrations in the aether. The resulting pattern in the distribution of aether channels the stream of corpuscles—*waves of aether guide particles of light*.

In the first few decades of the 1800s, the wave theory was reborn with a new analytic vigor. Resurrected principally by Thomas Young and Augustin Fresnel, the wave conception slowly gained acceptance. By supposing that light was a *transverse* disturbance (p. 372), they could understand everything that had been observed over the centuries. The particles were gone, light was once again a wave—presumably an elastic wave in the aethereal sea. But the story doesn't end here. {For more historical details click on **WAVES & PARTICLES** under **FURTHER DISCUSSIONS** on the **CD**.} 💿

22.2 Electromagnetic Waves

While this saga was being played out, and quite independent of it, the study of electricity and magnetism was reaching a high point of its own. The masterwork of the age was created by a good-humored, bright young man, one James Clerk Maxwell of Kirkcudbrightshire in the Lowlands of Scotland.

We saw earlier that *in the presence of a varying magnetic field, an emf will be induced in any closed path*; that's Faraday's Law (p. 710). Because there is an emf, any charges in the presence of this varying *B*-field would experience forces and that, in turn, is equivalent to the existence of a nonelectrostatic electric field (p. 719). This conceptual extension of Faraday's Law is today known as one of **Maxwell's Equations**. In essence, it says that **a time-varying B-field generates an E-field**.

Figure 22.2 depicts a localized upwardly directed increasing magnetic field $\vec{B}$ surrounded by an induced electric field $\vec{E}$. Three points are crucial here: $\vec{E}$ and $\vec{B}$ are everywhere perpendicular to each

James Clerk Maxwell (1831–1879).

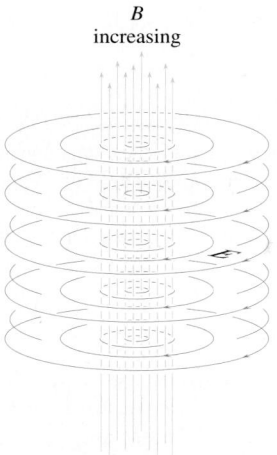

B
increasing

Figure 22.2 Imagine a wire loop surrounding the increasing *B*-field. A current would be induced such that the induced magnetic field $\vec{B}_I$ is downward, telling us the direction of the current and, therefore, the direction of $\vec{E}$.

other; the lines of $\vec{E}$, which do not now begin and end on charges, must therefore close on themselves; and $\vec{E}$ is not restricted to the region actually containing the flux—it extends beyond that.

The second equation that will be of special interest is Ampère's Circuital Law (p. 684), relating the amount of $\vec{B}$ parallel to a closed path *C* with the total current, $\sum I$, passing within the confines of *C*. This expression says that moving charges are the source of the magnetic field, and although that's true, it's not the whole truth. While charging or dis-

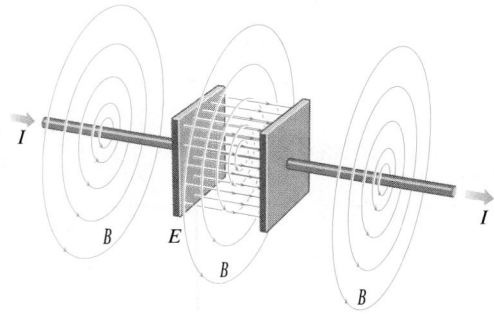

Figure 22.3 As a capacitor gets charged, a transient current exists, producing a *B*-field. In the gap where no actual current exists there is nonetheless a *B*-field. It's due to the time-varying *E*-field.

charging a capacitor, a magnetic field can be measured in the region between the plates, even though no actual current traverses the device (Fig. 22.3). To address this shortcoming Maxwell hypothesized the existence of another field-generating mechanism. He suggested that the time-varying *E*-field between the plates produces the *B*-field. This idea was totally

MAXWELL & ELECTROMAGNETIC THEORY

William Thomson (later to become Lord Kelvin) was, in the 1840s, the first to begin to mathematize the notion of *lines-of-force*. Faraday himself was ill at ease with mathematics, never having had any formal training in the subject. One day in 1846, Faraday was asked to step in at the last moment and give the evening lecture at the Royal Institution in place of Charles Wheatstone, who had had stagefright and run off minutes before. During the impromptu remarks that followed, Faraday conjectured that light might be a wave of some sort propagated along the lines-of-force. So it was, that Faraday's visions and Thomson's lovely though disconnected thoughts became the stimulus for Maxwell, who began to pursue his own researches within months of graduating from Cambridge in 1854. He drew together everything fundamental that was known about electricity and magnetism in a set of four equations. Two of these are credited to Gauss, the third is Faraday's Induction Law, and the fourth is Ampère's Circuital Law—from these, Maxwell built his theory.

A time-varying *E*-field is always accompanied by a *B*-field.

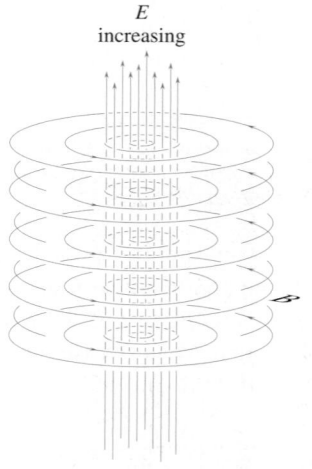

E increasing

Figure 22.4 The increasing *E*-field shown here has the effect of an upward current. The Right-Hand-Current Rule tells us the direction of the associated *B*-field.

new—it had never been proposed or observed prior to Maxwell. And it turns out to be true quite generally; **a time-varying $\vec{E}$-field generates a $\vec{B}$-field** (Fig. 22.4). It's the crucial creation that allows us to understand electromagnetic waves. {For a more detailed mathematical treatment click on **ELECTROMAGNETIC WAVES** under **FURTHER DISCUSSIONS** on the **CD**.}

Each form of radiant energy (radiowaves, microwaves, infrared, light, ultraviolet, X-rays, and γ-rays) is a web of oscillating electric and magnetic fields inducing one another. A fluctuating *E*-field creates a *B*-field perpendicular to itself, surrounding and extending beyond it. That magnetic field sweeping off to a point further on in space is varying there and so generates a perpendicular electric field that moves out and beyond that point as well, and so on: $\Delta E/\Delta t$ creating a perpendicular *B*, $\Delta B/\Delta t$ creating a perpendicular *E*, extending out indefinitely in an independent undulation of interlacing fields, an **electromagnetic wave**.

When an electromagnetic wave interacts with bulk matter, the *E*-field component has a much greater effect on charges than does the *B*-field. Accordingly, experiments have revealed that it is principally the electric field of the wave that effectuates vision, photochemistry, fluorescence, and so on. We shall usually explicitly consider only the *E*-field component, assuming that the *B*-field tags along—the two are inseparable.

22.3 Waveforms and Wavefronts

A *progressive* electromagnetic wave is a self-supporting, energy-carrying disturbance that travels free of its source (the light from the Sun sails through space, unleashed and on its own, for 8.3 minutes before arriving at Earth).

Armed with an *E*-field meter,* we could presumably measure the variations in an electric field as a wave swept by. Figure 22.5 attempts to show pictorially what might be seen as the meter-pointer wiggles, first positively then negatively, while the size of the *E*-field at that point in space fluctuates with the passing wave. The sizes of the drawn arrows correspond to the instantaneous values of the strength of the *E*-field—*nothing is actually displaced in space*. This phenomenon is not a wave rising and falling so many centimeters in the air like a rope. What's being pictured is a succession of readings in time, on a single

---

*Though these do exist for relatively low-frequency signals, a field meter that would function fast enough to follow the exceedingly rapid swings of the electric field associated with light is still beyond the capabilities of present technology.

Figure 22.5 The electric field of an electromagnetic wave. (a) If we observe the wave at a single, stationary location, we see an oscillating field as the wave passes. (b) If we observe the wave at many different locations all at the same time, we see the profile of the wave.

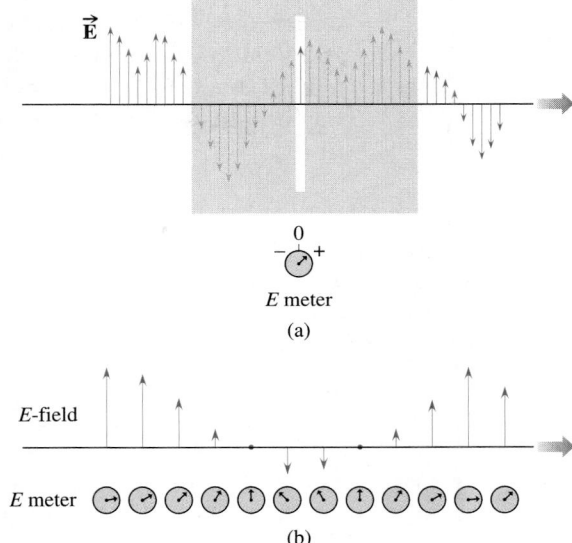

(a)

(b)

meter, at one point in space. Alternatively, Fig. 22.5*b* can also be thought of as showing the many readings from a whole row of meters at different points along a line in the direction of propagation of the wave, all at the same instant. The construct is of a disturbance whose "outline," or **profile**, is irregular.

Figure 22.6 depicts an electromagnetic wave whose profile is sinusoidal. (Take another look at Chapter 11 for a discussion of waveforms and the definitions of period T, wavelength λ, frequency f, and phase speed v.) Here, as with all periodic waves,

$$v = f\lambda \tag{11.1}$$

In vacuum, $v = c$, whereas in all other material media, v is usually less than c, though it can be greater under certain circumstances (p. 822).

In Fig. 22.6, along with the vertical E-field, is included the ever-present accompanying B-field. Every point on the disturbance experiences a sinusoidal oscillation in time, and the whole wave extends along the line-of-travel as an advancing sinusoid moving with a speed v. The E-field has an *amplitude* of E_0. It oscillates between $\pm E_0$, and so the profile of the wave can be represented mathematically as

$$E = E_0 \sin \frac{2\pi}{\lambda} x \tag{22.1}$$

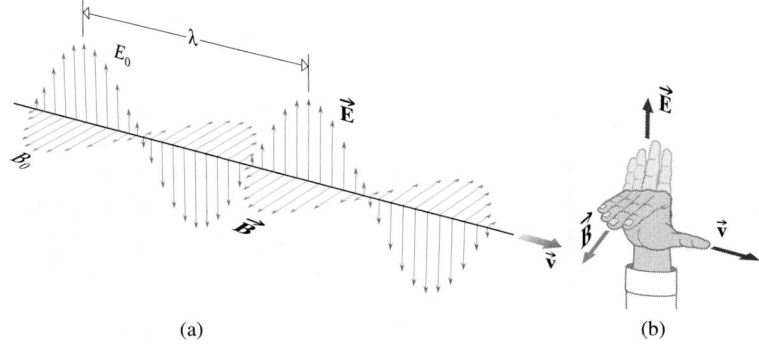

(a)

(b)

Figure 22.6 A harmonic electromagnetic wave. The coupled E- and B-fields are perpendicular to each other, and the propagation direction is given by $\vec{E} \times \vec{B}$.

very much like that of the mechanical wave of Eq. (11.2). To get a sense of the size of E_0, realize that bright sunlight on Earth might have an E-field amplitude of about 10 V/cm as compared with the tremendous value of 10^{10} V/cm that can occur in the focused beam of a high-power laser. The quantity $2\pi/\lambda$ arises so often in the analysis of waves that it's given its own name, the **propagation number**, and is symbolized by the letter k.

Example 22.1 **[I]** In New York City, WNYC broadcasts FM radio signals at 93.9 MHz. Assuming that the speed of propagation of these electromagnetic waves in air is negligibly different from c, please determine the corresponding wavelength.

Solution The problem is talking about frequency, speed, and wavelength, and that should remind you of one basic equation. (1) TRANSLATION—An electromagnetic wave of known frequency propagates in air; determine its wavelength. (2) GIVEN: $v = c$ and $f = 93.9$ MHz. FIND: λ. (3) PROBLEM TYPE—Electromagnetic wave/periodic. (4) PROCEDURE— Since the problem involves v, f, and λ, the one expression relat-

ing all of these, $v = f\lambda$, should come to mind. (5) CALCULATION—Rewriting Eq. (11.1),

$$\lambda = \frac{v}{f} = \frac{2.998 \times 10^8 \text{ m/s}}{93.9 \times 10^6 \text{ Hz}} = \boxed{3.19 \text{ m}}$$

Quick Check: $v = f\lambda \approx 300 \times 10^6$ m/s. Incidentally, TV sound and FM radio are VHF (very high frequency) signals ranging in wavelength from 1 m to 10 m.

[For more worked problems click on **WALK-THROUGHS** *under* **CHAPTER 22** *on the CD.]*

Example 22.2 **[I]** Write an expression for the profile of the E-field of a harmonic electromagnetic wave propagating in vacuum in the positive x-direction if its frequency is 600 THz (green light) and its amplitude is 8.00 V/cm.

Solution You're being asked to write an equation describing the shape of a harmonic wave, so think "sine function." (1) TRANSLATION—An electromagnetic wave of known frequency and amplitude propagates in a specified direction in vacuum; write an equation for it. (2) GIVEN: $v = c$, $E_0 = 800$ V/m, and $f = 600$ THz. FIND: The sinusoidal profile. (3) PROBLEM TYPE—Electromagnetic wave/periodic. (4) PROCEDURE—An expression in the form of Eq. (22.5) is what's needed. (5)

CALCULATION—But we will first have to find $k = 2\pi/\lambda$, and that means finding λ:

$$\lambda = \frac{v}{f} = \frac{2.998 \times 10^8 \text{ m/s}}{600 \times 10^{12} \text{ Hz}} = 500 \times 10^{-9} \text{ m} = 500 \text{ nm}$$

Hence, $k = 12.6 \times 10^6$ m^{-1} and so

$$\boxed{E = (800 \text{ V/m}) \sin(12.6 \times 10^6 \text{ m}^{-1})x}$$

Quick Check: $k = 2\pi/\lambda$ and $\lambda = 2\pi/k$; hence, $2\pi f/k = f\lambda = 2\pi(600 \text{ THz})/(12.6 \times 10^6 \text{ m}^{-1})$ should equal c, and it does.

Refer back to the harmonic electromagnetic wave traveling in vacuum (shown in Fig. 22.6) and notice that the direction of propagation corresponds to the direction of $\vec{E} \times \vec{B}$. The wave is *transverse*; $\vec{E}$ and $\vec{B}$ are perpendicular to each other and to the direction of propagation as well. Such **transverse electromagnetic waves**—so-called **TEM waves**—exist in vacuum, but things usually get much more complicated for waves traversing material media, where the fields may not always be totally transverse.

Electromagnetic waves, like sound waves, are three-dimensional. An ideal point source would radiate sinusoidal waves uniformly in all directions, as in Fig. 11.28. Here, *the surfaces of constant phase*—the **wavefronts**—are spherical. More practically, if we examine the light from a street lamp even a few hundred meters away, the portion of the wavefront intercepted by an eye or a small telescope will be essentially spherical. By allowing a spherical wave to expand out (Fig. 11.29) over a great distance, the wavefronts will get larger and flatter, ultimately resembling planes.

It might seem easier to just say that a wavefront corresponds to a surface over which the disturbance has some constant strength or magnitude. Indeed, quite often waves are

Figure 22.7 A plane harmonic electromagnetic wave. Notice how the fields change from one plane to the next. Over each plane both $\vec{\mathbf{E}}$ and $\vec{\mathbf{B}}$ are constant.

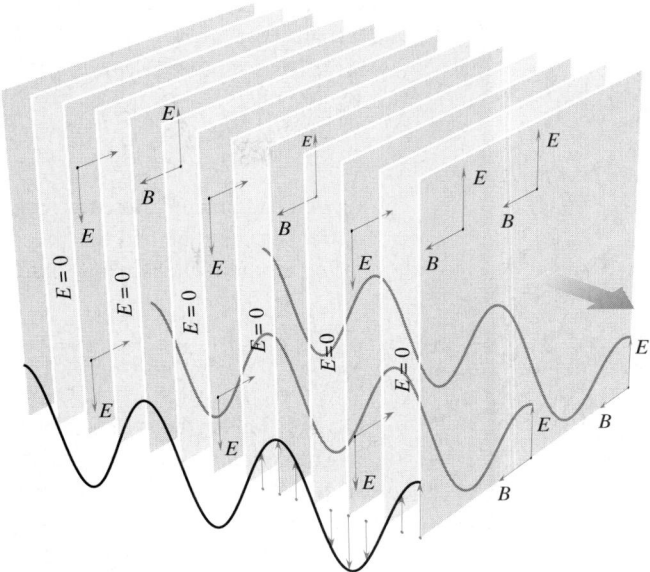

homogeneous; they have constant magnitudes over their wavefronts—but not always! A homogeneous harmonic plane wave traveling in free space can be envisioned as a moving stack of flat surfaces, like a deck of cards, over each of which E and B are constant, arranged so that as we go from one plane to the next, the fields change in a sinusoidal way (Fig. 22.7). The most popular form of laser light resembles plane waves that have a far greater electric field strength at the center (and are much brighter there) than at the edge of the beam—such a wave is *inhomogeneous*.

The Speed of Propagation: c

We can apply Maxwell's Equations to an electromagnetic pulse and arrive at an expression for the speed of the wave {To see how that's done click on **THE SPEED OF PROPAGATION** under **FURTHER DISCUSSIONS** on the **CD**} but it will suffice here just to examine the results. Accordingly, imagine a flat pulse of uniform magnetic field $\vec{\mathbf{B}}$ pointing in the z-direction laced with a uniform electric field $\vec{\mathbf{E}}$ pointing in the y-direction. The combination, resembling a rudimentary plane wave, is traveling through empty space in the x-direction at speed c (Fig. 22.8).

Envision an imaginary closed rectangular loop C of width L and indefinite length suspended in the xy-plane. As the B-field sweeps by, the magnetic flux in the region of space of the loop increases in the z-direction. That, in turn, induces an E-field in that region of space. The direction of this field is the direction of the induced current, which must be upward, so its induced B-field will oppose the increasing flux in the loop. In short, a traveling B-field generates an accompanying E-field, and only such coupled perpendicular fields propagate through space as a wave. Moreover, the two fields satisfy the expression

$$E = cB \tag{22.2}$$

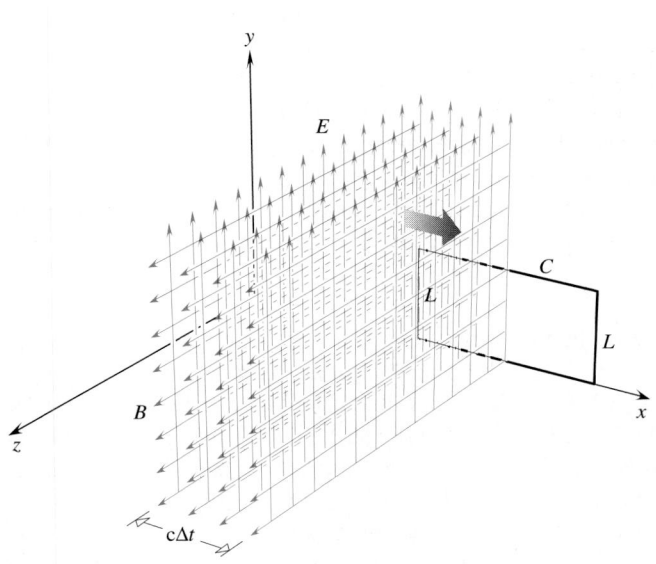

Figure 22.8 An electromagnetic pulse propagating past a loop. The change in magnetic flux through the loop induces a current in the loop and, therefore, an E-field as well.

Maxwell's own analysis was remarkably elegant. He began with the several laws of Electromagnetic Theory and combined them to form an expression that was known to describe all wave phenomena. And where there should have been a variable associated with the medium, there alternatively was E or B. And where there should have been the speed of the wave, there was $\sqrt{1/\varepsilon_0\mu_0}$; in other words

$$c = \frac{1}{\sqrt{\varepsilon_0\mu_0}} \qquad (22.3)$$

Maxwell had in hand these two constants (they had been determined in 1856 by Weber and Kohlrausch), and he had the speed of light (3.15×10^8 m/s) measured by Fizeau with a rotating toothed wheel in 1849. Substituting in the values of the constants,

$$c = \frac{1}{\sqrt{(8.85 \times 10^{-12}\ \text{C}^2/\text{N}\cdot\text{m}^2)(4\pi \times 10^{-7}\ \text{N}\cdot\text{s}^2/\text{C}^2)}} = 3.00 \times 10^8\ \text{m/s}$$

Amazing! This is the speed of light in vacuum.

The conclusion was inescapable: light is *"an electromagnetic disturbance in the form of waves"* propagated in the aether. And this description was true, as well, for "radiant heat, and other radiations if any" (light and "radiant heat"—that is, infrared—were known at the time, as was something called chemical rays, now referred to as ultraviolet). Maxwell's analysis, which forms the basis of classical Electromagnetic Theory, stands as one of the greatest theoretical achievements in the history of physics.

Today we envision electromagnetic waves propagating in the electromagnetic field, and all such waves travel in vacuum at exactly

$$c = 2.997\ 924\ 58 \times 10^8\ \text{m/s}$$

which is a tremendous speed. Light travels 1 ft in $\approx 10^{-9}$ s. As we'll see, this wave speed is a macroscopic manifestation of the fact that photons exist only at the speed c.

22.4 Energy and Irradiance

If we are to deal with light quantitatively, we have to measure the amount of radiant energy arriving on a surface—a photographic plate, a retina, whatever. It would be nice to determine the strength of the E- and B-fields directly, but that's technically impractical at the high frequencies ($\approx 10^{15}$ Hz) of light. The next best thing to find is the average amount of light energy flowing during a convenient interval of time. A surface is illuminated and we measure the amount of energy arriving per second on each square unit of area. The idea is that if all detectors are to get the same reading regardless of their size, we must measure energy-per-unit-area, and if they are all to get the same reading regardless of how long they take to make the measurement, we must measure energy-per-unit-area-per-unit-time. Earlier (p. 388), we called that quantity *intensity*, but in modern optics the *average amount of energy-per-unit-area-per-unit-time* is known as the **irradiance** (I) and it's specified in $\text{J/s}\cdot\text{m}^2$ or W/m^2. Notice that since *energy-per-unit-time* is power, irradiance is *power-per-unit-area*.

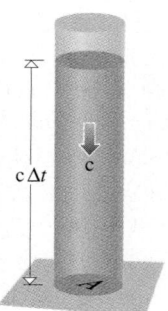

Figure 22.9 Traveling at a speed c, the column of light c Δt high will impinge on the plane in a time Δt. Thus, a volume of the beam equal to $Ac\,\Delta t$ will sweep down onto the plane in a time Δt.

Figure 22.9 shows a beam of light of cross-sectional area A impinging on a plane in vacuum. It travels at c, and in a time Δt all the light in the cylindrical section of length $(c\,\Delta t)$ and volume $V = (c\,\Delta t)A$ will sweep down onto the plane. If we knew the energy per unit volume u in the beam, then the amount of energy striking the surface in the time Δt would just be $uV = u(c\,\Delta t)A$. But we do know u from Eqs. (20.12) and (20.15)—it's the energy density in the E-field, $u_E = \frac{1}{2}\varepsilon_0 E^2$, plus the energy density in the B-field, $u_M = \frac{1}{2}B^2/\mu_0$. Before we press on, notice that for an electromagnetic wave $E = cB$ and $c = 1/\sqrt{\varepsilon_0\mu_0}$, and so $u_M = \frac{1}{2}E^2/c^2\mu_0 = \frac{1}{2}\varepsilon_0 E^2 = u_E$. Not surprisingly, the energy stored in the B-field of an electromagnetic wave equals the energy stored in the E-field. Thus, $u = \varepsilon_0 E^2$. Since the amount of energy arriving in a time Δt is $u(c\,\Delta t)A$, the energy-per-unit-area-per-unit-time is then $u(c\,\Delta t)A/A\,\Delta t = uc = c\varepsilon_0 E^2$.

Because E varies very rapidly, this quantity has to be averaged in time to obtain I:

$$I = [c\varepsilon_0 E^2]_{av} = c\varepsilon_0 [E^2]_{av}$$

The bracket $[\ ]_{av}$ just means the time average of what's inside. Here E is a time-varying sinusoidal function that we will write as $E = E_0 \sin \phi$, keeping in mind that ϕ varies in time. Thus

$$I = c\varepsilon_0 [E_0^2 \sin^2 \phi]_{av} = c\varepsilon_0 E_0^2 [\sin^2 \phi]_{av}$$

since $c\varepsilon_0 E_0^2$ is constant. We saw in Fig. 21.4 that when averaged over many oscillations $[\sin^2 \phi]_{av} = \frac{1}{2}$ and so

$$I = \frac{1}{2} c\varepsilon_0 E_0^2 \qquad (22.4)$$

The irradiance (or intensity), which we measure with a meter, is proportional to the square of the amplitude of the electric field of the wave.

Example 22.3 **[II]** A 1.0-mW laserbeam with a frequency of 4.74×10^{14} Hz has a cross-sectional area of 3.14×10^{-6} m². Determine (a) the energy arriving in 1.00 second on a screen intercepting the beam perpendicularly, (b) the irradiance, and (c) the amplitude of the electric field. Take the medium to be vacuum.

Solution This is a problem about the relationship between power, irradiance, and field amplitude, all of which we have discussed. (1) TRANSLATION—A laserbeam of known power, frequency, and cross-sectional area impinges on a surface; determine (a) the energy arriving in a specified time, (b) its irradiance, and (c) its electric field amplitude. (2) GIVEN: Beam-power P = 1.0×10^{-3} W, cross-sectional area $A = 3.14 \times 10^{-6}$ m², and $f = 4.74 \times 10^{14}$ Hz. FIND: (a) energy arriving in 1.00 s, (b) irradiance I, and (c) E_0. (3) PROBLEM TYPE—Electromagnetic wave/periodic/energy. (4) PROCEDURE—(a) The given beam-power P by definition equals the energy per unit time. (b) The irradiance is the energy-per-unit-area-per-

unit-time. (c) And we have an expression for I in terms of E_0. (5) CALCULATION—In $\Delta t = 1.00$ s, the energy that arrives is

$$P\Delta t = (1.0 \times 10^{-3} \text{ W})(1.00 \text{ s}) = \boxed{1.0 \times 10^{-3} \text{ J}}$$

(b) The irradiance or energy-per-unit-area-per-unit-time is

$$I = \frac{P}{A} = \frac{1.0 \times 10^{-3}\,\text{W}}{3.14 \times 10^{-6}\,\text{m}^2} = \boxed{3.2 \times 10^2 \text{ W/m}^2}$$

(c) Once we have I, we can get the amplitude from $I = \frac{1}{2} c\varepsilon_0 E_0^2$;

$$E_0^2 = \frac{2I}{c\varepsilon_0} = \frac{2(3.2 \times 10^2 \text{ W/m}^2)}{(3.00 \times 10^8 \text{ m/s})(8.85 \times 10^{-12} \text{ C}^2/\text{N}\cdot\text{m}^2)}$$

and $\boxed{E_0 = 0.49 \text{ kV/m}}$

Quick Check: $I \approx (1.0 \text{ mW})/(3 \times 10^{-6} \text{ m}^2) \approx 0.3 \text{ kW/m}^2$; $\frac{1}{2}c\varepsilon_0 \approx 1.3 \times 10^{-3}$ C²/N·m·s; $E_0^2 = I/\frac{1}{2}c\varepsilon_0 \approx [(0.3 \times 10^3)/(1.3 \times 10^{-3})]$ (V/m)² $\approx 0.2 \times 10^6$ (V/m)² and $E_0 \approx 0.5 \times 10^3$ V/m.

22.5 The Origins of EM Radiation

Though all forms of electromagnetic (EM) radiation share a single vacuum speed c, they do differ in frequency and wavelength. Still, there really is only one entity, one essence of electromagnetic "stuff." Maxwell's Equations, which are independent of frequency, do not suggest any fundamental differences in kind. Thus, it's natural enough to look for a common basic source-mechanism. What we find is that all the various types of radiant energy seem to have a common origin in that they are all associated somehow with *nonuniformly moving charges*. Of course, classically, we are dealing with waves in the electromagnetic field, and charge *is* that which gives rise to the field.

Free Charge

A stationary charge will have a constant *E*-field, no motional *B*-field, and hence, produce no radiation. (Where would the energy come from if it did radiate, and what would turn it on or off?) A uniformly moving charge has both an *E*- and a *B*-field, but these do not abruptly disengage, the motion is continuous, and, again, there is no radiation. If you moved along with the charge, the current would thereupon vanish; hence, *B* would vanish, and we would be back to the previous case of $v = 0$. That's reasonable since it would make no sense at all if the charge stopped radiating just because you started walking along next to it. Besides, if uniformly moving charges radiated, they would not move uniformly very long (not without an external agent supplying energy), and what then would happen to the Law of Inertia? That leaves *nonuniformly moving charges*, which, assuredly, will be accompanied by time-varying *E*- and *B*-fields.

We know, in general, that *free charges* (those not bound within an atom) emit electromagnetic radiation when accelerated. This much is true for charges sailing around in circles within a synchrotron, moving at a changing speed on a straight line in a linear accelerator, or simply oscillating back and forth in a radio antenna—**if charge accelerates, it radiates**. A portion of its kinetic energy is converted into radiant energy.

> **Accelerating charge** is the source of electromagnetic radiation.

The Dipole

Perhaps the simplest electromagnetic wave-producing mechanism to visualize is the oscillating dipole: two charges, one plus and one minus, vibrating to and fro along a straight line. The electric field lines, which begin and end on charges, close on themselves whenever the two charges overlap and thereby effectively vanish (Fig. 22.10). Moving charge constitutes a current (in this instance, an oscillating current), which is accompanied by an oscillating magnetic field whose closed circular lines-of-force are in planes at right angles to the motion. The *E*- and *B*-fields, which are in-phase, rising and falling together, are perpendicular to each other and to the propagation direction as well. Some distance from the dipole there will be developed an outgoing TEM wave. Keep in mind that this pattern is actually three-dimensional, roughly resembling a series of almost spherical concentric tires expanding radially outward.

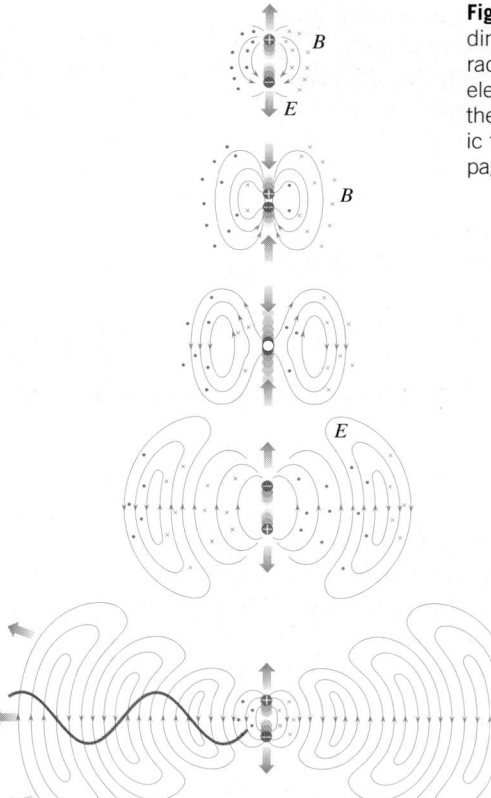

Figure 22.10 A cross-sectional view of the three-dimensional pattern of electromagnetic waves radiated by an oscillating dipole. The loops are electric field lines. The dots and crosses depict the emergence and exit of the circles of magnetic field, which are in planes perpendicular to the page and surround the dipole.

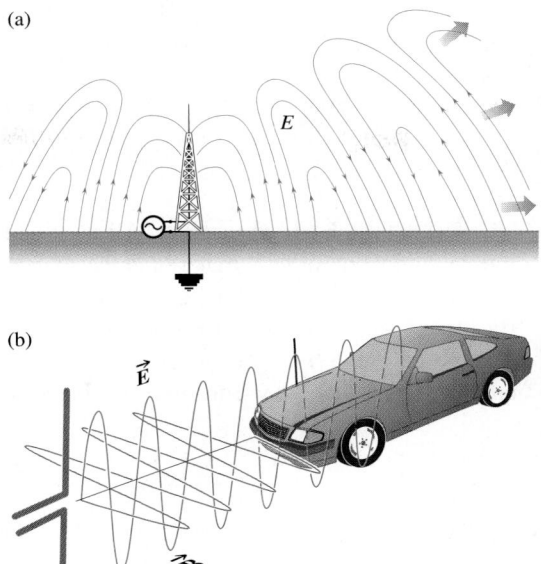

(a)

(b)

Figure 22.11 The *ground* wave generated by an AM antenna. The wave tends to hug the planet's surface, where most of the people with the radios and TVs are likely to be. Typically, this gives a commercial radio station a range of somewhere between 25 and 100 miles.

RADIO ANTENNAS

To receive a radio signal, one with a vertically oscillating *E*-field, we need only stick a straight length of wire up into the air, more or less parallel to *E*. The electric field strength of the incoming signal might be anywhere from microvolts per meter to millivolts per meter depending on where you are and what you want to receive. A signal with an amplitude of 1.0 mV/m reaching an antenna 2.0-m tall will induce a voltage having an amplitude of 2.0 mV. An old wire coat hanger will work adequately well as a replacement antenna for your car when you are driving around in the city, where the signals could be as strong as 10 mV/m..

It's an easy matter to attach an ac generator between two conducting rods and send currents oscillating up and down that transmitting "antenna." The result, Fig. 22.11, is a fairly standard AM radio tower. The antenna will function most efficiently if its length corresponds to the wavelength being transmitted or, more conveniently, to half that wavelength. The radiated wave is then formed at the dipole in sync with the oscillating current producing it. Unhappily, AM radiowaves are typically several hundred meters long. Accordingly, the antenna shown uses the conducting Earth so that it functions as if half of the $\frac{1}{2}\lambda$-dipole were buried in the ground, which allows us to build the thing only $\frac{1}{4}\lambda$ tall.

Every physicist thinks he knows what a photon is. I spent my life to find out what a photon is and I still don't know it.

ALBERT EINSTEIN

Photons

Einstein's Special Theory of Relativity revolutionized Classical Mechanics, completely removing the theoretical need for a stationary aether. Deprived of the necessity for an all-pervading elastic fiction, physicists simply had to get used to the idea that electromagnetic waves could propagate through an aetherless free space. The conceptual emphasis passed from aether to field—*light became an electromagnetic wave propagating in the electromagnetic field.* Aether vanished and field appeared as an entity in itself.

In 1905 Einstein shocked the scientific world with another brilliant insight. He boldly introduced a novel form of corpuscular theory that immediately explained several experimental problems that had developed since the late 1800s (Sect. 28.2). The energy carried by light (and all other forms of electromagnetic radiation) is not smoothly spread out across the wave but is somehow concentrated at points within it. Einstein asserted that light consists of massless quanta, "particles" of electromagnetic radiation.

22.6 Energy Quanta

Each *quantum of electromagnetic radiation*, or **photon**, as it came to be called in the 1920s, has an energy proportional to its frequency. The constant of proportionality, $h = 6.626 \times 10^{-34}$ J/Hz, or in units of electron volts, 4.136×10^{-15} eV/Hz, is known as Planck's

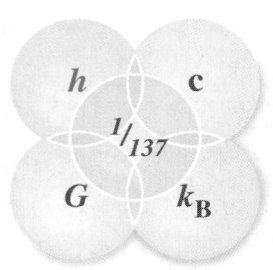

Figure 22.12 The atomic landscape is quantized; things (energy, angular moments, etc.) often come in tiny fixed amounts. And usually these amounts have something to do with Planck's Constant.

Constant (p. 261). As we'll come to see, this is one of those fundamental constants of Nature (Fig. 22.12). The energy of a photon of frequency f is then

$$E = hf \qquad (22.5)$$

The emission and absorption of light (i.e., the transfer of electromagnetic energy and momentum from, or to, a material object) takes place in a corpuscular way. Because photons have zero mass, each quantum of light carries very little energy. Even an ordinary flashlight beam must be thought of as a torrent of perhaps 10^{17} photons per second. In general, when we "see" light, what is recorded with eye, detector, or film is the average energy-per-unit-area-per-unit-time arriving at some surface.

Example 22.4 **[I]** A helium-neon laser puts out a beam of red light containing a very narrow range of frequencies centered at 4.74×10^{14} Hz. Determine the energy of each photon at that frequency.

Solution The energy of a photon only depends on its frequency. (1) TRANSLATION—A laserbeam has a known frequency; determine the energy of each photon. (2) GIVEN: $f = 4.74 \times 10^{14}$ Hz. FIND: Photon energy E. (3) PROBLEM TYPE—Electromagnetic radiation/photon energy. (4) PROCEDURE—

The energy of a photon is always given by $E = hf$. (5) CALCULATION—Using Planck's Constant in the form $h = 6.626 \times 10^{-34}$ J/Hz will provide an answer in joules:

$$E = hf = (6.626 \times 10^{-34} \text{ J/Hz})(4.74 \times 10^{14} \text{ Hz})$$

$$\boxed{E = 3.14 \times 10^{-19} \text{ J}}$$

Quick Check: Using h in eV/Hz, we get $E = 1.96$ eV and, since 1 eV $= 1.602 \times 10^{-19}$ J, $E = 3.14 \times 10^{-19}$ J. Photons of light have an energy of roughly 2 or 3 eV.

Light, and all other forms of electromagnetic radiation, interacting with matter in the processes of emission and absorption, behave like streams of particlelike concentrations of energy that exist only at the speed c and otherwise propagate in a wavelike fashion. We cannot say whether light *is* particle or wave; it seems to be both and so is likely neither. The "wee beasties" of the microworld, whether they be protons or photons, don't play our game of waves *or* grains—of "either/or." The electromagnetic wave (like the water wave made up of countless molecules) is an illusion of continuity. On the most fundamental level, there is no such thing as a continuous electromagnetic wave. Yet torrents of photons behave exactly as if they were dissolved into a smooth classical TEM wave.

Example 22.5 **[II]** Reconsider the laserbeam in Example 22.3. Taking the frequency of the light to be 4.74×10^{14} Hz, determine (a) the *photon flux*—the average number of photons per second impinging on a perpendicular flat target and (b) the *photon flux density*—the number of photons per second per unit area hitting that target.

Solution Each photon carries a specific amount of energy determined by its frequency. (1) TRANSLATION—A laserbeam of known power, frequency, and cross-sectional area impinges on a surface; determine (a) the photon flux and (b) the photon flux density. (2) GIVEN: beam-power $P = 1.0 \times 10^{-3}$ W, cross-sectional area $A = 3.14 \times 10^{-6}$ m^2, and $f = 4.74 \times 10^{14}$ Hz. FIND: (a) photon flux and (b) photon flux density. (3) PROBLEM TYPE—Electromagnetic radiation/photon energy. (4) PROCEDURE—(a) We have the power or energy per second,

1.0×10^{-3} W; hence, if we divide that quantity by the energy of each photon, we'll get the photon flux. (b) The photon flux density is the number of photons per second, per unit area. (5) CALCULATION—(a) From Example 22.4, $E = 3.14 \times 10^{-19}$ J;

$$\frac{P}{hf} = \frac{1.0 \times 10^{-3} \text{ W}}{3.14 \times 10^{-19} \text{ J}} = \boxed{3.2 \times 10^{15} \text{ photons/s}}$$

(b) The photon flux density is just the flux per unit area, or

$$\frac{3.18 \times 10^{15} \text{ photons/s}}{3.14 \times 10^{-6} \text{ m}^2} = \boxed{1.0 \times 10^{21} \text{ photons/s·m}^2}$$

Quick Check: The photon flux density (1.013×10^{21}) times the energy per photon equals 0.32 kW/m^2, the irradiance, which agrees with Example 22.3.

So here we have the latest picture of what seems to defy picturing: powerful in its ability to explain, wonderful in its subtlety—it is the latest theory of light, but probably not the last.

22.7 Atoms and Light

By far the most important mechanism for the emission and absorption of radiant energy—especially of light—is the *bound charge*, electrons confined within atoms. Much of the chemical and optical behavior of a substance is determined by only its outer electrons; the remainder of the cloud is formed into closed, essentially unresponsive shells around and tightly bound to the nucleus. Although it's not clear what exactly happens when an atom radiates, we do know with some certainty that light is emitted during changes in the outer charge distribution of the electron cloud (Sect. 28.5). It is the absorption and emission of light via these electrons that determine almost all optical phenomena in Nature.

Each electron is usually in the lowest possible energy state available to it, and the atom as a whole is in its **ground state**. There it will remain, indefinitely, if left undisturbed. Any mechanism that can pump energy into an atom, such as a collision with another atom, with a photon or an electron, will affect this situation. In addition to the ground state, there are specific well-defined higher energy levels, so-called **excited states**.

At low temperatures, atoms (and molecules) tend to be in their ground states; but as the temperature rises, more and more of them become excited through collisions. This process is indicative of a class of relatively gentle excitations (glow discharges, flames, sparks, etc.), which energize only the outermost unpaired valence electrons rather than the far more strongly bound inner ones. For the moment, we will concentrate on these outer-electron transitions, which give rise to the emission of light, infrared, and ultraviolet.

When just the right amount of energy (ΔE) is imparted to an atom (i.e., to a valence electron), it can respond by suddenly jumping from a lower to a higher energy level. Thus, the amount of energy that can be absorbed by an atom is quantized (i.e., limited to specific, well-defined amounts). This excitation of the atom is a short-lived resonance phenomenon. Usually, after a time of about 10^{-8} s or 10^{-9} s, the excited atom spontaneously relaxes back to a lower state, most often the ground state, losing the excitation energy along the way (Fig. 22.13). This transition can occur by the emission of light or (especially in dense materials) by conversion to thermal energy via interatomic collisions within the medium.

When the downward atomic transition is accompanied by the emission of light, the energy of the photon (hf) exactly matches the quantized energy decrease of the atom (ΔE). Consequently, $\Delta E = hf$, and there is a specific frequency associated with both the emitted photon and the atomic transition between the two states. An atom has many excited states, and these correspond to *resonant frequencies at which it very efficiently absorbs and emits energy.*

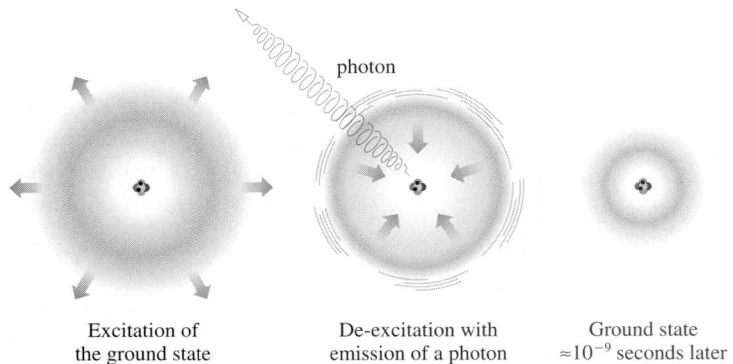

photon

| Excitation of | De-excitation with | Ground state |
| the ground state | emission of a photon | $\approx 10^{-9}$ seconds later |

Figure 22.13 A schematic representation of an atomic emission. (a) An atom is raised to an excited state. (b) With the emission of a photon, it (c) spontaneously drops back to the ground state.

Even though we don't know exactly what is going on during that 10^{-9} s, it is helpful to imagine the orbital electron somehow making the downward energy transition via a diminishing, oscillatory motion at the specific resonance frequency. Light can therefore be imagined as emitted in a short oscillatory pulse, or wavetrain, that is somehow representative of the radiated photon. Keep in mind that there are two intertwined models that have evolved to represent radiant energy: the particlelike and the wavelike—and light is both. Without the more complete development of Quantum Theory (which even then leaves a lot to be desired), we will have to irreverently talk about particles at one moment and waves at another.

Scattering and Absorption

The process whereby an atom absorbs a photon and emits another photon is known as **scattering.** The transmission of light through a windowpane, the reflection from a mirror or a face, the coloration of the sunset sky, are all governed by scattering.

We see most things not because they are self-luminous but because they absorb part of the incident ambient light and redirect some of the remainder toward our eyes. An atom (one on your cheek, for instance) can react to incoming light in essentially two different ways depending on the incident frequency (color) or, equivalently, on the incoming photon energy ($E = hf$). The atom can absorb the light, or alternatively, its ground state can scatter the light.

If the photon's energy matches that of one of the excited states, the atom will absorb it, making a quantum jump to that higher energy level. In the dense atomic clutter of solids and liquids, it's very likely that the excitation energy will be transferred, via collisions, to the random KE of the atoms rather than being reemitted when the atom returns to the ground state—the photon vanishes, its energy converted into thermal energy. This process (the taking up of a photon and its conversion into thermal energy) is called **dissipative absorption**. Most things in our environment have the colors they do because of selective dissipative absorption. Your skin dissipatively absorbs certain frequencies and scatters others, which is what gives it the color it has under white light illumination. A red apple appears red because it has a resonance in the blue and absorbs out the yellow-BLUE-green band, reflecting mostly red—what it doesn't "eat" it reflects.

In contrast to this resonant process, *ground-state*, or **nonresonant elastic scattering**, occurs for incoming light of other frequencies, that is, other than resonance frequencies. Even now, the light passing through your eyes is progressing via elastic scattering. Envision an atom in its lowest state and suppose that it interacts with a photon of frequency f, whose energy is too small to cause an excitation up into any of the higher states. Nonetheless, the electromagnetic field of the light can drive the electron cloud into oscillation. There can be no resulting atomic transitions; the atom will remain in its ground state while the cloud vibrates ever so slightly at the frequency of the incident light. The electron, once it begins to oscillate, is, of course, an accelerating charge and so will immediately begin to reemit light of that same frequency, f. This scattered light consists of a photon that sails off in some direction carrying the same amount of energy as did the incident photon. In effect, we are imagining the atom to resemble a little dipole oscillator. **When a material (like a piece of glass) is bathed in light, this almost-omnidirectional scattering gives each atom the appearance of being a tiny source of spherical wavelets.** As we'll see in the next chapter, it is precisely this nonresonant scattering that accounts for the transmission of light through all transparent material media and for the reflection of light from most surfaces.

The Electromagnetic-Photon Spectrum

The whole spread of radiant energy, which ranges in wavelength between zero and infinity, is referred to as the **electromagnetic spectrum**. It's usually subdivided into seven more or less distinct regions. These were delineated originally as much by historical circumstance

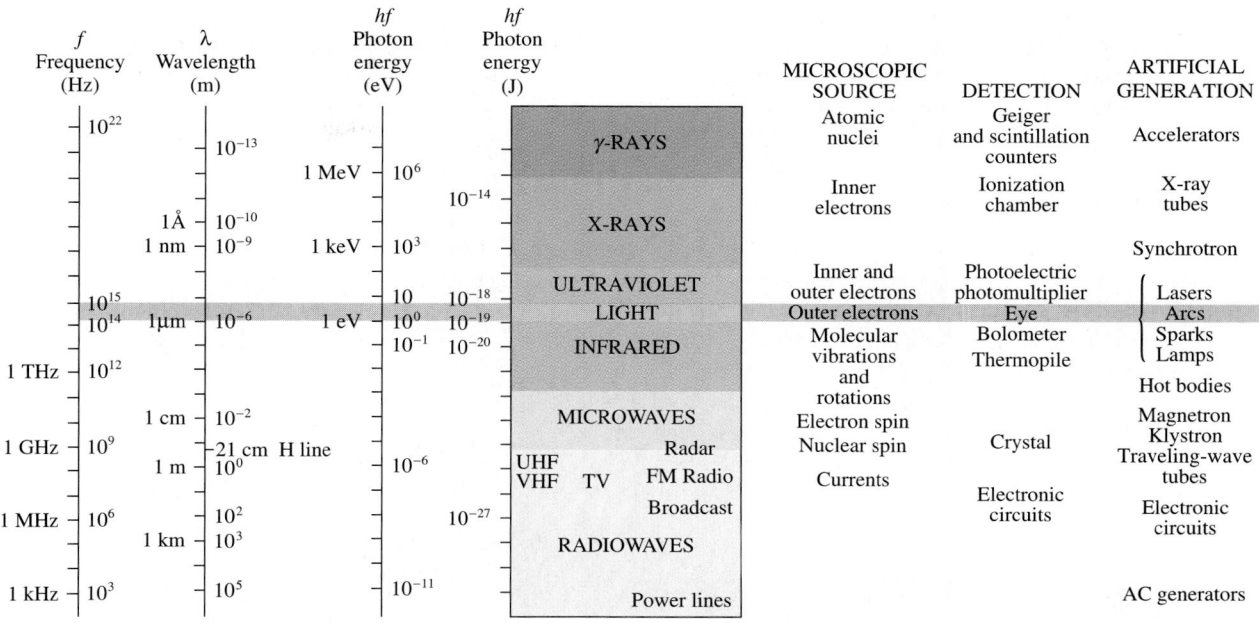

f Frequency (Hz)	λ Wavelength (m)	hf Photon energy (eV)	hf Photon energy (J)		MICROSCOPIC SOURCE	DETECTION	ARTIFICIAL GENERATION
10^{22}	10^{-13}			γ-RAYS	Atomic nuclei	Geiger and scintillation counters	Accelerators
		1 MeV 10^6	10^{-14}	X-RAYS	Inner electrons	Ionization chamber	X-ray tubes
	1Å 10^{-10}						Synchrotron
	1 nm 10^{-9}	1 keV 10^3		ULTRAVIOLET	Inner and outer electrons	Photoelectric photomultiplier	Lasers
10^{15}		10 10^{-18}		LIGHT	Outer electrons	Eye	Arcs
10^{14} 1μm	10^{-6}	1 eV 10^0 10^{-19}			Molecular vibrations and rotations	Bolometer Thermopile	Sparks Lamps
		10^{-1} 10^{-20}		INFRARED			Hot bodies
1 THz 10^{12}							
	1 cm 10^{-2}			MICROWAVES	Electron spin Nuclear spin	Crystal	Magnetron Klystron Traveling-wave tubes
1 GHz 10^9	21 cm H line 1 m 10^0	10^{-6}		UHF Radar VHF TV FM Radio	Currents	Electronic circuits	Electronic circuits
1 MHz 10^6	10^2 1 km 10^3	10^{-27}		Broadcast RADIOWAVES			
1 kHz 10^3	10^5	10^{-11}		Power lines			AC generators

Figure 22.14 The electromagnetic spectrum.

as by physical necessity, and so there tends to be a good bit of overlap in the categories. Needless to say, light was discovered first, then infrared (1800), ultraviolet (1801), radiowaves (1888), X-rays (1895), gamma rays (1900), and, finally, it was just a technical matter of filling in the microwaves, which was done in the 1930s, primarily with an eye toward radar (Fig. 22.14).

22.8 Radiowaves

Great, slowly rising and falling electromagnetic waves (more than 18 million miles long) have been measured impinging on the Earth, streaming in from the depths of space. These faint cosmic flutters are one extreme of the grouping known as **radiowaves**, which extends in wavelength down to about 0.3 meter.

AM radio transmissions in the United States are confined to the frequency band from 535 kHz to 1605 kHz (p. 760). The major failing of the AM technique is that it picks up extraneous noise sent out by everything from lightning and electric motors to car ignitions.

At 1.0 MHz, a radio frequency photon has an energy of 6.6×10^{-28} J, a very small quantity by any measure. In an ordinary flow of radio emission, there is an incredible number of photons, each individually so weak as to be essentially undetectable on its own. The result is an apparently continuous wavelike transport of energy. Radio photons have such

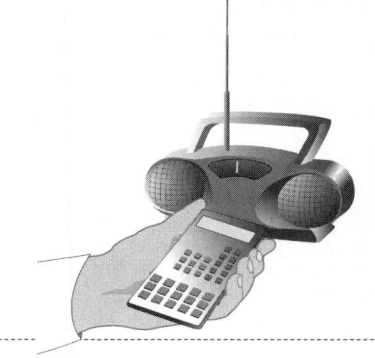

EXPLORING PHYSICS ON YOUR OWN

Impulse Radiowaves: Electronic calculators transmit radiowaves—not very powerfully and therefore not very far, but far enough to be picked up by a nearby AM radio. Turn the radio's dial to the low-frequency end between stations and punch in a beeping tune on your calculator (try it with a VCR remote control). You can even watch the effects of lightning—flashing white dashes running horizontally across your TV picture—during a storm. The picture is transmitted AM, the sound, FM. Or hold a telephone receiver next to your ear, push any button to get rid of the dial tone for a few seconds, and put it up against the body (CPU) of a computer. You'll hear the computer's radiated hum. Roll yourself in a ball around a portable radio and see what happens to the reception—are you transparent to radiowaves? Next, put some metal foil around the radio. Note the difference.

Example 22.6 [III] The AM receiver of Fig. 21.25 is tuned to accept a carrier frequency of 1.0 MHz. If the inductance in the tuning circuit has a value of $L = 300$ μH, (a) determine the value of the tuning capacitor. (b) What is the wavelength of the carrier wave? (c) What is the energy of a radio frequency carrier photon in electron volts?

Solution The circuit is tuned to resonate at the desired frequency. (1) TRANSLATION—A resonant circuit has a known frequency and inductance; determine (a) the capacitance, (b) the signal wavelength, and (c) the photon energy. (2) GIVEN: $f_0 = 1.0$ MHz and $L = 300$ μH. FIND: C, λ, and E. (4) PROCEDURE—(a) This is a resonant *L-C* circuit and so $f_0 = 1/(2\pi\sqrt{LC})$. (b) Having frequency and speed, wavelength follows from $c = f\lambda$. (c) Energy is determined from E $= hf$. (5) CALCULATION—(a) Start with the expression for f_0 and solve it for C

$$C = \frac{1}{4\pi^2 f_0^2 L} = \frac{1}{4(3.14)^2(1.0 \times 10^6 \text{ Hz})^2(300 \times 10^{-6} \text{ H})}$$

$$C = 84 \times 10^{-12} \text{ F} = \boxed{84 \text{ pF}}$$

(b) From $c = f\lambda$,

$$\lambda = \frac{c}{f} = \frac{3.0 \times 10^8 \text{ m/s}}{1.0 \times 10^6 \text{ Hz}} = \boxed{3.0 \times 10^2 \text{ m}}$$

(c) E $= hf = (6.63 \times 10^{-34} \text{ J/Hz})(1.0 \times 10^6 \text{ Hz}) = 6.63 \times 10^{-28}$ J. Since 1.00 J $= 6.24 \times 10^{18}$ eV,

$$\boxed{\text{E} = 4.1 \times 10^{-9} \text{ eV}}$$

Quick Check: Light has a frequency of about 500 THz and an energy of around 1 eV. These radio photons are roughly 0.5×10^9 lower in frequency and should be that much lower in energy; thus, E $= (1 \text{ eV})/(0.5 \times 10^9) = 2 \times 10^{-9}$ eV, which is the same order of magnitude as above.

low energies that they are not very likely to encounter atomic resonances, meaning that we can expect nonconductors (such as glass, bricks, and concrete) to be fairly transparent to radiowaves, while metals (with their free electrons) certainly are not. Incidentally, the human body most effectively serves as a conducting antenna in the range from about 30 MHz to 300 MHz, which is in the FM and VHF regions. Perhaps you've noticed as much while playing with an indoor TV antenna.

FM RADIO AND TV

Frequency modulation (FM) was introduced primarily to decrease the amount of noise and to extend the frequency content of the information transmitted. The range of FM is from about 88 MHz to 108 MHz, corresponding to wavelengths ranging from 3.4 m to 2.8 m. The carrier is made to vary in frequency (Fig. 22.15) in proportion to the amplitude of the audio message. Noise, which only affects the amplitude of the signal, is simply cropped off by the receiver since all the useful information is in the frequency distribution. FM signals, and therefore TV as well, are often transmitted with the *E*-field horizontal, which is why a typical TV receiving antenna has several $\frac{1}{2}\lambda$ horizontal bars.

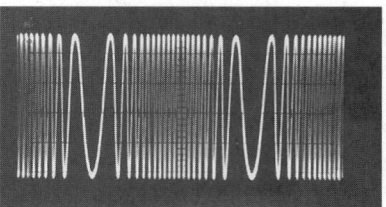
A frequency-modulated signal.

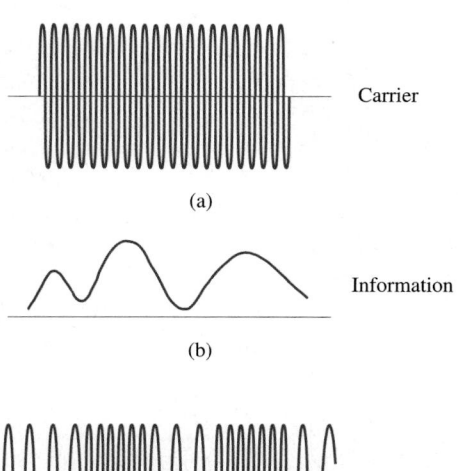

Figure 22.15 Frequency modulation (FM). (a) A constant-frequency, constant-amplitude carrier is modulated by (b) an information signal such that the amplitude of the information determines the frequency of the FM signal, as shown in (c).

A microwave (radar) image of the surface of Venus taken by the *Magellan* spacecraft in 1992.

22.9 Microwaves

The frequency region extending from about 10^9 Hz (1 GHz) up to roughly 3×10^{11} Hz is the domain of the **microwaves**. In wavelength, that corresponds to the range from approximately 30 cm to 1.0 mm. Electromagnetic radiation in the region from less than 1 cm to about 30 m can penetrate the Earth's atmosphere, and that makes microwaves especially useful for space-vehicle communications and radio astronomy. The ground state of the cesium atom consists of two very closely spaced levels, separated by only 4×10^{-5} eV. When the atom drops from one level to the other, the resulting microwave emission has a splendidly precise frequency of 9.192 631 77 $\times$ 10^9 Hz. This is the basis for the cesium clock, which is the present-day laboratory standard of frequency and time (p. 10).

To understand how the microwave oven works, recall that molecules can absorb and emit energy by altering the state of motion of their constituent atoms. The molecule can be made to vibrate and/or rotate; again, the energy associated with either such motion is quantized, and molecules therefore possess a number of rotational and vibrational energy levels in addition to those due to their electrons. Only when a molecule is polar will it experience forces via an electromagnetic wave that cause it to rotate into alignment with the changing E-field. Because molecules are massive and not able to swing around easily, we can anticipate low-frequency rotational resonances (infrared of 0.1 mm to microwave of 1 cm).

For example, the water molecule is polar—the hydrogen end is positive, the oxygen end is negative. When exposed to an alternating electromagnetic wave, it will swing around, trying to stay lined up with the E-field. Water molecules will very efficiently absorb microwave radiation at or near a resonant frequency, thereupon exhibiting large-amplitude oscillations. The oscillatory KE of these excited molecules is rapidly converted into thermal energy via collisions with other molecules.

After the Second World War, the people who were making military radar equipment had lots of powerful microwave generators and were looking for civilian applications; the microwave oven was an obvious one. As it turned out, the operating characteristics ($f =$ 2.45 GHz, $\lambda =$ 12.2 cm, P $\approx$ 1 kW, and $E_0 \approx$ 2 kV/m), though not ideal, were more than adequate for the job. Clearly, the thing to be heated has to contain water—a dry paper plate will remain quite cool. The diathermy machine, used to warm muscles and joints in order to relieve soreness, works on the same principle, at the same frequency.

Microwaves are now used for everything from carrying phone conversations and interstation TV to guiding planes and catching speeders (via radar) to studying the origins of the Universe and chatting with astronauts. Even though individual photon energies are small (in general, you want to avoid absorbing large numbers of them), people have been killed by massive exposure. In the United States, the supposedly safe limit to environmental microwave exposure is assumed to be 10 mW/cm^2, which some maintain is far too high.

Microwave antennae on the top of the Eiffel Tower in Paris.

A picture of a candy bar made using T-rays. The nuts, which were hidden beneath the chocolate, are visible as a result of refraction.

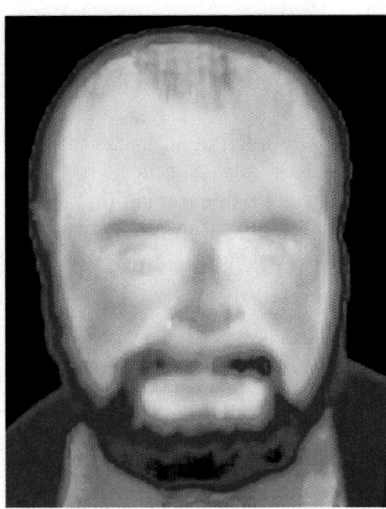

A portrait of the author taken using self-radiated IR. The beard is cool, and there seem to be no abnormal hot spots. The top of the mustache is warm from exhaled air.

22.10 Infrared

In 1800, the renowned astronomer (musician and deserter from the Hanoverian Foot Guards) Sir William Herschel made a surprising discovery. Using a prism, he had been studying the different amounts of "heat" conveyed by the various colors of sunlight. To his astonishment, he found that his thermometer registered its greatest increase just "beneath" the red region. Herschel rightly concluded that he was observing the effects of an invisible radiation, now called **infrared** (beneath the red).

The infrared (or IR) band merges with microwaves at around 300 GHz (1.0 mm) and extends to about 385 THz (780×10^{-9} m; i.e., 780 nm), where the photons have enough energy (1.6 eV) to break apart certain molecules. Most any material will radiate IR via thermal agitation of its molecules—just heat it up and it will pour forth IR. Infrared is copiously emitted from glowing coals and home radiators—roughly half the radiant energy from the Sun is IR, and an ordinary light bulb puts out far more IR than light. Like all warm-blooded creatures, we ourselves are IR emitters. The human body radiates very weakly, starting practically around 3000 nm; it peaks in the vicinity of 10 000 nm and trails off from there. This fact is exploited by some rather nasty "heat"-sensitive snakes (Crotalidae pit vipers and Boidae constrictors) that are active at night (p. 473).

In addition to rotating, a molecule can vibrate in several different modes. The corresponding vibrational emission and absorption spectra are, generally, in the infrared (1000 nm to 0.1 mm). Many molecules have vibrational and rotational resonances in the IR and are good absorbers, converting radiant energy into thermal energy, which is one reason IR is often misleadingly called "heat waves" (just put your face in the sunshine and you'll warm to the notion).

Heat lamps (1000 nm–2000 nm) used in physical therapy (and in bad restaurants) offer a more penetrating radiation than light. Typically, near-IR (i.e., near the visible) enters the human body to a depth of not much more than about 3 mm below the skin, regardless of the skin's color. If you've ever warmed yourself in sunlight streaming through a window, you already know that ordinary glass passes a good fraction of the incident near-IR; so do the cornea and lens of the human eye, and that's a good reason not to stare at the Sun, especially through cheap sunglasses that deceptively often pass much more IR than light.

There are photographic films sensitive to near-IR (<1300 nm) and TV systems that produce continuous infrared pictures known as *thermographs*; there are IR spy satellites that look out for rocket launches, IR satellites that look out for crop diseases, and IR satellites that look out into space; there are "heat-seeking" missiles that are guided by infrared and IR lasers and IR astronomical telescopes peering at the sky. You probably change stations on your TV with an IR-emitting remote control unit. Wherever subtle variations in temperature are of concern, from detecting brain tumors and breast cancer to simply spotting a lurking burglar, IR systems have found practical use.

22.11 Light

The narrow band of the spectrum that we humans "see" is often referred to as *light*. That's a rather inaccurate specification because we can "see" X-ray shadow patterns cast directly on the retina. Furthermore, many of us can see—if only poorly—into both the IR (up to roughly 1050 nm) and ultraviolet (down to about 312 nm). At those extremes, the sensitivity of the eye has dropped by a factor of about a thousand. Accordingly, let's fix the meaning of the word *light* to stand for that tiny range of the electromagnetic spectrum from 780 nm to 390 nm, one octave.

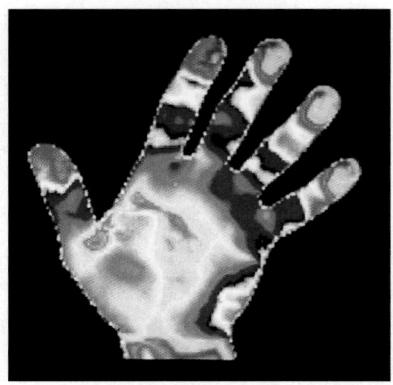

An infrared picture of a normal hand showing a typical temperature distribution. Abnormal blood flow becomes immediately obvious in IR images of this kind.

Newton was the first to realize that **white light** is actually a mixture of all the colors of the visible spectrum; the prism does not create color by changing white light to different degrees, as had been thought for centuries, but simply fans out the light, separating it into its constituent colors. Not surprisingly, the very concept of white light seems dependent on our perception of the Earth's daylight spectrum. The phenomenon of "daylight" changes from moment to moment and place to place, but it is nonetheless recognizable as a broad,

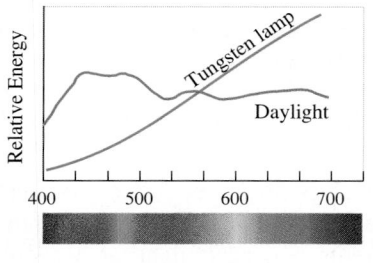

Figure 22.16 The distribution of the various frequencies in the light from a tungsten lamp and in sunlight.

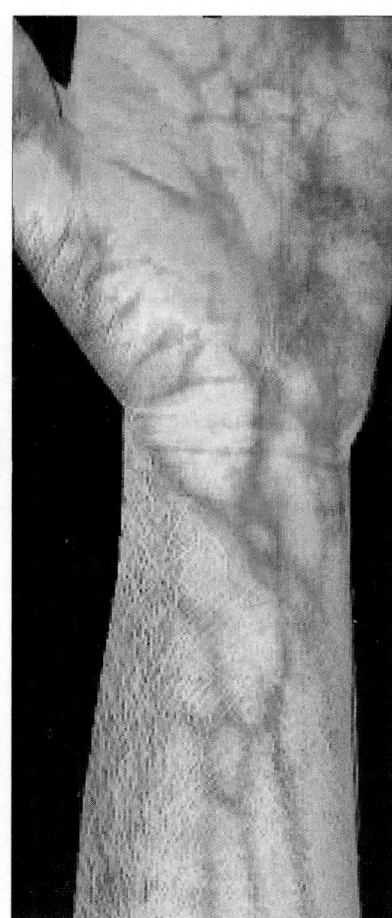

An arm viewed in a broad band of radiant energy extending from 468.5 nm (which is light) to 827.3 nm (which is near-infrared). The technique has many biomedical applications, among which is the early detection of skin cancer.

slightly wiggly frequency distribution that falls off more rapidly in the violet than in the red (Fig. 22.16).

What we perceive as whiteness is a wide mix of frequencies roughly in the same amounts as that of daylight, and that's what we mean when we talk about "white light." Be that as it may, lots of different distributions will still appear more or less "white." We recognize a piece of paper to be white whether it's seen indoors in incandescent light or outside in skylight, even though those whites are quite different. Indeed, many pairs of colored light beams (for example, 656 nm red and 492 nm cyan) will produce the sensation of whiteness, and the human eye cannot always distinguish one white from another—it cannot frequency analyze light into its harmonic components the way the ear can analyze sound (p. 385).

Colors themselves are the subjective human physiological and psychological responses primarily to the various frequency regions extending from about 384 THz for red, through orange, yellow, green, and blue, to violet at about 769 THz (Table 22.1). Thus, a harmonic wave of the single-frequency 508 THz will be perceived as yellow, as will 507 THz and 509 THz, and so on. Each such single-frequency, single-color light is spoken of as **monochromatic**. But to be monochromatic, the wavetrain must be perfectly sinusoidal, extending from $-\infty$ to $+\infty$. It can have no beginning and no end, and that's impossible. In reality, the best we can hope for is light having a narrow band of frequencies—**quasi-monochromatic light**.

Color corresponds to the human perception of photon energy or frequency. It is not a property of the light itself but a manifestation of the electrochemical sensing system: eye, nerves, brain. To be precise, we should not say "yellow light" but instead refer to light that is "seen to be yellow." Actually, a variety of different frequency mixtures can evoke the same color response from the eye-brain sensor. For example, a beam of red light overlapping a beam of green light will result, believe it or not, in the perception of *yellow* light, even though there are no frequencies present in the yellow band. The eye-brain averages the input and "sees" yellow! Furthermore, the nice reddish-blue or purple known as *magenta* does not even exist as a single frequency—it's not in the white-light spectrum.

The human eye, under daytime illumination levels, is most responsive to yellow-green (one reason that sodium-yellow street lights are so common and why yellow-tinted eyeglasses can be useful). Sensitivity gradually diminishes at both higher (blue, violet, ultraviolet) and lower (orange, red, IR) frequencies. It should be no surprise to learn that the solar spectrum also peaks at around 560 nm (2.2 eV) in the yellow-green so that there are usually plenty of the "right" photons flying around.

The wavelength of even red light at 0.000 000 780 m is rather small—780 nm is roughly 1/100 the thickness of this page. On an atomic scale the wavelengths of light are immense, several thousand times the size of an atom. If a uranium atom were enlarged to the size of a pea, a single wavelength of red light would then be about 54 feet long. This disparity is one of the crucial factors in determining the way light reflects off material objects. Remember that most of what we see is via reflected light. We make it a point not to look

Table 22.1

Approximate Frequency and Vacuum Wavelength Ranges for the Various Colors

Color	λ_0 (nm)	f (THz)*
Red	780–622	384–482
Orange	622–597	482–503
Yellow	597–577	503–520
Green	577–492	520–610
Blue	492–455	610–659
Violet	455–390	659–769

*1 terahertz (THz) = 10^{12} Hz,
1 nanometer (nm) = 10^{-9} m.

directly at bright, self-luminous objects even when they are around (apart from TV sets and campfires, which aren't very bright). Seeing the world almost exclusively in reflected light gives rise to a richly colored, shadow-filled, high-contrast, detailed picture—there's essentially no background noise.

In bright sunlight, where more than 10^{17} photons arrive each second on every square centimeter of a surface, the quantum nature of the process can easily be overlooked. Even so, light quanta are energetic enough ($hf \approx 1.6$ eV up to 3.2 eV) to produce effects on a distinctly individual basis. For instance, the human eye can detect as few as 10 photons impinging on it and perhaps as few as 1 arriving at the retina. Quanta of light can break up delicate chemical bonds, and so substances such as aspirin and wine are protected by dark bottles, and light-sensitive photographic films are commonplace. Premature infants sometimes develop jaundice, due to an excess of bilirubin in the blood, a condition that is successfully treated by exposing them to light. Blue-light photons have enough energy to dissociate the bilirubin molecule. Light is a major agency for the Sun → Earth transport of energy. Powered by sunlight, the process of photosynthesis results in the removal of upwards of 200 thousand million tons of carbon yearly from atmospheric carbon dioxide and the subsequent generation of complex organic molecules to the greater good of life on the planet.

22.12 Ultraviolet

A year after IR was discovered, J. Ritter (1801) found yet another invisible radiation. It was well known that white silver chloride would turn black, liberating metallic silver in the presence of light, especially blue light (this reaction was the precursor of photography). Ritter found that silver chloride did its little trick even more efficiently when exposed to the spectral region "beyond" the violet, where there was no visible radiation. This was the discovery of what came to be known as **ultraviolet** (UV) radiation, which corresponds to the range from about 8×10^{14} Hz to 2.4×10^{16} Hz. This is the so-called "black light," which is neither black nor light. Ultraviolet controls certain annoying dermatological conditions, tans the skin, and activates the synthesis of vitamin D within it.

At wavelengths of ≈ 300 nm and below, at the edge of the solar spectrum, UV can cause sunburn as well as tanning. Interestingly, this is roughly the energy needed (4 eV) to break a carbon-carbon bond. Passage through the atmosphere filters out much of this radiation, especially in the higher latitudes and at the low sun-angles that occur in winter and in the early and late parts of the day, even in summer. Someone in Chicago has to be a lot more determined to get tanned than someone in Florida. But then again, solar UV is the major cause of skin cancer in human beings. Our continued concern for the ozone (O_3) layer stems from the fact that this gaseous envelope absorbs (<320 nm) what would otherwise be a lethal stream of solar UV photons.

Ultraviolet in the wavelength range less than 300 nm will depolymerize nucleic acids and destroy proteins, both of which are strong absorbers, making UV quite incompatible with life on this planet. Extended exposure to UV, in time, causes wrinkles, liver spots, actinic keratosis (precancerous dark blotches), and finally cancer (80% of which is the curable form of basal-cell carcinoma). UV also inhibits the body's immune system, which may explain why some viral diseases, such as fever blisters and chicken pox, get more severe when exposed to sunshine.

Some materials reflect UV much as they reflect light, so long exposure while playing around on snow or water can be deceptively hazardous. The same is true for lying about on a beach on a totally overcast summer day, since water vapor passes a good deal of UV ($\approx 50\%$ in this case). In contrast, ordinary window glass invariably contains iron oxide contaminants, which make it quite opaque to near-UV. It's a waste of time to attempt to tan yourself behind such a window, no matter how warm it gets.

Humans don't see ultraviolet very well because the cornea absorbs it, especially at the shorter wavelengths, while the eye's lens absorbs most effectively beyond 300 nm.

An ultraviolet photo of Venus taken by *Mariner 10*.

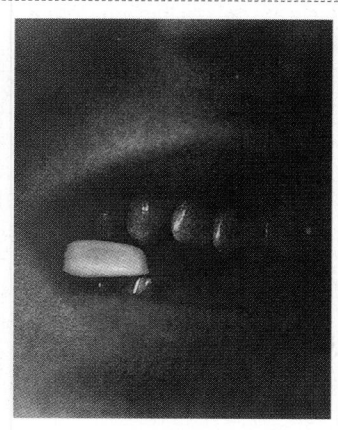

EXPLORING PHYSICS ON YOUR OWN

Fluorescence: There are a number of materials that can absorb a high-energy photon (UV) and almost immediately emit a lower-energy photon (light), in a process called fluorescence. Nowadays many packaged drinks (e.g., those made by Snapple and Crystal Light) have their caps sealed in place with a tamper proof overwrap of clear polyvinyl chloride. Happily, these plastics fluoresce rather beautifully. Hold one near a window so that it's bathed in solar UV and you'll see the plastic glow with a faint violet light. Now take it under an incandescent lamp—what happens? Try this at night with a mercury street light—the plastic will glow amazingly bright. When certain crystals are broken, a charge builds up on the separating faces and a spark leaps the gap. Francis Bacon discovered that sugar sparks when it's crushed. The flying electrons that constitute the spark excite N_2 molecules in the air, and they radiate, emitting a little blue light and a lot of invisible UV. The visual effect is enhanced when a fluorescing material like wintergreen oil is present to convert the UV into light. Go into a dark closet with a roll of wintergreen Lifesavers and a mirror. Crush one between your teeth and watch the blue-green sparks in the mirror.

Someone who has had a lens removed because of cataracts can see UV ($\lambda > 300$ nm). It now seems that in addition to insects such as honeybees, a fair number of creatures can visually respond to UV as well. Pigeons, for one, are quite capable of recognizing patterns illuminated by only UV and likely employ that ability to navigate by the Sun, even on overcast days.

An atom emits a UV photon when the electron makes a long jump down from a highly excited state. For example, the outermost electron of a sodium atom can be raised into higher and higher energy levels until it's simply torn loose altogether at 5.1 eV. The atom is then said to be ionized. Should it subsequently recombine with a free electron, the latter will quickly descend to the ground state, most likely in a series of jumps to ever lower levels, each resulting in the emission of a photon. If, however, the electron makes one long plunge to the ground state, a single 5.1-eV ultraviolet photon will result. Still more energetic UV photons can be generated when the inner electrons of an atom are excited.

Example 22.7 **[II]** Suppose that a singly ionized copper atom recombines with an electron. The ionization energy of copper is 7.72 eV. What is the shortest possible wavelength that can be emitted via the recombination?

Solution Think about Planck's Constant expressed in electron volts (p. 793). (1) TRANSLATION—A specified amount of energy is available as a photon; determine the corresponding wavelength. (2) GIVEN: E = 7.72 eV. FIND: The minimum λ. (3) PROBLEM TYPE—Electromagnetic radiation/photon energy/wavelength. (4) PROCEDURE—The maximum amount of energy available, corresponding to a transition directly to the ground state, equals 7.72 eV. This value, in turn, is associated with a maximum frequency given by E = hf and therefore a minimum wavelength. (5) CALCULATION—When you have frequency and want wavelength use $c = v = f\lambda$; therefore, E = $hf = hc/\lambda$ and

$$\lambda = \frac{hc}{E} = \frac{(4.136 \times 10^{-15} \text{ eV/Hz})(3.00 \times 10^8 \text{ m/s})}{7.72 \text{ eV}}$$

$$\boxed{\lambda = 161 \text{ nm}}$$

This, the most energetic radiation, is in the UV.

Quick Check: 161 nm is very roughly one-fourth the wavelength of light; so the energy of the photon should be approximately four times the energy of a light photon, which is roughly 1.5 eV. That value yields about 6 eV, which is close to the given value of 7.72 eV. Thus, 161 nm must be at least the right order-of-magnitude.

The unpaired valence electrons of isolated atoms are the source of much colored light. But when these atoms combine to form molecules or solids, those valence electrons can get paired up in the very process of forming the chemical bonds that hold the thing together. As a direct consequence, the electrons are often more tightly bound and their molecular-excited states are higher up, that is, in the UV. This gives rise to selective molecular UV absorption. Molecules in the atmosphere, such as N_2, O_2, CO_2, and H_2O, have such electronic resonances in the UV (which contributes to making the sky blue, p. 812).

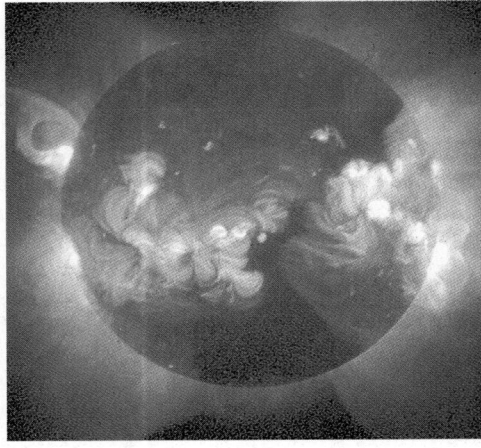

An X-ray photograph of the Sun. Modern focusing methods are producing detailed images of distant celestial sources. Orbiting X-ray telescopes have given us an exciting new view of the Universe.

Wilhelm Konrad Röntgen (1845–1923). As a result of his discovery of X-rays in 1895, Röntgen became the instant hero of his age and the first winner of the Nobel Prize in Physics.

Each UV quantum can carry enough energy (from 3.2 eV to 100 eV) to independently ionize an atom or rip apart a chemical bond. The particlelike aspects of radiant energy begin to become increasingly more evident as the frequency increases. At wavelengths less than around 290 nm, UV is germicidal; that is, it kills microorganisms.

22.13 X-rays

When Wilhelm Röntgen discovered X-rays on November 5, 1895, it was quite by accident (p. 973). Within months, the marvelous rays were at work everywhere. Without the slightest inkling of their inherently dangerous nature, X-rays were used for everything conceivable, from removing facial hair to examining luggage. Too often, the results of that cavalier attitude were horribly tragic. (And, of course, the Victorian ladies were well warned of the only danger anyone seems to have been concerned with: lurking X-ray peeping Toms.) Incredibly, X-rays were even being used as late as the 1950s to treat acne, though it's only recently that the victims began developing cancer, particularly of the thyroid.

Extending in frequency from roughly 2.4×10^{16} Hz to 5×10^{19} Hz, X-rays have exceedingly short wavelengths; most are smaller than the size of an atom. The individual photon energies (100 eV to 0.2 MeV) are so large that X-ray quanta can interact with matter one at a time in a clearly granular fashion, almost like bullets of energy. The primary mechanism for the production of this radiation is the rapid deceleration of high-speed charged particles.

Example 22.8 **[II]** An electron flies across an X-ray tube. Traversing a voltage difference of 1.00×10^4 V, it crashes into a metal target. Assuming it makes a rare head-on collision and comes to rest with the emission of a single photon, what is the wavelength of the radiation?

Solution If we know the voltage traversed, then we know the energy gained. (1) TRANSLATION—A specified amount of energy is available as a photon; determine the corresponding wavelength. (2) GIVEN: $V = 1.00 \times 10^4$ V and $q_e = 1.60 \times 10^{-19}$ C. FIND: λ. (3) PROBLEM TYPE—Electromagnetic radiation/photon energy/wavelength. (4) PROCEDURE—An electron that falls through a potential difference of V picks up an amount of energy Vq_e, which then appears as the photon energy. (5) CALCULATION—We can calculate the frequency using $E = hf =$

Vq_e. Once we have the frequency, use $c = v = f\lambda$ to get the wavelength, and therefore $E = hf = hc/\lambda$; hence

$$\lambda = \frac{hc}{E} = \frac{hc}{Vq_e} = \frac{(6.626 \times 10^{-34} \text{ J/Hz})(3.00 \times 10^8 \text{ m/s})}{(1.00 \times 10^4 \text{ V})(1.60 \times 10^{-19} \text{ C})}$$

and

$$\boxed{\lambda = 1.24 \times 10^{-10} \text{ m}}$$

Notice that the calculation could have been done a bit more easily using electron volts (Problem 73).

Quick Check: This value of λ is 0.12 nm, which is just about the size of an atom, and we know that's roughly the correct wavelength for X-rays.

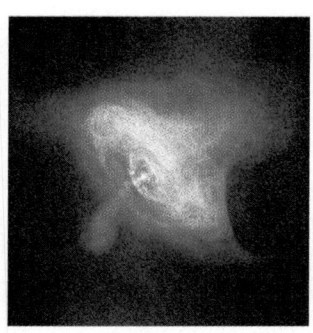

This amazingly detailed X-ray image of the Crab Nebula was recently taken by the orbiting Chandra X-Ray Observatory. The picture reveals the locations of the most energetic particles in the pulsar.

Diagnostic X-rays used in medicine have energies from 20 keV to 100 keV. Traditional medical film-radiography of this simple variety produces shadow castings. What arrives at the film is a crude mapping of the absorption that took place as the beam crossed all the interposed tissue.

In the 1970s, the marriage of the X-ray machine and the computer gave rise to a marvelous advance in X-ray technique known as *computed tomography*. Tomography (from the Greek *tomos*, or slice) is the process of creating an image of a cross-sectional region of a three-dimensional object—in this case, a person. The CT (or CAT) scan is done by rotating an X-ray source 360° around the patient and recording the transmission across the body at hundreds of different viewing angles. From that data, the computer then constructs a picture of the traversed region (see photo), producing a splendidly detailed representation showing all the soft tissue in that "slice" of the body.

22.14 Gamma Rays

Just as the atom is an interacting system of charged particles that exists only in well-defined configurations with well-defined energies, so, too, is the nucleus. Accordingly, the nucleus has a lowest-energy configuration, a ground state to which it will return once excited. A nucleus in one of its several well-defined excited states drops back to its ground state with the emission of a **gamma-ray** photon. Because the nuclear energy states are tightly bound, γ-ray energies range from keV to MeV.

There is no substantive difference between γ-rays and X-rays, other than the superficial historical one that the former were first seen to come from nuclei and the latter from atomic electrons—high-energy photons from a synchrotron come from neither. Thus, the distinction between X-rays and gamma rays vanished with the introduction of modern high-energy machines. Hundreds of modern hospitals are equipped with sophisticated devices such as linear accelerators and betatrons that provide multi-MeV electron beams that generate energetic photons for cancer treatment.

An optical image of the Crab Nebula. The light forming this picture comes from particles of intermediate energy. The filaments are due to hot gases at temperatures of tens of thousands of degrees.

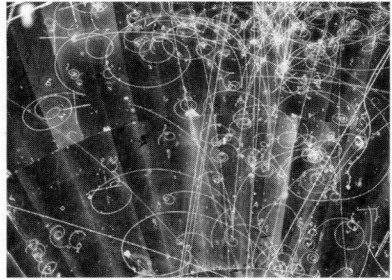

A number of events in this bubble chamber photo seem to occur in the middle of nowhere, all by themselves. But these are actually triggered by trackless gamma-ray photons. Because they are uncharged, gamma-rays don't leave a trail of bubbles and are therefore "invisible." Nonetheless they have a great deal of energy and are responsible for both the pair production (p.584) and Compton scattering (p. 535) evident in this photo.

Core Material & Study Guide

Light (and all other forms of radiant energy) is electromagnetic. A *progressive* wave is a self-sustaining, energy-carrying disturbance that travels free of its source (p. 786). The most important form of disturbance, from a theoretical perspective, is the **harmonic**, or sinusoidal, wave, which has a profile given by

$$E = E_0 \sin kx \qquad [22.1]$$

$k = 2\pi/\lambda$ is the **propagation number**. As with all waves

$$v = f\lambda \qquad [11.1]$$

Sections 22.1 (Waves and Particles), 22.2 (Electromagnetic Waves), and 22.3 (Waveforms and Wavefronts) provide basic introductory

material with few equations. To see how these are applied examine Examples 22.1 and 22.2. **Look at the CD Warm-Ups, study the Walk-Through Examples** 💿 **and then try the I-level problems.** For electromagnetic waves in vacuum

$$E = cB \qquad [22.2]$$

and

$$c = \frac{1}{\sqrt{\varepsilon_0 \mu_0}} \qquad [22.3]$$

where $\quad c = 2.997\ 924\ 58 \times 10^8$ m/s

These ideas are treated in the subsection called "The Speed of Propagation: c" (p. 789).

The energy per unit volume associated with the E- and B-fields of an electromagnetic wave are, respectively, $u_E = \frac{1}{2}\varepsilon_0 E^2$, and $u_M = \frac{1}{2}B^2/\mu_0$. The total energy density of a TEM wave is $u = \varepsilon_0 E^2$, from which it follows that the energy-per-unit-area-per-unit-time conveyed by a lightwave is its **irradiance** I, where

$$I = \frac{1}{2}c\varepsilon_0 E_0^2 \qquad [22.4]$$

This is a very practical notion and one you should be familiar with; study Section 22.4 (Energy and Irradiance) and review Example 22.3 so that you know how to do similar problems.

Light is wavelike, though unlike traditional waves, electromagnetic energy is not smoothly spread out across the wavefront but somehow concentrated within it. These energy concentrations are known as **photons**. The energy of a photon is

$$E = hf \qquad [22.5]$$

where $h = 6.626 \times 10^{-34}$ J/Hz (or 4.136×10^{-15} eV/Hz) is Planck's Constant. This is one of the great insights of the twentieth century; reread Section 22.6 (Energy Quanta) and go over Examples 22.4. and 22.5.

The **electromagnetic spectrum** ranges from radiowaves to microwaves, infrared, light, ultraviolet, X-rays, and gamma rays (p. 796).

Key Terms

wave theory	scattering
time-varying B-field	dissipative absorption
time-varying E-field	nonresonant elastic scattering
progressive wave	radiowaves
amplitude	microwaves
propagation number	infrared
wavefront	monochromatic
irradiance	quasimonochromatic
photons	ultraviolet
quanta	X-rays
ground state	gamma rays
excited states	

Discussion Questions

1. Why are fire engines sometimes painted yellow?

2. The Stanford Linear Accelerator imparts energies of up to 20 000 MeV to electrons, accelerating them along its straight two-mile course. Discuss the behavior of these particles as it pertains to electromagnetic radiation.

3. EXPLORING PHYSICS ON YOUR OWN: Does your hair dryer radiate radiowaves? How might you determine as much?

4. When NASA sends space probes to bodies beyond Earth, they routinely bathe the vehicles on the launchpad in ultraviolet radiation for several hours. Why? What's the significance of the so-called hole in the ozone layer?

5. People such as Ben Franklin rejected the corpuscular theory of light on the basis of momentum considerations. Devise an argument he might have used.

6. Is the region of space you are in right now crisscrossed by electromagnetic waves? In fact, are you permeated by them even as you read this question? Explain.

7. How does Newton's corpuscular theory of light compare with the modern quantum picture of light?

8. Central to Maxwell's theory of electromagnetic waves is the insight that the electric and magnetic fields generate one another. Explain. In vacuum, electromagnetic waves (including light) are said to be *transverse*. What does that mean?

9. How is it possible to cook a piece of meat on a paper plate in a microwave oven? What would happen to a dry cube of ice immediately after being placed in an operating microwave oven?

10. In his experiments (1780) on animal electricity, Luigi Galvani at one point tied a long wire to a dissected frog via a nerve and another wire from the feet was grounded down a well. Waiting for a thunderstorm, he observed the legs twitching in rhythm with the lightning. In what sense was this the first frog-radio? What was happening? Incidentally, H. Hertz (1880s) vainly attempted to use a frog to detect the TEM waves he was generating in his laboratory.

11. What is the physical meaning of the idea of whiteness? That is, when do we see a thing as being white? In what way is whiteness a property of the detector (that is, the human eyeball)?

12. One often hears terms such as "ultraviolet light" and occasionally even "infrared light" (although X-ray light and radio light are never spoken of). Why are such terms at best misleading?

13. Is a pulse a wave, or must a wave be something that "rises and falls" over and over again? What is it that waves in a lightwave?

14. In what regard is it better to say "light that is perceived to be yellow" rather than "yellow light"?

15. Figure Q15 shows the phenomenon of optical levitation. A tiny glass sphere one-thousandth of an inch in diameter is suspended in an upward laserbeam. What can you say about the ability of radiant energy to transfer momentum? Might it be possible to travel around in space not far from the Sun using a sail craft?

16. A few decades ago, some doctors were treating tonsillitis with X-rays. What would you expect would be a possible consequence of receiving such therapy?

Figure Q15

17. Explain the meaning of each of the Key Terms above.

The next four questions refer to Fig. MC1, which represents several different EM waves traveling in vacuum.

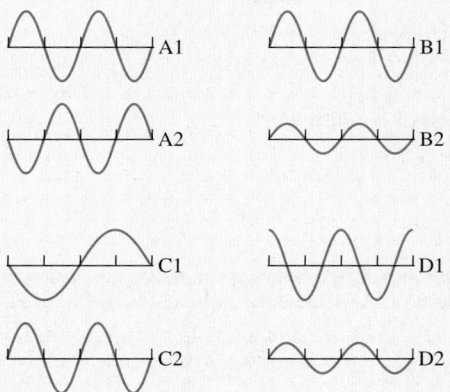

Figure MC1

1. Which two waves have different periods but the same amplitudes? (a) A1 and A2 (b) B1 and B2 (c) C1 and C2 (d) D1 and D2 (e) none of these.

2. Which two waves have the same frequencies but are 180° out-of-phase? (a) A1 and A2 (b) B1 and B2 (c) C1 and C2 (d) D1 and D2 (e) none of these.

3. Which two waves have the same frequencies, different amplitudes, and are in-phase? (a) A1 and A2 (b) B1 and B2 (c) C1 and C2 (d) D1 and D2 (e) none of these.

4. Which two waves have the same periods but are out-of phase by 90°? (a) A1 and A2 (b) B1 and B2 (c) C1 and C2 (d) D1 and D2 (e) none of these.

5. The source of electromagnetic waves is (a) a constant current (b) a charge moving only in circles (c) any accelerating charge (d) any accelerating particle (e) none of these.

6. The radio antenna for an AM station is a 75-m-high tower that is equivalent to $\frac{1}{4}\lambda$; another $\frac{1}{4}\lambda$ corresponds to the ground reflection. At what frequency does the station transmit? (a) 10 kHz (b) 75 kHz (c) 1 MHz (d) 300 kHz (e) none of these.

7. When sodium atoms are excited (for example, by heating salt in a flame) they emit bright yellow light at two wavelengths of 589.0 nm and 589.6 nm. That emission implies that the sodium atom has two nearby energy levels separated by only (a) 2 eV (b) 2×10^{-3} eV (c) 0.6 eV (d) 0.6 MeV (e) none of these.

8. In comparison to UV, light has (a) wavelengths that are shorter (b) frequencies that are higher (c) wavelengths that are equal (d) frequencies that are equal (e) none of these.

9. Dental X-ray photos are usually taken while operating the machine at about 50 000 V. The minimum wavelength of such radiation is (a) 0.025 nm (b) 0.25 nm (c) 2.5 nm (d) 25 nm (e) none of these.

10. Which statement is not true for a photon? (a) it is electromagnetic in nature (b) it always travels at the speed c independent of frequency (c) it possesses energy independent of frequency (d) it has wavelike properties (e) none of these.

11. An X-ray tube produces a beam of photons by bombarding a dense metal target with high-energy electrons. The resulting beam possesses a wide range of frequencies and is terminated at a maximum energy that (a) depends only on the target material and the temperature (b) depends on the voltage across the tube and is constant provided the voltage is constant (c) is not constant and changes even when the voltage is kept constant (d) depends on the tube length and the shielding (e) none of these.

12. Which of the following requires a physical medium to travel in? (a) lightwaves (b) radiowaves (c) sound waves (d) gamma rays (e) none of these.

13. The propagation number of a wave (a) varies inversely with wavelength (b) has to do with the number of pulses in a burst (c) varies inversely with frequency (d) depends on the speed of the wave (e) none of these.

14. Nowadays it is possible to directly measure, using electronic techniques, the frequencies of electromagnetic oscillations ranging up to 500 MHz. That corresponds to a wavelength of (a) 0.6 m (b) 0.6 cm (c) 6.0 m (d) 6.0 cm (e) none of these.

15. What is the frequency of 1-GeV gamma rays? (a) 2.4×10^{20} Hz (b) 1.5×10^{42} Hz (c) 3×10^{8} Hz (d) 2.4×10^{23} Hz (e) not enough information given.

16. For all practical purposes, the light from the following source can best be considered a plane wave (a) a street light overhead (b) a nearby desk lamp (c) a match held in the hand (d) a star in the constellation Orion (e) none of these.

17. An oscillating dipole is best described as (a) two poles moving in a circle (b) a single pole moving in two directions (c) two equal and opposite charges moving to and fro along a line (d) two like charges oscillating (e) none of these.

18. Light travels 1.000 m in vacuum in (a) 1.000 s (b) 3.336 $\times 10^{-19}$ s (c) 0.334 s (d) 3.336×10^{-9} s (e) none of these.

19. A typical AM radiowave is (a) 1.0-m long (b) 1.0-cm long (c) millions of meters long (d) hundreds of meters long (e) none of these.

20. Is there a relationship between the electric and magnetic fields of an electromagnetic wave in vacuum? (a) no (b) yes, $B = cE$ (c) yes, $E = cB$ (d) yes, $E = B/v$ (e) none of these.

21. An AM radio station transmits a signal whose electric field is received with a strength of 1.5 mV/m. If the antenna on a portable radio at that location is 0.75-m long and straight up in the air, the input voltage will be (a) 1.5 mV (b) 1.1 mV (c) 7.5 mV (d) 1.5 V (e) none of these.

22. A typical atomic transition lasts about (a) 10^{-5} s to 10^{-6} s (b) 1.0 s to 10.0 s (c) 0.01 s to 0.001 s (d) 10^{8} s to 10^{9} s (e) none of these.

23. In order of decreasing frequency, the entire electromagnetic spectrum is made up of (a) radiowaves, microwaves, IR, light, UV, and γ-rays (b) γ-rays, X-rays, UV, IR, microwaves, and radiowaves (c) radiowaves, microwaves, IR, light, X-rays, and γ-rays (d) light, IR, and UV (e) none of these.

24. The ground state of an atom is the state (a) with the highest energy (b) nearest the ground (c) with no energy (d) with the least energy (e) none of these.

25. The molecular rotational and vibrational energy states are primarily responsible for the emission of (a) radiowaves (b) light and UV (c) IR and microwaves (d) X-rays and gamma rays (e) none of these.

Suggestions on Problem Solving

1. Planck's Constant is given as either 6.6×10^{-34} J/Hz or 4.1×10^{-15} eV/Hz and, depending on how the data of the problem is provided, one or the other is the more convenient. Note that 1 Hz is 1 cycle per second, so the units of h are also J·s.

2. Many of the introductory wave problems make use of two equations ($v = \lambda f$ and $f = 1/T$) in endless variations. That was the case with sound, and it's true here as well.

3. Keep in mind that although $\lambda = c/f$, if we have a wavelength range $\Delta\lambda$, there will be a corresponding frequency range, but $\Delta\lambda \neq c/\Delta f$. We have to calculate $\lambda_1 = c/f_1$ and $\lambda_2 = c/f_2$ and then determine $\Delta\lambda = \lambda_2 - \lambda_1$ and $\Delta f = f_2 - f_1$.

4. When a charge q falls through a potential difference V, the energy change, qV, is in joules, *not* electron volts. That's a common point of confusion.

Problems ✚ Coordinated Problems ✚ Progressive Problems ✚ Solutions

STUDY GUIDE **1. Coordinated Problems:** The three problems within each magenta-colored grouping are solvable in similar ways. Note that the first of these always has a hint; moreover, its solution is provided in the back of the book. *Work out each of these sets; they'll strengthen technique and build confidence.* **2. Progressive Problems:** The problems introduced in blue unfold step-by-step carrying along the analysis in a more suggestive way than is customary. *Work out all of these; they'll guide you through the analytic process and help develop problem-solving skills.* **3. Worked-Out Solutions:** Studying worked-out solutions is an important part of learning how to solve problems. Accordingly, additional *solutions* to a number of model problems are given below. *Make sure you understand each of them before you go on to the next problem.* **4.** Also provided in the back of the book are the *Answers* to all odd-numbered problems, as well as worked-out *solutions* to those with boldface numbers. Problem numbers in italic indicate that a solution appears in the Student Solutions Manual.

SECTION 22.3: WAVEFORMS AND WAVEFRONTS

1. [I] THIS PROBLEM EXAMINES THE HARMONIC EM WAVE. Figure P1 represents the electric field of a sinusoidal electromagnetic wave propagating along the x-axis in vacuum. (a) What is the meaning of the term "amplitude" as applied to a wave (see p. 375)? (b) What is the value of the amplitude of the electric field for this wave? (c) Find the wavelength in meters. (d) What is this kind of electromagnetic radiation called? (e) What is the speed of the wave? (f) Determine the frequency of the wave.

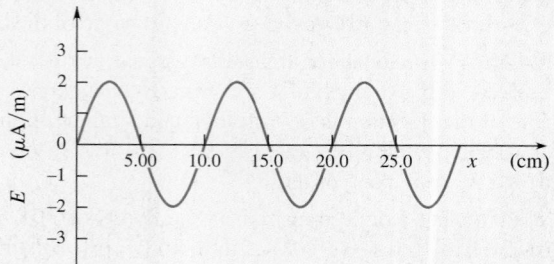

Figure P1

2. [I] THIS PROBLEM EXAMINES A SINUSOIDAL EM WAVE. A helium neon laser produces a lightbeam that has a wavelength of 632.8 nm in vacuum. (a) What is λ in meters? (b) At what speed does the beam propagate in vacuum.? That is, at what speed do the wavefronts travel? (c) Determine the frequency of the light. (d) What is the period of the EM wave?

3. [I] A lightwave has a wavelength of 500 nm. What is its propagation number? [*Hint: Go back to the definition of k.*]

4. [I] An infrared electromagnetic wave has a propagation number of 2000π m^{-1}. What is its wavelength?

5. [I] The propagation number of a harmonic electromagnetic wave is 6.283×10^{-4} m^{-1}. What is its wavelength?

6. [I] An electromagnetic wave has a profile given by

$$E = E_0 \sin kx$$

If the amplitude of the wave is 20.0 V/m, what is the size of the field at $x = 0$?

7. [I] An electromagnetic wave at $t = 0$ has the profile

$$E = E_0 \sin kx$$

Draw a plot of E versus x showing E at points $x = 0$, $\lambda/4$, $\lambda/2$, $3\lambda/4$, and λ. Remember that $k = 2\pi/\lambda$.

8. [I] Draw a plot of the function

$$E = E_0 \sin(kx - \pi/2)$$

and compare it to your result from Problem 7. How far has the profile advanced?

9. [I] The electric field of a stream of microwaves is given by

$$E = (20 \text{ V/m}) \cos \frac{2\pi}{1.00 \text{ mm}} [x - (3.00 \times 10^8 \text{ m/s})t]$$

What is its value at $x = 0$ and $t = 0$?

10. [I] The electric field of a TEM wave has the form

$$E = (5.0 \text{ V/m}) \sin k(x - vt)$$

What is the value of the E-field at $x = \lambda/4$ and $t = 0$?

11. [I] Make a sketch of the profile or shape of the wave $E = (10 \text{ V/m}) \sin[k(x - vt) + \varepsilon]$ when $\varepsilon = 0$ and again when $\varepsilon = \pi/2$. Set $t = 0$ and plot the curve for various values of x (namely, $x = 0$, $\lambda/4$, $\lambda/2$, $3\lambda/4$, λ), remembering that $k = 2\pi/\lambda$. What does the phase-shifted wave look like?

12. [I] Make a sketch of the wave

$$E = E_0 \sin k(x - vt)$$

at $t = 0$ given that $\lambda = 10$ m and $E_0 = 2.0$ V/m.

13. [I] THIS PROBLEM EXAMINES THE HARMONIC EM WAVE. Figure P13 represents the electric field of a sinusoidal electromagnetic wave propagating along the x-axis in vacuum. (a) What is the value of the amplitude of the wave? (b) What is its wavelength? (c) Determine the propagation number. (d) Compute the frequency of the wave. (e) What is the period (T) of the wave in seconds? (f) Consider the fields at points A, B, C, D, and E, and remember that

the wave advances to the right. At each point, will the field a moment later (say $T/100$) be larger or smaller, positive or negative?

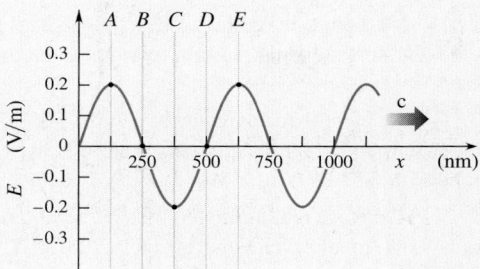

Figure P13

14. [I] Write an expression for the specific wave profile shown in Fig. P13 using the results of the previous problem.

> SOLUTION: The standard form of the sinusoidal profile is Eq. (22.1), namely, $E = E_0 \sin kx$. We know the amplitude and propagation number and so $E = (0.2 \text{ V/m}) \sin (12.6 \times 10^6 \text{ m}^{-1}) x$.

15. [I] In the article "The Longest Electromagnetic Waves" (*Sci. Amer.*, March 1962), J. R. Heirtzler described the detection of waves 18.6×10^6 miles in wavelength. What type of waves were they, and what was their period?

16. [I] How far does light travel in vacuum in 10^{-9} s?

17. [I] A tuning circuit in an FM radio is designed to pick up a station at 100 MHz. If the capacitance of the input circuit is 0.5 pF, how much is the inductance?

18. [I] In 1887, Heinrich Rudolf Hertz succeeded in generating and detecting long-wavelength electromagnetic waves. His transmitter was an induction coil (a device that converted low-voltage dc into high-voltage ac) attached to a loop of wire ending in a spark gap. Across the room, a wire loop with its own open gap served as the receiver. When the circuit dicharged, an oscillatory spark flashed across the transmitter's gap, and the event was almost immediately reported by a fainter spark at the distant receiver. If the frequency of the waves was 75 MHz, what was their wavelength?

19. [I] Determine the frequency of a 200-m-long AM radiowave, assuming the conditions to be that of vacuum. [*Hint: What is the relationship between v, f, and λ?*]

20. [I] On December 12, 1901, Marconi, using a 20-kW transmitter attached to a 200-ft antenna, sent electromagnetic signals (with a 1-km wavelength) across the Atlantic for the first time. Compute the frequency of the emission. Assume the speed in air is the same as that in vacuum.

21. [I] What is the vacuum wavelength of a 20.0-Hz electromagnetic wave?

22. [II] THIS PROBLEM EXAMINES THE HARMONIC EM WAVE. Figure P22 represents the electric field of an electromagnetic wave propagating along the positive *y*-axis in vacuum. (a) What is the value of the amplitude of the wave? (b) What is its wavelength? (c) Determine the propagation number. (d) Compute

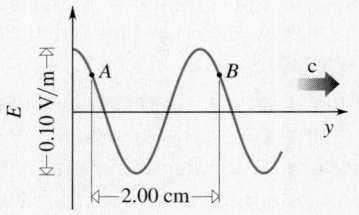

Figure P22

the frequency of the wave. (e) What is the period (T) of the wave in seconds? (f) Given that the field is 0.05 V/m at $y = 0$, write an expression for the profile of the wave.

23. [II] Figure P23 shows two electromagnetic waves in vacuum. Wave-1 travels to the right, wave-2 travels to the left. (a) If the frequency of wave-1 is 4.00 THz, what is the frequency and wavelength of wave-2? (b) What is the period of wave-1? (c) Write expressions for each wave in terms of k_1 and ω_1.

> SOLUTION: (a) As you can see from the diagram $\lambda_1 = 4\lambda_2$. Since all EM waves travel at the same speed, c, in vacuum, $f_2 = 4f_1 = 4(4.00 \text{ THz}) = 16.0$ THz. $c = \lambda_2 f_2$, hence $\lambda_2 = c/f_2 = (2.998 \times 10^8 \text{ m/s})/(16.0 \times 10^{12} \text{ Hz}) = 1.87 \times 10^{-5}$ m. (b) The period is 1 over the frequency, or 2.50×10^{-13} s. Since wave-1 travels to the right $E_1 = E_{01} \sin (k_1 x - \omega_1 t)$, and since wave-2 travels to the left $E_2 = E_{02} \sin (k_2 x + \omega_2 t) = E_{02} \sin 4(k_1 x + \omega_1 t)$.

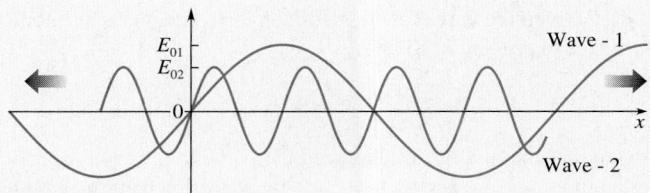

Figure P23

24. [II] Show that $\varepsilon_0 (\Delta \Phi_E / \Delta t)$ has the units of current.

25. [II] Show that the wave equation in Problem 10 can be written as

$$E = E_0 \sin (kx - \omega t)$$

26. [II] Show that the electric field of a progressive harmonic electromagnetic wave, as given in Problem 12, can be written as

$$E = E_0 \sin 2\pi f \left[\left(\frac{x}{v} \right) - t \right]$$

27. [II] What is the magnitude of the electric field of the wave given in SI units by $E = 20 \cos (kx - \omega t + \pi)$ at $x = 0$, when $t = 0$, $t = T/4$, and $t = T$?

28. [II] The electric field of an electromagnetic wave is given by

$$E = 2.0 \times 10^2 \sin [3.0 \times 10^6 \pi (x - 3.0 \times 10^8 t)]$$

where everything is in SI units. What is the wave's frequency? [*Hint: Compare this expression with the one in Problem 12.*]

29. [II] What is the amplitude of the magnetic field associated with the TEM wave in the previous problem?

30. [II] The antenna shown in Fig. P30 is called a folded dipole. How long should it be if it's to receive FM signals at 90 MHz?

31. [II] In 1982, workers at Bell Labs produced optical pulses lasting 30.0 femtoseconds. How many wavelengths of the 620-nm red light correspond to one of these little wavetrains?

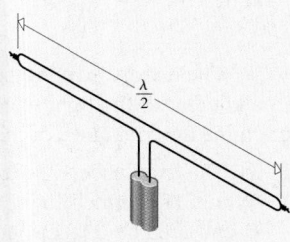

Figure P30

32. [II] Light with a frequency of 6.50×10^{14} Hz is traveling

through vacuum. How many of these waves (end-to-end) are there per centimeter?

33. [II] If light is emitted from an atom in little wavetrains (Fig. P33) lasting up to 10^{-8} s, how long, at most, is such a disturbance in space? If we approximate the wavelength as 500 nm, roughly how many waves long is the train?

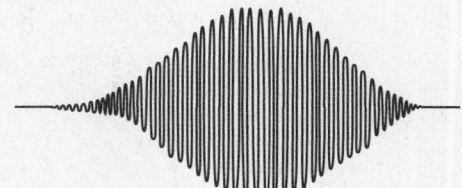

Figure P33

34. [III] Given the wave function for an electromagnetic harmonic disturbance expressed in SI units

$$E = 10^2 \sin \pi(3 \times 10^6 x - 9 \times 10^{14} t)$$

find (a) the amplitude, (b) the speed, (c) the frequency, (d) the wavelength, (e) the period, and (f) the direction of propagation.

35. [III] If the electric field of an electromagnetic wave traveling in vacuum, pointing in the y-direction, is given in SI units by

$$E_y = 200 \sin (1.00 \times 10^7 x - \omega t)$$

find (a) the vacuum wavelength and (b) the frequency. (c) In what direction does the wave progress?

36. [III] Using the results of Problem 35, write an expression for the accompanying B-field.

SECTION 22.4: ENERGY AND IRRADIANCE

37. [I] THIS PROBLEM DEALS WITH POWER AND IRRADIANCE. A 1.0-mW laser has a cross-sectional area of 0.10 cm². We want to determine the irradiance of the beam. (a) What physical quantity does 1.0 mW correspond to? (b) How many joules does the beam deliver in 2.00 s? (c) What is the beam's cross-sectional area in m²? (d) Write an expression relating the beam power and irradiance. (e) Determine the irradiance of the beam.

38. [I] The irradiance of a light beam is 500 W/m². How much power does it deliver when it impinges perpendicularly on a square photocell 1.00 cm by 1.00 cm? The entire cell is uniformly illuminated.

> **SOLUTION:** Irradiance is energy-per-unit-area-per-unit-time, or $I = P/A$. Let's first calculate the area in m², $A = (1.00 \times 10^{-2} \text{m})(1.00 \times 10^{-2} \text{m}) = 1.00 \times 10^{-4} \text{m}^2$. Hence P = IA = (500 W/m²)(1.00 × 10⁻⁴ m²) = 50.0 mW.

39. [I] The energy arriving per second on a 2.00-m² detector held perpendicular to the light from a bright star is about 2.4×10^{-9} J. Determine the irradiance of the radiation.

40. [I] The irradiance 1 m from a candle flame is just about 1.5×10^{-3} W/m². How much energy will arrive in 2.00 s on a disk having a 1.00-cm² area held as close to perpendicular as possible 1.00 m from the flame?

41. [I] The maximum irradiance I of solar radiation arriving on the lawn of the White House is 1.05×10^3 W/m². What maximum

energy will impinge each minute on a flat collector with a 1.00-square-meter area? [*Hint: Irradiance is energy-per-unit-area-per-unit-time.*]

42. [I] If a beam of light is to deliver 10.0 mJ to a perpendicular 1000-cm² surface every 1.00 s, what must be the irradiance of the beam?

43. [I] A detector receives energy via a 4.00-cm² aperture. If it is placed perpendicular to a broad beam of radiant energy with an irradiance of 500 kW/m², how much power will enter the meter?

44. [I] A surface is illuminated by a perpendicular beam that delivers 140 mJ/m² of radiant energy every 7.00 s. What is the irradiance of the beam?

45. [I] If the average irradiance of sunlight incident on a 10-m × 10-m collector is 800 W/m², how much power is arriving?

46. [I] Show that the irradiance of a harmonic electromagnetic wave can be written as

$$I = \frac{1}{2c\mu_0} E_0^2$$

47. [II] An electromagnetic wave is propagating through space. Write an expression for the total energy per unit volume of the wave in terms of the B-field and μ_0.

48. [II] Consider an electromagnetic wave propagating through space; write an expression for the instantaneous energy-per-unit-area-per-unit-time flowing along in terms of the B-field, μ_0, and c.

49. [II] A TEM wave given by

$$E = 2.0 \times 10^2 \sin [3 \times 10^6 \pi(x - 3.0 \times 10^8 t)]$$

where everything is in SI units, impinges on a perpendicular surface in space. What is the irradiance?

50. [II] A beam of TEM harmonic waves has an irradiance of 13.3 W/m². What is the amplitude of the electric field?

51. [II] A plane harmonic electromagnetic wave propagating in space has an irradiance of 50.0 W/m²; determine the amplitude of its B-field.

52. [II] A monochromatic point source radiates 100 W uniformly in all directions. What is the irradiance at a distance of 2.00 m? What is the value of the electric field amplitude at that point?

53. [III] A laser that emits pulses of UV lasting 2.00 ns has a beam diameter of 2.5 mm. If each burst contains an energy of 3.0 J, (a) what is the length in space of each pulse? (b) what is the average energy per unit volume (J/m³), the energy density, in one of these pulses?

54. [III] The *rms* electric field of a laserbeam in vacuum has a value 4.00 kV/m. If the beam has a cross-sectional area of 2.00×10^{-5} m², what's the average power it delivers?

SECTION 22.6: ENERGY QUANTA

55. [I] A photon of blue light has a frequency of 650×10^{12} Hz. What is its energy in joules?

56. [I] What is the energy of a 400-THz red photon?

57. [I] THIS PROBLEM WILL HELP US UNDERSTAND THE ENERGY OF A PHOTON. Two monochromatic EM waves are traveling in the positive x-direction in vacuum (Fig. P57). (a) What can you say, comparatively, about their wavelengths? (b) Express the frequency (f_1) of wave-1 in terms of the frequency (f_2) of wave-2. (c) Write an

expression relating photon energy and frequency. (d) If the photons that constitute wave-1 each have an energy of 1.0 eV, how much energy will the photons of wave-2 have?

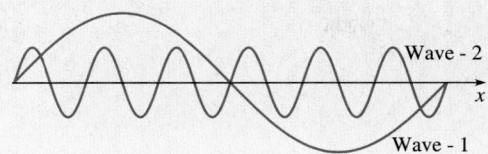

Figure P57

58. [I] A college radio station broadcasting at 91.1 MHz sends out a stream of electromagnetic radiation. What is the energy of an individual photon? Compute the answer in joules and eV.

> SOLUTION: 1 MHz = 1×10^6 Hz. E = hf = $(6.626 \times 10^{-34}$ J·s$)(91.1 \times 10^6$ Hz$)$ = 6.04×10^{-26} J and E = hf = $(4.136 \times 10^{-15}$ eV·s$)(91.1 \times 10^6$ Hz$)$ = 3.77×10^{-7} eV.

59. [I] A photon has an energy of 100×10^{-15} J; what is its frequency? What kind of electromagnetic radiation does that correspond to?

60. [I] What is the energy of a tangerine-colored photon where λ_0, the vacuum wavelength, equals 616 nm?

61. [I] Determine the vacuum wavelength, frequency, and energy in joules of a 2.00-eV photon. What type of radiation is it? [*Hint: Planck's Constant is given in eV/Hz.*]

62. [I] A photon has a wavelength in vacuum of 1.00 cm. What is its energy in eV?

63. [I] What is the energy in eV of a photon with a frequency of 1000 Hz?

64. [I] A continuous 1.0 mW laserbeam can be imagined as a stream of photons each having an energy of 2.00×10^{-19} J. When the beam is incident on a detector, how many photons arrive in 10.0 s?

65. [I] A beam of electromagnetic radiation has a cross-sectional area of 1.00×10^{-4} m². It is comprised of photons each of energy 2.0×10^{-19} J. If on average 4.00×10^{12} photons arrive on a target each second, what is the irradiance of the beam?

66. [I] It is found that a beam of radiant energy having a cross-sectional area of 1.00×10^{-6} m² delivers 2.00×10^{21} photons per second to a large perpendicular surface. What is the photon flux density?

67. [I] If each photon in Problem 66 has an energy of 6.0×10^{-15} J, what is the irradiance of the beam?

68. [I] A photon in vacuum has an energy of 12.4 eV. Determine its wavelength and identify the type of EM radiation it is.

> SOLUTION: We can find its frequency and from that its wavelength. E = 12.4 eV = hf = $(4.136 \times 10^{-15}$ eV·s$)f$; f = $(12.4$ eV$)/(4.136 \times 10^{-15}$ eV·s$)$ = 2.998×10^{15} Hz. λ = c/f = $(2.998 \times 10^8$ m/s$)/(2.998 \times 10^{15}$ Hz$)$ = 1.00×10^{-7} m. Radiant energy of wavelength 100 nm is UV.

69. [II] THIS PROBLEM DEALS WITH THE ENERGY OF A STREAM OF PHOTON. The antenna of a radio station beams out 50 megawatts of carrier wave at 103.1 MHz. We want to determine how many photos are emitted each second. (a) What is the frequency of each photon? (b) How much energy, in joules, is associated with each photon? (c) How much energy is radiated in joules per second? (d) How many photons are radiated per second?

70. [II] If 1.00×10^{15} photons of 450 THz red light are incident on a detector each second, what power does that correspond to?

71. [II] A laserbeam has a power of 100 mW and a cross-sectional area of 3.14×10^{-6} m². The frequency of the radiation is 0.500×10^{15} Hz. On average how many photons arrive on a perpendicular target receiving the entire cross section of the beam per second?

72. [II] A beam of yellow light (580 nm) with an irradiance of 1.00 mW/m² is incident on a target. How many photons arrive per square meter each second?

73. [II] An electron falls through a potential difference of 1.00×10^4 V. What is the maximum energy of a resulting X-ray photon? What is the minimum corresponding wavelength? Do this calculation in electron volts and then compare the results with that of Example 22.8.

74. [III] It takes an energy of 33 keV to remove the innermost electron from an atom of iodine. Show that iodine will be a powerful absorber of X-rays at a frequency of 8.0×10^{18} Hz.

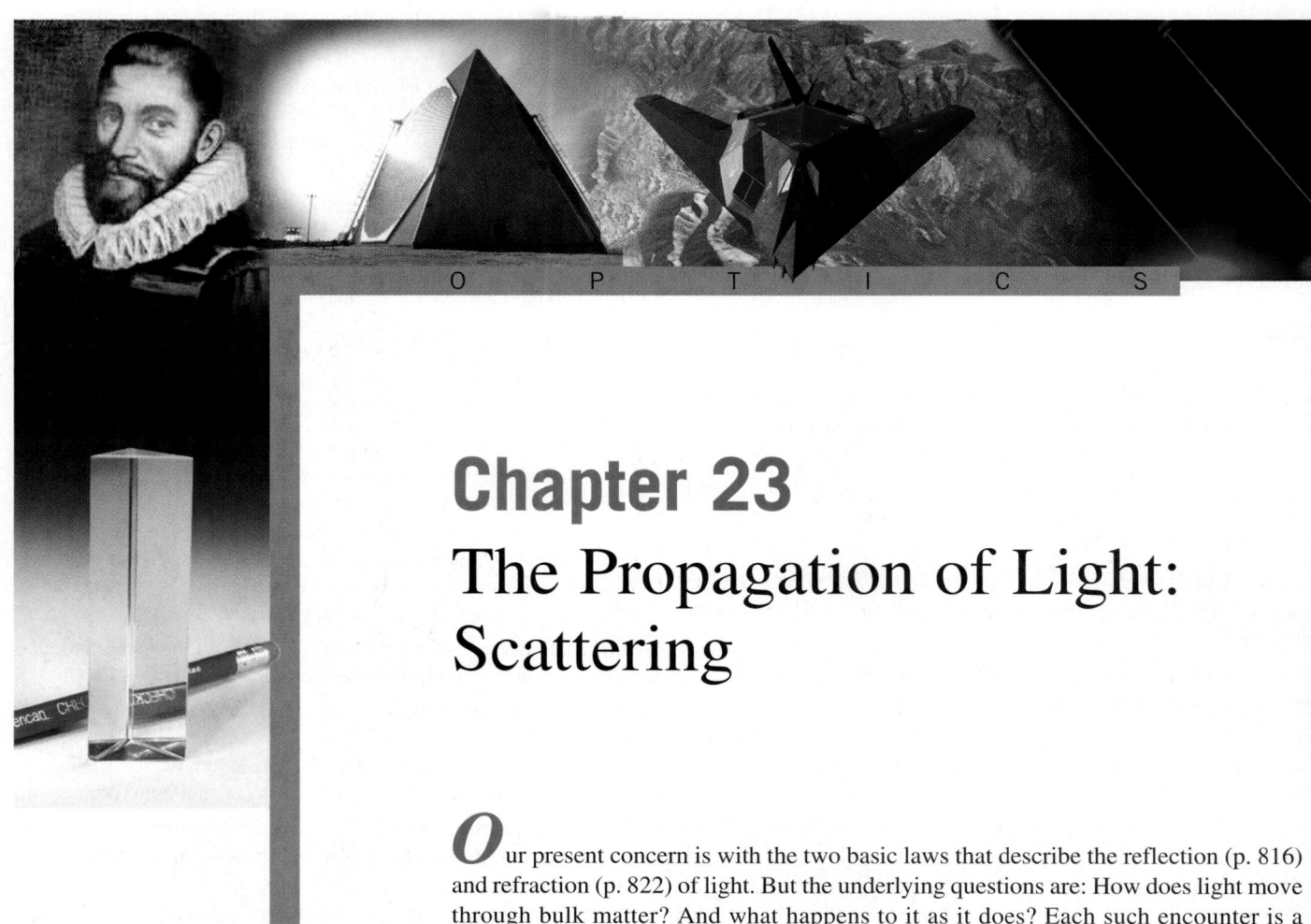

Chapter 23
The Propagation of Light: Scattering

*O*ur present concern is with the two basic laws that describe the reflection (p. 816) and refraction (p. 822) of light. But the underlying questions are: How does light move through bulk matter? And what happens to it as it does? Each such encounter is a stream of photons sailing through an array of atoms suspended in the void. The details of that marvelous journey determine why the sky is blue and blood is red, why your cornea is transparent and your hand opaque, why snow is white and rain is not.

Scattering

At its core, this chapter is about **scattering**, the absorption and prompt re-emission of electromagnetic radiant energy by atoms and molecules. The processes of reflection and refraction are macroscopic manifestations of scattering occurring on a submicroscopic level. Let's first consider the propagation of radiant energy through various homogeneous media.

23.1 Rayleigh Scattering: Blue Skies

Imagine a narrow beam of sunlight, a torrent of photons having a broad range of frequencies, advancing through empty space. As it progresses, the beam spreads out very slightly, but apart from that, all the energy continues forward at c. There is no scattering, and the beam cannot be seen from the side (the photons move straight ahead). Nor does the light tire or diminish in any way. When a star in a nearby galaxy 1.7×10^5 ly away was seen to explode in 1987, the flash of light that reached Earth had been sailing through space for 170 000 years. Photons are timeless.

Now, suppose we mix a wisp of air into the void—some molecules of nitrogen, oxygen, and so forth. Since these molecules have no resonances in the visible (pp. 796 and 803), no one of them can be raised into an excited state by absorbing a quantum of light and the gas is transparent. Instead, each molecule behaves as a little oscillator whose electron cloud can be driven into a ground-state vibration by an

Figure 23.1 A beam of light traversing a region of widely spaced molecules. The light laterally scattered is mostly blue, and that's why the sky is blue. The unscattered light, which is rich in red, is viewed only when the Sun is low in the sky at sunrise and sunset.

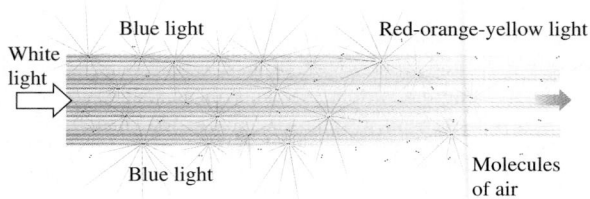

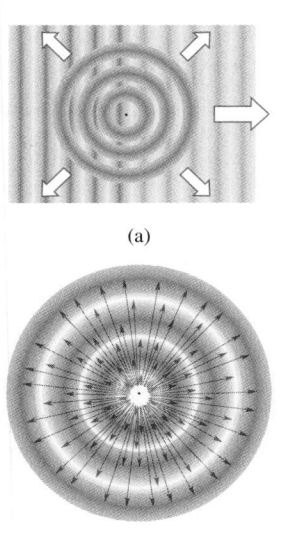

(a)

(b)

Figure 23.2 (a) Scattering of some of the energy out from a plane wave in the form of a spherical wavelet. (b) The process is continuous, and billions of photons per second stream out of the scattering atom in all directions as wave after wave sweeps over it.

The **sky is blue**, in part, because of Rayleigh scattering from air molecules in the tenuous upper atmosphere.

incoming photon. Immediately upon being set oscillating, the molecule initiates the re-emission of light. A photon is absorbed, another photon of the same frequency is emitted; the light is *elastically scattered*. The molecules are randomly oriented, and photons scatter out every which way (Fig. 23.1), over and over again as the atoms are excited and re-excited. Even when the light is fairly dim, the number of photons is immense, and it looks as if the molecules are each scattering little classical spherical wavelets (Fig. 23.2)—energy streams out in every direction. Still, the scattering process is quite weak and the gas tenuous, so the beam is very little attenuated unless it passes through a tremendous volume of air.

The amplitudes of these ground-state vibrations, and therefore the amplitudes of the scattered light, increase with frequency because the molecules all have electronic resonances in the UV. The closer the driving frequency is to a resonance, the more vigorously the oscillator responds (p. 356). So, violet light is strongly scattered laterally out of the beam, as is blue to a slightly lesser degree, as is green to a considerably lesser degree, as is yellow to a still lesser degree, and so on. The beam that traverses the gas will thus be richer in the red end of the spectrum, while the light scattered out (sunlight not having much violet in it in the first place) will be richer in blue. That, in part, is why the sky is blue.

Long before Quantum Mechanics, Lord Rayleigh (1871) analyzed scattered sunlight in terms of molecular oscillators and correctly concluded that the intensity of the scattered light was proportional to $1/\lambda^4$, which increases with f^4. Before this work, it was widely believed that the sky was blue because of scattering from minute dust particles. Since that time, *scattering involving particles smaller than a wavelength* has been referred to as **Rayleigh scattering**. A human's blue eyes, a blue jay's feathers, the blue-tailed skinks's blue tail, and the baboon's blue buttocks are all colored via Rayleigh scattering.

As we will see in a moment, a dense uniform substance will not appreciably scatter laterally, and that applies to much of the lower atmosphere. Something else beyond Rayleigh scattering must be contributing to making the lower dense regions of the sky blue. What happens in the atmosphere is that the thermal motion of the air results in rapidly changing *density fluctuations* on a local scale. And these momentary, fairly random fluctuations cause more molecules to be in one place than another and to radiate more in one direction than another. A theory of scattering from these fluctuations gives very much the same results as Rayleigh obtained for a tenuous gas.

Sunlight streaming into the atmosphere from one direction is scattered in all directions. Without an atmosphere, the daytime sky would be as black as the void of space, as black as is the Moon sky. With your back to the Sun, the light coming down toward your eyes from

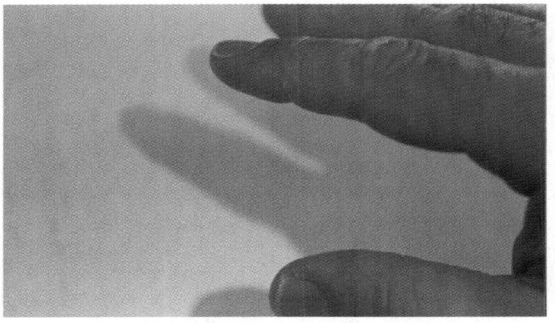

This photo of a white piece of paper was taken near the window of an airplane at 35 000 ft. Blue light scattered by the atmosphere streams in on the left through the window. Where the orange-white incandescent light from a reading lamp is shielded by the hand, the shadow (illuminated only via the window) is a deep blue. You can see the same effect on snow. When something blocks the direct sunlight the shadows are blue.

Example 23.1 **[I]** A beam of white light traverses a medium composed of randomly distributed particles that are each much smaller than a typical wavelength. Compare the amount of scattering occurring for the red (710 nm) component with that of the violet (400 nm) component.

Solution Because the particles are small, Rayleigh scattering applies. (1) TRANSLATION—Compare the amounts of light of two different wavelengths that are scattered from very small particles. (2) GIVEN: $\lambda_r = 710$ nm and $\lambda_v = 400$ nm. FIND: Compare their scattered intensities. (3) PROBLEM TYPE—Rayleigh scattering. (4) PROCEDURE—The degree of Rayleigh scattering is proportional to $1/\lambda^4$. (5) CALCULATION—The red

wave length $\lambda_r = [(710$ nm$)/(400$ nm$)]\lambda_v = 1.775\lambda_v$ and so

$$1/\lambda_r^4 = (1/1.775\lambda_v)^4$$

Consequently, violet is scattered $(1.775)^4 = \boxed{9.93}$ times more intensely than red.

Quick Check: We can check the order-of-magnitude of the answer by noting that red is very roughly twice the wavelength of violet, and the answer must be between $1^4 = 1$ and $2^4 = 16$. At least we pushed the right keys on the calculator.

[For more worked problems click on WALK-THROUGHS in CHAPTER 23 on the CD.]

Figure 23.3 Solar rays reach about 18° beyond the daytime terminator because of atmospheric scattering. Over this twilight band the skylight fades to the complete darkness of night.

Without an atmosphere to scatter sunlight, the Moon's sky is an eerie black.

the bright blue sky is all scattered. It's this scattering that's responsible for twilight on Earth. Were it not for the atmosphere and scattering, there would be a sharp transition from day to night (Fig. 23.3).

When the Sun is low over the horizon, its light passes through a great thickness of air (far more so than it does at noon). With the blue-end appreciably attenuated, the reds and yellows propagate along the line-of-sight from the Sun to produce Earth's familiar fiery sunsets.

Scattering and Interference

In dense media, there are a tremendous number of close-together atoms or molecules contributing an equally tremendous number of scattered electromagnetic wavelets. These wavelets overlap and interfere in a way that does not occur in a tenuous medium. As a rule, the denser the substance through which light advances, the less the lateral scattering.

The phenomenon of interference was touched on earlier (see The Superposition of Waves, p. 385) and will be treated in detail in Chapter 25; here, the basics suffice. **Interference** *is the superposition of two or more waves producing a resultant disturbance that is the sum of the overlapping wave contributions.* Figure 23.4 shows two harmonic waves of the same frequency and nearly the same amplitude traveling in the same direction. When they are in-phase, the resultant at every point is the sum of the two wave-height values (Fig. 23.4a). Crest falls upon crest, trough falls upon trough, and this extreme case is called *constructive interference*. The result is a wave of the same frequency with an amplitude equal to the sum of the constituent amplitudes. When the phase difference reaches 180°, trough falls upon crest, the waves tend to cancel, and we have the other extreme, called *destructive interference* (Fig. 23.4c). When the two constituent waves have the same amplitude, they completely cancel one another.

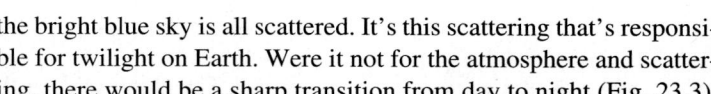

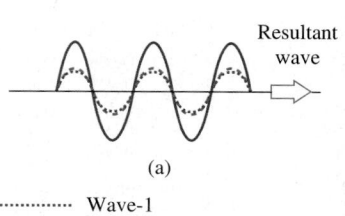

.... Wave-1
- - - - - - Wave-2
———— The sum of wave-1 and wave-2

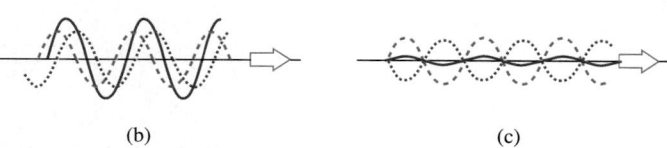

Figure 23.4 The superposition of two waves. These combine to form a resultant wave. (a) When the component waves are in-phase, we have *constructive interference*, and the resultant is large. (b) As the phase difference increases, the resultant decreases. (c) When the component waves are 180° out-of-phase, the resultant has its smallest value, and we have *destructive interference*.

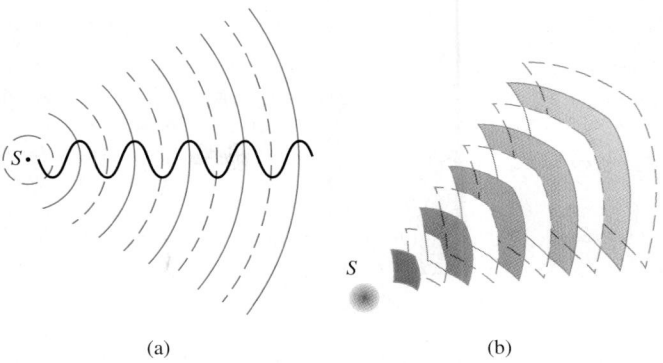

Figure 23.5 (a) Two-dimensional circular sinusoidal waves. (b) Three-dimensional segments of spherical waves streaming from a point source *S*. (See Fig. 11.28.)

In two dimensions the sinusoidal wave from a point source *S* can be represented as a series of circular wavefronts (Fig. 23.5*a*), which are drawn solid along crests and dashed along troughs. In three dimensions (Fig. 23.5*b*) the point source sends out a stream of spherical wavefronts. When the light from two such sources coexist in the same space, wherever the waves are in-phase (solid overlays solid and dashed overlays dashed), there will be constructive interference (Fig. 23.6). Similarly, wherever the waves are 180° out-of-phase (solid overlays dashed), there will be destructive interference.

The simple theory of Rayleigh scattering has the molecules randomly arrayed in space so that the phases of the wavelets scattered off to the side have no particular relationship to one another and there can be no sustained interference between them. That situation will occur when the separation between the molecular scatterers is roughly a wavelength or more, as it is in a tenuous gas. When there are many scatterers, the random hodgepodge of overlapping waves effectively averages away the interference. ***Random, widely spaced scatterers driven by an incident wave emit wavelets that are essentially independent of one another.*** Laterally scattered light, unimpeded by interference, streams out of the beam. And this is approximately the situation existing about 100 miles up in the Earth's tenuous high-altitude atmosphere, where a good deal of blue-light scattering takes place. The direction defined by the beam itself is special—**all the scattered wavelets add constructively with each other in the forward direction**, and what remains of the beam continues to advance. {For a detailed explanation of why that happens click on FORWARD SCATTERING under FURTHER DISCUSSIONS on the CD.} 💿

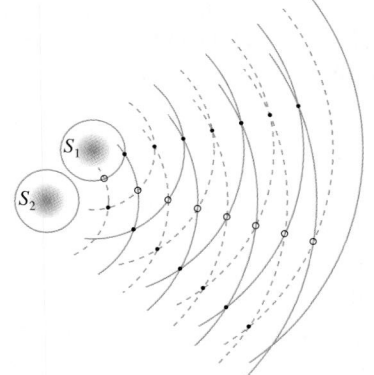

Figure 23.6 The overlapping of two circular harmonic waves to produce constructive (•) and destructive (○) interference.

The Transmission of Light through Dense Media

Now, suppose the amount of air in the region under consideration is increased. In fact, imagine that each little cube of air, one wavelength on a side, contains a great many molecules, whereupon it is said to have an appreciable *optical density*. At the wavelengths of light, the Earth's atmosphere at STP has about three million molecules in such a λ^3-cube. The wavelets ($\lambda \approx 500$ nm) radiated by sources so close together (≈ 3 nm) cannot properly be treated as random. The light beam effectively encounters a fairly uniform medium with no discontinuities to destroy the symmetry. Again the scattered wavelets interfere constructively in the forward direction (that much is independent of the arrangement of the molecules), but now because there are so many molecules so close together, destructive interference predominates in all other directions. For every atom scattering laterally there is always another atom a distance $\lambda/2$ away that scatters in the same direction; these waves will be 180° (i.e., $\lambda/2$) out-of-phase and cancel. *No light ends up scattered laterally or backwards*. {For a more detailed explanation of how that happens click on DENSE-MEDIA SCATTERING under FURTHER DISCUSSIONS on the CD.} 💿

The scattering phenomenon on a per-molecule basis is extremely weak. In order to have half its energy scattered, a beam of green light will have to traverse ≈ 150 km of

atmosphere. Since about 1000 times more molecules are in a given volume of liquid than in the same volume of vapor (at atmospheric pressure), we can expect to see an increase in scattering. Still, the liquid is a far more ordered state with much less pronounced density fluctuations, and that should suppress the nonforward scattering appreciably. Accordingly, an increased scattering, per unit volume, is observed in liquids, but it's more like 5 to 50 times as much rather than 1000 times. *Molecule for molecule*, liquids scatter substantially less than gases.

Transparent amorphous solids, such as glass and plastic, will also scatter light but very weakly. Good crystals, like quartz and mica, with their almost perfectly ordered structures, scatter even more faintly. Of course, imperfections of all sorts (dust and bubbles in the liquids; flaws and impurities in the solids) will serve as scatterers, and when these are small, as in the gem moonstone, the emerging light will be bluish.

Reflection

When a beam of light impinges on the surface of a transparent material, such as a sheet of glass, the wave "sees" a vast array of very closely spaced atoms that will somehow scatter it. Remember that the wave may be ≈ 500 nm long while the atoms and their separations (≈ 0.2 nm) are thousands of times smaller. In the case of transmission through a dense medium, the scattered wavelets cancel each other in all but the forward direction and just the ongoing beam is sustained. But that can only happen if there are no discontinuities, whereupon the scattering is uniform. This is not the case at an interface between two different transparent media (such as air and glass), which is a jolting discontinuity. When a beam of light strikes such an interface, some light is always scattered backward; we call this phenomenon **reflection**.

23.2 Internal and External Reflection

Imagine that light is traveling across a large block of glass (Fig. 23.7a). Now, suppose that the block is sheared in half perpendicular to the beam. The two segments are then separated, exposing the smooth flat surfaces depicted in Fig. 23.7b. Just before the cut was made, there was no lightwave traveling to the left inside the glass—we know the beam only advances. Whatever light was scattered back by one layer of atoms was canceled by the light scattered back by another layer of atoms a distance $\frac{1}{2}\lambda$ away. (Two such waves would be completely out-of-phase and would interfere destructively.) Now, however, there must be a wave (beam-I) moving to the left, reflected from the surface of the right-hand block. The implication is that the region of scatterers on and beneath the exposed surface of the right-hand block is now "unpaired," and the backward radiation they emit can no longer be canceled. The region of oscillators that was adjacent to these, prior to the cut, is now on the section of the glass that is to the left. When the two sections were together, these scatterers presumably also emitted wavelets in the backward direction that were 180° out-of-phase with, and canceled, beam-I. Now they produce reflected beam-II. Each molecule scatters

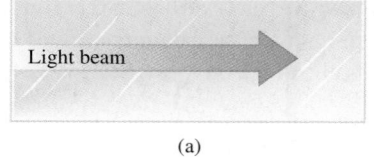

(a)

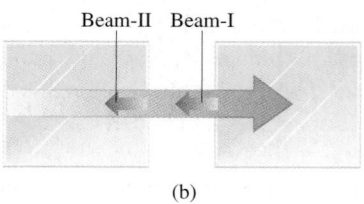

(b)

Figure 23.7 (a) A light beam traveling through a dense homogeneous medium such as glass. (b) If the block of glass is cut and parted, the light is reflected backward at the two new interfaces. Beam-I is externally reflected, and beam-II is internally reflected. When the two pieces are pressed back together, the two reflected beams cancel one another.

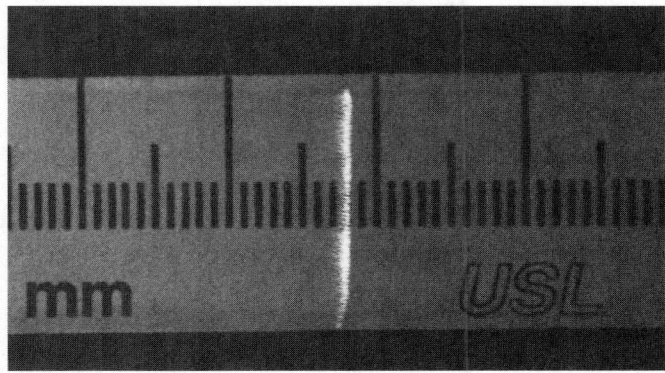

A plane wave pulse caught as it swept along the surface of a ruler (left to right). The pulse extended in time for a mere 300×10^{-15} s and so was only a fraction of a millimeter long.

light in the backward direction and, in principle, each and every molecule contributes to the reflected wave. Nonetheless, in practice, it is a thin layer ($\approx \frac{1}{2}\lambda$ deep) of unpaired atomic oscillators near the surface that is effectively responsible for the reflection. For an air-glass interface, about 4% of the energy of an incident beam falling perpendicularly *in* air *on* glass will be reflected straight back out by this layer of unpaired scatterers. And that's true whether the glass is 1.0 mm thick or 1.00 m thick.

Beam-I reflects off the right-hand block; because light was initially traveling from a less to a more optically dense medium, this is called **external reflection**. Since the same thing happens to the unpaired layer on the section that was moved to the left, it, too, reflects backward. With the beam incident perpendicularly *in* glass *on* air, 4% must again be reflected, this time as beam-II. This process is referred to as **internal reflection**. If the two glass regions are made to approach one another increasingly closely (so that we can imagine the gap to be a thin film of, say, air—p. 905), the reflected light will diminish until it ultimately vanishes as the two faces merge and disappear and the block becomes continuous again.

Remember this 180° *relative phase shift between internally and externally reflected light*—we will need it later on. This kind of reflection is of practical importance when you have a microscope with lots of compound lenses and perhaps a dozen or two interfaces each kicking back ≈4% of the incident light (just try looking through 20 or 30 layers of plastic food wrap). To overcome that difficulty, modern high-quality lenses are covered with antireflection coatings (p. 907).

We know from experience with the common mirror that white light is reflected as white—it certainly isn't blue. To see why, first realize that the layer of scatterers responsible for the reflection is very roughly $\lambda/2$ thick. Thus, the larger the wavelength, the deeper the layer contributing, and the more scatterers there are acting together. This tends to balance out the fact that each scatterer is less efficient as λ increases (remember $1/\lambda^4$). The combined result is that *the surface of a transparent medium reflects all wavelengths about equally and doesn't appear colored in any way*. That, as we will see, is why this page looks white under white-light illumination.

23.3 The Law of Reflection

When light arrives on a transparent surface, a portion of it is reflected and the remainder is transmitted. Figure 23.8 shows a beam composed of plane wavefronts impinging at some angle on the smooth flat surface of an optically dense medium (let it be glass). Assume that the surrounding medium is vacuum, and let's only concern ourselves with the reflected light. Follow one wavefront as it sweeps in and across the molecules on the surface. For the

Figure 23.8 A beam of plane waves incident on a distribution of molecules constituting a piece of clear glass or plastic. Part of the incident light is reflected and part refracted.

Vacuum

Incident beam

Reflected beam

Transmitted or refracted beam

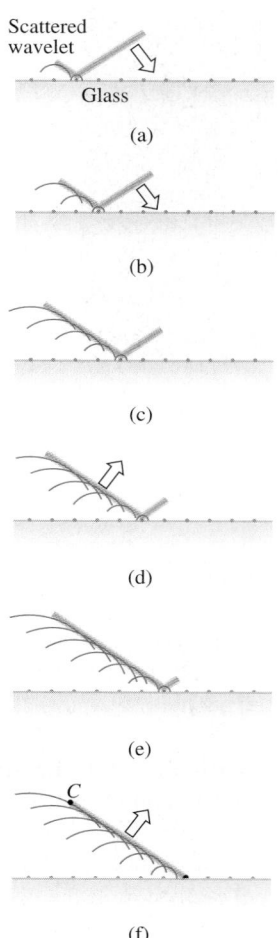

Figure 23.9 An incoming plane wave scattering off a layer of molecules. Because the wavelength is thousands of times larger than the molecular spacings, the reflection can be treated as occurring at the surface..

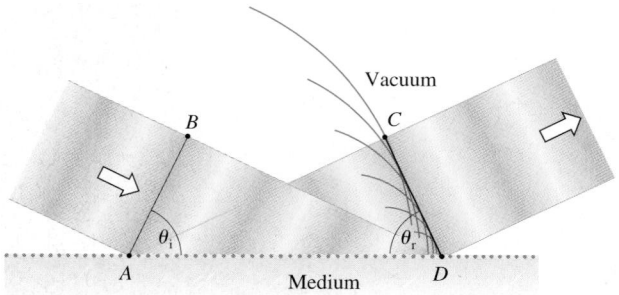

Figure 23.10 Wavefront geometry for reflection. The reflected wavefront $\overline{CD}$ is formed of waves scattered by the atoms on the surface from A to D. Just as the first wavelet arrives at C from A, the atom at D emits, and the wavefront along $\overline{CD}$ is completed.

sake of simplicity, in Fig. 23.9 we have omitted everything but one molecular layer at the interface. As the wavefront descends, it excites and re-excites, one scatterer after another, each of which reradiates a stream of photons that can be thought of as a hemispherical wavelet in the incident medium. Because the wavelength is so much greater than the separation between the molecules, the wavelets advance together and add constructively (crest overlapping crest, trough overlapping trough) in only one direction, and there is one well-defined **reflected** wave.

In Fig. 23.10, the line $\overline{AB}$ lies along an incoming wavefront while $\overline{CD}$ lies on an outgoing wavefront—in effect, $\overline{AB}$ transforms on reflection into $\overline{CD}$. With Fig. 23.9 in mind, we see that the wavelet emitted from A will arrive at C in-phase with the wavelet just being emitted from D (as it is stimulated by B), as long as the distances $\overline{AC}$ and $\overline{BD}$ are equal. If all the wavelets emitted from all the surface scatterers are to overlap in-phase and form a single reflected plane wave, it must be that $\overline{AC} = \overline{BD}$. Then, since the two triangles have a common hypotenuse

$$\frac{\sin \theta_{\mathrm{i}}}{\overline{BD}} = \frac{\sin \theta_{\mathrm{r}}}{\overline{AC}}$$

All the waves travel in the incident medium with the same speed v_{i}. It follows that in the time Δt it takes for point B on the wavefront to reach point D on the surface, the wavelet emitted from A reaches point C. In other words, $\overline{BD} = v_{\mathrm{i}} \Delta t = \overline{AC}$, and so from the above equation, $\sin \theta_{\mathrm{i}} = \sin \theta_{\mathrm{r}}$, which means that

$$\theta_{\mathrm{i}} = \theta_{\mathrm{r}} \tag{23.1}$$

The angle of incidence equals the angle of reflection. This equation is the first part of the **Law of Reflection**.

Drawing wavefronts can get things a bit cluttered, so we now introduce another con-

A F-117A Stealth fighter has an extremely small radar profile, that is, it returns very little of the incoming microwaves back to the station that sent them. That's accomplished mostly by constructing the aircraft with flat tiled-planes that use the Law of Reflection to scatter the radar waves away from their source. One wants to avoid $\theta_{\mathrm{i}} = \theta_{\mathrm{r}} \approx 0$.

By placing a pair of pins in front of a flat mirror and aligning their images with another pair of pins, you can easily verify that $\theta_{\mathrm{i}} = \theta_{\mathrm{r}}$.

It's tempting to think of reflection as the process wherein light "bounces off" a surface, but that's not what actually happens. A portion of the incoming light is scattered back into the incident medium by the atoms in and near the interface. And if that interface is smooth so that all the wavelets can stay together and interact in a coordinated way, $\theta_i = \theta_r$.

A modern phased-array radar system. The field of individual small antennas behaves very much like the atoms on a smooth surface. By introducing a proper phase shift between adjacent rows the antenna can "look" or transmit in any direction. A reflecting surface has a similar phase shift determined by θ_i as the incident wave sweeps over the array of atoms.

A **ray** is a line tracing the path of the flow of radiant energy.

venient scheme for visualizing the progression of light. The imagery of antiquity was in terms of straight-line streams of light, a notion that got into Latin as "radii" and reached English as "rays." **A ray is a line drawn in space corresponding to the direction of flow of radiant energy.** It is a mathematical device and not a physical entity. In a medium that's uniform (homogeneous), rays are straight. If the medium behaves in the same manner in every direction (isotropic), **the rays are perpendicular to the wavefronts**. Thus, for a point source emitting spherical waves, the rays, which are perpendicular to them, point radially outward from the source. Similarly, the rays associated with plane waves are all parallel (Fig. 23.11*a*). Rather than sketching bundles of rays, we can simply draw one incident ray and one reflected ray (Fig. 23.11*b*). *All the angles are measured from the perpendicular (or normal) to the surface*, and thus θ_i and θ_r have the same numerical values as before (Fig. 23.10).

The **Law of Reflection** states that the angle of incidence equals the angle of reflection. Moreover, the incident and reflected rays are in the plane-of-incidence.

The ancient Greeks knew the Law of Reflection—it can be deduced by observing the behavior of a flat mirror, and nowadays that observation can be done most simply with a flashlight or, even better, a laser. The second part of the Law of Reflection maintains that **the incident ray, the perpendicular to the surface, and the reflected ray all lie in a plane** called the **plane-of-incidence** (Fig. 23.11*b*)—this is a three-dimensional business. Try to hit some target in a room with a flashlight beam by reflecting it off a stationary mirror and the importance of this second part of the law becomes obvious!

When a beam is incident upon a reflecting surface that is smooth (one for which any irregularities are small compared to a wavelength), the light re-emitted by millions upon millions of atoms will combine to form a single well-defined beam in a process called **specular reflection**. If, on the other hand, the surface is rough, although the angle of incidence will equal the angle of reflection for each ray, the whole lot of rays will emerge

Figure 23.11 (a) We select one ray to represent the beam of plane waves. Now both the angle of incidence θ_i and the angle of reflection θ_r are measured from a perpendicular drawn to the reflecting surface. (b) The incident ray and the reflected ray define a plane, known as the *plane-of-incidence*, perpendicular to the reflecting surface.

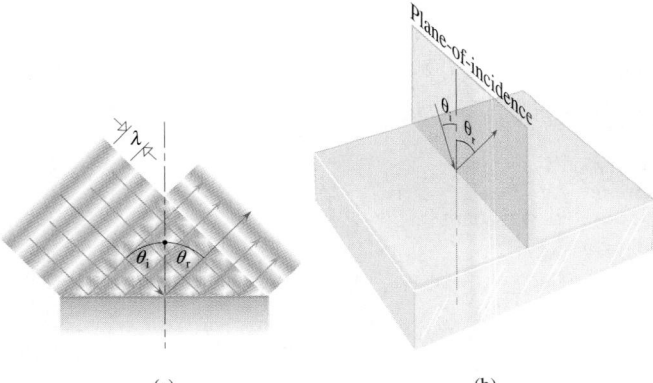

(a)   (b)

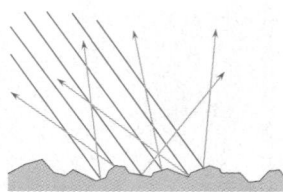

Figure 23.12 Diffuse reflection. When the surface roughness is large compared to λ, the scattering molecules are far apart, and there is no longer a direction in which all the wavelets will add constructively to produce a single reflected wave. For each ray $\theta_i = \theta_r$ but the normals point every which way. Energy goes off in a broad range of directions.

The cruiser *Aurora*, which played a key role in the Communist Revolution (1917), docked in St. Petersburg. Where the water is still, the reflection is specular. The image blurs where the water is rough.

Diffuse reflection. Here the surface is rough compared to λ, and the light (unable to interfere effectively) scatters in every direction. There is no well-defined reflected beam.

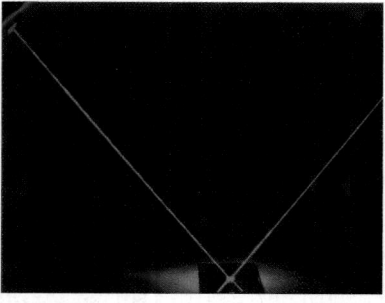

Specular reflection. A laserbeam reflected from a mirror in a well-defined beam. Note that the angle of incidence equals the angle of reflection. The laser is at the upper left in both photos.

every which way, constituting what is called **diffuse reflection** (Fig. 23.12). Both of these conditions are extremes—the reflecting behavior of most surfaces lies somewhere between them (see the photo). {Take a moment now to look at the **CD**. Go to **INTERACTIVE EXPLORATIONS** and click on **REFLECTION AND REFRACTION**. Just play with the reflection part; we'll come to refraction presently.}

The Plane Mirror

While standing in front of a plane (flat) mirror* (i.e., any polished smooth surface), every point on your body that is illuminated by some external source scatters light, some of which heads toward the mirror. Every molecule on your face that is driven by the incident *E*-field reradiates a more or less spherical wavelet or, if you prefer, sends out a cone of rays (Fig. 23.13). Some portion of these wavelets will reflect from the mirror and subsequently reach your eye (Fig. 23.14). But the eye-brain receiver, accustomed to straight-line propagation, will see the source as if it were behind the mirror. *The rays received by the observer diverge from the image point* and, as such, the image is said to be **virtual**—*it appears behind the mirror* and *cannot be projected onto a screen*. Every point on your hand, for instance, will have its corresponding image point behind the mirror, recreating the appearance of that hand (Fig. 23.15).

Figure 23.13 A person's face, like everything else we ordinarily see in reflected light, is covered with countless atomic scatterers.

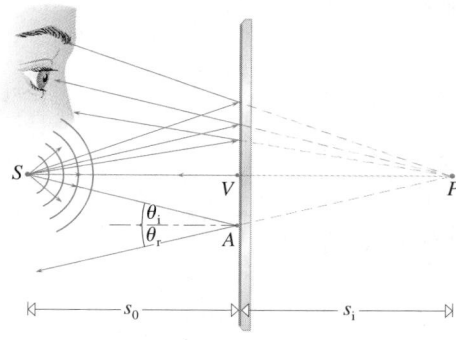

Figure 23.14 The light from an atom (*S*) on the face of someone looking into a mirror. Rays from *S* strike the mirror, $\theta_i = \theta_r$, and some reflect back to the viewer's eye. The eye-brain system apprehends the rays as if they were straight and being emitted from *P*, the image of *S* behind the mirror.

*Most mirrors in common use outside of the laboratory are back-silvered—they reflect light off a layer of metal behind the glass; that's done in order to protect the delicate reflecting coating. For simplicity, we'll draw only front-silvered mirrors of the type used in optics labs.

This tiny tiltable flat mirror (which is so small it can fit through the eye of a needle) is part of an array of thousands of such mirrors used to steer light beams in one of today's most important telecommunications devices.

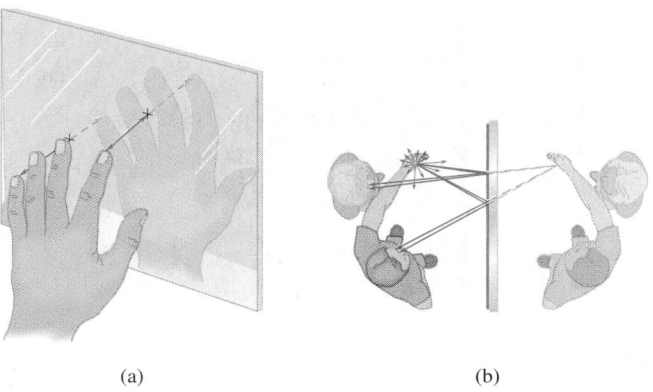

(a)   (b)

Figure 23.15 (a) The image of the left hand can be easily traced by using rays from it perpendicular to the mirror ($\theta_i = 0 = \theta_r$). They simply reflect back on themselves. A left hand becomes a life-size right hand. (b) The light from the young man's hand reflects both to his eye and the eye of the woman standing next to him. Both can therefore see the image of his hand behind the mirror.

Example 23.2 **[I]** Using Fig. 23.14, prove that the image in a flat mirror is located the same distance behind the mirror as the object is in front of it; that is, the **object distance** s_0 equals the **image distance** s_i.

Solution The central idea here is that $\theta_i 5 \theta_r$; the rest is geometry. (1) TRANSLATION—For a planar mirror, show that the object distance equals the image distance. (2) GIVEN: s_0 and s_i. FIND: Show that $s_0 = s_i$. (3) PROBLEM TYPE—Reflection/flat mirror. (4) PROCEDURE—Apply geometrical analysis to the triangles in Fig. 23.14. (5) CALCULATION—

The exterior angle of triangle SAP equals $\theta_i + \theta_r$, which equals the sum of the alternate interior angles of that triangle, namely, $\angle VSA + \angle VPA$. But $\angle VSA = \theta_i = \theta_r$ and therefore $\angle VSA = \angle VPA$, which means that triangles VAS and VAP are congruent by angle-angle-side. It follows that

$$s_0 = s_i$$

The image of the source is the same perpendicular distance behind the mirror as the object is in front of it.

The facts that the mirror is flat and that the angle of incidence equals the angle of reflection combine to produce an undistorted, right-side up, life-size image standing just as far behind the surface as the object is in front of it. Realize that the image of a left hand (the outline of which can be determined using rays falling perpendicularly on the mirror as in Fig. 23.15a) is a right hand. Smack your left hand up against a mirror (the reflected fingers must be directly against the actual fingers)—the image must be a right hand. *A single reflection changes a right-handed system into a left-handed one and vice versa.*

{At this point, open the **CD**, go to **INTERACTIVE EXPLORATIONS**, and click on **MIRRORS**. This is a lovely simulation and you'll be asked to come back to it later.} 💿

Example 23.3 **[II]** What is the length of the smallest vertical plane mirror in which you can see your entire standing body all at once, and how should it be positioned?

Solution You have to be able to see both the top of your head and the bottom of your foot. (1) TRANSLATION—Determine the

smallest planar mirror in which you can see your entire body. (2) GIVEN: $s_0 = s_i$. FIND: The minimum height of the mirror. (3) PROBLEM TYPE—Reflection/flat mirror. (4)PROCEDURE—Apply geometrical analysis to the triangles depicted in Fig. 23.16. (5) CALCULATION—A ray from your toe will enter your eye, striking point H somewhere such that $\angle DHC = \angle BHC$.

continued

Triangles *CHD* and *CHB* are congruent and so $\overline{GH} = \overline{HI} = \frac{1}{2}\overline{BD}$. Similarly, if you are to see the top of your head, $\overline{EF} = \overline{FG} = \frac{1}{2}\overline{AB}$. Thus, a mirror of length $\overline{FH}$ will do the job, where

$$\overline{FH} = \overline{FG} + \overline{GH} = \tfrac{1}{2}\overline{AB} + \tfrac{1}{2}\overline{BD} = \tfrac{1}{2}\overline{AD}$$

A mirror half your height with its upper edge lowered by half the distance between your eye and the top of your head serves nicely.

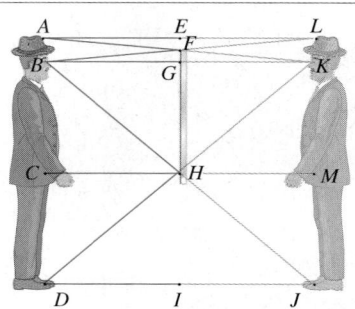

Figure 23.16 A determination of the smallest mirror ($\overline{FH}$) in which a person can see one's entire body.

STUDY GUIDE

There are lots of subtle ideas here, but basically you should know that as light propagates through a medium (e.g., glass) it scatters and rescatters off layer upon layer of atoms. The cumulative effect on the lightwave is to produce an overall speed v, which is usually less than c.

Refraction

The lightwave that crosses an interface and propagates into a transparent medium is called the **refracted wave**. And it has a remarkable property: despite the fact that **photons do not exist at any speed other than** c, the refracted electromagnetic wave never travels at c, except in the special case of vacuum. How could that be?

23.4 The Index of Refraction

When a beam of light enters a transparent material medium, it interacts with the atoms that constitute the glass, or plastic, or water, or whatever it happens to be. Some photons travel in the void between atoms, while others collide with the atoms and are absorbed and immediately re-emitted. The only photon paths are in the vacuum between atoms, and those photons travel at c. Ordinarily there will be billions upon billions of photons absorbed, re-emitted, and absorbed again and again as the beam makes its way through the medium. The refracted electromagnetic wave is the averaged macroscopic manifestation of this torrent of photons.

Because of the ongoing process of absorption and re-emission, electromagnetic waves propagate through material media at speeds other than c.

The crucial feature here is that the re-emitted photons constitute a wave—call it the secondary wave—that is necessarily somewhat out-of-phase with the incident wave. What happens is that the incident wave drives atoms into a ground-state vibration. But this vibration will not usually follow in step with the driver; there will be a phase difference, as we saw earlier when we studied Slinkies (p. 356). When the driving frequency is much below the atom's resonant frequency, there is only a slight phase difference. But that difference increases as the frequency of the incident beam increases toward the resonance. We needn't worry about all the details; what's of interest is that ***for an incident wave with a frequency below the atomic resonance, the secondary wave lags behind it, and for an incident wave above resonance, the secondary wave leads it.***

The refracted wave is the sum of what remains of the incident wave combined with the phase-shifted secondary wave. But adding a phase-shifted contribution to the incident wave has the effect of producing a refracted wave having a somewhat shifted phase. Figure 23.17 shows how the phase of the refracted wave can be shifted (advanced or retarded), and that's exactly equivalent to increasing or decreasing its average speed. Remember that the phase speed of any wave is the rate at which any point of constant phase, like a crest, moves. If its phase is retarded, the electromagnetic wave is effectively slowed.

Figure 23.17 If the secondary lags the incident wave, the resultant refracted wave will also lag it and vice versa.

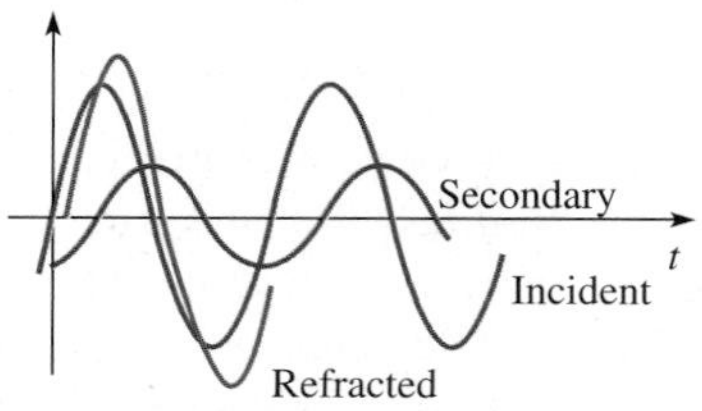

X-rays have frequencies in excess of the atomic resonances and so for them $v > c$. On the other hand, in transparent media, light frequencies are most often lower than the atomic resonances of the medium, and usually $v < c$. The dependence of the speed of propagation on the frequency of the radiant energy is known as **dispersion**; it's responsible for the

...

Table 23.1

Approximate Indices of Refraction of Various Substances*

Air	1.000 29
Ice	1.31
Water	1.333
Ethyl alcohol (C_2H_5OH)	1.36
Fused quartz (SiO_2)	1.458 4
Carbon tetrachloride (CCl_4)	1.46
Turpentine	1.472
Benzene (C_6H_6)	1.501
Plexiglass	1.51
Crown glass	1.52
Sodium chloride (NaCl)	1.544
Light flint glass	1.58
Polystyrene	1.59
Carbon disulfide (CS_2)	1.628
Dense flint glass	1.66
Lanthanum flint glass	1.80
Zircon ($ZrO_2 \cdot SiO_2$)	1.923
Fabulite ($SrTiO_3$)	2.409
Diamond (C)	2.417
Rutile (TiO_2)	2.907
Gallium phosphide	3.50

*Values vary with physical conditions—purity, pressure, etc. These correspond to a wavelength of 589 nm.

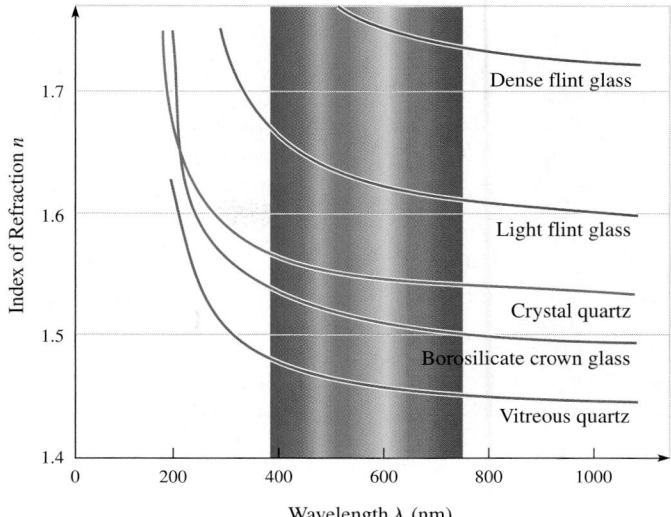

Figure 23.18 The wavelength dependence of the index of refraction for various materials. Transparent substances such as these usually have resonances in the UV (the region below 380 nm) and in the IR (far off to the right), which is why all of the curves here rapidly rise in the UV. That means that glass will dissipatively absorb UV. Can you guess why UV lamps are made of vitreous quartz rather than glass?

separation of white light into its constituent colors via a prism.* {For a more detailed treatment of all of this click on **REFRACTION** on the **CD**.}

The ratio of the speed of an electromagnetic wave in vacuum to that in a medium is defined as the **index of refraction** n; accordingly

$$n = \frac{c}{v} \tag{23.2}$$

This is also known as the *absolute index of refraction* (for reasons that will become clear on p.830). Keep in mind that n always varies somewhat with the frequency of the illumination. For most transparent materials, this variation, although significant, is not very large across the visible spectrum, and so it's common to use a single value of n (see Table 23.1) for such a substance illuminated with light of any frequency (Fig. 23.18). However, as f increases (and λ decreases), approaching an atomic resonance, the phase lag gets appreciably larger, v gets smaller, and n increases drastically (see the curves at around 200 nm in Fig. 23.18) . There is then a large increase in dissipative absorption as the amplitudes of the atomic oscillations increase. That's what happens to ordinary glass in the UV—glass eats UV—and it's why you can't get a tan behind a window that you can easily see through.

Example 23.4 [I] What is the apparent speed of light (589 nm) in diamond?

Solution Whenever you're asked for the speed of light in a medium, think of the index of refraction. (1) TRANSLATION —Determine the speed of light in a specified medium. (2) GIVEN: From Table 23.1, $n_d = 2.42$. FIND: v. (3) PROBLEM TYPE—Refraction/speed of light. (4) PROCEDURE—Use the definition of the index of refraction, the value of which we look up in the table. (5) CALCULATION—By definition $n = c/v$,

$$v = \frac{c}{n_d} = \frac{3.00 \times 10^8 \text{ m/s}}{2.42} = \boxed{1.24 \times 10^8 \text{ m/s}}$$

Quick Check: The index of vacuum is 1, and since diamond has an index of around 2.4, the speed in diamond is 2.4 times slower than $c \approx 3 \times 10^8$ m/s, or roughly 1×10^8 m/s.

*One need not fret about violating Relativity Theory with speeds in excess of c—these are phase speeds, there is no modulation and the waves carry no information. If they were modulated somehow, the signal would travel at a reduced rate known as the group *velocity*.

Looking down into a puddle (that's melting snow on the right) we see a reflection of the surrounding trees. At normal incidence water reflects about 2% of the light. As the viewing angle increases, here it's about 40°, that percentage increases. See Table 25.1 on p. 896.

At near normal incidence about 4% of the light is reflected back off each air-glass interface. Here because it's a lot brighter outside than inside the building, you have no trouble seeing the photographer.

Although n is appropriately called the index of refraction, it has a good deal to do with reflection as well. When light strikes the interface between two media, it's the difference between the indices of the two media that will determine how much of that light is transmitted and how much is reflected. As a rule, the greater the difference the more light is reflected. For example, at normal incidence diamond ($n_d = 2.4$) in air ($n_a = 1.0$) will reflect about 17%, glass ($n_g = 1.5$) in air will reflect about 4%, whereas water ($n_w = 1.3$) reflects only 2%. That's why people doing close work will often unthinkingly peer over their eyeglasses—otherwise they lose about 4% of the light at each interface. In addition, the greater the angle of incidence, the more light is reflected; at near glancing incidence a piece of "clear" glass will transmit little or no light.

23.5 Snell's Law

Let's now examine the transmitted beam in Fig. 23.8. Each atom radiates wavelets into the glass. These can be imagined as coalescing into a secondary wave that then recombines with the unscattered remainder of the incoming beam to form the refracted wave. However we visualize it, immediately on entering the transmitting medium, there is a single net field, a single net wave—the refracted, or transmitted, light. As we have seen, the repeated scattering in effect slows down the transmitted wave, which propagates with a speed $v_t < c$.

Fig. 23.19 picks up where we left off with Fig. 23.8. The diagram depicts several wavefronts, all shown at a single instant in time. Remember that each wavefront is a surface of constant phase and, to the degree that the phase of the net field is retarded by the transmitting medium, each wavefront is held back, as it were. The wavefronts bend as they cross the boundary because of the speed change. Alternatively, we can envision Fig. 23.19 as a multiple-exposure picture of a single wavefront showing it after successive equal intervals of time. Notice that in the time Δt, which it takes for point B on a wavefront (traveling at speed

Figure 23.19 The wave picture of refraction. The atoms in the region of the surface of the transmitting medium reradiate wavelets that combine constructively to form a refracted beam.

Willebrord Snel van Royen
(1580–1626)

v_i) to reach point D, the transmitted portion of that same wavefront (traveling at speed v_t) has reached point E. If the glass ($n_t = 1.5$) is immersed in an incident medium that is vacuum ($n_i = 1$) or air ($n_i = 1.000\ 3$) or anything else where $n_t > n_i$, $v_t < v_i$ and $\overline{AE} < \overline{BD}$, the **wavefront bends**. The refracted wavefront extends from E to D, making an angle with the interface of θ_t. As before, the two triangles ABD and AED share a common hypotenuse $\overline{AD}$ and so

$$\frac{\sin \theta_i}{\overline{BD}} = \frac{\sin \theta_t}{\overline{AE}}$$

where $\overline{BD} = v_i\,\Delta t$ and $\overline{AE} = v_t\Delta t$. Hence

$$\frac{\sin \theta_i}{v_i} = \frac{\sin \theta_t}{v_t}$$

Multiply both sides by c, and since $n_i = c/v_i$ and $n_t = c/v_t$

$$n_i \sin \theta_i = n_t \sin \theta_t \qquad (23.3)$$

This equation is the first portion of the **Law of Refraction**, also known as **Snell's Law** after the man who proposed it (1621), Willebrord Snel van Royen (1580–1626). Snel's analysis has been lost, but contemporary accounts follow the treatment shown in Fig. 23.20. What was found, purely through observation, was that the bending of rays could be quantified via the ratio of x_i to x_t which was constant for all θ_i, for any particular transparent medium in air. This constant was naturally enough called the *index of refraction* of the transmitting medium. In other words,

$$\frac{x_i}{x_t} \equiv n_t$$

and in air that's equivalent to Eq. (23.3). We now know that the Englishman Thomas Harriot had come to the same conclusion before 1601, but he kept it to himself.

At first, the indices of refraction were simply experimentally determined constants of the physical media. Later on, Newton was actually able to derive Snell's Law using his own corpuscular theory. By then, the significance of n as a measure of the speed of light was evident. Still later, Snell's Law was shown to be a natural consequence of Maxwell's Electromagnetic Theory.

Figure 23.21 is a ray diagram wherein all the angles are measured from the perpendicular. Along with Eq. (23.3), there goes the understanding that **the incident, reflected, and refracted rays all lie in the plane-of-incidence**. When $n_i < n_t$ (that is, when the light is initially traveling within the lower-index medium), it follows from Snell's Law that $\sin \theta_i >$

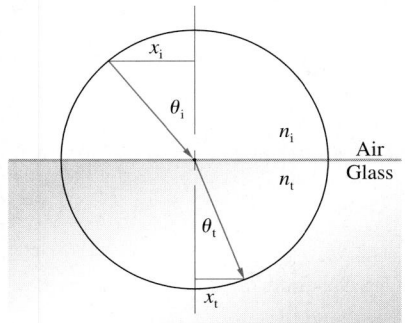

Figure 23.20 Descartes's arrangement for deriving the Law of Refraction.

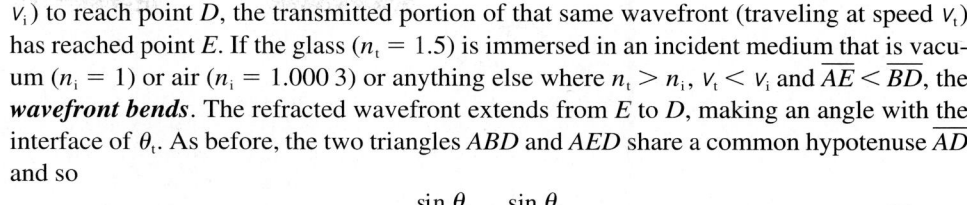

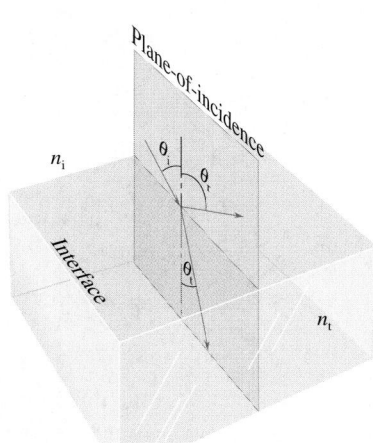

Figure 23.21 The incident, reflected, and transmitted beams all lie in the plane-of-incidence. The incident medium, containing the incoming beam, has an index of refraction n_i. The transmitting medium has an index n_t.

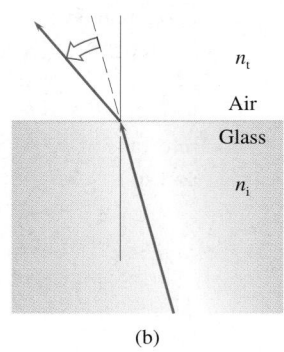

Figure 23.22 (a) When a beam of light enters a more optically dense medium, one with a greater index of refraction ($n_i < n_t$), it bends toward the perpendicular. (b) When a beam goes from a more dense to a less dense medium ($n_i > n_t$), it bends away from the perpendicular.

(a)   (b)

Any ray passing through a system can be reversed and consequently will renegotiate the system along the exact same path. In other words, if we simply reverse the arrowheads, the new rays so formed are physically correct.

$\sin \theta_t$, and since the sine function is everywhere positive between 0° and 90°, then $\theta_i > \theta_t$. Rather than going straight through, **a ray entering a higher index medium bends *toward* the normal** (Fig. 23.22*a*). The reverse is also true; that is, **upon entering a medium having a lower index, a ray, rather than going straight through, bends *away* from the normal** (Fig. 23.22*b*). Notice that this implies that the rays will traverse the same path going either way, into or out of either medium—the arrows can be reversed and the resulting picture is still true.

Example 23.5 **[II]** Suppose that the beam of light in Fig. 23.23 travels in air before it is incident on the glass plate ($n_g = 1.5$) at 60°. (a) At what angle will it be transmitted into the block? (b) Show that the beam emerges from the far side parallel to the original incoming light.

Solution This problem is about the bending of light at an interface, and that means Snell's Law. (1) TRANSLATION— Light is incident on a transparent plate at a known angle; (a) determine the transmission angle and (b) show that the emerging beam is parallel to the incident beam. (2) GIVEN: An air–glass interface ($n_g = 1.5$) and $\theta_i = 60°$. FIND: θ_{t1} and θ_{t2}, using Fig. 23.23. (3) PROBLEM TYPE—Refraction/ray angles / Snell's Law. (4) PROCEDURE—(a) Apply Snell's Law at the first interface. (5) CALCULATION:

$$n_a \sin \theta_{i1} = n_g \sin \theta_{t1}$$

and

$$\sin \theta_{t1} = \left(\frac{n_a}{n_g}\right) \sin \theta_{i1} = \left(\frac{1.0}{1.5}\right) \sin 60° = 0.577$$

Hence $\theta_{t1} = 35.3°$, and to two figures, $\boxed{\theta_{t1} = 35°}$. (b) Applying Snell's Law to the second interface where $\theta_{t1} = \theta_{i2}$,

$$n_g \sin \theta_{i2} = n_a \sin \theta_{t2}$$

and

$$\sin \theta_{t2} = \left(\frac{n_g}{n_a}\right) \sin \theta_{i2} = \frac{1.5}{1.0} \sin 35.3° = 0.866$$

Therefore $\boxed{\theta_{t2} = 60°}$, *and the light emerges parallel to the incident beam*.

Quick Check: Upon crossing the first interface, the ray bent toward the normal, which is what it should have done. It emerged parallel to the incident ray, which it must do if the interfaces are parallel, and that's a check of the calculation in itself.

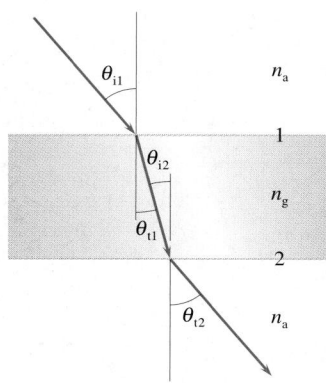

Figure 23.23 Light incident on a parallel plate of glass of index n_g immersed in air of index n_a.

Figure 23.19 illustrates the three important changes that occur in the beam traversing the interface. (1) It changes direction. Because the leading portion of the wavefront in the glass slows down, the part still in the air advances more rapidly, sweeping past and bend-

NONLINEAR OPTICS

In all the situations thus far treated, it was assumed that the reflected and refracted beams always had the same frequency as the incident beam; ordinary experience tells us this is a very reasonable assumption. Light of frequency f impinges on a medium and presumably drives the molecules into simple harmonic motion. That's certainly the case when the amplitude of the vibration is fairly small, as it is when the electric field driving the molecules is small. The E-field for bright sunlight is only about 1000 V/m (while the B-field is less than a tenth of the Earth's surface field). This isn't very large compared to the fields that keep a crystal together, which are of the order of 10^{11} V/m or just about the same magnitude as the cohesive field holding the electron in an atom. We can usually expect the oscillators to vibrate in simple harmonic motion so that the frequency will remain constant—the medium will ordinarily respond linearly. This will not be true, however, if the incident beam has an exceedingly large-amplitude E-field, as can be the case with a high-power laser. So driven, at some frequency f the medium can behave in a nonlinear fashion, resulting in reflection and refraction of harmonics ($2f$, $3f$, etc.) in addition to f. Nowadays, second-harmonic generators are available commercially; you shine red light (694.3 nm) into an appropriately oriented transparent nonlinear crystal (of, for example, potassium dihydrogen phosphate, KDP, or ammonium dihydrogen phosphate, ADP) and out will come a beam of UV (347.15 nm).

ing the wave toward the normal. (2) The beam in the glass has a broader cross section than the beam in the air; hence, the energy is spread thinner. (3) The wavelength decreases because the frequency is unchanged while the speed decreases, $\lambda = v/f$. This latter notion suggests that *the color aspect of light is better thought of as its frequency* (or energy, $E = hf$) than its wavelength, which changes with the medium through which the light moves. When we do talk about wavelengths and colors, we should always be referring to *vacuum wavelengths* (henceforth to be given as λ_0).

Example 23.6 [I] Someone is wearing a scarlet-red bathing suit that reflects light with a wavelength of primarily $\lambda_a = 629$ nm in air ($n_a \approx 1.00$). What is the corresponding wavelength in water ($n_w = 1.33$)? Do bathing suits change color under water? Explain.

Solution The frequency of light corresponds to its energy and that's fixed, but its wavelength is determined by the medium it's in. (1) TRANSLATION—Light having a known wavelength (in air) enters a medium (water) with a specified index; determine its new wavelength. (2) GIVEN: $\lambda_a = 629$ nm for $n_a = 1.00$. FIND: λ_w when $n_w = 1.33$. (3) PROBLEM TYPE—Refraction/wavelength. (4) PROCEDURE—The index of refrac-

tion tells us the speed; to find the wavelength in water we'll need that speed along with the frequency of the light ($v = f\lambda$), which we can get from its characteristics in air. (5) CALCULATION—$n_a = c/v_a = c/f\lambda_a$; hence, $f = c/n_a\lambda_a = 4.77 \times 10^{14}$ Hz. Using the same reasoning, in water

$$\lambda_w = \frac{c}{n_w f} = \frac{3.00 \times 10^8 \text{ m/s}}{1.33(4.77 \times 10^{14} \text{ s}^{-1})} = \boxed{473 \text{ nm}}$$

which, were it in vacuum, would be seen as blue. Of course, the suit still appears red in water. Color depends on photon energy.

Quick Check: Since wavelength decreases as index increases, $\lambda_w/\lambda_a = n_a/n_w$, and $\lambda_w = \lambda_a(1/1.33) = 473$ nm.

Notice that had we substituted ($\lambda_0 f$) for c in Example 23.6, we would have gotten

$$\lambda = \frac{\lambda_0}{n} \tag{23.4}$$

for the wavelength in any medium of index n.

The image of a pen seen through a thick block of clear plastic. The displacement of the image arises from the refraction of light toward the normal at the air-plastic interface. If this arrangement is set up with a narrow object (e.g., an illuminated slit) and the angels are carefully measured, you can confirm Snell's Law directly.

The fact that rays leaving a medium of higher index bend away from the perpendicular gives rise to the familiar effect in which distances within liquids appear foreshortened when viewed from above. Figure 23.24 shows a fish under water a real distance down d_R, which seems to be at an apparent depth d_A. These distances can be related in a particularly simple way if we limit the problem to one where the fish is not far away horizontally from the viewer; that is, x is small compared to the depth. Then θ_i and θ_t are both small. For small angles, the cosine is approximately 1.0 and the tangent then equals the sine:

$$\sin \theta_i \approx \tan \theta_i = \frac{x}{d_R} \quad \text{and} \quad \sin \theta_t \approx \tan \theta_t = \frac{x}{d_A}$$

It follows from Snell's Law that

$$\frac{n_i x}{d_R} = \frac{n_t x}{d_A}$$

and

$$\frac{n_i}{n_t} = \frac{d_R}{d_A}$$

The rays of light from the submerged portion of the pencil bend on leaving the water as they rise toward the viewer. The image of the part of the pencil in the water appears higher than it should be, just as the fish in Fig 23.24 seems to be higher than it actually is.

which is why a pencil seems to bend when it's dipped into water—the immersed end appears higher than it should in comparison to the portion remaining in the air.

{Take a break from the book and go to the **CD**. Open **INTERACTIVE EXPLORATIONS** and click on **REFLECTION AND REFRACTION**. Play with this simulation until you're sure you understand it completely.}

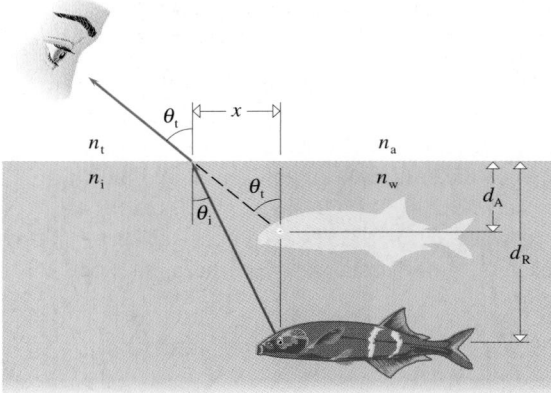

Figure 23.24 When light emerges from water into air ($n_w > n_a$), the ray bends away from the perpendicular. Anything below the surface appears higher in the liquid than it actually is, which makes spear fishing tougher than it might otherwise seem. Furthermore, the image will appear directly above the fish only when $\theta_t \approx 0$.

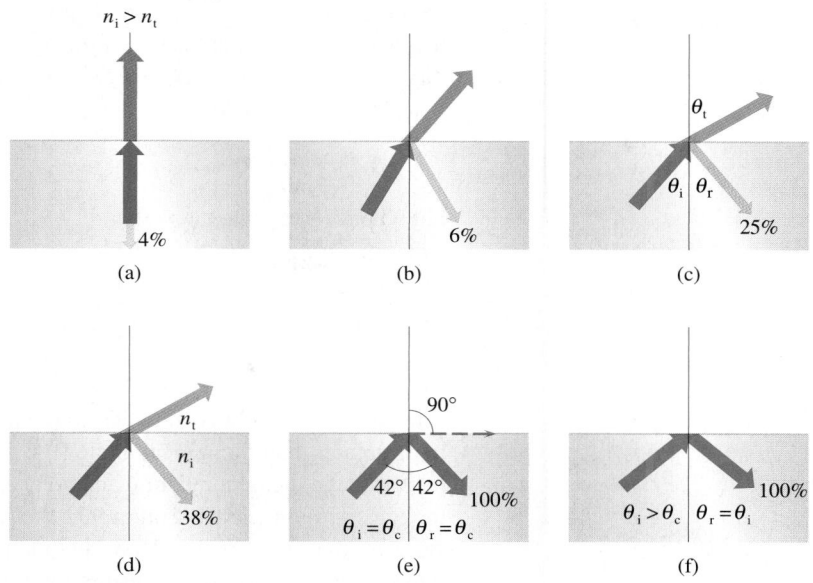

Figure 23.25 Internal reflection ($n_i > n_t$). As the incident angle increases, more and more light is reflected. (a) When $\theta_i = 0$, only about 4% of the light is reflected, while $\approx 96\%$ is transmitted.(b) (c) (d) As θ_i increases, more light is reflected and less transmitted. (e) At the critical angle ($\theta_i = \theta_c$), all the light is internally reflected. (f) At values of θ_i greater than θ_c, the incident beam continues to be totally internally reflected.

23.6 Total Internal Reflection

Often a beam of light originates in air (a low-index medium) and impinges on glass or water (a high-index medium). The reverse is also possible. When we look at a swimming fish, the light from the fish originates in the water (a high-index medium) and impinges on the air (a low-index medium). When light is incident on an interface where $n_i > n_t$, a curious and rather important thing happens. As the angle of incidence is made larger and larger (Fig. 23.25), the transmitted beam bends more and more away from the normal and toward the interface. As that occurs, the transmitted beam grows weaker, and the reflected beam—the beam traveling back into the higher-index medium—becomes stronger. As we've seen, that same sort of thing happens for external reflection, but there the decrease in transmission continues all the way up to $\theta_i = 90°$. Here however, when a particular incident angle is reached, known as the **critical angle** (θ_c), *all the light striking the interface will be reflected back; no light will be transmitted*. This **total internal reflection** continues for all incident angles greater than the critical angle. If, for example, $n_i = 2$ while $n_t = 1$, Snell's Law

Total internal reflection (at a glass-air interface). Note how the very bottom of this glass kettle is visible in totally reflected light. As a result we can't see the flame through the front part of the bottom of the kettle.

The prism behaves like a mirror and reflects a portion of the pencil (reversing the lettering on it). The operating process is total internal reflection.

maintains that $2 \sin \theta_i = \sin \theta_t$ and, provided $\sin \theta_i \leq \frac{1}{2}$, the equation can be satisfied. But for incident angles greater than this critical value of $\theta_i = 30°$, the $\sin \theta_t$ would have to be greater than 1, and that cannot be—there is no transmitted beam.

Snell's Law yields

$$\sin \theta_i = \frac{n_t}{n_i} \sin \theta_t$$

but when $\theta_i = \theta_c$, we know that $\theta_t = 90°$, whereupon $\sin 90° = 1$, and we get the defining equation

[Total internal reflection] $$\sin \theta_c = \frac{n_t}{n_i}$$ (23.5)

The ratio on the right is known as the *relative index of refraction*. For incident angles equal to or greater than the critical angle, 100% of the energy in the incoming beam is reflected back into the incident medium and no energy is transmitted. Recall that at the critical angle the transmission angle becomes 90°; the "transmitted" light propagates along the interface. Because that wave is limited to the boundary, its energy flows back and forth across the interface with no average transmission into the second medium.

Example 23.7 **[I]** Imagine a beam of light traveling in a block of glass for which $n_g = 1.56$. Now suppose the light comes to the end of the block and impinges on an air–glass interface. What is the minimum incident angle that will result in all the light being reflected back into the glass?

Solution The phrase "all the light reflected back" should immediately call to mind total internal reflection. (1) TRANSLATION—Light is internally reflected at a glass–air interface; determine the critical angle. (2) GIVEN: $n_i = n_g = 1.56$ and $n_t = n_a = 1.00$. (3) PROBLEM TYPE—Refraction / total

internal reflection / the critical angle. (4) PROCEDURE—Use the defining equation, $\sin \theta_c = n_t/n_i$. (5) CALCULATION:

$$\sin \theta_c = \frac{n_t}{n_i} = \frac{1.00}{1.56} = 0.641$$

and $$\boxed{\theta_c = 39.9°}$$

Quick Check: For $n_t = 1.5 = 3/2$, $\sin \theta_c = 2/3$ and $\theta_c = 42°$, so the preceding result is certainly reasonable.

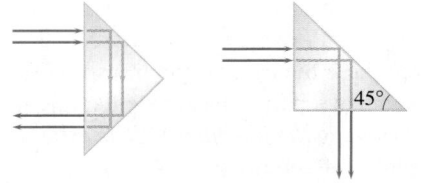

Figure 23.26 The rays strike the walls of these prisms at 45°, which is in excess of the critical angle ($36° \leq \theta_c \leq 43°$). The rays are then totally internally reflected.

The fact that the critical angle for air-glass typically ranges from 36° to 43° (depending on the kind of glass) is utilized in the reflecting prisms of Fig. 23.26. This is a convenient way to redirect a beam; and it's used in cameras, binoculars, telescopes, and a variety of other optical devices.

Fiberoptics

Easily the most important application of total internal reflection is **fiberoptics**. Techniques have evolved in recent times for efficiently conducting light and near-IR (1300 nm) energy along thin, transparent, dielectric fibers. Figure 23.27 depicts a glass (or plastic) fiber having a diameter of, say, 50 μm, just about the thickness of a human head-hair. Because of the small diameter, most of the light entering one face goes on to strike the cylindrical wall at a large angle and is subsequently totally internally reflected. This occurs over and over again, typically thousands of times per foot, as the beam propagates along the fiber.

All such thin filaments are quite flexible and, if not bent too tightly, will transmit light through twists and turns with relatively little loss. Bundles of free fibers whose ends are bound together and then polished form flexible light guides. If no attempt is made to keep the fibers in an ordered array so that the two end-faces are each a different jumble, the thing is called an *incoherent bundle*. These are light carriers. Easy to produce and inexpensive, they are used for remote light sensing and illumination. They conduct light, for example, from a conveniently positioned bulb to some otherwise inaccessible place such as an instru-

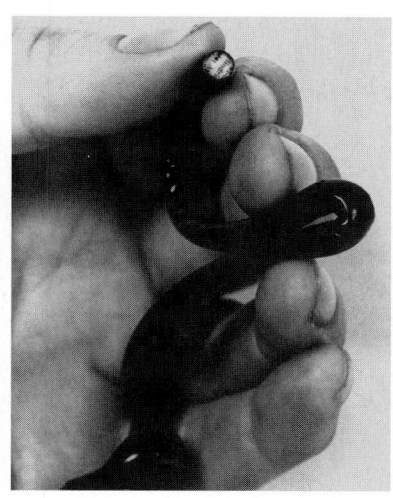

A bundle of thousands of carefully arranged thin glass fibers transmitting an image. One end of the bundle rests on a page of print, and the other end reveals the words below, despite the knots and bends.

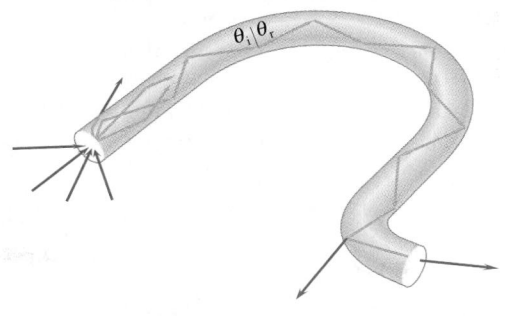

Figure 23.27 Light captured within a thin transparent fiber. Note that if a bend is too tight, $\theta_i < \theta_c$, and some light leaks out. Similarly, if a ray enters at too large an angle, it will leak out.

ment panel in an airplane or deep into a human body.

Conversely, when the fibers are arranged so that their terminations on both end-faces are exactly the same, the bundle is said to be *coherent*, and we have a flexible image carrier. Frequently, these bundles are tipped off with a small lens so that they need not be in contact with the surface being viewed. Today, it is commonplace to use fiberoptic apparatus to poke into all sorts of unlikely places from nuclear reactor cores and jet engines to stomachs and reproductive organs. When a device is used to examine internal body cavities, it's called an *endoscope*, and there are bronchoscopes, colonoscopes, gastroscopes, and so on. An additional incoherent bundle incorporated into the device usually supplies the illumination.

The world is now in the first stages of a new era of optical communications, of light flashing along fibers, replacing electricity moving in metal wires—not for transmitting power, but information. The much higher frequencies of light allow for an incredible increase in data-handling capacity. For example, using some sophisticated transmitting techniques, a pair of copper telephone wires can be made to carry up to about two dozen simultaneous conversations. To get a feel for how much information that is, consider the fact that a single ordinary TV transmission is equivalent to about 1300 simultaneous phone conversations, which, in turn, is roughly the equal of sending some 2500 typewritten pages

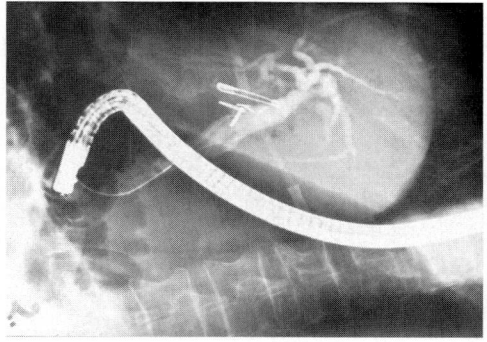

A colonoscope being used to examine a colon for cancer.

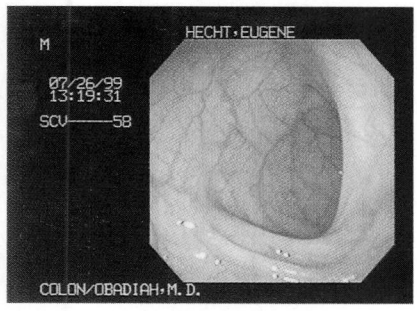

Remarkably detailed view as seen through a fiberoptic colonoscope.

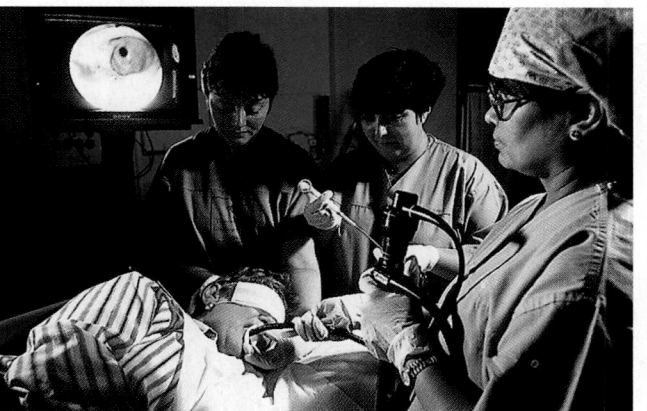

Most modern endoscopes have close-circuit TV links. Here doctors are performing a gastroscopy of a woman's stomach. The fiberoptic endoscope has been fed down through her throat.

each and every second! So, at present, it's quite impractical to attempt to send television over copper phone lines. By comparison, it's already possible to transmit far in excess of 12 000 simultaneous conversations over a single pair of fibers, which is more than nine TV channels—and this is only the beginning; the technology is in its infancy. Achieved capacities to date don't even begin to approach the theoretical limit. A pair of fibers will some-day connect your home to a vast network of communications and computer facilities that will make the era of the copper wire seem charmingly primitive.

The World of Color

We are now in a position to explore why the objects in our environment look the way they do. What makes paper white and coal black? Why is silver gray and grass green?

23.7 White, Black, and Gray

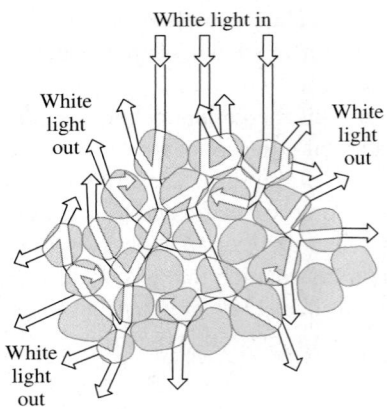

White light in

White light out

White light out

White light out

Figure 23.28 A number of transparent particles, which are large compared to the wavelengths of light. These might be cloth fibers, sugar crystals, talcum powder, or snowflakes. White light enters, is reflected and refracted out.

A reflecting surface is white when it diffusely scatters a broad range of frequencies under white-light illumination. This page is white. So are salt and snow. Clouds are white, as are powders and pills and cloth. Even the foam on a glass of beer is white, steam and soap bubbles and chalk and sugar and gray hair are white—and all this diversity shares a common structural similarity. All of these objects are composed of transparent grains or particles or fibers or bubbles that are large in size compared with the wavelength of light but otherwise quite small. Each is much too large to produce Rayleigh scattering, too large to act as a single oscillator preferentially scattering at the higher frequencies. Despite the common opinion that things white are opaque, each grain or fiber is actually transparent. There is no such thing as a single particle of white pigment that by itself is opaque white. Materials that have no atomic resonances in the visible are transparent to light of all frequencies—they do not appear colored because they do not absorb any range of colors.

Whiteness arises when the incident white light is reflected back out of the medium in all directions as if from countless point sources, from scatterers that show no preferential absorption. All colors of light come in, and all colors scatter out. When the medium is composed of many small transparent bodies, like the matted fibers of this page or the random jumble of sugar crystals on a spoon, each reflects some light, transmits some light, and reflects again (Fig. 23.28). And that process happens over and over, layer upon layer, with much of the light eventually reflected back into the general direction from which it came.

The fraction of light energy reflected at each surface depends on the difference between the indices of the two media—if there is no difference, the boundary essentially vanishes. *The closer the indices are to each other*, *the less light will be reflected* and the less luminous will be the surface. To make white paint, we need only mix a powdered transparent material, the pigment, with a clear vehicle (such as oil or acrylic). If the refractive indices are equal, the pigment simply vanishes and the paint is transparently useless. If the

EXPLORING PHYSICS ON YOUR OWN

Whiteness: Sprinkle a single layer of salt or sugar on this page and the print will still be legible, proving that the crystals are transparent. Crush a piece of clear glass and the grains, like grains of sand, will appear white. Water is clear, but snow and steam and clouds are white. Slightly lift a few pages of this book and peep under them—lots of light will be transmitted through the pages but, like a cloud, if the mass is thick enough, very little light will emerge and it will appear gray or black. The brightness of a white piece of paper arises from the difference between the index of the fibers and the index of the surrounding air—the bigger the difference, the more light will be reflected or scattered backward. The paper's whiteness will change tremendously if the fibers are immersed in a substance having an index between that of fiber and air, a substance like water. What happens to the whiteness of tissue paper or cloth when it becomes wet? Why do the colors of wet cloth seem so much more vivid than when it was dry?

indices differ appreciably, there will be strong reflections at the countless surfaces and the paint will be a brilliant white. A really splendid white surface can reflect up to 98% of the incident light. Because of the low index of air, small particles such as salt, talc, and sugar, as well as all kinds of undyed fibers, surrounded by air appear bright white.

A diffusely reflecting surface that absorbs somewhat uniformly right across the spectrum will reflect a bit less than a white surface and so appear matt gray. The less it reflects, the darker the gray, until it absorbs almost all the light and appears black. A surface that reflects perhaps 70% or 80%, but reflects specularly, will appear the familiar shiny gray of a typical metal. Metals possess tremendous numbers of free electrons that scatter light very effectively independent of frequency—the free electrons are not bound to the atoms and have no associated resonances. The amplitudes of the vibrations are an order-of-magnitude larger than they are for the bound electrons. The incident light cannot penetrate into the metal any more than a fraction of a wavelength or so before it's canceled completely. There is little or no refracted light—most of the energy is reflected out; the small remainder is absorbed.

If the metal is thin enough (only a few atom layers thick), it will transmit light and one can see through it. Look at a bright lamp through a CD. Nowadays, "two-way" mirrors that can be seen through are a common security device. A number of snack food products are packaged in shiny metal-coated plastic films. Hold one up to your eyes and look through it at a bright light. Note that the primary difference between a gray surface and a mirrored surface is one of diffuse versus specular reflection.

23.8 Colors

When the distribution of energy in a beam of light is fairly uniform across the spectrum, the light appears white; when it is not, the light usually appears colored. Figure 23.29 depicts typical frequency distributions for what would be perceived as red, green, and blue light. These curves show the dominant frequency regions, but beyond that there can be a great deal of variation in the distributions, and they will still provoke the eye-brain responses of red, green, and blue. Thomas Young, in the early 1800s, showed that a broad range of colors can be generated by mixing three beams of light, provided their frequencies were widely separated. When three such beams combine to produce white light, they are called **primary colors**. There is no single unique set of these primaries, nor do they have to be monochromatic. Since the widest range of colors can be created by mixing *light beams* of red (R), green (G), and blue (B), these tend to be the most commonly used. They are the three components (emitted by three phosphors) that generate the whole gamut of hues seen on a color TV set. Keep in mind that at the moment we are talking about mixing light beams and not paint pigments.

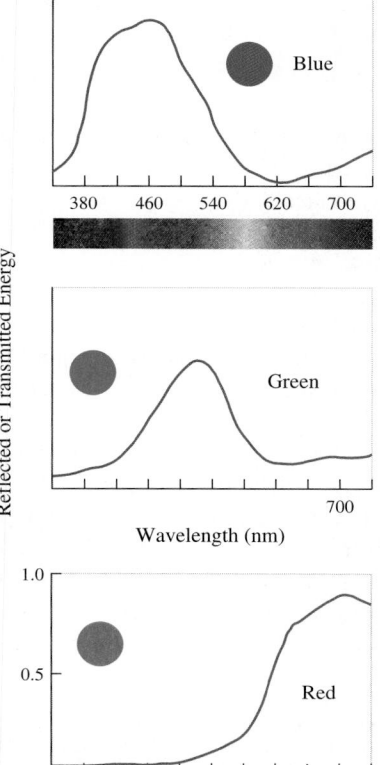

Figure 23.29 Reflection curves for blue, green, and red pigments are typical, but there is a great deal of variation within each color.

A white screen illuminated with red, blue, and green light appears white. Anyone standing between a lamp and the screen casts a shadow that has the complementary color.

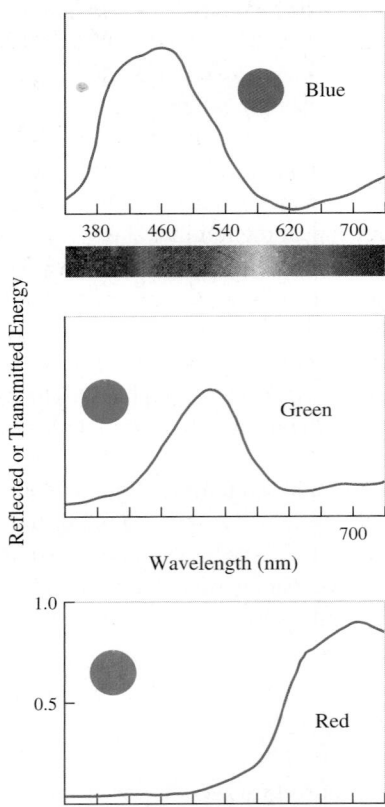

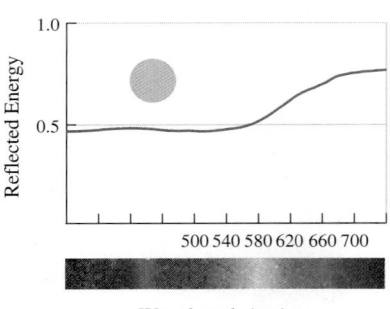

Figure 23.29 Reflection curves for blue, green, and red pigments are typical, but there is a great deal of variation within each color.

Figure 23.31 Spectral reflection of a pink pigment. The broad range of wavelengths produces a white background; adding the red peak yields pink. A shirt that reflects about half the light incident on it over most of the spectrum, but reflects even more in the red, looks pink.

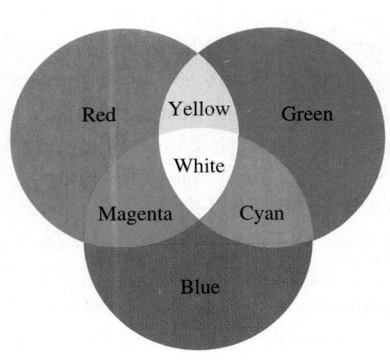

Figure 23.30 Three overlapping beams of colored light. A color TV set uses the same three primary light sources—red, green, and blue.

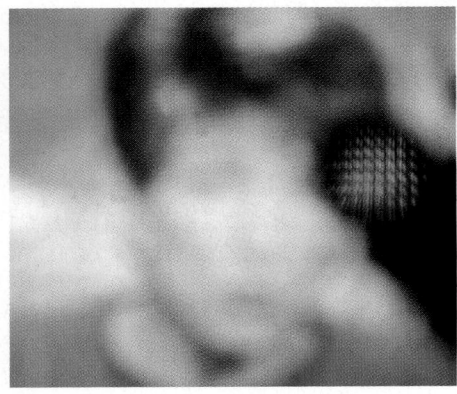

A color TV screen is made up of red, blue, and green regions that can be made visible with a magnifying glass. A hand-held lens on the right in the photo shows the three-color structure.

Figure 23.30 summarizes the result of overlapping these three primaries in a number of different combinations: red plus blue is *magenta* (M), a reddish blue; blue plus green is *cyan* (C), a bluish green; and, most surprising, red plus green is *yellow* (Y). And the sum of all the three primaries is white:

$$R + B + G = W$$

$$
\begin{aligned}
M + G &= W, && \text{since} && R + B = M \\
C + R &= W, && \text{since} && B + G = C \\
Y + B &= W, && \text{since} && R + G = Y
\end{aligned}
$$

Any two colored light beams that together produce white are said to be **complementary**, and the last three symbolic statements exemplify that situation. Now suppose we overlap a beam of magenta and a beam of yellow, represented symbolically by

$$M + Y = (R + B) + (R + G) = W + R$$

The result is pink, a combination of white and red. And that raises another point: we say that a color is **saturated**, that it is deep and intense, when it does not contain any white light. As can be seen in Fig. 23.31, pink is unsaturated red, that is, red superimposed on a background of white.

Imagine a piece of yellow stained glass, that is, glass having a resonance in the blue, which it strongly absorbs. Looking through it at a white-light source composed of red, green, and blue, the glass would absorb blue, passing red and green, which is yellow (Fig. 23.32). Yellow cloth, paper, dye, paint, and ink all selectively absorb blue and reflect what remains—yellow—and that's why they appear yellow. And if you peer at something that is a pure blue through a yellow filter, the object will appear black. Here, the filter or the paint colors the light yellow by removing blue, and we speak of the process as *subtractive* col-

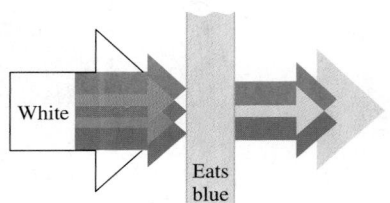

Figure 23.32 Yellow stained glass absorbs blue light, passing yellow. The incoming white beam corresponds to R + B + G. Subtracting blue yields R + G, which is yellow.

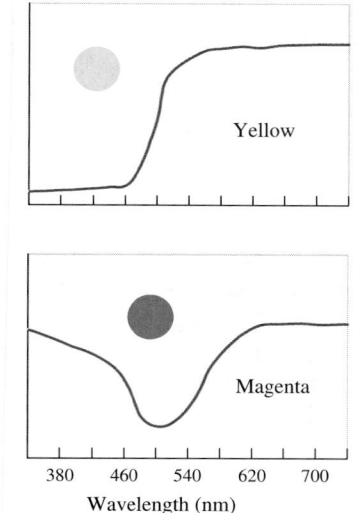

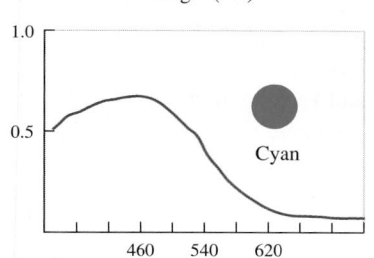

Figure 23.33 Transmission curves for colored filters. Or reflection curves for colored paints.

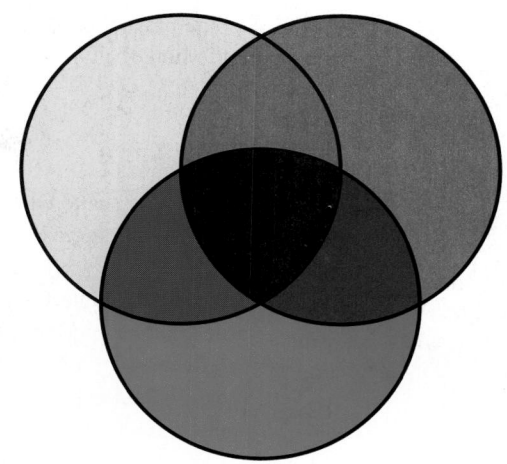

Figure 23.34 Filters of magenta, yellow, and cyan. These absorb certain frequencies and the process is called subtractive coloration. Where all three overlap, no light is passed.

oration. On the other hand, *additive* coloration can only result from the overlapping of light beams. Red light and green light make yellow; red paint and green paint make brown.

A wide range of colors (including red, green, and blue) can be produced by passing light through various combinations of magenta, cyan, and yellow filters (Fig. 23.33). Magenta, cyan, and yellow are the primary colors of subtractive mixing. (They are the primaries of the paint box, although they are very often mistakenly spoken of as red, blue, and yellow.) They are the basic colors of the dyes used to make photographs and the inks that printers use to print them (Fig. 23.34). To color paint or cloth or anything else, we need only surround the transparent grains or fibers with a colored filter, a dye layer. White light passes through the film of dye, loses a select portion of the spectrum due to dissipative absorption, and emerges with the complementary color. Ideally, if you mix all the subtractive primaries together (either by combining paints or by stacking filters), you get no color, no light, just blackness. Each removes a region of the spectrum, and together they absorb it all. Selective absorption is the principal mechanism that colors the world, from green grass to red roses to pink lips.

Core Material & Study Guide

SCATTERING

The fundamental mechanism underlying the propagation of light through matter is **scattering**—the absorption and re-emission of radiant energy. *A ray is a line drawn in space corresponding to the direction of flow of radiant energy* (p. 819). Section 23.1 (Rayleigh Scattering: Blue Skies) lays out all the basic notions and so should be read carefully in order to get the conceptual framework. There's only one relationship in this material that's useful for problem solving and it's applied in Example 23.1. Look at the **CD WARM-UPS** and **WALK-THROUGHS** 💿 and then try the I-level problems. Read the *Suggestions on Problem Solving* before you attempt to do any.

REFLECTION

When a beam of light reflects from a smooth surface

$$\theta_i = \theta_r \qquad [23.1]$$

The angle of incidence equals the angle of reflection. This equation is the first part of the **Law of Reflection**. The second part main-

tains that *the incident ray, the perpendicular to the surface, and the reflected ray all lie in a plane* called the **plane-of-incidence**. Sections 23.2 (Internal and External Reflection) and 23.3 (The Law of Reflection) elaborate the preceding ideas. Problems concerning plane mirrors are basic to the subject—see Examples 23.2 and 23.3.

REFRACTION

The ratio of the speed of an electromagnetic wave in vacuum to that in matter is the **index of refraction** n:

$$n = \frac{c}{v} \qquad [23.2]$$

This rather basic concept is discussed in Section 23.4 (The Index of Refraction) and explored in Example 23.4. The **Law of Refraction**, or **Snell's Law**, is

$$n_i \sin \theta_i = n_t \sin \theta_t \qquad [23.3]$$

where *the incident, reflected, and refracted rays all lie in the plane-of-incidence*. When light enters a refracting medium, its wavelength changes, and

$$\lambda = \frac{\lambda_0}{n} \qquad [23.4]$$

provides the wavelength in any medium of index n. Section 23.5 (Snell's Law) is a culmination of many ideas and quite important—it will be the basis for many problems.

When the light impinges on an interface where $n_i > n_t$, there will be a special incident angle, the **critical angle** (θ_c), at which all the incoming light will be reflected back and none will be transmitted. This **total internal reflection** continues for all incident angles greater than the critical angle defined by

[Total internal reflection] $$\sin\theta_c = \frac{n_t}{n_i} \qquad [23.5]$$

Reread Section 23.6 (Total Internal Reflection) and study Example 23.7.

THE WORLD OF COLOR

The widest range of colors can be created using a mix of light beams of red (R), green (G), and blue (B). Red plus blue is *magenta* (M); blue plus green is *cyan* (C); and red plus green is *yellow* (Y). And the sum of all three (p.833) is white.

Key Terms

elastically scattered	refraction
Rayleigh scattered	index of refraction
interference	Snell's Law
constructive interference	Law of Refraction
destructive interference	total internal reflection
reflection	fiberoptics
external reflection	incoherent bundle
internal reflection	coherent bundle
Law of Reflection	whiteness
angle of incidence	primary colors
angle of reflection	magenta, cyan, and yellow
plane-of-incidence	complementary
specular reflection	saturated
diffuse reflection	additive coloration
virtual image	subtractive coloration

Discussion Questions

1. (a) Describe the kind of reflection that is taking place off this page. (b) Why is it that glossy paints can be deeper and richer in color than flat paints? (c) Why do the colors on a wet watercolor painting look so much more vivid than when they dry?

2. In what regard can we say that both reflection and refraction are manifestations of scattering?

3. Shaving lather is often substituted for meringue in pie-throwing fests. Why do they look so similar?

4. Suppose you have the bad luck to own a car whose exhaust has a bluish tint to it. What can you say about the effluvium?

5. (a) What is the angle of reflection for a ray incident normally on a smooth surface? (b) What is the angle of refraction for a beam striking an air-glass interface perpendicularly?

6. Modern picture frames sometimes come with "nonreflecting" glass (i.e., glass whose surface has been deliberately made rough). What does it actually accomplish?

7. Is Venus in Velázquez's painting *The Toilet of Venus* (Fig. Q7) looking at her own face? Draw a ray diagram and in it drop a perpendicular from her nose to the plane of the mirror.

8. How is it that you can see your image in the surface of a polished *black* automobile? Explain what's happening. What allows you to see the gin in a glass, or the glass itself, for that matter?

9. The girl in Manet's painting, *The Bar at the Folies-Bergères* (Fig. Q9), is standing in front of a large plane mirror. We see reflected in it her back and the face of a man she seems to be talking to. From the Law of Reflection what, if anything, is amiss?

10. A properly cut diamond has a brilliant, lustrous appearance arising from the fact that most of the light entering the stone re-emerges. Explain how this happens and why diamond is especially well suited for the business.

Figure Q7 *The Toilet of Venus* by Diego Rodriguez de Silva y Velázquez—National Gallery, London.

Figure Q9 *The Bar at the Folies-Bergères* by Edouard Manet—Courtauld Institute Galleries, London.

11. Glass is often colored by adding to the silica, ions of the transition elements (Cr, Mn, Fe, Co, Ni, and Cu), usually in the form of oxides—all have a partially filled electron shell. Ordinary window glass contains, as contaminants, traces of iron oxides (in both ferrous and ferric forms) that have weak absorptions in the red and blue and a strong resonance in the near-UV. What radiation can you expect will traverse a sheet of glass? What color do you expect the glass to have if seen edge-on? Explain your answers.

12. Can a magenta light beam always be separated by a prism into red and blue? Similarly, can a cyan light beam always be separated into green and blue?

13. (a) What "color" will a red apple appear under red-light illumination? (b) What "color" will a cyan ink appear under the same conditions? (c) Why is a yellow filter often used when taking black-and-white photos of the clouds in the sky?

14. Which colored light appears the most saturated—the blue of the sky, the red of a traffic signal, or the pink of a rose?

15. (a) What "color" will a sheet of white paper appear under 650-nm illumination? (b) Given that you have a beam of cyan light shining on a white wall, what color beam should be added so that the reflected light is white?

16. Figure Q16 is a plot of the amount of light transmitted across the spectrum by a piece of stained glass. (a) What color does it appear? (b) Two beams of light are projected onto a white sheet of paper. If one passes through a cyan filter and the other through a yellow filter, what color results?

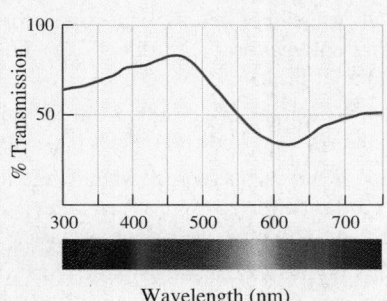

Figure Q16

17. (a) A beam of cyan light passes into a yellow filter. What color emerges? (b) A beam of yellow light passes into a magenta filter. What color emerges? (c) What color results when two beams of light, one cyan and one magenta, are made to overlap on a white screen? (d) White light passes through a cyan filter followed by a magenta filter. What color emerges?

18. Figure Q18 shows the transmission spectrum of a sheet of colored plastic. What color is it? Redraw the curve so that it corresponds to the frequency distribution reflected from an orange.

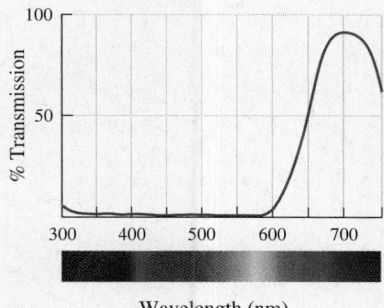

Figure Q18

19. Explain the meaning of each of the Key Terms on page 836.

Multiple Choice Questions

1. Standing on its surface looking upward, the Moon's sky is black because, unlike the Earth, the Moon   (a) has no streetlights   (b) is very cold   (c) has no atmosphere   (d) has no oceans to reflect sunlight   (e) none of these.

2. Cigarette smoke rising from the burning end has a bluish cast, while smoke exhaled is whitish. This difference happens because (a) the smoke contains a blue material that is left behind in the lungs   (b) the body changes the chemistry so that the exhaled smoke is white   (c) the smoke cools in the process and as a result turns white   (d) water droplets from the body surround the inhaled smoke particles and, because they are large, subsequently scatter white light   (e) none of these.

3. Take a piece of ordinary window glass and crush it into a fine powder; holding the mound in hand, it will   (a) appear black   (b) become green   (c) be as transparent as ever   (d) appear white (e) none of these.

4. If you have ever scratched a piece of clear plastic or the varnished finish on a tabletop, you are familiar with the white marks that result. They have that characteristic appearance due to   (a) changes in the chemistry of the surface   (b) melting of the surface so that it emits white light   (c) electron bombardment of the surface via static charge that causes fluorescence   (d) leaving behind a streak of powdered material that scatters a broad frequency band in the visible   (e) none of these.

5. A material is transparent to a certain band of frequencies of electromagnetic radiation. Accordingly, we can say that the atoms in the material   (a) are very small   (b) have resonances in that band   (c) are not interacting with each other   (d) are not close together   (e) none of these.

6. If the atmosphere were very much deeper than it is now, the Sun might appear   (a) blue at sunset   (b) red at noon   (c) green at sunrise   (d) violet at noon   (e) none of these.

7. From our knowledge of the atom, it seems reasonable that all types of matter must have resonances somewhere in the electromagnetic spectrum and therefore must, if the matter is dense enough,   (a) dissipatively absorb some radiant energy   (b) reflect across the spectrum   (c) absorb broadly across the spectrum   (d) be transparent at those resonances   (e) none of these.

8. When light is incident on a smooth surface, the angle of incidence ordinarily   (a) equals the angle of scattering   (b) doesn't equal the angle of reflection   (c) equals the angle of refraction   (d) equals the polarization angle   (e) none of these.

The next five questions refer to Fig. MC9 which shows a horizontal flat front-silvered mirror and a point source S.

9. If the angle β is 46°, what is the angle of incidence for the ray? (a) 46°   (b) 23°   (c) 92°   (d) 44°   (e) none of these.

10. If the angle α is 60°, what is the angle of reflection for the ray? (a) 90°   (b) 60°   (c) 30°   (d) 15°   (e) none of these.

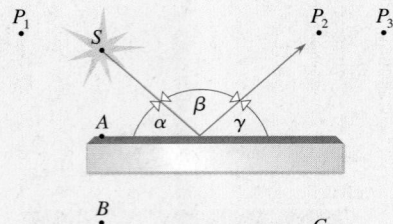

Figure MC9

11. An observer at P_2 looking into the mirror will see the point source S located at   (a) A   (b) B   (c) C   (d) D   (e) none of these.

12. An observer at P_3 looking into the mirror will   (a) see the point source located at A   (b) see the point source located at B   (c) see the point source located at C   (d) not see the point source at all   (e) none of these.

13. An observer at P_1 looking into the mirror will   (a) see the point source located at A   (b) see the point source located at B   (c) see the point source located at C   (d) not see the point source at all   (e) none of these.

14. Objects are placed in front of a flat mirror and subsequently moved around as their images are observed. In Fig. MC14 which curve best represents a plot of the resulting image size versus object size? And which best represents a plot of image distance versus object distance? (a) 1 and 2   (b) 2 and 4   (c) 3 and 2   (d) 4 and 4   (e) none of these.

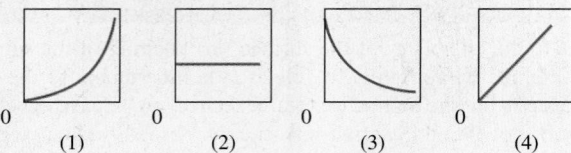

Figure MC14

15. When light is incident on an interface, the angle(s) of (a) the reflected and refracted beams depend on the wavelength (b) the reflected and refracted beams are independent of frequency (c) the refracted beam is independent of wavelength (d) the reflected beam is independent of frequency (e) none of these.

16. Compared to an object in front of it, the image in a plane mirror is always   (a) smaller   (b) virtual   (c) three times as far away   (d) distorted   (e) none of these.

17. A handful of white sandlike material is poured into a beaker of clear oil and vanishes from sight as it passes into the liquid. We can conclude that   (a) the material dissolved   (b) this phenomenon

is quite impossible and could not have happened   (c) the oil and material have mutually decomposed   (d) the oil has the same index of refraction as the material   (e) none of these.

18. A chicken is standing 1.0 m in front of a vertical plane mirror. A woman is standing 5.0 m from the mirror, behind and in line with the bird. How far from her will she see the image of the chicken? (a) 5.0 m   (b) 1.0 m   (c) 6.0 m   (d) 4.0 m   (e) none of these.

19. When a beam of light traveling in air enters a glass block, it ordinarily undergoes a change in   (a) speed only   (b) frequency only   (c) wavelength only   (d) speed and wavelength   (e) none of these.

20. The bending or refraction of a beam of light as it enters a medium that has a greater index of refraction than the incident medium is due to a change in its   (a) amplitude   (b) effective speed   (c) frequency   (d) period   (e) none of these.

21. What "color" will a yellow shirt appear under blue-light illumination?   (a) blue   (b) yellow   (c) white   (d) black   (e) none of these.

22. What combination of colored beams of light when overlapped on a white screen will produce black?   (a) R + B + G   (b) M + C + Y   (c) M + R + C + Y + G   (d) M + C + Y + R + B + G   (e) none of these.

23. A beam of yellow light passes through a cyan filter. The color that emerges is   (a) yellow   (b) blue   (c) green   (d) red   (e) none of these.

24. The color that results from the mixing of magenta and yellow paints is   (a) red   (b) blue   (c) green   (d) black   (e) brown.

25. What color emerges when white light passes through a magenta filter followed by a green filter?   (a) white   (b) yellow   (c) green   (d) blue   (e) none of these.

26. Figure MC26 shows three plots of the amount of light reflected over the visible region of the spectrum from a lemon, a ripe tomato, and a lettuce. These correspond, in turn, to curves   (a) 1, 3, and 2   (b) 3, 2, and 1   (c) 1, 2, and 3   (d) 2, 3, and 1   (e) 3, 1, and 2.

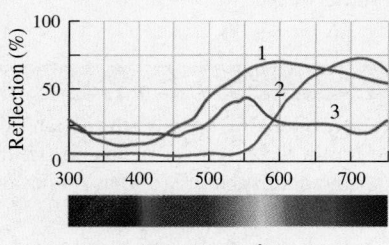

Figure MC26

27. A yellow surface is illuminated with magenta light. What color will it appear in reflected light?   (a) blue   (b) green   (c) yellow   (d) red   (e) none of these.

For more Multiple Choice Questions with answers click on WARM-UPS in CHAPTER 23 on the CD.

Suggestions on Problem Solving

1. The equation for the apparent depth of an object in a liquid (Fig. 23.24) can be remembered by keeping in mind that the ratio of indices equals the ratio of depths: smaller-to-larger (or possibly, larger-to-smaller). Thus, for example, when air is above water, it's the index of air to the index of water as the apparent depth is to the real depth.

2. An error commonly made when dealing with total internal reflection is to substitute the wrong values for the two indices—most often they are simply interchanged. Remember here $\sin\theta_c = n_t/n_i$ and $n_i > n_t$. The index ratio must be less than one.

3. When dealing with problems involving plane mirrors, always draw a diagram. The central notion is that the angle of incidence equals the angle of reflection—all the rest is geometry. Often it is useful to construct rays that touch the very edges of the mirror and thereby define the limits of the system.

4. When treating a problem that pertains to wavelengths, remember to distinguish between the vacuum wavelength and the wavelength in the medium ($\lambda_0 > \lambda$). Keep in mind that $n = \lambda_0/\lambda$.

Problems ✦ Coordinated Problems ✦ Progressive Problems ✦ Solutions

STUDY GUIDE **1. Coordinated Problems:** The three problems within each magenta-colored grouping are solvable in similar ways. Note that the first of these always has a hint; moreover, its solution is provided in the back of the book. *Work out each of these sets; they'll strengthen technique and build confidence.* **2. Progressive Problems:** The problems introduced in blue unfold step-by-step carrying along the analysis in a more suggestive way than is customary. *Work out all of these; they'll guide you through the analytic process and help develop problem-solving skills.* **3. Worked-Out Solutions:** Studying worked-out solutions is an important part of learning how to solve problems. Accordingly, additional *solutions* to a number of model problems are given below. *Make sure you understand each of them before you go on to the next problem.* **4.** Also provided in the back of the book are the *Answers* to all odd-numbered problems, as well as worked-out *solutions* to those with boldface numbers. Problem numbers in italic indicate that a solution appears in the Student Solutions Manual.

SECTION 23.1: RAYLEIGH SCATTERING: BLUE SKIES
SECTION 23.3: THE LAW OF REFLECTION

1. [I] Two beams of light, one red ($\lambda_r = 780$ nm) and one violet ($\lambda_v = 390$ nm), pass through several hundred meters of air. What is the ratio of the amount of scattering of red to violet?

2. [I] A beam of white light crosses a large volume occupied by a tenuous molecular gas mixture of mostly oxygen and nitrogen. Compare the amount of scattering occurring for the yellow (580 nm) component with that of the violet (400 nm) component.

3. [I] A laserbeam strikes a front-silvered mirror at an angle of $30°$ to the perpendicular. At what angle will it be reflected?

4. [I] A beam of parallel light strikes a flat polished metal surface perpendicularly. At what angle will light be reflected?

5. **[I]** Rays of light impinge on a smooth flat mirror at glancing incidence, making a tiny angle with respect to the surface. Approximately what are the angles of incidence and reflection?

6. [I] A narrow beam of light impinges on a smooth glass plate at $25°$ measured from the perpendicular to the surface. What is the angle between the incident and reflected beams?

7. [I] Two rays leave a point source and strike a front-silvered mirror, making angles with the normal of $30°$ and $40°$, respectively. What is the angle between the two reflected rays?

8. [I] A laserbeam is incident on a smooth surface at an angle of $62.5°$ measured up from the surface. A portion of the light is reflected and the remainder is transmitted. Determine the angle of reflection.

SOLUTION: From the Law of Reflection $\theta_i = \theta_r$, but these angles are measured from the normal. Accordingly, the incident angle is $90.0° - 62.5° = 27.5°$, and that's also the angle of reflection.

9. [I] THIS PROBLEM EXAMINES THE PHENOMENON OF REFLECTION. A man about 2.0 m tall stands erect in front of a vertical plane mirror. There is a laser pointer tucked behind his ear, and the beam strikes the mirror at $30°$ with respect to the normal and, reflecting off, illuminates his shoe. We want to find out how far he is from the mirror. (a) Draw a diagram. Does it matter how big the mirror is? (b) State the Law of Reflection. (c) If x is his perpendicular distance to the mirror, write an expression for the tangent of the angle of incidence at the mirror in terms of x. (d) Determine the approximate value of x.

10. [I] A very narrow beam of light from a laser is incident on a horizontal mirror at an angle of $58°$. The reflected beam strikes a wall at a spot 5.0 m away from the point of incidence where the beam hit the mirror. How far horizontally is the wall from that point of incidence?

11. [I] The tomb of FRED the Hero of Nod is a dark, closed chamber with a small hole in a wall 3.0 m up from the floor. Once a year, on FRED's birthday, a beam of sunlight enters via the hole, strikes a small polished gold disk on the floor 4.0 m from the wall, and reflects off it, lighting up a great diamond embedded in the forehead of a glorious statue of FRED, 20 m from the wall. Roughly how tall is the statue?

12. [I] There are a number of practical devices that use rotating plane mirrors. Show that if a mirror rotates through an angle α, the reflected beam will move through an angle of 2α.

13. **[I]** A woman can see an object clearly when it's held at a distance of 25 cm from her eyes. How far should she hold a flat mirror to see her own image clearly?

14. [I] A man is walking at 1.0 m/s directly toward a flat mirror. At what speed is his image approaching him?

15. [I] Two upright 1.00-m-tall plane mirrors are placed parallel to each other 2.8 m apart. The top of the mirror on the right is then moved back a little so that its surface tilts away from the other mirror at an angle of $10.0°$ off the vertical. A narrow laserbeam passes perpendicularly through a small hole in the very bottom of the mirror on the left. It subsequently strikes the tilted mirror from which it reflects. How many times will it reflect off the upright mirror on the left?

16. [I] Suppose both mirrors in Problem 15 were, say, 10 m high. What then would be the angle of reflection off the tilted mirror when the beam encounters it for the second time?

17. [I] Figure P17 shows an L-shaped object in front of a front-silvered plane mirror. Locate the images of points A, B, and C. Describe the resulting image completely. How does this compare with your experiences of looking into a plane mirror?

> **SOLUTION:** Make all measurements from the front reflecting surface. Drop a perpendicular from point-A to the mirror, and extend it behind (to the right of) the mirror. The image of A, namely A', appears on that line (8.0 cm + 10 cm) to the right of the reflecting surface. Similarly, B' is 18 cm to the right. Whereas C' is 10 cm to the right. The image is virtual, 12 cm tall and right-side-up, but it's turned around so that it faces toward the object (C' is closer than B'). That's why the image of your nose is always facing toward you in a mirror.

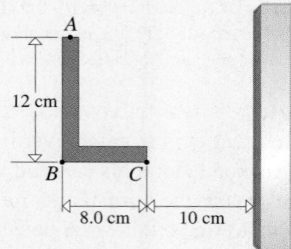

Figure P17

18. [I] THIS PROBLEM EXAMINES THE FORMATION OF AN IMAGE BY A PLANE MIRROR. The driver of a car looks into her rear-view mirror, which is flat, and sees the image of an SUV behind her. Her head is 60 cm from the mirror, almost in line with it, and the image appears to be 4.0 cm wide on the surface of the mirror. Assuming the approaching SUV is 2.4 m wide, we want to determine how far away it is. (a) Draw a ray diagram with the driver and the SUV on the same side of the mirror. Show two rays, one from each edge of the SUV, hitting the mirror (at the edges of the image), reflecting off, and converging to the driver's eye. (b) Now draw the image of the SUV behind the mirror so the rays appear to go from it in straight lines, directly to the driver's eye. (c) Using similar triangles involving the width of the SUV and its image, approximate the distance from the mirror to the SUV.

19. [I] A woman stands between a vertical mirror $\frac{1}{2}$ m tall and a distant tree whose height is H. She is 1.0 m from the middle of the mirror, and the tree is 11.0 m from the mirror. If she sees the tree just fill the mirror, how tall is the tree? [*Hint: The tree is behind her, and rays from its top and bottom get to her eye.*]

20. [I] A person stands 1.0 m from a window which is $\frac{1}{2}$ m tall and looking out sees a tree that is 11.0 m from the window. If she sees the tree just fill the window, how tall is the tree? Compare this geometry with that of the previous problem.

21. [I] An 8.00-m tall building is 10.0 m behind a person who is looking directly at the middle of a mirror 2.00 m in front of him. How tall should the mirror be if the image of the building is to completely fill it top-to-bottom?

22. [II] If 4.00% of the light energy incident normally on a microscope slide is reflected from the top surface, how much will be transmitted through two such slides?

23. [II] Suppose you are standing in front of a 4.0-ft-tall flat vertical mirror in which you can see some fraction of your body. What will happen to that fraction when you step farther from the mirror? Draw a ray diagram that proves your contention.

24. [II] THIS PROBLEM EXAMINES THE PHENOMENON OF REFLECTION. A kid holding a laser pointer stands a perpendicular distance d in front of a tall vertical plane mirror. The laser is at a height h, and points down making an angle θ with the horizontal. The narrow beam, traveling in a vertical plane, strikes the mirror at a height y and then (passing between her feet) hits the floor at a distance d behind the girl. We want to arrive at an expression for y in terms of h. (a) Draw a diagram. (b) What is the relationship between the angle of incidence at the mirror and θ? (c) What is the relationship between the angle of reflection at the mirror and θ? (d) What angle does the beam make with the ground? (e) We need an equation containing y and h. Using the fact that the beam strikes the mirror at a point (h − y) down from the laser, show that $\tan\theta = (h - y)/d$. (f) Using the fact that the beam strikes the mirror at a point y up from the ground, show that $\tan\theta = y/2d$. (g) Write an expression for y in terms of h.

25. [II] THIS PROBLEM APPLIES THE LAW OF REFLECTION. Figure P25 is a classic arrangement of two plane mirrors forming a wedge angle of α. A ray enters horizontally, makes four successive reflections, and then retraces its path, leaving the wedge along the same line as it entered. We want to find α. (a) From the physics what is the value of angle-1? Explain. (b) From the geometry what is the value of angle-2? Explain. (c) From the physics what is angle-3? (d) From the geometry find angles-4 and 5. (e) From the physics show that angle-6 = 2α. (f) From the geometry find angle-7 and show that angle-8 = 3α. (g) From the physics what is angle-9? (h) Prove that α = 22.5°.

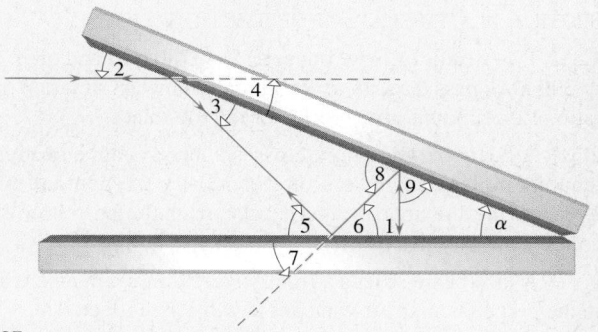

Figure P25

26. [II] A man 6 ft tall standing 10 ft from a 3-ft-high vertical mirror can see his entire image in it. If his eyes are 4 in. below the very top of his head, how high above the floor is the bottom edge of the mirror?

27. [II] We wish to set up an eye test in a rather modest-sized office and so must use a plane mirror to get the full required distance to the chart ($\overline{C_1 C_2}$) as shown in Fig. P27. If the observer is a distance d from the front-silvered mirror (which is H meters tall) and the chart is h meters in height, write an expression for the minimum value of H that will suffice, in terms of h, d, and the object distance s_0.

28. [II] Figure P28 shows a ray striking one of two perpendicular mirrors at angle θ_{i1} and then striking the other at θ_{i2}. Find a relationship between these two angles and describe the paths taken by the incoming and outgoing rays.

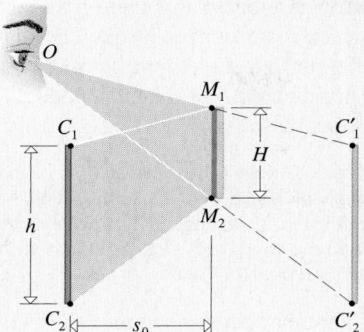

Figure P27

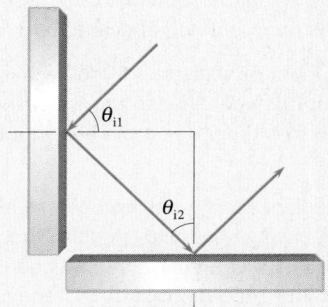

Figure P28

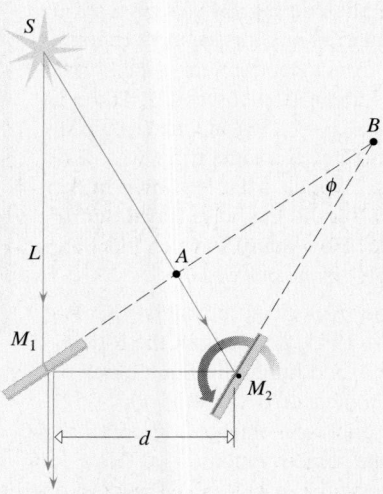

Figure P31

29. [II] Return to Fig. P28, but this time imagine there is a frog sitting between the mirrors looking at himself. Draw a ray diagram showing the locations of all the possible images. EXPLORING PHYSICS ON YOUR OWN: You should actually try this one with two ordinary mirrors held together. Look directly into the corner and hold up your right hand. Which hand will the image hold up? Notice where the seam is. Most people will see it run down the middle of their stronger eye. (A few with equal-strength eyes will see it down the middle of their faces.)

30. [II] Two vertical plane mirrors are brought together so that they make a wedge-angle of 35°, and a point source of light (S) is placed midway between them on a line bisecting the wedge-angle. Draw a simplified ray diagram locating all of the images of S by just using normal rays and making the object distance equal the image distance.

31. [III] Figure P31 shows an arrangement of mirrors forming a rangefinder for a camera. Mirror M_1 is partially silvered and fixed in position, while M_2 is totally silvered and rotates about a vertical axis. Looking into M_1, you would see two images, and by rotating M_2, they are made to exactly overlap. Since d is known accurately and the angle through which the second mirror is turned can be easily measured, a scale can be computed that displays the distance L directly on the camera. Write an expression for L in terms of ϕ and d, taking the tilt of the fixed mirror to be 45°.

32. [III] Most mirrors in common use are back-silvered. Suppose you stand 1.000 0 m from such a mirror, which is made of 5.00-mm-thick plate glass ($n = 1.50 = 3/2$). How far behind the front surface will your image appear? Draw a ray diagram.

SECTION 23.4: THE INDEX OF REFRACTION

33. [I] How far does yellow light travel in water in 1.00 s? [*Hint: How fast does it travel?*]

34. [I] What is the speed of yellow light in ethyl alcohol?

35. [I] What is the speed of a beam of light in diamond if the index of refraction is 2.42?

36. [I] If the wavelength of a lightwave in vacuum is 540 nm, what will it be in water, where $n = 1.33$?

37. [I] What should be the index of refraction of a medium if it is to reduce the speed of light by 10% as compared to its speed in vacuum?

38. [I] A beam of monochromatic light is traveling in a medium that has an index of refraction of 1.20. (a) If its frequency is 500 THz, what is its wavelength in the medium? (b) If light of a higher frequency enters the medium, will it experience the same index?

> SOLUTION: (a) To find λ we need to know the speed. Accordingly, $n = c/v$ and so $v = c/n = (2.998 \times 10^8 \text{ m/s})/1.20 = 2.498 \times 10^8$ m/s. Recall that $f\lambda = v$, which means $\lambda = v/f = 500$ nm. (b) No, all material media produce dispersion—the index will be greater.

39. [I] THIS PROBLEM WILL HELP US BETTER UNDERSTAND THE INDEX OF REFRACTION. Monochromatic light traveling in vacuum enters an optically dense medium where its speed is effectively halved. We want to determine the index of refraction of the medium. (a) What is the value of the speed of light in vacuum? (b) Compute the value of the speed of light in this medium. (c) State the definition of the index of refraction. (d) What is the value of the index of refraction in this case?

40. [I] If the speed of light (that is, the phase speed) in Fabulite (SrTiO₃) is 1.245×10^8 m/s, what is its index of refraction, to three significant figures?

41. [I] Lucite has an index of refraction of 1.49 whereas ice has an index of 1.31. In which does light travel faster?

42. [I] Carbon disulfide is a liquid that has an index of refraction of 1.63 as compared with PVC, which has an index of 1.54. In which does light travel faster, and what is the difference in the speeds?

43. [I] Dry air has an index of refraction of 1.000 293. By how much does a beam of light initially traveling in vacuum slow down upon entering air? Give your answer to three significant figures.

44. [II] A 500-nm lightwave propagating in vacuum impinges normally on a glass sheet of index 1.60. How many waves span the glass if it's 1.00 cm thick?

45. [II] The photo in Fig. P45 shows a pulse of green light (530 nm, the second harmonic of 1.06 μm from a neodymium-doped glass laser) lasting about 10 picoseconds. The cell contains water ($n \approx 1.36$) and the scale is in millimeters. The exposure time was also about 10 ps. Explain what is shown in the picture. How far did the pulse travel during the exposure? How many wavelengths long (of green light) is the pulse?

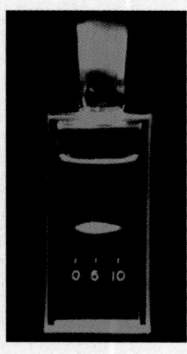

Figure P45

46. [II] The American physicist R. W. Wood (1868–1955) pointed out the following: a pulse of red light entering a block of glass (with a measured value of $n = 1.52$) 12.00 miles thick will emerge 1.80 miles ahead of a blue pulse that entered at the same moment. (a) How long will it take the red light to traverse the glass? (b) How fast does the blue pulse travel? (c) How much time will elapse between the emergence of the red and then the blue? Incidentally, Albert Michelson actually observed the effect in carbon disulfide.

47. [II] A lightwave propagates from point A to point B in vacuum. Suppose we introduce into its path a flat glass plate ($n_g = 1.50$) of thickness $L = 1.00$ mm. If the vacuum wavelength is 500 nm, how many waves span the space from A to B with and without the glass in place? What phase shift is introduced with the insertion of the plate?

48. [II] Yellow light from a sodium lamp ($\lambda_0 = 589$ nm) traverses a tank of glycerin (of index 1.47), which is 20.0 m long, in a time t_1. Now, if it takes a time t_2 for the light to pass through the same tank when filled with carbon disulfide (of index 1.63), determine the value of $t_2 - t_1$.

SECTION 23.5: SNELL'S LAW

49. [I] A beam of light impinges on an air-liquid interface at an angle of 55°. The refracted ray is observed to be transmitted at 40°. What is the refractive index of the liquid? [*Hint: Snell's Law relates the incident and transmitted rays via the associated indices and angles.*]

50. [I] If a narrow beam of light is incident on the surface of a tank of turpentine at 30°, at what angle will it be transmitted?

51. [I] A narrow beam of yellow light is incident in air on a salt crystal having an index of refraction of 1.544. If the beam makes an angle of 20.0° with the flat surface of the crystal, at what angle will it be transmitted?

52. [I] THIS PROBLEM IS ABOUT THE LAW OF REFRACTION. Figure P52 is a plot of the sine of the transmission angle versus the sine of the angle of incidence measured as light passed from air into some more optically dense medium. (a) Discuss the curve explaining its relationship to Snell's Law. (b) What is the value of the refractive index of the incident medium? (c) What is the significance of the slope of the line? (d) What is the numerical value of the slope of the line? (e) Guess at what the dense medium might be.

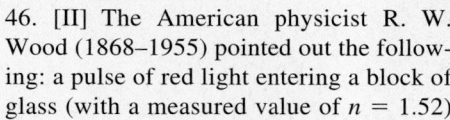

Figure P52

53. [I] A narrow laserbeam impinges on the smooth top of a sheet of plexiglass which is surrounded by air. The beam makes an angle of 30.0° with the surface. (a) A portion of the light is reflected. What is the angle of reflection? (b) The remainder of the light is transmitted. What is the transmission angle?

> SOLUTION: (a) The ray comes in at 30.0° w.r.t. the surface, so it has an incident angle measured from the normal of 60.0°. And that's also the angle of reflection. (b) Using Snell's Law, $n_i \sin \theta_i = n_t \sin \theta_t$ knowing that $n_i = 1.00$ and using Table 23.1, $\sin 60.0° = (1.51) \sin \theta_t$ and therefore $\sin \theta_t = 0.573$. Hence, $\theta_t = 35.0°$.

54. [I] A swimmer shines a beam of light in water up toward the surface. It strikes the air-water interface at 35°. At what angle will it emerge into the air?

55. [I] Prove that to someone looking straight down into a swimming pool, the water will appear to be 3/4 of its true depth.

56. [I] A pool of water is 3.00 m deep and 4.00 m wide. Someone lying with his face close to the water is looking for a coin that is directly across the pool on the bottom far edge. Will he be able to see it?

57. [I] Radiant energy can be conveniently converted from IR (1.06 μm) to UV using nonlinear crystals. First, a portion of the IR is converted into its second harmonic on passing through an appropriate crystal. Determine the resulting frequency. Then, these two frequencies are mixed in a second crystal to generate the third harmonic. Find the wavelength of the third harmonic.

58. [II] Figure P58 shows a semi-circular piece of clear test material surrounded by air. Determine the index of refraction of the sample. What is the speed of light in the material? This arrangement allows the ray to leave the material and the transmission angle to be easily measured. Explain why the emerging ray is unbent.

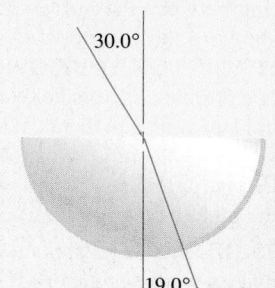

Figure P58

59. [II] We wish to determine the index of refraction of a liquid filling a small tank as shown in Fig. P59. By moving up and down, it is found that the back edge of the container is just visible at an angle of 20.0° above the horizontal. Find the index.

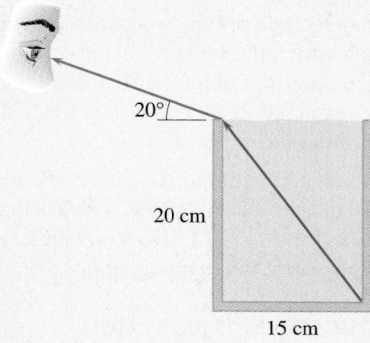

Figure P59

60. [II] A prism, *ABC*, is cut such that ∠ *BCA* = 90° and ∠*CBA* = 45°. What is the minimum value of its index of refraction if, while immersed in air, a beam traversing face *AC* is to be totally internally reflected from face *BC*?

61. [II] A beam of light impinges on the top surface of a 2.00-cm-thick parallel glass ($n = 1.50$) plate at an angle of 35°. How long is the actual path through the glass?

62. [II] THIS PROBLEM TREATS THE LAW OF REFRACTION. Imagine an aquarium filled with fresh water ($n_w = 1.33$) retained by flat thick glass ($n_g = 1.52$) walls. A ray of light from a fish impinges on the glass at 40.0° to the normal. Light is transmitted into the glass at an angle θ_{t1}, traverses the wall, and emerges at an angle θ_{t2}. We want to determine all the angles, especially the final transmission angle when the light enters air. (a) Draw a diagram. (b) What is the angle of reflection at the water-glass interface? (c) At what angle (θ_{t1}) is the ray transmitted into the glass? (d) What is the angle between the reflected and refracted rays at the water-glass interface? (e) What is the angle of incidence of the ray on the glass-air interface? (g) Determine the transmission angle (θ_{t2}) into the air.

63. [I] THIS PROBLEM IS ABOUT THE LAW OF REFRACTION. Figure P63 shows two parallel rays impinging on a glass ($n_g = 1.50$) prism. We wish to determine the angle (α) with which the rays converge. (a) Why don't the rays bend at the first interface? (b) Explain why the rays bend as they do at the second interface. (c) Use Snell's Law to compute the transmission angle (θ_{t2}) through the glass-air interface for each ray as it emerges from the prism. (d) By how much was each ray deviated from its original path? (e) Determine the angle α.

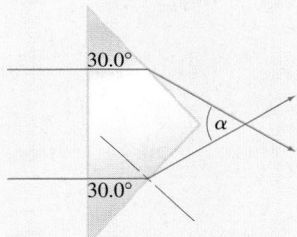

Figure P63

64. [II] Consider an air-glass interface with light incident in the air. If the index of refraction of the glass is 1.70, find the incident angle such that the transmission angle is to equal $\frac{1}{2}\theta_i$.

65. [II] Imagine that you focus a camera with a close-up bellows attachment on a letter printed on this page. Now, suppose the letter is covered with a 1.00-mm-thick microscope slide ($n = 1.55$). How high must the camera be raised in order to keep the letter in focus?

66. [III] Refer to Fig. 23.23. If the plate has a thickness τ, show that the emerging beam is laterally displaced by a perpendicular distance d from the incident beam where

$$d = \frac{\tau \sin(\theta_{i1} - \theta_{t1})}{\cos \theta_{t1}}$$

67. [III] A coin rests on the bottom of a tank of water ($n_w = 1.33$) 1.00 m deep. On top of the water floats a layer of benzene ($n_b = 1.50$), which is 20.0 cm thick. Looking down nearly perpendicularly, how far beneath the topmost surface does the coin appear? Draw a ray diagram.

68. [III] With the results of Problem 66 in mind, by how much would a beam, incident at 30°, be displaced on traversing a plate of glass 5.00 cm thick having an index of 1.60?

SECTION 23.6: TOTAL INTERNAL REFLECTION

69. [I] THIS PROBLEM TREATS TOTAL INTERNAL REFLECTION. A narrow beam of monochromatic light (of wavelength 589 nm) is traveling in an unknown medium. The beam impinges on a smooth flat interface between the medium and air. (a) What is the defining equation for the critical angle experienced at such an interface? (b) Keep in mind that the sine of an angle is always less than or equal to one. What does that tell you about the two indices in this case? (c) What is the value of the refractive index of air? (d) It is found that the sine of the critical angle for total internal reflection is 0.614 3. What physical quantity does that equal? (e) Determine the index of the unknown medium. (f) Guess at what the unknown medium might be.

70. [I] Two transparent media (*X* and *Y*) share a common smooth flat interface. A narrow laserbeam in medium-*X* travels at 2.00×10^8 m/s toward the interface. Medium-*Y* is known to have an absolute index of refraction of 1.36. With what angle of incidence, at the interface, will the beam be totally reflected back into medium-*X*?

> SOLUTION: This is a case of total internal reflection where the defining equation is $\sin \theta_c = n_t/n_i$. Here $n_i = n_X$ and $n_t = n_Y$. Having n_Y we must find n_X. Thus $n_X = c/v = 1.499$ and so $\sin \theta_c = n_t/n_i = n_Y/n_X = 1.36/1.499$ (which is properly less than 1) and so $\theta_c = \sin^{-1} 0.907\,3$ and $\theta_c = 65.1°$.

71. [I] A block of glass has an index of refraction of 1.540. Its upper surface is flat, smooth, and surrounded by air. A beam of light is projected upward through the block, and it makes an angle θ_i in the glass with the normal to the flat face. What minimum value must that angle have if no portion of the beam is to be transmitted into the air? [*Hint: This is an instance of total internal reflection with glass as the incident medium.*]

72. [I] A swimmer under the water shines a narrow beam of light up at the smooth surface. At what minimum angle, made with the normal to the interface, will the beam be completely reflected back into the water?

73. [I] A laserbeam traveling in ice impinges on the smooth flat interface separating the ice from the surrounding air. At what minimum incident angle, measured with respect to the normal, will the beam be totally reflected back into the ice?

74. [I] What is the critical angle for diamond in air? Give your answer to two significant figures. What, if anything, does the critical angle have to do with the luster of a well-cut diamond?

75. [I] A laserbeam inside a tank of benzene is directed up toward the interface at an angle of 45.0° with the normal. A gas is introduced above the liquid, and its pressure is increased until the laserbeam is totally reflected at the interface. What is the index of refraction of the gas at the moment?

76. [I] Using a block of a transparent, unknown material, it is found that a beam of light inside the material is totally internally reflected at the air-block interface at an angle of 48.0°. What is its index of refraction?

77. [I] What is the minimum incident angle for internal reflection at the interface between two materials of indices 1.33 and 2.40, respectively?

78. [II] A block of glass with an index of 1.55 is covered with a layer of water of index 1.33. For light traveling in the glass, what is the critical angle at the interface?

79. [II] A block made of a light crown glass, with an index of refraction of 1.501, is immersed in water. What is the critical angle of incidence if a ray is to be totally internally reflected within the glass? How will that change if the water is replaced by benzene?

80. [II] A prism made of Plexiglass is immersed in water. What is the critical angle of incidence if a ray is to be totally internally reflected within the prism? How will that change if the water is replaced by carbon disulfide?

81. [II] A fish looking straight upward toward the surface receives a cone of rays and sees a circle of light filled with the images of sky and ships and whatever else is up there (Fig. P81). This bright circular field is surrounded by darkness. Explain what is happening and compute the cone-angle.

Figure P81

82. [II] Suppose we took the same semicircular piece of material

that's in Fig. P58, and this time surrounded it by water and reversed the rays. In other words, shine a narrow beam of light upward and to the left radially; at what angle would the beam be totally internally reflected back down and to the left?

83. [II] THIS PROBLEM IS ABOUT REFRACTION AND TOTAL INTERNAL REFLECTION. Figure P83 shows a ray propagating through a flint glass block immersed in air. We want to find the minimum value of α that will cause the ray to reflect back into the block rather than emerging. (a) We'll need the index of the glass, which we can calculate from Snell's Law if we know α. Accordingly, determine the angle β from the geometry. (b) What is the value of α? (c) What is the transmission angle as the ray emerges from the block? (d) Determine the index of refraction of the glass. (e) By how much, at the least, must α be changed in order that the ray be totally internally reflected?

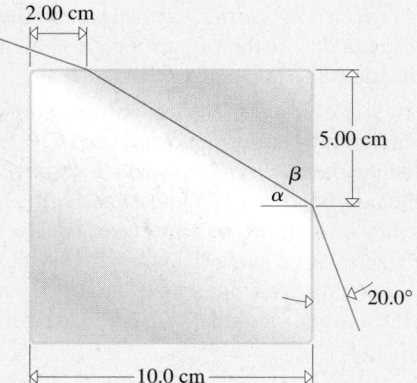

Figure P83

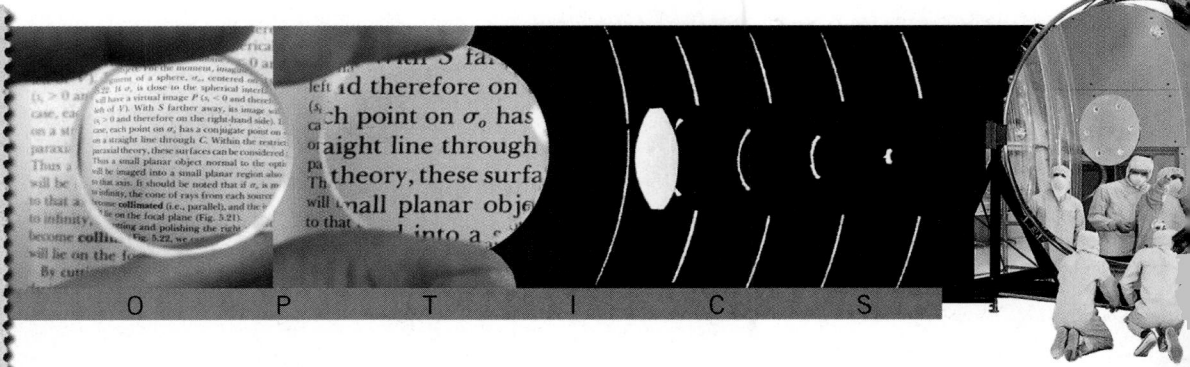

Chapter 24
Geometrical Optics and Instruments

An object that is either self-luminous or externally illuminated can be imagined to have a surface covered with radiating point sources. Your face in the sunshine is a distribution of countless atomic scatterers, each sending out little spherical wavelets (Fig. 23.13). The associated rays emanate radially outward in the direction of the energy flow (Fig. 24.1); they diverge from each point source *S*. Now suppose the object stands before some arrangement of reflecting and/or refracting surfaces constituting an *optical system* whose function it is to collect the light and cause it to *converge*. The energy in the diverging cone arrives at *P*, which is called the image of *S*. A billion points on a face contribute light to a billion points that make up the image of that face.

Real systems—cameras, telescopes, and eyeballs—cannot possibly collect *all* the light emitted from an object. When only a portion of any incident wavefront enters the system, there will inevitably be some deviation from straight-line propagation, a phenomenon called diffraction (p. 910). Even so, it is often reasonable to simply neglect diffraction and assume that all the rays travel in straight lines and go where they go via the Laws of Reflection and Refraction; we then have the idealized domain of **geometrical optics**.

Geometrical optics deals with the manipulation of wavefronts (or rays) by means of interposing reflecting and/or refracting objects, assuming straight-line propagation.

Lenses

The lens is no doubt the most widely used optical device, and that's not even considering the fact that we see the world through a pair of them. Human-made lenses date back at least to the *burning-glasses* of antiquity, which, as the name implies, were used to start fires long before the advent of matches—there's a reference to one of these in Aristophanes' play *The Clouds* (424 B.C.E.). From our perspective, *a lens is a refracting device (or discontinuity in the transmitting media) that reconfigures an incoming energy distribution*. And that much is true whether we are dealing with UV, lightwaves, IR, microwaves, radiowaves, or even sound waves.

The configuration of a lens is determined by the reshaping of the wavefront it is to perform. Point sources are basic, and so it is often desirable to convert diverging spherical waves into a beam of plane waves; flashlights, projectors, and searchlights all do this in order to keep the beam from spreading out and weakening as it progresses. In just the reverse, it is frequently necessary to collect incoming parallel rays and bring them together at a point, thereby focusing the energy, as is done with a burning-glass or a telescope lens.

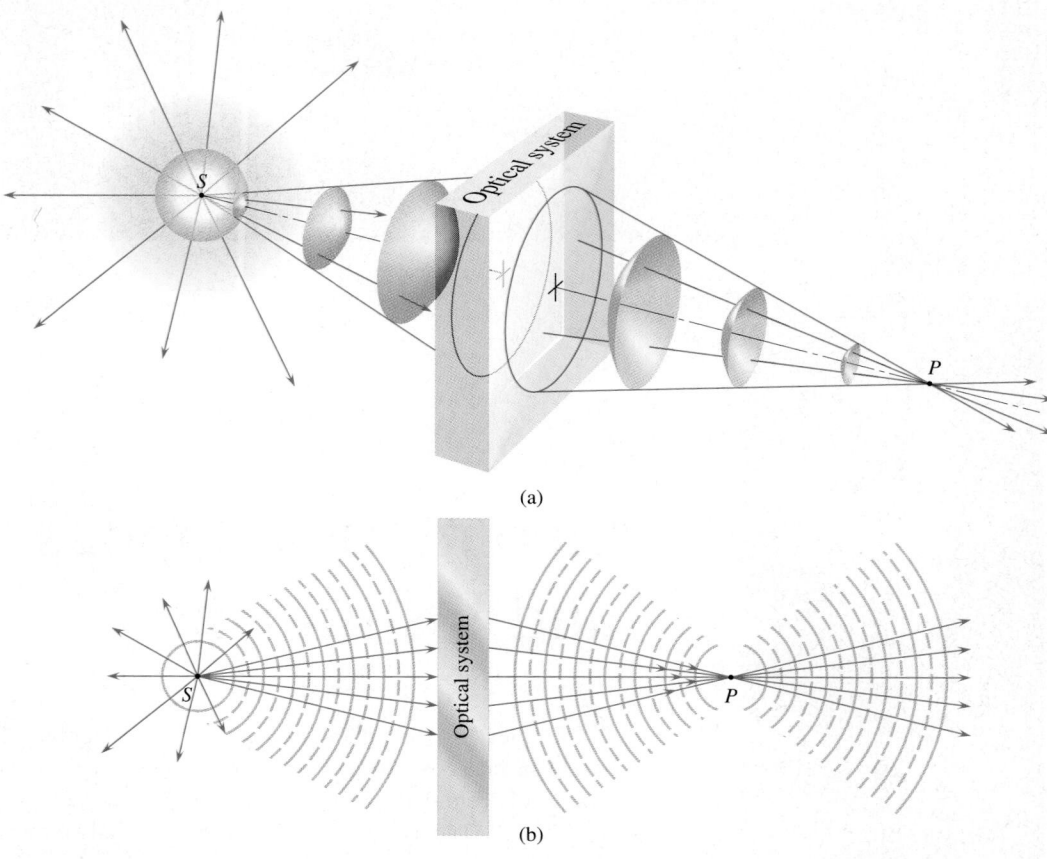

(a)

(b)

Figure 24.1 (a) A point source S sends out spherical waves. A cone of rays enters an optical system that inverts the wavefronts, causing them to converge on point P. (b) Rays diverge from S and a portion of them converge to P. If nothing stops the light at P, it continues on.

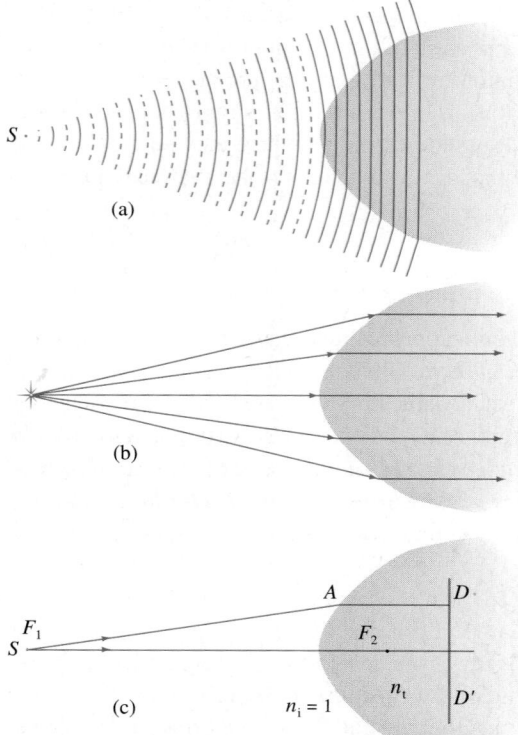

24.1 Aspherical Surfaces

We can begin to understand how a lens works by interposing in the path of the wave a transparent substance in which the wave's speed is different than it was initially. Fig. 24.2a shows a diverging spherical wave traveling in an incident medium of index n_i, impinging on the curved interface of a transmitting medium of index n_t. When $n_t > n_i$, the wave slows as it enters the new substance. The central area of the wavefront travels more slowly than its outer extremities, which are still moving through the incident medium. These extremities overtake the mid-region, continuously straightening out the wavefront. If the interface is properly configured, the spherical wavefront bends into a plane wave.

To find the required shape of the interface so that we can make such a device, refer to Fig. 24.2c, wherein point A can lie anywhere on the boundary. *One wavefront is transformed into another provided the paths along which the energy propagates are all equivalent, thereby maintaining the phase of the wavefront.* A little spherical surface of constant phase emitted from S must evolve into a flat surface of constant phase at $\overline{DD'}$. A detailed

Figure 24.2 Light striking a hyperbolic interface between air and glass. (a) Because the speed of the wave slows as it enters the glass, the wavefronts bend and gradually straighten out. (b) The rays become parallel. (c) The hyperbola is such that the optical path from S to A to D is the same no matter where A is.

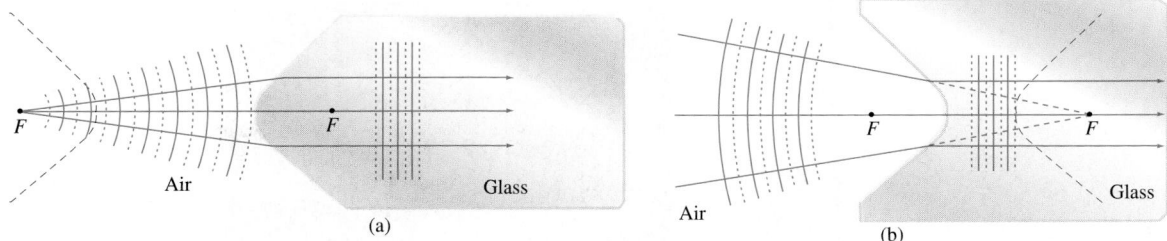

Figure 24.3 Hyperbolic surfaces have two foci, F. In (a), light originating at F passes into the glass in the form of plane waves. In (b), light converging toward F passes into the glass as plane waves. In both cases, the rays can be reversed so that they emerge into the air as either converging or diverging cones.

analysis, which can be viewed by clicking on ASPHERICAL SURFACES under FURTHER DISCUSSIONS on the CD ⊙ , shows that if the point A sweeps out a hyperbola, all the paths from S to A to P will be traveled in precisely the same amount of time—a *hyperboloidal* interface will transform a diverging spherical wave into a transmitted plane wave. Ellipsoidal and paraboloidal surfaces also do interesting things, but we'll limit our discussion here to hyperboloidal surfaces only (Fig. 24.3).

It's an easy matter now to construct lenses such that both the object and image points (or the incident and emerging light) will be outside of the medium of the lens. In Fig. 24.4a diverging incident spherical waves are made into plane waves at the first interface via the mechanism of Fig. 24.3a. These plane waves within the lens strike the back face perpendicularly and emerge unaltered; $\theta_i = 0$ and $\theta_t = 0$. And because the rays are reversible, *plane waves incoming from the right will converge to point F*, which is known as the **focal point** of the lens. Exposed to the parallel rays from the Sun, our rather sophisticated lens would serve nicely as a burning-glass.

In Fig. 24.4b, the plane waves within the lens are made to converge toward the axis by bending the second interface. Both of these lenses are thicker at their midpoints than at their edges and are therefore said to be **convex** (from the Latin *convexus*, meaning arched). Each lens causes the incoming rays to converge somewhat, to bend a bit more toward the central axis, and so they are referred to as **converging lenses**.

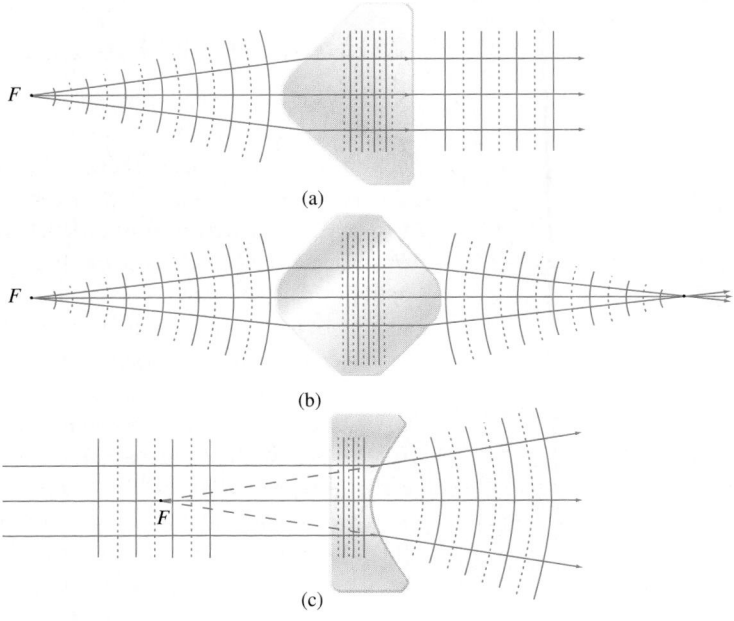

Figure 24.4 Several hyperbolic lenses.

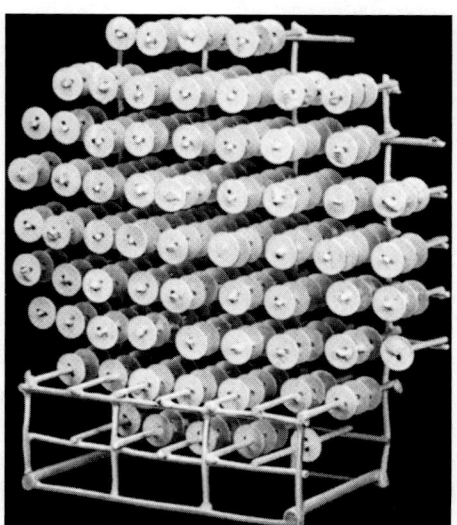

A lens for short-wavelength radiowaves. The disks serve to refract these waves much as rows of atoms refract light.

In contrast, a **concave** lens (from the Latin *concavus*, meaning hollow—and most easily remembered because it contains the word *cave*) is thinner in the middle than at the edges, as is evident in Fig. 24.4*c*. It causes the rays that enter as a parallel bundle to diverge. All such devices that turn rays outward away from the central axis (that in so doing add divergence to the beam) are called **diverging lenses**. In Fig. 24.4*c*, parallel rays enter from the left and, on emerging, seem to diverge from *F*; still, that point is taken as a focal point. *When a parallel bundle of axial rays passes through a converging lens, the point to which it converges (or when passing through a diverging lens, the point from which it diverges) is a focal point of the lens.*

> The parallel rays in a beam entering a converging lens along its central axis will all cross at a **focal point** beyond the lens.

Optical elements (mirrors and lenses) having at least one curved surface that's not spherical are referred to as *aspherical*. Although they are easy to understand and perform certain tasks exceedingly well, such elements are difficult to manufacture with great accuracy. Even so, many millions of aspherics have been made, and they can be found in instruments across the whole range of quality: in telescopes, projectors, cameras, and spy satellites.

24.2 Spherical Thin Lenses

In comparison to aspherics, spherical surfaces (actually, segments of spheres) are easy to fabricate. Unlike the hyperboloid, ellipsoid, or paraboloid, the sphere has no single central symmetry axis. All its diameters are alike, and the sphere can be generated by simply randomly rocking a rough-cut glass disk against an approximately spherical grinding tool. The two will wear each other away until both are perfectly spherical. The vast majority of lenses in use today have surfaces that are segments of spheres—despite the fact that those spherical surfaces are not ideal and will result in imaging errors known as *aberrations*. By using several components made of different materials to form compound lenses, these errors can be controlled quite well.

Because we cannot expect perfect imagery from spherical surfaces, it will be necessary to place limitations on the way a spherical lens can be used so that it behaves appropriately. We only allow the lens to receive rays that strike it *not far from the central axis and enter only at shallow angles*—rays of this kind are said to be **paraxial**. No matter how the rays are drawn (and they will often be depicted making large angles for the sake of clarity), the rays that actually enter the lens must be paraxial. To simplify matters further, we will only deal with **thin lenses**, that is, *lenses for which the radii of curvature of the surfaces are large compared to the thickness*. Such lenses are quite common—most telescope and eyeglass lenses are thin.

Figure 24.5 The radius drawn from C_1 is normal to the first surface; and as the ray enters the lens, it bends down *toward* that normal. The radius from C_2 is normal to the second surface; and as the ray emerges, since $n_l > n_a$, the ray bends down *away* from that normal. The lens causes the ray to converge toward the central axis and so we call this a *converging lens*.

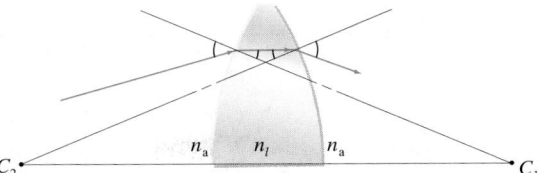

When the center of curvature of a surface is to the right of the surface, its **radius of curvature** is positive.

The distances of objects and images are usually measured from the lens and can be on either side of it. It's important to associate a specific sign with each such distance so that it can be properly manipulated algebraically. *As a rule, light enters from the left*, and Fig. 24.5 shows how a typical ray is twice bent toward the central axis as it traverses a convex spherical lens. In Fig. 24.6, we see two rays leaving the axial point source S and converging to the corresponding image point P. This is the basic geometry for which several possible sign conventions exist. We will take an **object distance** s_o to the *left* of the lens as positive and an **image distance** s_i to the *right* of the lens as positive. Furthermore, *the radius of curvature of a lens surface is positive when its center point C is to the right of the surface*. Here R_1, the radius of the first surface encountered, is positive while R_2, whose center C_2 is to the left, is negative. *All interfaces that bulge toward the left have positive radii, and all interfaces that bulge right have negative radii.*

The object and image distances, which is what we usually want to know, are related to the index of refraction of the lens and its radii by the formula

$$\frac{1}{s_o} + \frac{1}{s_i} = (n_l - 1)\left(\frac{1}{R_1} - \frac{1}{R_2}\right)$$ (24.1)

{For a complete derivation click on LENS EQUATION under FURTHER DISCUSSIONS on the CD.} 🔵 This is the famous **Thin-Lens Equation**, often referred to as the **Lensmaker's Formula** because the left side (which treats what is going on external to the lens) is

Figure 24.6 (a) The geometry of the thin lens. (b) The lens functions in three dimensions. Here several rays travel in a vertical plane.

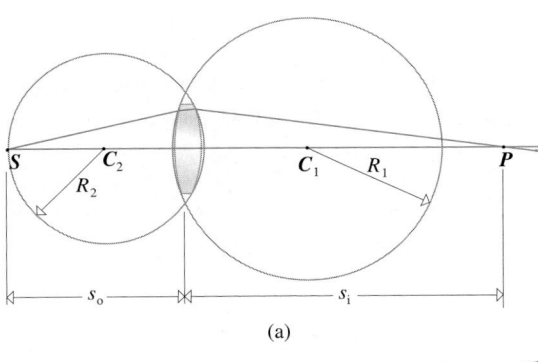

(a)

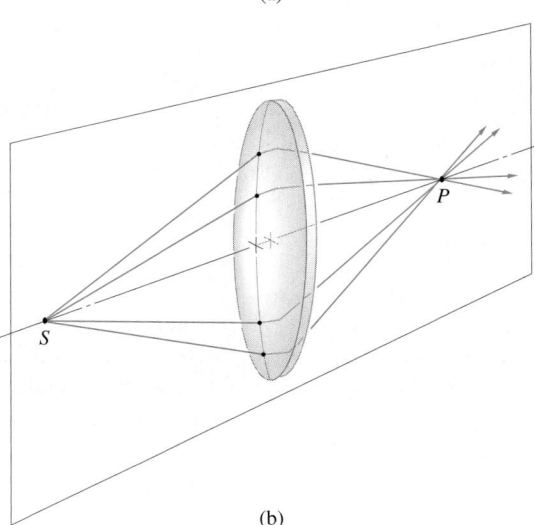

(b)

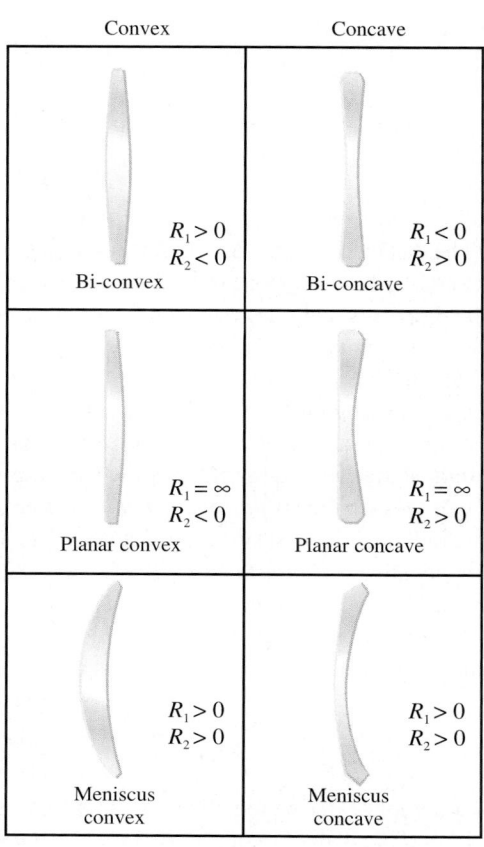

Figure 24.7 Cross sections of various centered spherical simple lenses. Take the surface on the left as number one since it's encountered first by light coming from the left.

Convex	Concave
$R_1 > 0$ $R_2 < 0$ Bi-convex	$R_1 < 0$ $R_2 > 0$ Bi-concave
$R_1 = \infty$ $R_2 < 0$ Planar convex	$R_1 = \infty$ $R_2 > 0$ Planar concave
$R_1 > 0$ $R_2 > 0$ Meniscus convex	$R_1 > 0$ $R_2 > 0$ Meniscus concave

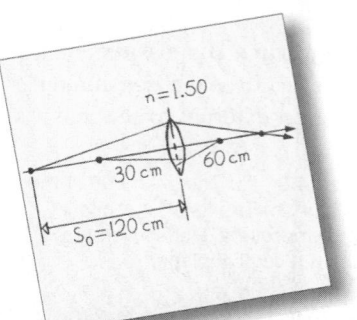

expressed on the right in terms of the physical variables that would have to be selected to fabricate the lens (Fig. 24.7).

Had the lens been immersed in a medium of index n_m rather than air, the Lensmaker's Formula would again result, but instead of the first term on the right being n_l, or equivalently $n_l/1$, it would now be n_l/n_m.

Example 24.1 **[I]** A small point of light lies on the central axis 120 cm to the left of a thin bi-convex lens (one curved outward on both sides) having radii of 60 cm and 30 cm. Given that the index of refraction of the lens is 1.50, find the location of the resulting image point. As far as the image distance is concerned, does it matter which way the lens is facing, that is, which surface is toward the object? Unless otherwise informed, always assume the surrounding medium is air.

Solution Whenever you are given the radii and index of a lens along with either the object or image distance, it's a good guess that the Lensmaker's Formula will be needed. (1) TRANSLATION—An object is placed at a specified distance from a converging thin lens whose index and radii are known; determine the image distance. (2) GIVEN: $s_o = 1.20$ m, $R_1 = 0.60$ m, $R_2 = -0.30$ m, and $n_l = 1.50$. FIND: s_i. (3) PROBLEM TYPE—Geometrical optics/thin converging lens. (4) PROCEDURE—Whenever you're given the radii of a lens (and its index of refraction), think of the right side of the Lensmaker's Formula. (5) CALCULATION—Because the lens is bi-convex, the second radius encountered by incoming rays from the left is

taken as a negative number. It follows from Eq. (24.1) that

$$\frac{1}{s_o} + \frac{1}{s_i} = (n_l - 1)\left(\frac{1}{R_1} - \frac{1}{R_2}\right)$$

$$\frac{1}{1.20 \text{ m}} + \frac{1}{s_i} = (1.50 - 1)\left(\frac{1}{0.60 \text{ m}} - \frac{1}{-0.30 \text{ m}}\right)$$

$$\frac{1}{s_i} = (0.50)\left(\frac{1}{0.20 \text{ m}}\right) - \frac{1}{1.20 \text{ m}} = \frac{2}{1.20 \text{ m}}$$

and $\boxed{s_i = 0.60 \text{ m}}$. The image distance is positive, and so the image lies to the right of the lens on the axis. Had we let $R_1 = 0.30$ m and $R_2 = -0.60$ m, nothing would have changed—*it doesn't matter which way a thin lens faces!*

Quick Check: The right side of the Lensmaker's Formula equals $(0.50)/(0.20 \text{ m})$, and that quantity should equal the left side of the formula. Hence, $1/(1.20 \text{ m}) + 1/(0.60 \text{ m})$ should equal $1/(0.40 \text{ m})$, and it does.

[For more worked problems click on WALK-THROUGHS in CHAPTER 24 on the CD.]

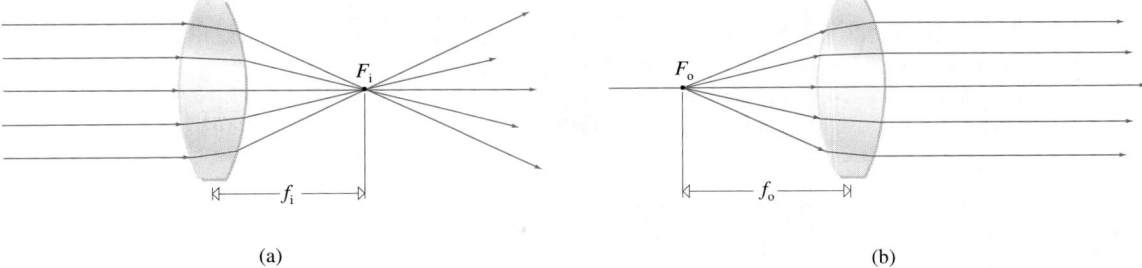

(a)   (b)

Figure 24.8   Focal points for a converging lens. (a) A parallel bundle of rays passing through a thin lens is brought to convergence at the image focal point F_i. (b) A point of light at the object focal point F_o emits light that emerges from the lens as a parallel beam.

24.3 Focal Points and Planes

Suppose that the object point S in Fig. 24.6 is moved far to the left. As $s_o \to \infty$, the rays enter the lens as a parallel bundle and are brought together at a specific image point known as the **image focal point**, F_i. The distance from the lens to this point is called the **image focal length**, f_i, where as $s_o \to \infty$, $s_i \to f_i$, as depicted in Fig. 24.8a. Similarly, the rays will emerge from the lens as a parallel bundle as $s_i \to \infty$ and the special object point for which this occurs is called the **object focal point**, F_o. The distance from the lens to this point is called the **object focal length**, f_o, where as $s_i \to \infty$, $s_o \to f_o$, as depicted in Fig. 24.8b. A thin lens surrounded by the same medium on both sides is a special case for which the image and object focal lengths are the same and the subscripts can be dropped altogether. Accordingly, go back to the Lensmaker's Formula and let $s_o \to \infty$, whereupon $(1/s_o) \to 0$ while $s_i \to f$, and we get

$$\frac{1}{f} = (n_l - 1)\left(\frac{1}{R_1} - \frac{1}{R_2}\right)$$

(24.2)

The same thing results as $s_i \to \infty$. The focal length of a lens is determined by its physical makeup and can be positive or negative. In Figs. 24.8a and b, $R_1 > 0$ and $R_2 < 0$; each focal length is positive. In Figs. 24.9a and b, $R_1 < 0$ and $R_2 > 0$, and each focal length is negative. Going by the sign of its focal length, we commonly refer to a converging lens as a **positive lens** and a diverging lens as a **negative lens**.

If for a bi-convex lens the radii are small (the lens has bulging faces), then the right side of the equation will be large, and the left side must also be large, meaning f must be small (the lens has a short focal length). *The more "bent" the surfaces are, the more they bend the rays.* On the contrary, when the surfaces are fairly flat (the radii are large), the focal length will be large. In fact, if the radii are equal and the lens is made of glass $(n_l \approx \frac{3}{2})$, it follows from Eq. (24.2) that $f \approx R$. *For bi-convex and bi-concave lenses, f is ordinarily of the order-of-magnitude of the smaller radius*, which is a good thing to keep in mind. This need not apply with meniscus lenses, where the magnitude of the focal length can be much greater than the radii.

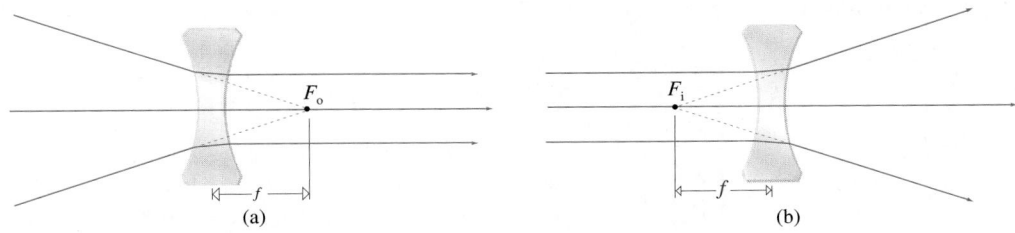

(a)   (b)

Figure 24.9   Focal points for a diverging lens. (a) Rays heading for F_o emerge parallel. (b) Rays entering parallel emerge as if from F_i.

Example 24.2 **[I]** Determine the focal length in air of a thin spherical planar-convex lens having a radius of curvature of 50 mm and an index of 1.50. What, if anything, would happen to the focal length if the lens is placed in a tank of water?

Solution Whenever you are given the radii and index of a lens (i.e., its physical specifications) it's a good guess that the right side of the Lensmaker's Formula will come into play. (1) TRANSLATION—Determine the focal length (in air and water) of a converging thin lens whose index and radii are known. (2) GIVEN: The fact that the first surface is flat means that it has an infinite radius of curvature, $R_1 = \infty$; $R_2 = -0.050$ m, and $n_l = 1.50$. FIND: f when $n_m = 1.00$ and 1.33. (3) PROBLEM TYPE—Geometrical optics/thin converging lens/focal length. (4) PROCEDURE—Whenever you're given the radii of a lens, think of the right side of the Lensmaker's Formula. (5) CALCULATION—The focal length of a thin lens is specified in Eq. (24.4),

$$\frac{1}{f} = (n_l - 1)\left(\frac{1}{R_1} - \frac{1}{R_2}\right) = (1.50 - 1)\left(\frac{1}{\infty} - \frac{1}{-0.050 \text{ m}}\right)$$

$$\frac{1}{f} = 0.50\left(0 + \frac{1}{0.050 \text{ m}}\right)$$

and $\boxed{f = +0.10 \text{ m}}$. When the lens is surrounded by a medium of index n_m rather than air, n_l must be replaced by $n_l/n_m = 1.50/1.33$. The effect is to reduce the lens's ability to bring the rays into convergence and to increase the focal length to $+0.39$ m.

Quick Check: f is positive as it should be for a convex lens. Moreover, $f = 2|R_2|$, which is also the right order-of-magnitude.

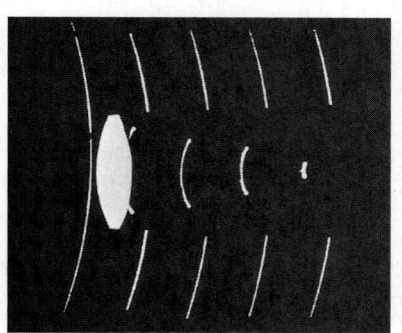

Wavefronts focused by a lens. The photo shows five exposures (each separated by 100×10^{-12} s) of a spherical pulse 10×10^{-12} s long as it swept over and through a converging lens. (The photo was made using a holographic technique.)

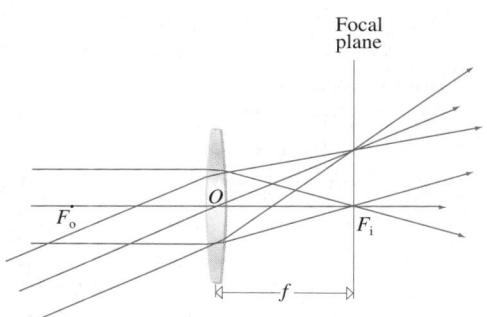

Focal plane

Figure 24.10 A ray headed for the center O of a thin lens passes straight through without bending. Ideally, all parallel bundles of rays focus on a plane, but actually only paraxial rays focus on that focal plane.

Table 24.1

Sign Convention for Spherical Refracting Surfaces and Thin Lenses*	
s_o, f_o	+ left of O
s_i, f_i	+ right of O
R	+ if C is right of O
y_o, y_i	+ above optical axis

*Light enters from the left.

It is especially convenient to draw a ray along the central axis of a lens because it strikes each surface perpendicularly and passes straight through undeviated (Fig. 24.10). Consider a tilted off-axis ray that enters the lens and emerges parallel to the incident direction. Such a ray passes through a fixed point on the axis known as the **optical center** of the lens, O. It actually enters, bends a little, and emerges almost parallel to its incident direction. But because the lens is thin, the lateral displacement of the emerging ray is negligible. We can assume that **any paraxial ray heading toward the center of any thin lens will pass through O undeviated and may be drawn as a straight line**. It is customary when treating a thin lens to simply place the point O at the geometric center of the lens (Table 24.1).

A bundle of parallel rays impinging on the lens at some angle with the central axis will be focused at a point on the central ray, the ray passing through O (Fig. 24.10). All such convergence points really lie on a curved surface, but if we limit the discussion to paraxial rays, that surface closely resembles a plane perpendicular to the central axis at the focal point. This is the **focal plane**, located a distance f from the lens given by Eq. (24.2).

24.4 Extended Imagery: Lenses

Any point on an extended object sends out light in a great many directions, and if some of that light enters a positive lens, it can be made to converge to an image point. In that way, point-by-point, the image is formed. Wherever two rays from an object point are made to converge, all rays from that point, passing through the lens, ideally converge. *Find where any two rays from any object point cross, and you have found the corresponding image*

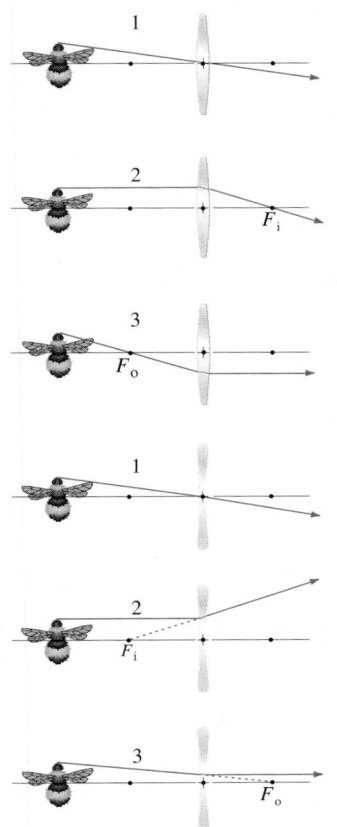

Figure 24.11 Tracing a few key rays through a positive and negative lens.

A **real image** is one formed by converging rays; it can be cast on a screen.

A **virtual image** is one formed by diverging rays; it cannot be cast on a screen.

point. And that's particularly easy to do using any two of the three special rays whose behavior we already know (Fig. 24.11). **Ray-1** heading for the center of any type of thin lens goes straight through. **Ray-2** entering a positive lens parallel to the central axis emerges passing through the image focus; a similar ray entering a negative lens emerges so that it can be extended back to pass through the image focus. **Ray-3** passing through the object focus of a positive lens emerges from the lens parallel to the central axis; a ray heading for the object focus of a negative lens emerges parallel to the central axis.

Since we know where these rays are going, we can simplify the drawings by constructing ray-paths with a single refraction taking place on a vertical line through the center of the lens. There really are two refractions, one at each face, but this way of proceeding will save us a lot of effort. Begin a ray diagram by setting down a horizontal axis and then sketching in a centered lens of arbitrary size and roughly the right shape. *The most important construction feature of the diagram is the location of the object and image focal points, which are at equal distances (f) on each side of the lens.* The vertical center-line, the focal length, and the location of the object determine the entire geometry. As a rule, when not drawing everything to scale, make the lens about the same size as the focal length. Should you inadvertently draw a lens too small to accept a particular ray, it can still properly be drawn refracted from the vertical center-line as if the lens were larger. At a distance of f on each side of the lens, put a mark on the axis to represent the focal points and then draw in another pair of points at distances of 2f.

Figure 24.12 depicts an object at some distance between f and 2f in front of a positive lens. To locate the resulting image, trace through the lens any two rays from each and every point on the flower—practically, a point on the very top and one on the bottom will do. From the top-most object point S, draw ray-1 straight through the center O of the lens. Now, we can either draw ray-2 or ray-3. If you like, draw all three; they, and indeed *all* rays entering the lens from that same object point, converge at the same conjugate image point P— that locates the top of the flower's image. The bottom of the image E can be found by doing the same thing over again for the bottom-most object point.

Here, **the rays converge toward the image**, and we say that the image is **real**. The light actually arrives at each such point and if there is a viewing screen there (as in a movie theater), the image would appear on it, which is why it's called a *real* image. By comparison, when an image is formed of light diverging from it (as was the case with the plane mirror, p. 820), we say that the image is **virtual**—it cannot be projected onto a screen, but it certainly can be viewed directly.

The rays in Fig. 24.12 don't just end on the image; they go on past it to diverge again. Moving the viewing screen toward or away from the lens intercepts a more or less blurred version of the flower, whose sharpest image exists at the one location in space where the rays converge to points. Unless the rays are intercepted (for example, by a film-plate), nothing of the image can be seen from the side. There is no transverse propagation—the image is not visible laterally as the flower itself *is*. The lens forms an image of the flower, but it

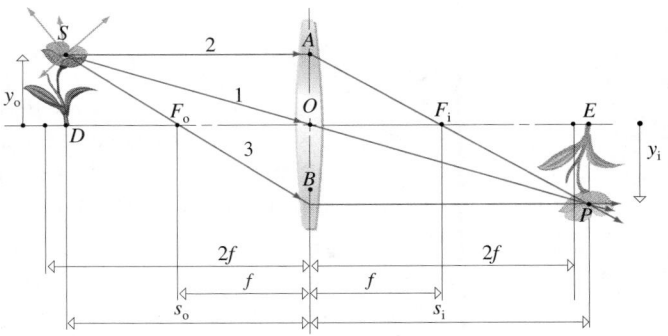

Figure 24.12 The geometry of image formation via a thin convex lens. **For simplicity, refractions take place at the *A-O-B* plane.**

A real image projected on the viewing screen of a 35-mm camera, much as the eye projects its image on the retina. Here a prism has been removed so that you can see the image directly.

does not re-create the original lightwave coming from the flower; something is lost in the transformation (p. 920.)

The ray diagram can provide us with an analytic relationship between the object and image distances and the focal length. In Fig. 24.12, triangles $F_i EP$ and $F_i AO$ are similar, all their angles are equal, and so

$$\frac{\overline{PE}}{\overline{AO}} = \frac{s_i - f}{f}$$

wherein $\overline{AO} = \overline{SD}$. Triangles SOD and POE are also similar and therefore

$$\frac{\overline{PE}}{\overline{SD}} = \frac{s_i}{s_o} \tag{24.3}$$

Setting these two equations equal and rearranging terms, we get

$$\frac{1}{s_o} + \frac{1}{s_i} = \frac{1}{f} \tag{24.4}$$

which is the famous **Gaussian Lens Equation** first derived by C. F. Gauss in 1841. It's also often called the **Thin Lens Equation**. Note that the same expression results on comparing Eqs. (24.1) and (24.2).

Example 24.3 **[I]** We wish to place an object 45 cm in front of a lens and have its image appear on a screen 90 cm behind the lens. What must be the focal length of the appropriate positive lens?

Solution This is a lens problem but we're not given its radii and index, and so the Thin Lens Equation should immediately come to mind. (1) TRANSLATION—Knowing the object and image distances, determine the focal length of a converging thin lens. (2) GIVEN: $s_o = 0.45$ m and $s_i = 0.90$ m. FIND: f. (3) PROBLEM TYPE—Geometrical optics/thin lens/focal length. (4) PROCEDURE—Whenever you're given or asked to find the object and image distances think of the Gaussian Lens Equation. (5) CALCULATION—The problem does not mention n or the radii of the lens, but we do have f and so the analysis

proceeds from Eq. (24.4);

$$\frac{1}{s_o} + \frac{1}{s_i} = \frac{1}{0.45 \text{ m}} + \frac{1}{0.90 \text{ m}} = \frac{1}{f}$$

Using a calculator, this can easily be evaluated term by term with the $(1/x)$ function

$$2.222 + 1.111 = \frac{1}{f}$$

and

$$\boxed{f = +0.30 \text{ m}}$$

Quick Check: Working without a calculator, recognize that Eq. (24.6) has the same form as that for two resistors adding in parallel;

$$f = \frac{s_i s_o}{s_i + s_o} = \frac{(0.45 \text{ m})(0.90 \text{ m})}{1.35 \text{ m}} = 0.30 \text{ m} \tag{24.5}$$

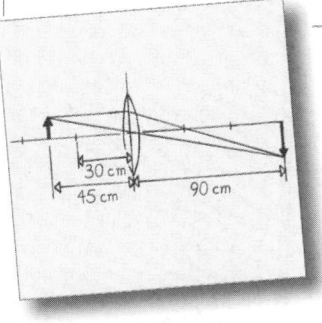

Magnification

The ratio of any transverse dimension of the image formed by an optical system to the corresponding dimension of the object is defined as the **transverse magnification** or, more often, the **magnification M_T**. In Fig. 24.12, the magnification is the height of the image divided by the height of the object:

$$M_T = \frac{y_i}{y_o} \tag{24.6}$$

Here y_i is below the central axis and is traditionally taken to be a negative number. The image is upside down, meaning that *the magnification is negative whenever the image is inverted and positive when the image is right-side-up*. Bear in mind that the magnification

The Thin Lens Equation is the ideal place to utilize the one-over key on your calculator. For example, suppose the object distance is 100 cm, the focal length is -20.0 cm, and we want the image distance. From the Thin Lens Equation

$$\frac{1}{s_i} = \frac{1}{f} - \frac{1}{s_o} = \frac{1}{-20.0 \text{ cm}} - \frac{1}{100 \text{ cm}}$$

and on a two-line calculator, s_i equals

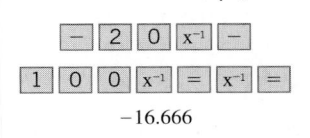

$$-16.666$$

Table 24.2

Meanings Associated with the Signs of Various Thin Lens Parameters

Quantity	Sign +	Sign −
s_o	Real object	Virtual object
s_i	Real image	Virtual image
f	Converging lens	Diverging lens
y_o	Right-side-up object	Inverted object
y_i	Right-side-up image	Inverted image
M_T	Right-side-up image	Inverted image

refers to the ratio of image size to object size and need not only correspond to enlargement ($|M_T| > 1$). The image can certainly be minified ($|M_T| < 1$) or life-size ($|M_T| = 1$) as well (Table 24.2).

From the similar triangles in Fig. 24.12, we got Eq. (24.3), which provides a convenient alternative statement for the magnification:

$$M_T = -\frac{s_i}{s_o} \qquad (24.7)$$

The minus sign here is necessary because both the object and image distances are positive and yet the image is inverted; that is, the magnification is negative.

Example 24.4 **[II]** The horse in Fig. 24.13 is 2.25 m tall, and it stands with its face 15.0 m from the plane of the thin lens, whose focal length is +3.00 m. (a) Determine the location of the image of the equine nose. (b) What is the magnification? (c) How tall is the image? (d) If the horse's tail is 17.5 m from the lens, how long—nose-to-tail—is the image of the beast?

Solution Given the object distance and focal length, the Thin Lens Equation should come to mind. (1) TRANSLATION —An object of known height and length is at a known distance from a lens with a specified focal length; determine (a) the image distance, (b) magnification, (c) image height, and (d) image length. (2) GIVEN: (a) $s_o = 15.0$ m, $f = +3.00$ m, $y_o = +2.25$ m; and (d) $s_o = 17.5$ m. FIND: s_i, front and rear; M_T and the image height. (3) PROBLEM TYPE—Geometrical optics / thin converging lens / magnification. (4) PROCEDURE—Since we have s_o and f, find s_i for the front and back of the horse using the Gaussian Lens Equation. (5) CALCULATION—The positive focal length means a converging lens. (a) From Eq. (24.4),

$$\frac{1}{s_o} + \frac{1}{s_i} = \frac{1}{15.0 \text{ m}} + \frac{1}{s_i} = \frac{1}{3.00 \text{ m}}$$

and $\boxed{s_i = +3.75 \text{ m}}$. (b) Computing the magnification from Eq. (24.9),

$$M_T = -\frac{s_i}{s_o} = -\frac{3.75 \text{ m}}{15.0 \text{ m}} = \boxed{-0.250}$$

(c) From the definition of magnification, Eq. (24.7), it follows that

$$y_i = M_T y_o = (-0.250)(2.25 \text{ m}) = \boxed{-0.563 \text{ m}}$$

where the minus sign tells us that the image is inverted. (d) Again from the Gaussian Equation, for the tail

$$\frac{1}{17.5 \text{ m}} + \frac{1}{s_i} = \frac{1}{3.00 \text{ m}}$$

and $s_i = +3.62$ m. The entire equine image is only $\boxed{0.13 \text{ m}}$ long.

Quick Check: Because the image distance is positive, the image is *real*. Because the magnification is negative, the image is *inverted*; and because the absolute value of the magnification is less than one, the image is *minified*. Moreover, the image is between 1 and 2 focal lengths from the lens. All of this matches Fig. 24.13, where the object distance exceeds $2f$.

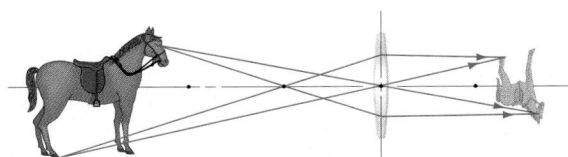

Figure 24.13 The image of a three-dimensional object is itself three-dimensional. The image exists in space, and any portion of it can be viewed on a screen even though it cannot be seen from the side as drawn here.

EXPLORING PHYSICS ON YOUR OWN

The Pinhole Camera: The fact that a small hole can cast a real image on a screen, very much like a lens, has been known for centuries. The geometry is shown in Fig. P39 (p. 884) and it's explored in Problem 39. I took the photo at the left on Polaroid film with a home-made pinhole camera (hole diameter 0.5 mm, film plane distance 25 cm, A.S.A. 3000, shutter speed 0.25 s). To see the effect, get a cardboard box, the bigger the better, but I used an empty cereal carton. Open it up and line one wall with white paper—that will be the screen. Seal the box with opaque tape so that it's light tight. Opposite the screen make a small round hole—start with one about the thickness of paper-clip wire. That will produce an inverted real image on the screen. To see it, cut a large peephole in the box facing the screen about a foot away from the pinhole so that your head doesn't block it. Because very little light will enter, you'll probably need to make your observations outside on a bright day; put a black cloth over your head to block extraneous light. Experiment with the hole-to-screen distance and the hole size. Try using your *camera obscura* on the Sun—you'll see a wonderful image and if you've got everything right, you might even see Sun spots. Incidentally, a tiny plane mirror will also work like a pinhole to produce real images.

It's easy to forget that the image exists in three-dimensional space. As a reminder, Fig. 24.13 depicts a horse standing in front of a very big lens (a small one would do, but the ray diagram would be awkward to draw). The resulting minified image is real, and it exists in an extended region of space (although we generally view it in transverse slices on a flat screen).

There are two important things to observe about the situation in Example 24.4 that apply to all real images formed by a positive lens. First, the image is evidently distorted, in that its length is reduced more than its height—the magnification transverse to the axis (M_T) is greater than the *longitudinal magnification* along the axial direction. This should not be surprising; the image of everything behind the horse, as far as the eye can see—out to infinity—must be compressed into the small space between the image-horse's rump and the lens. (In fact, as we will see shortly, it occupies even less than that, ending at F_i.)

The second important feature is that aside from being inverted, the image is oriented in an interesting way quite differently from that of the plane mirror. *The horse's nose, which is closer to the lens, is imaged farther away.* Were we to place a transverse observing screen far from the lens on the right and gradually move it to the left, closer in, we would first encounter the horse's face (with the saddle blurred). And then with the screen moved closer, the horse's head would be blurred, and the saddle would appear sharply "in focus." Nearest the lens, with horse and saddle fuzzy, only the tail might be clear.

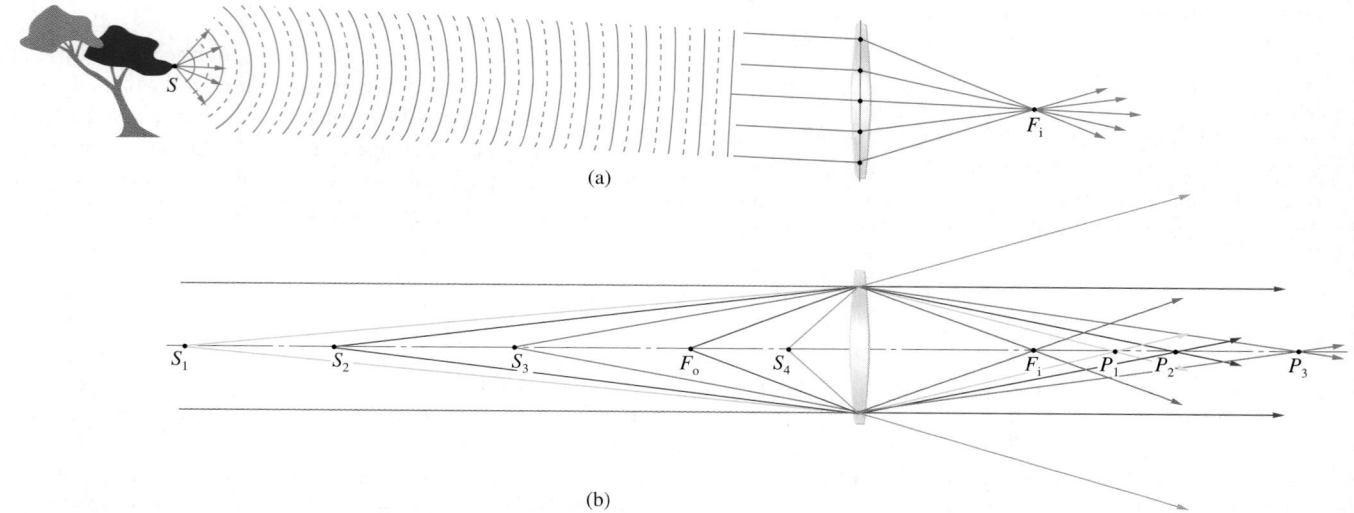

Figure 24.14 (a) Every point on a distant object emits spherical waves that gradually flatten out and ultimately transform into plane waves. (b) As the source moves closer, the rays (drawn yellow, then purple, then red) diverge more and the image point moves out away from the lens. The emerging rays no longer converge once the object reaches the focal point (green rays); nearer in still, they (pink rays) diverge.

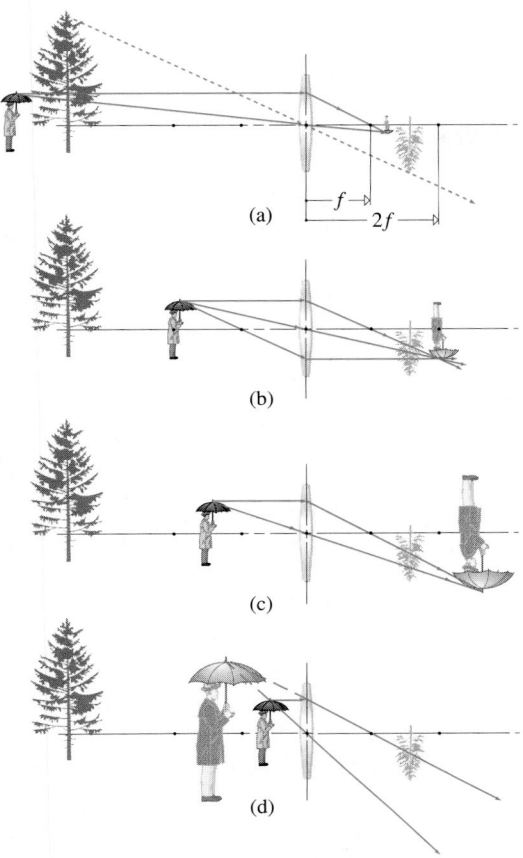

Figure 24.15 The operation of a thin positive lens.

A distant line of trees blocks sunlight from reaching this door. Nonetheless, tiny spaces between leaves act like pinholes, each casting a crude circular image of the Sun.

24.5 A Single Lens

We are now in a position to understand the entire range of behavior of a single convex or concave lens. To that end, suppose that a distant point source sends out a cone of light that is intercepted by a positive lens (Fig. 24.14). If the source is at infinity (i.e., so far away that it might just as well be infinity), rays coming from it entering the lens are essentially parallel and will be brought together at the focal point F_i. If the source point S_1 is closer, but still fairly far away, the cone of rays entering the lens is narrow and the rays come in at shallow angles to the surface of the lens. Because the rays do not diverge greatly, the lens bends each one into convergence and they arrive at point P_1. As the source moves closer, the entering rays diverge more and the resulting image point moves farther to the right. Finally, when the source point is at F_o, the rays are diverging so strongly that the lens can no longer bring them into convergence and they emerge parallel to the central axis. Moving the source point closer results in rays that diverge so much on entering the lens that they still diverge on leaving. The image point is now virtual—*there are no real images of objects closer in than f.*

A positive lens operates with three distinct regions of space where the object can exist. Suppose a man with an umbrella is standing near a tree somewhere in the most distant region, extending from ∞ to $2f$, as indicated in Fig. 24.15a. His real, inverted image will be formed on the right of the lens between f and $2f$; the farther away he is, the closer is his image to F_i (if he were at ∞, it would be at F_i). This situation corresponds to the way an **eye** or a **camera** works. The image on the retina is minified so that a panorama fits on the small screen; we can see a whole oak tree at once rather than one leaf at a time. *As the man walks toward the lens, his image grows in size and slowly moves away from the lens.* This domain, from ∞ to $2f$, is the first region of object space. It ends for the man at a distance of two focal lengths.

When he stands at $2f$, he is at the symmetry point of the system, where right and left sides (object and image) are identical. Figure 24.15b shows the situation: the triangles are now not just similar, they are congruent, and the real image is life-sized. This optical setup is that of a **photocopy machine**.

As the man continues to walk toward the lens, moving from $2f$ toward f, his initially life-sized image gradually grows while it moves away from the lens, advancing to the right from $2f$ toward ∞. The second region of object space, from $2f$ to f, transforms into the image space from $2f$ to ∞. Notice that everything thus far has happened in a smooth, continuous way. As he walks in from very far away, his real image walks out; starting small, it reaches life-size at $2f$ and continues to grow, becoming immense as it moves away. This arrangement is that of a **projector**. The slide or film is the object located between $2f$ and f, each point of which will have a corresponding image point on the picture screen. Thereon will appear a real, enlarged, upside-down image of the illuminated portion of the slide. The projector is focused by just changing the slide-to-lens distance. The inverted image is corrected by simply putting the slide or movie film in upside down.

It follows from Eq. (24.7) that the magnification approaches ∞ as the object nears f and s_i approaches ∞. When the object is exactly at the focal point, exactly at the boundary of the second region of object space, rays from each object point emerge from the lens parallel to one another. The image is no longer clearly observable on a screen, no matter how far away the screen is. Only a very large blur will be seen as if a real image were in focus somewhere far beyond the observing screen (namely, at ∞).

When the little man crosses beyond f into the third region of object space nearest the lens, he is so close and the rays so strongly diverging that no real image is possible. An eye or a camera looking into the lens could bring the emerging diverging rays together. What one would see is a virtual, right-side-up image much like the image formed by a plane mirror (Fig. 24.15d), except enlarged. With the object at f, the image is an immense blur; but if the little man leans back a bit, his image forms—large, real, and inverted; if he leans for-

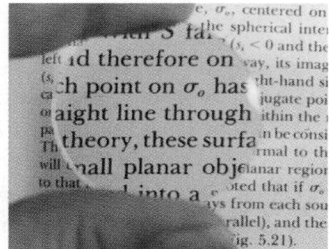

A positive lens serves as a magnifying glass when the object is closer to the lens than one focal length ($s_o < f$).

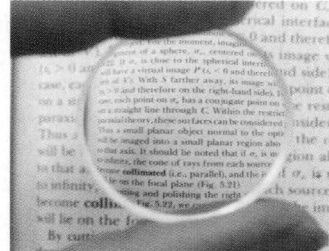

The minified, right-side-up, virtual image formed by a negative lens.

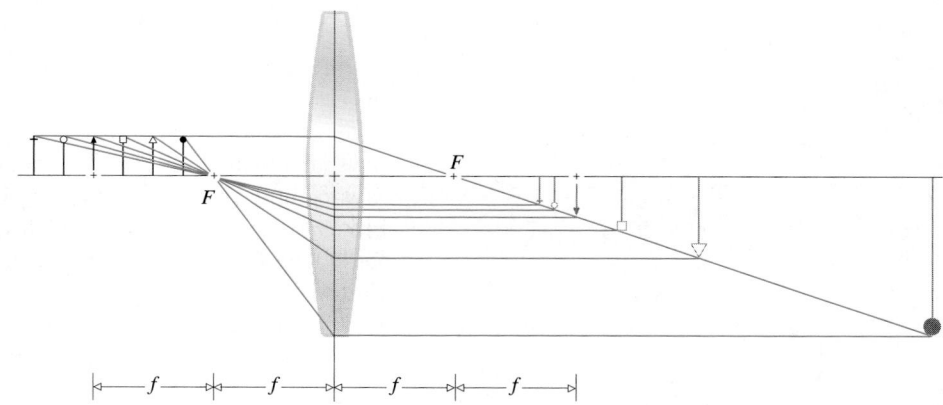

Figure 24.16 The number-2 ray entering the lens parallel to the central axis limits the image height. For simplicity, all refractions are drawn at the vertical mid-line of the lens.

ward, his image blurs, vanishes, and then reappears—large, virtual, and right-side-up. It flips over, but that discontinuity happens when he is at f and, reasonably enough, cannot be observed because the image is gone. As he walks closer to the lens, his image (as seen through the lens) will diminish ($M_T > 1$) until his face is flat up against the lens and he appears life-size. A positive lens operating on the light from an object located in this third region is known as a **magnifying glass**. Figure 24.16 shows how the number-2 ray entering the lens parallel to the axis fixes the height of the image.

The concave lens (Fig. 24.17) operates in only one way and so is much easier to keep track of. It produces *only* virtual, right-side-up, minified images no matter where the object is located. Rays diverging from any object point are made to diverge even more. An object pressed up against the lens appears very nearly life-size, but one more distant is minified correspondingly. In all cases, the image distance is negative, and the image always appears on the left side of the lens—on the same side the light enters from. Table 24.3 summarizes these conclusions for both types of lens.

{You should now click on **THE THIN LENS** under **INTERACTIVE EXPLORATIONS** on the **CD**. This excellent simulation makes it possible to see how a thin lens, concave or convex, forms a finite image. It's important that you spend some time working with this piece, it should be fun to play with and will certainly clarify the entire process of ray tracing and image formation.}

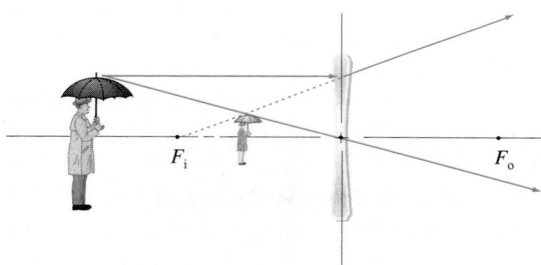

Figure 24.17 A concave lens forms a virtual, minified, right-side-up image.

Table 24.3

Images of Real Objects Formed by Thin Lenses

		Convex		
Object		**Image**		
Location	**Type**	**Location**	**Orientation**	**Relative size**
$\infty > s_o > 2f$	Real	$f < s_i < 2f$	Inverted	Minified
$s_o = 2f$	Real	$s_i = 2f$	Inverted	Same size
$f < s_o < 2f$	Real	$\infty > s_i > 2f$	Inverted	Magnified
$s_o = f$		$\pm \infty$		
$s_o < f$	Virtual	$|s_i| > s_o$	Right-side-up	Magnified

		Concave		
Object		**Image**		
Location	**Type**	**Location**	**Orientation**	**Relative size**
Anywhere	Virtual	$|s_i| < |f|$, $s_o > |s_i|$	Right-side-up	Minified

Example 24.5 **[II]** A 5.00-cm-tall matchstick is standing 10 cm from a thin concave lens whose focal length is −30 cm. Determine the location and size of the image and describe it. Draw an appropriate ray diagram.

Solution Given the object distance and focal length, the Thin Lens Equation should come to mind. (1) TRANSLATION —An object of known height is at a known distance from a negative lens with a specified focal length; describe the image and determine its (a) distance and (b) height. (2) GIVEN: $y_o = 0.0500$ m, $s_o = 0.10$ m, and $f = -0.30$ m. FIND: s_i and y_i. (3) PROBLEM TYPE—Geometrical optics/thin diverging lens/magnification. (4) PROCEDURE—Since we have s_o and f, find s_i using the Gaussian Lens Equation. (5) CALCULATION—(a) From Eq. (24.6),

$$\frac{1}{s_o} + \frac{1}{s_i} = \frac{1}{0.10 \text{ m}} + \frac{1}{s_i} = \frac{1}{-0.30 \text{ m}}$$

and $s_i = -1/13.3 = -0.075$ m $= \boxed{-7.5 \text{ cm}}$. The image distance is negative, meaning it is to the left of the lens, and therefore the image is virtual. (b) Using Eq. (24.7), we can compute the magnification and from that the size of the image via Eq. (24.6):

$$M_T = -\frac{s_i}{s_o} = -\frac{-0.075 \text{ m}}{0.10 \text{ m}} = +0.75$$

and so $\quad y_i = M_T y_o = 0.75(0.0500 \text{ m}) = \boxed{0.038 \text{ m}}$

Figure 24.18 is the corresponding ray diagram. Notice how ray-3 heading for the object focus off to the right emerges parallel to the axis, as in Fig. 24.9a, while ray-2 entering parallel to the axis appears, on emerging, to be coming from the image focus.

Quick Check: The image is right-side-up ($M_T > 0$), minified ($|M_T| < 1$), virtual ($s_i < 0$), and $s_o > |s_i|$, as Table 24.3 says it should be.

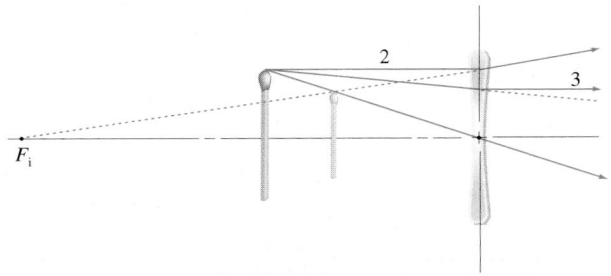

Figure 24.18 The image of a matchstick formed by a concave lens.

The Human Eye & The Camera

The eye (Fig. 24.19) is an almost spherical (24 mm long by about 22 mm across) jellylike mass contained within a tough, flexible shell, the *sclera*. With the exception of the front portion or *cornea*, which is transparent, the sclera is white and opaque. The cornea is the first and strongest converging element of the eye system. Most of the bending impressed on a bundle of rays by the eye takes place at the air-cornea interface. The reason you can't see

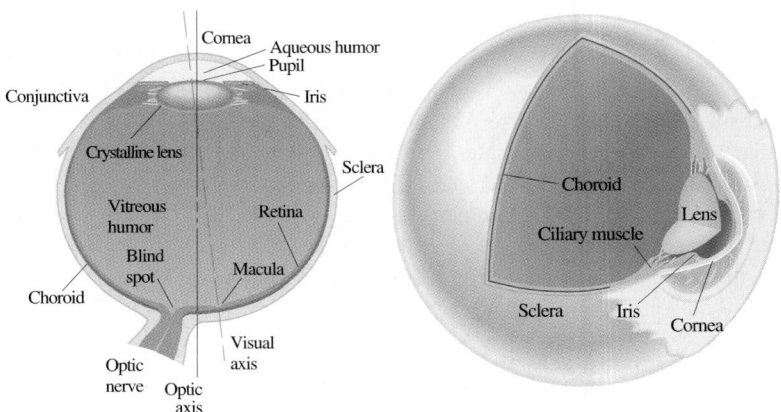

Figure 24.19 The human eye. The cornea is slightly flattened to control spherical aberration—the fact that rays from the edges of a lens usually do not focus at the same point as do rays near the center. Moreover, the spherical shape of the eyeball and retina eliminates problems associated with the fact that the image "plane" is not really flat. The retina contains 125 to 130 million photoreceptor cells.

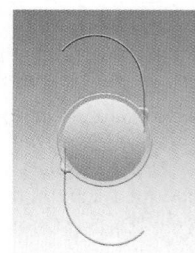

When the normally clear lens in the eye becomes cloudy, the condition is referred to as a **cataract**. The resulting haziness can have a devastating effect on vision. In extreme cases the crystalline lens is usually surgically removed. A small convex plastic lens (an **intraocular lens implant**) is then inserted in the eye to enhance its convergence. (The photo shows an enlarged image of this type of converging spherical lens, it's actually only about 6 mm in diameter.) Its use has all but eliminated the need for the thick "cataract eye glasses" that were once required after surgery.

very well under water ($n_w \approx 1.333$) is that the liquid's index is so close to that of the cornea ($n_C \approx 1.376$) that adequate refraction can no longer occur.

Having passed into the cornea, the light is only made slightly more convergent on emerging because it enters a chamber filled with a watery fluid known as the *aqueous humor* ($n_{ah} \approx 1.336$). Immersed in the aqueous is a variable diaphragm called the *iris*, which controls the amount of light entering the remaining portion of the eye by way of an aperture or *pupil*.

THE COMPOSITE LENS OF THE EYE

The cornea and crystalline lens can be treated as forming a double-element lens whose object focus is about 15.6 mm in front of the outer surface of the cornea and whose image focus is about 24.3 mm behind it on the retina. The combined lens has an optical center 17.1 mm in front of the retina, just at the rear edge of the crystalline lens. As a rule, $s_0 > 2f$.

Just behind the iris is the *crystalline lens*. The lens (9 mm in diameter and 4 mm thick) is a complex, layered fibrous mass surrounded by an elastic membrane. In structure, it is somewhat like a small transparent onion, formed of roughly 22 000 very fine layers. As a whole, the lens is quite pliable, albeit less so with age. Its index of refraction varies from about 1.406 at the inner core to roughly 1.386 at the less dense cortex. The crystalline lens provides the needed fine-focusing mechanism via changes in its shape.

Example 24.6 **[I]** As we saw earlier (p. 236), the Moon subtends an angle of about 0.009 rad as seen from Earth. Using the fact that the human eye has its optical center 17.1 mm in front of the retina, how big is the image of the Moon formed by the eye?

Solution This is essentially a geometry problem. (1) TRANSLATION—Determine the size of the retinal image of an object that subtends a known angle. (2) GIVEN: Subtended angle of 0.009 rad. FIND: y_i. (3) PROBLEM TYPE—Geometrical optics / converging lens. (4) PROCEDURE—See Fig. 24.20. Since the Moon subtends an angle of θ, it follows from the geometry that the image size is just $r\theta$. (5) CALCULATION—(a) Here $r = s_i$ and so

$$|y_i| = s_i\theta = (17.1 \text{ mm})(0.009 \text{ rad}) = 0.15 \text{ mm} = \boxed{0.2 \text{ mm}}$$

The absolute value is used because the image is inverted, and y_i is negative by convention. The fact that the diameter is 0.15 mm means that the retinal image of the face of the Moon is a dot smaller than the size of a period on this page!

Quick Check: $\theta = |y_i|/s_i \approx (15 \times 10^{-5}\text{ m})/(17 \times 10^{-3}\text{ m}) \approx 1 \times 10^{-2}$ rad.

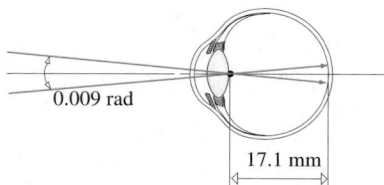

Figure 24.20 The angle subtended by the Moon on the retina.

Behind the lens is another chamber filled with a transparent gelatinous substance, the *vitreous humor* ($n_{vh} \approx 1.337$). A thin, delicate, transparent multilayer of cells (from 0.5 mm to 0.1 mm thick) covers about 65% of the interior surface of that chamber. This structure is

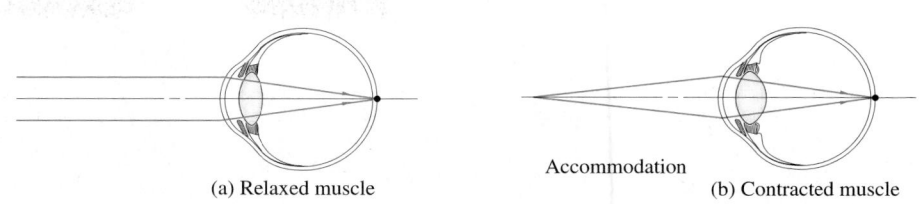

(a) Relaxed muscle   Accommodation   (b) Contracted muscle

Figure 24.21 The human eye focuses on an object by contracting the ciliary muscles around the edge of the crystalline lens, which relaxes the tension on the lens so that it contracts and bulges.

the light-sensitive *retina* (from the Latin *rete*, meaning net). The retina is the transducer that converts electromagnetic energy impinging on it into electrical nerve impulses that can be processed by the brain.

Accommodation. The fine focusing or **accommodation** performed by the human eye is carried out by the crystalline lens. Since the image distance for the eye is fixed, the only way we can see things clearly at different object distances is if the focal length is changed. The lens is suspended by ligaments that are connected to a circular yoke of muscles. Ordinarily, these are relaxed and elongated, the aperture they encompass is large, and in that state they pull back on the network of fine fibers holding the rim of the lens. This draws the pliable lens into a fairly flat configuration, increasing its radii of curvature (especially of the anterior surface), which increases its focal length. With the muscles completely relaxed, the light from an object at infinity (which is practically speaking anywhere beyond about 5 m) is focused on the retina (Fig. 24.21). Not all eyes will do that well; so the **far-point**—*that is, the point that is seen clearly by the unaccommodated eye*—is frequently closer in than infinity (or even 5 m).

As the object moves closer, the muscles contract, the aperture encircling the lens gets smaller, and the lens bulges a little under its own elastic forces. In so doing, the focal length decreases, keeping the image on the retina. *The closest point that can be clearly seen with maximum accommodation is the* **near-point.** The lens doesn't change thickness by much more than about 0.5 mm over the whole range. A 10-year-old with a flexible lens may have a near-point as close as 7 cm, but that will typically move out to 12 cm by age 25 and 28 cm at around age 45—still a workable distance (11 inches) for reading without any inconvenience. But by 50 years, the near-point jumps to about 40 cm; by 60 years, it's out to 100 cm; and by 70, it's at 400 cm.

The Camera. The camera works very much like the eye. Light enters via a lens, and an image is formed on a light-sensitive surface, which might be a photographic film or an electronic sensor such as a CCD (Fig. 24.22). Light passes into the camera by way of an adjustable hole, and the size of that aperture and the duration over which it stays open together determine the amount of energy allowed in. Unlike the human eye, most cameras are focused by moving the lens toward or away from the film plane. {For a more complete discussion of the physics click on **THE CAMERA** under **FURTHER DISCUSSIONS** on the **CD**.}

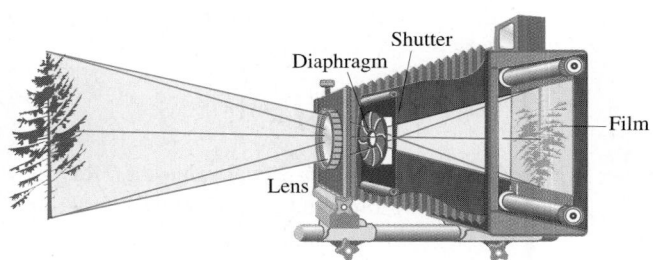

Figure 24.22 A large-format camera consists of a lens, followed by an adjustable diaphragm, a shutter that can open and close, and a sheet of film on which the image is formed.

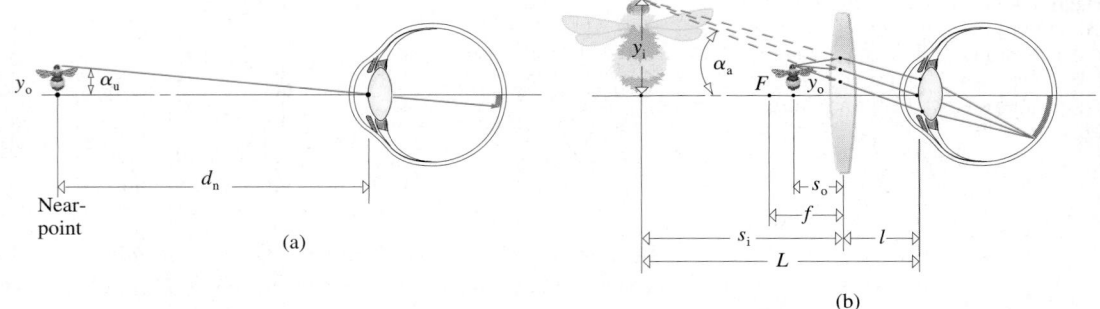

Figure 24.23 A magnifying glass. (a) An object is examined directly by placing it at the near-point. The retinal image is then as large as possible. (b) With a magnifying glass, the same object results in a much larger retinal image.

The Magnifying Glass   (Optional)

To examine an object in detail, you simply bring it nearer the eye so that the retinal image increases in size. That procedure can continue until the object is at the near-point, beyond which the eye can no longer provide adequate accommodation because the rays diverge too much (Fig. 24.14). To enlarge the object further, a single positive lens can be used to add convergence to the visual system, allowing the object to be brought still closer. A lens so used is a **magnifying glass**. Its function is ***to provide an image of a nearby object that is larger than that seen by the unaided eye***. It would be nice to have a right-side-up, magnified image where the rays entering the eye are not converging, and that's satisfied by placing the object within one focal length of the positive lens.

To deal with how large an object appears in some optical device, we must consider the size of its retinal image. Consequently, the *magnifying power*, or **angular magnification** M_A, of an instrument is defined as ***the ratio of the size of the retinal image formed by the device to the size of the retinal image formed by the unaided eye at normal viewing distance***. The latter is taken as the distance to the near-point d_n. In Fig. 24.23, M_A is equivalent to the ratio of the angles α_a (aided) and α_u (unaided):

$$M_A = \frac{\alpha_a}{\alpha_u} \qquad (24.8)$$

Being restricted to the paraxial region, $\tan \alpha_a = y_i/L$ is very small and therefore approximately equal to α_a while $\tan \alpha_u = y_o/d_n \approx \alpha_u$. Since $-s_i/s_o = y_i/y_o$,

$$M_A = \frac{y_i d_n}{y_o L} = -\frac{s_i d_n}{s_o L}$$

where both y_i and y_o are above the axis and positive. Taking all the previously unspecified distances in the diagram such as l, d_n, and L to be positive as indicated makes M_A positive as well. In the most commonly encountered application of the magnifying glass, the object is located at the focal point of the lens. In that case, the image is at infinity ($-s_i \approx L = \infty$; $s_o = f$) and

$$M_A = \frac{d_n}{f} \qquad (24.9)$$

for all practical values of l. Rays emerging from the lens are parallel and can be viewed with a relaxed eye, which is a very important practical consideration.

Single lens magnifiers are usually limited by their aberrations to powers of about 2× or 3×. The famous Sherlock Holmes reading glass is an example of the type. More complicated multielement magnifiers can be made in the range from 10× to 20×, and these are also used in microscopes and telescopes.

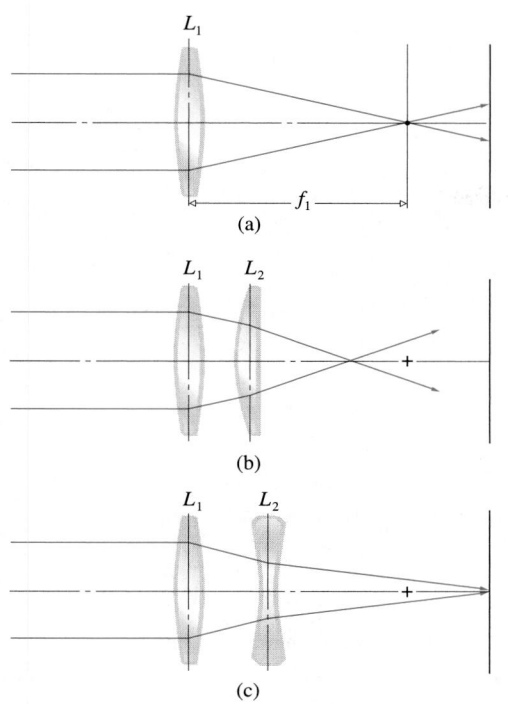

Figure 24.24 (a) The effect of placing a second lens, L_2, within the focal length of a positive lens, L_1. (b) When L_2 is positive, its presence adds convergence to the ray bundle. (c) When L_2 is negative, it adds divergence to the ray bundle.

24.6 Thin-Lens Combinations (Optional)

Optical systems are usually composed of several lenses. In the simple case of an incident parallel bundle of rays, adding a lens to an existing one simply changes the focal length (Fig. 24.24); more convergence (resulting from the addition of a positive lens) yields a shorter f, whereas less convergence (resulting from the addition of a negative lens) produces a longer combined f.

Figure 24.25 depicts two positive thin lenses L_1 and L_2 separated by a distance d, which here happens to be smaller than the focal length of either lens. The resulting image can be determined graphically by ray-tracing, using the following procedure. Imagine that L_2 is no longer there and find the image formed exclusively by L_1. This is easily done with ray-1 drawn through the center of L_1 and ray-2 entering parallel to the axis. These two rays fix point P' and the ***intermediate image***. If L_2 were in place, a ray through its center would be undeviated. Hence, construct ray-4 running backward from P' through O_2 to L_1 and back to S. This ray is one of the two crucial rays that will locate the final image since it passes properly through both lenses. The second necessary ray is just ray-3 through F_{o1}, which emerges from L_1 parallel to the central axis and therefore passes through F_{i2} on leaving the second lens. The point of intersection of these two rays determines the final image P of S. Notice that the intermediate image created by L_1 falls to the right of L_2 and can be thought of as a virtual object for L_2, which then goes on to create the final image.

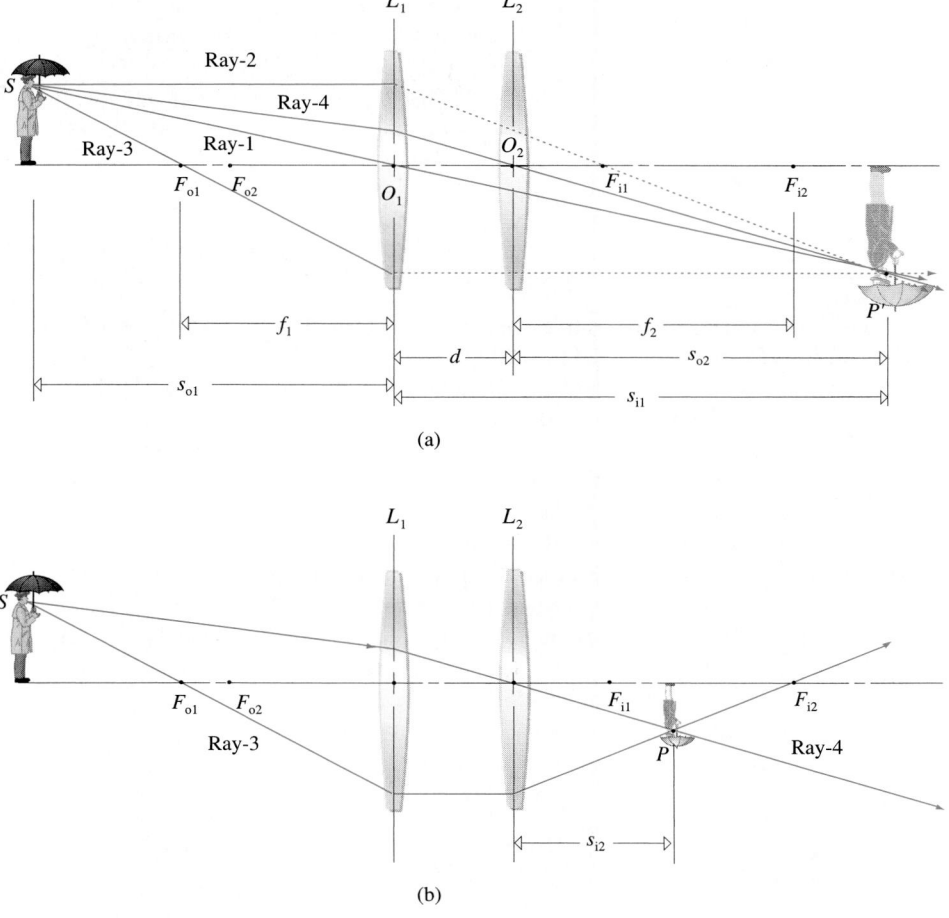

Figure 24.25 The image formed by two thin lenses. In (a) we imagine that lens L_2 is not there, and we locate the image formed by L_1. Ray-4 is found passing through point O_2. It will be unchanged when lens L_2 is put back (b) and, along with ray-3, locates the final image.

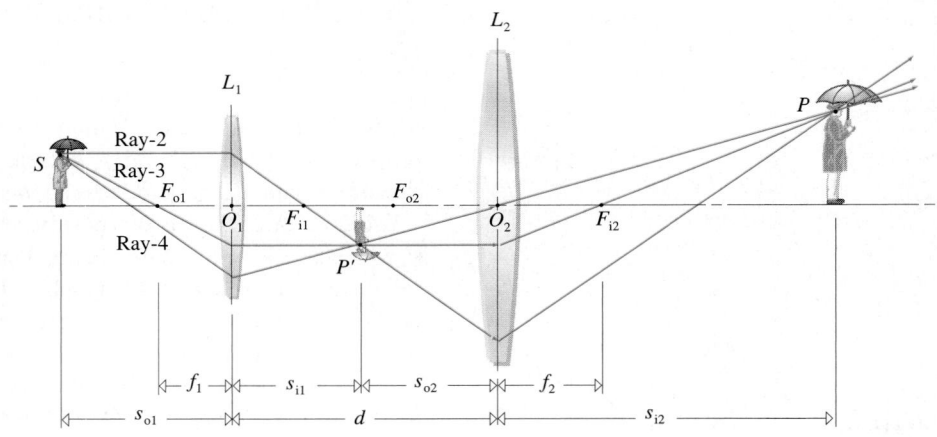

Figure 24.26 Two thin lenses separated by a distance $d > (f_1 + f_2)$. *Now a real intermediate image is formed at P', and rays from it can be traced through L_2 to produce the final image.*

Another example of the method is shown in Fig. 24.26 wherein the lenses are farther apart; in fact, $d > (f_1 + f_2)$. Again ray-1 and ray-2 fix the intermediate image at P'. And again ray-4 passing through P' and O_1 is drawn backward through L_1 to S. This ray and ray-3 once more locate the final image P. Here, the intermediate image is real, and we could have proceeded by simply treating it as the real object for the second lens, forgetting about the first lens. Any two convenient rays from P' through the second lens will suffice to locate the final image.

The system can be treated analytically in much the same way—that is, by finding the intermediate image formed by the first lens and letting that function as the object for the second lens.

Example 24.7 **[II]** Two positive lenses with focal lengths of 0.30 m and 0.50 m are separated by 0.20 m, as in Fig. 24.25. A red jellybean rests on the central axis 0.50 m in front of the first lens. Locate the resulting image with respect to the second lens.

Solution This is a two-lens problem, and that means the image formed by the first lens is the object for the second lens. (1) TRANSLATION—Two thin positive lenses with known focal lengths and separation are a specified distance from an object; determine the image-distance beyond the second lens. (2) GIVEN: $f_1 = +0.30$ m, $f_2 = +0.50$ m, $d = 0.20$ m, and $s_{o1} = 0.50$ m. FIND: s_{i2}. (3) PROBLEM TYPE—Geometrical optics / thin-lens combination. (4) PROCEDURE—Since we are dealing with focal lengths and object- and image-distances, apply the Gaussian Lens Equation to each lens. (5) CALCULATION—Applying the lens formula to L_1 only yields

$$\frac{1}{s_{i1}} = \frac{1}{f_1} - \frac{1}{s_{o1}} = \frac{1}{0.30 \text{ m}} - \frac{1}{0.50 \text{ m}} \quad (24.10)$$

and $s_{i1} = 0.75$ m. The intermediate image falls $(0.75 \text{ m} - 0.20 \text{ m}) = 0.55$ m to the right of the second lens. This value, then, is the object distance for L_2, and because it is to the right, it is negative ($s_{o2} = -0.55$ m). Again, the lens formula for the second lens with this object yields

$$\frac{1}{s_{i2}} = \frac{1}{f_2} - \frac{1}{s_{o2}} = \frac{1}{0.50 \text{ m}} - \frac{1}{-0.55 \text{ m}} \quad (24.11)$$

and $\boxed{s_{i2} = +0.26 \text{ m}}$. The image is real and to the right of the last lens.

Quick Check: The second lens adds more convergence, pulling the image in closer to the lens as if the object were farther away than it is. The image is real, inverted, and minified. See Example 24.8.

Example 24.8 **[III]** Considering two thin lenses separated by a distance d, derive an expression for s_{i2} in terms of s_{o1}, d, and the focal lengths.

Solution   This is a two-lens problem, and that means the image formed by the first lens is the object for the second lens. (1) TRANSLATION—Two thin positive lenses with known focal lengths and separation are a specified distance from an object; determine an expression for the image distance beyond the second lens.   (2) GIVEN: f_1, f_2, d, and s_{o1}. FIND: An expression for s_{i2}.   (3) PROBLEM TYPE—Geometrical optics/thin-lens combination.   (4) PROCEDURE—Since we are dealing with focal lengths and object- and image-distances, apply the Gaussian Lens Equation to each lens.   (5) CALCULATION—For L_1, using Eq. (24.10),

$$s_{i1} = \frac{s_{o1}f_1}{s_{o1} - f_1} \tag{24.12}$$

From Eq. (24.11), for L_2

$$s_{i2} = \frac{s_{o2}f_2}{s_{o2} - f_2}$$

Substituting in the fact that $s_{o2} = d - s_{i1}$, s_{i2} becomes

$$s_{i2} = \frac{(d - s_{i1})f_2}{(d - s_{i1} - f_2)}$$

Eliminating s_{i1} using Eq. (24.12), we have

$$s_{i2} = \frac{f_2 d - [f_2 s_{o1} f_1 / (s_{o1} - f_1)]}{d - f_2 - [s_{o1}f_1/(s_{o1} - f_1)]} \tag{24.13}$$

Quick Check: Using the numbers from Example 24.7, let's confirm that Eq. (24.13) produces the same value of the image distance:

$$s_{i2} = \frac{(0.50\text{ m})(0.20\text{ m}) - (0.50\text{ m})(0.50\text{ m})(0.30\text{ m})/(0.50\text{ m} - 0.30\text{ m})}{0.20\text{ m} - 0.50\text{ m} - (0.50\text{ m})(0.30\text{ m})/(0.50\text{ m} - 0.30\text{ m})}$$

$$s_{i2} = +0.26\text{ m}$$

Suppose that the individual lenses are now brought close enough to touch one another, as is often done in compound systems. When $d = 0$, in Eq. (24.13), we can find the focal length of the combination by letting $s_{o1} \to \infty$, whereupon $s_{i2} = f$. The terms with d vanish, $(s_{o1} - f) \to s_{o1}$ and

$$f = \frac{f_1 f_2}{f_1 + f_2}$$

or

[two lenses touching] $$\frac{1}{f} = \frac{1}{f_1} + \frac{1}{f_2} \tag{24.14}$$

The focal lengths add like resistors in parallel.

Eyeglasses   (Optional)

It is customary in physiological optics to speak about the **dioptric power** $\mathcal{D}$ of a lens, which is simply the *reciprocal of the focal length*. A lens possesses great power when it strongly bends rays, which happens when it has a *short* focal length. Power has the units of inverse meters, or *diopters* (D): $1\text{ m}^{-1} = 1$ D. For instance, a converging lens with a focal length of $+10$ m has a power of 0.10 D, while a diverging lens with a focal length of -2 m has a power of $-\frac{1}{2}$ D. It follows from Eq. (24.2) that

$$\mathcal{D} = (n_l - 1)\left(\frac{1}{R_1} - \frac{1}{R_2}\right) \tag{24.15}$$

The combined focal length of two lenses in contact is given by Eq. (24.14), and so their total power is the sum of the individual powers

$$\mathcal{D} = \mathcal{D}_1 + \mathcal{D}_2 \tag{24.16}$$

The human eye has a total power of roughly $+59$ D for the unaccommodated state (of which the cornea provides about $+43$ D). In the normal eye, that's just the refractive pow-

Example 24.9 **[I]** Two lenses with focal lengths of +0.100 m and −0.333 m are held close together on a common center-line. Compute both the focal length and the power of the combination..

Solution This is the special two-lens situation where the lenses touch. (1) TRANSLATION—Two thin lenses with known focal lengths are touching; determine the combined focal length and power. (2) GIVEN: $f_1 = +0.100$ m and $f_2 = -0.333$ m. FIND: f and $\mathcal{D}$. (3) PROBLEM TYPE—Geometrical optics /thin-lens combination/power. (4) PROCEDURE—We have an expression for the focal length of touching lenses. (5) CALCULATION—The focal length is obtained from Eq. (24.14)

$$\frac{1}{f} = \frac{1}{f_1} + \frac{1}{f_2} = \frac{1}{+0.100 \text{ m}} + \frac{1}{-0.333 \text{ m}}$$

and $\boxed{f = 0.143 \text{ m}}$. Equation (24.16) provides the power

$$\mathcal{D} = \mathcal{D}_1 + \mathcal{D}_2 = (10.0 \text{ D}) + (-3.0 \text{ D}) = \boxed{7.0 \text{ D}}$$

Quick Check: $1/f = 1/(0.143 \text{ m}) = 7.0$ D. A negative lens combined with a stronger positive lens yields a positive lens.

---

er needed to focus a parallel bundle of rays onto the retina. All too commonly, however, the image focus does not lie on the retina. This condition can arise either because of abnormal changes in the refracting mechanism (cornea, lens, and so on) or because of alterations in the length of the eyeball that upset the lens-retina distance. The latter is by far the more common cause. About 25% of the young adult population falls in the class of requiring as little as about ±0.5 D or less of eyeglass correction, and perhaps as many as 65% need only ±1.0 D or less.

Farsightedness, or *hyperopia*, is the defect that causes the image focus of the unaccommodated eye to fall behind the retina (Fig. 24.27). It is most often (perhaps 90% of the time) due to a shortening of the anteroposterior axis of the eye—the lens is too close to the retina. As a result, the image on the photoreceptors is formed of overlapping blotches of light and the picture is somewhat blurred. The *relaxed* hyperopic eye cannot bend the rays enough because it lacks the needed convergence and cannot see anything, near or far, clearly. But it can accommodate, thereby increasing its power and bringing into focus light from far away, which isn't very divergent to begin with (Fig. 24.27b). By accommodating, the farsighted eye can see clearly everything from infinity inward to some **near-point**. This near-point will be a lot farther away than it is in the normal eye. Any closer and the rays diverge too much; strain as the eye may, the image will be blurred.

To increase the power of the hyperopic visual system, a positive spectacle lens can be placed in front of the eye. This lens will allow the unaccommodated eye to see very distant objects clearly (Fig. 24.27d) while effectively pulling in the near-point so that it is at some close, convenient distance. Another way to appreciate this is to realize that a nearby object (closer in than the focal length of the corrective lens) results in a distant right-side-up, virtual image. That image is located farther out than is the eye's unaided near-point and so can be seen clearly. These spectacles will cast real images—try it if you happen to be hyperopic.

Nearsightedness or *myopia* is the condition where parallel rays are brought to a focus in front of the retina; the power of the eye's refractive system is too large for the anterior-posterior axial length (Fig. 24.28). The problem occurs primarily because the eye elongates or the cornea changes shape. It is a situation that usually becomes noticeable in the teens and then levels off in severity at about age 25 or so. Interestingly, myopia hardly exists in "primitive" populations, whereas it is exceedingly common in so-called "advanced" civilizations (there are perhaps 40 million myopes in the United States).

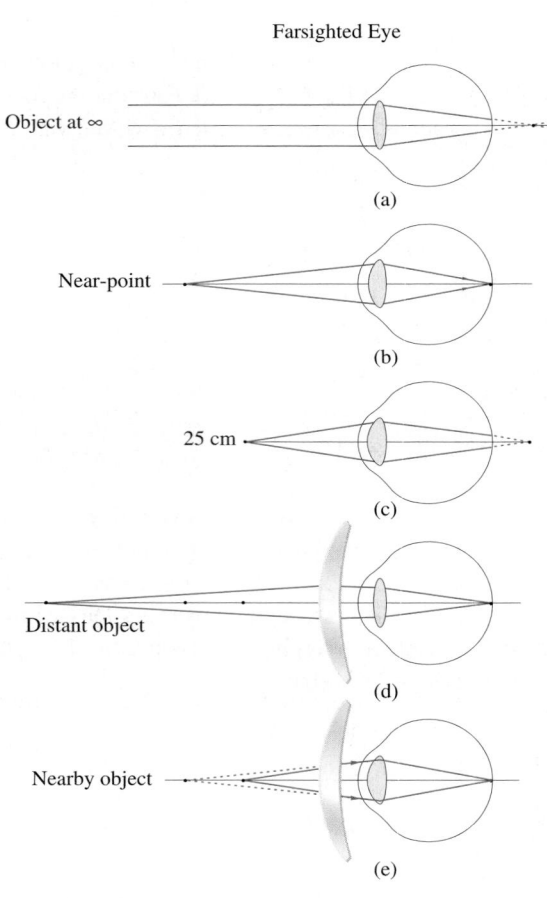

Figure 24.27 Correction of the farsighted eye (a), which focuses parallel light beyond the retina. (b) The near-point is now farther away than 25 cm, which is normal (c). By adding more convergence to the eye, via a positive eyeglass lens, distant objects can be viewed in a relaxed state (d). Moreover, objects at 25 cm, which were blurred, are now clearly seen at the near-point (e).

Example 24.10 **[III]** It is determined that a patient has a near-point at 50 cm. Approximate the eye to be 2.0 cm long. (a) Roughly how much power does the refracting system have when focused on an object at infinity? (b) When focused at 50 cm? (c) How much accommodation is required to see an object at a 50-cm distance? (d) What power must the eye have to see clearly an object at the standard near-point distance of 25 cm? (e) How much power should be added to the patient's vision system via reading glasses?

Solution The is a problem dealing with vision and so the mention of "near-point" and "power" should call to mind $\mathcal{D}$. (1) TRANSLATION—An eye has a known size and near-point; determine its (a) power when focused at infinity, (b) power when focused at near-point, (c) accommodation to go from (a) to (b), (d) power when focused at a 25-cm near-point, and (e) required correction to read at 25 cm. (2) GIVEN: Actual near-point is 50 cm, $s_i = 2.0$ cm, and standard near-point at 25 cm. FIND: (a) $\mathcal{D}_\infty$, (b) $\mathcal{D}_{50}$, (c) accommodation for 50-cm vision, (d) $\mathcal{D}_{25}$, and (e) prescription for correction. (3) PROBLEM TYPE—Geometrical optics/thin-lens combination/power/eyeglasses. (4) PROCEDURE—Use the definition of lens power. (5) CALCULATION—(a) When focused at infinity with the image falling on the retina 0.020 m beyond, the eye has a power of

$$\mathcal{D}_\infty = \frac{1}{f} = \frac{1}{\infty} + \frac{1}{0.020 \text{ m}} = \boxed{50 \text{ D}}$$

(b) Similarly, with $s_o = 0.50$ m and $s_i = 0.020$ m, the power is

$$\mathcal{D}_{50} = \frac{1}{f} = \frac{1}{0.50 \text{ m}} + \frac{1}{0.020 \text{ m}} = \boxed{52 \text{ D}}$$

(c) Thus, the eye adds 2 D of power via accommodation when changing focus from infinity to 0.50 m. This particular patient cannot provide any more than $\boxed{+2 \text{ D}}$ of accommodation. (d) When $s_o = 0.25$ m, the eye must bring to bear a power of

$$\mathcal{D}_{25} = \frac{1}{f} = \frac{1}{0.25 \text{ m}} + \frac{1}{0.020 \text{ m}} = \boxed{54 \text{ D}}$$

(e) Accordingly, this person is lacking $\boxed{+2 \text{ D}}$ of power that can be provided by correction lenses.

Quick Check: (a) $f = \infty(0.02 \text{ m})/(\infty + 0.02 \text{ m}) \approx 0.02$ m. (b) $f = (0.50 \text{ m})(0.02 \text{ m})/0.52 \text{ m} = 1.9$ cm. (d) $f = (0.25 \text{ m})(0.02 \text{ m})/0.27 \text{ m} = 1.85$ cm $= 1/(54 \text{ D})$.

Figure 24.28 Correction of the nearsighted eye (a), which focuses parallel light in front of the retina. By adding some divergence to the eye, via a negative eyeglass lens, distant objects can be viewed (d) in a relaxed state. The far-point (c) is brought in closer (e).

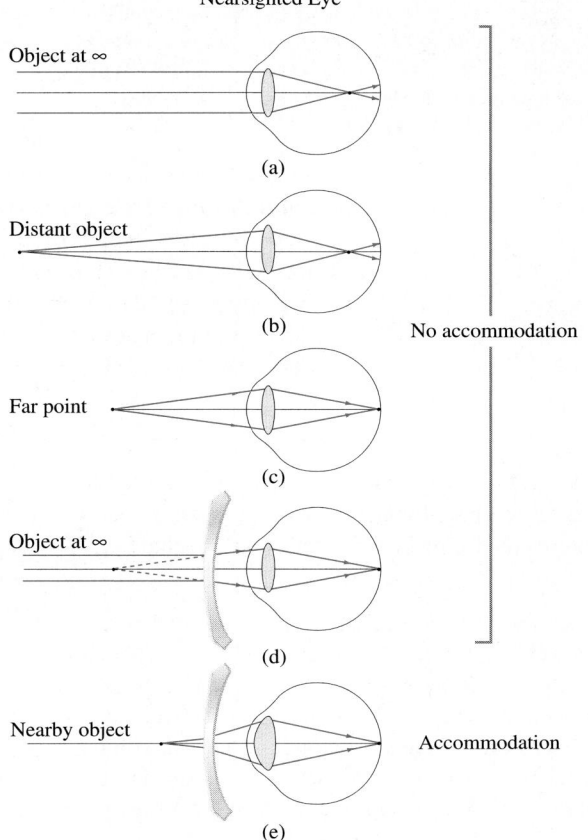

Nearsighted Eye

Object at ∞

(a)

Distant object

(b)

Far point

(c)

Object at ∞

(d)

Nearby object

(e)

No accommodation

Accommodation

Example 24.11 **[III]** An optometrist finds that a farsighted person has a near-point at 125 cm. What power contact lenses will be required if they are to effectively move that point inward to a more workable distance of 25 cm? Use the fact that if the object is imaged at the near-point, it can be seen clearly.

Solution The is a problem dealing with vision, and so the mention of "near-point" and "power" should call to mind $\mathcal{D}$. (1) TRANSLATION—An eye has a known near-point; determine the required contact lens correction to read at 25 cm. (2) GIVEN: s_o = 0.25 m and d_n = 1.25 m. FIND: $\mathcal{D}_c$. (3) PROBLEM TYPE— Geometrical optics/thin-lens combination/power/contact lens. (4) PROCEDURE—The eye can see things at its near-point clearly, so we want 1.25 m to be the image distance of the correction lens. This s_i must be on the left of the lens, and so it's

negative; the positive contact lens is being used as a magnifying glass. (5) CALCULATION—Accordingly, s_i = −1.25 m and s_o = 0.25 m. Both the focal length and power of the contact lens can be found via Eq. (24.4):

$$\frac{1}{s_o} + \frac{1}{s_i} = \frac{1}{f_c} = \mathcal{D}_c = \frac{1}{0.25\text{ m}} + \frac{1}{-1.25\text{ m}} = \boxed{+3.2\text{ D}}$$

The contact lens will form a virtual image of the book that appears at the near-point of the eye. By adding 3.2 D of power to the eye, the near-point of the corrected system becomes 0.25 m instead of 1.25 m.

Quick Check: s_o is less than f = 0.31 m; $1/(1/4) - 1/(5/4)$ = 4.0 D − 0.8 D = 3.2 D.

"FOR THE CONVENIENCE OF OLD MEN ..."

Spectacles were probably invented some time in the late thirteenth century, possibly in Italy or China. A Florentine manuscript (1299), which no longer exists, spoke of "spectacles recently invented for the convenience of old men whose sight has begun to fail." In 1804, Wollaston, recognizing that traditional (fairly flat bi-convex and concave) eyeglasses provided good vision only while looking through their centers, patented a new deeply curved lens. These were the forerunners of modern meniscus lenses that allow the turning eyeball to see through them from center to margin without significant distortion.

With the myopic eye, images of faraway objects fall in front of the retina. And that's true for object distances from infinity inward to the so-called **far-point**, where the rays diverge enough so that the image is finally right on the retina and clearly visible. It is the farthest point that can be seen sharply by the unaided myopic eye. Depending on the degree of the problem, the far-point can be very much closer in than infinity, and all objects beyond it in space appear blurred. Moreover, the near-point is also closer than normal, which is a convenience for doing detailed work because it provides a bit more magnification.

In effect, the myopic eye has too much convergence; its positive power is too great. To correct the symptoms, we need only place a negative lens in front of the eye. The image focus of the combined lens-eye system must fall on the retina. In other words, *parallel light is made to diverge just enough so that it appears to come from the far-point*, which can then be seen clearly by the unaccommodated eye. The intermediate image formed by the spectacle lens is at the far-point, which is the location of the object for the eye. *The far-point distance from the correction lens equals its focal length.* The eye views the right-side-up virtual images of all objects formed by the correction lens, and those images are located between its far- and near-points. The near-point also moves away a little, which is why myopes will often prefer to remove their spectacles when reading small print; they can then bring the material closer to the eye, increasing the magnification. If you are wearing glasses to correct myopia, try casting a real image with them—it can't be done.

Example 24.12 **[II]** A person sees objects beyond a distance of 2.0 m to be blurred, but otherwise everything seems fine. What kind of contact lenses, with what power, should they wear?

Solution This is a problem dealing with vision, and so the mention of "power" should call to mind $\mathcal{D}$. (1) TRANSLATION —An eye has a known finite far-point; determine the required contact lens correction. (2) GIVEN: A far-point of 2.0 m. FIND: $\mathcal{D}_c$. (3) PROBLEM TYPE—Geometrical optics/thin-lens combination/power/contact lens. (4) PROCEDURE—We want the image of an object at infinity to appear at the far-point of the eye

($s_o = \infty$, s_i = −2.0 m). Again, the far-point is on the left side of the lens, and therefore the image-distance is *negative*. Every distant object will then be imaged closer to the eye than the far-point and will be seen clearly. (5) CALCULATION—A lens must be added to the eye that has a power given by

$$\frac{1}{s_o} + \frac{1}{s_i} = \frac{1}{f} = \mathcal{D}_c = \frac{1}{\infty} + \frac{1}{-2.0\text{ m}} = \boxed{-\tfrac{1}{2}\text{ D}}$$

Quick Check: $f = \infty(-2.0\text{ m})/(\infty - 2.0\text{ m}) = -2.0\text{ m}$.

The Compound Microscope

The compound microscope goes the next step beyond the simple magnifier, providing still higher angular magnification (from 15× to around 1200×), achieved this time with a two-step arrangement. A simple version is illustrated in Fig. 24.29. The lens system closest to the object is the **objective**. It forms a real, inverted, magnified image of the object that is then viewed by the **eyepiece**. The latter is essentially a magnifying glass that looks at and enlarges the image created by the objective. Rays diverging from this intermediate image emerge from the eyepiece as a parallel bundle that can comfortably be viewed by a relaxed eye.

The problem is to examine an object that is close at hand; and since $M_T = -s_i/s_o$, the objective should have as small an s_o and as large an s_i as possible, which means, first, that the objective must be close to the object. Rearranging the lens equation yields $s_i = fs_o/(s_o - f)$, suggesting that the image distance will be appropriately large when $s_o \approx f$. *The objective must have a short focal length*, and the object must be positioned just beyond it so that the intermediate image is real. This image is then viewed by the eyepiece, which also must have a short focal length (f_E) since its magnification is $M_{AE} = d_n/f_E$. The intermediate image falls near the focal plane of the eyepiece so that the rays emerge parallel or almost so.

The eyepiece magnifies the intermediate image, which is a magnified version of the object; in other words, the total angular magnification of the system is the product of the magnifications of the objective (M_{TO}) and the eyepiece:

$$M_A = M_{TO}M_{AE}$$

Return to Fig. 24.12 and notice that, since triangles AOF_i and PEF_i are similar, $y_i/y_o = -(s_i - f)/f = M_T$. Applied to the objective, this expression becomes $M_{TO} = -L/f_O$, where the image distance minus the focal length of the objective is symbolized by L and is known as the **tube length**. Many manufacturers design their microscopes such that L is standardized at a length of about 160 mm. Using $M_{AE} = d_n/f_E$ and the fact that *it is customary to take the near-point d_n at 254 mm (10 in.)*, we have

$$M_A = -\frac{L}{f_O}\frac{d_n}{f_E} = \left(-\frac{160\,\text{mm}}{f_O}\right)\left(\frac{254\,\text{mm}}{f_E}\right) \tag{24.17}$$

where the focal lengths on the right are in *millimeters*. Because the intermediate image is real, it has to fit inside the tube. But the final image is virtual; it can be much larger than the tube diameter.

The barrel of an objective with a focal length of, say, 32 mm, is engraved with the mark 5×, indicating a *magnification* of 5 = 160 mm/32 mm. Combined with a 10× eyepiece ($f_E = 25.4$ mm), the microscope then has a magnification of 50×. The apparent size is 50 times the actual size.

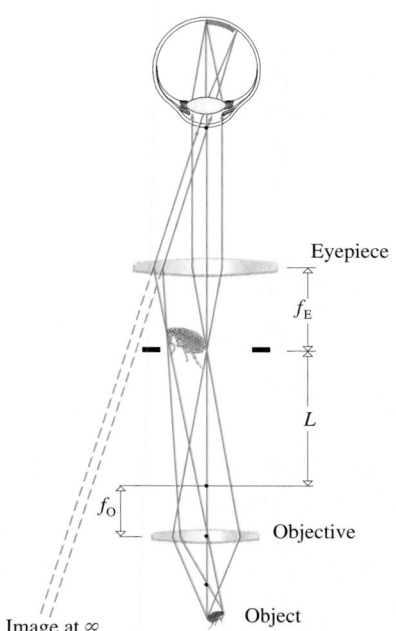

Figure 24.29 A rudimentary compound microscope. The objective forms a real magnified image of the object. That image is further magnified by the eyepiece. Because the intermediate image is at the focal point of the eyepiece, rays enter the eye in parallel bundles and the eye is relaxed. Notice the large retinal image, which is, of course, what we are after.

Example 24.13 **[II]** Suppose we wish to make a microscope (which can be used with a relaxed eye) out of two positive lenses both of focal length 25 mm. Assuming the object is positioned 27 mm from the objective, (a) how far apart should the lenses be and (b) what magnification can we expect?

Solution This is a two-lens problem specific to the microscope, so envision its configuration. (1) TRANSLATION—A microscope with two converging lenses of known focal length is to view an object at a specified distance; determine (a) the distance between the lenses and (b) the resulting magnification. (2) GIVEN: $f_O = f_E = 25$ mm and $s_o = 27$ mm. FIND: (a) lens separation and (b) M_A. (3) PROBLEM TYPE—Geometrical optics/thin-lens combination/ microscope. (4) PROCEDURE— (a) It follows from Fig. 24.29 that the lens separation is s_i plus f_E. (b) $M_A = M_{TO}M_{AE}$. (5) CALCULATION—(a) The intermedi-

continued

ate image distance is obtained from the lens formula applied to the objective

$$\frac{1}{27 \text{ mm}} + \frac{1}{s_i} = \frac{1}{25 \text{ mm}}$$

and $s_i = 3.38 \times 10^2$ mm. This value is the distance from the objective to the intermediate image, to which must be added the focal length of the eyepiece to get the lens separation;

$$3.38 \times 10^2 \text{ mm} + 25 \text{ mm} = \boxed{3.6 \times 10^2 \text{ mm}}$$

(b) We need $M_A = M_{TO} M_{AE}$, where $M_{TO} = -s_i/s_o = -(3.38 \times 10^2 \text{ mm})/(27 \text{ mm}) = -12.5\times$, and the eyepiece has a magnification of $d_n \mathscr{D}_E = (254 \text{ mm})(1/25 \text{ mm}) = 10.2\times$. Therefore, the total magnification is $M_A = M_{TO} M_{AE} = (-12.5)(10.2) = \boxed{-1.3 \times 10^2}$; the minus sign just means the image is inverted.

Quick Check: $L = s_i - f_o = 338 \text{ mm} - 25 \text{ mm} = 313 \text{ mm}$; $M_{TO} = -L/f_o = -(313 \text{ mm})/(25 \text{ mm}) = -12.5\times$; $M_{AE} = (254 \text{ mm})/f_E = 10.2\times$.

{It's a good idea at this point to take a break from the book and click on **THE MICROSCOPE** under **INTERACTIVE EXPLORATIONS** on the **CD**. That simulation allows you to study how a compound microscope forms a magnified image. You should spend some time working with this piece of cyberoptics; it's a great way to learn about the microscope.} 💿

The Refracting Telescope

The primary function of the telescope is to enlarge the image of a *distant* object. The device shown in Fig. 24.30 has an objective and an eyepiece just like a microscope but, because the job to be done is different, the structure is also different. The object is at a finite far distance from the device so that the intermediate image is located beyond the image focus of the objective. As with the microscope, this real, inverted image serves as the object for the eyepiece, which functions as a magnifier. The intermediate image is made to fall within one focal length (f_E) of the eyepiece so that the resulting final image it creates is virtual, enlarged, and remains inverted. In practice, *the position of the intermediate image is fixed, and only the eyepiece is moved in order to focus the instrument.*

The central piece of design information is that the object is far away; that is, the object distance for the objective is very large in comparison to all the other distances in the system. The Gaussian Lens Equation then tells us that, since $1/s_o \approx 0$, it follows that $s_i \approx f_o$. Unlike the microscope, where the intermediate image was magnified, here it must be *minified*. That might seem strange at first, but realize how awkward it would be to try to fit a larger-than-life real image of, say, the Moon inside the tube of a telescope. The important feature of the real image is that it is now so close that it can easily be examined with a magnifier.

Suppose the object focus of the eyepiece, which lies in front of it, overlaps the image focus of the objective, which lies behind it, as it does in Fig. 24.31. Then the separation between the two lenses equals the sum of their focal lengths. Parallel rays entering the

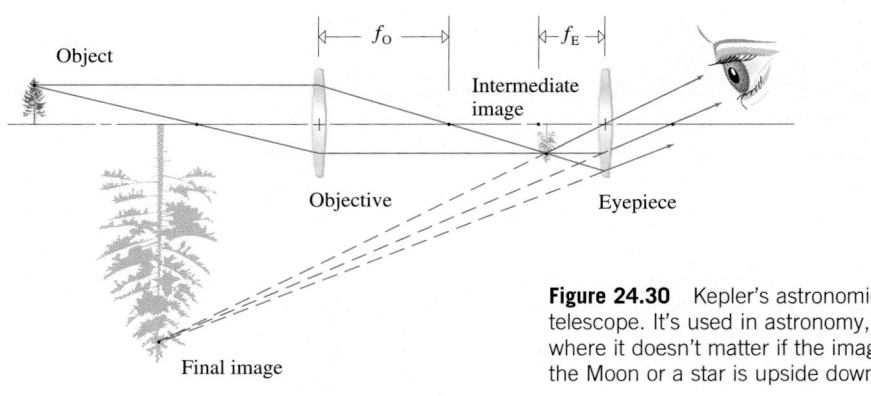

Figure 24.30 Kepler's astronomical telescope. It's used in astronomy, where it doesn't matter if the image of the Moon or a star is upside down.

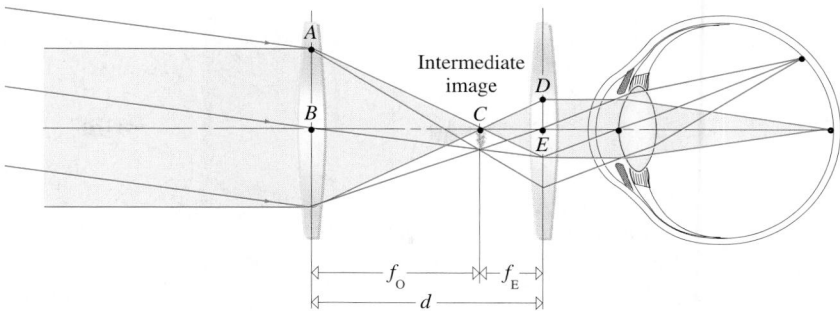

Figure 24.31 An astronomical telescope operating so that parallel light enters the objective and leaves the eyepiece.

scope from a very distant object exit in parallel bundles that can be viewed comfortably by the relaxed eye. That concern is important, and this arrangement is most often used.

In general, we want a minified intermediate image that is as large as is practical. Since $M_{TO} = -s_i/s_o \approx -f_O/s_o$, we need an objective lens with *as long a focal length as possible*. Again, the eyepiece views the intermediate image and magnifies it. Since the magnification varies inversely with f_E, the eyepiece should have a *short focal length*. In fact, the magnifying power of a telescope, adjusted so that parallel rays emerge, is

$$M_A = -\frac{f_O}{f_E}$$

This equation is the reason why high-power refracting telescopes usually have long tubes into which one inserts a short focal-length eyepiece. When you look through the back end of a telescope, everything appears minified. The roles of eyepiece and objective are reversed; and because their focal lengths are quite different, the effect is striking. Reversing the telescope reduces the image size by the same factor by which it was previously increased.

Mirrors

Mirror systems are finding increasingly more extensive and important applications, particularly in the infrared, ultraviolet, and X-ray regions of the spectrum. It is relatively easy to construct a reflecting device that performs satisfactorily across a broad range of frequencies; the same cannot be said for refracting systems. Today, mirrors play a significant role in all sorts of devices, from spy satellites and copy machines to cameras, microscopes, and lasers.

24.7 Curved Mirrors

Curved mirrors that form images very much like those of lenses have been known since the ancient Greeks. Fortunately, we have already developed much of the conceptual basis for analyzing curved mirrors and will be able to evolve the subject quickly without introducing many new ideas.

We again ask for the kind of surface that reshapes an incoming plane wave, this time via reflection, into an outgoing converging spherical wave. Figure 24.32 shows the geom-

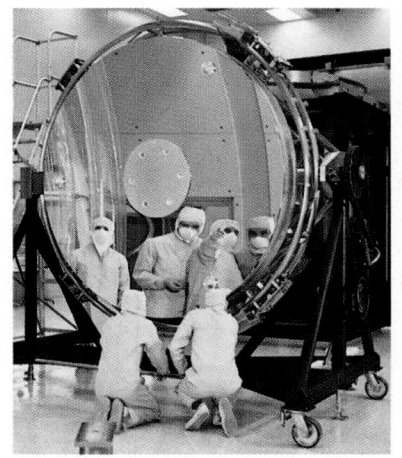

The 2.4-m-diameter primary mirror of the Hubble Space Telescope.

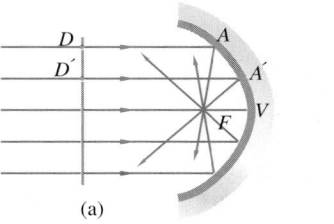

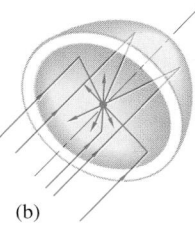

(a)   (b)

Figure 24.32 A paraboloidal mirror. (a) Parallel axial rays are brought to a focus at point F. (b) The configuration is three-dimensional and symmetric about the central axis.

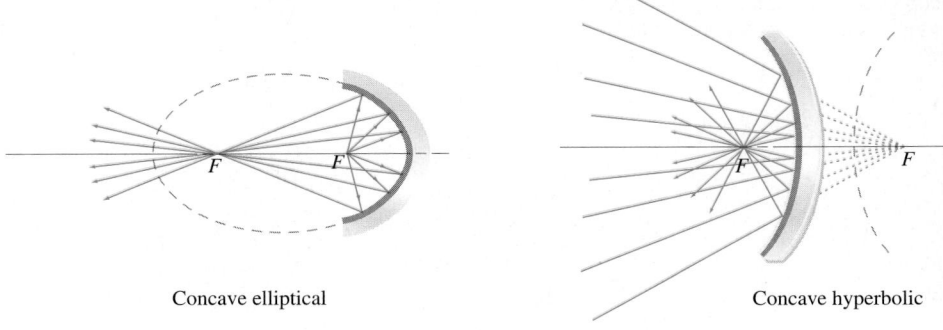

Figure 24.33 Two aspherical mirrors.

etry. For a plane wave to be bent into a converging sphere, the path length from any point D on the wavefront to point A directly opposite it on the surface, and thence to the fixed point F, must be constant. Descartes (1637) showed that the only surface that will satisfy these conditions is a parabola with its focus at F. {For a more complete treatment click on **ASPHERICAL MIRRORS** under **FURTHER DISCUSSIONS** on the **CD**.} 💿 A bundle of parallel rays reflecting off a paraboloidal mirror will be brought to a focus at F, a distance f from the **vertex** V of the mirror. Most of the world's older astronomical reflecting telescopes have energy-gathering mirrors that are paraboloidal.

Figure 24.33 depicts the behavior of several other aspherics. In recent years, these have been used to form images, via reflected X-rays, of a variety of phenomena from solar emission to laser-induced fusion. Today, the hyperboloid is the overwhelming choice for large telescopes, including the Hubble Space Telescope (p. 878).

Spherical Mirrors

A spherical mirror has no one particular symmetry axis, and that can be a great advantage, especially when the device is not movable. In many applications restricted to paraxial optics, spherical mirrors perform quite well. Problem 98 deals with the proof that a sphere and a paraboloid coincide in the region close to the symmetry axis provided that $|f| = |R|/2$, as shown in Fig. 24.34. Insofar as the rays are paraxial, they will encounter a region where sphere and paraboloid are nearly identical and the sphere will behave like the ideal mirror in Fig. 24.32.

Absolute values, $|f|$ and $|R|$, are used because we have not yet agreed upon signs for mirror quantities. Using the previous convention, R in Fig. 24.35 is negative because C is to the left of V. A parallel bundle of axial rays will converge to F, and so we take f to be positive, despite the fact that it is now measured to the left of the mirror (as compared with to the right of the lens). That being the case,

The 1000-ft radiotelescope at Arecibo, Puerto Rico, operates at 21 cm. The spherical bowl reflects radiant energy up to the focal-point detector suspended above it on cables attached to three towers.

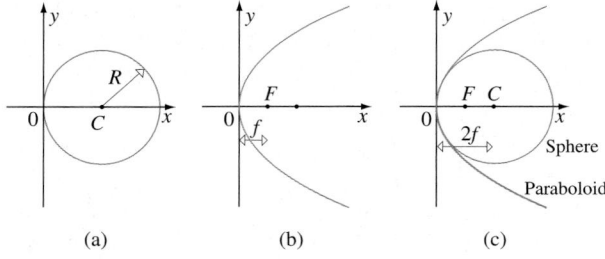

(a)   (b)   (c)

Figure 24.34 A sphere (a) and a paraboloid (b) coincide (c) in the region near the axis, such that $|R| = 2|f|$.

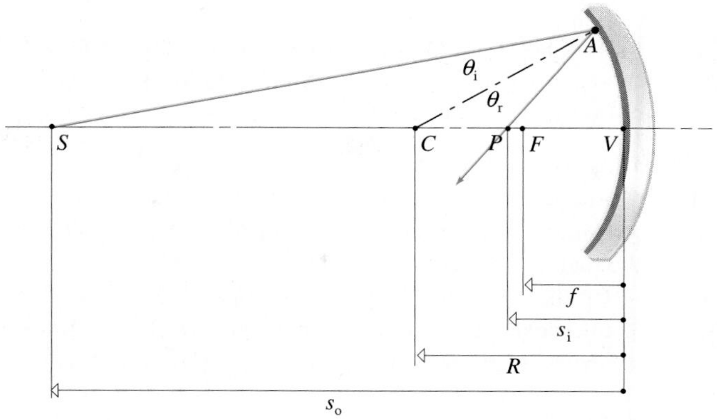

Figure 24.35 The geometry of a spherical mirror in the paraxial region.

[spherical mirror]

$$f = -\frac{R}{2}$$ (24.18)

The radius of curvature of a **concave mirror** is negative and since $f = -R/2$, its focal length is positive.

The focal length of a spherical mirror equals one-half its radius. We'll soon see that *the concave mirror behaves exactly like a convex lens, where the rays refracted to the right are instead reflected to the left.*

The image P of S in Fig. 24.35 is real, but it's *on the left of the mirror.* If we continue to require that real images have positive image distances, then we must henceforth take s_i to be *positive when left of the vertex*, exactly the same way that we took the focal length to be positive. These differences in the signs of f and s_i will be the only deviations from the convention for lenses, and with them all the appropriate equations will turn out the same as before. The image-formation geometry for the concave mirror is identical to that of the convex lens, and we need not rederive all the equations. Suffice it to say that paraxial rays in Fig. 24.35 obey the relationship

[spherical mirror]

$$\frac{1}{s_o} + \frac{1}{s_i} = -\frac{2}{R}$$ (24.19)

which is called the **mirror formula**. If $s_o \rightarrow \infty$, $s_i \rightarrow f_i = -R/2$; while if $s_i \rightarrow \infty$, $s_o \rightarrow f_o = -R/2$. In other words, both the image and object focal lengths equal f and

[spherical mirror]

$$\frac{1}{s_o} + \frac{1}{s_i} = \frac{1}{f}$$ (24.20)

Observe that f is positive for concave mirrors ($R < 0$) and negative for convex mirrors ($R > 0$). In the latter case, the image is formed behind the mirror in diverging light and cannot be projected upon a screen—it is virtual (Fig. 24.36a).

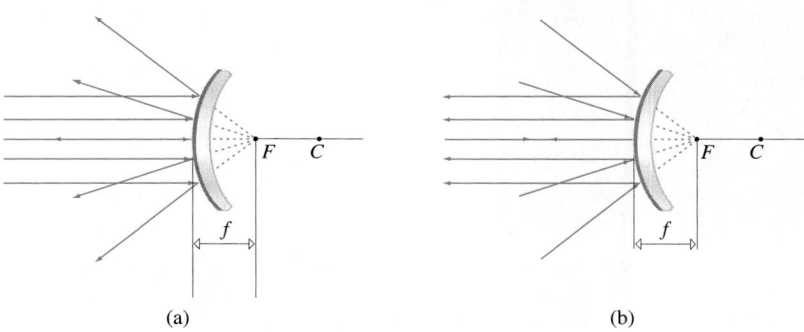

(a)   (b)

Figure 24.36 The focal point, F, of a convex spherical mirror.

Example 24.14 **[I]** A point source lies on the central axis 1.00 m in front of a concave spherical mirror having a radius of curvature of 20 cm. Locate and describe the resulting image.

Solution The words "concave spherical mirror" should bring to mind the mirror formula. (1) TRANSLATION—An object is at a specified distance from a concave spherical mirror of known radius; determine the image. (2) GIVEN: $s_o = 1.00$ m and mirror is concave ($R < 0$); therefore $R = -0.20$ m. FIND: s_i. (3) PROBLEM TYPE—Geometrical optics / curved mirrors / spherical. (4) PROCEDURE—We have s_o and R and need s_i, but we don't

have f. That suggests the mirror formula. (5) CALCULATION—From Eq. (24.19)

$$\frac{1}{s_i} = -\frac{2}{R} - \frac{1}{s_o} = -\frac{2}{-0.20 \text{ m}} - \frac{1}{1.00 \text{ m}}$$

and $\boxed{s_i = 0.11 \text{ m}}$. The image is real ($s_i > 0$).

Quick Check: $f = -R/2 = 0.10$ m $= s_o s_i / (s_o + s_i) = (1.00$ m$)(0.11$ m$)/(1.00$ m $+ 0.11$ m$) = 0.10$ m.

Extended Imagery: Spherical Mirrors

The remaining mirror properties are so similar to those of lenses that we need only mention them briefly without repeating the development. Accordingly, within the restrictions of paraxial theory, *any off-axis parallel bundle of rays will be focused to a point on the focal plane* a distance f from V. The image of any point on an object can again be located using any two easily drawn rays (Fig. 24.37). Unlike the lens, which is transparent and has separate object and image foci (one on each side), the mirror has only one focus. Table 24.4 summarizes the sign convention.

The ray diagram (Fig. 24.38) shows how the image of an extended object is created by a concave mirror. Because triangles *SDV* and *PEV* are similar (look at Fig. 24.12), their sides are proportional. Hence, taking distance measured down from the axis to be negative, $y_i/y_o = -s_i/s_o$, which, of course, is the transverse magnification M_T as defined earlier (p. 869). The striking similarity between the behavior of a concave mirror and a convex lens, on one hand, and a convex mirror and a concave lens, on the other, is evident on comparing Tables 24.3 and 24.5.

Just like a convex lens, a concave mirror can form a real image. Here the mirror in the foreground casts a real, minified, inverted image of the candle flame.

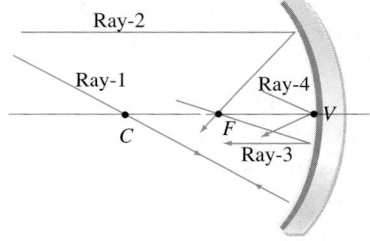

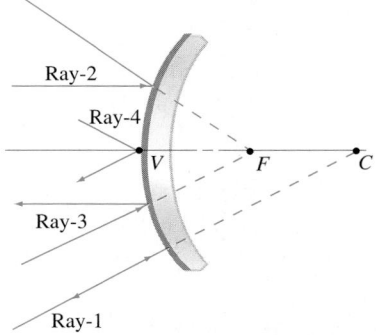

Figure 24.37 Four easily drawn rays. Ray-1 heads toward *C* and reflects back along itself. Ray-2 comes in parallel to the central axis and reflects toward (or away from) *F*. Ray-3 passes through (or heads toward) *F* and reflects off parallel to the axis. Ray-4 strikes point *V* and reflects such that $\theta_i = \theta_r$.

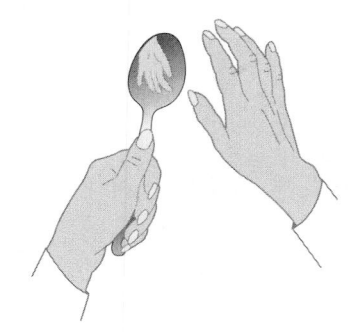

EXPLORING PHYSICS ON YOUR OWN

Spherical Mirrors: Get a variety of soup spoons and ladles that are spherical, or nearly so, and polish them up so that they're bright and shiny. Hold a finger straight up and bring it toward the concave surface. When the object distance is large the image will be inverted and minified. At two focal lengths it will be inverted and life-size. At the focal point the image will fill the bowl with an unrecognizable blur. Move your finger still closer and the image will reappear right-side-up and magnified. Depending on the spoon, there's likely to be lots of distortion. Now turn it around and explore the image properties of the convex side. How does it differ? Turn a lamp on in an otherwise dark room and see if you can project a real image of it onto a piece of paper using the concave surface—you'll have to tilt the spoon to the side a bit. This is best done with a shallow spoon having a large R and, therefore, a large f. Unless you've got a perfect spoon, the real image is going to be quite distorted, but it should still be recognizable.

Table 24.4

Sign Convention for Spherical Mirrors

	Sign	
Quantity	**+**	**−**
s_o	Left of V, real object	Right of V, virtual object
s_i	Left of V, real image	Right of V, virtual image
f	Concave mirror	Convex mirror
R	C right of V, convex	C left of V, concave
y_o	Right-side-up object	Inverted object
y_i	Right-side-up image	Inverted image
M_T	Right-side-up image	Inverted image

A convex spherical mirror forming a virtual, right-side-up, minified image.

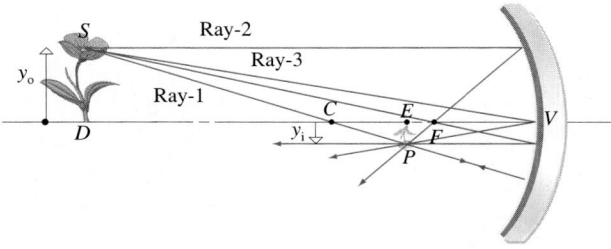

Figure 24.38 An extended image formed by a spherical concave mirror.

Table 24.5

Images of Real Objects Formed by Spherical Mirrors

		Concave		
Object		**Image**		
Location	Type	Location	Orientation	Relative size
$\infty > s_o > 2f$	Real	$f < s_i < 2f$	Inverted	Minified
$s_o = 2f$	Real	$s_i = 2f$	Inverted	Same size
$f < s_o < 2f$	Real	$\infty > s_i > 2f$	Inverted	Magnified
$s_o = f$		$\pm \infty$		
$s_o < f$	Virtual	$|s_i| > s_o$	Right-side-up	Magnified

		Convex		
Object		**Image**		
Location	Type	Location	Orientation	Relative size
Anywhere	Virtual	$|s_i| < |f|$, $s_o > |s_i|$	Right-side-up	Minified

Figure 24.39 The images formed by a concave spherical mirror.

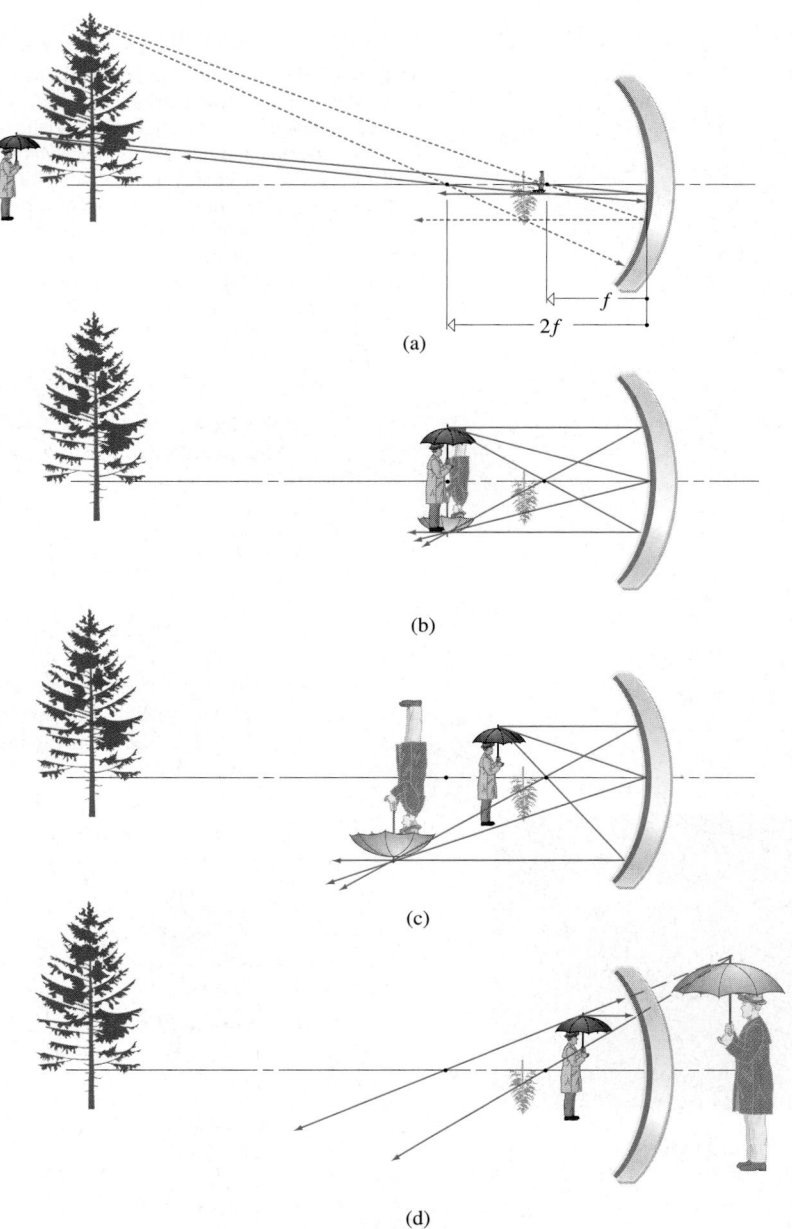

(a)

(b)

(c)

(d)

Figure 24.39 graphically illustrates the entire range of responses of the concave mirror. Except for the fact that the mirror folds the rays over (so that object and image space overlap when the image is real and do not when it's virtual), the diagram is the same as Fig. 24.15.

Example 24.15 **[II]** A youngster looking into the shiny convex back of a spoon with a spherical bowl sees herself reflected therein. The bowl has a radius of 3.00 cm, and her nose is 25.0 cm from its surface. Where will the image of her nose appear? Describe the image completely.

Solution This is a problem about a convex spherical mirror, and that means the mirror formula will likely be needed. (1) TRANSLATION—An object is at a specified distance from a convex spherical mirror of known radius; determine the image. (2) GIVEN: The mirror is convex ($R > 0$); therefore $R =$

continued

+0.030 m, and $s_o = 0.250$ m. FIND: s_i and M_T. (3) PROBLEM TYPE—Geometrical optics/curved mirrors/spherical. (4) PROCEDURE —We have s_o and R and need s_i, but we don't have f. That suggests the mirror formula. (5) CALCULATION:

$$\frac{1}{s_o} + \frac{1}{s_i} = -\frac{2}{R}$$

$$\frac{1}{0.250\text{ m}} + \frac{1}{s_i} = -\frac{2}{0.030\text{ m}}$$

and $\boxed{s_i = -0.014\text{ m}}$. The image distance is negative and so is to the right, behind the mirror. The magnification is

$$M_T = -\frac{s_i}{s_o} = -\frac{-0.014\,2\text{ m}}{0.250\text{ m}} = \boxed{+0.057}$$

The image is virtual, minified, and right-side-up.

Quick Check: $f = -R/2 = s_o s_i/(s_o + s_i) = -0.015\text{ m} = (0.250\text{ m})(-0.014\text{ m})/(0.250\text{ m} - 0.014\text{ m}) = -0.015\text{ m}$.

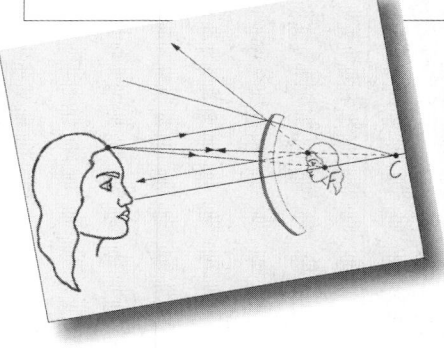

{Put the book aside for the moment and click on **MIRRORS** under **INTERACTIVE EXPLORATIONS** on the **CD**. This lovely simulation will allow you to study the behavior of both concave and convex mirrors. Spend some time working with this piece. It will be fun to play with and will definitely clarify the operation of mirrors.}

The Reflecting Telescope

In addition to providing magnification, serious astronomical telescopes must gather in as much light as possible because the objects being viewed are generally extremely faint. As with the camera, the energy entering the system is proportional to the diameter of the objective—the bigger, the better. But there is a real difficulty in making big lenses. The largest such instrument in the world is the 40-in.-diameter Yerkes refracting telescope in Wisconsin as compared to the 200-in. Palomar reflector in California. The problems are evident: a lens has to be perfectly transparent and free of internal flaws. A front-silvered mirror need not even be transparent. A lens can only be supported by its rim and may sag under its own weight; a mirror can be supported over its entire back. For these and other reasons (better frequency response, better aberration control, and so on), reflectors predominate in the domain of large telescopes, from observatories to spy satellites.

Invented by the Scotsman James Gregory in 1661, the reflecting telescope was first successfully constructed by Newton in 1668 (primarily because he had mistakenly concluded that the aberrations suffered by lenses were unavoidable). Two common reflectors are shown in Fig. 24.40. A plane mirror or prism brings the beam out to the side in the Newtonian version. The traditional Cassegrain arrangement uses a convex hyperboloidal secondary mirror to increase the effective focal length of a paraboloidal primary. A modern

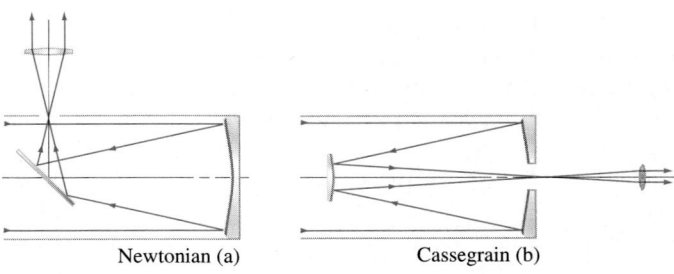

Newtonian (a) Cassegrain (b)

Figure 24.40 Reflecting telescopes.

Figure 24.41 The Hubble Space Telescope is 13 m long, about 5 m from the primary to the secondary, and has a mass of 11 600 kg.

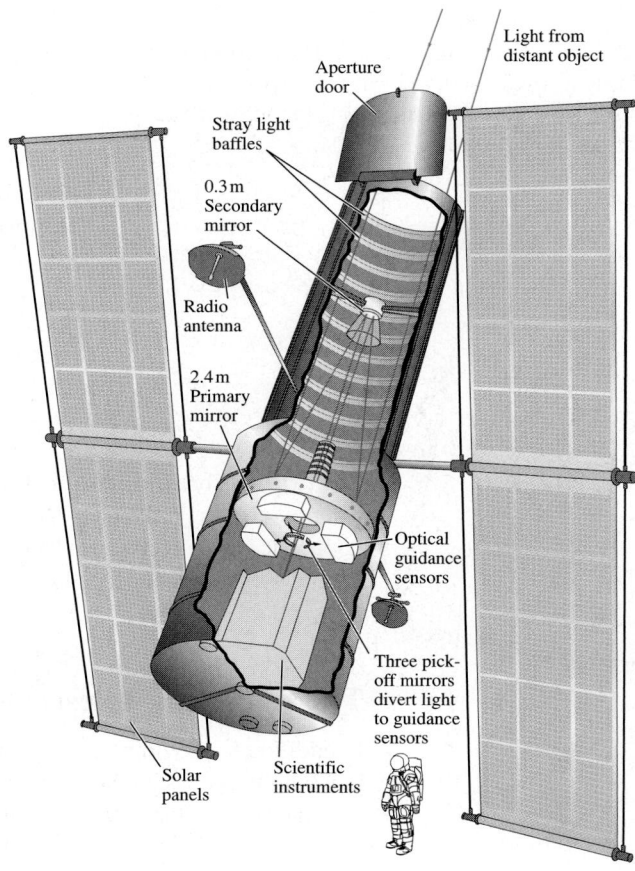

Cassegrain with a hyperboloidal primary and secondary is now the leading astronomical telescope configuration. The 2.4-m-diameter Hubble Space Telescope is one of these (Fig. 24.41).

Core Material & Study Guide

As the wavelength of the radiant energy being processed becomes vanishingly small, diffraction disappears, rectilinear propagation obtains, and we have the idealized domain of **geometrical optics**.

LENSES

For thin spherical lenses, there is the **Lensmaker's Formula**

$$\frac{1}{s_o} + \frac{1}{s_i} = (n_l - 1)\left(\frac{1}{R_1} - \frac{1}{R_2}\right) \quad [24.1]$$

Section 24.2 (Spherical Thin Lenses) considers this basic equation and lays out the associated conventions. Re-read Example 24.1, look at the **CD Warm-Ups** and **Walk-Throughs** 🔵, and then try the I-level problems. Read the *Suggestions on Problem Solving* before you begin.

The **focal length** of a lens f is given as

$$\frac{1}{f} = (n_l - 1)\left(\frac{1}{R_1} - \frac{1}{R_2}\right) \quad [24.2]$$

This very important idea is introduced in Section 24.3 (Focal Points and Planes) and explored in Example 24.2—make sure you understand it.

Comparing the preceding two expressions yields the **Gaussian**

Lens Equation

$$\frac{1}{s_o} + \frac{1}{s_i} = \frac{1}{f} \quad [24.4]$$

The ratio of any transverse dimension of the image to the corresponding dimension of the object is the **magnification**, M_T:

$$M_T = \frac{y_i}{y_o} \quad [24.6]$$

Alternatively, $\qquad M_T = -\dfrac{s_i}{s_o} \quad [24.7]$

The formation of images of extended objects is discussed in Section 24.4 (Extended Imagery: Lenses). Review Figs. 24.11 and 24.12, memorize Table 24.2, and be sure to master Examples 24.3 and 24.4. You must learn how to create simple ray diagrams—draw one for every problem you work on. It's good practice, when working a problem, to figure out the general characteristics of the image before making any calculations.

Section 24.5 (A Single Lens) pulls it all together, summarizing the operation of the lens and discussing its various applications. If the object is at the focal point of a **magnifying glass**, its image is at infinity and

$$M_A = \frac{d_n}{f} \qquad [24.9]$$

Study Fig. 24.15 and Table 24.3. Example 24.5 deals with a negative lens, and you should review it before going on.

When two lenses are being used together, the image of the first is the object of the second, and the analysis can be carried out one lens at a time (p. 863). The focal length of two thin lenses in contact is given by

$$\frac{1}{f} = \frac{1}{f_1} + \frac{1}{f_2} \qquad [24.14]$$

The **dioptric power** $\mathcal{D}$ of a lens is the reciprocal of the focal length, and its units are inverse meters or *diopters* (D). The total power of two lenses in contact is

$$\mathcal{D} = \mathcal{D}_1 + \mathcal{D}_2 \qquad [24.16]$$

Section 24.6 (Thin-Lens Combinations) treats multilens systems including eyes and eyeglasses, microscopes, and telescopes. This material should be reviewed before attempting any problems dealing with these devices.

MIRRORS

A spherical mirror behaves very much like a thin lens. It has a focal length given by

$$f = -\frac{R}{2} \qquad [24.18]$$

But we take f to be positive to the left of the mirror. The relationship between object distance, image distance, and focal length is represented by

$$\frac{1}{s_o} + \frac{1}{s_i} = -\frac{2}{R} \qquad [24.19]$$

and this is the **mirror formula**. These ideas are found in Section 24.7 (Curved Mirrors). Commit Table 24.4 to memory, study Table 24.5, and then review Fig. 24.39.

Key Terms

geometrical optics	real image
lens	virtual image
aspherical surfaces	transverse magnification
converging lens	longitudinal magnification
diverging lens	cornea
concave	crystalline lens
convex	retina
paraxial	accommodation
thin lens	far-point
object distance	near-point
image distance	angular magnification
Thin-Lens Equation	magnifying glass
Lensmaker's Formula	intermediate image
focal point	dioptric power
focal plane	farsightedness
positive lens	nearsightedness
negative lens	objective eyepiece
optical center	mirror formula

Discussion Questions

1. What happens to the focal length of a glass lens when it's taken from the air and placed in water? Explain your answer.

2. If a lens of glass surrounded by air is negative, what can be said about the identically shaped lens made of air surrounded by glass? A bubble in a glass of beer is a tiny lens. What kind?

3. Explain why it is that the focal length of a lens actually depends on the color of the light being transmitted.

4. How do goggles work to allow an underwater swimmer to see clearly?

5. What is a quick physical means of determining the approximate focal length of a converging lens? How might you use a known strong positive lens to find the focal length of a negative lens?

6. If a horse stands facing a positive lens, which part of the beast will be closest to the lens in the real image? in the virtual image? Draw the appropriate ray diagrams.

7. Imagine a converging glass lens in a chamber filled with a gas under a few atmospheres of pressure. Suppose the lens is illuminated by parallel light. What, if anything, will happen to the point at which the beam converges if the gas is gradually pumped out?

8. If a real image is formed by a large plane mirror, what can be said about the incoming rays? Incidentally, a tiny flat mirror can form a real image just as a pinhole does.

9. What is the focal length of a plane mirror? What does Eq. (24.19), the mirror formula, say about the image distance of such a device? What then is the magnification of a flat mirror according to the equations?

10. Figure Q10 depicts a hyperboloidal mirror and its accompanying geometry. Explain what is happening in the diagram. Describe the wavefronts before and after reflection from the hyperboloidal mirror.

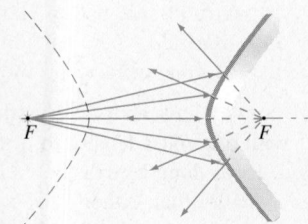

Figure Q10

11. Legend has it that Archimedes (ca. 287 B.C.E.–212 B.C.E.) burned the invading Roman fleet by focusing sunlight onto the sails. One version of the story has him on a mountainside lining up soldiers holding brightly

polished shields. How could he have arranged the soldiers to accomplish this task?

12. Suppose we take two positive lenses and place them in contact with one another. In what sense is the combination more powerful than either lens separately?

13. Figure Q13 is a diagram of an X-ray camera used for diagnosing the implosion of tiny laser-fusion targets (Sect. 30.10) at the Lawrence Livermore Laboratory in California. Explain how it works to form an image of the target. (You should be able to figure out what's happening, even though some aspects were not discussed in the text explicitly.)

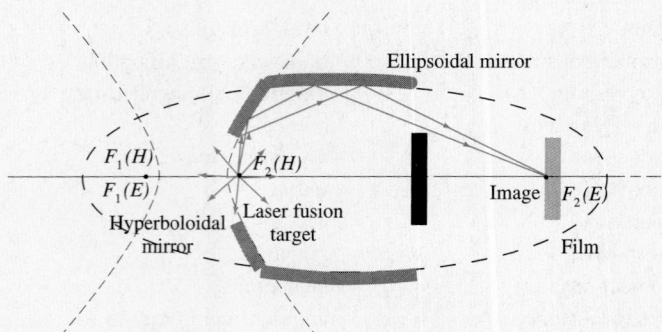

Figure Q13

14. A person with normal vision puts on a pair of eyeglasses made for a hyperope. What will things look like near and far with the eye relaxed? not relaxed?

15. Microscopes operating in the visible are limited in magnification to about $1200\times$. How big an object would produce an image seen in the eyepiece to be $\frac{1}{2}$ mm across? What does that suggest

about the limitations of the instrument? Will a $12\,000\times$ light microscope resolve any finer details of the object?

16. The spotlight in Fig. Q16 is a rather common, very simple one used in theaters in the United States. Explain what it does and how it works. Where is the filament located with respect to both the lens and the mirror? Why?

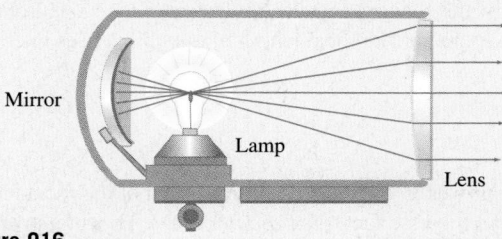

Figure Q16

17. Figure Q17 shows an ellipsoidal reflector spotlight, which is probably the most widely used form of the device. How does it work?

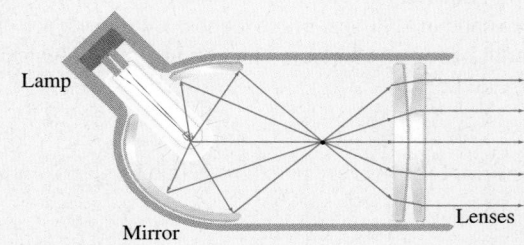

Figure Q17

18. Explain the meaning of each of the Key Terms on page 879.

Multiple Choice Questions

1. A converging lens is mounted on an optical bench, and a candle is placed in front of it so that an image of the flame appears on a screen $1.5f$ beyond the lens. The object is (a) between f and $2f$ (b) beyond $2f$ (c) closer than f (d) near infinity (e) none of these.

2. Which cannot be created by a negative lens? An image that is (a) virtual, right-side-up, and closer than f (b) right-side-up and smaller than life (c) minified, inverted, and virtual (d) minified, closer than f, and right-side-up (e) none of these.

3. Which cannot be formed by a positive lens? An image that is (a) virtual, right-side-up, and larger than life (b) virtual, inverted, and minified (c) real, inverted, and minified (d) real, inverted, and magnified (e) none of these.

4. A convex lens with a very small focal length is placed in contact with a convex lens with a very large focal length. The combined focal length will be (a) much larger than either (b) much smaller than either (c) approximately equal to the smaller (d) approximately equal to the larger (e) none of these.

5. A lens made of glass ($n = 1.50$) has a convex first surface with a radius of curvature of 2.0 m and a second concave surface with a radius of curvature of 1.0 m. Its focal length is (a) 3.0 m (b) $\frac{1}{2}$ m (c) $-\frac{1}{4}$ m (d) -4.0 m (e) none of these.

6. When an object is placed between the vertex and focal point of a negative lens, its image is (a) virtual, enlarged, and right-side-up (b) real, minified, and right-side-up (c) real, enlarged, and inverted (d) virtual, minified, and right-side-up (e) none of these.

7. A frog sitting 40 cm from a positive lens of focal length 20 cm hops out to 1.00 m. Its image on the right of the lens moves from (a) 40 cm to 1.00 m (b) 40 cm to 25 cm (c) 20 cm to 40 cm (d) 40 cm to 20 cm (e) none of these.

8. A bird is sitting on a perch peering into a small convex mirror in its cage. The image it sees is (a) always real (b) always virtual (c) always magnified (d) always inverted (e) none of these.

9. Suppose you make a calculation of the magnification of the image of a real object located somewhere in front of a convex spherical mirror and it turns out to be -1.5. You can then conclude that (a) the calculation was wrong (b) the image is smaller than life (c) the object was upside down (d) the image is larger than the object (e) none of these.

The next five questions refer to Fig. MC10, which depicts a concave spherical mirror whose center of curvature is at point C.

10. If a nonaxial ray traveling to the right, passes through C and

strikes the mirror, it will return (a) passing through point *A* (b) passing through point *B* (c) passing through point *C* (d) parallel to the central axis (e) none of these.

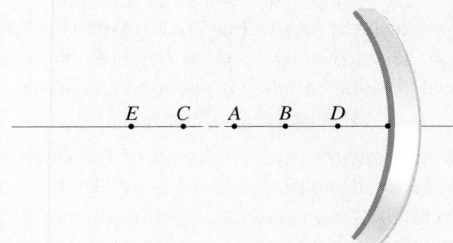

Figure MC10

11. If a nonaxial ray, traveling to the right, passes through *B* and strikes the mirror, it will return (a) passing through point *A* (b) parallel to the central axis (c) passing through point *C* (d) passing through point *B* (e) none of these.

12. Suppose a nonaxial ray, traveling to the right, passes through *A* and strikes the mirror at point *P*. If it makes an angle of 20° with respect to the line from *C* to *P*, what angle will the reflected ray make with respect to that same line? (a) not enough information is given (b) 40° (c) 90° (d) 20° (e) none of these.

13. Where should a point source be located if the reflected light is to emerge parallel to the central axis? (a) at point *A* (b) at point *C* (c) at point *B* (d) at point *D* (e) none of these.

14. If an object is placed at point *E*, its image will be (a) real, magnified, and inverted (b) virtual, magnified, and inverted (c) virtual, magnified, and right-side-up (d) real, magnified, and right-side-up (e) none of these.

15. A little mirror used by a dentist to see inside the mouth is a concave spherical device with a focal length of 25 mm. Held 2.0 cm from a tooth, it provides a magnification of (a) 0.5 (b) 10 (c) 5 (d) 0.1 (e) none of these.

16. Suppose that you carry out the analysis of a mirror system and find that the image distance is negative. This finding tells you that the image is (a) magnified (b) inverted (c) black and white (d) real (e) none of these.

17. Which cannot be formed by a concave spherical mirror? An image that is (a) real, inverted, and minified (b) virtual, right-side-up, and larger than life (c) magnified, upside down, and beyond 2*f* (d) life-size, real, and inverted (e) none of these.

18. A concave makeup mirror magnifies, by a factor of 2, the face of anyone looking into it from a distance of 25 cm. (The image is, of course, right-side-up.) The mirror's focal length must therefore be (a) 5 cm (b) 50 cm (c) 500 cm (d) 5000 cm (e) none of these.

19. The off-axis mirror forming a perfect point image depicted in

Fig. MC19 is (a) a segment of a paraboloid (b) planar (c) a segment of an ellipsoid (d) a segment of a hyperboloid (e) none of these.

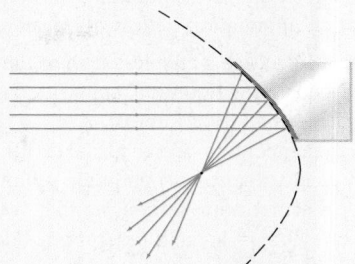

Figure MC19

20. Can a paraboloidal mirror be used to create a reflected diverging spherical wave? (a) no (b) yes, by sending a parallel beam into a concave mirror (c) yes, by sending a beam converging toward the focus of a convex mirror (d) yes, by sending a parallel beam toward a convex mirror (e) none of these.

21. George is nearsighted. George wears negative corrective lenses. The image focal point of either one of George's eyeglass lenses (a) lies at infinity (b) lies at his near-point (c) lies in the middle of his eye (d) lies at his far-point (e) none of these.

22. If the human eye, simply modeled, has a power of about 59 D, its focal length is (a) +17 mm (b) −5.9 mm (c) +5.9 mm (d) −17 mm (e) none of these.

23. Suppose you slip out the eyepiece from a microscope and replace it with one having twice the focal length. The magnification of the instrument will (a) double (b) halve (c) remain unchanged (d) quadruple (e) none of these.

24. Suppose we come upon a bag of lenses that are marked +1.0 D, −4.0 D, +10 D, +0.10 D, +95 D, and −0.50 D. Which would you use to build a microscope with the largest possible magnification? (a) +0.10 D as objective and +95 D as eyepiece (b) −4.0 D as objective and −0.50 D as eyepiece (c) +95 D as objective and +0.10 D as eyepiece (d) +10 D as objective and +0.10 D as eyepiece (e) none of these.

25. Suppose we come upon a bag of lenses that are marked +1.0 D, −4.0 D, +10 D, +0.10 D, +95 D, and −0.50 D. Which would you use to build an astronomical telescope with the largest possible magnification? (a) +0.10 D as objective and +95 D as eyepiece (b) −4.0 D as objective and −0.50 D as eyepiece (c) +95 D as objective and +0.10 D as eyepiece (d) +10 D as objective and +0.10 D as eyepiece (e) none of these.

For more Multiple Choice Questions with answers click on WARM-UPS in CHAPTER 24 on the CD.

Suggestions on Problem Solving

1. Read the problem carefully for *given* information that is subtly stated. If the image created by a single thin lens appears on a wall, the lens is positive, its focal length is positive, the image is real, the image distance is positive, the image is inverted, and the magnification and image height are both negative. All of that detail may *not* be explicitly spelled out. Make a sketch of the lens or mirror, the object

and image, putting everything in roughly where it belongs, with all the numbers included so that the setup can be visualized as a whole. Worry about the ray diagram later.

2. When working problems, keep in mind the sign convention as you substitute in the numbers. Read your diagram to confirm the signs. If you know that the image formed by the lens is inverted, y_i

must go in as a negative number, just as M_T must be negative. If that image is 3-m tall, $y_i = -3$ m. Watch out for the signs of the radii of curvature of lenses and mirrors—these are often messed up. The sign convention requires that the *light enter from the left*. If this is not the case in a particular situation, redraw the figure.

3. Several of the equations in this chapter, such as the Lensmaker's Formula and the Gaussian Lens Equation, are written in terms of the reciprocals of quantities that are of interest. Among the most common errors made is to neglect to invert the final result, for example, computing the numerical value for $1/f$ and then giving that number as your answer for f. Furthermore, remember that you *cannot* take an equation like $1/s_o + 1/s_i = 1/f$ and invert both sides!

4. In most cases, especially with only one lens or mirror involved, you should already have a good idea of what the image will be from Table 24.3 and/or Table 24.5. *Whenever possible, check the results of your calculations with reality, that is, with those tables.* If the

object is just a little bit farther from the positive lens than one focal length and the image is computed to appear roughly one focal length beyond the lens, then you should know something is terribly wrong!

5. When drawing ray diagrams, remember that objects nearer to a positive lens produce real images that are farther away. Also keep in mind that all images formed by a convex mirror lie between the vertex and the focal point behind the mirror. If you have a rough idea of what to expect, your ray diagrams will not go wild.

6. Notice that the angular magnification of both the compound microscope and the astronomical telescope are negative because the images are inverted. That means that you will have to be a little careful about signs when doing problems dealing with these devices. A $10\times$ telescope means $M_A = -f_O/f_E = -10$; both focal lengths are positive.

Problems ✦ Coordinated Problems ✦ Progressive Problems ✦ Solutions

SECTION 24.2: SPHERICAL THIN LENSES

SECTION 24.3: FOCAL POINTS AND PLANES

1. [I] A point source of light lies on the central axis of a bi-convex spherical lens whose radii of curvature are both measured to be 50 cm. The lens is made of glass with an index of refraction of 1.50, and the source is 1.00 m to the left of it. Where will the image be formed? [*Hint: When the radii and index are involved you should think of the Lensmaker's Formula.*]

2. [I] The radii of curvature of a bi-convex plastic spherical lens ($n = 1.50$) are measured to be 1.00 m and 3.00 m. If a tiny bright source is to be imaged 2.00 m behind the lens, where on the central axis should it be located?

3. [I] A point source 24.0 cm in front of a thin bi-convex spherical lens is imaged 24.0 cm behind the lens. If the lens is made of glass with an index of refraction of 1.50, what must be the radii of curvature of the lens's surfaces given that the lens is symmetrical (i.e., it's equi-convex)?

4. [I] A thin glass spherical lens ($n = 1.50$), which is fatter in the middle than at its edges, has one flat face and one face with a radius of curvature of 1.00 m. How far from the lens will it focus sunlight?

5. [I] A bi-convex spherical lens made of plastic ($n = 1.58$) has radii of curvature of $+1.00$ m and -1.00 m. What is its focal length?

6. [I] A glass ($n = 1.5$) bi-convex spherical thin lens has radii of curvature having magnitudes of 2.0 m and 5.0 m. Determine its focal length in air.

7. [I] A thin meniscus concave lens with an index of 1.50 has radii of 50.0 cm and 25.0 cm. Determine its focal length.

8. [I] A thin meniscus convex lens with an index of 1.50 has radii of 25.0 cm and 50.0 cm. Determine its focal length.

9. [I] We measure the radius of curvature of the face of a planar-convex lens to be 2.00 m and find its focal length to be 1.60 m. What is its index of refraction?

10. [II] When a thin spherical lens is used to focus the light from a distant star onto a screen, it's found that the screen must be placed 2.50 m from the lens in order for a sharp pointlike image to be formed. The lens is then placed 4.00 m from a tiny light-emitting diode; where should the screen now be located if it is to display the sharpest image of the diode? Is the image real or virtual?

11. [II] A point source is moved along the central axis of a thin spherical glass ($n = 3/2$) lens, and it is found that when the source is at 25.0 cm from the lens the emerging light forms a perfectly cylindrical beam traveling along the axis. Where will the image be when the source is moved out to 150 cm from the lens? Is the image real or virtual?

12. [II] Plane waves of light traveling along its central axis are incident on a thin meniscus convex spherical ($n = 3/2$) lens. The emerging light converges onto an axial point 100 cm from the lens. A point source is now placed on the axis 200 cm in front of the lens. Where will its image be located? Is the image real or virtual?

13. [II] Considering the lens in the previous problem, if the larger of the two radii of curvature is 100 cm, what is the value of the smaller one?

14. [II] A point source is placed on the central axis 50.0 cm from a planar convex thin spherical glass lens ($n = 1.50$). The light emerges from the lens as a beam parallel to the axis. What are the radii of curvature of the lens?

15. [III] Write an expression for the focal length (f_w) of a thin lens that is immersed in water ($n_w = \frac{4}{3}$) in terms of the focal length it had when it was surrounded by air (f_a). Take the index of the lens to be 1.5.

SECTION 24.4: EXTENDED IMAGERY: LENSES

SECTION 24.5: A SINGLE LENS

16. [I] THIS PROBLEM EXAMINES THE BASIC BEHAVIOR OF A THIN LENS. Using a thin positive lens having a focal length of 80.0 cm, a girl projects the image of the Sun onto a piece of paper. (a) What can you say about the object distance? (b) Compare the focal length and the object distance. At this point you should be able to describe the image in general terms. (c) Is it real or virtual? (d) Is it inverted or erect? (e) Is it magnified or minified? (f) Draw a ray diagram using an incident beam of essentially parallel light. (g) Calculate the position of the image.

17. [I] THIS PROBLEM EXAMINES THE BASIC BEHAVIOR OF A THIN POSITIVE LENS. A thin convex lens having a focal length of 100.0 cm is positioned in a vertical plane. A young man sitting 3.00 m from the lens faces it and smiles. We'd like to determine everything we can about the resulting image. (a) But first, what's the object distance? (b) Compare the focal length and the object distance. At this point you should be able to describe the image in general terms. (c) Can it be projected onto a screen? Is it real or virtual? (d) Is it inverted or erect? (e) Is it magnified or minified? (f) Draw a diagram using one ray parallel to the axis and one passing through the center of the lens. (g) Calculate the position of the image. (h) Which part of Fig. 24.15 does this situation correspond to?

18. [I] A flower is located 300 cm from the thin positive lens of a simple camera. The lens has a focal length of 50.0 mm and we want to determine everything we can about the image. Locate the image and describe it completely.

> SOLUTION: The object distance (300 cm) is much greater than the focal length (5.00 cm), and much greater than $2f$, hence the image is real, inverted, minified, and located near the focal plane.
>
> $$\frac{1}{s_o} + \frac{1}{s_i} = \frac{1}{f} \quad \text{and so} \quad \frac{1}{s_i} = \frac{1}{f} - \frac{1}{s_o} = \frac{1}{5.00 \text{ cm}} - \frac{1}{300 \text{ cm}} \quad \text{and}$$
>
> $$s_i = 300/59 = 5.1 \text{ cm}$$

19. [I] THIS PROBLEM EXAMINES THE BASIC BEHAVIOR OF A THIN NEGATIVE LENS. When a candle flame is placed 12.0 cm to the left of a thin concave lens, someone looking into the lens at the right sees an image that appears to be 4.8 cm to the left of the lens. We'd like to determine the focal length of the lens. (a) But first, what is the value of the object distance? (b) What is the image distance? Be careful about the signs! (c) Which equation depends on s_i and s_o, and can now provide f? (d) Compute the focal length. (e) What is the significance of the sign of the focal length? (f) Can the image be projected onto a screen? Is it real or virtual? (g) Draw a diagram using one ray from the top of the flame parallel to the axis and one passing through the center of the lens. (h) Is the image inverted or upright? (i) Which figure in the text most closely resembles this situation?

20. [I] A cat is 30.0 cm in front of a positive lens that has a 10.0-cm focal length. Locate and describe the image.

21. [I] A convex lens having a 60.0-cm focal length is placed 100.0 cm from a frog. Where will the image of the frog be located? Describe the image. [*Hint: The frog is between one and two focal lengths from the lens.*]

22. [I] Describe and locate the image of the flower formed when a thin converging lens having a focal length of 50.0 cm is positioned 100 cm from a potted plant.

23. [I] A light bulb is 80.0 mm in front of a positive lens with a focal length of 120 mm. Determine the location and type of image formed.

24. [I] A window on a spaceship in the Mongoian Royal Fleet is flat on one side and slightly concave on the other. As a result, it has a focal length of -10 m. What does the Universe look like through that porthole? A beacon light is 50 m from the window. How far away will it appear to someone looking out?

25. [I] A light bulb is 0.75 m in front of a thin positive lens that has a focal length of 0.25 m. (a) Completely describe the image. (b) Draw a ray diagram.

26. [I] A sharp image of a bus appears on a piece of paper held 2.5 m behind a positive eyeglass lens that has a focal length of 2.0 m. How far away is the vehicle?

27. [I] The lens in a camera has a focal length of 60.0 mm and is 100 mm from the film plane. How far away should the bug be that is having its picture taken?

28. [I] A 52.0-mm focal-length camera lens is focused on a very distant motorcycle heading toward the photographer. How far must the lens be advanced from the film plane if the driver is to be in focus when 5.00 m away?

29. [I] A lens used as a magnifying glass is found to provide an angular magnification of 2.0× when viewed by a normal eye in the relaxed state. What is its focal length?

30. [I] Holding the magnifying glass close to his eye, Dr. Watson brought the blood-stained scrap of cloth up toward it until his eye was relaxed and the image clear. If the glass has a focal length of 10 cm and the good doctor has a near-point of 25 cm, what was the magnification?

31. [I] A little magnifier marked 8× is for sale in a camera store. It is designed for examining photographs and allows light to enter from the sides through a clear cone-shaped stand that keeps the lens one focal length above whatever flat surface the thing sits on. How high above the surface should the lens be?

32. [II] THIS PROBLEM EXAMINES THE BEHAVIOR OF A THIN POSITIVE LENS. A thin convex lens having a focal length of 20.0 cm is held in a vertical plane. A screw 1.0 cm tall is placed upright on the axis of the lens, 10.0 cm to the left of it. We want to determine everything there is to know about the image. (a) What's the object distance? Compare the focal length and the object distance. At this point you should be able to describe the image in general terms. (b) Is it real or virtual? (c) Is it inverted or erect? (d) Is it magnified or minified? (e) Draw a diagram using one ray parallel to the axis and one passing through the center of the lens. Note that these rays diverge. (f) Calculate the position of the image. (g) What is the magnification? (h) What is the significance of the fact that the magnification is positive? (i) How big is the image?

33. [II] THIS PROBLEM EXAMINES THE BASIC BEHAVIOR OF A THIN

NEGATIVE LENS. A 3.0-cm tall butterfly lands 12.0 cm to the right of a thin lens, which is concave on one side and flat on the other. An entomologist looking into the left side of the lens sees the butterfly's image appearing 4.0 cm to the right of the lens. We'd like to study the image and determine the focal length of the lens. (a) But first, turn the whole thing around so the light enters from the LEFT. What are the values of the object and image distances? Be careful about the signs! (b) Compute the focal length. (c) What is the significance of the sign of the focal length? (d) Can the image be projected onto a screen? Is it real or virtual? (e) Draw a diagram using one ray from the top of the butterfly parallel to the axis, and one passing through the center of the lens. (f) Is the image inverted or upright? (g) Compute the magnification. (h) What is the significance of the sign of the magnification? (i) How big is the image?

34. [II] The image of a face is to be projected life-sized onto a screen via a bi-convex lens whose radii both equal 0.60 m. The lens is made of glass ($n = 1.5$), and the whole thing is taking place in air. (a) Compute the necessary location of the object. (b) How far must the lens be from the screen? (c) Draw a ray diagram.

35. [II] A grasshopper sitting 10 cm to the left of a convex lens sees its image on a screen 30 cm to the right of the lens. It then jumps 7.5 cm toward the lens. (a) Where will its image be now? (b) Describe the image in both instances. (c) Draw a ray diagram of the situation after the jump.

36. [II] Design the lens for a simple 35-mm slide projector that will cast an image on a screen 10 m away that is enlarged 100 times.

37. [II] A photographer wishes to take a picture of his pet chicken Fred, who happens to have a fine face 5.0 cm high. While standing 2.0 m away, he selects a lens that will fill the film (24 mm top-to-bottom) with Fred's poultry physiognomy. What lens should be used?

38. [II] Suppose you wanted to take a picture of the Moon (which has a diameter of 0.273 times that of Earth and is 3.84×10^8 m away) using a 35-mm camera with a normal lens having a 50-mm focal length. How large will the image be on the film? The diameter of the Earth is about 1.27×10^7 m.

39. [II] Using Fig. P39, show that the equation for the transverse magnification produced by a pinhole is the same as for a lens. What does this similarity tell us about the pinhole-film plane distance and the magnification?

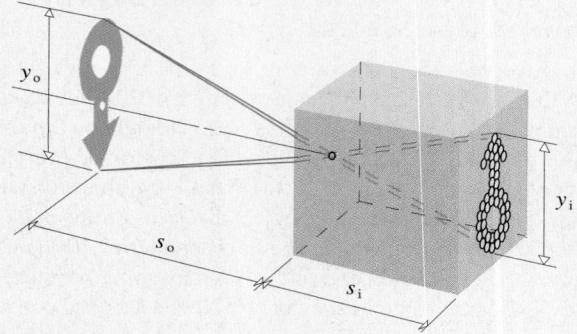

Figure P39

40. [II] A sewing needle is 3.00 cm tall and located in a plane perpendicular to the central axis of, and 60.0 cm from, a thin positive lens. The sharp image of the needle formed on a screen is 1.50 cm long. What is the focal length of the lens?

41. [III] The image projected by an equi-convex lens ($n = 1.50$) of a 5.0-cm-tall frog standing 0.60 m from a screen is to be 25 cm high. Compute the necessary radii of the lens.

42. [III] A thin double convex glass lens (with an index of 1.56) surrounded by air has a 10-cm focal length. If this lens is placed under water (having an index of 1.33) 100 cm beyond a small fish, where will the guppy's image be formed?

43. [III] A home-made TV projection system uses a large positive lens to cast the image of the screen onto a wall. The final picture is enlarged three times and, although rather dim, is nice and clear. If the lens has a focal length of 60 cm, what should be the distance between the screen and the wall? Why use a large lens? How should we mount the set with respect to the lens?

SECTION 24.6: THIN-LENS COMBINATIONS

44. [I] What is the power of a lens if it forms a real image on a screen 100 cm behind it of an object that is 100 cm in front of it?

45. [I] A person stands 20.0 m from a thin positive lens having a focal length of 0.50 m. Behind this lens a distance of 2.00 m, and centered on the same axis, is another positive thin lens having a focal length of 1.00 m. Where will the image of the person appear? Describe it. Draw a ray diagram. [*Hint: The intermediate image is the object for the second lens.*]

46. [I] Two positive thin lenses with focal lengths of 25.0 cm and 40.0 cm are on a common axis and separated by 130 cm. A 5.00-cm tall chess piece is placed 50.0 cm in front of the first lens. Where will its image appear? Draw a ray diagram.

47. [I] Sunlight traveling along the central axis of a two-lens system first impinges on a positive lens with a 50.0-cm focal length. The second positive lens has a focal length of 60.0 cm and is located 170 cm behind the first. Where should we place a screen if we are to see a clear image of the Sun? Draw a ray diagram.

48. [I] A horizontal negative lens has a power of -1.0 D. A clear liquid is poured into the concavity, and it is found that the combined power is now -0.75. What is the power of the liquid lens?

49. [I] What is the power of a thin equi-convex lens ($n = 3/2$) if both its radii of curvature are found to be 100 cm?

50. [I] A negative lens with a focal length of magnitude 15.0 cm is in contact with a positive lens having a focal length of 24.0 cm. What is the focal length of the combination?

51. [I] We wish to spread a narrow parallel laserbeam out into a diverging cone of light so that it makes a large blotch on a nearby screen. At our disposal are two positive lenses, each with a focal length of 20 cm. If one lens is placed in the beam and the resulting blotch is still too small, will it help any to put the other lens in right up against the first? What is the combined focal length?

52. [I] An optical system consists of three lenses with focal lengths of $+10$ cm, $+20$ cm, and $+5.0$ cm. The first and second of these are separated by 30 cm, the third and fourth by 5.0 cm. If parallel light is shone into the first lens, how far from the third will it be brought to a focus by the system?

53. [I] A rather expensive well-corrected (for aberrations) lens consists of three simple lenses of focal lengths $+10$ cm, -20 cm,

and +5.0 cm all glued together. What is the combined focal length? Can it form real images?

54. [I] A lens positioned 1.00 m from a light bulb produces a sharply focused image of the filament on a screen 33 cm beyond the lens. What is the refractive power of the lens?

55. [I] The near-point of a person's eye is 100 cm away rather than a more desirable 25.4 cm. What contact lens should be prescribed?

56. [I] The far-point of a person's eye is 100 cm away rather than a more desirable ∞. What contact lens should be prescribed?

57. [I] Someone wearing corrective contact lenses having a focal length of −5.0 m can see quite normally. Determine this person's unaided far-point.

58. [I] A compound microscope is made with a 20× objective and a 5× eyepiece. What is the total magnification of the device?

59. [I] Two lenses used to form a homemade microscope are mounted in a tube with a separation of 10.0 cm. If the objective has a focal length of 10 mm and the eyepiece has a focal length of 30 mm, what is the so-called tube length?

60. [I] Typically, a good eyepiece will have a focal length of approximately an inch. Suppose we have one with a focal length of 2.5 cm and we wish to make a 10× astronomical telescope with it. What objective shall we use, and how long will the scope end up?

61. [I] Suppose we wanted to make a telescope to look at the stars. At our disposal is an old eyeglass lens (+1.00 D) donated by hyperopic Aunt Jane and a little magnifier having a focal length of 3.0 cm that came with a stamp collection we got from someone. How long should the tube be for relaxed viewing? What magnification can be achieved?

62. [II] THIS PROBLEM EXAMINES A COMBINATION OF TWO THIN LENSES. A 1.0-cm tall candle flame is 15.0 cm to the left of a thin positive lens having a focal length of 10.0 cm. Twenty-five centimeters to the right of this lens is a second thin concave lens with a focal length of −7.5 cm. We'd like to determine the image generated by the system. (a) Determine the intermediate image distance as formed by the first lens. Note that the object is between one and two focal lengths. (b) What is the magnification produced by the first lens? (c) Describe the intermediate image completely. (d) Using the intermediate image as the object for the second lens, determine that object distance. What is its sign? (e) What is the final image distance? (e) Can the image be projected onto a screen? Is it real or virtual? (f) Draw a diagram. Use a ray from the top of the intermediate image that goes back through the center of the second lens, as in Fig. 24.25. (g) Is the final image inverted or upright? (h) Compute the magnification of the second lens. (i) Compute the magnification of the system. (j) How big is the final image?

63. [II] Two thin lenses are separated by 4.0 cm. The first lens on the left has a focal length of +5.0 cm, and the second (to the right of the first) has a focal length of −3.0 cm. A little light-emitting diode (LED) 0.30 cm tall is placed on the central axis 10.0 cm to the left of the positive lens. (a) Locate and describe the intermediate image. (b) Locate and describe the final image. (c) What is the total magnification of the system? (d) How big is the final image? (e) Draw a ray diagram.

SOLUTION: $1/s_{o1} + 1/s_{i1} = 1/f_1$ and so $1/s_{i1} = (1/5.0 \text{ cm}) − (1/10.0 \text{ cm})$ and $s_{i1} = 10$ cm. The magnification M_{T1} is −1.0. The inter-

mediate image is real ($s_{i1} > 0$), inverted ($M_T < 0$), and life size ($M_T = 1$). It is 6.0 cm behind the second lens and is a virtual object for that lens: $s_{o2} = −6.0$ cm. $1/s_{o2} + 1/s_{i2} = 1/f_2$ and so $1/s_{i2} = (1/−3.0 \text{ cm}) − (1/−6.0 \text{ cm})$ and $s_{i2} = −6.0$ cm. (c) The magnification M_{T2} is −1.0 and so $M_T = M_{T1}M_{T2} = 1.0$. The final image is virtual ($s_{i2} < 0$), upright ($M_T > 0$), and life size. (d) Final image is 0.30 cm tall. (e) Figure P63.

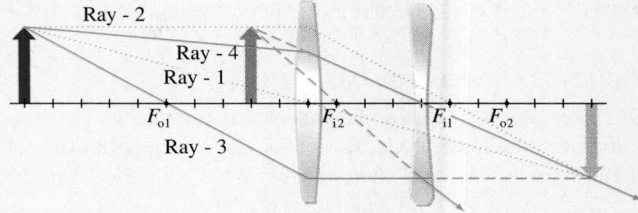

Figure P63

64. [II] A combination lens consists of a positive lens with a focal length of +0.30 m located 10 cm in front of a negative lens with a focal length of −0.20 m. Determine the location and magnification of the image of an object 30 cm in front of the first lens by finding the effect of each lens in turn.

65. [II] The 20-cm-diameter image of the face of a clock appears on a screen 1.0 m from a positive lens. A negative lens is then placed in the path 90 cm behind the first lens, and the image moves out an additional 10 cm beyond where it was. Determine the focal length of the second lens. How big is the final image?

66. [II] George has been nearsighted since he was eighteen; and now that he's fifty-five, his far-point has moved in to 3.00 m. But even more annoying, his near-point has recently migrated out to 45 cm. Assuming that both his eyes are the same, prescribe bifocals with the tops for distance and the bottoms for reading to be worn at 2.0 cm from the cornea. Incidentally, it was Ben Franklin who made the first pair of bifocals (by sawing two lenses in half).

67. [II] A person who is nearsighted has a far-point at 1.00 m and a near-point at 25.0 cm. (a) What corrective lens should this individual wear if it's to be mounted 15.0 mm from the eye? (b) Where is the near-point when glasses are worn?

68. [II] Suppose we have two lenses with focal lengths of 2.0 cm and 2.0 mm. How should they be arranged to make a microscope to view, with a relaxed eye, an object 2.5 mm from the front lens? What is the separation between lenses? What is the magnification of the device?

69. [II] A compound microscope formed of two lenses, a 20× objective and a 10× eyepiece, is adjusted for viewing by a relaxed eye. It has a standard tube length of 160 mm. (a) What is the total magnification of the device? (b) What is the focal length of each lens? (c) Compute the object distance.

70. [III] An equi-convex thin lens L_1 is placed in close contact with a thin negative lens L_2, whereupon the combination has a focal length of 0.50 m in air. If the lenses are made of glass with indices of 1.50 and 1.55, respectively, and if the focal length of L_2 is −0.50 m, compute the radius of each surface. Note that a convex surface of L_1 fits precisely against a concave surface of L_2.

71. [III] Mary Lou got her first pair of reading glasses (+2.0 D)

when she was forty-eight, in 1979. Now, in order to peruse her mail (still wearing those spectacles 2.0 cm down on her nose), she finds she must hold it 80 cm away from her eyes. Prescribe new glasses for her.

72. [III] A nearsighted person has a far-point of 200 cm. What power eyeglass lens should be worn 2.0 cm from the cornea? What contact lens is equivalent to this? Show that both have the same far-point. Verify that Eq. (24.3) agrees with your results.

SECTION 24.7: CURVED MIRRORS

73. [I] The curved surface of a planar concave spherical lens is measured to have a radius of 66.0 cm. It is covered with a thin film of metal and used as a front-silvered mirror. What is its focal length? [*Hint: Always have the light enter from the left.*]

74. [I] A convex mirror has a radius of curvature whose magnitude is 0.50 m. What is its focal length?

75. [I] A polished 50-cm-diameter steel ball in the hands of a statue reflects the scene around it. Determine the ball's focal length.

76. [I] We wish to convert light from a candle flame into a parallel beam using an inexpensive mirror 30 cm away from it. Design the appropriate arrangement and draw a ray diagram.

77. [I] If an object 200 cm from the vertex of a spherical concave mirror is imaged 400 cm in front of the mirror, what is the latter's focal length?

78. [I] A 15-cm-tall Teddy bear stands 60 cm from the vertex of a concave mirror having a radius of curvature of 60 cm. Determine the resulting image and describe it in detail. Draw a ray diagram illustrating the phenomenon.

79. [I] A display lamp having a bright 5.0-cm-long vertical filament is positioned 30 cm from a concave mirror that projects the bulb's image onto a wall 9.0 m from the vertex. (a) What is the radius of curvature of the mirror? (b) How big is the image?

80. [I] Suppose you had a spherical mirror with a 0.20-m radius. If you wished to project the image of a candle flame onto a piece of paper 1.10 m away, where should the candle be located? Describe the image, giving the magnification as well. Draw a ray diagram.

81. [I] If your nose is 20 cm from a convex spherical mirror having a radius of 100 cm, locate its image. What will that image look like? Draw a ray diagram. [*Hint: A convex mirror has a negative focal length.*]

82. [I] A polished metal sphere has a diameter of 80.0 cm. A mouse, whose face is 100 cm from the vertex of the spherical surface, looks directly at it. Locate and describe the mouse's image. Draw a ray diagram.

83. [I] While talking to someone who is wearing sunglasses you see yourself reflected in their lenses. Your face is 0.10 m from the glasses, which have a radius of curvature of 1.00 m. Locate and describe the image. Draw a ray diagram.

84. [I] The cornea of the human eye behaves like a little convex spherical mirror when seen from nearby. Suppose you look at yourself reflected in someone's eye a distance of 20 cm away. Taking the radius of curvature of the cornea to be 8.0 mm, what will your image look like? Where will it be located?

85. [II] THIS PROBLEM EXAMINES THE FUNCTIONING OF A CONCAVE

MIRROR A technician wants to project the image of a 4.0 mm tall light-emitting diode (LED) onto a screen using a spherical concave mirror. The screen is to be 90.0 cm to the left of the mirror and the diode is to be halfway between it and the mirror. Design the mirror and describe the image completely. (a) Since we don't already have it, the first thing to compute is the focal length. Is *f* positive or negative? At this point you should be able to describe the image in general terms. (b) Is it real or virtual? (c) Is it inverted or upright? (d) Is it magnified or minified? (e) Draw a ray diagram. (f) What is the magnification? (g) How tall is the image? (h) Determine the radius of curvature of the mirror.

86. [II] THIS PROBLEM EXAMINES THE FUNCTIONING OF A CONVEX MIRROR. An old convex spherical mirror, having a radius of curvature of 50.0 cm, hangs on a wall in a schoolroom. A kid 100 cm tall stands 200 cm away directly in front of the mirror. We want to study the resulting image seen by the youngster looking into the mirror. (a) Since we don't already have it, the first thing to compute is the focal length. Is *R* positive or negative? Is *f* positive or negative? At this point you should be able to describe the image in general terms. (b) Is it real or virtual? (c) Is it inverted or upright? (d) Is it magnified or minified? When would the image be essentially life size? Check your answers with Table 24.5. (e) Draw a ray diagram. (f) Calculate the position of the image. (g) What is the magnification? (h) How tall is the image?

87. [II] An Earth satellite in the shape of a 2.0-m ball is orbiting at an altitude of 500 km. It is being tracked by a telescope having a spherical concave mirror with a 1.0-m radius of curvature. How big is the image of the craft formed by the mirror?

88. [II] We wish to design an eye for a robot using a concave spherical mirror such that the image of a 1.0-m-tall object 10 m away fills its 1.0-cm-square photosensitive detector (which is movable for focusing purposes). Where should this detector be located with respect to the mirror? What should be the focal length of the mirror? Draw a ray diagram.

89. [II] You are herewith requested to design a little dentist's mirror to be fixed at the end of a shaft for use in someone's mouth. The requirements are (1) that the image be right-side-up as seen by the dentist and (2) that when held 1.5 cm from a tooth, the mirror should produce an image twice life-size.

90. [II] Prove that with a spherical mirror of radius *R*, an object at a distance s_o will result in an image that is magnified by an amount

$$M_T = \frac{R}{2s_o + R}$$

91. [II] With the results of Problem 90 in mind, suppose that we have a concave mirror with a radius of curvature of 60 cm, which is forming an image of an object 2.4 m away. What will be the resulting magnification? Describe the image. Draw a ray diagram.

92. [II] A keratometer is a device used to measure the radius of curvature of the cornea of the eye, which is useful information when fitting contact lenses. In effect, an illuminated object is placed a known distance from the eye, and the reflected image off the cornea is observed. The instrument allows the operator to measure the size of that virtual image. Suppose, then, that the magnification is found to be 0.037× when the object distance is set at 100 mm. What is the radius of curvature?

93. [II] Consider a spherical mirror. Show that the locations of the object and image are given by

$$s_o = \frac{f(M_T - 1)}{M_T} \quad \text{and} \quad s_i = -f(M_T - 1)$$

94. [II] Looking into the bowl of a spherical soup spoon, someone 25 cm away sees their image reflected with a magnification of -0.064. Determine the radius of curvature of the spoon.

95. [III] A large, upright, convex spherical mirror in an amusement park is facing a plane mirror 10.0 m away. A youngster 1.0 m tall standing midway between the two sees herself twice as tall in the plane mirror as in the spherical one. In other words, the angle subtended at the observer by the image in the plane mirror is twice the angle subtended by the image in the spherical mirror. What is the focal length of the latter?

96. [III] The telescope depicted in Fig. P96 consists of two spherical mirrors. The larger (which has a hole through its center) has a radius of curvature of 2.0 m and the smaller of 60 cm. How far from the smaller mirror should the film plane be located if the object is a star? What is the effective focal length of the system?

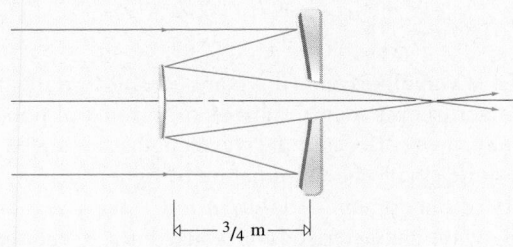

Figure P96

97. [III] Figure P97 shows the arrangement of a classic illusion that works surprisingly well. A person looks through a window into the box and sees a glowing bulb in a socket. The bulb is turned off and vanishes, leaving only the socket seen in roomlight entering the window. Explain what is happening. If the bulb-mirror distance is 1.0 m, compute the mirror's radius of curvature.

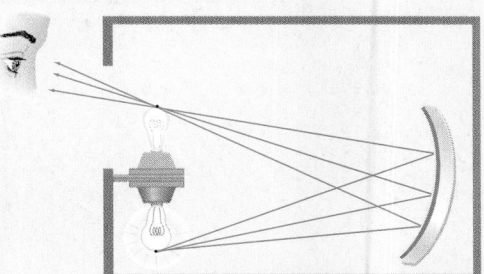

Figure P97

98. [III] Starting with the equation $y^2 + (x - R)^2 = R^2$ for a circle whose center is displaced by a distance R from the coordinate origin, solve for x and expand the appropriate term in a binomial series. Compare the first term with the equation of a parabola having its vertex at the origin, $y^2 = 4fx$.

Chapter 25
Physical Optics

*L*ight is electromagnetic, and it is wavelike, and (if we are careful to remember that what we usually observe are the macroscopic manifestations of torrents of photons) we can say that *light is an electromagnetic wave.* Because the wavelengths associated with the visible region of the spectrum are quite small, the wave nature of light can sometimes be neglected. Still, there are a variety of phenomena—*polarization*, *interference*, and *diffraction*—that reveal the underlying wave characteristics of radiant energy, and these form the study of **physical optics**.

Polarization

We can wiggle a length of rope up and down in a vertical plane while the wave progresses horizontally (Fig. 25.1). The disturbance remains in that fixed plane, and we say that the wave is plane-polarized. When viewed end-on, the vibration appears to be along a line perpendicular to the propagation direction, and the situation is also referred to as *linear polarization*. The plane in which such a transverse wave oscillates is the *plane of polarization*, and it certainly need not be vertical—any orientation is equally possible. When the $\vec{\mathbf{E}}$-field oscillates in a plane, the TEM (transverse electromagnetic) wave is **plane-polarized**.

STUDY GUIDE

Physical Optics is all about the behavior of light as a wave. The subject is historically divided into the study of the phenomena of polarization (p. 888), interference (p. 899), and diffraction (p. 910). Practically speaking, the discipline allows us to understand a variety of things from the inherent limitations of telescopes and microscopes to the colors of peacock feathers. But more fundamentally, it establishes that light, whatever it is, is wavelike.

Today in this era of lasers and computers, there isn't an area in all of science that doesn't benefit in some way from the insights of Physical Optics. Just look at the screen on your calculator: it works by polarizing light (p. 893)

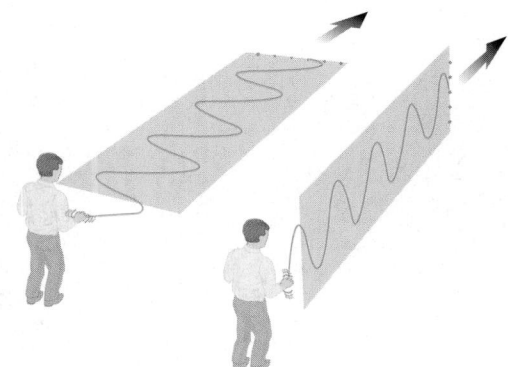

Figure 25.1 Plane-polarized waves. The rope oscillates in a plane. Notice how the end point vibrates along a line as the wave advances. As a result, the wave is also said to be linearly polarized.

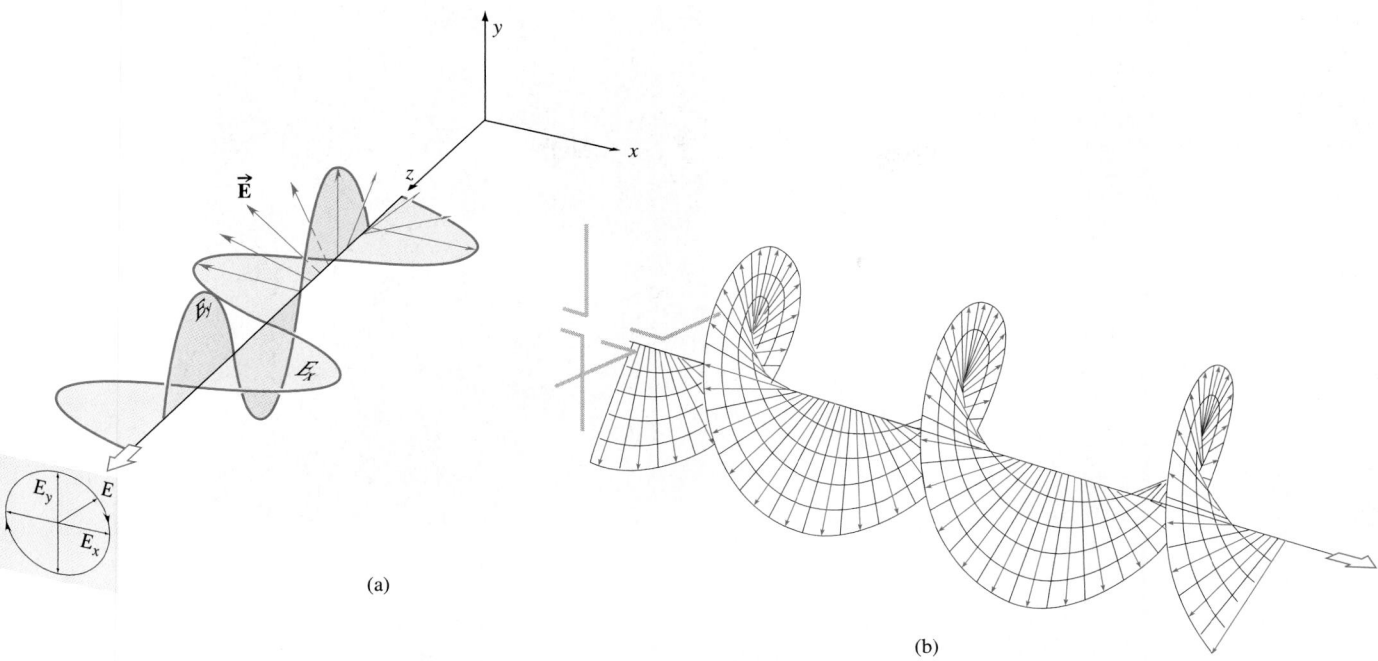

(a)

(b)

Figure 25.2 Right-circular light. As the wave advances, the electric field vector $\vec{\mathbf{E}}$ rotates clockwise once around per wavelength. The magnitude of $\vec{\mathbf{E}}$ is constant. The effect arises because the components of the field, $\vec{\mathbf{E}}_x$ and $\vec{\mathbf{E}}_y$, are out-of-phase by $\lambda/4$.

It's possible for the $\vec{\mathbf{E}}$-field to swing around in a circle, in which case the wave is said to be **circularly polarized**. With circular light, the electric-field vector remains constant in magnitude while it revolves once around with every advance of one wavelength (Fig. 25.2). Each sustained configuration is known as a *state of polarization*, and **plane-polarized** lightwaves are often referred to as $\mathscr{P}$**-state light**. {For more on the subject click on **Circular Light** under **Further Discussions** on the **CD**.}

25.1 Natural Light

The light coming from an ordinary (nonlaser) source is a polychromatic jumble of overlapping emissions from a tremendous number of more or less independent atoms. The stream of photons manifests itself on a macroscopic level as an electromagnetic wave. And so we

LIGHT IS A TRANSVERSE WAVE

A curious discovery made in 1669 led to one of the great insights into the nature of light. A newly found crystal, now called calcite, was observed to have the remarkable ability to produce double images of everything seen through it. That strange separation, or **polarization** of light, was first studied in depth by Huygens, who explained some aspects of it via his wave theory. In the early 1800s, Augustin Fresnel, joined by Arago, conducted a series of experiments to see if polarization had any effect on the process of interference. Their positive results were utterly bewildering because they, like Huygens, believed light was a longitudinal wave in the aether. For the next few years, Fresnel in France, allied with Thomas Young in England, wrestled with that stubborn problem until finally Young (in 1817) had the crucial revelation: **light is a transverse wave** (something Robert Hooke had suggested a century before).

Many animals can see variations in polarization just as we see variations in color. The pygmy octopus is one such creature—it's pictured here in a false color polarization image. The varying pattern of polarized light reflected from its surface suggests it might be "communicating" with other pygmy octopuses (the way birds display color).

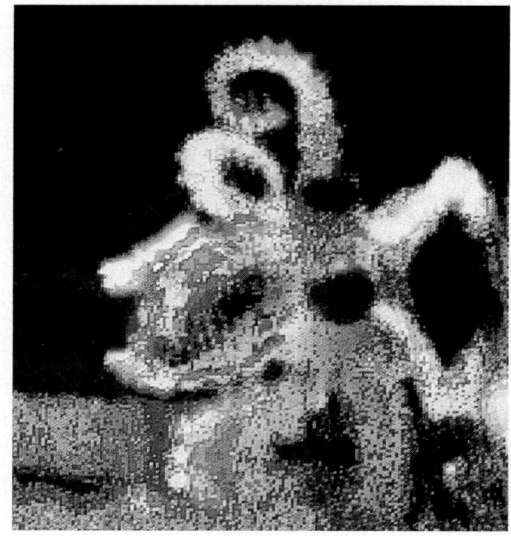

can imagine that each atom radiates a tiny polarized photon-wavetrain lasting less than $\approx 10^{-8}$ s, and these all arrive with no particular phase relationship or orientation, one to the other. The resultant field is the sum of all the separate wavetrain contributions at that location and is constantly changing as new photons arrive and others sweep by. Thus, the net optical field has some polarization at every point, at every instant, but that polarization varies exceedingly rapidly in a totally random fashion. Because the state of polarization of the light is sustained for less than $\approx 10^{-8}$ s, that state is essentially undetectable, and we refer to the light as **unpolarized** or **natural**. This is the case even though, strictly speaking, there is no such thing as unpolarized light. Often, the light we experience in the environment is a blend of polarized and unpolarized known as *partially polarized*. Ordinary sky light, the light reflected from objects, and the light transmitted through windows are all partially polarized.

There are several ways to conceptually visualize unpolarized light. One of the easiest is to think of it as a superposition of many differently oriented plane-polarized waves of different amplitudes, all changing rapidly and randomly (Fig. 25.3a). In time, the resultant E-field points every which way, swiftly jumping from one orientation and amplitude to another.

Any electric field vector can be resolved into two mutually perpendicular components

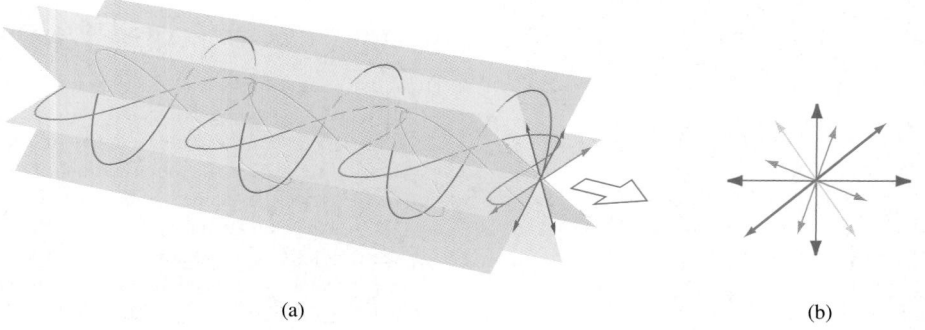

(a)   (b)

Figure 25.3 (a) Natural light is a jumble of random, rapidly changing fields having a range of frequencies. (b) Looking toward the source, one sees a resultant field oscillating, first in one direction, then another, each lasting for a fraction of a period before $\vec{\mathbf{E}}$ changes to a new random orientation.

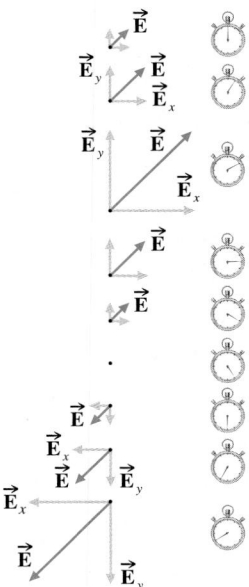

Figure 25.4 A sequence in time showing the oscillation of a linearly polarized light field. $\vec{E}$ is always along a 45° line in the first and third quadrants. The two component fields $\vec{E}_x$ and $\vec{E}_y$ are in-phase—when one is positive, the other is positive. As the wave sweeps past a fixed point in space, the field oscillates in time, as shown.

A pair of crossed Polaroids. The electric field passed by the first polarizer is perpendicular to the transmission axis of the second polarizer.

(which we take for convenience as horizontal and vertical). For example, if at some instant the electric field vector points up and to the right, it can be thought of as the resultant of two smaller constituent fields, one to the right and one up (Fig. 25.4). As the tilted field oscillates, it can be resolved into—and replaced by—the two perpendicular fields that oscillate at the same rate in-step with each other (Fig. 25.5).

The natural-light field depicted in Fig. 25.3b can similarly be resolved into horizontal and vertical components that change from moment to moment as the vector changes. The result at any instant is a net horizontal field and a net vertical field. ***We can think of the unpolarized jumble as equivalent to two independent, equal-amplitude $\mathcal{P}$-states varying rapidly and randomly along any two perpendicular axes.*** Each of these carries half the total irradiance (I_0) of the light.

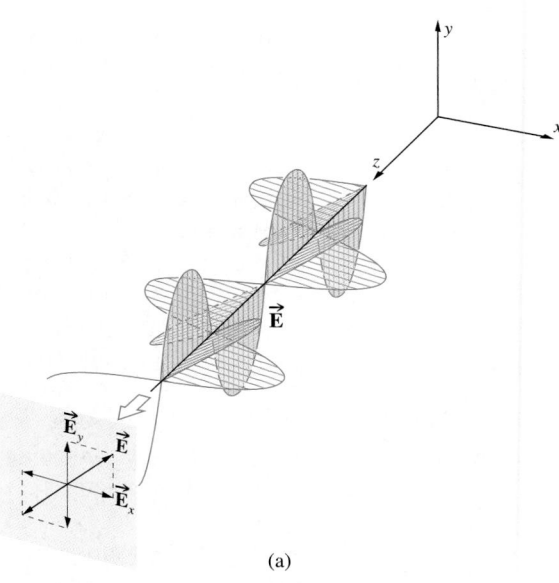

(a)

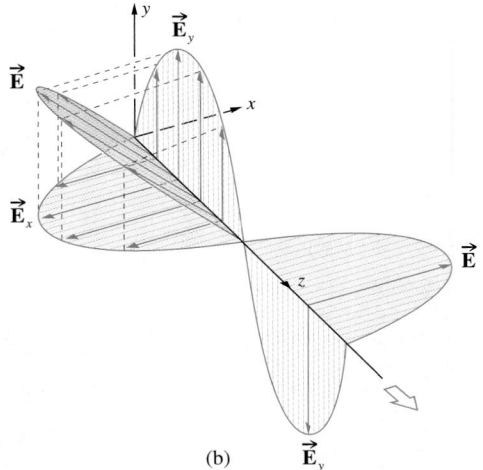

(b)

Figure 25.5 A linearly polarized wave. In Fig. 25.4, we examined the disturbance at a fixed point in space as time progressed. (a) Here, we examine the disturbance spread out in space at a given instant in time. (b) The resultant field at any instant is $\vec{E}$, the sum of $\vec{E}_x$ and $\vec{E}_y$.

Figure 25.6 Natural light incident on a linear polarizer. The transmission axis of the polarizer is tilted at an angle θ, and only the $\vec{\mathbf{E}}$-field parallel to θ is passed. In this case, the polarizer absorbs half the light that enters it.

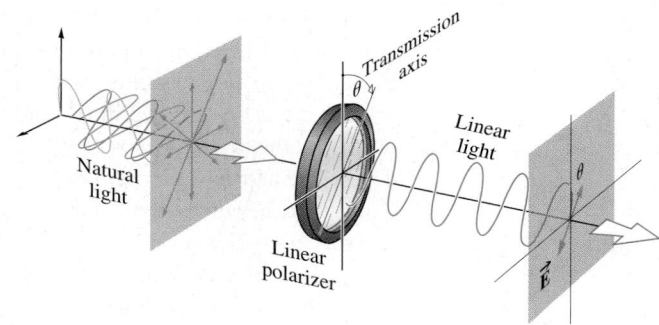

25.2 Polarizers

A device that takes an input of natural light and transforms it into an output of polarized light is a **polarizer**. Natural light is a superposition of two independent, perpendicular $\mathcal{P}$-states. Anything that separates these two equal-irradiance components, discarding one and passing on the other, will produce a beam of light that has its $\vec{\mathbf{E}}$-field fixed in a plane, a beam that is plane-polarized. Such a device is a *linear polarizer*. **Ideally, if natural light of irradiance I_0 is incident on a linear polarizer, $\mathcal{P}$-state light of irradiance $\frac{1}{2}I_0$ will be transmitted**.

Before we go on, we must establish a means of experimentally verifying that a device is in fact a linear polarizer. When natural light is incident on the ideal linear polarizer of Fig. 25.6, only $\mathcal{P}$-state (plane polarized) light emerges. It has an orientation parallel to a specific direction called the **transmission axis** of the polarizer. This is not literally a single axis but a direction—any line on the polarizer parallel to that axis is also a transmission axis. Only the component of the incident electric field parallel to the transmission axis will emerge as the transmitted light. An oscillating electric field $\vec{\mathbf{E}}$ tilted over at an angle θ will come out of the polarizer.

If the polarizer is revolved in its own plane about the central axis, changing θ in the process, the transmission axis will rotate along with it, as will the emerging $\vec{\mathbf{E}}$-field. Nonetheless, the reading of the detector that corresponds to the transmitted irradiance will be constant because of the symmetry of natural light—each transmitted $\mathcal{P}$-state carries the same average amount of energy. So overlooking any losses due to reflection, I_0 comes into the polarizer and $I_1 = \frac{1}{2}I_0$ emerges. The human eye cannot distinguish the various polarization states and will see only a constant irradiance (I_1) as the polarizer is revolved.

Now, suppose we introduce a second identical polarizer called an **analyzer**, whose transmission axis is vertical (Fig. 25.7). The first polarizer transmits a wave of amplitude E_{01} tilted over at an angle θ. Only the vertical component of that wave, namely,

Figure 25.7 A linear polarizer and analyzer—Malus's Law. Natural light enters the polarizer and emerges in a $\mathcal{P}$-state at an angle θ. The analyzer passes a fraction of that light, such that $E_{02} = E_{01} \cos \theta$. The irradiance of the natural light incident on the polarizer is I_0, and it, in turn, transmits an irradiance $I_1 = \frac{1}{2}I_0$. This is incident on the analyzer, and it transmits an irradiance $I = I_1 \cos^2 \theta$.

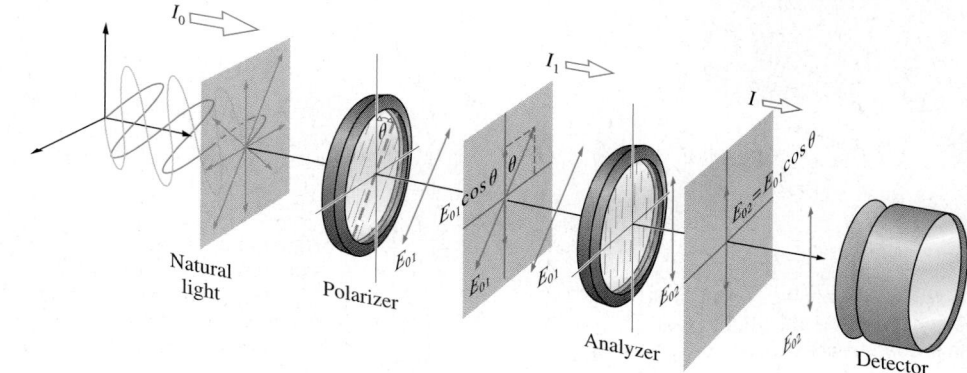

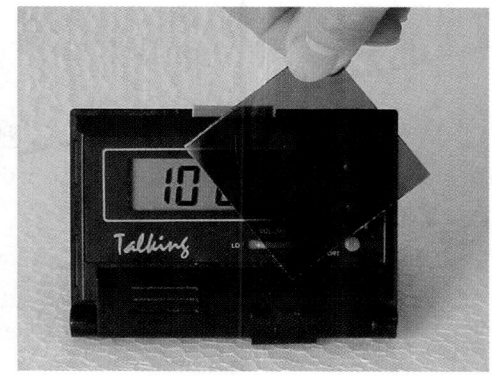

By rotating a linear polarizer in front of a liquid crystal display we can see the numbers appear and disappear. The light coming from the number display is linearly polarized. Try it with your calculator.

$E_{01} \cos \theta$, will be passed by the analyzer because its transmission axis is vertical. (Note that θ is the angle between the transmission axes of the polarizer and analyzer.) The irradiance reaching the detector is proportional to the square of the amplitude ($E_{02} = E_{01} \cos \theta$) of the wave coming out of the analyzer, the second polarizer,

[natural light] $I = \frac{1}{2}c\varepsilon_0 E_{02}^2 = \frac{1}{2}c\varepsilon_0 (E_{01} \cos \theta)^2 = (\frac{1}{2}c\varepsilon_0 E_{01}^2) \cos^2 \theta = I_1 \cos^2 \theta$

where you recall that the irradiance leaving the first polarizer is I_1. Slowly revolving the analyzer about the central axis varies θ, and the transmitted irradiance changes as the cosine squared; it varies from its maximum value ($I = I_1 = \frac{1}{2}I_0$ when $\theta = 0$ and the two transmission axes are parallel) to its minimum value ($I = 0$ when $\theta = 90°$ and the axes are said to be **crossed**). The expression

$$I = I_1 \cos^2\theta \qquad (25.1)$$

is known as **Malus's Law**, having first been published in 1809 by Étienne Malus, military engineer and captain in the army of Napoleon. Using the setup of Fig. 25.7 and Malus's Law, we can test polarizers to see if they are linear by simply revolving one in front of the other. In fact, if we analyze a beam of light through a revolving polarizer and find that it varies in accord with Eq. (25.1), going completely dark and then bright every 90°, the beam must be $\mathscr{P}$-state light. There are many kinds of linear polarizers, among them several that operate by the dissipative absorption of one of the two perpendicular $\mathscr{P}$-state components of the incident light. Of these, we will examine two types—wire grids and Polaroids—the first only because it explains the second.

> If I_0 is the amount of natural light incident on an ideal linear polarizer, $I_1 = \frac{1}{2}I_0$ will be transmitted. A second linear polarizer, with its transmission axis at θ w.r.t. the first, will then transmit a portion of this light (I) given by **Malus's Law**.

STUDY GUIDE

You can think of Eq. (25.1) as describing what happens to an incident beam of linear light of irradiance I_1 upon passing through a linear polarizer whose transmission axis makes an angle θ with the direction in which the light is polarized (i.e., the direction of the light's electric field).

Example 25.1 **[II]** A beam of unpolarized light with an irradiance of 1000 W/m² impinges on an ideal linear polarizer whose transmission axis is vertical. The light is studied using a second ideal linear polarizer, and it is found that the final emerging beam has an irradiance of 250 W/m². (a) How much light leaves the first polarizer? (b) What is the orientation of the second polarizer?

Solution The mention of "unpolarized light" and two successive "linear polarizers" should bring to mind Malus's Law. (1) TRANSLATION—Natural light of known incident irradiance passes through two linear polarizers and emerges with a specified irradiance; determine (a) the irradiance passing through the first polarizer and (b) the angle between the transmission axes. (2) GIVEN: $I_0 = 1000$ W/m² incident and $I = I_2 =$

continued

250 W/m². FIND: (a) I_1 and (b) θ. (3) PROBLEM TYPE—Physical optics/polarization/linear light. (4) PROCEDURE —(a) If natural light is incident with an irradiance I_0, half of that will be "eaten" by the first polarizer. (b) Use Malus's Law to determine θ. (5) CALCULATION—(a) The first polarizer must discard half the light if it is to pass only a $\mathscr{P}$-state:

$$I_1 = \tfrac{1}{2}I_0 = \boxed{500.0 \text{ W/m}^2}$$

(b) Because the second polarizer passes $I_2 = 250$ W/m² $= \tfrac{1}{2}I_1$, it follows from Malus's Law that

$\cos^2 \theta = \tfrac{1}{2}$. Consequently,

$$\cos \theta = \sqrt{\tfrac{1}{2}}$$

and $\boxed{\theta = 45.0°}$; the second polarizer is 45° from the vertical.

Quick Check: Substituting back into Eq. (25.1), $I_2 = (500$ W/m²$) \cos^2 45° = 250$ W/m².

{For more worked problems click on WALK-THROUGHS *in* CHAPTER 25 *on the CD.}*

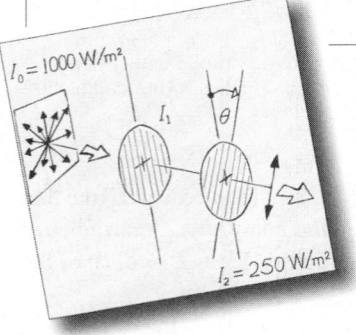

The Wire-Grid Polarizer

The **wire-grid polarizer** consists of a grid of parallel conducting wires each spaced less than a wavelength apart. The electric field of an incident unpolarized electromagnetic wave can again be imagined resolved into two independent, perpendicular components—one parallel to the wires and the other transverse to them (Fig. 25.8). The vertical component of the field effectively drives the free electrons within the wires. Oscillating, up-and-down currents result along the length of each conductor. These currents dissipate some of the wave's energy via joule heating, thereby diminishing the vertical field. At the same time, the accelerating charges reradiate like tiny dipoles. But these emissions are out-of-phase with the incoming wave, thus canceling most of the remaining vertical field in the forward direction. The reradiation in the backward direction constitutes the reflected wave and carries off the remainder of the energy. By contrast, the currents oscillating across the wires are very restricted, and there is hardly any effect on the transverse electric field, which passes through the grid almost the same as it arrived.

The transmission axis of the grid is perpendicular to the wires. It's an extremely common error to assume that the field somehow slips between the wires and is vertically polarized—it doesn't and it isn't! Such grids have been made to operate in the visible (one had 2160 wires per mm) and infrared, though they are much easier to fabricate for microwaves.

Polaroids

In 1928, Edwin H. Land, then a nineteen-year-old undergraduate at Harvard College, invented the first dichroic *sheet polarizer*. That was the humble start of the now-famous Polaroid Corporation. In 1938, Land devised the much improved *H-sheet Polaroid*, which is now easily the most widely used linear polarizer. A sheet of clear polyvinyl alcohol is heated and stretched, thereby aligning—in nearly parallel rows—its long-chain hydrocar-

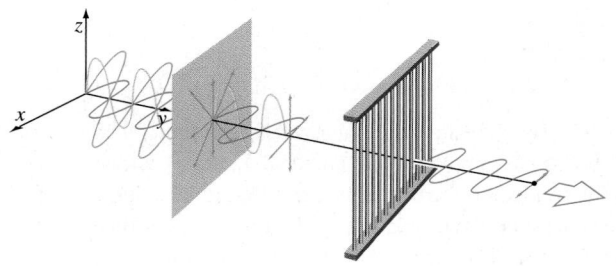

Figure 25.8 Natural light incident on a wire-grid polarizer. The unpolarized light can be represented as two uncorrelated $\mathscr{P}$-states, one horizontal and one vertical. The grid absorbs the vertical E-field and passes the horizontal $\mathscr{P}$-state.

bon molecules. The sheet is then dipped into a dye solution rich in iodine. The iodine atoms impregnate the plastic and attach to each polymeric molecule, coating it and effectively forming a conducting chain along it. The conduction electrons associated with the iodine can now move down each chain, from one atom to the next, as if the coated molecule were a long, thin microscopic wire. The result is essentially a wire-grid polarizer with its transmission axis transverse to the aligned molecules. In natural light, each sheet looks gray because it absorbs roughly half the incident light.

Example 25.2 **[II]** An unpolarized light beam of 800 W/m^2 is incident on an ideal pair of crossed linear polarizers. Now a third such polarizer is inserted between the other two with its transmission axis at 45° to that of each of the others. Determine the emerging irradiance (a) before and (b) after the insertion of the third polarizer and explain what's happening.

Solution The mention of "unpolarized light" and two successive "linear polarizers" should bring to mind Malus's Law. (1) TRANSLATION—Natural light of known incident irradiance passes through three linear polarizers where the front and back ones are crossed and the middle one is at 45°; determine the emerging irradiance (a) without and (b) with the middle filter. (2) GIVEN: Polarizers at $\theta = 0°$, $\theta = 90°$, and $\theta = 45°$ and $I_0 = 800$ W/m^2. FIND: (a) I_2 with two polaroids and (b) I_3 with three. (3) PROBLEM TYPE—Physical optics/polarization/linear polarizers. (4) PROCEDURE—(a) Before the third polarizer is inserted, ***no light passes through the crossed pair*** $\boxed{I_2 = 0}$. (b) It might seem that inserting yet another polarizer could have no effect, but that's wrong! If the

first filter has a vertical transmission axis, then $I_1 = \frac{1}{2}I_0 = 400$ W/m^2 of vertical $\mathscr{P}$-state light will emerge from it. (5) CALCULATION—(b) The second polarizer makes an angle of 45° with the first; by Malus's Law

$$I_2 = I_1 \cos^2 45° = 200 \text{ W/m}^2$$

The transmitted electric field is now tilted at 45°—*the presence of the second polarizer has changed the field reaching the third polarizer.* The field is at 45° with respect to the last polarizer, whose transmission axis is horizontal. Again from Malus's Law

$$I_3 = I_2 \cos^2 45° = \boxed{100 \text{ W/m}^2}$$

which is the amount of horizontal $\mathscr{P}$-state light finally emerging through all three filters.

Quick Check: $E_{01} \propto \sqrt{I_1}$; $E_{02} = E_{01} \cos 45°$; $I_2 \propto E_{02}^2 = E_{01}^2 \cos^2 45°$; $I_2 = I_1(1/\sqrt{2})^2 = \frac{1}{2}I_0(1/\sqrt{2})^2 = \frac{1}{4} 800 \text{ W/m}^2$.

25.3 Polarizing Processes

There are several important natural processes that produce polarized light. One of the most commonplace is the reflection from dielectric media. The glare spread across a window pane, a sheet of paper, or a classroom desk and the sheen on a balding head or a shiny nose are all partially polarized.

Polarization by Reflection

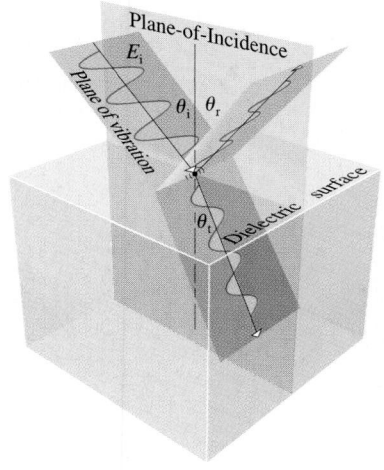

Figure 25.9 A wave with its *E*-field perpendicular to the plane-of-incidence reflecting and refracting at an interface. Electrons then oscillate perpendicular to the plane-of-incidence and reradiate light perpendicular to that plane.

We can understand what's happening in a simple way by again considering the electron-oscillator model of the atom, although the description is a little simplistic. Figure 25.9 shows an incident wave plane polarized *perpendicular to the plane-of-incidence*. The atoms in the surface are driven by the *E*-field parallel to the interface and they reradiate, producing the usual reflected and refracted waves, both of which are similarly polarized. On the other hand, if the incident *E*-field is linearly polarized *in the plane-of-incidence*, something rather different happens to the reflected wave. The atoms at and near the surface are again driven into oscillation by the *E*-field of the transmitted wave (Fig. 25.10), but this time that's not parallel to the interface. The oscillations are all perpendicular to the refracted ray, making a small angle θ with the reflected ray. Because these dipoles do not radiate along their oscillatory axes, the amount of light reflected is relatively small. In fact, if we could arrange things so that $\theta = 0$, whereupon the reflected and transmitted rays would be perpendicular ($\theta_r + \theta_t = 90°$), none of the incident linearly polarized light would be reflected. The special angle of incidence for which this occurs is designated by θ_p and is called

The **polarization angle** is the angle whose tangent is the ratio of the transmitting index to the incident index. At that angle reflected light is completely polarized.

the **polarization angle**, or **Brewster's angle**, where $\theta_p + \theta_t = 90°$. It follows from Snell's Law and the fact that $\theta_t = 90° - \theta_p$ that

$$n_i \sin \theta_p = n_t \sin \theta_t = n_t \sin(90° - \theta_p) = n_t \cos \theta_p \tag{25.2}$$

where shifting the sine by 90° makes it a cosine. Dividing both sides by $\cos \theta_p$ and again by n_i yields

$$\tan \theta_p = \frac{n_t}{n_i} \tag{25.3}$$

This equation is known as **Brewster's Law**, after the man who discovered it empirically, Sir David Brewster, the inventor of the kaleidoscope.

When an incoming unpolarized beam (I_i) strikes a dielectric surface (glass, plastic, water, etc.), it will have an equal amount of energy associated with electric fields parallel ($I_{i\parallel}$) and perpendicular ($I_{i\perp}$) to the plane-of-incidence (i.e., $I_{i\parallel} = I_{i\perp} = I_i/2$). But that will not be the case for the reflected light. Table 25.1 shows the percentages of reflected light at an air-water interface as θ_i increases. As you can see, most of the light is transmitted, unless the angle of incidence is quite large. What little is reflected is mostly light with its E-field perpendicular to the plane-of-incidence ($I_{r\perp}$). Notice how, as θ_i increases from 0, the portion of the reflected irradiance having its E-field parallel to the plane-of-incidence ($I_{r\parallel}$) first decreases to 0 at $\theta_i = 53°$, and then increases. That's Brewster's angle; at $\theta_i = 53°$ all the reflected light—albeit a mere 8%—is light with its E-field perpendicular to the plane-of-incidence (which makes it parallel to the reflecting surface).

When $\theta_i = \theta_p$, *only the component polarized normal to the plane-of-incidence will be reflected*: **the reflected light will be totally plane-polarized parallel to the interface** (Fig. 25.11). At any other angle, the reflected light will be partially polarized. This provides a handy way to find the transmission axis of a linear polarizer. Locate the reflected glare of some light source on a flat, horizontal, nonconducting surface (at an angle around 50°, which is a typical value for θ_p). Now, slowly revolving the polarizer, view the glare through it; when the reflected light essentially vanishes, the transmission axis is vertical (see the photos p 897).

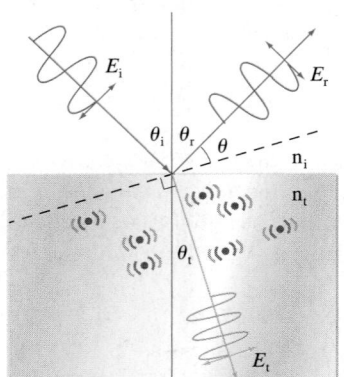

Figure 25.10 Electron-oscillators and Brewster's Law. Here, the incident E-field is in the plane-of-incidence. When $\theta = 0$, the electrons near the surface will oscillate parallel to what would ordinarily be the direction of the reflected beam. But they do not radiate along their axis of vibration, so there will be no reflected wave; $E_r = 0$.

Table 25.1

Amount of Light Reflected at an Air–Water Interface

θ_i (degrees)	Percent of light reflected perpendicular to the plane-of-incidence*	Percent of light reflected parallel to the plane-of-incidence†
0	2%	2%
30	3%	1%
45	5%	0.3%
53	8%	0%
60	11%	4%
75	31%	11%
90	100%	100%

* $I_{r\perp}/I_{i\perp}$

† $I_{r\parallel}/I_{i\parallel}$

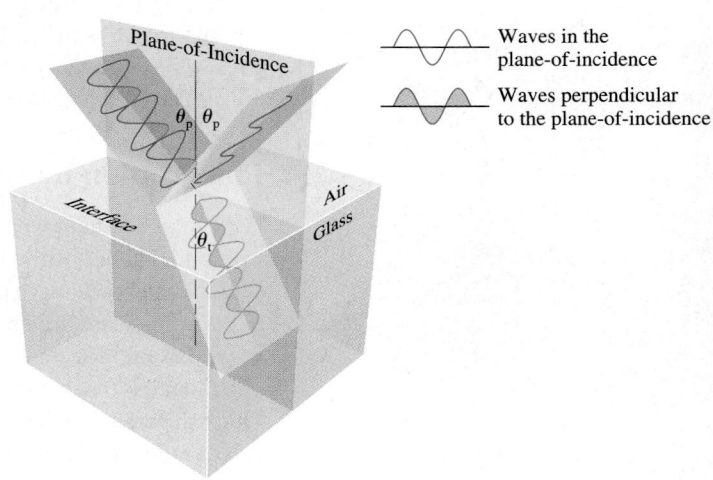

Figure 25.11 The polarization of light that occurs on reflection from a dielectric, such as glass, water, or plastic. At θ_p, the **reflected beam** is a $\mathcal{P}$-state perpendicular to the plane-of-incidence—it's completely polarized. The **transmitted beam** is strong in $\mathcal{P}$-state light parallel to the plane-of-incidence and weak in $\mathcal{P}$-state light perpendicular to the plane-of-incidence—it's partially polarized.

Light reflecting off the puddle is partially polarized. When viewed through a Polaroid filter whose transmission axis is parallel to the ground, the glare is passed and visible (a). When the Polaroid's transmission axis is perpendicular to the water's surface, most of the glare vanishes (b).

Example 25.3 **[I]** The image of a child reflects off a wet city street. At what angle should the reflection be viewed if it is to be seen in totally polarized, linear light? Give your answer to three significant figures.

Solution This problem deals with polarization by reflection and hence, Brewster's Law. (1) TRANSLATION—Determine the polarization angle at a known interface. (2) GIVEN: $n_i = 1.00$ and $n_t = 1.33$. FIND: θ_p. (3) PROBLEM TYPE—Physical optics/polarization/reflection. (4) PROCEDURE—The relation-ship that governs polarized reflections is Brewster's Law. (5) CALCULATION:

$$\tan \theta_p = \frac{n_t}{n_i} = \frac{1.33}{1.00} = 1.33$$

and so $\boxed{\theta_p = 53.1°}$.

Quick Check: $\tan 53.1° = 1.33$; at an air-water interface this must be greater than one.

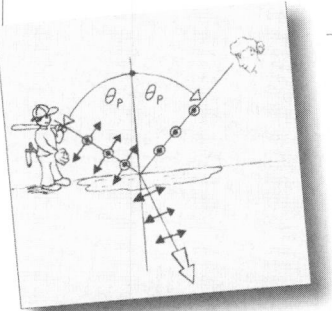

Polarization by Scattering

When light is scattered by molecular-sized particles, as in Rayleigh scattering (p. 812), the particles behave like tiny dipole oscillators. Because the dipole radiation pattern is not the same in all directions, the scattered light will be polarized. Figure 25.12*a* on page 898 depicts an incident vertically polarized plane wave driving a molecule, just as sunlight dri-ves the molecules of the atmosphere. Light is reradiated in all directions, except along the axis of vibration—straight up and straight down. In Fig. 25.12*b*, the incident beam is hori-zontally polarized, and again it is scattered into all directions, except along the axis of vibra-tion. Now, if the incident beam is unpolarized, it can be envisioned as equivalent to two perpendicular, uncorrelated $\mathscr{P}$-states. Figure 25.12*c* shows the scattering of such a distur-bance, and it's nothing more than the superposition of the two previous situations (a) and (b). Notice that the light coming off perpendicular to the original beam is completely lin-early polarized.

Birefringence

Many crystalline substances, such as ice, quartz, and calcite, have optical properties that are not the same in all directions. Because the atoms of these substances are not uniformly

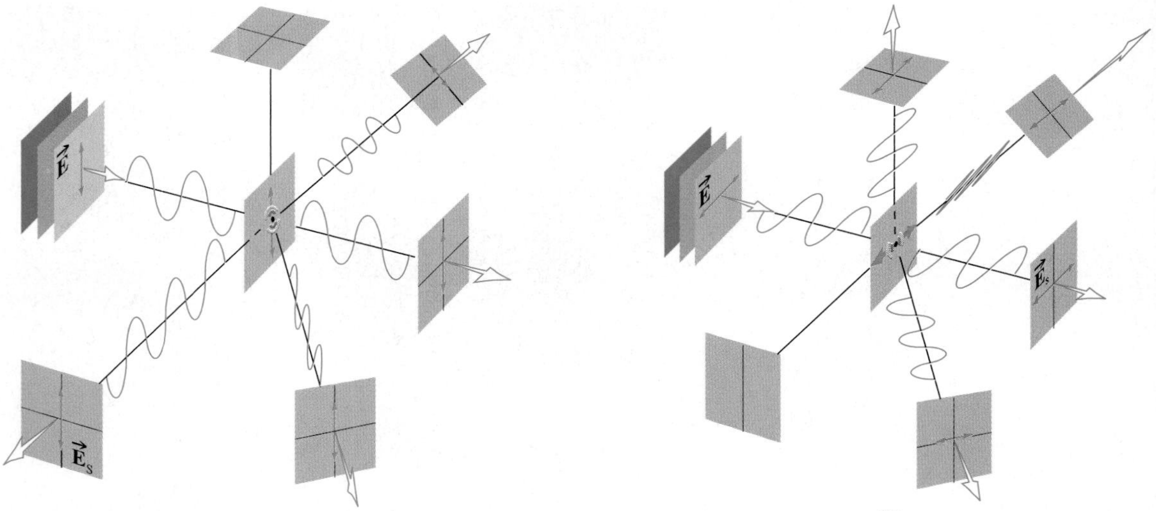

(a)

(b)

A piece of crumpled cellophane, which is birefringent, placed between crossed Polaroids. Depending on its thickness and the frequency of the light, the cellophane rotates the $\vec{\mathbf{E}}$-field by different amounts. Rotating one of the Polaroids will shift the colors to their complements.

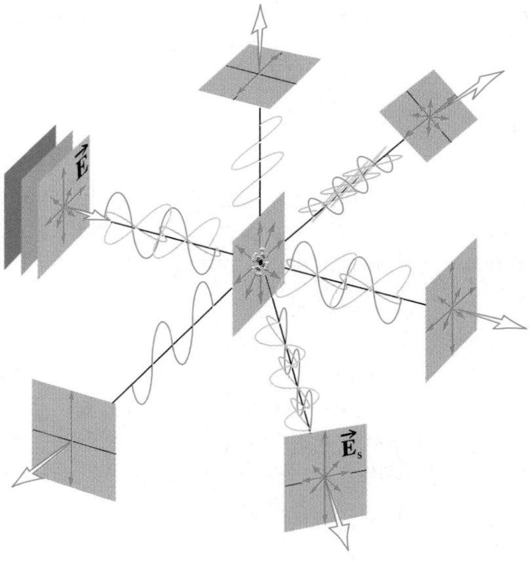

(c)

Figure 25.12 Scattering of polarized light by a molecule. Vertically polarized light scatters, as shown in (a); horizontally polarized light scatters as in (b). Since natural light (c) can be imagined as the sum of two perpendicular $\mathcal{P}$-states, it scatters as the superposition of (a) and (b). Notice how the light scattered perpendicular to the propagation direction in (c) is linearly polarized.

arrayed, the forces on their electron clouds are different in different directions. The response of the atomic oscillators depends on the direction of the incoming E-field. As a result, such crystals display two different indices of refraction. The word *refringence* was once used instead of our more modern term *refraction*, hence the name **birefringence**, meaning doubly refractive.

When unpolarized light enters a calcite crystal, each of the two constituent, independent, perpendicular $\mathcal{P}$-states encounters an essentially different medium, and each travels with a different speed. The two beams refract at different angles (according to Snell's Law) and therefore separate, produc-

Crystals of potassium chloride, calcium carbonate (calcite), and sodium chloride (table salt). Only the calcite produces a double image. It's because of this that calcite is said to be birefringent.

ing the double image discussed earlier on p. 889 (see the photo on p. 917). By viewing both images through a linear polarizer, we can easily see that the two are plane-polarized in perpendicular directions. Many human-made substances, such as certain transparent food wraps, clear plastic tapes, and plastic microscope slides, are birefringent because of the alignment of their long-chain molecules—the best of all of these is cellophane. Crumple up a sheet of cellophane and insert it between crossed Polaroids. A profusion of multicolored regions will appear, resulting from the variations in thickness of the birefringent material—the indices of refraction of all substances are frequency dependent.

Some transparent isotropic materials that ordinarily show no polarization effects can have their molecules shifted by an applied force, producing an anisotropy and what is known as *stress birefringence*. Most clear plastic implements—rulers, spoons, cups, etc.—have this stress birefringence frozen in as they solidify. Between crossed polarizers, they reveal the internal stress pattern. To study the stresses within a machine part or an architectural element, we need only make a model of the piece in plastic and examine it in polarized light. Glass lenses when stressed in manufacture or by improper mounting will also show birefringence. Federal regulations require that eyeglasses be heat-treated to remove internal stresses that make them more likely to break. A simple test with Polaroids is used to evaluate the annealing. The back windows of automobiles are often deliberately stressed so that they will shatter into small harmless pieces on impact; and that induced birefringence is often visible, even without polarizers, at Brewster's angle.

Interference

The basic idea of interference was introduced earlier: in the region of overlap, two waves of the same frequency can combine constructively or destructively depending on their relative phase (Fig. 23.4, p. 814). If two identical point sources S_1 and S_2 emit spherical waves with the same wavelength, we can expect that the space surrounding them will contain some pattern of interference. Assuming the two sources continue to pump out light in-step

A pair of crossed polarizers. The lower one is noticeably darker than the upper one, indicating that scattered sky light is partially polarized.

A piece of waxed paper between crossed polarizers. Light, having passed through the first filter, is plane-polarized, but the hodgepodge of fibers depolarizes the beam. Try it with an ordinary piece of wet paper.

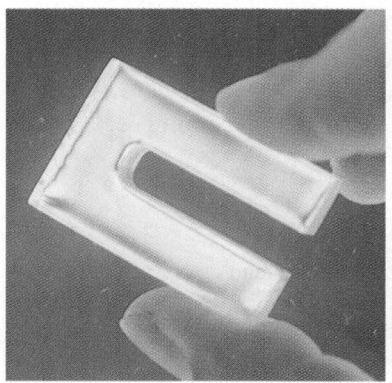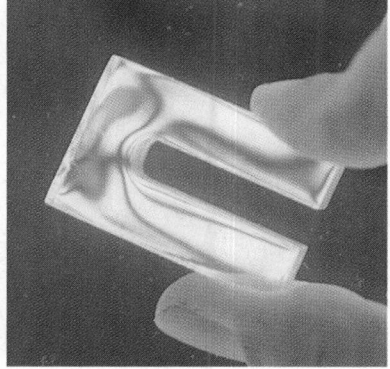

Stress within a material can cause shifts among its molecules, which then affect the transmission of polarized light. Here a piece of clear plastic between crossed Polaroids is being squeezed to increase that stress. Notice how dust, deliberately left between the crossed Polaroids scatters and depolarizes light, thereby becoming quite visible.

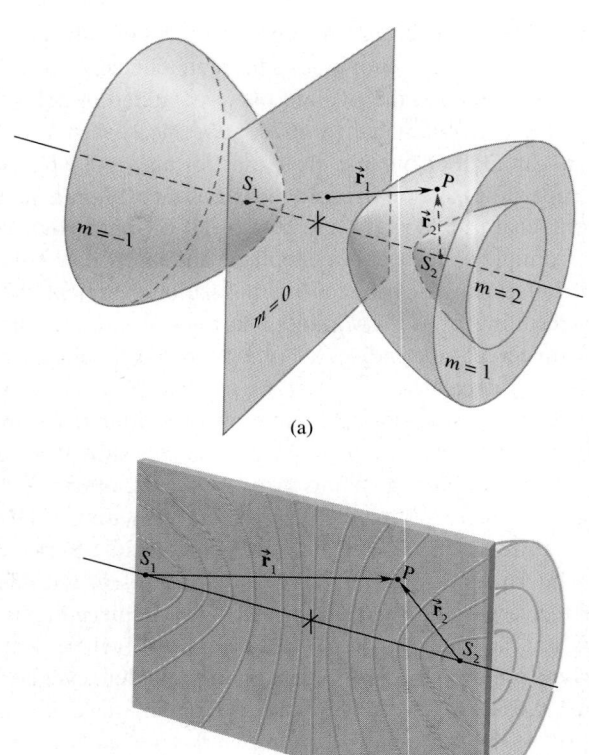

(a)

(b)

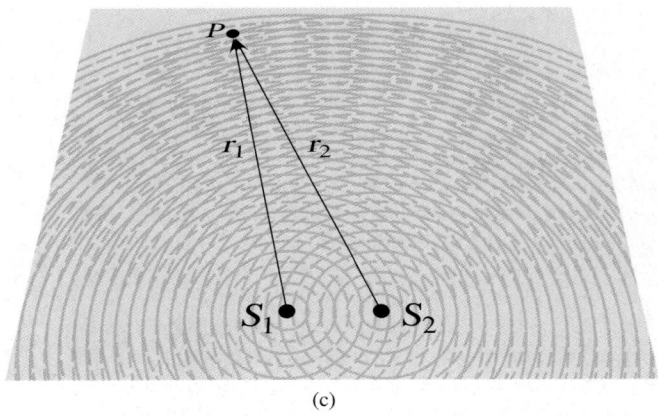

(c)

Figure 25.13 (a) Hyperboloidal surfaces of maximum irradiance for two point sources. The quantity m is positive where $r_1 > r_2$. (b) Here we see how the irradiance maxima are distributed on a plane containing S_1 and S_2. (c) Two point sources S_1 and S_2 sending out waves in-phase. Here we examine what's happening in a plane and draw the waves as circular. Where peak overlaps peak (two solid lines meet) and trough overlaps trough (two dashed lines meet), there is constructive interference and a maximum. Where peak overlaps trough (a solid line and a dashed line meet), there is destructive interference and a minimum.

with each other, the waves reaching an arbitrary point P in Fig. 25.13 will interfere in a stable observable fashion. Their relative phase (δ) at P will determine exactly how they interfere, and δ will depend on the index of the medium and the path length traveled by the waves. Taking the point sources to be in-phase, we will further assume that the index of refraction of the surrounding medium is one. Thus, if one route (r_1) is greater than the other (r_2) by half a wavelength ($r_1 - r_2 = \frac{1}{2}\lambda$), then $\delta = 180°$ and destructive interference occurs—the waves arrive out-of-step; trough overlays peak, there is cancellation, and P appears as a dark spot. The same thing would happen if the path difference were $\frac{3}{2}\lambda$ or $\frac{5}{2}\lambda$ or $\frac{7}{2}\lambda$, and so, quite generally, a **minimum** in the irradiance occurs when

[irradiance minima]
$$(r_1 - r_2) = m'\tfrac{1}{2}\lambda \tag{25.4}$$

where $m' = \pm 1, \pm 3, \pm 5, \dots$. Similarly, if the path difference corresponds to λ or 2λ or 3λ, the overlapping waves will arrive in-phase, $\delta = 0$ or, equivalently, 2π or 4π or 6π, etc. In general then, a **maximum** occurs when

[irradiance maxima]
$$(r_1 - r_2) = m\lambda \tag{25.5}$$

where $m = 0, \pm 1, \pm 2, \dots$.

Suppose that the right side of either of these two expressions is held fixed (that is, $m' = $ constant or $m = $ constant). The left side then represents all the possible locations of P for which $(r_1 - r_2) = $ constant. But if the difference in the distances from any point P to two fixed points S_1 and S_2 is a constant, as it is here, then the locations of P, which satisfy Eqs. (25.4) and (25.5), form a family of hyperbolas; that's the definition of a hyperbola.

In three dimensions, the two sources will be surrounded by a set of nested hyperboloidal surfaces, over each of which $(r_1 - r_2) = $ constant, and hence the irradiance is con-

Water waves in a ripple tank.

stant. Figure 25.13a shows several such surfaces representing irradiance maxima. Each of these is separated from the next by a hyperboloidal surface of minimum irradiance. Because the sources are in-phase, and because every point on the central plane is equidistant from both S_1 and S_2, that plane is a surface of maximum irradiance. The arrangement might call to mind the image of two stacks of skull caps, alternately white and black, getting larger and flatter as they approach one another. In any event, the interference pattern fills the space, and we need only put a screen almost anywhere in the vicinity to see it (Fig. 25.13b).

A vertical viewing screen placed at P, perpendicular to the plane of S_1 and S_2, shown in Fig. 25.13c, will be covered with vertical bright and dark bands known as **interference fringes**. You can even see the faint suggestion of fringes (Fig. 25.13c) in the plane containing P, S_1, and S_2, where the lines representing the circular waves overlap. In all interference patterns the energy, instead of being uniformly distributed, is redirected out of certain areas and into others. What is missing at minima appears at maxima: energy is conserved.

25.4 Young's Experiment

The two sources producing interference need not really be in-phase with each other. A somewhat shifted but otherwise identical pattern will occur even if there is some *initial phase difference* between the sources, as long as it remains constant. Such sources (which may or may not be in-step but are always marching together) are said to be **coherent**. Two ordinary sources, two light bulbs or candle flames, can maintain a constant relative phase for a time no greater than about 10^{-8} s; so the interference pattern they produce will randomly shift around in space at an exceedingly rapid rate, averaging out and making it quite impossible to observe. Until the advent of the laser, no two different sources could produce an observable interference pattern. Several decades ago interference was electronically detected using independent lasers, and recently (1993), fringe patterns produced by two separate lasers have even been photographed directly.

The main problem in producing interference is the sources: they must be *coherent*. And yet separate, independent, adequately coherent sources, other than the modern stabilized laser, don't exist! That dilemma was first solved two hundred years ago by Thomas Young in his classic double-slit experiment. He brilliantly took a single wavefront, split off from it two coherent portions, and had them interfere. In effect, he produced two coherent sources using a single wavefront. The technique is straightforward (Fig. 25.14a)—an opaque screen σ_a containing two identical small apertures is illuminated by a symmetrical wave that impinges on the two holes in the same way. That wave can be planar, spherical, or cylindrical provided that the phases of the two oscillating E-fields across the holes are identical. The light streaming from the apertures S_1 and S_2 pours out (diffracts) as if from two coherent identical sources.

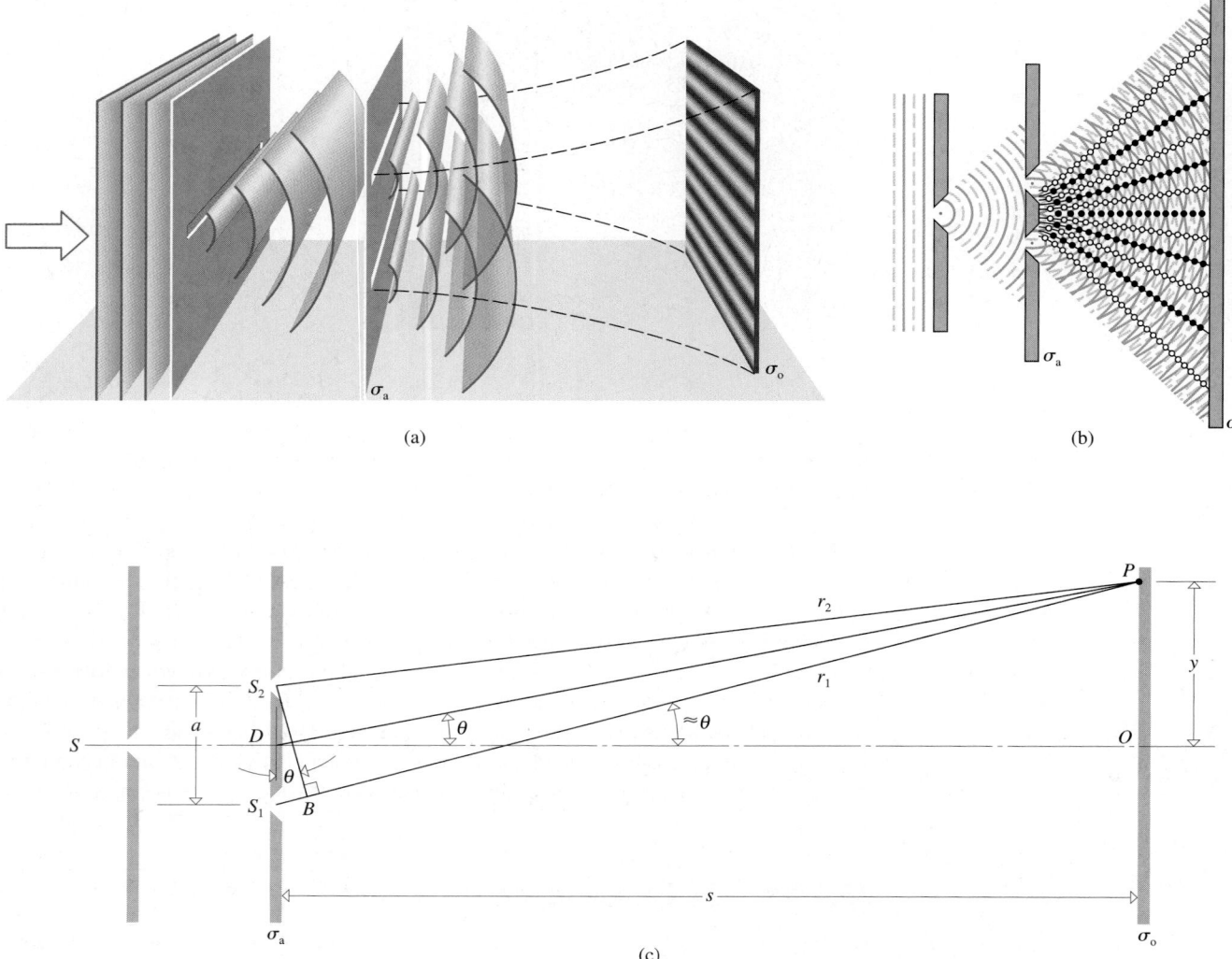

(a)

(b)

(c)

Figure 25.14 Young's Experiment. S_1 and S_2 are two coherent sources— they can be slits or holes (slits provide more light and a brighter pattern). The waves from them overlap, as in Fig. 25.13, and fill the space with fringes. The fringes that appear on a screen σ_o are horizontal in (a), (b), and (c). (b) is an edge-view of (a). With a laserbeam, σ_a can be illuminated directly without the need for the single slit at far left. (c) Usually $s \gg a$ and $s \gg y$, in which case angle $S_1 S_2 B \approx \theta$.

The apertures could be pinholes, in which case the setup closely resembles Fig. 25.13. In Fig. 25.14a, at every point where trough lies on trough (dashed line crosses dashed line) or peak overlaps peak (solid line crosses solid line), the waves are in-phase and a maximum or bright spot appears. The lines connecting all of these maxima are rather straight hyperbolas. Similarly, where trough overlaps peak (dashed line crosses solid line), the waves are out-of-phase and a minimum or dark spot appears. These, too, lie on fairly straight hyperbolas.

In order to pass a greater amount of light and get a brighter pattern, narrow slits are used rather than pinholes. An observation screen σ_o would then be covered with fairly straight dark and light bands running parallel to the slits, as in Fig. 25.14b. There we see plane waves impinging on a slit to produce a cylindrical wave, but a laserbeam could just as well illuminate σ_a with plane waves directly.

Generally, the slit separation a is very much smaller than the distance to the observation screen s. The former is usually less than a millimeter, and the latter is typically several

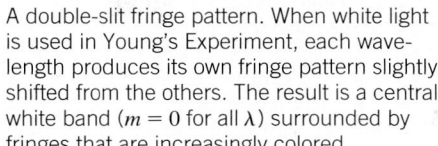 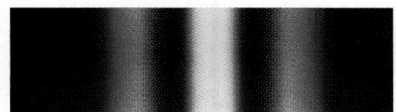

A double-slit fringe pattern. When white light is used in Young's Experiment, each wavelength produces its own fringe pattern slightly shifted from the others. The result is a central white band ($m = 0$ for all λ) surrounded by fringes that are increasingly colored.

A double-slit fringe pattern. When the separation between slits, a, is decreased, the distances between the fringes and the central axis, $y_m = sm\lambda/a$, increase. Similarly, the fringes broaden since $\Delta y = s\lambda/a$. All of this is easily seen with an ordinary long-filament display light bulb.

thousand times that. Consequently, some simplifying assumptions can be made in Fig. 25.14c. First, approximate the path-length difference $(r_1 - r_2)$ from the two apertures to any point P on the observation screen as the distance $\overline{S_1B}$ where $\overline{S_2B}$ is perpendicular to $\overline{S_1P}$. Then $a(\sin\theta) = \overline{S_1B}$ and

$$(r_1 - r_2) = a\sin\theta \qquad (25.6)$$

But θ is very small and so $\sin\theta \approx \theta$, in which case

$$(r_1 - r_2) \approx a\theta$$

Inasmuch as $\angle PDO = \theta$, $\tan\angle PDO = y/s \approx \theta$ and

$$(r_1 - r_2) \approx \frac{ay}{s}$$

Yet, for maxima

$$(r_1 - r_2) = m\lambda \qquad [25.5]$$

which means that the height of the mth bright band (counting the central one as the zeroth) above the axis is

[distances to maxima]
$$y_m \approx \frac{s}{a}m\lambda \qquad (25.7)$$

These maxima correspond to locations of P where, as shown in Fig. 25.15, the path difference is either 0 (where $m = 0$), $\pm\lambda$ (where $m = \pm1$), $\pm2\lambda$ (where $m = \pm2$), and so forth. The location of the fringes above or below the central axis is λ-dependent, which means

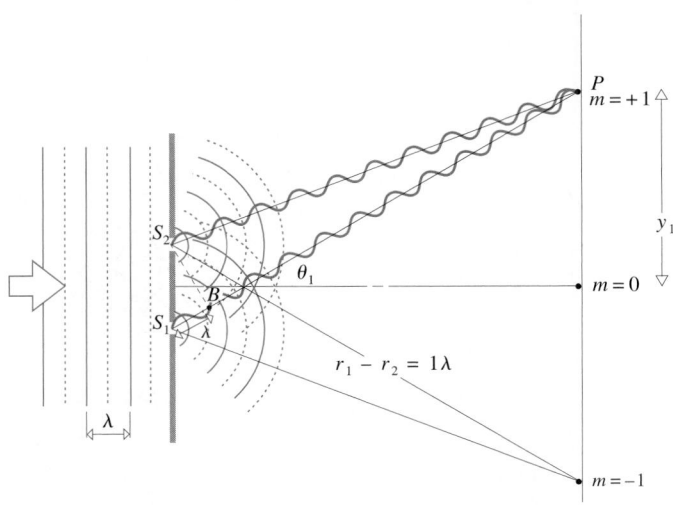

Figure 25.15 The plane waves from a laser illuminate the double-slit experiment. The $m = \pm1$ maxima occur where the optical path-length difference $(r_1 - r_2)$ equals $\pm\lambda$.

that with incident white light we can expect to see a white central band ($m = 0$) where all wavelengths overlap but that all other fringes will show some coloration.

The angular height of the mth maximum is obtained by comparing Eqs. (25.5) and (25.6):

$$a \sin \theta_m = m\lambda \tag{25.8}$$

or, since the angles are small, $\theta_m \approx m\lambda/a$.

The spacing between fringes on the screen, Δy, is just the difference between the locations of consecutive maxima. From Eq. (25.7) it follows that $\Delta y = y_{m+1} - y_m$ is

[adjacent fringe spacing]
$$\Delta y \approx \frac{s}{a}\lambda \tag{25.9}$$

For a particular wavelength of light, the spacing of the fringes is constant and inversely proportional to the slit separation: the wider apart the slits, the finer the fringes (Fig. 25.16).

A more complete analysis would show that under ideal circumstances the irradiance (I) of the light pattern on the plane of observation varies as $\cos^2 \theta$. Thus, these equally spaced, bands of light that gradually alternate from bright to black are often spoken of as *cosine-squared fringes*.

{For a discussion of coherence and how it affects Young's Experiment look at **COHERENCE** under **FURTHER DISCUSSIONS** on the **CD**. Before going on to the next section, click on **YOUNG'S EXPERIMENT** under **INTERACTIVE EXPLORATIONS** on the **CD**. This beautiful simulation makes it possible to study double-slit interference in a way that's particularly informative.}

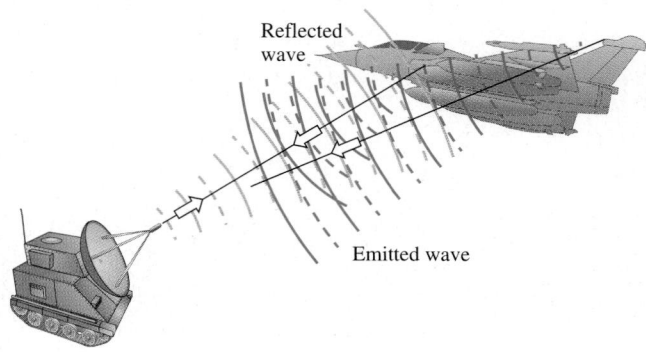

Figure 25.16 The French fighter *Rafale* uses active cancellation to confound radar detection. It sends out a signal that's one half a wavelength out-of-phase with the radar wave it reflects. The reflected and emitted waves cancel in the direction of the enemy receiver.

Example 25.4 **[III]** Red light from a He-Ne laser ($\lambda = 632.8$ nm) is incident on a screen containing two very narrow horizontal slits separated by 0.200 mm. A fringe pattern appears on a white piece of paper held 1.00 m away. (a) Approximately how far (in radians and mm) above and below the central axis are the first zeros of irradiance? (b) How far from the axis is the fifth bright band?

Solution The mention of "two narrow slits" should call up images of Young's Experiment. (1) TRANSLATION—For a double-slit experiment with known slit separation, wavelength, and screen distance, determine (a) the distance (angular and linear) to the first minimum and (b) the linear distance to the fifth maximum. (2) GIVEN: $\lambda = 632.8$ nm, $a = 0.200$ mm, and $s = 1.00$ m. FIND: (a) θ_1, $y_1(\text{min})$ and (b) $y_5(\text{max})$. (3) PROBLEM TYPE— Physical optics/interference/Young's Experiment. (4) PROCEDURE— In general, minima occur when $(r_1 - r_2) = m'\frac{1}{2}\lambda$ where $m' = \pm 1, \pm 3, \pm 5, \ldots$. Moreover, the angular distance θ is specified by $(r_1 - r_2) = a \sin \theta$. (5) CALCULATION—(a) The first minimum occurs when $m' = \pm 1$:

$$(r_1 - r_2) = \pm \tfrac{1}{2}\lambda$$

Hence,

$$a \sin \theta_1 = \pm \tfrac{1}{2}\lambda$$

and

$$\theta_1 \approx \frac{\pm \frac{1}{2}\lambda}{a} \approx \pm \frac{1}{2}\frac{(632.8 \times 10^{-9}\,\text{m})}{(0.200 \times 10^{-3}\,\text{m})}$$

$$\theta_1 \approx \pm 1.58 \times 10^{-3}\,\text{rad}$$

Therefore, $y_1(\text{min}) = s\theta_1 = (1.00\,\text{m})(\pm 1.58 \times 10^{-3}\,\text{rad}) = \boxed{\pm 1.58\,\text{mm}}$. (b) From Eq. (25.7)

$$y_5(\text{max}) \approx \frac{s5\lambda}{a} \approx \frac{(1.00\,\text{m})5(632.8 \times 10^{-9}\,\text{m})}{(0.200 \times 10^{-3}\,\text{m})}$$

$$y_5(\text{max}) \approx \pm 1.58 \times 10^{-2}\,\text{m}$$

Quick Check: The fringe irradiance varies as cosine-squared, and the answer to (a) is half a fringe width. The answer to (b) is a distance of 5 full fringes, and therefore it's 10 times larger.

25.5 Thin-Film Interference

In addition to the triumph of the double-slit experiment, Young was able to explain the colors arising from *thin films*: the colors of soap bubbles and oil slicks on a wet pavement, the rainbow of hues that appears on oxidized metal surfaces, and the iridescence of peacock feathers and mother-of-pearl.

When light impinges on the first surface of a transparent film, a portion of the incident energy re-emerges as a reflected wave while the remainder is transmitted. The refracted energy subsequently encounters the second surface where, again, a portion is reflected and the remainder transmitted out of the film (Fig. 25.17). The film has the effect of dividing the wave into three segments: one reflected from the top surface, one reflected from the bottom, and one transmitted. Rather than utilizing two separate regions of the incident wavefront (each with the same amplitude as the incoming wave) as was done in Young's Experiment, the film shears the entire wavefront. These two reflected waves come off in the same direction and can be made to overlap at some point on the viewer's retina. Because they travel different routes, depending on the film's thickness, they will ultimately interfere in some way that depends on that thickness.

The vivid iridescence of a peacock feather is due to interference of the light reflected from its complex layered surface.

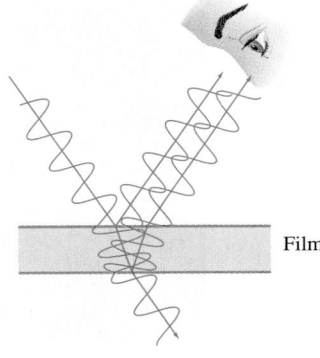

Film

Figure 25.17 Thin-film interference. Light reflected from the top and bottom of the film interferes to create a fringe pattern.

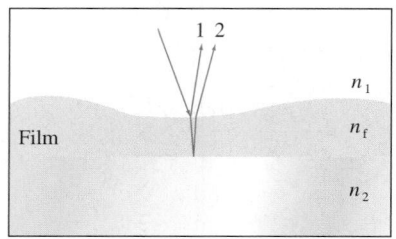

Figure 25.18 The reflection of light from the top and bottom of a thin film of index n_f.

A wedge-shaped film made of liquid dish-washing soap, showing interference fringes. The top part is drained thin. On the very top right a sliver of the film appears black, indicating that it's less than a quarter of a wavelength thick in that region.

The **optical path length** is the actual length of the path taken by light, multiplied by the index of refraction of the medium through which it travels. The optical path length divided by the vacuum wavelength (λ_0) equals the number of wavelengths corresponding to the route taken.

Consider a nonuniform oil film floating on a puddle so that it thins out to a thickness of only a few wavelengths (and the dark surface beneath absorbs any transmitted light, keeping it from scattering back to the viewer). Wherever the film is exactly the right thickness for the two waves of emerging red light to undergo constructive interference, the film will appear to reflect a spot of red light (and so on across the spectrum and across the film). These so-called *fringes of equal thickness* create a kind of colored topographical map of the film.

Let's analyze the situation for the simplified case of nearly perpendicularly incident light. Figure 25.18 shows a thin film of index n_f between media with indices n_1 and n_2. Ray-2 travels through the film, down and back up, crossing it twice. If the thickness at that point is d, then ray-2 traverses an additional path length of $2d$ before rejoining ray-1. The path multiplied by the index of refraction is called the *optical path length*—in this case, $2dn_f$. The optical path length divided by λ_0 is the number of wavelengths corresponding to the path. This means that ray-2 traveled an additional number of waves equal to $2dn_f/\lambda_0$ farther than ray-1. To see this, notice that the extra distance $\approx 2d$ in the film corresponds to an extra number of wavelengths equal to $2d/\lambda_f$, which, because $\lambda_f = \lambda_0/n_f$, equals $2dn_f/\lambda_0$. Since each wavelength is equivalent to a phase change of 2π rad, the two waves will have a relative phase difference δ of

$$\delta = \frac{4\pi dn_f}{\lambda_0} \tag{25.10}$$

When this quantity equals a whole number (m) multiple of 2π, the two waves will be back in-phase, and that particular wavelength of light will undergo constructive interference:

$$\delta = 2\pi m = \frac{4\pi dn_f}{\lambda_0}$$

Interference fringes due to an oil film on wet pavement. The dark roadway scatters very little light back up through the film and therefore doesn't wash out the pattern. These are fringes of equal thickness much like Newton's rings (see Fig. P74, p. 930).

Therefore, solving for d (Case-1),

Case-1 applies when the index of the film lies between the indices of the surrounding two media: $n_1 > n_f > n_2$ or $n_1 < n_f < n_2$

[maxima]
$$d = \frac{m\lambda_0}{2n_f} \qquad m = 1, 2, 3,\ldots \qquad (25.11)$$

Maxima in normally reflected light occur when the film thickness is a whole number multiple of half the wavelength ($\lambda_f/2$). **And in the same way, *minima occur when the thickness is an odd multiple of one-quarter of the wavelength*** ($\lambda_f/4$).

As the film becomes thicker, exceeding several wavelengths, it becomes possible for two different colors to have maxima at the same spot. For example, a film 1000-nm thick will produce maxima for light with wavelengths within it of both 400 nm and 500 nm. At still greater thicknesses, where many wavelengths can interfere constructively at the same time, the reflected color becomes increasingly unsaturated, the fringe contrast decreases, and the pattern ultimately vanishes. Monochromatic illumination will encounter no such limitations, and a sheet of window glass is still effectively a thin film for laser light.

Actually, Eq. (25.11) provides the thickness of the film for an interference maximum *only* when $n_1 > n_f > n_2$ or $n_1 < n_f < n_2$. In the first instance, the reflections are both *internal*; in the second, they're both *external*. As we saw earlier (p. 816), nearly normally incident light will undergo a relative phase shift of π rad between its internally and externally reflected components.* That's not relevant in either of the preceding cases, which is why Eq. (25.11) applies. Notice that as the film gets vanishingly thin, all wavelengths will interfere more or less constructively everywhere, to gradually produce the uniform reflection from the interface (between the two surrounding media) that must exist when the film disappears altogether.

One of the most important practical applications of these ideas is the *antireflection coating*. Whenever light reflects off an interface separating two dielectrics (such as glass and air), some fraction is reflected, and that can become very problematic in complex instruments where there might easily be a dozen or more such surfaces. As a solution, each optical element is coated with a thin, transparent, solid, dielectric film such that it does not reflect a specific range of wavelengths. Often that range is chosen to be in the yellow-green region of the spectrum, where the eye is most sensitive. Lenses coated for the yellow-green reflect in the blues and reds, giving the surface a familiar purple color. The simplest antireflection coating is a one-quarter of a wavelength ($\lambda_f/4$) thick layer of some hard transparent material like magnesium fluoride ($n_f = 1.38$), where in air on glass $n_a < n_f < n_g$.

An even more commonly occurring situation has $n_1 < n_f > n_2$ or $n_1 > n_f < n_2$, as with a soap film in air in the first case and an air film between two sheets of glass in the second. Now, one reflection is internal and the other is external, and there will be an additional $\pm\pi$ phase shift. Which sign we select is irrelevant; accordingly, using the minus

$$\delta = \frac{4\pi d n_f}{\lambda_0} - \pi$$

Constructive interference in reflected light now occurs when

$$\delta = 2\pi m = \frac{4\pi d n_f}{\lambda_0} - \pi$$

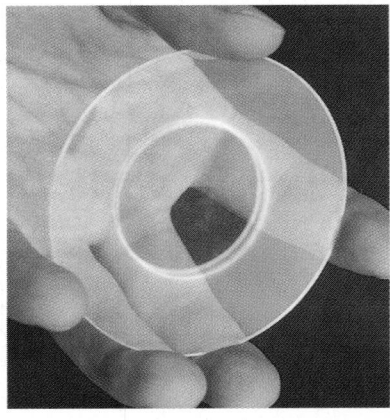

An antireflection coating in the shape of a circle, applied at the center of each side of a glass disk. The simplest antireflection coating is a Case-1 film one-quarter wavelength thick for the color whose reflection we wish to suppress.

*It's very often stated that a phase shift of π rad occurs when light is reflected off a higher index medium (external reflection) and no shift occurs when it's reflected off a lower index medium (internal reflection). Although that sounds appealingly simple, it's totally wrong! Still this incorrect notion, when applied naively to thin films, yields the correct overall phase shift. Actually, different phase shifts occur for the two electric field components parallel and perpendicular to the plane-of-incidence; moreover, those phase shifts vary with the angle of incidence. So the whole thing is rather complicated, but for $0 \leq \theta_i \leq \theta_p$, which is the way we generally look at thin films, there will be a relative phase shift of π rad between the internally and externally reflected waves, and that's the only thing we need to be concerned about here.

Interference from the thin air film between a convex lens and the flat sheet of glass it rests on. The illumination was quasimonochromatic, which is why the fringes aren't more colorful. These fringes were first studied in depth by Newton and are known as Newton's rings. See Problem 74, p. 930.

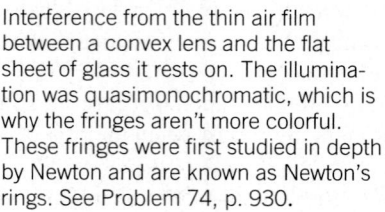

Case-2 applies when the index of the film is greater or less than the indices of the surrounding two media: $n_1 < n_f > n_2$ or $n_1 > n_f < n_2$.

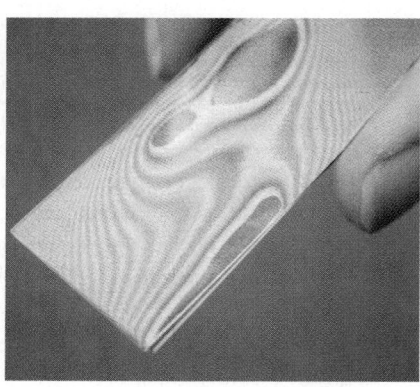

The interference pattern produced by a thin film of air between two clean microscope slides.

Therefore, solving for d (Case-2),

[maxima]
$$d = \frac{(m + \frac{1}{2})\lambda_0}{2n_f} \qquad m = 0, 1, 2,\ldots \qquad (25.12)$$

As the film becomes vanishingly thin ($d \to 0$, $d << \lambda_0$), the total phase difference, δ, approaches $-\pi$, there is total destructive interference, and the film appears uniformly black in reflected light and uniformly transparent in transmitted light. That's reasonable enough—since it must disappear altogether when $d = 0$, it should reflect less and less and transmit more and more as it gets there. Biological bimolecular lipid films (5-nm to 10-nm thick) appear black in reflected light, as does an ordinary liquid-soap film when it drains, and becomes a small fraction of a wavelength thick (see the photo on p. 906).

{The interference pattern produced by the circularly symmetric air film between a lens and a flat sheet of glass was first studied by Newton. It's examined in some detail in Problem 74. Before going on to the next section, take a look at **NEWTON'S RINGS** under **INTERACTIVE EXPLORATIONS** on the **CD**.}

Example 25.5 **[I]** A soap film in air has an index of refraction of 1.34. If a region of the film appears bright red ($\lambda_0 = 633$ nm) in normally reflected light, what is its minimum thickness there?

Solution The phrase "a soap film" tells us that this is likely to be a problem about thin-film interference. (1) TRANSLATION—A thin film of known index is illuminated by light with a specified wavelength; determine the minimum thickness for an interference maximum. (2) GIVEN: $\lambda_0 = 633$ nm and $n_f = 1.34$. FIND: d. (3) PROBLEM TYPE—Physical optics/interference/thin film. (4) PROCEDURE—This situation corresponds to Case-2, $1.00 < 1.34 > 1.00$. (5) CALCULATION—Using Eq. (25.12) with $m = 0$, which corresponds to the minimum thickness, we have

$$d = \frac{(m + \frac{1}{2})\lambda_0}{2n_f} = \frac{(0 + \frac{1}{2})(633 \text{ nm})}{2(1.34)} = \boxed{118 \text{ nm}}$$

Quick Check: The optical path-length difference is $2n_f d = 2(1.34)(118 \text{ nm}) = 3.16 \times 10^{-7}$ m; the number of waves this difference corresponds to is $2n_f d/\lambda_0 = 0.5$, which, correctly, is an odd whole number multiple of $\frac{1}{2}$.

25.6 The Michelson Interferometer

There are a number of practical devices known as **interferometers** that produce fringe patterns very much in the nature of the thin-film effects just considered. The most important of these, both historically and practically, is the *Michelson Interferometer*, shown in Fig. 25.19. Here, an extended source emits a wave that enters from the left. This could be

Many creatures in Nature produce their brilliant coloration via interference and/or diffraction.

In Fig. 25.17 the optical path-length depends on λ (i.e., the color of the light) and on the viewing angle. In a similar way the ink used to print the denominations on U.S. currency now contains structured particles that produce interference colors. Here the number 20 changes from black to green as the viewing angle changes.

an expanded laserbeam, the light from a discharge lamp, or an ordinary tungsten bulb. The wave is then sheared into two equal-amplitude parts by a beamsplitter (M_S), which is usually a half-silvered mirror. The two waves are subsequently reflected by mirrors M_1 and M_2 and returned to the beamsplitter. The wave from M_1 reflects toward the detector, and the wave from M_2 traverses M_S, passing on to the detector. The waves are reunited, and we can expect to see an interference pattern that depends on the path-length differences traversed as well as whatever phase shifts are introduced via reflections from the beamsplitter.

Since one beam passes through M_S three times and the other only once, there will be a good deal of difference in the optical path lengths, even when the two arms are of equal actual length. Moreover, the optical path length will depend on λ because of dispersion (p. 823) in the glass. The inclusion of a compensator plate, C, which is a duplicate of the beamsplitter (without the silvering), equals out the number of thicknesses of glass traversed and negates the effects of dispersion. With the compensator in place even white light can be

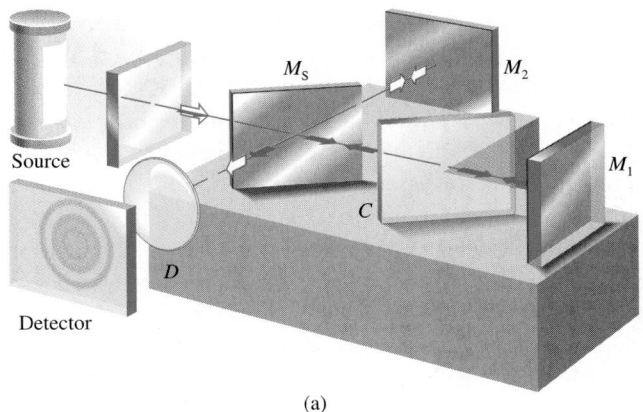

(a)

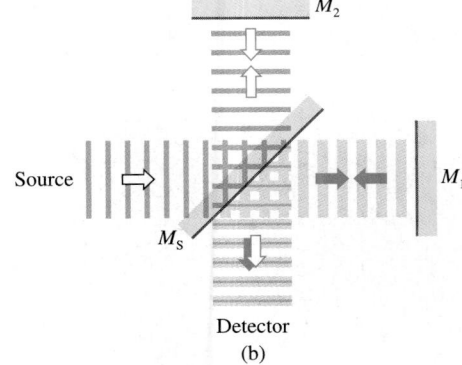

(b)

Figure 25.19 The Michelson Interferometer. (a) Light enters at the left and is split into two equal beams by the beamsplitter M_S. Part goes on to mirror M_1 and part to mirror M_S. These beams are reflected back to M_S, and both pass on to the detector. When M_1 and M_S are perpendicular, the fringes are circular and centered on the axis of the observer's eye. (b) is a simplified schematic representation of the wavefronts ignoring refraction.

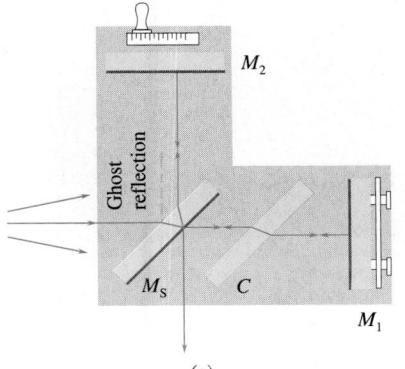

(c)

Wedge fringes in a Michelson Interferometer. The distortions are caused by a hot soldering-iron placed in one arm. It heated the air and produced local variations in the index of refraction. Before I inserted the iron, the fringes were all straight and parallel, as they are on the right in the photo.

The displacement of one of the mirrors in a **Michelson Interferometer** by a distance of $\Delta d = \lambda_0/2$ produces a shift of one (bright-dark) fringe pair.

used; without it, the illumination must be quasimonochromatic.

What we see when we look into the device at the position of the detector is the image of both mirrors superimposed. We see the beamsplitter and, at the same time, mirrored in it M_1 and, through it, M_2. In effect, we are looking at M_1 and M_2, one behind the other, with a layer of air between them whose thickness corresponds to the difference in the two path lengths. When one of the mirrors is moved a distance $\Delta d = \lambda_0/4$, the beam in that arm will traverse that distance twice and therefore change its path length by $2(\lambda_0/4)$. As a result, a dark band in the interference pattern will move over and a bright band will take its place. This is a shift of "half a fringe." Such *fringe shifts* are easily counted, and so if we displace one of the mirrors by a precisely measured distance, we can determine the wavelength of the light.

If M_2 and the image of M_1 are not parallel, fringes of equal thickness appear. Thus, when the mirrors are set to form a triangular-shaped air film, a system of parallel, equal-spaced straight fringes aligned with the edge of the wedge (corresponding to lines of equal film thickness) will be observed, as in the photo. And here we see another fundamental advantage of this kind of interferometer: there is space within the arms to which we have access. A hot soldering iron inserted into one of the arms will change the index of refraction of the air, shift the fringes, and reveal an otherwise invisible phenomenon. There are several variations of Michelson's instrument that exploit this marvelous ability to study the structure of transparent systems (such as gases, plasmas, and lenses).

Diffraction

The distinction between **diffraction** and **interference** is made for reasons that are more historical than physical. **Diffraction is a manifestation of interference.**

An object placed between a point source and a screen casts an intricate shadow made up of bright and dark regions quite unlike anything one might expect from the tenets of geometrical optics. This deviation from rectilinear propagation, known as **diffraction**, is *a general characteristic of all wave phenomena occurring whenever a portion of a wavefront is obstructed in any way.* When in the course of encountering an obstacle, either transparent or opaque, a region of the wavefront is altered in amplitude or phase, diffraction occurs.

Nowadays (because these fringes are characteristic of the objects that give rise to them, and because they can be analyzed by computer) all sorts of automatic processors utilize diffraction, for things as diverse as spotting tanks in aerial photos and checking fingerprints to measuring blood cell sizes.

The various unobstructed segments of a wavefront that propagate beyond an obstacle (be it sound or light) interfere to produce the particular energy-density distribution referred to as the diffraction pattern. There is no profound physical difference between interference and diffraction—in fact, the two are inseparable, though we tend to talk about them individually because that's the way the notions developed historically.

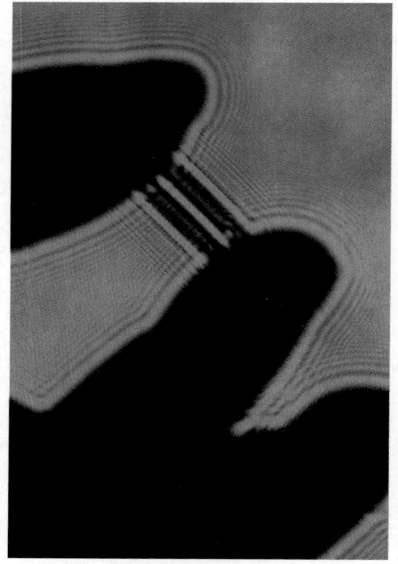

Diffraction pattern of a hand-held paper clip. This is simply the shadow cast on a wall using a He-Ne laserbeam (a distant point source would do as well).

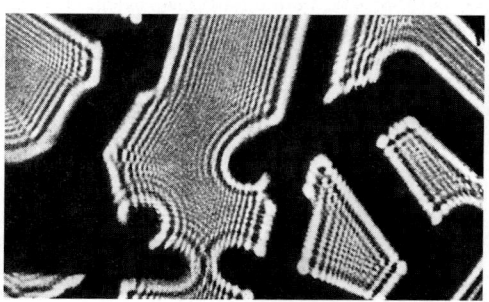

Fresnel diffraction of electrons by zinc oxide crystals. All subatomic particles (and even atoms themselves) manifest wavelike properties. Here electrons produce a diffraction pattern.

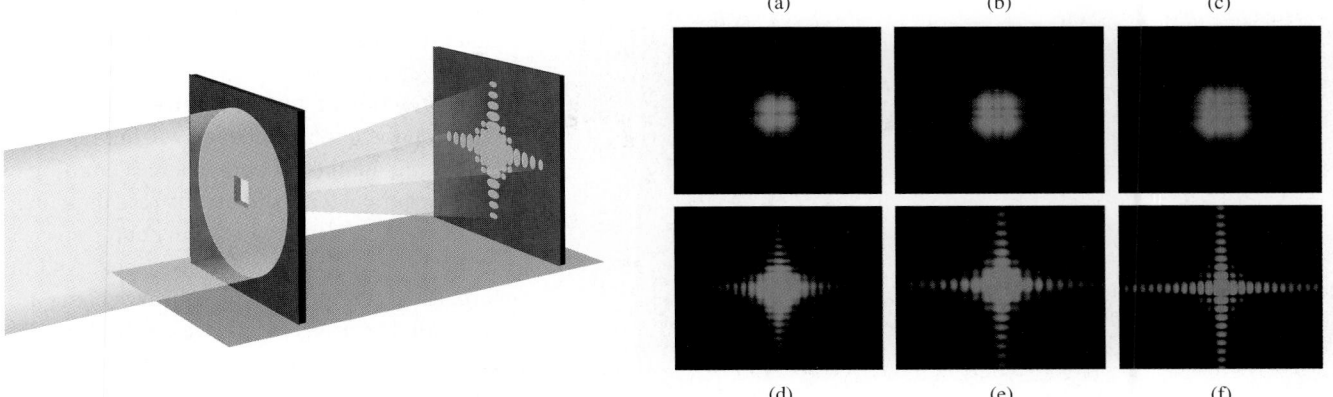

Figure 25.20 The shadow or diffraction patterns arising from a single square aperture. The hole was sequentially decreased, producing Fresnel diffraction (a), which gradually transformed (b), (c), (d), (e) into Fraunhofer diffraction (f). All apertures were illuminated by the plane waves from a He-Ne laser, and the viewing screen was fixed in place.

As far as the analysis is concerned, we recognize two distinct kinds of diffraction, though that, too, is more for practical reasons than fundamental ones. Figure 25.20 will help make the point. It shows a succession of diffraction patterns produced by a square hole illuminated by a laser. When the hole is large and the film very close to it by comparison, the pattern is complicated internally but the overall shape is recognizable—that's what is seen in (a), (b), and (c). But as the hole is made still smaller, *the pattern increasingly spreads out in directions perpendicular to the hole's edges* until it becomes quite unrecognizable. Finally, as in (f), a point will be reached where making the hole smaller simply makes the pattern grow larger without changing shape.

Since the effect depends on the relative size of the hole compared to its distance from the film, the same set of photos could just as well have been gotten by leaving the aperture fixed and moving the film plane farther and farther away. The first several pictures, from (a) to (e), represent what is called *near-field* or **Fresnel diffraction**, while (f) corresponds to *far-field* or **Fraunhofer diffraction**. The latter is a special case of the former, one that is comparatively easy to deal with mathematically. We will only consider *far-field diffraction where the incident light is planar and the film is essentially at infinity* (though 10 or 15 meters will generally do nicely in practice).

Consider a thin glass photographic plate that is completely blackened except for a small, clear aperture, perhaps in the shape of a little square. When this square is illuminated by a succession of normally incident plane waves, layer upon layer of atoms within the glass will scatter the light until the last sheet of atoms on the far side of the plate emits into the air in all directions. In effect, the light beyond the screen—the diffracted light—is the result of a tremendous number of point sources distributed uniformly across the aperture, all emitting secondary, more or less spherical, wavelets.

To make the point again from a different perspective, imagine the obstructing screen to be a perfect mirror, initially with no apertures. The incoming light reflects backward, and none is transmitted. Whatever radiation comes from the atoms of the mirror cancels the forward-moving wave and simply produces the reflected wave. Now, cut out the same hole as before and remove the square plug. The back-scattered radiation from the screen no longer completely cancels the wave, and light emerges from the region of the hole. Still, if the plug were reinserted, the light would be exactly canceled. As a first approximation, let's assume that the mutual interaction of all the atomic oscillators is negligible; that is, the atoms in the screen are unaffected by the removal of the atoms in the plug. The field in the region beyond the screen will be that which existed there prior to the removal of the plug, namely zero, minus the contribution of the plug alone. All of which suggests that the light emitted by all the atoms spread across the surface of the plug is essentially identical to the light emerging from the aperture.

The diffraction field can be pictured as arising from a set of oscillators distributed uniformly over the unobstructed area of the aperture. These imagined secondary point

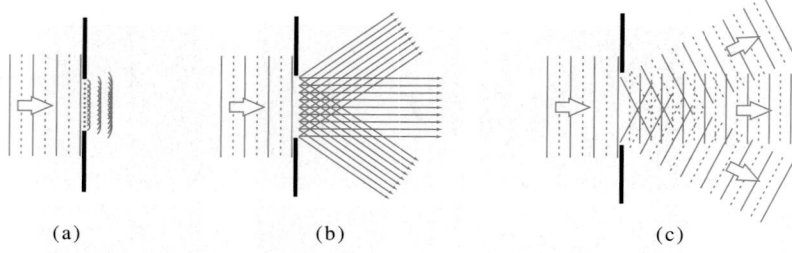

Figure 25.21 The diffraction of light through a narrow slit, represented by (a) spherical wavelets, (b) rays, and (c) plane waves.

sources emit wavelets beyond the obstruction that mutually interfere to create the diffraction pattern. Figure 25.21 illustrates how the notion is applied to the diffraction of a wave by an aperture. In (a), we envision the unobstructed wavefront replaced by a very large number of point sources emitting essentially spherical waves in the forward direction. This is equivalent (b) to rays traveling out from each secondary source-point in all directions. And this, in turn, is equivalent (c) to plane waves being diffracted in all directions. Since we are limiting the discussion to Fraunhofer diffraction, it is more convenient to use a few lenses to compress the setup rather than having to observe the pattern at infinity (Fig. 25.22).

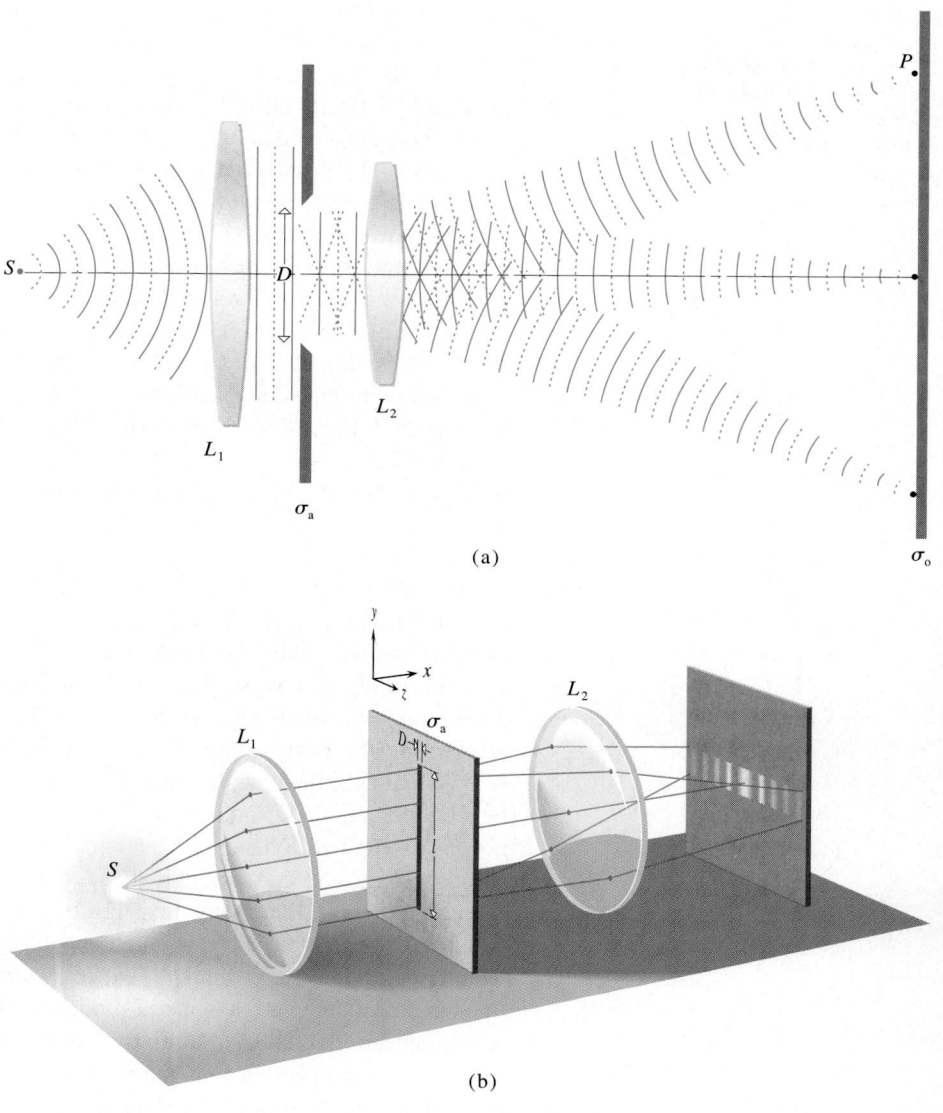

Figure 25.22 (a) The diffracted waves (seen in a horizontal plane) from the aperture are brought to a focus on a nearby screen by lens L_2. (b) When the aperture is a single slit, the diffraction pattern is a series of bands. With a point source S or a narrow laserbeam illuminating the slit, the fringe pattern is narrow as well.

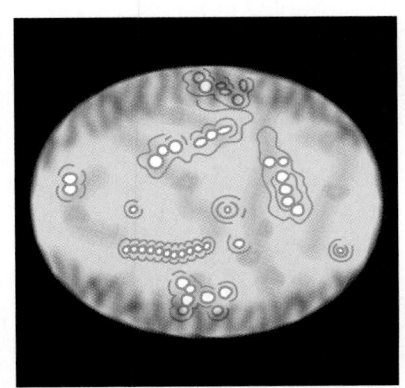

EXPLORING PHYSICS ON YOUR OWN

Diffraction & Seeing Your Eye from the Inside: Anyone who has ever walked at night in the rain wearing eyeglasses has probably seen the effect of diffraction. Just put a drop of water on a glass plate, hold it near your eye, and look through it at a distant streetlight, and you'll see a complex system of bright and dark diffraction fringes. Similarly, the amoeba-like floaters that can be seen within your own eye when you squint at a bright broad source are diffraction patterns of drifting cellular debris cast on your retina. Close one eye, and then squint as tight as you can at a light bulb (or the sky, or a computer screen close up) until you see the shadows of your lashes. Relax your focus and you'll see floaters. The circular dark boundaries on the right and left are the outlines of your own pupil. Cover that eye with a hand, wait a few seconds for your pupil to dilate, and then pull your hand away—what happens to the outline?

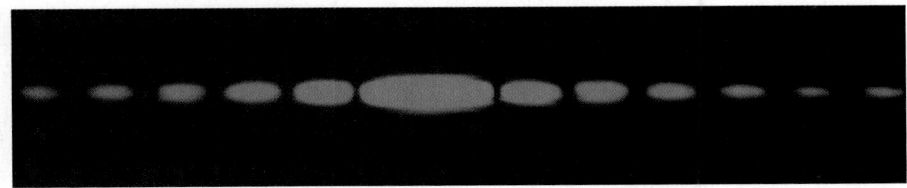

Diffraction pattern of a single vertical slit under narrow laserbeam illumination.

25.7 Single-Slit Diffraction

Suppose that the aperture in Fig. 25.22a is a long narrow slit of width D, perpendicular to the plane of the diagram. Under monochromatic plane-wave illumination, we envision every point in the aperture emitting rays in all directions. The light that continues to propagate directly forward (Fig. 25.23a) is the undiffracted beam, all the waves arrive on the viewing screen in-phase, and a central bright region is formed. Figure 25.23b shows the specific bundle of rays coming off at an angle θ_1 where the path-length difference between the rays from the very top and bottom, $D \sin \theta_1$, is made equal to one wavelength. A ray from the middle of the slit will then lag $\frac{1}{2}\lambda$ behind a ray from the top and exactly cancel it. Similarly, a ray from just below center will cancel a ray from just below the top and so on; all across the aperture ray-pairs will cancel, yielding a black minimum. The irradiance has dropped from its high central maximum to the first zero on either side (Fig. 25.24) at $\sin \theta_1 = \pm \lambda / D$.

As the angle increases further, some small fraction of the rays will again interfere constructively, and the irradiance will rise to form a subsidiary peak less than 5% of the central maximum. Increasing the angle still further produces another minimum (Fig. 25.23c) when

Figure 25.23 The zeros in the irradiance distribution produced by a single slit. (a) The central maximum is produced by undiffracted light. (b) The first pair of minima ($m' = \pm 1$) occurs when the waves from the edge and center of the slit are out-of-phase by $\lambda/2$. (c) The second pair of minima ($m' = \pm 2$) occurs when waves from each quarter of the slit cancel.

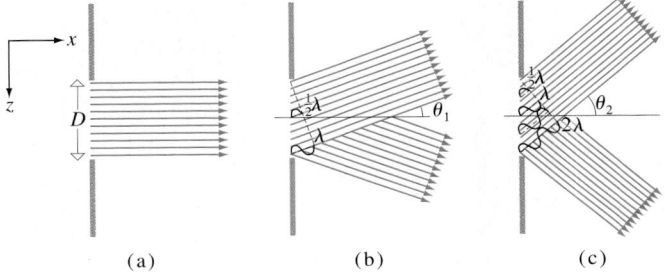

(a)   (b)   (c)

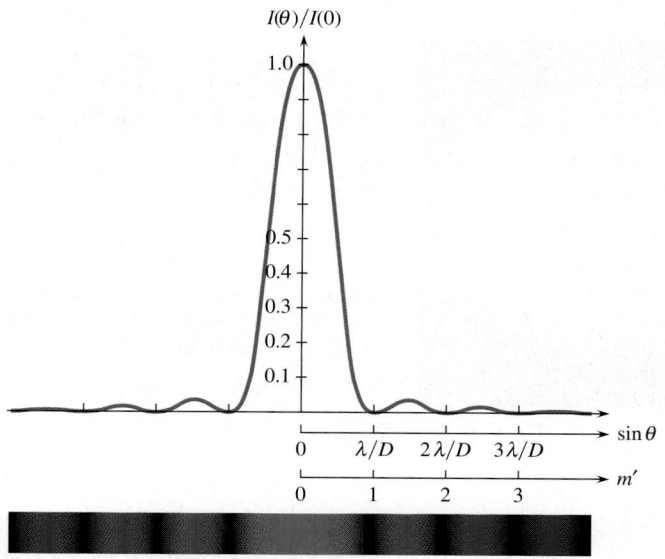

Figure 25.24 The Fraunhofer diffraction pattern of a single slit. This is a plot of the relative irradiance. Notice that the central maximum has twice the width ($I = 0$ to $I = 0$) of the little maxima.

$D \sin \theta_2 = 2\lambda$. Now, imagine the aperture divided into quarters. Ray by ray, the top quarter will cancel the one beneath it, and the next, the third quarter, will cancel the last quarter. Ray-pairs at the same locations in adjacent segments are $\lambda/2$ out-of-phase and destructively interfere. In general, then, zeros of irradiance will occur when

[minima]
$$D \sin \theta_{m'} = m'\lambda \tag{25.13}$$

where $m' = \pm 1, \pm 2, \pm 3, \dots$.

Example 25.6 **[II]** A single slit 0.10 mm wide is illuminated by plane waves from a helium-neon laser ($\lambda = 632.8$ nm). If the observing screen is 10 m away, determine the **width of the central maximum** (*defined as the distance between the centers of the adjoining minima*).

Solution The phrase "a single slit" tells us that this is very likely to be a problem about diffraction. (1) TRANSLATION— Light of known wavelength is incident on a slit of specified width; determine the width of the central maximum on a screen a known distance away. (2) GIVEN: $D = 1.0 \times 10^{-4}$ m, $\lambda = 632.8 \times 10^{-9}$ m, and $s = 10$ m. FIND: The width of the central maximum. (3) PROBLEM TYPE—Physical optics /far-field diffraction

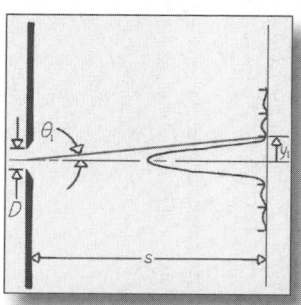

/ single slit. (4) PROCEDURE—The two minima bounding the central maximum correspond to $m' = \pm 1$ in Eq. (25.13). Find the angular width (θ_1) and from that the linear width. (5) CALCULATION—To calculate θ_1 begin with

$$\sin \theta_1 = \frac{\lambda}{D} = \frac{632.8 \times 10^{-9}\,\text{m}}{1.0 \times 10^{-4}\,\text{m}} = 632.8 \times 10^{-5}$$

and $\theta_1 = 0.36° = 6.3 \times 10^{-3}$ rad. The central band is $2\theta_1 = 2(6.3 \times 10^{-3}\,\text{rad})$ wide. Since the linear half-width is $s(\tan \theta_1)$, which in radians is approximately $s\theta_1$, for small θ_1, the overall width of the central fringe is

$$s2\theta_1 = (10\,\text{m})(0.012\,6\,\text{rad}) = \boxed{0.13\,\text{m}}$$

Quick Check: Using the preceding results, $2\theta_1 = (0.13\,\text{m})/s = 0.013 \approx 2 \sin \theta_1 = 2(632.8 \times 10^{-5}) = 0.013$. Alternatively, $\tan \theta_1 = y_1/s$; $2y_1 = 2s \tan \theta_1 = 2(10\,\text{m}) \tan 0.36° = 0.13$ m.

As the slit narrows and D gets smaller, the diffraction pattern spreads out and the central peak gets wider. Again, the light fans out perpendicularly against the edges of the aperture just as it did in Young's Experiment. In fact, during that experiment, very narrow slits are used and each produces a single-slit diffraction pattern with a very wide central bright band. These two diffraction patterns overlap to create the familiar double-slit fringes (Figs. 25.25a and b).

25.8 The Diffraction Grating

Something rather interesting develops as the number of parallel slits (each spaced a distance *a*) is increased beyond two (Fig. 25.25). Identical single-slit diffraction patterns again overlap to produce maxima at the same locations as in Young's Experiment, but now these principal peaks are narrower and faint subsidiary maxima appear between them. Three slits will produce one subsidiary maximum; four slits, two subsidiary maxima; and N slits, $(N-2)$ subsidiary peaks between successive principal bright bands. As N increases, the secondary peaks become more numerous and even fainter, with the effect that more of the diffracted light appears in the sharpened, widely spaced, principal bands. When there are thousands of slits, the bright bands are, for all practical purposes, separated by black regions where essentially no light arrives. It follows from Eq. (25.8) for Young's fringes that the principal maxima are to be found where

[principal maxima] $$a \sin \theta_m = m\lambda$$ [25.8]

and $m = 0, \pm 1, \pm 2, \dots$. A repetitive array of apertures or obstacles that alters the amplitude or phase of a wave is a **diffraction grating**, and Eq. (25.8) is known as the ***grating equation***. Master gratings with over 10 000 lines inscribed per centimeter are either ruled using a fine diamond point or generated holographically. Most gratings are plastic copies of master gratings, and they can be transparent *transmission gratings* or metal-coated *reflection gratings*.

Figure 25.25 Diffraction patterns for slit systems composed of (a) 1, (b) 2, (c) 3, (d) 4, and (e) 5 identical, equally spaced, vertical slits. Notice that the maxima for several slits are where they were in the two-slit case. The more apertures, the finer the maxima become and the more faint the fringes appear between them. When there are thousands of identical slits, the various order maxima are quite fine and they appear separated by black regions.

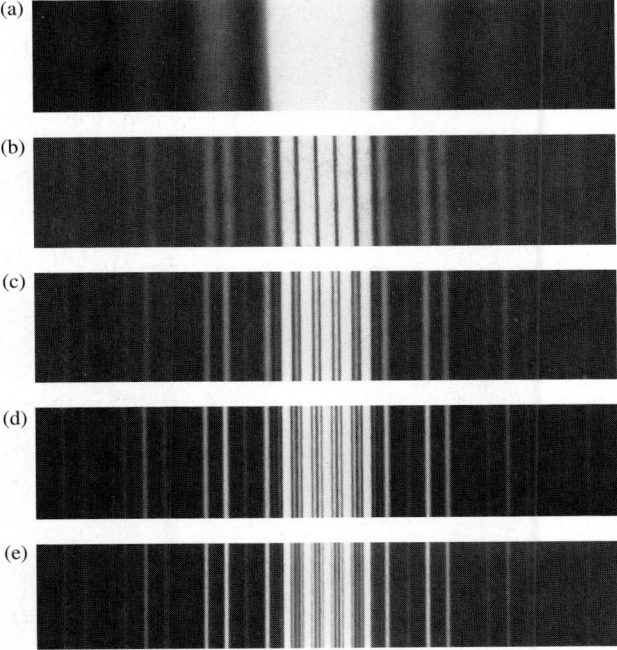

Example 25.7 **[I]** Light from a distant star enters a telescope and then passes through a diffraction grating. Each component of the emerging dispersed light is focused onto a curved strip of film. The grating is located on the central axis of the setup. It is found that a red beam (known as the hydrogen-α line) appears off axis, in the first-order ($m = \pm 1$) spectrum, at an angle of 25.93°. If the lines of the grating are separated by 1.50×10^{-6} m, determine the wavelength of that light.

Solution Here we're dealing with a diffraction grating and $a \sin \theta_m = m\lambda$ should immediately come to mind.

(1) TRANSLATION—Light passes through a diffraction grating having a known line separation and produces a first-order spectrum at a specified angle; determine the wavelength. (2) GIVEN: $\theta_1 = 25.93°$, $a = 1.50 \times 10^{-6}$ m, and $m = 1$. FIND: λ. (3) PROBLEM TYPE—Physical optics/diffraction/transmission grating. (4) PROCEDURE—Use the grating equation. (5) CALCULATION—From Eq. (25.8), we have

$$\lambda = \frac{a(\sin \theta_1)}{m} = \frac{(1.50 \times 10^{-6} \text{ m})(0.437\ 3)}{1} = \boxed{656 \text{ nm}}$$

Quick Check: Use this λ to calculate θ_1; $a \sin \theta_m = m\lambda$, $\theta_m = \sin^{-1} m\lambda/a = \sin^{-1}(0.437) = 25.9°$.

Notice in Eq. (25.8) that it is only because a is greater than λ that more than the single $m = 0$ wave exists. The smaller a is, the fewer are the values of m that will satisfy the formula. Moreover, if $\lambda > a$, since the sine cannot exceed one, only $m = 0$ is possible. As a result, when light scatters specularly from a smooth surface (p. 820), only one beam is reflected and *the angle of incidence equals the angle of reflection*. Because $\lambda \gg a$, where a is the spacing between atomic scatterers, only the $m = 0$ order can be sustained in ordinary reflection.

The source in the photo below is a hot solid filament that puts out light having a continuous range of frequencies. Each band of color emerging from the grating, that is, each order (here $m = \pm 1$ and $m = \pm 2$), consists of a continuous sweep from violet to blue to green to red. By contrast, light from a gaseous source, like a star, contains a number of distinctly different frequencies. These we can separate and examine, often so that they can reveal the identity of the emitting atoms. Accordingly, diffraction gratings are widely used to do chemical analysis in astronomy and elsewhere. Usually the light is passed through a vertical slit (see Fig. 27.15, p.982) and then through a grating (or in earlier times, a prism). Each wavelength appears as a separate vertical "line" of colored light clustered together and repeated in each of the several orders. These are the so-called *spectral lines*. Each atom emits its own distinctive group of light lines, like a fingerprint. The finer the spacing between the ridges, slits, or lines that constitute the grating, the more effective it will be at separating the spectral lines.

As we'll see, for X-rays all crystals are three-dimensional gratings (p. 974). The gemstone opal gets its shifting internal coloration from an ordered array of tiny silica (SiO_2) spheres that form a three-dimensional diffraction grating for light.

Light and UV passing through a transmission grating. The region on the left shows the visible spectrum; that on the right reveals the ultraviolet.

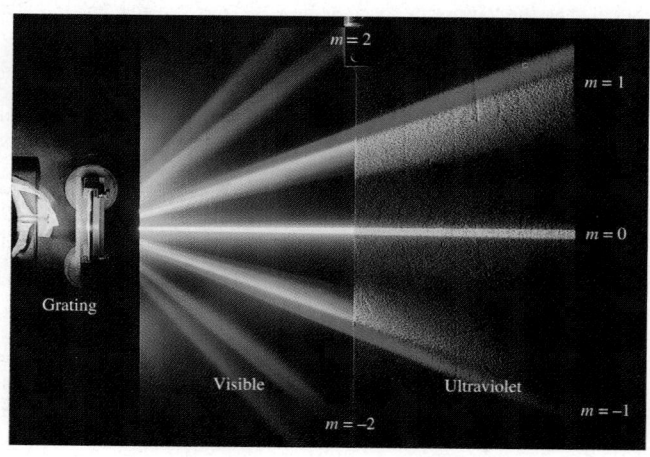

25.9 Circular Holes and Obstacles

When light from a distant point source (like a star or, more mundanely, an atom on a star's nose) is focused by a lens, the image formed is a small blotch rather than a perfect point. It must be so because the lens captures only a portion of the wavefront, and diffraction must occur—the blotch is the far-field diffraction pattern of the aperture of the lens. In the end, every image formed by a circular lens, whether in your eye or a TV camera, is composed of countless overlapping circular-aperture Fraunhofer diffraction patterns. And that alone makes this particular configuration of the greatest importance.

All Fraunhofer patterns, no matter how weirdly shaped the aperture, are symmetrical about the center point. Draw a line through that center, and the distribution of light outward along one-half of the line is the same as along the other. A circular hole must generate a circularly symmetric pattern. In fact, a circular aperture produces a circular disk of light, the so-called **Airy disk**, surrounded by a system of increasingly faint concentric rings. Actually, 84% of all the energy lies in the Airy disk. By the way, because there is so little energy in the higher order fringes, circular lenses are better image formers than rectangular lenses.

A hole with a diameter D followed by a lens of focal length f will generate a diffraction pattern on a screen (in the lens's focal plane) having an Airy disk of radius r_a that can be shown to be

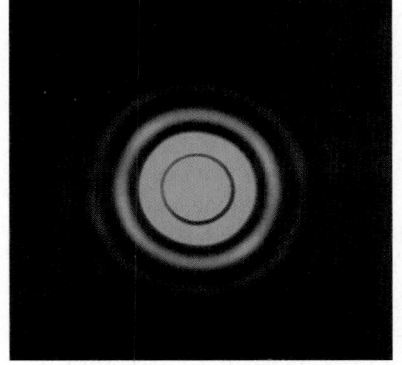

Airy rings (1.0-mm-hole diameter). I overexposed the central disk in order to make a few rings visible.

$$r_a \approx \frac{1.22 f \lambda}{D} \qquad (25.14)$$

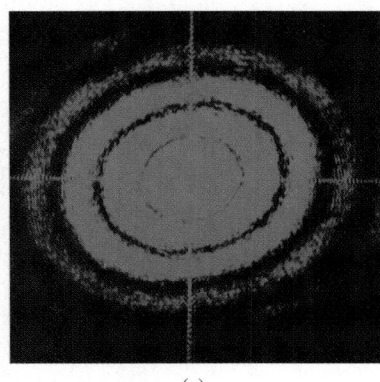 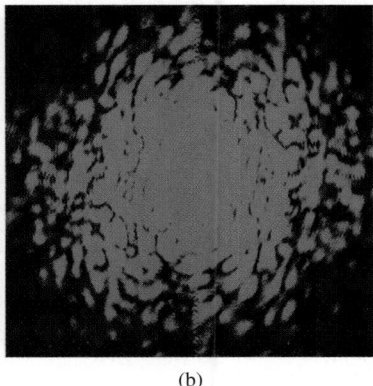

(a)   (b)

(a) The Fraunhofer diffraction pattern of a normal cervical cell. (b) The diffraction pattern of a malignant cervical cell is very different. Diffraction is being studied as a possible means of rapid automatic analysis of Pap tests for cancer.

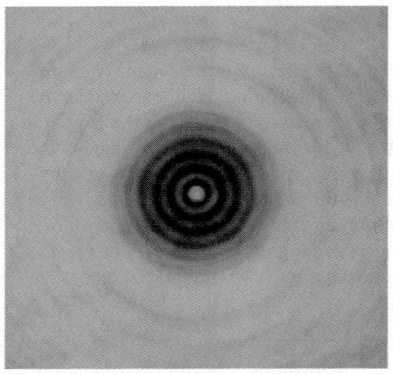

The shadow of an $\frac{1}{8}$-in.-diameter ball bearing illuminated with He-Ne laser light. To hold it in place, the ball bearing was glued to a microscope slide. The effect can also easily be seen with a halogen lamp as the point source.

THE ABSURD SPOT

In 1818, the young Fresnel entered his new wave theory in a competition sponsored by the French Academy. The judging committee consisted of such luminaries as Pierre Laplace, Jean Biot, Siméon Poisson, Dominique Arago, and Joseph Gay-Lussac. An ardent antagonist of the wave hypothesis, Poisson managed to deduce a remarkable and seemingly untenable conclusion from Fresnel's theory. He showed that the treatment predicted that a bright spot should appear at the very center of the shadow of a circular opaque obstacle; such an absurdity must surely disprove the entire theory! Arago, who was not one to accept anything on face value, went to the laboratory to learn the truth of Poisson's death blow. And there, wonder of wonders, at the center of the shadow, he saw the "absurd"—a bright spot of light (see the photo). Fresnel was right!

Alternatively, using the fact that $\theta_a \approx \sin \theta_a = r_a/f$, we can express the size of the central bright circle in terms of its angular half-width θ_a as

$$\theta_a \approx \frac{1.22\lambda}{D} \tag{25.15}$$

in radians.

Suppose we wish to examine two equal-irradiance, incoherent, point sources—two close-together stars seen through a telescope or two points on a virus viewed in a microscope. Because of diffraction, the images will spread out, and only if their angular separation exceeds this smearing will they appear distinct. As the two sources come closer together, their images come together, commingling into a single blend of fringes. Lord Rayleigh suggested that the criterion for *just being able to resolve the separate images* be that their center-to-center distance equal the radius of their Airy disks (Fig. 25.26). We can actually do a little better, but Rayleigh's criterion is both simple and standard. Equation (25.15) therefore specifies the **angular limit of resolution**. The resolution limit of the human eye is measured to be 2 or 3 arc minutes (1 arc minute equals 0.291 mrad). The *resolving power* of an image-forming system is the reciprocal of θ_a.

By decreasing the wavelength we can increase the resolving power of an instrument. Ultraviolet microscopes can "see" finer details than light microscopes (which at best can distinguish two points as distinct that are no closer than about 0.12×10^{-6} m). Similarly, electron microscopes with wavelengths equivalent to roughly 10^{-4} to 10^{-5} times that of light have a limit of resolution of about 0.5 nm. By increasing the diameter of the objective lens or mirror of a telescope it can collect more light, but the images it forms will be sharper as well. That's why spy satellites have large-diameter cameras.

{The above discussion is limited to *far-field diffraction* where, in effect, the size of the object causing the pattern is very small compared to the distance from it to the observing screen. As the object gets larger (or the screen comes closer), we get *near-field diffraction* which is more complicated and difficult to analyze. To put it all in perspective click on **DIFFRACTION** under **INTERACTIVE EXPLORATIONS** on the **CD**.}

Figure 25.26 Overlapping images of two point sources (separated by θ_a) that are just resolvable. The limit of resolution occurs when the center-to-center separation Δl equals the radius of the Airy disk (r_a).

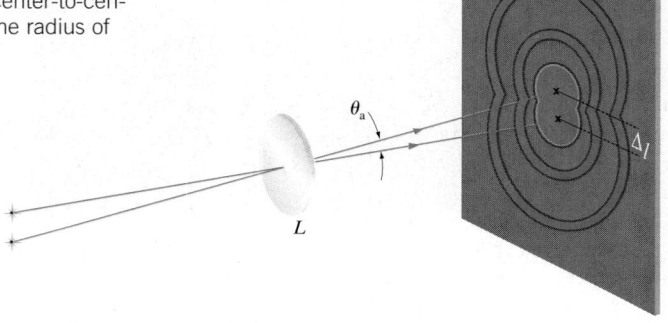

This picture is made up of only about 750 pixels. The subject is hard to see when the page is close to you because you can resolve the individual pixel squares. That's especially true when you hold the picture close to your face. To perceive it more clearly, decrease your ability to resolve each separate pixel: decrease D, the aperture of your eye (squint), or decrease the angular separation of the edges of each pixel (hold the picture farther away). If you do either, you should be able to make out the image of your humble author.

Example 25.8 **[II]** Compute the angular limit of resolution of the eye, assuming that it's determined only by diffraction, and that the medium in the eye is air. Take the pupil diameter to be wide open at 8.0 mm, and $\lambda_0 = 550$ nm. How far apart will two points be if they are just able to be distinguished at a distance of 25 cm from the eye? How does θ_a change with the presence of the vitreous humor ($n_{vh} = 1.337$)?

Solution The mere mention of "angular limit of resolution" tells us that we're going to have to use the defining equation for θ_a. (1) TRANSLATION—The aperture diameter of an instrument is known, as is the wavelength of the illumination; determine (a) its angular resolution and (b) the separation between two just-resolvable points at a specified distance. (2) GIVEN: $D = 8.0$ mm, $\lambda_0 = 550$ nm, $d_o = 25$ cm, and $n_{vh} = 1.337$. FIND: θ_a and y_o. (3) PROBLEM TYPE—Physical Optics / diffraction / resolution. (4) PROCEDURE—Use the definition of the limit of resolution. (5) CALCULATION—From Eq. (25.15)

$$\theta_a \approx \frac{1.22\lambda_0}{D} = \frac{1.22(550 \times 10^{-9} \text{ m})}{8.0 \times 10^{-3} \text{ m}} \approx \boxed{8.4 \times 10^{-5} \text{ rad}}$$

which is ≈ 0.1 mrad or 4.8×10^{-3} degrees. At a distance of 25 cm, this angle corresponds to a linear distance of $y_o = d_o\theta_a$ or

$$y_o = (0.25 \text{ m})\theta_a = \boxed{2.1 \times 10^{-2} \text{ mm}}$$

With the eye filled with liquid, $\lambda = \lambda_0/n_{vh} = 411$ nm and $\theta_a \approx 1.22\,\lambda/D = 6.3 \times 10^{-5}$ rad. By contrast the actual value of θ_a, which varies greatly from person to person, is limited by the size of the cone cells in the fovea and is only $\approx 0.05°$—about 10 times worse than the value we calculated.

Quick Check: $\theta_a D/\lambda_0$ should equal 1.22; therefore, $(0.084$ mrad$)D/\lambda_0 = (0.084$ mrad$)(8.0$ mm$)/(550$ nm$) \approx 1.2$.

A reflection hologram serving as a U.S. postage stamp. It's best viewed with a white light point source like the sun. The scene changes both its perspective and its color, as the viewing angle changes.

25.10 Holography

The technology of photography has been with us for a long time, and we're all accustomed to seeing the three-dimensional world flattened onto a scrapbook page. A photograph is a record of the energy per unit area per unit time that once impinged on each of its surface points, and the light it scatters (the light we see when we look at it) is nothing more than a reflection of that frozen irradiance distribution. It is not an accurate reproduction of the original light field that came from the subject but only a record of the square of the field's amplitude averaged over the exposure time. It reveals nothing about the phases of the waves that formed the image. On the other hand, if we could somehow reconstruct both the amplitude and phase of the wave coming from the object, the resulting light field would be indistinguishable from the original. One would then see the reformed image in perfect three-dimensionality, exactly as if the object were there before us, generating the wave.

The technique of *image reconstruction* was invented by Dennis Gabor sometime around 1948. His research, which won him the 1971 Nobel Prize in physics, led to a practical means of recording and playing back the complete wave emanating from an object. The process is now called **holography** (from the Greek *holos* meaning whole). The crucial insight is that both the amplitude and phase information can be captured in a coded form via interference and preserved photographically. When two monochromatic waves generate an interference pattern, the shape, contrast, and spacing of the fringes is a record of the physical characteristics of the overlapping waves themselves. If we wish to capture the vital features of an **object-wave**, we can cause it to interfere with a known, simply configured, **reference-wave**. The two should have roughly the same amplitude and, of course, be coherent. This last requirement kept Gabor's work in semi-unnoticed oblivion for about 15 years, until the laser was invented.

Dennis Gabor (1900–1979) Hungarian-born British physicist won the Nobel Prize in 1971.

Consider the simple case of two plane waves traveling in the same general direction (Fig. 25.27) with a photographic plate inserted in the region of overlap. The wedge-shaped film viewed in a Michelson Interferometer produces the same configuration of two tilted plane waves, and so we can anticipate a fringe pattern of parallel bands across the emulsion. To appreciate this more clearly, just lay a pencil along each wavefront and move them both in their respective propagation directions at the same speed. The point of overlap, the maximum, will travel to the right along a straight line—along the fringe. As the shape of the object-wave deviates increasingly from a simple plane, the fringe pattern becomes more modulated and complex. Thus, if the object-wave is the light reflected from someone's face, the fringe system will be an unrecognizable tumult of fine bright and dark regions. When developed, the silver atoms in the emulsion form a kind of three-dimensional diffraction grating, which is the hologram (Fig. 25.28).

All we need to do is illuminate a fine-grained film with both a scattered object-wave and a coherent reference-wave (Fig. 25.29). The developed film is a *transmission hologram*; and when it is illuminated from the back by a wave identical to the reference-wave (Fig. 25.29*b*), the original light field is reconstructed. The hologram during playback functions much like a diffraction grating that scatters the incoming plane waves into two off-axis beams. One of these gives rise to an extraneous real image, while the other corresponds to a duplicate of the original lightwave diverging from the object.

Figure 25.27 A schematic representation of the interference of two plane waves traveling toward the same side of a photographic film, thereby creating a transmission hologram. (For simplicity, refraction has been ignored.)

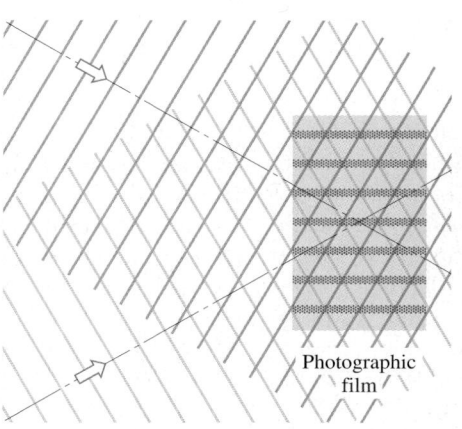

Photographic film

Figure 25.28 (a) A hologram. (b), (c), (d) Three different views photographed from the same holographic image generated by the hologram in (a). By moving your head (or in this case, a camera), you can see different regions of the scene. The lens in the hologram will magnify different objects depending on how you look through it.

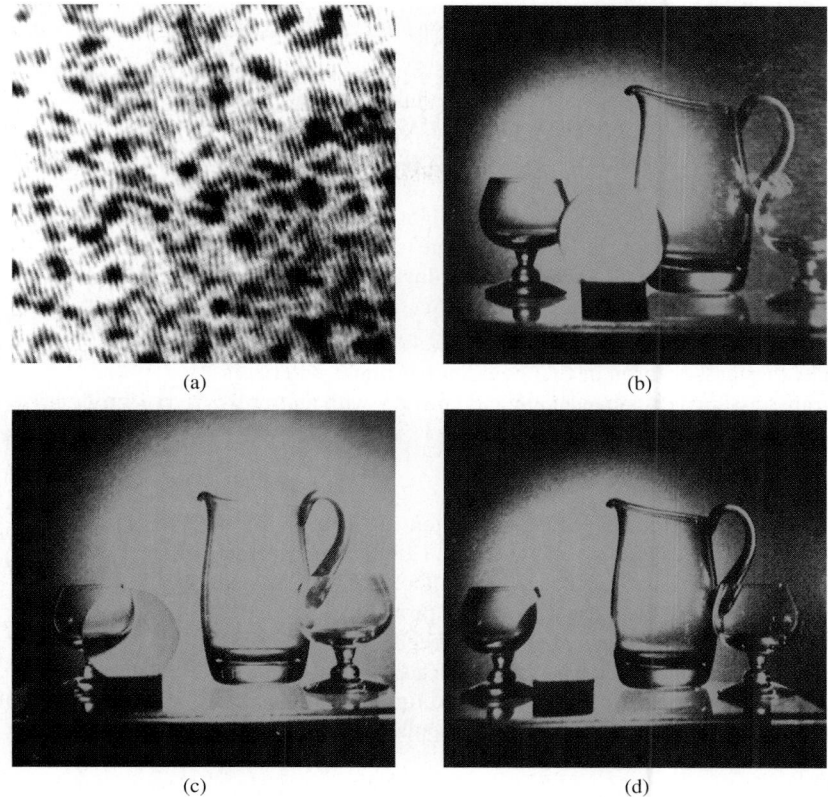

The light emerging from a hologram is identical to the light originally coming from the object (it's as if you were simply looking through a window). In the case of an extended scene, your eyes would have to refocus as you viewed different regions at various distances. And if you chose to photograph the vision, you would only have to point the camera through the "window," focus on the desired object, and shoot. Furthermore, since the light from any point on the subject spreads across much of the film plate, the information about that point is redundantly recorded all across the hologram. Consequently, just as you can peer out of a small opening in a window shade and see the entire scene beyond, you can peep through a small piece of the hologram and see everything as well. It's been suggested that the human brain functions in a similarly redundant way and that memory is, in that sense, holographic.

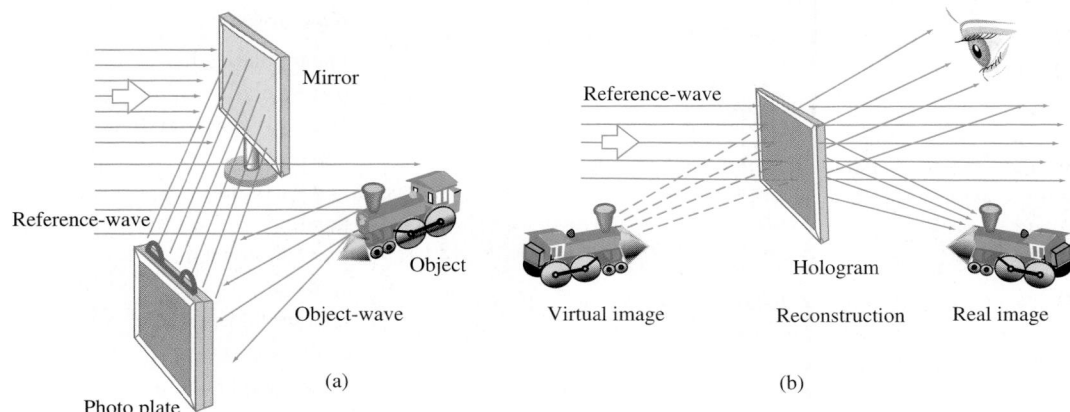

Figure 25.29 Holographic recording and reconstruction of an image. (a) The photo plate receives light from the object and a reference beam directly from the mirror. (b) To view the hologram, it is illuminated with a wave identical to the reference-wave.

POLARIZATION

When the electric field of a lightwave remains in a fixed plane, the wave is **plane-polarized**, or *linearly polarized*. With *circular light*, the electric-field vector remains constant in magnitude while it revolves once around with every advance of one wavelength (p. 889). Because the state of polarization of light from an ordinary source is sustained for less than 10^{-8} s, that state is essentially undetectable, and the light is said to be **unpolarized**, or **natural** (p. 889). These basic ideas are discussed in Section 25.1 (Natural Light).

Light that passes through two consecutive linear polarizers whose transmission axes make an angle θ emerges with an irradiance given by

$$I = I_1 \cos^2 \theta \qquad \text{[25.1]}$$

where I_1 leaves the first filter. This equation is **Malus's Law** (p. 893), and it's treated in Section 25.2 (Polarizers). Study Examples 25.1 and 25.2; go over all of the **CD WALK THROUGHS** 💿 , and then try the I-level problems. Read the *Suggestions on Problem Solving* before you begin.

When an unpolarized beam is incident on a surface at an angle θ_p, known as **Brewster's angle**, the reflected light will be totally plane-polarized parallel to the interface. The **polarization angle** is given by

$$\tan \theta_p = \frac{n_t}{n_i} \qquad \text{[25.3]}$$

Light can also be linearly polarized via scattering (p. 897). Reread Section 25.3 (Polarizing Processes) and make sure you understand Example 25.3.

INTERFERENCE

In a region of overlap, two waves of the same frequency can combine constructively or destructively, depending on their relative phase, to produce a redistribution of energy in that area—this is the essence of **interference**. The double-slit setup, known as Young's Experiment, is the premier demonstration of interference. A series of bright maxima appear (p. 901) symmetrically about the centerline at distances of

$$y_m \approx \frac{s}{a} m\lambda \qquad \text{[25.7]}$$

The maxima are displaced from the central axis by the angles θ_m such that

$$a \sin \theta_m = m\lambda \qquad \text{[25.8]}$$

or, since the angles are small, $\theta_m \approx m\lambda/a$. The spacing between fringes on the screen is

$$\Delta y \approx \frac{s}{a} \lambda \qquad \text{[25.9]}$$

These ideas are treated in Section 25.4 (Young's Experiment), which is one of the most important parts of the chapter. Example 25.4 is typical of the problems you may be asked to solve.

Thin transparent films generate fringe patterns when the two waves reflected from the front and back surfaces interfere (p. 905). Where the film is nonuniform, the pattern of so-called *fringes of equal thickness* creates a topographical map. When $n_1 > n_f > n_2$ or $n_1 < n_f < n_2$, reflected maxima occur where the thickness

is

$$d = \frac{m\lambda_0}{2n_f} \qquad m = 1, 2, 3,\dots \qquad \text{[25.11]}$$

An even more commonly occurring situation has $n_1 < n_f > n_2$ or $n_1 > n_f < n_2$. Now maxima occur where

$$d = \frac{(m + \frac{1}{2})\lambda_0}{2n_f} \qquad m = 0, 1, 2,\dots \qquad \text{[25.12]}$$

Section 25.5 (Thin-Film Interference) deals with these notions— study Example 25.5.

DIFFRACTION

The deviation of light from rectilinear propagation is **diffraction**. We concentrate on *Fraunhofer* diffraction, which effectively obtains *where the incident light is planar and the observation screen is at infinity*. In the case of a single slit of width D, zeros of irradiance will occur on both sides of the broad central maximum when

[minima] $$D \sin \theta_{m'} = m'\lambda \qquad \text{[25.13]}$$

where $m' = \pm 1, \pm 2, \pm 3,\dots$. If you wish to learn something about this phenomenon, study Section 25.7 (Single-Slit Diffraction) and go over Example 25.6.

Similarly, when many parallel slits, each separated by a distance a, are present, narrow principal maxima will appear at the same locations as Young's fringes. A multislit system is a **diffraction grating**, and the grating equation is

[maxima] $$a \sin \theta_m = m\lambda \qquad \text{[25.8]}$$

It's treated in Section 25.8 (The Diffraction Grating).

Section 25.9 (Circular Holes and Obstacles) considers the angular half-width of the Airy disk given, in radians, by

$$\theta_a \approx \frac{1.22\lambda}{D} \qquad \text{[25.15]}$$

This expression defines the angular limit of resolution of two closely spaced point sources, and Example 25.8 shows how it's applied.

Key Terms

linear polarization	interference
plane-polarized light	Young's Experiment
circularly polarized light	coherent
natural light	thin-film interference
unpolarized light	fringes of equal thickness
polarizer	optical path length
transmission axis	Michelson Interferometer
analyzer	diffraction
crossed linear polarizers	Fresnel diffraction
Malus's Law	Fraunhofer diffraction
wire-grid polarizer	diffraction grating
polarization angle	Airy disk
Brewster's angle	angular limit of resolution
Brewster's Law	hologram
birefringence	

Discussion Questions

1. Is it correct to say that ideally monochromatic light must be polarized? Explain.

2. Suppose that you had a beam of monochromatic light that you knew to be the in-phase superposition of two equal-amplitude, plane-polarized waves having their electric fields, respectively, horizontal and vertical. How might you verify this situation experimentally?

3. Can polychromatic light be polarized? In what sense is there no such thing as unpolarized liaght? Explain your answers.

4. Natural light is incident on a flat pane of glass at an angle of 45°. Describe the polarization of both the reflected and transmitted beams. How would that change if the light came in at Brewster's angle instead?

5. Suppose that you are looking through a linear polarizer, with its transmission axis vertical, at the surface of a pot of water, and at some angle of reflection find that the glare on the surface vanishes. What is happening? Would the vision change if you now replace the water with benzene, all else held constant? Explain.

6. What was the central problem that Young solved via his double-slit arrangement? How will the fringe pattern in his experiment change if the light source is gradually altered so that its coherence time diminishes? Describe what will happen to the fringe system in some region where the optical path-length difference from the two sources ultimately exceeds the coherence length.

7. An antireflection coating on glass has a thickness that corresponds to one-quarter of a wavelength of red light in the medium of the film. What color light will be reflected and what color transmitted when white light is incident normally? Assume $n_a > n_f > n_g$.

8. EXPLORING PHYSICS ON YOUR OWN: Put a few drops of liquid soap in a cup with a little water and shake it around, making a great froth of bubbles. After several seconds, the bubbles will begin to show bright rainbow color patterns. Why must you wait before the colors appear?

9. EXPLORING PHYSICS ON YOUR OWN: Take a very close look at a worn surface under direct sunlight. A fingernail, an old coin, or the roof of an old car will all show a very fine granular pattern of tiny colored dots. This phenomenon is the so-called *speckle effect*. What might be its cause? Notice how it seems to move as you slowly move your head parallel to the surface.

10. EXPLORING PHYSICS ON YOUR OWN: Fog up a piece of glass with your breath and look through the fogged plate in a darkened room at a white-light point source. You will see a system of concentric colored rings. Of what does this remind you, and what do you think is causing the effect? Figure Q10 shows a point source seen through a glass plate covered with a layer of transparent, spherical lycopodium spores.

11. Consider Young's Experiment where the apertures are two small circular holes, one next to the other, separated by a distance *a*, which is similarly small but still much larger than any of the incoming wavelengths. Assuming normally incident monochromatic plane-wave illumination, describe—qualitatively but in detail—the resulting fringe pattern on a very distant screen parallel to the plane of the apertures. Make a sketch of it.

12. Discuss the interrelationship between aperture size, illuminating wavelength, and *finite* distance to the observation screen (no

Figure Q10

lens after the hole) for near- and far-field diffraction. In other words, what will happen to the pattern on a screen as each of the above is varied?

13. A nearsighted person looking through an adjustable circular aperture at a distant object can improve his vision by making the hole a bit smaller than his pupil—people squint to improve acuity. Still, if the aperture is made less than about 1.0 mm in diameter, the image will again blur. What is happening?

14. Why do bright stars appear to the eye to be larger than faint stars?

15. EXPLORING PHYSICS ON YOUR OWN: Light a long-filament vertical light bulb. If you look at it from several meters away through a vertical slit cut in a 3 × 5 card, you will see a lovely fringe system. Explain what's happening.

16. Using a TV or computer screen as a convenient broad source, stand a few feet from it and peer at it through a narrow slit (for example, the space between two straight fingers). Hold your hand about five or six inches from your eye and form a slit between your fingers about a millimeter or less wide—you can rotate your hand slightly to effectively narrow the slit. Describe and explain the distribution of light within the slit.

17. Figure Q17 shows seven apertures and seven Fraunhofer diffraction patterns. Match the patterns with the apertures that produced them.

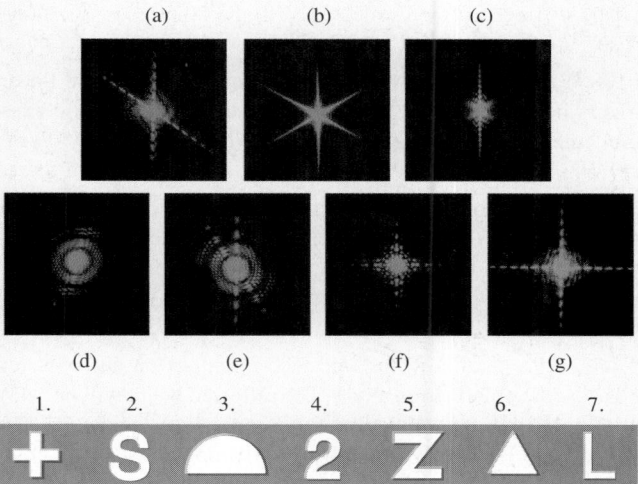

Figure Q17

18. Explain how the Fresnel double mirror works to produce interference fringes (Fig. Q18). What will the pattern look like?

19. Explain the meaning of each of the Key Terms on page 922.

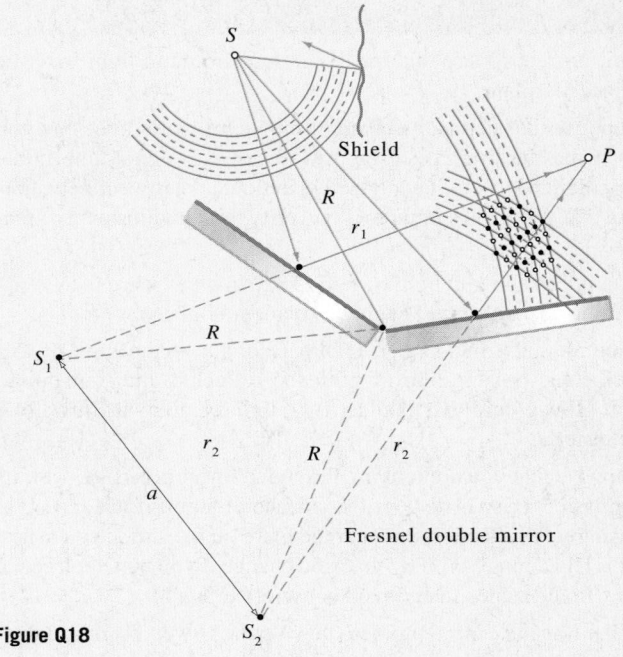

Figure Q18

Multiple Choice Questions

1. The irradiance of a polarized lightwave is   (a) a function of the amplitude (to the first power) of the electric field   (b) independent of the electric field   (c) a function of the square of the amplitude of the electric field   (d) a function of one over the amplitude of the electric field   (e) none of these.

2. Natural light is light that is   (a) only found in nature   (b) linearly polarized   (c) completely unpolarized   (d) polarized but rapidly changing   (e) none of these.

3. The electric field associated with circular light   (a) has a constant magnitude and rotates   (b) has an oscillating magnitude and rotates clockwise   (c) has a sinusoidal amplitude and rotates counterclockwise   (d) has a constant amplitude and oscillates in a fixed plane   (e) none of these.

4. Unlike transverse waves, longitudinal waves cannot   (a) interfere   (b) diffract   (c) be reflected   (d) be polarized   (e) none of these.

5. If 100 W/m^2 of natural white light impinges on two ideal linear polarizers held one behind the other with their transmission axes parallel, the amount of light emerging will be   (a) 100 W/m^2 (b) 50 W/m^2 (c) 25 W/m^2 (d) 200 W/m^2 (e) none of these.

6. Light linearly polarized in the plane-of-incidence impinges on the surface of a glass plate in air at Brewster's angle. The transmitted beam is   (a) nonexistent   (b) linearly polarized in the plane-of-incidence   (c) linearly polarized perpendicular to the plane-of-incidence   (d) partially polarized   (e) none of these.

7. $\mathcal{P}$-state light parallel to the interface impinges on an air-water boundary at the polarization angle. The reflected beam will be   (a) linearly polarized perpendicular to the plane-of-incidence   (b) nonexistent   (c) partially polarized   (d) unpolarized   (e) none of these.

8. For two sources to be coherent, it is both necessary and sufficient that each   (a) be exactly in step, that is, in-phase   (b) have exactly the same amplitude   (c) be monochromatic   (d) be linearly polarized   (e) none of these.

The next six questions refer to Fig. MC9 which shows two coherent in-phase monochromatic point sources S_1 and S_2 radiating EM waves into the space on the right. There is a flat screen a distance s from the sources. Keep in mind the similarity with Young's Experiment.

9. Destructive interference will occur at points   (a) A and C   (b) E and C   (c) B and A   (d) F and G   (e) none of these.

10. Constructive interference will occur at points   (a) A and B   (b) E and C   (c) B and G   (d) F and G   (e) none of these.

11. If only the phase of S_1 is shifted by 180°, constructive interference will occur at points   (a) A and G   (b) B and A   (c) F and C   (d) E and C   (e) none of these.

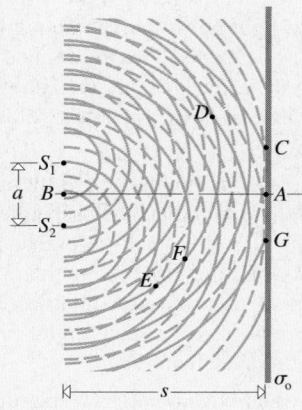

Figure MC9

12. With both sources in-phase, if distance a is reduced   (a) the circular fringes on the screen will move apart   (b) the straightline fringes on the screen will come closer together   (c) the circular fringes on the screen will come closer together   (d) the hyperbolic fringes on the screen will come closer together   (e) none of these.

13. With both sources in-phase, if distance s is increased   (a) the circular fringes on the screen will get larger and move apart.   (b) the straightline fringes on the screen will get smaller and come closer together   (c) the circular fringes on the screen will get smaller and come closer together   (d) the hyperbolic fringes on the screen will get larger and move apart   (e) none of these.

14. With both sources in-phase, if the frequency of both are increased the fringes on the screen will get   (a) larger and move apart.   (b) smaller and move apart.   (c) smaller and move closer together.   (d) larger and move closer together.   (e) none of these.

15. Coherent lightwaves never arise from   (a) two lasers   (b) two pinholes   (c) two candles   (d) two slits   (e) any of these.

16. The middle of the first-order maximum, adjacent to the central bright fringe in the double-slit experiment, corresponds to a point where the optical path length difference from the two apertures is equal to   (a) λ (b) 0   (c) $\frac{1}{2}\lambda$ (d) $\frac{1}{4}\lambda$ (e) none of these.

17. Suppose you take a ring-shaped wire and dip it in liquid soap so that a circular film forms. Now, if you place it horizontally on a platform, which is then spun about a central vertical axis, and look at it under white light,   (a) a series of straight colored fringes will appear   (b) you will see a uniformly bright surface   (c) a series

of concentric colored circular fringes surrounding a black central disk will appear   (d) a set of concentric circular colored bands surrounded by a black fringe will appear   (e) none of these.

18. A narrow slit of width D is illuminated by light coming from a monochromator so that the wavelength can be varied all across the visible spectrum. As λ decreases, the Fraunhofer diffraction pattern, viewed in the focal plane of a lens,   (a) shrinks with all the fringes getting narrower   (b) spreads out with all the fringes getting wider   (c) remains unchanged   (d) alters such that only the central maximum broadens   (e) none of these.

19. A small aperture is located some distance from an observation screen. As the wavelength of the illumination is gradually decreased, the far-field diffraction pattern seen on the screen   (a) remains unchanged   (b) changes to a near-field pattern   (c) changes to a Fraunhofer pattern   (d) stays as a far-field pattern but gradually expands   (e) none of these.

20. The effect of increasing the number of lines per centimeter of a grating is to   (a) increase the number of orders that can be seen   (b) allow for the use of longer wavelengths   (c) increase the spread of each spectral order   (d) produce no change in the diffracted light   (e) none of these.

21. A grating diffracts red light through an angle that is   (a) greater than the angle for blue light   (b) independent of frequency   (c) less than the angle for blue light   (d) the same as the angle for blue light   (e) none of these.

For more Multiple Choice Questions with answers click on Warm-Ups in Chapter 25 on the CD. 🔘

Suggestions on Problem Solving

1. A common error associated with calculations of the polarization angle is to invert the n_t/n_i—watch out for that. Another frequent oversight is to forget that a linear polarizer reduces the irradiance of incident *unpolarized* light by 50%. Don't fail to square the cosine in Malus's Law.

2. When dealing with interference in thin films, be sure to determine whether or not there is a relative phase shift, due to internal and external reflection, before making any numerical calculations.

3. Many of the equations from the analysis of Young's Experiment

carry over directly to the treatment of the diffraction grating. That means that Eqs. (25.7) through (25.9) apply to gratings, as does the expression $\theta_m = m\lambda/a$. Keep in mind that this last equation is always in radians!

4. The number density of lines on a grating is often given in lines per centimeter, *not* per meter, so upon inverting this factor to compute a, the line spacing, you must first convert to lines/m if a is to be in meters.

Problems ✦ Coordinated Problems ✦ Progressive Problems ✦ Solutions

SECTION 25.1: NATURAL LIGHT
SECTION 25.2: POLARIZERS

1. [I] What is the irradiance of a beam of linearly polarized electromagnetic radiation traveling through vacuum if the maximum value attained by its electric field is 1000 V/m?

2. [I] Plane-polarized light, oscillating in a vertical plane and having an energy flux density of 2000 W/m^2, impinges normally on a linear polarizer whose transmission axis is horizontal. What is the transmitted irradiance?

3. [I] A beam of natural light impinges perpendicularly on an ideal linear polarizer with a horizontal transmission axis. If the irradiance of the emerging beam is 1000 W/m², ignoring any losses due to reflections, what was the irradiance of the incident beam?

> **SOLUTION:** An ideal linear polarizer passes half of the natural light incident perpendicularly upon it. And it doesn't matter in what direction its transmission axis is in. Therefore if the emerging beam has an irradiance of 1000 W/m², the incident beam was 2000 W/m².

4. [I] THIS PROBLEM IS CONCERNED WITH LINEARLY POLARIZED LIGHT. Unpolarized light from a lamp passes through an ideal linear polarizer followed by an ideal analyzer. The analyzer is slowly rotated about its central axis, and the emerging irradiance is observed to vary from a maximum of 750 W/m² to a minimum of 0. (a) Draw a diagram. (b) Through how many degrees must the analyzer be rotated in order that the irradiance go from 750 W/m² to 0? (c) What was the value of the irradiance of the light reaching the analyzer? (d) What was the value of the irradiance of the light incident on the first polarizer?

5. [I] A 100-W/m² beam of linearly polarized light with its electric field vertical impinges perpendicularly on an ideal linear polarizer with a vertical transmission axis. Ignoring any losses due to reflections, what is the irradiance of the transmitted beam?

6. [I] Light from an ordinary tungsten bulb arrives perpendicularly at an ideal linear polarizer with a radiant flux density of 100 W/m². Overlooking any reflection, what is its corresponding value on emerging?

7. [I] A beam of natural light with an irradiance of 500 W/m² impinges on the first of two consecutive ideal linear polarizers whose transmission axes are 30.0° apart. How much light emerges from the two? [*Hint: Remember that a linear polarizer "eats" half of the unpolarized light that is incident on it.*]

8. [I] One thousand watts per meter squared of unpolarized light is incident normally on a pair of ideal linear polarizers whose transmission axes make an angle of 45° with one another. What is the irradiance emerging from the pair of filters?

9. [I] Two ideal linear polarizers are placed one behind the other with an angle of 60° between their transmission axes. If light with an irradiance of 1000 W/m² leaves the first and enters the second, how much will emerge from the second? What was the irradiance of the natural light incident on the first polarizer?

10. [I] Many substances such as sugar and insulin are *optically active*; that is, they rotate the plane of polarization in proportion to both the path length and the concentration of the solution. A glass vessel is placed between crossed linear polarizers, and 50% of the natural light incident on the first polarizer is transmitted through the second polarizer. By how much did the sugar solution in the cell rotate the light passed by the first polarizer? This sort of arrangement can be used to determine such things as the amount of sugar in urine.

11. [I] A beam of white light linearly polarized with its electric field vertical and having an irradiance of 160 W/m² is incident normally on a linear polarizer whose transmission axis is 30° above the horizontal. How much light is transmitted?

12. [I] 𝒫-state light aligned with its electric field vector at +40° from the vertical impinges on an ideal sheet polarizer whose transmission axis is at +10° from the vertical. What fraction of the incoming light emerges?

13. [II] THIS PROBLEM IS ABOUT POLARIZATION AND ENERGY FLOW. Suppose we place a sheet of an ideal Polaroid linear polarizer in front of a small tungsten bulb. An observer 200 cm away looks at the bulb through an identical Polaroid. Knowing that the bulb puts out 2.00 W of natural light we want to figure out what the observer will see. (a) What would be the irradiance at the observer if neither Polaroid was in place? (b) Is the light from the bulb polarized? (c) Find the irradiance at the observer when only the first Polaroid is in place. (d) If the transmission axes of the two Polaroids make an angle of 20.0°, how much energy per unit area per unit time will reach the observers eyes?

14. [II] THIS PROBLEM IS CONCERNED WITH THE LINEAR POLARIZER. An unpolarized beam of light is measured to have an irradiance of 1000 W/m². It impinges normally on an ideal Polaroid sheet linear polarizer. It then goes on to pass through another such Polaroid and emerges at 375 W/m². We want to find out the relative orientation of the two Polaroids. (a) Draw a sketch of the arrangement. (b) Assuming no reflection losses, what was the irradiance (I_1) exiting the first Polaroid? Explain. (c) Now, linearly polarized light of known irradiance (I_1) strikes the second Polaroid and emerges at 375 W/m². (d) What law describes this situation? Write it out. (e) Determine θ, the angle between the two transmission axes.

15. [II] Two ideal linear sheet polarizers are arranged with respect to the vertical with their transmission axes at 10° and 70°, respectively. If 𝒫-state light at 40° enters the first polarizer, what fraction of its irradiance will emerge?

16. [II] Four ideal linear polarizers are stacked one behind the other with the transmission axes of the first vertical, the second at 30°, the third at 60°, and the fourth at 90°. What fraction of the incident unpolarized light emerges?

17. [II] Two linear polarizers are placed one behind the other with their transmission axes parallel. The pair is illuminated by unpolarized light, and the emerging irradiance is measured. The first filter is then rotated until its transmission axis moves through 30°. What is the ratio of the irradiance now transmitted by the combination to that previously measured? [*Hint: Initially the irradiance transmitted by the first filter is transmitted by the second.*]

18. [II] Five hundred watts per square meter of unpolarized light impinges on two ideal linear polarizers that are placed one behind the other with their transmission axes both horizontal. One of the filters is then rotated through 45.0°. Determine the transmitted irradiance before and after the rotation.

19. [II] Eight hundred watts per square meter of unpolarized light impinges on two ideal linear polarizers that are placed one behind the other with their transmission axes making an angle of 90.0°. One of the filters is then rotated through 45.0°. Determine the transmitted irradiance before and after the rotation.

20. [II] Eight hundred watts per square meter of linearly polarized light with its electric field vertical impinges on an ideal linear polarizer with a vertical transmission axis. It is followed by an ideal analyzer with its transmission axis horizontal. One of the filters is then rotated through 45.0°. Determine the transmitted irradiance before and after the rotation.

21. [II] Two ideal linear polarizers are in place one behind the other. What angle should their transmission axes make if the incident unpolarized beam is to be reduced to 30% of its original irradiance?

22. [II] Linearly polarized light of irradiance I_0 is incident on a polarizer-analyzer pair. The electric field vector of the light makes an angle θ_1 with the transmission axis of the polarizer and an angle θ_2 with the transmission axis of the analyzer. Prove that the irradiance of the light emerging from the analyzer is given by

$$I_2 = I_0 \cos^2\theta_1 \cos^2(\theta_1 - \theta_2)$$

SECTION 25.3: POLARIZING PROCESSES

23. [I] The reflection of the sky coming off the surface of a pond ($n = 1.33$) is found to completely vanish when seen through a Polaroid filter. At what angle is the surface being examined? Give the answer to two significant figures. [*Hint: This is an air-water interface where $n_t > n_i$.*]

24. [I] At what angle, measured from the normal, should one view a piece of glass ($n = 1.50$) in order to see reflected light that is totally linearly polarized parallel to the interface?

25. [I] A narrow beam of natural light is incident on a sheet of dense flint glass having an index of refraction of 1.92. What is the angle of incidence if the reflected beam is found to be totally linearly polarized?

26. [I] What is the polarization angle for reflection of light from the surface of a piece of glass ($n_g = 1.55$) immersed in water ($n_w = 1.33$)?

27. [I] To determine the electrical characteristics of an opaque dielectric we might measure its index of refraction. Accordingly, a narrow beam of unpolarized light is found to be completely polarized when reflecting at the air-dielectric interface for an angle of incidence of 60.0°. Determine the index.

28. [I] To test a crystal believed to be diamond it is submerged in pure water and a narrow unpolarized light beam (589 nm) is reflected off its surface. Using a linear polarizer it is found that the reflected beam is totally polarized when it emerges into the water at 61.12°. Is it diamond? [*Hint: Use Table 23.1.*]

29. [I] A beam of light is reflected off the surface of a cup of benzene, and the light is examined with a linear sheet polarizer. It is found that when the central axis of the polarizer (that is, the perpendicular to the plane of the sheet) is tilted down from the vertical at an angle of 56.33°, the reflected light is completely passed, provided the transmission axis is parallel to the plane of the interface. From this information, compute the index of refraction of the liquid.

30. [II] A narrow beam of natural light shines onto a flat sheet of polystyrene ($n_p = 1.59$) 1.00 cm thick, and it is found that the reflected beam is linearly polarized. At what angle will the transmitted beam leave the sheet on its bottom side?

31. [I] A sample of a transparent material is immersed in water and a linearly polarized laserbeam is shined at it. The direction of the electric field of the beam is adjusted to lie in the plane-of-incidence. By varying the angle of incidence, it is found that at 61° the beam is entirely transmitted into the material. Determine the material's index of refraction and guess at what it might be.

> **SOLUTION:** With the E-field in the plane-of-incidence, a zero net reflection occurs at Brewster's angle so $\theta_i = \theta_p = 61°$. Accordingly,
>
> $$\tan\theta_p = \frac{n_t}{n_i} = \frac{n_t}{1.33}$$
>
> Thus $n_t = 1.33 \tan 61° = 2.4$, and the material is probably diamond.

32. [II] THIS PROBLEM IS CONCERNED WITH THE LINEAR POLARIZER. A narrow beam of unpolarized light is reflected off a puddle at 75°. A meter shows that the total irradiance reaching the puddle is 200 mW/m². A person wearing Polaroid sunglasses having a vertical transmission axis is viewing the reflected light. We want to learn about the light passing through her sunglasses. (a) What is the angle of incidence at the puddle? (b) What are the values of the incident irradiances (at the puddle), with E-fields both parallel and perpendicular to the plane-of-incidence? (c) Using Table 25.1, what are the reflected irradiances, with E-fields both parallel and perpendicular to the plane-of-incidence? (d) What net irradiance does the person receive without the glasses on? (e) What irradiance reaches the eye when the glasses are being worn? Assume them to be ideal polarizers.

33. [II] A narrow beam of unpolarized light shines onto a flat sheet of glass ($n_g = 1.55$) so that the beam strikes it at an angle of incidence θ_i. The transmitted beam is generally partially polarized with the component polarized in the plane-of-incidence being greater than the component perpendicular to it. Explain that observation. Determine the transmission angle θ_t within the glass such that the transmitted beam is most nearly linearly polarized in the plane-of-incidence.

34. [II] A sheet of glass ($n_g = 1.64$) 1.00 cm thick is held horizontally by a clamp. A narrow beam of light linearly polarized so that its electric field vector is perpendicular to the plane-of-incidence shines onto the glass. Ordinarily a transmitted beam emerges from the far side of the plate, but by varying the incident angle it is possible to make that transmitted beam vanish. Determine the value of θ_i for which this happens.

35. [II] A narrow beam of natural light shines through a linear polarizer whose transmission axis is perpendicular to the beam. A flat sheet of glass ($n_g = 1.60$) is placed so that the beam strikes it at an angle of incidence θ_i. The transmission axis of the polarizer, the incident beam, and the normal to the interface all lie in a common plane. Is there a value of θ_i for which no light is reflected at the air-glass interface? If so, determine the transmission angle θ_t within the glass of the refracted beam. Describe the polarization of that transmitted light.

36. [II] Light reflected from the top of a glass ($n_g = 1.65$) plate immersed in ethyl alcohol ($n_e = 1.36$) is found to be completely linearly polarized. At what angle will the partially polarized beam be transmitted into the plate?

37. [II] A flat glass plate is surrounded by air. A beam of light, linearly polarized so that its electric field vector is parallel to the plane-of-incidence, is shined onto the plate. The light is monitored as the incident angle is varied, and it is found that the downward beam in the glass is strongest at a transmission angle of θ_t. If the beam emerges into the air at the bottom of the plate at an angle of 60.0°, what is the index of refraction of the glass? Determine θ_t.

SECTION 25.4: YOUNG'S EXPERIMENT

38. [I] Two point sources S_1 and S_2 oscillating in-phase send out waves into the air at the same wavelength of 500 nm. Given that there is a minimum irradiance at a point P where the two waves overlap, what is the smallest corresponding path-length difference $(r_1 - r_2)$?

39. [I] THIS PROBLEM WILL HELP US LEARN ABOUT DOUBLE-SLIT INTERFERENCE. Two very narrow slits are illuminated perpendicularly by a He-Ne laser (632.8 nm). The resulting pattern is observed on a screen 6.7 m away. Suppose the separation between the central axis and the center of the first bright off-axis fringe is 21 mm. (a) Draw a sketch of the arrangement. (b) Figure P39 shows bright and dark bands corresponding to possible patterns of fringes made with either red or violet light. Which one best matches this setup? Explain. (c) What is the order of the fringe we are interested in? (d) Write an expression for the height above the axis of the *m*th bright fringe. (e) Watching out for the mix of units, determine the center-to-center separation of the slits. (f) If the wavelength of the light were reduced, which part of Figure P39 would now best match the observed pattern?

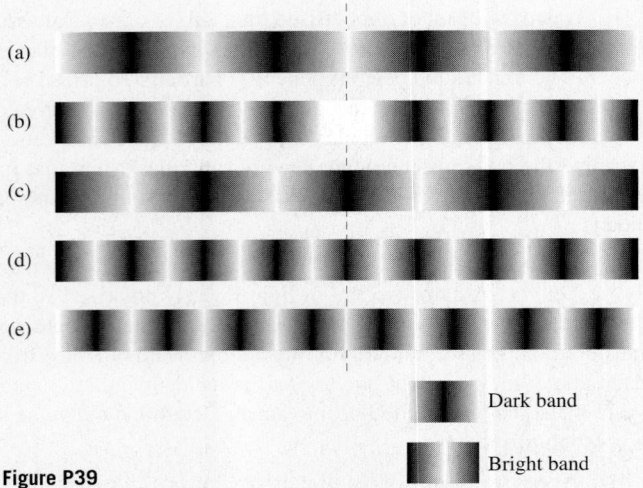

(a)

(b)

(c)

(d)

(e)

Dark band

Bright band

Figure P39

40. [I] A beam of plane waves of 400-nm violet light illuminates two narrow vertical slits in an opaque screen. The center-to-center separation of the slits is 0.20 mm, and the fringe pattern is observed on a wall 6.7 m away. (a) Describe the fringe pattern. Is it spread out vertically or horizontally? Figure P39 shows bright and dark bands corresponding to possible patterns of fringes made with either red or violet light. Which one best matches this setup? (b) What is the center-to-center separation between successive bright fringes? (c) If the light from either slit was phase shifted 180°, which pattern in Fig. P39 would best represent the new fringe system?

> SOLUTION: (a) The slits are vertical, so the fringes are vertical and spread out horizontally. This is a cosine-squared irradiance distribution, and it looks like both Fig. P39a and e, but since the wavelength is short, part e is the better choice. (b)
>
> $$\Delta y \approx \frac{s}{a}\lambda = \frac{6.7\text{ m}}{0.20 \times 10^{-3}\text{ m}} 400 \times 10^{-9}\text{ m} = 13\text{ mm}$$
>
> (c) The central fringe would now be a minimum, so Fig. P39d is the best choice.

41. [I] Blue light at 480 nm impinges perpendicularly on a pair of very narrow parallel slit apertures separated by 0.15 mm. A fringe pattern appears on a screen 2.00 m away. How far from the central axis on either side will we find the second-order bright bands?

42. [I] Red plane waves from a ruby laser (λ = 694.3 nm) in air impinge on two parallel slits in an opaque screen. A fringe pattern forms on a distant wall, and we see the fourth bright band 1.0° above the central axis. Calculate the separation between the slits.

43. [I] Parallel rays of blue light from an argon ion laser (λ = 487.99 nm) are incident on a screen containing two very narrow slits separated by 0.200 mm. A fringe pattern appears on a sheet of film held 1.00 m away. How far (in mm) from the central axis is the fourth irradiance maximum?

44. [I] A 3 × 5 card containing two pinholes, 0.08 mm in diameter separated center-to-center by 0.10 mm, is illuminated by red light from a He-Ne laser (λ = 632.82 nm). If the fringes on an observing screen are to be 10 mm apart, how far away should the screen be?

45. [I] Two parallel slits 0.100 mm apart are illuminated by plane waves of quasimonochromatic light, and it is found that the fifth bright fringe is at an angle of 1.20°. Determine the wavelength of the light.

46. [I] A collimated beam of light (λ = 550 nm) falls on a screen containing a pair of long, narrow slits separated by 0.10 mm. Determine the separation between the two third-order maxima on a screen 2.00 m from the apertures.

47. [II] Red light from a He-Ne laser (λ = 632.8 nm) is incident in air on a screen containing two very narrow horizontal slits separated by 0.100 mm. A fringe pattern appears on a screen 2.00 m away. How far (in mm) above and below the central axis are the first zeros of irradiance? [*Hint: Horizontal slits create a vertical fringe pattern composed of horizontal bands.*]

48. [II] Two vertical narrow slits separated by 0.20 mm are illuminated perpendicularly by 500-nm plane waves. If the centers of the first black bands in the interference pattern are 1.00 mm to the right and left of the central axis, how far away is the screen on which the fringes appear?

49. [II] A screen with two parallel narrow horizontal slits in it is 4.00 m from an observing screen. A beam of parallel rays of 450 nm light shines perpendicularly on the apertures, producing a pattern where the center-to-center separation between the first minima on either side of the central maximum is 0.500 cm. How far apart are the slits?

50. [II] White light falling on two long, narrow slits emerges and is observed on a distant screen. If red light (λ_1 = 780 nm) in the first-order fringe overlaps violet in the second-order fringe, what is the latter's wavelength?

51. [II] Considering the double-slit experiment, derive an equation for the distance $y_{m'}$ from the central axis to each irradiance *minimum*. [*Hint: Consider Eq. (25.4). The first dark bands on either side of the central maximum correspond to m' = ±1.*]

52. [II] A strip of photographic film is completely black except for two horizontal parallel slits, each 1.0 cm long by 0.10 mm wide, separated center-to-center by 0.50 mm. When illuminated by sunlight, all but the zeroth-order fringe show a spread of colors. If violet light appears on a screen 3.0 m away at a distance of 2.40 mm above and below the central axis in the first colored bands, what's its wavelength?

53. [II] Two 1.0-MHz radio antennas emitting in-phase are separated by 600 m along a north-south line. A radio placed 20 km east is equidistant from both transmitting antennas and picks up a fairly strong signal. How far north should that receiver be moved if it is again to detect a signal nearly as strong?

54. [II] Sunlight incident on a screen containing two long, narrow

slits 0.20 mm apart casts a pattern on a white sheet of paper 2.0 m beyond. What is the distance separating the violet ($\lambda = 400$ nm) in the first-order band from the red ($\lambda = 600$ nm) in the second order?

55. [II] Two stereo speakers are 5.00 m apart, up against the east-west wall of someone's living room. They have been wired carelessly so that when one cone is displaced forward by the driving signal, the other is displaced backward. What's the main disadvantage of this arrangement to someone sitting on the midway line equidistant from the speakers? How far due east should the listener move if she is sitting up against the opposite wall 10.0 m away and she wishes to hear a peak in the power level of a 1000-Hz sound? Take the speed of sound to be 346 m/s.

56. [III] With regard to Young's Experiment, derive a general expression for the shift in the vertical position of the *m*th *maximum* as a result of placing a thin parallel sheet of glass of index *n* and thickness *d* directly over one of the slits. Identify your assumptions.

57. [III] Plane waves of monochromatic light impinge at an angle θ_i on a screen containing two narrow slits separated by a distance *a*. Derive an equation for the angle measured from the central axis that locates the *m*th maximum.

SECTION 25.5: THIN-FILM INTERFERENCE

58. [I] A soap film has an index of refraction of 1.334. How thick is the film if one-half of a wavelength of red light (with a vacuum wavelength of 700 nm) extends from one surface across the film to the other surface?

59. [I] A thin film of ethyl alcohol ($n_f = 1.36$) is spread on a glass plate and viewed from directly above as the liquid evaporates. At a certain moment a point on the film appears bright red ($\lambda_0 = 700$ nm); what is the minimum thickness of the alcohol at that instant? [*Hint: Compare the index of refraction of the film to that of the media on either side of it. Is there a phase shift due to internal and external reflection?*]

60. [I] A film of oil ($n_f = 1.42$) 0.000 20 mm thick is spread on a glass plate and viewed from above at nearly 90° to the surface. What color will the film appear to be? Compute the vacuum wavelength.

61. [I] A thin film of turpentine ($n_f = 1.36$) covers a lanthanum flint glass plate ($n_g = 1.80$). What is the minimum thickness the film should have if it is to appear green ($\lambda_0 = 560$ nm) when viewed from above in white light?

62. [I] What is the minimum thickness a film of oil ($n_f = 1.42$) floating on water must have if it is to strongly reflect violet light ($\lambda_0 = 400$ nm) at near normal incidence?

63. [I] A thin film of ethyl alcohol ($n_e = 1.36$) spread on a flat glass plate and illuminated with white light shows a lovely color pattern in reflection. If a region of the film reflects only green light (500 nm) strongly, how thick is it?

64. [I] A glass camera lens with an index of 1.55 is to be coated with a cryolite film ($n \approx 1.30$) to decrease the reflection of normally incident green light ($\lambda_0 = 500$ nm). What minimum thickness should be deposited on the lens?

65. [I] A soap film of index 1.35 appears yellow (580 nm) when viewed from directly above. Compute several possible values of its thickness.

66. [II] A thin film with a refractive index of n_f lies between two media such that $n_1 < n_f > n_2$. Under white-light illumination, when viewed perpendicularly, when do reflection maxima occur? Show that wavelengths that satisfy the equation

$$d = m\lambda_0/2n_f$$

will correspond to transmitted maxima.

67. [II] A film ($n_f = 1.42$) of oil 250 nm thick floats on water. When illuminated from above by white light, what color light will the film reflect most brightly?

> SOLUTION: Since $n_w = 1.33$ this is an instance of Case-2 thin-film interference, where $n_1 < n_f > n_2$. Accordingly, using Eq. (25.12) $d = (m + \frac{1}{2})\lambda_0/2n_f$ and we have to determine both the order *m* and the wavelength. So we'll try a few different allowed values for *m*, namely 0, 1, 2, and 3. Leaving things in nanometers, for
>
> $$m = 0, \quad d = \tfrac{1}{2}\lambda_0/2n_f \quad \text{and} \quad \lambda_0 = 4n_f d = 1420 \text{ nm}$$
> $$m = 1, \quad d = 3\tfrac{1}{2}\lambda_0/2n_f \quad \text{and} \quad \lambda_0 = 4n_f d/3 = 473 \text{ nm}$$
> $$m = 2, \quad d = 5\tfrac{1}{2}\lambda_0/2n_f \quad \text{and} \quad \lambda_0 = 4n_f d/5 = 284 \text{ nm}$$
>
> The only suitable answer in the visible is 473 nm, which from Table 22.1 is blue.

68. [II] THIS PROBLEM EXPLORES THE BEHAVIOR OF AN ANTIREFLECTION COATING. Magnesium fluoride (MgF$_2$) is a hard, transparent, low-index (1.38) material that can be deposited in thin layers on glass (e.g., 1.52) lenses, mirrors, and prisms. We want to compute the thickness of such a film that will result in zero reflection for normally incident green light in the middle of the visible spectrum (550 nm). (a) Draw a sketch of the arrangement showing air, film, and glass. (b) Comparing all the indices, which of the two thin-film cases are we dealing with? Explain. (c) What does Eq. (25.11) provide? (d) Show that $d = m\lambda_f/4$ is the desired equation for the thickness of the nonreflecting film. (e) Watching out for the wavelength, compute the minimum thickness of the film. (f) Use Eq. (25.11) to prove that a film of this thickness does not result in any constructive interference (i.e., strong reflection) in the visible region.

69. [II] The water in a bowl is covered with a film of oil ($n_f = 1.42$) that is 500 nm thick. What visible wavelengths will not be present in the reflected light when the oil is illuminated by white light and viewed from straight above?

70. [II] It's an easy matter to see fringes of equal thickness in a wedge-shaped film (look at the photo at the top of p. 906). Take two flat sheets of glass, one on top of the other, and separate them with a piece of paper (Fig. P70). If the wedge angle is α and the illumination is near normal at a wavelength in the film of λ_f, derive an expression for the distance (x_m) measured from the apex *A* to the successive maxima.

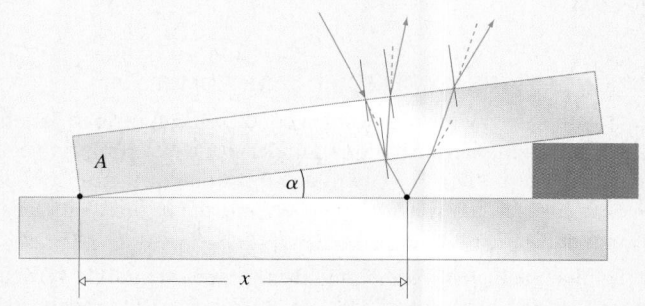

Figure P70

71. [II] Refer to the wedge-shaped film in Problem 70. What is the increase in film thickness as we move along from one maximum to the next? Derive an expression for the spacing Δx between consecutive bright fringes.

72. [II] With Fig. P70 in mind, suppose a wedge-shaped air film is made between two sheets of glass using a piece of paper 7.618×10^{-5} m thick as the spacer. If light of wavelength 550 nm comes down from directly above, determine the number of bright fringes that will be seen across the wedge.

73. [II] A soap film of index 1.340 has a region where it is 550.0 nm thick. Determine the vacuum wavelengths of the radiation that is not reflected when the film is illuminated from above with sunlight.

74. [III] Figure P74 shows a setup for examining the shape of a lens. The lens is placed on an optical flat and illuminated at normal incidence by quasimonochromatic light. The gap between the lens and optical flat constitutes a circularly symmetric, wedge-shaped air film. The amount of uniformity in the resulting concentric system of circular fringes, known as Newton's rings, is a measure of the degree of perfection (see photo on p. 908). Derive the expression

$$x_{\mathrm{m}} = [(m + \tfrac{1}{2})\lambda_{\mathrm{f}} R]^{\frac{1}{2}}$$

for the radius of the *m*th bright ring. Use the fact that $R \gg d$.

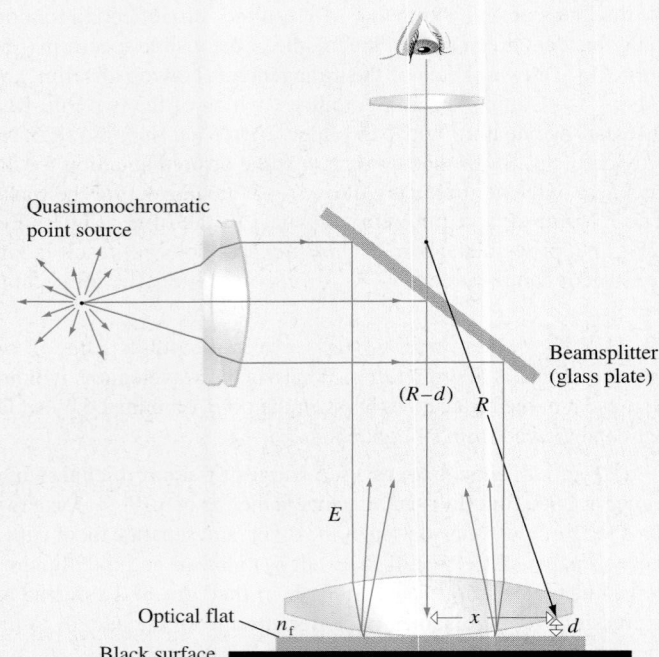

Quasimonochromatic point source

Beamsplitter (glass plate)

$(R-d)$ R

E

Optical flat n_{f} x d

Black surface

Figure P74

SECTION 25.6: THE MICHELSON INTERFEROMETER

75. [I] Suppose you are observing a concentric-ring fringe pattern in a Michelson Interferometer (one similar to the Newton's ring pattern shown on p. 908). By how much must one of the mirrors be moved to cause a dark ring to expand out into the position previously occupied by a bright ring?

76. [I] Imagine that you are observing a system of straight vertical fringes (see the photo on p. 909) in a Michelson Interferometer using light with a wavelength of 500 nm. By how much must one of

the mirrors be moved to cause a dark band to shift into the position previously occupied by an adjacent dark band?

77. [II] A Michelson Interferometer is illuminated with monochromatic light. One of its mirrors is then moved 2.53×10^{-5} m, and it is observed that 92 fringe-pairs, bright and dark, pass by in the process. Determine the wavelength of the incident beam.

78. [II] One of the mirrors of a Michelson Interferometer is moved, and 1000 fringe-pairs shift past the hairline in a viewing telescope during the process. If the device is illuminated with 500-nm light, how far was the mirror moved?

79. [III] Suppose we place a chamber 10.0 cm long (measured inside) with flat parallel windows in one arm of a Michelson Interferometer that is being illuminated by 600-nm light. If the refractive index of air is 1.000 29 and all the air is pumped out of the cell, how many fringe-pairs will shift by in the process?

SECTION 25.7: SINGLE-SLIT DIFFRACTION
SECTION 25.8: THE DIFFRACTION GRATING

80. [I] What happens to the single-slit diffraction pattern if we increase the wavelength of the light to 1.5 times its original value while simultaneously increasing the slit width to 1.5 times its original value?

81. [I] What happens to the linear width of the central maximum of the single-slit diffraction pattern if we increase the wavelength of the light and the slit width both by a multiplicative factor of two while doubling the distance from the aperture to the observation screen?

82. [I] Provided we look at only the first few fringes so that angular displacements are all small, what happens to the single-slit diffraction pattern if we decrease the slit width to half its original value keeping everything else as is?

83. [I] A parallel beam of microwaves impinges on a metal screen that contains a 20-cm-wide, long horizontal slit. A detector moving parallel to the screen locates the first minimum at an angle of 36.87° above the central axis. Determine the wavelength of the radiation. [*Hint: We only have one equation dealing with single-slit diffraction.*]

84. [I] A narrow vertical single slit (in air) is illuminated by IR from a He-Ne laser at 1152.2 nm, and it is found that the center of the first dark band lies at an angle of 4.2° off the central axis. Determine the width of the slit.

85. [I] A single slit 0.10 mm wide is illuminated (in air) by plane waves from a krypton ion laser ($\lambda = 461.9$ nm). Determine the angular position in degrees of the center of the first irradiance minimum with respect to the central axis.

86. [I] A narrow single slit (in air) is illuminated by IR from a He-Ne laser at 1152.2 nm, and it is found that the center of the tenth dark band lies at an angle of 6.2° off the central axis. Please determine the width of the slit.

87. [I] A single slit 0.10 mm wide is illuminated (in air) by plane waves from a krypton ion laser ($\lambda = 461.9$ nm). If the observing screen is 1.0 m away, determine the angular width of the central maximum as defined on p. 914.

88. [I] A transmission grating whose lines are separated by 3.0×10^{-6} m is illuminated by a narrow beam of red light ($\lambda = 694.3$ nm) from a ruby laser. Spots of light, on both sides of the undeflected beam, appear on a screen 2.0 m away. How far from the central axis is either of the two nearest spots?

89. [I] We wish to study several different gratings by holding them each near one eye and looking through each at an ordinary light bulb, blocked off by two pieces of cardboard so that it forms a slit source. What will happen to the locations of the various-order spectra as we go from 200 lines/cm to 400 lines/cm to 800 lines/cm?

90. [I] A diffraction grating with slits 0.60×10^{-3} cm apart is illuminated by light with a wavelength of 500 nm. At what angle will the third-order maximum appear?

91. [I] A diffraction grating produces a second-order spectrum of yellow light ($\lambda = 550$ nm) at 25°. Determine the spacing between lines.

92. [II] Plane waves with a wavelength of 500 nm are incident on a long, narrow aperture 0.10 mm wide. How wide will the central maximum be on a screen 3.00 m from the slits? (Use the fringe-width defined in Example 25.6, p. 914.)

93. [II] THIS PROBLEM WILL HELP US BETTER UNDERSTAND SINGLE-SLIT FRAUNHOFER DIFFRACTION. A beam of plane waves of frequency 4.738×10^{14} Hz impinges perpendicularly on an opaque screen containing a long 0.100 mm wide horizontal slit. We want to study the diffraction pattern appearing on a distant wall parallel to the plane of the slit and from it find the distance to the wall. (a) Draw a diagram of the setup. (b) Write down the equation that locates the minima in the diffraction pattern. How will the pattern be oriented? (c) This equation depends on λ (in air) so calculate it. (d) On the wall we measure the distance from the very center of the pattern to the center of the first dark band (y_1) to be 12.7 mm. Using the fact that for small angles $\sin \theta \approx \tan \theta$, find the approximate value of the sine of the angle (in radians) subtended at the slit (see Example 25.6, p. 914) by this distance. (e) What is the approximate angle in radians subtended? (f) Use that to compute the distance (s) to the wall.

94. [II] THIS PROBLEM TREATS SINGLE-SLIT FRAUNHOFER DIFFRACTION. Collimated light (i.e., plane waves) is passed through a blue filter making it quasimonochromatic. It then perpendicularly illuminates a long 0.400-mm wide vertical slit in an opaque screen. Wishing to determine the wavelength, we examine the diffraction pattern appearing on a wall 7.40 m from the plane of the slit. (a) Draw a diagram of the setup (see Example 25.6, p. 914). (b) Write down the equation that locates the minima in the diffraction pattern. How will the pattern be oriented? (c) Measuring from the central axis, we find that the distance to the center of the second dark band (on either side) is 17.0 mm. Using the fact that for small angles $\sin \theta \approx \tan \theta$, prove that

$$\lambda = \frac{D y_{m'}}{m' s}$$

(d) Determine the wavelength. (e) Which part of Fig. P39 best corresponds to the fringe pattern observed here?

95. [II] A beam of parallel rays of red light from a ruby laser ($\lambda = 694.3$ nm in air) incident perpendicularly on a single slit produces a central bright band that is 40 mm wide on an observing screen 3.50 m beyond. How wide is the slit? (See Problem 92 and use the fringe-width defined in Example 25.6, p. 914.)

96. [II] Consider Problem 52. At what angle would the tenth minimum appear if the entire arrangement were immersed in water ($n_w = 1.33$) rather than air ($n_a = 1.000\ 29$)?

97. [II] Bats use ultrahigh-frequency sound to reflect off targets. Estimate the typical frequency of this rodent sonar, presuming that bats eat moths. Take the speed of sound to be 330 m/s and a moth to be about 3.0 mm across. [*Hint: Remember that diffraction becomes pronounced as the wavelength becomes comparable to the aperture or obstacle size.*]

98. [II] White light falls normally on a transmission grating that contains 1000 lines per centimeter. At what angle will we find red light ($\lambda = 650$ nm) in the first-order spectrum?

99. [II] Suppose we shine light from a sodium lamp into a transmission grating that has 1.000×10^6 lines per meter. The light contains two strong yellow components with wavelengths of 589.592 3 nm and 588.995 3 nm. Determine the diffraction angles in the first-order spectrum measured from the center line for each of these emissions.

> **SOLUTION:** The equation for the grating, Eq. (25.8), is $a \sin \theta_m = m\lambda$. Here $m = 1$ and $\sin \theta_1 = 1\lambda/a$. The space between lines is $a = 1/(1.000 \times 10^6 \text{ lines/m}) = 1.000 \times 10^{-6}$ m and so
>
> $$\sin \theta_1 = \frac{1\lambda}{a} = \frac{589.592\ 3 \text{ nm}}{1.000 \times 10^{-6} \text{ m}} = 0.589\ 592\ 3$$
>
> and for the 589.592 3 nm line $\theta_1 = 36.13°$. Similarly,
>
> $$\sin \theta_1 = \frac{1\lambda}{a} = \frac{588.995\ 3 \text{ nm}}{1.000 \times 10^{-6} \text{ m}} = 0.588\ 995\ 3$$
>
> and for the 588.995 3 nm line $\theta_1 = 36.09°$.

100. [II] THIS PROBLEM CONCERNS THE DIFFRACTION GRATING. A gas discharge lamp containing hydrogen radiates a number of distinct wavelengths. We wish to examine the emissions at 434.046 nm and 410.173 nm. Accordingly, the light will be passed through a grating having 1000 lines per centimeter. The resulting diffraction pattern will be examined on a screen 2.00 m away. Looking at the first group of off-axis colored lines (i.e., the first order-spectrum), we should see the two lines corresponding to 434.046 nm and 410.173 nm. We want to predict their separation on the screen. (a) Draw a diagram of the setup. (b) Using the fact that for small angles $\sin \theta \approx \tan \theta$, prove that each line will be found at a perpendicular distance y_m from the central axis, where

$$y_m = \frac{m s \lambda}{a}$$

(c) If there are 1000 lines per cm on the grating, how many centimeters are there per line? That is, what's the space between grating lines? Now find the space in meters. (d) Determine the distance each spectral line is from the central axis. (e) What is their separation?

101. [II] Light from a sodium lamp has two strong yellow components at 589.592 3 nm and 588.995 3 nm. How far apart in the first-order spectrum will these two be on a screen 1.00 m from a grating having 10 000 lines/cm? (a) Assume Eq. (25.7) is accurate enough for our purposes. (b) Show that a more exact solution, arrived at using Eq. (25.8), yields a separation of 1.13 mm.

102. [II] Sunlight impinges on a transmission grating having 5000 lines per centimeter. Does the third-order spectrum overlap the

second-order spectrum? Take red to be 780 nm and violet to be 390 nm.

103. [III] Two long slits 0.10 mm wide, separated (center-to-center) by 0.20 mm, in an opaque screen are illuminated by light of wavelength 500 nm. How many Young's fringes will be seen within the central bright band on a screen 2.0 m away? Determine a general rule for this situation in terms of the number N where $a = ND$. [*Hint: Count fractions of fringes too.*]

104. [III] Light with a frequency of 4.0×10^{14} Hz is incident on a grating having 10 000 lines per centimeter. What is the highest order spectrum that can be seen with this device? Explain.

105. [III] A device used to measure the diffracted angle, and therefore the wavelength, of light passing through a grating is called a *spectrometer*. Suppose that such an instrument while in vacuum on Earth sends 500-nm light off at an angle of 20.0° in the first-order spectrum. By comparison, after landing on the planet Mongo, the same light is diffracted through 18.0°. Determine the index of refraction of the Mongoian atmosphere.

SECTION 25.9: CIRCULAR HOLES AND OBSTACLES

106. [I] A beam of parallel rays with a wavelength of 600 nm impinges perpendicularly on a converging lens 1.20×10^{-2} m in diameter. If the lens has a focal length of 0.50 m, what is the angular width of the central disk of the diffraction pattern appearing on a screen at the focal plane?

107. [I] Light from a very distant star enters a telescope that has an objective mirror 1.00 m in diameter with a focal length of 5.00 m. The image of the star is a tiny circular disk surrounded by a series of faint rings. What is the diameter of the disk in red light (650 nm)? [*Hint: We need twice the radius of the Airy disk.*]

108. [I] The light from a very distant star enters a convex lens that forms a real image of it on the lens's focal plane 1.50 m away. If the diameter of the lens is 0.50 m, how big will the central disk of the diffraction pattern be in 500-nm light?

109. [I] The second largest refracting telescope in the world is at the Lick Observatory. Its objective has a diameter of 0.914 m (i.e., 36 in.) and a focal length of 17.1 m (i.e., 56 ft). When it forms a real image of a far away star, that image is a tiny disk of light surrounded by very faint rings. What is the diameter of that disk in 580-nm yellow light?

110. [I] If you peeped through a 0.75-mm-diameter hole at an eye chart, you might notice a decrease in visual acuity. Compute the angular limit of resolution, assuming that it's determined only by diffraction; take λ within the eye to be 550 nm. Compare your results with the value of 1.7×10^{-4} rad that corresponds to a 4.0-mm pupil.

111. [II] The Mount Palomar telescope has a 508-cm-diameter objective mirror. Determine its angular limit of resolution at a wavelength of 550 nm, in radians, degrees, and seconds of arc.

112. [II] With Problem 111 in mind, how far apart must two objects be on the surface of the Moon if they are to be resolvable by the Palomar telescope? The Earth-Moon distance is 3.844×10^8 m; take $\lambda = 550$ nm.

113. [II] Compare the results of Problem 112 with the human eye; that is, how far apart must two objects be on the Moon if they are to be distinguished by eye? Assume a pupil diameter of 4.00 mm, and take λ within the eye to be 550 nm. In actuality, the eye will need a separation of about 10 times this amount.

114. [II] The Hubble Space Telescope has a diameter of 2.4 m. What is its diffraction-limited angular resolution at $\lambda = 500$ nm? It has been said that the Hubble can spot a firefly from a distance of about halfway around the world and that it can resolve two fireflies at the same distance provided they're separated by more than 5 m. Is that reasonable? The Earth has a diameter of 1.28×10^7 m.

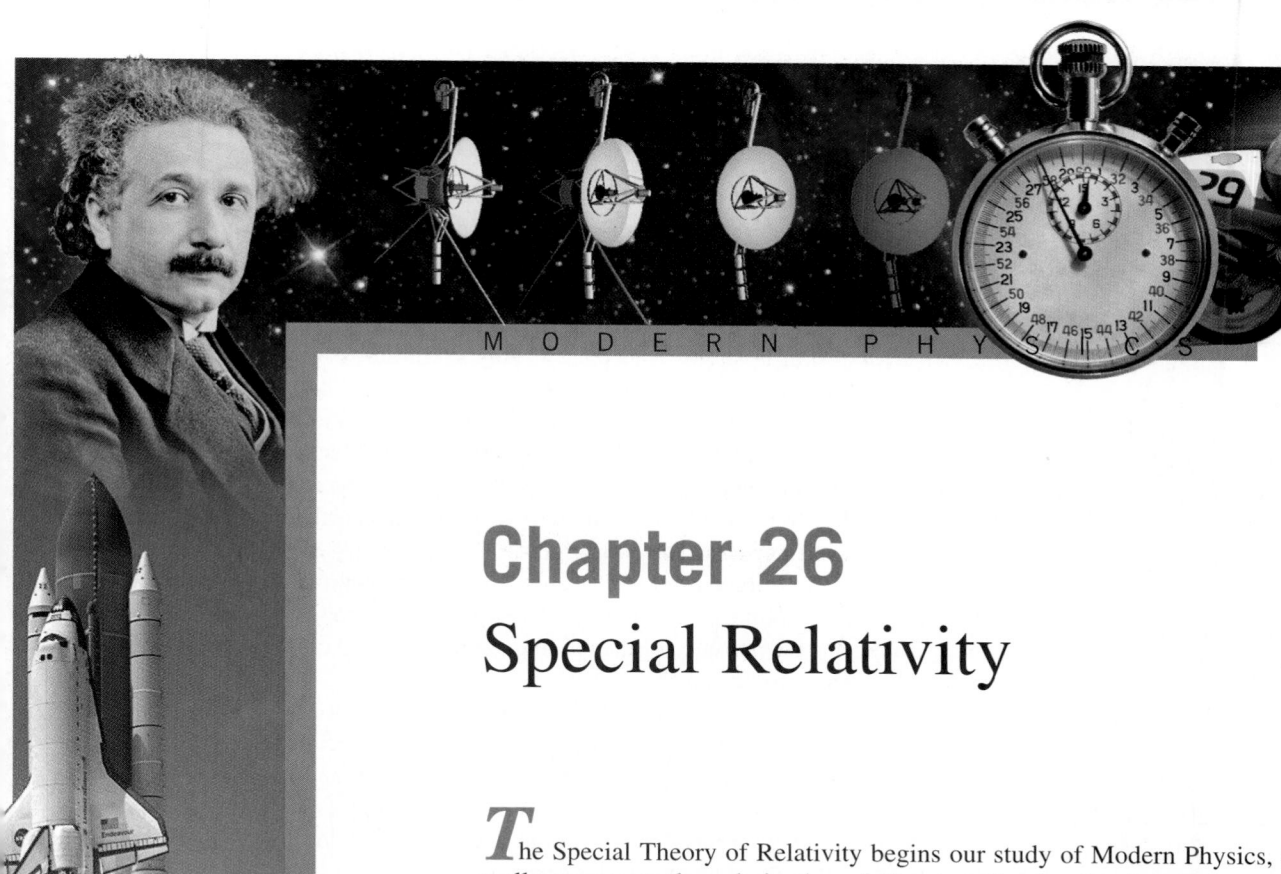

Chapter 26
Special Relativity

*T*he Special Theory of Relativity begins our study of Modern Physics, even though it really represents the culmination of Classical Physics. It arises from a rethinking of the essence of space and time. So profound are the ideas that they led to a flood of new conclusions, many of which were no less than shocking: time is relative and flows differently in systems that are in motion with respect to each other; the length of an object is not absolute but depends on its relative motion with regard to each particular observer; the speed of light, c, is invariant and, moreover, is the upper limit of all speeds in the Universe; the gravitational and electromagnetic interactions propagate at c rather than infinitely swiftly; except for Newton's Second Law (as he gave it, p. 93), most of the equations of classical dynamics are only approximations; mass and energy are equivalent in a way never conceived of before. These ideas and more follow from two simple postulates that Einstein introduced and then elegantly developed.

Before the Special Theory

The research that laid the groundwork for the Special Theory was concerned with the behavior of light. Ironically, Relativity has nothing to do with light per se, although it has a lot to do with the constancy of the speed of light.

Prior to Einstein's work in 1905, it was widely believed that space was filled with aether and light was an electromagnetically induced vibration of that invisible elastic "goop." The common wisdom was that aether was a transparent solid that could support transverse lightwaves and at the same time a perfect fluid that could be parted effortlessly by the wandering planets.

26.1 The Michelson–Morley Experiment

The aether, even though it doesn't exist, played a crucial part in the development of the Special Theory. At that time, the most important single observation was the Michelson–Morley experiment, which had failed to detect any evidence of the aether and which seemed to suggest the invariance of the speed of light. It is arguably one of the most important experiments ever performed but at the same time, it was a disappointing failure for the man who performed it.

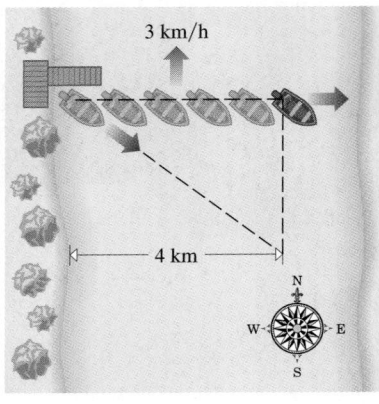

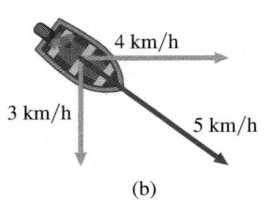

Figure 26.1 A boat traveling due east across a river flowing north at 3 km/h. It accomplishes this by having a southerly velocity component of 3 km/h in addition to its easterly component of 4 km/h.

James Clerk Maxwell, in 1879, the last year of his life (and the year Einstein was born), wrote a letter that was ultimately published. In it, he discussed a scheme for measuring the speed (v) at which the orbiting Earth plows through the aether. As the Earth moves, the ordinarily motionless aether presumably streams over the planet. The back-and-forth time of transit of a light beam in this aether wind should be different along or across the flow. To see this, imagine a wide river streaming north at 3 km/h and a motorboat capable of a single speed with respect to the water of 5 km/h (Fig. 26.1). Compare the transit times for two trips: 4 km upstream and back versus 4 km across stream and back. Going upstream, bucking the current, the boat makes 5 km/h − 3 km/h = 2 km/h and reaches its destination in 2 h, whereupon it turns around and returns at 5 km/h + 3 km/h = 8 km/h, arriving back after an additional $\frac{1}{2}$ h—the total trip taking $2\frac{1}{2}$ h. By contrast, to cross the river directly, the boat must be headed south a bit so that it has a southerly velocity component equal to the river's northerly 3 km/h (p. 42). The two north-south velocities cancel, and the boat travels due east. The 3-4-5 right-triangle geometry gives the boat a net easterly speed of 4 km/h. It travels out for 1 h and then returns (again heading somewhat south) after another 1 h—total transit time: 2 h. The time difference between the two journeys (here, $\frac{1}{2}$ h) reveals the presence of the moving medium. If v (the river's speed) were zero, the time difference would vanish.

Unfortunately, as Maxwell pointed out, in the case of an aether wind (instead of a river) flowing at v and a light beam (instead of a boat), the effect depends on $(v/c)^2$, which "is quite too small to be observed." Even with the aether streaming past the planet at the Earth's orbital speed of $v = 3 \times 10^4$ m/s (i.e., ≈66 000 mi/h), the quantity $(v/c)^2 \approx 10^{-8}$ and the transit-time difference for a 4-km-out-and-4-km-back journey is a mere tenth of a millionth of a millionth of a second. Not very promising.

Maxwell's letter came to the attention of a young instructor at the United States Naval Academy, Albert Abraham Michelson. Michelson took a leave of absence in 1880 to study in Europe and, while there, accepted Maxwell's challenge. In the past, several experimenters had used interferometers to examine the effects of motion on the transmission of light through the aether. But all of these efforts were limited in their precision; and though they failed to detect the aether, the results were unconvincing. Now Michelson designed an entirely new instrument that would allow him to make the decisive measurement precisely enough to reveal $(v/c)^2$ variations. His interferometer (funded by none other than Alexander Graham Bell) would surely detect an aether wind carrying along the lightwaves.

Figure 26.2 is a schematic drawing of the Michelson Interferometer; it's the same device we studied earlier (p. 909). The difference between the along-the-aether and the across-the-aether transit times, Δt, is

$$\Delta t \approx \frac{L}{c}\beta^2 \tag{26.1}$$

where $\beta = v/c \approx 10^{-4}$. One wave travels a little longer than the other, and they arrive a bit out-of-phase. Hence, the fringes are shifted from where they would have been were the Earth at rest (Fig. 26.3). Since we can't stop the Earth, this shift is not immediately apparent, but that can be rectified by rotating the entire interferometer while observing the fringes. This will gradually interchange the two arms so that the one that was along the aether wind is now across it and vice versa. The cross-stream wave will go from leading by Δt to lagging by Δt as it becomes an upstream-downstream wave. That change of 2 Δt in time is equivalent to an introduced path-length difference of c2 Δt, which causes the fringe pattern to move. A shift of one wavelength results in a displacement of one bright-dark fringe-pair, and the observer (following around as the device is rotated through 90°) should see $N = $ c2 $\Delta t/\lambda \approx 2L\beta^2/\lambda$ fringe-pairs move past a cross hair in a viewing telescope.

When Michelson first performed the experiment in 1881, N should have been about four one-hundredths of a fringe, but he saw no shift at all. Confident of his measurements, he published the startling result that "there is no displacement of the interference bands." Still, his so-called *null result* was not very persuasive. With the urging of Lord Rayleigh and Lord Kelvin, Michelson decided to redo the experiment in an improved fashion and set-

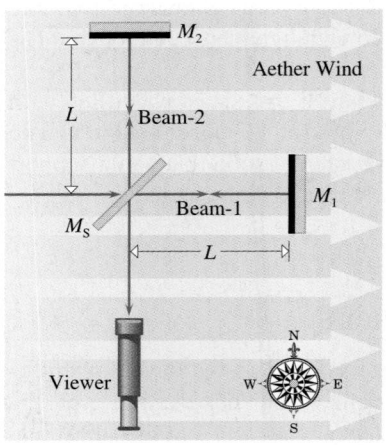

Figure 26.2 A simplified version of the Michelson Interferometer placed in the supposed aether wind.

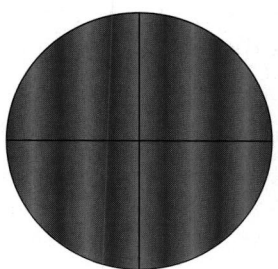

Figure 26.3 With the interferometer set up to produce straight fringes, we would see a pattern like this one in the viewing telescope. It occurs when M_1 and M_2 are perpendicular to the plane of the instrument but not to each other.

AETHER, DOES IT REALLY EXIST?

FitzGerald maintained that there was no fringe shift in Michelson's experiment because the arm of the interferometer aligned along the wind shrunk. And there's no easy way to measure that shrinkage because all rulers will likewise shrink. It was not until 1932, long after Einstein had put things right, that the notion of an aether-wind-induced contraction was laid to rest experimentally by R. Kennedy and E. Thorndike. In 1990, a modern laser version of their experiment accurate to within 70 parts per million confirmed their result.

Stimulated by Michelson's observation, physicists subsequently performed many elaborate optical, electrical, and magnetic experiments, all of which failed to detect the motion of the Earth with respect to the aether. In 1900, the brilliant French mathematician Jules Henri Poincaré wrote: "Our aether, does it really exist? I do not believe that more precise observations could ever reveal anything more than *relative* displacements."

The luminiferous aether, that is the only substance we are confident of in dynamics.... One thing we are sure of, and that is the reality and substantiality of the luminiferous aether.

LORD KELVIN
POPULAR LECTURES AND ADDRESSES (1891)

tle the matter. He was joined by E. W. Morley, and together they enlarged the apparatus to the point where the expected shift was now four-tenths of a fringe. Still no appreciable shift was observed—and this time the results could not be neglected—there was no detectable aether wind. ***The speed of light is not influenced by the motion of the Earth.***

A modern version of the Michelson–Morley experiment performed in 1979 used stable lasers to improve the precision by a factor of 4000; it, too, found no sign of an aether wind.

The Lorentz–FitzGerald Contraction

In 1892, G. FitzGerald proposed a rather imaginative hypothesis to get around the Michelson–Morley result and still keep the aether wind blowing. The idea was elaborated by Lorentz and is today known as the **Lorentz–FitzGerald Contraction**. However ad hoc the notion was, it would re-emerge over a decade later as a natural consequence of Einstein's Special Theory of Relativity. What FitzGerald proposed was that the aether wind exerts a pressure on a body moving through it and the body compresses slightly: *every object moving at a speed v contracts along the direction of motion by a factor equal to* $\sqrt{1 - \beta^2}$. If the $M_S M_1$ arm of the interferometer shrinks so that L becomes $L\sqrt{1 - \beta^2}$, $\Delta t = 0$ (see Problem 1). There is no such aether pressure, but strange as it may seem, there is a Lorentz–FitzGerald Contraction.

Example 26.1 **[I]** A particle is moving at a speed of 0.200 0c. Determine the value of $\gamma = 1/\sqrt{1 - \beta^2}$ at that speed. Redo the calculation for a speed of 0.002 0c. Don't worry about significant figures. This quantity (γ) will come up over and over again in Relativity.

Solution This problem is here just to familiarize you with γ. (1) TRANSLATION—An object is moving at a known speed; determine the corresponding value of γ. (2) GIVEN: Speeds of 0.200 0c and 0.002 0c. FIND: $\gamma = 1/\sqrt{1 - \beta^2}$. (3) PROBLEM TYPE—Relativity. (4) PROCEDURE—Straightforward substitution. (5) CALCULATION—$\beta = v/c = 0.200 0$,

$$\gamma = 1/\sqrt{1 - \beta^2} = 1/\sqrt{1 - 0.040 0} = 1/0.979 8 = \boxed{1.021}$$

Now for $v = 0.002 0c$, $\beta = 0.002 0$, $\beta^2 = 4.0 \times 10^{-6}$, $(1 - \beta^2) = (1 - 0.000 004 0) = 0.999 996 0$, and $\sqrt{1 - \beta^2} = 0.999 998 0$. Thus,

$$\gamma = 1/\sqrt{1 - \beta^2} = \boxed{1.000 002}$$

Quick Check: Knowing that we'll need it again (p. 954), we now derive a useful approximation. From the binomial expansion with $x = -\beta^2$ and $n = -\frac{1}{2}$, we get

continued

$$(1 + x)^n = 1 + nx + \frac{n(n-1)x^2}{2} + \cdots$$

[for small β] $\gamma \approx 1 + \beta^2/2$

All the terms beyond the second one can be dropped since β^2 is tiny, and therefore $x^2 = (-\beta^2)^2$ is minuscule. Hence, $1/\sqrt{1-\beta^2} \approx (1 + nx) \approx 1 + \beta^2/2 = 1 + 0.000\,004\,0/2 = 1.000\,002$. And all's well. The important result here is that

[For more worked problems click on **WALK-THROUGHS** *in* **CHAPTER 26** *on the* **CD.**]

How Comfortable the Old Ideas. The beginner will find it best to accept the aether theory, at least as a working hypothesis.... Even if future developments prove that the extreme relativists are right and that there is no aether, it is likely that the change will involve no serious readjustments so far as explanations of the ordinary phenomena are concerned.

A. A. KNOWLTON
PHYSICS FOR COLLEGE STUDENTS (1928)

The Special Theory of Relativity

Long after his billowing shock of hair had turned gray, Albert Einstein recalled how, when he was sixteen, he began to struggle with a troubling paradox. Light was understood to be electromagnetic, an intricate oscillatory web of linked time-varying electric and magnetic fields rippling through space. If we could travel out at speed c alongside a pulse of light, what would we see? Moving in imagination next to a wave peak, the disturbance would appear unchanging, motionless; yet electromagnetic theory does not allow such a situation. A stationary, nonvarying field-pulse cannot exist. In that regard, light is totally unlike a sound wave or even a stream of bullets, each of which can be followed and examined as if frozen in flight. Light is self-sustained by *change*. It is a thing of interwoven fields that, by alternation, generate each other—no variation in time, no existence. A stationary observer must see the pulse, even as the observer moving at c sees nothing—how strange. At sixteen, Einstein had recognized a profound dilemma, a conflict between Newton's mechanics, which allowed travel at lightspeed, and Maxwell's electrodynamics, which couldn't abide with it. One of these two formalisms was wrong.

26.2 The Two Postulates

At twenty-six (1905), Einstein published the *special* or *restricted* theory—restricted in the sense of specialized, since it pertained only to *uniform motion*. The theory assumed the validity of two *postulates* that he believed were correct but could not otherwise prove. He then set out to derive the physical implications of those postulates and, in the process, completely recast our understanding of space and time.

The Principle of Relativity

The first of the two postulates is called the ***Principle of Relativity***. It's a generalization of work done by Galileo and Newton. Both men recognized that uniform motion had no perceivable effect on mechanical systems. One can play pool or Ping Pong aboard a ship and never know the vessel is moving, regardless of the velocity, as long as it's constant. The woman juggling oranges in the lounge of a 747 can't tell from the behavior of the fruit in the air whether she is cruising at 600 mi/h or sitting at rest on the runway. This, then, is the *Classical Principle of Relativity*: *the laws of mechanics are the same for all observers in uniform motion.*

Newton had struggled with the distinction between absolute and relative motion. Was there something somewhere in the vast Universe that was totally stationary, something absolutely at rest from which all motion could be reckoned absolutely? "I hold space to be at rest" wrote Newton—space, unchanging and immovable, was his fixed reference frame. By the end of the 1880s, a motionless aether filled all space and provided a material backdrop for absolute rest throughout the Universe. A thing moves absolutely when it moves with respect to the aether. The aether had two theoretical functions that justified its existence in the face of untold contradictions: it provided the medium for lightwaves, and it was the signpost of absolute rest.

Albert Einstein (1879–1955).

HIDING RELATIVITY IN A DRAWER

In 1905, Einstein was an unknown clerk in the Bern Patent Office, Switzerland. Newly married, shy but friendly, the young man had plenty of time to think and create (reviewing patent applications wasn't very taxing). He often worked on his own ideas, hiding the calculations in a drawer whenever footsteps approached. That was the year he published five papers in the prestigious *Annalen der Physik*, representing three extraordinary new developments in physics. One of these was his first paper on Relativity, *"Zur Elektrodynamik bewegter Körper"* ("On the Electrodynamics of Moving Bodies").

The **First Postulate**: The laws of Nature—the ways in which things behave—are the same in all inertial systems regardless of their speeds.

The **Second Postulate**: The speed of light in vacuum is completely independent of the motion of the source emitting it.

Aether Persisted. We have learned too that radiant heat energy is believed to be transmitted by a medium called the aether. At the present time, some scientists believe that other aether waves produce various other effects.... It is possible, then, that light waves are aether waves.

CHARLES E. DULL
MODERN PHYSICS (1939),
A HIGH SCHOOL TEXT

We speak of a uniformly moving observer as an **inertial observer**—someone standing still in a train, plane, or rocket ship that is itself moving at a constant velocity. *A system that is moving at a constant velocity is an inertial system*. The name comes from the fact that the Law of Inertia holds in all inertial systems. A body at rest does not tend to stay at rest, nor does a body in motion tend to stay in uniform motion in a straight line if it's traveling in a system that's accelerating.

While in an inertial system, jet-plane commuters expect no novel experiences. Our battery-operated toothbrushes still hum along, computers compute, popcorn pops, cola tastes the same, and pizza smells the same. Experience suggests that not just the laws of mechanics are the same, but all the laws of physics are the same for inertial (i.e., uniformly moving) observers. All aspects of the physical environment are unaffected by uniform motion, and life goes on quite normally at 2 km/h or 2000 km/h. Einstein raised this conjecture to the status of his **First Postulate**, his **Principle of Relativity**:

All the laws of physics are the same for uniformly moving observers.

It is impossible, using experiments performed within inertial systems, to distinguish any one such system from any other; therefore, no experiment can establish whether a particular inertial system is uniformly moving or at "absolute rest." The concept of absolute rest thereby loses all significance; if it can never be determined, motion is relative, not absolute. The Principle of Relativity logically abolishes the concept of absolute rest and, along with it, the concept of absolute motion—**motion is relative**.

All the failed experiments that had vainly tried to measure the absolute motion of the Earth with respect to the aether had led Poincaré to the Principle of Relativity, and now Einstein embraced the same conclusion. He dismissed the aether wind, boldly dispensing with the concept of aether entirely:

The introduction of a "luminiferous aether" will prove to be superfluous, inasmuch as the view here to be developed will not require an "absolutely stationary space."

The Constancy of the Speed of Light

Einstein's **Second Postulate** is the **Principle of the Constancy of the Speed of Light**:

Light propagates in free space with a speed c that is independent of the motion of the source.

Now this much in itself is both orthodox and reasonable; after all, the speed of sound is independent of the motion of the source. A sound wave is launched into a medium, and the speed of the disturbance is only determined by the physical characteristics of the medium. The speed of the source is irrelevant. If light is a wave in the aether, this statement makes obvious sense.

Remember that the equations of Electromagnetic Theory—Maxwell's Equations—led to a wave equation that provided the speed of light in vacuum (p. 784). Assuming Maxwell's Equations are right and given the First Postulate, it must be that this same wave equation is applicable in all inertial systems—the vacuum speed of light measured on Earth or inside a rocket ship must be the same, independent of any uniform relative motion. In Maxwell's theory, the speed of light is a constant, not a motion-dependent variable. In other words, *the speed of light measured with respect to an inertial system must be the same for all such systems*.

The fact that the speed of light is independent of the motion of the source is not at all troublesome, but is it also independent of the motion of the detector (i.e., the observer)? Certainly, the speed of sound is not; if the detector rushes toward the source, moving with respect to the air, the measured speed of sound increases. Just imagine two identical ships headed toward a motionless sound-emitting buoy, one steaming along at full speed and the

Figure 26.4 Two spaceships, one at rest and the other moving at speed v with respect to a source S. Both have on-board light beams, which they measure to travel at c. Both must measure the light from S to travel at c as well.

other dead in the water. On both ships, the time it takes a blast of sound to sweep from bow to stern is measured, and the speed of the wave is computed in that inertial system. Clearly, for the ship moving toward the buoy, that time will be shorter (during the interval it takes the sound to traverse the ship, the stern will advance somewhat toward the pulse, shortening the effective length) and the wave speed will be determined to be faster.

Figure 26.4 depicts the same arrangement, this time with spaceships and lightwaves. Both ships receive light from the outside source S, and both compare the speed of that light to the speed of the light from their own sources. Ship-1 at rest with respect to the beacon must measure both its beam and the beacon's beam to have the same speed c. Ship-2 (moving at v with respect to S) must measure the beam from its own source to travel at c because of the Second Postulate, but it must also measure the light from the beacon to travel at c as well. If it didn't, that would mean that the two beams could be used together as a motion detector to establish that ship-2 was actually moving, which is not allowed by the First Postulate. The conclusion is an astonishing one: ***no matter how fast a light source moves toward or away from an inertial observer, and no matter how fast the observer moves toward or away from the source, the speed of the light passing from one to the other in vacuum will always be* c—the speed of light is constant**; it *is* absolute. The Michelson–Morley experiment ends with a null result because the speed of light along each arm is identically the same; there is no fringe shift because there is no transit-time difference, $\Delta t = 0$.

The two postulates separately are innocent enough. It's when we mix them together, when we demand that they both apply at once, that things seem to become fantastic. The paradox of Einstein's youth is now no longer a paradox—one simply cannot travel next to a light beam and catch up to it. The spaceship in Fig. 26.5 could be rushing either toward or away from the source at any speed you like, say, 99% of c, and still the inertial observer aboard it will measure the speed of the beam to be c! This is one of the premier conclusions of the analysis; its disturbingly "illogical" nature suggests that our familiar, comfortable understanding of space and time requires revision.

To our best knowledge, neutrinos and gravitons both also travel at c, and there must equally well be a *Principle of the Constancy of the Speed of Neutrinos*. The important thing is the speed c, not the photons doing the traveling—they just happen to be the most convenient, easily detected, c-speed probe.

SPACE & TIME

The link between space and time and c hardly intrudes in our everyday lives—it doesn't matter that the people you are talking to 10 feet away are also 10 thousandths of a millionth of a second away. Light takes about 10^{-9} s to travel 1 foot, and people are seen not as they are now but as they were 10×10^{-9} s ago. When you look out the window, the scene on your retina—the scene you see—is not all happening at the same moment; the view of the Sun in the sky shows it as it was 8.3 minutes before Uncle George smiled in the foreground. The stars in a photo of the night sky were not all there looking as they do at the same moment, even though the light from them arrived on the film at the same instant. For that matter, the stars you see might no longer exist even while you are "looking" at them: a star 100 light-years away appears as it was 100 years ago when the light you see now first left it. And if any of this seems strange, it is because you have always thought of the world as if you saw things the instant they happened, as if c = ∞, but it doesn't.

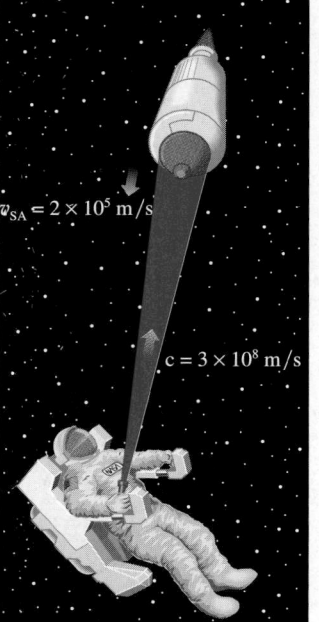

Figure 26.5 The speed of light in vacuum will be measured to be constant, regardless of the motion of either the source or the receiver. Here, a ship receives a light beam from an astronaut toward whom it is rushing at a speed v_{SA}. Regardless of the relative motion between the ship and the astronaut, a person on board the vessel will measure the light to arrive at c.

$v_{SA} = 2 \times 10^5$ m/s

$c = 3 \times 10^8$ m/s

Experimental Confirmations of the Second Postulate

At the time Einstein proposed the Second Postulate, there had been no direct experimental evidence to confirm it—it was at birth a purely logical conjecture. That situation has changed considerably over the intervening decades, and the extensive confirmation that now exists has transformed the Second Postulate into what might better be called the *Law of the Constancy of* c.

In a most convincing contemporary experiment, subatomic particles known as neutral pions (p. 1095) were produced at a tremendous speed of 99.98% of c. Free neutral pions have a mean life of only 8.7×10^{-17} s before naturally decaying—vanishing into gamma rays (p. 1098). So here we have a pulse of pions traveling at 0.999 8c, and in a short while each of them transforms into two photons. Detectors in the forward direction record bursts of photons, and their speeds can be determined over the 31-m flight path. Instead of finding a speed of 0.999 8c + c ≈ 2c, which would have been expected from Newtonian kinematics, researchers found a speed of $2.997\,7 \pm 0.000\,4 \times 10^8$ m/s, in excellent agreement with the standard value of c measured with a stationary source.

26.3 Simultaneity and Time

Einstein was troubled by the fact that, on one hand, the speed of light must be invariant and, on the other, such invariance violates the customary addition rules for velocities. He spent most of a year struggling with the problem. Then it came to him: "Time cannot be absolutely defined, and there is an inseparable relation between time and signal velocity. With this new concept, I could resolve all the difficulties …. Within five weeks the Special Theory of Relativity was completed."

Light is the swiftest instrumentality at our disposal for communication and for the perception of events (which are themselves the basis of time). Yet its speed is finite, and that shapes our understanding. If c were infinite, the distinction between Special Relativity and Newtonian theory would vanish. In fact, wherever c is so large in comparison with the relevant motions that it can be considered effectively infinite, Newtonian theory works very well. This is precisely why the discrepancies went unnoticed so long—we live, for the most part, in a comparatively slow-moving world. Running, driving, even spaceship-flying at thousands of miles per hour are all mere crawling compared to 2.998 $\times 10^8$ m/s (Table 26.1).

The National Institute of Standards and Technology (NIST) sends out radio signals with which we can set our clocks—"When you hear the tone it will be … BEEP." Yet the farther you are away from the transmitter, the later the BEEP will arrive. An observer in the Andromeda galaxy will have to wait 2 200 000 years for the BEEP to arrive marking *now*. That's presumably no problem since we can make the necessary corrections for communication-time lags, knowing distance and c. Still, what does *now* mean for us in regard to our friend on Andromeda? That is, what does it mean to say that

Table 26.1

Various Speeds and the Corresponding β and γ Values*

Object	Speed v	β v/c	γ $1/\sqrt{1 - (v/c)^2}$
Human walking	8 km/h	0.000 000 007	1.000 000 000
100-yard dash (max.)	10.0 m/s	0.000 000 033	1.000 000 000
Commercial automobile (max.)	62 m/s	0.000 000 21	1.000 000 000
Sound	333 m/s	0.000 001 11	1.000 000 000
SR-71 reconnaissance jet	980 m/s	0.000 003 27	1.000 000 000
Moon around Earth	1000 m/s	0.000 003 33	1.000 000 000
Apollo 10 (re-entry)	11.1 km/s	0.000 037	1.000 000 001
Escape speed (Earth)	11.2 km/s	0.000 037	1.000 000 001
Pioneer 10	14.4 km/s	0.000 048	1.000 000 001
Earth around Sun	29.6 km/s	0.000 099	1.000 000 005
Mercury orbital speed	47.9 km/s	0.000 16	1.000 000 013
Helios B solar probe	66.7 km/s	0.000 22	1.000 000 025
Earth-Sun around galaxy	2.1×10^5 m/s	0.000 70	1.000 000 245
Electrons in a TV tube	9×10^7 m/s	0.3	1.05
Muons at CERN	2.996×10^8 m/s	0.999 4	28.87
Electrons in Stanford Linear Accelerator (SLAC)	$2.997\,9 \times 10^8$ m/s	0.999 999 999 7	4×10^4

*To better show the behavior of β and γ, little concern is given to significant figures.

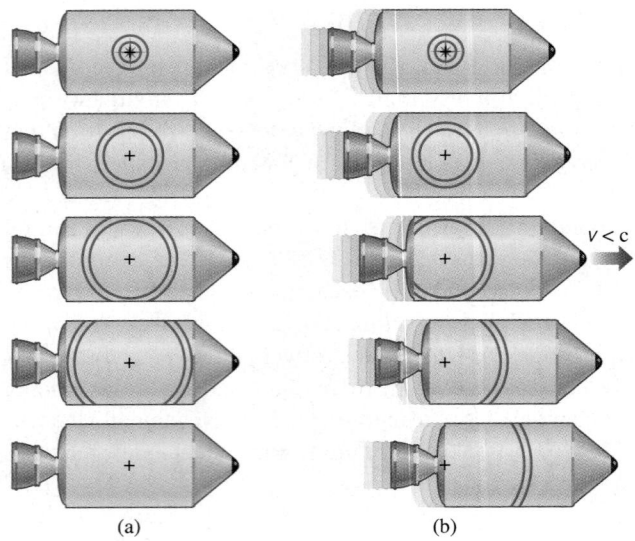

Figure 26.6 A flash of light in (a) a "stationary" ship and in (b) a "moving" ship as seen from outside. Or, equivalently, a flash in a ship seen (a) inside and (b) by an observer outside with respect to whom the ship is moving at a constant speed v.

two events occur simultaneously, one here and one there, now? To be sure, if two events occur simultaneously *at the same location*, there's no problem—if two comets are seen to blow up immediately on crashing, we know the two events occurred absolutely simultaneously. Every other inertial observer in the Universe will see the same thing. Difficulties arise when the events occur at separate locations; and the greater their separation in space, the longer the period of inconclusiveness in time.

Figure 26.6 illustrates the problem quite simply. Part (a) is the view of a pulse of light as seen sequentially by an observer at rest inside the ship. She sees the expanding lightwave strike the front and back walls simultaneously. Note that *once the light is launched into space, it's on its own and it must travel at c in all directions as seen by any and all inertial observers*. Part (b) is the view as seen by an observer outside with respect to whom the ship is uniformly moving. He sees the lightwave, traveling at c in all directions, but now the rear wall advances on the source and is struck first while the front wall recedes and is struck second—the events are *not simultaneous*. Notice that if the walls in the craft are mirrored, the waves will be reflected and meet in the center of the chamber in Fig. 26.6a. There's no ambiguity then; the same thing must happen in (a) and (b)—*events that occur simultaneously at one place in space must be simultaneous for any inertial observer*. What all of this suggests is that, like rest and motion, simultaneity is not absolute. You might say "who cares?" but if simultaneity, which is at the heart of the measurement of when events occur, is not absolute, time itself is not absolute.

Since that's really a big step to take, let's be a bit more rigorous about it. *Two spatially separated events are simultaneous if they are seen to occur at the same time by an observer located midway between the sites where the events happen.* This way, we will not have to introduce any corrections for different distances as was the case with the BEEPs from NIST. The question now is, do two events occurring simultaneously for one **midway inertial observer** occur simultaneously for all midway inertial observers regardless of their motion?

Imagine a rocket ship and a space station gliding past each other at any constant relative speed (Fig. 26.7). *We select the station commander*, a stubborn but honest fellow named Stan, *to be the one who will see the two events we are arranging as simultaneous*. Playing out his part, he switches on two explosive flares that just happen to be floating a few meters apart in space between him and the rocket. At the very moment they explode (as determined by him later), he finds himself lined up nose-to-nose with the rocket's pilot, a woman named Rosie no less willful than himself. Both are in the middle of their respective crafts. (It doesn't matter how these flares are themselves moving or how they got there. We are only interested in them as light sources that will conveniently leave dent marks on each hull.) According to Stan, points *A* and *A′* are adjacent to each other at the same moment that *B* and *B′* are adjacent—the distances between dents on the station and on the ship are the same.

Stan in the station is a midway observer, and the light from both flares, fore and aft, reaches him at the same instant (we arranged it that way). As far as he is concerned, the two explosions are unquestionably simultaneous. Naturally enough, looking out the window he takes himself to be at rest and sees the rocket ship moving off to his rear. As a result, he sees Rosie advance on flare-A and recede from flare-B. He sees the light from flare-A reach her first (Fig. 26.7*b*) before the light from flare-B arrives (Fig. 26.7*d*). He expects Rosie to see the two events as not having occurred simultaneously, but he knows that he's right.

Rosie is also a midway observer, and the two dents on the rocket ship from the explosions are permanent proof of that. She looks out her window and sees the station moving off to the rear of her rocket ship (Fig. 26.8). We have arranged for the two beams to arrive

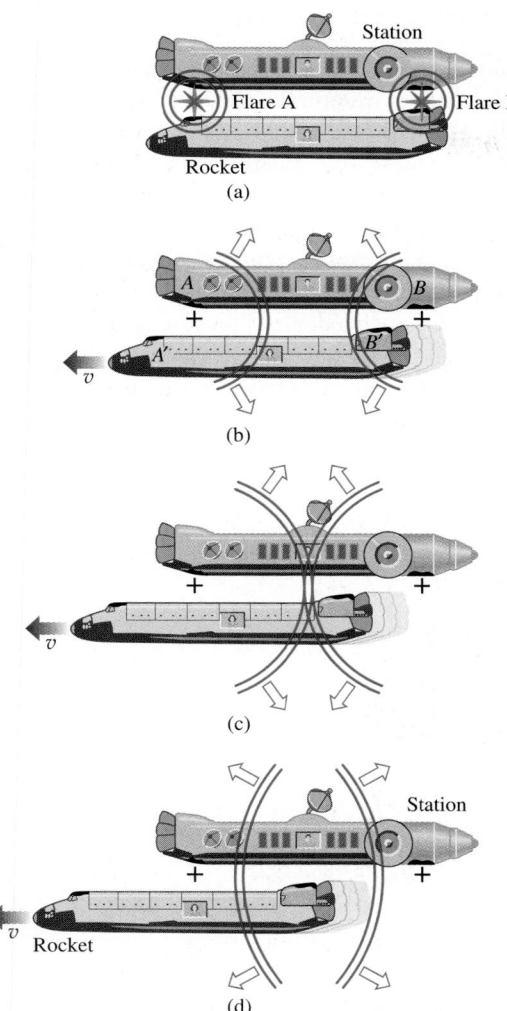

Figure 26.7 From the perspective of Stan, the Station Commander, the flares lit simultaneously and the rocket ship sailed off to the rear of his station. It follows from the Principle of Relativity that either inertial observer can assume that he or she is at rest and that it is the other person who is moving.

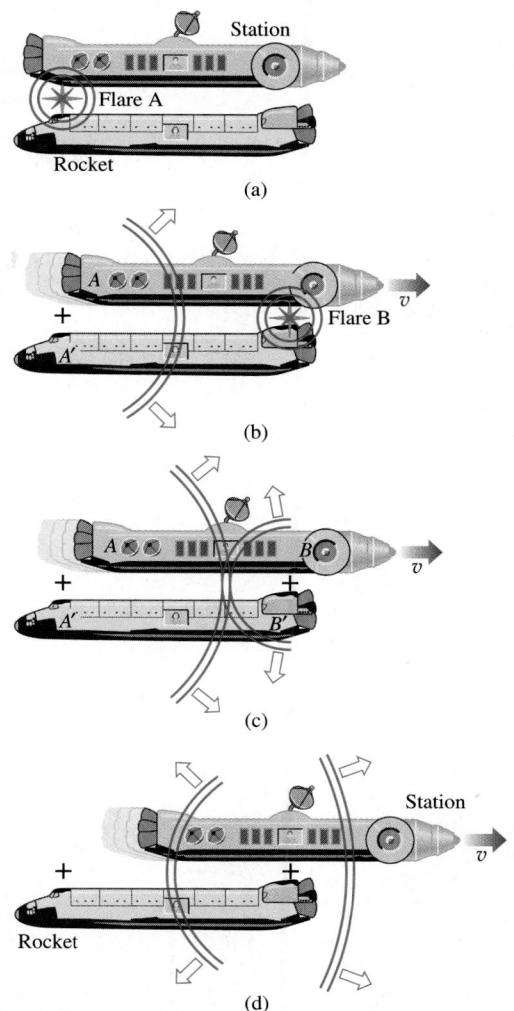

Figure 26.8 From the perspective of Rosie, the rocket's pilot, flare-A fires first and the station sails off to the rear of her ship. The drawing neglects the fact that the station must appear to Rosie to be somewhat shrunken, a point we'll come back to later.

STUDY GUIDE

The Special Theory of Relativity arises because c is finite. Here if c were infinite, Figs. 26.7 and 26.8 would be identical and both observers would see the flares fire simultaneously—simultaneity would be absolute.

As you work though the rest of this chapter, notice what would happen to each important conclusion if c became infinite.

together at Stan's location (Fig. 26.8*c*), and both observers see that happen because it takes place at a single point in space. Accordingly, Rosie must see the flash from flare-A (Fig. 26.8*b*), which she records as having fired first, and then the flash from flare-B, which she therefore knows must have fired second. Moreover, she observes that points *A* and *A*′ coincided before points *B* and *B*′ did. Understanding that the station is moving toward the later explosion and away from the earlier one, she anticipates Stan giving an erroneous report of simultaneity. She knows that the events she witnessed did not occur simultaneously, and all of the automatic detectors on board her ship will verify it.

If either the First or Second Postulate were wrong, we could determine who was really moving and would know which of the two observers was correct. But the discrepancies cannot be resolved—they are inherent in the nature of things; they are the reality. **There is no such thing as absolute simultaneity**—*events separated in space that are simultaneous for midway observers in one inertial system are not necessarily simultaneous for midway observers in any other inertial system.* Notice, too, that the lengths $\overline{AB}$ and $\overline{A'B'}$ which were seen by Stan in the station to be equal (Fig. 26.7), are not seen to be equal by Rosie in the rocket (Fig. 26.8); they will not agree on times or lengths, and both will be right.

26.4 The Hatter's Watch: Time Dilation

We reckon time by comparing simultaneous events—the moment the horse's nose touches the wire and the placement of the hands on a stopwatch. Evidently, if the front flare in Fig. 26.7 were a flashing digital clock, each observer would see the other flare explode at a different time. Despite the centuries of thought to the contrary, **time is relative**, not absolute. There is no universal time that dances over us all equally, unalterably—the Universe doesn't grind away according to a single silent drumbeat.

Suppose we construct a light-clock in which a pulse of light bounces back and forth between mirrors and some mechanism counts off the number of traversals, much like the familiar tick-tock. Figure 26.9a depicts such a device in a spaceship. With respect to the pilot on board, the clock is at rest, and she sees nothing extraordinary happening—everything aboard ship is normal. Quite the contrary is true according to an observer on the ground as the ship whisks by with a uniform relative velocity (Fig. 26.9b). He sees a light

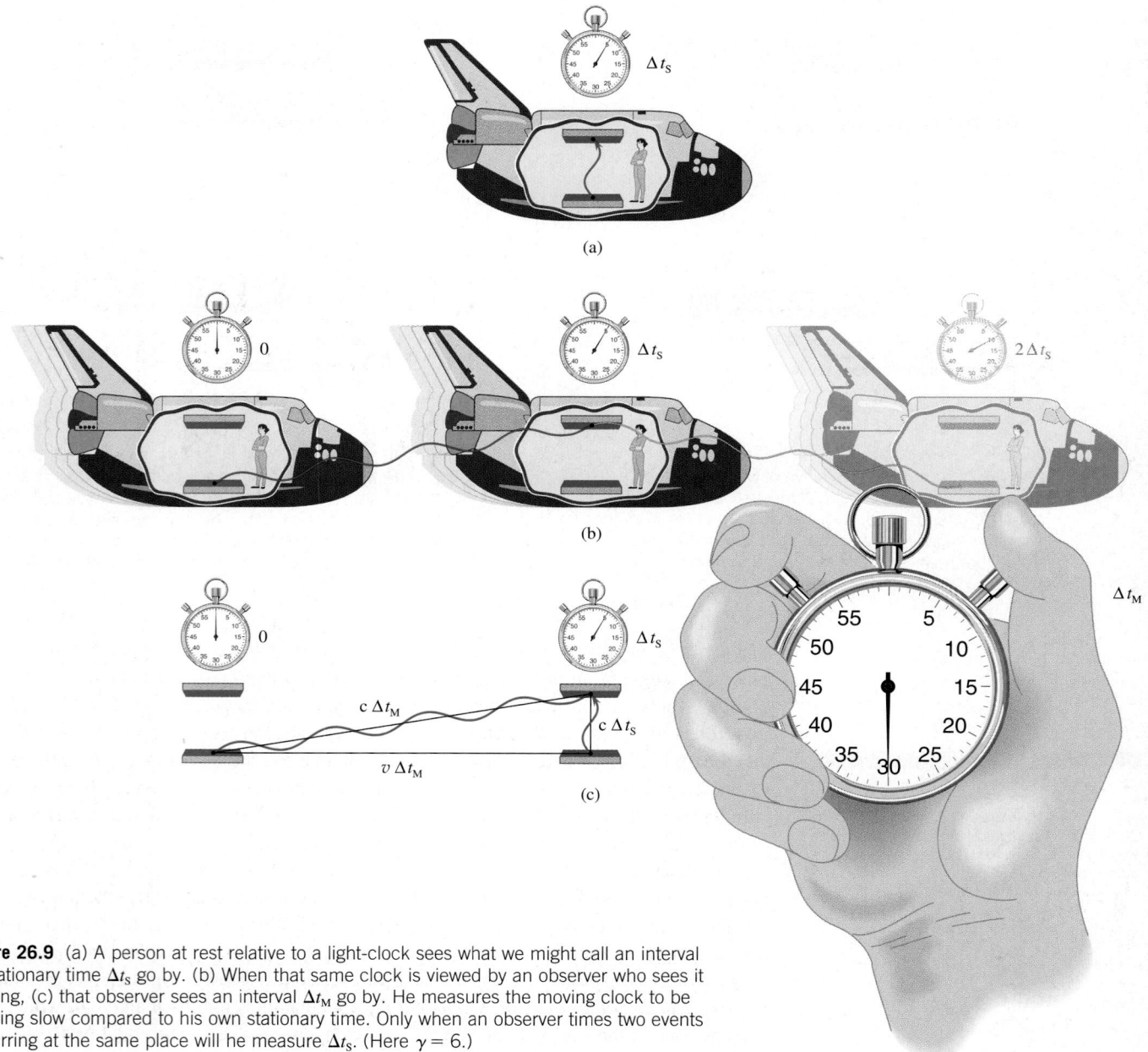

Figure 26.9 (a) A person at rest relative to a light-clock sees what we might call an interval of stationary time Δt_S go by. (b) When that same clock is viewed by an observer who sees it moving, (c) that observer sees an interval Δt_M go by. He measures the moving clock to be running slow compared to his own stationary time. Only when an observer times two events occurring at the same place will he measure Δt_S. (Here $\gamma = 6$.)

pulse in her clock behave as if it advanced diagonally as the vehicle carried the clock's mirrors past him. (Notice how the situation resembles the transverse arm of the Michelson Interferometer—indeed, the analysis will yield similar results.) While someone at rest with the clock sees the pulse travel to and fro between two stationary mirrors, the picture is very different when viewed from the ground. The pulse surely goes from one mirror to the other (both observers see the pulse strike either mirror—at such moments, the pulse and mirror are at the same point in space), but to the outside observer, the mirrors move along the flight path of the ship.

As seen by the pilot *who is stationary with respect to the clock*, the pulse travels a distance given by $c\,\Delta t_S$, where Δt_S is the elapsed interval on her clock (one tick's worth). To the observer on the ground with *respect to whom the ship is moving* at speed v, the pulse travels a longer diagonal path. Bound by the Second Postulate, the pulse (which must be seen by all inertial observers to progress at c) must take a longer interval of time, Δt_M, to traverse the longer distance, $c\,\Delta t_M$. During that interval, as perceived from the ground, the ship advances a distance $v\,\Delta t_M$. It follows from the Pythagorean Theorem that

$$(c\,\Delta t_M)^2 = (v\,\Delta t_M)^2 + (c\,\Delta t_S)^2$$

Hence

$$(c\,\Delta t_M)^2 - (v\,\Delta t_M)^2 = (c\,\Delta t_S)^2$$

and

$$(\Delta t_M)^2 = \frac{c^2(\Delta t_S)^2}{c^2 - v^2} = \frac{(\Delta t_S)^2}{1 - \frac{v^2}{c^2}}$$

$$\Delta t_M = \frac{\Delta t_S}{\sqrt{1 - v^2/c^2}} \qquad (26.2)$$

Inasmuch as $\sqrt{1 - \beta^2}$ must be less than 1, the time interval seen by the outside observer with respect to whom the clock is moving (Δt_M) must be greater than the corresponding time interval (Δt_S) seen by the inside observer with respect to whom the clock is stationary. The interval between ticks and tocks is longer for the observer who sees the clock moving than for the one who doesn't. Since $\gamma = 1/\sqrt{1 - \beta^2}$,

$$\Delta t_M = \gamma \Delta t_S \qquad (26.3)$$

with $\gamma > 1$. This slowing down of time in a system (as seen by an observer w.r.t. whom the system is moving) is known as **time dilation**, and it's usually a very small effect. A clock aboard a commercial plane flying at top speed for $\approx 70\,000$ years would lose about 1 s compared to a clock on the ground. Notice that when $c \gg v$, as is ordinarily the case, $\gamma \approx 1$, and $\Delta t_M \approx \Delta t_S$; that's why we're not troubled by time dilation in our everyday lives.

Time on a clock that is moving with respect to an observer is seen to run slower than time on a clock that is stationary with respect to that observer. And this is true for any clock (a wristwatch, a pendulum, a beating heart, a fertility cycle, or a dividing cell); all must slow down, all must match the light-clock. Otherwise we could easily learn from the difference who was actually moving, which is nonsense; no one is *actually* (or absolutely) moving—absolute motion violates the First Postulate. If the postulates are true, time is relative.

The duration of an event Δt_S, as measured by an inertial observer who sees the event begin and end in one place is called the **proper time**; it's always shorter (by a factor of γ^{-1}) than the corresponding interval Δt_M, as measured by an observer who sees the event begin and end at two different locations in his coordinate system.

STUDY GUIDE

The most important step in solving time dilation problems is correctly identifying the proper time (Δt_S). The proper time is the time interval between two events that occur at the same location in the particular reference system. When the interval is determined by an arbitrary inertial observer for whom the events occur at different locations, since $\gamma > 1$, $\Delta t_M > \Delta t_S$. The proper time is the shortest possible duration that will be observed.

The **proper time** is the time interval measured by someone who sees the event to occur at a fixed location in space. The proper time corresponds to the shortest interval that will be measured by any observer.

Example 26.2 **[I]** A college physics laboratory is under observation by aliens traveling on an asteroid. An undergrad seen measuring the period of a mass oscillating on a spring gets a value of 2.00 s. Given that the aliens are cruising by at a constant speed of 0.50c and that they have nothing better to do, what period will they determine? (Because they are moving rapidly, there can be substantial communication-time lags, so we assume they make any necessary corrections.)

Solution   We've got a time interval (occurring at a single location), measured by an inertial observer moving w.r.t. that location—all of which should bring to mind the concept of "time dilation." Who measures the proper time here?   (1) TRANSLATION—An event occurs during a known time interval at a fixed location in an inertial system. It is observed from a second inertial system moving at a known speed w.r.t. the location

of the event; determine the duration of the interval as perceived in the second system.   (2) GIVEN: $v = 0.50\,c$ and the **proper time** measured by the student, $\Delta t_S = 2.00$ s. FIND: The period determined by the aliens, Δt_M.   (3) PROBLEM TYPE—Relativity/time dilation.   (4) PROCEDURE—The period measured by an observer at rest with respect to the spring is $\Delta t_S = 2.00$ s and, moreover, $\Delta t_M = \gamma \Delta t_S$.   (5) CALCULATION—Here, $\sqrt{1 - \beta^2} = 0.866$ and so

$$\Delta t_M = \frac{1}{\sqrt{1 - \beta^2}} \Delta t_S = \boxed{2.3 \text{ s}}$$

Quick Check: The time interval is properly dilated; the aliens see the oscillations taking longer than the student does. Moreover, when $\beta = \frac{1}{2}$, Table 26.2 agrees with this value of γ.

> There is one single and invariable time, which flows in two movements in an identical and simultaneous manner.... Thus, in regard to movements which take place simultaneously, there is one and the same time, whether or no the movements are equal in rapidity....The time is absolutely the same for both.
>
> ARISTOTLE

> What then is time? If no one asks me, I know: if I wish to explain it to one that asketh, I know not.
>
> ST. AUGUSTINE

Time dilation has been measured. Certain nuclei vibrate and emit gamma-rays with very precise frequencies. When a sample of such a substance is heated, the gamma-ray frequency is reduced. The atoms move around more rapidly, and with respect to an observer at rest in the laboratory, their nuclear clocks run slower.

There have been many other experimental confirmations of time dilation over the years; the following is among the more interesting. Muons are subnuclear particles that are like heavy electrons. They are unstable, decaying into electrons and neutrinos. A muon at rest in the laboratory has a mean life of 2.2 μs, and this provides us with a convenient natural clock having a 2.2-μs tick-tock interval. In 1976, experimenters at the European Council for Nuclear Research (CERN) created a beam of muons traveling at 0.999 4 c. These were injected into a large doughnut-shaped storage ring, where they were confined by powerful magnets and circulated until they decayed. Although a typical muon might, on the basis of Newtonian theory, be expected to survive 14 or 15 trips around the ring, most muons actually made in excess of 400 orbits. Electron detectors surrounding the ring estab-

Table 26.2
Values of β, $1/\gamma$, and γ

β	$1/\gamma$	γ
v/c	$\sqrt{1 - (v/c)^2}$	$1/\sqrt{1 - (v/c)^2}$
0.000 000	1.000 00	1.000 000
0.100 000	0.994 987	1.005 038
0.200 000	0.979 796	1.020 621
0.300 000	0.953 939	1.048 285
0.400 000	0.916 515	1.091 089
0.500 000	0.866 025	1.154 701
0.600 000	0.800 000	1.250 000
0.700 000	0.714 143	1.400 280
0.800 000	0.600 000	1.666 667
0.900 000	0.435 890	2.294 157
0.990 000	0.141 067	7.088 812
0.999 000	0.044 710	22.366 27
0.999 900	0.014 142	70.712 45
0.999 990	0.004 472	223.607
0.999 999	0.001 414	707.107

To keep things neat, the proper number of significant figures has not been kept consistently.

lished that the rapidly moving muons had a mean life that was about 30 times longer than when they were at rest ($\gamma = 28.87$). Equation (26.2) was confirmed to an accuracy of 0.2%.

More recently (1985), fast-moving neon atoms (excited by a laser so that they emitted light of a precise frequency) were used to confirm the time dilation to within an accuracy of 40 parts per million. Not long ago (1999) a Boeing 747 made a transatlantic flight with a new atomic clock aboard. The clock lost 40 ns during the journey. {Go to the **CD**, click on **INTERACTIVE EXPLORATIONS** and open **THE RELATIVITY OF TIME**.}

EARTH TIME

Everyone at rest on the Earth more or less experiences the same Earth-time. Since none of us really rushes around very quickly, $c \gg v$, and quite generally $\Delta t_M \approx \Delta t_S$ for everyone on the planet, which is just what we expect. At its extremes, Relativity must yield the same results as our well-tested Classical Mechanics. As the Earth sails through space, it essentially has its own proper time. Anyone riding an asteroid past the planet will see our time running slower than their own time as seen on their "stationary" clocks. Someone else flashing by at a greater speed will see our "moving" clocks running still slower compared to their time. And the inverse is true: we will see their "moving" clocks run slow, as our own "stationary" clocks run at their normal rate. This reciprocity doesn't make the process less meaningful.

The **proper length** is the length measured by someone who is at rest w.r.t. the distance being measured. It corresponds to the largest value that will be measured by any observer.

26.5 Shrinking Alice: Length Contraction

Neither time nor space is absolute in this Universe where the speed of light is constant. The fall of absolute simultaneity takes with it both absolute time and absolute length (or distance). *Only if both ends of a moving rod can be located at exactly the same instant can its length be measured accurately, and that cannot be done absolutely.* It's not much good in finding the length of a rod to say that the front end lined up with the 3-m mark of a ruler at 1:00 P.M. and 2 seconds later the rear end was next to the 2-m mark. We must watch the alignment with the ruler at both ends (separated in space) simultaneously; different observers will not agree about that. If the flares in Fig. 26.7 are replaced by clocks, they can be synchronized to the same time by someone at rest with respect to them, but they will be seen to be unsynchronized by anyone moving relative to their inertial system. What is the distance between the exploding flares in Figs. 26.7 and 26.8? Remember: there were two different distances seen by the two observers just because they couldn't agree on whether the flares fired off simultaneously or not. A rod has one single **proper length** measured by any observer at rest with respect to it, but it can also have different shorter lengths measured by people who are in uniform motion with respect to it.

Imagine a meter stick in a rocket ship flying past you, and you also have a meter stick. *The sticks are aligned parallel to the relative velocity* and, along with appropriate clocks and sources, are used to measure the speed of light. As seen by Rosie in the rocket, a pulse of light travels the length of her ruler (L_S) in a certain proper time at speed c. As seen by you, with respect to whom that experiment is moving, the pulse in the rocket traverses a length L_M. You, looking into the rocket, see the pulse travel for fewer seconds on the rocket's clock, which you see running slow. You watch the pulse traverse the apparatus, in a room where time runs "slow," just as Rosie does, in a room where time runs "normally," and both of you must determine the speed to be c. For this outcome to be the case, the pulse in the rocket ship must travel along a path that to you is apparently shorter than it "ought to be" (shorter than L_S, shorter than 1 m) by a factor of γ; so that

$$L_M = \gamma^{-1} L_S \tag{26.4}$$

or

$$L_M = L_S \sqrt{1 - v^2/c^2} \tag{26.5}$$

There was a young fencer named Fisk,
Whose thrust was exceedingly brisk.
So fast was his action,
The Lorentz–FitzGerald contraction
Reduced his rapier to a disk.

ANONYMOUS

Pilot Rosie moving along with her apparatus sees everything in the ship as properly normal and measures c. You, the outside viewer, see the rocket ship *shrunk along the line of motion*, the experiment on board it shrunk, and everything in the cabin happening slowly. And you understand why Rosie found the pulse to travel at c (her time was running slow by a factor of γ, but her experiment was shrunk by a factor of $1/\gamma$). With your own stick and clock, your light pulses will also be measured to travel at c. And when Rosie views you on Earth, she sees everything of yours appropriately shrunk and your time running slow compared to her clock—the contractions and dilations are symmetrical via the First Postulate.

  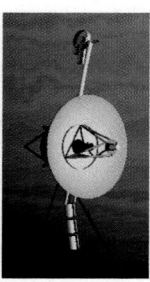 

Computer images of a space probe as seen by an observer traveling with respect to it at four different speeds: (a) 0, (b) 0.25c, (c) 0.5c, and (d) 0.75c. Moving away at high speed, the light is red-shifted via the Doppler Effect (p. 405).

The most important step in solving length contraction problems is correctly identifying the proper length (L_S). The **proper length** is the distance in space between two points when both points are at rest w.r.t. the observer. Imagine a ruler extending from one location to the other. The proper length of the ruler is the one measured by an observer who is at rest w.r.t. the ruler. When the length is determined by an arbitrary inertial observer (L_M) moving w.r.t. the ruler, since $\gamma > 1$, $L_S > L_M$. The proper length is the largest possible length that will be observed.

A moving observer measures an object to have a length (along the direction of motion) that is shorter than the length measured by an observer at rest with respect to the object (i.e., shorter than the proper length). This is the **length contraction**, and it applies only to the direction of motion; *transverse distances are unaltered*. Equation (26.5) is mathematically identical to the Lorentz–FitzGerald Contraction (p. 935), and it's often called by that name.

Example 26.3 **[I]** A flying saucer descending straight toward the Earth at 0.400 0c is first observed by an astronomer on the planet when it passes a satellite at an altitude of 3000 km. At that instant, what will be the ship's altitude as determined by its navigator?

Solution Imagine a ruler extending from the satellite to the Earth. There's an observer moving w.r.t. the ruler who is measuring its length. That should bring to mind the idea of "length contraction." Who measures the proper length, aliens or earthlings? (1) TRANSLATION—An observer moves at a known speed w.r.t. a specified distance; determine the corresponding distance measured by that observer. (2) GIVEN: the proper length as measured by someone on Earth, L_S = 3000 km and v = 0.400 0c. FIND: The measured altitude. (3) PROBLEM TYPE—

Relativity/length contraction. (4) PROCEDURE—The proper length, the height measured by an observer (the astronomer) w.r.t. whom the distance is stationary, is L_S = 3000 km. (5) CALCULATION—Using Eq. (26.5),

$$L_M = L_S\sqrt{1 - v^2/c^2}$$

$$L_M = (3000 \text{ km})\sqrt{1 - 0.160\,0}$$

$$\boxed{L_M = 2750 \text{ km}}$$

Quick Check: The speed is not very high, so we can expect a relatively small contraction. A glance at Table 26.2 confirms that $\sqrt{1 - v^2/c^2} = 0.92$.

There is no absolute distance between New York and Chicago; an atlas provides the proper distance as measured by people at rest with respect to the planet. But every traveler moving with regard to the surface sees his or her own version of that spatial interval, depending on his or her relative speed. Go fast enough, and London and San Francisco can be a meter or two apart. As for the reality of all of this (i.e., is the spaceship actually squashed?), the business is similar to what happens with the Doppler Effect. Run toward a source and the sound pitch increases. The wave itself doesn't physically change, but the perception of it certainly does change. What you hear and measure—the reality of the experience—most assuredly depends on how you move with respect to the source.

You might find it satisfying to know that the electromagnetic field seen by differently moving inertial observers will be different and that a detailed theory of the electromagnetic forces between the atoms within an object shows that the object must thereby contract by a factor of $1/\gamma$.

Example 26.4 **[I]** A starship (some time in the very distant future) is headed for a galaxy that, according to human astronomy texts, is 200 light-years away from Earth. Flying a direct course, the ship quickly reaches a cruising speed of 0.999c. What will be the Earth-galaxy distance as then determined by the navigator?

Solution   Imagine a ruler extending from the Earth to the galaxy. There's an observer moving w.r.t. this ruler who is measuring its length. That should bring to mind the idea of "length contraction." Who measures the proper distance, the earthbound astronomers or the crew of the starship?   (1) TRANSLATION—Determine the distance measured by an observer moving at a known speed if the proper length is specified. (2) GIVEN: the earthbound astronomers measure the proper distance (the length of the stationary ruler) $L_S = 200$ ly and $v =$ 0.999c. FIND: The navigator's distance.   (3) PROBLEM TYPE—Relativity/length contraction.   (4) PROCEDURE—The first thing to settle is which distances are which in Eq. (26.5).   (5) CALCULATION—L_M is the length as seen by someone moving with respect to the physical system in which the proper length is L_S. Thus,

$$L_M = L_S\sqrt{1 - v^2/c^2} = (200 \text{ ly})\sqrt{1 - (0.999)^2} = \boxed{8.94 \text{ ly}}$$

Quick Check: The speed is very high, so we can expect a considerable contraction. A glance at Table 26.2 confirms that $\sqrt{1 - v^2/c^2} = 0.044\,7$. At these great speeds, the length is just a few percent of the proper length. (Get a feeling for the numbers you can expect by looking over the table.)

26.6 Tweedledee & Tweedledum: The Twin Effect

Some day, we may have rocket ships that can attain speeds high enough to experience significant time dilations and length contractions. This feat requires engines that can continue to exert thrust for very long periods so that the craft can accelerate at a humanly bearable rate (1 g would be nice) for years. Such an achievement is far beyond our present capabilities and may be for centuries. Practicalities aside, suppose we have such a starship.

You and I meet at the launchpad, engage our identical stopwatches, shake hands, and you fly off to some star at a distance of 50.00 light-years, measured from Earth. Imagine a ruler extending from Earth to star. An earthbound observer sees the ruler at rest, and so 50.00 ly is the proper length, L_S. By contrast, an observer aboard the ship sees the ruler moving and measures its length to be L_M, which must be smaller than 50.00 ly.

The mission is to arrive at the star, plant the flag as it were, and promptly come home. To make the calculations simple, you quickly get the craft up to a wonderfully unrealistic cruising speed v of 0.999 8c and settle in for the journey. A glance out the window reveals to the passengers that the Earth–star distance is $L_M = L_S\sqrt{1 - v^2/c^2}$ (the length of our imaginary ruler seen from the ship).

The pilot of the craft sees the start of the trip to occur just outside her window on Earth. And some time later, the end of the trip, marked by the arrival of the star, also occurs just outside her window. From her perspective it's as if the ruler, with the Earth at one end and the star at the other, flew past her window while she simply waited at rest. For her the start and end of the journey happen at the same location in her coordinate system (her world), and the duration she observes on her clock is therefore the proper time (Δt_S).

Since we both agree on *our* relative speed (and neglecting the comparatively short times it takes to negotiate the several accelerations), the trip out takes a proper on-board time of $\Delta t_S = L_S\sqrt{1 - v^2/c^2}/v$. Thus, if T_S is the total round-trip flight time ($2\Delta t_S$) recorded by you in the spaceship, then

$$T_S = \frac{2L_S}{v}\sqrt{1 - v^2/c^2}$$

Similarly, if T_M is the total flight time I record while watching you through my telescope,

HOW THINGS LOOK AT HIGH SPEEDS

The question of what objects look like to a rapidly moving observer because of the length contraction is a complicated one. The semi-enlightened comic book renditions of things simply shrunk along the direction of motion—the long-skinny-people vision—is erroneous. Complications arise because light arriving at the retina (or on a piece of film) at some instant must have left different regions of the object at different times, depending on how far away they are. Three-dimensional objects will therefore seem to be twisted and distorted, the more so the greater v (see photo sequence on p. 946).

then the distance I see you travel is vT_M, but I know you traveled $2L_S$ and so $vT_M = 2L_S$ or

$$T_M = \frac{2L_S}{v}$$

If we substitute this expression into the preceding equation for T_S, we're right back to the time dilation equation.

As ever, when putting in the numbers, we could enter L_S in meters and v in meters per second, entering L_S in light-years is more convenient, especially when we have v in terms of c. And there's a little trick for doing just that. We'll replace the numerical value of L_S in light-years with its equivalent, the number-of-years-traveled-at-c multiplied by c (in whatever units are appropriate); that is,

$$1 \text{ light-year} = c(1 \text{ year})$$

Here $L_S = 50.00$ ly $= c(50.00$ y$)$. For this particular round trip, $T_M = 2L_S/v = 2(50.00$ ly$)/(0.9998c) = 2c(50.00$ y$)/(0.9998c) = 100.0$ y. Notice how the two c factors, regardless of their units, cancel, yielding T_M in years directly. Knowing that $\gamma = 50.00$, compare the fact that $T_M = 100.0$ y with $T_S = T_M/\gamma = 2.000$ y. I watch you travel 100.00 ly total (out and back) at a speed nearly that of light, taking a total of about 100 y. You see a journey contracted by a factor of $\gamma = 50.00$ and travel only 1.000 ly out and 1.000 ly back, taking about 2.000 y on your clock. I will greet you on the big day of your return in a wheelchair, having aged 100 y since you left, and you will saunter out having aged a mere 2 y.

Example 26.5 **[II]** The nearest galaxy to ours in all the Universe is the shapeless star-island known as the Magellanic Cloud (located by astronomers on Earth to be about 1.70×10^5 ly from midtown Manhattan). Assuming you could get up to a speed of 0.999 99c in a negligible amount of time (which is sheer nonsense), how long would you say the trip to that galaxy will take? Incidentally, the fastest anyone has ever gone is only around 0.000 037c.

Solution Imagine a ruler with N.Y. at one end and the galaxy at the other. You're an observer measuring the length of this ruler, w.r.t. which you are moving. That should bring to mind the idea of "length contraction." (1) TRANSLATION—Determine the time, as measured by an observer moving at a known speed, to travel a specified proper length. (2) GIVEN: $v = 0.999\,99c$ and the proper distance, as measured by the astronomers w.r.t. whom the imaginary ruler is stationary, namely, $L_S = 1.70 \times 10^5$ ly. FIND: The flight time on the traveler's clock, which is the proper time, Δt_S (see p. 943). (3) PROBLEM TYPE—Relativity/ length contraction/time traveled. (4) PROCEDURE —The first thing to settle is which distances are which in Eq. (26.5). You, the traveler, see a contracted distance $L_M = L_S\sqrt{1 - v^2/c^2}$,

which is to be traversed at a speed $v = 0.999\,99c$. (5) CALCULATION—L_M is the length as seen by someone moving with respect to the physical system in which the proper length is L_S. Your proper time is

$$\Delta t_S = \frac{L_S}{v}\sqrt{1 - v^2/c^2}$$

$$\Delta t_S = \frac{(1.70 \times 10^5 \text{ ly})(4.472\,12 \times 10^{-3})}{0.999\,99c}$$

and

$$\Delta t_S = \frac{760 \text{ ly}}{c} = \boxed{760 \text{ y}}$$

It would seem we're not likely to travel to other galaxies using the technology we have at hand.

Quick Check: Table 26.2 confirms the value of $\sqrt{1 - v^2/c^2}$. Multiplying 760 y by γ yields 1.7×10^5 y, which is the time in which we on Earth would see you make the 1.70×10^5-ly journey traveling at 0.999 99 c.

STUDY GUIDE

In Example 26.5 we have a situation where the observer who measures the proper time is not the same observer who measures the proper distance.

This **twin effect** has often been called the *twin paradox* (usually enunciated with one twin staying and one traveling), but it's no paradox at all. It may be a startling result, but it's quite understandable within the context of the two postulates. Even so, Einstein pointed out that because of the accelerations on the part of the traveler, the analysis should more properly be done using General rather than Special Relativity. Until now, the situations we have treated have been symmetrical; inertial observers in relative motion see each other's clocks run slow. Here, however, there are accelerations, and we know from the inertial forces that the traveler is moving and not the stay-at-home. That determination can be made in an isolated laboratory aboard the vessel. Those accelerations change the path of the traveler through space and time, impressing on him the burden of being out of step with all the friends left behind. In that sense, the time elapsed between two moments in a journey is route-dependent, just as is the distance traveled.

One of the most compelling confirmations of the reality of these conclusions was made in October 1971. Then, four exceedingly accurate cesium-beam atomic clocks were flown around the world twice, on regularly scheduled commercial jet flights. The idea was to compare the clocks with those at the U.S. Naval Observatory before and after they had circumnavigated the Earth. Because of the planet's spin, there were two trips around, first eastward and then westward. Things were complicated by the presence of gravity, which affects what's happening via the General Theory of Relativity. (The problem is that time speeds up as the gravitational potential decreases with altitude. This was in addition to the speed-dependent slowdown of time expected from the Special Theory.) In any event, the eastward clocks should have lost 40 ± 23 ns in the journey, and the westward-flying clocks should have gained 275 ± 21 ns. After $7600 was provided for airfare, it was found that, with respect to terrestrial reference standards, the eastward clocks lost 59 ± 10 ns and the westward clocks had gained 273 ± 7 ns—in breathtaking agreement with theory!

{For a brief treatment of the four-dimensional landscape of space and time click on **SPACE-TIME** under **FURTHER DISCUSSIONS** on the **CD**.}

26.7 Addition of Velocities

Classical theory maintains that velocities add vectorially (p. 40), and at everyday speeds that certainly seems to be true. Accordingly, consider Fig. 26.10, which shows a coordinate system S' moving with respect to another such system S. The scalar velocity of O' relative to O is $v_{\mathrm{O'O}}$. Now consider some object at point P moving in the space of both systems. For simplicity, we limit its motion to be in the $\pm x$-directions. Suppose that P travels at a scalar velocity relative to O of v_{PO} and at a scalar velocity relative to O' of $v_{\mathrm{PO'}}$. Classically, we learned that

$$v_{\mathrm{PO}} = v_{\mathrm{PO'}} + v_{\mathrm{O'O}} \tag{26.6}$$

for all objects in uniform motion. But this certainly couldn't be true for a photon, for which it follows from the Second Postulate that $|v_{\mathrm{PO}}| = |v_{\mathrm{PO'}}| = c$ regardless of $v_{\mathrm{O'O}}$. This dilemma can be sorted out by carefully applying the two postulates to the broader question of how coordinates in one system relate to those in another. We will skip the derivation—which isn't nearly as difficult as it is long—and simply state the one-*dimensional relativis-*

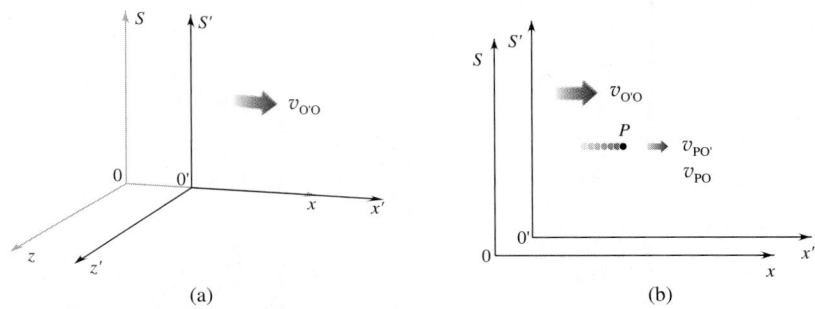

Figure 26.10 Two inertial systems S and S' moving with respect to each other at speed $v_{\mathrm{O'O}}$. (a) Here, S is at rest and S' advances in the x-direction. (b) A particle P moves at a speed v_{PO} relative to S and $v_{\mathrm{PO'}}$ relative to S'.

tic formula for the addition of velocities:

$$v_{PO} = \frac{v_{PO'} + v_{O'O}}{1 + \dfrac{v_{PO'}v_{O'O}}{c^2}} \tag{26.7}$$

Note that the individual scalar velocities can be either positive or negative. Equation (26.7) differs from Eq. (26.6) only because of the term $v_{PO'}v_{O'O}/c^2$. If either the object moves slowly with respect to S' (i.e., $|v_{PO'}| \ll c$) or S' moves slowly with respect to S (i.e., $|v_{O'O}| \ll c$), it follows that $|v_{PO'}v_{O'O}/c^2| \ll 1$. In that case, the relativistic expression for v_{PO} becomes identical to the classical one. That's why Eq. (26.6) was used for about two centuries before there was even a hint of a problem.

Now for the true test. Let's find the speed of light emitted from a moving source. Observer O', carrying a flashlight, is moving at a scalar velocity $v_{O'O}$ in the positive x-direction relative to observer O. A stream of photons is sent out in the positive x-direction at a speed $v_{PO'} = c$, and we want to find its speed v_{PO} with respect to O, namely,

$$v_{PO} = \frac{v_{PO'} + v_{O'O}}{1 + \dfrac{v_{PO'}v_{O'O}}{c^2}} = \frac{c + v_{O'O}}{1 + \dfrac{cv_{O'O}}{c^2}} = \frac{c(c + v_{O'O})}{c + v_{O'O}} = c$$

Wonderful! No matter what value $v_{O'O}$ has, each observer sees the speed of light to be c.

As another example, consider the rocket ship (S') in Fig. 26.11. It is flying away from us (S) at scalar velocity $v_{O'O} = \frac{1}{2}c$ when it emits a pulse of light. A traveler aboard the ship sees the pulse moving at $v_{PO'} = -c$; the minus sign shows that it's advancing in the negative x-direction. The pulse's scalar velocity with respect to us is

$$v_{PO} = \frac{-c + \frac{1}{2}c}{1 + \dfrac{(-c)(\frac{1}{2}c)}{c^2}} = -c$$

The light comes toward us at c even though it was emitted from a platform receding at a speed of $\frac{1}{2}$c.

The velocity transformations are such that **the speed of one entity with respect to another is always less than or equal to** c. As far as the two galaxies in Example 26.6 are concerned, they separate from each other at 0.96c and can intercommunicate directly (via light beams traveling at c). An important point here is that *any two entities, regardless of their motions, can interact*; they can transport energy (at speed c) from one to the other. Thus, even though the gravitational and electromagnetic interactions propagate at the finite speed c, two objects cannot outrun their mutual interactions.

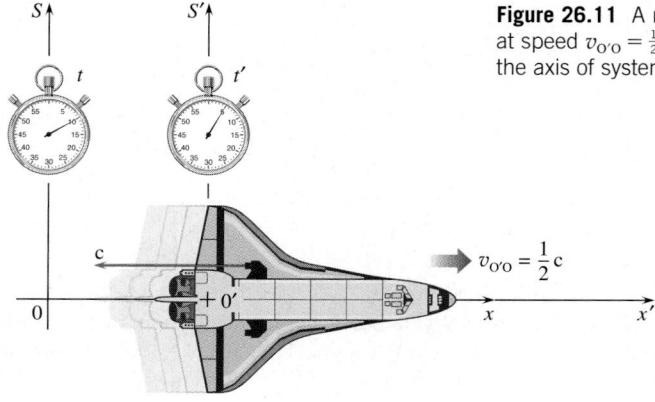

Figure 26.11 A rocket ship in system S' moving at speed $v_{O'O} = \frac{1}{2}c$ emits a pulse of light toward the axis of system S.

Example 26.6 **[II]** Two galaxies are speeding away from the Earth along a line in opposite directions, each with a speed of 0.75c with respect to the planet. At what speed are they moving apart with respect to each other?

Solution We want the "speed...w.r.t. each other." And that means the relativistic addition of velocities. (1) TRANSLATION —Two objects are moving away from a third object at known relativistic rates; determine the speeds of the two with respect to each other. (2) GIVEN: Speeds of each galaxy w.r.t. the Earth, namely, 0.75c. FIND: Relative speed. (3) PROBLEM TYPE— Relativity/addition of velocities. (4) PROCEDURE—First we have to assign the coordinate systems, and there are several equivalent possibilities. There are three moving bodies, and in Fig. 26.12 we let one galaxy be S, the Earth be S', and the other galaxy be P. Thus, S' moves forward with respect to S at $v_{O'O} = 3c/4$; P moves forward with respect to S' at $v_{PO'} = 3c/4$; and we want to find v_{PO}, the speed of P with respect to S. (5) CALCULATION—From Eq. (26.7)

$$v_{PO} = \frac{v_{PO'} + v_{O'O}}{1 + \dfrac{v_{PO'} v_{O'O}}{c^2}} = \frac{0.75c + 0.75c}{1 + \dfrac{(0.75c)(0.75c)}{c^2}} = \boxed{0.96c}$$

The galaxies separate at 96% of c.

Quick Check: This result is at least reasonable because the galaxies could not separate from each other at a speed equal to or greater than c since Eq. (26.7) says that even photons traveling in opposite directions ($v_{PO'} = $ c and $v_{O'O} = $ c) only separate at c.

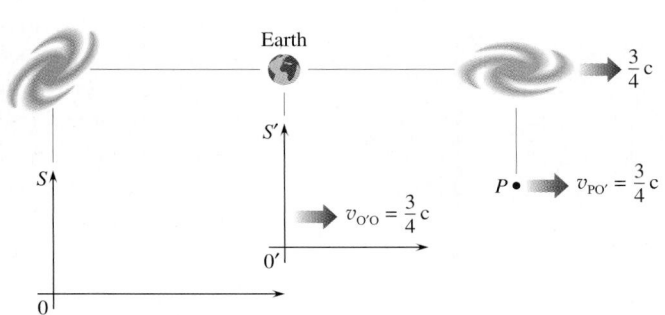

Figure 26.12 The galaxy on the right at point P is moving away from O' in S' at $v_{PO'} = \frac{3}{4}$c. With respect to the galaxy on the left (at rest in S), the Earth (at rest in S') is moving to the right at $v_{O'O} = \frac{3}{4}$c.

Relativistic Dynamics

Classical dynamics developed along two conceptual paths illuminated by the fundamental principles of Conservation of Momentum and Conservation of Energy. And so, faced with the necessity to reformulate dynamics, Einstein was guided by these same principles. Rather than attempt a complete derivation of relativistic dynamics, we limit our study to a few main results, their origins, and implications.

26.8 Relativistic Momentum

Both the energy and momentum of an object depend on the inertial reference system in which they are determined: the passenger holding a bowling ball on her lap flying to Paris "sees" it to have neither momentum nor kinetic energy. Since we now know, via Eq. (26.7), that the speed of an object with respect to different inertial frames has to be treated relativistically, it should be no surprise that both momentum (p) and energy (E) will have to be reformulated. Here, we are guided by our knowledge that the classical definition of momentum $\vec{p} = m\vec{v}$ works well at low speeds ($v << $ c) and must be an approximation of a more precise *relativistic momentum*. We can get a sense of how the momentum might be modified by examining the details of a simple collision.

Let's first establish that if momentum is to be conserved relativistically, it cannot be the classical notion given by $\vec{p} = m\vec{v}$. To that end, Fig. 26.13 on p. 952 depicts an elastic collision between two identical balls. There is a "stationary" frame S and a person therein, observer-1, holding ball-1. Passing nearby is a uniformly "moving" frame S' (traveling at speed $v_{O'O}$), in which there is a second person, observer-2, holding ball-2. Just at the right moment, both observers throw their balls vertically toward one another at identical speeds equal to u. Classically, the balls have momenta of $p_y = mu$ that are equal in magnitude and opposite in direction, and the initial net vertical momentum equals zero. After the head-on

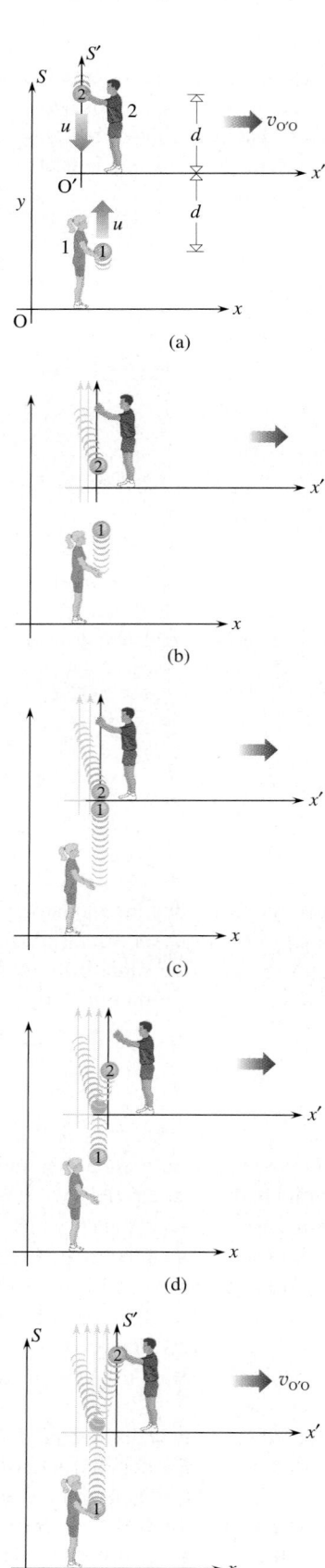

(a)

(b)

(c)

(d)

(e)

Figure 26.13 An elastic collision as viewed by observer-1 at rest in inertial system S. Observer-2 is at rest in system S', which moves relative to S at $v_{O'O}$. Observer-1 sees time running slow for ball-2 by a factor of γ, which depends on its net relative speed.

collision, the balls are both reversed; the net vertical momentum is still zero, and classical momentum is conserved.

Now, let's run through this situation again, taking into account that the relative speed of S' with respect to S is appreciable. Observer-2 sees his ball travel a vertical round-trip distance of $2d$ at a speed u in a time $2d/u$. The same is true of observer-1, who also sees her ball travel a vertical distance of $2d$ at a speed u in a time $2d/u$. But observer-1 sees ball-2 to be moving and therefore to be experiencing a time dilation (by a factor of γ). Thus, observer-1 must see ball-2 take a longer time to make the round trip from, and back to, the hand of observer-2. Observer-1 sees ball-2 travel a vertical distance of $2d$ in a longer time, and therefore at a slower speed, u/γ, than his own ball moves. And yet the collision reverses the motion of his ball—*classical momentum is not conserved*. So we need a new statement of what is conserved and, although we might guess that it looks like γmv, we certainly haven't proven it.

Nonetheless, the **relativistic linear momentum**, which is conserved, is indeed given by

$$\vec{\mathbf{p}} = \gamma m\vec{\mathbf{v}} = \frac{1}{\sqrt{1 - v^2/c^2}}\, m\vec{\mathbf{v}} \tag{26.8}$$

As v/c becomes negligible, $\beta \to 0$, $\gamma \to 1$, and $\vec{\mathbf{p}} \to m\vec{\mathbf{v}}$, which is the classical value. Figure 26.14 is a plot of p/mc—which classically equals β and relativistically equals $\gamma\beta$—against v/c. As the speed of a body increases beyond $\beta \approx \frac{1}{2}$, the relativistic momentum climbs away from the classical value, approaching infinity as $v/c \to 1$. It is the relativistic curve that is confirmed experimentally.

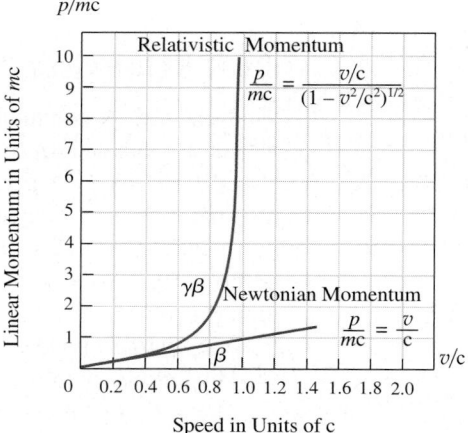

Figure 26.14 Linear momentum in units of mc plotted against speed in units of c; that is, p/mc versus v/c. Here, we compare relativistic momentum against Newtonian momentum. Observe that, although they both increase with v, the former approaches ∞ as v approaches c.

Example 26.7 [I] Electrons have a mass of 9.1094×10^{-31} kg. If one is traveling at $0.99c$, what is its momentum?

Solution Because of the high speed this is a relativistic momentum problem. (1) TRANSLATION—An object of known mass is traveling at a known relativistic speed ($v > 0.5c$); determine its momentum. (2) GIVEN: $m = 9.1094 \times 10^{-31}$ kg and $v = 0.99c$. FIND: p. (3) PROBLEM TYPE—Relativity/ momentum. (4) PROCEDURE—The momentum follows from Eq. (26.8), $p = \gamma m v$. (5) CALCULATION—First compute

$\gamma = 1/\sqrt{1 - v^2/c^2} = 7.0888$. Then

$p = \gamma m v = (7.0888)(9.1094 \times 10^{-31} \text{ kg})(2.9679 \times 10^8 \text{ m/s})$

and

$$p = 1.9165 \times 10^{-21} \text{ kg·m/s} = \boxed{1.9 \times 10^{-21} \text{ kg·m/s}}$$

Quick Check: From Table 26.2, γ looks okay at 7. The classical momentum is roughly $(9.1 \times 10^{-31})(3 \times 10^8) = 27 \times 10^{-23}$ kg·m/s, so our answer should be about $7(27 \times 10^{-23})$, which it is.

Though experiments have verified the validity of Eq. (26.8), the interpretation of $p = \gamma m v$ is unsettled at this moment. We can associate the γ with mass and suppose that there is a *relativistic mass* $m_R = \gamma m$ that is a function of speed. Then $p = m_R v$, which is nice. Einstein's early work seems ambivalent about the idea of a speed-dependent mass. Nonetheless, in 1948, he made it very clear that introducing the concept of relativistic mass was "not good." Experimentally, what we measure is a change in momentum with speed; there is no way to measure the mass of a moving body directly. Many physicists prefer to think of mass as speed-dependent. A growing number of others find it more appealing to hold that mass is an invariant property of matter like charge. Both views lead to the same measured results; the difference is a matter of interpretation. **We shall take m to be the invariant mass**—the speed-independent, reference frame–independent, almost Newtonian mass (m)—though that idea will need some more discussion (p. 957).

Newton's Second Law (for an average force) is, as ever,

$$\vec{F}_{av} = \frac{\Delta \vec{p}}{\Delta t} \qquad [4.1]$$

wherein $\vec{p}$ is the relativistic momentum given by Eq. (26.8). Now, suppose a constant force is applied to a body that is initially at rest. What happens to v? In that case, $F = p/t$, where p is the momentum after the elapse of a time t. Solving Eq. (26.8) for v^2, we obtain

$$v^2 = \frac{p^2/m^2}{1 + p^2/m^2c^2}$$

Dividing by c^2 and taking the square root gives

$$\beta = \frac{v}{c} = \frac{Ft/mc}{\sqrt{1 + (Ft/mc)^2}} \qquad (26.9)$$

Now consider the possible motion of an object. If the force and time are such that Ft is small and Ft/mc is therefore much less than 1, Ft/mc is negligible in the denominator and $\beta \approx Ft/mc$, whereupon $mv = Ft$, and we have the classical change-in-momentum-equals-impulse expression, Eq. (7.3). On the other hand, if Ft is large (as, for instance, if a large force is applied for a very long time), $(Ft/mc)^2 >> 1$ and $\sqrt{1 + (Ft/mc)^2} \to Ft/mc$, and therefore as $Ft \to \infty$, $\beta \to 1$. What this means is that **regardless of how great the force or how long it is applied, the body will neither reach nor exceed lightspeed. As a body moves faster and faster under the influence of a force, it takes longer and longer (because of the time dilation) for the speed to increase further.** In effect, an inertial observer will see the acceleration decrease (much as if the body's mass had actually increased).

As v approaches c, time slows [Eq. (26.3)] until at $v = c$, $\beta = 1$, $\sqrt{1 - \beta^2} = 0$,

$\gamma = \infty$, and a second stretches out into infinity. Nor is the Lorentz–FitzGerald Contraction any less strange when $v = c$ and everything shrinks to nothingness. For speeds beyond c, all this becomes unreal, literally and mathematically. Only one conclusion is evident: **the speed of light is an upper limit on the rate of propagation of objects that have mass**. Nothing with mass can be accelerated to c (Question 11). To be precise, this conclusion does not preclude the existence of particles that have somehow been created with speeds in excess of c. These fantastic hypothetical entities, known as *tachyons*, have never been observed and may well never be. They are a product of the theoretical school that maintains that whatever is not explicitly forbidden must exist. But since we are not possessed of all the theories we are ever going to have, such a notion seems a trifle premature.

> No object that has **mass** can be brought up to the speed of light; c is the **limiting speed**.

26.9 Relativistic Energy

We want now to arrive at an expression for the relativistic kinetic energy (KE) of a particle. The rigorous, though rather mathematically involved, way to proceed is to go back to the idea of the work done by a force in the process of changing the body's speed (in a zero-PE situation). Instead, let's take a less formal and easier approach. As we saw earlier (p. 936), when $\beta \ll 1$, $\gamma = 1/\sqrt{1 - \beta^2} \approx 1 + \beta^2/2$, and so $\gamma m \approx m(1 + \beta^2/2)$. Multiplying this by c^2 gives

$$\gamma mc^2 \approx mc^2 + \tfrac{1}{2}mv^2$$

The term on the far right is the familiar low-speed $KE = \tfrac{1}{2}mv^2$; thus,

[kinetic energy]
$$KE = \gamma mc^2 - mc^2 \qquad (26.10)$$

> Relativistic **kinetic energy** is given by $KE = mc^2(\gamma - 1)$

and the more rigorous analysis leads to this same expression for all speeds. This is the **relativistic kinetic energy**, and it's no longer simply $\tfrac{1}{2}mv^2$. (See Fig. 26.15 for a lovely experimental confirmation.) Consequently,

$$\gamma mc^2 = KE + mc^2$$

where each term has the units of energy. The second quantity on the right is independent of speed and is known as the **rest energy** (E_0). All of this suggests that we interpret the term

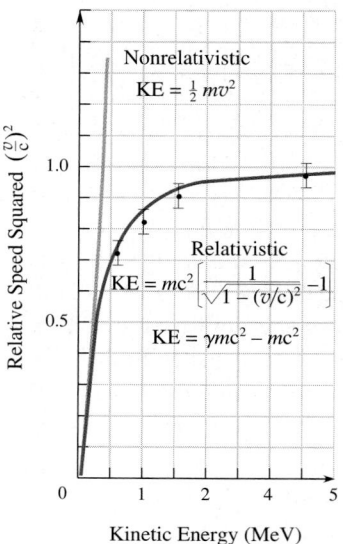

Figure 26.15 Here, electrons were given various values of kinetic energy by accelerating them across appropriate potential differences. Their speeds were then measured by finding the flight times over a fixed path. The dots represent the measured values, for which the vertical bars indicate the expected range of experimental error. Clearly, the results at high speeds confirm the validity of the relativistic expression for KE (the purple curve). The classical formula for KE corresponds to the straight dashed line, which is accurate at low speed. [Adopted from W. Bertozzi, *Am. J. Phys.* 32, 551 (1964).

on the left as the **total energy** (E) of the body, in which case

[total energy]
$$E = KE + E_0 \qquad (26.11)$$

and

[total energy]
$$E = \gamma m c^2 \qquad (26.12)$$

When the body is at rest ($\gamma = 1$), KE = 0, and the total energy (E $= \gamma m c^2$) equals the rest energy, whereupon E $= E_0$ and

[rest energy]
$$E_0 = mc^2 \qquad (26.13)$$

Example 26.8 **[II]** Every electron has a rest energy of 0.511 MeV. If one in particular is traveling at a speed of 0.900c, determine its total energy and its kinetic energy. Give your answers in MeV.

Solution Because of the high speed, this is a relativistic energy problem. (1) TRANSLATION—An object of known rest energy is traveling at a known relativistic speed ($v > 0.5c$); determine its total energy. (2) GIVEN: $E_0 = 0.511$ MeV and $v = 0.900c$. FIND: E and KE. (3) PROBLEM TYPE—Relativity/energy. (4) PROCEDURE—We have the rest energy and the speed and need the total energy: E $= \gamma m c^2$. (5) CALCULATION—We first compute γ:

$$\gamma = \frac{1}{\sqrt{1 - \beta^2}} = \frac{1}{\sqrt{1 - (0.900)^2}} = 2.294\,2$$

Then we realize that $mc^2 = 0.511$ MeV and so

$$E = \gamma m c^2 = 2.294\,2(0.511 \text{ MeV}) = \boxed{1.17 \text{ MeV}}$$

From Eq. (26.11)

$$KE = E - E_0 = 1.172 \text{ MeV} - 0.511 \text{ MeV} = \boxed{0.661 \text{ MeV}}$$

Quick Check: γ is greater than 1, which is a good sign, and it also checks with Table 26.2. As we will see in Problem 89, rest energy for a particle equals kinetic energy when $\beta = 0.866$; so this result where $\beta = 0.900$ is certainly the right magnitude.

This is perhaps the most famous outcome of the Special Theory. Confirmed experimentally, it stands beyond doubt. If there can be such a thing as the premier equation of the twentieth century, this is it. It has revolutionized Modern Physics and hurled us all into the age of nuclear weapons. Even so, the precise meaning of the relationship is still being argued in the scientific literature, primarily because we have not yet satisfactorily defined matter, mass, and energy. For example, is mass a congealed form of energy, or are mass and energy very different concepts that are simply proportional to each other and only interrelated via Eq. (26.13), just as F and a are interrelated by $F = ma$? Still, it has been established experimentally that matter possessing mass (e.g., electrons and positrons) can be transformed into massless electromagnetic radiation and vice versa.

MASS-ENERGY

When the PE of a system changes, the rest energy of the system changes, and its mass changes; $E_0 = mc^2$. Indeed the notion of potential energy, which was a necessary accounting device, can now be appreciated in a whole new way. Whenever energy is stored in a system—be it gravitational, elastic, chemical, nuclear, whatever—the mass of the system increases.

At the level of our discussion, it's probably best to ignore some of these subtleties and follow the most common usage.

Mass can be transformed into energy, and energy can be transformed into mass; hence,

$$1 \text{ kg} \leftrightarrow 8.987\,6 \times 10^{16} \text{ J}$$

$$1 \text{ kg} \leftrightarrow 5.609\,5 \times 10^{29} \text{ MeV}$$

There is one unified concept: **mass-energy**. Had we known *that* a few centuries ago, we would not now have joules, Btus, calories, and kilowatt-hours to fuss with: the kilogram would do for all forms of mass-energy. The constant c^2 is a scale factor that numerically

Table 26.3
Masses and Rest Energies for Some Particles and Atoms

Object	Mass (kg)	Rest energy (MeV)
Photon	0	0
Neutrino	0	0
Electron (or positron)	$9.109\,389\,7 \times 10^{-31}$	0.510 999
Proton	$1.672\,623\,1 \times 10^{-27}$	938.272
Neutron	$1.674\,929 \times 10^{-27}$	939.566
Muon	$1.883\,54 \times 10^{-28}$	105.659
Pion (+)	$2.416\,5 \times 10^{-28}$	135.56
Deuteron	$3.343\,584 \times 10^{-27}$	1857.612
Triton	$5.007\,357 \times 10^{-27}$	2808.920
Alpha	$6.644\,72 \times 10^{-27}$	3727.41
Hydrogen atom ($_1^1$H)	$1.673\,534 \times 10^{-27}$	938.783
Deuterium atom ($_1^2$H)	$3.344\,497 \times 10^{-27}$	1876.12
Tritium atom ($_1^3$H)	$5.008\,270 \times 10^{-27}$	2809.43
Helium-3 atom ($_2^3$He)	$5.008\,237 \times 10^{-27}$	2809.41
Helium atom ($_2^4$He)	$6.646\,482 \times 10^{-27}$	3728.40

relates mass to energy. It's an immense number, $\approx 9 \times 10^{16}$ m²/s², and even a tiny change in mass corresponds to an enormous change in energy. In Modern Physics, mass is often given in units of MeV/c² (i.e., million electron volts divided by c²); thus,

$$1 \text{ MeV/c} = 5.344\,29 \times 10^{-22} \text{ kg·m/s}$$

$$1 \text{ MeV/c}^2 = 1.782\,663 \times 10^{-30} \text{ kg}$$

whereupon, for example, the mass of an electron is 0.510 999 MeV/c² (Table 26.3). Here c² = 8.988 × 10¹⁶ m²/s²; we just don't carry out the division.

Example 26.9 [I] A 1.00-kg chicken is placed on the transporter of a fictitious starship, whereupon it is converted directly into electromagnetic energy—a process that is theoretically possible, though technologically quite beyond our poor powers—so that it could be "beamed" down to the galley. What is the equivalent energy of the chicken? How much is that in kilowatt-hours?

Solution This problem talks about converting mass into energy, and that should remind us of the defining equation for rest energy. (1) TRANSLATION—An object has a known mass; determine its rest energy. (2) GIVEN: $m = 1.00$ kg. FIND: E_0. (3) PROBLEM TYPE—Relativity/mass-energy/rest energy. (4)

PROCEDURE —We have the mass of the object and need its rest energy: $E_0 = mc^2$. (5) CALCULATION—From Eq. (26.13)

$$E_0 = mc^2 = (1.00 \text{ kg})(2.998 \times 10^8 \text{ m/s})^2 = \boxed{8.99 \times 10^{16} \text{ J}}$$

1 kilowatt-hour = 1000 J/s × 60 s/min × 60 min = 3.60 MJ; hence, dividing this into the value of energy determined above yields $\boxed{E_0 = 2.50 \times 10^{10} \text{ kW·h}}$. That's the equivalent of running ten 100-W bulbs for 2.5 × 10¹⁰ hours, about 3 million years.

Quick Check: Since 1 kg → 5.6 × 10²⁹ MeV and 1 MeV = 1.6 × 10⁻¹³ J, 1 kg → 9 × 10¹⁶ J.

In Chapter 6 we talked about Conservation of Mechanical Energy, and in Chapter 14 we generalized that concept to include thermal energy. The result was the First Law of Thermodynamics. Now we come to the final generalization: there is one all-encompassing law of **Conservation of Energy**:

The total energy of an isolated system always remains constant, although any portion of it can be converted from one form to another, including rest energy.

Example 26.10 **[II]** There are several fusion reactions that convert mass directly into energy and can power stars and drive hydrogen bombs. One such process fuses two nuclei of heavy hydrogen (deuterium) together, resulting in a still heavier hydrogen (tritium) nucleus, an ordinary hydrogen nucleus (proton), and the KE they fly off with. It's customary to write such a reaction in terms of the neutral atoms involved (neglecting the tiny amounts of energy holding the electrons to each atom, ≈ 10 eV). Thus,

$$^2_1\text{H} + ^2_1\text{H} \rightarrow ^3_1\text{H} + ^1_1\text{H} + \text{energy}$$

Determine the energy liberated per fusion.

Solution   This situation is only understandable in terms of Conservation of Mass-Energy. (1) TRANSLATION—Given a specific fusion reaction, determine the resulting energy generat-

ed. (2) GIVEN: The fusion reaction. FIND: The amount of energy liberated per fission. (3) PROBLEM TYPE—Relativity/Conservation of Energy. (4) PROCEDURE—The total energy on the right in the reaction equals the total energy on the left. In other words, the difference in the rest energy before and after is the liberated KE. (5) CALCULATION—From Table 26.3,

$$1876.12 \text{ MeV} + 1876.12 \text{ MeV} = 2809.43 \text{ MeV}$$
$$+ 938.783 \text{ MeV} + \text{energy}$$
$$\text{energy} = 4.03 \text{ MeV} = \boxed{6.45 \times 10^{-13} \text{ J}}$$

where 1 MeV = 1.602×10^{-13} J.

Quick Check: Nuclear reactions typically involve several MeV, so this result is the right order-of-magnitude.

ENERGY POSSESSES INERTIA

A single material particle at rest, by virtue of its very existence, has a rest energy. Similarly, a body composed of several particles possessing internal energy (thermal, potential, whatever), when taken as a whole, also has a net rest energy and a net mass. As Einstein put it, "the mass of a body is a measure of its energy-content." A hot apple pie has more rest energy and more mass than an otherwise identical cold apple pie. The greater mass is due purely to its greater energy content—**energy possesses inertia**. Thus, a flashlight emitting energy (ΔE) decreases in mass (by $\Delta E/c^2$), just as a plant absorbing that light gains in mass. A spring must weigh more after **elastic**-PE is stored in it than before. No one has ever measured these minuscule variations in mass, but there is ample confirmation elsewhere. In the final analysis, all the familiar occurrences that liberate energy—from burning marshmallows to exploding dynamite—transform a minute amount of mass into energy. Ultimately, that is the source of the reaction energy, and this is as true for chemical energy as it is for energy liberated by a nuclear weapon.

It is not always possible practically to convert a quantity of mass completely into energy, or vice versa, although both effects are now commonplace. Still, one single gram of mass is equivalent to 9×10^{13} J, which is enough energy to raise 200 000 000 kg of water from 0°C to 100°C. That corresponds to the peak power output of Boulder Dam operating for 19 hours, namely, 25 million kilowatt-hours. An exploding kilogram of TNT liberates about 5 million joules, which is certainly formidable, though it represents a mass loss of only about 6×10^{-11} kg—far too little to measure. Chemical reactions, which are at heart relatively weak electrical interactions, release energies of the order of a few eV. By comparison, nuclear transmutations involving the far more powerful *strong force* (p. 1062) correspond to mass changes of about 0.1%. This is roughly a million times more than the mass change in a chemical reaction (see Table 26.4) and, hence, directly observable.

The total energy can be written in terms of p without any explicit reference to v, and that provides some insights about photons. Starting with $E = \gamma mc^2$, square both sides and write it as

$$E^2 = \gamma^2 m^2 c^2 (c^2 + v^2 - v^2)$$

Table 26.4
The Fractional Change of Mass for Various Kinds of Processes

Process releasing energy $\Delta E = \Delta(mc^2)$	$\Delta m/m$
Chemical	$\approx 1.5 \times 10^{-8}\%$ to $\approx 10^{-7}\%$
Nuclear fission	$\approx 0.1\%$
Nuclear fusion	$\approx 0.6\%$
Electron-positron annihilation	100%
Neutral pion decay into two photons	100%

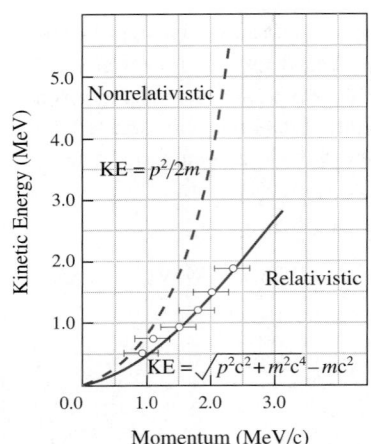

Figure 26.16 A comparison of the classical and relativistic relationships between KE and p. Momentum and energy are measured using electrons emitted via radioactive decay. Note how nicely the data fit the predicted relativistic curve. [Adopted from R. Kollarits, *Am. J. Phys.* 40, 1125 (1972).]

Now
$$E^2 = \gamma^2 m^2 c^2 (c^2 - v^2) + \gamma^2 m^2 c^2 v^2$$

Putting in the expression for γ, we have

$$E^2 = \frac{m^2 c^2 (c^2 - v^2)}{(1 - v^2/c^2)} + \gamma^2 m^2 c^2 v^2$$

and using $p = \gamma m v$, substitute in $p^2 = \gamma^2 m^2 v^2$ to get

$$E^2 = m^2 c^4 + p^2 c^2 \tag{26.14}$$

or

$$E^2 = E_0^2 + (pc)^2 \tag{26.15}$$

Figure 26.16 shows the kind of striking confirmation of these conclusions that, today, is available experimentally.

Notice that inasmuch as $E_0^2 = (mc^2)^2$ is a constant independent of any reference frame, $E^2 - (pc)^2$ must also be invariant, even though E and p separately depend on v and are thus relative quantities. We shall not explore the implications of this, other than to point out that the invariance of $E^2 - (pc)^2$ implies that **there is a single notion that ties together what until now were the two separate ideas of energy and momentum.**

Example 26.11 **[II]** A proton (of mass 938.3 MeV/c²) is accelerated across a potential difference of 202.0 MV, so that it has a kinetic energy of 202.0 MeV. Determine its total energy (in MeV) and momentum (in MeV/c). What is the speed of the particle?

Solution Because the potential difference is so large we can assume this is a case of relativistic dynamics. (1) TRANSLATION—A particle of known mass is given a specified KE; determine its total energy, momentum, and speed. (2) GIVEN: $m = 938.3$ MeV/c² and KE = 202.0 MeV. FIND: E, p, and v. (3) PROBLEM TYPE—Relativity/energy and momentum. (4) PROCEDURE—$E = KE + E_0$. (5) CALCULATION:

$$E = KE + mc^2 = 202.0 \text{ MeV} + (938.3 \text{ MeV/c}^2)c^2$$

$$\boxed{E = 1140 \text{ MeV}}$$

From Eq. (26.15)

$$p = \frac{\sqrt{E^2 - E_0^2}}{c} = \boxed{647.5 \text{ MeV/c}}$$

To find the speed, we use the fact that $E = \gamma E_0$, in which case $E = E_0 / \sqrt{1 - v^2/c^2}$ and so

$$v = c \sqrt{1 - \frac{E_0^2}{E^2}} = c \sqrt{1 - \left(\frac{938.3}{1140}\right)^2} = \boxed{0.568\,3\,c}$$

Quick Check: Starting with the momentum, $p = \gamma m v$ and $v = p/\gamma m = p/(E/c^2) = (647.5 \text{ MeV/c})/(1140 \text{ MeV/c}^2) = 0.568\,0\,c$.

Massless particles, like the photon, only exist traveling at the speed c.

In addition to the photon, physicists have found it necessary to deal with other zero-mass particles such as the several different neutrinos and the graviton. The existence of neutrinos is well confirmed experimentally (though it's still unclear if they are totally massless), whereas the graviton remains a matter of speculation. When $m = 0$ in Eq. (26.14)

[zero-mass particles]
$$E = pc \tag{26.16}$$

Moreover, because $E = \gamma mc^2$, $E/\gamma = mc^2$; and when $m = 0$, $E/\gamma = 0$. Since E is nonzero, $1/\gamma = 0 = \sqrt{1 - v^2/c^2}$, and it follows that for particles of zero mass $v = c$. **Particles of zero mass exist only at the speed c.**

Example 26.12 **[II]** A neutral pion has a mass 264 times larger than the electron mass and a rest energy of 135 MeV. It is unstable and decays into two oppositely directed γ-ray photons. Assume the pion to be at rest and determine the energy and momentum of the photons.

Solution This situation can be understood from the perspective of Conservation of Mass-Energy. (1) Translation—A particle of known rest energy and zero KE decays into two photons; determine the photons' energy and momentum. (2) Given: $E_0 = 135$ MeV. Find: The energy E and momentum p for the photons. (3) Problem Type—Relativity/massless particles/energy and momentum. (4) Procedure—Since the pion is at rest when it decays, initially $p = 0$, and the momenta

of the photons are equal and opposite. The total energy of the two photons (2E) must equal the energy of the pion. (5) Calculation—$E_0 = 135$ MeV $= 2E$, and for each photon $\boxed{E = 67.5 \text{ MeV}}$. From Eq. (26.16) for massless particles,

$$p = \frac{E}{c} = \boxed{67.5 \text{ MeV/c}}$$

and since 1 MeV/c $= 5.344\ 3 \times 10^{-22}$ kg·m/s, $p = 3.61 \times 10^{-20}$ kg·m/s.

Quick Check: $m = 264(9.11 \times 10^{-31} \text{ kg}) = 2.4 \times 10^{-28}$ kg; the photon energy is half the pion energy $E = \frac{1}{2}mc^2 = 1.08 \times 10^{-11}$ J $= 67.5$ MeV.

CAUSALITY & RELATIVITY

Though we shall not elaborate upon the argument, Relativity establishes that if an interaction is to be causal (in the sense that the act of initiation of an event must precede the event in time for all inertial observers), then the speed of propagation of the interaction (i.e., the signal) cannot exceed c. The uncovering of a radioactive source sets a distant counter clattering; the emitted γ-rays travel from source to detector at c, and the sequence of events is causal. The transport of information (matter-energy in a discriminated form) in a cause-and-effect sequence cannot occur at a speed in excess of c. That does not mean that there cannot be apparent motions that are greater than c. Shine a laserbeam at observer-*A* leaning on a very distant screen. Now quickly jerk the laser so that the light spot moves across the screen to observer-*B*. Clearly, if you shift the laser fast enough and if the target is far enough away, the light spot travels from *A* to *B* in excess of c (oscilloscopes actually perform this trick all the time). Still, the photons that get to *B* were never at *A*—new photons a little behind in the stream arrive at each successive point on the screen. Light never literally travels from *A* to *B*.

Relativity does not forbid noncausal interactions. In fact, the concept is at the center of a great deal of modern experimentation and interpretation as it relates to Quantum Mechanics. There are some wonderfully mind-boggling contemporary experiments that are challenging our familiar notions of reality.

Core Material & Study Guide

BEFORE THE SPECIAL THEORY
The Michelson–Morley experiment established that the motion of the Earth through the supposed surrounding aether could not be detected (p. 933).

THE SPECIAL THEORY OF RELATIVITY
The **Principle of Relativity**, the *First Postulate* of the Special Theory, states that *all the laws of physics are the same for nonaccelerating observers*. Einstein's *Second Postulate*, the **Principle of the Constancy of the Speed of Light**, states that *light propagates in free space at the invariant speed* c (p. 937). The basic ideas are put forth in Section 26.2 (The Two Postulates).

There is no such thing as **absolute simultaneity**—events separated in space that are simultaneous for midway observers in one inertial system are not necessarily simultaneous for midway observers in any other inertial system (p. 939). Although still not

quantitative, Section 26.3 (Simultaneity and Time) prepares the way for the next section, which is quantitative.

Time on a clock that is moving with respect to an observer runs slower than time on a clock that is stationary with respect to that observer; thus,

$$\Delta t_M = \frac{\Delta t_s}{\sqrt{1 - v^2/c^2}} \qquad [26.2]$$

This important equation is treated in Section 26.4 (The Hatter's Watch: Time Dilation). Make sure you understand Example 26.2 and look at the **CD Walk-Throughs**.

A moving observer sees an object to have a length (along the direction of motion) that is shorter than the **proper length** seen by an observer at rest with respect to the object; thus,

$$L_M = L_S\sqrt{1 - v^2/c^2} \qquad [26.5]$$

This Lorentz–FitzGerald Contraction is discussed in Section 26.5 (Shrinking Alice: Length Contraction), and it would be wise to study Examples 26.3 and 26.4.

The one-dimensional relativistic velocity transformation

$$v_{PO} = \frac{v_{PO'} + v_{O'O}}{1 + \dfrac{v_{PO'}v_{O'O}}{c^2}} \qquad [26.7]$$

is treated in Section 26.7 (Addition of Velocities). Review Example 26.6 and study Figs. 26.10 and 26.11.

RELATIVISTIC DYNAMICS

The relativistic momentum of a body of mass m is

$$\vec{p} = \gamma m\vec{v} = \frac{1}{\sqrt{1 - v^2/c^2}}\, m\vec{v} \qquad [26.8]$$

Here we take the approach that *m is not a function of v*. Study Section 26.8 (Relativistic Momentum) and reexamine Example 26.7.

The relativistic energy of a body is given by

[kinetic energy] $\qquad$ KE $= \gamma mc^2 - mc^2 \qquad [26.10]$

[total energy] $\qquad$ E $=$ KE $+ E_0 \qquad [26.11]$

[total energy] $\qquad$ E $= \gamma mc^2 \qquad [26.12]$

[rest energy] $\qquad$ $E_0 = mc^2 \qquad [26.13]$

The mass of a composite body changes if the total rest energy of its component parts changes (p. 957). In terms of momentum,

$$E^2 = E_0^2 + (pc)^2 \qquad [26.15]$$

For zero-mass particles

$$E = pc \qquad [26.16]$$

All of this is discussed in Section 26.9 (Relativistic Energy), and you must study Examples 26.8, 26.9, 26.10, 26.11, and 26.12 before attempting to do any problems.

Key Terms

aether	proper time
Michelson–Morley Experiment	proper length
Lorentz-FitzGerald Contraction	twin effect
Principle of Relativity	addition of velocities
inertial observer	relativistic momentum
First Postulate	relativistic mass
Second Postulate	relativistic energy
Law of the Constancy of c	relativistic kinetic energy
simultaneity	rest energy
time dilation	mass-energy

Discussion Questions

1. In terms of Classical Physics only, suppose you are sitting in a chair resting. (a) What will you see if you jump up, rush away at, say, a speed of 100c, spin around, and stop? (b) In a similar vein, how might you observe Lincoln's Gettysburg address? (c) Is all of this actually possible?

2. Prior to the Special Theory, Lorentz and others were faced with an interesting dilemma. Figure Q2 shows a beam of light and two observers moving along with it, one at a speed $v < c$ and the other at $v > c$. From a classical perspective, how would each see the light to move? What about the notion that the beam propagates in the direction of $\vec{E} \times \vec{B}$?

3. How might the history of physics have been different if c were 3×10^2 m/s rather than 3×10^8 m/s?

4. Could we connect two distant spacecraft by a long rigid rod and communicate faster than we might with light by just wiggling the ends of the rod? Explain.

5. Imagine an ideally rigid long pair of scissors. As the two blades are closed, the cutting point moves out away from the pivot. The

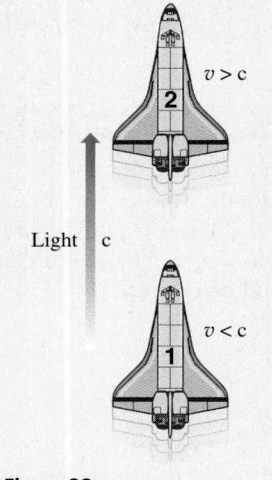

$v > c$

Light $\quad$ c

$v < c$

Figure Q2

more nearly parallel the blades become, the more rapidly that point travels away. If the scissors are long enough and closed rapidly enough, the contact point will travel out faster than c. An ax slammed almost horizontally onto an equally large horizontal sheet of wood suggests the same effect. Think of the point of overlap of two crossed laser-beams in the same plane that are moved toward being parallel. Does any of this violate Relativity? Explain. Can you think of another "thing" that moves faster than c?

6. Primary cosmic rays impacting on the upper atmosphere at altitudes of around 15 km create high-speed ($v = 0.999c$) muons. Despite their 1.5-μs half-life, about 200 muons are detected at ground level per square meter per second. That's surprising because half of them should disappear after 1.5 μs (or 450 m) and after 33.3 half-lives (15 km) a minuscule $10^{-8}\%$ of them should survive (much fewer than do). Explain what is happening from the perspective of the muon.

7. Suppose we are receiving a stream of photons of frequency f in vacuum and then decide to fly directly away from the source. Discuss what we will observe to happen to the photons' speed, momentum, total energy, and rest energy.

8. Astronaut Suzy inside her spaceship shines a flashlight beam on the front wall just as the craft passes the Earth. If the ship is traveling at or in excess of c, what would Suzy see? What would you on Earth see? Using the two postulates, what can you conclude?

9. As we have seen, when two parallel wires carry currents in the same direction, there is a "magnetic" attraction between them. Amazingly, it turns out that this effect is actually an electrostatic one and, more generally, that *magnetism is a relativistic manifestation of electrostatics*. There really is only one force—the electromagnetic force. Thus, envision two current-carrying wires and their moving electrons and stationary positive ions. What do things look like within wire-1 as seen by a moving electron in wire-2? Is there a Lorentz–FitzGerald Contraction, and, if so, what is its apparent effect on the positive and negative charge densities?

10. Newton said, "Absolute, true, and mathematical time, of itself, and from its own nature, flows equally without relation to anything external." Discuss the significance and validity of this statement.

11. Given that $E = \gamma mc^2$, what happens to a material object as it approaches the speed of light? What does this tell us about v compared to c?

12. The energy of a photon is entirely kinetic. Discuss the relationships between E, $p, f,$ and λ for photons. How does the Doppler Effect influence these considerations? Can a light beam exert pressure on a target?

13. Starting with Eq. (26.7), show that either $v_{PO'}/c = 1$ or $v_{O'O}/c = 1$ implies $v_{PO}/c = 1$. When is v_{PO} less than the classical value? When is it greater?

14. You may have learned in chemistry class somewhere along the line that if you "weighed" all the reactants before and after a chem-

ical reaction, the sum would be unchanged. That's the so-called Law of Conservation of Mass, and it's confirmed to an accuracy of one part per ten million ($1/10^7$). Inasmuch as Relativity shows us that most chemical reactions involve mass changes ($\Delta m/m$) of a few parts per ten thousand million ($\approx 2/10^{10}$), what can be said about the Law of Conservation of Mass?

15. Poincaré challenged the idea of immediate action-at-a-distance, although Newton had accepted it. From Poincaré's perspective, for example, a change in the gravitational field of the Sun would take a finite time to reach the Earth. Generalizing this assumption, show that nothing can travel at an infinite speed. Assuming there is an upper speed limit, discuss why it must be invariant with respect to all inertial reference frames. What does this suggest about the Michelson– Morley experiment and c?

16. You've no doubt heard in school the adage "Matter can neither be created nor destroyed" or its haunting corollary "Energy can neither be created nor destroyed." What can you say about these statements?

17. Here are two postulates: (1) All interactions propagate at a finite speed. (2) An object cannot interact with itself. Discuss how it follows that there must be an upper speed limit.

18. Show, through verbal argument, that if the First Postulate is true, the Law of Inertia must be true.

19. Explain the meaning of each of the Key Terms on page 960.

Multiple Choice Questions

1. The Michelson–Morley experiment established (a) that the Earth does not move with respect to the Sun (b) that the aether moves at c as the Earth travels in its orbit (c) that the aether is an elastic solid that streams over the Earth (d) that there is no observable aether wind at the surface of the Earth (e) none of these.

2. Simultaneity is (a) dilated (b) absolute (c) invariant (d) relative (e) none of these.

3. A clock is moving at a uniform velocity with respect to an observer. The latter, comparing things to her clock, reports that the time on the moving clock is (a) perfectly accurate (b) fast (c) slow (d) running backward (e) none of these.

4. According to the Lorentz–FitzGerald Contraction (a) a body contracts only when it accelerates (b) a body contracts along the direction of motion (c) the time it takes for a light clock to tick contracts (d) a body contracts transverse to the direction of its motion (e) none of these.

The next six questions refer to Fig. MC5, which shows four graphs plotted against $\beta = v/c$ on the horizontal axis.

5. If the vertical axis in each case was the length L_M, which of the graphs would best depict relativistic length contraction? (a) 1 (b) 2 (c) 3 (d) 4 (e) none of these.

6. If the vertical axis in each case was the time interval Δt_M, which of the graphs would best depict relativistic time dilation? (a) 1 (b) 2 (c) 3 (d) 4 (e) none of these.

7. If the vertical axis in each case was p, which of the graphs would best depict relativistic linear momentum? (a) 1 (b) 2 (c) 3 (d) 4 (e) none of these.

8. In a correct plot of $\Delta t_M/\Delta t_S$ versus β, what would be the value of the curve at $\beta = 0$? (a) 1 (b) 0 (c) Δt_S (d) ∞ (e) none of these.

9. In a correct plot of L_M/L_S versus β, what would be the value of the curve at $\beta = 1$? (a) L_S (b) 1 (c) 0 (d) ∞ (e) none of these.

10. In a correct plot of $\Delta t_M/\Delta t_S$ versus β, what would be the value of the curve at $v = c$? (a) 1 (b) 0 (c) Δt_S (d) ∞ (e) none of these.

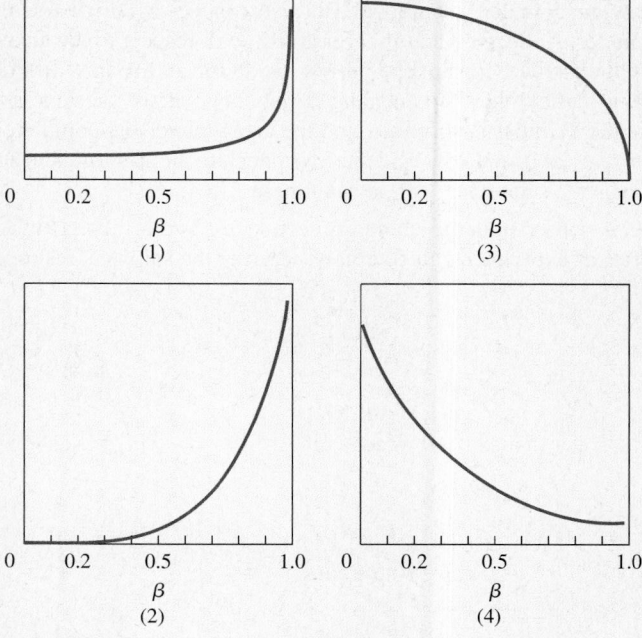

Figure MC5

11. How long will a vertical meter stick appear to someone moving horizontally with respect to it at a speed of 0.99c? (a) 99 cm (b) 100 cm (c) 0 cm (d) 0.01 cm (e) none of these.

12. Two inertial observers separating from each other at a speed of 99.99% of c each shine a light beam at one another. (a) each will see the light arrive at a speed of c (b) each will see the light arrive at a speed of 2c (c) the light will not reach either of them (d) the light will reach one at c and the other at 2c (e) none of these.

13. A spaceship observes an asteroid coming directly toward it at a speed v. The ship fires its engines and travels away along the line of approach. It measures the asteroid's (a) v and p to decrease, but E_0 is unaltered (b) v, p, and E to be unchanged (c) v only to decrease, p and E being unaltered (d) p and E to increase, leaving v and E_0 unaltered (e) none of these.

14. A long thin rocket takes off straight up from the ground. It is observed by three people: a man selling ice cream on the beach, a woman riding up an adjacent gantry elevator, and the pilot looking out a window. Referring to the rocket, ideally (a) all three see the same length (b) of the three, the pilot sees the shortest length (c) of the three, the woman sees the shortest length (d) of the three, the man with the ice cream sees the longest length (e) none of these.

15. An astronaut heading out toward a star at a constant high speed can determine that he is in motion by (a) the contraction of on-board meter sticks (b) the slowing down of time on his clocks (c) the increase of his mass (d) the speeding up of his heart (e) none of these.

16. An arrow 30 cm long is fired at very high speed directly toward a hollow cylinder 30 cm long. The arrow passes axially inside the cylinder. As seen by an observer at rest with respect to the cylinder, (a) the arrow is shrunken and fits inside for an instant (b) the arrow and cylinder are unchanged and stay exactly the same size (c) the cylinder is shrunken and the arrow never fits completely inside (d) the arrow and the cylinder are both shrunken and remain the same size (e) none of these.

17. An arrow 30 cm long is fired at very high speed directly toward a hollow cylinder 30 cm long. The arrow passes axially inside the cylinder. As seen by an observer at rest with respect to the arrow, (a) the arrow is shrunken and fits inside for an instant (b) the arrow and cylinder are unchanged and stay exactly the same size (c) the cylinder is shrunken and the arrow never fits completely inside (d) the arrow and the cylinder are both shrunken and remain the same size (e) none of these.

18. An observer in the laboratory section of a spacecraft performs a series of experiments to determine whether the ship is at rest or in uniform motion. He (a) can succeed by making very precise time measurements (b) can succeed by making very precise mass measurements (c) can succeed by making very precise length and time measurements (d) cannot succeed no matter what he does (e) none of these.

19. When a battery is drained in ordinary use, (a) its mass decreases because its internal rest energy decreases (b) its mass decreases, but its rest energy remains unchanged (c) its mass remains unchanged, but its internal rest energy decreases (d) its mass decreases because its internal rest energy increases (e) none of these.

20. When a photon is red-shifted via the Doppler Effect by having an observer moving away from the source, (a) its energy is unchanged (b) its momentum increases (c) its energy increases (d) both its momentum and energy decrease (e) none of these.

21. An arrow 30 cm long is fired at very high speed v directly toward a hollow cylinder 30 cm long. The arrow passes axially inside the cylinder. An observer moving parallel to the cylinder in the direction of the arrow at a speed $\frac{1}{2}v$ sees the arrow moving forward at the same speed he sees the cylinder moving backward. He observes that (a) the arrow is shrunken and fits inside for an instant (b) the arrow and cylinder are unchanged and exactly the same size (c) the cylinder is shrunken and the arrow never fits completely inside (d) the arrow and the cylinder are both shrunken and are the same size (e) none of these.

22. The proper time of a system is (a) the time measured on clocks at rest in that system (b) the time measured on clocks at rest in an inertial system moving properly with respect to the first system (c) the time measured on clocks moving uniformly in that system (d) the time that agrees with the National Institute of Standards and Technology (e) none of these.

23. A photon's proper time (a) is a function of its speed (b) changes slowly with respect to a moving observer (c) never changes (d) changes faster with respect to systems at rest (e) none of these.

24. Given that there are about 2.56×10^9 heartbeats in a statistically average lifetime of 70 years, people who are born and die on a spaceship moving at a constant speed of 0.600c can expect their hearts to beat a total of (a) $(0.600)(2.56 \times 10^9)$ times (b) 2.56×10^9 times (c) $(0.800)(2.56 \times 10^9)$ times (d) $(1.25)(2.56 \times 10^9)$ times (e) none of these.

For more Multiple Choice Questions with answers click on WARM-UPS in CHAPTER 26 on the CD.

Suggestions on Problem Solving

1. The factor $\gamma = 1/\sqrt{1 - \beta^2}$ is so ubiquitous in the work of this chapter that it merits a few words. It's common to forget either to square the β or to take the square root once having computed $1 - \beta^2$. Remember that $\gamma > 1$. A nice way to compute γ on an old-fashioned electronic calculator if v is given as a number (for example, $v = 2.43 \times 10^7$ m/s) with no reference to c is

$$v \;[\div]\; c \;[=]\; [x^2] \;[+/-]\; [+]\; 1 \;[=]\; [\sqrt{\ }\,] \;[1/x]$$

If you have v in terms of c (for example, $v = 0.9c$), then the calculation is more straightforward because you can enter v/c as a number (in this case, $v/c = 0.9$):

$$1 \;[-]\; v/c\; [x^2] \;[=]\; [\sqrt{\ }\,] \;[1/x]$$

Many calculators have constants stored in memory, and c is usually one of them.

2. When doing problems on time dilation and length contraction, the "hard part" is just figuring out which piece of information to call L_M and which L_S. For example: A kid on an asteroid holding a meter stick is going to measure the length of a space probe as it flies by at 0.9c parallel to the stick. As seen by the pilot, how long is the meter stick? "As seen by the pilot"—the stick moves past the probe at 0.9c, the stick is moving with respect to the frame we are interested in, the pilot's frame. Hence, the *proper length* of the stick measured by the kid is $L_S = 1$ m. We place ourselves in the frame of the pilot and see a moving stick of length L_M. The rest of the problem is just plug-in using Eq. (26.5).

Another useful approach is to *figure out who sees what shrunk or dilated before getting into the problem numerically.* In the preceding example, "As seen by the pilot" tells us whose perspective we are to determine. The pilot sees the ruler moving and *shrunk.* The answer must be less than 1 m. Keeping in mind that $\sqrt{1 - v^2/c^2} < 1$,

$$(\text{Length seen by pilot}) = (\text{length seen by kid})\sqrt{1 - v^2/c^2}$$

The stick has only one proper length (measured at rest), and that length is the longest possible measure.

3. There are clever things that can be done regarding units in momentum and energy problems. For example, given rest energy in MeV, if you are asked to find momentum, determine it in terms of MeV/c. You needn't convert MeV to joules. Thus, if $v = \frac{1}{2}c$, $p = \gamma m v = \gamma m \frac{1}{2}c$; *now multiply top and bottom by* c; $p = \gamma \frac{1}{2}mc^2/c = \frac{1}{2}\gamma E_0/c$ and you enter E_0 in MeV and get p in MeV/c.

Problems ✦ Coordinated Problems ✦ Progressive Problems ✦ Solutions

STUDY GUIDE   1. **Coordinated Problems:** The three problems within each magenta-colored grouping are solvable in similar ways. Note that the first of these always has a hint; moreover, its solution is provided in the back of the book. *Work out each of these sets; they'll strengthen technique and build confidence.* 2. **Progressive Problems:** The problems introduced in blue unfold step-by-step carrying along the analysis in a more suggestive way than is customary. *Work out all of these; they'll guide you through the analytic process and help develop problem-solving skills.* 3. **Worked-Out Solutions:** Studying worked-out solutions is an important part of learning how to solve problems. Accordingly, additional *solutions* to a number of model problems are given below. *Make sure you understand each of them before you go on to the next problem.* 4. Also provided in the back of the book are the *Answers* to all odd-numbered problems, as well as worked-out *solutions* to those with boldface numbers. Problem numbers in italic indicate that a solution appears in the Student Solutions Manual.

SECTION 26.1: THE MICHELSON–MORLEY EXPERIMENT
SECTION 26.4: THE HATTER'S WATCH: TIME DILATION

1. [I] In the Michelson–Morley experiment, the time it takes beam-1 to travel out and back, with and against the aether wind, is $t_\parallel$, and the time for beam-2 to go from M_S to M_2 to M_S is $t_\perp$, where

$$t_\parallel = \frac{2L}{c}\frac{1}{(1 - \beta^2)} \quad \text{and} \quad t_\perp = \frac{2L}{c}\frac{1}{\sqrt{1 - \beta^2}}$$

Suppose the $M_S M_1$ arm of the Michelson Interferometer shrinks as suggested by FitzGerald so that L becomes $L\sqrt{1 - \beta^2}$. Show that $\Delta t = t_\parallel - t_\perp$ then equals zero.

2. [I] A light-minute is often symbolically written as 1 c·min. Define the concept and give its value in SI units.

3. [I] Show that $\beta^2 = (\gamma^2 - 1)/\gamma^2$.

4. [I] An object in space is seen from Earth to be motionless. What is its value of γ? The pilot of a spacecraft sees the Earth moving toward him at 0.10c. What is the speed of the craft as seen from the Earth?

5. [I] Show that

$$v = c(1 - \gamma^{-2})^{1/2}$$

At what speed would you have to travel in order to have a γ value of 9.000?

> **SOLUTION:** $\gamma^{-1} = (1 - v^2/c^2)^{1/2}$ and so $\gamma^{-2} = (1 - v^2/c^2)$. Therefore $v^2/c^2 = (1 - \gamma^{-2})$, $v^2 = c^2(1 - \gamma^{-2})$ and $v = c(1 - \gamma^{-2})^{1/2}$. $v = c(1 - 9.000^{-2})^{1/2} = 0.993\,8c$.

6. [I] THIS PROBLEM IS CONCERNED WITH TIME DILATION. An astronomer on Earth looking through her telescope sees a spaceship traveling at 0.500c. She watches while 1.00 min goes by on the clock hanging on the wall of the spaceship. (a) How fast is the ship moving in m/s with respect to her? (b) What is the corresponding value of β? (c) What is the corresponding value of γ? (d) Write the equation for time dilation. (e) What is the value of the proper time interval? Explain. (f) How long will the 1.00-min observation take as shown on the clock in the observatory on Earth?

7. [I] THIS PROBLEM DEALS WITH TIME DILATION. Imagine a time in the future when spaceships roam the cosmos. Suppose that such a ship is detected traveling at 80.0% the speed of light with respect to the Earth. It sends out two omnidirectional electromagnetic signals separated by 10.00 min. We want to study the time between signals as recorded on Earth. (a) Who measures the proper time interval, observers on the Earth or on the ship? Explain. (b) What is the value of the proper time interval? (c) What is the value of β? (d) What is the value of γ? (e) Determine the interval separating the two signals on the Earth.

8. [I] A computer in a spacecraft takes 2.00 μs to make a calculation, as measured by a co-moving observer on board. Someone traveling at a relative speed of 0.998c looks in the window. How long will that person determine the calculation to take?

9. [I] According to a biology text, it takes 45 s for blood to circulate around the human body. How long on his rocket-borne clocks will it take blood to circulate around the body of an astronaut traveling such that with respect to Earth $\gamma = 2$? How long will a ground-based observer see the astronaut's circulation take? [*Hint: 45 s is the proper time.*]

10. [I] A spaceship passes Earth at a speed such that $\gamma = 100$. A creature aboard the vessel does the Mongoian greeting dance that takes it 20 s (on its clock—bought on a previous visit—which is why it reads in seconds) to perform. How long will the welcoming committee on Earth see the performance last?

11. [I] A subatomic particle traveling at a speed such that $\gamma = 10$ as observed by a lab technician is seen by her to live for 20 ns before decaying. How long would such a particle normally live when at rest?

12. [I] A stopwatch in the hands of a terrestrial observer is seen by a passerby in a flying saucer traveling relative to the planet with a γ of 2.00. If the earthling sees two events separated by a proper time interval of 60.0 s, how long will the interval be as seen by the traveler?

13. [I] The vaudeville team of Shultz and O'Hara, playing to a full house at Minsky's Burlesque, takes 60.0 s to tell their worst joke. How long will an observer in a rocket ship flying by at 0.995c have to wait to lip-read the punch line? Will the observer start laughing after seeing the audience respond? Assume that the observer is equidistant from the stage when he perceives the start and finish of the joke so that we needn't worry about communication-time lags.

14. [I] An astronaut on the way to Mars at 0.600c wishes to take a one-hour nap. She contacts Ground Control requesting a wake-up call. Neglecting any complications due to the time it takes the signal to reach the ship, how long should the flight controller let her sleep as measured on his clock?

15. [I] Suppose you look at the inhabitants of a passing asteroid and see all their clocks running such that 52.0 s goes by on them while 60.0 s elapse on your planetary clock. What is the relative speed of the asteroid and Earth?

16. [II] THIS PROBLEM DEALS WITH TIME DILATION. An astronaut flying by the Earth measures her pulse rate to be 80 beats per minute. The doctor in the ground control facility records her pulse to be 76 beats per minute. (a) What is the interval of time (in seconds) between beats as measured by the astronaut? (b) What is the interval of time (in seconds) between beats as measured by the doctor? (c) Which of these is the proper time interval? Explain. (d) What is the value of γ for the astronaut? (e) How fast is she traveling relative to the Earth?

17. [II] THIS PROBLEM DEALS WITH TIME DILATION. On a typical flight, the Space Shuttle might have an orbital speed of 17.0×10^3 mi/h. We want to prove that while traveling at that speed

$$\Delta t_M = \Delta t_S (1 + 3.21 \times 10^{-10})$$

(a) What is its speed in meters per second? (b) What is the corresponding value of β? (c) Using the results of Example 26.1, write an expression for γ when β is small. (d) Show that the above equation is true. (e) Compared to clocks on Earth does the Shuttle clock run fast or slow? Explain. Suppose 1.00 s went by on the Shuttle clock, how much time would pass on the Earth below? What is the value of the proper time? Don't worry about significant figures. (f) How long (in years) will it take for the Shuttle clock to be off 1.00 s from Earth time?

18. [II] Calculate the approximate time dilation that would be observed by someone on the ground watching a clock in a supersonic jet flying at 1800 mi/h. Use the binomial expansion (which is applicable at speeds less than around 0.3c). Try doing this calculation exactly—you'll see why the approximation is so useful.

19. [II] Two friends, Harvey (age 20) and Lisa (age 24), say their goodbyes. She boards a space shuttle for a round-trip flyby of the planet Mongo. Rapidly reaching cruising speed, she settles in for a long journey. Neglecting the short acceleration times (which is quite impractical), Lisa returns home 24 months later (as seen on her calendar). She is met by Harvey, who is 26 years old. At what speed was Lisa cruising?

20. [II] A subatomic particle when at rest typically exists for about 20 ns before it decays. A beam of such particles is created traveling at 0.80c with respect to the laboratory. How far might such a particle travel in the lab before it decayed?

21. [III] Consider the passengers aboard a jet plane cruising at 1000 km/h with respect to an observer on the ground. Neglecting all other effects (for example, gravity), how long would the flight have to last before the clock aboard the plane was 1.00 s behind a clock on the ground with which it had been synchronized?

22. [III] With Problem 21 in mind, consider a satellite in orbit or a plane flying around the Earth at a uniform speed. We wish to determine how long the flight must last for there to be a discrepancy of 1.00 s between clocks on the ground and in the air. Show that this will indeed be the case when the proper time on the moving craft is

$$\Delta t_S = \frac{1.00 \text{ s}}{\gamma - 1}$$

SECTION 26.5: SHRINKING ALICE: LENGTH CONTRACTION

SECTION 26.6: TWEEDLEDEE & TWEEDLEDUM: THE TWIN EFFECT

23. [I] If an electron is hurtling down the two-mile-long Stanford Linear Accelerator at 99.98% of c, how many meters long is the trip as seen by the electron? [*Hint: What is the proper length of the accelerator?*]

24. [I] A long pepperoni is fired out of a new secret weapon developed by the Mediterranean arm of NATO. As it flashes past the reviewing stand, its length is measured to be 44.0% of what it was before launch. How fast is it moving?

25. [I] How large would a 1.00-m-long Italian bread look to a hungry observer (with very sharp eyes) if he sees the bread to have a speed of 0.900 c in a direction parallel to the loaf?

26. [I] A 1.00-m-long rocket is to be fired past an observer who will locate both its ends simultaneously and determine its length. If that length is found to be 0.500 m, how fast was the rocket moving?

27. [I] A spacecraft passes a platform traveling parallel to it (at 0.600 c) along its full length. The pilot of the spacecraft determines the platform to be 400.0 m long. You witness this event from the platform lounge where you happen to be having a drink. What is the proper length of the platform?

28. [I] An astronaut is on the way directly to a star system along a straight line from Earth. Her instruments indicate that at the speed relative to the Earth at which she is traveling, her γ is 5/3 and the total distance she has to traverse is $(6.0c)(1.0 \text{ y})$. What is the corresponding proper distance shown on Earth-based instruments?

29. [II] A bright yellow line painted along the length of a rocket is seen to be 1.000 yd long by an observer who is standing on the Earth as the vehicle sails by. Knowing that the proper length of the line is actually 1.000 m, how fast is the rocket traveling?

30. [II] Calculate the percentage length contraction for a high-speed jet plane traveling at 3600 mi/h. Try doing it exactly and then use the binomial expansion (which is okay for speeds less than around 0.3 c).

31. [II] Two inertial observers pass each other at a relative speed v, and each notices a 10.0% length contraction of the other. Determine v.

32. [II] The navigator on a Federation starship traveling toward Earth along a path parallel to the Earth-Sun line sees the separation between these bodies to be 6.11 c·min. Consulting charts, the navigator finds the proper distance to be 8.33 c·min. How fast is the ship going?

33. [II] A 1.000-m bar of steel travels at a constant speed of 0.600 c with respect to an observer. The bar moves lengthwise along a line perpendicular to the observer's telescope. How long does it take the bar to fly past the cross hairs in the telescope?

34. [II] A starship passes a space station at a uniform 0.600 c. The pilot looking into the gym area sees a runner traveling perpendicular to the flight path. The runner covers the straight 100-m gym track in 10.0 s by his watch. What was the runner's average speed as seen by the pilot? [*Hint: How long is the track as seen by the pilot?*]

35. [II] The TV mast on a spaceship passing Earth is fixed at 21.0° to the line of flight by the communications officer. If the ship zips by at 0.852c, at what angle will someone on Earth see the mast?

36. [II] Mars is 80.00×10^6 km from Earth when a settler sends a laser-beam message back home. As fate would have it, just as the flash of light goes out, a spacecraft flies by on the way toward Earth at 0.500 0c. How long will someone on Earth see the ship take to make the journey? How long will the message take to arrive on Earth? How long will the crew of the ship say the trip took? How long will they say the message took to get to Earth?

37. [II] This problem is concerned with both time dilation and length contraction. According to its flight plan (filed in the year 2050) U.S.A.F. Shuttle-A10 was to travel at a fairly constant speed of 0.993 8c, so as to have a γ of 9.000, on the way to a planetary

system 9.000 ly away, as measured by astronomers on Earth. (a) What is the so-called proper length of the journey? (b) How far will the crew conclude they will have to travel? (c) Who measures the proper time? Explain. (d) About how long will the journey take from the crew's perspective (recall the discussion of how to express light years on p. 948)? (e) How long will the earthbound astronomers say the journey lasts?

38. [II] This problem deals with both time dilation and length contraction. The brightest star in the northern sky is Sirius which has been measured to be 8.58 ly from Earth. Far off in the future, a ship leaves our planet, quickly speeds up to 0.990c, and travels to Sirius at that speed. We want to determine how much time the journey will take for both the crew and the astronomers on Earth who are watching it unfold. (a) Draw a diagram. (b) What is the proper length of the journey? (c) The two central events are the arrival at Sirius and the departure from Earth. Does an astronomer see these events occur at the same location? Explain. (d) Does an astronaut see these events occur at the same location? Explain. (Reread p. 947 if necessary.) (e) Who measures the proper time (Δt_S)? (f) An astronomer sees the craft flying at 0.990c and traveling a distance of 8.58 ly. How long will the trip appear to take? (g) What is the value of γ? (h) As reported by an astronaut, how many years will the trip take, and how many light-years long will it be?

39. [II] Suppose you're an astronaut who is to travel to a planet in orbit around the fifth brightest star Vega, which is 25.32 ly from New Orleans. You'd like to get there via a journey taking 2.00 y (i.e., you only want to age 2.00 y along the way). Approximate the necessary cruising speed?

SOLUTION: Let's first find the distance as seen by the astronaut; that's $L_M = \gamma^{-1} L_S$, but we don't have γ. Still, if $v \approx c$, as it must be, the length of the trip as seen by the astronaut (L_M) has to be ≈ 2.00 ly. Which means that 25.32 ly $\approx \gamma(2.00 \text{ ly})$ and $\gamma \approx 12.66$. From Problem 5 above $v = c(1 - \gamma^{-2})^{1/2}$ and $v = c(1 - 12.66^{-2})^{1/2} \approx 0.997c$.

SECTION 26.7: ADDITION OF VELOCITIES

40. [I] The pilot in a spaceship traveling at 0.50c toward the star Canopus fires a laser pulse in the opposite direction toward some floating wreckage. (a) If the wreckage is at rest with respect to Canopus, at what speed does the pulse approach it? (b) What is the speed of the pulse with respect to the ship?

SOLUTION: (a) The speed of light in vacuum is c for all inertial observers, so the pulse approaches the wreckage at c, and it travels away from the spaceship at c.

41. [I] A spaceship traveling at 0.80c with respect to a permanent station fires a probe in the forward direction at a speed of 0.70c with respect to the craft. What is the speed of the probe with respect to the station? [*Hint: Let the probe be P and the ship be S'.*]

42. [I] Two particles are flying down a long evacuated tube. The first moves at 0.50c with respect to the tube, while the second moves in the same direction at 0.60c with respect to the first. What is the speed of the second as recorded by a detector fixed to the wall of the tube?

43. [I] Superman while traveling past the Daily Planet Building at 0.90c (as measured by Lois Lane in her office) throws a projectile at 0.40c, relative to himself, in the forward direction. How fast will Lois see the projectile move?

44. [I] A spaceship moving at a speed of 0.70c with respect to you at rest on Earth fires a rocket in the backward direction at 0.40c relative to the ship. What is the speed of the rocket with respect to you? What would be expected classically?

45. [I] A spaceship moving at a speed of 0.70c with respect to you at rest on Earth fires a rocket in the forward direction at 0.40c relative to the ship. What is the speed of the rocket with respect to you? What would be expected classically?

46. [I] A space probe is launched from a mother ship and sails away at 0.50c. It fires a laser in the direction of motion, and the beam moves off at a speed c with respect to the probe. The ship is traveling at 0.20c with respect to an observer on a nearby planet. What is the speed of the beam with respect to the ship?

47. [I] Two pulses of light are sent out in vacuum, one to the right and another colinearly to the left. What is the speed of either one with respect to the other? Show your work, please. What would be expected classically?

48. [II] Show that if the speed of an object in any one inertial frame is less than c, its speed in all other inertial frames (each moving with respect to the first at a speed less than c) will be less than c. [*Hint: Let $v_{PO'} = ac$ and $v_{O'O} = bc$, where a and b are < 1.*]

49. [II] A laserbeam fired from Earth is received by a spacecraft heading directly toward the source at a speed v. Show that a technician aboard the ship will measure the speed of the incoming beam to be c regardless of the value of v.

50. [II] Two spacecraft 1.0 km apart are traveling in opposite directions along a straight line that intersects the Earth so that one is approaching the Earth and the other is receding from it. An observer on the planet sees each ship to be moving at 0.80c. Determine the speed at which the two ships are moving with respect to one another.

51. [II] An electron traveling at 0.992c collides head-on with a positron traveling at 0.981c. Both speeds are measured with respect to the laboratory in which the experiment occurs. At what speed do they approach each other?

52. [II] THIS PROBLEM DEALS WITH RELATIVE MOTION. Two space probes are flying, one behind the other, toward the Earth. Probe-1 is moving at $v_1 = 0.500c$ and probe-2, in front of it, is moving at $v_2 = 0.600c$, both measured with respect to the Earth. We want to find the scalar velocity of probe-2 w.r.t. probe-1. (a) Draw a diagram. (b) Making use of Fig. 26.10, fix 0 in probe-1 and 0' in probe-2. Explain why $v_{PO} = -v_1$ and why $v_{PO'} = -v_2$. (c) Explain why the scalar velocity of probe-2 w.r.t. probe-1 is equivalent to $v_{O'O}$. (d) Using Eq. (26.7) prove that

$$v_{O'O} = \frac{v_2 - v_1}{1 - \frac{v_1 v_2}{c^2}}$$

(e) Compute the scalar velocity of probe-2 w.r.t. probe-1.

53. [III] THIS PROBLEM DEALS WITH RELATIVE MOTION AND LENGTH CONTRACTION. A federation space station detects the approach of a space cruiser closing at 0.500c along a collision course. While still 2.00×10^{-4} ly away, the cruiser launches a B-class surveillance device directly toward the station. The manufacturer's manual states that the device is 1.000 m long, but an observer on the cruiser who sees it fly by at a constant speed, records its length to be 752 cm. (a)

Draw a diagram. (b) What is the proper length of the surveillance device? (c) Knowing the contracted length as seen from the cruiser, what is the speed of the device w.r.t. the cruiser (v_{DC})? (d) Considering the addition of relativistic velocities (Fig. 26.10), fix 0 in the cruiser, 0' in the surveillance device, and P in the station. Show that in general

$$v_{PO'} = \frac{v_{PO} - v_{O'O}}{1 - \frac{v_{PO} v_{O'O}}{c^2}}$$

(e) Why does it make sense that $v_{PO'} = v_{SD}$ is negative? (f) What is the scalar velocity of the device w.r.t. the station ($v_{DS} = v_{O'P} = -v_{PO'}$)?

SECTION 26.8: RELATIVISTIC MOMENTUM

54. [I] A particle having a mass of 1.00 mg is traveling at a speed of 0.4c such that it has a γ of 1.09 with respect to an observer. What is its relativistic momentum as measured by that observer?

55. [I] A neutron ($m = 1.675 \times 10^{-27}$ kg) is traveling at 0.50c with respect to a "stationary" target. How much momentum does it have with respect to the target? [*Hint: At that speed you must use the relativistic momentum.*]

56. [I] A tiny space probe having a mass of 1.00 kg is seen to be moving in a straight line at a speed of 0.40c. What is its relativistic momentum?

57. [I] A proton ($m_p = 1.675 \times 10^{-27}$ kg) is moving at a speed of 3c/5 with respect to some inertial system. Determine its relativistic momentum as measured by an observer in that system.

58. [I] An electron ($m = 9.109 \times 10^{-31}$ kg) is traveling at 0.866c with respect to the face of a TV picture tube toward which it is heading. How much momentum does it have with respect to the tube?

59. [I] A particle of mass m travels at a speed of 0.600c; show that its relativistic momentum is given by the expression

$$p = mc3/4$$

60. [II] Show that as the momentum of a body of mass m becomes very large, its speed approaches c but does not exceed it.

61. [II] Suppose that an object traveling uniformly has a value of γ relative to an inertial observer. Derive an expression in terms of γ for the percent error introduced using the classical momentum (p_c) instead of the relativistic momentum (p), that is, $(p - p_c)/p$. What is this error for a speed of 0.600c?

62. [II] An electron ($m = 9.1095 \times 10^{-31}$ kg) has a momentum of 8.603×10^{-23} N·s. What's its speed?

63. [II] If the ratio of the momentum to the mass of a proton is 0.101 m/s, determine its speed. Compare this ratio to its classically determined speed. When will the relativistic speed approximate the classical speed?

64. [III] An object has an initial scalar velocity v and a corresponding momentum p. Show that if the momentum is to be doubled, the final scalar velocity must become

$$v_f = \frac{2vc}{\sqrt{c^2 + 3v^2}}$$

SECTION 26.9: RELATIVISTIC ENERGY

65. [I] What is the rest energy of a 1.00-g particle traveling at 3.00×10^4 m/s? [*Hint: Use the definition of rest energy and watch out for the units.*]

66. [I] A proton has a mass of $1.672\ 6 \times 10^{-27}$ kg. Determine its rest energy in joules and in MeV, and compare your results with Table 26.3 on p. 956.

67. [I] If the rest energy of a macroscopic particle is 200 MJ, how much mass does it have?

68. [I] If a tiny object with a mass of 0.0100 g is traveling at a speed such that $\beta = 0.700$ and $\gamma = 1.40$, determine its total energy.

69. [I] A hypothetical particle is traveling with respect to an observer such that its total energy is seen to be twice its rest energy (which is 1000 MeV). How fast is it moving? What is its kinetic energy?

70. [I] An electron with a rest energy of 0.511 MeV has a kinetic energy of 0.089 MeV. What is its total energy?

71. [I] The total energy of a muon is 106.7 MeV. If its rest energy is 105.7 MeV, what is its kinetic energy?

72. [I] A proton ($m = 1.672\ 6 \times 10^{-27}$ kg) moves with respect to the laboratory frame such that it has a $\gamma = 1.50$. What is its total energy in joules and electron volts?

73. [I] A particle of mass 1.000×10^{-6} kg is fired from a device such that it has a muzzle "velocity" of 0.200 0c with respect to an inertial observer. Determine the rest energy, total energy, and kinetic energy of the projectile with respect to the observer.

74. [I] An electron with a rest energy of 0.511 MeV travels with a γ of 5/3 with respect to a laboratory. What is its total energy?

75. [I] A deuteron is the heavy hydrogen nucleus $^2_1$H composed of a proton and a neutron. Its rest energy is 1875.6 MeV. How much energy must be supplied to rip it apart? That is, how much energy is liberated (as KE and as a γ-ray) when a deuteron is formed of a separate proton and neutron?

76. [I] A hypothetical particle has a rest energy of 1.500 MeV and a total energy of 3.000 MeV. What is its momentum?

77. [I] A clock, in the course of being wound, has 89.88 J of work done on its mainspring. By how much does its mass change?

78. [I] A proton and a neutron together form the nucleus of a deuterium atom that is completed with a single orbiting electron. The mass of such an atom is $3.344\ 5 \times 10^{-27}$ kg. When the particles come together, mass is transformed and the composite atom has less mass than the separate constituents; the equivalent energy difference is called the **binding energy**. Find the binding energy of deuterium in units of J and MeV to three figures.

79. [I] Compute the amount of energy liberated in the fusion reaction

$$^2H + {}^2H \rightarrow {}^3He + {}^1n + energy$$

Here, the neutron flies off with about three times the KE of the helium.

80. [I] Another important fusion reaction involving deuterium and tritium is

$$^2H + {}^3H \rightarrow {}^4He + {}^1n + energy$$

where the neutron flies off with about four times the KE of the helium. Determine the total amount of energy liberated.

81. [I] A photon is to have the same energy as a 1.000-MeV electron (that is, an electron with a KE of 1.000 MeV). What must be its frequency? [*Hint: That's total energy.*]

82. [II] An electron with rest energy 0.511 MeV travels at 0.80c with respect to the laboratory. What is its momentum in the lab frame in units of MeV/c?

83. [II] An object has a total energy equal to twice its rest energy. Show that $p = \sqrt{3}\ mc$.

84. [II] An electron has a kinetic energy of 0.200 0 MeV. Find its speed. Compare that with its classical speed.

85. [II] An electron at rest accelerates through a potential difference of 1.50 MV. Determine its speed on emerging, and compare it with the erroneous classical value.

86. [II] We wish to accelerate an electron from rest up to 0.990c. Through what potential difference must it pass?

87. [II] A proton with a rest energy of 938.3 MeV has a momentum of 100.0 MeV/c. How fast is it moving?

88. [II] Show that for an object moving at speed v, the ratio of the relativistic KE to the classical KE, call it Γ, is

$$\Gamma = 2(\gamma - 1)/\beta^2$$

89. [II] Given a particle whose rest energy equals its kinetic energy, determine its β value.

90. [II] With Problem 88 in mind, show that

$$\gamma = \tfrac{1}{4}\left(\Gamma + \sqrt{\Gamma^2 + 8\Gamma}\right)$$

91. [II] A photon is to have the same momentum as a 1.000-MeV electron (that is, an electron with a KE of 1.000 MeV). What must be its frequency?

Chapter 27

The Origins of Modern Physics

We now begin a sequence of five chapters dealing with the main concerns of Modern Physics. That discipline is a study of the submicroscopic world of atoms, and the particles that compose them, and the particles that compose those particles. It is essentially a creation of the twentieth century, but most of the work had its foundations in the late 1800s, a period when even the existence of the atom was not totally accepted. This chapter deals with the establishment of the atom as a real entity, and, beyond that, with the realization that it is a complex structure capable of coming apart into its constituent subatomic particles.

Subatomic Particles

The first subatomic particle to reveal itself was the electron, and that discovery was predicated on the study of electric currents in solids, liquids, and gases. We know now that charge is a property of substantial matter and that it is quantized; it comes in whole number multiples of a minimum amount corresponding to the magnitude of the charge on the electron. A little over a hundred years ago, physicists had no idea of what charge was—whether it was continuous or particulate, whether it was material or aethereal, like light.

27.1 The Quantum of Charge

Soon after the invention of the battery, Nicholson and Carlisle used one to decompose water via a process known as *electrolysis*. Their discovery was accidental; to improve the contact between a wire and their battery, they put some water on the connection and the liquid filled with gas bubbles. Thereafter, they inserted two wires into a beaker of water and passed a current through it. Bubbles of oxygen were released at the *anode*, and hydrogen was formed at the *cathode*. Humphry Davy and his assistant Michael Faraday used electrolysis to explore chemical interactions. When the conducting solution, the electrolyte, contains a dissolved salt of some metal like silver, that metal will be liberated at the cathode just as hydrogen is (Fig. 27.1). To account for the passage of current across the electrolyte,

An **ion** is, in effect, a charged atom: an atom that has picked up or lost orbital electrons.

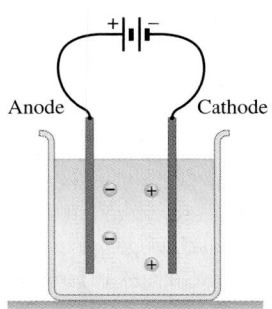

Figure 27.1 The passage of a current through an electrolyte. Negative ions move to the anode, positive ions move to the cathode.

Faraday supposed that there was a flow of charged particles, which he called **ions**. Although he did not speculate as to their nature, we know that *ions are atoms that have lost or gained one or more orbital electrons* and therefore carry a net charge.

Faraday measured the mass of monovalent silver deposited on the cathode and the net charge provided (i.e., current multiplied by the time it was applied) in the process. To liberate one *mole* of a single-valence element (singly charged ± ion), he found that a large quantity of charge F, now called a **faraday** and equal to 96 485 C, had to be provided. Remember that a mole of any element comprises $N_A = 6.022\ 136\ 7 \times 10^{23}$ atoms. For example, silver has an atomic mass of 107.87 units, and 1 faraday of charge liberates 6.02×10^{23} silver atoms with a net mass of 107.87 grams. Faraday teetered on the edge of electron theory, hinting that electricity was composed of particles of charge *e*. Given that there are N_A atoms per mole, and assuming each monovalent ion has a charge *e*, a faraday of charge must correspond to an amount

$$F = N_A e \tag{27.1}$$

The striking thing is that the same quantity F is found for all monovalent elements regardless of their chemical properties; $e = F/N_A$ is a constant. For bivalent ions like copper, a charge of 2F must be supplied to liberate a mole of Cu, and each ion carries a charge of $2e$. Though F was directly measurable, Avogadro's number was not determined for several decades and that left *e* unknown. Still, 1 faraday of charge is transported by 1 mole of monovalent substance. It follows that $\frac{1}{2}$ faraday is transported by $\frac{1}{2}$ mole, and so on down, *presumably* to the smallest unit of charge *e* associated with the smallest mass *m*, that is, the mass of a single atom of the material liberated. For example, it is found from electrolysis that the **charge-to-mass ratio** for singly ionized hydrogen (H^+) is

$$\left(\frac{e}{m}\right)_{H^+} = 9.58 \times 10^7\ C/kg$$

Hydrogen is the lightest of the elements, and this value turns out to be the largest charge-to-mass ratio of any element. Although neither *e* nor *m* was known at the time, the corresponding ratios could be determined experimentally.

Remarkably, in 1874, the Irish physicist George Stoney used Eq. (27.1) along with the best available values of F and Avogadro's number to determine the basic unit of charge. His estimate of *e* was too small by a factor of about 20, but even that was still quite good, all things considered.

Example 27.1 **[II]** During electrolysis, how much silver will be deposited by a current of 1.00 A applied for 1.00 s?

Solution Mention of the word "electrolysis" should call to mind the faraday. (1) TRANSLATION—A known current flows for a known time; determine the amount of a monovalent element it deposits. (2) GIVEN: Element silver, $I = 1.00$ A, and $t = 1.00$ s. FIND: *m*. (3) PROBLEM TYPE—Modern Physics/electrolysis. (4) PROCEDURE—The defining equation is $F = N_A e$, where F is the amount of charge that corresponds to the transfer of a mole. In this problem we need to find the amount of charge transferred and then compare that to F. (5) CALCULATION—We have the current and the time during which it flowed, so the net charge passed is

$$\Delta q = I\Delta t = (1.00\ A)(1.00\ s) = 1.00\ C$$

Since 1 faraday of charge liberates 1 mole of silver (that is, 107.9 g), 1.00 C of charge liberates an amount *m* where

$$\frac{96\,485\ C}{107.9\ g} = \frac{1.00\ C}{m}$$

and

$$\boxed{m = 1.12\ mg}$$

Quick Check: Here we are passing about 10^{-5} faraday, and that quantity should deposit about 10^{-5} of a mole of Ag, which equals 1.1 mg.

[For more worked problems click on **WALK-THROUGHS** *in* **CHAPTER 27** *on the* **CD***.]*

A so-called neon sign is just a long tube filled with an appropriate gas, across which there is a sustained high-voltage discharge.

The electrical matter consists of particles extremely subtle since it can permeate common matter, even the densest, with such freedom and ease as not to receive any appreciable resistance.

BENJAMIN FRANKLIN (CA. 1750)

By 1881, Helmholtz asserted that, "If we accept the hypothesis that the elementary substances are composed of atoms, we cannot avoid concluding that electricity also, positive as well as negative, is divided into definite elementary portions, which behave like atoms of electricity." For a while, the German literature referred to *e* as *das Helmholtzsche Elementarquantum* (the Helmholtz elementary quantity, or **quantum**). In 1891, Stoney christened that fundamental unit of charge the **electron**.

27.2 Cathode Rays: Particles of Charge

A totally different route to the electron began one dark night in 1675 when the French astronomer J. Picard noticed in amazement that the barometer he was swinging as he walked began to give off an eerie flickering light. The effect was duplicated in the laboratory by Hauksbee using an electrostatic generator to provoke a rarefied gas in a bottle into the mysterious luminosity. Thus began the study of electrical discharges in low-pressure gases.

Outstanding among the many scientists in the field during the 1870s was Sir William Crookes, a rather unorthodox fellow who believed he could communicate with the dead. A "Crookes tube" with two sealed electrodes is illustrated in Fig. 27.2. (It's the forerunner of all the blazing bar and motel signs—the neon uglies.) When the tube was connected to the terminals of a high-voltage source, a glowing beam spread down its length. Different gases glowed with different colors: mercury emitted greenish blue, while neon gave off a bright orange-red. By displacing the anode to the side (Fig. 27.3), Crookes verified that the emanations were actually streaming from the negative plate, not always making it to the anode. When these **cathode rays** struck the walls of the tube, the glass gave off a pale green fluorescent glow that betrayed the impact of the beam.

Objects that were inserted into the stream cast sharp shadows in the glow at the far end of the tube. This, too, suggested straight-line propagation. Crookes even put a little paddle wheel inside a tube and spun it around with the bombarding beam, which more and more appeared to be a stream of particles carrying energy and momentum. Later, he was able to deflect the beam with a magnet (p. 687) in the same way a current of negative charges would be deflected. What emerged from this experimentation was the "British" view that cathode rays were streams of negatively charged submicroscopic particles.

The opposing "German" view was most vigorously pressed by P. Lenard. Heinrich Hertz had discovered that cathode rays could pass through thin sheets of metal without leaving any holes, and that was enough to convince his assistant Lenard that the rays must be nonmaterial. Today, we know that bulk matter is mostly empty space and that some subatomic particles can sail through yards of concrete and steel as if they weren't there. To Lenard, cathode rays were without substance; they were waves like light—"phenomena in the aether."

There the matter stood, befogged by nationalism—corpuscle versus wave—a controversy that would last at least twenty years.

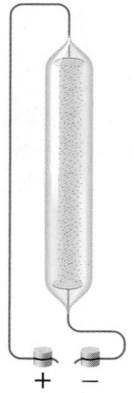

Figure 27.2 A cathode-ray or Crookes tube.

Figure 27.3 When the tube is bent, it becomes clear that the rays emanate from the negative terminal, or cathode.

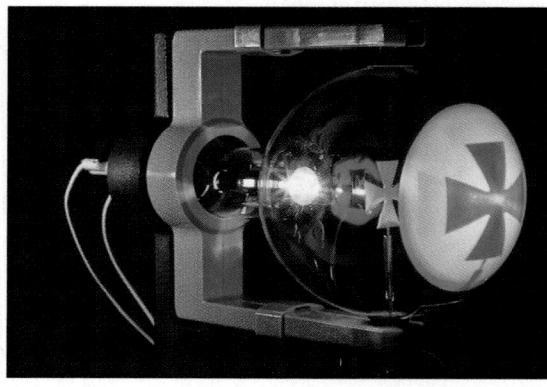

A cathode-ray tube. The metal cross intercepts the beam and casts a shadow on the face of the tube.

Discovering the Electron

When Joseph John Thomson began his research into the nature of cathode rays, he was the director of the famed Cavendish Laboratory at Cambridge University. In 1894, Thomson showed that cathode rays traveled much slower than light (another blow to the notion that they were "phenomena in the aether"). And then J. Perrin (1895) found that a metal obstacle located in the path of the beam acquired a negative charge. It remained for Thomson (his friends called him J. J.) to demonstrate that the cathode rays and the charge were one and the same.

To that end, Thomson constructed the device shown in Fig. 27.4, which is similar in many respects to a TV picture tube, a so-called CRT (cathode-ray tube). A high voltage between cathode (C) and anode (A) ionized the trace of residual gas, creating a cloud of positively charged atoms and negative electrons; the latter were the cathode rays. These rays were accelerated toward the positive anode reaching tremendous speeds (of as much as 60 000 mi/h). A small hole in the front of the anode allowed a narrow unobstructed stream to pass into the region beyond A. The speed v of the beam, which could be controlled via the cathode-anode voltage, was fairly uniform. Farther along the tube, the beam traveled between two metal plates (P) that were attached to a variable dc voltage, creating a vertical electric field $\vec{\mathbf{E}}$. Coils straddling the tube produced a controllable horizontal magnetic field $\vec{\mathbf{B}}$ crossing the same region.

The idea was to measure the charge-to-mass ratio in a two-step procedure. The beam was assumed to be composed of a stream of discrete particles of charge e and mass m_e that could be deflected in the vertical plane by either field. The first step was to determine v, which was done by adjusting E and B so that the forces produced were equal and opposite and the beam sailed through undeflected (the effect of gravity was negligible). This arrangement is the "velocity selector" we considered earlier (Problem 69, p. 706), where it was shown that there's no deflection of the beam provided

$$v = \frac{E}{B}$$

Next, the B-field was shut off, and the beam was deflected by the remaining E-field. The cathode-ray particles experienced a downward Coulomb force (eE) and so accelerated down uniformly, via Newton's Second Law (eE/m_e). They each "fell," much as a stone would fall in a gravity field, for all the time (t) that they were between the plates. Since the horizontal speed v is constant, the traversal time is $L/v = L/(E/B)$. We want the distance each particle drops (y) in the course of passing through the plates. Inasmuch as $y = \frac{1}{2}at^2$,

$$y = \frac{1}{2}\left(\frac{eE}{m_e}\right)\left(\frac{LB}{E}\right)^2$$

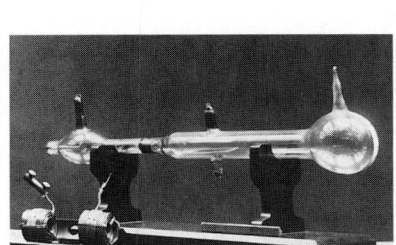

J. J. Thomson's original tube.

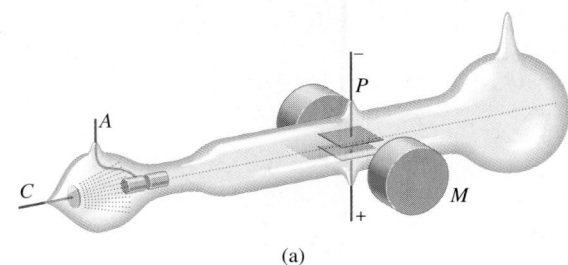

(a)

Figure 27.4 (a) J. J. Thomson's electron-beam device showing the horizontal electrical deflecting plates and the vertical magnetic coils. (b) Electrons experience a downward electric force and (with the B-field coming out of the page) an upward magnetic force.

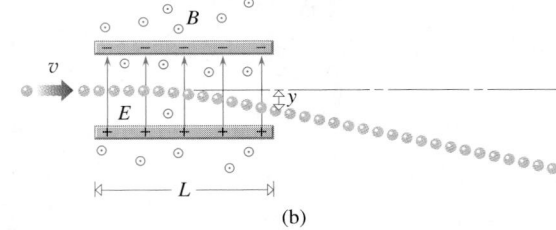

(b)

J. J. Thomson at the Cavendish Laboratory.

On Electrons. At first there were very few who believed in the existence of these bodies smaller than atoms. I was even told long afterwards by a distinguished physicist who had been present at my lecture at the Royal Institution that he thought I had been "pulling their legs."

J. J. THOMSON
ENGLISH PHYSICIST

Since y can be determined directly from the point where the beam hits the screen (using the geometry of the tube, p. 993), and because we know E from the plate voltage and B from the coil current, the only unknowns are e and m_e. Consequently, the electron's charge-to-mass ratio can be determined:

$$\frac{e}{m_e} = \frac{2yE}{L^2B^2} \tag{27.2}$$

The present-day accepted value of the ratio is

[electron] $$\frac{e}{m_e} = (1.758\,819\,62 \pm 0.000\,000\,53) \times 10^{11} \text{ C/kg}$$

Thomson suggested that the reason this ratio was so much larger (namely, 1836 times larger) than the corresponding ratio for hydrogen was that the mass of the electron was that much smaller than the mass of the H^+ ion. *The electron was a tiny fragment detached from a complex atom.* Not everyone in 1897 believed in atoms, and few were pleased with the notion of **subatomic particles**.

The Oil-Drop Experiment

Millikan's famous oil-drop experiment is pictured in Fig. 27.5. A fine mist of oil is squirted from an atomizer above a small hole in the top plate of what can be thought of as a large parallel-plate capacitor. A few droplets descend through the hole into the region occupied by a variable E-field. Lit from the side, each minute droplet shines like a tiny star when seen through a viewing telescope. Most of the droplets get negatively charged on passing through the nozzle, picking up some small but unknown number of electrons. Once a drop is caught sight of, the voltage on the plates is carefully varied, thereby controlling a downward E and slowing the fall until that sphere of oil is suspended motionless midair in a balance between gravitational (mg) and electrical (qE) forces. At that moment $qE = mg$, and if the mass of the droplet (m) were known, the charge that it carried (q) would be known. The next procedure determines m.

The E-field is shut off and the same droplet is watched as it again falls, soon reaching a constant terminal speed. By timing the motion of the droplet as it passes from one hairline down to another in the telescope, the terminal speed is measured. The theory of air resistance tells us that the terminal speed depends on the radius of the droplet as well as on a number of other factors, all of which are known. Thus, from the terminal speed we have the

Robert Andrews Millikan, American physicist (1868–1953).

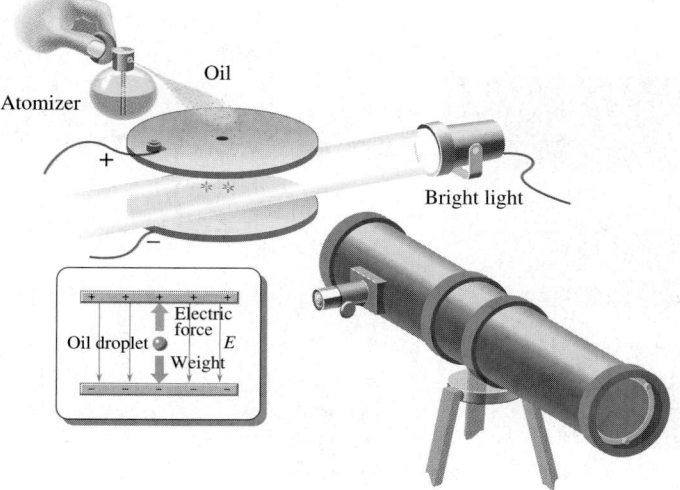

Figure 27.5 The Millikan oil-drop experiment. A mist of oil causes a few drops to enter between the plates of a capacitor. There, they experience a force due to the applied electric field.

J. J. AND THE DISCOVERY OF THE ELECTRON

J. J. put the finishing touches on what today is generally called his "discovery of the electron" with two further experiments in 1899. First, he repeated the preceding measurement using a totally different source of electrons, namely, the *photoelectric effect* (p. 1002), and got the same results for e/m_e. His second experiment was to measure e, again using a totally new procedure. His student C. T. R. Wilson (the inventor of the cloud chamber) had been working on the formation of clouds and found that droplets could form around charged particles. We shall not describe the methods used by Thomson and his colleagues other than to say that by 1901 he was able to arrive at a value for e that was only about 30% off. The work inspired a classic measurement by the American Robert Millikan, which we will examine.

radius, and from the radius we have the mass of the sphere. Millikan actually used a somewhat more complicated though less arduous approach, but the idea is the same. His graduate student H. Fletcher methodically found the net charge on thousands of droplets, one by one. Millikan and Fletcher were then able to show that the droplets carried whole number multiples of a basic charge $q_e = e$, which was ascribed to the electron itself. They arrived at an average value of e of 1.592×10^{-19} C. Today, the best value of this fundamental quantum of charge (Table 27.1) is

$$e = (1.602\ 177\ 33 \pm 0.000\ 000\ 49) \times 10^{-19} \text{ C}$$

All charged subatomic matter observed to date, whether positive or negative, carries a net charge that is an integer multiple of e.

Combining e with the charge-to-mass ratio provides another remarkable number—the mass of the electron:

$$m_e = (9.109\ 389\ 7 \pm 0.000\ 005\ 4) \times 10^{-31} \text{ kg}$$

Using Eq. (27.1) and the measured value of the faraday, Millikan computed Avogadro's number to be 6.062×10^{23} molecules per mole (as compared to the present-day value of 6.022×10^{23}—not bad at all).

Given a charge-to-mass ratio for a hydrogen ion of 9.58×10^7 C/kg and the value of e, we can make a rough estimate of the size of an atom—silver, for instance. The relative atomic mass of hydrogen is 1.008, whereas that of silver is 107.9. The mass of a hydrogen ion is e divided by the charge-to-mass ratio: $(1.602 \times 10^{-19}$ C$)/(9.58 \times 10^7$ C/kg$) = 1.67 \times 10^{-27}$ kg. A silver atom has a mass of $(107.9/1.008)(1.67 \times 10^{-27}$ kg$) = 1.79 \times 10^{-25}$ kg. The density of silver is 10.5×10^3 kg/m³, and so there must be $(10.5 \times 10^3$ kg/m³$)/(1.79 \times 10^{-25}$ kg/atom$) = 5.87 \times 10^{28}$ atoms/m³. The reciprocal of that is the number of cubic meters occupied per silver atom, namely, 1.70×10^{-29} m³/atom. Each atom sits in a little imaginary cube of this volume. Assuming the silver atoms are tightly packed spheres, the cube root of 1.70×10^{-29} m³ should be roughly the diameter of one such atom, that is, 0.26 nm.

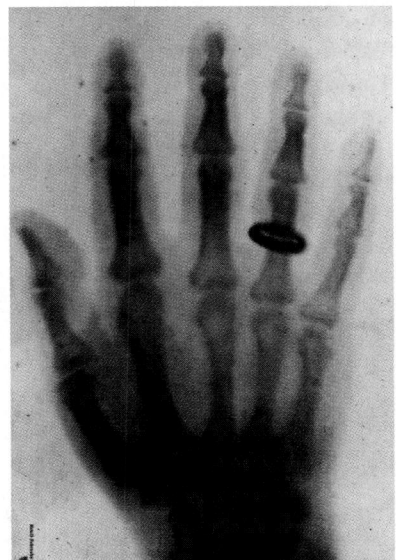

A very early X-ray photo taken by Professor Röntgen in Hamburg in January of 1896.

Table 27.1
Some Physical Characteristics of the Electron*

Mass m_e	$9.109\ 389\ 7(54) \times 10^{-31}$ kg
Charge e	$1.602\ 177\ 33(49) \times 10^{-19}$ C
Rest energy	$0.510\ 999\ 06(15)$ MeV
Charge-to-mass ratio e/m_e	$1.758\ 819\ 62(53) \times 10^{11}$ C/kg

*The parentheses show the uncertainty in the last two figures.

27.3 X-Rays

Wilhelm Conrad Röntgen was fifty-five in 1895, a well-respected professor of physics at Würzburg. He was a private person who conducted his research quietly, almost secretively, so much so that his students nicknamed him *der Unzugängliche*, the unapproachable one. He had become interested in the work done on cathode rays by Hertz and Lenard, especially the latter's study of the passage of the rays though an aluminum-foil window and thence outside beyond the vacuum tube into the air. Lenard had surrounded his apparatus with lead and iron sheet to shield it from light and electric fields; Röntgen also covered the

Van Dyck's painting of *Saint Rosalie Interceding for the Plague-Stricken of Palermo*, as viewed in reflected light and in transmitted X-rays. Because the X-rays pass through the painting to reach the photographic film placed behind it, they reveal information about what is beneath the surface. (See photos on p. 1079.)

tube, but luckily he used only thin black cardboard. In the darkened room, he noticed that one of his luminescent screens (a piece of paper coated with a barium salt) some distance away glowed brightly for no apparent reason. Astonished, he explored the effect until he was convinced that it was real. After a while, the cause seemed clear: some new invisible penetrating radiation was being emitted from the discharge tube. He called it **X-rays**, because *x* in mathematics represents an unknown.

Like light, X-rays expose photographic film, and because of their tremendous penetrating power, he was able to produce "photographs … of the shadows of the bones of the hand." Though the rays could ionize gases, they were similar to light in that they were not deflected by either electric or magnetic fields. To prove that X-rays were waves and not particles, Röntgen tried but failed to observe refraction, specular reflection, and polarization. Still, J. J. Thomson, among others, favored what we know to be the proper interpretation, namely, that X-rays are very short-wavelength electromagnetic radiation ($\lambda \approx 0.1$ nm). The correct wave picture was only slowly gaining acceptance when C. Barkla, in 1906, managed to observe the partial polarization of X-rays, thus establishing their transverse nature (see Discussion Question 7). Still, the wavelength of the radiation needed to be found.

X-Ray Diffraction

The question of the wavelength of X-rays was settled in 1912 by the German physicist Max von Laue (pronounced *fun lowùh*). He reasoned that X-rays could not be diffracted by ordinary gratings because their wavelengths were much too small. As a rule, if an array of scatterers is to produce an observable diffraction pattern, their spacing must be of the order of the wavelength of the wave. Von Laue proposed that the ordered array of atoms in a crystal (which at the time were correctly believed to be separated by about 0.1 nm to 0.2 nm) might serve to produce diffraction. The atoms would scatter X-rays in all directions; but because of the regularity of the array, the waves traveling in certain directions would interfere constructively (see the photo made by a two-dimensional diffraction grating on p. 917).

A crystal was placed in a narrow X-ray beam in front of a photographic plate (Fig. 27.6). If all went according to plan, the crystal would produce a number of off-

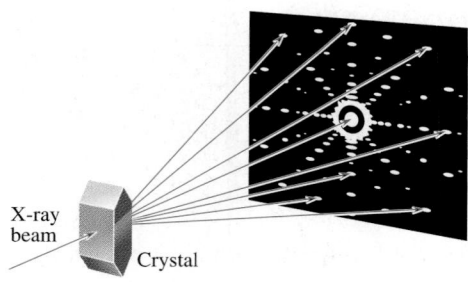

Figure 27.6 X-ray diffraction. A narrow X-ray beam enters a crystal, which diffracts it into many exiting beams. The process is akin to the operation of a three-dimensional transmission grating.

X-ray beam

Crystal

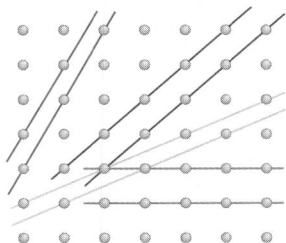

Figure 27.7 Several of the atomic planes off which X-rays may scatter. Notice that d is different in each case.

Although light coming in at any angle will be reflected, **the diffraction of X-rays** from a crystal takes place at only certain incident angles that satisfy the Bragg equation.

The **angle** between the transmitted beam and the diffracted beam is always 2θ, and it is this angle that is measured directly.

axis diffracted beams and an ordered pattern of dots would appear in the photo. The very first attempt was a resounding success, "proving" both the transverse nature of X-ray waves and the periodicity of the atoms in a crystal. The published accounts of this work caught the attention of W. H. Bragg, professor of physics at the University of Leeds, and his son, W. L. Bragg, a student at Cambridge. They devised a simpler diffraction process now known as Bragg scattering.

Figure 27.7 shows a regular array of atoms. We can imagine various sets of planes passing through the array in lots of different directions and containing different groups of atoms. Each such plane scatters rays in specific directions. For simplicity, let's deal with the set of horizontal planes. A parallel beam strikes the atoms and is scattered every which way. We have already seen that, for each plane, the reflected radiation will reinforce only in the direction such that it leaves the plane at the same angle at which it entered. That was what gave rise to the Law of Reflection (and it's true separately for each plane in the diagram). In Fig. 27.8, ray-1 and ray-2 correspond to waves that will exactly reinforce each other. The question now is, What is the condition for rays from the second plane (and all others as well) to reinforce those from the first plane?

The waves associated with ray-1 and ray-3 will be in-phase, provided that their path-length difference is a whole number of wavelengths. Ray-3 travels farther than ray-1 by a distance equal to $2d \sin \theta$. (Note that in X-ray work θ is measured up from the plane and not down from the normal as in Optics.) For special values of θ, namely, θ_m, $2d \sin \theta_m$ equals $m\lambda$, where m is a whole number (1, 2, 3, …). It follows that constructive interference (Fig. 27.8b) occurs between waves from the various planes only when

[maxima]
$$2d \sin \theta_m = m\lambda \qquad (27.3)$$

This formula has come to be known as the **Bragg equation**. Consistent with the equation (for a fixed d and λ), there may be several angles at which diffraction can occur, each corresponding to an integer value of m known as the *order*. Although the angle of incidence equals the angle of reflection as with light, *only those incident angles satisfying the Bragg equation will result in reflection (or diffraction)*, and even then the exiting beam will be quite weak.

Since $\sin \theta$ must be equal to or less than 1

$$\frac{m\lambda}{2d} = \sin \theta_m \leq 1$$

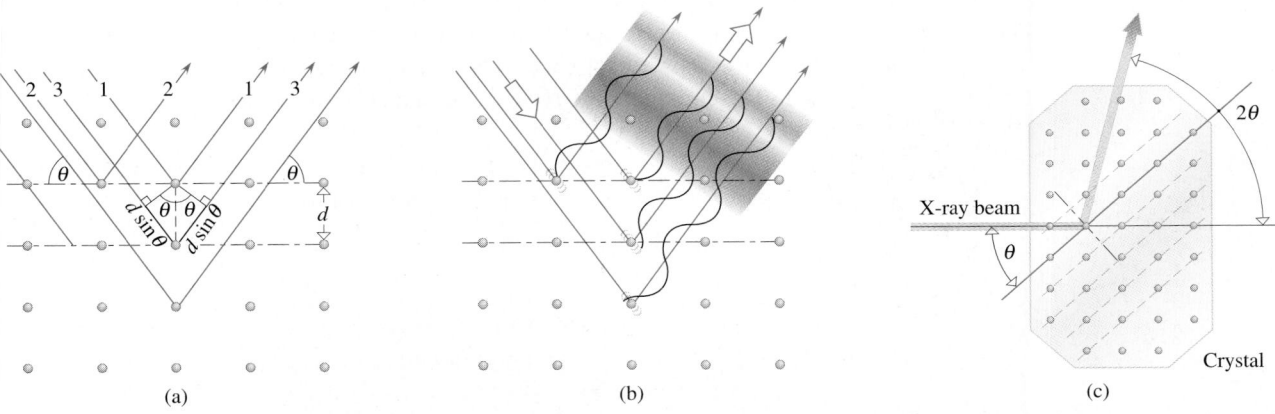

Figure 27.8 (a) The scattering of X-rays from the planes of a crystal. (b) Waves interfering constructively as they scatter off an array of atoms. (c) The geometry of the scattering. Practically, one knows the direction of the incident beam and can measure the scattered beam. Thus, 2θ is determined experimentally.

Hence with $m = 1$

$$\lambda \leq 2d$$

Generally, d, the distance between atomic planes, is at most 0.3 nm, which sets a practical upper limit of 0.6 nm on λ.

Example 27.2 **[I]** A beam of quasimonochromatic X-rays with a wavelength of 0.15 nm is incident on a crystal. What is the space between atomic planes if the first-order diffraction maximum occurs at $\theta_1 = 16.0°$?

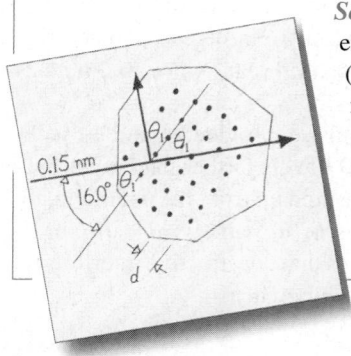
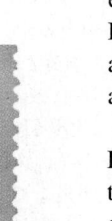

Solution Diffracting X-rays generally mean Bragg's equation. (1) TRANSLATION—A beam of X-rays of known wavelength is diffracted in a specified order at a known angle; determine the atomic spacing between the diffracting planes. (2) GIVEN:

$\lambda = 0.15$ nm, $m = 1$, and $\theta_1 = 16.0°$. FIND: d. (3) PROBLEM TYPE—Modern Physics/X-rays/Bragg's equation. (4) PROCEDURE—The problem treats X-ray scattering from a crystal and there's only one way to deal with it, namely via Bragg's analysis. (5) CALCULATION—We have λ, m, and θ and need d. From the Bragg equation

$$d = \frac{m\lambda}{2 \sin \theta_m} = \frac{(1)(0.15 \text{ nm})}{2 \sin 16.0°} = \boxed{0.27 \text{ nm}}$$

Quick Check: The answer is the right order-of-magnitude (i.e., a few tenths of a nanometer). $2d \sin \theta_m = 1.49 \times 10^{-10}$ m $= m\lambda$.

X-rays are generated when high-speed electrons crash into a material target and rapidly decelerate, thereupon emitting radiant energy. A modern X-ray tube (Fig. 27.9) boils electrons out of a hot metal filament in a process called *thermionic emission* (discovered by Edison while he was working with light bulbs). These electrons are then accelerated across a potential difference ranging from 10^4 V to 10^6 V, in vacuum, whereupon they impact on a metal anode, decelerate, and radiate.

X-ray diffraction has become a routine technique for determining crystal structure. Equally as important, it has become a powerful tool for studying molecules, especially in the case of large repeating configurations such as proteins and DNA. Indeed, it was with the guidance of X-ray data that J. D. Watson and F. H. C. Crick were able (1953) to figure out the double-helix structure of DNA.

Sir William Henry Bragg (1862–1942) and Sir William Lawrence Bragg (1890–1971).

27.4 The Discovery of Radioactivity

Röntgen's chance discovery of X-rays sent much of the scientific community into a flurry of activity. Henri Becquerel, in France, returned to his father's earlier work on fluorescent substances. These absorb and re-emit light, as does the pale green paint on the hands of a watch. The emission of X-rays had been related to the glowing region of the Crookes tube,

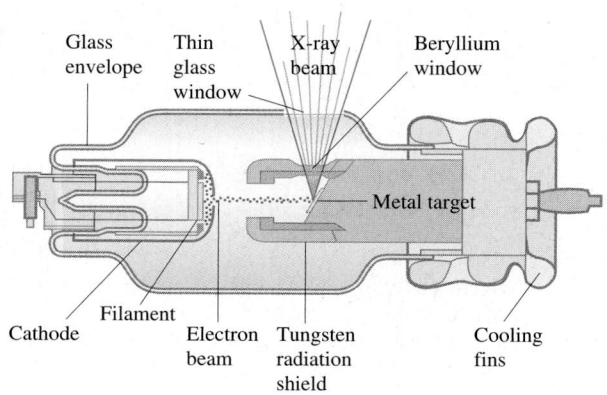

Figure 27.9 A modern X-ray tube. A beam of electrons crashes into a metal target, and X-rays are emitted.

Glass envelope · Thin glass window · X-ray beam · Beryllium window · Metal target · Cathode · Filament · Electron beam · Tungsten radiation shield · Cooling fins

Marie Sklodovska Curie (1867–1934). She is the only person to win Nobel prizes in physics and chemistry: in 1903 for her work on radioactivity and in 1911 for her discovery of two new elements.

TOO CLOSE TO THE FIREFLY

Marie received her Ph.D. at the Sorbonne on the basis of this epochal work. Within six months she, Pierre, and Becquerel shared the 1903 Nobel Prize in physics. Pierre was probably already suffering from radiation sickness (he used to keep the vials of radium in his pockets), and they were both too exhausted to even attend the ceremony in Stockholm. Three years later he was dead—run down in the street by a horsecart. In 1934, after a long illness, Marie Curie (the only person to ever win Nobel Prizes in both physics and chemistry) died of leukemia, a victim of many years of overexposure to radiation—too close to the firefly. Even the pages of her lab notebook were later found to be contaminated with radioactive fingerprints.

and Becquerel wondered whether his glowing materials might also send out X-rays. As fate would have it, he began to work with one of his father's compounds, potassium uranylsulfate—a *uranium* salt. He exposed the salt to sunlight, and it actually did emit radiation that could penetrate the heavy paper wrapping on a photographic plate and fog the film.

On a sunless day in the winter of 1896, Becquerel placed the fluorescent compound atop a wrapped film plate as usual but, this time, he put it aside in the darkness of a drawer to wait for clearer skies. A few days later, almost as an afterthought, he developed the plate; there, amazingly, was the blackened outline of the piece of uranium salt! Prior exposure to light was apparently quite unnecessary—the radiation was emitted without it, without the fluorescence. By accident, he had discovered what Madame Curie would later name **radioactivity**, the property of certain substances to give out, all by themselves, penetrating radiation.

Becquerel studied the radiation and reported several of its properties (a few of them erroneously), including that it could discharge a charged electroscope, as could X-rays. By 1897, he turned to other things. Unlike X-rays, these new emanations couldn't produce those exciting pictures of bones, and they were generally greeted with indifference. This limbo of neglect lasted for about a year and a half until the problem was taken up by a promising young student at the Sorbonne in search of a Ph.D. problem. Marie (Manya) Sklodovska Curie accepted the challenge and began her life's work.

Mme. Curie determined that thorium, like uranium, was radioactive. She and her husband, Pierre, discovered an element about 400 times more active than uranium and called it *polonium* (the name deriving from Marie's homeland, Poland). No one knew at the time that the radiation was essentially debris flung out from the nucleus of an unstable atom as it spontaneously readjusted itself. Even so, Marie was the first (1898) to report "that radioactivity is an atomic property." Working for 45 months under the most dreadful conditions in an abandoned, leaking wooden shed, the Curies finally isolated yet another radioactive element: *radium*. After four years of incredible labor, they had distilled from over a ton of the uranium ore a mere 1/30 ounce of the pure radium salt. But this new element was two million times more radioactive than uranium. Over the years, a single ounce of radium would spew out an amount of energy equal to that of burning ten tons of coal. That quietly blazing precious salt stays warm from its own internal heat and glows, self-luminous, like a firefly. The term *atomic energy* entered the vocabulary of science.

Alpha, Beta, and Gamma

Ernest Rutherford was twenty-four when he arrived at the Cavendish Laboratory in 1895. There, he began to study "Becquerel rays" and found that they were "complex, and that there are present at least two distinct types of radiation—one that is very readily absorbed, which will be termed for convenience α [**alpha**] radiation, and the other of a more penetrative character, which will be termed the β [**beta**] radiation." Only a short time before, Sir William Ramsay had discovered helium on Earth in a uranium-bearing mineral (and, along

Entitled "Radium," this lithograph appeared in the December 22, 1904, issue of *Vanity Fair*. It depicts the Curies, Pierre holding a sample of the glowing element radium and Marie standing properly behind him.

Ernest Rutherford (1871–1937) came to the Cavendish in 1895. Fresh from New Zealand, this unpolished colonial was the first of the new research students and the first to work with J. J. Thomson himself. Rutherford received the Nobel Prize in chemistry, strangely enough, in 1908.

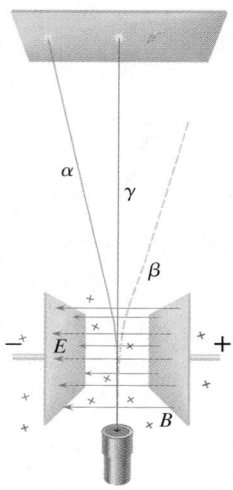

Figure 27.10 Alpha and beta rays, as charged particles, can be bent off course by both electric and magnetic means. Gamma rays, being electromagnetic radiant energy, are unaffected by these fields.

with F. Soddy, determined that it was liberated from radium). Thus, there was an early connection between α-rays and helium. (An alpha particle is a helium nucleus, two protons and two neutrons, but that would not be known for years.)

In 1903, Rutherford succeeded in deflecting α-rays using strong electric and magnetic fields, thus "proving" that they were positively charged particles (Fig. 27.10). Measurements of the charge-to-mass ratio of α-particles produced a value several thousand times smaller than that of the electron, again suggesting a large mass. Working with a young Ph.D., Hans Geiger (of Geiger counter fame, p. 985), Rutherford concluded that the α-particle carried a charge of $+2e$, that is, twice the positive fundamental charge (twice the charge of the hydrogen ion).

In a beautiful experiment (1908), Rutherford and Royds captured alpha particles (Fig. 27.11). Alpha-emitting radon gas was collected above mercury in a thin-walled tube. After a week, the alpha particles that had passed into the surrounding vacuum region were pushed up into a capillary tube by raising the mercury level. On electrically exciting the accumulated gas, they found it to produce the characteristic emission spectrum of helium. It was not yet clear what atoms were, but "the α-particle, after it has lost its positive charge [by gaining two orbital electrons], is a helium atom." **An alpha particle is the nucleus of the helium atom.**

The work on β-rays progressed quickly in a number of European laboratories: they were deflected by magnetic fields (1899), found to have a negative charge (1900), revealed to possess a charge-to-mass ratio very near that of cathode rays (1900), and finally were determined to have a mass equal to that of the electron (1902). **Beta rays are electrons.**

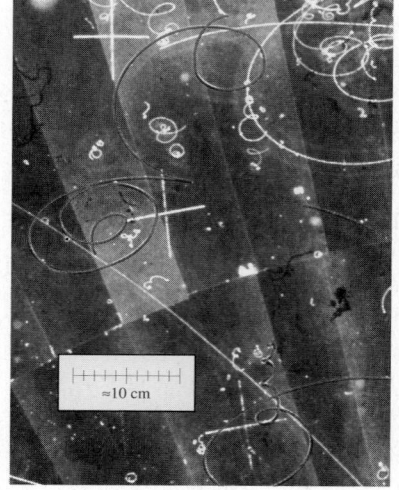

Each charged particle leaves a track of ionized atoms in its wake. Tiny bubbles form on these ions leaving a visible trail. Here $\vec{\mathbf{B}}$ is downward and that causes electrons (β-particles) to curve clockwise (red track), just as they do above in Fig. 27.10. Gamma rays (which don't leave tracks) are unaffected by the magnetic field and travel in straight lines. The blue positron track, at middle right, ends with a γ photon that invisibly travels in a straight line about 10 cm before it produces an electron-positron pair.

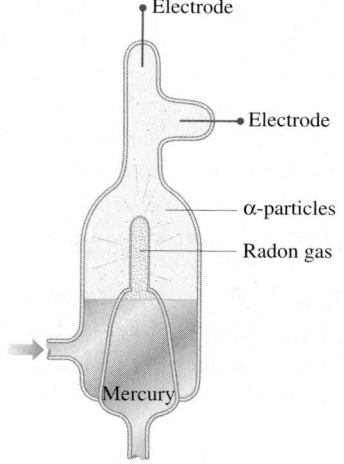

Electrode

Electrode

α-particles

Radon gas

Mercury

Figure 27.11 Radioactive radon in the thin-walled chamber emits α-particles, which are captured within the thick outer vessel. When the mercury level is raised, the alphas are forced into the discharge tube at the top. There, a helium spectrum is produced.

A smoke alarm. A small amount of a radioactive α-emitter ionizes the air between two parallel metal plates. A voltage across the plates causes the ions to drift, creating a current. The presence of smoke cuts off the current and triggers the alarm.

Table 27.2
Ionizing Radiation Sources in the United States

Sources	Percent
Human activity	Total 18%
Medical and dental X-rays	11
Nuclear medicine	4
Consumer products	3
Occupational	0.3
Fallout	<0.3
Nuclear fuel cycle	0.1
Miscellaneous	0.1
Natural sources	Total 82%
Radon	55
Internal body emission	11
Terrestrial minerals	8
Cosmic rays	8

P. Villard, in Paris, also studied the emissions from radium and, in 1900, reported the presence of highly penetrating rays. They were not even slightly deflected by strong magnetic fields, thus suggesting a kinship with light. Gradually, the evidence grew that this γ (**gamma**) radiation was electromagnetic—somewhat more energetic and shorter in wavelength, but otherwise identical to X-rays. The issue was settled in 1914 when Rutherford and Andrade succeeded in reflecting γ-rays off the surface of a crystal.

Gamma photons (typically 0.01 MeV to 10 MeV) are highly penetrating. They can be completely absorbed only after passing through several feet of concrete or about one to five centimeters of lead. More potent (and more dangerous) than X-rays, gamma rays have no trouble whisking through a human body and destroying molecules along the way. Beta particles (0.025 MeV to 3.2 MeV), traveling at roughly 25%c to 99%c, can sail through up to 15 meters of air. A millimeter or so of aluminum will block them totally, and they will not burrow very deeply into people (≈1 mm to 2 cm). They are about 100 times more penetrating than alphas, though far less so than gammas.

Alpha particles (4 MeV to 10 MeV) have a mass of 6.642×10^{-27} kg, roughly 7300 times that of an electron. They're ejected from atoms at formidable speeds (14×10^3 km/s to 22×10^3 km/s)—the alpha particles from radium are emitted at ≈15 190 km/s. Easily stopped, alphas will barely make it through a single sheet of paper, or 0.3 cm to 8.6 cm of air. Even so, breathing in radioactive dust can put alpha emitters inside the lungs, where they can be quite lethal. They are very highly ionizing (1000 times more so than β-rays). A 5-MeV alpha can create 40 000 ion pairs in the process of traversing 1 cm of dry air. As a result, alphas lose their energy rapidly and are especially hazardous to biological organisms (perhaps 20 times more damaging than beta or gamma radiation). Only very energetic alphas (>7.5 MeV) can penetrate human skin, and external radiation is usually not a problem (Table 27.2).

The Nuclear Atom

Today, we know beyond any doubt that an atom is composed of a central dense core of positive charge—the nucleus—surrounded by a fast-moving cloud of electrons. This nuclear model of the atom was advanced by Rutherford, but others before him paved the way. Earnshaw (1831) had argued that no system of charges interacting via an inverse-square-law force could be in a stable static equilibrium—if they're bound by a Coulomb force, *the particles that constitute the atom must be in motion*. It followed from the cathode-ray experiments that those particles "must" be electrons. J. Larmor (1900) proposed a vague scheme with *electrons attracted to a central positive charge*; soon after that, J. Perrin (1901) put forth a Solar System model with *orbiting electrons*. But as we'll see, there were serious problems with the stability of all such systems—*orbiting electrons are accelerating and so should radiate*, lose energy, and collapse into the positive center.

Lord Kelvin in 1902 proposed that the atom might be imagined as a jellylike sphere of positive charge, embedded throughout with an equal negative charge in the form of electrons, like the raisins in a glob of pudding. The following year J. J. made a thorough investigation of the stability of such a scheme, and as a result, it came to be known as the "Thomson atom." This ***raisin-pudding atom*** was widely considered, for want of anything better.

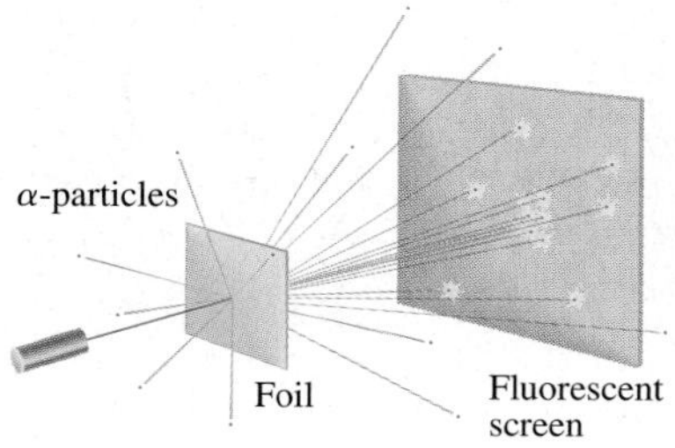

27.5 Rutherford Scattering

One day in early 1909, Geiger went to Rutherford to ask if his student, an undergraduate named Marsden, might be allowed "to begin a small research." Being of a similar mind, Rutherford suggested a simple scattering experiment, whose results he was sure he knew beforehand. Rutherford had been the first to scatter α-particles from matter, and Geiger had done some work in the area as well. The idea was to study the way alpha particles were deflected as they traversed a thin foil, and from that, possibly learn something about the hidden structure of the atoms of the target.

A few milligrams of a radium compound in a hollow lead tube served as the gun, shooting α-particles in a well-defined beam. The target, an exceedingly delicate gold foil about 0.000 06 cm (about 20 millionths of an inch) thick, corresponded to only ≈ 1000 layers of atoms. Alphas that easily penetrated the foil would slam into a distant zinc sulphide screen (Fig. 27.12). With each impact, the screen would give off a minute flash of light that could be seen in a totally dark room and counted with great effort. (The experiment is nerve-wracking, to say the least.) Positioning the detector straight down in the forward direction, the experimenters had found that very few alphas were even slightly deflected away from the original beam, which for the most part went right through undeviated. That finding was reasonable enough; after all, the alpha particles were very massive and moving exceedingly fast. They would presumably sail right through the tenuous positive pudding and could hardly be pulled aside appreciably by the minute electron raisins (Fig. 27.13). So, when Rutherford suggested that Marsden look for alphas scattered through large angles (greater than 90°), he knew the eager undergraduate simply wouldn't find any.

Two or three days later, Geiger rather excitedly rushed back to "Papa," as they called the great man in those days, to tell him the unbelievable news: "We have been able to get some of the alpha particles coming backwards." Rutherford recalled, "It was almost as incredible as if you fired a 15-inch shell [i.e., in diameter] at a piece of tissue paper and it came back and hit you." It took about two years for Rutherford to work out the details of a theory that would explain those extraordinary observations (Fig. 27.14).

It was clear from the start that each alpha particle could be blasted backward toward the source by a head-on collision with a concentrated, highly charged, positive, massive object—the **atomic nucleus**. We know from the study of elastic collisions (p. 221) that a projectile (the α-particle) will bounce back off a target (a nucleus) only when its mass is exceeded by that of the target (Fig. 7.14b, p. 222). This kind of collision must be a very rare event (the nucleus must be exceedingly small or else the undeviated penetration of the bulk of the beam would never occur). To be sure, the back-scattering happened only about once in every 10 000 or 20 000 impacts. Nonetheless, there were many millions of alphas being fired (a gram of radium undergoes about 4×10^{10} atomic disintegrations every second).

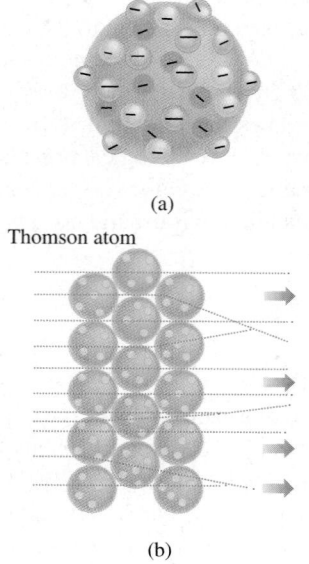

Figure 27.12 Rutherford's alpha particle scattering experiment.

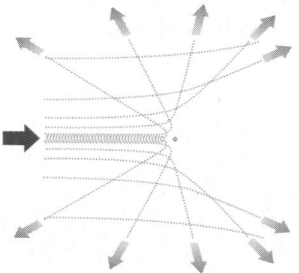

Figure 27.13 (a) The raisin-pudding atom—negative electrons embedded within and throughout a large positive fluff. In its simplest form, the electrons hover in circular patterns within fixed planes. (b) Bombarding α-rays should pass right through the raisin-pudding atom with very little deflection.

Figure 27.14 The scattering of a beam of α-particles by a positive massive nucleus.

"Papa" saw the connection with the arching flight of a comet in the Sun's attractive gravitational field. In time, he was able to derive an equation for the scattering that would occur in a repulsive interaction between a target and projectile when both were positively charged. The α-particles sail off in hyperbolic orbits (Fig. 27.14). One morning in 1911, Rutherford happily sauntered into Geiger's lab to share one of the great secrets of the Universe: ***Each atom consists of a tiny massive concentration of positive charge, the nucleus*** (a term he introduced in 1912), ***surrounded by a distribution of electrons***. Geiger immediately began to test the theoretical predictions—the specific dependence of the scattering on foil thickness, nuclear charge, alpha velocity, and so on. Within a year, his measurements would convincingly bear out the power of the image, however fuzzy: the nuclear atom had arrived.

The Size of the Nucleus

Atoms were known to be about 10^{-10} m across (p. 285). Now Rutherford's scattering experiments provided a means of approximating the size of the nucleus. Consider a head-on collision between an α-particle and a nucleus, which we suppose to contain a number of positive charges Z, each of magnitude e. The α-particle initially sails in with a kinetic energy of

$$\mathrm{KE}_\alpha = \tfrac{1}{2} m_\alpha v_\alpha^2$$

and it gradually slows down as it approaches closer and closer to the repelling nucleus. At any distance r from the nucleus, the electric potential is given by

$$V = \frac{k_0 Q}{r}$$

where $Q = Ze$, the nuclear charge (assumed to be a point, which it isn't). At any distance r, the alpha has an electric potential energy (gravity is negligible) equal to its charge $2e$ times V:

$$\mathrm{PE}_\alpha = \frac{k_0 (Ze)(2e)}{r}$$

This is only approximate since it neglects the cloud of orbital electrons (which will soon be penetrated anyway). The alpha will reach a point of closest approach to the nucleus at a distance $r = R$, where it momentarily stops (as if it had compressed a spring and is about to pop backward). At that point, its initial KE has been transformed into PE:

$$\mathrm{KE}_\alpha = \mathrm{PE}_\alpha$$

$$\tfrac{1}{2} m_\alpha v_\alpha^2 = \frac{k_0 2Ze^2}{R}$$

and

$$R = \frac{4 k_0 Z e^2}{m_\alpha v_\alpha^2} \qquad (27.4)$$

This distance represents an upper limit on the nuclear radius. Given $v_\alpha = 1.5 \times 10^7$ m/s and $m_\alpha = 6.64 \times 10^{-27}$ kg, we have to come up with a value of Z. At the time when Rutherford was doing these experiments, it was reasonable to guess that (as with the α-particles themselves) a nucleus would contain a number of charges equal to half the atomic weight; that is, $Z = \tfrac{1}{2} 197 \approx 99$. Actually, for gold, $Z = 79$, but that knowledge would have to wait until the work of Moseley (p. 1017). We can use either number; the choice will not affect things much. Hence,

$$R = \frac{4(9.0 \times 10^9 \text{ N·m}^2/\text{C}^2)(79)(1.60 \times 10^{-19} \text{ C})^2}{(6.64 \times 10^{-27} \text{ kg})(1.5 \times 10^7 \text{ m/s})^2} = 4.9 \times 10^{-14} \text{ m}$$

The nucleus is about ten thousand times smaller than the atom. If in imagination the nucleus of a typical atom is enlarged to the size of an apple (10 cm), the atom as a whole would be about 1 km across, most of it a vast emptiness.

Rutherford never satisfactorily addressed three outstanding issues concerning his model: the question of the stability of the atom, the necessity to explain atomic spectra, and the need to understand the Periodic Table within the context of atomic structure. That too would come later.

27.6 Atomic Spectra

In the late 1800s, there developed a body of research called *spectroscopy* that provided an accurate means of distinguishing between atomic species. The formulas describing the patterns of spectroscopic phenomena are of little interest to us in themselves. But they reflect, with extraordinary precision, the hidden structure of the atom. Ultimately, those formulas will serve as one of the most powerful tests of atomic theory (p. 1010).

Spectroscopy began with Newton and his experiments using white light and prisms, but it didn't become an analytic tool for exploring matter until the 1800s. A. J. Ångström, in 1853, used a discharge tube filled with various gases to study their spectra. Light from a specimen is made to pass through a slit in a screen and then through a prism or grating that separates the narrow beam into its constituent color bands, the **spectral lines** (Fig. 27.15). When a gas is excited, it emits specific wavelengths and we see colored lines on a black background. This is known as the **emission spectrum** (Fig. 27.16). Inversely, when white light passes through the same gas, the atoms absorb those same specific wavelengths. This is the **absorption spectrum**, and we see black lines on a smoothly varying, bright-colored background. Ångström first measured the wavelengths of the four bright visible emission lines of hydrogen. Because hydrogen is the simplest of all the elements, these data would play a special role in the later development of theoretical atomic physics. J. Plücker gave the lines the names appearing in Table 27.3 and, by 1858, had rightly suggested that the spectrum of any substance was a fingerprint that unambiguously specified its identity.

The very existence of spectral lines (and the knowledge that light was an oscillatory wave of some sort) suggested that atoms had a complex structure that could sustain many different internal vibrations. Maxwell pointed *that* fact out in 1875, but even before then, George Stoney noticed that the hydrogen spectral wavelengths formed simple ratios:

$$\text{H}_\alpha : \text{H}_\beta : \text{H}_\delta = \frac{1}{20} : \frac{1}{27} : \frac{1}{32}$$

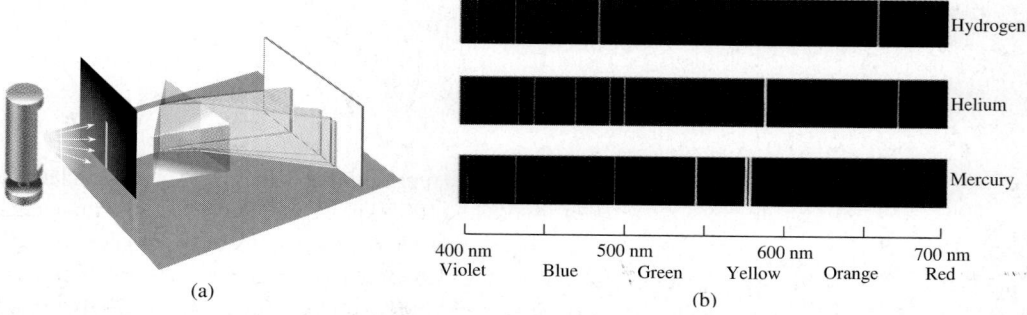

Figure 27.15 (a) The formation of spectral lines by a prism. (b) The emission spectra of hydrogen, helium, and mercury.

Table 27.3

Hydrogen Visible Spectra

Line	Wavelength (nm)	Color
H_α	656.28	red
H_β	486.13	blue-green
H_γ	434.05	violet
H_δ	410.12	violet

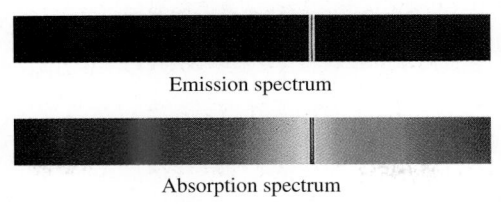

Figure 27.16 Emission and absorption spectra of sodium.

Emission spectrum

Absorption spectrum

Stoney talked about, "the 32nd, 27th, and 20th harmonics of a fundamental vibration" of the atom. In 1885, Johann Balmer, having investigated the notion of atomic harmonics, published a simple formula that rather remarkably yielded the observed hydrogen wavelengths. In its modern form, it is

$$\frac{1}{\lambda} = R\left[\frac{1}{2^2} - \frac{1}{n^2}\right] \qquad n = 3, 4, 5, \ldots \qquad (27.5)$$

The equation generates the wavelengths of the various *visible* lines—of what has come to be known as the **Balmer series**—when we substitute in turn, $n = 3$ or 4 or 5, and so on. When λ is expressed in meters, R, which is called the ***Rydberg constant***, is

$$R = 1.097\ 373\ 15 \times 10^7\ \text{m}^{-1}$$

which is quite close to the number Balmer suggested. Note that the series continues with the wavelengths getting shorter and shorter as $n \to \infty$ at the ***series limit*** (Problem 43); in time,

---

Example 27.3 **[I]** Use Balmer's formula to compute the wavelength of the red line in the hydrogen spectrum.

Solution Balmer's formula is something you just have to memorize. (1) TRANSLATION—Determine the wavelength of the first line in the Balmer series. (2) GIVEN: Balmer's formula. FIND: λ for red line. (3) PROBLEM TYPE—Modern Physics / spectra / Balmer series. (4) PROCEDURE—Just use the Balmer formula realizing that the larger n is, the smaller λ is. (5) CALCULATION—The red line has the longest wavelength and the

lowest n; namely, $n = 3$. Thus

$$\frac{1}{\lambda} = (1.097 \times 10^7\ \text{m}^{-1})\left[\frac{1}{4} - \frac{1}{9}\right]$$

and

$$\boxed{\lambda_\alpha = 656.3\ \text{nm}}$$

Quick Check: This λ is appropriate for red light and, of course, agrees with Table 27.3.

---

this too, was confirmed experimentally.

The first term in the brackets of Eq. (27.5) contains a denominator of 2^2, and Balmer was so sure of himself that he proposed (off the top of his head, as it were) that there might well be other sets of lines for 1^2, 3^2, 4^2, and so on. As luck would have it, he was essentially right and in time the ***Lyman series***

$$\frac{1}{\lambda} = R\left[\frac{1}{1^2} - \frac{1}{n^2}\right] \qquad n = 2, 3, 4, \ldots \qquad (27.6)$$

in the ultraviolet was observed—as were the ***Paschen series***

$$\frac{1}{\lambda} = R\left[\frac{1}{3^2} - \frac{1}{n^2}\right] \qquad n = 4, 5, 6, \ldots \qquad (27.7)$$

and two others (with $1/4^2$ and $1/5^2$) in the infrared (p. 1013).

It would take until 1913 before Niels Bohr could explain spectral lines in terms of atomic transitions between energy levels. We will deal with that discovery in the next chapter.

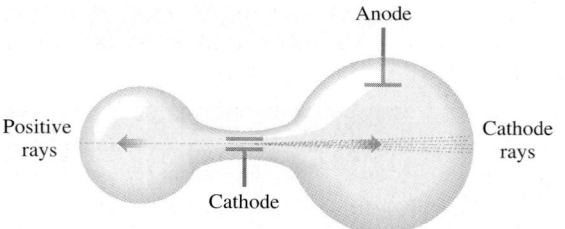

Figure 27.17 Cathode rays stream out toward the general location of the anode. Positive rays sail toward, through, and beyond the cathode.

27.7 The Proton

Recall the cathode-ray experiments in which a cathode and anode were placed in a tube containing a low-pressure gas. A high potential across the electrodes caused the gas to become ionized, and a beam of electrons accelerated away from the cathode out in the general direction of the anode (Fig. 27.17). It was noticed in 1886 that the glowing trail of a beam could also be projected in the opposite direction. The backward streaming of that emanation, along with the observation that it could be deflected by a magnetic field, was convincing evidence that these rays were positive particles; J. J. Thomson christened them *positive rays*. When their charge-to-mass ratio was measured, several rather low values were found (indicative of particles with large masses), on the order of those of the atoms themselves.

After the Rutherford model had been put forward, it was reasonable to suppose that positive rays were simply ionized atoms. Naturally enough, the e/m ratios depended on the particular trace gas in the tube. When Thomson performed the experiment with hydrogen (1907), he found two values of e/m, the larger for the light H^+ ion and the smaller for the ionized molecule H_2^+. The former compared nicely with the value arrived at via electrolysis. More and more, it seemed that the positive nuclear particle (the thing many people called the H-particle) was the hydrogen atom stripped of its single orbital electron.

The clincher came when Rutherford, pounding the nuclei of various elements with alpha particles, discovered H-particles among the flying fragments. The apparatus (Fig. 27.18) consisted of an evacuated chamber containing an alpha source. The chamber was sealed with a piece of foil just thick enough to stop all the alphas. Nitrogen gas was then introduced into the chamber (later on, fluorine and other elements were used; in all cases, the results were the same). Scintillations were observed, indicating the presence of a more penetrating radiation that proved to have the same range and charge as the H-particle. The range in air was easily determined by moving the screen, and it was indicative of the kind of radiation being dealt with. Apparently, this positive particle must be a basic component of many if not all of the elements. It wasn't until 1920 that Rutherford formally proposed the name **proton** (from the Greek *protos* for "first"), a word that had already quietly been in use for a dozen years.

With a mass of

$$m_p = 1.672\ 623\ 1(10) \times 10^{-27}\ \text{kg} = 938.272\ 31(28)\ \text{MeV}/\text{c}^2$$

Figure 27.18 Alpha particles crashing into nitrogen atoms liberate protons, which strike the fluorescent screen, causing flashes that can be seen with a microscope. The chamber is sealed with a thin foil that will pass protons but not alpha particles.

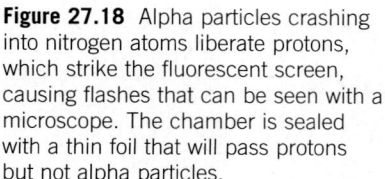

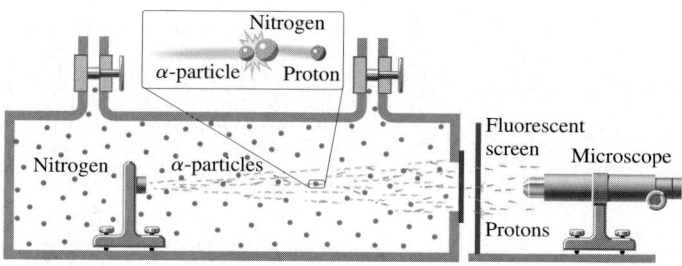

the proton (p) is 1836 times the mass of the electron (e). The proton carries a quantum of charge as does the electron; so these subatomic specks logically seem, if not twins, at least mates. Laboratory tests confirm that the proton and electron charges are indeed equal—in the most precise measurement ever carried out, the hydrogen atom has been found to be neutral to an accuracy of twenty-two decimal places. This equality of charge becomes even more remarkable when we learn that the proton is a complex structured thing; in fact, it remains one of the fundamental puzzles of Modern Physics (p. 1104).

27.8 The Neutron

We picture the hydrogen atom as a single nuclear proton coupled by an attractive Coulomb force to one orbital electron. The next element in the Periodic Table, helium, might then consist of two protons and two surrounding electrons; lithium might have three and three; and so on, all the way up to uranium with 92 protons and 92 electrons. Yet, the α-particle has a charge of $+2e$ and an e/m ratio of half that of the hydrogen nucleus; it has a mass of 4. And this is *not* quite right; something else must be happening in the nucleus. Such differences between **atomic number** (Z) and **atomic mass** (A) occur throughout the Table all the way up to uranium, which has 92 nuclear protons and a mass of 238. There is a good deal of mass unaccounted for when the uranium nucleus is considered to be a cluster of 92 protons. If everything is made up of the only two then-known particles—electrons and protons—we have a little problem here.

Rutherford, in 1920, proposed that all of these difficulties could be resolved by assuming the existence of tightly bound *proton-electron pairs*, neutral units that would add mass but not charge. The idea was wonderfully simple, but quite erroneous. Rutherford even suggested the name *neutron* for the pair-particle. And he promptly set a group of his "boys" to work to track down the neutral object; thus, James Chadwick began a 12-year search.

In 1930, it was discovered that when beryllium (Be) was bombarded with alpha particles, out streamed a flux of very energetic radiation that was assumed to be γ-emission, but that seemed a bit odd in some respects. It was soon determined that, as with γ-rays, the new radiation was both very penetrating and not deflected by a magnetic field. However, unlike γ-rays, it was not ionizing and could not discharge an electroscope. The Joliet-Curies, Irène (Mme Curie's daughter) and her husband, Frédéric Joliot-Curie, observed that the presence of the mystery emission (which they thought was electromagnetic) could be detected in a much enhanced way by making it impinge on a substance rich in hydrogen, that is, protons. When the beam struck a sheet of paraffin, protons were blasted out, and these could easily be picked up with a Geiger counter (Fig. 27.19). (The latter was simply a metal tube containing a low-pressure gas such as argon. A voltage of about 1000 V was applied between the central-wire anode and the tube. Ionizing radiation entered via a thin mica window and knocked electrons out of a few gas atoms. These rushed toward the anode, ionizing more atoms in the process. The resulting discharge created a current pulse that was amplified and sent to a counter.)

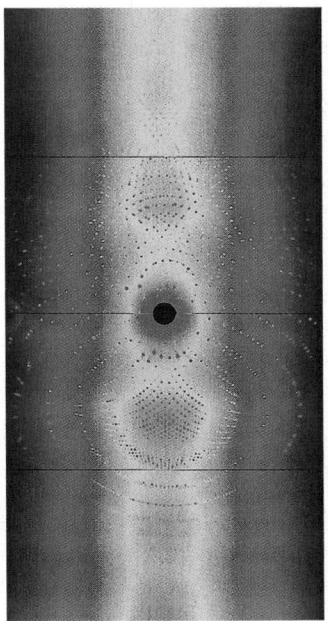

Neutron Laue diffractometry. This image was created using a crystal to scatter a beam of neutrons rather than X-rays (p. 974). Neutron diffraction is especially useful for determining the locations of hydrogen atoms. Since much of the structure of proteins is dominated by hydrogen, the technique is particularly suited for studying biological systems.

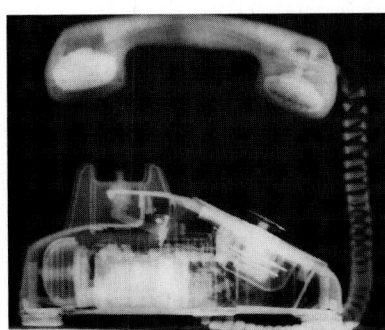

A radiograph of a telephone created with a beam of neutrons. These neutrophotos usually show considerably more of the subtle variations and details than do X-ray pictures, which they otherwise closely resemble.

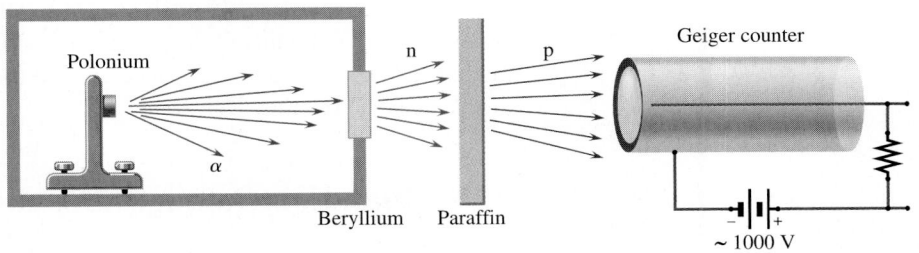

Figure 27.19 Polonium emits alphas, which knock neutrons out of the beryllium. These knock protons out of the paraffin.

Neutron beam cancer therapy at Fermi Lab. A laserbeam is used to align the target with the neutron beam that emerges from the square aperture.

Chadwick (1932) argued that a nonionizing particle capable of ejecting protons had to be the long-sought neutron (n). He could determine the speed of the protons from their range and was able to show from collision theory that all of his observations could be understood quantitatively by hypothesizing the presence of a neutral particle having about the mass of the proton. When the alpha bombardment of beryllium was finally understood, Chadwick wrote it out as

$$ {}_2^4\alpha + {}_4^9\mathrm{Be} \rightarrow {}_6^{12}\mathrm{C} + {}_0^1\mathrm{n} $$

(mass as the superscript, charge as the subscript). Beryllium absorbs an alpha and, with the emission of a neutron, is transformed into carbon.

Rutherford had been close (but no cigar): electrons don't live in the nucleus, and the neutron is as much a single particle as is the proton, though it is slightly more massive:

$$ m_\mathrm{n} = 1.674\,928\,6(10) \times 10^{-27}\ \mathrm{kg} = 939.565\,63(28)\ \mathrm{MeV}/c^2 $$

Very soon after the announcement of the discovery of the neutron, Werner Heisenberg proposed the now accepted neutron-proton model of the atomic nucleus. **All nuclei are composed exclusively of neutrons and protons**, the total number of which (A) determines the atomic mass. Thus, uranium has 92 protons (an atomic number Z of 92) and 146 neutrons, yielding an atomic mass (A) of 238—everything works out neatly, at last.

Once removed from the nucleus, the neutron is unstable, decaying into an electron, a proton, and a neutrino (actually an antineutrino, p. 1071) at a rate such that half of the neutrons present at any moment will decay within roughly 10.4 minutes. This process is an example of β-decay, and it explains how electrons can be ejected from a nucleus wherein they do not exist—they are a by-product of the transformation, created in the process. While nestled with the protons in the nucleus, the neutrons interact with (and seem to be more or less stabilized by) the others, so they don't decay in all nuclei. Moreover, neutrons and protons provide a glue (by way of the strong force) that helps keep the nucleus together. To underscore this intimate relationship (and because they are the exclusive components of all nuclei), the neutron and proton are called **nucleons**.

Neutrons are about 1.3 MeV/c^2 more massive than protons, and with this excess rest energy, it is the free neutron that decays "down" to the proton. Had the reverse been true, we could expect free protons to decay into neutrons. Happily for us, it's not the case, and hydrogen, water, and human beings are stable. (A proton can decay into a neutron but only in a fairly large nucleus that can provide the needed mass-energy.)

When nuclei decay, as when neutrons decay, the process is described statistically. We know that half the free neutrons in a box will decay after ≈10.4 minutes, but we have no idea which ones will be left. Each survivor seems identical, each seems unaffected by the passage of ≈10.4 min. The data suggests that the likelihood of a particular neutron decaying during any ≈10.4-minute interval is constant, no matter how many such intervals it has already survived. Neutrons don't age, and when they do decay, it is for no apparent reason. This strange result is the same one Rutherford observed for radioactive atoms, and it suggests that causality (and the accompanying predictability we are so fond of) has given way on some hidden level to the roulette wheel and the rule of probabilities. Today, it is widely believed that our inability to predict the individual behavior of a neutron or a nucleus is not the result of a correctable ignorance but is a fundamental manifestation of the quantum-mechanical nature of the Universe.

27.9 Radiation Damage: Dosimetry

When energetic radiation (generally in the form of positive ions, electrons and positrons, neutrons, and photons) passes through matter, it can have a variety of transforming effects. With enough exposure (that is, with the deposition of large amounts of energy), interatomic bonds are broken and atoms are literally displaced; metals become brittle, living cells die,

After exposure to 1.7×10^6 rads of gamma rays a Pyrex beaker turns brown and becomes brittle. Yet colonies of the bacterium **Deinococcus radiodurans** can survive such doses. These bacteria rapidly repair radiation-induced damage to their DNA. A colony of this sort could travel through space for millions of years and still remain viable in spite of the exposure to cosmic radiation.

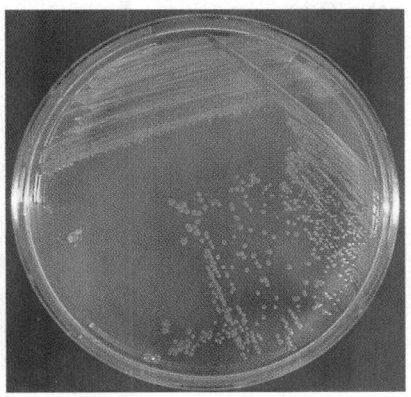

plastics decompose, and semiconductors (and their electronic progeny) deteriorate. Of course, this same ability to destroy can be harnessed to control cancer, sterilize food, introduce useful biological mutations, and produce sunglasses that darken automatically.

Among the most important effects of radiation on biological systems is the ionization of atoms and molecules. An atomic electron in the cloud surrounding the nucleus can be ripped away, leaving behind a positive ion. When that electron is subsequently captured by some other atom, a negative ion is created. The presence of these ion-pairs in living cells can alter the normal chemistry of the cell and have very deleterious effects, including the initiation of cancer.

The subject of ***dosimetry*** deals with the quantification and measurement of the "amount" of radiation, the so-called *dose* delivered to a system. There are four distinct radiation measurements that must be considered: the *source activity* (p. 1076), which deals with how much radiation is being generated at the place of origin; the *exposure*, which signifies the amount of ionizing radiation arriving at the system under study; the *absorbed dose*, which concerns the energy actually imparted to the system by the incident radiation; and the *biological equivalent dose*, which is associated with the effect the physical dose has on any particular biological system.

Exposure is defined specifically for photons with energies up to 3 MeV. It relates to the amount of ionization produced in a kilogram of dry air at STP. The earliest unit used was the *röntgen* (R), which corresponds to the generation of a specific number of ionization pairs (2.083×10^9 ion pairs per cm³ of air) or, equivalently, to the separation of a certain number of coulombs per kilogram of positive and negative charge: $1 \text{ R} = 2.58 \times 10^{-4}$ C/kg. Such a measure provides us with a sense of the ionizing intensity of the incident radiation. But it says nothing about the specific system being irradiated, which might even be totally transparent to the radiation.

The absorbed dose, measured in grays, is the energy per unit mass of absorber, given up by the radiation on ionizing the material.

The ***absorbed dose*** reflects the amount of the energy imparted to the system via the absorption of ionizing radiation incident upon it. If a one-roentgen photon beam impinges on soft biological tissue, an energy of approximately 0.01 J will be absorbed in each kilogram of material. On average a particle loses about 35 eV per ionization it produces. Accordingly, we define the *rad* (short for ***radiation absorbed dose***) such that

$$1 \text{ rad} = 0.01 \text{ J/kg}$$

This is probably still the most widely used unit of absorbed dose, although the gray (Gy) is now the official SI unit: $1 \text{ Gy} = 100 \text{ rad} = 1.0 \text{ J/kg}$. The röntgen pertains only to photons, whereas the rad and gray apply to all particles.

Every different kind of radiation is likely to have a somewhat different effect on a particular biological system (cell, tissue, organ, or organism); therefore it's not enough to simply say that the absorbed dose is so many rads. For instance, energetic neutrons are especially nasty at causing cataracts (the clouding of the lens in the eye that can lead to blindness). Thus, 1 rad of fast neutrons will inflict the same damage as 10 rads of X-rays.

Table 27.4
Quality Factors (*Q*)
for Several Kinds of Radiation

Radiation	Typical *Q*
Photons	1
Beta particles (>30 keV)	1
Beta particles (<30 keV)	2
Protons (1–10 MeV)	2
Neutrons (<0.02 MeV)	3–5
Fast neutrons (1–10 MeV)	10 (body)–20 (eyes)
Protons (1–10 MeV)	10 (body)–20 (eyes)
Alpha particles	10–20
Heavy ions	up to 20

That's because the secondary ionization caused by fast neutrons (colliding with other ionizing particles) is concentrated in dense tracks along the path, which causes more serious alterations of the tissue.

To begin to take account of these differences in the *relative biological effectiveness* of the various forms of radiation, a multiplicative **quality factor** (*Q*) was introduced. The damage from each form of radiation (which also depends on energy and the specifics of the target) is compared to that induced by 200-keV photons, taken as the standard; the resulting *Q* values are listed in Table 27.4. The **biological dose equivalent** was defined as the product of the absorbed dose *D* (in rads) times *Q*, and the corresponding unit was the rem (which stands for **rad equivalent man**). *One rem of any kind of radiation produces more or less the same amount of biological damage*, namely, that of 1 rad of 200-keV photons. Thus 1 rad of 200-keV photons (*Q* = 1) corresponds to 1 rem and generates the same cataract damage as about 0.05 rad of fast neutrons (*Q* = 20), which also corresponds to 1 rem.

The value of *Q* (which you will also see referred to as the *radiation weighting factor*) depends on the *linear energy transfer* (LET) of the charged particles. The LET corresponds to the spatial rate at which energy is imparted to the medium as the radiation progresses through it; namely, $d\mathrm{E}/dx$, and it can be measured directly. The equivalent dose (*H*) which is defined by

$$H = QD$$

where *D* is in grays, has as its SI unit the *sievert* (Sv); 1 Sv = 100 rem.

It is estimated that the average person receives an annual dose of radiation from all sources of about 0.40 rem (see Table 27.2). Of that perhaps 0.13 rem per year is the natural background due to cosmic rays and due to radioactivity in the immediate environment: in rocks, brick, soil, food, cigarette smoke, and so on. Beyond that, another roughly 0.07 rem per year is picked up via medical X-rays.

Animals can be classified according to how radiosensitive they are. That's usually done by specifying the $\mathrm{LD}_{50/30}$, that is, the single lethal whole-body dose that will result in death of 50% of the animals within a period of 30 days. (An animal surviving 30 days is not likely to expire soon thereafter.) Table 27.5 lists the $\mathrm{LD}_{50/30}$ for several species. The likelihood of death and the time it takes to occur both depend on the dose. The MST is the *mean survival time* for any whole-body dose. For humans, a whole-body dose of around 100 rem generally causes mild radiation sickness (Table 27.6). The few fatalities that occur have an MST of around 15 days. As the dose increases, the MST stays fairly constant, although the

Table 27.5
Radiosensitivities of Several Organisms

Species	$\mathrm{LD}_{50/30}$ (rem)
Dog	350
Mouse	400–600
Monkey	600
Man	500–700
Rat	600–1000
Frog	700
Newt	3000
Snail	8000–20 000
Virus	>100 000

Table 27.6
Immediate Effects of Whole-Body Single Exposure to Radiation—Adults

Dose (rem)	Effect
0–10	No observable short-term effects.
10–100	Slight blood changes, decrease in white cell count. Temporary sterility at 35 rem for women and 50 rem for men.
100–200	Significant reduction in blood platelets and white cells (temporary). Some nausea, vomiting.
200–500	Severe blood damage, nausea, vomiting, hemorrhage, hair loss, death in many cases.
500–2000	Malfunction of small intestine and blood systems. Death in less than 2 months in most cases, but survival possible if treated.

likelihood of demise gradually increases. Received in one dose, 250 rem will result in severe radiation sickness. A dose of 500 rem is likely to be lethal in 50% of the occurrences (with an MST that's still 15 days). All of these unfortunates succumb from damage to the bone marrow. As the dose increases further, the MST drops, reaching another plateau of three or four days at a dosage level of 1000 rem. Death now occurs in almost 100% of the untreated cases and is due to gastrointestinal damage, which manifests itself more rapidly than bone marrow deterioration. With doses beyond 10 000 rem or so, the MST drops to a few hours and death results from the destruction of the central nervous system. At still higher doses, death occurs almost simultaneously with the irradiation from the direct destruction of molecules vital to the support of life.

Core Material & Study Guide

SUBATOMIC PARTICLES
When F is a faraday of charge (96 485 C)

$$F = N_A e \qquad [27.1]$$

For single-valence elements, 9.65×10^4 C will liberate 1 mol, or equivalently, 9.65×10^7 C will liberate 1 kmol (p. 969). The **charge-to-mass ratio** for singly ionized hydrogen (H^+) is

$$\left(\frac{e}{m}\right)_{H^+} = 9.58 \times 10^7 \text{ C/kg}$$

These ideas are treated in Section 27.1 (The Quantum of Charge); study Example 27.1. Look at the **CD WALK-THROUGHS** .

The charge-to-mass ratio of the electron is

$$\frac{e}{m_e} = 1.758\ 8 \times 10^{11} \text{ C/kg}$$

where $e = 1.602\ 177\ 33 \times 10^{-19}$ C and $m_e = 9.109\ 389\ 7 \times 10^{-31}$ kg. Section 27.2 (Cathode Rays: Particles of Charge) discusses the classic research of J. J. Thomson and Robert Millikan—you should understand both of these landmark experiments.

The **Bragg equation** (p. 975) for X-ray diffraction is

$$2d \sin \theta_m = m\lambda \qquad [27.3]$$

Section 27.3 (X-rays) provides an important discussion of the origins of X-rays. It ends with a treatment of the Bragg equation, which is often the source of numerical problems (study Example 27.2).

THE NUCLEAR ATOM
Each atom consists of a tiny massive concentration of positive charge, the nucleus, surrounded by a distribution of electrons (p. 979). The nucleus is at least ten thousand times smaller than the atom. These are among the important conclusions described in Section 27.5 (Rutherford Scattering).

The **Balmer series** for the spectrum of hydrogen is

$$\frac{1}{\lambda} = R\left[\frac{1}{2^2} - \frac{1}{n^2}\right] \qquad n = 3, 4, 5,\ldots \qquad [27.5]$$

wherein $\qquad R = 1.097\ 373\ 15 \times 10^7$ m^{-1}

Section 27.6 (Atomic Spectra) introduces several ideas that will be significant in future chapters. Make sure you understand Balmer's equation and can follow the logic in Example 27.3.

The masses of the proton and neutron are

$$m_p = 1.672\ 623\ 1(10) \times 10^{-27} \text{ kg} = 938.272\ 31(28) \text{ MeV/c}^2$$

and

$$m_n = 1.674\ 928\ 6(10) \times 10^{-27} \text{ kg} = 939.565\ 63(28) \text{ MeV/c}^2$$

All nuclei are composed exclusively of neutrons and protons. Sections 27.7 (The Proton) and 27.8 (The Neutron) trace the discovery of these two extremely important particles.

Key Terms

electrolysis	atomic nucleus
faraday	spectral lines
charge-to-mass ratio	emission spectrum
electron	absorption spectrum
cathode rays	Balmer series
oil-drop experiment	Rydberg constant
X-rays	Lyman series
X-ray diffraction	Paschen series
Bragg equation	proton
thermionic emission	neutron
radioactivity	atomic number
radium	atomic mass
alpha, beta, and gamma	nucleon
nuclear atom	dosimetry
Rutherford scattering	absorbed dose

Discussion Questions

1. A molten bath of NaCl has a current passed through it. Chlorine gas bubbles off at the anode, and sodium is deposited at the cathode. Explain what's happening. Why did we start with molten NaCl rather than a water solution?

2. The chemical cell in Fig. Q2 depicts a silver nitrate solution, a silver anode, and a copper cathode. Explain what will happen when the switch is closed. How is it that the concentration of the solution remains unchanged? Describe the nature of the current both in the solution and in the external circuit. What, if anything, happens to the silver of the anode?

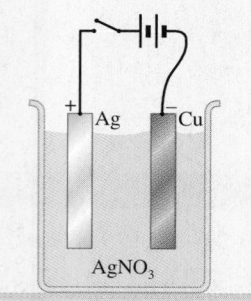

Figure Q2

3. Let's free our imaginations and speculate about the nature of the electron. Experiments to date have shown that if the electron has a "hard core," it's smaller than 10^{-16} cm. What are your feelings about what the electron "looks like," that is, size, shape, and charge distribution, for example? What questions immediately come to mind? Incidentally, no one knows what configuration it has or even if it has one. Nor do we know if it's finite or simply a point (whatever that means). We do believe that it is an elementary particle (that is, it has no smaller constituent parts). How does mass enter in your picture?

4. An equation for the mass M of material liberated at either electrode when a net charge of Q is passed across an electrolyte is

$$M = \frac{(Q)(\text{mass per mole})}{(F)(\text{valence})}$$

Explain how this expression comes about.

5. Beyond their dynamical parameters (that is, velocity, momentum, etc.), is it reasonable to assume that all electrons are identical?

6. Compare X-ray diffraction from a crystal with the reflection of light from a smooth surface. How do the two processes differ? Although we speak of the "reflection" of X-rays, that description is somewhat misleading. Why?

7. Using detectors that were not themselves sensitive to polarization, Barkla (1906) performed a crucial experiment that showed that X-rays could be polarized. His arrangement consisted of two blocks of carbon as depicted in Fig. Q7 and an incident unpolarized beam (entering at the left along the *x*-axis). From what you learned via Fig. 27.12, explain why the absence of scattered radiation from the second block outside the *xy*-plane proved the wave nature of the process.

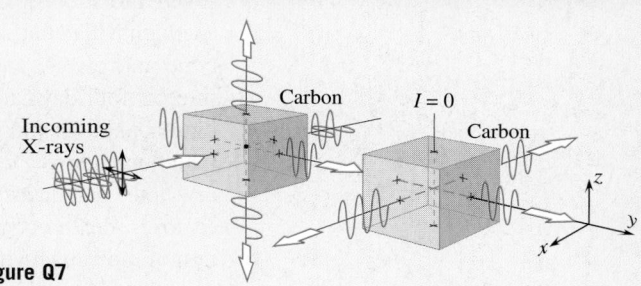

Figure Q7

8. Figure Q8 depicts a monochromatic beam of X-rays incident on a single crystal. Explain what is happening.

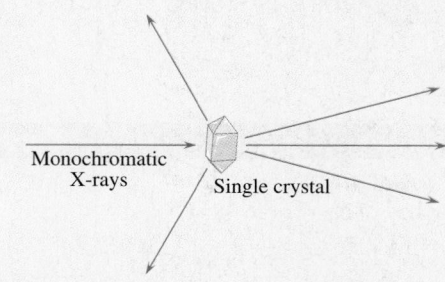

Figure Q8

9. Consider X-ray diffraction using both a single crystal and a chamber holding a monatomic gas. Figure Q9 shows the irradiance reaching a detector as a function of the angle from the beam axis. Explain the features of the two curves and why they are so different. What would the curve look like for a liquid or amorphous solid?

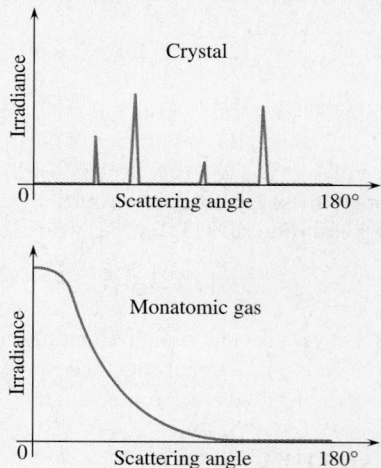

Figure Q9

10. Think of the nucleus as a uniformly charged positive sphere. Figure Q10*a* depicts the electric field for two different-sized spheres carrying the same total charge. Explain the diagram. Part (*b*) of the figure shows a head-on collision between each of these spheres and a small positive particle. Explain what's happening and relate it to Rutherford's analysis of alpha scattering.

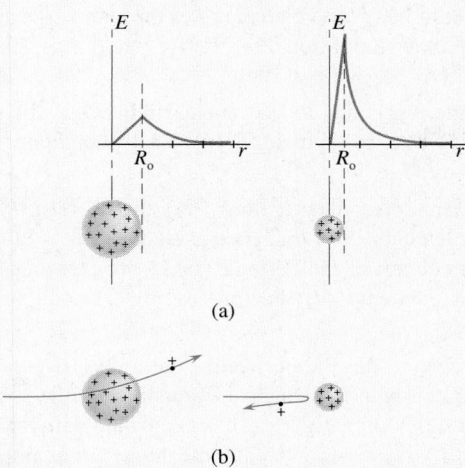

(a)

(b)

Figure Q10

11. Because of the form of Eqs. (27.5) through (27.7), it's useful to define $1/\lambda$ as the so-called *wave number*. It was noticed early on that the differences between certain wave numbers corresponded to other wave numbers. For example, the first four wave numbers in the Lyman series in units of $\times 10^3$ cm^{-1} are 82.3, 97.5, 102.8, and 105.3. Thus, the difference between the first two wave numbers in the Lyman series is the Hα wave number in the Balmer series, which is

$$97.5 - 82.3 = 15.2$$

Figure Q11 displays several of these remarkable relationships, and we see that the Hα line corresponds to what might be called a *transition* from the 3rd to the 2nd "level." **All the spectral lines can thus be seen as transitions from one level to another**, which is essentially the *Ritz Combination Principle*, first proposed in 1908. What is the significance of the lowest level (109 677), of the next

lowest level (27 419), and so on? What does all of this suggest about what might be happening in the atoms giving off the light? (The Bohr atom, treated in Sect. 28.5, p. 1010, is predicated on similar reasoning.)

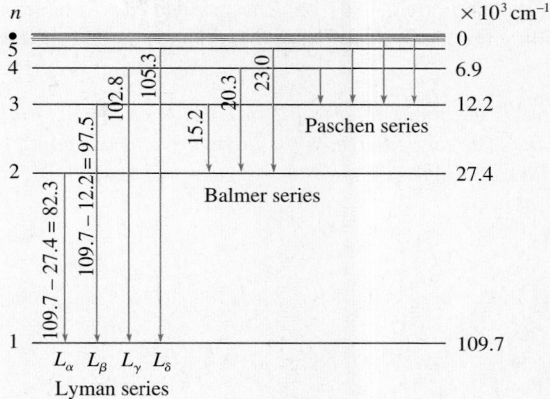

Figure Q11

12. The electron, proton, and neutron all have magnetic properties; they each have a magnetic moment (p. 694) and each behaves like a tiny bar magnet. The electron's magnetic moment is roughly 666 times stronger than the proton's, which is about 1.5 times stronger than the neutron's. Experiments show that the magnetic moments of ordinary nuclei are comparable to those of the proton. Does this relationship suggest anything about electrons residing in the nucleus of an atom? Explain.

13. Explain the meaning of each of the Key Terms on page 989.

Multiple Choice Questions

1. A typical atom has a diameter of roughly   (a) 0.2 mm   (b) 0.2 pm   (c) 0.2 m   (d) 0.2 nm   (e) none of these.

2. A typical nucleus has a diameter of roughly   (a) 1×10^{-14} mm   (b) 1×10^{-14} cm   (c) 1×10^{-14} m   (d) 1×10^{-14} nm   (e) none of these.

3. We learned from electrolysis that one atom of a univalent substance carries a charge   (a) equal to F   (b) equal to F/N_A (c) equal to $2F/N_A$ (d) equal to N_A/F (e) none of these.

4. We learned from electrolysis that one atom of a bivalent substance carries a charge   (a) equal to F   (b) equal to F/N_A (c) equal to $2F/N_A$ (d) equal to N_A/F (e) none of these.

5. The magnitude of the charge of the electron is   (a) equal to F   (b) equal to F/N_A (c) equal to $2F/N_A$ (d) equal to N_A/F (e) none of these.

6. When a monochromatic X-ray beam impinges on a crystal (λ and d are fixed), there may be several values of the   (a) angle θ at which diffraction will occur, and these correspond to different values of m (b) crystal spacing at which the reflected beam angle

will exceed the incident angle   (c) energy at which the reflected beams emerge at an angle in excess of 90°   (d) reflection angle for each incident angle, and these correspond to different values of f (e) none of these.

7. The largest Bragg angle through which a beam of X-rays can be bent is   (a) 0°   (b) 45°   (c) 90°   (d) 180°   (e) none of these.

8. The light emitted via the excitation of low-pressure hydrogen gas is   (a) a continuous spectrum   (b) made up of a discrete spectrum of bright lines   (c) a mix of bright spectral lines superimposed on a continuous bright background   (d) composed exclusively of X-rays   (e) none of these.

9. Naturally occurring radioactive atoms can spontaneously emit   (a) δ-, γ-, and ξ-rays   (b) α-, β-, and γ-rays   (c) N-rays   (d) X-rays   (e) none of these.

10. Neutrons are highly penetrating because they   (a) are very thin and can slide between atoms   (b) never strike an atom because they twist as they advance   (c) are neutral and do not lose energy via Coulomb interactions with the atoms   (d) always trav-

el at extremely high speeds in comparison to protons or electrons (e) none of these.

11. Gamma-ray emissions can (a) be distinguished from β-ray emissions using a magnetic field (b) not be distinguished from β-ray emissions (c) not be distinguished from α-ray emissions using a magnetic field (d) be distinguished from a neutron beam by bending the latter's path via a B-field (e) none of these.

The next six questions refer to Fig. MC12 which depicts Rutherford scattering. The angle between the incoming and out going paths is the scattering angle θ. The center-to-center distance b is called the impact parameter.

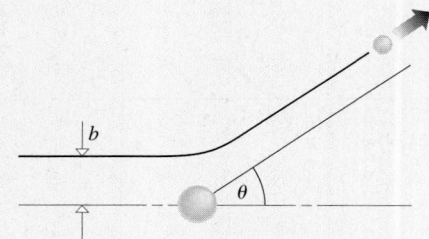

Figure MC12

12. The diagram pictures a high-speed (a) electron deflected by a neutron (b) proton deflected by a nucleus (c) nucleus deflected by a proton (d) α-particle deflected by a nucleus (e) none of these.

13. The scattered particles travel (a) elliptical paths (b) parabolic paths (c) cycloidal paths (d) hyperbolic paths (e) none of these.

14. Rutherford found that most incoming particles were scattered through an angle (a) $\theta = 0$ (b) $\theta = 45°$ (c) $\theta = 90°$ (d) $\theta = 180°$ (e) none of these.

15. When the impact parameter is zero we can expect (a) no scattering (b) $\theta = 45°$ (c) $\theta = 90°$ (d) $\theta = 180°$ (e) none of these.

16. Which of the following metal target foils would you think

would scatter more of the beam particles through large angles? (a) copper, $Z = 29$ (b) uranium, $Z = 92$ (c) gold, $Z = 79$ (d) aluminum, $Z = 13$ (e) none of these.

17. The scattering is due to (a) nuclear repulsion (b) the strong force (c) Coulomb repulsion (d) electromagnetic attraction (e) none of these.

18. An element emits spectral lines (a) that are exactly the same as all other elements (b) that are exactly the same as all other elements in its column of the Periodic Table (c) that are characteristic of that element (d) that are evenly spaced (e) none of these.

19. In reference to the Balmer series, the longest wavelength line (a) is associated with the smallest n, namely 1 (b) is associated with the smallest n, namely 3 (c) is associated with the largest n^2, namely 9 (d) is associated with the largest n, namely ∞ (e) none of these.

20. In reference to the Balmer series, the shortest wavelength line (a) is associated with the smallest n, namely 1 (b) is associated with the smallest n, namely 3 (c) is associated with the largest n^2, namely 9 (d) is associated with the largest n, namely ∞ (e) none of these.

21. The magnitude of the charge-to-mass ratio of the electron (a) is zero (b) is less than that for the proton (c) is greater than that for the proton (d) equals that of the neutron (e) none of these.

22. The proton and the electron (a) have the same size charge and mass (b) have the same mass but differ in charge by a factor of about 2000 (c) have the same size charge and differ in mass by a factor of about 2000 (d) differ in both charge and mass by a factor of about 2000 (e) none of these.

23. The neutron (a) has a mass slightly greater than that of the proton (b) has a charge slightly greater than that of the proton (c) has a mass slightly greater than that of the electron (d) has a mass slightly less than that of the proton (e) none of these.

For more Multiple Choice Questions with answers click on WARM-UPS in CHAPTER 27 on the CD. 💿

Suggestions on Problem Solving

1. Return to Fig. 27.8c and notice that the angle between the transmitted and diffracted beams is 2θ. It is this angle that is usually measured experimentally, and it is called the diffraction angle. Among the first things to determine in a given X-ray diffraction problem is which angle is being provided or asked for.

2. When dealing with the spectral series for hydrogen, the order is Lyman (1), Balmer (2), Paschen (3), Brackett (4), Pfund (5); these are the numbers that are squared in the first term of the denominators for $1/\lambda$ [that is, Eqs. (27.5)–(27.7)]. The first spectral line in any particular series is the next number up; for example, substitut-

ing $n = 3$ into Eq. (27.5) generates the first Balmer line. Remember to square those integers. Once again, we have one-over the quantity of interest (that is, $1/\lambda$), and it's common to forget to invert the result to get the final answer—be alert to this.

3. Remember that it's commonplace to specify the uniform electric field between parallel plates via $E = V/d$; that is, V and d are given, and it is assumed you therefore know E. Because d is small, it is often provided in centimeters; be sure to convert it into meters—failure to convert is a common error.

STUDY GUIDE   **1. Coordinated Problems:** The three problems within each magenta-colored grouping are solvable in similar ways. Note that the first of these always has a hint; moreover, its solution is provided in the back of the book. *Work out each of these sets; they'll strengthen technique and build confidence.* **2. Progressive Problems:** The problems introduced in blue unfold step-by-step carrying along the analysis in a more suggestive way than is customary. *Work out all of these; they'll guide you through the analytic process and help develop problem-solving skills.* **3. Worked-Out Solutions:** Studying worked-out solutions is an important part of learning how to solve problems. Accordingly, additional *solutions* to a number of model problems are given below. *Make sure you understand each of them before you go on to the next problem.* **4.** Also provided in the back of the book are the *Answers* to all odd-numbered problems, as well as worked-out *solutions* to those with boldface numbers. Problem numbers in italic indicate that a solution appears in the Student Solutions Manual.

SECTION 27.1: THE QUANTUM OF CHARGE

SECTION 27.2: CATHODE RAYS: PARTICLES OF CHARGE

1. [I] How many electrons would you get if you bought a gram of them? How does that compare with the total number of stars in the entire Universe ($\approx 10^{22}$)?

2. [I] Determine the total mass of all the electrons in 2.00 g of hydrogen (H_2) gas at STP.

3. [I] A current of 965 A passes through a molten solution of NaCl for 100.0 s. How much chlorine gas will be liberated? [*Hint: Chlorine is monovalent.*]

4. [I] How many atoms of a monovalent element will be deposited on an electrode if a current of 10.0 A flows through the appropriate electrolyte for 1000 s?

5. [I] Suppose that a substance is melted, appropriate electrodes are placed in the liquid, and a current of 100 A is made to flow across it. How long should the current be maintained if 60.2×10^{23} atoms of a monovalent element are to be deposited?

6. [I] For single-valence elements, verify that 1 faraday is equivalent to 9.65×10^7 C/kmole.

7. [I] What mass of metallic sodium will be deposited at the cathode if 50.0 A passes through a molten bath of NaCl for 5.00 minutes?

8. [I] How much charge must pass through a molten bath of NaCl in order to liberate 11.2 liters of chlorine gas (Cl_2) at STP?

9. [I] A current of 2.00 A is passed through molten $BaCl_2$ for 5.00 h. How much barium and chlorine will be liberated?

10. [I] A beam of electrons travels straight through a region occupied by an electric field of 1000 V/m perpendicular to a magnetic field of 20.0×10^{-2} T. If the electrons move perpendicular to both fields, how fast are they going?

11. [I] A tiny droplet of oil carries an extra charge of e and is held motionless between parallel plates 2.00 cm apart across which there is a potential difference of 60.0 kV. What is the mass of the droplet?

12. [I] In the Millikan oil-drop experiment, a droplet of mass 1.111 $\times 10^{-15}$ kg is held motionless by an electric field of 34.0 kV/m. How many extra electrons is it carrying?

13. [II] Imagine a block of pure metallic copper with a mass of 63.55 g and a density of 8.96 g/cm^3. Approximate the size of a copper atom.

14. [II] If 0.754 5 g of metallic silver are deposited onto an electrode using a current of 0.500 A for 22.5 minutes, what is the atomic mass of silver? The ions are singly charged. Refer to Discussion Question 2.

15. [II] Use Fig. P15 depicting Thomson's electron-beam apparatus to show that

$$\frac{e}{m_e} \approx \frac{E\theta}{B^2 L}$$

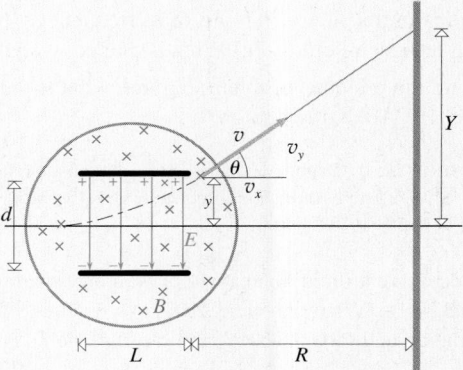

Figure P15

16. [II] If r is the radius of the path taken by an electron in the B-field region of Fig. P15, show that

$$\frac{e}{m_e} = \frac{E}{B^2 r}$$

and prove that this equation is equivalent to the result obtained in Problem 15.

17. [II] An electron in Thomson's apparatus moves under the influence of a B-field along a path with a radius of 15.00 cm. If an E-field of 20.0 kV/m makes the path straight and horizontal, find B.

18. [II] With the previous few problems in mind, knowing e/m for the electron, determine the strength of the B-field that will result in a deflection of 0.25 rad when $V = 180$ V, $L = 4.5$ cm, and the plates are separated by 1.66 cm.

19. [III] Use Fig. P15 depicting Thomson's electron-beam apparatus to show that

$$Y = \left(\frac{e}{m_e}\right) \frac{B^2 L}{2E} (L + 2R)$$

20. [III] An oil droplet carrying a net charge of Q and having a mass m falls in air at a steady vertical terminal speed between two vertical parallel plates separated by a distance d. When a potential difference V is applied across the plates, the droplet moves uniformly at an angle θ with the vertical. Show that

$$\tan \theta = \frac{VQ}{mgd}$$

21. [III] Rock salt (NaCl) is a cubic crystal with a molecular mass of 58.5 and a density of 2.16×10^3 kg/m³. Approximate the space between its atoms.

22. [III] An electron in a cathode-ray tube is emitted from a hot filament at a negligible speed, whereupon it accelerates across a potential difference of 1.00×10^6 V. Determine its final speed. [*Hint: The correct answer is about half the classical value.*]

SECTION 27.3: X-RAYS

23. [I] THIS PROBLEM EXAMINES X-RAY PHOTONS. Consider an X-ray photon having a frequency 5.20×10^{18} Hz. (a) Write the expression for the energy of the photon in terms of its frequency. (b) Determine the photon energy in joules. (c) What is that equivalent to in keV?

24. [I] With the previous problem in mind, what is the momentum of a 5.20×10^{18}-Hz X-ray photon?

> **SOLUTION:** The momentum and energy of a photon are related by Eq. (26.16); $E = pc$. From the previous problem $E = 3.4456 \times 10^{-15}$ J and so $p = E/c = 1.15 \times 10^{-23}$ kg·m/s.

25. [I] Successive atomic planes in a crystal are separated by a distance of 0.324 nm. What wavelength X-rays will be diffracted in the first order at an angle $\theta_1 = 16.0°$ from those planes?

26. [I] A beam of electrons in a 45-kV X-ray tube bombards the target, producing 750 W of thermal energy per second. Nonetheless, only 1.00% of the beam's energy ends up as X-rays. Compute the average rate at which electrons strike the target. [*Hint: Use the relationship between power, current, and voltage.*]

27. [I] A beam of X-rays having a wavelength of 0.090 nm impinges on an unknown crystal. The strongest reflection maximum ($m = 1$) is observed to occur at an angle $\theta = 25°$. What is the spacing between the atomic planes doing the scattering? [*Hint: X-rays diffract in accord with Bragg's equation.*]

28. [I] X-rays with a wavelength of 0.100 nm are incident on a set of atomic planes at an angle of 30.0°. Given that there is a strong first-order reflection at 30.0°, what is the spacing between the planes?

29. [I] A beam of 0.900 nm X-rays is strongly reflected in the first order from a crystal at an angle of 60.0° with respect to the incident axis of the beam. Accordingly, the reflection occurs at an angle of half that (Fig. 27.8c) with respect to the planes. What is the separation between the atomic planes?

30. [I] A crystal such as that in Fig. 27.8c can be used as a monochromator. It can select a single frequency (at a particular angle) out of a polychromatic incident beam. If the Bragg planes are known to be separated by 1.50 Å (that is, 0.150 nm), what wavelength will have a first-order peak at 30° from the beam axis?

31. [I] Determine the angle at which a narrow beam of monochromatic X-rays of wavelength 0.090 nm has a first-order reflection off a calcite crystal ($d = 0.303$ nm).

32. [I] A silver bromide crystal has atomic planes spaced by 0.288 nm. Given that an X-ray beam diffracts at an angle $2\theta = 35.0°$, producing a first-order maximum at a detector, what is the wavelength?

33. [I] X-rays reflect off a salt crystal whose atomic planes are 0.28 nm apart. If the first-order maximum is at 22° to the planes, what is the wavelength of the radiation?

34. [II] THIS PROBLEM EXPLORES THE PRODUCTION OF X-RAYS. A beam of electrons is accelerated across a high potential difference inside an X-ray tube. A continuous range of X-ray frequencies is produced as the electrons collide with the target atoms and emit radiation. The maximum frequency is generated when all of the electron's kinetic energy is transformed into an X-ray photon. Suppose the voltage difference is 38.0 kV. (a) How much energy does an electron pick up as it flies across a 38.0-kV potential difference? Give your answer in both keV and joules. (b) The beam of electrons is then shot at a metal target. Compute the maximum KE of the electrons hitting the target. (c) Compute the maximum energy of the photons. (d) Determine the maximum frequency and the minimum wavelength of the X-rays.

35. [II] THIS PROBLEM EXPLORES THE PRODUCTION OF X-RAYS. With the previous problem in mind, suppose an X-ray device generates a beam with a minimum wavelength of 0.0200 nm. We want to find the accelerating potential difference. (a) What is the maximum photon frequency emitted? (b) What is the maximum photon energy emitted? (c) What is the maximum KE of the electrons in the beam? (d) What is the potential difference across which the electrons accelerate?

36. [II] The continuous X-ray emission from a copper target impinges in a narrow beam on a calcite crystal with an atomic spacing of 0.303 nm. A detector picks up the first strong (first-order) maximum at 24.0° to the beam axis. What is the shortest wavelength present in the radiation?

37. [II] A narrow beam of X-rays of wavelength 0.200 nm impinges on a crystal as in Fig. 27.8c. The detector picks up the first-order maximum at an angle of 50.0° from the beam axis. What are the spacing and orientation of the atomic planes causing the reflection?

38. [II] X-rays of unknown wavelength are incident on a nickel single crystal having an interatomic separation of 0.215 nm, as shown in Fig. P38. The diffracted beam has a first-order maximum at an angle $\phi = 45°$. Determine the wavelength. [*Hint: First compute d. Note that the planes do not run diagonally across each little square of atoms.*]

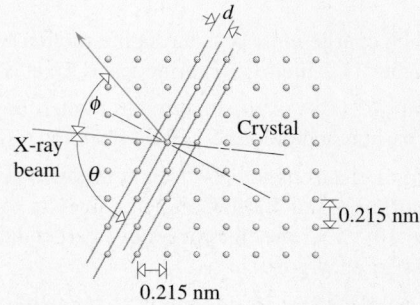

Figure P38

39. [II] A beam of 0.0200-nm X-rays impinges on a crystal having an array of Bragg planes separated by 1.22×10^{-10} m. What is the highest order reflected maximum?

40. [II] A polychromatic narrow beam of X-rays impinges on a crystal at 30° to a set of atomic planes. What wavelengths will be reflected at 30° if the atomic spacing is 0.0500 nm?

41. [III] An imperfect crystal has Bragg "planes" spaced on average by 0.100 nm, which, instead of being fixed everywhere, vary by

±0.001 nm from place to place. A polychromatic beam of X-rays diffracts from the crystal into a first-order peak at an angle of 30°. Determine the spread in the wavelengths of the emerging first-order beam at 30°.

SECTION 27.5: RUTHERFORD SCATTERING
SECTION 27.6: ATOMIC SPECTRA

42. [I] An alpha particle is fired at a sheet of gold with a kinetic energy of 8.00×10^{-13} J. Knowing that each gold atom contains 79 protons, about how close will the alphas come to the center of the nucleus?

43. [I] Compute (to four significant figures) the wavelength of the second line of the hydrogen Balmer series.

44. [I] Confirm Stoney's notion that the hydrogen spectral wavelengths formed simple ratios, such as $H_\alpha : H_\beta : = \frac{1}{20} : \frac{1}{27}$.

45. [I] Compute (to three significant figures) the frequency of the second line of the hydrogen Balmer series.

46. [I] What is the frequency (to four significant figures) of the third line in the Balmer series?

47. [I] Compute (to four significant figures) the wavelength that corresponds to the series limit of the Balmer series. [*Hint: n → ∞*]

48. [I] What is the shortest wavelength in the Lyman spectral series of hydrogen? Give your answer to four significant figures.

49. [I] What is the shortest wavelength in the Paschen spectral series of hydrogen? Give your answer to four significant figures.

50. [I] Compute (to four significant figures) the wavelength of the first line of the Paschen series.

51. [I] Determine (to four significant figures) the longest wavelength in the Balmer series for hydrogen.

52. [I] Compute (to four significant figures) the wavelength of the second line of the Paschen series.

53. [II] THIS PROBLEM IS CONCERNED WITH THE SIZE OF THE NUCLEUS. A uranium nucleus (like that in Fig. 30.4) is 14.8×10^{-15} m in diameter. We want to study the electrostatic characteristics of such a system. (a) How many protons are there in a typical uranium nucleus? (b) Write an expression for the potential outside a uniform sphere of charge. (c) Using that as a model, what is the electrical potential at a point just beyond the surface of a uranium nucleus? (d) What is the *electrical*-PE of an α-particle at a point near the face of the uranium nucleus? (e) How much is that in units of MeV? (f) Through what potential difference must such a particle, which was originally at rest, be accelerated if it's to have that much energy? Be careful how you divide energy in MeV by charge. You must use units of electron charge (e or q_e), not coulombs.

54. [II] THIS PROBLEM IS CONCERNED WITH THE SIZE OF THE NUCLEUS AND THE COULOMB INTERACTION IT PRODUCES. Imagine a monoener-

getic beam of alpha particles in which each one has a KE of 4.98 MeV. We want to shoot these at a thin foil of silver and know that in a head-on collision an α-particle will approach to a distance of 1.20×10^{-14} m from the center of a nucleus. (a) How many protons are there in the nucleus of a silver atom? (b) What energy in joules must the alpha possess? (c) How much is that in MeV? (d) If our beam is to do the job, how much additional energy will it need? (e) Through what potential difference should the beam be passed in order to boost up its energy appropriately? Be careful how you divide energy in MeV by charge. You must use units of electron charge (e or q_e), not coulombs.

55. [II] An alpha particle of mass 6.6×10^{-27} kg is fired at a speed of c/20 directly at an iron nucleus. About how close will it come to the center of the nucleus?

56. [II] What is the highest frequency of the Lyman series of hydrogen? Give your answer to four significant figures.

57. [II] If light is defined to correspond to the wavelength range from 390 nm to 780 nm, what is the shortest wavelength of the Balmer series that falls within that range?

58. [II] THIS PROBLEM DEALS WITH THE SPECTRAL LINES OF HYDROGEN. We want to examine the longest wavelength in the Paschen series. (a) Write an expression λ^{-1} corresponding to the Paschen series. (b) Show that as n gets larger, λ^{-1} gets larger, and λ gets smaller. (c) For what value of n will the frequency of the emitted UV have the largest possible value? (d) For what value of n will the wavelength of the emitted UV have the largest possible value? (e) What is the longest wavelength in the Paschen spectral series of hydrogen? Give your answer to four significant figures.

59. [II] THIS PROBLEM DEALS WITH THE SPECTRAL LINES OF HYDROGEN. Consider the electromagnetic radiation from hydrogen at 97.20 nm. (a) What kind of radiation is this (UV, IR, light, etc.)? (b) Which series would you guess it belongs to? (c) Compute $1/\lambda R$. (d) What's the significance of this being nearly equal to 1? (e) Write an expression λ^{-1} corresponding to the appropriate series. (f) Determine the value of n.

60. [II] What is the longest wavelength of the Lyman series of hydrogen? Give your answer to five significant figures.

61. [II] THIS PROBLEM DEALS WITH THE SPECTRAL LINES OF HYDROGEN. Suppose we have a device that can detect wavelengths of 390 nm or less and we need to determine which of the Balmer lines in the hydrogen spectrum can be detected. (a) Write an expression for λ^{-1} corresponding to the appropriate series. (b) Write an expression for $\lambda < 390$ nm. (c) Prove that

$$n > [(1/4) - 1/(390 \times 10^{-9} \text{ m})R]^{-1/2}$$

(d) What numerical value must n exceed?

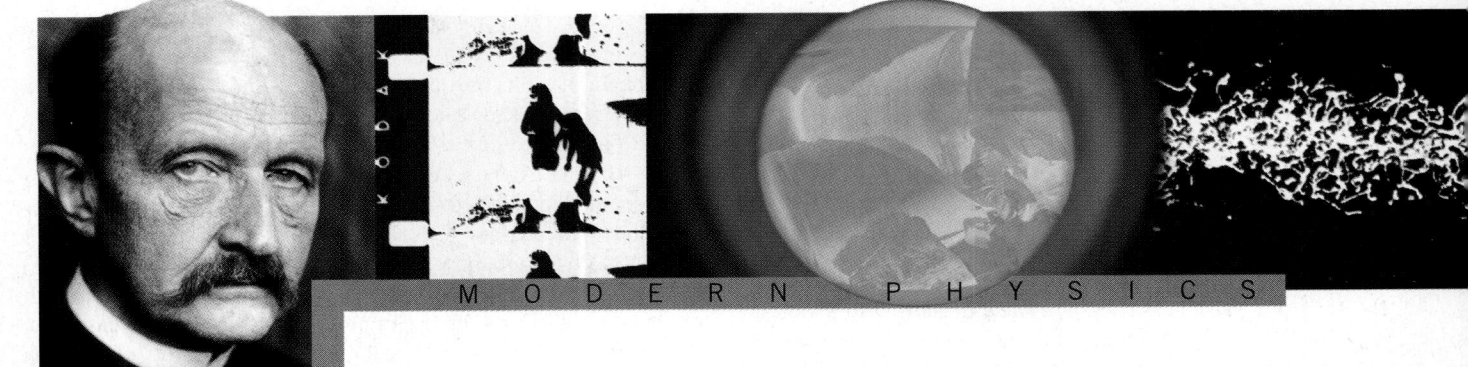

Chapter 28
The Evolution of Quantum Theory

Quantum Mechanics developed in two fairly distinct phases. First came the shocking basics: the appreciation of Planck's Constant, the quantum of action; the discovery of the quantization of energy and then, later, momentum; the conception of the photon; a crude but effective atomic model; and the wave-particle duality of matter. All these ideas form the body of knowledge known as the Old Quantum Theory. Building on that foundation, from 1925 to the present, physics moved into the contemporary, more theoretically abstract and mathematically driven stage of Quantum Mechanics, which will be discussed in Chapter 29.

Newton's Laws and the laws of Electromagnetic Theory give us a wonderful, reliable insight into the workings of the everyday world. But in the late 1800s, when we first began to really explore the atom, it became clear that these pillars of Classical Physics were inadequate to the task. The atomic domain is a much stranger place than was ever imagined.

This chapter deals with some of the discoveries and pivotal experiments that would usher in the new era of Quantum Mechanics (Chapter 29). Central among them was the revelation that the energy of a bound system, such as an atom, is quantized. And that light itself is quantized; all forms of radiant EM energy come in minute blasts called photons.

This chapter is essentially an introduction to early twentieth-century physics. It lays the groundwork for the great revolution in thought that would come in the 1920s.

The **emission coefficient** ε_λ is the emitted irradiance lying in a tiny wavelength range centered on λ.

The Old Quantum Theory

It shouldn't be surprising that if physics was to be turned upside down, it would be done while trying to figure out what *light* (i.e., radiant energy) was about. Quantum Mechanics had its earliest tentative beginnings in the theory of blackbody radiation, which itself began back in 1859. That year, Darwin published *The Origin of Species*, and it was the year that Gustav Robert Kirchhoff proffered an intellectual challenge that would lead to a revolution in physics.

28.1 Blackbody Radiation

Kirchhoff analyzed the way bodies in thermal equilibrium behave in the process of exchanging radiant energy. This *thermal radiation* (p. 473) is electromagnetic energy radiated by all objects, the source of which is their thermal energy. Suppose that the abilities of a body to emit and absorb electromagnetic energy are characterized by an **emission coefficient** ε_λ and an **absorption coefficient** α_λ. Epsilon is the energy per unit area per unit time emitted in a tiny wavelength range around λ (in units of $\mathrm{W/m^2/m}$); any energy-measuring device admits a range of wavelengths. Alpha is the fraction of the incident energy absorbed per unit area per unit time in that wavelength range; it's unitless. These coefficients depend

A red-hot furnace emits a spectrum of radiation that is independent of the furnace wall materials and is contingent only on the temperature.

on both the nature of the surface of the body (color, texture, etc.) and the wavelength—a body that emits or absorbs well at one wavelength may emit or absorb poorly at another.

Now, imagine an isolated chamber of some sort in thermal equilibrium at a fixed temperature T. Clearly, it would be filled with radiant energy at lots of different wavelengths—just think of a glowing electric furnace. Assume that there is some formula, or **distribution function I_λ**, that depends on T and tells us the intensity or amount of energy present at each wavelength. Put in numbers for T and λ, and the formula tells us the quantity of energy in the radiation within the cavity. Apparently, the *total* amount of energy at all wavelengths being absorbed by the walls versus the amount emitted by them must be the same, or else T will change. Kirchhoff argued that if the walls were made of different materials (which behave differently with T), that same balance would have to apply for *each* wavelength range individually. Thus, the energy absorbed at λ, namely, $\alpha_\lambda I_\lambda$, must equal the energy radiated, ε_λ, *and this is true for all materials no matter how different*. **Kirchhoff's Radiation Law** is thus

$$\frac{\varepsilon_\lambda}{\alpha_\lambda} = I_\lambda$$

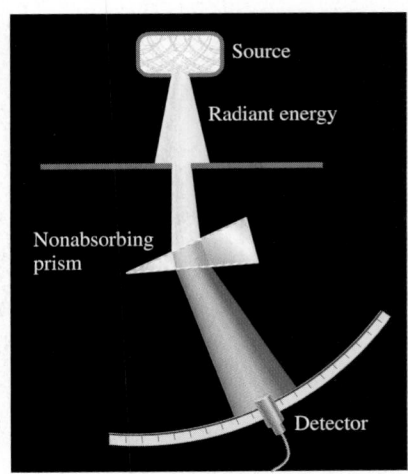

Figure 28.1 A schematic setup for measuring blackbody radiation.

wherein the distribution I_λ is a universal function that is the same for every type of cavity wall regardless of material, color, size, and shape and is only dependent on T and λ. That's extraordinary! Still, the famous British ceramist Thomas Wedgwood (1792) had long before noted that the objects in a fired kiln all turned glowing red together with the furnace walls regardless of their size, shape, or material constitution.

Although Kirchhoff did not provide the energy distribution function, he did point out that a perfectly absorbing body, one for which $\alpha_\lambda = 1$, will appear black and, in that special case, $I_\lambda = \varepsilon_\lambda$. The distribution function for a perfectly black object is the same as for an isolated chamber at the same temperature. This means that the radiant energy at equilibrium inside an isolated cavity is in every regard the same, "as if it came from a completely black body of the same temperature." ***The energy leaking from a small hole in the chamber should be identical to the radiation coming from a perfectly black object at the same temperature***; and that, as we will see, has important practical consequences.

Even though the scientific community accepted the challenge of determining I_λ, technical difficulties caused progress to be very slow. A simplified experimental setup is shown in Fig. 28.1. Radiant energy from a source passes through a slit and into a nonabsorbing prism. The wavelengths present are spread out into a continuous band that is sampled by a detector. Data must be extracted that are independent of the specifics of the detector. Thus, the best thing to plot is the radiant energy per unit time, which enters the detector per unit area (of the entrance window) per unit wavelength range (admitted by the detector). Figure 28.2 shows the kind of curves that were ultimately recorded (see also Fig. 13.7, p. 473), and each is a plot of I_λ at a specific temperature.

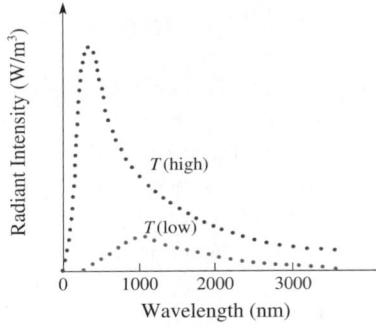

Figure 28.2 A plot of energy (J) per unit time (s), per unit area (m²), per unit wavelength interval (m), entering a detector successively centered at a number of different wavelengths. The source is a blackbody. Each curve represents I_λ at a given temperature.

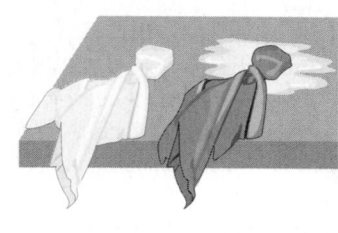

The Stefan-Boltzmann Law

In 1865, J. Tyndall published the result that the total energy emission of a heated platinum wire was 11.7 times as much operating at 1200°C (1473 K) as it was at 525°C (798 K). Amazingly, Josef Stefan (1879) noticed that the ratio of $(1473 \text{ K})^4$ to $(798 \text{ K})^4$ was 11.6, nearly 11.7, and he inferred that the rate at which energy is radiated is proportional to T^4. In this he was quite right (and quite lucky) because Tyndall's results were actually far from those of a blackbody. In any event, the conclusion was subsequently proven via a theoretical argument carried out by L. Boltzmann (1884). This was a traditional analysis of the radiation pressure exerted on a piston in a cylinder using the Laws of Thermodynamics and Kirchhoff's Law. The discussion progressed in much the same way one would treat a gas in a cylinder, but instead of atoms, the active agency was electromagnetic waves. The resulting **Stefan-Boltzmann Law** for blackbodies is

[blackbody radiation]
$$P = \sigma A T^4 \tag{28.1}$$

where P is the total power radiated at all wavelengths, A is the area of the radiating surface, T is the absolute temperature in kelvins, and σ is a universal constant now given as

$$\sigma = 5.67 \times 10^{-8} \text{ W/m}^2 \cdot \text{K}^4$$

The total area under any one of the blackbody-radiation curves of Fig. 28.2 for a specific T is the power per unit area, and from Eq. (28.1) that's just $P/A = \sigma T^4$.

No real object is a perfect blackbody; carbon black has an absorptivity of nearly one, but only at certain frequencies (obviously including the visible). Its absorptivity is much lower than that in the far infrared. Still, most objects resemble a blackbody (at least at certain temperatures and wavelengths)—you, for instance, are nearly a blackbody for infrared. Thus, it's useful to write a similar expression for ordinary objects. To that end, we introduce a factor called the **total emissivity** (ε), which relates the actual radiated power ($\varepsilon < 1$) to that of a blackbody for which $\varepsilon = 1$, at the same temperature,

[real object radiation]
$$P = \varepsilon \sigma A T^4 \tag{28.2}$$

Table 28.1 provides a few values of ε (at room temperature), where $0 < \varepsilon < 1$. Note that emissivity is unitless.

Suppose a body with a **total absorptivity** of α is placed in an enclosure such as a cavity or a room having an emissivity ε_e and a temperature T_e. The body will radiate at a rate $\varepsilon \sigma A T^4$ and absorb energy inside the enclosure at a rate $\alpha(\varepsilon_e \sigma A T_e^4)$. But at any equilibrium temperature between body and enclosure (i.e., $T = T_e$), these rates must be equal; hence, $\alpha \varepsilon_e = \varepsilon$, and that must be true for all temperatures. We can write an expression for the net power radiated (when $T > T_e$) or absorbed (when $T < T_e$) by the body:

[power radiated or absorbed]
$$P = \varepsilon \sigma A (T^4 - T_e^4) \tag{28.3}$$

Table 28.1

Representative Values of Total Emissivity*

Material	ε
Aluminum foil	0.02
Copper, polished	0.03
Copper, oxidized	0.5
Carbon	0.8
White paint, flat	0.87
Red brick	0.9
Concrete	0.94
Black paint, flat	0.94
Soot	0.95

*$T = 300$ K, room temperature.

Example 28.1 **[I]** A cube of rough steel 10 cm on a side is heated in a furnace to a temperature of 400 °C. Given that its total emissivity is 0.97, determine the rate at which it radiates energy from each face.

Solution The word "emissivity" and the fact that we have the temperature suggests the Stefan-Boltzmann Law. (1) TRANSLATION—Determine the rate at which an object of known surface area, temperature, and emissivity radiates electromagnetic energy. (2) GIVEN: $L = 10$ cm, $T = 400$ °C, and $\varepsilon = 0.97$. FIND: P. (3) PROBLEM TYPE—Blackbody radiation/power/real object radiation. (4) PROCEDURE—Because this is not an ideal blackbody, we use the modified form of the Stefan-Boltzmann Law, $P = \varepsilon\sigma AT^4$. (5) CALCULATION—With $T =$ 673 K and $A = 0.01$ m^2

$$P = \varepsilon\sigma AT^4$$

$$P = (0.97)(5.67 \times 10^{-8} \text{ W/m}^2\cdot\text{K}^4)(0.01 \text{ m}^2)(673 \text{ K})^4$$

$$\boxed{P = 1.1 \times 10^2 \text{ W}}$$

Quick Check: $P \approx (1)(6 \times 10^{-8} \text{ W/m}^2\cdot\text{K}^4)(10^{-2} \text{ m}^2)(7 \times 10^2)^4 \approx 1.4 \times 10^2$ W.

[For more worked problems click on **WALK-THROUGH EXAMPLES** *in* **CHAPTER 28** *on the CD.]*

In the cold north you'd expect people to have "white" skin and "white" hair ($\varepsilon < 1$) to minimize the emission of radiant energy, and just the opposite in the tropics.

All bodies not at zero kelvin radiate, and the fact that T is raised to the fourth power makes the radiation sensitive to temperature increases. If a body at 0 °C (273 K) is brought up to 100 °C (373 K), it radiates about 3.5 times the previous power. Raising the temperature raises the net power radiated; that's why it gets harder and harder to increase the temperature of an object. (Just try heating a steel spoon to 1300 °C.) Raising the temperature also shifts the distribution of energy among the various wavelengths present. When the filament of a light bulb "blows," the resistance, current, and temperature momentarily rise and the filament goes from its normal operating reddish-white color to a bright flash of blue-white.

Remember that whenever you see a temperature (T) in this section on Blackbody Radiation it must be given in kelvins.

The Wien Displacement Law

In 1893, Wilhelm "Willy" Wien derived what has come to be known as the **Displacement Law**. Each blackbody curve reaches a peak height at a value of wavelength (λ_p), which is particular to it and therefore to the absolute temperature T. At that wavelength, the blackbody radiates the most energy. Wien correctly arrived at the fact that

[peak-energy wavelength] $$\lambda_p T = \text{constant} \tag{28.4}$$

where the constant was found experimentally to be 0.002 898 m·K. The peak wavelength is inversely proportional to the temperature. *Raise the temperature, and the bulk of the radiation moves to shorter wavelengths and higher frequencies* (see the dashed curve in Fig. 28.3). As a glowing coal or a blazing star gets hotter, it goes from IR warm, to red-hot,

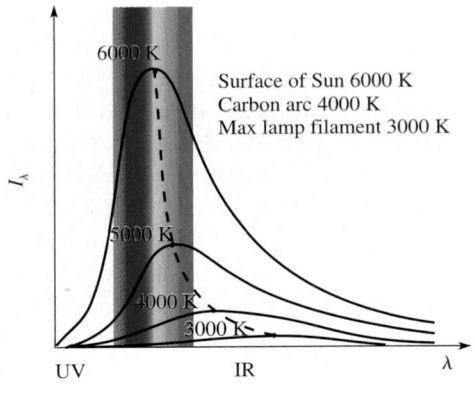

Figure 28.3 The amount of radiant energy emitted by a hot object at various wavelengths. Each curve peaks at a point where $\lambda_p T = $ constant, which is the Wien Displacement Law.

Surface of Sun 6000 K
Carbon arc 4000 K
Max lamp filament 3000 K

to blue-white. A person or a piece of wood, both only approximating blackbodies, radiates mostly in the infrared and would only begin to glow faintly in the visible at around 600 °C or 700 °C, long after either had decomposed. The bright cherry red of a chunk of hot iron sets in at around 1300 °C.

Example 28.2 **[I]** On the average, your skin temperature is about 33 °C. Assuming you radiate as does a blackbody at that temperature, at what wavelength do you emit the most energy per unit wavelength?

Solution The wavelength at peak energy is given by Wien's Displacement Law [Eq. (28.4)]. The Wien Displacement Law [Eq. (28.4)] is all about the peak wavelength radiated by a blackbody. (1) TRANSLATION—Determine the wavelength at which a blackbody of known temperature radiates a maximum amount of energy. (2) GIVEN: $T = 33$ °C. FIND: λ_{max}. (3) PROBLEM TYPE—Blackbody radiation/peak-energy wave-

length. (4) PROCEDURE—Wien's Displacement Law gives the wavelength at peak-energy for a blackbody. (5) CALCULATION—Using $\lambda_p T = $ constant,

$$\lambda_p = \frac{0.002\ 898\ \text{m·K}}{306\ \text{K}} = \boxed{9.5\ \mu\text{m}}$$

or

$$9.5 \times 10^3\ \text{nm}.$$

Quick Check: This result is well into the infrared and therefore reasonable. It's also independent of skin color, which is unrealistic; some of us radiate more than others.

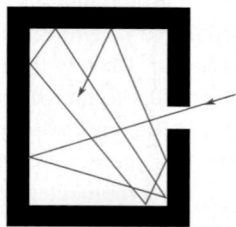

Figure 28.4 Energy entering a small hole in a chamber rattles around until it's absorbed. In reverse, the aperture in a heated enclosure appears as a blackbody source.

In 1899, researchers tremendously advanced the state of experimentation using, as a source of blackbody radiation, a small hole in a heated cavity (Fig. 28.4). Energy entering such an aperture reflects around inside until it's absorbed (the pupil of the eye appears black for precisely this reason). A near-perfect absorber is a near-perfect emitter, and the region of a small hole in an oven is a wonderful source of *blackbody radiation*.

Planck and His Energy Elements

Max Karl Ernst Ludwig Planck at forty-two was the reluctant father of Quantum Theory. Like so many other theoreticians at the turn of the twentieth century, he, too, was working on blackbody radiation. Planck would succeed not only in producing Kirchhoff's distribution function, but he would turn physics upside-down in the process. We cannot follow the details of his derivation—they are far too complicated and, besides, the original derivation is wrong (Einstein corrected it years later). Still, it had such a powerful impact that it's worth looking at some of the features that are right.

Planck knew that if an arbitrary distribution of energetic molecules were injected into a constant-temperature chamber, it would ultimately rearrange itself into the Maxwell-Boltzmann distribution of speeds as it inevitably reached equilibrium. Presumably, if an arbitrary distribution of radiant energy is injected into a constant-temperature cavity, it, too, will ultimately rearrange itself into the Kirchhoff distribution of energies as it inevitably reaches equilibrium.

In October of 1900, Planck produced a distribution formula that was based on the latest experimental results. This mathematical contrivance, concocted "by happy guesswork," fit all the data available. It contained two fundamental constants, one of which (*h*) would come to be known as **Planck's Constant**. That much by itself was quite a success, even if it didn't explain anything. Although Planck had no idea of it at the time, he was about to take a step that would inadvertently revolutionize our perception of the physical Universe.

Max Karl Ernst Ludwig Planck (1858–1947). In 1889 he became Professor of Physics at the University of Berlin. There, he soon took up the problem of blackbody radiation first raised by his old teacher Kirchhoff, who had recently died (1887) and whose former position Planck now held.

Naturally enough, Planck set out to construct a theoretical scheme that would logically lead to the equation he had already devised. He assumed that the radiation in a chamber interacted with simple microscopic oscillators of some unspecified type. These vibrated on the surfaces of the cavity walls, absorbing and re-emitting radiant energy independent of the material. (In fact, the atoms of the walls do exactly that. Because of their tightly packed configuration in the solid walls, the atoms interact with a huge number of their neighbors. That completely blurs their usual characteristic sharp resonance vibrations, allowing them to oscillate over a broad range of frequencies and emit a continuous spectrum.) Try as he

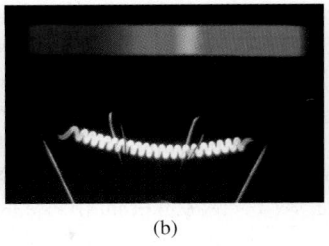

(a)   (b)   (c)

A hot filament and the spectra it emits. As the temperature rises from (a) to (b) to (c), the corresponding emission curves shift, as shown in Fig. 28.3. The peaks of the curves move toward the yellow, and the blue end of the spectrum increases in intensity as well. The result is that the filament shifts from cherry red to white-hot.

might, Planck was unsuccessful. At that time, he was a devotee of E. Mach, who had little regard for the reality of atoms, and yet the obstinate insolubility of the problem ultimately led Planck to "an act of desperation." He hesitantly turned to Boltzmann's "distasteful" statistical method, which had been designed to deal with the clouds of atoms that constitute a gas.

Boltzmann, the great proponent of the atom, and Planck were intellectual adversaries for a while. And now Planck was forced to use his rival's probabilistic formulation of entropy (p. 517), which—ironically—he would actually misapply. If Boltzmann's scheme for counting atoms was to be applied to something continuous, such as energy, some adjustments would have to be made in the procedure. Thus, according to Planck, the total energy of the oscillators had to be thought of, at least temporarily, as apportioned into "energy elements" so that they could be counted. These energy elements were given a value proportional to the frequency f of the resonators. Remember that he already had the formula he was after, and in it there appeared the term hf. Planck's constant

$$6.626\,075\,5 \times 10^{-34}\ \text{J} \cdot \text{s} \qquad \text{or} \qquad 4.135\,669\,2 \times 10^{-15}\ \text{eV} \cdot \text{s}$$

is a very small number, and so hf, which has the units of energy, is itself a very small quantity. Accordingly, he set the value of the energy elements equal to it:

$$\text{E} = hf \tag{28.5}$$

This was a statistical analysis, and counting was central. Still, when the method was applied as Boltzmann intended, it naturally smoothed out energy, making it continuous as usual. Again, we needn't worry about the details; the amazing thing was that Planck had stumbled on a hidden mystery of nature: **energy is quantized**—it actually comes in tiny bursts as given by Eq. (28.5), but he certainly didn't realize it then.

The equation that finally resulted is presented here only so that you can see the answer to Kirchhoff's challenge, namely,

$$I_\lambda = \frac{2\pi hc^2}{\lambda^5}\left[\frac{1}{e^{\frac{hc}{\lambda k_\text{B} T}} - 1}\right]$$

This is **Planck's Radiation Law**, and it fit blackbody data splendidly. Notice how the expression contains the speed of light, Boltzmann's Constant (p. 439), and Planck's Constant (h)—see Fig. 28.5. It bridges Electromagnetic Theory to the domain of the atom. In 1901, Planck used blackbody data to arrive at a numerical value of k_B. With that and the universal gas constant R (p. 438), he determined N_A. And using Avogadro's number, the faraday, and Eq. (27.1), he calculated the charge on the electron! Remarkably, his answer was only 2% too small (which was much better than the result J. J. Thomson had gotten), and it was arrived at eight years before Millikan's measurements.

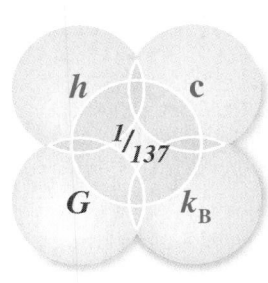

Figure 28.5 The Planck Radiation Law relates c, k_B, and h.

Although Eq. (28.5) represents a great departure from previous ideas, Planck did not mean to break with classical theory. It would have been unthinkable for him to even suggest that radiant energy was anything but continuous. "That energy is forced, at the outset, to remain together in certain quanta ...," Planck later remarked, "was purely a formal assumption and I really did not give it much thought." It was only around 1905, at the hands of a much bolder thinker, Albert Einstein, that we learned that the atomic oscillators were real and that their energies were quantized. Each oscillator could only exist with an energy that was a whole number (n) multiple of hf (a little like the *gravitational*-PE of someone walking up a flight of stairs). Moreover, **radiant energy itself is quantized**, existing in localized blasts of an amount $E = hf$.

EINSTEIN & THE PHOTOELECTRIC EFFECT

The concept of the energy quantum went unnoticed for almost five years. It languished until Einstein completely recast blackbody theory. He had published three brilliant papers between 1902 and 1904 that laid out the principles of the discipline known as *Statistical Mechanics*. Building on this work, Einstein in 1905 published a paper in which he showed that radiant energy behaved as if it were a collection of particles, light-quanta, each of energy hf. That same paper is most remembered for its analysis of the **Photoelectric Effect**.

28.2 Quantization of Energy: The Photoelectric Effect

The Photoelectric Effect was discovered by Heinrich Hertz, the man who had experimentally established the existence of electromagnetic waves. Ironically, the Photoelectric Effect, which he considered a *minor* observation, would ultimately lead to the overthrow of the classical understanding of electromagnetic waves. The Photoelectric Effect corresponds to the fact that *radiant energy (in the form of X-rays, ultraviolet, or light) impinging on various metals ejects electrons from their surfaces* (Fig. 28.6).

Classical theory suggests that the incident "light" arrives as an electromagnetic wave. If we use a uniform beam, its energy is presumably evenly spread across the entire wavefront (as it is with a water wave). The brighter the light, the greater its intensity, the larger the amplitudes of the *E*- and *B*-fields everywhere on a wavefront, and the more energy the wave delivers per second. These fields exert forces on the electrons in the metal and can even liberate some of them from the surface.

When the collector plate is made positive with respect to the emitter plate, *photoelectrons* easily traverse the tube, and that constitutes a *photocurrent* that is measurable with a microammeter. Classical theory predicts that increasing the brightness of the light beam provides more energy, thus liberating more electrons and increasing the photocurrent (Fig. 28.7). This proportionality between photocurrent and irradiance (brightness) was shown to hold across a broad range from a low, where the eye cannot even see the light, to a high, where it cannot bear it.

If the collector's positive potential is gradually decreased, we can expect the photocurrent to slowly decrease as well. Figure 28.8 illustrates the effect for monochromatic light. Some electrons make it to the collector even at zero potential difference; and the brighter the light, the more current there will be. Evidently, if the collector is made negative, it will repel photoelectrons, decreasing the current even further. At a negative potential—specific to each metal—known as the **stopping potential** (V_s), all the electrons hurled out of the metal are turned back, and the photocurrent goes to zero. The existence of a stopping potential characteristic of the metal, but *independent of intensity*, was troubling. It would seem

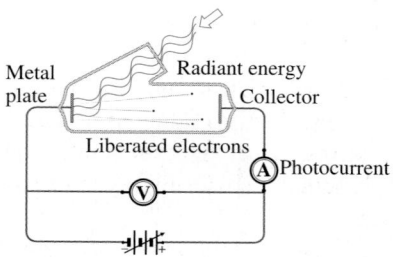

Figure 28.6 The Photoelectric Effect—radiant energy liberating electrons from a metal.

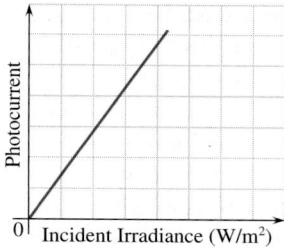

Figure 28.7 Photocurrent is directly proportional to the incident irradiance (that is, the energy per unit area per unit time).

Figure 28.8 The variation of photocurrent with collecting potential and light intensity. Different metals have different stopping potentials.

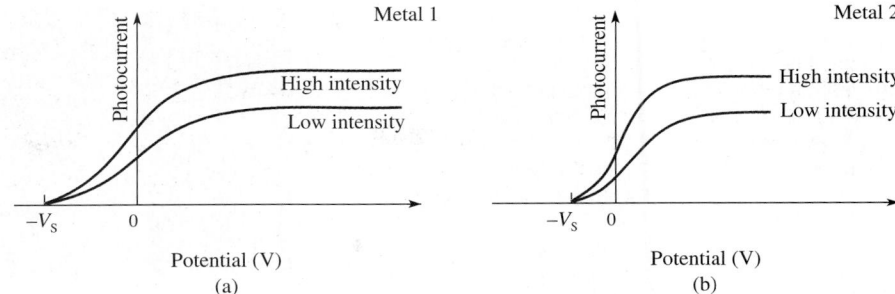

that V_S ought to depend on how bright, and therefore energetic, the light is. An electron has to traverse the region where the E-field is pushing against it and thus overcome a potential energy barrier in order to arrive at the collector. Reaching the maximum negative potential finally turns back even those electrons with a maximum KE, where

$$\mathrm{KE}_{\max} = eV_S \qquad (28.6)$$

The surprise is that $\mathrm{KE}_{\max}$ is independent of the light intensity. With a low light level and very little energy coming in, the electron must somehow store up what it needs to pop out of the metal, and yet at high light levels, it must refrain from absorbing any more energy than $\mathrm{KE}_{\max}$—very strange.

However disconcerting these experimental results proved, when attention was turned to the frequency of the radiation, things became utterly bizarre. For any given metal, there is a specific threshold frequency below which photoemission does not occur, no matter how intense the incident radiant energy. That simply should not be; provided enough energy (via a bright beam), electrons should come pouring out. Moreover, the astounding facts were established (1903) that *the photoelectron's maximum kinetic energy was independent of beam intensity and yet proportional to the frequency of the illumination.* Kinetic energy should only depend on the incident energy; it shouldn't have anything to do with frequency. How could frequency affect energy?

The time delay between the arrival of radiant energy and the emission of photoelectrons is now known to be less than 3×10^{-9} s; switching on even the dimmest source can "instantly" release electrons. That behavior is simply impossible within orthodox wave theory. Each incoming wave should spread uniformly over the surface of the illuminated metal, imparting only a minute amount of energy to each of the millions upon millions of atoms in its path. We can calculate, at least crudely, how long it would take an electron to absorb enough energy to attain the observed speeds (Problem 35). There is some ambiguity depending on how much absorbing "area" the electron presents to the beam, but all such calculations conclude that it will take from seconds to months of irradiation before an electron can be ejected—and yet there they were, flying off almost instantaneously. Something fundamental was obviously wrong with the accepted understanding of the way electromagnetic energy interacted with matter.

Remarkably, J. J. Thomson (1903) suggested that electromagnetic waves might well be radically different from other waves; perhaps a sort of concentration of radiant energy actually existed. After all, if one shone X-rays on a gas, only certain of the atoms, here and there, would be ionized, as if the beam had "hot spots" rather than being uniform. The notion was prophetic.

Einstein and the Photon

Planck had earlier introduced the concept of energy elements, and regardless of the inconsistencies, he had supposed that radiant energy was continuous and that it behaved like a classical wave. Now Einstein, in 1905, stepped into the muddle. Audaciously breaking with the long-standing traditional view, he proposed that **light itself is granular,**

A few frames of a motion picture showing the optical sound track at the right. Light passing through the track illuminates a photoelectric detector that converts the signal into a varying voltage.

A beam of X-rays enters a cloud chamber on the left. The tracks are made by electrons emitted via either the Photoelectric Effect (these tend to leave long tracks at large angles to the beam) or the Compton Effect (short tracks more in the forward direction). Although classically, the X-ray beam has its energy uniformly distributed along transverse wavefronts, the scattering seems discrete and random just as if the energy were concentrated in "hot spots."

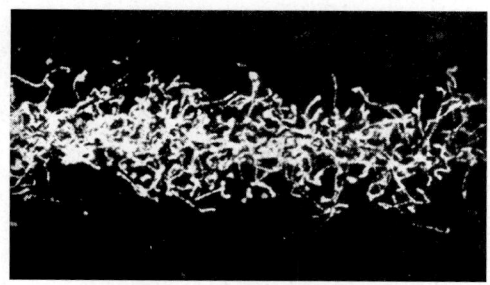

that it is actually composed of discrete bursts, particles of energy, or **photons**, as they came to be called.* The interaction between electromagnetic radiation and substantial matter occurs in jolts. Radiant energy is absorbed and emitted discontinuously because it itself is discontinuous!

Einstein's first paper on light-quanta, "On a Heuristic Point of View Concerning the Generation and Conversion of Light," was published in 1905 (before Relativity). *Heuristic* means something that serves as a guide in the solution of a problem but is otherwise itself unproved. In that spirit, he postulated that *every electromagnetic wave of frequency f is actually a stream of energy quanta, each with an energy*

$$E = hf = \frac{hc}{\lambda} \tag{28.7}$$

Example 28.3 **[II]** In the atomic domain, energy is often measured in electron volts. Accordingly, arrive at an expression for the energy of a light-quantum in eV when the wavelength is in nanometers. What is the energy of a quantum of 500-nm light?

Solution This problem is about the energy of a photon, and that should bring to mind $E = hf$. (1) TRANSLATION— Determine an expression for the energy of a photon in eV expressed in terms of wavelength in nanometers. (2) GIVEN: Generally λ in nm and particularly $\lambda = 500$ nm. FIND: E in eV. (3) PROBLEM TYPE—Quantization of energy/photons. (4) PROCEDURE—Use the defining relationship $E = hc/\lambda$.

(5) CALCULATION—With $h = 4.135\ 67 \times 10^{-15}$ eV·s and $c = 2.997\ 9 \times 10^{17}$ nm/s we get

$$E = \frac{hc}{\lambda} = \frac{1239.8 \text{ eV·nm}}{\lambda}$$

For $\lambda = 500$ nm, $\boxed{E = 2.48 \text{ eV}}$.

Quick Check: The numerator here can be remembered as 1234.5, which is accurate to about 0.4%. The energy of a 500-nm quantum is in keeping with the energies associated with valence-electron processes, a few eV.

Nobody knows what a photon "looks like," but it is both localized and wavy (it has a frequency). The higher the frequency, the greater the energy of the individual photons. The irradiance of a monochromatic beam, the energy per unit area per unit time, is determined by the number of photons in the stream. The brighter the beam, the more photons.

A photon colliding with an electron in a metal can vanish, imparting essentially all of its energy to the electron. *Here the electron cannot be totally free*: because of the demands of momentum conservation, momentum must be transferred to the metal atoms, which nonetheless only pick up a negligible amount of energy. Even in the faintest beam, a single adequately energetic photon can kick loose an electron, and this can happen as soon as the illumination is turned on—there is no time delay. Raising the irradiance (intensity) of the beam increases the number of photons and therefore proportionately increases the current (Fig. 28.7). It takes energy to bring an electron up to the surface of the metal and subsequently liberate it. If electrons were not bound to the metal, they would be escaping all the

*The word *photon* was coined by G. N. Lewis in 1926.

Table 28.2
Representative Work
Function Values

Metal	Work function (ϕ in eV)
Na	2.28
Co	3.90
Al	4.08
Pb	4.14
Zn	4.31
Fe	4.50
Cu	4.70
Ag	4.73
Pt	6.35

time, leaving it positively charged, which doesn't happen. For an electron already up near the surface, this liberation energy is a minimum called the **work function** ϕ (Table 28.2). A photon's energy goes into freeing the electron, and whatever is left appears as KE. When the electron is at the surface, the liberating energy is a minimum, and the electron takes on a maximum KE given by

$$hf = KE_{max} + \phi \qquad (28.8)$$

This wonderfully simple expression is known as **Einstein's Photoelectric Equation**, and it explains every aspect of the effect. Since $KE_{max} = hf - \phi$, increasing the intensity (i.e., irradiance) of the light leaves the maximum kinetic energy unchanged. Only by changing f is the KE_{max}, or equivalently, the stopping potential, changed for a given metal. The threshold frequency f_0 corresponds to the initiation of emission where $KE_{max} = 0$. Hence, $hf_0 = \phi$, and, below a frequency of $f_0 = \phi/h$, the photocurrent will be zero.

It follows from Eqs. (28.8) and (28.6) that

$$eV_S = hf - \phi$$

which has the familiar form of the equation of a straight line ($y = mx + b$). The theory predicted that for any and all metals, a plot of stopping-potential-multiplied-by-electron-charge (y) against frequency (x) will be a straight line of slope (m) equal to Planck's Constant with a y-intercept (b) equal to the negative of the work function. No such relationship was known prior to Einstein's paper of 1905 and, amazingly, exactly that relationship was subsequently observed in every respect! Not until 1914–1915 would Robert Millikan finally and conclusively determine that the Photoelectric Equation was in complete agreement with experiment. He had spent 10 years of meticulous labor intent on showing that Einstein was totally wrong, and in the end he had established "the exact validity" of the theory. And yet he was unwilling to accept the reality of the photon even as he accepted the Nobel Prize for his work. Figure 28.9 shows the kind of results Millikan got. Each metal produces a straight line that intersects the horizontal axis (where the voltage is zero) at its threshold frequency. And each intersects the vertical axis (where $f = 0$) at $-\phi$. Moreover, each

UNAPPRECIATED QUANTA

The idea of light-quanta was not received well at all, and it had very few early advocates. Paul Ehrenfest (who seems to have been independently thinking along similar lines in 1905), Max von Laue, and Johannes Stark (who in 1909 anticipated the Compton Effect, p. 1007) were the leading exceptions. Despite its clearly demonstrated theoretical power, it was so disconcerting to people educated in classical wave theory that it was especially slow to be accepted. When Planck recommended Einstein for membership in the Prussian Academy in 1913, he felt that, notwithstanding Einstein's demonstrated genius, he still had to apologize for him because "he may sometimes have missed the target in his speculations, as, for example, in his hypothesis of light-quanta."

line has a slope equal to Planck's Constant!

The preceding analysis is increasingly being called the *Single Photon Photoelectric Effect* because, nowadays, lasers can put out such a torrent of nearly monochromatic photons that it's possible to blast an electron with two or more light-quanta before it leaves the metal. The result is the *Multiple Photon Photoelectric Effect* (see Discussion Question 5).

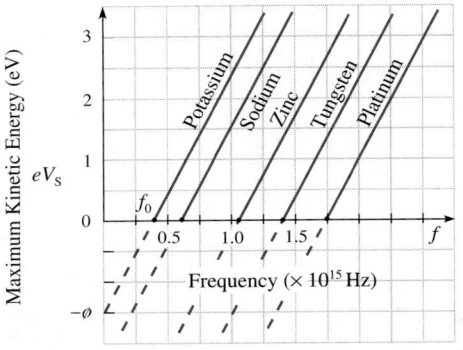

Figure 28.9 Each line follows the equation $eV_S = hf - \phi$. Thus, each intercepts the f-axis at f_0 and the energy axis at $-\phi$.

Most of the participants in the first Solvay Conference of 1911. Among the standees are Einstein (second from right), Planck (second from the left), and Rutherford (fourth from right). The young man next to him in the casual suit is Jeans. The lone woman is, of course, Madame Curie. J. J. Thomson missed the picture altogether.

Light propagates from one place to another as if it were a wave. Several centuries of work had established *that* with convincing credibility, and yet now it became equally as clear that ***light interacts with matter in the processes of absorption and emission as if it were a stream of particles***. This, then, is the so-called **wave-particle duality**, the schizophrenia of light (and, as we will see, of matter in general). Radiant energy appears and disappears in minute localized blasts and is seemingly transported via spreading waves.

Despite the general indifference, Einstein continued to use his heuristic principle to make a number of splendid predictions, one of which (1911) concerned the generation of X-ray quanta. It would be another 10 years before the strange and wonderful notion that light was no longer a simple TEM wave became so potent that it had to be accepted, even if not quite "understood." Einstein received the 1921 Nobel Prize "for his services to Theoretical Physics, and especially for his discovery of the law of the photoelectric effect."

28.3 Bremsstrahlung

X-rays are generated when high-speed electrons impacting on a dense target rapidly decelerate and thus radiate. The resulting broad continuous X-ray spectrum, illustrated in Fig. 28.10, is known as Bremsstrahlung (p. 973). The spikes in the curve arise from the atomic structure of the specific target, and we'll deal with them later in this chapter. The process of X-ray production is something of an inverted Photoelectric Effect; in come electrons, out goes radiant energy. And both can only occur in the presence of heavy atoms that can take up some momentum. Using the notion of light-quanta, Einstein predicted there would be a high-frequency (low-wavelength) limit on the radiation.

Imagine an incident electron arriving with an energy eV after being accelerated across a potential of V. As it slows down within the target, it radiates one or more photons. The maximum photon frequency (f_{max}) occurs when all the electron's kinetic energy is radiated as a single photon, whereupon

$$\mathrm{E} = hf_{max} = eV$$

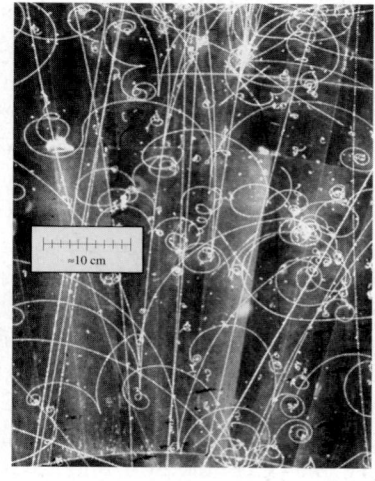

Particles like electrons and positrons have low mass and therefore little momentum. As a result their paths are easily bent. They lose some energy by ionizing atoms in the liquid medium, but they lose much more via bremsstrahlung. That's why the paths spiral. The amount of bremsstrahlung depends on the acceleration, which arises from the Coulomb forces between the particles and the nuclei of the medium. Low-mass particles have high accelerations and radiate copiously.

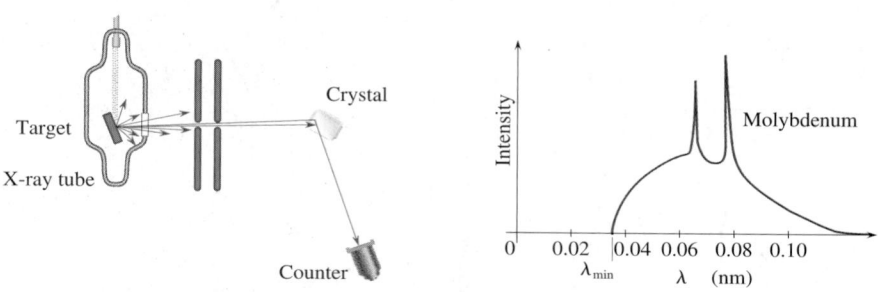

Figure 28.10 X-ray spectrum for a molybdenum target. The broad peak in intensity is due to Bremsstrahlung.

There will be a cut-off frequency of

$$f_{max} = \frac{eV}{h}$$

or, since $c = f\lambda$, a minimum wavelength of

$$\lambda_{min} = \frac{ch}{eV} \tag{28.9}$$

In other words, the electron cannot emit more energy than it has, and that maximum amount corresponds to the maximum frequency (minimum wavelength) end of the Bremsstrahlung curve.

Example 28.4 **[II]** Determine the smallest wavelength X-rays that can be emitted by an electron as it crashes into the metal mask on the front face of a color TV tube operating with an accelerating voltage of 20.0 kilovolts.

Solution The electron's energy is entirely converted into a photon. (1) TRANSLATION—Knowing the energy of an electron, determine the shortest wavelength (highest f) of X-rays that can be generated. (2) GIVEN: $V = 20.0$ kV. FIND: λ_{min}. (3) PROBLEM TYPE—Quantization of energy/photons/X-rays. (4) PROCEDURE—We want X-rays with the highest E and lowest λ. Use the defining relationship $\lambda_{min} = ch/eV$. (5) CALCULATION—It is easier here to work in eV than joules;

accordingly,

$$\lambda_{min} = \frac{ch}{eV} = \frac{(2.997\,9 \times 10^{8}\ \text{m/s})(4.135\,67 \times 10^{-15}\ \text{eV·s})}{20.0 \times 10^{3}\ \text{eV}}$$

Entering h in eV·s allows us to enter the denominator in eV. Finally

$$\boxed{\lambda_{min} = 0.062\,0\ \text{nm}}$$

Quick Check: An atom is about 0.1 nm across, and X-rays range in wavelength from there down to roughly 0.006 nm.

The existence of a maximum frequency was confirmed experimentally in 1915 by W. Duane and F. Hunt at Harvard. Using Eq. (28.8), they determined h to an accuracy of better than 4%.

28.4 The Compton Effect

Evidence of the reality of photons and the fact that they behave like particles with a well-defined energy and *momentum* was provided by the American Arthur H. Compton. He shone X-rays onto targets of low atomic number, like carbon. These have many loosely bound electrons that are essentially "free," and they scatter the radiant energy in a characteristic fashion. Classical theory is complicated by several factors, one of which is that the electrons recoil from the collision, and that introduces a Doppler Shift in the radiation they emit. Be that as it may, Compton reported in 1922 that the evidence at hand was in severe conflict with classical theory.

In 1923, Compton and Debye independently applied relativistic kinematics to the problem of a photon colliding with a "free" electron and thereby explained all the bewildering observations. Recall Eq. (28.16) for a massless particle, E = pc; hence, for photons for which E = hf,

$$p = \frac{E}{c} = \frac{hf}{c}$$

and

$$p = \frac{h}{\lambda} \tag{28.10}$$

Both the momentum and the energy of a high-frequency (short-wavelength) photon, such as an X-ray quantum, exceed that of a low-frequency (long-wavelength) photon, such as a microwave quantum.

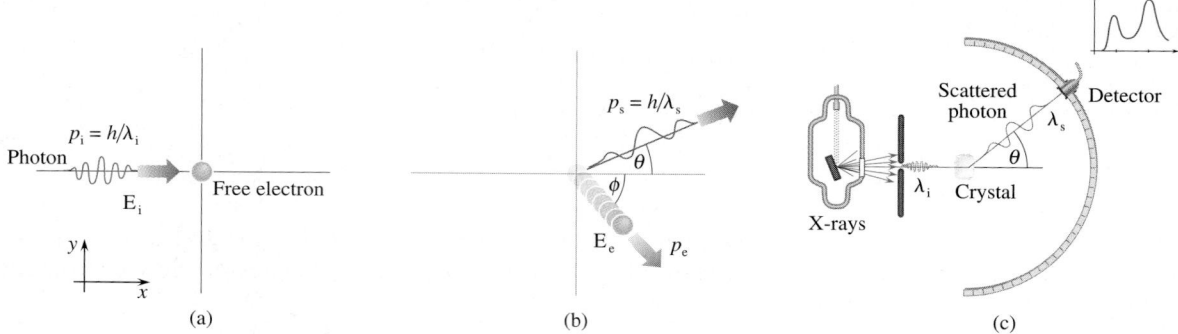

Figure 28.11 Compton scattering of X-rays. (a) A photon collides with a free electron and (b) both are scattered. (c) The scattered X-rays come off at an angle θ.

Imagine a beam of wavelength λ_i incident on a target. When photons collide with strongly bound electrons that stay put (and do not absorb energy), the photons are elastically scattered in all directions, and one sees this as radiation coming out with the incident wavelength unaltered. By contrast, picture an incident photon of momentum $\vec{p}_i$ (Fig. 28.11). If it strikes a free, essentially motionless, electron and imparts some of its momentum to that electron ($\vec{p}_e$), a new scattered photon with a longer wavelength and less momentum ($\vec{p}_s$) will come flying out at some angle θ.

This is an elastic-collision problem, and we analyze it just as we did earlier (p. 224). Applying Conservation of Momentum yields

$$\vec{p}_i = \vec{p}_s + \vec{p}_e \tag{28.11}$$

Similarly, the initial energy of the system is associated with the incident photon ($E_i = hf_i$) and the rest energy of the electron ($m_e c^2$). After the collision, the energy is that of the scattered photon ($E_s = hf_s$) and the moving electron (E_e). Conservation of Energy states that

$$E_i + m_e c^2 = E_s + E_e \tag{28.12}$$

Equations (28.11) and (28.12) can be combined with the relativistic expression [Eq. (26.15), p. 958] describing the energy of the electron

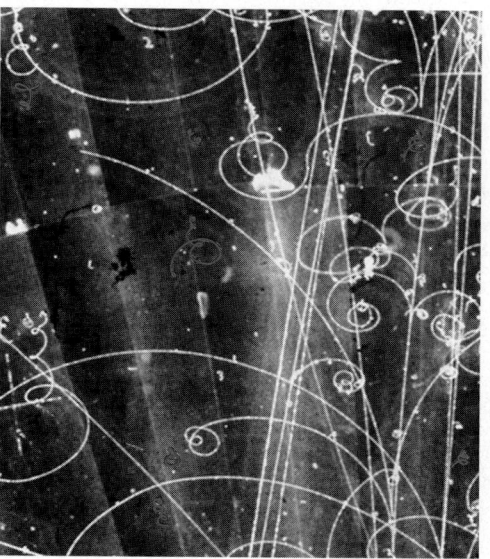

This (now familiar) bubble chamber photo shows the spiraling tracks of several lone electrons. Such trails are caused by electrons knocked out of atoms by trackless, high-energy photons (i.e., γ-rays). These are Compton electrons much like those depicted in Fig. 28.11.

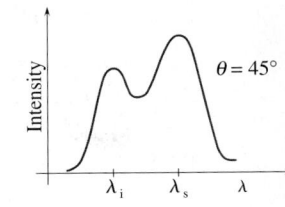
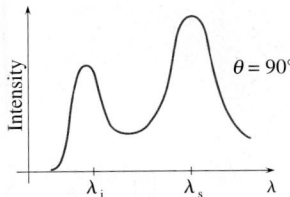

Figure 28.12 Scattered intensity at several angles.

$$E_e^2 = c^2 p_e^2 + m_e^2 c^4$$

to yield (see Problem 49) a formula for the increase in wavelength

$$\Delta\lambda = \lambda_s - \lambda_i = \frac{h}{m_e c}(1 - \cos\theta) \tag{28.13}$$

Figure 28.12 shows the sorts of results that are observed and that agree very nicely with Eq. (28.13). Note that the wavelength shift $\Delta\lambda$ only depends on the photon-scattering angle and is independent of the initial wavelength. It's also independent of the scattering material. The accuracy of the experiment was found to be within 1% by determining the multiplicative constant $(h/m_e c)$, which is called the **Compton Wavelength** and should theoretically equal 0.002 426 nm. In 1925, Bothe and Geiger, using counters, showed that the scattered electron and the scattered X-ray photon actually do fly off at the same time. More recent experiments (1950) verify that simultaneity to within 0.5 ns.

In the 1920s, the Compton Effect had the impact of eliminating the last vestiges of doubt: **whatever light is, it is a stream of particles capable of transferring energy and momentum that can also somehow behave as a wave**. Interestingly, though the contemporary consensus is that the Compton Effect is strong "proof" of the photon, people have nonetheless come up with alternative derivations in terms of the wave model—the business is far from settled.

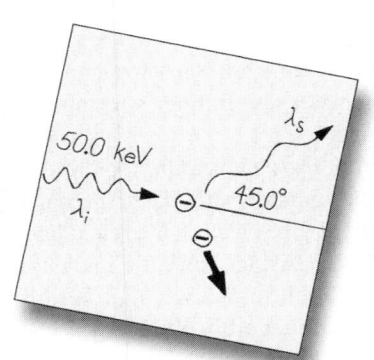

Example 28.5 A 50.0-keV photon is Compton-scattered by a quasi-free electron. If the scattered photon comes off at 45.0°, what is its wavelength?

Solution This problem treats the Compton Effect, and you'll just have to know the appropriate equation. (1) TRANSLATION—Knowing the energy and angle of a Compton-scattered photon, determine its wavelength. (2) GIVEN: $E_i = 50.0$ keV and $\theta = 45.0°$. FIND: λ_s. (3) PROBLEM TYPE—Quantization of energy/photons/Compton scattering. (4) PROCEDURE—Use the defining relationship: $\lambda_s - \lambda_i = h(1 - \cos\theta)/m_e c$. (5) CALCULATION—First compute λ_i using the results in Example 28.3, p. 1027:

$$\lambda_i = \frac{1240 \text{ eV·nm}}{50.0 \times 10^3 \text{ eV}} = 0.024\ 8 \text{ nm}$$

Since

$$\lambda_s - \lambda_i = \frac{h}{m_e c}(1 - \cos\theta)$$

we have

$$\lambda_s = (0.024\ 8 \text{ nm}) + (0.002\ 426 \text{ nm})(1 - 0.707)$$

$$\boxed{\lambda_s = 0.025\ 5 \text{ nm}}$$

Quick Check: $\Delta\lambda = \lambda_s - \lambda_i = 0.025\ 5$ nm $- 0.024\ 8$ nm $= 7 \times 10^{-4}$ nm as compared to Eq. (28.13), where $(0.002\ 426$ nm$)(1 - 0.707) = 7 \times 10^{-4}$ nm.

Atomic Theory

The atomic theory created by Niels Bohr was a brilliant accomplishment at the time. It has, however, been surpassed by the far more powerful formulations of Quantum Mechanics. Still, we will study the Bohr Theory because it provides the basic conceptual vocabulary for modern atomic physics and because many of its conclusions are essentially valid.

Niels Henrik David Bohr (1885–1962). Before Bohr left Denmark in 1943 with the Nazis on his trail, he dissolved in acid the gold Nobel Prize medals that Von Laue and Franck had given him for safekeeping. The bottle containing the gold solution was left on a shelf in Bohr's lab throughout the war. When he returned to Copenhagen, Bohr precipitated the gold and had the medals recast.

28.5 The Bohr Atom

Bohr began by guessing about the nature of the simplest atomic configuration, the single-electron atom; his first postulate assumed that *the electron sails around the nucleus in a circular orbit*. He then postulated that, unlike a planet, which can revolve permanently at any distance from the Sun, *atomic electrons exist in only certain stable, lasting orbits about the nucleus*. These are now called **stationary states**, and the one with the lowest energy is the **ground state**. *While in such a stationary state, the atom does not radiate.* Everyone knew that something was wrong with the planetary model of the atom; electrons orbiting the nucleus, accelerating, should continuously radiate. Losing energy, they should wind inward, finally crashing into the nucleus. And it's easy to calculate that this death spiral will take about a hundred-millionth of a second. Our very existence is therefore an embarrassment to such a theory, which ridiculously insists that all the atoms in the Universe should have long ago collapsed. Why doesn't Bohr's orbiting electron radiate? Here, Bohr was refreshingly outrageous. He simply insisted it didn't, and that was that. Of course, his stance implied that Maxwell's Electromagnetic Theory somehow reaches profound limitations on the atomic level, but he didn't bother with that.

Bohr was not the only one groping toward an orbital model. J. W. Nicholson, an English astrophysicist, was perhaps his chief rival. Like Maxwell long before him, Nicholson raised the question of there being a relationship between the structure of an atom and the kind of spectrum it emits. Bohr believed that any such interdependence must be exceedingly complicated, and at first he avoided the notion altogether. But when a friend, H. M. Hansen, returned to Copenhagen after studying spectroscopy and the two began to discuss Bohr's work, the issue of spectral colors immediately came up. Hansen suggested that spectra (p. 982) were not really so complex. "As soon as I saw Balmer's formula," Bohr recalled, "the whole thing was immediately clear to me."

Consider the simplest of all atoms, the hydrogen atom. It has a nuclear proton orbited by a single electron in its ground state. When the atom is appropriately stimulated (perhaps thermally via collisions, electrically, or even by absorbing light), the electron is excited into a higher energy orbit that is more distant from the nucleus. There it resides, ordinarily for about a nanosecond, before spontaneously descending to some inner orbit, ultimately dropping back to the ground state. During each drop (a moment when the theory goes impotent), the electron emits its excess energy as a burst of electromagnetic radiation—a photon. This is the now-famous **quantum jump**.

The picture is a little like a Greek amphitheater, in which a ball revolving around the lowest tier is boosted up to some higher ring, where it orbits for a while, only to drop back to ground level in either one large or many small descents. The sequence of Balmer's frequencies became a ladder of orbits, of **energy levels**. When the electron in an excited atom drops from an initial (E_i) to a final (E_f) energy level, it emits the difference as a quantum hf:

[$E_i > E_f$: energy emitted]

$$E_i - E_f = hf \qquad (28.14)$$

The larger the drop, the greater the photon frequency (Figs. 28.13 and 28.14).

By the time Bohr began his work, the notion that an oscillator could have quantized energy levels that were whole number multiples (n) of hf was widely accepted; $E = nhf$ was an essential feature of the new blackbody theory. Planck recognized that h had the units of energy multiplied by time (J·s) or, equivalently, of momentum multiplied by distance. Since that product, known as *action*, was important in classical mechanics, he called h the **quantum of action**. The idea that action was quantized would be generalized into a guiding principle. For example, for an electron in the nth circular orbit of radius r_n, the distance

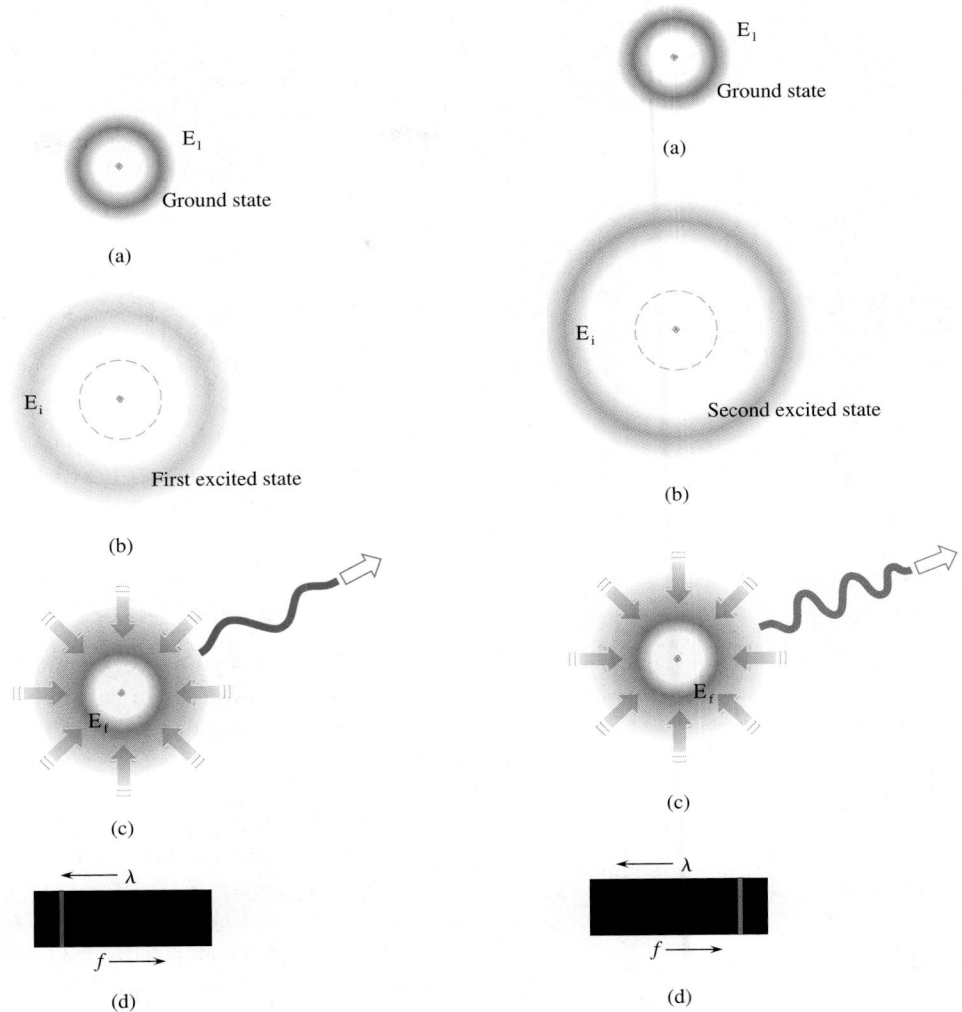

Figure 28.13 An atom in the first excited state drops back to the ground state with the emission of a long-wavelength, low-energy photon. The emission shows up as a long-wavelength, low-frequency spectral line.

Figure 28.14 An atom in the second excited state drops back to the ground state with the emission of a short-wavelength, high-energy photon.

traveled per revolution is $2\pi r_n$; and since $p_n = m_e v_n$, we might guess that the action (momentum times distance) for each orbit would equal a whole number multiple of h: $(m_e v_n)(2\pi r_n) = nh$. This immediately leads to $m_e v_n r_n = nh/2\pi$, and that we recognize (p. 259) as the angular momentum (L) of the orbiting electron. In any event, Nicholson (1912) suggested that "the angular momentum of an atom can only rise and fall by discrete amounts." A year later, Bohr adopted the same idea as his second postulate; that is,

$$L_n = m_e v_n r_n = n\frac{h}{2\pi} \qquad n = 1, 2, 3, \ldots \qquad (28.15)$$

Here, each successive value of n corresponds to a higher orbit of larger radius and lower speed. The term $h/2\pi$ comes up so frequently that P. A. M. Dirac gave it its own symbol, $\hbar$, and we refer to it as "h-bar":

$$\hbar = 1.05 \times 10^{-34} \text{ J·s}$$

Angular momentum is quantized in whole number multiples of $\hbar$.

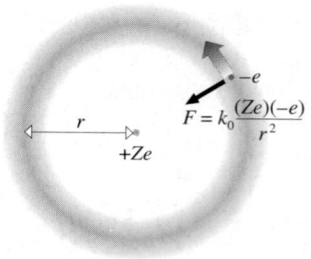

Figure 28.15 A single-electron atom showing the Coulomb interaction between electron and nucleus.

If we continue with this simple pictorial model (which we *will* have to abandon soon enough), we can derive an expression for the radii of the allowed orbits (Fig. 28.15) of a one-electron atom. Of course, hydrogen is the simplest such atom, but the theory works for such systems as singly ionized helium as well. The electron is attracted to a nucleus (which in general has Z protons and a net charge of $+Ze$) via a Coulomb interaction. Equating that force, $F_E = k_0(Ze)(e)/r^2$ with the centripetal force, $F_c = mv^2/r$, leads to

$$k_0 \frac{Ze^2}{r_n^2} = \frac{m_e v_n^2}{r_n} \tag{28.16}$$

which can be solved for

$$r_n = \frac{k_0 Ze^2}{m_e v_n^2}$$

The quantity v_n, which we do not know, can be eliminated from this expression by solving Eq. (28.15) for it. Whereupon $v_n = nh/2\pi m_e r_n$, and substituting this expression into the previous equation and simplifying (canceling a factor of r_n) gives

$$r_n = n^2 \frac{h^2}{m_e k_0 Ze^2} = n^2 r_1 \tag{28.17}$$

For hydrogen, $Z = 1$, and putting the appropriate numbers in Eq. (28.17) leads to $r_1 = 0.052\,917\,7$ nm. This is the radius of the ground state of the hydrogen atom, and it's called the **Bohr radius**. We know that an atom is about 0.1 nm across, so this result is very encouraging (Fig. 28.16). The theory suggests (it only deals directly with single-electron systems) that for heavier atoms (larger Z), the increased nuclear charge draws more strongly on the surrounding electrons, shrinking their orbits. Thus, atomic diameters are all close to the same size. Uranium, which is 238 times more massive than hydrogen, has a diameter only about 3 times as great.

The model Bohr devised pictured the discrete spectral lines resulting from transitions between discrete energy levels. Clearly, energy is the next thing to consider. The total energy of an orbital electron (E_n) is its KE plus its *electrical*-PE $= qV = -eV$. Here, V is due to the nucleus and so $V = k_0 q/r = k_0(Ze)/r$. Thus,

$$E_n = \tfrac{1}{2} m_e v_n^2 - \frac{k_0 Ze^2}{r_n}$$

which is the energy of the atom as a whole. It follows from Eq. (28.16) that $\tfrac{1}{2} m_e v_n^2 = k_0 Ze^2/2r_n$ and so

$$E_n = -\frac{k_0 Ze^2}{2 r_n}$$

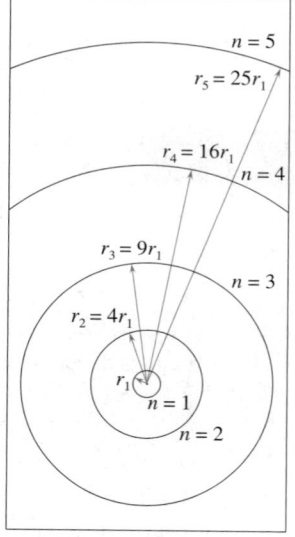

Figure 28.16 The first five Bohr orbits.

Notice that the total energy of the atom is negative, only becoming zero when the electron is removed to infinity, or the practical equivalent thereof (see Discussion Question 10). The dependence on r_n can be removed using Eq. (28.17), yielding

$$E_n = -\frac{2\pi^2 k_0^2 e^4 m_e}{h^2} \frac{Z^2}{n^2} \tag{28.18}$$

The integer n, which enumerates both the orbital radii and the energies, is known as the **principal quantum number**. Notice that as n gets larger, the radii increase as n^2, and the orbits get farther apart. On the other hand, the energy levels become less negative and approach zero as $n \to \infty$; they get closer together, ultimately becoming continuous when the energy is positive and the electron is unbound.

Equation (28.18) provides the energy of each orbit, and it can be simplified considerably by noting that the ground-state energy ($n = 1$) for hydrogen ($Z = 1$) turns out to equal

−13.6 eV. This value is the hydrogen atom's largest negative energy and its most tightly bound configuration. In that case

[for hydrogen]
$$E_n = -\frac{13.6\,\text{eV}}{n^2}$$
(28.19)

The **first excited state** ($n = 2$) of the hydrogen atom has an energy of $-(13.6/2^2)$ eV, or −3.40 eV. The minus sign tells us that that much energy must be added to raise the electron to the zero level. We can think of it as a kind of energy well with the electron usually down −13.6 eV, at the bottom. Accordingly, the **ionization energy** needed to remove the electron from the atom setting it free (to raise it up and out of the potential well) is 13.6 eV, and this agrees precisely with the measured value! Being able to derive that from basic principles is an amazing accomplishment.

Example 28.6 **[II]** How much energy must a hydrogen atom absorb if it is to be raised from the ground state to the first excited state? If that excitation energy is to come in the form of a photon, what must be its frequency?

Solution A photon is emitted when the atom descends from a higher to a lower energy level. (1) TRANSLATION—Determine the energy difference between the ground state and the first excited state of hydrogen. What is the frequency of a photon with that energy? (2) GIVEN: H atom in ground state. FIND: $E_2 - E_1$ and f. (3) PROBLEM TYPE—Bohr atom/energy levels/photon energy. (4) PROCEDURE—The difference between energy levels is the photon energy; $E_i - E_f = hf$. (5) CALCULATION—We already know that $E_1 = -13.6$ eV, $E_2 = $

−3.40 eV, and $E_2 > E_1$; hence, the atom must receive an energy of

$$E_2 - E_1 = (-3.4\,\text{eV}) - (-13.6\,\text{eV}) = \boxed{10.2\,\text{eV}}$$

if it's to be raised into its first excited state. A photon for which

$$\Delta E = hf = 10.2\,\text{eV}$$

should have a frequency of

$$f = \frac{(10.2\,\text{eV})}{(4.136 \times 10^{-15}\,\text{eV·s})} = \boxed{2.47 \times 10^{15}\,\text{Hz}}$$

Quick Check: $\lambda \approx 120$ nm, which is in the UV where it should be (see Fig. 28.17).

If the transition from an upper-excited state down to a lower state is accompanied by the emission of a photon as per Eq. (28.14), we should now be able to write an explicit expression for the resulting wavelength or, better yet, for $1/\lambda$. Since $hf = hc/\lambda$, rewrite Eq. (28.14) as

$$\frac{1}{\lambda} = \frac{1}{hc}(E_i - E_f)$$

or, using Eq. (28.18), as

$$\frac{1}{\lambda} = \frac{2\pi^2 k_0^2 e^4 m_e Z^2}{h^3 c}\left[\frac{1}{n_f^2} - \frac{1}{n_i^2}\right]$$
(28.20)

where $n_f < n_i$; the atom is initially in a higher state than the one it drops down to. Now compare this equation to the Balmer series ($Z = 1$), for which $n_i = n$ and $n_f = 2$; that is

$$\frac{1}{\lambda} = R\left[\frac{1}{2^2} - \frac{1}{n^2}\right] \qquad n = 3, 4, 5, \ldots$$
[27.5]

where all those constants in front of the bracketed term in Eq. (28.20) must equal RZ^2; recall that R is the Rydberg constant. When Bohr carried out the calculation in 1913 using the best values of the day, he got the measured R to within 1%—spectacular!

Each line in the Balmer series arises when the hydrogen atom in an excited state relaxes back in a single quantum jump to the first excited state ($n_f = 2$). Similarly, when the transition is down to the ground state ($n_f = 1$), the energies are greater and the resulting

Figure 28.17 (a) Energy levels of the hydrogen atom and the transitions between them corresponding to the emission of radiation. Shown are the Lyman, Balmer, and Paschen series. (b) The spectral lines corresponding to the Balmer series. (Take a look at Fig. Q11, p. 991.)

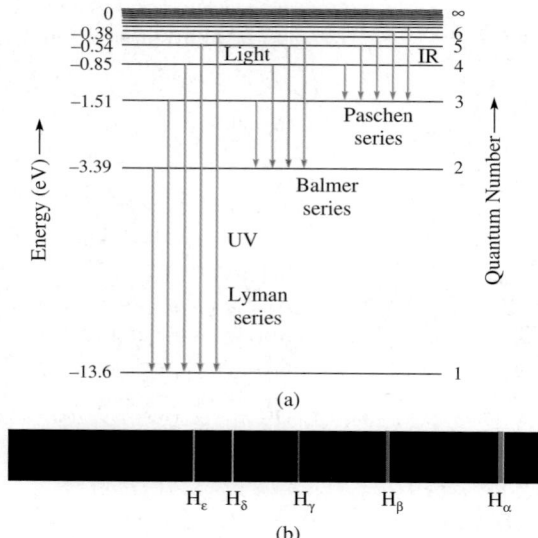

(a)

(b)

Lyman series, which wasn't discovered until 1916, is in the far-UV. The Paschen series (first observed in 1908) corresponds to $n_f = 3$. Brackett found a new series in 1922 in the IR for which $n_f = 4$, and Pfund, in 1924, located yet another in the IR for $n_f = 5$ (Fig. 28.17), all precisely as predicted by the Bohr Theory.

Energy Levels

An atom can be raised into any excited state by absorbing an amount of energy given by Eq. (28.14). This can happen, for instance, via collisions with other atoms or from bombardment by projectiles such as electrons (via an electric current) or photons. If an incoming photon does not have enough energy to raise the atom into its first excited state, the atom remains in its ground state, immediately elastically scattering away the energy. Thus, a visible photon with an energy of around 3 eV cannot raise hydrogen into its first excited state, and so hydrogen, normally in its ground state, is transparent to light; photons enter and leave the gas without any energy losses. Only if the photon (or any other impacting particle) has energy equal to, or in excess of, that needed for excitation will the atom "rise up" to a higher level. For example, consider a monatomic gas at room temperature. The atoms collide, but they don't give off light. Why not? According to Eq. (12.15), the average KE of such an atom is $\frac{3}{2}k_B T = \frac{3}{2}(1.38 \times 10^{-23} \text{ J/K})(300 \text{ K}) = 6.2 \times 10^{-21} \text{ J} = 0.04 \text{ eV}$. That's much too small to excite any of the atoms, and the collisions remain perfectly elastic (which is why the air doesn't glow on a warm day).

In recent times, it has become possible to examine the transitions made by single atoms from one specific energy level to another. For example, in 1986, an individual barium ion held, almost at rest, in an electromagnetic-field trap was stimulated using laserbeams of precise frequency. The experiments (see Discussion Question 13) directly confirmed the existence of quantum jumps between energy levels.

28.6 Stimulated Emission: The Laser

In a conventional light source, such as a tungsten lamp or a neon sign, energy is usually *pumped* into the reacting atoms via collisions with an electrical current. An excited atom drops back to its ground state *spontaneously*, without any external inducement, emitting a randomly directed photon. The atoms are all essentially independent, and each photon in the emitted stream bears no particular phase relationship to any other.

The first photograph of a solitary atom. The tiny blue-green dot at the center of this photo is a single barium ion. It was slowed or "cooled" by laserbeams and is suspended in a radio frequency trap. (The large, white-and-red structure surrounding the ion is part of the trap.) The ion absorbs energy from the laserbeam and re-emits at 493 nm.

Now, imagine a bunch of atoms somehow pumped up into an excited state. What happens next is not obvious. In 1917, Einstein pointed out that an excited atom can relax to a lower state via photon emission in two distinct ways. In one, the atom emits energy spontaneously; while in the other, it is triggered into emission by the presence of a photon of the proper frequency. The former process is known as **spontaneous emission**, the latter as **stimulated emission**. Suppose that the energy difference between an upper and lower energy state is $\Delta E = E_u - E_l$ such that if the transition down occurs, a photon hf_{ul} will be emitted. Let the atom be in the E_u-state with no concern as to how it got there. If the excited atom is exposed to a photon of frequency f_{ul}, it will immediately be stimulated into dropping down to the E_l state. A remarkable feature of the process is that *the emitted photon is in-phase with, has the polarization of, and propagates in the same direction as the stimulating radiation*. This is a manifestation of the basic nature of photons to cluster in the same state (p. 1039). The incident lightwave is thereby increased in irradiance (Fig. 28.18).

Since most atoms are usually in their ground states, any incoming light is far more likely to be absorbed than to produce stimulated emission. This raises an interesting point: what happens if a substantial number of atoms could be excited into an upper state, leaving the lower state almost empty? Such a condition is called a **population inversion**. An incident photon of the right frequency could then easily trigger an avalanche of stimulated photons—all in-phase, all perfectly in step. The initial triggering wave would continue to build as it swept across the active medium, as long as there were no dominant competitive processes (such as scattering) and provided the population inversion could be maintained. Moreover, the process could be enhanced by placing the active medium between two mirrors so that the lightwave passed back and forth through it many times. To allow some light to escape in the form of a beam, one of the mirrors could be made to be less than 100% reflecting; it would leak the beam (Fig. 28.19). In effect, energy (electrical, optical, chemical, whatever) would be pumped in to sustain the inversion, and a beam of light would be extracted. Such a device for **l**ight **a**mplification by **s**timulated **e**mission of **r**adiation is known as a **laser**.

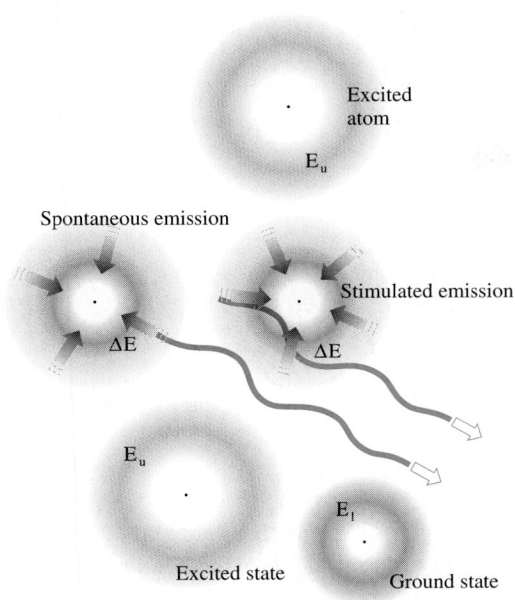

Figure 28.18 A population inversion: five atoms in a group where all but one are in an excited state. One atom (on the left) spontaneously returns to the ground state, emitting a photon, which causes a nearby atom to be stimulated into emitting an in-phase photon of its own.

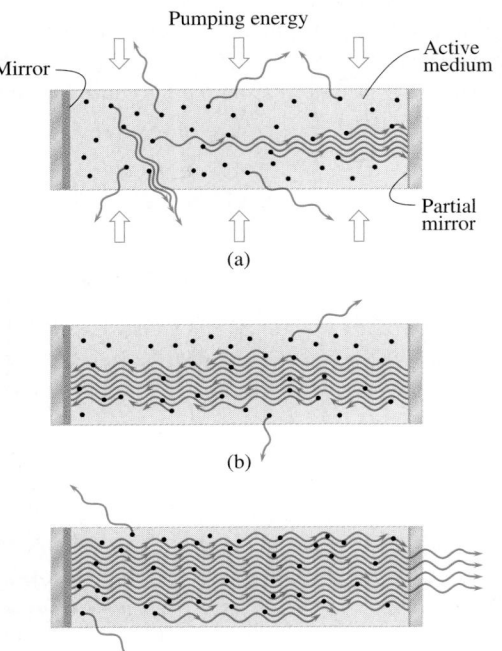

Figure 28.19 A schematic representation of a laser. An active medium with most of its atoms in an excited state is located between two mirrors—one totally reflecting, the other 99% reflecting. Atoms spontaneously emit. Only photons along the axis stay within the cavity and stimulate additional emission. As the wave sweeps back and forth across the active medium, the wave builds until a portion leaks out as the laserbeam. Energy can be pumped in continuously, in which case the beam can be continuous.

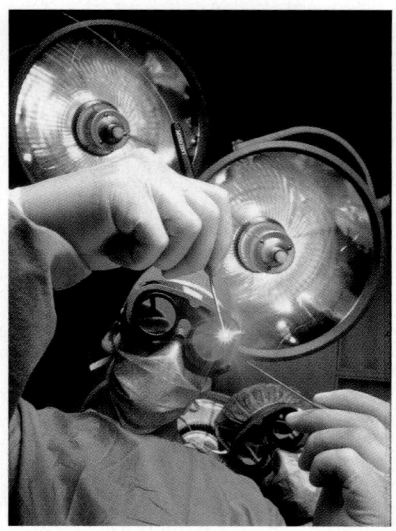

Surgery using a laser.

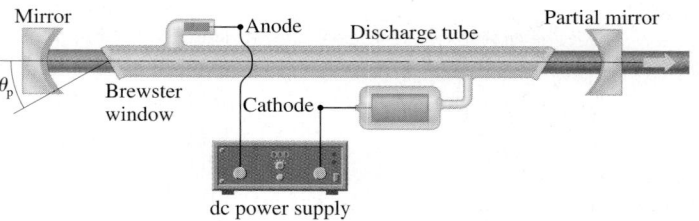

Figure 28.20 An early version of a helium-neon (He-Ne) continuous laser. The Brewster windows serve to polarize the beam.

The two mirrors constitute a **resonant cavity**. Only certain wavelengths that "fit" can set up standing waves in the cavity, and only those wavelengths will be sustained. Consequently, the laser produces light of unmatched spectral purity; it is quasimonochromatic and therefore has a very long coherence length. Moreover, only waves sweeping back and forth parallel to the axis can undergo many reflections and thereby build in intensity. Thus, the rays emerging from the cavity are very nearly parallel, and the laserbeam is highly directional.

The helium-neon (He-Ne) laser (Fig. 28.20) is still the most popular device of its kind. It puts out a continuous beam, usually of a few milliwatts of red light at 632.8 nm. The active medium, a mixture of about 10 Pa of neon (the active centers) to 100 Pa of helium, is placed in a gas-discharge tube. Pumping is accomplished through a high-voltage electrical discharge. What happens to the gases can be understood using the simplified energy-level diagram in Fig. 28.21. Many helium atoms are kicked up into a number of upper levels by the current. After dropping down, most accumulate in a long-lived state (E_{H2}), 20.61 eV above the ground level, from which there are no allowed radiative transitions.

The excited He atoms inelastically collide with and transfer energy to ground-state neon atoms, raising them into a long-lived energy state E_{N4}. This level is 20.66 eV above the ground level, the difference (0.05 eV) having been supplied from the KE of the colliding atoms. There then exists a population inversion, among the neon atoms, with respect to the lower E_{N3} level. Spontaneous photons initiate stimulated emission, and a chain reaction begins down from E_{N4} to E_{N3}. The result is the emission of bright red light at 632.8 nm. The E_{N3} level readily drains off to the E_{N2} level, sustaining the inversion. The latter is above the ground level, but that isn't significant—the important point is that lasing takes place between two upper levels. As a result, since the E_{N3} level is only sparsely occupied, the inversion is easy to maintain continuously, without having to half-empty the ground state.

Figure 28.21 Simplified He-Ne laser energy levels. Here the lasing transition $E_{N4} \rightarrow E_{N3}$ is not down to the ground state. E_{N3} quickly dumps to E_{N2}, so the population inversion between E_{N3} and E_{N2} is constantly sustained.

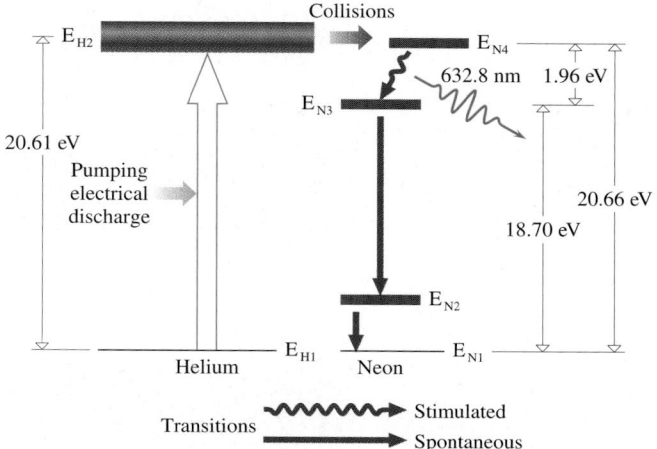

Example 28.7 [II] The helium-neon laser puts out a bright red beam at a wavelength of 632.8 nm. Please determine the difference in energy between the two states defining the transition.

Solution   A photon is emitted when the atom descends from a higher to a lower energy level. (1) TRANSLATION—Determine the energy difference between the two states when an atom emits a photon of a specified frequency. (2) GIVEN: $\lambda = 632.8$ nm. FIND: ΔE. (3) PROBLEM TYPE—Bohr atom/energy levels/photon energy. (4) PROCEDURE—The difference between energy levels is the photon energy: $\Delta E = E_i - E_f = hf$. (5) CALCULATION—The frequency follows from $c = f\lambda$

$$\Delta E = \frac{hc}{\lambda} = \frac{(6.626\ 1 \times 10^{-34}\ \text{J·s})(2.997\ 9 \times 10^8\ \text{m/s})}{632.8 \times 10^{-9}\ \text{m}}$$

and $\boxed{\Delta E = 3.139 \times 10^{-19}\ \text{J}} = 1.959\ \text{eV}$.

Quick Check: The order-of-magnitude of the answer, a few eV, is appropriate for outer electron transitions. The 1.96-eV transition corresponds to an energy in joules of $(1.96\ \text{eV})(1.602 \times 10^{-19}\ \text{J}) = 3.14 \times 10^{-19}$ J; since that value equals hf, $f = 4.738 \times 10^{14} = c/\lambda$ and $\lambda = 632.8$ nm.

28.7 Atomic Number

The **characteristic X-rays** that appear as sharp spikes superimposed upon the broad Bremsstrahlung spectrum (Fig. 28.10) come in groups much like the Balmer and Paschen series. Barkla (1911) named them the K, L, M, ... and so forth series, and today, the spikes in each group are labeled (recall the hydrogen lines) K_α, K_β, K_γ, and so on. Just as the Balmer lines are an indication of the hidden outer electron structure of the atom, these series are equally as significant for the inner electrons. Bohr discussed this matter with his friend Henry Moseley, and Moseley, in 1913, brilliantly found the pattern. After a tremendous solitary effort, he was able to devise an empirical equation describing the frequencies of the K_α lines in terms of the particular atomic structure of the target, that is, in terms of something he called the *atomic number*.

Each element has an integer atomic number, and it increases, one unit at a time, from hydrogen (1), to helium (2), to lithium (3) and beryllium (4), up to uranium (92), just like a place number marking consecutive positions in the Periodic Table (inside back cover). Moseley astutely concluded: "This quantity [atomic number] can only be the charge on the central positive nucleus." *The atomic number is the number of units of nuclear charge, the number of protons in the nucleus* (Z). The formula Moseley deduced was $f = C(Z - 1)^2$, where the constant C was subsequently shown to have the value $3cR/4$,

> The **atomic number** (Z) is the number of protons in the nucleus of the atom.

[for K_α lines]
$$f = \left(\frac{3cR}{4}\right)(Z - 1)^2 \tag{28.21}$$

Bohr conjectured that for all atoms, as the nuclear charge increased, the number of orbital electrons would also increase. One by one, they would build across the Periodic Table. Unhappily, the heavier atoms were beyond treating analytically with his simple theory. Despite that, W. Kossel (1914) suggested that *if an inner tightly bound electron of a heavy atom was removed from its orbit (e.g., by bombarding it with cathode rays), all the other whirling electrons would cascade downward until that vacancy was filled.* He called the innermost electron energy level K, the next L, and so on (Fig. 28.22), and that scheme led to a picture of electron transitions (much like those corresponding to the outer electrons) and the associated spectral lines. By comparison, these jumps were far greater in energy and so emitted X-rays. For example, while 7.4 eV will remove the outermost electron from a lead atom, it takes 88 keV to remove either one of the K electrons.

Equation (28.21) can be rewritten for wavelengths; it then takes on a familiar form

$$\frac{1}{\lambda} = R(Z - 1)^2 \left[\frac{1}{1^2} - \frac{1}{2^2} \right] \tag{28.22}$$

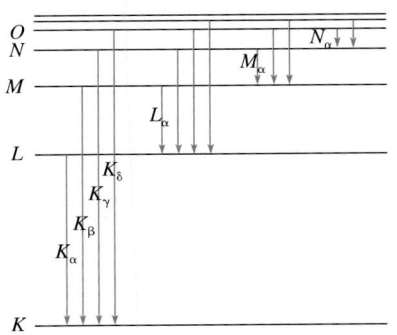

Figure 28.22 Inner electron energy levels and the transitions that give rise to X-ray emission.

This equation looks just like Eq. (28.20) except for the $(Z - 1)^2$ instead of Z^2. Ordinarily, the $n = 1$, K-level has two electrons in it, but now one has been removed, leaving one remaining. As an electron from an upper level plunges toward the hole, it sees an effective nuclear charge reduced (via Gauss's Law) by one negative unit because of the single nearby orbital electron; it sees $+e(Z - 1)$, not $+eZ$.

Example 28.8 **[III]** A target emits K_α radiation of wavelength λ. Show that

$$Z = 1 + \sqrt{\frac{4}{3\lambda R}}$$

Then suppose that $\lambda = 0.251\,0$ nm. What metal is being used?

Solution This problem is about X-ray emission. (1) TRANSLATION—Derive the specified expression for Z corresponding to K_α radiation. For a known wavelength, determine the emitting metal. (2) GIVEN: $\lambda = 0.251\,0$ nm. FIND: Prove above expression and find Z. (3) PROBLEM TYPE—X-ray emission/atomic spectra. (4) PROCEDURE—Equation (28.22) relates Z, R, and λ and is a good place to start. (5) CALCULATION:

$$\frac{1}{\lambda} = R(Z - 1)^2 \left[\frac{1}{1^2} - \frac{1}{2^2} \right] = \frac{R(Z - 1)^2 3}{4}$$

$$(Z - 1)^2 = \frac{4}{3\lambda R}$$

and

$$Z = 1 + \sqrt{\frac{4}{3\lambda R}}$$

Therefore,

$$Z = 1 + \left[\frac{4}{3(0.251\,0 \times 10^{-9}\,\text{m})(1.097\,373 \times 10^7\,\text{m}^{-1})} \right]^{1/2}$$

and $Z = 23.0$; therefore, the metal is $\boxed{\text{vanadium}}$.

Quick Check: Using Eq. (28.22) to check λ against $Z = 23.0$, it follows that $1/\lambda = R(22)^2 3/4$ and therefore $\lambda = 0.251$ nm.

The idea that the elements of the Periodic Table were to be ordered not by mass but by nuclear charge had been around for several years already. However, Moseley could directly observe the emitted X-rays and from them immediately tell whether or not the material under study was elemental and, if so, where in the Table it belonged! For instance, there was some controversy about argon (atomic weight 39.9) and potassium (atomic weight 39.1), since the former is an inert gas and should be located with the other inert gases (inside back cover), which would have it strangely preceding the lighter alkali metal. Moseley's X-rays revealed argon to have an Atomic Number of 18, whereas that of potassium was 19. Argon clearly belonged before potassium.

Unfortunately, Harry Moseley would solve no more of nature's mysteries; in 1915, at the age of twenty-eight, he was killed at Gallipoli in one of the most useless campaigns of World War I ("the war to end all wars").

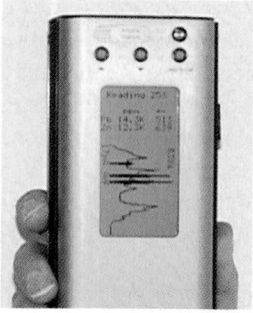

The source in a lead paint detector emits energetic electrons. These strike paint, causing the ejection of electrons from inner shells of atoms. The X-ray spectrum from the subsequent atomic transitions reveals the presence of lead.

Core Material & Study Guide

THE OLD QUANTUM THEORY

The total power radiated at all wavelengths by a blackbody is given by the **Stefan-Boltzmann Law**, namely,

$$P = \sigma A T^4 \qquad [28.1]$$

where A is the area, T is the absolute temperature in kelvins, and

$$\sigma = 5.670\,5 \times 10^{-8} \text{ W/m}^2 \cdot \text{K}^4$$

Given the **total emissivity** (ε), the power radiated by a surface is

$$P = \varepsilon \sigma A T^4 \qquad [28.2]$$

For a body in an enclosure, the net power radiated (when $T > T_e$) or absorbed (when $T < T_e$) is

$$P = \varepsilon \sigma A (T^4 - T_e^4) \qquad [28.3]$$

For blackbodies, **Wien's Displacement Law** is

$$\lambda_p T = \text{constant} \qquad [28.4]$$

where the constant is 0.002 898 m·K. **Planck's Radiation Law** is

$$I_\lambda = \frac{2\pi h c^2}{\lambda^5} \left[\frac{1}{e^{hc/\lambda k_B T} - 1} \right]$$

where $h = 6.626\,075\,5 \times 10^{-34}$ J·s (p. 1001). All of this material is discussed in Section 28.1 (Blackbody Radiation). Study it again and make sure you understand Examples 28.1 and 28.2. **Look at the CD Walk-Through Examples** and then try the I-level problems.

In the Photoelectric Effect, the stopping potential is given by

$$KE_{max} = eV_s \qquad [28.6]$$

According to Einstein, **light is a stream of energy quanta**, each with an energy

$$E = hf = \frac{hc}{\lambda} \qquad [28.7]$$

The **Photoelectric Equation** is then

$$hf = KE_{max} + \phi \qquad [28.8]$$

Reexamine Section 28.2 (Quantization of Energy: The Photoelectric Effect) and study Example 28.3.

The minimum wavelength of Bremsstrahlung (p. 1006) is given by

$$\lambda_{min} = \frac{ch}{eV} \qquad [28.9]$$

This equation is applied in Example 28.4 in Section 28.3 (Bremsstrahlung).

In the **Compton Effect**, a photon can scatter off a free electron with an accompanying wavelength shift given by

$$\Delta \lambda = \lambda_s - \lambda_i = \frac{h}{m_e c}(1 - \cos\theta) \qquad [28.13]$$

Reread Section 28.4 (The Compton Effect) and study Example 28.5 (it's fairly typical of the sort of thing you should know).

ATOMIC THEORY

According to Bohr, for a hydrogen atom, a photon is emitted when the atom makes a transition from one energy level to another; that is,

$$E_i - E_f = hf \qquad [28.14]$$

The resulting orbits have radii given by

$$r_n = n^2 \frac{h^2}{m_e k_0 Z e^2} = n^2 r_1 \qquad [28.17]$$

and energies given by

$$E_n = -\frac{2\pi^2 k_0^2 e^4 m_e}{h^2} \frac{Z^2}{n^2} \qquad [28.18]$$

These ideas are treated in Section 28.5 (The Bohr Atom)—review Example 28.6. These ideas are applied to the laser as discussed in Section 28.6 (Stimulated Emission: The Laser).

Moseley showed (p. 1017) that for X-ray emission

[for K_α lines] $\qquad f = \left(\frac{3cR}{4} \right)(Z - 1)^2 \qquad [28.21]$

Reread Section 28.7 (Atomic Number) and go over Example 28.8.

Key Terms

emission coefficient	bremsstrahlung
absorption coefficient	Compton Effect
distribution function	Compton Wavelength
Kirchhoff's Radiation Law	Bohr atom
Stefan-Boltzmann Law	ground state
total emissivity	quantum jump
total absorptivity	energy levels
Wien Displacement Law	Bohr radius
blackbody radiation	principal quantum number
Planck's Constant	first excited state
Planck's Radiation Law	ionization energy
Photoelectric Effect	spontaneous emission
stopping potential	stimulated emission
photon	population inversion
work function	laser
Photoelectric Equation	characteristic X-rays
wave-particle duality	atomic number

Discussion Questions

1. A candle or match flame usually has a distinct yellow color to it. One might guess that the color comes from sodium, but this is generally *not* the case. It's actually due to blackbody radiation from minute hot (≈ 2000 K) particles of soot. How might you confirm this fact? (Incidentally, the bluish light at the base of the flame is from CH, C_2, and CO_2 molecular emission.)

2. Not very many solid materials can be heated to incandescence, and even fewer liquids can. Discuss and explain this observation. How do the following substances behave regarding incandescence: iron, water, glass, carbon, plastic, gasoline, banana, soap, wood, ceramics?

3. In order to test the photon hypothesis, Walther Bothe conducted the following experiment (Fig. Q3). A low-intensity beam of X-rays was shone on a thin foil, which subsequently emitted X-rays (via X-ray fluorescence) toward two counters. The counts from the two detectors were recorded as they occurred, and no correlation in time was observed. What does that "prove"? Explain.

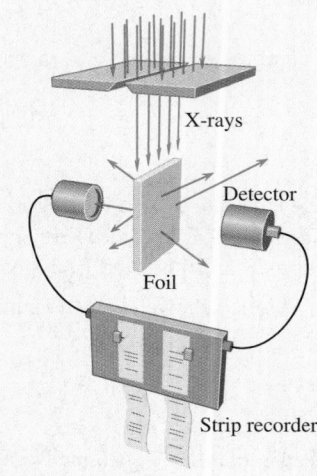

Figure Q3

4. Figure Q4 shows a sequence of drawings of three photons interacting with a metal. Explain each portion of the figure. Why does the liberated electron have a maximum kinetic energy?

5. With a bright source, such as a high-powered laser, it is possible to have a photoelectron absorb two or more photons before it leaves the metal (p. 1002). Given that an electron absorbs N identical photons of frequency f, how should Eq. (28.8) be modified, if at all? What will happen to the stopping potential as compared to the single photon effect? Does the work function change? Does the threshold frequency change? Explain your answers in detail.

6. Figure Q6 shows a current-versus-voltage curve for the Photoelectric Effect using polychromatic light. Explain every detail of the curve. Which frequency is higher?

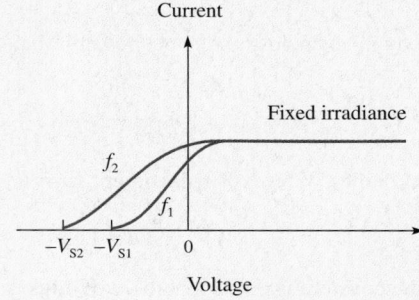

Figure Q6

7. Light absorbed by a substance may result in a chemical change of that substance via a process known as a *photochemical reaction* (which is why wine and aspirin are kept in dark-colored bottles). For each such reaction, there is a threshold frequency below which it will not occur. Assuming that the molecule must absorb an amount of energy that will trigger the process E_a, known as the *activation energy*, explain (as Einstein first did) the existence of a threshold frequency.

8. Figure Q8 shows some data for gamma-ray Compton scattering. Even though the gamma-ray wavelengths are much smaller than those of X-rays, the slope was found to be 2.4×10^{-12} m, in fine agreement with theory. Explain the curve in every regard.

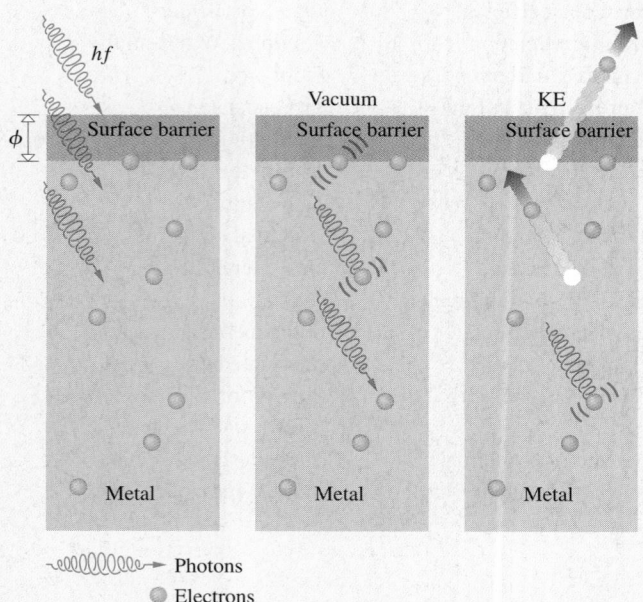

Figure Q4

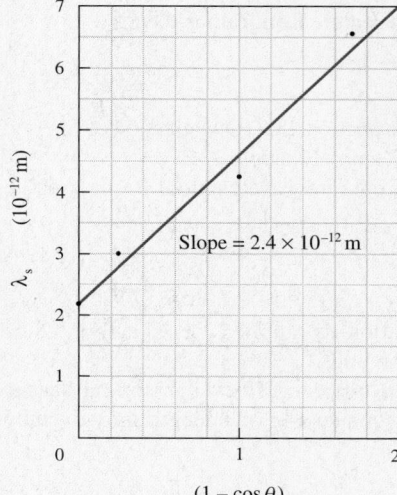

Figure Q8

9. A photon cannot be absorbed by a free electron—prove that statement by using Conservation of Energy and Momentum.

Assume the electron is initially at rest. First draw a diagram of the problem as seen from a coordinate system fixed with the center-of-mass—show it before and after the collision. Note that by definition the initial momentum in the center-of-mass system is zero. What is the final momentum? Write expressions for the initial and final energy. Now write the statement of Conservation of Energy (as seen from the center-of-mass). The fact that this implies that $m > \gamma m$ makes the initial assumption impossible.

10. Discuss the relationships between E, KE, and PE for any Bohr orbit of hydrogen. In the process explain Fig. Q10 and offer a definition of the binding energy of the electron. Which of these quantities is positive and which negative?

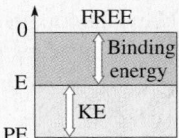

Figure Q10

11. Figure Q11 shows a hypothetical four-level laser. How does it work, and what are the advantages of the arrangement?

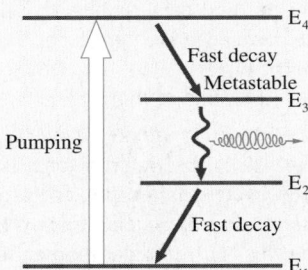

Figure Q11

12. It is theorized that the Universe was created roughly 17 thousand million years ago and is still expanding from that explosive moment. If that's the case, the original radiant fireball should have cooled considerably (via the Doppler Effect) as the Universe spread out to its present size. Less than a year after the so-called Big Bang, the radiation and matter should have reached equilibrium. By 300,000 years later, the radiation was red-shifted down to ≈ 4000 K and neutral atoms formed. Since then, the Universe has expanded about a thousandfold, and we can now expect isotropic cosmic background radiation with a blackbody temperature of about 3 K. Figure Q12 is a plot of brightness versus wavelength of the radiation detected by the *Cosmic Background Explorer* (*COBE*) satellite. The measured microwave spectrum corresponds to the black dots, and the solid line is the Planck blackbody curve for a temperature of 2.735 ± 0.06 K. What can you conclude from all of this?

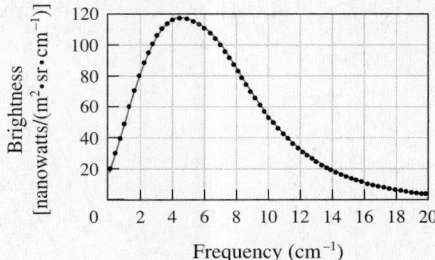

Figure Q12

13. In 1986, a single barium ion (and in a different experiment, a mercury ion) was held in a "trap" and excited by laser light (p. 1014). The ion has two energy levels above the ground state that were utilized. The first was an ordinary short-lived state, and when bathed in light of the proper frequency, the atom absorbed and emitted 100 million times per second, putting out a seemingly continuous stream of light that was easily observed in a microscope. The other higher state was metastable, and once excited, the atom took several seconds to make the transition back down to the ground state via the emission of a lone photon (which went undetected). When stimulated into the first state, the atom shone with a "continuous" light; but whenever it was twinked into the metastable state, the shining light blinked off only to blink back on seconds later. Explain what happened.

14. Explain the meaning of each of the Key Terms on page 1019.

Multiple Choice Questions

1. Pour hot water into a box whose outer walls are painted either black or white. Place thermal detectors at each face. Even though the entire surface of the box is at one temperature, (a) the black faces will be hotter (b) the white faces will be hotter (c) the black faces will radiate more energy (d) the white faces will radiate more energy (e) none of these.

2. It follows from the Stefan-Boltzmann Law that the energy per unit area per unit time radiated by a blackbody (a) depends on its total mass (b) depends on its color (c) depends on its temperature (d) depends on its total area (e) none of these.

3. When the absolute temperature of a blackbody is doubled, its total radiation (a) increases by a factor of 16 (b) stays constant (c) increases by a factor of 4 (d) decreases by a factor of 2 (e) none of these.

4. All blackbody radiation curves have a peak that (a) is shifted toward the longer wavelengths as the temperature goes up (b) is shifted toward the longer wavelengths as the temperature goes down (c) is shifted toward the higher frequencies as the temperature goes down (d) is not shifted as the temperature goes up (e) none of these.

5. Old-fashioned carbon filament incandescent light bulbs operated at around 2100 K. Modern tungsten lamps operate at around 2500 K. This more than triples the efficiency because (a) it's always better to have the strength of metal (b) at the higher temperature the blackbody curve puts more energy in the visible (c) the Stefan-Boltzmann Law tells us we will get more light out of the tungsten (d) we always get more energy from a metal than from a semiconductor like carbon (e) none of these.

6. For us humans the theoretically ideal incandescent light bulb would have a filament that operated at around (a) 20 °C (b) 98.6 °F (c) 2500 K (d) 6000 K (e) none of these.

7. In order to escape from the surface of a metal, an electron must overcome a potential barrier at the metal-vacuum interface, which requires (a) a high temperature $T = hf/k_B$ (b) an energy equal to ϕ (c) a stopping potential (d) a photon of energy $h\lambda$ (e) none of these.

The next nine questions relate to the Photoelectric Effect (Fig. MC8).

8. Which metal has the smallest threshold frequency (and, not surprisingly, is the most chemically active)? (a) potassium (b) sodium (c) zinc (d) platinum (e) none of these.

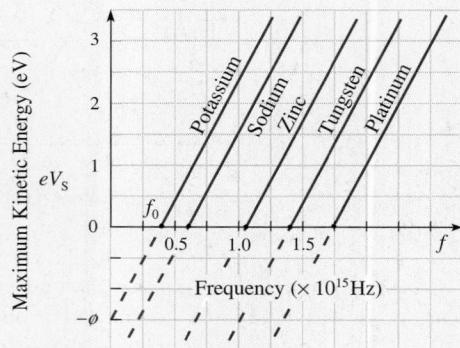

Figure MC8

9. Which metal has the largest work function (and, not surprisingly, is least chemically active)? (a) potassium (b) sodium (c) zinc (d) platinum (e) none of these.

10. Approximately, what is the work function for potassium (a) 1.0 eV (b) 1.5 eV (c) 2.0 eV (d) 2.0 J (e) none of these.

11. Below what frequency will the photocurrent be zero? (a) $f_0 = \phi/h$ (b) $f_0 = KE_{max}/h\phi$ (c) $f_0 = h/\phi$ (d) $f_0 = eV_s\phi/h$ (e) none of these.

12. Each straight-line plot has the same slope, and it's equal to (a) ϕ (b) KE_{max} (c) V_S (d) h (e) none of these.

13. With incident EM radiation such that $f > f_0$, if you wish to increase the number of photoelectrons emitted from any one of the metals in the diagram you need only (a) decrease the metal's work function (b) increase the wavelength of the radiation (c) decrease KE_{max} (d) increase f (e) none of these.

14. The work function of a metal depends on (a) the frequency of the incident radiation (b) the voltage applied (c) the maximum kinetic energy of the electrons (d) the stopping potential (e) none of these.

15. In the photoelectric effect, if $f > f_0$ and the irradiance of the incident beam is doubled, the photocurrent (a) is unchanged (b) decreases by a factor of 4 (c) doubles (d) is halved (e) none of these.

16. When dealing with photoelectrons, a plot of (a) e times the stopping potential against photon frequency is a straight line for each metal (b) photocurrent against potential is a straight line of slope h for each metal (c) photocurrent against incident intensity is a straight line of slope h for each metal (d) incident intensity against photon frequency is a straight line of slope ϕ for each metal (e) none of these.

17. When a photon is Compton-scattered, the (a) minimum shift $\Delta\lambda$ will equal twice the Compton wavelength (b) minimum shift $\Delta\lambda$ will equal h times twice the Compton wavelength (c) maximum shift $\Delta\lambda$ will equal h times twice the Compton wavelength (d) maximum shift $\Delta\lambda$ will equal twice the Compton wavelength (e) none of these.

18. Conservation of both energy and momentum leads to the conclusion that (a) only free electrons can absorb a photon (b) only bound electrons can absorb a photon (c) no electrons can absorb a photon (d) both free and bound electrons can absorb a photon (e) none of these.

19. As the hydrogen atom goes from one Bohr orbit to another, increasing in radius, the (a) electron's speed increases (b) electron's speed decreases (c) electron's speed remains unaltered (d) proton's speed increases (e) none of these.

20. The higher the principal quantum number of a Bohr orbit, the (a) smaller the orbit and the less negative the energy (b) larger the orbit and the more negative the energy (c) larger the orbit and the less positive the energy (d) larger the orbit and the more positive the energy (e) none of these.

21. The classical planetary model of the hydrogen atom is unacceptable because (a) the Coulomb force is too weak to hold the electron (b) the nucleus is too powerful to allow the electron to orbit it (c) the electron is accelerating and must radiate away its energy (d) the electron's angular momentum is all wrong to be in such an atom (e) none of these.

22. The greater the principal quantum number, the (a) closer are adjacent energy levels (b) farther apart are adjacent energy levels (c) more nearly constant are the separations of adjacent energy levels (d) more rapidly the separations between adjacent levels increase and decrease (e) none of these.

23. A population inversion means that (a) there are more atoms in one gas than in another (b) there are more atoms in some excited state than in a lower state (c) there are more states populated than unpopulated (d) the lower states are filled rather than the higher ones (e) none of these.

24. To pump a laser means to (a) lower all its electron states (b) remove some gas from it (c) put pressure on the active medium (d) excite its atoms (e) none of these.

For more Multiple Choice Questions with answers click on WARM-UPS in CHAPTER 28 on the CD.

Suggestions on Problem Solving

1. Some useful constants to keep in mind are

$$1\ eV = 1.602\ 177 \times 10^{-19}\ J$$

$$hc = 1.239\ 842 \times 10^{-6}\ eV \cdot m \approx 1240\ eV \cdot nm$$

$$h = 6.626\ 08 \times 10^{-34}\ J \cdot s = 4.135\ 67 \times 10^{-15}\ eV \cdot s$$

$$h = 1.054\ 573 \times 10^{-34}\ J \cdot s = 6.582\ 12 \times 10^{-16}\ eV \cdot s$$

2. In problems involving electrons, we might be given the energy in eV or, equivalently, told the potential across which an electron moves. It's then convenient to use $m_e = 0.511\ MeV/c^2$. The $1/c^2$ tells us we are dealing with mass, not energy (MeV). To get the electron's mass in kg, *multiply* $0.511\ MeV/c^2$ by 1.602×10^{-13} J/MeV and then *divide* by c^2 in $(m/s)^2$. The $1/c^2$ is not to be treated as if it were a unit like s in m/s, which we get rid of by multiplying by s. The quantities $1/c^2$ or $1/c$ are literal divisions; they're not carried out numerically because we anticipate that they will

soon cancel and it's convenient to leave them in this form, provided it doesn't confuse the issue.

3. When calculating things such as the minimum X-ray wavelength $\lambda_{min} = hc/eV$, the energy will often be encountered in electron volts, whereupon it's helpful to use $hc = 1.239\ 842 \times 10^{-6}$ eV·m.

4. Computations concerning the Compton Effect can be simplified a bit by utilizing the Compton wavelength: $(h/m_ec) = 0.002\ 426$ nm.

5. Many modern calculators have built-in constants like c, h, e, k_B, and m_e. It will often be easier to use these built-in numbers and let them determine the form of the units (that is, eV versus J, for example).

Problems ✦ Coordinated Problems ✦ Progressive Problems ✦ Solutions

SECTION 28.1: BLACKBODY RADIATION

1. [I] Suppose that a person has an average skin temperature of 33°C and a total naked area of 1.4 m². If the person's total emissivity is 97%, find the net power radiated per unit area, the irradiance, when the environment is room temperature. How much energy does that person radiate per second?

2. [I] An object resembling a blackbody is raised from a temperature of 100 K to 1000 K. By how much does the amount of energy it radiates increase?

3. [I] Assume you are a blackbody at 33 °C (external temperature). What is the wavelength at which you radiate most energy per unit wavelength?

4. [I] An object resembling a blackbody at room temperature (20 °C) radiates energy into the environment. What is the wavelength that carries away the most energy per unit wavelength?

5. [I] A class O blue-white star has a surface temperature of around 40×10^3 K. At what wavelength will it radiate the most energy per unit wavelength? [*Hint: Remember Willy Wien.*]

6. [I] Per unit wavelength, the Sun radiates the most energy at 470 nm. Taking it to be a blackbody, determine the temperature of its outer layer, the photosphere.

7. [I] At what wavelength will a stove at 1000 K radiate the greatest amount of energy per unit wavelength, assuming it to be a blackbody?

8. [II] If a blackbody radiates 500 J at 20.0 °C in the course of 5.00 h, how much will it radiate in that same amount of time when its temperature is raised to 1000 °C?

9. [II] A blackbody radiates 180 J at 20.0° in 8.00 h; what is its surface area?

10. [II] In Wien's Displacement Law the peak wavelength is inversely proportional to T; but because $f = c/\lambda$, the peak frequency is directly proportional to T. As a result, the frequency version of the law is

$$f_p = (58.8 \times 10^9\ Hz/K)T$$

The unit frequency interval is not equal to a constant times the unit wavelength interval. Show that $c \neq \lambda_p f_p$.

11. [II] THIS PROBLEM EXAMINES BLACKBODY RADIATION. The star Rigel (β Orionis) is a blue-white supergiant 50 times larger than the Sun. It radiates a maximum amount of energy per unit frequency interval, at a frequency (f_p) of 7.06×10^{14} Hz. (a) What does the fact that this frequency is almost in the ultraviolet suggest? Assuming the star radiates like a blackbody, we want to approximate its surface temperature. (b) With the previous problem in mind write an expression for f_p as a function of T. (c) Determine the surface temperature remembering that it's in kelvins. (d) Compare that to the surface temperature of the Sun (Table 12.1).

12. [II] Spica is the brightest star in the constellation Virgo, and it has a surface temperature of about 20×10^3 K. Review the previous two problems. (a) We measure the radiation using a detector that accepts a narrow frequency band, the center of which is swept across the spectrum. At what peak frequency (in the middle of the band) will the greatest amount of energy be received from the star. (b) At what wavelength will the energy be maximum as measured by a different device that admits a narrow wavelength range?

SOLUTION: (a) As we've seen in Problem 10 $f_p = (58.8 \times 10^9 \text{ Hz/K})T$ = $(58.8 \times 10^9 \text{ Hz/K})(20 \times 10^3 \text{ K}) = 1.176 \times 10^{15} \text{ Hz or } 1.2 \times 10^{15} \text{ Hz}$.
(b) Similarly from Eq. (28.4), $\lambda_p T = 0.002\,898 \text{ m·K}$ and $\lambda_p = 144.9 \text{ nm}$ or 1.4×10^2 nm.

13. [II] Determine the frequency at which a blackbody at 1000 °C will radiate the greatest amount of energy per unit frequency. [*Hint: Use the discussion of the previous problem.*]

14. [II] A blackbody is at a temperature of 6000 K. At what wavelength will it radiate the most energy per unit wavelength? At what frequency will it radiate the most energy per unit frequency? [*Hint: Reread the previous two problems.*]

15. [II] At what temperature will an object resembling a blackbody emit a maximum amount of energy per unit wavelength in the red end of the visible region of the spectrum ($\lambda = 650$ nm)?

SECTION 28.2: QUANTIZATION OF ENERGY:
THE PHOTOELECTRIC EFFECT

16. [I] What is the energy in joules of a 0.100-nm photon?

17. [I] What is the wavelength of a 2.50-eV photon?

18. [I] What is the energy in keV of a 0.20-nm photon?

19. [I] A typical light level for good reading corresponds to about 2×10^{13} photons per second per square centimeter. If these have an average wavelength of 550 nm, what is the corresponding irradiance?

20. [I] What is the threshold frequency that will result in photoemission from zinc?

21. [I] Photons of wavelength 650 nm liberate photoelectrons with negligible kinetic energy from a metal target. What is the threshold frequency of the metal?

22. [I] Determine the cut-off wavelength for the emission of photoelectrons from tungsten, which has a work function of 4.52 eV.

23. [I] Ultraviolet radiation of wavelength 200 nm is shone on a zinc target. What's the maximum kinetic energy of the emitted electrons?

24. [I] In the photoelectric effect, what frequency of radiant energy should be shone on iron to produce electrons with a maximum kinetic energy of 1.50 eV?

25. [I] THIS PROBLEM EXPLORES THE THRESHOLD FREQUENCY OF A METAL. A single crystal of any metal will have a work function that depends on the orientation of its atoms w.r.t. the surface being bombarded. (a) One such work function for a gold crystal is 5.47 eV, and we want to examine the corresponding threshold frequency. Write out Einstein's Photoelectric Equation. (b) Define the threshold frequency and explain what $KE_{max} = 0$ means. (c) Write an expression for the threshold frequency in terms of the work function. (d) Using h in units of eV·s compute f_0.

26. [II] THIS PROBLEM EXPLORES THE PHOTOELECTRIC EFFECT. The work function of a metal is known to be 3.1×10^{-19} J. (a) Write out Einstein's Photoelectric Equation. (b) If the metal is illuminated by a monochromatic beam, what minimum frequency must it have in oder to just begin to liberate electrons? What is the value of KE_{max} under these circumstances? (c) What is the corresponding wavelength? (d) Determine the maximum kinetic energy (in joules) of the photoelectrons when the illumination has a wavelength of 450 nm. (e) Express the maximum kinetic energy in eV.

(f) What is the value of the stopping potential under 450-nm illumination?

27. [II] THIS PROBLEM IS ABOUT THE PHOTOELECTRIC EFFECT. A polycrystalline sample of the metal cesium has a work function of 2.14 eV. White light in the frequency range from 384 THz to 769 THz is incident on the sample. (a) Define the threshold frequency and write an expression for it in terms of the work function. (b) Compute the threshold frequency for cesium. (c) What range of frequencies in the beam will produce photoelectrons? (d) Using h in units of eV·s, compute the maximum kinetic energy that will be imparted to the electrons in units of eV. How much is that in joules?

28. [II] THIS PROBLEM DEALS WITH THE PHOTOELECTRIC EFFECT. Ultraviolet radiation of wavelength 200 nm is shone on a metal target. Electrons are ejected from the metal with a wide range of speeds from essentially zero up to 1100 km/s. (a) Write out Einstein's Photoelectric Equation. (b) Determine the maximum kinetic energy of the electrons. (c) What is the frequency of the incident radiation? (d) Determine the energy in eV of the incident photons. (e) Using h in units of eV·s, compute the work function in eV.

29. [II] The current in a Photoelectric Effect experiment decreases to zero when the retarding voltage is raised to 1.25 V. What is the maximum speed of the electrons?

30. [II] Suppose that 60-nm radiant energy is incident on a metal whose work function is negligibly small. Compute the maximum speed of the liberated photoelectrons.

31. [II] Photons of wavelength 220 nm impact on a metal target and liberate electrons with kinetic energies ranging from 0 to 61×10^{-20} J. Determine the threshold frequency and wavelength. [*Hint: First calculate the work function.*]

32. [II] Ultraviolet radiation ($\lambda = 250$ nm) falls on a metal target and electrons are liberated. If the maximum kinetic energy of these electrons is 1.00×10^{-19} J, what is the lowest frequency electromagnetic radiation that will initiate a photocurrent on this target?

33. [II] A target is irradiated with UV at 240 nm. Electrons are emitted, and they have a broad range of kinetic energies from near zero all the way up to 12.0×10^{-20} J. The experiment is begun over, this time starting with incident light at 500 nm and increasing the frequency until a photocurrent just begins to flow. Determine that initiating photon frequency and the corresponding wavelength.

34. [II] In the Photoelectric Effect, the target has a stopping potential of 4.00 V when the incident energy has a momentum of 3.50×10^{-27} kg·m/s. What is the threshold frequency?

35. [III] Approximate the time it takes for an electron in the Photoelectric Effect to absorb enough energy (≈ 1 eV or 2 eV) from a classical electromagnetic wave to be liberated. First assume a He-Ne laser source of 10 W/m². Compare this result with the time corresponding to an irradiance of about 1 µW/m², which is detectable by sodium. Assume the electron in a sodium atom absorbs over an area equal in radius (0.10 nm) to its orbit (which should be giving it the benefit of the doubt).

SECTION 28.3: BREMSSTRAHLUNG
SECTION 28.4: THE COMPTON EFFECT

36. [I] What is the shortest-wavelength X-ray emanation that can be expected from a tube operating at a voltage of 30.0 kV?

37. [I] X-rays with a wavelength of 110 pm are scattered off free electrons at an angle of 20.0°. Find the change in λ. [*Hint: This is Compton scattering.*]

38. [I] A beam of X-rays shines onto a target and essentially free electrons and photons scatter away. If the scattered photons come off at an angle of 18.0° with respect to the incident-beam direction, what will be the shift in wavelength experienced by the X-rays?

39. [I] A sample is irradiated with X-rays of wavelength 0.100 nm. Some of this energy is Compton scattered at 45.0°; determine the wavelength of this radiation.

40. [I] A 60-keV photon is Compton-scattered off an electron at rest. In the process, a 25-keV photon comes sailing off at 35°. What is the kinetic energy of the scattered electron?

41. [I] At what scattering angle does the photon come off in the Compton Effect when the electron has a maximum kinetic energy?

42. [I] Compton shone 71.0-pm photons at a target containing loosely bound electrons and examined the scattered photon beam at 90.0°. What was the wavelength of the scattered photon?

43. [II] THIS PROBLEM EXPLORES THE PRODUCTION OF X-RAYS VIA BREMSSTRAHLUNG. A beam of electrons is accelerated across a high potential difference inside an X-ray tube. A continuous range of X-ray frequencies is produced as the electrons collide with the target atoms and emit radiation. (a) Under what circumstances do the electrons generate X-ray photons of maximum frequency? Suppose the voltage difference is 50.0 kV. (b) How much energy does an electron pick up as it flies across that potential difference? Give your answer in both keV and joules. (c) The beam of electrons is shot at a metal target. What is the maximum energy of the resulting photons. (d) Write an expression for the energy (E) of a photon in terms of its frequency. (e) Write an expression for the energy of a photon in terms of its wavelength. (f) Determine the minimum wavelength of the X-rays in the setup we're dealing with.

44. [II] An X-ray photon scatters from an essentially free electron and flies off at an angle of 39.5° with respect to the incident direction. By how much does the photon's wavelength change in the process?

> **SOLUTION:** This is an example of Compton Scattering and so, remembering that the Compton Wavelength is 0.002 426 nm, $\Delta\lambda = (h/m_e c)$ $(1 - \cos\theta) = (2.426 \times 10^{-12}$ m$)(1 - \cos 39.5°) = 5.54 \times 10^{-13}$ m.

45. **[II]** An electron traveling at 1.00×10^8 m/s is fired at a metal target. On impact, it rapidly decelerates to half that speed emitting a photon in the process. What is the wavelength of the photon?

46. [II] In the Compton experiment, derive an expression for the kinetic energy of the scattered electron in terms of the incident and scattered photon wavelengths.

47. [II] On scattering via the Compton Effect, a photon undergoes a *fractional wavelength change* $(\lambda_s - \lambda_i)/\lambda_i$ equal to 6.00%. If the incident photon has a wavelength of 0.020 0 nm, at what angle is the detector to the incident beam?

48. [II] X-ray photons of wavelength 0.220 0 nm are Compton-scattered at 45°. What is the energy of the scattered photons (in eV)?

49. [III] The Compton equation (28.13) is a formula in terms of photon parameters and the constant m_e. To derive it, start with Eq. (28.11) and write expressions for conservation of the x- and y-components of momentum. Show that

$$p_i^2 + p_s^2 - 2p_i p_s \cos\theta = p_e^2$$

Then, using Conservation of Energy, establish that

$$E_e = E_i + m_e c^2 - E_f$$

Substitute both of these formulas into the equation for the electron's energy

$$E_e^2 = c^2 p_e^2 + m_e^2 c^4$$

thereby getting an expression containing the electron's constant rest-energy and the energies and momenta of only the photons. Using $E = pc$ for photons, simplify, and Eq. (28.13) follows in short order.

SECTION 28.5: THE BOHR ATOM
SECTION 28.6: STIMULATED EMISSION: THE LASER
SECTION 28.7: ATOMIC NUMBER

50. [I] Determine (to three significant figures) the energy in electron volts and in joules of the first excited state of the hydrogen atom.

51. [I] Compute the energy in eV of the second excited state of a hydrogen atom (to three significant figures).

52. [I] With Problems 50 and 51 in mind, what is the energy (in joules) of the quantum of light given out when a hydrogen atom drops from its second to its first excited state?

53. **[I]** With the last problem in mind, what is the frequency and wavelength of the light given out when a hydrogen atom drops from its second to its first excited state?

54. [I] What is the value in eV of the $n = 5$ energy level of the hydrogen atom?

55. [I] What is the value in eV of the $n = 2$ energy level of the hydrogen atom?

56. [I] How much energy is liberated when a hydrogen atom in an $n = 5$ excited state drops down to the $n = 2$ state? Give your answer in eV and look at the two previous problems.

57. [I] Show that the units of action defined as either energy multiplied by time or momentum multiplied by position are equivalent.

58. [I] The carbon dioxide laser usually emits at 10.6 μm in the IR. Typically such a device puts out a continuous emission of from a few watts to several kilowatts. Determine the energy difference in eV between the two laser levels for this wavelength.

59. [I] What is the radius of the second excited state of the hydrogen atom according to the Bohr Theory? [*Hint: n = 1 is the ground state.*]

60. [I] Determine (to four significant figures) the radius of the first excited state of the hydrogen atom according to the Bohr theory.

61. [I] According to the Bohr theory the hydrogen atom in an excited state can have a radius of 0.846 68 nm. Which state is this?

62. [I] A certain variation of He-Ne laser operates between two levels that are 3.655×10^{-19} J apart. What color is the beam? Compute the wavelength.

63. [I] Compute the principal quantum number corresponding to a hydrogen atom with a 1-m orbital radius.

64. [I] The anode of an X-ray tube emits K_α radiation of wavelength 0.075 9 nm. What metal is the target made of?

65. [II] What is the speed of the electron in the ground state of the hydrogen atom according to Bohr?

66. [II] Determine the force holding the electron in orbit in the ground state of the hydrogen atom.

67. [II] THIS PROBLEM WILL HELP US BETTER UNDERSTAND THE BOHR THEORY OF THE ATOM. Suppose a monoenergetic beam of charged particles is shot into a chamber containing a low-temperature tenuous gas of atomic hydrogen. The atoms are all assumed to be in their ground states. The energy of the particles is gradually raised until the gas begins to give off light corresponding to a Balmer line. (a) What is the principal quantum number for the Balmer series? (b) What is the quantum number for the lowest energy level that the atom must be excited to if it is to drop back to $n = 2$ and emit a Balmer photon? (c) What is the energy of the lowest excited state that will do the trick? (d) Make a sketch of the energy levels. (e) Keeping in mind that the energy of the ground state is -13.6 eV, how much energy should the charged particles have in order to properly excite the atom?

68. [II] THIS PROBLEM UTILIZES THE BOHR THEORY OF THE ATOM. A hydrogen atom in the $n = 3$ excited state drops down to the ground state and emits a single photon. Because the photon flies off in some random but well-defined direction, the atom recoils in the opposite direction. The observation of this recoil is a strong testament to the photon picture of atomic emission. (a) Using Eq.

(28.25) along with the Rydberg constant, determine the wavelength of the radiation when the atom drops to the ground state. (b) What kind of EM radiation is this? (c) How much momentum does the photon have? (d) How much momentum is imparted to the atom?

69. [II] Compute the ground-state energy of singly ionized helium using the Bohr Theory.

70. [II] Use the Bohr Theory to prove that for hydrogen-like atoms

$$v_n = \frac{k_0 Z e^2}{n\hbar}$$

71. [II] Derive an expression for the frequency (f_n) of the electron revolving in the nth Bohr orbit of a hydrogen-like atom.

72. [II] With Problem 71 in mind, use the Bohr Theory to prove that

$$|E_n| = \tfrac{1}{2}nhf_n$$

for the various orbits (and not $E_n = nhf_n$).

73. [III] When a photon is absorbed by an atom in the process of excitation, the atom must recoil if it is to conserve momentum. Consider a photon able to raise a 939-MeV/c² hydrogen atom into its first excited state. Determine the recoil energy of the atom and notice that it is negligible.

74. [III] Compute the energy of the third excited state of the helium atom in the Bohr Theory and compare it to the first excited state of hydrogen.

Chapter 29
Quantum Mechanics

The pace of things intellectual picked up after the First World War (caution seemed less fitting in a battered world). The Jazz Era was largely a period of revolt and irreverence, and that selfsame turbulent mood gave form to the new physics that rose like a flash out of the Roaring Twenties. The hodgepodge of *ad hoc* ideas that constituted the Old Quantum Theory had gone about as far as it could. Quantum Mechanics emerged around 1925 as a complete theoretical structure that subsumed, and went far beyond, Classical Physics and the Old Quantum Theory.

Quantum Mechanics is far too mathematically complex for us to attempt to understand it here. Indeed, as Richard Feynman (1967) penetratingly remarked, "nobody understands quantum mechanics." But we can get a sense of it, see where it came from, learn a little about its strengths and weaknesses, and appreciate some of its accomplishments.

The Conceptual Basis of Quantum Mechanics

Amazingly, when physicists set about reformulating classical theory in the 1920s, they seemed at first to have come upon an overabundance of riches. Heisenberg created Matrix Mechanics, Dirac produced Transformation Theory, and Schrödinger devised Wave Mechanics, all almost simultaneously. Although their approaches were distinctive and their resulting formalisms seemed equally unique, it was soon shown that all three theories were essentially equivalent. These formulations are the basis of what is broadly called Quantum Mechanics. Our study will be cursory at best and limited to only Wave Mechanics.

29.1 De Broglie Waves

In the summer of 1923, the French aristocrat Prince Louis Victor Pierre Raymond de Broglie (pronounced to rhyme with *Troy*) proposed that the wave-particle duality (p. 1003) manifested by radiant energy might be a fundamental characteristic of *all* entities (for example, electrons, protons, and neutrons). Einstein had already shown that substantial matter (matter possessing mass) and radiant energy are interconvertible. Why

Prince Louis de Broglie (1892–1987) spent much of his life teaching physics at the Sorbonne and at the Institut Henri Poincaré, both in Paris.

should they not display similar properties? In particular, might not substantial matter have some sort of wave aspect associated with it? Suppose that Eq. (28.10), $p = h/\lambda$, which describes photons, applies to all matter. In that eventuality, with $p = mv$ for material entities, we have

$$\lambda = \frac{h}{mv} \quad (29.1)$$

Matter in motion has a wavelength. Not that de Broglie knew how to visualize such waves or even what was doing the waving. A particle moving with a constant momentum is associated with a monochromatic wave ($\lambda = h/p$) having a single frequency determined by its *total energy* ($f = E/h$) via Eq. (22.5). As with all waves, it moves with a phase speed given by $v_p = f\lambda$, which is usually different from, though related to, the speed of the particle (Problem 16).

A REMARKABLE ACCIDENT

However theoretical, the idea that electrons and protons were somehow wavelike, just as light was particlelike, brought a pleasing symmetry to Nature. This fascinating speculation was the basis of a Ph.D. thesis that de Broglie submitted to his physics professors in Paris. Still, the concept was considered outlandish, and a copy of the work was sent to Herr Professor Einstein for his opinion. The great man was enthusiastically supportive, and the degree was promptly granted. Six years later (1929), de Broglie received the Nobel Prize for the idea that particles are waves—but a remarkable accident happened first.

In April of 1925, the American C. J. Davisson was busy scattering electrons off a polycrystalline nickel target. While collaborating with L. Germer, an explosion rocked the lab, but after putting the experiment back together, it was clear that something very strange had happened: the data "completely changed." Unbeknownst to them, while cleaning up the target through prolonged heating, the nickel sample had reformed into a few large crystals.

The more important fundamental laws and facts of physical science have all been discovered, and these are now so firmly established that the possibility of their ever being supplanted in consequence of new discoveries is exceedingly remote. Our future discoveries must be looked for in the sixth place of decimals.

ALBERT MICHELSON (1899)
AMERICAN PHYSICIST

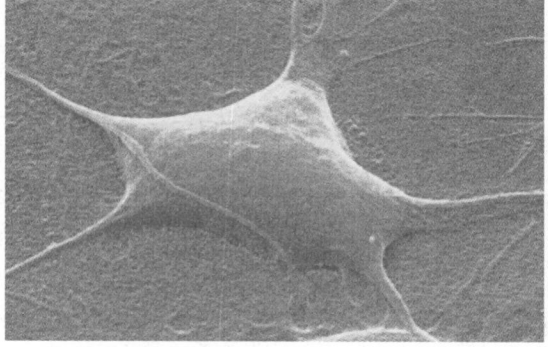

Electron micrograph of a neuron (see Fig. Q1). With an accelerating voltage of around 10^5 V, the electrons have a de Broglie wavelength of about 0.004 nm.

When they subsequently directed an electron beam at the target, they saw a diffraction pattern identical to the one produced by X-rays (p. 975), even though it's impossible for a stream of particles to do that. Another year passed before anyone realized that Davisson and Germer had verified de Broglie's hypothesis; Eq. (29.1) matched the data in every detail. ***Electrons with comparable momenta to those of X-ray quanta are diffracted as if they were waves of the same wavelength***.

De Broglie's wave-particle theory offered another provocative interpretation of the Bohr atom. The quantization of angular momentum, $L = nh$, means that $mvr = nh/2\pi$ and $mv = nh/2\pi r$. But if $mv = h/\lambda$, then $n\lambda = 2\pi r$; in other words, the circumference of each of Bohr's allowed circular orbits ($2\pi r$) exactly equals an integer multiple of the electron wavelength ($n\lambda$). The orbits are electron standing-wave configurations (Fig. 29.1).

Example 29.1 **[II]** Calculate the wavelength of an electron once it has been accelerated through a potential of 110 V and compare that to X-rays. Assume the speed is nonrelativistic.

Solution Whenever you're asked for the wavelength of a particle think de Broglie. (1) TRANSLATION—An electron is accelerated through a known potential; determine its wavelength. (2) GIVEN: $V = 110$ V. FIND: λ. (3) PROBLEM TYPE—Conservation of Energy/electric potential/de Broglie wavelength. (4) PROCEDURE—The defining equation is $p = h/\lambda$ or $\lambda = h/mv$. (5) CALCULATION—To get λ, we first need v. Since for the electron $KE_f = PE_i$, $\frac{1}{2}m_e v^2 = eV$

$$v = \sqrt{\frac{2eV}{m_e}} = 6.22 \times 10^6 \text{ m/s}$$

it follows that

$$\lambda = \frac{h}{m_e v} = 1.17 \times 10^{-10} \text{ m} = \boxed{0.117 \text{ nm}}$$

This is in the wavelength range of X-rays, just about the size of an atom.

Quick Check: We'll first derive a useful expression for λ: $KE = p^2/2m$, $\lambda = h/p = h/\sqrt{2m(KE)}$, multiplying top and bottom by c, and using the fact that $hc = 1239.8$ eV·nm (p. 1023), we have $\lambda = (1239.8 \text{ eV·nm})/\sqrt{2(mc^2)(KE)}$. For an electron, $mc^2 = 0.511$ MeV, and using the **kinetic energy in eV**, we have

[for electrons]
$$\lambda = \frac{1.226\,4}{\sqrt{KE}} \text{ nm} \qquad (29.2)$$

Here, where $KE = 110$ eV, $\lambda = 0.117$ nm.

{For more worked problems click on **WALK-THROUGHS** *in* **CHAPTER 29** *on the CD.}*

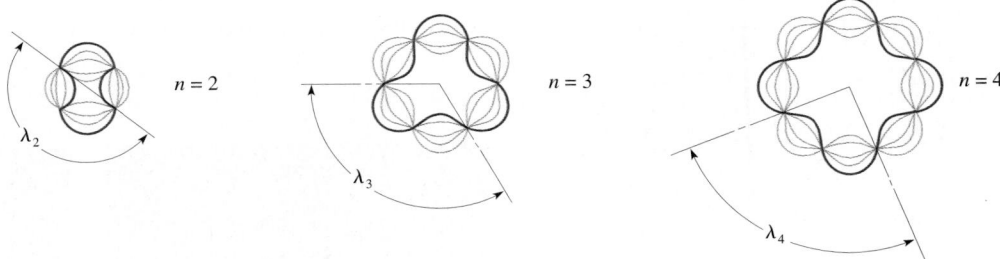

Figure 29.1 De Broglie standing waves for three excited states ($n = 2, 3, 4$) of hydrogen. Each Bohr orbit corresponds to a standing-wave configuration for the electron.

29.2 The Principle of Complementarity

When light propagates beyond an obstruction or through a hole, it casts a complex pattern in the process of diffracting, a pattern that can be analyzed in all its detail only if the light is taken as a wave. It is remarkable that researchers using all sorts of beams of matter, ranging from electrons and neutrons to potassium atoms, have now observed exactly the same complex interference and diffraction patterns. Classically, there is just no way to understand how beams of atoms can come together on a screen in clumps that exactly match optical interference patterns.

If the "logic" of Eq. (29.1) holds, and no one knows if it does, it applies equally well to oranges, baseballs, and fire engines as it does to electrons. The momentum of an ordinary macroscopic object in motion is so gigantic that λ is much too small to ever measure, and that even applies to something as small as a tiny dust mote gently falling through the air. Bullets may well have de Broglie wavelengths, though we'll never see them bending around obstacles. Unless you can pass a camel through the eye of a needle, observing diffraction (and thereby λ) is out of the question for structures made up of millions of atoms.

Microparticles (electrons, protons, photons, atoms, and so forth) propagate as if they were waves and exchange energy as if they were particles—that's the ***wave-particle duality***. When we measure the arrival of one of them at a detector, we always measure the energy it delivers to a single point, even though that point may be part of a pattern that could only be created by a wave. In any experimental event, such entities will do one or the other—they cannot simultaneously manifest the properties of wave and particle. The imagery of particle and wave are two aspects of the complete picture of the photon, and they do not compete; rather, they complement each other. That notion (which does not immediately make things any less disconcerting) was first mentioned by Bohr in 1925 and came to be known as the **Principle of Complementarity**. Energy multiplied by temporal period ($T = 1/f$) equals momentum multiplied by spatial period (λ):

$$E T = p\lambda = h$$

The particle notions of E and p complement the wave notions of T and λ.

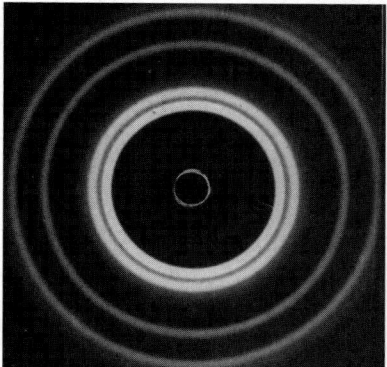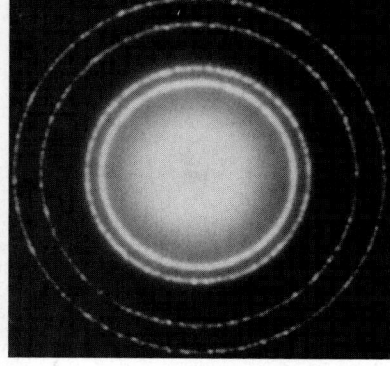

The concentric ring patterns produced when a beam of X-rays (left) and a beam of electrons (right) pass through the same thin polycrystalline aluminum foil. The two diffraction patterns are almost identical.

29.3 The Schrödinger Wave Equation

Erwin Schrödinger (1887–1961). In 1925, he became interested in de Broglie waves, and by the end of that year he derived a correct relativistic version of Wave Mechanics. Thinking he was in error (because, without the concept of spin, the theory does not predict fine structure), he abandoned it and published the more limited nonrelativistic wave equation.

Schrödinger's Equation is the equation of motion of de Broglie waves, ψ-waves, whatever they are.

In 1657, Pierre Fermat proposed the intriguing idea that light, in going from one place to another, traveled along the route that took the least time. In a single simple statement, Fermat seemed to have gotten an overview of optics. His successes (for example, the derivation of Snell's Law) stimulated a great deal of effort to supersede Newton's Laws with a similar formulation. The resulting dynamical *Principle of Least Action* was introduced, albeit in a confused fashion, by Maupertuis (1747) and formalized by William Rowan Hamilton about a hundred years later. The principle asserts that a dynamical system moves in such a way as to keep its action ($\sum p \Delta q$) a minimum. (Here p is the momentum and q is the corresponding, or *conjugate*, space variable, for example, p_x and x or L and θ.) The kinship between the propagation of light and the motion of a material particle is already there in this early work.

Hamilton found that there are mathematical surfaces over which the action of a system is constant and that these *surfaces of constant action* propagate in a completely analogous fashion to the way surfaces of constant phase (that is, wavefronts) propagate in optics. The same equations apply to both! The angular temporal frequency ($\omega = 2\pi f$) of the light corresponds to the energy (E) of the particle; the angular spatial frequency ($k = 2\pi/\lambda$) of the light corresponds to the momentum (p) of the particle. Hamilton maintained that Newtonian Mechanics corresponded to the ray representation of Geometrical Optics. But the latter is only an approximation of the complete description of Wave Optics. What it lacks is interference, and what Hamilton's ray-dynamics for material entities lacked was also interference. Of course, this all happened almost a hundred years before de Broglie introduced the notion of the interference of electrons (the electron wasn't even discovered yet). For us, the question almost leaps from the page: *Is Newtonian Mechanics an approximation of some as yet undiscovered Wave Mechanics?*

The Viennese physicist Erwin Schrödinger learned about de Broglie's wave hypothesis from a footnote in one of Einstein's papers—he was already quite familiar with Hamilton's work. Just before Christmas, 1925, he went off to the Swiss Alps for vacation, leaving his wife at home. He took along a copy of de Broglie's thesis and an old girlfriend (Erwin was a rather blatant philanderer). When he and the lady returned from their holiday two and a half weeks later, the great breakthrough had already been made. Starting essentially where Hamilton had left off, he produced an extraordinary formula now universally known as **Schrödinger's Wave Equation**.

THE SCHRÖDINGER WAVE EQUATION

The most general form of the Schrödinger Equation is three dimensional and depends on time, but the simpler one-dimensional, time-independent expression

$$\frac{1}{2m}(i\hbar)^2\frac{d^2\psi}{dx^2} + U\psi = E\psi$$

is fine for our purposes. This formula was literally created by Schrödinger and presented as a postulate because it works. The equation cannot be derived directly from Classical Theory, although it has its roots there. Mathematically, it's well beyond what we can deal with here, but because it's one of the most important intellectual accomplishments of the twentieth century, it's worth examining. During the derivation, the expression had a constant K in it. Upon applying the analysis to the orbital electron of the hydrogen atom, Schrödinger immediately got the Bohr energy levels, provided K was set equal to $\hbar$. Interestingly, the theory naturally produced the quantum number n, but (like both Heisenberg's and Dirac's previous analysis) $\hbar$ had to be inserted into the formalism.

Consider a particle, such as an electron in an atom, where m is its mass, U is its potential energy, and E is its total energy. Recalling that $p^2/2m = \text{KE}$, we can write the classical expression for the energy (KE + U = E) as Hamilton first wrote it, namely

$$\frac{1}{2m} p^2 + \text{U} = \text{E}$$

Schrödinger boldly turned this equation into a mathematical expression, a differential equation, that vaguely resembles the standard equation that describes all waves. In the process, momentum became a mathematical operator and what it operated on was a mathematical function ψ (Greek letter *psi*) referred to as the **wavefunction**. The Schrödinger Equation deals with the "motion" of ψ-waves, though it's not at all obvious what these actually represent. Still, the theory exquisitely describes so much of the behavior of the microworld that it must be taken very seriously even if we don't yet quite understand all of the physical processes lying beneath the mathematical machinery.

A particle constrained to a region by an interaction forms a *bound system*—for example, an electron in an atom. ***The quantization of the energy of a bound system is a consequence of the wave equation and certain restrictions placed on the wavefunction.*** Those "reasonable" restrictions simply require that ψ have only one value at each point, that it be continuous, and that it be finite everywhere. The quantum numbers that were introduced into the Bohr Theory in an ad hoc way now emerge, as Schrödinger put it, "in the same way as the integers specifying the number of modes in a vibrating string."

In general, ψ is complex; it contains $i = \sqrt{-1}$ and thus is a so-called imaginary quantity. Accordingly, **ψ is not directly measurable**; it has no physical existence. In other words, ψ does not carry energy, as do all physical waves, and cannot be detected first-hand via that energy. Still, it has the usual mathematical attributes of a wave: it has a frequency, amplitude, and phase; it obeys the superposition principle and so mathematically undergoes interference and diffraction. When two subatomic particles have identical physical characteristics (mass, charge, velocity, and so on), we say they are in the same *quantum state*, wherein they have identical wavefunctions. It is assumed that such particles are physically indistinguishable from one another.

New theories are rarely complete at the moment of presentation, and Schrödinger's was no exception. Although it proved to be remarkably effective, several features were quite troubling from the outset. In particular, what was the physical significance of ψ? In the early days, Schrödinger assumed that the negative charge of the atom was actually spread out in space around the nucleus and that ψ was related to the density of that charge-cloud. Curiously, although Schrödinger's theory became an immense success, the spread-out-charge interpretation of ψ was a disappointing failure.

Probability Waves

Max Born, who was a professor of physics at Göttingen, was troubled by the idea that ψ could undergo diffraction and be split up into separate portions—how could it characterize an electron that, of course, doesn't fragment? One can detect individual electrons with a Geiger counter and see their separate tracks in a cloud chamber. These are real manifestations of a highly localized entity.* In the spring of 1926, guided by a suggestion Einstein had made with regard to photons and the electric field, Born proposed what has now come to be the orthodox interpretation of ψ. He suggested that the ψ-wave associated with a particle was a probability function, an information wave that would tell us not where the particle is, but only where it is likely to be. Today, the first postulate of the orthodox interpretation of Quantum Mechanics is that **the state of a system is completely specified by its wavefunction.**

---

*Electrons interact with each other over distances through their electric fields. Thus, two colliding beams scatter electrons via this field interaction. If the electrons have some size, some extent in space, as the beams come closer and closer, there should be a deviation from pure field scatter. As of this writing, distances down to 10^{-16} cm have been probed with no sign of a hard core—if the electron has a size, it's extremely small.

Figure 29.2 Under exceedingly faint illumination, the pattern (each spot corresponding to one photon) seems random, but as the light level increases, the quantal character of the process gradually becomes obscured. The torrent of photons blends into a seemingly continuous flow of radiant energy. (See *Advances in Biological and Medical Physics* V, 1957, pp. 211–242.)

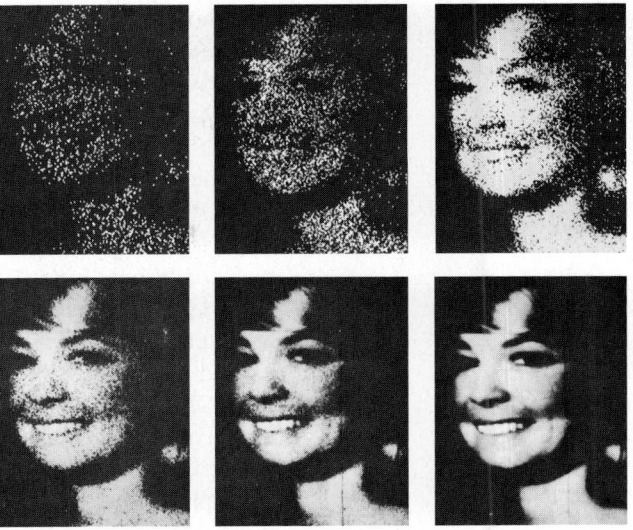

If we shine the image of some scene on a photon detector and then drastically lower the light level, we can watch the flash of each light-quantum arriving, one at a time. The recorded spots of light seem unpredictably random, but the image gradually builds until it is recognizable (Fig. 29.2). Under ordinary conditions, a great torrent of photons produces the picture almost all at once, obscuring the fundamental grainy nature of the process. Similarly, when light impinges on the double-slit arrangement of Young's Experiment (Fig. 29.3), it produces a flutter of flashes on the screen that blends into the familiar irradiance distribution of bright and dark bands. The irradiance (I) is proportional to the square of the amplitude of the electric-field wave (E_0^2) as per Eq. (22.4). The most likely place for a photon to arrive is where the pattern is brightest, where I is largest. Thus, the probability of a photon arriving at any particular location is given by the amplitude of the field squared. Naturally enough, Born suggested that if the experiment is carried out with electrons, or neutrons, or with any microparticle, the probability of finding such a particle at a specific location is given by the square of the amplitude of the wavefunction. Just as waves of electric fields interfere to produce complicated fringe patterns, wavefunctions interfere to produce much the same patterns.

Schrödinger's Equation is quite as deterministic as is Newton's; that is, once the wavefunction is known, its future form can be computed precisely. Nonetheless, *the wavefunction can only tell us the probability of finding a particle at a given point in space*. Determinism as it applies to measurable quantities vanishes in the microworld; what is observable is probabilistic. We never know where a particular electron will end up, but instead we can provide a picture of the likelihood of occurrence of all its various possibilities. Large numbers of particles respond as a group whose overall behavior can be accurately predicted statistically. Even if the individual is lost in the crowd, the crowd *en masse* does its thing in a well-defined fashion that can be measured, understood, and anticipated. If we shine a beam of electrons at a pair of tiny slits, we can calculate the resulting fringe pattern pre-

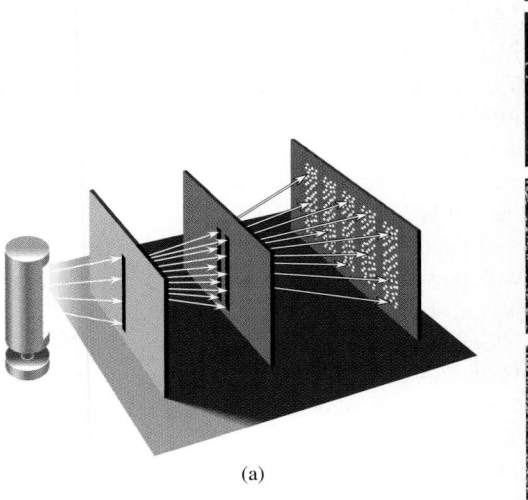

(a)

Figure 29.3 (a) Young's double-slit experiment. Flash by flash, photon by photon, the pattern gradually develops. (b) The double-slit interference pattern, but this time it's generated by a beam of electrons rather than photons. The number of electrons in each frame runs from 10 to 100 to 3000 to 20 000 and, finally, to 70 000.

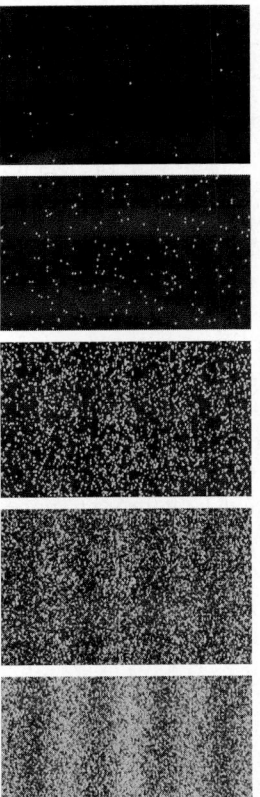

(b)

cisely even though we know nothing of the specific motion of any one particle. Similarly, an actuarial table can tell you the probability of dying for any person in your age group, but it obviously cannot reveal how long you, a particular person, will live.

Still, Quantum Mechanics as we currently understand it challenges us at almost every turn. Suppose we again shine a beam of photons, electrons, or neutrons on a double slit. Now lower the beam intensity to the point where only one microparticle enters the apparatus at a time. It will pass through one of the slits (or not) and cause a flash on the screen (or not) at some unpredictable spot. But if we wait long enough, perhaps a few months, flash by flash, the traditional fringe system will appear. It seems that *each photon, electron, or neutron interferes with itself* (whether it interferes *only* with itself is another question). And it follows that the wavefunction, which undergoes the interference, must extend over both apertures and pass through both apertures. In that sense, the individual photon passes through both slits.

Envision a laserbeam comprising a stream of photons. For what it's worth, one might picture the wavefunction of one such light-quantum as a thin, flat disk of progressing undulations—the photon has a probability of being anywhere within the confines of the beam, and so the wavefunction is defined by the beam diameter. Some would say that the photon was a flying disk, while others keep their wavefunctions purely mathematical and think of the photon as a localized concentration of energy. Still others don't worry about what a photon looks like since it can't be seen in transit anyway. It's even possible in the case of light with a broad classical wavefront (starlight, for example) to separate the slits in Young's Experiment by 10 or 20 meters and still observe interference. If we buy the ψ-wave picture, we must buy the idea that the photon's wavefunction then also extends over tens of meters.

The photon itself must somehow "see" both slits, however far apart they are, and yet arrive at the observation screen as a tiny localized blast of energy. If we cover either aperture, the interference vanishes. Indeed, if we place a particle detector behind either aperture so as to determine if the photon (or electron) has passed through it alone (as a particle might), the interference pattern also vanishes (see Discussion Question 5). And that is presumably true even if the beam is not obstructed as it is detected. Once the particle aspect is observed—once we know it has passed through one slit or the other—the wave aspect is obliterated. It's safe to say that many physicists are still not totally happy with the orthodox (or any other) explanation of these strange phenomena.

Just as surprising, if we set up several million identical double-slit experiments, shine a single identical photon (or electron or neutron) at each one at the same moment, and then superimpose the results from each, the total pattern should be the same fringe system as always. Each photon addresses the diffracting apertures individually and it progresses, guided by its wavefunction pattern, in a totally random way. Where on the theoretically computable, observably verifiable distribution it will land is indeterminate, but that it will land on the distribution is certain. This state of affairs is very peculiar! *It appears that identical particles in identical situations need not behave identically*—that's the essence of **quantum randomness**, which must be postulated in order to account for the observations. And again, not every physicist is happy with these weird goings on, though most are pragmatic enough to keep on cranking out amazing solutions to real physical problems.

If we determine ψ from Schrödinger's Equation for an electron moving in the Coulomb field of a proton, we get a whole range of potential electron locations, a cloud of probabilities. The most likely place to find the electron in the ground state is out at a distance identically equal to the Bohr radius, 0.052 9 nm (Fig. 29.4). There is no longer any reason to think of atomic electrons flying around in little orbits, though what they are doing from one moment to the next is a mystery. Figure 29.5 is a representation of the first few excited states of hydrogen. Keeping in mind that we are dealing with one rapidly moving electron, imagine a cloudlike afterimage around the nucleus, revealing where the electron has been and where it is likely to be.

The indeterministic substructure of the Universe was hinted at decades before Quantum Mechanics, when it was discovered that the decay of radioactive atoms was only

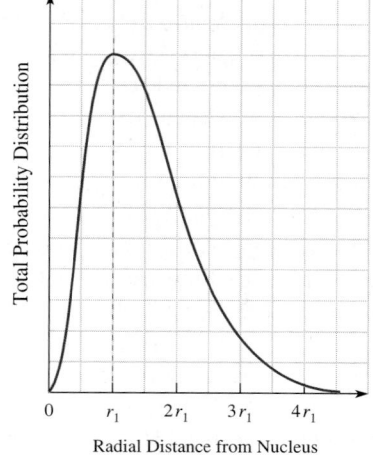

Figure 29.4 A plot of the probability of finding the electron at a distance from the nucleus of the hydrogen atom in its ground state. The most likely location is at the Bohr radius.

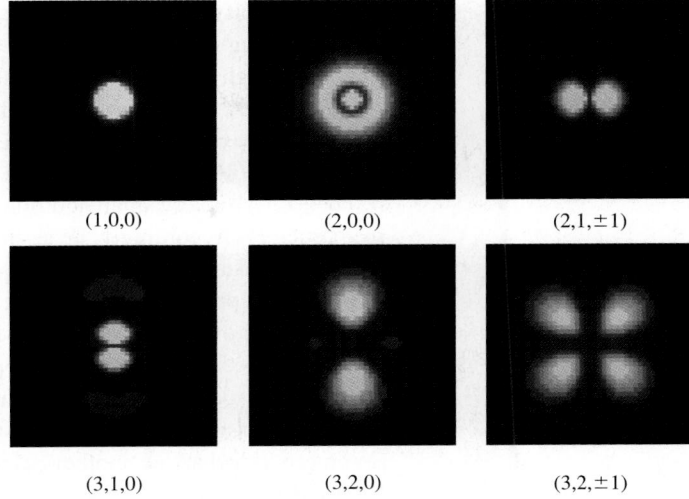

Figure 29.5 These are plots of the probability density, $|\psi|^2$, for the electron in the ground state and in several excited states of the hydrogen atom. Each picture is a slice of the three-dimensional electron cloud with the nucleus at the center. The brighter the region, the higher the probability of finding the electron. Each pattern corresponds to a specific set (p. 1039) of quantum numbers (n, l, m_l). For example, the ground state is specified by $(1, 0, 0)$.

predictable statistically. Generally, that realization was rationalized as simple classical ignorance, which we could hope to overcome, rather than profound quantum-mechanical indeterminism, which could never be eliminated. Even though it is now the consensus, the idea of a probabilistic world (wherein things like particle decay happen spontaneously, without apparent cause) does not sit well with all physicists. It certainly didn't sit well with Einstein, who complained that God does not play dice with the Universe.

Quantum Physics

Guided by challenging experimental observations and powered by a potent theoretical machinery, Quantum Mechanics developed rapidly in the 1920s and 1930s. Attempts to understand the Zeeman Effect stimulated the introduction of the fundamental concept of *spin*. Another revelation, the *Pauli Exclusion Principle*, helped to uncover the detailed structure of the atomic electron cloud. The *Uncertainty Principle*, a profound overview of the microworld, was formulated by Heisenberg. A relativistic quantum theory, *Quantum Electrodynamics*, was introduced by Dirac, and it predicted the existence of *antimatter*. All of these milestones are explored in the remainder of this chapter.

29.4 Quantum Numbers

Modern Quantum Mechanics generates a set of fundamental quantum numbers in a mathematical way that flows naturally from the theory of differential equations. These numbers describe the state of a quantum-mechanical system and therefore specify the wavefunction. The Schrödinger Equation has solutions that are energy values, and in one dimension, these are each associated with a value of the **principal quantum number** n. For a three-dimensional atomic system, there must be at least three quantum numbers (n, l, and m_l), and the

On the concept of quantum jumps. The whole idea of quantum jumps necessarily leads to nonsense… If we are still going to have to put up with these damn quantum jumps, I am sorry that I ever had anything to do with quantum theory.

ERWIN SCHRÖDINGER

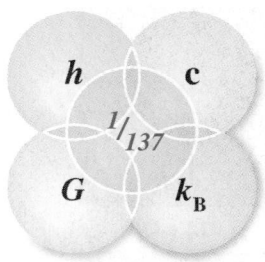

Figure 29.6 Although perhaps not a fundamental constant $\alpha \approx 1/137 \approx e^2/4\pi\varepsilon_0 hc$ arises over and over again in modern physics. (see Problems 20 to 25).

rules for specifying these numbers come directly from the mathematics. These same rules for quantizing dynamical systems were more or less anticipated by the Old Quantum Theory, though in an ad hoc fashion.

Ever since Michelson's work (1891), it was known that each of the hydrogen spectral lines actually comprised several lines, very close together. With that in mind, Arnold Sommerfeld (1915) generalized the Bohr Theory to encompass this **fine structure**. What he needed was a very small additional energy term that would produce the observed multiplet structure. Accordingly, he assumed the orbits were elliptical and would precess about the nucleus. Surprisingly enough, when this variation on Bohr's theme was framed relativistically, it produced the desired effect in full agreement with observation. Thus was the fine structure of hydrogen "explained." Sommerfeld's theory, which is now totally obsolete, was so accurate and visually appealing that the picture of interweaving orbital ellipses persists even today. (Just ask any youngster to draw an atom.)

Each possible motion of a system is known as a *degree of freedom,* and in effect, Sommerfeld had added another degree of freedom to the one circular orbital motion of the Bohr electron; elliptical orbits could swing around the nucleus. To the principal quantum number n, which essentially determined the orbital energy, he added a new quantum number—an important insight. To put things in a modern context, we call it the **orbital angular momentum quantum number**, symbolized by l and having integer values from 0 to $n - 1$ (i.e., $n \neq l$). This quantum number determines the magnitude of the total orbital angular momentum (L), fixing it in steps of multiples of $\hbar$. Here the modern theory departs dramatically from Bohr's treatment in that it allows for an $l = 0$, zero angular momentum state. That possibility does great mischief to any thoughts of *orbiting* electrons!

The crucial parameter in Sommerfeld's analysis is the dimensionless quantity

$$\alpha = \frac{e^2}{4\pi\varepsilon_0 hc} \approx \frac{1}{137} \qquad (29.3)$$

known as the **fine structure constant** (Fig. 29.6). Even though Sommerfeld's treatment is now only of historic interest, α continues to play a very significant role in the quantum theory of the atom. Indeed, it follows from Coulomb's Law that $e^2/4\pi\varepsilon_0$ is a measure of the strength of the electromagnetic interaction, as for example between an electron and a proton. In Quantum Electrodynamics α, which is called the *electromagnetic coupling strength,* is extremely important. Though not a fundamental constant, α relates two of the primary constants of electromagnetism (ε_0 and e) with the fundamental constants of Relativity (c) and Quantum Mechanics (h).

29.5 The Zeeman Effect

The next problem to demand another degree of freedom (and a new quantum number) was remotely necessitated by Faraday, who was convinced that magnetism should affect light. Following his lead, Pieter Zeeman (1896), who had the advantages of a powerful electromagnet and a fine diffraction grating, placed a sodium flame in a *B*-field. Whenever "the current was put on, the two *D* lines were distinctly widened"—the magnetic field influences the way atoms emit light. The **Zeeman Effect** was quickly explained classically: the orbiting electron is a tiny current loop that has a magnetic moment and experiences a torque. As a result, the angular momentum vector (drawn such that if you curl the fingers of the right hand in the direction of the motion, the thumb points in the direction of $\vec{L}$) revolves around $\vec{B}$, and that alters the energy slightly, producing a splitting of the spectral lines (Fig. 29.7).

Debye and Sommerfeld were able to interpret the Zeeman Effect within the Old Quantum Theory by associating a new quantum number with the magnetic moment of the orbital electron. Today, this number is written as m_l and is known as the **orbital magnetic quantum number**. The energy of a bar magnet, a dipole equivalent to a current loop, in a

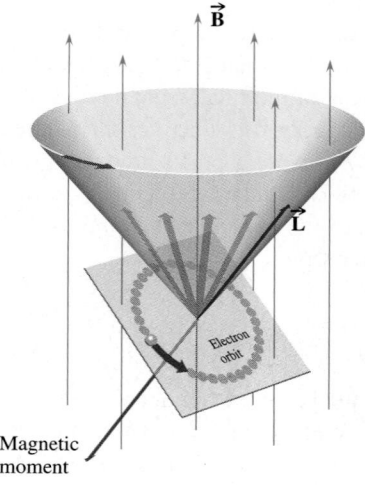

Figure 29.7 An orbiting electron creates a magnetic moment that revolves around $\vec{B}$.

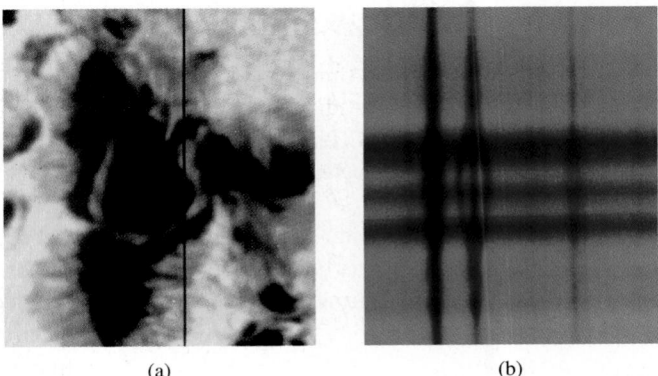

(a)   (b)

A sunspot. These dark blotches on the face of the Sun are associated with powerful localized magnetic fields. The thin black vertical
line in (a) corresponds to the location of the slit opening in a spec-
trograph that analyzes the light within that narrow region. In (b) we
see three spectral lines running vertically. Where the slit opening
passes into the strong magnetic field of the sunspot, the middle line
clearly splits (b) into three lines as a result of the Zeeman Effect. By
measuring the separations of these lines, the magnetic field ($\approx$4 T)
can be determined.

B-field depends on its orientation with respect to the field. Accordingly, consider the magnetic moment of an orbital electron pointing in the opposite direction of its angular momentum. When the electron is placed in a magnetic field, $\vec{\mathbf{L}}$ can only assume certain orientations
with respect to $\vec{\mathbf{B}}$. Sommerfeld called this *space quantization* ("spatial quantization" might
be better). In other words, if $\vec{\mathbf{B}}$ is in, say, the z-direction, L_z is quantized such that

$$L_z = m_l h$$

wherein m_l goes from $-l$ to 0 to $+l$ in integer steps. For example, when $l = 2$, m_l can be
$-2, -1, 0, +1$, or $+2$. And if $m_l = 0$, $L_z = 0$ and $\vec{\mathbf{L}}$ will be perpendicular to $\vec{\mathbf{B}}$. For states
other than the $n = 1$ ground state (for which $l = 0$ and $m_l = 0$), an applied magnetic field
will split the energy levels into 3, 5, and so on, sublevels whose separation depends on the
strength of the field (Fig. 29.8).

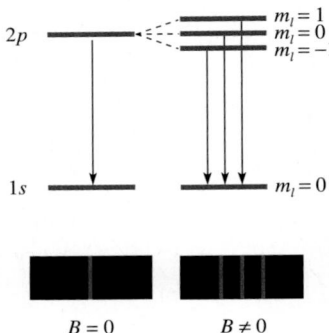

Figure 29.8 The splitting of energy levels in a magnetic field; the Zeeman Effect. The
application of a magnetic field causes the atom to have three close-together excited
states. The result is three emission lines where there ordinarily is only one.

29.6 Spin

Almost as soon as the problem of the Zeeman Effect was solved, it was complicated again. Although there were lines that behaved according to theory (the so-called Normal Zeeman Effect), it was discovered in 1897 that others, like those of sodium, actually split into a bewildering array of lines in a magnetic field (the Anomalous Zeeman Effect).

Classically, an object such as the Earth can have any value of spin angular momentum—the quantity has always been assumed to be continuous. But that's not possible with particles in the atomic domain. To match experimental results, the **spin angular momentum quantum number** (s) for electrons must have only one value, $s = \frac{1}{2}$; because of this,

SPINNING ELECTRONS

R. Kronig, from Columbia University, conceived the idea (1925) that the electron might be spinning about its own axis and therefore have a magnetic moment. Although Compton (1921) had already suggested that "the electron itself, spinning like a tiny gyroscope, is probably the ultimate magnetic particle," he never applied the idea. Kronig did, but there were relativistic problems with the notion of a finite spinning charged particle; when several colleagues (among them, Pauli and Heisenberg) rejected the hypothesis, he decided to let it go unpublished. Ironically, S. A. Goudsmit and G. E. Uhlenbeck together independently introduced the idea of electron spin after reading a paper by Pauli in which he showed the necessity for associating four quantum numbers with the electron. They, too, hesitated, but fortunately they were working under Ehrenfest who told them "that it was either highly important or nonsense" and, in any event, he had already sent their paper off to be published.

electrons are referred to as **spin-$\frac{1}{2}$ particles**. Spatial quantization again applies because the spin magnetic moment orients itself with regard to an applied B-field. The component of the spin angular momentum vector in the direction of $\vec{B}$ is quantized and equals $m_s\hbar$, where m_s is the **spin magnetic quantum number**. Moreover, m_s can only change by steps of 1 going from $+s$ to $-s$; in other words, $m_s = \pm\frac{1}{2}$. The component of spin angular momentum of an electron in the direction of B can only equal $\pm\frac{1}{2}\hbar$. We speak of **spin-up** $(+\frac{1}{2})$ and **spin-down** $(-\frac{1}{2})$ electrons. The electron's energy in the magnetic field depends on its spin orientation, and so all of the levels are split by the applied B-field, thereby producing the Anomalous Zeeman Effect.

The notion of spin is not supposed to be taken literally. Particles do have angular momentum—a beam of light can twist a torsion pendulum—yet there are serious relativistic problems associated with picturing particles as tiny revolving entities. The idea of the electron being a minute spinning charged sphere is appealing but very problematic. What is said instead is that it has an inherent spin angular momentum (without the need for revolving), but that concept doesn't completely satisfy those who like their angular momentum to involve a clear angular motion. Still, in 1928, Dirac demonstrated that a relativistic quantum theory of the electron naturally provides intrinsic spin via an additional quantum number (Table 29.1).

29.7 The Pauli Exclusion Principle

The quantum numbers specify the state of a system and are therefore extraordinarily important. As Pauli showed, the state of any atomic electron can be determined via four quantum numbers: n, l, m_l, and m_s. Furthermore, *no two atomic electrons can occupy the same state*; that is, no two electrons in the same atom can have the same four quantum numbers. In a generalized form, this idea, known as the **Pauli Exclusion Principle**, is a basic precept of nature applicable to a wide range of situations.

Table 29.1

Atomic Quantum Numbers		
Name	**Symbol**	**Values**
Principal quantum number	n	1, 2, 3, ...
Orbital angular momentum quantum number	l	0, 1, 2, ..., $(n-1)$
Orbital magnetic quantum number	m_l	0, ±1, ±2, ..., $\pm l$
Spin angular momentum quantum number	m_s	$\pm\frac{1}{2}$

We learn from Quantum Mechanics that the Exclusion Principle applies to particles that are known as **fermions**. These are particles (whether elementary or compound) that have spins that are odd integer multiples of $\frac{1}{2}$, particles such as the proton, neutron, and neutrino. It does not apply to the group of particles called **bosons**, which have zero or integer spins. Alpha particles, which are composed of even numbers of fermions, are bosons. Photons are spin-1 particles and therefore bosons.

The state of a photon is specified by its momentum and polarization. Thus, a planar monochromatic lightwave can be thought of as a stream of photons all in the same state—a circumstance that is possible because they are bosons. Indeed, bosons tend to accumulate in the lowest possible energy state. If that were the case for fermions, all atomic electrons would collect in the same lowest energy level and the atom would compress. Moreover, when atoms are brought very close to one another (so close that their wavefunctions overlap), the Exclusion Principle obtains, which is why two hydrogen atoms can form a molecule only when the atoms have spin-up and spin-down electrons, respectively.

To date, tests of the Exclusion Principle have never found it wanting; researchers have seen no violations of it down to an accuracy of less than 2 parts in 10^{26}.

29.8 Electron Shells

STUDY GUIDE

For a given shell (with a fixed n) the values of l go in steps of 1 from 0 to $(n-1)$. The values of m_l go in steps of 1 from $-l$ to $+l$, and there are a total of $(2l+1)$ of them. This is the number of orbitals in a subshell corresponding to a given orbital angular momentum quantum number. As a rule $n \geq (l+1)$.

During the 1920s, Bohr, Stoner, and others constructed a model of the electron structure of all the elements in the Periodic Table (inside back cover). Moseley's work provided a knowledge of the number of nuclear protons, and since the atom is neutral, that's the number of orbital electrons. It's no simple matter to sort out the electron structure (Fig. 29.9); for that analysis, chemical behavior and atomic spectra were the guides.

Electrons are ordered in shells and subshells about the various nuclei according to rules associated with their quantum numbers. Each shell corresponds to a specific value of n, and they are traditionally given letter names, where

$$\text{Shell} = K\ L\ M\ N\ O\ \ldots$$

$$n = 1\ 2\ 3\ 4\ 5\ \ldots$$

Remember that l ranges from $(n-1)$ to 0, that m_l goes from $-l$ to $+l$, and that m_s is either $+\frac{1}{2}$ or $-\frac{1}{2}$. Again, by tradition, the states of an electron are given letter names, where

$$\text{Subshell} = s\ p\ d\ f\ g\ h\ \ldots$$

$$l = 0\ 1\ 2\ 3\ 4\ 5\ \ldots$$

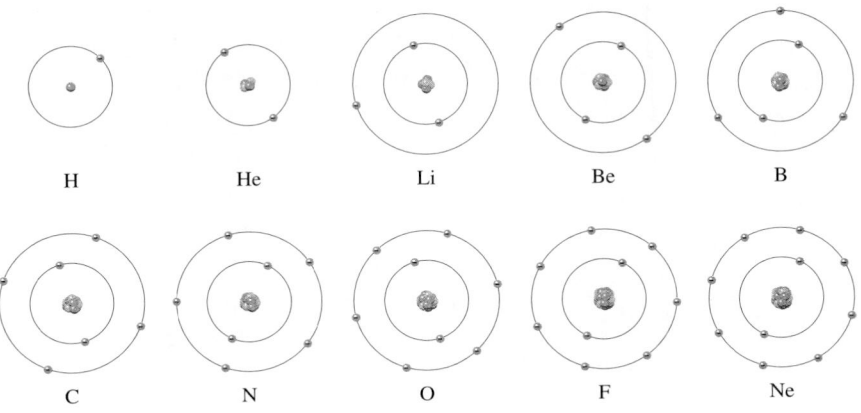

H   He   Li   Be   B

C   N   O   F   Ne

Figure 29.9 A schematic representation of the electronic structure of the first 10 atoms in the Periodic Table. (The sizes of the nuclei are tremendously exaggerated, as are the electrons.)

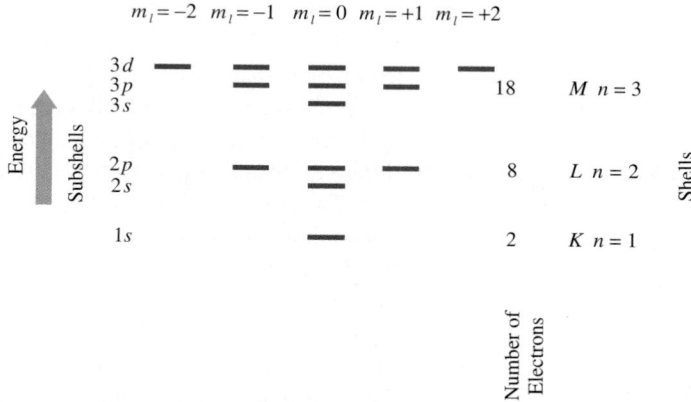

$m_l = -2 \quad m_l = -1 \quad m_l = 0 \quad m_l = +1 \quad m_l = +2$

Figure 29.10 Ordering of atomic energy levels in terms of quantum numbers n, l, m_l. For example, the $3d$ subshell corresponds to $n = 3$ and $l = 2$. It comprises five orbitals corresponding to $m_l = -2, -1, 0, +1,$ and $+2$.

A **shell** is a group of states that have the same principal quantum number. A **subshell** is a smaller group of states that has the same value of both n and l (Fig. 29.10). An **orbital** is specified by the three quantum numbers n, l, and m_l, and it can contain two electrons: one spin-up and one spin-down. (Each short horizontal line in Figs. 29.10 and 29.11 represents an orbital.) And a **state** is specified by all four quantum numbers and contains one electron, as per the Exclusion Principle.

Hydrogen ($Z = 1$) has one electron and is chemically active. Its one unpaired electron gives it a valence of one and requires that it enter into covalent bonds in which the electron is shared with another atom. In the ground state, $n = 1$, $l = 0$, $m_l = 0$, and $m_s = \pm\frac{1}{2}$ (Fig. 29.12). The electron configuration is said to be a $1s$ orbital ($n = 1$, $l = 0$) of the K-shell (Fig. 29.10). Once two hydrogen atoms have combined (H_2) by sharing their spin-up, spin-down electrons, there are no longer any unpaired charges and a third hydrogen cannot join the group (Table 29.2, p. 1041).

Helium ($Z = 2$) has two electrons, and because it's a stable *Noble Gas* (in the right-most column of the Periodic Table), we can conclude that two electrons must correspond to a completed system. Thus (even before the Exclusion Principle), it was presumed that the first (innermost) two electrons of any atom form a closed K-shell. In the ground state, the

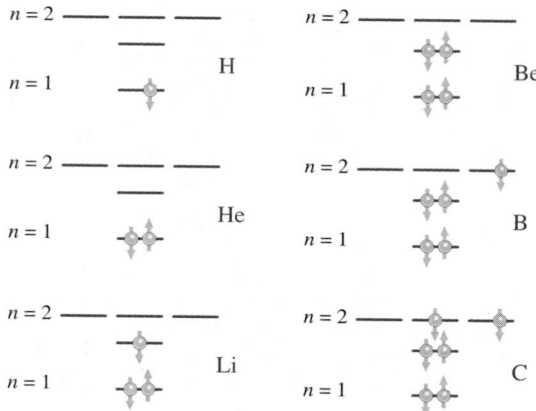

Figure 29.11 Distributions of electrons in the energy levels for the first six elements: hydrogen (H), helium (He), lithium (L), beryllium (Be), boron (B), and carbon (C).

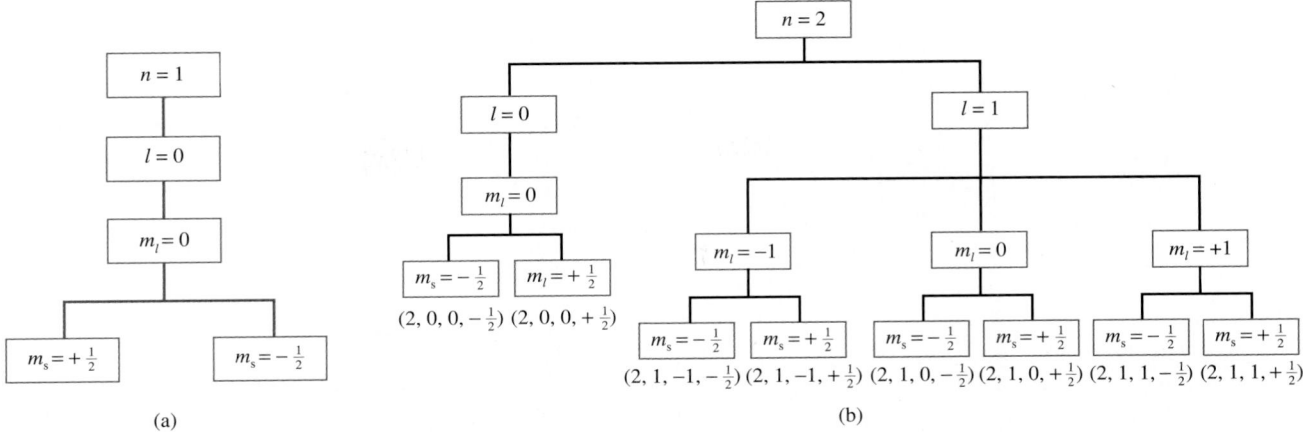

Figure 29.12 The quantum numbers and corresponding states (n, l, m_l, m_s) for the principal quantum numbers (a) $n = 1$ and (b) $n = 2$ for hydrogen.

quantum numbers (n, l, m_l, m_s) for one electron are $(1, 0, 0, -\frac{1}{2})$, and for the other they are $(1, 0, 0, +\frac{1}{2})$. The K-shell has only two states and is filled. The 2-electron configuration is $1s^2$, where the superscript is the number of electrons (Table 29.3, p. 1042). Both electrons are close to the nucleus, and the atom has a high ionization energy of 24.6 eV (Fig. 29.13); it holds tightly to its electrons and does not interact effectively with other atoms.

Lithium ($Z = 3$) is next. It has three electrons, two in the first closed K-shell and the third rather far off in the second, or L-shell. With no room in the $n = 1$ shell, the third electron goes into the $l = 0$, s-subshell of the $n = 2$ shell; (n, l, m_l, m_s) for the third electron corresponds to $(2, 0, 0, -\frac{1}{2})$. The overall configuration is $1s^2 \cdot 2s^1$.

Beryllium ($Z = 4$) has four electrons, the last of which fills the $l = 0$, s-subshell and has quantum numbers $(2, 0, 0, +\frac{1}{2})$. The four-electron ground-state configuration is $1s^2 \cdot 2s^2$; there are no unpaired electrons, and the valence is zero. As it happens, the energy difference between the $(2, 0, 0, \frac{1}{2})$ state and the $(2, 1, 0, \frac{1}{2})$ state is very small, and the beryllium atom

Table 29.2
Ground-State Configurations of Several Elements

Element	Symbol	Atomic number, Z	Electronic configuration
Hydrogen	H	1	$1s$
Helium	He	2	$1s^2 \cdot$
Lithium	Li	3	$1s^2 \cdot 2s$
Beryllium	Be	4	$1s^2 \cdot 2s^2$
Boron	B	5	$1s^2 \cdot 2s^2 2p$
Carbon	C	6	$1s^2 \cdot 2s^2 2p^2$
Nitrogen	N	7	$1s^2 \cdot 2s^2 2p^3$
Oxygen	O	8	$1s^2 \cdot 2s^2 2p^4$
Fluorine	F	9	$1s^2 \cdot 2s^2 2p^5$
Neon	Ne	10	$1s^2 \cdot 2s^2 2p^6 \cdot$
Sodium	Na	11	$1s^2 \cdot 2s^2 2p^6 \cdot 3s$
Magnesium	Mg	12	$1s^2 \cdot 2s^2 2p^6 \cdot 3s^2$
Aluminum	Al	13	$1s^2 \cdot 2s^2 2p^6 \cdot 3s^2 3p$
Silicon	Si	14	$1s^2 \cdot 2s^2 2p^6 \cdot 3s^2 3p^2$
Phosphorus	P	15	$1s^2 \cdot 2s^2 2p^6 \cdot 3s^2 3p^3$
Sulfur	S	16	$1s^2 \cdot 2s^2 2p^6 \cdot 3s^2 3p^4$
Chlorine	Cl	17	$1s^2 \cdot 2s^2 2p^6 \cdot 3s^2 3p^5$
Argon	Ar	18	$1s^2 \cdot 2s^2 2p^6 \cdot 3s^2 3p^6 \cdot$

Table 29.3

Electron Configurations of Some Atoms in Their Ground States

			K	L		M			N			
		n	1	2		3			4			
Z	Element	l	s	s	p	s	p	d	s	p	d	f
1	Hydrogen	H	1									
2	Helium	He	2									
3	Lithium	Li	2	1								
4	Beryllium	Be	2	2								
5	Boron	B	2	2	1							
6	Carbon	C	2	2	2							
7	Nitrogen	N	2	2	3							
8	Oxygen	O	2	2	4							
9	Fluorine	F	2	2	5							
10	Neon	Ne	2	2	6							
11	Sodium	Na	2	2	6	1						
12	Magnesium	Mg	2	2	6	2						
13	Aluminum	Al	2	2	6	2	1					
14	Silicon	Si	2	2	6	2	2					
15	Phosphorus	P	2	2	6	2	3					
16	Sulfur	S	2	2	6	2	4					
17	Chlorine	Cl	2	2	6	2	5					
18	Argon	Ar	2	2	6	2	6					
19	Potassium	K	2	2	6	2	6	·	1			
20	Calcium	Ca	2	2	6	2	6	·	2			
21	Scandium	Sc	2	2	6	2	6	1	2			
22	Titanium	Ti	2	2	6	2	6	2	2			
23	Vanadium	V	2	2	6	2	6	3	2			
24	Chromium	Cr	2	2	6	2	6	5	1			
25	Manganese	Mn	2	2	6	2	6	5	2			
26	Iron	Fe	2	2	6	2	6	6	2			
27	Cobalt	Co	2	2	6	2	6	7	2			
28	Nickel	Ni	2	2	6	2	6	8	2			
29	Copper	Cu	2	2	6	2	6	10	1			
30	Zinc	Zn	2	2	6	2	6	10	2			
31	Gallium	Ga	2	2	6	2	6	10	2	1		
32	Germanium	Ge	2	2	6	2	6	10	2	2		
33	Arsenic	As	2	2	6	2	6	10	2	3		

can easily go into the $1s^2 \cdot 2s^1 2p^1$ configuration; that is, one electron can be raised up into the p-subshell. Accordingly, it can have a valence of either 0 or 2, but nothing else (Fig. 29.11).

Boron ($Z = 5$) has five electrons, the last of which goes into the next open level in the p-subshell and has quantum numbers $(2, 1, +1, -\frac{1}{2})$. The five-electron ground-state configuration is $1s^2 \cdot 2s^2 2p^1$, and a valence of 1 is to be expected. Here again, the configuration

Figure 29.13 Ionization energy versus Atomic Number. Closed shells occur at $Z = 2, 10, 18, 36,$ and 54. An atom such as sodium has one electron beyond a closed shell ($Z = 11$), and it's shielded from the nucleus, so its binding energy is small.

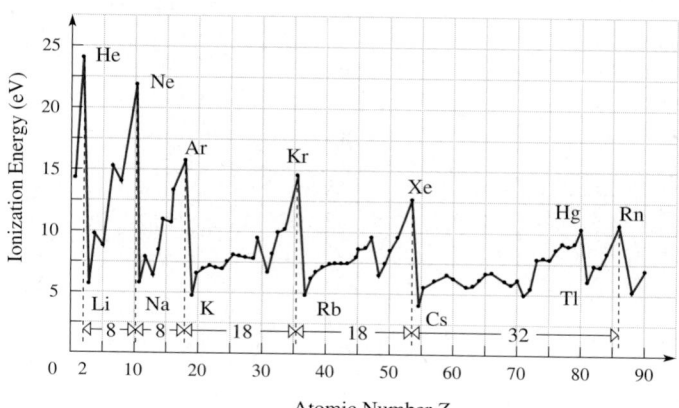

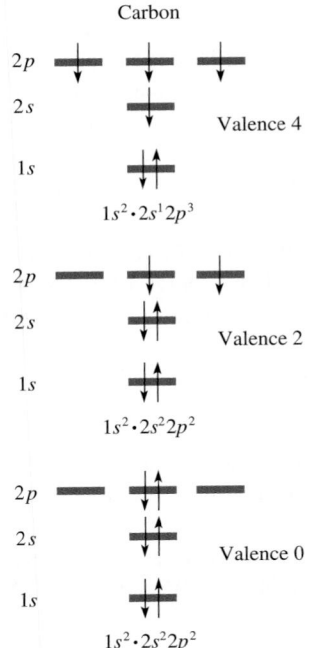

Carbon

$2p$

$2s$

Valence 4

$1s$

$1s^2 \cdot 2s^1 2p^3$

$2p$

$2s$

Valence 2

$1s$

$1s^2 \cdot 2s^2 2p^2$

$2p$

$2s$

Valence 0

$1s$

$1s^2 \cdot 2s^2 2p^2$

Figure 29.14 The distribution of electrons in carbon corresponding to valences of 0, 2, and 4. Each orbital fills up only when it contains two electrons. Accordingly, the valence corresponds to the number of single electrons that can be shared.

can be altered with the input of a small amount of energy, which is available to the atom via collisions even at room temperature. It then takes the form $1s^2 \cdot 2s^1 2p^2$.

Carbon ($Z = 6$) has six electrons, two in the $n = 1$ shell and four in the $n = 2$ shell. The last electron goes into the p-subshell and has quantum numbers $(2, 1, 0, -\frac{1}{2})$. As a rule, *electrons filling a subshell do not double up in an orbital until each orbital has one*. Because of their mutual repulsion, the electrons get as far apart as possible by going into different orbitals. The six-electron ground-state configuration is $1s^2 \cdot 2s^2 2p^2$. Thus, we would expect carbon (Fig. 29.11) to participate in covalent bonds and have a valence of 2. And yet it almost always displays a valence of 4. This occurs because it takes a mere ≈ 2 eV to break the $2s^2$ pair, raising one of the electrons into the empty orbital of the p-subshell. There are then four unpaired electrons, yielding a valence of 4. Figure 29.14 shows how carbon can have valences of 0, 2, or 4, but nothing else. It's this range of valences that allows carbon to share electrons, forming single, double, and triple bonds (for example, C—O, C=C, C=O, C=N, C≡C, C≡N), and thereby producing a wide range of complex molecules, including the ones that are the basis of all known life.

At the point where stable neon ($Z = 10$) is reached (Table 29.4), its ten electrons complete the first two shells, with two and eight electrons, respectively ($1s^2 \cdot 2s^2 2p^6$). Argon ($Z = 18$) is the next Noble Gas, and it fills the $3p$-subshell in the M-shell with a configuration of $1s^2 \cdot 2s^2 2p^6 \cdot 3s^2 3p^6$. After argon, the order of several subshells changes because, for example, the $4s$ level is a little less energetic than the $3d$, so it fills first (Fig. 29.15). Krypton ($Z = 36$) fills the M-shell and has a configuration of $1s^2 \cdot 2s^2 2p^6 \cdot 3s^2 3p^6 \cdot 3d^{10} 4s^2 4p^6$. And so on up the Periodic Table.

With few exceptions, *all the atoms in any one column of the Periodic Table have the same number of unpaired electrons*. The very active Alkali Metals all have only one such electron, which is easily given up or shared; the less active Alkaline Earths (such as magnesium and calcium) have two; the boron, carbon, nitrogen, and oxygen families have three, four, five, and six, respectively; and the highly reactive Halogens each contain an almost complete shell of seven electrons.

Table 29.4

Groups of Elements

Halogens	Number of electrons
Fluorine	9
Chlorine	17
Bromine	35
Iodine	53
Astatine	85

Noble Gases	Number of electrons
Helium	2
Neon	10
Argon	18
Krypton	36
Xenon	54
Radon	86

Alkali Metals	Number of electrons
Lithium	3
Sodium	11
Potassium	19
Rubidium	37
Cesium	55
Francium	87

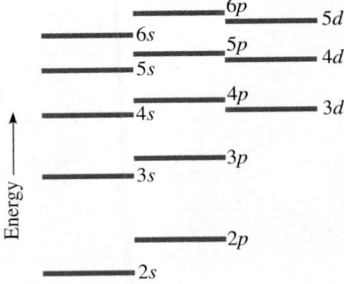

Figure 29.15 Relative positions of atomic orbitals.

Energy Levels

As you recall, there was only one quantum number (n) in the Bohr Theory of hydrogen, and it determined both energy and orbital angular momentum. That model maintained that the angular momentum of the hydrogen electron had only one value. By contrast, Quantum Mechanics correctly predicts a range of such values, associated with $l = 0, 1, 2, ..., (n - 1)$, for each n. Moreover, if we ignore some tiny internal magnetic and relativistic effects, Schrödinger's Equation produces the same result as Bohr's energy equation $E_n = (-13.6$ eV$)/n^2$; the energy levels of hydrogen depend only on n. That's not the case for mulielec-tron atoms; their energy levels depend on n and l. As a rule, when n increases (for a fixed l) the energy increases, and when l increases (for a fixed n), the energy also increases (Fig. 29.15).

Example 29.2 **[I]** Determine the ground-state electron configuration for $_{12}$Mg. Start with neon, for which $(1s^2 \cdot 2s^2 2p^6)$.

Solution This is a straightforward application of the electron shell scheme. (1) TRANSLATION—Knowing the ground state configuration for neon, determine it for a specified nearby element. (2) GIVEN: $_{12}$Mg. FIND: Its electron configuration. (3) PROBLEM TYPE—Pauli Exclusion Principle/quantum numbers/

electron shells. (4) PROCEDURE—Use all the rules we've learned for electron shells. (5) CALCULATION—Neon has 10 electrons $(1s^2 2s^2 2p^6)$ and a closed p-subshell. The $n = 2$ shell allows $l = 1$ or 0, so there is an s- and a p-subshell. The eleventh electron ($n = 3$) corresponds to $(1s^2 \cdot 2s^2 2p^6 \cdot 3s^1)$ and the twelfth to $(1s^2 \cdot 2s^2 2p^6 \cdot 3s^2)$.

Quick Check: Look at Table 29.2, p. 1041.

29.9 The Uncertainty Principle

Born's statistical interpretation suggests that we can compute only the likelihood of individual events. This "blurriness" of vision, which is inherent in the formulation of Quantum Mechanics, might seem peculiar. After all, if we know the precise location of a proton to begin with, as well as how fast and in what direction it's moving, why can't we follow it *exactly*? Perhaps that's not the right question, however; perhaps we should be bolder and ask whether or not we really can simultaneously measure both the position and velocity precisely in the first place. The product of position and momentum, or time and energy, brings us back to *action* and the quantum-of-action (h), which is at the heart of things quantum-mechanical. Questions about the measurability and definability of quantum concepts had occupied Werner Heisenberg in 1927 before he came up with the crucial insight known as the **Uncertainty Principle**.

Ordinarily, we observe the position of something by scattering something else off it: radar waves off a car; sound waves off a whale; lightwaves off your face; or the end of a cane off the curb. A stream of sunshine photons scattering from a flying baseball hardly affects its path, but one photon slamming into an electron drastically and unpredictably alters the electron's motion. Evidently, to measure anything in Atomland we will need a probe as well, but even the most delicate one possible will obtrude on the measurement. In order to observe the position of a microparticle precisely, we must—to cut down on diffraction—use a short-wavelength (and, therefore, a high-energy) probe. And that, in turn, will blast the particle away, making knowledge of its velocity and momentum even less precise. To approximate the effect, we use a photon of wavelength λ to locate a tiny object along the x-axis. Reasonably enough, assume the position can be measured to an accuracy of $\Delta x \approx \lambda$. At most, the photon can transfer all its momentum (h/λ) to the object, whose own momentum will then be uncertain by an amount $\Delta p_x \approx h/\lambda$. Hence

$$\Delta p_x \Delta x \approx h \qquad (29.4)$$

and this is a crude form of the Heisenberg Uncertainty Principle. It quantifies the "giveth

Werner Karl Heisenberg (1901–1976) was awarded the Nobel Prize for physics in 1932 for his discovery of the Uncertainty Principle.

and taketh away" interrelationship between so-called conjugate variables, such as position and momentum, and energy and time (p. 1030). Even though it's clear that the very act of measuring unavoidably obtrudes on the atomic level, most physicists now believe that the Uncertainty Principle has a yet more fundamental basis.

Consider a particle moving along the x-axis with a speed v and a momentum p_x. Momentum, according to de Broglie ($p = h/\lambda$), depends on wavelength, which, in turn, corresponds to an extension in space. To specify λ, we must think of observing a cycle of the wave in space. The wave-mechanical concept of a precisely determined momentum is thus in direct conceptual conflict with the notion of a precisely defined conjugate position (x). Similarly, consider a particle of energy E moving at a time t. Energy (E $= hf$) depends on frequency, which, in turn, corresponds to an extension in time. To specify f, we must think of observing a cycle of the wave in time. The wave-mechanical concept of a precisely determined energy is thus incompatible with the notion of a precisely defined conjugate time (t).

Suppose we set out to simultaneously determine a pair of conjugate variables, say, p_x and x. Having made a number of measurements of any physical quantity, we would find a spread in the data about an average value. Accordingly, we record a momentum spread of Δp_x and a position spread of Δx. Classically, we would expect to be able to simultaneously sharpen up these values, making them both more and more precise, with no inherent conceptual limitations keeping us from improving things. We need only measure v over a vanishingly small track (Δx) centered at x. But the wave nature of the particle makes that process much less straightforward. To fix the momentum, we must fix the wavelength with precision, and to do so will necessitate allowing it some extension in space, thereby blurring x.

This linkage between conjugate variables is at the center of Heisenberg's discovery of the Uncertainty Principle, which is

$$\Delta p_x \Delta x \geq \tfrac{1}{2}\hbar \tag{29.5}$$

The product of the simultaneous uncertainties in position and momentum is at best equal to $\tfrac{1}{2}\hbar$. And the conceptually related expression

$$\Delta E \Delta t \geq \tfrac{1}{2}\hbar \tag{29.6}$$

holds as well. We cannot simultaneously know both the momentum and the position to any accuracy we wish. Homing in on one conjugate variable decreases its uncertainty and increases the uncertainty in the other variable.

Consider the implications of the preceding conclusions as they relate to complementarity, that is, to the wave-particle duality. The Uncertainty Principle sees to it that we will never be able to experimentally resolve the either/or issue of whether a photon is a particle or a wave. Similarly, precisely measuring the position of an electron fixes it with a sharp spatial localization, whereupon it may correctly be considered a particle. Yet when $\Delta x = 0$, it follows that $\Delta p = \infty$, and the electron has zero wavelength ($p = h/\lambda$); it has no wavelike attributes. The basis of the wave-particle duality thus seems to be the conceptual conflict between position and momentum as reflected in the Uncertainty Principle, and that sets the agenda for much of Quantum Mechanics. If we cannot know precisely whether an electron's present, we will have to rely on statistical means to tell its future. The very existence of the quantum of action smears certainty into probability and blurs cause and effect so that reality is not quite as sharp as we once thought it was. *Just as the unique character of Relativity arises from the fact that c is finite and not infinite, the unique character of Quantum Mechanics arises from the fact that h is finite and not zero* (Fig. 29.6). The classical world behaves as if c $= \infty$ and $h = 0$. The real world behaves as if c $= 3 \times 10^8$ m/s and $h = 7 \times 10^{-34}$ J·s. It's only on the macroscopic scale, where h is relatively negligible, that energy and momentum seem continuous and the world appears to be classical. {For an interesting application of the Uncertainty Principle click on **Δ𝑝** AND **DIFFRACTION** under **FURTHER DISCUSSIONS** on the CD.} 🔘

Uncertainty and Nuclear Electrons

It was pointed out earlier (p. 985) that electrons do not reside in the nuclei of atoms, and a persuasive argument to that effect can be made using the Uncertainty Principle. An electron imprisoned in a nucleus must have a position uncertainty not greater than the size of the nucleus. The Uncertainty Principle then demands a corresponding indeterminacy in momentum, allowing values of p and therefore of energy quite inconsistent with observation.

Example 29.3 **[III]** Assume that an atomic nucleus is 1.0×10^{-14} m in diameter and use the Uncertainty Principle to calculate the kinetic energy of an electron confined to such a space. Compare that to the energy range for beta rays (0.025 MeV to 3.2 MeV). Recall that an electron has a rest energy of 0.511 MeV $= 8.19 \times 10^{-14}$ J.

Solution The Uncertainty Principle deals with Δp, so we'll have to come at KE via momentum. (1) TRANSLATION— Knowing the size of the nucleus, determine the KE of a particle of known mass confined to such a space. (2) GIVEN: Nuclear diameter of 1.0×10^{-14} m. FIND: KE. (3) PROBLEM TYPE— Uncertainty Principle/relativistic energy. (4) PROCEDURE— Compute Δp_x from the Uncertainty Principle via Δx; then get the total energy from $E^2 = E_0^2 + (pc)^2$, and the KE follows from that. (5) CALCULATION—The electron can be anywhere in the nucleus, and so $\Delta x \approx 10^{-14}$ m. The smallest uncertainty in momentum is

$$\Delta p_x \approx \frac{\frac{1}{2}\hbar}{\Delta x} = \frac{5.27 \times 10^{-35}\,\text{J}\cdot\text{s}}{1.0 \times 10^{-14}}\,\text{m} = 5.27 \times 10^{-21}\,\text{J}\cdot\text{s/m}$$

The electron's momentum must be at least as large as this result, and so its total energy is

$$E^2 = E_0^2 + (pc)^2 \qquad [26.15]$$

For an electron, $E_0 = 8.19 \times 10^{-14}$ J and

$$E^2 = 6.70 \times 10^{-27}\,\text{J}^2 + (5.27 \times 10^{-21}\,\text{J}\cdot\text{s/m})^2(2.998 \times 10^8\,\text{m/s})^2$$

Thus $\qquad E = 1.58 \times 10^{-12}$ J $= 9.87$ MeV

with $E_0 = 0.511$ MeV

$$KE = E - E_0 \approx \boxed{9.4 \text{ MeV}}$$

This is the smallest energy, and it's appreciably larger than the observed β-ray energies. That difference strongly suggests that there are ordinarily no electrons confined to the nucleus.

Quick Check: To redo the calculation in MeV, use $h = 4.14 \times 10^{-15}$ eV·s in the expression $\Delta p \approx \frac{1}{2}h/\Delta x$. Hence, $p = 0.0329$ eV·s/m and $p^2c^2 = 9.73 \times 10^{13}$ eV^2. Since $E_0 = 0.511$ MeV, it follows that $E^2 = E_0^2 + p^2c^2$ and $E \approx 9.9$ MeV. Here, where KE $= 110$ eV, $\lambda = 0.117$ nm.

{For more worked problems click on **WALK-THROUGHS** *in* **CHAPTER 29** *on the CD.}* 🔘

Electrons can only exist inside the nucleus if there is a tremendous force restraining them there. A large collapsing star can have such immense pressures at its core that orbital electrons are crushed into the nuclei to create a giant ball of neutrons—a neutron star.

Determinism

Ultimately, we do not even know if a particle simultaneously has a perfectly defined position and momentum. There are many physicists who believe that it does not. In any case, we could not measure it if it did—not because we're momentarily without the proper method, but because no such method is theoretically possible. Heisenberg's original German word for his concept was *Unbestimmtheitsprinzip*, which better translates as "Indeterminacy Principle." "Uncertainty" suggests that a particle has definite simultaneous values of conjugate variables whose determination is somewhat uncertain, which was not Heisenberg's belief, though not everyone agrees.

There was a time when scientists like famed Laplace idly fancied that if they knew the initial location and momentum of all the particles in the Universe, they could (theoretically

An intelligence which at a given moment knew all the forces that animate nature, and the respective positions of the beings that compose it, could condense into a single formula the movement of the greatest bodies of the Universe and that of the least atom: for such an intelligence nothing could be uncertain, the past and future would be before its eyes.

PIERRE SIMON, MARQUIS DE LAPLACE
(1749–1827)

at least) foretell the future and retell the past. (This, despite free will and passion, as if your next warm kiss had been set in motion some 15 thousand million years ago when the Universe began.) That's the ultimate in *determinism*, and however beyond our actual computational abilities, the logic still disturbs. Though Newtonian theory weaves a deterministic plot where every sigh results from the inexorable confluence of physical law, Quantum Mechanics happily puts a touch of chance in the Hatter's clockwork. The future of even one lone electron is quite beyond prediction; the Uncertainty Principle makes it impossible to precisely know its initial conditions. So kiss whomever you like; change your mind however you will; and rest assured that physics, at least, knows that *it* cannot hope to calculate the future.

29.10 QED and Antimatter

In 1928, things were still simple, ominously simple: atoms were composed only of electrons and protons; the neutron had not yet been discovered; and there were no other microparticles in the zoo except for the photon. That was the year P. A. M. Dirac published "The Relativistic Theory of the Electron," for which he would share the 1933 Nobel Prize with Schrödinger. It was Dirac's intention to reformulate Quantum Mechanics so that it was consistent with Special Relativity. A relativistic quantum theory would naturally mate with Maxwell's Equations and, as such, is now known as **Quantum Electrodynamics** and affectionately called QED. The resulting *Dirac equation* was a triumph: among other things, electron spin came out naturally, Sommerfeld's fine structure formula was derived, and the magnetic moment of the electron was calculated. However splendid, the theory seemed to be flawed by one small aspect, one enigmatic mathematical quirk: it had twice as many energy states as were presumably called for.

Dirac traced the problem back to the relationship $E^2 = c^2p^2 + m^2c^4$; as with all square roots, it has two solutions:

$$E = \pm c\sqrt{p^2 + m^2c^2}$$

Paul Adrien Maurice Dirac (1902–1984). He shared the 1933 Nobel Prize in physics with Schrödinger.

Yet the energy of a free electron cannot be negative. It made no sense to keep the negative solution—its implications were too bizarre to be real. As is customary in such cases, Dirac simply put aside the ugly twin as a mathematical aberration. We do that all the time in Classical Physics. It soon became clear, however, that the negative portion could not be overlooked without seriously weakening the whole theory. Now there was a dilemma—a richly potent theory embarrassed by what seemed a bit of technical minutia.

It took Dirac until the end of 1929 to come up with a bold solution. The negative-energy states are real: many electrons have presumably radiated their energy and descended down into these lowest possible states. Since the $\approx 10^{80}$ ordinary electrons in this Universe of ours are not observed to tumble into the oblivion of the negative Wonderland, it must be that all the negative states are occupied, in accordance with the Exclusion Principle. Wonderland has no vacancies. The situation was like a large hotel with half its floors below ground and half above. With most of the upper rooms empty, there could be lots of shifting around by the guests, but below ground level all the rooms were occupied. No one new could move down and in until someone moved up and out.

Instead of being empty, vacuum was now to be imagined as filled with an invisible multitude of unseen electrons sailing around with negative energies. If one of them were somehow removed, pulled up into the ordinary positive world, it would leave behind an empty hole, a bubble in the negative sea of electrons. Dirac recognized that this hole would appear to us as a positive charge and proposed that it was the only known positive particle, the proton. J. Robert Oppenheimer quickly showed that this proposal was untenable. The hole had to behave as if it were the mirror image of the electron. "A hole, if there is one would be a new kind of particle, unknown to experimental physics, having the same mass and opposite charge to the electron." So wrote Dirac in 1929. "We may call such a particle an anti-electron."

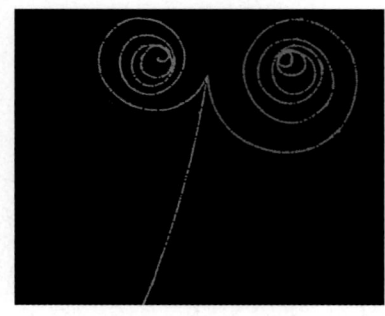

A subatomic event played out in a liquid-hydrogen bubble chamber. A gamma-ray photon (no track) descends from the top of the picture. It strikes an atom and knocks out an electron (long green track). The remainder of the photon's energy goes into creating an electron-positron pair. Because of the applied magnetic field, these two low-energy particles move along tight spirals.

This flight of creative imagination might have remained little more than a mathematical fantasy if not for a cosmic-ray study taking place at the California Institute of Technology. Carl Anderson, working with Millikan, was examining these particles (predominantly protons, possibly emitted from supernovas), which stream in on Earth from space. They were being observed using a cloud chamber placed in a uniform magnetic field so that positive and negative particles would follow oppositely curved paths. Among the thousands of photographs taken, Anderson noticed one that seemed curious. It clearly showed two oppositely curved tracks, one of which was certainly made by an electron; the other suggested an antielectron, but that conclusion was still "very radical at the time." Others had seen these positive tracks and either ignored them or dismissed them as "dirt." But Dirac's paper changed the world view, and the facts in the cloud chamber changed accordingly. By the summer of 1932, Anderson had clear evidence of the existence of the antielectron! It was Anderson who christened it the **positron**.

Today, it is commonplace to create electron-positron pairs—to have substantial matter materialize out of radiant energy. A blast of electromagnetic energy, a gamma ray with zero mass, can disappear and in its place an electron and a positron pop into existence. This ultimate alchemy occurs provided that the gamma-ray photon carries an energy equal to at least the total rest energy of the two particles (and provided there is a heavy object nearby to conserve momentum). Just as pair creation is possible, so is pair annihilation. A positron and *any* electron it approaches can come together and totally obliterate each other, vanishing in a puff of gamma rays (as if the electron had dropped back into the hole and disappeared).

Example 29.4 **[I]** Determine the minimum energy (in joules and MeV) that a gamma photon must have to create an electron-positron pair.

Solution When you talk about the creation or annihilation of particles, $E_0 = mc^2$ should come to mind. (1) TRANSLATION—How much energy is needed to create an electron-positron pair? (2) GIVEN: An electron-positron pair. FIND: Minimum energy to create. (3) PROBLEM TYPE—Antimatter/relativistic energy/rest energy. (4) PROCEDURE—Compute E_0 for the pair. (5) CALCULATION—The minimum energy, E, will equal the rest energy of the pair, name-

ly, $E_0 = 2m_e c^2$; therefore,

$$E = 2(9.11 \times 10^{-31}\ \text{kg})(3.00 \times 10^8\ \text{m/s})^2$$

and $\qquad E = 1.64 \times 10^{-13}\ \text{J} = \boxed{1.02\ \text{MeV}}$

Quick Check: The rest energies of both the electron and positron are 0.511 MeV. Moreover, since $E = hf$,

$$f = \frac{E}{h} = \frac{1.64 \times 10^{-13}\ \text{J}}{6.63 \times 10^{-34}\ \text{J·s}} = 2.47 \times 10^{20}\ \text{Hz}$$

This result corresponds to 0.001 2 nm, which is a gamma ray.

The proton is 1836 times more massive than the electron. Thus the creation of the **antiproton**, which would require that much more energy, could not be attempted until a new generation of accelerators was available. In 1955, the bevatron at the University of California, Berkeley, produced the first human-made antiproton—a negative speck with the same mass as the proton. Highly energetic beams of protons (p) and antiprotons ($\overline{\text{p}}$) can slam together, creating all sorts of new subnuclear particles (p. 1096). On rare occasions, there is a cancellation of the equal and opposite charges and the creation of a neutron and an **antineutron** (1956). The similarity of neutrons (n) and protons is underscored by the fact that protons can similarly interact with antineutrons ($\overline{\text{n}}$), as can antiprotons (often called p-bars) and neutrons.

Nowadays, antimatter is familiar in the laboratory; there are even artificial radioactive isotopes, such as ^{22}Na, that emit positrons on decaying and so serve as convenient sources. For example, by injecting someone with glucose laced with such a radioactive tracer, we

---

Despite its enormous practical success, quantum theory is so contrary to intuition that, even after 45 years, the experts themselves still do not agree what to make of it.

BRYCE DEWITT
PHYSICS TODAY (SEPTEMBER 1970)

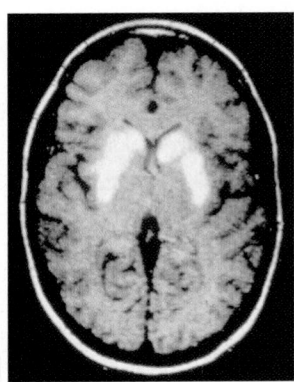
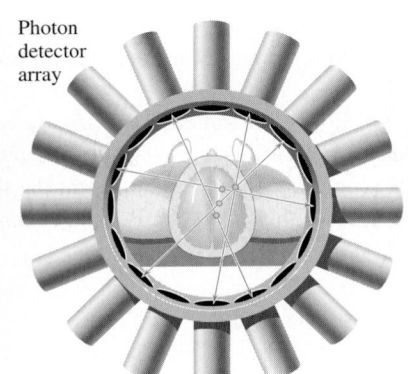

Photon
detector
array

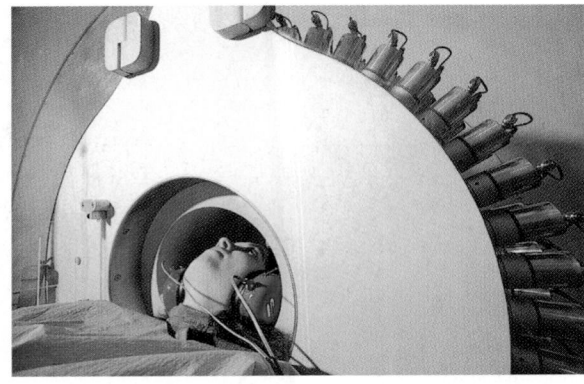

Figure 29.16 Positron Emission Tomography (PET). A patient receives an injection of some substance such as glucose that has been tagged with a radioactive element. The glucose soon goes to the brain, where it emits positrons. The positrons annihilate with electrons, producing pairs of gamma photons that are readily measured by a surrounding array of detectors. The images that result can reveal Alzheimer's disease, schizophrenia, and a wide range of other disorders of the brain.

A high-energy γ-ray invisibly entering from the bottom of the photo creates an electron-positron pair. The track of each of these particles has a large radius of curvature and therefore corresponds to a high speed. The positron (blue) subsequently collides with an electron in the medium. The two annihilate one another creating a photon with an energy of 265 ± 31 MeV, far exceeding the 1.02 MeV necessary for pair production. That photon travels without leaving a track for about 10 cm whereupon it materializes into an electron and a positron. In pair production, the nearby positive nucleus repels the positron while attracting the electron toward itself. As a result, the kinetic energies of the electron and positron are often very quickly not equal.

can get a picture of the metabolism of various regions of the brain (Fig. 29.16). Antinucleons have been joined together to make compound structures such as antideuterons and antialpha particles. Researchers have made *positronium*, a hydrogen-like atom composed of a positron and an electron bound together in a somewhat stable form. And an exotic atom called *antiprotonic hydrogen*, a proton and antiproton orbiting each other, was created in 1978. Protons, electrons, neutrons, neutrinos (indeed, all the material subnuclear particles) have their antimatter twins, their mirror-image annihilators.

Dirac's theory worked well; in its first approximation, it agreed nicely with the known experiments of the day. His prediction of the electron's magnetic moment was accurate to two decimal places. But there gradually developed some very troublesome features: an atomic electron should interact both with the field of the nucleus and, to some extent, with its own field. When attempts were made to add in appropriate contributions, the theory went wild, producing totally weird results—things took on infinite values. In 1947, Willis Lamb at Columbia University used microwave techniques to discover that two quantum states of hydrogen actually had slightly different energies, even though the Dirac theory predicted that they had to be *exactly* identical. This *Lamb shift* stimulated a modern-day reformulation of Quantum Electrodynamics, principally by Julian Schwinger and Richard Feynman of the United States and Sinitiro Tomonaga of Japan. They devised a mathematical scheme for making the theory workable known as *renormalization*. It got around the infinities, producing "normal" results.

Despite the fact that QED, the theory of the electron and photon, is certainly limited, it nonetheless is among the premier theoretical formalisms of physics. In the several instances where QED can be carried through, it provides an amazing degree of accuracy. For example, the magnetic moment of the electron as determined theoretically via QED has the numerical value 1.001 159 652 46, with an error of about 20 in the

Richard Phillips Feynman
(1918–1988).

last two digits. The accepted measured value is 1.001 159 652 193, with a possible error of about 10 in the last two digits. A theory that can do that will engender a great deal of faith in its practitioners.

As we'll see (p. 1100), the structure of QED is very special in a way that few people were concerned about before 1954; it's a *gauge theory*, derivable from a symmetry principle. In time, QED would become a model for all the theoretical advances in High Energy Physics, the great work of the second half of the twentieth century.

Core Material & Study Guide

THE CONCEPTUAL BASIS OF QUANTUM MECHANICS

According to de Broglie, the wavelength of a particle is

$$\lambda = \frac{h}{mv} \qquad [29.1]$$

and, using the kinetic energy in eV for nonrelativistic situations,

$$[\text{for electrons}] \qquad \lambda = \frac{1.226\,4}{\sqrt{\text{KE}}}\text{ nm} \qquad [29.2]$$

Material particles that are moving have momentum and therefore a wavelength. These ideas are treated in Section 29.1 (De Broglie Waves), and they form the basis for much of what follows. Study Example 29.1. Throughout this chapter examine the appropriate **WALK-THROUGH PROBLEMS on the CD**. Study each of the three-problem magenta groupings in the problem set and then try to do all the I-level questions.

Heisenberg, Schrödinger, and Dirac each formulated a quantum-mechanical theory, all of which are closely related. **Schrödinger's Equation** in simple form is

$$-\frac{\hbar^2}{2m}\frac{d^2\psi}{dx^2} + U\psi = E\psi$$

and it is the basis of Wave or Quantum Mechanics (p. 1031). The wavefunction, ψ, determines the probability of occurrences. Reread Section 29.3 (The Schrödinger Wave Equation). This material is too complicated for us to apply to problem solving.

QUANTUM PHYSICS

The dimensionless quantity

$$\alpha = \frac{e^2}{4\pi\varepsilon_0\hbar c} \approx \frac{1}{137} \qquad [29.3]$$

is known as the **fine structure constant**.

No two fermions in any system can occupy the same state, that is, have the same quantum numbers. This concept is the **Pauli Exclusion Principle** (p. 1038). The state of an atomic electron is

fixed by four *quantum numbers*: n, l, m_l, and m_s. Sections 29.4 through 29.8 lead to an important understanding of atomic structure. Make sure you understand Example 29.2.

One form of the **Heisenberg Uncertainty Principle** (p. 1044) is

$$\Delta p_x \Delta x \approx h \qquad [29.4]$$

or, more precisely,

$$\Delta p_x \Delta x \geq \tfrac{1}{2}\hbar \qquad [29.5]$$

The position and linear momentum of an object cannot both be measured with unlimited precision at the same time. Similarly

$$\Delta E \Delta t \geq \tfrac{1}{2}\hbar \qquad [29.6]$$

Key Terms

de Broglie wavelength	Pauli Exclusion Principle
Principle of Complementarity	fermion
Schrödinger Equation	boson
wavefunction	photon
quantum randomness	shell
principal quantum number	subshell
angular momentum quantum number	orbital
fine structure constant	Uncertainty Principle
Zeeman Effect	QED
magnetic quantum number	positron
spin quantum number	antiproton
spin magnetic quantum number	antineutron

Discussion Questions

1. Given that the resolving power of a microscope is proportional to the wavelength of the illumination, why might an electron microscope be appealing? Figure Q1 shows a transmission electron microscope invented by Ruska in 1935. How does it work? Incidentally, electrons traveling within the axially symmetric magnetic field inside the gap of a current-carrying coil tend to be brought inward to a focus. Compare this tendency with the behavior of X-rays. While an optical instrument might have a resolution of 200 nm or so, a modern electron microscope can resolve objects as small as about 0.1 nm. (The limit on resolution is actually set by the spherical aberration of the lenses and not λ.)

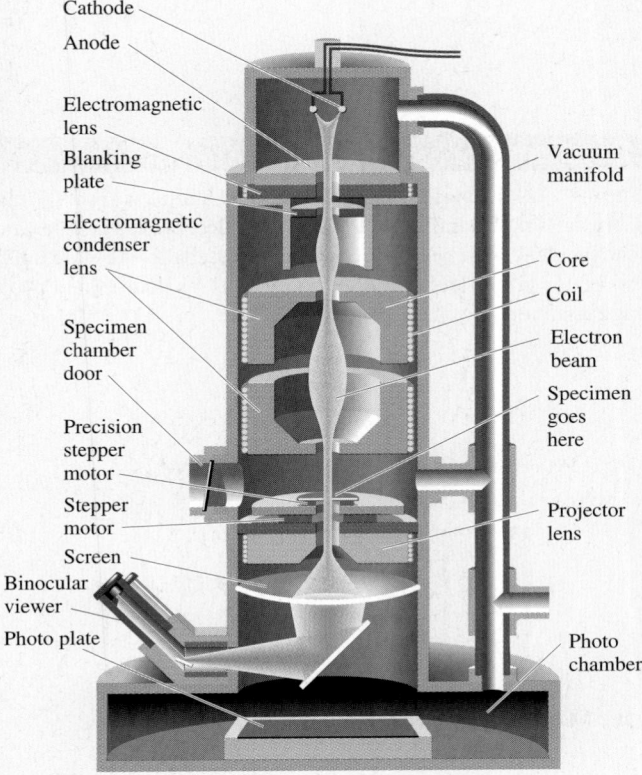

Figure Q1

2. A narrow, very sparse beam of monoenergetic protons is shone on an even narrower slit in an opaque screen. Make a set of drawings showing the pattern you would expect to observe after several increasing intervals of time. Explain your conclusions. If the beam were not monoenergetic, what would happen?

3. Draw a crude pictorial representation of the electron shell structure of sodium fluoride (NaF), and explain how the molecule is held together. Do the same for hydrogen chloride (HCl).

4. Considering the shell model of the atom, derive an expression for the number of electrons in the *n*th shell and also the number in the *l*th subshell. Explain your reasoning.

5. Imagine a screen with two narrow slits in it and suppose there is an incident beam of monochromatic electrons that encompasses both apertures (Fig. Q5). An interference pattern typical of Young's

Experiment will be observed on a distant surface. Now suppose that each slit is surrounded by a coil of wire such that, as an electron passes through the opening and then the coil, a current will be induced and the passage of the particle appropriately recorded. Use the arguments of Quantum Mechanics to describe what, if anything, will happen to the interference pattern once the switches on the coils are closed. If only one coil was used, would that change things?

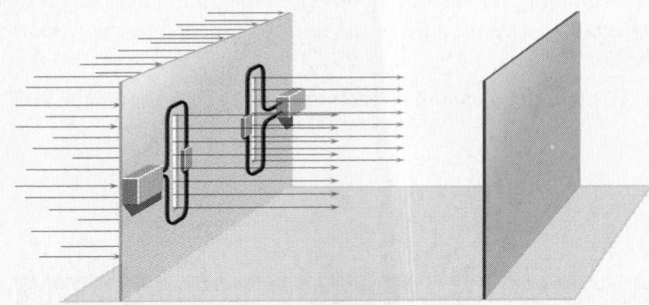

Figure Q5

6. Like photons, electrons have a wavelength, frequency, spin, and phase. Might it be possible to build a laserlike device that will produce tremendously intense beams of coherent electrons? Explain.

7. A far-reaching consequence of the Uncertainty Principle is that a particle confined to a region of space cannot have zero kinetic energy. The energy it does have is called the **zero-point energy**. Explain this statement and discuss how it might apply to the energy of a material at a temperature of absolute zero.

8. In 1929, Otto Stern shone a beam of neutral helium atoms (with a de Broglie wavelength of about 0.13 nm) at an angle to the surface of a crystal. What do you think he saw when he examined the pattern of the reflected helium atoms? Explain.

9. A reasonable way to represent a particle mathematically is to use a localized wavefunction like that of Fig. Q9. Such a pulse is known as a *wave packet*, and it's equivalent to the superposition of a number of monochromatic waves, as indicated in the figure. The tighter (shorter in space) the packet, the more numerous the needed monochromatic contributions (remember Fourier's analysis). In what sense does this description of the particle have built into it an uncertainty in energy and momentum? How might you make Δp zero? What then happens to Δx? What happens as $\Delta x \to 0$?

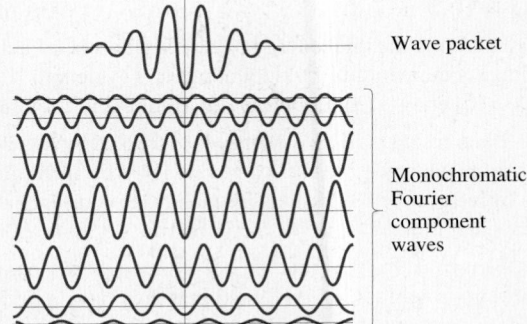

Wave packet

Monochromatic Fourier component waves

Figure Q9

10. Can a photon create a single electron? Explain your reasoning. What conclusions, if any, can you draw about the creation of particles in general?

11. Electron clouds are statistical distributions, and that makes the size of an atom somewhat ambiguous. Figure Q11 is nonetheless a summary of several atomic radii as determined, for example, from their spacings in solids. Explain the shape of the curve.

12. In what conceptual sense is a Bohr orbit of a hydrogen atom like a vibrating circular steel hoop? Does the central idea here in any way relate to the structure of the laser? [*Hint: Remember Kundt's tube?*]

13. Explain the meaning of each of the Key Terms on page 1050.

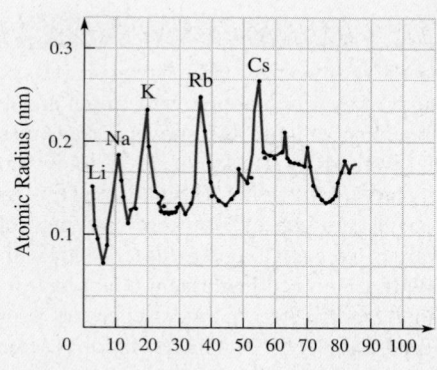

Figure Q11

Multiple Choice Questions

1. According to contemporary physics, wavelike behavior is a characteristic of   (a) all particles at rest relative to the observer   (b) all particles moving relative to the observer   (c) only moving charged particles   (d) only stationary charged particles   (e) none of these.

2. If Planck's Constant were 100 times larger than it is, the   (a) mass of a moving particle would be 100 times smaller   (b) momentum of a moving particle would be 100 times larger   (c) wavelength of a moving particle would be 100 times smaller   (d) wavelength of a moving particle would be unchanged   (e) none of these.

3. Doubling the momentum of a neutron   (a) decreases its energy   (b) doubles its energy   (c) doubles its wavelength   (d) halves its wavelength   (e) none of these.

4. In general, the speed of a material particle   (a) equals the phase speed of its de Broglie wave   (b) equals the phase speed of its ψ-wave   (c) does not equal the phase speed of its de Broglie wave   (d) equals c (e) none of these.

5. For a monochromatic photon, the phase speed in vacuum   (a) equals the phase speed of its de Broglie wave   (b) does not equal the phase speed of its ψ-wave   (c) does not equal the phase speed of its de Broglie wave   (d) equals the speed γc (e) none of these.

6. If the circumference of the first Bohr orbit is 3.3×10^{-10} m, what is the wavelength of the ground-state electron?   (a) 0   (b) ∞ (c) 3.3×10^{-10} m   (d) $(3.3 \times 10^{-10}$ m$)/h$ (e) none of these.

7. Doubling the total energy of a meson has the effect of   (a) doubling its momentum   (b) doubling its wavelength   (c) quartering its frequency   (d) doubling its frequency   (e) none of these.

8. If a moving particle's energy is increased by a factor of 10,   (a) its frequency increases by a factor of 10   (b) its frequency decreases by a factor of 10   (c) its frequency remains unchanged   (d) its wavelength increases by a factor of 10   (e) none of these.

9. Neutrons from a reactor are slowed down by passing them through graphite (Fig. MC9) so that they have a de Broglie wavelength of about 0.1 nm. The beam is directed at a crystal and the neutrons   (a) because they are uncharged and pass right through the crystal, totally unaffected by it   (b) reflect off at an angle equal to the incident angle, just like a stream of baseballs   (c) are totally absorbed   (d) reflect off in several beams, as determined by the Bragg equation   (e) none of these.

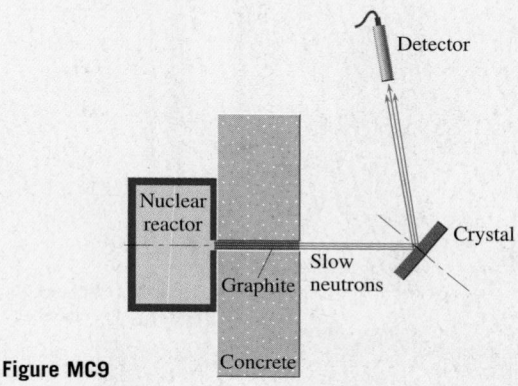

Figure MC9

10. It follows from the de Broglie hypothesis that   (a) $E = \omega h$ (b) $E = \omega h$ (c) $E = fh$ (d) $E = \lambda h$ (e) none of these.

11. It follows from the de Broglie hypothesis that   (a) $p = \omega h$ (b) $p = \lambda h$ (c) $p = kh$ (d) $p = \lambda h$ (e) none of these.

12. If ψ is the wavefunction for a particle, $|\psi|^2$ is proportional to   (a) the charge density of the particle   (b) the probability of finding the particle at a point in space   (c) the momentum of the particle at a point in space   (d) the energy of the particle at the point in space   (e) none of these.

13. The principal quantum number of the sixth excited state of hydrogen is   (a) 6   (b) 5   (c) 4   (d) 7   (e) none of these.

14. The maximum number of electrons that can be contained in the f-subshell of an atom is   (a) 14   (b) 10   (c) 6   (d) 12   (e) none of these.

15. The size of the space in which an electron is confined determines the uncertainty in its   (a) linear momentum   (b) maximum angu-

lar momentum   (c) spin angular momentum (d) its mean lifetime (e) none of these.

16. We might say that an electron in the ground state cannot radiate because   (a) if it did, its wavelength would increase and it could not fit in any smaller lower energy orbit   (b) it has no energy to radiate in the ground state   (c) it would decrease in wavelength

and lose momentum   (d) it would not conserve both energy and momentum   (e) none of these.

For more Multiple Choice Questions with answers click on WARM-UPS in CHAPTER 29 on the CD.

Suggestions on Problem Solving

1. Some useful constants to keep in mind are

$$1 \text{ eV} = 1.602\,177 \times 10^{-19} \text{ J}$$

$$h_c = 1.239\,842 \times 10^{-6} \text{ eV·m} \approx 1240 \text{ eV·nm}$$

$$h = 6.626\,08 \times 10^{-34} \text{ J·s} = 4.135\,67 \times 10^{-15} \text{ eV·s}$$

$$\hbar = 1.054\,573 \times 10^{-34} \text{ J·s} = 6.582\,12 \times 10^{-16} \text{ eV·s}$$

$$\tfrac{1}{2}\hbar = 3.29 \times 10^{-16} \text{ eV·s}$$

2. We often have to find such things as the wavelength of, say, an electron, and so need its momentum ($p = h/\lambda$). And here a decision has to be made as to whether to compute the classical value $p = \sqrt{(\text{KE})2m}$ or the relativistic value $p = \sqrt{(E^2 - E_0^2)}/c$. When $E_0 \gg \text{KE}$, we can use the classical formulation. For example, for

an electron with a KE of 0.1 MeV (as compared to its rest energy of 0.5 MeV), using the classical momentum in calculating wavelength introduces a 5% error.

3. When calculating the de Broglie wavelength ($\lambda = h/p$) and frequency ($f = E/h$) for anything other than a photon, remember that λf is the phase speed of the wave, *not* the speed of the particle and *not* c.

4. The following summary of levels and quantum numbers is handy to have:

Principal	$n = 1, 2, 3, 4, \ldots$
Orbital	$l = 0, 1, 2, 3, \ldots, n - 1$
Magnetic	$m_l = l, l - 1, \ldots, 0, \ldots, -l + 1, -l$
Spin	$m_s = -\tfrac{1}{2}, +\tfrac{1}{2}$

Problems ✦ Coordinated Problems ✦ Progressive Problems ✦ Solutions

STUDY GUIDE   **1. Coordinated Problems:** The three problems within each magenta-colored grouping are solvable in similar ways. Note that the first of these always has a hint; moreover, its solution is provided in the back of the book. *Work out each of these sets; they'll strengthen technique and build confidence.* **2. Progressive Problems:** The problems introduced in blue unfold step-by-step carrying along the analysis in a more suggestive way than is customary. *Work out all of these; they'll guide you through the analytic process and help develop problem-solving skills.* **3. Worked-Out Solutions:** Studying worked-out solutions is an important part of learning how to solve problems. Accordingly, additional *solutions* to a number of model problems are given below. *Make sure you understand each of them before you go on to the next problem.* **4.** Also provided in the back of the book are the *Answers* to all odd-numbered problems, as well as worked-out *solutions* to those with boldface numbers. Problem numbers in italic indicate that a solution appears in the Student Solutions Manual.

SECTION 29.1: DE BROGLIE WAVES

1. [I] Calculate the wavelength of a 60-kg person jogging along at 2.0 m/s.

2. [I] What is the de Broglie wavelength of a 10.00-g bullet traveling at 331 m/s?

3. [I] Determine the wavelength of an electron traveling at a speed of c/10.

4. [I] Determine the frequency of an electron traveling at a speed of c/10. Use relativistic considerations.

5. [I] A *thermal neutron* is one that is traveling at a speed comparable to that of a gas molecule at room temperature—in other words, one for which $(3/2)k_B T = \text{KE}$, and that (at 293 K) turns out to be 6.068×10^{-21} J. What is the wavelength of a thermal neutron?

6. [I] Use the results of Problems 3 and 4 to determine the phase speed of the de Broglie wave of an electron traveling at c/10. Note that v_p is greater than c.

7. [I] Suppose we are to build an electron microscope and we want it to operate at a wavelength of 0.10 nm. What accelerating voltage should be used?

8. [I] A tiny virus having a mass of 1.00×10^{-20} kg has a de Broglie wavelength of 660 nm. How fast is it traveling?

SOLUTION: The de Broglie wavelength is given via the momentum expression $p = h/\lambda$. The virus's momentum is $p = mv$ and so $v = h/\lambda m = (6.626 \times 10^{-34} \text{ J·s})/(660 \times 10^{-9} \text{ m})(1.00 \times 10^{-20} \text{ kg}) = 1.0 \times 10^{-7}$ m/s.

9. [II] THIS PROBLEM DEALS WITH THE DE BROGLIE WAVELENGTH. An electron which is initially at rest is accelerated across a potential difference (ΔV) of -100 V. (a) How much PE_E will the electron lose, and how much KE will it gain? (b) Explain the significance of the minus sign on the voltage difference. Since $E_0 \gg \text{KE}$, this problem does not require that we use relativistic dynamics. (c) Using Conservation of Energy, show that the momentum of the emerging electron is

$$m_e v = \sqrt{2m_e q_e \Delta V}$$

(d) Compute the momentum of the electron. (e) Determine the wavelength of the electron in nanometers after it's accelerated.

10. [II] What is the wavelength of an electron whose KE is 4.00

MeV? Since this energy is comparable with the rest energy, use relativistic arguments.

11. [II] What is the wavelength of an electron with a KE of 20 eV? [*Hint:* $E_0 \gg KE$, *so classical dynamics will do.*]

12. [II] A proton traveling in an evacuated tube descends through a potential difference of 100 V. Determine its final wavelength.

13. [II] A neutron is given a kinetic energy of 50 eV; determine its wavelength.

14. [II] Determine the kinetic energy of an electron that has a wavelength of 1.00 m.

15. [II] Consider a particle that has a large KE compared to its rest energy. Show that its wavelength is approximately equal to the wavelength of an equal-energy photon.

16. [II] If v_p is the phase speed of the de Broglie wave of a particle traveling at a speed v, prove that $v_p v = c^2$. Use Special Relativity.

17. [II] For a lightwave, the phase speed is given by $c = E/p$. Assume the same form for a de Broglie wave, and find a relationship between its phase speed (v_p) and the particle's speed v. Assume classical conditions; that is, let $E = KE$. Your result neglects the rest energy and will not compare very well with the relativistic expression $v v_p = c^2$.

18. [II] With Problem 15 in mind and assuming an electron in a hydrogen atom has an energy given by $E_n = -hf_n$, where f_n is the frequency of the electron in the nth "orbit," show that de Broglie's hypothesis leads to the same orbital energies as the Bohr Theory; namely,

$$E_n = -\frac{nhv_n}{2r_n}$$

provided $E = KE$.

19. [III] In the Davisson-Germer experiment, 54-V electrons were scattered from a nickel crystal (in the first order) at 65°. The spacing of the crystal's atomic planes was found, using X-rays, to be 0.091 nm. Show that these results conform to de Broglie's hypothesis.

SECTION 29.4: QUANTUM NUMBERS
SECTION 29.8: ELECTRON SHELLS

20. [I] Verify that the fine structure constant α actually equals $1/137$.

21. [I] Show that the fine structure constant α is dimensionless.

22. [I] In Problem 70 on p. 1026 we used the Bohr Theory to prove that for hydrogen-like atoms the speed of the single orbital electron in the *nth* state is given by

$$v_n = \frac{k_0 Z e^2}{nh}$$

Show that

$$v_n = \frac{Z}{n}\alpha c$$

where α is the fine structure constant.

23. [I] With the previous problem in mind, show that for the ground state of hydrogen

$$v_1 = \alpha c$$

where α is the fine structure constant.

24. [I] What is the ratio of the Bohr radius to the Compton wavelength in terms of α, the fine structure constant.

25. [I] Scattering experiments measure the so-called *classical radius of the electron*: $r_c = e^2/4\pi\varepsilon_0 m_e c^2$. Write an expression for the ratio of the Bohr radius to the classical electron radius, in terms of α, the fine structure constant.

26. [I] What are the values of the quantum numbers n and l for a $3d$ electron state?

27. [I] Consider the second excited state of the hydrogen atom. What are the values of the appropriate quantum numbers n, l, and m_l?

28. [I] Can two electrons in an atom have quantum number sets of $(2, 0, 0, +\frac{1}{2})$ and $(2, 0, 0, -\frac{1}{2})$, respectively? Explain.

29. [I] How many electrons can reside in the K, L, M, and N shells of an atom? See Problem 38.

30. [I] How many electrons can reside in each subshell of the M-shell of an atom?

31. [II] Using the Bohr Theory, determine the energy of a photon that would excite the electron in a hydrogen atom from the K-shell to the M-shell.

32. [II] Make a table of all of the allowed four quantum numbers for the first three shells of the hydrogen atom. How many electrons can each shell accommodate?

33. [II] An excited hydrogen atom with its electron in the O-shell drops down to the L-shell, emitting a photon in the process. Determine the energy of that photon using the Bohr Theory.

34. [II] What are the possible values of n, l, and m_l for a $5f$ atomic state? (First, check to see if it's allowed.)

35. [II] What are the possible values of n, l, and m_l for a $3f$ atomic state, and what can be said about that state?

36. [II] Determine the ground-state electron configuration for $_{17}$Cl.

37. [II] THIS PROBLEM DEALS WITH THE DISTRIBUTION OF ELECTRONS IN AN ATOM. We want to determine the ground-state electron configuration of phosphorus. (a) What is the atomic number of phosphorus? (b) How many electrons does a phosphorus atom have? (c) Considering Fig. 29.11, how many electrons occupy each orbital? Explain what this has to do with spin. (d) Referring to Fig. 29.10, how many *1s* orbitals are there? How many *2s* orbitals are there? How many *2p* orbitals are there? How many *3s* orbitals are there? How many *3p* orbitals are there? (e) How many electrons in total will it take to fill all the orbitals up to and including *3p*? (f) Write out the electron configuration for phosphorus.

38. [II] THIS PROBLEM DEALS WITH THE DISTRIBUTION OF ELECTRONS IN A MULTIELECTRON ATOM. With Problem 29 in mind, we want to determine a general expression for the number of electrons that can occupy a given shell in an atom. (a) Show that the number of values of m_l, for a given value of l, is $(2l + 1)$. (b) Here l goes, in steps of 1, from $l_{min} = 0$ to $l_{max} = (n - 1)$, and each such value has $(2l + 1)$ values of m_l. Show that for a given n the total number of m_l values (i.e., the number of orbitals) is $[1 + 3 + 5 + ... + (2n - 1)]$. (c) Explain why the number of electrons that can be accommodated in a shell of quantum number n (i.e., the number of states) is given by $2[1 + 3 + 5 + ... + (2n - 1)]$. (d) The sum in the brackets is

equal to n^2, and so the total number of electrons in the nth shell is $2n^2$. Verify that this is correct for $n = 3$.

39. [II] THIS PROBLEM DEALS WITH THE STRUCTURE OF ENERGY LEVELS. Consider a hydrogen atom in which an electron has an orbital angular momentum quantum number of 3. We want to determine the lowest shell (i.e., the smallest n) that can accommodate this subshell. (a) How many orbitals are in this subshell? (b) How many states are in this subshell? (c) Explain the statement $n \geq (l + 1)$. (d) What is the minimum value of n that specifies the lowest shell that can contain this subshell? (e) Using the Bohr Theory, determine the energy (in eV) of the electron in that shell.

SECTION 29.9: THE UNCERTAINTY PRINCIPLE

40. [I] The speed of an electron traveling along the x-axis is known, with an uncertainty of 2.0%, to be 100 m/s. At best, how accurately can we know its location; that is, what is the minimum uncertainty in its position?

> SOLUTION: The electron's momentum is $p_x = m_e v$, and the corresponding uncertainty is $\Delta p_x = 2.0\% p_x = 2.0\% m_e v = 0.02(9.11 \times 10^{-31}$ kg)(100 m/s) $= 1.822 \times 10^{-30}$ kg·m/s. The Uncertainty Principle $\Delta p_x \Delta x \geq \hbar/2$ yields $\Delta x \geq \hbar/2\Delta p_x$ and so at best $\Delta x = h/4\pi(1.822 \times 10^{-30}$ kg·m/s) = 2.9×10^{-5} m.

41. [I] Imagine a particle flying along the x-axis in a box of length L. What is the uncertainty in its momentum along the x-axis?

42. [I] A 100-g ball is confined to move in a 1.00-m-long frictionless tube lying along the x-axis. What is the minimum uncertainty in its speed? And how far will it move in a year at that speed?

43. [I] The position of a 0.001 00-kg particle along the x-axis is measured to be $1.243\ 7 \pm 0.000\ 5$ cm from the tip of a probe. What is the minimum uncertainty in its speed? [*Hint: That's $\pm 0.000\ 5 \times 10^{-2}$ m.*]

44. [I] Since a charged pi meson at rest exists on average for only 26 ns, its energy will not be able to be measured with unlimited precision. Determine the minimum uncertainty in the meson's rest energy.

45. [I] A typical excited atomic state has a lifetime of 10^{-8} s. Determine the minimum uncertainty in the energy of such a state in joules and electron-volts. This is the unavoidable "blurriness" of the energy level, and rather than a sharp line, it produces an emission band known as the *natural linewidth*. [*Hint: Take the lifetime to be the temporal uncertainty.*]

46. [I] The energy of a nuclear state is measured to within an uncertainty of 1.0 eV. What should be the corresponding lifetime of the state?

47. [I] Determine the minimum uncertainty in the energy of an atomic state if the state lasts for 0.10 μs?

48. [I] With Problem 45 in mind, determine the minimum uncer-

tainty in the frequency of the emitted photon when there is a transition down to the ground state.

49. [I] A rho meson at rest has a mean lifetime of 4.4×10^{-24} s and an energy of 765 MeV. Determine the minimum uncertainty in the energy and write it as a fraction of the rest energy.

50. [II] A 10.0-μg particle is traveling at 2.00 cm/s. Given that there is a 1.00% uncertainty in its speed, what is the least uncertainty in its position?

51. [II] A 1.0-kg projectile moving along a straight line in space is located with an accuracy of 1.00 mm. Determine the change in its speed resulting from the measurement.

52. [II] Prove that a 10-g beetle whose position is known to within 1.00 nm can move with a speed uncertainty of 1.00 nm per year and still not violate the Uncertainty Principle.

53. [II] A 10.0-g particle is traveling at 20.0 cm/s. If the uncertainty in its momentum is 1 part in 1000, determine the uncertainty in its position.

54. [II] Approximate the minimum kinetic energy of an electron confined to a region the size of an atom (0.10 nm).

55. [II] The position of an electron is measured within an uncertainty of 0.100 nm. What will be its minimum position uncertainty 2.00 s later?

56. [II] Angular momentum (L) and angle (θ), in radians, are conjugate variables. Accordingly, beginning with $\Delta p_x \Delta x \geq \frac{1}{2} \hbar$, apply it to circular motion and derive the uncertainty relationship for angular momentum.

57. [III] Gold is placed in an oven, melted, vaporized, and kept at a temperature of 1600 K. A parallel beam of atoms (each of mass 3.271×10^{-25} kg) emerges directly toward a circular hole of radius r in a screen (Fig. P57). Taking all the atoms to have the same speed, use the Uncertainty Principle to approximate the spot size formed by the gold on a screen that is 1.000 m away. Make whatever simplifying assumptions you need to.

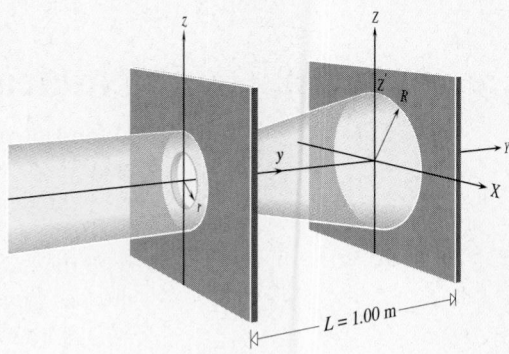

Figure P57

Chapter 30
Nuclear Physics

An atom, with a radius of about 10^{-10} m, consists of a cloud of electrons moving at great speeds around a positively charged nucleus comprised of protons and neutrons. That minuscule core, containing over 99.9% of the total mass, is roughly 10^4 times smaller than the atom as a whole. Like the atom, the nucleus is a bound system and can exist in a number of quantum states beyond its lowest energy ground state.

We now turn our attention to the atomic nucleus. Though crucial to the structure of the atom, it plays an unobtrusive role hidden deep below the swirling cloud of electrons. On Earth, the atomic nucleus reveals itself in the phenomenon of radioactivity and in nuclear weapons that unleash vast amounts of energy. On a much grander scale, nuclear fusion powers the stars (p. 1084).

Nuclear Structure

During the past several decades, experiments have provided reliable information about the size, shape, distribution of charge within, and magnetism of, the nucleus.

Three principal techniques are used: (1) the nucleus can be probed with high-energy (≈ 10 GeV), short-wavelength beams, usually of electrons; (2) orbital electrons interact with the nucleus, and the electromagnetic energy they emit provides information about nuclear structure; and (3) a beam of positive particles can be scattered inelastically, transferring some of its energy to the nucleus, which becomes excited. The gamma radiation emitted as the nucleus returns to the ground state provides data on the nuclear configuration. It has been found that the nucleus is a complex shifting structure made of rapidly moving parts. The first clues to its composition came from studying isotopes.

30.1 Isotopes: Birds of a Feather

All atoms of a particular element were supposed to be identical. That assertion had been central to atomic theory since before Dalton's time, but it was wrong. There *are* different kinds of every one of the elements, often six or seven distinct variations. Rutherford and F.

STUDY GUIDE

This chapter is a basic introduction to the atomic nucleus. It's a practical approach, so rather than spending much time on nuclear structure, we'll study the various nuclear transformations that can occur. The emphasis is on radioactivity—its origins and applications. But we'll also learn something about fission (p. 1079) and fusion (p. 1084).

Soddy traced the decay of several heavy, naturally occurring radioactive elements, such as uranium and thorium. These decayed, spewing out particles and continuously transforming into different elements until they ended up as lead. Along the way, elements appeared that were chemically identical to others but had different radioactive characteristics. Although ordinary lead has an atomic mass of 207.20 (we'll straighten out the units in a moment), lead present in uranium ore had a mass of only 206.05. Soddy named these variations of a given element **isotopes** (from the Greek *isos* meaning "same" and *topos* for "place"— having the same place in the Periodic Table).

In 1913, J. J. Thomson and F. W. Aston were the first to separate the isotopes of an element. They had put neon gas (atomic mass 20.2) into a positive ion tube and deflected the beam with charged plates and magnets to measure the atomic mass. At the face of the tube *two* distinct spots appeared. The conclusion was unmistakable: "Neon is not a single gas, but a mixture of two gases, one of which has an atomic weight [sic] of about 20, and the other of about 22."

A particular species of nucleus specified by a characteristic value of both **atomic number** (Z) and mass number, or **nucleon number** (A), is called a **nuclide**, and there are about 1500 nuclides. Every distinctly different nucleus is a specific nuclide. **Isotopes** of a given element are nuclides having the same Z (i.e., the same nuclear charge) but a different A; the total number of nucleons is different. The **neutron number** (N) is the number of neutrons in the nucleus. Since $A - Z = N$, it follows that *isotopes of an element differ from one another only in their value of N*. Letting X be the chemical symbol for any element, we can specify any nuclide by writing it as

$$^A_Z X_N$$

The name of an element is associated with a specific value of Z, a specific place in the Periodic Table. Any atom containing 10 protons is neon no matter what its nucleon number. Moreover, 10 protons and 10 neutrons make neon ($^{20}_{10}Ne_{10}$), just as 10 protons and 12 neutrons make a different form of neon ($^{22}_{10}Ne_{12}$). Apparently, this notation is redundant: $Z = 10$ means neon, and vice versa. The distinction is also often made by referring to neon-20, or Ne-20.

When found in our atmosphere, 90% of natural neon is of the first and lighter variety, while the other 10% is the heavier kind. The *chemical atomic mass* of an element referenced in the Periodic Table reflects the natural abundances of the isotopes in a typical sample. These values were arrived at before the knowledge of the existence of isotopes and correspond to a weighted average of all the isotopes naturally present in the environment. For neon, that average turns out to be 20.179 7.

In the early part of the nineteenth century, it made sense to determine relative atomic masses. Thus, hydrogen was set at 1, and all the other elements were measured with respect to it. Today, masses in the atomic domain are often specified in **unified atomic mass units** (u) where a neutral carbon atom ($^{12}_{6}C$) is defined to have a mass of precisely 12.000 000 u (see Table 30.1). Because of the practical link to the laboratory, with its accelerating elec-

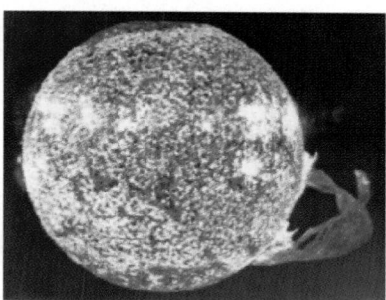

The Sun is a swirling sphere of plasma. At its center is a thermonuclear furnace driven by a gigantic fusion reaction.

Table 30.1

Atomic Mass of Some Representative Atoms

Element	Symbol	Mass (u)
Hydrogen	$^{1}_{1}\text{H}$	1.007 825
Deuterium	$^{2}_{1}\text{H (D)}$	2.014 102
Tritium*	$^{3}_{1}\text{H (T)}$	3.016 049
Helium	$^{3}_{2}\text{He}$	3.016 029
Helium	$^{4}_{2}\text{He}$	4.002 603
Lithium*	$^{5}_{3}\text{Li}$	5.012 54
Lithium	$^{6}_{3}\text{Li}$	6.015 121
Beryllium	$^{9}_{4}\text{Be}$	9.012 182
Nitrogen	$^{14}_{7}\text{N}$	14.003 074
Nitrogen	$^{15}_{7}\text{N}$	15.000 109
Nitrogen*	$^{16}_{7}\text{N}$	16.006 100
Oxygen	$^{16}_{8}\text{O}$	15.994 915
Oxygen	$^{17}_{8}\text{O}$	16.999 131
Oxygen	$^{18}_{8}\text{O}$	17.999 160
Lead	$^{204}_{82}\text{Pb}$	203.973 020
Lead*	$^{205}_{82}\text{Pb}$	204.974 458
Lead	$^{207}_{82}\text{Pb}$	206.975 872
Uranium*	$^{233}_{92}\text{U}$	233.039 628
Uranium*	$^{235}_{92}\text{U}$	235.043 924
Uranium*	$^{238}_{92}\text{U}$	238.050 785

*Radioactive. (Remember Table 26.3 p. 956).

tric fields, the unit MeV/c^2 is also widely used:

$$1 \text{ u} = 1.660\,540 \times 10^{-27} \text{ kg} = 931.494 \text{ MeV}/c^2$$

(For a derivation of this equivalence, see Problem 35.) As a rule, A for each isotope differs slightly, but significantly, from its mass in units of u. (The exception is ^{12}C.)

Example 30.1 **[II]** Show that the chemical atomic mass of neon should be about 20.18 u, given that ^{20}Ne and ^{22}Ne have natural abundances of about 90.51% and 9.22% and masses of 19.99 u and 21.99 u, respectively. (Your error will be due to the fact that we have overlooked trace amounts of ^{21}Ne.)

Solution The problem is based on the fact that any naturally occurring element is actually a mixture of isotopes. (1) TRANSLATION—Knowing the masses and relative abundances of the two isotopes of an element, determine its average atomic mass. (2) GIVEN: Isotopes with masses of 19.99 u and 21.99 u and abundances of 90.51% and 9.22%. FIND: The chemical atomic mass. (3) PROBLEM TYPE—Nuclear physics/isotopes/atomic mass. (4) PROCEDURE—We have to compute a

weighted average. (5) CALCULATION—The lighter neon is almost 10 times more abundant and therefore 10 times more influential in determining the chemical atomic mass of any natural sample than is the heavier isotope. The weighted average mass is therefore

$$90.51\% \ (19.99 \text{ u}) + 9.22\% \ (21.99 \text{ u})$$

or 18.09 u + 2.03 u = $\boxed{20.12 \text{ u}}$.

Quick Check: As already given, the actual value is 20.179 7 u.

[For more worked problems click on WALK-THROUGHS *in* CHAPTER 30 *on the CD.]*

Some 280 isotopes of the naturally occurring elements are stable and presumably will last for all time (Fig. 30.1). Around 1200 others (human-made and natural) are radioactive and transient. All the elements beyond uranium, from $Z = 93$ to ≈ 110, are produced in the laboratory and are radioactive; if they ever existed in abundance in nature, most were short-

Figure 30.1 A plot of *N* versus *Z* for stable and unstable nuclides. The so-called *line of stability* runs up through the gold-colored band, which represents the stable nuclei. Each integer value of *N* and *Z* lying on a colored region corresponds to a nuclide.

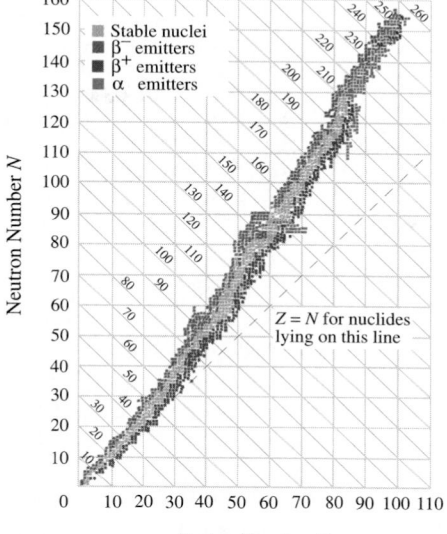

Hydrogen

Deuterium

Tritium

Figure 30.2 The three hydrogen isotopes.

lived enough to have decayed to unobservably low levels. Some elements, such as xenon and iodine, have more than a dozen known isotopes each. Nine of the xenon forms are stable, but iodine has only one stable isotope, as do aluminum and gold.

The three isotopes of hydrogen ($^1_1$H, $^2_1$H, $^3_1$H) are so different from each other and so important that, unlike all others, they have their own names (Fig. 30.2 and Table 30.2). Ordinary *hydrogen* with a single proton ($Z = 1$, $A = 1$) is the lightest and most common (99.985%). **Deuterium** (also written $^2_1$D) with an added neutron ($Z = 1$, $A = 2$) is quite rare (0.015%). For every 6500 ordinary hydrogen atoms in a natural sample, there is only 1 deuterium atom. Radioactive **tritium** ($^3_1$T) with still another neutron ($Z = 1$, $A = 3$) is by far the least abundant (for every 10^{18} atoms of ^{1}H there is 1 of ^{3}T). Since these isotopes differ enormously in mass, and since the nuclei must interact somewhat differently with the single orbital electron, it's not surprising that they are distinctive chemically. For example, living organisms respond differently to water formed of oxygen and deuterium, known as **heavy water**, than they do to the ordinary brew, which differs as well in its freezing and boiling points. Heavy-water ice cubes, though they look and taste ordinary enough, sink to the bottom of a glass of tap water.

Insofar as the nucleus has little influence on the outer electrons, isotopes behave identically chemically except for some minor effects. Different isotopes of the same element can be separated, but as a rule only by mechanical means. The fact that a uranium-based atomic bomb requires large amounts of ^{235}U, which must be separated from ^{238}U, makes their production quite difficult for all but the most technologically advanced nations.

30.2 Nuclear Size, Shape, and Spin

Experiments pioneered by R. Hofstadter in the 1950s have revealed that nuclei are fairly spherical, with most being slightly ellipsoidal, though they come elongated, flattened a little, and pear-shaped as well. Figure 30.3 is a plot of nuclear charge density measured out along a radial distance in femtometers. The same results hold for the mass-density, with

Table 30.2

Some Useful Nuclear Masses		
Particle	**Designation**	**Mass* (u)**
Neutron	n	1.008 665
Proton	p, ^{1}H	1.007 276
Deuteron	D, ^{2}H	2.013 553
Triton	T, ^{3}H	3.015 500
Helion	^{3}He	3.014 932
Alpha	α, ^{4}He	4.001 506

*1 u = 1.660 540 2 × 10^{-27} kg.

The 60-inch cyclotron at the Argonne National Laboratory. A beam of deuterons (the nuclei of deuterium) is streaming out at a speed of ≈28 000 miles per second.

Figure 30.3 (a) The charge density versus distance from the center of the nucleus for several nuclei. [Data from R. Hofstadter, *Annual Reviews of Nuclear Science* 7, 231 (1957).] (b) The radius of the germanium nucleus is about 4.9 fm.

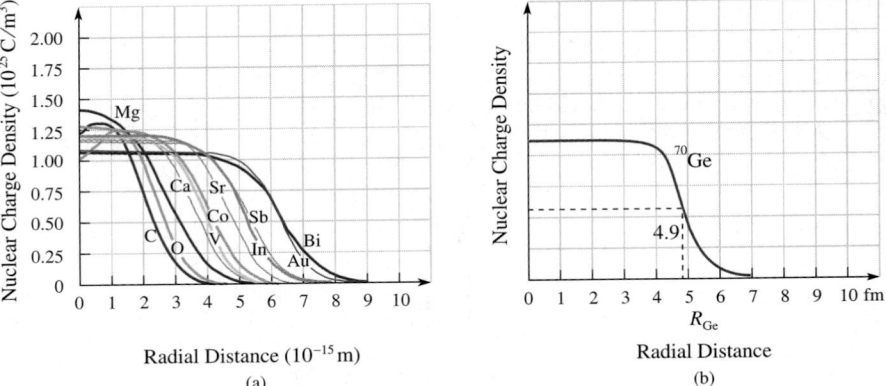

neutrons and protons being distributed in much the same way. Recall that 1 fm = 10^{-15} m; it's a distance that's convenient in the nuclear domain, just as the nanometer was convenient in the atomic domain. (Nowadays, 1 fm is often called a *fermi* in honor of the Italian-American physicist Enrico Fermi.) The density drops off gradually over an outer thickness of roughly 2.5 fm—the nucleus thins out across this one-nucleon-thick surface region.

The nuclear radius R is often taken as the distance from the center to the half-density point (Fig. 30.3b). Independent of A, the density of a nucleus is constant for much of its radius. That means that the number of nucleons contained in a nucleus (assumed to be spherical) simply depends on its volume $\frac{4}{3}\pi R^3$. Hence, $A \propto R^3$ and $R \propto A^{1/3}$. Using a proportionality constant R_0 to make this relationship into an equality, we have

$$R = R_0 A^{1/3} \tag{30.1}$$

R_0 turns out to be ≈ 1.2 fm (Fig. 30.4). This situation should remind us of a drop of water, which also has a volume proportionate to the number of molecules it contains.

Figure 30.4 (a) shows the relative sizes of helium and uranium nuclei. (b) gives a sense of the size difference between a nucleus and an atom.

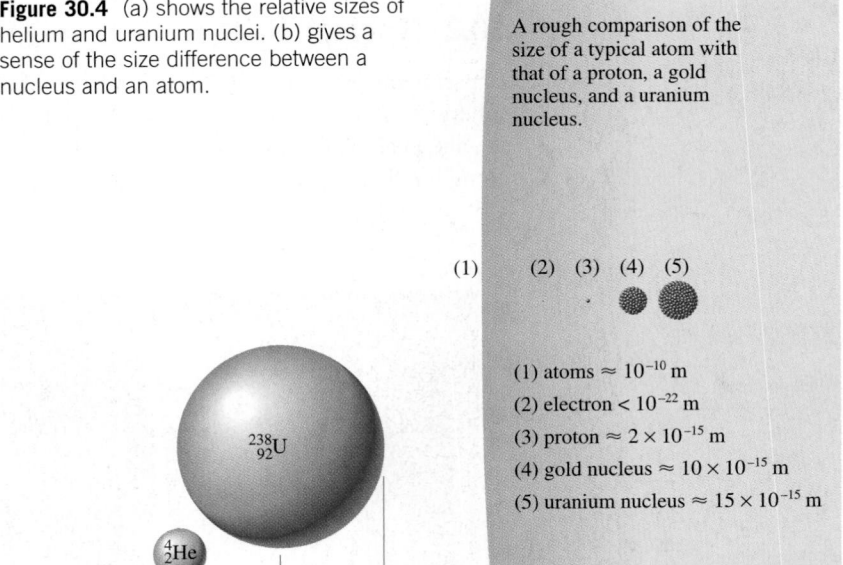

A rough comparison of the size of a typical atom with that of a proton, a gold nucleus, and a uranium nucleus.

(1) atoms $\approx 10^{-10}$ m

(2) electron $< 10^{-22}$ m

(3) proton $\approx 2 \times 10^{-15}$ m

(4) gold nucleus $\approx 10 \times 10^{-15}$ m

(5) uranium nucleus $\approx 15 \times 10^{-15}$ m

Example 30.2 **[I]** Determine the radius of the carbon nucleus ($A = 12.0$).

Solution We derived an expression for the radii of nuclei. (1) TRANSLATION—A nucleus has a specified nucleon number (A); determine its radius. (2) GIVEN: $A = 12.0$ u. FIND: R. (3) PROBLEM TYPE—Nuclear physics/nuclear radius. (4)

PROCEDURE—The defining equation is $R = R_0 A^{1/3}$. (5) CALCULATION—Using Eq. (30.1),

$$R = R_0 A^{1/3} = (1.2 \text{ fm}) A^{1/3} = (1.2 \text{ fm})(2.29) = \boxed{2.7 \text{ fm}}$$

Quick Check: Compare this result with Fig. 30.3.

The density of *nuclear matter*, as it's called, is the mass over the volume; and since A is equivalent to the mass in units of u, where $1 \text{ u} = 1.66 \times 10^{-27}$ kg.

$$\rho = \frac{m}{\frac{4}{3}\pi R^3} = \frac{A(1.66 \times 10^{-27} \text{ kg})}{\frac{4}{3}\pi R_0^3 A} = \frac{1.66 \times 10^{-27} \text{ kg}}{7.24 \times 10^{-45} \text{ m}^3} = 2.3 \times 10^{17} \text{ kg/m}^3$$

This value is tremendously large; by comparison, the density of water is a mere 10^3 kg/m^3. Nuclear matter in bulk does not exist on Earth. You would know if it did because a cubic inch of it would weigh about 4×10^9 tons. Still, it is likely that certain celestial objects, such as neutron stars, consist of nuclear matter. Each such object is essentially a gigantic nucleus with gravity forcing it together.

Neutrons and protons are fermions with spins of $\frac{1}{2}$ and angular momenta of $\frac{1}{2}\hbar$. The *Shell Model* (p. 1065) assumes that the nucleus has energy levels very like the atom. The Exclusion Principle requires that only two protons (one spin-up, one spin-down) and two neutrons (one spin-up, one spin-down) can occupy any level. The total spin of a nucleus is the sum of the spins of its parts—*odd-A nuclides are fermions, even-A nuclides are bosons*. Thus, for an even-even nucleus (even Z, even N), such as $_2^4$He, $_6^{12}$C, or $_8^{16}$O, the total spin is zero. The nucleus of deuterium, the **deuteron** ($_1^2$H), is typical of odd-odd nuclides; it has unfilled sublevels (one unpaired neutron and one unpaired proton) and a spin of 1. Even-odd and odd-even nuclides have one unpaired nucleon, and therefore total spins that are odd-integer multiples of $\frac{1}{2}$. Thus, the spin of $_3^7$Li (3 protons, 4 neutrons) arises from a group of two protons and two neutrons with spin zero plus the remaining 3 spin-$\frac{1}{2}$ nucleons, yielding a total of $\frac{3}{2}$.

The presence of spin suggests the possibility of a magnetic moment, which is certainly the case for a charged particle or an entity composed of charged particles. The magnetic moment of the proton is about 0.15% that of the electron and is in the same direction as its spin. This much is determined experimentally; as yet there is no complete theory of nuclear magnetism that accounts for the details. The magnetic moment of the proton ($+1.41 \times 10^{-26}$ J/T) is about three times larger than expected from its mass and the way the electron behaves. We can already anticipate that the proton will be a far more complicated entity than was first thought. The neutron, too, has a magnetic moment; it's smaller than the proton's and in the opposite direction to its own spin. The nonzero magnetic moment implies that the neutron, though it is neutral, has some internal nonuniform charge distribution. We now know that the proton and neutron are each composed of three still smaller charged quarks. A typical nucleus will have a net magnetic moment that depends on the number and arrangement of its nucleons.

By applying magnetic fields to an atom, it's possible to alter the spin state of the nucleus in a process discovered in 1946 called **nuclear magnetic resonance** (NMR). In the case of hydrogen, a proton can align its magnetic moment either parallel or antiparallel to an applied static B-field. The spin-up state has a slightly lower energy (ΔE) than the spin-down state. If a photon with an energy (hf) equal to the difference in energy between these two spin states (namely, ΔE) impinges on the proton, it can be absorbed. Thus, a pulse of radiofrequency electromagnetic energy of exactly the correct frequency can resonate the protons, exciting them into the higher energy state. The same is true of more complex

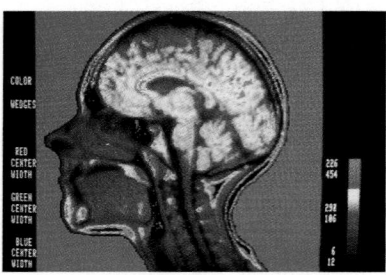

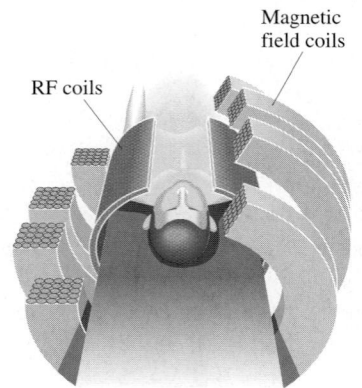

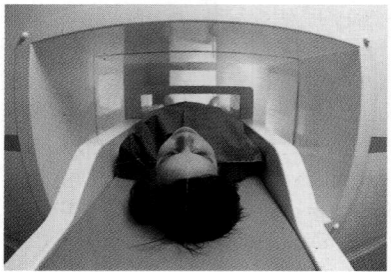

Figure 30.5 Magnetic resonance imaging (MRI). A human body is subjected to a powerful constant *nonuniform* magnetic field. The energy levels available to the hydrogen nuclei split in two (much like the Zeeman Effect), separating by $\Delta E \propto B$. A broadband radiofrequency (RF) pulse of electromagnetic energy then excites some of the nuclei ($hf = \Delta E \propto B$). When the nuclei subsequently relax, they emit RF photons ($f \propto B$), which are used to construct an image of the distribution of nuclei based on knowledge of the spatial configuration of B.

nuclei that have their characteristic resonant frequencies. When the nuclei relax back to their ground states, they each emit a photon of the same resonant frequency, which is easily detected. Either the absorption or re-emission of photons at different points in the specimen can be used to produce images of its internal structure. The human body is about 75% water and contains an abundance of hydrogen and, therefore, of protons. **Magnetic resonance imaging** (MRI) usually uses the NMR of protons to noninvasively form images of body tissue of living patients (Fig. 30.5).

30.3 The Nuclear Force

No sooner had the nucleus been revealed than there arose the obvious question: what holds it together? By 1925, there was a recognition of the need for a new type of force. A cluster of positively charged particles must repel one another with an electrostatic force (like charges repel). Moreover, we know from scattering experiments that the $1/r^2$ Coulomb force works right down to nuclear distances. A simple calculation of the repulsion between two protons separated by a distance that puts them just about in contact within a nucleus yields a value of around 50 N (about 11 lb). That interaction is enormous when considered in light of the tiny masses of the protons; no nuclei other than hydrogen should ever have formed, and those that exist should explode immediately. Evidently, there is another kind of force, the **nuclear force**, operating within the nucleus. **The nuclear force binds neutrons and protons together to form nuclei.** It had its conceptual origins in a proposed short-range, neutron-proton force first suggested (1932) by Heisenberg. Now known to have an effective range of only about 1 fm, the nuclear force is powerfully attractive, imparting potential energies to nucleons of as much as 100 MeV. It is repulsive at distances less than about 0.5 fm (two nucleons cannot occupy the same space), and it depends on the spins of the interacting particles.

The nuclear force is a manifestation of the more fundamental and less restricted *strong force*, which affects a whole class of particles, not just nucleons. The most potent of all known interactions, the strong force (at a proton-proton separation of 2 fm) is about 100 times stronger than the electromagnetic force and roughly 10^{34} times stronger than the gravitational force (Table 30.3). While it might take as much as 8 MeV to remove a nucleon from a nucleus, the electron in a hydrogen atom can be ionized with a mere 13.6 eV. For that reason, pound for pound, a nuclear reaction can liberate millions of times more energy than a chemical reaction. We will discuss the strong force at greater length in Chapter 31.

From experiments starting in 1936, we have found that the nuclear force exists between any two nucleons. The evidence for the n ↔ n attraction is inferential, but the p ↔ p and p ↔ n interactions can be measured, though indirectly, using beams of neutrons or protons scattered from a target consisting mostly of hydrogen (that is, protons). Electrons are completely immune to the nuclear force, which is why they were so effective at probing the nuclear charge distribution.

Table 30.3

The Four Forces of Nature*			
Force	**Interacts Between**	**Strength[1]**	**Effective Range**
Gravitational	All mass-energy[2]	10^{-34}	Unlimited
Weak	All material particles (quarks and leptons)	10^{-2}	$\approx 10^{-17}$ m
Electromagnetic	Electromagnetic charges	10^{2}	Unlimited
Strong	Many subnuclear particles (quarks and gluons)	10^{4}	$\approx 10^{-15}$ m

*At the temperatures that exist today, we see four distinct interactions. At much higher energies, these blend together (p. 1113).
[1]The strengths (in newtons) are for two protons separated, center-to-center, by 2 fm.
[2]Gravity and the strong force both act on their own field quanta.

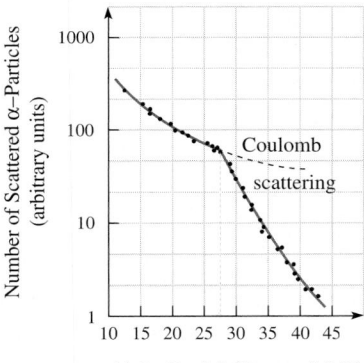

Figure 30.6 Alpha particles elastically scattered from a lead target. [Data from G. W. Farwell and H. E. Wegner, *Phys. Rev.* Vol. 95: 1212 (1954).] At around 27.5 MeV, the curve abruptly deviates from purely Coulomb scattering. The nuclear force now comes into play as the α-particles get close enough to the nucleus for that short-range force to be influential.

Rutherford's scattering experiments of 1913 established that, down to distances of roughly 10^{-14} m, α-particles interacted with nuclei via the Coulomb force. In 1919, he fired 5-MeV α-particles at low-Z nuclei, thereby minimizing the Coulomb repulsion. What he found (using hydrogen nuclei as targets) was that at distances of ≈ 3.5 fm the resulting scattering markedly deviated from that predicted by electrodynamics (Fig. 30.6).

At relatively large distances, protons repel one another via the Coulomb force, but there is a change at roughly 3 fm, where the interaction becomes increasingly attractive. A monoenergetic neutron beam colliding with protons shows little or no interaction to about 2 fm, whereupon there is again an increasingly strong attractive force. At very small separations, the interaction quickly becomes repulsive. The nuclear force is so powerfully repulsive at very small distances that nucleons rarely get closer to one another than about 0.4 fm. Estimates of the nucleon radius range from about 0.3 fm to 1 fm.

Figure 30.7*a* depicts a crude potential-energy curve for a proton interacting with a nucleus of radius R. At large distances, the curve is positive and Coulombic—an approaching proton experiences a repulsion. The positive *electrical*-PE increases as $1/r$. By contrast, a neutron does not "feel" any electrical force—it approaches the nucleus along the zero-PE axis in Fig. 30.7*b*. At the surface of the nucleus, a nucleon is tremendously influenced by the attractive nuclear force. The resulting negative potential energy confines the nucleon to the tiny region of the nucleus. Bound neutrons and protons rattle around inside the nucleus like submicroscopic bees flying in a well as much as 50 MeV deep.

The nuclear force is *saturable*: each bound nucleon interacts with only a few of its nearest neighbors. If a nucleon is added to a nucleus, it will not interact with all the other particles via the nuclear force. That means that the more massive nuclei should have nearly the same density as small nuclei, which is borne out by experiment (Fig. 30.3). Remember that atoms, held together by the $1/r^2$ Coulomb force, are nearly all the same size regardless of A. Adding more charge to both the nucleus and the electron cloud increases the long-range interaction, and that essentially compresses the atom. By contrast, nuclei get larger with A via Eq. (30.1). The nucleus will prove to be more similar in behavior to a droplet of water than it is to an atom.

The nuclear force performs as do the forces between molecules—both fall off rapidly with separation. Moreover, two water molecules also repel one another when they come very close together and their electron clouds start to overlap. Nucleons and water molecules, while shifting around, will tend to stay separated by a distance such that their mutual attraction is maximized. The separation between nucleons is comparable to the range of the nuclear force, and the surface of the nucleus is about that thick as well. As with a liquid, the nucleus also experiences surface tension; nucleons at the surface will be pulled inward, which is why nuclei are essentially spherical. In a sense, the nucleus is a droplet of nuclear matter.

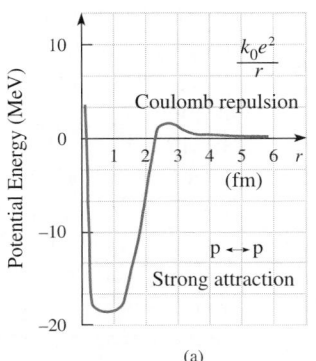

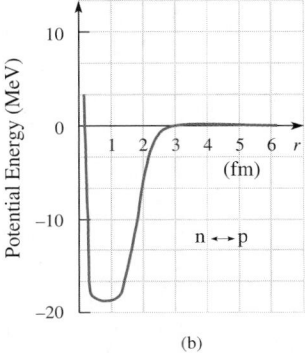

Figure 30.7 Approximate potential wells for (a) proton-proton and (b) neutron-proton interactions.

When a neutron and a proton come together to form a deuteron, they do it only if the two spins are parallel, suggesting yet another complication in the way the nuclear force works: it has a substantial spin dependence. When nucleon spins are antiparallel, the nuclear force between the two particles is about a factor of 2 weaker.

30.4 Nuclear Stability

Figure 30.1 reveals that stable light nuclei up to about $A = 20$ are almost all composed of equal numbers of neutrons and protons ($Z = N$). The hydrogen nucleus ($A = 1$) is a single proton and naturally stable. Protons repel one another with the Coulomb force, and it takes some doing to get two of them close enough together to stick via the nuclear force. There is evidence that the di-proton has been created, but it lasts for less than 10^{-18} s. On the other hand, neutrons experience the nuclear force while being immune to electrostatic repulsion; they therefore serve as a source of nuclear glue, but that's a bit simplistic, since the di-neutron is unstable. The next heavier nucleus ($A = 2$) results when a neutron clings to a proton with parallel spin, forming a stable deuteron ($Z = 1$, $N = 1$). Two protons can be held together with the inclusion of a neutron, thereby making stable ^3_2He. Two deuterons can combine to create helium ($Z = 2$, $N = 2$).

Both the spin and the magnetic moment of the α-particle are zero. Since the magnetic moments of the neutron and proton are different, that tells us that the two neutrons (spin-up and spin-down) pair together, as do the two protons (spin-up and spin-down). The corresponding closed stable system (the α-particle) plays an important role in the scheme of the nuclides (p. 1067). That's borne out by the observation that, while there are only four stable odd-odd nuclei (^2_1H, ^6_3Li, $^{10}_5\text{B}$, and $^{14}_7\text{N}$, for which $Z = N$), there are 160 stable even-even nuclei. Furthermore, when a nuclide is blasted with a nucleon, it's much more likely that an α-particle will be emitted than a deuteron.

Most of the radioactive nuclides in Fig. 30.1 are of the ***artificially induced*** variety, made in the laboratory by bombarding other nuclides. There are only about 20 naturally occurring radioactive isotopes in the range up to lead ($Z = 82$). Beyond that, all nuclides are radioactive, though bismuth decays so slowly it might as well be considered stable.

The Shell Model

Nuclides with certain numbers of neutrons or protons are especially stable and abundant. These so-called **magic numbers** with N or Z equaling

$$2, 8, 20, 28, 50, 82, 126$$

suggested a closed-shell structure reminiscent of the configurations of the Noble Gases in atomic theory (p. 1039). A nucleus with either N or Z magic is tightly bound. One with both N and Z magic is very tightly bound, highly stable, and abundant. Such is the case with $^4_2\text{He}_2$, $^{16}_8\text{O}_8$, $^{40}_{20}\text{Ca}_{20}$, $^{48}_{20}\text{Ca}_{28}$, and $^{208}_{82}\text{Pb}_{126}$, the five known nuclei with double magic (see Discussion Question 1).

A nucleon within the nucleus is surrounded by other nucleons, and the nuclear forces they exert on it in opposing directions should tend to cancel; an interior nucleon will experience little local influence via the nuclear force. Don't picture the nucleus as a static cluster of tightly packed cannon balls! Instead, imagine it to be more like a cloud of flying bees, each of which "feels" and attracts only its nearest neighbors. Although there is a complex tumult of motion, the swarm hangs together as a single entity. In fact, if the bees fly around in pairs, the analogy would be still better. In any event, we can assume that a nucleon "feels" an average force that retains it within the nucleus. The Uncertainty Principle requires that particles confined to a small region of space have a correspondingly large uncertainty in their motion and so at least a minimum kinetic energy.

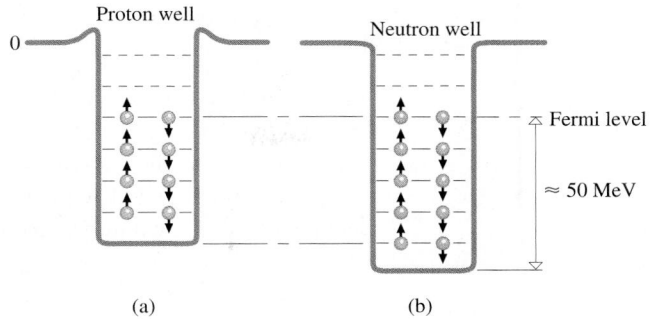

Figure 30.8 Potential-energy wells for (a) protons and (b) neutrons in the same nucleus. To bring the Fermi levels in line, protons can convert to neutrons by emitting positron-neutrino pairs. Note the unfilled higher energy levels available when the nucleus is excited.

Suppose that each nucleon moves around in a complex independent way within the nucleus. Each is restrained by the surface of the nucleus, as if in a potential-energy well. Figure 30.8 depicts a crude representation of the average potential-energy wells for a proton and neutron in a medium- to large-sized nucleus. The difference in shape and depth is due to the Coulomb force. The proton's well is not as deep because of the repulsion.

Each nucleon is accompanied by a de Broglie wave, a standing wave, fitting into the nucleus, just as the electron's wave fit into the atom. Each nucleon should correspond to a particular standing wave pattern or, equivalently, should occupy a particular quantum orbital. A group of energy levels that are close to one another is called a shell, and this scheme is the **shell model** of the nucleus. In the same way that an atomic level can contain at most two different electrons, each nuclear energy level can harbor two protons and two neutrons. No exclusion acts between neutrons and protons, and they can enter the same states. The levels fill up from the bottom. The highest occupied level corresponding to the greatest kinetic energy is called the **Fermi level**. Although we might expect a moving nucleon to undergo frequent collisions and be scattered every which way, the Exclusion Principle and the demand for filled levels will keep that from happening. There are very few available energy levels for a scattered nucleon to enter, and so they cannot easily be knocked off course.

Once in identical states, a proton and neutron experience a maximum attractive interaction, and so the simplest closed subshell system, that of the helium nucleus, is energetically favorable. When the energy levels were computed quantum mechanically, it was found that the nucleus had a system of closely spaced subshells. These come in groups constituting major shells that are themselves much farther apart energetically. M. Goeppert Mayer and J. H. D. Jensen further showed (1947) that the magic numbers corresponded to the number of states in the major shells. When the major shells are filled, the corresponding nuclei were especially stable. For example, a nucleus with 50 neutrons or 50 protons has a filled shell. Thus, tin ($Z = 50$), which is relatively abundant in nature, has 10 stable isotopes.

Nuclides off the **line of stability** that runs through the stable nuclei spontaneously decay, often changing neutrons to protons or vice versa and becoming more tightly bound and energetically stable. Observe how the line of stability in Fig. 30.1 increasingly bends toward the N-axis, indicating a larger percentage of neutrons. Because of the long range of the Coulomb force, the greater the number of protons, the more the nucleus tends to decay, and the more neutrons are needed to stabilize it. The progression of stable nuclei ends at lead; further increasing the percentage of neutrons will not keep things together. There are no pea-sized nuclei, and not until the mass becomes immense can gravity help to hold nuclear matter together on a large scale as in a neutron star.

Maria Goeppert Mayer (1906–1972). German-born American physicist shared the 1963 Nobel Prize in physics with J. H. D. Jensen.

Binding Energy

If we fire a neutron directly at a proton so that they come very close, the two can grab hold of each other via the strong force. Rammed together, the system will emit a burst of electromagnetic energy, a 2.224-MeV gamma-ray photon. Thus formed, the deuteron (2.013 553 u) has shed mass; its constituents have drawn tightly together and lost some of the plumpness they had apart. This transformed mass is known as the **mass defect** (Δm), and it indicates how tightly a nuclide is bound. In this case

$$m_{\text{p}} + m_{\text{n}} = (1.007\ 276\ \text{u}) + (1.008\ 665\ \text{u}) = 2.015\ 941\ \text{u}$$

whereas the deuteron mass (m_{d}) is only 2.013 553 u. The difference between the mass of the separate components and the combined nucleus is the mass defect; namely

$$\Delta m = 0.002\ 388\ \text{u}$$

Since 1 u is equal to 931.494 MeV/c^2, the mass defect corresponds to a **binding energy** ($E_{\text{B}} = \Delta mc^2$) of 2.224 MeV—exactly the energy ejected as a photon. In reverse, to split a deuteron into a neutron and a proton, 2.224 MeV (or 3.56×10^{-13} J) has to be supplied, for example, via a photon or a collision:

$$Q + {}_1^2\text{H} \rightarrow {}_1^1\text{H} + {}_0^1\text{n}$$

Of course, 2.224 MeV is an immense amount of energy on an atomic scale. It takes only 2×10^{-18} J, one hundred thousand times less energy, to pull the electron out of a deuterium atom, and even that is 10 times the energy released when the "burning" of a carbon atom forms CO_2.

Example 30.3 **[II]** When a neutron is removed from a ${}_{20}^{43}\text{Ca}$ atom (of mass 42.958 766 u), the latter will be transformed into a ${}_{20}^{42}\text{Ca}$ atom (of mass 41.958 618 u). What minimum energy must be provided to accomplish the removal?

Solution The difference in the masses before and after will have to be provided as energy if the transformation is to occur. (1) TRANSLATION—Determine the minimum amount of energy needed to remove a neutron from an atom with a known mass, thereby converting it into another atom of known mass. (2) GIVEN: Atomic masses before 42.958 766 u and after 41.958 618 u. FIND: The energy required to remove a neutron. (3) PROBLEM TYPE—Nuclear physics/binding energy. (4) PROCEDURE—First find the difference between the masses of the two nuclei. If that difference is less than the mass of a neutron, the deficiency (Δm) has to be made up by pumping energy in. (5) CALCULATION—We were given *atomic* masses because they

are what's generally tabulated rather than nuclear masses. That fact doesn't matter here since taking the difference cancels the mass of the atomic electrons:

$$(\text{mass of } {}^{43}\text{Ca}) - (\text{mass of } {}^{42}\text{Ca}) = 1.000\ 148\ \text{u}$$

Not surprisingly, this is less than the neutron mass (1.008 665 u); hence, the difference

$$\Delta m = (1.008\ 665\ \text{u}) - (1.000\ 148\ \text{u}) = 0.008\ 517\ \text{u}$$

(at the least) will have to be supplied as some form of energy. Consequently

$$E = (0.008\ 517\ \text{u})(931.494\ \text{MeV/u}) = \boxed{7.934\ \text{MeV}}$$

Quick Check: As we'll see presently, 8 MeV is typical for the binding-energy-per-nucleon (which is an average value); for heavy nuclei, our answer is reasonable. (See Problem 29.)

Planck first realized that **any bound system should have less mass than the sum of its constituents**, and in 1913, Langevin applied that insight to the nucleus. *The coming together of nucleons to form a nuclide is accompanied by a conversion of mass to energy.* The mass defect is equivalent to the total binding energy of the nuclide. Dividing that energy by the total number of nucleons yields an average measure of how strongly each nucleon is bound to the composite system, namely (E_{B}/A), the **binding-energy-per-nucleon**. (In a way, it's like our swarm of bees flying around in a well where we ask how much energy it will take on average to lift each of them up and out.)

A plot of binding-energy-per-nucleon for all the elements from hydrogen to uranium is given in Fig. 30.9. Hydrogen, with one proton, has no binding energy at all; from there, the curve rises to 1.1 MeV for the deuteron, 2.8 MeV for ^{3}H, 2.6 MeV for ^{3}He, then on up to a 7.074-MeV spike for tightly bound helium, falling back to 5.3 MeV for ^{6}Li, then rising to 5.6 MeV for ^{7}Li, and gradually continuing upward. The significance of the tightly bound α-particle structure is manifest by nuclides that can be thought of as comprising whole number multiples of ^{4}He. Thus, ^{8}Be (2 alphas), which decays almost immediately, nonetheless has a substantial binding energy of 7.06 MeV. Stable ^{12}C (3 alphas) and ^{16}O (4 alphas) have relatively high binding energies compared to their neighbors. Moreover, those nuclides, whose Z and A values correspond to whole number groupings of α-particles, also have zero spin and zero magnetic moment.

The binding-energy-per-nucleon curve reaches a maximum of 8.795 MeV at nickel-62, which is the most stable and most tightly bound of all nuclides. That fairly flat peak in the curve (around $Z = 60$) contributes to the abundance of iron in the Universe. From there on, the graph gradually drops, because of the Coulomb repulsion of the protons, all the way to uranium. Except for the lightest nuclei, E_B/A is fairly constant at about 8 MeV per nucleon. That this value is independent of A again suggests that the nucleus is held together by a short-range force. With a long-ranged force such as gravity, the size of the composite body is crucial; it's much more difficult to remove a 10-kg stone from the surface of the Earth than from the surface of the Moon. To the contrary, where the force is short-ranged, as with the intermolecular force holding water together, there is the same independence of quantity; it takes the same amount of energy to evaporate 10 kg of water from a kiddy pool as from an ocean.

The binding-energy-per-nucleon is the energy above and beyond its kinetic energy (the Fermi level) that must be added to a nucleon to remove it from the nucleus (Fig. 30.8). For small nuclei, the well depth is small because there are fewer nucleons acting, and therefore the binding energy is also small. For large nuclei, the electrostatic potential energy raises the bottom of the well and decreases the binding energy at the top.

Notice that if we select a nuclide way off on either side of the peak in the binding-energy-per-nucleon curve and alter its structure so as to move up the curve toward Ni, a very large amount of energy could be liberated. Thus, if two light nuclei (say, of hydrogen) could

Figure 30.9 Average binding-energy-per-nucleon for the nuclides that occur naturally (including short-lived ^{8}Be, the second spike near He) versus mass number.

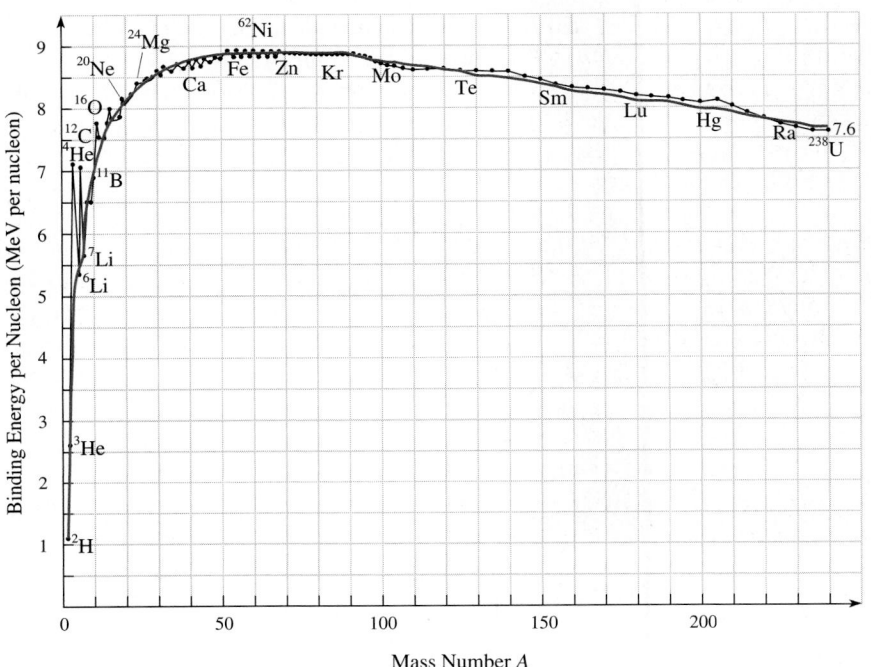

A self-luminous 1000 °C radioactive sphere. At its core are six ounces of plutonium (^{238}Pu) dioxide, clad in an iridium shell, surrounded by a graphite casing. The α-particles emitted in the decay are absorbed in the surrounding shell, where they impart thermal energy at a rate of about 100 W and will continue to do so for decades.

be joined together (that is, fused), the resulting nuclide would reside further up the curve and would have a greater binding-energy-per-nucleon. Each of its nucleons would be more tightly squeezed and individually less massive than it was prior to the union. The resulting mass defect would appear as liberated energy in the well-known process of ***nuclear fusion***. On the other hand, splitting a large nucleus (from the right side of the curve) into small fragments also transforms mass. The binding-energy-per-nucleon of the fragments is higher than it was for the original nucleus that broke up. This process of ***nuclear fission*** liberates copious amounts of energy. It's no accident that the key elements of the nuclear age, hydrogen and uranium, are the end points of the binding energy curve.

Nuclear Transformation

The transformation of one nuclide into another can take place either spontaneously or as a result of an external stimulus. The remainder of this chapter deals with a variety of nuclear transformations. We examine the spontaneous processes of alpha, beta, and gamma decay; introduce the weak force; study the mathematical description of the rate of radioactive decay; and explore the phenomenon of induced radioactivity. The discussion ends with an introduction to fission and fusion as applied to bombs and stars.

30.5 Radioactive Decay

As we saw earlier, nuclei spontaneously transform themselves into more energetically favorable configurations via alpha, beta, and gamma emission in a process called **radioactive decay**. Usually, when a nuclide resides above the line of stability in Fig. 30.1, it decays by emitting an electron, thereby transforming a neutron into a proton and moving downward and to the right, coming closer to the line. Similarly, when a nuclide resides beneath the line of stability, it emits a positron or, occasionally, an alpha, transforming to the left toward the line.

A given radioactive decay can be a single step in a long sequence of transformations from one unstable nuclide to another, ultimately ending in a stable form. There are three *naturally occurring* radioactive series beginning with ^{238}U, ^{235}U, and ^{232}Th (Table 30.4), all ending in different isotopes of lead. The series shown in Fig. 30.10 begins with the slowly decaying nuclide uranium-238. A little less than half of all the U-238 present at the formation of the Solar System some 5×10^9 years ago still remains. By contrast, radium-226 decays away fairly rapidly (half of it is transformed every 1600 years), and all of the Earth's original radium atoms have long ago disappeared. Yet, radium is still plentiful because, like so many other radioactive substances, it is continuously replenished (Fig. 30.10).

Alpha Decay

Alpha emission is rare in the light nuclides, although at around $Z = 60$ there is a small cluster of α-emitters. It is beyond $Z = 82$, where there are no stable nuclides, that alpha emission predominates. There, it drives the nuclides down in both Z and N, toward the line of stability. The emission of an α-particle ($^4_2$He, or $^4_2\alpha$) decreases N by 2, decreases Z by 2, and

The little Sojourner vehicle that drove around on the surface of Mars (1997) was warmed by one-tenth of an ounce of plutonium-238, which produced about 1W of thermal energy.

Table 30.4

The Three Natural Radioactive Series			
Series	Starting Nuclide	Half-Life (years)	Stable End Nuclide
^{235}U-Actinium	$^{235}_{92}$U	7.04×10^8	$^{207}_{82}$Pb
^{232}Thorium	$^{232}_{90}$Th	1.41×10^{10}	$^{208}_{82}$Pb
^{238}U-Radium	$^{238}_{92}$U	4.47×10^9	$^{206}_{82}$Pb

Figure 30.10 The naturally occurring decay series beginning with uranium-238.

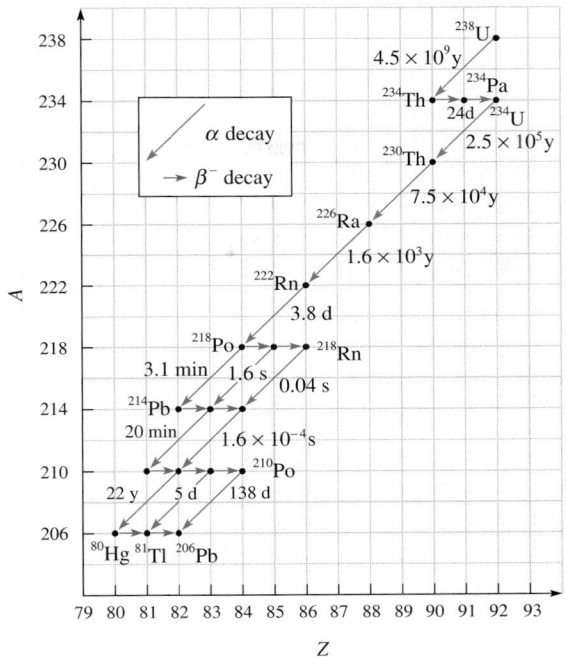

decreases A by 4. In general, the decay can be written as

[α-decay]
$$_{Z}^{A}X \rightarrow {}_{Z-2}^{A-4}Y + {}_{2}^{4}He + Q \qquad (30.2)$$

where X is called the **parent nucleus**, Y is the **daughter nucleus**, and Q is the **disintegration energy**. The rules of the game require that the process *conserve charge*, *nucleon number* (A), *angular and linear momentum*, and *mass-energy*. Alpha decay is a rearranging: it doesn't change the total number of nucleons present in the formula, nor does it alter the net charge, but it does alter the net mass. Whenever such a rearrangement happens by itself, energy is liberated in the amount Q such that

$$Q = (m_{X} - m_{Y} - m_{\alpha})c^{2} \qquad (30.3)$$

where m_{X}, m_{Y}, and m_{α} are the masses of the parent, daughter, and alpha, respectively. The disintegration energy (here given in joules) appears as the total kinetic energy of the daughter nuclide and the α-particle. A positive Q means that the process takes place spontaneously; a negative Q means it cannot. Using the binding energy data, it can be shown that for heavy nuclei, the average value of Q is around 5 MeV. Typical of the process is the *spontaneous decay* (Fig. 30.11) of uranium-238 into thorium-234:

$$_{92}^{238}U \rightarrow {}_{90}^{234}Th + {}_{2}^{4}\alpha + 4.3 \text{ MeV}$$

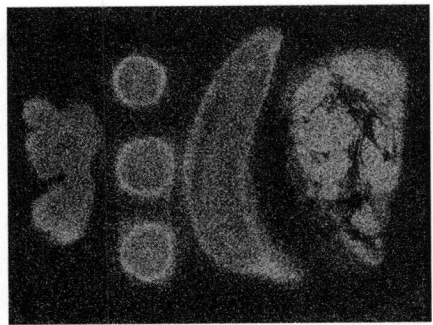

This image reveals the natural radioactivity of several foods. On the right is a piece of pork; the veins of fat are fairly free of radioisotopes and show up as dark regions. The banana, sliced longitudinally and across (in three pieces), is clearly more radioactive in its skin. On the left is a chunk of ginger. The most active agent here is the beta-decay of potassium-40.

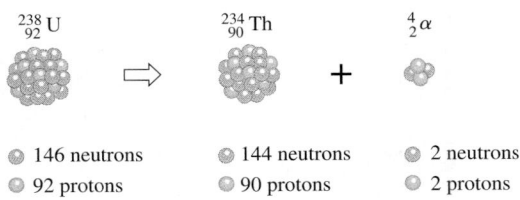

Figure 30.11 An unstable parent nuclide (U-238) decaying, via α emission, into a daughter nuclide (Th-234).

Example 30.4 **[II]** A too-common hazard in many homes is the radioactive gas radon-222. It's produced in the ground by the alpha decay of radium-226 and often leaks into basements. Write out the transformation equation and determine (to three figures) the total kinetic energy of the decay products in MeV. The masses of the radium, radon, and helium *atoms* are 226.025 406 u, 222.017 574 u, and 4.002 603 u, respectively. As a rule, most of the KE is carried off by the light particle, here, the alpha. Very little ($\approx$0.1 MeV) goes to the massive recoiling daughter.

Solution Radium decays into radon plus an energetic alpha particle. (1) TRANSLATION—Determine the kinetic energy of the resulting particles when radium alpha decays into radon. (2) GIVEN: $m_X = 226.025\ 406$ u, $m_Y = 222.017\ 574$ u, and $m_\alpha = 4.002\ 603$ u. FIND: KE. (3) PROBLEM TYPE—Nuclear physics /radioactive decay/alpha emission. (4) PROCEDURE—For α-alpha decay use $^A_Z X \rightarrow ^{A-4}_{Z-2}Y + ^4_2He + Q$. (5) CALCULATION—Here the parent is $^A_Z X = ^{226}_{88}Ra$, the daughter is $^{A-4}_{Z-2}Y = ^{222}_{86}Rn$, and

$$^A_Z X \rightarrow ^{A-4}_{Z-2}Y + ^4_2He + Q \qquad [30.2]$$

becomes

$$^{226}_{88}Ra \rightarrow ^{222}_{86}Rn + ^4_2He + Q$$

The KE equals Q, where

$$Q = (m_X - m_Y - m_\alpha)c^2 \qquad [30.3]$$

The atomic masses can be used because the electrons cancel. In MeV

$$Q = (226.025\ 406\ u - 222.017\ 574\ u - 4.002\ 603\ u)$$
$$\times (931.494\ MeV/u)$$

$$Q = (0.005\ 229\ u)(931.494\ MeV/u) = 4.870\ 8\ MeV$$

The net KE of the decay products is $\boxed{4.87\ MeV}$.

Quick Check: Our answer is roughly 5 MeV, as we know it should be.

Beta Decay and the Neutrino

Three distinct processes are each a form of *beta decay*. In **β^- decay**, an electron ($^0_{-1}e$) is emitted from a nucleus as a neutron transforms into a proton. In **β^+ decay**, a positron ($^0_{+1}e$) is emitted from a nucleus as a proton transforms into a neutron. In **electron capture**, one of the orbital electrons in an inner shell of the cloud is drawn into the nucleus, transforming a proton into a neutron.

By the end of the 1920s, experiments on electron beta decay had revealed some extremely puzzling aspects. A parent nucleus was assumed to decay into a daughter after the creation and immediate emission of an electron. One problem was an apparent violation of Conservation of Linear Momentum. When a parent nucleus (the best results are obtained with a monatomic gas) more or less at rest decays, the electron and recoiling daughter nucleus should move in opposite directions. That's the only way the net linear momentum before and after can be zero, and surprisingly, that didn't always happen. The basic process occurring was assumed to be the spontaneous transformation of a neutron into a proton and an electron, but each of these particles is a spin-$\frac{1}{2}$ fermion. The original neutron spin is $\frac{1}{2}$, whereas the resulting total proton-electron spin could be either 0 or 1. Clearly, this situation violated Conservation of Angular Momentum.

An even more confounding observation was made by Chadwick in 1914. Figure 30.12 shows the broad energy spectrum measured for emitted electrons from a typical beta emitter. Each parent atom is identical, as is each daughter, and so every escaping electron should be identical and have the same energy, but they do not. If the basic process corresponds to

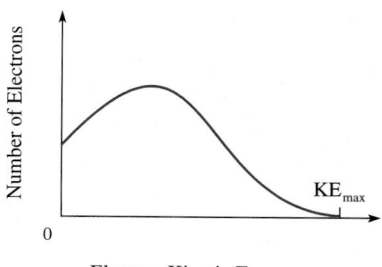

Figure 30.12 Energy spectrum of the electrons in a typical beta decay.

$n \rightarrow p + e^-$, we can expect a disintegration energy of

$$Q = (m_n - m_p - m_e)c^2$$

which equals 0.783 MeV (see Problem 72). Except for a very small amount of energy appearing as the recoil KE of the proton, Q ought to be exactly the KE of every emitted electron. In the case of a heavy nucleus, even less energy is given to the recoil. In other words, all electrons should appear with the maximum kinetic energy (KE_{max}) terminating the curve in Fig. 30.12 for each emitter. That energy (Q) is equivalent to the parent-daughter-electron mass difference. Experimentally, we find that all electrons are ejected with less than this energy. To explain that evident violation of Conservation of Energy, it was suggested that the β-particles lost energy via collisions while emerging from the atom. That suggestion was rejected by C. D. Ellis (who, incidentally, became a physicist only after being interned in a German prison camp with Chadwick during World War I) and W. A. Wooster. Using a thick lead-walled calorimeter, they proved that the total amount of energy deposited corresponded to the average energy of the spectrum, which is about 40% of what it ought to be (either energy was not conserved or 60% of it mysteriously escaped from the calorimeter).

Some physicists felt compelled to reject Conservation of Energy—Madame Curie and Niels Bohr were among those skeptics. Rather than accept the obvious facts that were there for all to see, Wolfgang Pauli remained steadfast. In 1930, he postulated that an invisible particle emitted along with each electron carried off precisely the amount of energy needed to sustain conservation. Sometime later, Enrico Fermi, while answering a question, almost jokingly referred to this phantom speck as a **neutrino** (which in Italian means "little neutral one"), and the name stuck. Today (for reasons that will become clear) we call it an **electron-antineutrino** ($\overline{\nu}_e$). Not surprisingly, Pauli proposed that it was uncharged and so immune from the electromagnetic force. Because this neutral particle could pass through the thick lead walls of the Ellis-Wooster calorimeter, it had to be unaffected by the strong force. To conserve angular momentum, Pauli argued that his phantom particle had a spin of $\frac{1}{2}$. Today, it's known that neutrinos and antineutrinos do have spin-$\frac{1}{2}$. When an electron and an antineutrino are created as a pair, their spins are oppositely directed.

The kinetic energies of some emitted electrons approach KE_{max}, meaning that, in the extreme, the accompanying antineutrino can have a total energy that is vanishingly small. This total energy equals its rest energy (mc^2) plus its kinetic energy. The only way that sum can be vanishingly small is if m is zero or nearly zero, while KE ≈ 0 as well. By 1933, Perrin had argued from the β-spectrum "that the neutrino has *zero intrinsic mass*." Like a photon, it would then travel at c; and like a photon with zero rest energy, it could carry ample kinetic energy to balance out the demands of conservation. Intrigued by these ideas, de Broglie (1934) suggested that, as with the electron, the neutrino should have a mirror image, the antineutrino.

Electron-beta decay arises when a neutron decays into a proton, an electron, and an electron-antineutrino,

$$n \rightarrow p + {}_{-1}^{0}e + \overline{\nu}_e \qquad (30.4)$$

In the case of a radioactive nuclide, the decay has the form

[electron β-decay] $${}_{Z}^{A}X \rightarrow {}_{Z+1}^{A}Y + {}_{-1}^{0}e + \overline{\nu}_e + Q \qquad (30.5)$$

where Q is the kinetic energy of all three particles emerging from the transformation. Similarly, positron-beta decay arises when a proton decays into a neutron, a positron, and

A stream of high-energy neutrinos (1 to 2 meters in diameter) traveling upward gave rise to much of what is pictured in this photo. The neutrino beam consists of bursts, each of which contains millions of neutrinos. Very few, if any, of these interact with the liquid. Some interact with the walls of the chamber or even farther upstream, producing highly penetrating muons and pions that cross the region of the photo.

an electron-neutrino:

$$p \rightarrow n + {}_{+1}^{0}e + \nu_e \qquad (30.6)$$

In the case of a radioactive nuclide, the decay has the form

[positron β-decay] $\qquad {}_{Z}^{A}X \rightarrow {}_{Z-1}^{A}Y + {}_{+1}^{0}e + \nu_e + Q \qquad (30.7)$

Electron-capture, the process wherein a parent nucleus absorbs one of its own orbital electrons, competes with β^+ decay; that is,

[electron capture] $\qquad {}_{Z}^{A}X + {}_{-1}^{0}e \rightarrow {}_{Z-1}^{A}Y + \nu_e + Q \qquad (30.8)$

For example, beryllium-7 captures an electron and transforms into lithium-7:

$${}_{4}^{7}Be + {}_{-1}^{0}e \rightarrow {}_{3}^{7}Li + \nu_e$$

The distinction between the electron-neutrino (ν_e) and the electron-antineutrino ($\overline{\nu}_e$) is that the former is left-handed (its spin angular momentum vector is antiparallel to its linear momentum vector) and the latter is right-handed (its angular and linear momentum vectors are parallel). ***The emission of a neutrino is equivalent in many regards to the absorption of an antineutrino, and vice versa.***

Because it doesn't "feel" either the strong or the electromagnetic force, a neutrino can sail through a thickness of over a light-year of solid lead before undergoing a single collision. In the 1930s, there was little hope of ever confirming their existence. Twenty-five years later, the feat was accomplished. Using a nuclear reactor (which is a copious source of antineutrinos, most arising from the decay of free neutrons), C. L. Cowan and F. Reines bombarded the protons in a huge quantity of water. If antineutrinos existed, the inverse process to that of Eq. (30.6)—add an antineutrino to both sides—should produce observable effects. Roughly 1 in 10^{12} antineutrinos struck a proton head-on, converting it into a neutron and a positron, both of which were then detected. The electron-antineutrino exists. More recently (1986), it has been determined that the neutrino mass is less than a mere 27 eV/c^2 (which is about 19 000 times less than the electron). There are actually three different types (or *flavors* of neutrinos, p. 1096). Results announced in April of 2002 strongly suggest that neutrinos can morph from one type to another. And to do that little trick they must have some mass, however small. The issue of a nonzero neutrino mass is crucial in astrophysics. Even if the neutrino corresponded to just a few eV, they are so numerous (billions pass through us every second) that much of the mass of the Universe would be invisible neutrinos. That hidden mass might well determine the future fate of the entire Universe.

SUPERNOVA 1987A

At 7:35 A.M. on February 23, 1987, counters on two huge particle detectors deep below the surface of the Earth, one near Cleveland and the other near Tokyo, began flashing for several seconds (p. 1114). The cause of that unprecedented event was revealed several hours later when the light from an exploding star in a nearby galaxy (170 000 ly away) arrived at Earth. The cataclysmic eruption of Supernova 1987A emitted a blast of neutrinos that immediately flew out into space at or very near the speed of light. Several hours later, the shock wave reached the surface of the star, and it began emitting a blaze of light a million times brighter than the Sun. The pulse of neutrinos and flood of photons had flown toward Earth for 170 000 years, arriving about two hours apart.

30.6 The Weak Force

In 1934, Fermi introduced a highly successful theory of beta decay predicated on the existence of a new interaction in nature, the **weak force**. The very fact that neutrinos were so penetrating suggested that they were uninfluenced by both the strong and the electromagnetic forces, and yet something (other than gravity) both blasted them into existence and promoted their occasional interaction with matter.

In its broadest meaning, a force is an agent of change; it may manifest itself as a push or a pull (a changer of motion), or it may be a changer of some aspect of the state of a system, a transformer. The electromagnetic force can transform mass-energy in an atomic system, thereby creating a photon (or it can transform a stick of dynamite into a puff of smoke). In the case of beta decay, it was necessary to conceive of a new force that could transmute a neutron into a proton, or vice versa.

Figure 30.13 The gravitational, electromagnetic, strong, and weak forces are the four fundamental interactions that effect all change.

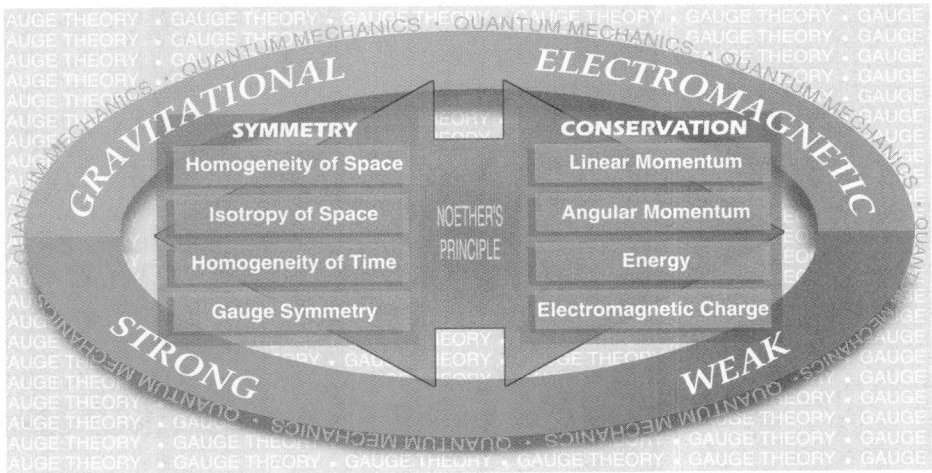

A free neutron has a half-life of 10.4 minutes (a very long time in Atomland), and that implies that the operative force is very weak—it's argued that *the stronger the force, the faster the processes it drives*. As we'll see in Chapter 31, there are hundreds of unstable exotic subnuclear particles. Some of these decay via the strong force (producing strongly interacting particles) and have lifetimes of only $\approx 10^{-23}$ s or so. By contrast, a decay resulting in a photon and so driven by the electromagnetic force might take $\approx 10^{-20}$ s. Weak decays involving neutrinos happen much more slowly, with lifetimes of perhaps $\approx 10^{-8}$ s.

The weak force is a million times more feeble than the strong force (Table 30.3), though it is immensely ($\approx 10^{32}$ times) more powerful than gravity. It reveals its presence in the macroscopic world primarily through radioactive transformation (there being hundreds of β-decaying nuclides) and also via the results of fission and fusion occasionally rather spectacularly in a supernova. Most subnuclear particles (the vast majority of them are unstable) interact via the weak force. The spontaneous transmutation of many subnuclear particles occurs through the weak force. It is extremely short-ranged, so much so that it was only in the 1980s that accurate determinations of its extent could be attempted. Beyond about 10^{-17} m (i.e., 10^{-2} fm), the weak force becomes exceedingly small. It is for the most part an inside-the-particles force rather than a between-the-particles force. Still, as we become more sophisticated experimentally, we are able to detect the minute influence of the weak force reaching from the nucleus to the electron cloud of an atom (Fig. 30.13).

30.7 Gamma Decay

After an alpha or beta decay, a daughter nucleus may momentarily end up in an excited state, that is, with a nucleon in a higher energy level than the ground state (Fig. 30.8). The nucleus then quickly relaxes, inevitably reaching the lowest available energy configuration. The energy difference (≈ 1 keV to ≈ 1 MeV) is emitted as one or more gamma photons (hf). As with an atom, it is possible to excite a nucleus by pumping in energy. This might be accomplished via the absorption of a gamma photon of the proper frequency, or it might happen as the result of a collision with a massive particle. For example, a slow-moving neutron can be absorbed by a ^{238}U nucleus, whereupon it becomes an excited ^{239}U* nucleus (the asterisk indicates an excited state). Returning to its ground state, it emits a gamma photon; that is,

$$_{0}^{1}\text{n} + {}^{238}\text{U} \rightarrow {}^{239}\text{U*} \rightarrow {}^{239}\text{U} + \gamma$$

When observed independent of its nuclear source, a gamma photon is identical in all respects to a photon of the same energy arising from any other means. It's only because most nuclear energy levels are widely separated that gamma rays are usually more energetic than photons emitted from atomic electron transitions.

Full-body, front and back, bone scans. Gamma-ray images of a patient three hours after being injected with 25 milli-curies of technetium-99, a γ-ray emitter with a half-life of 6 h.

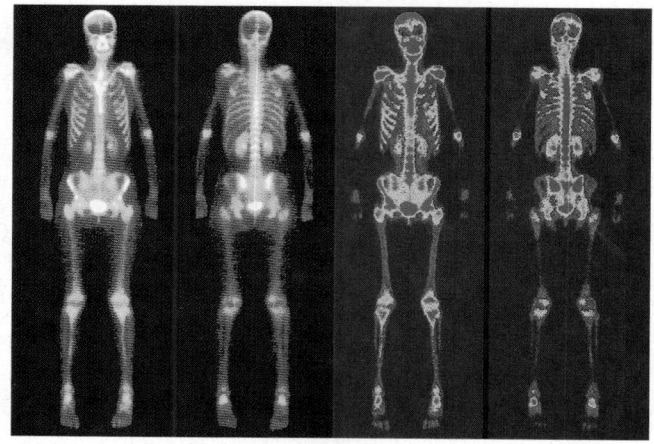

Example 30.5 **[II]** Nitrogen-12 beta decays into carbon-12. The latter, $^{12}C^*$, then emits a 4.43-MeV gamma photon during de-excitation. Compute the mass of the $^{12}C^*$ atom.

Solution This is a Conservation of Mass-Energy problem. (1) TRANSLATION—Beta decay produces an atom in an excited state that then emits a known photon; determine the net mass of the excited atom. (2) GIVEN: Gamma energy of 4.43 MeV and ^{12}C system. FIND: $^{12}C^*$ atomic mass. (3) PROBLEM TYPE—Nuclear physics/β-decay/mass-energy. (4) PROCEDURE—Find the equivalent mass of the photon and add that to the mass of ^{12}C. (5) CALCULATION—The mass of the ^{12}C atom is 12.000 000 u. The equivalent mass of the photon is gotten from $E = mc^2$ or the fact that 1.000 00 u = 931.494 MeV/c^2:

$$m_\gamma = \frac{4.43 \text{ MeV}}{931.494 \text{ MeV}/c^2} = 0.004\ 76 \text{ u}$$

The total mass of the excited carbon-12 atom is then

$$12.000\ 00 \text{ u} + 0.004\ 76 \text{ u} = \boxed{12.004\ 76 \text{ u}}$$

Quick Check: Let's redo the calculation using kilograms. (4.43 MeV) × (1.782 663 × 10^{-30} kg/MeV) = 7.897 197 × 10^{-30} kg; (12.000 00 u)(1.660 540 × 10^{-27} kg/u) = 1.992 648 × 10^{-26} kg; hence, the total mass is 1.993 438 × 10^{-26} kg, or 12.004 76 u.

30.8 Half-Life

While working with highly radioactive ^{220}Rn, Rutherford observed that the intensity of the emissions decreased with time in a precise and predictable way. The amount of radiation emerging from a sample of a radioactive element is virtually independent of the surrounding environment (that is, the chemical compound it's in, the temperature, pressure, etc.). The quantitative measure of radioactive intensity is the *number of disintegrations per second*, known also as the **decay rate**, or the **activity**. The SI unit of decay rate is the **becquerel** (Bq), which is *one disintegration per second*:

$$1 \text{ becquerel} = 1 \text{ Bq} = 1 \text{ decay/s}$$

Another, more commonly used unit of activity is the curie (Ci), named to honor Marie and Pierre Curie (p. 977). A single curie roughly corresponds to the activity of a gram of radium,

$$1 \text{ curie} = 1 \text{ Ci} = 3.7 \times 10^{10} \text{ decays}$$

That's a good deal of activity so you're more likely to encounter millicurie (1 mCi = 10^{-3} Ci) or microcurie (1 μCi = 10^{-6} Ci) samples.

THE QUANTUM-MECHANICAL MICROWORLD

Radioactivity is a quantum-mechanical phenomenon, and Einstein (1916) first realized that Eq. (30.13) was a manifestation of the inherent statistical nature of the process. It is now the common wisdom that an unstable nucleus decays spontaneously (just as an atom in an excited state emits a photon spontaneously); that there is no knowable external trigger that fires them off in a way that can be anticipated; and that identical atoms behave, all by themselves, in nonidentical ways that conform *en masse* to the rules of probability. It would seem that radioactivity provides a glimpse of the quantum-mechanical microworld playing havoc with classical cause and effect. An atom that is ten thousand years old is supposedly identical to an atom of the same species that is ten seconds old; from this moment on, one of these may live for ten thousand years and the other for ten seconds, and yet we don't know which will do what. Whether that ignorance is a profound one, forced on us by the probabilistic nature of the Universe, or just a practical one inflicted by our own limitations remains to be seen.

Marie Sklodovska Curie (1867-1934) and Pierre Curie (1859-1906).

Activity seems to be dependent only on the type and amount of radioactive material present at that moment, which is both a common and a very special behavior. Many processes unfold at a rate that depends on the amount of participating ingredient present at that instant (for example, the decay of the charge on a capacitor, Fig. 18.18).

Letting N stand for the number of radioactive atoms present at a given instant, $|\Delta N/\Delta t|$ represents the decay rate (R). This rate can be made independent of the size of the particular sample by dividing it by N. Thus, $|\Delta N/\Delta t|/N$ is the fraction of the atoms of a given species disintegrating per unit time, regardless of the sample size. This quantity has been found experimentally to be constant over a wide range of sources and for long periods of time. It's called the **decay constant** and is usually symbolized, rather unfortunately, by λ, which is *not* to be confused with wavelength (sorry about that):

[decay constant]
$$\lambda = \frac{|\Delta N/\Delta t|}{N} = \text{constant} \qquad (30.9)$$

The unit of λ is s^{-1}, and it should be thought of as the fractional disintegration rate.

Suppose a detector is set up next to the sample and the counting rate per convenient interval of time is recorded. Start at $t = 0$ with a rate R_0. From Eq. (30.9), $R_0 = \lambda N_0$ where N_0 is the number of atoms present initially. Figure 30.14 is a typical plot of R against time. The measured decay rate drops to $\frac{1}{2}R_0$ after a time interval Rutherford called the **half-life** ($t_{1/2}$). The known radioactive nuclides have half-lives ranging from roughly 10^{-22} s to about 10^{21} y. Upon waiting two half-lives, the decay rate will be one-quarter its original value; after three half-lives, it will be half of that or one-eighth its original value, and so on, with R approaching, but never reaching, zero. After n half-lives, the **decay rate** is

[decay rate or activity]
$$R = \left(\tfrac{1}{2}\right)^n R_0 \qquad (30.10a)$$

Since the decay rate (R) is proportional to the number of radioactive atoms (N) present, it follows that if m_0 is the initial radioactive mass, after n half-lives

[radioactive mass remaining]
$$m = \left(\tfrac{1}{2}\right)^n m_0 \qquad (30.10b)$$

Figure 30.14   The exponential decay of the activity R of a radioactive nuclide.

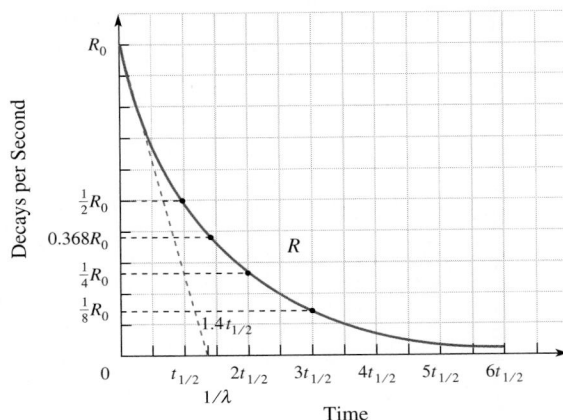

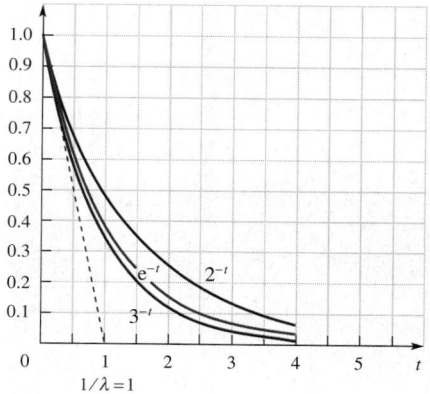

Figure 30.15 A comparison of 2^{-t}, e^{-t}, and 3^{-t}.

Each radioactive element has its own decay curve, some declining more steeply than others (just like the two curves in Fig. 20.22), but all will have the same basic shape. A straight line having the initial slope of each curve will intersect the time axis at the *time constant*, or **mean lifetime** (τ), at which moment $R = 0.368R_0$. The greater the activity ($R = |\Delta N/\Delta t|$), the greater λ is, the steeper the slope, and the smaller is τ. In fact,

[mean lifetime] $$\tau = 1/\lambda$$

The red curve in Fig. 30.15 is a replot of Fig. 30.14, with R_0 set equal to 1 and τ set equal to 1, the fundamental decay curve. Let's try to fit this nice smooth graph with a simple mathematical function: 2^{-t} or $1/2^t$ is too slow descending, and 3^{-t} is a little too fast. After some routine calculation, it can be determined that $(2.718\,281\,8\ldots)^{-t}$ fits the decay exceedingly well. It's traditional to call $2.718\,281\,8\ldots$ simply e (after the mathematician Euler), whereupon the basic decay curve is e^{-t}.

All decay curves drop to $1/e = 0.368$ of their initial values at $t = \tau = 1/\lambda$, and so the basic curve can be particularized to a specific decay rate by writing it as $1e^{-\lambda t}$. Moreover, letting the initial activity be some value R_0, the **decay rate** or **activity** for any radioactive element becomes

[decay rate or activity] $$R = R_0 e^{-\lambda t} \qquad (30.11)$$

At $t = 0$, $e^0 = 1$ and $R = R_0$, as it should be. Now, let's consider the half-life. When $t = t_{1/2}$, $R = R_0 e^{-\lambda t_{1/2}} = \frac{1}{2}R_0$, thus $e^{-\lambda t_{1/2}} = \frac{1}{2}$. Yet, as an electronic calculator can show, $e^{-0.693} = \frac{1}{2}$, and

$$\lambda t_{1/2} = 0.693 \qquad (30.12)$$

or $t_{1/2} = 0.693\tau$. The average value of the exponential is reached at $t = \tau$, which is why it's called the *mean lifetime*. If we know the half-life or the lifetime, we can calculate the decay constant or vice versa. A large decay constant means a short half-life.

It follows from Eqs. (30.9) and (30.11) that

[decay rate or activity] $$R = \lambda N_0 e^{-\lambda t} = \lambda N \qquad (30.13)$$

therefore, the **number of atoms remaining** after a time t is

$$N = N_0 e^{-\lambda t} \qquad (30.14)$$

Notice that both sides of this equation could be multiplied by the mass of the nuclide, thereby yielding an expression for the total mass of the remaining undecayed element.

Example 30.6 [III] Radium-226 is found to have a decay constant of 1.36×10^{-11} Bq. Determine its half-life in years. Given that the Curies had roughly 200 g of radium in 1898, how much of it will remain 100 years later?

Solution This is a Conservation of Mass-Energy problem. (1) TRANSLATION—Knowing the decay constant of an element, determine its half-life and how much of it remains after a specified time. (2) GIVEN: $\lambda = 1.36 \times 10^{-11}$ decays/s, $m_0 = 200$ g, and $t = 100$ y. FIND: $t_{1/2}$ and m. (3) PROBLEM TYPE—Nuclear physics/radioactive decay/half-life. (4) PROCEDURE—The decay constant and the half-life are related via $\lambda t_{1/2} = 0.693$. First compute $t_{1/2} = 0.693/\lambda$ and m using $N = N_0 e^{-\lambda t}$. (5) CALCULATION— Solving for $t_{1/2}$, we get

$$t_{1/2} = \frac{0.693}{1.36 \times 10^{-11} \text{ s}^{-1}} = 5.096 \times 10^{10} \text{ s}$$

$$\boxed{t_{1/2} = 1.62 \times 10^3 \text{ y}}$$

Inasmuch as $N = N_0 e^{-\lambda t}$, it follows that $m = m_0 e^{-\lambda t}$. Since $t = 100 \text{ y} = 3.15 \times 10^9$ s and $\lambda = 0.693/t_{1/2} = 1.36 \times 10^{-11} \text{ s}^{-1}$,

$$m = m_0 e^{-(1.36 \times 10^{-11} \text{ s}^{-1})(3.15 \times 10^9 \text{ s})}$$

$$m = (200 \text{ g})(0.958) = \boxed{192 \text{ g}}$$

Quick Check: The factor $(-\lambda t)$ to which the exponential is raised must be a pure number; its units must cancel as they do here. With a half-life of 1620 y, $\lambda = 0.693/(1620 \text{ y}) = 4.28 \times 10^{-4} \text{ y}^{-1}$ and $e^{-\lambda t} = e^{-0.042\,8} = 0.958$.

Potassium-40

The human body contains a fair amount of radioactive potassium, ^{40}K, which has a natural abundance of 0.011 9%. The so-called biochemical standard man contains 245 g of potassium (i.e., all naturally occurring isotopes) per 70 kg of body mass. Potassium-40, which has a half-life of 1.3×10^9 years, is a beta and gamma emitter; in fact, it's the most abundant gamma emitter found in the body. Even so, the beta radiation is more important, especially for ^{40}K deposited in the skeleton (see the photo on p. 1069). The doses from ^{40}K to the gonads, bones, and bone marrow are, respectively, 19, 11, and 11 millirem per year. To get a sense of what's happening inside you at this very moment, let's compute the number of radioactive disintegrations of ^{40}K that occur in the so-called standard (154-lb) body. The activity R follows from the fact that $R = \lambda N$ and $\lambda t_{1/2} = 0.693$. First, find the half-life in seconds: $t_{1/2} = (1.3 \times 10^9 \text{ y})(3.156 \times 10^7 \text{ s/y}) = 4.103 \times 10^{16}$ s. Then determine $\lambda = 0.693/t_{1/2} = 0.693/(4.103 \times 10^{16} \text{ s}) = 1.689 \times 10^{-17} \text{ s}^{-1}$. Now find N: the standard body contains 245 g of K, of which 0.011 9%, or 2.92×10^{-2} g, is ^{40}K. (The actual amount of potassium in the body varies from one individual to another, so this number can only be approximate.) Potassium has an average atomic mass of 39.100 u and hence the number of moles of ^{40}K in the body is $(2.92 \times 10^{-5} \text{ kg})/(39.100 \text{ kg/kmol}) = 7.46 \times 10^{-4}$ mol. Each mole contains $N_A = 6.022 \times 10^{23}$ atoms and so $N = (7.46 \times 10^{-4} \text{ mol})(6.022 \times 10^{23} \text{ atoms/mol}) = 4.49 \times 10^{20}$ atoms of ^{40}K. Therefore, $R = \lambda N = (1.689 \times 10^{-17} \text{ s}^{-1})(4.49 \times 10^{20} \text{ atoms}) = 7.6 \times 10^3$ disintegrations per second or 7.6 kBq.

Carbon-14

Carbon-14 is formed naturally (at a rate of about 7 to 10 kg per year) in the atmosphere as a result of cosmic-ray neutron bombardment of nitrogen:

$$^{14}\text{N} + \text{n} \rightarrow {}^{14}\text{C} + \text{p}$$

The half-life of radioactive carbon-14 is 5730 years (see Table 30.5), and it serves as an archeological clock. The ratio of the amount of ^{14}C to stable ^{12}C in the atmosphere is fairly constant at about 1.3×10^{-12}. Any living organism takes in both isotopes in that fixed proportion as it assimilates carbon. The resulting activity is about 0.25 Bq per gram of carbon. Once dead, the organism cannot take up any

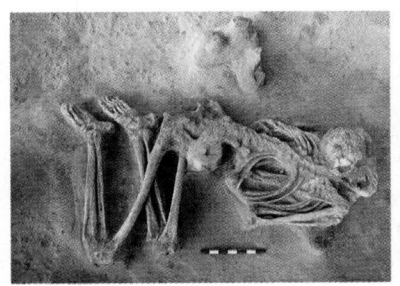

Archeological remains, like this skeleton, can be dated using carbon-14.

Table 30.5

Half-Lives of Some Radioisotopes		
Isotope	**Decay Mode**	**Half-Life**
Rubidium-87	e^-	4.7×10^{10} y
Uranium-238	α	4.5×10^9 y
Plutonium-239	α	2.4×10^4 y
Carbon-14	e^-	5730 y
Radium-226	α	1620 y
Strontium-90	e^-	28 y
Tritium-3	e^-	12.26 y
Cobalt-60	e^-	5.24 y
Iodine-131	e^-	8 d
Radon-222	α	3.82 d
Technetium-104	e^-	18 min
Fluorine-17	e^+	66 s
Polonium-213	α	4×10^{-6} s
Beryllium-8	α	1×10^{-16} s

A gamma-ray bone scan. Because of its short half-life, readily detected relatively low-energy (140.5 keV) photons, and absence of significant beta radiation, technetium-99m is the most commonly used isotope in nuclear medicine. 99m-Tc phosponates are often used in bone scanning to detect tumors. In this scan it's evident that cancer cells (red-yellow areas) have unfortunately migrated from their primary development site to other parts of the body.

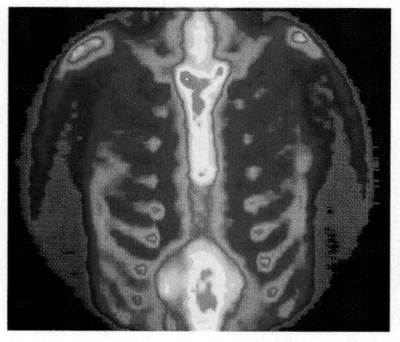

new carbon, the ^{14}C subsequently decays, and the ratio changes in a way that marks the passage of time. For instance, if a piece of charcoal from an ancient fireplace contains one-quarter of the original ratio of ^{14}C to ^{12}C, then a time of $2t_{1/2} = 11\,460$ years has passed since the wood died. Of course, atmospheric atomic weapons testing in the 20th century has changed the ratio of ^{14}C to stable ^{12}C. It's estimated that by the time the United States ended its tests in October of 1958 ($\approx$174 megatons later), it alone had released 25×10^{27} atoms of carbon-14 into the atmosphere, increasing the concentration to about 1.33 times its natural level.

30.9 Induced Radioactivity

Irène and Jean Frédéric Joliot-Curie were not having much luck in the early 1930s. Chadwick's announcement of his discovery of the neutron in 1932 was a shock. They had often produced these new particles in their own laboratory but had thought they were gamma rays. Hot on the trail of the positron, they lost that glory to Anderson in 1933. Still, their triumph (and the Nobel Prize in chemistry) would soon come (1934) out of a series of experiments bombarding light elements with α-particles. In particular, they transmuted aluminum into phosphorus:

$$^{4}_{2}\text{He} + {}^{27}_{13}\text{Al} \rightarrow {}^{31}_{15}\text{P}^{*} \rightarrow {}^{30}_{15}\text{P} + {}^{1}_{0}\text{n}$$

The α-particle was captured into the Al-27 nucleus, forming a highly unstable *compound nucleus*, P-31, that immediately decayed into P-30. The surprising thing was that the phosphorus went on emitting positrons even after the alpha bombardment ceased, as

$$^{30}_{15}\text{P} \rightarrow {}^{30}_{14}\text{Si} + {}^{0}_{+1}\text{e}$$

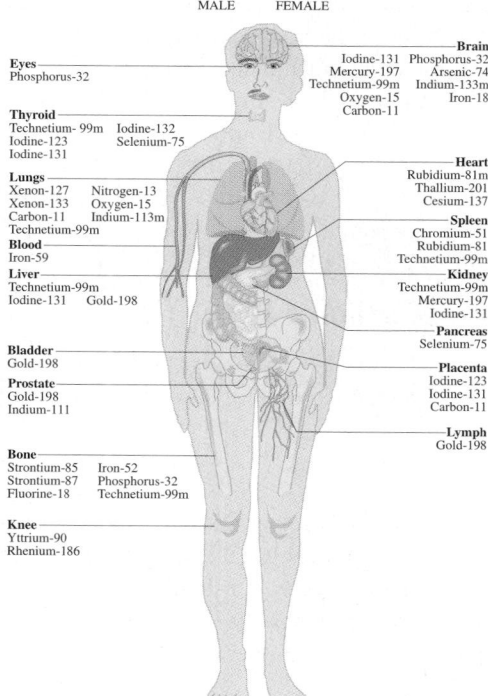

Figure 30.16 Radioisotopes, especially those with very short half-lives, are an important tool in nuclear medicine. They are used to affect anatomy (by destroying unwanted tissue), to influence physiology (by suppressing activity), and to contribute to diagnostics (by serving as tracers in the evaluation of organ function). Technetium-99m is the "workhorse" of nuclear medicine. It emits 0.14 MeV gamma photons that are easily detected, and it has a half-life of only six hours. Because of that short $t_{1/2}$, it's generated where it's going to be used from molybdenum-99. The nucleide is commonly called technetium-99m because it's a metastable state.

They had **artificially induced radioactivity** and, within a year, created a whole group of **radioisotopes**. The stage was set for our present-day medical, biological (Fig. 30.16), and industrial use of these materials—even then the implications were enormous.

Enrico Fermi in Rome was particularly fascinated by the work of the Joliot-Curies, and he set out to test the novel idea he had of using neutrons to create radioisotopes. Since these neutral bullets experienced no electrical repulsion, he rightly reasoned they would easily approach and enter a target nucleus. Within weeks, Fermi published his first positive results. Systematically, he bombarded every element he could get his hands on, all the way up to uranium. Along the way, he realized that neutrons that had been slowed down in their passage through certain materials were even more

Enrico Fermi (1901–1954), Italian-American physicist, was one of the leading figures in the development of the atomic bomb. After years of working with radiation, he died from cancer at the age of 53.

potent at instigating transmutations than were ordinary fast neutrons. These **thermal neutrons** (moving at the speeds of room temperature air molecules) spent more time in the vicinity of a nucleus as they traveled by and could be captured more effectively. "The Pope" (that was Fermi's nickname because of his strong advocacy of the "new faith" of Quantum Mechanics) concluded that the light hydrogen atoms (protons) in a sheet of paraffin would be very efficient at slowing the neutrons. Colliding objects that are about the same mass transfer the most momentum (p. 222). To everyone's amazement, inserting a sheet of paraffin into the neutron beam increased its ability to produce transmutations a hundredfold.

Uranium bombarded by thermal neutrons exhibited some surprising behavior in that it subsequently emitted beta rays. The natural conclusion was that a nucleus had swallowed up a neutron, emitted an electron, and ended up with an additional proton. Fermi believed that uranium had thus been transformed into a new element one box higher in the Periodic Table, the first *transuranic* element. Although that transformation does occasionally take place, the predominant effect was something far more spectacular: nuclear fission.

30.10 Fission and Fusion

Shortly after Fermi's announcement, Ida Noddack, a German chemist, published an alternative scenario that at the time, though it was correct, seemed absolutely crazy. She proposed that the incoming neutron had ruptured the uranium nucleus "into several big fragments which are really isotopes of already known elements." Everyone, including the world's foremost radiochemist, Professor Otto Hahn at the Kaiser Wilhelm Institute in Berlin, completely dismissed the idea as "impossible." After all, how could a small, low-energy neutron burst the largest known nucleus?

 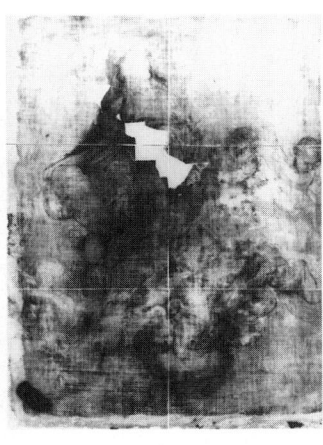 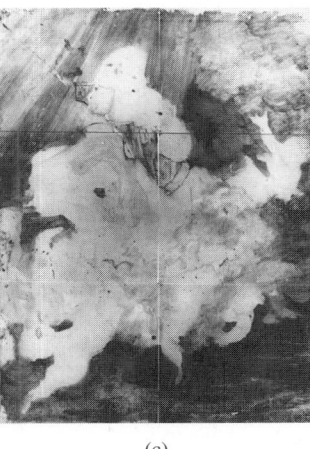

(a)   (b)   (c)

By bombarding a painting (a), in this case one by Van Dyck, with neutrons, different elements in the paint and varnish can be made radioactive. For example, manganese, once common in the dark brown pigment umber, absorbs neutrons with the subsequent emission of electrons. (b) A special film sensitive to electrons reveals the original layers of umber. The white region free of manganese is a modern repair. (c) Four days later, after the emission from the umber faded away, radiation from phosphorus in charcoal revealed a whole new aspect to the painting. Notice the inverted charcoal sketch of the head of a person, now clearly visible. (See p. 974.)

Lise Meitner (1878–1968).

Hahn and his colleague Lise Meitner, joined by F. Strassmann, repeated the Italian experiments with uranium. The chemical analysis was extremely difficult, and the results remained inconclusive, but that wasn't their only worry. Dr. Meitner was an Austrian Jew, and despite Planck's personal appeal to Hitler, she was dismissed from the Institute where she had worked for 30 years. That summer (1938), she fled to Stockholm. Just before Christmas, Hahn and Strassmann finally realized that they were dealing with isotopes of the light element barium. It was inescapable: the uranium nucleus had indeed split in two (Fig. 30.17). They dashed off a letter to the editor of the journal *Naturwissenschaften* and sent a copy to their exiled friend Meitner.

Just then Meitner was being visited by her nephew, a young physicist named Otto Frisch who had fled the Nazis and was working at Bohr's Institute in Copenhagen. With Hahn's letter in hand, the two went for a walk in the woods. Frisch was well versed in Bohr's new model, which maintained that the nucleus resembled a droplet of liquid bound by surface tension. They sat down on a tree stump and began making calculations on scraps of paper. A very large nucleus such as uranium with its many protons would experience an internal electrostatic repulsion that would all but cancel the cohesive surface tension. Even the slight nudge of an incoming neutron might cause it to oscillate and elongate (Fig. 30.18), whereupon the Coulomb repulsion would tend to increase the distortion until, swamped by the short-range strong force, the two pieces would snap into a separate existence and fly apart (all in $\approx 10^{-22}$ s).

Remembering the binding energy curve (p. 1067), Meitner could assume that the fragment nuclei might have a binding-energy-per-nucleon of about 8.5 MeV compared to about 7.6 MeV for a heavy nucleus. The energy released would be the difference $(8.5 - 7.6)$ MeV for each of the roughly 240 nucleons, or about 216 MeV. That's a lot of energy for a single atomic event (roughly a million times the amount liberated by the combustion of one molecule of gasoline). About 1% of the original nuclear mass would be converted into energy, most of it (168 MeV) going into the KE of the two fragments.

A few days later, Frisch caught up with Bohr as he was about to sail off to America to visit Einstein. "Oh, what fools we have been!" Bohr blurted out, "We ought to have seen

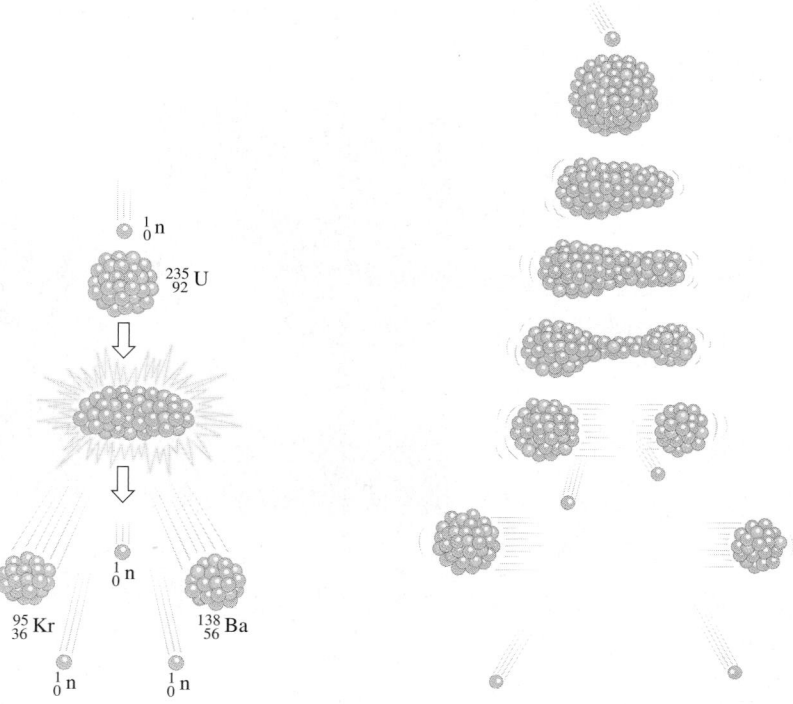

Figure 30.17 The fission of U-235.   **Figure 30.18** Nuclear fission.

that before." Hurriedly, Frisch and Meitner wrote a paper announcing their results, and borrowing a term from the biology of cell division, Frisch named the process **nuclear fission**.

Naturally occurring uranium is composed of three isotopes: U-234, U-235, and U-238. By far the most abundant at 99.27% is U-238, followed by U-235 at only 0.72% and a mere trace of U-234. On the basis of theoretical considerations, Bohr and his former student J. A. Wheeler concluded that it was the rare U-235 that had undergone slow-neutron fission. In fact, it's the only naturally occurring element that can do that little trick. (The common, and far more stable, isotope U-238 can be split, but only by fast-neutron bombardment.) A slow neutron captured by U-235 results in the highly unstable compound nucleus U-236, which immediately ruptures into two large pieces. U-236 can split in two in about forty different ways. One of the common fragment pairs is barium-138 and krypton-95:

$$\,^1_0\text{n} + \,^{235}_{92}\text{U} \rightarrow \,^{236}_{92}\text{U}^* \rightarrow \,^{141}_{56}\text{Ba} + \,^{92}_{36}\text{Kr} + 3\,^1_0\text{n}$$

The Chain Reaction: On the Way to the Bomb

In the 1930s, most people (including Einstein) shared Rutherford's view: "Anyone who expects a source of power from the transformation of these atoms is talking moonshine." Leo Szilard, a brash young Hungarian physicist, was incensed by Rutherford's remark, and in short order he conceived the concept of the **chain reaction**. Szilard speculated that if a nuclide could be found that, when struck by a neutron, would emit *two* or more neutrons, the resulting fission process could continue on its own, building up into a horrendous cascade (Fig. 30.19). Sensing that Europe would soon be engulfed in war, Szilard accepted a post at Columbia University and arrived in America in January 1939. Fermi came to Columbia that same month. His mother's antecedents were Jews, as was his wife, and so after stopping in Stockholm to pick up the 1938 Nobel Prize, the family continued on to New York.

Only weeks after Fermi's arrival, Bohr reached New York with the news of the discovery of fission. "The Pope" immediately appreciated the possibilities of making a nuclear bomb, but there were troubling questions that had to be resolved. Only if each fission event emitted an average of at least two neutrons could an explosive device be developed. In March, Szilard and Zinn determined experimentally that, on average, between two and three fast neutrons were indeed emitted per fission. A single such event liberates about 3.2 $\times 10^{-11}$ J, which is a great deal of energy in Atomland, but not very much on the human

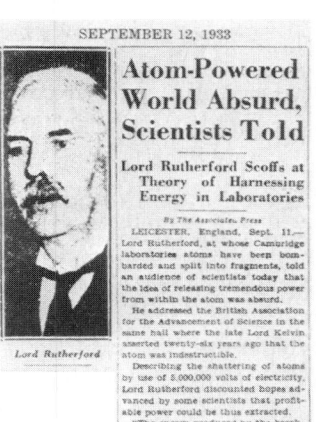

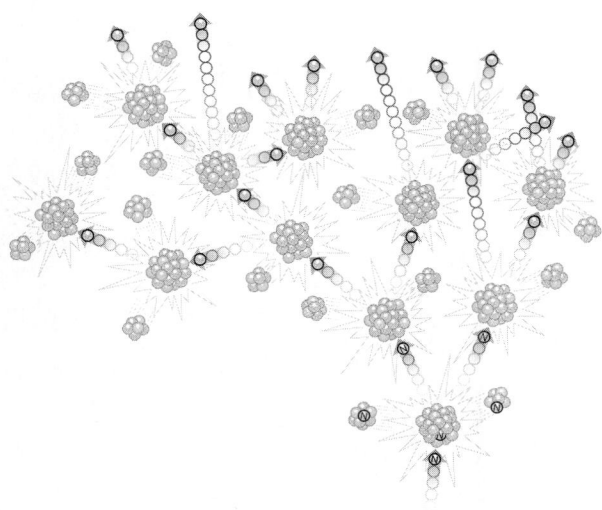

Figure 30.19 A branching chain reaction arising from uranium fission.

Enrico Fermi (1901-1954).

scale of things. That was why the chain reaction—and, in particular, a chain reaction that fanned out and grew—was so crucial if a vast amount of energy was to be liberated. If the first nucleus to split fired out two neutrons, those two could split two other nuclei in the second generation, and so on. Ideally, after, say, 80 generations, an incredible 1.2×10^{24} atoms would have shattered in a fraction of a millisecond. About 0.5 kg of uranium would vanish, releasing 3.8×10^{13} J, or the equivalent of roughly 10 kilotons of TNT.

When Hitler embargoed the export of uranium from Czechoslovakian mines, Szilard was convinced the Germans were developing nuclear weapons.* Since Szilard couldn't drive a car, he convinced Edward Teller to take him to Long Island in August 1939 to meet with Einstein. Szilard brought with him a letter he had written to President Roosevelt, which Einstein, who was already world famous, read, immediately understood, and signed. It advised "that extremely powerful bombs of a new type may thus be constructed." The all-out push to develop the atomic bomb began on December 6, 1941, just one day before Pearl Harbor was attacked by the Japanese.

Nuclear Reactors

A chunk of natural uranium is mostly U-238, and these nuclei don't easily split. Moreover, fast neutrons from a fissioning nucleus can be captured equally well by U-238 and U-235. Thus, pure natural uranium will not support a chain reaction. But U-238 has much less appetite for slow neutrons than does U-235, and that difference was exploited in the design of the first **fission reactor**, the controlled chain-reaction machine. If uranium were interspersed within a *moderator* (some light substance that slowed neutrons down), it would have almost the same effect as removing the U-238 altogether, leaving only the U-235 to interact. The most promising candidates for moderators were beryllium, heavy water (regular hydrogen absorbs too many neutrons), and carbon. Beryllium was crossed off the list because it is rare and toxic. Heavy water would have been perfect, but there were only a few quarts of it in the United States, and there wasn't time to prepare the several tons needed. The choice fell to carbon in the form of pure graphite.

Fermi and his group moved the top-secret operation to the University of Chicago. They set up shop in the squash courts (where the ceilings were more than 25 feet high) under the abandoned football stands of Stagg Field. There, they stacked 40 000 graphite bricks, into which were inserted chunks of uranium oxide. Several *control rods* of cadmium, which is a voracious absorber of neutrons, were built into the reactor to keep a leash on the chain reaction. On December 2, 1942, the cadmium rods were carefully withdrawn, and the world's first self-sustained controlled nuclear chain reaction occurred.

A depiction of the historic start-up of the world's first nuclear reactor under the direction of Fermi.

*Szilard was right. Germany had an active nuclear weapons program led by Heisenberg. Because he seems never to have actually believed in the possibility of making a bomb, Heisenberg's efforts were rather half-hearted. In 1943, the Japanese joined the race to build an atomic bomb. A lack of uranium made things difficult for them, and when a submarine from Hitler carrying two tons of it was sunk by the Allies, their project faltered.

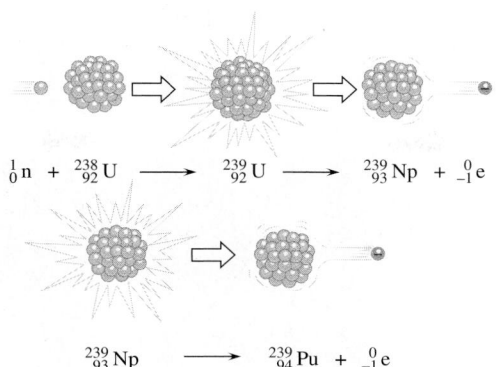

Figure 30.20 Plutonium-239 produced from uranium-238 by neutron bombardment.

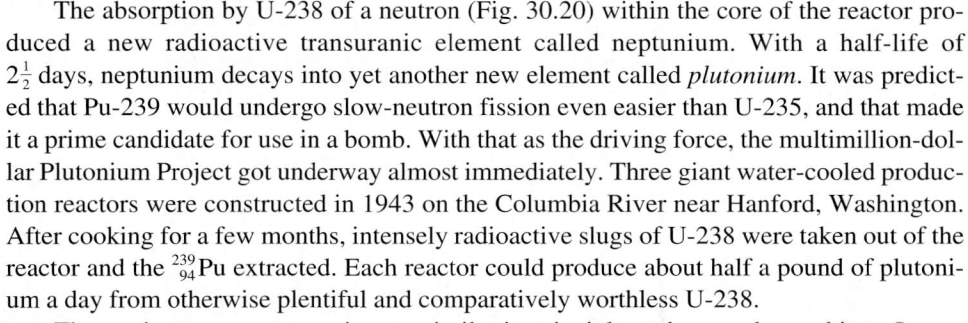

$$_0^1 n + {}_{92}^{238}U \longrightarrow {}_{92}^{239}U \longrightarrow {}_{93}^{239}Np + {}_{-1}^{0}e$$

$$_{93}^{239}Np \longrightarrow {}_{94}^{239}Pu + {}_{-1}^{0}e$$

Technicians loading the fuel into a modern reactor at a nuclear power plant.

The absorption by U-238 of a neutron (Fig. 30.20) within the core of the reactor produced a new radioactive transuranic element called neptunium. With a half-life of $2\frac{1}{2}$ days, neptunium decays into yet another new element called *plutonium*. It was predicted that Pu-239 would undergo slow-neutron fission even easier than U-235, and that made it a prime candidate for use in a bomb. With that as the driving force, the multimillion-dollar Plutonium Project got underway almost immediately. Three giant water-cooled production reactors were constructed in 1943 on the Columbia River near Hanford, Washington. After cooking for a few months, intensely radioactive slugs of U-238 were taken out of the reactor and the $_{94}^{239}Pu$ extracted. Each reactor could produce about half a pound of plutonium a day from otherwise plentiful and comparatively worthless U-238.

The modern power reactor is very similar in principle to these early machines. It usually uses natural uranium *enriched* to contain a few percent U-235, with ordinary water as the moderator (Fig. 30.21). The reactor is used essentially as a furnace to supply "heat" (via steam) that drives a traditional turbine and ultimately generates electricity. The kinetic energy of fission fragments and neutrons absorbed in the reactor transforms into copious amounts of thermal energy. A liquid coolant circulating around the fuel rods carries this energy out of the core. A modern 1000-million-watt reactor is charged with about 90 000 kilograms of fuel, a third of which is replenished each year. It's because a rather innocent-looking power reactor can generate plutonium that it becomes very difficult to control the spread of nuclear weapons (which is why the United States, in 1991, bombed Iraq's purported power reactors in the first few days of the Gulf War).

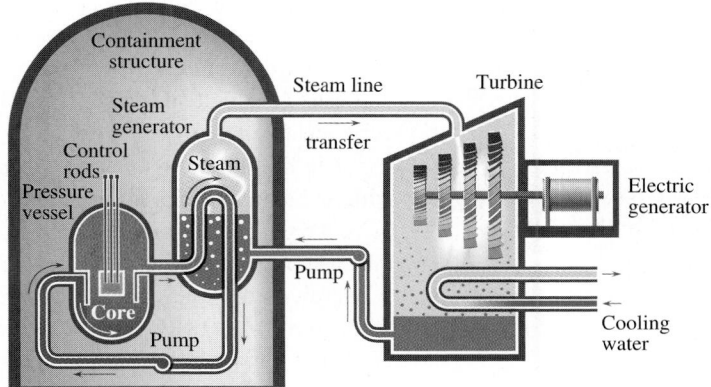

Figure 30.21 A pressurized water reactor system. A modern power reactor usually uses natural uranium enriched to contain a few percent U-235, with ordinary water as the moderator. The kinetic energy of the fission fragments and neutrons absorbed in the reactor transforms into thermal energy. This energy is carried out of the core by a circulating liquid coolant. The reactor is essentially a furnace to supply heat that drives a traditional turbine.

The article in the right column reports Einstein's opinion (1934) that attempts at loosing the "energy of the atom" would be fruitless—an opinion he soon changed.

The time will come when atomic energy will take the place of coal as a source of power.... I hope that the human race will not discover how to use this energy until it has brains enough to use it properly.

SIR OLIVER LODGE (1920)

J. R. Oppenheimer and General Groves in front of the remains of the vaporized 100-foot steel tower that had held the first atomic bomb (a plutonium device) ever exploded. The desert sand was seared into a glassy jade green crust. One wonders at the wisdom of visiting such a place so soon after an explosion.

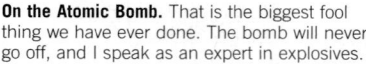

On the Atomic Bomb. That is the biggest fool thing we have ever done. The bomb will never go off, and I speak as an expert in explosives.

ADM. WILLIAM LEAHY
(1945)

The Atomic Bomb

The actual design of the bomb was carried out at Los Alamos, New Mexico, under the direction of J. Robert Oppenheimer ("Oppie" to almost everyone). The design team included, among many others, Bohr, Chadwick, Frisch, Fermi, and Feynman. General Groves, who headed the entire bomb program (code-named the Manhattan Project), once remarked, "At great expense, we have gathered here the largest collection of crackpots ever seen."

The principle underlying an A-bomb is simple. A small spherical chunk of fissionable material under bombardment by neutrons can support only a limited chain reaction; too many neutrons released from the fissioning nuclei escape from the sample through its surface. If a sphere of U-235 or Pu-239 is made larger, its volume increases faster than its surface area, more neutrons are produced within it, and proportionately fewer escape from its surface. There is therefore a *critical size* at which point the average number of neutrons from each fission that will cause another fission (rather than being lost somehow) equals one. Under such circumstances, the chain reaction is just initiated. Any increase in the amount of fissionable material beyond that produces a branching, rapidly growing chain reaction—such a mass is *supercritical*. A solid sphere of U-235 about the size of a softball, struck by a stray neutron, will roar into a furious, multimillion-degree plasma in a millionth of a second. That's all it takes: one chunk in excess of the *critical mass* (≈ 15 kg) and off it goes. Of course, the U-235 or Pu-239 has to be kept in a subcritical configuration so it can be safely transported. The hard part is to assemble a supercritical mass fast enough so that it doesn't just melt and fizzle. That was especially difficult to do with plutonium, which is very unstable.

The first bomb to be designed was a brute force device so simple it was used in war without having been previously tested. Two pieces of U-235, each too small to ignite on its own, were rapidly brought together into a *supercritical* mass. The bomb was little more than a short cannon that fired a subcritical projectile of U-235 into another subcritical target screwed into the end of the muzzle. On August 6, 1945, at $8:15\frac{1}{2}$ in the morning, it dropped from the belly of a B-29 over Hiroshima, Japan. Seconds later, 1500 feet above the ground, it burst into a 15-kiloton fireball.

Starfire: Thermonuclear Fusion

Stars and hydrogen bombs are both powered by a process called **thermonuclear fusion**. Figure 30.9 makes the point that fusing two light nuclei together produces a heavier nucleus higher up the binding-energy-per-nucleon curve. On bonding more tightly, a fraction of each nucleon's mass is converted to energy in accord with Einstein's $E_0 = mc^2$.

Inside the target chamber of the LLE laser-fusion device. Fusion reactions take place in a tiny target-sphere filled with deuterium/tritium and irradiated by the 30-kJ Omega laser.

Stars are born, live, and die—some explode and disperse; others, latecomers, rise out of their ashes. First-generation stars are composed of primordial hydrogen laced with helium. They are formed when a great interstellar cloud of gas slowly collapses under its own gravitational force, heating up and growing more dense as the material plunges inward. At densities 100 times that of water and temperatures of 15 to 20 million K, an almost-star is a seething plasma ball where bare nuclei rush about at tremendous speeds. Two protons, each with a charge $+e$, can overcome their Coulomb repulsion and come close enough ($r < 10^{-14}$ m) for the strong force to lock them together. That happens provided they approach one another with a kinetic energy that can override the repulsion, that is, a net KE equal to their electrical potential energy at a center-to-center separation r:

$$PE_E = k_0 \frac{e^2}{r} = (9 \times 10^9 \text{ N·m}^2/\text{C}^2) \frac{(1.6 \times 10^{-19} \text{ C})^2}{10^{-14} \text{ m}}$$

or
$$PE_E = 2.3 \times 10^{-14} \text{ J} = 0.14 \text{ MeV}$$

which follows from Eq. (16.7). A gas of protons with an average KE of 0.14 MeV has a temperature given by Eq. (12.15); accordingly

$$(KE)_{av} = \tfrac{3}{2} k_B T$$

and so
$$T = \frac{(0.67)(2.3 \times 10^{-14} \text{ J})}{1.38 \times 10^{-23} \text{ J/K}} = 1 \times 10^9 \text{ K}$$

Fusion can take place at temperatures roughly 50 times lower than this for two reasons. First, this value is an average KE, and many particles have much higher energies. Second, there are quantum-mechanical effects that allow the particles to penetrate the Coulomb barrier. In any event, high speeds and high temperatures are called for, which is why the process is known as *thermo*nuclear fusion. Once fusion begins, the furnace at the center of a star puffs it up with an outrushing of released energy, and the gravitational collapse stops; the plasma ball stabilizes and a star is born.

Many thermonuclear fusion processes can drive a star, depending on its physical circumstances (temperature, mass, available fuel, and so on). Each process generates energy out of matter, cooking up heavier elements along the way. Except for the hydrogen in your body (which goes back roughly 17 thousand million years to the first moments of creation), you are star dust, the ash created in the belly of some supernova long before our Sun was born. We believe the Sun (a second-generation star) is predominantly powered by a process known as the *proton-proton chain*. It begins with two protons coming together to form a

A nuclear explosion.

deuteron ($^2_1$H), a positron, and a neutrino:

$$^1_1\text{H} + ^1_1\text{H} \rightarrow ^2_1\text{H} + ^0_{+1}\text{e} + \nu_e$$

The deuteron then captures another proton, producing helium:

$$^2_1\text{H} + ^1_1\text{H} \rightarrow ^3_2\text{He} + \gamma$$

Finally, two helium-3 nuclei fuse in the transformation

$$^3_2\text{He} + ^3_2\text{He} \rightarrow ^4_2\text{He} + ^1_1\text{H} + ^1_1\text{H}$$

forming He-4 ash, two protons, and a great deal of energy. In effect, the total process is

$$4\text{p} \rightarrow \alpha + 2 \, ^0_{+1}\text{e} + 2\nu_e + \gamma$$

The liberated energy (6.2 MeV per proton) is carried off mostly by gamma rays, positrons, and neutrinos. The gamma photons are absorbed by the plasma. The positrons are annihilated when they come in contact with electrons generating more photons (which are also absorbed and add more energy). About 2% of the energy liberated comes off with the escaping neutrinos. The Sun unleashes 3.8×10^{26} J each second; each second it pours into space the energy equivalent of burning three million million million gallons of gasoline. Every second roughly 660 000 000 tons of hydrogen are transmuted into helium. In the process, about 4 600 000 tons (4.2×10^9 kg) of matter transform into radiant energy. And so it has raged for at least five thousand million years.

Core Material & Study Guide

NUCLEAR STRUCTURE

Masses in the atomic domain are often specified in *unified atomic mass units* (u) where a neutral $^{12}_6$C atom is defined to be precisely 12.000 000 u:

$$1 \text{ u} = 1.660\,540 \times 10^{-27} \text{ kg} = 931.494 \text{ MeV/c}^2$$

The nuclear radius R is

$$R = R_0 A^{1/3} \qquad \text{[30.1]}$$

R_0 turns out to be $\approx$1.2 fm (p. 1060). This important introductory material is considered in Sections 30.1 (Isotopes: Birds of a Feather), 30.2 (Nuclear Size, Shape, and Spin), and 30.3 (The Nuclear Force). The problems dealing with these ideas are quite straightforward. When nucleons combine to form a nucleus, an amount of mass Δm known as the **mass defect** is converted into **binding energy** $E_B = \Delta mc^2$. These significant ideas are discussed in Section 30.4 (Nuclear Stability) (p. 1064)—study Example 30.3.

NUCLEAR TRANSFORMATION

Alpha decay has the form

[alpha decay] $\qquad ^A_Z\text{X} \rightarrow ^{A-4}_{Z-2}\text{Y} + ^4_2\text{He} + Q \qquad$ [30.2]

where X is called the **parent nucleus**, Y is the **daughter nucleus**, and Q is the **disintegration energy** (p. 1069),

$$Q = (m_X - m_Y - m_\alpha)c^2 \qquad \text{[30.3]}$$

Here m_X, m_Y, and m_α are the masses of the parent, daughter, and

alpha, respectively. Electron-beta decay arises when (p. 1097)

$$\text{n} \rightarrow \text{p} + ^0_{-1}\text{e} + \bar{\nu}_e \qquad \text{[30.4]}$$

In the case of a radioactive nuclide, the decay has the form

[electron beta decay] $\qquad ^A_Z\text{X} \rightarrow ^A_{Z+1}\text{Y} + ^0_{-1}\text{e} + \bar{\nu}_e + Q \qquad$ [30.5]

Positron-beta decay arises when

$$\text{p} \rightarrow \text{n} + ^0_{+1}\text{e} + \nu_e \qquad \text{[30.6]}$$

In the case of a radioactive nuclide, the decay has the form

[positron beta decay] $\qquad ^A_Z\text{X} \rightarrow ^A_{Z-1}\text{Y} + ^0_{+1}\text{e} + \nu_e + Q \qquad$ [30.7]

Electron-capture, the process wherein a parent nucleus absorbs one of its own orbital electrons, competes with β^+ decay Electron-capture, the process wherein a parent nucleus absorbs one of its own orbital electrons, competes with β^+ decay (p. 1070):

[electron capture] $\qquad ^A_Z\text{X} + ^0_{-1}\text{e} \rightarrow ^A_{Z-1}\text{Y} + \nu_e + Q \qquad$ [30.8]

Reread Sections 30.5 (Radioactive Decay) and 30.7 (Gamma Decay) and study Examples 30.4 and 30.5.

The SI unit of decay rate is the *becquerel* (Bq), which is equivalent to one disintegration per second. After n half-lives, the **decay rate** R will be

[decay rate or activity] $\qquad R = (\tfrac{1}{2})^n R_0 \qquad$ [30.10a] and similarly,

[radioactive mass remaining] $m = \left(\frac{1}{2}\right)^n m_0$ [30.10b]

Moreover (p. 1076)

[decay rate or activity] $R = R_0 e^{-\lambda t}$ [30.11]

where λ is the **decay constant**. The **half-life** is related to λ by

$$\lambda t_{1/2} = 0.693$$ [30.12]

Since

[decay rate or activity] $R = \lambda N$ [30.13]

where N is the **number of atoms remaining**, it follows that

$$N = N_0 e^{-\lambda t}$$ [30.14]

All of this material about the rate of radioactive decay is treated in Section 30.8 (Half-Life). There are lots of problems on this stuff, so make sure you understand Example 30.6.

Key Terms

atomic number	electron capture
nucleon number	neutrino
nuclide	weak force
isotope	gamma decay
atomic mass unit	decay rate
deuterium	activity
tritium	becquerel
heavy water	decay constant
fermi	half-life
Shell Model	mean lifetime
NMR	induced radioactivity
MRI	thermal neutron
nuclear force	nuclear fission
mass defect	chain reaction
binding energy	nuclear reactor
alpha decay	critical mass
parent nucleus	nuclear fusion
beta decay	proton-proton chain

Discussion Questions

1. Figure Q1 is a plot of the excitation energy necessary to raise a nuclide with a given number of neutrons up to the next highest energy level. What's the significance of the data in light of the magic numbers?

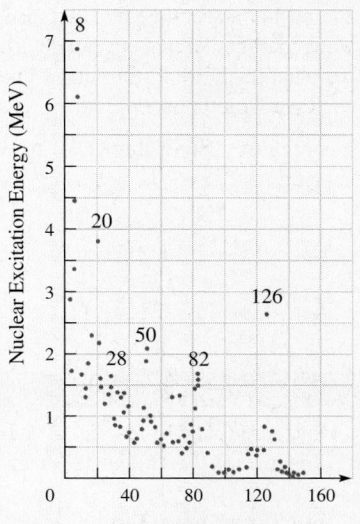

Figure Q1

2. The two nuclides ^{235}U and ^{239}Pu undergo fission via bombardment with neutrons of very little KE (even less than 0.01 eV). By contrast, ^{238}U and ^{232}Th will fission, provided the neutrons have kinetic energies in excess of 1 MeV. How is this fact used in a fission reactor? What does it suggest about fuels for bombs? Incidentally, it's been proposed that small atomic explosions could be used to propel a spaceship up to tremendous speeds. (Perhaps some good will come of this yet.)

3. Your standard H-bomb (what a thought) is essentially an ellipsoidal chunk of lithium-6 deuteride, an opalescent white solid, surrounded by an atom-bomb trigger. The whole thing is often encased in a U-238 shell, which initially reflects energy back onto the ^{6}LiD. The cheap U-238 also undergoes fission, allowing one to increase the output even more. (The Russians, with their big missiles and poor guidance systems, prefer large bombs.) The United States tested the first lithium-deuteride fission-fusion-fission device in 1953, ushering in a new era of human folly with a 15-megaton roar. Given the reactions

$$^2_1D + {}^3_1T \rightarrow {}^4_2He + n + 17.6 \text{ MeV}$$

and

$$^6_3Li + n \rightarrow {}^4_2He + {}^3_1T + energy$$

describe as much of what happens in an H-bomb as you can. Comment on the limitations, if any, on the mass of the U-238 shell. How might tritium be produced in a factory situation?

4. During World War II, the only large producer of heavy water in the world was the Norsk Hydro plant at Vemork, Norway. Toward the end of 1942, a group of Anglo-Norwegian commandos carried out two fairly successful raids on the facility. The plant was in Nazi hands, and the Nazis were shipping all of its output to Germany. In the fall of 1943, after the plant had been repaired, American planes attacked it again. At that point, it should have been clear that we knew and that they knew that we knew. Why did we attack the plant? In the United States, we often use enriched U-238 and ordinary water in reactors. By contrast, the Canadians generally use D_2O and natural uranium. Explain.

5. Figure Q5 depicts a vacuum chamber into which are pointed a number of powerful laserbeams. A tiny pellet of deuterium and tritium is dropped inside and blasted from all sides by the beams. This is a so-called *inertial confinement* reactor. What does it do and why? Knowing that neutrons are a likely by-product, see if you can figure out the formula for a possible nuclear reaction that might take place.

Figure Q5

6. Figure Q6 shows a Tokamak Fusion Reactor. A high-temperature deuterium-tritium plasma is confined by powerful magnets so that it stays suspended within the vacuum chamber. By pumping exceedingly large currents through the plasma, temperatures of tens of millions of degrees can be attained. What do you think this thing is supposed to do? What are the high temperatures for? Why is fusion power appealing?

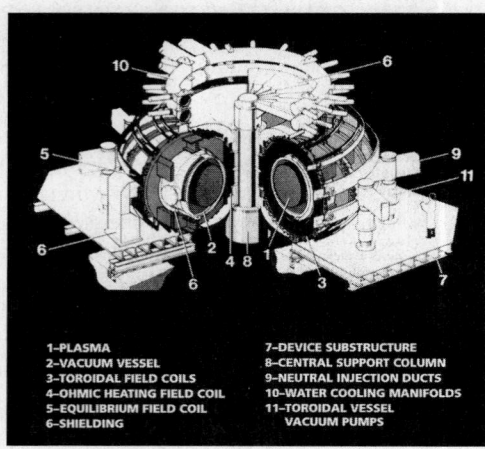

Figure Q6

7. Discuss the difference between the half-life and the mean life of a radioactive substance. Relate these ideas to the life of a human being. Given a typical sample of Americans, 90% will make it to age 30, 80% to age 50, 50% to age 68, 25% to age 77, 10% to age 84, 1% to age 92, 0.1% to age 97, 0.01% to age 100, and 0% to age 200.

8. Hans Bethe at Cornell proposed a high-temperature fusion process known as the *carbon-nitrogen cycle* that may well power some stars:

$$
\begin{array}{lr}
 & \text{MeV} \\
{}^{12}\text{C} + \text{p} \rightarrow {}^{13}\text{N} + \gamma & 2.0 \\
\quad {}^{13}\text{N} \rightarrow {}^{13}\text{C} + \text{e}^+ + \nu_\text{e} & 2.2 \\
{}^{13}\text{C} + \text{p} \rightarrow {}^{14}\text{N} + \gamma & 7.5 \\
{}^{14}\text{N} + \text{p} \rightarrow {}^{15}\text{O} + \gamma & 7.3 \\
\quad {}^{15}\text{O} \rightarrow {}^{15}\text{N} + \text{e}^+ + \nu_\text{e} & 2.7 \\
{}^{15}\text{N} + \text{p} \rightarrow {}^{12}\text{C} + {}^4\text{He} & \underline{5.0} \\
 & 26.7
\end{array}
$$

Explain what's happening. What form does the energy produced take? What is the net result of running through the cycle?

9. Return to Fig. 30.14, the exponential curve, which is a rather special function. For example, consider the slope of the curve. To do that qualitatively, imagine short line segments tangent to the curve at every point. What is the significance of the slope? Now plot the slope versus time. What is the shape of the curve?

10. Let's have a little fun and do a bit of speculating. What would happen to the Universe if the nuclear force vanished, *now*?

11. What would the Universe look like if the electromagnetic force vanished, *now*?

12. Suppose we produce the molecules of ${}^{12}_{6}\text{C}{}^{16}_{8}\text{O}$ and $({}^{14}_{7}\text{N})_2$. Then we singly ionize them (that is, remove one electron) and accelerate them inside a vacuum chamber so that we can determine their masses. Each molecule has a total of 14 protons and 14 neutrons. Explain in general terms what we would observe.

13. Explain the meaning of each of the Key Terms on page 1087.

Multiple Choice Questions

1. A nuclide is specified completely by   (a) its atomic number   (b) its proton number   (c) the number of nucleons it has   (d) giving both Z and A (e) none of these.

2. The quantity A is the   (a) proton number   (b) nucleon number   (c) neutron number   (d) atomic number   (e) none of these.

3. Isotopes are nuclides having   (a) the same A and Z values   (b) the same Z, but different values of A (c) the same A, but different values of Z (d) different values of both A and Z (e) none of these.

4. When a nucleus undergoes alpha decay, the daughter as compared to the parent has   (a) the same Z and an N reduced by 4   (b) Z reduced by 4 and A reduced by 2   (c) A reduced by 4 and Z reduced by 2   (d) Z increased by 2 and N reduced by 2   (e) none of these.

5. Beta particles are   (a) protons emitted from a nucleus   (b) electrons emitted from a nucleus   (c) neutrons emitted from the nucleus   (d) electrons emitted from the atomic cloud   (e) none of these.

6. The mass of a nucleus composed of several nucleons is   (a) always less than the sum of masses of its constituents   (b) sometimes less than the sum of masses of its constituents   (c) always more than the sum of masses of its constituents   (d) always equal to the sum of masses of its constituents   (e) none of these.

7. When a nucleus undergoes electron-beta decay, the daughter as compared to the parent has (a) the same Z and an N reduced by 1 (b) Z reduced by 1 and A reduced by 1 (c) A reduced by 1 and Z increased by 1 (d) Z increased by 1 and N reduced by 1 (e) none of these.

8. The nucleus $^{234}_{90}$Th β-decays into (a) $^{234}_{89}$Ac (b) $^{235}_{91}$Pa (c) $^{234}_{91}$Th (d) $^{234}_{91}$Pa (e) none of these.

9. The missing particle in the reaction

$$^{2}_{1}D + {}^{199}_{80}Hg \rightarrow {}^{197}_{79}Au + (?)$$

is a(n) (a) gamma (b) alpha (c) neutron (d) proton (e) electron.

10. The missing particle in the reaction

$$^{2}_{1}D + {}^{196}_{78}Pt \rightarrow {}^{197}_{79}Au + (?)$$

is a(n) (a) deuteron (b) alpha (c) neutron (d) proton (e) gamma.

11. The missing particle in the reaction

$$^{1}_{0}n + {}^{198}_{80}Hg \rightarrow {}^{197}_{79}Au + (?)$$

is a(n) (a) deuteron (b) alpha (c) neutron (d) proton (e) electron.

12. The missing particle in the reaction

$$^{1}_{0}n + {}^{196}_{78}Pt \rightarrow {}^{197}_{78}Pt + (?)$$

is a(n) (a) deuteron (b) alpha (c) neutron (d) proton (e) gamma.

13. The missing particle in the reaction

$$\gamma + {}^{198}_{80}Pt \rightarrow {}^{197}_{79}Au + (?)$$

is a(n) (a) electron (b) alpha (c) neutron (d) proton (e) gamma.

14. An antineutrino always accompanies (a) α decay (b) β^{-} decay (c) gamma emission (d) neutron emission (e) none of these.

15. In a nuclear reaction, the value of Q (a) is always negative (b) is positive in a spontaneous decay (c) is positive only for gamma emission (d) is always negative for proton emission (e) none of these.

16. The process whereby a nucleus splits into two very roughly equal parts is known as (a) fusion (b) beta decay (c) fission (d) nucleation (e) none of these.

17. A nuclide with a half-life of 10 days has a decay constant of (a) 0.069 decays/day (b) 6.9 decays/day (c) 1/10 decays/day (d) 10 decays/day (e) none of these.

18. Given that a freshly prepared nuclide has a half-life of 10 days, the percentage of it remaining after 30 days is (a) 30% (b) 10% (c) 12.5% (d) 72.5% (e) none of these.

19. As a naturally occurring radionuclide decays, (a) its λ increases (b) its τ decreases (c) its $t_{1/2}$ decreases (d) its R remains constant (e) none of these.

20. The Sun is powered by (a) nuclear beta decay (b) nuclear fission (c) thermonuclear fusion (d) nuclear gamma decay (e) none of these.

For more Multiple Choice Questions with answers click on WARM-UPS in CHAPTER 30 on the CD.

Suggestions on Problem Solving

1. Some useful numerical quantities for this chapter are

$1\ u = 1.660\ 540 \times 10^{-27}\ kg = 931.494\ MeV/c^2$

$1\ eV/c^2 = 1.782\ 663 \times 10^{-36}\ kg$

$1\ eV = 1.602\ 177 \times 10^{-19}\ J$

$m_e = 9.109\ 389\ 7(54) \times 10^{-31}\ kg$
$= 0.510\ 999\ 06(15)\ MeV/c^2 = 5.485\ 80 \times 10^{-4}\ u$

$m_p = 1.672\ 623\ 1(10) \times 10^{-27}\ kg$
$= 938.272\ 31(28)\ MeV/c^2 = 1.007\ 276\ u$

$m_n = 1.674\ 928\ 6(10) \times 10^{-27}\ kg$
$= 939.565\ 63(28)\ MeV/c^2 = 1.008\ 665\ u$

$m_\alpha = 4.001\ 506\ u$

2. Most tabulations of masses for the various isotopes are *atomic* mass listings, not nuclear mass. When you are given that the atomic mass for carbon-10 is 10.016 856 u, it is for the neutral atom, and it includes the six electrons. One way to handle this situation is to use atomic masses throughout any calculations (for example, of mass defect). Just *keep track of the electrons*. There will be a difference in the electron binding energies, but generally that's totally negligible. Accordingly, it may often be more convenient to use the mass of the hydrogen atom (1.007 825 u) in place of the proton mass—the difference is the electron.

3. A typical problem concerning radioactive decay might involve one or two of Eqs. (30.10) through (30.13). If you need the activity (or decay rate) R, that's Eq. (30.11). Any question about the mass, or fractional amount, number of atoms, or percentage of a substance remaining is addressed to Eq. (30.13). You may be given τ, λ, or $t_{1/2}$ and be required to find one of the others. Similarly, the equations for R and N are written in terms of λ, though you may be provided τ or $t_{1/2}$. *Don't forget the minus signs* in those two exponential expressions. Review APPENDIX A6 on exponentials and logarithms. You can be asked to solve for λ in either Eq. (30.11) or (30.13), and that will require undoing the exponential by taking the natural log. Watch the time units in the exponential equations: you can use λ in decays/year provided you enter t in years—just don't mix years with days or seconds.

4. The advent of the electronic calculator has made it just as easy to use Eq. (30.10), $R = (\frac{1}{2})^n R_0$, directly as to use Eq. (30.11), $R = R_0 e^{-\lambda t}$. Accordingly, you'll be able to check your work and are advised to always do so—everyone makes numerical errors, we just have to catch them.

5. To calculate $R = R_0 A^{1/3}$, use the $x^{1/y}$ key on your electronic calculator. Enter the number for A, which is x, hit the $x^{1/y}$ key, enter 3, and hit the = key.

Problems ✦ Coordinated Problems ✦ Progressive Problems ✦ Solutions

SECTION 30.1: ISOTOPES: BIRDS OF A FEATHER

1. [I] How many nucleons does the nuclide $^{111}_{50}$Sn possess?

2. [I] What is the proper symbol for a nucleus having 14 protons and 15 neutrons?

3. [I] How many neutrons and how many protons does the nuclide $^{15}_{7}$N possess?

4. [I] How many neutrons are in a ^{15}O nucleus?

5. [I] Identify each of the following nuclides: $^{211}_{87}$X, $^{202}_{82}$X, $^{105}_{47}$X, and $^{142}_{59}$X.

6. [I] What is the difference structurally between $^{183}_{76}$Os and $^{193}_{76}$Os?

7. [I] The stable isotope of nitrogen, $^{14}_{7}$N, has an atomic mass of 14.003 074 u. How much is that in MeV/c² and GeV/c²? Give the answers to six significant figures.

8. [I] Inasmuch as nuclear matter has a density of 2.3×10^{17} kg/m³, how many cubic meters of water would have to be compressed into a cubic centimeter to match that density?

9. [I] The mass of the lithium-6 isotope is 5.603 051 GeV/c². How much is that in kg? Give your answer to seven significant figures.

10. [I] There are two stable isotopes of chlorine: Cl-35 (with an atomic mass of 34.968 853 u) and Cl-37 (with an atomic mass of 36.965 903 u). They are found with relative abundances of 75.77% and 24.23%, respectively. Determine the atomic mass of chlorine as it would be listed in the Periodic Table.

11. [I] Niobium has several isotopes, but only Nb-93 is stable. Its atomic mass is $1.542\ 748 \times 10^{-25}$ kg. Determine its mass in unified atomic mass units and compare it with the atomic mass listed for niobium in the Periodic Table. Explain your results.

12. [II] Show that MeV/c² corresponds to mass by explicitly determining the equivalent in kg of 1.00 MeV/c².

13. [II] Bromine has at least two stable isotopes: Br-79 (with an atomic mass of 78.918 336 u) and Br-81 (with an atomic mass of 80.916 289 u). Given that the relative abundance of Br-79 is 50.7% and the Periodic Table lists its atomic mass as 79.909 u, is it likely to have any other long-lived isotopes? Explain via numerical analysis.

14. [II] Hydrogen has an atomic number $A = 1.0078$. Neglecting the electron-proton atomic binding energy, what is the rest energy of a hydrogen atom? Use this to determine a value for 1.00 u in MeV/c².

15. [II] Use Avogadro's Number and the definition of the atomic mass unit to show that 1 u = 1.66×10^{-27} kg.

SECTION 30.2: NUCLEAR SIZE, SHAPE, AND SPIN

16. [I] What is the radius of a gold nucleus? Gold has a nucleon number of 197.

17. [I] Determine the diameter of the uranium isotope ^{235}U. [*Hint: The diameter depends on the nucleon number.*]

18. [I] What is the radius of the oxygen (^{16}O) nucleus?

19. [I] Find the diameter of the nucleus of the isotope lead-208.

20. [I] The radius of a nucleus doubles whenever the number of nucleons increases by a multiplicative factor of how much?

21. [I] If a nucleus is determined to be 7.2 fm in diameter, what mass number does it correspond to?

22. [II] Approximate the ratio of the density of an atom to the density of its nucleus. Take the radius of the atom to be 0.05 nm and that of the nucleus to be 1.2 fm.

23. [II] How much bigger is the radius of the nucleus of U-238 than that of He-4?

24. [II] What is the atomic number of the nuclide that has a diameter one-quarter that of tellurium-128?

25. [II] Approximate the density of an atomic nucleus and of an atom, and compare the two.

SECTION 30.4: NUCLEAR STABILITY

26. [I] A boron atom ($^{10}_{5}$B) has a mass of 10.012 937 u. What is the mass defect of its nucleus?

27. [I] Using the results of Problem 26, determine the binding energy of the boron nucleus.

28. [I] Use Fig. 30.9 to approximate the binding-energy-per-nucleon for boron-10. What is the total binding energy of this nuclide? How does that compare with the computed value?

29. [I] Using the results of Problem 27, determine the average binding-energy-per-nucleon for boron-10. Compare your answer with that of Problem 28. [*Hint: The mass defect corresponds to the total binding energy.*]

30. [I] If the binding energy of $^{126}_{52}$Te is 1.066 GeV, what is the binding-energy-per-nucleon?

31. [I] If the binding-energy-per-nucleon for an alpha particle is 7.1 MeV, what is the total binding energy of the alpha? Take a look at Problem 38.

32. [I] Use Fig. 30.9 to determine an approximate value for the total binding energy of a U-238 nucleus.

33. [I] Find the binding energy of the last neutron in the nucleus of oxygen-16. The masses of the atoms ^{16}O and ^{15}O are 15.994 915 u and 15.003 065 u, respectively.

34. [I] Calculate the binding energy of the last neutron in $^{13}_{6}$C. The atomic mass of a carbon-13 atom is 13.003 355 u.

35. [II] THIS PROBLEM DEAL WITH THE UNITS OF MASS-ENERGY. We want to prove that 1 u = 931.494 MeV/c². (a) How much mass (in kg) does 1 u equal? (b) Using Einstein's equation, how much energy (in joules) does that amount of mass correspond to? In other words, what does (1 u)c² equal in joules? (c) Convert that number of joules to eV giving your answer to four significant figures. (d) Finally, express that value in MeV.

36. [II] THIS PROBLEM IS ABOUT BINDING ENERGY. Consider the isotope neon-20. We want to study its binding energy and total mass. (a) How many nucleons are there in neon-20? (b) Using the Periodic Table, how many protons are in this nuclide? (c) How many neutrons are in the nuclide? (d) Using Table 30.2, what is the net mass (in atomic mass units) of all the individual protons? (e) What is the net mass (in atomic mass units) of all the individual neutrons? (f) What is the total mass (in atomic mass units) of all the separate neutrons and protons? (g) Given that this nuclide has a binding energy of 160.647 MeV, express that in atomic mass units. (h) Determine the mass of the neon-20 nucleus.

37. [II] What minimum energy is needed to remove a neutron from a $^{40}_{20}$Ca atom (of mass 39.962 59 u) converting it to $^{39}_{20}$Ca (with a mass of 38.970 69)?

38. [II] Determine the binding-energy-per-nucleon of helium to four figures. [*Hint: Watch out for the electrons.*]

39. [II] Iron-54 has an atomic mass of 53.939 613 u. Determine its nuclear mass defect in atomic mass units. Find the binding-energy-per-nucleon in MeV (four figures will do).

40. [II] With Problem 39 in mind, find the binding-energy-per-nucleon for iron-55 (atomic mass 54.938 296 u) and compare results.

41. [II] The neutrons in an isotope tend to pair and thereby bind strongly. For example, compute the minimum energy needed to remove a neutron from calcium-41 (atomic mass 40.962 278 u) as compared to calcium-42 (atomic mass 41.958 618 u). The former has 21 neutrons, the latter 22. Calcium-40 has an atomic mass of 39.962 591 u.

42. [II] What is the minimum amount of energy necessary to remove a proton from the nucleus of a $^{42}_{20}$Ca atom, thereby converting it into a $^{41}_{19}$K atom? The former has a mass of 41.958 618 u, the latter 40.961 825 u, and a hydrogen atom has a mass of 1.007 825 u.

43. [II] Uranium-232 (with an atomic mass of 232.037 13 u) is radioactive. It emits an α-particle and transforms into thorium-228 (with an atomic mass of 228.028 73 u). Determine the KE available to the decay products. Assuming the U-232 was at rest, this KE must be shared by the Th-228 and the alpha so as to conserve momentum. [*Hint: The above are the masses of the atoms.*]

SECTION 30.5: RADIOACTIVE DECAY

SECTION 30.7: GAMMA DECAY

44. [I] THIS PROBLEM WILL HELP US BETTER UNDERSTAND ALPHA DECAY. As we saw in Example 30.4, radium decays into the radioactive gas radon, $^{222}_{86}$Rn. We want to study the subsequent decay of radon. (a) How many particles make up an alpha particle, and what are they? (b) Given that radon emits an alpha particle, how many protons would remain in the daughter nuclide after radon decays? (c) Into what element does radon transform in the process? (d) How many particles, neutrons and protons, are in the daughter nuclide? (e) Write out the decay equation.

45. [I] THIS PROBLEM WILL HELP US BETTER UNDERSTAND BETA DECAY. One of the most important isotopes used for radioactive dating is carbon-14. It decays by beta emission, and we want to study the process. (a) How many nucleons are there in carbon-14? (b) Of those how many are protons? (c) After beta emission occurs, how many nucleons remain in the daughter nuclide? (d) How many protons are in the daughter nuclide? (e) Identify the daugh-

ter nuclide. (f) Write out the decay equation and don't forget the antineutrino.

46. [I] Lead-214 undergoes electron-β decay. What is the daughter nuclide? Write out that decay reaction. Check your answer with Fig. 30.10.

> SOLUTION: According to the Periodic Table, lead has 82 protons. Losing a negative charge yields a daughter with one more proton, namely, 83 and that's bismuth. The total number of nucleons is still 214, so we end up with bismuth-214.

$$^{214}_{82}\text{Pb} \rightarrow ^{214}_{83}\text{Bi} + ^{0}_{-1}\text{e} + \bar{\nu}_e$$

47. [I] THIS PROBLEM WILL HELP US BETTER UNDERSTAND POSITRON DECAY. Bismuth-214 undergoes electron-β decay. (a) How many nucleons are there in bismuth-214? (b) Using the Periodic Table, how many protons are in the nucleus? (c) After beta emission occurs, how many nucleons remain in the daughter nuclide? (d) How many protons are in the daughter nuclide? (e) Identify the daughter nuclide. Check your answer with Fig. 30.10. (f) Write out the decay equation and don't forget the antineutrino.

48. [I] Rubidium-87 undergoes electron-β decay. What is the daughter nuclide? Write out that decay reaction.

49. [I] Samarium-147 decays via alpha emission. What is the resulting daughter nuclide? Write out that decay reaction.

50. [I] THIS PROBLEM WILL HELP US BETTER UNDERSTAND ALPHA DECAY. Astatine-218 undergoes alpha decay. (a) How many nucleons are there in ^{218}At? (b) Of those how many are protons? (c) How many are neutrons? (d) After alpha emission occurs, how many nucleons remain in the daughter nuclide? (e) How many protons are in the daughter nuclide? (f) Identify the daughter nuclide. (g) Write out the decay equation.

51. [I] THIS PROBLEM WILL HELP US BETTER UNDERSTAND POSITRON DECAY. Neon-19 undergoes positron decay. (a) How many nucleons are there in neon-19? (b) Of those how many are protons? (c) After positron emission occurs, how many nucleons remain in the daughter nuclide? (d) How many protons are in the daughter nuclide? (e) Identify the daughter nuclide. (f) Write out the decay equation and don't forget the neutrino.

52. [I] THIS PROBLEM WILL HELP US BETTER UNDERSTAND GAMMA DECAY. One of the most useful radioisotopes is technetium-99m (see Fig. 30.16). (a) How many nucleons are in technetium-99m? (b) Of those how many are protons? (c) Technetium-99m undergoes gamma decay. After gamma emission occurs, how many nucleons remain in the daughter nuclide? (d) How many protons are there in the daughter nuclide? (e) Identify the daughter nuclide. (f) Write out the decay equation.

53. [II] Polonium-210 is radioactive. Confirm that it is energetically possible for it to decay via alpha emission in the following way

$$^{210}_{84}\text{Po} \rightarrow ^{206}_{82}\text{Pb} + \alpha$$

What is the net kinetic energy of the decay particles, assuming the polonium nucleus to be at rest? The masses of the polonium and lead atoms are 209.982 848 u and 205.974 440 u, respectively. [*Hint: Watch out for the electrons.*]

54. [II] Can lead-206 (of atomic mass 205.974 440 u) spontaneously decay via alpha emission? Explain your answer completely. The atomic mass of mercury-202 is 201.970 617 u.

SECTION 30.8: HALF-LIFE

SECTION 30.9: INDUCED RADIOACTIVITY

SECTION 30.10: FISSION AND FUSION

55. [I] THIS PROBLEM EXAMINES THE DECAY RATE. Suppose you have a small sample of radium that has a decay rate of 10 mCi. We want to express that in SI units. (a) How many disintegration per second would take place if we had 1.0 Ci? (b) How many decays per second take place with our sample? (c) What is the activity of our sample in SI units?

56. [I] Iodine-131 has a half-life of 8.0 days. If we start with a 1.00-g sample, how much iodine will remain after 48 days?

> **SOLUTION:** How much (m) of the original $m_0 = 1.00$-g sample remains after $48/8 = 6.0$ half-lives?
>
> $$m = (\tfrac{1}{2})^n m_0 = (\tfrac{1}{2})^6 (1.00 \text{ g}) = 0.016 \text{ g}$$

57. [I] The isotope $^{226}_{88}$Ra has a half-life of 1622 years; what is its decay constant?

58. [I] What is the equivalent activity of one microcurie (μCi) in becquerels?

59. [I] A cobalt atom is at the core of the vitamin B_{12} molecule. Accordingly, radioactive Co-60 has been used as a tracer to study the B_{12} absorption defect in pernicious anemia. Cobalt-60 has a half-life of 5.3 years. What is its decay constant in disintegrations per second?

60. [I] With Problem 58 in mind, given that the level of radioactive activity in the human body is naturally 10 nCi, how many nuclear disintegrations occur per second inside you?

61. [I] A radioactive sample is studied for a period of 36 hours, during which time its decay rate decreases to one-eighth its original value. What is its half-life?

62. [I] In humans, iodine is readily taken up by the thyroid, which requires it in the making of the hormone thyroxine. To study metabolism and treat thyroid disease, the isotope $^{131}_{53}$I is often introduced into the body. Given that it has a half-life of eight days, what fraction of the original activity remains after eight weeks? Assume none is excreted. Take a look at Problem 64.

63. [I] Radium-226 emits an α-particle and decays into radon-222 with a half-life of 1620 years. What is its decay constant in decays/year?

64. [I] Radionuclides that have entered the human body are sometimes biologically excreted in an approximately exponential way. The effective half-life is then

$$1/t_{1/2}(\text{eff}) = 1/t_{1/2}(\text{bio}) + 1/t_{1/2}$$

Given that iodine-131 has a half-life of 8.0 d and a biological half-life of 138 d, what is its effective half-life? How much of it will remain after 8.0 weeks? (See Problem 62.)

65. [I] Given that radon has a half-life of 3.8 days, what is its mean life?

66. [I] Radon undergoes α decay with a half-life of 3.8 days. Given some original amount, how much of it will remain after 11.4 days?

67. [I] A radioactive sample has a decay constant of 7.69×10^{-3} decays/s. What is its average lifetime and its half-life?

68. [I] A sample of radioactive oxygen-15 has a half-life of 2.1 minutes and a decay rate of 5.5×10^{-3} decays/s. By how much will the amount of this isotope be diminished after 4.0 s?

69. [I] Protactinium-234 has a half-life of 1.18 minutes. If 1.00 mg of it is freshly created, what fraction will be left in 1.00 h?

70. [I] When boron is struck by an alpha, it is transmuted into nitrogen-13, the reaction being

$$^{10}_{5}\text{B} + ^{4}_{2}\text{He} \rightarrow ^{13}_{7}\text{N} + ^{1}_{0}\text{n}$$

which then decays (with a half-life of ≈ 10 minutes) via positron emission. Write out that decay reaction.

71. [I] Irène and Pierre Joliot-Curie bombarded a foil of aluminum with α-particles, transmuting some of the nuclei and producing neutrons in the process. Write out the transformation formula.

72. [I] Calculate the maximum kinetic energy available (in MeV) to the electron and electron-antineutrino created by the decay of a neutron.

73. [I] Consider the reaction in which lithium-7 is struck by a proton; that is,

$$\text{p} + ^{7}_{3}\text{Li} \rightarrow \alpha + \alpha$$

Compute the difference in mass before and after the collision. How much KE will the alphas have (in excess of the KE the proton delivered)? The masses of the ^{1}H, ^{7}Li, and ^{4}He *atoms* are 1.007 825 u, 7.016 003 u, and 4.002 603 u, respectively.

74. [I] THIS PROBLEM DEALS WITH INDUCED RADIOACTIVITY. As described in Section 30.10, fissionable Pu-239 was first produced by causing U-238 to absorb a neutron. (a) Write the equation for that reaction. (b) Show that the resulting nuclide is U-239? (c) The U-239 subsequently decayed via beta emission. Show that the resulting nuclide is Np-239 and write out the equation describing the process. What metal is this? (d) The Np-239 subsequently decayed via beta emission. Show that the resulting nuclide is Pu-239 and write out the equation describing the process. What metal is this?

75. [II] By bombarding beryllium with alphas, we get the reaction

$$^{9}_{4}\text{Be} + ^{4}_{2}\alpha \rightarrow ^{12}_{6}\text{C} + ^{1}_{0}\text{n} + Q$$

The mass of the Be atom is 9.012 182 u, and the mass of the He atom is 4.002 603 u. Please compute the value of Q in excess of the alpha's incoming KE.

76. [II] Given the reaction

$$^{6}_{3}\text{Li} + \text{p} \rightarrow \alpha + ^{3}_{2}\text{He} + Q$$

if $Q = 4.018\ 5$ MeV in excess of the proton's KE and the lithium is at rest, find the mass of the helium-3 atom.

77. [II] Determine the value of Q, which is required in order to initiate the reaction

$$\alpha + ^{14}\text{N} \rightarrow ^{17}\text{O} + \text{p}$$

This energy must come by way of the KE of the alpha, which actually has to be a bit higher than Q because some KE must be given to the other particles if momentum is to be conserved.

78. [II] What is the activity of 1.00 g of the radium isotope $^{226}_{88}$Ra, which has a half-life of 1622 years?

79. [II] Calculate the activity of one milligram of radon-222, which has a decay constant of 2.1×10^{-6} decays/s.

80. [II] Uranium-238 has a half-life of 4.5×10^9 years. The Earth was formed about 5×10^9 years ago. What fraction of the ^{238}U present then is still around today?

81. [II] Radioactive phosphorus-32 has been used to study bone metabolism and for the treatment of blood diseases. What is the activity of a sample of 5×10^{16} atoms if it has a half-life of 14.3 days?

82. [II] Carbon-11 is radioactive with a half-life of 20.4 minutes. If a sample initially has 1.0×10^{17} carbon-11 atoms in it, what is its activity after 10 minutes?

83. [II] Oxygen-15 has a half-life of 2.1 minutes. How many ^{15}O atoms are present in a source with an activity of 4.1 mCi? (See Problem 58.)

84. [II] Consider a sample of freshly cut wood that has been reduced to pure carbon. What is the activity per gram resulting from the decay of ^{14}C (half-life of 5730 y) in this material?

85. [II] Determine the maximum permissible concentration in the air of tritium in the workplace, given that the maximum permissible annual intake of tritium is 444 MBq. A typical person inhales ≈ 10 m^3 of air per working day. Use a 50 work-week year.

86. [II] A 50-g chunk of charcoal is found in the buried remains of an ancient city destroyed by invaders. The carbon-14 activity of the sample is 200 decays/min. Roughly when was the city destroyed (more accurately, when was the tree felled from which the charcoal came)? Refer back to Problem 84.

87. [II] Determine the amount of energy liberated in the fusion reaction

$$4p \rightarrow \alpha + 2\,{}_{+1}^{0}e + 2\nu_e + \gamma$$

The masses of the proton, electron, and alpha particle are 1.007 276 u, 0.000 548 580 u, and 4.001 506 u, respectively.

88. [II] Radioisotopes are sometimes used to produce electricity to power such things as interplanetary probes and pacemakers. Consider the radioactive nuclide Po-210, which has a half-life of ≈ 140 d, emitting 5.30-MeV alphas. If all the KE of the alphas is absorbed in a radioisotope thermal generator (known as an RTG, in the trade), what's the average thermal power (in watts) developed per gram during the first 140 days of operation?

89. [III] Show that $t_{1/2} = (\ln 2)/\lambda$.

90. [III] A radioactive sample shows a decrease by a factor of 10 in activity over a period of 5.0 minutes. What is its decay constant?

Chapter 31
High-Energy Physics

*A*s the twentieth century unfolded, the number of observed subnuclear particles gradually increased. The electron, proton, and neutron were joined by the muon, pion, positron, and so forth, until by the 1970s, several hundred distinct particles were identified. The obvious problem was to determine which of these tiny specks of matter was *elementary* in the sense of being a single homogeneous entity having no internal structure. It now seems certain that most of these subnuclear particles are clusters of two or more fundamental entities called *quarks* (p. 1104). These are the primary building blocks that fuse together via the strong force to create neutrons, protons, pions, and so forth. The complementary group of elementary entities, the leptons, do not experience the strong force and do not combine to form composite subnuclear particles. They exist only individually, and, to within the limits of our most powerful techniques, each (the electron, neutrino, and so on) appears to be a distinct structureless object.

Table 31.1 provides a representative sample of the more long-lived subnuclear particles. They are listed primarily according to their modes of interaction. All of them experience or *couple to* the gravitational interaction, which nonetheless plays a negligible role in particle physics because it's so feeble. The more dominant influences arise from the electromagnetic, strong, and weak interactions.

Elementary Particles

Our low-energy world of trees and rocks and politicians is made up almost entirely of three fundamental material particle types: the electron (a lepton), the u-quark, and the d-quark. Along with the neutrino, these constitute the *first generation* of matter, the ordinary stuff still remaining in the cool Universe as it presently exists. At higher temperatures and greater energies, we can produce more exotic forms of matter—the *second* and *third generations* (p. 1107). These highly unstable particles once existed in abundance in the blazing early moments of the primordial Universe (p. 1112).

STUDY GUIDE

This chapter treats our contemporary understanding of the very basic nature of matter. We now believe that all the stuff of the Universe from ants to galaxies is composed of countless combinations of identical particles, all of which are themselves formed of only a handful of fundamental entities. The theoretical picture that has evolved is called the Standard Model (p. 1114), and although it is wonderful in its ability to provide understanding and predictive power, it is surely not the whole truth. Still, there is good reason to believe that you and I, and everything else, are basically exquisitely organized, ever changing, transient clouds of timeless quarks (p. 1104) and leptons (p. 1095), abuzz with gluons (p. 1108), gravitons, and photons.

Table 31.1
Some Fairly Long-Lived Elementary Particles*

Category	Name	Particle	Antiparticle†	Mass (MeV/c²)	B	L_e	L_μ	L_τ	S	Lifetime (s)
LEPTONS	Electron	e^-	e^+	0.51100	0	±1	0	0	0	Stable
	Neutrino (e)	ν_e	$\bar\nu_e$	0 ($<14 \times 10^{-6}$)	0	±1	0	0	0	Stable
	Muon	μ^-	μ^+	105.658	0	0	±1	0	0	2.197×10^{-6}
	Neutrino (μ)	ν_μ	$\bar\nu_\mu$	0 (<0.25)	0	0	±1	0	0	Stable
	Tau	τ^-	τ^+	1784.2 ± 3.2	0	0	0	±1	0	$(2.9 \pm 1.2) \times 10^{-13}$
	Neutrino (τ)	ν_τ	$\bar\nu_\tau$	0 (<35)	0	0	0	±1	0	Stable
HADRONS										
Mesons	Pion	π^+	π^-	139.570	0	0	0	0	0	2.60×10^{-8}
		π^0	π^0	134.976	0	0	0	0	0	0.84×10^{-16}
	Kaon	K^+	K^-	493.677	0	0	0	0	±1	1.24×10^{-8}
		K^0	$\bar K^0$	497.677	0	0	0	0	±1	0.9×10^{-10}
Baryons	Proton	p	$\bar p$	938.3	±1	0	0	0	0	Stable
	Neutron	n	$\bar n$	939.6	±1	0	0	0	0	886.7
	Lambda	Λ^0	$\bar\Lambda^0$	1115.7	±1	0	0	0	∓1	2.6×10^{-10}
	Sigma	Σ^+	$\bar\Sigma^-$	1189.4	±1	0	0	0	∓1	0.80×10^{-10}
		Σ^0	$\bar\Sigma^0$	1192.6	±1	0	0	0	∓1	7.4×10^{-20}
		Σ^-	$\bar\Sigma^+$	1197.4	±1	0	0	0	∓1	1.5×10^{-10}
	Xi	Ξ^0	$\bar\Xi^0$	1315	±1	0	0	0	∓2	2.9×10^{-10}
		Ξ^-	Ξ^+	1321	±1	0	0	0	∓2	1.64×10^{-10}
	Omega	Ω^-	Ω^+	1672	±1	0	0	0	∓3	0.82×10^{-10}

*Wherever there are ± or ∓ symbols the top sign is for the particle and the bottom one is for the antiparticle.
†An antiparticle is denoted by the same symbol as the particle but with the opposite charge. When there is any ambiguity a bar is also placed over the symbol.

Today, we can just begin to recreate some of those incredibly violent conditions. In studying the subnuclear domain, we gain a view, however darkly, of the first moments of Creation.

31.1 Leptons

Leptons are the group of fermionic (half odd integer spin) *fundamental particles* that do not experience the strong force.

The first group of twelve particles in Table 31.1 constitutes the **leptons** and **antileptons** (from the Greek *leptos*, meaning "slight"). These are the particles that couple to the weak force, and if they are electrically charged, they react to the electromagnetic force as well. *All are immune to the strong force.* The most familiar and the lightest of the charged leptons is the electron. The muon (μ^-), found among the products of cosmic-ray bombardment (1936), was the first unstable subnuclear particle to be discovered. Muons are commonplace, rushing toward the surface of the planet in a flow of roughly 1 per cm² per minute. Even as you read these lines, muons are more or less harmlessly streaming downward through your body like a gentle penetrating cosmic rain. The muon decays (in about 2.2 μs via the weak interaction) into an electron and two neutrinos:

$$\mu^- \rightarrow e^- + \nu_\mu + \bar\nu_e$$

Otherwise, it behaves in every way like an overweight electron. The same can be said about the

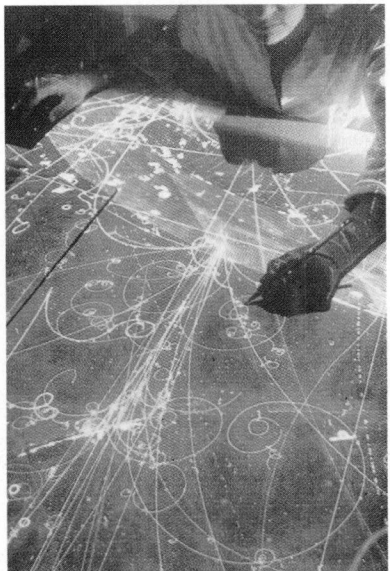
Subnuclear particle tracks left in a bubble chamber. (See Discussion Question 7.)

Table 31.2

Leptons*

Flavor	Charge	Mass (MeV/c^2)
e	-1	0.511
ν_e	0	$<14 \times 10^{-6}$
μ	-1	105.7
ν_μ	0	<0.25
τ	-1	1777
ν_τ	0	<20

Spin = 1/2

*Each particle has an antiparticle mirror image with the same mass and spin, but with opposite charge.

$$\begin{bmatrix} e \\ \nu_e \end{bmatrix} \quad \begin{bmatrix} \mu \\ \nu_\mu \end{bmatrix} \quad \begin{bmatrix} \tau \\ \nu_\tau \end{bmatrix}$$

tau (τ), although, since it was only discovered in the 1970s, we have less experience with it. The tau is almost twice as massive as the proton; nonetheless, it is believed to be truly elementary. Both muons and taus play no known significant role in the scheme of things, and their very existence is puzzling in a Universe that seems to have a preference for economy.

At our current limits of observation, leptons appear to be structureless, pointlike entities. All are spin-$\frac{1}{2}$ fermions. Each lepton having substantial mass is created in conjunction with its own variety of essentially massless neutrino. There are electron-neutrinos, muon-neutrinos, and tau-neutrinos. In total, there are probably more neutrinos than anything else—they outnumber electrons and protons by a factor of perhaps a thousand million. The Universe is awash with neutrinos, some of which are harmlessly passing through your body even now. Each lepton comes in both particle and antiparticle varieties, making a total of six electrically charged and six neutral leptons. These form three—electron, muon, and tau—particle-neutrino couplets: e and ν_e; μ and ν_μ; and τ and ν_τ (Table 31.2).

The systematic patterns of production and decay of leptons suggested that an underlying conservation law might be at work. Remember the special feature that leptons are created in couplets. It is believed that **lepton number** (L) is conserved in all presently attainable processes involving any members of this family. Each of the three subsets of leptons (electron-type, muon-type, and tau-type) conserves its own lepton number (L_e, L_μ, L_τ). In each subset, the particle (for example, e^-) has a lepton number (L_e) of $+1$, whereas the antiparticle (e^+) has a lepton number of -1, and the same is true of the associated neutrino and antineutrino.

The phenomenon of *electron-positron pair production* (p. 1048)

$$\gamma \rightarrow e^- + e^+$$

for which L_e: $0 \rightarrow (+1) + (-1)$

conserves lepton number as does the inverse phenomenon of *electron-positron annihilation*

$$e^- + e^+ \rightarrow \gamma + \gamma$$

for which L_e: $(+1) + (-1) \rightarrow 0 + 0$

The decay reaction

$$\mu^- \rightarrow e^- + \nu_\mu + \overline{\nu}_e$$

for which L_e: $0 \rightarrow (+1) + 0 + (-1)$

and L_μ: $1 \rightarrow 0 + (+1) + 0$

separately conserves both electron and muon-lepton numbers.

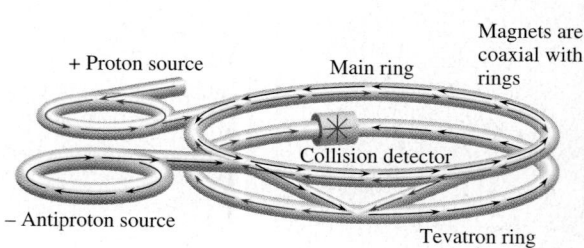

The world's most powerful particle accelerator at the Fermi National Accelerator Laboratory. When it actually has the funds to operate, it accelerates protons in one ring and antiprotons in the other. These beams are then made to collide with a total energy of $\approx$2000 GeV. Superconducting magnets force the particles to stay within the rings.

31.2 Hadrons

Hadrons are the strongly interacting composite particles. They are all fairly massive; ergo, the name, which derives from the Greek *hadros*, meaning "bulky." A succession of hadrons was discovered between 1947 and 1954 in cosmic-ray studies and at the Brookhaven Cosmotron in New York. That machine, a 3-GeV accelerator, made it possible for the first time to bring to bear the needed amounts of energy to create heavy particles via $E_0 = mc^2$. These conjured hadrons were christened the kaon (K: $\approx$500 MeV/c^2), the lambda (Λ: $\approx$1100 MeV/c^2), the sigma (Σ: $\approx$1200 MeV/c^2), and the xi (Ξ: $\approx$1300 MeV/c^2). All were observed via photographs of the several-centimeter-long tracks they left in visual detectors. These trails, marking their passage between creation and decay, corresponded to lifetimes of around 0.1 ns to 10 ns. In contrast to the twelve leptons, there are hundreds of hadrons, which increasingly suggested that hadrons might not be elementary. Every hadron couples to the strong, weak, and gravitational interactions; and some also couple to the electromagnetic interaction. All of them, with the possible exception of the proton, decay—*some rapidly, via the strong interaction; some less rapidly, via the electromagnetic interaction; and others still more slowly, via the weak interaction*. Hadrons form two distinct subgroups (defined by their spins): *mesons* and *baryons*.

Mesons

Mesons are bosonic (integer spin) hadrons composed of a quark and an antiquark (p. 1106).

The Greek root *meso* means "middle," and it was applied in 1939 to particles whose mass was between that of the electron and proton. Today, the term **meson** refers to hadrons that are bosons (they have spins of 0, 1, 2, ...). Table 31.1 lists only the several pions (or π-mesons) and kaons (or K-mesons). Although there are dozens of other known mesons, these serve our purposes for the time being.

The three members of the *pion* family (π^+, π^-, π^0) are the lightest of the mesons. Proton-proton collisions carrying enough energy create pions:

$$p + p \rightarrow p + p + \pi^+ + \pi^-$$

$$p + p \rightarrow p + n + \pi^+$$

and pion-proton collisions produce more of them, for example,

$$p + \pi^- \rightarrow p + \pi^- + \pi^0$$

$$p + \pi^- \rightarrow n + \pi^0$$

The purple and green trails that form a V reveal the spontaneous creation of a pair of particles. But these tracks are nearly straight and must have been created by high-speed particles far more massive than the electron and positron. The green track curves clockwise, indicating a negative charge. Unlike an electron, its trail ends in a burst of fragments (all shown in orange). Computer analysis reveals that this negative particle, which had a momentum of $\approx$15.5 GeV/c, was either a negative pion or an antiproton. Apparently, a neutral strange particle (a K^0 or Λ^0) entering from the bottom of the picture, and leaving no track, decayed into a $\pm$ partical pair thereby creating the V. Whether this was two pions or a proton-antiproton pair, we cannot know, since the results of the analysis were ambiguous.

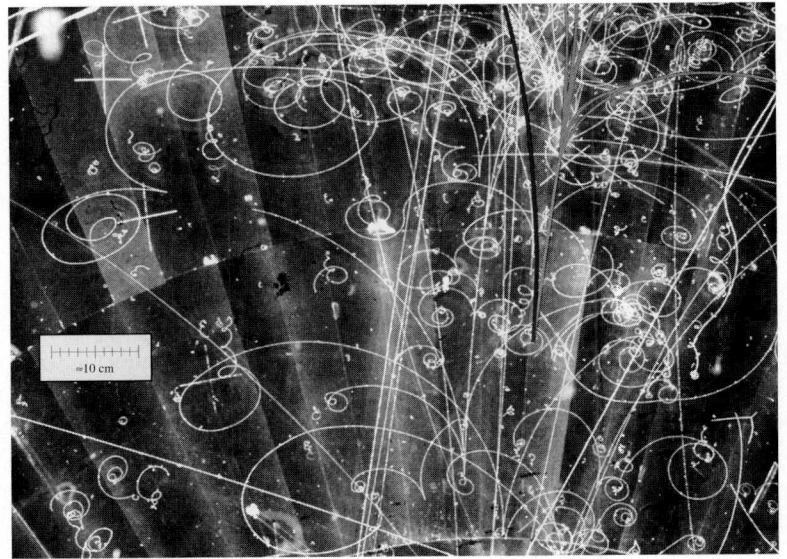

Each process conserves angular momentum (i.e., spin; **the pion has zero spin**), electromagnetic charge, and nucleon number (i.e., the number of nucleons). Whereas π^+ and π^- can be thought of as particle and antiparticle, neutral π^0 is its own antiparticle. The π^0 has a fleeting lifetime (0.87×10^{-16} s) usually decaying (via the electromagnetic interaction) into two photons

$$\pi^0 \rightarrow \gamma + \gamma$$

Notice that there is *no* conservation of meson number. The most common decay route for the charged pions is fixed by the fact that there are no hadrons lighter than themselves; they must decay into leptons (muons) via the weak interaction; thus

$$\pi^+ \rightarrow \mu^+ + \nu_\mu \qquad\qquad \pi^- \rightarrow \mu^- + \overline{\nu}_\mu$$

Pions play a very special role in nature: they make a major contribution to mediating the nuclear force between nucleons (p. 1062).

The kaon (K-meson) is produced in collisions via the strong interaction. There are four zero-spin kaons: K^+, K^-, K^0, and its antiparticle $\overline{K}^0$. They are unstable and undergo a number of different decay reactions. For example, each can decay into a pair of pions:

$$K^0 \rightarrow \pi^+ + \pi^-$$

$$\overline{K}^0 \rightarrow \pi^0 + \pi^0$$

$$K^+ \rightarrow \pi^+ + \pi^0$$

Two surprising things were observed here and with a group of other newly discovered particles as well. First, the reaction products were always created in twos, even though there was no known reason for it. For example, every time a K^0 was produced, a Λ^0 was also produced. Second, instead of a decay time typical of the strong interaction ($\approx 10^{-23}$ s), these transformations sauntered along with lifetimes from 0.1 ns to 10 ns. It was as if the particles, born out of the strong force, died by the weak force.

That mysterious behavior was denoted by calling these objects **strange particles**. To explain what was happening, Murray Gell-Mann, in the 1950s, introduced a new quantum number that reflects a new conserved quantity. Like electromagnetic charge or spin, he proposed that these particles uniquely possess a quality of *strangeness* (as if they carried a strangeness charge), and he assigned a numerical value of it (S) to each hadron, according to its observed behavior. *The strong and electromagnetic interactions both conserve strangeness. The weak interaction does not, and the strangeness may change, but by no more than one unit.* Pions and nucleons, which are not at all strange, have a strangeness of 0. The kaon (K^+ and K^0) has a strangeness of $+1$, whereas the strange baryon known as lambda has a strangeness of -1.

The strong interaction could create a pair of strange particles out of more ordinary matter (with zero strangeness), provided the resulting pair had canceling values of strangeness. Thus, a pion and a proton (net strangeness, 0) can interact strongly to produce a K-zero and a lambda-zero of strangeness $+1$ and -1, respectively: $\pi^- + p \rightarrow K^0 + \Lambda^0$. The logic forbids strange particles (like K^0), once created, from decaying to lighter nonstrange particles ($K^0 \rightarrow \pi^+ + \pi^-$) by the rapid route of the strong interaction. Strange particles can accomplish such decay only via the weak interaction and only slowly. This kind of ad hoc conceptual fine-tuning reveals the way physics develops: in the absence of a proven theoretical formalism, phenomenological relationships are derived, or guessed at, from observed patterns. The idea, however well it worked then, really makes sense only when viewed from the perspective of the quark theory (p. 1104). Strangeness simply depends on the number of strange and antistrange quarks composing the particles.

STUDY GUIDE

The weak force gives rise to beta decay and to all particle transformations in which strangeness changes.

Baryons are fermionic (half odd-integer spin) hadrons composed of three quarks (p. 1106). Antibaryons are composed of three antiquarks. All baryons have a mass equal to or greater than that of the proton.

Baryons

→ **Baryons** derive their name from the Greek *barys*, meaning "heavy." They are the heavy hadrons, but more importantly, they are all fermions. Neutrons and protons, the nucleons, are the most well-known baryons. Table 31.1 displays a representative selection of baryons. Generalizing from the concept of nucleon number, it was proposed that the number of baryons and antibaryons in a reaction is conserved, and so they are assigned a *baryon number* (B). For baryons $B = +1$, whereas antibaryons have $B = -1$, and all nonbaryons have $B = 0$ (see Discussion Question 11). This system has the effect of "explaining" why the proton does not decay into still lighter particles, even though such a decay is not otherwise in violation of any known principle. In all particle reactions (at the energies we can attain), the total baryon number must, it is assumed, be conserved. Thus, the strong reaction

$$\pi^- + p \rightarrow K^0 + \Lambda^0$$

conserves electromagnetic charge (zero on both sides), strangeness (zero on both sides), angular momentum ($\frac{1}{2}$ on both sides), and baryon number ($+1$ on both sides) and, therefore, can and does take place. The heavier strange baryons (for example, Λ, Σ, Ξ, Ω) decay to lighter baryons and ultimately to the proton.

Whereas there are conservation laws for leptons and baryons, there are no restrictions on the number of photons or mesons entering and leaving an encounter. These particles can appear and vanish at will. In fact, the photon and the pion have $L = S = B = 0$. In the case of the photon and the neutral pion, each is its own antiparticle. With all of these quantum numbers equaling zero, including the electromagnetic charge, it is impossible to distinguish particle from antiparticle. As we'll see presently, photons and pions play a very special role in Quantum Field Theory.

PARTICLES & FIELDS

Prior to the introduction of Quantum Mechanics, particles and fields were considered interrelated though distinct entities; particles possessed intrinsic features (e.g., mass and/or electromagnetic charge) that gave rise to external fields (e.g., gravitational and/or electromagnetic). Force fields emanated from particles and filled the surrounding space. They carried energy and were, in a sense, real continuous media that interconnected all interacting particles and mediated their interactions. Particles were composed of matter, fields were composed of energy. The force field was the nineteenth century's answer to the age-old mystery of action-at-a-distance. Still, particles that do not react to any force fields are unobservable and physically meaningless. Force fields that do not act upon any particles are equally without significance. The ideas of particle and field take meaning from their interrelationship.

Example 31.1 **[I]** The Ξ^- baryon, which has a relatively long lifetime of 1.64×10^{-10} s, most often decays in the following way:

$$\Xi^- \rightarrow \Lambda^0 + \pi^-$$

Determine what happens, as a result of this transformation, to the baryon number, charge, lepton number, and strangeness. What can you conclude?

Solution This problem deals with several elementary-particle quantum numbers, all of which are tabulated. (1) TRANSLATION —Knowing a particular particle transformation, determine the quantum numbers before and after. (2) GIVEN: $\Xi^- \rightarrow \Lambda^0 + \pi^- \rightarrow \Lambda^0 + \pi^-$. FIND: The values of B, Q, L, and S, before and after. (3) PROBLEM TYPE—Particle physics/conservation. (4) PROCEDURE—We have to get the quantum numbers we need from Table 31.1. (5) CALCULATION—Before the decay the baryon number for Ξ^- was $B = +1$. After the decay Λ^0 has a

value of $B = +1$ and π^- has a value of $B = 0$ and so

$$\Xi^- \rightarrow \Lambda^0 + \pi^-$$

B: $\qquad (+1) \rightarrow (+1) + (0)$

and baryon number is conserved. As for the charge

Q: $\qquad (-1) \rightarrow (0) + (-1)$

it too is conserved. The lepton numbers are all zero since none of the particles are leptons. Before the decay the strangeness number for Ξ^- was $S = -2$. After the decay Λ^0 has a value of $S = -1$ and π^- has a value of $S = 0$ and so

S: $\qquad (-2) \rightarrow (-1) + (0)$

The strangeness has changed by $+1$, which tells us that the decay was governed by the weak interaction.

Quick Check: The long lifetime is consistent with the weak interaction.

Figure 31.1 A Feynman diagram showing the scattering of two electrons via the exchange of a virtual photon.

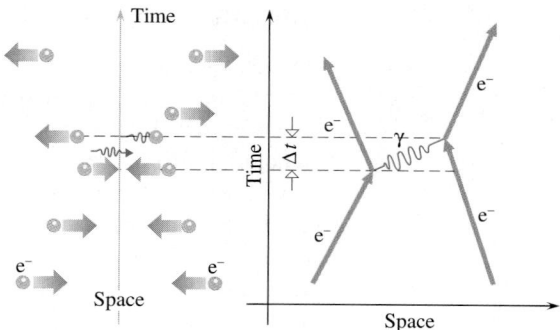

Quantum Field Theory

The concept of field began to change drastically with the introduction of Einstein's photon. The electromagnetic field does not, after all, have its energy continuously spread out in space. **The photon is the quantum of the electromagnetic field, and it carries the energy and momentum of the field.** The interaction of two charged particles corresponding to the electromagnetic force, transmitting energy and momentum from one to the other, must take place through the exchange of electromagnetic energy quanta—photons. Quantum Electrodynamics (QED), the theory of such interactions, was the first successful application of these ideas (p. 1047).

Figure 31.1 is a representation of two electrons as they undergo an elastic collision. It's called a **Feynman diagram**. Suppose each electron is initially traveling at the same speed. The electrons first approach and then recede from one another along a line in space that is projected upward in the increasing time direction in the diagram. The electron on the left emits a photon (the wiggly line), and for a moment (Δt) there are two electrons and one photon. The electron on the right absorbs the photon, and the interaction is momentarily over; other photons will subsequently go back and forth between the electrons. The average force is proportional to the rate of transfer of momentum mediated by the exchange of the photons. The measure of the probability of both emission and absorption of photons is the charge. The force must be proportional to both interacting charges (recall Coulomb's Law). Think of the repulsive interaction between two astronauts floating in space, throwing a ball back and forth (p. 218).

The electrons' exchange interaction is a quantum effect and cannot be visualized in classical terms. Still, repulsion by way of an exchange force can be thought of via the astronaut-ball analogy. However, attraction between an electron and a proton through exchange is unvisualizable, unless you resort to nonsense like the astronauts facing away from each other catching boomerangs backwards so that they're pushed together. Figure 31.2 depicts the attraction, but in a way that does not attempt to be faithful to the kinematics. Feynman diagrams are symbolic—they're computational devices in QED and are not concerned with accurately picturing the particle trajectories. Thus, the horizontal distances are of little significance, and the arrangement of Fig. 31.1 is often used for both attraction and repulsion. The important part is the interaction.

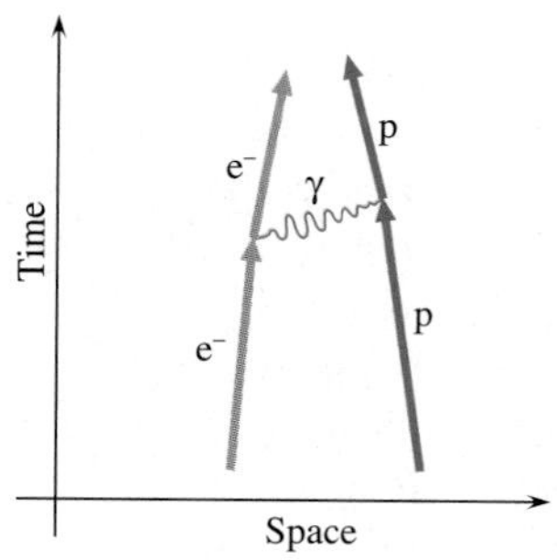

The collision in Fig. 31.1 is elastic; the energy of either electron is unchanged throughout, and yet during the time Δt, the system contains an additional amount of energy hf corresponding to the photon. For a time Δt, Conservation of Energy is seemingly violated! Can this situation be tolerated? One answer offered by modern physics is *yes*, provided it can never be observed. In other words, there is always some uncertainty (ΔE) in the measured value of the energy of a system. The Heisenberg Uncertainty Principle (p. 1044) tells us that

$$\Delta E \, \Delta t \geq \tfrac{1}{2}\hbar$$

Nonconservation of energy up to an amount ΔE will be hidden by the ever-present energy

Figure 31.2 A Feynman space-time diagram of the electron-proton interaction.

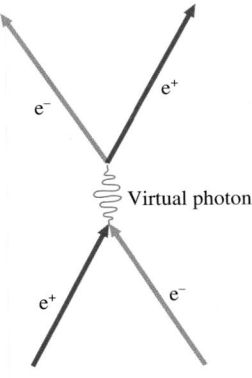

Figure 31.3 The annihilation and creation of electron-positron pairs.

uncertainty, provided the time available to make the observation (Δt) is restrictively small; namely,

$$\Delta t = \tfrac{1}{2} \frac{h}{\Delta E}$$

(If a moment of nonconservation is totally unobservable, is Conservation of Energy actually violated?) The energy uncertainty will exceed the photon energy hf if the photon exists for a time less than

$$\Delta t = \tfrac{1}{2} \frac{h}{hf} = \frac{1}{4\pi f}$$

This unobservable photon can travel a maximum distance of

$$R = c\Delta t = \frac{c}{4\pi f} \tag{31.1}$$

and since its frequency can be arbitrarily small, the range of the force transmitted by the massless photon is unlimited. Such unobservable exchange quanta are called **virtual photons**.

In contrast to *real* quanta, these *virtual* quanta are the messengers of the interaction. In the Feynman diagrams, they are the internal segments that begin and end within the figure. They effectively "tell" the material particles what's happening. A photon that is observable in the sense that it is detected by an eyeball or a Geiger counter is real enough. A photon that never leaves the region of interaction between charges (Fig. 31.3) and vanishes in the process of communicating the electromagnetic force—a photon that cannot therefore be seen by a detector (so that we never observe any violation of the basic conservation laws)—is a virtual photon. It need not obey the equation for the total energy of a real particle: $E^2 = m^2 c^4 + p^2 c^2$.

The distinction between real and virtual photons is not always obvious. At the extreme of short range and short existence, messenger photons can have tremendous energies (modern accelerators can generate virtual quanta of up to ≈ 100 GeV). At the other extreme, because the photon is massless and moves at the ultimate speed, there is no limitation on the time it can travel with its message (its clocks don't appreciate time anyway; for a discussion of the fact that photons are timeless click on **SPACE-TIME** under **FURTHER DISCUSSIONS** on the **CD**). The range of virtual photons is limitless, and the range of the electromagnetic interaction is limitless. If Δt is tiny, as it usually is in particle physics, there's no issue; if Δt is large, as it might be in astronomy, the imagery can get a little murky. The question of whether a particular photon that has traveled through space for a million years is real or virtual loses significance: ΔE is vanishingly small.

By the late 1920s, it was recognized that the known material particles (protons and electrons) could each be considered as the quantum of a specific particle field. From this perspective, there are electron fields, proton fields, and so on; the Universe is a set of quantized fields. Reality (a cup of coffee and a hamburger) is the totality of observable manifestations of field quanta. And from that perspective, the problem of the wave-particle duality no longer exists: matter is field.

When Fermi formulated his theory of the weak interaction in 1932, he founded it on the principles of QED. Not long after that, the Japanese physicist Hideki Yukawa proposed (1934) that the strong interaction was mediated by the exchange of a massive virtual boson. At the time, the only known particles were the electron, proton, and neutron. As we'll see, he was able to predict that the mass of this new messenger would be between m_e and m_n and thus it came to be called the meson. It has since been identified specifically as the π-meson, or pion. Strong (hadronic) interactions occur with equal strength between electrically positive, negative, and neutral particles; and so the idea was extended to include three exchange particles: one positive (π^+), one negative (π^-), and one neutral (π^0). It was proposed that virtual particles, emitted and absorbed, constantly fly back and forth between nucleons that, in turn, are transformed—the proton and neutron are two alternative states of the nucleon (Fig. 31.4). The *virtual meson* of mass m_π and rest energy $m_\pi c^2$ traveling at nearly the speed

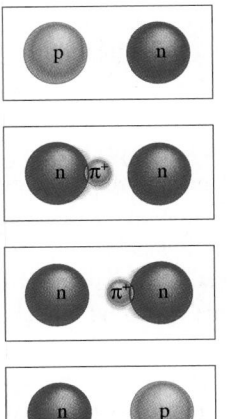

Figure 31.4 The attractive interaction between a neutron and proton through the exchange of a positive pion.

Figure 31.5 Nucleon-nucleon interactions through pion exchange.

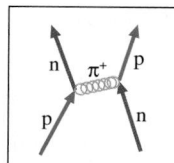 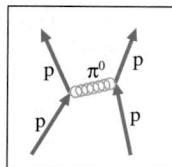 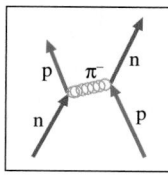

of light has a maximum range provided by Eq. (31.1), of

$$R \approx c\Delta t \approx c\left(\frac{\frac{1}{2}\hbar}{\Delta E}\right) \approx c\left(\frac{\frac{1}{2}\hbar}{m_\pi c^2}\right) \approx \frac{h}{4\pi m_\pi c}$$

Given the known range of the nuclear force (≈ 1 fm), the predicted particle mass comes out about $200m_e$ or ≈ 100 MeV/c^2. As a rule, **when enough energy is present such that $E_0 = mc^2$, a real particle corresponding to the virtual one can be created**. In 1934, the only source of sufficient energy was cosmic radiation. But not until 1947 was it possible to study high-energy cosmic-ray collisions using photographic emulsions. And only then was the Yukawa particle finally found: the **pion** was discovered (Fig. 31.5).

QUANTUM FIELDS

Contemporary Quantum Field Theory operates under several assumptions: (1) the essential reality is the set of quantum fields—nothing else exists; (2) these fields obey the rules of Special Relativity and Quantum Mechanics; (3) the intensity of a field at some location is a measure of the likelihood of finding an associated particle at that location; and (4) the field quanta interact as the fields interact. **All particle interactions (all forces) are mediated by field quanta.** We now believe that pion exchange between hadrons is actually a low-strength residual manifestation of a still more basic and more powerful interaction involving the exchange of gluons among quarks (p. 1108).

31.3 Gauge Theory

It is now widely accepted that all field theories that accurately portray nature must possess a particular type of mathematical structure known as **gauge symmetry**. The meaning of the term is subtle and depends on several subsidiary ideas. The discovery of the significance of gauge symmetry, one of the greatest of the century, was prompted by the realization that the General Theory of Relativity was gauge symmetric. That insight was underscored by the awareness that renormalized QED (p. 1047) happened also to be gauge symmetric. Today, it is believed that we have a mathematical test of the legitimacy of any new field theory; even better, we have a potent guide to formulating such a theory—that's an amazing thing to be able to say! Physicists maintain that the hidden structure of the Universe, whatever detailed form it takes, is gauge-invariant.

The philosophical insight that sprang from Relativity Theory was that all the laws of physics are the same for all observers—anyone anywhere must experience nature to behave in the same way. And yet a thousand scientists on a thousand planets across the Universe will surely have their own definitions and formulations, their own arbitrary sets of units, and arbitrary base levels for things such as zero speed, zero voltage, and zero potential energy. If one body of physical law rules the cosmos, different local constructs must be equivalent—the mindset of the scientist cannot alter the underlying physical reality. A correct theory must be expressible in different local languages in terms of local conceptions and still carry the same truth. All such theories should be translatable from one foreign mathematical language to any other.

The question of units isn't a problem: we can *transform* units from one system to another—an inch is as good as 2.54 centimeters. There are no completely natural units; all are dreamed up in the minds of scientists and cannot possibly affect physical law. Nor can the arbitrarily assigned base levels of relative concepts such as potential energy have any effect on law. There is no natural zero of voltage, and however it may be assigned, we can again transform from one level to another with no problem (that's why there are zero reset knobs on meters). This freedom to assign base levels of certain important physical quantities is a manifestation of an underlying symmetry.

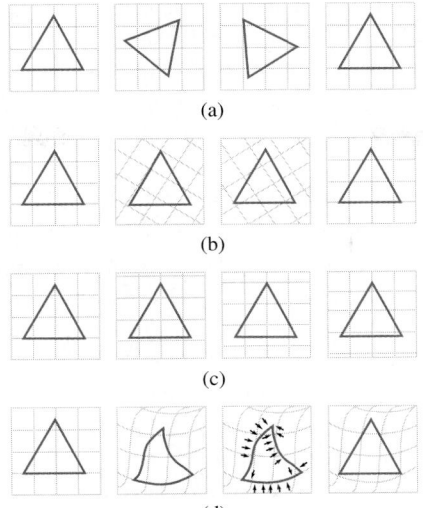

Figure 31.6 (a) The rotational symmetry of an equilateral triangle. (b) Rotating the entire coordinate system has no effect on the shape of the triangle. (c) Translating the entire coordinate system downward has no effect on the shape of the triangle. (d) Altering the coordinate system in a point-to-point, local way distorts the triangle. Even so, overlaying a force field, a gauge field, will reestablish the original configuration of the triangle.

In the broadest terms, a physical system possesses symmetry if something happens (whereupon there is a change of some kind) and no observable change in the final state of the system results. We've already talked about the geometrical symmetry of a snowflake (p. 198), which can be rotated through 60° and come up unaltered. Here, however, we are concerned with *abstract*, or *internal*, *symmetry* such that the alteration of some aspect of a system described by a set of equations does not change the solution of those equations. An equilateral triangle looks the same if we rotate it through 120° (Fig. 31.6*a*). The angle is continuously variable, and the symmetry operation is said to be *continuous*. By comparison, the triangle is also symmetric via a reflection through any one of its altitudes, but this is a jump, a discontinuous symmetry operation. Recall (p. 5) the discussion of **Noether's Principle**: *for every continuous symmetry, there is a corresponding conservation law and vice versa* (Fig. 31.7). Accordingly, we will only be concerned with continuous symmetries and indeed only with ones of a very restricted kind; namely, *local* rather than *global* symmetries. It is this last stricture that is the defining feature of gauge invariance.

Think of an object, a wire triangle, and imagine a mathematical coordinate grid extending in all directions against which the shape of the figure is to be described mathematically. Instead of an object, we might transform a physical quantity represented by a mathematical function (e.g., the potential), but the triangle is easier to visualize. Every point on the wire is transformed to a point on the coordinate grid. Rotating the grid (Fig. 31.6*b*)

Figure 31.7 For each of the conservation laws there is a corresponding symmetry: Conservation of Linear Momentum—Homogeneity of Space; Conservation of Angular Momentum—Isotropy of Space; Conservation of Energy—Homogeneity of Time; Conservation of Charge—Gauge Symmetry.

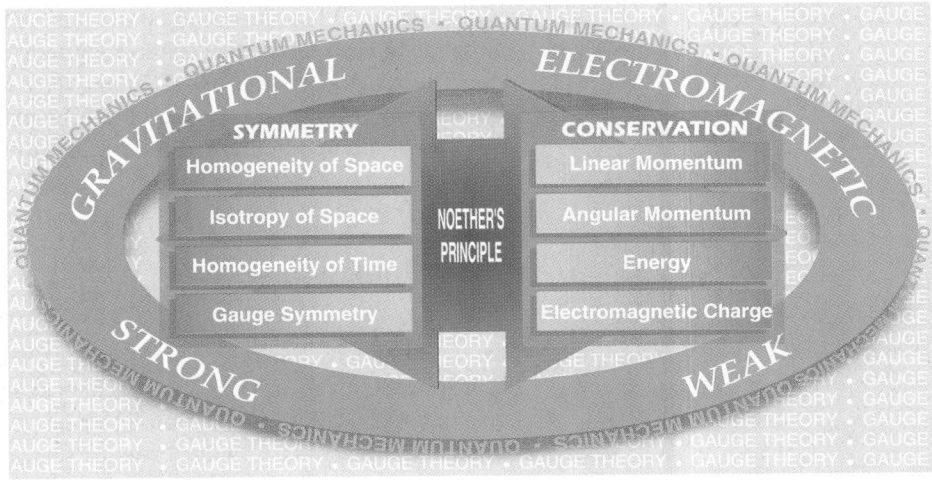

will not alter the lengths of the sides or the angles of the projected triangle, nor will translating the grid (Fig. 31.6c). These are global transformations; they happen everywhere identically and do not affect our system (the shape of the triangle). When a feature of the system is invariant under such transformation, we say the system has a ***global symmetry***.

To give the idea a more subtle application, consider yourself coasting around on a bicycle on top of a hill. It's a nice smooth mound, so the equipotential lines (the lines of equal height) are smooth concentric closed curves. If the entire hill is now transformed upward 100 m, you'll not notice a thing as you roll up and down its trails; assuming gravity is constant, the potential energy is independent of absolute height, and the forces on the bike only depend on the slopes of the surfaces. Similarly, the gravitational potential contour map, the potential field, is unchanged—this is a global symmetry.

By contrast, if the grid of Fig. 31.6d is altered independently from point to point so that it's twisted every which way, the shape of the triangle as formulated in this distorted space changes and the symmetry is lost. But suppose we bend the wire here and there, applying a force (call it a gauge force) at every point to counter the distortion and resymmetrize this local transformation. We then have imparted **local symmetry** in the presence of an added compensating **gauge force field**. Go back up the hill on that bike. With a local height transformation that changes the potential from point to point, the hillside will be distorted, covered with humps and potholes; the contour lines will be full of wiggles. But suppose we somehow introduce a resymmetrizing force field at every point that miraculously readjusts the net force field acting on you to its original configuration; we make a transformation of the potential field. You would then happily bike around as if nothing had changed. The compensating gauge field establishes a local symmetry, a gauge symmetry.

Alternatively, if we place a mapmaker at "every" point on the hill and ask that they all independently establish a potential for just the one location at which each resides, we'll again get contours that are a wiggly mess—the cartographers choose the zero altitudes wherever they like, making the height of each point quite unrelated to the next. Now, let each person make up a complete map (leaving out the numerical values of the contours): they'll all be identical—the hill is the hill. If we add the proper field, the gauge field, to the wiggly-mess field, the mess will be transformed into a correct contour map of the hill. *A correct theory must be independent of all the arbitrary definitions that went into it; it must be gauge-invariant.* It must be transformable from one measuring framework, or gauge, to another without affecting the conclusions of the formalism.

Classical electromagnetism is gauge-invariant; Conservation of Charge is associated with symmetry in the electromagnetic potentials (p. 582), and for that symmetry to be local, one must introduce a compensating gauge field that turns out to be Maxwell's familiar electromagnetic field. The conservation of electromagnetic charge inexorably leads to classical Electromagnetic Theory. This gauge invariance carries over into the quantized domain of QED. Here, the phase of the de Broglie wave of a charged particle is made locally symmetric. To accomplish that, a new compensating field interacting with the charged particles is required, and it turns out to be the photon field. **The photon is the quantum of the gauge field associated with electromagnetic charge.**

Both the gravitational and Coulomb force fields are long-range gauge fields. These are mediated by massless gauge quanta traveling at the speed of light (gravitons and photons). Moreover, the interaction is proportional to the source's quantum number, that is, its "charge" (mass, electromagnetic charge, etc.).

31.4 Quarks

The theoretical quandary produced by the proliferation of hadrons was addressed independently in 1963 by Murray Gell-Mann and George Zweig. They proposed that all known hadrons were composite particles, each made up of a cluster of two or more truly elementary particles. These, Gell-Mann lightheartedly called **quarks**, from the phrase "Three quarks for Muster Mark" in James Joyce's novel *Finnegans Wake*. How many quarks make

Table 31.3

Characteristics of Quarks and Antiquarks

Quarks

Flavor	Symbol	Charge	Spin	Baryon number	Strangeness	Charm	Bottomness	Topness
Up	u	$+\frac{2}{3}$e	$\frac{1}{2}h$	$\frac{1}{3}$	0	0	0	0
Down	d	$-\frac{1}{3}$e	$\frac{1}{2}h$	$\frac{1}{3}$	0	0	0	0
Strange	s	$-\frac{1}{3}$e	$\frac{1}{2}h$	$\frac{1}{3}$	-1	0	0	0
Charmed	c	$+\frac{2}{3}$e	$\frac{1}{2}h$	$\frac{1}{3}$	0	$+1$	0	0
Bottom	b	$-\frac{1}{3}$e	$\frac{1}{2}h$	$\frac{1}{3}$	0	0	$+1$	0
Top	t	$+\frac{2}{3}$e	$\frac{1}{2}h$	$\frac{1}{3}$	0	0	0	$+1$

Antiquarks

Flavor	Symbol	Charge	Spin	Baryon number	Strangeness	Charm	Bottomness	Topness
Up	$\bar{u}$	$-\frac{2}{3}$e	$\frac{1}{2}h$	$-\frac{1}{3}$	0	0	0	0
Down	$\bar{d}$	$+\frac{1}{3}$e	$\frac{1}{2}h$	$-\frac{1}{3}$	0	0	0	0
Strange	$\bar{s}$	$+\frac{1}{3}$e	$\frac{1}{2}h$	$-\frac{1}{3}$	$+1$	0	0	0
Charmed	$\bar{c}$	$-\frac{2}{3}$e	$\frac{1}{2}h$	$-\frac{1}{3}$	0	-1	0	0
Bottom	$\bar{b}$	$+\frac{1}{3}$e	$\frac{1}{2}h$	$-\frac{1}{3}$	0	0	-1	0
Top	$\bar{t}$	$-\frac{2}{3}$e	$\frac{1}{2}h$	$-\frac{1}{3}$	0	0	0	-1

STUDY GUIDE

Mesons are composed of a quark and an anti-quark. **Baryons** are composed of three quarks. **Antibaryons** are composed of three antiquarks.

up a baryon? Baryons are fermions, and if they have parts, the parts must be fermions—any cluster of bosons is a boson. Thus, all baryons must be made up of an odd number of quarks. The simplest thing is to assume quarks are spin-$\frac{1}{2}$ fermions. Accordingly, Gell-Mann and Zweig proposed that **all baryons are composed of three quarks and all antibaryons are composed of three antiquarks**. Each quark has a baryon number of $+1/3$ and each antiquark a baryon number of $-1/3$. Similarly, mesons are bosons and must have an even number of quarks. Moreover, they have zero baryon number, and so **all mesons are composed of one quark and one antiquark**.

To account for every then-known hadron, Gell-Mann and Zweig required three varieties, or **flavors**, of quark. With a whimsical flair, these quark types came to be called **up** (u), **down** (d), and **strange** (s). Their most off-putting characteristic was the requirement that quarks have fractional electromagnetic charge, either $\pm 1/3$ or $\pm 2/3$ of the fundamental electron charge (Table 31.3). The proton and neutron, the ordinary matter of our Universe, are quark clusters uud and udd (Fig. 31.8), which suggests that the u- and d-quarks are also ordinary matter rather than the exotic matter created in high-energy collisions. Table 31.4 shows the quark combinations forming a number of hadrons. Note that if a meson is composed of a quark and antiquark of the same type or flavor (as is π^0), the constituents can rapidly annihilate one another and the lifetime is very brief. **The strong interaction cannot transform flavor.** If the flavors are different, the quark and antiquark will eventually annihilate and the meson decay, but only via the weak interaction and only relatively slowly. **The weak interaction can transform flavor.**

Figure 31.8 The quark composition of the pion, proton, meson, and neutron. As we'll see, each hadron also contains a swarm of gluons (p. 1108) that bind the quarks together. These gluons contribute significantly to the hadron's physical characteristics. For example, much of a neutron's or proton's spin is carried by its gluons.

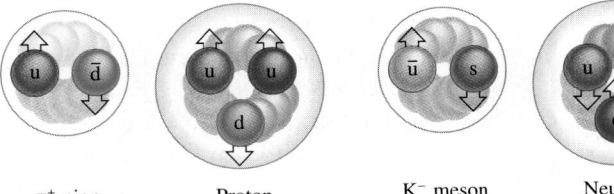

Example 31.2 **[I]** Table 31.3 lists several baryons that have the same quark content. These are distinguished by differences in quark spin. Thus the proton (p) corresponds to u(with spin-up) + u(with spin-up) + d(with spin-down), whereas the delta plus (Δ^+) corresponds to u(with spin-up) + u(with spin-up) + d(with spin-up). From their quark content, determine the charge and spin of each of these baryons. **Note that the spin designation of up and down is not to be taken literally, nor does it have anything to do with the name of the quarks.**

Solution This problem deals with quark attributes, all of which are tabulated. (1) TRANSLATION—Knowing the quark content of two baryons, determine their charges and spins. (2) GIVEN: A proton, uud (with spin-up, spin-up, and spin-down) and a delta plus, uud, (with spin-up, spin-up, and spin-up). FIND: Their charges and spins. (3) PROBLEM TYPE—Particle physics/quarks. (4) PROCEDURE—We have to add the individual contributions of the constituent quarks. (5) CALCULATION—According to Table 31.3 the u quark has a charge, in units of e, of $+\frac{2}{3}$ and the d quark has a charge of $-\frac{1}{3}$.

Hence for the proton the net charge is

$$\left(+\tfrac{2}{3}\right) + \left(+\tfrac{2}{3}\right) + \left(-\tfrac{1}{3}\right) = 1$$

and the same is true for the Δ^+, which is why it's called delta plus. According to Table 31.3 both the u and d quarks have a spin, in units of, of $\frac{1}{2}$. The net spin of the proton (spin-up, spin-up, spin-down) is

$$\left(+\tfrac{1}{2}\right) + \left(+\tfrac{1}{2}\right) + \left(-\tfrac{1}{2}\right) = \tfrac{1}{2}$$

Whereas the net spin of the delta plus (spin-up, spin-up, spin-up) is

$$\left(+\tfrac{1}{2}\right) + \left(+\tfrac{1}{2}\right) + \left(+\tfrac{1}{2}\right) = \tfrac{2}{3}$$

Incidentally, it takes more energy to align the three spins, and that energy manifests itself via $E_0 = mc^2$, as mass. The delta plus is more massive than the proton.

Quick Check: Clearly the proton has the right charge and spin.

Table 31.4

The Quark Composition of Several Hadrons*

Particle	Quarks
Mesons	
π^0	$u\bar{u}$, $d\bar{d}$ mix
π^+	$u\bar{d}$
π^-	$\bar{u}d$
η	$d\bar{d}$, $u\bar{u}$ mix
η'	$s\bar{s}$
K^0	$d\bar{s}$
$\bar{K}^0$	$\bar{d}s$
K^+	$u\bar{s}$
K^-	$\bar{u}s$
J/ψ	$c\bar{c}$
Υ	$b\bar{b}$
Baryons	
p	uud
n	udd
Δ^0	udd
Δ^{++}	uuu
Δ^+	uud
Δ^-	ddd
Σ^+	uus
Σ^-	dds
Σ^0	uds
Ξ^0	uss
Ξ^-	dss
Λ^0	uds
Ω^-	sss

*Where the quark compositions are the same, their spin alignments are different.

One might expect that blasting two protons together would produce a shower of quarks—they ought to be easy to generate and easy to identify, and yet their shyness proved a continuing embarrassment and impediment to the acceptance of the theory. No free quark has ever been observed. Accordingly, theoreticians have made a case for **quark confinement**, the notion that free quarks cannot exist, but that contention is not entirely convincing (it certainly wasn't in the early 1970s). Still, a series of experiments begun in 1969 at the Stanford Linear Accelerator Center (SLAC) in California and repeated using neutrinos at CERN gave the theory a needed boost. Probing with high-energy electrons (20 GeV), the SLAC group established that the proton and neutron were made up of three small, hard lumps of charge. The nucleon is composed of three pointlike charges that move around quite freely, like three bees inside a small balloon. But even that picture is now known to be an oversimplification—nucleons are much more complicated.

A theoretical objection to the quark model was raised in the early 1960s: there are several hadrons composed of the same flavor quarks, and their existence violates the Pauli Exclusion Principle. For example, the Δ^{++} baryon corresponds to uuu, three u-quarks (three fermions in the same state is a no no no). A way out of that quark quandary developed that has since proven to be remarkably fruitful for a number of other reasons. It was proposed that each flavor of quark (at the time u, d, and s) actually came in three varieties, or **colors**: red, green, and blue. Each quark carries one of three types of *color charge*. The u-quarks in Δ^{++}—namely ($u_R u_G u_B$)—were not really identical, and therefore there is no problem with the Exclusion Principle (Fig. 31.9).

Of course, nothing is colored in the ordinary sense of the word, but the analogy with light is convenient and makes the details easy to remember. Quarks carry color charges—they possess redness, blueness, or greenness. Antiquarks have anticolor charges of antired (think of it as cyan), antiblue (think of it as yellow), and antigreen (think of it as magenta). Red, blue, and green color are like positive electromagnetic charge; antired, antiblue, and antigreen anticolor are like negative electromagnetic charge—*it is these qualities that give rise to the strong force fields.*

Since no hadron displays an observable property that can be associated with color, it follows that color is an internal characteristic—**all hadrons are color neutral** (just as the neutron is electromagnetic-charge neutral). This means that either the total amount of each

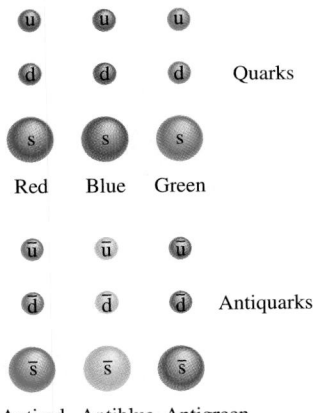

Figure 31.9 A few quarks and antiquarks, each coming in three colors or anticolors. Cyan is the complement of red so we take it as antired. Yellow is the complement of blue so we take it as antiblue. Magenta is the complement of green so we take it as antigreen.

color is zero (as with mesons that are color-anticolor pairs, for example, $q_R\overline{q_R}$) or that all the colors are present in equal amounts (as with baryons, $q_R q_B q_G$). The latter is analogous to the fact that red, blue, and green light beams add to make white. This scheme demands that no observable particle be composed of either two (qq) or four (qqqq) quarks—nor can free quarks exist. We might say that an atom is electromagnetically white and that an ion is colored. And because it attracts opposite charges, an ion has a higher energy and tends to become white by picking up an electron. Similarly, quarks tend to form white neutral composites.

As fate would have it, quark theory, which was none too popular, was revivified by a chance discovery. In 1974, two teams of scientists—one at Brookhaven National Laboratory under C. C. Ting and the other at SLAC under B. Richter—almost simultaneously discovered a new hadron, called J by one and ψ (psi) by the other. The J/ψ meson was three times more massive than the proton and, remarkably, lived for 10^{-20} s before decaying. That's 10^3 times longer than is normal for a hadron of that mass—there was some fundamentally new physics at work here. Moreover, the quark theory as it stood was no help; it was all filled up, there were no more particles that could be accounted for with three quarks.

The difficulty was soon settled by appealing to an idea suggested by Sheldon L. Glashow that had been around for some time but had found little support. Because there were then four known leptons, it had been proposed (on the grounds of natural symmetry among elementary particles) that there ought to be four quarks. The new addition had already been named **charm** (c), though until the discovery of J-psi, there was no reason to take it seriously. It is now accepted that J/ψ corresponds to the two-quark bound state ($c\overline{c}$). A whole clutch of very massive charmed mesons and baryons, particles containing the heavy charm quark, have since been discovered.

This story played itself out again in 1975 when the tau (τ) lepton was discovered. Assuming there was also a τ-neutrino to be found, there were then six leptons and four quarks. By 1977, a new meson, the upsilon, was observed. It is 75 times more massive than the pion. With little hesitation, a fifth heavy quark flavor—**bottom**, or *beauty* (b)—was conjured up, and the upsilon was recognized to be ($b\overline{b}$). The mesons ($b\overline{d}$) and ($b\overline{u}$) were found in 1983. Evidence of the discovery of the sixth heavy quark flavor, named **top**, or *truth* (t), was announced at the Fermi National Accelerator Laboratory in 1994. Experiments (1990) strongly suggest that nature probably cannot accommodate more than three kinds of neutrinos, and therefore there will be a total of six leptons and presumably no more than six quark flavors forming three generations of elementary particles (Fig. 31.10).

One might wonder what kind of interquark force is at work. Could some theoretical Coulomb-like interaction be devised that is proportional to color charge, one that would match all the diversity of strong reactions? In 1954, the first powerful step in that direction was taken by C. N. ("Frank") Yang and R. L. Mills. They produced the mathematical framework on which would ultimately rest modern Quantum Field Theory. Their work would make possible a satisfactory description in terms of the quantum of the gauge field of color: the gluon.

Family	Particle					Charge	Mass (GeV/c^2)
First Generation	Quarks	Up	•	•	•	$\frac{2}{3}$	0.003
		Down	•	•	•	$-\frac{1}{3}$	0.006
	Leptons	Electron	•			-1	5.11×10^{-4}
		e-neutrino	○			0	$< 1.4 \times 10^{-8}$
Second Generation	Quarks	Charm	•	•	•	$\frac{2}{3}$	1.3
		Strange	•	•	•	$-\frac{1}{3}$	0.1
	Leptons	Muon	•			-1	0.106
		μ-neutrino	○			0	$< 2.5 \times 10^{-4}$
Third Generation	Quarks	Top	●	●	●	$\frac{2}{3}$	≈ 173.8
		Bottom	●	●	●	$-\frac{1}{3}$	4.3
	Leptons	Tau	●			-1	1.78
		τ-neutrino	○			0	< 0.02

Figure 31.10 The three generations of elementary fermions (the associated antiparticles are not shown).

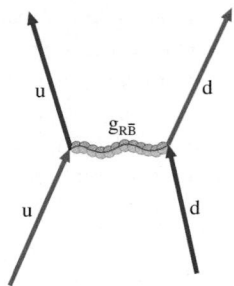

Figure 31.11 The interaction of an up-quark and a down-quark mediated by the exchange of a red-antiblue gluon.

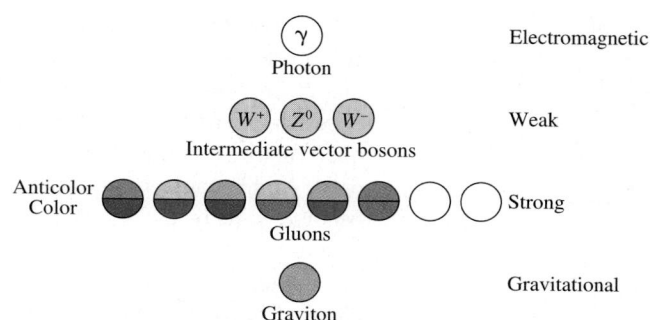

Figure 31.12 Boson force mediators.

Gluons are spin-1 bosons that, through their exchange, give rise to the strong or color force

31.5 Quantum Chromodynamics

▶ *The interaction between two point-like quarks is mediated by the exchange of a boson messenger amusingly called the* **gluon**. This is the basis of the fundamental *strong interaction*, or **color force** (Fig. 31.11). The quantum of the color field, the gluon, is a spin-1, electromagnetically neutral, massless particle referred to as a *vector boson*. Spin-1 particles are bosons that have wavefunctions in the form of a four-dimensional vector—hence, the name. Because the quarks come in three colors and the absorption or emission of a gluon can change the quark's color (though not its flavor), it turns out that there are eight possible different couplings and Color Gauge Theory postulates eight massless gluons (Fig. 31.12). These differ significantly from the photon, which is chargeless, in that six of them carry color charge. Each of these transports color and anticolor. Figure 31.13 shows how a red quark radiating a red-antigreen gluon loses red and has left behind green—it becomes charged green.

The strong interaction acts via color (for example, as with the meson depicted in Fig. 31.14). Gluon exchange also holds the hadrons together (Fig. 31.15) as composite entities (in twos and threes), but the transfer of individual gluons between separate hadrons is essentially precluded. Hadrons interacting with other hadrons experience the effects of the strong force mostly via the exchange of quark-antiquark composite particles (Yukawa's mesons). Figure 31.16 depicts the quark picture of the strong force proton-proton interaction. The creation and annihilation of quark pairs allows the transfer of a mediating pion, but the basic interaction is between quarks and gluons. The strong force holding the nucleus of an atom together is the rather feeble remnant of the interquark color force.

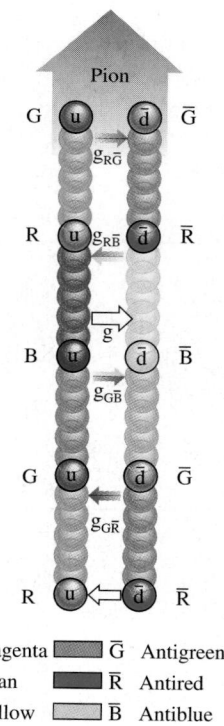

Magenta ▭ $\overline{G}$ Antigreen
Cyan ▭ $\overline{R}$ Antired
Yellow ▭ $\overline{B}$ Antiblue

Figure 31.14 The color force. A schematic representation of the interaction between the two quarks (u, $\overline{d}$) constituting a pion. The attraction is mediated by the exchange of color-charged and uncharged gluons.

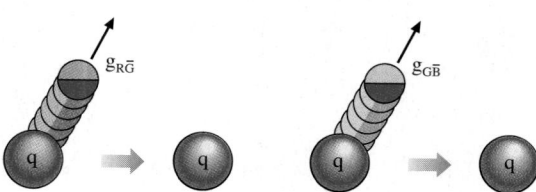

Figure 31.13 The transformation of a quark from one color to another. A $g_{R\overline{G}}$ gluon carries away red and antigreen (or magenta), leaving behind only green, and the quark becomes charged green. The emission of a $g_{G\overline{B}}$ gluon carries away green and antiblue (or yellow), leaving behind a blue quark.

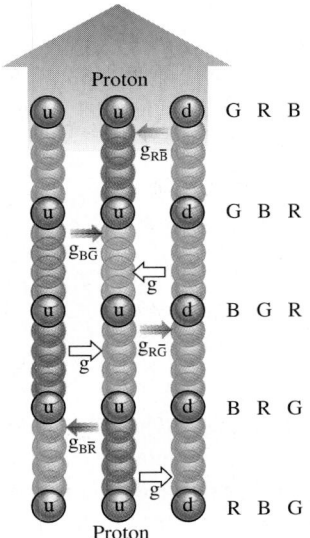

Figure 31.15 The color force. A schematic representation of the interaction between the three quarks (u, u, d) constituting a proton. The attraction is mediated by the exchange of color-charged and uncharged gluons.

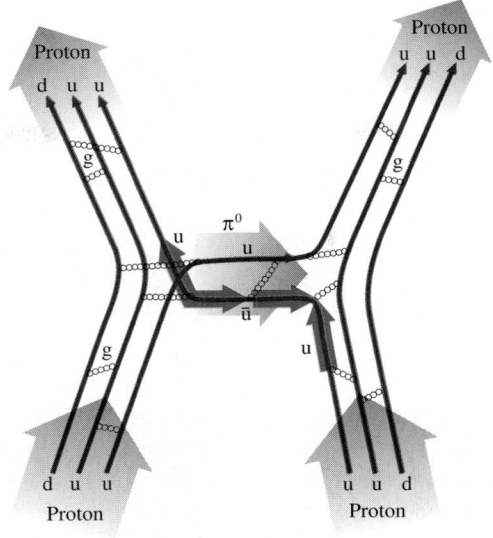

Figure 31.16 The strong interaction between protons. Two protons exchanging a neutral pion as understood via the quark model. The creation of one $u\bar{u}$ pair and the annihilation of another allows the transfer of a π^0. Gluons exchanged between quarks hold the hadrons together.

The gauge theory of the color field, the strong interaction, was rather colorfully christened **Quantum Chromodynamics** (QCD)—*chroma* in Greek means color—by Gell-Mann. This is a mathematically sophisticated, renormalizable gauge theory modeled after QED. To date, QCD has been successful in dealing with experimental findings, and although it's certainly incomplete, it may even be the right and true theory (or more likely, part thereof), but that remains to be seen.

IN SUMMARY, THEN

Quarks are bound to one another (via the strong or color force) to form hadrons through the exchange of gluons. Unlike the electromagnetic force, which is communicated by chargeless photons, most gluons carry color charge, and their absorption or emission changes the color of the participating quark. The strong interaction leaves the flavor (u, d, s, c, t, b) of a quark unaffected, although its color is generally altered. Only quark-antiquark pairs of the same flavor can annihilate each other in a strong interaction (that's not the case for a weak interaction). Moreover, quarks cannot decay via the strong force, although they can decay via the weak force.

31.6 The Electroweak Force

Fermi's nonrenormalizable theory of the weak interaction (1932) pictured the process of neutron decay at a single point in space-time (Fig. 31.17a). He had oversimplified things, avoiding the question of the carrier of his new force. Two years later, Yukawa suggested that the weak force (like the electromagnetic interaction of QED) was mediated by a massive messenger particle. After decades of neglect, the idea was revived by Julian Schwinger (1956), who attempted to describe the weak force in terms of Gauge Theory. He called the intermediary the W-particle (for *weak*). Like the photon, it had to have a spin of 1 if angular momentum was to be conserved; as a result, it came to be known as the *intermediate vector boson*. The extremely short range (≈ 0.01 fm) of the weak force required that its mediator be quite massive and quite small.

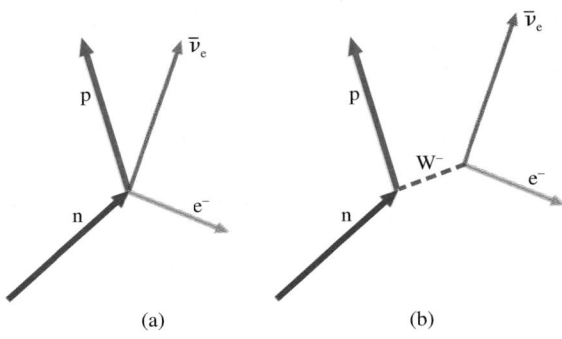

Figure 31.17 The decay of a neutron into a proton, electron, and electron-antineutrino. (a) The decay occurring at a single point. (b) The decay as mediated by a negative vector boson, which itself transforms into an electron and neutrino.

Figure 31.18 (a) Weak interactions as mediated by the three intermediate vector bosons. (b) A change in the charge of any of the particles creates a *charged current*. When, as in (c), there is no change in charge, we have a so-called *neutral current*. Note that if we slide the incoming neutrinos up so that they are created by the W⁻ or Z⁰, they become outgoing antineutrinos.

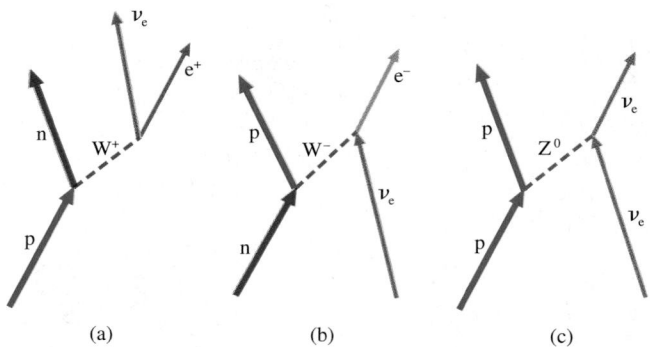

(a) (b) (c)

Beta decay results when a neutron transforms into a proton, an electron, and an antineutrino. That happens when a d-quark in the neutron changes ┈┈▶ into a u-quark, an electron, and an anti-neutrino through the action of the weak force.

Schwinger postulated the existence of two charged vector bosons. The neutron decays into a proton and a virtual W-particle, which must therefore carry a negative electromagnetic charge (Fig. 31.17*b*). The W⁻ then decays into an electron and an electron-antineutrino. Similarly, he argued that the creation of a positron and a neutrino must be the result of the decay of a positively electromagnetically charged W-particle, the W⁺(Fig. 31.18*a*). The emission or absorption of a charged intermediate vector boson by a fundamental particle results in the prompt transformation of that particle. That's something new among the forces we have studied. The emission or absorption of a photon alters the phase of the de Broglie wave, but it doesn't transform the emitting particle. Similarly, gluon exchange alters color, not flavor, so there's no change of particle type there either.

The weak force acts differently, depending on the handedness of the participants. The effect of the weak force on an electron changes depending on the particle's motion, that is, on the alignment or antialignment of its linear momentum and spin angular momentum. The weak force operates between fermions, and so leptons and quarks (i.e., left-handed particles and right-handed antiparticles) couple to it; they possess weak charge. The W's carry both electromagnetic charge and weak charge.

If, as a result of emitting a W⁻, a neutron (udd) is transformed into a proton (udu), it must be that a d-quark is transformed into a u-quark (Fig. 31.19). The weak interaction can alter the flavor of a quark without altering its color. In fact, the emission or absorption of a W± results in the prompt transformation of one type of lepton (for example, e⁻) into another (for example, $\bar{\nu}_e$), as shown in Fig. 31.20. *All the various nuclear decays that take place via the weak force involve quark flavor transformations.*

The fact that the W-particles are electromagnetically charged suggested to Schwinger that there is an intimate relationship between the weak and electromagnetic interactions; the W-particles carry both weak charge and electromagnetic charge. In the late 1950s, Schwinger turned to other concerns, but before he did, he asked one of his Ph.D. students, "Shelly" Glashow, to think about a possible connection between the weak and electromagnetic interactions. Years later (1961), Glashow published a seminal paper introducing a third messenger, the neutral Z boson, but few people paid it much attention. Like the W±,

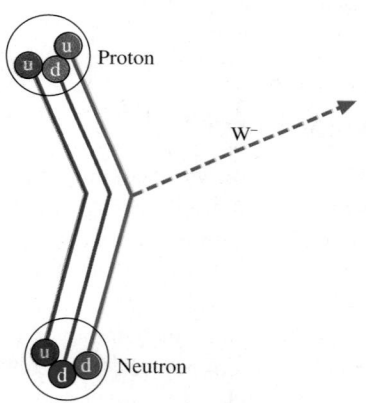

Figure 31.19 The transformation of a down-quark into an up-quark via the emission of a W⁻ intermediate vector boson. The result is the transformation of a neutron into a proton.

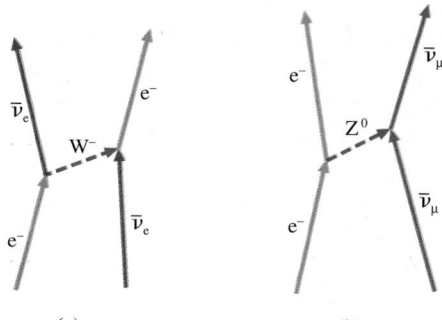

(a) (b)

Figure 31.20 (a) An electron scattering with an electron-antineutrino via the weak interaction mediated by a W⁻ boson. The W carries charge, and we see an electron transformed into a neutrino. (b) An e⁻ and $\bar{\nu}_\mu$ come in and go out, but now the scattering is mediated by a Z⁰ boson. The Z⁰ is neutral and there is no change in electromagnetic charge. Note how the interaction resembles that mediated by a photon.

Table 31.5
The Four Forces of Nature*

Force	Applicable to	Strength[1]	Effective range	Mediator
Gravity	All leptons and hadrons[2]	6×10^{-39}	Unlimited	Graviton
Weak	All leptons and hadrons	10^{-5}	$\approx 10^{-17}$ m	$W^{\pm}$ and Z^0
Electromagnetic	Charged leptons and hadrons	$1/137$	Unlimited	Photon γ
Strong	All hadrons	1	$\approx 10^{-15}$ m	Gluon g

*At the temperatures that exist today, we see four distinct interactions. At much higher energies, these blend together, presumably becoming one.
[1]Here is a slightly different scheme from the one we saw before for comparing the strenghts relative to the strong force as 1.
[2]Gravity and the strong force both act on their own field quanta.

these messengers were quite massive, and yet gauge bosons presumably had to be massless—that, along with the fact that Glashow's model was nonrenormalizable, left much to be desired. Amazingly, Steven Weinberg, an old high school rival of Glashow, was independently working on the same problem at the Massachusetts Institute of Technology. He devised a scheme (1967) by which the vector bosons could take on mass and still fit into a gauge theory, but again, no one, not even Glashow, seemed to notice it. A year later, Abdus Salam published a very similar unifying treatment that met with the same neglect. In 1969, a young graduate student in Holland, Gerhard 't Hooft, demonstrated how Yang-Mills gauge fields (p. 1107) could be made renormalizable, and the whole area of study came alive. The Glashow-Weinberg-Salam synthesis, the **Electroweak Theory**, won these three physicists the Nobel Prize in 1979. They had devised a quantum gauge theory that naturally resulted in four gauge quanta: the three massive vector bosons W^-, W^+ and a new neutral Z^0 (zee-zero), along with a fourth massless force-carrier, the familiar, though no less mysterious, photon (Table 31.5).

Gauge quanta are supposed to be massless. How did the theory produce massive gauge bosons? How did the symmetry demanded by Gauge Theory get suspended, or "broken"? The idea of a symmetry spontaneously being disrupted or broken is not new. Think of a sample of ferromagnetic material (p. 676) above its Curie temperature; its atoms have spins pointing every which way. A microscopic observer would encounter magnetic uniformity in all directions. Indeed, Maxwell's Equations are symmetric in space; there are no preferred directions for the theory. Yet, if the sample is cooled, that symmetry is spontaneously broken as the atoms align themselves arbitrarily into tiny domains—the energy of an aligned set of interacting magnets is less than the energy of a random grouping. The system (without any external asymmetrical influences) spontaneously breaks the symmetry of the laws of electromagnetism without violating those laws. In a quantum-mechanical Universe, where subatomic particles shift around constantly, this sort of energy-driven descent to the ground state is quite reasonable.

All of this cannot help but seem bizarre, and yet the Electroweak Theory made several striking predictions that were subsequently confirmed in every regard. First was the existence of the ***neutral-current process***, which is a weak interaction without any resulting change in electromagnetic charge. For example, the theory proposed that a very energetic neutrino might collide with a proton. Much of the neutrino's energy (≈ 100 GeV) could be given up to the creation of a neutral Z that could pass over to a quark in the proton (Fig. 31.21). The neutrino could scatter off physically unaltered while the energy transformed into a burst of quark-antiquark pairs that would form a spray of hadrons with a net electromagnetic charge of plus one. That bit of high-energy alchemy was confirmed at CERN and later at Fermilab in 1973.

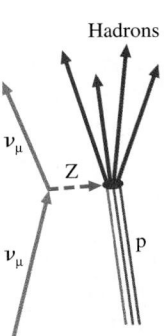

Figure 31.21 Inelastic scattering corresponding to a neutral current.

The Higgs boson is like a toilet. Everyone has one, but it's not necessarily an object of pride and beauty.

SHELDON LEE GLASHOW (1984)

THE HIGGS FIELD

The W- and Z-particles "should" be massless like the photon. According to Electroweak Theory, they acquire their mass as a result of the ad hoc presence of an additional field known (after the person who first studied it) as a *Higgs field*. The latter is a spinless, directionless scalar field (as yet undetected experimentally). Space, at the low temperature at which it exists today, is permeated by the Higgs field, which drags on the W- and Z-particles like cold, thick honey. In other words, vast numbers of hypothetical spin-0 Higgs bosons condense in the ground state, producing a classical field that acts on the Ws and Zs. No longer able to travel through space at lightspeed, the W and Z bosons behave like massive particles compared to the photon, which doesn't interact with the Higgs "honey" field. At high temperatures (that is to say, at high energies), the Higgs "honey" thins out and does not affect the W- and Z-particles. They supposedly become massless (at about 1000 GeV), thereby unifying in a blazing bliss with the ever-constant photon. The weak and the electromagnetic interactions are one and the same force, mediated by the exchange of these four field quanta.

Second, the theory predicted that the masses of the intermediate vector bosons would be 80 GeV/c^2 for $W^\pm$ and 90 GeV/c^2 for Z. These values were way beyond the energy of any existing accelerators, and the direct observation of the real incarnations of these bosons had to wait for the construction of the proton-antiproton collider at CERN. Finally, in 1983, a team of 130 physicists led by Carlo Rubbia and Simon Van der Meer (both of whom shared the 1984 Nobel Prize for their work) succeeded in producing and detecting the intermediate vector bosons. One out of roughly five million p-$\bar{\text{p}}$ collisions fused together a quark from a proton and an antiquark from an antiproton, creating a vector boson that then disintegrated in less than 10^{-24} s. From the tracks made by the debris, the researchers could unambiguously identify the bosons. The masses determined for $W^\pm$ (81 GeV/c^2) and Z^0 (91 GeV/c^2) were in splendid accord with the predictions.

A recent test of the Electroweak Theory used laserbeams to detect tiny distortions of heavy atoms such as cesium. The weak force between the nucleus and the orbital electrons of an atom has a minute but measurable effect. To date, all the results are in excellent agreement with theory. Still, the whole idea of the Higgs mechanism (which seems inelegant, but necessary) remains to be resolved, and until it is, few physicists will be completely content with the existing formalism.

31.7 GUTs and Beyond: The Creation of the Universe

Just as the electric and magnetic forces were unified by Maxwell's Theory into a single electromagnetic field, contemporary theory has produced a twofold unification, a merger of the weak and the electromagnetic fields into a single **electroweak field**. What we see as the separate weak and electromagnetic interactions is a result of the cool environment of today's Universe—the unity is hidden. Only in the largest machines ever made can we even begin to simulate the inferno of primordial creation. Our Universe, space and time, began some fifteen thousand million years ago. The interval from $\approx 10^{-34}$ s to $\approx 10^{-11}$ s after the initiation of Genesis (Time Zero) can be called the *electroweak epoch* (Fig. 31.22). Ending at an incredible temperature of around 3×10^{15} K, all there was was a dense seething soup of leptons, quarks, and massless gluons, photons, Ws and Zs. During the epoch (at temperatures above 3×10^{15} K), the weak force and the electromagnetic force shared the same inverse-square behavior and were of equal strength (Fig. 31.23). As the Universe cooled further, the electroweak symmetry shattered, Ws and Zs took on mass and vanished, and the cosmos became ruled by the Four Forces we know.

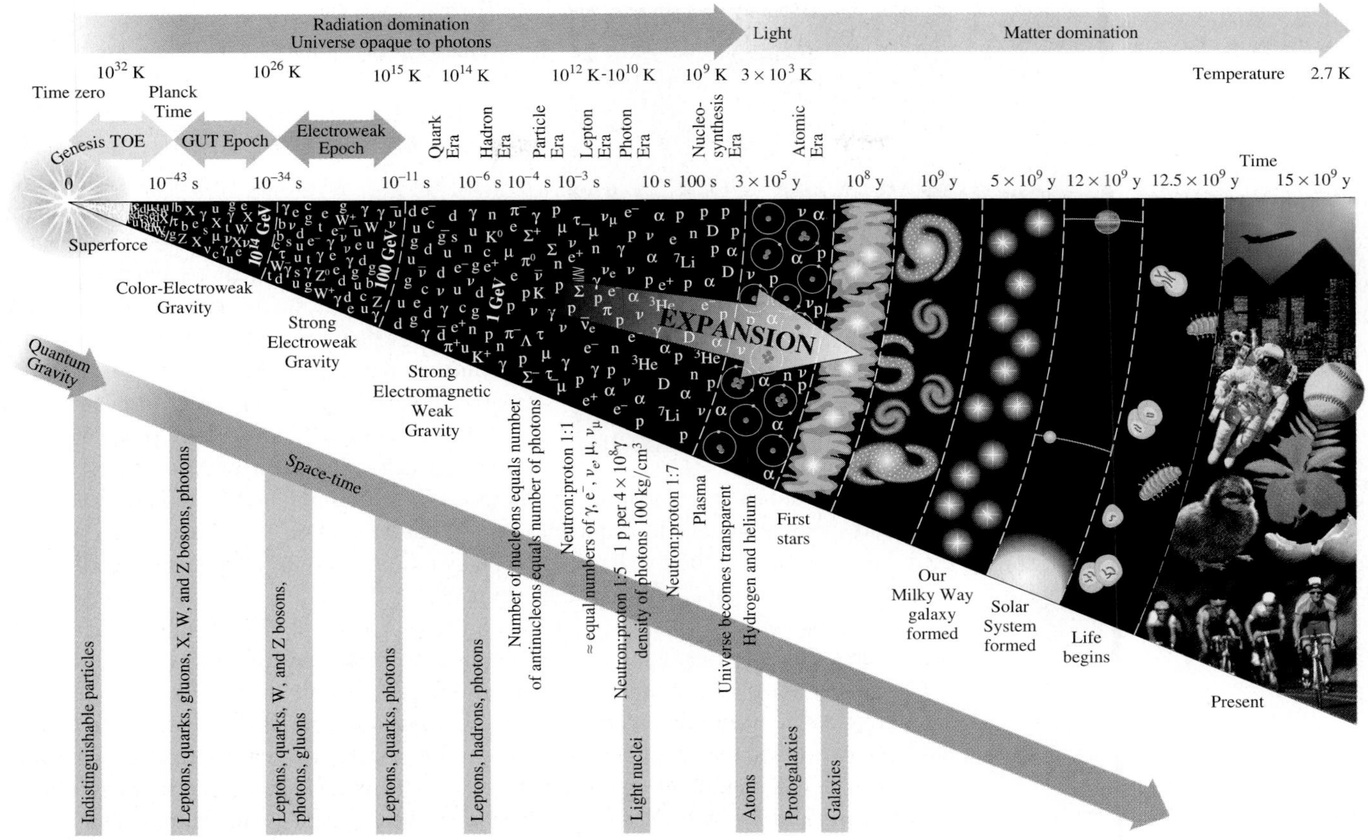

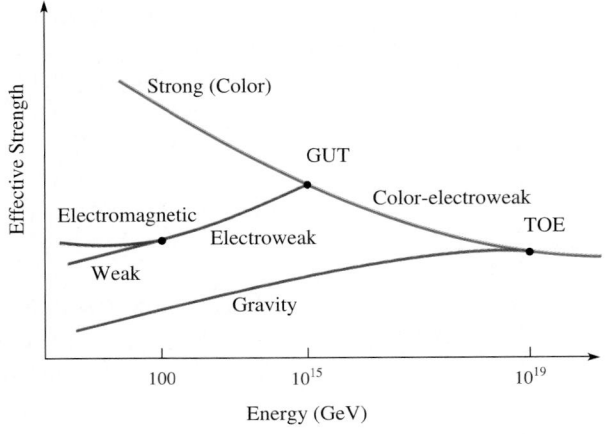

Figure 31.22 The creation of the Universe.

As the Universe continued to expand, it cooled to around 10^{14} K in less than its first microsecond. Free quarks and gluons vanished, coming together to form all the hadrons physicists have worked so hard to re-create. Then the heavy hadrons also annihilated, leaving (at $\approx 10^{-4}$ s) a dense quantum brew of leptons and light hadrons (protons, neutrons, pions, and so on). Neutrons and protons in almost equal number were transmuted into one another via weak interactions: $p + e^- \leftrightarrow n + \nu$ and $n + e^+ \leftrightarrow p + \bar{\nu}$. Because neutrons are heavier and require more energy to create, more protons resulted as the Universe cooled. After 1 s, the temperature was $\approx 10^{10}$ K. The unstable heavy leptons, the muons and taus, disappeared leaving neutrons, protons, electrons, positrons, and neutrinos—by now, there were five protons to every neutron.

Figure 31.23 The Four Forces coalesce as the energy (or temperature) of the interacting particles increases and they can approach closer together.

A 60 × 70 × 80-ft tank containing 8000 tons of ultra-pure water (a rich supply of protons) forms the body of a giant particle detector. Here a diver inspects the photo-multiplier tubes. Although designed to detect proton-decay debris predicted by GUTs, in February of 1987 it recorded a flux of neutrinos from supernova 1987A.

The world was pervaded by a fireball of radiant energy; photons ruled the cosmos for the next ≈ 100 s. The Universe continued to expand, the particles separated; collisions and nucleon transformations ceased. After several minutes, appreciable numbers of free neutrons decayed into protons. But, at around 100 s, the temperature had dropped to the point ($\approx 10^9$ K) where light nuclei (deuterons and α-particles) could form via fusion and remain bound—all was plasma. Alpha production stopped when the available neutrons were locked away. After a few hundred thousand years, the Universe cooled enough for atoms of hydrogen (75%) and helium (25%) to form without being blasted apart by electromagnetic radiation. The cosmos was then dominated by matter, and in a billion years or so, protogalaxies came into being, first-generation stars flashed on, lived, and died. Five thousand million years ago, the Sun was born, and the Earth came into being. A moment later life began, and you turned the page.

Today, we understand all the marvels of nature in terms of three primary gauge fields: the electroweak, the strong, and the gravitational fields. The first two, treated quantum mechanically via the Electroweak Theory and QCD, form the basis of what is generally referred to as the **Standard Model**. Driven by the aesthetic of *unification*, gauge theories have been proposed that attempt to unite the strong and electroweak interactions. These so-called **GUT**, or Grand Unification Theories, seek to find a relationship that will allow a threefold unification bringing together the electroweak gauge quanta and the gluons. The six quark flavors and the six lepton types then dissolve into a single lepton-quark fermion field. {For more on these ideas click on **GUT** under **FURTHER DISCUSSIONS** on the **CD**.}

The Universe came into existence as simple as it could be, becoming increasingly diverse only as it cooled enough for complexity to crystallize—for symmetries to spontaneously break. Already by $\approx 10^{-43}$ s beyond Time Zero, gravity had settled out from the once single *superforce* that had ruled from the beginning. Even more ambitious than GUTs are the attempts to bring a quantum theory of gravity (with its mediator, the graviton) into a fourfold unification. That ultimate goal is amusingly called the *Theory of Everything* (TOE). It is estimated that the effects of quantum gravity will become significant at a particle separation of 10^{-35} m, which corresponds to a phenomenal energy of 10^{19} GeV. This value is totally beyond direct experimental observation—our most powerful machines will not likely exceed 10^5 GeV for decades. Since gravity acts on both fermions and bosons, it would seem that the TOE should account for the transformation of these two particle types into one another.

During the TOE epoch prior to $\approx 10^{-43}$ s, the entire Universe was subnuclear in size and surely quantum-mechanical in disposition. The ambient temperature ($\approx 10^{32}$ K) and density ($\approx 10^{92}$ times greater than water) were unimaginable. Every point in the inferno could spontaneously become a Black Hole dropping through the fabric of the cosmos and then evaporating back. That disconnected seething foam of space and time was quantized. It is here in the primal whirlwind, even before space-time congealed into a gravitational continuum, that our physics is most challenged. It is here, in the distant reaches of imagination, that we are most bold. Yet who could doubt that the whirlwind was a gauge whirlwind?

All there is today, stars, wind, baseballs, lovers, all there is, was there in the foam of beginning. Shattered symmetries, epoch upon epoch, fanned out diversity. Quarkstuff, you and I, for a moment contemplating Creation—how naive, how wonderful.

Core Material & Study Guide

Leptons are the particles that couple to the weak force and, if they are electrically charged, to the electromagnetic force, but leptons *are immune to the strong force*. **Hadrons** are the strongly interacting composite particles that form two distinct subgroups: *mesons* and *baryons*. **Mesons** are bosons; **baryons** are fermions (p. 1097). On a nonfundamental level, the nuclear force arises from the exchange of mesons (p. 1100). With a mass *m*, these have a maximum range of

$$R \approx \frac{h}{4\pi m_\pi c}$$

All field theories must possess a particular type of mathematical structure known as gauge symmetry (p. 1102).

Baryons are composed of three quarks; antibaryons are composed of three antiquarks. Mesons are composed of one quark and one antiquark. There are six flavors of quark: *up* (u), *down* (d), *strange* (s), *charm* (c), *bottom* (b), and *top* (t) (p. 1104). Each flavor comes in three varieties or *colors*—red, blue, or green. Antiquarks have anticolor charges of antired, antiblue, and antigreen. Hadrons are color neutral (p. 1108). The interaction between quarks is mediated by the exchange of eight massless gluons (p. 1108). The gauge theory of the color field is called Quantum Chromodynamics (QCD).

The weak interaction is mediated by three *intermediate vector bosons*: W^+, W^-, and Z^0 (p. 1109). The weak interaction can alter the flavor of a quark without altering its color. *Electroweak Theory* combines the electromagnetic and weak forces (p. 1110). GUTs, or Grand Unification Theories, seek a threefold unification, bringing together the electroweak gauge quanta and the gluons.

Key Terms

leptons	charm
antileptons	bottom
lepton number	top
hadrons	gluon
mesons	color force
pion	vector bosons
strangeness	QCD
baryons	Electroweak Force
Feynman diagram	intermediate vector boson
virtual photon	neutral-current process
gauge symmetry	Higgs boson
quarks	electroweak epoch
antiquarks	Standard Model
quark flavors	superforce
up, down, & strange	GUT
quark confinement	TOE
color charge	

Discussion Questions

1. Figure Q1*a* is a drawing of the Stanford Linear Collider (SLC) showing its three-kilometer straightaway. The cathode fires two successive bunches of electrons. The damping rings condense the beams, which are focused down to a diameter of a few millionths of a meter later on. Discuss, in general terms, what this 100-GeV machine does and how it does it. What might this process have to do with the Z^0 intermediate vector boson? These are the heaviest known real particles and are of great interest because, among other reasons, they have many possible decay modes. Figure Q1*b* depicts the simplest linac, the drift-tube accelerator. How does it work? [*Hint: The particles spend the same amount of time drifting along inside each cylinder.*]

2. In many respects, the Z^0 behaves like a heavy photon. Discuss this notion using the diagrams of Fig. Q2. Here, the generic label *quark* or *lepton* means that any such particle coming in also sails out. Explain what's happening in each illustration. An interaction between quarks can take place via Zs or Ws. How do these differ?

3. The μ^- and μ^+ are antiparticles of each other, and because the μ^- decays to an e^-, we take it as the matter and the other as the antimatter. By contrast, π mesons are not so easy to categorize; π^0 seems to be its own antiparticle, but the π^+ and π^- always decay into a lepton and an antilepton. Discuss this situation as it relates to the quark picture. Is the classification of matter and antimatter unambiguous? Does it matter which pion we call the antiparticle?

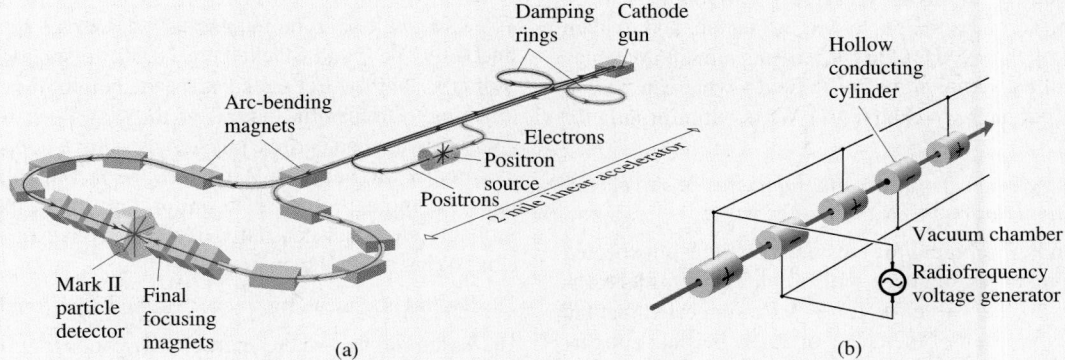

Figure Q1

(a)

(b)

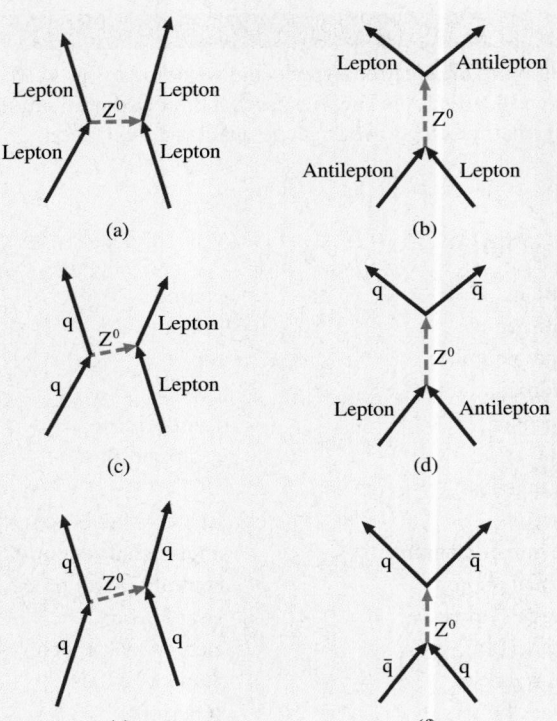

Figure Q2

(a) (b) (c) (d) (e) (f)

colliding particles had a total combined energy of about 3.6 GeV. Why is that significant?

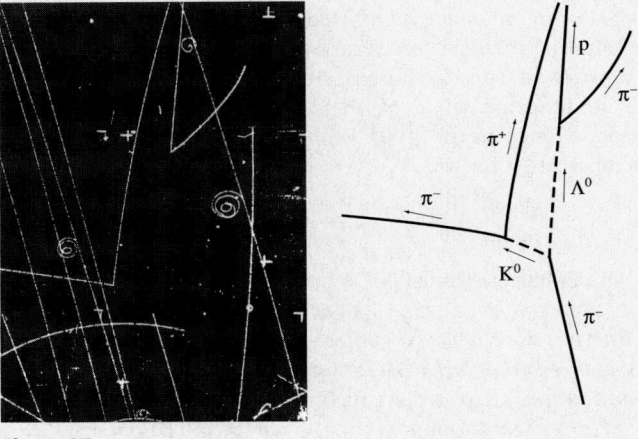

Figure Q7

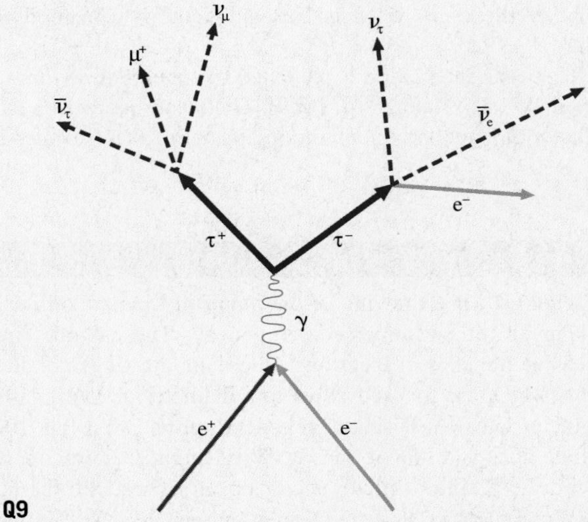

4. It is known that a Feynman diagram can be turned on its side to yield an entirely new physical insight, provided that particles moving backward in time are interpreted as antiparticles. Is this attribute true of Fig. Q2? Explain.

5. Every hadron interaction must be able to be analyzed in terms of the quark model. Accordingly, explain in detail how the reaction

$$\pi^- + p \rightarrow \Lambda^0 + K^0$$

takes place.

6. With Question 5 in mind, realize that the Λ^0 may rapidly decay into a negative pion and a proton while the K_0 may decay into a positive and a negative pion. Write out the complete reaction starting with the proton. In what sense do we end up with more matter than we start with? How is that manifest in the quark model? Where does the matter come from? Draw a Feynman diagram of the whole process.

7. Figure Q7 shows a bubble chamber photo of some particle tracks. The chamber is filled with supercooled liquid hydrogen. When a particle ionizes the hydrogen atoms, tiny bubbles form, revealing the path. Compare the photo and the sketch, and explain what happened. Why did a few of the participants remain invisible? The tracks are bent by a known magnetic field—what can we tell from the radius of curvature of each track? Write out a formula for what took place.

8. Is it correct to say that *every observable subnuclear event occurs via creation and annihilation*? Explain.

9. In 1975, Martin Perl reported the existence of some unexpected muon-electron events associated with high-energy electron-positron collisions: $e^+ + e^- \rightarrow e^- + \mu^+ + ?$. He suggested that the observations were the result of the decay of a new particle-antiparticle pair of very massive leptons he called *tau*. Discuss the reactions depicted in Fig. Q9. The effect first appeared when the two

Figure Q9

10. Neutrons decay when they are free—why might they be stable when bound to protons in a nucleus? [*Hint: Consider the situation in terms of energy.*]

11. The idea of baryon number as a conserved quantity was introduced by Weyl and Stückelberg (1930) to supposedly explain the proton's stability. Let's reconsider the notion. First, does it *explain* anything? Discuss the nature of the "law"—is it arbitrary? Is it applicable beyond particle physics, as is, for example, Conservation of Angular Momentum? What does the fact that the Universe contains far more protons than antiprotons suggest? How does the probable electrical neutrality of the Universe fit into this discussion? What about lepton number?

12. Explain the meaning of each of the Key Terms on page 1115.

13. Assuming that when a particle decays its fragments fly off with some kinetic energy, what does this imply about the stability of zero-mass particles?

Multiple Choice Questions

1. A proton and an electron that are close together   (a) will sometimes interact through the strong force   (b) will never interact through the gravity force   (c) will never interact through the weak force   (d) will never interact through the strong force   (e) none of these.

2. Mesons are   (a) always composed of either two u-quarks or two d-quarks   (b) always composed of two quarks of different flavor   (c) always composed of a quark and an antiquark   (d) never composed of a quark and an antiquark of the same flavor   (e) none of these.

3. A meson and a baryon that come very close together   (a) will only sometimes interact through the gravity force   (b) will never interact through the gravity force   (c) will never interact through the weak force   (d) will never interact through the strong force   (e) none of these.

4. Decays that last in the range from 10^{-20} s to 10^{-23} s are   (a) driven by the weak interaction   (b) driven by gravity   (c) driven by the strong force   (d) driven by the electromagnetic force   (e) none of these.

5. The reaction $e^- + p \rightarrow n + \bar{\nu}_e$ is   (a) possible because it conserves everything   (b) impossible because it does not conserve baryon number   (c) impossible because it does not conserve lepton number   (d) impossible because it does not conserve charge   (e) none of these.

6. From the perspective of the Standard Model, the reaction $p \rightarrow \pi^+ + \pi^0$ is   (a) possible because it conserves everything   (b) impossible because it does not conserve strangeness   (c) impossible because it does not conserve lepton number   (d) impossible because it does not conserve charge   (e) none of these.

7. From the perspective of the Standard Model, the reaction $p \rightarrow e^+ + \gamma$ is   (a) possible because it conserves everything   (b) impossible only because it does not conserve baryon number   (c) impossible because it does not conserve either baryon or lepton number   (d) impossible because it does not conserve charge   (e) none of these.

8. The decay $K^+ \rightarrow \gamma + \gamma$ is   (a) possible because it conserves everything   (b) impossible for several reasons, one being because it is electromagnetic and must conserve strangeness   (c) impossible for several reasons, one being because it is electromagnetic and does not conserve meson number   (d) impossible for several reasons, one being because it does not conserve baryon number   (e) none of these.

9. The reaction $\nu_e + n \rightarrow e^- + p$ is   (a) possible because it conserves everything   (b) impossible because it does not conserve baryon number   (c) impossible because it does not conserve lepton number   (d) impossible because it does not conserve charge   (e) none of these.

10. Which is true?   (a) only the weak interaction can change one flavor of quark into another   (b) only the strong interaction can change one flavor of quark into another   (c) both the strong and the weak interactions can change one flavor of quark into another   (d) only the gravitational interaction can change one flavor of quark into another   (e) none of these.

11. All strongly interacting material particles   (a) also couple to the weak force   (b) also couple to the electromagnetic force   (c) never couple to the weak force   (d) never couple to the electromagnetic force   (e) none of these.

12. The presence of a neutrino among the decay products of a particle   (a) tells us nothing about the force responsible   (b) suggests that the strong force was at work   (c) suggests that the electromagnetic force was at work   (d) tells us that the weak force was at work   (e) none of these.

13. We know that the electromagnetic force has controlled a particle decay when the decay products   (a) include at least one particle with electromagnetic charge   (b) include at least one intermediate vector boson   (c) have a zero net charge   (d) include at least one photon   (e) none of these.

14. Baryons are composed of   (a) two quarks and an antiquark   (b) two antiquarks and a quark   (c) a lepton and a quark   (d) three quarks   (e) none of these.

15. A quark can change flavor by   (a) absorkbing a W boson   (b) emitting a photon   (c) absorbing a gluon   (d) emitting an electron   (e) none of these.

16. The thing that holds a meson together is   (a) the exchange of W^- bosons   (b) the exchange of gluons   (c) the exchange of gauge bosons of the electromagnetic field   (d) the absorption and emission of quarks   (e) none of these.

17. Gluons can attract each other to form so-called *glueballs* because   (a) gluons are sticky   (b) gluons exchange photons   (c) gluons carry color charge   (d) gluons emit and absorb $W^\pm$ bosons   (e) none of these.

18. An antineutron can be distinguished from a neutron by   (a) its opposite electromagnetic charge   (b) its opposite magnetic moment   (c) its smaller mass   (d) its greater tendency to self-annihilate   (e) none of these.

19. Prior to 10^{-43} s after Time Zero, we believe that the Universe   (a) was the size of the Moon   (b) was ruled by a single superforce   (c) was in the so-called GUT era   (d) did not exist   (e) none of these.

20. We believe that a correct fundamental field theory must   (a) be simple   (b) predict that the proton is unstable   (c) have global symmetry   (d) be gauge-invariant   (e) none of these.

For more Multiple Choice Questions with answers click on Warm-Ups **in** Chapter 31 **on the CD.** 🔘

Suggestions on Problem Solving

1. Some useful numerical quantities for this chapter include

$$1 \text{ eV} = 1.602\ 177 \times 10^{-19} \text{ J}$$

$$hc = 1.239\ 842 \times 10^{-6} \text{ eV·m}$$

$$k_B = 1.380\ 66 \times 10^{-23} \text{J/K} = 8.6174 \times 10^{-5} \text{ eV/K}$$

2. How would you go about determining if the reaction

$$\mu^+ + \nu_\mu \rightarrow \pi^+$$

is possible? First, you might check the electromagnetic charge; $+1 + 0 \rightarrow +1$ (it's okay). Then check the spin; $\pm\frac{1}{2} + \pm\frac{1}{2} \rightarrow 0$ (it's okay). Then check the baryon number; $0 \rightarrow 0$ (it's okay). Then check the lepton number. There are only μ-type leptons; hence, on the left, we have -1 for the μ^+antimuon and $+1$ for the μ-neutrino, yielding a total of zero on both sides. These are not strange particles, so we needn't worry about S. Finally, checking the energy, the muon has less mass than the pion, and the neutrino can balance the formula by supplying the energy difference. Conclusion: the

reaction is allowed. Moreover, the inverse reaction

$$\pi^+ \rightarrow \mu^+ + \nu_\mu$$

must also be allowed.

3. Notice that adding the same particle or antiparticle to both sides doesn't change any of the requirements and must result in allowed processes as well. Thus, adding an antiparticle to both sides, we see that the reaction

$$\pi^+ + \bar{\nu}_\mu \rightarrow \mu^+ + \nu_\mu + \bar{\nu}_\mu$$

results in

$$\pi^+ + \bar{\nu}_\mu \rightarrow \mu^+$$

which is fine, provided the μ has the right amount of KE. This is equivalent to the notion that *carrying a particle from one side of the formula to the other changes it to an antiparticle* and vice versa. And *that's equivalent in a Feynman diagram to changing an incoming particle to an outgoing antiparticle* or vice versa.

4. Remember that **the production of strange particles via the strong force conserves strangeness**; only weak-force decays do not conserve strangeness.

Problems ✦ Coordinated Problems ✦ Progressive Problems ✦ Solutions

STUDY GUIDE   **1. Coordinated Problems:** The three problems within each magenta-colored grouping are solvable in similar ways. Note that the first of these always has a hint; moreover, its solution is provided in the back of the book. *Work out each of these sets; they'll strengthen technique and build confidence.* **2. Progressive Problems:** The problems introduced in blue unfold step-by-step carrying along the analysis in a more suggestive way than is customary. *Work out all of these; they'll guide you through the analytic process and help develop problem-solving skills.* **3. Worked-Out Solutions:** Studying worked-out solutions is an important part of learning how to solve problems. Accordingly, additional *solutions* to a number of model problems are given below. *Make sure you understand each of them before you go on to the next problem.* **4.** Also provided in the back of the book are the *Answers* to all odd-numbered problems, as well as worked-out *solutions* to those with boldface numbers. Problem numbers in italic indicate that a solution appears in the Student Solutions Manual.

ELEMENTARY PARTICLES

QUANTUM FIELD THEORY

1. [I] THIS PROBLEM DEALS WITH QUANTUM NUMBERS AND CONSERVATION. The positive kaon, which has a relatively long lifetime of 1.24×10^{-8} s, most often decays in the following way:

$$K^+ \rightarrow \mu^+ + \nu_\mu$$

Using the quantum numbers in Table 31.1, we want to verify that this decay is indeed possible. It will be allowed if baryon number, lepton number, mass-energy, and charge are all conserved. (a) Does strangeness also have to be conserved? (b) What is the total baryon number before and after the decay? (c) What is the total lepton number before and after the decay? (d) What is the total charge before and after the decay? (e) Is the mass of the K^+ greater than the mass of the decay products? (f) What is the strangeness before and after the decay? (g) Which force drives the transformation? Explain. How does this relate to the lifetime?

2. [I] THIS PROBLEM DEALS WITH QUANTUM NUMBERS AND CONSERVATION. One decay mode of the omega minus is given

$$\Omega^- \rightarrow \Xi^0 + \pi^-$$

Using the quantum numbers in Table 31.1, we want to verify that this decay is indeed possible. It will be allowed if baryon number, lepton number, mass-energy, and charge are all conserved. (a) Does strangeness also have to be conserved? (b) What is the total baryon number before and after the decay? (c) What is the total lepton number before and after the decay? (d) What is the total charge before and after the decay? (e) Is the mass of the Ω greater than the mass of the decay products? (f) What is the strangeness before and after the decay? (g) Which force drives the transformation? Explain.

3. [I] Knowing that Δ^* is a baryon with spin $\frac{3}{2}$ and 0 strangeness and that η and η' are mesons with 0 spin, 0 electromagnetic charge, and 0 strangeness, indicate which forces determine the following reactions: (a) $\eta \rightarrow \gamma + \gamma$ (b) $\Delta^* \rightarrow p + \pi$ (c) $\Lambda^0 \rightarrow \pi^- + p$ (d) $\eta' \rightarrow \eta + \pi + \pi$. Explain your answers fully.

4. [I] Determine the forces that are responsible for each of the following reactions: (a) $\pi^+ \rightarrow \mu^+ + \nu_\mu$ (b) $\pi^0 \rightarrow \gamma + \gamma$ (c) $\pi^- + p \rightarrow K^0 + \Lambda^0$. Explain your answers fully.

5. [I] State some of the conservation laws that are violated in each of the following reactions: (a) $\Lambda^0 \rightarrow \pi^+ + \pi^- + e^- + e^+$ (b) $\pi^+ \rightarrow \mu^+ + \gamma$ (c) $\Lambda^0 \rightarrow \pi^- + \pi^+ + p$ (d) $p + p \rightarrow \pi^+ + \Sigma^+$

6. [I] Give at least one conservation law violated in each of the following reactions: (a) $n + \nu_e \rightarrow e^+ + \pi^-$ (b) $p + \bar{\nu}_e + e^- \rightarrow \pi^0 + \gamma$ (c) $p + p \rightarrow \pi^+ + K^+ + n + n$ (d) $p + n \rightarrow K^- + K^+ + \pi^0$

7. [I] Does the reaction

$$p + n \rightarrow p + \bar{p} + p$$

occur and if not, why not?

8. [I] Is the reaction

$$\mu^- \rightarrow e^- + \bar{\nu}_e$$

possible and if not, why not?

9. [I] Is the reaction

$$\mu^- \rightarrow e^- + \nu_e + \nu_\mu$$

possible and if not, why not?

10. [I] Is the reaction

$$p + \pi^- \rightarrow p + \pi^- + \pi^0$$

possible and if not, why not? By what interaction could it proceed?

11. [I] Is the reaction

$$K^0 \rightarrow \pi^+ + \pi^- + \pi^0$$

possible and if not, why not? By which interaction will it proceed?

12. [I] Is the reaction

$$\pi^- + p \rightarrow K^0 + n$$

likely to be observed?

13. [I] Is the reaction

$$K^- \rightarrow \mu^- + \bar{\nu}_\mu$$

forbidden by any conservation law? How does it decay?

14. [I] Is the reaction

$$\pi^- + p \rightarrow K^+ + K^-$$

forbidden by any conservation law?

15. [I] Is the reaction

$$\pi^+ \rightarrow \mu^+ + \nu_\mu$$

forbidden by any conservation law? If not, how much energy will be released as KE as a result of the decay? (Assume the neutrino to have negligible mass and the pion to be at rest initially.)

16. [I] Determine the particle that is missing in each of the following reactions: (a) $p + (\ ) \rightarrow n + \mu^+$ (b) $p + p \rightarrow p + K^+ + (\ )$ (c) $p + p \rightarrow p + n + \pi^+ + (\ )$

17. [I] Consider the decay

$$\Lambda^0 \rightarrow \pi^- + p$$

which takes a rather long 10^{-10} s. Assuming the lambda to be at rest, what is the net KE of the resulting particles? What conservation law does the reaction not have to obey?

18. [I] If the reaction

$$\mu^- \rightarrow e^- + \bar{\nu}_e + \nu_\mu$$

can occur, what is the maximum energy carried by the two essentially massless neutrinos assuming the muon to be at rest?

19. [I] If a proton and an antiproton, both essentially at rest, were to completely annihilate each other, how much energy could be liberated?

20. [I] What is the maximum total kinetic energy shared by the pion and positron in the reaction

$$K^+ \rightarrow \pi^0 + e^+ + \nu_e$$

assuming the kaon to be at rest?

21. [I] Approximately how much energy must be provided to create a separated u-quark and $\bar{u}$-antiquark pair?

22. [I] What is the quark composition of an antiproton?

23. [I] If the mean lifetime of a proton were a mere 10^{30} y, how many protons would we have to put in a tank to observe an average of one decay per year?

24. [I] THIS PROBLEM DEALS WITH QUARK ATTRIBUTES. Table 31.3 lists several baryons that have the same quark content. These are distinguished by differences in quark spin. Thus the neutron (n) corresponds to u(spin-up) + d(spin-up) + d(spin-down), whereas the delta zero (Δ^0) corresponds to u(spin-up) + d(spin-up) + d(spin-up). We wish to determine the net charge and spin of each of these baryons. (a) What are the charges, in units of e, of each quark? (b) What is the net charge of the neutron? (c) What is the net charge of the delta zero? (d) Why do you think it's called delta *zero*? (e) What are the spins, in units of $\hbar$, of each quark in the neutron? (f) What is the net spin of the neutron? (g) What are the spins, in units of $\hbar$, of each quark in the delta zero? (h) What is the net spin of the delta zero?

25. [I] THIS PROBLEM DEALS WITH QUARK ATTRIBUTES. A positive pi-meson is composed of a quark u(spin-up) and an antiquark $\bar{d}$(spin-down), whereas a positive rho-meson is composed of a quark u(spin-up) and an antiquark $\bar{d}$(spin-up). We wish to determine the net charge and spin of each of these mesons. (a) What are the charges, in units of e, of each quark? Be careful with the antiparticle. (b) What is the net charge of the π-meson? (c) What is the net charge of the ρ-meson? (d) What are the spins, in units of $\hbar$, of each quark in the π-meson? (e) What is the net spin of the π-meson? (f) What are the spins, in units of $\hbar$, of each quark in the rho-meson? (g) What is the net spin of the ρ-meson?

26. [I] Suppose we decide to make a baryon using three quarks: u_R(spin-up) + u_G(spin-up) + u_G(spin-down). Are we likely to be successful? Explain.

> **SOLUTION:** No, the baryon has to be color neutral, which means $u_R u_G u_B$ or any permutation thereof.

27. [I] Suppose we decide to make a meson using two quarks: b_G(spin-up) + $\bar{u}_{\bar{G}}$(spin-down). Are we likely to be successful? Explain.

28. [II] Is the decay $\Xi^0 \rightarrow n + \pi^0$ forbidden and if so, why?

29. [II] Consider a mediating particle of mass m; derive an expression for its rest energy in terms of the range of the associated force.

30. [II] Two protons moving head-on toward each other at the same speed collide, causing the reaction

$$p + p \rightarrow p + p + \pi^0$$

Determine the minimum kinetic energy of each proton.

31. [II] Two protons moving toward one another at the same speed collide to produce the reaction

$$p + p \rightarrow n + p + \pi^+$$

What is the minimum amount of KE each must have for the reaction to take place?

32. [II] Consider the reaction

$$p + p \rightarrow p + \Lambda^0 + K^0 + \pi^+$$

Determine the minimum KE the protons must have if this event is to occur. For simplicity, we take the protons to be moving toward one another at the same speed.

33. [II] Determine the Q value of the reaction (i.e., the energy released)

$$\pi^- + p \rightarrow K^0 + \Lambda^0$$

Could this reaction take place if the pion and proton were moving slowly toward each other just before impact?

34. [II] How much energy would appear in the form of gamma rays if a positron with a KE of 10.0 MeV was to annihilate an electron with a KE of 20.0 MeV?

35. [II] A neutrino has an energy of 100 MeV. Assuming it has zero mass, what are its frequency, wavelength, and momentum? Use SI units.

36. [II] A D^+ meson has quantum numbers $B = 0$, $S = 0$, $C = +1$, and $Q = +1$ (and no topness or bottomness). What is its quark configuration?

37. [II] A D^0 meson has quantum numbers $B = 0$, $S = 0$, $C = +1$, and $Q = 0$ (and no topness or bottomness). What is its quark configuration?

38. [II] A K^- meson has quantum numbers $B = 0$, $S = -1$, $C = 0$, and $Q = -1$ (and no topness or bottomness). What is its quark configuration? What is the configuration of K^+? Explain your answer in detail.

39. [II] A Λ^0 baryon has quantum numbers $B = +1$, $S = -1$, $C = 0$, $Q = 0$, spin $= \frac{1}{2}$ (and no topness or bottomness). What is its quark configuration? Explain your answer in detail.

40. [II] Explain how, according to the quark model, this reaction takes place:

$$\pi^0 \rightarrow \gamma + \gamma$$

41. [II] Explain how, according to the quark model, this reaction

takes place

$$K^0 \rightarrow \pi^+ + \pi^-$$

(First write it out in terms of quarks and then discuss what happens.)

42. [II] Explain how, according to the quark model, this reaction takes place.

$$\Omega^- \rightarrow \Lambda^0 + K^-$$

(First write it out in terms of quarks and then discuss what happens.)

43. [II] Suppose an electron and a positron, each very nearly at rest, annihilate forming a single *virtual* photon. What would be its frequency?

44. [II] The energy at which the weak and electromagnetic forces unify into the electroweak force is around 100 GeV (Fig. 31.23). What temperature does that correspond to? (Check your answer against Fig. 31.22.)

45. [II] One GUT formulation supposes that quarks and leptons brought together at a distance of around 10^{-31} m will transform one into the other. Determine the mass of the gauge boson that would have such a range. Give your answer in GeV/c^2.

46. [II] Assume the body of a typical human being contains roughly 10^{28} protons. If the proton had a half-life of only 10^{10} years, what would be its disintegration rate in decays per second?

47. [II] With Problem 45 in mind, at what temperature will this unification occur? (Check your answer against Fig. 31.22.)

48. [II] GUTs predict that both neutrons and protons can decay in the nucleus via routes that do not conserve baryon number. Roughly how many nucleons are in a liter of water? If the mean lifetime of a nucleon was 10^{20} y, how many decays would occur in the liter of water per year? What does this result suggest about the lifetime of a proton?

Appendixes:
A Brief Mathematical Review

Appendix A   Algebra

Remember the following three logical rules:

1. The same quantity (constant or variable) can be added to or subtracted from both sides of an equation without changing the equality: $100 = 100$; $100 - 2 = 100 - 2$.

2. Both sides of an equation can be multiplied or divided by the same quantity (constant or variable) without changing the equality: $100 = 100$; $100/2 = 100/2$.

3. Both sides of an equation can be raised to the same power (squared or square rooted, cubed or cube rooted) without changing the equality: $100 = 100$; $\sqrt{100} = \sqrt{100}$.

Wherever possible, apply rule 1 first followed by rule 2. Isolate the variable to whatever power it is raised to and if that power is other than one, use rule 3 to make it one.

Let's solve $8x + 2 = 42$ for x. To get the x alone, first remove the 2 via rule 1, $8x = 42 - 2 = 40$. Next, use rule 2 to remove the 8; $x = 40/8 = 5$.

Now solve $5x^2 + 12 = 57$ for x. Apply rule 1: $5x^2 = 57 - 12 = 45$. Next isolate x^2 using rule 2 to divide both sides by 5: $x^2 = 9$. Now use rule 3, taking the square root of both sides: $x = \pm 3$. Note that there are two solutions: $+3$ and -3.

Solve $\frac{3}{4}t^2 - 6 = 0$ for t. Use rule 1 to move the 6, yielding $\frac{3}{4}t^2 = 6$. Now, using rule 2, multiply both sides by 4 to get $3t^2 = 24$. Use it again, dividing both sides by 3 to get $t^2 = 8$. Now, using rule 3, take the square root of both sides: $t = \sqrt{8}$.

Some Trouble Spots

1. Dividing by fractions can be troublesome. $8/\frac{1}{4}$ is *not* 2. *A fraction is not changed when the top and bottom are multiplied by the same quantity.* Here, multiply top and bottom by 4 to get $32/1$. Similarly, given $\frac{1}{4}/\frac{1}{2}$ multiply top and bottom by 2 to get $\frac{1}{2}/1 = \frac{1}{2}$.

2. $(a + b)^2$ is *not* equal to $a^2 + b^2$! $(a + b)^2 = (a + b)(a + b) = aa + ab + ba + bb = a^2 + 2ab + b^2$.

3. $\dfrac{1}{a} + \dfrac{1}{b}$ is *not* equal to $\dfrac{1}{a + b}$. We can only add terms with the same denominators. To get a common denominator (ab), multiply top and bottom by whatever is needed: $\dfrac{b}{ba} + \dfrac{a}{ab}$ and add. Thus

$$\frac{1}{a} + \frac{1}{b} = \frac{a + b}{ab}$$

4. Remember that $\sqrt{a}\sqrt{b} = \sqrt{ab}$; thus, $\sqrt{4}\sqrt{9} = \sqrt{36}$ or $2 \times 3 = 6$. Similarly, $\sqrt{8} = \sqrt{4}\sqrt{2} = 2\sqrt{2}$; hence, $\sqrt{25gt^2} = 5t\sqrt{g}$.

A-1   Exponents

Clearly, a raised to the *n*th *power* is a^n, where the *exponent n* can be any number, fractional or whole, positive or negative. Moreover, $a^2 \times a^3 = (a \times a) \times (a \times a \times a) = a^2 a^3 = a^5$;

$$(a^n)(a^m) = a^{n+m} \qquad \text{(A.1)}$$

Alternatively, if the base numbers (a and b) are different and the powers are the same, then

$$(a^n)(b^n) = (ab)^n \tag{A.2}$$

Now, for division: $2^3/2^2 = 8/4 = 2 = 2^1$; the two 2s on the bottom cancel two 2s on the top or, equivalently, the exponent on the bottom subtracts from that on the top. Thus

$$\frac{a^3}{a^2} = \frac{a \times a \times a}{a \times a} = a^{3-2} = a$$

As a rule

$$\frac{a^n}{a^m} = a^{n-m} \tag{A.3}$$

which suggests that

$$\frac{1}{a^m} = a^{-m} \tag{A.4}$$

Hence, $1/5 = 5^{-1}$, $1/3^2 = 3^{-2}$, and $1/4^{-2} = 4^2$. Inasmuch as $a/a = 1$, it follows from Eq. (A.3) that $a^{1-1} = a^0 = 1$. *Any quantity raised to the zero power is* 1. Fractional exponents correspond to *roots*;

$$a^{1/n} = \sqrt[n]{a} \tag{A.5}$$

Using Eq. (A.1), we have

$$\sqrt{a}\sqrt{a} = a^{\frac{1}{2}}a^{\frac{1}{2}} = a^1 = a$$

Since powers undo roots and vice versa, $(a^{1/n})^n = 1$. For example, $(25^{\frac{1}{2}})^2 = 5^2 = 25$. In general

$$(a^n)^m = a^{nm} \tag{A.6}$$

where n and m can be either whole or fractional numbers.$5t\sqrt{g}$.

A-2 Powers of Ten: Scientific Notation

Scientific notation is a shorthand way of writing numbers in terms of powers of ten:

$10^0 = 1$ $10^0 \ = 1$
$10^1 = 10$ $10^{-1} = 1/10 = 0.1$
$10^2 = (10 \times 10) = 100$ $10^{-2} = 1/(10 \times 10) = 0.01$
$10^3 = (10 \times 10 \times 10) = 1000$ $10^{-3} = 1/(10 \times 10 \times 10) = 0.001$
$10^4 = (10 \times 10 \times 10 \times 10) = 10\,000$ $10^{-4} = 1/(10 \times 10 \times 10 \times 10) = 0.000\,1$

and so forth. For positive exponents, *the number of zeros corresponds to the power of ten.*

Suppose we want to express a number greater than 1.0 given in ordinary form (such as the number of seconds in a year—31 560 000) in scientific notation. Start by indicating the decimal at its reference position (31 560 000.0). Then decide where you would like to move it to (for example, 31*560 000.0). Here the decimal is to be relocated six places to the left yielding 31.56, and multiplying by 10^6 will move it back six places to the right, where it started. Thus, $31\,560\,000 = 31.560 \times 1\,000\,000 = 31.56 \times 10^6$, or 3.156×10^7, or $0.315\,6 \times 10^8$. To write a number less than 1.0 in scientific notation (for example, 0.000 000 000 000 000 000 000 000 000 000 911 kg), again locate where you would like the decimal to appear (0.*000 000 000 000 000 000 000 000 000 000 9*11 kg). That's 31 places to the right, so multiplying 9.11 by 10^{-31} will shift the decimal back 31 places to the left where it started. The mass of an electron is 9.11×10^{-31} kg.

When multiplying or dividing numbers in scientific notation, process the numerical terms separately from the exponents; as

$$(1.1 \times 10^{12})(5.0 \times 10^{17}) = (1.1 \times 5.0)(10^{12} \times 10^{17}) = 5.5 \times 10^{29}$$

$$\frac{(1.1 \times 10^{12})}{(5.0 \times 10^{17})} = \frac{(1.1)}{(5.0)} \times \frac{(10^{12})}{(10^{17})} = 0.22 \times 10^{-5}$$

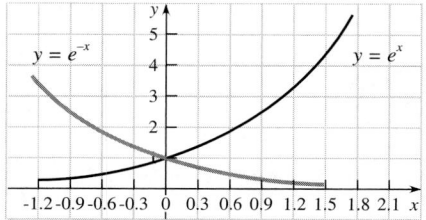

Figure A1

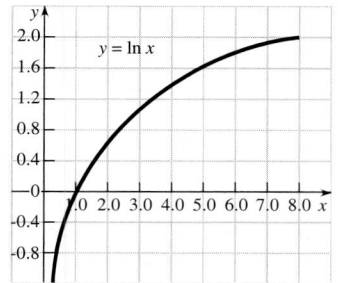

Figure A2

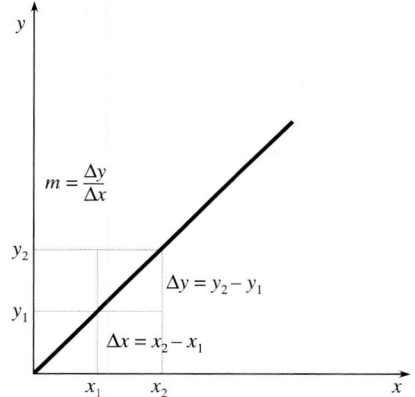

Figure A3

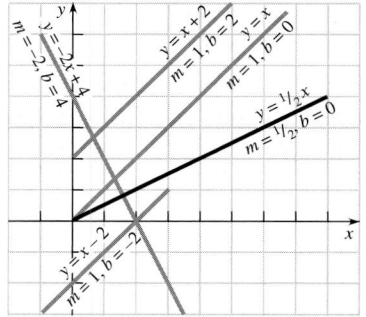

Figure A4

A-3   Logarithms

Suppose we have a positive number y expressed as a power of b where $b > 0$ and $b \neq 1$; accordingly $y = b^x$. Let's define x to be the *logarithm of y, to the base b*; thus

$$x = \log_b y$$

The logarithm x is the exponent, and the two equations are just different ways of saying the same thing. For example, if $2^3 = 8$, then $3 = \log_2 8$; if $8^{-4/3} = 1/16$, then $-4/3 = \log_8 1/16$. Accordingly,

$$\log(ab) = \log a + \log b$$

$$\log \frac{a}{b} = \log a - \log b$$

$$\log a^n = n \log a \qquad \log_b b = 1$$

$$\log 1 = 0$$

and

$$b^{\log_b a} = a$$

There are two widely used bases, 10 and $e = 2.718\ 281\ 828\dots$. The *common logarithms* arise when $y = 10^x$, whereupon $x = \log_{10} y$. The *natural logarithms* arise when $y = e^x$, whereupon $x = \ln y$ (it's customary to write $\ln y$ rather than $\log_e y$). The exponential function e^x (Fig. A1) and the natural log (Fig. A2) occur in a great many physical situations in which the rate-of-change of some quantity depends on that quantity. We can see this dependence in Fig. A1: the rate-of-change of the curve $y = e^x$ equals e^x. When the curve has small values, the slope is small; when the curve has large values, the slope is large. This property is very special and it means that if some quantity N changes in time exponentially, then $\Delta N / \Delta t \propto N$. It follows from the above that

$$\ln e^x = x \qquad \text{and} \qquad e^{\ln x} = x$$

A-4   Proportionalities and Equations

We frequently find that one physical quantity depends on another: the force exerted by a spring depends on the amount it's stretched; the current through a resistor depends on the voltage across it. Suppose the quantities A and B depend on one another such that doubling either doubles the other. They are said to be **directly proportional** to each other or, symbolically, $A \propto B$. On the other hand, it's possible that doubling one could halve the other, in which case $A \propto 1/B$, and they are **inversely proportional**. When two quantities are proportional, we can always turn the proportionality into an equality using an appropriate constant-of-proportionality. The circumference of a circle (C) is directly proportional to the diameter (D), where the constant of proportionality is π: $C = \pi D$.

When the variables in an equation occur only to the first power, it's called a **linear equation**; for example, $y = mx$, where m is a constant. Here, when $x = 0$, $y = 0$, and the curve (Fig. A3) is a straight line passing through the origin. For any two points on the line, $\Delta y / \Delta x$ is its **slope**, or tilt, and it equals m. If a constant (b) is added to y such that $y = mx + b$, the line shifts parallel to itself; b is the place where the line crosses the y-axis, the *y-intercept* (Fig. A4).

The Quadratic Equation

When an object is uniformly accelerating, it travels a distance s given by Eq. (3.9); namely, $s = v_i t + \frac{1}{2} a t^2$. This expression is typical of a class of equations that contains a variable (t) raised to the second power, and it's called a **quadratic equation**. The standard form has everything on the left set equal to zero on the right:

$$ax^2 + bx + c = 0$$

where a, b, and c are constants. This equation can be solved directly, but it's common practice to memorize the solution in the form

$$x_\pm = \frac{-b \pm \sqrt{b^2 - 4ac}}{2a}$$

where, provided $b^2 > 4ac$, there are two real solutions. For example, $2x^2 + 6x - 20 = 0$ corresponds to $a = 2$, $b = 6$, and $c = -20$; therefore

$$x_+ = \frac{-6 + \sqrt{36 - 4(2)(-20)}}{2(2)} = \frac{-6 + \sqrt{196}}{4} = +2$$

$$x_- = \frac{-6 - \sqrt{36 - 4(2)(-20)}}{2(2)} = \frac{-6 - \sqrt{196}}{4} = -5$$

Simultaneous Equations

For every unknown, we must have an independent equation. Situations with one and two unknowns are common throughout the discipline. Suppose there are two unknowns, x and y, and two equations $4x - 2y = 16$ and $3x + 4y = 23$. Lots of numbers satisfy each of these equations (for example, in the first one, $x = 4$, $y = 0$ or $x = 0$, $y = -8$). We want to solve these *simultaneously* so that the solutions satisfy both equations at the same time. There are two ways to accomplish this goal.

(1) *Solve either equation for either unknown in terms of the other unknown, and substitute that into the remaining equation.* From the first equation, $4x = 16 + 2y$, $x = 4 + \frac{1}{2}y$. Now put this result into the second equation so it has only one unknown $3(4 + \frac{1}{2}y) + 4y = 23$ and so $12 + 3y/2 + 4y = 23$, $3y/2 + 8y/2 = 23 - 12$, $11y/2 = 11$, $y/2 = 1$, $y = 2$. Put that back into the first equation: $4x - 2(2) = 16$, $4x = 20$, and $x = 5$.

(2) Alternatively, *multiply either or both of the equations by whatever numbers it takes to make the same coefficient appear for either one of the unknowns. Then add or subtract the two, thereby removing one unknown.* Here, to get rid of y, multiply the first equation $4x - 2y = 16$ by 2 getting

$$\begin{array}{r} 8x - 4y = 32 \\ 3x + 4y = 23 \\ \hline 11x \qquad = 55 \end{array}$$

and add

or $x = 5$. Substitute this result into either equation and solve for y.

A-5 Approximations

It's often desirable to approximate the solution to a problem, either as a quick check or because the exact solution is too elaborate to deal with. Newton's *Binomial Theorem*, expressed as

$$(a + b)^n = a^n + na^{n-1}b + \frac{n(n-1)}{2 \times 1}a^{n-2}b^2 + \frac{n(n-1)(n-2)}{3 \times 2 \times 1}a^{n-3}b^3 + \cdots \quad \text{(A.7)}$$

produces some very helpful approximations. In general, if n is a positive integer, the series ends in a finite number of terms, otherwise it has an infinite number. Expressions of the form $(1 + x)^n$ arise frequently and can be approximated using the Binomial Theorem ($a = 1$, $b = x$); thus

$$(1 + x)^n = 1 + nx + \frac{n(n-1)}{2}x^2 + \cdots \quad \text{(A.8)}$$

When x is very small ($x \ll 1$), the x^2 and higher terms will be negligibly small and

$$[x \ll 1] \qquad\qquad (1 + x)^n \approx 1 + nx$$

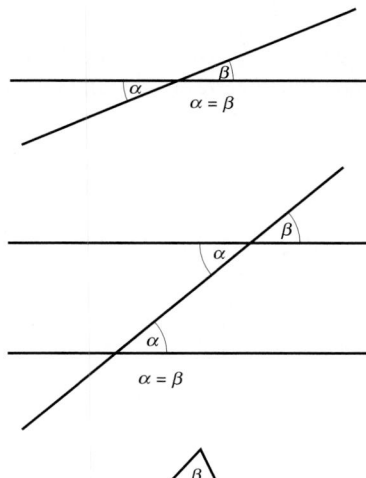

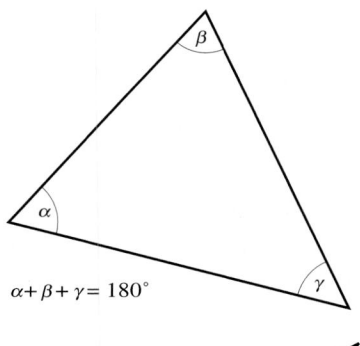

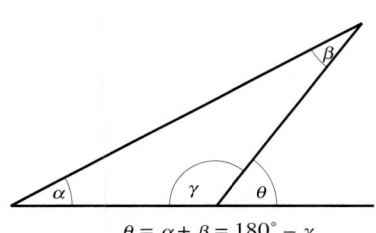

$\alpha = \beta$

$\alpha + \beta + \gamma = 180°$

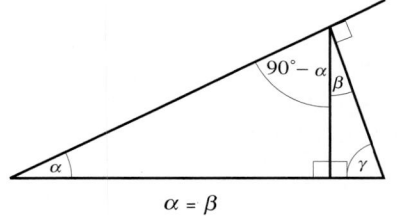

$\theta = \alpha + \beta = 180° - \gamma$

$90° - \alpha$

$\alpha = \beta$

Figure B1

with $n = 2$ $\qquad\qquad (1 + x)^2 \approx 1 + 2x$

with $n = 3$ $\qquad\qquad (1 + x)^3 \approx 1 + 3x$

with $n = \frac{1}{2}$ $\qquad\qquad (1 + x)^{\frac{1}{2}} = \sqrt{1 + x} \approx 1 + \frac{1}{2}x$

with $n = -\frac{1}{2}$ $\qquad\qquad (1 + x)^{-\frac{1}{2}} = \dfrac{1}{\sqrt{1 + x}} \approx 1 + -\frac{1}{2}x$

with $n = -1$ $\qquad\qquad (1 + x)^{-1} = \dfrac{1}{(1 + x)} \approx 1 - x$

What is the value of $a/(a + b)^2$ when $a >> b$? Since b is negligible compared to a, $(a + b)^2 \approx (a)^2$ and $a/(a + b)^2 \approx 1/a$. If $b >> a$, $(a + b)^2 \approx (b)^2$ and $a/(a + b)^2 \approx a/b^2$.

When a number is rounded off to the nearest power of 10, we say it's an **order-of-magnitude** figure. Thus, the order-of-magnitude of the acceleration due to gravity on Earth (9.81 m/s^2) is 10 m/s^2.

A-6   The Change in a Quantity

We will often encounter quantities that are formed of products such as $z = xy$. The question arises, what happens to z when either x or y or both change? The initial value of z is $z_i = xy$. Its final value, after a change in x of Δx, and a change in y of Δy, is

$$z_f = (x + \Delta x)(y + \Delta y)$$
$$z_f = xy + x\Delta y + y\Delta x + \Delta x \Delta y$$

Hence $\qquad\qquad \Delta z = z_f - z_i = x\Delta y + y\Delta x + \Delta x \Delta y$

When Δx and Δy are small, $\Delta x \, \Delta y$ is negligibly small and

$$\Delta z \approx x\Delta y + y\Delta x$$

This is an important relationship that we will make use of often.

Appendix B   Geometry

Figure B1 provides a number of useful relationships for the angles formed by intersecting lines and triangles. Of special interest is the Pythagorean Theorem, which relates the sides of any right triangle;

$$C = \sqrt{A^2 + B^2}$$

as in Fig. B2. There are a number of right triangles with whole number sides—the 3–4–5 and 5–12–13 are the most commonly encountered (Fig. B3).

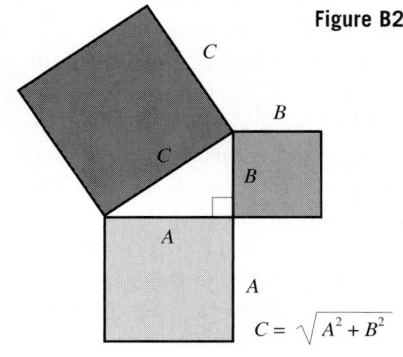

Figure B2

$C = \sqrt{A^2 + B^2}$

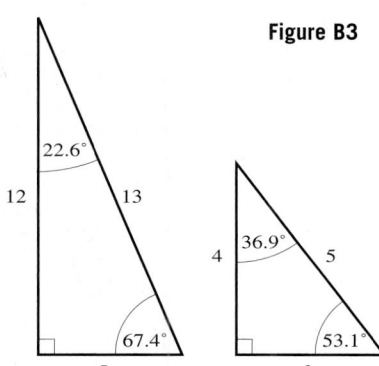

Figure B3

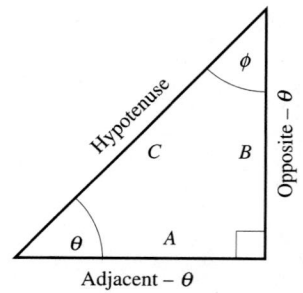

Figure C1

Appendix C   Trigonometry

Figure C1 shows a right triangle with an acute angle θ and the sides (A) adjacent and (B) opposite to it. The various possible ratios are given names, and electronic calculators are programmed to compute them. Knowing θ we can determine any ratio and vice versa. Thus $\sin \theta = opposite/hypotenuse$, $\cos \theta = adjacent/hypotenuse$, and $\tan \theta = opposite/adjacent$. Referring to Fig. C1, observe that $\sin \theta = B/C = \cos \phi$ and that $\cos \theta = A/C = \sin \phi$. Some useful relationships are

$$\tan \theta = \frac{\sin \theta}{\cos \theta}$$

$$\sin^2 \theta + \cos^2 \theta = 1 \qquad \text{and} \qquad \sin 2\theta = 2 \sin \theta \cos \theta$$

For all triangles (Fig. C2)

$$\text{Law of Sines} \quad \frac{\sin \alpha}{A} = \frac{\sin \beta}{B} = \frac{\sin \gamma}{C}$$

$$\text{Law of Cosines} \quad C^2 = A^2 + B^2 - 2AB \cos \gamma$$

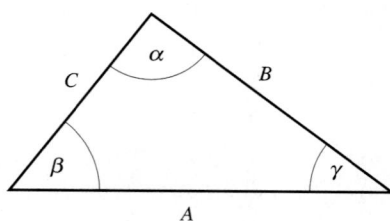

Figure C2

Appendix D   Vectors

Vectors are discussed at length in Chapter 2. Here, we'll go over some of the trouble spots often encountered when dealing with them. Figure 2.26 on p. 37 shows a vector $\vec{\mathbf{C}}$ in each of the four quadrants and the corresponding x- and y-components. In each case

$$C_x = |\vec{\mathbf{C}}| \cos \theta \quad \text{and} \quad C_y = |\vec{\mathbf{C}}| \sin \theta$$

where θ is measured either up or down from the x-axis and is always acute. Now, suppose we add several vectors along the x-axis and get $\vec{\mathbf{C}}_x$ and add several along the y-axis and get $\vec{\mathbf{C}}_y$. The next step in determining $\vec{\mathbf{C}}$ is to use $\tan \theta = C_y/C_x$ to find θ. It's best to write the tangent as $\tan \theta = |C_y|/|C_x|$, using the absolute values of the components so that $\tan \theta$ is always positive and θ is always between 0° and 90°. Thus, the inverse tangent of -1.732 is both 120° and 300°, lying in either the second or fourth quadrants. A calculator provides another form of the answer—namely, $-60°$ (equivalent to 300°). It neglects 120° altogether, which means that if we take the ratio of the components when finding the tangent and one or both of them are negative, there will be some ambiguity in the resulting angle. Use the ratio of absolute values; determine which quadrant the resultant is in by drawing C_x and C_y. Then measure $\theta < 90°$ up or down from the horizontal axis, as in Fig. 2.26.

Appendix E   Dimensions

Length is measured in units of feet, light-years, meters, and so forth. The type of unit is the *dimension* of the quantity; in this case, the dimension of length indicated as [L]. Area is a distance times another distance and so has the dimensions of [L][L] = [L^2]. We can assign dimensions to all the quantities in mechanics using the basic dimensions of *length* [L], *time* [T], and *mass* [M]. Thus, speed, which is measured in units of m/s, ft/s, mi/h, km/h, and so on, has the dimensions of length/time or [L]/[T].

The volume of a sphere of radius R is $\frac{4}{3}\pi R^3$, whereas that of a cube of side a is a^3. Each has the dimensions of volume [L^3]. This suggests that dimensional analysis can be used to check our equations. Consider the equation representing three physical quantities A, B, and C, as $C = A + B$. There are three rules. The first is: *dimensions can be treated as algebraic quantities*—added, subtracted, multiplied, and divided accordingly. Furthermore, A and B must have the same units (we can't add seconds to meters). Our second rule is: *only quantities having the same dimensions can be added or subtracted*. Whatever it might be, it's clear that C must have the same units as $A + B$. Our third rule is: *the dimensions on either side of an equal sign must be the same*. All physical equations must obey these rules.

Answer Section

Chapter 1

Answers to Odd-Numbered Multiple Choice Questions
1. c **3.** d **5.** a **7.** a **9.** b **11.** a **13.** b **15.** e **17.** c **19.** c **21.** b

Answers to Odd-Numbered Problems
1. $10\,000\,000\,000 = 10^{10}$ **3.** (a) 0.010 s (b) 0.001 s (c) 10 000 s (d) 100 000 000 s (e) 0.000 001 s **5.** 1.00×10^7 mm **7.** 269 m **9.** 76.2 mm **11.** 7.874 in. **13.** 5×10^2 nm **15.** 9.46×10^{15} m **17.** 4.0×10^{13} km **19.** 8.4×10^{10} bacteria **21.** 2×10^2 **23.** 25 marbles **25.** $t = 1/500$ s $= 2$ ms $= 2 \times 10^3 \,\mu$s $= 2 \times 10^6$ ns **27.** 4×10^9 reactions in 10 hours **29.** 0.393 7 in./cm **31.** 1.6×10^3 m, or 1.6 km **33.** 9.290×10^{-2} m^2 **35.** 2.37×10^{-4} m^3 **37.** 2.00 cm **39.** 1×10^{-7} g/particle; total mass shed in 50 years $\approx 0.3 \times 10^2$ kg **41.** 1.6×10^5 kg **43.** 1.000×10^5 g in scientific notation **45.** (a) 2 significant figures (b) 2 significant figures (c) 3 significant figures (d) 4 significant figures (e) 3 significant figures (f) 1 significant figure **47.** $\pi \approx 3$ (to one significant figure); ≈ 3.14 (to three significant figures); ≈ 3.142 (to four significant figures); $\approx 3.141\,6$ (to five significant figures) **49.** 1.000 liter = 1.000 $\times 10^{-3}$ m^3 **51.** $V = 2.2 \times 10^{19}$ m^3 **53.** 1.91 m^3 **55.** 1806.8 m **57.** 1.00 kg **59.** $\approx 1 \times 10^{30}$ to $\approx 2 \times 10^{31}$ **61.** $A \approx 2\pi r(r + h)$, $V \approx \pi r^2 h$, $N \approx 2 \times 10^{13}$ cells **63.** $\approx 3 \times 10^{-7}$ cm = 3 nm

Solutions to Selected Problems
3. (a) 0.010 s; (b) 0.001 s; (c) 10 000 s; (d) 100 000 000 s; (e) 0.000 001 s. **9.** 1.00 in. = 2.54 cm = 25.4 mm; 3.00 in. = 76.2 mm. **15.** $(5.88 \times 10^{12}$ mi$)(1.609$ km/mi$) = 9.46 \times 10^{15}$ m. **21.** 1 kg = 1000 g; (1000 g)/(5 g) = 200 = 2×10^2. **31.** (45 hairs/d)(365 1/4 d/y) = 16436 hairs/y; (16 436)(0.10 m) = 1.6×10^3 m. **35.** 1 liter = 1000 cm^3 = 10^{-3} m^3; hence, 1.00 cup = 237 $\times 10^{-6}$ m^3. **51.** $V = (4/3)\pi R_{\,c}^3 = 2.199\,1 \times 10^{19}$ m^3 = 2.2×10^{19} m^3. **55.** These add up to 1806.818 m,

and the least number of decimal places is one, so the answer is 1806.8 m.

Chapter 2

Answers to Odd-Numbered Multiple Choice Questions
1. d **3.** b **5.** d **7.** c **9.** c **11.** d **13.** a **15.** c **17.** e **19.** b **21.** c **23.** b

Answers to Odd-Numbered Problems
1. (a) 5.0×10^3 m (b) 9.00×10^2 s (c) 5.6 m/s **3.** 29.5 km/s **5.** 22 m/s **7.** 1×10^9 s **9.** 1.000 m/s = 3.600 km/h = 2.237 mi/h = 3.281 ft/s **11.** 125 km/h **13.** (a) 60 km/h, 20 km/h, (b) 30 km/h **15.** average speed = 9.80 km/h, average driving speed = 10 km/h **17.** 120 km/h **19.** 6.7 km/h **21.** 12 km/h **23.** (a) 0-10 s, 16-20 s (b) 20-34 s (c) 20 m/s **25.** (a) 1.0×10^{-9} s (b) 3.336×10^{-6} s **27.** 2.0 m; $v_{max} = 4.0$ m/s, $v_{min} = 0$; from 1.5 to 3.3 s and from 4.0 s to 5.0 s **29.** 15 m **31.** 0.33 m/s **33.** 1.7 km **35.** 4s **37.** $v = 16.0$ m/s, $l = 1.6 \times 10^3$ m **39.** proof **41.** 1.4 m **43.** 15 km **45.** 1005 steps, at 53.1° W of N **47.** $D = 9.99$ m **49.** (b) it's the resultant (c) 30.0 m **51.** zero, 36 km **53.** $\vec{s} = 24$ m–straight up **55.** $\vec{C} = 5$ units–north **57.** $\vec{C} = 5$ units –due N **59.** $\vec{s} = 63$ m–72° up from horizontal **61.** 36.1 m **63.** $s = 300$ m **65.** $v_{av} = 4.3$ m/s; $\vec{v}_{av} = 0.30$ m/s–south **67.** $\vec{v}_{av} = 2.84$ km/h– north; v_{av} cannot be determined **69.** $\theta_1 = 67.38°$; $\theta_2 = 22.62°$ **71.** 12 m **73.** 74 km/h **75.** (a) $v_x = 254$ m/s (b) $v_y = 159$ m/s (c) $v = 300$ m/s **77.** v = constant from 0 to 4.0 s, from 5.0 to 6.0 s, from 6.0 to 7.0 s, and from 7.0 to 8.0 s; $v = 1.0$ m/s at $t = 2.0$ s; $v = 0$ at $t = 6.5$ s; v changes its direction at around $t = 4.5$ s **79.** (a) zero (b) $\vec{v} = 0.4$ m/s–south **81.** (b) 224 m/s **83.** On the way up $v_{av} = 150$ m/s and $\vec{v}_{av} = 150$ m/s–upward; for the round trip $v_{av} = 150$ m/s and $\vec{v}_{av} = 0$ **85.** $\vec{v}_{av} = 2.1$ m/s–southeast **87.** (a) $\vec{v}_{av} = 2 \times 10^2$ m/s –north **89.** 100 s **91.** 1.0×10^3 m **93.** 15.0

m/s, 18.0 m/s, 0.50 s **95.** 10 m/s **97.** 224 m/s **99.** $v_{av} = 1.3 \times 10^3$ km/h **101.** $\theta = 11°$; $v = 10$ m/s **103.** (a) no (b) $\theta = 42°$ (c) 36 km/h **105.** 22 km/h, at 3° W of N **107.** (a) $v_{RS} = 1470$ km/h; $v_{RC} = 0$ (b) $v_{RC} = 2940$ km/h (c) eastward (d) Florida is closer to the equator.

Solutions to Selected Problems
3. $v_{av} = l/t = (885$ km$)/(30.0$ s$) = 29.5$ km/s. **9.** 1.000 m/s =

$$\frac{(1.00 \text{ m})/(1000 \text{ m/km})}{(1.000 \text{ s})(60.00 \text{ s/min})(60.00 \text{ min/h})} =$$

3.600 km/h; 1.000 m/s =

$$\frac{(1.000 \text{ m})(3.281 \text{ ft/m})/(5280 \text{ ft/mi})}{(1.000 \text{ s})/(60.00 \text{ s/min})(60.00 \text{ min/h})} =$$

2.237 mi/h; 1.000 m/s =

$$\frac{(1.000 \text{ m})(3.281 \text{ ft/m})}{(1.000 \text{ s})} = 3.281 \text{ ft/s}.$$

19. L is the one-way distance; $v_{av} = l/t = 2L/$ $[L/(10$ km/h$) + L/(5.0$ km/h$)] = 2/[1/(10$ km/h$) +1/(5.0$ km/h$)] = 20/(3.0$ km/h$) = 6.7$ km/h. **33.** Assuming the time for the light to reach you is negligible, $l = (3.3 \times 10^2$ m/s$)(5.0$ s$) = 1.7$ km. The error incurred by not including the time for light to travel is a few millimeters. **39.** $l = Ct^2$; $(l + \Delta l) = C(t + \Delta t)^2 = C[t^2 + 2t\Delta t + (\Delta t)^2]$; $\Delta l = Ct^2 + 2Ct\Delta t + C(\Delta t)^2 - 1$; $v_{av} = \Delta l/\Delta t = [2Ct\Delta t + C(\Delta t)^2\,]/\Delta t = 2Ct + C\Delta t$. Letting $\Delta t \to 0$, in $v_{av} = 2Ct + C\Delta t$, the second term vanishes and $v = 2Ct$. **43.** This is a 3-4-5 right triangle, so the displacement is 15 km. **51.** Zero, 36 km. **65.** $v_{av} = l/t = (43$ m$)/(10$ s$) = 4.3$ m/s; $\vec{v}_{av} = \vec{s}/t = (3.0$ m—south$)/(10$ s$) = 0.30$ m/s— south. **69.** $\tan \theta_1 = 12/5 = 2.400$, $\theta_1 = 67.38°$; $\tan \theta_2 = 5/12 = 0.416\,7$, $\theta_2 = 22.62°$. **71.** $\tan 25° = h/(25$ m$) = 0.466$; $h = 11.66$ m, or 12 m. **75.** (a) $v_x = v \cos \theta = (300$ m/s$) \cos 32.0° = 254$

m/s (b) $v_y = v \sin \theta = (300 \text{ m/s}) \sin 32.0° = 159$ m/s (c) $[v_x^2 + v_y^2]^{1/2} = v = [(254 \text{ m/s})^2 + (159$ m/s$)^2]^{1/2} = 300$ m/s.   **79.** (a) The slopes are zero, as are both velocities. (b) The slope is -2.0 m in 5.0 s or $v = (-2.0 \text{ m})/(5.0 \text{ s}) = -0.4$ m/s and so $\vec{v} = 0.4$ m/s—south.   **85.** From Eq. (2.6), $\vec{v}_{av} = (\vec{s}_f - \vec{s}_i)/\Delta t$; $\Delta \vec{s} = [(3.0 \text{ m})^2 + (3.0 \text{ m})^2]^{1/2}$ –SOUTHEAST $= 4.24$ m—SOUTHEAST; $\vec{v}_{av} = 2.1$ m/s –SOUTHEAST.   **87.** (a) $\vec{v}_{av} = (600 \text{ m})/(3 \text{ s})$ –NORTH $= 2 \times 10^2$ m/s– NORTH; (b) also 2×10^2 m/s–NORTH.   **89.** Each runs 500 m in time t at speed 5.00 m/s; $(500 \text{ m})/(5.00 \text{ m/s}) = 100$ s. Alternatively, together they cover the track at a speed of 5.00 m/s + 5.00 m/s = 10.0 m/s; hence, $(1000 \text{ m})/(10.0 \text{ m/s}) = 100$ s.   **95.** $\vec{v}_{MJ} = \vec{v}_{ME} + \vec{v}_{EJ}$ and $\vec{v}_{ME} \perp \vec{v}_{EJ}$; so $[v_{ME}^2 + v_{EJ}^2]^{1/2} = [(8.0$ m/s$)^2 + (6.0 \text{ m/s})^2]^{1/2} = 10$ m/s.   **99.** $\cos \theta = R_\theta/R$; $R_\theta = R \cos \theta = (6400 \text{ km}) \cos 41° = 4830$ km; the circumference is $2\pi R_\theta = 3.0 \times 10^4$ km; $v_{av} = (3.0 \times 10^4 \text{ km})/(24 \text{ h}) = 1.3 \times 10^3$ km/h, or 7.9×10^2 mi/h.   **103.** (a) No. Those velocity vectors cannot form a right triangle because the hypotenuse ($\vec{v}_{BW}$) must be larger than either of the other sides, and here $v_{BW} = v_{WE}$ (b) $\vec{v}_{BE}$ cuts directly across the river. Head at $\vec{v}_{BW}$ somewhat upstream at an angle θ such that $\vec{v}_{BE} = \vec{v}_{BW} + \vec{v}_{WE}$, $\sin \theta = v_{WE}/v_{BW} = (32 \text{ km/h})/(48 \text{ km/h}) = 0.667$; $\theta = 42°$. (c) $v_{BE} = [v_{BW}^2 - v_{WE}^2]^{1/2} = [(48 \text{ km})^2 - (32 \text{ km}/\text{h})^2]^{1/2} = 36$ km/h.   **107.** (a) $v_{RS} = 1470$ km/h; $\vec{v}_{RC} = \vec{v}_{RS} + \vec{v}_{SC} = 1470$ km/h–west + 1470 km/h–east = 0. (b) $\vec{v}_{RC} = 1470$ km/h–east + 1470 km/h–east = 2940 km/h–east. (c) Eastward. (d) Florida is closer to the equator and is moving faster than Maine.

Chapter 3

Answers to Odd-Numbered Multiple Choice Questions

1. b   **3.** d   **5.** a   **7.** a   **9.** b   **11.** d   **13.** e **15.** d   **17.** a   **19.** c   **21.** d

Answers to Odd-Numbered Problems

1. (a) average acceleration (b) $t = 0$ (c) zero   **3.** 10 m/s²   **5.** 0.17 m/s²–north   **7.** 0.23 m/s²   **9.** 18 m/s²   **11.** 9.67 m/s²–downward   **13.** (a) 12.0 m/s (b) zero (c) 20 s (d) no   **15.** $a_{av} = -8.0$ m/s², $v = 1.6$ m/s   **17.** The sprinter's speed increases with every step, while his acceleration is decreasing. $(a_{av})_1 : (a_{av})_2 : (a_{av})_3 = 3.0 : 1.2 : 0.8$ **19.** -17 km/s²   **21.** (a) instantaneous acceleration (b) no (c) zero (d) zero   **23.** 5.0 m/s²   **25.** $a = 0$ at $t = 3.0$ s; $a < 0$ at $t = 3.7$ s; $a > 0$ at $t = 1.1$ s; $a = 0$ at $t = 0.25$ s   **27.** The greatest accelerations are reached by cyclists (1), (2) and (3) at $t = 2.1$ s, 3.3 s, and 6.5 s, respectively.   29. -7.5 m/s²   **31.** 20 m/s²; at $t = 0$ and $t = 3.0$ s, $a = 0$ **33.** 3.5 m/s² **35.** (a) $a_{av} = 0$ (b) $a = -1.1$ m/s²   **37.** $a = A + 2Bt$ **39.** 25 m   **41.** 50.0 km/h   **43.** 8.0 m/s **45.** $v_{av} = 30.2$ m/s, $v_f = 60.5$ m/s.   **47.** (a) 3.70 m/s (b) 8.11 s (c) 0.617 m/s²   **49.** $v_{av} = 10$ m/s, $v = 7$ m/s   **51.** 64 m   **53.** (a) zero (b) 6.0 m/s² (c) 70 m/s   **55.** $v_f = 4.7$ m/s   **57.** $v_f = 5.9$ m/s **59.** $v_i = 48$ m/s or about 107 mi/h   **61.** $a = -0.9 \times 10^6$ m/s²   **63.** 0.2 s   **65.** $s_3 - s_2 = 1.6$ m

67. 244.5 m and they don't collide.   **69.** (a) 70 m (b) 90 m   **71.** 0.1×10^2 m   **73.** 10 m/s²   **75.** $a_{min} = 0.020$ m/s²   **77.** 74 m   **79.** 0.12 km   **81.** (a) zero (b) 20 m/s²   **83.** If your friend catches the ruler after it has dropped by s, then his or her reaction time t can be estimated from $s = \frac{1}{2}gt^2$.   **85.** 3.1 m/s; 10 ft/s; 7.0 mi/h   **87.** -9.8 m/s, -20 m/s, -49 m/s, -98 m/s; -4.9 m, -20 m, -0.12 km, -0.49 km   **89.** $v_f = 423$ m/s $= 947$ mi/h **91.** s versus t graph is a parabola.s versus t^2 graph is a straight line.   **93.** (a) 27.8 m/s (b) 128 m/s **95.** $v_i = 4.9$ m/s   **97.** 127.5 m   **99.** $a = -gs_{max}/s_a$ **101.** $t = (1.53 \pm 0.55)$ s   **103.** 6 m **105.** (a) zero (b) 1.00 s (c) 16.0 m   **107.** proof **109.** 20 m   **111.** 9.0 m   **113.** The equation must be wrong.   **115.** (a) 76.6 m/s and 64.3 m/s (b) 299 m (c) 1.44 s   **117.** 11 m/s   **119.** 24.6 m, $v_f = 50.8$ m/s   **121.** $v_f = 35.9$ m/s   **123.** proof **125.** $24.1t^4 - 255t^3 + 900t^2 - 400 = 0$, t in seconds

Solutions to Selected Problems

5. Taking north as positive, $a_{av} = \Delta v/\Delta t = [(+10$ m/s$) - (-10 \text{ m/s})]/(120 \text{ s}) = 0.17$ m/s²–north. **9.** $a_{av} = \Delta v/\Delta t = (244 \times 0.447 \ 0 \text{ m/s})/(6.2 \text{ s}) = 18$ m/s².   **19.** $a_{av} = \Delta v/\Delta t = (-60 \text{ km/h} \times 1000$ m/km $\div 3600 \text{ s/h})(1 \text{ s}/1000) = -17$ km/s². **31.** Draw a line on the curve from $t = 0$ to $t = 3$ s: the rise is $6(10 \text{ m/s})$; the run is 3.0 s; the slope is $(60 \text{ m/s})/(3.0 \text{ s}) = 20$ m/s².   **37.** $v = At + Bt^2$; ($v + \Delta v) = A(t + \Delta t) + B(t + \Delta t)^2 = At + A\Delta t + Bt^2 + B(\Delta t)^2 + B2t\Delta t$; $\Delta v = A\Delta t + B(\Delta t)^2 + 2Bt\Delta t$; $a_{av} = \Delta v/\Delta t = A + 2Bt + B\Delta t$; as $\Delta t \to 0$, $a = A + 2Bt$.   **55.** $v_f^2 = v_i^2 + 2as = (1.5 \text{ m/s})^2 + 2(1.0$ m/s²$)(10 \text{ m}) = 22$ m²/s²; $v_f = 4.7$ m/s.   **63.** $s = v_i t_R - v_i^2/2a$; $t_R = (23.3 \text{ m})/(16.7 \text{ m/s}) + (16.7$ m/s$)/2(-7 \text{ m/s}^2) = 0.2$ s.   **67.** To stop, the Express needs $s_E = -v_i^2/2a = -(26.67 \text{ m/s})^2/2(-4 \text{ m/s}^2) = 88.9$ m ; the Flyer needs $s_F = -v_i^2/2a = -(30.56 \text{ m/s})^2/2(-3 \text{ m/s}^2) = 155.6$ m; for a total of 244.5 m and they don't collide.   **73.** It covered 20 m in 2.0 s; $s = v_i t + \frac{1}{2}at^2$; the initial relative speed was zero; $a = 2s/t^2 = 10$ m/s².   **79.** His initial speed with respect to the train is 50 km/h, and he passes it as if it were standing still. Hence $s = v_i t + \frac{1}{2}at^2$, 500 m = (13.89 m/s)$t$ + $\frac{1}{2}(10 \text{ m/s}^2)t^2$, and using the quadratic formula, $t = [-(13.89 \text{ m/s}) \pm \sqrt{(13.89 \text{ m/s})^2 - 4(5.0 \text{ m/s}^2)(-500 \text{ m})]}/2(5.0$ m/s²$) = (-13.89 \text{ m/s} \pm 100.96 \text{ m/s})/(10 \text{ m/s}^2)$ $= 8.7$ s; the train travels $s = (13.89 \text{ m/s}) \times (8.7 \text{ s})$ $= 0.12$ km.   **89.** Taking down as positive, $v_f^2 = v_i^2 + 2as = 0 + 2(9.81 \text{ m/s}^2)(9144 \text{ m})$; $v_f = 423$ m/s $= 947$ mi/h.   **95.** $s = v_i t + \frac{1}{2}at^2$; $0 = v_i(1.0 \text{ s}) + \frac{1}{2}(-9.81 \text{ m/s}^2)(1.0 \text{ s})^2$; $v_i = 4.9$ m/s.   **99.** The lift-off speed is v_1, where $v_1^2 = 2as_a$. During the deceleration, $0 = v_1^2 + 2gs_{max}$ or $v_1^2 = -2gs_{max} = 2as_a$; $a = -gs_{max}/s_a$.   **103.** Down is positive; over his height $s = v_i t + \frac{1}{2}at^2$; 2 m = $v_i(0.20 \text{ s}) + \frac{1}{2}(9.81 \text{ m/s}^2)(0.20 \text{ s})^2$; $v_i = 9.02$ m/s at top of his head; at ground $v_f = v_i + gt = 9.02$ m/s + (9.81 m/s²)(0.20 s) = 10.98 m/s; for total fall, $v_f^2 = v_i^2 + 2as_B$; $(10.98 \text{ m/s})^2 = 0 + 2(9.81 \text{ m/s}^2)s_B$; $s_B = 6.1$ m.   **107.** $v_{ix} = v_i \cos \theta$, $v_{iy} = v_i \sin \theta$, hence Eq. (3.16) yields the desired expression.   **109.**

(Taking down as positive) $s_y = \frac{1}{2}gt^2 = \frac{1}{2}(9.81 \text{ m/s}^2)(2.0 \text{ s})^2 = 19.6$ m, or to two figures, 20 m.   **119.** To avoid the quadratic, use Eqs. (3.10) and (3.8) applied to the vertical motion, where $v_{iy} = (40.0$ m/s) $\sin 60.0° = 34.64$ m/s. Taking down as positive, $v_{fy}^2 = (34.64 \text{ m/s})^2 + 2(9.81 \text{ m/s}^2)(50.0 \text{ m})$; and $v_{fy} = 46.70$ m/s. $s_y = 50.0$ m $= \frac{1}{2}(v_{iy} + v_{fy})t = (40.67 \text{ m/s})t$ and $t = 1.23$ s. It strikes the water a distance downrange of $v_x t = (40.0 \text{ m/s})(\cos 60.0°)(1.23 \text{ s}) = 24.6$ m. It hits at a speed of $v_f = (v_{fx}^2 + v_{fy}^2)^{1/2} = 50.8$ m/s.   **125.** Using the Pythagorean Theorem, we obtain $(20 \text{ m})^2 = (v_x t)^2 + (v_{iy}t + \frac{1}{2}gt^2)^2$, $v_{ix} = v_i \cos \theta = (30 \text{ m/s}) \cos 60°$ and $v_{iy} = v_i \sin \theta = (30 \text{ m/s}) \sin 60°$. $24.1t^4 - 255t^3 + 900t^2 - 400 = 0$.

Chapter 4

Answers to Odd-Numbered Multiple Choice Questions

1. d   **3.** e   **5.** c   **7.** c   **9.** b   **11.** c   **13.** c **15.** d   **17.** c   **19.** d   **21.** a   **23.** a   **25.** a **27.** d   **29.** c

Answers to Odd-Numbered Problems

1. $\vec{v}_M = 10$ m/s-horizontal   **3.** 1.000 0 m   **5.** (a) 100 N (b) 100 N (c) 100 N   **7.** 100 N   **9.** 100 N on each person   **11.** $F = 298$ N at $\theta = +35°$ **13.** 0.12 kN at $\theta = +54°$ **15.** proof   **17.** 500 N at 77° up from x-axis   **19.** (b) 130° (c) 100 N   **21.** one due east; the other two, $\pm 60°$ **23.** 400 N–straight up

25.

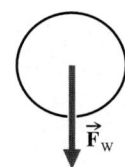
$\vec{F}_W$

27. 0.033 m/s²   **29.** -71 N   **31.** $F_f = 8.66$ N **33.** $a = 1.3$ m/s² at 23° N of E   **35.** 1.53 kN   **37.** 6.4×10^3 N   **39.** 1.0 kN   **41.** -13.9 kN   **43.** -0.12 MN   **45.** $\vec{a} = 13$ m/s²–upward   **47.** (a) 3.0 N (b) 0.510 kg (c) 5.9 m/s²   **49.** 200 N   **51.** 9.90 m/s²   **53.** 4.81 m/s²   **55.** $a = 1.7$ km/s², $F_{av} = 7.8 \times 10^{-4}$ N   **57.** $v_f = 12$ m/s   **59.** $\theta = 22°$ **61.** proof   **63.** 203 N   **65.** 98.1 N   **67.** $F_T = 300$ N, $F = 200$ N, downward   **69.** $a = 0$, $F_T = 98.1$ N   **71.** 49.1 N   **73.** (a) 3.0 N (c) no (e) $\frac{3}{8}g$ **75.** $F_{T1} = 1.00$ kN, $F_{T2} = 475$ N, $F_f = 125$ N   **77.** $a = 1.4$ m/s², $F = 1.3 \times 10^2$ N   **79.** $\vec{a} = 0.2$ m/s²–upward   **81.** $F_{T1} = 0.03$ kN, $F_{T3} = 0.02$ kN, $a = \frac{1}{4}g$, the tensions are internal to the 3-body system, the weights are external   **83.** $\mu_s = 0.53$ **85.** $a_{max} = \mu_s g$ **87.** 0.31   **89.** Crate slides toward back of truck   **91.** 20 m   **93.** 3 s   **95.** (a) 1.36 N (d) 0.173   **97.** (a) $a = 2 - \mu_k g$ (b) 40.0 N   **99.** $\mu_k = 0.070$ **101.** $\mu_k = 0.30$ **103.** (a) $a = 0$ (b) $a = 1$ m/s²   **105.** $\theta = 0.2 \times 10^2$ degrees   **107.** $\mu_k = 0.11$ **109.** 0.20 kN in the upper rope, 98 N in the lower one   **111.** $F_{Tb} = 0$, $F_{Tl} = 98$ N   **113.** $\theta = 38°$, $F_T = 16$ N   **115.** 39.7 N   **117.** $F_T = 50.0$ N, 100 N   **119.** 5.4 N   **121.** 6.7 N   **123.** $F_T = 3.1$ kN   **125.** 15.0°, from the vertical, 193 N   **127.** (b) 200 N (c) 48.2° (d) 224

N (e) $F_{T1} = 224$ N, $F_{T2} = 300$ N, $F_{T3} = 200$ N (f) 224 N   **131.** $F_{RH} = 866$ N, to the right; $F_{RV} = 1.50$ kN, upward   **133.** $F_R = 3.44$ kN upward; 44.3°   **135.** proof   **137.** $F_{RB} = 20.0$ N, $F_{RA} = F_{RC} = 17.3$ N, $F_{RD} = 30.0$ N

Solutions to Selected Problems

1. At maximum altitude, $v_y = 0$, hence $\vec{v}_M = 10$ m/s–horizontal.   **9.** The force is 100 N on each person.   **11.** $F_{1x} = (100$ N$)$ cos 45° = 70.7 N; $F_{2x} = (200$ N$)$ cos 30° = 173.2 N; $F_{1y} = (100$ N$)$ sin 45° = 70.7 N; $F_{2y} = (200$ N$)$ sin 30° = 100 N; $F^2 = (243.9$ N$)^2 + (170.7$ N$)^2$; $F = 298$ N; at $\theta = $ tan^{-1} 170.7/243.9 = +35°.   **17.** There is 90° between the two vectors, which form a 3-4-5 right triangle, so the resultant is 500 N at an angle of 36.9° + 40° up from the x-axis.   **23.** By symmetry, the horizontal components of the forces cancel, leaving only 4(200 N) cos 60° = 400 N acting straight up.   **29.** $F = ma = 0$; $F_f = -$F cos 45° = (100 N)(0.707) = -71 N.   **39.** Taking the initial direction as positive, $F_{av} = m\Delta v/\Delta t$; from Eq. (3.8), $s = \frac{1}{2}v_i \Delta t$, $\Delta t = 2(0.20$ m$)/(200$ m/s$) = 2.0$ ms; $F_{av} = (0.010$ kg$)(200$ m/s$)/(0.002\ 0$ s$) = 1.0$ kN.   **43.** 120 mi/h = 53.64 m/s; if v_T is his terminal speed and s_D the depth of the crater, Eq. (3.8) yields $\Delta t = 2s_D/v_T = 0.039\ 8$ s. $F_{av} = m\Delta v/\Delta t = (90$ kg$)(-53.64$ m/s$)/(0.039\ 8$ s$) = -0.12$ MN.   **49.** $F_{av} = ma_{av} = [(392.4$ N$)/g] \times 5.00$ m/s$^2 = 200$ N.   **55.** $s_R = 12$ in. $= 0.304\ 8$ m $= [2v_i^2/(9.8$ m/s$^2)]$ cos θ sin θ; $v_i = 1.729$ m/s; (a) $a = \Delta v/\Delta t = (1.729$ m/s$)(1.0 \times 10^{-3}$ s$) = 1.7$ km/s^2; (b) $F = ma = (4.5 \times 10^{-7}$ kg$) \times (1.73$ km/s$^2) = 7.8 \times 10^{-4}$ N.   **61.** $F_{av} = m \times \Delta v/\Delta t$; impact speed $v_1 = \sqrt{2gs_h}$. The duration of the collision follows from Eq. (3.8), $s_c = \frac{1}{2}v_1 \Delta t$, and so $\Delta t = 2s_c/v_1$; hence, $F_{av} = m\Delta v/\Delta t = mv_1(2\ s_c/v_1) = mv_1^2/2s_c = mgs_h/s_c$.   **75.** The tension in the front rope is 1.00 kN. For the first barge (taking the pull of the tug to be positive and to the right), $\overset{+}{\rightarrow}\Sigma F_{H1} = m_1 a = 1.00$ kN $- F_{T2} - F_f = (4.00 \times 10^3$ kg$)(0.100$ m/s$^2)$; for the second barge, $\overset{+}{\rightarrow}\Sigma F_{H2} = m_2 a = F_{T2} - F_f = (3.50 \times 10^3$ kg$)(0.100$ m/s$^2)$. Adding the two equations, $-2F_f = (7.50 \times 10^3$ kg$)(0.100$ m/s$^2) - 1.00$ kN; $F_f = +125$ N, opposing the motion; $F_{T2} = 475$ N.   **81.** For mass−1, $\overset{+}{\rightarrow\downarrow}\Sigma F_{V1} = m_1 a = F_{W1} - F_{T1}$; for mass−2, $\overset{+}{\rightarrow}\Sigma F_{H2} = m_2 a = F_{T1} - F_{T2}$; for mass−3, $+\uparrow\Sigma F_{V3} = m_3 a = F_{T2} - F_{W3}$; adding all three of these, $a(m_1 + m_2 + m_3) = F_{W1} - F_{W3} = g(m_1 - m_3)$; (b) $a = g(2$ kg$)/(8$ kg$) = \frac{1}{4}g$; (c) the tensions are internal to the 3−body system, the weights are external; (a) $m_1 a = m_1 g - F_{T1}$; $F_{T1} = m_1(g - a) = 29.4$ N $= 0.03$ kN. $m_3 a = F_{T3} - F_{W3}$, $F_{T3} = m_3(a + g) = 24.5$ N $= 0.02$ kN.   **85.** $F = ma$; $a = F_f/m = mg\mu_s/m = \mu_s g$.   **89.** For no slipping. $+\swarrow\Sigma F_\parallel = 0 = F_W$ sin $\theta - F_f($max$)$; and $\mu_s = $ tan θ, 0.3 = tan θ; $\theta = 16.7°$; hence, the crate slides toward the back of the truck.   **91.** $v_i = 50.0$ km/h = 13.89 m/s; to find the stopping distance, $v_f^2 = v_i^2 + 2as$; $s = -v_i^2/2a$; where $a < 0$ is determined from $F_f($max$) = \mu_s \mu g = ma$; hence, $a = \mu_s g$ and so $s = -v_i^2/2(-\mu_s g) = 20$ m.   **99.** $+\swarrow\Sigma F_\parallel = 0 = F_W$ sin $\theta - F_f$; tan $\theta = \mu_k = 0.070$. At $-10°$C, the water film that usually forms via melting under the skis would not.   **103.** (a) $F_H = F$ cos $\theta = (300$ N$)$

(0.866) = 259.8 N, which must exceed $F_f($max$) = 0.5[(50$ kg$)g + (300$ N$)$ sin 30°$] = 320$ N, but it does not, so there is no acceleration;   (b) $F_f($max$) = 256$ N—it moves! $\overset{+}{\rightarrow}\Sigma F_H = ma = 259.8$ N $- 0.3(640.3$ N$) = 67.7$ N and $a = 1.4$ m/s^2, or because the friction coefficients are known to only one figure, $a = 1$ m/s^2.   **107.** $v_f^2 = 2as$, $a = (8.33$ m/s$)^2/2(30.0$ m$) = 1.158$ m/s^2; $v_f = at$, $t = (8.33$ m/s$)/(1.158$ m/s$^2) = 7.196$ s; the box accelerates with respect to the truck at a_b, where $s_b = \frac{1}{2}a_b t^2 = 1.00$ m, and $a_b = 0.038\ 6$ m/s^2. If there were no friction, it would have accelerated back at 1.158 m/s^2; hence, $(1.158$ m/s$^2 - 0.038\ 6$ m/s$^2)m = F_f = \mu_k mg$, $\mu_k = 0.11$.   **123.** $+\uparrow\Sigma F_V = 2F_T$ sin 5.0° $- 533.8$ N $= 0$; $F_T = 3.1$ kN.   **131.** The net force exerted on the bar must vanish: $\Sigma\vec{F} = \vec{F}_T + \vec{F}_R + \vec{F}_W = 0$. Thus $\overset{+}{\rightarrow}\Sigma F_x = F_{Rx} - F_T$ cos $\theta = 0$, so $F_{Rx} = F_T$ cos $\theta = (1.00$ kN$)($cos 30.0°$) = 0.866$ kN $= 866$ N, to the right; and $+\uparrow\Sigma F_y = F_T$ sin $\theta + F_{Ry} - F_W = 0$, so $F_{Ry} = F_W - F_T$ sin $\theta = 2.00$ kN $- (1.00$ kN$)($sin 30.0°$) = 1.50$ kN, upward.   **137.** On the small ball, $\overset{+}{\rightarrow}\Sigma F_H = F_{RA} - F_{RB}$ cos 30° $= 0$ and $+\uparrow\Sigma F_V = F_{RB}$ sin 30.0° $- 1.00$ kN $= 0$; $F_{RB} = 2.00$ kN. $F_{RA} = 1.73$ kN $= F_{RC}$; $F_{RD} = $ the net weight $= 3.00$ kN.

Chapter 5

Answers to Odd-Numbered Multiple Choice Questions

1. d   **3.** e   **5.** a   **7.** b   **9.** c   **11.** d   **13.** c   **15.** c   **17.** d   **19.** c   **21.** b

Answers to Odd-Numbered Problems

1. $a_C = 0.500$ m/s^2 **3.** (a) 1.00 m (c) 0.707 m/s   **5.** 17 m/s   **7.** 13 m/s   **9.** $F_C = 6.6 \times 10^2$ N, friction   **11.** $F_C = 2.1$ kN   **13.** $a_C = 1.8 \times 10^6$ m/s^2 **15.** 0.12 kN   **17.** $F_N = m(v^2/r + g)$ **19.** $F_N($max$) = 2.10$ kN,   **21.** (a) 2.09 m/s (b) 4.39 m/s^2 **23.** 19.2 m/s   **25.** (a) 0.018 rotations per second   (b) $a_C = 4\pi^2 r/T^2$ **27.** 99.7 N   **29.** would be halved   **31.** $m = 1.22 \times 10^5$ kg   **33.** 4 times larger   **35.** 2.5×10^{16} N   **37.** 1.1 m   **39.** $F_E = 2.2 \times 10^{39} F_G$ **41.** $0.11\ M_\oplus$ **43.** 1.45 m/s^2 **45.** 4.4 m   **47.** $R = 1.82 \times 10^3$ km $= 0.286\ R_\oplus$ **49.** proof 51. $0.379\ g_0$ **53.** $g_n = 1 \times 10^{12}$ m/s^2 **55.** $F_G = GmMr/R^3$ **57.** $T = 1 \times 10^{-3}$ s   **59.** proof   **61.** 1.62 m/s^2 **63.** 1.65×10^3 m/s   **65.** proof   **67.** $T_2 = 2.8\ T_1$ **69.** 3.32×10^{-5} m/s^2 **71.** 5.9×10^{-3} m/s^2 **73.** 8.000 Earth-years   **75.** (a) 3.34×10^{24} km^3/Earth-years^2 (b) 0.616 Earth-years   **79.** 1.078 AU, 1.12 Earth-years   **81.** $g(r) = GM/2R^2$ **83.** 1.05×10^3 kg/m^3 **85.** 9.7 d, 1.440 km/s

Solutions to Selected Problems

5. $F_C = mv^2/r$, $v = [(20$ N$)(0.355\ 6$ m$)/(0.025$ kg$)]^{1/2} = 17$ m/s.   **11.** The distance traveled once around is $2\pi r$, hence $v = 2(2\pi r)/(1.0$ s$)$; $r = 1.828$ m; $v = 22.98$ m/s; $F_C = mv^2/r = (7.257$ kg$)v^2/r = 2096.8$ N, or 2.1 kN.   **13.** $a_C = v^2/r$; $= (0.100$ m$)$ sin 30° $+ 0.050$ m $= 0.10$ m; $v = 2\pi r(40\ 000)/60 = 418.88$ m/s; $a_C = 1.8 \times 10^6$ m/s^2.   **17.** $F_N = m(v^2/r + g$ cos $\theta)$; at the bottom, $\theta = 0$, cos $\theta = 1$; $F_N = m(v^2/r + g)$.   **25.**

(a) $a_C = v^2/r = 1.0g$; $v^2 = \frac{1}{2}(1500$ m$)g$; $v = 85.76$ m/s; the cylinder must go once around a distance $(2\pi r)$ in a time $(2\pi r)/v = T = 54.9$ s, or 55 s; (b) $a_C = v^2/r = (2\pi r/T)^2/r = 4\pi^2 r/T^2$, where the time to go once around is the same throughout the station. Hence, the simulated gravity varies linearly with r, decreasing with distance "up" from the floor.   **27.** $+\downarrow\Sigma F_V = ma_C = F_G - F_N$; F_N is equal and opposite to the effective weight; F_G is the actual gravitation force downward, namely, 100 N. $F_N = F_G - ma_C = F_G - (F_G/g)v^2/R_\oplus = 100$ N $- 0.345$ N $= 99.7$ N.   **31.** $F_G = Gm^2/r^2$; 1.00 N $= (6.67 \times 10^{-11}$ N·m$^2/$kg$^2)m^2/(1.00$ m$)^2$; $m = 1.22 \times 10^5$ kg.   **37.** $v_f^2 = 0 = v_i^2 - 2gs$; $s = v_i^2/2g$. $s_\oplus/s_\circ = g_\circ/g_\oplus$, $(1.00$ m$)/s_\circ = 0.88$, $s_\circ = 1.1$ m.   **41.** $g_\circ = GM_\circ/R_\circ^2$, $M_\circ = (3.7$ m/s$^2)(\frac{1}{2}6.8 \times 10^6$ m$)^2/(6.67 \times 10^{-11}$ N·m$^2/$kg$^2) = 6.4 \times 10^{23}$ kg $= 0.11\ M_\oplus$.   **45.** To find $g_\circ$, we need the mass; $M_\circ = \frac{4}{3}\pi r^3 \rho = 4.82 \times 10^{24}$ kg; $g_\circ = GM_\circ/R_\circ^2 = 8.79$ m/s^2; $s = \frac{1}{2}g_\circ t^2 = 4.4$ m.   **51.** $g_\circ = GM_\circ/R_\circ^2 = G(0.108\ M_\oplus)/R_\oplus^2(0.534)^2 = 0.379\ g_0$.   **55.** $F_G = GmM_r/r^2$, where $M_r = \frac{4}{3}\pi r^3 \rho$, and $\rho = M/(4\pi R^3/3)$; hence, $M_r = Mr^3/R^3$ and $F_G = GmMr/R^3$.   **59.** $g_p = GM/(R + h)^2 = GM/R^2(1 + h/R)^2 = (GM/R^2)(1 + h/R)^{-2}$. From the binomial expansion, the bracketed term becomes $1 + (-2)(1)h/R + \frac{1}{2}(-2) \times (-3)(1)(h/R)^2 + ...$, which is $\approx (1 - 2h/R)$ since the $(h/R)^2$ term is very small.   **61.** $g_\mathbb{C} \approx GM_\mathbb{C}(1 - 2h/R)/R^2 \approx GM_\mathbb{C}\ [1 - (200$ m$)/(1.74 \times 10^6$ m$)]/R_\mathbb{C}^2 \approx 0.999\ 885\ GM_\mathbb{C}/R_\mathbb{C}^2 \approx 1.62$ m/s^2. The difference is negligible.   **65.** $4\pi^2 r_\mathbb{D}^3/M_\oplus G = 9.90 \times 10^{-14}r_\mathbb{D}^3$; $T = 3.15 \times 10^{-7}\ r_\mathbb{D}^{3/2}$ **69.** $g_\mathbb{C} = GM_\mathbb{C}/r_{\oplus\mathbb{D}}^2 = 3.32 \times 10^{-5}$ m/s^2.   **73.** $T^2 = r_\mathbb{D}^3/C_\odot = (4.000$ AU$)^3/(1$ AU^3/Earth-years$^2) = 64.00$ Earth-years^2; $T = 8.000$ Earth-years.   **79.** $r_{\odot 1} = \frac{1}{2}(0.186$ AU $+ 1.97$ AU$) = 1.078$ AU. $r_{\odot 1}^3/T_1^2 = 1 = (1.078$ AU$)^3/T_1^2$; $T_1 = 1.12$ Earth-years.   **85.** From Eqs. (5.8) and (5.9), you find that $C_\oplus = GM_\oplus/4\pi^2 = 1.010 \times 10^{13}$ m^3/s$^2 = r_{\oplus s}^3/T_s^2 = (\frac{1}{2}3.844 \times 10^8$ m$)^3/T_s^2$ and $T_s = 0.838\ 5 \times 10^6$ s $= 9.7$ days; $v = 2\pi r_{\oplus s}/T_s = 1.440$ km/s.

Chapter 6

Answers to Odd-Numbered Multiple Choice Questions

1. b   **3.** a   **5.** a   **7.** a   **9.** c   **11.** c   **13.** a   **15.** a   **17.** c   **19.** c   **21.** d

Answers to Odd-Numbered Problems

1. 5.00 kJ   **3.** 4.4×10^5 J   **5.** 23 J   **7.** 800 J   **9.** 2.0 J   **11.** 4.00 N   **13.** 7.2×10^2 J, gravity   **15.** 20 MJ   **17.** 5.0 m   **19.** 4.35 kJ   **21.** 0.50 kN·m   **23.** 0.74 kJ   **25.** 1.2×10^2 J   **27.** (a) 0.20 kN·m (b) 2.5 N·m/cm^2 (c) 0.20 kN·m   **29.** 0.09 kJ   **31.** 26 J   **33.** 6.40×10^8 J   **35.** 197 J   **37.** (a) B-C, D-E, F-G, H-I   (b) A-B, C-D, E-F, G-H   (c) F-G   (d) at A, $v = 0$ and $a = 0$; at B, $v > 0$ and $a > 0$; at C, $v > 0$ and $a > 0$; at E, $v > 0$ and $a > 0$; at F, $v > 0$ and $a = 0$; at H, $v > 0$ and $a < 0$ **39.** 1×10^7 J   **41.** 0.29 kJ   **43.** 0.71 kJ for the shot, 0.15 kJ for the baseball; the shot is considerably more massive.   **45.** 0.50 kJ   **47.** 4.0 kJ   **49.** 0.11 kJ, 3.5 cm   **51.** 6×10^2 m   **53.** 17 km

55. (a) mg (c) mgh_1, 0, mgh_2 **57.** (a) $(h_1 - h_3)$
(c) $mg(h_1 - h_3)$ **59.** 1.2 MJ **61.** −0.78 MJ **63.**
(a) 0.10 kJ (b) no rope (c) 0.10 kJ **65.** (a) $W =$
mgh (b) $\frac{1}{2}Nmgh$ **67.** 1.5 kJ **69.** (a) 0.16 $g_\oplus$ or
1.6 m/s^2 (b) 1.6 × 10^5 J **71.** 1.92 kJ **73.** both
75. 5.1 m **77.** 4.85 × 10^3 J, h = 9.00 m, or 10.0
m above ground **79.** $v = \sqrt{2g(h_1 - h_2)}$ **81.**
−6.2 × 10^7 J, −3.1 × 10^7 J, −2.1 × 10^7 J, −1.6 ×
10^7 J, −1.2 × 10^7 J, −0.62 × 10^7 J, E = −1.0 ×
10^7 J **83.** 2.4 km/s **85.** v_{esc}^2(Moon) $/v_{esc}^2$(Earth)
= 0.046 **87.** 2121 m/s **89.** v equals the escape
speed. **91.** $v = [2(m_2 - m_1)gy/(m_1 + m_2)]^{1/2}$
93. (a) 3 × 10^2 W (b) 6 × 10^2 W **95.** 5.0 W
97. 10 m/s **99.** 1 × 10^{15} W **101.** 589 W
103. 5.3 s **105.** 1.2 × 10^2 W, 60 W/m^2 **107.**
74 W/m^2 **109.** 8 min

Solutions to Selected Problems

9. $W = Fl = (20\ \text{n})(0.10\ \text{m}) = 2.0$ J. **15.** The
thrust is constant at 10 × 10^3 N; hence, $W = (2.0$
× 10^3 m)(10^4 N) = 20 MJ. **21.** $\sin \theta = 5/13$, W
= $(F_W \sin \theta)\ s = (100\ \text{N})(5/13)(13\ \text{m}) = 500$
N·m. Straight up, $W = (100\ \text{N})(5.0\ \text{m}) = 0.50$
kN·m. **27.** (a) $W = (100\ \text{N})(2.0\ \text{m}) = 0.20$
kN·m; (b) (1.0 cm)(1.0 cm) = (10 N)(0.25 m) =
2.5 N·m/cm^2; (c) 10 cm × 8.0 cm = 80 cm^2; (80
cm2)(2.5 N·m/cm^2) = 0.20 kN·m. **39.** KE = $\frac{1}{2}$
$mv^2 = \frac{1}{2}(0.5 \times 10^{-3}\ \text{kg})(200 \times 10^3\ \text{m})^2 = 1 \times 10^7$
J. **51.** PE$_G$ = mgh; $h = (6 \times 10^9\ \text{J})/(10^6\ \text{kg})(9.8$
m/s^2) = 6 × 10^2 m. **69.** (a) $g_{\mathbb{C}}$ = $GM_{\mathbb{C}}/R_{\mathbb{C}}^2$ =
$[(1/100)/(1/4)^2]g_\oplus = 0.16\ g_\oplus$ (b) $\Delta PE = mg\Delta h$
= (1000 kg)(0.16)(9.8 m/s^2)(100 m) = 1.6 × 10^5
J. **73.** For one car, KE = $\frac{1}{2}mv^2 = \frac{1}{2}[(7.12$
kN)/(9.81 m/s^2)](26.67 m/s)^2 = 258 kN·m; and
for the other, KE = twice that; viz., 516 kN·m.
For one car on top of the mountain, PE = mgh =
(7.12 kN)(33.5 m) = 238.5 kN·m; and for the
other, PE = twice that; viz., 477 kN·m. Thus, both
cars have enough energy to coast over the top.
75. $\frac{1}{2}mv_f^2 + mgh_f = \frac{1}{2}mv_i^2 + mgh_i$; $mgh_f = \frac{1}{2}mv_i^2$;
$2gh_f = v_i^2$; $h_f = v_i^2/2g = 5.1$ m. **83.** v_{esc} =
$(2GM/R)^{1/2} = [2(6.67 \times 10^{-11}\ \text{N·m}^2/\text{kg}^2)(7.4 \times$
10^{22} kg)/(1.74 × 10^6 m)]$^{1/2}$ = 2.4 km/s. It can be
launched in any direction that does not intersect
the Moon. **91.** PE$_i$ = 0; E$_i$ = 0; hence, $\frac{1}{2}m_1v^2 +$
$m_1gy + \frac{1}{2}m_2v^2 - m_2gy = 0$, a moment after
release. $v = [2(m_2 - m_1)gy/(m_1 + m_2)]^{1/2}$. **97.** P
= Fv = 1.0 × 10^4 J/s = (1.0 × 10^3 N)v ; v = 10
m/s, or 22 mi/h. **105.** $\Delta W/\Delta t$ = (10^7 J)/ (24 h
× 3600 s/h) = 1.2 × 10^2 W. BMR = (90 W)/
(1.5 m^2) = 60 W/m^2. **109.** Pt = W = ΔKE = $\frac{1}{2}$
(2 × 10^3 kg)(268 m/s)^2 = 7.18 × 10^7 J; t =
8 min.

Chapter 7

**Answers to Odd-Numbered Multiple Choice
Questions**

1. a **3.** b **5.** b **7.** d **9.** b **11.** e **13.** c
15. a **17.** c **19.** a **21.** d

Answers to Odd-Numbered Problems

1. 4.8 × 10^3 kg·m/s–NORTH **3.** 37 m **5.** −10
N·s **7.** 11 kg·m/s **9.** unchanged, 10 kg·m/s
11. 0.012 s **13.** from 4.0 s to 5.0 s **15.** (a) it's
the impulse (b) 7.0 N·s (c) 14 m/s **17.** 221

kg·m/s; $\vec{F}_{av}$ = 2.21 × 10^4 N–upward **19.** $\vec{F}_{av}$ =
2.8 × 10^2 N–upward **21.** 6 kN **23.** 0.3 N **25.**
42 MN **27.** 83.9 kN **29.** 10.2 kg·m/s, at 156°
from the direction of the initial momentum **31.**
−2 MN **33.** (a) they're equal (b) it's the same
(c) it's the same (d) 18 kg·m/s (e) 9.0 m/s-WEST
35. 3.3 m/s–south **37.** 0.10 m/s–west **39.**
−0.12 m/s **41.** 2.01 × 10^{-2} m/s, forward **43.**
2.2 × 10^2 N, in the direction of the water jet **45.**
0.020 2 m/s, opposite to the snowball **47.** 18
m/s **49.** $v_{2f} = -m_1v_{1f}/m_2 = -1.00$ m/s **51.**
v_{Nf} = −0.100 m/s, v'_{Sf} = 0.198 m/s, v''_{Sf} = 0.400
m/s. **53.** (a) +10.0 kg·m/s, −15.0 kg·m/s (b)
−5.0 kg·m/s (c) −5.0 kg·m/s **55.** 10.2 m/s
57. 0.46 km/s **59.** Ball-1 stops, ball-2 moves
off at 20.0 m/s. **61.** 9.8 m/s, 4.7 × 10^5 J **63.**
0.070 m/s, 0.13 J **65.** 23.3 m/s **67.** v_{2f} = 15
m/s, and v_{1f} = −10 m/s **69.** $v_B = [(m_B + m_C)/$
$m_B]\sqrt{2gh}$ **71.** proof

Solutions to Selected Problems

5. Taking the direction of motion as positive, Δp
= $m\Delta v$ = (1.0 kg)(−10 m/s) = − 10 N·s **17.**
The speed of the youngster just before she touches
the floor is $v_i = \sqrt{2gh} = [2(9.81\ \text{m/s}^2)(1.00$
m)]$^{1/2}$ = 4.429 m/s, so her initial momentum is
$p_i = mv_i$ = (50.0 kg)(4.429 m/s) = 221 kg·m/s.
She is brought to a complete stop in Δt = 10.0
ms = 1.00 × 10^{-2} s while momentum changes from
p_i to zero; so from Eq. (7.2) $F_{av} = \Delta p/\Delta t$ = (0 −
$p_i)/\Delta t$ = −(−221 kg·m/s)/(1.00 × 10^{-2} s) =
2.21 × 10^4 N, upward. **23.** 2 oz = 2/16 lb; m =
(2/16 lb)(0.453 6 kg/lb) = 0.056 7 kg; F_{av} =
$m\Delta v/\Delta t$ = (0.056 7 kg) (0.50 m/s)/(0.1 s) = 0.284
N, or 0.3 N. **27.** Let's see what a is first, then t,
and then F. $v_f^2 = v_i^2 + 2as$; a = (69.5 m/s)^2/2
(914.4 m)= 2.64 m/s^2; $v_f = v_i + at$; t = (69.5 m/s)
/(2.64 m/s^2) = 26.3 s; m = 31 751 kg; $F = m\Delta v/$
Δt = (31 751 kg)(69.5 m/s)/(26.3 s) = 83.9 kN.
31. To find F_{av}, the force exerted on the gas, take
the direction of the exhaust as positive and use
$F_{av}\Delta t = m\Delta v$; F_{av} = (1000 kg) (2 km/s)/(1.000 s)
= 2MN. Thrust is −2 MN. **35.** The initial
momentum of the boat-person system is zero, $\vec{p}_i =$
$\vec{p}_f$ = 0, and $m_pv_{pf} + m_bv_{bf}$ = 0; taking motion north
as positive, (50 kg)(+10 m/s) = −(150 kg)v_{bf}; v_{bf}
= −3.3 m/s or 3.3 m/s–south. **47.** p_i =
(10 000 kg)(20 m/s) = 2.0 × 10^5 kg·m/s = p_f =
(10 000 kg + 900 kg)v_f; v_f = 18.3 m/s, or 18 m/s.
51. $p_{Ni} + p_{ai} = 0 = p_{Nf} + p_{af}$ = (100 kg)v_{Nf} +
(0.500 kg)(20.0 m/s); v_{Nf} = −0.100 m/s—he
moves in the opposite direction to the asteroid,
whose direction is taken as positive; she catches it
and so $p_{ai} = p'_{af} + p'_{Sf}$; (0.500 kg)(20.0 m/s) =
$v'_{Sf}(m_a + m_s) = v'_{Sf}(50.5$ kg); v'_{Sf} = 0.198 m/s. Now
she throws it back, $p''_{af} + p''_{Sf} = p''_{Si} + p''_{ai}$; (50.5 kg)
(0.198 m/s) = (0.500 kg)(−20.0 m/s) + (50.0
kg)v''_{Sf}; v''_{Sf} = 0.400 m/s. **57.** p_i = (1.00 kg)v_i + 0
= p_f = (91 kg)(5.0 m/s); v_i = 0.46 km/s. **63.**
Taking east as positive, p_i = (+8.0 kg)(0.15 m/s)
− (2.0 kg)(0.25 m/s) = 0.70 kg·m/s = p_f = (10.0
kg)v_f; v_f = 0.070 m/s. **67.** $p_i = m(15.0$ m/s) +

$m(-10\ \text{m/s}) = mv_{1f} + mv_{2f}$, +5.0 = $v_{1f} + v_{2f}$. But
$v_{2f} - v_{1f} = v_{1i} - v_{2i}$ = 25 m/s; hence, adding the
last two equations, $2v_{2f}$ = 30 m/s, v_{2f} = 15 m/s,
and v_{1f} = −10 m/s. **69.** $m_Bv_B = (m_B + m_C)v_C$.
After impact, $\frac{1}{2}(m_B + m_C)v_C^2 = (m_B + m_C)gh$, $v_C^2 =$
$2gh$, $v_B = [(m_B + m_C)/m_B]\sqrt{2gh}$.

Chapter 8

**Answers to Odd-Numbered Multiple Choice
Questions**

1. b **3.** d **5.** a **7.** a **9.** a **11.** c **13.** a
15. a **17.** d **19.** d **21.** c **23.** c **25.** d
27. e

Answers to Odd-Numbered Problems

1. 0.017 453 rad **3.** 180.00° **5.** $\pi/4$ rad **7.**
270 ° **9.** 31.8 rev **11.** 33.3 rad **13.** 3 × 10^5 m
in diameter **15.** (a) 5.00 m (b) proof **17.** 2.0 m
19. 69 rev **21.** 1.57 rad, 0.020 0 rad, 0.020 0
rad, 2.00 rad. **23.** 16 rad/s **25.** 3.14 × 10^4
rad/s **27.** v = 2.98 × 10^4 m/s **29.** 12.2 rad/s
31. (a) 0 (b) 12.5 rad/s^2 **33.** 105 rad/s^2 **35.**
4.2 s **37.** 33 rpm **39.** ω = 0.029 rad/s, 0.26
m/s^2 **41.** 0.1 m/s–upward **43.** $R_\perp$ = 4.85 ×
10^6 m, θ = 40.4° **45.** ω_4 = 1200 rpm; ω_5 = 400
rpm, $\omega_n = \omega_1N_1/N_n$ **47.** 1.4 m **49.** 0.75 m/s,
upward **51.** 2.3 m **53.** 150 rad **55.** 5.2 s, 27
rad **57.** 10.0 s, −2.00 rad/s^2 **59.** (a) 12.5 rad/s
(b) 62.8 rad (c) 5.02 s (d) 4.98 rad/s^2 **61.** 0.07,
0.2 m **63.** 2.7 m/s **65.** α = 9.82 rad/s^2, t =
16.0 s **67.** 8.7 rad/s^2, 25 rev **69.** τ = 50.0
N·m, counterclockwise around the north end of the
stick and clockwise round the south end **71.** 100
N·m **73.** 140 N **75.** (b) 40 N   (c) 80 N **77.**
F_{RB} = 6000 N, F_{RA} = 2000 N **79.** 11 N **81.** (b)
37° (c) 4.0 N (d) zero **83.** F_R = 355.9 N, x =
0.731 7 m **85.** 1.0 kg **87.** F_{RD} = 360 N, F_{RC} =
80.0 N, F_{RA} = 40 N **89.** F_N = 0.83 kN **91.** F_f
= 11 N **93.** F_{Ax} = −1.5 kN (to the left); F_{Ay} =
−0.50 kN (down) **95.** 1.0 m from either end
97. at the center of the + **99.** 10.0 cm to the
right of the large sphere **101.** midface in the glue
8 cm from the end **103.** $x_{cg} = y_{cg}$ = 6.3 cm **105.**
F_{T2} = 100 N, F_W = 173 N **107.** yes **109.** (a)
at the center, 5 units from either end (f) 98 N

111.

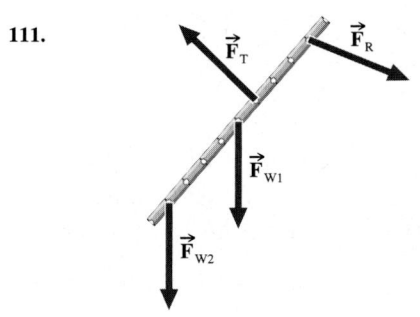

113. yes **115.** (a) at the center (b) to the right
up (e) 43° (f) 0.38 N (f) 3.6 N **117.** proof **119.**
(a) friction F_f and normal force $F_{RF} = F_{NF}$ (b) normal force $F_{RW} = F_{NW}$ (c) $F_f = F_{RW}$ **121.** F_{RG} =
0.31 kN at 74° up from the ground. **123.** 0.21
kN on each hand, 0.11 kN on each foot **125.** $x_{c.g.}$

$= F_{N2}h/(F_{N1} + F_{N2})$ **127.** $F_T = 90.4$ N **129.** MR^2 **131.** $\frac{1}{3}ml^2$ **133.** $\frac{3}{2}mR^2$ **135.** 426 rad/s^2 **137.** 1.50 N·m **139.** $a = mg/(m + I/R^2)$ **141.** $a = 1.5$ m/s^2 **143.** $a = \frac{1}{2}g \sin\theta = \frac{1}{4}g = 2.5$ m/s^2 **145.** $I \approx \frac{1}{3}Ml^2$ **147.** proof **149.** 493 J **151.** 4.47 rad/s **153.** $[M][L^2][T^{-2}]$ **155.** proof **157.** $KE_{total} = \frac{7}{10}mv^2$ **159.** proof **161.** 0.125 kg·m^2/s **163.** 450% decrease **165.** 3×10^{-3} kg·m^2/s **167.** 5.2×10^2 rev/s or 3.3×10^3 rad/s **169.** (a) 1.6×10^{-4} kg·m^2 (b) 8.0×10^{-2} kg·m^2 (c) $\approx$ 0.080 kg·m^2 (d) 0.010 kg·m /s (e) 4.0×10^{-3} kg·m^2/s (f) 5.0×10^{-2} rad/s **171.** 1.2 m/s, 1.8×10^4 kg·m^2/s, 24 m/s, 17×10^3 N

Solutions to Selected Problems

9. $200/2\pi = 31.8$ rev **3.** $\ell = r\theta = (3.8 \times 10^8$ m$)(8 \times 10^{-4}) = 3 \times 10^5$ m, which is the diameter. **21.** (ii) 90.0° = 1.57 rad; $\tan\alpha = 1/100$, $\alpha = 5.7 \times 10^{-1}$, $2\alpha = 1.15° = 0.0200$ rad; $\theta = \ell/r =$ (2.00 m)/(100 m) = 0.020 0 rad. (i) $\theta = \ell/r =$ 2.00 rad. **23.** 45.0° = 0.785 rad; $\omega = \Delta\theta/t =$ (0.785 rad)/(0.050 s) = 16 rad/s. **35.** $\omega_i = 200$ $(2\pi)/(60$ rad/s$) = 20.94$ rad/s; $0 = \omega_i + (-5$ rad/s$^2)t$; $t = 4.2$ s. **45.** 1500 rpm × 20 = $\omega_4 25$ $= \omega_5 75$; $\omega_4 = 1200$ rpm; $\omega_5 = 400$ rpm. $\omega_n = \omega_1 N_1/N_n$, where N is the number of teeth and n is the number of gears. If n is odd, the direction is unchanged. **47.** $a_C = R\omega^2$, $a_T = R(4.0$ rad/s$^2)$, $a = [a_C^2 + a_T^2]^{1/2}$, $(8.0$ m/s$^2)^2 = R^2\omega^4 + R^2(4.0$ rad/s$^2)^2$; $R^2 = 2.0$ m^2, $R = 1.4$ m. **49.** $\ell_3 = (0.24$ m$)\theta_1$, $\ell_5 = (0.12$ m$)\theta_1 = (0.20$ m$)\theta_2$, $\ell_4 = (0.15$ m$)\theta_2$, so $\theta_2 = 0.60\theta_1$ and $\ell_4 = (0.15$ m$) \times (0.60$ $\theta_1) = (0.90 \times 10^{-1})\theta_1 = 0.90 \times 10^{-1}$ $(\ell_3/0.24) =$ $0.375\,\ell_3$, $v_4 = 0.375v_3 = 0.375(2.0$ m/s$) = 0.75$ m/s, upward. **51.** $\omega = 2\pi(1800$ rpm$)/(60$ s/m$)$ $= 188.5$ rad/s, $v = R\omega = (0.12$ m$)(188.5$ rad/s$)$ $= 22.62$ m/s; $\ell = vt = (22.62$ m/s$)(0.10$ s$) = 2.3$ m. **55.** $\Delta\omega = -100$ rev/min $= -10.47$ rad/s; $t =$ $\Delta\omega/\alpha = 5.235$ s or 5.2 s; $\theta = -\omega_i^2/2\alpha = 27$ rad. **61.** $\omega^2 = 0 + 2\alpha\theta$, $(20 \times 2\pi/60)^2 = 2(5$ rad/s$^2)\theta$, $\theta = 0.4$ rad, or 0.07 turns. (0.4 rad)(0.5 m) = 0.2 m. **63.** $v = 2\pi R(100$ rpm$)/(60$ s/min$)$ $= 8.7$ ft/s $= 2.7$ m/s. **65.** $\omega_f^2 = \omega_i^2 + 2\alpha\theta$, $(1500 \times 2\pi/60)^2 = 2\alpha(200 \times 2\pi)$, $\alpha = 9.82$ rad/s^2; $\omega_f = \omega_i + \alpha t$, $t = (1500 \times 2\pi/60)/(9.82$ rad/s$^2) = 16.0$ s. **83.** $+\uparrow\sum F_V = 533.8$ N $-$ 889.66 N $+ F_R = 0$; $F_R = 355.9$ N; $(\!\circlearrowleft\!+\sum\tau_s =$ (355.9 N)(1.829 m) $-$ (889.66 N)$x = 0$; $x =$ 0.731 7 m from Selma. **89.** $\circlearrowleft\!+\sum\tau = (300$ N$)(0.25$ m$) - F_N(0.09$ m$) = 0$; $F_N = 0.83$ kN. **93.** Take the horizontal bar out and draw a free-body diagram; $\circlearrowleft\!+\sum\tau_A = (1000$ N$)(3.0$ m$) - (2.0$ m$)F_C \sin 45° = 0$; $F_C = 2121.6$ N. At A, the horizontal force is $F_{Ax} = -F_{Cx} = -1.5$ kN (force is to the left). $+\uparrow\sum F_y = F_C \sin 45° - 1000$ N $- F_{Ay} =$ 0; $F_{Ay} = -0.50$ kN (force is down). **103.** Taking the upright side as $y = 0$, $\sum mx = (0.50$ kg/m$)(0.25$ m$)(0.125$ m$) = 0.015\,6$ kg·m and x_{cg} $= (0.015\,6$ kg·m$)/(0.25$ kg$) = 0.063$ m, and, by symmetry, $y_{cg} = 0.063$ m. **113.** If each is of length L, the *c.g.* of the topmost block is at the edge of the one beneath it. Measured from the left end, for the top two blocks, $x_{cg} = (mL + mL/2)/$ $2m = 3L/4$, and that is located just above the edge of the third block. For the top three blocks, $x_{cg} =$ $(2mL + mL/2)/\,3m = 5L/6$. For the top four blocks, $x_{cg} = (3mL + mL/2)/4m = 7L/8$. The overhangs are $L/8$, $L/6$, and $L/4$, which is more than $L/2$. **121.** $+\uparrow\sum F_y = F_N - g(30$ kg$) = 0$; $F_N = 0.29$ kN. $\circlearrowleft\!+\sum\tau = 0 = F_N(8.0$ m$)\cos 60° -$ $F_f(8.0$ m$)\sin 60° -$ g(30 kg)(4.0 m) cos 60°; $F_f =$ 85 N, $F_{RG} = 0.31$ kN at 74° **125.** $\circlearrowleft\!+\sum\tau_1 =$ $F_W x_{cg} - F_{N2}h = 0$; $\circlearrowleft\!+\sum\tau_2 = F_{N1}h - F_W(h - x_{cg})$ $= 0$; $F_W = F_{N2}h/x_{cg}$; $F_{N1}h - F_{N2}h(h - x_{cg})/x_{cg} =$ 0; $x_{cg} = F_{N2}h/(F_{N1} + F_{N2})$. **127.** Taking O at the base of the wall, $\circlearrowleft\!+\sum\tau_o = (4.00$ m$)F_{N2} - (3.00$ m$)F_{N1} - F_W(2.00$ m$) = 0$; and $\overset{+}{\rightarrow}\sum F_x = F_T\cos\theta -$ $F_{N1} = 0$, where θ is the angle the rope makes with the ground. From the geometry (we are dealing here with a 3-4-5 right triangle), $\theta = 10.62°$; $+\uparrow\sum$ $F_y = F_{N2} - F_W - F_T \sin\theta = 0$. Eliminating F_{N1} from the first and second equations, $-(3.00$ m$)F_T$ $\cos\theta - F_W(2.00$ m$) + (4.00$ m$)F_{N2} = 0$; and substituting F_{N2} into the third equation, $(0.75\,F_T \times \cos$ $\theta + 0.50\,F_W) - F_W - F_T\sin\theta = 0$. Hence, $F_T(0.75\cos\theta - \sin\theta) = 0.50\,F_W$; $F_T = 90.4$ N. **135.** $\sum\tau = 50.0$ ft·lb $= I\alpha$, $I = mR^2 = (F_W/g)R^2$, 50.0 ft·lb $= [(3.22$ lb$)/(32.2$ ft/s$^2)] \times [(13$ in.$)/(12$ in./ft$)]^2\alpha$; $\alpha = 426$ rad/s^2. **141.** $+\!\downarrow\!\sum F_y$ $= ma = F_W - F_T$; $\circlearrowleft\!+\sum\tau = I\alpha = F_TR - 0.060$ N·m; $a = R\alpha$; $F_T = (I\alpha + 0.060$ N·m$)/R =$ $Ia/R^2 + (0.060$ N·m$)/R$; $F_W - Ia/R^2 - (0.060$ N·m$)R = ma$, $a(m + I/R^2) = Mg - (0.060$ N·m$)/R$, $a = 9.2/6.0 = 1.5$ m/s^2, as compared to 1.6 m/s^2 previously. **143.** $+\searrow\sum F_\parallel = F_W \times \sin\theta$ $- F_f = ma$; $\sum\tau_{cm} = F_fR = I_{cm}\alpha = mR^2\alpha$; $a =$ $R\alpha$; $F_fR = mRa$; $F_W = mg$; $a = \frac{1}{2}g \times \sin\theta = g/4$. **147.** $+\searrow\sum F_\parallel = mg\sin\theta - F_f = ma$; $\circlearrowleft\!+\sum\tau_{cm} =$ $F_fR = I\alpha$, $\alpha = a/R$, $I_{cm} = \frac{1}{2}mR^2$, $F_f = \frac{1}{2}ma$; mg $\sin\theta = (3/2)ma$, $a = (2/3)g\sin\theta$; $s = h/\sin\theta$, $v_f^2 = 2as = 2(2/3)(g\sin\theta)h/\sin\theta = (4/3)gh$. v_f $= 2\sqrt{gh/3}$. **163.** 450% decrease; $L_i = L_f$ **165.** $L = I\omega$; $I \approx mr^2 = (2 \times 10^{-3}$ kg$)(0.5$ m$)^2 = 5 \times$ 10^{-4} kg·m^2; $L \approx (5 \times 10^{-4}$ kg·m$^2)(2\pi$ rad/s$) = 3$ $\times 10^{-3}$ kg·m^2/s. **171.** $v = a_Tt = 1.0 \times 10^{-3}g(120$ s$) = 1.2$ m/s; $L = rmv = 1.8 \times 10^4$ kg·m^2/s. $L =$ $(5.0$ m$)(150$ kg$)v = 1.8 \times 10^4$ kg·m^2/s, $v = 24$ m/s, or 54 mi/h. $F_C = mv^2/r = 17 \times 10^3$ N, or 4 $\times 10^3$ lb!

Chapter 9

Answers to Odd-Numbered Multiple Choice Questions

1. d **3.** d **5.** d **7.** b **9.** a **11.** a **13.** a **15.** d **17.** b **19.** c **21.** c **23.** d

Answers to Odd-Numbered Problems

1. 6.02×10^{23} **3.** 6×10^7 kg/m^3 **5.** (a) 1.27 g/cm^3 (b) 1.27×10^3 kg/m^3 **7.** 49 kg **9.** 0.19 GJ **11.** 12.0×10^{23} **13.** 303 u **15.** 6×10^{77} **17.** 3.3×10^{28} molecules/m^3 **19.** 3.14 kg **21.** 0.46 nm **23.** 3.3 nm **25.** 3×10^4 Pa **27.** 50.7 N/cm^2 **29.** proof **31.** $P_G = 1.1 \times 10^8$ Pa = 1.6×10^4 psi **33.** 10.3 m **35.** 1.0×10^4 Pa **37.** (b) 3.04×10^6 Pa (b) 1.5×10^6 N **39.** 3.9×10^3 Pa **41.** -1.1×10^4 Pa **43.** 1.1×10^5 Pa **45.** (a) 2.4 kPa (b) yes (c) 1.1 kPa (d) 3.4 kPa **47.** 1.0×10^5 N **49.** 9.6×10^4 Pa **51.** proof **53.** 2×10^5 N **55.** 1.47×10^5 N **57.** (b) 78.6 kg **59.** 3.75 kN **61.** $P_o = 1.27 \times 10^6$ Pa, $F_i =$ 100 N **63.** 9.88×10^6 Pa **65.** $F_o = F_i(A_L/A_S) - m_Lg$ **67.** 4.9×10^4 W **69.** 2.9 kN **71.** $\Delta\rho = -1.5 \times 10^2$ kg/m^3 **73.** 10.5 **75.** 5.00 **77.** (a) 7.2×10^{-45} m^3 (b) $\rho = 2.3 \times 10^{17}$ kg/m^3 (c) 2.3×10^{14} **79.** 2.00% **81.** $V = 3.00 \times 10^{-8}$ m^3, sp.gr. = 21.5, it is probably platinum **83.** 10% visible above sea water, 8.0% visible above fresh water **85.** (a) 0.390 g/cm^3 (b) yes (c) 3.83 N (d) 3.83 N (e) 11.6 N (f) no **87.** 72.7 kg **89.** 4.6 kN, 0.15 m **91.** 16×10^{-3} m^3/s **93.** $\Delta P = -27$ kPa **95.** $\Delta P = -8.8$ kPa **97.** $\sqrt{2P_G/\rho}$ **99.** 5.10 m **101.** $2\sqrt{yh}$ **103.** (a) 2.83×10^7 cm^3 (b) 5.07 cm^3 (c) 1.55×10^3 cm^3/s (d) 5.08 h **105.** 12 m/s **107.** (b) 31.3 m/s (c) 5.54×10^5 Pa **109.** $v_2 = 2\sqrt{g}$ **111.** $v =$ $\sqrt{2\Delta P/\rho}$ **113.** $v_p = \{2g\Delta y/[(A_p^2/A_t^2) - 1]\}^{1/2}$ **115.** 82 m/s **117.** $v_2 = \sqrt{2gh}$, $v_3 = [2g(h + Y)]^{1/2}$ **119.** $v_2 = \sqrt{2gH}$ **121.** $y = h\sin^2\theta$ **123.** proof

Solutions to Selected Problems

1. 6.02×10^{23} **7.** (13 gallons)(3.785 liters/gallon) = 49.2 liters; 1 liter = 1000 cm^3; hence, the human body contains $\approx 49 \times 10^3$ cm^3, or $\approx 49 \times 10^3$ g, or 49 kg. **11.** $C_{12}H_{22}O_{11}$: 12(12) + 22(1) + 11(16) = 342; hence, 684 g is 2 gram-moles and, therefore 12.0×10^{23} molecules. **15.** $(10^{33}$ g per star$)(10^{11}$ stars per galaxy$)(10^{11}$ galaxies$)/(10$ g per mole) = number of moles = 1×10^{54} moles; $(10^{54}$ moles$)(6.02 \times 10^{23}$ atoms/mole$) = 6$ $\times 10^{77}$ atoms. **19.** The mass of a molecule is 18 u, of which 2 u is hydrogen; that is, 11.1%, hence, 11.1% of 62.4 lb is 6.93 lb → 3.14 kg. **23.** 22.4 liters contain 6.0×10^{23} molecules; hence, the volume "occupied" by each one is $(22.4 \times 10^3$ cm$^3)(10^{-6}$ m^3/cm$^3)/(6.0 \times 10^{23}) = 3.73 \times 10^{-26}$ m^3; and the cube root of this is 3.3×10^{-9} m, or 3.3 nm, which is roughly 10 times the nucleus-to-nucleus separation in a solid. **27.** $P_i = 5(1.013$ $\times 10^5$ N/m$^2) = 5.07 \times 10^5$ N/m$^2 = (5.07 \times 10^5$ N$)/(10^4$ cm$^2) = 50.7$ N/cm^2. **31.** $P = \rho gh =$ $(1.025 \times 10^3$ kg/m$^3) \times (9.8$ m/s$^2)(11 \times 10^3$ m$) =$ 1.1×10^8 Pa, or 1.6×10^4 psi. **35.** (3.0 in.) (2.54 cm/in.) = 7.62 cm; $\Delta P = \rho g\Delta h = (13.6 \times$ 10^3 kg/m$^3)(9.8$ m/s$^2) \times (7.62 \times 10^{-2}$ m$) = 1.0 \times$ 10^4 Pa. **39.** $P = \rho gh = (1.0 \times 10^3$ kg/ m$^3)(9.8$ m/s$^2)(0.40$ m$) = 3.9 \times 10^3$ Pa. **43.** At the level of the lower surface, $P = P_A + \rho g\Delta h = (1.013 \times$ 10^5 Pa$) + (1.00 \times 10^3$ kg/m$^3)(9.8$ m/s$^2)(0.61$ m$)$ $= 1.1 \times 10^5$ Pa, or 1.1 atm. **51.** The area of the left face is $A_a = L(L\sin\phi)$, the area of the top face is $A_c = L^2$, and that of the bottom face is $A_b =$ $L^2\cos\phi$. So the corresponding forces are $F_a =$ $P_aL^2\sin\phi$, $F_c = P_cL^2$, and $F_b = P_bL^2\cos\phi$. Taking the sum of the horizontal forces equal to zero, we have $F_a = F_c\sin\phi$; $P_aL^2\sin\phi = P_cL^2$ $\sin\phi$. Thus, $P_a = P_c$. The weight of the fluid wedge is the volume, $\frac{1}{2}\,L(L\sin\phi)(L\cos\phi)$, times ρg; hence, the sum of the vertical forces yields F_b $= F_c\cos\phi + F_W$ or $P_bL^2\cos\phi = P_cL^2\cos\phi +$ $\frac{1}{2}\rho gL^3\sin\phi\cos\phi$ and $P_b = P_c + \frac{1}{2}\rho gL\sin\phi$. As $L \to 0$, we get $P_b = P_c$, which equals P_a. **53.** Because the pressure varies uniformly with depth, the average pressure on the wall is $\frac{1}{2}$ the pressure at the bottom. From Problem 9.25, $P_1 = 3 \times 10^4$ Pa; hence, $F = (1.5 \times 10^4 \text{ Pa})(5 \text{ m} \times 3 \text{ m}) = 2 \times 10^5$ N. **59.** Apply Pascal's Principle: $F_o/F_i = A_o/A_i$. Here $A_o = \pi D_o^2/4$ and $A_i = \pi D_i^2/4$, where $D_o = 20.00$ cm and $D_i = 8.00$ cm. Also, $F_i = 600$ N. Thus the output force exerted on the larger piston is $F_o = F_i(A_o/A_i) = F_i(D_o/D_i)^2 = (600 \text{ N})(20.00 \text{ cm}/8.00 \text{ cm})^2 = 3.75$ kN. **73.** From Eq. (9.10), $(10.0 \text{ g})/(0.952 \text{ g}) = 10.5$, and it's likely silver. **81.** From Eq. (9.10), $gm/(gm - gm_r) = (6.327 \times 10^{-3} \text{ N})/(6.327 \times 10^{-3} \text{ N} - 6.033 \times 10^{-3} \text{ N}) = 21.5$; platinum; weight of water displaced is $(6.327 \times 10^{-3} \text{ N} - 6.033 \times 10^{-3} \text{ N}) = 2.94 \times 10^{-4}$ N, equivalent to 3.00×10^{-5} kg, or 3.00×10^{-8} m³. **87.** mass of air displaced = mass of helium + mass of balloon; $V(1.29 \text{ kg/m}^3) = V(0.178 \text{ kg/m}^3) + 454$ kg; $V = 408.3$ m³, for a mass of $\rho V = (0.178 \text{ kg/m}^3) \times (408.3 \text{ m}^3) = 72.7$ kg. **93.** $P_1 - P_2 = \rho g(y_2 - y_1) = (0.68 \times 10^3 \text{ kg/m}^3)(9.8 \text{ m/s}^2)(-4.0 \text{ m}) = -27$ kPa. $P_2 > P_1$. **97.** $v = \sqrt{2gh}$, $P_G = \rho gh$; hence, $v = \sqrt{2P_G/\rho}$. **105.** Consider a slug of gas of volume $V = AL$, where the length $L = v \Delta t$. The mass per unit time (Δt) flowing is 1.0 kg/s = $V\rho/\Delta t = Av\Delta t\rho/\Delta t = \frac{1}{4}\pi D^2\rho v$; $v = 4(1.0$ kg/s$)/\pi D^2\rho = 12$ m/s. **117.** At the surface and the spigot using gauge pressure, $0 + \rho g(h + Y) + 0 = 0 + \rho g Y + \frac{1}{2}\rho v_2^2$, hence, $v_2 = \sqrt{2gh}$. Then between points 2 and 3, $0 + \rho g Y + \frac{1}{2}\rho(2gh) = 0 + 0 + \frac{1}{2}\rho v_3^2$; $v_3 = [2g(h + Y)]^{1/2}$. These are just what we expect if the water free-falls. **121.** $P_2 + \frac{1}{2}\rho v_2^2 + \rho g y_2 = P_3 + \frac{1}{2}\rho v_3^2 + \rho g y_3$; $0 + \frac{1}{2}\rho v_2^2 + 0 = 0 + \frac{1}{2}\rho v_3^2 + \rho g y$; since there is no horizontal acceleration, $v_3 = v_2 \cos\theta$, $y = \frac{1}{2}(v_2^2 - v_2^2 \cos^2\theta)/g$, but $\cos^2\theta + \sin^2\theta = 1$; hence, $y = (\frac{1}{2}v_2^2 \times \sin^2\theta)/g$ and, since $v_2^2 = 2gh$, $y = h \sin^2\theta$ and it cannot exceed h. **123.** Imagine a flow tube from some distant point 1 in front of the plane at the level of the top of the wing; then $P_1 + \frac{1}{2}\rho v_1^2 + \rho g y_1 = P_\alpha + \frac{1}{2}\rho v_\alpha^2 + \rho g y_\alpha$. Now do the same thing for a tube starting at point 2 just below 1 and running to the bottom of the wing as $P_2 + \frac{1}{2}\rho v_2^2 + \rho g y_2 = P_\beta + \frac{1}{2}\rho v_\beta^2 + \rho g y_\beta$. Since $P_1 \approx P_2$, $v_1 \approx v_2$, $y_1 \approx y_2$, $P_\alpha + \frac{1}{2}\rho v_\alpha^2 + \rho g y_\alpha = P_\beta + \frac{1}{2}\rho v_\beta^2 + \rho g y_\beta$, and with $y_\alpha \approx y_\beta$, $F_L = A(P_\beta - P_\alpha) = \frac{1}{2}\rho(v_\alpha^2 - v_\beta^2)A$.

Chapter 10

Answers to Odd-Numbered Multiple Choice Questions

1. a **3.** d **5.** c **7.** d **9.** c **11.** b **13.** b **15.** a **17.** c **19.** b **21.** b **23.** a **25.** b **27.** c **29.** d **31.** a **33.** c

Answers to Odd-Numbered Problems

1. (a) no (b) 15×10^{-2} m (c) 21 N **3.** 1.0 kN/m **5.** 2.00 cm **7.** 1.00 J **9.** 643 N/m **11.** (a) 78.4 N (b) yes (c) 78.4 mm (d) 3.07 J **13.** 5.0×10^2 N/m **15.** 0.025 m **17.** 4.24 cm **19.** they're equal **21.** $\sigma' = 4\sigma$ **23.** 2.50×10^7 Pa **25.** 2 MPa **27.** 0.15% **29.** 0.399 8 m **31.** same **33.** 1×10^{-2} m **35.** (b) 0.11 GPa (d) 2.0×10^{-3} **37.** 29.5 kN **39.** 0.22% **41.** 0.911 MN **43.** 0.765 mm **45.** proof **47.** (a) 0.50 MPa (c) 2.5×10^{-6} **49.** 2.5×10^{-4}% **51.** 0.65 kN **53.** 5.0×10^{-6} cm **55.** 25 GPa **57.** $\Delta L_1 = 2\Delta L_2$

59. (a) 1.01×10^8 Pa (d) 0.11% **61.** 2.03 cm, 0.173% **63.** proof **65.** 1.8 s **67.** $f = 1.3$ Hz, $\omega = 2.6\pi$ rad/s = 8.2 rad/s **69.** 4.9×10^2 m/s² **71.** $T = 4.4$ s **73.** (a) 5.0 m (b) 0.064 Hz (c) 0.10 rad (d) 3.1 m **75.** proof **77.** $\epsilon = \pi$ **79.** $T = 2\pi\sqrt{bp/\rho_w g}$ **81.** proof **83.** $z = +0.50$ m, $z = -0.50$ m, $z = 0$ **85.** proof **87.** $k = 2.6$ N/m **89.** (a) 1.70 kg (b) 3.0 Hz (c) 6.0×10^2 N/m **91.** $A = [(mv_1^2 + kx_1^2)/k]^{1/2}$ **93.** $k = 785$ N/m, $f = 3.15$ Hz **95.** $f_0 = (1/2\pi)[(k_1 + k_2)/m]^{1/2}$ **97.** $y = (0.10 \text{ m})\sin(4.4t - \frac{\pi}{2})$, or $y = -(0.10 \text{ m})\cos 4.4t$, where t is in seconds **99.** 3.5 Hz **101.** $g = 4\pi^2\Delta L/T^2$ **103.** (a) 24.3 N/m (b) 7.02 J (c) 7.02 J (d) -0.534 m (e) 3.46 J (f) 3.56 J (g) 2.98 m/s **105.** $T = 2\pi[m/k]^{1/2}$ **107.** f is halved **109.** 0.158 Hz **111.** $L = 24.8$ m **113.** 9.7 m **115.** $f_c \approx 0.41 f_\theta$ **117.** proof **119.** (c) they're equal (d) $h = \frac{1}{2}v^2/g$ (e) 0.46 m **121.** 1.2T **123.** 98.3 cm **125.** (c) $T = 2\pi r[L/GM_\oplus]^{1/2}$

Solutions to Selected Problems

3. $F = ks$; $k = F/s = (50 \text{ N})/(0.05 \text{ m}) = 1.0$ kN/m. **13.** $k = F/s = (10 \text{ N})/(2.0 \times 10^{-2} \text{ m}) = 5.0 \times 10^2$ N/m. **25.** $\sigma = F/A = (200 \text{ N})/(10^{-4} \text{ m}^2) = 2 \times 10^6$ N/m² = 2 MPa. **29.** $\Delta L/L_0 = \varepsilon$; $\Delta L = (0.400 \text{ 0 m})(5.000 \times 10^{-4}) = 2.000 \times 10^{-4}$ m and $L = L_0 - \Delta L = 0.399$ 8 m. **33.** $Y = (F/A)/(\Delta L/L_0)$; $\Delta L = (F/A)(Y/L_0) = FL_0/YA$; the mike is already on the rod, and so only the weight of the *sumatori* enters; $\Delta L = (9.8 \text{ m/s}^2)(181.4 \text{ kg})(4 \text{ m})/(200 \times 10^9 \text{ Pa})\pi(\frac{1}{2}2 \times 10^{-3} \text{ m})^2 = 1 \times 10^{-2}$ m. **39.** $Y = 50$ GPa = $(110 \text{ MPa})/\varepsilon_R$; $\varepsilon_R = (110 \text{ MPa})/(50 \text{ GPa}) = 2.2 \times 10^{-3} = 0.22\%$. **43.** $F/A < 435$ MPa; $(200 \text{ N})/(435 \text{ MPa}) < A$; $A > 4.598 \times 10^{-7}$ m²; $\pi R^2 > 4.598 \times 10^{-7}$ m²; minimum diameter is 0.756 mm. **45.** $PE_e = F^2\frac{1}{2}L_0/2AY + F^2\frac{1}{2}L_0/2(2A)Y = 3F^2L_0/8AY$. **49.** $Y = \sigma/\varepsilon = (F/A)/(\Delta L/L_0)$; $\varepsilon = \sigma/Y = (0.50 \text{ MPa})/(200 \text{ GPa}) = 2.5 \times 10^{-6}$ or equivalently 2.5×10^{-4}%. **63.** $k = YA/L_0$, $Y = (F/A)/(\Delta L/L_0) = \sigma_R(\Delta L/L_0)$; $PE_e = \frac{1}{2}k(\Delta L)^2 = \frac{1}{2}(YA/L_0)$ $(\sigma_R L_0/Y)^2 = \frac{1}{2}AL_0\sigma_R^2/Y$. **65.** $T = 1/f = 1/[(33\frac{1}{3}$ rpm$)/(60 \text{ s/min})] = 1.8$ s. **69.** $a_0 = -A(2\pi f)^2$; dropping the sign, $a_0 = (0.005 \text{ m})4\pi^2(50 \text{ Hz})^2 = 4.9 \times 10^2$ m/s². **73.** By comparison with Eq. (10.13), (a) $A = 5.0$ m (b) $\omega = 2\pi f = 0.40$ rad/s; $f = 0.064$ Hz (c) $\epsilon = 0.10$ rad (d) $x = 3.1$ m. **79.** The additional buoyant force if the block is pushed down a distance y is $-acy\rho_w g$, which is the restoring force $-ky$, so the motion is SHM, since $F \propto -y$ and $k = ac\rho_w g$. Here, $m = abc\rho$, so $T = 2\pi[(abc\rho)/(ac\rho_w g)]^{1/2} = 2\pi[(bp)/(\rho_w g)]^{1/2}$. **85.** $x = L\cos\phi + R\cos\theta$; $\cos\phi = 1 - \sin^2\phi$, but $L\sin\phi = R\sin\theta$; $x = R\cos\theta + L[1 - (R/L)^2\sin^2\theta]^{1/2}$. Although the first term corresponds to SHM, the presence of the second term negates that. **87.** $f = (1/2\pi)\sqrt{k/m}$; $k = (2\pi f)^2 m = 2.6$ N/m. **95.** Since $x = x_1 = x_2$ and $F = F_1 + F_2$; hence, $kx = k_1 x_1 + k_2 x_2$ and $k = k_1 + k_2$. Now, simply substitute k into Eq. (10.19). **101.** $k\Delta L = mg$; hence, $m/k = \Delta L/g$ and, $T = 2\pi\sqrt{\Delta L/g}$; hence, $g = 4\pi^2\Delta L/T^2$. **111.** From Eq. (10.24), $T = 2\pi \times \sqrt{L/g}$; hence, $L = gT^2/4\pi^2 = 24.8$ m. **121.** From Eq. (10.24), $T = 2\pi$

$\sqrt{L/g}$, and if $L' = L + 50\%L = 1.50L$, $T' = 2\pi [L'/g]^{1/2} = \sqrt{1.50} T = 1.2 T$.

Chapter 11

Answers to Odd-Numbered Multiple Choice Questions

1. d **3.** b **5.** c **7.** c **9.** d **11.** b **13.** d **15.** c **17.** e **19.** d **21.** d **23.** b **25.** c **27.** a **29.** b **31.** c

Answers to Odd-Numbered Problems

1. 5×10^2 nm **3.** $\lambda = 3.40$ m **5.** proof **7.** $A = 1.2$ cm, $\lambda = 10.0$ cm, $f = 20.0$ Hz **9.** $A = 2.5$ cm, y-axis, $\lambda = 78.9$ m **11.** (a) 2 mm (b) 50 Hz (c) 80 cm (d) 40 m/s (e) 0.020 s **13.** +3.8 cm **15.** 12 m **17.** proof **19.** proof **21.** v is doubled **23.** 1.3×10^2 m/s **25.** 14.8 g **27.** 1.08×10^3 N

29.

31. 6.3 m/s² **33.** 100 m/s, 314 rad/s, 0.020 0 s, 2.00 m **35.** (e) 1.3 g **37.** $v = 1.0 \times 10^2$ m/s; $F_T = 3.2 \times 10^2$ N when v is doubled **39.** $F_T = 1.9 \times 10^2$ N **41.** $F_T = g\Delta y(m/L)$, $v_{max} = \sqrt{gL}$, $v = [g\Delta y + MgL/m]^{1/2}$ **43.** $y = -0.7 A$, $y = -0.7 A$ **45.** 360 GPa **47.** 1.4 km/s **49.** 1×10^3 m/s **51.** 40 ms **53.** 1.43 km/s **55.** $T = 2.27$ ms, $\lambda = 0.780$ m **57.** 2.91 **59.** 3.3×10^{-4} m. **61.** 1.9 ms **63.** $t = 2.0$ ms, $\lambda = 51$ m **65.** 0.973 79 **67.** 2.5×10^{-5} W **69.** 1.1 mW **71.** v increases by a factor of $\sqrt{2}$ **73.** 0.79 m **75.** $I = 4.0 \times 10^{-2}$ W/m², E = 4.0 μJ **79.** (b) displacement and pressure are $\frac{1}{4}\lambda$ out-of-phase **81.** 30 dB **83.** 40 dB **85.** 10 dB **87.** 0 dB **89.** 4.8 dB **91.** (d) $I = 1.6 \times 10^{-7}$ W/m² **93.** 5.0×10^{-5} W/m² **95.** 90 dB **97.** (c) $I = 1.0$ W/m² (d) $\beta = 1.2 \times 10^2$ dB **99.** 87 dB, 75 dB **101.** $R = 2.2$ km **103.** 0.20 km **105.** $\beta = 20\log_{10}(P/P_0)$ **107.** 25.8 dB **109.** 996 Hz or 1004 Hz **111.** 1.0 Hz **113.** $\lambda = 2.0$ m **115.** $\lambda = 2N$ m **117.** $\lambda_6 = 0.17$ m **119.** $f_1 = 0.10$ kHz **121.** proof **123.** (a) 266 Hz or 254 Hz (b) 274 Hz or 266 Hz (c) 266 Hz **125.** air, $\lambda = 1.4$ m; string, 66 cm **127.** 2 Hz **129.** 1.76 kHz **131.** 210 N **133.** 1.8 kHz **135.** $2A\cos[\frac{1}{2}(\omega_1 - \omega_2)t]$ is a continually modulated amplitude **137.** (a) 1.10 m (b) 1.00 m (c) 1.20 m **39.** 1016 Hz **141.** 466 Hz **143.** 0.782 m/s **145.** 62 kHz **147.** $f_0 - f_s = 2v_t f_s/(v - v_t)$

Solutions to Selected Problems

3. $v = f\lambda$; $\lambda = (1498 \text{ m/s})/(440 \text{ Hz}) = 3.40$ m. **7.** Compare $y = (1.2 \text{ cm})\sin[2\pi x/(10.0 \text{ cm})]$ to $y = A\sin(2\pi x/\lambda)$; $A = 1.2$ cm, $\lambda = 10.0$ cm, $f = v/\lambda = 20.0$ Hz **15.** The round-trip distance is $2x$

$= vt = (330\ \text{m/s})(70\ \text{ms}) = 23.1\ \text{m}, x = 12\ \text{m}$. **19.** Using the trigonometric identity $\sin(\alpha \mp \beta) = \sin\alpha\cos\beta \mp \cos\alpha\sin\beta$, where $\alpha = k(x - vt)$ and $\beta = \mp\frac{1}{2}\pi$, we get $y = \mp A\cos\frac{2\pi}{\lambda}(x - vt)$. **23.** $v = [F_T/(m/L)]^{1/2} = [(10\ \text{N})/(0.59 \times 10^{-3}\ \text{kg/m})]^{1/2} = 1.3 \times 10^2\ \text{m/s}$. **27.** $F_T = v^2(m/L) = 1.08 \times 10^3\ \text{N}$. **33.** $v = [F_T/(m/L)]^{1/2} = 100\ \text{m/s}$; $\omega = 2\pi f = 314\ \text{rad/s}$; $v = f\lambda$, $\lambda = 2.00\ \text{m}$; $T = 1/f = 0.020\,0\ \text{s}$. **39.** $m = F_W/g = (0.20\ \text{N})/(9.81\ \text{m/s}^2) = 0.020\,4\ \text{kg}$; $m/L = 1.7 \times 10^{-3}\ \text{kg/m}$; $v = [F_T/(m/L)]^{1/2}$; $F_T = v^2 m/L = 1.9 \times 10^2\ \text{N}$, or 42 lb. **43.** The phase is $2\pi[(t/0.01) - (5.0/40)] = 2\pi[t/0.01) - (1/8)] = (200\pi t - \frac{1}{4}\pi)$; hence, $y = A\sin(200\pi t - \frac{1}{4}\pi)$ is the displacement of the bead. At $t = 0$, $y = A\sin(-\frac{1}{4}\pi) = -0.7\,A$, and at $t = 0.01$ s, which is one period later, $y = A\sin(2\pi - \frac{1}{4}\pi) = -0.7\,A$. **47.** $v = \sqrt{B/\rho} = 1.4\ \text{km/s}$. **55.** $T = 1/f = 1.440\ \text{Hz} = 2.27\ \text{ms}$. $\lambda = v/f = (343\ \text{m/s})/(440\ \text{Hz}) = 0.780\ \text{m}$. **61.** $\tau = \frac{1}{2}T = 1/2f = 1/2(261.6\ \text{Hz}) = 1.9\ \text{ms}$. **71.** $v = [1.4P/\rho]^{1/2}$; the volume doubles and $\rho_i = 2\rho_f$; $v_f = \sqrt{2}v_i$. **79.** (a) $\text{N/m}^2 = (\text{N/m}^2) \times (1/\text{m})(\text{m})$; (b) the displacement and pressure are $\frac{1}{4}\lambda$, or 90°, out-of-phase; (c) $P_0 = \frac{2\pi}{\lambda}BA$; $v = [B/\rho]^{1/2}$; $P_0 = \frac{2\pi}{\lambda}v^2\rho A$. The Bulk Modulus relates pressure change to volume change in the gas, and that volume change, in turn, is related to the displacement of the molecules. P is a pressure change from ambient, and so we can expect it to be associated with B and displacement. **85.** Ten times the power means 10 times the intensity; therefore $\Delta\beta = 10\log_{10}10 = 10\ \text{dB}$. **89.** $\Delta\beta = 10\log_{10}3 = 4.8\ \text{dB}$. **99.** $\beta = 10\log_{10}(56 \times 10^{-5}\ \text{W/m}^2)/(10^{-12}\ \text{W/m}^2) = 87\ \text{dB}$. The intensity drops with the inverse square of the distance; $(5.0\ \text{m})^2/(20\ \text{m})^2 = I/(56 \times 10^{-5}\ \text{W/m}^2)$; at 20.0 m, $I = 3.5 \times 10^{-5}\ \text{W/m}^2$; and there $\beta = 75\ \text{dB}$. **103.** -40 dB means a change of -4 bel or 10^{-4}; hence, from the Inverse Square Law, $R^2/(2.0\ \text{m})^2 = 1/10^{-4}$ of whatever the original intensity was; $R = 0.20\ \text{km}$. **109.** The beat frequency is 4 Hz; hence, the wire is vibrating at *either* 996 Hz or 1004 Hz. **117.** $\frac{1}{2}\lambda = L = \frac{1}{2}\ \text{m}$, $\lambda = 1.0\ \text{m}$; $\lambda_6 = 1/6\ \text{m}$. **121.** $f_N = (N/2L)[F_T/(m/L)]^{1/2} = (\frac{1}{2}N)[F_T/L^2(m/L)]^{1/2}$, which equals the desired expression. **133.** $v = \sqrt{Y/\rho} = 3516\ \text{m/s}$; $v = f\lambda$, $\lambda = 2L = 2.00\ \text{m}$; $f = v/2L = (3516\ \text{m/s})/(2.00\ \text{m}) = 1.8\ \text{kHz}$. **135.** $y = y_1 + y_2 = A\sin(\omega_1 t) + A\sin(\omega_2 t)$. Using the identity $\sin\alpha + \sin\beta = 2\sin\frac{1}{2}(\alpha + \beta)\cos\frac{1}{2}(\alpha - \beta)$, we get the desired result. Remember $\omega_1 \approx \omega_2$; there is a relatively high-frequency sinusoidal carrier oscillating at the average frequency $\frac{1}{2}(\omega_1 + \omega_2)$, and this is multiplied by an amplitude term $2A\cos[\frac{1}{2}(\omega_1 - \omega_2)t]$. This amplitude varies slowly in time, and so the modulated carrier has maxima whenever $\cos[\frac{1}{2}(\omega_1 - \omega_2)t] = \pm 1$; hence, beats occur at twice that frequency, namely, at $(\omega_1 - \omega_2)$ or $(f_1 - f_2)$. **143.** $f_o = f_s(v + v_o)/(v + v_s)$; here $v_s = 0$ and $v_o = v(f_o - f_s)/f_s = (344\ \text{m/s})(1/440) = 0.782\ \text{m/s}$.

Chapter 12

Answers to Odd-Numbered Multiple Choice Questions

1. b **3.** a **5.** c **7.** c **9.** c **11.** b **13.** a **15.** b **17.** d **19.** a **21.** a **23.** b **25.** e **27.** a

Answers to Odd-Numbered Problems

1. 37.0 °C **3.** Fig. P3 is a plot of the Celsius temperature T_C versus the Fahrenheit temperature T_F **5.** $T_{C1} = -17.8$ °C **7.** -270.42 °C **9.** -215 °C **11.** 9.6×10^3 °F **13.** 55.6 K **15.** 42 K **17.** 10×10^6 °C $\approx 18 \times 10^6$ °F **19.** 927 °F **21.** proof **23.** 2×10^{-4} m **25.** $\Delta L = 25 \times 10^{-6}$ m **27.** 5.0 mm **29.** 0.54 m **31.** (a) $\Delta L = \alpha L_0 \Delta T$ (b) they're equal (c) 36 K (d) 1.0 mm **33.** 1001 cm^2 **35.** (a) 3.14×10^{-4} m^2 (b) 2.00×10^{-2} m (c) 0.007×10^{-4} m^2 **37.** (a) heat the pipe (b) $\Delta D = \alpha D_0 \Delta T$ (c) 66.2 °C (d) 86 °C **39.** 10.002 m **41.** 0.51 cm^3 **43.** $\Delta H = -1.5$ mm **45.** 0.79×10^{-6} m^3 **47.** 7.9×10^3 kg/m^3 **49.** 2.006 s, ran fast **51.** proof **53.** 9.3 MPa **55.** (a) $V = CT$ (b) faster (c) out (d) $V_2 = 2V_1$ **57.** (a) PV/T = const (b) slower (c) no (d) $V_2 = V_1$ **59.** 1.0 MPa **61.** Gauge pressure is 1.1 atm **63.** $T_f = 199$ K **65.** 1.1×10^3 cm^3 **67.** 2.992×10^{-26} kg **69.** 3.01×10^{26} molecules, 500 moles **71.** 1.1 m^3 **75.** (a) V/T = const (b) we measure volume at constant pressure (c) capillary tube is open, mercury doesn't move, P is constant (d) 1 atm (e) 328 K **77.** $V_f = (1 + \rho gh/P_A)V_i$ **79.** (a) $\rho_{Hg}gh$ (b) $\rho_{Hg}gH + \rho_{Hg}gh$ (c) $V = LA$ (d) PV = const., PL = const., P = const.$(1/L)$ **81.** 189 cm^3 **83.** 4.3 atm **85.** 174 K **87.** $N = 2.7 \times 10^{19}$ **89.** 8.21 mol **91.** 3.1×10^5 Pa **93.** (a) $PV = Nk_BT$ (b) 347 K (c) 4.0×10^{-3} m^3 (d) 0.303×10^6 Pa (e) 2.5×10^{23} molecules **95.** 1.2×10^{22} **97.** 0.33 g **99.** 43 kg **101.** 3.0 m/s **103.** 478.03 m/s **105.** 6.1×10^{-21} J **107.** (a) $\text{KE}_{av} = \frac{3}{2}k_BT$ (b) it would double (c) 34 K (d) 307 °C (e) same **109.** 76 kJ **111.** 0.147 MJ **113.** $v_{rms} = \sqrt{3P/\rho}$ **115.** 1.6×10^7 J **117.** 0.427%

Solutions to Selected Problems

1. 37.0°C **17.** 10 000 000 K $\rightarrow$ 10 000 273°C $\approx 10 \times 10^6$ °C $\approx 18 \times 10^6$ °F. **25.** $\alpha = 25 \times 10^{-6}\ \text{K}^{-1}$; hence, $\Delta L = 25 \times 10^{-6}$ m. **29.** $\Delta L = \alpha L_0 \Delta T = (12 \times 10^{-6}\ \text{K}^{-1})(1280\ \text{m})(35\ \text{K}) = 0.54$ m. **39.** $\Delta L = \alpha L_0 \Delta T = (12.15 \times 10^{-6}\ \text{K}^{-1})(10.000\ \text{m})(16.0\ \text{K}) = 1.9 \times 10^{-3}$ m. $L = 10.002$ m. 2 mm is an appreciable error, but it's not likely to trouble a carpenter. **43.** $\Delta V = \beta V_0 \Delta T = (182 \times 10^{-6}\ \text{K}^{-1})(800.0 \times 10^{-6}\ \text{m}^3)(-95\ \text{K}) = -1.383 \times 10^{-5}\ \text{m}^3$ for the mercury. $\Delta V = \beta V_0 \Delta T = (26 \times 10^{-6}\ \text{K}^{-1})(800.0 \times 10^{-6}\ \text{m}^3)(-95\ \text{K}) = -1.98 \times 10^{-6}\ \text{m}^3$ for the glass. Hence, the total $\Delta V = -1.185 \times 10^{-5}\ \text{m}^3$. $\Delta V = A\Delta H = \pi r^2 \Delta H$; $\Delta H = -1.5$ mm. **47.** $\rho_i = m/V_i$; $\rho_f = m/(V_i + \Delta V)$; $\rho_i V_i = \rho_f(V_i + \Delta V) = \rho_f V_i(1 + \beta\Delta T)$; $\rho_i/(1 + \beta\Delta T) = \rho_f = (7.85 \times 10^3\ \text{kg/m}^3)/[(1 + (36 \times 10^{-6}\ \text{K}^{-1})(-20\ \text{K})] = 7.9 \times 10^3\ \text{kg/m}^3$. **51.** $\rho_0 = m/V_0$; $\rho = m/(V_0 + \Delta V)$; $\rho_0 V_0 = \rho(V_0 + \Delta V) = \rho V_0(1 + \beta\Delta T)$; $\rho_0/(1 + \beta\Delta T) = \rho$: from the Binomial Theorem, $(1 + x)^n \approx (1 + nx)$ for $x << 1$; here, $n = -1$, and so $\rho \approx \rho_0(1 - \beta\Delta T)$. **53.** Stress $= Y\alpha\Delta T = (25\ \text{GPa})(10 \times 10^{-6}\ \text{K}^{-1})(37\ \text{K}) = 9.3$ MPa. Concrete is strong in compression so it's more likely to buckle than break. **59.** $P_fV_f = P_iV_i$; $P_f = 10\,P_i = 1.0$ MPa. **63.** No change in P; $V_i/T_i = V_f/T_f$; $(300\ \text{cm}^3)/(298\ \text{K}) = (200\ \text{cm}^3)/T_f$; $T_f = 199$ K. **67.** 2H $\rightarrow$ 2 (1.008 u); O $\rightarrow$ 15.999 u; (18.015 u)(1.660 6 $\times 10^{-27}$ kg/u) $= 2.992 \times 10^{-26}$ kg. **81.** To find the pressure on the gas, 3.90 cm Hg = 5.199 6 kPa, P_i = (101.05 − 5.199 6) kPa = 95.85 kPa; $P_iV_i/T_i = P_fV_f/T_f$; $(95.85\ \text{kPa})(214\ \text{cm}^3)/(293.15\ \text{K}) = (101.32\ \text{kPa})V_f/(273\ \text{K})$; $V_f = 189\ \text{cm}^3$. **85.** $PV = nRT$; $(202.6 \times 10^3\ \text{Pa})(10.0 \times 10^{-3}\ \text{m}^3) = (1.40\ \text{mol})(8.314\ \text{J/mol}\cdot\text{K})T$; $T = 174$ K. **95.** $PV = Nk_BT$; $T = 98.6$°F $= 37$°C $\rightarrow 310$ K; $PV/k_BT = (101.46\ \text{kPa})(500 \times 10^{-6}\ \text{m}^3)/(1.380\,7 \times 10^{-23}\ \text{J/K})(310\ \text{K}) = N = 1.2 \times 10^{22}$ molecules. **99.** $n = PV/RT = (19.4 \times 10^3\ \text{Pa})(2000\ \text{m}^3)/(8.314\ \text{J/mol}\cdot\text{K})(218\ \text{K}) = 21.4 \times 10^3$ moles. Each mole has a mass of 0.002 kg; hence, the total mass is $(0.002\ \text{kg/mol})(21.4 \times 10^3\ \text{mol}) = 43$ kg. **101.** v_{av} = (1.0 m/s + 2.0 m/s + 3.0 m/s + 4.0 m/s + 5.0 m/s)/5 = 15/5 m/s = 3.0 m/s. **109.** $\text{KE}_{av} = (3/2)k_BT = (3/2)(1.380\,662 \times 10^{-23}\ \text{J/K})(273.15\ \text{K}) = 5.66 \times 10^{-21}$ J/molecule. 0.50 m^3 = 500 liter; (500 liter)/(22.4 liter/mol) = 22.3 mol; (22.3 mol)(6.022 $\times 10^{23}$ molecules/mol)(5.66 $\times 10^{-21}$ J/molecule) = 76 kJ **117.** $v_{rms} = \sqrt{3k_BT/m}$; hence, the ratio of the *rms*-speeds varies as the inverse of the square roots of the masses. The masses are 238 u + 6(18.998 u) = 351.99 u and 235 u + 6(18.998 u) = 348.99 u. $(v^{235}_{rms} - v^{238}_{rms})/v^{235}_{rms} = [(348.99\ \text{u})^{-1/2} - (351.99\ \text{u})^{-1/2}]/(348.99\ \text{u})^{-1/2} = 0.427\%$.

Chapter 13

Answers to Odd-Numbered Multiple Choice Questions

1. e **3.** c **5.** b **7.** a **9.** e **11.** c **13.** c **15.** b **17.** a **19.** c **21.** a **23.** b **25.** b **27.** c

Answers to Odd-Numbered Problems

1. $Q = 2.0$ kcal **3.** 0.20 kcal. **5.** 17 K **7.** 63 kJ **9.** $T_f = 60.0$ °C **11.** (a) 0.27 kJ (b) 1.10×10^4 J (c) 2.62 kcal (d) 0.38 K **13.** 0.2 Mcal $\approx$ 0.9 MJ **15.** $W = 4.15\ \text{J} \approx 1$ cal **17.** T_f = 26.9 °C **19.** $T_f = 58$ °C **21.** 17.6 MJ **23.** 427 m **25.** 1.9 h **27.** 0.68 kJ **29.** 0.68 kg **31.** 0.12 GJ **33.** (a) 510 J/s (b) 2.1 kJ (c) 4.1 s (d) 5.5 min **35.** (a) 4.186 kJ/kg·K, 0.39 kJ/kg·K (b) 6.4×10^2 kJ (c) 57 kJ (d) 1.7×10^2 kcal **37.** 0.11 kg **39.** 0.9 kg **41.** +0.43 K **43.** T_f = 72 °C **45.** 1.1×10^2 kJ or 27 kcal **47.** 0.74 MJ **49.** $c = 0.65$ kJ/kg·K **51.** 1.8 h **53.** 13 g **55.** $c = 0.44$ kJ/kg·K **57.** proof **59.** 1380 °C **61.** 3.1×10^5 J **63.** 41.0 kJ **65.** 0.13 kg **67.** 3.5 MJ **69.** 0.230 kg vaporizes **71.** *b* **73.** 0.17 MJ **75.** 3.5 kW **77.** (a) 0 °C (b) 15.6 J (c) 15.6 J (d) 4.67×10^{-5} kg **79.** $T_f = 24$ °C **81.** 2.4 h **83.** 7.45×10^{-20} J/molecule **85.** 0.068 kg **87.** 353 m/s **89.** 1×10^2 g **91.** (a) 2.21×10^6 kg (b) 16.7 °C (c) 1.54×10^8 kJ (d) 6.49 $\times 10^4$ kg/min (e) 1.7×10^4 gallons/min **93.** T_f = 0 °C **95.** 0 °C **97.** 0.36 MJ, 960.8 °C (or 1234 K) **99.** 3.9 kJ **101.** 2.1×10^5 J **103.** 2.3 m **105.** 1.6%

Solutions to Selected Problems

1. $Q = cm\Delta T = (1.00\ \text{cal/g}\cdot\text{C°})(100\ \text{g})(20\ \text{°C}) = 2000\ \text{cal} = 2.0$ kcal **7.** 15 kcal = 63 kJ. **13.**

(7 kcal/g)(30 g) = 210 kcal = 879 kJ, or to 1 significant figure, 9×10^5 J.   **19.** $Q/t = cm\Delta T/t$; 1.5 liters/min = 1.5 kg/min = 0.025 kg/s; 4000 W = (4186 J/kg·K)(0.025 kg/s)ΔT; ΔT = 38 K; T_f = 58 °C.   **25.** $Q = cm\Delta T$ = (4186 J/kg·K) (1.00 kg)(80 K) = 334 880 J; (334 880 J)/(50 J/s) = 6697.6 s = 1.9 h.   **27.** $Q = cm\Delta T$ = (0.060) (1.0 kcal/kg)(0.030 kg)(−90 K) = 0.162 kcal = 162 cal = 0.68 kJ.   **31.** $m = (3785 \times 10^{-6}$ m^3) (0.68 × 10³ kg/m³) = 2.574 kg; (2.574 kg) (48 MJ/kg) = 124 MJ = 1.2×10^8 J.   **39.** −(910 J/kg·K) × m(−173 K) = (4186 J/kg·K)(4.95 kg) (7.00 K); m(157 430 J/kg) = 145 044.9 J; m = 0.921 kg.   **47.** 150 lb → 68.04 kg; $Q = cm\Delta T$; ΔT = 33.89 °C − 37 °C = −3.11 K; Q = (3474.4 J/ kg·K)(68.04 kg)(3.11 K) = 0.74 MJ.   **51.** The average specific heat capacity of the body is 3.5 kJ/kg·K (0.83 kcal/kg·C°), and normal temperature is 37°C. $Q = cm\Delta T$ = (3.5 kJ/kg·K)(70 kg) (6 K) = 1470 kJ will raise the temperature 6 K. Since 200 kcal/h = 837.2 kJ/h, it will take (1470 kJ)/(837.2 kJ/h) = 1.8 h.   **55.** The tension in the rope is g(6 kg) = 58.8 N, and this is the force exerted (via friction) on the rope by the rod, and it does it over a distance of $2\pi R(240)$ = 22.62 m. The work thus done is (58.8 N)(22.62 m) = 1330 J, and this result equals the heat−in: $Q = cm\Delta T$; 1330 J = c(0.250 kg)(12.0 K); c = 443 J/kg·K, which is what we might expect for some sort of steel.   **57.** In the first set-up, $Pt = c_w m\Delta T + P_1 t$ and, in the second, $P't = c_w m'\Delta T + P_1 t$; note that the mean temperature is the same in both cases, and so P_1 is the same. Thus, solve for P_1 in both equations and set them equal. $P − c_w m\Delta T/t = P' − c_w m'\Delta T/t$; and $(P − P')t = c_w m\Delta T(m − m')$ **65.** $−m$(334 000 J/kg) = (0.50 kg)(4186 J/kg·K) (0 °C − 20 °C); m = 0.13 kg.   **67.** L_v = 2336 × 10³ J/kg; $Q = mL_v$ = (1.5 kg)(2336 × 10³ J/kg) = 3.5 MJ.   **73.** $Q = cm\Delta T + mL_v$ = (138 J/kg·K) (0.50 kg)(630 K − 240 K) + (0.50 kg)(296 × 10³ J/kg) = 26 910 J + 148 000 = 0.17 MJ   **83.** 1 mol = 18 g; at STP there are 6.02 × 10²³ molecules/mol; to vaporize 1 mol requires $Q = mL_v$ = (0.018 kg)(2492 kJ/kg) = 44.86 kJ. Hence, per molecule the energy is (44.86 × 10³ J)/(6.02 × 10²³ molecules/mol) = 7.45 × 10⁻²⁰ J.   **87.** $Q = mL_f$; mL_f = (22.9 × 10³ J/kg)m; to which must be added Q = $cm\Delta T$ = (130 J/kg·K)m(327 K − 23 K) = (39 520 J/kg)m; the total is then (62 420 J/kg)m = $\frac{1}{2}mv^2$; v = 353 m/s.   **93.** $−Q_{out} = Q_{in}$; if it dropped to zero, the calorimeter would provide an amount of heat equal to (1.00 kcal/kg·K)(0.398 kg)(5.1 K) + (0.093 kcal/kg·K)(0.102 kg)(5.1 K) = 2.078 kcal; the amount of heat needed to melt the ice is (0.040 5 kg)(80 kcal/kg) = 3.24 kcal; hence, not all the ice melts, and T_f = 0 °C.   **101.** $Q/t = \lambda_T A\Delta T/d$ = (18 Cal·cm/m²·h·C°)(1.4 m²) × (4°C)/(2.0 cm) = 50.4 Cal/h = 0.014 kcal/s = 0.059 kJ/s = 0.059 kW. Therefore, 2.1 × 10⁵ J per hour.   **103.** $Q/t = k_{TB}A\Delta T/d_B = k_{TA}A\Delta T/d_A$; $k_{TB}/d_B = k_{TA}/d_A$; $k_{TB}/k_{TA} = 0.60/0.026 = 23 = d_B/d_A$; d_B = 2.3 m.

Chapter 14

Answers to Odd-Numbered Multiple Choice Questions

1. b   **3.** d   **5.** d   **7.** b   **9.** e   **11.** b   **13.** c **15.** b   **17.** b   **19.** c   **21.** a   **23.** d   **25.** b

Answers to Odd-Numbered Problems

1. +250 J   **3.** −450 J   **5.** +500 J   **7.** (a) V increases (b) V = 1.6 × 10⁻² m³ (c) W = −0.256 kJ, done by gas (d) ΔU = 144 J, increased   **9.** (a) P increases (b) ΔU = 4.0 kJ, increased   **11.** 519 J **13.** $W = -\frac{1}{4}P_i V_i$ **15.** W = 2.03 MJ, T will decrease   **17.** +12 J   **19.** $W = -\frac{1}{2}P_i V_i$ **21.** (a) W = 0, ΔV = 0 (b) take heat-out (c) 29.5 kJ (d) $T_B > T_A$ (e) $U_B > U_A$, ideal gas (f) −0.080 MJ, work done by gas (g) 29.5 kJ   **23.** W = −9.8 kJ, ΔU = +0.2 kJ   **25.** −4.8 kJ.   **27.** 0.22 g   **29.** 7.47% **31.** (a) ΔT = 0 (b) ΔU = 0 (c) removed (d) −10.5 kJ   **33.** −4.21 kJ   **35.** 702 J   **37.** 0.989 MJ   **39.** (a) 1/2 (b) 1/2 (c) 2.00 atm (d) on (e) $W = -P_i V_i \ln(V_f/V_i)$ (f) 42 J   **41.** 2.2 liters **43.** $W = -nRT \ln(P_i/P_f)$ **45.** proof   **47.** (a) 0.038 4 MPa (b) 207 K   **49.** T_f = −78 °C   **51.** proof   **53.** T_i/T_f = 0.49   **55.** 22 J   **57.** 5.5% **59.** 22.1%   **61.** (a) zero; back to starting state (b) 0.32 kJ (c) −0.32 kJ, work-out (d) 16% **63.** 24.6%   **65.** 4.23%   **67.** 3.5 kW   **69.** 0.11 J **71.** (a) Eq. (14.11) (b) 26.5% (c) raise T_H (d) T_H = 1.10 × 10³ K   **73.** impossible, 435 kJ   **75.** 48.9 MJ   **77.** (a) 72% (b) 28% (c) friction, exhaust, radiation, convection, conduction (d) 0.800 kJ   **79.** T_H = 547 K   **81.** e = 62.5%, irreversible   **83.** 399 W   **85.** Q_H/W_i = 6.5, pay for 1 unit of work and get 6.5 units of heat   **87.** 0.69 W **89.** +2.68 J/K   **91.** +19.0 J/K   **93.** (a) Eq. (14.17) (b) 1.33 × 10⁶ J (c) 273.15 K (d) 4.89 × 10³ J/K   **95.** see Student Solution Manual **97.** +75.6 J/K   **99.** ΔS = +1.22 kJ/K for melting, ΔS = +6.06 kJ/K for vaporization   **101.** +14 J/K   **103.** (a) 0 (b) $P_i/2$ (c) 0; isothermal (d) $|W| = |Q|$ (e) $W = -nRT \ln 2$ **105.** W_o = 6.0 MJ, e = 67%   **107.** −2.71 J/K.   **109.** +5.8 J/K

Solutions to Selected Problems

1. $\Delta U = Q + W$ = (+500 J) + (−250 J) = +250 J.   **5.** Since ΔV = 0, W = 0, $Q = \Delta U$ = +500 J. **13.** $W = \frac{1}{2}(\frac{1}{2}P_i)(2V_i − V_i) = \frac{1}{4}P_i V_i$.   **17.** The work done is the area under the curve, which is $-\frac{1}{2}$(3.0 MPa)(2.0 × 10⁻⁶ m³) − (4.0 × 10⁻⁶ m³)(2.0 MPa) − (1.0 MPa)(1.0 × 10⁻⁶ m³) = −3.0 J − 8.0 J − 1.0 J = −12 J, and this is negative, since the gas expanded and it did work on the surroundings. **23.** Isobaric process: $W = −P\Delta V = −PAh$ = −(0.30 × 10⁶ Pa)(0.50 m²)(0.065 m) = −9.8 kJ; $\Delta U = Q + W$ = 10.0 kJ − 9.8 kJ = 0.2 kJ.   **33.** Use Eq. (14.5) and the Ideal Gas Law: $W = −nRT \ln(V_f/V_i)$ = $−P_i V_i \ln(V_f/V_i)$ = −[(10.0 atm)(1.013 × 10⁵ Pa/atm)(6000 × 10⁻⁶ m³) ln[(12 000 cm³/6000 cm³)] = −4.21 × 10³ J = −4.21 kJ.   **43.** $W = −nRT \ln(V_f/V_i)$, but $P_i V_i = P_f V_f$; hence, $W = −nRT \ln(P_i/P_f)$.   **49.** $P_i V_i^\gamma = P_f V_f^\gamma$, where we assume the rapid expansion means an adiabatic process; final pressure is 1 atm; we don't have ini-

tial volume, so let's work per−unit−volume, (4.5 atm)(1 m³) = (1 atm) V_f^γ; V_f = 4.5 $^{1/\gamma}$ = 2.93 m³ per m³, where γ = 1.4. To find T_f, use Ideal Gas Law; $P_i V_i/T_i = P_f V_f/T_f$; (4.5 atm)(1 m³)/(300 K) = (1.0 atm)(2.93 m³)/T_f; T_f = 195 K, or −78 °C. Notice that we didn't need to know the actual initial volume.   **73.** The maximum efficiency requires the temperatures in kelvins; T_H → 144 °F = 62.2 °C or 355.2 K; T_L → 5 °F = −15 °C or 258 K; e_c = 1 − (258/335.2) = 23%. As compared to the general expression for efficiency, e = 26 kJ)/(102 kJ) = 25.4%. The engine exceeds the efficiency of a Carnot engine and is therefore impossible.   **79.** e_c = 1 − T_L/T_H = 42.2% and T_L = 273.4 K, T_H = 547 K.   **85.** η = heat-in/ work-in = Q_L/W_i = 5.5. Moreover, $Q_H = Q_L + W_i$ = 5.5 $W_i + W_i$ = 6.5 W_i; hence, the ratio we want Q_H/W_i = 6.5. You pay for 1 unit of work and get 6.5 units of heat into the house, which compares very well with electrical heating (6.5 times better), where 100% of the electrical energy is converted into heat and you have to pay for each unit of energy that comes into the house.   **89.** ΔS = (1.00 kJ)/(373 K) = 2.68 J/K.   **97.** $\Delta S = Q/T$ = mL_f/T = (0.500 kg)(205 kJ/kg)/(1356 K) = +75.6 J/K.   **101.** Since the masses are equal, ΔT = ±4.0 K; thus, $Q = mc\Delta T$ = (20 kg)(4.186 kJ/ kg·K)(±4.0 K) = ±334.88 kJ. Average temperatures are 38 °C and 34 °C; ΔS = (−334.88 kJ)/ (311 K) + (+334.88 kJ)/(307 K) = (−1.077 kJ/ K) + (+1.091 kJ/K) = +0.014 kJ/K.   **109.** From Problem 108, $\Delta S = nR \times \ln(V_f/V_i)$ = (1.0 mol)(8.314 4 J/mol·K) ln 2 = 5.8 J/K.

Chapter 15

Answers to Odd-Numbered Multiple Choice Questions

1. e   **3.** a   **5.** c   **7.** c   **9.** b   **11.** c   **13.** a **15.** a   **17.** d   **19.** b   **21.** c   **23.** d   **25.** e

Answers to Odd-Numbered Problems

1. 5.7 × 10⁻¹⁸ kg   **3.** (a) charges combine (b) +4.0 μC (c) $Q/2$ (d) +2.0 μC   **5.** (a) repulsive (b) $10q_e$ (d) 5.77 × 10⁻²¹ N   **7.** ±1.1 × 10⁻⁵ C   **9.** 2.3 N   **11.** ± 5.3 μC   **13.** 1.52 × 10⁻¹⁴ m   **15.** zero   **17.** 7.9 μN   **19.** zero   **21.** 0.58 N   **23.** 1.7 × 10⁻¹⁶ N toward the third electron   **25.** 0.288 N attractive, 3.60 × 10⁻² N repulsive   **27.** 8.3 N, ϕ = 49°   **29.** 6.5 N, ϕ = 74° **31.** 0.11 μC   **33.** q = −0.343Q, a = 0.414, where Q and q are separated by ad **35.** 5.0 × 10⁻¹⁰ N/C **37.** 8.0 × 10² N/C   **39.** 0.10 N/C due east   **41.** 8.00 × 10⁻⁵ N due east   **43.** 5.6 × 10⁻¹¹ N/C straight down   **45.** (a) 2$\vec{E}$ (b) $\vec{E}$/4 (b) $\vec{E}$/2 **47.** 5.1 × 10¹¹ N/C   **49.** 7.8 × 10⁴ N/C   **51.** 6.4 × 10² N/C straight upward   **53.** 50.0 N/C, 36.9° N of E   **55.** 6.00 × 10⁻¹² C   **57.** 2.6 × 10¹⁵ m/s² **59.** 9.6 × 10¹⁰ m/s²   **61.** (a) − (b) F_E, F_T, F_w (e) 10.0 × 10⁻³ N (g) 1.04 × 10⁻⁵ C   **63.** (a) upward (b) positive (c) F_E and F_w (f) 8.7 × 10⁻¹⁰ C   **65.** $E = \sigma/\varepsilon_0$ **67.** $E = Q/2\pi RL\varepsilon_0 K_e$ **69.** 1.2 × 10⁹ N/C out perpendicularly from the sheet in the ± z-direction   **71.** 4.4 nC   **73.** outside the sphere $E = Q/4\pi\varepsilon_0 r^2$, radially outward if $Q > 0$ and radially inward if $Q < 0$; inside the sphere $E = 0$ **75.** $E = \sigma/\varepsilon_0$ **77.** $E = R\sigma/r\varepsilon$ **79.** $E = Q/4\pi\varepsilon_0 r^2$ within the space, zero beyond the shells **81.** $E = rQ/4\pi\varepsilon_0 R^3$ **83.** 1.4×10^{21} N/C

Solutions to Selected Problems

1. $N = (10^{-6}\ \text{C})/(-1.6 \times 10^{-19}\ \text{C}) = 6.2 \times 10^{12}$ electrons. Each has a mass of 9.109×10^{-31} kg; hence, the mass increase is 5.7×10^{-18} kg. **9.** $F = kqq/r^2 = (9.0 \times 10^9\ \text{N·m}^2/\text{C}^2)(1.6 \times 10^{-19}\ \text{C})^2/(1.0 \times 10^{-14}\ \text{m})^2 = 2.3$ N. **15.** By symme-try, the net force is zero. **21.** $F_3 = F_{31} + F_{32} = (9 \times 10^9\ \text{N·m}^2/\text{C}^2)(-12.5 \times 10^{-6}\ \text{C})(-10.0 \times 10^{-6}\ \text{C})/(3.0\ \text{m})^2 + (9 \times 10^9\ \text{N·m}^2/\text{C}^2)(-5.0 \times 10^{-6}\ \text{C})(-10.0 \times 10^{-6}\ \text{C})/(1.0\ \text{m})^2 = 0.125\ \text{N} + 0.450\ \text{N} = 0.58$ N, in the increasing x-direction. **27.** See Fig. AN 27. $F_{12} = kq_{,1}q_{,2}/r_{12}^2 = 3.6$ N; $F_{13} = kq_1q_3/r_{13}^2 = 4.5$ N; $F_{14} = kq_{,1}q_{,4}/r_{14}^2 = 1.8$ N; $F_{x1} = (1.8\ \text{N}) + (4.5\ \text{N})4/5 = 5.4$ N; $F_{y1} = -(3.6\ \text{N}) - (4.5\ \text{N})3/5 = -6.3$ N; $F_1 = (5.4^2 + 6.3^2)^{1/2} = 8.3$ N; $\phi = \tan^{-1}(6.3/5.4) = 49°$. **33.** The repulsive

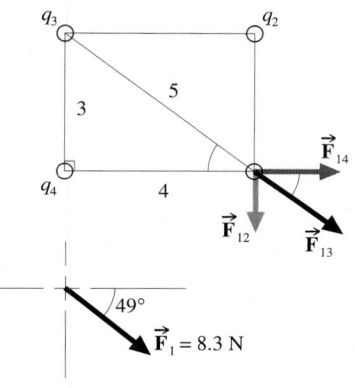

Fig. AN 27

force between the two known charges is $F = kQ2Q/d^2$; put a negative charge $-q$ on the line connecting the two positive charges and between them at a distance ad from $+Q$ and $(1 - a)d$ from $+2Q$. The attraction between $+Q$ and $-q$ will cancel the repulsive force on $+Q$ when $k2Q^2/d^2 = kqQ/(ad)^2$ or when $a^2 = q/2Q$; similarly $+2Q$ will be in equilibrium when $k2Q^2/d^2 = kq2Q/[(1 - a)d]^2$, that is, $(1 - a)^2q/Q$; hence, $(1 - \sqrt{q/2Q}) = \sqrt{q/Q}$; $1 - \sqrt{q}/\sqrt{2Q} = \sqrt{q}/\sqrt{Q}$; $1 = \sqrt{q}(1 + \sqrt{2})/(\sqrt{2}\sqrt{Q})$; $q = 2Q/(1 + \sqrt{2})^2 = -0.343Q$ and $a = 0.414$. **39.** make east $+$, $E = F/q = (-2.0\ \text{nN})/(-20\ \text{nC}) = (1/10)$ N/C; points due east. **47.** $E = kq_e/r^2 = (9.0 \times 10^9\ \text{N·m}^2/\text{C}^2)(1.6 \times 10^{-19}\ \text{C})/(5.3 \times 10^{-11}\ \text{m})^2 = 5.1 \times 10^{11}$ N/C. **51.** See Fig. AN 51. The horizontal field cancels, leaving only twice the vertical con-tribution from each charge: $E = 2(kq_{,1}/r^2)\cos\theta = 2[(9.0 \times 10^9\ \text{N·m}^2/\text{C}^2)(+50 \times 10^{-9}\ \text{C})/(1.0\ \text{m})^2](\cos 45°) = 6.4 \times 10^2$ N/C, straight upward. **59.** $F = qE = ma$; $a = qE/m = (1.6 \times 10^{-19}\ \text{C})(20 \times 10^3\ \text{N/C})/(3.35 \times 10^{-26}\ \text{kg}) = 9.6 \times$

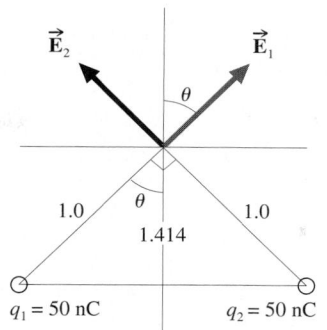

Fig. AN 51

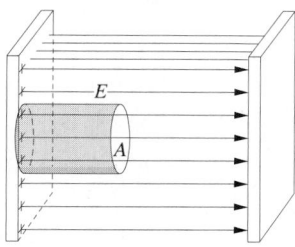

Fig. AN 65

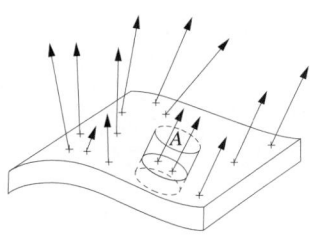

Fig. AN 75

10^{10} m/s². **65.** See Fig. AN 65. The field is par-allel to the curved side and perpendicular to the endface; since $E = 0$ inside the metal, $EA = \sigma A/\varepsilon_0$ and $E = \sigma/\varepsilon_0$. **75.** See Fig. AN 75. Use a very short perpendicular Gaussian cylindrical sur-face. Assume that E and σ are constant over the tiny area of the cylinder (both may vary from region to region, depending on the shape of the conductor). Embed one endface within the conduc-tor, where $E = 0$. Hence, $EA = \sigma A/\varepsilon_0$ and $E = \sigma/\varepsilon_0$. **79.** See Fig. AN 79. Surround the inner sphere with a spherical Gaussian surface of radius r ranging in the gap between the shells. The field is radial and $EA = E4\pi r^2 = Q/\varepsilon_0$; hence, $E = Q/4\pi r^2\varepsilon_0$, within the space. Now draw a Gaussian surface encompassing the large sphere anywhere beyond it. Again, if there is a field, by symmetry it would have to be radial, but now the net charge enclosed is zero and the field is therefore zero. **83.** $E = kq_e/r^2 = (9 \times 10^9\ \text{N·m}^2/\text{C}^2) \times (1.6 \times 10^{-19}\ \text{C})/(1.0 \times 10^{-15}\ \text{m})^2 = 1.4 \times 10^{21}$ N/C.

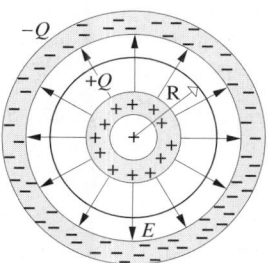

Fig. AN 79

Chapter 16

Answers to Odd-Numbered Multiple Choice Questions

1. b **3.** d **5.** b **7.** b **9.** c **11.** b **13.** d **15.** d **17.** d **19.** c **21.** d **23.** b **25.** b

Answers to Odd-Numbered Problems

1. (a) 6.0 V (b) increase (c) 1.8×10^{-4} J (d) 1.8×10^{-4} J (b) by external force **3.** 3.00 V **5.** -3.50×10^3 V **7.** (a) $+1.602 \times 10^{-16}$ C (b) 3.20×10^{-16} J (c) increase, 10 V (d) gain (e) W done on sphere, (f) 1.60×10^{-15} J **9.** -2.50 μJ **11.** -1.00 V **13.** ± 0.100 V **15.** $\Delta\text{KE} = +500\ \text{eV}$, $\Delta\text{PE}_E = -500$ eV **17.** $\Delta\text{KE} = 4.8 \times 10^{-15}$ J, 1.0×10^8 m/s **19.** (a) 2.58×10^{-4} N (b) 3.87×10^{-5} J (c) 3.87×10^{-5} J (d) 1.94 V (e) 98.1 V **21.** 5.58×10^{-12} V **23.** -0.500 μJ, -10.0 V **25.** 11 MV/m **27.** $V = 0$ **29.** 1.11 nC **31.** $\Delta V = 213$ kV; the closer detector is at the higher potential. **33.** -1.24 μC **35.** 1.4×10^{-17} J **37.** 1.3×10^2 V **39.** 5.6×10^4 V **41.** 8.5×10^{-17} J **43.** 539 V **45.** -44.9 V **47.** -3.6×10^5 V **51.** (a) 0.089 9 m (b) 8.00 kV (c) 8.00×10^{-8} C (d) 7.88×10^{-7} C/m² **53.** $+0.23$ J **55.** $V = 1.80$ kV **57.** (a) $+1.00$ μC (b) $\sqrt{5}$ m (c) 4.02 kV (d) 8.04 kV (e) 3.22×10^{-2} J **59.** (a) 20.0×10^{-2} m, 5.0×10^{-2} m (b) -45.0 kV (c) potentials are equal (d) -0.20 μC, -0.80 μC **61.** $V = -8.3$ kV, $E = 0$ **63.** 0.15 nC **65.** 1×10^{-5} C **67.** $R = 90$ mm **69.** 20.0 V **71.** 1.1 nF **73.** 25 nF **75.** 5.0 μF **77.** (a) 12 V (b) $E_0/10$ (c) $10C_0$ (d) 1.2 V **79.** 8.9×10^{-10} F **81.** 2.5 pF **83.** 0.01 F/m² **85.** 7×10^{-4} C/m² **87.** proof **89.** 2.0 nF **91.** 3.0 μF **93.** 10 pF in series, 110 pF in parallel **95.** 0.99 μF **97.** 10 pF **99.** 14 μF **101.** (a) it's in parallel (b) 12.0 μF (c) 24.0 V (d) 288 μC (e) 192 μC, 96.0 μC **103.** (a) 8.0 μF (b) 4.0 μF (c) 24.0 μF (d) 12.0 μF (e) 8.0 μF (f) 96 μC **105.** 9.0 pF **107.** 13.0 μF **109.** 3.6×10^{-10} C **111.** 1.0 V **113.** 24 nF **115.** 1.44 mJ **117.** 10.0 μJ **119.** 60.0 μC **121.** 1.4 nJ **123.** 1.5 nJ/m **125.** 2.2 nJ **127.** 1.4 nJ

Solutions to Selected Problems

3. From Eq. (16.3) $\Delta V = \Delta W/q_o = 60.0\ \text{nJ}/ 20.0$ nC = 3.00 V. **9.** $\Delta W = \Delta\text{PE}_E = q\Delta V = (-25.0 \times 10^{-9}\ \text{C})(100\ \text{V}) = -25.0 \times 10^{-7}$ J. This is the work done against the field. The negative sign means the charge has work done on it by the field. It moves from 0 to 100 V and so in the opposite direction to the field. That's the direction a negative charge moves in spontaneously, propelled by the field.   **27.** A charge of $-Q$ is induced on the inner surface of the outer sphere. At a point outside, the potential is equivalent to that due to a point charge of $+Q$ and a point charge of $-Q$, both at the center. Hence, at P, $V = 0$.   **31.** Use Eq. (16.7), with $q = +25.0\ \mu C$; so $V_1 = k_0 q/r_1 = (8.98755 \times 10^9\ \text{N·m}^2/\text{C}^2)(25.0 \times 10^{-6}\ \text{C})/(1.00\ \text{m}) = 2.24 \times 10^5\ \text{V}$ and $V_2 = k_0 q/r_2 = (8.98755 \times 10^9\ \text{N·m}^2/\text{C}^2)(25.0 \times 10^{-6}\ \text{C})/(20.0\ \text{m}) = 1.12 \times 10^4\ \text{V}$. $\Delta V = V_1 - V_2 = 2.13 \times 10^5\ \text{V}$, with the first (closer) detector at the higher voltage.   **35.** At the surface, $V = k_0 Q/R = (8.99 \times 10^9\ \text{N·m}^2/\text{C}^2)(1.00 \times 10^{-9}\ \text{C})/(0.10\ \text{m}) = 90\ \text{V}$. $W = q_e \Delta V = (1.6 \times 10^{-19}\ \text{C})(90\ \text{V}) = 1.4 \times 10^{-17}\ \text{J}$.   **43.** Each charge is $q_\_ = +30.0\ \text{nC}$, so $V = 2kq_\_/r = 2(8.98755 \times 10^9\ \text{N·m}^2/\text{C}^2)(30.0 \times 10^{-9}\ \text{C})/(1.00\ \text{m}) = 539\ \text{V}$.   **53.** The potential at the point is $V = k_0 Q_1/r_1 + k_0 Q_2/r_2 = (8.99 \times 10^9\ \text{N·m}^2/\text{C}^2)(+10\ \mu\text{C})/(4.0\ \text{m}) + (8.99 \times 10^9\ \text{N·m}^2/\text{C}^2)(-25\ \mu\text{C})/(5.0\ \text{m}) = -22.5\ \text{kV}$, $W = q\Delta V = (-10\ \mu\text{C})(-22.5\ \text{kV} - 0) = +0.23\ \text{J}$.   **61.** Each face has a diagonal of $\sqrt{2}$ m, and so there is a tilted rectangle $\sqrt{2}$ m by 1.0 m passing through the center and having four spheres at its corners. The diagonal of that square is $\sqrt{3}$ m long and half of that, $\frac{1}{2}\sqrt{3}$, is the corner-to-center distance. Hence, $V = 8kQ/r = -8.3\ \text{kV}$. And $E = 0$.   **63.** $Q = CV = (100 \times 10^{-12}\ \text{F})(1.5\ \text{V}) = 1.5 \times 10^{-10}\ \text{C}$.   **75.** $C = Q/V = (20\ \mu\text{C})/(4.0\ \text{V}) = 5.0\ \mu\text{F}$.   **81.** $C = \varepsilon A/d$; $A = \pi R^2 = \pi(0.25 \times 10^{-2}\ \text{m})^2 = 1.96 \times 10^{-5}\ \text{m}^2$; C $= (4.8)(8.85 \times 10^{-12}\ \text{F/m})(1.96 \times 10^{-5}\ \text{m}^2)/(0.33 \times 10^{-3}\ \text{m}) = 2.5\ \text{pF}$.   **83.** The membrane is like a rolled-up parallel-plate capacitor, so $C = \varepsilon A/d$; $C/A = \varepsilon/d = 7\varepsilon_0/d = 7(8.85 \times 10^{-12}\ \text{C}^2/\text{N·m}^2)/(6\ \text{nm}) = 0.01\ \text{F/m}^2$.   **89.** $1/(3.0\ \text{nF}) + 1/(6.0\ \text{nF}) = 1/C$; C = 2.0 nF.   **97.** The 12-pF and 4.0-pF capacitors are in series and so equivalent to 3.0 pF, which, in turn, is in parallel with 7.0 pF. Hence, $C = 10$ pF.   **107.** The two 6.0-μF capacitors are in series (yielding 3.0 μF), as are the two 5.0-μF capacitors (yielding 2.5 μF). Now everything is in parallel, so 3.0 μF + 7.5 μF + 2.5 μF = 13.0 μF.   **109.** The 9.0-pF capacitor is in parallel with the battery, and so the voltage is 12 V. All the capacitors are in parallel; hence, $C = 30$ pF. $Q = CV = (30\ \text{pF})(12\ \text{V}) = 3.6 \times 10^{-10}\ \text{C}$.   **117.** $\text{PE}_E = \frac{1}{2}Q^2/C = 10.0\ \mu\text{J}$ **123.** $\Delta \text{PE}_E/L = \frac{1}{2} CV^2/L = \frac{1}{2}(3 \times 10^{-7}\ \text{F/m})(0.1\ \text{V})^2 = 1.5\ \text{nJ/m}$, or to one significant figures, 2 nJ/m.   **127.** Figure AN 127 shows successive simplifications af the circuit. The equivalent capacitance is 20 pF; hence, $\text{PE}_E = \frac{1}{2}CV^2 = \frac{1}{2}(20\ \text{pF})(12\ \text{V})^2 = 1.4\ \text{nJ}$.

Chapter 17

Answers to Odd-Numbered Multiple Choice Questions

1. d   **3.** d   **5.** c   **7.** d   **9.** c   **11.** e   **13.** b   **15.** b   **17.** a   **19.** c   **21.** d   **23.** c   **25.** a

Answers to Odd-Numbered Problems

1. 10 μA   **3.** 0.011 MC   **5.** 96.5 A   **7.** 3.6 $\times$ 10^2 C   **9.** 6.2×10^{14} **11.** 120 C   **13.** 1.5×10^3

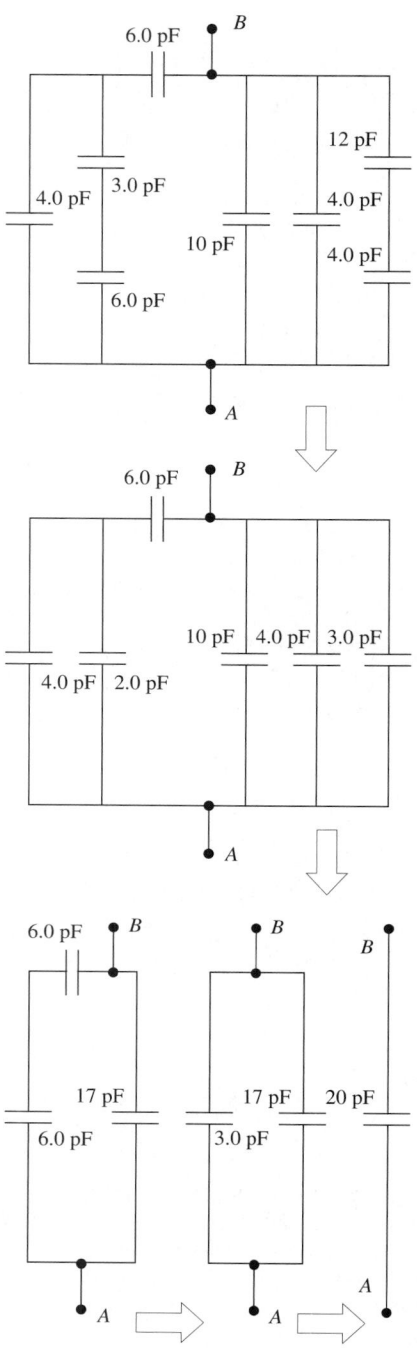

Fig. AN 127

cells   **15.** 36 kC   **17.** (a) +12 V   (b) 0 V   (c) +12 V   (d) 12 V   (e) B (f) counterclockwise   **19.** 27 h   **21.** (a) A and B   (b) A and E   (c) A, D, and C   (d) A and D   (e) A and F; when the headlights are on   **23.** 20 cells in series form a row; there are 20 rows in parallel   **25.** We need 3 big cells in series to get the voltage; two such rows in parallel handling 4.0 A each would do the job. We have 4 big cells and must make a substitute for 2 more; each big cell can be duplicated by putting 4 small cells in parallel. Total: 4 big, 8 small.   **27.**

0.11 MC   **31.** (a) 0   (b) +10 V   (c) +10 V   (d) 10 V   (e) A (f) $V = IR$ (g) 5 Ω **33.** 60 mA   **35.** 2.50 V   **37.** (a) 56 Ω (b) 1.5 V   (c) $V = IR$ (d) 27 mA   **39.** 10 Ω **41.** $50 \times 10^9\ \Omega$ **43.** 50 mA   **45.** 100 V, 0.10 mA   **47.** 100 V, 25 Ω **49.** if it's 1 m on a side   **51.** (a) ρ, L, A (b) $r = 5.0 \times 10^{-6}$ m, $L = 10 \times 10^{-3}$ m   (c) $A = 10 \times 10^{-11}$ m^2 (d) $0.06 \times 10^9\ \Omega$ (e) small cross-sectional area   (f) $6 \times 10^9\ \Omega/\text{m}$ **53. proof**   **55.** $1.6 \times 10^{-8}\ \Omega\cdot\text{m}$ **57.** 1.9 mm   **59.** 0.12 Ω **61.** 0.10 Ω **63.** (a) $L = 2.10$ m   (b) ρ, L, A (c) $R = \rho L/A$ (d) $3.6 \times 10^{-4}\ \Omega$ (e) 12 V   (f) 3.4×10^4 A   (g) No, I is very big, will kill battery   **65.** 1×10^3 °C   **67.** 26 Ω **69.** 220 °C   **71.** $\Delta R/R_0 = 2.7$, poor: 12   **73.** 0.1 A   **75.** 0.1 W   **77.** 1.7 A   **79.** 0.2 A   **81.** 2.2 kW   **83.** (a) 0.30 kC   (b) 3.3×10^4 J   (c) 55 W   (d) 55 W   (e) 0.22 kΩ   **85.** 87 kW   **87.** (a) $2.0 \times 10^{-4}\ \Omega$ (b) 2.4×10^{-5} V   (c) 2.9×10^{-7} W   **89.** 6.0 kW, 5.2 kW   **91.** 30 W   **93.** 1.3 W, 12 Ω **95.** 12 W   **97.** 2.8 kW, 17 A

Solutions to Selected Problems

1. $I = \Delta Q/\Delta t = (10\ \mu\text{C})/(1.0\ \text{s}) = 10\ \mu\text{A}$.   **9.** $I = \Delta Q/\Delta t$; $\Delta Q = (1.0\ \text{mA})(0.10\ \text{s}) = 0.10\ \text{mC}$. The number of protons is $(0.10 \times 10^{-3}\ \text{C})/(1.602 \times 10^{-19}\ \text{C}) = 6.2 \times 10^{14}$.   **13.** (220 V)/(0.15 V per cell) = 1466.7 cells = 1.5×10^3 cells. Salt water, because of the presence of ions, is a much better conductor.   **19.** Number of coulombs passing per second is $(I/A)A = (1.0\ \text{MA/m}^2) \times (1.00 \times 10^{-6}\ \text{m}^2) = 1.0\ \text{C/s}$; or 6.24×10^{18} electrons/s; hence, $(6.02 \times 10^{23}$ electrons$)/(6.24 \times 10^{18}$ electrons/s$) = 96\,507\ \text{s} = 27\ \text{h}$.   **23.** First, do the voltage, (9.0 V)/(0.45 V) = 20; we need 20 cells in series forming one row. To get the current, we must have (440 mA)/(22 mA) = 20 such rows in parallel.   **33.** $V = IR$; (9.0 V)/(150 Ω) = 0.060 A   **39.** $R = V/I = (100\ \text{V})/(10\ \text{A}) = 10\ \Omega$.   **49.** If the cube is 1 m on a side, $R = \rho$.   **55.** $R = \rho L/A$; $\rho = AR/L = 1.6 \times 10^{-8}\ \Omega\cdot\text{m}$.   **59.** $R = 0.10\ \Omega + (0.10\ \Omega)(0.005\ 0\ \text{K}^{-1})(30\ \text{K}) = 0.12\ \Omega$.   **67.** $R_f = R_0(1 + \alpha_0 \Delta T) = (10\ \Omega)(1 + 0.003\ 9\ \text{K}^{-1} \times 400\ \text{K}) = 25.6\ \Omega = 26\ \Omega$.   **75.** $\text{P} = IV = (16 \times 10^{-3}\ \text{A})(9\ \text{V}) = 0.144\ \text{W} = 0.1\ \text{W}$.   **85.** The current entering the factory must be $I = \text{P}/V = (45 \times 10^3\ \text{W})/(110\ \text{V}) = 409.09$ A; the total power loss in the cables is $I^2 R = (409.09\ \text{A})^2(2 \times 0.25\ \Omega/\text{mile} \times 0.50\ \text{mile}) = 41.84$ kW; hence, the net power supplied must be 45 kW + 41.8 kW = 87 kW.   **89.** P = (12 A)(500 V) = 6.0 kW; 86%(6.0 kW) = 5.2 kW = 6.9 hp   **93.** P = IV = (1/3 A)(3.9 V) = 1.3 W. P = I^2R = 1.3 W = (1/3 A)^2R; $R = 12\ \Omega$.

Chapter 18

Answers to Odd-Numbered Multiple Choice Questions

1. c   **3.** d   **5.** d   **7.** d   **9.** c   **11.** c   **13.** a   **15.** b   **17.** d   **19.** c   **21.** a   **23.** b   **25.** b

Answers to Odd-Numbered Problems

1. 1.5 A   **3.** 1.48 V   **5.** 1 Ω **7.** 75 mW   **9.** 7.9 W, 8.8 V   **11.** 12.2 V   **13.** 0.1 Ω **15.** (a) 50 Ω (b) 20 Ω (c) 80 Ω (d) 40 Ω (e) 150 Ω **17.** 0.60 A before 1.5 A after; goes out   **19.** 0.15

A, P doubles from 1.8 W to 3.6 W **21.** 1.0 Ω **23.**
5.00 A **25.** Proof **27.** 1.2 V **29.** 2 Ω **31.** 5.5
Ω **33.** 5.0 Ω **35.** 5 Ω **37.** 9 A through 17 Ω, 6
A through 3 Ω, 3 A through 4 Ω and 2 Ω **39.** 1.0
A through 4.0 Ω, 0.67 A through 3 Ω, 0.33 A
through 6.0 Ω **41.** 10 Ω **43.** (a) 20 Ω (c) 5-J-
K-6 (d) shorts out the rest of the circuit (e) 0 Ω
45. (a) 300 Ω (c) shorts out the rest of the circuit
to the right (d) 150 Ω **47.** R_e = 1.8 Ω **49.** 9 A
through 17 Ω and 23 Ω, 0.9 A through 9 Ω, 8 A
through left 1 Ω, 2 A through 3 Ω and 1 Ω, 5 A
through 2 Ω, 2 A through 4 Ω and 1 Ω **51.** (a)
100 Ω (b) 12 V (c) clockwise (d) 0.12 A (e)
starting at A and going clockwise, 1.2 V, 2.4 V, 7.2
V, 1.2 V (f) 8.4 V (g) 8.4 V (h) $V_D > V_B$ **53.**
4 Ω, 95 W **55.** 24 W, C is 1.5 V above A **57.** 27
W **59.** 0.89 A, 8.6 V **61.** (a) 2600 Ω (b) 2400
Ω (c) 200 Ω (d) ∞ (e) 1200 Ω (f) 0 (g)
0.10 A (h) 24 V (i) 24 V **63.** 4.0 mA **65.** R_s
= 0.050 Ω **67.** R_s = 0.100 Ω **69.** 5.0 ms **71.**
4.0 s **73.** 6.0 μA, 2.0 s **75.** 10 μA, 3.7 μA **77.**
0.12 A, 0.60 ms **79.** 1.6 μA **81.** 0.541 μA **83.**
3 A, 3 A, 6 A **85.** I_1 = 2 A, I_2 = (2/3) A, I_3 =
(4/3) A **87.** +4 V **89.** 32 V **91.** I_1 = 2.0 A
right, I_2 = 1.0 A left, I_3 = 1.0 A left **93.** 8.0 μC
95. R_1 = 46 Ω, $\mathscr{E}_1$ = 14 V **97.** 28 Ω **99.** (a) zero
(b) 36 μC **101.** R = 4.0 Ω, V = 30 V, I_1 = 8.0 A,
I_2 = I_6 = 2.0 A, I_3 = 1.0 A, I_4 = I_5 = 5.0 A **103.**
A at +14 V; E at +29 V; F at +12 V; D, C, B, and
G at zero **105.** (42/5) Ω

Solutions to Selected Problems

3. From Eq. (18.1) $V = \mathscr{E} - Ir$ = 1.50 V − (0.50
A)(0.05 Ω) = 1.475 V ≈ 1.48 V. **19.** R_e = 80 Ω;
$I = V/R_e$ = 0.15 A. After S is closed R_e = 40 Ω,
and I = 0.30 A. P = IV, so the power doubles from
1.8 W to 3.6 W. **23.** Each lamp draws I = P/V =
(100 W)/(100 V) = 1.00 A; hence, the source must
provide 5.00 A. **29.** 6 Ω in parallel with 3 Ω is
2 Ω. **35.** The only resistance between A and B is
5 Ω. **39.** The 3.0-Ω and 6.0-Ω resistors in parallel equal 2.0 Ω, and that's in series with the 4.0-Ω
resistor; hence, R_e = 6.0 Ω. $I = V/R_e$ = (6.0 V)/
(6.0 Ω) = 1.0 A. This current passes through the
4.0-Ω resistor and splits 6 parts out of 9, that is,
(6/9) A, through the 3.0-Ω resistor and (3/9) A
through the 6.0-Ω resistor. **47.** 3 Ω and 9 Ω are in
parallel, and that's in parallel with the 18-Ω resistor
yielding 2 Ω, which is in parallel with 3 Ω and 6 Ω
for a total of 1 Ω; hence, R_e = 1.8 Ω. **55.** See
Fig. AN 55. The equivalent resistance is 6.0 Ω; the
battery provides 2.0 A and 24 W. Working backwards, the currents split as shown. Going from A to
B, there is a rise of $\frac{1}{2}$ V, and from B to C, another
rise of 1 V; hence, C is $1\frac{1}{2}$ V above A. **59.** See
Fig. AN 59. The equivalent resistance is 10.14 Ω.
The current from the battery is 0.89 A, and terminal
voltage is 9.0 V − (0.89 A)(0.50 V) = 8.6 V. **65.**
(50.0 Ω)(0.001 0 I) = (0.999 0 I)R_s; R_s = 0.050 Ω.
67. The coil carries 1.00 mA when the meter carries 1.00 A, and so the shunt carries a maximum of
0.999 A; (1.00 × 10⁻³ A)r = (0.999 A)R_s, R_s =
0.100 Ω. **73.** $I_i = V/R$ = (12.0 V)/(2.0 MΩ) =
6.0 μA; RC = (2.0 MΩ)(1.0 μF) = 2.0 s. **77.**

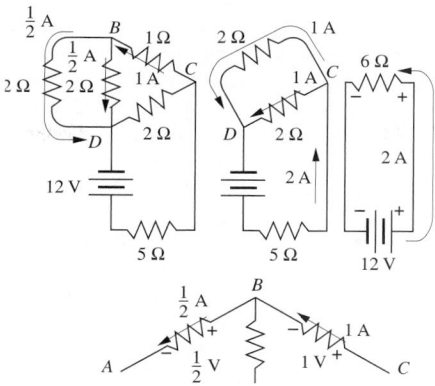

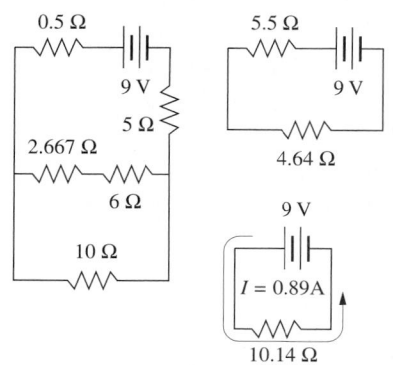

Fig. AN 55

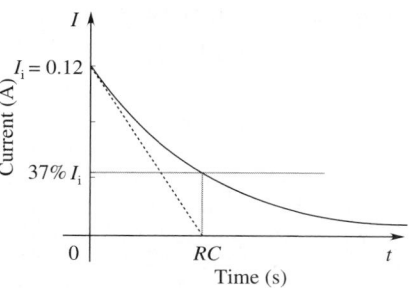

Fig. AN 59

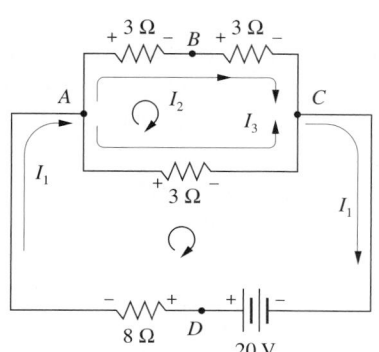

Fig. AN 77

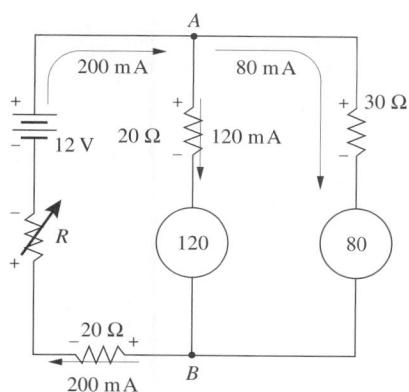

Fig. AN 97

See Fig. AN 77. $I_i = V/R$ = (12 V)/(100 Ω) =
0.12 A; RC = (100 Ω)(6.0 μF) = 0.60 ms. **85.**
See Fig. AN 85. At A, $I_1 = I_2 + I_3$ around loop A-
B-C-D-A going clockwise; $-3I_2 - 3I_1 + 20$ V −
$8I_1 = 0$. around loop A-B-C-A, clockwise: $-3I_2 -$
$3I_1 + 3I_3 = 0$, around loop A-C-D-A, clockwise:
$-3I_3 + 20$ V − $8I_1 = 0$. Hence, I_1 = 2 A, I_2 =
(2/3) A, I_3 = (4/3) A. **89.** The drop across the
top branch is 16 V; with that across the middle
branch the current through it must be 1.0 A; hence,
the current in the bottom branch, from the Node
Rule, is 1.0 A. The drop across the two bottom
resistors is 16 V; hence, the voltmeter must read 32
V. **93.** The equivalent resistance in the steady
state is 12 Ω; hence, the current is 1.0 A. The voltage across C is (1.0 A)(4.0 Ω) = 4.0 V, and since
CV = Q, Q = 8.0 μC. **97.** See Fig. AN 97. I_1 =
200 mA, V_{AB} = +2.4 V; hence, around the first
loop clockwise: −(0.200 A)R + 12 V − 2.4 V − 4
V = 0, R = 28 Ω. **105.** Place a source of, say, 18
V, across the group. Then solve for the six branch
currents: here, (6/7) A goes through both 12-Ω
resistors, (3/7) A goes through the central 6-Ω
resistor, and 97 A passes through each outer 6-Ω
resistor. Thus, the source provides (15/7) A, and
so the equivalent resistance is (42/5) Ω, independent of the source.

Chapter 19

Answers to Odd-Numbered Multiple Choice Questions

1. d **3.** c **5.** c **7.** b **9.** b **11.** a **13.** a
15. c **17.** d **19.** a **21.** b **23.** c **25.** c
27. a

Answers to Odd-Numbered Problems

1. (a) this is correct, (b) field lines must penetrate
the bar, (c) field lines must penetrate the bar,
(d) field should enter south pole of magnet on the
right. **3.** proof **5.** 4.0 × 10⁻⁶ T **7.** 4.0 mT **9.**
0.10 mT **11.** 0.010 m **13.** 63 μT **15.** 0.63 nT
17. 0.20 A **19.** 2.5 mT **21.** 3.98 × 10⁴ turns
23. 1.3 × 10⁻¹¹ T, clockwise looking toward the
source **25.** (b) 0.98 m (c) $B = \mu_0 I / 2\pi r$ (d)
4.5 × 10⁻⁷ T and 2.2 × 10⁻⁷ T (e) 2.2 × 10⁻⁷ T—
NORTH **27.** 500 A **29.** 2.4 μT **31.** 31° W of
N **33.** $B_z \approx \mu_0 IR^2/2z^3$ **35.** $I \approx 6 \times 10^9$ A **37.**

proof **39.** 0 **41.** 1.00×10^{-5} T **43.** $B = 0$ **45.** $B = \mu_0 NI / 2\pi r$, same for single wire **47.** proof **49.** proof **51.** $B = \mu_0 i$ inside, $B = 0$ outside **53.** $+x$-direction **55.** 30° above the $-x$-direction **57.** 1.8×10^{-11} N **59.** (a) 2.99 T (b) $+1.60 \times 10^{-19}$ C (d) 4.39×10^6 m/s—DOWN **61.** proof **63.** 1.2 T **65.** proof **67.** 1.8×10^7 m/s **71.** 3.8 $\times 10^{18}$ m/s^2 **73.** 1.01×10^{-2} N **75.** circle, $R = $ 14 mm **77.** $p = qBR$ **79.** proof **81.** 0.5 m/s **83.** $f = qB/2\pi m$ **85.** $-y$-direction **87.** 64 $\times 10^{-3}$ N **89.** 1.00 A **91.** 3.5 N **93.** 0.31 mN **95.** 0.019 N·m **97.** 1.1 N, upward **99.** $\tau_f = 0.75$ τ_i **101.** 2.0×10^{-5} N, repulsive **103.** (b) 25 $\times 10^{-3}$ kg/m (c) $(F_M/l) = \mu_0 I_1 I_2 / 2\pi d$ (f) 39 A.

Solutions to Selected Problems

5. From Eq. (19.2), $B = (1.26 \times 10^{-6}$ T·m/A) $\times$ (10 A)/2π(0.50 m) = 4.0×10^{-6} T. **9.** $V = IR$; $I = (12$ V)/(1.2 Ω) = 10 A; $B = \mu_0 I/2\pi r = 0.10$ mT. **15.** $B_z = \mu_0 I/2R = 0.63$ nT. **19.** $V = IR$, $I = 0.20$ A; $B_z \approx \mu_0 nI = 2.5$ mT. **27.** $B = \mu_0 I/2\pi r$; $I = 500$ A. **31.** $B = \mu_0 I/2\pi r = 0.30 \times 10^{-4}$ T, west—due to the current; the net field is at an angle of $\tan^{-1} 0.30/0.50$, or 31° west of north. **35.** We take a single circular equatorial current loop circulating with a radius of 2.3×10^3 km—it would be more realistic to assume several such loops distributed over the spherical core. We know that the field drops off as in Fig. 19.19b and the formula given. At a distance of $2.78R$, it's down by a factor of about 26. $B = \mu_0 I/2R$; 0.6×10^{-4} T = $(1.26 \times 10^{-6}$ T·m/A)$I/2(2.3 \times 10^6$ m); $I = 0.22 \times 10^9$ A, but this distance is to be 6.3×10^6 m away along the z-axis, so the current should be roughly 26 times larger, or $I \approx 6 \times 10^9$ A. Since there is a south magnetic pole in the north, the current circulates from east to west. **41.** Apply Ampère's Law to a circular loop of radius R, concentric with the pipe and in a plane perpendicular to it: $\sum B_\parallel \Delta l = \sum B \Delta l = B \sum \Delta l = (2\pi R)B = \mu_0 I$, where I is the current carried by the pipe. So $B = \mu_0 I/2\pi R = (4\pi \times 10^{-7}$ T·m/A) $\times$ (50.0 A)/2π(1.00 m) = 1.00×10^{-5} T. **45.** Using Ampère's Law around a circular path of radius R within the torus, $\sum B_\parallel \Delta l = \mu_0 \sum I$. If we take the field to be axial (a solenoid bent around on itself), $B 2\pi r = \mu_0 NI$; $B = \mu_0 NI/2\pi r$. The field varies inversely with r—the density of the windings is not the same on the inner and outer surfaces. For a single wire, $B = \mu_0 NI/2\pi r$, which is identical to the coil. **49.** See Fig. AN 49. For a finite sheet, the field lines would loop around the sheet counterclockwise looking into the current (along the minus z-axis). The sheet is infinite, so there cannot be a preferred direction other than parallel to it—no $\pm y$-component to the field. Moreover,

there's no reason for the field to be any different at different distances from the sheet because, being infinite, the sheet "looks" the same as seen from any point above it. In effect, the notion of "distance from" loses its traditional meaning in this idealized situation. Going around the Ampèrean loop in the figure, $\sum B_\parallel \Delta l = \mu_0 \sum I$; $Bl + Bl = \mu_0 il$; and $B = \frac{1}{2}\mu_0 i$. **53.** In the $+x$-direction. **65.** Using $F = qvB$; $B = F/qv$; 1 T = 1 N/C·m/s = 1 (N·m)s/C·m^2 = 1 (J/C)s/m^2 = 1 V·s/m^2. **69.** $qE = qvB$; $v = E/B$. **71.** $F = qvB \sin\theta = ma$; $a = (1.60 \times 10^{-19}$ C)(5.0 $\times 10^6$ m/s)(5.0 T)0.866/$(9.11 \times 10^{-31}$ kg) = 3.8×10^{18} m/s^2. **77.** $qvB = mv^2/R$; $qB = mv/R$; $p = qBR$. **81.** $V = vBl$; $v = V/Bl = (1.0 \ \mu$V)/ (0.50 mT)(4 mm) = 0.5 m/s. **83.** $qvB = mv^2/R$, $qB = mv/R$, $v = qRB/m$; if f is the frequency, $1/f$ is the period, which equals $2\pi R/v$; hence, $f = v/2\pi R$ and $f = qB/2\pi m$. **89.** $F = IlB$; $I = F/lB = 1.00$ A. **95.** $\tau = NIAB \sin\phi = (20)(1.5$ A)(1.3 $\times 10^{-3}$ m^2)(0.90 T)(0.529 9) = 0.019 N·m. **99.** $\tau = NIAB$; hence, $\tau_i/\tau_f = I_i/I_f = I_i/0.75 I_i$; $\tau_f = 0.75 \tau_i$.

Chapter 20

Answers to Odd-Numbered Multiple Choice Questions

1. e **3.** c **5.** b **7.** a **9.** a **11.** a **13.** c **15.** c **17.** b **19.** d **21.** d **23.** b **25.** b

Answers to Odd-Numbered Problems

1. 3.0 μWb **3.** 1.2 Wb/m^2 **5.** 0.1 T, south **7.** 0.01 T, west **9.** 0.50 V **11.** 75 ms **13.** 8.0 V **15.** (a) 4.0 V (b) 2.0 A **17.** -59 mV **19.** -3.5 mV, counterclockwise **21.** 4.1 μC **23.** $B = RK\theta/NA$ **25.** 0.83 mT **27.** 1.4 mV **29.** 0.60 V **31.** 0.12 V, the east end is positive. **33.** bulb would not light **35.** 90 μV **37.** $I = Blv/R$, constant for a time l/v, first counterclockwise, later clockwise **39.** P = $(Blv)^2/R$ **41.** 31 m **43.** 11.8 T **45.** 1.6 V **47.** 39 V **49.** 100 V, 60.000 Hz **51.** 21 mT **53.** 12 mV **55.** $\mathcal{E} = (1.3$ kV) sin (200 $\times \pi t$), t in s **57.** (a) T·m/A (b) T·m^2 (c) H = T·m^2/A **59.** $L = 0.26$ H **61.** 80.0 μWb **63.** 0.80 mWb **65.** $N = 2.5 \times 10^3$ **67.** 8.4×10^2 **69.** 5.0 H **71.** 2.5 H **73.** 0.15 s **75.** 6.0 A **77.** proof **79.** (a) 0.50 s (b) L/R (c) 2.0 Ω (d) 6.0 A (e) 0 **81.** proof **83.** 2.00 kA/s **85.** 0.30 H **87.** (a) 6.0 A (b) 12 ms (c) 3.8 A **89.** (a) 0 (b) 1.0 A (c) 0.20 ms (d) 1.0 A, positive (e) 2 (f) 0.40 ms **91.** (a) 2.0 A (b) 4.0 s (c) $\approx$20 s **93.** $\Delta I/\Delta t = (V - IR)/L$; 2.4 kA/s, 0.89 kA/s, never reaches 1.0 A **95.** 80 mJ **97.** 6.3 A **99.** 1.2 H **101.** 79 mJ/m **103.** 7×10^{16} J, 7×10^8 gallons of gasoline

Solutions to Selected Problems

1. $\Phi_M = B_\perp A = (1.2$ mT)(0.002 5 m^2) = 3.0 μWb. **5.** The final field minus the initial field is the change in the field, and that's 0.1 T, south. **11.** From Eq. (20.3), $\Delta t = 150(0.030$ Wb)/(60 V) = 75 ms. **19.** From Eq. (20.1), $N\Delta\Phi_M = 1(0.500$ T)(cos 30°)(1.6 $\times 10^{-3}$ m^2) = 6.928×10^{-4} T·m^2; from Eq. (20.3), emf = 3.5 mV. Counter-clockwise. **23.** $I = V/R$; $\Delta Q = N\Delta\Phi_M/R = NBA/R = K\theta$, $B = RK\theta/NA$. **29.** $\mathcal{E} = vBl = (200$ m/s)

$(0.050$ mT$)(60$ m$) = 0.60$ V. **37.** As soon as the leading edge of the loop starts cutting field lines, there is an emf = Blv, which produces a counterclockwise current $I = Blv/R$. The emf is constant (for a time l/v) until the trailing edge of the loop enters the field. It produces an oppositely directed emf (the flux no longer changes), and the net emf goes to zero. It remains zero until the leading edge emerges from the field at a time $3l/v$ after it entered, whereupon the emf jumps to Blv in the opposite direction. A current $I = Blv/R$ appears clockwise and remains constant for a time l/v. **41.** P = IV, $V = \mathcal{E} = vBl$, so P/$I = vBl$, and $l = $ P/IvB = 5.0 W/[$(200 \times 10^{-3}$ A)(2.00 m/s)(0.400 T)] = 31 m. **47.** For each length $\mathcal{E} = vBl = (22$ m/s)(0.35 T)(0.10 m) = 0.77 V. Since there are 2 lengths per turn and 25 turns, 50(0.77 V) = 39 V. **53.** $\mathcal{E} = \frac{1}{2}Br^2 2\pi f = \frac{1}{2}(0.050$ mT cos 40°)(5.0 m)$^2 2\pi(240/60)$ = 12 mV. **59.** $N\Phi_M = LI$; $L = (500)(0.002 \ 0$ Wb)/(3.8 A) = 0.26 H. **65.** $N\Phi_M = LI$; $N = (2.5$ H)(1.80 A)/(1.80 mWb) = 2.5×10^3. **69.** $\mathcal{E} = -L \Delta I/\Delta t$, $L = -(10$ V)/(2.0 A/s) = 5.0 H (forget the sign, since we don't know if the current is increasing or decreasing). **73.** $\mathcal{E} = -L \Delta I/\Delta t$; $\Delta t = (1.5$ H)(10 A)/(100 V) = 0.15 s. **77.** From Eq. (20.9), $L \approx \mu N^2 A/l = An^2 l\mu$ where An^2 is a constant, so ΔL depends on the change in μl. Once the shaft is inserted, part of the region of the coil (d long) changes from μ_0 to μ; hence, $\mu d + \mu_0(l - d)$ is the new value, $\mu_0 l$ the original value. The difference $d(\mu - \mu_0)$ is what we want; hence $\Delta L \approx d(\mu - \mu_0)N^2 A/l^2$. **81.** $1 - 0.368 = 0.632$. **83.** The current is zero at $t = 0$, when Eq. (20.11) reads $V = L\Delta I/\Delta t = |\mathcal{E}| = 20.0$ V, and so $\Delta I/\Delta t = |\mathcal{E}|/L = (20.0$ V)/(10.0 $\times 10^{-3}$ H) = 2.00 kA/s. **87.** (a) $I = V/R = (150$ V)/(25 V) = 6.0 A; (b) $L/R = (300$ mH)/(25 Ω) = 12 ms; (c) 0.632(6.0 A) = 3.8 A. **93.** From Eq. (20.11), $\Delta I/\Delta t = (V - IR)/L$, which (in the limit as $\Delta t \to 0$) is also the instantaneous rate of change of current, or the slope of the curve in Fig. 20.22 at any instant. At $t = 0$, $I = 0$ and $\Delta I/\Delta t = V/L = 2.4$ kA/s. At $t = L/R$, $I = 0.63V/R$ and $\Delta I/\Delta t = (0.37$ V)/L = 0.89 kA/s. The maximum current is V/R = 0.60 A, so it never reaches 1.0 A. **95.** Use Eq. (20.13): PE$_M = \frac{1}{2}LI^2 = \frac{1}{2}(40.0$ mH)(2.0 A)^2 = 80 mJ.

Chapter 21

Answers to Odd-Numbered Multiple Choice Questions

1. b **3.** d **5.** e **7.** d **9.** c **11.** c **13.** b **15.** e **17.** a **19.** b

Answers to Odd-Numbered Problems

1. (a) zero (b) 84.9 V **3.** $i = (14$A$) \sin 2\pi(50$ Hz$)t$ **5.** 141 V **7.** 0.833 A **9.** (b) 1.58 kΩ (c) 69.6 mA (d) 69.6 V and 40.4 V (e) 110.0 V. **11.** 113 W **13.** 0.13 W **15.** $f = 100$ Hz **17.** 618 mA **19.** The 50-Ω resistor is probably shorted internally. **21.** 20 A, 120 V **23.** 15 W, 240 W **25.** P$_m = I_m^2 R = 2$P$_{av}$ **27.** (b) $X_L = 2\pi fL$ (c) 43 Ω (e) 4.7 A **29.** $f = 11$ Hz **31.** 1.3 H **33.** 94 Ω **35.** proof **37.** 2.12 A **39.** 22.0 V **41.** (a) 12.0×10^{-12} F (b) 100×10^3 Hz (c) 133 kΩ (e) 45.2 μA **43.** 2.00 Ω **45.** 1.3 kΩ **47.** 22 μF

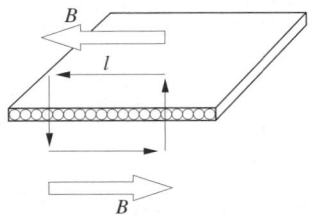

Fig. AN 49

49. 44.4 mA **51.** 1.81 A **53.** (a) Something is wrong with it. (b) New capacitor is shorted internally. **55.** $10.1\ \Omega$ **57.** $159\ k\Omega$ **59.** $219\ \Omega$ **61.** $1.30\ k\Omega$ **63.** $\theta = 26.6°$, voltage leads current **65.** 0.25 A **67.** $\theta = -63°$ **69.** $R = 1.7\ k\Omega$ **71.** 500 Ω **73.** 5.1 nF **75.** 52.8 mH **77.** $1.6\ \mu F$ **79.** 5.0 A, $\theta = +51°$ **81.** $V_L = 302$ V, $V = 447$ V **83.** 27 μF **85.** proof **87.** (b) $X_L = 245\ \Omega$ and $X_C = 549$ Ω (c) $304\ \Omega$ (d) $369\ \Omega$ (e) 20.0 V (f) 0.054 2 A (g) $V_{Rm} = 11.4$ V, $V_{Lm} = 13.3$ V, $V_{Cm} = 29.7$ V (h) No **89.** (a) 70.7 V (b) 1.18 A (c) 27.8 W **91.** $Z = 448\ \Omega$, $\theta = 70.4°$, $I = 268$ mA, $f_0 = 26.9$ Hz, at resonance $I = 800$ mA **93.** (a) $44.7\ \Omega$ (b) 63.4° (c) 0.159 H (d) 1.00 A (e) 2.00 A (f) 2.24 A **95.** 50 turns **97.** 100:1 **99.** 0.12 kV **101.** $V_s = 30$ V, $I_s = 4.0$ A **103.** (a) 20:1 (b) $I_p = 0.11$ A **105.** 98.3% **107.** 20:1

Solutions to Selected Problems

3. $I_m = 1.414(10\ A) = 14$ A; $i = (14\ A)\sin 2\pi \times (50\ Hz)t$. **7.** $P_{av} = 100$ W $= IV$; $I = (100\ W)/(120\ V) = 0.833$ A. **11.** $P_{av} = IV = (1.50\ A) \times (75.0\ V) = 113$ W. **15.** $v = V_m\,0.951$; $0.951 = \sin 2\pi f(0.002\ 00\ s)$; $f = 100$ Hz. **21.** $1/R_e = 1/(10\ \Omega) + 1/(15\ \Omega)$, $R_e = 6.0\ \Omega$; $I = V/R_e = 20$ A. The voltmeter reads 120 V. **29.** $X_L = 2\pi fL$; $f = (10\ \Omega)/2\pi(0.15\ H) = 10.6$ Hz ≈ 11 Hz. **35.** $X_L = 2\pi fL$; $V = -L\Delta i/\Delta t$, so $L = -V\Delta t/\Delta i$ and 1 H $= 1$ V·s/A; hence, reactance has units of $(1/s)(H) = V/A = \Omega$. **43.** $X_C = 1/2\pi fC$; $X_{Ci}/X_{Cf} = f_f/f_i = 100$; hence, the final reactance is 2.00 Ω. **53.** (a) $X_c = 1/2\pi fC = 132.6\ \Omega$; hence, $I = V/X_C = (120\ V)/(132.6\ \Omega) = 0.90$ A. The current should be about 0.9 A and, instead, the ammeter reads 0.25 A. something is wrong with that capacitor. (b) The fuse would blow if the new capacitor was shorted internally—it would short out the first capacitor. **55.** $Z = (R^2 + X^2)^{1/2} = 10.1\ \Omega$. **59.** $Z = (R^2 + X^2)^{1/2} = 219\ \Omega$. **63.** $\tan \theta = X/R$; $\theta = 26.6°$, and voltage leads current. **67.** $\tan \theta = (200\ \Omega)/(100\ \Omega) = 2.0$; $\theta = -63°$, this is a capacitive reactance. **71.** $Z = 2500\ \Omega - 2000\ \Omega = 500\ \Omega$. **73.** $C = 1/4\pi^2 f_0^2 L = 5.1 \times 10^{-9}$ F—such values are not unusual. **79.** $2\pi f = 1257\ s^{-1}$; $X_L = 2\pi fL = 37.7\ \Omega$; $Z = (R^2 + X^2)^{1/2} = 48.18\ \Omega$; $I = V/Z = 5.0$ A; $\tan \theta = X_L/R = 1.257$, $\theta = 51°$, voltage leads current. **81.** $2\pi f = 377$ s^{-1}; $X_L = 2\pi fL = 150.8\ \Omega$; $Z = (R^2 + X^2)^{1/2} = 238.7\ \Omega$; $I = V_R/R = (370\ V)/(185\ \Omega) = 2.00$ A. $V_L = IX_L = 302$ V. $\tan \theta = X_L/R$; $\theta = 39.18°$; $V = V_R/\cos \theta = 477$ V. **85.** $Z = (R^2 + X_C^2)^{1/2}$; $I = V_i/Z$; $V_0 = IX_C$; $V_0 = V_iX_C/Z$; $V_0/V_i = X_C/Z = (1/2\pi fC)/(R^2 + X_C^2)^{1/2} = 1/[1 + (2\pi fRC)^2]^{1/2}$. **97.** 100:1. **103.** (a) $V_pN_s = V_sN_p$; $N_p/N_s = V_p/V_s = 20:1$; (b) $2\pi f = 377$ s^{-1}; $X_L = 2\pi fL = 1.131$ $k\Omega$; $I_p = V/X_L = 0.11$ A. **107.** $Z_p = V_p/I_p$; $Z_s = V_s/I_s$; $Z_p/Z_s = (V_p/V_s)(I_s/I_p)$ but, as we saw earlier, $V_pN_s = V_sN_p$ and $N_pI_p = N_sI_s$; hence, $Z_p = Z_s(N_p/N_s)^2$. $\sqrt{Z_p/Z_s} = N_p/N_s = 20:1$.

Chapter 22

Answers to Odd-Numbered Multiple Choice Questions

1. c **3.** b **5.** c **7.** b **9.** a **11.** b **13.** a **15.** d **17.** c **19.** d **21.** b **23.** e **25.** c

Answers to Odd-Numbered Problems

1. (a) half the peak-to-peak (b) $2\ \mu V/m$ (c) 0.100 m (d) microwave (e) 2.998×10^8 m/s (f) 3.00×10^9 Hz **3.** 12.6×10^6 m^{-1} **5.** 10.00 km **7.** $E = 0$ at $x = 0$, $E = E_0$ at $x = \lambda/4$, $E = 0$ at $x = \lambda/2$, etc. **9.** 20 V/m **11.** When $\varepsilon = 0$, $E = 0$ at $x = 0$, $E = 10$ V/m at $x = \lambda/4$, $E = 0$ at $x = \lambda/2$, etc. When $\varepsilon = \pi/2$, $E = 10$ V/m at $x = 0$, $E = 0$ at $x = \lambda/4$, $E = -10$ V/m at $x = \lambda/2$, etc. **13.** (a) 0.2 V/m (b) 500 nm (c) 12.6×10^6 m^{-1} (d) 6.00×10^{14} Hz (e) 1.67×10^{-15} s **15.** radiowaves, 100 s **17.** 5 μH **19.** 1.50 MHz **21.** 1.50×10^7 m **23.** (a) $f_2 = 16.0$ THz, $\lambda_2 = 1.87 \times 10^{-5}$ m. (b) 2.50×10^{-13} s (c) $E_1 = E_{01}\sin(k_1x - \omega_1t)$ and $E_2 = E_{02}\sin 4(k_1x + \omega_1t)$. **25.** proof **27.** -20 V/m, 0, -20 V/m **29.** $B_0 = 6.7 \times 10^{-7}$ T **31.** 14.5 **33.** 3 m, 6×10^6 wavelengths **35.** (a) 628 nm (b) 4.77×10^{14} Hz (c) positive x-direction **37.** (a) power (b) 2.0×10^{-3} J (c) 0.10×10^{-4} m^2 (e) 100 W/m^2 **39.** 1.2×10^{-9} J/m^2·s **41.** 6.30×10^4 J **43.** 200 W **45.** 80 kW **47.** B^2/μ_0 **49.** 53 W/m^2 **51.** 6.47×10^{-7} T **53.** (a) 0.600 m (b) 1.0×10^6 J/m^3 **55.** 4.31×10^{-19} J **57.** $\lambda_1 = 6\lambda_2$ (b) $f_1 = (1/6)f_2$ (c) $E = hf$ (d) 6.0 eV **59.** 1.51×10^{20} Hz, γ-rays **61.** 620 nm, 484 THz, 3.20×10^{-19} J, orange-red light **63.** 4.136×10^{-12} eV **65.** 8.0 mW/m^2 **67.** 1.2×10^{13} W/m^2 **69.** (a) 103.1 MHz (b) 6.83×10^{-26} J (c) 50×10^6 J/s (d) 7.3×10^{32} photon/s **71.** 3.02×10^{17} photons/s **73.** $E_{max} = 1.00 \times 10^4$ eV, $\lambda_{min} = 1.24 \times 10^{-10}$ m

Solutions to Selected Problems

3. $k = 2\pi/\lambda = 2\pi/(500\ nm) = 12.6 \times 10^6$ m^{-1} **7.** See Fig. AN 7. At $x = \lambda/4$, $kx = (2\pi/\lambda)(\lambda/4) = \pi/2$. **11.** See Fig. AN 11. $E = 10\sin(kx + \varepsilon)$ mark off the axis in intervals of $\lambda/4$. Notice that shifting the phase by 90° results in a cosine function. **19.** $f = c/\lambda = (3.00 \times 10^8\ m/s)/(200\ m) = 1.50 \times 10^6$ Hz. **27.** $E(0,0) = 20\cos(0 - 0 + \pi) = -20$; $E(0, T/4) = 20\cos(0 - 2\pi fT/4 + \pi) = 20\cos(+\pi/2) = 0$; $E(0, T) = 20\cos(0 - 2\pi fT + \pi) = 20\cos(-\pi) = -20$ V/m. **31.** $\Delta\lambda = c\ \Delta t = (3.00 \times 10^8\ m/s)(30.0 \times 10^{-15}\ s) = 9.00 \times 10^{-6}$ m. Now dividing this result by the wavelength yields the number of waves: $(9.00 \times 10^{-6}\ m)/(620 \times 10^{-9}\ m) = 14.5$. **35.** (a) $k = 1.00 \times 10^7 = 2\pi/\lambda$, $\lambda = 628$ nm. (b) $f = c/\lambda = (2.998 \times 10^8$ m/s)/(628 \times 10^{-9}$ m) $= 4.77 \times 10^{14}$ Hz. (c) It moves in the positive x-direction. **41.** Irradiance is energy divided by area divided by time: $E/At = I$, $E = (1.05 \times 10^3$ W/m^2)(1.00 m^2)(60.0 s) $= 6.30 \times 10^4$ J. **53.** (a) $l = c\Delta t = (3.00 \times 10^8$ m/s)(2.00 \times 10^{-9}$ s) $= 0.600$ m. (b) The volume of one pulse is $(0.600\ m)(\pi R^2) = 2.945 \times 10^{-6}$ m^3; hence, $(3.0\ J)/(2.945 \times 10^{-6}\ m^3) = 1.0 \times 10^6$ J/m^3. **61.** 1 eV $= 1.602 \times 10^{-19}$ J; $E = hf = 3.20 \times 10^{-19}$ J; $f = 484$ THz; $\lambda_0 = c/(484\ Thz) = 620$ nm; orange−red light.

Fig. AN 7

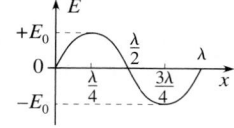

Fig. AN 11

Chapter 23

Answers to Odd-Numbered Multiple Choice Questions

1. c **3.** d **5.** e **7.** a **9.** b **11.** b **13.** d **15.** d **17.** d **19.** d **21.** d **23.** c **25.** e **27.** d

Answers to Odd-Numbered Problems

1. 6.25% **3.** 30° **5.** $\approx 90°$ **7.** 10° **9.** (a) no (d) 1.7 m **11.** 12 m **13.** 12.5 cm **15.** It misses the mirror. **17.** virtual, 12 cm tall, upright **19.** 6.0 m **21.** 1.14 m **23.** unchanged **25.** (a) 90° (b) α (c) α (d) α, 2α (f) 2α (g) 3α **27.** $H = hd/(d + s_0)$ **29.** See solution below. **31.** $L = d/\tan 2\phi$ **33.** 2.25×10^8 m **35.** 1.24×10^8 m/s **37.** 1.1 **39.** (b) 1.499×10^8 m/s (d) 2.00 **41.** ice **43.** 8.78×10^4 m/s **45.** 2.2 mm, 4.2×10^3 waves **47.** $(\overline{AB} - L)/\lambda_0$, $(\overline{AB}/\lambda_0) + L(1/\lambda - 1/\lambda_0)$, 2000π **49.** 1.3 **51.** $\theta_t = 37.5°$ **53.** (a) 60.0° (b) 35.0° **55.** Proof **57.** 566 THz, 353 nm **59.** $n_i = 1.6$ **61.** 2.2 cm **63.** (a) $\theta_i = 0°$ (b) they bend away from the normal (c) 48.6° (d) 18.6° (e) 37.2° **65.** 0.36 mm **67.** 0.885 m **69.** (b) $n_t < n_i$ (c) $n_t = 1$ (d) $1/n_i$ (e) 1.628 (f) carbon disulfide **71.** 40.49° **73.** $\theta_c = 49.8°$ **75.** 1.06 **77.** $\theta_c = 33.7°$ **79.** $\theta_c = 62.63°$ for water, 90.00° for benzene **81.** 97.2° **83.** (a) 58.0° (b) 32.0° (c) 70.0° (d) 1.77 (e) +2.3°

Solutions to Selected Problems

1. $(1/\lambda_r)^4/(1/\lambda_v)^4 = (\lambda_v/\lambda_r)^4 = 6.25\%$. **11.** The statue is 16 m from the point of incidence, and since the ray-triangles are similar, 4m:16 m as 3m:Y and $Y = 12$ m. **19.** See Fig. AN 19. $h = \frac{1}{2}$ m, $d = 1.0$ m, $D = 11.0$ m, find H, $h/d = H/(d + D)$, $h = H/(1.0 + D)$, $\frac{1}{2}(1.0 + D) = H = \frac{1}{2}(1.0 + 11.0) = 6.0$ m. **23.** You still see the same fraction of your total body. **29.** See Fig. AN 29. There will be two images arising from single reflections from each of the single mirrors, and there will be a third image due to the double reflections, one from each mirror. The central image is not reversed in its handedness—you hold up a right hand, and diagonally across from it, the image will hold up its right hand. **31.** See Fig. AN 31. Construct a normal at M_2; the angle it makes with $\overline{SM_2}$ call α. $\angle M_1AM_2 = 45° + 2\phi$, hence (45° +

Fig. AN 19

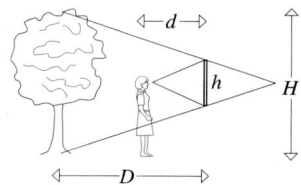

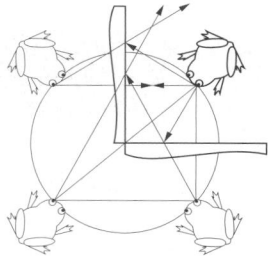

Fig. AN 29

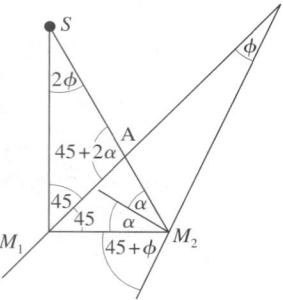

Fig. AN 31

$2\alpha) + (45° + 2\phi) = 180°$, and so $\phi = 45° - \alpha$ and $\angle M_1 SM_2 = 90° - 2\alpha = 2\phi$, thus $\tan 2\phi = d/L$; hence, $L = d/\tan 2\phi$. **33.** $l = vt = (c/n)t = (3.00 \times 10^8 \text{ m/s})(1.00 \text{ s})/1.333 = 2.25 \times 10^8 \text{ m}$. **45.** The photo shows the trail of the pulse as it moves through the water during the 10-ps exposure time. $v = c/n = (3.0 \times 10^8 \text{ m/s})/ 1.36 = 2.2 \times 10^8 \text{ m/s}$; the pulse moves $v \Delta t = (2.2 \times 10^8 \text{ m/s})(10 \times 10^{-12} \text{ s}) = 2.2 \text{ mm}$ during the exposure and is about 2.2 mm long to begin with. $(2.2 \times 10^{-3} \text{ m})/(530 \times 10^{-9} \text{ m}) = 4.2 \times 10^3$ waves. **47.** The number of waves in vacuum is $\overline{AB}/\lambda_0$. With the glass in place, there are $(\overline{AB} - L)/\lambda_0$ waves in vacuum and an additional L/λ waves in glass for a total of $(\overline{AB}/\lambda_0) + L(1/\lambda - 1/\lambda_0)$. The difference in number is $L(1/\lambda - 1/\lambda_0)$, giving a phase shift $\Delta\phi$ of 2π for each wave; hence, $2\pi L(1/\lambda - 1/\lambda_0) = 2\pi L(n/\lambda_0 - 1/\lambda_0) = 2\pi L/2\lambda_0 = 2000\pi$. **49.** $1.00 \sin 55° = n \sin 40°; n = 1.27$ or 1.3. **57.** $\lambda = (1.06 \text{ }\mu\text{m})/2 = 0.530 \text{ }\mu\text{m} = 530 \text{ nm}$, hence $f = 566 \text{ THz}$. $(1.06 \text{ }\mu\text{m})/3 = 0.353 \text{ }\mu\text{m} = 353 \text{ nm}$ in the near-UV. **59.** $\tan\theta_i = 15/20$, $\theta_i = 36.87°$; from Snell's Law $n_i \sin 36.87° = 1.00 \sin 70°$, $n_i = 1.6$. **65.** The glass will change the depth of the object from d_R to d_A, where $d_A/d_R = 1.00/1.55$; but $d_R = 1.00 \text{ mm}$; hence, $d_A = 0.645 \text{ mm}$, and the camera must be raised $1.00 \text{ mm} - 0.645 \text{ mm} = 0.36 \text{ mm}$. **67.** See Fig. AN 67. $d_{A1}/d_{R1} = 1.50/1.33$; $d_{R1} = 1.00 \text{ m}$; $d_{A1} = 1.1278 \text{ m}$: $d_{R2} = d_{A1} + 0.20 \text{ m}$; $d_{A2}/d_{R2} = 1.00/1.50$; $d_{A2} = (1.3278)(1.00/1.50) = 0.885 \text{ m}$. **71.** For total internal reflection $\theta_i \geq \theta_c$, where $\sin\theta_c = n_t/n_i = 1.000/1.540 = 0.649\,35$, $\theta_c = \sin^{-1}(0.649\,35) = 40.49°$. So $\theta_i \geq 40.49°$.

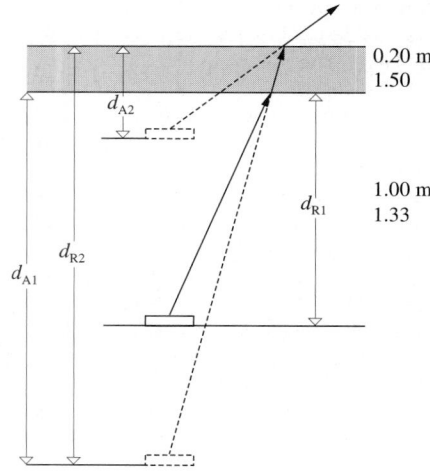

Fig. AN 67

Chapter 24

Answers to Odd-Numbered Multiple Choice Questions

1. b **3.** b **5.** d **7.** b **9.** a **11.** b **13.** c **15.** c **17.** e **19.** a **21.** d **23.** d **25.** b

Answers to Odd-Numbered Problems

1. 1.0 m to the right of the lens **3.** 0.12 m **5.** 0.86 m **7.** −1.0 m **9.** 2.25 **11.** 30.0 cm beyond lens, real **13.** 33.3 cm **15.** $f_w = 4f_a$ **17.** (a) 3.00 m (b) $f = 100.0 \text{ cm} = s_o/3$ (c) yes, real (d) inverted (e) minified **19.** (a) +12 cm (b) −4.8 cm (d) −8.0 cm (e) concave lens (f) no, virtual (h) upright **21.** $s_i = +150 \text{ cm}$ **23.** $s_i = -240$ mm, virtual **25.** $s_i = +0.38 \text{ m}$, real, inverted, and minified **27.** $s_o = 150 \text{ mm}$ **29.** 0.13 m **31.** 3 cm **33.** (a) +12.0 cm and −4.0 cm (b) $f = -6.0$ cm (c) negative len (d) no, virtual (f) upright (g) +1/3 (h) image is upright (i) 1.0 cm **35.** (a) −3.8 cm (b) 1st—real, inverted, magnified; after jump—virtual, right-side-up, magnified **37.** $f = 6.5 \times 10^2 \text{ mm}$ **39.** The greater is s_i, the greater is M_T. **41.** $R = 0.083 \text{ m}$ **43.** 3.2 m, set should be upside down. **45.** 3.1 m to the right of the second lens, real, right-side-up, and minified **47.** 120 cm behind the second lens **49.** +1.00 D **51.** $f = 10$ cm, will help **53.** $f = +4.0 \text{ cm}$, yes **55.** $\mathcal{D} = +2.94 \text{ D}$ **57.** 5.0 m **59.** $L = 6.0 \text{ cm}$ **61.** 1.03 m, 33× **63.** (a) $s_{i1} = 10 \text{ cm}$, the intermediate image is real, inverted, and life size (b) $s_{i2} = -6.0 \text{ cm}$, the final image is virtual, upright, and life size (c) The magnification is 1.0. (d) life size, 0.30 cm tall. **65.** $f = -20 \text{ cm}$, 40 cm in diameter. **67.** (a) $\mathcal{D} = -1.02 \text{ D}$ (b) $s_o = 30.9 \text{ cm}$ **69.** (a) 200× (b) $f_E = 25.4 \text{ mm}$, $f_o = 8.0 \text{ mm}$ (c) $s_o = 8.4 \text{ mm}$ **71.** $\mathcal{D} = +5.0 \text{ D}$ **73.** 33.0 cm **75.** −13 cm **77.** +1.33 m **79.** (a) −0.58 m (b) 1.5 m **81.** virtual, right-side-up, minified (0.7 times), 0.14 m behind mirror **83.** 8.3 cm behind the glasses, virtual, right-side-up, and minified **85.** (a) positive, $f = +30.0 \text{ cm}$ (b) real (c) inverted

(d) magnified (f) −2.0 (g) 8.0 mm (h) −60 cm **87.** $2.0 \times 10^{-6} \text{ m}$ **89.** $R = -0.060 \text{ m}$ **91.** $M_T = -0.14$, real, inverted, and minified **93.** proof **95.** $f = -2.5 \text{ m}$ **97.** −1.0 m

Solutions to Selected Problems

1. From Eq. (24.2), $1/f = (n_l - 1)(1/R_1 - 1/R_2) = (1.50 - 1)[1/(0.50 \text{ m}) - 1/(-0.50 \text{ m})]$, $f = 0.50$ m. Now apply Eq. (24.4), $1/s_o + 1/s_i = 1/f$, so $s_i = s_o f/(s_o - f) = (1.00 \text{ m})(0.50 \text{ m})/(1.00 \text{ m} - 0.50 \text{ m}) = 1.0 \text{ m}$, i.e., the image is located 1.0 m to the right of the lens. **5.** $1/f = 0.58(2 \text{ m}^{-1}); f = 0.86$ m. **9.** $1/f = (n - 1)\{1/(200 \text{ m}) - 1/\infty\}$; $1/(1.60 \text{ m}) = (0.500 \text{ m}^{-1})(n - 1)$; $n = 2.25$. **15.** From Eq. (24.2), $1/f_w = (n_l/n_w - 1)/[(n_l - 1)f_a] = (1.5/1.33 - 1)/(1.5 - 1)f_a = 0.125/0.5f_a$; $f_w = 4f_a$. **21.** $1/s_o + 1/s_i = 1/f$ so $s_i = s_o f/(s_o - f) = (100.0 \text{ cm})(60.0 \text{ cm})/(100.0 \text{ cm} - 60.0 \text{ cm}) = 150$ cm, the image is beyond $2f$, it is real, inverted, and minified. **25.** Given the focal length and the object-distance, we want the image-distance that follows from the Gaussian lens formula; $1/s_o + 1/s_i = 1/f$; $1/s_i = 1/f - 1/s_o = 2.66 \text{ m}^{-1}$; $s_i = 0.38 \text{ m}$. The object is three focal lengths away from the lens and the image is real, inverted, and minified. **31.** For the standard observer $M_A = (0.254 \text{ m})\mathcal{D} = 8\times$; $\mathcal{D} = 31.5 \text{ D}$ and $f = 0.03 \text{ m}$. The lens is 3 cm above the surface. **37.** $M_T = y_i/y_o = -s_i/s_o$; $s_i = (-2.0 \text{ m}) \times (-0.024 \text{ m})/(0.05 \text{ m}) = 0.96 \text{ m}$. $1/(2.0 \text{ m}) + 1/(0.96 \text{ m}) = 1/f$; $f = 6.5 \times 10^2 \text{ m}$. **45.** For the first lens $s_{o1} = 20.0 \text{ m}$ and $f_1 = 0.50 \text{ m}$, $1/f_1 = 1/s_{o1} + 1/s_{i1}$, so $s_{i1} = s_{o1} f_1/(s_{o1} - f_1) = (20.0 \text{ m})(0.50 \text{ m})/(20.0 \text{ m} - 0.50 \text{ m}) = 0.512\,8 \text{ m}$, and $s_{o2} = 2.00 \text{ m} - 0.512\,8 \text{ m} = 1.487$ m. Thus for the final image $s_{i2} = s_{o2} f_2/(s_{o2} - f_2) = (1.487 \text{ m})(1.00 \text{ m})/(1.487 \text{ m} - 1.00 \text{ m}) = 3.1 \text{ m}$, i.e., it is 3.1 m behind the second lens, real, right-side-up, and minified. **53.** Applying Eq. (24.14) to two lenses at a time, $1/f = 1/10 + 1/(-20) + 1/5.0$; $f = +4.0 \text{ cm}$. Since it's positive, it can form real images. **57.** The lens makes an object at a distance s_o, corresponding to infinity, appear to be at the far-point (s_i), where it can be seen clearly. $1/(-5.0 \text{ m}) = 1/\infty + 1/s_i$; $s_i = -5.0 \text{ m}$. **61.** Putting the long focal length $(+1.00 \text{ m})$ first, we'll need a tube of length $f_o + f_E = 1.00 \text{ m} + 0.030 \text{ m} = 1.03 \text{ m}$. $M_A = -(1.00 \text{ m})/(0.030 \text{ m})$ that's 33×. **65.** For the second lens $s_o = -10 \text{ cm}$ and $s_i = +20$ cm; hence, $1/f = 1/(-10 \text{ cm}) + 1/(20 \text{ cm})$; $f = -20 \text{ cm}$. $M_T = -(20 \text{ cm})/(-10 \text{ cm}) = +2$, and the image is 40 cm in diameter. **69.** (a) $20 \times 10 = 200\times$ (b) $f_E = (254 \text{ mm})/M_{AE} = 25.4 \text{ mm}$, $f_o = -(160 \text{ mm})/(-20) = 8.0 \text{ mm}$. (c) $1/s_o = 1/f_o - 1/s_i = 1/(8.0 \text{ mm}) - 1/(L + f_o) = 1/(8.0 \text{ mm}) - 1/(168 \text{ mm})$; $s_o = 8.4 \text{ mm}$. **71.** With her glasses on, $2.0 \text{ D} = 1/(0.80 \text{ m} - 0.020 \text{ m}) + 1/s_i$, which gives the distance from the eyeglass plane at which she can see clearly; that is, $(s_i + 2.0 \text{ cm})$ is her new near-point. The prescribed focal length of her new spectacles must satisfy the expression $1/f = 1/(0.254 \text{ m} - 0.020 \text{ m}) + 1/s_i$. Combining these two equations; $1/f - 4.274 \text{ D} = 2.0 \text{ D} - 1.282 \text{ D}$; and $\mathcal{D} = 5.0 \text{ D}$. **77.** $1/(2.00 \text{ m}) + 1/(4.00 \text{ m}) = 1/f$; $f = 1.33 \text{ m}$. **81.** $1/(0.20 \text{ m})$

$+ 1/s_i = -2/(1.00 \text{ m})$; $s_i = -0.14$ m. $M_T = -(-0.14 \text{ m})/(0.20 \text{ m}) = +0.7$. The image is virtual, minified (0.7 times life-size), and 0.14 m behind the mirror. **89.** For the image to be magnified and erect, the mirror must be concave and the image virtual; $s_o = 0.015$ m; $M_T = 2 = -s_i/(+0.015 \text{ m})$; $s_i = -0.030$ m and, hence, $1/f = 1/(0.015 \text{ m}) - 1/(0.030 \text{ m})$; $f = 0.030$ m and $f = -R/2$; $R = -0.060$ m. **93.** Take another look at your solution of Problem 80; $M_T(s_o - f) = -f$; hence, $s_o = (-f + fM_T)/M_T$, which equals what we want. Furthermore, $s_i = fs_o/(s_o - f)$, and on substituting in for s_o, we do indeed get the desired expression. **97.** The mirror projects a real, inverted, life-size image onto the empty socket. $s_o = 2f = s_i = 1.0$ m $= 2(-R/2)$; $R = 1.0$ m.

Chapter 25

Answers to Odd-Numbered Multiple Choice Questions

1. c **3.** a **5.** b **7.** a **9.** b **11.** d **13.** d **15.** c **17.** c **19.** b **21.** a

Answers to Odd-Numbered Problems

1. 1327 W/m^2 **3.** 2000 W/m^2 **5.** 100 W/m^2 **7.** 188 W/m^2 **9.** 2.5×10^2 W/m^2, 2000 W/m^2 **11.** 40 W/m^2 **13.** (a) 3.98×10^{-2} W/m^2 (b) no (c) 1.99×10^{-2} W/m^2 (d) 1.76×10^{-2} W/m^2 **15.** 0.19 **17.** 0.75 **19.** 0 W/m^2 before rotation, 200 W/m^2 afterwards **21.** 39° **23.** 53° **25.** 62.5° **27.** 1.73 **29.** 1.502 **31.** $n_t = 2.4$ **33.** 32.8° **35.** 32.0° **37.** 1.73, 30.0° **39.** (b) red is long wavelength so wide fringes (c) 1 (e) $a = 0.20$ mm (f) e **41.** 13 mm **43.** 9.76 mm **45.** 419 nm **47.** ± 6.33 mm **49.** 0.360 mm **51.** $y_{m'} = sm'\frac{1}{2}\lambda/a$ for odd m' **53.** 10 km **55.** ≈ 0.35 m **57.** $\theta_m = \sin^{-1}(\sin\theta_i + m\lambda/a)$ **59.** 257 nm **61.** 206 nm **63.** 184 nm or 368 nm **65.** 107 nm, 322 nm, 537 nm, etc. **67.** 473 nm **69.** Using Table 22.1, 710 nm and 473 nm **71.** $\Delta x = \lambda_f/2\alpha$ **73.** 1474 nm, 737.0 nm, 491.3 nm, 368.5 nm **75.** $\frac{1}{4}\lambda$ **77.** 550 nm **79.** 96.6 **81.** It doubles **83.** 12 cm **85.** 0.26° **87.** 0.53° **89.** Δy doubles each time. **91.** 2.6 μm **93.** (b) horizontal bands spread vertically (c) 6.328×10^{-7} m (d) 6.328×10^{-3} rad (e) 6.328×10^{-3} rad (f) 2.01 m **95.** 0.12 mm **97.** 11×10^4 Hz **99.** $\theta_1 = 36.09°$ **101.** 0.597 mm **103.** 4 fringes, $2N$ **105.** 1.11 **107.** 7.93 μm **109.** 26.5 μm **111.** 1.32×10^{-7} rad, 7.57×10^{-6} degrees, 2.72×10^{-2} s **113.** 64.5 km

Solutions to Selected Problems

1. $I = \frac{1}{2}c\varepsilon_0 E_0^2 = \frac{1}{2}(2.998 \times 10^8 \text{ m/s})(8.854 \times 10^{-12} \text{ C}^2/\text{N·m}^2)(1000 \text{ V/m})^2$; hence, $I = 1327$ W/m^2. **7.** $I = I_1 \cos^2\theta = (250 \text{ W/m}^2) \cos^2 30 = 188$ W/m^2. **17.** Before rotation, the emerging irradiance is $I_1 = \frac{1}{2}I_0$. After one of the filters is rotated through $\theta = 30°$, $I = I_1 \cos^2\theta = \frac{1}{2}I_0 \cos^2 30°$; $I/I_1 = \cos^2 30° = 0.75$. **29.** $\tan\theta_p = n_t/n_i$; $n_t = \tan 56.33° = 1.501$. **41.** Use Eq. (25.7): $y_m \approx (s/a)m\lambda$, with $s = 2.00$ m, $a = 0.15$ mm, $m = \pm 2$, and $\lambda = 480$ nm: $|y_m| \approx |sm\lambda/a| = |(2.00 \text{ m})(\pm 2)(480 \times 10^{-9} \text{ m})/(0.15 \times 10^{-3} \text{ m})| = 13$ mm. **43.** $y_4 = s4\lambda/a = (1.00 \text{ m})4(487.99 \times 10^{-9} \text{ m})/$

$(0.200 \times 10^{-3} \text{ m}) = 9.76 \times 10^{-3}$ m. **47.** According to Eq. (25.4), the first minimum occurs when $m' = \pm 1$. Hence, $a \sin\theta_1 = \pm\frac{1}{2}\lambda$; $\theta_1 \approx \pm\frac{1}{2}\lambda/a = \pm\frac{1}{2}(632.8 \times 10^{-9} \text{ m})/(0.100 \times 10^{-3} \text{ m}) = \pm 3.16 \times 10^{-3}$ rad, since $y = s\theta_1 = (2.00 \text{ m})(\pm 3.16 \times 10^{-3} \text{ rad}) = 66.33$ mm. **53.** From Eq. (25.7), $y_1 = s\lambda/a = sc/fa = (20 \times 10^3 \text{ m}) \times (3.0 \times 10^8 \text{ m/s})/(1.0 \times 10^6 \text{ Hz})(600 \text{ m}) = 1.0 \times 10^4$ m. **59.** Since $n_1 < n_f < n_2$, Eq. (25.11) applies: $d = m\lambda_0/2n_f$, where $m = 1, 2, 3, \ldots$. For d_{min} take $m = 1$, so $d_{min} = m\lambda_0/2n_f = (1)(700 \text{ nm})/2(1.36) = 257$ nm. **65.** There will be a 180° phase shift due to the reflections, and so the film must be a quarter-wavelength thick or odd multiples thereof, $d = (m + \frac{1}{2})\lambda_0/2n_f$; for $m = 0$, $d = \lambda_0/4n_f = (580 \text{ nm})/4(1.35) = 107$ nm; for $m = 1$, $d = 322$ nm; for $m = 2$, $d = 537$ nm, etc. **71.** The film increases half a wavelength, $\lambda_f/2$. Using the results of Problem 60, we obtain $x_{m+1} - x_m = \Delta x = \lambda_f/2\alpha$. **73.** There will be a relative phase shift on reflection of $\frac{1}{2}\lambda$; hence, the longest wave not reflected is 550.0 nm $= \frac{1}{2}\lambda_f$, or $\lambda_f = 1100$ nm, and so $\lambda_0 = (1100 \text{ nm})1.340 = 1474$ nm in the IR. When 550.0 nm $= \lambda_f$, $\lambda_0 = (550.0 \text{ nm})1.340 = 737.0$ nm. Similarly, when 550.0 nm $= 3\lambda_f/2 = 366.6$ nm or $\lambda_0 = 491.3$ nm, there is no reflection. And, finally, when $2\lambda_f = 550.0$ nm, $\lambda_f = 275.0$ nm; and $\lambda_0 = 368.5$ nm in the UV, which is not reflected. **77.** Each arm is traversed twice, so a motion of $\lambda/2$ causes a single fringe-pain to shift past; hence, $92\lambda/2 = 2.53 \times 10^{-5}$ m and $\lambda = 550$ nm. **79.** If the cell has a length L and is traversed twice, then the change in optical path length with and without air is $2Ln_a - 2L$, and the number of waves changed is $2L(n_a - 1)/\lambda_0$, each corresponding to a fringe-pair, and so there are 96.6 of them. **83.** From Eq. (25.13), $\lambda = (20 \text{ cm}) \sin 36.87° = 12$ cm. **91.** From Eq. (25.8), $a = 2(550 \times 10^{-9} \text{ m})/0.422\ 6 = 2.6 \times 10^{-6}$ m. **97.** The wavelength should be less than the size of the target or else diffraction will be troublesome. $v = f\lambda$, $f > (330 \text{ m/s})/(0.003\ 0 \text{ m}) = 11 \times 10^4$ Hz. **103.** The angular width of the diffraction maximum is obtained from Eq. (25.13), $\sin\theta_1 = 1\lambda/D = (500 \times 10^{-9} \text{ m})/(0.10 \times 10^{-3} \text{ m}) = 0.005$; $\theta_1 = 0.286\ 5°$, and twice this value is $0.573\ 0°$. The angular width of one of Young's fringes is $\theta = \lambda/a = (500 \times 10^{-9} \text{ m})/(0.20 \times 10^{-3} \text{ m}) = 2.5 \times 10^3$ rad $= 0.143\ 2°$; hence, 4.0 fringes will exist. and that might have been obvious from the fact that $a = 2D$ and the number of fringes is $2N$. **107.** diameter $= 2r_a \approx 2 \times 1.22 f\lambda/D = 2(1.22)(5.00 \text{ m})(650 \times 10^{-9} \text{ m})/(1.00 \text{ m}) = 7.93$ μm. **113.** $\theta_a = 1.22\lambda/D = 1.22(550 \times 10^{-9} \text{ m})/(4.00 \times 10^{-3} \text{ m}) = 1.68 \times 10^{-4}$ rad; hence, $L/(3.844 \times 10^8 \text{ m}) = 1.68 \times 10^{-4}$ rad; $L = 64.5$ km.

Chapter 26

Answers to Odd-Numbered Multiple Choice Questions

1. d **3.** c **5.** c **7.** b **9.** c **11.** b **13.** a **15.** e **17.** c **19.** a **21.** d **23.** c

Answers to Odd-Numbered Problems

1. proof **3.** proof **5.** proof **7.** (a) on the spaceship (b) 10.00 min. (c) 0.80 (d) 1.67 (e) 16.7 min. **9.** 0.9×10^2 s, Earth-based observer; 0.5×10^2 s, astronaut **11.** 2.0 ns **13.** 601 s **15.** 0.499 c **17.** (a) 7.60×10^3 m/s (b) 2.53×10^{-5} (c) $\gamma = 1 + \beta^2/2$ (e) shuttle clock runs slow (f) 98.6 y **19.** $v = 0.9$ c **21.** 73.8×10^3 y **23.** 64.37 m **25.** 0.436 m **27.** 500 m **29.** $v = 0.404\ 8$ c **31.** 0.436 c **33.** 4.45 ns **35.** 36.2° **37.** (a) 9.000 ly (b) 1.000 ly (c) the ship's crew (d) 1.006 y (e) 9.054 y **39.** ≈ 0.997c **41.** 0.96 c **43.** 0.96 c **45.** relativistically 0.86 c, classically 1.1 c **47.** relativistically c, classically 2 c **49.** proof **51.** 0.999 9 c $\approx$ c **53.** (b) 1.000 m (c) $v_{DC} = 0.659$c (f) $v_{DS} = +0.872$c **55.** 2.9×10^{-19} kg·m/s **57.** 3.77×10^{-19} kg·m/s **59.** $3mc/4$ **61.** $(\gamma - 1)/\gamma$, 20% **63.** $v = 0.101$ m/s for both, when $\gamma \approx 1$ **65.** 8.99×10^{13} J **67.** 2.23×10^{-9} kg **69.** KE = 1000 MeV, $v = 0.866\ 0$ c **71.** 1.0 MeV **73.** $E_0 = 8.988 \times 10^{10}$ J, E $= 9.173 \times 10^{10}$ J, KE $= 1.85 \times 10^9$ J **75.** 2.2 MeV **77.** 1.000×10^{-15} kg **79.** 3.26 MeV **81.** $f = 3.654 \times 10^{20}$ Hz **83.** proof **85.** $v = 0.967$ c, $v = 2.42$ c (classically) **87.** 0.106 0 c **89.** 0.866 **91.** 3.438×10^{20} Hz

Solutions to Selected Problems

1. $t_\| = (2L/c)[1/(1 - \beta^2)] \rightarrow 2L(1 - \beta^2)^{1/2}/c(1 - \beta^2) = (2L/c)[1/(1 - \beta^2)^{1/2}]$. **9.** $\Delta t_M = \Delta t_S/(1 - v^2/c^2)^{1/2} = 2(45 \text{ s}) = 90$ s (i.e., 0.9×10^2 s) for the Earth-based observer; 45 s (i.e., 0.5×10^2 s) for the astronaut. **15.** $\Delta t_M = \Delta t_S/(1 - v^2/c^2)^{1/2}$; 60.0 s $= (52.0 \text{ s})/(1 - v^2/c^2)^{1/2}$; $(1 - v^2/c^2)^{1/2} = 0.866\ 7$, $v = 0.499$ c. **21.** $\gamma \approx 1 + \frac{1}{2}\beta^2 = 1 + \frac{1}{2}8.585 \times 10^{-13}$; $\Delta t_M = \gamma\Delta t_S = (1 + \frac{1}{2}8.585 \times 10^{-13})\Delta t_S = \Delta t_S + (\frac{1}{2}8.585 \times 10^{-13})\Delta t_S$, and we want $(\frac{1}{2}8.585 \times 10^{-13})\Delta t_S = 1.00$ s; hence, $\Delta t_S = 2.330 \times 10^{12}$ s $\approx 73.8 \times 10^3$ y. **23.** $L_M = L_S (1 - v^2/c^2)^{1/2} = (2 \text{ mi})(5280 \text{ ft/mi})(0.020\ 00) = 211.2$ ft $= 64.37$ m. **31.** $L_M = 0.900 L_S = L_S/\gamma$, $\gamma = 1/0.900 = 1.111$, $1.234\ 5 = 1/(1 - \beta^2)$, $\beta^2 = (\gamma^2 - 1)/\gamma^2 = 0.190$, $v = 0.436$ c. **41.** Let the probe be P, the ship be S', and the station be S. Then $v_{O'O} = 0.80$ c, and $v_{PO'} = 0.70$ c. So from Eq. (26.7) $v_{PO} = (v_{PO'} + v_{O'O})/(1 + v_{PO'} v_{O'O}/c^2) = (0.70 \text{ c} + 0.80 \text{ c})/[1 + (0.70\text{c})(0.80 \text{ c})/c^2] = 0.96$ c. **47.** Let S' be the frame of our laboratory, such that one photon moves right at $v_{PO'} = $ c. The other photon in S moves to the left; that is, S' moves right ($+x$-direction) with respect to S at $v_{O'O} = $ c. The speed of one photon with respect to the other is $v_{PO} = (c + c)/(1 + cc/c^2) = $ c. Classically: 2 c. **51.** Put the electron at rest in S, and have the proton move in S', which is approaching S at -0.992 c (i.e., moving left). The proton travels at -0.981 c (also moving left). $v_{PO} = [(-0.981\text{c}) + (-0.992 \text{c})]/[1 + (0.981\text{c})(0.992 \text{ c})/c^2] = -1.973$ c/1.973 2 $\approx -$c. **55.** $p = \gamma mv = 1.155(1.675 \times 10^{-27} \text{ kg}) \times (0.50 \text{ c}) = 2.9 \times 10^{-19}$ kg·m/s. **61.** $(\gamma mv - mv)/\gamma mv = (\gamma - 1)/\gamma$, $\gamma = 1.25$, $(\gamma - 1)/\gamma = 20\%$. **65.** Use Eq. (26.13): $E_0 = mc^2 = (1.00 \times 10^{-3} \text{ kg})(2.998 \times 10^8 \text{ m/s})^2 = 8.99 \times 10^{13}$ J. **69.** E $= $ KE $+ E_0 = 2E_0$, KE $= E_0 = 1000$ MeV; E $= 2000$ MeV $= \gamma mc^2 = \gamma(1000 \text{ MeV})$, $\gamma = 2 = 1/(1 - \beta^2)^{1/2}$, $v = 0.866\ 0$ c. **73.** $E_0 = mc^2 = 8.988 \times 10^{10}$ J; E $= \gamma mc^2 = 1.021(8.988 \times 10^{10}$ J)

$= 9.173 \times 10^{10}$ J; E = KE + E_0, KE = E − E_0 = 1.85×10^9 J. **75.** 938.272 MeV + 939.566 MeV = 1877.83 MeV; this result minus 1875.6 MeV is 2.2 MeV. **79.** 2(1876.12 MeV) − 2809.41 MeV − 939.566 MeV = 3.26 MeV. **85.** KE = 1.50 MeV, E = (1.50 MeV + 0.511 MeV) = γmc^2 = $\gamma E_0 = \gamma(0.511$ MeV$)$, $\gamma = 3.935$, $\beta^2 = (\gamma^2 - 1)/\gamma^2$ = 0.935, $v = 0.967$ c. $KE_c = \frac{1}{2}mv^2 = 2.40 \times 10^{-13}$ J, $v = 2.4$ c. **89.** KE = E_0, E = KE + E_0 = 2E_0, $\gamma mc^2 = 2mc^2$, $\gamma = 2 = 1/(1 - \beta^2)^{1/2}$, $4 - 1 = 4\beta^2$, $\beta = \sqrt{3}/2 = 0.866$. **91.** $E^2 = E_0^2 + (pc)^2$, $(pc)^2$ = $(1.000$ MeV + 0.511 MeV$)^2$ − $(0.511$ MeV$)^2$, p = 1.422 MeV/c, which is the momentum of the electron. For the photon E = pc = 1.422 MeV = hf, $f = 3.438 \times 10^{20}$ Hz.

Chapter 27

Answers to Odd-Numbered Multiple Choice Questions

1. d **3.** b **5.** b **7.** c **9.** b **11.** a **13.** d **15.** d **17.** c **19.** b **21.** c **23.** a

Answers to Odd-Numbered Problems

1. $1.097\ 8 \times 10^{27}$ **3.** 35.5 g of Cl **5.** 9.65×10^3 s **7.** 3.57 g **9.** Ba, 25.6 g; Cl, 13.2 g **11.** 4.90×10^{-14} kg **13.** 2.28×10^{-10} m **17.** B = 8.71 $\times 10^{-4}$ T **19.** proof **21.** 0.282 nm **23.** (b) 3.45×10^{-15} J (c) 21.5 keV **25.** 0.179 nm **27.** 0.11 nm **29.** 0.900 nm **31.** $\theta = 8.5°$ **33.** 0.21 nm **35.** (a) 1.50×10^{19} Hz (b) 9.93×10^{-15} J (c) 9.93×10^{-15} J (d) 62.0 kV **37.** 0.237 nm at 25.0° **39.** 12 **41.** $\Delta\lambda = \pm 0.001$ nm **43.** λ = 486.0 nm **45.** 617 THz **47.** $\lambda_\infty = 364.5$ nm **49.** 820.1 nm **51.** $\lambda_\alpha = 656.1$ nm **53.** (a) 92 (c) 1.79×10^7 V (d) 5.74×10^{-12} J (e) 35.8 MeV (e) 17.9 MV **55.** 1.6×10^{-14} m **57.** 397 nm **59.** (a) UV (b) Lyman (c) 0.937 5 (f) 4 **61.** (d) 7.8

Solutions to Selected Problems

3. (965 A)(100.0 s) = 96.5 kC = 1 faraday; hence, 1 mole of each is liberated; 35.5 g of Cl. **9.** The valence of barium is 2, and that of chlorine is 1. Hence, for Ba, m = (2.00 A)(5.00 h)(3600 s/h)(137 g)/(96.5 kC)2 = 25.6 g; for Cl, m = (2.00 A)(5.00 h)(3600 s/h)(35.5 g)/(96.5 kC)1 = 13.2 g. **15.** $\tan\theta \approx \theta = v_y/v_x = a_y t/v_x$ = $(eE/m_e)t/(E/B)$ where $t = L/v_x$; hence, $\theta \approx eLB^2/Em_e$. **19.** Y = y + R $\tan\theta$; $\tan\theta = v_y/v_x$ = eLB^2/Em_e. Using Eq. (27.2), we have y = $eL^2B^2/2m_eE$ and Y = $(e/m_e)(B^2L/2E)(L + 2R)$. **27.** d = 1(0.090 nm)/(2 sin 25°) = 0.11 nm. **33.** λ = 2(0.28 nm) sin 22° = 0.21 nm. **37.** d = 1λ/(2 sin 25°) = 0.237 nm at 25.0°. **39.** The largest angle is 90°; 2(1.22 × 10^{-10} m) sin 90° = m(0.020 0 × 10^{-9} m); m = 12.2, so to nearest integer m = 12. **41.** mλ = 2d sin θ; m Δλ = 2Δd sin θ; Δλ = ±0.001 nm. **47.** 1/λ = (1.097 373 × 10^7 m^{-1})(1/4 − 1/∞); and λ_∞ = 364.5 nm. **51.** 1/λ = (1.097 373 × 10^7 m^{-1})(1/4 − 1/9); and λ_α = 656.1 nm. **57.** 1/(390 nm) = (1.097 × 10^7 m^{-1})(1/2^2 − 1/n^2); and n = 7.8, so we must use n = 7; 1/λ = (1.097 × 10^7 m^{-1})(1/2^2 − 1/7^2), and λ = 397 nm.

Chapter 28

Answers to Odd-Numbered Multiple Choice Questions

1. c **3.** a **5.** b **7.** b **9.** d **11.** a **13.** e **15.** c **17.** d **19.** b **21.** c **23.** b

Answers to Odd-Numbered Problems

1. I = 77 W/m^2, P = 0.11 kW **3.** 9.4 μm, IR **5.** 72 nm **7.** 2.898 μm **9.** 1.49×10^{-5} m^2 **11.** (a) must be a hot star (b) $f_p = (58.8 \times 10^9$ Hz/K$)T$ (c) 12 000 K **13.** 74.9 THz **15.** T = 4.46×10^3 K **17.** 496 nm **19.** 0.07 W/m^2 **21.** 4.61×10^{14} Hz **23.** 1.89 eV **25.** (b) when $KE_{max} = 0$, $V_{elect} = 0$ (d) 1.32×10^{15} Hz **27.** (b) 5.17×10^{14} Hz (c) 517 THz to 769 THz (d) 1.04 eV, 1.67×10^{-19} J **29.** 0.663×10^6 m/s **31.** 4.4×10^{14} Hz, 6.8 × 10^{-7} m **33.** $f = 1.07 \times 10^{15}$ Hz, λ = 281 nm **35.** ≈58 d **37.** 1.46×10^{-4} nm **39.** 0.099 nm **41.** 180° **43.** (a) $KE_{max} = hf$ (b) 50.0 keV, 8.01 × 10^{-15} J (c) 50.0 keV (e) E = hc/λ (f) 0.024 8 nm **45.** Classically, 5.82×10^{-11} m; relativistically, 5.22×10^{-11} m **47.** 59.6° **51.** −1.51 eV **53.** $f = 4.57 \times 10^{14}$ Hz, λ = 656 nm **55.** −3.40 eV **59.** 476 pm **61.** n = 4 (the third excited state) **63.** n = 137 467 **65.** 2.19×10^6 m/s **67.** (a) n = 2 (b) n = 3 (c) −1.51 eV (e) 12.1 eV **69.** −54.4 eV **71.** $f_n = m_e k^2 Z^2 e^4 / 2\pi n^3 (h/2\pi)^3$ **73.** 5.54×10^{-8} eV

Solutions to Selected Problems

1. P/A = $\varepsilon\sigma(T^4 - T_e^4)$ = (0.97)(5.670 × 10^{-8} W/m^2·K^4)(306^4 − 293^4) = I = 77 W/m^2. P = 0.11 kW. **5.** From Eq. (28.4), $\lambda_p T$ = 0.002 898 m·K; λ_{max} = 72 nm; $f_{max} = 4.1 \times 10^{15}$ Hz. **15.** Taking $\lambda_p \approx$ 650 nm, $\lambda_p T$ = 0.002 898 m·K; T = 4.46×10^3 K. **17.** E = hf = hc/λ, λ = hc/E = 496 nm. **21.** KE = 0 is the threshold condition; f_0 = c/λ = 4.61×10^{14} Hz. **31.** In order to use hf_0 = ϕ, we first need to find ϕ; hf = $KE_{max} + \phi$; ϕ = 2.929×10^{19} J ; $f_0 = \phi/h = 4.4 \times 10^{14}$ Hz ; λ_0 = 6.8×10^{-7} m. **37.** Δλ = $\lambda_s - \lambda_i$ = $(h/m_e c)$ (1 − cos θ) = (0.002 426 nm)(1 − cos 20°) = 1.46 × 10^{-4} nm. **47.** (0.060 0)(20.0 pm) = (2.426 pm)(1 − cos θ); cos θ = 0.505; θ = 59.6°. **49.** For x-component, $p_i - p_s$ cos θ = p_e cos φ; for y-component, p_s sin θ = p_e sin φ; square and add these two, noting that cos$^2\phi$ + sin$^2\phi$ = 1; (a) $p_i^2 + p_s^2 - 2p_i p_s$ cos θ = p_e^2. For the electron (b), $E_e^2 = c^2 p_e^2 + m_e^2 c^4$. Conservation of Energy is (c) $E_i + m_e c^2 = E_s + E_e$. Now rewrite this equation as $E_e = E_i + m_e c^2 - E_s$ and square it and plug it into (b); also multiply (a) by c^2 and plug it into (b) to get $(E_i + m_e c^2 - E_s)^2$ = $c^2(p_i^2 + p_s^2 - 2p_i p_s$ cos θ) + $m_e^2 c^4$. Using E = pc for photons and simplifying, this equation becomes $1/E_s − 1/E_i = (1/m_e c^2)(1 − $ cos θ), and since E = hf = hc/λ for photons

$$\lambda_s - \lambda_i = \frac{h}{m_e c}(1 - \cos\theta) \qquad [28.13]$$

51. n = 1 is the ground state, n = 2 the first excited state, n = 3 the second excited state; E_3 = −(13.6 eV)/3^2 = −1.51 eV. **59.** n = 3; from Eq.

(28.17), r = 3^2(0.052 9 nm) = 476 pm. **69.** From Eqs. (28.18) and (28.19), $E_1 = (-13.6$ eV$)Z^2$ = −54.4 eV. **73.** ΔE = (−3.40 eV) − (−13.6 eV) = 10.2 eV; the photon of momentum p = E/c, imparts that momentum to the atom; the atom's recoil energy is then $p^2/2M$ = $E^2/2Mc^2$ = (10.2 eV)^2/2(939 × 10^6 eV/c^2)c^2 = 5.54×10^{-8} eV.

Chapter 29

Answers to Odd-Numbered Multiple Choice Questions

1. b **3.** d **5.** a **7.** d **9.** d **11.** c **13.** d **15.** a

Answers to Odd-Numbered Problems

1. 5.5×10^{-36} m **3.** 2.4×10^{-11} m **5.** 0.147 nm **7.** 150 V **9.** (a) lose 100 eV of PE_E and gain 100 eV of KE (b) voltage is negative as is charge (d) 5.402×10^{-24} kg·m/s (e) 0.123 nm **11.** 0.27 nm **13.** 0.004 0 nm **17.** $v_p = \frac{1}{2}v$ **19.** proof **21.** proof **23.** proof **25.** $r_1/r_c = \alpha^{-2}$ **27.** l = 0, m_l = 0; l = 1, m_l = −1, 0, +1; l = 2, m_l = −2, −1, 0, +1, +2 **29.** K has 2, L has 8, M has 18, N has 32. **31.** 12.1 eV **33.** 2.86 eV **35.** cannot exist **37.** (a) 15 (b) 15 (c) two electrons per orbital (d) 1, 1, 3, 1, 3 (e) 18 (f) $1s^2 \cdot 2s^2 2p^6 \cdot 3s^2 3p^3$ **39.** (a) 7 (b) 14 (c) l goes from 0 to (n − 1) (d) 4 (e) −0.85 eV **41.** $(h/2\pi)/2L$ **43.** 5×10^{-27} m/s **45.** 5×10^{-27} J, 3×10^{-8} eV **47.** 5.3×10^{-28} J. **49.** 75 MeV, 9.8% **51.** 5.3×10^{-32} m/s **53.** 2.64×10^{-29} m **55.** 1.16×10^6 m **57.** R = r + (1.79 × 10^{-13} m)/r

Solutions to Selected Problems

1. p = h/λ, λ = h/p = h/mv = 5.5×10^{-36} m. **5.** λ = h/p = h/$\sqrt{2m\ KE}$ = 0.147 nm. **11.** E_0 >> KE; hence, classical form is okay; p = $\sqrt{2m_e}$ $\overline{KE}$, KE = (20 eV)(1.60 × 10^{-19} J/eV); p = $2.414\ 6 \times 10^{-24}$ kg·m/s; λ = h/p = 2.7×10^{-10} m = 0.27 nm. **17.** v_p = E/p = $(p^2/2m)/p$ = p/2m; $v_p = \frac{1}{2}v$. **27.** n = 3; n − 1 = 2; hence, l = 2, 1, or 0. m_l goes from −1 to +1; that is, when l = 0, m_l = 0 ; l = 1, m_l = −1, 0, +1 ; l = 2, m_l = −2, −1, +1, +2 **33.** From Eq. (28.19), ΔE = (−12.6 eV)/5^2 − (−13.6 eV)/ 2^2 = 2.86 eV. **41.** Δx = L, $\Delta p_x \approx \frac{1}{2}(h/2\pi)/\Delta x = \frac{1}{2}(h/2\pi)/L$. **45.** Δt = 10 ns; ΔE = $\frac{1}{2}(h/2\pi)/\Delta t = (5.272\ 9 \times 10^{-35}$ J·s)/(10 ns) = 5×10^{-27} J = 3×10^{-8} eV. **55.** $\Delta z\ \Delta p_z \geq \frac{1}{2}(h/2\pi)$; $m \Delta v_z \approx \frac{1}{2}(h/2\pi)/\Delta z$; $\Delta v_z \approx$ 0.578 8 × 10^6 m/s. After 2.00 s, Δz = $\Delta v_z t$ = 1.16×10^6 m. **57.** Δz = 2r; $\Delta p_z \approx \frac{1}{2}(h/2\pi)/\Delta z \approx \frac{1}{2}(h/2\pi)/2r$; $m \Delta v_z \approx \frac{1}{2}(h/2\pi)/2r$, $v_z \approx \frac{1}{2}(h/2\pi)/2mr$; in the time τ it takes to reach the observation screen, the beam spreads a distance Z′ = $v_z \tau$. To find τ, we find v_y ; KE = $(3/2)k_B T$; $\frac{1}{2}mv_y^2 = (3/2)k_B(1600$ K); $v_y^2 = 3k_B(1600$ K$)/m$; v_y = 450.1 m/s. τ = (1.000 m)/ (450.1 m/s) = 2.221 ms. Z′ = $[\frac{1}{2}(h/2\pi)/2mr] \times$ (2.221 ms) = (1.79 × 10^{-13} m)/r; and R = r + Z′ = r + (1.79 × 10^{-13} m)/r.

Chapter 30

Answers to Odd-Numbered Multiple Choice Questions

1. d **3.** b **5.** b **7.** d **9.** b **11.** a **13.** d
15. b **17.** a **19.** e

Answers to Odd-Numbered Problems

1. 111 **3.** 7 protons, 8 neutrons **5.** Fr, Pb, Ag, Pr
7. 13 043.8 MeV/c^2, 13.043 8 GeV/c^2 **9.** 9.988
352 × 10^{-27} kg **11.** 92.906 38 u **13.** no **17.** 15
fm **19.** 14 fm **21.** 27 **23.** 3.9 times **25.** ρ_{nucleus}
≈ 10^{17} kg/m^3, ρ_{atom} ≈ 10^5 kg/m^3, $\rho_{\text{nucleus}}/\rho_{\text{atom}}$ ≈
10^{12} **27.** 64.749 MeV **29.** 6.474 9 MeV/nucleon
31. 28 MeV **33.** 15.663 MeV **35.** (b)
1.492 419 × 10^{-10} J (c) 931.5 MeV **37.** 15.62
MeV **39.** 8.736 MeV **41.** Ca-41, 8.36 MeV;
Ca-42, 11.48 MeV **43.** 5.40 MeV **45.** (a) 14
(b) 6 (c) 14 (d) 7 (e) $^{14}_{7}$N (f) $^{14}_{6}$C → $^{14}_{7}$N + $^{0}_{-1}$e
+ $\bar{\nu}_e$ **47.** (a) 214 (b) 83 (c) 214 (d) 84
(e) polonium (f) $^{214}_{83}$Bi → $^{214}_{84}$Po + $^{0}_{-1}$e + $\bar{\nu}_e$ **49.** $^{147}_{62}$
Sm → $^{143}_{60}$Nd + $^{4}_{2}\alpha$ **51.** (a) 19 (b) 10 (c) 19
(d) 9 (e) fluorine (f) $^{19}_{10}$Ne → $^{19}_{9}$F + $^{0}_{+1}$e + ν_e **53.**
the net KE is 5.407 MeV **55.** (a) 3.7 × 10^{10} (b)
3.7 × 10^8 (c) 3.7 × 10^8 Bq **57.** 1.35 × 10^{-11} s^{-1}
59. 4.1 × 10^{-9} s^{-1} **61.** $t_{1/2}$ = 12 h **63.** 4.28 ×
10^{-4} decays/y **65.** 5.5 d **67.** 90.1 s **69.** N/N_0
= 4.97 × 10^{-16} **71.** $^{27}_{13}$Al + $^{4}_{2}\alpha$ → $^{30}_{15}$P + $^{1}_{0}$n **73.**
17.346 MeV **75.** 5.700 74 MeV **77.** 1.191
MeV **79.** 5.7 × 10^{12} Bq **81.** 3 × 10^{10} Bq **83.**
2.8 × 10^{10} atoms **85.** 0.18 MBq/m^3 **87.** 24.685
4 MeV **89.** proof

Solutions to Selected Problems

5. Fr, Pb, Ag, and Pr. **9.** (5.603 051 GeV/c^2) ×
(1.782 663 × 10^{-36} kg/eV)(1.000 000 × 10^9
eV/GeV) = 9.988 352 × 10^{-27} kg. **13.** R =
$R_0 A^{1/3}$ = (1.2 fm)(6.17) = 7.4 fm; 2R = 15 fm.
15. Mass of C-12 atom is (12 kg)/(6.022 × 10^{26}
atoms/kmole) = 1.99 × 10^{-26} kg; dividing by 12
yields 1 u = 1.66 × 10^{-27} kg. **29.** If E$_B$ =
64.749 MeV, E$_B$/A = 6.474 9 MeV/nucleon. **33.**

Δm = 1.008 665 u + 15.003 065 u − 15.994 915 u
= 0.016 815 u → 15.663 MeV. **41.** For Ca-41:
(40.962 278 u) − (1.008 665 u) − (39.962 591 u)
= −0.008 978 u. For Ca-42: (41.958 618 u) −
(1.008 665 u) − (40.962 278 u) = −0.012 325 u.
One must add 8.36 MeV to remove a neutron from
Ca-41 and 11.48 MeV to remove one from Ca-42.
49. $^{147}_{62}$Sm → $^{4}_{2}\alpha$ + X; X = $^{143}_{60}$X, which is neodymi-
um-143. **59.** λ = 0.693/$t_{1/2}$ = 0.693/(167 × 10^6
s) = 4.1 × 10^{-9} s^{-1}. **63.** λ = 0.693/$t_{1/2}$ = 4.28 ×
10^{-4} y^{-1}. **69.** λ = 0.693/(1.18 min) = 0.587
min^{-1}. N/N_0 = exp[(−0.587 min^{-1})(60.0 min)] =
4.97 × 10^{-16}. **77.** Using atomic masses; initial −
final mass; (4.002 603 u) + (14.003 074 u) −
(16.999 131 u) − (1.007 825 u) = −0.001 279 u; α
must supply 1.191 MeV. **81.** $t_{1/2}$ = 1.236 × 10^6 s;
from Eq. (30.12), λ = 5.609 × 10^{-7} s^{-1}; R_0 = λN_0
= 3 × 10^{10} Bq. **85.** (444 MBq)/(50 weeks)(5
d/week) × (10 m^3/d) = 0.18 MBq/m^3. **87.**
4(1.007 276 u) − (4.001 506 u) − 2(0.000 548 580
u) = 0.026 501 u; or 24.685 4 MeV. **89.** $N = N_0$
exp(−λt); $N_0/2 = N_0$exp(−$\lambda t_{1/2}$); exp($\lambda t_{1/2}$) = 2; ln
2 = $\lambda t_{1/2}$.

Chapter 31

Answers to Odd-Numbered Multiple Choice Questions

1. d **3.** e **5.** c **7.** c **9.** a **11.** a **13.** d
15. a **17.** c **19.** b

Answers to Odd-Numbered Problems

1. (a) no (b) B: 0 → 0 + 0 (c) L: 0 → −1 + 1 =
0 (d) Q: +1 → +1 (e) yes (f) S: +1 → 0 + 0
(g) weak force because ΔS = +1 **3.** (a) electromagnetic (b) strong (c) weak (d) strong **5.**
(a) baryon number and strangeness (b) lepton
number (c) charge and strangeness (d) strangeness and baryon number **7.** no **9.** no, electron-
lepton number **11.** yes **13.** no, weakly **15.** not
forbidden, 33.9 MeV **17.** 37.7 MeV, strangeness

19. 1877 MeV **21.** 0.660 GeV **23.** 10^{30} **25.** (a)
+2/3, +1/3 (b) +1 (c) +1 (d) +1/2, −1/2
(e) 0 (f) +1/2, +1/2 (g) 1 **27.** yes, this is col-
or neutral **29.** $E_0 = hc/4\pi R$ **31.** 70.45 MeV
33. −535.4 MeV, no—must have appreciable KE
35. 2.42 × 10^{22} Hz, 1.24 × 10^{-14} m, 5.34 × 10^{-20}
kg·m/s **37.** c$\bar{\text{u}}$ **39.** Λ^0 = uds **41.** d$\bar{\text{s}}$ → u$\bar{\text{d}}$ + d$\bar{\text{u}}$
43. 2.47 × 10^{20} Hz **45.** ≈10^{15} GeV/c^2 **47.** 8 ×
10^{27} K

Solutions to Selected Problems

3. (a) Only the electromagnetic force affects photons. (b) Since there's no strangeness, this must
be a strong-force decay. (c) Λ^0 has strangeness, so
this must be a weak-force decay. (d) With no
strangeness, this must be a strong-force decay. **7.**
No. It does not conserve baryon number—the
p$\bar{\text{p}}$ pair can cancel, leaving only one nucleon on the
right. Also, it doesn't conserve angular momentum.
11. Yes. It conserves electromagnetic charge, spin,
and B. There is no meson number to worry about.
Because of the strangeness of K^0, its decay will be
controlled by the weak force. **15.** Not forbidden.
(139.6 MeV) − (105.7 MeV) = 33.9 MeV. **19.**
2(m_pc^2) = 2(938.3 MeV) = 1877 MeV. **23.** If 1
proton takes an average of 10^{30} y to decay, in a
group of N = 10^{30} protons one should decay within
1 y. τ = 10^{30} y = 1/λ; λ = 10^{-30} fractional
decays/y; λN = 1. **31.** 2(938.3 MeV) − (939.6
MeV) − (938.3 MeV) − (139.6 MeV) = −140.9
MeV; each proton must bring in 70.45 MeV. **35.**
E = hf; f = E/h = 2.417 99 × 10^{22} Hz = 2.42 ×
10^{22} hz; λ = c/f = 1.239 8 × 10^{-14} m = 1.24 ×
10^{-14} m; p = h/λ = 5.34 × 10^{-20} kg·m/s. **39.**
It's made up of three quarks (no antiquarks). To get
the strangeness, we need S = −1 or one s-quark
that has a Q = −1/3, so we must add a net charge
of +1/3. One u and one d will do it: Λ^0 = uds.
43. E = hf = 2m_ec^2 = 1.637 × 10^{-13} J; f = 2.47 ×
10^{20} Hz. **47.** E = (3/2)$k_B T$; T = 2E/3k_B = 8 ×
10^{27} K.

Credits

Photo Credits

Chapter 1: **Opener**, EH, Courtesy of the Bryn Mawr College Library, EH, Fermilab, Neutrino Experiment E 632, NASA; **4,** Fermilab, Neutrino Experiement E 632; **5,** The Bryn Mawr College Archives; **7** (top); **10** (top); **11** (top); NASA; **7** (second); **10** (second); NASA; **7** (third); **10** (third); NASA; **7** (bottom); **10** (bottom), **11** (bottom), NASA; **9,** Courtesy of the National Institute of Standards and Technology; **10** (top right), EH; **12,** EH; **13,** Fotomas Index, London; **15** (top), EH; **17,** J. Anthony Tyson/AT&T Bell Labs

Chapter 2: **Opener, p. 20,** David Madison/Tony Stone Images; EH; **21,** David Madison/Tony Stone Images; **22,** EH; **25,** Fermilab, Neutrino Experiment E 632; **28** (top), EH, **28** (bottom), Courtesy of The Cooke Corporation, Auburn Hills, MI; **39;** Courtesy of the Canadian Forces

Chapter 3: **Opener, p. 53,** NASA, Biblioteca Ambosiana, Milan Italy; The Stock Market; Dawson Jones/Stock Boston; **62,** Biblioteca Ambosiana, Milan, Italy; **63,** Kaz Sheekey—Skydiving Videographer; **68,** Runk/Schoenberger, Grant Heilman Photography, Inc.; **70,** Dawson Jones/Stock Boston; **73,** U.S. Army; **74,** AP/Wide World Photos; **77,** NASA

Chapter 4: **Opener, p. 85,** The Burndy Library, Dibner Institute for the History of Science and Technology, Cambridge, Massachusetts, ©FPG/PNI, EH, EH, NASA; **86** (top), The Burndy Library, Dibner Institute for the History of Science and Technology, Cambridge, Massachusetts; **86** (bottom), Fermilab, Neutrino Experiment E 632; **87** (top), NASA; **87** (bottom), Gamma Liaison/Roger Voillet; **89,** EH; **94,** EH; **98,** Michael J. Martin, **100,** EH; **102,** @FPG/PNI; **103,** McDonough/Focus on Sports; **105,** EH; **108** (left), Kellar Autumn and Ed Florance, Lewis & Clark College; **108** (right), Kellar Autumn, Lewis & Clark College; **109,** EH; **112,** EH; **113,** EH; **114** (left), EH; **114** (right), EH; **115,** EH; **117,** Canadian Forces Photo, **118,** EH; **120,** EH; **121,** Armed Forces Institute of Pathology; **122,** EH; **123,** DeGolyer Library, Southern Methodist University; **126,** U.S. Air Force; **132,** EH; **133,** EH; **135,** EH; **136,** EH; **137,** EH; **138,** EH

Chapter 5: **Opener, p. 141,** James Sugar/Black Star; Burndy Library, Focus on Sports, Six Flags Great Adventure, Jet Propulsion Laboratory; **143,** Focus on Sports; **146,** Fermilab, Neutrino Experiment E 632; **147** (top) Six Flags Great Adventure; **147** (bottom), SKA; **150** (top), NASA; **150** (bottom), James Sugar/Black Star; **151,** Central Scientific Company; **154,** James Sugar/Black Star; **156,** The Burndy Library, Dibner Institute for the History of Science and Technology, Cambridge, Massachusetts; **157,** Uwe Fink/Department of Planetary Sciences, University of Arizona; **159,** Jet Propulsion Laboratory; **161** (top) NASA; **161** (bottom), NASA; **162,** Dr. Raymond E. Arvidson/Washington University; **163,** EH

Chapter 6: **Opener, p. 171,** © Harold & Esther Edgerton Foundation, 2002, courtesy of Palm Press Inc., Smithsonian Institution, Tom Herbst/Max-Planck-Institute fur Astromomie, Heidelberg; **180** (top), Smithsonian Institution; **181,** Fermilab, Neutrino Experiment E 632; **186** (left), EH; **186** (right), © Harold & Esther Edgerton Foundation, 2002, courtesy of Palm Press Inc.; **188,** Brian Sprulock/Getty Images; **190,** © Harold & Esther Edgerton Foundation, 2002, courtesy of Palm Press Inc; **193,** Tom Herbst/Max-Planck-Institute für Astronomie, Heidelberg; **195,** James Sugar/Black Star; **197,** © David Madison 2001; **201,** Brooks/Cole

Chapter 7: **Opener, p. 209,** Physics Curriculum and Instruction, Rick Stewart/Getty Images, Courtesy of The Cooke Corporation, Auburn Hills, MI; **210** (top), Physics Curriculum and Instruction; **212,** Hewlett-Packard; **215** (left), Erik Anderson/Stock Boston; **215** (right), Rick Stewart/Getty Images; **219** (top), © Loren M. Winters; **219** (bottom), EH; **223** (top), Fermilab, Neutrino Experiment E 632; **223** (bottom), John Gaps III/AP/Wide World Photos; **225** (top left), Courtesy of Oxford University Press; **225** (top right), Central Scientific Company; **225** (bottom), Berenice Abbott/ Commerce Graphics Ltd., Inc.

Art Credits

Index

A

Aberrations, 848
Absolute gravitational acceleration (g_0), 152, 154, 163
Absolute pressure (P), 295
Absolute rest, 937
Absolute simultaneity, 941, 959
Absolute temperature (T), 426, 506, 507, 510
Absolute temperature scale, 426
Absolute zero, 425–429, 446
Absorbed dose, 987
Absorption, 796
    process of, 377–378
Absorption spectrum, 982, 983
ac. *See* Alternating current
AC capacitive circuit, 751
AC generator, 721–722
AC series circuits, behavior of, 757
AC shaded-pole induction motor, 745
AC voltages, 722
Acceleration ($\vec{a}$), 53–75
    angular (α), 239–243
    average (a_{av}), 53–55, 76
    centripetal (a_C), 142–143, 149, 163, 239, 348
    concept of, 53–56
    constant, 56–62, 76, 241–242
    down-plane, 104–106
    due to gravity (g), 63–65, 149, 154, 257
    equations of constant, 60–62
    in free-fall, 63–64
    instantaneous ($\vec{a}$), 55–56
    physiological effects of, 104
    relationship to force and mass, 94–106
    SI unit of, 54
    in simple harmonic motion, 348–349
    tangential, 56
    uniform, 56–62, 76
    vertical, 103–104, 160
Accommodation, of the human eye, 861
Acoustic power, 387–388
Acoustic waves, 379
    intensity-level of (β), 393–395
Acoustics, 382–410
    hearing sound, 390–392
    intensity-level of (β), 393–395
    sound waves, 382–410
    superposition of waves, 385

Action-at-a-distance, 162, 544, 546, 670
Activity (R), defined, 1076
Acute angles, sine and cosine functions of, 34
Addition
    significant digits in, 12–13
    of vectors, 30–31
Additive coloration, 834
Adhesion, role in producing friction, 113–115
Adiabat, 500
Adiabatic change, 500–502
Adiabatic demagnetization, 429
Adiabatic process, 494, [502, CD]
Aerodynamic effects, analysis of, 88
Aether, 783, 784
Air
    as an insulator, 533
    wavelengths in, 374
Air capacitor, 589
Air columns, standing waves in, 401–404
Air friction, 63, 76, 87
Airfoils, 311–312
Airy, George B., 546
Airy disk, 917, 922
al-Biruni, 300
Algebra, 13
    of vectors, 30–32
Alhazen, 237
Alkali metals, 1043
Alkaline earths, 1043
Alpha decay, 1068–1070, 1086
Alpha particles, 984
    in helium atom, 978
    scattering experiment with, 980–982
Alpha radiation, 977–978
Alternating current (ac), 743–772
    capacitance and, 750–753
    inductance and, 748–750
    power (P_{av}), 757
    resistance and, 745–748
    R-L-C ac networks, 753–764
Alternating-current ampere, 746
Alternator, 721
al-Zarqali, 155
AM radio, 760–761
    transmissions, 793
Amber, 529–530
Ammeter, 615, 626, 644–646
    placement of, 638

Amontons, G., 425
Amorphous solids, 288
Ampere (A), 6, 607, 625
    as unit of alternating current, 746
Ampère, A. M., 680, 695, 696
    on electromagnetic induction, 709
    on magnetism, 677
Ampère's Circuital Law, 785
Ampère's Currents, 680, [684, CD]
Ampère's Law, 684–686, [684, CD], 696, 785
Amp-hour rating, 612
Amplifiers, electronic, 771–772
Amplitude and Speed, [390, CD]
Amplitude modulation (AM), 760
Amplitude of SHM, 346
Amplitude of the oscillation, 350
Analog meters, 644–645
Analyzer, 892
Anderson, Carl, 1048
Andromeda galaxy, 237
Angle of flight, determining, 41
Angles
    in degrees and radians, 235
    measurement of, 234–236
Angström, A. J., 18, 982
Angular acceleration (α), 239–243
Angular displacement (θ), 234–237, 265
Angular frequency (ω), 345
Angular limit of resolution, 918
Angular magnification, 862
Angular momentum (L), 259–265, 266
    on atomic and subatomic levels, 260
    conservation of, 6, 260–265, 266
    orbital, 263
    quantization of, 1011
    similarity to linear momentum, 259
    symmetry and, 264–265
Angular momentum vector ($\vec{L}$), 259
Angular speed (ω), 236–239
    determining, 241–242
    instantaneous, 237–238
Anisotropic space, 265
Anode, 613, 968
Anomalous Zeeman Effect, 1038
Antennas, radio, 793
Antibaryons, 1105
Anticolor charges, 1107
Antileptons, 1095
Antimatter, 1035, 1047–1050

Antineutron, 1048
Antinodes, 398, 411
Antiparallel vectors, 31, 32, 40
Antiproton ($\bar{p}$), 1048
Antiprotonic hydrogen, 1049
Antiquarks ($\bar{q}$), 1105, 1115
Antireflection coating, 907
Aorta, 305–306
Apollo space mission, 62, 86, 159, 162, 200–201
Apparent power, 757, 758, 772
Approximations, 15
Arago, Dominique, F. J., 684, 709
    on magnetism, 677
Arc (inverse) tangent function, 36, 41
Archimedes of Syracuse, 148
    Buoyancy Principle of, 298–299
    Law of the Lever, 243
Arc-length (l), 236, 265
Area, computing, 9, 12
Area-under-the-curve analysis, 59
Argonne National Laboratory, 1059
Argument of a function, 13
Aristotle, 62
    aether conception of, 783
    definition of speed, 20
    ideas on sustained motion, 85
    Law of the Lever, 243
    on time, 944
Armature, 694
Arteriosclerosis, 311, 312
Ascending objects, effect of g on, 66
Aspherical mirrors, 871–872, [872, CD]
Aspherical optical elements, 848
Aspherical surfaces, 846–848, [847, CD]
Aston, F. W., 1057
Astronomical telescope, 870–871
Astronomical units (AU), 156
Atmosphere, density fluctuations in, 813
Atmosphere (atm), 294–295, 316
Atmospheric pressure, 293–295
Atomic bomb, 1084
Atomic clocks, 10, 949
Atomic emission, schematic of, 795
Atomic energy, 977, 1014
Atomic mass units (u), 1057, 1086
    table of, 1058
Atomic masses (u), 284, 985
    of nuclides, 1058
Atomic nucleus, 3, 4, 980, 1056–1086
    neutron-proton model of, 985
Atomic number (Z), 985, 1017–1018, 1057
Atomic particles, angular momentum of, 260
Atomic quantum numbers, 1038
Atomic theory, 3, [466, CD], 1009–1018
    Bohr atom, 1009–1014
    stimulated emission, 1014–1017
Atomism, 283–284, [284, CD]
Atoms, 3–4
    as elastic systems, 284

electromagnetic interaction of, 109, 114
    ground state of, 795
    light and, 795–796
    magnetism and, 673–674
    masses and rest energies for, 956
    matter and, 283–290
    neutrality of, 531
    polarized, 536
    rotational motion within, 234
    size of, 285–287
    in solids and liquids, 287–289
Attack angle, 311–312
Attwood's machine, 134
Augustine, Saint, 1085
    on time, 944
Auroras, 690, 691
    and radiation belts, [690, CD]
Average acceleration, 53–55, 76
Average angular acceleration, 239
Average angular speed, 236–238
Average current, 607
Average deceleration, calculating, 100–101
Average density, 153
Average dissipated power, 757
Average induced emf, 712, 726
Average power, 195
Average speed, 20–23, 44
Average velocity, 32–33, 44
Avicenna, 87
Avogadro, Amadeo, 284–285
Avogadro's number (N_A), 284–285, 969, 973, 1001
Axis of rotation, 259–260

B

Back-emf, 725–727, 731
Bacon, Francis, 441, 803
Balance, stability and, 251–253
Balanced steelyard, 244
Ballistic motion, 87
Ballistic projectiles, shape of the path of, 68–72
Ballot, 405
Balmer, Johann, 983
Balmer series, 983, 989, 1013
Banking angle, 144–146
Bar magnet, force field around, 669
Barkla, C., 1017
Barometer, 290, 293–294
Bars, 118
Baryon number, 1099
Baryons, 1099, 1105, 1115
Baseball, 220
Basilar membrane, 390
Basketball, 220
Batteries, 609–614, [612, CD], 626
    capacitors across, 589
    potential difference in, 574
    terminal voltage of, 637–638
Bay Bridge, San Francisco, 433

Beat frequency, 396
Beats, 395–397
Becquerel (Bq), 1074, 1086
Becquerel, Henri, discovery of radioactivity by, 976–977
Bel, 393
Bell, Alexander Graham, 393, 934
Benedetti, G. B., 87, 141
Bennet, A., 534
Berkeley, George, 88
Bernoulli, Daniel, 306, 442
    on electric force, 537
Bernoulli's Equation, 315
Bernoulli's Theorem, 306–309, 316
Beryllium, electron structure of, 1041
Beta decay, 1070–1072
Beta radiation, 977–978
Beta rays, as electrons, 978
B-fields, calculating, 685–686
Big Bang, 2, 162, [162, CD], 410
Binding energy, 1066–1068, 1086
Binding-energy-per-nucleon, 1066–1067
Biot, Jean, 678, 680
Birefringence, 898–899
Black, Joseph, 459, 460, 462, 463
Black holes, 2, 195
Blackbody radiation, 996–1002
Blake, William, 424
Blood, speed of, 306
Blood pressure, 314–315
Bodies, rigid, 246–248
Bohr, Niels, 983, 1009–1014
    atomic theory, 1009
    beta decay and, 1071
    on electron shells, 1039–1044
    energy equation, 1044
    nuclear fission and, 1080–1081
    principle of complementarity, 1030
Bohr atom, 1009–1014
Bohr radius (r_1), 1012, 1034
Boiling, 471–473
Boiling-point curve, 441
Boltzmann, Ludwig, 442, 443, 516–518, 998
Boltzmann's Constant (k_B), 439, 446, 518, 1001
Bomb calorimeter, 466
Bones
    broken, 338, 341, 342
    as compression members, 123
    flexibility of, 333
Born, Max, 1032–1033
Boron, electron structure of, 1042–1043
Bosons, 1039, 1115
Bostock, John, 611
Bottom quark, 1107
Bound charge, 795
Boundary layer, 304
Boyle, Robert, 64, 283
Boyle's Law, 435–436, [436, CD], 438, 442, 446, 500

Bragg, W. H., 975
Bragg, W. L., 975
Bragg equation, 975, 989
Brahe, Tycho, 155
Brain
    equipotentials in, 578
    magnetic particles in, 670
Bramah, 297
Branch, defined, 638
Branching chain reaction, 1081
Breaking point, 333
Breaking strength, 120, 338
Bremsstrahlung, 1006–1007
Brewster, David, 896
Brewster's angle (θ_p), 896, 922
Brewster's Law, 896
Bridge circuit, 650
British Royal Society, 219
British thermal unit (Btu), 461
Brittle materials, 333, [333, CD], 338
Brookhaven Cosmotron, 1097
Brushes, 694
Bubble chamber, 1095
Bulk Modulus (B), 341–342, 359, 379–380
Buller, A. H. Reginald, 943
Bullialdus, Ismael, on gravity, 148
Buoyancy Principle, 298–299
Buoyant force (F_B), [299, CD]
Buridan, Jean, 93
Byron, Lord, 154

C

Caloric, 784
Caloric theory, 456–457
Calorie (cal), 459–462
Calorimeter, 464–465
Calorimetry, 463–466
Camera, structure of, 861
Camera obscura, 856
Cameras, 861, [861, CD]
Capacitance (C), 584–595
    ac and, 750–753
    capacitors in combination, 588–592
    of a cow, 585
    energy in capacitors, 592–595
    parallel plate capacitor, 586–588
    of a person, 585
Capacitive reactance (X_C), 752, 772
Capacitor banks, 594
Capacitors, 584–595, 626
    air, 589
    as charge storers, 586
    charged, 594–595
    in combination, 588–592
    dc and, 752
    energy in, 592–595
    parallel plate, 586–588
    radio, 587
Capillaries, 305–306
Carbon, electron structure of, 1043

Carbon-14
    dating with, 1077–1078
    half-life of, 1077
Carbon-filament lamps, 743
Cardiovascular system, 315
Carnot, Nicholas Léonard Sadi, 504, 512
Carnot Cycle, 504–507, 510
Carnot efficiency, 507
Carnot engine, 504, 506–507, 515, 519
Carrier, 374
Carrier wave, 760
Carus, Titus Lucretius, 63
Cassegrain reflecting telescope, 877
CAT scan, 805
Cathode, 613, 968
Cathode rays, 984
Cathode-ray tube, 554, 573, 970–972
    electron beam in, 690
Cauchy, Augustin Louis, 334
    stress and strain, 334
Cavendish, Lord Henry, 150
Celestial bodies, escape speeds from,
        193–194
Cells, electrically responsive, 548, 609
Celsius, Anders, 424
Celsius scale, 424
Center-of-gravity, 249–255
    degrees of equilibrium and, 251–252
    locating, 249–251
Center-of-mass, 257–258, [257, CD], 266
Center-seeking forces, 142–146
Center-to-center gravitational interaction,
        147
Centigrade scale, 424
Centimeter, 8, 9
Central generating stations, 743
Centrifuge, 146
Centripetal acceleration (a_C), 142–143,
        [143, CD], 163, 239, 348
    computing, 143, 240
    computing magnitude of, 143
    of the Moon, 149
Centripetal force (F_C), 141–146, 163
    center-seeking forces, 142–146
    centripetal acceleration, 142–143
    circular motion, 145–146
    defined, 142
Cesium clock, 10, 799, 949
Cgs system, 9
    units of force, mass, and acceleration in,
        95
Chadwick, James, 985, 986, 1070, 1071,
        1084
Chain reaction, 1081–1082
Change
    adiabatic, 500–502
    energy as a measure of, 171
    as a manifestation of energy, 714
    in quantity, notation for, 26
Change of state, 467–473
    boiling, 471–473

    melting and freezing, 467–469
    vaporization, 469–471
Characteristic X-rays, 1017
Charge (q), 530, 558, 561
    conservation of, 531, 582–584
    electromagnetic, 529–537
    free, 791–792
    interaction between, 540
    mobile carriers, 608
    particles of, 970–973
    potential and distribution of, 581–582
    potential of several, 579–582
    quantization of, 530
    quantum of, 969–970
    storing, 584–586
    units of, 539–540
Charge conductors, 535
Charge densities, 559
Charge per unit length (λ), 558
Charged conductors, 535
Charge-field-charge concept, 546
Chargeless objects, electrifying, 531
Charge-to-mass ratio, 969, 989
Charles, Jacques Alexandre César, 436
Charles's Law, 436–437, 446
Charm quark, 1107
Chemical atomic mass, 1057
Chemical energy, 172
Chemical-PE, 187, 191
Chisov, I. M., 64
Choke coil, 749–750
Circuits, 635–653
    ammeters and voltmeters, 644–647
    defined, 589
    domestic, 762–764
    impedance of, 756, 757, 772
    network analysis, 649–653
    parallel, 588–590, 595
    power factor of, 757
    R-C, 647–649
    reactance of, 756, 757, 772
    resistors in series and parallel, 639–645
    series, 590–592, 595, 753–761
    sources and internal resistance, 636–638
Circular Light, [889, CD]
Circular motion, 145–146
Circular polarization, 889
Circulatory system, human, 305–306,
        314–315
Classical field theory, 730
Classical physics, 2
Classical Principle of Relativity, 936–937
Clausius, Rudolf, 511, 512, 514–515
Clausius formulation, 511–512
Clausius statement of the Second Law, 512,
        520
Cochlea, 390
Cockroft, J. D., 191
Coefficient of friction, 108, 123
Coefficient of kinetic friction (μ_k), 111–
        114, 123

Coefficient of performance, 508, 519
Coefficient of rolling friction (μ_r), 113, 123
Coefficient of static friction (μ_s), 108, 123
Coefficients of volumetric expansion (β), 432
Coherence, 901, [904, CD]
Coherent bundle, 831
Cold welding, 115
Colinear vectors, 54
Collagen, 336, 337
Collector, 769
Collision time, 219
Collisions, 219–226
    elastic, 221–224
    force-time curves for, 214–215
    inelastic, 219–221
    in two dimensions, 224–227
Color, 832–835
    frequency and wavelength ranges for, 801
    human perception of, 800–801
Color charge, 1106, 1108–1109
Color force, 1108–1109
Colored light, 833–835
    source of, 803
Combustion
    heats of, 467
    internal, [507, CD]
Common emitter amplifier, 771
Commutator, 694, 722
Complementarity, Principle of, 1030
Complementary colors, 834
Composite solids, 288
Compound microscope, 869–870
Compounds, 284
Compression, 116
Compression waves, 379–380, [379, CD]
Compton, Arthur H., 1007
Compton effect, 1007–1009, 1019
Compton Wavelength, 1009
Computed tomography (CT), 805
Concave lenses, 848, 859
Concurrent force systems, 118–124
Concurrent Forces, [119, CD]
Condensation, 379
Condenser, 584
Conduction, 476–479
Conduction band, 766
Conductors, 532–537, 619
    fields and, 554–557, [679, CD]
    immersed in an E-field, 555
Conjugate variables, 1031
Conservation laws, 6
Conservation of Angular Momentum, 6, 198, 260–265, 266
    isotropy of space and, 264–265
Conservation of Charge, 531, 582–584
    Law of, 531, 651
Conservation of Energy, Law of, 6, 171, 187, 198–199, 219, 221, 306, 491–502, 956
    and gravitons, 162

homogeneity of space and, 226
Conservation of Linear Momentum, Law of, 6, 198, 216–218, 226, 227
    origin of concept, 216
Conservation of Mass-Energy, 956
Conservation of Mechanical Energy, 187–199
Conservation of Momentum, Law of, 198, 209, 219, 220
Conservative force, 186
Constancy of the speed of light, 937–938
Constant acceleration, 56–62, 76
    calculating, 60–62
    equations of, 60–62, 76, 241–242
    Galileo's experiments with, 64–65
Constant angular speed, determining, 239
Constant speed, 24–25
Constant velocity, 85–87
Constant-a equations, 60–62
Constant-volume gas thermometer, 425
Constructive interference, 814, 815, 900
Contact forces, 5, 102–105
Continuity Equation, 305–306, 310, 311, 316, 557
Convection, 475–476
Convection currents, 475–476
Converging lenses, 847
    focal points for, 851
Conversion factors, 24
Convex lenses, 847
Copernicus, Nicolaus, 147
Coplanar concurrent forces, 124
Coriolis, Gaspard, 172, 180
Cosine function, 13, 34
Cosmic force, 155–162
    effectively weightless, 160–162
    gravitational field, 162
    laws of planetary motion, 155–157
    Newton's proof of Kepler's laws, 156–157
    satellite orbits, 157–160
Coulomb (C), 540, 541
    defined, 696
Coulomb, Charles, 538, 539
Coulomb force, 569, 1063, 1065
Coulomb interactions, summation of, 540
Coulomb scattering, 1063
Coulomb's Law, 538–544, 545, 557, 561, 678
Countercurrent exchange, 477
Coupled motions, 106–107
Covalent bonding, 765, 766
Cowan, C. L., 1072
Crab Nebula, pulsar in, 261
Crick, F. H. C., 976
Critical angle (θ_c), 829, 836
Critical constants, 445
Critical field, 677
Critical mass, 1084
Critical size, 1084
Critical temperature (T_c), 621, 626
Critically damped system, 355

Crookes, William, 970
Crookes tube, 970
Cross product, 688, [688, CD]
Crossed axes, polarizes, 893
Crystalline lens, 860
Crystalline solids, 288
Crystallization, of water, 434–435
Crystals, [288, CD]
Cubic crystal form, 288
Cubits, 7
Curie, Marie, 977, 1071, 1074
Curie, Pierre, 676, 977, 1074
Curie (Ci), 1074
Curie temperature, 676
Current (I), 607
    direct, 612
    eddy, 724
    effective, 746, 772
    fields and, 677–684
    gain, 771
    instantaneous, 746, 751
    maximum, 746
    steady-state, 612
    as transporter of energy, 623
Current balance, 696
Current loop, 680–681
    torque on, 692–694
Current-carrying coil, 681, 719
    field of, 679
    magnetic field lines produced by, 677, 678
Current-carrying wire, 686
    force on, 691
Curved mirrors, 871–877
Curvilinear motion, equations of, 241, 242
Curvilinear translation, 235
Cyclic processes, 503–510
    absolute temperature and, 510
Carnot Cycle, 504–507
    internal combustion, [507, CD]
    refrigeration machines and heat pumps, 507–510, [508, CD]
Cyclotron, 1059
Cylinder, determining acceleration of, 256

D

da Vinci, Leonardo, 97, 115, 244, 372
    on wave theory, 826
Dalton, John, 284, 441
Damped harmonic motion, 350
Damped motion, 354–356
Damping, 354–356
Dashpot, 378
Data, collection of, 1
Daughter nucleus, 1069, 1086
Davisson, C. J., 1028–1029
Davy, Humphrey, 546, 968
Daylight, phenomenon of, 800–801
dc. See Direct current
de Broglie, Louis Victor Pierre Ramond, Prince, 491, 1027–1028

antineutrinos and, 1071
de la Tour, 396
Decay constant (λ), 1075, 1087
Decay rate (R), 1074–1076, 1086
Deceleration, 54
  calculating, 100–101
Decibel, 393
Decimal measurement, 7
Deflection yoke, 690
Degrees of freedom, 1036
Delta notation, 26
Dense media, light transmission through, 815–816
Dense media scattering, 815–816, [815, CD]
Density (ρ), 285–287
  Earth's, 152–154
  pressure-energy, 308
  relative, 300
Depletion layer, 767
Depolarization, 611
Derivative, 28, [28, CD]
Descartes, René, 93, 180, 187, 669, 825, 872
  Conservation of Linear Momentum, 209
Descending objects, effect of g on, 66
Destructive interference, 814, 815, 900
Determinism, 1046–1047
Deuterium, 1059
Deuteron, 1061
Dewitt, Bryce, 1048
Diamagnetism, 673
Diastolic pressure, 314
Diathermy machine, 799
Diderot, Denis, on dimensions of space-time, 949
Dielectric constant (K_e), 550
Dielectrics, 533, 536
  in a parallel plate capacitor, 588
Diffraction, 910–921, [918, CD]
  circular holes, obstacles and, 917–919
  diffraction grating and, 915–917
  Fraunhofer, 911, 917, 922
  Fresnel, 911
  holography and, 920–921
  single-slit, 913–915
  X-ray, 974–976
Diffuse reflection, 819, 820
Diffusion of gases, [446, CD]
Digital meters, 645
Diode laser, 768
Diodes, 767–768, 773
Diopters (D), 865, 879
Dioptric power ($\mathscr{D}$), 865, 879
Dip poles, 672
Dipole, 553, 668
  field, 552
  radiation, 792–793
Dirac, P. A. M., 1011, 1047
  on Quantum Electrodynamics, 1047
Dirac equation, 1048
Direct current (dc), 607–625

capacitors and, 752
cell, 626
generator, 722–724
motor, 694, 722, 724
resistance, 614–625
Discharge rate (J), 305–306
Disintegration energy, 1069, 1086
Dispersion, 823
Displacement, distance vs., 29
Displacement Law, 999–1000
Displacement (θ), 29, 32, 33
  angular, 234–237, 265
  in simple harmonic motion, 346, 347
Displacement triangle, 35
Displacement vector ($\bar{s}$), 29
Dissipative absorption, 796
Distance
  displacement vs., 29
  varying of force with, 185–186
Distance-time graph, 24–25
Distributed force, 333
Distribution function, 997
Diverging lenses, 848
  focal points for, 851
Division, significant figures in, 11–12
DNA structure, 976, 987
Dolphins, and ultrasound, 381
Domains, magnetic, 674–676
Domestic circuits, 762–764
Doping, 766
Doppler, Johann, 405
Doppler Effect, 404–410, 946
Doppler shift, 408–410, 1007
Dosimetry, 986–989
Double-slit experiment, 901–904, 1033–1034
Down quarks, 1094, 1105
Down-plane acceleration, 104–106
Drag force, 63
Drift velocity, electric fields and, 613–614, [614, CD]
Dry cell, 611
  electric field of, 613–614
  flow of charge through, 612
Dry friction, 107
Duane, W., 1007
Ductile materials, 333, 337
Dufay, C., 530
Dull, Charles E., 937
Dulong and Petit's Law, [466, CD]
Dynamics, 85, 97
  fluid, 290, 303–315
  Newton's Laws and, 253
  of rotation, 253–257
Dynamos, 720, 724
Dyne, 539

E

$E = mc^2$, 191–192, 955, 1066
Ear, structure of, 390
Earnshaw, 979

Earth
  angular momentum of, 260
  asphericity of, 154
  charge stored by, 544
  composition of atmosphere of, 289
  as a conductor, 578
  density of, 152–154
  gravitational interaction with objects, 101–102
  gyroscopic stability of, [264, CD]
  magnetism of, 672
  rotation of, 145
  spin rate of, 265
Earth satellite, 157
Earth-Moon system, 265
Earthquake compression waves, 380
Eddy currents, 724, [724, CD]
Edison, Thomas, 976
  incandescent lamp and, 743
Effective current, 746, 772
Effective voltage, 746, 772
Effective weight, 102–105, 160
Effective weightlessness, 160–162
E-field, amplitude of, 787
E-field meter, 786
Ehrenfest, Paul, 1005
Eiffel Tower, 799
Einstein, Albert, 2, 3, 16, 187
  on Ampère's hypothesis, [684, CD]
  on atomic energy, 1082, 1084
  on blackbody theory, 1002
  on Conservation of Energy, 187, 957
  on corpuscular theory, 793
  on Electrodynamics of Moving Bodies, 937
  on Electromagnetic Theory, 560
  on electromotive force, 719
  General Theory of Relativity of, 2, 147, 355
  on light-quanta, 1004
  Photoelectric Effect and, 1002
  on photon, 793, 1003–1006
  Quantum Mechanics, 1050
  Special Theory of Relativity of, 936–950
  on speed-dependent mass, 953–954
  on stimulated emission, 1015
Elastic collisions, 221–224
Elastic energy (PE_e), 172, 186, 187
Elastic force, 5, 389
Elastic force (spring) constant (k), 330
Elastic limit, 330
Elastic materials, 332–333
Elastic restoring force, 349–352
Elastic shear strain, 335
Elastic stress per unit strain, 340, 359
Elasticity, 329–344
  fatigue and, 338–339
  Hooke's Law and, 329–334
  strain and, 335–336
  strength and, 336–338
  stress and, 334
Elastic-PE (PE_e), 187, 190, 331–332, 491

Elastin, 336
Electric battery, first, 609
Electric car, 645
Electric current (*I*), 607–614, 625
   batteries, 609–614
   electric fields and drift velocity,
     613–614
Electric field (*E*), 544–560, [554, CD],
   560
   calculating, 547–549
   of a charged conductor, 555
   conductors and, 554–557
   defined, 546
   of a distribution of charge, 558
   drift velocity and, 613–614
   energy density of (u_E), 730
   flux of, 557
   Gauss's Law and, 557–560
   induced, 719–720
   lines, 546
   magnitude of, 575
   of a parallel-plate capacitor, 554
   of a point-charge, 547–549
   Quantum Field Theory and, 560
   rationalized units and, 549–550
   shielding against, 555–556
   table of values, 548
   between two charged plates, 560
   of two equal positive point-charges, 550,
     551
Electric field lines, 545, [554, CD], 558
Electric force, 529, 537–544, [538, CD]
   Coulomb's Law, 538–544
   electrostatic induction, 535–537
Electric motor, 694
Electric potential (*V*), 569–584, 595
   Conservation of Charge and, 582–584
   distribution of charge and, 581–582
   equipotentials and, 576–579
   of a point-charge, 575–576
   of several charges, 579–582
   in a uniform field, 573–575
Electric power, consumption of, 763
Electrical capacity, 584
Electrical energy, 172
Electrical-PE (PE_E), 187, 491, 569
Electrocardiogram, 345, 623
Electrodes, 610
Electrodynamics, 677–686, 696
   Ampère's hypothesis, 684–686
   currents and fields, 677–684
Electrolysis, 968–969
Electrolytes, 609
Electromagnet, creation of, 683
Electromagnetic charge, 529–537
   insulators and conductors of, 532–537
   positive and negative, 530–532
   as quantized and conserved, 530–531
   by rubbing, 531–532
   transfer of, 532
Electromagnetic force, 4–5, 103, 287

Electromagnetic induction, 709–730
   electromotive force (emf) and, 710–720
   generators and, 720–724
   self-induction, 725–730
Electromagnetic interaction, of atoms, 103,
   107, 114
Electromagnetic radiation (EM), origins of,
   791–793
Electromagnetic spectrum, 797, 806
Electromagnetic Theory, 2
Electromagnetic waves, 784–786, [786, CD]
   electric field of, 786–787
Electromagnetic-PE, 333
Electromagnetic-photon spectrum, 796–805
   gamma rays, 805
   infrared radiation, 800
   light, 800–802
   microwaves, 799
   radiowaves, 797–798
   ultraviolet radiation, 802–804
   X-rays, 804–805
Electromagnetism, 667
Electromotive force (emf), 610, 625
   induced, 710–720
   motional, 716–720
   polarity of, 714, 715
Electromotive series, 610
Electron capture, 1070, 1086
Electron drift velocity, [614, CD]
Electron shells, 1039–1044
Electron spin, 1038
Electron volt (eV), 572
Electron-antineutrino, 1071
Electronics
   pn-junction and diodes, 767–768
   semiconductors, 765–772
   transistors, 769–771
Electron-oscillators, 895, 896
Electron-positron pairs, 1048, 1101
Electrons (e⁻)
   affinities for, 531
   angular momentum of, 260
   beam of, 608
   behavior of, 673
   beta rays as, 978
   charge on (q_e), 530–531, 561
   as dipole magnets, 669
   discovery of, 971–972
   fields of, 673
   as fundamental unit of charge, 970
   J. J. Thomson on, 973
   magnetic moment of, 1049
   physical characteristics of, 973
   spin of, 1038
   transfer of, 532
   valence, 765, 766
   value of charge of, 540
Electroscope, 534–536
   inducing a charge on, 535–536
Electrostatic generator, 535, 555, 581
Electrostatic induction, 535–537

Electrostatics, 529–560, 569–595
   capacitance, 584–595
   defined, 529
   electric field, 544–560
   electric force, 537–544
   electric potential, 569–584
   electromagnetic charge, 529–537
   energy, 569
Electroweak epoch, 1112
Electroweak field, 1112
Electroweak force, 5, 1109–1112
Electroweak Theory, 1111, 1115
Elemental semiconductor, 766
Elementary particles, 3, 4, 1095–1099
   angular momentum of, 260
   chemical atomic mass of, 1057
   hadrons, 1095, 1097–1099
   leptons, 1094–1096
Elements, 284
Ellis, C. D., 1071
Ellis-Wooster calorimeter, 1071
Emission coefficient, 996
Emission spectrum, 982, 983
Emission theory, 783
Emitter, 769
Empire State Building, 3, 345, 346
Empty space, isotropic nature of, 265
Endoscope, 831
End-pivoted lever, 251
Endurance (fatigue) limit, 338–339
Energy, 171–199
   applying conservation of, 192–193
   atomic, 977
   chemical, 172
   elastic potential, 331–332
   gravitational, 172
   internal (*U*), 186–187, 492–493, 497
   kinetic (KE), 172, 180–182, 186
   mass and, 190–192
   mechanical, 180–199, 188–190
   mechanical change in, 172, 180–187
   moduli and, 340–342
   nuclear, 172
   potential (PE), 183–187
   symmetry and, 198–199
   thermal, 172, 186, 456–479
   of a wave, 375
   zero-point, 429
Energy conservation, 6, 171, 198–199, 306,
   491–502
   binding, 1066–1068
   in capacitors, 592–595
   current as transporter of, 623
   electrostatics and, 569–595
   irradiance and, 790–791
   in magnetic field, 730
   mechanical, 188–190
   Newton's Third Law and, 199
   as possessing inertia, 957
   power and, 623–625
   quantization of, 1002–1006

relativistic, 572, 954–959
symmetry and, 198–199
total, 956
Energy content, 466
Energy density, 730
Energy efficiency, 505–507
Energy elements, Planck on, 1000–1002
Energy levels, 1010, 1044
atomic, 1014
Energy quanta, 793–795
Energy transfer, 171–179, 189
power (P), 195–197
work (W), 172–179
Engine thrust, calculating, 216
Engines, 503–510
Entropic neutrality, 514
Entropy (S), 513–520
as probability, 517–519
Entropy Principle, 514
Epitaxial layers, 769
Equations, 13
Equilibrium, 85, 115–123, 243–244
First Condition of, 116
force triangle corresponding to, 121
rotational, 243–244
Second Condition of, 246–248
stable, 251
static, 85
of systems, 85, 493
thermal, 462
translational, 115
Equilibrium of extended bodies, 249–251
center-of-gravity, 249–255
stability and balance, 251–253
Equilibrium vapor pressure, 440
Equipotential surfaces, 577
Equipotential volume, 578
Equipotentials, 576–579, [579, CD]
in the brain, 578
Equivalence Principle, 355
Equivalent resistance, 639, 654
Escape speed (v_{esc}), 193–194, 199
from surfaces of celestial bodies,
193–194
Euler, Leonhard, 94, 343
European Council for Nuclear Research
(CERN), 944
Evaporation, 469
Exact measurement, 11–13
Excited states, 795
Exclusion Principle, 1047, 1061, 1065
Exponentials, 11, 647–648, 1075
Exposition of the System of the World
(Laplace), 195
Exposure, radiation, 987
Extended bodies, equilibrium of, 249–251
Extended imagery
lenses and, 852–856
mirrors and, 874–877
External force, 96–97, 217, 218
weight as an, 103

External reflection, 816–817
Eye, structure of, 859–860
Eyeglasses, 865–868

F

$\vec{F} = m\vec{a}$, 94–101
Fahrenheit, G. D., 424
Fahrenheit temperature scale, 424
Farad (F), 584
Faraday (F), 969
Faraday, Michael, 545, 546
on diamagnetism, 673
electric generator and, 720
on electromagnetic fields, 545
on electromagnetic induction, 709–710,
761
ice-pail experiment, 556–557
on lines-of-force, 545, 785
use of electrolysis, 968–969
Faraday cage, 555, 556
Faraday's Induction Law, 710–716, [713,
CD], 731
Faraday's Law, 784
Far-field diffraction, 911
Far-point, 861, 868
Farsightedness, 866
Fat tank, 300
Fatigue, 338–339
Fermat, Pierre, 1031
Fermi (fm), 1060
Fermi, Enrico, 1060
and atomic bomb, 1084
beta decay, theory of, 1071
neutrino and, 1071
nuclear reactor and, 1082
radioisotopes, creation of, 1079
Fermi level, 1065
Fermi National Accelerator Laboratory,
1096
Fermions, 1039, 1115
Ferrite-core inductor, 727–730
Ferrites, eddy currents and, 725
Ferromagnetic substances, 674
Ferromagnetism, 674
Feynman, Richard P., 89, 172, 1047, 1049
on atomic bomb, 1084
and Quantum Electrodynamics, 1049
on Quantum Mechanics, 1047
Feynman diagram, 1100, 1101
Fiberoptics, 830–832
coherent bundle, 831
incoherent bundle, 831
Field lines, 551–554, 671, 672
cutting, 718–719
for a positive charge, 576, 577
Field of force, 544–545
Field particles, 162
Field Theory, 162
Fields
conductors and, 554–557

currents and, 677–684
Fine structure, 1036
Fine structure constant (α), 1036, 1050
Firing angle, of projectiles, 72–73
First Condition of Equilibrium, 116, 124,
247
First harmonic, 386
First Law of Thermodynamics, 491–502,
519, 956
First Postulate of Relativity, 937, 959
Fission, nuclear, 1079–1084
Fission reactor, 1082–1083
FitzGerald, G., 935
Fizeau, 790
Flavors of quarks, 1105
Fletcher, H., 973
Floating, 300–301
Floating circuits, 623
Fluid dynamics, 290, 303–315
airfoils and, 311–312
Bernoulli's Equation and, 306–309
blood pressure and, 314–315
continuity equation and, 305–306
fluid flow and, 303–305
turbulent flow, 304–305
vortices and, 313, 314
Fluid statics, 290–302
buoyant force and, 298–299
gases and, 289–290
hydrostatic pressure and, 290–296
liquids and, 289
Pascal's Principle and, 296–298
specific gravity and, 300
surface tension and, 301–302
Fluids, 290–302
flow of, 557
force exerted by, 291
Flux
defined, 557
of the electric field, 557
of the magnetic field, 712, 731
Flux density, 712
Flux linkage, 713
FM transmissions, 798
Focal length, (f), 871, 878
Focal plane, 852
Focal points, 847, 851–852
Food, energy provided by, 460
Foot-pounds, 173
Force ($\vec{F}$), 88–93, 171
buoyant, 298–299
center-seeking, 142–146
centripetal, 141
conservative, 186, 199
constant external, 172
cosmic, 155–162
defined, 4, 88–90, 171
elastic, 5, 389
electromagnetic, 4–5
external, 96–97, 217, 218
frictional (F_f), 102, 108, 177–178

gravitational, 3–5, 89, 101–106, 147, 154
  impulse of, 211
  instantaneous, 212–213
  interaction and, 96–97
  interatomic, 287
  internal, 186, 217, 258
  kinetic-friction, 111–113
  line-of-action of, 90, 244
  measuring, 89
  moment of a, 244–245
  moment-arm of, 244–245
  on a moving charge, 687–690
  nonworking, 181
  normal, 102–104, 144–146
  reaction, 102, 248
  relationship to change in motion, 85
  relationship to mass and acceleration, 94–107
  restoring, 330, 349–352
  static-friction, 110–111
  strong, 5
  tensile, 90, 100
  varying, 212–215
  varying with distance, 185–186
  as a vector, 89–93
  weak, 5, 1072–1073
  on wires, 691–696, [691, CD]
Force field, 162
Force law, 540
Force of friction, 102
Force per unit area (pressure), 114
Force systems
  concurrent, 118–124, [119, CD]
  coplanar, 124
  parallel and colinear, 117–118
Force triangle, 92, 120, 121
Force vectors, 89–92, 545
Force-carrying particles, 147
Forced oscillation, 356–357
Force-multiplier, 117–118
Forces, nonconcurrent, 246–248
Force-time curve, 213–215
Force-time graph, 213
Forward biased diode, 767
Forward scattering, [815, CD]
Fosbury Flop, 251, 252
Four Forces, 4–5, 88–90, 94, 171, 560, 1062, 1111, 1112
  energy transformation and, 491
  field particles and, 560
Fourier, Baron de, 386, 479, 615
  on waveforms, 386
Fourier analysis, 386
Fourier's Conduction Law, 479
Franklin (unit of charge), 539
Franklin, Benjamin, 530
  on electric force, 537–538
  on electrical matter, 970
  parallel plate capacitor and, 586
Fraunhofer diffraction, 911, 917, 922

Free charge, 791–792
Free particles, trajectory of, 689–690
Free space
  permeability of, (μ_0), 676
  permittivity of, (ε_0), 550, 561
  propagation of light in, 937
Free-body diagram, 98–100, 103
Free-fall, 62–76, 88
  acceleration in, 63–65
  orbiting astronauts in, 160–161
Freeze drying, 441
Freezing, 467–469
Frequency, 344–345
Frequency (f)
  angular (ω), 345
  resonant (f_0) 356–357, 758–759
  wave, 374
Frequency modulation (FM), 798
Frequency spectrum, 386
Fresnel, Augustin Jean, 784
  on polarization, 889
  wave theory of, 918
Fresnel diffraction, 911
Friction, 5, 107–115
  air, 63, 76, 87
  causes of, 113–115
  cold welding, 114
  contact area, 109
  as a driving force, 110
  dry, 107
  force of (F_f), 102, 108, 177–178
  heating by, 457
  kinetic, 107, 111–114, 177–178
  motion and, 107–115
  Newton's experiments with, 85–88
  rolling, 113
  static, 108–114
  stick-slip, 359
Friction coefficients, 108, 111–114
Friction electric machine, 585
Friction force, 177–178
Fringes of equal thickness, 910, 922
Frisch, Otto, 1080, 1081
Full-scale current, 646
Full-wave rectifier, 768, 769
Fundamental (frequency), 386, 400
Fusion, 467
  nuclear, 1084–1086

G

g (acceleration due to gravity), 63–65
  effect of Earth's rotation on, 154
  and free-fall, 65
  and weight, 89
G (Universal Gravitational Constant), determining, 150–151
$\vec{g}$ (vector acceleration due to gravity), 87, 102
Gabor, Dennis, 920
Galaxy, 4, 15

Gale, Henry G., 784
Galilei, Galileo, 54, 63, 64, 68, 71, 85, 86, 88, 93, 352–353
  acceleration experiments, 64–65
  ballistics and, 85–86
  discourse on falling bodies, 64
  energy and, 171
  law of inertia and, 85–88
  pendulum, 352–353
  projectile motion, 68
  thermoscope of, 423
  thought experiment, 76
  waves, 400
Galvani, Luigi, 609
Galvanometer, 646
Gamma decay, 1073–1074
Gamma radiation, 805, 979
Gamma rays, 805, 987
Gas Laws, 435–446
  Ideal Gas Law, 438–441
  kinetic theory and, 441–446
  liquefaction, [441, CD]
  phase diagrams and, 440–441
Gases, 283–315
  diffusion of, [446, CD]
  ideal, 497–502
  light scattering of, 815–816
  molecule speed in, 444–446
  Newton's experiments on, 96
Gauge blocks, 115
Gauge force field, 1104
Gauge pressure, 295–296
Gauge symmetry, 1102, 1104
Gauge Theory, 1050, 1102–1104
  importance of, 582
Gauss, Friedrich, 557, 854
Gaussian Lens Equation, 854, 859, 870, 878
Gaussian surfaces, 557–559
Gauss's Law, 557–561, [558, CD], 1018
Gay-Lussac, Joseph Louis, 437
Gay-Lussac's Law, 437, 446
Geiger, Hans, 978, 980–981
Gell-Mann, Murray, 1098, 1104, 1105
Gemini spacecraft, 212
General Theory of Relativity, 2, 147, 355
Generators, 720–724, 731
  ac generator, 721–722
  dc generator, 722–724
  eddy currents and, 724
  maximum voltage of, 756
Genesis, 3
Geomagnetic poles, 672
Geometric center, as center-of-mass, 250, 257–258
Geometrical optics, 845–877
  aspherical surfaces, 846–847
  extended imagery, 852–856
  focal points, 851
Gaussian Lens Equation, 854
  lenses, 845–871
  magnification (M_T), 854

mirrors, 871–877
sign convention, 852, 855
spherical thin lenses, 848
thin lens combinations, 863–871
Geostationary (geosynchronous) orbit, 160
Gilbert, William, 669
on gravity, 147, 148
on magnetism, 669, 698
versorium and, 562
Gillet, J. A., 785
Glancing collision, 224
Glashow, Sheldon L., 1107, 1110–1111
Global symmetry, 1104
Gluons, 1108
Goddard, Robert H., 197, 216
Goeppert Mayer, Maria, 1065
Gold-leaf electroscope, 534
Golf ball, 220
Goudsmit, S. A., 1038
Gram (g), 9, 10
Gramme dynamos, 724
Gram-molecular mass, 284
Grand Unification Theories (GUTs), 1112–1114, [1114, CD]
Graphs, 14
Grating equation, 915
Gravimeter, 154
Gravitation, Law of Universal, 147–152
Gravitational acceleration, [65, CD]
Gravitational attraction, computing, 151
Gravitational Constant, Universal (G), 150–152
Gravitational energy, 172
Gravitational fields, 162, 163
action upon planets, 265
Gravitational force (FG), 3–5, 89, 101–106
center-seeking, 147
Gravitational mass, [186, CD], 355
Gravitational-PE (PE$_G$), 184–186, 188–190, 192–194, 308, 344, 491, 569, [575, CD]
Graviton, 162
Gravity
acceleration due to (g), 63–65
artificial, 146
center of, 249–251
effect on gases, 290
effect on motion, 87
hydrostatic pressure and, 291–292
Kepler's Laws and, [155, CD]
mass and, 93–94
Newtonian theory of, 147–162
relationship to planetary motion, 155
specific, 300
of a sphere, [154, CD]
stability and, 252
terrestrial, 152–154
varies with distance, 185–186
vector acceleration due to, 87
weight and, 9
Gravity of a sphere, [539, CD]

Gray, Stephen, 533, 562
Gregory, James, 877
Ground, 535
Ground state, 765, 796, 1010
Ground wave, 793
Ground-state configurations, 1041
Group velocity, 823
Gyros, [264, CD]
Gyroscopic stability, [264, CD]

H

Hadrons, 1095, 1097–1099, 1106–1107, 1115
quark composition of, 1106–1107
Hahn, Otto, 1079–1080
Haldane, J. B. S., 64
Half-life ($t_{1/2}$), 1074–1078, 1087
Half-wave rectifier, 768
Halley's comet, 158
Halogens, 1043
Hamilton, William Rowan, 1031
Hanger, 117
Hansen, H. M., 1010
Harmonic electromagnetic wave, 787, 788
Harmonic motion, 344–359
damping and, 354–356
elastic restoring force and, 349–352
forcing and resonance and, 356–357
self-excited vibrations and, 357–358
simple, 344–349
Harmonic oscillator, 348–350
Harmonic waves, 375–376, 411, 806
Harmonics, 386
Hauksbee, Francis, 585, 970
Hauksbee's Influence Machines, 585
Hearing, 390–392
Heart activity, equipotentials due to, 580
Heat (Q), 456, 459
change of state and, 467–473
defined, 459
mechanical equivalent of, 462
quantity of, 459–460
specific, 463–464
temperature and, 457–459
thermal energy and, 456–479
work and internal energy and, 492–493
Heat engine, 504, 519
Heat of fusion (L_f), 467–468, 480
Heat of vaporization (L_v), 468, 470, 480
Heat pumps, 507–510
Heat-as-matter theory, 457
Heats of combustion, 467
Heaviside, Oliver, 549
Heavy water, 1059
Heisenberg, Werner, 674, 986, 1044, 1062
and atomic bomb, 1082
on ferromagnetism, 674
Heisenberg Uncertainty Principle, 1044–1047, 1050, 1100
and diffraction, [1046, CD]

Helium
discovery of, 978
electron structure of, 1040–1041
liquid, 428–429, 621
Helium atom, nucleus of, 978
Helmholtz, Hermann von, 490
on atoms, 970
on sound, 396
Henry (H), 725
Henry, Joseph, 683, 684, 725
on electromagnetic induction, 710
electromagnetism and, 683
Hermann, Jacob, 94
Herschel, William, on infrared radiation, 800
Hertz, Heinrich R., 345, 970
generation of electromagnetic waves, 809
on Photoelectric Effect, 1002
Hertz (Hz), 345
Higgs boson, 1111
Higgs field, 1112
High-energy physics, 1094–1114
elementary particles, 1095–1099
Quantum Field Theory, 1100–1114
High-temperature superconductors, 621
"High-tension" wires, 745
High-voltage ac, 744
Hip joints, rotational inertia of, 255
Hofstadter, R., 1059
Holography, 920–921
Homogeneity of space, 226
Homogeneous waves, 789
Hooke, Robert, 283, 783–784
on elasticity, 329–330
on gravity, 147–149, 151
wave theory of, 783–784
Hooke's Law, 14, 329–334, 349, 359
Horizontal balance, computing, 243
Horizontal force, computing, 99, 119, 120
Horizontal speed, of projectiles, 68–75
Horrocks, Jeremiah, 158
Horsepower (hp), 196
"Hot" wires, 764
H-particles, 984
H-sheet Polaroid, 894
Hubble Space Telescope, 473, 871, 872, 877
Human body
capacitance of, 585
center-of-gravity of, 249–253
as a conducting antenna, 797
effect of electricity on, 764
effect of radiation on, 987–989
fluid pressures in, 295
as a fluid-dynamical system, 290
head, concurrent force system and, 247
levers and, 243
radiation by, 800
temperature of, 423–424
Human bones, torques and breaking angles for, 342

Hume, David, 460
Hunt, F., 1007
Huygens, Christiaan, 141, 180, 187, 306, 355
    on collisions, 220
    wave theory of, 889
Hydraulic machines, 296–298
Hydrodynamic systems, 302–303
Hydrogen
    antiprotonic, 1049
    charge-to-mass ratio of, 969
    electron structure of, 1039
    isotopes of, 1059
    liquid, 428
    molecular, 284
    structure of, 1036
Hydrogen atom, energy levels of, 1014
Hydrogen visible spectra, 983
Hydrostatic pressure, 290–296
Hyperbolic surfaces, 846–848
Hyperopia, 866
Hypotheses, 1–2

I

IC microprocessor chip, 765
Ideal Gas Law, 438–441, 446–447
Ideal gases, 497–502
    isothermal expansion of, 497–500
    temperature of, 444
Ideal liquids, 289
Ideal voltage source, 639
Ideal voltmeter, 637
Image distance (s_i), 849
Image focal length (f_i), 851
Image focal point, 851
Image reconstruction, 920
Impact, force-time graph for, 213
Impact time, 212, 214
Impedance (Z), 749, 756, 757, 772
Impedance triangle, 756, 757
Impulse
    defined, 211
    interaction and, 216
    momentum change and, 210–212
Impulse-momentum, 215
Impurity semiconductors, 766
Incident angle, (θ_i), 829–830
Inclined plane, 104–106
Inclines
    friction on, 109–110
    rolling down, 259
Incoherent fiber bundle, 831
Index of refraction (n), 822–824, 835
    table of, 823
Induced emf (electromotive force), 710–720
    Faraday's Induction Law, 710–716
    fields and, 719–720
    Lenz's Law, 714–716
    motional emf, 716–720

Induced radioactivity, 1078–1079
Inductance (L), 725–727, 731
    ac and, 748–750
Induction, electrostatic, 535–537
Induction motor, 744
Inductive reactance (X_L), 749, 772
Inductor, defined, 727
Inelastic collisions, 219–221
Inertia
    defined, 88
    moment of (I), 253–255, 262
    Newton's Law of, 85–97, 253
    rotational, 253–257
Inertial mass (m), 253–254, 355
Inertial observer, 937
Inertial reference frame (system), 937
Inflation, [162, CD]
Infrared electromagnetic radiation, 474
Infrared (IR) pictures, 800
Infrared radiation, 800
Infrasound, 381–382
Inhomogeneous waves, 789
Initial phase, [346, CD]
Ink jet printer, E-field in, 560
Inner ear, 390
Input current, 771
Input transducers, 635
Instantaneous acceleration, 55–56
Instantaneous angular acceleration, 239, 265
Instantaneous angular speed, 237–238, 265
Instantaneous current, 746, 751
Instantaneous force, 212–213
    time (t) vs., 213, 214
Instantaneous power, 195, 746
Instantaneous speed, 26–28, 44
Instantaneous velocity, 32–33, 44
Instantaneous voltage, 745, 749, 751
Insulators, 532–537, 619
Intensity (I)
    of a sound wave, 411
    wavefronts and, 387–388
Intensity-level (β), 411
    of an acoustic wave, 393–395
Interacting systems, work and, 174–175
Interaction
    Newton's Law of, 85, 96–97, 103–105, 110
    in rockets, 215–216, 216
Interaction pairs, 96, 97, 103, 247
Interatomic binding, 429–431
Interference, 899–910, 922
    coherence and, 901, [904, CD]
    Michelson Interferometer and, 909–910
    scattering and, 814–815
    thin film, 905–908
    Young's Experiment with, 901–905
Interference fringes, 901
Interferometers, 909–910
Intermediate vector bosons, 1109, 1112, 1115
Internal combustion, [507, CD]

Internal energy, 186–187, 492
    as a measure of state, 497
    work and heat and, 492–493
Internal forces, 97, 216, 258
Internal reflection, 816–817, 829–832
Internal resistance (r), 636
    of a battery, 638
International Bureau of Weights and Measures, 9
Intrinsic semiconductor, 766
Invariant mass, 953
Inverse (arc) tangent function, 36, 41
Inverse relationships, 14
Inverse Square Law, 388
Ionization, 287
Ionization energy, 1013
Ionizing radiation, sources of, 979
Ions, 533, 969
Iron-core transformer, 761
Irradiance (I), 790–791, 806
    minimum and maximum in the, 900–901
Isentropy, 515
Isobaric process, 494–496
Isolated systems, 490, 493
    momentum changes in, 217
Isothermal change, 497–500
Isothermal process, 494, 500
Isothermal processes, [500, CD]
Isothermals, 440
Isotopes, 1056–1059
    half-lives of, 1077
Isovolumic (isochoric) process, 493

J

Jammer, Max, 92, 209
Jensen, J. H. D., 1065
Joliot-Curie, Irène, induced radioactivity and, 985, 1078
Joliot-Curie, Jean Frédéric, induced radioactivity and, 985, 1078
Joule, James Prescott, 173, 461, 491, 624
Joule heat, 624
Joules, 173, 181, 199, 462
Junction diodes, 767–768
Jupiter, Great Red Spot on, 313

K

Kaons, 1095, 1097, 1098
Kelvin (K), 6, 426, 446
Kelvin, Lord (William Thomson), 180, 426, 491, 510
    on aether, 934
    on atoms, 979
    on lines-of-force, 785
Kelvin-Planck statement, 511, 512, 520
Kepler, Johannes, 147, 148, 155
    astronomical telescope of, 870

on gravity, 147, 148
  Laws of Planetary Motion, 155–157, 163, [264, CD]
Keplerian constant, computing, 156
Kevlar, [338, CD]
Kidney, 381
Kilocalorie (kcal), 461, 462
Kilogram (kg), 6, 9–10
Kilogram Number 20, 10
Kilomole, 284
Kinematic parameters, linear and angular, 239
Kinematics, 20–44
  acceleration, 53–75
  relative motion, 39–43
  of rotation, 234–243
  speed, 20–28
  velocity, 28–38
Kinetic energy (KE), 172, 180–182, 186, 190–191, 199
  origin of concept, 180
  as a relative quantity, 182
Kinetic friction, 107, 111–114, 177–178
Kinetic Theory, 441–446, 492
  Caloric Theory vs., 456–457
Kinetic Theory of Gases, 447
King Kong, 361
Kirchhoff, Gustav Robert, 650
  on thermal radiation, 996–997
Kirchhoff's Radiation Law, 997
Kirchhoff's Rules, 650–653, 728, 748
Knowlton, A. A., 936
Kohlrausch, 790
Kossel, W., 1017
Kronig, R., 1038

L

Lamb, Willis, 1049
Lamb shift, 1049
Lambert, J. H., 425
Laminar flow, 303
Land, Edwin H., 894
Laplace, Pierre Simon de, 195, 1046
  speed of sound, 389
Larmor, J., 979
Lasers, 1014–1017
Latent heat of fusion, 467–468
Launch speed, computing, 159
Law of Conservation of Energy, 6, 171, 187, 198–199, 219, 306, 491–502, 650, 714
Law of Conservation of Momentum, 199, 209, 219, 220
Law of Inertia, 85–88, 96–97, 123, 792, 937
  rotational equivalent of, 253–255
Law of Interaction, 85, 96–97
  friction and, 110
  vertical acceleration and, 103–105

Law of Reflection, 817–822, 835
Law of Refraction, 825, 835
Law of Sines, 44, 92
Law of the Constancy of c, 939
Law of the Lever, 243
Law of Universal Gravitation, 147–152, 163
  asphericity of Earth, 154
  confirmation of $1/r^2$, 148–149
  evolution of, 147–148
  gravitational constant, 150–152
  product of the masses, 148
  terrestrial gravity, 152–154
Lawrence Livermore National Laboratory, 594
Laws, defined, 1
Laws of Planetary Motion, 155–157, 163, [264, CD]
L-C-R ac networks. *See R-L-C ac networks*
Leacock, Stephen, 531
Lead paint detector, 1018
Lead storage cell, 611
Leahy, William, 1084
Least Action, Principle of, 1031
Leclanché, 611
Leibniz, Gottfried Wilhelm von, 180, [457, CD]
Lenard, P., 970
Length, measurement of, 7–8
Length contraction, 945–947
Lens equation, [849, CD]
Lenses, 845–871
  aspherical surfaces and, 846–848
  extended imagery and, 852–856
  focal points and planes and, 851–852
  magnification and, 854–856
  single, 857–862
  spherical thin lenses, 848–850
  thin-lens combinations, 863–871
Lensmaker's Formula, 849, 878
Lenz, Heinrich Friedrich Emil, 714
Lenz's Law, 714–716, 719, 725, 731
Lepton number (*L*), 1096, 1099
Leptons, 3, 1095–1096, 1115
  angular momentum of, 260
Leucippus, 283
Levers, 243–244
  end-pivoted, 251
Lewis, G. N., 1004
Leyden jar, 585–586, 610
Libration point, 164
Lift, 311–312
Light, 783–805
  atoms and, 795–796
  behavior of, 1009, 1019
  elastic scattering of, 813
  electromagnetic radiation, 794
  electromagnetic waves, 784–786, 888
  electromagnetic-photon spectrum, 796–805
  energy quanta, 793–795
  granularity of, 1003–1004

irradiance and, 790–791
  as massless quanta, 793
  measurement of the speed of, 9
  natural, 889–891
  nature of, 783–796
  propagation in free space, 937
  propagation of, 812–835
  quasimonochromatic, 801
  scattering and absorption, 796
  speed of, 789, 790, 935, 937–938
  transmission through dense media, 815–816
  as a transverse wave, 889
  unpolarized, 890
  vacuum speed of, 790
  waveforms and wavefronts, 786–789
  waves and particles, 783–784
  white, 800–801
Light rays, 819
Light waves, sources of, 784
Light-emitting diodes (LEDs), 768
Light-quanta, 1005
Line of stability, 1059, 1065
Line strengths, 120
Line voltage, 744
Linear acceleration, link to angular acceleration, 239–240
Linear charge density, 559
Linear energy transfer (LET), 988
Linear equations, 14
Linear expansion, temperature coefficient of (α), 429–431
Linear momentum ($\vec{p}$), 209–210, 226–227
  conservation of, 6, 216–218, 226, 227
  rotational equivalent of, 259
  symmetry and, 226–227
Linear polarization, 888, 922
Linear polarizer, 891, 892, 893
Linear speed, determining, 239–240
Line-of-action, 90, 244–256
  center-of-gravity and, 249
Lines-of-force, 545, 785
Liquefaction of gases, [441, CD]
Liquid-gas friction, 109
Liquid-hydrogen bubble chamber, 584
Liquids, 283–315
  compression waves in, 379–380
  ideal, 289
  light scattering of, 816
  polar and nonpolar, 542
  surface tension of, 301–302
  volumetric expansion of, 432–435
Lithium, electron structure of, 1041
Load, 103
  relationship to pressure, 114–115
  skeletal, 116
  supported by ropes, 117–118
Load resistor, 768
Local symmetry, 1104
Lodestone, 667, 668
Lodge, Sir Oliver, 1084

Logarithmic dependence, 499
Longitudinal elastic wave, 379
Longitudinal magnification, 856
Longitudinal waves, 372–373
Loop, defined, 638
Loop equations, 652–653
Loop Rule, 650–651, 654, 728
Lorentz-FitzGerald Contraction, 935
Loudness, sensation of, 392–393
Low-temperature superconductivity, 621
Low-voltage ac, 744
Lucretius, 63, 671
Lyman series, 983, 1014

M

Mach, E., 88, 1001
Machines
    hydraulic, 296–298
    perpetual-motion, 493, 511
Mach's Principle, 88, 180
Magalotti, 531
Magic numbers, 1064
Maglev (magnetically levitated train), 715
Magnetic dipole moment, 694, [694, CD]
Magnetic domains, 674–676
Magnetic field (B)
    energy in, 730
    flux of (Φ_M), 712, 731
    induced, 719–720
    straight conductor, [679, CD]
    table of, 671
Magnetic flux (Φ_M), 712
Magnetic force (F_M), 687–696
Magnetic materials, permeability of,
    678–679
Magnetic moment, 694
Magnetic poles, 668–670
Magnetic resonance imaging (MRI), 1062
Magnetic torques, 692–694
Magnetic-PE (PE$_M$), 187
Magnetism, 148, 667–696
    on atomic level, 673–674
    of a current loop, 680–681
    diamagnetism, 673
    of the Earth, 672
    electrodynamics, 677–686
    ferromagnetism, 674
    magnetic force, 687–690
    magnets and magnetic field, 670–677
    paramagnetism, 674
    of a solenoid, 681–682, 686
Magnetite, 667
Magnetite crystal, 670
Magnetotactic bacterium, 670
Magnification, 854–856, 878
    angular, 862
Magnifying glasses, 858, 862, 878
Magnitude, 29
    of angular momentum, 259–260
    of centripetal acceleration, 143
    of instantaneous velocity vector, 33
    of normal force, 105

    of a reaction force, 248
    of velocity, 43, 55
Malus, Etienne, 893
Malus's Law, 892, 893, 922
Manhattan Project, 1084
Manometer, 296, 315
Maricourt, Peter de, 668
Mariner 2, 46
Marx, Karl, 107
Mass (m), 9, 93
    atomic standard for, 9–10
    center of, 257–258, 266
    determining, 212
    energy and, 190–192
    gravitational, 355
    inertial, 93, 253–254, 355
    measurement of, 9–10
    Newton's Second Law and, 93–95
    operational definition of, 95
    relationship to force and acceleration,
      94–95, 98–107
Mass damper, 357
Mass defect, 1066, 1086
Mass flux, 306
Mass-energy, 954, 955
    conservation of, 957
Masses
    comparison of, 89
    electromagnetic interaction and,
      529–530
    as field, 1101
    product of, 148
    wavelength of, 1028
Materials, elastic, 332–333
Mathematical symbols, 13
Matter
    atoms and, 283–290
    center-to-center attraction in, 147
    changes of state of, 467–473
    density of, 285–287
    energy as a property of, 171
    interaction of, 4
    states of, 287–290
    tendency to sustain rotational motion,
      261
    thermal properties of, 423–446
Maupertuis, 1031
Maximum acceleration, computing, 63
Maximum current, 746
Maximum emf (electromotive force), calcu-
    lating, 723
Maximum force, computing, 214
Maximum Power Transfer Theorem, [644,
    CD]
Maximum range, 72–74
Maximum voltage, 746
Maxwell, James Clerk, 1, 442, 443, 785
    Electromagnetic Theory of, 825
    on measuring the aether, 934
Maxwell-Boltzmann distributions, 446,
    1000
Maxwell's Equations, 784, 937
Mayer, Julius Robert, 461, 490

Mayer, Maria Goeppert, 1065
Mean free path, 445
Mean lifetime (τ), 1076
Mean-Speed Theorem, 58–60, 593
Measurement, 6–15
    exact, 11–13
    of force, 89
    of length, 7–8
    of mass, 94
    of mass and weight, 9–10
    significant figures in, 11–13
    of time, 10
Mechanical change in energy, 171
Mechanical energy, 180–199
    conservation of, 187–199
    defined, 188
    internal energy, 186–187
    kinetic energy, 180–182
    potential energy, 183–187
Mechanical equivalent of heat, 462
Mechanical waves, 371–410
    characteristics of, 341–376
    compression waves, 379–380
    transverse waves, 376–377
Medicine, nuclear, 1078
Medicine, use of X-rays in, 805
Megohms, 617
Meissner Effect, 677
Meitner, Lise, 1080–1081
Melting, 467–469
Melting points, [430, CD]
Melting-point curve, 441
Mendel, Gregor, 16
Meniscus lenses, 868
Mercury
    density of, 434
    superconducting state of, 621
Mercury barometer, 290, 293–294
Mercury cell, 611
Mercury manometer, 296, 315
Mersenne, Father, 396
    speed of sound, 389
    waves on a string, 400
Mesons, 1095, 1097–1099, 1101, 1105,
    1115
    kaon, 1095, 1098
    pion, 1095, 1097–1098
    quark composition of, 1105, 1106
Messenger particles, 560
Metal fatigue, 339, [339, CD]
Metals
    conductivity of, 476, 477, 533
    electromotive series for, 610
    electrostatic induction and, 536
    resistivity of, 618
    scattering by, 833
Meter (m), 6, 7
    defined, 9
Method of mixing, 463
Metric System, 7–8
Michelson, Albert Abraham, 934, 1028
Michelson Interferometer, 909–910, 920,
    934, 943

Michelson-Morley experiment, 933–935, 959
Microcoulomb, 541
Microfarad, 584
Microscope, [870, CD]
  compound, 869–870
  electron, 1028, 1051
Microstates, 517–518
Microwave images, 799
Microwaves, 799
Middle ear, 390
Millikan, Robert, 784, 973, 1005
Millikan oil-drop experiment, 972–973
Mills, R. L., 1107
Mirror formula, 873, 879
Mirrors, 871–877, 879
  aspherical, 871–872, [872, CD]
  curved, 871–877
  extended imagery and, 874–877
  paraboloidal, 871, 872
  plane, 821–822
  spherical, 872–877, 879
Mobile charge carriers, 608
Modern monopoles, [669, CD]
Modern Physics, 3–4
  nuclear atom, 979–989
  subatomic particles, 968–986
Moduli, 340–342
  shear and bulk, 341–342
  Young's modulus (*Y*), 340
Modulus of Elasticity (*Y*), 340
Mole, 284, 316
Molecules, 284
  mobility of, 291
  speed in gases, 444–446
Molecules per cubic meter, 288
Moment of a force, 244
Moment of inertia (*I*), 253–255, 259
  calculating, 253–255
  of a point-mass, 254
  spin rate and, 262
Moment-arm, 244–245
  for torques, 249
Momentum, 85, 93–95, 209–218
  angular (*L*), 259–266
  Conservation of Linear, 111–113, 216–218, 227
  homogeneity of space and, 226
  linear (*p*), 93–94, 209–210, 226–227
  relativistic, 951–954
  time rate-of-change of, 93–94, 211, 443
  transfer of, 219
Momentum change (Δ$\vec{p}$), 93–94, 210–212
  impulse and, 210–212
  in systems, 217
Monatomic gases, 501
Monochromatic light, 801
Monolithic IC, 771–772
Monopoles, magnetic, 668
Moon
  average orbital speed of, 22
  centripetal acceleration of, 163

diameter of, 235
Morgan, J. P., 744
Morley, E. W., 935
Morley, E. W., 935
Morphing Neutrinos, 1073
Motion, 20–44
  in an *E*-field, [571, CD]
  around center-of-mass, 257–258
  ballistic, 85–88
  changes in, 53
  circular, 142–143
  coupled, 106–107
  curvilinear, 141–146
  friction and, 107–115
  harmonic, 344–359
  one-dimensional, 39
  periodic, 344
  power and, 196
  projectile, 68–75
  relative, 39–43, 937
  repetitive, 344
  resistance to, 63
  resistance to a change in, 88–89
  rotational, 234–265
  straight-line, 32, 211
  translational vs. rotational, 234
  uniform, 39, 85–87
  uniformly accelerated, 56–62
  vibrational, 344
Motion variables, constant-a equations for, 60–62
Motional emf (electromotive force), 716–720, 731
Motion-induced vortices, 359
Motions, coupled, 106–107
Motor, dc, 694
Moving bodies, center-of-mass of, 257–258
Moving coil galvanometer, 646
Multiple Photon Photoelectric Effect, 1005
Multiplication
  significant figures in, 11–12
  of a vector by a scalar, 31–32
Muons, 944, 1095
Muscles
  fibers, 177
  force exerted by, 250–251
  functioning of, 177
  tissue strength, 338
  torque, 245–246
Musculoskeletal system
  structure of, 123
  support in, 116
mv (product of mass and speed), 93
Myopia, 866–868

N

National Aeronautic and Space Administration (NASA), 161
National Institute of Standards and Technology (NIST), 10, 939
Natural angular frequency, 345
Natural convection, 475–476

Natural light, 889–891, 922
Natural logarithm, 499
Natural radioactive series, 1068
Navier, Claude, 339
Near-point, 861, 866
Nearsightedness, 866–868
Necking, 338
Negative acceleration, 55, 60
  uniform, 61
Negative charge, 530–532
Negative lenses, 851
Negative magnification, 859
  computing, 543
Net applied force, 90
Net force, 98
  computing, 91–92, 543
Net potential, determining, 581
Net torque, 248
Network, equivalent capacitance of, 592
Network analysis, 649–653
Neutral-current process, 1111
Neutrality, electrical, 530–532
Neutrinos (ν), 283, 1096
  beta decay and, 1070–1072
  morphing, 1073
Neutron star, 261
Neutrons (n), 1106
  angular momentum of, 260
  decay of, 1109
  interaction with protons, 1101
Neutrophotos, 985
Newton (N), 89, 123
  defined, 89
Newton, Isaac, 85, 86, 171, 374
  confirmation of $1/r^2$, 148–149
  curvilinear motion, 141
  death mask, 94
  definition of force, 89
  definition of quantity-of-motion, 93
  g as a function of altitude, 65
  Law of Inertia of, 85–88
  light theory of, 784
  momentum and, 85
  speed of sound, 389
  on wave diffraction, 784
  wave speed, 374
Newtonian Mechanics, 2, 171
Newtonian reflecting telescope, 877
Newtonian Theory of Gravity, 141–162
  cosmic force, 155, 163
  Law of Universal Gravitation, 147–152, 163
Newton-meter, 173, 199
Newton's First Law, 85, 86
Newtons per coulomb (N/C), 546
Newton's Second Law, 92–93, 123, 150, 210–211
  alternative formulation of (*F* = *ma*), 94–95, 98–99
  relationship of weight and mass in, 102
  rotational equivalent of, 261
Newton's Third Law, 96–97, 123, 539
Niagara Falls, 743

Nicholson, J. W., 1010, 1011
Nickel-cadmium cell, 611
Nitrogen, liquid, 289, 428
Noble gases, 1040–1041, 1043, 1064
Noddack, Ida, 1079
Node equations, 651–653
Node Rule, 651, 654
Nodes, 397
  defined, 638
  electric circuits, 590
Noether's Principle, 198, 226, 264, 582, 1103
Noether, Amalie, 5, 6, 198
Noise, 395
Nollet, Abbé, 586
Nonconcurrent forces, 246–248
Nonconductors, 532–537
Nonisolated systems, 493
Nonohmic electrical devices, 616
Nonresonant elastic scattering, 796
Nonuniform magnetic field, 1062
Nonworking force, 181
Normal compressive strain (ε), 335, 359
Normal force ($\vec{\mathbf{F}}_N$), 102–104
  relationship to centripetal force, 144–146
Normal tensile strain (ε), 335, 359
Normal Zeeman Effect, 1038
Norman, Robert, 676
North pole, 668
  npn transistor, 769, 771, 773
  n-type semiconductor, 766
Nuclear atom, 979–989
  atomic spectra, 982–983
  neutron, 985–986
  proton, 984–985
  Rutherford scattering, 980–982
Nuclear Coulomb barrier, 981
Nuclear electrons, uncertainty and, 1046
Nuclear energy, 171
Nuclear fission, 1068, 1079–1084
Nuclear force, 1062–1064
  nucleus formation and, 1062
Nuclear fusion, 1068, 1084–1086
Nuclear magnetic resonance (NMR), 1061
Nuclear medicine, 1078
Nuclear physics, 1056–1086
Nuclear reactors, 1082–1083
Nuclear stability, 1064–1068
Nuclear structure, 1056–1068
Nuclear transformation, 1068–1086
Nuclear-PE, 187
Nucleon number (A), 1057
Nucleon-nucleon interactions, 1102
Nucleons, 986
Nucleus, 3–5
  composition of, 986
  Coulomb interaction with electron, 1012
  radius of, 1060
  shell model of, 1064–1065
  size, shape, and spin of, 1059–1062
  size of, 981–982
  stability of, 1064–1068

Nuclides, 1057
  atomic mass of, 1058
  binding-energy-per-nucleon for, 1066

O

Object distance (s_o), 849
Object focal length (f_o), 851
Object focal point, 851
Object size, relationship to terminal speed, 63
Objects, point-masses in, 258
Object-wave, 920
Oersted, Hans Christian, 667, 677, 687
Ohm, Georg Simon, 614–615
  on resistance, 615–616
Ohmic circuit element, 616
Ohm-meters, 618
Ohm's Law, 14, 16, 615–617, 624, 625, 746, 747, 756
Oil-drop experiment, 972–973
Old Quantum Theory, 996–1009
Onnes, H. Kamerlingh, 621
Oppenheimer, J. Robert, 1084
  on antimatter, 1047
  and atomic bomb, 1084
Optical center, 852
Optical density, 815
Optical path length (O.P.L.), 906
Optics, 783
  geometrical, 845–877
  physical, 888–921
Orbital, 1040
Orbital angular momentum quantum number (l), 1036
Orbital magnetic quantum number (m_l), 1036
Orbital speed, relationship to orbit size, 157–158
Orbiting planet, gravitational field acting on, [264, CD]
Orbits, [159, CD]
  of satellites, 157–160
Order, disorder and, 516–517
Order-of-magnitude, 22
Oresme, Nicolas, 59
Oscillating dipole, 792
Oscillating spring, 349–352
Oscillation, 329, 344–359, [351, CD]
  forced, 356–357
Oscillators, 745
Ounce, 10
Outer ear, 390
Outlets, two- and three-prong, 763, 764
Output current, 771
Overdamping, 356
Overtones, 391–392
Ozone, 290

P

$\vec{\mathbf{p}}$ (linear momentum), 209
Papin, Denis, 472
Parabola, 14

Parabolic Trajectories, 68, [69, CD], 71, 75
Parallel circuit, 595
  and capacitors, 588–590
  and cells, 611–613
  and resistors, 640
Parallel force systems, 117–118
Parallel resistance, 639–645
Parallel vectors, 31, 32, 40
Parallel wires, current in, 695–696
Parallel-Axis Theorem, 280
Parallelogram method, 30, 34, 37
Parallel-plate capacitor, 560, 586–588
  charged, 587
Paramagnetic substances, 674
Paraxial rays, 848
Parent nucleus, 1069, 1086
Partial polarization, 890
Particle accelerator, 1096, 1097
Particle interactions, mediation by field quanta, 1102
Particle theory of light, 783
Particles
  angular momentum of, 259–260
  charged, motion in electrical field, [579, CD]
  determining speed of, 958
  elementary, 3, 4, 260
  force-carrying, 147
  masses and rest energies for, 956
  messenger, 560
  moment of inertia of, 254
  relativistic kinetic energy of, 954–959
  subatomic, 968–986
Particles of zero-mass, 958
Pascal, (Pa), 334
Pascal, Blaise, 290, 296
Pascal's Principle, 296–298, 316
Paschen series, 983, 1014
Pauli, Wolfgang, 1071
  beta decay and, 1071
Exclusion Principle, 1035, 1038–1039, 1050, 1106
  on spin, 1038
Peak altitude (s_p), 71–72
Peak-to-peak voltage, 746
Pendulum, 352–354
  period of, 352
Perfect diamagnetism, 676–677
Period of a cycle (T), 344, 350–351
Periodic motion, 344
Periodic Table, 1057. *See also inside back cover*
Periodic waves, 374
Permanent magnet, 667–669
  magnetic field of, 672
Permeability (μ), 678
  of free space (μ_0), 676
  property of, 676–677
Permittivity (ε), 550, 561
  of free space (ε_0), 550, 561
  table of values, 550
Perpendicular components of vectors, 34–43
  summation of two vectors using, 36

Perpetual-motion machine, 493
 of the second kind, 511
Perrin, J., 971, 979
Phase, 346, 375
 initial, [346, CD]
Phase diagrams, 440–441
Phase differences, 748
Phasor, 754–755
Phasor addition, 754
Photocopy machine, 531, 857
Photocurrent, 1002, 1003
Photoelectric effect, 973, 1002–1006
 work function (ϕ), 1005
Photoelectric Equation, Einstein's, 1005
Photoelectrons, 1002
Photon flux, 794
Photon flux density, 794
Photons, 3, 793, 806, 1004, 1006
 Einstein on, 1003–1006
 gamma-ray, 805
 as quantum of the electromagnetic field,
  1100
 as quantum of the gauge field, 1104
 virtual, [560, CD], 1100, 1101
Physical optics, 888–921
 diffraction, 910–921
 interference, 899–910
 polarization, 888–899
Physical quantities, symbols for, 13
Physics
 Classical, 2
 defined, 1
 high-energy, 1094–1114
 language of, 13–15
 Modern, 3–4
 nuclear, 1056–1086
 quantification function of, 1
 testing of, 2
 thermal, 423–446
Picard, J., 970
Picofarad, 584
Pioneer 10, 86, 194
Pions, 1095, 1097–1098, 1102
 quark composition of, 1105
Pitch, 390–391
Pith balls, 533
Planck, Max, K. E. L.
 on energy elements, 1000–1002
 on quantum theory, 1001–1002
Planck's Constant (h), 260, 794, 1000,
  1028, 1030, 1044, 1045
 quantum of action, 1010
Planck's Radiation Law, 1001, 1019
Plane harmonic electromagnetic wave, 789
Plane mirrors, 821–822
Plane of polarization, 888
Plane-of-incidence, 819, 826, 835, 895, 896
Plane-polarized waves, 922
Planes, focal points and, 851–852
Planetary motion, 158, [264, CD]
 Kepler's laws of, 155–157, 163, [264,
  CD]
Planetary spin, effect on g, 65, 155

Planets
 elliptical orbits of, 155, 159
 gravitational field acting on, [264, CD]
 gravitational-PE and, 193
Planimeter, 179
Plasma, 287
Plato, 13, 441, 456, 668
Plücker, J., 982
Plutarch, 117
Plutonium, 1068, 1069, 1083
Plutonium Project, 1083
Plutonium-239, production of, 1083
pn-junction, 767–768, 773
pnp transistor, 769, 771, 773
Poincaré, Jules Henri, 173, 935
Point-charges
 electric field of, 547–549
 force exerted on a test-charge, 546
 magnitude of electric field of, 547–549
 potential of, 575–576
Point-mass, 249
 angular momentum of, 259–260
 moment of inertia of a, 254
Poise, 244
Polarity, of emf (electromotive force), 714,
  715
Polarization, of light, 888–899
 Malus's Law, 892, 893
 natural light and, 889–891
 polarizers and, 892–895
 processes of, 895–899
 by reflection, 895–897
Polarization angle (θ_p), 896, 922
Polarization of dielectrics, 588
Polarized atom, 536
Polarized light, scattering of, 898
Polaroids, 892, 894–895
Poles, 668–670
Pollio, Marcus Vitruvius, 382
Polycrystalline solids, 287, 288
Polyphase ac, 745
Pope, Alexander, 1047, 1048
Population inversion, 1015
Positive charge, 530–532
 direction of flow of, 608
 transfer of, 532
Positive lens, 851, 857, 859
Positive magnification, 854
Positive rays, 984
Positron (e^+), 1048
Positron Emission Tomography (PET),
  1049
Positron-beta decay, 1086
Positronium, 1049
Potassium-40, 1069, 1077
Potential, 570–571, 595
 electric, 569–584
 of a point-charge, 575–576
 in a uniform field, 573–575
Potential difference, 571–575, 589, 595
 in batteries, 574
 proton falling through, 572
 table of, 572

Potential energy (PE), 183–187, 199
 elastic (PE_e), 331–332
 electrical (PE_E), 569
 gravitational (PE_G), 184–186, 191, 344
 mass and, 191
Potential-energy wells, 1065
Potentiometer, 623, 626
Pound (lb), 90
Power (P), 195–197, 199
 apparent, 757, 758, 772
 electrical energy and, 623–625
 equivalencies, 196
 factor, 757
 instantaneous, 195, 746
 maximum transfer of, [644, CD]
 real dissipated, 757
Pressure, force per unit area (P), 114–115,
  290, 316
 absolute, 295
 in a container, 292–293
 gauge, 295–296
 in human body, 295
 hydrostatic, 290–296
Pressure cooker, 472
Prevost, P., 473
Primary charge distribution, 544
Primary colors, 833
Principal quantum number, 1012, 1035
Principia (of Newton), 86, 784
Principle of Complementarity, 1030
Principle of Least Action, 1031
Principle of Relativity, 936–937, 959
Principle of the Constancy of the Speed of
  Neutrinos, 938
Profile of a wave, 787
Progressive electromagnetic wave, 786
Progressive (traveling) waves, 372, 397,
  805
Projected horizontal distance, 72–73
Projectile motion, 68–75
Projectiles
 ballistic, 68–75
 effect of air friction on, 75
 range, 72–75
Proof-plane, 537
Propagation, speed of, light (c), 789–790,
  [789, CD]
Propagation number (k), 788, 805
Propagation of Sound, [410, CD]
Proper length, 945
Proper time, 943
Proton decay, 1114
Proton-electron pairs, 985
Proton-proton chain, 1085–1086
Protons (p)
 angular momentum of, 260
 charge on, 530–531
 interaction with neutrons, 1101
 quark composition of, 1105
p-type semiconductor, 767
Pulleys, 117–118
Pulsars, 2, 261
Pure tone, 382, 390

Pythagorean Theorem, 29, 36, 41, 348, 443, 943

Q

Quadratic equations, 14
Quality factor (Q), 988
Quanta, 3, 970
  energy, 793–795
  field, 1102
  light, 795, 1005
Quantity
  change in, 26
  measuring, 1
Quantity-of-angular-motion, 259
Quantity-of-matter, 9
Quantity-of-motion, 93, 215
Quantization, 4
Quantum Chromodynamics (QCD), 1108–1109, 1115
Quantum Electrodynamics (QED), 1035, 1047–1050
Quantum Field Theory, 162, 219, 560, 1100–1114
  creation of the Universe, 1112–1114
  electroweak force, 1109–1112
  gauge theory, 1102–1104
  quarks, 1104–1107
Quantum jump, 1010
Quantum Mechanics, 3–4, 260, 560, 959, 1027–1050
  de Broglie waves, 1027–1029
  Principle of Complementarity, 1030
  Schrödinger Wave Equation, 1031–1035
Quantum numbers, 1044, 1050
Quantum of action (h), 1010
Quantum physics, 1035–1050
  electron shells, 1039–1044
Pauli Exclusion Principle, 1038–1039
Quantum Electrodynamics and antimatter, 1047–1050
  quantum numbers, 1035–1036
  spin, 1038
  Uncertainty Principle, 1044–1047
  Zeeman Effect, 1036–1037
Quantum randomness, 1034
Quantum state, 1032
Quantum theory, 796, 996–1018
  atomic theory, 1009–1018
  origins of, 996–1009
Quark confinement, 1106
Quarks, 3, 4, 531, 1094, 1104–1107, 1115
  angular momentum of, 260
  table of, 1105
  transformation of, 5
Quasi-crystalline solids, 288
Quasimonochromatic light, 801

R

Racquetball, 220
rad (radiation absorbed dose), 987, 988

Radial acceleration, 144
Radial force, 146
Radian (rad), 235, 236
Radiant energy, 172, 458, 783–805
  electromagnetic-photon spectrum, 796–805
  nature of light, 783–796
  quantized, 1002
Radiation, effect on humans, 986–989
Radiation, thermal, 473–474
Radiation belts, [690, CD]
Radiation sickness, 988–989
Radio
  AM, 760–761
  circuitry in, 618
Radio receiver, 760–761
Radioactive decay, 5, 1068–1072
  measurement via, 958
Radioactivity
  discovery of, 976–979
  induced, 1078–1079
  natural, 1069
Radiowaves, 797–798
Radium, 977
Radius of curvature, 143, 849
Radon, 979, 1070
Raisin-pudding atom, 980
Ramsay, William, 977
Range, 72–75
Rankine, W., 187
Rarefaction, 379, 402
Rationalized units, 549–550
Rayleigh, Lord, 392, 813, 918
Rayleigh scattering, 812–814
R-C circuits, 647–649
Reactance (X), 749
  capacitive, 752, 772
  of a circuit, 756, 757, 772
  inductive, 749, 772
Reaction force, 102, 248
  magnitude of, 248
Reactor, nuclear, 1082–1083
Real capacitors, [591, CD]
Real dissipated power, 757
Real gases, [441, CD]
Real images, 853
Real particles, creation from virtual particles, 1102
Recoil speed, calculating, 218
Rectification, 768, 769, 773
Rectilinear translation, 235
Reference circle, 346
Reflected waves, 377–378
Reflecting telescope, 877
Reflection
  defined, 816
  diffuse, 819, 820
  external, 816–817
  internal, 816–817
  Law of, 817–822, 835
  of light waves, 816–822
  polarization by, 895–897

of sound waves, 377–378
  specular, 820
  total internal, 829–832, 836
Reflection gratings, 915
Reflections on the Motive Power of Heat (Carnot), 504
Refracted waves, 822
Refracting telescope, 870–871
Refraction, 822–832, [823, CD]
  index of (n), 823
  Law of, 825, 835
  of lightwaves, 822–832
Refrigeration machines, 507–510, [508, CD], 519
Reines, F., 1072
Relative atomic masses, 284
Relative density, 300
Relative motion, 39–43
  velocity and, 39
Relative permittivity (K_e), 550
Relativistic dynamics, 951–959
  energy, 954–959
  momentum, 951–954
Relativistic linear momentum, 952
Relativity, Theory of, 2, 39, 147, 355. *See also* Special Theory of Relativity
  energy and mass, 190–192
  Lorentz-FitzGerald Contraction, 935, 946
  Michaelson-Morley Experiment, 933–935
  postulates, the two, 936–939
  relativistic dynamics, 951–959
  simultaneity, 939–941
  table of, β and γ, 939
  time dilation, 942–945
  twin effect, 947–949
rem (rad equivalent man), 988
Renaldini, C., 424
Renormalization theory, 1049
Repetitive motion, 344
Resilience, 361
Resistance (R), 608, 614–625, 626
  ac and, 745–748
  energy and power, 623–625
  equivalent (R_e), 639, 654
  internal (r), 636
  movement against, 172
  power dissipated in, 624, 626
  resistivity, 618–621
  superconductivity, 620–621
  voltage drops and rises, 622–623
Resistivity (ρ)
  table of, 618
  temperature dependence of, 619–620
Resistors, 616–618, 626
  in series and parallel, 639–645
Resolution, angular limit of, 918
Resolving power, 918
Resonance, 356–357, 772–773
  ac, series, 758–760
Resonance phenomenon, 356–357

Resonant cavity, 1016
Resonant frequency (f_0), 356, 758–759
Rest, uniform motion and, 86
Rest energy (E_0), 190, 955, 960
Restoring force, 330
    elastic, 349–352
Resultant vector, 29, 44
    computing magnitude of, 36
    finding, 37
Reverse biased diode, 768
Reverse potential, 750
Reversible engines, 504, 505
Reversible processes, 500, 519, 520
Revolving-armature machine, 721
Revolving-field machine, 721
Rey, J., 423
Reynolds, O., 303
Right-circular polarization, 889
Right-Hand-Current Rule, 677–678, 680,
    688, 693, 694, 786
Rigid bodies, 246–248
    center-of-mass in, 257–258
Ripple, 768
Ritter, J., discovery of ultraviolet radiation
    by, 802
R-L circuit, 728–729
R-L-C ac networks, 753–764
    domestic circuits and, 762–764
    series circuits, 753–761
    transformer and, 761–762
rms current, 747
Roberval, 148
Robison, J., 396
Rockets, 215–216
Rolfe, W. J., 785
Rolling friction, 113
Roman balance, 244
Röntgen, Wilhelm Konrad, 804, 973–974
Ropes
    breaking strength, table of, 120
    determining tension in, 106–107,
        119–121, 256
    pulleys and, 117–118
    weights on, 100
Rotating systems, dynamical equation of,
    256
Rotation
    dynamics of, 253–257
    kinematics of, 234–243
Rotational equilibrium, 243–244
    rigid bodies, 246–248
    torque, 244–245
Rotational inertia, 253–257
Rotational kinetic energy, 258–260
Rotational motion, 234–265
    resistance to change in, 254
Rowland, H. A., 460
Royds, alpha experiment, 978
Rubbia, Carlo, 1112
Rubbing, charging by, 531–532
Rubbing machines, 581
Rumford, Count. *See* Thompson, Benjamin

Russell, Bertrand, 1, 971
Rutherford, Ernest, 977–978
    on atomic power, 1081, 1082
    on protons, 984
    scattering, 980–982
Rydberg constant, 983

S

s (scalar value of displacement), 66
Salam, Abdus, 1111
Salt (NaCl), 288
Satellite System, Tethered, 717
Satellites, orbits of, 157–160
Saturable nuclear force, 1063
Saturated color, 834
Saturated vapor pressure, 440
Saturn, ring system of, 159
Savart, Félix, 678, 680
Scalar, multiplication of a vector by, 31, 32
Scalar components, 34–38, 43
Scalar quantity, 28
Scattering, 796, 812–835, 835
    color and, 832–835
    dense media and, 815–816
    elastic, 813
    interference and, 814–815
    nonresonant elastic, 796
    polarization by, 897
    Rayleigh scattering, 812–814
    reflection and, 816–822
    refraction and, 822–832
    Rutherford, 980–982
Schrödinger Wave Equation, 1031–1035,
    1044, 1050
Schrödinger, Erwin, 5, 1031
    on quantum jump, 1035
Schwinger, Julian, 1109
Scientific calculator, 13
Scientific vocabulary, 13–15
Scotch yoke, 345, 346
Scott, David R., 62
Search coil, 719–720
Sears Tower, 64, 153, 196
Second Condition of Equilibrium, 246–248
Second harmonic, 289
Second Law of Thermodynamics, 506,
    510–519
    Clausius formulation and, 511–512
    entropy and, 513–519
    order and disorder and, 516–517
Second Postulate of Relativity, 937, 959
Secondary waves, 823
Second-Law equations, coupled, 106–107
Seconds (s), 6, 9, 10
Seismic waves, 373
Self-excited vibrations, 357–358
Self-field, 588
Self-induced emf, determining, 726
Self-induction, 725–730
    back-emf and, 726–727

energy in magnetic field and, 730
    R-L circuit and, 728–729
Semiconductors, 619, 765–772
Series circuit, 595
    and capacitors, 590–592
    and electric cells, 611–613
    and resistors, 639–640
    R-L-C circuit, 753–761
Series limit, 983
Series resonance, 758–760
Series *R-L-C* circuit, 753–761
Shakespeare, William, 150
    on time, 943
Sharks, 548
Shear Modulus, 341–342, 359
Shear strain (ϵ_s), 335, 359
Shear stress (σ_s), 334
Sheet polarizer, 894
Shell, defined, 1040
Shell model, 1061, 1064–1065
Shelley, P.B., 518
Shock absorbers, 330, 331, 355
Short circuit, 594, 612
Shunt resistor, 646
SI (Système International), 6, 16
    units of force, mass, and acceleration in,
        95
Significant figures, 11–13
Silicon, doping of, 766
Silicon atoms, 765–766
Simon, Pierre, Marquis de Laplace, 1046
Simple cubic structure, 288
Simple harmonic motion (SHM), 344–349
Simple harmonic oscillator, 349, [351, CD]
    acceleration, 348
    displacement, 346, 347
    velocity, 346–348
Simple pendulum, 352–354
Sine function, 13, 34
    of vectors, 34
Single Photon Photoelectric Effect, 1005
Single-phase system, 744
Single-slit diffraction, 913–915
Sink of a field, 551, 557
Sinusoidal sound wave, 384
Sinusoidal voltage, 746
Sinusoidal wave, 805
Sinusoidal waveforms, 387
Skin, human, 335
Slip, 333
Slope of a line, 25
Smith, Adam, 460
Snell's Law, 824–828, 835, 898
Socrates, 667
Soddy, F., 490, 978, 1056–1057
Sojourner, Martian, 1068
Solar system, Brahe's model of, 155
Solenoids, 681–684
Solid-gas friction, 109
Solid-liquid friction, 109
Solids, 283–315
    atoms and matter, 283–290

band structure of, 765–766
classification of, 288
crystalline, 288
elasticity and, 329–344
light scattering of, 816
moduli and energy and, 340–342
volumetric expansion of, 432–435
Solid-solid friction, 109
Solutions, checking, 15
Solvay Conference, 1005
Sommerfeld, Arnold, 1036
Sonograms, 381
Sound, 382–410
acoustics, 382–410
hearing of, 390–392
speed of, 389–390
wavelengths, 382
Sound waves, 382–410
beats, 395–397
Doppler Effect, 404–410
standing waves, 397–404
Sound-level, 411
Source of a field, 551, 557
South pole, 668
Space
defining, 10
homogeneity of, 226
Space quantization, 1037
Space & time, [10, CD]
Space-time, [949, CD]
Spafford, H. G., 784
Sparks, 533
Spatial symmetry, 226
Special Theory of Relativity, 2, 39,
933–959
addition of velocities, 949–951
effect on Classical Mechanics, 793
energy and mass, 190–192
groundwork for, 933–936
length contraction, 945–947
magnetism and, 677
and Michelson-Morley experiment,
933–935
postulates of, 936–939
relativistic dynamics, 951–959
simultaneity and time, 939–941
time dilation, 942–945
twin effect, 947–949
Specific gravity, 300
Specific heat capacity (c), 463–466, [464,
CD], 480
Spectral lines, 982
Spectroscopy, 982
Specular reflection, 820
Speed (v), 21
angular, 241
average, 20–23
banking angle for, 145
of compression waves, 379–380
constant, 24–25
instantaneous, 26–28, 44
mean, 58–60

of sound, 389–390
table of, 21
terminal, 63
of waves, 374–375
Speed conversion table, 24
Speed of efflux, 308
Speed of light (c), 9, 790, 937–938
characteristics of, 954
constancy of, 959
Speedometer, 26–28
Speed-time graph, 25, 56, 59
Sphere, gravity of, [154, CD]
Spherical mirrors, 872–877
Spherical thin lenses, 848–850
Spherical waves, amplitude of, 388
Spin
concept of, 1035, 1038
defined, 673
Spin angular momentum quantum number,
1038
Spin magnetic quantum number, 1038
Spin rate, moment of inertia and, 262
Spin-down electrons, 1038
Spin-up electrons, 1038
Split-ring commutator, 723
Spontaneous emission, 1015
Spring, oscillating, 349–352
mass of, 353
Spring (elastic force) constant, 330
Spring scale, for measuring force, 90
SQUID, 675, 678
Stability
balance and, 251–253
gyroscopic, [264, CD]
Stable equilibrium, 251
determining, 252
Stagnation point, 327
Standard Model, 1114
Standard temperature and pressure (STP),
284–285
Standards of measure, 6
Standing (stationary) waves, 397–404
and the laser, 1016
Standing still, physics of, 102–103
Stanford Linear Accelerator Center (SLAC),
1106
Staphylococcus bacterium, 295
Stark, Johannes, 1005
Stars, thermonuclear fusion in, 1084–1086
State, defined, 494–497, 1040
State of polarization, 889
Static equilibrium, 85, 555
Static friction, 108–114
moving via, 111–113
Static-friction coefficient (μ_s), 108–109
Statics, 85, 97
applied to hip dislocations, 121
fluid, 290–302
Stationary states, 1010
Statistical mechanics, 1002
Steady-state current, 612
Steady-state motion, 303

Steam, 472
Steam engine, 504
Steelyard, 244
Stefan, Joseph, 998
Stefan-Boltzmann Law, 998–999, 1019
Step-down transformer, 761
Step-up transformer, 761
Stevin, Simon, 93, 292
Stick-slip friction, 359
Stimulated emission, 1014–1017
and the laser, 1014–1017
Stoney, George, 969, 982, 983
Stopping distance, calculating, [61, CD]
Stopping potential, 1002
Straight-line motion
in a fixed direction, 32
infinite, 85–87
Strain (ε), 335–336, 359
Strain gauges, 636
Strange particles, 1098
Strange quarks, 1105
Strassmann, F., 1080
Strato, 63
Streamlines, 303
Strength, 336–338, 359
breaking, 338
of materials, [338, CD]
tensile, 337
ultimate, 337
yield, 337
Stress birefringence, 899
Stress (σ), 334, 359
Stress-strain diagram, 336, 337
Strings, standing waves on, 398–400
Strong force, 5, 89, 957, 1062
Strong interaction, 1105
Structures
stability of, 251–253
study of, 116
total tension, 100
Sturgeon, W., 683
Subatomic matter, net charge of, 973
Subatomic particles, 968–986
cathode rays and, 970–973
charges of, 530–531
conservation of momentum in, 219
discovery of radioactivity, 976–979
quantum of charge, 969–970
X-rays, 974–976
Sublimation, 441
Sublimation curve, 441
Subnuclear particle tracks, 1095
Subshell, defined, 1040
Subtraction
significant digits in, 12–13
of vectors, 31
Subtractive coloration, 834
thermonuclear fusion and, 1085
Sum-of-the-forces, 247
Sum-of-the-torques, 247
Sun
attractive action on planets, 147, 148

diameter of, 236
distance to, 11
Superconducting elements, critical temperatures of, 621
Superconducting magnets, 685
Superconducting quantum interference device (SQUID), 675
Superconductivity, 620–621, [621, CD], 626
Superconductors, 620–621, 676–677
Supercooled state, 435
Supercooling, 435
Supercritical mass, 1084
Superfluids, 429
Superheated steam, 472
Supernova 1987A, 87, 1072, 1114
Superposition Principle, 385
Surface charge density (σ), 559
Surface tension, [302, CD]
Surfaces of constant action, 1031
Surfaces of constant phase, 387
Symbols, mathematical, 13
Symmetry, [6, CD]
    angular momentum and, 264–265
    energy conservation and, 198–199
    global, 1103
    linear momentum and, 226
    local, 1104
Syncom II, 160
Système International, 6, 16
    units of force, mass, and acceleration in, 95
Systems
    binding energy in, 1066–1068
    concurrent force, 118–123
    equilibrium in, 85
    isolated, 217, 490
    momentum changes in, 217
    nonconcurrent forces, 246–248
    order and disorder in, 516–517
    parallel and colinear force, 117–118
Systolic pressure, 314
Szilard, Leo, 1081, 1082

T

Tachyons, 954
Tacoma Narrows Bridge, 358, 359
Tait, P. G., 376
Tangent, 13
Tangent function, 36
Tangential acceleration, 239, 240, 253
Tangential orbital speed, 158
Tau lepton, 1096
Telescope
    reflecting, 877
    refracting, 870–871
Temperature (T), 424–429, 480
    absolute, 426, 507, 510
    critical, 621, 626
    Curie, 676
    defined, 458

heat and, 457–459
range of, 427–429
thermodynamic temperature and absolute zero, 425–429
Temperature coefficient of linear expansion (α), 430, 432, 446
    table of, 430
Temperature coefficient of resistivity (α_0), 619, 626
Temperature coefficient of volume expansion (β), 432–435, 446
    table of, 432
Temperature gradient, 478
Temperature map, 458
Temporal period (T), 374
Tendon, 338
Tensile force ($\vec{F}_T$), 90, 100
Tensile (ultimate) strength, 337
Tension (F_T), 90
    on weighted ropes, 100
Terminal speed, 63, 117
Terminal voltage, 637
Terrestrial gravity, 152–154
Tesla (T), 670, 671, 696
Tesla, Nikola, 670
    ac induction motor and, 744
    polyphase ac and, 744–745
Test-charge, 544
Tethered Satellite System, 717
Theories, defined, 1
Theory of Everything (TOE), 1114
Theory of Gravitation, 2
Thermal conductivity, 476, 480
Thermal energy, 172, 186, 456–479
    caloric vs. kinetic, 456–457
    change of state and, 467–473
    heat and temperature, 457–459
    mechanical equivalent of heat, 462
    quantity of heat, 459–460
    specific heat capacity, 463–466
    Zeroth Law, 462
Thermal energy transfer, 473–479
    conduction, 476–479
    convection, 475–476
    radiation, 473–474
Thermal equilibrium, 462
Thermal neutrons, 1079
Thermal physics, 423–446
    gas laws, 435–446
    temperature, 424–429
    thermal expansion, 429–435
Thermal radiation, 473–474, 996–997
Thermal reservoir, 497
Thermionic emission, 976
Thermodynamic absolute temperature, 425
Thermodynamic temperature (T), 425–429, 446
Thermodynamics, 2, 490–519
    cyclic processes and, 503–510
    First Law of, 491–502, 519, 956
    Second Law of, 506, 510–519
    Zeroth Law of, 462, [462, CD], 480

Thermographs, 800
Thermometers, 423–425
Thermonuclear fusion, 1084–1086
Thermoscope, 423
Thin film interference, 905–908
Thin lenses, combinations of, 863–871
Thin positive lens, 857
Thin-Lens Equation, 849
Third harmonic, 386
Thompson, Benjamin, 456–457
Thompson, Francis, 162
Thomson, G. P., 1030
Thomson, J. J., 971
    on electromagnetic waves, 1003
    separation of isotopes by, 1057
Thomson, William. *See* Kelvin, Lord
"Thomson atom," 980
't Hooft, Gerhard, 1111
Thorndike, E., 935
Three Laws of Planetary Motion, 155–157, 163, [264, CD]
Three-coil dc generator, 724
Three-phase ac, 745
Thrust, calculating, 216
Tidal motions, 261
Timbre, 391–392
Time (t)
    defining, 10
    instantaneous force vs., 213, 214
    measurement of, 10
    momentum change and, 211
    with respect to an observer, 945
    simultaneity and, 939–941
Time constant, 647, 728, 1076
Time dilation, 942–945
Time rate-of-change
    of momentum, 93–94, 211
    of velocity, 53
Time-varying E-field, 719–720
Tip-to-tail method, 30–31, 36
Tokamak Fusion Reactor, 1088
Tomonaga, Sinitiro, 1049
Top quark, 1107
Torque (τ), 244–257, 265, 266
    center-of-gravity and, 249
    on a current loop, 692–694
    in a gravitational field, 257
    as rotational equivalent of force, 259
    rotational motion and, 253
    as time-rate-of-change of angular momentum, 260–261
Torque vector ($\vec{\tau}$), [246, CD]
Torr, 294
Torricelli, Evangelista, barometer and, 290, 293–294
Torricellian vacuum, 293
Torricelli's Result, 308–301, 316
Torsion balance, Coulomb's, 538
Total absorptivity, 998
Total emissivity, 998, 999, 1019
Total energy, 955
Total impulse, 214

Total internal reflection, 829–832, 836
Total tension structures, 100
Total-flight time, 72, 76
Toughness, 361
Transformation of energy, 491
Transformers, 761–762, [762, CD]
    turn ratio, 761
Transistors, 769–771
Translational equilibrium, 115–123, 247
    concurrent force systems, 118–123
    parallel and colinear force systems, 117–118
Translational motion, 234
Transmission axis, 892
Transmission gratings, 915
Transmission hologram, 920
Transmitted waves, 377–378
Transuranic element, 1079
Transverse electromagnetic (TEM) waves, 788
Transverse magnification (M_T), 854, 878
Transverse waves, 372–373, 376–377, 411
Traveling waves, 372
Treadmill scheme, 39
Triboelectric sequence, 532
Triple point, 441
Triple point of water, 426
Tritium, 1059
Trusses, 123
Tube length, microscope, 869
Tuned-mass damper, 357
Tungsten filament, 619
Turbulent flow, 304–305
    breathing, 304
Turn ratio, 761
Twin "paradox," 948
Two New Sciences (Galileo), 329
Two-coil dc generator, 723
Two-force systems, equilibrium of, 118
Tyndall, J., 998

U

u-quark, 1094, 1105
U-235, fission of, 1080
Uhlenbeck, G. E., 1038
Ultimate (tensile) strength, 337
Ultrasonic levitation, 398
Ultrasound, 381–382
Ultraviolet radiation (UV), 790, 802–804
Uncertainty, nuclear electrons and, 1046
Uncertainty Principle, 1035, 1044–1047, 1050, 1064
Underdamped system, 354
Unequal-arm balance, 244
Unified atomic mass units (u), 284, 1057, 1086
    table of, 1058
Uniform acceleration, 56–62, 76
    equations of, 60–62
    of falling bodies, 65

problems concerning, 61–62
Uniform electric field, 595
    potential in, 573–575
Uniform motion, rest and, 39, 86
Uniform negative acceleration, 61
Uniform speed, 24
Unit conversions, 24
Unit prefixes, 8
Units of charge, 539–540
    rationalized, 549–550
Units of measure, 6
    rationalized, 549–550
Universal gas constant, 439, 446
Universal Gravitation, Law of, 147–152, 163
Universal Gravitational Constant (G), 150–152
Universe
    age of, 194
    Aristotle's theory of, 68
    constant net charge of, 531
    creation of, [162, CD], 1112–1114
    expansion of, 194
    indeterministic nature of, 1035
    spatial symmetry in, 226
    temperature range in, 427–429
    total momentum of, 217
    understanding, 1
Unpolarized light, 890
Unstable equilibrium, 252
Up quarks, 1094, 1105
U.S. Customary Units, 8, 16, 24

V

Vacuum, 293, [293, CD]
    propagation of electromagnetic waves in, 790
    speed of light in, 161, 790, 937, 938
Vacuum pump, [293, CD]
Vacuum wavelengths, 827
Valence band, 765–766
Valence electrons, 765–766
    as source of colored light, 803
Van de Graaff, Robert J., 562
Van de Graaff generator, 556, 562
Van der Meer, Simon, 1112
    van Helmont, Jan Baptista, 289
    van Musschenbroek, P., 585
Vapor pressure, equilibrium, 440
Vaporization, 469–471
    heats of, 468
Vapors, condensation of, 440
Variable capacitor, 626
Variable dc voltage source, 626
Variable resistor, 626
Vector boson, 1108
Vector force field, 545
Vector quantity, 29, 44, 90
Vectors, 29
    addition of, 30–31, 44, 90

algebra of, 30–32
colinear, 54
components of, 34–38, [37, CD]
displacement vector ($\vec{s}$), 29
force, 89–93
magnitude, 29
mathematics of, 30–32
reversing direction of, 40–41
scalar components of, 70
subtraction of, 31
Velocity ($\vec{v}$), 28–38, [33, CD]
    addition and, 949–951
    angular, 236–239
    average, 32, 44
    instantaneous, 32–33, 44
    relative motion and, 39
    relativistic addition of, 949–950
    in simple harmonic motion, 346–348
    time rate-of-change of, 53
    of waves, 374–375
Velocity triangle, 142–143
Velocity vectors, 33
    adding, 39
Velocity-time curve, 59
Venturi Effect, 311, [311, CD], 312
Venus, ultraviolet photo of, 802
Versorium, 562
Vertical acceleration, 103–104, 160 [160, CD]
    gravity and, 68
Vertical arcs, [146, CD]
Vertical displacement, 35
    work and, 177
Vertical reaction force, 248
Vertical speed, of projectiles, 71–72
Vibrational motion, 344
Vibrations, self-excited, 357–358
Vibratory periodic motion, 344
Villard, P., 979
Virtual images, 821–822, 853, 857
Virtual mesons, 1101
Virtual particles, 94, 560
    creation of real particles from, 1102
Virtual photons, [560, CD], 1100, 1101
Visco-elastic dampers, 356
Viscosity, 289
Viscous damping, 354
Visual perspective, 237
Voices, frequency ranges of, 383, 391, 394
Volta, Alessandro, 571, 584, 585, 609
    on capacitance, 584
    on measuring potentials, 586
Voltage (V), 569, 571
    determining, 644
    drops and rises in, 622–623
    effective, 746, 772
    of household appliances, 645
    instantaneous, 745, 751
    line, 744
    maximum, 746
    peak-to-peak (V_{pp}), 746
    rms (V_{eff}), 747

Voltage divider, 623
Voltage gain, 772
Voltaic pile, 609
Voltaic wet cell, 609, 610
Voltaire, 153
Voltmeter, 626, 646–647
    placement of, 638
Volts per meter, 571
Volume (V), 9
Volume charge density (ρ), 559
Volume expansion, temperature coefficient
    of, 432–435
Volume flux (J), 305–306
Vomit Comet, 161
von Guericke, Otto, electrostatic generator
    and, 581
von Karman vortex street, 313
von Kleist, G., 585
von Laue, Max, 974, 1005
Vonnegut, Kurt, Jr., 949
Vortices, 313, 314
    motion-induced, 359

W

Wallis, John, 220
Walls, stability of, 251–253
Water
    absorption of microwave radiation by,
        799
    heavy, 1059
    phase diagram for, 440
    polarized molecules in, 542
    specific heat of, 463–466
Water calorimeter, 464
Water reactor system, 1083
Watson, J. D., 976
Watt (W), 196, 199, 626
Watt, James, 196, 460
Wave theory of light, 783–784
Waveforms, 373–375, 385–386, 786–789
Wavefront, 387
    intensity and, 393–395
    wavefronts and intensity, 387–388
Wavefunction, 1032
    state of a system and, 1032
Wavelength, 374
    relationship to transmitting medium, 378
Wave-particle duality, 1006, 1030
Wavepulse, 373
Waves, 371–410
    acoustic, 379, 393–395
    circularly polarized, 889
    compression, 379–380
    de Broglie, 1027–1029

diffraction of, 910–921
    energy of, 375
    finding velocity of, 374–375
    harmonic, 375–376, 411
    linearly polarized, 891
    mechanical, 371–410
    phase speed of, 822–823
    plane-polarized, 888, 922
    probability, 1032–1035
    reflection, absorption, and transmission
        of, 377–378
    refracted, 822–823
    speed of, 411
    on strings, 376–377, [376, CD]
    transverse, 372–373
    ultrasonic, 381
Waves and particles, [784, CD]
Wavetrain, 373
Weak force, 5, 89, 1072–1073
    theory of, 5, 1109
Weak interaction, 1105, 1109–1112
Weber (Wb), 712, 731
Weber, Ernst, 397, 790
Weber, Wilhelm, 397, 712
Wedgwood, Thomas, 997
Weight ($\vec{F}_W$), 9
    effect on equilibrium, 249–250
    effective, 102–105, 160
    gravity and, 101–106
    measurement of, 9–10
    on ropes, 100
Weight-dependent fall, notion of, 63
Weightlessness, 160–162
Weight-mass correspondence, 10
Weinberg, Steven, 1111
Weisskopf, Victor, 290
Wells, H. G., 950
Westinghouse, George, 744
White light, 800–801, 832–833
White noise, 395
Width of the central maximum, 914
Wien, Wilhelm, 999
Wien's Displacement Law, 999–1000, 1019
Wilson, C. T. R., 973
Wilson, E. T. S., 191
Wire-grid polarizers, 894
Wires, forces on, 691–696, [691, CD]
Wollaston, 868
Wooster, W. A., 1071
Work (W), 172–179, 199
    along a straight line, 172–173
    change in kinetic energy and, 180–182
    changing forces and, 178–179
    defined, 172, 458
    First Law of Thermodynamics and,
        494–497

    in general, 174–175
    and gravity, [177, CD]
    heat and internal energy and, 492–493
    internal energy sources and, 186–187
    overcoming friction, 177–178
    overcoming gravity, 176–177
    power and, 195–197
    vertical displacement and, 177
Work diagram, 179
Work function (ϕ), 1005
World Trade Center, 336, 345, 351, 356
    collapse of, 342–344
W-particles, 1109, 1112
Wrenches, 244–245
w.r.t. (with respect to), 39

X

X-ray spectrum, 1006–1007
X-ray tube, 976
X-rays, 804–805
    Compton scattering of, 1008
    emissions from lead paint, 1018

Y

Yang, C. N., 1107
Yield point, 333
Yield strength, 336–337
Young, Thomas, 180, 302, 339, 385, 784
    on color, 833
    energy and, 180
    on light waves, 889
    superposition principle, 385
    on surface tension, 302
Young's Experiment, 901–905, 922, 1033
Young's Modulus, 340, [340, CD], 343, 359
Yukawa, Hideki, 1101, 1108

Z

Z boson, 1110–1111
Zeeman, Pieter, 1036
Zeeman Effect, 1036–1037, 1062
"Zero-g," 160
Zero-mass particles, determining energy and
    momentum of, 959
Zero-point energy, 429, 1051
Zeroth Law of Thermodynamics, 462, [462,
    CD], 480
Zweig, George, 1104, 1105

USEFUL PHYSICAL DATA

Standard Temperature and Pressure (STP)		$0\ °C = 273.15\ K$ $1\ atm \equiv 101.325\ kPa$
Water		
Density, relative (4°C)	ρ_w	$1.000 \times 10^3\ kg/m^3$
Heat of fusion	L_f	$333.7\ kJ/kg$
Heat of vaporization	L_v	$2259\ kJ/kg$
Specific heat capacity	c	$4.186\ kJ/kg \cdot K$
Index of refraction	n_w	1.33
Standard acceleration due to Earth's gravity		$9.806\,65\ m/s^2$
Speed of sound in air (20°C)		$343\ m/s$
Speed of sound in air (STP)		$331\ m/s$
Density of dry air (STP)		$1.29\ kg/m^3$
Molecular mass of air		$28.98\ g/mol$

ASTROPHYSICAL DATA

	Earth	Moon	Sun
Mass	$5.975 \times 10^{24}\ kg$	$7.35 \times 10^{22}\ kg$	$1.987 \times 10^{30}\ kg$
Mean radius	$6.371 \times 10^6\ m$	$1.74 \times 10^6\ m$	$6.96 \times 10^8\ m$
Mean density	$5.52 \times 10^3\ kg/m^3$	$3.33 \times 10^3\ kg/m^3$	$1.41 \times 10^3\ kg/m^3$
Orbital period about galactic center			$200 \times 10^6\ y$
Orbital period	365 d 5 h 48 min	27.3 d	
Mean distance from Sun	$1.50 \times 10^{11}\ m$		
Mean distance from Earth		$3.85 \times 10^8\ m$	
Surface gravitational acceleration	$9.81\ m/s^2$	$1.62\ m/s^2$	$274\ m/s^2$
Surface pressure	$1.013 \times 10^5\ Pa$		
Magnetic moment	$8.0 \times 10^{22}\ A \cdot m^2$		
Surface temperature	$\approx 287\ K$	125 K–375 K	$5.8 \times 10^3\ K$
Power output			$3.85 \times 10^{26}\ W$
Period of rotation	23 h 56 min 4.1 s	27.3 d	~26 d

THE GREEK ALPHABET

Alpha	A	α	Nu	N	ν
Beta	B	β	Xi	Ξ	ξ
Gamma	Γ	γ	Omicron	O	o
Delta	Δ	δ	Pi	Π	π
Epsilon	E	ϵ	Rho	P	ρ
Zeta	Z	ζ	Sigma	Σ	σ
Eta	H	η	Tau	T	τ
Theta	Θ	θ	Upsilon	Y	υ
Iota	I	ι	Phi	Φ	ϕ
Kappa	K	κ	Chi	X	χ
Lambda	Λ	λ	Psi	Ψ	ψ
Mu	M	μ	Omega	Ω	ω